Nanotechnology 2010: Advanced Materials, CNTs, Particles, Films and Composites

Technical Proceedings of the 2010 NSTI Nanotechnology Conference and Expo

Nanotech
Conference & Expo 2010

*An Interdisciplinary Integrative Forum on
Nanotechnology, Biotechnology and Microtechnology*

**June 21-24, 2010
Anaheim Convention Center
Anaheim, California, U.S.A.
www.nsti.org**

NSTI
Nano Science and
Technology Institute

CRC Press
Taylor & Francis Group

Nanotechnology 2010: Advanced Materials, CNTs, Particles, Films and Composites

Technical Proceedings of the 2010 NSTI Nanotechnology Conference and Expo

NSTI-Nanotech 2010, Vol. 1

Nanotech
Conference & Expo 2010

An Interdisciplinary Integrative Forum on Nanotechnology, Biotechnology and Microtechnology

June 21-24, 2010
Anaheim, California, U.S.A.
www.nsti.org

NSTI Nanotech 2010 Joint Meeting

The 2010 NSTI Nanotechnology Conference and Expo includes:
2010 NSTI Bio Nano Conference and Expo, Bio Nanotech 2010
2010 Microtechnology Conference and Expo, Microtech 2010
9th Workshop on Compact Modeling, WCM 2010
NSTI Nanotech Ventures 2010
2010 TechConnect Summit
Clean Technology 2010

NSTI Nanotech 2010 Proceedings Editors:

Matthew Laudon
mlaudon@nsti.org

Bart Romanowicz
bfr@nsti.org

Nano Science and Technology Institute
Boston • Geneva • San Francisco

Nano Science and Technology Institute
696 San Ramon Valley Blvd., Ste. 423
Danville, CA 94526
U.S.A.

The papers in this book comprise the proceedings of the 2010 NSTI Nanotechnology Conference and Expo, Nanotech 2010, Anaheim, California, June 21-24 2010. They reflect the authors' opinions and, in the interests of timely dissemination, are published as presented and without change. Their inclusion in this publication does not necessarily constitute endorsement by the editors, Nano Science and Technology Institute, or the sponsors. Dedicated to our knowledgeable network of willing and forgiving friends.

ISBN 978-1-4398-3401-5
ISBN 978-1-4398-3418-3 (Volumes 1-3 SET)
ISBN 978-1-4398-3421-3 (TechConnect World 2010 Proceedings: Nanotech, Clean Technology, Microtech, Bio Nanotech Proceedings DVD)

Additional copies may be ordered from:

CRC Press
Taylor & Francis Group
an informa business
www.taylorandfrancisgroup.com

6000 Broken Sound Parkway, NW
Suite 300, Boca Raton, FL 33487
270 Madison Avenue
New York, NY 10016
2 Park Square, Milton Park
Abingdon, Oxon OX14 4RN, UK

Printed in the United States of America

Nanotech 2010 Proceeding Editors

VOLUME EDITORS

Matthew Laudon, NSTI, USA

Bart Romanowicz, NSTI, USA

NANOTECH 2010 CONFERENCE CHAIRS

Wade Adams, Rice University, USA

Mansoor M. Amiji, Northeastern University, USA

Wolfgang S. Bacsa, University of Toulouse, FR

Fiona Case, Case Scientific, NSTI Soft Nanotech Coordinator, USA

Elena Gaura, Coventry University, UK

Srinivas Iyer, Los Alamos National Laboratory, USA

Sha-Chelle Manning, NSTI, USA

Clayton Teague, NNCO, USA

Wolfgang Windl, Ohio State University, USA

Xing Zhou, Nanyang Technological University, SG

TOPICAL EDITORS

FABRICATION, CHARACTERIZATION & TOOLS

Computational Methods, Simulation, Software Tools

Keith Glassford, Nanostellar, Inc., USA

Michael Makowski, PPG Industries R&D, USA

Nano Fabrication

Jason Haaheim, NanoInk, Inc., USA

Guy A. DeRose, California Institute of Technology, USA

Aaron Stein, Brookhaven National Laboratory, USA

Nanoscale Materials Characterization

Pierre Panine, Xenocs SA, FR

Greg Haugstad, University of Minnesota, USA

Micro and Nano Reliability

Jürgen Keller, AMIC GmbH, DE

Bernd Michel, Berlin and Chemnitz Fraunhofer IZM and ENAS, DE

ADVANCED MATERIALS

Carbon Nano Structures and Devices

Wolfgang S. Bacsa, University of Toulouse, FR

Robert Hauge, Rice University, USA

Philip G. Collins, University of California, Irvine, USA

Nanoparticle Synthesis and Application

Sotiris E. Pratsinis, Swiss Federal Institute of Technology, CH

Bio Nano Materials

Darrin J. Pochan, University of Delaware, USA

Composite Materials

Thomas E. Twardowski, Twardowski Scientific, USA

Page McAndrew, Arkema Inc., USA

Tinh Nguyen, National Institute of Standards and Technology, USA

Soft Nanotech and Colloids

Fiona Case, Case Scientific, NSTI Soft Nanotech Coordinator, USA

Nanostructured Coatings, Surfaces & Films

Jan Sumerel, Dimatix, Inc., USA

Nanostructured Materials and Devices

Fiona Case, Case Scientific, NSTI Soft Nanotech Coordinator, USA

MICROSYSTEMS & ELECTRONICS

Nano Electronics & Photonics

Bob Doering, Texas Instruments, USA

Frank Register, University of Texas, USA

Peter Burke, University of California, Irvine, USA

Eric M. Vogel, The University of Texas at Dallas, USA

Luigi Colombo, Texas Instruments Inc., USA

Sensing Systems and Wireless Networks

Elena Gaura, Coventry University, UK

Inkjet Design, Materials and Fabrication

Elena Gaura, Coventry University, UK

Inkjet Design, Materials & Fab

Chris Menzel, Dimatix, Inc., USA

Jan Sumerel, Dimatix, Inc., USA

Micro & Nano Fluidics

Daniel Attinger, Columbia University, USA

Hang Lu, Georgia Institute of Technology, USA

Steffen Hardt, Leibniz Universität Hannover, DE

Workshop on Compact Modeling

Xing Zhou, Nanyang Technological University, SG

MEDICAL & BIOTECH

Biosensors, Diagnostics and Imaging

Srinivas Iyer, Los Alamos National Laboratory, USA

Cancer Nanotechnology

Mansoor M. Amiji, Northeastern Univ., USA

Vladimir Torchilin, Northeastern Univ., USA

Nano Medical Sciences

K. Kostarelos, University of London, UK

Nano Particles in Imaging Technologies

J. Manuel Perez, University of Central Florida, USA

Neurology Nanotech

Gabriel A. Silva, Univ. of California, San Diego, USA

Sarah Tao, The Charles Stark Draper Laboratory, Inc., USA

Materials for Drug and Gene Delivery

K. Kostarelos, University of London, UK

Phage Nanobiotechnology

Valery A. Petrenko, Auburn University, USA

ENERGY & ENVIRONMENT

Water Technologies

Nitin Parekh, Palo Alto Research Center, USA

Armin R. Völkel, Palo Alto Research Center, USA

Robert C. Renner, Awwa Research Foundation, USA

Nanotech for Oil and Gas

Sean Murphy, Univ. of Texas, USA

Lewis Norman, Halliburton, USA

Jay Kipper, Univ. of Texas, USA

Howard Schmidt, Rice Univ., USA

Photovoltaics

Loucas Tsakalakos, General Electric, USA

Matt Law, University of California, Irvine, USA

Tinh Nguyen, NIST/Industry Polymer Interphases Consortium, USA

Dean M. DeLongchamp, National Institute of Standards and Technology, USA

Power Generation and Storage

Wade Adams, Rice University, USA

Sponsors

Advanced Energy Consortium
Buchanan Ingersoll & Rooney PC
Caltech/MIT Enterprise Forum
Clean Technology and Sustainable Industries
Organization (CTSI)
CMP: Circuits Multi-Projets
Day Pitney LLP
Food and Drug Administration (FDA)
Greenberg Traurig
Houston Technology Center
Italian Trade Commission
Jackson Walker L.L.P.
Lockheed Martin
Nano Science and Technology Institute (NSTI)

nano tech Japan
NanoEurope Fair & Conference
NanoInk, Inc.
NanoTecNexus
National Cancer Institute (NCI)
National Institutes of Health (NIH)
Research in Germany Land of Ideas
Richard Smalley Institute for Nanoscale Science
and Technology
Small Times
Taylor & Francis Group LLC - CRC Press
TechConnect
Texas Nanotechnology Initiative
Transducers 2009

Supporting Organizations

Antenna Systems & Technology
Asia Pacific Nanotechnology Forum (APNF)
AZoNano.com
Battery Power Products & Technology
Berkeley Nanotechnology Club
BioExecutive International
Biolexis.com
BioOptics World
Biophotonics International
BioProcess International
BioTechniques
Biotechnology Industry Organization (BIO)
BusinessWeek
Drug & Market Development Publications
EurekAlert! / AAAS
Foresight Institute
Fuel Cell Magazine
GlobalSpec
IEEE CiSE
IEEE San Francisco Bay Area Nanotechnology
Council
Innovation Europe
Innovation UK
Institute of Physics
Journal of Experimental Medicine
KCI Investing
Laser Focus World
Materials Today
MEMS Investor Journal
Micronews
Microsystem Technologies - Micro- and
Nanosystems   mst|news
NANO Magazine
Nano Today
Nano World News

NanoBiotech News
Nanomedicine: Nanotechnology, Biology and
Medicine
NanoNow!
nanoparticles.org
NanoReg
NanoSPRINT
NanoTechCafe
Nanotechnology Law & Business
Nanotechnology Now
Nanotechnology.org.il
nanotechweb.org
NanoTecNexus
nanotimes
Nanovip.com
Nanowerk
Nature
Nature Publishing Group
Photonics Media
Photonics Spectra
PhysOrg.com
PR Newswire
R&D Magazine
Science Magazine
Solid State Technology
Springer
Springer Micro and Nano Fluidics
TechConnect News
Technology Transfer Tactics
The Journal of BioLaw & Business (JB&B)
The Journal of Cell Biology
The Real Nanotech Investor
The Scientist
tinytechjobs
Understanding Nanotechnology

Table of Contents

Carbon Nano Structures & Devices

Nanostructured Coatings, Surfaces & Films

Composite Materials

Polymer Nanotechnology

NSTI Nanotech 2010 Program Committee

TECHNICAL PROGRAM CO-CHAIRS

Matthew Laudon, NSTI , USA
Bart Romanowicz, NSTI, USA

NANOTECH 2010 CONFERENCE CHAIRS

Wade Adams, Rice University, USA
Mansoor M. Amiji, Northeastern University, USA
Wolfgang S. Bacsa, University of Toulouse, FR
Fiona Case, Case Scientific, NSTI Soft Nanotech Coordinator, USA
Lynn Foster, BPT Pharma, USA
Elena Gaura, Coventry University, UK
Srinivas Iyer, Los Alamos National Laboratory, USA
Sha-Chelle Manning, NSTI, USA
Clayton Teague, NNCO, USA
Wolfgang Windl, The Ohio State University, USA
Xing Zhou, Nanyang Technological University, SG

TOPICAL AND REGIONAL SCIENTIFIC ADVISORS AND CHAIRS

Computational Methods, Simulation, Software Tools

Keith Glassford, Nanostellar, Inc., USA
Michael Makowski, PPG Industries R&D, USA

Nano Fabrication

Jason Haaheim, NanoInk, Inc., USA
Guy A. DeRose, California Institute of Technology, USA
Aaron Stein, Brookhaven National Laboratory, USA

Nanoscale Materials Characterization

Pierre Panine, Xenocs SA, FR
Greg Haugstad, University of Minnesota, USA

Micro and Nano Reliability

Jürgen Keller, AMIC GmbH, DE

Bernd Michel, Berlin and Chemnitz Fraunhofer IZM and ENAS, DE

ADVANCED MATERIALS

Carbon Nano Structures and Devices

Wolfgang S. Bacsa, University of Toulouse, FR
Robert Hauge, Rice University, USA
Philip G. Collins, University of California, Irvine, USA

Nanoparticle Synthesis and Application

Sotiris E. Pratsinis, Swiss Federal Institute of Technology, CH

Bio Nano Materials

Darrin J. Pochan, University of Delaware, USA

Composite Materials

Thomas E. Twardowski, Twardowski Scientific, USA
Page McAndrew, Arkema Inc., USA
Tinh Nguyen, National Institute of Standards and Technology, USA

Soft Nanotech and Colloids

Fiona Case, Case Scientific, NSTI Soft Nanotech Coordinator, USA

Nanostructured Coatings, Surfaces & Films

Jan Sumerel, Dimatix, Inc., USA

Nanostructured Materials and Devices

Fiona Case, Case Scientific, NSTI Soft Nanotech Coordinator, USA

MICROSYSTEMS & ELECTRONICS

Nano Electronics & Photonics

Bob Doering, Texas Instruments, USA
Frank Register, University of Texas, USA
Peter Burke, University of California, Irvine, USA
Eric M. Vogel, The University of Texas at Dallas, USA
Luigi Colombo, Texas Instruments Inc., USA

Sensing Systems and Wireless Networks

Elena Gaura, Coventry University, UK

Inkjet Design, Materials and Fabrication

Elena Gaura, Coventry University, UK

Inkjet Design, Materials & Fab

Chris Menzel, Dimatix, Inc., USA
Jan Sumerel, Dimatix, Inc., USA

Micro & Nano Fluidics

Daniel Attinger, Columbia University, USA
Hang Lu, Georgia Institute of Technology, USA
Steffen Hardt, Leibniz Universität Hannover, DE

Workshop on Compact Modeling

Xing Zhou, Nanyang Technological University, SG

MEDICAL & BIOTECH

Biosensors, Diagnostics and Imaging

Srinivas Iyer, Los Alamos National Laboratory, USA

Cancer Nanotechnology

Mansoor M. Amiji, Northeastern Univ., USA
Vladimir Torchilin, Northeastern Univ., USA

Nano Medical Sciences

K. Kostarelos, University of London, UK
Nano Particles in Imaging Technologies
J. Manuel Perez, University of Central Florida, USA

Neurology Nanotech

Gabriel A. Silva, Univ. of California, San Diego, USA
Sarah Tao, The Charles Stark Draper Laboratory, Inc., USA

Materials for Drug and Gene Delivery

K. Kostarelos, University of London, UK

Phage Nanobiotechnology

Valery A. Petrenko, Auburn University, USA

ENERGY & ENVIRONMENT

Water Technologies

Nitin Parekh, Palo Alto Research Center, USA
Armin R. Völkel, Palo Alto Research Center, USA
Robert C. Renner, Awwa Research Foundation, USA

Nanotech for Oil and Gas

Sean Murphy, Univ. of Texas, USA
Lewis Norman, Halliburton, USA
Jay Kipper, Univ. of Texas, USA
Howard Schmidt, Rice Univ., USA

Photovoltaics

Loucas Tsakalakos, General Electric, USA
Matt Law, University of California, Irvine, USA
Tinh Nguyen, NIST/Industry Polymer Interphases Consortium, USA
Dean M. DeLongchamp, National Institute of Standards and Technology, USA

Power Generation and Storage

Wade Adams, Rice University, USA

NANOTECH PHYSICAL SCIENCES COMMITTEE

Wade Adams, Rice University, USA
M.P. Anantram, NASA Ames Research Center, USA
Phaedon Avouris, IBM, USA
Wolfgang S. Bacsa, University of Toulouse, FR
Gregory S. Blackman, DuPont, USA
Alexander M. Bratkovsky, Hewlett-Packard, USA
Roberto Car, Princeton University, USA
Fiona Case, Case Scientific, USA
Seth B. Darling, Argonne National Laboratory, USA
Alex Demkov, University of Texas at Austin, USA
David K. Ferry, Arizona State University, USA
Toshio Fukuda, Nagoya University, JP
Keith Glassford, Nanostellar, Inc., USA
Sharon Glotzer, University of Michigan, USA
William Goddard, California Institute of Technology, USA
Gerhard Goldbeck-Wood, Accelrys, Inc., UK
Jay T. Groves, University of California at Berkeley, USA
James R. Heath, California Institute of Technology, USA
Christian Joachim, CEMES-CNRS, FR
Jürgen Keller, Fraunhofer Institut, Berlin, DE
Anantha Krishnan, Lawrence Livermore National Lab, USA
Kristen Kulinowski, Rice University, USA
Shenggao Liu, ChevronTexaco, USA
Lutz Mädler, University of Bremen, DE
Chris Menzel, FUJIFILM Dimatix, USA
Arben Merkoçi, Catalan Institute of Nanotechnology, ES
Meyya Meyyappan, National Aeronautics and Space Agency, USA
Martin Michel, Nestlé, CH
Tinh Nguyen, NIST, USA
Sokrates Pantelides, Vanderbilt University, USA
Philip Pincus, University of California at Santa Barbara, USA
Joachim Piprek, NUSOD Institute, USA

Sotiris E. Pratsinis, ETH Zürich, CH
Serge Prudhomme, University of Texas at Austin, USA
Nick Quirke, University College Dublin, Ireland
PVM Rao, IIT Delhi, India
Mark Reed, Yale University, USA
Philippe Renaud, Swiss Federal Institute of Technology of Lausanne, CH
Doug Resnick, Molecular Imprints, USA
Mihail Roco, National Science Foundation, USA
Robert Rudd, Lawrence Livermore National Laboratory, USA
Brent Segal, Nantero, USA
Douglas Smith, University of San Diego, USA
Donald C. Sundberg, University of New Hampshire, USA
Clayton Teague, National Nanotechnology Coordination Office, USA
Loucas Tsakalakos, GE Global Research, USA
Wolfgang Windl, Ohio State University, USA
Xiaoguang Zhang, Oakridge National Laboratory, USA

NANOTECH LIFE SCIENCES COMMITTEE

Mansoor M. Amiji, Northeastern University, USA
Stephen H. Bryant, National Institute of Health, USA
Fred Cohen, University of California, San Francisco, USA
Tejal Desai, University of California, San Francisco, USA
Robert S. Eisenberg, Rush Medical Center, Chicago, USA
Mauro Ferrari, University of Texas Health Science Center at Houston, USA
Andreas Hieke, GEMIO Technologies, Inc., USA
Sorin Istrail, Brown University, USA
Srinivas Iyer, Los Alamos National Laboratory, USA
Brian Korgel, University of Texas-Austin, USA
Jeff Lockwood, Novartis, USA
Hang Lu, Georgia Institute of Technology, USA
Atul Parikh, University of California, Davis, USA
Andrzej Przekwas, CFD Research Corporation, USA
George Robillard, BioMade Corporation, NL
Rafal Romanowicz, NSTI, CH
Gabriel A. Silva, University of California, San Diego, USA
Srinivas Sridhar, Northeastern University, USA
Sarah Tao, The Charles Stark Draper Laboratory, Inc., USA
Tom Terwilliger, Los Alamos National Laboratory, USA
Vladimir Torchilin, Northeastern University, USA
Thomas J. Webster, Brown University, USA
Steven T. Wereley, Purdue University, USA

NANOTECH MICROSYSTEMS COMMITTEE

Narayan R. Aluru, University of Illinois Urbana-Champaign, USA
Daniel Attinger, Columbia University, USA
Supriyo Bandyopadhyay, Virginia Commonwealth University, USA
Stephen F. Bart, Analog Devices, Inc., USA
Bum-Kyoo Choi, Sogang University, KR
Bernard Courtois, TIMA-CMP, FR
Peter Cousseau, Honeywell, USA
Mark da Silva, Analog Devices, Inc., USA
Robert W. Dutton, Stanford University, USA
Gary K. Fedder, Carnegie Mellon University, USA
Edward P. Furlani, Eastman Kodak Company, USA
Elena Gaura, Coventry University, UK
Steffen Hardt, TU Darmstadt, DE

Michael Judy, Analog Devices, USA
Jan G. Korvink, University of Freiburg, DE
Mark E. Law, University of Florida, USA
Mary-Ann Maher, SoftMEMS, USA
Kazunori Matsuda, Tokushima Bunri University, JP
Tamal Mukherjee, Carnegie Mellon University, USA
Andrzej Napieralski, Technical University of Lodz, PL
Ruth Pachter, Air Force Research Laboratory, USA
Marcel D. Profirescu, Technical University of Bucharest, RO
Marta Rencz, Technical University of Budapest, HU
Siegfried Selberherr, Technical University of Vienna, AT
Sudhama Shastri, California Micro Devices, USA
Dragica Vasilesca, Arizona State University, USA
Gerhard Wachutka, Technical University of Münich, DE
Jacob White, Massachusetts Institute of Technology, USA
Thomas Wiegele, Goodrich, USA
Andreas Wild, Freescale Semiconductor, FR
Cy Wilson, North American Space Agency, USA
Xing Zhou, Nanyang Technological University, SG

NANO FABRICATION COMMITTEE

Adekunle Adeyeye, National University of Singapore, SG
Ronald S. Besser, Stevens Institute of Technology, USA
Gregory R. Bogart, Symphony Acoustics, USA
Charles Kin P. Cheung, NIST, USA
Zheng Cui, Rutherford Appleton Laboratory, UK
Guy A. DeRose, California Institute of Technology, USA
Majeed Foad, Applied Materials, USA
Zhixiong Guo, Rutgers University, USA
Takamaro Kikkawa, Hiroshima University, JP
Jungsang Kim, Duke University, USA
Uma Krishnamoorthy, Sandia National Laboratory, USA
Andres H. La Rosa, Portland State University, USA
Warren Y.C. Lai, New Jersey Institute of Technology, USA
Pawitter Mangat, Motorola, USA
Vivian Ng, National University of Singapore, SG
Leonidas E. Ocola, Argonne National Laboratory, USA
Sang Hyun Oh, University of Minnesota, USA
Nag Patibandla, Applied Materials, USA
Stanley Pau, University of Arizona, USA
John A. Rogers, University of Illinois at Urbana-Champaign, USA
Nicolaas F. de Rooij, Université de Neuchâtel, CH
Aaron Stein, Brookhaven National Laboratory, USA
Rajesh Swaminathan, Applied Materials, USA
Vijay R. Tirumala, Cabot Corporation, USA
Gary Wiederrecht, Argonne National Laboratory, USA
Eric Witherspoon, Applied Materials, USA

NSTI Nanotech 2010 Proceedings Topics

Nanotechnology 2010: Advanced Materials, CNTs, Particles, Films and Composites,

NSTI-Nanotech 2010, Vol. 1, ISBN: 978-1-4398-3401-5:

1./ Nanoscale Materials Characterization
2./ Carbon Nano Structures & Devices
3./ Nanoparticle Synthesis & Applications
4./ Nanostructured Coatings, Surfaces & Films
5./ Composite Materials
6./ Polymer Nanotechnology
7./ Soft Nanotech

Nanotechnology 2010: Electronics, Devices, Fabrication, MEMS, Fluidics and Computational,

NSTI-Nanotech 2010, Vol. 2, ISBN: 978-1-4398-3402-2:

1./ Electronics & Photonics
2./ Nanostructured Materials & Devices
3./ Micro & Nano Reliability
4./ NanoFab: Manufacture, Instrumentation
5./ MEMS Fab: Design, Manufacture, Instrumentation
6./ MEMS & NEMS Devices
7./ Sensing Systems & Wireless Networks
8./ Micro & Nano Fluidics
9./ Inkjet Design, Materials & Fabrication
10./ Computational Methods, Simulation & Software Tools
11./ Compact Modeling

Nanotechnology 2010: Bio Sensors, Instruments, Medical, Environment and Energy,

NSTI-Nanotech 2010, Vol. 3, ISBN: 978-1-4398-3415-2:

1./ Bio Sensors, Diagnostics & Imaging
2./ Bio Analytical Instrumentation
3./ Nano for Biotech, Interfaces & Tissues
4./ Bio Nano Materials
5./ Drug & Gene Delivery
6./ Cancer Nanotechnology
7./ Nano Medical Sciences
8./ Environment, Health & Safety
9./ Water, Oil, Gas & Bio Energy
10./ Photovoltaics, Solar, Lighting & Displays
11./ Power Generation & Energy Storage

Nanotechnology 2010, Vol. 1-3, ISBN: 978-1-4398-3418-3 (hardcopy)

TechConnect World 2010 Proceedings: Nanotech, Clean Technology, Microtech, Bio Nanotech Proceedings DVD, ISBN: 978-1-4398-3421-3

Advanced Nanoscale Elastic Property Measurement by Contact-Resonance Atomic Force Microscopy

G. Stan and R. F. Cook

Nanomechanical Properties Group, National Institute of Standards and Technology, Gaithersburg, MD, USA, gheorghe.stan@nist.gov, robert.cook@nist.gov

ABSTRACT

In atomic force microscopy (AFM)-based techniques, material information (topography, mechanical, electrical, magnetic, etc.) is retrieved from the nanoscale interaction between the AFM tip and the material probed. As the size of the apex of the AFM tip is comparable with the size of the investigated structures, it is very important to recognize the tip contribution to the quantities measured. In particular, this is necessary when nanoscale elastic properties are probed by contact-resonance AFM (CR-AFM), in which case, the tip contribution needs to be considered in both the topographical and contact mechanical responses. By combining topography and mechanical response, an advanced characterization of non-flat nanoscale structures by CR-AFM has been accomplished. Two major applications will be discussed here: (1) self-correlation of the topography image and contact stiffness map for a granular Au film and (2) point measurements across oxidized Si nanowires.

Keywords: nanoscale elastic modulus, contact resonance, atomic force microscopy, nanowires

1 INTRODUCTION

Invented two decades ago [1], atomic force microscopy now provides simple and direct access to nanoscale investigations of material surfaces. Besides the common topography mapping, over the years innovative AFM modes have advanced and differentiated AFM nanoscale characterization by exploiting the sensitive detection of probe-sample interactions: In the class of dynamic AFM modes, contact resonance atomic force microscopy (CR-AFM) [2] [3] distinguishes itself as a sensitive technique for quantitative measurements of the elastic responses of materials at the nanoscale. CR-AFM has been used successfully to measure the elastic properties of a large variety of materials and structures: piezoelectric ceramics [4], metal films [5], diamond-like carbon films [6], glass-fiber-polymer matrix composites [7], clay minerals [8], polycrystalline materials [9], and nanostructures (nanobelts [10], nanowires (NWs) [11], and nanotubes [12]). In the past few years, we have improved and developed the capabilities of CR-AFM in various ways: established clear criteria for assessing the precision and

accuracy in CR-AFM measurements [13], introduced a load-dependent CR-AFM protocol [14] suitable for elastic modulus measurements on compliant materials, and tested the suitability of CR-AFM measurements on various materials with elastic moduli in the range of few GPa to hundreds of GPa. In the new load-dependent CR-AFM approach, the deflection and resonance frequency shift of the AFM cantilever-probe are recorded simultaneously as the probe is gradually brought in and out of contact. Another valuable CR-AFM application has been the mapping of elastic properties at the nanoscale of granular nanocrystalline surfaces [15] and microstructures [16]. By combining the CR-AFM point measurements with AFM scanning capability, the elastic modulus was mapped with better than 10 nm spatial resolution. To determine the elastic modulus from CR-AFM measurements on non-flat surfaces, both topography and contact stiffness maps were self-consistently correlated. We have also extended CR-AFM applicability to quantitative elastic modulus measurements on various one-dimensional nanostructures (NWs [11] [17] and nanotubes [12]). From such CR-AFM measurements, not only were the mechanical properties of nanostructures probed but also intriguing characteristics of the nanoscale structure-mechanical properties relationship were revealed. In the case of ZnO NWs [11] and Te NWs [17], the elastic moduli were found to be strongly size dependent for diameters less than 100 nm. From recent CR-AFM measurements on oxidized Si NWs [18], the effect of the compressive stress at the Si-SiO$_2$ interface was revealed in the diameter dependence of the elastic moduli of these NWs. In this work we will review some advances in using the topographical information to retrieve correctly the local elastic response of nanostructures probed by CR-AFM. This is necessary when non-flat surfaces are imaged by CR-AFM and the local nanoscale elastic response has to be amended for the continuous change in the contact geometry during scanning. The example here will be for a granular Au surface, with local changes in the contact spring couplings imposed by the topographical variations. A precise determination of the size of the structure measured is critical also in establishing the correct contact geometry in CR-AFM point-measurements across one-dimensional structures; as an example, the CR-AFM

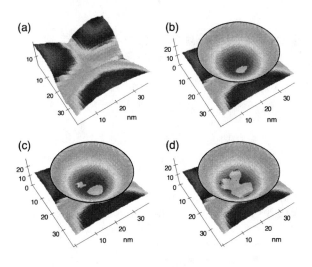

Figure 1: Progressive contact intersection between the apex of the AFM tip of radius 34.4 nm and a triple junction of a granular Au surface as the applied load was increased from (a) 0, to (b) 100 nN, (c) 250 nN, and (d) 350 nN, respectively.

analysis will be shown here for elastic modulus measurements of oxidized Si NWs.

2 CR-AFM Application To Image Nanoscale-Rough Surfaces

In CR-AFM, a small-amplitude oscillation that vibrates the tip-sample contact excites acoustic waves in the sample and the resonance state of the clamped-coupled cantilever system changes accordingly. The elastic response of the material is retrieved then from the variation experienced by the resonance frequency of the AFM cantilever. Besides the dynamical analysis that converts directly the measured resonance frequencies into the tip-sample contact stiffness, the final step consists of extracting the elastic modulus of the sample from the calculated contact stiffness through a contact mechanics model that adequately describes the contact formed between tip and sample. On non-flat surface topographies, variations in the contact geometry need to be known a priori and incorporated into the contact model used. In the case of a complex contact geometry it could be relevant for not only the normal component but also the lateral component of the tip-sample coupling. Ultimately, the normal and lateral contact couplings alter the dynamics of the cantilever and their effect is observed in additional changes of the contact resonance frequencies of the cantilever. Detail reviews of CR-AFM operation, measurement analysis, and applications can be found in Refs. [19], [20], and [21].

On nanoscale-rough surfaces, the contact geometry changes not only in size by also through the number of contacts established between the scanning tip and adjacent contacted material, varying from single-asperity contact (SAC) to multiple-asperity contact (MAC) [15]. Such an example is shown in Fig. 1 where the apex of the AFM tip was brought gradually into contact with the intergrain region formed between three adjacent Au grains. As the applied load was increased progressively from 0 to 350 nN, the contact evolved from SAC to MAC. The contacted grains act as individual springs with the load distributed between their summits. In such cases of complicated contact geometries, the actual topography, free of tip artifacts, needs to be used for a correct interpretation of the contact mechanics contained in the CR-AFM measurements.

Two scan images, topography and contact resonance frequency (see Figs. 2a and 2b), were acquired over the same area of about 600 nm^2 of a granular Au film. The contact resonance frequency map (Fig. 2b) was reconstructed from the frequency spectra acquired at every point in the scan (125 x 125 pixels) by a lock-in amplifier detection technique [15]. Subsequently, the scanning tunneling microscopy (STM) mode of the same AFM used for CR-AFM was chosen to re-scan the same investigated area and provide high-resolution details of the scanned topographical features (Fig. 2a). A very good feature correspondence can be observed between the two images and they can be self-consistently correlated. Using the information from STM and CR-AFM maps, at every location in the scan, the contact reconstruction is processed for the tip radius and applied load used in CR-AFM scan. Thus, the tip and surface were simulated to come into contact by gradually increasing the load until it reached the value of the applied load. Once the setpoint was reached, the contact geometry was observed to be either from a SAC or MAC. We point out that this is an advanced approach in tip-sample contact reconstruction as it includes the local elastic deformation of the tip and surface.

With the self-consistent consideration of the elastic properties of the contacted surface, the maps for the contact area and elastic modulus are obtained (see Figs. 2c and 2d). Good correlations and intriguing details are observed between the four maps. Nominally, a slight reduction in the elastic modulus in the intergrain regions is expected as a result of the orientation mismatch between adjacent grains. On contrary, in Fig. 2b, the contact resonance frequency is observed to increase in the intergrains regions. The explanation is that, as can be seen in Fig. 2c, the contact area increases in the intergrain regions due to MAC formation. This in turn determines an increase in the contact resonance frequency, but its proper consideration provides the correct calculation for the elastic modulus. Indeed, the elastic modulus (Fig. 2d) decreases consistently in the intergrain regions. In addition to that, small variations in the contact area

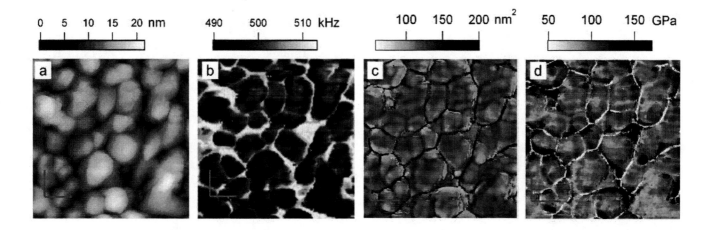

Figure 2: (a) Topography and (b) contact-resonance frequency map, acquired over the same area of a granular Au film, are self-consistent correlated to calculate (c) the contact area and (d) elastic modulus maps. The horizontal and vertical scale bars are 100 nm in every image.

and elastic modulus are also observed within grain regions. As can be seen in Fig. 2d, there are still unresolved regions (grain boundaries with sharp ripples) where the contact geometry is not described properly by the Hertzian contact model used; in such cases finite element analysis could be a viable alternative.

3 CR-AFM Application To One-Dimensional Structures

The measurement of mechanical properties of one-dimensional structures is important both for understanding the material-structure relationship at the nanoscale and practical applications. Among the multitude of nanostructures, Si NWs have distinctively remarkable electrical, mechanical, optical, and thermoelectric properties that can be exploited in building-blocks of various nanodevices. From CR-AFM measurements of the elastic modulus of as-grown and oxidized Si NWs, we have found that the mechanical properties of oxidized Si NWs can be slightly tuned by the stress developed at the Si-SiO_2 interface during the oxidation process.

CR-AFM measurements were performed on three types of Si NWs: as-grown Si NWs, Si NWs oxidized at 900 °C for 5 min, and Si NWs oxidized at 1000 °C for 5 min. As in the case of the granular topography analyzed above, the correct consideration of the contact geometry is necessary in converting the measured contact resonance frequencies into the elastic modulus. The measurements, as shown in Fig. 3a, were made along a path perpendicular to the NW probed, starting on the substrate, passing over the NW, and ending on the substrate. The contact geometry during these measurements changes from sphere-to-flat on the substrate to sphere-on-cylinder on the NW, with the normal applied force varying across the NW. The parameters needed to resolve these contact

geometries were extracted from regular AFM scans encompassing the NW. At every location, both the deflection of the cantilever and its resonance frequency were recorded as the tip was brought gradually in and out of contact (refer to Fig. 3a). The measured signals were combined to determine the force dependence of the contact stiffness over the range of the applied forces (less than 50 nN). The Hertzian contact mechanics model was then used to fit the load-dependent contact stiffness measured on each NW and determined the radial elastic moduli of the NWs probed.

In the investigated range of NW radii, between 15 nm and 55 nm, for as-grown Si NWs (orientated along $< 112 >$ direction) and fully oxidized Si NWs (those oxidized at 1000 °C for 5 min) no significant variations were observed in their elastic modulus as a function of radius. Their measured elastic moduli, around 160 GPa for as-grown Si NWs and 75 GPa for fully oxidized Si NWs (dotted lines in Fig. 3b), are close to the known values of bulk Si and SiO_2, respectively. Contrary to that, a clear radius dependence was observed for the elastic modulus of Si NWs oxidized at 900 °C for 5 min. With the exact knowledge of the Si core-SiO_2 structure geometry, the elastic modulus of these NWs can be calculated from a core-shell model [11]. However, a good data fit [18] was obtained with a modified core-shell model that includes the contribution from the mechanically modified Si-SiO_2 interface (refer to the gray curve in Fig. 3b with the interface stress specified on the graph). This interface stress originates from the difference between the volumes of consumed Si and newly formed SiO2. As the oxide is progressively formed at the Si-SiO_2 interface, it is pushed outwards and a radial compressive stress is developed at the interface and decays towards the free surface of the oxide [22]. From the radius dependence

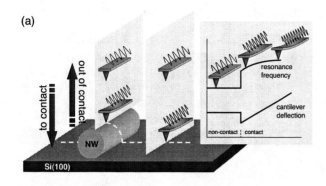

(a)

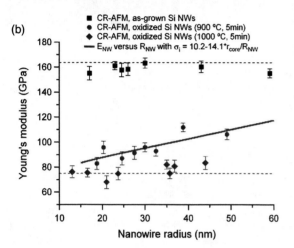

(b)

Figure 3: (a) Load-dependent CR-AFM measurements were performed along a direction perpendicular to an investigated NW. At points along this path (dotted trace), on the substrate as well as over the NW, the resonance frequency and deflection of the cantilever were both recorded as the probe was gradually brought in and out of contact with the sample tested. The resonance frequency changes as a function of material probed (substrate or NW), contact geometry, and applied force. (b) Radius dependence of the elastic modulus of as-grown Si NWs, Si NWs oxidized at 900 °C for 5 min, and Si NWs at 1000 °C for 5 min.

of the elastic modulus determined in our CR-AFM measurements, the interface stress is predicted to decrease with the increase in the NW radius and this is because, as the NW radius increases, the oxide shell thickness becomes thicker and a less deformed oxide is accommodated.

In this application, from CR-AFM measurements performed on the outer surface of oxidized Si NWs, we were able to probe the effect of the buried Si-SiO$_2$ interface on the elastic modulus of these nanostructures. This demonstrates the application of using CR-AFM in probing the mechanical properties of subsurface interfaces.

REFERENCES

[1] G. Binning, C. F. Quate, and C. Gerber, Phys. Rev. Lett. 56, 930, 1986.

[2] U. Rabe, K. Janser, and W. Arnold, Rev. Sci. Instrum., 67, 3281, 1996.

[3] K. Yamanaka and S. Nakano, Jpn. J. Appl. Phys., 35, 3787, 1996.

[4] U. Rabe, M. Kopycinska, S. Hirsekorn, J. Munoz Saldana, G. A. Schneider, W. Arnold, J. Phys. D: Appl. Phys, 35, 2621 (2002).

[5] M. Kopycinska-Muller, R. H. Geiss, J. Muller, D. C. Hurley, Nanotechnology, 16, 703, 2005.

[6] D. Passeri, A. Bettucci, M. Germano, M. Rossi, A. Alippi, Appl. Phys. Lett., 88, 121910, 2006.

[7] D. C. Hurley, M. Kopycinska-Muller, A. B. Kos, R. H. Geiss, Adv. Eng. Mater., 7, 713, 2005.

[8] M. Prasad, M. Kopycinska, U. Rabe, W. Arnold, Geophys. Res. Lett., 29, 31, 2002.

[9] A. Kumar, U. Rabe, S. Hirsekorn, W. Arnold, Appl. Phys. Lett., 92, 183106, 2008.

[10] Y. G. Zheng, R. E. Geer, K. Dovidenko, M. Kopycinska-Muller, D. C. Hurley, J. Appl. Phys., 100, 124308, 2006.

[11] G. Stan, C. V. Ciobanu, P. M. Parthangal, and R. F. Cook, Nano Lett., 7, 3691, 2007.

[12] G. Stan, C. V. Ciobanu, T. P. Thayer, G. T. Wang, J. R. Creighton, K. P. Purushotham, L. A. Bendersky, and R. F. Cook, Nanotechnology, 20, 035706, 2009.

[13] G. Stan and W. Price, Rev. Sci. Instrum., 77, 103707, 2006.

[14] G. Stan, S. W. King, and R. F. Cook, J. Mater. Res., 24, 2960, 2009.

[15] G. Stan and R. F. Cook, Nanotechnology, 19, 235701, 2008.

[16] G. Stan, S. Krylyuk, A. V. Davydov, M. D. Vaudin, L. A. Bendersky, and R. F. Cook, Ultramicroscopy, 109, 929, 2009.

[17] G. Stan, S. Krylyuk, A. V. Davydov, M. Vaudin, L. A. Bendersky, and R. F. Cook, Appl. Phys. Lett., 92, 241908, 2008.

[18] G. Stan, S. Krylyuk, A. V. Davydov, and R. F. Cook, submitted, 2010.

[19] U. Rabe, in "Applied Scanning Probe Methods II", pp 37-90, ed. B. Bhushan and H. Fuchs, Berlin, Springer, 2006.

[20] D. C. Hurley, in "Applied Scanning Probe Methods IX", pp. 97-138, ed. B. Bhushan and H. Fuchs, Berlin, Springer, 2009.

[21] G. Stan and R. F. Cook, in "Scanning Probe Microscopy in Nanoscience and Nanotechnology", pp. 571-611, ed. B. Bhushan, Berlin, Springer, 2010.

[22] D. B. Kao, J. P. McVittie, W. D. Nix, and K. Saraswat, IEEE Trans. Electron. Devices, ED-35, 25, 1988.

Nanoscale Infrared Spectroscopy with the Atomic Force Microscope

Craig B. Prater, Debra Cook, Roshan Shetty and Kevin Kjoller*

*Anasys Instruments
121 Gray Ave. Suite 100 Santa Barbara, CA, US, kevin@anasysinstruments.com

ABSTRACT

The ability to unambiguously identify arbitrary material at the nanoscale using a scanning probe microscope has long been a major research goal of probe microscopy. While the atomic force microscope (AFM) has the ability to measure a range of material properties including mechanical, electrical, magnetic and thermal, the technique has lacked the robust ability to characterize and identify unknown materials. Infrared spectroscopy is a benchmark technique routinely used in a broad range of sciences to characterize and identify materials on the basis of specific vibrational resonances of chemical bonds. We will discuss a new technique (AFM-IR) which integrates the capabilities of AFM and infrared spectroscopy to allow chemical characterization on the micro and nanoscale. We will share the details of the measurement technique including application examples on polymer composites and blends.

Keywords: AFM-IR, nanoscale IR spectroscopy, AFM, infrared spectroscopy, photothermal induced resonance, nanothermal analysis

1 INTRODUCTION

A number of AFM probe-based techniques have been developed to beat the diffraction limit of conventional IR measurements, including near field optical techniques.[1-5] The near field techniques make use of either small apertures which significantly limit the throughput of the light in the IR range or an "apertureless" scattering technique. The "apertureless" techniques suffers from significant background signals and weak scattering signals requiring long integration times and typically limiting it to select samples with ideal properties such as a strong phonon resonance. Another technique to achieve high resolution IR spectroscopy is based on measuring the local temperature rise from spectral absorption through the use of temperature-sensing probes integrated with conventional Fourier Transform IR (FTIR) spectrometers.[6-7]. The thermal technique, although demonstrating high resolution on select samples, more typically has resolution similar to traditional IR microscopy. This is due to the time scale at which the measurement occurs and the thermal diffusion within the material limiting the resolution to the several micron scale.

To our knowledge, however, none of these techniques provide readily interpretable broadband IR spectroscopy with nanoscale resolution.

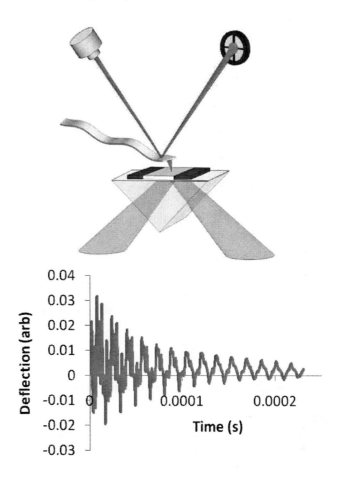

Figure 1. (Top) Schematic diagram of the AFM-IR technique for nanoscale IR spectroscopy. (Bottom) An example cantilever ringdown after excitation by sample IR absorption and rapid thermal expansion.

2 TECHNIQUE

AFM-IR[8-10] uses the tip of an AFM to measure the local photothermal expansion resulting from IR light incident upon a sample. This technique enables the ability to first map the topography of the sample using the contact mode AFM, then obtain a high quality IR spectrum at a selected point in an AFM image, automatically map spectra at an array of points on a sample or create an image of the local absorption of the sample at a fixed wavenumber to enable chemical mapping

Previous work with the AFM-IR technique used a free electron laser as the source of IR radiation. We have employed a pulsed tunable IR source with an AFM. The tunable IR source can be adjusted in terms of the output wavenumber to continuously cover the range from 1200 to 3600 cm⁻¹, with lower power also available from 1000-1200 cm⁻¹ This range covers a major portion of the mid-IR spectrum including important CH, NH and CO bands, as well as the carbonyl and amide I/II bands allowing for extensive characterization of different samples. The pulse length is in the nanosecond range and the repetition rate of the source is 1 kHz.

The pulse of infrared light from the tunable source is directed under a region of a thin sample that is placed on the surface of an IR transparent prism. This configuration causes most of the illumination to reflect back into the prism as the angle of illumination is selected to cause total internal reflection. Thus minimal IR illumination is incident on the AFM probe. When the source is tuned to an absorption band of the sample, the absorbed radiation heats up the local area, resulting in a rapid thermal expansion of the sample. This rapid thermal expansion rapidly deflections the cantilever and excites resonant oscillation modes of the cantilever as shown in Figure 1. The ringdown can be analyzed using Fourier techniques or other modes of analysis to determine the amplitude of the oscillation. The amplitude of this ringdown is proportional to the strength of IR absorption under the AFM tip.

In addition to the IR absorption, Fourier analysis can be used to measure the frequencies of cantilever oscillation in the ringdown which are the contact resonances of the AFM probe in contact with the sample surface. These contact resonances are dependent on the contact stiffness of the sample directly under the AFM tip.[11,12] While the current measurements just provide qualitative values of the stiffness and viscosity, recent progress has been made in extracting quantitative measurements of elastic and viscous properties of materials.[13]

This allows complementary and simultaneous mapping of mechanical properties through measurements of the ringdown resonant frequencies. By combining these measurements with the topography and IR absorption, the sample under study can be characterized for a number of its material properties on the nanoscale. In addition the thermal properties of the sample, such as the melt or glass transition temperature can be determined locally by using nanoscale thermal analysis.

3 RESULTS

The AFM-IR technique works best with thin samples that have sufficient thermal expansion to allow a measurable response. In addition, if the sample is too thick the IR illumination will be absorbed through the thickness of the sample and not reach the surface of the sample to excite the probe. This means the samples need to have a thickness range from a couple of microns down to approximately 100 nm. The samples also need to be placed on the surface of the IR transparent prism. There are a number of ways to accomplish this but the two primary ways are to use ultramicrotomy or to spin or drop cast a

sample out of a solvent solution. Using ultramicrotomy we target sample thicknesses between 100 nm and 1000 nm and then we transfer the sections to the prism surface. We have used many styles of cantilevers, including standard contact mode cantilevers, nanothermal analysis cantilevers, and novel enhanced resonance probes.[14]

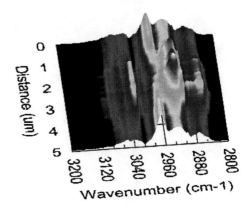

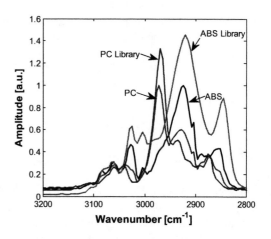

Figure 2. (Top) Spectral line map over a polymer blend containing acrylonitrile-butadiene styrene (ABS) and polycarbonate (PC). (Bottom) Example spectra measured by AFM-IR compared to conventional Fourier Transform Infrared (FTIR) library spectra.

We have used AFM-IR to measure and map a variety of polymer and biological samples. Figure 2 shows measurements performed on a blend of polycarbonate (PC) and acrylonitrile butadiene styrene (ABS). These materials are blended to create a high impact material which provides improved properties including high flow for molding, heat resistance and toughness. Polymer blends like these are used in many applications including automotive and consumer electronics. A number of AFM-IR spectra across the 2800-3200 cm⁻¹ range, which includes the C-H stretch, were made along a line across the sample. The separation between each spectra was ~300 nm which allowed measurements in both the domain and matrix within the material. The top portion of Figure 2 displays a line spectral map which is an image of the IR absorption as a function of wavenumber and position on the sample. Local chemical variations on the sub micron scale are clearly resolved via the change in the position of the CH peak between 2920 and 2970 cm^{-1} as the tip moves from the PC to the ABS domains.

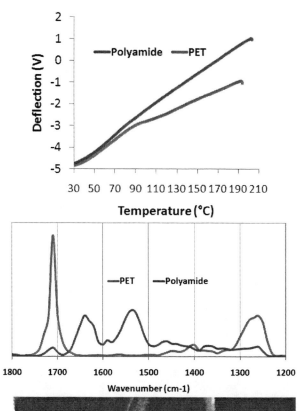

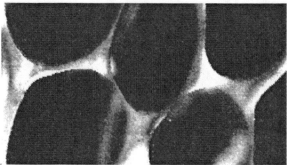

Figure 3. Nano thermal measurements (top), AFM-IR spectra (middle) and a chemical map (bottom) of a PET-nylon composite sample.

Figure 3 demonstrates the benefits of the nano thermal analysis and nano IR spectroscopy on a polymer composite sample. The sample is a microtomed section of a PET fiber bundle embedded in a polymeric matrix, originally misidentified by the material provider as polylactic acid. The sample was initially measured using the thermal analysis. This showed the characteristic glass and melt temperatures of PET on the fibers but did not show the expected transition for the matrix material had it been polylactic acid. The sample was then measured using the

nano IR spectroscopy. Again the PET fibers showed the characteristic absorption bands for PET including a strong carbonyl peak. The matrix had absorption peaks at the Amide I and Amide II. Spectra were exported to BioRad KnowItAll software which suggested that the matrix material was a polyamide (nylon) material. This correlated well with the thermal measurements which showed a melt transition at ~200° C. It was later independently confirmed that the AFM-IR material identification was correct. In addition to measuring spectra at select locations, the sample surface can be mapped in terms of the absorption at a select wavenumber. An example of this is shown in bottom portion of Figure 3. The IR source was tuned to 1640 cm^{-1}, which is the Amide I band, while the probe was scanned across the sample surface over an area of 30 x 15 um. The matrix appears mostly light indicating a stronger absorption in the nylon.

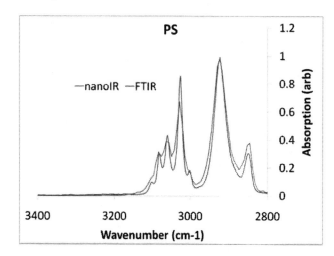

Figure 4. The AFM-IR technique provides spectra of nanoscale areas that have a good correlation to FTIR measurements on bulk samples. This plot shows a comparison of the AFM-IR spectra of polystyrene with a reference spectra measured by FTIR.

One of the primary benefits of the AFM-IR technique is that the technique provides rich spectra that are well correlated to traditional FTIR spectra. Figure 4 shows a comparison between measurements made on a polystyrene sample with the library spectrum from traditional FTIR. Despite the fact that the AFM-IR measurements are performed on length scales orders of magnitude smaller that bulk FTIR, the measurements exhibited generally good correlation between FTIR and AFM-IR. This enables the identification of the material in contact with the probe by importing the AFM-IR spectra into a FTIR database library.

4 CONCLUSION

The AFM-IR technique allows the measurement of the IR absorption of a sample at the nanoscale. In addition, the topography, mechanical and thermal properties can also be measured on the same sample allowing a more complete characterization of the sample. The broad spectral range and the direct correlation of the signal to the IR absorption makes possible the comparison of the AFM-IR spectra with the extensive library of FTIR spectra allowing for the identification of unknown samples.

REFERENCES

1. F. Zenhausern, Y. Martin and H. K. Wickramasinghe, Science **269**, 1083-1085 (1995).
2. B. Knoll and F. Keilmann, Nature **399**, 134-137 (1999).
3. M. S. Anderson, Applied Physics Letters **76**, 3130-3132 (2000).
4. M. Brehm, T. Taubner, R. Hillenbrand and F. Keilmann, Nano Letters **6**, 1307-1310 (2006).
5. M. M. Qazilbash, M. Brehm, B. G. Chae, P. C. Ho, G. O. Andreev, B. J. Kim, S. J. Yun, A. V. Balatsky, M. B. Maple, F. Keilmann, H. T. Kim and D. N. Basov, Science **318**, 1750-1753 (2007).
6. A. Hammiche, L. Bozec, M. J. German, J. M. Chalmers, N. J. Everall, G. Poulter, M. Reading, D. B. Grandy, F. L. Martin and H. M. Pollock, Spectroscopy **19**, 20-42 (2004).
7. A. Hammiche, L. Bozec, M. J. German, J. M. Chalmers, N. J. Everall, G. Poulter, M. Reading, D. B. Grandy, F. L. Martin and H. M. Pollock, Spectroscopy **19**, 20 (2004).
8. A. Dazzi, R. Prazeres, E. Glotin and J. M. Ortega, Optics Letters **30**, 2388-2390 (2005).
9. A. Dazzi, R. Prazeres, F. Glotin and J. M. Ortega, Ultramicroscopy **107**, 1194-1200 (2007).
10. A. Dazzi, R. Prazeres, F. Glotin, J. M. Ortega, M. Al-Sawaftah and M. de Frutos, Ultramicroscopy **108**, 635-641 (2008).
11. Yamanaka, K. and S. Nakano, Applied Physics A: Materials Science & Processing 66(0): S313-S317, 1998.
12. Rabe, U., et al., Applied Physics A: Materials Science & Processing 66(0): S277-S282, 1998.
13. Yuya, P.A., et al., Journal of Applied Physics 104(7): 074916-7, 2008.
14. J. R. Felts, K. Kjoller, D. Cook, C. B. Prater and W. P. King, IEEE MEMS 2010 **submitted**.

Subsurface characterization of high dielectric nanostructures in low dielectric polymer matrix using Electric Force Microscopy and Scanning Electron Microscopy

Minhua Zhao[1], Bin Ming[2], Andras E. Vladar[2], Paul Stutzman[1], Guodong Chen[1], Xiaohong Gu[1], Cheol Park[3], Y.C. Jean[4], and Tinh Nguyen[1]

[1,2]National Institute of Standards and Technology, Gaithersburg, MD 20899, [1]Materials and Construction Research Division, [2]Precision Engineering Division, [3]National Institute of Aerospace, Hampton, [4]Department of Chemistry, University of Missouri, Kansas City

ABSTRACT

Scanning Probe Microscopy (SPM) and Scanning Electron Microscopy (SEM) are two powerful techniques for surface characterization of nanostructures. With the development of nanotechnology, there is a growing need to nondestructively characterize nanostructures at the *subsurface* of materials, such as dispersion of carbon nanotubes (CNTs) in polymer composites, which is beyond the capabilities of conventional SPM and SEM techniques. Here we report subsurface characterization of CNTs in polyimide composites using Electric Force Microscopy (EFM), a special type of SPM based on long-range electrostatic interactions; and Poly-Transparent SEM (PT-SEM) based on the charge contrast at the interface of CNTs and polymers. The effectiveness of these techniques is demonstrated by subsurface imaging of polyimide nano-composites containing different concentrations of CNTs using EFM technique alone and combined EFM and PT-SEM techniques at the same location of the sample. Key experimental parameters are studied and a new contrast mechanism for EFM and SEM subsurface imaging of CNTs in polymer matrix is proposed. Furthermore, these techniques can be universally applied to characterize high dielectric nanostructures in low dielectric matrices, with a broad range of applications in nanotechnology.

Keywords: electric force microscopy, scanning electron microscopy, nanocomposites, subsurface, carbon nanotube.

1 INTRODUCTION

Scanning Probe Microscopy (SPM) and Scanning Electron Microscopy (SEM) are two complementary techniques for high resolution surface characterization. SPM is a branch of microscopy that forms images of surfaces using a physical probe that scans the specimen. An image of the surface is obtained by mechanically moving the probe in a raster scan of the specimen, line by line, and recording the probe-surface interaction as a function of position. The main advantage of a SPM is that samples can be observed in air or liquid environments without the need for partial vacuum. In comparison, SEM is a type of electron microscopy that images the sample surface by scanning it with a high-energy beam of electrons in a raster

scan pattern. The main advantages of SEM over SPM include: (i) much faster scan speed and much larger scan area; (ii) compositional information through x-ray energy-dispersive spectroscopy. With the advent of nanotechnology, there is also a growing need for high resolution subsurface, e.g., understanding the dispersion of carbon nanotubes (CNTs) in a polymer matrix is critical for improving the performance of nanocomposites. While conventional SPM and SEM are mainly used for surface characterization, two specific techniques named Electric Force Microscopy (EFM) [1] and Poly-Transparent SEM (PT-SEM) [2], are promising for high resolution subsurface characterization. EFM is a special type of SPM technique based on long-range electrostatic interactions between a probe and a sample for the characterization of local electronic and/or dielectric properties of insulating, conducting and semi-conducting materials. For instance, EFM was used for subsurface imaging of CNTs dispersed in polymer composites [3]. PT-SEM was also applied to provide a quantitative assessment of CNT dispersion in a polymer matrix through the use of an electric field or charge contrast [2]. In this study, subsurface imaging of polyimide nano-composites containing different concentrations of CNTs is conducted using EFM alone and combined EFM and PT-SEM techniques at the same location in the sample. Key experimental parameters are studied and a new contrast mechanism for subsurface imaging of CNTs in polymer matrix is proposed.

2 EXPERIMENTS*

Single wall carbon nanotube (SWCNT)-polyimide nanocomposite films were prepared by in-situ polymerization as described previously [4]. The purified laser ablated (LA) and high-pressure carbon monoxide decomposition (HIPCO) SWCNTs were purchased from Rice University and Carbon Nanotechnologies, Inc., respectively. The polymer matrix was a transparent polyimide (CP2). As-received anhydrous dimethyl formamide (DMF, Fisher Scientific) was used as a solvent. A series of SWCNT-polyimide nanocomposite films with SWCNT concentrations at mass fractions of (0, 0.05, 0.2, 0.5, 2 and 10) % were prepared as follows. The SWCNT-poly(amic acid) solution was cast onto a glass plate, dried in a dry air-flowing chamber and peeled from the glass substrate. Subsequently, the dried tack-free film was thermally imidized in a nitrogen-circulating oven to obtain a solvent-free SWCNT-polyimide film having a thickness between 25 μm and 65 μm. The smoother interface side (glass-film interface) was used for EFM and SEM imaging.

Figure 1 is a schematic of the EFM setup for subsurface imaging of nanofillers in polymer films based on two-pass scanning techniques. During the first pass, the topographic image was acquired in normal tapping mode using a Veeco Dimension 3100 Atomic Force Microscopy (AFM) equipped with Nanoscope IV controller. The second pass raised the conductive AFM probe above the sample to a fixed distance and re-scanned the surface with a bias voltage applied between the probe and AFM stage, following the previous-recorded topography to maintain a constant tip-sample separation. The amplitude and phase of the probe during the second pass were recorded as EFM signals. The bias voltage applied to the AFM probes ranged from -12 V to +12 V and the lift height was kept at 20 nm. The EFM imaging was carried out using the NIST-patented (US 6.490.913 B1) environmental chamber that was custom-designed and fabricated for the Dimension 3100 AFM.

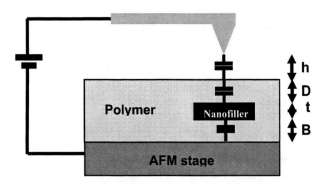

Figure 1: Schematic of EFM setup for subsurface characterization of nanofillers in a free-standing polymer composite film.

The capacitive electrostatic force between the probe and sample is given by Equation (1).

$$F = \frac{1}{2}\frac{dC}{d\lambda}V_{tip}^2 \qquad (1)$$

where F is the electrostatic force, C is the capacitance between the tip and sample, λ is the tip-sample distance, and V_{tip} is the bias voltage on the tip. The change in phase shift, $\Delta\Phi$, of the vibrating AFM probe defined by Equation (2) is recorded pixel by pixel to form an EFM image during the second pass under the approximation of a harmonic oscillator [5, 6] using a conventional capacitive coupling model:

$$\Delta\Phi = \Phi - \Phi_0 \approx tan(\Phi - \Phi_0) \approx \frac{Q}{k}F' = \frac{Q}{2k}(C_1''(z) - C_2''(z))V_{tip}^2 \qquad (2)$$

where Φ is the phase shift when the probe is over the nanofiller (Figure 1), Φ_0 is the phase shift when the probe is over the unfilled film, Q is the quality factor of the AFM

probe, and k is the spring constant of the AFM probe. Assuming a plane-capacitor model [5], Equation (3) describes the second derivative of the tip-nanofiller capacitance, $C_2''(z)$, and Equation (4) describes the second derivative of the tip-polymer film capacitance, $C_1''(z)$,

$$C_2''(z) = 8\pi\varepsilon_0\left[ln(\frac{1+cos\theta}{1-cos\theta})\right]^{-2}\frac{1}{\{h+(D+B)/\varepsilon_{film}+t/\varepsilon_{filler}\}^3} \qquad (3)$$

$$C_1''(z) = 8\pi\varepsilon_0\left[ln(\frac{1+cos\theta}{1-cos\theta})\right]^{-2}\frac{1}{\{h+(D+B+t)/\varepsilon_{film}\}^3} \qquad (4)$$

where ε_0 is the permittivity constant, θ is the AFM tip cone half angle, h is the distance between tip apex and top surface of the film, D is the depth of the nanofiller located from the film top surface, z ($=h+D$) is the tip-nanofiller distance, B is the distance of the nanofiller to the bottom of the film, t is the thickness of the nanofiller, ε_{filler} is the permittivity of the filler, and ε_{film} is the permittivity of the unfilled polymer matrix. According to the Equations 2, 3 and 4 based on the capacitive coupling model, EFM phase contrast of nanofillers can only be revealed when the dielectric constant of the nanofillers is sufficiently different from that of the polymer matrix. However, we will demonstrate later that capacitive coupling model is not adequate to describe the EFM contrast mechanism of nanofillers embedded in polymer matrix.

The same location in the region characterized by EFM was also studied using a FEI Helios NanoLab 600 Dual-Beam system. In this experiment, the beam accelerating voltage was varied between 1 kV, 2 kV, 4 kV, 6 kV, 8 kV and 15 kV, while the beam current was maintained at 43 pA, with the sample at 4.3 mm working distance. Collected SEM images had a horizontal width of 1024 pixels and a height of 768 pixels. A beam dwell time of 40 μs was used for each image pixel. These parameters were used throughout the study to obtain image-to-image consistency.

3 RESULTS AND DISCUSSIONS

3.1 EFM for subsurface characterization

Figure 2 shows subsurface EFM images of SWCNT-polyimide nanocomposite films with different SWCNT loadings. The left, middle and right columns of Figure 2 are height, AFM phase and EFM phase images, respectively. Neither the height nor the conventional AFM phase images show the dispersion of SWCNTs in the film. However, as the SWCNT loadings are increased from 0.05 % to 10 % in the film, more curved line features with dark phase contrast are revealed in the EFM images. Moreover, the control sample CNT-free polyimide film does not have any line features with negative contrast (not shown). Because the dielectric constant of SWCNTs is much higher than that of the polyimide ($\cong$ 4), it is reasonable to obtain negative EFM contrast when SWCNTs are dispersed in the polyimide matrix [3].

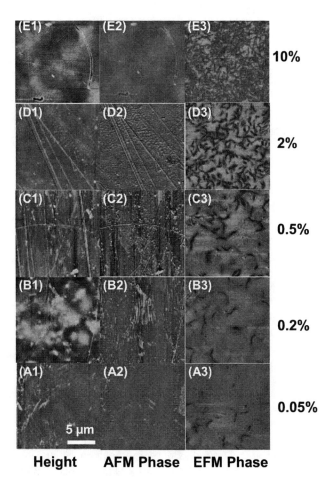

10%

2%

0.5%

0.2%

0.05%

5 µm

Height   AFM Phase   EFM Phase

Figure 2: EFM subsurface imaging of SWCNT (HIPCO) - polyimide composites with SWCNT mass fraction of (A) 0.05 %, (B) 0.2 %, (C) 0.5 %, (D) 2 % and (E) 10 %. Left column: Height, scale 50 nm. Middle column: AFM phase, scale 40°. Right column: EFM phase, scale 40°.

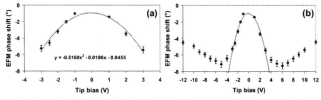

Figure 3: Effect of the bias voltage on the EFM phase signal. (a) Relationship of EFM phase signal to the bias voltage in the range of -3 V to +3V. The line is a theoretical fitting by parabolic law. (b) Relationship of EFM phase signal to the bias voltage in the range of -12 V to +12 V. The line is the same parabolic fitting in (a). Uncertainty bars represent one standard deviation.

A key parameter for EFM subsurface imaging is the applied voltage between the probe and sample. The characteristic parabolic shape of phase shift vs. tip bias is observed at low bias range (Figure 3a), which conforms to the conventional model of capacitive coupling between the probe and sample as defined by Equation (2). However, at a higher bias, the voltage dependence of EFM phase shift is no longer parabolic as shown in Figure 3b. More interestingly, the EFM phase shift is even reduced with the increase of bias voltage beyond a certain threshold, which is dependent on the sample and lift height. These experimental findings cannot be explained by the conventional capacitive coupling defined by Equation (2). Hence, a new mechanism to explain the voltage dependence of EFM phase contrast is required. Since most of the previous EFM work dealt with conductive samples, there is little understanding of EFM contrast mechanisms between high dielectric constant nanostructures (e.g., CNTs) and a low dielectric constant matrix (e.g., polymer film). When a biased tip is placed very close to a low dielectric film, charge injection into the film is possible [7]. The effect of charge injection on the EFM signal [8, 9] can be accounted for using Equations (5) & (6),

$$F_{Electrostatic} = F_{Capacitive} + F_{Columbic} = \frac{V^2}{2}\frac{\partial C}{\partial z} - \frac{Q_s C V}{4\pi\varepsilon_0 z^2} - \frac{Q_s^2}{4\pi\varepsilon_0 z^2} \quad (5)$$

$$\Delta\phi \propto \frac{\partial F}{\partial z} = \frac{V^2}{2}\frac{\partial^2 C}{\partial z^2} + \frac{Q_s V}{2\pi\varepsilon_0 z^2}[\frac{C}{z} - \frac{1}{2}\frac{\partial C}{\partial z}] + \frac{Q_s^2}{2\pi\varepsilon_0 z^3} \quad (6)$$

where V is the bias voltage, C is the capacitance between the tip and sample, z is the tip-sample distance, Q_s is charge injected into the sample and ε_0 is the permittivity constant. Electrostatic interactions between the probe and sample are comprised of capacitive and columbic coupling. The first term in Equation (5) accounts for capacitive coupling, while the others are related to columbic coupling. If the injected charge Q_s and/or tip-sample capacitance C is voltage dependent [10, 11], the relationship between EFM phase shift $\Delta\Phi$ and tip bias V will no longer be parabolic according to Equation (6), which can explain why the EFM contrast could even be reduced at large tip bias as shown in Figure 3b. Furthermore, the EFM contrast (top row) of CNTs in polymer composites changed from dark to bright with an increase of the tip bias voltage as shown in Figure 4. Interestingly, similar inversion of SEM contrast was observed on CNTs when the SEM acceleration voltage was increased, which will be shown later.

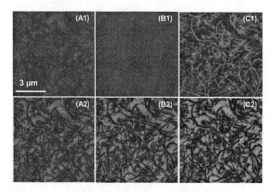

Figure 4: Inversion of EFM contrast of CNTs in 10 % SWCNT (HIPCO)-polyimide composite film by bias

voltage at lift height of 20 nm. Top row: EFM amplitude signal with 100 mV scale. Bottom row: EFM phase signal with 20° scale. (A1)&(A2): -3 V bias; (B1)&(B2): -4 V bias; (C1)&(C2): -5 V bias.

3.2 Combined EFM and PT-SEM

Figure 5 shows the combined EFM and PT-SEM subsurface imaging of CNT-polyimide composites at the same location. While conventional SPM height image (Fig.5a) could not detect the subsurface CNTs, both EFM (Fig.5b) and PT-SEM (Figs.5c-5d) can reveal the CNTs embedded below the surface. Furthermore, the PT-SEM contrast of CNTs was reversed from dark to bright with the increase of acceleration voltage, similar to what was observed in EFM imaging (Fig. 4). It is hypothesized that the contrast mechanism of subsurface imaging of CNTs by PT-SEM is related to the charge accumulated or dissipated at the interface between the CNTs and polymer matrix. The similarity of PT-SEM and EFM contrast reverse implies that there is a common mechanism for both techniques, in which the interface of CNT/polymer may play an important role in charge accumulation and dissipation during subsurface imaging. It is worth mentioning that PT-SEM only works when the conductivity of the sample is within a certain range ($\sim 10^{-7}$-10^{-1} S/m) and no conductive coating is applied at the surface. Furthermore, more subsurface CNTs are revealed at higher SEM acceleration voltage (Figs.5c & d), which is due to deeper electron-sample interaction region at higher acceleration voltage. More research is going on to verify this mechanism postulation.

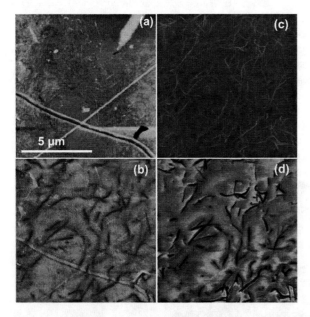

Figure 5: Combined EFM and PT-SEM subsurface imaging at the same location of SWCNT(LA)-polyimide composite film. (a) AFM height scale 50 nm, (b) EFM phase scale 20 °, (c) SEM image at acceleration voltage of 8 kV, (d) SEM image at acceleration voltage of 1 kV.

4 CONCLUSIONS

Both EFM and PT-SEM have been successfully applied to characterize the CNTs dispersed in the subsurface regions of a nanocomposite. A new contrast mechanism based on interface charge accumulation and dissipation has been proposed to account for the experimental findings. These techniques can be universally applied to characterize high dielectric nanostructures embedded in low dielectric matrix, with a broad range of applications in nanotechnology. Further optimization of these techniques is ongoing.

REFERENCES

1. Zhao, M.H., et al., Nanotechnology. **19**, 235704, 2008.
2. Lillehei, P.T., et al., Nanotechnology. **20**, 325708, 2009.
3. Zhao, M.H., et al., Nanotechnology. **submitted**, 2010.
4. Park, C., et al., Chemical Physics Letters. **364**, 303, 2002.
5. Bockrath, M., et al., Nano Letters. **2**, 187, 2002.
6. Staii, C., A.T. Johnson, and N.J. Pinto, Nano Letters. **4**, 859, 2004.
7. Terris, B.D., et al., Journal of Vacuum Science & Technology a-Vacuum Surfaces and Films. **8**, 374, 1990.
8. Baird, P.J.S., J.R. Bowler, and G.C. Stevens. *Quantitative methods for non-contact electrostatic force microscopy.* **163**, 381, 1999.
9. Jespersen, T.S. and J. Nygard, Nano Letters. **5**, 1838, 2005.
10. Dragoman, D. and M. Dragoman, Journal of Applied Physics. **101**, 3, 2007.
11. Heo, J.S. and M. Bockrath, Nano Letters. **5**, 853, 2005.

Characterization of Metallic Nanowires by Combining Atomic Force Microscope (AFM) and Scanning Electron Microscope (SEM)

E. Celik[*] and E. Madenci[**]

[*]The University of Arizona, Tucson, AZ, 85721, USA, emrah@email.arizona.edu
[**]The University of Arizona, Tucson, AZ, 85721, USA madenci@email.arizona.edu

ABSTRACT

A new experimental method is introduced in order to characterize the mechanical properties of metallic nanowires. An accurate mechanical characterization of nanowires requires simultaneous imaging using scanning electron microscope (SEM) and mechanical testing with an atomic force microscope (AFM). In this study, an AFM is located inside an SEM chamber in order to establish the visibility of the nanowires. The tip of the AFM cantilever is utilized to bend and break the nanowires. Nanowire specimens are prepared by electroplating of metal ions into the nano-pores of the alumina membranes. Mechanical properties are extracted by using existing analytical formulations along with the experimental force versus bending displacement response. Preliminary results revealed that copper nanowires have unique mechanical properties compared to their bulk counterparts.

Keywords: atomic force microscope, nanomechanics, nanowire, scanning electron microscope

1 INTRODUCTION

Nanowires offer broad range of applications due to their unique properties especially in the fields of nano-electronics and computing technology. Mechanical characterization of nanowires is necessary in order to establish the feasibility of replacement of current electrical, mechanical and optical devices with nanowire integrated devices based on the reliability requirement. Although the characterization of electrical and optical properties of different types of nanowires has been studied extensively, mechanical characterization of nanowires still poses challenges to many existing testing and measurement techniques because of their small length scale. Manipulation and installation of individual nanowires in a test setup is rather challenging. Therefore, new mechanical characterization methods must be developed to quantify the mechanical properties of these nanostructures.

Existing nanowire mechanical characterization techniques in the literature can be grouped into three major categories; resonation of nanowires in an electron microscope, mechanical characterization via axial loading and bending of nanowires using atomic force microscope.

The first technique involves the oscillation of nanowires inside an electron microscope by applying alternating voltage and examining the corresponding natural frequencies of nanowires [1-3]. Frequencies are then related to mechanical properties. Although this method is easy to perform, it is limited only to elastic properties of nanowires. Therefore, the yield and failure stresses, deformation and defect initiation mechanisms cannot be obtained. Mechanical characterization of nanowires has been performed by applying axial loading as well [4, 5]. In this method, nanowires or nanotubes are clamped in between two cantilevers and subjected to tension or compression loads in electron microscope. This technique requires significant amount of specimen preparation time and is not practical for testing of high number of nanowire specimens. The last group of mechanical characterization techniques involves bending of nanowires by using the tip of a commercially available AFM cantilever and this technique has been used successfully by several researchers [6-8]. However, commercial AFMs used in these studies are not capable of real time imaging of nanowires during testing. The correct alignment of nanowires according to AFM test probe poses a challenge due to the lack of nanowire visibility during testing. Also, deformation mechanism of nanowires cannot be investigated during testing. Therefore, there is a need for a new testing technique for mechanical characterization of nanowires that provides coexistence of two components of experimental analysis which are imaging and mechanical testing. A custom AFM placed in SEM for improved visibility can satisfy this requirement.

This study presents a new experimental test setup for bending test of nanowires involving a custom AFM and a SEM. The AFM is placed inside the SEM chamber; this high magnification imaging capability allows for correct alignment and placement of the specimens. Metallic nanowires with different sizes are fabricated using electroplating. Bending deformation measurements of nanowires concern different materials in different sizes.

2 MATERIALS AND METHODS

2.1 Specimen Preparation

Metallic nanowire specimens are prepared by electro-deposition of metal ions into nanopores of commercially available Whatman alumina membranes. Up to date, only copper nanowires have been grown in porous membranes with 200 nm pore diameter with this technique. A thin layer of metal is sputtered on one side of the alumina membrane as a contact layer onto which the nanowires grow. During electroplating, copper ions existing in electrolyte solution (0.5 M CuSO4) are deposited into the nanopores of the alumina membrane cathode. For the electroplating of copper, cathode voltage is set to -0.1 mV relative to the reference electrode and deposition duration is varied between 5 to 30 minutes which results in the deposition of nanowires in different lengths. After the nanowires are electroplated into the alumina membrane, the membrane is dissolved in 3M NaOH solution for 1 hour. Then, nanowires attached to the bottom metal layer are washed with distilled water and dissolved in toluene in ultrasonic bath. The vibration in the bath breaks the bonds between the nanowires and the thin metal layer and a homogenous distribution of nanowires in toluene is obtained. Nanowires dissolved in toluene solution are then spread over a silicon substrate. The substrate involves pre-etched trenches with the thickness range of 5 μm to 10 μm and the depth of 5 μm. After evaporation of toluene, some of the nanowires are suspended over the trenches and used for mechanical testing. The clamping of nanowires suspending on the trenches can be obtained by conformal deposition of a thin metal layer over the nanowires. The deposited thin metal film provides mechanical support for the mechanical testing of nanowires. After the nanowire specimens are prepared and clamped over the trenches of silicon substrate, they are tested using a custom AFM located inside SEM chamber.

2.2 Experimental Setup

The experimental setup combines the mechanical testing ability of AFM with imaging capability of SEM. The concept of locating a custom made AFM inside a SEM chamber is shown in Fig. 1.a. The detailed view of the custom AFM is given in Fig. 1.b. Main components of the setup are numbered in Fig. 1.b as follows:

1.	AFM head	5.	Piezoactuator
2.	Micromotor	6.	Mirrors
3.	Laser diode	7.	Manual control knobs
4.	Photodiode		

Electronic components are connected to a computer via National Instruments (NI) data acquisition boards. NI's Labview software is utilized in order to acquire photodiode signal and to control micromotor, laser diode and piezoactuator from outside the SEM chamber. Laser beam is used to monitor the deflection of the AFM cantilever during mechanical testing. In Fig. 1.b, the path of the laser beam from laser source to the photodetector is sketched. The laser beam is reflected by two mirrors before reaching the AFM cantilever. The third reflection of the laser occurs at the back of the cantilever and the reflected beam is

directed to the final mirror before reaching to the quadrant photodiode.

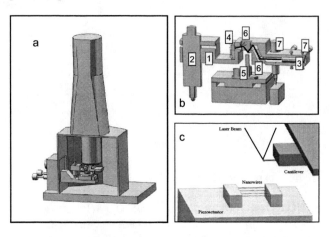

Figure 1: Experimental setup for mechanical testing of nanowires a) Location of AFM in SEM chamber, b) Custom AFM design, c) Detail of cantilever and nanowires

The tip of the AFM cantilever is positioned at the center of the nanowire located on the piezoactuator with respect to the nanowires. At this stage of the test, AFM cantilever and the nanowires are brought side by side without touching each other. After the vertical alignment is completed, the piezoactuator and the sample are moved towards the AFM tip, perpendicular to the nanowire axis establishing physical contact and causing the bending of the nanowires at the contact point by AFM tip. The lateral force acting between the nanowire and the AFM tip results in twisting of the cantilever and therefore changes the position of the laser beam on photodiode. The application of further load causes the nanowire to fracture and the laser beam goes back to original non-contact position on photodiode.

The change in the deflection of the AFM cantilever and the position of the laser beam on photodiode leads to change of the photodiode voltage which is converted into the data of force versus bending displacement of the nanowires in order to extract mechanical properties. The force exerted on nanowires can be obtained by multiplying the lateral deflection and the lateral stiffness of the AFM cantilever as given by Eq. (1).

$$F = k \times \delta_{cantilever} \qquad (1)$$

where k is lateral stiffness and $\delta_{cantilever}$ denotes the lateral deflection of the AFM cantilever. The lateral stiffness of the cantilever is obtained by using the dimensions and the shear modulus of the AFM cantilever. In order to obtain bending displacement of the nanowire at the contact point, deflection of the cantilever is subtracted from the total piezo displacement as follows;

$$\delta_{bending} = \delta_{piezo} - \delta_{cantilever} \qquad (2)$$

where $\delta_{bending}$, and δ_{piezo} represent bending displacement, and piezo displacement respectively. By using Eqs. (1) and (2), force versus bending displacement data can be obtained experimentally and these data are compared against the analytical formulations/models in order to extract mechanical properties of the nanowires.

In linear beam bending theory, the force acting on the mid-section of a double clamped beam is related to the bending displacement at the same location as follows;

$$F = \frac{192EI}{L^3} \delta_{bending} \qquad (3)$$

where E, I, L are Young's modulus, moment of inertia and length of the clamped portion of the beam. Therefore, the force is assumed to change linearly as a function of bending displacement. However, if the deformation of the nanowire beam is much larger than the diameter of the beam, the axial force acting on the beam leads to nonlinearity in force versus displacement variation. In this case, force can be defined by the approximate solution to the force/bending displacement expression according to [9].

$$F = \frac{192EI}{L^3} \delta_{bending} \left(1 + \frac{A}{24I} \delta^2_{bending} \right) \qquad (4)$$

The elastic limit of the nanowires can be estimated by using the linear and nonlinear analytical equations described above. The plastic zone is achieved after which the force versus bending displacement curve starts deviating from the predicted responses defined by Eqs. (3) and (4). The force at the deviation location is defined as the yielding force and at this point the yield stress can be described as follows according to [10].

$$\sigma_y = \frac{F_y L}{4\pi r^2 \delta_y} \qquad (5)$$

where r denotes the nanowire radius and F_y and σ_y are yielding force and yield stress respectively. δ_y is the bending displacement at the yield point. Similarly ultimate stress can be calculated by the equation described by [10].

$$\sigma_u = \frac{F_u L}{4\pi r^2 \delta_u} \qquad (6)$$

where and F_u, δ_u and σ_u are force, bending displacement and the ultimate stress at the failure point, respectively.

3 RESULTS

Bending displacement versus force relations obtained in four independent tests on copper nanowires are given in Fig. 2. According to these results, the force increases nonlinearly and decreases rapidly after the fracture of nanowires. The fracture points where the maximum force is achieved are marked by the arrows in the figure.

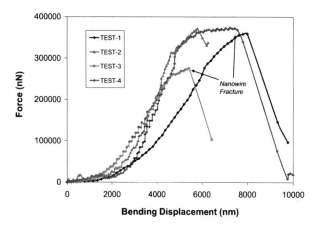

Figure 2: Force versus bending displacement variation.

In order to extract elastic modulus, force-bending displacement response of the nanowires obtained experimentally was compared against the predicted response described by Eq. (4). The elastic modulus values were obtained by matching the experimental and the predicted values. An example of the comparison between two data is given in Fig. 3. Force versus bending displacement data shows a remarkable agreement with the analytical model until the nanowire starts yielding, and the two responses deviate from each other.

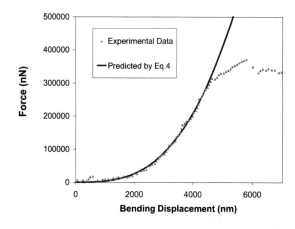

Figure 3: Comparison of experimental data and predictions.

The elastic modulus values for copper nanowires with 200 nm diameter were obtained between 2 to 3.5 GPa with an average of 2.7GPa. The elastic modulus of bulk copper is given as 110 GPa in the literature. The elastic moduli of the copper nanowires were observed to be significantly lower than that of the bulk material.

	Averaged Measured Value	Literature Value for Bulk Copper
Elastic Modulus (GPa)	2.7	110
Yield Stress (GPa)	2.3	0.05-0.3
Ultimate Stress (GPa)	2.8	0.23-0.38

Table 1: Averages of the mechanical properties obtained for 200 nm copper nanowires.

The yield stress values were calculated by using the Eq. (5) in four different tests and the results are given in Table 1. The yield stress of copper nanowires with 200 nm diameter was obtained between 1.01 and 3.09 GPA with an average value of 2.3 GPa. Average value of yield stress obtained in four tests is nearly an order of magnitude larger than the yield stress of the bulk copper which is given in literature within the range of 55-300 MPa.

The ultimate stress values of the nanowires were calculated by using Eq. (6) and the experimental data. The ultimate stress of copper nanowires was obtained between 2.32 and 3.31 GPA with the average value of 2.83 GPa (Table 1). The ultimate stress of bulk copper is given in the literature in the range of 230-380 MPa. The average value of ultimate stress obtained in this study is also an order of magnitude larger than the ultimate stress of bulk copper.

4 CONCLUSIONS

The preliminary results indicate that the elastic modulus (2.7 GPa) for copper nanowires with 200 nm diameter is significantly lower than that of the bulk material (110 GPa). This difference in elastic moduli is unexpected; therefore, it needs further investigation. Both the yield and ultimate strengths of the nanowires are an order of magnitude higher than those of the bulk copper. Higher mechanical strength can be explained due to reduced defect density at the nanoscale compared to the bulk scale [9]. High strength and durability of these nanowires in addition to their size advantages of copper nanowires will make them excellent candidates for many future applications.

REFERENCES

[1] Z. L. Wang, "Mechanical properties of nanowires and nanobelts", Dekker Encyclopedia of Nanoscience and Nanotechnology, 1773-85, 2004.

[2] Y. Huang, X. Bai and Y. Zhang, "In situ mechanical properties of individual ZnO nanowires and the mass measurement of nanoparticles", Journal of Physics: Condensed Matter, 18, 179-84, 2006.

[3] C. Q. Chen , Y. Shi, Y. S. Zhang, J. Zhu, and Y. J. Yan, "Size dependence of Young's modulus in ZnO nanowires", Physical Review Letters, 96, 075505, 2006.

[4] M. F. Yu, O. Lourie, M. J. Dyer, K. Moloni, T. F. Kelly, R. S. Ruoff, "Strength and Breaking Mechanism of Multiwalled Carbon Nanotubes Under Tensile Load", Science, 287, 637-40, 2000.

[5] C. H. Lin, M. Chang, X. Li and J. R. Deka, "Measurement of Mechanical Properties of Boron Nanowire Using Nanomanipulation System", Proceedings of the 35th International MATADOR Conference, 275-278, 2007.

[6] S. Paulo, J. Bokor, R. T. Howe, R. He, P. Yang, D. Gao, C. Carraro, and R. Maboudian, "Mechanical elasticity of single and double clamped silicon nanobeams fabricated by the vapor-liquid-solid method", Applied Physics Letters, 87, 053111, 2005.

[7] H. Ni, X. Li, and H. Gao, "Elastic modulus of amorphous SiO2 nanowires", Applied Physics Letters, 88, 043108, 2006.

[8] B. Wu, A. Heidelberg and J.J. Boland, "Mechanical properties of ultrahigh-strength gold nanowires", Nature Materials, 4, 525-9, 2005.

[9] A. Heidelberg, L. T. Ngo, B. Wu, M. A. Phillips., S. Sharma, T. I. Kamins, J. E. Sader, and J. J. Boland, "A Generalized Description of the Elastic Properties of Nanowires", Nano Letters, 6, 1101-6, 2006.

[10] D. Almecija, D. Blond, J. E. Sader, J. N. Coleman, and J. J. Boland, "Mechanical properties of individual electrospun polymer-nanotube composite nanofibers, Carbon, 47, 2253-8, 2008.

Imaging three-dimensional microstructures fabricated by two-photon polymerization using CARS microscopy

T. Baldacchini[*], R. Zadoyan

[*]Technology and Applications Center, Newport Corporation
1791 Deere Avenue, Irvine, CA 92606
tommaso.baldacchini@newport.com

ABSTRACT

Two-photon polymerization (TPP) is an enabling technology for the fabrication of three-dimensional microstructures with sub-micron resolution. Because of its unique set of advantages in respect of other microfabrication techniques such as photolithography, TPP has been employed in the last ten years for the manufacturing of devices in as diverse fields as microfluidics, microelectronics, photonics, and bioengineering. In this article, we exploit the imaging capabilities of CARS microscopy as a diagnostic tool for microstructures fabricated by TPP.

Keywords: photopolymerization, two-photon absorption, microfabrication, vibrational imaging.

1 INTRODUCTION

Among the arsenal of unconventional microfabrication techniques now available to scientists, two-photon polymerization (TPP) is unique for the fabrication of three-dimensional microstructures [1]. It is capable of producing geometries with no topological constrains and with resolution smaller than 100 nm. Because of its versatility, TPP has been employed to create devices previously impossible to manufacture with conventional microfabrication procedures such as photolithography. For example, by taking advantage of specific chemical moieties in the materials used in TPP, three-dimensional microstructures coated with conductors and semiconductors were fabricated delivering devices with distinctive and practical features for uses in microelectronics and nanophotonics [2, 3]. Furthermore, TPP is being used with promising results in biomedical applications by producing devices for drug delivery and tissue engineering [4].

TPP offers a unique combination of advantages: when fabricating three-dimensional microstructures, no topological constrains apply; sacrificial layers are not required in order to fabricate movable parts; by using laser intensities just above the polymerization intensity threshold, sub-diffraction-limited resolution can be attained. Although these features have been exploited successfully in diverse applications, there are still some fundamental questions that need to be answered. For example, how does the laser or sample scanning pattern used to fabricate a specific microstructure influence its mechanical properties? How can we optimize experimental conditions to minimize microfabrication time while maintaining structural integrity? At which dimensions do we have to consider the properties of the polymeric microstructures different from those of the bulk polymer?

Recently we have shown that coherent anti-Stokes Raman scattering (CARS) microscopy can be used to image microstructures fabricated by TPP and, more importantly, we have shown that some of the aforementioned questions can be answered by using this imaging technique [5, 6]. CARS microscopy is a nonlinear imaging technique capable of providing contrast with chemical selectivity [7]. Furthermore, because of the nonlinear optical nature of its light absorption, CARS microscopy allows for three-dimensional imaging.

Among a variety of approaches for performing CARS microscopy, we have chosen multiplex CARS (mCARS) microscopy [8]. The reason is that the experimental setup of mCARS shares the same laser source required for performing TPP. Thus, integrating TPP and mCARS microscopy in one experimental setup give us the opportunity to quickly investigate the process of TPP in situ.

2 MICROFABRICATION AND IMAGING

2.1 Two-Photon Polymerization

Two-photon polymerization is based on the nonlinear interaction of light with a photosensitive material (resin). Typically, near-infrared photons are used to induce two-photon absorption in molecules (photoinitiators) in the resin capable of starting a polymerization process. Since most photoinitiators posses small two-photon cross-sections, strong focusing lenses and ultrashort pulsed lasers are employed to increase the probability of this unlikely event. Under these conditions two-photon absorption and subsequent polymerization can be localized in a volume as small as a femtoliter. The ability of TPP to spatially confine matter transformation is the key step for three-dimensional microfabrication. Indeed, if the resin can be solidified by two-photon polymerization, complex geometries can be patterned with high precision by means of accurate positioning of the laser focal point. As long as the solubility properties of the solidified and unsolidified resin are different, the non polymerized material can be washed away to leave the freestanding structures.

2.2 Coherent anti-Stokes Raman scattering microscopy

Coherent anti-Stokes Raman scattering microscopy provides three-dimensional imaging with contrast based on molecular vibration. In order to generate CARS, two laser beams are focused in the sample, a pump beam at frequency ω_p and a Stokes beam at frequency ω_s. Vibrational sensitivity is obtained when the difference frequency $\omega_p - \omega_s$ matches the frequency of a Raman active vibrational mode of the sample. The anti-Stokes signal at $\omega_{as} = 2\omega_p - \omega_s$ is coherently driven by the two laser beams, resulting in coherent radiation. Traditionally, CARS microscopy has been performed by using two synchronized picosecond lasers as sources of pump and Stokes beams. In this way, imaging with high spectral resolution can be obtained. To acquire images in different spectral regions of the Raman spectrum of the investigated sample, the output wavelengths of the two lasers have to been adjusted accordingly. Alternatively, mCARS microscopy allows gathering the entire vibrational spectrum of the sample at once. In mCARS microscopy, the pump pulse has a narrow bandwidth which defines the spectral resolution of the technique. The Stokes pulse is spectrally broad, usually in the femtosecond regime. Application of the pump and Stokes beams excites simultaneously multiple Raman transitions within the bandwidth of the Stokes pulse. Thus, in a single shot the entire CARS spectrum of the sample exited states is generated.

3 EXPERIMENTAL SETUP

The experimental setup consists of three major parts (Figure 1a). A femtosecond oscillator (800 nm, 100 fs, 80 MHz), a unit for the generation of a pump and Stokes beams, and a microscope equipped with a laser scanning system. The microscope is completed with a detection scheme for collecting forward propagating signal by means of a photomultiplier tube (PMT). A detailed schematic for the unit which generates the pump and Stokes beams is shown in Figure 1b.

The broadband CARS unit is based a photonic-crystal fiber (PCF) which has a zero-dispersion wavelength at 800 nm. A portion of the laser output is used to generate supercontinuum in the photonic-crystal fiber. The near-infrared part of the supercontinuum (longer than 800 nm) is used as Stokes pulse. The remaining part of the laser output is spectrally narrowed to 3 nm by a bandpass filter and is used as pump beam. Pump and Stokes beams are then overlapped both in space and in time with the aid of a razor edge long pass filter (RELP) and a delay line. Each arm of the setup has a variable attenuator which allows independent control of intensity and polarization. The collinearly propagating beams are directed into the

microscope where a pair of galvanometric mirrors is used to scan the sample. One focusing objective is used for both imaging and fabrication (40x, 0.75 NA).

TPP is performed by removing the 3 nm bandpass filter from the path of the pump beam and by blocking with a shutter the Stokes beam. Three-dimensional structures are patterned with the aid of the galvos and a piezo-focusing stage. The sample consists of an acrylic-based photoresist that undergo radical polymerization when irradiated by femtosecond laser source centered at 800 nm [9]. After fabrication, the microstructures are revealed by washing away the unsolidified portion of the sample with an organic solvent.

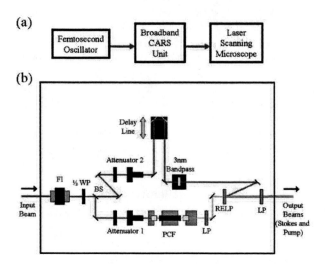

Figure 1 – (a) Block diagram of the experimental setup employed for both TPP and mCARS microscopy, and (b) schematic drawing of the broadband mCARS unit. FI: Faraday isolator; ½ WP: half-wave plate; BS: beam splitter; PCF: photonic crystal fiber; LP: long-pass filter; RELP: razor edge long pass filter.

4 RESULTS AND DISCUSSIONS

Scanning electron microscopy (SEM) images of representative three-dimensional microstructures fabricated by TPP are shown in Figure 2. The truncated cone with square cross-section in Figure 2(a) was created by stacking in the z direction square patterns with smaller and smaller diameters. A separation between the different stacks of 3 μm was selected and a total height of 60 μm was reached. The laser average power (as measured before the shutter), stages velocity, and excitation wavelength used for the fabrication process were 10 mW, 10 μm/s, and 775nm, respectively. A more complex microstructure is depicted in Figure 2(b). Two towers were fabricated one inside the other; a larger one with a hexagonal base and a smaller one with a square base. At the top of each microstructure, two freestanding beams connect adjacent sides. Since the outside tower is 40 μm tall while the insider tower is 20 μm tall, there is no overlap among the two sets of beams. The experimental conditions used for the fabrication of the microstructure in Figure 2(a) were repeated for the fabrication of the microstructure in Figure 2(b).

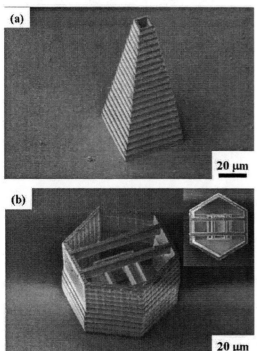

Figure 2 – SEM images of microstructures fabricated by TPP. The samples were tilted 45° in order to reveal the tree-dimensional nature of the microstructures. A top view of the microstructure in (b) is shown in the inset. The polymeric beams that run across the towers in Figure 2(b) are 4 µm wide.

In order to assess the capabilities of the broadband CARS unit, a test sample was fabricated by TPP. First a set of four walls 35 µm long and 20 µm tall were fabricated. Then, a cross-like structure 20 µm tall was patterned within the walls. Finally, the cube was closed by fabricating a thin roof on top of it. A schematic diagram of the microstructure is depicted in Figure 3(a); it consists of a cube with four chambers filled with unpolymerized photoresist. A scanning electron microscope of the microstructure is shown in Figure 3(b).

Since the photoresist used in this experiment is rich in carbon-hydrogen (C-H) bonds, we placed in front of the PMT used in collecting forward CARS signal a 10 nm bandpass filter centered at 650 nm. This spectral range corresponds to the CARS signal originating from the C-H stretching mode (~ 3000 cm^{-1}). Thus, when imaging the structure shown in Figure 3(b) with broadband CARS microscopy, the signal intensity is correlated with the local density of C-H bonds.

Figure 3(c) is a xy cross-section CARS image of the cube recorded at a height of 18 µm from the substrate. Stronger signal is observed from the walls and cross-like

structure than from the inside of the pockets that contain unpolymerized material. Furthermore, because of the proximity of the scanned area with the roof of the microstructure, part of the signal in image shown in Figure 3(c) arises from the roof itself. Upon TPP, the photoresist degree of cross-linking increases enormously. Thus, the density of the polymerized photoresist is higher than the density of the unpolymerized photoresist. Broadband CARS microscopy has revealed this material properties change with high spatial resolution, rapidly, and by using the same experimental setup employed for TPP.

Figure 3 – (a) Schematic drawing of the test microstructure fabricated by TPP. (b) SEM image of the microstructure with the sample tilted at an angle. Inset: top view of the microstructure. (c) mCARS image of the microstructure relative to its C-H stretching modes (LUT shown on far right). The scale bar in (b) and (c) are 25 µm and 20 µm, respectively.

5 CONCLUSION

We have demonstrated the use of broadband CARS imaging for the microscopic characterization of three-dimensional microstructures fabricated by TPP. Both microfabrication and imaging were performed on the same experimental setup allowing for fast in situ monitoring of TPP. This result demonstrates the diagnostic potential of CARS microscopy in analyzing written patterns in the native photoresist in real time.

REFERENCES

[1] LaFratta, C. N.; Fourkas, J. T.; Baldacchini, T.; Farrer, R. A.; *Angew. Chem. Int. Ed.* **2007**, *46*, 6238.

[2] Farrer, R. A.; LaFratta, C. N.; Praino, J.; Naughton, M. J.; Saleh, B. E. A.; Teich, M. C.; Fourkas, J. T. *J. Am. Chem. Soc.* **2006**, *128*, 1796.

[3] Ledermann, A.; Cadematiri, L.; Hermatschweiler, M.; Toninelli, C.; Ozin, G. A.; Wiersma, D. S.; Wegener, M.; von Freymann, G. *Nat. Mater.* **2006**, *5*, 942.

[4] Tayalia P.; Mendonca, C. R.; Baldacchini, T.; Mooney, D. J.; Mazur, E. *Adv. Mater.* **2008**, *20*, 4494.

[5] Baldacchini, T.; Zimmerley, M.; Potma, O. E.; Zadoyan, R. *Proc. SPIE* **2009**, *7201*, 72010Q-1.

[6] Baldacchini T.; Zimmerley, M.; Kuo, C.-H.; Potma, E. O.; Zadoyan, R. *J. Phys. Chem. B* **2009**, *113*, 12663.

[7] Evans, C. L.; Xie, S. *Annu. Rev. Anal. Chem.* **2008**, *1*, 883.

[8] von Vacano, B.; Meyer, L.; Motzkus, M. *J. Raman Spectrosc.* **2007**, *38*, 916.

[9] Baldacchini, T.; LaFratta, C. N.; Farrer, R. A.; Teich, M. C.; Saleh, B. E. A.; Naughton, M. J.; Fourkas, J. T. *J Appl. Phys.* **2004**, *95*, 6072.

Individual NW Chemical Composition Depth Profiling and Surface Oxidation Analyzed by Auger Electron Spectroscopy

U. Givan*, J. Hammond**, D. Paul** and F. Patolsky*

*Tel-Aviv University, Tel-Aviv, Israel, urigivan@post.tau.ac.il.
** Physical Electronics Inc., 18725 Lake Drive East, Chanhassen, MN, USA, 55317, jhammond@phi.com
** Physical Electronics Inc., 18725 Lake Drive East, Chanhassen, MN, USA,55317, dpaul@phi.com
*Tel-Aviv University, Tel-Aviv, Israel, fernando@post.tau.ac.il.

ABSTRACT

Nanowires (NWs), as well as other nanoscale structures, are in the heart of extensive study over the last few decades for both scientific and application interests. NWs physical properties are profoundly dependent on their surface chemical composition and size, prompting the need for the challenging characterization of an **individual** NW. Recent instrumentation advances in Field Emission Scanning Auger combined with sputter ion depth profiling allow the analysis of the lateral and depth elemental distributions of individual NWs.

In this study, a series of CVD-VLS grown, SiGeNWs and phosphorus doped SiNWs were characterized by Scanning Auger with a sub 10 nm primary electron beam. The lateral concentrations of oxide, silicon, germanium and phosphorus were obtained as a function of the NW's diameter showing the nature and depth of the surface oxidation and the Si/Ge ratio radial homogeneity in SiGeNWs. In addition, initial results of depth dependent phosphorus doping concentrations were obtained.
Keywords: "scanning Auger microscopy" "doping" "depth profiling" "silicon nanowire" "silicon germanium nanowire"

1 INTRODUCTION

The rapidly growing interest in novel nanostructure based applications requires the integration of complicated heterostructures and superlattices leading to new fabrication and analytical characterization challenges. Specifically, the study of the interplay between fabrication processes and the resulting nanostructure size dependent properties requires the use of analytical tools with nanometer scale probing capabilities. For example, doped silicon nanowires (SiNWs) based FET devices are being evaluated as promising candidates for replacing traditional MOSFETs, prompting the requirement to understanding their unique electronic properties. Nevertheless, only recently has an appropriate analytical tool lead to the correlation of the inhomogeneous radial dopant concentration to the doped SiNWs high resistivity [1-2].

Auger electron spectroscopy (AES) is one of the most widely used analytical techniques for high spatial resolution analysis of the chemical composition of solid surfaces. Its surface sensitivity of 0.5 nm-2 nm combined with sub-micron SEM and Auger spatial resolution and the ability to detect and quantify all elements above helium has led to the scanning Auger microscope (SAM) becoming a standard analytical tool in both research laboratories and production support laboratories in the microelectronics industry. The development of new nm scale electron probe beams combined with ion beam sputter depth profiling has made the new generation of Scanning Auger Nanoprobe a promising analytical tool for emerging research in nanotechnology.

In this study the elemental and chemical depth compositions of individual NWs were analyzed by a nanoprobe SAM. Previous direct analytical measurements of the depth distributions on NWs have only been reported with the infrequently used atom probe tomography (APT) [1]. Mostly indirect measurement techniques have also been extensively employed in recent years which showed that the dopant profiles were inhomogeneous [1, 3-7]. The dopant profiles in this study are in agreement with previous studies and demonstrate that the nanoprobe SAM is a simple alternative for the complicated APT technique and can be used as a direct measurement technique compared to previous indirect measurement techniques. The chemical composition depth profiling of SiGeNWs is also reported. Previous research has studied the miscibility of Si and Ge in SiGeNWs employing Raman spectroscopy to detect Ge segregation [8-9]. The present findings show no signs of inhomogeneity in the Si/Ge ratio and are in agreement with TEM-EDS measurements.

2 EXPERIMENTAL

2.1 Si$_{(1-x)}$Ge$_x$NWs and SiNWs Growth

The growth procedures for phosphorus doped SiNWs and Si(1-x)Gex NWs have previously been described in detail [10]. In summary, all NWs produced for this study were grown by the VLS technique in a hot-wall CVD set-up, with 20 nm gold nanoparticles (AuNPs) as the catalyst, high-purity silane (SiH4) and germane (10% GeH4 in H2 carrier) as precursor and dopant gases, and hydrogen as the carrier gas. Phosphine (1000 ppm in H2 carrier) served as the N-doping gas. Silicon substrates were dipped in a solution of poly-L-lysine, washed with deionized distilled water (DDW), and dried in a stream of dry nitrogen. After drying, the substrates were dipped in a diluted solution of AuNPs, washed and dried as described above. Finally, organic traces were removed by oxygen plasma. Substrates were then placed at the center of a horizontal quartz tube in a hot-wall furnace, which was then evacuated and purged in a stream of argon and hydrogen. Flow rates were determined with the use of mass-flow controllers (MFC), and the pressure in the growth chamber was computer-controlled via a throttle valve that limited the flow to the vacuum pump. In order to suppress radial growth, the following growth parameters were chosen for Si(1-x)Gex NWs: growth temperature - 3500C; inlet-gas ratio (GeH4/(GeH4+SiH4)) - 0.02. The inlet-gas flows were: 10%GeH4 -1 sccm (standard centimeter cube per minute at STP), SiH4 - 5sccm, H2 -100 sccm. The growth parameters of the 1/2000 phosphorus doped SiNWs were: growth temperature - 5000C. The inlet-gas flows were: SiH4 – 4 sccm, H2 -20 sccm. Phosphorus doping was carried out by adding a PH3 flow of 2 sccm during growth. Growth times varied from 20 to 5 minutes, resulting in 35 to 30 µm long NWs. The growth process was ended by purging the sample with argon and hydrogen while it was cooling.

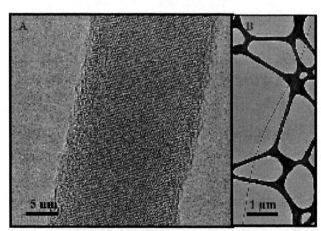

Figure 1: TEM images of slightly tapered typical 1/2000 phosphorus doped SiNW (B), with 20nm diameter (A).

2.2 Sample Preparation

NWs were removed from the growth substrate by sonication in isopropanol. The NWs containing solution was dispersed onto GaAs substrates for SAM measurements or onto lacey carbon-coated copper grids for TEM analysis. Prior to NWs deposition, E-beam lithography and thermal evaporation were used to pattern the GaAs substrates with 30 µm Au crosses grid. After NWs deposition the pattern was used for samples mapping via dark field optical microscopy.

2.3 Auger measurements

The Auger data was acquired with a Physical Electronics PHI 700Xi Scanning Auger equipped with a 25 kV Schottky field emission electron gun and a coaxial Cylindrical Mirror Analyzer (CMA). The base pressure of the PHI 700Xi is < 6.7 x 10^{-8} Pa (<5 x 10^{-10} Torr). All secondary electron images, Auger spectra and Auger depth profiles were acquired with a 20 kV electron beam with an incident current on the sample of 5 nA with a probe diameter of 9.5 nm. The Auger data was acquired at 0.5% CMA energy resolution for elemental identification and for calculations of surface atomic concentration. All atomic concentration calculations used dN(E)/dE peak to peak height measurements. High energy resolution spectra of Ge LMM spectra and Si KLL spectra on a SiGeNW were acquired at 0.1% CMA energy resolution and displayed in the N(E) mode after a polynomial background subtraction.

Depth profiling of the SiGeNWs and the phosphorous doped SiNWs was performed using an alternating Auger data acquisition with 500 eV Argon ion sputtering. Each sputtering cycle was for 20 seconds resulting in a sputter rate of 1 nm per cycle, as calibrated on a standard SiO$_2$ thin film. A computer controlled image registration of the beam position was used to maintain precise electron beam alignment during the high sensitivity depth profiling. The presence of the low concentration P dopant was verified by the exact phosphorous LMM N(E) peak position and the atomic concentration values as a function of depth were calculated using the dN(E)/dE peak to peak heights of the P LMM and Si KLL spectra and PHI supplied sensitivity factors.

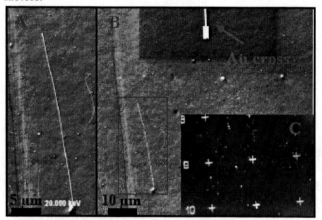

Figure 2: The SEM capacity of the PHI700 enables detection and focusing the e-gun on individual NW without undesired carbon coating. "in-situ" SEM image of individual SiNW: (A) – a 30 µm SiNW on GaAs substrate (the enclosed area in B). (B) – lower magnification SEM image showing the SiNW and the Au cross ending, the inset (c)- Dark field optical image of patterned GaAs substrate after NWs deposition .

3 RESULTS AND DISCUSSION

3.1 Si$_{(1-X)}$Ge$_X$NWs

The Si$_{(1-X)}$Ge$_X$NW morphology, crystalline structure and chemical composition were studied with SEM, TEM and EDS prior to the scanning Auger in depth chemical analysis. The NWs were found to be: 20-30 nm diameter, up to 30 μm length, single crystalline structure with a smooth, thin amorphous Si shell layer. EDS results have shown the NWs to be 45% Si and 55% Ge (Si$_{0.45}$Ge$_{0.55}$NWs). Auger scans of a 30 nm thick Si$_{0.45}$Ge$_{0.55}$NW revealed the following elements: Si, Ge, Ga, As, O and C. Comparing the scans taken on the Si$_{0.45}$Ge$_{0.55}$NW to scans taken on the GaAs substrate has shown similar carbon percentages. The carbon percentages of both the substrate and the Si$_{0.45}$Ge$_{0.55}$NW decreased after each round of Ar$^+$ etching, leading us to conclude that the carbon contamination covered the entire sample and was eliminated after a few nm of etching. In contrast, the oxygen percentages on the Si$_{0.45}$Ge$_{0.55}$NW were found to be higher than those found on the substrate and substantially decreased only after 3-5 nm of etching. This observation was reinforced by the broadening of the Si peak attributed to SiO which disappeared after the same etching depth. From both observations we conclude that the measured oxygen originated in ~5 nm width of the amorphous shell. Figure 3A shows the differential (dN(E)/dE) scan results over the entire energy range taken after 1 nm etching with observable peaks from all of the above mentioned elements.

The Si peak broadening is shown in figure 3B. The same broadening was observed in both the as received and after 4 nm etching Auger scans. After additional etching the broadening was no longer observed. The Si/Ge atomic concentrations as a function of sputter depth, presented in figure 3C, were obtained by further etching and Auger analyses in 1 nm intervals. The Si/Ge ratios are similar to the TEM-EDS results throughout almost the entire NW's width. A 7% increase in the Si percentage was observed after ca. 15 nm of etching which seems to exceed the measurement noise level (ca. 2%).

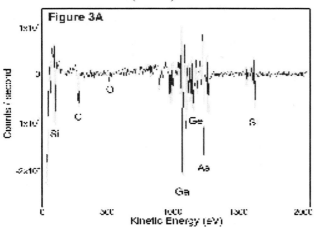

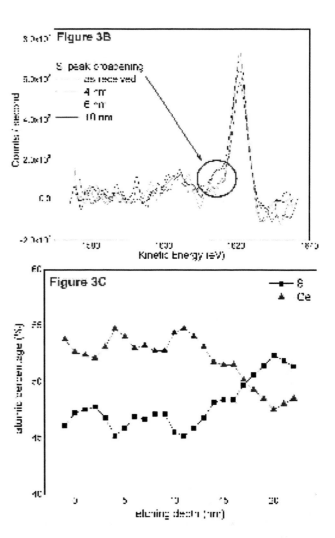

Figure 3: Auger measurments of Si$_{0.45}$Ge$_{0.55}$NW. (A)- Differential mode Auger scan (prior to etching). (B)- The Si Auger peak at various depths. (C)- compositional depth profiling.

3.2 Phosphorus doped SiNWs

Figure 4A shows the differential mode Auger scan results of the entire energy range taken from a 1/2000 phosphorus doped SiNW before etching. Ga, As, Si, O, C and P elemental peaks are readily observable. The calculated atomic percentage values of Ga and As were similar to those measured of SiGeNWs. Comparable reduction of carbon by the Ar$^+$ sputtering was observed for both the SiNW and the GaAs wafer suggesting that the surface contamination is easily eliminated. The oxygen concentration was substantially reduced after 5-6 nm of etching in a similar manner observed with the SiGeNWs. Surprisingly high phosphorus concentrations close to 1% were observed at the SiNW's surface. The phosphorus concentration was dramatically reduced with further etching. The phosphorus detectability limit is determined by the signal to noise limit of the observable Auger phosphorous LMM peak and was found to be ~ 1/1250. Nevertheless, we believe that lower concentrations may be detected with further improvements of the measurement conditions.

The radial distribution of phosphorus dopant concentration is presented in figure 4B. The phosphorous atomic concentration depth profile shows a 1/120 phosphorus to silicon ratio on the NW's surface which decreases to about 1/350 after 8 nm etching. The high dopant ratios (as compared to the 1/2000 phosphorus to silicon inlet gases ratio) might be the outcome of the specific NW's growth conditions- in this case the NW was grown at 500^0C and was slightly tapered.

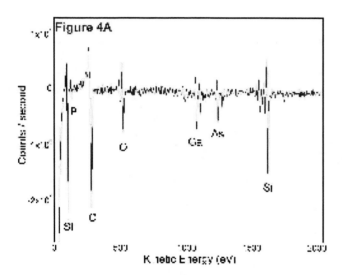

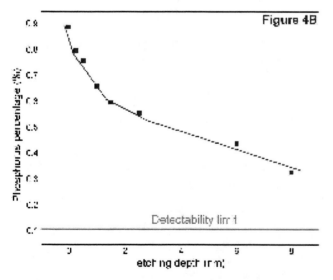

Figure 4. Scanning Auger chemical analysis of a phosphorus doped SiNW (A)- full energy range scan in differential mode. (B)- Depth profiling reveals the non-homogenous radial phosphorus doping distribution.

4   SUMMARY

In conclusion, successful analysis of the lateral and depth elemental distributions of individual NWs was achieved employing a nanoprobe field emission scanning Auger combined with low voltage sputter ion depth profiling.

The surface of $Si_{0.45}Ge_{0.55}$NWs was studied and found to be composed of 4-6nm SiO layer with no evidence for GeO. The $Si_{0.45}Ge_{0.55}$NWs in depth chemical analysis was obtained by further etching and the atomic fractions were in agreement with previously obtained TEM-EDS results. The compositional depth profiling revealed the Si/Ge ratio to be constant throughout the first 15nm with some 5 nm cyclic diversity of 2%. An increase of 7% in the Si atomic percentage was found closer to the $Si_{0.45}Ge_{0.55}$NW core. Further study is needed to verify both findings (the cyclic behavior and the Si rich core).

The nanoprobe scanning Auger was found to be relevant for depth compositional analysis of individual phosphorus doped SiNWs. The dopant concentration distribution profile was found to be rapidly decreasing with depth, starting with over 15 fold higher concentration values at the NW's surface than the expected values from the inlet gas ratio (1/2000). The inhomogeneous distribution is in agreement with previous direct measurements (carried out with APT), indirect measurements and theoretical predictions. The detectability limit was found to be ~1/1250. We believe this value could be substantially decreased by further improvement of technical measurement conditions.

REFERENCES

[1] D. E. Perea, E et al, *Nature Nanotechnology*, 4, 315, 2009.
[2] V. Schmidt, et al, *Adv. Mater.*, 21, 2681, 2009.
[3] J. E. Allen, et al, *Adv. Mater.*, 21, 1, 2009.
[4] E. C. Garnett, et al, *Nature Nanotechnology,* 4, 311, 2009.
[5] E. Koren, et al, *Nano Lett.*, 10, 1163, 2010.
[6] P. Xie, Y. et al, *Proc. Natl. Acad. Sci. USA* 106, 15254, 2009.
[7] E. Koren, et al, *Appl. Phys. Lett.* 95, 92105, 2009.
[8] Q. Lu, et al, *J. Phys. Chem. C, 112*, 3209, 2008.
[9] C. Nishimura, et al, *Appl. Phys. Lett.* 93, 203101, 2008.
[10] U. Givan, F. Patolsky, *Nano Lett.*, 9, 1775. 2009.

Advances in Helium Ion Microscopy
For High Resolution Imaging, Analysis and Materials Modification

William B. Thompson, John Notte, Larry Scipioni,
Mohan Ananth, Lewis Stern, Sybren Sijbrandij and Colin Sanford

Carl Zeiss SMT,
One Corporation Way, Peabody, MA 01960
b.thompson@smt.zeiss.com

ABSTRACT

The rapidly increasing use of materials like carbon, ceramics and soft organic materials in nanotechnology causes a number of challenges for materials imaging, characterization, analysis and modification. With its 'Top Down' 0.35 nm resolution and its ability to obtain surface sensitive images of uncoated insulating specimens, the helium ion microscope (HIM) is showing improved usefulness in addressing some of these imaging, characterization and analysis challenges. With the addition of controlled dose patterning, the HIM is further capable of modifying materials such as graphene, silicon, and magnesium oxide with dimensions at or below 10 nm. These modifications can be accomplished by either the physical processes associated with a pure ion/material interaction or by the chemical processes of ion induced etching and deposition. In what follows, we provide an overview of the machine architecture, a summary of the more important HIM material interactions and cover the system's current level of imaging, analysis and material modification performance.

Keywords: Helium Ion Microscope, Sub-nanometer imaging, Rutherford Backscatter Spectroscopy, Nanometer patterning, Graphene

1 INTRODUCTION

Currently, the Orion helium ion microscope (HIM) can be operated in two main imaging modes; ion induced secondary electron (SE) mode and Rutherford backscatter imaging (RBI) mode. This paper will provide an overview of the microscope's ion source, its ion optics, the system architecture, the fundamentals of the SE and RBI imaging modes and some material analysis and modification related examples.

A helium ion microscope SE mode image differs from an SEM image in that it provides high resolution, surface sensitive information, on both conductive and insulating samples without the need for any conductive coating, unlike a conventional SEM. HIM RBI mode, on the other hand, contains considerable material and channeling contrast image content.

Recently integrated with the microscope is a Rutherford Backscatter (RBS) detector for materials analysis. The detector is a modified silicon diode drift detector with a collected pulse height that is proportional to the energy of the backscattered helium ion.

The high resolution beam induced modification of materials is an important nanofabrication technique with applications in material science, semiconductor processing, circuit edit, basic physics research and NEMS, for example.

The removal of material can be accomplished in the helium ion microscope through either sputtering; ion impact material loss induced by radiolysis and thermal effects or by ion beam induced etching.

The addition of material can only be accomplished by ion beam induced chemistry. To accommodate our customers, we have added a Gas Injection System (GIS) to the HIM to provide the user with the opportunity to perform ion induced chemistry research. Although both the etched and deposited dimensions of geometries generated with the helium ion beam are greater than the 0.35 nm beam diameter, these geometries are typically in the sub 10 nm regime and generally much smaller than those obtained by other lithographic patterning methods.

In addition to describing the machine architecture in the succeeding sections, we will cover these three main HIM application topics of imaging, RBS and HIM initiated material modification.

2 THE HIM ION SOURCE, OPTICS, SYSTEM BASICS AND SECONDARY ELECTRON IMAGING

2.1 The HIM Ion Source

The system's high resolution is the result of the virtual source for the helium ion probe being subatomic. The ion source consists of a single crystal metal needle sharpened to a three atom pyramid that sits at the apex of the needle. The needle is immersed in a helium atmosphere held at a temperature of approximately 77 degrees K at a voltage that is variable from 10 to 35 keV. Ions launched from one of the three atoms are sent down the column to become the scanned HIM probe.

2.2 The HIM Optics

The microscope electrostatic optics contains a gun lens, deflectors and a final or objective lens. The ratio of source size to probe size is approximately 1:1, i.e. the column magnification is around 1. Hence in the absence of aberrations, the scanned probe would be of order atomic dimensions. The actual probe size, at 35 keV, is limited by chromatic aberrations to 0.2 nm. As a consequence of this probe diameter, SE mode resolution of the microscope is around 0.35 nm as is evident in the gold on carbon image of figure 1.

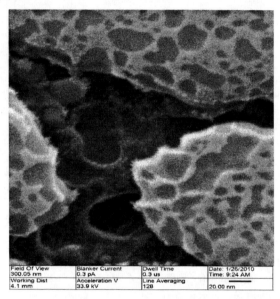

| Field Of View 300.05 nm | Blanker Current 0.3 pA | Dwell Time 0.3 us | Date: 1/26/2010 Time: 9:24 AM |
| Working Dist 4.1 mm | Acceleration V 33.9 kV | Line Averaging 128 | 20.00 nm |

Figure 1: A 300 nanometer field of view, HIM SE, image of a gold on carbon sample with 0.35 nm resolution.

The high source brightness and low column demagnification gives the microscope an unusually large, 2.5 μm, depth of field. For the beam diameters available with this optics, beam currents in the 0.5 to 10 pA range are easily attainable.

2.3 The Helium Ion Microscope System

The design goal for the SE imaging mode of the microscope was TEM resolution without the need of elaborate sample preparation. For this reason, the work chamber module is supported by a double tiered vibration isolation system mounted on a granite table. Sample loading, stage motion and the user interface are very similar to that of conventional SEM. Figure 2 shows the overall helium ion microscope system.

For a more extensive description of the HIM source, optics and system one is referred to a previous microscope publication [1].

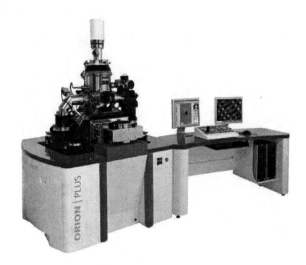

Figure 2: The Orion Plus Helium Ion Microscope

2.4 Secondary Electron Imaging

In many respects the microscope behaves and is operated in the same manner as an SEM. The primary difference being that the scanned probe is helium ions rather than electrons. When the high energy helium ions are scanned over a sample surface, they generate secondary electrons, backscattered ions and photons. The present microscope configuration uses an Everhart-Thornley (ET) secondary electron detector to form an SE image. A microchannel plate (MCP) is used to acquire a backscattered ion, or RBI image. This RBI image is generated by the video signal output of the MCP produced when the Rutherford backscattered helium from the sample impinges on the MCP detector. Both the SE and the RBI images can be acquired simultaneously and provide significantly different image information.

The energy transferred by the helium to the electrons of a sample is less than 10 eV and the energy of the electrons that escape from the sample is around 2 eV. As a consequence of these low energies and the short mean free path of the electrons in most materials, the only electrons that can escape from the sample have originated near the sample surface. HIM SE imaging is therefore a very surface sensitive imaging technique. The low SE electron energy also means that small changes in surface potential and work function will readily modify the image contrast. One implication of this effect is that low doping concentration changes in silicon are often detectable.

Any insulating or dielectric material imaged with a positive ion beam will invariably charge positively. This is not the case in an SEM in which samples can charge both negatively and positively. The ion induced positive surface charge can be neutralized in the HIM with low energy flood electrons. The helium ion microscope has been equipped with a low energy electron flood gun that permits those samples that might conceivably charge up to be imaged, without sample coating, at full beam energy and resolution. An example of a charge neutralized sample is shown in figure 3.

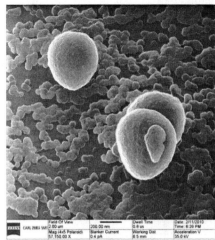

Figure 3: A 2 micron field of view, charge neutralized, HIM SE, image of hydroxyapatite on PLLA.

3 RUTHERFORD BACKSCATTER IMAGING AND SPECTROSCOPY

3.1 Rutherford Backscatter Imaging (RBI)

In the RBI mode, in order to generate a video signal, the helium ion microscope uses a micro-channel plate with a voltage output that is proportional to the number of backscattered helium atoms and ions entering the detector in a given time.

RBI Material Contrast

The differential scattering Rutherford backscatter cross-section is quadratically dependent both on the atomic number of the target material and the reciprocal (1/E) of the incident helium energy. In other words, the helium backscatter yield in any Rutherford backscatter image (RBI) increases as the sample target atomic number goes up and as the helium energy goes down. This means that in a sample which includes, for example, carbon, aluminum, copper and gold elemental materials, the regions of carbon will be black while those of gold will be white. This gray level variation simply means that more helium is backscattered into the RBI detector by gold than is back scattered into the RBI detector by carbon. To some degree the amount of white in any given RBI pixel is a measure of the Z2 of the individual pixel. Such a gray level variation could aid the operator viewing an RBI image to quickly categorize a defect particle as either a high, medium or low atomic number material without having to resort to RBS spectroscopy.

RBI imaging mode has little topographical contrast, the RBI image contrast is generated purely by variations in sample material and not by topographic or sample voltage variations. Since both the SE image and the RBI image can be acquired simultaneously in the helium ion microscope, the RBI imaging mode will aid the user in quickly discriminating between a material variation and a sample edge or voltage artifact. In summary, SE mode imaging will give resolution as well as material and topographical contrast information to the operator while RBI mode will give material contrast information.

RBI Channeling Contrast

Incident helium ions that are well aligned to a crystal grain axis, such as the silicon <110> axis, have a backscatter yield that is 20% of the backscatter yield from a randomly oriented crystal axis. Similarly, a copper grain with an axis well aligned to the incident helium ion microscope axis will appear black in an RBI image whereas a misaligned grain will appear white. This channeling contrast is significantly greater in RBI imaging than in an SEM or HIM SE imaging mode and can be useful in determining the granularity and grain orientation of crystalline samples.

3.2 Rutherford Backscatter Spectroscopy

The energy of a helium ion backscattered from a sample is a function of the mass of the scattering target nucleus. The more massive the scattering nucleus, the more energetic the helium ion scattered back from the target. This is the principle underlying an entire analytical discipline, Rutherford Backscattering Spectroscopy, RBS. As the incident helium ion enters the specimen, it interacts primarily with electrons which have relatively small effect on the ion's trajectory, but will deplete energy at a steady rate, dE/ds. For example, the rate of energy loss for a 30 keV helium ion in iron is about 100 eV / nm. At some depth beneath the surface, the helium ion (perhaps neutralized at this point) can undergo a back scattering event with a nucleus of the sample. The actual scattering phenomenon is well described by the straightforward physics of a two body collision in which energy and momentum are conserved There is an instantaneous drop of the ion's energy after the collision. This drop in energy is most significant for target atoms with low atomic masses (e.g. carbon and silicon versus tungsten and gold). Subsequent to the back scattering event, interactions with electrons are again responsible for energy losses of the scattered ion along its path length, dE/ds. A spectrum of the energy of helium scattered from a gold-on-silicon target acquired with this SDD spectrometer is shown in figure 4. This spectrum was fit with the RBS peak interpretation routine, SIMNRA, developed at the Max Planck Institute [2,3]. From this gold on silicon spectra fit, the thickness of the gold film was determined to be 10 ±1 nm. In addition to film thickness measurements, the spectrometer has the ability to perform material analysis on relatively small particles and to determine the elemental ratio of alloys of known composites to better than a percent.

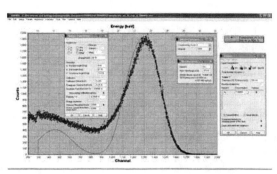

Figure 4: HIM Gold on silicon RBS Spectrum

4 HIM MATERIAL MODIFICATION AND LITHOGRAPHY

4.1 Ion Induced Material Removal

When finely focused energetic ions interact with the atoms of the sample to be modified, they can initiate three basic mechanisms that will cause the material to be removed; sputtering, radiolysis and thermal evaporation. Sputtering and radiolysis are a total dose dependent effect while thermal evaporation is a dose rate effect. Radiolysis, the breaking of chemical bonds that leaves the remaining sample constituents volatile, is a removal effect associated only with materials having weak chemical bonds. All of these phenomena are initiated by local doses that exceed the imaging dosages by several orders of magnitude.

There has been considerable interest in the electrical properties of nanometer scale graphene ribbons. Most of the graphene patterning techniques induce impurities or damage to the ribbons that alter their electrical properties. HIM induced patterning of free standing graphene films has shown considerable promise.

An example of HIM work done by Dan Pickard at the National University of Singapore is shown in figure 5.

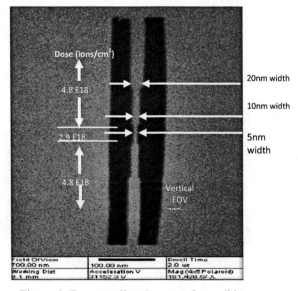

Figure 5: Free standing 5nm graphene ribbon.

4.2 Ion Induced Etching and Deposition

With the addition of a Gas injection System (GIS) to the microscope it became possible to introduce halogenated compounds, such as xenon diflouride, as well as tungsten hexacarbonyl, organo-metalic platinum compounds and TEOS to do ion induced etching, metal deposition and insulator deposition work.

An example of the HIM induced deposition of a 6000nm high, 30 nm diameter, pillar is shown in figure 6.

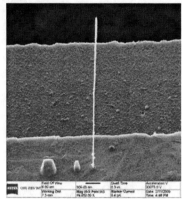

Figure 6: Free standing 6 μm pillar 30 nm in diameter.

4.3 HIM Lithography

Recent work in the HIM patterning of PMMA and HSQ resists has demonstrated dense arrays of 5nm dots and closely spaced 10 nm 'L' bars. The sensitivity, in ion/cm^2, required to expose the PMMA and HSQ resists is nearly an order of magnitude lower than that of comparable e-beam sensitivities. While these results are promising, it is believed that by the proper parameter optimization, they can be made even better.

5 SUMMARY

The helium ion microscope has demonstrated resolution on bulk samples that is superior to other particle beam imaging techniques. Insulating samples need not be coated in order to see their surface detail at high resolution.

Applications such as RBS analysis and sample patterning and modification are only now being explored and seem to be showing interesting and promising results.

REFERENCES

[1] J. Morgan, J. Notte, R. Hill, B. Ward, An Introduction to the Helium Ion Microscope, Microscopy Today 16, No 4, p. 24 (2006)
[2] M. Mayer, SIMNRA Computer Code, Max Planck Institute, Garching, Germany
[3] W.B. Thompson, J. Notte, S. Sijbrandij, L. Scipioni, and. L. Stern, The Helium Ion Microscope Operation, Imaging and Materials Analysis Fundamentals, The Vacuum and Surface Science Conference of Asia and Australia, (2008) p. 62

Atomic scale dopant metrology in an individual silicon nanowire by atom probe tomography

W. H. Chen[*], R. Lardé[*], E. Cadel[*], T. Xu[**], J. P. Nys[**], B. Grandidier[**], D. Stiévenard[**] and P. Pareige[*]

[*]Groupe de Physique des Matériaux, Université et INSA de Rouen, UMR CNRS 6634, Av. de l'université, BP 12, 76801 Saint Etienne du Rouvray, France, philippe.pareige@univ-rouen.fr
[**]Institut d'Electronique, de Microélectronique et de Nanotechnologie. UMR CNRS 8520, Département ISEN, 41 bd Vauban, 59046 Lille Cedex, France

ABSTRACT

In this work, the p-type silicon nanowires (SiNWs) are grown by the Chemical Vapor Deposition (CVD) method using gold as catalyst droplet, silane as precursor and diborane as dopant reactant and are analyzed at the atomic scale using the three dimensional laser assisted Atom Probe Tomography (APT). This paper reports the preparation of SiNWs and their observation. A three dimensional dopant distribution and its accurate concentration are determined. The possible dopant incorporation pathway into an individual SiNW is also discussed.

Keywords: silicon nanowire, dopant, atom probe

1 INTRODUCTION

One dimensional nanostructure can play an important role for the future applications. Among them, the carbon nanotubes (CNTs) [1] and the silicon nanowires [2] attract an extreme attention because of their relative easy synthesis and potential properties. By controlling their synthesis, CNTs and SiNWs can both be candidate for ultimate nanoelectric devices. However, in comparison to CNTs, it is easier to control the electrical properties of SiNWs. Thus, a lot of efforts and progresses have been made on SiNWs in different domains in the last few years, such as electronics [3, 4], bio or molecular sensors [5, 6], modeling [7, 8] and energy devices [9, 10].

SiNWs can typically be grown by vapor liquid solid (VLS) mechanism. CVD is a the method commonly used [11]. A typical SiNW growth kinetic is schematically illustrated in Fig.1. This process is governed by three different configurations of the Si atoms: Si atom holder (Si^H the substrate), Si atom provider (Si^P, the precursor) and Si atom receiver (Si^R, the supersaturated binary Au-Si alloys). Si^H can be chosen as Si [111], Si [110] or Ge [111] etc...; Si^P can be chosen as SiH_4, Si_2H_6 or $SiCl_4$ etc... and Si^R can be chosen as a mixing with Au, Cu or Fe etc... Before the SiNWs growth, Si^R exists on Si^H as droplets. The Si^P is then introduced into the chamber while the substrate is kept to high temperature (e.g. 500 °C). To that point, three steps are necessary for Si atoms to contribute the SiNW growth : *i)* the production of Si atoms via the decomposition of Si^P; *ii)* the incorporation of Si atoms into Si^R and *iii)* the crystallization of Si atoms at SiNW and Si^R interface.

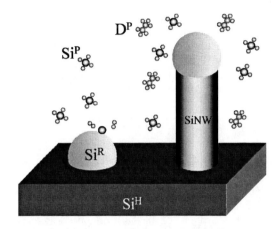

Figure 1. Schematic view of VLS mechanism system for SiNWs growth. This system includes three parts for SiNWs without dopant: Si atoms holder (Si^H), Si atoms receiver (Si^R) and Si atoms provider (Si^P). After the Si^R is deposited on Si^H as a droplet, the Si atoms coming from the decomposition of Si^P can incorporate into Si^R and then crystallize at Si^R. Crystallization at SiNW interface also occurs depending on experimental growth condition. For doped SiNWs, a fourth part, namely the dopant provider (D^P) either for p-type or n-type, should be added into the system.

The doping of SiNWs is an important issue for their electrical applications. In the CVD process, the doping process can be easily realized by the addition of dopant provider (D^P) in Si^P. For example, D^P can be chosen as B_2H_6 and PH_3 for a p-type and n-type SiNWs respectively. The different doping concentration can be realized by changing the ratio (gas pressure) between Si^P and D^P. The p-n heterostructure can also be grown by tuning the D^P level during the SiNWs growth [12]. The relation between the Si^P/D^P ratio and the accurate incorporated dopant atomic concentration should be established for having a controllable doping process. However, the control of the doping process is still very difficult because of the lack of information or measurement on dopant level in an individual SiNW. It is also very important to know through which pathway the dopant atoms can incorporate into SiNW. In this work, we use a laser assisted three-dimensional APT to characterize an individual p-type SiNW. APT is a three dimensional high resolution analytical microscope that can map the distribution of atoms in materials. The principle of the APT is based on the field evaporation of atoms. In order to realize this, an electrical potential of several kilovolts is applied to the sample, prepared in the form of a tip whose diameter should be smaller than 100 nm. Femtosecond laser pulses are superimposed to the high voltage in the case of semiconductor materials such as SiNWs. It must be noted that a SiNW is already in the form of a tip. This combination (voltage and laser pulses) gives rise to the field ionization of surface atoms. The laser pulse frequency triggers the evaporation of ions from the surface of the specimen towards a position sensitive detector. Measuring the time of flight allows the identification of the chemical nature of evaporated ions (time of flight mass spectrometry) and their impact positions on the detector allows to calculate their initial position on the surface (projection microscope). These information are used for the three dimensional reconstruction of the material at the atomic scale. The information about the concentration and distribution of dopant (e.g. B) in an individual SiNW can thus be obtained using APT.

2 EXPERIMENT

2.1 SiNWs growth

The choice of Si^R is important for the SiNWs growth. In theory, all the elements that can form eutectic liquid with Si can be used as Si^R. In this work, Au was chosen for SiNW growth because of its relatively low eutectic temperature and its easy preparation in the form of Au droplets. Several methods can be used to obtain the catalyst droplet, such as colloids [13] or evaporation [14]. The Au droplets, in our work, are growing on the Si substrate during the gold evaporation of massive Au. This method allows to control effectively and precisely the catalyst droplet number density which directly influences the SiNW surface number density. For Au droplets deposition, a 7×7 n-type doped Si (111) substrate is prepared. It should be pointed that there is a relation between the SiNWs growth direction and the Si substrate orientation. In order to facilitate the manipulation of the SiNWs, they were grown on the top surface of Si-micro-pillars. These Si-micro-pillars are prepared from Si substrate using a deep reactive ion etch process. Then the Si-micro-pillars are placed in an ultra-high vacuum chamber (10^{-10} mbar) for the Au droplets deposition. A chosen average diameter and droplet number density can be obtained by controlling, according to Ostwald ripening effect [15], the substrate temperature. In order to have a series of Si-micro-pillars with different droplet number densities, three different temperatures (400°C, 430°C and 460°C) at a fix evaporation rate (0.025 monolayer/s) were used (Fig. 2). As expected, the Au droplet number density increases with the decrease of the deposition temperature. The average diameter for 400°C, 430°C and 460°C are 40nm, 55nm and 90nm respectively. For consecutive APT analyses, a small Au droplet diameter and a weak Au droplet number density are needed. Only the Si-micro-pillars in Fig.2 (b) can be used for example.

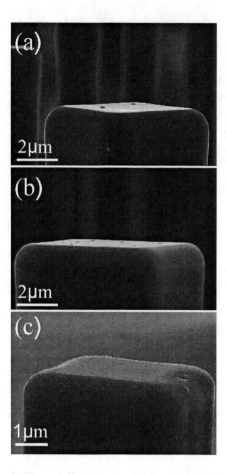

Figure 2. Three different series of Si-micro-pillars with different Au droplet number densities. The evaporation pressure is fixed at 10^{-10} mbar and temperature ranges between: (a) 460 °C; (b) 430 °C and (c) 400 °C. The Au droplet number density increases with the decrease of the deposition temperature.

The SiNWs can be grown on the Si-micro-pillars using the CVD method. In this work, SiH_4 as Si^P and B_2H_6 as p-type D^P were chosen. The carrier gas is chosen to be hydrogen which is also the by-product of SiH_4. The SiH_4 gas is diluted with H_2 in the ratio 50:49. The flow rate of SiH_4 is 50 sccm and silane partial pressure is kept at 0.4 mbar during the NWs growth. The flow rate of B_2H_6 is 1 sccm. The growth temperature and time are 500 °C and 30 min. respectively. It should be noted that this method allows SiNWs with diameter of 50-100nm to be grown perpendicularly to the Si (111) substrate, namely the SiNW growth direction is [111]. However it is shown, as reported in the literature, that when the diameter of SiNW decrease down to 20nm, the [111] growth direction may turn to [112] or [110] because of the surface energy competition between Au/SiNW interface and SiNW sidewall [16]. Taking advantage of this phenomenon, a suitable SiNW sample for APT, namely, a single SiNW perpendicular to the pillar surface, can be grown. As shown in Fig. 3, a SiNW is located vertically on the top of Si micro pillar. The top end of SiNW is illustrated in the inset part and the Au droplet can be clearly seen. A tapering of SiNW can also be seen because of the diffusion of gold atom during SiNW growth [17].

Figure 3. A p-type SiNW on a Si micro pillar fabricated by CVD method. The inset is a zoom of SiNW end and a gold droplet can be clearly seen on the top of SiNW.

2.2   APT sample preparation

For the laser assisted APT, the analyzed sample is normally connected to a W tip, it self included in a steel tube. Thus, after the growth of SiNW on Si micro pillar, an effective four-step *in-situ* process is needed to prepare the APT sample (Fig. 4 (a)).

i) Preparing a W tip using regular electro-polishing;
ii) Choosing a suitable Si micro pillar by using a micro tweezers (Kleindiek, nanotechnik, Germany);
iii) Cutting the Si pillar using a Focused Ion Beam (FIB)/Scanning Electron Microscope (SEM) dual beam system;
iv) Welding the Si pillar and W tip by using a Gas Injection System (GIS).

It should be noted here that a great attention should be paid to SiNW during the step iii) for cutting Si pillar because of the possibility of Ga atoms contamination. A Si micro pillar on a W tip can be prepared with success using this four-step process (Fig. 4 (b)).

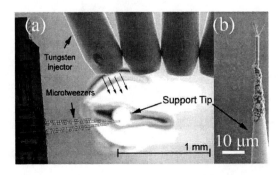

Figure 4. (a) A four-step process for transferring a Si micro pillar to a W tip.

3   RESULTS AND DISCUSSION

The result of the analysis of a p-type SiNW by APT is presented and discussed in this part. The analysis is realized in an ultra high vacuum (10^{-10} mbar) and low temperature condition (80 K). The reason to choose a low temperature is to decrease the atom vibration. The wavelength of the femtosecond laser is chosen to be green (515 nm). The laser power is chosen to be 20 mw. A high electrical potential (10kV) is also applied to the sample (tip/wire). Fig. 5 is the mass-to-charge state ratio spectrum. As indicated on Fig.5, isotopes of different elements (e.g. Si, B and H) can be clearly distinguished. Hydrogen exists in the chamber as a background gas even under ultra high vacuum. Thus, during the evaporation process, hydrogen has an opportunity to interact with the sharp tip (e.g. SiNW), and thus be detected in the spectrum. Both B isotopes are also detected. the three Si isotopes are also detected. As shown in Fig. 5 the large peak tail following the Si peaks is due to thermal effect during laser evaporation.

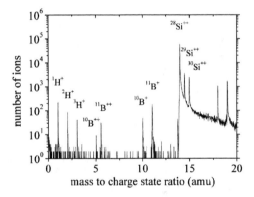

Figure 5. The mass to charge state ratio spectrum of SiNW analyzed by APT. Two types of B isotopes peaks are clearly shown in this spectrum.

The data set obtained after evaporation of an individual SiNW can be reconstructed in a three dimensional view using our APT data treatment software. One of the advantages of APT is that a combination of high spatial resolution (depth < 0.1nm and lateral < 1nm) and accurate concentration (< 100ppm) can be obtained at the same time compared to other techniques. Fig. 6 is a three dimensional reconstruction of an individual SiNW. The reconstruction volume is $16 \times 16 \times 58$ nm^3. The matrix and dots represent individual Si and B atoms in SiNW respectively. For clarity of the image, only 20% of Si atoms have been represented in this volume. The B concentration can be calculated and is equal to 1.4×10^{20} B/cm^3 with a detection limit of 5×10^{18} at/cm^3. The B atoms are homogeneously distributed in the core of the SiNW as shown in Fig. 6. It should be noted here that the APT reconstruction volume is representative of the core of the SiNW. The information about the SiNW sidewall surface is missing during the evaporation process. The first conclusion here is that under this SiNWs' growth condition, B atoms can easily be introduced into SiNW (core). Now, the B atoms can integrate the SiNWs by two main pathways:

i) Through the Au droplet during SiNWs growth. The B atoms dissociate at the droplet interface. A complete description of this process needs to take into account the modification of the binary alloys (Au/Si) towards a ternary alloy (Si/Au/B) with different characteristic such as eutectic composition and temperature.

ii) By the lateral surface of the SiNW. Here again a complete description of the phenomena needs to consider the possible mechanisms: 1) diffusion (Fick's law) from the sidewall surface into the SiNW during growth or 2) deposition of B on the sidewall surface during the lateral growth of the SiNW. Works are in progress to work on these hypotheses.

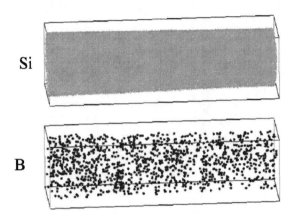

Figure 6. The three dimensional reconstruction of an individual SiNW analyzed by APT. The analyzed volume is $16 \times 16 \times 58$ nm^3. The matrix and dots represent Si and B atoms in the SiNW respectively. Only 20% Si atoms are shown in the reconstruction for clarity. B atoms distribute uniformly in SiNW core.

To conclude, we have successfully prepared the optimized Au droplets size and number density for SiNWs growth on Si micro pillars by using an *in-situ* evaporation/deposition method. Taking advantage of this pre-growth preparation, an individual p-type SiNW is grown on a Si micro pillar by CVD method using Au, SiH$_4$ and B$_2$H$_4$ as Si atoms receiver, Si atoms provider and B atoms provider respectively. We have shown that B atoms can be located in SiNW core with a relatively high concentration (1.4×10^{20} B/cm^3) according to our APT results. It indicates that B atoms can effectively incorporate into SiNW following the precursor/dopant ratio. The incorporation mechanism of the dopant into the nanowire is under studies.

Acknowledgement

This work was supported partly by the DGA (Direction Générale de l'Armement) under the contract REI - N°2008.34.0031.

[1] S. Iijima, Nature, 354, 56, 1991.

[2] A. M. Morales, C. M. Lieber, SCIENCE, 279, 208, 1998.

[3] D. Wang, B. Sheriff, M. McAlpine, J. Heath, Nano Research, 1, 9, 2008.

[4] J.-P. Colinge, C.-W. Lee, A. Afzalian, N. D. Akhavan, R. Yan, I. Ferain, P. Razavi, B. O'Neill, A. Blake, M. White, A.-M. Kelleher, B. McCarthy, R. Murphy, Nat Nano. 5, 225, 2010.

[5] D. R. Kim, C. H. Lee, X. Zheng, Nano Lett. 9, 1984, 2009.

[6] J. W. Lee, D. Jang, G. T. Kim, M. Mouis, G. Ghibaudo, J. Appl. Phys. 107, 044501, 2010.

[7] W. H. Chen, R. Lardé, E. Cadel, T. Xu, B. Grandidier, J. P. Nys, D. Stiévenard, P. Pareige, J. Appl. Phys. 107, 2010, in press.

[8] V. Schmidt, J. V. Wittemann, U. Gosele, Chem. Rev. 110, 361, 2010.

[9] B. Tian, T. J. Kempa, C. M. Lieber, Chem. Soc. Rev. 38, 16, 2009.

[10] A. I. Hochbaum, P. Yang, Chem. Rev. 110, 527, 2010.

[11] R. S. Wagner, W. C. Ellis, Appl. Phys. Lett. 4, 89, 1964

[12] L. Wei, X. Ping, C. M. Lieber, Electron Devices, IEEE Transactions 55, 2859, 2008.

[13] J. H. Woodruff, J. B. Ratchford, I. A. Goldthorpe, P. C. McIntyre, Chidsey, Nano Lett. 7, 1637, 2007.

[14] T. Xu, J. P. Nys, B. Grandidier, D. Stiévenard, Y. Coffinier, R. Boukherroub, R. Larde, E. Cadel, P. Pareige, J. Vac. Sci. Technol B, 26, 1960, 2008.

[15] W. Ostwald, Engelmann: Leipzig, Germany, Vol. 2, 1896.

[16] Y. Wu, Y. Cui, L. Huynh, C. J. Barrelet, D. C. Bell, C. M. Lieber, Nano Lett. 4, 433, 2004.

[17] T. Xu, J. P. Nys, A. Addad, O. I. Lebedev, A. Urbieta, B. Salhi, M. Berthe, B. Grandidier, D. Stiévenard, Phys. Rev. B, 81, 115403, 2010.

In-situ Surface Characterization of Nano- and Microparticles by Optical Second Harmonic Generation

B. Schürer,[1,3] S. Wunderlich,[2,3] U. Peschel[2,3] and W. Peukert[1,3]

[1]Institute of Particle Technology, University of Erlangen-Nuremberg, Erlangen, Germany
[2]Max Planck Institute for the Science of Light, Erlangen, Germany
[3]Cluster of Excellence - Engineering of Advanced Materials and Erlangen Graduate School in Advanced Optical Technologies (SAOT), University of Erlangen-Nuremberg, Erlangen, Germany

ABSTRACT

Nonlinear optical spectroscopic techniques, namely Second Harmonic Generation (SHG) spectroscopy and Sum Frequency Generation (SFG), are highly interface sensitive and can be used to obtain information exclusively from the surface of particles. Second Harmonic Generation (SHG) has proven to be a versatile technique that can be used to probe different kinds of surface properties of dispersed particle systems in in-situ. Prominent examples are the direct study of molecular adsorption at the surface of particles and the determination of surface potentials of nano- and microparticles. However, in most of these studies well-defined spherical and monodisperse particles with low number densities were used. Here we demonstrate the application of the SHG technique for the study of more complex and technically relevant particle systems. We discuss the influence of particle concentration and particle size distribution. Moreover, we apply a novel nonlinear Mie Model that is capable to simulate the SH scattering from the surface of spherical particles and we show that the model can be used to extract information about molecules at the particle liquid interface.

Keywords: surface characterization, nonlinear Mie model, Second Harmonic Generation, optics, nanoparticles

1 INTRODUCTION

The macroscopic properties of particulate systems are mostly controlled by microscopic properties of the particle interfaces. Therefore in nanotechnology it is not only necessary to get information on size and shape of particles but also on additional surface properties as charge, surface structure and interface molecules.

There are several optical techniques as static and dynamic light scattering or image analysis which can be used to get information on size and shape and number densities of particles in suspensions and powders and which can also be used for online measurements during production and processing of particles. Methods for the in-line characterization of particle surfaces are rare. Most surface analytical techniques cannot be used directly in the process since they require often high vacuum, special preparation techniques or long measuring times.

Nonlinear optics opens the possibility to study embedded surfaces in-situ and in real time. Second Harmonic Generation (SHG) Spectroscopy is a well established technique for the investigation of planar interfaces [1–3]. Recently, it has also been applied to study the surface properties of colloidal interfaces [4]. Since then a lot of different particulate systems have been studies by nonlinear optical spectroscopy [5–15]. The Second Harmonic light is scattered at half the wavelength of the light which is used for excitation. In contrast to linear optical methods, where the separation of bulk and surface signals is often difficult if not impossible, second order nonlinear effects are exclusively surface sensitive. The signal generation depends on the nonlinear susceptibility $\chi^{(2)}$ of the system which vanishes for centrosymmetric materials. Interfaces between two such materials represent a symmetry break and do therefore possess a nonlinear surface susceptibility, which depends on the number of surface molecules, their orientation and local electro-magnetic fields. The method can in principle be applied to any optically accessible interface between two centrosymmetric bulk media. However, a factor that can hinder the application of SHG is that many molecules and interfaces have only a weak nonlinear susceptibility $\chi^{(2)}$ and therefore their Second Harmonic (SH) signal can be too low to be detected [12]. The particle sizes, particle concentration as well as the absorption of the media and the particles also have a strong effect on the total detectable Second Harmonic signal intensity [16].

For the application of SHG to more complex and technically relevant systems the influence of the detection geometry and the dependence of the SH scattering process on particle concentration and particle size distribution have to be elucidated. Therefore we have performed systematic studies with mixtures of monodisperse polystyrene particles in aqueous suspension and have compared the generated signal intensities quantitatively [17]. The applicability of the SHG technique for the study of technical relevant particles is demonstrated by experiments with ground TiO_2 particles.

Further, we have used angle-resolved SHG to derive additional information about surface molecules that are not accessible with conventional experimental transmission setups. A new nonlinear Mie theory is applied to simulate the SH scattering from individual molecules with arbitrary orientation that are located at the surface of spherical particles. Using this nonlinear Mie model we are able to derive additional information about molecules at the particle liquid interface [18].

2 EXPERIMENTAL SECTION

The experimental transmission setup used to study the influence of particle size distribution and to derive the surface potential of TiO_2 particles is shown in Figure 1.

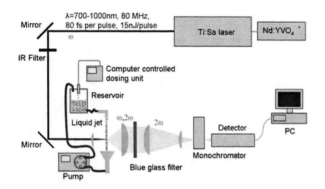

Figure 1: Schematic SHG transmission setup.

For the excitation of the SHG signals a pulsed Ti:Sa femtosecond laser with a pulse length of about 80 fs is used. The wavelength can be varied between 700 and 1000 nm. The beam is focused into the sample where the SHG signal is generated. With a second lens after the probe the scattered light is collected. A blue color filter that transmits only light shorter than 650 nm is used to block the transmitted and linear scattered laser light. The filtered light is then focused into the slit of a monochromator for a further separation of the SHG signal from other background radiation as two photon fluorescence. The SHG light is detected by a photomultiplier tube and recorded by a computer. For the measurements we use either a quartz cuvette with an inner diameter of 2 mm of a flow cell, consisting of a reservoir that contains the sample and an outlet tubing. A laminar liquid jet is formed which is collected with a funnel and is pumped back to the reservoir by a peristaltic pump. The SH signal is generated by a laser beam the laser beam that is focused inside the laminar jet. By the continuous flow of the suspension local heating of the sample and temporal variations of the solid concentration due to sedimentation are avoided. A computer controlled dosing unit was used to systematically vary the electrolyte concentration of the suspension by adding a 100 mM NaCl solution.

The experimental setup for the angle-resolved SHG measurements is shown in Fig. 2.

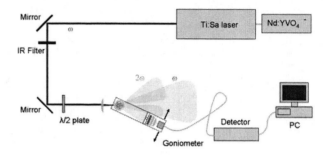

Figure 2: Schematic SHG setup for angle resolved experiments.

In contrast to the transmission setup, the optics for the detection of the SH light are mounted on a computer controlled pivot arm. For the angle resolved measurements a cylindrical cuvettes with an inner diameter of 3.5 mm is used. The angular resolution of about 1.4° of the goniometer setup is confined by a spherical aperture in front of the detector head.

3 RESULTS AND DISCUSSION

3.1 Influence of Particle Concentration and Particle Size Distribution

In Fig. 3 the characteristic dependence of the detected SH intensity in forward direction on the particle number density for monodisperse 770 nm polystyrene particles is shown. At low number densities, the SH signal increases linearly with the number of the particles. At a number density of about $1.25 \cdot 10^9$ particles / ml a maximum occurs and at even higher particle concentrations the SH signal in transmission decreases. In principle this decline could be caused by two different processes. The first is scattering and absorption of the laser light and the generated SH light by the particles in the suspension. The second process is interference of the SH signals from individual particles at high particle densities.

When the laser beam passes through the cuvette a part of the light is scattered and absorbed by particles in the suspension. The total intensity of the generated SH signal is proportional to the square of the intensity of the incoming laser beam. Thus, most of the SH signal is generated in the focal region which is schematically indicated in Fig. 4. Laser light that is absorbed before the focus region as well as light that is scattered in a way that it not reaches the focus region causes a decrease of the generated SH signal. On the other hand, SH light that is generated at the surface of the particles in the focus region is also scattered and absorbed by particles on the way to the detector. In a transmission setup the light is typically collected with a large solid angle of 40° - 50°. Thus the intensity of the generated SH signal that reaches the detector depends on the detection geometry as well as on the angular dependent Mie scattering by the particles.

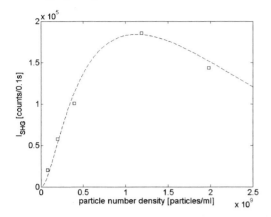

Figure 3: Dependence of the detected SH intensity in forward direction on the particle number density of 770 nm polystyrene particles with surface adsorbed malachite green

However, the decrease in the transmitted SH intensity with higher particles densities could also be caused be interference. The coherence length of the SHG process is in the order of micrometers. If the distance between individual particles is smaller than the coherence length for SHG, than the individual signals from individual particles do not simply add up, but interfere with each other and a nonlinear dependence between the generated SH signal and the particle density is expected. If this is the case, the total SH signal from particle samples with a wide size distribution can not be described as sum of the SH signals from the individual particles.

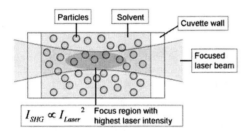

Figure 4: Schematic illustration of the focused laser beam inside the particle suspension

To determine the dominating mechanism that causes the decrease of the SHG signal in transmission at high particle concentrations, different samples of monodisperse polystyrene (PS) particles were mixed together and the intensities of the transmitted SH light within a solid angle about 48° were measured. The diameter of the PS particles used for these experiments was between 50 nm and 1.1 μm. The experiments were performed at different particles densities with solid concentrations ranging from 0.01 to 0.5 w/v%. For all samples a nonlinear relation between SH signal and particle density was observed at high particle concentrations. However, we found a linear dependence of the total SH signal intensity on the mixing ratio of smaller and larger polystyrene particles as long as the extinction of the samples was comparable for all samples. From our result we conclude, that the decrease of the SH signal at high particle densities is mainly caused by absorption and scattering. This suggests, that the total SH signal intensity can also be used for the quantitative study of surface processes of polydisperse and high concentrated particle samples.

3.2 Study of technical particle systems

To demonstrate the applicability of SH light scattering for the quantitative study of technical relevant particles we have analyzed the SH signal from anatase TiO_2 particles with a mean diameter $x_{50,3}$ of 370 nm in water at different electrolyte concentrations. In Fig. 5 a scanning electron microscope (SEM) image of a dry particle sample on a silicon wafer is shown.

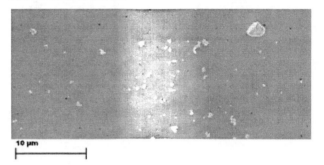

Figure 5: SEM image of a dry TiO_2 particle sample

As the particles were produced by a grinding process the particles are non-spherical and exhibit a wide size distribution as can be seen from the representative probe in Fig. 5. The SH signal from these particles consists of two source terms that interfere with each other. One of these is attributed to a contribution from the bulk of the particles. The magnitude of this term is not depending on the electrolyte concentration of the surrounding liquid. The other term is attributed to a layer of oriented interfacial water. The static electric field of the negative charges at the particle surface induces the orientation of the dipolar water molecules near the particle-liquid interface. In the presence of ions in the suspension an electrical double layer is formed and the negative surface charge is screened by positive counter ions. Thus, the penetration depth of the static electric field and the degree of orientation of the water molecules depends on the electrolyte concentration [19]. This correlation can be used to estimate the SH surface potential of the particles by fitting the dependence of the SH signal intensity on the electrolyte concentration with the Gouy Chapman double layer model [20]. In titration experiments with TiO_2 particles at different NaCl concentrations we found a SH surface potential of -67 mV at 10 mM NaCl and neutral pH. Using laser Doppler electrophoresis a zetapotential of -43 mV was measured for the same sample. This discrepancy between the SH surface potential and the zetapotential can be explained by the difference in the measurement principles. The zetapotential is a measure for the potential in the slipping plane whereas the SH surface potential is a measure for the orientation of the interfacial water layer which is closer to the particle surface than the slipping plane. Thus SHG measurements provide a potential that can be expected to be closer to the real surface potential than the commonly used zetapotential.

3.3 Application of Nonlinear Mie theory

In the previous mentioned applications of SH light scattering only the total SH signal intensity within a large solid angle was measured. However, it has been shown that SH light scattering from spherical particles exhibits distinct angular scattering profiles that differ significantly from the profiles of linear light scattering. We have developed a nonlinear Mie model that is capable to calculate SH light scattering from molecules at the surface of spherical particles [18]. The model can be applied to quantitatively analyze the nonlinear response of the interfacial molecule layer, allows us to derive information about the density and order of the surface molecules and to determine the dominating components of the surface susceptibility.

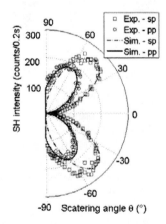

Figure.6: Comparison of experimental and simulated SH scattering profiles for a 200 nm PS particle with adsorbed MG at the surface

The experimental scattering profiles from a suspension of 200 nm PS particles with surface adsorbed MG molecules as well as respective simulations are shown in Fig. 6. A good agreement between simulations and experiment is found indicating that the nonlinear Mie model can be applied for the quantitative analysis of angular-resolved SH experiments with different types of colloidal particles.

4 CONCLUSION

We could show that the SH light scattering can be applied for the in situ study of surface properties of polydisperse particles in aqueous suspension. Furthermore a nonlinear Mie theory was applied to quantitatively analyze the angular SH scattering profiles from suspensions of spherical particles.

ACKNOWLEDGMENT

The authors gratefully acknowledge the funding of the German Research Council (DFG), which, within the framework of its `Excellence Initiative´ supports the Cluster of Excellence `Engineering of Advanced Materials´ (www.eam.uni-erlangen.de) and the Erlangen Graduate School in Advanced Optical Technologies (SAOT) at the University of Erlangen-Nuremberg.

REFERENCES

[1] Y.R. Shen, Appl. Phys. Sci. 93, 12104, 1996.
[2] Y.R. Shen, Solid State Commun 102, 221, 1997.
[3] K.B. Eisenthal, Chem. Rev. 96, 1343, 1996.
[4] H. Wang, E.C.Y. Yan, E. Borguet, K.B. Eisenthal, Chem. Phys. Lett. 259, 15, 1996.
[5] A. Srivastava, K.B. Eisenthal, Chemical Physics Letters 292, 345, 1998.
[6] E.C.Y. Yan, K.B. Eisenthal, J. Phys. Chem. B 103, 6056, 1999.
[7] E.C.Y. Yan, K.B. Eisenthal, Biophys. J. 79, 898, 2000.
[8] H. Wang, T. Troxler, A.G. Yeh, H.L. Dai, Langmuir 16, 2475, 2000.
[9] H.M. Eckenrode, H.L. Dai, Langmuir 20, 9202, 2004.
[10] S.H. Jen, H.L. Dai, J. Phys. Chem. B 110, 23000, 2006.
[11] S.H. Jen, H.L. Dai, G. Gonella, J. Phys. Chem. C 114, 4302, 2010.
[12] S.H. Jen, G. Gonella, H.L. Dai, J. Phys. Chem A 113, 4758, 2009.
[13] L. Schneider, W. Peukert, Part. & Part. Syst. Char. 23, 351, 2007.
[14] L.M. Tomalino, A. Voronov, A. Kohut, W. Peukert, J. Phys. Chem. B 112, 6338, 2008.
[15] K.B. Eisenthal, Chem. Rev. 106, 1462, 2006.
[16] L. Schneider, H.J. Schmid, W. Peukert, Appl. Phys. B 87, 333, 2007.
[17] B. Schürer, W. Peukert, Particul. Sci. Technol., accepted.
[18] B. Schürer, C. Sauerbeck, W. Peukert, S. Wunderlich, U. Peschel, in preparation.
[19] S. Ong, X. Zhao, K.B. Eisenthal, Chem. Phys. Lett. 191, 327, 1992.
[20] E.C.Y. Yan, Y. Liu, K.B. Eisenthal, J. Phys. Chem. B 102, 6331, 1998.

Size and Count of Nanoparticles by Scattering and Fluorescence Nanoparticle Tracking Analysis (NTA)

D. Griffiths*, P. Hole**, J. Smith**, A. Malloy, and B. Carr**

*NanoSight USA, 3027 Madeira Ave., Costa Mesa, CA 92626, USA duncan.griffiths@nanosight.com
**NanoSight Ltd., Amesbury, Wiltshire, SP4 7RT, UK Patrick.hole@nanosight.co.uk

ABSTRACT

A novel addition to a technique for the analysis of nanoparticles in a suspension is described. The Nanoparticle Tracking Analysis (NTA) technique sizes individual nanoparticles, based on their Brownian motion. NTA allows nanoparticles to be sized on a particle-by-particle basis, resulting in a higher resolution analysis and therefore a better understanding of polydispersity than ensemble methods (such as dynamic light scattering, DLS) and it also yields directly a count/concentration measurement. Analysis of scattering intensity is a recent development allowing sub-populations of nanoparticles with varying scattering characteristics to be resolved in a complex mixture.

Now this technique has been extended to the analysis and differentiation of fluorescently labeled nanoparticles. With the appropriate wavelength lasers and optical filters, the technique has been shown to be able to differentiate between sub-populations in a heterogeneous mixture.

Keywords: fluorescence, aggregate, nanoparticle, characterization, sizing.

1 INTRODUCTION

The analysis of nanoparticle properties is an increasingly important requirement in a wide range of applications areas and size analysis is usually carried out by either electron microscopy or dynamic light scattering (DLS). Both techniques suffer from disadvantages; the former requiring significant cost and sample preparation, the latter frequently generating only a population average size, which itself can be heavily weighted towards larger particles within the population.

A new method of microscopically visualizing individual nanoparticles in a suspension, called Nanoparticle Tracking Analysis (NTA), allows their Brownian motion to be analyzed and from which the particle size distribution profile (and changes therein in time) can be obtained on a particle-by-particle basis [1-3]. The technique offers significant advantages over traditional light scattering techniques (such as DLS- and SLS-based systems) for the characterization of polydispersed populations of nano-scale particles. Independent of particle density or refractive index, NTA dynamically tracks individual particles within the range of 10 - 1,000nm and provides size distributions along with a real-time view of the nanoparticles being measured.

This technique also provides a measurement of particle count within the measured volume. By knowing the interrogated volume, this particle count can be converted to a total concentration measurement.

Additionally, the technique works equally well whether the light from the particle is scattered or fluorescence. This allows sizing of counting of either naturally fluorescent materials or of particles that have been tagged with fluorophores.

2 MEASUREMENT METHODOLOGY

A small (250µl) sample of liquid containing particles at a concentration in the range 10^6-10^{10} particles/ml is introduced into the scattering cell through which a finely focused laser beam (approximately 40mW at wavelengths appropriate to the fluorophore of interest) is passed. Particles within the path of the beam are observed via a microscope-based system (NanoSight LM10 or NS500) onto which is fitted a CCD camera.

The motion of the particles in the field of view (approx. 100 x100 µm) is recorded (at 30 frames per second) and the subsequent video analyzed. Each and every particle visible in the image is individually but simultaneously tracked from frame to frame and the average mean square displacement determined by the analytical program. From this can be obtained the particle's diffusion coefficient. Results are displayed as a sphere-equivalent, hydrodynamic diameter particle distribution profile. The only information required to be input is the temperature of the liquid under analysis and the viscosity (at that temperature) of the solvent in which the nanoparticles are suspended. Otherwise the technique is one of the few analytical techniques which is absolute and therefore requires no calibration. Results can be obtained in typically 30-60 seconds and displayed in a variety of formats.

The minimum particle size detectable under scattering mode depends on the particle refractive index but for highly efficient scatterers, such as colloidal silver, 10nm particles can be detected and analyzed. For weakly scattering (e.g. biological) particles, the minimum detectable size may only be 30-50nm. For fluorescence measurements, the minimum detectable size also depends significantly on the specific fluorophore, incident wavelength, and number of fluorphores per particle.

The upper size limit to this technique is defined by the point at which a particle becomes so large (>1000nm) that Brownian motion becomes too limited to be able to track accurately. This will vary with particle type and solvent viscosity but in normal (e.g. aqueous) applications is approximately 800-1000nm. See www.nanosight.com for details.

3 SIZE DETERMINATION BY NANOPARTICLE TRACKING ANALYSIS

Brownian motion in a Newtonian fluid is governed by the Stokes-Einstein equation. Whilst the motion clearly occurs in three dimensions, NTA observes motion only in two dimensions. It is possible to determine the diffusion coefficient from measuring the mean squared displacement of a particle in the two observed dimensions;

$$\overline{(x,y)}^2 = \frac{4TK_B t}{3\pi\eta d}$$

where the first term is the mean squared displacement, T, is temperature, K_B is Boltzmann's constant, t is the time period (here given by $1/framerate$), η is viscosity and d is the hydrodynamic diameter.

By tracking the centers of the particles the mean squared displacement for each and every particle is calculated. This process is depicted below in figure 1. By recording a video of particles (fig 1a), tracking them (fig 1b) and compiling the resulting sizes a particle size distribution is established (fig 1c).

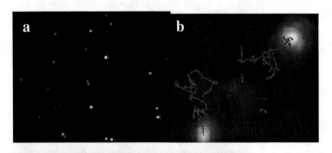

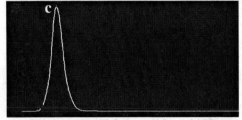

Fig 1. a) A still from a video of 100nm polystyrene calibration particles, b) showing only some (for clarity) of the Brownian motion trajectories analysed and c) subsequent particle size distribution.

4 INTENSITY DISTRIBUTION ANALYSIS

In addition to measuring the particle size for each and every particle, it is possible to extract further information about the particles. The intensity of the light scattered is strongly dependent on the particle size. Therefore simultaneously measuring the intensity of light scattered along with the particle size can further be used to gain more information about the particles. This can either infer information about the size distribution at higher resolution than diffusion rates alone or give an indication of the different particle materials used.

The results shown in Fig 3 were obtained from an analysis of a mixture of 200 and 300nm latex beads (overlaid with the normal particle size distribution plot, 3b) and shows that the two populations can be well resolved from each other. Furthermore, because the technique analyses particles on an individual basis and can collect information on their relative brightness as well as their size (measured dynamically) these two data can be combined to give an intensity v size plot (Fig 3c). This capability shares many features in common with conventional flow cytometry but is unique in this deeply sub-micron size range. [4]

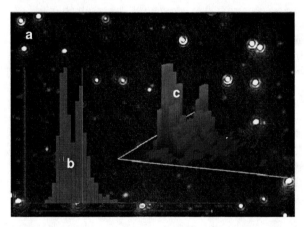

Fig. 3 A mixture of 200nm and 300nm particles; a) still image, overlaid with b) analysis plot and c) 3D number v. relative intensity v. diameter plot.

Figures 4 shows a heterogeneous mix of 50nm gold nanoparticles and 100nm polystyrene latex). Note how the intensity distributions demonstrate that the 50nm (gold) particles scatter at higher intensity than the 100nm (polystyrene latex particles). This example uniquely demonstrates the ability of the technique to differentiate species not just on their hydrodynamic size but also by their scattering ability.

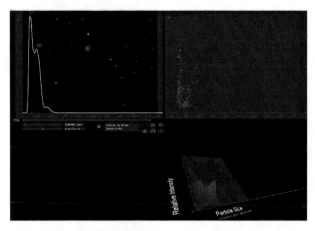

Fig. 4: Particle size distribution, intensity scatter plot, and intensity map for a 50nm gold, 100nm polystyrene mix.

5 FLUORESCENCE MEASUREMENTS

Extension of the NTA technique allows fluorescent nanoparticles to be individually tracked in real-time from which labeled particle size and concentration can be determined. Under light scatter mode, the total number of particles can be measured and subsequently compared to the concentration of labeled particles when measuring in fluorescence mode.

The NanoSight LM10 system uses either a 405nm (blue) or 532nm (green) laser source to excite suitable fluorophores whose fluorescence can then be determined using matched 430nm and 560nm long-pass filters respectively.

For example, a mixture of 100nm fluorescent (Fluoresbrite™, PolySciences Inc.) and 400nm non-fluorescent calibration polystyrene particles was measured under scattered light (Figure 5a) and through an optical fluorescence filter (Figure 5b). Under scattered light, both fluorescent and non-fluorescent particles were observed, sized and counted, while under the fluorescence filter only 100nm fluorescence particles could be visualized.

Note that it was also possible to retain concentration information on the fluorescently labeled nanoparticles for comparative labeling efficiency purposes.

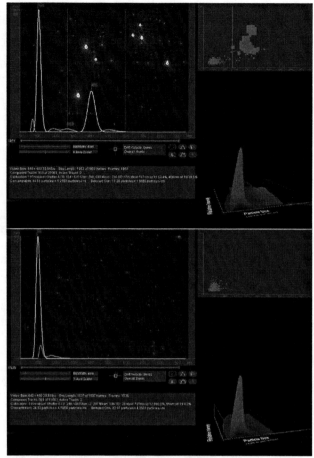

Fig. 5: Particle Size Distribution profiles (yellow graph) of a mixture of 100nm fluorescent and 400nm non-fluorescent polystyrene particles analyzed under a) scatter mode and b) fluorescent (optically filtered) mode.

In the following example an approx. 50:50 mixture of fluorescently labelled (Fluoresbrite) 100nm and unlabelled 100nm polystyrene beads were analyzed under light scatter mode (red line and top image) and when fluorescently filtered (white line and bottom image). The size result was the same for both results, but the difference in the y-axis shows the ability to accurately measure concentration differences.

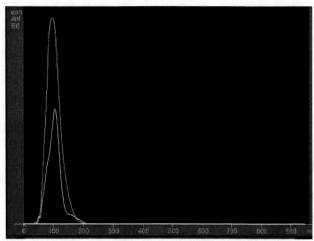

Fig. 6 Mixture of 100nm polystyrene and fluorescent polystyrene, showing difference in concentration in fluorescence measurement mode.

Semiconductor nanocrystals have recently emerged as a powerful and attractive alternative to conventional fluorescent labels due to their great chemical and optical stability and ease of use. Now commercially available as pre-functionalized kits with a choice of emission wavelengths, these interesting materials are rapidly gaining in popularity in the biosciences. While conventionally restricted to being imaged when immobilized (i.e. visualized by long exposure microscopy or when used to multiply label larger structures (e.g. Cellular structures)), NanoSight's new fluorescent versions of their LM Series instrument allow, for the first time, quantum dots to be visualised, sized and counted when unbound and moving freely under Brownian motion in liquids.

The following example is an analysis of a suspension of Invitrogen's non-functionalised QD655 QDot® nanocrystals in an aqueous buffer. Excited by NanoSight's 405nm (blue) laser and detected through a suitable filter, these 655nm emitting QDot® structures are visualized, sized and counted on an individual basis in less than 60 seconds.

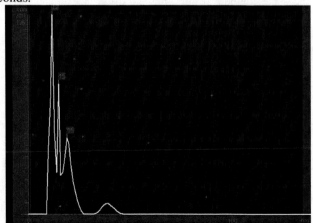

Fig. 7 Measurement of individual fluorescent quantum dot particles.

5 CONCLUSION

The NTA technique is a robust and direct method for characterizing particle size distributions and concentrations for a wide range of particulate materials. It represents an attractive alternative or complement to higher cost and more complex methods of nanoparticle analysis such as light scattering or electron microscopy that are currently employed. The technique uniquely allows the user a simple and direct qualitative view of the sample under analysis (perhaps to validate data obtained from other techniques) and from which an independent quantitative estimation of sample size, size distribution and concentration can be immediately obtained [5-7]. In addition, fluorescence experiments can be conducted to isolate either fluorescing materials or fluorescently-tagged populations in a heterogeneous mixture.

REFERENCES

[1] CARR, R (2005), NanoParticle Detection and Analysis, 27th International Fine Particles Research Institute (IFPRI) Meeting, June 25-29th, 2006, SANTA BARBARA, USA

[2] CARR, R., DIAPER, T. and BARRETT, E. (2005) NanoParticle Tracking Analysis – The Halo system. Abs Proc 9th Particulate Systems Analysis Conference 2005, September, Stratford-upon-Avon, UK

[3] WARREN, J and CARR, R (2006) 'A Novel Method for Characterisation of Nanoparticle Dispersions', Nanoparticelle: sintesi e caratterizzazione, Politecnico di Torino, Italy 8 March 2006

[4] van der SCHOOT, A (2007) "Sizing of nanoparticles by visualising and simultaneously tracking the Brownian motion of nanoparticles separately within a suspension" ChinaNANO2007 –Abs. International Conference on Nanoscience & Technology, June 4, 2007, Beijing, China.

[5] GHONAIM, H., LI, S., SOLTAN, M.K., POURZAND, C. and BLAGBROUGH, I. S. (2007) "Chain Length Modulation in Symmetrical Lipopolyamines and the effect on Nanoparticle Formulations for Gene Delivery", British Pharmaceutical Conference, Manchester, 10th Sept, 2007

[6] SAVEYN, H., De BAETS, B., HOLE, P., SMITH J. and VAN DER MEEREN, P. (2008) Accurate particle size distribution determination by Nanoparticle Tracking Analysis based on 2-D Brownian dynamics simulation in Abs PSA2008, Stratford on Avon, September, 2008

[7] MONTES-BURGOS, I., SALVATI, A., LYNCH, I. and DAWSON K. (2007) "Characterization techniques for nanoparticle dispersion" at European Science Foundation (ESF) Research Conference on Probing Interactions between Nanoparticles/Biomaterials and Biological Systems, Sant Feliu de Guixols, Spain, 3 November 2007

Conformation of pH responsive alternating copolymers and self-assembly in nanoarchitectures

A.S.W. Chan, M.N. Groves, X. Li and C. Malardier-Jugroot[*]

[*] Department of Chemistry & Chemical Engineering,
Royal Military College of Canada, Kingston, ON, Canada,
anita.chan@rmc.ca, cecile.malardier-jugroot@rmc.ca

ABSTRACT

In this paper, we discuss the formation of distinct nanoarchitectures from Poly(isobutylene-*alt*-maleic acid) (IMA) copolymer chains in water, and the dependence of pH on the self-assembly process. An experimental approach was chosen to investigate and characterize the association of IMA chains at different pH values. A change in structure at different environments was observed using Dynamic Light Scattering (DLS), Small Angle Neutron Scattering (SANS) and Transmission Electron Microscopy (TEM). DLS results reveal an association occurring at neutral pH whereas the copolymer was unable to from any observable structure at low and high pH. From the fitting of SANS pattern, it is suggested that IMA copolymer at pH 7 forms double layer sheets in which the outer surfaces are hydrophilic and the center gap is hydrophobic. In addition, TEM images of a dried sample showed sheet-like structures associated in layers.

Keywords: poly(isobutyl-*alt*-maleic anhydride), alternating copolymer, self-assembled nanoarchitecture

1 INTRODUCTION

The most widely used materials for the production of self-assembled nanomaterials are amphiphilic blockcopolymers. Example applications are drug delivery [1] and nanowire templates [2]. The association, shapes and properties of block copolymers have been extensively characterized [1]. On the other hand, the association of alternating copolymers has been rarely studied, and are predicted to have several major advantages over their block counterparts [3]. Nanoscale structures formed by alternating copolymers have proven very efficient for the solubilization of small hydrophobic molecules in water for applications in storage, delivery and nanoelectronics [4]. These copolymers can be structurally very close to biological systems with hydrophilic monomers alternating with hydrophobic monomers along the chain giving rise to complex and well-defined nanoarchitechtures [5]. In order to design new nanomaterials, it is therefore essential to understand the fundamental properties of the polymer assembly and the factors influencing it.

Our research focuses on the characterization of the association and the properties in aqueous solution of nanostructures composed of amphiphilic alternating copolymers. They are composed of a pH sensitive hydrophilic group, maleic anhydride alternating with a hydrophobic group. We will present the characterization of the self-assembly mechanism of poly(isobutyl-*alt*-maleic anhydride) (IMA) by a combination of experimental methods. It has been well observed that IMA forms stable linear chains [6] at neutral pH only due to strong electrostatic interactions between the sodium cations and the hydrophilic group of the copolymer. These linear chains (Figure 1) allows for the formation of nanoarchitecture by minimizing the entropic cost of association [3].

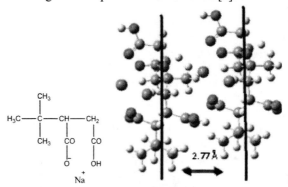

Figure 1. (Left) Representation of IMA monomer at pH 7, (Right) Representation of the interaction between two copolymer chains at pH 7.

Different structures were observed using Dynamic light scattering (DLS) at different pH environments. Small Angle Neutron Scattering (SANS) and Transmission Electron Microscopy (TEM) were used to characterize the nanoarchitecture at neutral pH, where the association among the polymer chains was predicted to occur [6].

2 METHODS

2.1 Sample Preparation

The IMA polymer was purchased from Aldrich Co. and has molecular weights of 6,000 (~85% purity). Similar to our previous work on poly(isobutylene-alt-maleic anhydride) in aqueous solution [6], low concentrated (0.05% wt in Milli-Q water) amphiphilic copolymer nanoarchitecture solutions of IMA were prepared and then sonicated for a few hours (6 hours to 1 day). The pH of the solution was adjusted accordingly by stoichiometric addition of sodium hydroxide. The solutions were centrifuged at 14,500 rpm for 10 minutes prior to analysis.

High concentrated samples (2% wt in ultra pure water) were produced by adding sodium hydroxide to the IMA solution to increase the pH to 12, and then by adding accordingly hydrogen chloride to adjust to neutral pH. This method allows for a higher concentration due to a higher solubility of IMA at high pH; however the concentration of salt in the solution and its ionic strength are much greater. The solutions were centrifuged at 14,500 rpm for 10 minutes prior to analysis.

2.2 Nanoarchitecture Characterization

The radii of gyration of the copolymer nanoarchitectures as a function of pH were measured in solution using DLS (Brookhaven BI-200SM Research Light Scattering) equipped with a 632.8nm He-Ne laser. The measurements were taken at a 90° angle. The samples were filtered using Titan 2 HPLC Filters with 0.45μm regenerated cellulose membranes.

The transmission electron microscopy (TEM) images were carried out on at Brockhouse Institute for Materials Research Canadian Centre for Electron Microscopy at McMaster University in Hamilton, Ontario. The images of the nanostructure sample after drying were obtained using FEI Titan 80-300 Cubed High-Resolution Transmission Electron Microscope equipped with Gatan 866 model spectrometer optics.

SANS experiments were carried out on the NG3 30-m Small Angle Neutron Scattering Instrument at the NIST Center for Neutron Research in Gaithersburg, Maryland. The incident wavelength was 6 Å. Three sample-to-detector distances, 1.00, 6.00 and 13 m, were employed, covering a q-range of 0.005 to 0.40 Å^{-1}. The samples were loaded in quartz cells. The sample aperture was 10*10 mm^2. The experiment was carried out using hydrogenated polymer in deuterated water as the solvent to increase the contrast. The scattering pattern was reduced and analysed using the Igor Macros package [7]. The averaged data were corrected for empty cell and background. Rigid Cylinder and Diluted Lamellar form factors were chosen for the fitting of the scattering patterns. The scattering intensity is described by eq 1 for rigid cylinders [8]:

$$I(q) = \varphi P(q) \tag{1}$$

$$P(q) = \frac{scale}{V_{cyl}} \int_0^{\frac{\pi}{2}} f^2(q, \alpha) \sin \alpha \, d\alpha$$

$$f(q, \alpha) = \frac{2(\rho_{cyl} - \rho_{solv})V_{cyl}j_0(qH \cos \alpha)(J_1(qr \sin \alpha))}{(qr \sin \alpha)}$$

$$V_{cyl} = \pi r^2 L \quad \text{and} \quad j_0(x) = \frac{\sin(x)}{x}$$

where φ is the particle volume fraction. J$_1$(x) is the first order Bessel function. Alpha is defined as the angle between the cylinder axis and the scattering vector, q. The integral over alpha averages the form factor over all possible orientations of the cylinder with respect to q. The returned value is in units of [cm^{-1}], on absolute scale.

For diluted Lamellar structures, the scattering intensity is given by eq 2 [9-10]:

$$I(q) = \frac{2\pi P(q)}{\delta q^2}$$

$$P(q) = \frac{2\Delta\rho^2}{q^2 \left[1 - \cos\left[(q\delta)e^{-\frac{q^2\sigma^2}{2}}\right]\right]} \tag{2}$$

where δ is the bilayer thickness and σ = variation in bilayer thickness = δ*polydispersity

3 RESULTS AND DISCUSSION

The pH values were chosen relative to the degree of hydrolysis of the maleic acid groups in the copolymer after titration analysis. In an aqueous solution of poly(isobutylene-*alt*-maleic anhydride) at pH 3, the maleic anhydride ring is hydrolyzed to give two acid groups. With the increase of pH, one acid group is hydrolysed and a sodium counterion stabilizes the site at pH 7. The second acid group is also hydrolysed when the pH of the solution is further increased to pH 12.

Theoretical analysis was performed to determine the most stable structure of the copolymer at different pH value to predict the structure effect of the pH on the polymer [7]. The fully optimized structures reveal a linear conformation at neutral pH due to the electrostatic interaction between the hydrophilic regions of the copolymer and the sodium counterion. From a rigid 2-D scan using AMBER and another optimization at DFT-6-31++G(d,p), the association of two chains revealed that two IMA chains are about 2.77Å apart (Figure 1). This linearity is only observable at neutral pH; hence we expect to see an observable structure at pH 7. At low and high pH, the copolymer chain takes a bent shape. As a result, associations are less likely to form due to the large entropic cost that would derive from this association.

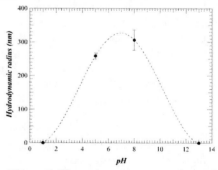

Figure 2. The points represent values of the hydrodynamic radius of 0.05 wt%. IMA at varying pH obtained from DLS. Values at pH 1 and pH 13 are below the resolution of the instrument. N.B.: the dotted line is an eye guide and not an actual fit.

Following the characterization using theoretical methods, the hydrodynamic radius was measured using DLS (Figure 2). The resolution of the instrument was between 20 nm to a few hundreds of nm. The results confirmed the association at neutral pH obtained from the theoretical models. Indeed, an association occurs at neutral pH and allows the polymer chain to form stable structures. At low or high pH values, the polymer chain is unable to form a structure large enough to be detected by DLS. The values for pH 1 and pH 13 are not zero, but they are below the resolution of the instrument. It can be concluded that the linear association predicted by the simulations would form larger nanostructures. The association can only occur at neutral pH. A similar pH dependence for the association of amphiphilic copolymers was observed for SMA [3] which revealed a self-assembly into nanotubes [11].

In order to analyze the association of the chains, the copolymer was characterized using HR-TEM images taken of a 2%wt IMA sample after drying (Figure 3). This image shows the formation of liner structures of width 5Å. The obtained d-spacing is 2.86 Å which is comparable to the value obtained from the simulations. The image shows very uniform and well-ordered structures which are unusual in polymer sample since the entropic cost associated with such linear structures would be very high. In this case, the copolymer chains are able to remain linear even after the removal of water which drives the self-assembly. This linearity is an optimistic sign that IMA nanostructures are indeed an excellent template to synthesize well ordered nanostructures.

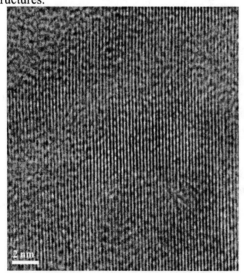

Figure 3. HR-TEM image after drying at scale 2 nm. The high resolution image shows lattice fringes. The measure the d-spacing of uniform structure is ~2.9Å.

Finally, SANS, a key technique for the study in complex fluids and polymers, provides detailed information on the structure of the nanostructure *in situ*. The experiment was performed in D_2O using hydrogenated IMA to maximize the contrast between the solvent and the polymer. The scattering curve corresponds to the IMA structure,

IMA-IMA association, and chain interactions. The scattering-length density (SLD) contrast is caused by the D_2O-IMA scattering length contrast (Table 1) [12]. The scattering profile obtained for a 2 wt % solution at pH 7 is shown in Figure 5.

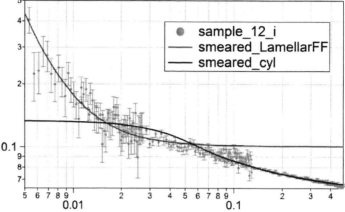

Figure 5. Neutron scattering pattern of 2% wt IMA solution at pH 7. The pattern was fitted using two different form factors: rigid cylinder for high Q values and diluted Lamellar at low Q values. There is a noticeable break between the form factors.

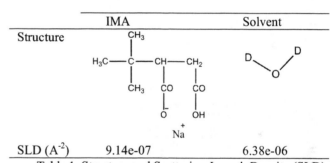

	IMA	Solvent
Structure		
SLD (A^{-2})	9.14e-07	6.38e-06

Table 1. Structure and Scattering-Length Density (SLD) of IMA and D2O

At low Q range ($0.005-0.01Å^{-1}$), there is a decaying slope. At mid Q range ($0.055Å^{-1}$), a noticeable break appears. This break is not the result of the connectivity of data between two detector positions. As the data is split between three detector positions (1, 6 and 13m from the sample), if the data was not properly reduced, breaks would appear at Q range of 0.02 and 0.1 $Å^{-1}$. As this break is at not around these points, it was concluded that is part of the scattering pattern.

Several different models were fitted to this data set, and a sum of two different models was required to characterize the scattering pattern for the entire range. The Dilute Lamellar Form Factor [9-10] and the Cylinder Form Factor [8] were use to describe the pattern at low ($0.005-0.055Å^{-1}$) and at high ($0.055-0.500Å^{-1}$) Q ranges respectively. With the use of this particular method, we are able to describe the structure at all level of resolution of the copolymer sample such as the uniform linear copolymer chains and the formation of nanoscale sheets that were observed using HR- and Cryo-TEM respectively.

At high Q range, the Cylinder Form Factor was chosen because of our assumption that the copolymer would behave like a rigid linear chain [6]. The resulting fit yields a radius of 2.063±0.002Å and a length of 87.37±0.04Å. The scales of these values indicate that the model is representing a single copolymer chain. The height of the chain found using this model is similar to that found in the simulation (4.2Å). The model did indeed fit the range of the scattering pattern in which would describe the linear chains.

At low Q range, where the scattering pattern describes the larger system, the above model would not accurately fit. Referring to the HR-TEM image (figure 3) which showed that individual linear copolymer chains would interact with each other and form a sheet, we were led to select the dilute Lamellar Form Factor which resulted in a better fit. Indeed, while keeping the SLD of the solvent and the copolymer (Table 1) fixed, the thickness of the sheets was determined to be 2.096±0.005Å. This value is similar to the width of one IMA copolymer chain, which suggests that the model is describing only one bi-layer of the amphiphilic sheet. It was well observed that the isobutylene groups are very hydrophobic. The self-assembly in water suggests that the sheets would form with the hydrophilic groups pointing outward whereas the hydrophobic region would face inwards and interact with the hydrophobic region of another sheet (Figure 6). This behaviour is analogous to a phospholipids' bi-layer sheet.

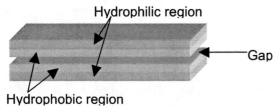

Figure 6. Representation of the copolymer bi-layer.

4 CONCLUSION

An experimental approach was chosen to investigate and characterize the association of poly(isobutylene-*alt*-maleic acid) chains at different pH values. A change in structure at different pHs was observed using DLS and SANS. It is suggested that IMA copolymers at pH 7 form double layer sheets composed of well-order uniform copolymer chains; the surfaces are hydrophilic and the center gap is hydrophobic. In addition, TEM images of a dried sample showed layers of sheet-like structure. HR-TEM images show a uniform formation of well-ordered wire structures 2.8Å apart which confirms the results of the scattering pattern fitting.

These results are analogous the theoretical predictions [6]. IMA is able to self-assemble into observable nanoarchitectures in water at neutral pH by adopting a linear structure, which would decrease the entropic cost of association. This linearity was only observable at neutral pH; which was confirmed by experimental characterization; hence we expect to see an observable structure at pH 7. At low and high pH, the copolymer chain takes a bent shape. As a result, associations are less likely to form due to the large entropic cost that would derive from this association.

We also observed the formation of IMA polydispersed spheres at low pH when the second preparation method was used: with an increase amount of salt. It is suggested that IMA can self-assemble into different nanoarchitectures at neutral or low pH. It is possible that, with enough ionic interactions present, the copolymer chains can reach the entropic barrier and will remain linear, thus leading to different forms of nanoarchitecutres. In the case of high pH, that barrier will not be attained; hence, no observable formation of a nanostructure would occur.

The complexity of poly(isobutylene-*alt*-maleic acid) nanostructure is vastly interesting. Unlike other polymer, its most energetically favored structure is linear once an energy barrier is reached. Once that is done, it is able to form different multi-scale nanostructures depending on the pH. Controlling the formation of these structures would have great implications to the application of bio-compatible nanosystems,

REFERENCES

[1] Zhang, Zhang, Wang, Liu, Chen and Liu, "Self-Assembled Nanostructures," Klumer Academic/Plenum Publishers, New York, 2003.

[2] E. J. W. Crossland, *et al.*, Soft Matter, 3, 94–98, 2007.

[3] C. Malardier-Jugroot, T.G.M. van de Ven and M.A. Whitehead, J. Phys. Chem. B., 109, 7022-7032, 2005.

[4] J. Xiang, W. Lu, Y. Hu, Y. Wu, H. Yan and C.M. Lieber, Nature, 441, 489-493, 2006

[5] D. C. Sherrington and K. A. Taskinen, Chem.Soc. Rev., 30, 83–93, 2001.

[6] A.S.W. Chan, M.N. Groves and C. Malardier-Jugroot, Molecular Simulations, accepted November 2009.

[7] S.R. Kline, "Reduction and Analysis of SANS and USANS Data using Igor Pro", J Appl. Cryst. 39(6), 895, 2006.

[8] A. Guinier and G. Fournet, "Small-Angle Scattering of X-Rays", John Wiley and Sons, New York, 1955.

[9] Nallet, Laversanne, and Roux, J. Phys. II France, 3, 487-502, 1993.

[10] Nallet, Laversanne, and Roux, J. Phys. Chem. B, 105, 11081-11088, 2001.

[11] C. Malardier-Jugroot, T.G.M. van de Ven, T. Cosgrove, R.M. Richardson, M.A. Whitehead, Langmuir, 21, 10179-10187, 2005.

[12] http://www.ncnr.nist.gov/resources/sldcalc.html, last accessed April 12, 2010.

Strong room-temperature photoluminescence of β-crystalline Ta$_2$O$_5$ nanobrick arrays

Yuan-Ron Ma, Jin-Han Lin, Wei-Der Ho, and Rupesh S. Devan

Department of Physics, National Dong Hwa University, Hualien 97401, Taiwan, Republic of China,
ronma@mail.ndhu.edu.tw

ABSTRACT

We analyzed the structural and photoluminescence properties of a new morphological form, stacking Ta$_2$O$_5$ nanobrick arrays that were synthesized by hot filament metal-oxide vapor deposition. Field-emission scanning electron microscopy showed the stacking Ta$_2$O$_5$ nanobricks to be arranged in a large-area array, on average ~20.7 nm wide. X-ray diffractometery showed the stacking Ta$_2$O$_5$ nanobricks to be the orthorhombic (β) phase, assigned to the space group P2$_1$2$_1$2. Photoluminescence spectra showed very strong green-light emissions, which emerged from the trap-levels of the oxygen vacancies within the Ta$_2$O$_5$ bandgap. One of the trap-levels, called the midgap state, can provide a very intense 566.3 nm emission with 44.0 % and 69.8 % of the intensity of the incident 486.8 and phonon-assisted 499.5 nm lasers, respectively, a strong indication that the stacking Ta$_2$O$_5$ nanobricks are good room-temperature visible-light emitters.

Keywords: Ta$_2$O$_5$, nanobricks, x-ray diffraction (XRD), photoluminescence (PL), room-temperature.

1 INTRODUCTION

Over the past few decades, researchers have been engaged in the study of metal-oxide thin-films and bulk materials. The high dielectric constant values as well as the steady thermal and chemical properties [1-5] of Ta$_2$O$_5$ thin films, as compared to those of SiO$_2$ thin films [2-4,6-7], make them good candidate materials for capacitors to be integrated into memory devices and for advanced electronic packing applications. Ta$_2$O$_5$ films also offer a wide-range of refractive-index adjustments [8] for integrated optical circuits, high-quality antireflection properties for solar cells [9-10], and better ionic conductivity for electrochromic devices (such as the chromogenic glazing in windows and large-scale information displays [11-12]). In addition, Ta$_2$O$_5$ is a good thermochromic material; its optical properties can be altered by simply elevating the temperature [13]. Hence, Ta$_2$O$_5$ films have potential for use in thermochromic 'smart windows', display devices [14-16], and many other similar applications necessitating electrochromic capabilities [17-18].

The crystalline structures of Ta$_2$O$_5$ are complicated, but the structures are based only on a network of TaO$_6$ octahedra and TaO$_7$ pentagonal bipyramids [19-20] with shared oxygen atoms. Specifically, Ta$_2$O$_5$ consists of only two polyhedral building blocks, TaO$_6$ octahedra and TaO$_7$ pentagonal bipyramids. This is similar to SiO$_2$, where amorphous or crystalline structures of SiO$_2$ are composed of only one type of building block, the SiO$_4$ tetrahedron. Oxygen (O) is easily volatilized in the complicated structures of Ta$_2$O$_5$, thereby yielding many types of oxygen vacancies. Note that although oxygen vacancies are the most predominant type of defect in Ta$_2$O$_5$, they are located at only two sites, the in-plane and the cap [19-20]. Each oxygen vacancy leaves one or two dangling bonds close to the vacancy location and donates two electrons. The dangling bonds can give rise to electronic states within the bandgap, which are able to trap excited electrons. This creates many trap-levels (electronic states) within the bandgap of Ta$_2$O$_5$. There is a large variety of in-plane and cap oxygen-vacancy sites that occur at the complex interfacial intersections of TaO$_6$ octahedra and the TaO$_7$ pentagonal bipyramids, and more complex interfacial intersections can supply more trap-levels. When the trapped electrons are released from the trap-levels to the ground state, varying wavelengths of visible light are emitted. Therefore, the presence of oxygen vacancies is expected to make Ta$_2$O$_5$ an *n*-type semiconductor and is important to play a strong role in the photoluminescence (PL) emissions of Ta$_2$O$_5$.

In this study, we synthesized large-scale arrays of stacking Ta$_2$O$_5$ nanobricks via hot-filament metal-oxide vapor deposition. The development of the hot-filament metal-oxide vapor deposition technique is very significant, because it can be used to synthesize a variety of nanostructures with different morphologies and crystalline traits [19,21-24] This development has opened up a new field in nanotechnology leading to the discovery of the dependence of electrical, thermal transport and mechanical properties on dimensionality and size reduction. The structural morphology and size distribution of the stacking Ta$_2$O$_5$ nanobricks were examined through field-emission scanning electron microscopy (FESEM). X-ray diffractometery (XRD) showed the crystalline structure of the stacking Ta$_2$O$_5$ nanobricks. PL spectroscopy is a well-developed optical tool for the inspection of semiconductor materials and devices for impurities and defects. Our room-temperature PL spectra showed very strong green-light emissions. For example, the intensity of the green-light emission of 566.3 nm, which emerged from the midgap trap-level, was about 44.0 % and 69.8 % that of the incident 486.8 and phonon-assisted 499.5 nm lasers, respectively.

The results are a good indication that the stacking Ta_2O_5 nanobricks are effective visible light emitters at room temperature, consequently confirming the viability of a large-scale array of these stacking Ta_2O_5 nanobricks as a potential candidate material for the fabrication of optoelectronic nanodevices, such as light emitting diodes (LEDs) and laser diodes (LDs).

2 EXPERIMENTAL METHODS

Large-scale arrays of stacking Ta_2O_5 nanobricks were synthesized using the hot filament metal-oxide vapor deposition technique. Clean tantalum (Ta) wires (99.98 % pure) with a diameter of 1 mm were passed through a pure graphite disc fixed to two supporting Cu electrodes mounted in a vacuum chamber. Once the pressure of the vacuum chamber was pumped down to 1×10^{-2} Torr, the Ta wire was heated to ~1800 °C for 10 min to generate hot tantalum vapor. When the hot tantalum vapor encountered the residual oxygen (or leaking air), a metal-oxide vapor arose which condensed into nanobricks on the cold Si wafer substrates. The substrates were placed on a graphite disc holder (~ 3 mm above the Ta-wires). Characterization of the as-synthesized large-area arrays of stacking Ta_2O_5 nanobricks thus produced was carried out with a field emission scanning electron microscope (JEOL JSM-6500F). Structural analysis of the stacking Ta_2O_5 nanobricks was carried out using an x-ray diffractometer (Philips X'Pert PRO) with Cu Kα radiations (λ=1.541 Å). The PL properties of the stacking Ta_2O_5 nanobricks were studied by scanning near-field optical microscopy (WITec, Alpha300 S) with an excitation laser of 486.8 nm (CVI Melles Griot 43 series, Ar ion laser, model 35-LAL-415-208).

3 RESULTS AND DISCUSSION

The surface morphology of the large-scale array of stacking Ta_2O_5 nanobricks can be seen in the FESEM image in Fig. 1(A). The stacking Ta_2O_5 nanobrick array is very compact, but its textural boundaries are clearly visible. The array contains ~1630 nanobricks per micrometer square. Various widths of the stacking Ta_2O_5 nanobricks all are smaller than 50 nm. The XRD patterns reveal details of the crystalline structures. Fig. 1(B) displays the XRD spectra of the stacking Ta_2O_5 nanobricks. From the diffraction peaks it can be seen that the large-area array of the stacking Ta_2O_5 nanobricks consists of only the orthorhombic (β) phase. The orthorhombic (β) phase is assigned to the space group $P2_12_12$ with lattice constants of $a = 0.6198$ nm, $b = 4.029$ nm, $c = 0.3880$ nm and $\alpha = \beta = \gamma = 90°$ (JCPDS 25-0922). The diffraction peaks (indicated by β) at 2θ = 22.9°, 28.2°, 28.5°, 28.7°, 28.8°, 28.9°, 37.1°, 37.2°, 38.4°, 44.7°, 59.8°, 60.1°, 60.4°, and 61.4°, respectively correspond to the (001), (1 11 0), (141), (200), (210), (081), (201), (211), (1 12 1), (2 15 0) or (340), (2 20 1), (410), (420), and (2 22 1) lattice planes.

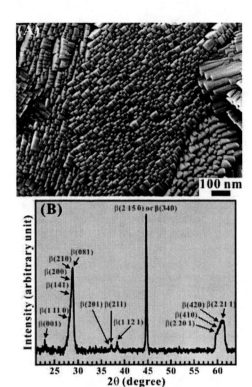

Fig.1. (A) FESEM image and (B) XRD spectrum of large-scale arrays of stacking Ta_2O_5 nanobricks

In previous experimental studies [25-33], trap-levels and shallow centers of oxygen vacancies within the bandgap of Ta_2O_5 have been observed. However, some theoretical studies [19-20,34] predict more trap-levels and shallow centers to be created within the bandgap. Therefore, it is of great interest to inspect the PL performance of the stacking Ta_2O_5 nanobricks. PL spectroscopy is a powerful optical tool that can be used to probe the impurities and defects in semiconductor materials. Fig. 2 shows the strong room-temperature PL spectra of the large-area array of stacking Ta_2O_5 nanobricks. The two sharp peaks at 486.8 and 499.5 nm (labeled by **1** and **2**, respectively) in Fig. 2(A) correspond to the incident and phonon-assisted lasers. Their full widths at half maxima (FWHMs) are only 0.9 nm. The very narrow FWHMs verify that the two sharp peaks represent laser emissions. As known, we synthesized the stacking Ta_2O_5 nanobricks on Si substrates. When the incident 486.8 nm laser impacts on the Si substrate, strong Si Raman scattering (or phonon effect) occurs. Here, some of the incident 486.8 nm laser (labeled by **1**) was absorbed by Ta_2O_5 samples, but some passed through the Ta_2O_5 samples and impacted on the Si substrate. Thus, the incident laser can be adsorbed to induce multiple lasers, which are called phonon-assisted lasers. The phonon-assisted 499.5 nm lasers (labeled by **2**) have intense intensity, because it corresponds to the Si Raman band of ~520 cm^{-1}.

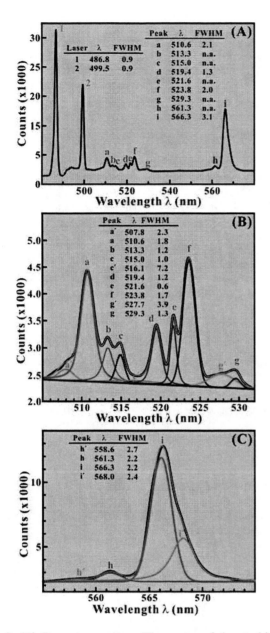

Fig. 2. (A) Room-temperature PL spectra of the stacking Ta$_2$O$_5$ nanobricks. The PL spectra in (B) and (C) are decomposed via partial Gaussian and Lorentz curve fittings.

In addition, there are two emission spectral bands located at green-light wavelengths ranging from 510 to 530 nm and from 560 to 570 nm. The first emission band consists of seven green-light emissions of 510.6 (labeled by **a**), 513.3 (labeled by **b**), 515.0 (labeled by **c**), 519.4 (labeled by **d**), 521.6 (labeled by **e**), 523.8 (labeled by **f**), and 529.3 (labeled by **g**). The second emission band contains two green-light emissions of 561.3 (labeled by **h**), and 566.3 (labeled by **i**) nm. Only the four **a**, **d**, **f**, and **i** peaks have indisputable FWHMs of 2.1, 1.3, 2.0, and 3.1 nm. Apparently, the **i** peak receives strong intensity, which is 44.0 % and 69.8 % that of the laser **1** and **2** peaks,

indicating that the stacking Ta$_2$O$_5$ nanobrick array can generate very strong green-light emissions. The remaining **a**, **d**, and **f** peaks also have very high intensities, which are 15.1 % and 24.0 %, 11.9 % and 18.9 %, and % 15.8 and 25.0 % those of the lasers **1** and **2** peaks, respectively. The first emission band can be decomposed with partial Gaussian and Lorentz curve fittings. The results of decomposition are shown in Fig. 2(B). After decomposition, there are three more peaks found at 507.8 (labeled by **a′**), 516.1 (labeled by **c′**), and 527.7 (labeled by **g′**) nm with the FWHMs of 2.3, 7.2, and 3.9 nm, respectively. The **a**, **b**, **c**, **d**, **e**, **f**, and **g** peaks reach their genuine FWHMs of 1.8, 1.2, 1.0, 1.2, 0.6, 1.7, 1.3 nm. The second emission band is also decomposed using partial Gaussian and Lorentz curve fittings, with the decomposition results being shown in Fig. 2(C). After decomposition, there are two more peaks discovered at 558.6 (labeled by **h′**) and 568.0 (labeled by **i′**) nm with FWHMs of 2.7 and 2.4 nm. The **h** and **i** peaks reach their authentic FWHMs of 2.2 nm. In brief, the PL results demonstrate that the large-area array of the stacking Ta$_2$O$_5$ nanobricks can provide numerous and various trap-levels within the bandgap, due to variations in the number of types of oxygen vacancies on the surfaces of the large-area array.

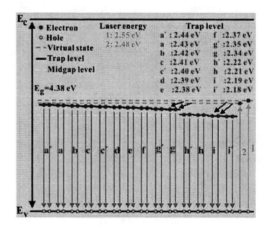

Fig. 3. Schematic diagram of the trap-level emissions of the stacking Ta$_2$O$_5$ nanobricks.

Fig. 3 shows a schematic diagram of the room-temperature trap-level emissions of the stacking Ta$_2$O$_5$ nanobrick array. The schematic representation of the various trap-levels within the bandgap of Ta$_2$O$_5$ helps to explain the room-temperature PL results in Fig. 2. The **a′**, **a**, **b**, **c**, **c′**, **d**, **e**, **f**, **g**, **g′**, **h′**, **h**, **i**, and **i′** peaks correspond to oxygen vacancy trap-levels of 2.44, 2.43, 2.42, 2.41, 2.40, 2.39, 2.38, 2.37, 2.35, 2.34, 2.22, 2.21, 2.19 and 2.18 eV, respectively. Electrons in the valence band are excited by two laser lights with wavelengths of 486.8 and 499.5 nm to the virtual states of 2.55 and 2.48 eV. The excited electrons undergo nonradiative transitions to the varying trap-levels, implying that at room temperature they are naturally trapped by the dangling bonds around the oxygen vacancies. When the trapped electrons return to the valence band, they emit photons (i.e., PL) to release photonic energy during the radiative transitions. Since the **i** peak is large, the trap level of 2.19 eV corresponds to the midgap state [25-33]. The midgap state implies that the energy bandgap of the stacking Ta_2O_5 nanobrick array is ~ 4.38 eV, which agrees with previous results [27,35-36]. The stacking Ta_2O_5 nanobrick array has a wide bandgap of ~4.38 eV making it more suitable for ultraviolet (UV) and blue optoelectronic applications. The stacking Ta_2O_5 nanobrick arrays not only provide green-light emissions, but also UV-light and blue-light emissions. The next stage will be the fabrication of electroluminescent nanodevices from them, such as light emitting diodes (LEDs) and laser diodes (LDs), when suitable *p*-type materials are found for the *n*-type stacking Ta_2O_5 nanobrick array.

4 CONCLUSION

We scrutinized the structural and electronic properties of a large-area stacking Ta_2O_5 nanobrick array synthesized using hot filament metal vapor deposition. The stacking Ta_2O_5 nanobricks were about 20.7 nm wide, and were composed of the orthorhombic (β) phase only at room temperature. The stacking Ta_2O_5 nanobrick arrays provided very strong green-light emissions at room temperature, which emerged from the numerous oxygen vacancies of the large-area array surfaces. The numerous oxygen of the stacking Ta_2O_5 nanobrick large-scale array arrangement make it a strong candidate for optoelectronic nanodevices, such as light emitting diodes (LEDs) and laser diodes (LDs).

5 ACKNOWLEDGEMENT

The authors would like to thank the National Science Council of the Republic of China for their financial support of this research under Contract Nos. NSC-97-2112-M- 259-005-MY2 and NSC-96-2811-M-259-004.

REFERENCES

[1]/ S. C. Sun and T. F. Chen, *IEEE Electron. Device. Lett.* 1996, **17**, 355-357.

[2]/ C. Chaneliere, J. L. Autran, R. A. B. Devine and B. Balland, *Mater. Sci. Eng. R-Rep.* 1998, **22**, 269-322.

[3]/ J. P. Chang, M. L. Steigerwald, R. M. Fleming, R. L. Opila and G. B. Alers, *Appl. Phys. Lett.* 1999, **74**, 3705-3707.

[4]/ G. Lucovsky, G. B. Rayner and R. S. Johnson, *Microelectron. Reliab.* 2001, **41**, 937-945.

[5]/ S. D. Kim, *Curr. Appl. Phys.* 2007, **7**, 124-134

[6]/ M. S. Mattsson, G. A. Niklasson, K. Forsgren and A. Hårsta. *J. Appl. Phys.* 1999, **85**, 2185-2191.

[7]/ S. Zafar, A. Kumar, E. Gusev and E. Cartier, *IEEE Trans. Device Mater. Reliab.* 2005, **5**, 45-64.

[8]/ H. Terui, and M. Kobayashi, *Appl. Phys. Lett.* 1978, **32**, 666-668.

[9]/ D. Bouhafs, A. Moussi, A. Chikouche and J. M. Ruiz, *Sol. Energy. Mater. Sol. Cells.* 1998, **52** 79-93.

[10]/M. Cid, N. Stem, C. Brunetti, A. F. Beloto and C. A. S. Ramos, *Surf. Coat. Technol.* 1998, **106**, 117-120.

[11]/M. J. Duggan, T. Saito and T. Niwa, *Solid. State. Ion.* 1993, **62**, 15-20.

[12]/C. Corbella, M. Vives, A. Pinyol, I. Porqueras, C. Person and E. Bertran, *Solid. State. Ion.* 2003, **165**, 15-22.

[13]/R. S. Devan, W.-D. Ho, J.-H. Lin, S. Y. Wu, Y.-R. Ma, P.-C. Lee and Y. Liou, *Cryst. Growth. Design.* 2008, **8**, 4465-4468.

[14]/C. Sella, M. Maaza, O. Nemraoui, J. Lafait, N. Renard and Y. Sampeur, *Surf. Coat. Technol.* 1998, **98**, 1477-1482.

[15]/N. Ozer and C. M. Lampert, *Sol. Energy. Mater. Sol. Cells.* 1998, **54**, 147-156.

[16]/C. C. Liao, F. R. Chen and J. J. Kai, *Sol. Energy. Mater. Sol. Cells.* 2007, **91**, 1282-1288.

[17]/J. S. Hale, M. DeVries, B. Dworak and J. A. Woollam, *Thin Solid Films* 1998, **313-314**, 205-209.

[18]/H. Yoshimura and N. Koshida, *Appl. Phys. Lett.* 2006, **88**, 093509.

[19]/H. Sawada and K. Kawakami, *J. Appl. Phys.* 1999, **86**, 956-959.

[20]/R. Ramprasad, *J. Appl. Phys.* 2003, **94**, 5609-5612.

[21]/Y.-R. Ma, C.-M. Lin, C.-L. Yeh and R.-T. Huang, *J. Vac. Sci. Technol. B* 2005, **23**, 2141-2145.

[22]/L. Kumari, Y.-R. Ma, C.-C. Tsai, Y.-W. Lin, S. Y. Wu, K.-W. Cheng and Y. Liou, *Nanotechnology* 2007, **18**, 115717.

[23]/L. Kumari, J.-H. Lin and Y. -R. Ma, *Nanotechnology* 2007, **18**, 295605.

[24]/L. Kumari, J.-H. Lin and Y. -R. Ma, *J. Phys. Condens. Matter.* 2007, **19**, 406204.

[25]/J. H. III Thomas, *Appl. Phys. Lett.* 1973, **22**, 406-408.

[26]/J. H. III Thomas, *J. Appl. Phys.* 1974, **45**, 835-842.

[27]/J. H. III. Thomas, *J. Appl. Phys.* 1974, **45**, 5349-5355.

[28]/S. Seki, T. Unagami and B. Tsujiyama, *J. Vac. Sci. Technol. A* 1983, **1**, 1825-1830.

[29]/W. S. Lau, L. Zhong, A. Lee, C. H. See, T. Han, N. P. Sandler and T. C. Chong, *Appl. Phys. Lett.* 1997, **71**, 500-502.

[30]/W. S. Lau, L. L. Leong, T. J. Han, and N. P. Sandler, *Appl. Phys. Lett.* 2003, **83**, 2835-2837.

[31]/M. M. Zhu, Z. J. Zhang and W. Miao, *Appl. Phys. Lett.* 2006, **89**, 021915.

[32]/H. Shin, S. Y. Park, S. T. Bae, S. Lee, K. S. Hong and H. S. Jung, *J. Appl. Phys.* 2008, **104**, 116108.

[33]/W. Miao, M. M. Zhu, Z. C. Li and Z. Zhang, *J. Mater. Trans.* 2008, **49**, 2288-2291.

[34]/B. R. Sahu and L. Kleinman, *Phys. Rev. B* 2004, **69**, 165202.

[35]/Y. L. Chueh, L. J. Chou and Z. L. Wang, *Angew. Chem-Int. Edit.* 2006, **45**, 7773-7778.

[36]/J. Robertson, *J. Vac. Sc. Technol. B* 2000, **18**, 1785-1791.

Nano-scale TEM Studies of the Passive Film and Chloride-induced Depassivation of Carbon Steel in Concrete

P. Ghods[*], O. B. Isgor[**], J. Li[***], G. J. C. Carpenter[****], G. McRae[*****] and G. P. Gu[******]

[*] Carleton University, Ottawa, ON, Canada, pghods@connect.carleton.ca
[**] Carleton University, Ottawa, ON, Canada, oisgor@carleton.ca
[***] CANMET Materials Technology Laboratories, Ottawa, ON, Canada, jili@nrcan.gc.ca
[****] CANMET Materials Technology Laboratories, Ottawa, ON, Canada, gcarpent@nrcan.gc.ca
[*****] Carleton University, Ottawa, ON, Canada, gmcrae@mae.carleton.ca
[******] CANMET Materials Technology Laboratories, Ottawa, ON, Canada, ggu@nrcan.gc.ca

ABSTRACT

In the highly alkaline environment of concrete, carbon steel rebar is protected against corrosion by a passive iron oxide film. The partial or complete loss of the protective oxide film, known as depassivation, leads to increased rates of rebar corrosion. Understanding the compositional and morphological characteristics of the passive film on rebar, and how it depassivates, is key to mitigate problems associated with corrosion of rebar in concrete. In this study, nano-scale TEM techniques have been used to characterize the passive films on carbon steel in surrogate concrete pore solutions. In chloride-free solutions, oxide films on the samples had a relatively uniform thickness that ranged from 5 to 8 nm, however, after exposure to chloride in amounts greater than typical depassivation thresholds, the thickness of the films became more variable and thinner on average. TEM dark-field images showed the oxide to be comprised of crystalline nano-particles with widths less than 8 nm.

Keywords: TEM, passive film, steel rebar, corrosion, chloride.

1 INTRODUCTION

Understanding the depassivation mechanism of carbon steel rebar in concrete due to chloride attack is key to the development of methods to mitigate problems associated with chloride-induced corrosion initiation in reinforced concrete structures. Although a number of researchers have proposed general mechanistic models [1-3], they do not explain experimental observations regarding the depassivation of carbon steel in highly alkaline environments such as concrete pore solution; therefore, additional studies are needed.

The investigation of passive films and depassivation processes is challenging and requires the use of nano-scale techniques. So far, only a few studies using some of these techniques have been carried out to study iron or steel passivity in highly alkaline environments [4-7]; and, to our knowledge, transmission electron microscopy (TEM) has not been used to investigate the passivity and depassivation of carbon steel in concrete pore solutions.

In this study, (bright-field and dark-field) TEM imaging and energy-dispersive X-ray spectroscopy (EDS) were used as nano-scale techniques to study the passive films on carbon steel that were grown in saturated calcium hydroxide (CH) solution.

2 EXPERIMENTAL SETUP
2.1 Preparation of specimens

TEM samples were made from the cross sections of 10 mm long rebar specimens cut from as-received deformed carbon steel rebar with 10 mm nominal diameter. The chemical composition of the steel is given in Table 1. Two rebar samples were mounted in low-shrinkage epoxy under vacuum. The cross sections of each cold-mounted specimen (see Fig. 1) was subsequently ground and polished using water-free polishing media and finished with 1 μm diamond suspension. After polishing, all cold-mounted specimens were kept in anhydrous alcohol, to avoid exposure to air, until they were transferred to the beakers containing CH solution (pH 12.5) where the passive films were formed on the cross sections of the rebar specimens. The CH solution was prepared by dissolving analytical grade calcium hydroxide in distilled water. The specimens were kept in the solutions for 15 days to allow the formation of stable passive films on the cross sections [8]. After passivation, one specimen was taken out from the solution; this specimen (CH-0) was used to investigate the passive films before exposure to chlorides. Following this, chloride was added to 0.45 M, which is beyond typical chloride thresholds [9]. After two weeks of immersion, the second specimen (CH-Cl) was taken out of the solution.

Figure 1: A cold-mounted rebar specimen that was used in the FIB-TEM study.

Element	Weight %
C	0.26
Si	0.27
Mn	1.10
Cr	0.05
Ni	0.07
Mo	<0.01
Cu	0.21
Al	<0.005
Nb	<0.01
V	<0.005
Ti	<0.005
B	<0.0005
P	0.01
S	0.03
W	<0.01
Sn	0.021
Co	0.01
Zr	<0.01
Fe	Balance

Table 1: Elemental composition of the rebar used in the TEM investigation

2.2 FIB sampling

Each specimen was transferred to the focused ion beam (FIB) microscope. A layer of tungsten (W) was sputtered on a ~200 μm by ~50 μm area of interest on the gold-covered cross section of the rebar specimen. The FIB H-bar lift-out method was used to extract the TEM sample from the gold and tungsten covered cross section of the rebar specimen. A gallium ion beam was used to make two trenches on the surface of the specimen. The edges of the area between the two trenches were then cut in order to release the sample from the bulk material. The released sample was placed on a copper TEM grid using a micro-manipulator. The middle part of the sample was milled to ~100 nm thickness using a gallium ion beam (Fig. 2) so that it was electron transparent.

2.3 TEM microscopy

The prepared TEM samples were examined in a transmission electron microscope (TEM) (Model: Philips CM20FEG) equipped with a Schottky field emission gun operated at 200 kV. Conventional bright-field TEM imaging was used to visualize the passive film. Dark-field TEM imaging was also used to determine the amorphous and crystalline phases. Chemical analysis was performed using a thin-window EDS detector from Oxford Instruments equipped with an INCA system analyzer. EDS analyses were obtained in scanning transmission electron microscopy (STEM) mode using a probe diameter of 1~2 nm.

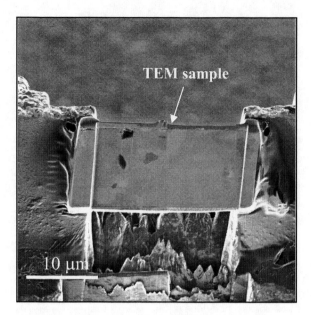

Figure 2: Process of thinning the TEM sample in the FIB microscope using the Gallium ion beam

3 RESULTS AND DISCUSSIONS

TEM micrograph of the CH-0 sample exposed to chloride-free CH passivating solution is shown in Fig. 3. A thin oxide film between the steel substrate and the gold layer is visible in the micrograph image. The oxide film has relatively uniform thickness. From this image, it is difficult to extract a good estimate of the thickness of the film, mainly due to the fact that the TEM samples were approximately 100 nm thick, and the layer that is visible in Fig. 3 represents the two dimensional projection of the oxide films across the depth of the TEM samples. Therefore, it is expected that the oxide film is thinner than the thickest part of the film in the TEM micrographs given in Fig. 3. The thickness of the oxide film can be estimated to range from 5 to 8 nm.

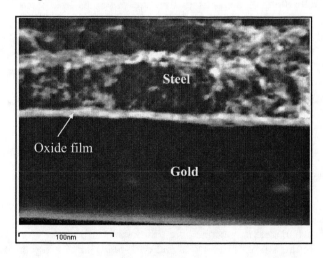

Figure 3: An TEM image of the CH-0 sample passivated in chloride-free solution (the oxide film thickness: 5 to 8 nm).

EDS analysis was carried out to investigate the elemental composition of the oxide films. Typical EDS spectrum obtained at the midpoint of oxide films grown in the CH solution is presented in Fig. 4. EDS analyses of different locations along the oxide film provided similar spectra. Oxygen and iron peaks in the spectra are the indication of the presence of iron oxides. Trace amounts of manganese, silicon and carbon in the EDS spectra originate from the steel substrate. Calcium was also detected in the spectra; this can be attributed to the presence of calcium in the structure of oxide film.

EDS analyses across the width of the oxide films provided information on the film composition with thickness. Results for the oxide film on the CH-0 sample are presented in Fig.5. Figure 5(a) shows the line along which EDS spectra were measured. Figure 5(b) shows the relative atomic percentages of the major elements along the line. The ratio of iron-to-oxygen decreases in the direction of the free surface (i.e., Gold layer).

TEM micrographs of the CH-Cl sample, which was exposed to the CH solution plus 0.45 M chloride, are shown in Fig. 6. Figure 6(a) shows that the thickness of the oxide film on the CH-Cl sample is thinner and more variable when compared with the thicker films on the CH-0 samples (see Fig. 3). On average, the thickness of the film on the CH-Cl samples is ~3 nm, and ranges between 2 nm and 10 nm, whereas the CH-0 films are more uniform and in the range of 5-8 nm, as described previously. Fig. 6(b) shows part of the oxide film where the thickness is < ~3 nm.

Further examination of the CH-Cl sample also showed the presence of pits, with depths ranging from 30 nm to 100 nm, on the steel surface. In Fig. 7, two pits are shown on the surface of the CH-Cl sample on which the oxide film is partially destroyed.

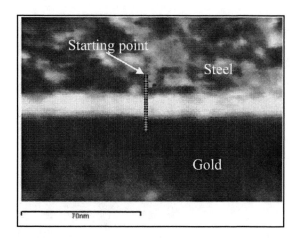

(a)

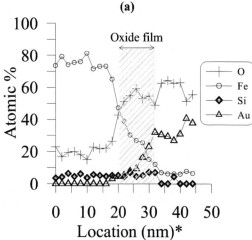

* Distance form the starting point shown in Fig. 5(a)

(b)

Figure 5: EDS analyses conducted on the CH-0 sample across the oxide film perpendicular to the steel surface: (a) TEM image showing the locations of the analyses; (b) EDS results corresponding to the points shown in the TEM image.

Additional evidence on the nature of the oxide film was obtained from dark-field TEM images. A dark-field TEM image taken of the oxide on the CH-0 sample is presented in Fig. 8. The bright phases, which are as large as 8 nm, are crystalline particles that compose the nanostructure of the oxide film.

4 CONCLUSIONS

The oxide films were 5 to 8 nm thick on rebar immersed for 15 days in chloride-free saturated calcium hydroxide 'simulated concrete pore' solution. Oxide films were mainly composed of nano-crystalline particles in various sizes up to 8 nm. After exposure to chloride in amounts larger than depassivation thresholds, iron oxides on the steel were no longer uniform and the average thickness of the oxide film was smaller than that observed before chloride exposure. In addition, some pit initiation sites were observed on the surface of steel.

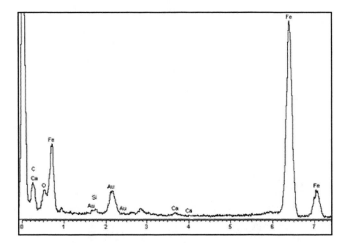

Figure 4: Typical EDS spectrum of an oxide film formed in the chloride-free solution (CH-0) obtained at the midpoint through the film.

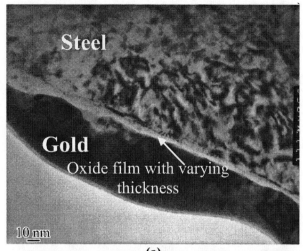

(a)

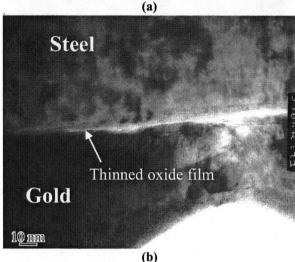

(b)

Figure 6: Typical TEM images of the oxide film from different parts of the CH-Cl sample showing: (a) non-uniform oxide thickness; (b) thinned oxide film.

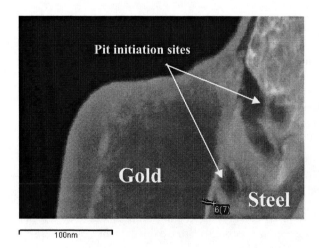

Figure 7: An TEM image of the CH-Cl sample showing two pits.

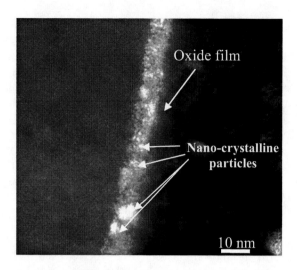

Figure 8: TEM dark-field image of the oxide film on the CH-0 sample.

ACKNOWLEDGMENTS

We would like to thank Ms. Catherine Bibby for her skilled assistance in the TEM investigation. This research was supported by a grant from the Natural Sciences and Engineering Research Council (NSERC) of Canada along with technical and financial support of CANMET-MTL laboratories, through the Resource for Innovation of Engineered Materials program. Both contributions are gratefully acknowledged.

REFERENCES

[1] L. F.Lin, C. Y.Chao and D. D. Macdonald, Journal of the Electrochemical Society, 128, 1194, 1981.

[2] J. Kruger, USA-Japan Seminar, NACE, Houston TX, 91, 1976.

[3] M. G. Alvarez and J. R. Galvele, Corrosion Science, 24, 1, 27, 1984.

[4] T. Zakroczymski, C. J. Fan and Z. Szklarska-Smialowska, Journal of the Electrochemical Society, 132, 2862, 1985.

[5] J. M. Sarver and Z. Szklarska-Smialowska, Localized Corrosion: Proceedings of an International Symposium Honoring Professor Marcel Pourbaix, 84, 328, 1984.

[6] S. Joiret, M. Keddam, X. R. Nóvoa, M. C. Pérez, C. Rangel and H. Takenouti, Cement and Concrete Composites, 24, 1, 75, 2002.

[7] P. Schmuki, M. Büchler, S. Virtanen, H. S. Isaacs, M. P. Ryan, and H. Böhni, Journal of the Electrochemical Society, 146, 6 , 2097, 1999.

[8] P. Ghods, O. B. Isgor, G. McRae and T. Miller, Cement and Concrete Composites, 31, 1, 2, 2009.

[9] P. Ghods, O. B. Isgor, G. A. McRae and G. P. Gu, Corrosion Science, doi:10.1016/j.corsci.2010.02.016.

Enhanced ferromagnetism and weak polarization of multiferroic BiFeO$_3$ synthesized by sol-gel auto-combustion method

GeMing Wang[*], JianMei Xu[**], De Zhang[***]

[*]Engineering Research Center of Nano-Geomaterials of Ministry of Education, China University of Geosciences, Room 419, China, wanggeming2006@sohu.com
[**]Engineering Research Center of Nano-Geomaterials of Ministry of Education, China University of Geosciences, Room 419, China, xjmchina2009@sohu.com
[***]Engineering Research Center of Nano-Geomaterials of Ministry of Education, China University of Geosciences, Room 219, China,zhd8638@163.com

ABSTRACT

BiFeO$_3$ powders and ceramics were prepared by sol-gel auto-combustion method. Optimal conditions for the synthesis of pure BiFeO$_3$ powders were experimentally obtained by X-ray diffraction patterns analysis. Enhanced room magnetic property with saturated magnetization (m_s) of 2.5emu/g and 0.02emu/g was observed respectively both in powders calcined at 500ºC (3h) and in ceramics sintered at 830ºC (3h). The latter also exhibited distinct ferroelectricity features with a considerable spontaneous polarization ($2P_s$=13.2µC/cm^2) at a low electric field of 6KV/cm, but failed to saturate due to its large leakage current.

Key words: ferroelectrics, magnetic property, sol-gel auto-combustion method; X-ray techniques

1.INTRODUCTION

Multiferroic materials which exhibit a coexistence of (anti)ferromagnetism, (anti)ferroelectricity, or (anti)ferroelasticity, has become the focus of research due to both their fascinating fundamental physics and great potential of multifunctional application.But multiferroic materials are rare as natural resources. As an interesting candidate of the few multiferroic materials, single-phase BiFeO$_3$ (BFO) is a rhombohedrally distorted perovskite crystallized in space group $R3C$, possessing ferroelectricity order (T_C=1083K) coupled with antiferromagnetism order (T_N=643K) at room temperature [1]. Recent, work on density functional calculations predicated that BFO would represent a macroscopic magnetization of about 0.9emu/g owing to a residual moment from a canted spin structure [2], as proved in the observation about BFO single-crystal and thin films [3-4]. However for bulk BFO, M-H loops are only observed at very low temperature. On the other hand, a relatively large leakage current density in BFO ceramics may also make it difficult to obtain a well-saturated hysteresis loop [4].

In our present work, we investigated Sol-gel auto-combustion method and obtained the optimal technical parameters to manufacture pure BFO. The structure, magnetic and ferroelectricity of BFO at room temperature will be discussed.

2.EXPERIMENTAL

BFO powders and ceramics were prepared via sol-gel auto-combustion method. The phase identification was examined by using X-ray diffractometer (XRD, Philips X'pert PW3373/10). Microstructure of ceramics was investigated by Scanning electric microscope (SEM, FEI Quanta 2000). Magnetic hysteresis (M-H) measurements were carried out at room temperature by using vibrating sample magnetometer with the maximum magnetic field of 12kOe (VSM, Nanjing Univeristy, BHV-55). Ceramics were polished to ~0.3mm and pasted by silver on both surfaces as electrodes, and then Ferroelectric hysteresis (P-E) measurements at room temperature were measured by using ferroelectric loop trace at a frequency of 50Hz(Huazhong University of Science and Technology, ZT-1).

3. RESULTS AND DISCUSSION

Fig.1 shows the influence of precursor's pH values on BFO powders' structure.

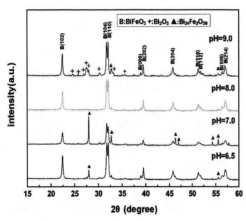

Fig.1: XRD patterns of BFO powders obtained by different precursor solution's pH

It is found that all the samples have single-phase BFO pervoskite crystal structure. For pH=9.0, it is observed that powders exist a few impurities peaks corresponding to $Bi_{24}Fe_2O_{39}$ (JCPDS NO.42-0201) and Bi_2O_3 (JCPDS No: 50-1088). The onset of Bi_2O_3 may be contributed to high pH value to make Bi_2O_3 separate from sol-gel during heating and calcining process. For pH=8.0, pure BFO phase is found and the powders' average size is 112nm (measured by zeta potential & particle sizing device, Macvern 2S90), indicating that it is viable to fabricate the pure and smaller BFO powders when pH value is 8.0.

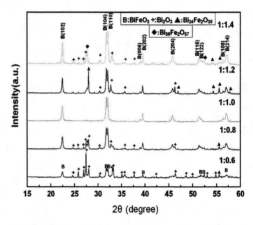

Fig.2: XRD patterns of BFO powders obtained by different metal cations/ citrate acid

Fig.2 demonstrates the effect of the proportion of iron cation to citrate acid on BFO powders' structure. For the proportion of 1:0.6 and 1:0.8, the powders can be defined as the mixture of Bi_2O_3 and BFO. Until up to 1:1, all the XRD patterns were indexed with BFO structure. However with further increasing in the proportion (1:1.2, 1:1.4), impurities such as $Bi_{24}Fe_2O_{39}$, Bi_2O_3 or $Bi_{36}Fe_2O_{57}$ occurred.

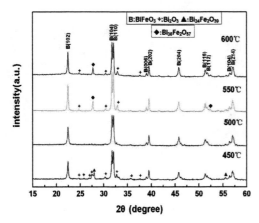

Fig.3: XRD patterns of BFO powders calcined at different temperature for 3h

Fig. 3 represents XRD patterns of powders calcined at different temperatures. It can be seen that BFO was obtained above 450°C. While for powders calcined at 450°C, 550°C and 600°C, there were impurity phases of Bi_2O_3, $Bi_{24}Fe_2O_{39}$ and $Bi_{36}Fe_2O_{57}$.

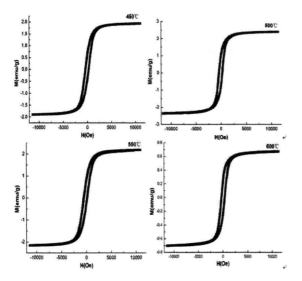

Fig.4: Room temperature magnetic hysteresis loops of BFO powders calcined at different temperature (3h).

Fig.4 shows the magnetic hysteresis loops of BFO powders calcined at different temperatures, measured by vibrating sample magnetometer at room temperature. All the calcined powders show weak ferromagnetic ordering at room temperature, similar to thin films[3]. The values of the saturated magnetization (m_s) are respectively 2.0, 2.5, 2.1 and 0.7 emu/g for 450°C, 500°C, 550°C and 600°C. The corresponding values of the coercive field (H_c) are almost the same (≈400Oe).

Fig.5 shows the magnetic M versus applied fields H of BFO ceramic calcined at 830°C (3h) at room temperature by vibrating sample magnetometer.A weak net magnetization of 0.02emu/g is much better than previous work [4] and comparable to some lanthanum-modified BFO ceramics.

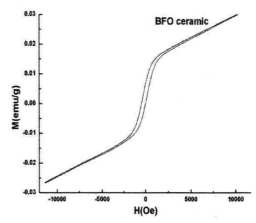

Fig.5 Room temperature magnetic hysteresis loops of multiferroic properties of BFO ceramic sintered at 830°C

The hysteresis loops together with a small remnant magnetization and coercive field are achieved in our samples which confirmed that BFO is antiferromagnet and weak intrinsic ferromagnetism. The powders (500°C) arecorresponding to pure phase (from fig.3) and smaller average size (112nm) and hence the largest M_s. At 600°C, the M_s decreased largely, indicating higher calcined temperature hampers the release of potential ferromagnetism. Magnetic properties of our samples both powders and ceramic are much better than those in other reports and theory prediction [2]. The oxygen vacancies and valence fluctuation (the coexistence of Fe^{2+} and Fe^{3+}) may lead to the structure distortion in the pervoskite with the canting of antiferromagnetically ordered Fe-O-Fe chains of spin, resulting in suppressing the spin spiral and hence releasing potential weak ferromagnetic.

The ferroelectricity polarization hysteresis loops are presented in Fig.6.for BFO sample sintered at 830 °C for 3h.The spontaneous polarization (P_s), remnant polarization (P_r) and the coercive field (E_c) are about $6.6\mu C/cm^2$, $2.4\mu C/cm^2$,and 2.0KV/cm, respectively. The P-E curve of our samples shows distinct ferroelectricity features and

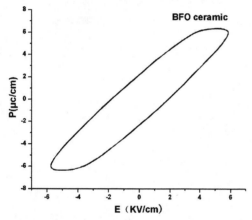

Fig.6: Room temperature P-E curves of multiferroic properties of BFO ceramic sintered at 830°C

exhibits a considerable Ps even under a low applied electric field (less than 6KV/cm) but without saturation. However, as the electric field increased to 7KV/cm, breakdown occurred, which may be mainly due to the ceramics' large leakage current and relatively low resistivity which restrains ferroelectricity property.

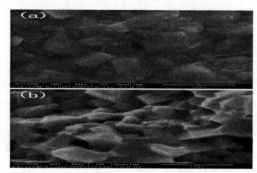

Fig.7: Scanning electron micrographs of BFO ceramics sintered at 830°C (a) natural surface (b)

SEM photograph (Fig.7) reveals that the grain size in ceramics is submicronmeter with approximately less than 1μm and randomly oriented, and the grain growth is homogeneous.

4.CONCLUSIONS

In summary, sol-gel auto-combustion method was an effective way to prepare high quality BFO powders and ceramics. Optimal conditions for synthesizing pure BFO powders with average size of ~112nm are as follows: precursor solution's pH 8.0, mental cations/ citrate acid ratio 1:1 and calcined temperature 500°C (3h). Enhanced room ferromagnetic with M_s of 2.5emu/g and 0.02emu/g was observed in BFO powders and ceramics. Meanwhile, The P-E curve of ceramic shows distinct ferroelectricity features with $2P_s$ of $13.2\mu C/cm^2$, $2P_r$ of $4.8\mu C/cm^2$, Hc of 2.0KV/cm under a low restricted applied electric 6KV/cm.

REFERENCES

[1] N. A. Hill. Phys. Chem. B 104(2000), 6694-6709.

[2] Kubel .F and Schimid.H. Acta Crystallogr. B .46(1990), 698-702.

[3] Wang.J, J.B.Neaton, H.Zheng, V.Nagarajan, S.B.Ogale and B.Liu. Science.299 (2003), 1719-1722.

[4]M.Mahesh Appl.phys.lett.Vol.76, No.19, 2000

Size Distribution Determination of Nanoparticles and Nanosized Pores by Small-Angle X-Ray Scattering on a Multi-Purpose X-ray Diffractometer Platform

J. Bolze[*], S. Rekhi[**], K. Macchiarola[**] and B. Litteer[**]

[*] PANalytical B.V., Almelo, The Netherlands, joerg.bolze@panalytical.com
[**] PANalytical Inc., Westborough, MA, USA, kathy.macchiarola@panalytical.com

ABSTRACT

A multi-purpose X-ray diffractometer platform was configured for small-angle X-ray scattering (SAXS) measurements. This technique was applied for nanoparticle and pore size distribution determination in nanopowders, colloidal dispersions, nanocomposites and porous materials in a range of 1-100 nm. It is extremely versatile and yields truly ensemble-averaged results from a large sample volume, does not require knowledge about the refractive index or any other physical properties, and it is applicable to crystalline and amorphous materials alike. Furthermore sample preparation is minimal. Advanced data analysis software allows to reveal complex multimodal size distributions with unrivaled resolution. Data acquisition and analysis can be automated for routine production control. The instrument platform used can be reconfigured within minutes for a multitude of complementary X-ray analytical techniques, such as powder diffraction, thin film analysis, and computed tomography.

Keywords: X-ray, SAXS, size distribution, nanocomposite

1 INTRODUCTION

Nanoscaled materials are becoming increasingly important in various applications such as coatings, paints, cosmetics, ceramics, polymers, catalysis and drug delivery. Even at relatively low concentrations such additives may be highly effective in delivering significant improvements in the desired field of application. The properties and performance characteristics of nanoparticles are to a large extent determined by their size distribution, specific surface area, and degree of dispersion. Similarly, with porous materials containing nanosized pores, as used e.g. in delivery systems, separation processes or catalysis, the pore size distribution and specific surface area are key quality parameters. It is thus essential to control and quantify these aforementioned properties.

With most available experimental methods this task becomes increasingly challenging with decreasing sizes below 100 nm. As an example, laser diffraction is a well established and powerful sizing method for particles in the range of micrometers down to 100 nm. However, with particles smaller than 100 nm and the wavelength of laser light being significantly longer, this technique is becoming less suitable. For nanomaterials characterization it is thus logical to make use of methods that are based on electromagnetic radiation of a shorter wavelength. X-rays having a wavelength of the order of 1 Å are widely used in diffraction experiments for the analysis of crystalline materials with atomic resolution. Powder diffraction allows for qualitative and quantitative analysis of crystalline phases, as well as for the determination of crystallite size.

When extending such wide-angle X-ray diffraction (XRD) measurements down to very small angles, one is probing the electron density distribution on increasingly larger length scales in the range of nanometers. The small-angle X-ray scattering (SAXS) technique [1-2] is thus ideally suited for the structural characterization of nanoscaled materials, and among others, allows for nanoparticle and pore size analysis and specific surface area determination. It is important to note that crystallite size and particle size are usually not the same. XRD and SAXS thus yield complementary structural information.

In this paper we present how SAXS measurements can be performed on a multi-purpose X-ray diffractometer platform. Application examples for nanopowders, colloids, nanocomposites and porous materials will be given. The advantages of SAXS as compared to other available measurement techniques will also be discussed.

2 THEORY

For a homogeneous spherical particle the scattering intensity I as a function of the scattering vector q ($q = 4\pi / \lambda \sin(\theta)$ where λ is the wavelength of radiation and 2θ the scattering angle) exhibits a steep decay that is modulated by distinct oscillations. The oscillation frequency is inversely related to the size of the particle (see simulation in fig. 1, top). A well-known analogous case from optics are the interference fringes that are observed on a screen when laser light passes a pinhole.

For an ensemble of spheres of different sizes, in first approximation, the scattering intensities of all fractions of particles simply add up. It may thus be readily understood that with increasing width of the size distribution the pronounced oscillations that are observed in the scattering curve of a single particle become more and more smeared out (see fig. 1, bottom). This is a result of the summation of different oscillation frequencies that are related to the various size fractions present in the sample.

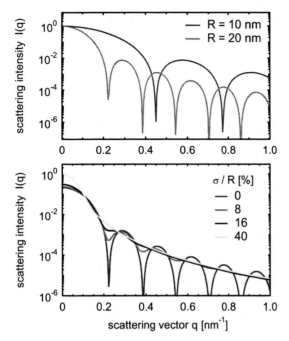

Figure 1: Simulated SAXS data from spherical particles with a distinct radius R (top) and with a certain size distribution width σ centered around R = 20 nm (bottom)

An indirect Fourier transformation procedure that searches for the frequency components in experimental SAXS data may thus be used to retrieve the particle or pore size distribution. The specific surface area from nanoparticles can then be readily determined from an integral over the size distribution function, given that the surface-to volume ratio of spheres is 3/R and by taking into account the mass density of the particles. More details can be found in [1-2] and in further references given therein.

3 EXPERIMENTAL DETAILS

Setups for SAXS measurements require a narrow, highly collimated and intense X-ray beam, the effective suppression of any parasitic scattering, and a detector with a high linearity range. The objective is to measure the scattered intensities in the immediate vicinity of the direct beam, typically down to 0.1 deg and below. The smallest accessible scattering angle determines the upper limit of the structural feature size (e.g. particle diameter) that can be studied.

A multi-purpose X-ray diffractometer platform (Empyrean, PANalytical), which can be configured for a variety of applications by choosing the optimal combination of optical modules, sample stages and detectors, was used. Typical applications on this platform range from the identification of unknown crystalline phases and quantification of mixtures, to the determination of residual stress and preferred orientation of crystallites in bulk materials as well as in thin films. It may also be configured for computed X-ray tomography. The system makes use of

pre-aligned fast interchangeable X-ray (PreFIX) modules. Thus there is no need for realignment when changing between different setups.

Using this platform, SAXS measurements were done in a transmission geometry (see fig. 2) using Cu Kα radiation (wavelength λ = 0.154 nm) from a long fine focus X-ray tube powered by a high voltage X-ray generator at 45 kV and 40 mA. A well-collimated, intense, monochromatic X-ray beam was obtained by using an X-ray mirror. The size of the incident beam at the sample position was set to approx. 20 mm x 50 μm by using a combination of low-angle slits and a beam mask. The samples were placed in the center of the diffractometer; a reflection-transmission stage was used for solids, and a capillary holder for liquid samples. An automatic sample changer allowed to measure multiple solid samples in a batch. On the diffracted beam side a narrow anti-scatter slit was used to reduce parasitic scattering effects and a solid-state pixel detector (PIXcel[3D], PANalytical) to sequentially measure the scattered intensities as a function of the scattering angle 2θ. The latter is defined as the angle between the incident and the scattered beam.

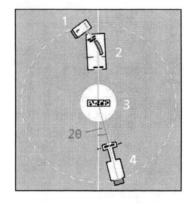

Figure 2: Schematics and photograph of experimental SAXS setup: 1: X-ray tube, 2: beam defining slits and X-ray mirror, 3: sample stage, 4: anti-scatter slit and detector.

Solid samples were placed in circular transmission holders between thin X-ray transparent polymer foils (see fig. 3). Liquid samples were confined in quartz capillaries with a diameter of 1 mm. As SAXS is a non-destructive and non-intrusive method, the samples can be recovered from the sample holders after a measurement.

Figure 3: Sample preparations (left to right): empty sample holder, nanopowder and polymer nanocomposite.

SAXS data were acquired by measuring the scattered intensities as a function of 2θ - typically up to 5 deg, with a step size of 0.01 deg. The background signal was measured using an empty sample holder or the dispersion medium (in case of nanoparticle dispersions). The profile of the direct beam (transmitted through the sample) was included in each measurement for the determination of the absorption factor of a given sample. The smallest accessible scattering angle was around 0.07 deg. Typical measurement times were 2-20 minutes for solid samples, and 30 minutes up to several hours for dilute liquid samples.

4 DATA ANALYSIS

SAXS data analysis was done using a commercial software package (EasySAXS, PANalytical) that was developed in cooperation with the European Molecular Biology Laboratory (EMBL, Germany). It is partly based on the ATSAS software suite [3]. The software performs primary data handling steps, such as absorption correction, background-subtraction, and conversion of the scattering angle 2θ to the scattering vector q. Determination of nanoparticle or pore size distribution was done by an indirect integral transform technique that relates experimental SAXS data to the volume distribution function $Dv(R)$ using a regularization procedure [2-3].

Whereas the nanoparticles or pores are assumed to be (quasi-)spherical, no further assumptions about the shape and modality of the size distribution curve are made. By using this general approach, rather complex, multi-modal size distributions can also be revealed. This is a significant advantage of the indirect transformation technique over direct model fitting procedures.

5 EXPERIMENTAL RESULTS

5.1 Nanopowder

A commercial sample of a titania nanopowder with photocatalytic properties was investigated. A quick XRD scan revealed the presence of 95% of anatase and 5% of rutile. Due to the effective collimation used in the experimental SAXS setup, the background quickly decreases at lowest angles and then remains on a very low level. The characteristic hump that is observed in the steeply decaying, background-corrected scattering curve indicates the presence of nanoparticles with a rather well-defined particle size (fig. 4). A strong upturn of the scattering intensity occurs towards lowest angles, where the largest dimensions are being probed. This may be attributed to the presence of micron-sized, loose agglomerates of primary particles. For the determination of the size distribution of the primary particles, this region was thus excluded from further data analysis.

The result from size distribution analysis (fig. 4) indicates that the sample contains particles having a radius

predominantly in the range of 15-75 Å, with a volume-average of 48 Å and a relative standard deviation (size polydispersity) of 24%. The specific surface area of the particles was determined to 189 m^2/g. These results are in good agreement with those obtained from the same samples using TEM and the gas adsorption (BET) method. Note that these SAXS results were obtained within 5 minutes only. Due to the non-avoidable agglomeration effects in nanopowders, the dynamic light scattering method does not allow to determine the size of the primary particles from such a sample. Instead only a signal from the large agglomerates is observable.

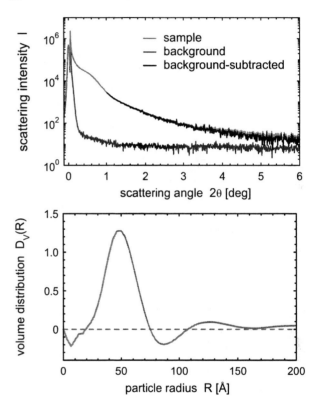

Figure 4: SAXS data from a titania nanopowder (top) and the determined volume distribution (bottom)

5.2 Colloidal Dispersion

An aqueous dispersion of silica at a concentration of 0.8 Vol.% was measured and analyzed with SAXS. This was a control experiment in which three samples of colloidal silica were mixed in an equal ratio. The goal was to demonstrate the superior capability of SAXS to analyze complex, multimodal size distributions.

In spite of the very low concentration, a pronounced excess signal from the nanoparticles over the background could be observed down to as low as 0.07 deg 2θ (fig. 5). Data analysis was performed without any presumption with respect to the shape of the size distribution curve. The result from SAXS indeed clearly revealed a trimodal distribution in which the three fractions only have a small difference in size (namely 5, 8 and 14 nm). Furthermore, the 1:1:1

mixing ratio that was used for sample preparation is correctly found back in a deconvolution of the size distribution curve. The SAXS data also show clear evidence for monodispersed, non-aggregated particles.

For analyzing multimodal size distributions of particles of such small size as in this example, SAXS is the most accurate experimental method. With dynamic light scattering only a monomodal, broad distribution would be observed (fig. 5). Whereas with SAXS and electron microscopy the actual, hard core particle radius is determined, hydrodynamic methods yield effective radii that include also the hydration shell.

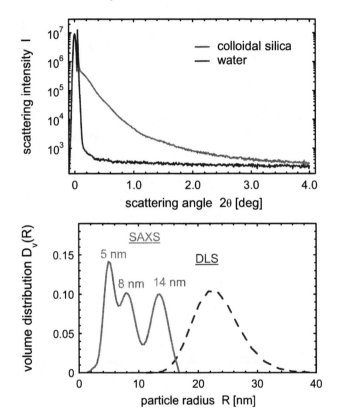

Figure 5: SAXS data from 0.8% colloidal silica (top) and the determined trimodal volume distribution (bottom), together with a typical result from dynamic light scattering.

5.3 Polymer Nanocomposite

PMMA was modified with 3 wt.% of silica nanoparticles in a melt compounding process to yield a nanocomposite with improved thermal and mechanical properties [4] (sample and TEM courtesy C. Beckers, Inst. Henri Tudor , Luxembourg). SAXS data from this sample allowed to determine the nanoparticle size distribution and to confirm that the surface-functionalized particles are well dispersed in the polymer matrix, which is crucial for the application properties (fig. 6). These findings are in good agreement with observations from TEM. However SAXS has the important advantage of quickly delivering such information over a relatively large volume of the sample, and with minimal sample preparation.

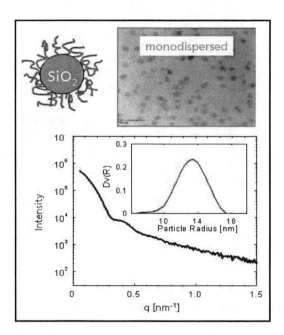

Figure 6: SAXS results from a PMMA-silica nanocomposite and a comparison with TEM.

5.4 Porous Materials

Pore size distribution analysis results obtained with SAXS from two silica samples with a random porous structure are shown in fig. 7. The measurement time per sample was only 5 minutes. The data were analyzed assuming spherical pores and the distributions correlate well with results from the mercury intrusion method.

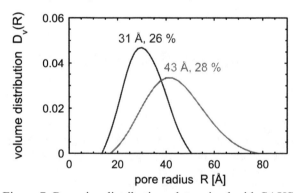

Figure 7: Pore size distributions determined with SAXS

REFERENCES

[1] O. Glatter & O. Kratky, "Small Angle X-Ray Scattering", Academic Press, London (1982).
[2] L. A. Feigin & D. I. Svergun, "Structure Analysis by Small-Angle X-ray and Neutron Scattering", Plenum Press, New York (1987).
[3] P. V. Konarev, M. V. Petoukhov, V. V. Volkov & D. I. Svergun, J. Appl. Cryst. 39, 277-286, 2006.
[4] S. Etienne, C. Beckers et al., J. Therm. Anal. and Calor. 87, 101-104, 2007.

Effect of local curvature on the interaction between hydrophobic surfaces

P. M. Hansson[*], A. Swerin[*], E. Thormann[**], P. M. Claesson[*,**], J. Schoelkopf[***] and P. A. C. Gane[***,****]

[*] YKI, Institute for Surface Chemistry, Box 5607, SE-114 86 Stockholm, Sweden, petra.hansson@yki.se [**] Department of Chemistry, Surface and Corrosion Science, Royal Institute of Technology, SE-100 44 Stockholm, Sweden, [***] Omya Development AG, CH-4665 Oftringen, Switzerland, [****] Aalto University, Faculty of Chemistry and Materials Sciences, Department of Forest Products Technology, P.O. Box 16300, FI-00076 Aalto, Finland

ABSTRACT

Surface structure, including roughness and chemical heterogeneities, is known to be of importance for the surface interaction forces observed between hydrophobic and superhydrophobic surfaces. In this study, silica particles have been used to prepare structured particulated surfaces with a controlled roughness using Langmuir-Blodgett deposition or a drop coating technique. The surfaces were characterized by SEM and AFM. The AFM colloidal probe technique was employed for probing the interaction between a micro-size hydrophobic particle and hydrophobic surfaces with a sintered and silanized nano-sized silica particle monolayer. These measurements indicate that the adhesion force is increased by a decrease in particle size. Larger roughness gives larger crevices on the surface and more air/vapor accumulation but the capillary growth is impaired since the three-phase line (solid-liquid-air) has to move over a longer distance with an increase in the size of the surface features.

Keywords: surface forces, AFM, capillary force, hydrophobic interaction, surface roughness

1 INTRODUCTION

The first direct measurements of the attractive interaction between hydrophobic macroscopic surfaces was reported in 1982 and the study showed that long-range forces that are considerably stronger than the van der Waals force dominate the interaction [1]. It was later suggested that cavitation or bridging bubbles between the surfaces could be responsible for the long-range nature of the attractive force observed on bringing the surfaces together [2-4]. The hydrophobicity of the surfaces involved is not solely responsible for the range and magnitude of the interaction between hydrophobic surfaces in aqueous solutions. Observations have shown that surface roughness and chemical heterogeneities are of vital importance for the long-range attractive forces. Air pockets created within the rough and heterogeneous surface are likely to be the reason for that [5-6].

Particulated surfaces with a densely packed and well-ordered structure can give valuable information about how the bubble/cavity induced attractive forces are influenced by topography and roughness. Recently, Wallqvist et al.[7], reported how the adhesive and long-range capillary forces between slightly hydrophobic surfaces are influenced by the surface topography of disordered particulated surfaces. Langmuir-Blodgett (LB) deposition is well-known to be capable of preparing films with a highly ordered structure and controlled roughness. Previously, hexagonally close-packed films with silica particles of sizes 0.5, 1.0 and 1.5 µm have been created by Tsai et al [8].

This study aims at gaining information about how the range and magnitude of the hydrophobic interaction are affected by the surface roughness of the interacting surface by measuring forces between a colloidal probe and a surface coated with particles of the same size range as the probe as well as particles more than 100 times smaller than the probe. Focus has been on the preparations of the structured surfaces and characterization of particles from which the surfaces are created.

2 EXPERIMENTAL

2.1 Materials

Silica particles of 30, 90, 200, 800 and 4000 nm in diameter were purchased from G. Kisker GmbH Germany. The cationic surfactant hexadecyltrimetyhylammonium bromide (HTMAB, Sigma, UK) was used to physically modify the particle surfaces. To hydrophobize the particle surfaces, dodecyl-dimethyl-chlorosilane (Sigma, UK) was used. Polystyrene particles with a radius of 5 µm (G. Kisker GmbH,Germany) were used as probes in AFM force measurements.

2.2 Surface preparation

Modification of the silica particles to be used for LB deposition was done following a protocol developed by Lee et al and Tsai et al [8-9]. Water was removed from the silica particle suspensions by careful heating in an oven. Then the particles were redispersed in 2-propanol by sonication for 2 h followed by addition of 1000 ppm HTMAB and sonication for another hour. The dispersion was left to equilibrate for 24 h followed by removal of the solvent. Finally, chloroform was added to give a particle concentration of 20 mg/mL.

The deposition of particles and monolayer experiments were conducted in a Langmuir trough (KSV Instruments, Finland). The particles were spread on a clean water subphase by using a microsyringe (705N, Hamilton Co, USA) followed by a 15 min waiting period to let the solvent evaporate. The amount of particles added to the trough differed with respect to particle size and an optimal amount was decided through surface pressure-area (π-A) isotherms. For all types of experiments; π-A isotherms and film deposition, the monolayer at the air-water interface was compressed at a speed of 10 mm/min. Glass microscope slides thoroughly cleaned in ethanol and water with dimensions of 15 mm x 15 mm x 1 mm were used as substrates for film deposition. The particles were deposited at a speed of 1 mm/min in the upstroke direction at the selected surface pressure.

Drop coating was performed by putting a droplet of the aqueous silica particle stock solution on a glass microscope slide and letting it dry.

In order to obtain stable and organics free particulated films, the deposited films were heated in a furnace for 5 min at 650 °C. The sintered surfaces were cleaned in a plasma cleaner (PDC-3XG, Harrick Scientific, USA) for 15 min. In the next step, the surfaces were exposed to dodecyl-dimethyl-chlorosilane in gas phase by placing them in a desiccator next to a few droplets of the silane for 15 hours. Finally, the surfaces were rinsed consecutively in chloroform, heptane, 2-propanol, ethanol and Milli-Q water.

2.3 Methods

Surface modified silica particles were characterized by measuring their size with dynamic light scattering (Zetasizer Nano, Malvern Instruments, UK). Water contact angles were measured using a dynamic contact angle and absorption tester (FibroDAT 1100, FibroSystems AB, Sweden). Images of the surfaces were recorded using an Atomic force microscope (AFM, Nanoscope III, Digital Instruments, USA) in tapping mode with non-contact silicon cantilevers (NCH-VS2-W, Nanoworld, Switzerland) or by using a Scanning electron microscope (SEM, XL30, Philips FEI, USA). The forces between the hydrophobized surfaces and a hydrophobic colloidal probe were measured using the same AFM as employed for imaging equipped with a liquid cell. The cantilevers (NSC12/tipless/No Al, Mikromasch, Estonia) were calibrated using the method proposed by Sader et al.[10] Measurements were performed in aqueous 10 mM NaCl.

3 RESULTS AND DISCUSSION

The size of the silica particles both before and after modification with the cationic surfactant was measured and gave an indication that the smaller particles, 30 and 90 nm, tend to aggregate during the drying process and are impossible to disperse by using ultrasonication after they have aggregated (Figure 1). However, a filter with a pore size of 500 nm could be used to remove most of the aggregates.

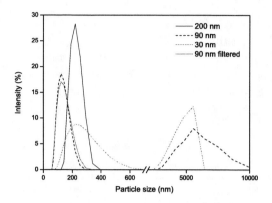

Figure 1: Size measurements of silica particle suspensions using dynamic light scattering.

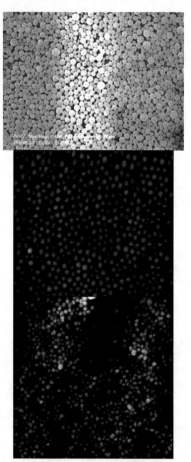

Figure 2: Particulated surfaces A) 4000 nm particles deposited by LB, recorded by SEM, B) 200 nm particles deposited by LB, recorded by AFM (5x5 μm^2) C) 30 nm particles deposited by drop coating, recorded by AFM (2x2 μm^2)

The images displayed in Figure 2A and 2B illustrate Langmuir-Blodgett (LB) deposited surfaces. LB deposition of 30 and 90 nm particles gave surfaces that contained large particle aggregates.

According to the previously described particle size measurements, the aggregates are present already before spreading of the particles in the Langmuir trough. Particle filtration before LB deposition resulted in films without any apparent aggregates. Drop coated surfaces with 30 nm particles are shown in Figure 2C which displays a well-ordered surface. However, the surface coverage tends not to be complete and large holes without particles were sometimes visible on the surface.

Contact angle measurements of silanized particulated surfaces showed that the macroscopic contact angles of LB deposited films were 120-125°, compared to a clean hydrophobized glass slide which exhibited a contact angle of 108° (Figure 3). As expected, an increase in surface roughness due to the deposited particles gives an increased hydrophobicity of the surfaces. However, the particle size has an insignificant effect on the contact angle. As mentioned above, due to lower surface coverage on the drop coated surfaces, the contact angle is slightly lower than those exhibited by the LB surfaces.

Figure 3: Water contact angles on A) Flat silanized glass surface, 108°, B) LB film of 4000 nm particles 122°, C) LB film of 200 nm particles 124°, D) Drop coated film of 90 nm particles 116°.

The force measurements presented here were done by measuring interactions between a polystyrene colloidal probe and a LB deposited particulated surface. Figure 4 displays force curves measured on 30, 90 and 200 nm particle surfaces. These measurements indicate that the adhesion force is increased by a decrease in particle size. This is in line with measurements reported, indicating the strong influence of air/vapour cavities contributing to capillary forces between the surfaces [7]. In the present study, the measurements were extended to larger particles sizes and to higher macroscopic contact angles. The interaction forces are larger when the nanoscale roughness is smaller. There is more air/vapour accumulation due to larger crevices on rougher surfaces but the capillary growth is impaired since the three-phase line (solid-liquid-air) has to move over a longer distance with an increase in the size of the surface features.

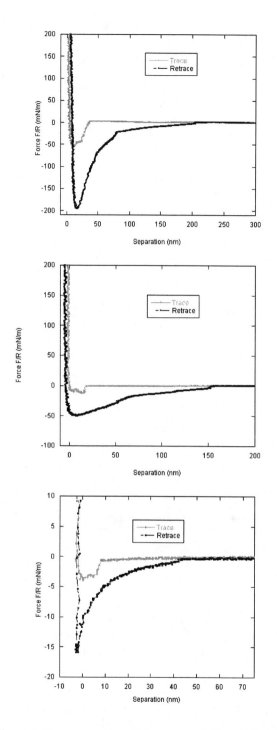

Figure 4: Force measurements on approach (jump-in) and separation (adhesion) between a polystyrene probe and LB deposited hydrophobic particulated surface. Particle size for the surfaces: A) 30 nm, B) 90 nm and C) 200 nm.

4 CONCLUSIONS

Silica particles of diameters 30, 90, 200, 800 and 4000 nm were surface-modified with a cationic surfactant and used for preparation of structured particulated monolayer films by using Langmuir-Blodgett deposition or drop coating. The smaller particles displayed a higher tendency to aggregate during the process of surface modification.

After hydrophobization of the particulated surfaces, forces between them and a hydrophobic colloidal probe were measured. The first results indicate strong interaction forces between hydrophobic particulated surfaces and hydrophobic probe due to capillary forces and a significant influence of the size of the surface features.

REFERENCES

[1] Israelachvili, J. & Pashley, R. "The Hydrophobic Interaction Is Long-Range, Decaying Exponentially With Distance." Nature, 300, 341-342, 1982.

[2] Christenson, H. K. & Claesson, P. M. "Cavitation And The Interaction Between Macroscopic Hydrophobic Surfaces." Science, 239, 390-392, 1988.

[3] Parker, J. L., Claesson, P. M. & Attard, P. "Bubbles, Cavities, And The Long-Ranged Attraction Between Hydrophobic Surfaces." J. Phys. Chem. 98, 8468-8480, 1994.

[4] Carambassis, A., Jonker, L. C., Attard, P. & Rutland, M. W. "Forces measured between hydrophobic surfaces due to a submicroscopic bridging bubble." Physical Review Letters, 80, 5357-5360, 1998.

[5] Hato, M. "Attractive forces between surfaces of controlled "hydrophobicity" across water: A possible range of "hydrophobic interactions" between macroscopic hydrophobic surfaces across water." J. Phys. Chem. 100, 18530-18538, 1996.

[6] Nguyen, A. V., Nalaskowski, J., Miller, J. D. & Butt, H. J. "Attraction between hydrophobic surfaces studied by atomic force microscopy." Int. J. Miner. Process., 72, 215-225, 2003.

[7] Wallqvist, V., Claesson, P.M., Swerin, A., Östlund, C., Schoelkopf, J., Gane, P.A.C.. "Influence of Surface Topography on Adhesive and Long-Range Capillary Forces between Hydrophobic Surfaces in Water." Langmuir, 25, 9197-9207, 2009.

[8] Tsai, P. S., Yang, Y. M. & Lee, Y. L. "Fabrication of hydrophobic surfaces by coupling of Langmuir-Blodgett deposition and a self-assembled monolayer." Langmuir, 22, 5660-5665, 2006.

[9] Lee, Y. L., Du, Z. C., Lin, W. X. & Yang, Y. M. "Monolayer behavior of silica particles at air/water interface: A comparison between chemical and physical modifications of surface." J. Colloid Interface Sci., 296, 233-241, 2006.

[10] Sader, J. E., Chon, J. W. M. & Mulvaney, P. "Calibration of rectangular atomic force microscope cantilevers." Review Of Scientific Instruments, 70, 3967-3969,1999.

Structural Elucidation of Doxorubicin-Loaded Liposomes by Atomic Force Microscopy in air and water

B. Tian[a], W. T. Al-Jamal[a], M. Stuart[b] and K. Kostarelos[a]*

[a] Nanomedicine Laboratory, Centre for Drug Delivery Research,
The School of Pharmacy, University of London, London WC1N 1AX, United Kingdom
[b] Physical Organic Chemistry Unit, Stratingh Institute, University of Groningen, Nijenborgh 4, 9747 AG Groningen, The Netherlands
bowen.tian@pharmacy.ac.uk; kostas.kostarelos@pharmacy.ac.uk

ABSTRACT

In this study, we are reporting the structural elucidation of doxorubicin-loaded liposomes (L-Dox) by atomic force microscopy (AFM) in water and air. Our results showed that liposome vesicles ruptured on the mica surface and resulted in Dox crystals exposed outside in water, enabling direct imaging by AFM. The height of such crystals was determined ranging from10 nm to 30 nm by the cross-section analysis. The first experimental proof of the hard nature of Dox crystals was confirmed in phase image by AFM. Consistent results were obtained in air scan when vesicles were dried. In conclusion, Dox crystals formed inside liposomes following encapsulation were visualized by AFM in water and air for the first time.

Keywords: Doxil, doxorubicin crystal, AFM, cancer, liposomes

Atomic force microscopy (AFM) is a powerful imaging technique to image and characterize nanoparticles, for example liposomes.[1,2] AFM has many advantages over electron microscopy since it does not require special sample processing and can be operated both in air and water, plus offering three-dimensional visualization.[1,2] AFM have been used to elucidate liposome-based structures in air, such as lipid-enveloped adenovirus[3] and functionalized QD-liposome hybrids.[4] AFM imaging of liposomes under fluid conditions is a very challenging task mainly due to their soft nature. So far, only several studies have attempted to image liposome vesicles in fluid, using tapping mode by AFM.[5-8] Very recently, two studies have pioneered to image more complicated liposome-based delivery systems in water, such as iron oxide-encapsulated liposomes and action-containing liposomes in fluid, using tapping-mode AFM.[9,10] Both studies have pointed out that the rupture of liposomes on mica in fluid resulted in the exposure of encapsulated entities outside.

Liposomes are one of the most-developed nanometer-scale drug delivery systems.[11] Liposomes encapsulating cytotoxic drugs have been widely used for the treatment of various tumors.[11-13] Liposomes (smaller than 200 nm in mean diameter) can preferentially accumulate in tumors *in vivo* due to enhanced permeation and retention effect (EPR).[11-14] Furthermore, the attachment of targeting ligands to the liposome surface facilitate their cellular uptake [15], resulting in a significant improvement in cancer therapy.

Doxil is a clinically used, FDA-approved liposomal product against cancer. Anticancer drug Dox is loaded into liposomses using the osmotic gradient technique, with high loading efficiency achieved over 95%.[16] Structural elucidation of L-Dox by cryo-EM showed that Dox formed crystal-like structures inside of liposomes.[17-19] Since 1992, cryo-EM has been the only technique used to visualize the Dox crystals inside of liposomes. The formation of such crystals has been reported due to the fact that Dox precipitates with sulphate inside of liposomes.[17] Li and co-workers pioneered to investigate physical state of Dox crystals inside of liposomes using cryo-EM and circular dichroism spectra, concluding that Dox crystals account for 99% of Dox molecules loaded inside of liposomes.[20] The formation of such crystals also lead to stable encapsulation of Dox inside of liposomes.[16] However, the physical properties of Dox crystals have to be fully studied yet and their residing inside of liposomes poses many difficulties.

In this study, we attempted to elucidate the structure of of liposomes encapsulating Dox by AFM. We first optimized the experimental conditions for both AFM in air and water to image PEGylated liposomes without leading to fusion between adjacent vesicles. Liposomes were deposited on mica surface and scanned using the tapping mode by AFM to avoid liposome damage. Our results showed that liposome vesicle deposition onto the mica surface resulted in Dox crystal exposure in fluid. The hard nature of Dox crystals was clearly shown in phase images where the height of such crystals was determined to range between 10nm to 30nm. The structure of liposomes scanned in air showed consistent results with that of in water scan. However, we found that the structure of liposomes in air was more dependent on the deposition time and the drying process applied. In conclusion, Dox crystals formed inside liposomes following encapsulation were visualized by AFM in water and air for the first time. These optimized conditions in our study allow the structural elucidation of individualized liposome vesicles and indicate that AFM can be applied as a routine technique for the structural characterisation of soft nanoparticles such as hydrated lipid bilayer vesicles.

Reference

(1)   Spyratou, E.; Mourelatou, E. A.; Makropoulou, M.; Demetzos, C. *Expert Opinion on Drug Delivery* **2009**, *6*, 305-317.

(2)   Ruozi, B.; Tosi, G.; Leo, E.; Vandelli, M. A. *Talanta* **2007**, *73*, 12-22.

(3)   Singh, R.; Tian, B. W.; Kostarelos, K. *Faseb Journal* **2008**, *22*, 3389-3402.

(4)   All-Jamal, W. T.; Al-Jamal, K. T.; Tian, B.; Cakebread, A.; Halket, J. M.; Kostarelos, K. *Molecular Pharmaceutics* **2009**, *6*, 520-530.

(5)   Solletti, J. M.; Botreau, M.; Sommer, F.; Brunat, W. L.; Kasas, S.; Duc, T. M.; Celio, M. R. *Langmuir* **1996**, *12*, 5379-5386.

(6)   Ruozi, B.; Tosi, G.; Forni, F.; Fresta, M.; Vandelli, M. A. *European Journal of Pharmaceutical Sciences* **2005**, *25*, 81-89.

(7)   Thomson, N. H.; Collin, I.; Davies, M. C.; Palin, K.; Parkins, D.; Roberts, C. J.; Tendler, S. J. B.; Williams, P. M. *Langmuir* **2000**, *16*, 4813-4818.

(8)   Liang, X. M.; Mao, G. Z.; Ng, K. Y. S. *Journal of Colloid and Interface Science* **2004**, *278*, 53-62.

(9)   Li, S. L.; Palmer, A. F. *Langmuir* **2004**, *20*, 7917-7925.

(10)   Dagata, J. A.; Farkas, N.; Dennis, C. L.; Shull, R. D.; Hackley, V. A.; Yang, C.; Pirollo, K. F.; Chang, E. H. *Nanotechnology* **2008**, *19*, -.

(11)   Torchilin, V. P. *Nature Reviews Drug Discovery* **2005**, *4*, 145-160.

(12)   T Al-Jamal, W.; Kostarelos, K. *Nanomedicine* **2007**, *2*, 85-98.

(13)   Torchilin, V. P. *AAPS J* **2007**, *9*, E128-47.

(14)   Maeda, H.; Wu, J.; Sawa, T.; Matsumura, Y.; Hori, K. *Journal of Controlled Release* **2000**, *65*, 271-284.

(15)   Kirpotin, D. B.; Drummond, D. C.; Shao, Y.; Shalaby, M. R.; Hong, K. L.; Nielsen, U. B.; Marks, J. D.; Benz, C. C.; Park, J. W. *Cancer Research* **2006**, *66*, 6732-6740.

(16)   Haran, G.; Cohen, R.; Bar, L. K.; Barenholz, Y. *Biochimica Et Biophysica Acta* **1993**, *1151*, 201-215.

(17)   Lasic, D. D.; Frederik, P. M.; Stuart, M. C. A.; Barenholz, Y.; Mcintosh, T. J. *Febs Letters* **1992**, *312*, 255-258.

(18)   Lasic, D. D. *Nature* **1996**, *381*, 630-630.

(19)   Lengyel, J. S.; Milne, J. L. S.; Subramaniam, S. *Nanomedicine* **2008**, *3*, 125-131.

(20)   Li, X. G.; Hirsh, D. J.; Cabral-Lilly, D.; Zirkel, A.; Gruner, S. M.; Janoff, A. S.; Perkins, W. R. *Biochimica Et Biophysica Acta-Biomembranes* **1998**, *1415*, 23-40.

Hydrophobic Carbon Nanotube AFM Probes for High Resolution Imaging of Biological Materials

Jae-Hyeok Lee, Najeeb C.K., Jingbo Chang, Won-Seok Kang,
Gwang-Hyeon Nam and Jae-Ho Kim*

* Department of Molecular Science and Technology,
Ajou University, Suwon 443-749, Republic of Korea, jhkim@ajou.ac.kr

ABSTRACT

Carbon nanotube (CNT) tipped atomic force microscope (AFM) probes have shown a significant potential for obtaining high resolution imaging of nanostructure and biological materials. In this work, we report a simple method to fabricate single-walled carbon nanotube (SWNT) nano-probes for atomic force microscopy (AFM) using the Langmuir-Blodgett (LB) technique. Moreover, to obtain hydrophobic property of SWNT-AFM probes with a capability of high resolution imaging in atmospheric and high relative humidity conditions, the SWNT-AFM probe was modified through vapor deposition of trichloro[1H, 1H, 2H, 2H,-perfluorotyl]silane (FTOS). Using this FTOS-modified SWNT-AFM probe, a high resolution image of plasmid DNA standard sample was acquired at 80~90 % RH condition. The measured dimension of plasmid DNA had a middle width of 4~5 nm (true width 2nm). We demonstrate that the FTOS-modified SWNT-AFM probes have extremely high resolution imaging capability, which allows to measure higher resolution images compared with conventional AFM probe.

Keywords: Atomic force microscope, Carbon nanotube, Surface modification, High resolution images.

1 INTRODUCTION

Since their discovery in 1991, carbon nanotubes [1] have attracted much attention because of their unique properties and wide variety of applications, for examples, probe tips for scanning probe microscopy [2-4], nanoelectronic devices [5], field emission displays [6], hydrogen storage [7], and batteries [8-9]. Among these applications, the SWNT-AFM probes offer important advantages of high resolution imaging over conventional silicon probes due to their small tube diameter, high aspect ratio, stiffness, reversible elastic buckling and chemical stability. Since the first report of SWNT-AFM probe by Dai et al., [10] various fabrication methods of SWNT-AFM probe have been developed; for example, mechanical assembly of nanotube tips, applying an electric or magnetic field attraction [11] and direct growth of nanotubes onto AFM tip by catalytic Chemical Vapor Deposition (CVD)

[12]. In our previous work [13], we reported that the SWNT-AFM probe was well formed by the Langmuir-Blodgett technique. The purpose of this paper is to investigate the effect of hydrophobic property of SWNT-AFM probe for high resolution AFM imaging at high humidity conditions. In this work, to induced hydrophobicity, we modified the SWNT-AFM probe through vapor deposition of a trichloro[1H, 1H, 2H, 2H,-perfluorotyl]silane (FTOS). We assume that if SWNT-AFM probe modified with hydrophobic materials it would reduce the water meniscus effect between the tip and sample. Moreover, we evaluated image resolution using the FTOS-modified SWNT-AFM probes and compared with a conventional high resolution AFM probe by utilizing the standard samples of a deoxyribonucleic acid (DNA). The FTOS-modified SWNT AFM probe showed excellent imaging capability and stability, allowing higher-resolution images compared with conventional AFM probes.

2 EXPERIMENTAL

2.1 Materials and Methods

SWNTs synthesized by the arc discharge method were purchased from Iljin Nanotech Co. (Seoul, Korea). Here, we used thiophenyl (-SH)-modified SWNTs (SWNT-SHs) solution for fabricating SWNT-AFM probes. In a typical experiment, purified SWNTs were shortened and carboxylated, by chemical oxidation, in a mixture of concentrated sulfuric and nitric acids (3:1, v/v, 98 % and 70 %) under ultrasonication (Cole-Palmer, 55 kHz) at 70 ℃ for 4h. The reaction mixture was filtered through an alumina filter (Whatman, England, pore size = 0.2 μm). The remains left on the filter were then washed by deionized water until the filtrate pH became nearly neutral. The resulting filtrate was dispersed by sonic agitation for 1 h in 100 ml of aqueous Triton X-100 surfactant solution (3 wt %). Immediately after sonication, the sample was centrifuged at 6000 rpm for 4 h. The supernatant was then carefully decanted. The resulting carboxylated SWNT were then reacted with 4-aminothiophenol in the presence of 1-[2-(dimethylamino)propyl]-3-ethylcarbodiimide hydrochlorid and N-hydroxysuccin-imide SWNTs to obtain

SWNT-SHs. The SWNTs obtained were dissolved in chloroform.

2.2 Manufacture of SWNT-AFM Probes

Commercially available AFM probes (M2N, Inc., Korea) were immerged into the water surface. The SWNTs solution (~3 ml) was spread by carefully casting minute droplets on the air/water interface of the LB trough (KSV 10002, KSV instruments). After the solvent evaporated, the hydrophobic SWNTs bundles remained on the air/water interface. The bundles were then compressed by moving the barriers at a speed of 4 mm/min until the surface pressure isotherm recorded 50mN/m. A vertical dipping method was employed to transfer the SWNTs onto the commercial AFM probes surface with a deposition rate 1 mm/min at a surface pressure of 50 mN/m. The SWNT modified probes were dried under the air atmosphere for 30 min. Figure 1. is schematic diagrams of the fabrication of individual SWNTs onto the tip of a conventional AFM probe using LB technique.

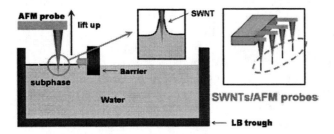

Figure 1. Schematic diagrams of the SWNT-AFM probes fabrication method using the LB technique.

2.3 Surface Modification of SWNT-AFM Probes

To obtain hydrophobic property of SWNT-AFM with capability of high resolution imaging in high RH (80~90%) condition, we modified the SWNT-AFM probe through vapor deposition of FTOS (trichloro[1H, 1H, 2H, 2H,-perfluorotyl]silane). First, the SWNT-AFM probes were cleaned and hydroxylated simultaneously by a UV/ozone cleaning for 5 minute. The light source was an excimer lamp with $\lambda = 172$ and 10 mWcm^2 (Ushio Electric, UER20-172V). After UV/ozon cleaning the SWNT-AFM probes were functionalized by FTOS. The vapor deposition procedure involves placing a SWNT-AFM probe in a reaction vessel saturated with FTOS vapor for 30 second.

2.4 Characterization and Instrumentation

The attachment of SWNTs onto the end of the tip was confirmed by scanning electron microscopy (SEM, LEO SUPRA 55, Carl Zeiss NTS GmgH) operated at 10 kV. All AFM images were recorded in non-contact mode on a XE-100 AFM system (PSIA, Inc., Korea) in air and in high RH (80~90%) condition. SWNTs-AFM probes were vibrated

with a resonance frequency at approximately 310 kHz. SWNT Langmuir-films and DNA plasmid pGem7zf+ (3000 b. p. linearized with the SmaI endonuclease deposited on freshly cleaved mica, NT-MDT, Rusia) were used as standard samples.

3 RESULTS AND DISCUSSION

3.1 Evaluation of Hydrophobic SWNTs LB Films and SWNTs-AFM Probes

In our previous report, we have demonstrated a simple and effective method for transferring SWNT on AFM tip for high resolution imaging using the LB technique. To study the effect on the wettability of the SWNT LB film by FTOS functionalization, dynamic contact angles were measured by using contact angle analyzer. The contact angles of the FTOS modified SWNT LB film was compared with that of the bare silicon wafer and the thiophenyl-modified SWNTs LB film. Figure 2. shows that the typical contact angle measurements with a series of bare silicon wafer and thiophenyl-modified SWNTs LB film after FTOS coating. The contact angles were measured with a water droplet of 3 µL. The substrate was silicon wafer. Figure 2. shows a consistent increase of contact angle from 31° (bare silicon wafer) to 66° (SWNT LB film) and then to 109°, when SWNT LB film was functionalized. This clearly suggests that FTOS can be used as a suitable reagent for modifying the hydrophobic behavior of SWNT LB film. It was shown that such approach results in smooth and stable monolayer silane film with hydrophobicity. The mechanism of interaction between the FTOS and SWNT LB film can be attributed to the lowering of interfacial free energy which is evident from the change in the contact angles of the functionalized SWNT LB film.

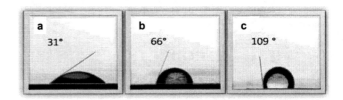

Figure 2. Contact angle measurements with a series of (a) bare silicon wafer (b) SWNT LB film (c) after FTOS coating on SWNT LB film. The contact angles were measured with a water droplet of 3 µL.

We also employed the same method of functionalization in SWNT-AFM probe mentioned above. To obtain hydrophobic property of SWNT-AFM with capability of high resolution imaging in atmospheric and high humidity environments, we modified the SWNT-AFM probe through vapor deposition of FTOS. Figure 3. demonstrate the SEM images of SWNT-AFM probes assembled by the LB technique. Figure 3. (a) represent SEM image of the bare silicon AFM probe. Figure 3. (b) shows SEM image of

SWNT-AFM probe obtained from a single LB process as shown in Figure 1. and Figure 3. (C) indicates the SWNT-AFM probe modified with FTOS through vapor deposition method. From this data, it is observed that the surface morphology of FTOS-modified SWNT-AFM probe was similar in comparison with non treated SWNT-AFM probe.

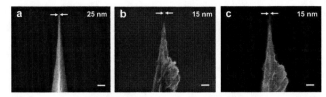

Figure 3. SEM images of SWNT-AFM probes assembled by the LB technique. (a) SEM image of the bare silicon AFM probe. (b) SEM image of SWNT-AFM probe obtained from a single LB process. (C) The SWNT-AFM probe modified with FTOS through vapor deposition method (scale bar : 50 nm)

3.2　High Resolution Images of DNA

To acquire FTOS-modified SWNT-AFM probe with the capability of high resolution imaging in air, we evaluated the FTOS-modified SWNT-AFM probe and compared to conventional high resolution AFM probe by utilizing the standard sample of a plasmid DNA. Figure 4. represent AFM images and cross-section profiles of a plasmid DNA deposited on a freshly cleaved mica surface, which was obtained by using a conventional AFM probe (Fig. 4 (a)) and an FTOS modified SWNT-AFM probe (Fig. 4 (b)). Figure 4 (a) shows the radius of curvatures of the DNA was 29~30 nm.

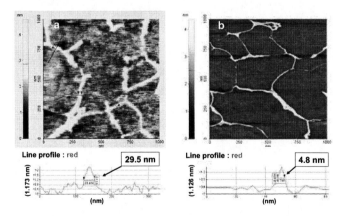

Figure 4. AFM images and cross-section profiles of a plasmid DNA obtained by using a conventional AFM probe (a) and an FTOS modified SWNT-AFM probe (b). (a) The radius of curvatures of the DNA was 29~30 nm. The result suggests that the radius of a conventional AFM probe is 27~28 nm and (b) the radii of curvatures of the DNA was measured to be 4~5 nm, indicating the radius of an FTOS modified SWNT-AFM probe is 2~3 nm.

The result suggests that the radius of a conventional AFM probe is 27~28 nm and Fig. 4 (b) indicate the radii of curvatures of the DNA was measured to be 4~5 nm, indicating the radius of an FTOS modified SWNT-AFM probe is 2~3 nm. This indicated that FTOS-modified SWNT-AFM probe have extremely high resolution imaging capability, which allows to measure higher resolution images compared with conventional AFM probe. Although the non-treated SWNT-AFM probe also has a high resolution imaging property in atmospheric condition, [13] the FTOS-modified SWNT AFM probe would be more better for measurement in atmospheric as well as in 80~90 % RH condition. It is due to the fact that a continuous and homogeneous SWNT film leads to AFM probe coated with densely packed, hierarchical SWNT arrays, which are mechanically stable because of the strong interaction between SWNT bundles. Also, the FTOS-modified SWNT-AFM probe has a high hydrophobicity so it reduces the water meniscus effect between the tip and the sample.

4　CONCLUSION

We have developed an LB process which can be used to fabricate SWNT-modified nanoprobes for high-resolution AFM measurement. A simple dipping process of conventional tips into a uniaxially aligned SWNT Langmuir monolayer results in the formation of well-oriented, robust SWNT nanoprobes that maintain their shape and direction even after successive AFM measurements. Moreover, to obtain hydrophobic property of SWNT-AFM with capability of high resolution imaging in atmospheric and aqueous environments, we modified the SWNT-AFM probe through vapor deposition of FTOS. Using this FTOS modified SWNT-AFM probe, we demonstrated the high resolution imaging of a plasmid DNA standard sample with a middle width of 4~5 nm (true width 2nm). This indicated that FTOS-modified SWNT-AFM probe have extremely high resolution imaging capability, which allows to measure higher resolution images compared with conventional AFM probe.

5　ACKNOWLEDGMENT

This research was supported by a grant (06K1401-00411) from Center for Nanoscale Mechatronics & Manufacturing, one of the 21st Century Frontier Research Programs supported by Ministry of Science and Technology, KOREA. In addition, we thank BK21 program of the molecular science and technology in Ajou University and M2N, Inc for providing AFM tips.

REFERENCES

[1] S. Iijima, T. Ichihashi, Nature (London) 363, 603, 1993.

[2] C V. Nguyen, C. So, R M. Stevens, Y. Li, L. Delzeit, P, Sarrazin and M J. Meyyappan. Phys. Chem. B 108, 2816, 2004.

[3] J H. Hafner, C L. Cheung and C M. Lieber, Nature (London) 398, 761, 1999.

[4] H. Nishijima, S. Kamo, S. Akita, Y. Nakayama, K I. Hohmura, S H. Yoshimura and K, Takeyasu, Appl. Phys. Lett. 74, 4016, 1999.

[5] A. Bachtold, P. Hadley, T. Nakanishi, C. Dekker, Science 294, 1371, 2001.

[6] J A. Misewich, R. Martel, P. Avouris, J C. Tsang, S. Heinze and J. Tersoff, Science 300, 783, 2003.

[7] Q H. Wang, A A. Setlur, J M. Lauerhaas, J Y. Dai, E W. Seelig, Appl. Phys. Lett. 72, 2912, 1998.

[8] J, Li, Y J. Lu, Q. Ye, M. Cinke, A. Javey, O. Wang, H J. Dai, S. Peng and K J. Cho, Nano Lett. 3, 929, 2003.

[9] M. Trojanowicz, Trends in Analytical Chemistry, 25, 480, 2006.

[10] H. Dai, J H. Hafner, A G. Rinzler, D T. Colbert and R E. Smalley, Nature (London) 384, 147, (1996).

[11] R. Stevens, C. Nguyen, A. Cassell, L. Delzeit, M. Meyyappan and J. Han Appl. Phys. Lett. 77, 3453. 2000.

[12] Q. Ye, A M. Cassell, H. Liu, K-J. Chao, J. Han and M. Meyyappan, Nano Lett. 4, 1301, 2004.

[13] Lee. Jae-Hyeok, Kang. Won-Seok, Choi. Bung Sam, Choi. Sung-Wok and Kim. Jae-Ho, Ultramicroscopy 108, 1163, 2008.

Dynamics of intermediate to late stage concentration fluctuations during free diffusion experiments

A. Oprisan, S.A. Oprisan, and A. Teklu

College of Charleston, SC, USA, oprisana@cofc.edu

ABSTRACT

Two experiments were performed using shadowgraph visualization technique in order to study the dynamics of concentration fluctuations. Our experimental setup includes an objective attached to the CCD camera to increase the field of view. Using two colloidal suspensions, one with gold and the other with silica colloid, we extracted both the structure factors and the correlation time during the intermediate to late stages of concentration fluctuations during a free diffusion experiment. The temporal evolution of fluctuations is qualitatively investigated using recursive plots and spatial-temporal sections of fluctuating images. Our experiment reveals significant differences between concentration fluctuations induced in silica and gold colloids. The difference is explained not only in terms of the particle size but due to the possible plasmonic interaction between gold nanoparticle and the incident light.

Keywords: concentration fluctuations, nanocolloids, free diffusion

1 INTRODUCTION

Equilibrium fluctuations in fluids, and binary mixture were investigated in the past both on Earth and under microgravity conditions using different light scattering techniques [1, 2, 3, 4]. Recent experiments have shown that non-equilibrium fluctuations exists during a free diffusion process in liquid mixture and are caused by a coupling between velocity fluctuations and the macroscopic concentration gradient established between two liquids [5, 6]. It has been shown that gravity suppressed the long wavelength fluctuation below a certain cutoff value. As a result, some experiments were carried out in microgravity to investigate the interplay between the gravity and diffusion in fluctuating interface of miscible fluids.

In this paper, we present experimental results regarding the intermediate to late stage of free-diffusion in silica and gold nanocolloids. We were interested in establishing a time frame for the concentration fluctuations. We definitely know that large scale giant fluctuations are present in recordings done during the late stages at 50000 s or 100000 s [6, 7], but when do they actually develop? Our experimental setup include an objective focused on the free diffusion layer that in order to observe and analyze the evolution of fluctuations. Our image processing technique used sets of 1000 images [8]. However, we repeated our analyses for sets that were 250 s apart to capture any possible change in the structure factor and investigated the evolution of fluctuations.

2 EXPERIMENTAL SET-UP

We studied the evolution of the wavelength fluctuations in silica and gold colloids. Our optical setup of a free-diffusion experiment is schematically represented in Fig. 1. A water-soluble colloid was confined in a Hellma 120-OS-20 cylindrical glass cell, 2.0 cm diameter and 2.0 cm height with plane-parallel optical glass windows [7, 8]. The interface between the two fluids was perpendicular to the gravitational field. A denser colloid (silica or gold) was slowly injected using a bent needle at the bottom of the cell completely filled with degassed water. The two fluids are initially separated by a sharp horizontal interface localized almost in the middle of the diffusion cell. A parallel beam of light from a 10 mW He-Ne laser, with a wavelength of 635.5 nm, was collimated and passed vertically upward through the diffusion cell. A progressive scan SONY CCD camera, with sensor area of 5.8 mm x 4.92 mm and a pixel resolution of 4.65 μm, and a Video Lens-10X objective were used for image recording.

We used two different liquids mixture: 1) silica colloids with the average diameter of particles of 200 nm, and 2) gold nano colloids with the average nano-particle of 20 nm. Our results showed that the density concentration fluctuations causing local perturbation in the intensity of transmitted and scattered light cover both gravitational and diffusion effects. Using our experimental data, we were able to analyze the structure factor and their corresponding power spectra for different wave vectors and the correlation time from intermediate to late stage.

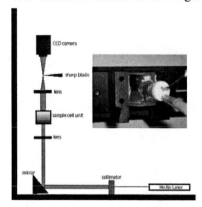

Figure 1. Schematic representation of the experimental setup (left) and a picture of the sample cell unit (right) used to study the giant fluctuations determined by diffusion in a mixture of silica (bottom) and water

3 METHOD

We applied a dynamic light scattering algorithm to study intermediate to late stage concentration fluctuations in a free diffusion experiment in silica and gold colloids. The full description of the experimental set-up is described elsewhere [8]. The first step in applying the dynamic algorithm was to normalize each image by diving each image to its spatial average (noted < >), in order to reduce the effect of variable light intensity [7, 8]:

$$i(\boldsymbol{x},t) = I(\boldsymbol{x},t)/<I(x, t)>, \qquad (1)$$

where $I(x,t)$ is the two-dimensional image intensity cropped from a snapshot of the sample cell unit, and $<I(x, t)>$ is the spatial average over all pixel positions in the image $I(x,t)$. The second step is to find the fluctuation image, which is defined as the difference between two normalized images separated by a delay time Δt:

$$\delta i(\boldsymbol{x},t,\Delta t) = i(\boldsymbol{x},t+\Delta t) - i(\boldsymbol{x},t). \qquad (2)$$

The third step is to calculate the spatial two-dimensional Fast Fourier Transform (FFT2) for each fluctuation image in order to find the power spectrum of each fluctuating image:

$$\delta i(\boldsymbol{q},t,\Delta t) = FFT2(\delta i(\boldsymbol{x},t,\Delta t)). \qquad (3)$$

The fourth step consists in computing the time-dependent structure function $c_m(\boldsymbol{q},t,\Delta t)$, which is the temporal average of all power spectra in a set of images with a fix delay time Δt :

$$c_m(\boldsymbol{q},t,\Delta t) = \overline{\left| \delta i(\boldsymbol{q},t,\Delta t) \right|^2}_t . \qquad (4)$$

One of the characteristics of the power spectra is the azimuthally symmetry and using this property we computed the radial average of structure function $C_m(q,t,\Delta t)$.

$$C_m(q,t,\Delta t) = \overline{c_m(\boldsymbol{q},t,\Delta t)}_{|q|} . \qquad (5)$$

The structure function $C_m(q,t,\Delta t)$ is related to the sample structure factor $S(q,t)$, the temporal autocorrelation function $G(q,t,\Delta t)$, and the background noise $B(q,t)$ as follows [24]:

$$C_m(q,t,\Delta t) = 2\{[S(q,t)T(q)]\cdot[1 - G(q,t,\Delta t)] + B(q,t)\}, \qquad (6)$$

where the shadowgraph transfer function is $T(q) = 4sin^2(q^2z/2k)$, with z the distance from the sample to the CCD, q the wave vector and k the vacuum wave vector of the incident light. The correlation function specific for thermal fluctuations is given by [9]:

$$G(q,t,\Delta t) = exp(-\Delta t/\tau(q,t)), \qquad (7)$$

where $\tau(q,t)$ is the correlation time of thermal fluctuations. By fitting the radial average of the time-dependent structure function $C_m(q,t,\Delta t)$ with the equations (6) and (7), we determined $S(q,t)T(q)$, $\tau(q,t)$ and $B(q,t)$. The two signals, $S(q,t)$ and $T(q)$, obtained with this method cannot be separated using image processing techniques and we will refer to $S(q,t)T(q)$ as the structure factor.

The correlation time is defined as the lifetime of the thermal fluctuations. The correlation time of non-equilibrium fluctuations is given by: [8, 10]

$$\tau_c(q,t) = \frac{q^2}{D(q^4 + q_c^4)}. \qquad (8)$$

4 RESULTS

4.1. Silica nanocolloids

We expect a very slow relaxation of nonequilibrium fluctuations for silica due to large nanoparticle sizes. The structure factor versus wave vector (Fig. 2) shows power law dependence for large (170 a.u. $<q<230$ a.u.) wave vectors. The exponent of the power law decreases from -4.1, which indicated giant concentration fluctuations at the beginning of the relaxation process, down to -2.5 after 100 hours (Fig. 2).

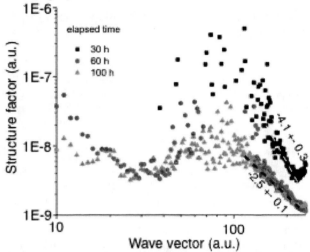

Figure 2. Structure factor for silica at 100 hours elapsed time has a power law exponent of -2.5 +- 0.1 and Chi^2/DoF = 1.1549E-18, and a coefficient of confidence R^2 = 0.85. After 60 h, the exponent is -2.6 +-0.1, Chi^2/DoF = 1.7086E-19, R^2 = 0.88. At 30 h, the exponent of the power law is is -4.1 +- 0.3, Chi^2/DoF = 2.2357E-18, R^2=0.73.

The correlation (delay) time has the characteristic shape described by Eq. (8). For large wave numbers (q > 100 a.u.), the exponent of the power law is around -2.0 with small variations. This is a characteristic signature of concentration fluctuations under gravitational relaxation regime.

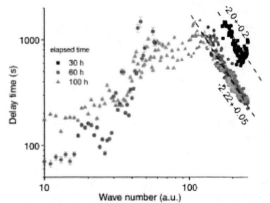

Figure 3. After 100 hours, the diffusion coefficient is D = 5.4952E-8+- 1.3825E-9 cm²/s, qc = 93+-1 a.u. with R^2 = 0.78. Power law exponent of delay time for large q is -2.22+-0.05, with R^2 = 0.93. After 60 h, the diffusion coefficient is D = 4.984E-8 +- 1.1814E-9 cm²/s, qc = 87+-1 a.u., R^2 = 0.8. For 30 h D = 2.4601E-8 +-4.6227E-10

cm^2/s, qc = 13.24249 +- 0.37315, R^2 = 0.64. After 30h, the power law exponent for large q is -2.0+-0.2.

4.2. Gold nanocolloids

Gold particles used in this experiment were one order of magnitude smaller than silica. As a result, the expected relaxation time is much shorter for this nanocolloid. The structure factor (Fig. 4) has the same shape as for silica (Fig. 2). The exponent of the power law decreases from -3.6 at the beginning of the experiment to almost -2 after 40 hours or recording. Large power law exponents are signatures of giant concentration fluctuations and a -2 exponent is a signature a Brownian fluctuations.

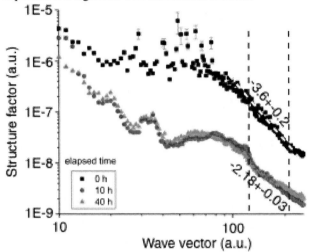

Figure 4. After 40 h, the slope is -2.18+-0.03 with R^2 = 0.98. After 10 h, the exponent of power law is -2.73+-0.04, R = 0.98. Initially the exponent of power law was -3.6+-0.2, R = 0.87.

The correlation time follows expected concentration fluctuation Eq. (8). The power law exponent for large wave vectors is -1.54. However, as diffusion progresses, the correlation time becomes almost independent of the wave vector.

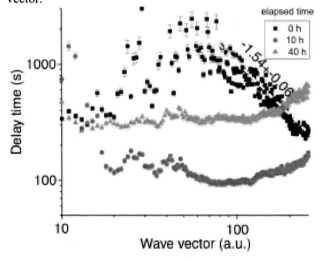

Figure 5. The correlation (delay) time for gold nanocolloids. Initially, the power law exponent for large wave numbers is -1.54+-0.06 wit R^2 = 0.85. The coefficient of diffusion was measured as 9.4139E-8+-3.6048E-9 cm^2/s, the critical wave number was 58+-1 a.u. with R^2=0.60.

5 CONCLUSIONS

It has been shown that the structure factor $S(q)$ is the ideal tool in describing fluctuations in a fluid stressed by temperature or concentration gradient. $S(q)$ measurements suggested that giant fluctuations are long-range correlated [1, 3, 7, 8] and they are proportional to the gradient that generated them [5].

We performed free diffusion experiments using 200 nm silica colloids and 20 nm gold colloids to investigate the contribution of different processes (viscosity, gravitational field, and concentration gradients) on the fluctuations of light intensity (Fig. 1).

Our results showed that the power spectra have characteristic rings determined by the shadowgraph transfer function $T(q)= 4 \sin^2(q^2z/2k)$, where k is the vacuum wave number for incident light, z is the distance between the CCD camera, the fluctuating interface, $q = 2nk \sin(\theta/2)$, is the scattering wave vector, and θ is the scattering angle. We used a FFT algorithm to compute the power spectra with 512 discrete values for wave numbers in arbitrary units (a.u.). Since the power spectra present azimuthal symmetry, averages over constant wave number values were computed in order to extract the scalar azimuthal structure function $C_m(q, t, dt)$, which is the power spectrum when plotted versus wave vector for fixed delay time dt. The power spectrum measures the intensity of fluctuations in the wave vector domain for a fixed value of delay time dt. The same information is conveyed by the scalar structure function $C_m(q, t, dt)$ in the temporal domain of delay times dt for fixed values of the wave vector q. The advantages of using the scalar structure function $C_m(q, t, dt)$ over the more traditional correlation function allowed us to a) obtain both the structure factors and the correlation time with only one sweep of the data, and b) eliminated any slow drifts in the recorded signal [8, 9]. We used a dynamic method to extract the structure factor $S(q)T(q)$, the correlation time $\tau(q)$, and the background noise $B(q)$ from fluctuation images.

Based on a detailed analysis of the slopes of the structure factor $S(q)T(q)$, we were able to observe giant concentration-induced fluctuations signaled by a slope of about -4 for during the early stages of the relation process (Figs. 2 and 4). During the early stages of the experiment, both for silica (Fig. 3) and gold (Fig. 5), the correlation time has a convex shape well approximated by classical theory encapsulated in Eq. (8). By fitting the correlation time with Eq. (8) we estimated the diffusion coefficient D for both nanocolloids and the corresponding critical wave vector at which the fluctuation due to viscosity (small wave vectors) is replaced by large gravitation-driven fluctuations. We identified at least two possible causes for the large variability of the exponent of the power law for correlation time: 1) at low wave vectors, the azimuthal average is performed over a very small number of configurations leading to large errors and 2) the maximum point of the correlation time plot, which determines the critical wave vector, shifted over time towards smaller values according to the power law $q_c \propto t^{1/8}$ derived from (3). Our results are in good agreement both with theoretical predictions [7] and with other experiments [10]. For silica, giant concentration fluctuations are still observed even after 30 hours as indicated by the large value of the power law exponent in the structure factor (Fig. 2). However, for gold nanocolloids, even after 1 hour the power law exponent is down to -3.6 and converges very fast to -2, which indicates Brownian motion. At the same time, the correlation time of concentration fluctuations is almost independent of the wave vector (Fig. 5).

ACKNOWLEDGMENTS

This research was supported by a R&D grant from the College of Charleston to A. Oprisan and S. A. Oprisan. We are grateful to Dr. Fabrizio Croccolo for fruitful discussions both on experimental details of shadowgraph setup and the numerical implementation of the dynamic method.

REFERENCES

[1] D. Brogioli, A. Vailati, and M. Giglio, "Universal behavior of nonequilibrium fluctuations in free diffusion processes," Phys. Rev. E 61, R1-R4 (2000).

[2] P. Guenoun, F. Perrot, and D. Beysens, "Microscopic observation of order-parameter fluctuations in critical binary fluids: Morphology, self-similarity, and fractal dimension," Phys. Rev. Lett. 63, 1152-1155 (1989).

[3] S. Mazzoni, R. Cerbino, A. Vailati, and M. Giglio, "Fluctuations in Diffusion Processes in Microgravity", Ann. N.Y. Acad. Sci. 1077, 351–364 (2006).

[4] P. N. Segre, R. W. Gammon, and J. V. Sengers, "Light-scattering measurements of nonequilibrium fluctuations in a liquid mixture," Phys. Rev. E 47, 1026-1034 (1993).

[5] A. Vailati and M. Giglio, "Giant fluctuations in a free diffusion process," Nature 390, 262-265 (1997).

[6] F. Croccolo, D. Brogioli, A. Vailati, M. Giglio, and D. S. Cannell, "Effect of gravity on the dynamics of nonequilibrium fluctuations in a free diffusion experiment," Ann. N.Y. Acad. Sci. 1077, 365–379 (2006).

[7] F. Croccolo, Ph.D. Dissertation, University of Milano, 2005; F. Croccolo, D. Brogioli, A. Vailati, M. Giglio, and D. S. Cannell, "Use of dynamic schlieren interferometry to study fluctuations during free diffusion," Appl. Opt. 45, 2166-2173 (2006).

[8] A. Oprisan, S.A. Oprisan, and A. Teklu, "Experimental study of non-equilibrium fluctuations during free diffusion in nano-colloids using microscopic techniques", Applied Optics, Vol. 49, No. 2, (2010).

[9] F. Croccolo, D. Brogioli, A. Vailati, M. Giglio, and D. S. Cannell, Nondiffusive decay of gradient-driven fluctuations in a free-diffusion process, Phys. Rev. E, 76, (2007).

[10] R. Cerbino, S. Mazzoni, A. Vailati and M. Giglio Scaling behavior for the onset of convection in a colloidal suspension, Phys. Rev. Lett. 94, 64501, (2005).

Effects of Non-radiative energy transfer processes on the visible upconversion in ZrO$_2$:Yb^{3+}/Er^{3+} nanocrystals

O. Meza,* L.A. Díaz-Torres,* P. Salas,** E. De la Rosa,*, D. Solís*

* Centro de Investigaciones en Óptica A.C., León, Gto. 37150 México, ditlacio@cio.mx.
** Centro de Física Aplicada y Tecnología Avanzada, Universidad Nacional Autónoma de México, A.P.
1-1010, Querétaro, Qro. 76000, México, psalas@fata.unam.mx.

ABSTRACT

The luminescence dynamics of both near infrared (NIR) and visible (VIS) emission of Yb and Er in nanocrystalline ZrO2 is studied. It is found that upconversion emission is dominated by back transfer processes over direct Yb to Er energy transfer processes. A microscopic rate equation model was developed to fit both NIR and VIS fluorescence decay trends simultaneously, and to quantify the different interaction parameters responsible of the involved energy transfer processes. The microscopic model takes in to account the crystalline phase as well as the size of nanocrystals.

Keywords: Lifetime, Nanocrystals, Ytterbium, Erbium, Zirconium Oxide.

1 INTRODUCTION

Infrared-to-visible upconversion luminescence in nanocrystals has attracted considerable attention due its potential in applications as color displays and up-converted lasers [1-3]. In recent years many nanocrystalline materials have been proposed has efficient upconverters, among these, nanosized ZrO$_2$ has proven to be a best candidate for visible ligth generation. Many trivalent rare earths have been studied for Infrared-to-visible conversion [4]. Particularly, Erbium has been extensively studied in optical communications systems and Infrared-to-visible conversion processes [5]. The Er^{3+} ion has a relatively low absorption cross-section for the transitions in the near-infrared NIR region around 1000 nm [6]. Furthermore, the Yb^{3+} ion exhibits a higher absorption cross-section in this region and then, codoping with Yb^{3+} has proven to be a successful alternative for the upconversion process [7].

One of most important characteristics of a material is the fluorescence lifetime [8].The fluorescence lifetime is difficult to predict because of its dependence of several parameters, such as crystalline phase, material host, crystal size and codoping ions. There are many theoretical models explaining those observations, Forster was pioneer in this work, continued later by Dexter [9], Inokuti and Hirayama [10], whose main purpose was to estimate the rates for energy-transfer processes. In spite that those models have been successfully applied to several materials, these models have not improved to overcome an important fact and main assumption: All ions are subject to a sum of interactions over a continuous distribution of luminescence trap centers [11]. This is certainly not the case in crystalline medium, where the ions are distributed in specific positions in the host lattice. Therefore, interactions among ions present a discrete distribution and for nanocrystals this property becomes quite important. In this work we present the fluorescence characterization of Yb/Er codoped nanocrystalline ZrO$_2$ with dopant concentrations. An estimation of the upconversion parameters (red and green emission) and their corresponding efficiencies are calculated and discussed by solving the microscopic rate equations that govern the dymanics of the fluorescence decays subject to different Non-Radiative Energy Transfer processes among Yb and Er ions [12].

2 EXPERIMENTAL CHARACTERIZATIONS

Nanocrystals were prepared by using the sol-gel method. We synthesize four samples (1%)Yb/(1%)Er, (2%)Yb/(0.1%)Er , (2%)Yb/(0.2%)Er and (2%)Yb/(0.5%)Er. The samples were obtained using zirconium n-propoxide (ZP) 70% purity, erbium nitrate (Er(NO)3)3 · 5H$_2$O) and ytterbium chloride (YbCl$_3$ · 6H$_2$O) 99.99% purity as precursors. All samples were aged at 65 °C for 15 ~ 18 h. Afterwards the samples were dried at 300 °C for 2 h, dehydrated (stabilized) at 500 °C for 2 h and annealed at 1000 °C for 5 h applying increment ramps of temperature. A detailed explanation of sample preparation was reported elsewhere [13].

The crystalline structure of the samples was investigated by X-ray diffraction (XRD) using a SIEMENS D-500 equipment provided with a Cu tube with Kα radiation at 1.5405 Å, scanning in the 20º to 70º 2θ. The average particle sizes obtained from XRD patterns using the Scherer equation were between 60 and 80 nm. Crystalline structures were refined with the Rietveld technique by using DBWS-9411 and FULLPROF-V3.5d codes, peak profiles modeled with a pseudo-Voigt function contained average crystallite size as one of its characteristic parameters .Transmission electron microscopy (TEM) was performed in JEM-2200FS transmission electron microscope with accelerating voltage of 200 kV.

For the photoluminescence (PL) characterization a CW semiconductor laser diode centered at 970 nm was used as a pump source. The fluorescence emission was analyzed with a monochromator Acton Pro 500i and a R955 photomultiplier tube (Hamamatsu) connected to a mode-locking amplifier SR860 (Stanford). Fluorescence lifetime was measured using an SR540 chopper (Stanford) with a monochromator and photomultiplier connected to a LeCroy digital Oscilloscope.

3 THEORETICAL MODEL

The microscopic origin for non-radiative energy transfer processes can be visualized as an interaction between an excited ion, the donor D, and another not excited ion, the acceptor A, with an absorption transition resonant with the de-excitation of the first one [12]. For simplicity we represent this process by $\left(D_{FS}^{IS} \rightarrow A_{ES}^{IS}\right)_{ji}$. This notation indicates the j-th donor ion D_j has a non radiative relaxation from an initial state IS to a final state FS, and the freed energy excites the i-th acceptor ion Ai from an initial state IS up to an excited state ES. The transfer probabilities are due to dipole-dipole, can be written as [14]

$$W\left(D_0^1 \rightarrow A_1^0\right)_{ji} = \frac{1}{\tau_{D_0^1}} \left(\frac{R_{06}\left(D_0^1 \rightarrow A_1^0\right)}{R_{D_i A_j}}\right)^6 \quad (1)$$

where $R_{0S}\left(D_0^1 \rightarrow A_1^0\right)$ defines the critical distance of the non radiative transfer. τ_{FS}^{IS} is the radiative lifetime of a donor relaxing radiatively from the initial state IS to a final state FS. $R_{D_i A_j}$ is the distance between the j-th Donor and the i-th Acceptor. In general it is assumed a crystalline host, and in consequence the D and A ions are uniformly distributed in the possible crystalline sites within the crystal lattice. To study the non radiative energy transfer from the Yb^{3+} ions to the Er^{3+} ions, we have to consider the donor ions (Yb^{3+}) are two level ions whereas the acceptors ions (Er^{3+}) are five level entities (see Figure 1). Such active ions will be uniformly distributed at the available crystallite sites within the nanocrystallite. N_D and N_A are the total number of active ions inside of the $ZrO_2:Yb^{3+}/Er^{3+}$ nanocrystal. Under this assumption, we represent the probability of the j-th D ion to remain in its excited energy level n (n = 1) at time t by $P_{D_j}^n(t)$; and the probability of the i-th acceptor A to remain in its excited energy level m (m = 1, 2,...,5) at time t by $P_{A_j}^m(t)$. This means that we need to write $N_D + 5N_A$ ordinary differential equations (GETME):

$$\frac{dP_{D_i}^1}{dt} = -\frac{1}{\tau_{D_0^1}} P_{D_i}^1 - P_{D_i}^1 \sum_{j=1}^{NA} W\left(D_0^1 \rightarrow A_2^0\right)_{ij} P_{A_j}^0 \quad (2a)$$

$$- P_{D_i}^1 \sum_{j=1}^{NA} W\left(D_0^1 \rightarrow A_3^1\right)_{ij} P_{A_j}^1 - P_{D_i}^1 \sum_{j=1}^{NA} W\left(D_0^1 \rightarrow A_4^2\right)_{ij} P_{A_j}^2$$

$$\frac{dP_{A_i}^1}{dt} = -\frac{1}{\tau_{A_0^1}} P_{A_i}^1 + \frac{1}{\tau_{A_1^2}} P_{A_i}^2 - P_{A_i}^1 \sum_{j=1}^{ND} W\left(D_0^1 \rightarrow A_3^1\right)_{ij} P_{D_i}^1 \quad (2b)$$

$$\frac{dP_{A_i}^2}{dt} = -\frac{1}{\tau_{A_0^2}} P_{A_i}^2 - \frac{1}{\tau_{A_1^2}} P_{A_i}^2 + P_{A_i}^0 \sum_{j=1}^{ND} W\left(D_0^1 \rightarrow A_2^0\right)_{ij} P_{D_i}^1 \quad (2c)$$

$$- P_{A_i}^2 \sum_{j=1}^{ND} W\left(D_0^1 \rightarrow A_4^2\right)_{ij} P_{D_i}^1$$

$$\frac{dP_{A_i}^3}{dt} = -\frac{1}{\tau_{A_0^3}} P_{A_i}^3 + \frac{1}{\tau_{A_3^4}} P_{A_i}^4 + P_{A_i}^1 \sum_{j=1}^{ND} W\left(D_0^1 \rightarrow A_3^1\right)_{ij} P_{D_i}^1 \quad (2d)$$

$$\frac{dP_{A_i}^4}{dt} = -\frac{1}{\tau_{A_0^4}} P_{A_i}^4 - \frac{1}{\tau_{A_3^4}} P_{A_i}^4 + P_{A_i}^2 \sum_{j=1}^{ND} W\left(D_0^1 \rightarrow A_4^2\right)_{ij} P_{D_i}^1 \quad (2e)$$

Where $Yb(^2F_{5/2}, ^2F_{7/2}) \rightarrow Er(^4I_{15/2}, ^4I_{11/2})$, $Yb(^2F_{5/2}, ^2F_{7/2}) \rightarrow Er(^4I_{13/2}, ^4F_{9/2})$ and $Yb(^2F_{5/2}, ^2F_{7/2}) \rightarrow Er(^4I_{11/2}, ^2H_{11/2}+^4S_{3/2})$ transition were denoted as $W\left(Yb_0^1 \rightarrow Er_2^0\right)$, $W\left(Yb_0^1 \rightarrow Er_3^1\right)$ and $W\left(Yb_0^1 \rightarrow Er_4^2\right)$, and $\tau\left(Yb_0^1\right)$ and $\tau\left(Er_0^{IS}\right)$ are the free ion lifetimes for Yb^{3+} and Er^{3+} respectively from the level IS to ground state, when no energy processes are present (see Table). And $1/\tau\left(Er_{ES}^{IS}\right)$ are the relaxation rates of the Er^{3+} ions from its upper state IS (Initial State) to the next lower (Final State) state FS=IS-1, for IS=4,3,2. These radiative relaxation rates were obtained from Judd-Ofelt calculation done with the reported Judd-Ofelt parameters reported in [15].

By solving GETME, we are able to predict at any time all the individual populations of all the excited states of all the active ions within the nanocrystallite. For this model the only free parameters are the $W\left(D_0^1 \rightarrow A_2^0\right)$, $W\left(D_0^1 \rightarrow A_3^1\right)$ and $W\left(D_0^1 \rightarrow A_4^2\right)$ parameters corresponds to direct energy transfer $Yb(^2F_{5/2}, ^2F_{7/2}) \rightarrow Er(^4I_{15/2}, ^4I_{11/2})$ and to upconversion processes $Yb(^2F_{5/2}, ^2F_{7/2}) \rightarrow Er(^4I_{13/2}, ^4F_{9/2})$ and $Yb(^2F_{5/2}, ^2F^{7/2}) \rightarrow Er(^4I_{11/2}, ^2H_{11/2} - ^4S_{3/2})$ respectively.

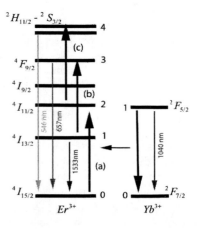

Figure 1: Energy levels diagram for the Yb^{3+}/Er^{3+} system. (a) direct energy transfer rate $Yb(^2F_{5/2}, ^2F_{7/2}) \rightarrow Er(^4I_{15/2}, ^4I_{11/2})$ and (c,b) upconversion processes $Yb(_2F_{5/2}, ^2F_{7/2}) \rightarrow Er(^4I_{13/2}, ^4F_{9/2})$ and $Yb(^2F_{5/2}, ^2F_{7/2}) \rightarrow Er(^4I_{11/2}, ^2H_{11/2} - ^4S_{3/2})$.

Table 1: Free ion relaxation rates for ZrO2:Yb,Er

Transition	Value (s⁻¹)	Transition	Value (s⁻¹)
Yb ($^2F_{5/2} \rightarrow {}^2F7/2$)	1052.632	Er($^4F_{9/2} \rightarrow 4I15/2$)	1142.857
Er($^4I_{13/2} \rightarrow {}^4I15/2$)	147.058	Er($^4S_{3/2} \rightarrow {}^4I_{15/2}$)	1190.476
Er($^4I_{11/2} \rightarrow {}^4I15/2$)	86.956	Er($^4I_9/_2 \rightarrow 4I_{11}/2$)	12974.948
Er($^4I_{11/2} \rightarrow 4I13/2$)	58.719	Er($^4F_{9/2} \rightarrow 4I9/2$)	15986.433
Er($^4I_{9/2} \rightarrow 4I15/2$)	115.340	Er($^4S_{3/2} \rightarrow 4F9/2$)	14452.316

Sample	Critical distance (Å)		
Yb/Er%	$R(D_0^1 \rightarrow A_2^0)$	$R(D_0^1 \rightarrow A_3^1)$	$R(D_0^1 \rightarrow A_5^2)$
1/1%	6,010	5,996	5,996
2/0.1%	6,010	5,996	5,996
2/0.2%	10,575	8,929	5,913
2/0.5%	10,851	6,6118	7,951

The macroscopic normalized fluorescence emission of the excited state ES (ES=1..4) of species X (X= D or A) from an aggregate of k nanocrystallites is proportional to [12],

$$\varphi_X^{Es}(t;C) = \frac{\sum_{i=1}^{N_X} P_X^{Es}(t;C)}{N_X} \quad (3)$$

Where $\varphi_X^{Es}(t;C)$ is a function of time (t) and dopant concentration (C), in order to solve the GETME by a 4th order Runge-Kutta numerical method, and then we use the

$$\varphi_X^{Es}(t;C) = \frac{\sum_{i=1}^{N_X} P_X^{Es}(t;C)}{N_X}$$

Eq.) for each excited level. To find $W(D_0^1 \rightarrow A_2^0)$, $W(D_0^1 \rightarrow A_3^1)$ and $W(D_0^1 \rightarrow A_4^2)$ parameters we fit the experimental lifetime decay $\varphi_X^{Es}(t;C)$ with Eq. (3) for all dopant concentration for any time, that is to say the square difference between the model solution and experimental measurement for each dopant concentration (C), energy level (Es), ion type (X) and any time (t_i) need to be minimal:

$$f = \sum_C \sum_{Es} \sum_{X=A,D} \sum_{i=1}^{Nt} (\log \varphi_X^{Es}(t_i;C) - \log \Phi_X^{Es}(t_i;C))^2$$
$$(4)$$

To find minimum of constrained nonlinear multivariable function f we use the Trust-Region-Reflective Optimization method. [15].

4 RESULTS AND DISCUSSIONS

Figure 2 shows the VIS and NIR spectra emission integration as a function of dopant concentration. The Er($^4I_{13/2} \rightarrow {}^4I_{11/2}$), Yb($^2F_{5/2} \rightarrow {}^2F_{7/2}$), Er($^4F_{9/2} \rightarrow {}^4I_{15/2}$) and Er($^2H_{11/2} \rightarrow {}^4S_{3/2} \rightarrow {}^4I_{15/2}$) emissions were observed among 1400-1620 nm, 1000-1100 nm, 640-690 nm and 510-580 nm respectively.

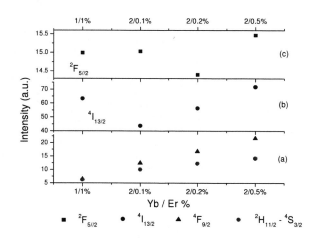

Figure 2. Emission intensity of ZrO2:Yb3+/Er3+ doped with Yb(1%),Er(1%), Yb(2%),Er(0.1%), Yb(2%),Er(0.2%) and Yb(2%),Er(0.5%).

The visible emission increases with the dopant concentration increment, nevertheless the red emission it is more intense than the green emission. The Er($^4I_{13/2} \rightarrow {}^4I_{11/2}$) emission for the same Yb concentration increases with the Er concentration increment. The Yb($^2F_{5/2} \rightarrow {}^2F_{7/2}$) emission does not show dependency with dopant concentration for the studied samples. Our simulations and experimental lifetime curves are shown in Figure 3. The effective lifetime [10] is obtained from the experimental fluorescence decay by the expression: $\tau_{eff} = \int I(t)dt / I(t=0)$. Our model, applied to all crystal samples, places the dopant ions randomly into the Zr sites of the lattice. With such a random placement of dopants, the total transfer rate is then calculated according to the assumed interaction (see eq. 1). Solving the eq. (4) for all samples, we estimate the critical distances (see table 1). This difference is due principally to the change of crystalline phase with dopant concentration. In addition we found that the Er($^4I_{11/2} \rightarrow {}^4I_{13/2}$) = (58.719 + 249.281) s⁻¹ transfer rate is greater than that estimated by the Judd-Ofelt parameters. Then, the most probably is that increment is promoted partly by the phonon energy introduced by the presence of some impurities such as H₂O (470cm⁻¹). For most of the wet method synthesized nanocrystals there is always some contamination with high phonon energy centers, such as OH⁻ at 3400 cm⁻¹ and CO⁻ at 1500 cm⁻¹, on the surface of nanocrystals, which can induce efficient nonradiative relaxation processes.

5 CONCLUSIONS

In summary, we have measured emission and lifetime curves of the fluorescence decay of Er(4I15/2), Yb($^2F_{5/2}$), Er(4F9/2) and Er($^2H_{11/2} \rightarrow {}^4S_{3/2}$) energy levels in ZrO₂:Yb/Er. The crystalline phase of ZrO2: Yb3+, Er3+ nanophosphor is determined by the Yb³⁺ and Er³⁺ concentration and affected the emission properties of the nanocrystal.

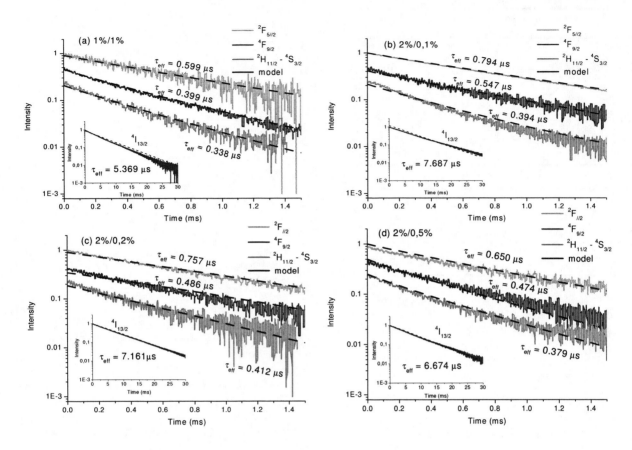

Figure 3: Experimental and simulated fluorescence decays of the ZrO_2:Yb,Er doped with (a) Yb(1%),Er(1%), (b) Yb(2%),Er(0.1%), (c) Yb(2%),Er(0.2%) and (d) Yb(2%),Er(0.5%).

The energy exchange between Yb^{3+} and Er^{3+} ions in the nanocrystal can be characterized by a dipole-dipole interaction. The critical distances depend of the dopant concentration due to the crystalline phase change. In addition the Er($^4I_{11/2} \rightarrow {}^4I_{13/2}$) transfer rate is greater than that estimated by the Judd-Ofelt parameters due to the action of impurities. Our interpretation is further supported by the application of solving General Energy Transfer Master Equations that govern the excitation dynamics of Yb and Er ions in a nanocrystal.

References

[1] H. Eilers, Materials Letters, 60 (2006) 214-217.

[2] R.N. Bhargava, J Lumin, 72-74 (1997) 46-48.

[3] A. Capobianco, F. Vetrone, T. D'Alesio, G. Tessari, A. Speghini, M. Bettinelli, Phys. Chem. Chem. Phys., 2 (2000) 3203-3207.

[4] G. Chen, G. Somesfalean, Y. Liu, Z. Zhang, Q. Sun, F. Wang, Physical Review B (Condensed Matter and Materials Physics), 75 (2007) 195204.

[5] D. Khoptyar, S. Sergeyev, B. Jaskorzynska, J. Opt. Soc. Am. B, 22 (2005) 582-590.

[6] E. Cantelar, J.A. Sanz-Garcia, F. Cusso, Journal of Crystal Growth, 205 (1999) 196-201.

[7] N. Yamada, S. Shionoya, T. Kushida, J. Phys. Soc. Jpn, 32 (1972) 1577-9.

[8] J.R. Lakowicz, Principles of Fluorescence Spectroscopy, Third Edition ed., Springer, 2006.

[9] D.L. Dexter, The Journal of Chemical Physics, 21 (1953) 836-850.

[10] M. Inokuti, F. Hirayama, The Journal of Chemical Physics, 43 (1965) 1978-1989.

[11] S.O. Vásquez, Physical Review B, 60 (1999) 8575.

[12] L.A. Díaz-Torres, O. Barbosa-García, C.W. Struck, R.A. McFarlane, J. Lumin, 78 (1998) 69-80.

[13] E. De la Rosa-Cruz, L.A. Diaz-Torres, R.A. Rodriguez-Rojas, M.A. Meneses-Nava , Appl. Phys. Lett, 83 (2003) 4903-4905.

[15] O. Meza, L.A. Diaz-Torres, P. Salas, E.D.l. Rosa, C. Angeles-Chavez, D. Solis, Journal of Nano Research, 5 (2009) 121-.130

[16] R.I. Merino, V.M. Orera, R. Cases, M.A. Chamarro, Journal of Physics, (1991) 8491.

[17] T.F. Coleman, Y. Li, An Interior Trust Region Approach for Nonlinear Minimization Subject to Bounds, in, Cornell University, 1993.

Enzymatic Synthesis: Acyl Ascorbate and Its Industrial Use as a Food Additive

R. Sharma[1], Y. Pathak[2]

[1]Center of Nanobiotechnology, TCC and Florida State University, Tallahassee, FL 32304;
[**]Department of Pharmacy, Sullivan University, Louisville, KY 40205

ABSTRACT

Acyl ascorbates were synthesized through the condensation of various fatty acids with L-ascorbic acid using immobilized lipase in a water-soluble organic solvent, and their properties as food additive were examined. The optimal conditions, which were the type of organic solvent, reaction temperature, the initial concentrations of substrates and the molar ratio of fatty acid to ascorbic acid, for the enzymatic synthesis in a batch reaction were determined. The continuous production of acyl ascorbate was carried out using a continuous stirred tank reactor (CSTR) and plug flow reactor (PFR) at 50°C, and each productivity was ca. 6.0×10 for CSTR and 1.9×10^3 g/(L-reactor·d) for PFR for at least 11 days, respectively. The temperature dependences of the solubility of acyl ascorbate in both soybean oil and water could be expressed by the van't Hoff equation, and the dissolution enthalpy, ΔH, values for the soybean oil and water were ca. 20 and 90 kJ/mol, respectively, irrespective of the acyl chain length. The decomposition kinetics of saturated acyl ascorbate in an aqueous solution and air was empirically expressed by the Weibull equation, and the rate constant, k, was estimated. The activation energy, E, for the rate constant for the decomposition in both systems depended on the acyl chain length. The surface tensions of acyl ascorbates in an aqueous solution were measured by the Wilhelmy method, and the critical micelle concentration (CMC) and the residual area per molecule were calculated. The CMC values were independent of temperature but dependent on the pH. The effect of pH of aqueous phase on the stability of O/W emulsion prepared using acyl ascorbate as an emulsifier was examined, and the high stability at pHs 5 and 6 was ascribed to the largely negative surface-charge of droplets in the emulsion. The addition of saturated acyl ascorbate, whose acyl chain length was from 8 to 16, lengthened the induction period for the oxidation of linoleic acid in a bulk and microcapsule with maltodextrin as a wall material. The oxidative stability in bulk system increased with increasing the acyl chain length, whereas that in the microcapsule was the highest at the acyl chain length of 10. The esterification of various polyunsaturated fatty acids, such as linoleic, α- and γ-linolenic, dihomo-γ-linolenic, arachidonic, eicosapentaenoic, docosahexaenoic and conjugated linoleic acids with ascorbic acid and subsequent microencapsulation significantly improved their oxidative stability.

Keywords: *Acyl ascorbate; Immobilized lipase; Continuous production; Amphiphilic food additive; Antioxidative emulsifier*

1 INTRODUCTION

Lipid oxidation is a complicated process. Antioxidants prolong the oxidation such as BHA (butylated hydroxyanisole), BHT (butylated hydroxytoluene), propyl gallate, vitamin C and tocopherols (vitamin E), inhibit free radical chain reactions. Derivatives of ascorbic acid acylated with a long-chain fatty acid such as palmitic acid are common additives in foods rich in lipid. 6-O-palmytoyl ascorbate shows antitumor activity and metastasis-inhibitory effects. Polyunsaturated fatty acid (n-6 PUFA) have advantage as a substrate for the synthesis of acyl ascorbate along with other physiological functions such as antithrombotic, cholesterol depressant and antiallergenic properties, as a precursor of prostaglandins, leukotrienes and related compounds in inflammation and immunity. But they suffer as 6-O-palmitoyl and stearoyl ascorbates are insoluble in water than the shorter acyl chains ascorbates. Enzymatic vitamin C esters is other advantage than a chemical method because of the simplicity of its reaction process and its high regioselectivity. Examples are synthesis sugar and fatty acids esterification by lipase, 6-O-Palmitoyl ascorbate using lipase shown in Figure 1. Enzymatic synthesis of 6-O-acyl ascorbates using immobilized lipase in a batch reaction system a CSTR or PFR, solubility in water - oil conditions (Section 3), emulsifier property and antioxidative ability of acyl ascorbate in a lipid microencapsulation are evaluated (Sections 4, 5).

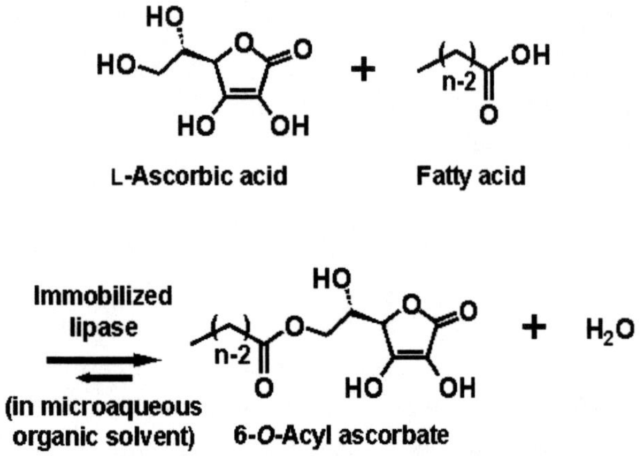

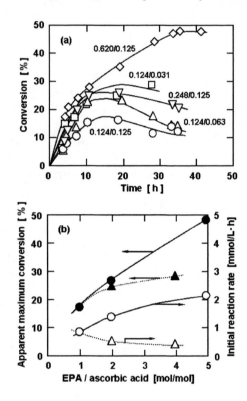

Figure 1. Scheme of condensation reaction of ascorbic acid with fatty acid using lipases from various origins in a microaqueous organic solvent with low water content.

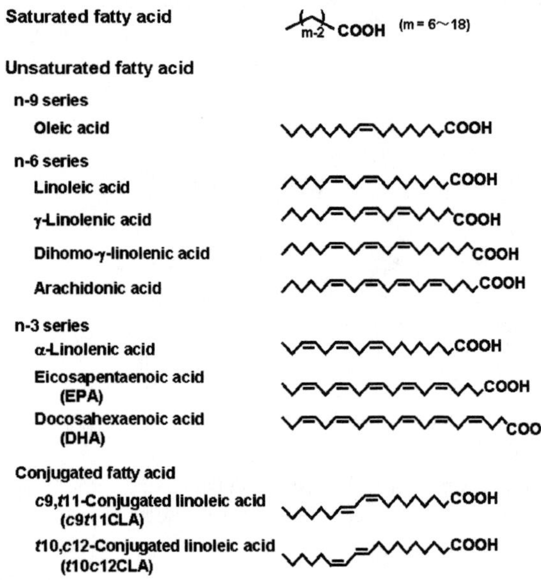

Saturated fatty acid $\underset{m-2}{\bigwedge}$COOH (m = 6~18)

Unsaturated fatty acid

n-9 series

Oleic acid

n-6 series

Linoleic acid

γ-Linolenic acid

Dihomo-γ-linolenic acid

Arachidonic acid

n-3 series

α-Linolenic acid

Eicosapentaenoic acid (EPA)

Docosahexaenoic acid (DHA)

Conjugated fatty acid

c9,t11-Conjugated linoleic acid (c9t11CLA)

t10,c12-Conjugated linoleic acid (t10c12CLA)

Figure 2. Molecular structures of fatty acids used for the synthesis of acyl ascorbate.

2 CONDITIONS FOR ENZYMATIC ACYL ASCORBATE SYNTHESIS : BATCH REACTION

The condensation of L-ascorbic acid and lauric acid in dehydrated acetonitrile by lipases (Chirazyme®L-2, Novozym®435, lipase types PS,QL,PL,A,F,M,MY,OF immobilized on macroporous acrylic resin can catalyze the condensation in the medium. Optimal conditions are 60°C. Examples: eicosapentaenoic acid and ascorbic acid xondensation is shown in Figure 3(a). Major factors: water content of reaction medium; molar ratio of eicosapentaenoic acid : ascorbic acid = 0.129 mmol:0.125 mmol; transient changes in the conversion (see Figure 4a and 4b). The equilibrium constant, K_C, is defined by

$$K_C = \frac{C_{P_e} C_{W_e}}{C_{A_e} C_{F_e}}$$

where C is the concentration and the subscripts A, F, P, and W represent ascorbic acid, lauric acid, lauroyl ascorbate, and water, respectively. The subscript e indicates the equilibrium. The values of C_{P_e} and C_{W_e} were experimentally determined.

Figure 3. Lipase-catalyzed condensation of eicosapentaenoic acid (EPA) and ascorbic acid at their various molar ratios in acetone at 55°C.

Figure 3(a) and 4(a), show the formation rate, conversion (equilibrium conversion).

3 CONTINUOUS PRODUCTION OF ACYL ASCORBATE USING A CSTR OR PFR

As illustration, continuous stirred tank reactor (CSTR) system is shown in Figure 5(a) An immobilized lipase (3.0 g by dry weight), Chirazyme®L-2 C2, is shown packed into the stainless steel basket. The volume of the solvent in the reactor, mass of ascorbic acid, fatty acid (decanoic, lauric or myristic acid) added with right solvent concn, reactor size, flow rate, water-bath temperature and magnetic stirrer speed are very crucial factors in this plug flow reactor system (PFR) is shown in Figure 5(b) for Chirazyme®L-2 C2 particles. The dependence mean residence times in the reactor, τ, and the concentrations of the decanoyl, lauroyl and myristoyl ascorbates in the effluent shown in Figure 6 with productivity of lauroyl ascorbate, decanoyl and myristoyl ascorbate in Figure 7.

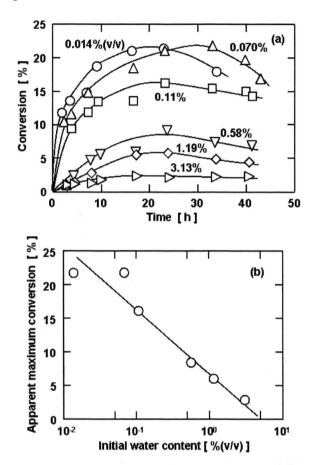

Figure 4. Effect of initial acetone water content on the synthesis of 6-O-eicosapentaenoyl ascorbate through lipase-catalyzed condensation.

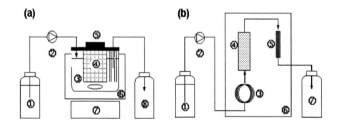

Figure 5 (a) Scheme of CSTR for the continuous synthesis of saturated acyl ascorbates. 1: feed reservoir, 2: pump, 3: reactor, 4: basket packed with immobilized lipase, 5: lid, 6: water-bath, 7: magnetic stirrer, 8: effluent reservoir;

(b) Scheme of a PFR; 1: feed reservoir, 2: pump, 3: pre-heating coil, 4: column packed with ascorbic acid, 5: column packed with immobilized lipase, 6: thermo-regulated chamber, 7: effluent reservoir.

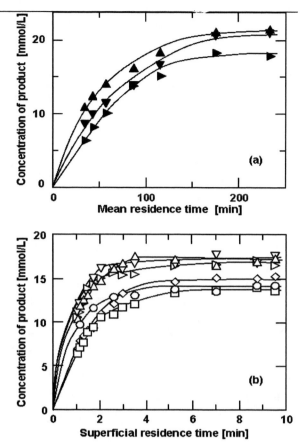

Figure 6. (a) Outcome of mean residence time, τ, and the concentration of (▲) decanoyl, (▼) lauroyl or (▶) myristoyl ascorbate in the CSTR at 50°C. (b) Relationship between the superficial residence time, τ, and the concentration of (○) arachidonoyl, (□) oleoyl, (◇) linoleoyl, (△) decanoyl, (▽) lauroyl or (▷) myristoyl ascorbate using the PFR at 50°C. In both reactor systems, the fatty acid concentration in the feed was 200 mmol/L.

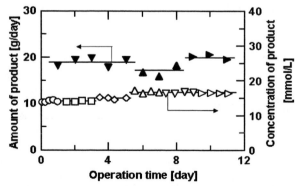

Figure 7. Continuous production of (▲, △) decanoyl, (▼, ▽) lauroyl or (▶, ▷) myristoyl, (○) arachidonoyl, (□) oleoyl, and (◇) linoleoyl ascorbate using the CSTR (closed symbols) or PFR (open symbols) with immobilized lipase, Chirazyme® L-2 C2, at 50°C.

4 SOLUBILITY OF ACYL ASCORBATE IN WATER AND OIL

The solubilities of the saturated acyl ascorbates in water or soybean oil show specific behavior. The acylation of ascorbic acid improves its solubility in soybean oil but decreased the solubility in water. The temperature dependence of the solubilities, S, of the decanoyl, lauroyl, myristoyl, palmitoyl and stearoyl ascorbates in water or soybean oil (Figure 8) could be expressed by the following van't Hoff equation:

$$\frac{dS}{d\left(1/T\right)} = -\frac{\Delta H}{R}$$

where ΔH is the dissolution enthalpy, R is the gas constant, and T is the absolute temperature. The dependence of the solubility in water on the acyl chain length of the acyl ascorbate was much stronger than that of the solubility in soybean oil. The plots in Figure 8 shows a straight line for each acyl ascorbate. The dissolution enthalpy, ΔH, was evaluated from the slope. The relationship between the acyl chain length and ΔH for the solubilization of the saturated acyl ascorbates in soybean oil or water is unique.

5. APPLICATIONS : ACYL ASCORBATE FOR LIPID MICROENCAPSULATION

Microencapsulation of a lipid with a wall material is a promising technology in the food and other industries. Microencapsulation of polyunsaturated fatty acid or its acylglycerol suppresses or retards lipid oxidation.

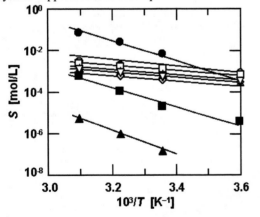

Figure 8. van't Hoff plots for the solubilities, S, of ($\bigcirc$, $\bullet$) decanoyl, ($\square$, $\blacksquare$) lauroyl, ($\triangle$, $\blacktriangle$) myristoyl, ($\triangledown$) palmitoyl and ($\diamond$) stearoyl ascorbates in soybean oil (open symbols) and water (closed symbols).

Microencapsulation is emulsification of lipid core with a polysaccharide wall followed by drying of the emulsions(Spray-dried microcapsules). As illustration, linoleic acid can be mixed with acyl ascorbates and microencapsulated with maltodextrin or gum arabic by spray-

drying with monitoring antioxidative ability of the ascorbates toward the encapsulated linoleic acid(see Figure 9). Microencapsulation needs optimal conditions spray-dryer flow rate of 3.0 kg/h, centrifugal atomizer at ca. 3×10^4 rpm, temperatures of air 200°C and 100 - 110°C, flow rate of air was ca. 7.5 m^3/min, relative humidity 12%,dark at 37°C.

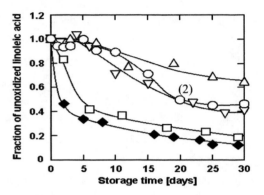

Figure 9. Effect of the addition of various saturated acyl ascorbates to linoleic acid on the oxidation of linoleic acid encapsulated with maltodextrin; ($\diamond$) no addition, ($\square$) octanoyl, ($\triangle$) decanoyl, ($\triangledown$) lauroyl and ($\bigcirc$) palmitoyl ascorbate. The molar ratio of the saturated acyl ascorbate to linoleic acid was 0.1. Temperature 37°C and a relative humidity 12%.

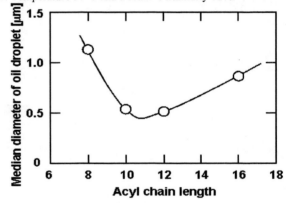

Figure 10. Example of oil droplets in emulsions prepared with various saturated acyl ascorbates depend on radii.

The oxidative stability of the microencapsulated arachidonoyl ascorbate and arachidonic acid is crucial at 37°C and 12% relative humidity.

6 CONCLUSION

Acyl ascorbate is synthesized through lipase-catalyzed-condensation of ascorbic acid with fatty acid. Enzymatic synthesis of acyl ascorbate is more advantageous than a chemical method The application of acyl ascorbate for microencapsulation of a lipid as antioxidative emulsifier is useful technology for suppressing or retarding the oxidation of polyunsaturated fatty acid.

7 REFERENCE

http://www.scribd.com/doc/25799862/Biotechnology-and-Nanotechnology/

Method Of Enhancing Polyaniline Conductivity Using Different Oxidizing Agent As Dopant

M.A.Iqbal and Saima Manzoor

Department of Physics, University of the Punjab, Lahore, Pakistan, iqbalma@yahoo.com

ABSTRACT

The synthesis and characterization of Polyaniline pellets has been done by using oxidizing agent Ammonium per oxy disulphate (APS). Samples of different molar ratios of Aniline Hydrochloride to oxidative agents have been analyzed. In these samples variation in the conductivity due to variation in the molar ratio is noticed, that the conductivity of Polyaniline pellets first increases up to the molar ratio of 4:1 and then start decreasing. The highest value of conductivity among the above mentioned three oxidizing agents is observed for APS i.e. 4.7 S/cm. Further Polyaniline emeraldine base film has been casted by dissolving it in Dimethylsulfoxide (DMSO). For all the samples, pellets as well as the film, variation in conductivity, XRD, and FTIR spectral variation has been studied. The thermal analysis was done for the powder form of polyaniline Emeraldine base by T.G.A and D.S.C methods indicate cross linking, melting and the decomposition of the material.

Keywords: Polyaniline, APS, DMSO, thermal degradation, conductivity, FTIR, TGA.

1. INTRODUCTION

Now a day's researchers show a great concern in the development of polymers whose strength of electrical properties can be customized for different practical application. Polyaniline is one of the conducting polymers whose electrical conductivity is enhanced upon doping of different oxidizing agents for example, Ammonium peroxy disulphate, hydrogen per oxide etc. Different conducting polymers such as poly pyrrole and poly acetylene cannot be converted into operable forms due to the infusible properties. The advance study in this field revealed that the polyaniline is the one which executes this application. [1] Such as stability and the relatively high conductivity in the doped state. The conductivity is referred to macromolecular structure and morphology and the acid used for protonation, and the stability is linked to the extrinsic factors for instance atmosphere and the temperature.

The process for controlling the conductivity is demonstrated by the slow deprotonation of the PAni salt to PAni base, which actually produces the changes at molecular level. [2] So the secondary dopants for the post polymerization can increase the conducting properties of polyaniline. Emeraldine salt films are made by their reaction with one of the solvents which are, DMSO, DMF, NMP and chloroform. But these show relatively decrease in conductivity typically ranges from 0.1-1 S/cm. So this form is converted into salt form by the acidic doping. The chemical reaction for the conversion is given as below.

The polyaniline formed through the oxidative polymerization of the monomer i.e. aniline, conducts electrically [3-4]. It is found that polymer processing is Hygroscopic in nature [5-8]. Thermal stability and temperature (Tg) are very important.

2. EXPERIMENTAL

The procedure of chemical polymerization of aniline given by the IUPAC was adopted. Ratio of 0.2 M aniline hydrochloride was oxidized to prepare Polyaniline with the variation in the molar ratio of oxidizing agent ammonium peroxy disulphate. The sample obtained was in the pallet form. Each pellet was pressed at 800 MPa. The conductivity measurements were done to know about the ratio having the best conductivity.

Further for the preparation of emeraldine base film, 0.045 M Ammonium per oxy disulphate was used. The aqueous solution of ammonium peroxy disulphate and aniline hydrochloride were prepared and left for half an hour. After this the ammonium per sulphate solution was added drop wise in Aniline hydrochloride solution. This was kept for 24 hours and then filtered.

The precipitates were washed with distilled water and 1.5 N HCl, until the remaining became colourless. The precipitates were dehydrated in oven for 24 hours at 80 ^{0}C.

Then the emeraldine salt was converted into base by dissolving it in DMSO (Dimethyl sulfoxide). Firstly this PAni powder was put in 100 ml of 5% wt ammonia solution which was prepared from the concentrated reagent solution. After 12 hours, powder was washed with distilled water until the filtrate become colourless.

Filtrate was dried for 48 hrs at 50-60 ^{0}C. Then 0.6 g of EB was dissolved in 50 ml of DMSO and stirred for 5 hrs using magnetic stirrer. The dark blue solution was filtered and used to cast the film on glass strip, dried and then peeled off from the glass by immersing it in water. Then film was pressed between the glossy weighing papers and dried in air for few days.

The results showed that the conductivity of base is very low, so the film was dipped in 1N aqueous solution of HCl. For the complete doping the film was immersed in solution for half day.

An FTIR spectrum was taken SHIMADZU 8400S, XRD to confirm the sample prepared, conductivity measurements were made. And the T.G.A and D.S.C analysis was done of the powder form of emeraldine base.

3. RESULT AND DISCUSSION

3.1 Conductivity measurement

Conductivity of the various oxidants to monomer ratio was measured using the four point probe method and it was found that the best ratio having the highest concentration is 4:1 and the value of conductivity was 4.7 S/cm. And the conductivity of the polyaniline (emeraldine base) film was of the order of 10^{-10} S/cm. After doping in HCl, the value of the conductivity was found to be 11.7 S/cm.

3.2 XRD Analysis:

It was done just to find the structure of PANI base sample. It was done by using Regaku Geiger Flex D-MAX/A diffractometer of Company.

The value of the operating voltage was 35 kV and current was 225 mA. The source of radiation used was copper.

The diffraction patterns were taken counts verses incident angle (2θ). The graph is shown in the figure 1. It is obvious from the graph that the chains are perturbed. The polyaniline EB shows the peaks at almost 2 = 27.7^0.

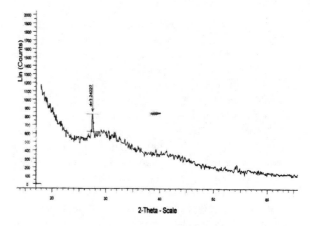

Fig.1: X-Ray diffraction of polyaniline base.

3.3 FTIR Analysis:

Figure 2 represents the spectra of PANI base film.

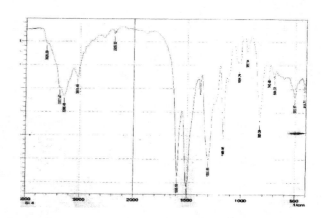

Fig. 2 FTIR of PAni base film.

The spectrum exhibits the 9 main absorption peaks at 3377.47, 3037, 1595, 1504, 1305.85, 1168.90, 1024.24, 950.94 and 833.28 /cm. the peaks at 3377.47is allotted to N-H(H free). The band corresponding to 3037 is due to C-H aromatic ring vibrations. The peaks at 1595 and 1504 /cm are attributed to C=C (Q) and C=C (benzenic). The peak at 1305.85 is assigned to C-N from aromatic amines. The band at 1168.90/cm represents N=Q=N. And the peaks 1024.24, 950.94 and 833.28 /cm are assigned to aromatic ring. The C-Cl stretching peaks originates in the region 590 to 700 /cm. [9]

The absorption characteristics of the PANI film are found to match with the literature [10].

3.4 Thermal analysis

3.4.1 Thermal Gravimetric Analysis (TGA)

The main points or the levels at which the weight losses of polyaniline base occurs are two, no matter whatever the environment is, when the experimental work is carried out. The first one is at smaller temperature at which the dehydration and possible evaporation of other unknown minor molecules occurs.

The second weight loss is the more important one which is actually at high temperatures and tells us about the degradation of the polymer.

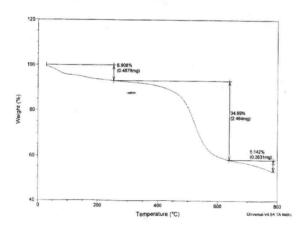

Fig.3 TGA curve of polyaniline base powder.

TGA curve is shown the figure 3, So in our case the initial weight loss of 6.908 % was observed up to the temperature 350 ^{0}C due to the moisture contents and due to the other solvents present. The major weight loss was observed between the temperature 400 to 650 ^{0}C, which was about 34.89 %. After this the further decomposition was 5.142 % observed up to 800 ^{0}C. This almost completed the polymer decomposition. So, it can be said that polyaniline base was thermally stable at least to 300 ^{0}C.

The weight loss detected after 350^{0}C is due to volatile and evaporable products leading from the oxidation of polymer. There are some important factors on which thermal stability depends, such as,

acid used for doping purpose, state of oxidation, heat treatment environment and the conditions of preparation. So the Polyaniline base has the steady and stable behavior among all form of polyaniline. [11]

3.4.2 Differential Scanning Calorimeter (DSC):

DSC is the commonly used technique to determine the transition temperature, cross linking, melting and the decomposition of the material.

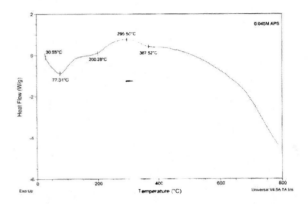

Fig.4: Differential Scanning Calorimeter trace of pani base powder.

Figure represents the DSC curve exhibiting the total heat flow and thermal conversions. The peak for endothermic reaction is there from 30.55^{0}C to almost 500^{0}C. and the exothermic peak is from 200 to 360^{0}C was found, which is similar to the work already reported about the moisture contents of polyaniline base powder [12]. The glass transition temperature is 30.55^{0}C.

The first exothermic peak representing the evaporation of water is logical in accordance with the TGA results. The next exothermic peak may be for the chemical process associated with the recrystallization [13] or the cross linking reaction [14]

4. CONCLUSION

The method of producing the polyaniline salt and base from the monomer, electrical, structural and the thermal stability of the samples was done. It is concluded that the best ratio of oxidant to monomer is 4:1. The increase in conductivity from base to salt form is in the order of 10 $^{-11}$ to 10 S/cm. Thermal analysis curves indicate, temperatures corresponding to the exothermic and endothermic peaks illustrates the decomposition of the polyaniline.

REFERENCES

[1] 12 J. A. Conklin, S.-C. Huang, S.-M. Huang, T. Wen and R. Kaner, Macromolecules, 28,6522 1995.

[2] W. R. Salaneck and A. Hilger, Science and applications of conducting Polymers, ed. IOP publishing Ltd.

[3] Prokes J, TrochoVa M, Hlavata D, Stejeskal J, Polym. Degrad. Stab., 78: 393-401,2002.

[4] K.G.Meoh,E.T.Kang, S.H. Khor and K.L.Tan, Polym. Degrad. And Stab. 107, 1990.

[5] V.G.Kulkarni, Thermochim, Acta, 188, 265, 1991.

[6] Y.Wei and K.F. Husueh, J. Polym. Sci.: Part A: Polym. Chem., 27,4351, 1989.

[7] S. H. O. Chan, P. K. H. Ho, E. Khor, M. M. Tan, K. L. Tan, B. T. G. Tan and Y. K. Lim, Synthetic Metals, 31, 95, 1989.

[8] 13 E. S. Matveeva, R. Diaz Calleja and V. P. Parkhutik, Synthetic Metals, 72, 105, 1995.

[9]] P.S.Rao, J.Anand, S.Palaniappan and, D.N. Sathyanarayana, European Polymer Journal, 36, 915-921, 2000.

[10] P. C. Rodrigues, M. P. Cantao, et al. Polyaniline/lignin blends: FTIR, MEV and electrochemical characterization, European Polymer Journal 38, 2213-2217, 2002

[11] R. Ansari and M.B.Keivani, Polyaniline conducting electroactive Polymer: thermal and environmental stability studies. E-Journal of Chemistry, 3, 202-217, 2006.

[12] A. J. Conklin, S. Huang, T. Wen, B. R. Kaner, Macromolecules 28, 6522, 1995.

[13] J. -C. LaCriox, A. F. Diaz, J. Electrochem, Soc, 135, 1457, 1988.

[14] Y. Wei, K. F. Hsueh. J. Polym. Sci, Part A 27, 4351, 1989.

Synthesis of Magnetically Separable Photocatalyst—

TiO₂/SiO₂ coated Mn-Zn Ferrite—and Its Photocatalytic Study

K. Laohhasurayotin[*], S. Pookboonmee and D. Viboonratanasri

National Nanotechnology Center, National Science and Technology Development Agency,
111, Thailand Science Park, Pathumthani, 12120, THAILAND
*kritapas@nanotec.or.th

ABSTRACT

Magnetically separable photocatalysts are of much interest since they can be used for waste management application and with recyclable advantage. Several TiO_2 incorporated with magnetite have been synthesized but their preparation methods are not convenient. Our aim of this work is to minimize the complication of synthesis procedures and reduce the use of chemicals that were not actually converted to the composite, i.e., surfactant, emulsifier, and stabilizers. With our process, a manganese-zinc ferrite nanoparticle could be prepared by simple co-precipitation of three metal ions, Mn^{2+}, Zn^{2+}, and Fe^{2+} and the resulting nanoparticles were coated with silica and titania through sol-gel methods. The rate of base addition and reagent mixing were found to be crucial for the size of the forming magnetic particles. Transmission electron microscopy was used to reveal the morphology of magnetic core and silica as well as titania coatings. Vibrating sample magnetometer was used to measure the superparamagnetism of the composite, performing good saturating magnetization as high as 10-30 emu/g. In addition, the amorphous silica interlayer was found to be advantageous in cancelling the synergistic effect between the two metal oxides. This was observed by monitoring the improved photo-decomposition of methylene blue with UV-Vis spectrophotometer, indicating the possibility in environmental remediation.

Keywords: Titanium dioxide, photocatalyst, manganese-zinc ferrite, superparamagnetic, co-precipitation

1 INTRODUCTION

Titanium dioxide (TiO_2) is known to possess the decomposition ability on many organic and inorganic pollutants due to its generation of two strong oxidant species, $\cdot OH$ and O_2^- [1]. It is cheap, stable, non-toxic, and especially easy to be prepared via simple chemical process, so-called hydrolysis. However, its usual size in nanometer range leads itself to the difficulty in handling or storing although the suspending particles are in the most effective form. The immobilized TiO_2 on many substrates have been developed by many researchers, but loss of surface area

limits the photocatalytic activity [2-3]. TiO_2 coated on magnetic materials, such as magnetite or ferrite, in a form of core-shell particle, provides a better solution because it preserves high surface area and gains ability to be re-collected by external magnetic source. A suitable magnetic particle employed as a supporting core for titanium dioxide depositing can be prepared by several methods suggested in the literatures, such as co-precipitation, citrate precursor, hydrothermal, and microemulsion. The magnetic nanoparticle, which is mainly formed as soft ferrite, MFe_2O_4 where M = Fe, Mn, Zn, Ni, etc., can be synthesized based on Schikorr's reaction [4].

$$M^{2+} + 2Fe^{2+} + 6OH^- + \frac{1}{2}O_2 \rightarrow MFe_2O_4 + 3H_2O \text{--------}(1)$$

The crystalline structure of MFe_2O_4 is mostly formed as inverse spinel. It usually has large magnetic polarization and high electrical resistivity. However, the photocatalytic and magnetic properties of the composite can be less effective. This is due to the electron migration in the electronic states between the photocatalyst and magnetic ferrite. To solve this problem, it has been suggested to allow the presence of silica layer between the contact point of the two semiconductors [5].

Recent publications reported by other research groups describe the preparation method of these magnetically separable photocatalysts with their results [6-8]. We carried out the synthesis of TiO_2/SiO_2 coated Mn-Zn ferrite by using three preparation steps including co-precipitation of Mn-Zn ferrite, hydrolysis coating of silica, and hydrolysis coating of titanium dioxide. It was found that without a complex condition provided by stabilizer or emulsifier, the nanoparticles of Mn-Zn ferrite could be well-formed with narrow size distribution by co-precipitation methods. This was resulted from the homogeneous growth of nano-magnetic particles controlled by rate of reagent addition and reaction mixing. The photocatalytic reaction of the as-prepared catalyst was observed. The superparamagnetic property of the magnetic core enhanced the photocatalyst to be able for recovering and redispersing after use. The improvement of the photocatalysis is being investigated.

2 EXPERIMENTAL

2.1 Materials

All chemicals are of analytical grade and used without further purification. $MnSO_4 \cdot H_2O$ and NH_4OH (30%) were purchased from Carlo Erba. $FeSO_4 \cdot 7H_2O$ was purchased from Fischer. $ZnSO_4 \cdot 7H_2O$ was purchased from Rankem. Tetraethyl orthosilicate (TEOS) and titanium (IV) tert-butoxide (TBOT) were purchased from Fluka. Methylene blue (MB, $C_{16}H_{18}N_3CIS \cdot 2H_2O$) was purchased from Unilab. Absolute ethanol and acetone were purchased from Merck.

2.2 Characterization methods

Scanning electron microscopic (SEM) and energy dispersive X-ray spectroscopic (EDS) measurement of the samples were obtained from HITACHI S-3400 microscope and EMAX Horiba, respectively. The transmission electron microscopy (TEM) was recorded by JEOL JEM 2010. X-ray diffraction patterns were obtained from X-ray diffractometer JEOL JDX 3530. Saturated magnetization (M_s) and hysteresis curve were recorded by Lakeshore 7404 Vibrating sample magnetometer.

2.3 Preparation of $Mn_{0.5}Zn_{0.5}Fe_2O_4$ (MZF)

A 1:1:1 mixture (30 mL) of 0.057 M Mn^{2+}, 0.057 M Zn^{2+}, and 0.23 M Fe^{2+} was added with 1.78 M NH_4OH solution (15 mL) while vigorously stirring. The resulting green solution was then heated in an oil bath up to 90-100 °C for 1 h. The mixture with dark brown precipitate was added with DI water (20 mL). The precipitate was washed with DI water and acetone then separated by centrifugation.

2.4 Preparation of SiO_2-coated $Mn_{0.5}Zn_{0.5}Fe_2O_4$ (S-MZF)

Typical method for the coating of silica on MZF was done by mixing MZF (180 mg) with 1:5 mixed solution of DI water and ethanol (30 mL). After dispersion by ultrasonicator, TEOS (30 μL) and 30 % aqueous ammonia (40 μL) were added and the hydrolysis was allowed to proceed for further 10 min. The other equal portions of these two reagents were added and this adding step was repeated until 450 μL total volume of TEOS was used. The resulting mixture was placed in a sonicator bath for 45 min. The resulting mixture was washed with water and ethanol using centrifugation. The collected precipitate was dried at room temperature.

2.5 Preparation of TiO_2/SiO_2-coated $Mn_{0.5}Zn_{0.5}Fe_2O_4$ (TS-MZF)

Typical method for the coating of titania on S-MZF (or MZF) was done by mixing S-MZF (150 mg) with 1:14

mixed solution of DI water and ethanol (75 mL) in a sonicator bath. This mixture was added with TBOT (0.5 mL) in absolute ethanol (10 mL). The resulting mixture was stirred in an oil bath at room temperature, subsequently heated up to 90 °C, and further maintained at this temperature for 2 h. A condenser was also used in order to prevent the vaporization of the solvent. The product was washed with DI water and ethanol, and then separated by centrifugation. The composite was received after drying at 60 °C for 24 h and calcined at 500 °C for 2 h.

2.6 Photocatalytic reaction

An amount of catalyst (12 mg; MZF, TS-MZF, and T-MZF)) was dispersed in a 60 mL of 5 ppm MB aqueous solution. The adsorption/desorption of MB molecules on photocatalyst surface was allowed until reaching equilibrium at 18 h in the dark. To conduct the experiment, each 3 mL of the catalyst suspensions was irradiated by 20 W of 366 nm low pressure UV lamp. At 10 min. interval, the catalyst was separated by centrifugation, and the MB removal was analyzed by monitoring the decrease of UV absorbance at 664 nm.

3 RESULTS AND DISCUSSION

3.1 Formation of magnetic nanoparticles by co-precipitation method

The chemical co-precipitation of MZF has been known to involve two sequencing steps of hydrolysis and oxidation of alkaline $Fe(OH)_2$ in the presence of Mn^{2+} and Zn^{2+} at elevated temperature based on Schikorr's reaction to form inversed spinel structure. In this case,

$$xMn^{2+} + yZn^{2+} + (3F - x - y)Fe^{2+} + 6OH^- + \frac{1}{2}O_2 \rightarrow$$
$$M_xZn_yFe_{3-x-y}O_4 + 3H_2O \quad \text{----(2)}$$

Because of its simplicity, co-precipitation is one of the most preferable methods for the preparation of magnetic particles. However, the large polydispersity of the formed particles is always a critical problem. Four MZF particles shown in Table 1 were prepared by co-precipitation method

Table 1: Reaction conditions and particle sizes of MZFs

Sample code	Addition type	Dispersion type	Particle size (nm)
MZF1	One shot	Stirred	8.7±2
MZF2	Dropwise	Stirred	67.5±12
MZF3	Dropwise	Homogenized	10.8±2
MZF4	One shot	Stirred	49.4±14

using different mixing rate as well as the rate of reagent (base) addition. Suggested by Auzan [9], the growth of

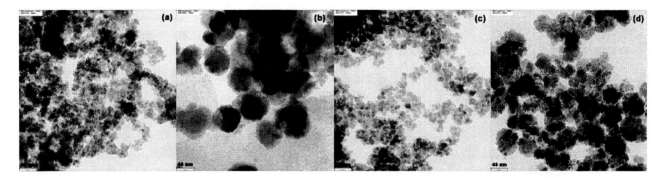

Figure 1: TEM images of (a) MZF1, (b) MZF2, (c) MZF3, (d) MZF4

magnetic nanoparticles by co-precipitation method can be controlled by two main practices. The addition of base reagent, which leads to the formation of nanoparticles, should be as fast as possible. Also, the mixing of the reaction solution should be at a very high rate. These will lead the reaction condition to approach the homogeneity that is essential for the uniform size nanoparticles with less aggregation. In our experiment, the two addition techniques, one shot and dropwise, of base solution were comparatively carried out. Figure 1a-d shows the TEM images of MZF samples prepared by the conditions in Table 1. It was obviously seen that the fast addition of NH₄OH resulted in small particles (MZF1) while the slow addition resulted in greater particle size (MZF2). However, the influence of homogeneity due to the mixing speed was more important than the initially formed particles as found that the homogenized reaction provided narrow-distributed particles with non-aggregated character in MZF3. Furthermore, MZF4 confirmed that when the volume of reaction increased, resembling the loss of homogeneity, the magnetic particles were formed variously in size and shape.

The superparamagnetic properties of the MZF samples were analyzed at room temperature by VSM. Figure 2 shows the hysteresis curves of MZF1, MZF3, and MZF4. The saturation magnetization of MZF3 (M_s = 44.5 emu/g) was higher than the other two samples ((M_s = 35.0, 32.9 emu/g for MZF1 and MZF4, respectively). This may be due to the uniformity of MZF3 particles formed in homogenized condition, contrarily to the stirred conditions that allowed the variation in size and thus crystallinity of MZF 1 and 4. Interestingly, the saturation magnetizations of these two were not different considerably, indicating of the complete formation of magnetic inducible materials by our synthetic design. In addition, the superparamagnetic property of MZF was confirmed as the hysteresis loop was not at all observed.

3.2 Coating of silica and titania on MZF

Silica coating on MZF was based on the intention to inhibit the electronic interaction between the core MZF and the TiO_2 layer. Figure 3 shows the TEM image of S-MZF that clearly demonstrate the coating layer of amorphous silica. The thickness of silica layer was approximately 10 nm on each S-MZF particle measured from the surface to its core size.

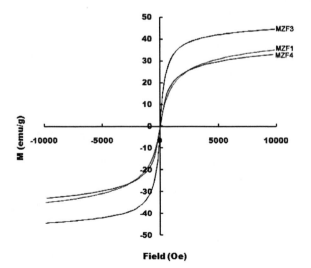

Figure 2: Magnetization measurement of MZFs at room Temperature

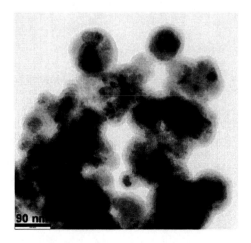

Figure 3: TEM images of S-MZF

Titania coating on S-MZF was confirmed by the TEM image shown in Figure 4. Due to a fast hydrolysis rate of TBOT, the TiO₂ crystals were formed as a rough layer of small granules covering entirely on the S-MZF particles. The X-ray diffraction patterns of the three samples are shown in Figure 4, confirming that each step of coating provides a new layer of the desired material.

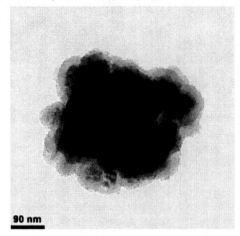

Figure 4: TEM image of TS-MZF

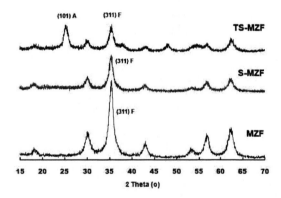

Figure 5: XRD pattern of the uncoated and coated MZFs

3.3 Photocatalytic activity

Figure 6 shows the photocatalytic activity of the prepared catalysts on methylene blue solution. After saturating small amount of samples with MB solution in dark place, the concentration decreased as the MB was adsorbed by the catalyst surface. When the composites were exposed to the UV radiation, the additional decrease of MB absorption was observed only in TS-MZF where as little or no change was found in the rest of the catalysts. This indicated that the photocatalysis was present in the three-layer composite, and thus the silica interlayers behave genuinely as an electron-transfer inhibitor of the two semiconductors. A similar photo-activity of T-MZF and MZF was attributed to the absence of silica interlayer.

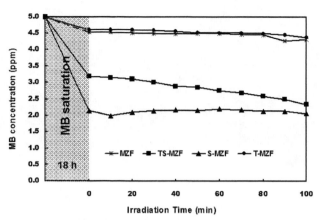

Figure 6: Decomposition of methylene blue by the prepared catalysts

4   CONCLUSION

A magnetically separable photocatalyst, TiO₂/SiO₂ coated Mn-Zn ferrite, was prepared and its properties were thoroughly analyzed. Via a simple co-precipitation method without involving preserved condition by stabilizer, monodisperse nanoparticles could be formed using proper mixing of reagents. The superparamagnetism of the particles was observed with rather high value reaching up to 40 emu/g. Activation of the titania/silica coated Mn-Zn ferrite by UV light was apparently observed which was certainly improved by silica interlayer between two semiconductors. These nanoparticles are promising in environmental remediation.

ACKNOWLEDGEMENT
This work was supported by National Nanotechnology Center, National Science and Technology Development Agency, Thailand (Grant no. P00-60061).

REFERENCES
[1] S. Banerjee, J. Gopal, P. Muraleedharan, A.K. Tyagi, B. Raj, *Curr. Sci.* 90, 1378, 2006.
[2] I.J. Ochuma, O.O. Osibo, R.P. Fishwick, S. Pollington, A. Wagland, J. Wood, J.M. Winterbottom, *Catal. Today* 128, 100, 2007.
[3] S. Horikoshi, N. Watanabe, H. Onishi, H. Hidaka, N. Serpone, *Appl. Catal. B.* 37, 117, 2002.
[4] G. Schikorr, *Z. Allg. Chem.*, 212, 33, 1938.
[5] D. Beydoun, R. Amal, G.K.C. Low, S.J. McEvoy, *Mol. Cat. A: Chem.* 180, 193, 2002.
[6] X. F. Song, L. Gao, *J. Am. Ceram. Soc.* 90, 4015, 2007.
[7] S. Xu, W. Shangguan, J. Yuan, M. Chen, J. Shi, Z. Jiang, *Nanotechnology* 19, 095606, 2008.
[8] C. Wang, L. Yin, L. Zhang, L. Kang, X. Wang, R. Gao, *J. Phys. Chem. C.* 113, 4008, 2009.
[9] E. Auzan, D. Zin, E. Blums, R. Massart, *J. Mater. Sci.* 34, 1253, 1999.

Effect of Compounding Method on the Mechanical Properties of Polypropylene-Clay Nanocomposites

J.W. Nader[*], A. Kiziltas[*, **], D.J. Gardner[*] and J. Parent[*]

[*]AEWC Advanced Structures& Composites Center, University of Maine, Orono, ME, USA
[**]Department of Forest Industry Engineering, Faculty of Forestry, University of Bartin, Bartin, Turkey

ABSTRACT

This study investigates the effects of direct melt compounding, masterbatch dilution and compatibilizer addition on the mechanical, thermal and morphological properties of organoclay nanocomposites. Nanoclay reinforced polypropylene composites manufactured using melt mixing with and without adding polypropylene-graft-maleic anhydride (PP-g-MA). These composites were compared with nanoclay reinforced polypropylene materials with similar compounds diluted from commercial masterbatches. All compounded materials were prepared at 2, 4, 6, 8, and 10 wt% nanoclay. Series of tensile, flexural, and impact tests were conducted to evaluate mechanical properties.

Key words: polypropylene, mechanical properties, nanocomposites, masterbatch dilution

1. Introduction

During the last decade, nanoreinforced polymer composite materials have received great attention both in academia and industry due to their superior mechanical properties, and relatively low cost. Due to their high aspect ratio, nano-particles/fibers provide better mechanical properties by enhancing the bonding between the matrix and the filler, and therefore providing a better load transfer between the matrix and the reinforcement [1-15]. Clay is one of the most studied nanofillers among researchers because of it is availability, relatively easy surface modification and low price among nanofillers [1, 2]. Clay is naturally a hydrophilic material and increasingly used for reinforcement of polymeric materials [1-3]. Hydrophilic structure makes clay difficult to exfoliate in a polymer matrix especially polyolefin based matrices because of their strong hydrophobicity. This contraction is overcome by the surface treatment of clay to render its surface more hydrophobic and by using adhesion promoters to make the matrix and clay more compatible [1-3]. It is very well documented that the mechanical properties of PP/nanoclay nanocomposite is directly related to the degree of exfoliation of the nanoclay particles [4-7]. Prior studies have extensively focused on studying the clay exfoliation and reported successful attempts in achieving full exfoliation of clay particles by using a mixture of stearyl ammonium-exchanged montmorillonite, maleated

polypropylene (MAPP), and PP homopolymer [2, 4-15]. In recent years, specific efforts have been devoted to the development of melt-compounded polypropylene (PP) based Polymer nanocomposites with organically modified fillers in which coupling agents, such as polypropylene-graft-maleic anhydride (PP-g-MA), were used to improve both the nano-fillers' dispersion and particle–matrix adhesion [16, 17]. Most studies have shown that Young's modulus and tensile strength increase in nanocomposites as compared to neat PP [2-19]. However, in some cases, the impact resistance of nanocomposites decreased when a coupling agent is used [2, 8, 12, 15].

Melt compounding is one of the most used techniques to produce nanocomposites especially at the industrial scales [20, 21]. Melt compounding has mainly concentrated on twin-screw extrusion, however several work has been reported using laboratory scale batch mixers and microextruders. Batch mixers are conventionally utilized in preparation of masterbatches, but twin-screw extruder is very common and increasingly used nanocomposite formation process [1, 2]. Despite the importance of this topic, few studies has focused on investigating the effect of compounding method, screw rotation and masterbatch dilution on the mechanical and thermal properties of nanocomposites [2, 15, 20-22]. Previous studies [1, 2] have shown that using masterbatches lead to more effective dispersive mixing, which is more uniform and show better exfoliated structure, and enables more precise control of filler concentration in the final compound in comparison with direct mixing .

In this study we seek to investigate the effects of direct melt compounding, masterbatch dilution and compatibilizer addition on the mechanical of nanoclay reinforced PP matrix.

2. Experimental Procedure

The polymer used in this study was polypropylene impact copolymer (FHR Polypropylene AP5135-HS) from Flint Hills Resources, LP. It has a melt flow index of 35 g/10 min and a density of $0.9 g/cm^3$. The polypropylene-graft-maleic anhydride (PP-g-MA) used had a molecular weight equal to 9100 and was purchased from Sigma-Aldrich. The nanoclay Nanomer I.44P was supplied by Nanocore and it has a quaternary ammonium chemistry based surface modification and an average particle size 15-20 micrometer. The commercial masterbatches nanoMax®-

PP-HiST were supplied by Nanocore. The nanoclay content in the nanoMax®-PP-HiST masterbatches was 50%. %. Note that the commercial masterbatches contain additives that are kept trade secret. Nanoreinforced PP composites were prepared by melt mixing PP pellets with nano-fillers using a C.W. Brabender minicompounder. Prior to compounding, the PP and nano materials were oven dried at 70°C for 12 hours. The in-house made compounds were prepared by direct melt mixing. The screw speed, the barrel temperature, and the blending time were kept constant during the manufacturing process at 60rpm, 180°C, and 20 min, respectively. PP/nanoclay material was produced with and without adding a compatibilizer (PP-g-MA), and with varying the nanofillers loading from 2 wt. % to 10 wt. % using an increment of 2 %. Note that in the case of using a compatibilizer the ratio of the nanofillers to the PP-g-MA was always kept constant at 2:1. The commercial masterbatches (nanoMax®-PP-HiST) containing 50% nanoclay was diluted to achieve 2, 4, 6, 8, and 10 wt %. nanoclay. The nanoMax®-PP-HiST pellets and the PP pellets were premixed using a speed mixer and simultaneously introduced in the hopper of the C.W. Brabender minicompounder. Adding 4wt%, 8wt%, 12wt%, 16wt% and 20wt% commercial masterbatch result in 2, 4, 6, 8, and 10 wt %. All compounded materials were cut into pellets and then oven dried at 70°C for 12 hours before being injection molded to make mechanical testing coupons. A summary of all compounded materials is shown in Table 1.

3. Evaluation of Nanocomposites

All the tension tests were conducted according to ASTM D 638-03. The tensile behaviors of composites were measured using an Instron 8801 with a 10 kN load cell. All the tension tests were tested at a rate of 0.2 in/min. The flexure tests were conducted according to ASTM D 790-03, using an Instron 8801 with a 4.48 N load cell. The support span was 50 mm. and tests were run at a test speed of 0.05 in/min. The impact tests were conducted according to ASTM D 256-06. The notches were added using a NotchVIS machine manufactured by Ceast.

4. Results and Discussion

Tensile modulus of elasticity and ultimate strength values of all compounded materials are shown in Figure 1 and 2, respectively. The test results showed that the tensile elastic modulus consistently increases when increasing the nanoclay loading in PP-C, PP-C-MAPP, and PP-COMMB. The maximum increase in the MOE was around 60% and it was achieved by adding 6% of nanoclay (Nanomer I.44P) powder to the PP matrix (PP-C-6). Moreover, the results show that adding PP-g-MA to the PP matrix (PP-C-MAPP) leads to lower MOE values when compared to PP-C and PP-COMMB, as shown in Figure 1. This reduction in performance is probably attributed to the macromolecular

modification of the PP matrix caused by the addition if PP-g-MA, a low molecular weight and high grafting content. Examining the ultimate strength data it can be seen that there was not significant improvements in the ultimate strength of PP-C and PP-MAPP-C when adding nanoclay. In contrast, there was clear improvement in the ultimate strength of PP-COMMB which constantly increased when increasing the nanoclay loading. It is important to note that adding more than 2wt% nanoclay resulted in a reduction in ultimate strength, as shown in Figure 2. The maximum increase in the ultimate strength was around 11% and it was achieved by adding 4wt. % nanoclay using commercial masterbatches (PP-COMMB-4).

The flexural modulus of elasticity and ultimate flexural strength of nano-reinforced PP composite are shown in Figure 3 and 4, respectively. The test results show that the elastic modulus consistently increases when increasing the wt% of nanoclay fillers for all types of fillers, as shown in Fig. 3. The maximum increase in the elastic modulus was around 45% and it was seen in PP-COMMB-10. The ultimate flexural strength decreased when adding nanoclay in both PP-C and PP-MAPP-C nanocomposites, and it increased when adding nanoclay to PP-COMMB. For instance adding 4wt% nanoclay resulted in 10% and 12% decrease in ultimate strength or PP-C and PP-MAPP-C, respectively and a 20% increase in PP-COMMB. The maximum increase in flexural strength was about 16% and was obtained when adding 10%wt in PP-COMMB composites, as shown in Figure 4. Similar to the tensile strength results, adding PP-g-MA resulted in lower flexural modulus and ultimate flexural strength. The reduction in performance can be also attributed to the macromolecular modification of the PP matrix caused by the addition of PP-g-MA, as previously explained. In summary, it was shown that PP-COMMB-2 and PP-COMMB-4 nanocomposites exhibit superior tensile and flexural properties as compared to neat PP, PP-C, and PP-MAPP-C nanocomposites.

The impact test results for notched specimens are shown in Figure 5. The impact test results show that the energy absorption of pure PP consistently decreases as the filler content increases. The decrease in the notched impact strength was not surprising since the matrix becomes more crack sensitive by adding fillers. Notched impact behavior is generally controlled to a greater extent by factors affecting the propagation of fractures initiated at predominating stress concentration at the notched tip.

5. Conclusions

In this study, the effects of filler loadings, surface treatments and compatibilizers on the tensile, flexural and impact properties were investigated. The most promising results were shown in PP-COMMB-2 and PP-COMMB-4 nanocomposites which exhibited superior tensile and flexural properties as compared to neat PP, PP-C, and PP-MAPP-C nanocomposites. Adding PP-g-MA compatibilizer to the PP matrix resulted in lower mechanical properties compared to PP-C and PP-COMMB. The notched impact strengths were lower when adding nanoclay to the PP matrix for PP-C, PP-MAPP, and PP-COMMB. It seems that adding 2 wt% of nanoclay to PP matrix, increases the modulus of elasticity and ultimate strength, of the PP matrix. In general, PP-COMMB nanocomposites show better mechanical properties in comparison with PP-C and PP-C-MAPP nanocomposites.

6. Acknowledgements

The authors would also like to thank U.S. Army Corps of Engineers, Engineer R&D center for financial support.

7. References

[1] Etalaaho P., Nevalainen K., Suihkonen R., Vuorinen J. and Jarvelap P. Journal of Applied Polymer Science, 114: (2) 978-992, 2009.

[2] Etelaaho P., Nevalainen K., Suihkonen R., Vuorinen J., Hanhi K. and Jarvela P. Polymer Engineering and Science, 49: (7) 1438-1446, 2009.

[3] Lertwimolnun W. and Vergnes B. Polymer, 46: (10) 3462-3471, 2005.

[4] Hasegawa N., Okamoto H., Kato M. and Usuki A. Journal of Applied Polymer Science, 78: (11) 1918-1922, 2000.

[5] Hasegawa N., Okamoto H., Kawasumi M., Kato M., Tsukigase A. and Usuki A. Macromolecular Materials and Engineering, 280: (1) 76-79, 2000.

[6] Kawasumi M., Hasegawa N., Kato M., Usuki A. and Okada A. Macromolecules 30: 6333, 1997.

[7] Nam P., Maiti P., Okamoto M., Kotaka T., Hasegawa N and Usuki A. Polymer, 42: (23) 9633-9640, 2001.

[8] Dennis H., Hunter D., Chang D., Kim S., White J., Cho J. and Paul D. Polymer, 42: (23) 9513-9522, 2001.

[9] Hasegawa N., Kawasumi M., Kato M., Usuki A. and Okada AJournal of Applied Polymer Science, 67: (1) 87-92, 1998.

[10] Kato M., Usuki A. and Okada A. Journal of Applied Polymer Science, 66: (9) 1781-1785, 1997.

[11] Kim K. N., Kimm H. and Lee J. W. Polymer Engineering and Science, 41: (11) 1963-1969, 2004.

[12] Lan T. and Qian D. Q. In Proceedings of Additives', 2000.

[13] Lee J. W., Lim Y. T. and Park O O. Polymer Bulletin, 45: (2) 191-198, 2000.

[14] Manias E., Touny A., Wu L., Strawhecker K., Lu B. and Chung T. C. Chemistry of Materials, 13: (10) 3516-3523, 2001.

[15] Reichert P., Nitz H., Stefan K., Brandsch R., Thomann R. and Mulhaupt R. Macromolecular Materials and Engineering, 275: (1) 8-17, 2000.

[16] Marchant D. and Jayaraman K. Industrial and Engineering Chemistry Research, 41: (25) 6402-6408, 2002.

[17] Yang J. L., Zhang Z. and Zhang H. Composites Science and Technology, 65: (15-16) 2374-2379, 2005.

[18] Bureau M. N., Ton-That M. T. and Perrin-Sarazin F. Engineering Fracture Mechanics, 73: (16) 2360-2374, 2006.

[19] Saminathan K., Selvakumar P. and Bhatnagar N. Polymer Testing, 27: (3) 296-307, 2008.

[20] Kim D. H., Fasulo P. D., Rodgers W. R. and Paul D. R. Polymer, 48: (18) 5308-5323, 2007.

[21] Ton-That M. T., Perrin-Sarazin F., Cole K. C., Bureau M. N. and Denault J. Polymer Engineering and Science, 44: (7) 1212-1219, 2004.

[22] Treece M. A., Zhang W., Moffitt R. D. and Oberhauser J. P. Polymer Engineering and Science, 47: (6) 898-911, 2007.

Nanocomposites	Abbreviation	Method
PP+ clay 2wt%	PP-C-2	Direct melt compounding
PP+ clay 4wt%	PP-C-4	Direct melt compounding
PP+ clay 6wt%	PP-C-6	Direct melt compounding
PP+ clay 8wt%	PP-C-8	Direct melt compounding
PP+ clay 10wt%	PP-C-10	Direct melt compounding
PP+ clay 2wt% + MAPP 1wt%	PP-C-MAPP-2	Direct melt compounding
PP+ clay 4wt% + MAPP 2wt%	PP-C-MAPP-4	Direct melt compounding
PP+ clay 6wt% + MAPP 3wt%	PP-C-MAPP-6	Direct melt compounding
PP+ clay 8wt% + MAPP 4wt%	PP-C-MAPP-8	Direct melt compounding
PP+ clay 10wt% + MAPP 5wt%	PP-C-MAPP-10	Direct melt compounding
PP+ nanoMax®-PP-HiST 4wt%	PP-COMMB-2	masterbatch dilution
PP+ nanoMax®-PP-HiST 8wt%	PP-COMMB-4	Masterbatch dilution
PP+ nanoMax®-PP-HiST 12wt%	PP-COMMB-6	Masterbatch dilution
PP+ nanoMax®-PP-HiST 16wt%	PP-COMMB-8	Masterbatch dilution
PP+ nanoMax®-PP-HiST 20wt%	PP-COMMB-10	Masterbatch dilution

Table 1: Composition and processing history of the nanoclay reinforced PP composites

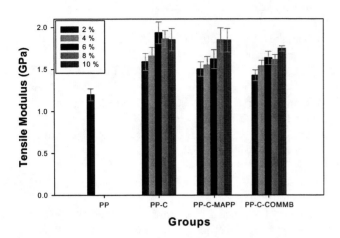

Figure. 1. Tensile modulus of elasticity as a function of the filler weight percent

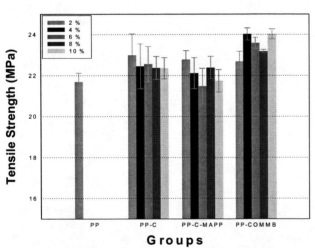

Figure. 2. Tensile strength as a function of the filler weight percent.

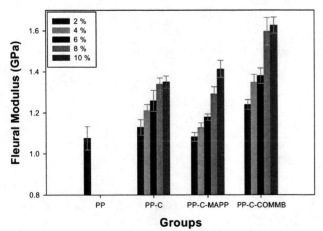

Figure 3. Flexural modulus of elasticity as a function of the filler weight percent.

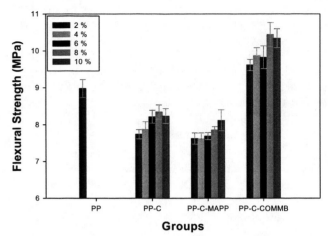

Figure 4. Flexural strength as a function of the filler weight percent

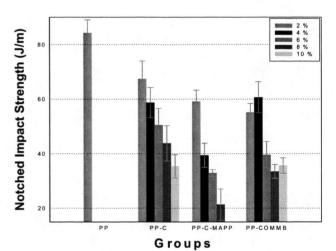

Figure 5. Notched impact strength as a function of the filler weight percent.

High Strain Rate Compression Behavior of Conventional and CNF-Filled-Nanophased Glass/Polyester Composites

M. K. Hossain[*], M. E. Hossain[**], M. Hosur[***] and S. Jeelani[****]

[*]Assistant Professor, hossainm@tuskegee.edu
[**]Graduate Student, Muhammad.Hossain@tuskegee.edu
[***]Associate Professor, hosur@tuskegee.edu
[****]Professor of ME, Director of T-CAM & Ph. D Program in Materials Science and Engineering, and Vice President for Research and Sponsored Program at Tuskegee University, jeelanis@tuskegee.edu
Center for Advanced Materials (T-CAM), Tuskegee University
101 Chappie James Center, Tuskegee, AL 36088, USA

ABSTRACT

The engineering materials are subjected to high strain rate loading in different structural applications and it is deemed necessary to characterize these materials under this loading situations. In this study, a high intensity ultrasonic liquid processor was used to infuse carbon nanofibers (CNFs) into polyester matrix that was then mixed with hardener using a high speed mechanical agitator. The trapped air and reaction volatiles were removed from the mixture using a high vacuum. The conventional and CNFs filled glass/polyester composite laminates using 35 plies of woven E-glass were processed with vacuum assistant resin transfer molding (VARTM). Quasi-static compression tests performed on the unfilled, 0.1 wt. %, 0.2 wt. %, and 0.3 wt% CNF filled Glass/polyester composites, respectively, showed increase in modulus and strength with increasing loading percentage of CNFs up to 0.2 wt%. The better dispersion of 0.2 wt% of CNFs into polyester matrix was observed by scanning electron microscopy (SEM). The high strain rate compressive behavior of this new material system was studied using a Split Hopkinson Pressure Bar (SHPB) setup at strain rates of 550, 700, and 800 s^{-1}. Results showed that the dynamic mechanical strength for both conventional and nanophased glass reinforced polyester increases with increasing strain rates. Fracture morphology of compression tested specimens examined under scanning electron microscope (SEM) revealed fiber kinking and shear fracture at low strain rates, with global delamination, interfacial separation, matrix cracking, fiber breakage and matrix-fiber debonding dominating at high strain rate failure regime.

Keywords: Polyester, CNFs, VARTM, Split Hopkinson Pressure Bar, High Strain Rate.

1 INTRODUCTION

Fiber reinforced polymer matrix composites (FRPC) have become attractive structural materials in aerospace industry, marine, armor, automobile, railways, civil engineering structures, sport goods [1] due to their high specific strength and specific stiffness to weight ratios, fatigue properties, and corrosion resistance. Depending on the service conditions, many of these structures are prone to dynamic loadings such as high impact loads. Hence, it is necessary to have a good knowledge on the high strain rate compression behavior of polymer composites reinforced with nanoparticles, such as carbon nanptubes (CNT), carbon nanofibers (CNF), clay nanoplatelets, and other nanoparticles. Hayes et al. [2] have shown the ability to incorporate nanosized alumina structures in the matrix and interlayer regions of prepreg based carbon/epoxy composites.

Ochola et al. [3] studied the strain rate sensitivity of both carbon fiber reinforced polymer (CFRP) and glass fiber reinforced polymer (GFRP) at strain rates of 10^{-3} and 450 s^{-1}. They documented the dynamic strength for GFRP increases with increasing strain rates and the strain to failure for both composites decrease with increasing strain rates. Evora and Shukla [4] investigated the dynamic response of polyester/TiO$_2$ nanocomposites under high strain rate (2000 s^{-1}) compression loading and obtained results showed that the addition of nanoparticles contributed to a moderate stiffening effect, but there was no significant effect on ultimate strength. Guo and Li [5] also concluded the compressive strength of SiO$_2$ filled epoxy nanocomposites was higher than that of pure epoxy at the high strain rate. Montiel and Williams [6] reported the dynamic behavior of AS4 graphite/PEEK cross-plied composites laminates at strain rate upto 8 s^{-1} using a drop tower assembly and documented 42 and 25% increased in strength and strain to failure, respectively, over the static values. Hosur et al. [7] studied the high strain compression response of affordable woven carbon/epoxy composites and found considerable increase in dynamic compression peak stress as compared to static loading; whereas the strain at peak stress was lower by 35–65%. Guden et al. [8] found the modulus and failure strength of woven glass fiber/ SC-15 composites to be strain rate sensitive and the failure of the composite occurred through matrix fracture and delamination.

In this study, glass/polyester nano-composites with 0.1wt.%, 0.2 wt.% and 0.3 wt.% CNFs loading and conventional ones were fabricated. Quasi-static compression tests showed the improvement in properties of nanophased composites compare to the conventional ones. Both conventional and nanophased composites were characterized using Split Hopkinson Pressure Bar (SHPB) setup and fracture morphology was examined using SEM. The experimental results of the new material system obtained were used to assess the influence of CNF on the properties of glass/polyester nanocomposites.

2 EXPERIMENTAL

2.1 Materials Selection

Commercially available B-440 premium polyester resin and styrene from US Composite, heat treated PR-24 CNF from Pyrograf Inc., and woven E-glass fiber from fiberglasssite.com were considered as matrix, nanoparticle, and reinforcement, respectively, in this current study because of their good property values and low cost. Polyester resin contains two-part: part-A (polyester resin) and accelerator part-B (MEKP- methyl ethyl ketone peroxide).

2.2 Resin Preparation

Ultrasonic cavitation technique is one of the most efficient means to disperse nanoparticles into a polymer [9]. In this study, sonication was performed using a high intensity ultrasonic irradiation (Ti-horn, 20 kHz Sonics Vibra Cell, Sonics Mandmaterials, Inc, USA) for 90 minutes, adding 0.1, 0.2, and 0.3wt.% CNF with corresponding percent polyester resin with 10 wt% styrene in a glass beaker. To remove the bubbles, high vacuum was applied using Brand Tech Vacuum system for about 90-120 minutes. Once the bubbles were completely removed from the mixer, 0.7 wt% catalyst was mixed with the mixer using a high-speed mechanical stirrer.

2.3 Composite Fabrication

Both conventional and nanophased E-glass/CNF-polyester composites were manufactured by VARTM process. The panel was cured for about 12-15 hours at room temperature. The room temperature cured material was taken out from the vacuum bagging and trimmed, and test samples were machined according to ASTM standard. They were thermally post cured at 110 °C for 3 hours in a mechanical convection oven.

2.4 Test Procedure

Quasi-static tests were performed on the servo-hydraulic MTS testing unit according to ASTM D695 to determine the ultimate strength and young modulus of the glass/polyester composites. The machines were run under the displacement control mode at a crosshead speed of 1.27 mm/min and tests were performed at room temperature.

For high strain rate testing, a modified SHPB test setup was used in this study. Typical setup of modified SHPB is shown schematically in Figure 1. Details of the setup and stress reversal technique are discussed by Hosur et al. [10]. During high strain rate loading, specimen is sandwiched between the incident bar and the transmission bar. Petroleum jelly is applied at surfaces of the specimen in contact with the bars to reduce the effect of friction.

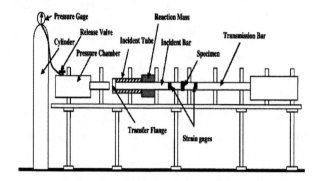

Figure 1: Schematic of compression Split Hopkinson Pressure Bar setup

The transient strain history was recorded from the strain gages mounted on the incident and the transmission bars. Two gages were mounted diametrically opposite to each other on each bar to eliminate recording of any bending strains. The data was acquired using a high-speed data acquisition card with Gagescope V3.42 software at a sampling rate of 2 MHz. The stress–strain relation was developed based on one dimensional elastic bar-wave theory for a pulse propagating in a uniform bar, which was initially unstrained and at rest before the pulse arrives. For this data analysis, VuPoint signal analysis software was used.

Fractography of neat and nanocomposite samples was studied using a Field Emission Scanning Electron Microscope (FE-SEM Hitachi S-900) JEOL JSM 5800). An accelerating voltage was applied to accomplish desired magnification. Fracture morphology of quasi-static tested samples was examined using SEM.

3 RESULTS AND DISCUSSIONS
3.1 Quasi-static Tests

Quasi-static tests were performed on the Neat, 0.1, 0.2, and 0.3 wt.% CNF-filled glass reinforced polyester nanocomposites (CNF-GRPC) to evaluate their compression stiffness and strength. Their typical stress-strain behaviors are shown in Figure 2. It is clear from these stress-strain curves that all the samples of CNF-filled nanophased E-glass/polyester composites showing considerable nonlinearity before reaching the maximum stress. However, more or less ductility was observed in each type of laminate sample but no obvious yield point was found.

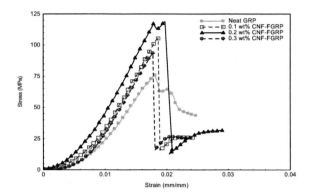

Figure 2: Stress-strain curves of GRPC and CNF-GRPC

Five samples were tested for each condition and the average properties obtained from these tests. It is evident that for 0.2 wt.% CNF loaded GRPC the compressive strength and modulus increased by about 43 and 71%, respectively, as compared to the neat GRPC samples. Quasi-static compression results are given in Table 1.

GRP and wt% CNF-FGRP	Max. Stress (MPa)	% Change	Modulus (GPA)	% Change
Neat	80.72 ± 6.27	------	3.78 ± 0.25	------
0.1wt%	106.98 ± 1.65	32.50	5.52 ± 0.48	46.03
0.2wt%	**115.71 ± 3.25**	**43.35**	**6.45 ± 0.27**	**70.64**
0.3wt%	95.4 ± 2.75	18.19	5.14 ± 0.44	35.98

Table 1: Quasi-static results of GRPC and CNF-GRPC

3.2 High Strain Rate Tests

The High strain rate tests were performed on 35 ply GRPC and CNF-GRPC at three different strain rates of 550, 700 and 800 s^{-1}, respectively. The transient data for each sample tested under high strain rate data was recorded and stored. The data is triggered at the point when the initial compressive pulse reaches the location of the strain gage on the incident bar. The strain rate versus time and stress versus time data are stored in separate files. To plot the dynamic stress–strain curve, it is important to synchronize the two pulses. The starting time is selected from the transmitted pulse at the instant when it starts deviating from zero and the ending time is selected as the time when the transmitted pulse flattens out. The portion of the reflected pulse is chosen for the corresponding time range and integrated to get the strain versus time data. Strain versus time and stress versus time data are superimposed by choosing stress for the y-axis and strain for the x-axis to obtain stress–strain curve. The data for dynamic tests is summarized in Table 2, which gives the peak stress, the slope of stress–strain curves, and the average values. To determine the modulus (slope of stress–strain curve), linear portion of the curve is zoomed in using Easyplot graphic software. The zoomed in portion is then fitted with a linear curve. Slope of the linear fit equation gives the stiffness of the sample. Figures 3 and 4 illustrated dynamic stress-strain curves of conventional and nanophased GRPC at different strain rates at room temperature.

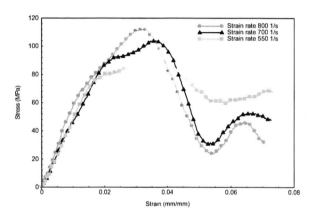

Figure 3: Typical stress –strain curve for neat GRPC

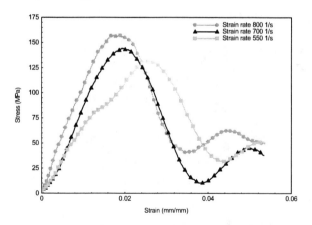

Figure 4: Stress-strain curve for 0.2 wt.% CNF-GRPC

The average modulus in both conventional and nanophased composites increases as the strain rate increases from quasi-static to high strain rates. The average maximum stress within the strain rate regimes also increases with increasing strain rate from quasi-static value to high strain rates. But the strain at failure slightly decreases with increasing strain rates and addition of CNFs.

Composites	Strain rates, 1/s	550	700	800
Neat GRPC	Max. Stress, MPa	92	103	110
	Modulus, GPa	4.87	4.93	5.05
0.2 wt.% CNF -GRPC	Max. Stress, MPa	126	140	152
	Modulus, GPa	7.01	8.85	10

Table 2: Summarized results from dynamic tests

Also at higher strain rates (550, 700, 800 s⁻¹), both the failure stress and modulus increases nonlinearly with increasing strain rates in both conventional and nanophased GRPC. The strain rate sensitivity on failure stress showed in Figure 5.

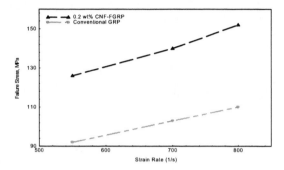

Figure 5: Failure stress vs. strain rate of GRPC

3.3   Fracture Morphology

Fracture morphology of quasi-static tested samples was studied using SEM. The SEM micrographs of the fractured surfaces of GRPC and CNF-GRPC are illustrated in Figures 6 and 7. Comparative analyses of the fractured surfaces of

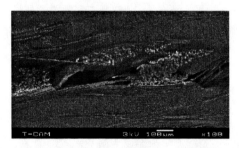

Figure 6: Fracture of 0.2% CNF-GRPC

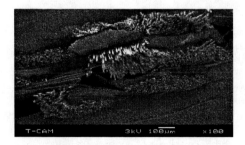

Figure 7: Fracture of conventional GRPC

GRPC and CNF-GRPC showed that the CNF-GRPC exhibited less region of the fiber–matrix separation, matrix breaking due to better addition in presence of CNFs and the conventional GRPC showed huge region of fiber-matrix debonding, matrix breaking, fiber breaking, and spalling of the matrix.

4   CONCLUSIONS

Compression responses of plain weave GRPC and GRP-CNF-GRPC were carried out at high strain rates. The significant conclusions drawn from the investigation are given below.

1. 0.2% CNF-GRPC exhibited the 43 and 71% improvement of compression strength and flexural modulus, respectively.
2. High strain rate test also illustrated that dynamic responses of laminate showed a strong influence of CNFs addition with the GRPC. In both the cases, there was also a considerable increase in the peak stress and stiffness at high strain rate loading as compared to lower strain rate loading.
3. Both failure stress and modulus increased nonlinearly with increasing strain rates in both conventional and nanophased GRPC.
4. SEM studies revealed the better adhesion of fiber – matrix in the CNF-GRPC due to the uniform dispersion of the CNFs and showed less damage.

REFERENCES

[1] Millerschin, E., Compos. Fab., 40-42, 1999.
[2] Hayes, B. S., Nobelen, M., Dharia, A. K. and Seferis J. C., 33ʳᵈ International SAMPE Symposium and Exhibition, Seattle, WA, 2001.
[3] Ochola, R.O., Marcus, K., Nurick, G. N. and Franz, T., Compos. Struct., 63, 455-467, 2004.
[4] Evora, V., M. F., and Shukla, A., Mat. Sci. and Eng. A, 361, 358-366, 2003
[5] Guo, Y. and Li, Y., Mater. Sci. and Eng. A, 458, 330-335, 2007.
[6] Monteal, D. M. and Williams, C. J., ASTM STP 1120(10), 54–65, 1992.
[7] Hosur, M.V., Alexander, J., Jeelani, S., Vaidya, U.K. and Mayer, A. , J. Reinf. Plast. Compos, 22(3), 271–296, 2003.
[8] Guden, M., Yildirim, U. and Hall, I. W., Polymer Testing, 23, 719-725, 2004
[9] Eskin, G.I., Ultrasonic Sonochemistry, 8 (3), 319-325, 2001.
[10] Hosur, M. V., Alexander, J., Vaidya, U.K. and Jeelani, S., Compos. Struct. 52, 405–417, 2001.

In-situ Techniques in Optical Spectroscopy for the Analysis of Carbon Nanotubes

G. Freihofer, B. Mohan and S. Raghavan

University of Central Florida
Mechanical, Materials and Aerospace Engineering
4000, Central Florida Blvd, Orlando, Florida, sraghava@mail.ucf.edu

ABSTRACT

In-situ temperature studies of the Raman tangential (G-band) and D modes are reported here on single-walled carbon nanotube engineered paper samples. The expansive temperature range was achieved through a temperature stage coupled with the Raman microprobe. The in-situ arrangement allowed for a large temperature variation while spectral data acquisitions were collected. Changes in the characteristics of the peaks were analyzed using a novel deconvolution and fitting algorithm featuring a pseudo-voigt function for the fitting. Results show the variation of spectral parameters over the temperature range and shows specifically the increase in D/G ratio with laser excitation is unaffected by temperature within the range.

Keywords: in-situ, optical spectroscopy, carbon nanotubes

1 INTRODUCTION

Optical spectroscopy has shown promise as a fundamental characterization tool for carbon nanotubes (CNTs). With the potential of CNTs being utilized for aerospace applications[1]–[3], in-situ techniques including temperature and strain measurements with spectroscopy, are gaining significance. The G and D bands have been known to exhibit changes in spectral parameters due to variation in physical parameters [4]–[8], axial stress [9]–[11] and hydrostatic pressure [12], [13].
Temperature induced shifts of the raman peaks have been observed specifically as RBM downshifts, and attributed mainly to the temperature-induced softening of the intra- and intertubular bonds [14]. While RBM temperature effects have been shown to vary when individual CNTs are probed as opposed to bundles, the G bands show no observable differences between individual and bundled CNT studies [15]. In both cases, generally a downshift of wavenumber as well as diminishing intensity is observed with increasing temperature, however variations of diameters, chiralities and dispersion in bulk samples studied under temperature makes this behavior specific to the CNT composite studied.
In-situ experimental studies will provide valuable information to advance design and tailoring of CNT re-

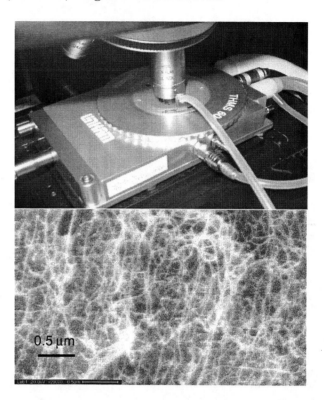

Figure 1: In-situ temperature studies using a temperature stage coupled with Raman spectrometer on CNT engineered paper (SEM image shown)

inforcements in composites for maximum strength or damping. Development of these in-situ techniques allow for the study of the effects of external parameters on the spectra of CNTs. The results from these in-situ studies can further be used to develop non-invasive measurement capabilities to establish the integrity of CNT reinforced materials. The temperature and laser excitation studies presented here provide quantitative behavior of these bands under environmental conditions. These results will be used with future in-situ load studies for the development of calibrated strain effects on the bands.

2 EXPERIMENT

The sample used for the study was single-walled CNT (SWCNT) engineered paper composed of 1.8nm CNTs

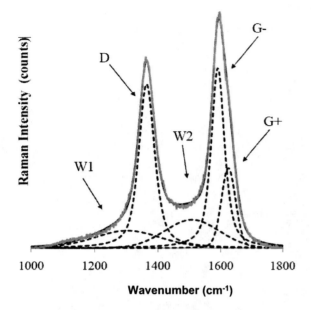

Figure 2: Deconvolution of the Raman bands using 5 peaks, including the G and D peaks. W modes refer to the wide peaks

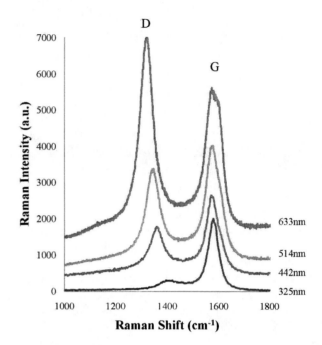

Figure 3: Variation of the Raman G and D bands at 88K with varying laser excitation

within 100nm diameter bundles estimated using scanning electron microscopy. The details behind the manufacturing of this sample are presented in earlier literature [3]. This material was selected due to good dispersal of the single-walled nanotubes in the composite with high loading of CNTs and their potential for use as structural materials. For a bulk sample of this nature, the raman spectra represents a large number of both semiconducting and metallic SWCNTs. The sample was mounted in a temperature controlled stage on a silver, liquid nitrogen-cooled sample holder and the temperature was varied from 88K to 388K at a rate of 4K/s. The Raman measurements were made at 10K increments using a confocal Raman microscope (Renishaw inVia) using an excitation wavelength of 514nm. Effects of variation in laser excitation was also studied at extreme temperatures. The laser excitations used included 325 nm, 442nm, 514 nm and 633 nm. The resolution of the grating used was 1800 g/mm.

3 ANALYSIS & RESULTS

3.1 Deconvolution method

Deconvolution and fitting of the Raman bands are required in order to correctly study the behavior of the individual peaks with temperature. Prior to this, pre-processing of the data was required whereby the baseline was removed from the entire spectrum using a 4th order polynomial. The range of deconvolution is extended from the endpoints or zero slope locations of the data at 1000 and $1800cm^{-1}$. The Raman bands were

deconvoluted in a systematic procedure using a pseudo-Voigt formulation where the optimization for the curve fitting was performed using a genetic-algorithm based code that is described in detail in literature [16], [17]. The number of peaks chosen to be fitted to the spectra was five based on previous studies on the effective number of resolvable peaks that would provide a reasonable goodness of fit. The initial guesses were obtained using gradient-based fitting methods. Fig. 2 shows the results of the fitting using the methodology described.

3.2 Results of temperature and laser excitation studies

The G and D bands showed a strong dependence upon excitation wavelengths with higher D intensities as excitation wavelengths increased towards near IR and this is illustrated in Fig. 3. While the G-band dominated at 325nm, the D band was significant at 633nm. This variation in intensity ratio was plotted in Fig 4 and found to be linear.

4 Discussion & Conclusions

A dispersion of the D band occurs in the excitation experiment due to a double resonance process that takes place in most carbonaceous materials [18], [19]. The magnitude of the D band dispersion for this CNT engineered paper was $-49.5cm^{-1}/eV$ which was predicted by the large amounts of sp^2 carbons in this composite. A very different type of dispersion takes place for the G band, that can be described as a splitting into G^+ and

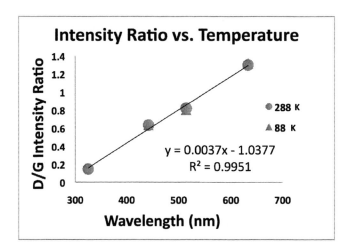

Figure 4: Intensity ratio of the D and G band with varying excitation wavelength at limiting temperatures

G^- as wavelengths approach near IR values and this is seen in Fig. 3. This is explained by the fact that as the excitation wavelength increases, the nanotubes that are resonated, have smaller diameters, and an increase in curvature [7]. The transition energies of the CNT are very sensitive [20], and this explains the large differences in the optical spectra for the discrete laser wavelengths used. In addition, the results show a small shift in the peak position of the D and G band with excitation wavelength. The change in the intensity ratio of the D and G bands were established at low temperature for variable excitation laser wavelengths as shown in Fig. 3. Importantly, the slope of this change was found to remain unchanged at limiting temperatures, indicating this to be a temperature independent relation within the range. The findings are significant to the development of optical spectroscopy techniques as characterization tools for CNTs and CNT reinforced composites. Further studies for in-situ strain measurements are in progress for bulk and individual CNTs.

5 Acknowledgements

Use of the Center for Nanoscale Materials was supported by the U. S. Departmentof Energy, Office of Science, Office of Basic Energy Sciences, under Contract No. DE-AC02-06CH11357. Dr. Gou Jihua (UCF) is acknowledged for providing the CNT Engineered paper samples for our experiments and the images of the samples. The genetic algorithm codes used in our spectral analysis programs were courtesy of Dr William Crossley (Purdue University).

REFERENCES

[1] Baughman, R. H., Zakhidov, A. A., and de Heer, W. A., "Carbon Nanotubesthe Route Toward Applications," *Science's Compass*, Vol. 297, 2002, pp. 787–792.

[2] Zhao, Z., Gou, J., and Khanp, A., "Processing and structure of carbon nanofiber paper," *Journal of Nanomaterials*, Vol. 2009, 2009, pp. 1–7.

[3] Gou, J., "Single-walled nanotube buckypaper and nanocomposite," *Polymer International*, Vol. 55, 2006, pp. 1283–1288.

[4] Pimenta, M. A., Marucci, A., Empedocles, S. A., Bawendi, M. G., Hanlon, E. B., Rao, A. M., Eklund, P., Smalley, R. E., Dresselhaus, G., and Dresselhaus, M. S., "Raman modes of metallic carbon nanotubes," *Physical Review B*, Vol. 58, 1998, pp. R16016–R16019.

[5] Rao, A. M., Richter, E., Bandow, S., Chase, B., Eklund, P. C., Williams, K. A., Fang, S., Subbaswamy, K. R., Menon, M., Thess, A., Smalley, R. E., Dresselhaus, G., and Dresselhaus, M. S., "Diameter-Selective Raman Scattering from Vibrational Modes in Carbon Nanotubes," *Science*, Vol. 275, 1997, pp. 187–197.

[6] Jorio, A., Pimenta, M. A., Filho, A. G. S., Saito, R., Dresselhaus, G., and Dresselhaus, M. S., "Characterizing carbon nanotube samples with resonance Raman scattering," *New Journal of Physics*, Vol. 5, 2003, pp. 139.

[7] Kasuya, A., Sasaki, Y., Saito, Y., Tohji, K., and Nishina, Y., "Evidence for Size-Dependent Discrete Dispersions in Single-Wall Nanotubes," *Physical Review Letters*, Vol. 78, 1997, pp. 4434–4437.

[8] Kahn, D. and Lu, J. P., "Vibrational modes of carbon nanotubes and nanoropes," *Physical Review B*, Vol. 60, 1999, pp. 6535–6540.

[9] Kumar, R. and Cronin, S. B., "Raman scattering of carbon nanotube bundles under axial strain and strain-induced debundling," *Physical Review B*, Vol. 75, 2007, pp. 155421.

[10] Lee, S. W., Jeong, G.-H., and Campbell, E. E. B., "In situ Raman Measurements of Suspended Individual Single-Walled Carbon Nanotubes under Strain," *Nano Letters*, Vol. 7(9), 2007, pp. 2590–2595.

[11] Cronin, S. B., Swan, A. K., Unlu, M. S., Goldberg, B. B., Dresselhaus, M. S., and Tinkham, M., "Resonant Raman spectroscopy of individual metallic and semiconducting single-wall carbon nanotubes under uniaxial strain," *Physical Review B*, Vol. 72, 2005, pp. 035425.

[12] Merlen, A., Bendiab, N., Toulemonde, P., Aouizerat, A., Miguel, A. S., Sauvajol, J. L., Montagnac, G., and Petit, H. C. P., "Resonant Raman spectroscopy of single-wall carbon nanotubes under pressure," *Physical Review B*, Vol. 72, 2005, pp. 035409.

[13] Venkateswaran, U. D., Rao, A. M., Richter, E.,

Menon, M., Rinzler, A., Smalley, R. E., and Eklund, P. C., "Probing the single-wall carbon nanotube bundle: Raman scattering under high pressure," *Physical Review B*, Vol. 59, 1999, pp. 10928.

[14] Raravikar, N. R., Keblinski, P., Rao, A. M., Dresselhaus, M. S., Schadler, L. S., and Ajayan, P. M., "Temperature dependence of radial breathing mode Raman frequency of single-walled carbon nanotubes," *Physcal Review*, Vol. 66, 2002, pp. 235424.

[15] Zhou, Z., Dou, X., Ci, L., Song, L., Liu, D., Gao, Y., Wang, J., Liu, L., Zhou, W., Xie, S., and Wan, D., "Temperature Dependence of the Raman Spectra of Individual Carbon Nanotubes," *Journal of Physical Chemistry B*, Vol. 110, 2005, pp. 1206–1209.

[16] Raghavan, S., Imbrie, P., and Crossley, W. A., "The spectral analysis of R lines and vibronic sidebands in the emission spectrum of ruby using genetic algorithms," *Applied Spectroscopy*, Vol. 62, 2008, pp. 759–765.

[17] Raghavan, S. and Imbrie, P., "The Development of Photo-stimulated Luminescence Spectroscopy for 3D Stress Measurements in the Thermally Grown Oxide Layer of Thermal Barrier Coatings," *Proceedings of the Materials Science and Technology 2007 conference*, 2007.

[18] Thomsen, C. and Reich, S., "Double Resonant Raman Scattering in Graphite," *Physical Review Letters*, Vol. 85, 2000, pp. 5214.

[19] Saito, R., Jorio, A., Filho, A. G. S., Dresselhaus, G., Dresselhaus, M. S., and Pimenta, M. A., "Probing Phonon Dispersion Relations of Graphite by Double Resonance Raman Scattering," *Physical Review Letters*, Vol. 88, 2002, pp. 027401.

[20] Telg, H., Maultzsch, J., Reich, S., Hennrich, F., and homsen, C. T., "Chirality Distribution and Transition Energies of Carbon Nanotubes," *Physical Review Letters*, Vol. 93, 2004, pp. 177401-1–4.

NSTI-Nanotech 2010, www.nsti.org, ISBN 978-1-4398-3401-5 Vol. 1, 2010

Synthesis and Characterizations of nano-sized Barium Hexa Ferrites using Sol-Gel Methods

M.A.Iqbal, Amna Mir and Shezad Alam*

Physics Department, University of the Punjab, Lahore, Pakistan, iqbalma@yahoo.com

*PCSIR Labs, Lahore, Pakistan

ABSTRACT

Nanosized Barium Hexaferrites were synthesized using sol gel auto combustion method. Samples of different composition with varying molar ratio of metal nitrates and citric acid i.e. (*MN\CA* = 0.5, 1, 1.5, 2, 3) were prepared. Pellets of powders of each composition were sintered at 1100 °C for 1 hr. All samples were structurally characterized by X-ray diffractometer. Lattice parameter, volume of unit cell and particle size was determined. It has been found that size of the particle increases by decrease in the citric acid content. It reconfirmed the particle size calculated by debey Sherrer formula. Resistivities of specimens indicate semiconducting behavior. Measured values of resistivity are well in agreement with standard Barium ferrites.

Keywords: Barium Hexaferrite, Sol gel method, particle size, debey Sherrer, conductivity, XRD,

1. INTRODUCTION

Barium Hexaferrite is hard magnetic material with quite large magnetocrystalline, anisotropy, high curie temperature, large saturation magnetization, good chemical stability and high coercivity, low density and low cost make barium hexaferrite the most commercialized permanent magnet. Because of its incredible magnetic properties barium hexaferrite has potential for being used as high density magnetic and magneto-optic recording media [1-2]. Nanosized barium hexa ferrite are of great importance because of its application as MLIC (multilayer chip inductors [3] which are important components of camcorders, pagers and notebook computer [4].

Preparation methods of barium Hexaferrite are also of great importance because homogeneity, particle size and material's magnetic properties are strongly determined by synthesis method.Hexaferrite can be prepared by classical ceramic method but magnetic particle so obtained are large and results in multi-domains structure, ball milling can reduce size of particles but it results in non homogeneous mixtures and lattice strains also appears [5]. In order to achieve homogeneity various techniques such as hydrothermal copreciptation [6], ion exchange resin method [7], glass crystallization [8], sol gel etc has been adopted .to achieve homogeneity and nano sized particles. In sol gel technique *MN/CA* molar ratio plays a very significant role on formation of Barium Hexaferrite. The objective of this paper is to examine the influence of *MN/CA* molar ratio on the phase formation, particle size and microstructure of Barium Hexaferrite prepared by sol gel combustion method.

2. EXPERIMENTAL PROCEDURE

Sol gel auto combustion method was used to prepare nanosized barium hexa ferrite ($BaFe_{12}O_{19}$) powders. The starting materials were barium nitrate $Ba(NO_3)_2$, iron nitrate $[Fe(NO_3)_2.9H_2O]$, citric acid $C_6H_8O_7$ and liquor ammonia[NH_3OH],all of analytic purity. Measured stoichiometric amounts of $[Fe(NO_3)_2.9H_2O]$ and $Ba(NO_3)_2$ in the molar ratio of 1:1, where disslolved in a minimum amount of deionized water to get a clear solution. To chelate ions Ba^{2+} and Fe^{3+} in the solution the citric acid $[C_6H_8O_7]$ was added in to prepared aqueous solution. Then liquid ammonia was slowly added to neutralize solution to PH 7. By heating the neutralized solution at 100 °C on a hot plate with continuous stirring it was evaporated to dryness. While water is evaporated, the solution became thick and finally a very thick brown gel formed. Temperature was increased up to 200 °C which lead to explosion of gel. The dried gel burnt in a self propagating combustion manner until all gel were completely burnt out to form a loose powder, during combustion, exothermic redox reaction associated with fuel oxidation and nitrate decomposition took place. Sol gel method used energy produced during this reaction. A large amount of gases such as N_2, H_2o, Co, No and CO_2 were also evolved during formation of particle ashes contain the oxide product after only a few minutes. As a final point this as burnt powders were calcined in air at 850 °C for 1 hour with a heating rate of 10 °C/ min.

Five samples with different molar ratios of metals nitrates to citric acid (*MN/CA*) of 0.5, 1, 1.5, 2, 3 were prepared and marked as A, B, C, D, E. Powders of each composition was pressed help of die and hydraulic press (model OGAWA-STIKICO) by applying load of 7 ton before sintering. Two pellets of each composition with diameter of 16.131 mm of each pellet were made.

2.1 Sintering

These pellets of each composition were sintered at 850 °C in digital control furnace for 1 hour. Heat treatment was followed by natural cooling in the furnace. These pellets are again sintered at 1100 °C for 1 hour . Phase identification of the sintered pellets at different sintering temperature was performed in a Rigaku Geigerflex D-MAX/A Diffractormeter using Cu-Kα radiation. Scherrer's formula is used to calculate Crystallite sizes

$$D = K\lambda / B \cos\theta \qquad (1)$$

Where D is crystallite size in nm, K is shape factor, B is is full width half maximum of highest peak, is is wave length of X-rays and is Bragg's diffraction angle.

Lattice parameters are calculated by analytical method of indexing. Microstructure features such as crystalline size was examined by S-3400N scanning electron microscope.

3 RESULTS AND DISCUSSION

3.1 XRD analysis

The influence of molar ratio of / on the XRD patterns of synthesis of $BaFe_{12}O_{19}$ can be shown in fig(1-3). The X-ray diffraction patterns of samples with *MN/CA* varying with 0.5, 1, and 1..5, 2, 3 calcined at 850°C for 1 hour. The result of XRD analysis of samples indicates the formation of different crystalline phases with hexagonal type structure.

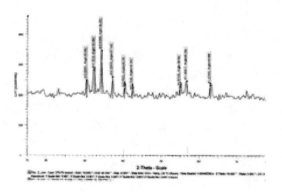

Fig 1: XRD of sample C

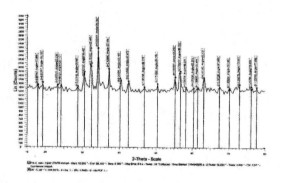

Fig 2: XRD of sample E

Analysis of XRD results of ferrite pellets, calcined at 850°C,with *MN/CA = 0.5, 1, 2, 3* reveals the co-existence of antiferromagnetic phase ($BaFe_2O_4$), hexaferrite ($BaFe_{12}O_{19}$) and other secondary phases i-e. $BaFeO_{3-x}$, $Ba_2Fe_6O_{11}$,

- Fe_2O_3 $Ba_2Fe_6O_{14}$ and $BaFeO_3$. These secondary phases other than BaM indicate that citric acid has not completely chelated iron and barium ions and distribution of ions is inhomogeneous. It is supposed that - Fe_2O_3 is vital phase to form Barium hexaferrite because it is the cubic spinel having chemical formula $Fe[Fe_{5/3} \quad _{1/3}]O_4$, stands for cation hole [10]. Its structure is analogous to 'S' block in barium ferrite so it can be easily transformed in $BaFe_{12}O_{19}$.

XRD graph of ferrite pellet with molar ratio (/) = 1.5 shows that all peaks belong to barium hexaferrite and single phase Barium hexaferrite is formed.

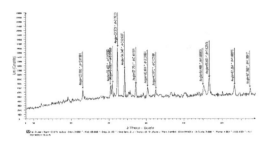

Fig 3: XRD pattern of sample 'c' sintered at 1100 C

Fig. 3 shows the XRD patterns of ferrite pellet of sample 'c' sintered at 1100 C for 1 hour. From XRD patterns it is clearly evident that single phase barium hexaferrite is formed at this temperature. Actually due to decomposition of carbon chains in citrates adjacent iron and barium ions can come into contact more easily and form a crystal lattice of barium hexa ferrite. Generally it can be said that increasing amount of citric acid results in decrease of intermediate phases. A. Mali A.Ataie has also reported the same fact.

3.2 Analysis of crystallite size and lattice parameter

Amount of citric acid is linked with the crystallite size because of nucleation and growth. Therefore molar ratio MN/CA took effects on crystallite size. The crystallite size of Barium hexaferrite calcined at 850C is calculated with the help of Scherrer formula and data of XRD.

Also lattice parameters of each sample is calculated by analytical method and it is observed that the lattice parameter varies with varying MN/CA ratio. Calculated values of crystallite size and lattice parameter are listed in table 1.

Table 1: XRD parameters of Barium hexaferrite

No.	MN/CA Ratio	Crystallite size D (nm)	a A^0	C A^0	Volume V
1	0.5	18.41	5.41	22.83	578.65
2	1.0	36.039	5.32	22.94	562.25
3	1.5	55.274	5.86	22.98	683.38
4	2.0	58.57	5.86	22.88	680.40
5	3.0	73.153	5.74	22.78	649.97

It can be seen that crystallite size decreases with increasing citric acid contents. This fact was also reported in literature [9, 10]. Figure 4 exhibits variation of lattice parameter 'a' and 'c' with varying molar ratios.

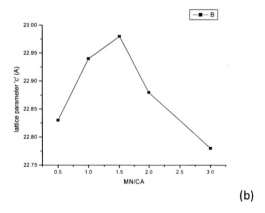

(b)

Fig 4: variation of lattice parameter with MN/CA

Crystallite size of the sample sintered at 1100 C is also calculated by virtue of Scherrer formula and XRD data. Lattice parameters are calculated by analytical method . Calculated values of crystallite size and lattice parameter are listed in table 2.

Table: 2 parameters of barium hexaferrite at 1100 C

No.	*MN/CA*	Crystallite size D (nm)	a (A)	c (A)
1	0.5	38.2	5.88	23.2
2	1.0	41.1	5.86	23.01
3	1.5	42	5.89	23.4
4	2.0	40	5.87	23.33
5	3.0	40	5.88	23.40

Figue 7 shows various crystallite size with increasing molar ratio of metal nitrate to citric acid.

Figure 7: Plot of *MN/CA* vs partical size

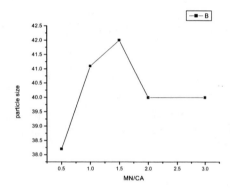

This graph shows increasing citric acid contents results in decrease of crystallite size.

4 CONCLUSION

Nano size barium hexaferrites are obtained after sintering at 1100 C. Crystallite size is greatly influenced by citric acid contents. SEM analysis reveals the presence of rod like and spherical shaped grains with grain size in the range of 200-500 nm. Electric resistivity of all samples is found to be very high. Its variation with applied voltage indicates its semiconducting behaviour.

REFERENCES

[1] I.Zaquine, H. Benazizi and J.C Mage, J . Appl. Phys,64,5822,1988.

[2] D.E Speliostis, IEEE Trans. Magn. Mag- 23, 25,1987.

[3] V.R.Caffarena,T.Ogasawara et al. , Revista Materria 13, 374-378, 2008

[4] H.Zhang, J.Zhou,et al. Mater, Lett,56, 397-403, 200

[5] W. Zhong, Weiping Ding, et al. J.Magn. Magn. Mater.168, 196-202, 1997.

[6] D. Mishra, S. Anand , R. K. Panda and R. P. Das, Material chemistry and physics, 86, 132-136,2004.

[7] W. Zhong, Weiping Ding, et al. J. of App phy. 85, 5552-5554, 1999.

[8] L.Rezlescu, E. Rezlescu, et al. J.Magn. Magn Mater. 193, 288-290, 1993.

[9] Yawowen Li, Qin Wang, Hua Yang, Current applied physics 9, 1375-1380, 2009.

[10] A.Mali and A.Ataie, Ceramic international vol. 30, 1979-1983, 2004.

Fabrication of bulk Al-TiB₂ nanocomposite by spark plasma sintering of mechanically alloyed powder

Z. Sadeghian[*], B. Lotfi[*], M. H. Enayati[**] and P. Beiss[***]

[*]Materials Engineering Department, Faculty of Engineering,
Shahid Chamran University, Ahvaz, Iran, z.sadeghian@scu.ac.ir
[**] Isfahan University of Technology, Isfahan, Iran, ena78@cc.iut.ac.ir
[***] RWTH Aachen University, Aachen, Germany, p.beiss@iwm.rwth-aachen.de

ABSTRACT

Production of Al-TiB₂ nanocomposite powder was studied by using mechanical alloying (MA). The target was to obtain Al-20 wt. % TiB₂ metal matrix nanocomposite by mechanical alloying of pure Ti, B and Al powder mixture. A double step process was used to prevent the formation of undesirable phases like Al₃Ti intermetallic compound, which has been described in our previous papers. The resultant powder was consolidated by spark plasma sintering (SPS). The structural characteristics of powder particles and sintered samples were studied by X-ray diffractometry (XRD), scanning electron microscopy (SEM) and transmission electron microscopy (TEM). Hardness measurements were conducted on the cross section of powder particles and sintered sample.

Keywords: aluminum matrix nanocomposite, mechanical alloying, in situ TiB₂, spark plasma sintering

1 INTRODUCTION

TiB₂ is known as a suitable reinforcing phase for Al-base composites because of its thermodynamic stability. It also presents a high modulus, excellent refractory properties and a high resistance to plastic deformation even at high temperatures [1-3]. In the last decade in situ metal matrix composites (MMCs) have been developed, in which the reinforcements are synthesized in the metal matrix during fabrication, by chemically reacting the constituents. In comparison to traditionally fabricated materials, in situ MMCs undergo less degradation at high temperatures. This is because of their superior thermodynamic stability, stronger interface bonding resulting from the clean reinforcement-matrix interfaces and finer and more uniform distribution [4]. Several techniques including exothermic dispersion, reactive-gas injection, reactive sintering, reactive milling and mechanical alloying have been developed to fabricate in situ MMCs [4-5]. MA as a solid state powder processing method has been extensively used to fabricate in situ ceramic particle reinforced MMCs. MA has an advantage over other in situ fabrication routes as it is capable of producing nanostructured composite powder with high uniformity [5].

In previous investigations it has been reported that during the in situ synthesis of Al-TiB₂ composites form different starting powder mixtures, Al₃Ti intermetallic compound is also formed, which is brittle and has been reported to considerably reduce the fatigue life of composites [6-7]. Therefore it is of interest to eliminate the formation of Al₃Ti intermetallic compound.

A challenge in processing of nanostructured materials is that long time exposure at high temperature sintering, often results in severe grain growth. Several consolidation techniques such as hot-pressing, shock consolidation, sintering with the application of ac currents, pulsed electric current and spark plasma sintering (SPS) are introduced to overcome these difficulties [8]. SPS is a newly developed rapid sintering technique with a great potential for achieving fast densification results with minimal grain growth in a short sintering time. It is proven that enhanced sinterability of powders subjected to SPS mainly associated with particle surface activation and increased diffusion rates on the contact zones caused by applied pulse current [9].

In this study, Al-20 wt. %TiB₂ nanostructured composite powder was successfully synthesized by double step mechanical alloying without the formation of unwanted phases. Consolidation of the powder was conducted via SPS technique and microstructure the powder and sintered bulk nanocomposite was also investigated.

2 EXPERIMENTAL PROCEDURE

Elemental Al, Ti and B powders were used as starting materials. The aluminum powder with an average particle size of 63 μm was supplied from ECKA Granulate, Velden. The titanium powder with a particle size of 40-60 μm was obtained from GKN Sinter Metal Filters, Radevormwald, and the Merck boron powder had a particle size of about 2 μm. Powder mixtures were milled by a Fritsch planetary ball mill with a rotating speed of 360 rpm. The ball to powder weight ratio was chosen to be 10 and the diameter of the chromium steel balls was 15 mm. The hard chromium steel vial was evacuated and filled with argon to prevent oxidation during the mechanical alloying process. To avoid severe adhesion of aluminum powder to the balls and the vial, 1 wt. % zinc stearate was added to the mixture as the controlling agent.

A double step process was used to obtain an in situ Al-TiB$_2$ metal matrix powder. In the first step Ti and B powders were milled with the composition of Al- 62.01 wt. % Ti-27.99 wt. % B. This powder was then used in further milling with additional aluminum powder to achieve the Al- 20 wt. % TiB$_2$.

Powders were sintered by a FCT HP D 250 spark plasma sintering (SPS) equipment, at Fraunhofer Institute for Manufacturing and Advanced Materials, Powder Metallurgy and Composite Materials, Franhofer, Germany. Four sets of variables were examined from which one was chosen for consolidation of the powder. Powder was placed into a graphite die with a diameter of 40 mm and compacted under argon atmosphere with an applied pressure of 35 MPa. The whole process lasted 600 sec and the maximum temperature was 550 C with the dwell time of 0 sec. The changes of the SPS parameters during the consolidation process are shown in Fig. 1. Compacted disks of 40 mm diameter and 5 mm thickness were obtained from the SPS. The density of the SPS materials was determined using Archimedes' principle.

Investigation of the structural changes during mechanical alloying and after sintering was conducted by a SEIFERT 30033 PTS diffractometer employing monochromatic Cu K$_\alpha$ radiation (λ= 0.15406 nm). XRD scans were performed with a step size of 0.05° in 2θ and a dwell time per step of 20 s. The cross-sectional microstructures of the powder particles and the sintered samples were studied by a LEO scanning electron microscope (SEM). Microhardness examination of Al-TiB$_2$ powders was conducted by a Leco testing machine while the hardness of compacted samples was measured by a Zwick microhardness machine.

3 RESULTS AND DISSCUSION

Fig. 2 shows the X-ray diffraction patterns of the Al- 62 wt.% Ti- 28 wt.% B and Al- 20 wt.% TiB$_2$ powders after different milling times. After 10 hours of milling the peaks related to TiB$_2$ ceramic phase are obvious, as well as Al and Ti, showing the gradual formation of TiB$_2$ (Fig. 2-b). After 20 hours of milling (Fig. 2-c) no evidence of remaining titanium or formation of undesired phases could be detected on the XRD pattern. The lack of Al peaks on XRD pattern can be due to the several effects including high strain level induced in the Al lattice, nanosized Al grains and lower x-ray scattering intensity of Al compared to the other constituents. Milling of the powder, resulted from first step, together with aluminum to achieve Al- 20 wt.% TiB$_2$ composition resulted in no clear structural change in the powder (Fig. 2-d). Detailed explanations of the milling process and results have been presented elsewhere [10-11]. The grain size of the aluminum matrix and TiB$_2$ was obtained from XRD analysis using Williamson–Hall (WH) equation [12]. In the resulting powder from the double step mechanical alloying, the grain

size of both TiB$_2$ particles and aluminum matrix was measured to be about 15 nm.

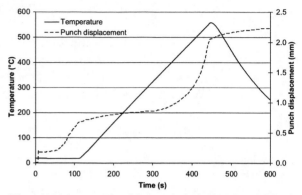

Figure 1: An example of the evolution of SPS conditions during the consolidation process.

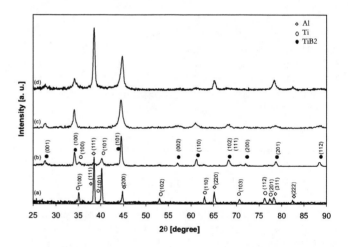

Figure 2: XRD patterns of Al- 62 wt.% Ti- 28 wt.% B powder mixture: (a) as-received, (b) as-milled for 10 h, (c) as-milled for 20 h and (d) final Al- 20 wt.% TiB$_2$ powder obtained from double step mechanical alloying.

Fig. 3-a shows the cross sectional SEM micrograph of an Al-20wt.%TiB$_2$ particle obtained from the third series. A fine and uniform distribution of TiB$_2$ particles within the Al matrix is obtained with many particles being too small to be measured on SEM. A higher magnification of the final Al- 20 wt.% TiB$_2$ composite powder obtained from STEM is presented in Fig. 3-b. The approximate average TiB$_2$ particle size measured using image analysis software was about 90 nm, ranging from 10 nm to maximum 1 μm. The mean particle size seems to be smaller than that of previous studies on in situ Al-TiB$_2$ composite produced by MA [6]. After 40 hours of MA the TiB$_2$ particles and aluminum matrix achieved a mean grain size of about 20 nm, according to According to the dark-field image (Fig. 4).

This is in good agreement with the values obtained from the WH equation.

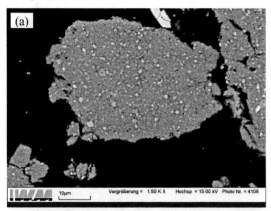

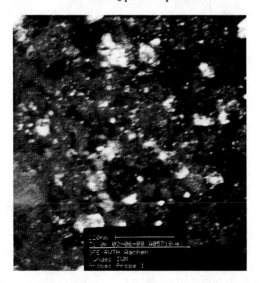

Figure 3: a) Typical cross sectional SEM micrograph of Al-20 wt.% TiB$_2$ powder particles, b) STEM image of an Al-20 wt.% TiB$_2$ powder particle.

Figure 4: Dark field image of Al- 20 wt.% TiB$_2$ powder particles obtained from double step MA.

Final powder from the double step mechanical alloying consisting 20 wt.% TiB$_2$ was sintered using SPS technique.

According to the XRD results, after SPS at 550 ºC no obvious structural change occurred in the consolidated Al-TiB$_2$ nanocomposite (Fig. 5). It can be concluded that by the double stage synthesis process, the formation of undesirable compounds even after exposure to sintering temperature, is prevented. Lu et al. reported that during the high temperature exposure of as milled Al–Ti–B powder mixture, Al$_3$Ti phase is formed in the aluminum matrix along with TiB$_2$ [7].

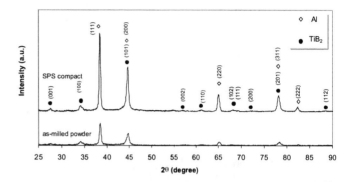

Figure 5: XRD pattern of the SPSed 20 wt.% TiB$_2$ nanocomposite compared to the as MAed powder.

Fig. 6 shows the SEM micrograph of SPSed Al-20wt.%TiB$_2$ nanocomposite.The size of TiB$_2$ particles in the sintered sample was measured from about 50 nm to 1 µm, which has not increased obviously in comparison with the as MAed powder. From TEM examination of the sintered samples (Fig. 7) the mean grain size of phases in the composite was measured about 25 nm. It can be concluded that the in situ synthesized Al-TiB$_2$ nanocomposite powder has a desirable stability to grain growth at high temperatures.

The density of the sintered sample was determined 2.8 g. cm^{-3} by using Archimedes' principle, which is 95.2 % of theoric density and porosity was measured 3.8 %. Microhardness measurements showed an almost large decrease in the hardness form powder to sintered state. Hardness measurement of the samples showed a large decrease from powder to sintered sample. Microhardness of the powder was measured about 480 VHN, while the hardness of sintered sample was about 206 VHN. Reduction in hardness of the SPSed sample in comparison with the as-MAed powder can be related to the recovery, and relief of work hardening effects during sintering process. However hardness of the compacted sample in this study is about two times higher than that reported in previous findings [6]. This is proposed to be the consequence of nanostructure composite obtained in the present study.

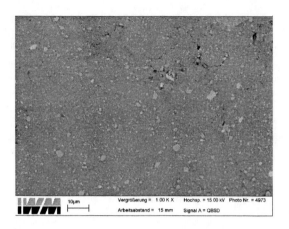

Figure 6: Cross sectional SEM micrograph of the SPSed Al- 20 wt.% TiB$_2$ nanocomposite

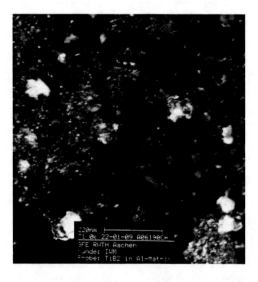

Figure 7: Dark field images of (**a**) Al- 20 wt.% TiB$_2$ powder particles and (**b**) sintered sample obtained from TEM experiments

CONCLUSIONS

Al–TiB$_2$ nanocomposite powder was synthesized by a double-step MA process of elemental powders. TEM observations showed formation of TiB$_2$ particles with a mean size of about 90 nm and uniform distribution in the Al matrix after a total milling time of 40 h. No traces of undesirable phases such as titanium aluminides were observed even after consolidation of the powder by SPS. The fabricated powder showed a suitable stability against grain growth during the sintering process.

ACKNOWLEDGEMENT

Z. Sadeghian's sincere thanks go to the Deutscher Akademischer Austauschdienst (DAAD) for supporting the stay in Germany.

REFERENCES

[1] S. C. Tjong, and K. C. Lau, "Properties and abrasive wear of TiB$_2$/Al-4%Cu composites produced by hot isostatic pressing", Composites Science and Technology, Vol. 59, pp. 2005-2013, 1999.

[2] Z. Y. Chen, Y. Y. Chen, G. Y. Shu, D. Li And Y. Liu, "Microstructure and properties of in situ Al/TiB$_2$ composite fabricated by in-melt reaction method", Metallurgical and Materials Transactions A, Vol. 31A, pp. 1959-1964, 2000.

[3] H. Yi, N. Ma, X. Li, Y. Zhang and H. Wang, "High-temperature machanics properties of in situ TiB$_{2p}$ reinforced Al-Si alloy composites", Materials science and Engineering A, Vol. 419, pp. 12-17, 2006.

[4] S. C. Tjong, and Z. Y. Ma, "Microstructural and mechanical characteristics of in situ metal matrix composites", Materials Science and Engineering, Vol. 29, pp. 49-113, 2000.

[5] C. Suryanarayana, "Mechanical alloying and milling", Progress in Materials Science, 2001, 46, pp. 1-184

[6] L. Lü, M. O. Lai, Y. Su, H. L. Teo, and C., F. Feng, "In situ TiB$_2$ reinforced Al alloy composite", Scripta Materialia, Vol. 45, pp. 1017-1023, 2001.

[7] l.Lu, H. Lai and Y. Wang, "Synthesis of titanium diboride TiB$_2$ and Ti-Al-B metal matrix composites", Journal of Materials Science, Vol.35, pp. 241-248, 2000.

[8] V. Viswanathan, "Challenges and advances in nanocomposite processing techniques", Materials Science and Engineering R, Vol. 54, pp. 121-285, 2006.

[9] V. Mamedov, "Spark plasma sintering as advanced PM sintering method", Powder Metallurgy, Vol. 45, No. 4, pp. 322-328, 2002.

[10] Z. Sadeghian, M. H. Enayati and P. Beiss, "Characterization of in situ Al-TiB$_2$ nanocomposite powder synthesized by mechanical alloying", Powder Metallurgy, In Press.

[11] Z. Sadeghian, M. H. Enayati and P. Beiss, "In situ production of Al–TiB$_2$ nanocomposite by double-step mechanical alloying", Journal of Materials Science, Vol.44, pp. 2566-2572, 2009.

[12] C. Suryanarayana, Norton MG (1998) X-ray diffraction: a practical approach. Plenum Press, New York.

Fabrication and Thermo-Mechanical Characterization of CNF-Filled Polyester and E-glass/Polyester Nanophased Composites

M. K. Hossain[*], M. E. Hossain[**], M. Hosur[***], and S. Jeelani[****]

Assistant Professor, hossainm@tuskegee.edu
[**]Graduate Student, Muhammad.Hossain@tuskegee.edu
[***]Associate Professor, hosur@tuskegee.edu
[****]Professor of ME, Director of T-CAM & Ph. D Program in Materials Science and Engineering, and
Vice President for Research and Sponsored Programs at Tuskegee University, jeelanis@tuskegee.edu
Center for Advanced Materials (T-CAM), Tuskegee University
101 Chappie James Center, Tuskegee, AL 36088, USA

ABSTRACT

A high intensity ultrasonic liquid processor was used to infuse CNFs into the polyester matrix which was then mixed with catalyst using a high speed mechanical agitator. Results showed the significant improvement of mechanical properties in the dispersion of CNFs in sonication over the mechanical mixing, magnetic stirring and thinky mixing methods. Flexure tests performed on the neat polyester (NP), 0.1wt.%, 0.2 wt.%, 0.3 wt.% and 0.4 wt.% CNF-filled polyester (CNF-FP) showed 86% and 16% increase in flexural strength and modulus, respectively, compared to the unfilled polyester with increasing loading percentage of CNFs up to 0.2%. Both conventional and nanophased glass reinforced polyester composites (GRPC) were fabricated using vacuum assisted resin transfer molding (VARTM). Similar increasing trend in the mechanical properties was observed in these nanophased GRPC. Dynamic mechanical analysis (DMA) studies indicated an increase of about 35% and 92% in the storage modulus, and slight increasing trend of glass-transition-temperature (T_g) in the CNF-FP and CNF-filled-GRPC (CNF-GRPC), respectively, as compared to the conventional ones. Thermo-gravimetric analysis (TGA) results showed the insignificant improvement in the decomposition temperature with addition of CNFs in these materials systems. Thermo-mechanical Analysis (TMA) of this new material system studied, and results of which illustrated the coefficient of thermal expansion (CTE) of polyester decreased with an infusion of CNFs into the polyester.

Keywords: Polyester, CNFs, VARTM, Nanophashed composites, Thermo-mechanical properties.

1 INTRODUCTION

Fiber reinforced polymer matrix composites due to their high specific strength and specific stiffness to weight ratios have become attractive structural materials in aerospace industry, marine, armor, automobile, railways, civil engineering structures, sport goods etc. [1]. Extensive research are currently underway to develop the nanophased fiber reinforced polymer matrix composites. Hence, it is very important to have a sound knowledge on the thermal and mechanical behavior of these advanced materials tailored with nanoparticles, such as carbon nanotubes (CNTs), carbon nanofibers (CNFs), clay nanoplatelets, and other nanoparticles. Hayes et al. [2] have shown the ability to incorporate nanosized alumina structures in the matrix and interlayer regions of prepreg based carbon/epoxy composites. Contribution made by the interphase modification adding low nanofiller provides possibilities to enhance performances of GRPC. It was documented that a small percentage of strong fillers can significantly improve the mechanical, thermal, and barrier properties of the pure polymer matrix [3].

Timmerman et al. [4] studied the influence of nanoclay on the carbon fiber/matrix composites under thermal cyclic loading and reported that the transverse cracking in symmetric carbon fiber/epoxy laminates was significantly reduced when nanoparticle fillers were used. Pervin et al. [5] evaluated the thermal and mechanical properties of carbon nanofiber reinforced SC-15 epoxy and documented the significant improvement in the thermal and mechanical properties of the material system. Mahfuz et al. [6] found the higher tensile strength and modulus by about 15–17% and increased thermal stability and crystallinity of the carbon nanoparticles/whiskers reinforced polyethylene system as compared to the neat polyethylene control samples. Chowdhury et al. [7] investigated the flexural and thermo-mechanical properties of woven carbon/nanoclay-epoxy laminates and noticed improvement in the thermal properties including glass transition temperature and storage modulus. Pramoda et al. [8] showed a good dispersion of graphite throughout the PMMA matrix and the addition of graphite into the matrix led to the significant improvement in the thermal stability and the storage modulus of the PMMA nanocomposites. The coefficient of thermal expansion (CTE) of the PMMA/GNCs decreases with an increase in the graphite content. Tognana et al. [9] summarized the coefficient of thermal expansion (CTE) experimental results of epoxy matrix particulate composite as a function of filler content in both glassy and rubbery state. Morales et al. [10] used high speed dispersion mixer to disperse CNFs into matrix and observed improvement in flexural and tensile strengths with addition of 1 wt. % CNFs in the glass/polyester composite.

In this study, nanophased polyester and glass/polyester composites were fabricated tailoring with 0.1wt.%, 0.2 wt.%, 0.3 wt.% and 0.4 wt.% CNFs. Flexure results showed enhanced properties in nanophased composites compared to the conventional ones. Both nanophased polyester and glass/polyester composites were characterized using dynamic mechanical analysis (DMA), thermo-gravimetric analysis (TGA), and thermo-mechanical Analysis (TMA).

2 EXPERIMENTAL

2.1 Materials Selection

Commercially available B-440 premium polyester resin and styrene from US Composite, heat treated PR-24 CNFs from Pyrograf Inc., and woven E-glass fiber from fiberglasssite.com were selected as matrix, nanoparticle, and reinforcement, respectively, in this current study because of their good property values and low cost. Polyester resin contains two-part: part-A (polyester resin) and accelerator part-B (MEKP- methyl ethyl ketone peroxide).

2.2 Resin Preparation

Ultrasonic cavitation technique is one of the most efficient means to disperse nanoparticles into a polymer [11]. In this study, sonication was performed using a high intensity ultrasonic irradiation (Ti-horn, 20 kHz Sonics Vibra Cell, Sonics Mandmaterials, Inc, USA) for 60, 90, and 120 minutes, respectively, adding 0.1, 0.2, 0.3, and 0.4 wt.% CNFs with corresponding percent polyester resin and 10 wt.% styrene in a glass beaker. The mixing process was carried out in a pulse mode of 30 sec. on/15 sec. off at amplitude of 50%. Continuous external cooling was employed by submerging the beaker in an ice-bath to avoid temperature rise during the sonication process. To remove the bubbles, high vacuum was applied using Brand Tech Vacuum system for about 90-120 minutes. Once the bubbles were completely removed from the mixer, 0.7 wt.% catalyst was mixed with the mixer using a high-speed mechanical stirrer for about 2-3 minutes and vacuum was again applied for about 6-8 minutes to degasify the bubbles produced during the accelerator mixing. The whole mixing system is shown in Figure 1.

a.   b.   c.

Figure 1: a. Sonication, b. Degasification, c. Mechanical mixing

2.3 Composite Fabrication

Both conventional GRPC and CNF-GRPC were manufactured by VARTM method. The panel was cured for about 12-15 hours at room temperature. The room temperature cured material was taken out from the vacuum bagging and trimmed, and test samples were machined according to ASTM standard. They were thermally post cured at 110 °C for 3 hours in a mechanical convection oven.

2.4 Test Procedure

Flexural tests under three-point bend configuration were performed on the Zwick Roell testing unit according to ASTM D790-02 to determine the ultimate strength and young modulus of the polymer nanocomposites and its laminates. The machine was run under the displacement control mode at a crosshead speed of 2.0 mm/min [5] and tests were performed at room temperature.

Storage modulus, glass transition temperature (T_g), and loss factor, tanδ, of the fully cured samples were obtained from a TA instruments Q 800 operating in the dual cantilever mode at a heating rate of 5 °C/min from 30 °C to 175 °C and an oscillation frequency of 1 Hz. The specimens were cut by a diamond cutter in the form of rectangular bars of nominal dimensions 3 mm×60 mm×12 mm. The test was carried out according to ASTM D4065-01 [12]. Thermo-gravimetric analysis (TGA) was conducted with a TA Instruments Q 500, which was fitted to a nitrogen purge gas. The temperature was increased from ambient to 600 °C at a rate of 10 °C/min. The coefficient of thermal expansion (CTE) were obtained from a TA instruments Q 400 operating in the expansion mode at a heating rate of 5 °C/min from 30 to 175 °C. The nominal thickness of these samples was 3.5 mm.

3 RESULTS AND DISCUSSIONS
3.1 Flexural Properties

Flexure tests were performed on the NP, 0.1, 0.2, 0.3, and o.4 wt.% CNF-FP nanocomposites and its laminates to evaluate their bulk stiffness and strength. Their typical stress-strain behaviors are shown in Figure 2. It is clear from these stress-strain curves that all the samples of CNF-FP composites failed immediately reaching to their maximum values showing significant improvement in the mechanical properties up to 0.2 wt.% of CNFs loading, beyond that there was a decreasing trend. Similar trend was found for the CNF-GRPC showing considerable nonlinearity before reaching the maximum stress. However, more or less ductility was observed in each type of laminate sample but no obvious yield point was found.

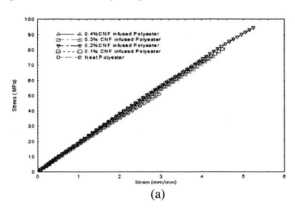

(a)

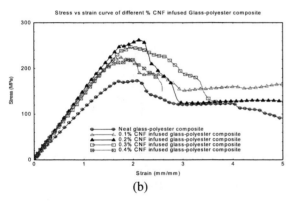

(b)

Figure 2: Stress-strain curves of (a) NP & CNF-FP (b) GRPC and CNF-GRPC

Five samples were tested for each condition and the average properties obtained from these tests. It is evident that for 0.2 wt.% loading the flexural strength and modulus were increased by about 86% and 16%, respectively, as compared to the NP samples. Similar trend was observed for the CNF-GRPC.

3.2 Dynamic Mechanical Analysis (DMA)

The storage modulus and loss factor, tan δ, of polyester and CNF-FP plotted as a function of temperature from DMA are shown in Figures 3 and 4, respectively. It is observed from the analysis that the storage modulus was increased and peak height of the loss factor was decreased with the addition of CNFs. The addition of 0.2 wt.% of CNFs infused polyester showed the maximum improvement of 35% in the storage modulus at room temperature and increasing trend on T_g of this material.

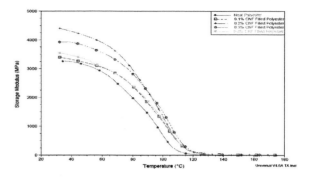

Figure 3: Storage modulus-temparature curve

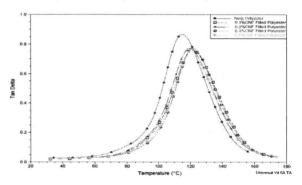

Figure 4: Loss factor-temperature curve

In case of glass/polyester composites it is observed that the addition of 0.2 wt.% of CNFs infused glass/polyester showed the maximum improvement of 92 % in the storage modulus at room temperature and also increasing trend of Tg. The DMA results are given in Table 1.

Type of Composites	Storage Modulus (MPa)	% Change	Glass Transition Temp.(°C)
GRPC	22067 ± 1375	-------	116.4 ± 0.21
0.1%	32550 ± 1245	47.5	118.4 ± 0.8
0.2%	42393 ± 1074	92.1	118.6 ± 0.2
0.3%	31942 ± 1023	44.8	119.1 ± 0.2
0.4%	25780 ± 1336	16.8	119.0 ± 0.3

Table 1: DMA results for GRPC and CNF-GRPC

3.3 TGA Results

TGA responses of GRPC and CNF-GRPC have been shown in Figure 5 as a function of temperature. It is clear from the plots that the decomposition temperature was slightly improved and overall weight loss was less for the 0.2 wt.% CNF-FP sample compared to all other samples.

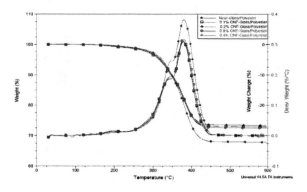

Figure 5: TGA of GRPC and CNF-GRPC

3.4 TMA Results

The coefficient of thermal expansion (CTE) of polyester decreased with an infusion of CNFs into polyester. This result agreed well with the DMA and also TGA results. The TMA results of NP and CNF-FP showed in Figure 6 and Table 2, respectively.

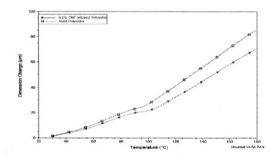

Figure 6: Dimension change with temperature both NP and CNF-FP

Type	Coefficient of Thermal Expansion	
	Below Tg (40--80°C)	Above Tg (120--170°C)
NP	106 μm/(m°C)	201 μm/(m°C)
0.2% CNF-FP	104 μm/(m°C)	196 μm/(m°C)

Table 2: CTE of NP and CNF-FP

The reduction of CTE was observed in the CNF-FP due to the better dispersion of CNFs in the polyester and the reduced segmental motion of the polyester matrix. [13]

4 CONCLUSIONS

A significant improvement in the properties of engineering structural polymeric composite materials can be made by tailoring a small amount of organic or inorganic fillers at a nanoscale level. Carbon nanofibers (CNFs) fillers have been shown to be effective reinforcements in the neat polyester resin to enhance the flexure behavior of this resin system. The neat, CNF-filled polyester, conventional and nanophased composites samples were manufactured and tested under the flexural and thermo-mechanical loading conditions. The significant conclusions drawn from the investigation are given below.

1. 0.2 wt.% CNFs and 90 minutes sonication showed the maximum 86 and 16% improvement in the flexural strength and modulus, respectively, over the mechanical mixing method.
2. 0.2 wt.% CNF-GRPC exhibited the 49 and 31% improvement in the flexural strength and modulus, respectively.
3. DMA results illustrated an increase of 35% in the storage modulus and insignificant improvement in the T_g of 0.2 wt.% CNF-FP.
4. DMA evaluation indicated the 92% improvement in the storage modulus and insignificant effect on the T_g of the 0.2 wt.% CNF-GRPC.
5. TGA analysis showed that addition of CNFs has insignificant effect in the decomposition temperature.
6. The CTE of polyester decreased with addition of CNFs into polyester.

REFERENCES

[1] Millerschin, E., Compos. Fab., 40-42, 1999.
[2] Hayes, B. S., Nobelen, M., Dharia, A. K. and Seferis J. C., 33rd International SAMPE Symposium and Exhibition, Seattle, WA, 2001.
[3] Adams, W. W. and Kumar, S., "Conventional and Molecular Composites - Past, Present, and Future in Ultrastructure Processing of Advanced Materials", John Wiley, 343- 359, 1992.
[4] Timmerman, J. F., Hayes, B. and Seferis J. C., Compos. Sci. and Tech., 62, 1249-1258, 2002.
[5] Pervin, F., Zhou, Y., Rangari, V. J., and Jeelani, S., Mater. and Sci. Eng. A, 405, 246-253, 2005.
[6] Mahfuz, H., Adnan, A., Rangari, V. K., Compos. Part A: Appl. Sci. and Eng., 35(5), 519-527, 2004.
[7] Chowdhury, F. H., Hosur, M. V., and Jeelani, S., Mater. Sci. and Eng. A, 421, 298-306, 2006.
[8] Pramoda, K.P., Linh, N.T.T., Tang, P.S., Tjiu, W.C., Goh, S.H. and He, C.B., Compos. Sci. and Tech., 70, 578-583, 2010.
[9] Tognana, S., Salgueiro, W., Somoza, A., Pomarico, J.A. and Ranea-Sandoval, H.F., Mater. Sci. and Eng. B, 157, 26-31, 2009.
[10] Morales, G., Barrena, M.I., Gómez de Salazar, J.M., Merino, C. and Rodríguez, D., Compos. Struct., 92, 1416–1422, 2010.
[11] Eskin, G.I., Ultrasonic Sonochemistry, 8 (3), 319-325, 2001.
[12] Annual Book of ASTM Standards, D 4065-01, "Standard Practice for Determining and Reporting Dynamic Mechanical Properties of Plastics", 2002.
[13] Pramoda, K.P., Linh, N.T.T., Tang, P.S., Tjiu, W.C., Goh, S.H. and He, C.B., Compos. Sci. and Tech., 70, 578-583, 2010.

The Effects of POSS on the Interlaminar Shear Strength of Marine Composites under Various Environmental Conditions

H. Mahfuz*, F. Powell, R. Granata, and M. Hosur**

*Florida Atlantic University, Boca Raton, FL 33431
{hmahfuz, fpowell1, rgranata}@fau.edu
**Tuskegee University, Tuskegee, AL, 36088, hosur@tuskegee.edu

ABSTRACT

The fiber/matrix (F/M) interface of carbon/vinyl ester composites has been modified by surface treating the carbon fiber with polyhedral oligomeric silsesquioxane (POSS). The objective was to improve the durability of the F/M interface in various environments. Two POSS systems with different functionalities, namely octaisobutyl and trisilanolphenyl, have been investigated. Carbon fibers were soaked in a sonicated dispersion of POSS in a compatible solvent, then fabricated into layered composites using Derakane 8084 vinyl ester resin. Treated and untreated composite samples were immersed in three different environments: seawater at room temperature (SWRT), seawater at 40°C (SW40), and in 85% relative humidity at 50°C (HM50). After six weeks, POSS modified samples reduced water absorption by 20-32% and there was no degradation of the interlaminar shear strength.

Keywords: nanocomposite, POSS, carbon fiber, vinyl ester, and F/M interface

1 INTRODUCTION

Increasing maintenance costs, fuel consumption, and warfare vulnerabilities of conventional steel vessels has raised Naval interest in the technological advances in composite fabrication [1]. Initial design studies indicate that the high specific modulus and strength, low weight, and corrosion resistance properties of composites decreases the life-cycle cost by 7% and the structural weight by 50% depending on the type of ship [2-4]. However, the durability of composites in moist environments has been a limiting factor in capitalizing on the structural advantages of composite vessels.

Several studies have been conducted investigating the effects of water on the mechanical properties of carbon/vinyl ester composites [5-8]. The mechanical and thermal analysis of all these studies indicated that the primary area of failure was the F/M interface due to water degradation. Structurally, poor interfacing of the F/M compromises the overall performance of the material system. Studies by Langston [9] and Verghese et at. [10] have shown that fiber surface modification improves the interface properties of carbon/vinyl ester components.

However, an improvement across the range of mechanical, thermal and hygrothermal behavior of composites was not observed.

To address this issue, we are proposing the introduction of nanotechnology to the composite material system. POSS is an inorganic-organic hybrid nanosilica engineered to strengthen composites, as well as, promote hydrophobicity, which is ideal for marine applications. POSS are nano sized inorganic cage based structures of silica comprised of eight silicon atoms linked to twelve oxygen atoms. The silicon atoms can be linked to almost any unreactive and/or reactive organic groups creating the innovation of a hybrid inorganic-organic composition. It is therefore expected that if POSS particles, with appropriate functional groups, can be installed on the fiber surface, they would provide a stable and strong F/M interface. Recent studies by Zhang [11] have demonstrated 25-27% improvement of the interlaminar shear strength with various POSS products. Earlier work in this study showed that POSS improved the shear strength by 17-25%, impact by 38%, tensile properties by 8-10% and glass transition temperature by 8% in dry conditions[12].

In this investigation, composites treated with POSS systems, octaisobutyl and trisilanolphenyl, have been exposed to various moist environments. Over a period of six weeks the rate of water absorption was monitored in seawater at room temperature (SWRT), seawater at 40°C (SW40), and in 85% relative humidity at 50°C (HM50). Short beam shear testing was conducted on samples before and after exposure to determine the interlaminar shear strength (ILSS) of moisture exposed composites.

2 EXPERIMENTATION

2.1 Fabrication

Prior to surface treatment, unidirectional carbon fibers were soaked in acetone (Sigma Aldrich) and dried to remove the sizing. A liquid media was produced by de-agglomerating POSS (Hybrid Plastics Inc.) in a compatible solvent (hexane or ethanol) using homogenization followed by 90 minutes of sonication. The concentration for each POSS system was 1.0 wt% and 0.2 wt%, respectively, for octaisobutyl and trisilanolphenyl. Carbon fibers (Vectorply, Inc.) were first soaked for three hours in the liquid media dispersed with POSS. Once soaking was complete, fibers were dried in an oven to evaporate the solvent. POSS-coated fibers were then used with vinyl ester resin, Derakane 8084, (J.I. Plastics) to fabricate layered composites following the standard wet lay-up and compression molding method. The resin was cured at ambient temperature for 24 hrs.

2.2 Short Beam Shear

To determine the interlaminar shear strength (ILSS), at least 5 samples from each composite were tested by short beam shear. In accordance with ASTM standard D2344 the sample span/thickness ratio was maintained at 4 to ensure shear failure: 30 mm (length) × 10 mm (width) × 5 mm (thickness). Testing was conducted in ambient conditions on the Z050 Zwick/Roell machine at 1.3mm/min.

2.3 Hygrothermal Conditioning

The specimens were prepared according to ASTM D2344 [45] requirements: 30 mm × 10 mm × 5 mm, in order to conduct interlaminar shear strength testing of the moisture conditioned F/M interface. Twelve specimens from each type of panel were weighed under dry conditions prior to immersion in saltwater at room temperature, saltwater at 40°C, and an 85% humidity chamber at 50°C. For six weeks the rate of water absorption was observed by weighing each specimen to the milligram and taking the average.

After environmental exposure, specimens were tested by short beam shear and compared to the shear strength values under dry conditions. To maintain the exposed condition of the specimens, samples from the saltwater environment were transferred in Ziploc® bags containing saltwater from the testing environment. The humidity samples were transferred in portable humidity chambers established by mixing potassium bromide with distilled water in a sealed container [13]. For comparison, specimens were dried in ambient conditions and intermittently weighed to the milligram until the composite weight was consistent. Five samples from each category were tested under short beam shear.

3 RESULTS

3.1 Mechanical Analysis

The ILSS property of treated and untreated composites were compared to study the affects of POSS in dry conditions. Figure 1 shows that introduction of POSS the F/M interface improved the composite shear strength. Individually, POSS Octa improved the shear strength by 17%; however, POSS TriS increased to 25% as seen in Table 2. This difference is attributed to the respective chemical structures of the POSS systems. TriS is functionalized with silanol to promote chemical bonding

specifically with vinyl ester; whereas, Octa is not functionalized.

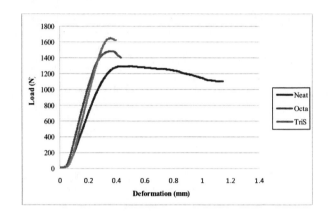

Figure 1. Short beam shear testing results

Panel Type	Carbon Fiber wt.%	Shear Strength (MPa)	Gain/Loss (%)
Neat	---	18.38 ± 0.72	---
Octa	1.0%	21.51 ± 0.73	17.03
TriS	0.2%	23.06 ± 1.82	25.48

Table 1. Comparison of interlaminar shear strength

3.2 Moisture Analysis

Having demonstrated that POSS improves mechanical performance in dry conditions, composites were exposed to moist environments. For six weeks, composite samples were immersed in three different environments: saltwater at room temperature (SWRT), saltwater at 40°C (SW40), and in 85% relative humidity at 50°C (HM50).

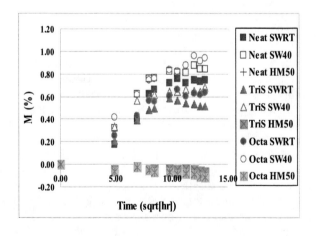

Figure 2: Percentage of weight change

| Condition | Percentage of Weight Change (%) | | | | |
	Neat	Octa	G/L (%)	TriS	G/L (%)
SWRT	0.75	0.64	14.67	0.51	32.00
SW40	0.85	0.94	-10.59	0.68	20.00
HM50	-0.08	-0.11	-37.50	-0.06	25.00

*G/L is Gain/Loss

Table 2: Comparison of weight change

The moisture absorption of the composites shows that the effects of Octa and TriS improves the composite durability as observed in dry conditions. Figure 2 shows that the composites absorbed water at differing rates in saltwater; however, the carbon/vinyl ester composites exhibited negligible weight change in the humid environment. In the 40°C saltwater environment, the composites absorbed water at a higher rate suggesting that temperature is an influencing factor. Initially, Octa sizing reduced the absorption rate in the saltwater environment at room temperature by 15%. However, Table 2 shows that in 40°C saltwater water absorption increases by 11% indicating that Octa is no longer reinforcing the F/M interface. On the contrary, introduction of TriS consistently reduced water absorption by 20% at 40°C and by 32% at room temperature.

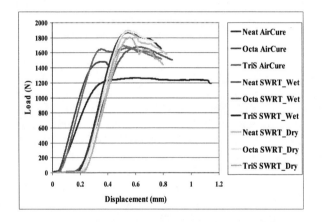

Figure 3: Comparison of seawater at room

| Specimen | Shear Strength (MPa) | | |
	SWRT	SW40	HM50
Neat	26.08 ± 0.49	29.29 ± 1.35	31.66 ± 1.41
Octa	24.30 ± 0.81	28.02 ± 0.80	28.76 ± 0.32
G/L (%)	-0.29	14.98	18.01
TriS	26.60 ±1.58	28.64 ± 1.56	32.15 ± 1.91
G/L (%)	9.15	17.52	31.92

*G/L is Gain/LOSS

Table 3: Comparison of shear strength of wet moisture exposed composites

| Specimen | Shear Strength (MPa) | | |
	SWRT	SW40	HM50
Neat	28.03 ± 1.12	29.58 ± 1.07	31.97 ± 0.99
Octa	26.19 ± 1.15	28.98 ± 1.03	30.03 ± 1.69
G/L (%)	7.47	18.92	23.23
TriS	26.06 ± 1.20	30.30 ± 2.16	31.33 ± 1.13
G/L (%)	6.93	24.33	28.56

*G/L is Gain/LOSS

Table 4: Comparison of shear strength of dried moisture exposed composites

Short beam shear testing of the composites indicated no degradation from moisture exposure, instead improvement of the F/M interface (Fig. 3). Under all three conditions: saltwater at temperature, saltwater at 40°C, and 85% humidity at 50°C, the shear strength of the wet and dry tested composites improved for both POSS systems. Table 3 shows that the shear strength values of the moist tested composites increased with the increase in temperature of the environment. Composites treated with Octa improved from approximately 0-18% in shear strength across the environments and TriS sizing gained 9-32%. This trend was also observed in Table 4 with composites dried after testing, where Octa improved from 7-23% in shear strength and TriS increased 7-29%. Even the shear strength property of the control samples increased from 8-14 MPa for both moist and dry testing. This behavior suggests that the varying temperature of the three aqueous environments is affecting the mechanical performance of the composites.

Based on this observation, ambient cured specimens were post cured for 2 hrs at 90°C as specified by Ashland, Inc., the manufacturer of Derakane 8084. Shear strength of the post cured composites was 33 MPa, which is nearly equivalent to 31-32 MPa shear strength of humidity exposed composites indicating that that composite were curing in the moist environments. A study conducted by Herzog et al. [14] with reinforced Derakane 8084 vinyl ester resin, demonstrated that curing the composites at room temperature as specified by the manufacturer does not achieve the advertised mechanical properties. The composites are not fully cured, which lowers the composites mechanical performance. For accurate mechanical characterization, post curing is required for complete crosslinking of the resin.

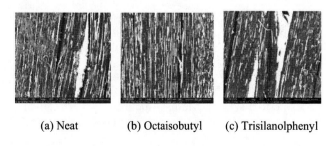

(a) Neat   (b) Octaisobutyl   (c) Trisilanolphenyl

Figure 4: Shear failure of composites exposed to saltwater at room temperature

Figure 4 shows SEM images of shear failure from composites that were dried after exposure to saltwater at room temperature conditions. Microscopic examination of the post fractured composites was required due to the overall reduction of failure propagation. After moisture conditioning, the shear failure seems to have reduced to a predominately intralaminar failure.

4 CONCLUSION

The following can be summarized from the above investigation:

1. Investigation of POSS systems, Octa and TriS, have shown that the nanosilica chemical structure influences the particle effectiveness. POSS systems, like TriS, functionalized with vinyl ester compatible silanol, promotes better adhesion with carbon fiber than unfunctionalized Octa.

2. In dry conditions, the composite shear strength property improved by 17-25% with POSS surface treatment of the carbon fibers.

3. Environmental exposure experiments have revealed that water uptake during a six week period of time was minimal. Although water absorption was insignificant at less than 1%, it has been shown that composites with trisilanolphenyl performed the best under each of the exposure conditions reducing absorption by 20-32% compared to neat specimens.

4. Short beam shear testing of samples after exposure to environmental conditions have shown that there is no degradation in interlaminar shear strength. Instead there was improvement of about 7-32%. This was true with neat as well as POSS reinforced composites. This observation suggests that while under environmental exposure, the resin was in fact curing rather than deteriorating within the six week time, especially at elevated temperature.

5 REFERENCES

[1] Mouritz, A.P., Gellert, E., Burchill, P., and Challis, K., "Review of Advanced Composite Structures for Naval Ship and Submarines," Composite Structures, 53, 21-41, 2001.

[2] Shivakumar K.N., Swaminathan G., and Sharpe M., "Carbon/Vinyl Ester Composites for Enhanced Performance in Marine Applications," Journal of Reinforced Plastics and Composites, 25, 1101 – 1116, 2006

[3] Makinen K., Hellbratt S.E., and Olsson K.A., "The Development of Sandwich Structures for Naval Vessels During Years," Mechanics of Sandwich Structures, Kluwer Academic Publishers: Netherlands, 1988.

[4] Goublat P, and Mayes S., "Comparative Analysis of Metal and Composite Materials for the Primary Structure of Patrol Boat," Journal of Naval Engineering, 108(3), 387-397, 1996.

[5] Kootsookos A. and Mouritz A.P., "Seawater Durability of Glass- and Carbon-Polymer Composites," Composites Science and Technology, 64, 1503-1511, 2004.

[6] Rivera J. and Karbhari V.M., "Cold-Temperature and Simultaneous Aqueous Environment Related Degradation of Carbon/Vinyl Ester Composites," Composites: Part B Engineering, 33, 17-24, 2002.

[7] Acha B.A., Carlsson L.A., and Granata R.D., "Moisture Effects on Carbon/Vinyl Ester by Dynamic Mechanical Analysis," ECCOMAS Thematic Conference on Mechanical Response of Composites, Porto Portugal, 12-14 September 2007.

[8] Ramirez F.A., "Evaluation of Water Degradation of Polymer Matrix Composites by Micromechanical and Macromechanical Tests," Thesis Florida Atlantic University, 2008.

[9] Langston, T.A., "The Effects of Nitric Acid and Silane Surface Treatments on Carbon Fibers and Carbon/Vinyl Ester Composites Before and After Seawater Exposure" Dissertation Florida Atlantic University 2008.

[10] Verghese K.N.E., Broyles N.S., Lesko J.J., Davis R.M., and Riffle J.S., "Pultruded Carbon Fiber/Vinyl Ester Composites Processed with Different Fiber Sizing Agents. Part II: Enviro-Mechanical Durability," Journal of Materials in Civil Engineering, 17(3), 335-342, 2005.

[11] Zhang X., Huang Y., WangT., and Liu L., "Effects of Polyhedral Oligomeric SIlsesquioxane Coatings on the Interface and Impact Properties of Carbon Fiber/Polyarylacetylene Composites," Journal of Applied Polymer Science, 2006, 102; 5202-5211

[12] H. Mahfuz, F. Powell, R. Granata, and M. Hosur, "Treatment of Carbon Fibers with POSS and Enhancement in Mechanical Properties of Composites," SAMPE, Seattle, WA, May 17-20, 2010.

[13] Spencer, H.M., "Laboratory Methods for Maintaining Constant Humidity," International Critical Tables of Numerical Data, Physics, Chemistry and Technology, 7, 67-68, 1930.

[14] Herzog, B., Gardner, D.J., Lopez-Anido, R., and Goodell, B., "Glass-Transition Temperature Based on Dynamic Mechanical Thermal Analysis Techniques as an Indicator of the Adhesive Performance of Vinyl Ester Resin," Journal of Applied Polymer Science, 97, 2221-2229, 2005.

Volume Grain Analysis in Organic Thin Film Semiconductors

T. Gredig*, D. Bergman, K. P. Gentry

Department of Physics & Astronomy, California State University Long Beach,
1250 Bellflower Blvd., Long Beach, CA 90840-3901, USA
*tgredig@csulb.edu

ABSTRACT

The grain structure of organic thin films is quantitatively studied using atomic force microscopy images. The grain morphology affects the charge transport of organic polycrystalline thin film devices due to charges at grain boundaries. Thus, the grain growth plays a crucial role for the optimization of organic thin film devices. Iron phthalocyanine, a planar small molecule, is thermally evaporated and grain distributions for thin films are measured. The nominal grain size diameter can be controlled between 30 - 200 nm for samples deposited between room temperature and 260 °C. The phthalocyanine grains are elongated and can be approximated by an ellipsoid. The grain size distributions are analyzed with a watershed algorithm that provides the grain area and volume. The grain volume distribution can be fit to a lognormal distribution. The minor and major axes distribution of the areal ellipse differ and best fit a normal and lognormal distribution, respectively. The anisotropic growth is attributed to the asymmetry of the small molecule.

Keywords: grain size distribution, atomic force microscopy, phthalocyanine, morphology, thin film, grain volume

1 INTRODUCTION

Polycrystalline organic thin films have many technological applications, such as gas sensors, flexible photovoltaics, and organic light emitting diodes.[1] Recent advances have shown that the charge transport of organic semiconductors shows a variety of new concepts due to the interaction of the π-electronic structure and its thin film morphology.[2]–[4] Therefore, a quantitative understanding of the film growth and grain structure is pertinent to improve device fabrication. Furthermore, it allows for a better understanding of intrinsic characteristics such as bias stress, trap levels, defects, adsorption rates and other properties of π-conjugated materials. In organic field-effect transistors the first monolayers form the charge transport channel.[5] Recently, quantitative analyses using atomic force microscopy in pentacene thin films have shown diffusion limited aggregation for the growth of the first few layers.[6] Also,

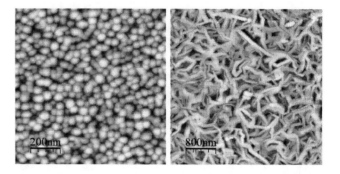

Figure 1: Atomic force microscopy images for iron phthalocyanine thin films deposited at room temperature (left) and 200 °C (right). The $1 \times 1\mu m^2$ image shows small rounded grains at room temperature, whereas the $4 \times 4\mu m^2$ image shows elongated grains growing in random directions at high deposition temperatures.

polycrystalline silicon films can be fabricated by thermal annealing from amorphous precursors. The crystallite area distributions are measured with transmission electron microscopy and reveal lognormal-like distributions.[7] This process can be understood with a random nucleation and growth model and yields analytical solutions for kinetics and grain size distributions.[8], [9] Such models and simulations provide important insight in the time-dependent growth process. Yet, these examples and most other experimental data are limited to two-dimensional grain area distributions ignoring the full details of the crystallite structure. From a theoretical point of view, important model and simulation verification comes from the precise three-dimensional structure.[10] In the following a quantitative analysis of the three-dimensional grain volume of phthalocyanine thin films is presented based on an atomic force microscopy analysis that includes height information.

2 EXPERIMENT

In order to study grain size distributions, thin films of iron phthalocyanine (FePc) were deposited by thermal evaporation in vacuum. Metallo-phthalocyanines (M-$C_{32}N_8H_{16}$) are the archetype of planar and π-conjugated small molecules.[11] The base pressure of the evaporation chamber was below $7 \cdot 10^{-7}$ mbar. The FePc powder

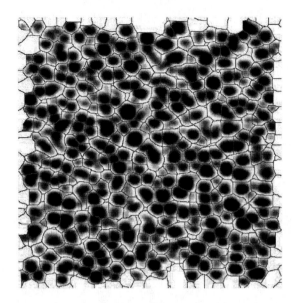

Figure 2: Watershed processed atomic force microscopy image of an iron phthalocyanine thin film deposited at room temperature. The image is $1 \times 1\mu m^2$ in size. The partial edge grains are removed from the statistics using the information from the bounding box. The height information from the original AFM image is used to infer the grain volume for each grain outlined in the image.

was purchased from Sigma-Aldrich and then purified in three cycles using a thermal gradient vacuum tube furnace. The substrate surface is carefully cleaned before all depositions. The deposition rate was between 0.3 and 0.4 Å/s. X-ray diffraction data of the FePc thin films show a pronounced peak at $2\theta = 6.8^o$ for films deposited on both silicon and sapphire substrates and confirm the polycrystalline nature of the thin film.[12] In particular, the (200) peak indicates that the plane of the FePc molecule is perpendicular to the substrate surface; i.e. the unit cell's b-axis is parallel to the substrate.[13] The thin films are subsequently analyzed using atomic force microscopy (AFM) employing a NanoScope III MultiMode^{TM} in tapping mode.[14]

3 RESULTS

The grain size and surface morphology of FePc thin films are strongly influenced by the deposition temperature. From the atomic force microscopy images, small rounded grains are measured in samples deposited at room temperature (Fig. 1). The nominal grain size diameter ranges from 20 - 50 nm and only a small asymmetry is apparent. However, it should be noted that the stacking order implies that the major axis contains four to five times more molecules as compared to the minor axis. This is due to the planarity of the molecule. At elevated deposition temperatures, the grain size dramatically increases for nominally equal films as shown

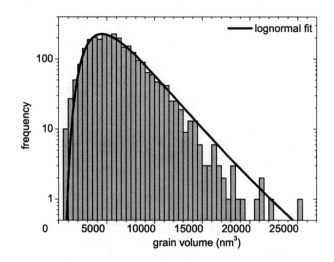

Figure 3: The grain volume is plotted for an iron phthalocyanine thin deposited at room temperature. The distribution is fit with three parameters to the lognormal function (line), where the mean is 5676 nm^3 and the standard deviation 0.49.

in Fig. 1. The mean grain size increases by almost one order of magnitude such that there are 650 molecules on average in a grain along the major axis and 70 molecules along the minor axis. The shape of the grain is strongly elliptical, as the major axis grows much faster than the minor axis with deposition temperature. From the line trace of the atomic force microscopy image, individual plateaux appear that are spaced 1.3 nm corresponding to the molecule's size.[15]

Since the resolution of atomic force microscopy (AFM) images generally limits the number of grains to at most a few hundred per image, several AFM images were combined to produce statistics of several thousand grains. The grain boundaries are determined with a watershed algorithm[16] and partial grains at the edge of the image are removed from the analysis, see Fig. 2. For each grain, the perimeter, area, and bounding or Zingg box are determined to build distributions. Using the explicit AFM height information, each grain is integrated over its area to sum up the total grain volume under the assumption that no smaller grains are hidden underneath. This assumption appears to be mostly true for thin films with the thickness less than the mean grain size diameter. It should be noted that the grain volume is systematically overestimated due to the finite AFM cantilever tip size and the tip-sample interaction. In this fashion, the distribution of more than 2200 grains for an iron phthalocyanine thin film deposited at room temperature is summarized in Fig. 3. The lognormal distribution has a lower reduced χ^2 of 86 compared to 211 for a best fit to a normal distribution. The asymmetry of the grains is further analyzed by fitting an ellipse of equal area to each grain. The major axis has a

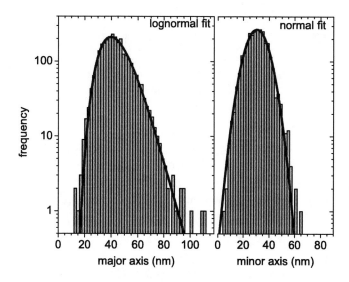

Figure 4: The grain is fit to an ellipse and the distributions of the major and minor axis are fit with 3 parameters each. The lognormal distribution describes the major axis length better and yields a mean of 42.3 nm and a standard deviation $\sigma = 0.25$. For the minor axis a normal distribution has a lower reduced χ^2 value with a mean of 30.5 nm and a distribution width of $\sigma = 16.5$ nm.

mean value of 42 nm versus the minor axis length, which is only 30 nm. However, the circularity decreases dramatically with deposition temperature as more kinetic energy allows the molecules to diffuse farther and form longer chains evolving into the characteristic needle-like shapes seen in micrographs.

The separated distributions of the major and minor axes differ; the major axis fits better to the lognormal distribution, whereas the minor axis grain distribution has a lower reduced χ^2 for the normal distribution as demonstrated in Fig. 4. This is evidence for the asymmetric growth of grains and crystallites in phthalocyanine thin films.

4 CONCLUSIONS

The distributions of iron phthalocyanine grains of polycrystalline thin films are studied quantitatively. A watershed-based algorithm is employed to determine grain boundaries of atomic force microscopy images. The AFM height information is integrated to determine not only the grain area, but importantly the grain volume. In a simple model, the grains are evaluated as ellipsoids having three axes. The major and minor axis in the plane are best described with a lognormal and normal grain distribution respectively. The grain volume fits well the lognormal distribution. This analysis provides detailed experimental data to improve the understanding of grain growth, grain boundaries, and volume.

5 ACKNOWLEDGEMENTS

This research was supported by the National Science Foundation in the Division of Materials Research (Grant No. DMR-0847552) and the Undergraduate Research Provost Award at the California State University, Long Beach. The authors would like to thank Dr. José de la Venta for performing x-ray diffraction measurements and Prof. Andreas Bill for many insightful discussions.

REFERENCES

[1] R. D. Yang, T. Gredig, C. N. Colesniuc, J. Park, I. K. Schuller, W. C. Trogler, and A. C. Kummel, Appl. Phys. Lett. **90**, 263506 (2007).

[2] V. Coropceanu, J. Cornil, D. A. da Silva Filho, Y. Olivier, R. Silbey, and J.-L. Brédas, Chem. Rev. **107**, 926 (2007).

[3] J. Locklin and Z. Bao, Anal. Bioanal. Chem. **384**, 336 (2006).

[4] T. Gredig, I. N. Krivorotov, and E. D. Dahlberg, Phys. Rev. B **74**, 094431 (2006).

[5] G. Horowitz, Adv. Mat. **10**, 366 (1998).

[6] B. Stadlober, U. Haas, H. Maresch, and A. Haase, Phys. Rev. B **74**, 165302 (2006).

[7] R. B. Bergmann, F. G. Shi, H. J. Queisser, and J. Krinke, App. Surf. Sci. **123-124**, 376 (1998).

[8] R. B. Bergmann and A. Bill, J. Cryst. Growth **310**, 3135 (2008).

[9] A. V. Teran, A. Bill, and R. B. Bergmann, Phys. Rev. B **81**, 075319 (2010).

[10] M. Hillert, Mat. Sci. Forum **204-206**, 3 (1996).

[11] T. Gredig, K. P. Gentry, C. N. Colesniuc, and I. K. Schuller, J. Mat. Sci. **ICAM 2009**, 1573 (2010).

[12] G. Liu, T. Gredig, and I. K. Schuller, Europhys. Lett. **83**, 56001 (2008).

[13] H. Peisert, T. Schwieger, J. M. Auerhammer, M. Knupfer, M. S. Golden, J. Fink, P. R. Bressler, and M. Mast, J. Appl. Phys. **90**, 466 (2001).

[14] I. Horcas, R. Fernández, J. M. Gómez-Rodríguez, J. Colchero, J. Gómez-Herrero, and A. M. Baro, Rev. Sci. Instr. **78**, 013705 (2007).

[15] K. P. Gentry, T. Gredig, and I. K. Schuller, Phys. Rev. B **80**, 174118 (2009).

[16] L. Vincent and P. Soille, IEEE Trans. Patt. Anal. Mach. Intel. **13**, 583 (1991).

Characterization of an electrical powering nano-generator based on aluminum doped ZnO thin films

N. Muñoz Aguirre[*], J. E. Rivera López[*], L. Martínez Pérez[**], R. Rivera-Blas[*], L.W. Rodríguez-Alvarado[*], R. Valderrabano[*] and P.Tamayo Meza[*]

[*]Instituto Politécnico Nacional. Sección de Estudios de Posgrado e Investigación. Escuela Superior de Ingeniería Mecánica y Eléctrica Unidad Azcapotzalco. Av. Granjas, N°682, Colonia Santa Catarina. Del. Azcapotzalco, CP.02550, México, DF. México, ptamayom@ipn.mx
[**] Unidad Profesional Interdisciplinaria en Ingeniería y Tecnologías Avanzadas del Instituto Politécnico Nacional, Av. IPN No. 2580, Col. Barrio La Laguna Ticomán, C.P. 07340, México D.F. México, buhomartin@yahoo.com

ABSTRACT

The conversion of mechanical to electrical powering of a nano-generator which consists of a no biased conductive atomic force microscope tip that is pressing over the surface on an aluminum doped ZnO thin film is showed. The measured voltage only induced by the atomic force of the tip-ZnO thin film interaction is of the order of tenths of volts.

Keywords: Piezoelectric ZnO thin films, Atomic Force Microscopy, c-AFM, nano-generator.

1 INTRODUCTION

It was well demonstrated the conversion of nanoscale mechanical energy into electrical energy by means of piezoelectric zinc oxide nanowires arrays using a conductive atomic force microscope tip in contact mode [1]. The coupling of piezoelectric and semiconducting properties in zinc oxide creates a strain field and charge separation across the nanowires as a result of its bending. In this work, it is demonstrated the conversion of nanoscale mechanical into electrical powering by means of a conductive atomic force microscope tip that is pressing over the surface of aluminum doped ZnO thin film. No external voltage to the conductive tip was applied in any stage of the AFM experiment.

2 EXPERIMENTAL

Aluminum doped ZnO thin films by the water-mist-assisted spray pyrolysis technique [2] were synthesized. An ultrasonic generator operated at 0.8MHz was used for producing the spray of source solution. A 0.3032 M solution of zinc acetylacetonate (from Sigma Aldrich) dissolved in N,N-dimethylformamide (N,N-DMF, from Mallinckrodt) was prepared and 10 % of molar concentration of acetylacetonate of aluminum was added to this chemical solution. High purity air with flows of approximately 10 l/min was used as the carrier gas of the

mist solution to the top of the substrate. The synthesis time was ten minutes. A molten tin bath was the substrate heating. Corning 7059 glasses were used as substrates. A stream of water mist was simultaneously supplied to the spray of the source solution. The depositions were carried out at substrate temperatures of 500 °C. X-ray diffraction (XRD) measurements were made using a D-5000 Siemens Diffractometer and reported in reference [3]. The experimental configuration of the electrical powering nano-generator is shown in Figure 1. As it can be observed, for the characterization of the nano-generator, the conductive AFM method in contact mode [3,4,5] was used. The experiments were carried on a JSPM-5200 microscope from JEOL Ltd. The used tip was the OSCM-PT model from Veeco Ltd. Which have a spring constant in the range of $k =0.5$-4.4 N/m. Because of a fresh tip for the measurements was used, the higher value equal to $k =4.4$ N/m was taking for obtain the tip-sample force.

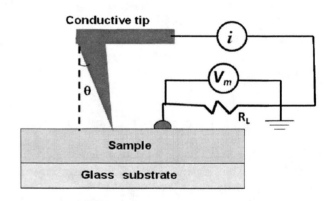

Figure 1: Scheme of the nano- generator showing an external load resistance R_L. Here θ is the angle of the cantilever surface with the horizontal level which is of the order of 15 degree. The distance between the tip and the silver paint contact was 1 mm.

3 RESULTS

When the tip is scanning and pressing the sample surface a voltage V_m is measured across the external load resistance $R_L=100$ MΩ which was connected in series to the resistance of the thin film (See Figure 1). The voltage was measured using an acquisition data card CDAQ model NI-9172 module NI-9205 from National Instruments Inc. controlled by means of the LabView software. The results are shown in Figure 2. In this figure, it is shown the generation of about 10 mV when the tip is scanning in contact mode on the surface of the thin film. It is possible increase the generated voltage increasing the tip-sample force as for example vertically pressing on the sample surface. Figure 4 shows the result of this experiment got up to 200 mV of generated voltage.

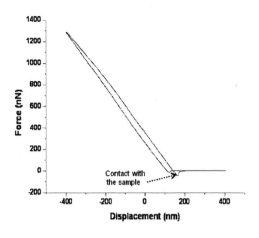

a)

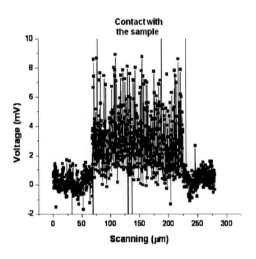

a)

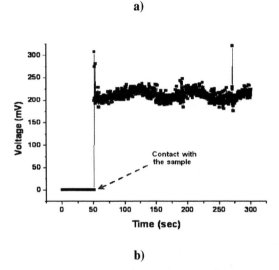

b)

Figure 3: a) Force curve showing the contact point with the sample. The movement of the tip is vertical towards the sample surface. b) When the tip is pressing the surface of the thin film, a high generated voltage is taking place; for a tip-sample force about of 1200 nN a voltage of about 200 mV is generated.

4 CONCLUSIONS

In a briefly conclusion, the simple experimental configuration of the nano-generator shown in Figure 1 has potential applications in the development of nanodevices which involves conversion of mechanical energy using piezoelectric aluminum doped ZnO thin films.

Acknowledgements: Work supported by Instituto Politécnico Nacional from México with the CGPI projects number SIP-20101303, SIP-20100627. Lilia Martinez Perez Acknowledge the Scanning Probe Microscopy laboratory facility from ESIME-Unidad Azcapotzalco also from Instituto Politecnico Nacional, Mexico.

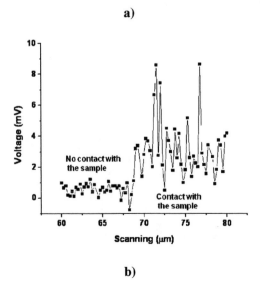

b)

Figure 2: a) Voltage output (V_m) across the external load resistance (R_L) as a function of the scanning position. b) Close up in the interval 60-80 μm. The velocity of the scanning was 0.97 μm/sec.

REFERENCES

[1] Zhong Lin Wang and Jinhui Song, *Science* 14 April 2006:Vol. **312**. no. 5771, pp. 242 – 246.

[2] L. Martínez Pérez, M. Aguilar-Frutis, O. Zelaya-Angel, and N. Muñoz Aguirre, Physica Status Solidi a, **203, No. 10**, (2006) 2411-2417.

[3] N. Muñoz Aguirre, J.E. Rivera-López, L. Martínez Pérez and P. Tamayo Meza, Synthesis and Atomic Force Microscopy contact current images of aluminium doped ZnO thin films, AEM-NANOMAT09 Proceedings at Materials Science Forum Vol. **644** (2010), 109-112.

[4] Liam S. C. Pingree, Obadiah G. Reid, David S. Ginger, *Adv. Mater.*, **21, No 1**, (2009)19-28.

[5] C. Ionescu-Zanetti, A. Mechler, S. A. Carter, R. Lal, *Adv. Mater.,* **16**, **No 5,** (2004) 385-389.

NSTI-Nanotech 2010, www.nsti.org, ISBN 978-1-4398-3401-5 Vol. 1, 2010

Growth and optical properties of high-density InN nanodots

W. C. Ke[*], C. Y. Kao[*] and B.-R. Huang[**]

[*]Department of Mechanical Engineering, Yuan Ze University, Chung-Li, Taiwan, R.O.C.,
wcke@saturn.yzu.edu.tw
[**]Graduate Institute of Electro-Optical Engineering, National Taiwan University of Science and
Technology, Taiwan, R.O.C., huangbr@mail.ntust.edu.tw

ABSTRACT

The alternative supply of source precursor growth (i.e. pulsed mode, PM) of InN/GaN nanodots by metal organic chemical vapor deposition (MOCVD) was investigated. The InN nanodots density of up to $\sim 5 \times 10^{10}$ cm^{-2} at a growth temperature of 550 ℃ was achieved, which is relatively higher than those ($\sim 8.5 \times 10^{9}$ cm^{-2}) by conventional mode (CM) method. The kinetic mechanism indicated that the diffusion activation energy at growth temperature region from 600 to 700℃ was 2.65 and 1.25 eV respectively for the PM and CM method. The high diffusion activation energy due to the high NH$_3$ flow rate generated more reactive nitrogen adatoms on the growth surface, and is believed to be the main reason for the growth of high density InN nanodots. In addition, the PL intensity of InN nanodots that grown by PM method exhibit more intense signal than CM method. It was implied superior emission property for the PM growth of InN nanodots

Keywords: *InN nanodots, pulsed mode, conventional mode, diffusion activation energy, PL.*

1 INTRODUCTION

Since the InN band gap energy was revised to ~0.7 eV, this discovery has provided new opportunities for further applications of III-nitride based optoelectronic devices. Recently there has been substantial interest in InGaN as a new thin-film solar cells material because of its wide tunable energy band gaps, which cover almost the entire solar spectrum. However, to date there are only a few reports on InGaN based solar cells due to some issues (phase separation, doping issue and poor quality material etc.) which are difficult to overcome in the commercial MOCVD system. In the past couple of years, the use of nanostructures in solar cells has offered the potential for a higher level of efficiency by using new physical concepts. For example, the benefit of multiple exciton generation (MEG) has been demonstrated in colloidal suspensions of PbSe, PbS, PbTe, CdSe, and GaAs quantum dots (QDs). A maximum theoretical efficiency of 42% has been predicted for single-junction devices employing MEG-active absorbers.

Numerous research efforts have been directed toward techniques based on the principle of self-assembly for the fabrication of nitride QDs. The growth of GaN QDs on AlN using the commonly used Stranski-Krastanow (S-K) growth mode was not reported until 1997 by Daudin et al. using MBE, and more recently by Miyamura et al. using MOVPE. Otherwise, antisurfacant method has been used to grow GaN QDs on AlGaN ternary. Despite the numerous studies on (In)GaN QDs, the published reports on InN QDs, particularly on specific sample preparation procedure, are still quite limited.

Up to now, only a few papers have been reported about the growth of InN QDs by conventional MBE [1-2] and MOCVD [3-5]. This paper presents a feasible method for preparing high density InN nanodots on a GaN surface. Preliminary results indicate that by alternating the source precursor method, the so-called pulsed mode (PM) during the MOVPE epitaxial growth, high density InN nanodots could be achieved on a GaN epilayer. This method has been proven to be a simple yet effective way for preparing QDs structures in the InN material system and may have the potential to be used in the fabrication of infrared InN-based solar cells, light-emitting diodes, laser diodes and detector devices. In addition, result of the anomalous temperature dependence of optical emission in high density InN nanodots was of great interest and was investigated as well.

2 EXPERIMENT

The InN nanodots were grown on GaN/sapphire (0001) at a temperature varying between 550 to 750℃ by the PM method using trimethylgallium (TMGa), trimethylindium (TMIn) and ammonia (NH$_3$) as the source materials. The gas flow sequence for the PM method, basically consists of four steps: 20-sec TMIn+NH$_3$ growth step, 20-sec NH$_3$ source step with 10-sec purge steps in between. During the growth step, the mole flow rates of TMIn and NH$_3$ are 1.53×10^{1} and 4.46×10^{5} μmole/min, respectively. It is worth noting that during the 20-sec NH$_3$ source step an amount of NH$_3$ (8.04×10^{5} μmole/min) is introduced intentionally to achieve high density InN nanodots. In the present study the InN nanodots were grown by the CM method, in which the TMIn and NH$_3$ flow rates are kept constant at 1.53×10^{1} and 4.46×10^{5} μmole/min, respectively. The total growth time for the InN nanodots was 2 min, which is equal to the total time of six cycles of growth steps in the PM method. An NT-MDT Solver HV atomic force microscopy (AFM) system was used to perform the morphology measurement. Photoluminescence (PL) measurements were performed by using the 488-nm line of an argon-ion laser as an excitation source. The PL signals were analyzed by a 0.5-m monochomator and detected by a cooled InGaAs photodiode with a cut-off wavelength at 2.05 µm.

3. RESULTS AND DISCUSSION

In order to understand the high density of InN nanodots grown by PM method, we performed a series experiment that changed the growth temperature from 550 to 750 ℃. The InN nanodots density as a function of reciprocal temperature is shown in Fig. 1. As anticipated, the dots density depends greatly on the substrate temperature. There are two distinct regions in our dots density curve, divided by a temperature of ~ 700℃. As can be seen in the Fig. 1, the dot density is reduced gradually from 1.4×10^{10} to 5.1×10^{8} cm^{-2} as the temperature is increased from 600 to 700 ℃, which then tends to drop sharply with further increasing temperature and eventually become zero, i.e. no dots growth, as the substrate temperature is beyond 750℃. For temperature higher than 700℃, the dots density drops drastically. This dot density effect is contributed by dots coarsening, driven by the desire of the system to reduce the boundary free energy by transforming small dots into large ones. The process involves both migrations of adatoms across terraces and the evaporation of atoms from dots. It is therefore that the dots density should become a much steeper function of temperature.

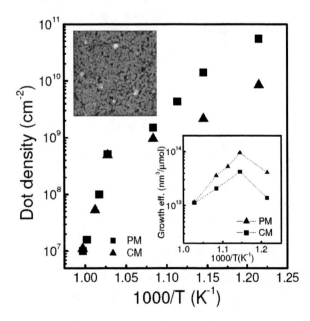

Fig. 1 The resulted Arrhenius plot of InN nanodots density as a function of reciprocal temperature. The insert shows

the growth efficiency of the InN nanodots as a function of 1000/T for the CM and PM methods.

By referring to the island nucleation mechanism proposed by Robison et al., we learn that the dot density at low temperature is governed by the diffusion capability of adatom, while that at high temperatures is determined predominately by the re-evaporation rate of adatoms, hence the binding energy of adatom to the adsorbed site. The respective characteristic equations are $N_S \propto N_{0L}\exp(E_d/3kT)$ for low temperature where N_s is dot density, N_{0L} pre-exponential parameters. The E_d activation energy of diffusing In adatom to the adsorbed sites were 2.65 eV for InN nanodots growth on GaN film by PM growth technique. The value of E_d is 1.25 eV for InN nanodots growth on GaN film by the CM method. Because of abundant supply of NH_3 in NH_3 step (8.04×10^5 µmole/min), result large density of nitrogen dangling bonds on the growing interface, and form the diffusion barrier of the In adatoms in the next growth step in pulsed mode growth scheme. This explains the high E_d diffusion activation energy of In adatom in pulsed mode as compared to conventional MOCVD method. In the high temperature region, the desorption of In adatoms from growing surface is effectively suppressed by the high NH_3 supply in pulsed growth mode. For InN dot growth, O. Briot et al. [3] also conducted similar experiments in their study. They found the activation energy of diffusion In adatom E_d is 4.19 ± 0.53 eV which is much larger than our results for InN nanodots prepared either by pulsed mode or conventional MOCVD method. Since they performed InN dot grown at V/III of ~30000, much higher than the V/III ratio (~20000) used in our study, this may be the reason causing the discrepancy in diffusion activation energy in the InN dot growth study.

The growth efficiency of these two kinds of growth methods is shown in the insert of Fig.2. The growth efficiency is defined by the InN nanodots volume divided by the mole flow rate of TMIn. Each InN nanodots volume was estimated by using the dome structure. The volume of the nanodots was calculated by taking the volume of a single nanodot and multiplying it by the nanodot density for a 1 cm^2 area. For example, the total volume of nanodots was 1.47×10^{15} nm^3 and 6.48×10^{14} nm^3 respectively using the PM and CM method at 600℃ for a 1 cm^2 area. It must be noted that the growth efficiency curves show a maximum value at the growth temperature of 600 ℃. It is known that the NH_3 cracking efficiency increases with the increase in growth temperature. Thus, the increase of the growth efficiency with the increase in growth temperature from 550 to 600 ℃ was believed to be relative to the increase in NH_3 cracking efficiency. However, the growth efficiency decreased with the increase in growth temperature from 600 to 700 ℃. This was evident from the increase in dissociation of the InN nanodots. In addition, the growth efficiency for the PM method was about twice as large as that of the CM method. The higher growth efficiency for the PM method was believed to be due to the higher concentration of nitrogen atom participates into the growth of InN nanodots. Despite of the dot density, the other dots parameters that concern us are height and diameter. From AFM measurements (in Fig. 2), the InN nanodots size listed as diameter/height were 190/35 and 180/18 nm, respectively for the CM and PM technique.

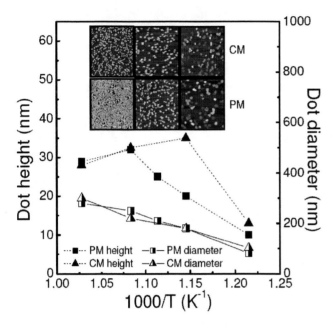

Fig. 2. Dependence of average height and average diameter of InN nanodots on growth temperature by CM and PM method. Insert shows the AFM images of CM and PM method at 600, 650 and 700℃.

The 17K PL spectra of the InN nanodots grown by the CM and PM methods are shown in Figs. 3 (a) and (b), respectively. It should be noted that these samples for PL measurement were grown at 600 ℃ without any GaN capping layer. The PL spectrum for the InN nanodots grown by the CM method shows a peak energy at 0.8 eV, with full width at half maximum (FWHM) of 87 meV. However, for the InN nanodots grown by the PM method, the PL peak energy was observed at 0.82 eV with FWHM of 143 meV. This emission energy of InN nanodots is higher than the reported bandgap energy of 0.69 eV [15], indicating a strong Burstein-Moss effect due to the presence of high electron concentration in the InN nanodots. Since the dot size is still too large to produce a pronounced quantum size effect, any differences in the peak energy between two samples are due mainly to variations of electron concentration in the InN nanodots. In addition, Fu et al. provides a convenient formula to determine the free-electron concentration in InN films by PL measurement [6]. The relationship between the FWHM and the free-electron concentration can be well described by the empirical formula [7-8] The free electron concentration were

estimated $\sim 7\times 10^{18}$ cm^{-3} and $\sim 2\times 10^{19}$ cm^{-3}, respectively for the CM and PM growth of InN nanodots.

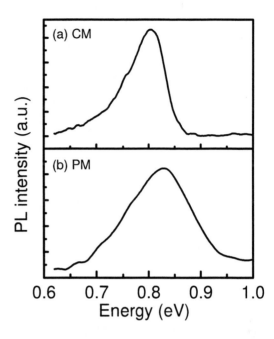

Fig. 3 The 17 K PL spectra of the InN nanodots grown by the CM and PM growth methods, respectively.

4. CONCLUSION

The high density InN/GaN nanodots ($\sim 5\times 10^{10}$ cm^{-2}) were achieved by the PM MOCVD growth technique at a temperature of 550℃. The high density InN nanodots were achieved mainly as a result of the high NH$_3$ flow rate in the NH$_3$ source step, resulting in a high density of nitrogen adatoms on the growing surface, and forming a diffusion barrier in the next growth step of the PM growth method. The higher PL peak energy and FWHM indicate that the free carrier concentration is higher for InN nanodots grown by the PM method. It is believed that the high carrier concentration due to the high In vacancy (V_{In}) in the InN nanodots is the main reason causing an anomalous temperature dependence of the PL peak energy for high density InN nanodots. In addition, the higher thermal activation energy of the InN nanodots indicates a stronger localization of carriers in the PM grown InN nanodots.

REFERENCES

[1] Y. F. Ng et al., Appl. Phys. Lett., 81, 3960, 2002.
[2] A. Yoshikawa et al., Appl. Phys. Lett., 86, 153115, 2005.
[3] O. Briot et al., Appl. Phys. Lett., 83, 2919, 2003.
[4] W. C. Ke et al., Appl. Phys. Lett., 88, 191913, 2006.
[5] W. C. Ke et al., Appl. Phys. Lett., 89, 263117, 2006.
[6] S. P. Fu et al., Semicond. Sci. Technol., 21, 244, 2006.
[7] R. M. Sieg et al., J. Appl. Phys., 80, 448, 1996.

[8] H. Q. Zheng et al., J. Appl. Phys., 87, 7988, 2000.

 NSTI-Nanotech 2010, www.nsti.org, ISBN 978-1-4398-3401-5 Vol. 1, 2010

Theoretical Studies in Epitaxial Stabilization of ZnO-based Light-emitting Semiconductors

Po-Liang Liu[1*], Yu-Jin Siao[1], and Ming-Hsien Lee[2]

[1] Graduate Institute of Precision Engineering, National Chung Hsing University, Taichung, Taiwan 402, Republic of China

[2] Department of Physics, Tamkang University, Tamsui, Taipei, Taiwan 251, Republic of China
E-mail addresses: pliu@dragon.nchu.edu.tw

ABSTRACT

We present first-principle calculations on the strained $Zn_{1-x}(Be,Mg,Al)_xO$ alloy systems. The epitaxial softening of (0001)-oriented $Zn_{1-x}(Be,Mg,Al)_xO$ strained layer lattices are investigated, and the ZnO strained layers could be stabilized with adding the aluminium compositions. A detailed analysis of the preferred (epitaxial) orientation of ZnO strained layer superlattices is examined, and $<2\bar{1}\bar{1}0>$-oriented ZnO is more stable than (0001)- or $(10\bar{1}0)$-oriented ZnO. Using the harmonic elasticity theory, we find the $q_{harm}([0001])$ is the highest and the $q_{harm}([2\bar{1}\bar{1}0])$ is much softer than $<0001>$ or $<10\bar{1}0>$. In addition, the slightly softening of $<11\bar{2}l>$ in $Zn_{1-x}(Be,Mg,Al)_xO$ strained layer lattices is observed. The findings agree with previously reported experimental results.

Keywords: LED, ZnO, first-principle calculation, epitaxy.

I. INTRODUCTION

The device quality ZnO films are expected to next generation materials applied in the field of photo-electronic devices. It is partly due to the widely tunable lattice constant for being a new class of templates and partly due to the direct tunable bandgaps. Thus, they have the potential to develop inexpensive lasers and nanophotonics covered a wide wavelengths range in the blue, green and UV lighting. Light-emitting semiconductor devices rely on materials that possess single-crystal epilayers due to high quality intrinsic epilayers having high carrier mobility and low residual carrier concentration. Unfortunately, most of epitaxial ZnO and related compound semiconductors belong to nanostructures in nature having incoherent grain boundaries, such as a nanowire, nanorod, nanowall, or nanobelt. Epilayer growth process of device quality ZnO was limited success even after many years of research. A major problem encountered in the single-crystal approach is routinely grown on the highly mismatched heteroepitaxial substrates, such as Si, sapphire, and glass substrates [1-3]. The fabrication of $Zn_{1-x}(Be,Mg,Al)_xO$ makes it possible to de-couple strain and band gap engineering to achieve unique device structures that lead to novel photonic devices based on Group II-VI materials [4-11]. These systems include strain-engineered direct gap diodes and multi-quantum well lasers, photodetectors, emitters and modulators grown on $Zn_{1-x}(Be,Mg,Al)_xO$ buffered sapphire or virtual substrates.

To expand our insight into the stress/strain distributions in and around coherently strained ZnO-based group-II-VI $Zn_{1-x}(Be,Mg,Al)_xO$ alloy systems we performed ab initio calculations to elucidate the epitaxial stabilization. Our calculations using density functional theory (DFT) methods based on the generalized gradient approximation (GGA) proceed as follows. First of all, we construct $Zn_{1-x}(Be,Mg,Al)_xO$ strained layer models, and their atomic structures are optimized by full relaxation to zero force positions. Second, we calculate the epitaxial softening of (0001)-oriented $Zn_{1-x}(Be,Mg,Al)_xO$ models. Third, we calculate the epitaxial softening of (0001)-, $(10\bar{1}0)$-, and $(2\bar{1}\bar{1}0)$-oriented ZnO strained layer superlattices. Finally, we employ the harmonic elasticity theory to study the harmonic behavior of the $Zn_{1-x}(Be,Mg,Al)_xO$. Above calculations enable us to determine the preferred growth direction, the effect of dopant constituents on the epitaxial stabilization of the $Zn_{1-x}(Be,Mg,Al)_xO$ alloy systems, and the compressive or tensile relaxation in and around strained $Zn_{1-x}(Be,Mg,Al)_xO$ alloy systems.

II. COMPUTATIONAL METHODOLOGY AND BACKGROUND

A fundamental quantity which influences the epitaxial deformation behavior of thin films is the epitaxial stabilization (also commonly referred to as the epitaxial softening) which is defined as the ratio between the epitaxial increase in energy ΔE^{epi} due to bi-axial deformation, and the hydrostatic increase in energy ΔE^{bulk} due to tri-axial deformation. Formally, the epitaxial stabilization is defined by the dimensionless parameter [12]:

$$q(a,\hat{G}) = \Delta E^{epi}(a,\hat{G}) / \Delta E^{bulk}(a). \qquad (1)$$

Here $\hat{G}$ is the growth direction and a is the in-plane lattice constant.

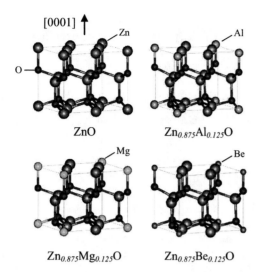

FIG. 1. Atomistic representations of ZnO, $Zn_{0.875}Al_{0.125}O$, $Zn_{0.875}Mg_{0.125}O$, and $Zn_{0.875}Be_{0.125}O$ models. The atoms are represented by spheres: Zn (light gray), O (red), Al (pink), Mg (light green) and Be (orange).

The epitaxial stabilization of the (0001)-oriented $Zn_{1-x}(Be,Mg,Al)_xO$ strained layer superlattices was studied in detail by conducting first-principles DFT-GGA calculations using the Vienna Ab Initio Simulation Package (VASP) [13], which utilizes a plane wave basis to solve the electronic structure. We employ ultrasoft pseudopotentials [14] and a plane-wave basis to treat the electronic structure within the PW 91 GGA [15,16]. We took into account 16-atom crystallographic unit cells with the compositions of ZnO, $Zn_{0.875}Be_{0.125}O$, $Zn_{0.875}Mg_{0.125}O$, and $Zn_{0.875}Al_{0.125}O$. All atomic positions were optimized by full relaxation to zero force positions. The convergence of electronic properties was achieved with a kinetic energy cutoff of 396 eV and 18 irreducible k-points generated within the Brillouin zone (BZ), yielding the optimized structures shown in the four models depicted in Fig. 1.

To investigate the systematic change in the preferred growth directions of ZnO systems, we examined (0001)-, $(10\bar{1}0)$-, and $(2\bar{1}\bar{1}0)$-oriented ZnO strained layer superlattices using the same computational parameters, i.e., a kinetic energy cutoff of 396 eV and 18 irreducible k points. The three

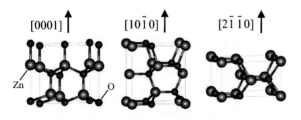

FIG.2. Schematic illustration of (0001)-, $(10\bar{1}0)$-, and $(2\bar{1}\bar{1}0)$-oriented ZnO strained layer superlattices

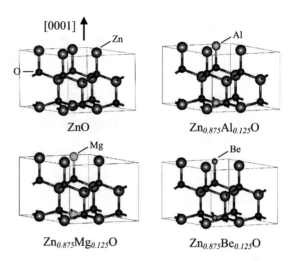

FIG. 3. Ball and stick structural representations of the four $Zn_{1-x}(Be,Mg,Al)_xO$ strained layer models considered in this study

models treated in the present work are optimized and described in Fig. 2.

$Zn_{1-x}(Be,Mg,Al)_xO$ are hexagonal crystals with six independent elastic stiffnesses C_{11}, C_{33}, C_{44}, C_{66}, C_{12}, and C_{13}. These elastic stiffnesses can be calculated through polynomial fitting to an energy-strain relation for the hexagonal crystals deformed using an appropriate strain tensor and carried out using the CAmbridge Serial Total Energy Package (CASTEP) [17]. Thus, to calculate the elastic stiffnesses of $Zn_{1-x}(Be,Mg,Al)_xO$ systems we adopt the $Zn_{1-x}(Be,Mg,Al)_xO$ crystallographic unit cell containing 8 formula unit (f.u.) of ZnO. The supercell consists of a 2×2×1 basic hexagonal primitive cell, which atomic positions were optimized by full relaxation to zero force positions as shown in Fig. 3. A 4×4×3 Monkhorst Pack (MP) grid in the first BZ and a kinetic energy cutoff of 300 eV were chosen to yield converged elastic stiffnesses through strain-stress theory implemented in the CASTEP.

III. RESULTS AND DISCUSSIONS

Epitaxial models with a variable and controllable range of compositions were constructed by using bulk crystalline configurations. The bulk strain energy is due to the (tri-axial) hydrostatically deformation and the epitaxial strain energy is due to the (bi-axial) epitaxially deformation. The bulk and epitaxial strain energies of all models are compared. Fig. 4 shows $q(a,[0001])$ in ZnO, $Zn_{0.875}Be_{0.125}O$, $Zn_{0.875}Mg_{0.125}O$, and $Zn_{0.875}Al_{0.125}O$ strained layer lattices over the wide strain range from -2% to 2%, indicating that hexagonal $Zn_{0.875}Al_{0.125}O$ is significantly softer than others in a film state. It is apparent that ZnO-based group-II-VI $Zn_{1-x}(Be,Mg,Al)_xO$ alloy systems become stiffer upon dilation (tensile strain). Our findings also indicate that the ZnO strained layers could be stabilized with adding the aluminum composition, which Al

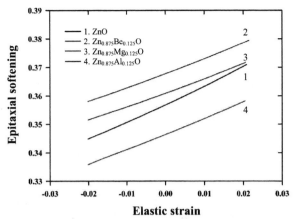

FIG. 4. $q(a_f,\hat{G})$ *vs* the elastic strain for (0001)-oriented ZnO (line 1), $Zn_{0.875}Be_{0.125}O$ (line 2), $Zn_{0.875}Mg_{0.125}O$ (line 3), and $Zn_{0.875}Al_{0.125}O$ (line 4) strained layer lattices.

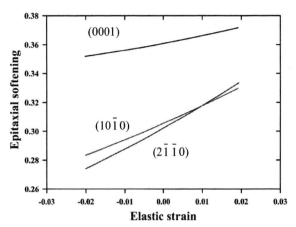

FIG. 5. The calculated epitaxial softening functions $q(a,\hat{G})$ versus epitaxial strain for (0001)-, $(10\bar{1}0)$- and $(2\bar{1}\bar{1}0)$-oriented ZnO.

is at substitutional sites in the underlying Zn lattice.

Furthermore, we are going to analyze the epitaxial softening of (0001)-, $(10\bar{1}0)$-, and $(2\bar{1}\bar{1}0)$-oriented ZnO strained layer superlattices for providing further insight into the stress/strain distributions in and around coherently strained ZnO-based systems. The results presented in Figure 5 show the elastically softest direction is $<2\bar{1}\bar{1}0>$ and the hardest direction is $<0001>$. The results of our calculations are consistent with reported experimental observations that $<2\bar{1}\bar{1}0>$-oriented ZnO is more stable than (0001)- or $(10\bar{1}0)$-oriented ZnO [18].

To further elucidate the behavior of longitudinal and shear modes under pressure we report the elastic constants (C_{11}, C_{33}, C_{44}, C_{66}, C_{12}, and C_{13}) obtained by calculating the change in energy with strain for six different strain configurations, as shown in Table 1. The pronounced softening of the longitudinal modulus C_{11} and C_{33} is found with adding the aluminum composition to ZnO. In addition, the table lists the bulk modulus K calculated from the fitting of total energy versus volume to the equation of state of a third-order Birch-Murnaghan. The relation $K = \frac{2}{9}(C_{11}+C_{12}+2C_{13}+0.5C_{33})$ provides a useful internal consistency check for calculated elastic stiffnesses. Using the C_{66}/C_{44} ratio as a shear-mode

elastic anisotropy criterion, we found that the ZnO and $Zn_{0.875}Al_{0.125}O$ have the weak elastic anisotropy (or isotropic elastic properties). The dopant compositions such as elemental Be and Mg have a direct influence on the elastic anisotropy.

The harmonic behavior of the $Zn_{1-x}(Be,Mg,Al)_xO$ is not well known. For solving this mystery, we employ the hexagonal harmonics elasticity theory [19]. As shown in Fig. 6, the $q_{harm}([0001])$ is the highest and the $q_{harm}([2\bar{1}\bar{1}0])$ is much softer than $<0001>$ or $<10\bar{1}0>$. Apart from $<2\bar{1}\bar{1}0>$, the $q_{harm}([11\bar{2}l])$ is the only other smaller value in $Zn_{1-x}(Be,Mg,Al)_xO$ layer lattices, where l is an arbitrary value. Our result is consistent with the experimental observations that the $(11\bar{2}3)$ faces of three-dimensional ZnO pyramids grow faster than (0001) and $(10\bar{1}0)$ faces [20].

IV. CONCLUSIONS

The epitaxial softening of $Zn_{1-x}(Be,Mg,Al)_xO$ strained layer lattices was studied by first-principles DFT-GGA calculations. We found that the ZnO strained layers could be stabilized with adding the aluminum composition. Dopant compositions such as elemental Be and Mg have a direct

TABLE 1. Summary of calculated elastic module for ZnO, $Zn_{0.875}Al_{0.125}O$, $Zn_{0.875}Mg_{0.125}O$, and $Zn_{0.875}Be_{0.125}O$.

	C_{11} (GPa)	C_{33} (GPa)	C_{44} (GPa)	C_{66} (GPa)	C_{12} (GPa)	C_{13} (GPa)	K (GPa)	C_{66}/C_{44}
ZnO	174.90	186.55	40.89	38.96	75.10	62.96	105.69	0.95
$Zn_{0.875}Al_{0.125}O$	168.85	173.13	37.32	37.37	75.36	62.02	98.27	1.00
$Zn_{0.875}Mg_{0.125}O$	186.20	192.15	36.63	51.48	88.09	72.28	110.41	1.41
$Zn_{0.875}Be_{0.125}O$	156.69	198.72	34.06	47.09	69.90	48.75	92.46	1.38

influence on the elastic anisotropy. The elastically softest direction is $<2\bar{1}\bar{1}0>$ and the hardest direction is $<0001>$. According to the harmonic elasticity theory, the $<11\bar{2}l>$ direction such as $<11\bar{2}3>$ is the other soft direction in $Zn_{1-x}(Be,Mg,Al)_xO$ layer lattices

ACKNOWLEDGMENTS

This work was supported by the National Science Council under Contract No. NSC98-2112-M-005-005-MY3, by the Ministry of Economic Affairs, Taiwan, Republic of China with Grant No. 97-EC-17-A-07-S1-097, and in part by the Ministry of Education, Taiwan, R.O.C. under the ATU plan. Computational studies were performed using the resources of the National Center for High-performance Computing.

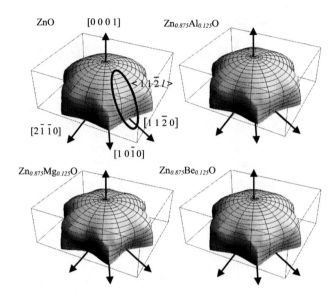

FIG. 6. 3D plot of epitaxial softening q_{harm} of the four $Zn_{1-x}(Be,Mg,Al)_xO$ strained layer models.

REFERENCES

[1] Y. Chen, D. M. Bagnall, H. Koh, K. Park, K. Hiraga, Z. Zhu, and T. Yao, J. Appl. Phys. 84, 3912 (1998).

[2] W. Xu, Z. Ye, T. Zhou, B. Zhao, L. Zhu, and J. Huang, J. Cryst. Growth 265, 133 (2004).

[3] H. W. Lee, S. P. Lau , Y. G. Wang, B. K. Tay and H. H. Hng, Thin Solid Films 458, 15 (2004).

[4] A. Tsukazaki, A. Ohtomo, T. Onuma, M. Ohtani, T. Makino, M. Sumiya, K. Ohtani, S. F. Chichibu, S. Fuke, Y. Segawa, H. Ohno, H. Koinuma and M. Kawasaki, Nat. Mater. 4, 42 (2005).

[5] T. P. Smith, W. J. Mecouch, P. Q. Miraglia, A. M. Roskowski, P. J. Hartlieb, and R. F. Davis, J. Cryst. Growth 257, 255 (2003).

[6] H. Xu, K. Ohtani, M. Yamao, and H. Ohno, Appl. Phys. Lett. 89, 71918 (2006).

[7] K. Koike, K. Hama, I. Nakashima, S. Sasa, M. Inoue and M. Yano, Jpn. J. Appl. Phys. 44, 3822 (2005).

[8] H. Tampo, H. Shibata, K. Maejima, A. Yamada, K. Matsubara, P. Fons, S. Niki, T. Tainaka, Y. Chiba, and H. Kanie, Appl. Phys. Lett. 91, 261907 (2007).

[9] Y. R. Ryu, J. A. Lubguban, T. S. Lee, H. W. White, T. S. Jeong, C. J. Youn, and B. J. Kim, Appl. Phys. Lett. 90, 131115 (2007).

[10] Y. Ryu, T. S. Lee, J. A. Lubguban, H. W. White, B. J. Kim, Y. S. Park, and C. J. Youn, Appl. Phys. Lett. 88, 241108 (2006).

[11] Y. Luo, J. Bian, J. Sun, H. Liang, and W. Liu, J. Mater. Process. Technol. 189, 473 (2007).

[12] V. Ozolins, C. Wolverton, and A. Zunger, Appl. Phys. Lett. 72, 427 (1998).

[13] G. Kresse and J. Furthmüller, Phys. Rev. B 54, 11169 (1996); G. Kresse and J. Furthmüller, Comput. Mater. Sci. 6, 15 (1996); G. Kresse and J. Hafner, J. Phys. Condens. Matter 6, 8245 (1994); G. Kresse and J. Joubert, Phys. Rev. B 59, 1758 (1999).

[14] D. Vanderbilt, Phys. Rev. B 41, 7892 (1990).

[15] J. P. Perdew and Y. Wang, Phys. Rev. B 45, 13244 (1992).

[16] J. P. Perdew, J. A. Chevary, S. H. Vosko. K. A. Jackson, M. R. Petersen, and C. Fiolhais, Phys. Rev. B 46, 6671 (1992).

[17] M. C. Payne, M. P. Teter, D. C. Allan, T. A. Arials, and J. D. Joannopoulos. Rev. Mod. Phys. 64, 1045 (1992); V. Milman, B. Winkler, J. A. White, C. J. Pickard, M. C. Payne, E. V. Akhmatskaya, and R. H. Nobes, Int. J. Quantum Chem. 77, 895 (2000).

[18] X. Y. Kong and Z. L. Wang, Nano Lett. 3, 1625, (2003).

[19] V. Ozoliņš, C. Wolverton, and Alex Zunger, Appl. Phys. Lett. 72, 427 (1998).

[20] J. B. Baxter, F. Wu, and E. S. Aydil, Appl. Phys. Lett. 83, 3797 (2003).

Magnetic properties of sputtered IrMn/CoFeB nanometric bilayers for exchange biased magnetic tunnel junctions

Himanshu Fulara, M. Raju, Sujeet Chaudhary, Subhash C. Kashyap and D. K. Pandya

Thin Film Laboratory, Indian institute of Technology Delhi, New Delhi-110016 (India)

ABSTRACT

In the present work, magnetic properties of sputtered IrMn (y nm)/CoFeB (x nm) bilayers were studied systematically as a function of annealing temperature, applied magnetic field and thickness of individual layers. The MOKE loops of the A- and B-series bilayers with x =7, 10, 15 and 20nm and y =5, 10, 15 and 20nm were obtained for both, as-deposited as well as magnetically-annealed films. The coercivity, as inferred from these loops, was found to increase by nearly three times with magnetic annealing. The coercivity values estimated from current in plane AMR measurements at room temperature match with the MOKE results.

Keywords: IrMn, MOKE, Thin Films, Exchange bias, MTJ

1 INTRODUCTION

Ultrathin magnetic multilayers find application in spintronic devices including spin valves and magnetic tunnel junctions (MTJs). In a spin valve [1], typically two ferromagnetic (FM) layers are separated by a metallic layer, while in an MTJ [2] the metallic layer is replaced by an insulator film also called as barrier. The electrical resistance of MTJs depends on the relative orientation of the magnetization of the FM layers. In order to obtain well separated magnetic states, an antiferromagnetic (AF) layer is deposited juxtaposed to one of the FM layers. Due to the exchange interaction at the AF/FM interface, the FM layer is pinned and the reversal of the magnetization (M) in this layer will occur at a high applied magnetic field, and hence the name pinned FM layer. The magnetization of the other so-called free layer FM layer can be reversed with a small applied magnetic field [3]. The exchange bias (EB) is manifested by a shift in magnetic hysteresis loop, enhanced coercivity, and asymmetry in the magnetization reversal process. Reported high interfacial exchange energy (J_k or H_{ex}), low coercivity (H_c), high blocking temperature (T_B), and good thermal stability of $Ir_{20}Mn_{80}$ thin films make it most suitable AF layer [1]. Magneto-Optical Kerr Effect (MOKE) is an effective technique to characterize such magnetic structures.

The aim of the present research is to study the effect of thickness of FM and AF layers on the coercivity of IrMn/CoFeB,and hence to pin the bottom FM layer in CoFeB/MgO/CoFeB magnetic tunnel junctions (MTJs). In this paper, we report systematic study of magnetic properties such as coercivity (H_C) and magnetoresistance (MR) of the $Ir_{22}Mn_{78}$(IrMn)/ $Co_{20}Fe_{60}B_{20}$ (CoFeB) bottom-configuration system (i.e., AF layer below the FM layered) as a function of annealing temperature, direction of applied magnetic field and thicknesses of individual layers.

2 EXPERIMENTAL DETAILS

The bottom configuration $Ir_{22}Mn_{78}$/ $Co_{20}Fe_{60}B_{20}$ bilayer system was sputter deposited at room temperature onto an oxidised Si(100) substrate. Two sets of bilayers: (A) IrMn(10 nm)/CoFeB(x nm), where $x = 7, 10, 15$ and 20 nm and (B) Si(100)/IrMn(y nm)/CoFeB(10 nm), where $y = 5, 10, 15$ and 20 nm, were prepared. The IrMn layers were deposited by DC magnetron sputtering at a working pressure of 1×10^{-2} Torr, and the CoFeB layers were deposited by ion-beam sputtering at a working pressure of 1.3×10^{-4} Torr. The layer thicknesses were estimated by using X-ray reflectivity (XRR) measurements. For each set of bilayers, we have three different sample type, namely (a) as-grown samples deposited at substrate temperature (T_S) of 25°C (RT); (b) Magnetically annealed samples, i.e., which are subjected to post-deposition annealing in presence of magnetic field of 3 kOe at a temperature (T_A) of 350°C for 30 min in a vacuum of 4×10^{-6} torr, followed by field cooling to RT; (c) the same procedure as in (b) except that T_A was 420°C.

The magnetic properties of the IrMn/CoFeB system were investigated at room temperature by Magneto-Optical Kerr Effect (MOKE). MOKE M-H hysteretic loops were obtained in the transverse geometry, with H in the film plane and perpendicular to the laser-beam incident plane. The magneto resistance (MR) of few samples was measured at room temperature in a four terminal geometry.

3 RESULTS AND DISCUSSION

The MOKE M-H loops obtained for both, as-deposited as well as magnetically-annealed IrMn (10 nm)/CoFeB (x nm) bilayers are presented in Fig. 1 for selected thicknesses (x=7, 10, 15 and 20 nm). We observe a smooth and relatively less sharp magnetization reversal in M(H) for as-deposited as compared that observed in magnetically annealed bilayers. The loops in latter cases were almost of square shape. The H_C obtained from these MOKE loops shows a peak and then decreases to its value as x increases above 10 nm (i.e., greater than AF layer thickness) in as-deposited as well as magnetic annealed samples.

This may be interpreted by considering the random field model [4] as described below.

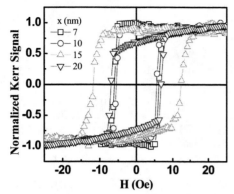

Fig. 1a: MOKE loops obtained for IrMn(10 nm)/CoFeB(x nm) as deposited bilayers.

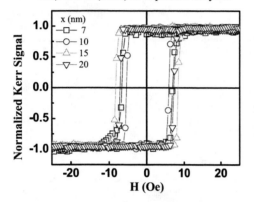

Fig. 1b: MOKE loops obtained for IrMn(10 nm)/CoFeB(x nm) bilayers magnetic annealed at 350°C

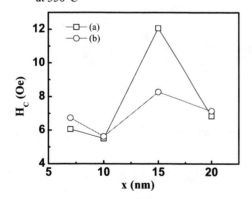

Fig. 1c: Dependence of H_C on the FM layer thickness for Si/IrMn(10 nm)/CoFeB(x nm) (a) as deposited, and (b) magnetic annealed (350°C).

At lower film thickness, the FM layer is expected to break into large number of domains, possibly of size much smaller than grain size. The random field arising due to inhomogeneities present in the AF grains can better act upon the FM layer and can delay the magnetisation reversal of the FM layer resulting in higher H_C. We believe that this is the case at an optimum thickness of 15 nm for the present

CoFeB/IrMn system, where the additive grain boundary pinning contributions causing further increase in H_C are at maximum. Since the ferromagnetic coupling between two neighbouring ferromagnetic blocks is proportional to the thickness of the FM layer, thicker FM layer produce larger domain sizes in the FM layer. For larger domains, the net coercive force (random fields from the underlying AF grains) acting on the FM layer is averaged out. Understandably, this would lead to small H_C as observed in the present case.

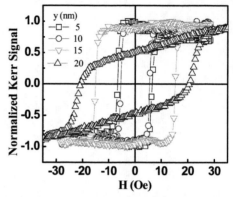

Fig. 2a: MOKE loops obtained for IrMn(y nm)/CoFeB(10 nm) as deposited bilayers

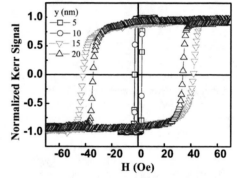

Fig. 2b: MOKE loops obtained for IrMn(y nm)/CoFeB(10 nm) bilayers magnetic annealed at 350°C

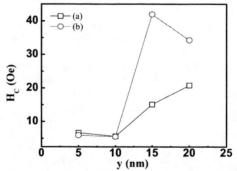

Fig. 2c: Dependence of H_C on the FM layer thickness for Si/IrMn(y nm)/CoFeB(10 nm) (a) as deposited, and (b) magnetic annealed (350°C).

Figure 2 shows the y-dependence of H_C for the IrMn(y nm)/CoFeB(10 nm) system. The higher H_C observed CoFeB film deposited on for thicker AF layer (i.e., y=15 and 20) in both as deposited and magnetically annealed samples can be attributed to the effective interaction of the random field in the thicker AF layer with the FM domains, since on increasing the AF film thickness the number of bulk inhomogeneities in the AF layer is increased leading to more effective pinning/biasing of the FM spins. The precise nature of the inhomogeneities in the AF layer (e.g., structural, thickness or compositional randomness), is unclear [5].

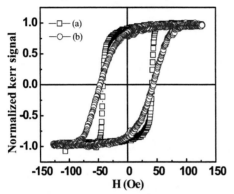

Fig. 3: MOKE MH loops for magnetic annealed Si/IrMn(15 nm)/CoFeB(10 nm), H applied in film plane but (a) parallel (b) perpendicular to the direction of annealing field (3kOe).

Figure 3 Shows the MOKE MH-loops recorded on the same magnetically annealed Si/IrMn(15 nm)/CoFeB(10 nm) sample by applying magnetic field (in film-plane) in a direction parallel (Fig 3a) and perpendicular (Fig 3b) to the direction along which 3 kOe field was aligned during magnetic annealing. It was observed that M-H loop measured along hard axis (normal to the direction of annealing field) deviates from the loop measured along the easy axis (in the direction of annealing field). This clearly shows that magnetic annealing conditions have induced anisotropy in these bilayers.

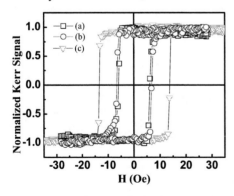

Fig. 4a: MOKE loops obtained for Si/IrMn(10 nm)/CoFeB(7 nm) under three different conditions: as-deposited at RT (a), magnetic annealed at 350°C (b) and 420°C (c).

Subsequently, the effect of varying the annealing temperature in presence of magnetic field was

also investigated in both the series, i.e., Si(100)/IrMn(y nm)/CoFeB(x nm), and Si(100)/IrMn(10 nm)/CoFeB(x nm). It was found that higher annealing temperature always resulted in increase of H_C (See Figs. 4a, 4b and Figs. 4c & 4d). Also, the increase was found to be more when thickness of IrMn layer was increased further to 20 nm.

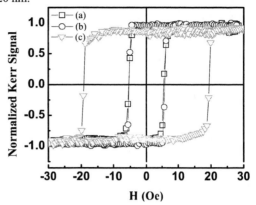

Fig. 4b: MOKE loops obtained for Si/IrMn(10 nm)/CoFeB(10 nm) under three different conditions: as-deposited (a), magnetic annealed at 350°C (b) and 420°C (c).

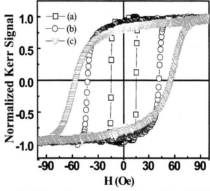

Fig. 4c: MOKE loops obtained for Si/IrMn(15 nm)/CoFeB(10 nm) under three different conditions: as-deposited at RT (a), magnetic annealed at 350°C (b) and 420°C (c).

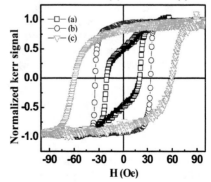

Fig. 4d: MOKE loops obtained for Si/IrMn(20 nm)/CoFeB(10 m) under three different conditions: as-deposited at RT (a), magnetic annealed at 350°C (b) and 420°C (c).

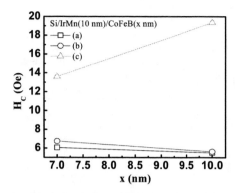

Fig. 5a: Variation in H_C with the FM layer thickness x under different annealing temperatures as-deposited at RT (a), magnetic annealed at 350°C (b) and 420°C (c).

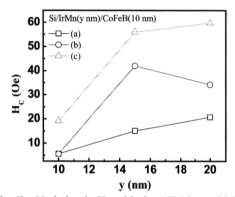

Fig. 5b: Variation in H_C with the AFM layer thickness y under different annealing temperatures as-deposited at RT (a), magnetic annealed at 350°C (b) and 420°C (c).

Figures 5a and 5b summarises the results of magnetic annealing temperature for two series samples. The substantial increase in H_C, irresepective of the thickness of the FM/AF layer, with increase in annealing temperature from 350°C to 420°C is likely to be due to the fact that the exchange bias effects get more pronounced as the magnetic annealing temperature is increased across the Neil temperature (T_N) of the underlying AF layer. In the present case, T_N of the IrMn layer is ~410-420°C, and hence magnetic annealing at 420°C shifted the magnetisation-reversal to higher field resulting in higher H_C~60Oe. We further stress here that other pinning mechanisms such as in homogeneities in the AFM-FM coupling or thermal fluctuations may also in part account for the observed higher H_C [1].

Figure 6 show the current-in-plane magnetoresistance (MR) measurements recorded at room temperature for the magnetically annealed (350°C) IrMn(5nm)/CoFeB(10nm) bilayer sample. The MR data has been recorded by applying magnetic field normal to the measuring current. The H_C values obtained from these measurements on different samples agree with that obtained by MOKE M-H results.

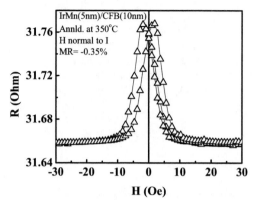

Fig. 6: MR curve (RT) of the magnetic annealed sample IrMn(5 nm)/CoFeB(10 nm).

4 CONCLUSIONS

We performed MOKE studies on IrMn/CoFeB bottom configuration systems with variable CoFeB or IrMn thickness, as a function of annealing temperature and applied magnetic field. The study reveals that higher IrMn thicknesses are needed to increase the coercivity of CoFeB films. Tunability H_C of FM electrode in IrMn/CoFeB exchanged bias systems can be realized by IrMn thickness. Further higher annealing temperature helps in increasing the coercivity by nearly three times compared to as deposited case. Coercivity values obtained from the MR measurements agrees with the coercivity obtained from MOKE measurements. The study promised that compared to Fe/MgO/Fe system, the coercivity values of top and bottom FM electrodes can be easily independently tuned to widely different H_C values in Si/IrMn(y nm)/CoFeB(x nm)/MgO(t nm)/CoFeB(z nm) TMR structures.

REFERENCES

[1] J. Nogues and I. K. Schuller, "Exchange bias", J. Magn. Mag. Matter.,Vol. 192, pp.203-232, 1999.

[2] J. S. Moodera, L. R. Kinder, T. M. Wong, and R. Meservey, "Large Magnetoresistance at Room Temperature in Ferromagnetic Thin Film Tunnel Junctions", Phys. Rev. Lett. Vol. 74, 3273, 1995.

[3] J.M. Teixeira, A.M. Pereira, J. Ventura, R.F.A. Silva, F. Carpinteiro, J.P. Araujo, J.B. Sousa, M. Rickart, S. Cardoso, R. Ferreira, P.P. Freitas, "Magnetic characterization of MnPt/CoFe bilayers using the MOKE technique", Vacuum, Vol. 82, pp.1486-1488, 2008.

[4] Z. Li and S. Zhang, "Coercive mechanisms in ferromagnetic-antiferromagnetic bilayers", Phys. Rev. B, Vol. 61, R14 897, 2000

[5] R. Jungblut, R. Coehoorn, M. T. Johnson, J. aan de Stegge, and A. Reinders, "Orientational dependence of the exchange biasing in molecular-beam-epitaxy-grown $Ni_{80}Fe_{20}/Fe_{50}Mn_{50}$ bilayers", J. Appl, Phys., Vol. 75, pp. 6659-6664, 1995.

Synthesis and Analysis of ZnO Nanorods using Molecular Beam Epitaxy

M. Asghar[*], A. Ali[*], F. Iqbal[*], Jon Merkert[**], A. S Gerges[***], M.-A. Hasan[***], M. Y. A. Raja[***]

[*]Department of Physics The Islamia University of Bahawalpur 63100, Pakistan
[**]University of North Carolina, Department of Chemistry, Charlotte, NC 28223, USA
[***]University of North Carolina, Department of Physics and Optical Sciences, Charlotte NC 28223, USA; Contact Emails: mhashmi@uncc.edu and raja@uncc.edu

Abstract

This paper presents a useful study on the synthesis and characterization of ZnO nanorods using (MBE). ZnO layers of thickness ≤500nm were grown on p-type silicon wafer Si (100) under conditions: substrate temperature $300^{\circ}C$, temperature of the Zn-Knudsen cell ~$286^{\circ}C \pm 0.5$, pressure of the chamber ~$4\times10^{-4} \pm 0.5\times10^{-4}$ mbar and oxygen plasma was generated by RF power supply operated at 300W. The as-grown ZnO layer was characterized by means of Fourier transform infrared (FTIR), scanning electron microscopy (SEM), Raman scattering and x-ray diffraction (XRD) and I-V/C-V measurements. FTIR displays a clear Zn-O bond at 406 cm⁻¹ excitation, large area SEM image demonstrates well defined nonorods of length ~ 100 μm while energy dispersive analysis x-ray spectrum (EDAX) reveals Zn-O at% as 56:44 that means the stiochiometric ratio is less than 1. Raman scattering exhibits nonpolar modes of Zn and O (sub-lattices) at 109 and 436 cm⁻¹ respectively which clearly justifies the presence of nanocrystals/nanoparticles and in addition it also yields A1(LO) mode (polar) associated with ZnO. XRD pattern shows that the preferable direction of the growth of ZnO is along (0002) direction however additional peaks (weak) are also present. From the existing literature, useful discussion has been carried out to justify the obtained results and henceforth interesting conclusion has been drawn for device quality material.

Keywords: MBE; ZnO nanorods; FTIR; SEM; XRD; Raman scattering

1 Introduction

ZnO nanorods have potential as one of the most important candidates for next generation opto- and nanoelectronics [1]. However, synthesis of high quality and stable ZnO nanorods remains a challenge thus far [2].

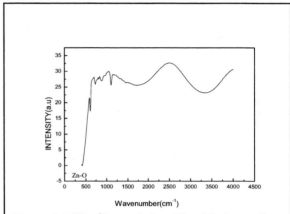

Figure 1 FTIR (Transmission %) of ZnO on Si to evidence Zn-O and Si-Si vibrational modes at well established frequencies *i.e.* 406 and 611 cm⁻¹.

Similarly fabrication of thermally stable and rectifying metal/nanorod-ZnO and Ohmic contacts also pose an uphill task. Furthermore, due to the presence of huge amount of interstitials and/or vacancies, ZnO has

intrinsically, n-type conductivity; as a result, it is difficult to dope it with p-type dopants [3]. Henceforth as way around such challenges, ZnO is being grown on friendly p-type substrates where GaN is more prominent. But, GaN itself is a costly material and therefore inexpensive templates are attracting the attention of the researchers for this goal. Various growth techniques are being attempted to grow improved quality ZnO nanorods but each of the method leaves defects in the form of vacancies and/or

interstitials in the layer. However, molecular beam epitaxy (MBE) because of its kinetic mode promises a better control and quality growth of the material.

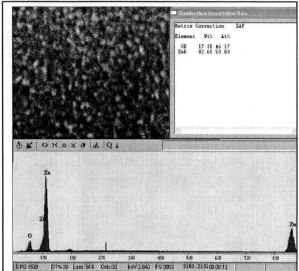

Figure 2 Self explanatory energy dispersive analysis x-rays (EDAX) of the as-grown ZnO/Si sample. The compositional ratio (At%) of Zn and O are listed on the right (upper) block ~ 54:46, which mean the growth is Zn-rich.

In continuation with our preliminary results presented in the 'International Nanotechnology Conference for Industries INCI 2009 held at King Saud University Saudi Arabia [4], this paper presents a study on the synthesis and analysis of ZnO nanorods using molecular beam epitaxy (MBE). ZnO layers of thickness ~1μm were grown on p-type silicon wafer Si(100) under conditions: substrate temperature 300 - 430°C, temperature of the Zn-Knudsen cell 300°C, pressure of the chamber 4×10^{-4} mbar and oxygen plasma was generated by RF power supply operated at 300W. The as-grown ZnO layers were characterized using Fourier transform infrared (FTIR), energy dispersive analysis x-rays (EDAX), atomic force microscopy (AFM) Raman and x-ray diffraction (XRD). FTIR displays a clear Zn-O bond at excitation 406 cm^{-1}, large area SEM image demonstrates well defined nanorods erected upwards while energy dispersive spectrum (EDAX) reveals Zn-O At% as 54:46 that means the stiochiometric ratio is less than 1. AFM confirms growth of nanorods however with the increase of thickness of the ZnO layers the rods are merged into bigger grains. XRD pattern shows that the preferable direction of the growth of ZnO is along (0002) direction which justifies the hexagonal structure of ZnO. Contrary to typical intrinsic carrier concentration of ZnO, the low value of the same (~ 10^{15} cm^{-3}) our sample reciprocates to the more quality ZnO layer. From the existing literature, useful discussion has been carried out to justify the obtained results and

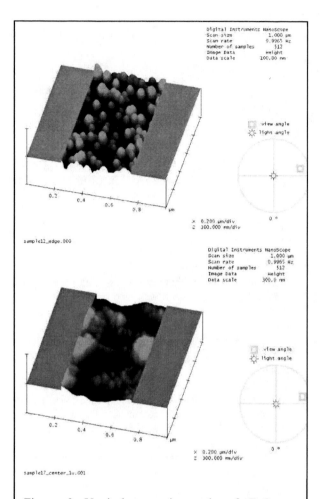

Figure 3 Vertical nano-size rods of ZnO are demonstrated in AFM 3D image (upper) and the merging of rods due to high growth temperature in (lower). The associated data such as scanning area and etc. with the measurement are depicted therein.

henceforth interesting conclusion has been drawn for device quality material. Experimental detail, results and discussion, and concluding remarks are described in sections II, III and IV, respectively.

2 Experimental

ZnO layers were grown on *p*-type silicon wafer Si(100) under conditions: substrate temperature 300°C - 430 °C, temperature of the Zn-Knudsen cell 300°C (Zn beads of purity 99.9999 were filled therein), pressure of the chamber during the growth was ~ (1-4) × 10^{-4} mbar and oxygen plasma was generated by RF power supply operated at 300W. Characterizations of the ZnO films were carried out with the help of following instruments/equipment: Tencor Alpha Step 200

Profilometer for Ellipsometry, Thermo Nicolet Nexus 870 FT-IR for FTIR measurement, PANalytical X'pert x-ray diffractometer (XRD) with Cu Kα radiation source for XRD measurements, JEOLJSM-5910LV for SEM, Horiba Raman Spectrometer for Raman and 72AD Boonton Capacitance meter for C-V measurements and Keithley 617 Programmable Electrometer for I-V characteristics. For I-V and C-V measurements, Ohmic contacts on ZnO/Si (Ag/Ni 2000:200Å)/Al (2000Å) were deposited under standard conditions.

3 Results and Discussion

Ellipsometery of the as-grown layers at four different points of the wafer covered by the substrate holder were used to measure the thicknesses of the layer which were estimated to be in the range of 0.95 – 7.45 μm. The variation in the layer thickness is related with substrate temperature (300 – 430°C). Figure 1 demonstrates transmission FTIR spectra of as-deposited samples, where we can see three prominent peaks at 404, 611 and 1150 cm^{-1}. Peak appearing at 404 cm^{-1} is related well established excitation of vibrating Zn–O bond [5] and those at 611 and 1150 cm^{-1} are due to Si–Si and C–O bonds, respectively. SEM of the ZnO layer shows plane view of vertically aligned nanorods of ZnO (c.f. left upper figure 2) and energy dispersive analysis x-rays (EDAX)

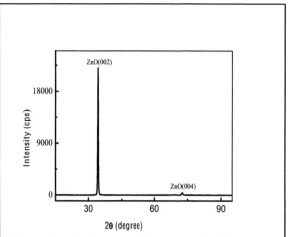

Figure 5 Single crystalline growth display of Wurtzite ZnO along 0002 direction. Si (004) peak is too small to appear in the XRD large y-scale spectrum. This observation is attributed to fairly thicker ZnO layer on Si substrate.

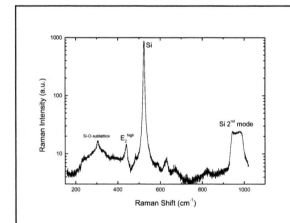

Figure 4 Typical Raman scattering due to ZnO/Si sample witnesses the non-poplar mode of ZnO which in way, justifies the presence on ZnO nano-crystallites in the layer.

results indicate that Zn-O At% is as 54:46 that means the stiochiometric ratio is less than 1 and the growth is Zn-

rich (see figure 2). As a matter of fact, the excess Zn contents are supposed to occupy O sites or interstitials to provide n-type conductivity of the layer (detail comes later). Figure 3 (upper and lower) display the AFM 3D pictures due to AFM measurements corresponding to thinner to thicker (0.95 to 7.45 μm) ZnO as grown layers respectively, where we can see that nanorods of size nearly 30 nm are vertically aligned but their grain size is increased due to merging process and eventually larger plateau like surface is developed. The plane view of the surface (1μm × 100nm) from AFM supports the SEM micrograph (not shown here). Figure 4 demonstrates Raman scattering exhibiting nonpolar mode of ZnO at 436 cm^{-1} which clearly justifies the presence of nanocrystals/nanoparticles and in addition it also yields A1(LO) mode (polar) associated with ZnO. Figure 5 depicts the XRD data of the ZnO/Si sample, where we can see that the preferential growth direction of ZnO is along (0002) *i.e.* along *c*-axis, in other word, this also means that Wurtzite ZnO has been grown on diamond Si. Since XRD spectra show only two peaks originating from ZnO: (0002), (0004) and Si (004) peak is too small to appear in the XRD large y-scale spectrum [6]. This observation is attributed to fairly thicker ZnO layer on Si substrate. Moreover, the single crystalline growth of hexagonal ZnO along 0002 direction is confirmed as well.

Typical C^{-2}-V measurements on n-ZnO-pSi are shown in figure 6 which reveals a one-sided junction and the depth profile of the ZnO layer indicates that at the interface free electron will be depleted and therefore the profile ascends towards the bulk (neutral region). The free carrier concentration measured from these data is 1±0.5 x 10^{15} cm^{-3}, the source of n-type conductivity is certainly due to the excess of Zn contents as pointed out by EDAX data. Since the *p*-Si used for the substrate has hole concentration 1×10^{17} cm^{-3}, this mean that about 90% of the depletion region should be in ZnO ($w_D \approx N_D/N_A$). This observation strengthens *Type II* heterojunction of ZnO/Si device as explained in figure 6. In this way, it is simply clear that we have succeeded to produce *p-n* junction on ZnO and Si, further studies will ascertain the optical properties of the device, reported elsewhere [7].

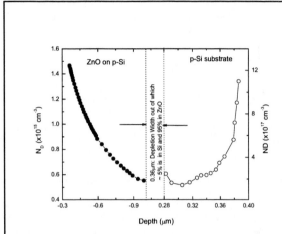

Figure 5 Typical *C-V* characteristics to justify successful donor/acceptor concentration depth profiles of *n*-ZnO/*p*-Si measured at room temperature.

3 Concluding Remarks

ZnO was grown on *p*-Si (1 0 0) by means of MBE in order to characterize the properties of the *p-n* junction. Prior to making Ohmic contacts on the two materials, the as-grown layer was characterized using FTIR, SEM/EDAX, AFM Raman and XRD. FTIR showed a well defined Zn-O bond, AFM demonstrates the formation of nanorods and their integration subject to the thickness of the ZnO layers and XRD revealed single crystalline Wurtzite structure of ZnO along 0002 direction. Raman scattering exhibits nonpolar mode of ZnO (sub-lattices) 436 cm^{-1} respectively which clearly justifies the presence of nanocrystals/nanoparticles. *C-V* measurements confirm the presence of *Type II* heterojunction between ZnO and Si where about 90% of the depletion width comes from ZnO.

Acknowledgements

The authors, M. Asghar, in particular, are thankful to Fulbright Organization (CIES) for financial support, Department of Electrical and Computer Engineering and Department of Physics and Optical Sciences at UNC-Charlotte and the Charlotte Research Institute UNC-Charlotte for providing research facility to carry out the activity. We are also especially grateful John Hudak and Alec Martin, the Faculty Associates for helping in experimental work.

REFERENCES

[1] S. P. Lau, H.Y. Yang, S.F. Yu, H.D. Li, M. Tanemura, T. Okita, H. Hatano, H.H. Hng, Appl. Phys. Lett., 87, 013104, 2005.

[2] D. C. Look, Mater. Sci. Eng. 80, 383, 2001.

[3] K. Nomura, H. Ohta, K. Ueda, T. Kamiya, M. Hirano, and H. Hosono, Science 300, 1269, 2003.

[4] M. Asghar, H. Noor, M-A. Hasan, and M. Y. A. Raja, INCI 2009 (International Nanotechnology Conference for Industries), King Saud University Riyadh, Saudi Arabia, April 05-07, 2009.

[5] M. Andres-Verges, A. Mifsud, C. J. Serna, J. Chem. Soc., Faraday Trans. 86, 959, 1990.

[6] A. Umar, S. H. Kim, E.–K. Suh, and Y.B. Hahn. Chem. Phys. Lett. 110, 440, 2007.

[7] K. Mehmood , *et al* (manuscript under progress).

Novel Nanoparticles Dispersing Beads Mill with Ultra Small Beads and its Application

M. Inkyo

Kotobuki Industries Co., Ltd, Ohashi Gyoen Eki-Bldg. 2F, 1-8-1 Shinjuku-ku,
Tokyo 160-0022, Japan, inkyo.m@kotobuki-ind.jp, www.kotobuki-ind.jp

ABSTRACT

Kotobuki Industries has developed a new type of beads mill which has successfully solved many problems related to nanoparticle dispersion such as reaglomeration and damage to the crystal structure of nanoparticles. The Ultra Apex Mill uses centrifugation technology which enables the use of ultra small beads with a diameter of less than 0.05mm for the first time in the world. Now 0.015mm beads are available. This technology has pioneered practical applications for nanoparticles in various areas, such as composition materials for LCDs, ink-jet printing, ceramic condensers and cosmetics.

Fig.1 UAM-5

Keywords: beads mill, dispersing, dispersion, nano, fine particle, nanoparticles, beads

1 STRUCTURE and OPERATION

The Ultra Apex Mill (UAM) consists of a vessel with a cooling jacket and rotor. The bead separator is located at the top of the rotor and rotor pins are located on the lower section of the rotor. The separator has the function of separating the beads using centrifugal force. The rotor pins are used to agitate the beads. Both the rotor and the vessel are sealed with mechanical seals. The dispersion of nanoparticles occurs during operation as the beads are agitated

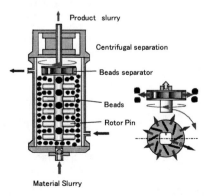

Fig.2 Structure of UAM

by the rotor pins inside the mill chamber while the material slurry is fed into the mill chamber from bottom. The slurry is dispersed by the impulse from the beads inside mill. Finally the milled slurry will exit the mill chamber at the top after passing through the separator to remove the beads. The difference in density will allow the milled slurry to pass through the separator while the beads are kept inside the mill chamber.

2 FEATURES

2.1 Utilizing centrifugal force for bead separation

The use of ultra small beads (0.015mm~0.1mm size) can only be realized using centrifugal force for bead separation. Despite the probability, bead leakage is rare and easily controlled.

2.2 Realizing nano particle dispersion

Nano particle dispersion is achieved by using ultra small beads. Dispersing speed is very fast and attained particle sizes will also become much smaller. Dispersion down to primary nano particle size is now possible.

2.3 Feed pressure is very low

Pressure loss is kept very low when using centrifugal separation. Feed pressure of the UAM is around one tenth when compared with conventional mills. This allows continuous, stable operation.

2.4 No need to pre-disperse a material slurry

Clearance of all internal passages is set to a minimum of 5mm. This eliminates the need to pre-disperse the material slurry and also contributes to the stable operation of the mill.

2.5 Simple Structure

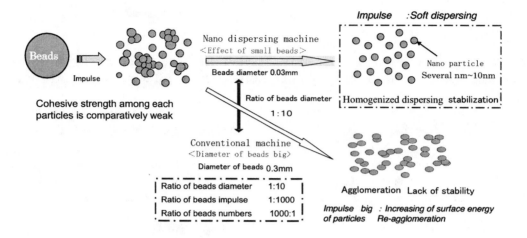

Fig.3 Principle of Nano particle dispersion

The UAM structure is simple. So, disassembly, cleaning and maintenance is very easy. The vertical type mill offers many operational advantages to conventional mills.

3 PRINCIPLE of NANO PARTICLE DISPERSION

Why is the Ultra Apex Mill superior in comparison with conventional machines in the nano dispersing field? The reason is simply because the Ultra Apex Mill can use the smallest beads in the world.

When dispersing agglomerated nano particles, the beads impulse is very low because the beads are very small. There is little influence on the nano particle. The surface energy of the nano particles hardly increases and the property is not likely to change. As a result, stable nano particles can be achieved without re-cohesion. On the other hand, conventional machines cannot use small beads. So, the impulse of the beads is very high and the result is strong influence on the nano particles. The surface energy increases and the properties are changed. As a result, stable nano particles cannot be achieved. Here, let us compare the difference of the dispersing performance between 0.03mm beads and 0.3mm beads.

The rate of beads diameter is one-tenth. Rate of beads impulse is one-1000th. Rate of beads number is 1000 times. That means the Ultra Apex Mill has one-1000th impulse, and is 1000 times more likely to make contact with the nano particles. It proves to be very effective. See Fig.3

4 The difference of structure between UAM and Conventional Mill

The Ultra Apex Mill is able to use very small beads because the bead separation is accomplished utilizing centrifugal force. Conventional mills are not likely to achieve stable operation because of clogging issues with the screens used for beads separation. See Fig.4.

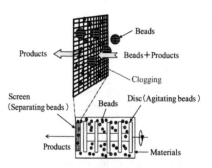

Fig.4 Conventional beads mill with screen

5 PRACTICAL EXAMPLE

5.1 Dispersion of TiO2

We would like to demonstrate the high-dispersing performance of the UAM using data from an actual experiment and then compare it with a conventional beads mill. The UAM used 0.03mm beads and the conventional beads mill used 0.3mm beads. As a result, there was the big difference in performance. The UAM achieved a primary particle size of 15nm and the slurry was transparent. On the other hand, in the conventional beads mill the particle size of product attained was 150nm and it was re-cohesive. You can see the result from TEM photograph too. See Fig.5

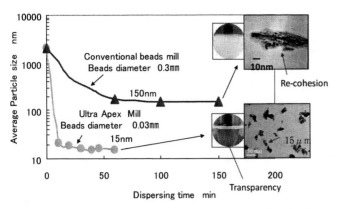

Fig.5 The effect for particle size as related to beads size

5.2 Dispersion of Organic pigment

First, the peripheral speed of the rotor-pins was set to 11m/sec. Under this condition, the particle size remained unchanged after reducing to 70nm. Next, raising the peripheral speed to 13m/sec immediately caused the particle size reduced to 10nm. Finally, the peripheral speed was increased to 14m/sec. Due to the increased peripheral speed, agglomeration of the particles occurred. See Fig.6.

Fig.7 shows the change of the particle size distribution. Particle size distribution is very sharp at optimum dispersing time.

Judging from these results, you can see the optimal rotation speed exist even if using ultra small beads.

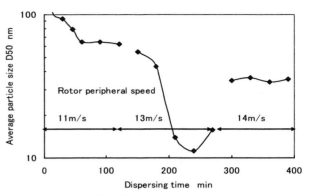

Fig.6 The effect of rotation speed of rotor

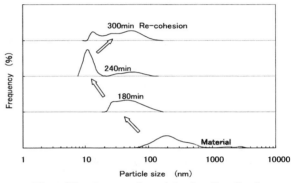

Fig.7 The change of particle size distribution

6/ APPLICATION

Material	Application	Purpose of dispersing	Attained particle size nm
Organic pigment	LCD color filter	Transparency, Contrast-up, Grain size refinement	20
	Ink jet printer ink	Grain size refinement	30
ZnO	Cosmetics : UV cutting	Transparency , Grain size refinement	10
Al₂O₃	Silicon wafer polishing	Grain size refinement, Sharpness of PSD	15
	Hard coating	Grade-up of hardness	10
ZrO₂	Anti-reflection coating	High refractive index, Grain size refinement,	15
SiO₂	Nano-composite	Strength up, Transparency	10
	Abrasive finishing	Grain size refinement, sharpness of particles size	10
ITO, ATO	Transparent electrode	Conductivity, Transparency	30
	Heat absorbing film	Transparency	30
	Anti-reflection coating	Transparency	30
TiO₂	Photocatalyst	Keeping crystal form, Grain size refinement	10
	Cosmetics : UV cutting	Transparency	10
BaTiO₃	MLCC (Ceramic condenser)	Keeping crystal form, Primary particle size	20, 200
Nano diamond	Polishing	Sharpness of PSD, Grain size refinement	8
Carbon nano horn	Fuel cell catalyst	Leveling	5
Cu paste	Print base wiring	Keeping primary particle form	200
Ni	Print base wiring	Keeping primary particle form	200
Nano-Ag	Transparent electrode	Conductivity, Transparency	10

SUMMARY

The Ultra Apex Mill has realized a breakthrough in conventional dispersing limits. Particle size limitation in the nano-fields has become much smaller by using 0.015 ~ 0.03mm dia. beads. Re-agglomeration is prevented and new products can be developed from common materials. The basic concept is that the impulse power of the small beads is very low which allows the crystal shape of the material to be maintained during milling. This allows the UAM to create new products with special properties that are not attainable with conventional machines. This breakthrough technology has attracted much attention in the nano technology field. As a result, Kotobuki Industries has supplied over 400 machines to various universities, research institutes and private companies around the world.

REFERRENCES

1) M. Inkyo, et al.: "Experimental investigation of nanoparticle dispersion by beads milling with centrifugal bead separation", J. Colloid and Interface Sci., **304** (2006)

2) M. Inkyo, et al.: "Beads Mill-Assisted Synthesis of Poly Methyl Methacrylate (PMMA)-TiO2 Nanoparticle Composite", Ind. Eng. Chem. Res., **47**, 2597-2604 (2008)

CONTACT

Kotobuki Industries Co., Ltd, Bellevue, WA, USA, Thomas Hagarty, tom@kotobuki-usa.com

Marubeni Techno-Systems America, Inc., Newport Beach, CA, USA, Tadahito (Tod) Arai, tadahito.arai@mts-america.com

Evaluation of Tools for Nanoparticles-modulated Hyperthermia by Atomic Force Microscopy

I.V.Reshetov*, V.D.Reva**, V.I.Chissov*, O.V.Matorin*, A.V.Korizky* and S.S.Sukharev**

* P.A.Hertzen Moscow Research Oncological Institute, 2-nd Botkinsky proezd 3, Moscow 125248, Russian Federation, hertz@portal.ru
** IPK FMBA of Russsia, Volokolamskoye shosse, 30, Moscow, Russia 123182

ABSTRACT

In 2004-2005 authors carried out the experiments [1] which make it possible to expect increase in the effectiveness of the tumor treatment by the method of percutaneous radiofrequency ablation (RFA): the tissue-damaging effect of RFA-induced hyperthermia does increase in the presence of nanoparticles.

This article told how Atomic-Force Microscopy (AFM) can be used for evaluation and characterization of nanoparticles supposed to be the modulators of hyperthermia and for studying of dynamics of an electrodes surface change.

Keywords: AFM, hyperthermia, percutaneous radiofrequency ablation, nanoparticles

1 AIM

Any acting nanoparticles, including ones, supposed to be used as hyperthermia modulators, must be introduced into an organism in the form of medicinal formulation, which should ensure:

1./ Stable concentration to guarantee the dosage control.
2./ Protection from seizure and neutralization by the natural shielding forces of organism.
3./ For the nanoparticles the special importance acquires the need of averting the formation of conglomerations.

Besides, it is necessary to evaluate the dynamics of a change in the expensive electrodes, used for the hyperthermia induction, surface properties.

Some of the enumerated properties can be evaluated by measurements with the aid of the AFM: sizes and concentration of nanoparticles, their capability to form the conglomerations, the dynamics of a change in the electrode surface properties.

The aims of the work were: the development of the methods of this estimation and their approbation based on several configurations of hyperthermia-modulating tools.

2 MATERIALS AND METHODS

AFM-system NTEGRA Prima, NT-MDT Co., Zelenograd, Russia.

Particles of rhenium sulfide and particles of silver with sizes of 80 and 120 nanometers, dispersed in different gels. Electrodes and temperature-sensitive elements for the universal complex for the RFA called "Metatom-2" manufactured by "Technosvet", Russia.

Development or selection of the specimen preparation methods

The suspension of nanoparticles in viscous fluid or gel was placed by thin layer to the microscope slide and was investigated by the AFM in the semicontact regime. It turned out that preparations on the basis of dextran and glue formed the steady structures, which could be repeatedly scanned with the reproducible results.

The estimation of the regular flat structures sizes

This was accomplished by repeated measurements of their linear dimensions (special function of regular NTEGRA-system software). The second method of evaluating the sizes of flat formations consisted in the measurement of the linear dimensions "of the disagreement zones" or, in other words, the zones of the artifacts appearance. The fact is that before scanning the special semiautomatic procedure selecting the parameters of the Cantilever fluctuations is carried out. With a substantial change of the surface properties during scanning parameters selected did not ensure shaping of the correct relief of surface (appearance of the artifacts in the form of pyramids, figure 1).

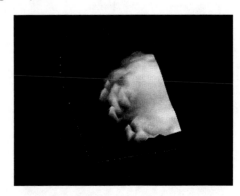

Figure 1: Example of "disagreement zone". Such sections treated as the zones of the protrusion of the solid nanoparticles above the surface of viscous phase.

Obviously, it is impossible for such sections to determine the height of the peaks, since the mentioned pyramids reflect not the components of surface, but the internal processes of the scanning system auto-oscillations.

Estimation of the vertical sizes of formations

The real height of formations is not equal to the height $h1$ of the highest point of formation above the level of colloid. For its estimation it is necessary to determine the thickness of colloid $h2$ and to add it to $h1$. We have found the areas with destroyed colloid (figure 2),

Figure 2: Area of destroyed colloid suitable for determination of colloid level.

where $h2$-estimation is possible in the sections (figure 3).

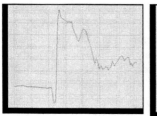 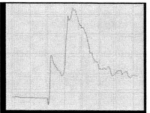

Height - about 1 micrometer   Height – about 0,6 micrometer

Figure 3: Determination of colloid thickness.

Estimation of the internal structure of formation

Characteristic vertical and horizontal dimensions must be close to each other for the solid structures, while for the conglomerations the horizontal sizes are expected to predominate above the vertical (figure 4).

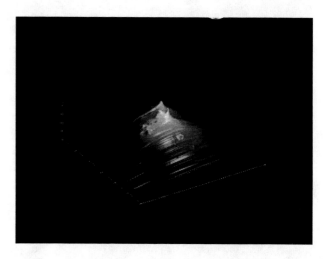

Figure 4: Example of conglomeration. Horizontal sizes are about 20 micrometers, vertical sizes – about 2 micrometers.

Estimation of the distances between the formations.

The minimum distance between the formations correlates with their sizes and can be determined by regular software of the NTEGRA system during image processing.

3   ESTIMATION OF ELECTRODE SURFACE DYNAMICS

The presence of two sets of the parameters, which characterize its form, is the special feature of object of study (the electrode).

The first one defines the long tube with diameter 1-2 millimeters, being seen by the naked eye its surface appears to be smooth.

The NTEGRA system does not have equipment for fastening of this type of specimens (and this should be considered as a deficiency in the context of the conducted study). Tis a the reason why during each scanning the larger or smaller (always random) inclination of the subject of studies relative to the plane in which cantilever moves has been demonstrated. The circumstances named lead to the appearance of the regular artifacts (illustrated in figure 5) during each scanning.

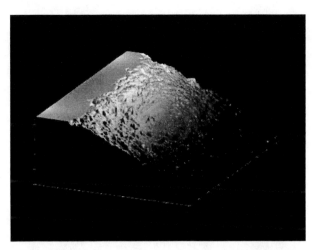

Figure 5: Regular artifacts: 1. Plane above illustrates the exit of the scanning system beyond the limits of linear interval because of too large difference in the heights of the different points of specimen. 2. The decrease of flat surface width occurs because of nonstandard fastening of model (inclination in the direction of the surface expansion).

The second set of the parameters characterizes the fine structure of the electrode surface, whose change under the influence of repeated cycles of application - sterilization is the object of study. The image of inclined fragment in figure 4 is apparently realistic taking into account the difference of vertical and horizontal scales. However, it is easy to note that the thin details, supposed to be the object of study, are located not on the plane, but on the complex surface, which is the part of the tube of electrode.

Software tools of NTEGRA system can convert complex surface into the plane. In particular, figure 6 shows the surface from figure 5, scanned in the regime "Subtraction of 4th order curve".

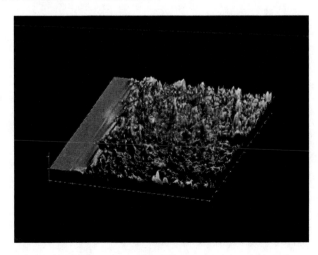

Figure 6: The surface from figure 4, scanned in the regime "Subtraction of 4th order curve".

After scanning specimens in the regime mentioned the following parameters which characterize the fine structure of the surface were calculated:

- Rq - Root mean square roughness, ISO 4287/1, nm;
- Ra - Average roughness, ISO 4287/1, nm;
- Rz ten point height, ISO 4287/1, nm. The parameter expresses surface roughness by the selected five maximal heights and hollows, nm.

Tables 1 and 2 present the obtained characteristics of two examples of electrodes, obtained in three stages. First stage - immediately after removing from factory package (columns 1 in both tables), second stage - immediately after the first use for the RFA - induction (columns 2 in both tables), third stage - after the first sterilization of electrodes (columns 3 in both tables).

	Stages of lifecycle of electrode		
	1	2	3
Rq	66,99	123,91	67,29
Ra	51,3	91,81	46,80
Rz	948,25	2057,79	599,05

Table 1: Surface dynamics of the first electrode.

	Stages of lifecycle of electrode		
	1	2	3
Rq	28,79	444,40	48,47
Ra	21,0	346,01	35,37
Rz	164,61	2237,32	412,50

Table 2: Surface dynamics of the second electrode.

The substantial increase in all the roughness parameters was noted after use of electrodes.

- All parameters, which characterize the roughness of the surface of electrode, increased after its use for RFA-induction.
- After electrode sterilization the parameters of roughness of the first electrode surface practically returned to the initial values (prior to the beginning of use). The roughness parameters of the second electrode also substantially decreased but remain above initial.
- It is possible to guess that the elements, which determined the high roughness of the surface of electrode, appeared after the first use, disappeared as a result of sterilization (fragments of the carbonized tissues?).
- If the previous thesis is correct, then the roughness of the surface of electrode after sterilization can characterize the quality of the sterilization

4 CONCLUSIONS

In the work are proposed the methods of evaluating the sizes of the nanoparticles, dispersed in the viscous carriers, and also the estimations of the internal structure of larger formations.

The electrode surface dynamics discovered in our studies has obvious interpretation and corresponds to the expectations.

Methods are approbated based on several configurations of hyperthermia-modulating tools.

REFERENCES

[1] V.I. Chissov, I.V. Reshetov, P.N. Luskinovich, A.A Shepelev, G.D. Efremov, O.V.Matorin, S.S. Sukharev. The Evolution of Views of in vivo Use of Nanoparticles for Tumor Treatment. Technical Proceedings of the 2005 Nanotechnology Conference and Trade Show, Nanotech2005, vol. 1, Pages 31-34

Laser-Induced Breakdown Spectroscopy (LIBS) for on-line monitoring and workplace surveillance of nanoparticle production processes

Tanguy Amodeo*, Christophe Dutouquet*, Jean Paul Dufour**
Olivier Le Bihan* and Emeric Fréjafon*

*INERIS, parc technologique ALATA BP2, F-60550 VERNEUIL EN HALATTE, France
tanguy.amodeo@ineris.fr
**CILAS, 8 avenue BUFFON, F-45063 Orléans, France, dufour@cilas.com

ABSTRACT

Research studies up to technical optimisations were realized on laser / plasma / particle interactions in order to further achieve LIBS measurements with optimum efficiency. Polydisperse and monodisperse flows of salt and metallic particles with sizes ranging from 40 nm up to few μm produced by two different particle generators were introduced inside a cell where they were vaporized by the laser induced plasma for LIBS analysis purposes. Time-resolved emission spectroscopy measurements were carried out and the influence on the LIBS signal of parameters such as chemical nature of particles, their concentration, laser wavelength, laser energy, kind of background gas was investigated. Then, calibration curves and limit of detection have been investigated for a wide range of metallic particles (Ti, Al, Cu) [1,2]. These results allowed to make a first assessment of LIBS potentialities for manufactured nanoparticle detection in workplace.

Keywords: LIBS, laser, plasma, spectroscopy, on-line monitoring, workplace surveillance, nanoparticles,

1 INTODUCTION

Today, nanotechnology is a growing field of research and nanoparticle-based material production is expected to soar in years to come. With the emergence of nanoparticle-based materials, concerns regarding health and safety have been brought to the fore.

Little is known yet about potential health effects nanoparticles might induce. Due to their small sizes, nanoparticles are more likely to get past biological barriers and penetrate much deeper into organs (e.g. the lungs) than their micrometric counterparts. Inhalation, ingestion and skin absorption are possible ways of penetration through the human body. The potential dangers of the engineered nanoparticles stem from their unique physical and chemical properties conferring them specific properties in the way they interact with biological systems. Nanoparticle characteristics such as chemical composition, surface area, solubility may influence toxicity [3].

In consequence, there is a matter of urgency to assess the risks associated to nanoparticle exposure. Though research is still currently under way to secure nanoparticle production processes, the risk of accidental release is not to be neglected. Thus, there is an urgent need for the manufacturers to have at their command a tool enabling leak detection in-situ and in real time so as to protect workers from potential exposure. Several reports and articles [4] have emphasized the lack of appropriate metrology tools for workplace surveillance. At present time, several expensive and sophisticated off-line systems exist such as microscopic techniques (STEM, STM or PEEM) giving structural information that can be combined with electron and X-ray diffraction for information on compositions (RHEED, LEED, XRD). There are also in-situ spectroscopy (ATR, Raman, GD-OES) or X-rays photoelectron spectroscopy (XPS), Auger electron spectroscopy (AES), photoelectron emission spectroscopy and photon correlation spectroscopy (PCS) but most of them suffer of high detection limit and might depend on nanoparticle shape or size.

Consequently, there is currently no device that directly measures worker exposure to, and the toxicity of, engineered nanoparticles, with monitoring being limited to expensive laboratory analysis of ambient air. As a consequence, most of the available tools do not allow differentiating manufactured nanoparticles from those of background air, thereby rendering targeted nanoparticle detection arduous. Such problem may be addressed by chemically identifying nanoparticles. To achieve this goal, the LIBS (Laser-Induced breakdown Spectroscopy) technique was deemed as a potential candidate.

2 LIBS PRINCIPLE

LIBS measurements consist in focusing a powerful laser pulse on a material which elemental composition is to be determined. At the focus spot, the matter whatever its state be (solid, liquid, gas, aerosol) is strongly heated resulting thereby in the generation of a hot and luminous ionized gas called plasma. Elemental composition of the irradiated target is then determined through plasma analysis by optical emission spectroscopy.

LIBS displays advantages of great interest for industrial applications. Detection of all the elements is possible at various pressure conditions. In addition, samples do not need preliminary preparation making LIBS eligible for on-line monitoring. It is a multielemental analysis technique with simultaneous detection of all the elements contained in the sample. Being all optical, it is not intrusive requiring only optical access to the sample and thus allowing in-situ measurements. LIBS is potentially fast and suitable for real-time monitoring, its speed depending on the number of laser shots required to obtain reliable results. Remote or stand-off analyses are even possible, as needed.

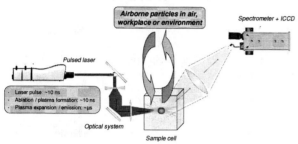

Figure 1 : description of the LIBS principle, case of nanoparticle detection in air.

These qualities are advantages over other techniques as the LIBS analyzer is intended to be operated on the production sites: no sample preparation required, in-situ and real time detection and survey for a wide range of on-site application.

3 RESULTS

Polydisperse flows of nanoparticles of various elemental compositions with median and maximum sizes inferior to 100 and 300 nm respectively produced using two different nanoparticle generators (a TSI atomizer and a PALAS spark generator) were introduced in the LIBS cell for detection limit determination. The LIBS results were expressed in terms of mass concentration, thereby facilitating comparison with exposure limit values issued by HSL (Health and Safety Laboratory, UK) or INRS (French Institute for Health and Safety at Work, France). So far, concentrations such as 4 milligrams per cubic meter for fine TiO_2 corresponding to the respirable fraction for a long term exposure limit (8-hour reference period) are advised for worker protection (see table 1, exposure limits values).

In order to evaluate the limits of detection, the nanoparticle sizes and total mass concentrations were lowered until detection was no longer possible. Experimental parameters as energy, laser wavelength, time delay, gate width and number of accumulation were optimized for each element.
As major results, polydisperse nanoparticles with diameter smaller than 300 nm were detected leading to a detection limit corresponding to an estimated mass ranging from a few micrograms per cubic meters (Sodium) to few tens (copper) up to a few hundreds (aluminum) of microgram

per cubic meters. The repeatability assessed during these measurements was found to be no more than 25 % for the whole campaign of experiments.

Monodisperse particles of sodium and copper with diameter ranging from 40 nm to few µm have been generated in order to prove that LIBS intensity is linear as a function of mass concentration indicating that no particular particle size effects take place under our experimental conditions, namely considering particles smaller than 2,5 µm which is in accordance with the inhalable fraction.

Thus, most LIBS response signals were found below the occupational threshold limit values. However, one could argue that these exposure limits might be revised downwards in the near future as they were not specifically released for nanoparticles.

No specific regulation has hitherto been enforced regarding exposure to nanoparticles. As a consequence, these will probably be lowered in the near future.

Materials	LIBS Detection limit (µg/m^3)	Exposure Limit Values (µg/m^3)	Guideline EU-OSHA « Nano : 0,066 »
Al	250	4000	264
Cu	50	1000	66
Ti	280	4000	264
Ca	10	4000	264
Mg	10	4000	264
Cd	300	30	1,98
Cr	45	500	33
Fe	200	5000	330
Ni	250	100	6,6
Si	100	4000	264

Table 1 : LIBS detection limits for some nanomaterials, actual associated occupational exposure limits and an example of future occupational exposure limits guidelines dedicated to nanoparticles (EU-OSHA)

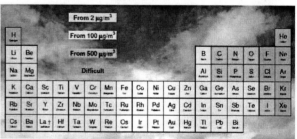

Table 2 : overview of LIBS expected detection limits for all elements

The above tables show detection limits obtained for some compounds, actual exposure limit and an extrapolation to future exposure limit assuming that specific surface is the key parameter in term of toxicity assessment.

4 CONCLUSION AND PERSPECTIVES

The results obtained by INERIS show a strong improvement of the continuous monitoring of aerosol containing engineered nano and ultra-fine particles. The sensitivity and detection limits are below the actual recommended level in the micro range and could be compatible with future « nano » legislation.

This is a first and very important step towards legislation and effective control of airborne particles concentrations in air, workplaces and environment.

CILAS & INERIS join forces to make this innovative technology available for nano-community faced with exposure risk. <u>Thanks to a collaborative program, we are working on a commercially available instrument.</u>

The major advantages of this product are the following:
-Multi element detection
-Continuous measurement in real time of the concentration
-Distinction between the background natural NP & engineered NP
-Easy to use, reliable, affordable

Figure 2: Safe-Air, description of the 'airborne real time selective particle detector'

REFERENCES

[1] T. Amodeo, C. Dutouquet, F. Tenegal, B. Guizard, H. Maskrot, O. Le Bihan, E. Frejafon, On-line monitoring of composite nanoparticles synthesized in a pre-industrial laser pyrolysis reactor using Laser-Induced Breakdown Spectroscopy, Spectrochimica. Acta, Part B 63 (2008) 1183-1190

[2] T. Amodeo, C. Dutouquet, O. Le Bihan, M. Attoui and E. Frejafon, On-line determination of nanometric and sub-micrometric particle physicochemical characteristics using spectral imaging-aided Laser-Induced Breakdown Spectroscopy coupled with a Scanning Mobility Particle Sizer, Spectrochimica Acta Part B 64 (2009) 1141–1152

[3] Approaches to safe nanotechnology: an information exchange with NIOSH, version 1.1, July 2006

[4] A. Franco, S.F. Hansen, S.I. Olsen, L. Butti, Limits and prospects of the "incremental approach" and the European legislation on the management of risks related to nanomaterials, Regulatory Toxicology and Pharmacology 48 (2007) 171-183

[5] List of approved workplace exposure limits (as consolidated with amendments October 2007), EH40/2005 Workplace exposure limits, Health and Safety Laboratory (HSL)

[6] Valeurs limites d'exposition professionnelle aux agents chimiques en France, ED984 june 2008, INRS (Institut National de Recherche et de Sécurité, French Institute for Health and Safety at work, France)

Application of Nanoindentation for Constituent Phases Testing in Ceramic-Metal Composites

Irina Hussainova[*], Iwona Jasiuk[**] and Medhat Hussainov[***]

[*]Tallinn University of Technology, Department of Materials Technology, Estonia, irhus@staff.ttu.ee
[**]University of Illinois, Department of Mechanical Science, Urbana, IL, USA, ijasiuk@illinois.edu
[***]University of Tartu, Institute of Physics, Estonia, medhat.hussainov@ut.ee

ABSTRACT

In this study, mechanical properties of the constituents of three-phase composite based on chromium carbides were measured by means of nanoindentation. Different loading scenarios were applied to different phases testing. The semi-ellipse method for accounting of pile-ups was applied to determine the hardness and modulus of elasticity of the constituents. After the contact area re-consideration, the properties of phases show good agreement with those obtained for the bulk form of the materials. Difference in values that are commonly referred and obtained during the test may point to the formation of new phases during sintering.

Keywords: nanoindentation, composite, hardness, pile-up, modulus of elasticity

1 INTRODUCTION

Many structural materials represent multiphase heterogeneous composites to tailor the properties needed for a specific application. Composite materials with well-controlled microstructures allow obtaining unique combinations of mechanical and/or thermal properties of a final product. Therefore, development of advanced composites requires a detailed knowledge of their constituents' properties as they are often somewhat different from the materials at their bulk form. Evaluation of in situ properties involves measurements in small volumes that can be accomplished by nanoindentation tests. Their main advantage is continues monitoring both the load and displacement of an indenter with high accuracy. Based on the analysis of loading – unloading data, elastic moduli (E) and hardness (H) as well as their distribution may be successfully determined [1].

The hardness of a material is usually defined as a load divided by a residual projected area. The correct determination of contact area is, therefore, of crucial importance for mechanical properties evaluation because of pile-ups appearing at the indent sides support part of the load and lead to overestimation of E and H.

In the present study, the nanoindentation techniques of different loading scenarios have been applied to chromium carbide based and nickel – chromium alloy bonded cermets to evaluate the micromechanical properties of the constituent phases. The aim of this paper is to report the experimental data on modulus of elasticity and hardness of cermets' constituents taking into consideration the indentations size effects and uncertainties in the projected area calculation due to pile – ups. Three-phase Cr_3C_2 – Cr_7C_2 – $CrNi_3$ composite, Figure 1, was chosen as a model material to study possibility of distinguishing between ceramic phases of quite similar properties (Cr_3C_2 – Cr_7C_2) as well as dissimilar constituents (carbides – binder alloy). Grain size of the carbides was in the range between 1 and 15 microns.

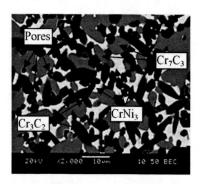

Figure 1: SEM image of the chromium carbide – based composite tested in the present study.

2 EXPERIMENTAL PROCEDURE

Prior to the testing, the samples were polished with a 0.25 µm diamond paste at a final step to obtain a smooth surface of nanometer-scale roughness.

A Hysitron Triboscope Nanomechanical Testing Instrument attached to a Veeco Nanoscope IV atomic force microscope (AFM) with a scanning probe microscope (SPM) controller was used to indent the composite by using a three-sided, pyramid-shaped diamond probe tip (Berkovich tip). The tests were load-controlled and conducted at room temperature. The area for testing was located by AFM imaging, and indentations marks were imaged by AFM after testing. In this study, two load scenarios were prescribed: (1) single – indent mode defined by loading segment duration of 10 s, a holding period at maximal load of 5 s, and unloading segment duration of 10 s; and (2) multi – indent mode defined by multiple step loading and which consisted of 10 loading segments, each up to a partial load and followed by a hold, and, respectively, 10 unloading segments as indicated in Figure 2. During nanoindentation tests the load on specimen and the depth of penetration were recorded and care was taken to properly calibrate for the range of the penetration depths recoded in both hard and soft phases. Because a small amount of cobalt and, therefore, small value of "mean free path", only scenario 1 with the maximum load of 2000 µN was applied during testing of the binder phase. Loading up to 8000 µN was used in scenario 2.

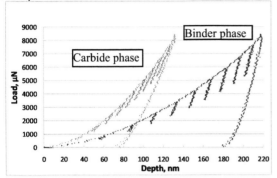

Figure 2: Load – displacement curves obtained during nanoindentation tests via scenario 2.

The analyzed area was firstly detected with the help of scanning electron microscope (SEM) JEOL 6060LV, Figure 3, guided by Vickers indents marks. This procedure is necessary to identify the phases despite the fact that it is time consuming. More than 20 indentations were performed for each phase and values of hardness and reduced moduli averaged. Figure 3 also represents the STM image of one of the tested zones with well recognizable indentation sites.

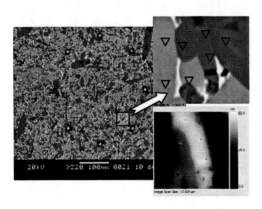

Figure 3: SEM and SPM images of the tested area.

The well-developed method for H and E determination based on the Oliver – Pharr approach [1], implying that contact stiffness of a material, S, can be experimentally obtained from the slope of unloading part of a load – displacement curve, was applied at the present study. For the Berkovich tip the contact area, A, is calculated as $A = 24.56h_c^2$, where h_c is a contact depth. Based on the theory of rigid contact, the reduced modulus can be calculated as following:

$$E_r = \frac{S\sqrt{\pi}}{2\beta\sqrt{A}} = \left[\frac{(1-v^2)}{E} + \frac{(1-v_i^2)}{E_i}\right]^{-1} , \qquad (1)$$

where E_i and v_i are Young's modulus and Poisson's ratio of the indenter (for diamond $E_i = 1141$ GPa and $v_i = 0.07$).

To evaluate the true contact area, the AFM imaging was incorporated into measurements and simple, yet effective method proposed in [2] was applied for H and E re-calculations. The basic idea is approximating the area of the pile-up contact perimeter as a semi-ellipse and including this pile-up area in the analysis [2].

3 RESULTS AND DISCUSSION

Figure 2 shows the representative load – displacement curves for different phases. Once a contact depth is known, the resulting radius of circle of contact is determined by simple geometry and used to determine the mean contact pressure or hardness H and reduced elastic modulus E_r (Eq. 1). Then, elastic modules are converted from reduced modules assuming a Poisson's ratio of 0.21 for both carbides and 0.3 for nickel alloy.

Analysis of the results obtained through the scenario 1 has revealed two well distinguished groups of parameters: one of them is ranged around a mean value of 5 (±2.7) GPa for H and 230 (±27) GPa for E_r while another one is grouped around 25 (±9) GPa for H and 360 (±50) GPa for E_r, Figure 4.

Quite large data scattering reflects the uncertainty in the spatial placement of indentations as under AFM the constituent phases are not clearly recognizable. The imprints in carbides subjected to a low load were not distinguished under SEM, also. Figure 4c shows the schematic diagram of possible indentation sites. Pores, presence of different phases in the nearest proximity of the tested area of material, interface and so on, all these features influence the results obtained.

To distinguish the carbide phases the higher load under scenario 2 was applied. Analysis of the data obtained from testing revealed three ranges of parameters for the studied cermet. The first one is grouped around the mean values of 28 (±3) GPa for H and 390 (±20) GPa for E_r; while another group exhibits values of about 22 (±2.5) GPa for H and 320 (±11) GPa for E_r. Undoubtedly, these parameters describe the carbides in the cermet materials. The third group corresponded to 5.5 (±3.6) GPa for H and 250 (±43) GPa for E_r. Figure 5 displays the modulus and hardness of the constituent phases measured at 20 arbitrary selected points in each phase.

The residual imprints left by indent at 8000 µN are well visible under SEM and two carbide phases can be clearly indicated. At this load scattering of the data obtained for binder phase is much larger.

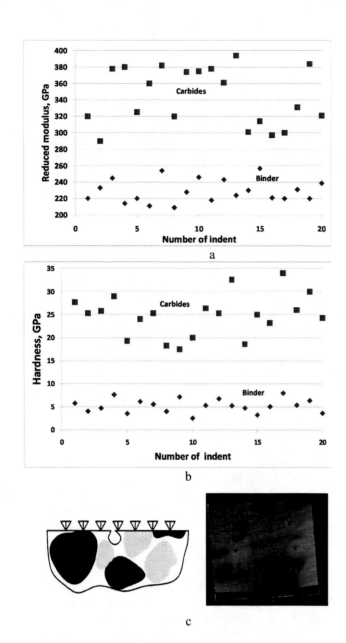

a

b

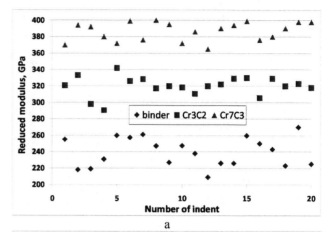

a

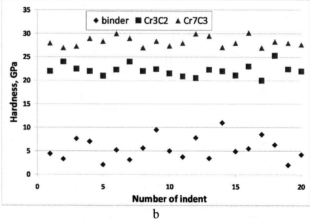

b

Figure 5: (a) – Reduced modulus of the constituents; and (b) – hardness as measured via scenario 2.

Figure 4: (a) – Reduced modulus of the constituents; and (b) – hardness as measured via scenario 1; (c) – schematic view of possible indentation sites.

The hardness measured by the nanoindentation in the binder metal is much higher than that in the bulk alloy. The high hardness of the binder metal may be due to the solid solution of Cr and C in nickel during sintering, and to the surface preparation (polishing). Somewhat higher than expected hardness of both types of carbides may be attributed, to some extent, to the residual stresses present within the specimens as a result of processing (mismatch in coefficients of thermal expansion between phases) leaving carbides in compression [3]. However, the true values of hardness and modulus may only be obtained by taking the pile-ups into consideration.

That may be explained by low metal content leading to the strong "proximity effect" due to violation of the conditions $h/D \leq 1/10$, where h is a penetration depth and D is the radius of an indented particle. The neighboring matrix affects the measured values and test gives somewhat "joint" properties of the neighboring phases. A special care should be taken when the binder phase is tested. In the case when scenario 1 was applied, the contact depth was around 300 nm that is close to 1/10 of alloy islands between hard grains. In contrast, the properties of carbides have larger scattering when measured at low load.

Examination of the indentation sites indicates the pile-ups around the indents, Figure 6. Line profiles along a median of one of the indents are made with the help of AFM. The procedure for calculation of the contact area due to pile-ups is described in details elsewhere [2]. The present study follows the algorithm of the pile-up contact width determination as the width of the semi-ellipse projected by the squeezed out material as it shown in Figure 6b.

Assuming that the projected contact area, determined at contact depth, h_c, traces an equilateral triangle of side b, then for a perfect Berkovich tip, $A = 24.5\ h_c^2 = 0.433\ b^2$ and $b = 7.531\ h_c$. At each indentation, a_i for three sides may be different and, therefore, the total contact area can be calculated as $A = A_{impr} + A_{ad} = A_{impr} + 5.915h_c \sum a_i$, where A_{impr} is the area of imprint derived from the Oliver–Pharr method and A_{ad} is the total pile-up contact area.

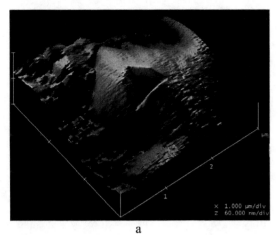

a

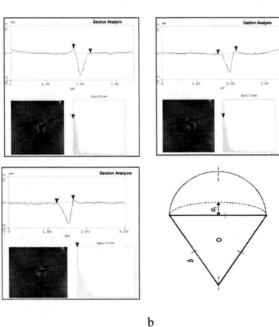

b

Figure 6: (a) - 3D image showing pile-ups around a nanoindent; (b) – lateral images of an imprint and cross-sectional profiles for a real contact area calculation; and schematic approximation of the indent.

Moduli of elasticity and hardness values for the constituents as reported at the published literature [4] as well as determined before and after area correction are summarized in Table 1.

Values of H before pile-up corrections were higher than the upper limits given for these materials in the literature. After re-consideration of the contact areas a very good agreement with published data was found. Binder $CrNi_3$ solid solution shows the modulus of elasticity quite similar to that reported for bulk nickel-chrome solution.

Phase	Modulus of elasticity, GPa		
	Bulk [4]	Measured	Calculated
$CrNi_3$	$210 - 255$	230 ± 27	216.3 ± 10.5
Cr_3C_2	$373 - 386$	320 ± 11	314.6 ± 8.8
Cr_7C_3	$340 - 400$	390 ± 20	340.3 ± 4.9
Hardness, GPa			
$CrNi_3$	$0.7 - 3$	5 ± 2.7	2.5 ± 1.0
Cr_3C_2	$10.2 - 18$	22 ± 2.5	17.8 ± 1.6
Cr_7C_3	$16 - 20$	28 ± 3.0	22.5 ± 1.5

Table 1: Modulus of elasticity and hardness of the composite constituents.

However, the modulus measured for both modifications of chromium carbides is smaller than it could be expected. It is very possible that the solubility of nickel in carbide and its influence on the carbide properties results in the formation of carbide structures that differ from structures of commonly referred ones. EDS analysis of the elemental distribution indicates the presence of high amount of free carbon throughout the structure, solubility of nickel in hexagonal chromium carbide Cr_7C_3 and spurious phases formation.

4 CONCLUSIONS

The data analysis and re-calculation of the projected area are of primary importance for evaluation of the mechanical properties (E and H) extracted from the nanoindentation tests on the constituent phases of composites. All constituents show some degree of plastic deformation forming the pile – ups around nanoindents. The semi-ellipse method for accounting of pile – ups has been applied to multiphase material and after the contact area re-consideration, the hardness and module of phases show good agreement with published data. Difference in values that are commonly referred and obtained during the test may point to the formation of new phases during sintering and, therefore, to the possible different behavior of the bulk composite under loading from that as it may be expected at the design stage. Data extracted from nanoindentation may serve as the inputs for materials design and modeling.

5 ACKNOWLEDGEMENTS

Estonian Science Foundation under grants No 8211, JD120 and T062 are acknowledged for financial support of this research.

REFERENCES
[1] W.C. Oliver, G. Pharr, J.Mater.Res., 7, 1564, 1996.
[2] K.O. Kese, Z.C. Li, Scipta Materialia, 55, 699, 2006.
[3] I.Hussainova, I. Jasiuk, M. Sardela, M. Antonov, Wear, 267, 152, 2009.
[4] MatWeb. Searchable database of materials sheets. Obtained through Internet: www.matweb.com (accessed 04/08/2010).

Oxygen vacancies induced ferromagnetism in undoped and Cr doped SnO$_2$ nanowires

Li Zhang[*], Shihui Ge[*], Yalu Zuo[*] and Jing Qing[*]

[*] Key Laboratory for Magnetism and Magnitic Materials of Ministry of Education,
Lanzhou University, Lanzhou 730000, P.R.China, gesh@ lzu.edu.cn

ABSTRACT

We presented detailed characterizations of the ferromagnetic properties of undoped and Cr-doped SnO$_2$ nanowires synthesized by chemical vapor deposition. All the samples exhibit room-temperature ferromagnetism. Our experiments reveal that, in addition to the contribution of surface oxygen vacancies, Cr doping into SnO$_2$ plays an important role in tuning the ferromagnetism of SnO$_2$ nanostructures. Doping the SnO$_2$ nanowires with 1.8at.% Cr enhances the magnetization by 28%. We applied the bound magnetic polaron model to successfully explain the enhancement of ferromagnetism.

Keywords: chemical vapor deposition (CVD), nanostructure, ferromagnetism, oxygen vacancies, bound magnetic polaron (BMP)

1 INTRODUCTION

Following the first discovery in Co-doped anatase TiO$_2$ thin film [1], significant work has shown that $3d$ transition-metal (TM)-doped semiconducting and insulating oxides such as ZnO, TiO$_2$ and SnO$_2$ [2-4], exhibit ferromagnetism (FM) above room temperature (RT), whose origin was primarily attributed to the doping. However, the finding of unexpected FM in HfO$_2$ thin films [5] has excited the revisit of the real role that a doping may play in tailoring the magnetic properties. The observations of FM in various pristine oxides such as TiO$_2$, In$_2$O$_3$, ZnO confirmed that magnetism is certainly possible in pure semiconducting and insulating oxides, where FM is attributed to the oxygen vacancies (OVs) or cation defects [6-8]. Recently, we have reported ferromagnetic properties in undoped SnO$_2$ nanowires prepared by chemical vapor deposition (CVD), and showed that the observed FM is likely due to the OVs at the surface [9]. However, the role of TM doping still remains unclear in these systems and there currently exist considerable debates in the literature [10-12]. In this paper, we present a detailed characterization of the magnetic properties of pristine and Cr-doped SnO$_2$ nanowire to investigate the role of a TM doping in tailoring FM. Unlike many other TM, Cr itself is antiferromagnetic and would not induce an extrinsic FM even if Cr clustering occurs. Moreover, trivalent Cr^{3+} ions exhibit $3d^3$ high-spin configuration, which helps to generate large magnetic moments in the host semiconductors [13]. Therefore, our system provides an excellent platform to conveniently investigate the origin of intrinsic FM.

2 EXPERIMENTAL METHODS

Our samples were synthesized with CVD method. To grow Cr-doped SnO$_2$ nanowires, highly pure Sn (99.9%) and CrCl$_3$ (99.9%) powders were mixed and ground thoroughly in a mortar. The mixture was loaded at a ceramic boat, which was located at the center of a quartz tube. The Si (100) substrates coated with 5 nm Au layers served as the catalyst positioned in the boat at an appropriate distance away from the source materials. The quartz tube was evacuated to a pressure of less than 2Pa and then heated to 900°C. A stable argon flow with a rate of 100 SCCM (SCCM denotes cubic centimeter per minuter at STP) was introduced into the quartz tube as the carrier gas during heating and cooling process. At 900°C, oxygen was introduced and the flux rate of argon was modulated to 30 SCCM; the process took 30 minutes, and the white material was formed on the surface of the substrates. In our work, Sn$_{1-x}$Cr$_x$O$_{2-\delta}$ nanowires with x=0.018, 0.048 and 0.072 were obtained by modulating the weight ratio of the Sn and CrCl$_3$ powders. For the synthesis of undoped SnO$_2$ nanowires, all the experimental conditions remained the same except that CrCl$_3$ was not used in the source powder.

The morphology and structure of the samples were characterized by scanning electron microscope (SEM) (Hitachi S-4800) and X-ray diffraction (XRD) (Philips X' Pert model). Further microstructural characterization was performed using high-resolution transmission electron microscope (HRTEM) (JEOL, JEM-2010 at 200 kV) imaging and selected area electron diffraction (SAED) pattern analysis. The magnetic properties of the samples were measured by a Quantum Design magnetometer (MPMS XL SQUID) from 5 to 305 K and a vibrating sample magnetometer (VSM) (LakeShore 7304 model) at RT. The chemical bonding states and the compositions of the products were determined by x-ray photoelectron spectroscopy (XPS) (a VG Scientific ESCALAB-210 spectrometer). Micro photoluminescence (PL) measurement were carried out at RT using a He-Cd laser with a wavelength of 325 nm as the excitation source.

Throughout all the experimental steps taken during synthesis and measurements, only high pure reagent, gas source and alumina boat were used. All the tools used for handing the samples were nonferromagnetic and plastic based to avoid any unintentional contact with metals. Moreover, all the Si (100) substrates uncoated or coated with Au layers were measured under the same sequences as for the samples to examine the possible contamination contributing to the observed FM. All the substrates showed the expected diamagnetic behavior.

3 RESULTS AND DISCUSSION

Figure 1a shows the SEM images of the undoped SnO_2 sample. Clearly, all the products are in the form of nanowires, direction-free on the substrate with a high density. The SnO_2 nanowires are homogeneous with diameters around 50nm (Fig. 1a, inset) and length greater than 20 μm. Compared with the undoped SnO_2 nanowires, Cr doping does not change the morphology and all the products remain as nanowires. However, although each individual nanowire for all samples has a uniform diameter along its length (which reaches 15 μm), the nanowire diameters scatter between 50 and 100 nm (as seen in Fig. 1b for the $Sn_{0.982}Cr_{0.018}O_{2-\delta}$ nanowires), indicating the possible influence of Cr doping on the nanowire width. Figure 1c shows a catalyst cap on the tip of a nanowire with a size close to that of the nanowire, which evidences that the growth mechanism of SnO_2 nanowires in our experiment is the vapor–liquid–solid (VLS) mechanism [14].

The structural characteristics of the $Sn_{1-x}Cr_xO_{2-\delta}$ nanowires were investigated in detail by using TEM and HRTEM. Fig. 1d shows a typical TEM bright-field image of $Sn_{0.982}Cr_{0.018}O_{2-\delta}$ nanowires with a width ~ 70 nm, indicating the diameter uniformity along the nanowire length. A ripple-like contrast observed in the TEM image is due to the strain resulting from the bending of the nanowires. The HRTEM image shows that the nanowire is structurally uniform, and the clear lattice fringes illustrate that the nanowire is a single crystalline (Fig. 1e). The interplanar spacing is about 0.343 nm, which corresponds to the {110} plane of the rutile crystalline SnO_2, implying that the nanowires grow along <110> direction. The inset shows the SAED pattern that can be indexed on a tetragonal cell with lattice parameters of a=4.737 Å and c=3.185 Å, which further confirms that the nanowire is a single-crystal rutile SnO_2. No secondary phase or clusters were observed, verifying that Cr has been doped into SnO_2 lattice. The undoped SnO_2 nanowires have the same structures as Cr-doped SnO_2 nanowires.

Figure 1f shows XRD patterns of undoped SnO_2 nanowires and $Sn_{1-x}Cr_xO_{2-\delta}$ nanowires with x = 0.018, 0.048 and 0.072. For each sample, all observed peaks can be indexed with the rutile-type tetragonal structure of SnO_2 (JCPDS 21-1250), which are consistent with the standard values for bulk SnO_2. No characteristic peaks of impurities, such as other forms of tin oxides, pure Sn or Cr oxides, were observed within the detection sensitivity. Compared with the undoped SnO_2 nanowires, the diffraction peak position shifts to a larger angle with increased Cr content,

revealing a possible change in lattice parameters. Such a change is quantified in Fig. 1g, where the lattice parameter a decreases monotonously with increased Cr content. The shrinkage of lattice constant after doping can be attributed to the substitution of the smaller Cr^{3+} ions (0.65 Å) onto the Sn^{4+} (0.69 Å) lattice sites [15], which indicates that Cr ions are incorporated into the SnO_2 matrix.

Figure 1. Low-magnification SEM images for (a) SnO_2 and (b) $Sn_{0.982}Cr_{0.018}O_{2-\delta}$ nanowires. (c) The end of an individual $Sn_{0.982}Cr_{0.018}O_{2-\delta}$ nanowire. (d) TEM image of $Sn_{0.982}Cr_{0.018}O_{2-\delta}$ nanowires. (e) HRTEM image and corresponding SAED pattern (inset) of $Sn_{0.982}Cr_{0.018}O_{2-\delta}$ nanowires. (f) XRD patterns of SnO_2 and $Sn_{1-x}Cr_xO_{2-\delta}$ nanowires with different x. (g) Evolution of the lattice parameters a and c as a function of Cr content x.

Further evidence for the purity and composition of the products were obtained by XPS, showing that no impurities were detected in all the samples within the detection limit. Two representative XPS spectra of the undoped SnO_2 and $Sn_{0.982}Cr_{0.018}O_{2-\delta}$ nanowires are shown in Fig. 2, where Figs. 2a –2c are the survey spectra, the Sn 2p, and O 1s core-level spectrum, respectively. The double spectral lines of Sn 3d appear at the binding energy of 486.6 eV (Sn $3d_{5/2}$) and 495 eV (Sn $3d_{3/2}$) with a spin-orbit splitting of 8.4 eV, which coincides with the findings for the Sn^{4+} bound to oxygen in the SnO_2 matrix [16]. The peak of Sn $3d_{5/2}$ shows only one symmetric component without a shoulder peak, indicating the absence of Sn^{2+} ions. On the other hand, an asymmetric O 1s peak is observed for the sample, which has a shoulder at the higher binding energy side. By fitting with Gaussian distributions, we obtained two peaks at 531 and 533 eV. The dominant one is located at 531 eV, which corresponds to the O 1s core peak of O^{2-} bound to Sn^{4+}, and normally assigned as the low binding energy component (LBEC). The peak at 533 eV is referred to as the high binding energy component (HBEC) [17]. We will discuss them in detail below. Moreover, as shown in the inset of Fig. 2a, the Cr $2p_{3/2}$ peak was detected at 577.5 eV for the $Sn_{0.982}Cr_{0.018}O_{2-\delta}$ sample, which matches that of Cr^{3+} ions

(577.2 eV) [18,19]. It confirms that Cr ions in our samples have a chemical valence of 3+.

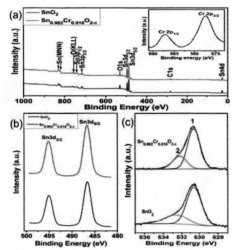

Figure 2. XPS spectra for SnO_2 and $Sn_{0.982}Cr_{0.018}O_{2-\delta}$ nanowires: (a) the survey spectra, (b)Sn $3d$, (c)O1s. The inset of (a) is the detail scan of Cr $2p_{3/2}$ and $2p_{1/2}$ peaks for $Sn_{0.982}Cr_{0.018}O_{2-\delta}$.

Figure 3a shows the magnetic hysteresis (M-H) curves of all samples measured at RT, where the contributions from the diamagnetism of Si substrate and the paramagnetism (PM) of samples have been subtracted. All the samples are clearly ferromagnetic at RT and their loops have small coercivities in the range from 80 to 120 Oe. Note that the observed FM should be intrinsic in all samples, since the measurements above (HRTEM, SAED, XRD and XPS) did not detect the Sn-based binary, ternary oxide candidates, pure Sn clusters or Cr secondary phase.

A large special magnetization M of 0.081 emu/g is observed for the undoped SnO_2 nanowires, much larger than that of SnO_2 nanoparticles (<0.002emu/g) in Ref. [20]. Note that $Sn_{0.982}Cr_{0.018}O_{2-\delta}$ nanowires exhibits M of 0.104emu/g, larger than that of undoped SnO_2 nanowires, which decreases with further increased dopant concentration x (Fig. 3, inset). Clearly, doping with 1.8% of Cr can certainly enhance the magnetization of the pristine SnO_2 nanowires, implying additional contributions brought by Cr doping. Given that Cr^{3+} ions enter the SnO_2 lattice and substitute for Sn^{4+} ions, it infers that Cr doping generates additional OVs which might be related to the FM enhancement.

Such additional OV generation is confirmed by micro PL measurement, since OVs in SnO_2 samples dominantly determine their optical emission around 500-600 nm [21]. As shown in Fig. 4, the intensity of 554 nm emission of $Sn_{0.982}Cr_{0.018}O_{2-\delta}$ is much stronger than that of pure SnO_2, indicating that the much larger OVs concentration were generated due to the Cr doping.

However, it turns out that these additional OVs are generated at the inner of the the samples rather than at the surfaces. This can be seen clearly on XPS given in Fig. 2c,

where the relative intensity of HBEC peak to LBEC peak determines the quantity of surface OVs [22]. As shown in Fig. 2c, the relative intensity of HBEC peak to the LBEC peak is 0.362 and 0.358 for the SnO_2 and $Sn_{0.982}Cr_{0.018}O_{2-\delta}$ samples, respectively. The nearly same contribution of the HBEC peak for the two samples indicates that the Cr doping does not induce more surface OVs. Note that, as the penetration depth of Mg Kα XPS is several nanometers, the measurement of XPS indicates the surface proporties of samples, whereas the PL measurement provides the optical properties of the whole sample. The difference between the PL and XPS measurement on our samples indicates that the substitution of Cr^{3+} for Sn^{4+} does not create OVs on the surface, but at the inner of samples. As surface OVs contribute to the FM of the pure SnO_2 nanowires [9], our observations imply that Cr doping does not tune the FM of the pure SnO_2 nanowires by introducing surface OVs.

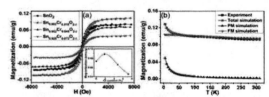

Figure 3. (a) Magnetic hysteresis loops of $Sn_{1-x}Cr_x O_{2-\delta}$ nanowires with x=0–0.072 measured at RT, the inset shows the V content dependence of the magnetization.
(b) Temperature dependence of the magnetization for $Sn_{0.982}Cr_{0.018}O_{2-\delta}$ nanowires.

In this case, we prefer to explain the enhanced FM with the Cr doping in the framework of the bound magnetic polaron (BMP) model. Recent studies have shown a direct dependence of ferromagnetic ordering on the OVs concentration in some dilute magnetic semiconductor systems [2-4], where A BMP mediated exchange interaction was suggested as responsible for magnetic ordering. A BMP is an oxygen vacancy with a trapped electron. This trapped electron occupies an extended orbital state that overlaps with the d shells of several TM ions nearby. Since the OVs arise naturally to ensure charge neutrality when a trivalent ion is substituted in SnO_2 and the Cr^{3+} is in $3d^3$ spin state, they would have the unoccupied minority spin orbitals available for exchange with the trapped electron. The trapped electron will align in an antiparallel configuration with the individual dopant ion spins. This leads to an effective ferromagnetic coupling between coupled dopant ions. The greater density of OVs yields a greater overall volume occupied by BMPs, thus increasing their probability of overlapping more Cr ions into the ferromagnetic domains and enhancing FM. On the other hand, our experiment shows that the magnetization drops with increasing Cr content for the samples with x > 0.018 (Fig. 3, inset). This behavior can be explained that the further reduction will result from any antiferromagnetic Cr^{3+}–O^{2-}–Cr^{3+} superexchange interaction taken place within the nearest Cr^{3+} ions through O^{2-} ions.

NSTI-Nanotech 2010, www.nsti.org, ISBN 978-1-4398-3401-5 Vol. 1, 2010

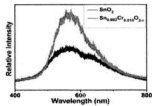

Figure 4. RT micro PL spectra of SnO_2 and $Sn_{0.982}Cr_{0.018}O_{2-\delta}$ nanowires taken at RT.

To further confirm the coexistence of ferromagnetic and antiferromagnetic coupling, the temperature dependence of magnetization was measured from 5 to 300 K in a 5 kOe field for $Sn_{0.982}Cr_{0.018}O_{2-\delta}$ nanowires (Fig. 3b). With increased temperature, the curve shows a rapid decay of magnetization at low temperature region, after which the magnetization declines gradually and does not go to zero at 300 K, indicating a paramagnetic phase mixed with a ferromagnetic phase for the sample. The PM comes from the magnetic moments of isolated Cr ions which are not bounded into magnetic polarons and hence do not exhibit ferromagnetic behavior. Based on these analyses, we can describe the M-T curve as a combination of the $T^{3/2}$ law of saturation magnetization for FM phase and a Curie–Weiss law for PM phase, given by the following equation: [23]

$$M = M_0(1 - BT^{3/2}) + CH/(T - P) \qquad (1)$$

where M_0 is the saturation magnetization at T = 0K, B is a constant, C is the curie constant, P is the Curie–Weiss temperature, and H is the applied magnetic field during measurement. In Fig. 3b, the ferromagnetic part, the paramagnetic part, and the total Ms-T curve simulated by equation (1) are plotted separately. It is seen that the simulated Ms-T curve agrees very well with the experimental results. Fitting the experimental data, we obtained P = -7.9. The negative P value implies the weak antiferromagnetic interaction existing in the samples, which is likely to be responsible for the magnetization decrease with increasing Cr content.

4 CONCLUSIONS

We have synthesized undoped SnO_2 and Cr-doped SnO_2 nanowires with RT FM using a catalyzer-assisted CVD method. 1.8 at.% Cr doping significantly enhances the magnetization, implying that a TM doping does play an important role in introducing FM. We suggest that the in intrinsic FM arises from (i) the surface OVs in the SnO_2 lattice, (ii) the ferromagnetic coupling between the unpaired $3d$ electrons of Cr as a result of hybridization between the $2p$ states of O with the $3d$ state of the dopant (i.e. BMP model). Therefore, both the OVs and the Cr dopants are supposed to be equally important to induce FM of SnO_2.

REFERENCES

[1] Y. Matsumoto, M. Murakami, T. Shono, T. Hasegawa, T. Fukumura, M. Kawasaki, P. Ahmet, T. Chikyow et al., Science, 291, 854, 2001.

[2] M. Venkatesan, C.B. Fitzgerald, J.G. Lunney, J.M.D. Coey, Phys. Rev. Lett. 93, 177206, 2004.

[3] N.H. Hong, J. Sakai, W. Prellier, A. Hassini, A. Ruyter, F. Gervais, Phys. Rev. B 70, 195204, 2004.

[4] S.B. Ogale, R.J. Choudhary, J.P. Buban, S.E. Lofland, S.R. Shinde, S.N. Kale, V.N. Kulkarni, J. Higgins et al., Phys. Rev. Lett. 91, 77205, 2003.

[5] M. Venkatesan, C.B. Fitzgerald, J.M.D. Coey, Nature 430, 630, 2004.

[6] N.H. Hong, J. Sakai, N. Poirot, V. Brizé, Phys. Rev. B 73, 132404, 2006.

[7] S.D. Yoon, Y. Chen, A. Yang, T.L. Goodrich, X. Zuo, D.A. Arena, K. Ziemer, C. Vittoria, V.G. Harris, J. Phys.: Condens. Matter 18, L355, 2006.

[8] N.H. Hong, J. Sakai, V. Brizé, J. Phys.: Condens. Matter 19, 036219, 2007.

[9] L. Zhang, S.H. Ge, H.X. Zhang, Y.L. Zuo, J. Nanosci. Nanotechnol. 10, 2010, in publish.

[10] N.H. Hong, E. Chikoidze, Y. Dumont, Physica B 404, 3978, 2009.

[11] Q.Y. Xu, H. Schmidt, S.Q. Zhou, K. Potzger, M. Helm, H. Hochmuth, M. Lorenz et al., Appl. Phys. Lett. 92, 082508, 2008.

[12] M. Kapilashrami, J. Xu, V. Ström, K.V. Rao, L. Belova, Appl. Phys. Lett. 95, 033104, 2009.

[13] G.Z. Xing, J.B. Yi, D.D. Wang, L. Liao, T. Yu, Z.X. Shen, C.H.A. Huan, T.C. Sum, J. Ding, T. Wu, Phys. Rev. B 79, 174406, 2009.

[14] X.F. Duan, C.M. Lieber, Adv. Mater.12, 298, 2000.

[15] H. Takeuchi, M. Arakawa, J. Phys. Soc. Jpn. 52, 279, 1983.

[16] L. Yan, J.S. Pan, C.K. Ong, Mater. Sci. Eng. B 128, 34, 2006.

[17] G. Tyuliev, S. Angelov, Appl. Surf. Sci., 32, 381, 1988.

[18] C. Xu, M. Hassel, H. Kuhlenbeck, H.J. Freund, Surf. Sci. 258, 23, 1991.

[19] N.J.C. Ingle, R.H. Hammond, M.R. Beasley, J. Appl. Phys. 89, 4631, 2001.

[20] A. Sundaresan, R. Bhargavi, N. Rangarajan, U. Siddesh, C.N.R. Rao, Phys. Rev. B 74, 161306, 2006.

[21] J.H. He, T.H. Wu, C.L. Hsin, K.M. Li, L.J. Chen, Y.L. Chueh et al., Small 2, 116, 2006.

[22] M. Naeem, S.K. Hasanain, M. Kobayashi, Y. Ishida, A. Fujimori, S. Buzby, S.I. Shah, Nanotechnology 17, 2675, 2006.

[23] S.H. Liu, H.S. Hsu, C.R. Lin, C.S. Lue, J.C.A. Huang, Appl. Phys. Lett. 90, 222505, 2007.

Evaluation of Uncertainty in Nanoparticle Size Measurement by

Differential Mobility Analysis

Chih-Min Lin, Ta-Chang Yu, Shih-Hsiang Lai, Hsin-Chia Ho, Han Fu Weng, Chao-Jung Chen*

Center for Measurement Standards, Industrial Technology Research Institute
321, Sec. 2, Kuang Fu Rd., Hsinchu, 30011, Taiwan, R. O. C., Chao-Jung.Chen@itri.org.tw

ABSTRACT

This paper presents the measurement results and the uncertainty analysis for four batches of polystyrene latex (PSL) spheres with nominal sizes of 20 nm, 100 nm, 300 nm and 500 nm using the measurement system which differential mobility analysis (DMA) method is applied. The possible error sources are evaluated based on the DMA method and the measurement procedures. The resulting expanded uncertainties are 1.3 nm for the 20 nm particles and 13 nm for the 500 nm particles at 95% confidence interval.

Key words: nanoparticle size, measurement uncertainty, Differential Mobility Analysis.

1. INTRODUCTION

Nanoparticles have found their unique advantages and immediate applications in numerous industrial, commercial, and consumer products. The success of the nanoparticle applications rely first-hand on the advancements of nanometrology to support the increasing demands in measurement accuracy. One measurement parameters of great interests on nanoparticles is their sizes, or the more precisely defined term, diameters. However, existing techniques for nanoparticle size characterizations have resulted in performance inconsistency due to lack of standardization of nanoparticle measurements [1]. Therefore, it is crucial and necessary to establish measurement standards to accurately and consistently determine the sizes of the nanoparticles in nanometer range. A measurement system for particle size, based on the differential mobility analyzer (DMA), has been developed in Center for Measurement Standards of Industrial Technology Research Institute (CMS, ITRI) to obtain particle diameters with high accuracy and small uncertainty. The DMA is calibrated by the primary standard measurement system "electro- gravitational aerosol balance (EAB)", which was also developed in ITRI [2].

2. MEASUREMENT SYSTEM

The DMA, widely used in various applications, is capable of accurate size measurement with highly-resolved size-classification of aerosol particles. Basically, this equipment has two major applications: one is to measure the particle size distribution; the second is to sort out narrow-ranged particle size.

The approach used in this study is to sort aerosol particles according to their electrical mobility, Z, which is defined as the ratio of the velocity to the electric field. The electrical mobility is related to the particle diameter, D, and is given by

$$Z = \frac{v}{E} = \frac{q \cdot C}{3\pi \cdot \eta \cdot D} \quad (1)$$

Where v is the velocity of the particles in the electric field, E is the electric field applied to the particles, q is the charges carried by the particles, C is the slip correction factor determined by the properties or aerosol carrier gas, and η is the gas viscosity.

As shown in figure 1, differential mobility analyzer is an axially symmetric structure, which contains a cylindrical outer sheath of radius r_1 and a central cylindrical electrode of radius r_2. The central cylindrical electrode is connected to high voltage and the outer sheath is connected to ground to form a radiated high voltage electric field. When the particle is within this electric field, the electric field strength for any point of distance r to the center will have electric field strength of:

$$E = \frac{V}{r \cdot \ln(r_1/r_2)} \quad (2)$$

Where V is the voltage of the central cylindrical electrode.

From equation (1) and equation (2), we can obtain the radial velocity of particles in this system is v_E:

$$v_E = \frac{dr}{dt} = Z \cdot E = \frac{Z \cdot V}{r \cdot \ln(r_1/r_2)} \quad (3)$$

Where t is the time. And the axial velocity of this

particle is v_R:

$$v_R = \frac{dx}{dt} = u(r) \qquad (4)$$

Where $u(r)$ is the radial velocity. From equation (3) and equation (4), we can obtain:

$$\int_0^L dx = \int_{r_1}^{r_2} \frac{r \cdot \ln(r_1/r_2) \cdot u(r)}{Z \cdot V} dr$$
$$= \frac{\ln(r_1/r_2)}{2\pi \cdot Z \cdot V} \int_{r_1}^{r_2} 2\pi \cdot r \cdot u(r) dr \qquad (5)$$

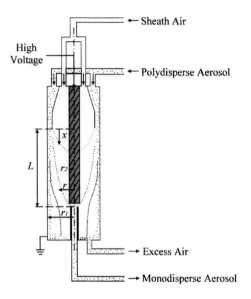

Figure 1: Differential mobility analyzer used in this work.

Equation (5) is used to describe the particle's behavior from the outer top of differential mobility analyzer ($r = r_1$ and $x = 0$) to the central bottom ($r = r_2$ and $x = L$) affected by electrostatic force and dragging force. When polydisperse aerosol flow rate (Q_a) is equal to monodisperse aerosol flow rate (Q_s), from equation (5), we can obtain:

$$Z = \frac{Q_c \cdot \ln(r_1/r_2)}{2\pi \cdot L \cdot V} \qquad (6)$$

Where Q_c is sheath air flow rate. Therefore, from equation (1) and equation (6), we can obtain the particle size D:

$$D = \frac{2q \cdot L \cdot V \cdot C}{3\eta \cdot Q_c \cdot \ln(r_1/r_2)} \qquad (7)$$

3. SYSTEM EVALUATION

Nanoparticle size measurement system using differential mobility analysis, through polystyrene particles, uses comparison method to be traced to EAB nanoparticle

size calibration system. Meanwhile, based on ISO/IEC Guide 98-3:2008[3], each of the error source and standard uncertainty is analyzed. Finally, the expanded uncertainty is calculated. The polystyrene particles and flow rate setup of differential mobility analyzer for system evaluation are shown in table 1.

3.1. Mathematical model

We can obtain the calibration values of reference particles by EAB nanoparticle size calibration system. Meanwhile, when the geometrical parameters (r_1 , r_2 , L) and sheath air flow rate (Q_c) of differential mobility analyzer are all of constant values, the mobility of particle (Z) is only related to voltage (V) from equation (6). Therefore, under the same parameters setup, when nanoparticles size measurement system using differential mobility analysis is used respectively to perform the mobility measurement of reference particles and particles to be tested, the calibration equation of comparison method can be:

$$Z_t = \frac{\tilde{Z}_t}{\tilde{Z}_r} \times Z_r \qquad (8)$$

Where Z_t is the mobility of particles to be tested, Z_r is the mobility corresponding to the traced particle size of reference particles (D_r). $\tilde{Z}_t$ and $\tilde{Z}_r$ is the mobility obtained by the measurement system for particles to be tested and reference particles, respectively. Meanwhile, from equation (1), we can obtain particle size measurement values of particles to be tested and reference particles of $\tilde{D}_t$ and $\tilde{D}_r$, respectively. Therefore, particle size calibration value (D_t) of particles to be tested can be obtained from equation (1) and equation (8):

$$D_t = (\frac{\tilde{D}_t}{\tilde{D}_r} \times D_r) \times (\frac{\tilde{C}_r}{\tilde{C}_t} \times \frac{C_t}{C_r}) \qquad (9)$$

Where C_t , C_r , $\tilde{C}_t$ and $\tilde{C}_r$ are the slip correction factors corresponding to each particle size D_t , D_r , $\tilde{D}_t$ and $\tilde{D}_r$, respectively.

3.2. Uncertainty evaluation

Let the measured Y is the particle size calibration value D_t of particles to be tested, and we combine all the slip correction factors of equation (9) as C_C , then we can obtain the measuring equation:

$$Y = C_C(\frac{\tilde{D}_t}{\tilde{D}_r} \times D_r + e_r) + e_C \qquad (10)$$

Where C_C and D_r are constant values, and their variances are e_C and e_r , respectively, and the estimated mean values of e_C and e_r are 0. Therefore, the sources of the measurement uncertainty include the uncertainty source caused by the measurement process by particles to be tested and reference particles, and the combined standard uncertainty of the measuring system is:

$$u_c^2(Y) = C_C^2 [(\frac{\partial D}{\partial \widetilde{D}_t})^2 \cdot u^2(\widetilde{D}_t) + (\frac{\partial D}{\partial \widetilde{D}_r})^2 \cdot u^2(\widetilde{D}_r)$$
$$+ u^2(e_r)] + u^2(e_C) = C_C^2 \cdot u^2(D) + u^2(e_C) \quad (11)$$

where $u(D)$ is the standard uncertainty generated during the measurement process. The sources of uncertainty are analyzed as following:

(a) Standard uncertainties caused by particles to be tested during the measurement process $u(\widetilde{D}_t)$:

The sources of uncertainty caused by particles to be tested during the measurement process include environmental parameters (temperature and atmospheric pressure), system setup (flow rate and voltage) and the stability of the particles to be tested, etc. The standard uncertainties are shown in table 2.

The uncertainty of five repeated measurements which are done on each particles to be tested are A type standard uncertainty. For B type standard uncertainty, the uncertainty might caused by the equipment resolution is considered, which includes the resolution of flow rate (0.1 L/min) and voltage (0.1 V). The uncertainty caused by the resolution of applied voltage can be neglected by the calculation.

(b) Standard uncertainty caused by reference particles during the measurement process $u(\widetilde{D}_r)$:

It is the same as the evaluation method for the uncertainty caused by particles to be tested during the measurement process. The standard uncertainties caused by reference particles during the measurement process are shown in table 3.

(c) Standard uncertainty of the particle size calibration value of reference particles $u(e_r)$:

The calibration values of reference particles by EAB nanoparticle size calibration system are used as constant value and each of the traced value is shown in table 1. Meanwhile, the relative uncertainty of each tracked value is supposed as 5 %, and we can obtain the standard uncertainty of 0.602 nm, sensitivity coefficient of 1, component of uncertainty of 0.602 nm, and degree of freedom of 200.

(d) Standard uncertainty caused by the measurement process $u(D)$:

The standard uncertainties caused during the measurement process include measurement uncertainty and traced uncertainty, etc., which are shown in equation (12). The calculation results of the standard uncertainty of each particles to be tested are summarized in table 4.

$$u^2(D) = (\frac{\partial D}{\partial \widetilde{D}_t})^2 u^2(\widetilde{D}_t) + (\frac{\partial D}{\partial \widetilde{D}_r})^2 u^2(\widetilde{D}_r) + u^2(e_r) \quad (12)$$

(e) Standard uncertainty generated by the variance of slip correction factor $u(e_C)$:

The slip correction factor representative equation will change the final measuring value of the particles to be tested and the value of slip correction factor. Therefore, seven kinds of slip correction factor representative equations as proposed in literature [4] are used to obtain respective slip correction factor and variance as corresponded to each particle size correction value. Meanwhile, B type evaluation method is used to obtain the standard uncertainty as shown in table 5.

From the above uncertainty source analysis and equation (11), we can obtain the $u_c(D_t)$ (the combined standard uncertainty) of each of the particles to be tested. The uncertainty evaluation result of this measurement system on each particles to be tested is shown in table 6. The U (the expanded uncertainty) is obtained from the multiplication of combined standard uncertainty by the coverage factor (k) at 95% confidence interval. The coverage factor can be obtained from effective degree of freedom (v_{eff}) [5].

4. CONCLUSION

From the uncertainty evaluation results of the measurement of particles to be tested by DMA method, the mean particle sizes of the measurement result of particles to be tested all fall within the calibration report values. Meanwhile, the expanded uncertainties for measurement range from 20 nm to 500nm are from 1.3 nm to 13 nm, respectively. Therefore, the measurement capability of this measurement system can be recognized. The completion of the setup of nanoparticle size measurement system using differential mobility analysis will provide the traceability system of domestic nanoparticle size standard.

REFERENCES

[1] B. Scarlett, "Standardization of Nanoparticle Measurements," J. Nanoparticle Res., 2, 1, 2000.
[2] Chih-Min Lin, Ta-Chang Yu, Shan-Peng Pan, Han-Fu Weng, and Chao-Jung Chen, "Evaluation of Uncertainty in Nanoparticle Size Measurement by Electro- gravitational Aerosol Balance," Nanotech Conference & Expo 2009, Texas, U.S.A., 2009.
[3] ISO/IEC Guide 98-3:2008, Uncertainty of measurement — Part 3: Guide to the expression of uncertainty in measurement (GUM: 1995).
[4] M. D. Allen and O. G. Raabe, "Slip Correction Measurements of Spherical Solid Aerosol Particles in an Improved Millikan Apparatus," Aerosol Sci. Technol. 4, 269, 1985.
[5] K. A. Brownlee, Statistical Theory and Methodology in Science and Engineering, John Wiley and Sons, New York, pp. 236, 1960.

Table 1: Polystyrene particles and flow rate setup of differential mobility analyzer for system evaluation.

Number	Particles to be tested	Value of calibration report (nm)		Reference particles	Value of calibration report (nm)		Flow rate setup (L/min)	
		Average particle size	Uncertainty		Average particle size	Uncertainty	Q_c	Q_a
	DUKE-3020A	21	1.5	CMS-100	109.0	1.3	10.0	1.0
	JSR-SC-0100-D	100	3	CMS-100	109.0	1.3	10.0	1.0
	JSR-SC-032-S	309	9	CMS-300	284.9	1.3	6.0	6.0
	JSR-SC-051-S	506	12	CMS-500	532.4	1.3	3.0	3.0

Table 2: Standard uncertainty caused by the measurement process of the particles to be tested.

Number	Measured value of particle size	Repro-ducibility measure-ment	Error caused by flow rate resolution	Standard uncertainty	Sensitivity coefficient	Component of uncertainty	Degree of free-dom
DUKE-3020A	20.45 nm	0.15 nm	0.045 nm	0.157 nm	0.960	0.151 nm	4
JSR-SC-0100-D	101.21 nm	0.07 nm	0.222 nm	0.233 nm	0.960	0.223 nm	136
JSR-SC-032-S	306.39 nm	0.13 nm	1.213 nm	1.220 nm	0.999	1.219 nm	203
JSR-SC-051-S	504.38 nm	1.29 nm	4.307 nm	4.495 nm	0.997	4.478 nm	152

Table 3: Standard uncertainty caused by the measurement process of the reference particles.

Number	Measured value of particle size	Repro-ducibility measurement	Error caused by flow rate resolution	Standard uncertainty	Sensitivity coefficient	Component of uncertainty	Degree of freedom
CMS-100	113.51 nm	0.03 nm	0.249 nm	0.251 nm	0.173	0.043 nm	199
CMS-100	113.46 nm	0.03 nm	0.249 nm	0.255 nm	0.857	0.218 nm	191
CMS-300	284.69 nm	0.30 nm	1.127 nm	1.166 nm	1.075	1.254 nm	167
CMS-500	531.42 nm	1.23 nm	4.538 nm	4.700 nm	0.946	4.444 nm	168

Table 4: Standard uncertainty caused by the measurement process.

Number	D	$u(\tilde{D}_t)$	$u(\tilde{D}_r)$	$u(e_r)$	$u(D)$	Degree of freedom
DUKE-3020A	19.64 nm	0.151 nm	0.043 nm	0.602 nm	0.622 nm	199
JSR-SC-0100-D	97.23 nm	0.223 nm	0.218 nm	0.602 nm	0.678 nm	191
JSR-SC-032-S	306.08 nm	1.219 nm	1.254 nm	0.602 nm	1.849 nm	167
JSR-SC-051-S	502.46 nm	4.479 nm	4.444 nm	0.602 nm	6.338 nm	168

Table 5: Standard uncertainty caused by slip correction factor variance.

Number	D_t	C_c	D	ΔC_c	ΔY	$u(e_c)$	Degree of freedom
DUKE-3020A	19.74 nm	1.0052	19.64 nm	1.06E-03	0.0208 nm	0.006 nm	6
JSR-SC-0100-D	97.30 nm	1.0007	97.23 nm	2.96E-04	0.0288 nm	0.008 nm	6
JSR-SC-032-S	306.08 nm	1.0000	306.08 nm	1.70E-06	0.0005 nm	0.000 nm	6
JSR-SC-051-S	502.48 nm	1.0000	502.46 nm	2.37E-05	0.0119 nm	0.003 nm	6

Table 6: Expanded uncertainty of particles to be tested.

Number	D_t	$u(D)$	$u(e_c)$	$u_c(D_t)$	v_{eff}	U ($k = 2.07$)
DUKE-3020A	19.7 nm	0.626 nm	0.006 nm	0.630 nm	45	1.3 nm
JSR-SC-0100-D	97.3 nm	0.679 nm	0.008 nm	0.680 nm	221	1.4 nm
JSR-SC-032-S	306.1 nm	1.857 nm	0.000 nm	1.857 nm	60	3.9 nm
JSR-SC-051-S	502.5 nm	6.397 nm	0.003 nm	6.397 nm	23	13 nm

Nano Crystalline Cellulosic Nematic monocrystals optical properties

D. Simon[*], Y. Kadiri[*], G. Picard[*], F. Ghozayel[**] and J.-D. Lebreux[**]

[*]Ahuntsic College, Physics Department,
9155 St-Hubert, Montreal H2M 1Y8, Qc, Canada, dominique.simon@collegeahuntsic.qc.ca
[**]The Quebec Institute of Graphic Communications
999 Emile-Journault East, Montreal H2M 2E2, Qc, Canada, fghozayel@icgq.qc.ca

ABSTRACT

Optically active solid films were produced by leaving to evaporate Nano Crystalline Cellulose (NCC) liquid dispersions. A NCC is a nanorod around 5 x 10 x 150 nm^3. It is real nanocrystals, as each cellulose making molecules are distributed inside a square lattice. If the aqueous dispersion is left to evaporate, microscopic liquid crystals appear very quickly and grow. The final result is a solid film showing vivid colors ranging from blue to red. In some conditions, it also becomes transparent, its optical activity being in the IR spectral region. It is also demonstrated that the solid film is in fact a mosaic made of several mono crystals. In good conditions, each crystal reaches the millimeter to centimeter scale. It is believed that twists in their crystalline nanostructure combined with their surface electric charges are responsible for their auto organization in bulk volume into liquid crystals. This generally accepted statement is however questioned in this report.

Keywords: nano crystalline cellulose, cellulose nanocrystal, self-organization, optical and structural properties

1 INTRODUCTION

The interest on NCC's optical properties is growing quickly. Since it is possible to generate colors with only natural and recyclable materials, like wood fibers, the application potential is highly interesting for the printing industry.

In order to follow the generation of color from the aqueous dispersion, new equipment was introduced in this study by our team. This is the laser light diffraction optical set-up. Thus, the liquid crystallization kinetics and the lattice dimensions in water were measured from just after shaking the dispersion for a better homogeneity to equilibrium. To better understand the role of electric charges in this process, the ionic concentration was measured and controlled. Although a powerful sonicator was initially used to better disperse the dilution, it also became a key player in the understanding of the crystallization process.

The optical activities of the polycrystalline solid film were observed with various optical set-ups and microscopes. The observations are converging toward the conclusion that the NCC nanorods in the aqueous suspension self organize into loosely organized thin layers. The layers are roughly parallel one to the other, with water in between. Upon drying, these liquid formations are shrinking gradually by the removal of water to multiple solid plates, stacked one on top of the other.

This research is very promising. This material coming from the forest, is easy to produce and abundant. Mass applications in several optical components may be possible, very respectful of the environment.

An interesting potential commercial application of NCC is to make colored films or active displays without using pigments. Since the material used in this work comes from the Canadian forest, it can be said NCC are environmental friendly. However, to better exploit any chromatic effect from NCC, it is obvious that the self organization mechanism present in the aqueous phase has to be better understood and controlled.

Indeed, several theories already exist (1,2). They all rest on the ability of the NCC to self orient, due to a combination of electric charges on the NCC surface, and twists in the NCC nanocrystals structure, a dislocation that should be present on all NCC, in an equal manner. These defects would provide the nanorods an asymmetric electric cloud around its main body, resulting with a helical structure.

Although quite satisfactory from the point of view of a conceptual model that explains most chromatic observations, this model is revisited. Nonobstantly, the most important point remains the color control and the industrial application.

2 MATERIAL AND METHOD

In order to better understand the organization structure, and the influence of various parameters, several instruments of various types were used. Also, our manipulations were either on the aqueous dispersion, or after evaporation, on the solid thin film.

The liquid dispersion was received from the local pulp and paper industries (FP Innovation, Paprican Division, Pointe-Claire, P. Québec, Canada). The NCC nanorods were prepared by exposing the cellulosic dispersion to sulfuric acid. The chemical reaction digests the amorphous component faster than the crystalline nanostructure. At some point, mostly nanostructures are remaining and the reaction is stopped. Therefore, the nanocrystals are separated from the rest and dispersed in water. The chemical process has charged the nanorods with electric negative charges.

The preparation was delivered to our laboratories free from any other components for instance salts or fungicides. The initial concentration is around 4% w/v. Then, the dispersion is first hand shaken, and then sonicated for 5 minutes with a Sonics Vibra cell ultrasonic (130 W, 20 kHz). It is centrifuged to remove remaining large aggregates, and forced through a 400 nm filter to further reduce the undesired formations. The dispersion is therefore passed through a resin bed (MTO-Dowex Marathon MR-3, SUPELCO Analytical). From this point, this is the NCC working preparation. It can be modified by adding sodium chloride to study the impact of ionic screening on the colloidal suspension.

The dispersion is then analyzed with a spectrophotometer. Assuming that the NCC dimensions are small compared to the wavelength, the Rayleigh diffusion is used to monitor the NCC size. Although the size distribution of nanorods is only estimated, and directly measured by AFM, it quickly reveals any changes in the average size for instance after an energetic sonication.

2.1 Preparation of thin solid disks

The NCC dispersion was deposited in Petri dish. The dish was placed in heavy Pyrex desiccators, itself placed onto foam. The assembly was installed on the shelf. If left to evaporate at 20 °C, fan like crystals are seen. If placed in a cold chamber kept at 4 °C, cubic crystals are seen. The final results are therefore a matter of crystallization conditions.

2.2 Laser diffraction

A new and simple optical set up was introduced. This is a He-Ne laser and a 5 mm glass cylinder containing NCC dispersed in water. The laser beam was focused at the center of the cylinder with a 250 mm focal converging lens. The resulting pattern was projected onto a screen 3 meters away. Typically, when only water was present, the result is a simple bright spot at the center of the screen with no other feature. With either latex, glass or other one hundred nanometer scale colloidal dispersions, a large quantity of speckles was seen moving at random around the central spot, without apparent slowing down after 10 minutes. With NCC dispersions, the initial rapid motion is followed by an apparent full stop within one or two minutes, depending of the initial concentration. The next motion is very slow, several orders of magnitude below the initial speed.

As the motion slows down, a ring of diffraction appears around the central spot. It reveals the formation of liquid crystals in the aqueous dispersion. Then later, a second ring appears wider than the first one, revealing a second diffraction order, thus a larger organization. As the time passes by, more rings appear, and they become brighter.

This indicates a higher degree of organization with time. The time required for the first ring formation is within one minute.

2.3 Atomic force microscopy

The NanoSurf easyScan 2 model of Atomic Force Microscope (AFM) was used. It is equipped with image software for a better contrast. The resolution limit allows the visualization of the molecular network. We can therefore easily see an individual nanorod, and the resulting organization after evaporation.

2.4 Scanning electron microscopy

In order to have a view of the stacking of plates one on top of the other, the solid thin films were broken and seen on the thickness side by a Scanning Electron Microscope (SEM). The images were then contrasted for a better analysis.

2.5 Optical microscopy

The resulting thin solid films, after evaporation, were observed with an optical microscope. Pictures were taken with a commercial 10 Mega pixel digital camera. Polarizers were also introduced in the optics, to better observe individual single crystals embedded in the films. Also, the films were broken into pieces. Then, single crystals were isolated and observed individually.

2.6 Sonication

A powerful sonication was used to disperse the NCC in the aqueous volume. However, it became apparent that the effect of sonication was more subtle. Usually, the sonicator was used at full power, for a given number of seconds. The instrument is able to give us the quantity of energy invested.

3 OBSERVATIONS

At the figure 1, an AFM image of a diluted aqueous NCC dispersion sonicated for 10 minutes and left to evaporate on a glass plate is showed. The glass plate was previously cleaned with acid and rinsed with Millipore deionized water. Observations showed nanorods chains connected by thin bridges apparently left over by the chemical reaction. From several pictures, the length distribution of each individual nanorod was established, and an average value was found. This wide size distribution is very common.

An interesting fact is that energetic sonication does not influence NCC morphology. This conclusion is confirmed by the light scattering measurement.

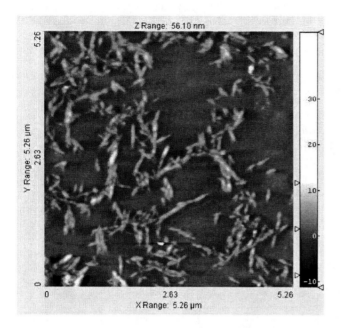

Figure 1: AFM picture of NCC after an energetic sonication. The nanorod chains are apparently intact.

A laser beam, after passing through a NCC aqueous dispersion, shows diffraction pattern. Right after shaking, only the central spot appears with moving speckles around it. After one minute, a steady circular ring appears. This shows that great quantities of small liquid crystals are formed, beginning almost immediately after shaking. Three minutes later, the circle becomes a hexagon (see above). This shows that the nanorod liquid crystals are growing into larger organization, proven by the appearance of new concentric rings. If this dispersion is shaken again, all patterns disappear, leaving again the central spot with speckles. Therefore, the process is reversible.

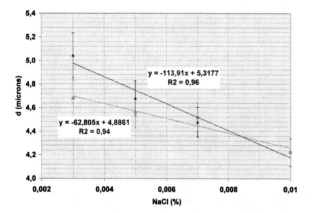

Figure 2: The liquid crystal lattice distance related with the salt concentration in neutral (square) and acidic (round) dispersions. A He-Ne laser beam passing through a NCC dispersion (5.53 % w/v) produces diffraction rings.

The crystal lattice distance d can be estimated by the following equation (for the cylindrical container):

$$d = \frac{m\lambda L}{nR} \qquad (1)$$

m is the diffraction order, L distance between the NCC dispersion and the screen, n the refraction index, R the ring horizontal radius, and λ the laser wavelength.

Once left to dry in a Petri dish, the preparations are showing bright colors, from blue to red. Pictures of Figure 3 are illustrating this effect. The films were also observed by optical microscopy as shown in Figure 4. Using crossed polarizers, square crystals formation were clearly observed. As well, the fingerprint pattern was also seen. Interestingly, this square pattern is one centimeter long. This was due to a very slow thinning of the NCC films.

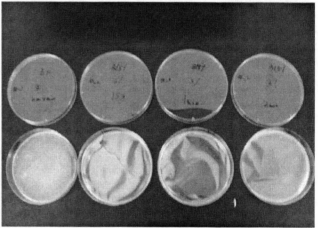

Figure 3: Petri dishes after the evaporation of NCC dispersions. The thin solid films final colors are due to the rotation of the light electric field. The reflection spectrum is complementary to the absorption spectrum.

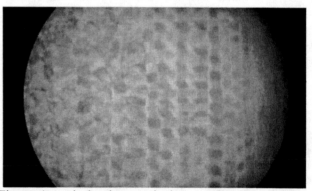

Figure 4: optical microscopic image of NCC solid thin films. Note the square formations.

These thin films were analyzed by spectroscopy. The peak wavelength is clearly dependent on the salt concentration. Increasing the salt concentration, the peak went from red to blue. Interestingly, the rate of chromatic shift is very similar to the one calculated on figure 2. This means that once formed, the liquid crystal keeps its structure while drying up to solid film. The lattice distance in the aqueous dispersion compared to the solid state is about 12 to 1.

AFM images of the solid film show a multilamellar system (Figure 5).

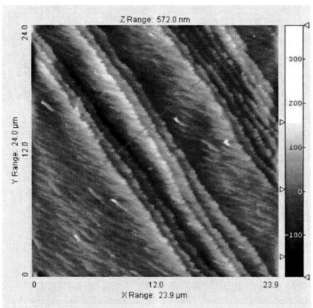

Figure 5: AFM picture of a NCC solid thin film. The lamellar structure can be easily seen. The general orientation of NCC fibers is also clearly visible. Each lamella is in the order of 100 nm thick.

The same system could be seen by SEM, and optical microscopy. The distance center-to-center is well below the micrometer. The orientation of the fibers is the same from plane to plane. The solid films spectra will have to be considered from the logic of the Bragg reflexion of a multilayer system.

4 CONCLUSION

Our experiments demonstrated that NCC self-organize in multilayers. Through AFM, SEM and optical observations, the research team established that the orientation of nanorod remained the same from one layer to the subsequent one. A possible interpretation is that our diffraction pattern of the aqueous dispersion comes from NCC liquid lamella stacked up after drying.

Our built model (3) verifies well this behavior. This behavior is mainly due to the fact that the real nanorods are not so identical, rather different in size, shape, and charge.

Instead of a long helical structure, a stratification of nanorods is observed. We keep studying these organizations, and a publication with a more detailed analysis will follow.

Nevertheless, the main point is that NCCs are capable to produce bright colors using wood and water. The following steps are to improve the contrast and better control the color opacity and wavelength.

REFERENCES

[1] Y. Habibi, L. A. Lucia and O. J. Rojas, Cellulose Nanocrystals: Chemistry, Self-Assembly, and Applications, Chem. Rev., 2010.

[2] D. Simon, Y. Kadiri and G. Picard, Nano cellulose crystallites: optical, photonic and electro-magnetic properties, NSTI, 2008.

[3] Y. Kadiri, D. Simon, G. Picard, F. Ghozayel and J.-D. Lebreux, Nano Crystalline Cellulose – Structural and Optical Properties by Molecular Dynamic Simulations NSTI, 2010.

INTERFACIAL EFFECT IN Fe/ InGaAs MAGNETISM

N. Berrouachedi[*], S. Rioual[**], B. Lescop[**], B. Rouvellou[**], J. Langlois[**], **M'**. Bouslama[*], Z. Lounis[*], A. Abdellaoui[*]

[*]*Laboratoire matériaux (LABMAT), ENSET d'Oran BP 1523 Oran Mnaouar, 31000 Oran, Algeria.*
Berr_naima@yahoo.fr
[**]*Laboratoire de Magnétisme de Bretagne (FRE 3117), Université de Bretagne Occidentale, UFR Sciences et Techniques, 6, av. Le Gorgeu 29285 Brest Cedex, France.*
rioual@univ-brest.fr
rioual@univ-brest.fr

ABSTRAT

Fe layers 4, 8, 20 and 60nm-thick, were epitaxially deposited by Molecular-Beam-Epitaxy on InGaAs/InP(100) wafers. We found that significant changes in cristallinity and microstructure occur with increasing the Fe thickness from the AFM measurement. Also, we have used both complementary techniques: VSM and FMR measurements to display how change the magnetic anisotropies as a function of thickness Fe layer. A strong in-plan uniaxial magnetic anisotropy with magnetic axes along the both major crystallographic directions is present at Fe/InGaAs/InP(100) interface for the 4-nm-thick Fe layer. The 8-nm-thick Fe film have a cubic anisotropy that is mixed with an interface induced uniaxial anisotropy. The magnetic anisotropy of the bulk Fe dominates for 20-nm-thick Fe film. On the other hand, in our FMR experiments, we observed unconventional two-picks FMR spectra for the thinner sample as compared to the bulk. This anomalous peaks, allowing one to discriminate between various in plane magnetic anisotropies. Moreover, after the thermal treatment a drastic change of the hysteresis loop has generated and the disappearance of the peak associated to the Fe (110) plane. In addition to this, we have performed a depth profile of thermally diffused Fe/InGaAs/InP using phoemission spectroscopy combined with Ar+ ion sputtering. We found that Fe ion was thermally diffused into the deep region of the wafer and the strong segregation of metallic indium at the surface under the UHV annealing treatment.

Keywords: Fe/InGaAs/InP; AFM, VSM and FMR; uniaxial and cubic magnetic anisotropy, DRX and XPS, annealing treatment

1. INTROCUCTION

Over the last decades, the characterization of Ferromagnet / Semiconductor interfaces has been a topic of great interest because of the possible future development of spintronics, e.g., spin injection through the ferromagnetic thin film to the semiconductor [1]. Particularly, the Fe/GaAs system has been extensively studied [2] due to its ferromagnetism at room temperature and the small lattice mismatch (1.3 %) between the two materials. The numerous investigations performed on such a system have revealed the crucial role played by the interface, particularly on the magnetic properties of iron. Indeed, elaboration of ultrathin films of iron has evidenced a reduction of its magnetization with respect to the bulk, and thus suggested the existence of a magnetically-deficient region at the interface [3]. Moreover, uniaxial magnetic anisotropy has been reported for iron ultrathin films. Its origin has been explained, at first, by Krebs *et al.* [4] by the dangling bond at GaAs surface. However, more recently, such an anisotropy has been observed for different GaAs surface reconstructions and consequently interpreted by magnetoelastic effects due to the lattice mismatch at the interface [5].

Another promising candidate for spintronic is the Fe/$In_{0.5}Ga_{0.5}As$ system. However, contrarily to Fe/GaAs, only few studies have been reported [6,7]. Magnetization reversal investigations [7] have demonstrated the presence of a uniaxial anisotropy for an iron thickness of less than 6 nm as well as a reduction of the magnetization at low iron coverage. Investigations of the interface after 1-hour annealing at temperatures in the range 350 - 450°C were performed by X-Ray Diffraction and Mosbauer techniques [6] and led to the identification of metallic indium, Fe-As compounds and $Fe_3Ga_{2-x}A_x$ (x = 0.2 – 0.3). The presence of indium at the interface is of high interest since it is expected to modify its magnetic properties with respect to GaAs. This is of first relevance since the indium atoms are known to strongly diffuse, even at moderate temperatures, and to not make alloy with iron. In the present paper, we focus on the characterization of Fe / $In_{0.5}Ga_{0.5}As$ / InP (001) heterostructure. Iron layers with different thicknesses were epitaxially deposited. The magnetic properties of the as-deposited samples will be studied and explained by a competition between interfacial and bulk iron magnetism. To point out the main moving species at the interface, thermal treatment of the heterostructures will then structurally and magnetically be investigated.

2. EXPERIMENTAL DETAILS

Iron films with thicknesses from 4 to 60 nm were grown by Molecular Beam Epitaxy (MBE) deposition onto oriented semiconductor $In_{0.5}Ga_{0.5}As$ / InP(100) substrates. $In_{0.5}Ga_{0.5}As$ was grown by metal-organic chemical vapor deposition (MOCVD) onto an InP (100) substrate within the *Laboratoire des Multimatériaux et Interfaces*. Because of the mismatch agreement (a = 0.5868 nm) between InGaAs layer and InP(001) substrate, there is a limited amount of constraints. Iron growth was achieved in a UHV chamber at a base pressure of 5 x 10^{-10} Torr. InGaAs substrates were preliminary degreased with acetone, and the native oxide was removed by Ar^+ spttering (800 eV). The Fe deposition (4, 8, 20 and 60 nm) was performed with a Knudsen cell, and the substrate temperature was 100 ± 10 °C. The deposition rate was calibrated with a quartz microbalance and was fixed at 2.0 Å / min

The deposition of 4, 8, 20 and 60 nm of iron was magnetically characterized by Ferromagnetic Resonance (FMR) and Vibrating Sample Magnetometer (VSM) experiments. FMR measurements were made with a standard spectrometer ELEXYS 500-BRUKER operating in X band at 9.7 GHz under application of the field in the sample plane. The morphology and the surface roughness of the samples were controlled through Atomic Force microscopy (AFM) experiments. The eventual diffusion of species was studied by both X-Ray Diffraction (XRD) and X-ray photoelectron spectroscopy (XPS) depth profile; the latter permits one to investigate the different layers after controlled Ar^+ etching [8].

3. RESULTS AND DISCUSSION

Figures 1(a) and 1(b) show the morphologies of 4- and 20-nm-thick iron layers. They clearly highlight differences in roughness and morphology. The 4-nm-thick Fe film looks flat, whereas granular structures features are observed for the higher thickness. The root mean square is 0.4nm for the first sample and 1.4 nm for the other. These differences in roughness and morphology are induced by the difficulty to growth epitaxial 3D films; they likely affect the magnetic properties, i.e. magnetic moments, magnetic anisotropy or pinning of domain wall.

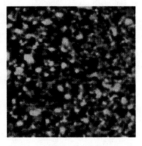

Fig 1. Atomic force microscope (AFM) images of 4-nm-thick (a) and 20-nm-thick (b) Fe layer on InGaAs (001) substrate. Scan area was 1 µm x 1 µm.

Figure 2 shows the magnetization reversal for 4 different iron thicknesses at room temperature. It displays clearly an increase of the coercivity field, H_C, with iron thickness. This behavior seems thus be correlated to the roughness increase observed by AFM. However, since the magnetization reversal is a rather complex mechanism, the relationship between coercivity and roughness for a rather thick film is not a simple monotonic relationship [9]. A similar effect was recently been reported on Fe/GaAs (001) [10]. The comparison of the different pictures highlights a second magnetic effect evidenced through the occurrence of a uniaxial anisotropy when the thickness is 4 nm: it is clearly evidenced in Fig. 2(a) by the existence of magnetic easy and hard axis. Increase in thickness causes a

reduction of the uniaxial anisotropy (Fig. 2(b)) prior to its complete disappearance (Figs. 2c and 2d). Iron layer being essentially single domain by nature, this observation confirms the competition between the uniaxial anisotropy related to the first layers and the anisotropy associated to the Fe bulk as described by Richomme et al. [7]. The shape of the hysteresis loops is unchanged, and the squareness is almost constant from 8 nm to higher coverage. For the 4 nm sample, the coercive field, H_C, in the easy axis was measured at 15 ± 2 Oe. For comparison, the H_C value for a similar iron layer deposited on Si (100) wafer under the same conditions was measured and found equal to 6 ± 1 Oe. This low value highlights the strong influence exerted by the interface on H_C for the lowest thickness sample. Its observed enhancement in Fe/InGaAs might be due to species diffusion and / or alloy formation as already reported by Haque et al. [11] for Nickel thin films deposited on Si and GaAs substrates.

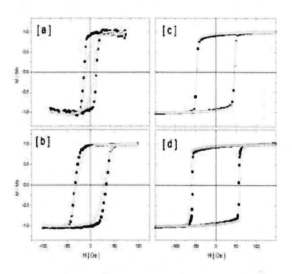

Fig. 2. Hysteresis loops for Fe films of different thicknesses (a: 4-, b: 8-, c: 20- and d: 60-nm) deposited on InGaAs (001). Both cycles (hard o, easy ■ axis) were perpendicularly acquired. Magnetization was normalized to M_S at saturation.

To elucidate this point, the samples were characterized by FMR, a technique highly sensitive to interfaces. Figure 3 illustrates the results of measurements and shows that, at high iron coverage (8 and 20 nm), only one peak is observed. It corresponds to the uniform mode due to the precession of layer magnetization around the applied field. The observed increase of the FMR peak width at higher iron thicknesses is correlated with the number of inhomogeneities in the layer due, for example, to the increase of the roughness. Decreasing the iron thickness makes appear several modes: the 4-nm spectrum shows a double peak structure. These structures are explained by a uniform mode in relation with the iron layer at low field and to an interface mode, respectively. They constitute evidence of the presence of two magnetic iron species at the interfaces. Furthermore, the shift towards high fields for the two peaks observed at 4 nm is explained by the reduction of the magnetization at low coverage.

As shown by the interfacial FMR mode and by the enhancement of H_C for the 4 nm sample, the interaction between the substrate and the firsts iron layers is a key parameter in the magnetic properties of the bilayer. To investigate the mobility of ion during the deposition and to thermal annealing at 250°C for 90 minutes in an Ultra

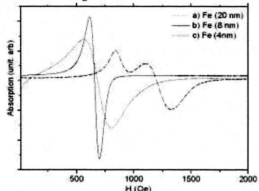

Fig. 3. FMR Spectra of Fe (t) / InGaAs (001): t = 20 nm (a); t = 8 nm (b) and t = 4 nm (c)

high Vacuum chamber. Figure 4 presents the magnetization reversal of a 60-nm-thick Fe layer without and with thermal annealing. It shows a drastic change of the hysteresis loop generated by the thermal treatment; the coercivity value being about 650 Oe. Moreover, the cycle shape is no longer a square, and discontinuities appear. The observed decrease of the saturation magnetization by a factor of 3 suggests the creation of non magnetic alloys. Similarly, no ferromagnetic signature was observed after the thermal treatment of the 20 nm layer. These results prove the strong interdiffusion process in the innerlayers in contrast to the indication of only a weak interdiffusion below 350°C reported by Monterverde and al. [6].

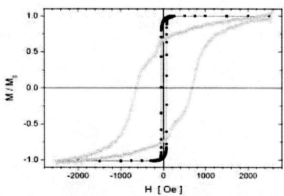

Fig. 4. Hysteresis loops for 60-nm-thick Fe layer deposited on InGaAs (001). The black and open dots correspond to data before and after annealing, respectively.

The 60-nm-thick iron sample was studied by XRD before and after annealing. The θ-2θ XRD pattern is dominated by the (002) and (004) reflections from the substrate InGaAs / InP (001). Figure 5 presents a restricted (40° to 48°) XRD pattern located around the Fe (110) peak located at 44.68°. One should note, after the annealing treatment, the disappearance of the peak associated to the Fe (110) plane. This is consistent with the study by Montverde *et*

al. [6] who also reported on the disappearance of the Fe(110) at 400°C. However, in contrast to this report, one

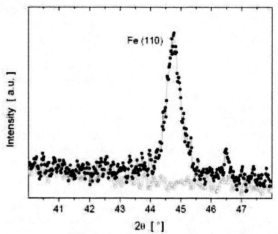

Fig. 5. θ/2θ diffraction pattern obtained for a 60-nm-thick Fe film deposited on InGaAs (001). The black and open dots correspond to data before and after annealing, respectively.

should note, here, after the annealing at low temperature, the lack of the additional peaks attributed to the presence of Indium and Fe_2As. This suggests a subsequent process consisting in a strong migration of atoms followed with crystallization at higher temperature. To gain more insight into the interdiffusion process, XPS measurements were made after different ion bombardment times, *i.e.* at different depths. To extract the depth profile composition, the Ga(3p), As(3p), P(2p), In ($3p_{3/2}$) and Fe(3p) spectra were fitted and subsequently corrected by their atomic

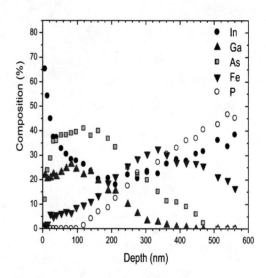

Fig. 6. XPS depth profile of Fe (60 nm) / $In_{0.5}Ga_{0.5}As$ (190 nm) / InP after *thermal annealing*

atomic sensitivity factors and sputtering yields, which were determined from SRIMS simulations. On Fig. 6, it is worth noting the strong interdiffusion of Iron within the whole system with a shallow maximum at about the 300-observation, at a depth of about 5 nm, of an outerlayer nm depth. This explains the reduction of the magnetization as well as the increase of the coercitive field. Besides, the region mainly composed of indium is indicative of the strong segregation of metallic indium at the surface under surface.under the UHV annealing treatment. Between 5 and 200 nm, Fig. 6 also shows an increase of the arsenic content and a decrease of the indium one, which lead to nearly the original InGaAs composition. The observation of P atoms released from the InP substrate from 120 nm indicates a migration towards the

4. CONCLUSION

The epitaxial growth of Fe films of different thicknesses (up to 60 nm) was achieved on InGaAs/InP(100) substrates by Molecular beam Epitaxy. FMR and VSM techniques provided evidence of the strong influence of the interface on the magnetic properties. Among them, a uniaxial magnetic anisotropy as well as an FMR interface mode were both revealed for the 4-nm Fe/InGaAs/InP(100) film. The thermal annealing of iron deposited at moderate temperature (250°C) in an Ultra High Vacuum chamber highlighted the strong mobility of atoms, which is a directly correlated dramatic change in magnetic properties. It results in both a diffusion of iron atoms within the multilayers and a strong segregation of indium atoms towards the surface. Consequently, even low temperature annealing has to be taken into account for technical applications.

ACKNOWLEDGMENTS

The authors wish to thank L. Auvray (LMI–UMR 5615) for the InGaAs / InP (001) elaboration as well as P. Elies and N. Kervarec for AFM and FMR measurements.

REFERENCES

[1]/ S. Datta and B. Das, Appl. Phys. Lett. 56, 665 (1990).

[2]/ Y. Chye, V. Huard, M.E. White and P.M. Petroff, Appl. Phys. Lett. 80, 449 (2002).

[3]/ G.A. Prinz and J.J. Krebs, App. Phys. Lett. 39, 397 (1981).

[4]/ J.J. Krebs, B.T. Jonker and A.Prinz, J. Appl. Phys. 61, 2569.(1987)

[5]/ O. Thomas, Q. Shen, P. Shieffer, N. Tournerie, and B. Lépine, Phys. Rev. Lett 90, 17205 (2003).

[6]/ F. Monteverde, A. Michel, A. Fnidiki and J.P. Eymery, Eur. Phys. J. 21, 179 (2003).

[7]/ F. Richomme, A. Fnidiki and J.P. Eymery, J. Appl. Phys. 97, 23902 (2005) 1.

[8]/ M. Salou, S. Rioual, J. Ben Youssef, D.T. Dekadjevi, S.P. Pogossian, P. Jonnard, K Le Guen, G. Gamblin, B. Rouvellou, Surf. Interf. Anal. 40, 1318. (2008)

[9]/ Y.-P. Zhao, R.M. Gamache, G.-C. Wang, T.-M. Lu, G. Palasantzas and J. Th. M. De Bosson, J. Appl. Phys. 89, 1325 (2001).

[10]J. Islam, Y. Yamamoto, E. Shikoh, A. Fujiwara and H. Hori, J. Magn. Magn. Mater. 320, 571 (2008).

[11]S.A. Haque, A. Matsuo, Y. Yamamoto and H. Hori, Physica B 325, 259 (2003).

Bundling of double wall carbon nanotubes and Raman G band

P. Puech[*], S. Nanot[**], B. Raquet[**], J.M. Broto[**], M. Millot[**], A.W. Anwar[*], E. Flahaut[***]
and W.S. Bacsa[*]

[*]CEMES, UPS, CNRS, Université de Toulouse
23 rue Jeanne Marvig, Toulouse 31055, France, wolfgang.bacsa@cemes.fr
[**]LNCMI, UPS, INSA, CNRS, Université de Toulouse
143 av. de Rangeuil, 31077 Toulouse, France
[***]CIRIMAT, UPS, CNRS, Université de Toulouse
118 route de Narbonne, 31062 Toulouse, France

ABSTRACT

We find that the Raman G band for individual double wall carbon nanotubes is split when the two walls are closely spaced or strongly coupled. We show that agglomeration of tubes into bundles leads to line broadening and overlapping of the G bands from the inner and outer tubes. This demonstrates that the observed broadening is not due to defects in the tube walls. We identify the tube helicity of an individual double wall tube. The consequence of our finding is that mapping of the Raman G band allows us to access the dispersion state of carbon nanotubes in a composite matrix.

Keywords: nanotubes, Raman spectroscopy, composites, dispersion

1 INTRODUCTION

The dispersion of carbon nanotubes in a polymer composite is a key challenge. The control of the dispersion state of carbon nanotubes (CNTs) allows reducing the percolation threshold. The formation of the percolation network in composites is used to evacuate electrical charges or for reinforcement. The percolation thresholds in CNT composites varies by more than two orders of magnitude (0.01-1w%) depending on tube type used [1]. The composition of carbon nanotube samples itself varies on a number of process parameters which are often not well controlled. As a result the agglomeration state and exact composition differs considerably. When mixing the tubes in the composite several additional process parameters play an important role in the final dispersion state of the tubes. To take advantage of the one dimensional nature of carbon nanotubes, it is essential to control their dispersion state to guarantee low percolation threshold. In double wall carbon nanotubes (DWs), the two walls are subject to a different environment. While the inner tube is surrounded by the outer tube, the outer tube is surrounded by the matrix or neighbouring DWs. As a result the interaction of the tube walls differs for the inner and outer tube which results in different spectroscopic positions of the Raman G band for inner and outer tubes.

DWs can be grown either by conversion of peadpods [2] or through the catalytic chemical vapor deposition method (CCVD) [3]. The wall spacing is typically smaller for CCVD [4] grown DWs while DWs grown from peapods show larger inter wall spacing [5]. This difference can be explained by the different growth conditions of the DWs. In the case of DWs grown with the CCV method, the inner and outer tubes are formed at the same time while in the case of peapod conversion, the inner tube is formed in the presence of a already present outer tube. Raman spectra of RBMs are consistent with this difference [4, 5]. The G band line shape also shows clear differences for the two types of DWs due to changes in the tube coupling. The pressure transmission on the inner tube is delaying for DWs grown from peapods as compard for DWs grown with the CCVD method [6]. The two types of DWs show also differences in the G'2D band. The G'2D band for DWs grown form peapods has two separated contributions due to inner and outer tubes while the G'2D band for DWs grown with the CCVD method shows one band with a shoulder similar to what is observed for graphite. We are here using DWs grown with CCVD and with small wall spacing.

To establish the difference in the Raman G band between individual and bundled DWs we have performed measurements on individual DWs and on isolated bundles of DWs on silicon oxide and compared the spectroscopic band positions and line broadening as a function of excitation wavelength.

2 EXPERIMENTAL

Raman spectra were acquired with a T64000 spectrometer from Horiba Jobin-Yvon. The laser power was measured after the objective. Polarization has been selected along the tube axis according to the scanning force microscopy images of the tube. We find that laser heating is less important for individual DWs in contact with the substrate or connected to the metal electrode. We use 1 mW with an objective of magnitude 100. We find that bundles of DWs are more sensitive to laser power. To reduce heating effects on DW bundles we use a low laser power, typically 0.1 mW and an objective of magnitude 40 is used to increase the focal spot size (2 μm) decreasing the power density. All spectra have been recorded by integrating the signal from the scattered light for 100-500 seconds. We were able to increase the laser power up to 3 mW before the sample deteriorated.

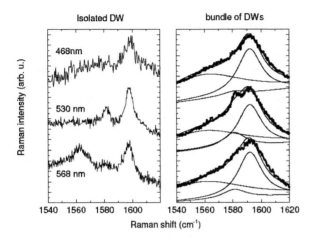

FIG 1: G band of individual DW (left side) and bundles of DWs (right side) using three excitation energies.

Fig. 1 shows the Raman G band region of the individual DW and bundled DWs excited with three different excitation wavelengths. By changing the excitation wavelength we find the resonance which is specific to the tube structure and this allows us to assign the tube structure of the inner and outer tube. The spectra from the isolated DW are narrower and contain up to three spectral bands. Bundles of DWs show a larger and asymmetric G band. The spectra of the DW bundles have been fitted by three Lorentzian line shapes using fixed spectral position and taking the relative intensities as a free parameter [6]. The half width at have maxima (HWHM) is considerable smaller for individual DWs. The G band of the inner and outer tube is 4±1 cm−1 which is the same HWHM observed for isolated SWs or for single layer graphene [17]. The HWHM in general depend on the number of defects and it has been shown that for SWs that the HWHM varies with applied voltage with a minimum value of 4 cm^{-1} [18, 19]. For bundles of DWs the HWHM is 10 cm^{-1}. The larger HWHM for bundles can be explained by the interaction with neighbouring tubes leading to heterogeneous line broadening. For comparison, pyrolytic graphite has a similar HWHM (7 cm^{-1}) [20].

For the individual DW, the spectral position of the G band of the inner tube is at 1580±2 cm^{-1} and for the outer tube at 1597±2 cm^{-1}. The spectral position of the inner tube is consistent with the extrapolated spectral position deduced from high pressure experiments [21] and when studying the

influence by chemical doping of DWs [15]. The spectral position of the G band for the outer tube is high compared to the G band in single wall tubes. When changing the excitation energy, we can see considerable changes in the intensity of the inner tube reflecting changes in the resonance condition. At 568 nm an additional band is observed at 1563 cm^{-1} which has been previously observed and identified as being associated with electronic coupling with the environment and the outer tube [6].

To identify the tube structure of the inner and outer tube we use the electronic transition energies from the Kataura plot as reported by Araujo et al [13]. We can then estimate the transition energies for E^S_{11}, E^S_{22} and E^M_{11} where S denotes semiconducting tubes and M metallic tubes. For the E^S_{33} and E^S_{44} or E^S_{55} and E^S_{66}, we use for $\gamma_p = 0.305$ for unbound excitonic states as suggested by the authors where γ_p is a correction associated to exciton localisation. For the E^M_{22} transition, we use $\gamma_p = 0.305$ which fits well with the experimental results of Sfeir et al [22] reducing the error to less than 0.1 eV for the corresponding transition energies.

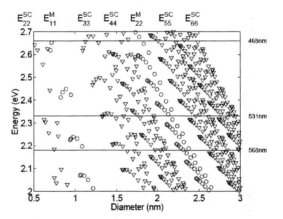

FIG. 2: Optical transition energies from Araujo et al [12] indicating the 3 different excitation wavelengths. The circles show the transition energies for metallic and the triangles show the transition energies for semiconducting tubes.

From the diameter histogram of our DW sample as determined by TEM, we consider 3 configurations with increasing diameter: M@S, S@M and S@S. We use the following notation for the four combinations of tubes for DWs: M@M, S@M, M@S, S@S (inner@outer). From height estimations using scanning force microscopy, we conclude that the M@S configuration with an outer diameter of 1.7 nm and an inner one of 1 nm is the most likely configuration for the DW investigated here. A S@S configuration would imply an outer diameter of 2.7 nm and an inner diameter of 2 nm. This configuration can be excluded by our preliminary electronic transport measurements which show a small on/off ratio in the source-drain current vs. gate voltage dependence characteristic for metallic tubes [23]. From the gate voltage at the largest variation of the source drain current we find that the semiconducting tube is p doped.

We note that for a similar DW, M@S configuration and similar diameter, Villalpando-Paez et al [14] found a G band frequency of 1591 cm^{-1} without separating between contributions from the inner and the outer tube although the tube diameters are similar even so the samples have been grown using the CCVD method. Interestingly no splitting is observed here for the G'2D band while Villalpando-Paez et al [14] observe a clear splitting of the G'2D as in the case for DWs grown from the peapod method. This shows that the two tubes of DWs grown with the CCVD method can be either strongly coupled or can be decoupled as in the case for DWS grown with the peapod method. It is clear that the parameters used in the CCVD method play an important role in the inter wall spacing. In our case, the inner and outer tubes are strongly coupled, leading to a clear difference in the spectral positions. The G band position associated with the outer tube of the individual DW is high. We attribute this to charge transfer due to the proximity of the palladium contact.

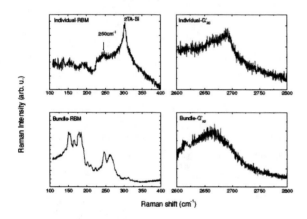

FIG 3: Frequency range of RBM and G'2D band of individual DW on SiO2 (Top) and DW bundles (below).

Fig. 3 shows the radial breathing mode (RBM) region and the second order D band (G'2D band) of the individual DW and bundled DWs using 531 nm excitation. The absence of sharp RBM bands in the case of individual tube indicates that the excitation does not coincide with the resonance maximum with neither of the two tubes. For bundles we observe several RBMs in the 140-180 cm^{-1} and 240-270 cm^{-1} spectral range. For the individual DW we observe a RBM band at 250 cm^{-1} consistent with an inner tube of 0.94 nm diameter and a less intense spectral band at 150 cm^{-1} consistent with the RBM of the outer tube (diameter: 1.61 nm). We use the constants for the determination of the diameter from the RBM frequency as reported by Telg et al

[7]. The G'2D band for isolated tubes is less broad then those for bundles.

By varying the laser power from 1 mW to 3 mW for the individual DW on SiO2 using 568 nm excitation, we observe changes in the G band of the outer tube and the disappearance of the additional spectral band at 1560 cm^{-1}. It has been previously found that this additional band is correlated with the shift of the G band of the outer tube with chemical doping [6]. The G band of the outer tube shifts to higher frequency and broadens with laser power while the G band of the inner tube remains at the same spectral position. A temperature increase, however, is known to shift the G band to lower frequency and broadening the band uniformly. The up shift and non-uniform broadening can be associated to stronger interaction with the substrate leading to a larger doping. Doping can lead to an up shift of the G band and can increase the HWHM for semi-conducting tubes [24]. This is compatible with a inner metallic and outer semiconducting tube.

3 DISCUSSION AND CONCLUSION

Electrical conductance measurements have shown that the signal associated to the outer tube can up shift or down shift and shows hysteresis when scanning the applied voltage [16]. As the tube lies on an insulating SiO2 layer, the Fermi level of the outer tube can be influenced when illuminated through its interaction with its environment. By increasing the laser power at 568 nm close to the resonance for the E_{33} transition energy one increases the population of delocalised excitons on the outer tube. The presence of the substrate and the increased population of excitons in the outer tube modifies the electronic band structure. The outer tube is in contact with the substrate and the palladium contact. This is correlated with the disappearance with the additional band observed at 1560 cm^{-1}. The G band of the inner tube is not influenced by the laser power increase and shows that the inner tube can be used as a reference.

We find that coupled individual DWs show narrow and separated G bands corresponding to the inner and outer tube. The line-widths of the G band of the inner and outer tubes are comparable to what has been reported for individual SWs and single layer graphene. Bundling broadens the G band considerably. Using height estimation from scanning force microscopy, excitation wavelength and preliminary transport measurements, we can identify the tube configuration to be M@S. An increase of the laser power at 568nm leads to a preferential modification of the outer tube which is correlated with the disappearance of the additional band at 1560 cm^{-1} associated with the outer tube. Double wall carbon nanotubes with closely spaced walls are thus suitable to determine the dispersion state in a carbon nanotube polymer composite. Splitting of the Raman G band can be used as an indication of the presence of individual tubes as compared to broader G bands indicating the presence of bundled tubes.

REFERENCES

[1] W. Bauhofer et al Composite Science and Technology 69 (2009) 1486

[2] S. Bandow, M. Takizawa, K. Hirahara, M. Yudasaka, and S. Iijima, Chem. Phys. Lett. 337, 48 (2001)

[3] E. Flahaut, R. Bacsa, A. Peigney and Ch. Laurent, Chemical Communication 12, 1442 (2003)

[4] R. R. Bacsa, A. Peigney, Ch. Laurent, P. Puech, and W. S. Bacsa, Phys. Rev. B 65, 161404 (2002)

[5] R. Pfeiffer, F. Simon, H. Kuzmany, and V. N. Popov, Phys. Rev. B 72, 161404 (2005)

[6] P. Puech, A. Ghandour, A. Sapelkin, C. Tinguely, E. Flahaut, D.J. Dunstan and W. Bacsa, Phys. Rev. B 78, 045413 (2008)

[7] A. M. Rao, E. Richter, S. Bandow, B. Chase, P. C. Eklund et al., Science 275, 187 (1997)

[8] H. Telg, J. Maultzsch, S. Reich, F. Hennrich, and C. Thomsen, Phys. Rev. Lett. 93, 177401 (2004)

[9] C. Fantini, A. Jorio, M. Souza, M. S. Strano et al., Phys. Rev. Lett. 93, 147406 (2004)

[10] I.C. Gerber, P. Puech, A. Gannouni, and W. Bacsa, Phys. Rev. B79, 075423 (2009)

[11] H. Kataura, Y. Kumazawa, Y. Maniwa, I Umezu, S. Suzuki, Y. Ohtsuka and Y. Achiba, Sythetic Metals 103, 2555 (1999)

[12] S.M. Bachilo, M.S. Strano, C. Kittrell, R.H. Hauge, R.E. Smalley, and R.B. Weisman, Science 298, 2361 (2002)

[13] P.T. Araujo, S.K. Doorn, S. Kilina, S. Tretiak, E. Einarsson, S. Maruyama, H. Chacham, M.A. Pimenta and A. Jorio, Phys. Rev. Lett. 98, 067401 (2007)

[14] F. Villalpando-Paez, H. Son, D. Nezich, Y.P. Hsieh, J. Kong, Y.A. Kim, D. Shimamoto, H. Muramatsu, T. Hayashi, M. Endo, M. Terrones, and M.S. Dresselhaus, Nanoletters 8, 3879 (2008)

[15] G. Chen, S. Bandow, E.R. Margine, C. Nisoli, A.N. Kolmogorov, V.H. Crespi, R. Gupta, G.U. Sumanasekera, S. Iijima,and P.C. Eklund, Phys. Rev. Lett. 90, 257403 (2003)

[16] S. Yuan, Q. Zhang, Y. You, Z.X. Shen, D. Shimamoto and M. Endo, Nano Lett.9, 383, (2009)

[17] A. Das, S. Pisana, B. Chakraborty, S. Piscanec, S.K. Saha, U.V. Waghmare, K.S. Novoselov, H.R. Krishnamurthy, A.K. Geim, A.C. Ferrari, and A.K. Sood, Nature Nanotechn. 3, 210 (2008)

[18] J.C. Tsang, M. Freitag, V. Perebeinos, J. Liu and Ph. Avouris, Nature Nanotechn. (2007)

[19] A. Jorio, C. Fantini, M.S.S. Dantas, M.A. Pimenta, A.G. Souza Filho, Ge. G. Samsonidze, V.W. Brar, G. Dresselhaus, M. S. Dresselhaus, A. K. Swan, M. S. Unlu, B. B. Goldberg, and R. Saito, Phys. Rev. B 66, 115411 (2002)

[20] M. Hanfland, H. Beister, and K. Syassen, Phys. Rev. B 39, 12598 (1989)

[21] P. Puech, H. Hubel, D. Dunstan, R.R. Bacsa, C. Laurent, W.S. Bacsa, Phys. Rev. Lett. 93, 095506 (2004)

[22] M.Y. Sfeir, T. Beetz, F. Wang, L. Huang, X.M. Henry Huang, M. Huang, J. Hone, S. O'Brien, J.A. Msewich, T.F. Heinz, L. Wu Y. Zhu, and L.E. Brus, Science 312, 554 (2006)

[23] S. Wang, X. L. Liang, Q. Chen, Z. Y. Zhang, and L.-M. Peng, J. Phys. Chem. B 109, 17361 (2005)

[24] A. Das and A.K. Sood, Phys. Rev. B 79, 235429 (2009)

Carbon-based nanocomposite EDLC supercapacitors

C. Lei and C. Lekakou

Faculty of Engineering and Physical Sciences, University of Surrey,
Guildford, Surrey GU2 7XH, UK, C.Lekakou@surrey.ac.uk

ABSTRACT

EDLC cells based on activated carbon as host material electrodes were fabricated and tested. The MWCNT additive was added into the host material to form a nanocomposite and the effect of the MWCNTs concentration on improving the performance of the EDLC cells was explored. 1M TEABF4-PC solution was used as organic electrolyte. The nanocomposites functioned differently in improving the specific power and energy density of EDLC cells., where impedance spectroscopy, cyclic voltammetry and galvanostatic charge-discharge measurements were performed to characterize the EDLC cells.

Keywords: Supercapacitors, EDLC, carbon, multicall carbon nanotubes

1 INTRODUCTION

A supercapacitor known as electrochemical double layer capacitor (EDLC) is an electric device with power and energy densities between those of a traditional capacitor and battery [1]. As the demand for energy storage devices with a high power and a long durability increases, the supercapacitor becomes more and more important. The EDLC supercapacitor differs from a traditional capacitor in that its electrodes are composed of a porous conductor such as activated carbon, which has a huge surface area, and it accumulates and keeps the charges on the thin layer of electrode/electrolyte interface via electrostatic force or non-faradic effect, so that it possesses a huge capacitance (> 100F/g), and has higher power and much longer recycle lifetime (>100000 cycles) than rechargeable batteries [2]. However, up to now the energy density of EDLC supercapacitors was not so high [1,2].

Carbonaceous materials such as carbon aerogel [3,4], powder [5,6] and carbon fiber are the most commonly used materials as electrodes in EDLC devices, as carbon can have high surface area, is chemically and thermally stable, of relatively low cost and environmentally friendly. A method to increase EDLC performance includes creating novel carbon nanocomposite electrodes [7], which increasing the conductivity and surface area of the electrodes.

In this report, a series of carbon based symmetric electrodes for EDLCs were explored, using a commercial, activated carbon powder as the basic active material and carbon nanotubes as a conductive filler. The performances of organic electrolyte based on LiPF6 or Et4NBF4 in mixed solvents were also explored, which have an electrochemical window of more than 3V [8,9]. A VersaSTAT MC analyser was used to carry out impedance spectroscopy, cyclic voltammetry and galvanostatic charge-discharge tests on the devices.

2 EXPERIMENTAL DETAILS

2.1 Carbon electrodes

Activated carbon powder (AC) with a surface area of 1000m²/g was used as the host electrode material. Multiwall carbon nanotube powder (MWCNT) was added without modification, in a series of investigations at a weight percentage of 0.15 wt%, 1 wt% and 7 wt%, respectively. PVDF (Poly(vinylidene fluoride)) was added as a binder at 5 wt.% of total solid component. The carbonaceous film had an areal density of 4~5 mg/cm².

2.2 Fabrication of capacitors

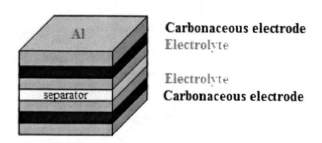

Figure 1. Diagram of an EDLC supercapacitor cell

Figure 1 presents a diagram of the type of supercapacitor cells fabricated in this study. The electrolyte was 1M Et4NBF4 (Tetraethylammonium tetrafluoroborate or TEABF4) in PC (propylene carbonate). Both the salt and solvent were from Sigma-Aldrich, Et4NBF4 was 99% pure, PC was anhydrous, 99.7% pure.

To assemble a capacitor cell, two rectangle strips of the carbonaceous material coated Al were cut out and combined with the carbon side face to face, a separator was soaked with electrolyte solution and was sandwiched in between. Thus a symmetric electrode EDLC cell was formed, with the electrodes overlap area defined as the working area, which was fixed to 2cm² in all the devices.

2.3 Electrochemical characterization of capacitors

The performance of different EDLC cells was determined by impedance spectroscopy, cyclic voltammetry and galvanostatic charge-discharge tests, using a VersaSTAT MC analyser. During the electrochemical tests, the EDLC cells were put under a pressure of 0.5Kg/cm² to ensure a good electrical contact of carbon-Al, and a fixed gap between the two electrodes. The devices were characterized in a 2-terminal configuration like in most commercially packaged electrical capacitor devices. No reference electrode was used in the tests.

3 RESULTS AND DISCUSSION

3.1 Impedance spectroscopy analysis

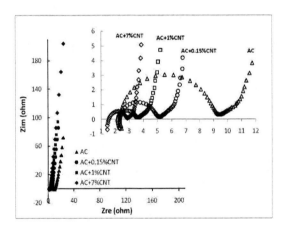

Figure 2. (a) Nyquist plots for the EDLC cells with AC, AC+0.15%CNT, AC+1%CNT and AC+7%CNT electrodes at a dc bias of 0.2 V, sinusoidal signal of 20 mV over the frequency range from 1000 kHz to 0.01 Hz. Z_{re}: real impedance. Z_{im}: imaginary impedance. Inset shows an enlarged scale graph in the high frequency region.

The Electrochemical Impedance Spectroscopy (EIS) measurements were carried out at a DC bias of 0.2 V with sinusoidal signal of 20 mV over the frequency range from 1MHz to 10mHz. Figure 2 shows the corresponding Nyquist plot for four devices with AC, AC+0.15%CNT, AC+1%CNT and AC+7%CNT carbonaceous electrodes. In the region of high frequency to medium frequencies, only small amounts of charge complexes can overcome the activation energy to migrate with the alternating potential. The Z_{im}-Z_{re} semicircle is developed by the moving charge-complexes close to the Helmholtz plane and is represented by an interfacial contact capacitance (Cc) and charge

transfer resistance (Rc) [10]. The semicircle has 2 intersection points on the real axis, the Zre value at the left end of the semicircle represents the equivalent series resistance (ESR) of the capacitor, which is the combination of the carbon-Al contact resistance, the bulk solution resistance of the electrolyte and the resistance of the electrode material itself: ESR values are similar for all electrodes except for the AC+7%CNT electrode which has a very low ESR. The diameter of the semi-circle represents the charge transfer resistance Rc: they are 7.1Ω, 2.8Ω, 1.4Ω and 1.2Ω for the AC, AC+0.15%CNT, AC+1%CNT and AC+7%CNT electrodes, respectively. This indicates that the conductive MWCNT can reduce the charge transfer resistance. On the right side of the semicircle, a ~45° Warbury region was observed, which is the result of the ion diffusion into the bulk of the electrodes, all of which have a pore size distribution [11]. At lower frequency range, the curve is almost vertical, indicating the capacitive nature of the device.

3.2 Cyclic voltametry analysis

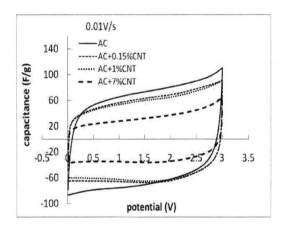

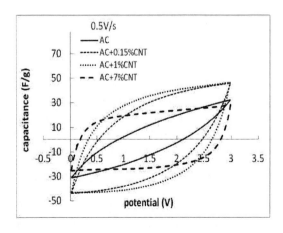

Figure 3. Cyclic voltammograms of EDLC cells with AC, AC+0.15%CNT, AC+1%CNT and AC+7%CNT electrodes using 1M TEABF₄-PC electrolyte at two extreme scan rates: (a) 0.01V/s, (b) 0.5V/s

The cyclic voltammetry (CV) response of the three devices was conducted between 0 and 3V at a range of scan rate from 0.01V/s to 0.5V/s, and the results are presented in Figure 3. For each cell, there are two typical CV curves at two extreme scan rates, from which a normalized specific capacitance, C, can be derived according to the following equation:

$$C = \frac{i}{mv}$$

where i is the current, v is the scan rate, m is the mass of one electrode (the current collectors not included). The curves are regular and symmetric to the 0 horizontal axis, with a Faradaic fraction [12] of ~0.05 in the 0-3 V scanned region, indicating a non-redox or non-faradic property of the cell.

At the low scan rate, the capacitances of all AC+CNT electrodes are lower than the AC only electrode, the higher the CNT percentage the lower the capacitance is. This can be justified by the fact that at the low charge/discharge rate, all surface are of AC is used in which case the high energy storage capacity of the AC carbonaceous material is confirmed. On the other hand, at the high scan rate, the capacitance of AC+CNT is higher than the AC electrode, and the CV shape is more like rectangle, indicating the fast charge-discharge process in the AC+CNT electrodes. In the fast charge-discharge case, only the AC+CNT electrodes are effective, where the presence of CNTs brings this charge efficiency. It can be concluded that the MWCNT have formed a continuous electrically conductive network, which might short the electric circuit in the gaps and some pores so that charges cannot diffuse into deep gaps and pores, as MWCNTs are more conductive and the unmodified CNT is lower in surface area than AC. As a result, the total capacitance is reduced.

3.3   Galvanostatic charge-discharge analysis

In order to examine the device rate performance and the power and energy density relationship, galvanostatic charge-discharge cycling between 0 and 3V at various current densities was carried out and it is shown in Figure 4. Two typical plots of constant current charge-discharge are presented at two extreme current densities. At low the current density of 2 mA, the performance of the four electrodes is in the order of AC>AC+0.15%CNT>AC+1%CNT>AC+7%CNT, which is consistent with their CV behaviour at low scan rate. At the high current density of 50 mA, the performance of the CNT containing AC electrodes is better than that of the AC alone. The IR drop at the turning point of charge-discharge indicates the energy loss due to the internal DC resistance; the data show that the IR drop changes with CNT content

according to the following sequence: AC<AC+0.15%CNT<AC+1%CNT<AC+7%CNT.

Two key factors of a supercapacitor, namely its gravimetric power and energy density, are evaluated from the discharge curves of the galvanostatic cycles at different current densities according to [13]:

$$E = I \int \frac{Vdt}{2m}, \qquad P = \frac{E}{\Delta t_d}$$

where I is the current density, m is the mass of one electrode (the current collectors not included), Δt_d is the time elapsed during discharge. From this, it can be concluded that the addition of 0.15% and 1% conductive CNT can improve the capacitor performance especially in the high power region, but at 7%CNT content the energy density was greatly reduced.

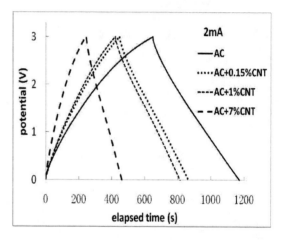

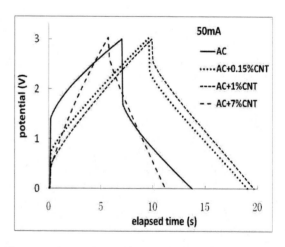

Figure 4. Galvanostatic charge-discharge curve of EDLC cells with AC, AC+0.15%CNT, AC+1%CNT and AC+7%CNT electrodes using 1M TEABF$_4$-PC electrolyte at two extreme constant current densities: (a) 2 mA, (b) 50mA.

4 CONCLUSIONS

Electrochemical analysis of the EDLC cells shows that activated carbon is suitable material for supercapacitor electrodes, whereas a small amount of conductive MWCNT can improve its performance through the reduction of charge transfer resistance in the electrode/electrolyte interface. At higher CNT content, the specific energy density is reduced, as the unmodified CNT is lower in surface area than AC. The MWCNTs might form a highly conducting network in the bulk carbon electrode and thus prevent charges from diffusing into the deep bulk of carbon electrode.

ACKNOWLEDGMENTS

This research was supported by the Engineering and Physical Sciences Research Council (EPSRC). We would like to thank Professor Joe Keddie's laboratory in the Physics Department at the University of Surrey for the courtesy of facility usage.

REFERENCES

[1] Y. Zhang, H. Feng, X. Wu, L. Wang, A. Zhang, T. Xia, H. Dong, X. Li, L. Zhang, Int. J. Hydrogen Energy 34(2009)4889

[2] L. L. Zhang and X. S. Zhao, Chem. Soc. Rev. 38(2009)2520

[3] J. Li, X. Wang, Q. Huang, S. Gamboa, P.J. Sebastian, J. Power Sources 158(2006)784.

[4] B. Fang, L. Binder, J. Power Sources 163(2006)616.

[5] V. Khomenko, E. Raymundo-Pi~nero, F. B´eguin, J. Power Sources 177(2008)643

[6] C.-C. Hu, C.-C. Wang, F.-C. Wu, R.-L. Tseng, Electrochim. Acta 52(2007)2498.

[7] T. Bordjiba, M. Mohamedi, and L.H. Dao, Adv. Mater. 20(2008)815

[8] A. Laheäär, H. Kurig, A. Jänes, E. Lust, Electrochim. Acta 54(2009)4587

[9] F.-Q. Li, Y.-Q. Lai, Z.-A. Zhang, H.-Q. Gao, J. Yang, Acta Phys. -Chim. Sin. 24(2008)1302

[10] K. Hung, C. Masarapu, T. Ko, B. Wei, J. Power Sources 193(2009)944.

[11] C. Masarapu, H.F. Zeng, K.H. Hung, and B. Wei, ACS Nano 3(2009)2199

[12] D. Moosbauer, S. Jordan, F. Wudy, S.S. Zhang, M. Schmidt and H.J. Gores1, Acta Chim. Slov. 56(2009)218

[13] C. Arbizzani, M. Mastragostino, F. Soavi, J. Power Sources 100(2000)164

Carbon Nanotube Field Effect Transistors using Printed Semiconducting Tubes

N. Rouhi[*], D. Jain[*], K. Zand[*] and P. J. Burke[*]

[*]Department of Electrical Engineering & Computer Science, University of California-Irvine, California, USA, pburke@uci.edu

ABSTRACT

Purified, all-semiconducting nanotubes offer great promise for a variety of applications in RF and microwave electronics. It has been showed theoretically that carbon nanotube electronic devices have the potential of going into THz regime since 2004 [1]. In this work, we present device performance of thin-film transistors fabricated using a carbon nanotube spin-coating method that is economical and lends itself to mass manufacturing of nanotube electronics.

Keywords: carbon nanotube, high mobility, printed electronics

1   INTRODUCTION

Carbon nanotube devices have been speculated as having potentially THz cutoff frequencies since 2004 [1]. Since then, there have been many efforts to make this dream come true. However, only recently have the device cutoff frequencies crossed into the GHz range, as reviewed in [2]. To date, no RF devices with all-semiconducting nanotubes have been demonstrated. Both theoretical and experimental results show that having purified semiconducting tubes, as the channel of the transistor, will properly lead to desired Radio Frequency results. Study on different methods, such as dielectrophoresis, to deposit nanotubes, rather than in-place growth, have been a major focus of nanoelectronics recently [2]. Recent works also demonstrated a method of depositing self-sorted nanotube networks which can be used in fabrication of nanotube transistors [2].

In this work, we present progress towards RF devices using all semiconducting nanotubes as the starting material. A spin-on process, which is compatible with standard semiconductor processing, is presented.

The device performance is measured in an RF compatible electrode geometry. In addition, the DC measurements demonstrate high mobility (more than 30 cm^2/V-s) and good on/off ratio (in the range of more than 10^3 and 10^4 in some cases) achievements.

2   DEVICE FABRICATION

Devices reported here were fabricated using a solution enriched up to 90% in semiconducting single walled carbon nanotubes (Isonanotubes-S, semiconducting purity 90+, diameter range – 1.2-1.7 nm, length range 300 nm to 5 micron). These solutions were made using density gradient centrifugation process for the separation of nanotubes with different chiralities (n,m indexes) [3]. In the process, nanotubes were first dispersed in deionized water using ionic surfactants and then iodixanol, a non-ionic, water-soluble iodine derivative was added to the solution to be used as a density gradient medium prior to the separation based on centrifugation. For our experiments, we used nanotube's solutions as they were received without any further processing.

Prior to the deposition of nanotubes, Si/SiO2 wafers were first modified with APTES (Figure 1). An amine terminated self-assembled monolayer was formed after this modification to assist with the solution based deposition process. For this modification wafers with thin oxide layers were first treated with hot piranha solution for 1 hour and then washed thoroughly with DI water. Piranha treatment introduces –OH groups on the oxide surface. For forming a thin monolayer, piranha treated wafers were dipped into 1% APTES solution (v/v) in isopropanol. After an hour wafers were washed several times with isopropanol to remove excess APTES from the wafer surface and then they were dried by blowing air. Nanotube solution was then either spin-coated or a 20 µL nanotube solution was poured onto per sq cm of wafer. After letting it dry for an hour, wafers were washed with plenty of deionized water and then air dried prior to observing them on SEM (Figure 2).

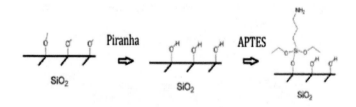

Figure 1: APTES treatment protocol.

Following the nanotube deposition, we patterned the wafer to deposit source and drain electrodes. The minimum

NSTI-Nanotech 2010, www.nsti.org, ISBN 978-1-4398-3401-5 Vol. 1, 2010

feature size on the mask is 5μm and the patterning process was done using the conventional photolithography by the means of MA-6 mask aligner.

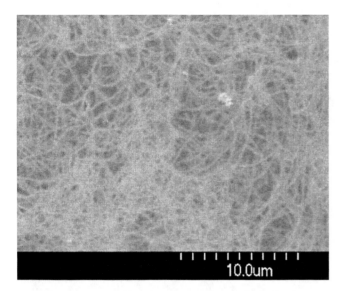

Figure 2: SEM image of nanotubes after deposition.

Gate lengths of 5~100μm (5μm and 10~100μm with 10μm increments) was designed to study the effect of gate length on the device characteristics. In all the devices, gate width is fixed and 200μm. After the patterning step, source and drain electrodes (30 nm Pd/ 100 nm Au) were deposited using e-beam evaporation followed by lift-off process. Palladium is used as the contact metal to nanotubes to make sure that the nanotube-metal contact is a resistive contact rather than Schottky. Devices were fabricated on Si substrate with 300nm high quality Silicon-dioxide cap and the Si wafer was used as the back gate. Figure 3 shows an SEM image of a fabricated device.

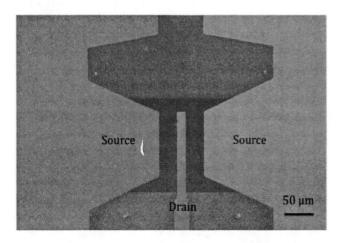

Figure 3: SEM image of fabricated device.

This design will allow for top-gate, three-terminal RF probing in the next generation of devices.

3 ELECTRICAL CHARACTERIZATION

The dc measurement data were extracted from devices with gate lengths of 5~100 μm and gate width of 200 μm. Figure 4 and 5 depict the current-voltage characteristic of one of the devices under DC bias.

The I_D-V_D curve is linear for V_D between 1 V and -1 V, indicating good ohmic contact between nanotube and electrodes as a result of depositing Palladium for the first layer on top of the nanotubes (Figure 4).

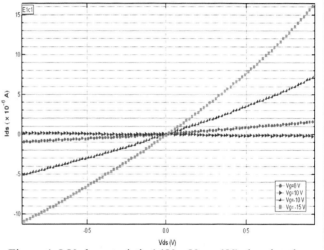

Figure 4: I-V characteristic (-1V < V_{DS}< 1V) showing the ohmic contact between electrodes and nanotube

By applying more negative V_D the devices clearly show saturation behavior (Figure 5).

The on/off ratio is more than 1000 in almost all devices. A reasonable assumption states that OFF current is roughly corresponding to the metallic nanotubes while both semiconducting and metallic tubes contribute in the ON current. Since we are using purified all-semiconducting (90% semiconducting) tubes in the channel, we are expecting high on/off ratio, which is true in our devices. In addition, the devices have a great p-type I-V curve response to the change in the back-gate voltage, which also is a result of forming the channel with semiconducting tubes. Gate voltage has been changed in the range of -10V to 10V with 2V increments.

Using the conventional I-V equations for MOS devices, we calculated the mobility by assuming the gate capacitance to be only the oxide capacitance.

$$\mu = \frac{L}{WCV_{ds}} \cdot \frac{dI_d}{dV_g}$$

L is the length of the gate and W is the width that is fixed (200nm).

Mobilities more than 30 cm^2/V-s was calculated using the above method and curve fitting the I_D-V_D characteristic while, taking the nanotube-substrate capacitance into account will make the total capacitance be less than the oxide capacitance therefore, resulting in higher mobility. These devices clearly show potential for RF performance when suitably optimized.

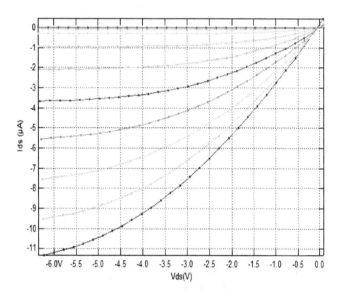

Figure 5: DC characterization showing the saturation behavior. (Vg varies from -10V to 10V with 2volts increment)

4   RESULTS AND CONCLUSION

In summary we reported thin film transistors fabricate using semiconductor-enriched carbon nanotubes. The APTES chemical treatment of the surface also allowed selective deposition of semiconducting nanotubes on the surface.

The dc electrical measurements show a great improvement in terms of mobility, current density and on/off ratio compared to previous works. Mobility of more than 30 cm^2/V-s was obtained. Besides, on/off ratio of more than 1000 and perfect response to gate voltage (p-type I-V curve) will bring us to the conclusion that all-semiconducting nanotubes were deposited between the source and drain electrodes.

Pushing this technology into the microwave (GHz) range will require shorter gate lengths and higher mobilities. Since these are some of the first spin-on, all semiconducting nanotube devices ever made, these initial results are indeed quite promising for RF and microwave electronics.

ACKNOWLEDGMENT

This work was funded by the National Science Foundation, the Army Research Office, the Office of Naval Research, Northrop Grumman and the Korean National Science Foundation (KOSEF) World Class University (WCU) program.

REFERENCES

[1] P. J. Burke, "AC performance of nanoelectronics: towards a ballistic THz nanotube transistor" Solid State Electronics, vol. 48, pp. 1981-1986, 2004.

[2] C. Rutherglen, D. Jain, P. Burke, "Nanotube Electronics for RF Applications" Nature Nanotechnology, vol. 4, No. 11, 2009.

[3] M. Hersam, "Progress towards monodisperse single-walled carbon nanotubes", Nature Nanotechnology, vol. 3, no. 7, pp. 387-394, May2008.

Gels of Carbon Nanotubes and A Nonionic Surfactant and High-quality Carbon Nanotube Films Prepared by Gel Coating

Xiaoguang Mei and Jianyong Ouyang*

Department of Materials Science and Engineering, Faculty of Engineering, National University of Singapore, Sinagpore 117574, mseoj@nus.edu.sg

ABSTRACT

This paper reports gels of carbon nanotubes (CNTs) with a nonionic surfactant, polyoxyethylene(12) tridecyl ether (POETE), and high-quality CNT films prepared by gel coating. Mixtures of multi-walled CNTs (MWCNTs) and POETE that was a fluid at room temperature became gels after mechanical grinding. The heavily entangled MWCNTs debundled during the grinding and dispersed with fewer bundles in POETE. The mechanism for the gel formation was studied by the dynamic mechanical measurements and scanning electron microscopy. The results suggest that the formation of the CNT/POETE gel is the result of the physical-crosslinking CNT networks, mediated by the van der Waals interaction between CNTs and the nonionic surfactant. The CNT/POETE gels were electrically conductive and could be processed into conductive CNT films by coating the CNT/POETE gels on a substrate by a doctor blade and subsequent heating. POETE was removed during the heating, while the heating did not degrade the CNTs. The CNT films had a good adhesion to substrate.

Keywords: carbon nanotube, gel, film, coating, surfactant

1 INTRODUCTION

CNTs have been attracting strong attention owning to their fascinating one-dimensional structure and unique mechanical, electrical and thermal properties [1-6]. CNTs are promising in various important applications. Though great progress has been made in CNTs, their application is severely impeded by their poor processibility, since they do not melt and usually have quite low solubility in solvents. Many methods have been developed to improve the dispersion capability of CNTs in solvents, including covalent sidewall modification, polymer wrapping, modification through π-π interaction with aromatic molecules, and surfactants [7-9]. However, these methods require an ultrasonication of CNTs under a high power or a chemical reaction. These methods could alter the physical properties and bring severe defects to CNTs. Moreover, they are still far away from processing of CNTs in large scale, because only in a small amount of CNTs can be dispersed by these methods. To efficiently process CNTs is still an important research topic. A simple and efficient method is strongly required. The method reported by Fukushima et al. on bucky gels of CNTs and ionic liquids opens a possibility to process CNTs in large scale [10,11]. The bucky gels were prepared by mechanical grinding, and CNTs debundled during the mechanical grinding.

Here, we report the preparation of gels of CNTs and a nonionic surfactant, POETE through mechanical grinding [12]. Conductive CNT films were prepared by coating the gels on substrates by doctor blade and removing the nonionic surfactant by subsequent heating. The CNT films have good adhesion to the substrate and can be produced in large scale.

2 EXPERIMENTAL

MWCNTs produced by chemical vapor deposition (CVD) were purchased from Chengdu Organic Chemicals Co. Ltd. run by the Chinese academy of Sciences. The specifications were: average outer diameter of 8–15 nm, average length of 50 μm, and purity higher than 95 %. POETE was purchased from Sigma-Aldrich. All the materials were used without further purification.

The CNT/POETE gels were prepared by mechanical grinding. 10 mg CNTs were mixed with 2 ml POETE in an agate mortar. The mixture turned into a sticky paste after ground for 1 hr. The paste was centrifuged with a Profuge 14D at 10,000 rpm for 30 min. The mixture separated into two phases. The top transparent liquid phase was decanted from the containers after centrifugation, and the bottom phase was collected.

The CNT/POETE gels were coated on glass substrates through a doctor blade. The gels turned into CNT films on glass substrate after heating at 250 °C for 15 min and subsequently at 475 °C for another 5 min in air.

The rheological properties of the gels were studied using a Rheometric Scientific Advanced Rheometric Expansion System. Scanning electron microscopic (SEM) images were obtained with a Hitachi S-4100 scanning electron microscope. Transmission electron microscopic (TEM) images were taken with a JOEL JEM 2010F transmission electron microscope equipped with a field emission gun. The MWCNT/POETE gel was dispersed in deionized (DI) water, and was sonicated in a Branson 1510 ultrasonic bath for 5 min. Then, the MWCNTs were put on a copper grid with carbon film, purchased from Electron Microscopy Sciences, for the TEM measurements. The conductivitives of the gels and the CNT films were measured by the four-point probe technique with a Keithley

2400 source/meter. The conductivities of the CNT/POETE gels were measured by placing the gel on a glass slide attached with 4 copper foils which served as the 4 probes. The conductivity of the CNT film was tested on a square CNT film on a glass substrate with the van der Pauw four-point probe technique. The thickness of the gels for the conductivity study was measured by a micrometer caliper, while the thickness of the CNT films was determined by cross-sectional scanning electron microscopy (SEM). The thermal gravimetric analyses (TGA) of POETE and MWCNTs were carried out in air with a Du Pont 2950 TGA. The samples were put at 120 °C for 10 min. Then, they were heated to decomposition at a heating rate of 10 °C min^{-1}. The Raman spectra were acquired in a back scattering configuration using a Jobin Yvon Horiba LabRam HR800 Raman system with a 514.5 nm Ar$^+$ laser as the excitation source. The laser power was 2 mW, and the resolution was 2.4 cm^{-1}. The accumulation time for each sample was 900 seconds.

3 RESULTS AND DISCUSSION

3.1 Gels of MWCNTs and POETE

Surfactants have been frequently used to disperse CNTs in solvents, particularly water [9]. POETE as a nonionic surfactant exhibited good wettability for MWCNTs, when they were mixed together. The mixture of MWCNTs and POETE turned to a soft, black, and sticky paste after mechanically ground for about 1 hr. Excess POETE was used to have a better dispersion of CNTs. The excess POETE was removed by the subsequent centrifugation of the paste. The paste separated into two phases in the centrifuge tube after the centrifugation. The bottom phase was a black gel, while the top was transparent POETE. Fig. 1a shows a picture of an upside-down centrifuge tube containing the two phases. The black gel remained on the bottom of the centrifuge tube and did not fall down under gravity. Since the gels of CNTs and ionic liquids are called bucky gels of ionic liquids [10], the gels of CNTs and a nonionic surfactant can be named as bucky gels of the nonionic surfactant as well. The weight percentage of MWCNTs in the MWCNT/POETE gel was determined after removing the excess POETE. It was 1.5 % by weight, that is, the maximum gelling efficiency was 66 g POETE for 1 g MWCNTs. The MWCNT/POETE gel could be extruded from a needle and did not break down under gravity. (Fig. 1b) This research work focused on the CNT/POETE gels with maximum gelling efficiency since they had the best CNT dispersion.

The MWCNT/POETE gels were stable at room temperature. They did not show any remarkable change after several weeks. The gels still kept the solid-like shape under heating up to 200 °C. They also did not shrivel even in vacuum of 10^{-4} mbar. The stability under heating or in vacuum is related to the low volatility of POETE. The gels were electrically conductive. Their conductivity was about

4×10^{-3} S cm^{-1} as determined by the four-point probe technique. The electrical conduction can be attributed to MWCNTs since the liquid component, POETE, in the gel is non-conductive.

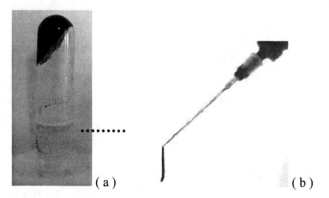

(a) (b)

Fig. 1 Photographs of a MWCNT/POETE gel. (a) Phase-separation behavior of the gel (upper phase) and excess POETE (lower phase), observed after centrifugation of a ground mixture of MWCNTs and POETE. The dotted line indicates the surface of the liquid phase. (b) Extrusion of the gel from a needle.

Fig. 2 shows the SEM image of a MWCNT/POETE gel. The as–received MWCNTs, which were heavily entangled, and some of them bundled together, became separated and showed less bundles in the gel. This indicates that the MWCNTs debundled during the grinding, just similar to the grinding of CNTs in ionic liquid [10].

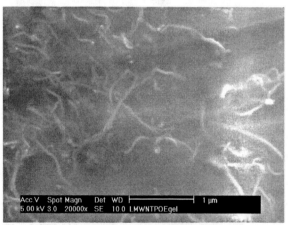

Fig. 2 SEM image of a MWCNT/POETE gel.

The rheological properties of the MWCNT/POETE gels were characterized as well (Fig. 3). The dynamic storage modulus (G') is almost flat over a wide range of the angular frequency (ω) from 0.1 to 100 rad s^{-1} at the amplitude of applied strain (γ) being 1 % or 5 %, while G" does not show any sign of relaxation. G' is always higher than G" at a certain angular frequency. This dynamic mechanical behavior is similar to that of CNT/ionic surfactant gels or CNT/surfactant suspensions obtained by ultracentrifugation of solution dispersed with surfactant and CNTs [10,11,13-

15]. These results indicate that the MWCNT/POETE gel prepared by the mechanical grinding has permanent networks and therefore behaves as a gel.

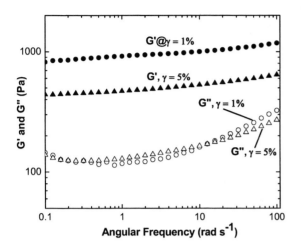

Fig. 3 Angular frequency (ω) dependencies of dynamic storage (G', solid symbols) and loss moduli (G", open symbols) of a MWCNT/POE gel at room temperature. Applied strain amplitudes (γ) were 1 % (circles) and 5 % (triangles).

The dispersion of MWCNTs in POETE is related to the van der Waals interaction between MWCNTs and POETE, since POETE is a nonionic surfactant and surfactants have been frequently used to disperse CNTs in solvents [9]. This interaction helps MWCNT debundle when ground, and it stabilizes MWCNTs in POETE. Thus, the networks in the MWCNT/POETE gel are formed by MWCNTs and mediated by the van der Waals force between MWCNTs and POETE. The conductive property of the MWCNT/POETE gel is due to the MWCNT networks in the gel.

The formation of the MWCNT/POETE gel is related to the length of the MWCNTs. Short MWCNTs, which were about 1 μm in length, were also used for the gel preparation through a similar process. The maximum gelling efficiency was down to 32.3 g POETE for 1 g short MWCNTs. The lower gelling efficiency of the short MWCNTs compared with long MWCNTs also implies that the networks in the MWCNT/POTET gels are formed by the MWCNTs. More short MWCNTs are needed to form the networks.

3.3 CNT films prepared from CNT/POETE gels

The formation of a homogeneous CNT film in a large area is important for the applications of CNTs in many aspects. The MWCNT/POETE gels could be used for the fabrication of CNT films through coating. POETE could be removed by heating. TGA results of POETE, MWCNTs and SWCNTs in air indicate that the decomposition of POETE starts at about 150 °C and finishes at 250 °C (Fig. 4). During the decomposition of POETE, the MWCNTs or SWCNTs are quite stable. Hence, the heating to remove POETE does not degrade MWCNTs and SWCNTs.

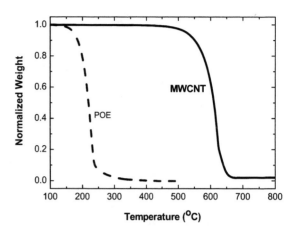

Fig. 4 TGA results of POETE (dashed line) and MWCNTs (solid line).

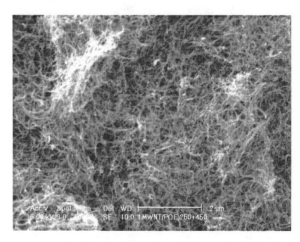

Fig. 5 SEM image of a MWCNT film. The MWCNT film was prepared from a MWCNT/POETE gel.

MWCNT films were prepared by coating the MWCNT/POETE gel on a glass substrate by a doctor blade. Then, POETE was removed by heating. A two-stage heating process in air was needed to obtain a homogeneous CNT film: at 250 °C for 15 min, subsequently at 475 °C for another 5 min. The first heating at 250 °C is important for homogeneity of the CNT film, while the following heating at 475 °C can completely decompose POETE in a short time. The MWCNT film had a conductivity of about 13 S cm^{-1} and good adhesion to the substrate. It did not detach from the substrate under strong gas flow or when immersed into water. Fig. 5 shows the SEM image of a MWCNT film on glass. The SEM image of the MWCNT film indicates that MWCNTs entangle with each other.

The fabrication of CNT film using CNT/POETE gels is an efficient way thanks to the simple dispersion process, dispersion of high amount of CNTs in POETE, simple coating, and easy removal of POETE. This method which does not require solvent and an ultracentrifugation of solution is different from that reported by Dan et al [16]. Dan et al. demonstrated fabrication of CNT films in large scale by rod-coating suspensions of CNTs and surfactants obtained through the ultracentrifugation of aqueous solution dispersed with surfactants and CNTs. Moreover, the CNT films prepared by our method have good adhesion to the substrate, since they are fabricated through coating and the surfactant is removed by heating. This is different from the methods of preparing the CNT film through filtration of solution dispersed with CNTs or a solid-state drawing process [2,4]. The CNT films obtained by these methods have to be transferred to the desirable substrate, so that they may have a problem of adhesion to the substrate.

Finally, it is worth to point out that our method to disperse CNTs in POETE by mechanical grinding is different from the conventional methods to disperse CNTs in solution of surfactants, in which an ultrasonication process is needed [9,16-18]. No solvent is used in our method, and our method can disperse CNTs in a much higher concentration than the conventional ways.

4 CONCLUSIONS

Bucky gels were formed by mechanically grinding mixtures of MWCNTs and POETE which was a nonionic surfactant and a fluid at room temperature. The MWCNTs debundle during the grinding and form networks through the physical crosslinking in the MWCNT/POETE gels. The gels have good stability up to 200 $^{\circ}$C or in vacuum. Homogeneous CNT films could be fabricated from the MWCNT/POETE gels through coating, and POETE could be removed by heating. The MWCNT films fabricated through this method had a conductivity of about 13 S cm^{-1} and good adhesion to the substrate. This opens a new avenue to efficiently fabricate CNT films in large scale.

ACKNOWLEDGEMENT

This research work was financially supported by the Ministry of Education, Singapore with project number: RG-284-001-223.

REFERENCES

[1] M. S. Dresselhaus, G. Dresselhaus and P. Avouris, Carbon nanotubes: synthesis, structure, properties, and applications, New York, Springer, 2000.

[2] Z. C.Wu, Z. H. Chen, X. Du, J. M. Logan, J. Sippel, M. Nikolou, K. Kamaras, J. R. Reynolds, D. B. Tanner, A. F. Hebard, and A. G. Rinzler, Science, 305, 1273, 2004.

[3] G. Gruner, J. Mater. Chem., 16, 3533, 2006.

[4] M. Zhang, S. Fang, A. A. Zakhidov, S. B. Lee,A. E. Aliev, C. D. Williams, K. R. Atkinson, and R. H. Baughman, Science, 309, 1215, 2005.

[5] R. H. Baughman, A. Z. Zakhidov, and W. A. de Heer, Science, 287, 787, 2002.

[6] B. Fan, X. Mei, K. Sun, and J. Ouyang, Appl. Phys. Lett., 93, 143103, 2008.

[7] A. Hirsch, Angew. Chem. Int. Ed., 41, 1853, 2002.

[8] M. J. O'Connell, P. Boul, L. M. Ericson, C. Huffman, Y. Wang, and E. Haroz, Chem. Phys. Lett., 342, 265, 2001.

[9] M. F. Islam, E. Rojas, D. M. Bergey, A. T. Johnson, and A. G. Yodh, Nano Lett., 3, 269, 2003.

[10] T. Fukushima, A. Kosaka, Y. Ishimura, T. Yamamoto, T. Takigawa, N. Ishii, and T. Aida, Science, 300, 2072, 2003.

[11] T. Fukushima and T. Aida, Chem. Eur. J., 13, 5048, 2007.

[12] X. Mei and J. Ouyang, Carbon, 2010, 48, 293.

[13] L. A. Hough, M. F. Islam, P. A. Janmey, and A. G. Yodh, Phys. Rev. Lett., 93, 168102, 2004.

[14] L. A. Hough, M. F. Islam, B. Hammouda, A. G. Yodh, and P. A. Heiney, Nano Lett. , 6, 313, 2006.

[15] E. K. Hobbie and D. J. Fry, Phys. Rev. Lett., 97, 036101, 2006.

[16] B. Dan, G. C. Irvin, and M. Pasquali, ACS Nano, 3, 835, 2009.

[17] V. C. Moore, M. S. Strano, E. H. Haroz, R. H. Hauge, R. E. Smalley, J. Schmidt, and Y. Talmon, Nano Lett., 3, 1379, 2003.

[18] Z. Sun, V. Nicolosi, D. Rickard, S. D. Bergin, D. Aherne and J. N. Coleman, J. Phys. Chem. C, 112, 10692, 2008.

An Amplification Strategy to Label Free Opiate Drug Detection using Liquid-Gated Carbon Nanotubes Transistor

J.N. Tey[*], I.P.M. Wijaya[**], S. Gandhi[***], J. Wei[*], I. Rodriguez[**], C.R. Suri[***] and S.G. Mhaisalkar[****]

[*]Singapore Institute of Manufacturing Technology, 71 Nanyang Drive, Singapore 638075,
jntey@simtech.a-star.edu.sg, jwei@simtech.a-star.edu.sg
[**]Institute of Materials Research and Engineering, Singapore,
ip-mahendra@imre.a-star.edu.sg, i-rodriguez@imre.a-star.edu.sg
[***]Institute of Microbial Technology (CSIR), Chandirgargh, India
sonugandhi@imtech.res.in, raman@imtech.res.in
[****]Nanyang Technological University, Singapore, Subodh@ntu.edu

ABSTRACT

The proliferation of narcotic analgesic drug, heroin, has necessitated the development for on-site drug diagnosis. Carbon nanotubes (CNTs), which are renowned for its good biocompatibility and excellent electrical sensitivity, has been an area of active research in the past decade for its applications in biosensing. In this paper, a CNT based field effect transistor (CNTFET) was employed for the direct electrical detection of 6-monoacetylmoprhine. An indirect detection scheme was adopted to enable the detection of morphine antibodies. By incorporating gold nanoparticles to the morphine antibodies, an improvement of the detection limit by two orders of magnitude from 130 fg/ml to ~1 fg/ml was achieved. The mechanism of signal enhancement effect was attributed to the electrostatic gating contribution from the negatively charged gold nanoparticle in addition to the antibodies themselves.

Keywords: carbon nanotube biosensor, carbon nanotube transistor, opiate, real time detection, signal amplification, label-free

1 INTRODUCTION

The potential of carbon nanotube (CNTs) in biosensing application stems from their unique inherent properties such as small size, high strength, high electrical and thermal conductivity, high specific surface area, and good biocompatibility [1, 2]. The extremely high surface-to-volume ratio results in most of the constituent atoms residing on the surface. Thus, binding of biomolecules to the nanotubes leads to the depletion or accumulation of charge carriers throughout the entire nanostructure, making even the slightest disturbance from their proxy surrounding sufficient to alter their electrical properties; an advantage over the conventional planar sensor device [3].

Among the various device configurations and detection methodologies incorporating CNTs in their sensing platforms, liquid-gated field effect transistors (LG-FETs) based architectures outperform the conventional electrochemical measurements by their ability in providing label-free, direct electronic read-out and real-time detection. To date, the sensing capability of CNTFETs has been demonstrated on a variety of molecules from different applications [4, 5], and has been envisaged to complement the current golden standard analytical methods such as ELISA for laboratory uses, point-of-care diagnostic and on-site sample analysis tool.

In this paper, a signal amplification scheme for opiate drug detection was reported using the LG-CNTFET platform. A heroin metabolite, 6-monoacetylmorphine (MAM), was used in the study. Heroin is a naturally occurring substance extracted from unripe seeds or capsules of *Papaver somniferum* (poppy plant) [6]. Upon ingestion, heroin is metabolized sequentially to MAM, morphine, and morphine glucuronide [7]. These drug molecules are potent narcotic analgesics and are often abused as recreational drugs. There is an urgent need to develop a simple and sensitive method to monitor these opiate drugs. As of today, the typical tests to detect these opiate include: electrochemical methods, high-pressure liquid chromatography, mass spectroscopy, and fluorescence immunoassays [6]. Sensitive as they may be, these methods are usually time-consuming, expensive and not adoptable for on-site analysis. Very often, an electrochemical or magnetic tag is required to perform the measurement, inflating the number of the process steps.

Using our previously developed, novel PDMS based, all CNTs LG-FET [8], we have demonstrated the detection of small morphine metabolite through indirect immunoassay approach. The device fabrication offers advantages such as scalability and simplicity, and obviates the requirement for clean room facility, therefore can be applied for disposable, on-site screening purpose.

2 EXPERIMENTAL PROCEDURES

Carboxylated single-walled CNT powder was purchased from Cheaptubes. Inc. PDMS (Sylgard 184) was purchased from, Dow Corning, Inc. 6-Monoacetylmorphine (MAM) was purchased from Cerilliant Analytical Reference Standards Sodium dodecyl sulfate (SDS), tween-20, poly (ethylene glycol) (PEG), skim milk, bovine serum albumin (BSA), Complete Freund's adjuvant (CFA), Incomplete Freund's adjuvant (IFA), sodium dihydrogen phosphate (NaH_2PO_4) and disodium hydrogen phosphate (Na_2HPO_4) were purchased from Sigma Aldrich. 1-ethyl-3-(3-dimethylaminopropyl) carbodiimide hydrochloride (EDC), and N-hydroxysulfosuccinimide (sulfo-NHS) were purchased from Pierce Chemicals. Protein-A Sepharose, were procured from Amersham Biosciences, India.

2.1 Biomolecules synthesis

Synthesis of MAM-BSA conjugates

Synthesis of hapten, derivatization was done by refluxing the reaction mixture for 3 h at 90 °C in an inert nitrogen atmosphere, containing 3 μM of MAM, 24 μM of chloroacetic acid, 45 μM sodium hydroxide and 30 μM acetonitrile. The presence of the –COOH group was confirmed by a thin layer chromatograph (TLC) and infra-red (IR) spectroscopy [6]. The derivatized hapten (MAM-COOH) was used for the conjugation with BSA (carrier protein) using carbodiimide coupling chemistry. For the activation, 50 μM MAM–COOH, 75 μM EDC and 75 μM sulfo-NHS were mixed and incubated for 1 h at room temperature, followed by overnight incubation at 4 °C, and centrifuged for 10 min at 10,000×g to remove the urea precipitate. For the conjugation of activated hapten with BSA, 30 μM of activated hapten was mixed with 0.15 μM (10 mg) of BSA to prepare the molar ratio of 100:1.

Generation of morphine antibodies (Mor-Ab)

The morphine antibodies (Mor-Abs) were raised against MAM-BSA conjugate in young six to eight weeks old New Zealand white rabbits. The rabbits were immunized subcutaneously with 250 μg of MAM-BSA mixed with equal volume of Freund's complete adjuvant at the time of the first booster followed by Freund's incomplete adjuvant in subsequent booster doses. The rabbits were bled after 5th day of each booster and blood was collected, serum precipitated and antibodies (IgG) were purified using a Protein A sepharose column. The fractions were then dialyzed against PBS and the IgG concentration was determined at 280 nm and stored at -20°C until used.

Synthesis of gold labled Mor-Ab (Au-Mor-Ab)

Monodispersed (30 nm) colloidal gold was prepared by a modified Frens method. A 200 mL solution of 0.01% tetrachloroauric acid in Milli-Q water was brought to boiling. 4 mL sodium citrate solution (1% w/v) was added to the boiling gold chloride solution. The solution was allowed to boil for 10 min until it developed the typical bright wine red color of colloidal gold. The average particles size of colloidal gold was determined using a transmission electron microscope (Hitachi Model H-7500) operated at 120 kV. The average particle size was estimated to be approximately 30 ± 4 nm.

For the preparation of the antibody gold conjugate, 90 μg of Mor-Ab was prepared in a 20 mM phosphate buffer, pH 7.4 and added drop-by-drop into 1 ml colloidal gold solution ([Au] = 2.4×10^{-4} mol/L) under mild stirring condition. The pH of the colloidal gold solution was maintained at 7.4 by addition of 10 mM Na_2CO_3 before adding the antibody. The mixture was incubated overnight at 4°C and centrifuged at 12000 rpm for 30 min to remove the unconjugated antibodies from the solution. The pellet obtained was washed three times with 10 mM Tris (pH 8.0) containing 3% BSA under centrifugation at 12000 rpm for 30 min to remove traces of unconjugated antibodies. The pellet was resuspended in 2 ml of phosphate buffer (20 mM, pH 7.4) and stored at 4°C before its use. The final concentration of colloidal gold in the antibody-gold conjugate solution was 4.8×10^{-4} mol/L. A Hitachi 2800 UV-vis spectrophotometer was used to measure the absorbance of gold nanoparticles and Au-Mor-Ab.

2.2 Immunosensor preparation

The fabrication process of flexible, all CNT/PDMS LGFET was described in details in the previous publication [9]. Briefly, only two materials are required in this fabrication route: SWCNT and PDMS. A thin SWCNT random network, formed from suspension through vacuum filtration method [10], was transfer-printed from an alumina filter onto a blank PDMS substrate to form the semiconducting layer, while another thick, metallic-like SWCNT random network prepared from the same methodology was transfer printed onto another PDMS substrate (with an integrated microfluidic channel) as the electrode material. The stamping of CNT on the PDMS with the defined microchannel results in the auto-separation and formation of source-drain pads on each side of the microchannel. Lamination of the two PDMS substrates face-to-face encloses the microfluidic channel and completes the fabrication process.

After the completion of device fabrication, the carboxylic functional groups on the SWCNT at the active channel region were activated with EDC and sulfo-NHS for 1 hour at room temperature, rinsed with 50 mM PB solution (pH 9.5), followed by 10 μg/ml MAM-BSA injection and overnight incubation at 4°C. Excess unbound molecules were removed by rinsing the microchannel with copious of PB solution, and the device was ready for target antibodies detection.

2.3 Electrical measurement

Electrical measurement of the CNT-LGFET was performed using a home-built LabView system with the testing protocol similar to the reported literature [11, 12]. For real time monitoring, a liquid gate potential (V_G) at -0.5 V was applied to the electrolyte through reference electrode (3M KCl, FLEXREF, World Precision Instruments) and a small drain bias (V_D) of 10 mV applied over the source and drain electrodes to obtain the kinetic response at respective sensing steps.

3 RESULTS & DISCUSSION

The simplest method to detect the MAM or MAM-BSA would be a direct immunoassay: using Mor-Ab as a probe molecule immobilized onto the CNT network (see Fig. 1(i)). However, a charge screening phenomenon by the ions present in the electrolyte diminishes the electrical signal originated from the biomolecular interaction to be detected by the underlying CNT network.

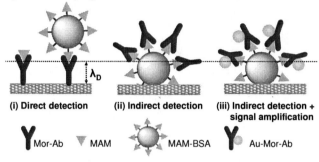

Figure 1: Illustration showing the configurations for different detection schemes.

The length which characterizes this screening effect is represented by the Debye length (λ_D) (Equation (1)). In this equation, the term I represents the ionic strength (mol/m^3), ε_0 is the permittivity of free space, ε_r is the dielectric constant, k represents the Boltzmann's constant, T is the absolute temperature in Kelvin, N_A is Avogadro's number and e is the elementary charge. The inverse relationship between λ_D and the I highlights the importance of ionic-strength of the electrolyte in modulating charge screening effect and hence influences the sensitivity in LGFET.

$$\lambda_D = \sqrt{\frac{\varepsilon_r \varepsilon_0 kT}{2z^2 e^2 I}} \qquad (1)$$

For a 50 mM PB solution used in this study, the λ_D was estimated to be ~3.5 nm. Given the usual size of an antibody (10-15 nm) [13], the immunocomplex formation in a direct immunoassay approach is certainly out of the λ_D distance. One simple alternative is to reduce the concentration of the electrolyte. However, it comes at the expense of reduced binding efficiency. Therefore, an indirect immunoassay approach (Fig. 1(ii)) is proposed wherein the MAM-BSA is used as the receptor for antibody detection. The hapten-carrier conjugation is engaged so that the amine-rich BSA could be bound covalently to the carboxylic groups on the CNT through carbodiimide coupling chemistry, while the MAMs are conjugated to BSA through established protocol to preserve its activity towards the corresponding antibody [6].

The indirect detection scheme allows a fraction of immunocomplexes to fall within the λ_D distance, hence increases the signal level for Mor-Ab detection, as illustrated in Fig. 1(ii). Fig 2 shows the real time data for Mor-Ab detection in a logarithmic serial dilution manner from 10 fg/ml, 100 fg/ml, 1 pg/ml, 100 pg/ml, 1 ng/ml, 10 ng/ml, 100 ng/ml, 1 µg/ml to 10 µg/ml (labeled from 1 to 9 respectively), along with the corresponding calibration plot which is based on triplicate kinetic measurements. The electrical current increment upon the immunocomplex formation was attributed to the electrostatic gating mechanism, where the negatively charged Mor-Ab induces positive doping in the CNTs, raising the overall hole carrier concentration in the p-type CNT transistor, and leading to the increment of overall conductance [8].

To determine the limit of detection (LOD), the calibration plot in Fig. 2(b) was first fitted with a log-linear regression equation at the linear regime from 10 fg/ml to 100 ng/ml. The fitting yields a R^2 of 0.981, indicating good fit with the underlying experimental data. By using Equation (2) and taking the 3σ criterion [14], the LOD for Mor-Ab was estimated to be 130 fg/ml. The term σ in the equation is the standard deviation of the fitted line with the underlying data point; S is the sensitivity calculated from the slope of the linear regime; and C is the curve fitting constant.

$$LOD = (3\sigma / S) + C \qquad (2)$$

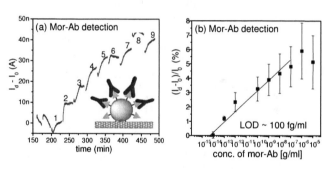

Figure 2: Indirect detection scheme where MAM-BSA is used as the receptor for Mor-Ab detection. (a) Kinetic measurement plot (b) Concentration plot

To further enhance the signal, an amplification strategy was employed by tagging the gold nanoparticles (AuNPs) to the Mor-Ab. AuNPs have been widely used in electrochemical based biosensing as labels for molecule detection, carriers for other electroactive labels, or signal enhancement tools [15]. It was chosen in our study because of its simple geometry and uniform charge distribution. AuNPs typically carry a negative charge as a result of the citrate ion absorption during the nanoparticles synthesis through the citrate reduction method [15]. From the LG-FET perspective, if the sensing mechanism is dominated by electrostatic gating, the addition of AuNPs to the existing detection system should further induce more positive charge carriers into the channel, resulting in larger signal increment. We verified this alternative route of signal enhancement through an offline experiment to confirm effect of direct AuNPs interaction with CNT network (Fig. 3). Within the pH window tested, increased conductance signal were observed upon the injection of 50 nM AuNPs solution into the microchannel.

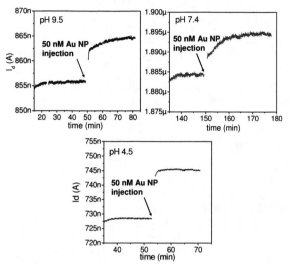

Figure 3: Direct detection of AuNPs on CNT [16].

Fig.4 shows the kinetic measurement with different Au-Mor-Ab concentrations: (1)1 fg/ml, (2)10 fg/ml, (3)100 fg/ml, (4)1 pg/ml, (5)100 pg/ml, (6)1 ng/ml, (7)10 ng/ml, (8) 100 ng/ml,(9) 1 µg/ml and (10) 10 µg/ml, and the corresponding calibration plot based on triplicate kinetic measurements. The estimated LOD was about 1 fg/ml, two orders of magnitude more sensitive than the previous scheme.

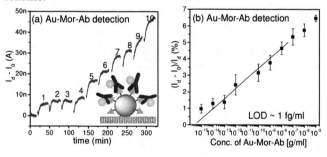

Figure 4: Indirect detection scheme where MAM-BSA is used as the receptor for Au-Mor-Ab detection. (a) Kinetic measurement plot (b) Concentration plot

4   CONCLUSIONS

We have demonstrated that by adopting the indirect detection approach using the MAM-BSA as the receptor for Mor-Ab detection, a LOD of ~130 fg/ml was achieved. This limit was further improved down to ~ 1 fg/ml by incorporating AuNPs into the sensing system. The signal amplification was a result of the increased carrier concentration owing to the electrostatic gating influence

from the negatively charged AuNPs and the MAM-BSA: Mor-Ab interaction.

It may be noted that the detection of Au-Mor-Ab is only the first step toward the detection of heroin family, since ultimately the analytes of interest is free MAM molecules instead of its antibody counterpart. A combination of competitive immunoassay with indirect detection approach should therefore be employed for the ultimate MAM detection [16].

REFERENCES

[1] N. Sinha, J. Ma, J.T.W. Yeow, Journal of Nanoscience and Nanotechnology, 6, 573, 2006.

[2] M.P. Anantram, F. Leonard, Reports on Progress in Physics, 69(3), 507, 2006.

[3] Y. Cui, Q. Wei, H. Park, C.M. Lieber, Science, 293(5533), 1289, 2001.

[4] K. Bradley, J.-C.P. Gabriel, M. Briman, A. Star, G. Grüner, Phys. Rev. Lett. J1 - PRL, 91(21), 218301, 2003.

[5] R.J. Chen, S. Bangsaruntip, K.A. Drouvalakis, N. Wong Shi Kam, M. Shim, Y. Li, W. Kim, P.J. Utz, H. Dai, PNAS, 100(9), 4984, 2003.

[6] S. Gandhi, P. Sharma, N. Capalash, R.S. Verma, C.R. Suri, Analytical and Bioanalytical Chemistry, 392, 215, 2008.

[7] P.P. Dillon, B.M. Manning, S.J. Daly, A.J. Killard, R. O'Kennedy, Journal of Immunological Methods, 276(1-2), 151, 2003.

[8] J.N. Tey, I.P.M. Wijaya, Z. Wang, W.H. Goh, A. Palaniappan, S.G. Mhaisalkar, I. Rodriguez, S. Dunham, J.A. Rogers, Applied Physics Letters, 94(1), 013107, 2009.

[9] J.N. Tey, I.P.M. Wijaya, Z. Wang, W.H. Goh, A. Palaniappan, S.G. Mhaisalkar, I. Rodriguez, S. Dunham, J.A. Rogers, Appl. Phys. Lett., 94, 013107, 2009.

[10] Y. Zhou, L. Hu, G. Gruner, Appl. Phys. Lett., 88, 123109, 2006.

[11] I. Heller, A.M. Janssens, J. Mannik, E.D. Minot, S.G. Lemay, C. Dekker, Nano Lett., 8(2), 591, 2008.

[12] E.D. Minot, A.M. Janssens, I. Heller, H.A. Heering, C. Dekker, S.G. Lemay, Appl. Phys. Lett., 91(9), 093507, 2007.

[13] N.H. Thomson, Journal of Microscopy, 217(3), 193, 2005.

[14] L. Torsi, G.M. Farinola, F. Marinelli, M.C. Tanese, O.H. Omar, L. Valli, F. Babudri, F. Palmisano, P.G. Zambonin, F. Naso, Nature Materials, 7(5), 412, 2008.

[15] M.T. Castañeda, S. Alegret, A. Merkoçi, Electroanalysis, 19(7-8), 743, 2007.

[16] J.N. Tey, S. Gandhi, I.P.M. Wijaya, A.Palaniappan, J. Wei, I. Rodriguez, C.R. Suri, S.G. Mhaisalkar, Small, Accepted, 2010.

RNA Functionalized Carbon Nanotube for Chemical Sensing

M. Chen[*], S. S. Datta[**], S. M Khamis[***], J. E. Fischer[****] and A. T. Johnson[*****]

University of Pennsylvania, Philadelphia, PA, USA
[*]Current Address: Simmons College, Boston, MA, USA, michelle.chen@simmons.edu
[**]Current Address: Harvard University, Cambridge, MA, USA, ssdatta@fas.harvard.edu
[***]Current Address: Nanosense, Inc., Redwood City, CA, USA, skhamis@evolvedmachines.com
[****]fischer@seas.upenn.edu
[*****]cjohnson@physics.upnn.edu

ABSTRACT

We demonstrate a versatile class of nanoscale chemical sensors based on single stranded RNAs (ssRNAs) as the chemical recognition sites and single-walled carbon nanotube field effect transistors (SWNT-FETs) as the electronic readout components. The sensor responses differ in sign and magnitude depending on both the type of gaseous analyte and the sequence of ssRNA being used. Such rapid response, sensitivity to large variety of analytes, self-regenerating ability, and reproducibility make ssRNA-functionalized SWNT-FETs promising building blocks in developing large arrays of sensors for electronic olfaction and disease diagnosis.

Keywords: carbon nanotubes, field effect transistors, chemical sensors, RNA functionalization

1 INTRODUCTION

Semiconducting single-walled carbon nanotubes (SWNTs) have electronic states that lie on the one-dimensional carbon cage structure, making them exceedingly sensitive to environmental stimuli. Bare and polymer-coated SWNTs are found to be sensitive to various gases [1-6]. However, SWNT functionalized with biomolecular complexes [7-11] could detect species that otherwise would have only weak interaction with unmodified nanotubes. These derivatized SWNTs and semiconducting nanowires [12-14] are attractive chemical and molecular sensors due to their high sensitivity, fast response time, and compatibility with array fabrication [15].

Recently, bridging the science of SWNTs with biology has evolved into a thriving research initiative [12] Nucleic acid biopolymers' affinity to analytes can be specifically engineered [16,17]. For example, high throughput screening is used to select films of dye-labeled single stranded DNA (ssDNA) for use as gas sensors with fluorescent readout [19,20]. SsDNA, as well as ssRNA, has high affinity to SWNT due to an attractive $\pi-\pi$ stacking interaction with the nanotube surface [17]. Even with such interaction the electronic integrity of the SWNT is preserved. These facts motivate the exploration of ssDNA/ssRNA functionalized SWNT hybrid nanostructures as electronic gas sensors.

Nucleic acids (DNA) and ribonucleic acid (RNA) are the genetic materials of all living organisms. RNA, in particular, relays the information stored in DNA to synthesize proteins. Although, the structure of RNA and DNA only slightly differs in their sugar and bases, RNA is much more reactive to chemical and biological species than DNA. In biological research, RNAs have been used as catalysts and mediators for a variety of organic and inorganic reactions [21-26]. Recently, short interfering RNAs have been coupled to carbon nnanotubes as drug delivery systems for RNA interference to silence genes [27]. Long RNA strands have been bound to carbon nanotubes for fluorescent and Raman spectroscopic studies [18]. However, there are still many interesting science and application for RNA-SWNT complex that have yet been explored.

In light of our work on DNA-functionalized carbon nanotube for chemical sensing [16], we aim to improve the sensitivity and specificity of our sensors by functionalization of SWNTs with molecules that interact more strongly with a larger variety of species. Single stranded RNA is a good candidate for such functionalization because it, like DNA, also binds to the nanotubes via $\pi-\pi$ stacking interaction. RNA can be a more superior chemical recognition component than DNA due to its higher reactivity to more chemicals. In this report, we discuss the RNA-functionalized SWNT-FETs as chemical sensors

2 EXPERIMENTAL DETAILS

2.1 Device Fabrication

SWNTs are grown on SiO_2/Si substrate (oxide thickness ~ 200-400 nm) by catalytical chemical vapor deposition with $Fe_2(NO_3)_3$ as catalyst. Field effect transistors (FETs) are fabricated by writing source-drain contacts to the nanotubes using electron beam lithography followed by thermal evaporation of chrome and gold. The degenerately doped silicon substrate is used as the backgate. Source-drain current I_{SD} is measured as a function of gate voltage. Only devices with individual p-type semiconducting SWNT with ON/OFF ratio >1000 are selected for the electrical experiments.

2.2 RNA Functionalization

Two sequences of single stranded RNA (ssRNA) that correspond to prior DNA experiments [16] are used so the data in the two systems can be directly compared:

RNA S1
5' GAG UCU GUG GAG GAG GUA GUC 3'

RNA S2
5' CUU CUG UCU UGA UGU UUG UCA AAC 3'

Single stranded RNA is functionalized onto carbon nanotube surfaces via non-covalent $\pi-\pi$ stacking interaction, similar to that of DNA [??]. For both the sequences used, application of single-stranded RNA causes the threshold voltage to decrease by 3-6 volts in the characterization of source-drain current vs. backgate voltage. This corresponds to a hole density decrease of roughly 400/mm, assuming a backgate capacitance (25 aF/mm) that is typical for this device geometry [29]. Furthermore, the functionalization of single stranded RNA typically decrease the ON-state current of SWNT-FET by ~ 10-20%. This suggests weak carrier scattering by the RNA molecular coating.

2.3 Sensing Setup

Five gaseous analytes are measured: methanol, propionic acid, trimethylamine (TMA), dinitrotoluene (DNT; 50 mg/mL in dipropylene glycol), and dimethyl methylphosphonate (DMMP; a simulant to the nerve agent sarin [30]. Saturated vapor of each chemical is stored in a reservoir. Air is pumped through the reservoir and delivers about 3% of the saturated vapor concentration at 0.1 mL/sec to the devices. The air or air/analyte mixture was directed toward the sample through a 2 ± 0.1 mm diameter nozzle positioned 6 ± 1 mm above the sample surface. Valves can be controlled so that diluted analyte and air can be applied to the sample alternately for a set time (~ 50 seconds), after which the flow reverted to plain air. For each sample, source-drain current (I) vs. gate voltage is characterized and air/analyte experiments are done on both the bare and single stranded RNA functionalized SWNT-FETs. The analyte-induced changes in the source-drain current (I) are measured at $V_B = 100$ mV and $V_G = 0$V.

3 RESULTS AND DISCUSSIONS

3.1 AFM Analysis

Figure 2 shows the AFM image of a ssRNA (S1) functionalized SWNT. By observing along the length of the nanotube, the ssRNAs appear to "pearl" up on the nanotube, as more brighter/higher structures are seen along the nanotubes. A linescan (data not shown) over the RNA "pearls" along the length of the tube indicates that the height increase due to the RNA is 1.2 ± 0.2 nm. This height increase is in agreement with the value in the DNA case [16].

The pearling of RNA on SWNT is not seen in the case of single stranded DNA. DNA forms a rather uniform thin layer on the nanotube [16]. This can be reasoned by the fact that single stranded RNA is more prone to self-hybridization and formation of secondary structures than DNA, while the single stranded DNA tends to maintain linear structure. Also, note that the pearling of RNA occurs mostly on the nanotube, and not much on the SiO$_2$ surface. This is different from the pronounced coverage of DNA on the substrate [16]. The elevated structures seen on the sample surface in Figure 2 are residual resist from e-beam lithography process. These residual resists can be removed through an annealing process that is discussed above.

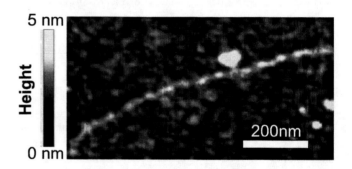

Figure 1: SWNT functionalized with RAN S1 shows pearling along the length of the tube.

3.2 Sensing Response to Chemical Vapors

We electrically detect various chemical vapors by monitoring the current change in the FET device as a function of time when the RNA functionalized carbon nanotube is exposed to the gas/air cycles. Two sequences of RNA and five analytes are tested. The two RNA strands used in this experiments have the same base sequences as the corresponding DNA strands used in the previous experiment [16]. The gases used for evaluating RNA-SWNT sensors are the same as the previous DNA-SWNT experiment, allowing direct comparison of sensing response in the two systems.

Table 1 lists the percent change in the devices' current for all combinations of RNA vs. analytes when the device is exposed to chemical vapors. It is evident from our data that bare SWNTs do not show response to any of the analytes except TMA. However, the response to the analytes is significantly enhanced after the same device is coated with single stranded RNA. Therefore, we conclude that the single stranded RNA enhances the binding affinity of the analytes and increases the sensor's response.

Analyte	Bare SWNT	SWNT + RNA S1	SWNT + RNA S2
PA	0±1	+45±10	+40±5
TMA	-9±2	-40±10	-45±5
Methanol	0±1	+32±5	+20±5
DMMP	0±1	+30±5	-35±10
DNT	0±1	-20±5	-70±10

Table 1. Summary of gas sensing results for ssRNA functionalized SWNT-FETs to five chemical analytes. Each data value (percent current change) is the mean of 5-10 samples, with standard deviation as the uncertainty.

Comparing the responses of SWNTs functionalized with RNA S1 and RNA S2 on the same gas, the two columns cannot be related by a simple multiplicative factor. This suggests that each RNA sequence is a unique chemical recognition component for SWNT-FET sensors.

Different RNA sequence can also have opposite response to the same chemical analyte. In particular, the sign of response to DMMP is opposite for RNA S1 and S2 devices functionalized devices. RNA S1 device shows a 30% current increase with exposure to DMMP (Figure 2), but RNA S2 device shows a 35% decrease in the current with DMMP exposure (Table 1). Although the detailed origin for the difference in the sign of sensing response is not clearly understood, it can be reasonably assumed that the different shapes in secondary structures of RNA S1 and RNA S2 elicit different interactions even with the same chemical analyte.

From the sensing response to DMMP as shown in Figure 2, it is intriguing that the sensor responds very rapidly (< 300 ms) to DMMP, but recovers slower when DMMP is removed. Similar sensing characteristics of fast response and slow recovery are observed when RNA-SWNTs are exposed to TMA and to methanol (data not shown). Such results may suggest that there exist some similarities in the interaction mechanism between RNA and these three chemical analytes: DMMP, TMA, and methanol.

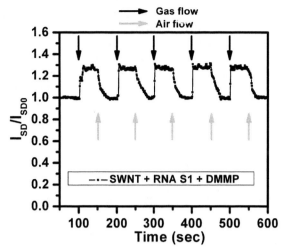

Figure 2. RNA S1 functionalized carbon nanotube shows 30% increase in the current when exposed to DMMP.

The RNA-functionalized SWNT sensors show very rapid response and recovery time to the chemical analytes. Figure 3 shows the nearly rectangular wave-like response characteristics to propionic acid on a RNA (S2) functionalized SWNT. When propionic acid (PA) is introduced (as indicated by the black arrows) the sensor responds rapidly (within 1 second) with 40% increase in the current, indicating that the PA is de-protonated in the residual water near the nanotube, forming a highly negatively charged entity near the nanotube. After flowing air replaces the PA (as shown by the gray arrows), the device current rapidly comes back to baseline, indicating that the device can quickly refresh itself once the analyte is removed.

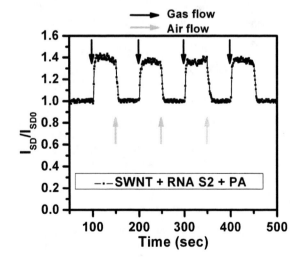

Figure 3. Sensor response to propionic acid (PA) on RNA S2 functionalized SWNT-FET. The sensor responds and recovers very rapidly.

The sign and the magnitude of the sensing response depend on the RNA sequence being functionalized and the chemical analyte being detected (Table 1). The sensor can respond and regenerate very quickly, in the order of seconds (Figure 3). The sensing response is reproducible for more than 50 cycles. Therefore, the RNA-functionalized SWNT-FETs show promise as specific and efficient nano-scaled sensor component for realization of an electronic nose.

An electronic nose is inspired by biological olfactory systems where thousands of different odor receptors, each responsive to many different odorants, perform molecular identification and analysis. The different receptor component in an electronic nose can be achieved by functionalizing SWNT-FETs with different sequences of single stranded RNA. The appropriate RNA sequence, each with distinct response to a collection of chemicals, can be selected by high throughput screening of RNAs' chemical versatility, and nucleic acid engineering. When combined with multiplexed arrays fabrication, a large electronic olfaction array made of single-stranded RNA functionalized SWNT-FETs can be realized.

4   CONCLUSIONS

We demonstrate that single stranded RNA-functionalized SWNT-FETs behave as specific, sensitive, and reproducible gas sensors, with fast response and recovery time. SsRNAs serve as effective chemical recognition sites on SWNT-FET sensors due to their strong interactions with chemical vapors. The sign and magnitude of sensing response for depends on the specific sequence of the ssRNA. Our results suggest that RNA-functionalized SWNT-FETs show great promise for development into a specific and versatile nanosensor for a large variety of gases and molecules.

ACKNOWLEDGEMENTS

This work was partially supported by the JSTO DTRA as well as the Army Research Office Grant #W911NF-06-1-0462; by the Nano/Bio Interface Center through the National Science Foundation under contract NSEC DMR-0425780; and by the Nanotechnology Institute of the Commonwealth of Pennsylvania (MC and ATCJ).

REFERENCES

[1] J. Kong, N.R. Franklin, C.W. Zhou, M.G. Chapline, S. Peng, K.J. Cho and H.J. Dai, Science, 287, 622 2000.

[2] P.G. Collins, K. Bradley, M. Ishigami and A. Zettl,Science, 287, 1801, 2000.

[3] K. Bradley, J.-C. P. Gabriel, A. Star and G. Grüner, Appl. Rhys. Lett., 83, 3821 2003.

[4] J.P. Novak, E.S. Snow, E.J. Houser, D. Park, J.L. Stepnowski and R.A. McGill, Appl. Phys. Lett., 83, 4026, 2003.

[5] J. Kong and H. Dai, J. Phys. Chem. B, 105, 2890, 2001.

[6] K. Bradley, J.-C. P. Gabriel, M. Briman, A. Star and G. Grüner, Phys. Rev. Lett., 91, 218301, 2003.

[7] A. Star, T.-R. Han, J.-C. Gabriel, K. Bradley, and G. Grüner, Nano. Lett., 3, 1421, 2003.

[8] K. Bradley, M. Briman, A. Star and G. Grüner, Nano Lett. 4, 253,2004.

[9] E.S. Snow, F.K. Perkins, E.J. Houser, S.C. Badescu and T.L. Reinecke, Science, 307, 1942, 2005.

[10] A. Star, J.-C. Gabriel, K. Bradley and G. Grüner, Nano. Lett., 3, 459, 2003.

[11] A. Star, T.-R. Han, V. Joshi, J.-C.P. Gabriel and G. Grüner, Adv. Mater., 16, 2049, 2004.

[12] S.S. Wong, E. Joselevich, A.T. Woolley, C.L. Cheung and C.M. Lieber, Nature, 394, 52, 1998.

[13] K. A. Williams, P. T. M. Veenhuizen, B. G. de la Torre et al., Nature, 420, 761, 2003.

[14] R.J. Chen, S. Bangasaruntip, K.A. Drouvalakis, N.W.S. Kam, M. Shim, Y. Li, W. Kim, P.J. Utz and H. Dai., Pro. Natl. Acad. Sci. U.S.A., 100, 4984, 2003.

[15] P.W. Baron, S. Baik, D.A. Heller and M.S. Strano, Nat. Mater., 4, 86, 2005.

[16] C. Staii, M. Chen, A. T. Johnson, Jr. and A. Gelperin, Nano Letters, 5, 1774, 2005.

[17] M. Zheng, A. Jagota, E.D. Semke, B.A. Diner, R.S. McLean, S.R. Lustig, R.E. Richardson and N.G. Tassi, Nature Materials, 2, 338, 2003.

[18] R. Rao, J. Lee, Q. Lu, G. Keskar, K.O. Freedman, W.C. Floyd, A.M. Rao and P.C. Ke, Appl. Phys. Lett., 85, 4228, 2004.

[19] J.E. White and J.S. Kauer, US Patent 2004/0101851

[20] J.E. White, L.B. Williams, M.S. Atkisson and J.S. Kauer, Assoc. Chemoreception Sciences, XXVI Annual Meeting Abstracts, 32, 2004.

[21] M. Illangasekare, G. Sanches, T. Nickles, and M. Yarus, Science, 267, 643, 1995.

[22] P.A. Lohse and W. Szostak, Nature, 381, 442, 1995.

[23] T.M. Tarasow, S.L. Tarasow and B.E. Eaton, Nature 389, 54, 1997.

[24] B.L. Zhang and T.R. Cech, Nature, 390, 96, 1997.

[25] L. Gugliotti, D.L Feldheim and B.E. Eaton, Science, 304, 850, 2004.

[26] A. Fraser, Nature, 428, 275, 2004.

[27] N.W.S. Kam, Z. Liu and H. Dai. J. Am. Chem. Soc., 2005.

[28] M. Freitag, A.T. Johnson, S.V. Kalinin and D.A. Bonnell, Phys. Rev. Lett., 89, 216801, 2002.

[29] M. Radosaviljevic, M. Freitag, K.V. Thadani and A.T. Johnson, Nano Lett., 2, 761, 2002.

[30] A. R. Hopkins and N. S. Lewis, Anal. Chem., 73, 884, 2001.

Incorporation of Plasma Modified Multi-Walled Carbon Nanotubes into TiO$_2$ Photoanode for Dye-Sensitized Solar Cells (DSSC)

Y.F. Chan*, C.C. Wang**, C.Y. Chen*

*National Cheng Kung University, 70101, Tainan, Taiwan, n38987107@mail.ncku.edu.tw
**Southern Taiwan University, 71005, Tainan, Taiwan, ccwang@mail.stut.edu.tw

ABSTRACT

Dye-sensitized solar cells (DSSC) are fabricated with photoanode comprises of nanocomposite material of TiO$_2$ and multi-walled carbon nanotubes (MWNT) treated *in situ* by plasma and subsequent functionalization by grafting of maleic anhydride (MA). The functionalized MWNTs enable the nanotubes to disperse homogenously among the TiO$_2$ nanoparticles and become an integral part of the nanocomposites. As a result, MWNTs-MA/TiO$_2$ nanocomposites presented obvious improvements in photocurrent and solar power conversion efficiency as the CNT provides efficient electron transfer through the film as well as the charge collection at the surface of counter electrode due to its excellent conductivity. At an optimum incorporation of 0.075wt.% MWNTs into the working electrode, the photo conversion efficiency of the DSSC was improved by more than 30%.

Keywords: TiO$_2$, MWNTs, DSSC, nanocomposites

1 INTRODUCTION

Numerous studies related to the development of dye-sensitized solar cells (DSSC) have emerged ever since it was first reported by Grätzel in 1991 [1]. Meanwhile, carbon nanotubes (CNTs) have attracted massive amount of attention in various applications due to their unique properties, inclusive of excellent electrochemical stability, low resistivity and high surface area [2-5]. Efforts have been made to incorporate CNTs into component of DSSCs, for example, the working and counter electrode, in order to improve the cell performance. The presence of CNTs in TiO$_2$ matrix [6] or ionic liquid electrolyte [7] shows limited improvement in the power conversion efficiency. The underlying problems can be attributed to the difficulty to obtain homogeneously dispersed CNT nanocomposites arising from the non-reactive nature of the CNT surface and the unavoidable bundle formation due to van der Waals attraction during synthesis [8-10]. Most published works have focused on chemical modification of the CNT to improve dispersion through functionalization, oxidization, melt mixing, and compounding [11-13]. In order to improve the dispersion of CNT in titania nanoparticles, we use "grafting to" method or functionalization of carbon nanotubes by making use of plasma [14-17]. Here we report a simple method to improve the DSSCs performance by the incorporation of multi-walled carbon nanotubes (MWNTs), of which the surface was grafted with maleic acid (MA) via plasma treatment, into the front electrode using a direct mixing method. The morphology of TiO$_2$-MWNTs film was observed by Scanning Electron Microscope (SEM) and Transmisison Electron Microscpe (TEM). The presence of MWNTs-MA into the film was confirmed by the Raman Spectroscopy and the photochemical performance of the solar cells fabricated with these nanocompomsites was examined.

2 EXPERIMENTAL

2.1 Materials

Multi-walled carbon nanotubes (MWNTs) used in this work were purchased from the Industrial Technology Research Institute in Taiwan and were synthesized by ethylene CVD using Al$_2$O$_3$ supported Fe$_2$O$_3$ catalysts (purity >95%). Prior to plasma modification, the MWNTs were purified by sonicating in ethanol for 1 hr to remove the catalysts and then diluting with deionized water by filtration through a filter paper. Maleic anhydride (MA) (Aldrich Co.), ethanol (99.99%) (J.T. Baker) and the TiO$_2$ powder (P25 degussa) were used as received.

2.2 Preparation of MWNTs Grafted Maleic Anhydride (MWNTs-MA) by Plasma Treatment.

The plasma treatment (Ar, 50W, 13.56MHz, 10min) was carried out in a stainless steel reactor (6×10^3 cm^3) at a vacuum level of about 10^{-4} torr. The electrodes were made of circular Cu plates and kept at a distance of 3cm. The MWNTs were spread evenly inside the chamber. After the plasma treatment, a large number of radicals were generated on the surface of the MWNTs, and the maleic anhydride (MA) with a concentration of 0.1 M dissolved in dimethylacetamide (DMAC) was immediately injected into the reactor to graft onto the MWNTs at room temperature for 3 h. After the grafting polymerization, the MWNTs grafted maleic anhydride, MWNTs-MA, was washed by DMAC repeatedly and subsequently separated by filtration. The sediments were dried in a vacuum oven at 80 °C overnight to get rid of the residual solvent. The resulting products obtained were the functionalized MWNTs, or known as MWNTs-MA for abbreviation.

2.3 Preparation of the MWNTs-MA/TiO$_2$ Nanocomposites

The MWNTs-MA was immersed in 0.1N HNO$_3$ and ultrasonicated for 3 hours to achieve maximum dispersion. TiO$_2$ powder was added into the solution, and the mixture was stirred vigorously for 3 h before adding the surfactant to form the paste. The paste of MWNTs-MA/TiO$_2$ was now ready for subsequent DSSC fabrication.

2.4 DSSC Fabrication

The MWNTs-MA/TiO$_2$ composite films were obtained after annealed at 350°C for 0.5 h in air. When cooled to 100°C, the MWNTs-MA/TiO$_2$ films were immediately soaked in an ethanol solution of ruthenium complex dye, N3 (cis-di(thiocynate)- bis(2,2'-bipyridyl, -4,4'-dicarboxylate) ruthenium (II); ruthenium 535, Solaronix SA, Switzerland). Pt-coated FTO glass substrates were used as the counter electrode. The acetonitrile solution of LiI, I$_2$, 1-methyl-3-propy-limidazolium iodide, and 4-tert-butylpyridine was used as an electrolyte. In the DSSC photocurrent performance test, a 100W/cm^2 Xe lamp (AM1.5) was served as a light source in conjunction with a potentiostat/galvanostat (PGSTAT 30, Autolab, EcoChemie, Netherlands). FT-IR spectra were recorded at room temperature in the range 700 – 4000 cm^{-1} using a Bio-Rad FT-IR system coupled with a computer.

3 RESULTS AND DISCUSSION

Fourier transform infrared (FT-IR) was used to characterize the functional groups on the surface of the MWNTs. Figure 1 shows the FTIR of MWNTs treated with and without the plasma treatment followed by MA grafting. Characteristic bands at 1719 and 1735cm^{-1} that represents the C=O and O-C=O groups from MA can be clearly seen from the one of MWNTs- MA. This indicates that the maleic acid has been successfully grafted onto MWNTs after the Ar plasma treatment that created lot of free radicals inside the chamber. The carboxyl groups attached on the tube wall of the MWNTs grafted with MA (MWNTs-MA) enable the bonding with Ti when mixed into TiO$_2$ matrix through the O-C=O opening and dehydration reactions among the groups. The electrodes were fabricated with nanocomposite obtained by a direct mixing method of MWNTs-MA into TiO$_2$ (Degussa P25) with aid of organic surfactants (Triton X-100)

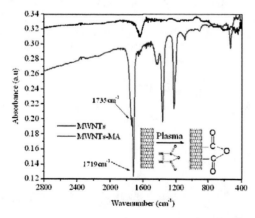

Figure 1: FT-IR for MWNTs before and after the plasma treatment followed by grafting with MA (maleic anhydride)

Figure 2 illustrates the snapshot of the MWNTs-MA nancomposites prepared from direct blending method via maximum ultrasonication and mechanical mixing. Without the CNTs addition, the nanocomposite appear as white color, which is solely the nanocrystalline titania. As the amount of CNTs added increased, the color of the mixture solution approachs greysi color, which was primarily attributed to the all-black CNTs. It can also been seen that the CNTs were well mixed with the titania, wherein no aggregation is observed.

Figure 2: Snapshot of nanocomposite paste of MWNTs-MA/TiO$_2$ of various CNTs weight percent

Figure 3 displays the SEM and TEM images of the MWNTs-MA/TiO$_2$. Both SEM and TEM images illustrate the well dispersion of the MWNTs-MA among the nanocrystals of TiO$_2$. Since the amount added into the TiO$_2$ matrix was relatively small, CNTs could not be identified from the XRD data obtained, as shown in Figure 4(a). The TiO$_2$ after annealed showed mainly the anatase phase. Therefore, Raman spectroscopy was used and it confirmed the presence of TiO$_2$ and CNTs in the prepared front electrodes. Three Raman peaks at approximately 395, 518, and 642 cm^{-1} indicate the anatase form of TiO$_2$, while the other two peaks at approximately 1348 and 1624 cm^{-1} represent the D and G bands of MWNTs, respectively (Figure 4 (b)).

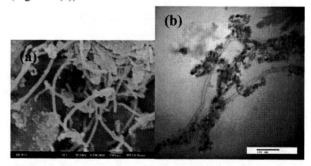

Figure 3: (a) SEM, and (b) TEM images of the nanocomposite material of functionalized multi-walled nanotubes (MWNTs-MA) and TiO$_2$

J-V characteristics of DSSCs fabricated with MWNTs-MA electrodes were exhibited in Table 1. The photovoltaic performance was measured under the AM 1.5 illumination standard with an active area of 0.25 cm^2. The solar cell based on 0.075 wt.% MWNTs-MA in TiO$_2$ working electrode displayed the greatest light harvesting ability with a fill factor (FF) of 61%, short-circuit current (J_{sc}) of 14.7 mA/cm^2, open-circuit voltage (V_{oc}) of 0.77 V, and an overall light to electricity conversion efficiency 6.9%. Open-circuit voltage and fill factor remain nearly the same for all conductive CNTs concentration. Compared to a cell fabricated with conventional TiO$_2$ film, TiO$_2$ electrode incorporated with 0.075 wt.% MWNTs-MA shows an enhancement of short-circuit photocurrent and overall conversion efficiency of 25% and 32%, respectively.

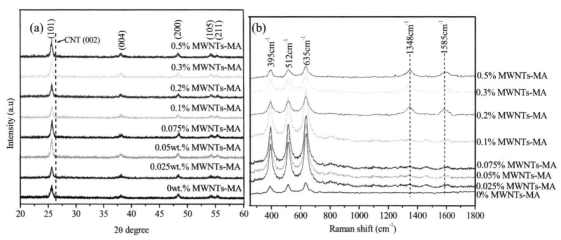

Figure 4: (a) XRD pattern, and (b) Raman spectra of TiO_2 and MWNTs-MA/TiO_2 nanocomposite

Weight percent of MWNTs-MA in TiO_2 (wt.%)	V_{oc} (V)	J_{sc} (mA/cm^2)	FF	Efficiency, η (%)
0.000	0.77	11.8	0.58	5.26
0.025	0.76	11.3	0.57	5.31
0.050	0.81	14.5	0.58	6.82
0.075	0.77	14.7	0.61	6.93
0.100	0.76	13.1	0.66	6.52
0.200	0.80	10.8	0.70	6.02
0.300	0.79	11.0	0.60	5.17
0.500	0.80	10.4	0.63	5.28

Table 1: J-V characteristics of the MWNTs-MA/TiO_2 photoanode of the DSSCs. V_{oc}, J_{sc} and FF are the open circuit voltage, the short-circuit current, and the fill-factor of the DSSCs, respectively.

Further increase in the amount of MWNTs-MA resulted in a gradual decrease, owing to and losses in optical transparency, decrease of electron injection to the conduction band of TiO_2 particles since the CNTs in black absorb some of the light illuminated. As a consequence, reduced of short circuit current, followed by degradation of the DSSCs performance was observed.

4 CONCLUSION

Improvement of the power conversion efficiency of DSSCs was demonstrated by the incorporation of plasma-modified MWNTs in the TiO_2 photoanode using a direct mixing method. At the best conditions, the MWNTs-MA/TiO_2 DSSCs attained in the short-circuit current density of 14.7 mA/cm^2, leading to the overall power conversion efficiency of 6.93% under the AM 1.5 illumination standard with an active area of 0.25 cm^2. An excess amount of MWNTs-MA in the electrode led to degradation of the DSSC performance due to optical transparency losses as well as agglomeration of the CNTs. The plasma modification process that is more environmental-friendly can serve as an alternative to replace the conventional wet-chemical

method that might alter their unique characteristics by shortening and damaging the tubes. Future work should focus on further optimization of the MWNT DSSC fabrication processes and detailed studies of the mechanisms for the solar power conversion efficiency enhancement.

REFERENCES

[1] B. O'Regan and M. Grätzel, "A low-cost, high-efficiency solar cell based on dye-sensitized colloidal TiO_2 films", Nature, 353, 737-740, 1991.

[2] S. Ijima, "Helical microtubules of graphitic carbon", Nature, 354, 56-58,1991.

[3] J. Wang, "Carbon-nanotube based electrochemical biosensors: A review", Electroanalysis, 17, 7-14, 2005.

[4] A. Kay and M. Grätzel, "Low cost photovoltaic modules based on dye sensitized nanocrystalline titanium dioxide and carbon powder", Solar Energy Materials and Solar Cells, 44, 99-117, 1996.

[5] H.J. Kim, D.Y. Lee, B.K. Koo, W.J. Lee, and J.S. Song, "Preparation of CNT electrodes for enhanced electrochemical properties". Proceedings of the KIEEME Annual

Conference, 17, 1090, 2004.

[6] P. Vincent, A. Brioude, C. Journet, S. Rabaste, , S.T. Purcell, J. Le Brusq and J.C. Plenet, "Inclusion of carbon nanotubes in a TiO sol-gel matrix", Journal of Non-Crystalline Solids, 311, 130-137, 2002,.

[7] N.G. Park, J. van de Lagemaat, and A.J. Frank, "Comparison of dye-sensitized rutile- and anatase-based TiO_2 solar cells", Journal of Physical Chemistry B, 104, 8989-8994, 2000.

[8] C. Park, Z. Ounaies, K.A. Watson, R.E. Crooks, J.J. Smith and S.E. Lowther et al., "Dispersion of single wall carbon nanotubes by in situ polymerization under sonication", Chem Physics Letters, 364, 303–308, 2002.

[9] Z. Ounaies, C. Park, K.E. Wise, E.J. Siochi and J.S. Harrison, "Electrical properties of single wall carbon nanotube reinforced polyimide composites", Composites Science and Technology, 63, 1637-1646, 2003.

[10] A. Yu, H. Hu, E. Bekyarova, M.E. Itkis, J. Gao and B. Zhao et al., "Incorporation of highly dispersed single-walled carbon nanotubes in a polyimide matrix", Composites Science and Technology, 66, 1190-1197, 2006.

[11] M.S.P. Shaffer, and A.H. Windle, "Fabrication and characterization of carbon nanotube/poly (vinyl alcohol) composites", Advanced Materials, 11, 937–941, 1999.

[12] J. Sandler, M.P.S. Shaffer, T. Prasse, W. Bauhofer, K. Schulte and A.H. Windel, "Development of a dispersion process for carbon nanotubes in an epoxy matrix and the resulting electrical properties", Polymer, 40, 5967–5971, 1999.

[13] E. Kymakis, I. Alexandou and G.A.J. Amaratunga, "Single-walled carbon nanotube–polymer composites: electrical, optical and structural investigation", Synthetic Metals, 127, 59–62, 2002.

[14] H. Li, F. Cheng, A.M. Duft and A. Adronov, "Functionalization of single-walled carbon nanotubes with well-defined polystyrene by "click" coupling", Journal of American Chemical Society, 127, 14518-14524, 2005.

[15] S. Qin, D. Qin, W.T. Ford, J.E. Herrera, D.E. Resasco and S.M. Bachilo et al., "Solubilization and purification of single-wall carbon nanotubes in water by in situ radical polymerization of sodium 4-styrenesulfonate", Macromolecules, 37, 3965-3967, 2004.

[16] Y.Z. You, C.Y. Hong and C.Y. Pan, "Functionalization of carbon nanotubes with well-defined functional polymers via thiol-coupling reaction", Macromolecular Rapid Communications, 27, 2001-2006, 2002.

[17] C.C. Tseng, C.C. Wang and C.Y. Chen, "Modification of polypropylene fibers by plasma and preparation of hybrid luminescent and rodlike CdS nanocrystals/polypropylene fiber", Journal of Nanoscience Nanotechnology, 6, 1–7, 2006.

NSTI-Nanotech 2010, www.nsti.org, ISBN 978-1-4398-3401-5 Vol. 1, 2010

Large Area, Directionally Aligned Single-Walled Carbon Nanotube Films by Self-assembly and Compressed Sliding Methods

C. K. Najeeb, J. -H. Lee, J. Chang and J.-H. Kim*

*Department of Molecular Science and Technology,
Ajou University, Suwon 443-749, Republic of Korea, jhkim@ajou.ac.kr

ABSTRACT

We demonstrate the fabrication of well-aligned array of surface modified single-walled carbon nanotubes (SWNTs) and nano silver decorated SWNTs from their surfactant stabilized dispersions by self-assembly on solid substrates in a large area. The shape, rigidity and surface charge of the nanomaterials are found to be the key factors determining the alignment by self-assembly. We also demonstrate preparation of highly transparent conductive thin films (800±50 $\Omega/\square$, 95-97 %T at 550 nm) of directionally aligned SWNTs from high concentration aqueous dispersion of SWNT by compressed sliding of a thin liquid film between two glass plates. Evaluation of thin films by SEM, AFM, optical and electrical characterization demonstrates that SWNTs are directionally aligned with high density. Thin films of directionally aligned SWNTs can be used for photovoltaic and other optoelectronic application such as in touch screens, antistatic coatings, flat panel displays, optical communication devices, and solar cells.

Keywords: Carbon nanotube, Surface modification, Self-assembly, Sliding, Alignment.

1 INTRODUCTION

Carbon nanotubes (CNTs) and nano crystals are among the most promising materials predicted to impact future nanotechnology owing to their unique structural and electronic properties [1, 2]. Single-walled carbon nanotubes (SWNTs) have drawn great interests for application in a wide range of potential nano-devices owing to their exceptional mechanical, electrical, optical and thermal properties [3]. Functionalization of CNTs with various chemical groups and nanocrystals can further enhance the properties of CNTs. Alignment of CNTs from as produced materials is an important challenge for achieving novel or enhanced physical properties. Aligned assembly of carbon nanotube arrays have shown excellent electrical, opto-electronic and electromechanical properties compared to that of disordered network film [4]. Therefore, most of the realistic applications of SWNTs would require directional alignment of nanotubes as thin films or patterns on a substrate to explore their unique out standing properties.

There are several different methods available for the alignment by pre or post synthesize processing of CNTs including chemical vapor deposition (CVD), dielectrophoresis, drawing polymer–CNT composite film, drop drying CNT film, Langmuir–Blodgett deposition, application of magnetic fields and electrospinning [5,6]. Large area crack alignments have been obtained from CNT aqueous dispersions by self organization of the surfactants on drying or due to the cracking mechanism in case of composites with polymer [7,8]. Recently, Liu et al. reported transfer and aligned assembly of large area CVD grown CNT films from growth substrates to receiver substrate by contact transfer and sliding [9]. In this contribution, we present two simple methods for the aligned assembly of SWNTs via self assembly of surface modified SWNTs and compressed sliding a thin liquid film of high concentration aqueous surfactant dispersion between two flat substrates. Surface modification of SWNTs by sodiumdeoxycholate (SDC), a bile acid salt with large and rigid hydrophobic moiety of a steroid skeleton, rendered a rigid shape and the further treatment with anionic surfactant sodiumdodecylbenzene sulfonate (SDBS) provided negatively charged surface. The self assembly of SWNTs and SWNT-Ag hybrid are driven by the shape, rigidity and surface charge. Compressed sliding of the SWNT liquid film between parallel surfaces causes mechanical stretching of the film, which lead to the formation of ultrathin film with directional orientation of nanotubes. The directionally oriented nanotube thin films can be used for applications such as chemical-gas sensor, bio sensors, catalysts, flexible displays, touch screens and solar cells.

2 EXPERIMENTAL

2.1 Self-assembly of Surface Modified SWNTs

An aqueous dispersion of CNT was prepared by ultra-sonic agitation of 20 mg SWNT (arc discharge, Iljin Nanotech Co., Korea) in 50 mL of 1 wt% SDC (Aldrich) solution. The dispersion was centrifuged at 6000 rpm for 30 min to remove larger bundles and impurities. To reduce the surface tension and also to generate negative charge on SWNTs, the dispersion was mixed with 0.1 wt% aqueous solution of SDBS in the ratio 1:2 or 1:4 (v/v). Thin films were prepared on cleaned glass and silicon substrates by spin coating, dip coating and drop deposition methods. A schematic illustration for self assembly in drop dried film of SWNT-SDC/SDBS is shown below.

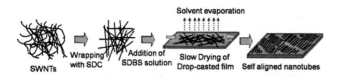

Scheme 1. Fabrication of self assembled CNT array by drop drying

2.2 Directional Alignment of SWNTs by Sliding Method

High concentration SWNT ink was prepared by sonic agitation of 25 mg SWNTs immersed in 20 mL of 0.08% SDBS aqueous solution by high power ultrasound sonication for 15 min. The dispersion was then centrifuged at 6000 rpm for 30 min to remove larger bundles and impurities. The concentration of SWNT ink was ~1mg/mL. To prepare ultrathin aligned transparent films, 50 μL ink was placed at the center of a cleaned glass substrate and placed another substrate over it and pressed gently. Excess CNT ink was wiped off from the sides of substrate and, then it was slided one over the other at constant speed with pressing. The films were allowed to dry at room temperature under atmospheric conditions. A schematic illustration for the film fabrication by sliding method is depicted below. For annealing, the films were heated in hot air oven at 90 °C for 1h.

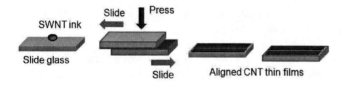

Scheme 2. Fabrication of directionally aligned thin film of SWNT by compressed sliding.

3   RESULTS AND DISCUSSION

3.1   Characterization and Evaluation of Aligned SWNT Films Prepared by Self-assembly of Surface Modified CNTs.

Scanning electron microscopic (SEM) images of SWNT thin films deposited on Si substrate is shown in Figure 1. The SEM image of SDC modified SWNTs (Fig. 1(a)) illustrates that the nanotubes are well dispersed and possess a rigid rod like structure. The large and rigid hydrophobic moiety of a steroid skeleton in the SDC molecule imparts rod-like rigid shape to originally flexible nanotubes [10]. Figure 1(b) and (c) show SEM images of aligned SWNT

thin films prepared by drop drying from SWNT-SDC dispersion mixed SDBS solution in the ratio 1:2 and 1:4, respectively. The film with SWNT-SDC: SDBS in the ratio of 1:4 demonstrated high degree of alignment as well as high packing density compared to that of film with 1:2 ratios. Figure 1(d) displays densely aligned assembly of SWNT-Ag hybrid dispersed in SDBS solution. The self assembly of nanomaterials from surfactant dispersion is based on their shape, aspect ratio, surface charge and capillary action during the drying process [11]. The surface modification of SWNTs with SDC rendered a rigid shape and the subsequent treatment with anionic surfactant provided negatively charged surfaces. Similarly, decoration of SWNT surface with metal nanoparticles gave more rigid shape and the surfactant adsorbed on the surface provided a charged surface. Additionally, the surfactant helps uniform spreading of nanotubes by lowering surface tension and also reduces rupture of film by dewetting. The thin films of SWNT-SDC/SDBS (ratio, 1:4) prepared by spin coating and dip coating also demonstrated aligned assembly of nanotube (Fig. 1(e) and (f)), but the degree of alignment and packing density were relatively low compared to drop deposited films.

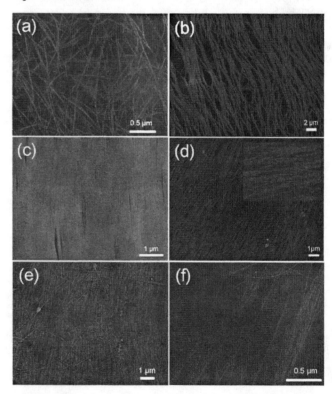

Figure 1. SEM images of (a) SWNT-SDC (b) SWNT-SDC/SDBS (ratio 1:2) (c) SWNT-SDC/SDBS (ratio 1:4) and (d) SWNT-Ag hybrid/SDBS thin films prepared by drop deposition. (e) & (f) thin film of SWNT-SDC/SDBS (ratio 1:4) prepared by dip coating and spin coating methods, respectively. Inset of Fig.1 (d) shows Ag nano particles decorated on SWNT surface.

Polarized spectroscopic techniques, including polarized Raman spectroscopy and / or polarized absorption spectroscopy have been shown to useful techniques for alignment characterization due to the strong anisotropic absorption of polarized radiation by SWNTs [12]. The optical absorption spectra of the sample films were measured at different rotation angles from -90 to 90 degree and the order of alignment was investigated as a function of wavelength. The intensity absorbance peak corresponding to S_{11} transition showed noticeable change with rotation angle. Figure 2 illustrates the angular dependence of absolute absorption for thin films corresponding to S_{11} transition at a wavelength of 1010 nm. The remarkable change in intensity of absorption band with rotation angle provides spectroscopic evidence for the directional alignment of the nanotubes.

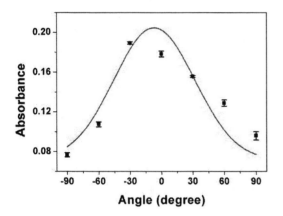

Figure 2. Angular dependence of S_{11} absorption peaks of aligned SWNT film

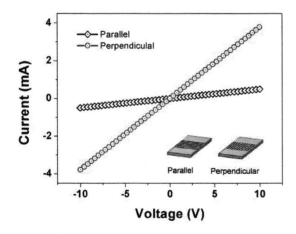

Figure 3. Current-voltage characteristic of SWNT film in which CNTs are aligned parallel and perpendicular to the electrodes.

Figure 3 shows the I-V characteristics of SWNT film corresponding to the electric field applied parallel and perpendicular to the SWNT alignment direction measured by depositing Au electrodes at 1 cm apart. The current was very low for parallel oriented nanotubes compared with that of CNTs oriented perpendicular direction against the electrodes. It implies to macroscopic anisotropic electrical transport in aligned nanotube films. Briefly, these results illustrate high order directional alignment and anisotropic behavior of self assembled nanotube arrays.

3.2 Characterization and Evaluation of Aligned SWNT Films Prepared by Sliding method

The SEM and AFM images in Figure 4 show that the nanotubes are horizontally aligned in the plane of the substrate by sliding the CNT liquid film in between two parallel substrates. The CNTs are in direct contact with each other though their ends, which make the easy transport of charge carriers along the aligned SWNT direction. The thickness of film measured by AFM line profile was 10±2 nm, which corresponds to the bundle diameter, illustrates the formation of monolayer SWNT array.

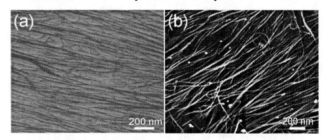

Figue 4. a) SEM and b) AFM images of horizontally aligned SWNT thin films prepared by sliding CNT liquid film between two glass slides.

When an external pressure is applied in perpendicular direction on sample ink drop placed in between two parallel flat substrates, a thin liquid film of nanotubes with monolayer thickness was formed owing to the reduction of gap between the parallel walls to minimum possible spacing. The mechanical stress exerted on the nanotube liquid film at such a small gap would be very huge and which make the original random network nanotubes in the film to orient in different directions. A one-dimensional flow of solution is developed when a pair of substrates is moved one over the other in opposite direction with uniform velocity under constant stress, which is believed to facilitate the aligned architecture nanotubes along the direction of sliding [13]. It is quite similar to mechanical stretching of solid SWNT film.

The surfactant molecule has a crucial role in this process to form uniform aligned SWNT film because it helps uniform spreading of nanotubes by lowering surface tension and also renders them to adhere firmly on the substrates. Experiments with surfactant free CNT dispersions failed to form uniform aligned film due to poor adhesion on the substrate and dewetting of the liquid film. To overcome the issue of film rupture, the dewetting velocity (V_{dewet}) must be low so that the liquid film can dry before deformation of the film occurred by dewetting. The dewetting can be effectively avoided if drying time of liquid film is much shorter than the dewetting time, $t_{dewet} = L/V_{dewet}$, where L is the characteristic length scale of the film [14]. Consequently, the dewetting can be minimized by lowering surface tension and by properly reducing the drying time of coated film. The surfactant SDBS satisfies these conditions. Since the thickness of film was nearly 10 nm, it quickly dried under atmospheric conditions.

Figure 5 shows angular dependence of optical absorption corresponding to S_{11} transition. The anisotropic optical absorption property demonstrates directional alignment of nanotubes in the plane of the substrate.

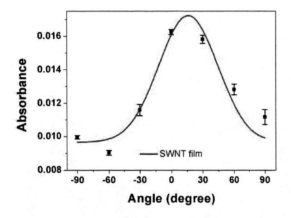

Figure 5. Angular dependence of S_{11} absorption peaks of aligned SWNT film

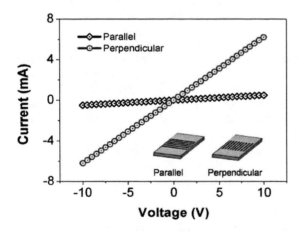

Figure 6. Current-voltage characteristic of SWNT film in which CNTs are aligned parallel and perpendicular to the electrodes.

The current-voltage characteristic of film in parallel and perpendicular directions of nanotube alignment direction (Fig.6) further illustrated directional orientation of nanotubes in the film prepared by sliding method. The ratio of resistivity for parallel ($R_{\parallel}$) to perpendicular ($R_{\perp}$) direction of SWNT film is 12.6, which shows the electrical anisotropic property. The number of inter-tube junctions in aligned nanotube film is very less compared to that in network film. Thin films prepared on glass substrate by sliding method demonstrated 95-97%T at 550 nm and sheet resistance of 800±50 $\Omega/\square$.

4 SUMMARY

In summary, we have fabricated large area directionally aligned thin films of CNTs by two simple methods. First, self assembled aligned array of surface modified SWNTs was fabricated by drop drying, spin coating and dip coating methods. We attribute that the rigidity of the nanostructure and surface charge are crucial factors for the alignment of SWNTs by self assembly. Second, compressed sliding of SWNT film between parallel substrates provided ultra-thin transparent conductive films with high degree of alignment of CNTs in the plane of the substrate. Directionally aligned monolayer of SWNTs can be used for transparent conductors, touch panel, field emission display applications.

5 REFERENCES

[1] P. M. Ajayan, Chem. Rev. 99, 1787, 1999.
[2] S. Eustis and M. A. El-Sayed, Chem. Soc. Rev. 35, 209, 2006.
[3] J. Bernholc, D. Brenner, M. B. Nardelli V. Meunier and C. Roland, Annu. Rev. Mater. Res. 32, 347, 2002.
[4] L. Dai, A. Patil, X. Gong, Z. Guo, L. Liu, Y. Liu and D. Zhu, ChemPhysChem 4, 1150, 2003.
[5] J. H. Lee, W. S. Kang, G. H. Nam, S. W. Choi and J. H. Kim, Journal of Nanoscience and Nanotechnology 9, 7080, 2009.
[6] K. Iakoubovskii, Cent. Eur. J. Phys. 7, 645, 2009.
[7] L. Huang, X.Cui, G. Dukovic and S. P. O'Brien, Nanotechnology 15, 1450, 2004.
[8] T. J. Simmons, D. Hashim, R. Vajtai and P. M. Ajayan, J. Am. Chem. Soc. 129, 10088, 2007.
[9] H. Liu, D. Takagi, S. Chiashi and Y. Homma, ACS Nano 4, 933, 2010.
[10] R. Pomponio, R. Gotti, M. Hudaib and V. Cavrini, J. Chromatogr. A 945, 239, 2002.
[11] N. R. Jana, Angew. Chem. Int. Ed. 43, 1536, 2004.
[12] C. L. Pint, Y. Q. Xu, S. Moghazy, T. Cherukuri, N. T. Alvarez, E. H. Haroz, S. Mahzooni, S. K. Doorn, J. Kono, M. Pasquali and R. H. Hauge, ACS nano, 4, 1131, 2010.
[13] G. Barnes and K. B. MacGregor, Phys. Plasmas 6, 3030, 1999.
[14] B. Dan, G. C. Irvin and M. Pasquali, ACS nano, 3, 835, 2009.

Characterizing the Oxygen Reduction Reaction for the Proton Exchange Fuel Cell with with Novel Nanotechnology-based Catalyst Supports

M.N. Groves*, C. Malardier-Jugroot** and M. Jugroot***

* Royal Military College of Canada, Kingston, ON, Canada, michael.groves@rmc.ca

** Royal Military College of Canada, Kingston, ON, Canada, cecile.malardier-jugroot@rmc.ca

*** Royal Military College of Canada, Kingston, ON, Canada, manish.jugroot@rmc.ca

Abstract

Using the ΔG of adsorption of O_2 gas, three doped catalyst supports were selected for a complete optimization of all the steps of the oxygen reduction reaction as well as their transition states using density functional theory. A single oxygen doped substrate, a 3.79Å N-N doped substrate, and a hemoglobin-like substrate are all compared to the undoped graphene case. It was found that the transition states were not significant so the energy required to dissociate water from the catalyst and the difficulty to create hydrogen peroxide were used to compare the four systems. Given these considerations, the 3.79Å N-N doped system showed the most positive characteristics versus the undoped case.

Keywords: density functional theory, proton exchange membrane fuel cell, doped catalyst support

1 Introduction

Commercialization of the proton exchange membrane fuel cell (PEMFC) is still not feasible due to many factors. One significant obstacle yet to over come is that the oxygen reduction reaction (ORR) on the cathode is still very inefficient resulting in a large Pt catalyst loading making the overall unit prohibitively expensive[1]. By changing the catalyst support there have been both experimental and theoretical evidence that the activity and durability of the catalyst changes. For instance, it has been shown experimentally that nitrogen doping of a carbon catalyst support can increase the durability and activity of a Pt catalyst[2], [3]. Computationally, the authors have demonstrated that the closer and more numerous nitrogen substitutions are to the carbon which binds to the Pt atom, the stronger the binding energy. This is due to the fact that the nitrogen atoms modify the bonds in the support around the carbon atom and allows it to use deeper orbitals to interact with the Pt. This was demonstrated for both graphene and single walled carbon nanotubes[4], [5].

In the literature, a combination of experimental and theoretical studies have shown that measuring the change in Gibbs free energy of gas adsorption (ΔG_{ad}) is a good indicator of activity of a catalyst[6]. As the $|\Delta G|$ of gas adsorption on the catalyst decreases the activity of the catalyst increases. Density functional theory (DFT) will be used to calculate the ΔG_{ad} of O_2 gas on a variety cathode systems. Once that is complete, the surfaces with a ΔG_{ad} of O_2 gas closest to zero will have their zero point energy (ZPE) calculated for every reaction step along the ORR along with examining the transition state energies of each state according to the following ORR reaction mechanism:

$$O_2 + * \rightarrow O_2^*$$
$$O_2^* + (H^+ + e^-) \rightarrow HO_2^*$$
$$HO_2^* + (H^+ + e^-) \rightarrow H_2O + O^*$$
$$O^* + (H^+ + e^-) \rightarrow HO^*$$
$$HO^* + (H^+ + e^-) \rightarrow H_2O^*$$
$$H_2O^* \rightarrow H_2O + *$$

where the * indicates a reaction site.

2 Computational Procedure

All calculations were performed using Gaussian 03[7] either on the High Performance Computing Virtual Laboratory (HPCVL) using revision C.02 or on a native Mac OSX running revision E.01. DFT was used to calculate both ΔG_{ad} data and the ZPE values. Due to the presence of large atoms, an effective core potential (ECP) was used to reduce the computational cost of each calculation. To increase the precision of these results, the GenECP basis set was employed so that an ECP could be mixed with the split valence basis set in the same system. It has been shown that using the GenECP basis set so that 6-31G(d) can be used for light elements such as hydrogen, carbon, nitrogen and oxygen while the ECP SDD is used for heavier elements from the first transition metal proved to be more precise than simply applying the ECP Lanl2DZ to the entire system assuming that the B3LYP hybrid functional [8] was used on both cases [9]. As a result, the combination of the 6-31G(d) split valence basis set [10], [11] was used for all non-transition metal elements, and the ECP SDD [12], [13] was used for all transition metal elements in all calculations unless otherwise noted. For all hydrogens being adsorbed to form water the 6-31G(d,p) basis set was used

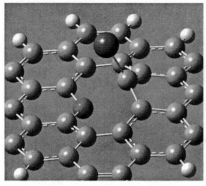
(a) Single oxygen doped surface.

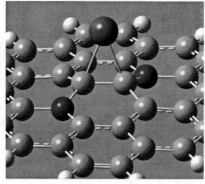
(b) 3.79Å N-N doped surface.

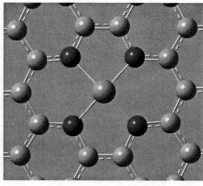
(c) Hemoglobin-like surface.

Figure 1: The three doped substrates selected to have each reaction intermediate optimized in the ORR. Hydrogen is in white, carbon is in grey, nitrogen is in dark blue, oxygen is in red, Platinum is in light blue, and iron is in slate blue.

To determine which substrates would be used when calculating the ZPE for each reaction step as shown above, a ΔG calculation of O_2 gas absorption was performed on many graphene systems. This includes all the nitrogen doped systems examined in [4], single doped berellym, boron, and oxygen doped graphene substrates as well as the pure carbon surface. This pure carbon surface is formed by using 42 carbon atoms in a sheet. The outer carbon atoms are all terminated with hydrogen atoms. Any dopants to the system are exchanged in the lattice for a carbon atom. These surfaces were all geometrically optimized before adding a lone Pt atom which represents the catalyst and optimized again. From this optimization a value for the Gibbs free energy is evaluated. A second optimization is performed on an O_2 molecule and its Gibbs free energy is evaluated. These two values add to be the Gibbs free energy of the reactants. The Gibbs free energy of the products is calculated after a geometry optimization where O_2 is added to the already optimized Pt/surface structure. From this the ΔG_{ad} is calculated.

Evaluating the ZPE of each reaction step meant performing a geometry optimization for each molecule in each step and adding up the results. It is also important to account for everything not listed in each step. For instance, in the first step, there are still four protons and four electrons, evaluated as four single hydrogen atoms, to include in the total ZPE value for the products and the reactants for that step. This is to ensure that each step stays consistent in terms of total energy with every other step. To determine the transition states, first a fixed scan is performed where only the atom being adsorbed on the surface is moved to characterize the position of where the saddle point. If a maximum is found during the scan then it becomes the initial guess in a QST3 optimization [14] to determine the transition state. These two calculations are performed using the

smaller 3-21G basis set on the surface while the transition metal still uses SDD. The oxygen being adsorbed still had 6-31G(d) applied while the adsorbed hydrogen used 6-31G(d,p) in order to still retain the effects the d and p orbitals potentially contribute to the process respectively. Once this was completed, the system was optimized again using the full basis set outlined above.

3 Results and Discussion

As previously mentioned, many substrates were evaluated using ΔG_{ad} of O_2 to select for the systems to be studied step by step. Three surfaces, an oxygen doped graphene system, the 3.79Å two nitrogen doped graphene system, and a hemoglobin-like system, all shown in figure 1 were selected to be compared to the undoped case for the full ORR. The ΔG of O_2 adsorption for these four systems can be found in figure 2. The oxygen doped system represents the best single dopant system measured, while the 3.79Å two nitrogen doped system represents the best multiple doped with a single element system. The hemoglobin-like system was selected due to its novelty in that iron is a much cheaper element than platinum. If its reaction kinetics are somewhat comparable to the other three cases then more catalyst sites can be created to offset its limited activity.

The results from the zero point energy optimizations for each reaction intermediate is shown in figure 3. Every step is exothermic except for when water dissociates from the surface. Finding the transition states for each of these steps proved to be difficult. For the cases when they were found they only confirmed that there was not any significant transition state. One example of the potential energy surface can be found in figure 4. It shows the dissociation of O_2 gas onto the oxygen doped catalyst structure. Each reaction coordinate reflects a fixed position along the path that the two oxygen atoms are thought to take between being adsorbed

Energy (kJ/mol)	1st H_2O	2nd H_2O
Undoped	8.5	25.2
3.79Å N-N	11.0	14.4
Hemo	9.7	23.5
O doped	8.5	22.0

Table 1: The energy, in kJ/mol, necessary to dissociate each water molecule from the catalyst.

onto the platinum and as an O_2 molecule well above the platinum. After the initial interaction between the O_2 molecule and the platinum atom, there is a small transition state. The minimum before the transition state can not be found via optimization as it is too shallow so and is not considered significant. Similar results are found for every other adsorption and desorbtion step. Since no significant transition states have so far been found, the rate determining step seems to be the water dissociation steps as they are endothermic.

The change in energy required for these two water dissociation steps are outlined in table 1. Based on this it would seem that the 3.79Å N-N doped substrate would be the most favourable cathode surface. Even though it requires the largest energy gain to shed the first water molecule, it only requires about half the energy to lose the second water relative to all the others.

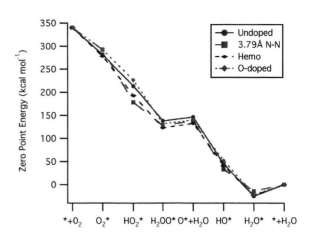

Figure 3: The zero point energy optimization results for the four systems analyzed for the ORR.

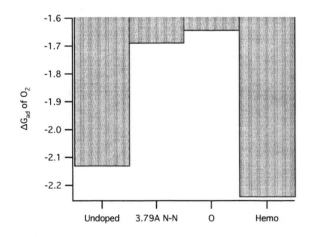

Figure 2: The ΔG of O_2 adsorption for the oxygen doped, 3.79Å two nitrogen doped, hemoglobin-like, and undoped substrates.

An important side reaction which also requires some attention is the water/hydrogen peroxide reaction that can occur instead of dissociating the first water from the catalyst site. The transfer of one of the protons between the two oxygen molecules when adsorbed onto the platinum catalyst does have a measureable transition state and can be seen in figure 5.

This result shows that all three substrate combina-

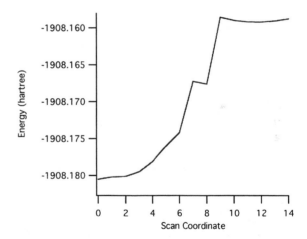

Figure 4: An example of a potential energy scan of the dissociation of an O_2 molecule along a hypothesized reaction pathway on the oxygen doped surface. The extremes in the coordinate domain represent O_2 adsorption (O-O distance is 1.38Å) and O_2 gas when it is not interacting with the platinum. The pathway involves the two oxygen molecules rotating around the platinum atom until they are 1.21Å from each other. They then move away from the platinum together in 0.02Å increments.

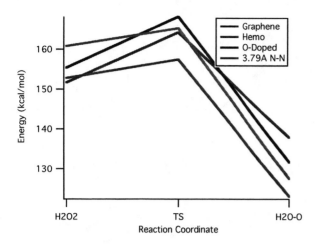

Figure 5: The optimized transition state energies as well as the products and reactant energies for the water/hydrogen peroxide reaction. The reference energy is the same as in figure 3.

tions are an improvement over the undoped surface in that they all require more energy to overcome the transition state energy to form hydrogen peroxide, and require at most the same amount of energy to overcome the transition state to form water. It would seem that the 3.79Å N-N doped substrate is again the best suited since it requires the least amount of energy to go from hydrogen peroxide to water, and the most energy for the reverse reaction. In all four cases, the dissociation of water requires less energy than the activation energy which would form hydrogen peroxide. The difference, however, is not large so it would not be inconceivable that the ORR would be slowed down by this side reaction which makes the large activation energy to form hydrogen peroxide and a small activation energy to form water desirable.

What is positive to see is that the other two catalyst support combinations also show stronger results when compared to the undoped case. This is important since creating this particular nitrogen doped arrangement might prove to be difficult given that different surface-platinum interactions can occur when the nitrogen atoms are moved closer or further away. The distribution of nitogen dopants may not be easily controlled creating regions of the catalyst with different activities. The simpler oxygen doped system would be easier to create since it does not rely on the precise positioning of dopants. The hemoglobin-like system while complex would be cheaper to create since it uses iron as the main catalyst. The lower success rate in creating catalyst sites could be offset by the ability to cost-effectively create many more than a platinum based catalyst.

4   Conclusion

Of the three novel catalyst supports analyzed, the 3.79Å N-N doped substrate had the most favourable characteristics. The transition states of all the reaction intermediates seem to not be significant so the small energy to dissociate water and difficulty to create hydrogen peroxide were used as the primany indicators of efficiency. The other two supports also had positive characteristics compared to the undoped version which is important since they might prove to be simpler to fabricate.

REFERENCES

[1] B. Wang, J. Power Sources **152**, 1 (2005).

[2] G. Wu, D. Li, C. Dai, and D. Wang, Langmuir **24**, 3566 (2008).

[3] X. Lepro, E. Terres, Y. Vega-Cantu, F. J. Rodriguez-Macias, H. Muramatsu, Y. A. Kim, T. Hayahsi, M. Endo, T. R. Miguel, and M. Terrones, Chem. Phys. Lett. **463**, 124 (2008).

[4] M. N. Groves, A. S. W. Chan, C. Malardier-Jugroot, and M. Jugroot, Chem. Phys. Lett. **481**, 214 (2009).

[5] M. N. Groves, A. S. W. Chan, C. Malardier-Jugroot, and M. Jugroot, Mol. Simulat. (2009), accepted November 13.

[6] J. Greely, T. Jaramillo, J. Bonde, I. Chorkendorff, and J. Norskov, Nature Mater. **5**, 909 (2006).

[7] M. J. Frisch *et al.*, Gaussian 03, Gaussian, Inc., Wallingford, CT (2004).

[8] A. D. Becke, J. Chem. Phys. **98**, 5648 (1993).

[9] M. Bühl and H. Kabrede, J. Chem. Theory Comput. **2**, 1282 (2009).

[10] W. Hehre, R. Ditchfield, and J. Pople, J. Chem. Phys **56**, 2257 (1972).

[11] M. Frisch, J. Pople, and J. Binkley, J. Chem. Phys **80**, 3265 (1984).

[12] M. Dolg, U. Wedig, H. Stoll, and H. Preuss, J. Chem. Phys **86**, 866 (1987).

[13] D. Andrae, U. HäuBermann, M. Dolg, H. Stoll, and H. PreuB, J. Chem. Phys **77**, 123 (1990).

[14] C. Peng, P. Ayala, H. Schlegel, and M. Frisch, J. Comp. Chem. **17**, 49 (1996).

Influence of silica precursor on the cobalt incorporation on mesoporous silica MCM-41 used for the synthesis of single wall carbon nanotubes

Frank C. Ramírez, Betty L. López, Luis F. Giraldo*

*Grupo de investigación Ciencia de los Materiales, Instituto de Química, Universidad de Antioquia, calle 62 # 52-59
Lab. 310, Medellín, Colombia*
luis.giraldo@gmail.com

SUMMARY

In this work several silica precursors have been used for synthesis of MCM-41 mesoporous silica, i.e. Cab-O-Sil, sodium silicate and TEOS with incorporation of cobalt in-situ during the synthesis of the molecular sieves. The materials have been characterized by TPR, nitrogen adsorption and XRD. The synthesized catalysts were employed in the chemical vapor deposition of methane at 800 °C for 30 minutes in order to obtain single wall carbon nanotubes. The reaction products were characterized by RAMAN, TGA and TEM. The results show an important effect of synthesis pH and silica precursors on the final selectivity and yield to SWCNTs. Although all the precursors gave way to uniform mesoporous silica type MCM-41, the selectivity to SWCNTs was higher for the catalysts prepared with Cab-O-Sil, followed for the silica prepared with silicate and the catalyst prepared with TEOS shows the lowest activity to SWCNTs. This behavior is related with the TPR profile for each catalyst. It showed that cobalt ions are reduced at higher temperatures for the support prepared with Cab-O-Sil as precursor. The higher is the reduction temperature the inner is the cobalt species into the silica walls, then sinterization process is less feasible, giving way to single wall carbon nanotubes with more uniform diameters and high yield.

Keywords: carbon nanotubes, mesoporous support, MCM-41, catalysis, silica precursor.

1 INTRODUCTION

Many efforts are currently being carried out to develop catalysts with high selectivity and yield to single walled carbon nanotubes (SWCNTs)[1, 2], i.e. carbon nanotubes with uniform and predesigned diameters. Even though parameters as temperature of reaction, pressure and hydrocarbon source can affect this selectivity, the main focus for those proposes has to be the catalysts design in CCVD process[3]. The catalysts consist of an active metal and the support[1, 4]. The mesoporous silica are widely used as a support due to the high surface area, and pore volume[5, 6]; However, these parameters are not enough to ensure an adequate dispersion of metal and small cluster size after reduction process, therefore it is necessary to pay attention to the interaction among the metal species and silica precursors during the catalysts preparation. The electrostatic and covalent interaction between the metal and silica play an important role in the final properties of the catalysts. If the interaction of cobalt ions with the silica monomers is enhanced during the sol-gel formation, the polymerization of silica over the surfactant micelles will include cobalt species incrusted in the silica layer, this species might be in a tetrahedral way or not depending on the concentration and electrostatic repulsion with the silica.

2 EXPERIMENTAL

Materials: Colloidal silica Cab-O-Sil, tetramethylammonium silicate ($TMASiO_2$) and tetraethylortosilicate used as a silica sources and cetyltrimethylammonium bromide (CTMABr), ammonium solution was purchased from Sigma-Aldrich. The ion interchange resin Ambersep 900 (OH) was from Alfa-Aesar. Sodium silicate (15-20 wt% SiO_2), acetic acid and cobalt nitrate were from Merck.

Synthesis of Co-MCM-41/Cab-O-Sil: 2.5g of Cab-O-Sil and 10 g of $TMASiO_2$ were dissolved in 50 mL deionized water and mixed for 30 minutes under magnetic stirring. After that, the cobalt salt at 3.0 wt% in Co was added with two drops of antifoam A[7]. 27.89 g of CTMAOH (Obtained by ion exchange) surfactant solution at 20 wt % were added to the previous solution, then pH was adjusted to 11.5 using acetic acid. The gel was heated in an autoclave at 100 °C for 3 days. After that the gel was filtrated and washed with deionized water and dried in an oven at 80 °C overnight. In order to obtain the mesoporous catalyst, the silica was calcined at 540 °C for 3 hours in air.

Synthesis of Co-MCM-4/sodium silicate: 2.8 g of CTMABr were dissolved in 10 mL of HCl 1.0 M and 15.0 mL of deionized water by means of magnetic stirring. After that, the cobalt precursor was added with Co 3.0 wt % by stirring for 30 minutes. Separately 8.0 g of Sodium silicate were dissolved in 30.0 mL of water, this solution was added drop by drop the former solution. The pH is adjusted to 10.0 using ammonium solution[8]. The hydrothermal treatment and calcinations were carried out with the same procedure described before.

Synthesis of Co-MCM-41/TEOS: 2.8 g of CTMABr were dissolved in 60 mL of deionized water with magnetic stirring. After that, 8.0 mL of ammonium solution (28-30 % wt) were added. The cobalt precursor was introduced with Co 3.0 wt % by stirring for 30 minutes. Five grams of TEOS were added drop by drop to the former solution adjusting the final pH to 11.5 with ammonium solution[9]. The same hydrothermal and calcinations treatment were used.

Previous studies made in our research group and by other researchers showed that the selected pH for each synthesis gave way to the mesoporous silica MCM-41 with structural regularity and good metal incorporation[10].

Carbon nanotubes synthesis: Two hundred mg of each catalyst were placed in a quartz tube with a fritted disc of 25x5 mm using a vertical oven at 700 °C for 30 minutes in flowing hydrogen and nitrogen mixture of 50:150 sccm respectively at 1 atm. After reduction, the hydrogen is displaced with nitrogen and the carbon nanotubes were synthesized by conversion at 800 °C for 30 minutes with methane and nitrogen mixture 50:150 sccm. The oven is cooled in a nitrogen flow and the material is recovered. The Raman spectra of nanotubes were recorded in a Horiba Laser Raman Spectrophotometer model LabRAM HR with an excitation radiation at 736 cm^{-1}.

CHARACTERIZATION

The nitrogen adsorption-desorption studies were carried out at -196 °C in a Micromeritics ASAP 2010 to examine the mesoporous properties of the catalysts. The catalysts were preheated at 250 °C for 2 hours in vacuum.

The X-ray analyses were performed at room temperature in a Siemens D5000 equipped with a Cu-Kα source (λ= 1.54Å), with a 0.048 step.

The thermogravimetric analyses were carried out in air from room temperature up to 800 °C at 10 °C/minute.

The reduction temperature analyses were made in an AutoChem II 2920 Micromeritic with TCD detector. Fifty milligrams of each catalyst are preheated up to 250 °C for 1 h in argon flow in order to eliminate any absorbed gas. The catalyst is heated again in a flow of 5% of hydrogen in argon from room temperature up to 1000 °C and hold at this temperature for 1h more in order to ensure the complete cobalt reduction.

3 RESULTS AND DISCUSSIONS

3.1 Catalyst characterization

Figure 1 shows the nitrogen adsorption isotherms for all the catalysts employed in the study. All the catalyst show a defined mesoporous adsorption profiles with isotherm type IV with large surface area ranging from 970 to 990 m^{2}g^{-1} determined with BET model, and BJH pore size distribution with uniform range from 2.6-2.7 nm for all the samples.

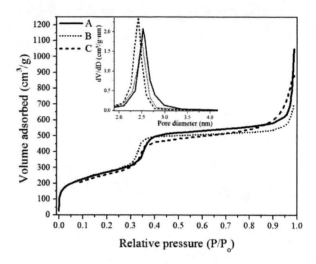

Figure 1. Adsorption isotherm of nitrogen at 77 K for the catalyst Co-MCM-41 prepared using Cab-O-Sil (A) sodium silicate (B) and TEOS (C) as precursors.

Silica precursor	BET area (m^2/g)	PD (nm)	a$_0$ (nm)	WT (nm)
TEOS	990	2.43	4.91	2.45
Sodium silicate	975	2.29	5.21	2.54
Cab-O-Sil	978	2.53	4.50	1.84

Table 1. XRD and nitrogen physisorption summary. WT = wall thickness; PD = Pore diameter; a$_0$ = Cell parameter

As it is summarized in table 1, there are not significant differences among the surface areas for all the catalysts, but catalyst synthesized by using Cab-O-Sil shows the higher pore diameter distribution and textural porosity.

X-ray diffraction experiments show diffractograms with a typical profiles corresponding to the reflections (100), (110) y (200) for a MCM-41 silica structure (Figure 2). The incorporation of the cobalt into the silica walls has not a significant effect in the structural regularity in relation to the pure silica. However, catalysts prepare using TEOS as precursor showed the highest structural ordering due to the reflection at the plane (210) and higher intensity in the diffractogram. The wall thicknesses for the mesoporous catalysts were determined as the differences among the cell parameter (a$_0$) from de XDR results and BJH pore size distribution. As a result, the catalyst prepared using sodium silicate as precursor has the biggest wall thickness, which could be related to the pH synthesis. A reduction in the pH synthesis promotes the polycondensation of silica species.

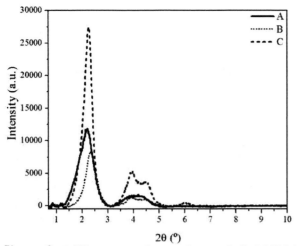

Figure 3. Diffractorams for catalysts of Co-MCM-41 prepared using Cab-O-Sil (A), Sodium silicate (B) and TEOS (C) as precursors.

The temperature programmed reduction (TPR) profiles are shown in figure 3. According to the TPR profiles, the catalyst prepared with sodium silicate as precursor shows the highest reduction temperature which could be related with his large wall thickness, in this case cobalt ions can be in an inner place far from the surface, for that reason more temperature is required to promote the reduction and migration of cobalt clusters. In the catalyst prepared with TEOS as silica precursor more heterogeneous species are on the silica with cobalt outside the walls as it is shown by the lower reduction peak about 500 °C. According to the TPR, the catalyst prepared with Cab-O-Sil shows the most homogeneous distribution of cobalt on the silica walls with a maximum temperature reduction at 750 °C.

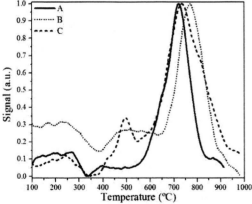

Figure 3. TPR profile for the catalyst Co-MCM-41 prepared using Cab-O-Sil (A) sodium silicate (B) and TEOS (C) as precursors.

During the reduction pretreatment in the carbon nanotubes synthesis the catalyst is heated below the maximum reduction temperature, thus the sintering process is not completed because the cobalt atoms that leave the silica wall will stay anchored with the cobalt cations that remain into the silica wall by electrostatic interaction[11].

2.1 Catalysts performance

Figure 4 and 5 show the Raman spectra of the SWNT growth on the different catalysts. The three most important signals in the spectra are the radial breathing mode (RBM) that gives information about the tubes diameter distribution, and the D band at about 1350 cm^{-1} that is related with the presence of amorphous carbon and structural disorder of the graphite sheet. The G band about 1570 cm^{-1} is related to the graphite sheet order thus the relative intensity D/G gives information about the order of the SWNT[12], i.e. the lower the ratio D/G the highest is the order in the SWNT structure. According to the RBM the single carbon nanotubes obtained have diameters from 0.8 up to 1.5 nm.

In our results it is clear that the precursor plays an important role in the selectivity to the formation of SWNT. The catalyst prepared with Cab-O-Sil as precursor showed the highest selectivity to single wall carbon nanotubes compared to the others supports prepared with sodium silicate and TEOS as precursors, despite of the highest structural regularity the catalyst prepared using TEOS precursor, for the last case the deposition of SWNT was only 1.2 wt% according to the thermogravimetrical analysis (TGA). On the other hand sodium silicate gave the most disordered carbon nanotubes in all the samples due to his highest D/G intensities ratio as it is shown in figure 4. This catalyst gave only 1.6 wt% of carbon deposition, showing low selectivity to SWCNTs, This effect could be related to sodium effect on the cobalt incorporation. Some studies have demonstrated that the presence of sodium during the metal incorporation could displace metal cations to the silica surface.

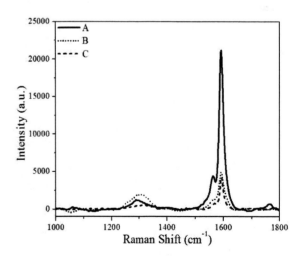

Figure 4. RAMAN spectra for D and G bands of the carbon synthesized on Co-MCM-41 prepared using Cab-O-Sil (A) sodium silicate (B) and TEOS (C) as precursors.

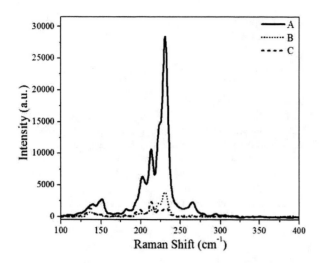

Figure 5. RAMAN spectra for the RBM region of the carbon synthesized on Co-MCM-41 prepared using Cab-O-Sil (A) sodium silicate (B) and TEOS (C) as precursors.

The Cab-O-Sil gave the highest selectivity and deposition, about 3.8 wt% under the reaction conditions. Even when all the samples have almost the same physical properties, with almost the same surface area and pore distribution, the performance is quite different for all the catalysts prepared with the precursors, therefore is not only important a high surface area and uniform pore size distribution, but also the interaction between the silica and cobalt precursors.

ACKNOWLEDGEMENTS

We are grateful to CODI, Universidad de Antioquia, Medellín for the financial support and Colciencias, Bogotá, Colombia.

REFERENCES

[1]   G. L. Haller, "New catalytic concepts from new materials: understanding catalysis from a fundamental perspective, past, present, and future", Journal of Catalysis, 216, 12-22, 2003.

[2]   S. Lim, et al., "Improved synthesis of highly ordered Co-MCM-41", Microporous and Mesoporous Materials, 101, 200-06, 2007.

[3]   J. E. Herrera, et al., "Relationship between the Structure/Composition of Co-Mo Catalysts and Their Ability to Produce Single-Walled Carbon Nanotubes by CO Disproportionation", Journal of Catalysis, 204, 129-45, 2001.

[4]   A.-C. Dupuis, "The catalyst in the CCVD of carbon nanotubes--a review", Progress in Materials Science, 50, 929-61, 2005.

[5]   J. Liu, et al., "Pore size control of mesoporous silicas from mixtures of sodium silicate and TEOS", Microporous and Mesoporous Materials 106, 62–67, 2007.

[6]   L. F. Giraldo, et al., "Mesoporous Silica Applications", Macromol. Symp., 258, 129–41, 2007.

[7]   Y. Chen, et al., "Synthesis of uniform diameter single-wall carbon nanotubes in Co-MCM-41: effects of the catalyst prereduction and nanotube growth temperatures", Journal of Catalysis, 225, 453-65, 2004.

[8]   L. F. Giraldo; M. Echeverri, and B. L. López, "Reinforcement of Polyamide 6 with Nanoparticles", Macromol. Symp., 258, 119–28, 2007.

[9]   M. Grün, et al., "Novel pathways for the preparation of mesoporous MCM-41 materials: control of porosity and morphology", Microporous and Mesoporous Materials, 27, 207-16, 1999.

[10]   S. Lim, et al., "The effect of synthesis solution pH on the physicochemical properties of Co substituted MCM-41", Topics in Catalysis, 34, 2005.

[11]   S. Lim, et al., "Evidence for anchoring and partial occlusion of metallic clusters on the pore walls of MCM-41 and effect on the stability of the metallic clusters", Catalysis Today, 123, 122-32, 2007.

[12]   M. S. Dresselhaus, et al., "Raman spectroscopy of carbon nanotubes", Physics Reports, 409, 47-99, 2005.

Spin Dependent Transport in Carbon Nanostructures

B. Montanari[b], and N. M. Harrison[a,b]

[a]Thomas Young Centre, Department of Chemistry,
Imperial College London, Exhibition Road, London, SW7 2AZ, United Kingdom
[b]Computational Science and Engineering Department, STFC Rutherford Appleton Laboratory,
Didcot, Oxon, OX11 0QX, United Kingdom
barbara.montanari@stfc.ac.uk, nicholas.harrison@imperial.ac.uk

ABSTRACT

Keywords: graphene, transport, spintronics, theory, magnetism

1 INTRODUCTION

For the past fifty years progress in the miniaturisation of silicon devices has produced an exponential growth in performance but quantum limits are now being reached as transistor gate lengths approach 5nm. The exploitation of electron spin rather than charge for information storage and manipulation (*spintronics*) has the potential for efficient operation on the sub 1nm length scale and thus to facilitate another generation of devices. In order to achieve this a readily available spin injection material is required; ideally, a material that is a semiconductor, ferromagnetic at room temperature, with a highly tunable band gap and magnetic coupling. The controlled synthesis of such a material has not yet been achieved.

The occurence of ferromagnetism in purely *sp*-bonded systems provides a challenge to theoretical physics and an opportunity to generate a new class of materials that may be of great value for spintronics.

In the current work hybrid exchange density functional theory is used to demonstrate that local spin moment formation can occur in graphene sheets at both localised and extended defects. In addition, the localised moments are strongly coupled via a spin polarisation of the sheet which, under particular circumstances, is sufficient to produce a ferromagnetic state stable at ambient temperatures. The resultant optical energy gap and magnetic coupling are sensitive to the scale of the nanostructure, the arrangement and concentration of defects; this suggests a mechanism for the tuning of magnetic and electronic properties for specific device applications.

2 METHODOLOGY

Spin localisation at local and extended defects and long range magnetic coupling are studied here using hybrid exchange density functional theory (B3LYP [1]) which significantly extends the reliability of the widely used local and gradient corrected approximations to density functional theory in strongly interacting systems [2].

In particular the energy gap and the tendency to electron localisation in extended systems are well described. All calculations are performed using the CRYSTAL implementation [3] in which the all electron periodic crystalline wavefunctions are expanded as a linear combination of atom centred Gaussian orbitals (LCAO) with s, p, d, or f symmetry. This approach is particularly suitable for use with hybrid functionals as the Fock exchange can be computed reliably and efficiently. Basis sets of double valence quality (6-21G* for C and 6-31G* for H) are used. A reciprocal space sampling on a Pack-Monkhorst grid of shrinking factor 6 is adopted which is sufficient to converge the total energy to within 10^{-4} eV per unit cell. The energies of various magnetic states have been generated by constraining the initial spin density matrix and then seeking stationary points of the self consistent field procedure.

The graphene sheet is a planar lattice of carbon atoms arranged as edge sharing hexagons which is defined by a unit cell containing two atoms each of which is a member of a distinct sublattice; an arrangement referred to as a bipartite lattice. The bonding of the sheet is sp^2 with each carbon atom σ bonded to three neighbours in the plane and contributing one electron to a delocalised, half filled, π-electron band.

In the present work extended defects are studied by considering ribbons of graphene cut from this sheet and local defects by considering carbon vacancies. In both cases the key issue is the behaviour of the π-electrons and so σ bonds are, where sterically possible, simply saturated by hydrogen atoms. These defects are chosen as representative examples and their behaviour is here taken to be representative of the many different types of defect that can be generated in graphene nanostructures.

The optimised structure of the sheets has a lattice constant of 2.460 Å, 0.04% smaller than the observed lattice constant of natural graphite and corresponding to a C–C bond length of 1.420 Å.

3 GRAPHENE RIBBONS

A graphene ribbon was obtained by cutting a graphene sheet along two parallel zig-zag lines and saturating the σ-bonds of the edges with H atoms, figure 1. The ribbon is periodic in the x direction only; the periodic unit cell is delimited by dashed lines in the figure. The width of the ribbon along the non periodic dimension y, is defined here by the number, N, of *trans*–polyacetylene-like rows of carbon atoms that run along x.

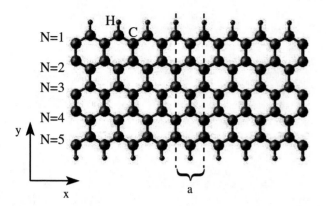

Figure 1: (color online) A mono-hydrogenated ribbon of width N=5 along y. The system is periodic only along x and the dashed lines delimit the periodic unit cell of length a.

The spin density of two stable magnetic states have been found; an *antiferromagnetic* (AF) state in which the spin localised on the edges is anti-parallel, Fig 2(a), and a *ferromagnetic* state in which it is parallel, Fig. 2(b).

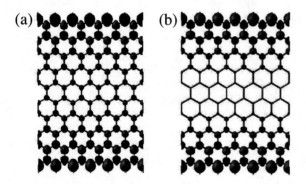

Figure 2: (color online) Isovalue surfaces of the spin density for the antiferromagnetic case (a) and ferromagnetic case(b). The red surfaces represents spin up density and the blue surface spin down density. The range of isovalues is [-0.28:0.28] $\mu_B/Å^3$ in case (a) and [-0.09:0.28] $\mu_B/Å^3$ in case (b).

The energy difference between the non-magnetic and the antiferromagnetic states, plotted in Fig. 3, is a measure of the strength of the magnetic instability. The

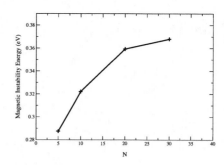

Figure 3: The energy difference (per unit cell) between the non-magnetic and antiferromagnetic case as a function of N.

stabilisation is larger for wider ribbons and converges to about 0.38 eV per unit cell at $N \sim 30$.

In the ground state AF configuration opposite spin polarisation on neighboring sites localises the electrons in singlet pairing between the two sublattices and a band gap opens at $k = \frac{2}{3}\frac{\pi}{a}$, where the valence and conduction bands of graphene are degenerate. The electronic ground state of a mono-hydrogenated ribbon of finite width is therefore insulating. As the width increases, the band gap tends to zero following an algebraic decay of the type 1/N. The band gap vanishes, within room temperature thermal energy, when N≃400, i.e., for an 80 nm wide ribbon. This length scale could be reached by the nanotechonolgy industry in the future with the synthesis of single, relatively thin sheets of nanographite [4].

4 ARRAYS of POINT DEFECTS

The representative point defect adopted here is a carbon vacancy in which, due to steric hindrance, two of the three dangling σ bonds are saturated with H-atoms. It is expected that the remaining dangling σ bond has no significant role to play in the conclusions drawn here.

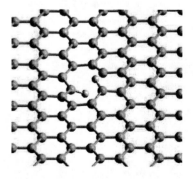

Figure 4: The structure of the defect considered in the present study.

Forming a periodic array of these defects reveals the localisation of spin at the defect and the mechanism for spin coupling between the defects. The resultant spin density for a defects spacing of 20 Å is displayed in figure 5.

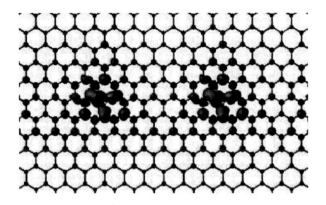

Figure 5: (color on line) Isovalue surfaces of the spin density of the graphene sheet with a defect separation of 20 Å. The red (at 0.021 μ_B/Å) and blue (at -0.021 μ_B/Å)) isosurfaces represent the majority and minority spin densities, respectively.

The majority spin density within the defect is concentrated on the carbon atoms surrounding the vacancy and their second neighbours. The operation of the spin alternation rule, observed within graphene ribbons, in the spin polarised lattice is clearly visible [5], [4].

The electronic band structure of this state is shown in figure 6. The computed ground state is semiconducting. The presence of the defects and the magnetic ordering break the symmetry of the graphene π system, opening band gaps of 0.51 and 0.55 eV in the majority and minority spin band structures, respectively. The energy gap of the majority states is shifted upwards by about 0.20 eV with respect to the energy gap of the minority states, and the shape of the bands around the Fermi energy is markedly spin dependent. A strong spin asymmetry in the conductivity of the sheet is therefore to be expected.

Calculations at a variety of defect separations establish that the band gap scales as L^{-2} [5]. This decay law is consistent with that computed for graphene ribbons where the edges are line defects at which the spin moments localise and the band gap scales as L^{-1} with the ribbon width [4]

Although creating the particular defect structure studied here is not currently possible it is notable that there are now a number of techniques under development for generating and controlling the density of defects in a graphene sheet including the use of electron beams [6].

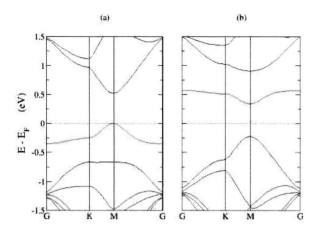

Figure 6: (color on line) The electronic energy bands of the majority (a) and minority (b) spin states of the ferromagnetic ground state in a defective graphene sheet with a defect separation of 20 Å plotted with respect to the Fermi energy (E_F). The rather flat impurity bands near the Fermi energy, which are associated with the states localised at the defect, are indicated in red.

5   CONCLUSIONS

In conclusion, first principles calculations have been used to establish that the electronic structure of graphene ribbons with zig-zag edges is unstable with respect to a magnetic polarisation of the edge states. The calculated magnetic interaction of the edge states is remarkably long ranged. A band gap in the bulk of the ribbon exists in the antiferromagnetically alligned ground state with the size of the gap also displaying a very long range dependence on the ribbon width closing, at room temperature for ribbon widths exceeding 80nm. The mechanism for the opening of the band gap and the long range magnetic coupling is a spin alternation instablity of the bipartite graphene lattice.

The introduction of a periodic array of point defects also generates localised magnetic moments and, through a very similar mechanism, can produce a semiconducting ferromagnetic state at room temperature for defect separations up to 20 Å and thus for defect concentrations as low as 10^{13}/cm^2. The energy gap and magnetic coupling depend strongly on defect concentration. We conclude that a doped or defective graphene sheet is a very promising material with an in built mechanism for tailoring properties for a variety of spintronics applications.

These results have direct implications for the control of the spin dependent conductance in graphitic nanoribbons using suitably modulated magnetic fields.

REFERENCES

[1] Becke AD. A New Mixing of Hartree-Fock and Local Density-Functional Theories. J Chem Phys. 1993;98:1372.

[2] Muscat J, Wander A, Harrison NM. On the prediction of band gaps from hybrid functional theory. Chem Phys Letts. 2001;342:397.

[3] Dovesi R, Saunders VR, Roetti C, Orlando R, Zicovich-Wilson CM, Pascale F, et al. CRYSTAL 2006 Users's Manual. University of Torino; 2007.

[4] Pisani L, Chan JA, Montanari B, Harrison NM. Electronic Structure and Magnetic Properties of Graphitic Ribbons. Phys Rev B. 2007;75:064418.

[5] Pisani L, Montanari B, Harrison NM. A Defective Graphene Phase Predicted to be a Room Temperature Ferromagnetic Semiconductor. New J Phys. 2008;10:033002.

[6] Warner JH, Rummeli MH, Ge L, Gemming T, Montanari B, Harrison NM, et al. Structural Transformations in Graphene Studied with High Spatial and Temporal Resolution. Nature Nanotech. 2009;4:500.

Effect of Position and Orientation of Stone-Wales Defects on the Fracture Behavior of a Zigzag Single-Walled Carbon Nanotube

Keka. Talukdar[1], Apurba Krishna Mitra[2]

[1, 2]Department of Physics
National Institute of Technology, Durgapur-713209, India.
Email : [1]keka.talukdar@yahoo.co.in,   [2]akmrecdgp@yahoo.com

ABSTRACT

Molecular dynamics simulation has been carried out to study the mechanical properties of a 42.59 Å long zigzag (10, 0) single-walled carbon nanotube with an increasing number of Stone-Wales defects, by varying their relative position and orientation. Brenner bond order potential has been employed for energy minimization. Changes in tensile failure stress, maximum strain and fracture behavior are reported in the present work. Maximum reduction of the tensile strength is observed for overlapping defects. Ductility of the tube is little affected by the presence of defect in the tube structure. Breaking patterns when modeled show variation in their initiation positions. The zigzag tube shows necking before failure with some deformations in other places.

Keywords: Carbon nanotube, Stone-Wales defect, Mechanical property, Molecular dynamics simulation

1. INTRODUCTION

Fabrication of super strong composite materials can be realized in practice by the use of carbon nanotubes(CNTs) as reinforcing fibres. The choice of CNTs in these fields is found to be very much beneficial for their tremendous high strength and low density. But the prediction of their mechanical properties [1-3] often ends with some uncertainties especially due to some unavoidable defects produced in them during their production, purification or fictionalization. Stone-Wales (SW) defects [4], vacancies, pentagons, heptagons, lattice-trapped states, ad-dimers etc. are many types of defects that can appear in the CNT structure. Influence of defects can be observed in the mechanical properties of CNTs as well as electronic or magnetic properties. However, the effect of Stone-Wales defects have been investigated by many researchers [5-7] to obtain a clear understanding of their influence. A SW defect is produced by 90^0 rotation of a C-C bond and thus producing two pentagons and two heptagons. No matter what the process or potential adopted, reduction of failure strength and failure strain was reported by most of the authors. Adopting Brenner's bond order potential, Pozrikidis [8] has shown that inclined, axial and circumferential defect orientations have a strong influence on the mechanical response of   zigzag and armchair single-walled carbon nanotubes (SWCNTs). Recently an attempt to explain the reason of scattering in data of the mechanical properties of CNTs has been made by Tunvir et al. [9] with Morse potential. To investigate the interference effect of spatial arrangements between two neighboring vacancy and SW defects with respect to the loading direction, their relative distances as well as their local orientations have been varied and the results compared. But they have considered the defects in the middle of a (10, 10) SWCNT and in only one side of the tube. The influence of odd and even number of SW defects are investigated in a recent study [10] by the present authors.

As the SWCNTs have cylindrical geometry, defects may lie on its surface in a particular spatial configuration. Keeping this in mind, more than one defect are considered at different positions of a zigzag (10, 0) SWCNT with varying separating distances and different angular orientations to simulate them in atomic scale. Tersoff-Brenner [11] potential is employed to represent the interaction of the C-C bond, which can explain adequately the creation and destruction of bonds in a CNT structure. Defects are distributed in all parts of the tube and moreover, diagonal, overlapping and neighboring defects are taken along with many other possible arrangements. Fracture modes are modeled and compared, with a detailed study of the mechanical responses of the tubes viz., Young's modulus, failure strength and ductility.

2. THEORETICAL METHOD

A zigzag (10, 0) SWCNT of length 49.19 Å is considered in this study. The basic pattern of 40 atoms is repeated in space to form the tube which contains 400 atoms as a whole. Berendsen thermostat [12] is used to allow small changes in the velocities of the atoms such that the temperature of the system reaches an equilibrium value close to 300K. The tube is stretched in the axial direction keeping the other end fixed. By stretching in small strain increments, the equilibrium potential energy is calculated by simulation, at first in absence of any defect, and then with 1, 2, 3 and 4 defects at different positions of the tube.

Stretch is applied along the axis of the tube. No other constraint is set up other than the condition that the force is zero in any other direction.

Stress is calculated from the energy-strain curve as $\sigma = 1/V \, (dE/d\varepsilon)$ where σ is the longitudinal stress, V the volume of the tube, ε the strain and E the strain energy of the tube. Volume of the tube is calculated as $V = 2\pi r \delta r l$ where r is the inner radius of the tube, δr its wall thickness and l the length of the tube. We have taken δr as 0.34 nm, which has been the standard value, used by most of the authors. To calculate stress from the energy-strain curve we have used a linear relationship for the elastic region and appropriate non-linear polynomial fit for the segments of high strain deformation regions. Young's modulus is found from the slope of the linear portion of the stress-strain curve. By changing the (z, r, θ) values, defects are created at different positions of the same tube. θ is changed by 90^0 from its first value to form two oppositely oriented defects. To produce two diagonal defects, θ is changed by 45^0 and z is adjusted to the desired value.

3. RESULTS AND DISCUSSION

3.1 Mechanical characteristics of a defect- free tube

In our calculation, the failure strength of a zigzag SWCNT is 115.4 GPa and maximum strain before failure is 18%. Young's modulus of the perfect tube is found to be 1.06 TPa [Fig 1]. Our calculation matches with the experimental values of failure stress by Demczyk [13] which is 150±45 GPa. Young's modulus is also close to the experimental value of 1.28 TPa by Wong et al. [14] and 1.25 TPa by Krishan et al.[15]. Maximum strain can be compared with the results of quantum mechanical calculations [6, 16] though 10-13% maximum strain and failure stress between 13-52 GPa were observed by Yu et al. [17]. For a pristine tube. several SW rotations are observed at strain above 10%.

3.2: Effect of a single defect

Young's modulus of the SWCNT is reduced by 24% by the inclusion of one defect. But ductility remains almost same. In maximum cases stress-strain curves are straight lines up to 7-8% strain. Strain energy (E_d) increases when a defect is introduced in the tube structure. Fig 1 below shows the stress-strain curves of a (10, 0) SWCNT with a single defect at different positions. Stress-strain curves [Fig 1] show the variation of the mechanical behavior of the tube mainly in the non-linear portion after inclusion of a single defect at different places. Failure stress is reduced to a maximum value of 103.59 GPa from its initial value. The results are tabulated in Table 1.

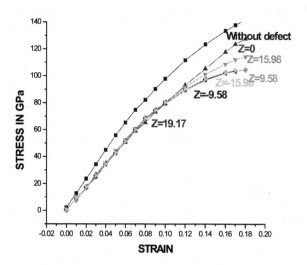

Fig 1: Stress-strain curves for a (10, 0) SWCNT with a single defect at different positions

3.3 Effect of a couple of defects

Interesting results are obtained with different combinations of two defects. Results are shown in Fig 2.

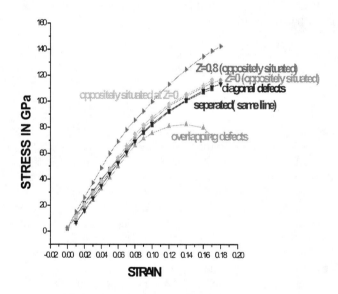

Fig. 2: Stress-strain curves for a (10,0) SWCNT with two defects at different positions

Tensile strength and Y value here are higher than the result of inclusion of one defect except for two diagonally situated defects and when the defects are situated on the same line separated by a distance. Defect-defect interaction [18] is different in each case. Repulsive interaction between defects is said to occur when failure strength is reduced greatly. But when this reduction is not much, we can say that the defects attract each other.

216

3.4 Influence of three defects

Different combinations of three defects are taken to study the effect of odd number of defects. An aggregation of three diagonal defects in a zigzag tube reduces the strength of the tube to 89.58 GPa.[Fig 3]. Reduction of strength is also remarkable when the defects are close and almost on the same line (orientation of one defect differs only by 9^0). Like two defects on the same line, three such defects exactly on the same line have no such pronounced effect on the tensile strength of the tube. The middle defect here is influenced in a complex manner by the other two defects. In each case, different results are obtained [Table 1]. For the defects that are oriented by some angle with respect to each other, the curvature plays a significant role in decreasing the strength. For defects on the same line, mainly the separation between the defects is important.

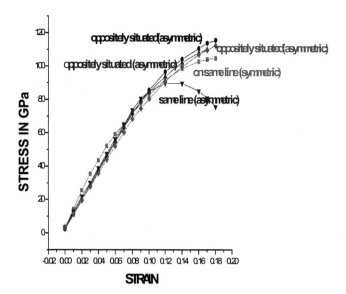

Fig 4: Stress-Strain curves of a (10,0) SWNT with 4 defects at different positions

3.6 Modeling of fracture behavior

Necking is observed before failure for the pristine tube. Fracture of a defect- free tube is shown in fig 5(a). Breaking of the tube with one or more defects sometimes [Fig 5(c, e, f)] show deformation of the tube in other places. With three diagonal defects [Fig 5(d)]the fracture of the tube is associated with several breaking of bonds. Fracture in no case is sharp. With four defects [Fig 5(f)] the tube does not break from any defect site.

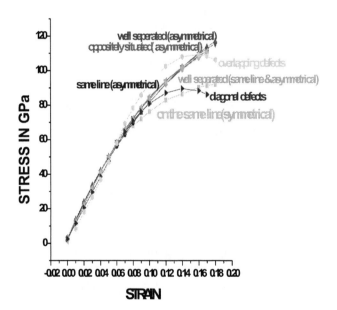

Fig 3: Stress-strain curves for a (10, 0) SWCNT with 3 defects at different positions

3.5 Results of inclusion of four defects

When four defects are placed asymmetrically on the same line in a zigzag (10, 0) SWCNT, the failure strength is very much reduced. 93.81 GPa is the lowest strength of the tube with four defects.

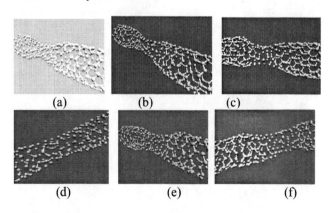

Fig 5: Breaking of a (10, 0) SWNT **(a)** without defect **(b)** with a single defect at (z, r, θ) position (15.98, 3.87, 9) **(c)** with 2 defects at (-1.06,3.87,-9) & (1.06,3.87,27) **(d)** with 3 defects at (0.00,3.91,90), (4.26,3.91,126) & (3.91, 3.87, 171) **(e)** (0.00,3.91,90), (-8.52,3.91,-90) & (-17.04.3.89, 90) and **(f)** with 4 defects at (-4.26, 3.91, 90), (12.78, 3.91, 90), (4.26,3.91,90) & (-12.78, 3.91,90)

	(z, r, θ) Position of defects	Y.M.[*] (TPa)	T.S.[**] (GPa)
1 D E F E C T	(0, 3.91, 18)	.832	105.34
	(9.58, 3.87, 99	.859	104.66
	(-9.58, 3.87, 99)	.847	103.59
	(15.98, 3.87, 9)	.824	116.67
	(-15.98, 3.87, 9)	.846	104.14
	(19.17, 3.91, 36)	.821	127.71
	(-19.17, 3.91, 36)	.829	106.67
2 D E F E C T S	(-4.26,3.91,54), (4.26,3.91,54)	.891	109.44
	(-4.26,3.91,90),(4.26,3.91,90)	.873	116.38
	(0,3.91,90), (8.52,3.91,-90)	1.00	141.7
	(-1.06,3.87,-9), (1.06,3.87,27)	.849	82.31
	(2.13.3.91,0), (0.00,3.91,90)	.893	113.5
	(0,3.91,-18), (,3.91,162)	.900	116.58
	(-3.19,3.87,27), (3.19,3.87,27)	.832	85.87
3 D E F E C T S	(-4.26,3.91,99), (0.00,3.91,90),(4.26,3.91,90)	.904	92.19
	(0.00,3.91,90), (-4.26,3.91,90),(8.52,3.91,90)	.865	111.41
	(-4.26,3.91,-90), (4.26,3.91,90),(17.04,3.91,90)	.95	117.02
	(0.00,3.91,90), (-8.52,3.91,-90),(-17.04.3.89,90)	.94	115.6
	(0.00,3.91,90), (-8.52,3.91,90),(-17.04,3.91,90)	.924	110.36
	(0.00,3.91,90), (8.52,3.91,90)(17.04,3.91,90)	.946	111.28
	(0.00,3.91,90),(4.26,3.91,126) ,(3.91,3.87,171)	.892	89.58
	(0.00,3.91,90), (3.91,3.87,99),(7.45,3.87,99)	.898	109.28
4 D E F E C T S	(-4.26,3.91,90), (12.78,3.91,90),(3.04, 2.65, 99), (7.45, 2.7, 90)	.74	93.81
	(-15.52, 2.65, 88), (-7.45, 2.73, 90),(7.45, 2.7, 90), (14.30, 2.65, 88)	.758	145.22
	(-15.52, 2.65, 88), (-5.02, 2.73, 97),(5.02, 2.73, -97), (15.52, 2.65, -88)	.756	88.09
	(-19.93, 2.7, 97), (-9.89,2.73, 83),(0.00,3.91,90), (4.26,3.91,90)	.792	98.92
	(-6.24, 2.7, 94), (-2.59, 2.73, 105)(0.61, 2.65, 92), (3.8, 2.73, 79)	.814	86.81

Table 1: Young's Modulus and Tensile strength of a (10, 0) SWCNT with defects at different positions
[*]Y.M.-Young's modulus, [**] T.S.- Tensile Strength

5. CONCLUSIONS

The mechanical properties of a (10, 0) SWCNT are influenced by the presence of SW defects. Their number, position, orientation and their arrangement are responsible for the changes in mechanical characteristics. The ductility of the zigzag tube is not influenced by the inclusion of defects. Results with more than one defect show defect-defect correlation. Neighboring defects, especially a combination of overlapping defects reduces the failure strength of the tube significantly. An increased number of defects do not show any marked influence on the mechanical properties of the nanotubes.

6. REFERENCES

[1] B.I. Yakobson , C.J. Brabec and J. Bernholc, Phys. Rev. Lett. **76**, 2511, 1996.
[2] B.I. Yakobson , Appl. Phys. Lett. **72**, 918,1998.
[3] K.M. Liew , C.H. Wong , X.Q. He , M.J. Tan and S.A. Meguid, Phys. Rev. B **69**, 115429,2004.
[4] A.J. Stone and D.J. Wales, Chem. Phys. Lett. **128,** 501,1986
[5] T. Belytschko, S.P. Xiao , G.C. Schatz and R. Ruoff, Phys Rev. B 65, 235430, 2002.
[6] D. Troya, S.L. Mielke and G.C. Schatz. Chem. Phys. Lett. **382**, 133, 2003.
[7] N. Chandra, S. Namilae and C.Shet, Phys. Rev. B **69**, 094101, 2004.
[8] C. Pozrikidis, Arch. Appl. Mech. **79**, 113, 2009
[9] K.Tunvir, A.Kim and H. Nahm, Nanotechnology **19**, 065703, 2008.
[10] K. Talukdar and A.K. Mitra. Composite structures. **92**, 1701, 2010.
[11] D.W. Brenner, Phys. Rev. **B 42**, 9458, 1990.
[12] H.J.C. Berendsen, J.P.M. Postma, W.F. van Gunsteren, A. DiNola and J. R. Haak, J. Chem. Phys. **81,** 3684, 1984.
[13] B.G. Demczyk , Materials Science and Engineering **A334**, 173, 2002.
[14] E.W. Wong, P.E. Sheehan and C.M. Lieber. Science, **277**, 1971, 1997.
[15] A. Krishnan , E. Dujardin, T.W. Ebbesen, P.N. Yianilos and M.M.J. Treacy, Phys. Rev **B 58**, 14013, 1998.
[16] T. Ozaki et al. Phys. Rev. Lett. 84, 1712, 2000.
[17] M.F. Yu , B.S. Files, S. Arepalli and R.S. Ruoff, Phys Rev Lett. **84**, 5552, 2000.
[18] Ge. G. Samsonidze , G.G. Samsonidze and B.I. Yakobson. Comput. Mater. Sci. **23**, 62, 2002.

Controlled Dispersion of Carbon Nanotubes Wrapped in Amphiphilic Block Copolymers: Elaboration of Polymer Nanocomposite

C. Semaan, M. Schappacher, A. Soum

Laboratoire de Chimie des Polymères Organiques- UMR 5629 -
Université Bordeaux I - IPB – CNRS
ENSCBP, 16 Avenue Pey-Berland, 33607 Pessac, France, semaan@enscbp.fr

ABSTRACT

In this work, we aim to prepare polymer nanocomposites by the dispersion of multiwall carbon nanotube (MWCNT) in aqueous solution using amphiphilic block copolymers. First, we report our success in wrapping MWCNT with amphiphilic block copolymers in aqueous solution. We selected poly(ethylene oxide) (PEO) as the hydrophilic block because of its strong affinity for water and its easiness synthesis, while for hydrophobic block we used one of the following polymers: polyethylene, poly(propylene oxide), polythiophene. The dispersions were characterized by different techniques including optical microscopy (OM) and transmission electron microscopy (TEM), along with UV-visible adsorption and dynamic light scattering. Of particular interest, are our findings related to nanocomposites composed of the PEO or poly(methyl-methacrylate)-wrapped MWCNT. We investigated through of their rheological and electrical (percolation) behavior, about the structure and the dispersion of the nanoparticles in the polymer matrix. In terms of the rheological behavior, both dynamic modulus G' and complex viscosity η^* increased especially at low frequencies as the CNT loading increases. We observed two rheological percolations. Electrical measurements showed one electrical percolation.

Keywords: multiwall carbon nanotube, amphiphilic block copolymers, polymer wrapping, nanocomposite, rheological measurement.

1 INTRODUCTION

The next generation of high-performance nanocomposites would certainly benefit from both the numerous properties of carbon nanotubes (CNT) and the processability of polymers [1]. However, the scope of CNT applications in practical devices has been hampered by poor dispersion and weak interfacial bonding with polymer matrices [2]. Therefore, effective methods for CNT dispersion and consequently for dispersion characterization are required for industrial applications.

In recent years, many groups worked on the dispersion of CNT using amphiphilic bloc copolymers [3; 4]. In this paper, we describe a noncovalent process for surface functionalization of MWCNT using amphiphilic block copolymers.

Our work shed the light on elaborating and characterization of nanocomposite with a well dispersed wrapped MWCNT inside to benefit from all the properties for the CNT [5].

2 EXPERIMENTAL

The MWCNT used in this study are Graphistrength® supplied by Arkema. The MWCNT synthetized by the chemical vapor deposition have an outer diameter within the range 12-20 nm and an initial average length between 1 and 10 μm.

2.1 Aqueous solution

CNT was stored in a wet form (90% of water) and mixed with an aqueous solution of amphiphilic block copolymers. The mixture were sonificated for 15 minutes (15 Watts) using an ultrasonic probe (Vibra Cell model 75186). To evaluate the solutions obtained with different concentrations of CNT and copolymers different techniques at different stages were used. First, we observe dispersions with optical microscopy and analyzed their adsorption using Varian® Cary 3E between 200 and 800 nm. The most effectively dispersed system was further characterized by TEM and dynamic light scattering. The TEM images were obtained by on HITACHI H7650. Dynamic light scattering measurements were performed to determine the dimension (length, diameter) of CNT in suspension according to the model describe by Badaire and Coll [6].

2.2 Nanocomposite

The dispersions MWCNT-copolymer were used to elaborate nanocomposites with a polymer matrix made of

POE (evaporation technique) or PMMA (melt mixing technique). Viscoelastic properties of nanocomposites were assessed with a strain controlled parallel-plate rheometer (AR 2000) equipped with aluminum disks of 25 mm in diameter. Measurements were performed with oscillatory shear method at 120°C and 230°C. In the linear viscoelastic domain a dynamic strain sweep was carried out to determine both dynamic moduli (G', G'') and complex viscosity (η*). The dielectric measurements were performed using a frequency response analysis system consisting of Solartron 1250, at room temperature; in a frequency domain ranging from 0.01 to 1000 Hz. Typically the sample thickness and diameter were respectively around 0.9 and 10 mm.

3 RESULTS AND DISCUSSION

3.1 Aqueous dispersion characterization

On the base of collected data, we could establish the influence on CNT dispersion quality of the molar mass of copolymers, the nature of the hydrophobic block and the length of hydrophilic block. For example, as shown in table 1, the decreased of molar mass of the copolymers corresponds to a significant improvement in dispersion. Further evidence for the dispersion state of the different solutions was provided by UV- visible adsorption as shown in figure 1; homogeneous dispersion absorb more because of a higher number of MWCNT dispersed in aqueous solution and the almost total absence of aggregate.

Block copolymers	Molar mass	Molar % of blocks	Dispersion state	
			1 week after	3 month after
PPO-POE	5200	27-73	No aggregation	
	10200	28-72	Presence of aggregation	

Table 1: Molar mass effect of block copolymers

Transmission electron microscopy (TEM) allows the polymer adsorbed onto the nanotubes to be observed. Figure 2 shows that in the presence of PPO-POE (Mn 5200), the average MWCNT diameter increases from 18nm to 25nm and that CNT are coated with an amorphous layer of copolymers.
Systems stabilized by PPO-POE are perfectly homogenous and can be characterized using depolarized light scattering. Considering the nanotubes as rigid rods and using the Broersma equations [7], we measure the depolarized component of the scattered light using dilute suspensions. From the corresponding diffusion coefficients, the average length L and the average diameter d can be calculated. We find L = 250 nm and d = 40 nm. The calculated diameter is larger than the diameter deduced from TEM observations on dried materials. This difference is not surprising since the polymer in solution is expected to be swollen. In addition, dynamic light scattering measures the

hydrodynamic diameter which is always to some extent larger than the actual one.

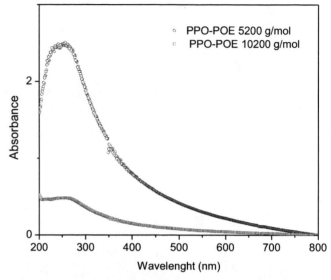

Figure 1: UV-visible absorption spectrum of PPO-POE wrapped MWCNT in water solution.

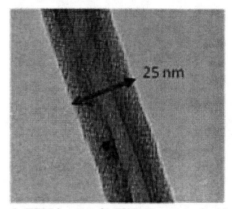

Figure 2: TEM image of MWCNT wrapped with PPO-POE, (average diameter 25 nm)

3.2 Polymer nanocomposite characterization

Rheological behavior of wrapped - MWCNT

Melt rheology is a powerful technique for studying dispersion of MWCNT in polymer matrix. The frequency dependence of the complex viscosity η* is shown in log-log plots in figure 3 for nanocomposite made with PEO and MWCNT wrapped with 10 wt. % PE-POE.

As the CNT loading increases, η* increases especially at low frequencies. In fact, we observe a progression from liquid-like to solid-like behavior, the Newtonian region disappeared and only the shear-thinning region remained [8]. The progression is attributed to a presence of a well organized network which is called percolation network of MWCNT [9; 10]. In order to observe clearly the rheological transitions the reduced viscosity η^*_r, was plotted versus wrapped MWCNT content. η^*_r is defined as:

$$\eta_r^* = \frac{\eta^*}{\eta_0^*}$$

Where η^*_0 is the zero shear viscosity of the pure copolymer mixing polymer and calculated at 0.1 rad/ (figure 4). The plot clearly indicates two rheological percolations, the first one at 0.085% and the second one at 2%.

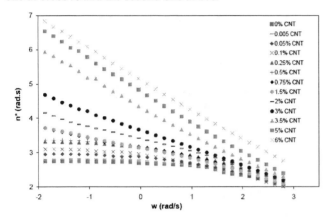

Figure 3: the complex viscosity η^* as a function of frequency for different PE-POE wrapped MWCNT dispersed wrapped in PEO polymer matrix.

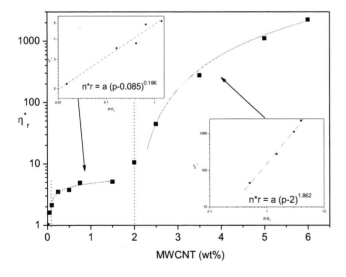

Figure 4: The reduce viscosity η^*r versus MWCNT concentration calculated at 0.1 rad/s for PE-POE-MWCNT/POE nanocomposites.

Electrical behavior of wrapped MWCNT-

To understand the impact of MWCNT on the nanocomposite microstructure, we compare rheological behavior with electrical conductivity variation. The concentration of MWCNT in nanomposite has a strong impact on the electrical conductivity [11]. Figure 5 shows a percolation behavior with only one electrical percolation a 2%.

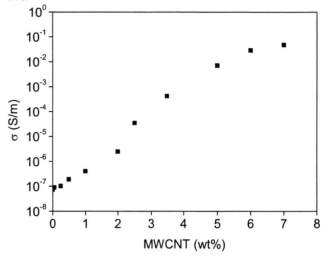

Figure 5: The conductivity σ versus MWCNT concentration for PE-POE-MWCNT/POE nanocomposites.

Discussion

The presence of two rheological percolations and only one electrical percolation can be explained by the presence of different types of interaction in the nanocomposite and as a result of the quality dispersion. On one hand, the viscosity increases with increasing the concentration of MWCNT in two phases. First, at low concentration, particle-polymer interactions predominate. Then, at higher concentration the particle-particle interactions appears (figure 6). As a result, we observe two rheological percolations. On the other hand, the change in conductivity (conduction) will not occur until we reached the particle-particle interaction domain; this is why we observe only one percolation threshold.

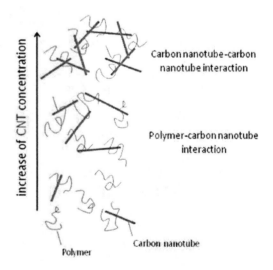

Figure 6: Schematic representation of MWCNT dispersion in polymer nanocomposites

Wrapping effect on nanocomposite

As shown in figure 6, the value of reduce viscosity of nanocomposite made with MWCNT without wrapping is greater, in the particle-particle interaction domain, than that of wrapped MWCNT. This can be explained by the level of dispersion of MWCNT in the PEO matrix. A better dispersion of MWCNT in the polymer matrix leads to a decrease of viscosity. Moreover, the tube-tube connections (frictions) are decreased by the copolymer layer.

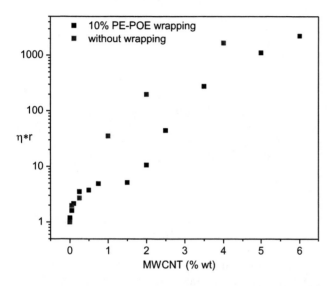

Figure 3: Comparison of the reduce viscosity η*r versus MWCNT concentration with or without copolymer wrapping

4   CONCLUSION

In summary, we have shown the efficiency of different amphiphilic block copolymers to wrap MWCNT. PEO-PE, PEO-PPO and POE-PT block copolymers were the most effective for the dispersion of MWCNT in aqueous phase. The wrapped MWCNT were well dispersed in water with PEO block exposed to the aqueous phase and PE or PPO or PT adsorbed to CNT. We also create PEO-wrapped MWCNT nanocomposites. Rheological and electrical analyses show the existence of two rheological percolations and one electrical percolation.

REFERENCES

[1] Y. Lin, Meziani, M.J., Sun, Y.-P., Functionalized carbon nanotubes for polymeric nanocomposites. Journal of Materials Chemistry 17 (2007) 1143-1148.

[2] P.M. Ajayan, Nanotubes from Carbon. Chemical Reviews 99 (1999) 1787-1800.

[3] C. Richard, F. Balavoine, P. Schultz, T.W. Ebbesen, C. Mioskowski, Supramolecular Self-Assembly of Lipid Derivatives on Carbon Nanotubes. Science 300 (2003) 775-778.

[4] C. Park, S. Lee, J.H. Lee, J. Lim, S.C. Lee, M. Park, S.-S. Lee, J. Kim, C.R. Park, C. Kim, Controlled assembly of carbon nanotubes encapsulated with amphiphilic block copolymer. Carbon 45 (2007) 2072-2078.

[5] E.W. Wong, P.E. Sheehan, C.M. Lieber, Nanobeam Mechanics: Elasticity, Strength, and Toughness of Nanorods and Nanotubes. Science 277 (1997) 1971-1975.

[6] S. Badaire, P. Poulin, M. Maugey, C. Zakri, In Situ Measurements of Nanotube Dimensions in Suspensions by Depolarized Dynamic Light Scattering. Langmuir 20 (2004) 10367-10370.

[7] S. Broersma, Rotational Diffusion Constant of a Cylindrical Particle. The Journal of Chemical Physics 32 (1960) 1626-1631.

[8] Y.S. Song, Rheological characterization of carbon nanotubes/poly(ethylene oxide) composites Rheologica Acta 46 (2006) 231-238.

[9] W. Defeng, W. Liang, S. Yurong, Z. Ming, Rheological properties and crystallization behavior of multi-walled carbon nanotube/poly(caprolactone) composites. Journal of Polymer Science Part B: Polymer Physics 45 (2007) 3137-3147.

[10] G. Galgali, C. Ramesh, A. Lele, A Rheological Study on the Kinetics of Hybrid Formation in Polypropylene Nanocomposites. Macromolecules 34 (2001) 852-858.

[11] W. Bauhofer, J.Z. Kovacs, A review and analysis of electrical percolation in carbon nanotube polymer composites. Composites Science and Technology 69 (2009) 1486-1498.

Catalyst-Free Synthesis of Carbon Nanotubes with Low-Intensity Lasers

V. Krishna[1], B. Koopman[1, 2] and B. Moudgil[1, 3]

[1]Particle Engineering Research Center,
[2]Department of Environmental Engineering Sciences,
[3]Department of Materials Science and Engineering,
University of Florida, PO Box 116135, Gainesville, FL 32611, USA
vkrishna@perc.ufl.edu

ABSTRACT

We have discovered a novel method of synthesizing carbon nanostructures such as single-walled carbon nanotubes (SWNTs), multi-walled carbon nanotubes (MWNTs) and carbon onions without catalyst. The method uses low-intensity continuous-wave laser irradiation of functionalized fullerenes at ambient temperature and pressure. Formation of SWNTs vs. MWNTs can be controlled by varying the wavelength and dose of laser irradiation. Complete graphitization of MWNTs can be achieved. Potential applications include *in situ* synthesis of CNTs, doped CNTs for bioimaging or hydrogen storage and electrical connects.

Keywords: Fullerenes, Carbon nanotubes, Catalyst, Laser, Onions

1 INTRODUCTION

Carbon nanotubes have unique optical, electrical and mechanical properties. Since Ijima's report in 1991, carbon nanotubes have been synthesized by various techniques such as arc-discharge, laser ablation, chemical vapor deposition and templated synthesis [1, 2]. The arc-discharge and laser ablation methods usually use graphite as the source of carbon, whereas in chemical vapor deposition and templated synthesis the source of carbon is carbon monoxide or other hydrocarbons [1]. Carbon nanotubes have also been synthesized by thermal treatment of carbon nanostructures. Thermal treatment and annealing at 800–1200°C result in single-walled carbon nanotubes (SWNTs). Annealing at a temperature of 1600–2000°C results in coalescence of SWNTs [3]. Above 2200°C, multi-walled carbon nanotubes (MWNTs) begin to form, while SWNTs dissappear. Only MWNTs are present at temperatures higher than 2400°C [4].

SWNTs and double-walled carbon nanotubes (DWNTs) have been synthesized by coalescence of pristine fullerenes with thermal or laser treatments. In the case of thermal treatment, the temperature is around 1200°C [5]. Guan et al., have achieved coalescence of fullerenes at 550°C by doping the SWNTs with iodine to lower the energy barrier

for coalescence of fullerenes present inside SWNTs [6]. To our knowledge, MWNTs have not previously been synthesized by coalescence of either pristine or functionalized fullerenes.

Coalescence of pristine fullerenes by laser treatment to form SWNTs or DWNTs requires intensities in the range of 1–3 MW/cm^2 [7, 8]. Although oxygen-free conditions are generally employed for coalescence of pristine fullerenes by laser, Manfredini et al. have reported that oxidative conditions are required for coalescence of fullerenes by laser irradiation [9]. Formation of carbon schwarzites by laser irradiation of pristine fullerenes has also been reported [10]. Formation of carbon nantoubes by laser irradiation of either pristine or functionalized fullerenes at room temperature and in absence of catalyst has not been reported.

Recently, we reported our discovery on optical heating, ignition and transformation of functionalized fullerenes. [11]. Functionalized fullerenes glow under laser irradiation in an oxygen-free environment (e.g., vacuum or nitrogen atmosphere). The white-light emission is blackbody radiation and the emission spectrum corresponds to a 4500 K emission temperature. We have also discovered that in an oxygen-free environment, laser irradiation of functionalized fullerenes results in extensive molecular reconstruction of functionalized fullerenes, leading to formation of carbon nanotubes, carbon onions and carbon schwarzites. We are the first group to synthesize carbon nanotubes, onions and schwarzites by laser irradiation of fullerenes in absence of catalyst and at room temperature and pressure [11].

2 MATERIALS & METHODS

Optical transformation experiments were carried out with polyhydroxy fullerenes (PHF) and carboxy fullerenes (CF) [11]. PHF or CF was coated on a silica wafer (1x1 cm^2). The sample was equilibrated with argon in a glovebox. After 24 hours of equilibration, the PHF film was exposed to a 785 nm laser with an optical fiber (400µm diameter). The output power of the laser was varied from 0.5–1.5 W. The irradiated area was approximately 1 mm in diameter. The functionalized fullerene samples glowed under laser irradiation with emission intensity increasing with higher incident laser intensity. The sample was irradiated at different locations for up to 30 seconds at any single spot. After completion of the experiment, the sample was placed in a Petri dish and the dish was sealed with Parafilm while still within the argon atmosphere of the glove-box. The black spots were scraped from the film and collected for imaging by high resolution transmission electron microscopy (HR-TEM).

3   RESULTS & DISCUSSION

Laser irradiation of functionalized fullerene coated silica substrate was marked with white and black spots. The white spots are believed to be due to melting and recrystallization of silica. HR-TEM (up to 500,000x magnification) revealed that the black spots contained numerous single-walled carbon nanotubes (SWNTs), multi-walled carbon nanotubes (MWNTs), carbon onions, carbon schwarzites and other carbon nanostructures (Figures 1 & 2). The MWNTs ranged from 2 nm to greater than 1000 nm in length. Higher resolution images revealed that the walls of the MWNTs are graphitized. Complete graphitization of MWNTs occurs above 2400°C [1]. Some of the MWNTs were welded together. Tour et al. have patented the use of electromagnetic irradiation for welding of carbon nanotubes [12]. Our observations indicate that the size, shape, number of walls, graphitization and alignment of nanotubes varies. Control of these properties should be possible through manipulation of synthesis parameters such as nature of functionalized fullerenes, sample preparation (e.g., matrix, film thickness), and intensity and duration of irradiation.

Laser irradiation induced conversion of functionalized fullerenes to carbon nanostructures has many potential applications, some of which are listed below.

In-situ synthesis of CNTs in polymer matrix

Carbon nanotubes (CNTs) have been incorporated in polymer matrices to improve the mechanical, electrical and optical properties of the polymer. A major limitation associated with current techniques is low solubility of CNTs and therefore low dispersability. Functionalized fullerenes are soluble in polar as well as non-polar solvents, thereby overcoming the dispersion issue. *In-situ* synthesis of CNTs, presently unique to our discovery, can be controlled for size, type, graphitization, dispersion, alignment and precise positioning of nanotubes. CNT-polymer composites have a wide range of applications, including bullet-proof materials, conducting polymers, high-mechanical strength materials and transparent display panels. Additionally, irradiation of functionalized fullerenes in a polymer matrix can lead to formation of welded CNTs, which can further improve properties of polymer composites.

Hydrogen storage

Carbon nanotubes are being widely researched as potential candidates for hydrogen storage because of their ability to adsorb hydrogen as well as their high surface area. According to our discovery, MWNTs synthesized in a hydrogen atmosphere can provide a route for trapping hydrogen inside MWNTs.

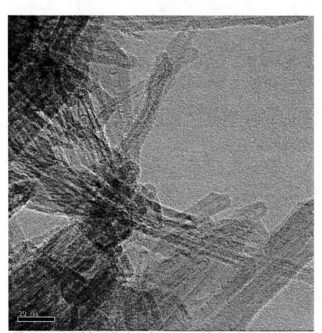

Figure 1: Multi-walled carbon nanotubes synthesized by laser irradiation of functionalized fullerenes at room temperature and pressure and in absence of catalyst.

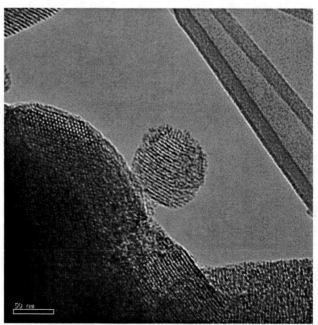

Figure 2: Carbon nanostructures synthesized by laser irradiation of functionalized fullerenes at room temperature and pressure and in absence of catalyst.

Electronics

Carbon nanotubes are being extensively researched for various electronic applications, including electrical connects, field emission devices and displays. One of the major hurdles for successful application of CNTs in these areas is the low dispersability of CNTs and the presence of impurities, such as metal catalysts. *In-situ* synthesis of CNTs with functionalized fullerenes can overcome these barriers. For example, CNTs can be synthesized in place to act as connects between electrical components. This can be accomplished by first coating the connect area with functionalized fullerenes dissolved in an appropriate solvent and then scanning the connect area with a narrow-beam-width laser in an oxygen-free environment. Functionalized fullerenes that remain after CNT synthesis can be removed by laser irradiation in the presence of oxygen. The solvent compatibility of functionalized fullerenes is a major advantage for coating them on or incorporating them within various substrates. A transparent substrate (e.g., glass, polymer) coated on one side with functionalized fullerenes can be irradiated to form an aligned and uniform matrix of CNTs, which can be used as a panel for displays or for other transparent electronic devices. Fabrication of other features (e.g., electrodes) in micro/nano electromechanical systems can also be easily achieved through *in-situ* synthesis of CNTs using the disclosed technology.

Doped CNTs

Elements or isotopes that impart novel capabilities (e.g., detection, therapeutics) can be incorporated within the core or walls of CNTs using suitable functionalized fullerenes. These modified CNTs can be used as sensors for detection of biological or chemical agents. For example, irradiation of $Gd@C_{60}(OH)_x$ in oxygen-free environment can produce Gd doped CNTs, which can be used as contrast agents for magnetic resonance imaging (MRI). The modified CNTs can also be applied for therapeutic properties such as drug delivery.

4 REFERENCES

1. Delpeux-Ouldriane, S., K. Szostak, E. Frackowiak, and F. Beguin, *Annealing of template nanotubes to well-graphitized multi-walled carbon nanotubes.* Carbon, 2006. **44**(4): p. 814-818.

2. Wang, W.Z., J.Y. Huang, D.Z. Wang, and Z.F. Ren, *Low-temperature hydrothermal synthesis of multiwall carbon nanotubes.* Carbon, 2005. **43**(6): p. 1328-1331.

3. Terrones, M., H. Terrones, F. Banhart, J.C. Charlier, and P.M. Ajayan, *Coalescence of single-walled carbon nanotubes.* Science, 2000. **288**(5469): p. 1226-1229.

4. Metenier, K., S. Bonnamy, F. Beguin, C. Journet, P. Bernier, M.L. de La Chapelle, O. Chauvet, and S. Lefrant, *Coalescence of single-walled carbon nanotubes and formation of multi-walled carbon nanotubes under high-temperature treatments.* Carbon, 2002. **40**(10): p. 1765-1773.

5. Bandow, S., M. Takizawa, K. Hirahara, M. Yudasaka, and S. Iijima, *Raman scattering study of double-wall carbon nanotubes derived from the chains of fullerenes in single-wall carbon nanotubes.* Chemical Physics Letters, 2001. **337**(1-3): p. 48-54.

6. Guan, L.H., K. Suenaga, T. Okazaki, Z. Shi, Z. Gu, and S. Iijima, *Coalescence of C-60 molecules assisted by doped iodine inside carbon nanotubes.* Journal of the American Chemical Society, 2007. **129**(29): p. 8954-+.

7. Kramberger, C., A. Waske, K. Biedermann, T. Pichler, T. Gemming, B. Buchner, and H. Kataura, *Tailoring carbon nanostructures via temperature and laser irradiation.* Chemical Physics Letters, 2005. **407**(4-6): p. 254-259.

8. Milani, P. and M. Manfredini, *Laser-Induced Coalescence of C-60 in the Solid-State.* Journal of Physical Chemistry, 1995. **99**(43): p. 16119-16120.

9. Manfredini, M., C.E. Bottani, and P. Milani, *Raman-Scattering Characterization of Amorphous-Carbon from Photothermal Oxidation of Fullerite.* Journal of Applied Physics, 1995. **78**(10): p. 5945-5952.

10. Spadoni, S., L. Colombo, P. Milani, and G. Benedek, *Routes to carbon schwarzites from fullerene fragments.* Europhysics Letters, 1997. **39**(3): p. 269-274.

11. Krishna, V., N. Stevens, B. Koopman, and B. Moudgil, *Optical heating and rapid transformation of functionalized fullerenes.* Nature Nanotechnology, 2010. **10.1038/NNano.2010.35**.

12. Tour, J.M., C.A. Dyke, J.J. Stephenson, and B.I. Yakobson, *Use of microwaves to crosslink carbon nanotubes to facilitate modifcation,* in *USPTO.* 2004, William Harsh Rice University: U.S.A.

Direct Growth of Carbon Nanotubes on a Plastic Substrate

A. V. Vasenkov[*], D. Carnahan[**], D. Esposito[**], and I. Izrailit[***]

[*]CFD Research Corporation, 215 Wynn Drive, Huntsville, AL 35805, avv@cfdrc.com
[**]NanoLab, Inc., 55 Chapel Street, Newton, MA 02458
[***]Security Technology Systems, Inc., 197 Main St., North Reading, MA 01864

ABSTRACT

The inability to grow carbon nanotubes (CNT) on temperature-sensitive substrates such as plastic tapes has impeded progress in CNT electronics. We report a novel method that minimizes substrate temperature allows growth on plastic tapes by selectively delivering energy via high radio frequency heating to carbon atoms for their catalytic assembly into individual CNTs. Direct growth of CNTs on a flexible Kapton substrate is demonstrated. A mechanism of selective heating is also discussed.

Keywords: carbon nanotube synthesis, flexible devices, high radio frequency heating, low temperature growth, CNT electronic devices

1 INTRODUCTION

Carbon nanotubes (CNTs) have the potential to play a central role in nanotechnology due to their unprecedented molecular-scale electronic and mechanical properties. [1, 2] CNT applications can be grouped into two general categories. The first category includes applications in which CNTs are used as a filler addition in composite materials, hydrogen storages, fuel cells, super-capacitors, or membranes. Applications in this category do not have strict requirements to the spatial arrangement and alignment of every single CNT, but rather rely on the properties of many mixed CNTs. The second category involves potential electronic applications including CNT-based sensors, displays and memories which could benefit the most from the unique properties of individual CNTs [1, 2]. In those applications, each CNT is used as individual functional element of a device which imposes strict requirements to the spatial arrangement and alignment of each CNT, as well as to the properties of a contact between CNTs and device surface. To address these needs, significant progress was made in the past in the development of growth methods for the synthesis of free-standing aligned CNTs with durable contact to a substrate [3].

CNT synthesis is a catalytic process requiring a high growth temperature of 700°C [4]. Such a high temperature makes problematic the direct growth of CNTs on temperature-sensitive substrates including plastic tapes. In an attempt to tackle this critical issue, a few CNT post-growth manipulation methods have been reported. For example, a method involving the initial growth of patterns of CNTs on silicon surfaces using a thermal catalytic chemical vapor deposition (CCVD) process, and the consequent transfer of these CNTs onto a desirable substrate by contact printing method was reported in [5]. In a similar effort, a mechanical transfer technique capable of controlling the positions of CNTs with micron-scale precision was reported in [6]. More recently, DNA-based method for the organization of individual CNTs in nano-patterned surfaces has been proposed in [7]. The major limitation of existing post-growth methods is in their inability to scale effectively because of difficulties in manipulating tiny CNTs.

Some success has been reported in reducing growth temperature of CNT synthesis via a plasma-assisted CCVD process [8, 9]. In such an approach, the energy necessary for CNT growth is provided by the plasma rather than by an external heating source. Although the results of published studies are encouraging, deterministic growth of well-graphitized CNTs at substrate temperatures below 700°C has not been established yet. Also, in view of results reported in [10], a further investigation is needed to explain how plasma can be sustained at low temperature during CNT growth.

In this paper, we report on a new scalable method of CNT growth in which energy, required for endothermic catalytic reactions used to assembly carbon atoms into CNT, is delivered via high radio frequency (RF) heating. RF treatment of micro- and nanoscale materials attracts a growing interest of researchers worldwide due to its ability for a selective heating or melting [11, 12]. Typically, RF heating is produced by a microwave source operating at a frequency of 2.45 GHz with a power in excess of 1 kW, which is sufficient to increase the temperature of a nano-scale object only by a few tens of degrees. For example, DNA-coated metal particles of a few nanometers in size were successfully heated to a temperature of 70°C [13]. We demonstrated that a more efficient RF heating can be achieved at higher frequencies since RF absorption power is proportional to the frequency of RF waves. The high selectivity of heating during CNT growth was achieved due to the efficient absorption of RF waves in dielectric TiO_x powder loosely packed on the plastic substrate.

2 CNT GROWTH VIA SELECTIVE RF HEATING

CNT growth via selective RF heating is schematically shown in Fig. 1 and can be outlined as follows. A Kapton tape was patterned by air-brush spraying iron-based catalyst nanoparticles suspended in ethanol, mounted on a resistively heated substrate holder in the vacuum chamber, and covered with a thin coating of TiO_x powder. The substrate holder was made from aluminum oxide which was absorptive to RF waves. Consequently, we detected no reflected waves from the substrate holder.

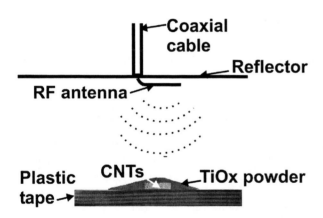

Figure 1. Schematic of direct growth of CNTs on plastic tape via selective heating. RF power is dominantly absorbed inside TiO_x powder. Hydrocarbon gas is heated while diffusing between loosely packed TiO_x particles and decomposes on the tape surface patterned by catalytic nanoparticles. Resulting carbon atoms assemble into CNT via endothermic catalytic reactions. After CNT growth, TiO_x powder is washed off.

Once the CNT growth chamber was evacuated, the resistive heater was heated to a temperature of 290°C. 40 W of 18 GHz RF power was then feed through a coaxial cable into an antenna. The antenna was suspended above the flexible Kapton tape and used to transmit RF waves for selective heating of TiO_x powder. A reflector was installed atop of antenna to minimize RF wave leakage. The RF

heating resulted in a temperature gradient of 360°C across TiO_x - Kapton.

The typical heating time required to reach a steady temperature of 650°C inside the TiO_x layer was 7 minutes, as measured using an infrared radiation pyrometer, to avoid interference with RF waves. The TiO_x powder was loosely packed so that processing C_2H_2/NH_3 gas molecules were heated while diffusing between TiO_x particles. This heating promoted C_2H_2 decomposition on the surface of catalyst and assembly of resulting carbon atoms into CNT via endothermic catalytic reactions. No visible-light illumination was detected, indicating the absence of gas breakdown and plasma discharge formation during CNT growth. This finding is important, as it points to compatibility of our growth technique and plasma-assisted growth methods of free standing aligned CNTs [3].

After CNT growth, the TiO_x powder remained loose and was easily removed using air flow. In contrast, CNTs stuck to the Kapton tape and could only be removed by abrasion or sonication in ethanol. The typical scanning electron transmission microscope image of deposited CNTs on overall intact Kapton tape is shown in Fig. 2. The formation of some bubbles on the surface of the tape was detected indicating that the temperature on the tape surface was near the Kapton glass transition temperature between 360°C and 410°C.

A transmission electron microscope was used to probe the structure of the grown CNTs. CNTs were found to have diameters in the range of 5-15 nanometers, and variable graphitization because of non-optimized growth conditions as shown in Fig. 3. CNTs with diameter >10 nanometers are individual tubes, while smaller CNTs tend to bundle together due to van der Waals attractions.

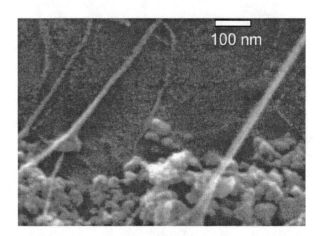

Figure 2. A typical scanning electron transmission microscope image of deposited CNTs on overall intact Kapton tape. Some bubbling on the surface of the tape was detected indicating that the surface temperature was near the glass transition temperature of Kapton.

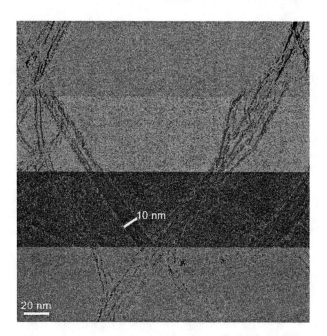

Figure 3. The transmission electron microscope image shows nanotubes of 5-15 nm in diameter, with variable graphitization because of non-optimized growth conditions.

3 HIGH SELECTIVITY OF HEATING

The high selectivity of heating in our method is due to the efficient absorption of RF waves in dielectric TiO$_x$ powder loosely packed on top of the plastic substrate. RF heating of dielectric materials is a volumetric heating in contrast to RF surface heating of conductive materials [11, 12, 14]. The mechanism of this heating involves induced dipole rotations, and is most commonly observable in the microwave oven where it operates most efficiently on liquid water. The power dissipated in a dielectric material due to dielectric heating is given by [12, 14]:

$$P_{RF} = 2\pi f \varepsilon_0 \varepsilon' \tan(\delta) E^2 \qquad (1)$$

Here, f is the frequency of RF waves, ε_0 is the vacuum dielectric permittivity, ε' is the relative dielectric permittivity, and $\tan(\delta)$ is the dielectric loss tangent. Considering that electric field E inside the TiO$_x$ dielectric sphere is given by $E = 3E_0/(\varepsilon + 2)$ [14], where E$_0$ is the electric field in vacuum and $\varepsilon = \varepsilon'(1 - j\tan(\delta))$ is the complex dielectric permittivity of TiO$_x$ powder, Eq. (1) can be rewritten as

$$P_{RF} = 18\pi f \varepsilon_0 K E_0^{\,2} \qquad (2)$$

where K is a parameter proportional to $\tan(\delta)$, and inversely proportional to ε'. In deriving Eq. (2), we considered that $\varepsilon \gg 2$, and $\tan(\delta)^2 \ll 1$. The values of parameter K are not very well known in a frequency range above 2.45 GHz. However, it was demonstrated that both an increase in dielectric permittivity with decreasing oxygen fraction x of TiO$_x$ and a near linear increase of loss tangent with increasing frequency can substantially increase K [15, 16]. To estimate K in our CNT growth experiments, we solved the Eq. (2) in conjunction with the thermal conductivity equation given by

$$-k\frac{\partial^2 T}{\partial x^2} = P_{RF}, \qquad (3)$$

where, κ is the thermal conductivity. The left side of Eq. (3) accounts for the heat loss due to the heat transfer from TiO$_x$ powder layer to Kapton tape and surrounding gas, and the right side represents the heat gain due to RF power absorption. Two Dirichlet boundary conditions at the bottom of the Kapton tape (T=T$_1$), and in the gas phase far from the TiOx particles T=T$_2$ were used in solving Eq. (3). We assumed that T$_1$=288°C is the resistive heater temperature, and T$_2$ =25°C. By solving Eqs. (2-3) and by fitting the computed profile to the measured referenced temperature at the gas/TiOx boundary, we found that K in our CNT growth experiments was ~ 0.2. This value is significantly larger than $K \sim 1 \times 10^{-5}$ typically reported for TiO$_{x=2}$ at 2.45 GHz [17]. A computed temperature profile inside Kapton and TiO$_x$ layers is shown in Fig. 4. A 100°C temperature gradient within the Kapton tape explain the formation of some bubbles on top of Kapton shown in Fig. 2. Optimization of growth conditions could lead to a slightly lower temperature on top of Kapton tape and eliminate the observed bubble formation. A 200°C temperature gradient between TiO$_x$ powder and Kapton was due to the selective RF heating.

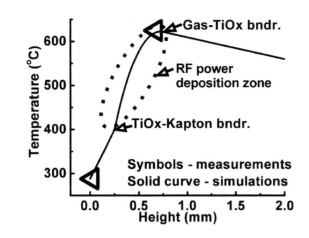

Figure 4. Computed temperature profile across Kapton - TiO$_x$-gas during CNT synthesis. A large temperature gradient across Kapton-TiO$_x$ was due to the selective heating.

4 SUMMARY

The presented results clearly show the feasibility of CNT growth on temperature-sensitive substrates such as polymer plastic tapes, aluminum foil, and doped semiconductor structures. In addition, the absence of gas breakdown and plasma formation during CNT growth permits the integration of reported technique with low-temperature plasma-assisted growth methods of free-standing aligned CNTs. Finally, our method can be easily scaled up in contrast to previously reported post growth methods of CNT manipulations. Consequently, reported method offers a unique route towards direct assemble of next generation devices such as CNT-based flexible e-paper displays, field emission sources for high-power defense electronics, bio and chemical sensors, and interconnects in computer chips.

ACKNOWLEDGMENT

This work was supported by the National Science Foundation (SBIR Award # 0724878).

REFERENCES

[1] J. Robertson, Materials Today, 2007. **10**: p. 36.
[2] *Carbon Nanotube Electronics*. Series: Integrated Circuits and Systems, ed. A.K. Javey, Jing (Eds.). 2009: Springer US.
[3] A.V. Melechko, V.I. Merkulov, T.E. McKnight, M.A. Guillorn, K.L. Klein, D.H. Lowndes, and M.L. Simpson, J. Appl. Phys., 2005. **97**: p. 041301-1.
[4] P.M. Ajayan and O.Z. Zhou, in *Carbon Nanotubes*, M. S. Dresselhaus, G. Dresselhaus, and P. Avouris, Editors. 2001. p. 391.
[5] A. Kumar, V.L. Pushparaj, S. Kar, O. Nalamasu, and P.M. Ajayan, Appl. Phys. Lett., 2006. **89**: p. 163120.
[6] X. Ming, H. Huang, R. Caldwell, L. Huang, S.C. Jun, M. Huang, M.Y. Sfeir, S.P. O'Brien, and J. Hone, Nano Lett., 2005. **5**: p. 1515.
[7] R.J. Kershner, L.D. Bozano, C.M. Micheel, A.H. Hung, A.R. Fornof, J.N. Cha, C.T. Rettner, M. Bersani, J. Frommer, P.W.K. Rothemund, and G.M. Wallraff, Nature Nanotech., 2009. **4**: p. 557.
[8] B.O. Boskovic, V.B. Golovko, M. Cantoro, B. Kleinsorge, A.T.H. Chuang, C. Ducati, S. Hofmann, J. Robertson, and B.F.G. Johnson, Carbon, 2005. **43**: p. 2643-2648.
[9] S. Hofmann, G. Csanyi, A.C. Ferrari, M.C. Payne, and J. Robertson, Phys. Rev. Lett., 2005. **95**: p. 1.
[10] K.B.K. Teo, D. Hash, G. Lacerda, and N.L. Rupesinghe, Nano Letters, 2004. **4**: p. 921.
[11] M. Tanaka, H. Kono, and K. Maruyama, Phys Rev. B, 2009. **79**: p. 104420.
[12] A.M. Hasna. *Composite Dielectric Heating and Drying: The Computation Process*. in *Proceedings of the World Congress on Engineering 2009 WCE 2009, July 1 - 3, 2009, London, U.K.* 2009, Vol. I.
[13] K. Hamad-Schifferli, J.J. Schwartz, A.T. Santos, S. Zhang, and J.M. Jacobson, Nature, 2002. **415**: p. 152.
[14] K.I. Rybakov, V.E. Semenov, S.V. Egorov, A.G. Eremeev, I.V. Plotnikov, and Y.V. Bykov, J. Appl. Phys., 2006. **99**: p. 023506.
[15] B. Vaidhyanathan, D.K. Agrawal, and R. Roy, J. Amer. Ceram. Soc., 2004. **87**: p. 834.
[16] J. Sheen, R. Guo, A.S. Bhalla, and L.E. Cross, Ferroelectric Letters, 2008. **35**: p. 79.
[17] S.A. Suvorov, I.A. Turkin, L.N. Printsev, and A.V. Smirnov, Refractories and Industrial Ceramics, 2000. **41**: p. 295.

Electric-Arc Plasma Installation and Method for Preparing Nanodispersed Carbon Structures

G. Vissokov, D. Garlanov, K. Zaharieva, S. Rakovsky

Institute of Catalysis, Bulgarian Academy of Sciences, "Acad. G. Bonchev" Str., Bl. 11,
1113 Sofia, Bulgaria, Fax/Tel: (+359 2) 971 29 67, vissokov@abv.bg

ABSTRACT

An electric-arc plasma installation operated in the "hidden" anode arrngement is constructed and used for the pereparation of carbon nanostructures. A device and a method were developed to produce nanomaterials in the form of nano-particles that are stable, polymorphic, head modified crystal structures of any type of elecrtically conducting materials, in particular of graphite, and possess unique physical, chemical and mechanical properties. The operating technological parameters of the contracted plasma arc were deternined in view of obtaining nanosised carbon structures withe given shape (nano-tubes, fullerenes, clusters, disc-like shapes, leaf-shaped etc.), size and structure.

The method, the divice and the nanomaterials produced can find application in the plasma-chemical industry, the nanotechnologies, the space technologies, the medicine and for fabrication of composite materials.

Keywords: thermal plasma method, carbon nanostructures, properties

1 INTRODUCTION

The high energy parameters of low-temperature plasma (LTP), the high rates of the vaporization and condensation processes, combined with the possibilities for automation, optimization and modeling of the plasma-chemical processes, determine the practical expediency of employing LTP for nano-dispersed powders (NDPs) preparation. Thanks to the exceptional plasma properties (high temperature and energy density, presence of a large number of charged particles, etc.) and to the specific gas-dynamic and thermo-physical conditions existing in the plasma-chemical reactor (PCR), all physical and chemical processes involved in the plasma treatment of any initial material (heating, vaporization, thermal dissociation, chemical interactions, nucleation and condensed-phase particle growth) take place at very high rates and, therefore, for a very short time ($10^{-2} - 10^{-3}$ s). As a result, the desired products obtained are in most cases characterized by a very small particle size (1 – 100 nm) and high number of defects in the crystal. In many cases (high-melting point nitrides, oxides, carbides, etc.), the plasma-chemical (PC) treatment is the only possible technique for preparation of NDPs [1, 2].

At present, carbon nanostructures are regarded as artificially composed structures with nanometer size. Their properties are the subject of both theoretical and experimental investigations; and are expected to find a very wide range of possible applications. The history of carbon nanostructures begins in 1985, when the Buckminster fullerene C60 was discovered by Kroto [3]. Since then, the number of discovered structures has been rapidly increasing. Examples of them are: the nanotubes discovered by Ijima [4], the family of fullerenes C70, C76, C84, C60 in a crystalline form, carbon nanocones, carbon nanohorns, nanoscale carbon toroidal structures and helicoidal tubes, etc. These carbon structures could be single-walled or multi-walled; they may have zero, positive or even negative Gaussian curvature (Schwarzites). Recently, a few types of similar non-carbon structures were discovered, such as boron nitride nanotubes, molybdenum disulfide or tungsten disulfide structures and even silicon nanotubes.

The aim of the study reported here is to use a modified technique of interaction between the material treated and a contacted plasma arc; the latter is transferred directly on a fixed or moving graphite electrode in the "hidden" anode arrangement thereby making it possible to achieve temperatures about 6 000 K in and above the interaction zone. The electrode material is thus vaporized and, following quenching and fixing, nanosized carbon structures are obtained.

2 EXPERIMENTAL

A schematic of a plasma-chemical installation for production of carbon nanostructures operating under the "buried" anode arrangement is shown in Figs. 1 and 2 [5]. The plasma arc (38) generated by the plasma torch (1) (see Fig. 1) has mass-averaged temperature exceeding 10×10^3 K.

The arc is strongly contracted, intensive and with high speed of motion of the current-carrying plasma stream. The plasma arc is transferred directly on a cylindrical graphite electrode (anode - 8) in the "buried" anode arrangement. The plasma arc and the plasma stream envelops entirely the anode contact area formed (40). As a result of this frontal attack with high heat power density (about 2 kW/mm^2), a temperature of about 6 000 K is reached in "hidden" anode contact area and above it. This results in intensive explosive vaporization of the graphite, which sublimates into a solid aerosol phase. The vaporization process takes place continuously, as the vaporized material is replaced be new one by means of moving the electrode in a direction opposite to the plasma stream. The electrode diameter is smaller than that of the positive column of the contracted plasma arc formed by the plasma torch nozzle. Thus, complete "burying" is ensured of the anode contact area within the moving plasma stream and a temperature of 6 000 K is permanently maintained in this area throughout the entire vaporization cycle.

A part of the plasma arc and the plasma stream after the contact area move into a closed space within the chamber. It is formed by two hollow open cones located within a closed space (41), the latter being connected through holes to another multi-chamber space used for collection of the nanomateriales produced. The open cones are coaxial with respect to one another and to the electrode (8); their walls, whose geometry can be varied, are formed by a moving fluid. The walls quench, mix and support the saturated plasma-aerosol carbon phase produced through vaporization of the electrode material. This phase is carried by the plasma stream, which leads to the initiation of homogeneous crystallization of carbon nuclei followed by quenching at a given moment. The latter interrupts the crystallization process and ensures the formation of heat-modified stable polymorphic nanosized crystalline carbon particles. If the graphite electrode is replaced by an electrode of another electrically conducting material, one will obtain nanoparticles of that material.

Throughout the vaporization cycle, the plasma arc remains with constant length and very narrow temperature profile. Thus, the installation allows one to perform continuous and complete transformation of the entire initial material into a plasma-aerosol phase.

1 - water-cooled elongated cylindrical three-body plasma torch; 2 - insulating air-tight coupler; 3 - chamber; 4 - holes for removing the products; 5 - two-body vaporizer; 6 - fixed electric power supply; 7 - rotating planetary head; 8 - electrically conducting anode of the plasma torch; 9 - electrical motor; 10 - plasma torch nozzle; 11 - gap; 12 - stream twisting device; 13 - tangential channels; 14 - holes for feeding quenching fluid; 15 - protective lid; 16 - tapered hole; 17 - cylindrical hole; 18 - holder; 19 - supporting metal disk; 20 - stream twisting device; 20' - tangential channels in the stream twisting device; 21 - gap; 22 - holes; 23 - internal body; 24 - spring holders; 25 - current feeding brushes; 26 - current feeding cylindrical graphite bushings; 27 - metal cylinder; 28 - silicon rubber seal; 29 - cap; 30 - metal bottom; 31 - holes for quenching fluid; 32 - outside housing; 33 - cooling gap; 34 - holes; 35 - holes for cooling liquid; 36 - flange; 37 - fixing plate; 38 - plasma arc; 39 - saturated plasma-gas-aerosol phase; 40 - anode contact area; 41 - volume enclosed by the moving fluid walls; 42 - body; 43 - inter-electrode gap; 44 - vapor-gas mixture; 45 - moving conical fluid wall; 46 - mixed gas-plasma stream; 47 - lower open conical space; 48 - upper open conical space;

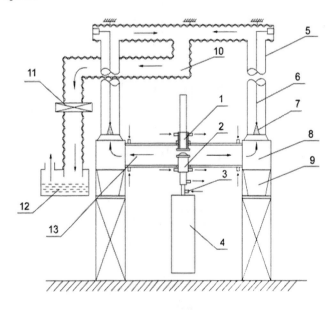

Figure 2: Schematic of the plasma installation for production of nanosized carbon structures

1 - plasma torch coupled to a quenching device; 2 - vaporizer coupled to an electric power supply; 3 - electric power supply; 4 - magazine-type electrode-feeding device; 5 - electrostatic filter; 6 - electrostatic filter cathode; 7 - counterweight of the electrostatic filter cathode; 8 - chamber-electrostatic filter coupler; 9 - bunker collecting the nanodispersed material; 10 - pipes for the stream after the electrostatic filter; 11 - suction fan; 12 - water filter; 13 - quenching chamber.

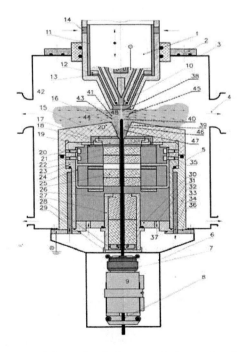

Figure 1: Schematic of plasma unit for production of nanosized carbon structures

3 RESULTS AND DISCUSSION

The nanostructures produced were studied by a Hitachi S570 TEM at magnification as denoted on each microphotograph, depending on the size of the structures and the preparatory techniques used. The TEM images show spheres of very small size (about 10 nm) (Figures $3^a \div 3^s$), as well as tubular structures clogged at both ends by spherical formations.

The samples thus observed can be described as follows: the presence is seen of crushed graphite (Figs. 3^f, 3^l, 3^m), of spherical formations of various size (Figures. 3^a, 3^g, 3°), of clusters (Figures. 3^k, 3^n), of elongated structures (Figures. 3^h, 3^j, 3^p) resembling tubes or threads (Figures. 3^h, 3^j, 3^p). Some of the spherical formations have holes, so that one can assume that they are hollow. The TEM images show spheres of very small size (about 10 nm), as well as tubular structures clogged at both ends by spherical formations (Figures. 3^j, 3^k, 3^p).

Figure 3^a

Figure 3^b

Figure 3^c

Figure 3^d

Figure 3^e

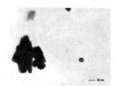

Figure 3^f

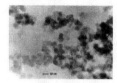

Figure 3^g

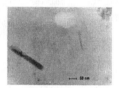

Figure 3^h

Figure 3^i

Figure 3^j

Figure 3^k

Figure 3^l

Figure 3^m

Figure 3^n

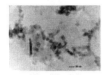

Figure 3°

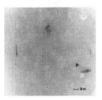

Figure 3^p

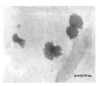

Figure 3^q

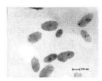

Figure 3^r

Figure 3^s

Figure 3: TEM images of synthesized nanodispersed carbon structures

The TEM images of a sample extracted from water by using toluene show particles of elongated or disc-like shape. Some of them resemble leaves with embedded small black particles of various geometry. One can also see conglomerates of complex shape consisting of spherical and elongated particles. The TEM image at magnification of 4×10^3 shows that their size is about 50 nm. At magnifications of 180 000 - 200 000, one can see that the discs are of variable density, decreasing to the periphery with a sharply defined borderline; the discs have sizes of about 10 - 25 nm. Spherical particles are also present in this image, which are also of variable density and sizes of about 10 - 25 nm. One can further observe tubular structures with channels in the middle. The outer diameter of the tubes is 10 nm, the inner, 2 nm, and the overall length is 50 nm.

At a magnification of 2×10^5, one can see a mesh-like structure containing spherical dot formations with sizes of about 5 - 10 nm. Needle-like and triangular formations are also present that exhibit multi-layered structures. The results of the X-ray phase analysis show two clearly seen lines (d = 2,81 Å and d = 3,00 Å), which describe the carbon nanostructures.

4 CONCLUSIONS

1. The application of a technique employing a contracted plasma arc transferred directly on a moving cylindrical graphite electrode (anode) in a "buried" anode arrangement allowed the authors to produce nanosized carbon structures of various size and texture. The high energy parameters of the contracted plasma arc determine the efficiency of the plasma process in what concerns the graphite material vaporization.

2. The operating technological parameters of the contracted plasma arc were determined in view of obtaining nanosized carbon structures with given shape, size and structure.

3. Collection of the nanosized carbon structures by means of a quenching water cone is efficient and environmentally appropriate.

REFERENCES

[1] G.Vissokov, Plasmananotechnologies. Nanopowders – preparation, properties and application ("St. Iv. Rilski" Publishing House, Sofia), 303, 2005.
[2] G. Vissokov, P. Pirgov, Ultradispersed powders (plasma-chemical preparation and properties) (Polyprint, Sofia), 395, 1998, (in Bulgarian).
[3] H.W. Kroto et al., Nature 318, 162, 1985.
[4] Ijima S., Nature (London) 354, 56, 1991.
[5] D. Garlanov, P. Stefanov, G. Vissokov and Tzv. Tzvetanov, BG Patent № 65 887, 20.07.2005.

Large area graphene sheets synthesized on a bi-metallic catalyst system

E. Dervishi,[*1] Z. Li,[*] J. Shyaka,[*] F. Watanabe,[*] A. Biswas,[**] J. L. Umwungeri,[*] A. Courte,[***] A. R. Biris,[****] and A. S. Biris[*1]

[*]Nanotechnology Center University of Arkansas at Little Rock, Little Rock, AR-72204, U.S.A.
[**]Center for Nanoscience and Technology (NDnano), Department of Electrical Engineering, University of Notre Dame, Notre Dame, IN 46556, U.S.A.
[***]Ecole d'Ingenieurs du CESI-EIA, La Couronne, France
[****]National Institute for Research and Development of Isotopic and Molecular Technologies, Cluj-Napoca, Romania.

ABSTRACT

Large area graphene sheets (several micro meters) were synthesized on a bimetallic catalyst system utilizing acetylene as the hydrocarbon source. After a simple purification, graphene sheets were thoroughly characterized by several microscopy techniques. It was observed that as the hydrocarbon rate increased from 4.5 to 8 ml/min, the diameter of the graphene sheets increased from a few hundred nanometers to a few micrometers. Therefore, by varying the hydrocarbon flow rate, we were able to control the size and the layer number of the CVD grown graphene sheets. This is a substrate free synthesis which makes it very attractive and flexible for a scaled up graphene production.

Keywords: large area, graphene, chemical vapor deposition, scale up production, catalyst system.

1 INTRODUCTION

Graphene, a monolayer of graphite, is one of the allotropes of pure carbon [1]. This 2D material has exceptional electrical and mechanical properties making it an excellent candidate for a wide range of applications [2,3,4,5]. Single-wall carbon nanotubes are formed by rolling a graphene sheet into a hollow seamless cylinder. Several groups around the world are utilizing graphene in various areas such as nanoelectronic, solar cell devices, sensor, batteries, and transparent electrodes, just to name a few [6,7,8,9]. Due to its extraordinary properties, graphene has also been incorporated into other materials to create nanocomposites with novel and enhanced properties [10]. Since most of the applications require large area graphene sheets, development of an efficient and large scale synthesis has been explored in the recent years. Current production methods of graphene films include mechanical or thermal exfoliation of graphite, chemical approaches such as chemical vapor deposition and epitaxial growth on various substrates [11,12,13,14]. Furthermore, large area graphene sheets are synthesized on thin film substrates such as copper or nickel foils, which depending on their size limit the dimensions of the desired product [7,15,16]. Although a lot of research has been conducted, fabrication of large area graphene still remains a challenge [17]. Synthesizing graphene sheets on particular substrates makes it difficult to scale up and might not be very practical for many applications [18]. Another disadvantage of using different type of thin film substrates to synthesize graphene is the difficulty of transferring the as-produced product without any damage, onto other surfaces for characterization or other functions. Chemical vapor deposition (CVD) is a well known technique previously used for carbon nanotube synthesis [19,20]. This is a very attractive method because the reaction parameters can be adjusted in order to control the growth and morphological properties of carbon nanostructures. In addition, this method has the potential to scale up for a large production of carbon materials. Previously we have reported synthesis of graphene nanosheets by utilizing the radio-frequency CVD technique [21]. In this work, we optimized the growth parameters in order to synthesize large area graphene sheets on transition metal nano-particles. Since the graphene sheets are not synthesized on any particular films or substrates, they can be deposited virtually on any surface (with little to no damage) making our method flexible for a wide range of applications. The purified product can be solubilized and deposited on various surfaces by using several methods such as membrane filtration, spin coating, airbrushing etc. [22,23]. In this manuscript we present a practical and scalable substrate free method for production of large area graphene sheets.

2 EXPERIMENTAL DETAILS

The Fe-Co/MgO catalyst system with a stoichiometric composition of 2.5:2.5:95 wt.%, was prepared by the impregnation method [21]. Graphene sheets were

---

[1] Corresponding authors: exdervishi@ualr.edu and asbiris@ualr.edu

NSTI-Nanotech 2010, www.nsti.org, ISBN 978-1-4398-3401-5 Vol. 1, 2010

synthesized via RF-cCVD on the MgO supported Fe-Co bi-metallic catalyst system utilizing acetylene as a hydrocarbon source. Approximately 100 mg of the catalyst was uniformly spread into a thin layer on a graphite susceptor and placed in the center of a quartz tube with an inner diameter of 1 inch. The tube was purged with Argon (the carrier gas) for 10 minutes at 150 ml/min. Next, the RF generator was powered on and used for heating the graphite boat to the desired synthesis temperature. Once the temperature reached 1000 °C, acetylene was introduced for 30 minutes. The flow rate of the hydrocarbon was varied between at 4.5-8 ml/min. At the end of the synthesis, the system was cooled down under the presence of Argon for 10 minutes. The as-produced carbon nanostructures were purified in one step by washing with a diluted hydrochloric acid solution under sonication.

2.1 Characterization Techniques

Transmission Electron Microscopy (TEM) images were collected on a field emission JEM-2100F TEM (JEOL INC.), equipped with a CCD camera. The acceleration voltage was 100 kV for the graphene analysis. The carbon nanostructures were homogeneously dispersed in 2-propanol under ultra-sonication for 30 minutes. Next, a few drops of the suspension were deposited on the TEM grid, dried, and evacuated before analysis. Scanning Electron Microscopy (SEM) images were obtained using a JEOL 7000F high-resolution scanning electron microscope. The final products were mounted on aluminum pins with double-sided carbon tape and their corresponding SEM images were obtained. Scanning Tunneling Microscopy (STM) images were obtained from a homebuilt beetle-style STM with RHK technology SPM1000 controller and XMP software with Axon CV4 current amplifier. STM imaging was performed in dry N_2 at room temperature. A drop of graphene solution (flakes of graphene dispersed in iso-propanol) was deposited on an Au/mica substrate and allowed to dry in air. Atomic Force Microscopy (AFM) measurements were performed using a Veeco Dimension 3100 instrument with a scan range of 90 μm for the "x/y" direction and 6 μm for the "z" direction. After the graphene sheets were individually dispersed into 2-propanol solution, a few drops of the final solution were pipetted onto Si substrates. Next, the substrates were air dried and placed directly under the AFM tip for morphology analysis.

3 RESULTS AND DISCUSSIONS

Large area graphene sheets were synthesized via RF-cCVD method, utilizing acetylene as a hydrocarbon source. As the flow rate of acetylene was varied from 4.5 to 6 and 8 ml/min, the size of the graphene sheets increased from a few hundred nm to a few μm. We have previously reported that when the flow rate of acetylene was set at 4.5 ml/min, graphene sheets with diameter of around 100 nm were synthesized [21]. For this work, the flow rate of acetylene

was increased to 6 and 8 ml/min and the as-produced graphene sheets were characterized by several microscopy techniques. Figs. 1(a) and (b) show the SEM images of graphene sheet synthesized on the Fe-Co metal nano-particles utilizing acetylene at 6 ml/min. The diameter of the graphene sheets varies between 30-50 μm. These semi-transparent sheets demonstrate that few-layer graphene were synthesized at both hydrocarbon rates. EDAX analysis were also performed (not-shown here) to confirm the presence of carbon structures.

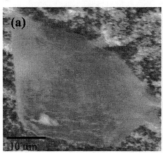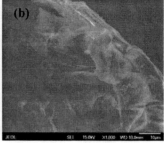

Figures 1(a) and (b): SEM images of graphene sheets synthesized on the bi-metallic catalyst system utilizing acetylene at 6 ml/min.

Similar results were also noticed as the flow rate of acetylene was increased to 8 ml/min. SEM images on Figs. 2 (a) and (b) show that graphene sheets with large diameters were synthesized. In this case there is no significant difference in size when compared to the results obtained from the lower hydrocarbon rate (6 ml/min). The SEM images also revealed that the large scale graphene sheets are very flaky. This could be due to the high rate of hydrocarbon which contributes to synthesis of graphene sheets with non-uniform layers.

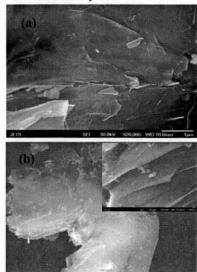

Figures 2(a) and (b): SEM images of graphene sheets synthesized on the bi-metallic catalyst system utilizing acetylene at 8 ml/min. The scale for the inset image in (b) is 100 nm.

TEM analyses were performed on the RF-cCVD grown graphene sheets utilizing acetylene at various flow rates. Fig. 3(a) shows the TEM image of the graphene synthesized with acetylene at 6 ml/min. Fig 3 (b) shows a high magnification TEM image indicating the layers of the graphene sheets overlaid on the TEM grid. It was observed that most of the graphene sheets have between 6-8 layers. We are currently working on modifying the synthesis conditions to lower the number of layers and increase the diameter of the graphene sheets.

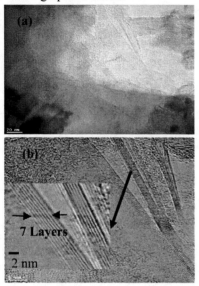

Figures 3 (a) and (b): TEM image of graphene sheets synthesized utilizing acetylene at 6 ml/min. The inset image in (b) shows a high magnification TEM image indicating the number of graphene layers.

Figs. 4 (a) and (b) present the low and high magnification TEM images of the graphene sheets synthesized with high hydrocarbon flow rate. As seen in Figure 4 (a), the simple acid treatment did not fully remove the metal catalyst particles entirely, which can be possibly explained by the entrapment of these metallic nano-clusters between the graphitic sheets. As a result a more sophisticated purification process needs to be used in order to fully remove the non-graphitic impurities from the samples.

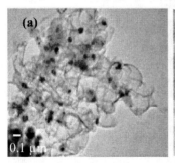
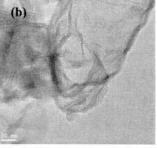

Figure 4 (a): Low magnification TEM image of graphene sheets synthesized utilizing acetylene at 8 ml/min. (b) High magnification TEM image of the CVD grown graphene sheets.

AFM measurements of the graphene sheets synthesized at two different hydrocarbon rates, were performed for thickness analysis (Figs (a)-(f)). It was noticed that as the flow rate of the hydrocarbon increased from 6 to 8 ml/min, so did the number of graphene layers. The graphene sheets synthesized with acetylene at 6 ml/min were found to have between 5 to 7 layers. This is in good agreements with the microscopy measurements shown in Fig. 3. On the other hand, the samples synthesized with the highest flow rate were seen to have a larger number of layers (between 12-21).

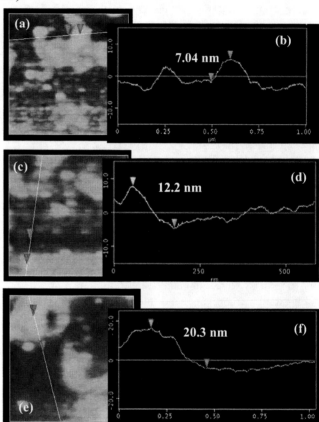

Figure 5(a): 5 μm × 5 μm AFM scan of graphene sheets overlaid onto a Si surface synthesized with acetylene at 6 ml/min. (b) The corresponding height image indicating the presence of graphene sheets with 5-7 layers. (c) and (e) AFM image scans of the graphene sheets synthesized utilizing acetylene at 8 ml/min. Their corresponding height images are shown in (d) and (f) respectively.

STM imaging was done on graphene flakes collected on atomically flat Au/mica substrates. Figs. 6 (a) and (b) show the STM topographic images of graphene sheets that were synthesized at different hydrocarbon flow rates. In bulk graphite, the carbon atoms on the surface are not equivalent. This results in an asymmetry of the surface atom electronic environment that produces a threefold symmetry pattern. Our FFT analysis is consistent with the threefold symmetry pattern (not shown here). The dark regions in the image correspond to vacancy islands that are one atomic layer deep pits in the Au{111} surface; a characteristic due to growth conditions [24].

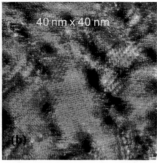

Figures 6 (a) and (b): STM image of the graphene flakes synthesized with acetylene at 6 ml/min and 8 ml/min respectively.

During the CVD process, hydrocarbon decomposes in the presence of a catalyst system and the carbon atoms start rearranging themselves into various carbon nanostructures. Graphene synthesis at high temperatures is not fully understood yet. It is believed that at high temperatures (1000 °C or higher) the catalyst nano-particles start melting, generating a better wetting property which is more suitable for the formation of the graphene structure [25]. Another theory is that magnetic metal atoms aid to the so called "cutting" process of graphene and unzipping CNTs [26]. It was noticed that as the flow rate of the hydrocarbon increased, the diameter as well as the layer number of the graphene sheets increased. This can be explained by the concentration of acetylene on the catalyst nano-particles. Similar phenomenon was also observed by others [16]. A low hydrocarbon flow rate leads to a low carbon accumulation onto the metal nano-particles, synthesizing graphene sheets with small diameters and very few layers.

In conclusions, this is an economical and attractive method which can be scaled up to synthesize large area graphene sheets. By increasing the flow rate of acetylene from 4.5-8 ml/min, the size of the graphene sheets increased from nm to μm range.

4 ACKNOWLEDGMENT

This research was partially supported by the DOE (Grant No. DE-FG 36-06 GO 86072). Also the financial support from Arkansas Science and Technology Authority (ASTA) grant # 08-CAT-03 is highly appreciated.

5 REFERENCES

[1] A. K. Geim and K. S. Novoselov, Nature Mater., 6, 183-191, 2007.
[2] Y. Zhang, J. W. Tan, H. L. Stormer, and P. Kim, Nature, 438, 201-204, 2005.
[3] C. Lee, X. Wei, J. W. Kysar, and J. Home, Science, 321, 385-388, 2008.
[4] M. Y. Han, B. Ozyilmaz, Y. B. Zhang and P. Kim, Phys. Rev. Lett., 98, 206805, 2007.
[5] C. E. Lee, X. Wei, J. W. Kysar, and J. Hone, Science, 321, 385, 2008.
[6] F. Schedin, A. K. Geim, S. V. Morozov, E. W. Hill, P. Blake, M. I. Katsnelson, and K. S. Novoselov, Nat. Mat., 6, 662, 2007.
[7] E. Yoo, J. Kim, E. Hosono, H. Zhou, T. Kudo, and I. Honma, Nano Lett., 8, 2277, 2008.
[8] X. Wang, L. Zhi, K. Mullen, Nano Lett., 8, 323, 2008.
[9] N. Tombros, C. Jozsa, M. Popinciuc, H. T. Jonkman, and B. J. van Wees, Nature, 448, 571-574, 2007.
[10] S. Stankovich, D. A. Dikin, G. H. B. Dommett, K. M. Kohlhaas, E. J. Zomney, E. A. Stach, R. D. Piner, and S. T. Nguyen, Nature, 442, 282-286, 2006.
[11] S. Kim, J. Nah, I. Jo, D. Shahrjerdi, L. Colombo, Z. Yao, E. Tutuc, and S. K. Banerjee, Appl. Phys. Lett., 94, 062107, 2009.
[12] X. Li, W. Cai, J. An, S. Kim, J. Nah, D. Yang, R. Pines, A. Velamakanni, I. Jung, E. Ttuc, S. K. Banerjee, L. Colombo, and R. S. Rouff, Science, 324, 2009.
[13] P. W. Sutter, J. I. Flege, and E. A. Sutter, Nature Mater., 7, 406-411, 2008.
[14] G. Eda, G. Fanchini, and M. Chhowalla, Nature Nanotechnology, 3, 2008
[15] X. Fan, W. Peng, Y. Li, X. Li, S. C. Wang, G. Zhang, and F. Zhang, Adv. Mater., 20, 4490, 2008.
[16] A. Reina, S. Thiele, X. Jia, S. Bhaviripudi, M. S. Dresselhaus, J. A. Schaefer, and J. Kong, Nano Res., 2, 509-516, 2009.
[17] V. Lee, L. Whittaker,C. Jaye, K. M. Baroudi, D. A. Fischer, and S. Banerjee, Chem. Mater. 2009.
[18] K. S. Kim, Y. Zhao, H. Jang, S. Y. Lee, J. M. Kim, K. S. Kim, J. Ahn, P. Kim, J. Choi, and B. H. Hong, Nature, 457, 2009.
[19] E. Dervishi, Z. Li, A. R. Biris, D. Lupu, S. Trigwell, and A. S. Biris Chem. Mater., 19, 2, 179–184, 2007.
[20] A. R. Biris, A. S. Biris, D. Lupu, S. Trigwell, E. Dervishi, Z. Rahman,and P. Marginean, Chem. Phys. Lett., 429, 204, 2006.
[21] E. Dervishi, Z. Li, F. Watanabe, A. Biswas, Y. Xu, A. R. Biris, V. Saini and A. S. Biris, *Chem. Commun.*, 4061-4063, 2009.
[22] D. A. Dikin, S. Stankovich, E. J. Zimney, R. D. Piner, G. H. B. Dommett, G. Evmenko, S. T. Nguyen, and R.S. Rouff, Nature, 448, 457, 2007.
[23] D. Li, M. B. Muller, S. Gilje, R.B. Kaner, and G. G. Wallace, Nat. Nanotehcnology, 3, 2008.
[24] C. Vericat, M. E. Vela and R. C. Salvarezza, Phys. Chem. Chem. Phys., 7, 3258, 2005.
[25] R. B. Little, *Journal Cluster Sci.,* 14, 2, 135, 2003.
[26]L. Ci, L. Song, D. Jariwala, A. Laura Eli´as, W. Gao, M. Terrones, and P. M. Ajayan, *Adv. Mater.,* 21, 1, 2009.

Is Aluminium Suitable Buffer Layer for Growth of Carbon Nanowalls?

P. Alizadeh Eslami[*], M. Ghoranneviss[**] and S. Nasiri Laheghi[**]

*Applied Chemistry Department, Science Faculty, Islamic Azad University,
Tabriz Branch, Tabriz, Iran, sp.alizadeh@iaut.ac.ir
**Plasma Physics Research Centre, Islamic Azad University, Science and Research branch,
Tehran, Iran, ghoranneviss@gmail.com; sp.nasiri@gmail.com

ABSTRACT

This work concentrates on the growth of carbon nanowalls (CNWs) using thermal chemical vapour deposition (T-CVD) technique on a glass substrate coated with Fe nanocatalyst and Al thin film as a buffer layer. A combination of $C_2H_2/NH_3/H_2$ renders the growth of carbon nanostructures at atmospheric pressure. Parameters affecting the growth of the CNWs such as C_2H_2 flow ratio, deposition time, and temperature are investigated. Surface morphologies of thin films are observed using AFM technique, while the thicknesses were measured by RBS technique. The samples utilized are characterized by the SEM, TEM, and Raman spectroscopy techniques. Results show forming some of maze like carbon nanostructures TCVD employing $C_2H_2/NH_3/H_2$ (20:80:100 sccm) for 45 minutes at atmospheric pressure, and 600°C substrate temperature. One of the important conclusions is that the an increase of the temperature from 600°C to 700°C causes the formation of the CNTs and the CNWs on the substrate.

Keywords: carbon nanowalls, glass substrate, thermal CVD, Fe nanocatalyst, Al buffer layer

1 INTRODUCTION

Self-assembled carbon nanostructures like carbon nanotubes (CNTs) [1] and nanofibers [2] have attracted much attention for several applications, such as gas storage, membranes for electrochemical energy storage, and field emitters [3,4]. In addition, two-dimensional graphitic carbon structures, commonly referred to as carbon nanowalls (CNWs) or nanoflakes, are found attractive both as far as their structures, and their applications due to their high surface to volume ratios. Moreover, these structures have a great potential to act as metallic clusters supporting elements for electrochemical fuel cells. The morphologies of the metallic clusters supports have much effect on the chemical activity of the metallic clusters themselves [5]. Also, the carbon nanowalls very sharp and thin edges being perpendicular to the substrates can act as field emitters [6]. In another work, a transmission electron microscopy study has revealed that the CNWs consisting of a few graphene sheets were stacked together, like thin graphite flakes [7]. Wu, et al. [8] have shown that CNWs can grow vertically on a substrate with catalyst using a microwave plasma enhanced chemical vapor deposition (PE-CVD). Hiramatsu, et al. [9] have shown that CNWs can grow vertically on a substrate without catalyst using radio-frequency (rf) PECVD with a H radical injection. Recently, Escobar and co-workers [10] reported the effects of acetylene pressure on nanotubes characteristics by CVD on Fe nano-particles at 180 torr. Very recently, Alizadeh et al. [11] and ghoranneviss et al. [12] have reported the growth of CNWs on a Fe catalyzed substrate using thermal chemical vapor deposition (TCVD) in atmospheric pressure.

In this paper, a glass substrate is coated with Fe nanocatalyst and Al thin film as a buffer layer by low temperature plasma. Carbon nanostructures (two-dimensional carbon nanostructures and CNTs) are grown on the Fe-Al-glass substrate by TCVD using acetylene (C_2H_2) as a carbon sources, and a number of diluting gases such as hydrogen (H_2) and ammonia (NH_3). Effective parameters such as the C_2H_2 flow ratios, time, and CNWs temperature growth are to be studied. The samples are to be analyzed for characterization of the surface morphology and the roughness parameters of the substrate, alignments, distributions, and carbon nanostructures types. These characters are measured by atomic force microscopy (AFM), Rutherford back scattering (RBS) spectrometer, scanning electron microscope (SEM), Raman spectroscopy and transmission electron microscopy (TEM).

2 EXPERIMENTAL SET UP

2.1 Substrate Preparations

The glass substrate is coated with Al thin film as a buffer layer in first step, and then Fe nanocatalyst coated on the Al buffer layer in second step. Al thin film with 1 nm thickness and Fe nanocatalyst with 2 nm thicknesses were deposited on the microscopic glass slides (5 mm × 5 mm × 0.9 mm) by low temperature plasma. The instrument used in this experiment was a direct current magnetron sputtering (DC-MS) device. This instrument consisted of a cylindrical glass tube and two Al or Fe co-axial cylinders used as a cathode (the inner one: 30 mm diameter and 200 mm length) and an anode (the outer one: 100 mm diameter and 200 mm length) in its chamber. Argon (Ar) was selected as a sputtering gas, and the operation pressure was set at 0.03 torr. The substrates were cleaned by ultrasonic vibration with acetone, ethanol, and deionized water to remove all the contaminants. For Al and Fe Coating, a uniform magnetic field of 400 Gauss was induced from the outside. The applied voltage between the anode and the cathode was about 900V which produced a 120mA current. The microscope glass slide used as a substrate was fixed on the anode. The distance between the target and the substrate was measured to be about 30mm. The surface morphologies of the films obtained in this manner, were observed using the AFM (Auto probe PC, Park Scientific Instrument, USA). Film thickness was measured using the RBS technique, using He^+ ion beam of 10 μm in diameter with 2.0 MeV energy. A typical RBS spectrum of Fe thin film is given in Fig. 1.

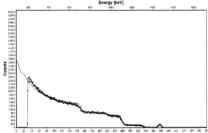

Fig.1. Rutherford back scattering (RBS) curve shows that that the thicknesses of Fe thin film coated on glass substrate is 2 nm.

2.2 Growth of Carbon Nanostructures

A thermal chemical vapour deposition (T-CVD) system was used in this study for growing the carbon nanostructures. This system consists of a horizontal quartz tube used as a reaction chamber, and a furnace surrounding the reaction chamber. The Fe-Al-glass substrate was placed on the ceramic boat, and was pushed into the reaction chamber at the centre of the furnace. To avoid the catalyst oxidation, argon (Ar) gas at 200 standard cubic centimetres per minute (sccm) was fed into the chamber of the CVD, and the chamber temperature was increased to 350°C. For etching the substrate, the Ar gas valve was closed, and the H_2 gas valve was opened (100 sccm) until the temperature was reached 500 °C. The time taken for the etching was 30 minutes. A combination of $C_2H_2/NH_3/H_2$ with 20:80:100 (sccm) flow ratios were fed into the reaction chamber. The reaction time and the temperature for the growth were 45 minutes, and 600°C, respectively. In the other experiments, the conditions were changed by changing the gas flow, the time, and the temperature from 20 to 48 sccm, 45 to 60 minutes, and 600°C to 700°C, respectively, one change at a time. The carbon nanostructures obtained using this technique were analyzed using the SEM (XL30, 15-30.kV, Philips Company), and the Raman spectroscopy (Thermo Nicolet Crop. Madison, Wisconsin 53711, USA with 1064 nm wave length). TEM (CM120, 120 kV, Philips Company, with a W cathode).

3 RESULTS AND DISCUSSIONS

Surface morphology of the Fe nanolayer on the Al-glass was observed using the AFM, and the configuration is shown in Fig.2. The average roughness of the Fe-glass is measured to be 2.64Å.

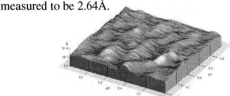

Fig.2. AFM analysis of Fe nanolayer deposited on the Al-glass substrate by D.C magnetron sputtering.

The SEM micrographs in Fig. 3 show the morphologies of the carbon nanostructures grown on the Fe nanocatalyst coated on the Al-glass substrate. Maze like carbon nanostructures were obtained by the T-CVD by applying the $C_2H_2/NH_3/H_2$ mixture (20:80:100 sccm) for 45 minutes at atmospheric pressure, and at 600°C. Fig. 3(a) shows the SEM image of these nanostructures. The image shows that the two-dimensional carbon nanosheets were grown vertically on the substrate. The thin two-dimensional carbon nanosheets formed have thicknesses of less than 28nm. Fig 3(b) shows the SEM image of the high density no uniform carbon nanostructures grown by TCVD employing the $C_2H_2/NH_3/H_2$ (20:80:100 sccm) mixture for 60 minutes at atmospheric pressure and 600°C substrate temperature. Fig 3(c) shows that an increasing in the C_2H_2 flow rate caused to grown high density no uniform carbon nanostructures on the substrate, too. Fig. 3 (d) shows the SEM image of the CNTs and CNWs obtained on the substrate by the TCVD employing $C_2H_2/NH_3/H_2$ (20:80:100 sccm) mixture for 45 minutes at atmospheric pressure and 700°C temperature. This SEM image shows that the CNTs with 61nm in diameter were formed on the CNWs with 24nm in thicknesses. Growing of the CNWs in this manner has randomly orientation. However, average of distances between the adjacent CNWs were grown, by TCVD employing the $C_2H_2/NH_3/H_2$ (20:80:100 sccm) mixture for 45 minutes at atmospheric pressure and 700°C substrate temperature, is 86 nm (fig. 3d). The limitations of the process for growing CNWs on a Fe coated Al-glass were analyzed through some parameter studies. The parameters in this case were temperature, time, and gas flow ratio. The results of these parameters studies show that the best CNWs growth condition takes place at 700°C, using $C_2H_2/NH_3/H_2$ flow ratios of 20:80:100, and a 45 minutes period. Since the experimental runs occur at atmospheric pressure, to minimize the unwanted reactions and ionization of the species included in the process, a high flow rate of Ar gas is fed into the reaction chamber. In other hand, the results of these parameters studies show that the maze like carbon nanostructures were grown by using $C_2H_2/NH_3/H_2$ flow ratios of 20:80:100, at 600°C temperature, and a 45 minutes growth time. One of important characterizations of ideal CNWs is high ratio for length/thicknesses ratio. So, investigation of this amount show that, the length/thicknesses ratio for CNWs were grown, by TCVD employing the $C_2H_2/NH_3/H_2$ (20:80:100 sccm) mixture for

45 minutes at atmospheric pressure and 700°C substrate temperature, is about 14. These results show that, increasing the substrate temperature from 600 to 700°C causes to obtain the better structures of the CNWs in this manner.

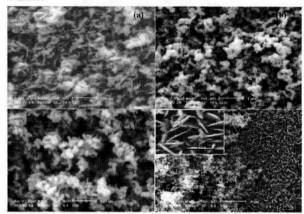

Fig.3. SEM images of the carbon nanostructures grown on the glass substrate coated with the Fe nanocatalyst on the Al buffer layer with conditions: (a) $C_2H_2/NH_3/H_2$ (20:80:100 sccm) for 45min. and 600°C; (b) 60 min., (c) 48 sccm, and (d) 700°C.

The Raman spectra of the carbon nanostructures grown on the Fe glass substrate coated with Fe nanocatalyst on the Al nanolayer as a buffer layer, for two conditions are shown in Fig. 5. Generally, for the carbon nanowalls, three Raman bands can be identified. The D-band around 1320 cm^{-1}, the G-band around 1580 cm^{-1}, and the D'-band around 1610 cm^{-1}. As shown in Fig. 5(a), the Raman spectrums for the two-dimensional carbon nanosheets were found to have a G-band peak at 1580 cm^{-1}, indicating the formation of a graphitized structure and a D-band peak at 1310 cm^{-1}, corresponding to the disorder-induced phonon mode. The G and D peaks are comparable in their intensities. It should be noted that the G-band peak is accompanied by a shoulder peak at 1610 cm^{-1} (D'-band). This shoulder peak is associated with the finite-size graphite crystals [13]. The strong G-band peak suggests a more nanocrystal configuration structure, and a presence of graphene edges and defects such as distortion, voids, and strains in the graphitic lattices which are among the prevalent features of the CNWs. The crystallite size "La" stems in I_D/I_G intensity ratio. The empirical formula "$La=44/(I_D/I_G)$" has been successfully used for Raman spectra taken with the excitation energies around 514.5 nm [14], although the I_D/I_G intensity ratio strictly depends on the Raman excitation energy [15]. Therefore, the above formula was also employed for Raman spectra with 1064 nm excitation energy in the present experiments. From these estimations, it is found that the average crystallite size "La" for the CNWs obtained in this study is 11.67 Å. Fig. 5(b) shows three Raman bands. The D-band is around 1320 cm^{-1}, the G-band is around 1580 cm^{-1}, and the D'-band is around

1607 cm^{-1}. These images show that the carbon nanowalls (CNWs) were grown vertically on the substrate. With increasing the temperature to 700°C CNWs were grown completely. The average crystallite size, "La", in the CNWs were obtained in the present experiment is 49.43 Å. Also, fig. 5(b) shows the Raman spectrums of the CNTs were grown on the CNWs. In this present work it is found that an optimum $NH_3/H_2/C_2H_2$ gas ratio is 4:5:1. The experimental runs indicate that the presence of a catalyst on a buffer layer is a necessary condition for the growth, since no CNWs could be synthesized in the absence of a Fe catalyst and Al buffer layer. In this work a qualitative analysis of the carbon coating is performed with Raman spectroscopy. Each spectrum shows three characteristic carbon bands, i.e., the disorder induced the D-band around 1350 cm^{-1}, the G-band around 1580 cm^{-1} (that is related to in-plane sp^2 vibrations), and the D'-band around 1610 cm^{-1} [15-17]. For a qualitative analysis of the graphite-like structures, the intensity ratio of the D-band is compared with that of the G-band. This ratio here is denoted as the R-value, "$R = I_D/I_G$" [16]. CNWs usually suffer from defects, and are well-known for their relatively large R-values, exceeding unity, while pure CNTs and highly ordered graphite can have R-values close to zero [16,17]. The Raman spectra of the CNWs showed the R = 3.77 and 0.89 for Figures 3a and 3d, respectively. These results indicate that the CNWs are somewhat more in order with respect to their sp^2 structures. In this present work the equation "$La=44/(I_D/I_G)$" [13] is implemented to investigate the crystallite size of the CNWs.

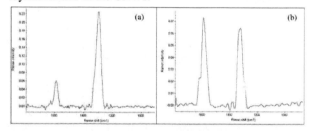

Fig.4. Raman spectrum of the carbon nanostructures grown on the glass substrate coated with the Fe nanocatalyst on the Al buffer layer with conditions: (a) $C_2H_2/NH_3/H_2$ (20:80:100 sccm) for 45min. (b) 700°C.

These results show that the increase of temperature from 600°C to 700°C causes formation of the CNTs and the CNWs on the substrate. Fig. 3(d) shows the SEM image of the CNTs with 61nm in diameter and CNWs obtained on the substrate. These CNTs characterized by TEM. For the TEM studies, a small amount of the CNTs was put down on a grid. Both the morphologies and the CNTs size are estimated by the TEM micrographs. Fig. 6 shows TEM micrograph of CNTs obtained by TCVD on the glass substrate at 700°C temperature. As it can be seen, the CNTs exhibit a non uniform diameters and high length/diameter ratios. The various CNTs show in the TEM micrograph with different diameters multi wall carbon nanotubes (MWCNTs). These CNTs have helical configuration. Length and minimum diameter of the CNTs are amount 550 nm and 9 nm, respectively. Maximum diameter of the MWCNTs is 25 nm. Some of metal catalyst particles detected that encapsulated in the nanotube and/or at the tip of the nanotubes.

Fig.6. TEM image of CNTs were grown by increasing the growth temperature from 600ºC to 700ºC. This image shows that the CNTs are collective of MWCNTs and helical configuration. Minimum diameter of CNTs is 9 nm and maximum diameter is 25 nm.

4 CONCLUSIONS

In this paper, the carbon nanowalls, and carbon nanotubes were synthesized by the T-CVD system at atmospheric pressure. These carbon nanostructures were produced on the glass where Fe thin film as the magnetic catalyst and Al thin film as the buffer layer were deposited by a DC magnetron sputtering. C_2H_2 was used as carbon sources. H_2 and NH_3 were applied as diluting gases, and the following concluding remarks were reached:

a) The SEM characterization procedure shows that the two-dimensional carbon nanosheets were grown on the Fe catalyst coated on the Al buffer layer.

b) Increases in the C_2H_2 flow ratio, and time duration of the reaction cause the producing no uniform carbon nanostructures on the substrate.

c) An increase of the temperature from 600°C to 700°C causes an increase in the crystallite size (*La*) of the CNWs and the formation of the CNTs on the substrate.

REFERENCES

[1] C. Bower, W. Zhu, S. Jin and O. Zhou, Plasma-induced alignment of carbon nanotubes, Appl. Phys. Lett., 77, 830-832, 2000.

[2] J.B.O. Caughman, L.R. Baylor, M.A. Gulliorn, V.I. Merkulov, D.H. Lowndes and L.F. Allard, Growth of vertically aligned carbon nanofibers by low-pressure inductively coupled plasma-enhanced chemical vapor deposition, Appl. Phys. Lett., 83, 1207-1209, 2003.

[3] A.C. Dillon, K.M. Jones, T.A. Bekkedahl, C.H. Kiang, D.S. Bethune and M.J. Heben, Storage of hydrogen in single-walled carbon nanotubes, Nature (London), 386, 377-379, 1997.

[4] G. Che, B.B. Lakshmi, E.R. Fisher and C.R. Martin, Carbon nanotubule membranes and possible applications to electrochemical energy storage and production, Nature (London), 393, 346-347, 1998.

[5] L. Giorgi, Th.D. Makris, , R. Giorgi, N. Lisi and E. Salernitano, Electrochemical properties of carbon nanowalls synthesized by HF-CVD, Sens. Actuators B, 126, 144–152, 2007.

[6] C.C. Chen, C.Fu. Chen, I.H. Lee and C.L. Lin, Fabrication of high surface area graphitic nanoflakes on carbon nanotubes templates, Diamond Relat. Mater., 14, 1897-1900, 2005.

[7] Y. Wu, B. Yang, B. Zong, H. Sun, Z. Shen and Y. Feng, Carbon nanowalls and related materials, J Mater Chem., 14, 469–477, 2004.

[8] Y. Wu, P. Qiao, T. Chong and Z. Shen, Carbon nanowalls grown by microwave plasma enhanced chemical vapor deposition, Adv. Mater., 14, 64-67, 2002.

[9] M. Hiramatsu, K. Shiji, H. Amano and M. Hori, Fabrication of vertically aligned carbon nanowalls using capacitively coupled plasma-enhanced chemical vapor deposition assisted by hydrogen radical injection, Appl. Phys. Lett., 84, 4708-4710, 2004.

[10] M. Escobar, M.S. Moreno, R.J. Candal, M.C. Marchi, A. Caso, P.I. Polosecki, G.H. Rubiolo and S. Goyanes, Synthesis of carbon nanotubes by CVD: Effect of acetylene pressure on nanotubes characteristics, Appl. Surf. Sci., 254, 251-256, 2007.

[11] P. Alizadeh Eslami, M. Ghoranneviss, Sh. Moradi, P. Abroomand Azar, S. Abedini Khorrami and S. Nasiri Laheghi, Growth of carbon nanowalls by thermal CVD on magnetron sputtered Fe thin film, Fullerene, nanotubes, and carbon nanostructures, XX (2010) XX.

[12] M. Ghoranneviss, P. Alizadeh Eslami and S. Nasiri Laheghi, US patent pub. No.: US 2009/0274610 A1

[13] S. Kurita, A. Yoshimura, H. Kawamoto, T. Vchida, and H. Nakai, Raman spectra of carbon nanowalls grown by plasma-enhanced chemical vapor deposition, J.Appl.phys., 7, 104320-104325, 2005

[14] F. Tuinstra and J.L. Koenig, Raman spectrum of graphite, J. Chem. Phys., 53, 1126-1130, 1970.

[15] P. Nikolaev, M.J. Bronikowski, R.K., Bradley, F. Rohmund, D.T. Colbert, K.A. Smith and R.E. Smalley, Gas phase growth of single-walled carbon nanotubes from CO, Chem. Phys. Lett., 313, 91-97. 1999.

[16] Z.H. Ni, H.M. Fan, Y.P. Feng, Z.X. Shen, B.J. Yang and Y.H. Wu, Raman spectroscopic investigation of carbon nanowalls, J. Chem. Phys., 124, 204703–20478, 2006.

[17] J. Kastner, T. Pichler, H. Kuzmany, S. Curran, W. Blau, D.N. Weldon, M. Delamesiere, S. Draper and H. Zandbergen, Resonance Raman and infrared spectroscopy of carbon nanotubes, Chem. Phys. Lett., 221, 53–58, 1994.

Fabrication of Nanoporous Ultrananocrystalline Diamond Membranes

Olga V Makarova[*], Ralu Divan[**], Nicolaie Moldovan[***] and Cha-Mei Tang[*]

[*]Creatv MicroTech Inc., 2242 West Harrison St., Chicago IL, 60612 olga@creatvmicrotech.com
[**] Argonne National Laboratory, 9700 South Cass Av., Argonne IL, 60439 divan@aps.anl.gov
[***] Advanced Diamond Technologies Inc., 429 B Weber Road #286, Romeoville, IL, 60446
moldovan@thindiamond.com

ABSTRACT

Nanoporous membranes have a wide range of applications in many fields, including medical diagnostics, drug delivery, and hemodialysis. Ultrananocrystalline diamond (UNCD®) coatings are becoming more and more significant in medical applications because of the highest degree of biocompatibility, unmatched by other materials. The 100-nm- and 200-nm-diameter pores have been fabricated in 1-μm-thick UNCD film on silicon wafers using e-beam lithography, reactive ion etching and laser writing.

Keywords: ultrananocrystalline diamond, RIE, e-beam lithography, nanonporous diamond membrane

1 INTRODUCTION

Nanoporous membranes engineered to mimic natural filtration systems can be used in smart implantable drug delivery systems, bioartificial organs, and other novel nano-enabled medical devices [1]. Some of the key properties required by these membranes are narrow and controlled pore size distribution, high pores density and low thickness to enable high flux, as well as mechanical and chemical stability. Nanoporous silicon [2] and silicon nitride membranes [3] have been used in drug delivery systems [4]; silicon nanopillars [5] were used for capturing circulating tumor cells in blood. Interestingly, irrespective of the regular nature of pore geometry, blocking of pores by proteins or cell debris is still a major problem. Biocompatibility and anti-biofouling are central issues when membranes are used as interfaces in implantable devices.

Ultrananocrystalline diamond films are becoming more and more significant in medical applications because of the highest degree of biocompatibility and anti-biofouling, unmatched by other materials [6]. The porous diamond membranes may open new opportunities in implant medicine.

Here we report results on high porosity, high-aspect-ratio ultrananocrystalline membranes fabricated using e-beam lithography, reactive ion etching and optical lithography.

2 EXPERIMENTAL

The nanoporous membranes were fabricated from 1-μm-thick UNCD films (Aqua 25 UNCD®, Advanced Diamond Technologies). These UNCD films have the hardness, Young's modulus, and other extreme properties of natural diamond, as well as smooth (~7 nm rms) surface, and a very low internal stress, which is essential for membrane fabrication [7]. The nanoporous UNCD membrane fabrication followed the sequence shown in Figure 1, as detailed in the following subsections. The optical masks for alignment marks and membrane were fabricated with a LW405 MicroTech laser writer system, and the array of holes was patterned with a Raith 150 e-beam lithography equipment.

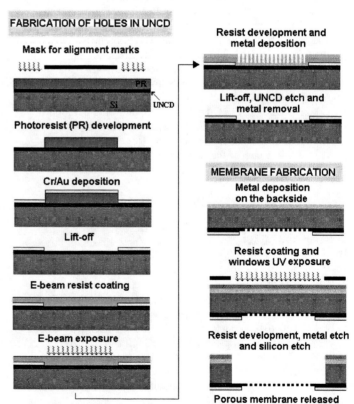

Figure 1. Illustration of the fabrication of holes and UNCD membrane.

The front-to-back side alignment and optical lithography were performed with a Karl Suss MA6 aligner. Metal depositions were performed with a Lesker PVD-250 electron-beam evaporator with a Sigma deposition controller.

2.1 Fabrication of holes in UNCD

The UNCD surface was treated prior to resist coating with hexamethyldisilazane vapors for 20 min in a YES-3/5TA priming oven (Yield Engineering Systems) at temperature of 150°C and 1Torr. This step improved the resist adhesion significantly.

The alignment marks were patterned separate from the main arrays of holes by direct write optical lithography, to avoid time-consuming large area exposures with the e-beam lithography tool. 0.5-μm-thick positive resist (S1805) was used (spin coated at 3000 rpm and baked on a hot plate at 115°C for 1 min). The alignment marks were fabricated by lift off, using a Cr/Au (5 nm/20 nm) layer, providing good contrast in scanning electron microscopy (SEM) mode for alignment in the e-beam lithography tool. These marks allowed a 3-points alignment in the e-beam writer and for backside optical lithography alignment.

The 25 nm films were evaporated at room temperature, with a base pressure of 2×10^{-8} Torr at a deposition rate of 2 Å/s. The lift-off step was performed in 1165 remover (Rohm and Haas) at 70°C for 30 min.

After cleaning with piranha solution for 10 min the samples were treated again in the priming oven. Negative e-beam resist ma-N 2405 (Micro Resist Technology GmbH, Germany) diluted 1:1 in anisole was spin coated at 3000 rpm to obtain ~200 nm-thick layer, followed by baking on a hot plate at 100°C for 90 sec, e-beam exposure, and then development in ma-D 533S developer (Micro Resist Technology GmbH, Germany) for 12 s. Resist exposure was performed using a Raith 150 e-beam system, an acceleration voltage of 20 keV, a beam current of 250 pA, and a dose of 300 mC/cm^2. The pattern consisted of arrays of 100- and 200-nm-diameter circles with a separation distance of 400 nm between the circle centers. The usage of negative resist for this process is essential in reducing the exposure time and the negative side wall angle after development facilitates the lift-off task.

To obtain a hard mask for pattern transfer in UNCD, after resist development, a ~5-nm-Ti and ~55-nm-thick Ni layers were e-beam evaporated using the Leskar system at a deposition rate of 3 Å/s. The proper choice of the metal for the lift-off mask is very important. It was shown [8] that Al can be used as hard mask for etching UNCD, but the sputtering yield of Al is high and metal re-deposition on UNCD surfaces can lead to a significant slow-down of the etching process and roughing of the etched surfaces. For this reason, and because Ni was shown to be a durable mask in inductively coupled plasma reactive ion etching (ICP-RIE) of nanometer size holes in GaN [9], it has been chosen as a hard mask for UNCD etch. Lift-off of the Ni

was done at 100°C in 1165 Remover for 3 hours, and an ultrasonic agitation was used for 90 s at the end of this step.

The UNCD layer was etched by ICP- RIE using oxygen plasma in a PlasmaLab 100 System (Oxford Instruments). The etching recipe used O_2 =50 sccm (sccm denotes standard cubic centimeter per minute at standard temperature and pressure, STP), SF_6 = 0.5 sccm, pressure 9 mTorr, temperature 20°C, ICP power 2700 W, and RIE power 300 W [10]. At these parameters, UNCD etching rate on large open surfaces was found to be ~500 nm/min and Ni etching rate was ~ 20 nm/min. After UNCD etching, the hard mask was removed in Cr etch solution (Transene). One wafer was cut at 90° angle using a dicing saw (Thermocarbon TCAR 864-1 Programmable Dicing Saw), and the cross-section of the UNCD was examined using SEM of an FEI Nova NanoLab microscope, to certify the complete through-etching.

2.2 Membrane fabrication

A 5-nm-thick Ti and 250-nm-thick Ni layers were deposited on the back side of the silicon wafers with patterned UNCD layer. Positive resist S1818 of 1.5-μm-thick layer was spin coated at 3000 rpm and baked on a hot plate at 115°C for 1 min. The resist was exposed with the second optical mask for window fabrication using backside alignment method with MA6 mask aligner. After development for 30 s in 351 developer, diluted 1 to 3, the metals were wet etched. The Si etching for membrane fabrication was done in PlasmaLab 100 System ICP chamber using: 50 sccm CHF_3, 7 sccm SF_6, pressure 15 mTorr, RF power 30W, ICP power 1200W at 20°C.

The UNCD membranes contain 100-nm- and 200-nm-diameter pores forming arrays stretching over a 5 mm x 5 mm area, with a silicon frame of 15 mm x 15 mm (Figure 2).

Figure 2. Photograph of the UNCD membranes on Si wafer.

3 RESULTS AND DISSCUSSION

A micrograph of the sample after lift-off with 100-nm-diameter openings in Ni layer is shown in Figure 3.

We noticed some residual spots on the metal layer, which could be attributed to resist residue or to non-uniformity of metal coating. They didn't affect on performing the next step of UNCD etching.

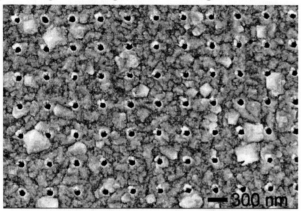

Figure 3. SEM image of the sample after lift-off.

For the used etching parameters, the etching rate of Ni was found to be ~ 12 nm/min, so 4 min is a max possible etching time for using ~ 55- nm-thick Ni mask. The thickness of Ni layer was limited by e-beam resist thickness in order to obtain ~ 100-nm-diameter circles and perform then lift-off.

Using 200-nm-diameter openings in Ni mask, ~ 1 μm-deep pores have been obtained in UNCD after 4 min-etching (Figure 4).

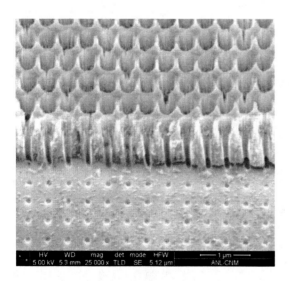

Figure 4 Cross-sectional SEM image of UNCD film that was etched using ~200-nm-diameter openings in Ni mask.

The Ni mask was completely etched in 4 min and this caused an increase in pore sizes at the surface. The holes were etched through, down to the Si surface where the hole pattern was slightly transferred into the Si due to the fluorine content of the plasma mixture. For the same etching time the 100-nm-diameter openings were not etched through, as shown in Figure 5.

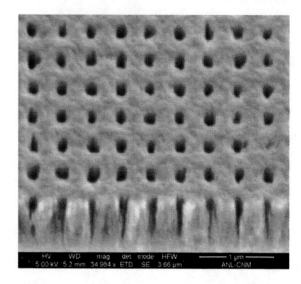

Figure 5. Cross-sectional SEM image of UNCD film with 100-nm-diameter pores, which were not etched completely. through openings in Ni mask.

A longer etching time is required to obtain etched through ~100-nm-diamter pores but thicker layer of Ni is needed. Based on these results, we did similar experiments with a thicker Ni layer (80 nm) pushing the limits of the lift-off step for a 200-nm-thick resist layer. With this Ni thickness it was possible to etch through ~100-nm and ~200-nm-diameter holes in ~1-μm UNCD layer. In Figure 6 SEM images of the membrane with 200-nm (left) and 100-nm-diameter (right) pores are shown.

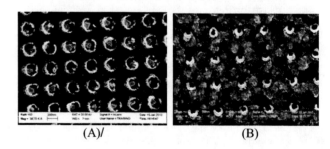

(A)/   (B)

Figure 6. SEM image of the membrane with 200-nm (A) and 100-nm-diameter (B) pores.

Thin UNCD membrane has very low conductivity, and we could perform SEM metrology only after coating the membrane with a conductive polymer ESPACER 300 (Showa Denko, Japan). Even with ESPACER the sample was charging a lot, and it is likely that the residue-like features around the holes in the SEM images can be attributed to the low conductivity of UNCD thin film because the cleaning with piranha solution didn't change the quality of the images.

4 CONCLUSION

High porosity (~ 50%), high-aspect-ratio (up to 10:1) UNCD membranes with 100-nm and 200-nm-diameter pores were fabricated using e-beam lithography, reactive ion etching and optical lithography. Ni mask showed good durability in the RIE-ICP etching of diamond film. Pore diameters were limited by the thickness of the Ni mask. Further experiments will investigate the usage of SiO_2 as hard mask for UNCD pore etching.

Use of the Center for Nanoscale Materials, Argonne National Laboratory was supported by the U. S. Department of Energy, Office of Science, Office of Basic Energy Sciences, under Contract No. DE-AC02-06CH11357

5 REFERENCES

[1] S.P. Adiga, C. Jin, L.A. Curtiss, N.A. Monteiro-Riviere, R.J. Narayan, WIREs nanomedicine and nanotechnology, 1, 568-581, 2009

[2] T.A. Desai, T. West, M. Cohen, T. Boiarski, A. Rampersaud, Adv. Drug Deliv Rev., 56 (11), 1661-1673, 2004.

[3] H. D. Tong, H. V. Jansen, V. J. Gadgil, C. G. Bostan, E. Berenschot, C. J. M. v. Rijn and M. Elwenspoek, Nano Lett., 4, 283-287, 2004.

[4] S.L. Tao, T.A. Desay, Adv. Drug Deliv Rev., 55 (3), 315-328, 2003.

[5] S. Wang, H. Wang, J. Jiao, K.-J. Chen, G.E. Owens, K.-i. Kamei, J. Sun, D. J. Sherman, C. P. Behrenbruch, H. Wu, H.-R. Tseng, Angew. Chem. Int. Ed., 48, 8970–8973, 2009.

[6] Ultrananocrystalline diamond, (eds. O.A. Shenderova, D.M. Gruen), William Andrew, Norwich, UK 2006.

[7] S. Jiao, A. Sumant, M.A. Kirk, D.M. Gruen, A.R. Krauss, O. Auciello, J.Appl.Phys. 90, 118, 2002.

[8] N. Moldovan, O. Auciello, A.V. Sumant, J. A. Carlisle, R. Divan, D. M. Gruen, A. R. Krauss, D. C. Mancini, A. Jayatissa, J. Tucek, SPIE, Vol. 4557, 288- 298, 2001.

[9] D.S. Hsu, C.S. Kim, C.R. eddy, R.T. Holm, R.L. Henry, J.A. Casey, V.A. Shamamian, A. Rosenberg, J.Vac. Sci. Technol. B 23, 1611-1614, 2005.

[10] N. Moldovan, R. Divan, H. Zeng, J.A. Carlisle, J. Vac.Sci.Technol. B 27(6), 3125-3131, 2009.

Quantum Electrodynamics of Nanosystems

Samina S. Masood

Physics Department, University of Houston Clear Lake

2700 Bay Area Blvd, Houston, TX 77058, USA; masood@uhcl.edu

ABSTRACT

Quantum description of mulitiparticle nano-systems is studied in a hot and dense electromagnetic medium. We use renormalization techniques of quantum field theory to show that the electromagnetic properties like electric permittivity and magnetic permeability depend on the temperature and density of the media. Casimir force also depends upon the physical properties of the medium and becomes a function of these parameters within the nano-systems. We discuss the effect of the Casimir force on the nanosystems in terms of temperature and density of the system. We present carbon nanotubes and biomolecules as examples.

Keywords: *Nanosystems, Casimir Force, Protein Molecules, Nanotubes,*

1. INTRODUCTION

The electromagnetic description of chemical processes and its application to biochemistry is used for a long time to understand the complicated structures and growth of living cells. Water being the most abundant solvent plays a big role in the growth and survival of life. A majority of the chemical reactions involve polar molecules dissolved in water. In this process, they release H^+ or OH^- that can affect the acidic properties of the medium. The change in pH value affects the cell growth [1-4] directly. However, the lipid walls may develop Casimir [5] type force, if it becomes within an appropriate distance with the non-polar molecules. Not only that, the entanglement in the carbon nanotubes the hydrophobic part of the protein molecules may also develop accidental Casimir force that can lead to unexpected behavior.

Casimir force is known to depend on the electromagnetic properties of the medium. It is simply proportional to the electric permittivity of the medium [6]. It has also been calculated that temperature [7] affects the Casimir force. Protein densities [8] are believed to be between (1.22-1.47) g/cc. The local densities at certain ends of protein molecule may be large enough to include the density effects. Therefore the vacuum polarization and other radiative corrections are included in this study. We show the dependence of Casimir force on temperature and density using the renormalization scheme of quantum electrodynamics (QED). The brief details of the calculations are given in the next section. We devote section 3 to discuss QED of Nanosystems. In the last section we will present some of the applications of these results to Nanosystems.

2. MEDIUM EFFECTS

We consider Nanosystems as multi-particle QED systems in a theromodynamic equilibrium background. It is believed that the QED effects are non-ignorable on nanoscales and the perturbative effects of QED can add up to sizeable corrections. Renormalization techniques are a way to extract finite perturbative contribution to physically measureable quantities in quantum field theory.

Second quantization in QED leads to the calculation of vacuum polarization at nanoscales to prove renormalization of QED as a gauge theory and remove the unwanted singularities to prove it as a physical theory. Electromagnetic properties of the medium including the electric permittivity ϵ_E and the magnetic permeability μ_B are calculated from the vacuum polarization of photon. We have previously shown [9-15] that the vacuum polarization at the one loop level does not affect the QED coupling at low temperatures. However we calculate nonzero effects to ϵ_E and μ_B at extremely low temperatures if the chemical potential μ is larger than temperature.

We showed [9] that the electromagnetic properties are modified as a function of temperature and chemical potential of the media regardless of their values at the higher loop level though the contribution may be too small to be worried about. It basically happens that the longitudinal and the transverse components of the polarization tensor are not the same in this situation.

The one loop radiative corrections to the vacuum polarization tensor are proportional to the coupling constant α_R which can be renormalized in hot and dense media to pick up the effect of temperature and density on the QED coupling constant at low temperature and small values of density are calculated [9] as,

$$\alpha_R = \alpha(T = 0)(1 + \frac{\alpha^2 T^2}{6m^2}).$$

At high temperatures (more than millions of Kelvin) the one-loop corrections are non-negligible but those temperatures are relevant to the very early universe or the stellar cores and are out of the scope of this paper. However, the ϵ_E and the μ_B of a highly dense medium are modified with temperature and chemical potential. It has pronounced effects at large value of density or chemical potential μ. Much larger values of μ correspond to local densities of large organic molecules like proteins that may be much larger than the average protein densities [8]. The mass of electron itself is modified in the highly dense background of organic molecules at nanoscales that is almost a local effect in that region. The change in the electron mass m has been calculated [10] as a function of chemical potential as

$$\frac{\delta m}{m} \simeq -\frac{3\alpha}{\pi} \ln\frac{\mu}{m} + \frac{\alpha}{\pi} \left[1 - \frac{\mu^2}{m^2}\right]\left[3\frac{m^2}{\mu^2} + 2\frac{p^2}{\mu^2} - 1\right] .$$

This extremely large local density will have an associated local magnetic field that will couple with the charged particles of the system including electrons, polar atoms and the molecule through the magnetic moment [11,12] depending on the strength of the magnetic field.

$$\mu_a^{el} \simeq \frac{\alpha}{2\pi} - \frac{2\alpha\pi}{9}\frac{T^2}{m^2} - \frac{4\alpha}{3\pi m^2}\left[I_1' - \frac{3m^2}{2}I_2' - \frac{m^4}{8}I_3'\right]$$

Whereas I' are complicated functions of chemical potentials. The details of these functions can be found in Ref. [12]. A very simple form of these integrals at $T \ll \mu$ is given by

$$I_1' = \frac{\mu^2}{2}\left[1 - \frac{m^2}{\mu^2}\right] ,$$

$$I_2' = \ln(\mu/m) ,$$

$$I_3' = \frac{1}{2m^2}(1 - m^2/\mu^2) .$$

The magnetic moment of other charged particles like atoms or molecules may not get sizeable corrections from background and their intrinsic magnetic dipole moment may be ignorable also.

The high local density inside the large and complicated organic molecules will affect the vacuum polarization in the non-polar portion of molecules. The local values of electric permittivity ϵ_E and the magnetic permeability μ_B of certain regions may be expressed [10] in terms of the plasma frequencies of photons ω and the wave-vector k. The details of these calculations can also be found in Ref. [10]. Local values of these quantities are the ones which may fluctuate between very high and very low values significantly due to the free energy of the system.

The ϵ_E and μ_B are evaluated as functions of chemical potential of the system as,

$$\epsilon_E(K,\mu,T) \simeq 1 - \frac{2\alpha}{\pi^2 K^2 k}\left\{\left[k - 2w \ln\frac{w-k}{w+k}\right]\mathcal{I}_1 \right.$$
$$\left. -2A_1\mathcal{I}_2 + \frac{1}{2}A_2\mathcal{I}_3\right\}\left[1 - \frac{w^2}{k^2}\right]$$

And

$$\frac{1}{\mu_B(K,\mu,T)} \simeq 1 + \frac{2\alpha}{\pi k^3 K^2}\left\{\left[\frac{k^2}{2} + w^2\right]\left[1 - \frac{w^2}{k^2}\right]\left[2w \ln\frac{w-k}{w+k}\mathcal{I}_1 - 2A_1\mathcal{I}_2 - \frac{1}{2}A_2\mathcal{I}_3\right]\right.$$
$$\left. - \frac{k^3K^2 + k^4 m}{4}\mathcal{I}_2 - \frac{m^2}{2}\left[\frac{k^5}{12} + \frac{(3w^2 + k^2)k^3K^2}{24m^2} + \frac{m^2k^2}{8k}(5w^2 - 3k^2)\right]\mathcal{I}_3\right\}$$

This difference arises because of the fact that the vacuum polarization tensor indicates different polarization along the line of separation of conductors and the perpendicular components. π_L is along the direction of separation of two conductors and π_T is perpendicular to the separation of conductors. The longitudinal and transverse components of the vacuum polarization tensor are affected differently by the temperature and densities of the medium and are expressed as a function of T and μ of the system are given as

$$\Pi_L(0,\mathbf{k}) \equiv K_L^2 \simeq \frac{e^2}{2\pi^2}\left[\mathcal{I}_1 + 2m^2\mathcal{I}_2 - \frac{9m^4}{2}\mathcal{I}_3\right] ,$$

$$\Pi_T(0,\mathbf{k}) \equiv K_T^2 \simeq \frac{e^2}{2\pi^2}(m^2\mathcal{I}_2 - \frac{17}{4}m^4\mathcal{I}_3) ,$$

Whereas,

$$\mathcal{I}_1(T,\mu) = \frac{1}{2}\int E\,dE\{n_F(E+\mu) + n_F(E-\mu)\}$$
$$= \frac{1}{2}\left[\frac{\mu^2}{2}\left[1 - \frac{m^2}{\mu^2}\right] + \frac{1}{\beta}\{a(m\beta,\mu) - a'(m\beta,\mu)\}\right.$$
$$\left. - \frac{1}{\beta^2}\{c(m\beta,\mu) + c'(m\beta,\mu)\}\right] ,$$

$$\mathcal{I}_2(T,\mu) = \frac{1}{2}\int \frac{dE}{E}\{n_F(E+\mu) + n_F(E-\mu)\}$$
$$= \frac{1}{2}\left[\ln\frac{\mu}{m} + b(m\beta,\mu) + b'(m\beta,\mu)\right] ,$$

$$\mathcal{I}_3(T,\mu) = \frac{1}{2}\int \frac{dE}{E^3}\{n_F(E+\mu) + n_F(E-\mu)\}$$
$$= \frac{1}{2}\left[\frac{1}{2m^2}\left[1 - \frac{m^2}{\mu^2}\right] - \frac{1}{4\mu^2} + \frac{1}{m^2}\{n_f(\mu+m) + n_F(\mu-m)\}\right.$$
$$\left. + \frac{\beta}{m}\left\{\frac{e^{-\beta(\mu+m)}}{[1+e^{\beta(\mu+m)}]^2} + \frac{e^{-\beta(\mu-m)}}{[1+e^{\beta(\mu-m)}]^2}\right\} + d(m\beta,\mu) + d'(m\beta,\mu)\right]$$

At the expected local densities of organic molecules they can simply be written as (for i = 1,2, or 3)

$$\mathcal{J}_i = \tfrac{1}{2}I_i'$$

Using π_L and $\pi_{T,}$, the real and imaginary parts of the dielectric function and the absorption coefficient of carbon Nanotubes can also be shown to be a function of temperature and density. In the real physical Nanotubes, the structural defects and the un-expected entanglement may be due to these probabilistic effects. Casimir effect may be a big cause of this entanglement which is as random and difficult to explain as the Casimir force is.

This difference between the longitudinal and the transverse polarization is a possible source of Casimir type force between two conducting plates. This force may be attractive or repulsive depending on the possibility of facing like or unlike charges. This kind of force can exist between neutral ends of molecules may locally act as small conductors.

3./ OTHER QED EFFECTS

The basic electromagnetic processes that control the dynamics of polar and non-polar molecules in water based fluid in living cells and determine the growth and the survival probability of life. Two basic interactions of solute molecules in water solvent are *hydrophilic* (water-loving) and *hydrophobic* (water-fearing) interactions. The study of these interactions with this approach may resolve some basic issues of chemical synthesis and even may lead to methods to develop more efficient energy sources.

Generally, hydrophilic molecules carry polar and/or ionized functional groups whereas hydrophobic molecules are electrically neutral with symmetric carbon chains. When non-polar molecules like lipids are mixed with water, they move away from water molecules and end-up gathering at the surface of water through hydrophobic interaction. Another good example of hydrophobic interaction is the addition of oil to water that gives two distinct layers of non-polar and polar liquid. Large molecules with a polar/ionized region and one end and a non-polar region at the other end have both hydrophilic and hydrophobic characteristics. The increasing solubility of these molecules in water is believed to be due to the cluster formation on the hydrophilic region and the hydrophobic pockets on the other side. The physics of these interactions is not so well-understood. We are using quantum electrodynamics (QED) to properly describe these hydrophobic and hydrophilic interactions. Even the oral consumption and/or inhaling of chemical components like alcohol or other charged species will change the electrodynamics of living bodies. It may even effect the magnetic moments of molecules.

The effect of electromagnetic radiation as a source of external energy can also be expected to have tremendous effects on human body. The quantum mechanical uncertainties and the probability theory may explain the causes of gene mutation [16] that can better describe the cause of developing cancer in human bodies. It is so much dependent on the dosage and location that the radiation therapy is used to control the cancer cell growth as the radiation provides energy which shields the cell nutrition and may reduce the growth. The affect on the pH value of the nutrient also plays a role in the cell growth.

Another type of QED effect is possible due to the inhomogeneous magnetic fields which may induce Eddy currents in different places of electrolytes that serve as nutrient and effect the cell growth. These currents are extremely small and ignorable individually, however for large molecules they may add up to sizeable effect. Especially because they may cause the molecular movements in preferred directions and effect the cell mechanism, its consumption of charged radicals like calcium ions or potassium ions and ultimately effect the cell growth. Analytical calculations of these currents are too cumbersome even the existing computational models cannot handle it. So the best way to study the importance of these effects is the experimental study which can lead to the theoretical explanation of those results. These Eddy currents can change the membrane potential and even the channel potentials. External weak magnetic fields will couple with the magnetic moment of molecules at different level depending on the strength and may give measurable effects. We have been working on some very simple experiments to just check the plausibility of this approach.

4./ DISCUSSION OF RESULTS

The temperature contributions to the QED parameters are not important for Nanosystems at all so we do not need to study them in detail. However, we need to use the thermodynamic principals to quantify the chemical potential effects in Nanosystems.

Casimir force may be the accidently generated force that may lead to the unexpected and unexplained behavior of the protein folding and carbon Nanotubes. Not only that, dense localities of these molecules like hydrophobic portion of protein molecules may have high densities and chemical potentials instantaneously to generate stronger than expected Casimir force.

The dependence of electric permittivity and magnetic permeability on temperature and chemical potential changes the electromagnetics of thermodynamic media. This is a possible reason of change of dielectric properties of the medium as well as the displacement vectors in the long conducting chains of organic molecules or some portions of carbon Nanotubes or protein molecules.

In carbon nanotubes, instantaneous Casimir force gives a possible explanation of synthetic defaults like entanglement of carbon Nanotubes, etc.

We are mainly proposing to use basic concepts of perturbative QED such as dipole moment of molecules, their interaction with electromagnetic fields, eddy currents produced through the variable magnetic fields and especially the possible Casimir forces on the non-polar end of the large molecule and the Lamb shift.

REFERENCES :

1. See for example: S. Humez et al.`The role of intracellular pH in cell growth arrest induced by ATP' *Am J Physiol Cell Physiol* 287: C1733-C1746, (2004).

2. M. Mataragas, `Influence of nutrients on growth and bacteriocin production by *Leuconostoc mesenteroides* L124 and *Lactobacillus curvatus* L442' , A *ntonie van Leeuwenhoek* **85:** 191–198, (2004).

3. H. `Umashankar, et. al., Influence of nutrients on cell growth and xanthan production by Xanthomonas campestris' Bioprocess Engineering 14 307-309 (1996).

4. Dan Zilberstein, et al;, Escherichia coli Intracellular pH, Membrane Potential, and Cell Growth, J. of Bacteria, , 246-252 (1984); Sarah Sundelacruz, Michael Levin and David L. Kaplan, `Role of Membrane Potential in the Regulation of Cell Proliferation and Differentiation, Stem Cell Rev and Rep 5: 231–246 (2009).

5. Casimir, H. G. B. "On the attraction between two perfectly conducting plates." *Proc. Con. Ned. Akad. van Wetensch* B51 (7): 793-796 (1948). Lamoreaux, S. K. "Demonstration of the Casimir force in the 0.6 to 6 mm range." *Physical review Letters* 78 (1): 5-8 (1997).

6. K. A. Milton, Invited Lectures at 17th Symposium on Theoretical Physics, Seoul National University, Korea, June 29 - July 1, (1998); http://arxiv.org/PS_cache/hep-th/pdf/9901/9901011v1.pdf; and references therein.

7. Simen Å. Ellingsen, Iver Brevik, Kimball A. Milton,` Casimir effect at nonzero temperature for wedges and cylinders', Phys.Rev.D81: 065031, (2010). And references therein.

8. Pieter Rein ten Wolde and Daan Frenkel,` Enhancement of Protein Crystal Nucleation by Critical Density Fluctuations' SCIENCE VOL. 277 Spt. 1797O9.; H. Fischer, Igor Polikarpov and Aldo F.Craievich, `Average protein density is a molecular-weight dependent function', Protein Sci. 2004 October; 13(10): 2825–2828 and several other papers.

9. **Samina Masood** and Mahnaz Haseeb,`Second Order Corrections to QED Coupling at Low Temperatures` *Int. J. Mod. Phys.*, A23 (4709-4719), 2008.

10. **Samina Masood**, Renormalization of QED in Superdense Media', Phys. Rev. D47. (648-652) 1993. And references therein.

11. K.Ahmed and **Samina Masood**, `Vacuum Polarization Effects at Finite Temperature and Density in QED', *Ann. of Phys. (*N.Y), 207 , (460-473), 1991.

12. **Samina Masood**, `Photon mass in the classical limit of finite temperature and density', Phys. Rev. D 44, 3943–3948 (1991).

13. **Samina Masood**, `Study of Casimir Effect in Nanosystems', BAPS.2010. MAR.S1.254

14. **Samina Masood**, ` Casimir Effect in Carbon Nanotubes' A talk presented at the First International Symposium on Nanotechnology, energy and space, Oct. 2009 Houston.

15. Phu Nuygen, Mike Cabrera, Channing Moeller and **Samina Masood,** `Casimir Effect and its Applications to Biophysics' ,BAPS.2009. TSF.D1. 8. (Fall APS meeting of Texas section, Oct. 2009)

16. Bertram J "The molecular biology of cancer". *Mol. Aspects Med.* **21** (6): 167–223(2000).

Electron Irradiation Induced Structural Changes in Nickel Nanorods Encapsulated in Carbon Nanotubes

Neha Kulshrestha, K. S. Hazra, S. Roy, R. Bajpai, D. R. Mohapatra and D. S. Misra

Department of Physics, Indian Institute of Technology Bombay,
Powai, Mumbai-400076, India, nirdesh@phy.iitb.ac.in

ABSTRACT

Nickel nanorods encapsulated inside the multiwalled carbon nanotubes were prepared using microwave plasma chemical vapor deposition (MPCVD) technique. The as prepared tubes were investigated by scanning electron microcopy (SEM), x-ray diffraction (XRD) and high resolution transmission electron microscopy (HRTEM). HRTEM study of these tubes exhibits long, good crystalline nickel filled CNTs with well-graphitized walls. Different orientations of filled metal with respect to graphitic walls have been observed and the orientations were found to get affected by the electron accelerating voltages of 200 KeV. We report that by deliberate e-irradiation at specific sites the crystalline planes of nickel can change their orientation. With continuous irradiation for 10-15 minutes the nickel starts evolving out of the tube. We propose that metal at the tip of the tube may be exposed by employing e-beam irradiation and it may be useful to study the true electrical properties of both the multiwalled carbon nanotubes and Ni nanorod inside the tube.

Keywords: Ni nanorods, carbon nanotubes, electroplating, electron irradiation.

1 INTRODUCTION

Carbon nanotubes are the most investigated nanostructures in the research area of nanoscience and nanotechnology. These peerless nanostructures are well known for their excellent structural, electrical and mechanical properties and find their proposed applications in almost all fields varying from basic science to advance chip technology [1,2,3] and space exploration [4], however very few realized applications such as in composites, probe tips and field emission displays exist at present [5,6]. On the other hand nanoscale magnetic materials also find significant interest in the wake of their revolutionary applications in magnetic memory devices. While the stability of nanoscale materials is a critical issue, incorporation of these two important nanomaterials is inevitably interesting for both the pedagogical and experimental viewpoint. The nanotube walls protect the magnetic material against oxidation and other damages and the process of filling and interaction of filled magnetic metal with the walls of carbon nanotube itself provide insight for the various phenomena in low dimensional physics.

CVD technique has been recognized as an easy method for the synthesis of ferromagnetic metal (such as Ni, Co or Fe) nanowires or nanorods inside nanotubes as the large amount of same metal which is used as a catalyst for CNT growth can be filled in situ through controlled experimental conditions [7,8].

Besides the important and useful applications of ferromagnetic metal filled carbon nanotubes, as suggested by the experimental studies on magnetic properties of these structures, such as nanomagnets [9] and high density magnetic recording media [10,11], engineering of these nanostructures by electron irradiation is another interesting aspect. Electron beam irradiation has been employed to get atomically sharp tips and soldering through the filled metal [12].The high pressure caused by e-beam irradiation in the core of closed shell carbon structures has been found to induce the phase transition and deformation of filling material, suggesting their use as nanoextruders or high pressure cylinders [13].

In this paper we report that the electrons in HRTEM at accelerating voltage of 200 KeV can be used to study and alter the structure of both the Ni nanorod and the nanotube encapsulating them. We propose that the nanorod can be exposed at any desired site by deliberate e-beam irradiation for their further important applications such as for contact formation for electrical measurement purpose.

2 EXPERIMENTAL

Commercially available copper foils of thickness about 0.2 mm are used as substrates for the growth of nanotubes filled with Ni nanorods. These copper substrates were thoroughly cleaned and coated with nickel layers of approximately 5μm thickness by electroplating technique. The details of nickel electroplating process can be found elsewhere [14]. However no boric acid has been used to balance the pH of the nickel sulphate and nickel chloride solution. For the synthesis of these Ni nanorods encapsulating tubes the ammonia plasma treatment has been carried out at a pressure of 10 torr for 15 seconds with ammonia flow rate of 180 sccm. The plasma treatment temperature has been kept 350 ± 10^0C. The treatment of the nickel layer by ammonia (NH_3) plasma converts the nickel layer into the nanoclusters of 50-60 nm size and facilitates the filling of nickel inside the MWNTs in the subsequent growth. Feedstock gas methane (CH_4) decomposes over Ni clusters in hydrogen plasma created at the pressure of 40 torr and 800^0C temperature. The flow rates of CH_4 and H_2 have been kept 8 sccm and 40 sccm respectively. In the growth time of 4-5 minutes a good black deposition becomes visible thorough the viewport in the MPCVD chamber and the methane flow has been stopped while H_2 flow still continuing up to 400^0C in order to ensure the less amorphous carbon deposition during growth.

The diameter and length of the nickel filled tubes and hence the nanorods inside them can be controlled by the ammonia plasma treatment time and temperature and their synthesis time in MPCVD. For the ammonia plasma treatment at 350 ^0C and synthesis time about 5 minutes the diameter and length of the tube were obtained to be 20-50 nm and 1 µm-3µm respectively.

3 CHARACTERIZATION

The XRD pattern of the as grown tubes is shown in figure 1, with an SEM image in inset. The high intensity peaks are of different polycrystalline (hkl) planes of Cu and Ni, as indexed by using joint committee on power diffraction Standards (JCPDS) file numbers and 03-1005 (Cu), 04-0850 (Ni) for copper and nickel respectively. One clear peak corresponding to (002) reflection plane of HCP graphite (file no.75-2078, C) is also visible. Other graphite peak corresponding to (010) plane is also present in the pattern but not clearly visible due to the high intensities of the peak corresponding to Cu (111) and Ni (111). The graphite peaks inform the growth of carbon nanotubes on Ni electroplated Cu foil.

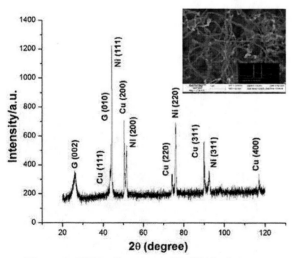

Figure 1: XRD of as grown Ni filled tubes.

The surface morphology of as grown tubes on Ni electroplated copper substrate is shown in figure 2. It has been found that large density of tubes grows in big clusters.

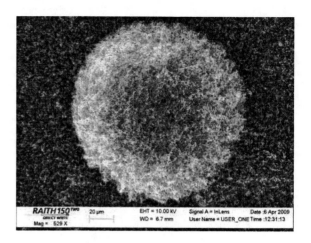

Figure 2: SEM image of Ni filled carbon nanotubes

During the optimization of growth parameters for best filling inside the tube it has been observed that the quality and density of the metal filled carbon nanotubes can be improved if the Cu foils are made rough using some sharp pin hammer before the Ni layer electroplating. And on these rough substrates the tubes deposit in a form of flower like clusters as is shown in the figure 2. We suggest that surface roughness facilitates the well separated Ni nanoclusters during the plasma treatment of Ni layer. The nanotubes grow densely over these nanoclusters rather than the other sites at the substrate.

However SEM images of the as grown Ni encapsulated tubes can show merely the morphology of tubes in bulk on the substrate and transmission electron microscopy (TEM) imaging is needed in order to see the filling inside the tube. For the TEM investigation the as grown tubes were detached from the substrate and ultrasonicated for 15 minutes in isopropyl alcohol. Few drops from the solution containing carbon soot which includes carbon nanotubes, catalytic nanoparticles and some amorphous carbon have been dispersed on conventional copper grids used for TEM analysis. The same technique has been used to prepare sample for HRTEM analysis. In the TEM image shown in figure 3 the Ni nanorod is clearly visible inside the multiwall carbon nanotube with outer diameter of about 50 nm and the inner diameter which is the diameter of the nanorod also is about 30 nm.

4 STRUCTURAL STUDY IN HRTEM

High resolution images of Ni filled tubes were obtained using JEOL- JEM 2100 field emission electron microscope. HRTEM study of these tubes exhibits long, good crystalline nickel filled CNTs with well-graphitized walls as shown in figure 4. It has been observed that the nickel filling inside might be single or poly crystalline as shown in figure 5 and 6 respectively. We have also observed that the Ni planes can have different orientations with respect to graphitic walls. In some extreme cases they may be parallel to the tube walls (figure 5) and in some they were perfectly perpendicular with respect to the graphitic walls (not shown here). It is new observation as in the earlier reported works different planes of Ni or Ni alloys were always found to be inclined at some angle with respect to nanotubes walls [15, 16]. However in reference 17 the fcc Ni (111) planes were reported parallel to tube axis which is contrary to reference 15 where the fcc Ni (111) planes were said to always have an angle of 39.6^0 with respect to the tube axis. For The particular tube shown in the figure 5, the lattice fringes of the Ni nanorod encapsulated inside 9 graphitic walls is visible. The spacing between observed Ni lattice planes is 0.20 nm. This d-spacing corresponds to the Ni (111) planes with fcc structure. These (111) Ni planes are parallel to the tube walls and hence the nanotube axis, maintaining the same orientation and spacing throughout the tube. This indicates the single crystalline nature of filling. While in figure 6, the randomly oriented different planes of filling indicate the polycrystalline nature of Ni.

We observed that electron beam irradiation affects the orientation of Ni planes. The graphitic walls of CNT, which are more sensitive to e-beam irradiation get distorted and sometime disappear. For 10-15 minute time of e-beam irradiation, a prominent change in the Ni planes becomes visible as shown in figure 7. With continuous irradiation on particular sites the metal of that site starts evolving out of the tube.

In another interesting observation we found that, owing to the combined effect of temperature and pressure conditions at the time of growth and e-beam irradiation during TEM imaging, nickel rod may come out from the tube by breaking the walls of the tube as shown in figure 9. This type of structures are useful for making the contacts for electrical measurement in both the side wall geometry and top contact geometry as the better electrical coupling between shells of nanotubes and contacting metal can be achieved.

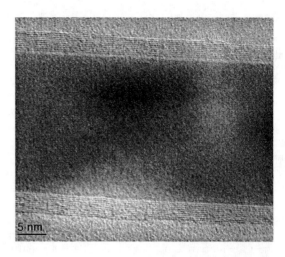

Figure 4: HRTEM image showing graphitic walls of the tube.

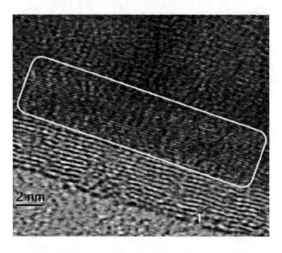

Figure 5: Single crystalline Ni planes parallel to the tube walls.

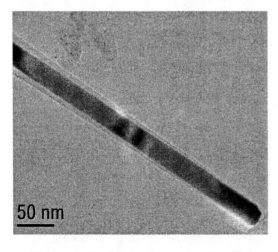

Figure 3: TEM image of Ni nanorod inside MWCNT.

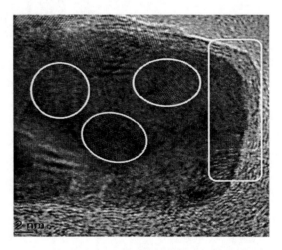

Figure 6: Polycrystalline nature of filled Ni in another tube.

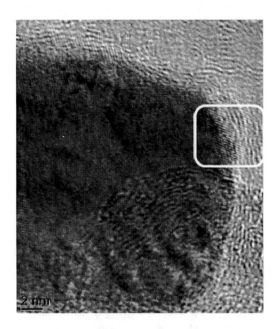

Figure 7: Evolution of Ni planes as a result
of electron irradiation.

Figure 8: Ni nanorod evolving out from the
tube.

5 CONCLUSION

Nickel nanorods encapsulated inside multiwalled carbon
nanotubes have been synthesized by MPCVD technique.
Good, uniform rod shaped filling has been obtained using
the process parameters described in the present work.
Different orientations of the filling metal inside the tube
have been observed and found to get affected by electron
beam irradiation. Electron irradiation studies done in situ in
HRTEM reveal the structural changes both in multiwall
structure of the carbon nanotube and the Ni nanorod inside
the tube.

Our results of filling evolution during e-beam irradiation
may provide insight for various phenomena related to the
interaction of filling metal and nanotube walls. The so
obtained CNTs with exposed metal at the tip of the tube
may prove ideal for contact formation to study the electrical
properties of these intriguing multiwalled structures.

ACKNOWLEDGEMENT

We would like to acknowledge Sophisticated Analytical
Instrument Facility (SAIF), IIT Bombay for HRTEM
imaging and Centre for Excellence in Nanoelectronics
(CEN), IIT Bombay for SEM imaging.

REFERENCES

[1] K. Banerjee and N. Srivastava, Proc. 43rdDAC, San
Francisco, CA, 809-814, 2006.
[2] F. Kreupl et al., Proc. IEEE International Electron
Device Meeting (IEDM), 683-686, 2004.
[3] W.I. Milne et al., interconnect Technology Conference,
IEEE, 105-107, 2008.
[4] Mark S. Avnet, space policy, 22, 133-139, 2006.
[5] P. Sarrazin et al., Copyright ©JCPDS-International
Centre for Diffraction Data, Advances in X-ray Analysis,
47, 232-239, 2004.
[6] N. S. Lee et al., Diamond and Related Materials, 10,
265-270, 2001.
[7] P. K. Tyagi et al., Thin Solid Films, 469, 127– 130,
2004.
[8] H. Li et al., J. Alloys and Compounds, 465, 51-55,
2008.
[9] J. W. Jang et al., Phys. Stat. Sol. (b) 241, No. 7, 1605–
1608, 2004.
[10] X. X. Zhang et al., J. Magn. Magn. Mater. 231, L9-
L12, 2001.
[11] D. C. Li et al., Chem. Phys. Letts., 316, 349–355,
2000.
[12] A. Misra and C. Daraio, Adv. Mater. , 21, 2305–2308,
2009.
 [13] L. Sun et al., Science, 312, 1199-1202, 2006.

[14] M. K. Singh et al, Chem. Phys. Letts., 354, 331–336,
2002.

[15] P. K. Tyagi et al. Appl. Phys. Letts., 86, 253110, 2005.

[16] H.Q. Wu et al, J. Mater. Chem., 12, 1919–1921, 2002.

[17] B. K. Pradhan et al., Chem. Commun., 1317–1318,
1999.

Tuning the electronic and chemical surface properties of graphane combining alkali deposition and low-energy ion irradiation

B. Heim[*], C.N. Taylor[*] and J.P. Allain[*],[**]

*School of Nuclear Engineering, Purdue University,
400 Central Drive, West Lafayette, IN 47907, allain@purdue.edu
**Birck Nanotechnology Center, Purdue University, West Lafayette, IN

ABSTRACT

The presence of defects in graphene due to thin-film processing can lead to alterations of its electronic properties. Tuning of graphene's electronic properties could be achieved through control of thermal deposition of alkaline dopants (e.g. lithium) combined with low-energy deuterium ion irradiation. In this case defect generation may be controlled enhancing tunability. Using X-Ray photoelectron spectroscopy (XPS) the functionality of lithiated polycrystalline graphite and lithited highly ordered polycrystalline graphite (HOPG) reveal distinct correlations between the fluency of deuterium atoms on the surface. Chemical functionalities of Li-O-D and Li-C-D are observed on the O1s and C1s XPS spectra as deuterium fluency increases. Bulk HOPG and graphite study of deuterium chemical functionality on lithiated graphite lay the foundation elucidating the synergistic effects of doping of lithium and deuterium irradiation have on graphene and hydrogenated graphene (graphane) property tunability.

Keywords: graphane, graphene, alkali metal, ion irradiation, surface chemistry,

1 INTRODUCTION

Single-layer graphene has unique electronic properties with promise for post-silicon nano-electronics [1]. The presence of defects originating from thin-film processing can lead to variation of graphene's electronic properties in addition to interaction with substrate and environmental conditions. Similar modifications are possible for carbon nanotubes. In particular, the chemical state of the surface plays a major role in the functionality of carbon nanotubes of various geometries (e.g. single-wall carbon nanotubes, few-wall carbon nanotubes, etc...) [2].

Chemical modification can lead to substantial changes in graphene's or CNT's electronic properties; such as reversible hydrogenation or alkali adsorption at the graphene or CNT surface [3]. The balance between metallic and covalent bonding in CNTs can also be dramatically influenced by the chemical state at the surface and interface of these materials. Tailoring of graphene's electronic properties could also, in principle, be achieved by controlled introduction of defects using low-energy ion irradiation [4]. This paper presents unique *in-situ* measurements that examine the combination of thermal deposition of alkali adsorbates (namely lithium) combined with low-energy deuterium irradiation and its effect on chemical and electronic properties of graphane (hydrogenated graphene) by operating at sub and near-damage threshold ion-beam energy regimes (i.e. $E <$ 100-500 eV/amu). Alkali absorbates have been found to uniquely tailor the electronic band structure of graphene [5]. To elucidate on the fundamental mechanisms that govern alkali-metal interactions with polycrystalline graphite-based systems, experiments depositing lithium coatings on graphite surfaces irradiated with deuterium are conducted. The hydrogen isotope is selected to minimize interference with chemical breakdown of water leading to contributions of hydrogen on the graphitic surface. Lithium is selected as the alkali metal due to its nearly free-electron like electronic properties.

The chemical functionality of lithium with oxgyen and carbon in polycrystalline graphite has been recently studied [6,7]. However the chemical functionality and intercalation of lithium in graphane and graphene is undocumented. Furthermore, *in-situ* surface characterization of lithiated graphitic surface irradiated by D ions on both polycrystalline and highly-oriented pyrolytic graphite is also undocumented. X-ray photoelectron spectroscopy (XPS) results show that the surface chemistry of a Li conditioned graphite sample after D ion bombardment is fundamentally different from that of non-Li conditioned graphite with hydrogenation followed by sp, sp2 and sp3 hybridizations [6]. Instead of simple LiD bonding as seen in pure liquid Li, graphite introduces additional complexities with XPS measuring bonding functionalities between Li-O-D (manifest at 533.0 ± 0.5 eV in the photoelectron spectra) and Li-C-D (manifest at 291.2 ± 0.5 eV) shown in Fig. 1. XPS spectra show Li-O-D and Li-C-D peaks become "saturated" with D at a fluence between 3.8 and 5.2 x 10^{17} cm^{-2}. Controlled experiments indicate D binds to O and C in the presence of Li by means of dipole interactions between Li atoms and the graphite matrix. XPS analysis and TDS (thermal desorption spectroscopy) show that deuterium has at least two bonding states in lithiated graphite: 1) weak bonding due to dipole interaction and 2) covalent bonds to C and O. This paper presents novel *in-situ* D-irradiation experiments on lithium-treated polycrystalline graphite and HOPG surfaces. Data includes in particular the carbon and oxygen states for lithiated graphite as a function of D irradiation dose.

2 EXPERIMENTAL SETUP

The following experiments were performed at OMICRON surface science facility at Birck Nanotechnology Center located at Purdue University. This facility consists of two vacuum chambers, one designated for characterization of samples, and the other for preparation of thermal and energetic particle interactions with material surfaces. The characterization chamber consists of a vacuum in the range of 4.0×10^{-11} Torr. Samples are characterized using X-ray photoelectron spectroscopy (XPS), using an Al K-alpha source (1486.6eV). The preparation chamber is held at 2.0×10^{-9} Torr that includes a Residual Gas Analyzer (RGA) with a quadrupole mass spectrometer (QMS) used for thermal desorption spectroscopy (TDS). Juxtaposed to the sample surface is a quartz crystal microbalance (QCM) that calibrates *in-situ* the amount of lithium flux from a nearby lithium evaporator. In addition a broad beam ion-beam source that delivers energetic deuterium beams with energies between 250-1000 eV/amu is also used. The beam currents average about 5-10 μA on a 1-cm^2 area on the sample.

Samples of HOPG and ATJ graphite used for the following experiments consisted of a 1-cm^2 area. Thermal lithium was evaporated on sample surfaces to a desirable thickness of 2-μm. The thickness was selected based on the known intercalation properties of lithium on graphite. This thickness ensured that an active lithium layer could last through the measurement of the XPS spectra, which ranged from a few minutes to about half an hour. The QCM measured the deposition rate of lithium in Å/s and corresponded to the expose time of the surface to equate to 2-μm of lithium on the surface. The sample surface was then transferred in vacuum for characterization using XPS. Irradiation with deuterium ions was conducted at an angle of 30-degrees to surface normal and a flux of 2×10^{13} ions cm^{-2}s^{-1}). Sequential irradiations were performed and repeated to record XPS spectra at corresponding fluencies of deuterium.

3 RESULTS AND DISCUSSION

3.1 Li intercalation on the surface

Lithium intercalation and chemical functionality in a carbon matrix has been well documented [8,6]. Past work has identified the peak locations and chemical functionalities of XPS spectra for O1s, C1s, and Li 1s binding energies. Literature reports the functionalities of the peak locations in the O1s spectrum to be 530.2 eV relating to Li$_2$O$_2$ [9] and the other peak located around 532

eV related to the presence of sub-oxides and hydroxides [10]. Documented XPS spectra yielded a substrate of normal polycrystalline substrate, and didn't explore the effects on Highly Ordered Polycrystalline Graphite (HOPG).

3.2 Lithium-deuterium interaction on polycrystalline graphite

Deuterium ion bombardment on lithium-treated graphite has been shown using XPS to be fundamentally different than that of non-Li graphite due to hydrogenation followed by sp, sp2 and sp3 hybridizations [11]. The interaction of lithium and the graphitic matrix introduces a chemical state whereby when exposed to energetic deuterium, lithium binds they hydrogen isotope via indirect interactions with carbon and oxygen atoms. Moreover, direct interaction of Li and D atoms do occur, however at much higher heats of formation, near 5-6 eV.

Results of deuterium irradiation on 2-μm lithium treated polycrystalline graphite surface are shown in Figs 1 and 2. The data shows the results of multiple XPS scan taken in-situ as a function of D fluence. The resulting data from O1s spectra from Fig 1 shows two dominant peaks located around 533eV and 530eV, the chemical functionality of (Li-O-D) correlates to the 533eV peak position, where the Li$_2$O$_2$ remains the chemical functionality of the peak at 530eV. Correlation to the effect of D interactions can be seen as the fluence of D2 ions to the surface is increased. The peak associated with Li-O-D gains relative intensity over the Li$_2$O$_2$ as the fluence of D2 ions is increased. This direct relationship between the D2 ions and the relative peak intensity can be further justified when looking into the C1s spectrum in Fig 2.

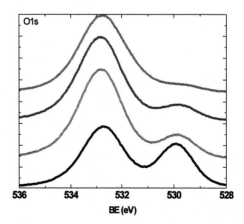

Figure 1: O1s Spectrum for 2-μm Lithium Evaporated Polycrystalline Graphite with corresponding D2 fluence. Black = 2×10^{15} D$^+$/cm^2, Red = 1.8×10^{16} D$^+$/cm^2, Blue = 4.0×10^{16} D$^+$/cm^2, Green = 1×10^{17} D$^+$/cm^2

In the polycrystalline graphite matrix the original C1s spectrum yields a single line at 284.5 eV associated directly with the graphitic peak of carbon. After 2-μm lithium evaporation the graphite peak is still apparent and no new peaks are developed. However after irradiation of the sample with D2 ions a peak at 291eV begins to develop. This peak is not associated with lithium carbonate (292 eV) due to lack of sample exposure to oxygen or atmosphere. [12]. The 291eV relative intensity begins to increase with D fluence and is eventually saturated and equal to the relative intensity of the graphite peak around 1×10^{17} D$^+$/cm^2.

Fig 3 and 4 show the same relative peak intensity as a function of the deuterium fluency on the surface. The C1s in Fig 4 again illustrates peak saturation at 1^{17} D$^+$/cm^2 D2 fluency. This result illlustrates that the surface structure does not play a major role on the chemical functionality of Li-C-D and Li-O-D interactions. This results is consistent with those found by Dahn et al.where thin lithium layers on polycrystalline graphite showed similar effects on D binding compared to crystalline graphite.

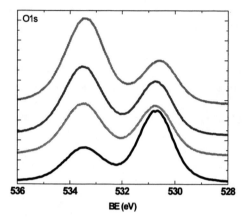

Figure 3: O1s spectra for 2-μm lithium evaporated HOPG with corresponding D2 fluency. Black= 2.2×10^{15} D$^+$/cm^2, Red = 2×10^{16} D$^+$/cm^2, Blue = 3.6×10^{16} D$^+$/cm^2, Green = 1.1×10^{17} D$^+$/cm^2

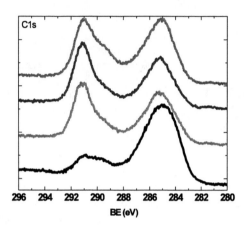

Figure 2: C1s Spectrum for 2-μm Lithium Evaporated polycrystalline graphite with corresponding D2 fluency. Black = 2×10^{15} D$^+$/cm^2, Red = 1.8×10^{16} D$^+$/cm^2, Blue = 4.0×10^{16} D$^+$/cm^2, Green = 1×10^{17} D$^+$/cm^2.

Such peak parameters as a function of the D2 fluency lead to the chemical functionality of the peak at 291ev and involve Li-C-D interactions. Since the chemical state can depend on surface organization, HOPG experiments were determined critical to these studies. In addition, work on HOPG would be closer to the behavior of graphane.

3.3 Lithium-deuterium interaction on HOPG

A 0.5mm×10mm×10mm sample underwent irradiation conditions identical to the ATJ graphite samples and was exposed to the same fluence of D2 singly-charged ions. Fig 3 and 4 show XPS data from the O1s and C1s regions. Compared to the ATJ graphite peaks spectra there is a slight shift in each of the spectra, 0.5eV, this is due to surface charging on the HOPG samples and the inability to properly ground the sample. These experiments will be repeated in a HR-XPS, monochromic, and charge neutralizing surface science facility in the future.

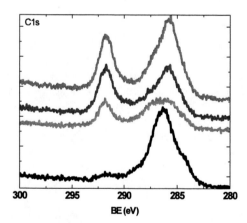

Figure 4: C1s Spectrum for 2-μm lithium evaporated HOPG with corresponding D2 fluency. Black= 2.2×10^{15} D$^+$/cm^2, Red = 2×10^{16} D$^+$/cm^2, Blue = 3.6×10^{16} D$^+$/cm^2, Green = 1.1×10^{17} D$^+$/cm^2

4 CONCLUSION

The chemical functionality of deuterium to lithium and oxygen yield distinct binding states and associated peak structure from XPS. Both HOPG and polycrystalline graphite samples reveal relative peak intensity from XPS O1s and C1s correlation directly with the fluency of D2 ions. XPS spectrum also relates the chemical saturation of the relative peak intensity's around $1x10^{17}$ D^+/cm^2. Carbon's role in the chemical functionality of lithium and deuterium has been observed as well with the depends on D2 irradiation.

5 FUTURE WORK

Modification of not only bulk highly-ordered and polycrystalline graphite, but the functionality of single layer of graphite (graphene) with lithium yields promise for post-silicon nano-electronics. Understanding the change on the electronic and chemical properties of HOPG and graphene need the use of angle-resolved photoelectron spectroscopy (ARPES) coupled to *in-situ* ultraviolet photoelectron spectroscopy (UPS) and XPS. ARPES can map the electronic band structure of graphane doped with alkali metal (e.g. lithium or potassium) and thus the effect low energy heavy ion irradiations have on graphene electronic properties. Further analysis will identify the irradiation conditions that minimize defect production while tuning the electronic properties of graphene and graphene combined with alkali-metal doping.

REFERENCES

[1] K. S. Novoselov et al., Science, 306 (2004) 666.

[2] Y. Hou et al. ACS Nano, 3 (5) (2009) 1057-1062

[3] D.C. Elias et al. Science Vol. 323, 2009, 610;

[4] A. Gruneis and D.V. Vyalikh Phys. Rev. B 77, 193401, 2008

[5] L. Tapasztó et al Phys. Rev. B 78 (2008) 233407.

[6] S.S. Harilal et al. / Applied Surface Science 255 (2009) 8539–8543

[7] J.P. Allain et al. / Journal of Nuclear Materials 390–391 (2009) 942–946

[8] J.R Dahn et al./Science 270 (1995) 590-593

[9] M.V Landau. A. Gutman, M. Herskowitz, R. Shuker, Y. Bitton, D. Mogilyansky, J. Mol Catal. A: Chem 176 (2001) 127

[10] M.Towe, P. Reinke, P Oelhafen, J. Nucl Mater. 290 (2001) 153

[11] P.S. Krstic, et al., New J. of Phys. 9, (2007) 219.

[12] D. Ensling et al./Applied Surface Science 255 (2008) 2517-2523

Fabrication and characterization of optoelectronic device using CdSe nanocrystal quantum dots/single-walled carbon nanotubes heterostructure

Hyung Cheoul Shim[*,**], Sohee Jeong[**], Soohyun Kim[*†] and Chang-Soo Han[**††]

[*] Korea Advanced Institute of Science and Technology (KAIST), 373-1, Guseong-dong, Yuseong-gu, Daejeon 305-701, Korea, scafos@kaist.ac.kr (H.C. Shim), soohyun@kaist.ac.kr (S. Kim)
[**] Korea Institute of Machinery and Materials (KIMM), 171 Jang-dong, Yuseong-gu, Daejeon, 305-343, Korea, sjoeng@kimm.re.kr (S. Jeong), cshan@kimm.re.kr (C. Han)

ABSTRACT

The ability to transport extracted carriers from the development of NQDs (nanocrystal quantum dots) based optoelectronic sensors have drawn considerable attention due to NQDs' unique optical properties. Coupling of NQDs to 1-D nanostructures such as SWNTs (single-walled carbon nanotubes) is expected to produce a composite material which facilitated selective wave length absorption, charge transfer to 1-D nanostructures, and efficient electron transport. In this paper, we fabricated the optoelectric device based on cadmium selenide (CdSe) nanocrystal quantum dots (NQDs)/single-walled carbon nanotubes (SWNTs) heterostructure using dieletrophoretic force. The efficient charge transfer phenomena from CdSe to SWNT make CdSe-Pyridine (*py*)-SWNT unique heterostructures for novel optoelectric device. We monitored the electrical signatures from NQDs decorated SWNTs upon excitation and were able to elucidate the carrier transfer mechanism. The conductivity of CdSe-*py*-SWNT was increased when it was exposed at ultra violet (UV) lamp, and showed a function of wavelength of incident light.

Keywords: Nanocrystal quantum dot (NQD), single-walled carbon nanotube (SWNT), charge transfer

1 INRODUCTION

NQDs (nanocrystal quantum dots) have been attracted considerable attention due to their semiconducting properties [1]. In addition, the band-gap energy of NQDs can be tuned using varying their size. These unique properties of NQDs make them have wide applications including bio-medical applications, electronic and optical devices [2,3]. Especially, in optical component based on NQDs, it is very important that how to detect electronically the carriers from NQDs when they were photo-excited. In this work, the novel typed optical sensor was fabricated using hybrid nano-materials which were comprised of NQDs and SWNTs (single-walled carbon nanotubes) that transport the carriers from the NQDs. The SWNTs can be the best candidate for the transport materials because of their nano-scaled size and ballistic transport properties [4].

2 RESULTS AND DISCUSSIONS

Fig. 1 shows the schematic diagram of device fabrication procedure. First, the CdSe (cadmium selenide) as the NQDs having average size of 2.8 nm were synthesized by conventional method [5], and covered with pyridine ligand instead of surfactant for non-covalent coupling with single-walled carbon nanotubes. As shown in TEM (transmission electron microscope) image of **Fig. 2(a)**, we can check the CdSe were coupled well with SWNTs using pyridine linking. Next, the CdSe-*py*(pyridine)-SWNTs nano hybrid materials were aligned between Cr/Au microelectrodes using AC (alternative current) dielectrophoretic force as shown in Fig. 1, and we also confirm the aligned CdSe-*py*-SWNTs between microelectrodes through the FESEM (field emission scanning electron microscope) image of **Fig. 2(b)**.

The 100 mV of bias voltage was applied to the optical sensor, and the UV (ultra-violet) lamp having wavelength of 365 nm and power of ~5 mW was used for the light source. The change of resistance value was monitored by Ohm meter (Fluke 189) in a real time (See **Fig. 3**). The conductance of optical device was increased when they exposed at UV lamp was on as shown in *Fig. 4*. However, the conductance was decreased and reached at their original resistance value when the UV lamp was off. The position of conduction band on CdSe (~3.5 eV) is higher than that of SWNTs (~4.8 eV), the electrons can be transferred thermokinetically from the CdSe to SWNTs when they were photo-excited. For this reason, the conductance of optical device can be changed based on the transferred carriers from NQDs to SWNTs. Moreover, the more carriers can be generated using excitation frequency with higher energy. Therefore, the change in resistance values with an excitation frequency of 490 nm is much higher than when the optical sensor was excited with wavelength of 550 nm. In this typed of optical device that composed of CdSe and SWNTs, the detection of selective wavelength is possible due to tunable band gap energy of CdSe. Furthermore the speed of device can be enhanced by using 1-D (one-dimensional) structure of SWNTs that transport the carriers with theoretically no loss.

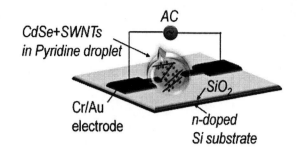

Fig. 1 Schematics of the dielectrophoretic assembly of CdSe-*py*-SWNT optoelectronic devices

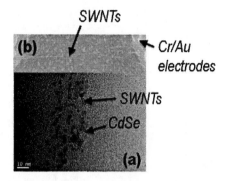

Fig. 2 (a) TEM image of CdSe-*py*-SWNTs, (b) FESEM image of CdSe-*py*-SWNTs between microelectrodes

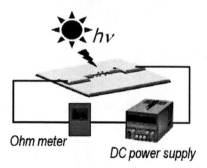

Fig. 3 Schematics of the measurement on CdSe-*py*-SWNT optoelectronic devices with ohm meter

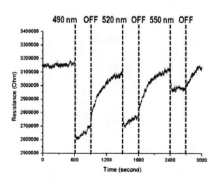

Fig. 4 The changes in resistance *vs* time of CdSe-*py*-SWNT optoelectronic device *via* different wave length illumination.

ACKNOWLEDGEMENT

This research was supported by the 21C Frontier Research Program from the Center for Nanoscale Mechatronics and Manufacturing, and the second stage of the Brain Korea 21 Project. S. Jeong acknowledges a financial support by Nano R&D program through the Korea Science and Engineering Foundation funded by the Ministry of Education, Science and Technology (Grant 2009-0083219). This research was also supported partially the Ministry of Knowledge Economy (MKE) and Korea Institute for Advancement in Technology (KIAT) through the Workforce Development Program in Strategic Technology.

REFERENCES

[1] Alivisatos, A. P., "Semiconductor Clusters, Nanocrystals, and Quantum Dots," *Science*, Vol. 271, pp. 933~937, 1996.

[2] Michalet, X., Pinaud, F. F., Bentolila, L. A., Tsay, J. M., Doose, S., Li, J. J., Sundaresan, G., Wu, A. M., Gambhir, S. S., Weiss, S., "Quantum Dots for Live Cells, in Vivo Imaging, and Diagnostics," *Science*, Vol. 307, pp. 538~544, 2005.

[3] Bhattacharya, P., Ghosh, S., Stiff-Roberts, A. D., "Quantum dot optoelectronic devices," *Annu. Rev. Mater. Res.*, Vol. 34, pp. 1~40, 2004.

[4] Avouris, P., Appenzeller, J., Martel, R., Wind, S. J., "Carbon nanotube electronics," *Proc. IEEE*, Vol. 91, pp. 1772, 2003.

[5] Olek, M., Bulsgen, T., Hilgendorff, M., Giersig, M., "Quantum Dot Modified Multiwall Carbon Nanotubes," *J. Phys. Chem. B.*, Vol. 110, pp. 12901, 2006.

EFFECT OF SOURCE, SURFACTANT, AND DEPOSITION PROCESS ON ELECTRONIC PROPERTIES OF NANOTUBE ARRAYS

Dheeraj Jain, Nima Rouhi, Christopher Rutherglen, Peter J. Burke
Integrated Nanosystems Research Facility
Department of Electrical Engineering and Computer Science
University of California, Irvine, CA 92697, USA

ABSTRACT

Carbon nanotubes are currently being sought as a potential candidate in the device research area. However, a scalable manufacturing technology for nanotube devices with appropriate yield, reproducibility, and performance in various metrics (mobility, on/off ratio, cost, chemical sensitivity, sensor specificity, etc.) is currently lacking. This is, in part, due to the variety of synthesis and deposition techniques and challenges due to the inherent differences in nanotube physical properties. All known synthesis methods for single walled carbon nanotubes result in a mixture of chiralities and diameters, resulting in heterogeneous electrical properties, particularly a mixture of semiconducting and metallic nanotubes. The chemistry community has lead a successful effort aimed at sorting and purifying nanotubes in solution post-synthesis[1]. In solution, as produced single walled carbon nanotubes exist as bundles. In order to de-bundle and sort the nanotubes, various surfactants are typically used. At present, it is not known how these surfactants affect the electronic properties of nanotube arrays and films made from depositing nanotubes using various techniques. While a complete understanding of the physical processes involved in these complex manufacturing steps is still lacking, in this work we have made an attempt to make a comprehensive study and report a raw set of empirical guidelines to guide future technology development in this area.

Keywords: Carbon nanotubes, surfactants.

We obtained carbon nanotubes from several different sources differing in manufacturing process used with a variety of average properties such as length, diameter, and chirality. We then used several common surfactants to disperse each of these nanotubes and then deposited them on Si wafers from their aqueous solutions using dielectrophoresis[2]. While most studies of purified nanotubes use spectroscopic techniques to determine nanotube purity and composition, these studies are useless for the electronics industry unless they can be directly correlated with transport measurements. Thus, in this study, transport measurements were performed to compare and determine the effect of different surfactants, deposition processes, and synthesis processes on nanotubes nanotubes synthesized using CVD, CoMoCAT, laser ablation, and HiPCO.

From our experiments, we found when a gap width close to the average nanotube length was used, it was possible to deposit enough well aligned nanotubes from all sources to have dc resistance values lower than 100 W for a width of 100 mm. For the effective dispersion of nanotubes, sodium cholate (SC) was most suitable among others despite of source (or synthetic process) of nanotubes. An average 2 hours of sonication in 1% surfactant solution

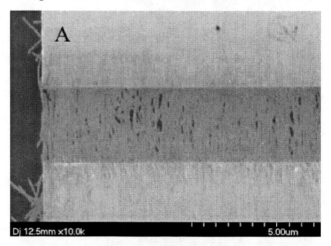

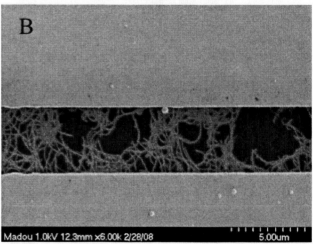

Figure 1. Nanotubes deposited on Si wafers using different frequencies for electric field direct alignment (dielectrophoresis). A) 25 MHz, yielding well-aligned arrays and B) 2 MHz, yielding poorly aligned arrays. Statistical properties of devices yielded reasonably low scatter in the electronic properties from device to device.

(w/v in DI water) in an ultrasonic bath @ 40 watts, followed by repeated centrifugation for up to 6 hours yielded fairly clean nanotubes solutions to be used for device fabrication. DEP parameters such as frequency and field strength play an important role for effective deposition of nanotubes on wafers. Using a dilute solution and giving more time for the deposition or using a comparatively concentrated solution with less time yielded devices with similar properties. Also using any amplitude above 3 V did not make much difference on device characteristics. In our observations, we also found the frequency used for DEP to be the most important factor having a direct effect on device properties. As shown in Figure 1, using a frequency higher than 10 MHz resulted in far better alignment of nanotubes deposited on a wafer. In fact using 25 MHz resulted devices with well-aligned nanotubes between the electrodes despite the source of the nanotubes used. Also Figure 2 shows the dependence of device resistance on electrode gap length. These studies lay important groundwork for useful application of nanotube electronics using purified, mono-disperse nanotubes in solution as the starting material for a diverse array of heterogeneously integrated systems.

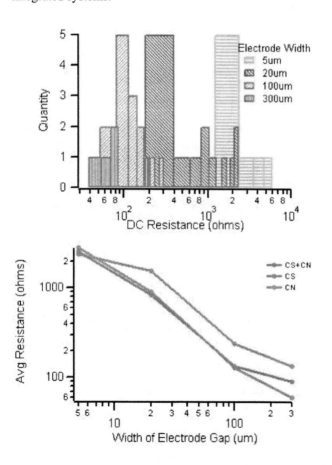

Figure 2. The results of deposition from two different manufacturers (Carbon Solutions CS, and Carbon Nanotechnologies, CN), gave similar results for the resistance at different electrode widths. D)

TABLE 1
SUMMARY OF RESULTS

SWNT Sources	Electrode spacing	Avg. length after deposition (in SEM)	DC resistance (1000's of SWNTs in parallel)	On/off ratio
Cheap Tubes	3 µm	> 3 µm	< 50 Ω	2-8%
SWeNT	1 µm	< 1 µm	50-100 Ω	2-6%
CSI	1 µm	<1.5 µm	100-200 Ω	<1%
CNI/Unidym	1 µm	<1 µm	100-500 Ω	<1%
Las Alamos	1 µm	<1.5 µm	120-150 Ω	<1%
Nanointegris (semi-enriched, 90%)	1 µm	<1.5 µm	150-180 Ω	<1%

ACKNOWLEDGMENTS

This work was supported by the NSF, ONR, ARO, the WCU program, and a gift from Northrop Grumman.

REFERENCES

[1.] M. Hersam, "Progress Towards Monodisperse Single-Walled Carbon Nanotubes", *Nature Nanotechnology*, **3**, 387, (2008).

[2.] C. Rutherglen, D. Jain and P. Burke, "Rf Resistance and Inductance of Massively Parallel Single Walled Carbon Nanotubes: Direct, Broadband Measurements and near Perfect 50 Omega Impedance Matching", *Applied Physics Letters*, **93**, 083119-3, (2008).

Novel Electrostatic Digital Printing Device Based on Single Walled Carbon Nanotube Film

M. Kanungo and K. Y. Law*

Xerox Research Center Webster, Xerox Innovation Group
Webster, NY, USA, klaw@xeroxlabs.com

ABSTRACT

Carbon nanotube thin films because of their unique optical and electronic properties coupled with the flexibility and easy patternability are promising candidates for application in low cost, large area flexible displays and optoelectronics. In this work, a novel printing device comprising a single-walled carbon nanotube (SWCNT) thin film and a molecularly doped polymer layer of arylamine in polycarbonate is reported. Results show that hole injection from the SWCNT thin film to the arylamine layer occurs under the influence of an electric field. The efficiency of the hole injection process was found to be dependent of the surface conductivity of the SWCNT film and the strength of the electric field across the bilayer device. The use of this hole injection process to generate electrostatic latent images for novel digital printing application is discussed.

Keywords: carbon nanotube films, printing device, electrostatic latent image, electron transfer, hole injection

1 INTRODUCTION

The basic steps in the xerographic (electrophotographic) process involve charge and imagewise discharge of the photoconductor to create the electrostatic latent image, followed by development of the latent image with toner electrostatically and then image fixation with heat and pressure [1]. Xerography has become a multi-billion dollar industry and the technology is applying to almost all color laser printers ranging from low speed printers in home office to high speed color presses in print shop. It is interesting to note that, other than the image file input through the laser ROS, which is digital, the entire printing process is still in an analog mode. In this work, we report a study of a new electron transfer reaction from a molecularly doped layer of an aryl amine in polycarbonate to a thin film of single walled carbon nanotube. The net result of this electron transfer reaction is a hole injection from the CNT film to the arylamine charge transport layer. This hole injection process triggers the discharge of the bilayer device process. In addition, this hole injection process is shown to be sensitive to the applied electric field as well as the conductivity of the CNT film. The mechanism for the charge-discharge process of the bilayer is proposed. The use of the bilayer device to digitize the xerographic printing process is discussed.

2 EXPERIMENTAL

Materials. Single walled carbon nanotube films of surface resistivity ranging from ~ 100 ohms/sq to ~ 5000 ohms/sq) on Mylar substrate were prepared using the procedure as described by Weeks et. al. [2]. Hole transporting molecule, TPD [3] and the polycarbonate polymer binder PCZ200 were obtained from internal source (structures in Fig. 1). All coating solvents (methylene chloride, tetrahyfrofuran and toluene) were analyzed reagent grade from Fischer and were used as received.

Devices. The bilayer devices studied in this work was fabricated by simply coating a solution containing TPD and PCZ200 in a mixed solvent of tetrahydrofuran and toluene (70:30 in ratio) over the carbon nanotube film (on Mylar) on a lab draw-down coater using a 5 mil draw bar. A typical coating solution consisted of ~ 14% of solid. The concentration of TPD in the charge transport layer (CTL) was at 40% by wt. The thickness of the CTL was typically ~ 20 μm and was controlled by the solid concentration of the coating solution as well as the wet gap of the draw bar. The resulting bilayer device was air dried for 0.5 hour followed by vacuum drying at 100°C for 2 hours before electrical evaluation.

Measurements and Techniques. The surface resistivity of the carbon nanotube films were measured by a four probe point method using a Keithley 237 high voltage source measure unit. The charge-discharge characteristics of the bilayer device were performed on an in-house static scanner. Gold dot was evaporated on the CTL for the electrical contact. A schematic description of the apparatus is shown in Figure 1. Typically the bilayer devices were charged by the HV corona device and the surface potential were monitored using an electrostatic voltmeter (ESV). Since the bilayer device was "static" throughout the measurement, the charging and monitoring of the surface potential was controlled electronically through the electric circuit within the static scanner, typically there was a ~ 0.1 s delay between charging and monitoring.

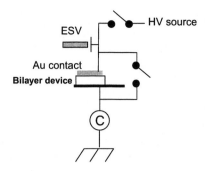

Fig. 1 A schematic of the "static" scanner.

3 RESULTS AND DISCUSSION

3.1 Device Configuration and Electrical Characterization

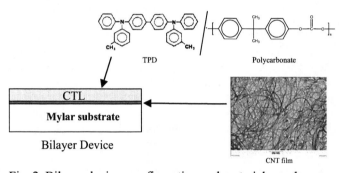

Fig. 2 Bilayer device: configuration and materials used.

Figure 2 shows the configuration of the bilayer device and the materials used in this work. The device comprises a molecularly doped charge transport layer (CTL) made of hole transport molecule TPD in polycarbonate [3] over a thin film of pure single walled carbon nanotube on a Mylar substrate. The charge and discharge characteristic of the bilayer device was studied on an in-house static scanner (Fig. 1). Figure 3a shows the surface potential curves obtained from a typical CNT bilayer device. Figure 3b shows the surface potential curves from a controlled bilayer device where the CNT film is replaced by a Ti/Zr metal layer. By comparing with the control, the result indicates that the CNT bilayer device is charge capacitively. Unlike the control, the CNT bilayer device undergoes rapid discharge as soon as the electric field across the bilayer device is established.

Based on the results in Figure 3, the discharge of the CNT bilayer device can be schematically summarized in Figure 4. As soon as the bilayer is charged up and an electrical field is established, the field facilitates an electron-transfer reaction from TPD to the CNT film (or hole injection in the opposite direction). This is followed by a series of isoenergetic electron-transfer across the CTL, leading to nearly total discharge of the bilayer device [3].

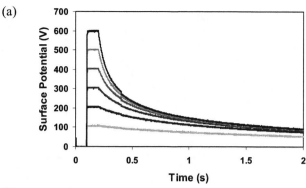

(a)

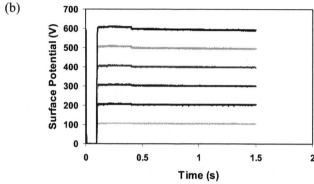

(b)

Fig. 3 (a) Typical charge – discharge curves for a CNT bilayer device, and (b) charge – discharge curves for a controlled bilayer device.

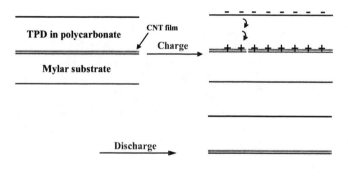

Fig. 4 Schematic illustration of the charge – discharge process.

3.2 Effect of Surface Resistivity of Carbon Nanotube Films on Discharge Rate

The rate of the dark discharge was found to be sensitive to the conductivity of the CNT film for a common CTL (40% TDP in polycarbonate, ~ 18 μm thick). Figures 5a and 5b depict the discharge curves for two bilayer devices with CNT films of different surface resistivity. The result shows that the higher the conductivity of the CNT film, the faster the discharge rate and the more sensitive the device is.

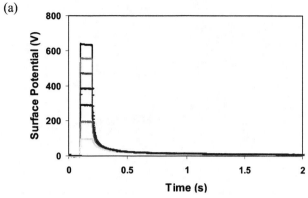

(a)

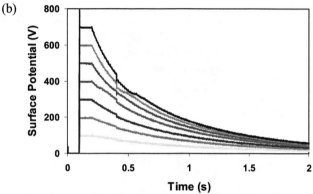

(b)

Fig. 5 Charge – discharge curves for bilayer devices with CNT films of different surface resistivity (a) 250 Ohm/Sq and (b) 5000 Ohm/Sq.

3.3 Effect of Electric Field on Discharge Rate

From the discharge curves in Figures 5a and 5b as well as similar plots obtained from bilayer devices of different CNT surface resistivity, one can analyze the initial discharge rates at different surface potential for these devices. Plots of the initial discharge rate (dV/dt) of these bilayer devices at different surface potentials are given in Figure 6.

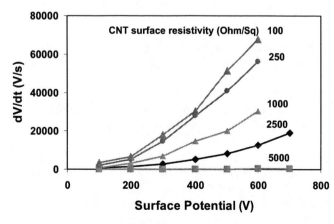

Fig. 6 Plot of initial discharge rate as a function of surface potential for bilayer devices with varying CNT surface resistivity.

The results indicate that (1) the discharge rate is highly sensitive to the surface resistivity of the CNT film, the higher the conductivity, the faster the discharge rate. (2) The discharge rate increases as the surface potential increases. Since the thickness of all these devices are held to be constant, ~ 18 μm, the result indicates that there is a strong electric field effect on the discharge rate, the higher the electric field across the device, the faster the discharge rate. Another important feature in Figure 6 is the apparent existence of a threshold electric field for the discharge rate. Specifically, dV/dt becomes negligible when the surface potential is < 200 V or when the electric field across the device is $< \sim 10$ V/μm.

The conclusions on the field and the threshold effects on dV/dt are substantiated by a second set of experiments where we vary the electrical field systematically by varying the CTL thickness.

3.4 Proposed Discharge Mechanism

The results in Figures 5 and 6 clearly suggest that the discharge rate of the CNT bilayer device is very sensitive to the electric field. There also appears to be a threshold field to trigger the discharge process. We hypothesize that, under the influence of the electrical field, "bounded" electron-hole (e-h) pairs are generated in the bilayer device. At fields weaker than the threshold field, the e-h pairs recombine and no discharge occurs. On the other hand, when the field across the bilayer device is higher than the threshold field, the e-h pairs dissociate. This is followed by an electron transfer from the arylamine CTL to the CNT film. In other words, hole injection from the CNT film to the CTL results. The injected holes then migrate across the CTL, discharging the bilayer device. This hypothesis is supported by the observation that the discharge process is facilitated when the conductivity of the CNT film is high. Presumably, the bounded e-h pair can also dissociate under weak field when the CNT film is conductive. The mechanistic picture of the e-h pair generation and dissociation is summarized in Figure 7.

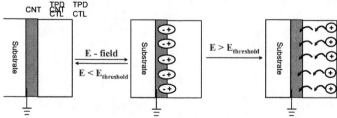

Fig. 7 Proposed discharge mechanism for the CNT bilayer device.

4 CONCLUDING REMARKS

In this work, we report a new electric field induced electron transfer reaction between an arylamine hole transport molecule, TPD, and single walled carbon nanotube thin film in bilayer device. Evidence is provided that bounded e-h pairs are formed under the influence of an electric field. At field strength lower than the threshold field, the e-h pairs recombine harmlessly. At higher electric fields, the e-h pairs dissociate, and this is followed by a hole injection reaction and a series of isoenergetic electron transfer across the CTL, leading to total discharge of the bilayer device. In separate experiments, we demonstrate that we can successfully develop the latent electrostatic images from the bilayer device with toner electrostatically. Since CNT films are known to be patternable, we suggest that, we should be able to digitize the xerographic process when we couple the discharge process of the present bilayer device with an addressable TFT backplane.

5 ACKNOWLEDGEMENTS

The authors thank Dr. Markus Silvestri for the use of the static scanner and Dr. Yuanjia Zhang for the support of the printing experiments.

REFERENCES

[1] For reviews about materials and the xerographic process, please see, D. Weiss and M. A. Abkowitz, Chem. Rev., 110, 479, 2010; also K. Y. Law, Chem. Rev., 93, 449, 1993.

[2] C. Weeks, P. Glatkowski and D. Britz, MicroNano, 12, 1, 2007.

[3] D. M. Pai, J. F. Yanus and M. Stolka, J. Phys. Chem., 88, 4714, 1984.

A Dual-Frequency Wearable MWCNT Ink-Based Spiral Microstrip Antenna

Taha A. Elwi*, Daniel G. Rucker*, Hussain M. Al-Rizzo**, Haider R. Khaleel*, Enkeleda Dervishi***, and Alexandru S. Biris***

*Department of Applied Science, University of Arkansas at Little Rock, Little Rock, AR 72204, USA, Phone: (501)682-6990, Fax: (501)569-8698
**Department of Systems Engineering, University of Arkansas at Little Rock, Little Rock, AR 72204, USA, Phone: (501)371-7615, Fax: (501)569-8698
***Nanotechnology Center, Department of Applied Science, University of Arkansas at Little Rock, Little Rock, AR 72204, USA, Phone: (501)551-9067, Fax: (501)569-8020

ABSTRACT

In this paper, we present a dual-frequency, square spiral microstrip patch antenna constructed from Multi-Walled Carbon Nanotubes (MWCNT) ink deposited on a thin flexible FR4 substrate using a cold printing process. The proposed antenna operates at 1.2276 GHz for wearable Global Positioning System applications and 2.47 GHz for wearable Multichannel Multipoint Distribution Services. The MWCNT ink, possessing an electrical conductivity of 2.2×10^4 S/m and a relative permittivity of 5-j1, is used as the conductor patch. CST Microwave Studio (CST MWS) was utilized to evaluate the return loss, matching impedance, resonant frequency, bandwidth, and far-field radiation patters. The antenna gain is found to be -12.5 dBi and -4.25 at 1.2276 GHz and 2.47 GHz, respectively. The corresponding gain values when copper is used for the patch are -12.05 dBi and -4.01dBi. Simulations and measurements based on MWCNT ink and copper are presented and compared. An excellent agreement was found between the measured and simulated results.

Keywords: Conductive ink, CST MWS, Square spiral antenna

1 INTRODUCTION

Rapid developments wireless communications coupled with recent progress in material exploration resulted in a growing interest to apply nanostructure materials for microwave devices. The need for developing new materials that provide wireless devices with novel properties, in particular wearable microstrip antennas, is subtle for applications where regular conductors may fail such as mining, fire, chemical or for biological degrading environments. Nanoscale materials are the most promising candidates since they have proven to provide new properties that were not possible for traditional bulk material. Electrically conductive ink based on nanoscale material is one such example. We have developed MWCNTs suspended in an adhesive solvent to create printable ink that dries quickly, adheres homogenously, and possess the characteristics of a good conductor at microwave frequencies. For the last several decades, conductive inks have stimulated the consideration and interest of many researchers due to their wide applications in microelectronics[1]. These materials have numerous advantages over traditional conductive bulk material such as lower mass density and high resistivity against corrosion. Moreover, the electromagnetic constitutive parameters can be controlled by chemical manipulation using different functional groups, density of doping, or mixing with different hosting[2, 3]. The MWCNT ink is painted using an ink jet cold printing process, which provides a low cost manufacturing process to manufacture passive microwave printed circuit boards such as transmission lines and microstrip antennas and is characterized by high resistance against corrosion. In this paper, we present a microstrip square spiral antenna constructed from MWCNT ink for wearable applications. The antenna design is conducted using CST MWS[4]. Measurements and simulation results are presented for both MWCNT and copper patches. The DC electrical conductivity is measured by the four-probe technique, while the permittivity measurement is conducted over the frequency range from 300 kHz to 8.5 GHz using an Agilent 85070B dielectric probe.

2 MATERIAL CHARACTERIZATION

In this section, we present the measured constitutive parameters of the MWCNT ink in terms of electrical conductivity and relative permittivity. The electrical conductivity measurement has been performed using the four-probe technique and it is found in the range of 2.2×10^4 S/m. The measurement of the real and imaginary parts of the complex relative permittivity is conducted using the Agilent 85070B dielectric probe kit and the 300 kHz - 8.5 GHz ENA series Network Analyzer. The measured real and imaginary parts of the relative permittivity are 5±0.1 and 1±0.05, respectively.

3 ANTENNA DESIGN AND CHARACTERIZATION

The CST model for the MWCNT antenna is presented in this section using the measured constitutive parameters presented in Section 2. The thickness of the MWCNT based patch is 0.5 mm. The patch is shaped as a square spiral trace truncated horizontally on a lossy FR4 substrate with a relative permittivity of 2.5, a loss tangent of 0.018, and a thickness of 0.5 mm. A thin copper layer, 35 μm in thickness is used as the ground plane. The 50 Ω feeding point is fixed at $X_p = -17.5$ mm and $Y_p = 17.5$ mm from the substrate's center as seen in Fig. 1. The MWCNT patch is positioning at the center of the substrate as shown in Fig. 1.

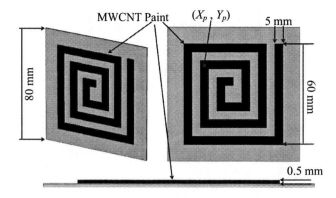

Fig. 1 The numerical model of the square spiral microstrip antenna.

Comparison of the simulated S_{11} spectrum for the MWCNT patch and the copper patch is presented in Fig. 2 (a) and (b). Fig. 2(a) shows that the return loss for the MWCNT patch is -12 dB at 1.2276 GHz, while the return loss for copper is -13 dB at 1.25 GHz. In Fig. 2(b), the MWCNT patch shows a resonance mode at 2.47 GHz with a matching return loss of -27 dB, while the copper antenna is matched with -13 dB at 2.53 GHz. However, the position of the feed has been shifted for the copper patch to the point at X_p = -7.5 mm, Y_p = -7.5 mm to obtain an optimum match to the 50 Ω feeding point. This shift in feeding position is attributted to the electrical conductivity of the MWCNT patch being less than that of copper. The -10 dB bandwidth of the MWCNT patch at both resonating frequency is about 3%, which is higher than the copper patch, which is about 1.9%. This is due to the relatively higher loss of the

MWCNT patch. The resonant frequancies for the two modes of the MWCNT patch are shifted by about 45 MHz with respect to the copper patch because of the relative permittivity of the MWCNT patch. In Fig. 2(c) and (d), the absolute gain versus frequency is presented for the MWCNT- and copper- based antennas, for both the first and second modes, respectively. The peak gain of the MWCNT patch is about -12.5 dBi and -4.25 dBi at 1.2276 GHz and 2.47 GHz, respectively, which are compared to the copper case in which the gain is about -12.05 dBi and -4.01 dBi at both 1.25 GHz and 2.53 GHz, respectively. This slight decrease in gain is caused by the conductor and dielectric losses of the MWCNT patch. The radiation patterns for both MWCNT and copper patches at the two operating frequencies are presented in Fig. 3. In Fig. 3(a) and (b), the radiation patterns of both antennas are illustrated in the $\theta = 0°$ and $90°$ planes at 1.2276 GHz for MWCNT patch and 1.25 GHz for copper. The radaition patterns of the MWCNT and copper at 2.47 GHz and 2.53 are presented in the $\theta = 0°$ and $90°$ planes, respectively, in Fig. 3(c) and (d). By inspecting Fig. 3, it is clear that the degradation in conductivity of the MWCNT ink with respect to copper does not significantly affect the shape of the far-field radiation patterns. In Fig. 4, we present the power dispated by the patch due to conductor and dielectric losses at the two resonance modes. In Fig. 4(a) and (b), the conductor and dilectric losses for the MWCNT patch is higher than the conductor losses of copper as presented in Fig. 4(c) and (d). The radiation efficiency at both resonance frequencies of the MWCNT were found to be 2% and 16%, while copper was found about 2.2% and 18%.

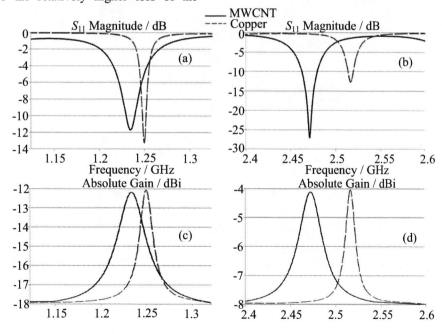

Fig. 2 Numerical results for the MWCNT and copper patches at the targeted frequancies (a) S_{11} versus frequency of MWCNT shows resonance at 1.2276 GHz and copper shows resonance at 1.25 GHz, (b) S_{11} versus frequency of MWCNT shows resonance at 2.47 GHz and copper shows resonance at 2.53 GHz, (c) absolute gain versus frequency of MWCNT shows -12.5 dBi at 1.2276 GHz and copper shows -12.05 at 1.25 GHz, and (d) absolute gain versus frequency of MWCNT shows -4.25 dBi at 2.47 GHz and copper shows -4.01 at 2.53 GHz.

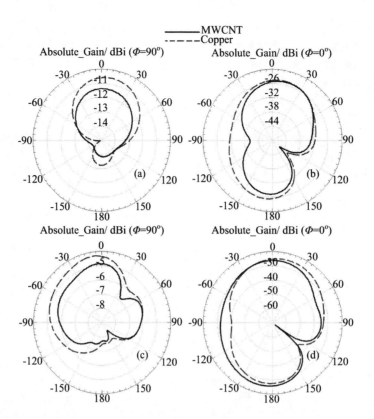

Fig. 3 Radiation patterns for the MWCNT and copper patches at the targeted frequancies (a) radiation patterns at $\theta = 0^\circ$ of MWCNT at 1.2276 GHz and copper at 1.25 GHz, (b) radiation patterns at $\theta = 90^\circ$ of MWCNT at 1.2276 GHz and copper at 1.25 GHz, (c) radiation patterns at $\theta = 0^\circ$ of MWCNT at 2.47 GHz and copper at 2.53 GHz, and (d) radiation patterns at $\theta = 90^\circ$ of MWCNT at 2.47 GHz and copper at 2.53 GHz.

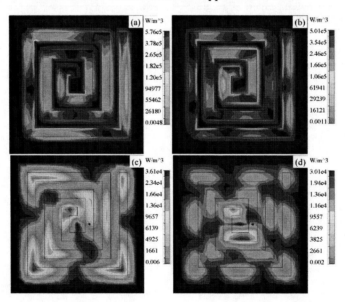

Fig. 4 Power dissipated by the conductor patch (a) at 1.2276 GHz for MWCNT, (b) at 2.47 GHz for MWCNT, (c) at 1.25 GHz for copper, (d) at 2.53 GHz for copper.

4   FABRICATION PROCESS

MWCNTs were synthesized on the Fe-Co/CaCO$_3$ catalyst system by chemical vapor deposition approach using Radio Frequency heating method as presented by Dervishi et. al.[1]. About 100 mg of the catalyst was uniformly spread into a thin layer on a graphite susceptor and placed in the center of a quartz tube with inner diameter of 1 inch. The quartz tube was horizontally positioned at the center of the RF generator. Next, the system was purged with nitrogen at 200 ml/min for 10 minutes, and when the temperature reached around 720 °C, acetylene was introduced at 3.3 ml/min for 30 minutes. The as-produced MWCNTs were purified in one simple step using diluted hydrochloric acid solution and sonication. A conductive ink composition consisting of purified MWCNTs is dispersed in sodium cholate (NaCh) aqueous solution (CNT: sodium cholate 1:1 wt., 5 mg/L). An inkjet printer based on a single inkjet head controlled by a piezoelectric actuation process is involved to drop the MWCNT ink onto a flexible FR4 substrate. The flexible FR4 substrate is treated by oxygen plasma for 2 minutes to increase the hydrophilicity of the polymeric films. The plasma treatment process is carried out at 200 W inside a cylindrical reactor connected to a rotary pump in series with an RF power source vacuumed at 2×10^{-3} Torr before introducing oxygen inside the reactor. The MWCNT patch is fed through an N - type female SMA coaxial cable to the 50 Ω matching point by silver paste, which dries at room temperature and adheres easily to the MWCNT structure. Fig. 5(a) shows the manufactured prototype of the MWCNT antenna. The copper patch has been deposited on the FR 4 substrate and the N - type female SMA connector is soldered to the 50 Ω matching point as shown in Fig. 5(b).

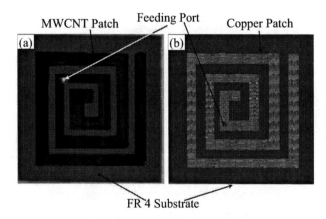

Fig. 5 Front veiw of the MWCNT and copper antennas, (a) MWCNT patch, (b) Copper patch.

5 EXPERIMENTAL MEASUREMENTS

The MWCNT and copper spiral patch antennas are tested and characterized by measuring the resonant frequency, return loss, and bandwidth as shown in Fig. 6 using an Agilent E5071B network analyzer. The -10 dB return loss bandwidth is 3.2% at the two resonant frequencies. On other hand, the copper patch shows a bandwidth of 1.8%, at the two resonant frequencies. The measured simulated results of the S_{11} of both cases and resonance frequencies are presented and compared in Fig. 6; an excellent agreement has been achieved between the measured and simulated results.

6 CONCLUSION

In this paper, we presented a square spiral microstrip antenna based on MWCNT conductive ink with an electrical conductivity of 2.2×10^4 S/m and a relative permittivity of 5-j1. The antenna performance is investigated numerically using CST MWS. The MWCNT antenna performance in terms of S_{11} spectrum was validated experimentally. In addition, the identical antenna

geometry based on copper has been tested numerically and experimentally. The antenna performance is compared to the copper case and an excellent agreement has been achieved between experimental and simulated results. The comparison pointed out that the bandwidth of the MWCNT patch is about 3%, which is wider than antenna made from copper. The resonant frequencies of the MWCNT patch are shifted by about 45 MHz with respect to the copper patch. The gain and far-field radiation patterns of the MWCNT are found not to be significantly different compared to the copper patch. Finally, the MWCNT antenna offers a larger gain bandwidth product compared to the copper-based antenna.

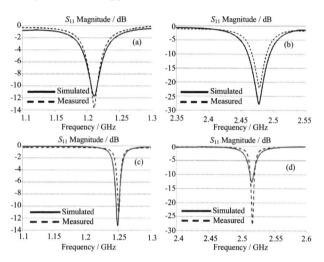

Fig. 6 Measured versus simulated S_{11} spectrum for both the MWCNT and copper patches (a) MWCNT at 1.2276 GHz, (b) MWCNT at 2.47 GHz, (c) copper at 1.25 GHz, and (d) copper at 2.53 GHz.

REFERENCES

1. Dervishi, E.; Li, Z.; Biris, A. R.; Lupu, D.; Trigwell, S.; Biris, A. S., "Morphology of multi-walled carbon nanotubes affected by the thermal stability of the catalyst system," *Chemistry of Materials*, vol. 19 no. 2, pp. 179-184, Apr. 2007.
2. Z. Li, H. Kandel, E. Dervishi, V. Saini, and A. Biris, "Does the wall number of carbon nanotubes matter as conductive transparent material?," *Applied Physics Letter*, vol. 91, no. 5, pp. 12 - 20, Aug. 2007.
3. A. Santos and T. Leite, "Morphology, thermal expansion, and electrical conductivity of multi-walled carbon nanotube/ epoxy composites," *Journal of Applied Polymer Science*, vol. 108, Issue 2, pp. 979 - 986, Jan. 2008.
4. CST MICROWAVE STUDIO® (CST MWS) is a specialist tool for the 3D EM simulation of high frequency components. http://www.cst.com/.

Acknowledgments

This research was funded in part by the National Science Foundation Grants: EPS-0701890 and CNS-0619069

VERTICALLY ALIGNED CARBON NANOTUBE ARRAYS ASSEMBLED ON GLASSY CARBON ELECTRODE

Hua-Zhang Zhao[*,a,b], Juan-Juan Sun[a,b] and Gang Shi[a,b]

[a]Department of Environmental Engineering, Peking University, Beijing 100871, China,

[b]The Key Laboratory of Water and Sediment Sciences, Ministry of Education, Beijing 100871, China

ABSTRACT: Carbon nanotubes have been widely utilized in electrode modification for their fascinating electrochemical catalytic properties. Given that the tube ends are the reactive sites of carbon nanotubes, modifying electrode with up-right carbon nanotubes is particularly desired. However so far, the well-organized standing carbon nanotubes on carbon electrode can only be realized by in situ nanotube growth which is very complicated and hard to manage. Herein, we present a new strategy to modify carbon electrode surface with vertically aligned carbon nanotubes. The single –walled carbon nanotubes were first chemically oxidized and carboxyl-derivatized at the open ends. The underlying glassy carbon electrode was introduced with nitrophenyl group via the chemical reduction of the corresponding nitrobenzenediazonium tetrafluoroborate salt.Then aminophenyl groups on electrode were obtained by reduction of nitrophenyl group in alkaline iron ammonium sulfate solution. Finally the nanotubes were attached through acylation-amidation. X-ray photoelectron spectroscopy and cyclic voltammetry were employed to characterize the two reduction reactions. Tapping mode atomic force microscopy (AFM) images clearly show that nanotubes have been successfully vertically organized on the glassy carbon electrode surface. Electrochemical behavior of glucose oxidase on the assembled electrode was measured to test the electrode performance.

INTRODUCTION

Bioelectrochemistry is a new interdisciplinary field, which combines biotechnology with electrochemical science and focuses on the structural organization and the electron transter functions of biointerfaces (Chen, Wang et al. 2007). Among the bioelectrochemical systems, biofuel cells are gaining more and more attention. The bioelectrochemical systems such as microbial fuel cell (MFC) and microbial electrolysis cell (Mele, Cardos et al. 1979; Murphy 2006) are of fundamental interest in view of their potentials in energy generation. To scale up the devices towards actual industrial application, particular emphasis is directed to the exploration of bioelectrochemical design of electrode interface, especially to those where direct electron transfer can be obtained on electrode surface, this is rather critical for the performance of entire system(Sun, Zhao et al. 2010).

In this respect, carbon nanotubes (CNTs) have came into spotlight, because CNTs have the characteristics of high surface-to-volume ratio, high mechanical strength, ultra-light weight, rich electronic properties, and excellent chemical and thermal stability. Recent studies demonstrated that CNTs can enhance the electrochemical reactivity and the electron transfer rates of biomolecules, accumulate important biomolecules, and alleviate surface fouling effects (Guiseppi-Elie, Lei et al. 2002; Lim, Cirigliano et al. 2007; Rivas, Rubianes et al. 2007). To take advantage of these remarkable properties of CNTs in bioelectrochemical applications, the CNTs need to be properly functionalized and immobilized.

In addition, Studies have demonstrated that the vertical alignment of CNTs on electrode surface is in favor of the direct electron transfer between enzyme/bacteria and electrode(Liu, Chou et al. 2005; Gooding, Chou et al. 2007). However, how to stably manufacture CNT based nanostructures with appropriate volume and orientation control to realize efficient transport of electrons, and to successfully apply them to practical application are still deemed of critical importance.

Given that the tube ends are the reactive sites of carbon nanotubes, and taking into consideration of the extraordinary electrical conductivity in the direction along the tube axis, modifying electrode

NSTI-Nanotech 2010, www.nsti.org, ISBN 978-1-4398-3401-5 Vol. 1, 2010

with up-right carbon nanotubes is particularly desired. So far, the well-organized standing carbon nanotube on carbon electrode has only been realized by in situ nanotube growth which is very complicated and hard to manage. Several other methods for general vertical alignment of CNTs were also used, for example, by using magnetic fields (Tian, Park et al. 2009), but the strict manufactural conditions have confined their applications and made them not suitable for large quantity production or modification.

Except in-situ growth, another way to control orientation of CNTs is mostly conducted by Au-S bond on gold electrode surface (Gooding, Wibowo et al. 2003), which restrict the rang of underlying electrode material. Carbon is an attractive bio-benign electrode material and would be an suitable choice for undelying electrode. Modifying carbon electrode surface in a way that allows efficient electron transfer between the electrode and the redox protein or bacteria is of great importance in the development of enzyme biofuel and microbial fuel cell. So we present another free-radical reaction based modification strategy, which can be also applied to carbon electrode surface and will support large quantities of manufacture.

MATERIALS AND METHODS

Chemicals. SWCNTs (Shenzhen Nanotech Port Co. Ltd., China) were purified and functionalized through a well-established way with slight modification. It is worth noting that the SWCNTs thus prepared were functionalized with carboxyl groups and could be dispersed in both water and dimethylformamide(DMF). 4-nitrobenzenediazonium tetrafluoroborate and N,N'-dicyclohexylcarbodiimide (DCC) were from J&K，All other reagents were purchased from Sinopharm Chemical Reagent, Beijing Co. Ltd. (Beijing, China), and were of highest grade available and used without further purification.

Instruments. X-ray photoelectron spectroscopy (XPS) measurements were performed on an AXIS-Ultra electron spectrometer from Kratos Analytical. The data were taken using monochromatic Al Ka radiation (225W, 15 Ma, 15 kV) and to compensate for surface charge effects, binding energies were calibrated using C 1s

hydrocarbon peak at 284.80 eV. The film topography was observed by tapping mode atomic force microscopy (AFM) imaging in respect to CNT structure and orientation.

Oxidation of Carbon Nanotube. Following a reported procedure (Sun, Zhao et al. 2010), the as-received SWCNTs (95%, 10–20nmdiameters) were purified and oxidized prior to use. Firstly the SWCNT oxidation procedure was as follows. First, the SWCNTs were sonicated in a mixture of H2SO4–HNO3 (volume ratio: 3:1) for 4 h, neutralized with a large amount of ultrapure water, filtered with a Millipore membrane (pore size 0.22 lm), and finally dried at 60 oC overnight. The ultrapure water was obtained from a Millipore-MilliQ system with a resistivity of 18M cm. The carboxylated and shortened SWCNTs were prepared and could be well-dispersed in water after ultrasonication for extended periods of time.

Introduction of amino groups onto electrode surface. Glassy carbon (GC) plates were used as substrate electrode. Prior to surface modification, the GC electrode was polished carefully to a mirror with alumina slurries and then washed ultrasonically in distilled water and ethanol for a few minutes, respectively. The 4-nitrophenyl groups were introduced by using the chemically activated one-electron reduction of the corresponding 4-nitrobenzenediazonium salt. The hypophosphorous acid (50% v/v aqueous solution). First, the GC plates were immersed into 10ml 50 mM aqueous solution of 4-nitrobenzenediazonium tetrafluoroborate at 5 ^{o}C, 20 ^{o}C and heated to 40 ^{o}C; then 20ml of hypophosphorous acid was slowly added and stayed for 60min. The resulting 4-nitrophenyl modified carbon electrode was washed with acetonitrile, acetone and water to remove any unreacted species. Next, the GC electrode with 4-nitrophenyl group was reduced in 25ml of 10 mM aqueous ferrous ammonium sulfate solution, and 20ml of 2N alcoholic sodium hydroxide solution was added. The resulting 4-aminophenyl GC electrode was then immersed in hydrochloric acid (37%) to remove any Fe(III) precipitates.

The vertical alignment of CNTs on GC

electrode. The shortened SWNT was dispersed in dimethylformamide (DMF) to form a well dispersed solution of 0.1mg/ml SWNTs, together with 0.5 mg/ml dicyclohexyl carbodiimide (DCC) to convert the carboxyl groups at the ends of the shortened SWNT into active carbodiimide esters. The 4-aminophenyl group modified GC electrode was placed in the nanotube solution for 24 hours to allow the amines at the electrode surface formed amide bonds with the end of the CNT tubes.

RESULTS AND DISCUSSION

It is known that the poor solubility of CNTs in most solvents has been standing in the way of facile fabrication of homogeneous and stable CNTs film or composite layer. Versatile modification approaches have been employed for this issue; including most commonly used the covalent attachment of chemical groups through reactions onto the Π-conjugated skeleton of CNT, and the non-covalent adsorption or wrapping of various functional molecules like cationic /anionic surfactant attachment. Here, in order to protect the electron structure of CNTs to the utmost extent, the concentrated acid oxidation procedure has been taken to add the negative charge carboxylic groups to CNTs to make them disperse stably and homogeneously in solution, and simultaneously remove metal catalyst and amorphous carbon left from synthesis process.

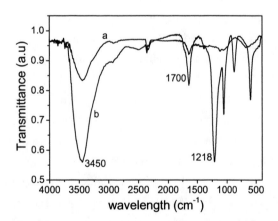

Figure 1. the Mid-IR spectrum of CNTs before and after oxidation.

CNTs were pretreated with mixed acid following the described procedure. After the oxidation process the CNTs were slightly shortened,

which makes it easier for the abundantly negative charged CNTs to stay uniformly dispersed in aqueous suspension for a long period of time (more than 2 weeks). The mid-IR spectrum of primary SWCNTs and oxidized SWCNTs are displayed in Fig. 2. For the concentrated acid-treated CNTs, an absorption peak at 1218 cm^{-1} is observed and can be ascribed to the stretching vibration of C-OH. In addition, two drastically enhanced absorption peaks at 3454 cm^{-1} and 1708 cm^{-1} are clearly displayed, which are typically attributed to the O-H stretching vibration and C=O carbonyl vibrations of carboxylic and carboxylate groups respectively, indicating a successful introduction of –COOH to the end or sidewalls of the SWCNTs.

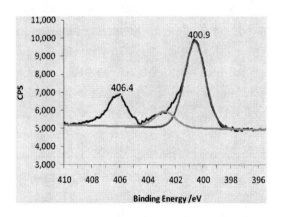

Figure 2. the N1s region of XPS results for 4-aminophenyl group modified GC electrode surface.

To confirm the introduction of amino groups onto GC surface and successful reduction of nitrophenyl groups, XPS was performed on the 4-aminophenyl group modified GC electrode surface. As shown in Figure 3, the major peak exhibited at 400.9 eV was ascribed to aniline (398.3–403.5 eV) which demonstrated the presence of amino group. However the peak is observed at 406.4eV corresponding to a nitrogen atom in an aromatic nitro group (N1s emission for nitrobenzene itself is 405.9 eV)(Masheter, Wildgoose et al. 2007), implying a not complete reduction of nitro groups. The ratio of peak areas indicates that the conversion of nitro to amino groups using ferrous hydroxide is ca. 64%.

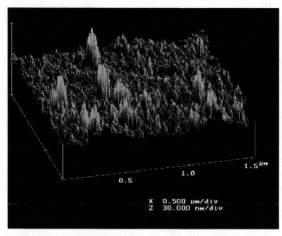

Figure 3. the AFM images of vertically aligned SWCNTs on GC electrode surface.

The result of shortened SWNT aligned normal to the electrode surface was shown in Figure 4. The AFM image clearly shows that the tubes assemble on the surface were in bundles and with different length, which is in accord to other research result(Gooding, Wibowo et al. 2003), and may be caused by the aggregation tendency of SWCNT and would be alleviated by shortening the reaction time.

CONCLUSIONS

In summary, with the proposed process, we have shown that shortened SWNTs can be aligned normal to carbon electrode by self-assembly.

REFERENCES

Chen, D., G. Wang, et al. (2007). "Interfacial bioelectrochemistry: Fabrication, properties and applications of functional nanostructured biointerfaces." Journal of Physical Chemistry C 111(6): 2351-2367.

Gooding, J. J., A. Chou, et al. (2007). "The effects of the lengths and orientations of single-walled carbon nanotubes on the electrochemistry of nanotube-modified electrodes." Electrochemistry Communications 9(7): 1677-1683.

Gooding, J. J., R. Wibowo, et al. (2003). "Protein Electrochemistry Using Aligned Carbon Nanotube Arrays." Journal of the American Chemical Society 125: 9006-9007.

Gooding, J. J., R. Wibowo, et al. (2003). "Protein electrochemistry using aligned carbon nanotube arrays." Journal of the American Chemical Society 125(30): 9006-9007.

Guiseppi-Elie, A., C. H. Lei, et al. (2002). "Direct electron transfer of glucose oxidase on carbon nanotubes." Nanotechnology 13(5): 559-564.

Lim, J., N. Cirigliano, et al. (2007). "Direct electron transfer in nanostructured sol-gel electrodes containing bilirubin oxidase." Physical Chemistry Chemical Physics 9(15): 1809-1814.

Liu, J. Q., A. Chou, et al. (2005). "Achieving direct electrical connection to glucose oxidase using aligned single walled carbon nanotube arrays." Electroanalysis 17(1): 38-46.

Masheter, A. T., G. G. Wildgoose, et al. (2007). "A facile method of modifying graphite powder with aminophenyl groups in bulk quantities." Journal of Materials Chemistry 17(29): 3008-3014.

Mele, M., M. J. Cardos, et al. (1979). "A Biofuel Cell as a Bioelectrochemical Sensor of Glucose-Oxidation." Anales De La Asociacion Quimica Argentina 67(4): 125-138.

Murphy, L. (2006). "Biosensors and bioelectrochemistry." Current Opinion in Chemical Biology 10(2): 177-184.

Rivas, G. A., M. D. Rubianes, et al. (2007). "Carbon nanotubes for electrochemical biosensing." Talanta 74(3): 291-307.

Sun, J. J., H. Z. Zhao, et al. (2010). "A novel layer-by-layer self-assembled carbon nanotube-based anode: Preparation, characterization, and application in microbial fuel cell." Electrochimica Acta 55(9): 3041-3047.

Tian, Y., J. G. Park, et al. (2009). "The fabrication of single-walled carbon nanotube/polyelectrolyte multilayer composites by layer-by-layer assembly and magnetic field assisted alignment." Nanotechnology 20(33).

Carbon Nanotubes and Graphene for Solar Cells

Zhongrui Li,[1] Enkeleda Dervishi, Viney Saini, Shawn Bourdo, Liqiu Zheng, Tito Viswanathan, Alexandru S. Biris[1]

Nanotechnology Center and Chemistry Department, University of Arkansas at Little Rock, AR, USA,

ABSTRACT

Carbon nanotubes and graphene represent attractive materials for photovoltaic devices. Single wall carbon nanotubes (SWNTs) and graphene layers can be directly configured as energy conversion materials to fabricate thin-film solar cells, serving as both photogeneration sites and charge carriers collecting/transport layers. SWNTs can be modified into either p-type conductor through chemical doping (like thionyl chloride, or just exposure to air) or n-type conductor through polymer functionalization. The solar cells consist of either a semitransparent thin film of p-type nanotubes (functionalized graphene) deposited on an n-type silicon wafer or a semitransparent n-type SWNT thin film on p-type substrate to create high-density p-n heterojunctions between nanotubes (graphene) and silicon substrate to favor charge separation and extract electrons and holes. The high aspect ratios and large surface area of carbon nano-structured materials could be beneficial to exciton dissociation and charge carrier transport thus improving the power conversion efficiency.

Keywords: photovoltaic devices, single wall carbon nanotube, graphene

1 INTRODUCTION

Semiconducting single wall carbon nanotubes (SWNTs) are potentially ideal material for photovoltaic applications thanks to their superior properties such as a wide range of direct bandgaps matching the solar spectrum [1], strong photoabsorption [2,3] from infrared to ultraviolet, and high carrier mobility [4] and reduced carrier transport scattering [5]. A standard oxidative purification process is known to induce p-type charge-transfer doping of SWNTs [6], so it is easy to fabricate p-type SWNT solar cells. It is possible to fabricate n-type SWNTs by doping low work function metals (like Al) onto the tubes [7] or through nitrogen doping of metallic SWNTs [8]. Graphene has shown remarkable photonic properties, such as high transparency and wide absorption spectral range, as well as outstanding electronic and mechanical properties, so extensive interest has developed toward applying it to optoelectronic devices. Unlike SWNTs, which contain 1/3 metallic species, narrow graphene nanoribbons (<10 nm) are semiconducting due to their edge effects [9]. Charge carriers in graphene behave as massless relativistic particles (Dirac fermions) and exhibit ballistic transport on the sub-micrometer scale at room temperature. Graphene combines high electron mobility with atomic thickness. Theoretic studies show that the substitutional doping can modulate the band structure of graphene [10], leading to a metal-semiconductor transition [11]. In this work, we designed and tested simple photovoltaic devices based on the high density heterojunctions formed between either acid purified or polymer functionalized SWNTs or graphene and different type semiconductor substrates.

2 EXPERIMENTAL

The CoMoCat SWNTs were utilized as photovoltaic media [12]. (1) To fabricate SWNT/n-Si device, the purified SWNTs were first dissolved in dimethylformamide (DMF, 0.5 mg/mL) under sonication, and the uniform solution was directly sprayed on n-type silicon wafers by airbrushing. A silicon wafer with a window of predeposited insulating layer was placed on a heating platform and heated up to 150 $^{\circ}$C in order to evaporate the DMF solvent from the film. (2) To make PEI-SWNT/p-Si cells, the purified SWNTs were submerged in a 10 wt% solution of polyethylene imine (PEI, average molecular weight ~2000) in methanol overnight, followed by rinsing with methanol. This removed PEI nonspecifically adsorbed on the sample surface, leaving nearly a monolayer of PEI irreversibly adsorbed on tubes as described below. The PEI-functionalized SWNTs (PEI-SWNT) were dispersed in dimethylformamide (DMF, 0.5 mg/mL) under sonication, and the uniform solution was directly airbrushed onto p-type silicon substrate (p-Si) which was heated up to 150 $^{\circ}$C. (3) For the preparation of acid treated graphene/n-Si device, graphene (1-2 layers, Angstrom Materials LLC) was sonicated at 40 $^{\circ}$C in 120 mL of a 3:1 mixture of H_2SO_4 and HNO_3 for 6 h, and the dispersion was allowed to stand for 4 days. After repeated washing with water, the oxidized grapheme was filtered, and further dispersed in $SOCl_2$ at 70 $^{\circ}$C for 2 days. At the end of reaction the excess $SOCl_2$ was removed by distillation. The product was dissolved in DMF and airbrushed onto a silicon wafer with a window of pre-deposited insulating layer under heating.

3. RESULTS AND DISCUSSION

3.1   p-type SWNT/n-type Si solar cells

---

[1] Corresponding authors: zxli3@ualr.edu, asbiris@ualr.edu

The randomly distributed networks of SWNTs containing many well-aligned tubes alternating with empty regions (Figure 1a, inset) are coated on the Si substrates, showing structural irregularity. In this work optical transmission was used to characterize the thickness of various thin SWNT networks prepared on a glass substrate. The film thickness can be controlled to reach hundreds of nanometers by using different number of airbrushing strokes. The SWNT films show sheet resistance varying from 250 to 3900 Ω/sq for SWNT films with transmittance of 51-86% at wavelength of 550 nm. Thus, the SWNT films can be directly used as a transparent conductive layer for solar cells, which can simplify the fabrication process of the SWNT solar cell.

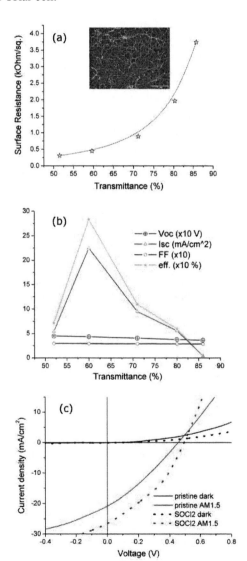

Figure 1: (a) Optoelectronic performance of SWNT films with different thickness at wavelength of 550 nm. Inset (bottom) shows the top view SEM image of a device. (b) Summary of short-circuit current (J_{sc}), open-circuit voltage (V_{oc}), fill factor (f), and efficiency (η) for SWNT cells with

different thickness, showing maximum current and efficiency of the device with transmittance of about 60%. (c) The effect of the thionyl chloride (SOCl$_2$) treatment on the photovoltaic performance of SWNT/n-Si heterogeneous junctions.

The flexibility of the fabrication process allows the construction cell devices containing SWNT films with different thickness by controlling the airbrushing process. We have made cells with tailored SWNT coating thicknesses roughly ranging from 30 to about 270 nm (corresponding to 86-51% transmittance). All the cells show relatively stable voltages (V_{oc}) ranging from 0.37 to 0.45 V, but quite different current densities (J_{sc}) (Figure 2b). Increasing SWNT network density can increase the total area of SWNT-Si junctions in the device, which could enhance the current generation/transport. The cell with transmittance of ~60% shows the highest J_{sc} (21.8 mA/cm^2) and η (2.7%). At further increased film thickness, the values of J_{sc} and η decrease significantly. For the thicker coating (T < 60%), the SWNTs at top layers are suspending on the underneath layers and cannot touch the Si substrate to form junctions. Also, thicker films are less transparent, preventing more incident light from reaching the substrate. The devices have the potential to be further optimized, for example, by tailoring the bundle size (dispersion) and the semiconducting to metallic tubes ratio of SWNTs, so that power conversion efficiency can be further improved.

The excellent photovoltaic performance of the SWNT solar cell originates from the numerous heterojunctions created at the interface between the SWNT film and the n-Si [13]. Acid purified SWNTs usually behave as p-type semiconductors [14]. When fully expanded on a planar Si substrate, there will be numerous p-n junctions formed due to close contact between SWNTs and underlying n-Si. The I-V curve in the dark actually shows a typical diode behavior, further confirming the existence of p-n junction of this SWNTs-on-Si configuration. The rectifying behavior in the dark and the photocurrent phenomenon under illumination of the SWNT/n-Si heterojunction cell can be explained by the energy band structure of this heterostructure. Since the band gaps (E_g) of our Si and SWNTs are 1.1 and <2.7 eV, respectively, an asymmetrical energy barrier would be formed at the junction interface. The holes and electrons generated in both sides of the heterojunction are collected effectively due to the large built-in electric field at the junction, where electrons are directed to the n-type Si region and holes are transported through the SWNTs, and thus yield the photocurrent. The high J_{sc} of the solar cells suggests that the presence of high density p-n junctions significantly enhanced the generation and transport of charge carriers from both SWNTs and silicon under light irradiation. Nanotubes act mainly as hole collectors and conductors in polymer/nanotube solar cells [15]. In our SWNT/n-Si devices, SWNTs might have participated in the photogeneration process as well as charge transport, and their high mobility ensures much enhanced efficiency compared with polymer composite cell structures. The high aspect ratios and large surface area of nanotubes can help exciton dissociation and charge carrier transport, accordingly the power conversion efficiency gets improved.

To enhance the performance of the solar cells, we carried out chemical doping using $SOCl_2$, a liquid organic solvent with remarkable reactivity toward the SWNTs surfaces. The $SOCl_2$ treatment involved dripping about 3 droplets of pure $SOCl_2$ onto the SWNT films followed by drying in air. The chemical attachments of functionals to the SWNTs are in the form of acyl chloride groups. Figure 1c demonstrates the effect of the $SOCl_2$-treatment on the photovoltaic and electrical properties of a SWNT/n-Si device. After the $SOCl_2$ treatment, the J_{sc} jumps onto 26.5 mA/cm^2 from 20.7 mA/cm^2, and the V_{oc} slightly increased to 0.49 V from 0.46 V, and the f is raised up to 0.35 from 0.29, as a consequence, the $SOCl_2$ post treatment leads to 59% increase in power conversion. Initial tests have shown a power conversion efficiency of about 4.5%.

The conductivity of the pristine SWNT films can be significantly increased by $SOCl_2$-treatment [16]. As a strong oxidizing agent, $SOCl_2$ exhibits remarkable electron-withdrawing ability when adsorbed onto the SWNT surface. The significant charge transfer induced by $SOCl_2$ could also enable Fermi level shifting into the van Hove singularity region of SWNTs, resulting in a substantial increase in the density of states near the Fermi level. With this functioning technique, typical sheet conductance of the SWNT films can be improved about 5-10 folds. In our configuration, SWNTs act as photogenerator as well as charge transporter. Semiconducting tubes would benefit photogeneration process but have poor conductivity; on the contrary, metallic tubes contribute more charge transportation but less photogeneration. Since $SOCl_2$ treatment can significantly improve the conductivity of semiconducting SWNT film, metallic tubes play less important role in the photoconversion of the $SOCl_2$-enhanced SWNT/n-Si device.

3.2 n-type SWNT/p-type Si

Figure 2 shows the current-voltage characteristics of a typical PEI-SWNT/p-Si solar cell, in which the SWNT film has about 30% transmittance at wavelength of 550 nm, in dark and under white light illumination (AM1.5, ~1000 W/m^2). A rectifying behavior is observed in PEI-SWNT/p-Si: the current is blocked when the PEI-SWNT is biased oppositely but starts to increase while the polarity is switched to the negative. Therefore the device is at forward (reverse) bias when the voltage applied to PEI-SWNT is negative (positive). Under illumination, the blend device shows an open-circuit voltage V_{oc} of about 0.42 V, short-circuit current density J_{sc} of 0.45 A/m^2, and fill factor of 0.16. The above results lead to several important conclusions: (1) PEI irreversibly adsorbs onto the sidewalls of SWNTs, and the adsorbed PEI is capable of and responsible for n-doping of SWNTs. The high density of electron donating amine functionalities in PEI brings about significant n-doping to a level where the adverse effect of p-doping by O_2 adsorption is overcome. (2) There will be numerous p-n junctions formed due to close contact between PEI-SWNTs and underlying p-Si substrate. Photons with energies larger than the tubes' energy gaps or/and the silicon energy gap can be absorbed and converted into electron-hole pairs (excitons), which were subsequently diffused to the depletion region of SWNT-Si junctions and separated. (3) The PEI-SWNT film can be directly used as a transparent conductive layer for solar cells, which can simplify the fabricating process of the CNT solar cell.

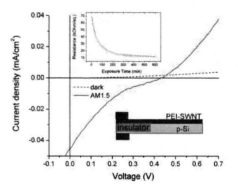

Figure 2: Current-voltage plot of a typical PEI-SWNT/p-Si device in dark (solid blue) and under illumination (doted red) before and after 12 h UV irradiation. The inset (bottom) displays the schematic diagram of a PEI-SWNT/p-Si solar cell. The inset (top) shows the resistance change of a PEI-SWNT film coated on a glass plate as a function of exposure time to UV (maximum intensity at 254 nm) irradiation in N_2 atmosphere.

Exposure of SWNTs to ultraviolet (UV) light results in efficient molecular photodesorption from SWNTs.[17] Photodesorption of oxygen from the PEI-coated SWNTs renders nanotubes to higher n-doping levels. Figure 2 inset (top) shows that the resistance of a PEI-SWNT film decreases under UV illumination. This is due to an increase in electron density upon photoremoval of oxygen adsorbed on the nanotube prior to PEI functionalization. The increased n-doping is also clearly observed in the evolution of I-V characteristics of the PEI-SWNT/p-Si solar cell. Although the open circuit voltage basically has no change, the short circuit current and the fill factor increase to 0.51 A/m^2 and 0.27, respectively, upon photoremoval of oxygen.

3.3 Functionalized graphene/n-type Si

Figure 3 shows the I-V characteristics of a typical graphene/n-Si solar cell, in which the functionalized graphene film, in dark and under white light illumination (AM1.5, ~100 mW/cm^2). The device shows an evident p-n junction behavior in the dark, where the reversed current density is very low (< 0.1 $\mu A/cm^2$, over 400 times lower than the forward current density) when the bias voltage sweeps from -1.0 to 0 V compared to forward current (~13 $\mu A/cm^2$ at 1.0 V). Under illumination, the I-V curve shifts downward, with an open-circuit voltage V_{oc} and short-circuit current density J_{sc} of about 0.28 V and 0.03 mA/cm^2, respectively. The graphene sheet represents an extreme dish-like-shape material; it has a 2D structure and a high aspect ratio. Such a structure of graphene can effectively facilitate the desired nanometer-scale donor/acceptor interface formation for charge separation and charge transfer. In general, optimal photocurrent generation requires a tradeoff between a maximization of the interfacial area between donor and acceptor and an optimized percolation through the film avoiding charge-carrier recombination losses during the charge transport [18]. The former maximizes the generation of charge carriers, while the percolation directly improves the charge transport properties. Acid functionalized grapheme layers likely behave as p-type semiconductors. When fully expanded on a planar Si substrate, there will be numerous p-n junctions formed due to close contact between graphenes and underlying n-Si. The holes and electrons generated in both sides of the heterojunction are collected effectively due to the large built-in electric field at the junction, where electrons are directed to the n-type Si region and holes are transported through the graphenes, and thus yield the photocurrent.

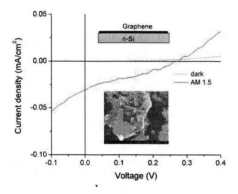

Figure 3: Current-voltage plot of a functionalized-G/n-Si device under dark and illumination. Inset (top) displays schematic diagram of a functionalized graphene/n-Si solar cell. Inset (bottom) shows the top view SEM image of a device.

4. CONCLUSION

In brief, the solar cells consist of a semi-transparent thin film of the p/n-type nanotubes/graphene conformally deposited on an n/p-type crystalline Si substrate to create high-density p-n heterojunctions between carbon nanostructure materials and silicon substrates to favor charge separation and extract electrons and holes. The carbon nanomaterials were serving as both photogeneration sites and a charge carriers collecting/transport layer. It is expected that after complete development, pure carbon cells may become more economical than silicon-based solar cells, and the cost would be extremely low and, unlike silicon, it is highly chemically and environmentally stable.

ACKNOWLEDGEMENTS

This research was partially supported by the DOE (Grant No. DE-FG 36-06 GO 86072). Also the financial support from Arkansas Science and Technology Authority (ASTA) grant # 08-CAT-03 is highly appreciated.

REFERENCES

[1] A. Hagen, T. Hertel, Nano Lett. 3, 383, 2003.
[2] T. G. Pedersen, Phys. Rev. B 67, 073401, 2003.
[3] T. W. O. Odom, J. L. Huang, P. Kim, C. M. Lieber, Nature 391, 62, 1998.
[4] M. S. Fuhrer, B. M. Kim, T. Du1rkop, T. Brintlinger, Nano Lett. 2, 755, 2002.
[5] M. Freitag, V. Perebeinos, J. Chen, A. Stein, J. C. Tsang, J. A. Misewich, R. Martel, P. Avouris, Nano Lett. 4, 1063, 2004.
[6] K. Bradley, J.C. P. Gabriel, G. Grüner, Nano Lett. 3, 1353, 2003.
[7] H.-S. Kim, B.-K. Kim, J.-J. Kim, J.-O. Lee, N. Park, Appl. Phys. Lett. 91, 153113, 2007.
[8] V. Krstic, G. L. J. A. Rikken, P. Bernier, S. Roth, and M. Glerup, Europhys. Lett. 77, 37001, 2007.
[9] X. L. Li, X. R. Wang, L. Zhang, S. W. Lee, H. J. Dai, Science 319, 1229, 2008.
[10] T. B. Martins, R. H. Miwa, A. J. R. Silva, A. Fazzio, Phys. Rev. Lett. 98, 196803, 2007.
[11] F. Cervantes-Sodi, G. Csanyi, S. Piscanec, A. C. Ferrari, Phys. Rev. B 77, 165427, 2008.
[12] W. E. Alvarez, F. Pompeo, J. E. Herrera, L. Balzano, D. E. Resasco, Chem. Mater. 14, 1853, 2002.
[13] J. Wei, Y. Jia, Q. Shu, Z. Gu, K. Wang, D. Zhuang, G. Zhang, Z. Wang, J. Luo, A. Cao, D. Wu, Nano Lett. 7(8), 2317, 2007.
[14] T. Shimada, T. Sugai, Y. Ohno, S. Kishimoto, T. Mizutani, H. Yoshida, T. Okazaki, H. Shinohara, Appl. Phys. Lett. 84, 2412, 2004.
[15] A. D. Pasquier, H. E. Unalan, A. Kanwal, S. Miller, M. Chhowalla, Appl. Phys. Lett. 87, 203511, 2005.
[16] B. B. Parekh, G. Fanchini, G. Eda, M. Chhowalla, Appl. Phys. Lett. 90, 121913, 2007.
[17] R. J. Chen, N. R. Franklin, J.Kong, J. Cao, T. W. Tombler, Y. Zhang, H. Dai, Appl. Phys. Lett. 79, 2258, 2001.
[18] L. H. Nguyen, H. Hoppe, T. Erb, S. Gunes, G. Gobsch, N. S. Sariciftci, Adv. Funct. Mater. 17, 1071,2007.

Determination of cefixime in aqueous solution using multiwalled carbon nanotubes solid- phase extraction cartridge

M. Jafari[*], S.F. Aghamiri[**] and G.R. Khaghani[***]

[*] Biotechnology Engineering Department, Engineering Faculty, University of Isfahan, Isfahan, Iran,
maryamjafari659@gmail.com
[**] Chemical Engineering Department, Engineering Faculty, University of Isfahan, Isfahan, Iran,
aghamiri@petr.ui.ac.ir
[***] Research and development Department, Farabi pharmaceutical company, Isfahan, Iran,
gholamrezakhaghani@yahoo.com

ABSTRACT

One of the most important chemical properties of carbon nanotubes is having a hydrophobic surface, and because of that many compounds and materials can be retained on the surface of these nano structures. In this study, by using this feature of carbon nanotubes, their ability in adsorption of cefixime at trace amount from aqueous solution has been tested. Samples were strongly adsorbed by carbon nanotubes and acceptable recoveries were obtained. Analyses of samples were done by high performance liquid chromatography and UV. Limit of detection of cefixime under optimum conditions is ranged between 0.15-0.2痢/ml. recoveries of spiked sample analysis ranged from 86.8% to 88.4%.

Keywords: carbon nanotubes, solid-phase extraction, high performance liquid chromatography, cefixime

1 INTRODUCTION

Today all over the world there are lots of pharmaceutical company which produce antibiotics and are causing the water and waste water be polluted by companies wastes. In this case the risk of antibiotic entrance to the underground water and microorganisms' resistance will occur [1]. To protect the environment and human health efficient facilities of sample preparation and sample preconcentration for accurate analysis seems necessary.

Solid-phase-extraction (SPE) is an efficient sample preparation method that is usually used to prepare liquid samples. SPE is suitable for sample extraction, concentration and clean up. They are accessible in wide range of chemistries, adsorbents and sizes. So it has become a common preconcentration method in environmental analytical application recently. Selecting the most suitable product for each application and sample is important. Today different kind of new adsorbent like carbon nanotubes are tested to be used in this extraction packages [2-9].

One group of nanostructures which have been surprisingly in the centre of scientists' attention in recent two decades is carbon nanotubes (CNTs). According to their unique physical and chemical properties, these materials from their discovery time in 1991 by Iijima [11] till now have been used in different fields of application such as development of sensors and biosensors, nano probes, drug delivery, nano electronic, gas separation and etc [10-14].

CNTs are considered as hollow graphitic cylinders that have one (single walled carbon nanotubes, SWCNTs) or more (multi walled carbon nanotubes, MWCNTs) graphene layers. The length of these tubes can range from hundreds of nano meters to some micrometers and their diameter for single and multi wall CNTs comes between 0.2-2 nm and 2-100 nm, respectively [7].

High specific area and hydrophobic surface are two specifications of CNTs that make them capable sorbents for retaining vast amounts of compounds on their surface [5, 7]. Extraction of amino acids, proteins, tetracyclins, sulphonamides, phenolic compounds, several phthalate esters, chlorophenols, fungicides, prometryn and cephalosporins are some examples of adsorbed materials by CNTs [2-5, 7, 8, 15 and 16].

In the present study, the ability of CNTs for determination of cefixime in aqueous solution has been examined. The solid-phase extraction (SPE) cartridge was self-made in our lab and was packed with MWCNTs. Percentages of remained analytes in the sample were measured by using high performance liquid chromatography and ultra violet (HPLC-UV). At the end, comparing studies between CNTs and silica gel (Si) was performed.

2 EXPERINENTAL

2.1 Materials

Cefixime was the target analyte and was taken from Farabi pharmaceutical company (Isfahan, Iran). Chemical structure of the selected compound was shown in table (1) Acetonitrile (ACN), methanol (MeOH), HPLC grade where obtained from Merck (Darmstadt, Germany). 500 $\mu g \mu l^{-1}$

standard stock solution of each sample was prepared in deionized water and standard solutions were made by diluting (500μgl-1) stock solution. Sodium hydroxide and Hydrochloric acid were purchased from Merck (Darmstadt, Germany). TLC- silica gel 60 GF which have particle size of 15 μm (Merck, Darmstadt, Germany) was used as comparing sorbent. MWCNTs with an average external diameter of 5-20 nm and apparent density of 150- 350 mg/cm3were provided by Plasmachem GmbH (Berlin, Germany).

Chemical structure	(chemical structure of cefixime)
Chemical formula	$C_{16}H_{15}N_5O_7S_2$
Chemical weight	g/mol 453/452

Table 1: Chemical structure of cefixime

2.2 Chromatographic conditions

Chromatographic experiments were performed by using Alliance HPLC system (Waters, USA) included 515 HPLC pumps, an in-line connected degasser, a 717 plus automatic sample injector, a column compartment and a UV detector in which wave length used to detect cefixime was set at 254nm . The analytical column was included a C_{18} column (4*125mm: particle size, 5μm) for cefixime. Isocratic separations carried out using (pH 6.5, adjusting by tetra butyl ammonium hydroxide/ACN (3:1)). The flow rate for this antibiotic was 1.5 ml min^{-1}.

2.3 Solid- phase extraction cartridge

The cartridge was made of a 10 ml glass syringe which was packed with 100 mg of sorbent and sorbent was retained by two porous stainless steel disks and cork as frits. Disk`s pores size was 50 μm, so we were forced to use a little cork, not to let CNTs washed out. MWCNTs were purified by 1M hydrochloric acid (sonicated about 2hr) and washed with water till the sorbent was neutral [5].

3   RESULTS AND DISCUSSION

3.1   Desorption conditions

Compounds which have amino group or hydroxyl group were recovered with difficulty and our target analyte belongs to this category [6]. Thus we faced problems for recovery of samples. To catch better recoveries, some tests have been done to optimize the desorption conditions, including composition and volume of eluents.

3.1.1 Composition of eluent

To find out the most capable solvent to wash cefixime from sorbent surface MeOH and CAN were chosen .eluting of cefixime with these eluents was not efficient and satisfactory results were not observed with mixture of MeOH and ACN at studied different ratios. After applying different mixtures of MeOH, ACN, TBAH and water, it was discovered that the most effective solvent for washing cefixime out of CNTs is TBAH/ACN (3:1), so it was accepted as .chosen eluent.

3.1.2 Eluent volume

To find out the most suitable amounts of eluent in order to get the best recoveries, different volumes of eluent from 2 to 5ml were examined three times and outcomes are shown in Fig1.

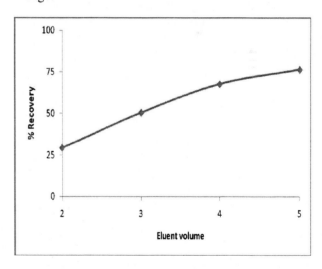

Fig 1: Effect of eluent volume on the % recovery

3.2   Effect of solution pH

pH-value plays an important role in the extraction of organic compounds in environmental samples because the pH-value of the sample solution determined the existing state of analytes and the analytes can only be adsorbed in molecule form, so pH of sample solution determined the extraction efficiency of the target analyte [8]. In this study, the effect of pH was investigated in the range of 2-7. Higher pH was not examined because Cephalexin was instable in alkaline medium [5]. According to results shown in Fig 2 pH=5 is the optimum pH of the sample solution.

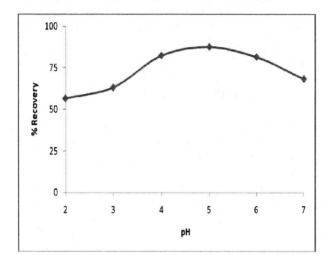

Fig 2: Effect of pH

3.3 Sample volume

To investigate the effect of sample`s volume, fixed amount of analyte was solved in different volumes up to 100ml. After passing the sample over CNTs, CNTs were recovered with optimum eluent volume and sample pH which were determined in previous steps. As it is shown in Fig3 the less the concentration is, the more the recovery is obtained.

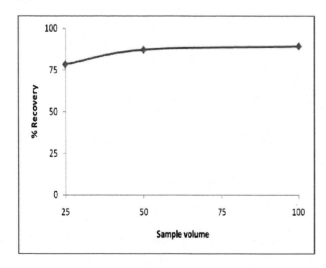

Fig 3: Effect of sample volume

4 COMPARISON STUDY

To evaluate the ability of CNTs as a new sorbent for solid-phase extraction cartridge this nanostructure was compared with one common sorbent, named Si. Much more satisfying results were gained for CNTs in comparison to Si. The following figure show the comparison results obviously.

5 ANALYTICAL PERFORMANCE

Under the optimized conditions, the analytical performance was examined with standard solution. Table 2 shows the analytical features of the proposed method. There is a linear correlation between peak area and concentration.

Table 2: Linear equation, Correlation coefficient, Detection limit

Analyte	cefixime
Linear equation	y=10.57x-1456.4
Correlation coefficient	0.9957
Detection Limit	$0.2\mu g/ml$

The method was applied to analysis of tap water samples and the results are shown in table 3.

Table 3: Analysis of spiked water samples

Analyte	cefixime
Added $(\mu g/l)$	2.5
Found $(\mu g/l)$	2.19
Recovery (%)	87.6
RSD (%)(n=3)	0.90

6 CONCLUSION

In this research, the ability of CNTs as a new sorbent of SPE was tested. Acceptable results proved that these nano materials could be a good alternative for traditional sorbents. Using small amount of sorbent is the positive point. The results showed that small amount of eluent are sufficient and when our samples were diluted better recoveries were concluded.

REFERENCES

[1] R.G. Finch, D. Greenwood, S.R. Norrby, R.J. Whitley, Antibiotic and chemotherapy: anti-infective agents and their use in therapy, Elsevier Health Sciences, 2003.
[2] Y. Cai, S. Mou, Y. Lu, J. Chromatoger. A, 1081, 245-247, 2005.
[3] Y.Q. Cai, G.B. Jiang, J.F. Liu, Q.X. Zhou, Anal. Chim. Acta, 494,149-156, 2003.

[4] G.Z. Fang, J.X. He, S. Wang, J. Chromatoger. A, , 1127, 12-17, 2006.

[5] H. Niu, Y. Cai, Y. Shi, F. Wei, J. Liu, S. Mou, G. Jiang, Anal. Chim. Acta, 594, 81-92, 2007.

[6] L.M. Ravelo-Pérez, A.V. Herrera-Herrera, J. Hernánddez-Borges, M. Rodríguez-Delgado, J. Chromatoger. A, ,2009..

[7] B. Suarez, B. Santos, B.M. Simonet, S. Cardenas, M. Valcarcel, J. Chromatoger. A, 1175, 127-132, 2007.

[8] Q. Zhou, J. Xiao, Y. Ding, Anal. Chim. Acta, 602 , 223-228, 2007.

[9] F. Bruno, R. Curini, A. Di Corcia, M. Nazzari, R. Samperi, J. Agric. Food Chem, 49 , 3463, 2001.

[10] K.M. Manesh, H.T. Kim, P. Santhosh, A.I. Gopalan, K.P. Lee, Biosci. Bioelectron, 23 ,771, 2008.

[11] D.R.S. Jeykumari, S.S. Narayanan, Biosci. Bioelectron, 23 , 1686, 2008.

[12] I. Singh, A.K. Rehni, P. Kumar, M. Kumar, H.Y. Aboul-Enein, Fullerenes, Nanotubes, Carbon Nanostruct, 17 , 361, 2009.

[13] S. Kim, T.W. Pechar, E. Marand, Desalination, 192 , 330, 2006.

[14] S. Polizu, M. Maugey, S. Poulin, P. Poulin, L.H. Yahia, Appl. Surf. Sci, 252 , 6750, 2006.

[15] X. Xie, L. Gao, J. Sun, Colloids. Surf. A, 308 , 54, 2006.

[16] T. Roman, W.A. Dió, H. Nakanishi, H. Kasai, Thin Solid Films, 509 , 218, 2006.

Wafer Level Assembly of Single-Walled Carbon Nanotube Arrays with Precise Positioning

J. Cao, A. Arun, K. Lister, D. Acquaviva, J. Bhandari, A. M. Ionescu

Ecole Polytechnique Fédérale de Lausanne, Nanolab
CH-1015 Lausanne, Switzerland, ji.cao@epfl.ch

ABSTRACT

A novel wafer level assembly method to fabricate Single-Walled Carbon Nanotube (SWCNT) arrays with precise positioning of the individual SWCNT by dielectrophoresis (DEP) is proposed and validated. Unlike previous CNT assembly performed directly between the guiding electrode pairs, in this work, by means of patterned trenches in some unique "sandwich"-like resist layers, super-aligned SWCNT arrays with controllable density and location can be achieved on various electrode dimensions and geometries. We demonstrate very high densities (~100 million/cm^2) of doubly clamped suspended SWCNT arrays, which is two orders of magnitude higher than earlier works. This self-assembled technique could enable a precise positioning with deep sub-micron control at the wafer level for the fabrication of future NEMS devices, such as SWCNT resonator arrays for RF or sensing applications.

Keywords: SWCNT array, assembly, precise positioning, dielectrophoresis, SWCNT resonator

1 INTRODUCTION

Single-walled carbon nanotubes (SWCNTs) have been intensively studied as a potential building block for future Nano-Electrical-Mechanical-System (NEMS) [1], owing to their unique mechanical and electrical properties [2], e.g., high Young's modulus and good conductivity. Despite significant research in this field, the controllable and precise integration of SWCNTs into underlying silicon NEMS circuitry at the wafer level still remains a great challenge for the engineering applications of CNTs [3].

Two methods of manipulating CNTs have attracted significant research interests: the directed CNT growth onto a desired location [4] and the deposition of CNTs by dielectrophoresis (DEP) between guiding electrodes [5]. The dielectrophoresis technique was first brought into SWCNT manipulation by Krupke *et al.* [6, 7]. Dielectrophoresis has already proven its potential for the fabrication of CNT-integrated microchips for relatively simply structured NEMS, in which two opposing electrodes build the interface for the connection between CNT and NEMS. Many efforts have been made to develop this method till now. The simultaneous and site-selective deposition of single SWCNT bundles onto a large number

of contacts was reported [8]. Recently, A. Vijayaraghavan et al. presented directed and precisely assembled SWCNT arrays by applying dielectrophoresis to capacitively coupled electrodes [9]. Although inspiring progress has been made, present approaches still lack the desired controllability in the arbitrary positioning of SWCNTs, the higher density, the uniformity over large scales and the flexibility to complex structural configurations on substrates, all of which are essential for complex CNT based NEMS.

To address this issue, we propose a novel approach of fabricating well-aligned assembled dense SWCNT arrays with precise positioning of individual SWCNT with high uniformity and reproducibility simply by dielectrophoresis. Instead of performing CNT deposition directly on electrodes as reported earlier [6-8], both single and highly dense parallel SWCNT arrays can be assembled uniformly on electrodes with flexible dimensions and geometry by means of defining patterned trenches in the "Sandwich"-like resist layers. This is of great importance since various applications have different criteria for the desired device characteristics, e.g., some demanding high density while others requiring high current drives. The proposed method gives access to a wider range of electronic device structures based on SWCNTs.

Meanwhile, clamped-clamped SWCNT arrays with very high densities (beyond 100 million devices per square centimeter) are successfully fabricated; this density is about two orders of magnitude higher than the previous work [8], with a yield of at least 85%. This versatile assembly strategy also leads the way to a wafer-level uniform fabrication of future NEMS RF devices. Two typical super-aligned SWCNT resonator arrays for RF or sensing applications are demonstrated.

2 DESIGN AND FABRICATION

The SWCNT suspension used in the experiment is prepared by dispersing commercially available SWCNTs in deionized (DI) water with Sodium Cholate Hydrate (1%wt) as surfactant. The assembly of SWCNT was performed on a test wafer in a simple experimental setup under ambient conditions with CNT suspension following the process flow described in Figure 1.

First, Chromium/Platinum guiding electrodes were patterned by a metal lift-off process on a (100) Si substrate protected by 500nm dry oxide (Figure 1a). The substrate was then coated with a thin photoresist layer and a thin e-

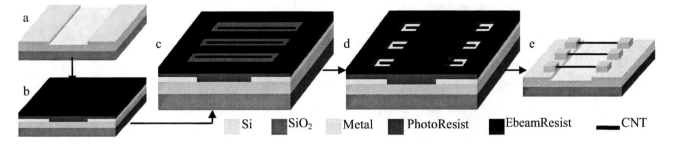

Figure 1: Bottom-up integration scheme used to fabricate suspended SWCNT arrays. a), Guiding electrodes are patterned. b), A photoresist layer and an e-beam resist layer are coated. c), Dielectrophoresis is applied to preferentially align single SWCNTs in trenches patterned in the e-beamresist layer. The photoresist thickness determines the SWCNT suspended height. d), Metal clamp windows are defined in a second e-beamresist layer. e), Metal clamps are deposited around the SWCNT tips. The resists are dissolved to suspend the clamped SWCNTs and lift off misaligned ones (unclamped).

beamresist layer (Figure 1b). Arrays of trenches with a typical width of 100nm were transferred to the e-beamresist layer by e-beam lithography, exposing the photoresist layer.

Then, dielectrophoresis [10] is applied to deposit SWCNTs in the trenches (Figure 1c). SWCNTs are attracted to the sample surface and aligned along the electric-field gradient by long-range dielectrophoretic forces. By stronger local electric field and capacitive forces in the trenches, SWCNTs were further preferentially aligned and centered in the trenches. The SWCNTs were retained in the trenches by capillary forces produced during evaporation of the suspending liquid within each trench. The metal clamp windows were defined on a second e-beamresist layer by e-beam lithography and by dissolving the resists on the unmasked metal electrodes region (Figure 1d). The final metal clamps were deposited by another metal lift-off. Lastly, the double clamped SWCNTs were released by stripping the resists and by using a critical point dryer (Figure 1e). Misaligned SWCNTs outside the trenches were lifted off at the same time.

3 RESULTS AND DISCUSSION

3.1 Experimental results

The alignment of SWCNTs was imaged by Scanning Electron Microscopy (SEM). The standard electrical transport characterization of the clamped-clamped SWCNTs is carried out by means of the Cascade probing system and the Agilent 4156C analyzer. During electrical transport measurements, one Cr/Pt electrode served as the source and the other as the drain. The Si substrate was used as the gate electrode.

Figure 2a shows an individual SWCNT or a small SWCNT bundle (diameter<4nm) placed precisely in a pre-defined 100nm wide trench on the resist layers with nanometer precision by means of this self-assembly method. The trapped SWCNT or the small SWCNT bundle is fixed

by a pair of Pt clamps after a second metal lift-off (Figure 2c).

The dielectrophoresis duration decides the amount of SWCNTs in each trench. The longer the dielectrophoresis is performed, the more SWCNTs will be attacted per trench. With the same SWCNT suspension concentration, by applying V_{ac} for additional 25s, one more SWCNT was deposited in each trench (Figure 2b). Figure 2d corresponds to the 2 SWCNTs / trench case after metal clamp deposition.

Furthermore, the higher the SWCNT suspension concertration is, the more SWCNTs will be trapped per trench with the same dielectrophoresis duration and applied voltage. Thus, it is more controllable to undertake the SWCNT assembly by relatively lower SWCNT suspension concertration.

Simply by adjusting the number and interval of the trenches on top of the guiding electrodes, uniform assembly of SWCNT devices with an arbitrary density is feasible at the wafer level.

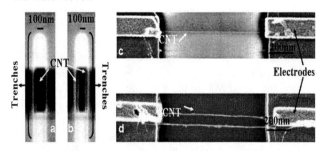

Figure 2: SEM pictuures of: 1.SWCNTs in a trench before (a,b) and after clamp deposition (c,d); 2.Effects of dielectrophoresis duration on the amount of SWCNTs aligned by a pair of electrodes.
a, c: One SWCNT deposited after 45s (8 V_{pp}, 5 MHz);
b, d: Two SWCNTs deposited after 70s (8 V_{pp} , 5 MHz).

Figure 3 show the representatives of large-scale aligned arrays with individual SWCNTs brigding each metal clamp

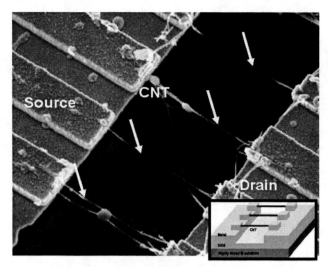

Figure 3: Four adjacent (1μm spacing) highly doped substrate bottom gate SWCNT resonators (device schematic shown in insert) shown as the representative of the whole array, with each electrode pair bridged by mostly one SWCNT or a small SWCNT bundle.

pair with no misalignment. We observe high yields (over 85%) of the well-aligned metal clamp pairs spanned by single SWCNT or single small SWCNT bundle in arrays. Even higher yields and better quality are expected by using more optimized and uniform SWCNTs and adjusting the dielectrophoresis duration with proper suspension concerntration.

High density of SWCNT arrays was achieved with arbitrary metal clamp spacing (500nm in Figure 3). According to the dimensions of the electrode gap between each metal clamp (2um), the width of a metal clamp (100nm) and the interval between adjacent metal clamp pairs (1um), a ultra-high density of more than 100 million SWCNT devices per square centimeter was demonstrated, two orders of magnitude denser than the previous work. In fact, the metal clamp interval is even arbitrary, which makes the potential array density higher and this techinque more flexible for NEMS applications.

The ability to uniformly assemble highly dense parallel arrays of SWCNTs on substrates in large-scale meets various application criteria for device characteristics. The proposed technique enables a robust way for fabricating and exploring a broad field of SWCNT electronic devices.

3.2 I-V characteristics

To confirm that the devices fabricated by the proposed method are electrically active and robustly assembled, we used the Cascade probing systems to contact the individual electrode pairs (100μm x 50μm) and the Agilent 4156C analyzer to measure the electronic transport through the individually accessible SWCNT-electrode pairs. All

SWCNTs on the sample were suspended by BHF etching for 1 min and dried by a critical point drier (CPD).

Over 60 pairs of bridged electrodes pairs were measured and all of them turned out to be functioning devices, indicating the reliability of our self-assembly technique. The smaller electrodes (~150nm x 3μm) were too small to be probed by our probing systems for electrical characterization. Figure 4 shows source-drain current (I_{DS}) vs. voltage (V_{DS}) of a semiconducting SWCNT with gate grounded. Semiconducting SWCNTs show nonlinear I-V curve due to the Schottky barrier at the CNT-metal contact and and its dependence on the substrate bias (acting as a transistor gate).

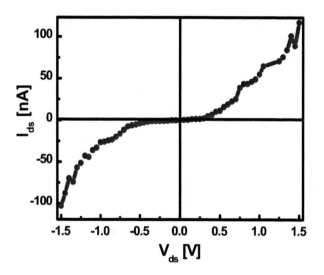

Figure 4: Nonlinear I-V characteristics of the semiconducting SWCNT in Figure 3, due to the Schottky barrier at the CNT-metal contact and dependence on substrate bias.

It is proved that typical contact resistances of our SWCNT bridging devices range from 0.5-3MΩ, which means the contact between SWCNT and evaporated metal clamp is of good quality. Improved contacts can be achieved by annealing the devices at 200°C for around 2h to improve the interfaces between the CNTs and the metal.

3.3 SWCNT resonator arrays

The proposed assembly approach enables the fabrication of ultra-dense well-aligned SWCNT NEMS devices for RF and sensing applications, such as clamped-clamped SWCNT resonator arrays. Two types of doubly clamped SWCNT resonator arrays with different gate configurations were demonstrated.

Figure 3 shows four adjacent well-aligned SWCNT resonators with highly doped Si substrate as back gate, as representatives of the whole dense array. Figure 5 shows four neighboring self-aligned SWCNT resonators with metal bottom gate, as representatives of the whole parallel

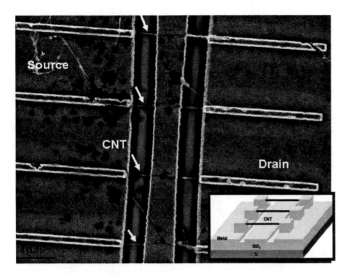

Figure 5: Four adjacent (1.5μm spacing) metal bottom gate SWCNT resonators (device schematic shown in insert) shown as the representative of the whole array, with each electrode pair bridged by one SWCNT or a small SWCNT bundle.

array. Most of the resonators are spanned by a single SWCNT or a small SWCNT bundle. As the proposed process can be carried out at die or wafer level, a yield of ~85% is observed in both SWCNT resonator arrays.

4 CONCLUSIONS

In conclusion, we have proposed a novel wafer level self-assembled technique for fabricating both single SWCNT and ultra dense SWCNT arrays with precise positioning and with individual and reproducible clamps by dielectrophoresis. Combing the four mechanisms: long range and local electric-field forces, capillary forces and SWCNT lift-off, high-yield positioning of SWCNT arrays with ultra-high density (>100 million SWCNT devices/cm^2), two orders of magnitude higher than previous reports, were successfully realized. The SWCNTs are proved to have reliable contacts to the clamping electrodes. In addition, arrays of two typical SWCNT resonators were fabricated, exhibiting the potential of our method for integrating SWCNTs for NEMS applications. Also, the method described in this report is very versatile as it is fully compatible with silicon-based microelectronics fabrication technologies and a variety of pre- and post-processing micro-fabrication techniques. At the wafer level, this approach allows for the self-aligned fabrication of large-scaled individually addressable SWCNT devices for future RF and sensing commercial applications.

ACKNOWLEDGMENT

This work was supported by the Swiss Nanotera project CABTURES and by the FP6 IST-028158 project NANORF (Hybrid Carbon Nanotube – CMOS RF Microsystems). The authors would like to thank CMI-EPFL for technology support and to Professor Christofer Hierold from ETH Zurich for many useful discussions.

REFERENCES

[1] Z. Hou, D. Xu and B. Cai, Appl. Phys. Lett., 89, 213502, 2006.

[2] T. Schwamb, T.-Y. Choi, N. Schirmer, N. R. Bieri, B. Burg, J. Tharian, U. Sennhauser, and D. Poulikakos, Nano Lett., 7, 12, 2007.

[3] T. Schwamb, T.-Y. Choi, N. Schirmer, N. R. Bieri, B. Burg, J. Tharian, U. Sennhauser, and D. Poulikakos, Nano Lett., 7, 12, 2007.

[4] Y. G. Zhang, A. L. Chang, J. Cao, Q. Wang, W. Kim, Y. M. Li, N. Morris, E.; Kong, J. Yenilmez, and H. Dai, J. Appl. Phys. Lett., 79, 19, 2001.

[5] J. Y. Chung, K. H. Lee, J. H. Lee, and R. S. Ruoff, Langmuir, 20, 8, 2004.

[6] R. Krupke, F. Hennrich, H. Löhneysen, and M. Kappes, Science, 301, 344, 2003.

[7] R. Krupke, F. Hennrich, M. Kappes, and H. Löhneysen, Nano Lett., 4, 8, 2004.

[8] R. Krupke, F. Hennrich, H. B. Weber, M. M. Kappes, and H. V. Löhneysen, Nano Lett., 3, 8, 2003.

[9] A. Vijayaraghavan, S. Blatt, D. Weissenberger, M. Oron-Carl, F. Hennrich, D. Gerthsen, H. Hahn, and R. Krupke, Nano Lett., 7, 6, 2007.

[10] A. Le Louarn, F. Kapche, J.-M. Bethoux, H. Happy, G. Dambrine, V. Derycke, P. Chenevier, N. Izard, M. F. Goffman, and J.-P. Bourgoin, Appl. Phys. Lett., 90, 233108, 2007.

100nm RANGE SHORT TUBE PRODUCTS OF NANO CARBON FROM SOLID PHASE SYNTHESIS PROCESS WITHOUT SHORTENING EFFORT

K. C. Nguyen.*, V. V. Tran*, AT.N. Mai*, M.C. Bui*, H. D. Dang**

* Saigon Hi Tech Park Research Laboratories (SHTP LABS)
** TH LLC

ABSTRACT

We have studied a solid phase synthetic process of carbon nano tube using full *solid raw materials* blended with the *tube control additives (TCA) and metal catalyst*. It is discovered from this synthetic process that TCA is the key component producing tube shape of nano carbon products from the process. It is also observed that relatively short tube length products (60 - 100nm) can be obtained without any effort to shorten them. The small scale of the raw material precursor is believed to be the key of the short tube products when a suitable baking condition (sudden heating or gradual heating) is practiced. Short tube of nano carbon products is desired for multiple applications, firstly, improving polarity and compatibility with other media in nano composites , then tip for AFM, electronic device,

Keywords: *60-100nm sort tube, shortening, sudden heating, Solid phase synthesis, solid raw materials, tube control additives (TCA),*

1 INTRODUCTION

The carbon nanotube (CNT) products from catalytic growth process, generally, are long tube ranging from um to mm. The in situ products are very hard to use for multiple applications, especially, for dispersing in certain medium to form nano composite. A lot of efforts had been made to shorten carbon nano tube products. The shortening of carbon nano tube had been known by thermally assisted field evaporation [1] , by electron bombardment [4] (100nm short tube) for AFM tip applications , by ball milling [2], by E beam for transistor applications (20nm short nanotube) [3], by steam [5] , by oxidation with sulfuric acid [6], by electric arc [7] for tube functionalization . The post shortening process easily causes the tube damage.

Recent years, we have investigated the solid phase process to produce CNT using precursor as admix of solid carbon source, tube control additive (TCA), metal catalyst and found that the TCA plays a significant role in providing tube shape for nano carbon products. The solid phase process, also provides several advantages over the conventional gas phase catalytic support growth process for examples, a) good mix between carbon source (CS) and metallic catalyst source (MS) for better control of CS/MS ratio and thus, better control of compatibility between CS and MS, better control of CS adsorption on suitable amount of MS b) products having better uniformity in terms of tube length and tube diameter. The reaction can occur in a very simple oven containing non oxidizing gas feeder, heater and temperature controller. The technology had been published in the proceeding of Nanotech 2007, 2008, 2009.

As continuing efforts to develop single walled nanotube (SWNT) with solid phase process, we had investigated several factors including the scale and shape of raw material sources. We discovered that very short, small tubes were able to achieve when much finer particle of carbon source is used with specific baking condition.

2 REACTOR

In the present study, the CNT was prepared by the pyrolysis of solid CS in a furnace equipped with high heat resistant ceramic materials including oven cover, heat resistant layer, heat resistant ceramic tube, coil heater, heat controller and a Pyrex glass reactor tube. The Pyrex glass tube is connected with 3 neck connector in one end where suitable inert gases can be fed in or the air can be succeed out to form unoxidizing environment in the reaction chamber . Another connector neck can be used to remove the gaseous waste from the reactor. The gas feeding is controlled by a gas flow meter supplied by Kobold, USA. The reactor chamber is a Pyrex glass tube with variable diameter for different quantity of solid raw material feeding. The larger diameter the higher product throughput. The present furnace can provide up to 3 kg CNT product for each reaction batch taking place between 1 to 3 hrs. The heating system (heater and controller) can provide a well controlled temperature to the reactor chamber up to 1200C.

3 MATERIAL SET

Fig. 1A Effect of raw material size scale on the shape of solid phase nano tube (SPNT) product

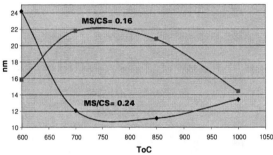

DIAMETER vs BAKING TEMP (oC)

Fig. 1B indicates the effect of baking temperature on the diameter of the SPNT products fabricated with sudden heating process having different MS/CS ratio

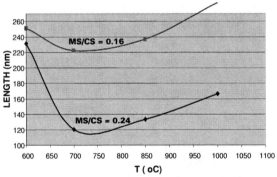

Fig 1C shows the relationship between baking temperature and tube length in a sudden heating mode

Solid carbon source is selected from flammable solid having high C content such as natural tree products

Catalysts are selected from the compounds containing transition metal elements of Ni, Fe, Co

TCA are proprietary of the present study, which is selected from wide range of low molecular weight and high molecular weight molecules having specific functionalities such as –OH, -NR,

The full solid precursor of the solid phase process is prepared by mixing powder of solid carbon source with catalyst and TCA in solvents followed with the evaporations of solvent.

4 PROCESS

The dried full solid precursors were carefully weighed into a quartz (Pyrex) tube then the air in the tube was evacuated followed with the purging of unoxidizing gases (N2 or Ar). The reactor tube must be inserted into the furnace where the heating occurs to form carbon nano tube products. There is several heating process which were practiced in the present study. In the ***sudden heating process***, the quartz tube filled with precursor was inserted into a preheated chamber and keeps heating temperature constant during baking process. On the other hand, the quartz tube containing the raw material is inserted into the reactor fist set at room temperature the slowly raise the reactor to the set temperature and keep it constantly during baking time . This is called as ***gradual heating process***. This process does show the effect on controlling the tube length as well as tube diameter.

At the reaction end, the heat was slowly removed under unoxidizing environment and the product was taken out at normal atmosphere environment.

5 TESTING

I) FE-SEM measurement was carried out using Hitachi S4800.

ii) *TEM measurement* was carried out with Hitachi TEM

iii) *XRD measurement* was carried out with Siemens 5000

IV) *Raman Spectroscopy* measurement was done with Model LabRam-1B (Jobin Yvon)

6 RESULTS

Effect of raw material size scale As indicated in Fig1.A a normal wood dust (CS) having average particle size of 500nm long and 150nm wide was treated with long alkyl chain alcohol at high temperature (180C) for 4 hours yielding ultra fine wood fibril with average width of 30-50nm , 3-5X reduction in size . For comparison purpose, these materials were used to formulate the precursor containing TCA and catalyst and baked the same way (850C, 1 hr,). As result, the normal wood dust yields products having average diameter D= 50nm with average length > 1000nm, while the ultra fine wood dust gives rise to product having average diameter of 10nm and average length of 100nm. So, changing the size scale of the carbon source would be the way to narrow down the tube size. Fig 1B indicates the effect of baking temperature on the average diameter for two different MS/CS ratios. It should be noted that (MS= metallic catalyst source, CS= carbon source) the (MS/CS) ratio does show some effect on tube diameter in *the sudden heating mode*. The MS/CS ratio = 0.24 shows smaller diameter tube in a scale of 2X compared to smaller MS/CS ratio=0.16.

Fig 1C shows the relationship between baking temperature and tube length in *a sudden heating mode* for two different (MS/CS) ratios. Again, the larger (MS/CS) exhibits shorter length compare to the larger one.

In another study, we calculated the volume V (nm^3) of the tube based on the formulae V= πR^2 L. The value of V indicates how small the tube is. Fig 1D exhibits the

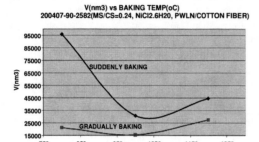

V(nm3) vs BAKING TEMP(oC)
200407-90-2582(MS/CS=0.24, NiCl2.6H20, PWLN/COTTON FIBER)

Fig. 1 D Relationship between baking temperature and tube volume (nm3) in two different heating modes: sudden and gradual

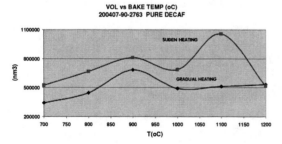

VOL vs BAKE TEMP (oC)
200407-90-2763 PURE DECAF

Fig. 1E shows the relationship between baking temperature and tube volume (nm^3) for coffee precursor baked at different mode of heating: sudden and gradual

relationship between V (nm3) and baking temperature for

comparing the effect of *sudden heating mode* and *gradual heating mode*. The precursor is made out of *cotton* as CS, PWLN as TCA and NiCl2 as MS. It is found that sudden bake process gives rise to much large size tube (volume) than gradual bake no matter what the baking time is the same for both.

Fig. 1E shows the effect of baking temperature on tube volume for two different baking modes using *coffee precursor*. Again, the similar results was obtained; sudden heating mode gives larger volume tube (bigger size) than gradual heating mode.

7 CONCLUSIONS

It can be concluded that in the solid phase synthesis of CNT utilizing TCA

1. Reduction of size scale of the raw material can reduce the diameter, the length and the overall tube volume at well

2. Larger MS/CS ratio gives rise to smaller diameter and length

3. Gradual heating mode always gives smaller tube than sudden heating mode

REFERENCES

[1] Seiji Akita and Yoshikazu Nakayama, Jpn. J. Appl. Phys. 41 (2002) pp. 4887-4889
[2] N. Pierard[a], A. Fonseca[a], Z. Konya[a], I. Willems[a], G. Van Tendeloo[b] and J. B.Nagy[a], Letters, Volume, 16 February 2001, Pages 1-8
[3] R. V. Seidel,* A. P. Graham, J. Kretz, B. Rajasekharan, G. S. Duesberg, M. Liebau, E. Unger, F. Kreupl, and W. Hoenlein, *Nano Lett.*, 2005, 5 (1), pp 147–150
[4] Seiji Akita and Yoshikazu Nakayama, Jpn. J. Appl. Phys. 41 (2002) pp. 4887-4889
[5] Ballesteros B, Tobias G, Shao L, Pellicer E, Nogués J, Mendoza E, Green ML., Small. 2008 Sep; 4(9):1501-6.
[6] FORREST Gavin A.; ALEXANDER Andrew J.; Journal of physical chemistry. C ISSN 1932-7447; 2007, vol. 111, n°29, pp. 10792-10798 [7 page(s) (article)]
[7] M.A. Hamon, H. Hui, P.Bhowmik, H.M.E. Itkis, R.C. Hardon, Appl. Phys. A74, 333-338(2002)

SYNTHESIS AND CHARACTERIZATION OF SINGLE WALL CARBON NANOHORNS PRODUCED BY DIRECT VAPORIZATION OF GRAPHITE

Cesare Pagura*, Simona Barison*, Simone Battiston*, Mauro Schiavon**

* Institute for Energetics and Interphases, National Council of Research (CNR-IENI)
C.so Stati Uniti, 4, 35127 Padova, Italy, c.pagura@ieni.cnr.it
**Carbonium srl, Via Penghe 28, 35100 Selvazzano (PD), Italy, info@carbonium.it

ABSTRACT

A new method is described for large-scale production of Single Wall Carbon Nanohorns (SWCNH). In a prototype reactor (a 30 kW plant for an estimated production of about 100 g/h of soot) optimal thermodynamic conditions that favour nanohorn formation were determined. The characterization of collected soot was performed by several techniques: thermogravimetric analyses (TGA), Raman spectroscopy, and transmission/secondary electron microscopies (HRTEM - SEM). The soot analyses revealed conspicuous presence of good-quality carbon nanostructure, mainly SWCNHs, with small amounts of graphene foils and amorphous phase material.

The first results demonstrated the excellent capability of the new approach that can be easily scaled up to tens the present scale, potentially becoming a powerful method to bring SWCNHs into every day applications

Keywords: carbon, nanotubes, nanohorns, graphene

INTRODUCTION

Single Wall Carbon Nanohorns [1,2] (SWCNH) represent one of the most interesting carbon nanostructures belonging to the thrived nanotube family.

A key characteristic is their tendency to group together and form aggregates (spherical clusters or bundles) like dahlia flowers or buds, with overall diameters of tens/hundreds nm. Advantage of this "self assembling" characteristic is not only the very large surface area, but also an easy permeation of gases and liquids inside their structure.

In spite their tiny tubular structure, SWCNHs maintain many of the typical properties of carbon nanotubes: easiness of functionalization and good electrical and thermal conductivities. Nowadays, SWCNHs therefore represent more than a promise for a very wide range of possible uses, e.g.: methanol fuel cells support; efficient hydrogen storage; super-capacitors; generation of hydrogen by decomposition or steam-reforming of methane; composites for lubrication. Moreover, recent literature [3,4] focussed the important role of SWCNH for photovoltaic applications, since they are considered to be ideal for electron-acceptors hybrid systems.

In addition, many applications in life science were recently reported, e.g. ranging from drug delivery to NIR laser-driven functional CNH complexes used as antiviral materials. Regarding SWCNH toxicity, a very important issue for social sustainability of nanotechnologies, it is widely confirmed that, in experiments using mice and rats, the cytotoxicity was negligibly small [5,6]. Also SWCNH-based hybrid materials and nanofluids with promising properties were reported [7,8].

Not still commercially available, SWCNHs can be produced in laboratory-scale quantities without catalysts, a key feature since they are very difficult to remove, and with high purity starting from graphite by laser ablation/vaporization processes [1,9] or by AC or DC arc discharge [10,11].

For relevant amount, the only known industrial-scale process uses an advanced graphite vaporization method, consisting in spotting a powerful CO_2 laser on a rotating cylindrical rod [12].

With all the foreseen potential market applications, an enormous impact is expected if SWCNHs could become a low-cost raw material thanks to diffusion of methods for massive production.

In this framework, a new method, based on heating of graphite rods by induction of very intense, high frequency, eddy currents (a patented method already tested successfully for fullerenes and carbon nanotubes production [13]) was tailored for mass production of SWCNHs and presented here.

EXPERIMENTAL

To produce carbon nanostructures like fullerenes, nanotubes or nanohorns starting from solid precursors (graphite), a very high specific energy (~60 kJ/g) supply is necessary for the vaporization/atomization of precursor. Arc discharge and laser ablation reach such critical energy density, but discharge methods, being limited in the maximum current compatible with a controlled arc current, cannot be easily scaled up and the laser ablation technique requires a very high power and costly CO_2 laser for real mass productions (hg/h of soot), although it is the only actually used.

The process of heating a conductors by electromagnetic induction, where eddy currents are generated within the object itself giving rise to Joule heating, is well known and consolidated for industrial application. One of the most interesting properties of heating by eddy currents is the relevant power density that is possible to concentrate on workpieces, that can reach several kW·cm⁻², sufficient to reach the temperature (3000-3200°C) necessary to sustain the evaporation of graphite and permits the successive condensation of carbon nanostructures. Moreover, this kind of plant can be scaled up till to MW scale, even increasing the overall efficiency.

Recently, the patented reactor [13] was realized to demonstrate feasibility of induction heating for CNTs production, where a synergic heating of both solid and plasma phases reaches the necessary thermodynamic conditions for the NTs synthesis. A new 30 kW power plant was specifically set up for SWCNH fabrication for soot production rates around 0.1 Kg/h. A detailed description of the apparatus and typical operational conditions is given in the cited patent.

The reactor chamber, cooled by water circulation, is pumped down by primary and turbomolecular vacuum pumps (Edwards and Pfeiffer Vacuum), equipped with a (Air Liquide) gas streaming inlet, controlled by a mass flow controller (MKS). A cooled finger acts as collector to harvest soot production. Since production occurs in dynamic gas flow condition, an exit port drain the excess gas, that is filtered (0.1 micron mesh, Donaldson) and recycled in the process. A quartz window allows the process to be observed externally. No direct temperature control is possible in the process, but off-line measures by an optical pyrometer through the quartz window assured that temperatures over 3000°C on graphite surface were reached during heating.

Soots were produced under diverse experimental conditions and with different kind of commercial graphite (SGL Carbon Gbmh). Experiments to optimize the flux and the fluidodynamic behavior in presence of extreme temperature gradient are still in progress.

RESULTS

All the mentioned SWCNH production methods tends to produce some quantities of wrapped or unwrapped graphene foils, amorphous carbon and graphite sub-micron particles, aka "giant graphite balls" (GGBs) [14]. Thermo-Gravimetric Analysis (TGA) in air or in oxygen flux is a very efficient tool for determining the overall quality of soot material, providing information on the different carbon structures, due to differences in their decomposition temperatures.

Fig.1 (red lines) represents typical TGA curves (800°C max, in air flux, 5°C/min ramp, performed with a SDT Q600 - TA Instruments) of a production batch of soot "as prepared". In accord with literature ([7,14]), in the derivative curves different phases are well observable: an amorphous phase peak at 400°C, followed by peaks related to SWCNHs, and almost two peaks ascribable to residual

graphitic structures as in [7] (GGBs and graphene). The residual is quite low, below 4%, compatible with the purity of graphite used in these experiments. For comparison, neat TGA curves of used graphite are reported (blue lines) in the same figure.

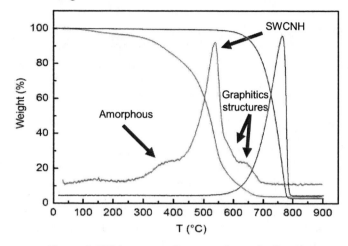

Figure 1: TGA curves of untreated soot (red) and pure graphite (blu).

TEM is the most accredited technique to evaluate the soot, even if, visualizing less than few pg of sample, cannot be assertive for overall production quality. Fig. 2 shows TEM images of single particles found in the soot produced by the reported method: both typical assemblies of nanohorns (bud-like and dhalia-like) are present. Fig.3 represents a direct comparison of HR-TEM and SEM images, showing agglomerates of dhalia- and bud-like nanohorns.

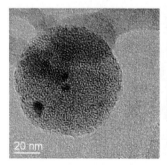 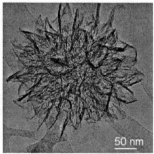

Figure 2: TEM images of single nanohorn particles in the untreated soot: both bud-like and typical Dhalia-like structures are present

The other acknowledged technique able to evidence the presence of SWCNHs is the Raman spectroscopy. Raman spectrum in fig. 4 shows typical bands observed for SWCNH, from two samples of soot respectively collected from the cool finger inside the reactor and from the residual dust from the chamber walls, with a high intensity D band

Figure 3: Comparison of TEM (left) and SEM (right) images of SWCNH

at 1314 cm^{-1} and the G band at 1594 cm^{-1} with intensity comparable to that of the D band. The spectrum also shows the G' band (overtone of the D band) at 2620 cm^{-1} and a characteristic band at 2895 cm^{-1}.

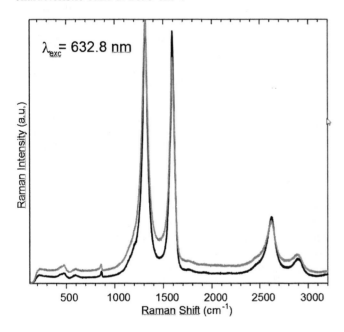

Figure 4: Raman spectra of soot sampled in two zones of rector chamber. Typical bands for SWCNH are observed in both samples

CONCLUSION

The presented method demonstrated its remarkable potentiality in large-scale production of SWCNHs with good purity. The plant can be scaled up to tens the present scale. Furthermore, by changing the reactor parameters and injecting suitable catalyst (Fe or Co organometallic compounds) or adding other gases to carrier gas (like nitrogen or diborane), the same plant could be able to produce N-doped or B-doped SWCNHs or different kind of carbon nanomaterial, i.e. single wall nanotubes.

REFERENCES

[1] Iijima S., Yudasaka M. et al., "Nano-aggregates of single-walled graphitic carbon nano-horns", Chemical Physics Letters, 1999;309(3-4):165-70.

[2] Iijima S., "Carbon nanotubes: past, present, and future", Physica B: Condensed Matter, 2002;323(1-4):1-5.

[3] Cioffi C., Campidelli S. et al., "Synthesis, characterization, and photoinduced electron transfer in functionalized single wall carbon nanohorns", Journal of the American Chemical Society, 2007;129(13):3938-45.

[4] Pagona G. et al., "Covalent functionalization of carbon nanohorns with porphyrins: Nanohybrid formation and photoinduced electron and energy transfer", Advanced Functional Materials, 2007; 17(10):1705-11.

[5] Miyawaki J., Yudasaka M., Azami T., Kubo Y., Iijima S., "Toxicity of Single-Walled Carbon Nanohorns", ACS Nano, 2008;2(2):213-26.

[6] Lynch R.M. et al., "Assessing the pulmonary toxicity of single-walled carbon nanohorns", Nanotoxicology, 2007;1(2):157-66.

[7] Battiston, S. et al. "Growth of titanium dioxide nanopetals induced by single wall carbon nanohorns" Carbon. 2010, Article in Press

[8] Sani, E. et al., "Carbon nanohorns-based nanofluids as direct sunlight absorbers" (2010) Optics Express, 18 (5), pp. 5179-5187.

[9] Azami T. et al. "Production of small single-wall carbon nanohorns by CO2 laser ablation of graphite in Ne-gas atmosphere" Carbon, 2007;45(6):1364-7.

[10] Gattia D.M., Vittori Antisari M., Marazzi R., "AC arc discharge synthesis of single-walled nanohorns and highly convoluted graphene sheets", Nanotechnology, 2007;18(25).

[11] Gattia M., Antisari M.V., Marazzi R., Pilloni L., Contini V., Montone A., "Arc-discharge synthesis of carbon nanohorns and multiwalled carbon nanotubes", Materials Science Forum 2006; 2006. p. 23-8.

[12] Azami T., Kasuya D., Yuge R., Yudasaka M., Iijima S., Yoshitake T. et al., "Large-scale production of single-wall carbon nanohorns with high purity", Journal of Physical Chemistry C, 2008;112(5):1330-4.

[13] Schiavon M., "Device and method for production of carbon nanotubes, fullerene and their derivatives" U.S. Pat. 7,125,525 - EP 1428794 2006.

[14] Fan J., Yudasaka M., Kasuya D., Azami T., Yuge R., Imai H., et al., "Micrometer-sized graphitic balls produced together with single-wall carbon nanohorns", Journal of Physical Chemistry B, 2005;109(21):10756-9.

Effect of Strain on the Energy Band Gaps of (13,0) and (15,0) Carbon Nanotubes

G. Dereli, N. Vardar and O. Eyecioğlu

Department of Physics, Yildiz Technical University, İstanbul, Turkey
gdereli@yildiz.edu.tr, necativardar@gmail.com, oeyeci@yildiz.edu.tr

ABSTRACT

The electronic energy band gap is a basic property of all semiconductors since it is responsible for the electrical transport and optical properties. Control of the size of the energy band gap is important in optimizing electronic devices. In this study, we have shown that the application of tensile strain modifies the size of the band gaps in semiconducting Single Walled Carbon Nanotubes (SWCNTs) and opens a band gap in metallic SWCNTs. We simulated (13,0) and (15,0) SWCNTs under various compressive and tensile strain at 300K. Pristine (13,0) SWCNT is a semiconductor with a band gap of 0.44 eV and (15,0) SWCNT is metallic. The strain was applied as both compression and stretching along the z-direction. The density of states are obtained in real space using a parallel Order N, Tight Binding Molecular Dynamic (O(N) TBMD) simulation code designed by Dereli et. al. [1-3].The energy band gaps of (13,0) and (15,0) SWCNTs demonstrate different behaviors with the presence of strain. For (13,0) tube, energy band gap decreases with negative strain values (compression) and the onset of semiconductor-metal transition occurs at a negative strain value of %8. For (15,0) SWCNT, application of positive and negative strains opens up the band gap causing metal - semiconductor transitions.

Keywords: Single-walled carbon nanotubes, Order-N Tight-Binding Molecular Dynamics simulations, strain, electronic energy band gap modifications.

1 INTRODUCTION

Single Walled Carbon Nanotubes (SWCNTs) display a variety of electronic structures depending on their chirality. The structure of a SWCNT is specified by a chiral vector $(\vec{C}_h = (n, m))$. Depending on the chirality vector, SWCNTs are classified into three types, namely, armchair *(n,n)*, zigzag *(n,0)* and chiral *(n,m)*. The armchair SWCNTs are always metallic, and the zigzag SWCNTs are metallic only when n is a multiple of 3.

Control of the size of the energy band gap is important in optimizing such devices as p-n junctions, transistors, photodiodes and laser works. In the present work, we have shown that the presence of positive and negative strain modifies the size of the energy band gaps in semiconducting SWCNT s and opens a band gap in metallic SWCNTs.

We demonstrated the modification of the electronic band gap with strain on (13,0) and (15,0) SWCNTs. At 300K and zero strain, (13,0) SWCNT is a semiconductor with a band gap of 0.44 eV and (15,0) SWCNT is metallic. The strain was applied as both compression and tension along the z direction. The results are obtained using a parallel Order N, Tight Binding Molecular Dynamic (O(N) TBMD) simulation code designed by Dereli et. al [1-3]. This code is used in SWCNTs simulations successfully and the details of the technique can be followed in [4-6].

O(N) TBMD calculates the band structure energy in real space and makes the approximation that only the local environment contributes to the bonding, and hence the band energy of each atom. In this case, the run time would be linearly scaled with respect to the number of atoms.

We simulated (13,0) and (15,0) SWCNTs under various compressive and tensile strain values. Positive strain values correspond to tensile and the negative strain values to compressive strain. Strain values are chosen to stay in the plastic region for these SWCNTs. Total energy and Fermi energy levels are obtained as functions of strain. Band gap is calculated from the behavior of electronic density of states near the Fermi energy level for each strain values.

2 METHOD

Our simulations are performed using Order N, Tight Binding Molecular Dynamic (O(N)-TBMD) simulation technique for the canonical (NVT) ensemble. Inter-atomic forces needed to move atoms are calculated from TBMD Hamiltonian by means of the Hellman-Feynman theorem. The many body Hamiltonian (H_{tot}) is reduced to Hamiltonian (h) of one electron moving in the average field due to other electron and ion cores:

$$H_{tot} = K_i + K_e + U_{ee} + U_{ie} + U_{ii} \tag{1}$$
$$h = K + U_{ie} + U_{ee} \tag{2}$$

The corresponding Schrödinger's equation is written as:

$$h|\psi_n\rangle = \varepsilon_n|\psi_n\rangle \tag{3}$$

where $|\psi_n\rangle$ are eigenstates corresponding to the eigenvalues ε_n. In TB formalism, $|\psi_n\rangle$ are taken as linear combinations of atomic orbitals $|\phi_{l\alpha}\rangle$ (l:quantum number index, α: labels the ions)

$$|\psi_n\rangle = \sum_{l\alpha} C_{l\alpha}^n |\phi_{l\alpha}\rangle \qquad (4)$$

Since the $|\phi_{l\alpha}\rangle$ basis set is not orthogonal, one may use instead the orthogonal set of atomic orbitals known as Löwdin orbitals $|\varphi_{l\alpha}\rangle$ 'n terms of which the secular equation becomes

$$\sum_{l'\beta} \left(\langle \varphi_{l'\beta} | h | \varphi_{l\alpha} \rangle - \varepsilon_n \delta_{ll'} \delta_{\alpha\beta} \right) C_{l'\alpha}^n = 0. \qquad (5)$$

To solve this secular equation with as small number of TB parameters as possible, some approximations are made [7]: (i) Minimal basis set is used in secular equation. In this study we use sp^3 basis set for SWCNTs. (ii) the hopping integrals corresponding to interatomic distance larger than a suitable cutoff are not considered. (iii) only two-center integrals are taken into account.

The key objects of the secular equation are the matrix elements $\langle \varphi_{l'\beta} | h | \varphi_{l\alpha} \rangle$ known as hopping integrals of one electron Hamiltonian. The equilibrium hopping integrals $\langle \varphi_{l'\beta} | h | \varphi_{l\alpha} \rangle = h_{ll'}^{(0)}$ are generalized as

$$h_{ll'}(r_{\alpha\beta}) = h_{ll'}^{(0)} f_{ll'}(r_{\alpha\beta}) \qquad (6)$$

where $f_{ll'}(r_{\alpha\beta})$ is known as the scaling function for atoms which are not in their equilibrium positions .

The solution of the secular equation determines the single electron eigenvalues ε_n. The total energy of a system of ion cores and valence electron is obtained as

$$U_{tot} = 2 \sum_n \varepsilon_n f(\varepsilon_n, T) + U_{ii} - U_{ee} = U_{bs} + U_{rep} \qquad (7)$$

where $f(\varepsilon_n, T)$ is the Fermi-Dirac distribution function, $-U_{ee}$ corrects the double counting of the electron-electron interaction in the first term. The sum of all the single particle energies is commonly called the band structure energy and denoted by

$$U_{bs} = 2 \sum_n \varepsilon_n f(\varepsilon_n, T) . \qquad (8)$$

The factor of 2 in the first term comes from spin degeneracy. The effective repulsive potential is expressed as a sum of suitable short ranged two-body potentials [8]

$$U_{rep} = U_{ii} - U_{ee} = \sum_{\alpha,\beta>\alpha} \Phi(r_{\alpha\beta}) \qquad (9)$$

where $\Phi(r_{\alpha\beta})$ is a pairwise potential between atoms α and β and $r_{\alpha\beta}$ is a pairwise distance between atoms. The functional form suggested by Goodwin, Skinner and Pettifor is used for pair wise potentials in [8,9] so that the TBMD Hamiltonian and corresponding forces are expressed as:

$$H_{TBMD} = \sum_\alpha \frac{p_\alpha^2}{2m_\alpha} + \sum_n \varepsilon_n f(\varepsilon_n, T) + U_{rep} \qquad (10)$$

$$\vec{F}_{tot,\alpha} = - \sum_n \left\langle \psi_n \left| \frac{\partial H}{\partial \vec{r}_\alpha} \right| \psi_n \right\rangle f(\varepsilon_n, T) - \frac{\partial U_{rep}}{\partial \vec{r}_\alpha} . \qquad (11)$$

The repulsive force term can be computed analytically since repulsive potential U_{rep} is known as a function of the interatomic distances:

$$F_{rep} = - \frac{\partial U_{rep}}{\partial r_\alpha} = - \frac{\partial}{\partial r} \sum_\alpha f\left(\sum_j \phi(r_{\alpha\beta}) \right) . \qquad (12)$$

Attractive force can be computed only numerically through Hellman-Feynman theorem. Hellman-Feynman contribution to the net force is given by

$$\sum_n \left\langle \psi_n \left| \frac{\partial H}{\partial \vec{r}_\alpha} \right| \psi_n \right\rangle f(\varepsilon_n, T) =$$
$$\sum_n f(\varepsilon_n, T) \sum_{l\alpha} \sum_{l'\beta} C_{l'\alpha}^n \frac{\partial H_{ll'\alpha\beta}(r_{\alpha\beta})}{\partial \vec{r}_\alpha} C_{l\alpha}^n . \qquad (13)$$

In order to determine the band structure energy and Hellman- Feynman forces, we need to know the full spectrum of eigenvalues ε_n and the corresponding eigenvectors $C_{l\alpha}^n$. So we must diagonalize the TBMD Hamiltonian matrix at every time step of the simulation. The standard diagonalization of the TB matrix of dimension nNxnN (N is the number of the atoms in the system and n is the number of valence electrons per atom) requires computation time that scales as $O(N^3)$ and dominates the overall computational workload of the TBMD simulations. On the other hand by using an O(N) algorithm, it is possible to reduce the scale of the computation time from N^3 to N. The recent O(N) methods solve for the band energy in real space and make the approximation that only local environment contributes to the bonding and hence bond energy of each atom[10]. One of the O(N) methods commonly used to carry out quantum calculations is the divide and conquer (DAC) approach [11]. The key idea of this approach is to provide a description of a large system in terms of contributions from subsystems. The system is split into sub systems and each subsystem is solved independently. Each subsystem comes with its own Hamiltonian matrix, constructed relative to the corresponding buffer. The effect of the buffer size on O(N) TBMD is very important. We calculated the buffer size as 4.8A for a (15,0) SWCNT and 4.6 A for a (13,0) SWCNT. The cubodial box size is taken as the distance between two cross-sectional layers along the uniaxial direction (1.229A) and periodic boundary condition (PBC) is applied along the z direction only. Interaction cut off distance is chosen as 2.6 A. The simulations are performed using parallel O(N) TBMD simulation method on distributed memory system consisting of a Linux cluster with 16 PC's. The simulation code was written in Fortran77 programming language by G. Dereli et. al. [1-3]. Dynamic memory allocation is used for all dynamic variables. Code is designed according to SIMD taxonomy.

3 RESULTS AND DISCUSSION

Snapshots of the simulated SWCNTs can be seen in Figure 1. Strain was applied along the axial (z) direction and SWCNTs are allowed to relax along the other directions (x and y). The axial strain is obtained from $\varepsilon = (l - l_0)/l_0$ where l_0 is the equilibrium length in the axial direction for unstrained SWCNT and the l is the corresponding length in the strained SWCNT. Positive strain values correspond to tension and negative strain values to compression. Simulations are performed at room temperature and the periodic boundary condition is applied along the z direction. Velocity Verlet algorithm along with the canonical ensemble molecular dynamics (NVT) is used. Our simulation procedure is as follows:

(i) The tube is simulated at 300K for a 3000 MD steps of run time with a time step of 1 fs. This eliminates the possibility of the system to be trapped in a metastable state. We wait for the total energy to reach the equilibrium state.

(ii) Next, uniaxial strain is applied to the tubes. We further simulated the deformed tube structure (the under uniaxial strain) for another 2000 MD steps. Figure 2, shows the variation of the total energy of strained SWCNTs during MD simulations for the several strain values corresponding to compression and tension. First 3000 MD steps show equilibrium of the unstrained SWCNTs and the next 2000 MD steps show the variation of the total energy of the SWCNTs during the simulations under applied strain. In the figure, it is seen that (13,0) and (15,0) SWCNTs are able to sustain their structural stability under applied strain in the elastic limit. In Figure 3, we present the change in the density of states of (13,0) and (15,0) SWCNTs under strain. Energy band gap is calculated from the behavior of eDOS around the fermi level. Strain is applied in the plastic region of $\% - 8 \leq \varepsilon \leq \%8$.

Effect of uniaxial strain on the energy band gaps of (13,0) and (15,0) SWCNTs are given in Figure 4. The energy band gaps of (13,0) and (15,0) SWCNTs respond differently to the presence of uniaxial strain. At zero strain (13,0) SWCNT is a semiconductor with an energy band gap of 0.44 eV . Application of negative strains closes the band gap causing semiconductor-metal transitions. For (13,0) SWCNT, energy band gap decreases with negative strain (compression) and the onset of semiconductor-metal transition occurs at the negative strain value of %8. On the other hand, positive strain increases the energy band gap of (13,0)SWCNT. For (15,0) SWCNT energy band gap increases both with the presence of positive and negative strain. Applications of positive and negative strains open up the band gap causing metal - semiconductor transitions in (15,0) SWCNT.

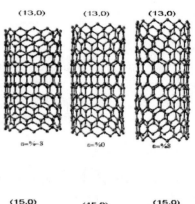

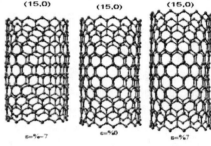

Figure 1: Snapshots of simulated SWCNTs

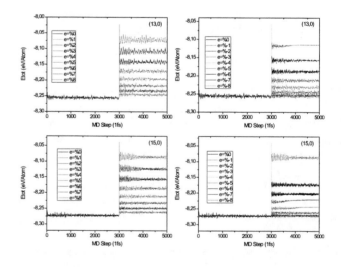

Figure 2: Variations of total energy with strain

4 CONCLUSION

In summary, the electronic structures of deformed (13,0) and (15,0) SWCNTs under strain are simulated in real space using Order N Tight-Binding Molecular Dynamics simulation technique. Our studies show that the energy band gaps of these zig-zag SWCNTs respond differently to the presence of uniaxial strain. The zigzag SWCNTs are generally semiconducting and they are metallic only when n is a multiple of 3. (15,0) SWCNT is a zig-zag nanotube and the application of external strain brings it back to semiconducting state. For (13,0) tube, electronic band gap decreases with compression and the onset of semiconductor-metal transition occurs at negative strain value of %8. What determines if a SWCNT is metallic or not is whether the quantized components of the electronic wave vectors along the circumferential direction intersects or not the graphene's Fermi surface during folding. For example, $(n,0)$ zigzag SWCNTs are metallic only when n is a multiple of three, because only in this case do the quantized k vectors cross the vertices of the hexagonal Brillouin zone of the graphene sheet. Our study shows the potential of a tunable energy band gap of SWCNTs through the application of positive and negative uniaxial strain. The findings of this paper will be helpful in the construction of SWCNT based nano-electronic devices.

The research reported here is supported through the Yildiz Technical University Research Fund Projects YTU 29-01-01-YL02 and YTU 24-01-01-04. The simulations are performed at the Carbon Nanotubes Simulation Laboratory, Department of Physics, Yildiz Technical University, Istanbul, Turkey

(*http://www.yildiz/edu/tr/~gdereli/lab_homepage/index.html*)

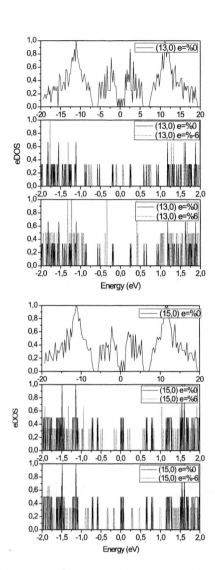

Figure 3: Density of states of (13,0) and (15,0) SWCNTs under strain

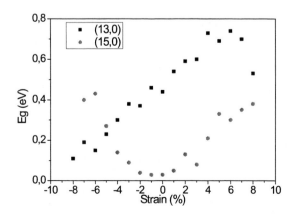

Figure 4: Variations of energy band gaps with strain

REFERENCES

[1] G. Dereli, C. Özdoğan, Phys. Rev. B67, 0354416, 2003.
[2] G. Dereli, C. Özdoğan, Phys. Rev. B67, 0354415, 2003.
[3] C. Özdoğan, G. Dereli, T. Çağın, Comput. Phys. Commun., 148, 188-205, 2002.
[4] G. Dereli, B.Süngü, Phys. Rev. B75, 184104, 2007.
[5] G. Dereli, B.Süngü, C. Özdoğan, Nanotechnology 18, 245704, 2007.
[6] B.Onat, M.Konuk, S.Durukanoğlu and G. Dereli, Nanotechnology 20 ,075707, 2009.
[7] L.Colombo, Comput. Mater. Sci., 12, 278, 1998.
[8] C. H. Xu. C. Z. Wang, C. T. Chang. K. M. Ho, J. Phys.: Cond. Matt. 4, 6047, 1992.
[9] L. Goodwin, A. J. Skinner and D. G. Pettifor, Europhys. Lett. 9, 701, 1989.
[10] P. Ordejon, Comput. Mater. Sci., 12, 157, 1998.
[11] W. Yang, Phys. Rev. Lett. 66, 1438, 1991.

Ab-initio investigation of Graphane

N. Virk[1], B. Montanari[2], N. M. Harrison[2, 3]

[1]Department of Chemistry, Imperial College London, London, SW7 2AZ, UK,
nsv09@imperial.ac.uk

[2] STFC Rutherford Appleton Laboratory, Didcot, Oxford, OX11 0QX, UK,
barbara.montanari@stfc.ac.uk

[3] Department of Chemistry, The Thomas Young Centre, Imperial College London, London,
SW7 2AZ, UK, nicholas.harrison@imperial.ac.uk, web: http://www.ch.imperial.ac.uk/harrison

ABSTRACT

First-principles calculations employing the hybrid exchange functional B3LYP are used to determine the structural parameters, electronic structure and associated properties of a graphene sheet, graphane chair and graphane boat confirmations. Graphane is found to have a band gap corresponding to an electronic insulator at room temperature. The relative performance of the B3LYP functional in comparison to calculations employing generalized gradient approximations (PBE) is also assessed. The B3LYP functional finds the band gap to be almost double that predicted by PBE methods. The choice of functional is found to be insignificant in the determination of the structural parameters

Keywords: dft, hybrid-exchange, graphane, electronic structure,

1 INTRODUCTION

Graphene is a two dimensionally periodic sheet of carbon atoms arranged in an hexagonal lattice. The recent development of relatively simple techniques for isolating such sheets by exfoliation from graphite[1] has lead to a significant increase in interest in the control of mechanical and electronic properties. Graphene has been proposed for use in a wide variety of applications[2].

An area of graphene research that shows great promise is the altering of its structure and composition to create graphene derivatives which have distinct electronic and magnetic properties comparative to the parent material. The graphene lattice can be considered as a single chemical entity, which could potentially react with specific adsorbates resulting in particular arrangements of patterns. The control of such patterns could give the ability to tune its properties. The simplest example of an adsorbate is the hydrogen atom and consequently the mechanisms behind hydrogen adsorption have been extensively researched[3,4,5]. The possible existence and stability of a structure comprised of hydrogen atoms periodically bound to each carbon atom in the graphene lattice was theoretically predicted in 2007[6]. This structure was termed graphane and is thought to have two favorable confirmations; the chair and boat conformers (figure 1). An

interesting feature of graphane is the significant changes in the electronic structure of the graphene sheet that are predicted to be induced by periodic hydrogenation. Hydrogenation of graphene, a semimetal, to form graphane leads to an opening of a band gap and a thus a transition to an insulating state[6]. Graphane was recently synthesized by exposing an annealed graphene sample to cold hydrogen plasma for two hours[7]. This example demonstrates the potential of chemically altering graphene to create functional derivatives. In the case of graphane this functionality arises via the introduction of a band gap, which has far reaching consequences, for example in prospective applications in the electronics industry. The example also highlights the importance of the predictive power of electronic structure methods, which have a particularly important role to play in graphene research as properties are very sensitive to composition and long range structure, and experimental control of these factors is particularly challenging[7].

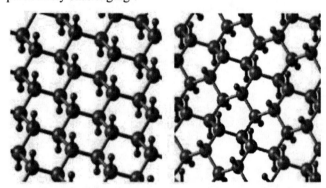

Figure 1-Graphane Chair (left) and Boat (right) confirmations. Green atoms represent carbon atoms of graphene lattice and the black atoms correspond to adsorbed hydrogen

From a theoretical perspective the predominant method utilized in the quantum mechanical simulation of periodic systems such as graphene is density functional theory (DFT) as it provides a tractable theory of the ground state energetics while not relying on empirical parameterization. For example the work by Sofo et al and other more recent theoretical studies[6, 8, 9] investigating the material have all employed DFT. A crucial component of a DFT calculation

NSTI-Nanotech 2010, www.nsti.org, ISBN 978-1-4398-3401-5 Vol. 1, 2010

is the choice of functional employed to treat electronic exchange and correlation. All the above mentioned studies employed functionals based on the generalized gradient approximation (PBE)[10]. The purpose of this work is to perform similar calculations using DFT, with a particular focus on computing the electronic structure and associated properties, along with the structural parameters, of the material. For this reason, in contrast to previous investigations, the functional of choice for this work is the hybrid exchange density functional in the B3LYP form[11, 12, 13]. The mixing of nonlocal and semilocal exchange present in hybrid exchange functional overcomes the major flaws of the local spin-denisty approximation (LSDA)[14] and, more pertinently, generalized gradient approximations (PBE) in the prediction of electronic and magnetic ground states for strongly correlated systems [15]. Specifically the B3LYP functional is thought to predict binding energies, geometries and frequencies systematically more reliably and with substantially improved performance compared to PBE methods[16]. Furthermore it is expected that band gaps calculated using B3LYP will be more accurate than those derived from PBE calculations, where the non analytic dependence of the effective potential[17] on the density often leads to calculated gaps which are less than half their actual value[15].

2 METHODOLOGY

The first-principles calculations in this work have been performed using the hybrid exchange density functional B3LYP in the CRYSTAL package[18]. In CRYSTAL the crystalline wavefunctions are expanded as a liner combination of atom centred Gaussian orbitals (LCAO) with s, p, d or f symmetry. The calculations reported here all electron, i.e. with no shape approximation to the ionic potential or electron charge density.

Basis sets of double valence quality (6-21G* for C and 6-31 G* for H) are used. A reciprocal space sampling on a Monkhorst-Pack grid of shrinking factor equal to 24 is adopted. The Gaussian overlap criteria which control the truncation of the Coulomb and exchange series in direct space are set to 10^{-7}, 10^{-7}, 10^{-7}, 10^{-7} and 10^{-14}. Typically linear mixing of 80% and an Anderson second-order mixing are used to guide the convergence of the SCF procedure

3 RESULTS AND DISCUSSION

Calculations were performed on graphene sheet, graphane chair and graphane boat confirmations. The geometries of all three structures were fully optimized. The respective structural and electronic data are presented in table 1.

Table 1-Structural parameters, total energy, formation energy and band gaps of three graphene and graphane

	Graphene Sheet	Graphane Chair	Graphane Boat
$d_{C-C}(\text{Å})$	1.42	1.54	(a) 1.538 (b) 1.569
$d_{C-H}(\text{Å})$	0.00	1.11	1.10
Lattice Parameter (Å)	2.46	2.54	(a) 2.53 (b) 4.32
Total Energy (per unit cell) (eV)	-2,071.49 (per C_2)	-2,104.09 (per C_2H_4)	-2,103.86 (per C_2H_4)
Formation energy (per unit cell)		-0.70	-0.48
Band Gap (eV)	0.00	6.93	6.84

Firstly focusing on the structural data, optimization of the graphene sheet gave a lattice constant value of 2.46 Å with a corresponding C-C bond length of 1.42 Å. These optimized parameters were subsequently used as a starting geometry for the two graphane confirmations, with the initial C-H bond length set to 1.1 Å. In the graphane chair confirmation each carbon atom has a hydrogen atom bound to the opposite side of the graphene sheet in relation to it's 3 nearest neighbors. While in the graphane boat confirmation this only applies to two nearest neighbors, with the remaining nearest neighbor having a hydrogen atom bound to the same side of the graphene sheet (figure 1). This difference has a notable effect on their respective structural parameters. After optimization the calculated C-C bond length of the graphane chair confirmation was 1.54 Å, which is significantly larger than that of the graphene sheet. This is due to the fact that each carbon in graphane is attached to a hydrogen atom and hence is sp^3 hybridized, in comparison to the non-hydrogenated graphene sheet where all the atoms are sp^2 hybridized. This graphane C-C bond length is comparable to that in the sp^3 carbon allotrope diamond which also has a C-C bond length of 1.54Å. In the boat confirmation H-H repulsion between neighboring carbon atoms with hydrogens bound to the same side of the graphene sheet leads to an extension of that particular C-C bond length leading to two distinct C-C bond lengths, this also breaks the hexagonal symmetry and lattice parameter values. As the (b) lattice parameter (table 1) lies parallel to the extended C-C bond it is extended relative to (a) and hence there are two different lattice parameter values.

In comparison to similar theoretical studies[6, 8, 9] all of which employed GGA functionals, there are some slight differences in the structural parameters but no qualitative changes. This indicates that the detailed treatment of electronic exchange and correlation does not have a significant effect on structural properties. The small remaining differences could arise from the numerical details of the calculations. For a quantitative comparison the data above (table 1) should be compared to table I in ref 6, 8, 9.

Comparing total energies (table 1), the graphane boat is a higher energy confirmation than the graphane chair. This instability is due to the aforementioned unfavourable repulsion between hydrogen atoms bound to adjacent carbon atoms on the same side of the lattice. The formation energy of each graphane confirmation was also calculated, this was defined as the total energy of a graphane structure minus the total energy of a plain graphene sheet minus the total energy of an individual hydrogen molecule. Following this the difference between the formation energies of each confirmation was computed, thus enabling a direct comparison of the calculated energies with those of other relevant theoretical investigations[6, 8]. The calculated difference between the formation energy of the graphane chair confirmation and the graphane boat confirmation was found to be 0.225 eV per C_2H_2, i.e. per unit cell, in favour of the chair confirmation. Sofo et al[6] calculated what they term to be the "binding energy" difference between the two confirmations, which we assume is the same as what is defined as the formation energy in this study. Their binding energy difference is given as 0.055 eV/atom, also in favour of the chair confirmation. Assuming by units per atom they are referring to the energy required to bind a single hydrogen atom to a single carbon atom, and considering our value is defined per C_2H_2, in order to make them directly comparable our calculated formation energy difference was halved to yield a value of approximately 0.1125 eV/atom. This is close to being exactly twice as large as the Sofo et al value. Initially this was believed to be a direct consequence of the choice of functional, as this study employed a B3LYP functional while Sofo et al a PBE functional. In order to test whether this was case, we repeated the calculations on the chair and boat confirmations but using a PBE as opposed to a B3LYP functional and computed the formation energy difference just as before. The formation energy difference between the two confirmations using the PBE functional was computed to be 0.222 eV per C_2H_2 in favour of the chair confirmation; which is almost exactly the same as the value that was calculated originally using the B3LYP functional. This therefore means that the observed difference is not a consequence of the functional used.

Looking at graphene's electronic structure, it is termed a zero gap semiconductor because the highest occupied and lowest unoccupied bands become degenerate at a single point in k-space, the K point of the hexagonal lattices first Brillouin zone, and yet it still has a zero density of states at E_F (figure 2). This can be explained by the fact that graphene has two inequivalent carbon atoms per unit cell (denoted C1 and C2 in figure 2). As each of these atoms is sp^2 hybridised, they both have p_z orbitals perpendicular to the lattice plane, which combine to form the π interactions in graphene. In the periodic sheet the bands of those p_z orbitals combine to form the π and π^* bonding and antibonding bands. As each band can hold up to two electrons and two electrons are donated from the two p_z orbitals per unit cell, the lower band is full and the upper band is empty. As a result the density of states at E_F is zero.

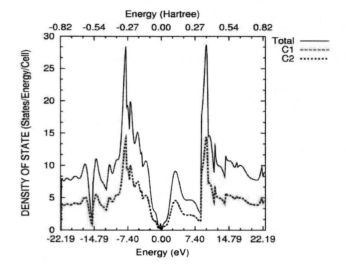

Figure 2-Total and projected density of states of graphene sheet. Projection is onto two inequivalent carbon atoms of the unit cell termed C1 and C2.

Hydrogenation of the graphene sheet to form graphane, results in a change in hybridization from sp^2 to sp^3 for each carbon atom. This therefore removes all π interactions and importantly the π conducting bands, thus opening an energy gap. The computed DOS displaying this band gap, with projections onto the carbon and hydrogen atoms of the unit cell is shown below (figure 3). As the electronic structures of both confirmations are qualitatively the same, only the DOS of the boat confirmation is presented. As the DOS shows the top of the valence band is comprised mainly of carbon p orbitals, while the bottom of the conduction band is more populated by hydrogen s orbitals (figure 3).

The calculated values of the band gaps (E_g) are displayed in table 1, the chair confirmation has a band gap of 6.93 eV, while the boat confirmation has band gap of 6.84 eV. This contrasts with the study by Sofo et al[6] which found the chair confirmation to have an E_g of 3.5 eV while the boat confirmation has an E_g of 3.7eV.Two points that stand out are that firstly the boat confirmation has a higher E_g than the chair, the opposite is true in this work. Secondly and more importantly the band gaps are much reduced in comparison to those calculated in the current work. This also true of the E_g values computed by Samarakoon et al[8] which found the band gaps to be 3.5 eV for both graphane confirmations. A probable reason for such a significant discrepancy is the choice of functional employed in the respective calculations. Both Sofo et al[6] and Samarakoon et al[8] use PBE functionals, while this investigation utilizes the B3LYP functional. In DFT the potential is computed from the total density and as each electron contributes to this density, this means that self interaction of an electron with itself is always included. Band widths are usually overestimated and band gaps underestimated in PBE calculations for this reason. In the hybrid exchange functional the mixing of non-local Fock exchange corrects for the self interaction, and thus provides a much more reliable first order description of band gaps than PBE theory. It is therefore likely that the graphane structures described here will behave as electronic insulators at room temperature rather than semiconductors as previously predicted.

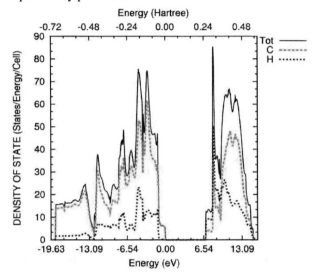

Figure 3-Total and projected density of states of graphane boat confirmation. Projection onto C and H

4 CONCLUSION

First-principles calculations using the hybrid exchange density functional B3LYP have been used to compute the electronic structure and band gaps of two confirmations of graphane. The size of these gaps, 6.93 eV and 6.94, indicate that graphane is electrically insulating at room temperature. This is contrary to similar investigations[6, 8] which through the size of their computed band gaps seem to indicate that graphane will be semiconducting at room temperature. As both 6, 8 treated the electronic exchange and correlation terms using PBE functionals, the observed results corroborate with the idea that the inclusion of the fock exchange when computing band gaps corrects for self interaction present in PBE functionals. The relative performance of the B3LYP hybrid exchange functional against PBE functionals in determining structural parameters and energies was assessed. Results seem to indicate that detailed treatment of electronic exchange and correlation does not have a significant effect on structural properties. Possible future work could involve assessing performance of B3LYP on different hydrogen adsorbed graphane structures, for example graphone. The

performance of B3LYP in treating magnetic graphene based systems could also be investigated.

5 REFERENCES

1. K. S. Novoselov, A. K. Geim, S. V. Morozov, D. Jiang, Y. Zhang, S. V. Dubonos, I. V. Grigorieva and A. A. Firsov, *Science,* 2004, **306**, 666-669.

2. A. K. Geim, *Science,* 2009, **324**, 1530-1534.

3. D. W. Boukhvalov, M. I. Katsnelson and A. I. Lichtenstein, 2008, **77**, 7.

4. S. Casolo, O. M. Lovvik, R. Martinazzo and G. F. Tantardini, 2009, **130**.

5. J. Kerwin and B. Jackson, 2008, **128**.

6 . J. O. Sofo, A. S. Chaudhari and G. D. Barber, *Phys. Rev. B,* 2007, **75**, 4.

7. D. C. Elias, R. R. Nair, T. M. G. Mohiuddin, S. V. Morozov, P. Blake, M. P. Halsall, A. C. Ferrari, D. W. Boukhvalov, M. I. Katsnelson, A. K. Geim and K. S. Novoselov, 2009, **323**, 610-613.

8. D. K. Samarakoon and X. Q. Wang, 2009, **3**, 4017-4022.

9. M. Z. S. Flores, P. A. S. Autreto, S. B. Legoas and D. S. Galvao, 2009, **20**, 465704.

10. J. P. Perdew, K. Burke, and M. Ernzerhof, Phys. Rev. Lett., 1996, 77, 3865.

11. A. D. Becke, Phys. Rev. A, 1988, **38**, 3098.

12. A. D. Becke, J. Chem. Phys., 1993 **98**, 5648.

13. C. Lee, W. Yang, and R. G. Parr, Phys. Rev. B, 1988 **37**, 785.

14. U. von Barth and L. Hedin, J. Phys. C, 1972, 5, 1629,

15. J. Muscat, A. Wander, N. M. Harrison, Chem. Phys. Lett, 2001, **342**, 397

16. N. M. Harrison, "An Introduction to Density Functional Theory" Computational Materials Science, 187, Ed. C. R. A. Catlow and E. Kotomin (NATO Science Series III, IOS Press, 2003).

17 J. P. Perdew, M. Levy, Phys. Rev. Lett, 1983, **5**, 1884

18. V. R. Saunders, R. Dovesi, C. Roetti, R. Orlando, C. M. Zicovich-Wilson, N. M. Harrison, K. Doll, B. Civalleri, I. J. Bush, Ph. D'Arco, and M. Llunell, *CRYSTAL2003 User's Manual* University of Torino, Torino, 2003

19. M.Catti, A.Pavese, R.Dovesi, V.C.Saunders, Physical Review B, 1993, **47**, 9189

Effect of Titanium on the Growth and Field Emission Properties of MPECVD Grown Multiwalled Carbon Nanotubes

Himani Sharma*, A. K. Shukla, V. D. Vankar*

Thin Film Laboratory, Department of Physics, Indian Institute of Technology Delhi, New Delhi-110016, India

ABSTRACT

The growth behavior and enhanced field emission properties of multiwalled carbon nanotubes were investigated using Ti as an underlayer and over layer. Ti of 10 nm thickness was deposited underneath Fe catalyst and their interaction behavior was observed for growth of carbon nanotubes. It was observed that underlayer deposition of Ti improves the growth of CNTs by acting as a barrier. In the other set of studies, Ti film of extremely small thickness (15 Å) was post deposited over CNTs to investigate their field emission behavior. The turn on and threshold field values of Ti modified CNTs were found to be low ~ 0.8 V/μm and 2.65 V/μm, respectively, as compared to pure CNTs. HRTEM studies confirmed that Ti nanoclusters on post deposition adsorbs on the edges and walls of CNTs. This improved adhesion and bonding with CNTs is due to the unfilled d-shell of Ti, resulting in the modification of CNTs structure.

Keywords: Carbon nanotubes (CNTs);; Micro Raman spectroscopy; High Resolution Transmission electron microscopy (HRTEM); Field Emission.

1 INTRODUCTION

Carbon nanotubes (CNTs) are of special interest due to their promising applications in field emission displays [1-2], nanoelectronics [3-4], hydrogen sensing [5] and fuel cells [6]. Both the electrical properties and geometrical aspects (alignment, diameter, length, etc) are critical factors in device fabrication. The electronic properties of the CNTs differ considerably and depend upon the defect structure, aspect ratio, configuration etc. To grow CNTs of desired characteristics is a real challenge. Thus, to grow CNT of desired characteristics, various metal underlayers have been used.

It is widely accepted that Fe, Ni and Co and their alloys or compounds show the highest catalytic activity for the growth of CNTs [7]. In addition to this, underlayer deposition is also employed for the growth of CNTs. Various metals (Pt, Mo, Ti, Al, and Au) have been used as underlayers for the growth of CNTs. The underlayer metal must not react with the catalyst to form alloys at CNT growth temperatures, which may poison the catalyst in the reaction [8].

On the other hand top coating on CNTs also leads to modification in their properties. Composite materials made of metals or metal oxides and CNT have gained much importance, as it results in the modification in the electrical, structural and mechanical properties of carbon nanotubes through orbital hybridization and thus results in the passivation of defect sites. Recently, various studies have been carried out to modify the electrical and structural properties of CNTs, by coating of various metals, oxides and fluorides such as Ru, Pd, ZnO, RbF, CrO_3 [9,10,11,12].

In this paper we report the results of experiments concerning the MPECVD growth of Fe and Ti-Fe catalyzed multiwalled carbon nanotubes. Ti was used as under layer to study the growth of CNTs. In other set of studies, Ti was used as top layer to investigate the field emission behavior of CNTs.

2 EXPERIMENTAL DETAILS

In the first set of studies, Ti was used an underlayer. MWCNT samples were synthesized using tubular microwave plasma enhanced chemical vapor deposition (MPECVD) [13]. Ti of 10nm thickness was pre deposited on p-type Si wafer that acted as an underlayer. Fe of 10 nm thickness was deposited on Ti coated Si wafer by thermal evaporation at the base pressure of 3×10^{-6} torr. The samples were pretreated in the presence of hydrogen and argon plasma at 720^0C for 15 minutes at 2.5 torr. The microwave power of 550 W was applied to generate the plasma at 1 torr. CNTs were grown on passing acetylene over Ti-Fe catalysts. The reactant gases were a mixture of hydrogen and acetylene with a ratio of 1:5 at a pressure of 5 torr.

In the second set of studies, Ti was used as a top layer. Pristine MWCNTs were grown by the above mentioned method using 10 nm thick Fe as a catalyst. As deposited CNTs were named as A. In the second set, CNTs were modified by Ti coating 15 Å thicknesses. The Ti decorated CNTs were named as B.

The surface morphology of MWCNTs were characterized by scanning electron microscopy (SEM: ZEISS EVO 50) operating at 20 kV accelerating voltage by secondary electron imaging. Carbon films were scratched from the substrate and ultrasonicated in acetone at least for 15 minutes so that CNTs could disperse properly. Few drops of the suspension were then transferred on to carbon coated copper grid. Their microstructures were analyzed by transmission electron microscope (TEM: Philips CM 12) and high resolution TEM (HRTEM: Technai G^2 20 S-Twin model) operating at 200 kV. Energy dispersive analysis of X-rays (EDAX) was carried out for the elemental analysis. Micro Raman studies (Micro-Raman T64000 Jobin Yvon triple monochromator system) of the samples were carried out at an excitation wavelength of 514.5 nm to analyze the structure of the samples. Field Emission measurements were carried using a diode set up in the vacuum chamber at 2×10^{-6} torr. CNTs were used as cathode and stainless steel plate was used as anode. Emission current density was determined by dividing current with the area of the sample.

3 RESULTS AND DISCUSSION

3.1 Ti as an underlayer

The results are discussed when Ti was used as an underlayer.

3.1.1 Surface Morphology and Structure of CNTs

First, we study the effect of catalyst (Fe) film on the growth of CNTs. As seen in Fig 1(a), the size of the particles varied from 20-100 nm. Breaking of pure Fe film into nanoparticles is due to the continuous collisions of active ions in the plasma with the film and due to heating in the plasma; nanoparticles of Fe agglomerate and become isolated bigger nanoclusters on which he growth of CNTs take place. Figure 1(b) shows the SEM image of Ti-Fe nanoparticles, formed after plasma treatment.

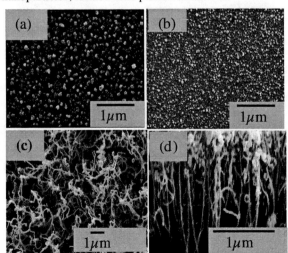

Figure1: SEM images showing (a) Fe nanoparticles (b) Ti-Fe nanoparticles (c) CNTs grown on Fe catalyst (d) CNTs grown on Ti-Fe catalyst.

The thickness of Ti as an underlayer and Fe as catalyst film was kept 10 nm.

Fe deposited on Si forms iron silicide that hinders the growth of CNTs. In order to prevent the iron silicide formation upon heating the samples to 700 °C, a layer of Ti

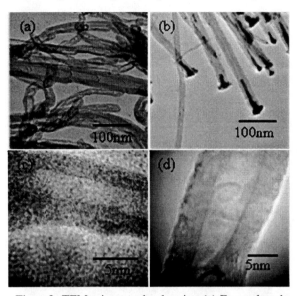

Figure2: TEM micrographs showing (a) Fe catalyzed CNTs. (b) Ti-Fe catalyzed CNTs. HRTEM images showing planes of (c) Fe catalyzed and (d) Ti-Fe catalyzed CNTs.

was pre deposited onto the Si wafer prior to the deposition of Fe film. Ti underlayer acts as a buffer layer and prevents the formation of iron silicide and thus optimize the growth of CNTs. Two intermetallics are formed between Fe and Ti depending on the concentration of particular metal. In our studies, CNTs grow on the nanoparticles of Fe rich phase. Further no growth of CNTs could be seen on Ti covered substrates without Fe. According to Sun et.al, there is not much strong interaction occurs between Fe and Ti underlayer [8]. This leads to the growth of CNTs on Ti-Fe catalyst.

The surface morphology of Fe based CNTs and Ti-Fe based carbon films can be seen in fig 1(b) and 1(c). It can be inferred that CNTs grown on Ti-Fe catalyst are much denser than Fe catalyzed films. Also, CNTs are vertically aligned through this combination of catalysts.

TEM images, in Fig. 2 shows the Fe and Ti-Fe catalyzed CNTs. Fe catalyzed CNTs have bamboo shaped, having diameter in the range of 20-30 nm, as seen in Fig. 2(a and c). From Fig. 2(b and d), Ti-Fe catalyzed CNTs are stacked up nanostructures having diameter 30-40 nm.

3.2 Ti as top layer

The results are discussed when Ti was used as a top layer.

3.2.1 Microstructure of pristine and Ti coated MWCNTs

Fig. 3 (a and b) shows the HRTEM images of pristine and Ti coated CNTs. HRTEM studies of Ti coated MWCNTs were carried out to observe the effect of defect passivation. In the present study, it is observed that due to very thin coating of Ti (15Å), Ti nanoparticles gets adsorbed on the CNT walls, because of the presence of certain defects. The defects reduce the cohesive energy of Ti to a negative value which improves its adhesion of Ti to the CNT walls [14]. Incorporation of Ti overcomes the defects and thus enhances the emission and crystallinity of CNTs.

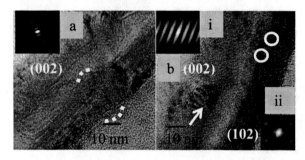

Figure3 HRTEM images showing (a) defect planes in bare MWCNTs and (b) Ti decorated MWCNTs, where Ti nanoparticles have been encircled. Inset: (a) MWCNT planes (b) (i) MWCNT planes; (ii) Ti planes.

3.2.2 Micro Raman spectroscopy of pristine and Ti coated MWCNTs

Fig. 4(a) shows the first-order Raman spectra of pristine and Ti coated CNT samples (A and B). Two major peaks observed at 1355 cm^{-1} and 1584 cm^{-1} are referred to as D and G-band [15, 16]. In addition, a shoulder peak, referred to as D^*, is evident towards the higher wave number side of the G-band peak, located at 1625 cm^{-1}.

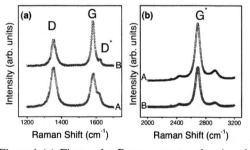

Figure4 (a) First-order Raman spectra showing the increasing order of crystallinity of samples A and B where (A) pristine CNTs (B) Ti coated CNTs of thickness 15Å. (b) Second- order Raman spectra showing decreasing D^* mode of samples A and B.

The D-band corresponds to the defects and structural disorders in polycrystalline graphite. The G-band corresponds to the tangential vibrations of carbon atoms. The intensity ratio of the D band to G band (I_D/I_G) is used as one of the parameter to ascertain the degree of crystallanity of CNT samples. The I_D/I_G ratios are 1.24, and 0.57 for the samples of A and B respectively. It is clear that Raman spectra exhibit a decreasing trend in the I_D/I_G ratio of Ti coated CNTs. This trend with increasing Ti thickness corresponds to the increased crystallanity of CNTs and decreased sp^3 bonded carbon. It was found that Ti coated CNTs were more crystalline than the pristine CNT samples. This was possibly due to passivation of defect sites and resulting changes in local order. The peaks at 2689 and 2691 cm^{-1} are referred to as G^* mode of the pristine and Ti coated CNTs, respectively, that corresponds to the second-order overtone of D mode, seen in Fig. 4(b).

3.2.4 Field Emission Studies

The field emission behaviour of pristine and Ti coated CNTs was investigated using diode setup. The emission characteristics were studied at a distance of 200 μm between the cathode and anode.

Field emission behaviour was analyzed using Fowler-Nordheim (F-N) equation [17]:

$$J = (A\beta^2 E^2 / \varphi) \exp(-B\varphi^{3/2} / \beta E) \quad (1)$$

where, J is the emission current density; φ the work function, E is the electric field and β the enhancement factor.

The F-N plot is obtained by plotting between ln (J/E^2) and 1/E. The enhancement factor was calculated from the slope of F-N plot by considering the work function of CNTs as 5 eV [18, 19]. Fig. 5 (a and b) shows the emission current density versus electric field (J vs. E) and Fowler- Nordheim plots (F-N) plots for the samples A and B respectively. A significant change in field emission behaviour of pristine and Ti coated CNTs were observed, as seen from Fig. 5 and table 3. The turn on field at 10 μA/cm^2, as seen in Fig.5(a) is found to be 1.8, and 1.4 V/μm for samples A and B. The threshold field at 1mA/cm^2 for the same samples was found to be 4.0, 2.1 V/μm (Fig. 5(b)). The field emission properties of Ti coated CNTs were enhanced as compared to pristine CNTs. This is due to the lower work function of Ti (4.29 eV) as compared to CNTs (5.0 eV). As very thin layer (15 Å) of Ti is coated on CNTs, Ti nanolusters adhere on the tip and edges of MWCNTs, thus shifting the Fermi level towards the higher energy side. The presence of Ti nanoparticles is expected to reduce the local surface work function of CNT film leading to the enhancement in the electron emission. This leads to the increase in the field enhancement factor. Consequently, the density of states increases near the Fermi level resulting in the enhancement of tunnelling current.

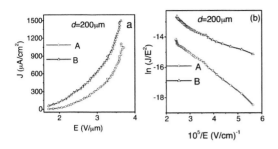

Figure5 (a) J vs. E plot for samples A and B at the distance of 200 μm (b) corresponding F-N plot. d is the distance between the cathode and anode. The turn on and threshold field values are taken at 10 μA/cm^2 and 1mA/cm^2.

4 CONCLUSIONS

In conclusion, we have reported the role of Ti on the growth and enhanced field emission properties of multiwalled CNTs. Ti, as an under layer improves the growth of CNTs without poisoning the catalysis, and as top layer improves the emission characteristics and crystallanity of CNTs as compared to pristine CNTs. It was observed that Ti coated CNTs have low function and higher field enhancement factor as compared to pristine CNTs. The enhancement in the field emission properties of Ti coated CNTs was due to the shifting of the Fermi level towards the higher energy side that lead to the increase in the density of states near the Fermi level resulting in the enhancement of tunnelling current.

ACKKNOWLEDGEMENTS

One of the authors (H.S.) is thankful to director IIT Delhi for providing a research scholarship. Authors are thankful to Mr. Rajkumar, Mr. Vivek and Mr. Kapil for their assistance in field emission measurements and micro Raman spectroscopy. Authors are also grateful to Dr. C. Singh and Mr. D.C. Sharma for carrying out scanning electron microscopy.

REFERENCES

[1] N. Lee, D. Chung, I. T. Han, J. H. Kang, Y. S. Choi, H. Y. Kim, S.H. Park, Y.W. Jin, W.K. Yi, M.J. Yun, J. E. Jung, C.J. Lee, J.H. You, S. H. Jo, C. G. Lee, J. M. Kim, Diamond and Related Materials, 10, 265, 2001.

[2] A. A. Talin, K. A. Dean, J. E. Jaskie, Solid-State Electronics, 45, 963, 2001.

[3] V. Q. Nguyen, D. H. Nguyen, A. Myungchan, C. Yousuk, K. Dojin, Nanotechnology, 18, 345201, 2007.

[4] L. M. Paul, Physics World, 3, 1, 2000.

[5] J. S. Oakley, H. T. Wang, B. S. Kang, Z. Wu, R. Fan, A. G. Rinzler, S. J. Pearton, Nanotechnology, 16, 2218, 2005.

[6] M. Endo, T. Hayashi, Y. A. Kim, M. Terrones, M. S. Dresselhaus, R. Soc., 362, 2223, 2004.

[7] S. Esconjauregui, C. M. Whelan, K. Maex, Carbon, 47, 659, 2009.

[8] X. Sun, K. Li, R. Wu, P. Wilhite, T. Saito, Jing Gao, Cary Y Yang, 21, 045201, 2010.

[9] C. Liu, K. S. Kim, J. Baek, Y. Cho, S. Han, S. W. Kim, N. K. Min, Y. Choi, J. U. Kim, C. J. Lee, Carbon, 47, 1158, 2009.

[10] D. S. Kim, S. M. Lee, R. Scholz, M. Knez, U. Gösele, J. Fallert, H. Kalt, M. Zacharias, Appl. Phys. Lett., 93, 1031081, 2008.

[11] S. Chhoker, S. K. Arora, P. Srivastava, V.D. Vankar, J. Nanosci. Nanotechnol. 8, 4309, 2008.

[12] S.F. Lee, Y.P. Chang, Li.Y. Lee, New Carbon Materials, 23, 104, 2008.

[13] S. K. Srivastava, A. K. Shukla, V. D. Vankar, V. Kumar, Thin Solid Films, 492, 124, 2005.

[14] F. Y. Meng, L. G. Zhou, S. Q. Shi, R. Yang, Letters to the editor/Carbon, 41, 2009, 2002.

[15] J. Maultzsch, S. Reich, C. Thomsen, Phys. Rev. B, 65, 233402, 2002.

[16] M.S. Dresselhaus, G. Dresselhaus, R. Satio, A. Jorio, Physics Reports, 1, 2004 .

[17] F.T. Chuang, P.Y. Chen, T.C. Cheng, C.H. Chien, B.J. Li, Nanotechnology, 18, 395702, 2007.

[18] E. Titus, M.K. Singh, G. Cabral, R.P. Babu, W.J. Blau, J. Gracio, Diamond and Related Materials, 18, 967, 2009.

[19] J.S. Lee, J.S. Suh, Bulletin Korean Chemical Society, 24, 1827, 2003.

* Authors for correspondence - himanitiet427@gmail.com, vdvankar@physics.iitd.ernet.in, Ph: (91) 11- 26596117, Fax: (91) 11-26581114

Violet-Blue Emission from TiO₂/SWNT Hybrid Synthesized Following a Very Simple Technique

R. Paul, P. Kumbhakar and A.K. Mitra[*]

Department of Physics, National Institute of Technology Durgapur
Durgapur-713209, West Bengal, India
*akmrecdgp@yahoo.com

ABSTRACT

Modification of CNT surfaces by hybridization with other nanostructures can drastically change their physical properties and transform them into useful functional nanomaterials. We report here a chemical precipitation process to synthesize a heterostructure of single walled carbon nanotubes (SWNT) and TiO₂ nanoparticles of average size 5 nm. The hybrid structure is characterized by high resolution transmission electron microscopy (HRTEM), scanning electron microscopy (SEM), energy-dispersive X-ray analysis (EDAX), powder X-ray diffractometry (XRD) and Raman spectroscopy. It is clearly revealed that titania nanocrystals of anatase phase and of nearly uniform size are attached to the surfaces of SWNT bundles. The UV-vis absorption study shows a blue shift of 16 nm in the absorbance peak position of the hybrid structure with respect to the pristine SWNT. The photoluminescence study shows violet-blue emission in the range 325-500 nm with peak emission at around 400 nm.

Keywords: SWNT, titania, nanohybrid, photoluminescence

1 INTRODUCTION

The carbon nanotube (CNT) since its discovery in the year 1991[1], has been continuously drawing attention of the researchers due to its fascinating physical and chemical properties [2]. The modification of the CNTs by surface functionalization and hybridization with other nanostructures can drastically change their physical properties to transform them into functional nanomaterials with desired properties and can find several novel applications in useful devices. Hybridizing semiconductor nanoparticles on CNT surfaces can modify their optical absorption and luminescence properties significantly and such composites have been used to tailor light-emitting diodes [3], to organize sensor systems and to fabricate electrochromic devices [4]. TiO₂ is one of the most investigated semiconductor which finds demanding applications related to semiconductor photocatlysts [5-7], optical devices [8], gas sensors [9] and solar cells [10]. The photocurrent or photochemical efficiency of TiO₂ is greatly influenced by its crystal structure, particle size, surface area and porosity [11]. The design and development of highly efficient photocatalytic composites have attracted the interest owing to their potential application for the degradation of toxic organic dyes and industrial effluents. CNTs can be used as catalytic support materials due to their high aspect ratio and their ability to disperse catalytically active metal particles [12]. There are several techniques to decorate CNT walls with TiO₂. Here we report one of the easiest methods to prepare TiO₂/SWNT hybrid. The hybrid material showed violet-blue emission at excitation wavelength of 250 nm and may find applications in sensor based devices. Also, the hybrid structure may render it useful for catalysis applications.

2 EXPERIMENTAL DETAILS

We procured SWNTs (1-2 nm outer diameter, length: 1-3 μm and purity > 95%) from Chengdu Organic Chemicals Co., Ltd. Chinese Academy of Sciences and further purified them by high temperature oxidation, acid treatment, ultrasonication, pH control and filtration [13]. To obtain TiO₂/SWNT hybrid structure, 30 mg of purified SWNT was taken in 20 ml of TiCl₃ and stirred using a magnetic stirrer (REMI 2MLH) for 10 min. 4.8M of NH₄OH solution was then added drop wise to the solution containing SWNT till pH became 7 and again stirred for 18 h using magnetic stirrer. The product colloidal solution was centrifuged for 15 min at rpm 6000 at 9°C. The precipitate was then washed thoroughly with de-ionized water followed by 2-propanol and left for drying at room temperature. We also synthesized TiO₂ nanoparticles following similar method. For our experiments, chemicals from Merck (GR grade) were used at room temperature and without further purification.

The prepared samples were then characterized for their nanostructural as well as compositional properties. The hybrid structure was revealed in the HRTEM (JEOL JEM 2100, operating voltage 200 KV) micrograph. SEM and EDAX spectroscopy (HITACHI S 3000N) were used for compositional analysis of the samples. XRD patterns were obtained using Philips PANalytical X-Pert Pro diffractometer. Raman spectroscopy was performed using TRIAX550 JY Horiba USA (provided with edge filter and a CCD detector). Argon ion laser of wavelength 488 nm was used as excitation source.

To study the optical properties, the dried samples were dispersed separately in sodium dodecyl sulfate (SDS) solution and their optical absorbance spectra were observed using UV-visible (HITACHI U-3010) spectrophotometer. Photoluminescence (PL) spectrum of the samples were studied using FL spectrofluorimeter (HITACHI F-2500) over a wide range of excitation wavelength from 220-400 nm.

3 RESULTS AND DISCUSSIONS

The HRTEM micrograph of TiO$_2$/SWNT hybrid is shown in figure 1. The micrograph clearly reveals that TiO$_2$ nanocrystals adhered uniformly onto the SWNT surfaces. The inset shows the particle size histogram of the TiO$_2$ nanocrystallites. The average size of TiO$_2$ nanocrystals is found to be 5 nm as obtained by fitting the particle size distribution from HRTEM micrograph.

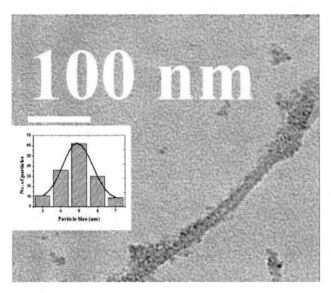

Figure 1: HRTEM micrograph of TiO$_2$/SWNT hybrid; inset shows the particle size distribution of TiO$_2$ nanocrystals.

Figure 2 shows the EDAX spectrum of TiO$_2$/SWNT hybrid.

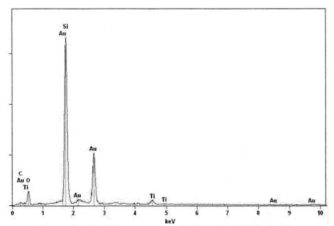

Figure 2: EDAX spectrum of TiO$_2$/SWNT hybrid

From the spectrum, the presence of titanium (Ti) and oxygen (O) with carbon (C) of SWNT are confirmed.

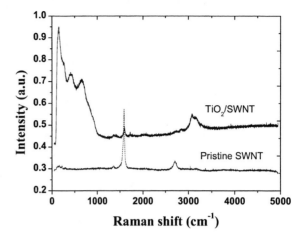

Figure 3: Raman spectrum of pristine SWNT and TiO$_2$/SWNT hybrid.

Figure 3 shows the Raman spectrum of pristine SWNT and that of TiO$_2$/SWNT hybrid. Additional peaks in the range 150-700 cm^{-1} of the hybrid sample are found which indicate that radial breathing mode (RBM) of SWNT has been influenced on decorating SWNT walls with TiO$_2$ nanocrystals. There is an up-shift by 12 cm^{-1} in the position of G band indicating charge transfer to SWNTs from TiO$_2$ nanocrystals. This type of up-shift of G band upon modification of CNT surfaces has been reported earlier by several authors [14, 15]. I_D/I_G for pristine SWCNT is 0.546 while that for TiO$_2$/SWNT hybrid has been found to be 0.939. This increase in I_D/I_G factor indicates that TiO$_2$ nanocrystals adhered to the SWCNT surfaces via chemical bonding. Figure 4 shows the XRD pattern of pristine SWNT and TiO$_2$/SWNT hybrid. For pristine SWNT, the peaks centered at 26°, 42° and 44° correspond to (002),

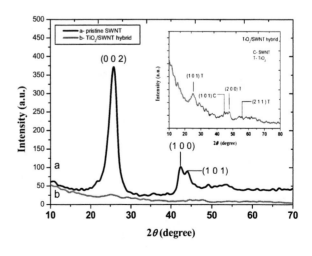

Figure 4: XRD pattern of pristine SWNT and TiO$_2$/SWNT hybrid

(100) and (101) reflections of graphitic planes from the SWNTs respectively (JCPDS card no. 75-1621). The XRD pattern of TiO$_2$/SWNT heterostructure gives the peak assigned to (101), (200) and (211) planes of the anatase phase of TiO$_2$. These are in good agreement with the reported data for TiO$_2$ (JCPDS card no. 21-1272). Moreover, the hybrid structure also shows the (101) reflection from SWNT surface. The inset of figure 4 shows the magnified image of the XRD pattern of the hybrid structure. The absorption spectra of pristine SWNT and TiO$_2$/SWNT nanostructure is shown in figure 5. The absorption peak position blue shifted by 16 nm from 264 nm to 248 nm on decorating SWNT walls with TiO$_2$ nanocrystals and this may also affect the catalytic properties of TiO$_2$/SWNT nanostructure [16], making their possible use in catalysis applications.

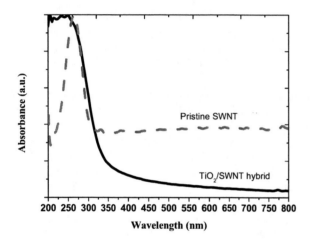

Figure 5: Absorption spectra of pristine SWNT and TiO$_2$/SWNT hybrid.

Figure 6 shows the photoluminescence spectrum of pristine SWNT and TiO$_2$/SWNT hybrid at 250 nm excitation

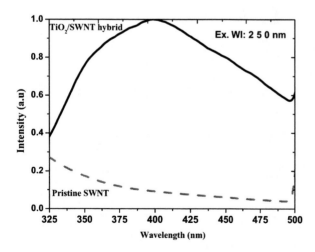

Figure 6: Photoluminescence spectra of pristine SWNT and TiO$_2$/SWNT hybrid at 250 nm excitation wavelength.

wavelength. The hybrid structure showed violet-blue emission when excited by UV radiation of wavelength 220-260 nm. Beyond 260 nm excitation wavelength, no considerable photoluminescence has been observed. The spectrum covered a region from 325-500 nm with the peak position at around 400 nm. The pristine SWNT sample showed no luminescence in this region. The photoluminescence emission of the hybrid structure is attributed to the charge transfer between titania nanocrystals and carbon nanotubes.

4 CONCLUSIONS

We reported a simple wet chemical process to synthesize a nanohybrid structure of SWNT and TiO$_2$ nanoparticles. Titania nanocrystals of average size of 5 nm uniformly decorated the walls of the SWNT bundles and the composite exhibited distinct optical properties, distinguished from individual components. A broad luminescence in the visible region in the range of 325-500 nm was observed which could be attributed to the charge transfer between the titania nanocrystals and SWNTs. This hybrid structure can find application as an useful optical material.

ACKNOWLEDGEMENTS

Authors acknowledge with thanks the support from Dr. D. Sukul of Department of Chemistry, NIT Durgapur, for PL study. Authors are grateful to NIT Durgapur and Government of India for financial support and one of the authors (R.Paul) is grateful for providing the Institute Research Fellowship. We extend our thanks to Dr. R. Mitra and Dr. A. Roy, IIT Kharagpur, for making available the HRTEM and Raman Spectroscopy facilities, respectively.

REFERENCES

[1] S. Iijima, Nature, 354, 56, 1991
[2] M.S. Dresselhaus, G. Dresselhaus, P. Avouris, "Nanotubes: synthesis, structure, properties and application", vol. 80 Springer Berlin, 2001.
[3] N. Tessler, V. Medvedev, M. Kazes, S.H. Kan, U. Banin, Science, 295, 1506, 2002.
[4] C. Bechinger, S. Ferrer, A. Zaban, J. Sprague, B.A. Gregg, Nature, 383, 608, 1996.
[5] M. Anderson, L. Osterlund, S. Ljungstrom, A. Palmqvist, J. Phys. Chem. B, 414, 238, 2002.
[6] Y.M. Sung, J.K. Lee, Cryst. Growth Des., 4, 737, 2004.
[7] Y.M. Sung, J.K. Lee, W.S. Chae, Cryst. Growth Des., 6, 805, 2006.
[8] D. Appell, Nature, 419, 553, 2002.
[9] A. Rothschild, A. Levakon, Y. Shapira, N. Ashkenasy, Y. Komem, Surf. Sci., 456, 532, 2003.

[10] N.G. Park, J. Van de Lagemaat, A.J. Frank, J. Phys. Chem. B, 104, 8989, 2000.

[11] J.C. Lee, K.S. Park, T.G. Kim, H.J. Choi, Y.M. Sung, Nanotechnology, 17, 4317, 2006.

[12] K. Byrappa, A.S. Dayananda, C.P. Sajan, B. Basavalinga, M.B. Shayan, K. Soga, M. Yoshimura, J. Mater Sci., 43, 2348, 2008.

[13] A. Suri, A.K. Chakraborty, K.S. Coleman, Chem. Mater., 3, 1705, 2008.

[14] A. M. Rao, P.C. Eklund, S. Bandow, A. Thess, R.E. Smalley, Nature, 388, 257, 1997.

[15] G.U. Sumanasekera, J.L. Allen, S.L. Fang, A.L. Loper, A.M. Rao, P.C. Eklund, J. Phys. Chem. B, 103, 4292, 1999.

[16] N. Satoh, T. Nakashima, K. Kamikura, K. Amamoto, Nat. Nanotechnol., 3, 106, 2008.

Novel interdigital actuators and sensors based on highly overlapped branched carbon nanotubes

S. Darbari, Y. Abdi, A. Ebrahimi and S. Mohajerzadeh

Nano-electronic Center of Excellence,
Thin Film and Nano-Electronic Lab,
School of Electrical and Computer Eng,
University of Tehran, Tehran, Iran, mohajer@ut.ac.ir

ABSTRACT

We report a novel interdigital sensor and actuator device based on branched treelike carbon nanotubes (CNT) on silicon-based membranes with an ultra high capacitance value. The presence of treelike CNTs leads to a high overlap between interdigital fingers as well as high electron emission thanks to many nanometric branches attached to their central stem. The fabricated B-CNT-based device resulted in a high capacitance value in comparison to typical devices, leading to a stronger electromechanical coupling behavior. The electro-mechanical behavior of the device has been investigated both with electron emission and capacitive characteristic of the sensor, which confirm superior performance of the investigated devices over silicon-based devices.

Keywords: Carbon nanotubes, branched carbon nanostructures, micro electromechanical actuator, sensor.

I. INTRODUCTION

Exceptional electrical and mechanical properties of carbon nanotubes have attracted the attention of many researchers in different fields. Their unique characteristics such as small size, high stiffness, flexibility and strength [1], high electrical and thermal conductivity as well as their exceptional electromechanical characteristics [2] propound great applications in nano-technology, as well as offering an absolute structure to study nano-cosmos. The incorporation of nano-structures in micro and nano-electromechanical systems has made great enhancement in the performance of such devices [3]. As examples of carbon nanotube-based devices are nano tweezers and grippers [4], nano switches and nano relays [5] and CNT-based bearing for nano-rotational devices [6]. The high aspect ratio of CNTs has exemplified them as excellent electron emitters suitable for field emission displays [7] and nano lithography [8].

We have recently reported the realization of novel branched carbon nanotubes on silicon substrates by means of a plasma enhanced chemical vapor deposition [9]. In this paper, we report for the first time, the application of branched nano structures in silicon-based electrostatic actuator-sensor systems to realize a considerable improvement in the devices.

II. FABRICATION PROCESS

The growth of CNTs is achieved using plasma enhanced CVD method on patterned structures. Nickel is used as the seed layer for the CNT growth and can be patterned using precision photolithography with features around 0.8μm. Once the initial growth is achieved, the formation of branched treelike structures is feasible by applying a sequential treatment and growth steps on the previously grown CNTs. Since a "tip-growth" mode is dominant, the Ni seed is present at the very top side of the vertical CNTs. After the initial growth, individual CNTs are coated with an amorphous carbon layer to encapsulate Ni on top of them. The sample is then exposed to hydrogen plasma to leach the nickel from the tip side. The treated sample is then subjected to a subsequent growth of nanotubes in the same reactor. The second growth is achieved from the tiny nickel seeds just at the very tip of the already grown CNTs. Figure 1 shows in detail, the evolution of branched CNTs on silicon substrates.

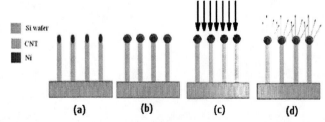

Fig. 1: The fabrication process of branched nanostructures. (a) The growth of vertical CNTs and (b) their encapsulating by a TiO_2 layer during an intermediate treatment step. (c) The hydrogenation step to expose the encapsulated Ni at the very top of the CNT, and (d) the subsequent growth to realize branched carbon nano structures.

To realize a sensor/actuator system a Si-based membrane is needed. Figure 2 depicts schematically the formation of the interdigital (I/D) structures on such a membrane and the subsequent growth and actuation of B-CNTs. A Si membrane (2-4μm thick) is obtained by standard backside micromachining in KOH solution, followed by deposition of Si₃N₄ on the backside. After doping the Si membrane, a 9-nm thick layer of Ni is deposited on the front side of the membrane and it is patterned to form proper interdigital structures on the membrane. A deep reactive ion-etching step is needed to remove Si from unwanted areas and to reach the bottom Si₃N₄ layer. Backside silicon-nitride layer behaves as the ultimate membrane for the sensing/actuation device. The growth of branched treelike CNTs is then achieved only on the parts, which have nickel remaining from previous steps (parts 4 to 6 of figure 2).

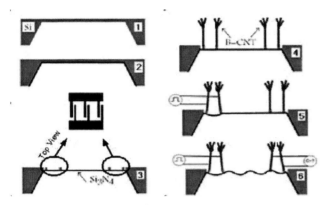

Fig. 2: Formation of sensor/actuator system by B-CNTs, (1) Si membrane, (2) back-side Si₃N₄ coating, (3) interdigital structures on top, (4) growth of B-CNT on digital structures, (5) actuation is possible by applying external ac voltage, (6) sensing the plane wave by measuring the capacitance of the opposite fingers.

III.   RESULT AND DISCUSSION

The value of the capacitance has been measured at different stages of the fabrication. For a simple interdigital structure without CNTs, we observe a value of 0.2pF while this value rises to 20 pF for the branched CNTs, thanks to a heavy overlap between parallel lines. By applying a pulsed signal to one interdigital part, the electrical oscillation leads to a mechanical resonance which can be propagated through the membrane and reach the other I/D side. Figures 3 collects several SEM images pertaining to vertical CNTs grown on closely patterned parallel lines of catalyst layer (left image) that have been converted to B-CNTs in next parts of the figure. There is a heavy overlap between parallel lines of B-CNTs, displayed apparently when the lines are placed with a little spacing as SEM images of figure 3. Figure 4 shows the device structure prior and after the B-CNTs have been grown. From inset, one can deduce there is a heavy overlap between neighboring lines.

Fig. 3: A collection of SEM images corresponding to various stages of B-CNT formation on linear and dotted arrays. The left-most image shows the growth of normal vertical CNTs on parallel lines whereas other images correspond to the formation of branched treelike structures with one or two subsequent growth steps.

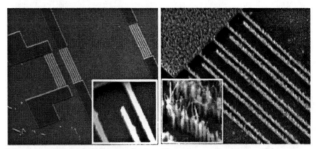

Fig. 4: The left image shows the sensor/actuator system on silicon membrane prior to the growth of CNTs. Inset magnifies individual fingers. The image at right shows the interdigital structure after the B-CNTs have been grown. Inset B-CNTs.

The sensing/actuating characteristic of the system has been investigated by applying a pulse generator to the actuator and the value of capacitance has been affected and measured at the opposite I/D side. For this purpose a 1MΩ resistor has been placed in series with the I/D capacitor and its voltage has been measured.

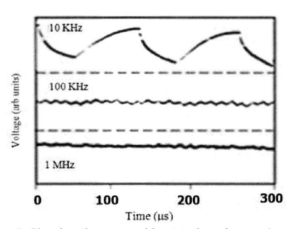

Fig. 5: Signal at the sensor side at various frequencies of actuation. Top curve shows the response to an actuation frequency of 10 KHz with actuation amplitude of 1V. Raising the frequency to 1MHz lowers the signal dramatically.

Figure 5 shows the results of the sensing operation in response to the actuation from the first I/D actuator. The actuation caused by the source signal leads to a mechanical vibration of the membrane and in turn a change in capacitance value of the other I/D capacitance in the opposite side of the membrane. The alternative capacitance value causes the fluctuation of the measured voltage drop over the resistance, which has been recorded in figure 5, for three different actuation frequencies.

As seen by raising the stimulation frequency to values of 100 KHz, the sensed signal vanishes. The optimum frequency is found to be around 10 KHz. Similar experiment has been carried out at a frequency of 10 KHz but with varying the amplitude of the actuator signal (see figure 6). Apart from sensor/actuator capability, such structure can be used as a source for efficient electron emission in a controlled manner.

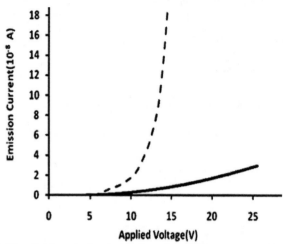

Fig. 7: The field emission behavior of B-CNTs measured in vacuum (dashed line) as opposed to normal CNTs (solid line) further confirming the superiority of the branched structure.

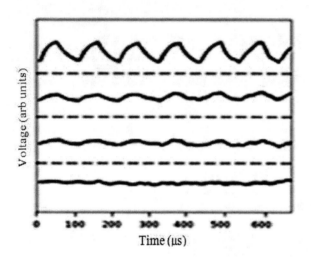

Fig. 6: Signal at the sensor side at various intensities of the actuator part. As observed a nearly linear response is observed between the applied voltage and the measured signal. The frequency of actuation is set at 10 KHz.

The field emission behavior of branched CNTs has been thoroughly examined and part of the results are presented in Figure 7, showing the superior performance of branched CNTs as opposed to vertical regular CNTs. In an attempt to observe the controllable field emission by means of the electromechanical actuation, the setup of Figure 8(a) is proposed where the anode electrode has a small surface and it is placed, with a narrow air gap, against the B-CNT holding substrate. By applying proper signal to one of the I/D parts, a mechanical wave propagates through the membrane which in turn vibrates the emitter part. The current measured at the anode electrode is significantly altered by the vibration of emitter, which itself has been imposed by the actuator I/D (part b). We believe such structure can be used for the fabrication of high sensitivity electro-mechanical pressure and acceleration sensors not achievable using standard silicon technology.

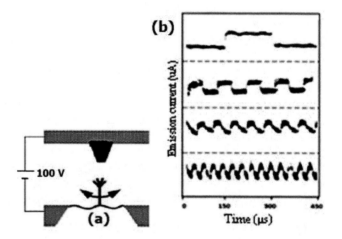

Fig. 8: (a) The schematic of the emission measurement. Top substrate is connected to negative and B-CNT-holding substrate is connected to the positive voltage.
(b) Signal (emission current) at the anode side due to actuation at different frequencies.

Figure 9 illustrates the response of investigated emission current of the structure to a pressure difference of about 1 atmosphere applied to the backside of the upholding membrane, at time intervals of about 10 seconds. The considerable increment in emission current in response to the mechanical stimulation confirms the potential to be applied as high sensitive pressure and acceleration sensors. Further investigation on the application of such devices in low frequency seismologic sensors is being pursued.

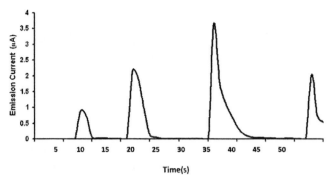

Fig. 9: The response of the emission current to the applied pneumatic pressure with step-like values with time intervals of 10 S.

IV. CONCLUSION

The fabricated structures show the efficacy of branched carbon nanotubes to realize high sensitivity sensors and actuators on silicon-based membranes. The heavy overlapping of neighboring lines, which is not possible using standard silicon technology, leads to a high value of capacitance and subsequently the actuation has become possible. Such structures are suitable devices for high sensitivity vibration sensors.

Authors wish to acknowledge the financial support of the Research Council of the University of Tehran.

REFERENCES

[1] M. M. Treacy, T.W. Ebbesen, J.M. Gibson, "Exceptionally high Young's modulus observed for individual carbon nanotubes" Nature, vol. 381, June 1996, pp. 678–680.

[2] T.Tombler, C. Zhou, L. Alexeyev, J. Kong, H. Dai, W. Liu, C. Jayanthi, M. Tang, S.Y.Wu," Reversible electromechanical characteristics of carbon nanotubes under local-probe manipulation", Nature, vol. 405, June 2000, pp. 769–772.

[3] M. H. Huang, C. A. Zorman, M. Mehregany, and M. L. Roukes, "Nanoelectromechanical systems: Nanodevice motion at microwave frequencies" , Nature (London), vol. 421, Jan. 2003, pp. 496-498.

[4] P. Kim, C.M. Lieber, "Nanotube Nanotweezers" Science, vol. 286, Dec.1999, pp. 2148–2150.

[5] J. E. Jang, S. N. Cha, Y. Choi, and Gehan A. J. Amaratunga, D. J. Kang, D. G. Hasko, J. E. Jung and J. M. Kim, "Nanoelectromechanical switches with vertically aligned carbon nanotubes", Applied Physics Letters, vol. 87, Oct. 2005, pp. 1631141-3.

[6] A.M.Fennimore, T.D. Yuzvinsky, W.Q. Han, M.S. Fuhrer, J.Cumings and A.Zettl, "Rotational actuators based on carbon nanotubes.", Nature, vol. 424, July 2003, pp. 408–410.

[7] W. B. Choi, D. S. Chung, J. H. Kang, H. Y. Kim, Y. W. Jin, I. T. Han, Y. H. Lee, J. E. Jung, N. S. Lee, G. S. Park and J. M. Kimd "Fully sealed, high-brightness carbon-nanotube field-emission display", Appl. Phys. Lett, vol. 75, Nov.1999, pp. 3129-3131.

[8] Y. Abdi, S. Mohajerzadeh, H. Hoseinzadegan, and J. Koohsorkhi, "Carbon nanostructures on silicon substrates suitable for nanolithography", Appl. Phys. Lett., vol. 88, 2006, pp.1-3.

[9] Y. Abdi, A. Ebrahimi, S. Mohajerzadeh and M. Fathipour, "High sensitivity interdigited capacitive sensors using branched treelike carbon nanotubes on silicon membranes", Appl. Phys. Lett., vol. 94, Apr. 2009, pp. 1735071-3.

[10] A. Sammak, S. Azimi, N. Izadi, B. Khadem Hosseinieh, and S. Mohajerzadeh, "Deep vertical etching of silicon wafers using a hydrogenation assisted reactive ion etching.", J. Microelectromech. Syst. Vol.16, Aug.2007, pp. 912-918.

Computer Simulations and Study of Bridge-Like Radiation Defects in the Carbon Nano-Structures in Composite Materials

A M Ilyin[*] and G W Beall[**]

[*]Physical Department of Kazakh National University, 96a, Tole-bi str., Almaty,Kazakhstan, e-mail: ilyinar@mail.ru
[**]Center of Nanophase Research, Texas State University San Marcos, Department of Chemistry and Biochemistry , 601 University Drive, TX, USA

ABSTRACT

We perform computer simulation and study of a special type of bridge-like defects in carbon nanostructures, which can be produced under fast electrons and ions irradiation. The binding energy and structure characteristics of the defects in carbon nanotubes and in graphene have been determined using known semiempirical extended Hückel techniques.

Keywords: computer simulation, carbon nanotube, graphene, bridge-like defect, radiation effects

1 INTRODUCTION

Carbon nanotubes and graphene attract great attention of scientists and engineers because their unique mechanical and physical properties. In particular, single-walled carbon nanotubes (SWNT) and multi-walled carbon nanotubes (MWNT) can be used in production of composites. These composites can be based on metal, ceramic or polymer matrices filled with carbon nanotubes as elements of reinforcement. Many of the good physical and electrical properties can be exploited by incorporating the nanotubes into some form of the matrix material [1-4]. Obviously the main goal of using nanotubes in making composites is using their extremely high mechanical properties in combination with low density. The same is true of the very high electrical conductivity of especially the SWNT where normally insulating polymers can be rendered conducting with very low weight percent of the nanotubes. This situation unfortunately is not true of the very high thermal conductivity of the carbon nanotubes. Composites produced from carbon nanotubes only yield thermal conductivities following the rule of mixtures. One of the important and unfortunately poorly understood factor is being a way of bonding of a matrix with carbon nanostructures. It should be taking into account that the sp^2 electron structure often results in very low binding energy between CNT's surface and atoms of many elements. This results in poor interfacial bonding of the nanotubes or nanostructures fragments that can slide in the matrix, or relatively to each other under stressed condition. It results in decreasing the part of CNT's surface, transferring stress between matrix and reinforcement elements under stressed state of the composite. This therefore limits rather severely the amount of ultimate strength present in the nanotubes that can effectively be transferred to the matrix. It is reasonable to suppose that defects in such structures might improve situation by linking nanoelements to each other allowing them to act as cross-links which will increase the stiffness of the composite. In general, investigations of radiation effects in carbon nanostructures are of great interest at present [5-7]. It is not yet well understood which kinds of stable atomic complex defects exist in carbon structures and it is not easy to observe such defects directly and to make interpretation of measurements [8]. In this situation computer simulation of radiation defects in carbon nanostructures becomes of great importance [9-12]. It is posited, that some types of radiation defects can act as a route to additional chemical bonds, promoting the strengthening of composite materials consisting of high concentration of carbon nanostructures. In this paper we consider results of simulation and investigation of structural and energetic properties of some possible configurations of defects, which can be produced by fast particles irradiation of a nanocomposite, involving carbon nanostructures. We used for calculations the well known semiempirical atomic basis extended Hückel method.

2 SIMULATION

Further we consider different types of possible radiation defects in carbon nanostructures rising under irradiation. We imply that all carbon species will be introduced into some matrices and focus our attention on possible effects of binding carbon nanospecies   in stiff   clusters for the limitation of relative displacement and stiffening   the materials.

Figure 1 presents a possible bonding configuration built between two parallel SWNT (6,6) with a bridge-like bond, which can be produced under ion-, fast electrons or neutron irradiation. We considered a case, when vacancies were produced in both carbon nanotubes in such a way, that vacancies were faced each other and the interstitial atom i was placed between them. The calculation was performed with using energy minimization of the defect zone.

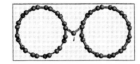

Figure 1. A bridge-like defect linking   carbon nanotubes

After minimization procedure the binding energy of the configuration was obtained as large as 3.9 eV for every bond of bridge-like configuration

Figure 2. Electron charge distribution for the configuration presented in Fig.1

Figure 2   presents the electron charge distribution for the defect in Fig.1 by the electron density equals 1.4 $el/Å^3$. It should be noticed, that it is a typical electron charge density for carbon nanostructures , corresponding to covalent bonds. So that it proves a fast bridge-like bond, rising between two nanotubes.

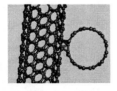

Figure 3. The linking defects between crossing nanotubes

Figure 3 presents the linking defects between two crossing nanotubes with total binding energy equal to 9.3 eV.

Figure 4. Atomic configuration of an inner bridge-like defect in a two-walled carbon nanotube. The interstitial atoms here and below are distinguished by light.

Figure 4 presents a bridge-like defect   which can be produced under irradiation of a system, involving multi-walled carbon nanotubes. In this case we performed the calculation   for the two-walled nanotube with the inner tube (5,0) and the outer tube (14,0).   Diameters of CNT's (5,0) and (14,0) are equal to 3.9 Å   and 10.9 Å accordingly. The inside between for this two-walled nanotube equals to 3.5 Å and is close to graphite interlayer distance.

In our simulation the vacancies in both inner and outer nanotubes were faced each other and the interstitial carbon atom i was placed symmetrically between the vacancies. The configuration, arising after using the energy minimization, is shown in Figure 4 . One can see that some atoms, closest to vacancies have moved slightly into the gap between the nanotubes. This movement is necessitated to facilitate the creation of the bonding bridge between the inner and outer nanotubes. For example, the distance between atoms 1-2 equals 2.49 Å , between i - 1 : 1.56 Å , i - 2: 1.49 Å . In this case the angle between bonds is very near to 180°. The total binding energy of the i atom equals -4.7 eV.

We proposed, that the open ends of   carbon nanotubes in principle can also serve as bonding sites for knocked out atoms.   A possible configuration with a bridge-like single end-bond, based on the interstitial i rising between the external and the inner tubes is presented in Figure 5.   After the relaxation of the structure was performed, the binding energy of the defect was equal to 10.0 eV.

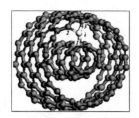

Figure 5. A possible atomic configuration of a single bridge-like defect between the ends of a two-walled carbon nanotube (5,0) / (14,0). The interstitial is marked by light. Lengths of the bonds: i-1: 1.31 Å ; i-2, i-3 : 1.41 Å .

The rather unusual nanotubular object is presented in Figure 6. It involves 5 bridge-like end-bonds between the external and the inner tubes. Such defect results not only in bonding, but also in closing the end gap of two-walled carbon nanotubes.

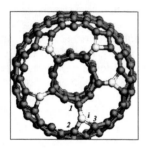

Figure 6. View of the atomic configuration of the complex bridge-like defects. Lengths of bonds: $i-1$: 1.30 Å , $i-2$, $i-3$: 1.39 Å . Distances between interstitial atoms vary approximately from 3.5 to 4.1 Å .

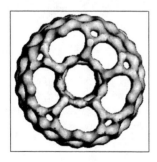

Figure 7. Electron charge distribution for the defect presented in Fig.6.

One can see in Fig.7 the electron charge distribution for the defect presented in Fig.6 by the electron density equals to 1.4 el/Å^3 . The calculations were made also for bilayer graphene fragment, in the usual graphite-like stacking configuration presented in Figure 8. The coupled atom pair removed by creating the vacancy pair is marked by black.

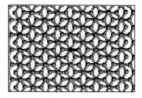

Figure 8. Graphite-like bilayer graphene used for making coupled vacancies

The interstitial atom i was placed between them. After that energy minimization procedure was used to obtain a configuration with minimum value of the total energy of the defect volume. During the minimization the edge atoms of graphenes were fixed in order to account the size effect of a larger graphene sheet. One can see from Figure 9 that after relaxation the center parts of graphenes near the defect significantly entered into the gap. The distances between atoms 1-2 and 3-4 turned out equal nearly 1.43 Å , that approximately are the distances in ideal graphene structure and the arising of additional bonds between graphenes was expected.

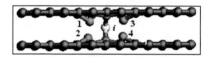

Figure 9. The bridge-like defect in the central part of two-layer graphene

The distribution of the electron charge presented in Figure 10 by the charge density 1.4 el/Å^3 proves existing of two additional covalent bonds between graphene layers together with primary one, created by the interstitial atom i. The total binding energy for this defect configuration was calculated as large as -11.3 eV.

Figure 10. The distribution of the electron charge in the configuration presented in figure 9.

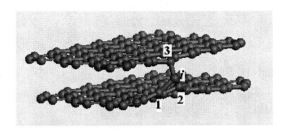

Figure 11. The bridge-like defect based on an interstitial atom i at the end of bilayer graphene.

Result of modeling of end-bridge-like defect for graphene sheet is presented in Fig.11. One can see a significant deformation of graphene fragments near the defect zone. The binding energy of the interstitial at this configuration was equal to -9.3 eV.

3 CONCLUSION

Some possible types of stable bridge-like radiation defects, which can originate under fast particles irradiation in typical carbon nanostructures were simulated and calculated using known semiempirical extended Hückel techniques.

Results of calculation illustrate possible linked configurations for SWNT, MWNT, graphene. The binding energy and structure characteristics of defects have been determined. A new type of defect configuration based on bridge-like bonds between ends of MWNT is presented. Radiation production of such types of defects can be useful for application in materials science, particularly in R&D of new materials with using the carbon nanostructures as reinforcement elements, for stiffening the nanostructures and modifying physical characteristics like electrical- and heat conductivity.

REFERENCES

[1] P.J.F.Harris, Inter. Mater. Rev., V.49, 1,31, 2004.
[2] P.M.Ajayan, L.S. Schadler, C. Giannaris, & A. Rubio, Adv. Mater. 12, 750, 2000.
[3] Y.Hu, O.A.Shenderova, Z. Hu, C.W. Padgett, &D.W. Brenner. Rep. Prog. Phys. 69, 1847, 2006.
[4] T.Laha, Y.Liu and A.Agarwal, Journal of Nanoscience and Nanotechnology,V. 7,1, 2007.
[5] A.Hashimoto, K.Suenaga, A.Gloter, K.Urita & S.Iijima, Nature 430, 870, 2004.
[6] A. V. Krasheninnikov, K.Nordland, M.Sirvio, E.Salonen, and J.Keinonen, Phys.Rev. B.63, 245405, 2001.
[7] A. V. Krasheninnikov, & F.Banhart, Nat. Mater. 6, 723, 2007.
[8] D.Teweldebrhan and A.A.Balandin, Appl.Phys.Lett., 94, 013101, 2009.
[9] A.M.Ilyin, E.A.Daineko and G.W.Beall, Physica E, 42, 67, 2009.
[10] A M Ilyin, G.W.Beall and I.A.Tsyganov. J.Comput. & Theoret. Nanoscience. V.7, 2010 (to be published).
[11] M.G.Futa and P.C.Kelires,Appl.Phys.Lett. 86, 191916, 2005.
[12] O.V.Yazyev, I. Tavernelli, U. Rothlisberger, & L.Helm , Phys. Rev. B 75, 115418, 2007.

Rapid Determination of Purity of Carbon Nanotubes

J. Hodkiewicz[*] and M. Wall[**]

[*]Thermo Fisher Scientific, 5225 Verona Road
Madison, WI 53711, USA, joe.hodkiewicz@thermofisher.com
[**]Thermo Fisher Scientific, Madison, WI, USA, mark.wall@thermofisher.com

ABSTRACT

Raman spectroscopy can provide for quick screening measurements of the carbonaceous purity of carbon nanotube samples. Carbonaceous purity is a critical parameter for many applications of carbon nanotubes. Raman spectroscopy has been reported many times in the literature as a technique which can be used to quickly measure carbonaceous purity although there are few instances in which details are provided on how the analysis is conducted or where limitations to the analysis are described. The goal of this work is to demonstrate the effectiveness and the speed of the technique for this measurement, as well as to help the reader understand the limitations of the analysis and in which situations it can be applied effectively and in which situations it would not be appropriate.

Keywords: raman, d-band, g-band, nanotube, purity

1 EXPERIMENTAL

All Raman microscopy characterization was performed with a Thermo Scientific DXR Raman instrument configured with a 633 nm excitation laser. The instrument was operated using the Thermo Scientific OMNIC 8 software suite.

2 RESULTS AND DISCUSSION

2.1 Measuring Purity

There are numerous publications which discuss using Raman spectroscopy to measure purity or quality of carbon nanotubes through the use of the ratio of the intensity of the D-band (~1350 cm^{-1}) to the intensity of the G-band (~1582 cm^{-1}) as the metric. In some cases variations on this basic scheme are used in which the intensity of the 2D-band (~2700 cm^{-1}) is used in place of the intensity of the D-band. In most common application of this technique the metric is used only to compare against a reference standard rather than to provide an actual quantitative measure of purity, although there are a few examples in which this metric has been used to provide an actual measurement of purity by preparing a set of calibration standards and referencing the samples to the calibration set. In most of the cases though, this metric is only used to compare to a known good reference standard. If the D/G intensity ratio in the sample is greater than the same ratio in the reference then the material is flagged for closer examination.

2.2 Speed of Analysis

The primary reason why Raman spectroscopy is used for this analysis is the speed of analysis. Both the measurement time and sample preparation time are very fast compared with most other techniques being used for this type of characterization. Typical measurement times range from as little as 5 seconds for high density samples to as much as a few minutes for loosely packed powder samples. The samples are run neat. Typically a small quantity of the material is placed on a glass slide and compressed with another glass slide to increase the density. In some cases, where the material is already in solution, the solution may be cast onto a glass slide and allowed to dry. Usually sample preparation takes only a few minutes.

2.3 Limitations of the Analysis

While the analysis is very quick and convenient, there are limitations to the analysis and it is important to understand these limitations before putting the method into practice. The first step to understanding the limitations of the analysis is to understand the origin of these bands and what is actually represented by them. The G-band is the primary Raman active mode in graphite and it provides a good representation of the sp^2 bonded carbon that is present in perfect planar sheets. The D-band originates from edge configurations in graphene where the planar sheet

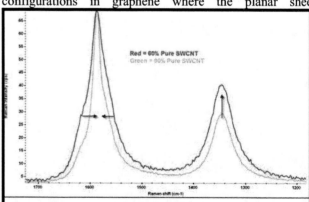

Figure 1. Comparison of the Raman spectra of 60% and 90% pure single-wall carbon nanotubes.

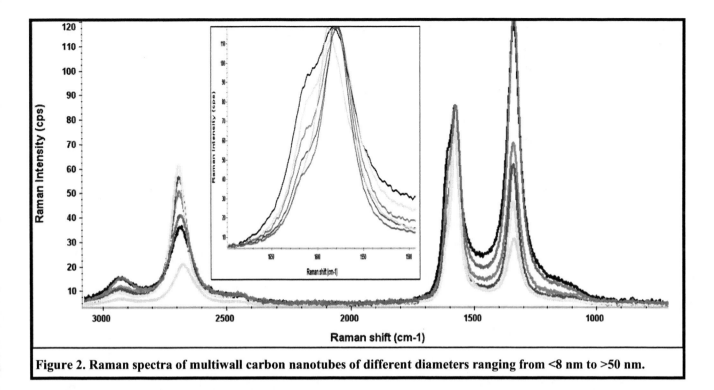

Figure 2. Raman spectra of multiwall carbon nanotubes of different diameters ranging from <8 nm to >50 nm.

configuration is distorted. You will also see some sp2 bonded amorphous carbon contributing to this band. As such the conditions that contribute to a higher intensity D-band are a higher concentration of amorphous material, nanotubes with a higher frequency of defects, shorter tubes where the edge configuration represents a higher percentage of the overall material, and multiwall tubes with a higher number of outer layers. Figure 1 provides an example of how purity can impact this ratio and figure 2 provides and example of how the number of layers in the tube can impact the ratio. With so many different attributes of the nanotube contributing to this band, it is not really possible to put a purity number on any random sample of nanotubes based on this metric alone. As I have already mentioned, people have reported that they have been able to successfully assign a purity value based on this approach, but in those cases they have confined their analysis to a highly uniform production stream where much of the potential variability is tightly controlled. If your samples do not fall into this category then it is probably not realistic to expect high accuracy purity measurements. However, despite the limitations, this is still a very useful industrial measurement and it is being used routinely in a great number of facilities around the world today. The reason for this is that if you are monitoring production line quality or the quality of incoming materials, most of the attributes that contribute to the D-band represent something that is undesirable in the product, so even if you can not immediately pinpoint which attribute has changed, the fact that there has been a change in the D/G band ratio can be a good indicator that there is a problem of some sort. The speed of the analysis also makes it very attractive as a quick screening methodology.

3 SUMMARY

Raman is a very powerful and valuable technique that can be of great benefit to characterization of carbon nanomaterials. The ratio of the Raman intensity of the D-band to the intensity of the G-band provides a quick method to screen materials for carbonaceous quality although you have to be careful when trying to extract more precise information using this method. Every lab that is monitoring the quality of Carbon nanomaterials will benefit from having access to Raman instrumentation.

Introduction to Raman Spectroscopy as a Characterization Tool for Carbon Nanotubes, Graphene, and other Carbon Nanostructures

J. Hodkiewicz[*] and M. Wall[**]

[*]Thermo Fisher Scientific, 5225 Verona Road
Madison, WI 53711, USA, joe.hodkiewicz@thermofisher.com
[**]Thermo Fisher Scientific, Madison, WI, USA, mark.wall@thermofisher.com

ABSTRACT

While there have been many publications produced in which Raman spectroscopy has been used to characterize various aspects of carbon nanostructures, there have only been a few works published which provide an overview of what the technique can offer to the field of carbon nanostructures, and works which discuss optimization of analysis parameters and the variation that is observed as those parameters are varied are still lacking. This work will provide an overview of the information that Raman spectroscopy can provide on carbon nanostructures with emphasis placed on carbon nanotubes and graphene. This work will go further to demonstrate the impact that varying key analysis parameters such as laser power and laser excitation will have on the measurements. The goal of this work is to provide people who are looking to characterize carbon nanostructures with an understanding of what Raman spectroscopy can offer and with some insight on how to optimize measurement conditions.

Keywords: raman, d-band, g-band, nanotube, graphene

1 EXPERIMENTAL

All Raman microscopy characterization was performed with a Thermo Scientific DXR Raman microscope. All bulk Raman spectroscopy characterization was performed with a Thermo Scientific DXR SmartRaman spectrometer. Both instruments were configured with 532 nm, 633 nm, and 780 nm lasers and data is labeled with the excitation laser that was used.

2 RESULTS AND DISCUSSION

2.1 Raman Highly Sensitive to Morphology

Raman spectroscopy is one technique which is particularly well suited to molecular morphology characterization of carbon materials. Every band in the Raman spectrum corresponds directly to a specific vibrational frequency of a bond within the molecule. The vibrational frequency and hence the position of the Raman band is very sensitive to the orientation of the bands and weight of the atoms at either end of the bond. Figure 1 shows and example in which the Raman spectrum of

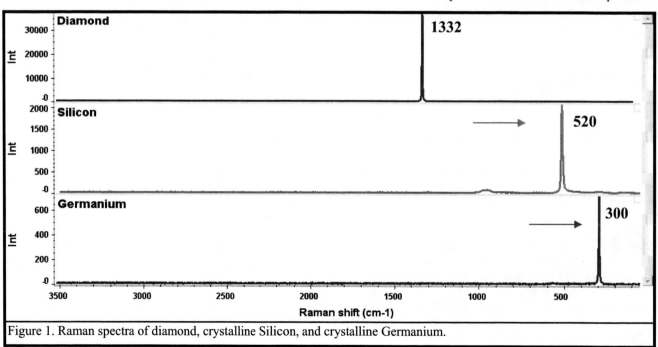

Figure 1. Raman spectra of diamond, crystalline Silicon, and crystalline Germanium.

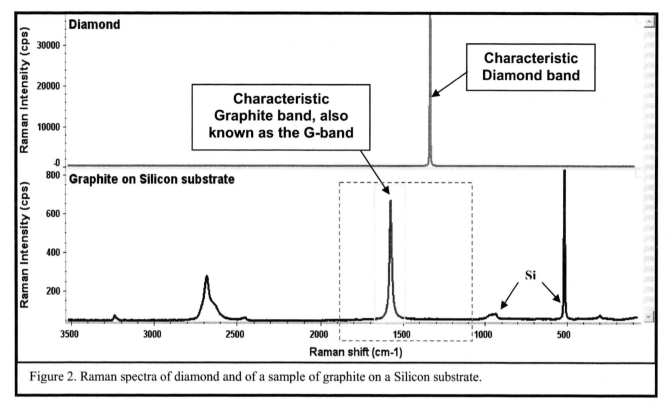

Figure 2. Raman spectra of diamond and of a sample of graphite on a Silicon substrate.

diamond is compared to the Raman spectra of crystalline Silicon and Germanium. These spectra show us several things. First, note that in the case of Diamond, where the material consists of highly uniform C-C bonds in a tetrahedral crystal structure that the Raman spectrum is very simple. It consists of only a single band because all of the bonds in the crystal are of the same orientation and strength resulting in a single vibrational frequency. We also see that the spectrum of Diamond is easily distinguished from the spectra of Silicon and Germanium by the frequency (cm^{-1} position) of the band even though they share the same tetrahedral crystal configuration. The heavier atoms of Silicon and Germanium slow the vibrational frequency and shift the corresponding Raman band to lower frequency as well.

Similarly in figure 2, when we compare the Raman spectra of two Carbon allotropes – diamond and graphite, again we can easily distinguish the two materials by their Raman spectrum even though both are composed entirely of C-C bonds. The graphite spectrum has several bands in the spectrum and the main band has shifted from 1332 cm^{-1} in diamond to 1582 cm^{-1} in graphite. The reason for this is that graphite is composed of sp^2 bonded carbon in planar sheets in which the bond energy of the sp^2 bonds is higher than the sp^3 bonds of diamond. The higher energy of the sp^2 bonds in graphite pushes the vibrational frequency of the bonds and hence the frequency of the band in the Raman spectrum to higher frequency. The 1582 cm^{-1} band of graphite is known as the G band. The presence of additional bands in the graphite spectrum indicate that there are some Carbon bonds with different bond energies in the graphite sample and this is in fact the case, as graphite is not quite as uniform in structure as diamond.

2.2 What Raman Can Reveal About More Complex Carbon Structures

We will start out by taking a look at fullerenes. Fullerenes are essentially hollow carbon shells of various sizes. The most well known of these is a 60-carbon unit called Buckminster fullerene or C60. There are many other fullerenes, from a few to many hundreds of carbon atoms. Figure 3 compares the Raman spectra of C60 and C70. The main feature in the C60 spectrum is a relatively sharp line at around 1462 cm^{-1}, known as the pentagonal pinch mode. This tells us several things. Firstly, it tells us that C60 is composed of sp^2 bonded Carbon. The sharpness of the band also tells us that the bonds are for the most part very uniform in nature. In fact, the carbon atoms in C60 are equivalent and indistinguishable. In contrast, the spectrum of C70 is littered with numerous bands. This is due to a

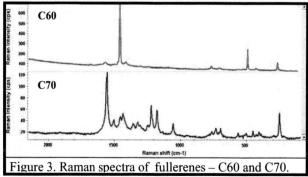

Figure 3. Raman spectra of fullerenes – C60 and C70.

reduction in molecular symmetry which results in more Raman bands being active. Additionally Raman can also be very sensitive to doping and stress due to temperature or pressure

Next we will take a look at graphene. Graphene is the fundamental building block of many important carbon materials including graphite. Graphite consists of stacks of sp^2 bonded planar graphene sheets. When comparing Raman spectra of graphene and graphite, at first glance the spectra look very similar. This is not too surprising as graphite is just stacked graphene. However, there are some significant differences. The most obvious difference is that the band at 2700 cm^{-1}, which is known as the G' band, is much more intense than the G band in graphene compared to graphite. Figure 4 allows us to take a closer look at the G' band of these two materials where we can see that both the shape of the band and the position are different and both tell us something. The peak shift in graphite is a result of interactions between the stacked graphene layers which has a tendency to shift the bands to higher frequency. The G' band in a single layer graphene spectrum fits to a single band whereas curve fitting reveals a multitude of underlying bands in the graphite spectrum. These bands are a result of the different interlayer interactions that occur at different depths within the graphene.

Lastly, we will take a look at carbon nanotubes. Carbon nanotubes are essentially rolled up graphene sheets that have been sealed to form hollow tubes. Single-wall carbon nanotubes (SWCNT) are cylindrical tubes with a single outer wall with diameters that are usually only 1 – 2 nm. There are also double-wall carbon nanotubes (DWCNT) which have a second layer of graphene wrapped around an inner single-wall carbon nanotube. These are a subset of the larger category of multi-wall carbon nanotubes (MWCNT) that have many layers of graphene wrapped around the core tube. Due to their unique mechanical, electrical and thermal properties, Carbon nanotubes are one of the most active areas in the field of carbon nanotechnology today.

The Raman spectrum of a SWCNT bears a lot of similarity to graphene, which is not too surprising as it is simply a rolled up sheet of graphene. Figure 5 shows us a Raman spectrum of a SWCNT in which we can see well defined G and G' bands as there are in graphene and graphite. We also see a prominent band around 1350 cm^{-1}. This band is known as the D band. The D band originates

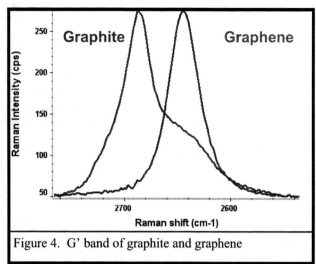

Figure 4. G' band of graphite and graphene

from a hybridized vibrational mode associated with graphene edges and it indicates the presence of some disorder to the graphene structure. This band is often referred to as the disorder band or the defect band and its intensity relative to that of the G band is often used as a measure of the quality with nanotubes. There is also a series of bands appearing at the low frequency end of the spectrum known as Radial Breathing Mode or RBM bands. The RBM bands are unique to SWCNTs and as their name suggests, correspond to the expansion and contraction of the tubes. The frequency of these bands can be correlated to the diameter of SWCNTs and they can provide important information on their aggregation state.

MWCNTs have very similar spectra to those of SWCNTs. The primary differences are the lack of RBM modes in MWCNTs and a much more prominent D band in MWCNTS. The RBM modes are not present because the outer tubes restrict the breathing mode. The more prominent D band in MWCNTs is to be expected to a certain extent given the multilayer configuration and indicates more disorder in the structure.

2.3 Recording a Raman Measurement

It is relatively easy to collect a Raman spectrum of a carbon nanomaterial. The materials are generally recorded neat under standard atmospheric conditions which make for very easy sampling and provide a lot of flexibility as to what can be sampled. Although if samples are in the form of a loose powder, then the powder is usually compressed to provide a denser sample. This can be done by simply pressing a small amount of sample between two microscope slides. If you measure the powders loose, the density is often too low to get a measureable spectrum. Many times CNTs are cast onto slides within a surfactant. This works quite well as the density is generally much higher than the powdered form of the material although it is worth noting that there may be some bands present from the surfactant. These can be easily identified by measuring a blank sample of the surfactant though. Films are typically measured in

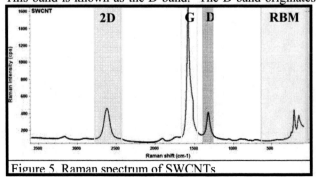

Figure 5. Raman spectrum of SWCNTs

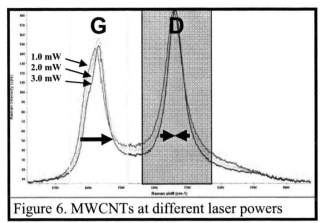

Figure 6. MWCNTs at different laser powers

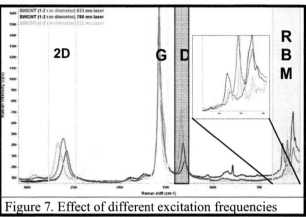

Figure 7. Effect of different excitation frequencies

their native state on whatever substrate they happen to be on.

In terms of time of analysis, Raman measurements are generally short compared to other techniques. The measurement time typically ranges anywhere from a few seconds to five minutes depending primarily on the density of the sample and on the laser power at the sample. The denser a sample the shorter the necessary measurement time will be and the higher the laser power the shorter the necessary measurement time will be. Note that when we talk about laser power – it is the laser power at sample and really the optical density of the laser power at the sample that is most critical. If you have a lot of laser energy dispersed over a large area it will not have the same impact as a lesser amount of energy tightly focused on the area of interest. It is also worth noting that many Carbon nanomaterials are sensitive to laser power, so the amount of laser power you can use is often limited by the samples and it is important to have good control of the laser power to insure reproducible measurements.

It turns out that it is especially important that you have good laser power control when measuring carbon samples. This is for two reasons. The first is that some carbon materials are very sensitive to laser power and even low laser powers can damage the sample. It is important to understand the laser power tolerance of your samples before you start your experiments for this reason. The second reason that good laser power is important is that even when tolerant of the laser power, carbon nanotube samples will exhibit significant spectral differences depending on the laser power, so it is important that exactly the same laser power be used. Figure 6 provides an example of this. In this example you can see that the D-band contracts in width and the G-band shifts to lower frequency as you increase laser power. This is a reversible effect, but it is still important to control laser power tightly to avoid misinterpreting effects such as this.

We have not discussed this so far, but excitation laser frequency can also have a dramatic effect on the spectrum. Figure 7 provides us with an example of this. The reason for these differences is that several of the bands in the Raman spectrum of carbon nanotubes exhibit what is known as resonant dispersive behavior. Meaning that there are a number of very weak bands present, but the bands that are visible are those that are resonant at the excitation frequency. These bands are greatly enhanced and since each excitation frequency enhances different bands, you get significant differences when comparing the spectra obtained with different excitation lasers. For some applications it may be desirable to characterize the samples with multiple excitation frequencies, but often it is sufficient to use only one provided that you always use the same excitation frequency.

3 SUMMARY

Raman is a very powerful and valuable technique that can be of great benefit to characterization of carbon nanomaterials. Raman is particularly well suited to detect small changes in structural morphology of Carbon nanomaterials making it an indispensible tool for many material scientists working with carbon nanostructures. Raman instruments are very fast and provide a great deal of flexibility in samples that can be accommodated. Every lab that is characterizing Carbon nanomaterials will benefit from having access to Raman instrumentation.

CNTs and Silver coated CNTs provide immune responses against RSV

S. Boyoglu[1], L. Adiani[1], C. Boyoglu[1], M. Mishra[1], V. Rangari[2], V. Dennis[1], S.R. Singh[1]

Center for Nanobiotechnology Research, Alabama State University, Montgomery, AL, USA; [2]Tuskegee University; Tuskegee, AL, USA

ABSTRACT

Nanobiotechnology and its potential to be used in treating and diagnosing general and viral infections have allowed us to step into a new realm, when dealing with Respiratory synctical virus (RSV). RSV has long been considered a major respiratory tract pathogen of children causing lower respiratory tract infections and pneumonia. In the present study, Silver coated CNTs and CNTs were evaluated for their cytotoxicity on HEp-2, and Vero using MTT assays. The cells were incubated with various concentrations of the nanoparticles and assessed for the cell viability at 24, 48, and 72 hours. The results show that at 1.25 µg/ml the cell viability is at 85 % which decreased to about 65 % at 5µg/ml after 48 h post-transfection. HEp-2 and Vero cells were infected with RSV in presence and absence of nanoparticles and inhibition effect of the nanoparticles on RSV was evaluated using immunofluorescence. Additionally, the characteristics of the nanoparticles were evaluated using AFM, TEM. Moreover, the immunological response of the nanoparticles is currently being analyzed using HEp-2 cell line.

Keywords: Respiratory syncytical virus, Carbon Nanotubes (CNTs), Silver coated CNTs, AFM

1. INTRODUCTION

Respiratory Syncytial Virus (RSV) belonging to the Paramyxovirus family is one of the most common causes of upper and lower respiratory tract infections. It is a leading cause of bronchiolitis and pneumonia worldwide, affecting children under the age of five, the immunocompromised, and the elderly with a high mortality rate. The RSV genome is composed of a single stranded negative sense RNA of 15,200 nucleotides encoding 11 proteins [1]. Among these 11 proteins only the fusion (F), which allows membrane penetration and fusion of the virus with the host cell, and the attachment (G) glycoproteins have been studied for their potential use as a vaccine. Nanoparticles such as CNTs and Silver coated CNTs have been applied to determine the inhibition of RSV.

A concern with using nanoparticles is their cytotoxic effect. Therefore, prior to applying these nanoparticles to inhibit RSV, their cytotoxicity must be determined. The aim of this present study was to assess the cytotoxicity of the CNTs and Silver coated CNTs followed by their inhibition of RSV. Subsequently, we studied the nanoparticles effect on the gene expressions.

2. MATERIALS AND METHODS

2.1 Sonochemical Synthesis of Silver coated CNTs

Multi walled carbon nanotubes (CNTs) were purchased from Nanostructured & Amorphous Materials. These CNTs are < 95% pure (stock #1205YJ) with a density of 0.04-0.05 g/cm3. CNTs (10 mg) were added to 60 ml dimethylformamide and magnetically stirred for 30 minutes. Then 500 mg of silver (I) acetate (Sigma Aldrich 98+ %) was added to the solution and stirred again for 10 minutes. This mixture was irradiated with a high intensity ultrasonic horn (Ti-horn, 20 kHz, 100 W/cm2) under argon gas at 13° C external cooling temperature for 72 h. The product was further washed with distilled water 3-4 times followed by drying with absolute alcohol in a vacuum overnight.

2.3 Transmission Electron Microscopy (TEM) analysis

High resolution TEM (HRTEM) was performed using a JOEL-2010 machine. Pictures were taken at 50,000-125,000 magnifications.

2.4 Atomic Force Microscopy (AFM) analysis

AFM pictures were obtained using the NANOSCOPE-R2 AFM (Pacific Nanotechnology, Santa Clara, CA, USA). CNTs and Silver coated CNTs were placed onto a slide, dried and visualized under the microscope. Close contact mode and standard silicon cantilevers (Pacific Nanotechnology, Santa Clara, CA, USA) 450 μm in length and 20 μm in width were employed for imaging. The cantilever oscillation frequency was tuned to the resonance frequency of approximately 256 kHz. The set point voltage was adjusted for optimum image quality. Both height and phase information were recorded at a scan rate of 0.5 Hz, and in 512 x 512 pixel format.

2.2 Cell cytotoxicity of nanoparticles

The cytotoxicity of CNTs and Silver coated CNTs on HEp-2 cells was determined using the MTT dye reduction assay in Vero and HEp-2 cells. Cells were seeded in a 96-well plate at a density of 2.0×10^{4} cells/well and incubated overnight at 37 °C. Then the cells were incubated in 100 μl serum free medium containing selected amounts (0.25, 0.5, 1.25, 2.5, 5 and 10 μg/ml) of CNTs and Silver coated CNTs. After 24, 48 and 72 h, the medium was removed and the cells were rinsed twice with sterile PBS. Next, 10ul of MTT (5mg/ml) solution was added into each well and allowed to react for 4 h at 37 °C. DMSO (150 μl) was added to each well and the plate was incubated for 30 min at room temperature. Absorbance at 490 nm was measured with an ELISA plate reader

2.3 Cell and Virus

HEp-2 cells were purchased from American Type Culture collection (ATCC, Manassas, VA; CCL-23) and were propagated by standard methods using Minimum Essential Medium (MEM) supplemented with 10% Fetal Bovine Serum (FBS), 2 mM L-Glutamine, 75 U/ ml Penicillin, 100 μg/ml Kanamycin and 75 μg/ml Streptomycin.

Human RSV Long strain was purchased from ATCC (VR# 26). Virulent RSV stocks were prepared and propagated in HEp-2 cells. RSV with multiplicity of infection (m.o.i) of 4:1 was added to the flask and virus adsorption was carried out for 1 h at 37°C in a humidified atmosphere with 5% CO_2. MEM supplemented with 2% FBS and 2 mM L-Glutamine were added to the flask and infection of cells was observed for 72 h. RSV infected cells were harvested and cell suspensions were subjected to 2 freeze-thaw cycles at -80°C followed by centrifugation at 3,000 x g at 4°C to remove cellular debris. The viral stock was aliquoted and stored at -80°C or liquid nitrogen until further use. Viral titer of the prepared stock as determined by plaque assay revealed a titer of 10^{6} PFU / ml.

2.4 Immunofluoresence

Once the cytotoxicity of the CNTs and Silver coated CNTs were determined in both HEp-2 and Vero cell lines, we were able to concentrate on determining the inhibition of RSV using the nanoparticles. The nanoparticle samples at concentrations of 0.25, 0.50, 1.25, 2.5, and 5μg/ml were mixed with 100 PFU of RSV and added to the cells. After plating the cells in an 8-chamber slide and incubation for 24 h, RSV and desired concentrations of CNTs and Silver coated CNTs were added to each chamber and incubated for 48 h. This was followed by fixing the cells with 10%TCA for 15 minutes and washed with 70%, 90%, and absolute alcohol for 5 minutes each. Washed cells were blocked using 0.5ml of 3% dry milk powder in PBS for 30 minutes. Polyclonal goat anti-RSV antibody (Chemicon) in antibody buffer (2% dry milk in PBS) were added to the cells, and incubated for 1 h at room temperature. The cells were washed three times again with 1X PBS followed by the addition of the secondary antibody, rabbit anti-goat Ig (H+L) (Southern Biotechnology). Following secondary antibody incubation, the cells were stained with DAPI.Cells were visualized with a Nikon fluorescent microscope (Model Ti-U Phase, Lewisville, TX).

2.5 Gene Expression of HEp-2 cells Treated with Nanoparticles

HEp- 2 cells were plated (1×10^{6} cells in MEM-10) in a six well plate and incubated for 24 h to get 80-90% confluency. The nanoparticles at 2.5μg/ml concentrations were mixed with RSV (100 PFU) and incubated for 30 minutes at room temperature. The RSV, nanoparicles alone and nanoparticle+RSV mixtures (100 μl) were added to respective wells and the plates were rocked gently. The cells were incubated for 2 hours and then the media was aspirated from the 12 well plates and added more MEM -10 media and incubated for 16hr,24hr and 48hr at 37°C under 5% CO2 in a humid atmosphere.

2.6. Measurement of cytokine concentrations

Cytokine ELISAs were performed as previously described [2]. Concentrations of IL-8 and IL-6 cytokines were quantified in cell cultures supernatants using Opti-EIA sets (BD-PharMingen, San Diego, CA) according to the manufacturer's instructions.

3. RESULTS

3.1 TEM and AFM analysis

Nanoparticles were characterized in order to analyze the size and shape microscopically using TEM and AFM. TEM and AFM analysis revealed that the CNTs 10-20 nm in diameter and ~15 μm in length and Silver coated CNTs had a size of ~10 nm with nanotubes ~ 30 nm (Figure 1).

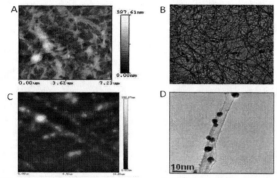

Figure 1: a) AFM 2D image of CNTs b) TEM images of CNTs c) AFM images of Silver coated CNTs d) TEM image of Silver coated CNTs. CNTs- 10-20nm in diameter and 0.5-20mm in length. Silver nanoparticles attached to CNTs are 5-10nm in diameter.

3.2. Cytotoxicity analysis

The percent viability of both cell lines was determined using the MTT Assay. The results show that at 1.25 μg/ml CNTs, the HEp-2 cell viability was 85 % which decreased to about 70 % at 5μg/ml after 48 and 72h post infection (Figure 2a). For the Vero cell lines, cell viability was 85% even at 5μg/ml after 48 and 72h post infection.(Figure 2b).Moreover, our results showed that Silver coated CNTs were more toxic then CNTs at the concentration of 5 μg/ml (cell viability was about 60%) after 72 h post-infection in HEp-2 cell lines. For the Vero cell lines, cell viability decreased to about 50% at 5μg/ml after 72h post infection. .

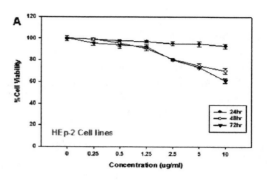

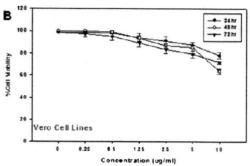

Figure 2: MTT cytotoxicity assay of CNTs with HEp-2 cells (a) and Vero cells (b).

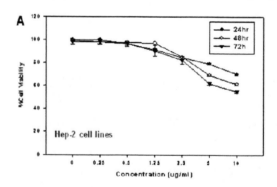

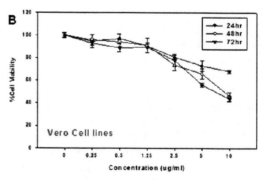

Figure 3: MTT cytotoxicity assay of Silver coated CNTs with HEp-2 cells (a) and Vero cells (b).

3.3. Immuno-Fluorescence Studies

Nanoparticles were mixed with 100 PFU of RSV and added to the cells to determine the inhibition of infection. Reduction in cytopathic effects (syncytia) in HEp-2 cells was observed to determine inhibition of RSV infection by CNTs and Silver coated CNTs. The cells infected with RSV showed marked syncytia formation (data not shown) and extensive immunofluroscence (Figure 4). Cells infected with RSV mixed with CNTs and Silver coated CNTs nanoparticles clearly show significant reduction in RSV infection compared to the cells infected with RSV alone (Figure 4a-b).

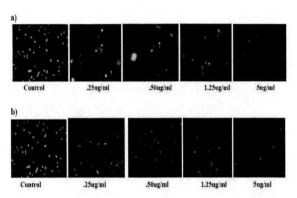

Figure 4: RSV inhibition by immunofluroscence. **Control:** Cells infected with RSV a) Cells infected with RSV mixed with Silver coated CNTs (b) cells infected with RSV mixed with CNTs

3.4. Gene Expression of HEp-2 cells Treated with Nanoparticles

Figure 5 shows the gene expression of HEp-2 cells treated with Nanoparticles, RSV and RSV mixed with Nanoparticles for 24 h and 48h.It was evident that IL-6 and IL-8 up-regulated after treatment with RSV+ Silver coated CNTs. For the same period, the expression of these genes in the control of HEp-2 cells and RSV groups were relatively lower. Our results also showed that when we increased the incubation time from 24h to 48h, the gene expression was also increased. Moreover, the Silver coated CNTs also up-regulated IL-8 expression relatively higher compared to HEp-2 cells and CNT groups (Figure 5). In conclusion, we can say that the Silver coated CNTs might activate some pathways or inhibit RSV affect so it increased some of the gene up-regulations.

The results of the present study show that CNTs and Silver coated CNTs can inhibit RSV infection. This approach may prove valuable in developing a therapeutic regime for RSV, and perhaps other dangerous infections including HIV.

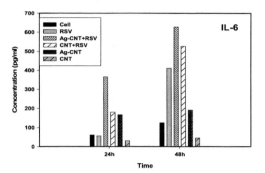

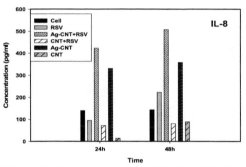

Figure 5: Gene expression of IL-6 and IL-8 in HEp-2 cells incubated with culture medium containing nanoparticles at 2.5µg/ml, RSV and RSV mixed with nanoparticles for 24 h and 48h.

REFERENCES

[1]. Hacking D, Hull J.Respiratory syncytial virus--viral biology and the host response. J Infect. 2002 Jul;45(1):18-2

[2] Ganapamo F, Dennis VA, Philipp MT. Differential acquired immune responsiveness to bacterial lipoproteins in Lyme disease-resistant and -susceptible mouse strains. Eur J Immunol 2003;33:1934–40. 4.

[3]. Sun Lova, Singh A., Vig K., Pillai S. R., Singh S. R.2008. J. Biomed. Nanotech. 4, 1.

Morphology Changes and Defects in Carbon Nanoparticles Due to Different Atom Arrangements

R Sharma[*], S.Kwon[**], A Sharma[***]

[*]Innovations And Solutions Inc.USA,3945 W Pensacola Street,#98, Tallahassee, Florida 32304 USA
[**]Biological Engineering Department, Utah State University, Logan, Utah USA
[***]Nanotechnology Lab, Electrical Engineering Department, Maharana Pratap A&T University, Udaipur, Rajasthan 313001, India

ABSTRACT

Motivation: Pentagon and heptagon fullerene carbon nanostructures (CNT) have potentials in superlight weight industries. CNT have structural defects caused by irregular atom arrangement during CNT growth. For better use, it needs modifications to remove defects. To accomplish it, investigations were made to identify bad atom arrangement. Findings: Present study highlighted the following: 1. The arrangement of carbon atoms in CNT rings was complex and different from common 6 carbon atom ring in nature; 2.Structures with incorporated n-gons are a new kind of nanostructure composed of parts equal in atom arrangement joined with suitable carbon connection bonds; 3.Still a lot of bio-industrial and superlight-weight interest in these CNT structures is not focused because of n-gons structural properties are unknown in the entire CNT complex; 4. The carbon atom arrangement in n-gons may answer the troubleshooting the defects of lightweight CNT complexes as elements in nanoscale devices and composite materials; 5.The present nanotechnology development allows the complicated CNT nanostructures with limited known chemical properties in experimental conditions by using self-assemble processes and atom-by-atom manipulation in CNT; 6. The industrial demand is increasing for more specific superlight weight CNT nanocomposites with atomic properties suited to environmental climatic resistance with utrastrength in more complicated aero-industrial applications. Conclusion: To meet these industrial applications, the arrangement of n-gons in CNT structures become complicated with new applications tailoring with their modified arrangement in CNT structures. It also provides more functionality necessary through incorporation of different n-gons and their combinations. It also inspired the continuous and further study of the n-gon role, with the exception of pentagons and heptagons, on carbon nanoparticle properties.
Key words: CNT, nanomaterials, polygon

1 INTRODUCTION

Carbon materials are in the most stable energetic situation if their atoms are connected via sp3 bonds and build 6 atom rings (n-gons with n=6), which, on their side, construct a honeycomb-like network.

Fig 1:A general sketch shows different carbon nanotubes.

2 PENTAGON-HEPTAGON FULLERENES

Most favored emergence of n-gons with n = 5 and 7, explains honeycomb structure of the observed carbon nanoparticles. The simultaneous emerged 5/7 defects such as pentaheptite, defects oppositely situated in the nanoparticle structure and the repetitive pair and carbon network deformation. The rings of *n* atoms (*n*-gon), for *n* = 1 up to infinity, and the changes in the network due to an *n*-gonal defect. Fig. 2 illustrates the honeycomb network that corresponds to the *n*-gons, when *n* = 1, 2, 3, 4 and 5. The bird's-eye view of transformed network due to an *n*-gonal defect is shown in Fig. 2b for *n* = 3, 4 and 5.

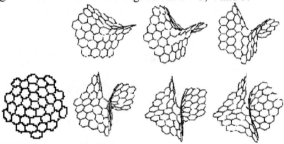

Fig 2: Honeycomb (left panel) and n-gonal defects of CNT

The rings of *n* atoms (*n*-gon), for *n* = 1 up to infinity, and the changes in the network due to an *n*-gonal defect, are widely discussed and explained in [47]. Fig. 2 illustrates the honeycomb network that corresponds to the *n*-gons, when *n* = 1, 2, 3, 4 and 5. The bird's-eye view of transformed network due to an *n*-gonal defect is shown in Fig. 2b for *n* = 3, 4 and 5. CNT can have defects similar to graphene (incorporation of carbon rings with n ≠ 6). Defects change nanotube geometry, e.g., their diameter and chirality and through that their electronic properties without introducing any impurities. It gives the possible electronic properties via defect changing. Defect arrangement is stable in time and due to that electronic properties will stay constant for the whole time of tube existence. This is illustrated in Fig. 3 for 3 orientations (Fig. 3a) of the tube axe to the carbon-carbon bonds. The nanotubes will be (6,0), (6,2) and (6,6) and will have diverse diameters (see Fig. 3b,c). Their hexagons are arranged in a different way according to the nanotube axe

(Fig. 3b). Those 3 kind of CNT are known as zigzag (when n = 0), chiral (different from its mirror image and m ≠ n), and armchair (when m = n). That means that each possible pair of coordinates (m,n) corresponds to a nanotube which will have unlike orientation as compared to all the others and, hence, of different diameter.

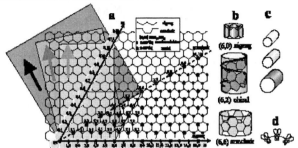

Fig 3: A coordinate system used for explaining carbon network and carbon structure properties. Three type of CNT (zigzag, armchair and chiral) are exhibited with their maps (Fig. 3a), their ring arrangement in accordance with tube axis (Fig. 3b), tube diameter (Fig. 3c) and a sketch of the electron orbital deformations by some bending (Fig. 3d).

• The pentagon and the heptagon are the frequently arising defects and can margin single or simultaneously in the nanotube structure. The single pentagon makes the circumference to decrease as the axis is extended behind the defect. On the contrary, the heptagon causes increasing of the tube circumference and the diameter. Merging of an isolated pentagon in the nanotube and transformation of the nanotube to a cone-like structure through that is illustrated in Fig. 4a. The tube is divided in two through the pentagon and the maps of the both parts are exposed right hands and left hands at the cut line DF of the pentagon. The exposition is over the carbon honeycomb to show how such a structure can be obtained via shearing the marked area and joining the line AB with A1B1 and CD with C1D1. In the case of the heptagon in Fig. 4b that is made analogically.

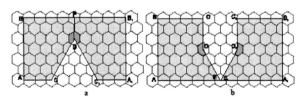

Fig 4: (a) a map of CNT with a pentagonal defect and (b) a map of CNT with a heptagonal defect.

• On the contrary, the defects caused by an isolated pentagon and a heptagon perturb the electronic properties of the hexagonal network significantly. If the symmetry axis of the aniline structure pair is nonparallel to the tube axis it changes chirality of a nanotube by one unit as shown in Fig 5. The heptagon splits the rows and as a result after the defect merging the row number 5 will be connected to number 4b instead to 5a after the map rolling in a nanotube. The pentagons arise frequently in pair with heptagons and they are responsible for the most of changes in the nanotube

structures. The diameter extension becomes significant if the pentagon and the heptagon in the pair are separated by hexagons because each hexagon added between the pentagon and heptagon changes the nanotube circumstance by one unit as shown in Fig 4b. These pairs also rule the electronic behavior around the Fermi level. The pentagon and the heptagon in the pair can be isolated one to other or to have one joined edge as in the aniline structure. The effect of the 5/7 pair on the growing structure depends on its place and orientation to the network direction. When the pentagon is attached to a heptagon, the pair creates only a topological change, but no net disclination, which may be treated as a single local defect. The defect can be only a small local deformation in the width of the nanotube or a small change in the helicity, depending on its orientation in the hexagonal network.

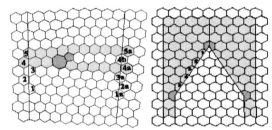

Fig 5: (left panel) A change of tube chirality by 5/7 defect nonparallel to tube axis. (Right panel) A change of tube circumference as a result of pentagon heptagon separation via 5 hexagons.

• The 5/7 pairs can be more than one in a junction zone. The pair can show different arrangements X-shape, T-shape, Y-shape, zig-zag, arm chair, double- and triple-strand and super coiled supercell e_x and e_y (determined in the honeycomb), L and M, and the vectors a and b defining nanoassemblies. In such case, polygonal defects other than pentagons and heptagons may play certain roles. The presence of up to 3 pairs in zigzag nanotubes in different configurations show several defects on the electronic properties of the system. The nanotube systems are (12,0); (11,0); (10,0) and (9,0). Those are selected because their electronic properties are expected to differ radically (the junction (12,0)-(9,0) is expected to be a metallic one, the two others are prototypes of a metal-semiconductor junction).

• A new class of layered carbon material, consisting of ordered arrangements of pentagons, hexagons, and heptagons is known as Haeckelites. They can be rectangular, hexagonal or oblique, surrounded by six-membered rings. They are more stable than C60 and can thus be regarded as energetically viable. All these structures are predicted to be metallic, exhibiting a high density of states at the Fermi energy, and to possess high stiffness. H5,6,7 and R5,7 haeckelites are predicted to be flat, while O5,6,7 is corrugated. H5,6,7 is found to be the most stable structure, possessing an energy of 0.304 eV per atom. All

Haeckelite nanotubes are metallic, independently of chirality and diameter.

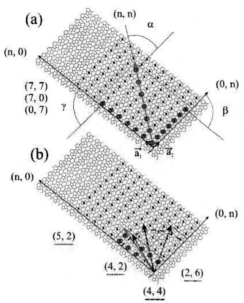

Fig 6 shows regular tiling of the plane with hexagons and azulene units in proportion 3:2.

• For illustration,'unit cell' is defined by the primitive vectors aW1 and aW2. The bonds shared by two pentagons have been highlighted by dots. In Fig 6(a), a 7x7 cell illustrates the wrapping vectors (7,0), (0,7), and (7,7) along the special directions (n,0), (0,n), and (n,n). The pentagons are encircled and rows are indicated by the thin solid lines defining the angles α, β, and γ. These rows can generate helical patterns along the structure. In Fig 6(b), the wrapping vectors (4,2), (5,2), (4,4), and (2,6) are highlighted by encircling. In the unit cell (2,6), all stressors are symmetrically placed around the wrapping vector. The dashed-line circle corresponds to a critical nanotube radius of 0.45 nm above which some stressors can dip inward toward the structure.

• The buckyballs have two aligned double bonds, each one in the center of a 5/6/6/5 pyracylene. This orientation conveniently permits standard polymerization by C60-pairs coalesce into the sections of (5,5) tubes either numbers 0 and 1 are the separate and the dimerized buckyballs.

• The nanopeapods are a family of carbon nanostructures where the condition for buckyballs coalescence is found to play significant role. Nanopeapods consist of an array of fullerene molecules (inner peas) and a single-wall carbon nanotube (SWCNT) (an outer pod), and all components are separated from the others by the van der Waals distance. Stable defects incorporated already in the nanostructures can also influence the growth of multi wall carbon nanotubes (MWCNT) growth. Each layer acts as a template for the sequent. During the simultaneous growth the atoms

which achieve opened layer ends are under the influence on the attracting forces from the neighbor layers and at suitable conditions can connect both layers, seamed them in one closed surface as shown in Fig 7. Fullerenes start to coalesce at 800ºC and complete transformation to a single-wall nanotube at 1200ºC. fullerenes start to coalesce at 800ºC and complete transformation to a single-wall nanotube at 1200ºC.

Fig 7: Sketch of nanotubes is shown with shells seamed one to other at the end of the tube.

• The defects and the electronic states initiated through that depend on tube type and size. The change in the rings of nanotubes (with different carbon network orientation to their axis) by adsorption of carbon dimers is shown in Fig 8.

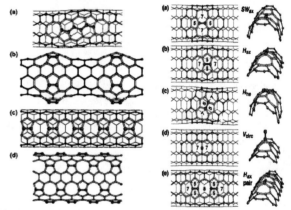

Fig 8(on left) Structure of multiple C-2 adsorbates.
Fig on right) Bonding structure of carbon dimer defects on a sample armchair (5,5) nanotube.

3 LIMITATIONS

Pentagons and heptagons emerged in the carbon nanostructures (CNT) as nano scale defects caused by irregular atom arrangement during growth. From this point of view, the induced investigations were attempted to pinpoint the reasons for bad atom arrangement and how those can be managed. From another point of view, the arrangement of carbon atoms in rings different from 6 carbon atom ring is also normal in nature, and in this case the structures with incorporated n-gons are to be treated as a new kind of nanostructure composed of parts equal in atom arrangement joined with suitable carbon connection bonds. There still remains lot of interest in those CNT structures, not focused before more on n-gons but on their structural properties of the entire CNT complex. Regardless of how the differences in carbon atom arrangement are studied or viewed, they still remain an interesting and a very important subject for study because of the utility of nanoparticles as elements in nanoscale devices and composite materials.

4 FUTURE PROSPECTS

The continuous progress in nano-science and nanotechnology stimulates the creation of nanostructures with new properties and functions. The level of nanotechnology development at the present moment allows complicated nanostructures constructed and their planned properties to be realized via suitable conditions by using self-assemble processes and atom-by-atom manipulation. The more increasing demand for specific CNT particle properties increased the more complicated applications. As the structures become complicated their more frequent applications tailoring with their structures and functions are necessary through incorporation of different n-gons and their combinations. It inspired the continuous and further study of the n-gon role, with the exception of pentagons and heptagons, on carbon nanoparticle properties.

5 REFERENCES

[i]

[i] http://www.scribd.com/doc/18663985/Nano-Materials-Lectures/   pages 352-85.

Gas-Phase Synthesis of Highly-Specific Nanoparticles on the Pilot-Plant Scale

T. Huelser*, S.M. Schnurre*, H. Wiggers*,**, C. Schulz*,**

*IUTA, Institute for Energy and Environmental Technology, 47229 Duisburg, Germany, huelser@iuta.de
**IVG, Institute for Combustion and Gas Dynamics, University of Duisburg-Essen and CeNIDE, Center
for NanoIntegration Duisburg-Essen, 47057 Duisburg, Germany, christof.schulz@uni-due.de

ABSTRACT

Highly-specific nanoparticles provide promising properties for current and future applications as the size dependence of their properties allows to tailor materials for specific applications [1-3]. The synthesis of particles in the size regime of a few nanometers follows two different routes, the wet-chemical synthesis route that usually yields materials grown by thermodynamic control and the gas-phase synthesis processes, which allow a kinetic control of nanoparticle formation. The second route is favored for the formation of doped nanoparticles and nanocomposites. However, industrial nanoparticle production is frequently carried out by cost-effective flame processes with high throughput [4]. The quality of the resulting materials is limited and often agglomerates with a broad size distribution are formed. Highly specific nanomaterials are often available in minute quantities and therefore, subsequent processing steps to generate nanoparticulate composites cannot be studied.

Keywords Nanoparticles, Particle Synthesis, Silicon, Pilot-Plant

1 PILOT PLANT SET UP

We designed a unique pilot-plant scale particle synthesis facility enabling for three different synthesis routes using either a hot-wall reactor (HWR), a flame reactor (FLR) or a plasma reactor (PLR). In these reactors the energy required for precursor decomposition is provided by either an electrical heat source, a flame, or a microwave-supported plasma. Figure 1 shows a schematic and a photo of the pilot plant. The pilot plant exhibits a base area of about 60 m^2 and an overall height of 7.2 m, containing a multi level lab with a floor space of about 180 m^2.

The HWR and the FLR are located on the third level. In these reactors the process gases and the precursor gases are injected at the top of reactors via complex nozzles. After the precursor decomposition and the particle formation, the particle-laden hot gases are delivered to the second level. While precursor materials in the FLR are being burned by a natural gas flame under the presence of oxygen, the precursors in the hot-wall reactor are decomposed by convective heating at temperatures up to 1100°C. The microwave-supported plasma reactor is located on the ground level of the pilot plant. In this reactor all gases are injected at the

bottom via a nozzle into the reaction zone, where the energy delivered by a microwave feeds a plasma and keeps it burning for the duration of synthesis. In all synthesis reactors the overall gas flow with adjustable mixtures of inert process gases and gaseous precursors can reach values up to 200 l/min. The pumping unit is located on the ground level, while the exhaust gas post-combustion system is situated in an intermediate level above the pumps.

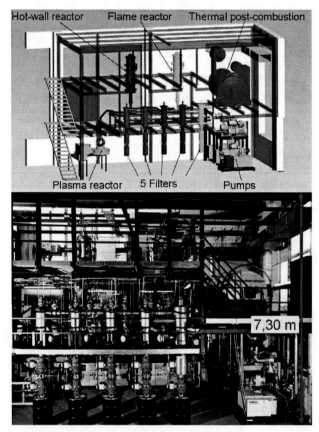

Fig. 1: Schematic (top) and image (down) of 3-level nano-particle synthesis plant: Filter system in front of the image on the ground level; reactors on the first and second level

Figure 2 a-c shows the different types of gas-phase reactors in more detail. All reactors provide several ports that can be equipped with quartz windows to allow for on-line laser diagnostics of the synthesis process and the formation of particles. In each of the independently operating reactor systems the particle-laden gases are delivered to the filter system, which is located on the second level of the pilot plant area.

The PLR and HWR can operate in a continuous mode, due to their individual double filter system. The particles are filtered alternately by one of the filter cartridges, while the other one can be released from collected particles and prepared for further operation. Additionally, this system enables to separate particles, which are generated in the first minutes of production process from those particles, which are synthesized later under stable operating conditions.

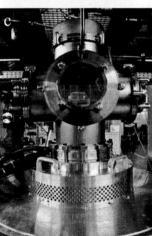

Fig. 2: Magnified view of the three reactors:
a) hot-wall reactor (HWR),
b) flame reactor (FLR),
c) plasma reactor (PLR) during synthesis.

1.1 On-line diagnostics

For an *in-situ* analysis of the synthesis process and the particles morphology highly sophisticated on-line diagnostic measurement set-ups are required. In recent years, laser diagnostics has proved to be a very powerful tool for the

on-line analysis of combustion processes [5]. Strategies that have been developed for combustion diagnostics can often be modified for applications in other reactive gas-phase flow systems. Consequently, the pilot-plant is equipped with on-line laser diagnostic techniques and on-line gas analysis systems. Laser-induced incandescence (LII) can be used for the monitoring of the particle and aggregate size during different stages of the formation processes and as well for on-line monitoring of the final particle size, which allows adjusting and varying the product parameters during synthesis [6]. Using a particle mass spectrometer (PMS) during the gas phase synthesis the size of generated particles can be monitored on-line [7]. Furthermore, for the analysis of the gas and flame temperatures within the FLR a laser induced fluorescence (LIF) system is installed. The obtained data are used as input for simulations of the combustion process.

The overall chemical process of the decomposition of precursor materials is continuously monitored by a quadrupol mass spectrometer (QMS) to ensure the complete conversion of the chemical starting materials.

1.2 Off-line diagnostics

Several standard off-line diagnostics methods are available at IUTA and associated research groups. Typically, electron microscopy techniques (SEM and TEM) are used to characterize the particle morphology and the state of agglomeration, while high-resolution transmission electron microscopy (HR-TEM) is performed on selected materials to study details of the material structure. The received on-line and off-line measurement results are compared to BET measurements, which yields off-line results of specific surfaces and thus particle sizes within one day. The structure and the crystallinity of the generated nanoparticulate powder are usually observed by X-Ray Diffraction (XRD) measurements. Using Scherrer's equation the crystallite size of the multi-domain particle ensemble is calculated and compared to HR-TEM investigations.

2 SYNTHESIS OF SILICON NANOPAR-TICLES

In this paper we present two synthesis routes for silicon nanoparticles. First, the generation process in the PLR will be visualized by images and second, results of the synthesis of nanoparticulate silicon in the HWR will be shown. Both reactor types enable for the synthesis of non-oxide materials, due to the fact, that gaseous precursor materials and process gases without oxygen content can be used.

2.1 Microwave-Supported Plasma Reactor

In this reactor all gases are injected at the bottom via a complex nozzle into the reaction chamber, while the energy delivered by microwave radiation at 915 MHz ignites a plasma and keeps it burning for the duration of synthesis.

For the synthesis of silicon nanoparticles in the microwave-supported plasma reactor, first, a mixture of Hydrogen and Argon is injected into the reaction zone to ignite and stabilize the plasma before the synthesis. After the stabilization process, Monosilane (SiH_4) as precursor is additionally injected into the system. This step directly changes the color of the plasma emission from deep blue to orange-red. Figure 3 shows images of the plasma without and with the presence of silicon nanoparticles in the system.

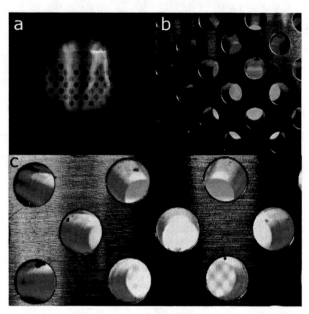

Fig.3: a and **b**: Hydrogen/Argon Plasma in the reaction chamber of the microwave-supported plasma reactor. **c**: Hydrogen/Argon with Silane during the synthesis of silicon nanoparticles

2.2   Si nanoparticles from the hot-wall reactor

The silicon nanoparticles are synthesized by thermal decomposition of SiH_4 in a hot-wall reactor at temperatures of 1000°C, while the process pressure is varied in the range from 15 to 100 kPa. The resulting material is characterized directly after the synthesis using specific surface measurements (BET). The measurements reveal an increase of the specific surface with decreasing synthesis pressure. Si particles generated at pressure of 100 kPa exhibit a specific surface of 21 m^2/g corresponding to a particle diameter of 120 nm, while BET measurements on particles synthesized at 15 kPa reveal a specific surface area of 44 m^2/g corresponding to particle diameter of about 60 nm. Figure 4 shows the particle diameter as a function of the synthesis pressure.

To characterize the structure of the materials XRD measurements are performed. Figure 5 shows the received diffraction pattern, which proofs the presence of a silicon crystallite structure, due to Bragg angles, which are in agreement with the bulk values. The position of the (111)

maximum is calculated by a Lorentzian fit. Using the fitting results we calculated a crystallite size of 13 nm within the nanoparticles. Furthermore, a shift of the maxima of the Bragg reflexes to smaller angles compared to the bulk values can be observed. From this displacement a decrease of the Si lattice constant can be estimated.

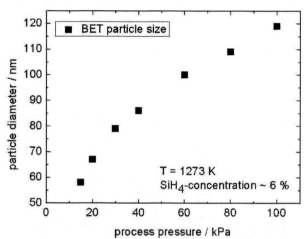

Fig 4: Si particle size as a function of the process pressure for particles generated at 1273 K with a total precursor concentration of 6 % Silane in a Hydrogen/Nitrogen mixture.

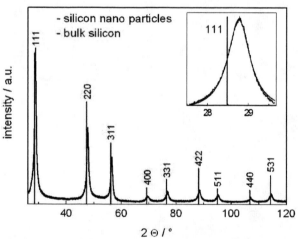

Fig 5: X-ray diffraction measurement of Si nanoparticles. The red peaks indicate the position of the Bragg angles for bulk material. The intensity of these peaks is normalized to the maximum of the (111) reflex. The inset shows a magnification of the (111) reflex and the Lorentzian fit (red).

HRTEM and TEM analysis confirm the crystalline structure throughout the particles, an oxygen content of ~5 atom% in the sample, but do not indicate the presence of a distinctive SiO_2 shell. Furthermore, HR-TEM investigations reveal a crystallite size of 12 nm, which is in good agreement with XRD analysis. Figure 6 shows a TEM image of Si nanoparticles. Typically, the particles are agglomerated and the resulting particle assemblies exhibit diameters of up to 1 µm and more.

Scanning electron microscopy (SEM) investigation is performed on the nanoparticulate powder material. For SEM analysis the particles are dispersed in acetone and droplet of the suspension is put on a p-doped silicon substrate.

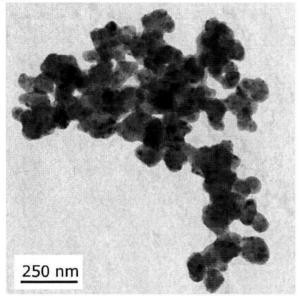

Fig. 6: TEM image of Silicon nanoparticles

Fig 7: SEM image of Silicon nanoparticles

Figure 7 shows a SEM image of the Si nanoparticles. The analysis reveals a homogenous thin film of particles with mainly distinct grain boundaries. The size distribution of the particles is in the range of $\sigma = 1.3$, as it is expected for the hot wall synthesis route.

3 HANDLING OF NANOPARTICULATE POWDER

The production rate of nanoparticulate material in the pilot-plant can be increased up to the kg/h range. Therefore, the handling of the generated powder material must be optimized to fulfill pre-industrial standards and scientific re-

quirements. Additionally, a continuous extraction of particles from the filter system under inert conditions during the synthesis must be possible. Therefore, a polyethylene (PE) hose, which is welded at the bottom, is rolled on and fixed at an outlet branch below the filter. The powder is released from the filter cartridge and falls into a lock, where the material is purged with gaseous nitrogen. After purging and opening the lock, the particles fall into the bottom-welded PE hose. After the collection of particles, the hose is sealed by welding above the collected powder. To separate the welded nanoparticulate powder, the PE hose is cut directly in the joint. With this method we create flexible bags with nanoparticles without nanoparticle exposure during the synthesis and the filling procedure.

ACKNOWLEDGEMENTS

This work was funded by the EU Ziel-2-Programm (NRW) and the Bundesministerium für Wirtschaft und Technologie. We gratefully acknowledge the support of NPPT, University of Duisburg-Essen, R. Theissmann, CeNIDE for TEM imaging and B. Stahlmecke, IUTA for SEM imaging.

REFERENCES

[1] F. E. Kruis et al., J. Aerosol Sci. **29**, 511-535 (1998).
[2] K. Wegner et al., Powder Technology **150**, 117-122 (2005).
[3] T. Hülser et al., Sens. & Act. B **109**, 13-18 (2005).
[4] P. Roth, Proc. Combust. Inst. **31**, 1773-1788 (2007).
[5] C. Schulz, Z. Phys. Chem. **219**, 509-554 (2005).
[6] A.V. Eremin et al., Appl. Phys. B **83**, 449-454 (2006).
[7] P. Ifeacho et al., Proc. Combust. Inst **31**, 1805-1812 (2007).

Gas-phase synthesis of silica nanoparticles in a hybrid microwave-plasma hot-wall reactor

A. Abdali[1], B. Moritz[1], H. Wiggers[1,2] and C. Schulz[1,2]

[1]IVG, University of Duisburg-Essen, Lotharstr. 1, 47057 Duisburg, Germany, ali.abdali@uni-due.de
[2]CeNIDE, Center for Nanointegration Duisburg-Essen, 47057 Duisburg, Germany

ABSTRACT

The synthesis of silica nanoparticles from tetraethoxysilane (TEOS) was investigated in a combination of a microwave-plasma reactor with a subsequent hot-wall furnace. This combination was used to provide variable temperature/time profiles and therefore to investigate the growth of silica particles in detail. Particle properties such as size, size distribution, shape, morphology, and agglomeration were investigated via transmission electron microscopy (TEM). The samples were collected on TEM grids in-situ from the reactor and ex-situ from a filter device. The particles collected directly from the gas flow in the reactor are non-agglomerated and have a spherical morphology. However, those obtained from the filter are spherical in shape but have a high degree of agglomeration.

Keywords: TEOS, silica particles, gas-phase synthesis, plasma reactor

1 INTRODUCTION

The synthesis and characterization of silica nanoparticles has received attention in recent years due to their large range of practical applications. They are widely used as fillers in plastics and coatings to improve material properties such as hardness, tensile strength, abrasion resistance, and thermal stability. In particular, applications with a demand for high transparency require silica particles with a specific size, morphology, and surface coating [1,2]. Therefore, the synthesis of particles with highly defined particle size, size distribution, and morphology is desired. Most of the silica powder is produced by gas-phase processes, most frequently using flame reactors and spray pyrolysis. However, these methods require exact knowledge of the kinetics of precursor decomposition [3,4], particle formation and particle growth. Tetraethoxysilane (TEOS) as a halide free and inexpensive precursor material is subject to growing interest for particle formation from the gas phase. Various studies have been conducted on the synthesis of silica nanoparticles from the gas phase using different methods. Ahn et al. and Jang have investigated the synthesis of silica nanoparticles from tetraethoxysilane (TEOS) in diffusion flames [5,6], while Goortani et al. have worked on the synthesis of SiO_2 nanoparticles from quartz in a RF plasma reactor [7]. They all found that, depending on temperature and precursor concentration, silica nanoparticles with different morphologies and sizes arise along the reaction coordinate ranging from spherical, spatially separated particles to large agglomerates and aggregates. As Ahn et al. have shown, an initially high temperature seems to support the formation of non-agglomerated, spherical particles, and Tsantilis et al. calculated that fast cooling results in soft agglomerates while in systems with low cooling rates hard agglomerates are formed [8]. To investigate the gas-phase synthesis of silica, experiments are usually performed at atmospheric pressure. To stretch the spatial axis for a detailed analysis of particle formation and growth, our experiments were conducted at reduced pressure (30 to 70 mbar). This allows for a highly resolved sampling of materials with different reaction times from the reactor. Because the focus of our experiments is to investigate the formation of non-agglomerated, spherical particles, a microwave-supported plasma reactor that enables for steep temperature gradients was chosen for the experiments. Additionally, it allows for high temperatures in the initial reaction step and enables for varying the gas composition over a large range.

2 EXPERIMENTAL

The hybrid reactor which consists of a microwave plasma reactor and a subsequent hot-wall furnace has been constructed for gas-phase synthesis of high purity silica nanoparticles using TEOS as the precursor. A microwave-induced plasma heats the injected gas mixture consisting of gaseous TEOS, O_2, and Ar within a few microseconds to high temperatures. This initiates precursor decomposition and chemical reactions followed by particle formation. Behind the plasma typically fast cooling is observed. To achieve longer and variable residence times at high temperature we combine the plasma reactor with a hot-wall furnace (see Fig. 1).

In order to obtain reproducible synthesis conditions, the current apparatus allows the control of gas flow rates, gas composition, pressure, and the temperature of the hot-wall furnace. The reactor pressure can be varied between 20 and 1000 mbar using a control valve. The reactor pressure is varied in order to achieve different residence times. Based on the ideal gas law, the residence time is proportional to pressure for a constant gas flow rate and temperature. The microwave generator in use has a maximum output power of 2 kW and a frequency of 2.45 GHz. In all experiments the power was fixed at 200 W.

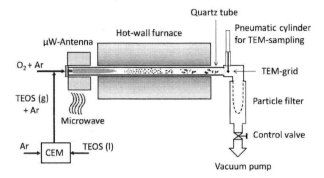

Figure 1: Simplified schematic of the hybrid microwave-plasma hot-wall reactor.

The required gaseous TEOS (in a mixture with Argon as the carrier gas) was supplied using a controlled evaporator mixer (CEM) from Bronkhorst. The CEM system supplies gas mixtures that can feed up to 13 g/h of TEOS vapor into the reactor despite the poor vapor pressure of TEOS (2.0 mbar at 20°C). The tube that connects the CEM and the reactor inlet is heated to 80°C to prevent condensation of TEOS. An additional gas mixture of Argon and Oxygen, as oxidizing reagent, is added to the TEOS/Argon mixture prior to entering the reactor. The purity of the gases used is 99.999% for Argon and Oxygen, and 98% for TEOS.

The particles produced are filtered from the gas using a cellulose extraction thimble. The size, shape, and morphology of the product were analyzed by transmission electron microscopy (TEM) using a Philips CM12 microscope. The TEM samples were either collected from the filter or directly from the reactor chamber by thermophoretic deposition onto TEM grids using a double acting pneumatic cylinder that limits the sampling time to an exactly defined time. For our experiments we used sampling times of 0.5 s.

In addition, material sampled from the filter device was also used for BET analysis (isothermal Brunauer-Emmett-Teller nitrogen adsorption). The primary average BET particle diameter was calculated based on the measured surface area assuming monodisperse and spherical particles. In order to calculate the equivalent diameter, the following equation was used:

$$D = \frac{6 \times 10^6}{\rho \, SSA} \qquad (1)$$

where ρ is the bulk density of silica (2200 kg/m³), SSA is the specific surface area in m²/g, and D is the diameter of the nanoparticles in nm.

3 RESULTS AND DISCUSION

The growth of the particles was systematically investigated by changing different parameters which are expected to influence the particle size as well as the particle morphology. The particle size was investigated at furnace temperatures that were varied between 200 and 800°C, whilst keeping the gas flow rates constant (Argon: 2.1 l/min, Oxygen: 1.0 l/min, TEOS: 3 g/h (equivalent to 1700 ppm)). A representative TEM image of silica nanoparticles obtained by thermophoretic sampling from the reactor chamber is shown in figure 2. In all cases the particles were non-agglomerated and had an almost spherical morphology.

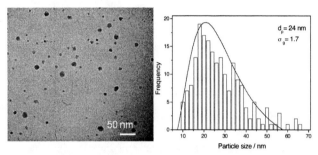

Figure 2: TEM-image of silica nanoparticles sampled from the reactor chamber (left); histogram and fitted log-normal size distribution (right)

The diameter of about 200 silica nanoparticles, observed on the different TEM grids, was measured and used to generate a histogram such as the one shown in figure 2. A log-normal particle size distribution was fitted to the histogram to calculate the mean particle diameter. These values were used to create the graphs that show the dependencies between experimental parameters and particle size (Figs. 3 and 4).

In figure 3 the red profile shows that for furnace temperatures up to 400°C the mean particle size increases as expected, with the maximum particle size of 28 nm. However, the particle sizes decrease when the furnace temperature exceeds 400°C. We attribute this to two reasons: As it is expected from plasma synthesis of nanoparticles, Coulomb repulsion plays an important role in preventing agglomeration and at high furnace temperatures the particles are likely to gain a higher charge, thus hindering particle collisions and growth due to Coulomb repulsion [9]. Thus, further particle growth would be increasingly prevented with increasing the particle charge. Additionally, the residence time decreases with increasing furnace temperature because of the reduced density at fixed pressure and gas flow rate. The influence of decreasing residence times on the particle size at high temperatures was investigated. For different temperatures the reactor pressure was varied in order to keep the residence time fixed at 164 ms. The results of these tests are shown in figure 3 (black profile). In figure 3 it can be seen that the particle sizes are bigger at the longer residence time of 164 ms. However, the particle size again decreases above furnace temperatures of 400°C due to the Coulomb repulsion.

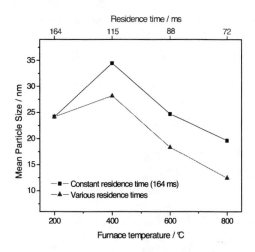

Figure 3: Influence of the furnace temperature on particle size. (▲) various residence times at constant pressure; (■) constant residence time at various pressures

To investigate whether the residence time within the reactor influences the particle growth, the furnace temperature was kept at 800°C whilst the residence time was increased from 72 to 126 ms by raising the reactor pressure from 40 to 70 mbar. As can be seen from figure 4, the mean particle diameter depends very strongly, and almost linearly, on the residence time in this case.

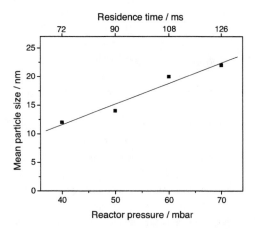

Figure 4: Influence of reactor pressure on the particle size, furnace temperature 800°C.

These results clearly demonstrate that we were able to adjust the size of silica nanoparticles, prepared by gas-phase synthesis, via two independent parameters, namely the furnace temperature and the residence time in the hot zone of the reactor.

All samples received from the filter or by thermophoretic sampling from the cold walls downstream of the reactor always showed a very high degree of particle agglomeration (see Fig. 5 left). The findings discussed above demonstrate that the particles in the gas phase are well separated (as it was shown earlier in figure 2), while hard agglomerates are found in samples taken from the filter and from the cold reactor walls. The highest agglomeration index was obtained with TEOS concentration of 3700 ppm, reactor pressure of 40 mbar and furnace temperature of 400°C. Nevertheless, the size distribution of the primary silica nanoparticle that form the large aggregates on the filter followed the same log-normal size distribution function as the non-agglomerated particles sampled from the reactor (see Fig. 5 right).

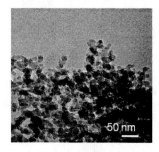 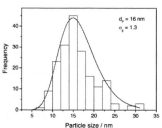

Figure 5: TEM-image of the collected particles from the filter (left); histogram and fitted log-normal size distribution (right)

The BET particle size of the samples from the filter, as shown in figure 5, was 24 nm which is much bigger than that found in the material from the reactor (see Fig. 5 right, mean particle size is 16 nm). The high degree of agglomeration reduces the specific surface area (SSA) and therefore shows larger particle size. The aggregation of formerly spherical, non-agglomerated particles may be attributed to the surface chemistry of the freshly prepared silica nanoparticles. Particle adhesion, especially in the nanosized regime, originates from solid-solid interaction forces (van der Waals forces), capillary bridging, and long-range electrostatic (Coulomb) forces that create strong particle-particle interaction [10]. It is found that materials like inorganic oxides that contain hydroxyl groups on their surface tend to build new, stable chemical bonds in the presence of water [11]. For this reason a particle sample from the filter has been investigated with Fourier Transform Infrared (FTIR) spectroscopy.

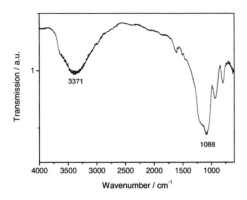

Figure 6: FTIR-Transmission spectrum of SiO$_2$ nanoparticles.

Figure 6 shows the FTIR spectrum of silica nanoparticles. In the spectrum, a broad peak around 1088 cm^{-1} appears, which can be assigned to stretching vibration of Si-O-Si. The peak is broad due to the amorphous nature of silicon oxide. Additionally, the signature from OH vibration around 1700 cm^{-1} may be due to some physisorbed water on the nanoparticle surface. The FTIR spectrum also shows a strong signal at about 3371 cm^{-1} indicating silanol groups that might be responsible for the highly active particle surface. As a result, we propose that silica nanoparticles that touch each other will sinter together due to condensation of silanol groups on their surface. In this case, temperature and humidity play an important role in the formation of silica aggregates.

4   SUMMARY

The gas-phase synthesis of silica nanoparticles was carried out in a hybrid microwave-plasma hot-wall reactor. The particles were collected by thermophoretic deposition on TEM grids directly from within the reactor chamber. TEM analysis reveals that in all cases the particles sampled from the reactor chamber are non-agglomerated, spherical in shape and have a wide size distribution. The influence of the main parameters such as hot-wall furnace temperature and residence time was investigated. The results show that short residence times yield small particles whereas their size increases at longer residence times. In contrast, the analysis of particles from the filter shows large hard agglomerates. When varying the furnace temperature between 200 and 800°C the maximum agglomeration index of the particles sampled in a filter device downstream of the reactor was observed at 400°C. The strong aggregation of the particle powder sampled from the filter is attributed to surface chemistry subsequent to the particle formation process in the gas phase.

ACKNOWLEDGEMENTS

Financial support of the German Research Foundation (DFG PAK 75/2 Gasdynamically induced nanoparticle synthesis) is gratefully acknowledged.

REFERENCES

[1]   A. Gutsch, H. Mühlenweg, M. Krämer, Small **1**, 30 (2005).

[2]   S. Tsantilis, S. E. Pratsinis, J. Aerosol Sci. **35**, 405 (2004).

[3]   A. Abdali, M. Fikri, H. Wiggers, C. Schulz, 26th Intern. Symp. Shock Waves, **1**, 781 (2009).

[4]   J. Herzler, J.A. Manion, W. Tsang, J. Phys. Chem. A **101**, 5500 (1997).

[5]   H. D. Jang, Aerosol Sci. Technol. **30**, 477 (1999).

[6]   K. H. Ahn, C. H. Jung, M. Choi, J. S. Lee, J. Nanopart. Res. **3**, 161 (2001).

[7]   B. M. Goortani, N. Mendoza, P. Proulx, Int. J. Chem. Reactor Eng. **4**, 1 (2006).

[8]   S. Tsantilis, S.E. Pratsinis, Langmuir **20**, 5933 (2004).

[9]   S. Vemury, S. E. Pratsinis, Appl. Phys. Lett. **66**, 3275 (1995).

[10]   J.A.S. Cleaver, J.W.G. Tyrrell, KONA Powder and Particle **22**, 9 (2004).

[11]   M.S. Kaliszewski, A.H. Heuer, J. Am. Ceram. Soc. **73**, 1504 (1990).

Ionic Liquids - Novel Tenside like Materials for size controlled Preparation of Nanoparticles and safe-to-handle Nanoparticle Dispersions

Dr. Thomas J. S. Schubert*, Dr. Tom F. Beyersdorff**, Dr. Frank Stiemke*

*IoLiTec Ionic Liquids Technologies GmbH, Heilbronn, Germany
**IoLiTec Ionic Liquids Technologies Inc., Tuscaloosa, AL, USA

ABSTRACT

Ultra small Ag and Ru nanoparticles with a narrow size distribution have been synthesized in ionic liquids using reductive, thermal or photochemical conditions. The influence of the ionic liquid on the size of the nanoparticles has been investigated. In addition stable dispersions of different kinds of MWCNT in water have been prepared by the addition of ionic liquids as novel tenside like materials to a mixture of the MWCNT in water. These dispersions have been characterized by Photon Cross Correlation Spectroscopy (PCCS).

Keywords: ionic liquids, nanoparticles, dispersions

1 INTRODUCTION

Ionic Liquids (ILs) are a class of materials, consisting entirely of ions, which are liquid at unusual low temperatures[1]. Typical structural motifs combine organic cations with inorganic, or, more rarely, organic anions. The lower symmetry of the cations or anions and the delocalisation of the charge over larger parts of the ions by resonance are mainly responsible for the low melting points of ionic liquids. The manifold combinations of cations and anions provide a large number of liquid materials with (tunable) unique properties such as high conductivity, high thermal stability, negligible vapor pressure, non-flammability, good solubility of many inorganic precursor salts as a consequence of the tunable polarity of ILs and their tenside like character. As a consequence, this unique combination of properties makes ILs – among many other different applications - the media of choice for synthesis and dispersing nanoparticles[2].

2 SYNTHESIS OF NANO PARTICLES

The controlled and reproducible synthesis of metal nanoparticles (MNPs) is of great interest since many properties of MNPs relate directly to the size of these materials. Especially transition metal MNPs are of interest due to their application in many different areas of sciences, including catalysis or chemical sensing. MNPs can be synthesized in different ways including reduction of metal salts with hydrogen gas, electrochemical reduction or photochemical reduction.

A problem that occurs during the reduction of metal salts MX_n with hydrogen gas is the formation of the corresponding acids HX, which leads to a destabilization of the MNPs and clustering. In order to prevent the MNPs from clustering a base has to be added to scavenge the HX byproducts.

Janiak *et al.* demonstrated the influence of the size of the ionic liquid ions on the size and size distribution of silver nanoparticles synthesized by the reduction of a silver salt AgX with hydrogen gas in an ionic liquid and an alkylimidazole base as scavenger[3]. Different Ag precursors were dissolved in dried ILs and reacted with hydrogen (4atm, 85°C) in the presence of n-Butylimidazole in a stainless steel reactor.

Selected results for these experiments are summarized in Table 1.

Table 1: Ag nanoparticle size and distribution from different precursors and ILs

Ionic Liquid	Scavenger	Silver Precursor	AgNP median (min-max diameter/nm)	Standard deviation
BMIM BF4	BuIm	AgBF4	2.8 (1.2-4.7)	0.8
BMIM PF6	BuIm	AgPF6	4.4 (2.0-9.8)	1.3
BMIM OTf	BuIm	AgOTf	8.7 (3.5-18.4)	3.4

From the results in Table 1 it is obvious, that the size of the anion in the ionic liquid has a dramatic influence on the particle size and the size distribution of the synthesized Ag nanoparticles. All synthesized Ag-IL-dispersions were stable under Argon over a period of at least 3 days.

In addition Janiak *et al.* synthesized other transition metal nanoparticles such as iron, ruthenium or osmium in ionic liquids by photochemical and thermal procedures of the respective metal carbonyl precursors[4].

The synthesis was performed by heating a mixture or suspension of the metal carbonyl precursors in BMIM BF4 under argon up to 250°C for several hours or by irradiation at 200-450nm for 15 minutes.

Selected results of these experiments are summarized in Table 2.

Table 2: Nanoparticle Size and Distribution in BMIM BF$_4$

Ionic Liquid	Metal carbonyl (wt % in IL)	Product	TEM median diam/nm (standard deviation σ)c
BMIM BF4	Fe2(CO)9 (0.2)	Fe	5.2 (±1.6)
BMIM BF4	Ru3(CO)12 (0.2)	Ru	1.6 (±0.4)
BMIM BF4	Ru3(CO)12 (0.08)	Ru	2.0 (±0.5)
BMIM BF4	Os3(CO)12 (0.2)	Os	2.5 (±0.4)

Extremely small and uniform MNPs were received as stable dispersions in the ionic liquid in all cases. The nanoparticles produced by photolysis were somewhat larger than those produced by thermolysis due to faster decomposition and growth of the particles in the ionic liquid.

3 STABILIZATION OF NANOMATERIALS IN IL

Dispersions in general are interesting for the easy and the even more important safe handling of nanoparticles, also with the background of uncertainties concerning their specific toxicity. It is e.g. assumed that some particles can cause cancer like asbestos. Due to this reason it is crucial to avoid any type of nanoparticle contamination. Stable IL-based dispersions of these materials will overcome these drawbacks: they are not dusty, non-volatile, non-inflammable and can be designed to be biodegradable.

IOLITEC *nanomaterials* succeeded in dispersing MWCNTs with same size in diameter but different length (5-15 μm and 1-2 μm) in water by using a suitable IL and treatment with ultrasound. The stabilizing effect of the IL is extremely obvious if the dispersions are stressed by centrifugal forces. If no or an unsuitable additive was added, the dispersion will collapse even at low centrifugal force impact and sediments will be obtained.

Samples of prepared dispersion were analyzed by Photon Cross Correlation Spectroscopy (PCCS)[5] which allows analysis of particles in the range of 1 nm – 1 μm using the principle of dynamic light scattering. Furthermore, measurements can be conducted often without any dilution or even at opaque samples. This is possible due to the 3D-measurement setup with two independent crossbred lasers and detectors allowing to detect the decay of light intensity and elimination of multiple scattered light. At the same time information about particle size and stability of the dispersion can be obtained from the measured correlation functions of the decays of the light intensity.

Decays of several repeated individual measurements for long and short MWCNTs with same diameter have been measured.

The sample of dispersed long MWCNTs (0.5%wt in water) is a very stable dispersion even over a period of 24 hours..

The stability analysis of the second sample (short MWCNTs, 0.5%wt in water) results in statistical distributed correlation functions. In this case a stabilizing effect of a suitable ionic liquid as dispersing additive (a few mol% based on particle concentration) can be seen and a very stable dispersion is obtained.

4 CONCLUSIONS

Ionic liquids are suitable solvents for the size controlled synthesis of metal nanoparticles. By variations of the anions different particle sizes and size distributions can be obtained. The dispersion of nanomaterials by using ionic liquids as additives leads to very stable, easy and safe-to-handle dispersions. This procedure is not limited to CNTs or water, but is rather applicable to different other particles or solvents, also the stability measurement by PCCS.

REFERENCES

[1] P. Wasserscheid, T. Welton, *Ionic Liquids in Synthesis*, Wiley-VCH, Weinheim, 2007.

[2] T. J. S. Schubert, *Nachr. Chem.* **2005**, *53*, 1222.

[3] a) E. Redel, R. Thomann, C. Janiak, *Inorg.Chem.* **2008**, *47*, 14-16. b) C. Janiak, E. Redel, M. Klingele, T. F. Beyersdorff, T. J. S. Schubert, DE102007038879.

[4] a) J. Kraemer, E. Redel, R. Thomann, C. Janiak, *Organometallics* **2008**, *27*, 1976-1978. b) C. Janiak, E. Redel, M. Klingele, T. F. Beyersdorff, T. J. S. Schubert, DE102007045878.

[5] Sympatec GmbH, www.sympatec.com

Highly Conducting Nanosized Monodispersed Antimony-Doped Tin Oxide Particles Synthesized via Nonaqueous Sol-Gel Method

Vesna Müller,[*] Jiri Rathousky,[**] Matthias Rasp,[*] Goran Štefanić,[***] Sebastian Günther,[*] Markus Niederberger,[****] and Dina Fattakhova-Rohlfing[*]

[*]Department of Chemistry and Biochemistry, University of Munich (LMU), Butenandtstrasse 5-13, 81377 Munich, Germany, Dina.Fattakhova@cup.uni-muenchen.de
[**]J. Heyrovsky Institute of Physical Chemistry, Academy of Sciences of the Czech Republic, Dolejskova 3, 182 23 Prague 8, Czech Republic, jiri.rathousky@jh-inst.cas.cz
[***]Division of Materials Chemistry, Ruđer Bošković Institute, P.O. Box 180, HR-10002 Zagreb, Croatia, stefanic@irb.hr
[****]Department of Materials, ETH Zürich, Wolfgang-Pauli-Strasse 10, 8093 Zürich, Switzerland, markus.niederberger@mat.ethz.ch

ABSTRACT

Conducting antimony-doped tin oxide (ATO) nanoparticles were prepared by a nonaqueous solution route, using benzyl alcohol as both the oxygen source and the solvent, and tin tetrachloride and Sb(III)/Sb(V) compounds as tin and antimony sources, respectively. This reaction produced non-agglomerated crystalline particles 3-4 nm in size, which can be easily redispersed in high concentrations in a variety of solvents to form stable transparent colloidal solutions without any stabilizing agents. The introduction of Sb dopant dramatically increased the particle conductivity, which reaches a maximum of $1 \cdot 10^{-4}$ S cm^{-1} for 4 % of Sb. Annealing in air at 500 °C further improves the conductivity to $2 \cdot 10^{2}$ S cm^{-1}.

Keywords: TCO, antimony-doped tin oxide, nanoparticles, conductivity, nonaqueous sol-gel synthesis

1 INTRODUCTION

Due to the unique combination of transparency in the visible range and high electric conductivity, transparent conducting oxides (TCOs) are widely used as electrodes in thin layer transparent devices such as solar cells, flat-panel displays, smart windows or electronics and chemical sensors. [1,2] Very recently, the focus of scientific attention has shifted from thin films to crystalline TCO nanoparticles. The use of nanocrystals opens the way to the deposition of conductive coatings by wet chemical methods at mild conditions, replacing expensive vacuum-based techniques and enabling coating temperature-sensitive plastic substrates. Moreover, such colloidal dispersions are of great interest for ink-jet printing technologies and are very attractive as primary building units for assembling nanostructured materials with defined porous architectures.[3]

Antimony-doped tin oxide (ATO) is a well-known transparent semiconductor with a large band gap of over 3.6 eV, providing transparency in the visible range and high n-type electric conductivity, which makes it a valuable alternative to much more rare and expensive indium tin oxide (ITO). ATO nanoparticles can be prepared by both chemical and physical methods, the former providing a much better control over the particle size and their stabilization in solution than the physical ones. In the present study we are reporting a nonaqueous-solution route to preparation of ATO nanoparticles with very small particle size, narrow size distribution and especially high electric conductivity.[4] Special emphasis was laid on good self-assembling properties of nanoparticles, as they are intended as suitable building blocks for the assembly of more complex nanoarchitectures. [5] The good electric properties of conducting oxides require an optimum doping level, a uniform distribution of the doping atom within the host lattice without phase separation and without any surface enrichment. Therefore, we have prepared ATO particles containing Sb in a wide range of concentrations, focusing on the effect of the nature and relative concentration of the metal oxide precursors and processing conditions on the structure, morphology and conductivity properties of the obtained nanoparticles. This synthetic approach provides ATO nanoparticles with exceptional properties such as small size, facile redispersibility in various solvents and high electric conductivity.

2 RESULTS AND DISCUSSION

In the synthesis of the ATO nanoparticles, anhydrous benzyl alcohol was used both as a solvent and as an oxygen source, and tin tetrachloride as the tin precursor. In order to test the influence of the nature of the antimony precursor on the doping efficiency and solubility of Sb in the tin oxide lattice, several commercially available antimony compounds with different antimony valence states and various ligands were used, namely antimony (III) chloride, antimony (V) chloride, antimony (III) acetate and antimony (III) ethoxide. The rate of the particle formation and the particle yield greatly depend on the reaction temperature. While the reaction completion requires 20 h at 100 °C, only 2 h are sufficient if the temperature is increased to 150 °C. On the contrary, the nature of Sb precursor only marginally influences the mean particle size and its distribution, and the degree of agglomeration. For all the used precursors, the TEM images show practically non-agglomerated or only loosely agglomerated highly crystalline nanoparticles, as proven by lattice fringes in HRTEM and STEM (Fig. 1). The average size of the particles is about 3 nm and 4 nm if synthesized at 100 and 150 °C, respectively. Independently of the ligand nature, all the Sb(III) precursors provide similar non-agglomerated particles exhibiting narrow particle size distribution. The use of precursors more reactive towards solvolysis, such as of $SbCl_5$, leads to some acceleration of the particle formation reaction, slight increase in the particle size and an increased amount of agglomerated nanoparticles.

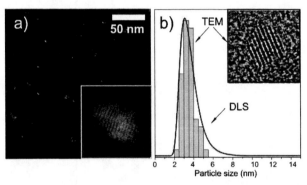

Figure 1. Morphology, size and dispersibility of 10 % ATO nanoparticles prepared using Sb(ac)$_3$ at 150 °C: (a) STEM-HAADF image. The inset 6 x 6 nm in size show high resolution STEM-HAADF images of a single nanoparticle. (b) Size distribution of nanoparticles determined from HRTEM images (gray bars) and from DLS measurement (red line) of a colloidal dispersion in EtOH of the same nanoparticles. The inset 6 x 6 nm in size shows high resolution TEM image of a single nanoparticle.

With regard to the properties of obtained ATO particles, the solubility limit of Sb dopant in the SnO_2 lattice and the influence of the dopant concentration on the structure, crystallinity, size and the conductivity of the formed ATO particles are of primary importance. Therefore, we prepared the particles containing from 0 to 50 mol % of Sb using Sb(ac)$_3$ as antimony source. Synthesis without any addition of antimony results in the formation of white particles. The color of the obtained particles changes to brownish-green when an Sb source is added to the reaction mixture. Besides the change in color, an increase in Sb content leads to the acceleration of the particle formation and an increase in the

particle yield, which is especially pronounced at lower reaction temperatures.

The particles can be easily redispersed at high concentration of about 3-8 wt% in different solvents without the addition of any surface-stabilizing agent, such as water, ethanol or THF, to form stable transparent colloidal solutions. The particles in the solution are practically non-agglomerated and show a narrow size distribution, as determined by DLS (Fig. 1b). The dispersibility of the antimony-doped nanoparticles is much better than that of the pristine SnO_2 ones. Moreover, it is worse in more polar solvents such as water, but is greatly improved by the addition of a few drops of concentrated hydrochloric acid. Such a striking difference could originate from the different surface charge of the particles with the varying Sb content, and points out the significant role of electrostatic effects in the colloid's stabilization.

The XRD patterns of ATO nanoparticles with an Sb content up to 30 mol% show the presence of only one crystalline phase structurally closely related to cassiterite, which indicates that the SnO_2 lattice can accommodate up to 30 mol% of Sb atoms without significant changes in the structure. The first indication of a second crystal phase ($Sb_4O_5Cl_2$) appears only at a very high Sb concentration of 40 mol%. Solid solubility limit of antimony ions in the cassiterite lattice is approximately 38 mol%, as estimated from the dependence of the diffraction line intensities of the $Sb_4O_5Cl_2$ phase on the initial content of Sb, and by extrapolation to the zero intensity. With increasing antimony content peaks corresponding to the cassiterite structure become broader as a result of the decrease in particle size from 4.3 nm for the undoped sample to 4.1 nm and 4.0 nm for samples with 10 mol% and 20 mol% of antimony, respectively, in good agreement with the results obtained by TEM. The incorporation of Sb ions causes an asymmetric distortion of the cassiterite lattice due to the difference in the ionic radii of the Sn^{4+} ion (0.69 Å in coordination 6) and antimony ions (0.76 Å and 0.61 Å in coordination 6 for Sb^{3+} ion and Sb^{5+} ion, respectively). Estimated values of unit-cell parameters of cassiterite (0% ATO) and cassiterite-type solid solutions (10 % ATO and 20 % ATO) obtained from the Le Bail and Rietveld refinements of XRD patterns show that the increase in the antimony content cause a linear decrease of parameter a and a linear increase of parameter c.

The phase composition of the ATO particles was further studied by Raman spectroscopy, which is able to deliver structural information on amorphous and poorly crystalline samples and is very sensitive to the presence of even minor amounts of crystalline phases. The typical bulk vibration modes predicted by group theory for the SnO_2 rutile structure, namely A_{1g} (630 cm^{-1}), B_{2g} (775 cm^{-1}) and E_g (477 cm^{-1}), are observed for the undoped SnO_2 sample although with low intensity, probably due to the small particle size. Moreover, the spectra of the samples containing up to 40 mol% of antimony show the bands corresponding to the rutile type of SnO_2. Position of these bands shifts with the increasing antimony content, indicating the influence of the replacement of Sn with Sb within the lattice. A_{1g} and B_{2g} modes corresponding to the vibrations perpendicular to the c axis decrease in intensity and shift to lower wave numbers with increasing antimony content, while the E_g mode corresponding to the vibration in direction of c axis follows the opposite trend. This continuous change in the vibration modes of the doped oxides is in good agreement with the XRD data, which also prove the slight asymmetric distortion of the crystalline SnO_2 lattice due to the incorporation of Sb ions. No bands corresponding to antimony oxides were found for the samples containing up to 40% Sb. Strong bands of the antimony (III) oxide were observed only for the sample containing 50 mol% Sb. All the spectra of ATO particles feature an intense band at 333 cm^{-1} and bands between 500 and 570 cm^{-1}. They were often observed for nanosized tin oxides, being frequently assigned to the surface vibration modes of tin oxide. The intensity of these bands changes with increasing antimony content, which can be related to vibration modes of incorporated antimony atoms inside the cassiterite lattice.

photoelectron spectroscopy (XPS). The overview spectra (Fig. 2a) obtained for the ATO samples demonstrate the strong peaks corresponding to Sn (Fig. 2b) and to oxygen and antimony (Fig. 2c). The total Sb content determined by the surface-sensitive XPS method is practically the same as that obtained by the bulk-sensitive energy-dispersion X-ray spectroscopy EDX method, indicating that particle composition is homogeneous throughout the complete particle without any enrichment of the surface of the nanoparticles with Sb atoms. In addition, as we have found all the particles independently of the Sb content contain ca. 60% of Sb^{5+} and 40% of Sb^{3+}, the effective donor Sb^{5+} concentration exceeds that of the acceptor by 20 % (effective n-type doping).

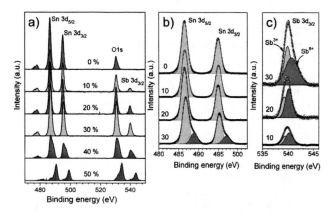

Figure 2. X-ray photoelectron spectra (XPS) of the as prepared ATO nanoparticles with varying Sb content synthesized at 150 °C using Sb(ac)$_3$ as an antimony source: overview spectra for the energy region between 460 and 560 eV (a) and the sections corresponding to the Sn$_{3d}$ doublet (b) and Sb 3d$_{3/2}$ antimony peaks (c). The solid lines correspond to the experimental spectra, and the open circuits and the grey shaded areas are the fitted curves and the Gaussian peak fits of the individual elements, respectively.

Conductivity of the ATO nanoparticles depends to a large extent on the antimony content and its oxidation state in the tin oxide lattice. While Sb^{5+} ions act as electron donors forming a shallow donor level close to the conduction band of SnO_2, Sb^{3+} behaves as an electron acceptor. If both oxidation states coexist, which is often observed for ATO materials, the resistivity is given by the ratio of Sb^{5+} and Sb^{3+} sites, which was analyzed by X-ray

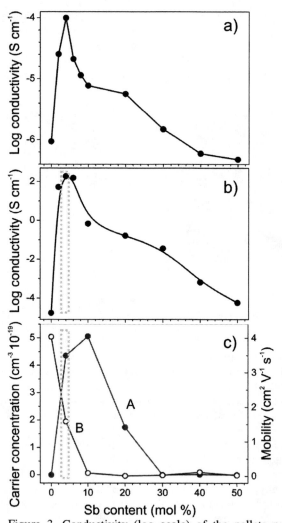

Figure 3. Conductivity (log scale) of the pellets pressed from ATO nanoparticles synthesized at 150 °C containing different Sb concentration using Sb(ac)$_3$ as antimony precursor: as prepared (a) and after calcination in air at 500 °C (b). For the latter, corresponding charge carrier concentration (A) and charge carrier mobility (B) are shown in (c).

Electrical conductivity of the ATO nanoparticles measured in pressed pellets strongly depends on the Sb content. The conductivity of the pure SnO_2 sample is very low, being only $9 \cdot 10^{-7}$ S cm^{-1}. The conductivity dramatically increases due to the introduction of small amounts of Sb, reaching a maximum for the 4 % ATO nanoparticles. Further increase in Sb content leads to a deterioration of the electrical conductivity, samples containing more than 20 mol % exhibiting conductivity comparable to or even worse than the undoped SnO_2 (Fig. 3a). To the best of our knowledge, the obtained conductivity of $1 \cdot 10^{-4}$ S cm^{-1} is the highest ever reported for non-annealed ATO particles only a few nm in size.

The pellets conductivity further increases by four orders of magnitude after their calcination at 500 $^{\circ}$C in air (Fig. 3b). The thermal treatment mainly leads to particle sintering and removal of organic impurities, providing good electrical contact between them, and to some extent to an increase in the particle size. The calcined pellets exhibit a similar effect of the Sb content on the conductivity, but at a much higher conductivity level.

The dependence of the charge carrier concentration on the Sb content in our ATO particles exhibits a pronounced maximum for the Sb concentration of 4-10 % (Fig. 3c), which is attributed to the formation of a shallow donor level close to the conduction band of SnO_2. For even higher Sb contents the carrier concentration decreases, as the increased disorder raises the effective activation energy of the donor. Even more strongly than the charge carrier concentration, the carrier mobility is affected by the variation of the dopant concentration, rapidly decreasing with the introduction of rather small amounts of antimony. As the XRD, Raman and TEM experiments have shown, the particle size decreases and the lattice disorder increases with the increasing Sb content. This contributes to the increased effects of grain boundary scattering, and ionized impurity and optical lattice scattering mechanisms, which are the reasons of the deteriorated charge carrier mobility in doped samples. Consequently, the best performance of the sample with 4 mol% of antimony is the result of the optimum combination of the high charge carrier concentration and high mobility inside the lattice.

3 CONCLUSIONS

Conducting ATO nanoparticles were prepared by a nonaqueous solution route, using benzyl alcohol, tin tetrachloride and a suitable Sb(III) compound. The non-agglomerated particles 3-4 nm in size are easy to redisperse in a variety of solvents to form stable transparent colloidal solutions without any stabilizing agents. The increase in antimony content causes a continuous decrease in particles size and a slight asymmetric distortion of the cassiterite SnO_2 lattice. The particle composition is homogeneous throughout the whole volume, without any enrichment of the surface with Sb atoms. Introduction of Sb dopant dramatically increases the particle conductivity, which

reaches a maximum for 4 % of Sb being more than two orders of magnitude higher than that of the pristine SnO_2 nanoparticles. The conductivity further improves to ca. $2 \cdot 10^2$ S cm^{-1} due to particle sintering during annealing in air at 500 $^{\circ}$C.

The combination of exceptional virtues makes the developed ATO nanoparticles an ideal material for electrostatic applications or as conducting additive for both organic and inorganic composites. As such, colloidal solutions of conducting nanoparticles are of special interest for direct printing of patterned electrodes. Above all, the reported ATO particles are attractive building blocks for assembling transparent conducting 3-D architectures, which offer favorable extra properties in comparison to those of traditional conducting layers. The high crystallinity of the nanoparticles serving as building blocks enables to obtain the fully crystalline inorganic frameworks with sufficient electric conductivity and high thermal stability already at temperatures as low as 300 $^{\circ}$C without any need for the elaborate post-synthetic treatment. [5] Owing to the open and accessible character of porosity, the high surface area and the uniform pore size, the obtained mesoporous frameworks are the ideal host materials for the accommodation of functional redox moieties. The high potential of the obtained mesoporous layers as nanostructured transparent electrodes with the high surface area was demonstrated on an example of ferrocene molecules which were covalently immobilized in the conducting matrix, showing the greatly enhanced electrochemical response proportional to the electrode surface area. [5]

ACKNOWLEDGEMENT

The authors are grateful to the German Research Foundation (DFG, grant No. FA 839/1-1) and the Grant Agency of the Czech Republic (grant No. 104/08/0435-1) for the financial support.

REFERENCES

[1] D. Fattakhova Rohlfing, T. Brezesinski, J. Rathouský, A. Feldhoff, T. Oekermann, M. Wark, B. Smarsly, Adv. Mater. 18, 2980, 2006.

[2]. D. Fattakhova-Rohlfing, T., B. Smarsly, J. Rathouský, Superlattices and Microstructures 44, 686, 2008.

[3] J.M. Szeifert, D. Fattakhova-Rohlfing, D. Georgiadou, V. Kalousek, J. Rathousky, D. Kuang, S. Wenger, S.M. Zakeeruddin, M. Grätzel, T. Bein, Chem. Mater. 21, 1260, 2009.

[4] V. Müller, M. Rasp, G. Štefanić, J. Ba, S. Günther, J. Rathousky, M. Niederberger, D. Fattakhova-Rohlfing, Chem. Mater. 21, 5229, 2009.

[5] V. Müller, M. Rasp, J. Rathousky, M. Niederberger, D. Fattakhova-Rohlfing:, Small, 2010, 6, 633.

Stabilization Mechanism of ZnO Quantum Dots

D. Segets[*], R. Marczak[*], S. Schäfer[*] and W. Peukert[*]

[*]Institute of Particle Technology, FAU Erlangen-Nuremberg
Cauerstr. 4, D-91058 Erlangen, Germany, D.Segets@lfg.uni-erlangen.de

ABSTRACT

The current work addresses the understanding of the stabilization of small ZnO quantum dots in ethanolic suspension. Thereby different degrees of purification were investigated. The results revealed that not only the well-known ζ-potential as a representative for the surface charge determines the colloidal stability but also the surface coverage of acetate groups bound to the particle surface. Those acetate groups act as molecular spacers between the nanoparticles and prevent agglomeration. Regarding DLVO-calculations using a core-shell model, not only the height of the maximum but also the depth of the primary minimum has to be considered to see if a colloidal suspension is stable or not.

Keywords: ZnO, quantum dot, purification, stability, core-shell model

1 INTRODUCTION

ZnO semiconductor quantum dots have attracted considerable attention during the past ten years due to their promising electro-optical properties. Besides fundamental research on the particle synthesis itself,[1] especially the nanoparticles optimization regarding electronic devices or solar cells is increasingly in the focus of interest. However, for the successful incorporation of nanoparticles into electronic devices, stable suspensions and high solid concentrations are required. At first glance, the stabilization of small nanoparticles seems to be rather challenging as the collision events between the particles are supposed to be very frequent due to the strong influence of Brownian motion in this size regime. In contrast, experiments have revealed that ZnO particles around 5 nm in diameter can be stored over months at room temperature without noticeable aggregation. This is recently explained in the literature using an extended core-shell model of the well-known DLVO-theory which considers not only the surface charge of the nanoparticles but also a protecting shell around them.[2,6] Thus, the current work shows that for a successful stabilization of nanoparticles not only the height of the maximum but also the depth of the primary minimum has to be considered.

2 EXPERIMENTAL

2.1 Synthesis of ZnO nanoparticles

All chemicals were analytical grade reagents used without further purification. ZnO nanoparticles were obtained using a preparation procedure adapted from Spanhel et al.[3] and Meulenkamp.[4] In the first step 1.1 g of zinc acetate dihydrate (ACS Grade, 98.0%, VWR, Germany) was dissolved in 50 mL of boiling ethanol (99.98%, VWR, Germany) at atmospheric pressure. The obtained zinc acetate solution was cooled down to 25 °C. A white powder of anhydrous zinc acetate precipitated close to room temperature. In the second step lithium hydroxide (98%, VWR, Germany) was dissolved in 50 mL of boiling ethanol and also cooled down to the synthesis temperature. Then, the lithium hydroxide solution was added dropwise to the cold zinc acetate solution under vigorous stirring. The reaction mixture became transparent as the zinc acetate reactant was consumed during the ZnO particle formation. The particles were grown in their mother liquors to achieve different particle sizes before the purification was started by adding an access of n-heptane.

2.2 Purification of ZnO nanoparticles

To purify the so-prepared ZnO nanoparticles in order to remove the reaction byproduct lithium acetate, the precipitation-redispersion procedure introduced by Meulenkamp[4] was performed. Thereby, the unwanted byproduct was removed by reversible flocculation of the ZnO nanoparticles with the addition of n-heptane with a fivefold excess of the antisolvent to the ZnO colloid. White ZnO flocculates precipitated immediately after the addition of heptane. Afterwards, the supernatant containing the lithium acetate was separated from the white ZnO precipitate by centrifugation and decantation. The suspension was investigated after one, two and three times of washing.

For the optical characterization by using UV–Vis absorption spectroscopy and DLS, the ZnO flocculates were easily redispersed in ethanol using sonication. For the compositional characterization by using TGA, the ZnO flocculates were dried under nitrogen for about 5 min at room temperature in order to obtain a solid powder. The detailed specification of the measuring devices can be found in the literature.[2d]

3 RESULTS AND DISCUSSION

3.1 Particle size distributions from absorption spectra

A model for the calculation of particle size distributions (PSD's) of colloidal ZnO nanoparticles using their absorption spectra is presented. Using literature values for the optical properties of bulk ZnO which are correlated with the measurement wavelengths by a tight binding model (TBM) from the literature,[5] an algorithm deconvolutes the absorption spectra into contributions from the distinct size fractions. Thereby an excellent agreement between the size distribution determined from TEM-images and the calculated PSD's is found.[1]

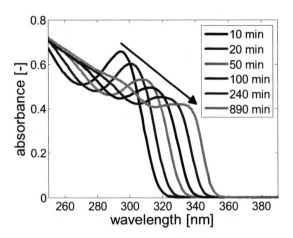

Figure 1: Absorption spectra of ZnO during ripening used for the validation of the algorithm.[1]

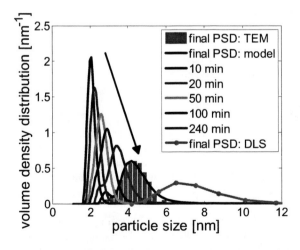

Figure 2: Comparison between PSD's calculated using the algorithm, optical image analysis of TEM-micrographs and dynamic light scattering (DLS).[1]

3.2 Temperature influence on ripening

To investigate the temperature influence on the ripening behavior of the particles the suspensions were put into a thermostated cuvette and the change of the optical properties was monitored with time. The final particle size distributions for T = 25 °C and T = 35 °C obtained from the absorption spectra are shown in Figure 3.[6]

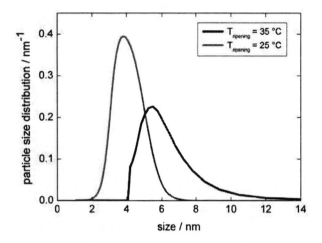

Figure 3: Particle size distributions after ripening at 25 °C and 35 °C, respectively, obtained from absorption spectroscopy.[6]

3.3 Washing

For the washing the procedure of precipitation and redispersion was performed one, two and three times. After each washing step the ζ-potential was determined and the amount of acetate molecules bound to the particle surface was measured using thermogravimetric analysis (TGA). The results are shown in Figure 4.

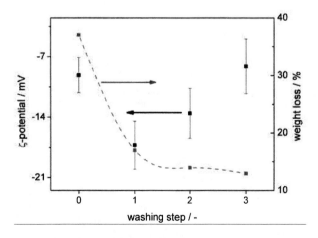

Figure 4: ζ-potential and conductivity obtained by DLS and corresponding weight losses obtained by TGA after zero, one, two and three washing steps.[6]

Thereby it becomes clear that the amount of the ζ-potential increases after the first washing as ions are removed from the solvent which are leading to a compression of the double layer. However, with subsequent washing, the amount of the ζ-potential decreases again as then the acetate ions directly bound to the particle surface are washed away. This removal of acetate can be monitored using TGA where the weight loss significantly decreases from nearly 40% for the unwashed sample down to 13% for the samples after three times of washing. However, after washing the powders could be easily redispersed without changing the original particle size distribution.[6]

Regarding the three-times washed sample it is highly unstable. This is leading to a white precipitate of agglomerates obtained without the addition of heptane immediately after the washing procedure has finished (see Figure 6).

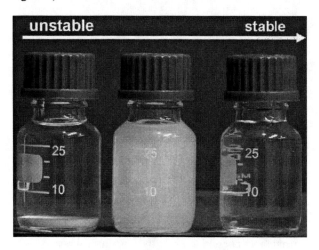

Figure 6: Evaluation of stability.

For a deeper understanding of the mechanisms which are leading to stable or unstable suspensions, calculations based on the DLVO-theory using a Core-Shell model were performed.

3.4 Core-Shell model

As shown in the previous section the stability of the nanoparticles can not be explained solely by electrostatics. Therefore Figure 7 shows the devolution of the interaction potential of a ZnO suspension calculated with a core-shell model with and without the consideration of acetate molecules bound to the particle surface. It becomes clear that for a proper interpretation of the stability not only the height of the maximum has to be evaluated but also the depth of the primary minimum. Thereby, the interaction potentials for the unwashed suspension as well as for the samples after one and two cycles of washing, were calculated under consideration of a protecting shell of acetate ions around the particle surface. They show neither a maximum nor a noticeable minimum.

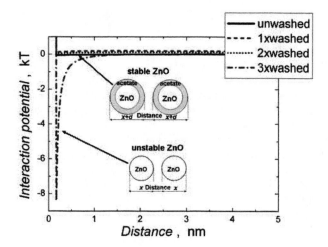

Figure 7: Comparison of the interaction energy of 5 nm ZnO nanoparticles before washing (solid line), after one (dashed line), two (dotted line) and three (dash-dotted line) washing cycles.[6]

In contrast, the suspension after three times of washing which was calculated with a hard-sphere model shows a comparatively deep primary minimum.[6] Thus, when two particles approach down to the minimum contact distance at 0.165 nm they "feel" a remarkable force sticking them together and agglomeration of the suspension is observed.

However, this holds for the assumption of a complete removal of acetate ions bound to the particle surface. For a detailed understanding of all influencing factors on particle stabilization being present at the particle surface, a more general approach has to be found which independently compares the influence of electrostatics, surface coverage and attractive forces.[6]

4 CONCLUSION

In the present work the stabilization of ZnO quantum dots was investigated. Thereby the particle size distributions were calculated from the absorption spectra, the ζ-potentials were determined for different washing cycles and the weight loss of acetate molecules bound to the particle surface was measured with TGA. For a clear identification of stable conditions under consideration of electrostatics as well as the protecting acetate coverage around the particles DLVO calculations using a Core-Shell model were performed. They revealed that for understanding colloidal stability not only the height of the maximum but also the depth of the primary minimum has to be considered.

However, for the future a more general approach has to be found. Thereby the different influencing factors on nanoparticle stabilization, namely electrostatics, surface coverage and attractive forces have to be considered independently.

5   ACKNOWLEDGMENTS

The authors would like to thank the German Research Council (DFG) for their financial support (Projects PE427/18-2) which supports within the framework of its Excellence Initiative the Cluster of Excellence 'Engineering of Advanced Materials' (www.eam.uni-erlangen.de) at the University of Erlangen-Nuremberg.

REFERENCES

[1]   (a) Segets, D.; Gradl, J.; Klupp Taylor, R.; Vassilev, V.; Peukert, W., *ACS Nano, 3*, 1703-1710, **2009**; (b) Segets, D.; Martinez Tomalino, L.; Gradl, J.; Peukert, W., *J. Phys. Chem. C, 113*, 11995, **2009**.

[2]   (a) Kim, B.-S.; Lobaskin, V.; Tsekov, R.; Vinogradova, O. I., *J. Chem. Phys., 126*, 244901, **2007**; (b) Sabin, J.; Prieto, G.; Sennato, S.; Ruso, J. M.; Angelini, R.; Bordi, F.; Sarmiento, F., *Phys. Rev. E, 74*, 031913, **2006**; (c) Yadav, P. S.; Dupre, D.; Tadmor, R.; Park, J. S.; Katoshevski, D., *Surf. Sci., 601*, 4582, **2007**;

[3]   Spanhel, L.; Anderson, M. A., *J. Am. Chem. Soc., 113* (8), 2826, **1991**.

[4]   Meulenkamp, E. A., Synthesis and Growth of ZnO Nanoparticles. *J. Phys. Chem. B, 102*, 5566-5572, **1998**.

[5]   Viswanatha, R.; Sapra, S.; Satpati, B.; Satyam, P. V.; Dev, B. N.; Sarma, D. D., *J. Mater. Chem., 14*, 661-668, **2004**.

[6]   Marczak, R.; Segets, D.; Voigt, M.; Peukert, W., *Advanced Powder Technology, 21*, 41, **2010**.

The use of PEO--PPO--PEO block copolymers in the synthesis of ZnO via the hydrothermal method

M. Young[*] and X. Bao

[*]Loughborough University, Leicestershire, LE11 3TQ
UK, m.i.young@lboro.ac.uk

ABSTRACT

Zinc oxide particles were produced by aqueous precipitation in hydrothermal conditions (140°C for 6 hours). The influence of surfactant and base to control morphology was investigated. With a change in surfactant, rod-like bundles of different size were obtained, depending on the core and corona size. By changing the base different complexes were formed which interacted differently with the surfactant present (L64). Platelets, rod-like and elliptical particles were obtained.

Keywords: Zinc Oxide, Hydrothermal, Pluronic, PEO-PPO-PEO

1 INTRODUCTION

Zinc oxide (ZnO) is a widely used semiconductor, which has a band gap of 3.37eV and an exciton binding energy of 59meV. ZnO has a wide range of uses in industry as gas-sensors, varistors and UV light emitters. This is due to its piezoelectric, pyroelectric, semi-conducting and optical properties.[1]

Two distinct routes can be used to synthesise ZnO: gas phase and solution phase. Gas phase synthesis uses a closed gaseous environment and high temperatures from 500°C to 1500°C, including vapour transport, physical vapour deposition and chemical vapour deposition. Due to the high temperature and the need for a controlled environment these techniques can be expensive. Solution phase synthesis requires the synthesis of the ZnO to be carried out in a solution. Some of the methods used are hydrothermal, solvothermal, precipitation and microemulsion. Unlike the gaseous methods, the simplicity and low cost of solution synthesis make the methods a more attractive route for the fabrication of ZnO.[1]

Self-assembly, in particular the formation of micelles, represents a facile approach to the control of morphology of oxides. Non-ionic surfactants such as Pluronics (PEO-PPO-PEO), having a range of ordered microdomain morphologies along with their low cost, low toxicity and bio-degradation make them an ideal tool for designed synthesis.[2]

In this paper the influence of different chain lengths of Pluronic surfactants and different starting precursors on the particle size and morphology of ZnO synthesised were reported.

2 EXPERIMENTAL

2.1 Materials

All chemicals used in this study were analytical grade and used without further purification. Acetone, ethanol and iso-propanol were used as supplied. Distilled water was used throughout the experiments. The surfactants used were three types of Pluronic: L64 ($PEO_{13}PPO_{30}PEO_{13}$ M_w=2900), F68 ($PEO_{80}PPO_{30}PEO_{80}$ M_w = 8400) and P123 ($PEO_{13}PPO_{80}PEO_{13}$ M_w = 5750). All surfactants were purchased from Sigma-Aldrich.

2.2 Synthesis of zinc precursor solutions

ZnO was prepared by the precipitation. One of the reactions is given as follows:

$$Zn(CH_3CO_2)_2 + 2NaOH \rightarrow ZnO + 2Na(CH_3CO_2)_2 + H_2O$$

For a typical preparation, 7.2×10^{-4}M surfactant (Pluronic L64, F68, P123) was dissolved in 40ml H_2O [Sol A] for 1 hour. This solution was then mixed with another solution of 2×10^{-3}M zinc acetate dihydrate dissolved in 16ml H_2O [Sol B]. After completion of the addition of Sol A to Sol B, the surfactant/salt solution [Sol C] was mixed for 30 minutes. Finally 2×10^{-2}M base (NaOH, TMAH, DEA, NH$_4$OH) was dissolved in 4ml H_2O was added drop wise using a peristaltic pump to Sol C and left to mix for 30 minutes. All the reactions were carried out at ph = 10. 20ml of the solution was transferred to a Teflon cup and placed in a stainless steel autoclave, which was hydrothermally treated. The remaining solution was washed for characterization.

2.3 Hydrothermal treatment

The hydrothermal treatment of the nanoparticles was carried out in an autoclave at 140°C for 6 hours. All samples were washed three times with distilled water and iso-propanol between centrifugal cycles of 5 minutes at 3000rpm.

Figure 2: TEM micrographs of hydrothermally grown ZnO particles synthesised with
NaOH and a different surfactant [A] No surfactant [B] L64 [C] F68 [D] P123

2.4 Characterization

The size and morphology of the particles were investigated using Transmission Electron Microscopy (TEM) (JEM 100CX, JEOL Ltd, Tokyo Japan) and Field Emission Gun Scanning Electron Microscopy (FEGSEM). The XRD patterns were recorded on a Bruker D8 X-ray using CuKα radiation. For TEM, a copper grid coated with carbon film was covered with a drop of the washed zinc oxide solution. For FEGSEM and XRD, the solution was cast onto a clean glass slide and dried for 4 hours at 60°C.

3 RESULTS AND DISCUSSION

3.1 The effect of surfactants

The effect of the change of non-ionic surfactants on the particle shape and size of zinc oxide particles after hydrothermal heat treatment are shown in Figure 1. It can be seen that without surfactant the ZnO formed into thin platelets. With the addition of surfactant rod-like structures were formed in all cases. In the case of the L64 and F68 surfactants, bundles of rods were formed with sizes of 1μm and 2μm respectively. For the addition of P123, singular rods were formed, ranging in length from between 1 – 2μm. Figure 2 shows the XRD results for pre and post –

hydrothermal treatment of the solution with P123 as the surfactant. The definitive peaks within the 30 – 40 degree range indicated the formation of ZnO after hydrothermal treatment.[3,4]

It was reported that the growth velocity of the ZnO particles and the micelle dimension of the surfactant controlled the size and arm lengths.[3] It is possible to control the growth velocity by the use of a peristaltic pump and a constant heat treatment.

The formation of the ZnO begins with the addition of NaOH, which reacts with $Zn(CH_3CO_2)_2$ to form

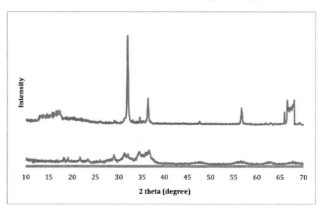

Figure 2: XRD patterns for pre hydrothermal
(bottom) and post hydrothermal treatment (top).

Figure 3: FEGSEM micrograph of hydrothermally grown ZnO with NaOH and P123

Zn(OH)$_2$. It is known that triblock copolymers such as Pluronics, can spontaneiously assembly into aggregates, such as micelles, vesicles or lyotropic mesophases in a given circumstance.[4] In this case it is considered that the surfactants form micelles.[3] In the aqueous solution, the surfactant forms a hydrophobic core consisting of PPO and a hydrophilic corona (or tails) with the PEO blocks. It is thought that surface of the PPO could be adsorbed onto the surface of the Zn(OH)$_2$, while the PEO blocks are in constant contact with H$_2$O. This could mean that during the hydrothermal treatment, the ZnO nuclei formed from the Zn(OH)$_2$ were restricted to the hydrophobic PPO core. During the heat treatment, the ZnO should grow in its preferential (0001) direction, resulting in the formation of rods. The difference in bundle size may be explained by the difference in aggregation size of the two surfactants. F68 (PEO$_{80}$PPO$_{30}$PEO$_{80}$) and L64 (PEO$_{13}$PPO$_{30}$PEO$_{13}$) share the same core size, but different corona size. This increase in corona size, whould lead to an increase in micelle size, leading to the bigger rod size, as shown in Fig. 1c

However when P123 (PEO$_{13}$PPO$_{80}$PEO$_{13}$) was used similar rod like structure, but with less aggregated bundles were observed (Figure 3) which is possible due to the change in core size. With a much bigger core size, the adsorbed Zn(OH)$_2$ were not confined to a small area unlike in L64 and F68. This means that instead of a core packed with aggregated ZnO crystallites, they were spread around the core. This may well lead to undisturbed growth to form more defined crystals.

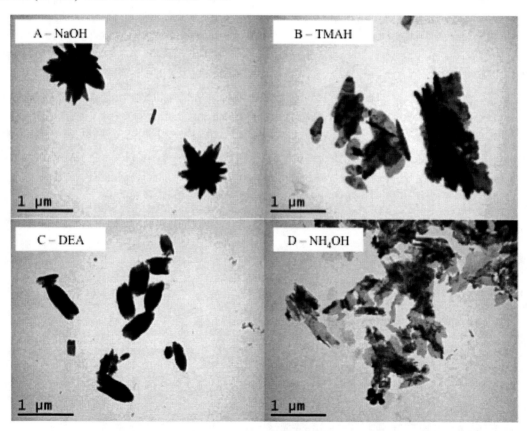

Figure 4: TEM micrographs of hydrothermally grown ZnO particles synthesised with Pluronic L64 and a different base [A] NaOH [B] TMAH [C] DEA [D] NH$_4$OH

3.2 The effect of base

Figure 4 shows the typical TEM micrographs of ZnO particles synthesised with different bases and L64 surfactant. As shown previously, the addition of NaOH results in the formation of bundles of rod-like structures. With the addition of TMAH (Fig. 4b) and NH_4OH (Fig 4d) a platelet structure is formed. In both cases the platelets were between 500nm – 1µm in length. The shape of the platelets were not uniform. By using DEA (Fig 4c) as the base, hollow elliptical shaped particles were obtained. A more detailed FEGSEM image is shown in Figure 5. The sizes of the particles are between 750nm – 1µm in length and a width of approximately 200-300nm.

These results suggest that different bases have an effect on the morphology of ZnO formed which is probably due to the change in complexes formed when the $Zn(CH_3CO_2)_2$ was reacted with the base. By using TMAH and NH_4OH, the results show a similar morphology (platelets) to that of no surfactant (Fig. 1a). This would suggest that there is no interaction between them and the surfactant.

The most interesting result is when DEA was used as a base in which hollow elliptical particles were formed. It is suggested that a Zn-DEA complex was produced in solution instead of $Zn(OH)_2$. With the increase in pressure and temperature, Zn-DEA complex began to hydrolyze into ZnO crystallites. The formation of the hollow particles was maybe due to the ZnO crystallites beginning to grow on the outside surface of the micelle instead of inside the core like before. During hydrothermal treatment, the process of Ostwald Ripening increased the size of the ZnO crystallites and a much larger hollow particle was formed.[4]

Figure 5: FEGSEM micrograph of hydrothermally grown ZnO with DEA and L64

4 CONCLUSIONS

In summary, we demonstrated the preparation of a variety of different morphologies of ZnO, by a simple change of surfactant and base. It was found that the by using a Pluronic surfactant, rod like structures were formed. By changing the size of the surfactant, in particular the core size, it was possible to change the size and aggregation of the rods. By changing the choice of base used in the reaction with $Zn(CH_3CO_2)_2$ the new complexes formed would also behave differently with the surfactants (L64). Though TMAH and NH4OH did not interact with the surfactant, the addition of DEA caused hollow elliptical particles to be formed. Ongoing investigations are focusing on the effect of ionic surfactants, heat treatment changes and on the doping of the zinc. It is expected that these variables will lead to more interesting structures, with a potential to be used in such applications as photovoltaic devices, gas sensors and catalysts.

REFERENCES

[1] S. Baruah and J. Dutta. Hydrothermal growth of ZnO nanostructures. Science and Technology of Advanced Materials, 10, 1-18, 2009

[2] Wan et al. Designed synthesis of mesoporous solids via nonionic-surfactant-templating approach. Chemical Communications, 9, 897-926 2007

[3] Zhang and Mu. Hydrothermal synthesis of ZnO nanobundles controlled by PEO–PPO–PEO block copolymers. Journal of colloid and interface science, 307, 79-82, 2007

[4] Tao et al. Controllable preparation of ZnO hollow microspheres by self-assembled block copolymer. Colloids and Surfaces A: Physicochemical and Engineering, 330, 67-71, 2008

Au, Ag and Au-Ag nanoparticles: microwave-assisted synthesis in water and applications in ceramic and catalysis

M. Blosi[*], S. Albonetti[**], F. Gatti[**], M. Dondi[*], A. Migliori[***], L. Ortolani[***], V. Morandi[***], G. Baldi[****]

[*] ISTEC-CNR, Institute of Science and Technology for Ceramics, National Research Council, Via Granarolo 64, 48018, Faenza (Italy), magda.blosi@istec.cnr.it
[**] Department of Industrial Chemistry and Materials, Bologna University, Viale del Risorgimento 4, 40136 Bologna (Italy), stalbone@fci.unibo.it
[***] IMM-CNR Bologna - Via Gobetti 101, 40129 Bologna (Italy), morandi@bo.imm.cnr.it
[****] CERICOL, Via Pietramarina 123, 50053 Sovigliana Vinci (Italy), baldig@colorobbia.it

ABSTRACT

A simple, microwave-assisted route for producing Au/Ag concentred sols by glucose reduction in water was developed. Ag-Au bimetallic nanoparticles stabilized by polyvinylpyrrolidone (PVP) were characterized and their catalytic activity was studied in the reduction of 4-nitrophenol (4-NP) with NaBH$_4$. Moreover they were tested as red ceramic inks for ink-jet printing technology. Since the research was developed in collaboration with a company interested in the large scale production of the suspensions, some fundamental properties for an industrial scale up were developed: high metal concentration, long time stability. The Au-Ag core-shell structures were prepared by a two-step process. Particle size-control and colloidal stability were achieved thanks to the accurate reaction optimization, combined with microwave heating, that allows the intensification of process even on large scale production. Prepared Au, Ag and Au/Ag nanocrystals acted as effective catalyst for the reduction of 4-NP.

Keywords: Au, Ag, microwave, bimetallic sols, nanoparticles, water media.

1 INTRODUCTION

In recent years, colloidal suspensions of different metals have found applications in various fields, catalysis [1], cancer therapy, thermal nanofluids and also as special inks [2-4]. In particular, alloy and core-shell metallic nanoparticles have received special attention due to the possibility of tuning the optical and electronic properties over a broad range by simply varying the composition [5]. For example, current technology of ceramic inks, consisting in the synthesis of gold nanoparticles or their in situ formation in glazes from Au precursors, can take advantage from the use of AuAg core-shell structure of nanoparticles modulating colour and reducing the cost. Moreover metallic nanoparticles show high catalytic activity in different reactions because a large fraction of the catalytically active metal sites is exposed to the reactants.

2 EXPERIMENTAL PROCEDURE

2.1 Bimetallic Synthesis

All the chemical reagents used in this experiment were analytical grade (Sigma Aldrich). The Au-Ag core-shell structures were prepared by a two-step process, both Au and Ag sols were used as core, acting as seed for the shell nucleation (Scheme 1). Au-core and Ag-core sols were synthesized separately by using the reduction of HAuCl$_4$ or AgNO$_3$ by glucose in alkaline water. PVP-coated metal seeds (nPVP/nMetal = 5.5) were synthesized in 5 minutes following a patented procedure [6] by using microwave heating. The reactions were carried out in few minutes at low temperature, 70°C for Ag or 90°C for Au. In order to control the particle size, the glucose amount and the solution pH were carefully optimized for each metal (1<nGlucose/nMetal<2; 1<nNaOH/nMetal<8). In Table 1 the characteristics corresponding to the synthesized samples are shown. Core metals are in brackets, Ag/Au molar ratio and total metal concentration are reported

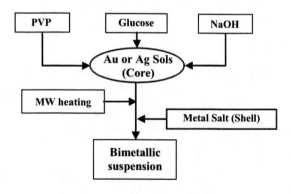

Scheme 1 – Synthesis flow chart

SAMPLE	%Au (mol/mol)	%Ag (mol/mol)	Ag/Au (mol/mol)	Metals Conc [mM]
Au	100	0	-	11
(Au)Ag01	67	33	0.5	15
(Au)Ag02	40	60	1.5	8.3
(Au)Ag03	18	82	4.5	6.1
Ag	0	100	-	50
(Ag)Au01	67	33	0.5	15
(Ag)Au02	40	60	1.5	8.3
(Ag)Au03	18	82	4.5	6.1

Table 1 – Prepared bimetallic nanoparticles

2.2 Apparatus

The microwave system used is a Milestone MicroSYNTH plus, whose reaction chamber is provided with magnetic stirring, reflux system and an optical fiber temperature controller. The microwave power is generated by 2 x 800 W magnetrons, with frequency 2.45 GHz. In order to respect the scheduled heating ramps, the power is continuously supplied and automatically modulated by a software; for each ramp only the maximum deliverable power can be imposed. The power is automatically changed by the instrument to follow the temperature profile.

2.3 Catalytic reaction

The catalytic reduction of 4-nitrophenol by $NaBH_4$ was studied at room temperature (25°C) in a standard quartz cuvette with 1 cm path length and about 3 mL volume. Due to their lowest dimension, only the Au-Core samples were considered for the catalytic characterization. The prepared samples were properly diluted with distilled water in order to achieve a metal concentration of 1.1×10^{-2} mM. Thus 10 ml of diluted suspensions were mixed with 5 mL of a 4-nitrophenol solution (9.0×10^{-2} mM) and with 1 mL of a $NaBH_4$ aqueous solution (0.72 M). An aliquot of the solution was poured into the quartz cuvette and the absorption spectra were collected by a Lambda 35 spectrophotometer (Perkin Elmer, USA) in the range between 250 and 500 nm. The rate constants of the reduction process were determined through measuring the change in absorbance at 400 nm, corresponding to 4-NP, as a function of time.

2.4 Ceramic application

Prepared nanosuspensions were properly diluted and applied on porcelain stoneware throughout an aerospray gun, taking care that each tiles received the same quantities of pigment. Ceramic tiles were fired at 1150°C for 10 minutes.

2.3 Analytical characterization

UV-VIS extinction spectra were measured with a Lambda 35 spectrophotometer (Perkin Elmer, UK), using a quartz cuvette as sample-holder. Samples for UV-VIS spectroscopy were prepared by diluting the as prepared colloidal suspension with water in order to get into the cuvette the same metal concentration for every sample. Particle size distribution, based on hydrodynamic diameter, was evaluated by Nano S (Malvern, UK), a dynamic light scattering analyzer (DLS), taking care to dilute and pour it in a polystyrene cuvette before measurement. Hydrodynamic diameter includes the coordination sphere and the species adsorbed on the particle surface such as stabilizers, surfactants and so forth. Unreacted metal cations were detected by ICP-AES analysis in order to determine the reaction yield. Samples were prepared as follow: 50 ml of synthesized colloid was poured into a semi-permeable osmotic membrane (Visking tube), which was submerged in a de-ionized water bath. Osmotic pressure caused the exchange of unreacted cations into the external water and the water entrance into the tube. After three hours, the equilibrium was attained and the external liquid underwent ICP (Liberty 200, Varian, Australia) quantitative analyses. Suspensions were dropped and dried on a copper grid, then observed by high resolution transmission electron microscopy (HRTEM) (Tecnai F20) and by the STEM mode with microanalysis EDX. The colour of ceramics was measured by diffuse reflectance spectroscopy (Miniscan MSXP4000, Hunter Lab, Reston, USA) in the 400-700 nm range (illuminant D65, observer 10°) taking a white glazed tile as reference.

3 RESULTS AND DISCUSSION

3.1 Nanoparticles synthesis

Since particles composition influences their optical properties, the UV-VIS spectra collected for bimetallic structure showed a clear shift with respect to single metal spectra. In fact, while the pure Au nanoparticles solution has a characteristic resonance peak at 520 nm, Au-Ag samples evidenced a clear blue-shifting of the plasmon resonance band increasing the silver content. Figure 1 shows the shift of UV-VIS spectra collected for the Ag-core series. The plasmon resonance shifting confirmed the formation of multi-component nanostructures, in fact for a physical mixing two distinct bands would be observed: at 400 nm for silver and at 520 nm for gold (Fig. 2).

The prepared sols show excellent stability up to several months of storage, indicating that no nanoparticles aggregation occurred.

Hydrodynamic diameter (HD) measured by DLS (Fig. 3) evidenced that with respect to Ag nanoparticles, Au nanoparticles have a lower HD diameter and samples containing Ag exhibited a progressive increase in particle size. Furthermore, it was clearly observed that the diameter of the particle used as seed influenced the final core-shell size, in fact the size of the Au-Core particles was lower than Ag-Core samples.

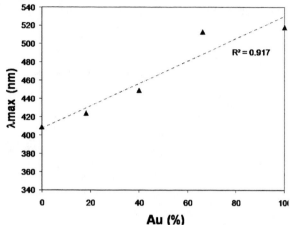

Figure 1 - Surface plasmon resonance shift in Ag-core series depending on particle composition.

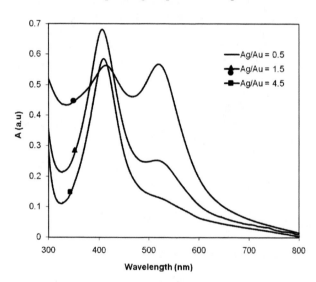

Figure 2 - Extinction spectra of Au-Ag physical mixing

Since DLS values correspond to the hydrodynamic diameter, HRTEM analysis generally indicated a lower nanoparticle diameter (Fig. 4a), highlighting the difference between hydrodynamic diameter, comprehensive of coordination sphere, and real size. However, HRTEM indicated that nanoparticles are typically spherical and polycrystalline and confirmed that samples containing higher shell element concentration exhibited a slight increase in the diameter, indicating that reduction/deposition rather than other process dominates the coating process.

EDX-STEM line scanning across the nanoparticles was carried out to analyze the distribution of the chemical composition. For example in Au core systems, in the samples with Ag/Au ratio of 1.5 or 4.5, the maximum of the Au signal in the center of the nanoparticles, supports the formation of Au-core Ag-shell nanostructures (Fig. 4b). On

the contrary, samples with low amount of Ag exhibit a typical Au-Ag alloy behaviour.

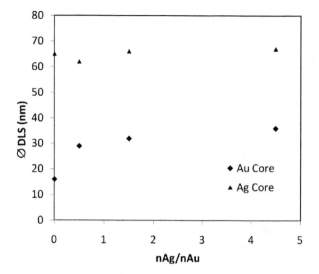

Figure 3 – Mean hydrodynamic diameter measured by DLS for Au-Core and Ag-Core series.

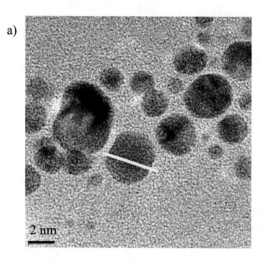

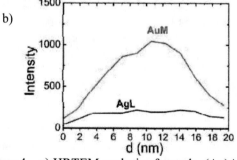

Figure 4 – a) HRTEM analysis of sample (Au)Ag02; b) EDX analysis of a particle profile

ICP analysis on external water containing the unreacted cations confirmed for all samples reaction yields higher than 98%, indicating that the so-optimized synthesis allows the complete reduction of precursors. This synthesis method has been extended to the preparation of bimetallic nanocluster of Au-Cu, both as core-shell structure and as alloy.

3.2 Catalytic behavior

Synthesized nanocrystal of Au, Ag and AuAg act as effective catalyst for the hydrogenation of p-nitrophenol in the presence of $NaBH_4$.

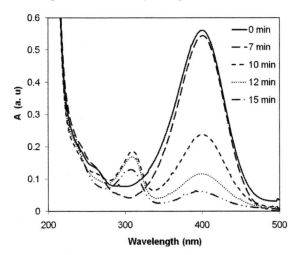

4- Nitrophenol (4-NP)   4- Aminophenol (4-AP)

Since $NaBH_4$ was added in large excess, the reduction rates can be regarded as being independent by its concentration. The peak of 4-NP at 400 nm decreased, while at 290 nm a new peak, assigned to 4-AP, appeared (Fig. 5). The pseudo-first-order rate constants reported in table 2, were calculated by the decreasing of the band of 4-NP, considering the slope of the ln(At/A0) function of time. The final kinetic constants resulted similar to literature values. Increasing the silver amount, a lower catalytic activity was observed, confirming the lower activity of Ag in this reaction.

Figure 5 – UV-Vis spectra of 4-NP reduction for sample AuAg 03.

SAMPLE	Ag/Au (mol/mol)	Kinetic Constant (s^{-1})
Au	-	1.75×10^{-2}
(Au)Ag01	0.5	0.71×10^{-2}
(Au)Ag02	1.5	0.48×10^{-2}
(Au)Ag03	4.5	0.31×10^{-2}
Ag	-	0.17×10^{-2}

Table 2 – Calculated kinetic constants for Au-Core samples.

3.3 Colouring performance

Preliminary tests showed that the red component of coating colour is directly correlated with gold concentration. Bimetallic nanopigments behave similarly to Au nanoparticles. Their colouring perfomance, in fact, is nearly the same of gold nanoparticles given the same Au concentration. Ag-Core samples are characterized by a better colouring performance with respect to Au-Core ones. In particular, sample (Ag)Au02 exhibited the best behaviour, as its performance is practically the same of nanopigments with higher gold concentration.

However from preliminary testes it was noticed that silver seems not to affect significantly the colour quality and despite the deep yellow shade of Ag nanoparticles suspensions, little contribution is given to the pigment-bearing coatings in terms of yellow component (b*).

3.4 Conclusions

Au-Ag stable bimetallic nanoparticles were synthesized by a microwave assisted route, by using an eco-friendly method that provided a total reaction yield. Both UV-VIS spectroscopy and microscopy data confirmed the formation of bimetallic nanostructures, in the form of core-shell or alloy. Catalytic tests showed that prepared nanostructure of Au, Ag and AuAg act as effective catalyst in the hydrogenation of p-nitrophenol, resulting in agreement with literature data, but with a decreased activity for higher silver content.

The best colouring performance was observed for Ag-Core samples, further analysis are ongoing to explain the reasons of this behavior.

REFERENCES

[1] W. Hou, N. A. Dehm, R. W.J. Scott, J. Catal. 253, 22, 2008.

[2] P. K. Jain, I. H. El-Sayed, M. A. El-Sayed, NanoToday, 2, 18, 2007.

[3] M. S. Liu, M. C. Lin, C.Y. Tsai, C. C. Wang, Int. Journ. Heat Mass Trans. 49, 3028, 2006.

[4] D. Gardini, M. Dondi, A. L. Costa, F. Matteucci, M. Blosi, C. Galassi, G. Baldi, E. Cinotti, Jour. Nanosc. Nanotech. 8, 1979, 2008.

[5] S. Link, Z. L. Wang, M. A. El-Sayed, J. Phys. Chem. B, 103, 3529, 1999.

[6] M. Blosi, S. Albonetti, M. Dondi, G. Baldi, A. Barzanti PCT/EP2010/052534 (March 2010).

In-Situ Quick XAFS Study on the Formation Mechanism of Cu Nanoparticles Synthesized in Aqueous Phase

Shun Nishimura, Atsushi Takagaki, Shinya Maenosono* and Kohki Ebitani*

*School of Materials Science, Japan Advanced Institute of Science and Technology,
1-1 Asahidai, Nomi, Ishikawa, 923-1292, Japan,
shinya@jaist.ac.jp (S.M.), ebitani@jaist.ac.jp (K.E.).

ABSTRACT

The formation process of Cu nanoparticles (NPs) capped by poly(N-vinyl-2-pyrrolidone) (PVP) using a wet-chemical reduction method was observed with in-situ time-resolved X-ray adsorption fine structure (XAFS) method in combination with other analytical techniques. XAFS analysis directly indicated that the $Cu(OH)_2$ and Cu^+ species were formed as intermediates, and their life-time were extended by adding PVP capping agent. The morphology analysis and indicator method also supported that both reticulated $Cu(OH)_2$ structure and Cu^+ intermediate remained for a long time with an increase in PVP capping agent. Our results suggested that the PVP capping agent plays an important role to the reaction kinetic resulting from extending the life-time of intermediates such as $Cu(OH)_2$ and Cu^+ species.

Keywords: in-situ XAFS, Cu nanoparticles, formation mechanism, role of capping agent, reaction kinetic

1 INTRODUCTION

Wet-chemical reduction method has been used extensively for the synthesis of metal nanoparticles (NPs) because it makes highly dispersed NPs with narrow size distribution by a simple procedure. Many researchers already reported that the preparation conditions like temperature, pH, concentration, kinds of precursor, capping agents, additive sources, solvent, etc. can change the morphologies and properties of synthesized NPs.[1,2] There were many reports to reveal the mechanism during the NPs formation, however, it was still remained to investigate dynamic changing directly with time resolution because of using ex-situ analysis (e.g., sampling, quench…).[3,4]

Recently, with the development of in-situ observation techniques, the mechanisms for the highly dispersed NPs formation were gradually clarified with time variation. Harada et al. investigated the kinetics of Rh, Pd and Ag NPs prepared by photoreduction method, and they found that the induction periods occurred during reduction except for the case of Pd.[5,6] Very recently, we succeeded in the observation of the formation process of Cu NPs synthesized with chemical reduction method, and suggested the poly(N-vinyl-2-pyrrolidone) (PVP) capping agent acts as not only protector from overgrowth and aggregation, but also as a control agent over the reaction kinetics of NP formation.[7]

In this study, we discuss the relationships between the concentration of PVP and the reaction kinetics during the Cu NPs formation process in more detail using in-situ X-ray adsorption fine structure (XAFS) method in combination with UV-vis spectroscopy, indicator method, TEM measurement, and so on.

2 EXPERIMENT

The synthesis of Cu NPs with wet-chemical reduction method was performed according to the previous reports[8] with some modifications. In this procedure, the reducing agent (NaBH4/NaOH) is injected into the aqueous solution containinng Cu(II) acetate ($Cu(OAc)_2$) and PVP under stirring in N_2 atmosphere. The molar ratio of the mixture was $Cu(OAc)_2/PVP/NaBH_4/NaOH/H_2O = 1.0/x/1.0/1.0/3.7$ (x = 0.0033, 0.033, and 0.165). By using this procedure, highly dispersed Cu NPs were obtained.[7] A plastic cell (PMMA, $1 \times 1 \times 4.5$ cm^3) was used as a reactor.

The fractions for each Cu species during Cu NPs formation process were characterized by in-situ XAFS at the Cu-K edge (8.98 keV) with quick scanning in fluorescence mode (SPring-8, BL01B1, proposal No. 2008B1328 and 2010A1598). The beam size was approximately 1×5 mm^2, and each scan was carried out for every 68 seconds with interval. To evaluate time profiles of the fraction for each Cu species, the time-resolved X-ray absorption near-edge structure (XANE) spectra were fitted by a linear combination of the XANES spectra reference ($Cu(OAc)_2$ aq, $Cu(OH)_2$, Cu_2O, and Cu metal).[9-11]

Morphology was analyzed by transmission electron microscopy (TEM) using a Hitachi H-7100 operated at 100 kV. Liquid samples were extracted by pipet, dropped onto a carbon-coated copper grid, and dried in vacuo before operation with TEM.

Sodium bicinchoninic acid (BCA) was used as a color indicator method for detecting Cu^+ species using the absorption peak at 562 nm. UV-vis absorption spectroscopy was carried out on a Perkim Elmer Lambda 35 UV-vis spectrometer in the range of 400-800 nm.

The zeta potential was measured with a Zetasizer Nano ZS ZEN 3600 (Malvern, UK) under alkaline conditions controlled by NaOH aqueous solution.

3 RESULTS AND DISCUSSIONS

The color of the solution immediately turned from light blue (absorption peak at ca. 750 nm) to an emerald green solution (ca. 650 nm) in the first step, then gradually changed to brownish-red (adsorption peak was extended to all wavelengths) for all samples. Moreover, the ratio of color change was also retarded for increasing PVP. These changes implied that the PVP capping agent effected to the reaction kinetics.

Figure 1 shows the time variation of the fraction for each Cu species under conditions of various amounts of PVP, that calculated by using the time-resolved XANES spectra with a linear combination method. The transformations were as follows: the precursor of $Cu(OAc)_2$ drastically transformed to the $Cu(OH)_2$ in the first step, then $Cu(OH)_2$ gradually changed to Cu^0 through the Cu^+ species. All samples indicated the same pathway, however, the life-time of $Cu(OH)_2$ was clearly different. When small amounts of PVP was used (x = 0.0033), $Cu(OH)_2$ could only keep within ca. 4 min reaction. But when the case of more large amount of PVP (x = 0.033, and 0.165), the life-time of the $Cu(OH)_2$ became ca. 5 min and above 5 min, respectively. Additionally, the formation of Cu^+ species also became slower with following the life-time of $Cu(OH)_2$. From the view point of the life-time of Cu^+ species behavior, it was indicated that the fraction of species existing with the peak during formation process under such strong reduction atmosphere. These results suggested the PVP capping agent was associated with slower reaction kinetics by elongating the life-time of the intermediates of $Cu(OH)_2$ and Cu^+ species.

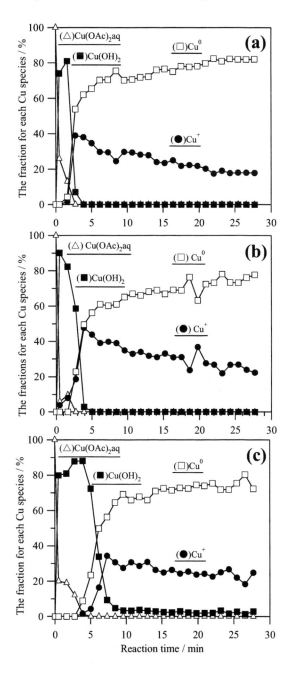

Figure 1. The time variation of the fraction for each Cu species at the condition of various amount of PVP, x = 0.0033 (a), 0.033 (b), and 0.165 (c), respectively.

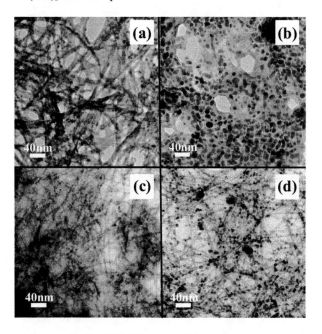

Figure 2. The TEM images of the solution after 2 min (left side) and 15 min (right side) reaction in the case of adding different amounts of PVP, x= 0.0033 (a, b), and 0.165 (c, d).

To compare the morphology change of $Cu(OH)_2$, the TEM images are shown in Figure 2. After 2 min reaction, the reticulated structure was obtained in both conditions with size differences. Using small amount of PVP (Fig. 2(a)) caused larger reticulated structure, and a large amount caused smaller ones (Fig. 2(c)). In addition, the large reticulated structures were consumed and disappeared (Fig. 2(b)), however the small reticulated structure remained after 15 min reaction (Fig. 2(d)). Generally, the morphology of $Cu(OH)_2$ prepared with simple pH control is known as a nanostrand.[12-14] In addition, the concentration of capping agent to metal precursor, relating to the average size of synthesized NPs was reported.[15-18] Therefore, it is possible that the PVP capping agent acts as a stabilizer to the $Cu(OH)_2$ intermediate, and the reaction kinetics become slower.

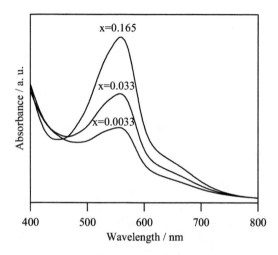

Figure 3. The UV-vis spectra after the addition of BCA in the case of varying the amount of PVP after 30 min reaction.

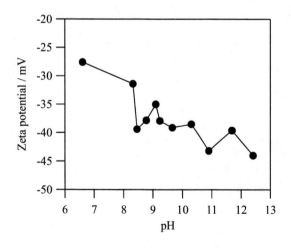

Figure 4. The zeta potential of PVP as a function of various pH conditions controlled with NaOH.

To verify the presence of the Cu^+ species despite a strong reducing atmosphere, an indicator method with BCA was performed. BCA is a well known reagent possessing the capacity to selectively-form an intense purple complex (adsorption peak at 562 nm) with Cu^+ ion.[19-21] Figure 3 shows the UV-vis spectra after the addition of BCA to the reaction solution after 30 min reaction. Every sample shows a sharp absorbance peaks at 562 nm, this indicates the stable Cu^+ species exist during Cu NPs formation process despite the PVP amount. Furthermore, the intensity of the specific peak was progressively increased by the PVP amount. It suggests the PVP could make the life-time of Cu^+ species longer. To consider this assumption, the zeta potential of PVP as a function of pH variation was graphically-illustrated in Figure 4. Under alkaline condition similar to the reaction conditions, PVP with a negative charge was observed.

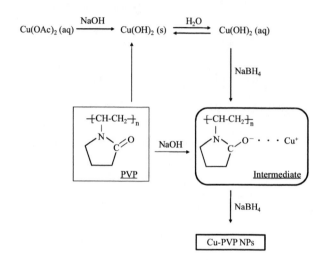

Scheme 1: Reaction pathway of Cu NPs formation

Scheme 1 illustrates the reaction pathway of Cu NPs formation using liquid chemical reaction. First, the precursor $Cu(OAc)_2$ reacted with NaOH (containing with $NaBH_4$ reductant), and reticulated $Cu(OH)_2$ was formed. This reaction occurs immediately (within a few minutes). Secondly, the slightly-soluble $Cu(OH)_2$ gradually reduced to Cu^0 by strong reductant of $NaBH_4$ through stable Cu^+ intermediate states. Finally, after consumption of almost all of $Cu(OH)_2$, the Cu^+ intermediates transformed to Cu^0 step by step. The PVP capping agent acts as a stabilizer to not only the synthesized Cu NPs but also intermediate such as $Cu(OH)_2$ and Cu^+ species. This proposed pathway agrees with the above experimental observations.

4 CONCLUSIONS

In summary, we succeeded to reveal the Cu NPs formation process under various amounts of PVP capping agents using the *in-situ* time-resolved XAFS observation in combination with several analytical techniques. It was observed that the life-times of $Cu(OH)_2$ and Cu^+ species were gradually extended with increasing the PVP capping agent. From investigation on the morphology changs of $Cu(OH)_2$ against the amount of PVP and zeta potential under alkaline condition, it is suggested that PVP capping agent can potentially coordinate with $Cu(OH)_2$ and Cu^+ species. These results indicated that the PVP capping agent plays an important role not only to stabilize the synthesized NPs, but also to contol the reaction kinetics by acting as the stabilizer of intermediates.

REFERENCES

[1] Y. Xia, Y. Xiong, B. Lim, S. E. Skrabalak, Angew. Chem. Int. Ed. 48, 60, 2009.

[2] B. Wiley, Y. Sun, B. Mayers, Y. Xia, Chem. Eur. J. 11, 454, 2005.

[3] J. Polte, T. T. Ahner, F. Delissen, S. Sokolox, F. Emmerling, A. F. Thunemann, R. Kraehnert, J. Am. Chem. Soc. 132, 1296, 2010.

[4] Y. Song, E. E. Doomes, J. Prindle, R. Tittsworth, J. Hormes, C. S. S. R. Kumar, J. Phys. Chem. B 109, 9330, 2005.

[5] M. Harada, Y. Inada, Langmuir 25, 6049, 2009.

[6] M. Harada, Y. Inada, M. Nomura, J. Colloid Interface Sci. 337, 427, 2009.

[7] S. Nishimura, A. Takagaki, S. Maenosono, K. Ebitani, Langmuir 26, 4473, 2010.

[8] H. Hirai, H. Wakabayashi, M. Komiyama, Bull. Chem. Soc. Jpn. 59, 357, 1986.

[9] T. Shido, A. Yamaguchi, Y. Inada, K. Asakura, M. Nomura, Y. Iwasawa, Top. Catal. 18, 53, 2002.

[10] J. Stotzel, D. Lutzenkirchen-Hecht, R. Frahm, B. Kimmerle, A. Baiker, M. Nachtegaal, M. J. Beier, J-D. Grunwaldt, J. Phys. Conf. ser. 190, 012153, 2009.

[11] G. Silversmit, H. Poleman, V. Balcaen, P. M. Heynderickx, M. Olea, S. Nikitenko, W. Bras, P. F. Smet, D. Poelman, R. D. Gryse, M-F. Reniers, G. B. Marin, J. Phys. Chem. Solids 70, 1274, 2009.

[12] Y. H. Luo, J. Huang, J. Jin, X. Peng, W. Schmitt, I. Ichinose, Chem. Mater. 18, 1795, 2006.

[13] W. Wang, C. Lan, Y. Li, K. Hong, G. Wang, Chem. Phys. Lett. 366, 220, 2002

[14] X. Wen, W. Zhang, S. Yang, Z. R. Dai, Z. L. Wang, Nano Lett. 2, 1397, 2002.

[15] R. Sardar, J. W. Park, J. S. Shumaker-Parry, Langmuir 23, 11883, 2007.

[16] J. Ohyama, Y. Hitomi, Y. Higuchi, T. Tanaka, Top. Catal. 52, 852, 2009.

[17] S. Giuffride, L. L. Costanzo, G. Ventimiglia, C. Bongiorno, J. Nanopart Res. 10, 1183, 2008.

[18] S. Kapoor, T. Mukherjee, Chem. Phys. Lett. 370, 83, 2003.

[19] P. K. Smith, R. I. Krohn, G. T. Hermanson, A. K. Mallia, F. H. Gartner, M. D. Provenzano, E. K. Fujimoto, N. M. Goeke, B. J. Olson, D. C. Klenk, Anal. Biochem. 150, 76, 1985.

[20] H. D. Hill, J. G. Straka, Anal. Biochem. 170, 203, 1988.

[21] K. J. Wiechelman, R. D. Braun, J. D. Fizpatrick, Anal. Biochem. 175, 231, 1988.

Hydrothermal synthesis and photoluminescent characterization of nanocrystalline BaZrO3

L.A. Diaz-Torres*,[a], P.Salas [b], V.M. Castaño[b], J. Oliva [a], E. De la Rosa [a],

a Centro de Investigaciones en Optica A.C., Leon, Gto. 37150 Mexico.

b Centro de Física Aplicada y Tecnología Avanzada, Universidad Nacional Autónoma de México, A. P. 1-1010, Querétaro 76000, Mexico.

* corresponding author: ditlacio@cio.mx.

ABSTRACT

Uniform well faceted sub-microcrystalline BaZrO$_3$ particles have been obtained by a facile hydrothermal reaction. It is found that both morphology and photoluminescence properties depend on reaction time, complexation surfactant, solvent water to ethanol ratio, as well as the coprecipitation agent. Band gap estimations suggest that blue-green emission intensity, under ultraviolet radiation excitation, depends on the degree of crystal lattice disorder. Such disorder degree on turn depends strongly on the coprecipitating agent and post hydrothermal processing annealing time. Morphology depends on coprecipitant agent molarities and reaction times.

Keywords. nanocrystals, Barium zirconate, Hydrothermal method.

1 INTRODUCTION

Since the phenomenon of high temperature protonic conduction in strontium ceramic was discovered by Iwahara et al. in the early 1980s [1], acceptor-doped A^{2+}B^{4+}O3 oxides with the perovskite-type structure have been paid much attention because of their potential applications in solid oxide fuel cells, hydrogen sensors, steam electrolysis and catalyst [2]. Recently, more and more investigations were focused on their luminescent properties [3]. Perovskite-type oxides phosphors are very stable and can steadily work in various environments, which is a merit for application. Moreover, Perovskite-type oxides phosphors have been found to be potential candidate in field emission display (FED) and plasma display panel (PDP) devices because they are sufficiently conductive to release electric charges stored on the phosphor particle surfaces [3]. Thus a great deal of perovskite-type oxides phosphors, such as A^{2+}B^{4+}O3 (A = Ca, Sr, Ba; B=Ti, Zr, Si, Hf, etc.) [4–6] have been prepared and their luminescent properties were also investigated carefully. Among these, BaZrO3 (BZO) has found interest in a number of applications including as a refractory material, as a structural material in ceramic container crucibles, as components of thermal barrier coatings, as a substrate in the synthesis of superconductors, as pinning centers in superconducting cables, as gate insulator materials, and as high-temperature microwave dielectrics [7-9]. It has been experimentally challenging though to generate uniform, monodisperse particles with reproducible reliability along with simultaneous control over product phase, size, and morphology [10-14]. It has even been difficult to eliminate impurities associated with the products [15]. In this manuscript, we report on the successful preparation of reasonably pure, crystalline, single-phase BaZrO3 submicron regular particles through a simple hydrothermal synthesis method using a alkali salts as precipitants. It is also presented the dependence of morphology and photoluminescence properties on synthesis parameters such as coprecipitant, alkali precipitant concentration, reaction time, and alcohol to water solvent ratio.

2 EXPERIMENTAL CHARACTERIZATIONS

BZO nanoparticles were obtained by hydrothermal process. All chemicals were analytical grade and were used as received. Barium nitrate (Ba(NO3)2), zirconyl chloride octahydrate (ZrOCl2 8H2O), were used as the starting materials for barium zirconate nanoparticles and sodium hydroxide (NaOH) or potassium hydroxide (KOH) were used as the precipitating agents. Cetyl-trimethylammonium-bromide (CTAB) or Pluronic 123 (P123) were used as the surfactants. In a typical procedure, barium nitrate (4.53 g), zirconyl chloride (5.58 g) and CTAB (1.9 g) were dissolved in a solution of ethanol-water at room temperature applying vigorous stirring for 30 min. Under strong stirring, sodium hydroxide was added and stirred again for 1 h at room temperature. The hydrothermal reactions were carried out in a Teflon autoclave (500 ml in total capacity) under autogenous pressure at 100°C for different reaction times, Rt = 2, 6, 12, and 24 h. The precipitate was then washed with distilled water and dried in an oven at 100 °C for 15 h. various water to ethanol solvent ratios, SR, were used with a fixed water volume of 100 ml. Also CTAB was substituted by Pluronic 123, in both surfactant cases the surfactant to BZO molar ratio was 0.3:1; the precipitants (NaOH or KOH) were used with concentration of 1M. Table 1 resumes the different synthesis parameters and the corresponding sample labeling.

The X-ray diffraction patterns of annealed samples were measured in a θ- θ Bruker D-8 Advance diffractometer having the Bragg – Brentano geometry, Cu Kα radiation, a Ni 0.5% Cu-Kβ filter in the secondary beam and a 1-dimensional position sensitive silicon strip detector (Bruker, Lynxeye). Diffraction intensity as a function of the angle 2θ was measured between 20 ° and 110º with a 2θ step of 0.01946º and a counting time of 53 s per point. All observed peaks were indexed in correspondence with JCPDS 6-0399 standard for cubic perovskite phase. The morphology of nanocrystals was investigated by Scanning electron Microscopy (SEM) with a JEOL XL30 microscope The optical absorption spectra were obtained with a Perkin-Elmer UV–VIS–NIR Lambda 900 spectrophotometer in diffuse reflectance mode using a 1.5 in integrating sphere (Labsphere Co.). The photoluminescence characterization measurements were performed with a conventional fluorescence setup, the excitation source was a 100W Xe lamp, and the excitation spectrograph, SP300i spectrograph (Acton Research), was setup so the excitation beam had a bandwidth of around 5 nm. The fluorescent emission from the sample was focused onto a SP-500i spectrograph (Acton Research) and detected by a photo multiplier tube R955 (Hamamatsu) connected to a SR830 DSP lock-in amplifier (Stanford Research). All photoluminescence measurements were done at room temperature, and a low band pass filter (cut off at 420 nm) was placed before the emission spectrograph to prevent spurious excitation and its harmonics to reach the detector.

3 RESULTS AND DISCUSSIONS

All synthesized BZO samples, but sample BZ18, have XRD patterns with main characteristic peaks that are in correspondence with JCPDS 6-0399 standard for pure perovskite cubic phase of BZO, see figure 1. The XRD pattern of sample BZ18 presents an amorphous envelope with three main peaks at 24.08°, 30.04°, and 34.34° that correspond to monoclinic zirconia and some other secondary peaks that might correspond to BZO. Samples BZ18, BZ12, and BZ11 represent the BZO crystalline structure evolution with increasing reaction time. From that it is observed that for reaction times of 2h or lower the formed nanoparticles are mainly amorphous precursos of BZO with a small amount of monoclinic zirconia. For reaction times 6h or lager the formed nanoparticles are fully crystallized in the cubic BZO phase, with a small segregation of monoclinic zirconia as is evidenced by the small peak still present at 24.08°. By comparing the BZ11 XRD pattern with the patterns of samples BZ15, BZ16, and BZ22 it is clear that the effects of substituting P123 by CTAB, or KOH by NaOH, or of the change in solvent ratio, are negligible as long as the reaction time is equal or larger than 6h.

SEM micrographs shown that the morphology of the secondary particles depends strongly on reaction time, see figure 2. For reaction times smaller that 2 h the amorphous

Table 1: Synthesis parameters

Sample	Surfactant	Solvent ratio [water : ETOH]	Precipit ant	Reaction Time [h]
BZ11	CTAB	4:1	NaOH	24
BZ12	CTAB	4:1	NaOH	6
BZ13	CTAB	4:2	NaOH	24
BZ15	CTAB	4:4	NaOH	24
BZ16	P123	4:1	NaOH	24
BZ17	CTAB	4:1	NaOH	12
BZ18	CTAB	4:1	NaOH	2
BZ20	CTAB	4:4	KOH	2
BZ21	CTAB	4:4	NaOH	2
BZ22	CTAB	4:4	KOH	24
BZ23	P123	4:4	KOH	24

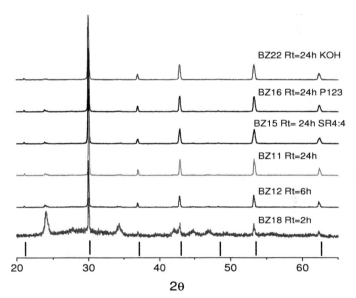

Figure 1. XRD patterns of different BaZrO3 samples as synthesized. See table 1 for the particular synthesis parameters. Bars at the bottom correspond to the JCPDS60399 standard for cibic BaZrO3.

particles are irregular in shape and present a huge size dispersion, see figure 2a. By increasing the reaction time up to 6 h, two main morphologies become present: long needles and spherical like particles, being the needles the dominant morphology. As reaction time increases up to 24 h, sample BZ11, the particles become regular well faceted truncated dodecahedra with an almost monodisperse size distribution around 1.5μm. However there are also many small irregular particles. When the amount of ETOH is increased, with respect to the one used for sample BZ11, to a water to ETOH solvent ratio of 4:4, regular well faceted particles are still observed in sample BZ15 (figure 2d).

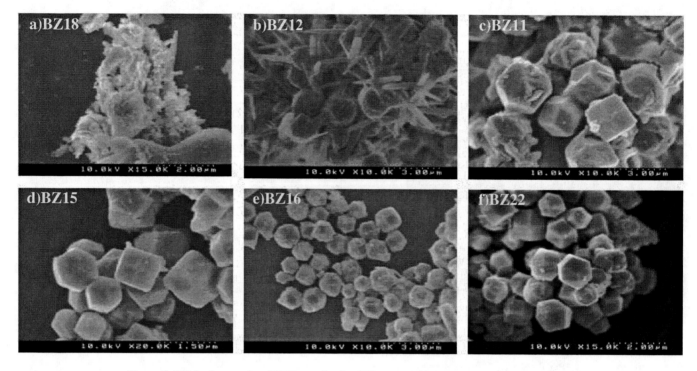

Figure 2. SEM micrographs of BZO samples for different synthesis parameters. See Table 1.

Particles in sample BZ15 are smaller (around 1μm) than the ones in BZ11, also less monodisperse, and a notable fact is that the sample looks cleaner, that is the amount of irregular smaller particles has been greatly reduced with respect to the BZ11 sample. The effect of substituting the surfactant CTAB by P123 results in the formation of smaller and rounder particles, almost sphere like with average sizes of 700 nm, see micrograph of sample BZ16 in figure 2f. On the other side, substitution of the precipitant NaOH by KOH results again in well faceted regular particles of around 600 nm, but in this case there are also a considerable amount of smaller irregular particles, in addition particles in sample BZ22 have huge size dispersion. In summary, the best synthesis conditions in order to obtain clean particles of regular morphology correspond to the ones used for sample BZ15.

Reflectance and wavelength data were converted to Kubelka-Munk [7] (k/s) units and energy in electron volts (eV), respectively. These data were combined in a Tauc plot [8] and the linear region was extrapolated in order to determinate the band gap of the samples in the energy axis, see figure 3. The estimated band gap values were are in the range between 3.907 eV for sample BZ18 and 4.679 eV for sample BZ23. These values are within the range already reported in the literature for pure BZO, from 3.8 to 4.8 eV [9]. The variations on band gap with increasing time suggest that as reaction time increases the crystallinity of the products increases, which is in correspondence with the evolution observed for the main diffraction peak in the XRD patterns in figure 1. The increasing crystallinity might be correlated to a reduction of disorder (increasing band gap) due to the reduction of oxygen vacancies [9]. This also might be correlated with the reduction of the particle sizes as reaction time increases, and the aggregation of smaller irregular particles the larger well faceted particles as observed in figure 2. The estimated band gap values are wide enough to allow transparency for visible and infrared photonic applications.

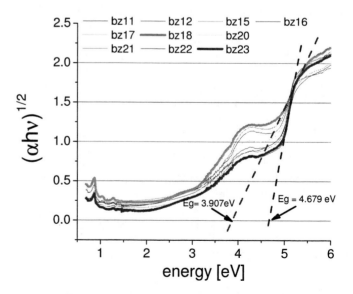

Figure 3. Absorbance spectra of BZO samples in Kubelka-Munk units.

PL spectra have been used widely to examine the efficiency of charge transfer, migration and to understand the behavior of the electron-hole pairs in semiconductors. The BZO samples have a UV to visible broad-band emission (120 nm half width) with to main peaks at about 504 and 552 nm, see figure 4. The excitation spectrum within the region of 220–400 nm has two maximums at about 267 nm and 380 nm, see figure 4a. Their luminescence is mainly caused by the charge transfer transition between the ligand (O^{2-}) and the metal (Zr^{4+}), which belongs to an intrinsic luminescence caused by annihilation of a self-trapped exaction (STE). The excitation peak at 267 nm is caused by the ZrO6–Vo defect center, which yields the emission similar to that of ZrO6 [2, 10]. The observed Stokes blue shift in the emission of BZO respect to previous reported values around 584 nm [10] implies a strong electron-lattice interaction and excitation relaxation in the semiconductor that is related to de increasing degree of order within the crystal lattice. On the other side, the excitation peak at 380 nm might be related to the existence of structural defects within the zirconium octahedral sites, which are an indication of the degree of order-disorder within the BZO crystallites [10]. This might be in agreement with the fact that the PL emission under excitation at 380 nm does not contribute greatly to the 552 nm emission peak.

4 CONCLUSIONS

In summary, we have obtained BZO submicron crystals with well faceted particle morphologies by means of a simple hydrothermal method. The PL properties indicate that in spite of having good level order there are many structural defects that reduce the band gap and promote luminescence centers localized at an excitation energy of 380 nm. It is found that the better synthesis conditions correspond to sample BZ15: water to ethanol solvent ratio 4:4, surfactant CTAB with a molar ratio to BZO of 0.3:1, NaOH at 1M as coprecipitant, and a reaction time of 24 h for the hydrothermal reaction at 100°C under autogenous pressure.

References

[1] H. Iwahara, T. Esaka, H. Uchida, N. Maeda, Solid State Ionics 3/4 (1991) 359.
[2] H. Iwahara, Y. Asakura, K. Katahria, M. Tanaka, Solid State Ionics 168 (2004) 229.
[3] Y. Pan, Q. Su, H. Xu, T. Chen, W. Ge, C. Yang, M. Wu, J. Solid State Chem. 174 (2003) 69–73.
[4] Z. Lu, L. Chen, Y. Tang, Y. Li, J. Alloy Compd. 387 (2005) L1–L4.
[5] N.J. Cockroft, S.H. Lee, J.C. Wright, Phys. Rev. B 44 (1991) 4117.
[6] W.Y. Jia, W.L. Xu, I. Rivera, A. Perez, F. Fernandez, Solid State Commun. 126 (2003) 153.
[7] P. Kubelka and F. Munk, Z. Tech. Phys. (Leipzig) 12, (1931) 593.
[8] J. Tauc, R. Grigorovici and A. Vancu, Phys. Stat. Sol. 15, (1966) 627.
[9] Cavalcante, L. S.; Longo, V. M.; Zampieri, M.; Espinosa, J. W. M.; Pizani, P. S.; Sambrano, J. R.; Varela, J. A.; Longo, E.; Simones, M. L.; Paskocimas, C. A. J. Appl. Phy. 103, (2008) art No. 063527.
[10] M.L. Moreira, J. Andres, J.A. Varela, E. Longo, Cryst. Growth and Design 9, (2009) 833-839.

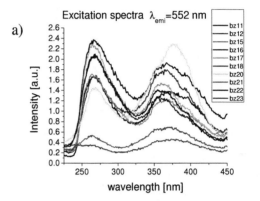

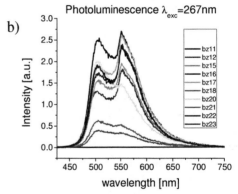

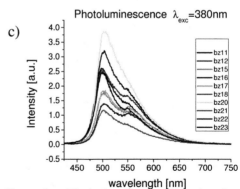

Figure 4. a)Excitation (λ_{em} = 552 nm) and emission spectra at b)λ_{exc} = 267 nm and c)λ_{exc} = 380 nm , of the BZO samples at room temperature.

Synthesis of Sulfated and Carbon-Coated Al₂O₃ Aerogels and Their Activity in Dehydrochlorination of 1-Chlorobutane and (2-Chloroethyl)ethyl Sulfide

A. F. Bedilo[*] A. M. Volodin[*] and K. J. Klabunde[**]

[*]Boreskov Institute of Catalysis, Novosibirsk 630090, Russia, abedilo@bk.ru
[**]Department of Chemistry, Kansas State University, Manhattan, KS, USA, kenjk@k-state.edu

ABSTRACT

The effect of gel modification with organic and inorganic acids, and β-diketones on the textural and adsorption properties of nanocrystalline alumina synthesized by the modified aerogel procedure was studied. The highest specific surface areas obtained exceeded 1200 m^2/g, being 4-5 times higher than those of typically used alumina materials. The activity of high surface area nanocrystalline alumina materials in dehydrochlorination of (2-chloroethyl)ethyl sulfide (2-CEES) and 1-chlorobutane compared favorably with that of other nanocrystalline metal oxides. The temperature of 1-chlorobutane decomposition could be decreased by ca. 100°C in comparison with MgO aerogel extensively studied earlier while maintaining 98% selectivity to butenes. Sulfated alumina aerogels had high surface areas ca. 600 m^2/g after calcination at 600°C and showed excellent acidic properties typical for sulfated alumina combined with exceptionally high surface area.

Keywords: Al₂O₃, aerogel, dehydrochlorination, destructive sorption, sulfated alumina

1 INTRODUCION

Aerogels are materials obtained by a sol-gel process followed by supercritical drying of the latter. Under supercritical conditions, the liquid-vapor interface that produces the collapse of the initial gel framework during conventional drying is eliminated. The obtained materials have high surface areas and pore volumes.

Various nanocrystalline aerogel-prepared (AP) metal oxides prepared by the modified aerogel procedure are known to have high surface areas, small crystallite sizes, unusual morphology, and enhanced adsorption properties and have been extensively studied for their use as destructive adsorbents [1-4] and decontamination of vegetative bacterial cells and spores [5] and various catalytic applications [6-9].

The use of these nanocrystalline metal oxides is limited under conditions where liquid water or water vapor is present due to their tendency to adsorb water, and thereby be partially deactivated toward adsorption of the target pollutants. We have earlier shown that coating of a destructive sorbent with an intelligent carbon layer significantly improves the stability of such nanocrystalline material in the presence of water and the possible time of its storage in air in comparison with not coated nanoscale oxide without a considerable loss of activity [4].

In the current publication we discuss the effect of various organic and acidic modifying agents on the textural properties of alumina aerogels and their performance in destructive adsorption of chlorinated hydrocarbons.

2 EXPERIMENTAL

For preparation of modified Al₂O₃ aerogels, aluminum isopropoxide (Aldrich) was dissolved in a desired amount of ethanol and/or other solvent. The modifying agents and hydrolysis water were dissolved in 10 ml of the same solvent each. Then, the solution of the modifying agent was quickly poured into the reaction vessel with aluminum isopropoxide solution and the resulting mixture was stirred for 10 min. Finally, a stoichiometric amount of hydrolysis water dissolved in same solvent was added. The reaction vessel was continuously maintained under nitrogen flow during the whole reaction. The gel formed after adding water was stirred overnight with a magnetic stir bar unless a solid gel was obtained.

The obtained gel was placed into an autoclave (Parr Instruments) equipped with a stirrer and heated in nitrogen atmosphere to 260°C with 1.3 °C/min heating rate. The final pressure and temperature were varied to study their effect on the properties of the obtained aerogels. After reaching the desired temperature the solvents were vented and the autoclave was purged with nitrogen for 20 min.

The resulting materials were calcined under vacuum at 500°C to obtain carbon-coated Al₂O₃ aerogels. The sulfated Al₂O₃ aerogels were calcined at 600°C in air.

Dehydrochlorination of 1-chlorobutane was performed in a flow reactor. 99% 1-chlorobutane (Aldrich) used in the experiments was introduced into the reactor by saturation of the argon flow with C₄H₉Cl vapor at room temperature. 10 μl injections of C₄H₉Cl were used. The volume flow rate was about 2 l/h, the catalyst loading was equal to 0.02 g. The product composition after the reactor was analyzed by gas chromatography. Prior to each experiment the catalyst was activated in an argon flow at 500°C for 1 h for removal of adsorbed water.

Experiments on decomposition of (2-chloroethyl)ethyl sulfide (2-CEES) were carried out in a 25 ml three-necked flask under argon with stirring. Ten milliliters pentane, 15 ml decane as internal standard, and 15 ml 2-CEES were placed in the flask. Then 200 mg of powder sorbent was added and the slurry stirred at a constant rate by means of a Teflon stir bar. Samples were extracted periodically and analyzed by gas chromatography using a Varian 3600 Gas Chromatograph equipped with a flame ionization detector.

3 RESULTS AND DISCUSSION

3.1 Synthesis of C/Al_2O_3 and SO_4^{2-}/Al_2O_3 aerogels

Organic groups used for modification of typical metal alkoxide precursors in the sol-gel process can serve two different purposes: i) to control the reaction rates of the reactants as well as the homogeneity and microstructure of derived gels being degraded during subsequent calcination to give purely inorganic materials; or ii) to modify or functionalize the oxide material. The latter approach could be applied for synthesis of oxide nanocrystals coated with carbonaceous structures. Nanostructures of this type can be created very effectively by controlled pyrolysis of organically modified aerogels. In this case organic groups are already located at the surface of the oxide, while their relatively homogeneous distribution can result in a higher number of well-distributed nucleation centers during the pyrolysis leading to smaller carbon particles.

Substituting standard isopropanol used a co-solvent with ethanol proved to increase the surface area of the resulting material (compare samples Al-11 and Al-22, Al-12 and Al-24 in Table 1). Therefore, the rest of samples were prepared using ethanol.

Unlike AP-MgO synthesis where the addition of toluene (or other aromatic solvent) has a major role and increases the resulting surface area by a factor of 2-2.5 [4], here in has a minor positive effect increasing the resulting surface area by 15-20% (compare samples Al-11 and Al-12, Al-22 and Al-24). Still, the highest specific surface areas exceeding 1200 m^2/g that are 4-5 times higher than those of typically used alumina materials were obtained for the samples prepared with an ethanol-toluene mixture used as the solvent.

Therefore, we have carefully investigated the possibility to go to truly supercritical conditions by either eliminating toluene or substituting it with fluorobenzene. In the first case, standard drying temperature of 265°C is above critical for ethanol. The critical temperature of fluorobenzene is 287°C. So, we can go above it if we increase the drying temperature to 295°C. Note that we cannot go above toluene critical temperature with our equipment.

Neither approach is perfect. Aluminum isopropoxide used as a starting material has only moderate solvability in ethanol or other alcohols. So, in this case, we have to deal with a solution containing a precipitate. Increasing the drying temperature generally has itself a negative effect on the surface area of the resulting material.

In order to prepare carbon-coated alumina, we tried to use MAA as a modifying agent. It has a β-diketonate fragment that provides strong binding to the surface, which

should survive hydrolysis and high-temperature drying and remain on the surface of dried aerogels. In addition, it has a terminal double bond that can be involved in polymerization reactions. By adding intermediate amounts of MAA we manage to deposit larger amounts of organic groups on the surface and improve the surface area of the material as well (Table 1).

Sample	Solvent/modifying agent	S.A. m^2/g	Pore volume, cc/g
Al-11	Isopropanol/toluene	1180	1.89
Al-22	Ethanol/toluene	1290	1.99
Al-12	Isopropanol	980	0.88
Al-24	Ethanol	1000	4.46
Al-25	Ethanol/0.5 mol/mol MAA	1090	2.80
Al-28	Ethanol/1.0 mol/mol MAA	1040	1.75
Al-27	Ethanol/1.5 mol/mol MAA	990	1.08
Al-29	Ethanol/2.0 mol/mol MAA	80	0.25
Al-33	Ethanol/fluorobenzene dried at 295°C	1040	2.56
Al-35	Ethanol/fluorobenzene/ 1.0 mol/mol MAA dried at 265°C	1280	0.75
Al-36	Ethanol/fluorobenzene/ 1.0 mol/mol MAA dried at 295°C	1130	2.31
Al-S51	Ethanol/15wt.% H_2SO_4	1050	1.58

Table 1. Textural characteristics of Al_2O_3 aerogels

We have earlier shown that resorcinol could be used as an efficient modifying agent for synthesis of carbon-coated MgO aerogels [4]. In the case of alumina the resorcinol addition resulted in a continuous decrease of the surface area and pore volume of the obtained aerogels (Fig. 1).

After activation under vacuum at 500°C the surface area of the alumina aerogels without the organic modifying agents was typically between 600 and 700 m^2/g. The surface areas of the carbon-coated aerogels was typically somewhat lower, about 400-500 m^2/g. HRTEM data (Fig. 2) clear shows that the carbon-coated are nanocrystalline with typical dimensions of the alumina nanoparticles synthesized by this method of several nanometers covered with few layer of graphitic carbon. XRD data did not show any clear peaks indicating that the aerogels even after calcination remained amorphous to x-rays.

Sulfated alumina aerogels were synthesized by a similar procedure with sulfuric acid used as a modifiying agent. Earlier we showed that this approach yields sulfated zirconia catalysts active in isomerization of n-butane [6]. The introduction of sulfuric acid to the solution before gelation results in a substantial decrease of the pore volume, whereas the high surface area of the Al_2O_3 aerogels is preserved. The surface areas of the sulfated alumina aerogels after calcination at 600°C usually required to make active acid catalysts was about 600 m^2/g. This value appears to the highest ever reported for sulfated alumina catalysts and is 2-3 times higher than those reported in the literature.

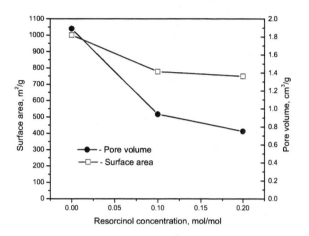

Figure 1. Effect of resorcinol on the textural parameters of modified Al_2O_3 aerogels.

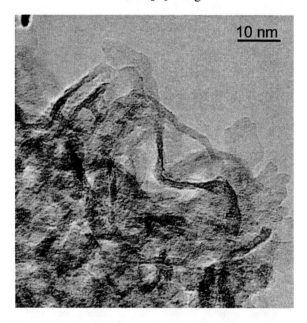

Figure 2. HRTEM image of modified alumina aerogel

3.2 Dehydrochlorination of 1-chlorobutane

The catalytic properties of the synthesized Al_2O_3 aerogels were compared in the dehydrochlorination of 1-chlorobutane. Earlier we have extensively studied this reaction for AP-MgO [8] and C/MgO aerogels [4]. We found that the activity of high surface area nanocrystalline alumina materials in dehydrochlorination of 1-chlorobutane

compared favorably with that of other nanocrystalline metal oxides. The temperature of 1-chlorobutane decomposition could be decreased by ca. 100°C in comparison with MgO aerogel (Figure 3) while maintaining 98% selectivity to butenes.

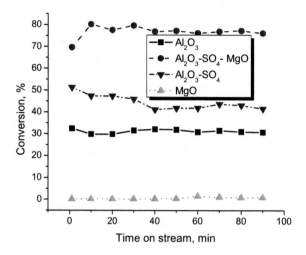

Figure 3. 1-Chlorobutane conversion at 250°C over aerogel nanocrystalline MgO, Al_2O_3, sulfated Al_2O_3 and mixed sulfated Al_2O_3 – MgO samples.

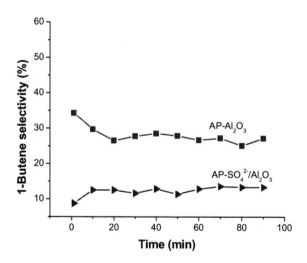

Figure 4. Selectivity to 1-butene during 1-chlorobutane dehydrochlorination at 250°C over alumina and sulfated alumina aerogels.

Apparently, the higher activity of alumina in this reaction is related to its higher acidity combined with high surface area. Sulfated alumina has even higher activity due to its higher acidity. In fact, even more active catalysts were prepared by cogelation of MgO and sulfated alumina in 1:1 ratio. The more acidic catalysts were characterized by lower selectivity to 1-butene (Fig. 4), indicating that the reaction follows E2 mechanism with easy abstraction of the chloride ion with the formation of the carbocation. The latter is than subjected to isomerization to form more stable secondary carbocations.

3.3 Destructive adsorption of (2-chloroethyl)ethyl sulfide

2-CEES is a mustard gas mimic commonly used to characterize the performance of materials developed for destruction of poisonous chlorinated compounds. We found that alumina and carbon-coated alumina aerogels chemisorbed 2-CEES at room temperature from pentane solution even more efficiently than MgO aerogels characterized in detail before [10] (Table 2).

Material	S.A., m^2/g	Conversion (molec./nm^2)	Reaction rate in first 5 min (mmol/g/min)
AP-MgO	660	0.65	0.020
C-MgO	370	1.02	0.037
AP-Al_2O_3	750	0.65	0.050
C/Al_2O_3	510	0.75	0.040
SO_4^{2-}-Al_2O_3	520	0.96	0.145
Act. carbon	1450	0.11	0.001

Table 2. 2-CEES conversion at room temperature over various sorbents.

Sulfated alumina aerogels showed even better performance in the destructive sorption. 2-CEES conversion rate in pentane solution at room temperature over sulfated aerogel alumina was 3 times higher than on Al_2O_3 aerogel and 8 times higher than on MgO aerogel. This reaction was exceptionally fast over all sulfated oxides, both aerogel and conventionally prepared, so almost maximum conversion was reached in the first 2 minutes. Furthermore, the high surface area of the developed sulfated alumina aerogels made it possible to achieve the absolute highest conversion. To prove that 2-CEES is destructively chemisorbed on the surface of sulfated alumina aerogels, we added excess water after the adsorption was complete. Almost no 2-CEES was extracted into the liquid phase, proving its efficient destruction.

The obtained results prove that, like 1-chlorobutane dehydrochlorination, 2-CEES destructive sorption is also more efficient over acidic materials than over basic ones. Most likely, these reactions follow the same E2 elimination mechanism initiated by chloride ion abstraction by very strong acid sites.

4 CONCLUSIONS

We have demonstrated that the textural parameters of alumina aerogels can be effectively controlled by selection of the solvents and modifying agents. Carbon-coated Al_2O_3 with permeable carbon coating can be prepared by adding modifying agents with bulky organic groups during gelation followed calcination under vacuum. The highest specific surface areas exceeding 1200 m^2/g that are 4-5 times higher than those of typically used alumina materials were obtained for the samples prepared with an ethanol-toluene mixture used as the solvent. Typical dimensions of the alumina nanoparticles synthesized by this method did not exceed few nanometers.

The activity of high surface area nanocrystalline alumina materials in dehydrochlorination of (2-chloroethyl)ethyl sulfide (2-CEES) and 1-chlorobutane compared favorably with that of other nanocrystalline metal oxides. For example, the temperature of 1-chlorobutane decomposition could be decreased by ca. 100°C in comparison with MgO aerogel while maintaining 98% selectivity to butenes. Sulfated alumina aerogels had high surface areas ca. 600 m^2/g after calcination at 600°C and showed excellent acidic properties typical for sulfated alumina. For instance, (2-chloroethyl)ethyl sulfide conversion rate in pentane solution at room temperature over sulfated aerogel alumina was 3 times higher than on Al_2O_3 aerogel and 8 times higher than on MgO aerogel.

Acknowledgment. This study was supported by the Russian Foundation for Basic Research (Grant 10-03-00806) and the US Civilian Research and Development Foundation (Grant RUE1-2893-NO-07).

REFERENCES

[1]. K. J. Klabunde, J. Stark, O. Koper, C. Mohs, D. G. Park, S. Decker, Y. Jiang, I. Lagadic and D. J. Zhang, J. Phys. Chem., 100, 12142, 1996.

[2]. A. Khaleel, P. N. Kapoor and K. J. Klabunde,. Nanostruct. Mater., 11, 459, 1999.

[3]. R. Richards, W. F. Li, S. Decker, C. Davidson, O. Koper, V. Zaikovski, A. Volodin, T. Rieker and K. J. Klabunde, J. Am. Chem. Soc., 122, 4921, 2000

[4]. A. F. Bedilo, M. J. Sigel, O. B. Koper, M. S. Melgunov and K. J. Klabunde, J. Mater. Chem., 12, 3599, 2002.

[5]. P. K. Stoimenov, R. L. Klinger, G. L. Marchin and K. J. Klabunde, Langmuir, 18, 6679, 2002.

[6]. A. F. Bedilo and K. J. Klabunde, Nanostruct. Mater., 8, 119, 1997.

[7]. V. V. Chesnokov, A. F. Bedilo, D. S. Heroux, I. V. Mishakov and K. J. Klabunde, J. Catal., 218, 438, 2003.

[8]. I. V. Mishakov, A. F. Bedilo, R. M. Richards, V. V. Chesnokov, A. M. Volodin, V. I. Zaikovskii, R. A. Buyanov and K. J. Klabunde, J. Catal., 206, 40, 2002.

[9]. I. V. Mishakov, A. A. Vedyagin, A. F. Bedilo, V. I. Zailovskii and K. J. Klabunde, Catal. Today, 144, 278, 2009.

[10]. R. M. Narske, K. J. Klabunde and S. Fultz, Langmuir, 18, 4819, 2002.

Sol-gel Synthesis of Ferrite Foam Materials for H$_2$ Generation from Water-Splitting Reaction

R. Bhosale, R. Shende* and J. Puszynski

Department of Chemical and Biological Engineering,
South Dakota School of Mines & Technology,
Rapid City, SD 57701 USA, Rajesh.Shende@sdsmt.edu

ABSTRACT

This paper reports the synthesis of Ni-ferrite and Sn-doped Ni-ferrite using sol-gel technique for H$_2$ generation from water-splitting reaction. In this synthesis, salts of Ni, Sn and Fe were sonicated in ethanol and propylene oxide was added to achieve the gel formation. In the case of Sn-doped Ni-ferrite, polymer microspheres were added prior to the gel formation. As-synthesized gels were aged, dried and then fired rapidly upto 1000°C in a resistive heating furnace or industrial scale microwave furnace and quenched in N$_2$ environment. The calcined powder was characterized using powder x-ray diffraction, scanning electron microscopy and BET surface area analyzer. Sol-gel derived Ni-ferrite was then used to investigate H$_2$ from water-splitting reaction at 700°C.

Keywords: Ni-ferrite, Sn-doped Ni-ferrite, sol-gel synthesis, ferrite foam, thermochemical water-splitting.

1 INTRODUCTION

Phase pure ferrites such as NiFe$_2$O$_4$ [1], ZnFe$_2$O$_4$ [2], MnFe$_2$O$_4$ [3] and doped ferrites, for instance, Ni$_x$Mn$_y$Fe$_2$O$_4$ [4], Zn$_x$Mn$_y$Fe$_2$O$_4$ [5] have application in H$_2$ generation from thermochemical water-splitting reaction. These ferrites are generally prepared by solid-state mixing, oxidation of aqueous metal hydroxide suspension, self-propagation high temperature synthesis (SHS) and aerosol-spray pyrolysis methods [1,4,6,7]. Although beneficial, there have been reports indicating that these methods sometimes result in chemical heterogeneity, non-stoichiometry, coarser particles and lower specific surface area (SSA). Sol-gel synthesis involving molecular or atomic level mixing and networking of chemical components significantly improves the chemical homogeneity, stoichiometry, phase purity and powder characteristics [8-11]. Because of these advantages, the sol-gel technique is an attractive synthesis approach available for the synthesis of different ferrite materials [12].

A two-step thermochemical water-splitting process seems to be thermodynamically more feasible as it may produce H$_2$ at significantly lower temperature. In the first step, the metal oxide is reduced at elevated temperatures by releasing O$_2$ according to the following reaction (1):

$$MO_{oxidized} \longrightarrow MO_{reduced} + \tfrac{1}{2} O_2 \qquad (1)$$

In the second step, the reduced metal oxide is oxidized at lower temperature by taking O$_2$ from water and producing H$_2$ via water-splitting reaction (2):

$$MO_{reduced} + H_2O \longrightarrow MO_{oxidized} + H_2 \qquad (2)$$

Reaction (1) is endothermic (activation/regeneration step) whereas reaction (2) is exothermic (water-splitting step). The combination of one activation/regeneration step and one water-splitting step can be termed as one thermochemical cycle. Typically, the water-splitting reaction requires lower temperatures as compared to that of regeneration step which takes place at relatively higher temperatures, making redox materials to undergo thermal cycling [13]. As a result, potential problems such as grain growth, thermal stress, and eventually spalling of redox material coating may occur in a thermal reactor [14]. This may further clog the reactor and the process lines, which may results in increasing the pressure drop and thereby creating unsafe processing conditions.

The above issues can be addressed if porous or foam ferrite materials are employed for thermochemical water-splitting reaction. These porous materials will have higher SSA which may provide higher number of active sites for water-splitting reactions resulting into higher yields of H$_2$. Also, the packed-bed reactors loaded with porous ferrites with different morphologies will provide lower pressure drop and minimize resistance for gas transport. Porous ferrite materials can be synthesized using a surfactant templating assisted sol-gel method. In this method the ionic/non-ionic surfactants can be added to the precursor solution prior the gel formation to achieve mesoporous structure with pore size in the range of 2-50 nm. To prepare the ferrite foam material, the precursor solution can be mixed with polymer microspheres (EXPANCEL® from Akzo Nobel) prior the gel formation. Expancel is a free flowing powder of spherical, expanded particles. The interior spherical void contains hydrocarbon gases and the shell is made up of a thermoplastic polymer. When gels containing Expancel microspheres are fired rapidly, foam-like materials with higher SSA can be realized.

Among several ferrites investigated so far, Ni-ferrite was reported to be the most promising material for H$_2$ production [14]. Fresno et al. [15] reported average H$_2$ production of 18.02 ml/g/cycle (water-splitting at 1000°C and regeneration at 1500°C) using Ni-ferrite, which was higher than the H$_2$ produced by other ferrites. Bhosale et al.

[16] reported that the sol-gel derived $NiFe_2O_4$ (calcined in air) was capable of producing a large volume of H_2 at lower water-splitting and regeneration temperatures of $700^{\circ}C$ and $900^{\circ}C$, respectively.

In this paper, Ni-ferrite and Sn-doped Ni-ferrite were synthesized using sol-gel technique. To prepare ferrite foam materials, gels were mixed with Expancel microspheres and heated rapidly upto $1000^{\circ}C$ in N_2 environment. The ferrite materials were used to investigate H_2 production from water-splitting reaction at $700^{\circ}C$.

2 EXPERIMENTAL

2.1 Synthesis of Ferrites

In the synthesis, chloride salts of Ni, Sn, and Fe were added in ethanol and sonicated to obtain a clear solution. To this solution, a predetermined amount of propylene oxide (PO) was added and the solution was set aside undisturbed at room temperature until a gel was observed. Gel formation was observed in 5–10 min. To prepare the ferrite foam material, the clear solution was mixed with polymer microspheres (20 wt% of the gel) prior to the gel formation. The gels were aged for 24 h at room temperature, dried at $100^{\circ}C$, heated rapidly upto $1000^{\circ}C$ in a resistive heating furnace or industrial scale microwave furnace (VIS 300-01 B, 5kW CPI Company) and finally quenched in N_2 environment.

2.2 Characterization of ferrites

The calcined powders were characterized using X-ray diffraction ($10^{\circ} \leq 2\theta \leq 80^{\circ}$ and scanning speed of 2° per minute), scanning electron microscopy (SEM) and BET surface area analyzer.

2.3 Hydrogen generation set-up

The H_2 generation ability of the sol-gel synthesized Ni-ferrite was investigated using the experimental set-up shown in Figure 1. It consists of a tubular packed bed Inconel reactor enclosed in a vertical split furnace (Carbolite Inc., USA). The temperature of the furnace was controlled and regulated precisely with the in-built PID controller. Accurately weighed 5.0 g of Ni-ferrite was packed in the middle section of the tubular reactor with the help of raschig rings and quartz wool. A horizontal tube furnace was provided with a stainless steel tube and it was used as a water vaporizing/preheating set-up. Distilled water from the reservoir was fed to the vaporizing/preheating furnace using a metering pump (Fluid Metering Inc., USA) at the flow rate of 1 ml/min. Water vaporized and the preheated steam obtained was fed to the packed bed reactor with N_2 as a carrier gas. A mass flow meter for N_2 (AALBORG Inc., USA) was mounted on the feed line to control the gas flow rate; a flow rate of about 100 ml/min was randomly selected. The exit gas stream from the packed bed reactor was first cooled down

in a water bath and finally fed to the moisture adsorption column containing anhydrous calcium sulfate purchased from W.A. Hammond Drierite Company Inc., OH, USA. The exit gas stream (mixture of N_2 and H_2) from the adsorption column was continuously monitored using an online H_2 sensor. The sensor provided H_2 concentration in vol% at an interval of 1 sec. The effluent gas stream was burned continuously. The water supply was shut down when the online H_2 sensor indicated a minimum of 0.95 (vol%) concentration level.

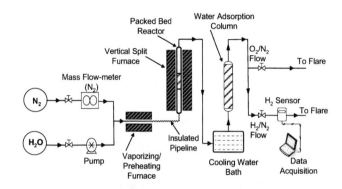

Figure 1: Thermochemical water-splitting reactor set-up for H_2 generation [14].

3 RESULTS AND DISCUSSION

Ni-ferrite gel was synthesized as per the procedure described in Section 2.1. The gels were dried and heated rapidly upto $1000^{\circ}C$ in a tube furnace under N_2 environment. The compositional purity of the calcined powder was determined using the powder x-ray diffraction. The XRD pattern presented in Figure 2a indicates a phase composition containing primarily $Ni_{0.4}Fe_{2.6}O_4$ mixed with metallic Ni. The SEM image of the calcined $Ni_{0.4}Fe_{2.6}O_4$ particles is shown in Figure 2b, which indicates faceted grains and broad particle size distribution as both smaller ($<1\mu m$) and larger ($>2\mu m$) grains were observed. The SSA of the calcined $Ni_{0.4}Fe_{2.6}O_4$ particles was 0.9 m^2/g.

The H_2 generation ability of $Ni_{0.4}Fe_{2.6}O_4$ was investigated by performing four thermochemical cycles (water-splitting at $700^{\circ}C$ and regeneration at $900^{\circ}C$) using the experimental set-up shown in Figure 1. The integrated H_2 volume profile presented in Figure 3 indicates that the amount of H_2 generated by $Ni_{0.4}Fe_{2.6}O_4$ in the 1st water-splitting step at $700^{\circ}C$ was 18.84 ml/g. The 2nd, 3rd and 4th water-splitting steps produced 6, 4.5 and 3.7 ml/g of H_2. It is believed that a major portion of H_2 was routed from water-splitting reaction by the $Ni_{0.4}Fe_{2.6}O_4$; however, it is also possible that Ni might have undergone redox transition as $Ni^0 \longrightarrow Ni^{2+}$ and contributed for additional H_2. A reverse transition of $Ni^{2+} \longrightarrow Ni^0$, however, may not be possible under the experimental condition employed here as higher water-splitting temperatures have been reported by other investigators for Ni/NiO [17] redox pair in solar thermal reactor for H_2 generation. Overall, an average of 8.37 ml of H_2/g/cycle was observed at the water-splitting temperature of 700°C. After performing four thermochemical cycles, the ferrite material was analyzed using the BET surface area analyzer. The SSA for the reacted material was 0.39 m²/g. Also, the reacted ferrite was analyzed using powdered X-ray diffraction. We observed a change in phase composition as well as reduction in metallic Ni content.

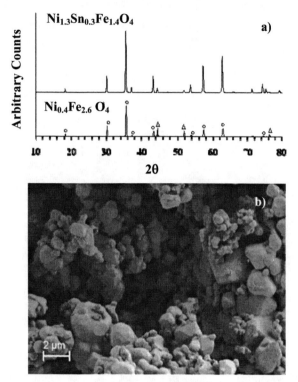

Figure 2: a) XRD patterns of $Ni_{0.4}Fe_{2.6}O_4$ and $Ni_{1.3}Sn_{0.3}Fe_{1.4}O_4$ calcined at 1000°C in N_2 environment and b) SEM image of calcined $Ni_{0.4}Fe_{2.6}O_4$ (SSA= 0.9 m²/g).

Doping of the ferrites with several metals should increase the O_2 vacancies as compared with the one without any dopant [12]. Using Ni and Mn doped-ferrites, the temperature needed for water-splitting was decreased to 627°C [18]. Thus, it appears that doping is effective in reducing the water-splitting temperature.

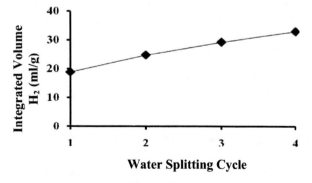

Figure 3: Integrated volume of H_2 generated via water-splitting reaction carried out using $Ni_{0.4}Fe_{2.6}O_4$ at 700°C, in four thermochemical cycles.

Abanades et. al. [19] reported that the complete hydrolysis of Sn is possible at < 550°C. It reflects that Sn could be used as a doping agent in Ni-ferrite, which may lower down the water-splitting temperature. Therefore, Sn-doped Ni-ferrite gel was synthesized using the synthesis method outlined in Section 2.1. The gel was dried at 100°C and rapidly fired upto 1000°C in N_2 environment. The powder x-ray diffraction pattern of the calcined powder as presented in Figure 2a, which represents a phase pure $Ni_{1.3}Sn_{0.3}Fe_{1.4}O_4$. The SEM image as shown in Figure 4 indicates faceted grains and broad particle size distribution. The BET surface area of calcined $Ni_{1.3}Sn_{0.3}Fe_{1.4}O_4$ was about 1 m²/g.

The calcined $Ni_{1.3}Sn_{0.3}Fe_{1.4}O_4$ was used to investigate the H_2 production from water-splitting reaction at 700°C. The preliminary results indicate that this doped ferrite has a great potential to produce H_2 from water-splitting reaction. The experiments are in progress to quantify H_2 production.

Figure 4: SEM image of $Ni_{1.3}Sn_{0.3}Fe_{1.4}O_4$ calcined at 1000°C in N_2 environment (SSA=1 m²/g).

It was observed that the calcination of ferrites in N_2 environment results in lower SSA. As mentioned earlier, higher SSA may be advantageous in water-splitting applications because it will provide more number of active sites for water-splitting reactions. If the ferrite is synthesized as a porous material, higher SSA can be achieved. Therefore, during synthesis of Sn-doped Ni-ferrite, about 20 wt% polymer microspheres were added in the precursor solution prior to the gel formation. A well mixed, aged and dried Sn-doped Ni-ferrite gel containing these microspheres was fired rapidly upto 1000°C in N_2 environment in a resistive heating furnace and industrial scale microwave furnace. The SEM images of the two calcined powders are presented in Figure 5a and 5b.

Figure 5a clearly indicates a foam-like morphology with pore size ranging from 50-500 nm. Critical evaluation of few samples under higher resolution showed the presence of the mesopores (2-50 nm) as well. Figure 5b shows fibrous foam-like morphology. After rigorous analysis, we observed that the few fibers in the foam have nano-tubular structure. The BET SSA of these ferrite foams was 156 m²/g, which is significantly higher than that of previously reported ferrites.

Figure 5: SEM images of Sn-doped Ni-ferrite calcined rapidly at 1000°C in N_2 in a) resistive heating furnace and b) industrial scale microwave furnace.

The investigations are in progress to examine the effectiveness of sol-gel synthesized doped ferrites and ferrite foam materials for H_2 generation using thermochemical water-splitting reactions.

4 CONCLUSIONS

Ni-ferrite and Sn-doped Ni-ferrite were successfully synthesized using sol-gel method involving preparation of a solution containing Ni, Sn and Fe precursors followed by gel formation using propylene oxide and subsequent calcination in N_2 environment. Sol-gel synthesized $Ni_{0.4}Fe_{2.6}O_4$ was used for H_2 production in four consecutive thermochemical cycles where water-splitting was performed at 700°C and the regeneration was carried at 900°C. In four consecutive thermochemical cycles, $Ni_{0.4}Fe_{2.6}O_4$ generated about 8.37 ml H_2/g per cycle. Sol-gel technique involving the addition of polymer microspheres in the Sn-doped Ni-ferrite gel produced foam-like or fibrous foam-like morphology. The SSA of this foam-like material was greater than 150 m^2/g, which is higher than that of the ferrite materials conventionally used for thermochemical water-splitting reaction. Due to such a higher SSA it is believed that the foam-like material will have tremendous potential for generating H_2 from thermochemical water-splitting reaction.

ACKNOWLEDGEMENT

The authors gratefully acknowledge the financial support by the National Science Foundation, Grant No. CBET-0756214.

REFERENCES

[1] M. Kojima, T. Sano, Y. Wada, T. Yamamoto, M. Tsuji and Y. Tamaura, J. Phys. Chem. Solids. 57, 1757, 1996.

[2] K. Go, S. Son and S. Kim, International Journal of Hydrogen Energy. 33, 5986, 2008.

[3] Y. Tamaura, Y. Ueda, J. Matsunami, N. Hasegawa, M. Nezuka, T. Sano and M. Tsuji, Solar Energy. 65, 55, 1999.

[4] Y. Tamaura, N. Kojima, T. Sano, Y. Ueda, N. Hasegawa, and M. Tsuji, International Journal of Hydrogen Energy. 23, 1185, 1998.

[5] M. Inoue, N. Hasegawa, R. Uehara, N. Gokon, H. Kaneko and Y. Tamaura, Solar Energy. 76, 309, 2004.

[6] Y. Tamaura, N. Hasegawa, M. Kojima, Y. Ueda, H. Amano and M. Tsuji. Energy, 23, 879, 1998.

[7] C. Agrafiotis, M. Roeb, L. Konstandopoulos, V. Zaspalis, C. Sattler, P. Stobbe and A. Steele, Solar Energy. 79, 409, 2005.

[8] D. Chen and X. He, Materials Research Bulletin. 36, 1369, 2001.

[9] X. Huang and Z. Chen, Journal of Magnetism and Magnetic Materials. 292, 79, 2005.

[10] S. Ebrahimi and J. Azadmanjiri, Journal of Nan-crystalline solids. 353, 802, 2007.

[11] M. George, A. John, S. Nair, P. Joy and M. Anantharaman, Journal of Magnetism and Magnetic Materials. 302, 190, 2006.

[12] R. Shende, J. Puszynski, M. Opoku and R. Bhosale, NSTI-Nanotech 2009. 1, 201, 2009.

[13] T. Sano, N. Hasegawa, M. Tsuji and Y. Tamaura, J. Mater. Chem..6, 605, 1996.

[14] R. Bhosale, R. Shende and J. Puszynski, Journal of Energy & Power Engineering. (In Press)

[15] F. Fresno, R. Fernandez-Saavedea, M. Gomez-Mancebo, A. Vidal, M. Sanchez, M. Rucandio, A. Quejido and M. Romero, International Journal of Hydrogen Energy. 34, 2918, 2009.

[16] R. Bhosale, R. Shende and J. Puszynski, AICHE conference proceedings. 2009.

[17] T. Rager, J. Golczewski, Z. Phys. Chem.. 219, 235, 2005.

[18] Y. Tamaura, M. Kojima, Y. Ueda, N. Hasegawa and M. Tsuji, International Journal of Hydrogen Energy. 23, 1185, 1998.

[19] S. Abanades, P. Charvin, F. Lemont and G. Flamant, International Journal of Hydrogen Energy. 33, 6021, 2008.

The Mechanochemical Formation of Functionalized Semiconductor Nanoparticles for Electronic and Superhydrophobic Surface Applications

B.S. Mitchell[*] and M.J. Fink[**]

[*]Department of Chemical and Biomolecular Engineering
Tulane University, New Orleans, LA, USA, brian@tulane.edu
[**]Department of Chemistry
Tulane University, New Orleans, LA, USA, fink@tulane.edu

ABSTRACT

A rapid, direct and efficient method for the fabrication of functionalized semiconductor nanoparticles is described. The mechanochemical method involves the simultaneous top-down formation of nanoparticles using high energy ball milling (HEBM) and reaction with a liquid medium to passivate (functionalize) the surface of nanoparticles as they are formed. In one embodiment of this process, elemental silicon is ball-milled in the presence of a reactive hydrocarbon, such as an n-alkyne. As the silicon fractures during the mechanical attrition, the formation of reactive surface species leads to the reaction with the surrounding organic medium and the creation of direct Si-C bonds. As the particles further fracture into the nano-regime and become sufficiently functionalized with organic molecules, they become soluble in the parent liquor and can be easily removed in a post-milling centrifugation step. Examples of how this method are being adapted to other semiconductors such as germanium, and other reactive organic compounds such as alkenes, alcohols, aldehydes and carboxylic acids are given. Furthermore, the adaptation of this technique to yield water-soluble analogues is elucidated. The importance of these functionalized semiconductor nanoparticles is illustrated through a brief description of the opto-electronic properties, including photoluminescence. Finally, the use of "waste" material from processing for the production of superhydrophobic films and surfaces is described. The importance of processing parameters and continuous operation to scale-up are discussed.

Keywords: mechanochemistry, nanoparticle, silicon, germanium, superhydrophobic.

1 INTRODUCTION

Fluorescent nanoparticles continue to garner tremendous interest as materials for use in optoelectronic devices [1-3] and as fluorescent biomarkers [4,5]. In semiconductor nanoparticles, such as silicon, their photoluminescent emission potentially arises from both excited states due to quantum confinement and from surface states. Although many routes to the synthesis of alkyl passivated silicon nanoparticles have been developed [6-17], most require the use of highly reactive and corrosive chemicals or the modification of nanoparticles with reactive silicon surfaces. The recent introduction of a new mechanochemical method for the production of passivated semiconductor nanoparticles [18-21] has shown this technique to be rapid, efficient and flexible. In this article, we give an overview of the types of semiconductor nanoparticles that can be produced via this method, the resulting surface functionalities, and their potential applications.

2 MATERIALS AND METHODS

2.1 Nanoparticle Passivation and Production

The production of semiconductor nanoparticles functionalized with alkyl substituents via a mechanochemical method has been described previously [18]. In brief, the semiconductor to be functionalized is placed in a high-energy ball mill (HEBM) vial along with milling balls and reactive organic compound, such as an alkyne. During the comminution process, which typically occurs over a 4-12 hour period, the semiconductor decreases in size due to particle fracture caused by impacts, and is functionalized through direct covalent bond formation between the semiconductor surface and the reactive organic compound. The process yields a solution of dissolved and functionalized nanoparticles and a "sediment" phase of larger and partially functionalized semiconductor material. These two phases are separated by centrifugation. The supernatant phase contains the nanoparticles, which can be isolated by solvent evaporation. These nanoparticles can then be dissolved in a variety of solvents for further characterization.

The process is flexible enough to accommodate a variety of metals, alloys, and semiconductors (both doped and undoped). It can also be used with a variety of functionalizing solvents such as alkynes, alkenes, aldehydes, carboxylic acids and alcohols [19]. The nanoparticles produced in this work are from either undoped silicon or germanium, and are functionalized with hydrocarbons of varying polarity.

2.2 Characterization Methods

Nuclear magnetic resonance spectroscopy (NMR) was conducted using a Bruker 300 Avance instrument. The static and dynamic contact angles were measured with a Standard Goniometer (ramé-hart Model 250) with DROPimage Advanced v2.4 and a droplet volume of 7.5 µl. The excitation-emission spectra and photoluminescence data from the nanoparticles were obtained using a Varian Cary Eclipse spectrofluorimeter. Particles were dissolved in heptane. Transmission electron microscopy (TEM) was performed using a JEOL 2011 TEM with an accelerating voltage of 200 kV.

3 RESULTS AND DISCUSSION

3.1 Alkyl-Passivated Silicon Nanoparticles

The first nanoparticles to be functionalized via this mechanochemical route were silicon nanoparticles passivated with 1-octyne [18]. These nanoparticles exhibited strong blue photoluminescence as a result of quantum confinement effects and the passivating layer. Detailed NMR studies showed that the 1-octyne was bound to the surface of silicon nanoparticles via two primary covalent bonding mechanisms: a single Si-C bond involving radical surface silicon species; and a bridging bond involving two Si-C bonds across surface double-bonded silicon species. This technique was extended to the passivation of silicon nanoparticles with octanol, octaldehye, octanoic acid, and octane [19]. These analogues also exhibited photoluminescence, but with varying intensities. The chain length of the passivating molecule can also be varied. Carbon chains of six to 12 units have been attached to silicon nanoparticles via this technique. All exhibit photoluminescence with characteristic Stokes shifts (see example of hexyne-capped Si nanopoarticles in Fig. 1).

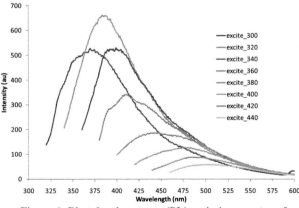

Figure 1 Photoluminescence (PL) emission spectra of fractionated hexyne-capped silicon nanoparticles at various excitation wavelengths.

The technique is now being extended to bi-functional alkyl chains. For example, milling in 1,7 octadiene results in not only a covalent bond with the silicon surface atoms, but also a terminal double bond (see Fig. 2) which can be utilized for further reactions with surface chain termini. Similar bifunctional molecules such as octadiyne, chlorooctyne, and valeric/pimelic acid can be utilized to produce nanoparticles capped with triples bonds, –Cl, and –COOH groups at the chain termini, respectively.

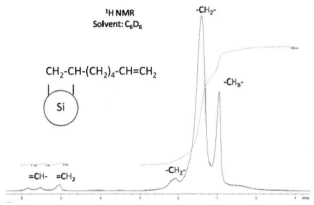

Figure 2 Proton NMR of silicon nanoparticles functionalized with 1,7 octadiene.

3.2 Alkenyl-Passivated Germanium Nanoparticles

Reactions analogous to those conducted with silicon can be carried out with germanium as the base semiconductor [21]. As with silicon, passivated nanoparticles in the ~5 nm regime are formed (see Fig. 3). There is a particle size distribution that results with this technique, but filtering and chromatographic techniques can be employed to narrow the size distribution.

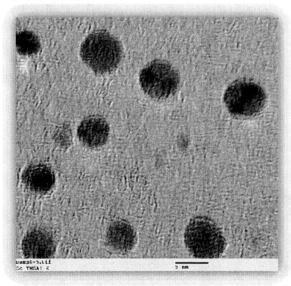

Figure 3 TEM photomicrographs of alkenyl-passivated germanium nanoparticles.

3.3 Water-Soluble Nanoparticles

Water-soluble semiconductor nanoparticles can be made by manipulating the passivating agent and its chain-end chemistry. There are several such chemical routes currently being studied. By utilizing the bi-functional passivating molecules referred to in section 3.1., water-soluble nanoparticles can be made through click chemistry approaches. In one such embodiment, chloro-terminated alkyl chains can be converted to primary amines via Gabriel [22] (see Fig. 4) or Williamson ether [23] synthesis routes.

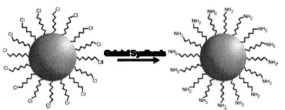

Figure 4 Synthetic route to formation of amine-capped silicon nanoparticles.

3.4 Superhydrophobic Films

Material that is not sufficiently functionalized or is too large to be solubilized in the surround reactive liquid can be separated via centrifugation and is referred to as "sediment." The sediment phase still contains a large quantity of passivated silicon, albeit larger in diameter with less surface coverage than the nanoparticles in solution. These functionalized particles are nevertheless useful for the production of superhydrophobic surface. Films of alkyl-passivated silicon nanoparticles are easily produced from the sediment phase by spreading a concentrated suspension on glass slides and evaporating the solvent. The resulting films have been shown to have high water contact angles (> 150°, see Fig. 5) [20]. Their contact angle hysteresis is still too high for superhydrophobic applications, and their mechanical integrity must be improved, but initial results are encouraging that the sediment phase can also lead to useful products.

Figure 5 Static contact angle of water on passivated silicon particle film.

4 CONCLUSIONS

A rapid, direct and efficient method for the fabrication of functionalized semiconductor nanoparticles is described. The method has been demonstrated for both silicon and germanium nanoparticles, with surface passivating species comprising alkynes, alkenes, alcohols, aldehydes, carboxylic acids of varying chain length (6-12 carbon units), and their bi- and multi-functional analogues. All nanoparticles produced via this technique exhibit some photoluminescence, and can be fractionated using filtration and chromatographic separation techniques.

5 ACKNOWLEDGMENTS

Financial support for this project was provided by a grant from the National Science Foundation (CMMI-0726943) and NASA (NNX08AP04A), administered by the Tulane Institute of Macromolecular Engineering and Science.. The contributions of Steffen Hallmann, Tapas Purkait, and Luigi Verdoni to this paper are gratefully acknowledged.

REFERENCES

[1] S. Pillai, K. Catchpole, T. Trupke, G. Zhang, J. Zhao, M. Green, Appl. Phys. Lett., 88, 161102, 2006.

[2] L. Raniero, S. Zhang, H. Aguas, I. Ferreira, R. Igreja, E. Fortunato, R. Martins, Thin Solid Films, 487, 170, 2005

[3] O. Boyraz, B. Jalili, Optics Express, 13, 796, 2005.

[4] D. R. Larson, W. R. Zipfel, R. M. Williams, S. W. Clark, M. P. Bruchez, F. W. Wise, W. W. Webb, Science, 300, 1434, 2003.

[5] L. Wang, V. Reipa, J. Blasic, Bioconjugate Chem. 15, 409, 2004.

[6] L.H. Lie et. al. Electroanal. Chem, 538, 183, 2002,

[7] M.H. Neyfeh et. al. Appl. Phys. Lett., 78, 1131, 2001.

[8] B. Sweryda-Krawiec et. al., J. Phys. Chem. B, 103, 9524, 1999.

[9] R.S. Carter, S.J. Harley, P.P. Power, M.P. Augustine Chem. Mat., 17, 2932, 2005,

[10] J. Zou, R.K. Baldwin, K.A. Pettigrew, S.M. Kauzlarich, Nano Letters, 4, 1181, 2004.

[11] R.D. Tilley, J.H. Warner, K. Yamamato, I. Matsui, H. Fujumori, Chem. Commun., 1833, 2005.

[12] K. Pettigrew, Q. Liu, P. Power, S. Kauzlarich. Chem. Mater., 15, 4005, 2003.

[13] D. Neiner, H.W. Chiu, S.M. Kauzlarich J. Am. Chem. Soc., 128, 11016, 2006.

[14] Z.F. Ding et. al. Science, 296, 1293, 2002.

[15] X.G. Li, Y.Q. He, M.T. Swihart,, Langmuir, 20, 4720, 2004.

[16] Z. F. Li, E. Ruckenstein Nano Lett. 4, 1463, 2004.

[17] J.D. Holmes, K.A. Ziegler, R.C. Doty, L.E. Pell, K.P. Johnston, B.A. Korgel J. Am. Chem. Soc., 123, 3743, 2001.

[18] A. Heintz, M. Fink and B.S. Mitchell, Adv. Mat., 19, 3984, 2007.

[19] A.S. Heintz, A.S., M.J. Fink, and B.S. Mitchell, App. Organometallic Chem., 24, 236, 2010.

[20] Hallmann, S., M.J. Fink, and B.S. Mitchell, J. Coll. Int. Sci., accepted, 2010.

[21] Purkait, T., B.S. Mitchell and M.J Fink, to be submitted, 2010.

[22] S. Gabriel; Ber. 20, 2224, 1887.

[23] A. Williamson, Phil. Mag., 37, 350, 1850.

Electrical Properties of Functionalized Silicon Nanoparticles

S. Hartner*, A. Gupta* and H. Wiggers**

*Institute for Combustion and Gasdynamics (IVG), University of Duisburg-Essen,
47048 Duisburg, Germany, sonja.hartner@uni-due.de, anoop.gupta@uni-due.de
**Institute for Combustion and Gasdynamics (IVG) and CeNIDE, University of Duisburg-Essen
47048 Duisburg, Germany, hartmut.wiggers@uni-due.de

ABSTRACT

Silicon nanoparticles are a promising, non-toxic material with respect to printed electronics and quantum size effects. As silicon nanoparticles tend to form a native oxide shell when handled in air, a stable surface functionalization is required to make them applicable for semiconductor purposes. The most favorable way is the stabilization by means of alkylation and hydrosilylation. Therefore, the natural oxide of silicon nanoparticles was removed by etching the particles with hydrofluoric acid. In order to prevent the particles from re-oxidation, the surface of freshly etched particles was functionalized with different alkenes via thermal alkylation. The electrical conductivity of as-prepared, freshly etched, and functionalized Si-NPs was measured using impedance spectroscopy in the temperature range between 323 K and 673 K. The electrical properties of silicon nanoparticles stabilized with n-alkenes ranging from C_6 to C_{18} are investigated.

Keywords: silicon nanoparticles, functionalization, electrical conductivity

1 Introduction

The utilization of silicon nanoparticles (Si-NPs) for printable electronics like solar cells, sensor devices, light emitters and transistors requires highly stable, (semi)conducting materials. While bulk silicon is one of the key materials for microelectronics, it is not applicable for printed devices. Additionally, it is an indirect band gap semiconductor [1] which limits its utilization for optoelectronic applications. Decreasing the size of the particles below 10 nm results in an increased optical activity with luminescence properties approaching the visible [2]. Nevertheless, as-prepared silicon nanoparticles suffer from poor resistivity against oxidation resulting in poor optical and electronic properties. By removing the oxide layer with hydrofluoric acid, light emission as well as electrical conductivity can be improved but when leaving them in air, the oxide layer appears again limiting conductivity and photoluminescence [3,4]. To avoid the re-oxidation of freshly etched silicon nanoparticles, they must be covered with a stable shell e.g. consisting of organic molecules [5].

In this paper, we report on the electrical properties of silicon nanoparticles. A comparison between as-prepared, freshly etched hydrogen terminated, and silicon nanoparticles functionalized with different n-alkenes is given.

2 EXPERIMENTAL DETAIL

Phosphorous-doped as well as undoped Si-NPs with a mean particle diameter of about 50 nm were synthesized in a microwave plasma reactor by thermal decomposition of silane in the presence of hydrogen, argon and phosphine. The etching of Si-NPs was carried out using hydrofluoric acid, which removes the surface oxides and terminates the silicon surface with hydrogen. The quantitative removal of oxygen was verified by FT-IR spectroscopy. In order to prevent the particles from re-oxidation, the surface of freshly etched particles was functionalized with different alkenes (hexene (C_6), decene (C_{10}), dodecene (C_{12}), teradecene (C_{14}) and octadecene (C_{18})) via thermal alkylation as described [5].

Transmission electron microscopy (TEM) was performed to observe any change in size or morphology. It is found that – despite the removal of the native oxide shell – neither particle size nor particle morphology has changed significantly.

As-prepared, hydrogen-terminated and n-alkene-terminated particle powder were pressed into small discs with 5 mm in diameter applying a force of 1.02 GPa (see figure 1a and 1b).

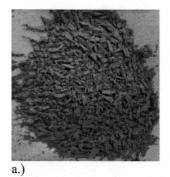 

a.)　　　　　b.)

Figure 1a: Dried Silicon nanoparticle powder after synthesis and functionalization;
b: Mechanically stable disc for electrical measurements prepared from silicon nanopowder.

The mechanically stable discs were placed between two polished platinum electrodes inside the measurement cell for the electrical measurements (see figure 2).

The electrical conductivity of as-prepared, freshly etched, and functionalized Si-NPs was measured in hydrogen atmosphere using impedance spectroscopy in the temperature range between 323 K and 673 K using an impedance spectrometer (HP4192A) in the frequency range between 10 Hz and 10 MHz with 20 frequency points per decade. For a stepwise measurement of the temperature-dependent conductivity, heating ramps with a step size of 25 K were applied and the impedance was measured at each temperature point after the temperature has stabilized. After reaching 673 K the sample was cooled to 323 K and the measurements were repeated two times. Due to annealing effects occurring during the first heating cycle, these measurements usually differ from cycle two and three which are almost identical. Therefore, the data from the third cycle were used for further analysis

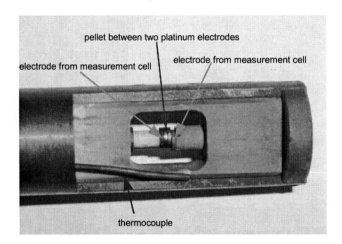

Figure 2: Compacted particles placed between two platinum electrodes of the measurement cell.

3   RESULTS AND DISCUSSION

3.1   Structural Properties

Transmission Electron Microscopy (TEM) images of as-prepared silicon nanoparticles show agglomerated and highly crystalline silicon particles with an oxide shell of about one to two nanometers in thickness. After etching and functionalization the particles are still crystalline but no oxide shell was observable and the agglomeration was broken resulting in single particles.

The XRD measurements analyzed with Rietveld refinement [6] show crystalline silicon nanoparticles with a diamond cubic structure. Before and after functionalization the crystallinity, the type of crystal lattice and the size of the crystals do not change. In figure 3, a Scanning Electron Microscopy (SEM) image of the surface of compacted sili-

con nanoparticles is shown illustrating the existence of porous silicon nanoparticle network.

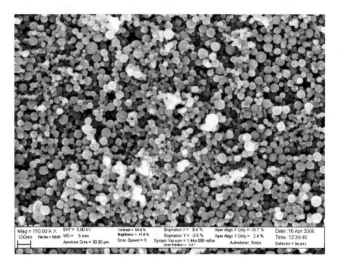

Figure 3: SEM picture of compacted silicon nanoparticles.

3.2   Electrical Properties

The conductivity of as-prepared silicon nanoparticles was measured in hydrogen atmosphere at different temperatures and compared to those, who were etched and functionalized. The resistivity of as-prepared Si-NPs decreases going to higher temperature which is typical for semiconducting materials and the activation energy is found to be in the range of 1.2 eV.

The freshly etched Si-NPs showed a very large increase in conductivity (about 4 orders of magnitude for measurements performed at 323 K) compared to the respective as-prepared samples (see fig. 4). With surface functionalization, the conductivity changes depending on the length of the alkenes used for stabilization.

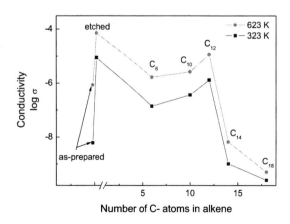

Figure 4: DC-conductivity of as-prepared, etched and functionalized silicon nanoparticles measured at 323 and 623 K.

Freshly etched as well as the functionalized samples show ntc conduction behavior as it is typical for semiconducting materials. Interestingly, conductivity increases from C_6 to C_{12} and is higher compared to the as-prepared materials. In contrast, samples functionalized with C_{14} and C_{18} show very poor conductivity. Particles functionalized with C_6 - C_{10} show a lower conductivity than C_{12} but still a better one than C_{14} and C_{18}. While FTIR-spectroscopy indicated that surface functionalization with C_6 - C_{10} is not very stable due to a creeping re-oxidation, dodecene (C_{12})-terminated nanoparticles showed the highest conductivity, even after storage in ambient conditions for half a year. We attribute the poor conductivity of C_{14} and C_{18} terminated Si-NPs to a large tunneling barrier caused by a thick organic layer which is expected to consist of well-aligned hydrocarbon chains perpendicular to the nanoparticle surface.

From our measurements we conclude that the conductivity of silicon nanoparticle ensembles is dominated by inter-particle transport rather than by the transport properties within the silicon core itsself. Therefore, engineering of the particle surface is the most important challenge for a useful application of silicon nanoparticles in printable electronics e.g. ink jet printed films.

REFERENCES

[1] C. Meier, A. Gondorf, S. Lüttjohann and A. Lorke "Silicon nanoparticles: Absorption, emission, and the nature of the electronic bandgap" J. Appl. Phys., 101, 103112, 2007

[2] L. T. Canham, "Silicon quantum wire array fabrication by electrochemical and chemical dissolution of wafers" Appl. Phys. Lett., 57, 1046, 1990

[3] A. Puzder, A. J. Williamson, J. C. Grossman and G. Galli, "Computational Studies of the Optical emission of silicon nanocrystals" J. Am. Chem. Soc., 125, 2786, 2003

[4] A. R. Stegner, R. N. Pereira, K. Klein, R. Lechner, R. Dietmueller, M. S. Brandt, M. Stutzmann and H. Wiggers, "Electronic Transport in Phosphorus-Doped Silicon Nanocrystal Networks" Phys. Rev. Lett., 100, 026803, 2008

[5] A. Gupta, M. T. Swihart and H. Wiggers, "Luminescent Colloidal Dispersion of Silicon Quantum Dots from Microwave Plasma Synthesis: Exploring the Photoluminescence Behavior Across the Visible Spectrum" Adv. Funct. Mater., 19, 696, 2009

[6] L. Lutterotti, M. Bortolotti, G. Ischia, I. Lonardelli and H. R. Wenk, "Rietveld texture analysis from diffraction images" Z. Kristallogr. Suppl., 26, 125, 2007

Precipitant molar concentration effect on the upconversion emission in BaZrO3:Er,Yb nanocrystalline phosphor

L.A. Diaz-Torres*,a, P.Salas b, V.M. Castaño b, J. Oliva a, C. Angeles-Chavez c .
a Centro de Investigaciones en Optica A.C., Leon, Gto. 37150 Mexico.
b Centro de Física Aplicada y Tecnología Avanzada, Universidad Nacional Autónoma de México, A. P. 1-1010, Querétaro 76000, Mexico.
C 4 Instituto Mexicano del Petróleo, A. P. 14-805, Cd. De México, D.F., 07730 México.

* corresponding author: ditlacio@cio.mx.

ABSTRACT

Strong green and red upconversion emissions of Er3+ in nanocrystalline BaZrO3:Yb,Er is observed. Powder samples were obtained by a facile hydrothermal process at 100°C under different NaOH molar concentrations and annealed in air at 1000°C for 24 h. Band gap and luminescence depend on precipitant molar concentration. Well faceted single phase BaZrO3 secondary particles were obtained for precipitant concentrations of 1.0 M whereas for other precipitant concentrations ZrO2 segregation was observed.

Keywords
Hydrothermal synthesis, barium zirconate, upconversion, Ytterbium, Erbium

1 INTRODUCTION

Upconversion (UPC) is the generation of visible or UV light by excitation with larger wavelengths, usually NIR, of trivalent rare-earth (RE) ions supported into a solid-state host. It is one of the nonlineal processes most widely investigated in the two last decades, as well as the trivalent RE ions in different host matrix in order to generate such optical process [1-3]. Practical UPC phosphors from ultra-violet to green spectral range have a wide range of applications as high-density optical storage, color displays, optical fiber communications, and recently as biological labels for medical imaging [4-6]. UPC of Er3+ doped host sensitized with Yb3+ has been widely studied in matrices such as GGG, NaYF4, ZrO2, LaF3, Y2O3 and YAG [7-11], and recently in BaZrO3 [12]. The excited state of the Yb3+ ion $(10,000cm^{-1})$ has a broad absorption band between 800 and 1100 nm and a much higher absorption cross-section than the 4I11/2 excited state of Er3+. The large spectral overlap between Yb3+ emission (2F5/2→2F7/2) and Er3+ absorption (4I15/2→4I11/2) results in an efficient resonant energy transfer (ET) from Yb3+ to Er3+ in Yb3+-Er3+-codoped systems. It is well known that in the UPC emission of RE ions, the highest phonon frequencies of the host lattice are responsible for nonradiative relaxation, resulting in a reduction of quantum efficiency. Then, matrices with low phonon energy are much more favorable in promoting the upconversion process. Although Fluorides have the best upconversion conversion efficiencies on account of very low phonon energies, they lack chemical stability. In most cases Fluorides are hydroscopic and that poses a serious drawback for biological applications and luminescent systems subject to humid environments. The oxides, on the contrary, are quite resistant to corrosive environments and have good optical properties. However, most perovskites have temperature phase transformations that lead to anisotropies in their refraction index, and this causes non linear optical phenomena such as second harmonic generation [1]. For some high energy upconversion applications this becomes a limiting factor since the upconversion generated heat can lead to such induced anisotropies limiting the range of pump power. Barium zirconate $(BaZrO_3)$ is a very promising refractory structural material with a very high melting point (2600 ºC) and a low chemical reactivity towards corrosive compounds like the ones used in the fabrication of high temperature superconductors. Among the cubic oxide perovskites, $BaZrO_3$ (BZO) is the only one that does not follow phase transitions over the range from 1600 K down to 4 K [13]. Recently, columnar inclusions of BZO nanocrystallites were found to enhance the transport currents in YBCO superconductors [14]. Besides, $BaZrO_3$ has attracted attention as a promising candidate for protonic electrolytes in Fuel cell applications [15]. In this paper, we report on the effects of the NaOH precipitant concentration, in the hydrothermal synthesis, on the morphology and upconversion emission of BZO of nanocrystalline particles. Green and red visible upconversion in Yb3+-Er3+ codoped BZO is explored under 978 nm excitation.

2 EXPERIMENTAL

Yb^{3+}–Er^{3+} codoped BZO microparticles were obtained by a facile hydrothermal process. RE content was fixed to 1 mol % Yb_2O_3 and 1 mol % Er_2O_3. All chemicals were analytical grade from Sigma Aldrich Inc. and were used as received. Barium nitrate ($Ba(NO_3)_2$), zirconyl chloride octahydrate ($ZrOCl_2$ $8H_2O$), erbium nitrate ($Er(NO_3)_3$ $5H_2O$) and ytterbium nitrate ($Yb(NO_3)_3$ $5H_2O$) were used as the starting materials for the Yb^{3+}–Er^{3+} co-doped BZO samples. Sodium hydroxide (NaOH) was used as the precipitating agent. Cetyl-trimethyl-ammonium-bromide (CTAB) was used as the surfactant. In a typical procedure, barium nitrate (4.53 g), zirconyl chloride (5.58 g), erbium nitrate (0.16 g), ytterbium nitrate (0.32 g) and CTAB (1.9 g) were dissolved in a 1:1 solution of ethanol-water at room temperature applying vigorous stirring for 1 h. Under strong stirring NaOH was added at three different molar concentrations: 0.1 M, 1.0 M, and 2.0 M. The hydrothermal reactions were carried out in a Teflon autoclave (500 ml in total capacity) under autogenous pressure at 100 °C for 24 h. The precipitate was then washed with distilled water and dried in an oven at 100 °C for 15 hr, and then annealed at 1000 °C for 3 h, with a heating rate of 3 °C/min.

The X-ray diffraction (XRD) patterns of as prepared and annealed samples were obtained using a SIEMENS D-500 equipment provided with a Cu tube with Kα radiation at 1.5405 Å, scanning in the 20° to 70° 2θ range with increments of 0.02° and a sweep time of 1 s. The morphology of the nanocrystalline particles was analyzed by scanning electron microscopy (SEM) using a Hitachi S-4500II operating at 25 kV. For absorption spectra, 7 mm diameter pellets were obtained by pressing the powder samples at 15 Ton pressure. Visible absorption spectra measurements were carried out in a Lambda 900 Perkin Elmer spectrophotometer in the reflectance mode with a 1 inch integrating sphere from Labsphere Co. The photoluminescence characterization was performed under 978 nm excitation form a tunable OPO system (MOPO from Spectra Physics) pumped by the third harmonic of a 10 ns Nd:YAG pulsed laser. The fluorescence emission was analyzed with a Acton Pro 500i spectrograph and a Hamamatsu photomultiplier tube R7400U-20 connected to a SR860 lock-in amplifier (Stanford Research). All photoluminescence measurements were done at room temperature. Special care was taken to maintain the alignment of the set up in order to compare the intensity of the upconversion signal produced by different samples.

3   RESULTS AND DISCUSSIONS

Secondary particles of regular shape and low dispersion in size were obtained for NaOH concentrations of 1 and 2 M, see SEM micrographs a) and b) in figure 1. Whereas for a NaOH concentration of 0.5 M the obtained particles were irregular in shape and with sizes ranging from nanoparticles to a few microns, see SEM micrograph c) in figure 1. These secondary particles are nanocrystalline aggregates of smaller nanoparticles as can be observed in figure 1b, and in figure 1S in supplementary information. The XRD patterns shown in figure 1 for the three NaOH precipitant concentrations suggest a good degree of crystallinity. For all samples the main peaks in the XRD patterns are in agreement with the JCPDS60399 standard

for cubic BaZrO3, marked with stars. However for the 0.5 and 2.0M NaOH there are other well resolved peaks that correspond to a monoclinic zirconia phase, marked by arrows in figure 1. This suggests deviations in NaOH precipitant molar concentrations from 1.0 M promote the segregation of zirconia phase. This is in agreement with the fact that pH control is a critical parameter in hydrothermal synthesis. Thus, a concentration of 1 M of NaOH leads to a estequiometric precipitation of Ba and Zr hydroxide precursors which after subsequent thermal treatments result in estequiometric BaZrO3. When the molar concentrations of NaOH are higher than 1 M a faster formation of zirconium oxide precursors leads to its segregation from the BZO phase. The same happens to a greater extent for NaOH molar concentrations lower than 1 M. In fact, for the 0.5 M NaOH concentration, such great non-estequiometric balance in Zr and Ba hydroxide precursors inhibits the aggregation of the well faceted secondary particles observed for 1.0 and 2.0 M NaOH. This might be due to the excess of ZrO_2 nanoparticles which form faster than the $BaZrO_2$ nanoparticles, and then during the annealing process the excess of ZrO_2 nanoparticles inhibits the coalescence of the BaZrO3 nanoparticles.

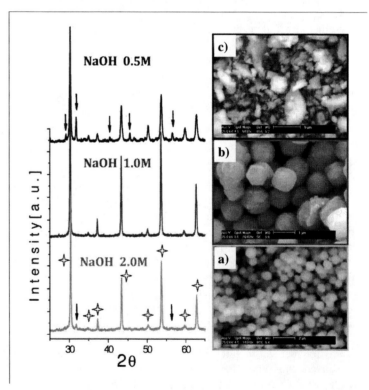

Figure 1. XRD patterns of BaZrO3 samples for different NaOH molar concentrations, and annealed at 1000°C. Bars in micrographs correspond to a) 5 μm, b) 1 μm, and c) 2 μm. as synthesized. Arrows correspond to main peaks of monoclinic zirconia. Stars correspond to the mail peaks of the JCPDS60399 standard for cubic BaZrO3.

Room Temperature (RT) absorption spectra for the three BZO samples annealed at 1000 °C are shown in figure 2. All bands in the spectra have been identified and correspond to the transitions from the ground state of Er (or Yb) ions to the final state of Er (or Yb) ions as indicated in figure 2. As expected, the main difference between these spectra is in the range between 800 to 1000 nm where Yb ions have its well known absorption band, of its only exited state $^4F_{5/2}$, superimposed on the $^4I_{11/2}$ absorption of the Er ions. The broad band centered at 1482 nm, that is superimposed with the absorption band of the 4I13/2 Er state, indicates the presence of residual OH anions that haves been kept occluded within the BZO crystallites. One might note that such band is quite strong in the sample with smaller NaOH molar concentration, which is also the one with an irregular morphology and a high particle size dispersion. Bandgap estimations, from the absorbance spectra in Kubelka-Munk units [16] has depicted in the inset of figure 2, were in the range between 3.75 and 4.57 eV in agreement with the reported values for BZO [17]. It is observed that the band gap decreases as NaOH molar concentration decreases. The band gap for the lower NaOH concentration is smaller than the expected value of 4.8 eV for well ordered bulk BZO [17], that suggest that this sample has a lot of defects, might be due to the high segregation of zirconia nanoparticles. The other two samples have bandgap values closer to the bulk value, and their lower bandgap are assumed to be a consequence of its nanocrystalline nature.

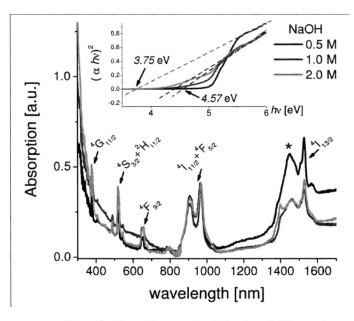

Figure 2. Absorption spectra of the three BZO samples annealed at 1000°C. Inset shows the absorbance Kubelka-Munk units.

Under NIR excitation at 978 nm upconverted Er3+ emissions in the green ($^2H_{11/2} + {}^4S_{3/2} \rightarrow {}^4I_{15/2}$) and red ($^4F_{9/2} \rightarrow {}^4I_{15/2}$) regions are observed for all synthesized samples, see figure 3. Both upconversion emission have similar structures, but different red to green intensity ratios. The higher red to green ratio corresponds to the 0.5 M NaOH sample which is the one with the greater segregation of

zirconia nanoparticles. The following red to green ratio corresponds to the 2.0 M NaOH that has less zirconia segregation than the 0.5 M NaOH sample, and the smaller red to green ratio corresponds to the 1.0 M NaOH sample that has no measurable zirconia segregation. This is in agreement with previous reports that reduction on nanocrystals size in Er-Yb codoped phosphors leads to an increasing red to green ratio [2,4,7]. This suggests that as NaOH concentration increases in our BZO samples the nanocrystalline secondary particles have smaller sizes, and that as NaOH concentrations reduces from 1.0 M the amount of zirconia nanocrystals increases as well.

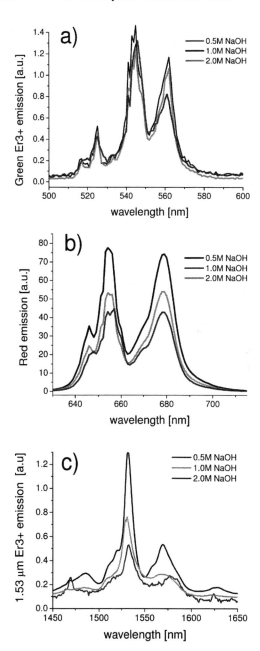

Figure 3. Emissions of Er3+ ions under 978 nm. a) and b) correspond to the Er3+ green and red upconversion emissions, respectively; and c) corresponds to the 1.53 mm emission from Er3+ ions.

Near infrared emission at 1.53μm from the $(^2I_{13/2} \rightarrow {}^4I_{15/2})$ transition of Er3+ ions is also observed for all samples, see figure 3c. It is observed that for the samples with zirconia segregation as NaOH molar concentration increases the 1.53μm emission decreases, in agreement with the fact that the overall amount of zirconia phase is decreasing. In addition the 1.53μm emission from the 1.0 M NaOH concentration is the weaker one since it has almost no zirconia content. The fact that the 1.53μm band structure for this latter sample is quite different from the band structure of the two former samples suggest that the Er3+ ions have been substituted in the Zr4+ sites within the BZO cubic phase. Besides, the lower level of emission supports the fact that the Er3+ ions are substituted in sites of lower symmetry [7]. In summary, segregation of zirconia might affect considerably the red and NIR emissions of BZO nanocrystalline particles, and such segregation strongly depends on the molar concentration of the precipitant used in the hydrothermal synthesis.

4 CONCLUSIONS

In summary, we have obtained BZO nanocrystalline secondary well faceted particles. Morphologies and phase composition as well as upconversion and NIR emissions depend strongly on the molar concentration of the precipitant used in the hydrothermal synthesis. Thus there is a compromise between morphology and phase versus overall emission intensities, both in the visible and near infrared regions. The best morphology corresponds to a NaOH molar concentration of 1.0 M, whereas the best emission properties correspond to a 0.5 M concentration of NaOH.

5 REFERENCES

[1] Boulon, G., Journal of Alloys and Compounds 451 (2008) 1–11.

[2] Chen, G. et al., *Nanotechnology* 20 (38), art. no. 385704.

[3] Auzel, F., *Chemical Reviews* 104 (1), pp. 139-173.

[4] Chen, G., et al., Phys. Rev. B 75, art. No. 195204.

[5] Jiang, C. et al., *Nanotechnology* 20 (15), art. no. 155101.

[6] Xu, C.T. et al., *Applied Physics Letters* 93 (17), art. no. 171103.

[7] Vetrone, F. et al., *Journal of Physical Chemistry B* 107 (39), pp. 10747-10752.

[8] Zhang, Q. et al, *Journal of Colloid and Interface Science* 336 (1), pp. 171-175.

[9] Salas, P. et al, *Optical Materials* 27 (7), pp. 1295-1300.

[10] Qu, Y. et al., *Materials Letters* 63 (15), pp. 1285-1288.

[11] Martinez A. et al., *Microelectronics Journal* 39 (3-4), pp. 551-555.

[12] Hinojosa, S. et al., *Optical Materials* 27 (12), pp. 1839-1844.

[13] P.G. Sundell, M.E. Björketun, G. Wahnström, Phys Rev B 73, 104112 (2006).

[14] S. Kang, A. Goyal, J. Li, A.A. Gapud, P.M. Martin, L. Heatherly, J.R. Thompson, D.K. Christen, F.A. List, M. Paranthaman, DF Lee, Science 311(31), 1911 (2006).

[15] A. S. Patnaik, A.V. Virka, J Electrochem Soc, 153(7), A1397 (2006).

[16] P. Kubelka and F. Munk, Z. Tech. Phys. (Leipzig) 12, (1931) 593.

[17] Cavalcante, L. S.; Longo, V. M.; Zampieri, M.; Espinosa, J. W. M.; Pizani, P. S.; Sambrano, J. R.; Varela, J. A.; Longo, E.; Simones, M. L.; Paskocimas, C. A. J. Appl. Phy. 103, (2008) art No. 063527.

[18]

[19]

Ageing Mechanism of Gold and Silver Nanorods in Aqueous Solution

D. Segets[*], C. Damm[*], B. Vieweg[**], G. Yang[**], E. Spiecker[**] and W. Peukert[*]

[*]Institute of Particle Technology, FAU Erlangen-Nuremberg
Cauerstr. 4, D-91058 Erlangen, Germany, D.Segets@lfg.uni-erlangen.de
[**]Department of Materials Science and Engineering, WW7, FAU Erlangen-Nuremberg
Cauerstr. 6, D-91058 Erlangen, Germany, Erdmann.Spiecker@ww.de

ABSTRACT

The current work addresses the understanding of the shape transformation of silver and gold nanorods towards more spherical nanostructures in aqueous suspension at ambient temperature. The shape transformation process was characterised using UV/Vis absorption spectroscopy, scanning electron microscopy (SEM), anodic stripping voltammetry and high resolution transmission electron microscopy (HRTEM). The increase in the particle number density during ageing indicates that the rods break spontaneously in smaller particles by surface defect induced ripening.

Keywords: silver nanorods, gold nanorods, optical characterization, ageing, shape stability

1 INTRODUCTION

Optical properties of noble metal nanoparticles including gold or silver strongly depend on particle size and shape.[1] Triggered by their interesting plasmonic properties, an increasing number of synthesis routes are leading to nanorods, nanowires and platelets, as it was recently reported in the literature.[2] These anisotropic particles typically show defective crystal stuctures.[3] Due to their face-centered cubic (fcc) crystal lattice, the equilibrium shape of noble metals usually is close to spherical. Thus, strongly anisotropic shapes may be thermodynamically unstable. Therefore, the long-term stabilization of rods and wires should be addressed in dependency of the related defects. In comparison to the well-known stabilization of spherical nanoparticles against agglomeration,[4] shape stability is a comparatively new field which is addressed in the following.

2 EXPERIMENTAL

2.1 Materials and nanoparticle preparation

The silver and gold nanorods were prepared by reduction of silver nitrate and chloro gold(III)-acid, respectively, in CTAB micelles in the presence of seed particles.[2c-e, 5] The glassware used for every particle synthesis was carefully rinsed with Millipore water prior to use. For the preparation of silver seed particles 0.25 ml of a 0.01 molar aqueous solution of silver nitrate were mixed with 8.9 ml of water. To this solution 0.25 ml of a 0.01 molar aqueous solution of trisodium citrate and 0.3 ml of a freshly prepared ice-cold 0.01 molar aqueous solution of sodium borohydride were added under vigorous stirring for 1 min. The seed particles have been stored at ambient temperature for 2 h prior to use. For the preparation of silver nanorods 0.25 ml of a 0.01 molar aqueous solution of silver nitrate were added to 10 ml of a 0.1 molar aqueous CTAB solution. To this solution 0.5 ml of a 0.1 molar aqueous solution of ascorbic acid and 0.25 ml of the silver seed suspension were added under gently shaking. Finally, 0.1 ml of a 1 molar sodium hydroxide solution was added to this solution resulting in the formation of sodium ascorbate acting as a reducing agent for silver ions. 10 min after addition of the sodium hydroxide solution the silver particles were purified by centrifugation (6.000 rpm, 30 minutes) and subsequent washing two times with Millipore water.

For the preparation of gold seed particles 0.2 ml of a 0.01 molar aqueous solution of chloro gold(III)-acid were mixed with 7.5 ml of a 0.1 molar aqueous CTAB solution. To this solution 0.6 ml of a freshly prepared ice-cold 0.01 molar aqueous solution of sodium borohydride were added under vigorous stirring for 1 min. The seed particles have been stored at ambient temperature for 2 h prior to use. For the preparation of gold nanorods 0.4 ml of 0.01 molar aqueous chloro gold(III)-acid were added to 9.5 ml of a 0.1 molar aqueous CTAB solution. To this solution 60 µl of a 0.01 molar aqueous silver nitrate solution and 64 µl of a 0.1 molar aqueous solution of ascorbic acid were added under gently shaking. After addition of the ascorbic acid solution the brownish solution became colourless because the Au(III) ions are reduced to Au(I). Finally, 8.5 µl of the gold seed suspension were added to this solution resulting in reduction of Au(I) ions to rod-like gold nanoparticles. The suspension of the gold nanorods was stored at ambient temperature for 1 h and purified by centrifugation (14.000 rpm, 10 min) and washing two times with Millipore water.

2.2 UV/Vis absorption spectroscopy

The extinction spectra of the suspensions of the silver nanorods were recorded in the range from 900 nm to 300 nm using a UV/Vis absorption spectrophotometer "Cary 100 Scan" (Varian). Disposable PMMA cuvettes (optical path: 1 cm) were used for absorption spectroscopy.

2.3 Anodic stripping voltammetry

A device consisting of a glassy carbon working electrode, a silver/silver chloride reference electrode and a platinum wire counter electrode supplied by Metrohm was used for anodic stripping voltammetry experiments. The electrolyte was a 0.1 M aqueous KNO_3 solution.

2.4 Electron microscopy

A scanning electron microscope (SEM) "Gemini Ultra 55" (Zeiss) was used to investigate the morphology of the silver nanoparticles. For the preparation of samples for SEM-investigation one droplet of the suspension of the nanoparticles was deposited onto a silicon wafer. The coated wafers were dried at ambient temperature.

Additionally, the shape transformation was investigated using a transmission electron microscope (TEM) "CM 300 LaB6/Ultratwin" (Philips). For the measurement 7 µl of suspension were deposited onto a carbon-coated copper grid and dried at ambient temperature.

3 RESULTS AND DISCUSSION

3.1 Stability of rod-like nanoparticles

Figure 1 shows the SEM micrographs of the silver nanorods immediately after synthesis (A), after 2 h (B), 24 h (C) and after 17 days (C) of ageing at 25 °C.

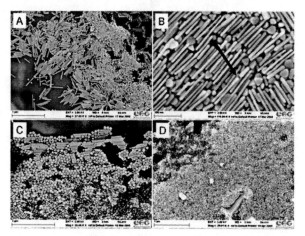

Figure 1: SEM micrographs of silver nanorods immediately after synthesis (A), after 2 h of ageing at 25 °C (B), after 24 h of ageing at 25 °C (C) and after 17 days of ageing at 25 °C (D), respectively.

It can be observed that the nanorods become shorter and significantly thicker with time and that the amount of spherical nanoparticles increases significantly. Moreover, as the optical properties of silver nanoparticles depend strongly on their size and shape, the ageing can also be observed by UV/Vis absorption spectroscopy. Regarding the extinction spectrum measured immediately after synthesis, two peaks are observed. One peak around 495 nm which refers to the longitudinal surface plasmon resonance (SPR) and one at 415 nm which refers to the transverse SPR of the initial rods.

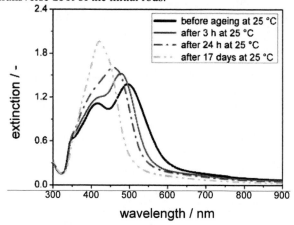

Figure 2: Extinction spectra of aqueous suspensions of rod-like silver particles as a function of ageing time at 25 °C.

During ageing, the shortening of the rods is reflected by the shift of the longitudinal SPR towards shorter wavelengths until after 17 days of ageing at 25 °C only one peak at 415 nm and a small shoulder which can be ascribed to the small and thick "nanobuns" marked in Figure 1D is obtained. For a more quantitative insight the aspect ratio distributions of the silver nanoparticles are derived from the SEM images. Before ageing a mean aspect ratio of about 14 is calculated whereas after 17 days of ageing the aspect ratio decreases down to 3. However, regarding the SEM micrographs in Figure 1A and 1D the intuitional assumption that one nanorod is leading to one spherical particle seems not to be true as the particles become smaller.

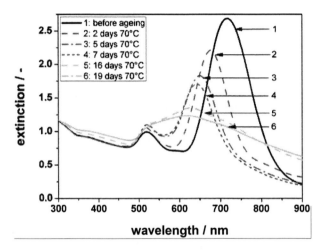

Figure 3: Extinction spectra of aqueous suspensions of rod-like gold particles as a function of ageing time at 70 °C.

In comparison, the gold nanorods show a much larger stability then the silver nanorods as the shape transformation process only starts at elevated temperatures, e.g. at 70 °C. However, a similar change in the absorption behaviour is observed as already seen from the silver nanoparticles. The peak at larger wavelengths of the original, bimodal absorption spectrum shifts towards smaller wavelengths and vanishes.

To obtain a more quantitative insight the suspensions were investigated by anodic stripping voltammetry. With this method dissolved silver ions can be detected down to a lower concentration of 5 µg/l. This corresponds to about 0.006% of the overall silver concentration. However, no silver ions could be detected in the liquid phase after 1 h, 3 h, 24 h or 17 days of ageing at 25 °C, respectively. Thus, the whole amount of silver must be situated in the solid phase and no permanent dissolution of solid can take place. To see how the particle number density changes during ageing, the number density distribution of the volume equivalent sphere is calculated (see Figure 4).

Accordingly, the following assumption has to be fulfilled.

$$m_{solid}^A = V_{solid}^A \cdot \rho_{Ag} \overset{!}{=} m_{solid}^B = V_{solid}^B \cdot \rho_{Ag} \qquad (1)$$

The index A refers to the suspension immediately after synthesis and the index B to the suspension after ageing.

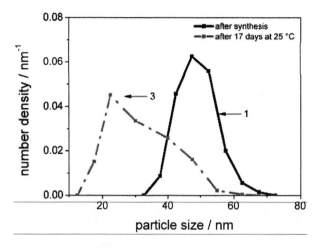

Figure 4: Number density distributions of the silver particles after synthesis (1, black line) and after 17 days (2, orange line) at 25 °C derived from the volume density distributions of the volume equivalent spheres.

Cancelling the density of silver ρ_{Ag} and replacing the overall particle volume by the corresponding number density distributions $q_0^A(x)$ and $q_0^B(x)$, multiplied by the absolute number of particles N^A and N^B and by the volume of each particle size fraction $\frac{\pi}{6}x_i^3$, the following equation 2 is derived.

$$\underbrace{\sum_i \left(x_i^3 \cdot q_0^A(x_i) \cdot \Delta x_i\right) \cdot N^A}_{M_{3,0}^A} \overset{!}{=} \underbrace{\sum_i \left(x_i^3 \cdot q_0^B(x_i) \cdot \Delta x_i\right) \cdot N^B}_{M_{3,0}^B}$$

$$(2)$$

The ratio of the particle number density before and after ageing goes reciprocal with the ratio of the third moment of their number density distributions. As the calculation reveals a moment ratio of around 1/3 the number density increases during ageing and the particles must decay.

3.2 Investigation of the nanorods by TEM

From the previous section it can be concluded that there is material transport from surfaces of higher energy to surfaces of lower energy. Different surface energies can be either an intrinsic phenomenon, e.g. regarding gold in the fcc crystal structure, the 111 surface possesses a slightly smaller surface energy than the 100 or the 110 surface, respectively.[6] On the other hand, e.g. regarding silver in the fcc crystal structure, all surfaces, 111, 100 and 110 have about the same intrinsic surface energies.[6] Different surface energies can only be obtained by different surface curvatures, i.e. by varying the particle diameter, or by defects at the particle surface. A careful analysis of the SEM-micrographs of the silver particles shows that during early stages of ripening the diameter of the rods becomes locally smaller (Fig. 1b, particles marked with an arrow). Also a TEM-micrograph of aged gold nanorods (Fig. 5, particles marked with an arrow) reveals that some of the particles got a shape like a "dog-bone".

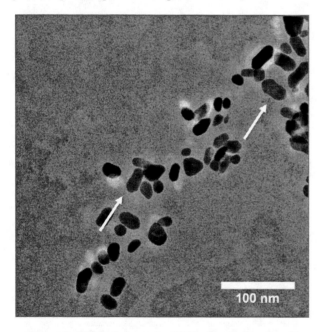

Figure 5: TEM micrograph of "dog-bone" shaped gold nanoparticles after 19 days of ripening at 70 °C.

Consequently, material must have been removed from other areas of the rod leading to a decrease of the rod diameter there. In general, during ripening the rod diameter becomes more and more inhomogeneous until it starts to decay. The calculation above even shows that the particle number increases during ripening. Thus, the rods break in smaller particles. This effect probably depends on the internal five-fold defect structure which leads to defects at the surface which in turn act as "nucleation" centres for fracture during ripening.

4 CONCLUSION

We have demonstrated that during the ageing of aqueous suspensions of rod-like silver and gold nanoparticles at 25 °C or 70 °C, respectively the aspect ratio decreases with time whereas the absolute particle number increases. This was evidenced by anodic stripping voltammetry and general calculations on the particle size distribution. Microscopic investigations showed the appearance of so-called dog-bone structures which are seen to be transition states during the particle decay. Regarding the increased interest in anisotropic nanoparticles during the past years also a control of their long-term stability will be one of the great challenges in nanotechnology.

5 ACKNOWLEDGEMENTS

The authors would like to thank the German Research Council (DFG) for their financial support (Projects PE427/18-2) which supports within the framework of its Excellence Initiative the Cluster of Excellence 'Engineering of Advanced Materials' (www.eam.uni-erlangen.de) at the University of Erlangen-Nuremberg. Furthermore, we thank Dr. Jochen Schmidt and Stefan Romeis for helpful discussion.

REFERENCES

[1] Gans, R., *Ann. Phys, 342*, 881, **1912**.
[2] (a) Huang, M. H.; Choudrey, A.; Yang, P., *Chem. Commun., 12*, 1063; (b) Nikoobakht **2000**; B.; El-Sayed, M. A., *Chem. Mater., 15*, 1957, **2003**; (c) Sau, T. K.; Murphy, C. J., *Langmuir, 20*, 6414, **2004**; (d) Murphy, C. J.; Gole, A. M.; Hunyadi, S. E.; Orendorff, C. J., *Inorg. Chem., 45*, 7544, **2006**; (e) Jana, N. R.; Gearheart, L.; Murphy, C. J., *J. Phys. Chem. B, 105*, 4065, **2001**.
[3] Xia, Y.; Xiong, Y.; Lim, B.; Skrabalak, S., *Angew. Chem. Int. Ed., 48*, 60, **2009**.
[4] (a) Reindl, A.; Peukert, W., *J. Coll. Int. Sci., 325*, 173-178, **2008**; (b) Marczak, R.; Segets, D.; Voigt, M.; Peukert, W., Optimum between purification and colloidal stability of ZnO nanoparticles. *Advanced Powder Technology, DOI: 10.1016/j.apt.2009.10.005*, **2009**.
[5] (a) Jana, N. R.; Gearheart, L.; Murphy, C. J., *Chem. Commun., 7*, 617, **2001**; (b) Murphy, C. J.; Sau, T. K.; Gole, A. M.; Orendorff, C. J., *MRS Bulletin, 30*, 349, **2005**; (c) Sau, T. K.; Murphy, C. J., *Langmuir, 21*, 2923, **2005**.
[6] Vitos, L.; Ruban, A. V.; Skriver, H. L.; Kollár, J., *Surface Science, 411*, 186-202, **1998**.

Durability of Pt/MWCNT Nanocatalyst in High Temperature H$_3$PO$_4$/PBI PEMFC

F. Buazar[*,**], K. Almoor[*], Q. Li[*], J.O. Jensen[*], N.J. Bjerrum[*] and T. Steenberg[***]

[*] Energy and Materials Science Group, Department of Chemistry, Building 207,
Technical University of Denmark, DK-2800 Lyngby, Denmark, foadboazar@yahoo.com
[**]Department of Marine Chemistry, Khoramshahr Marine Science and Technology University,
P.O. Box 669, Khoramshahr, Iran
[***] Danish Power Systems Aps, Raadhusvej 59, DK 2920 Charlottenlund, Denmark, ts@daposy.dk

ABSTRACT

The durability of Pt/MWCNT nanocatalyst was scrutinized by a series of constant life tests (11000 cycle operation) of single proton exchange membrane fuel cell (PEMFC) with phosphoric acid doped PBI (H$_3$PO$_4$/PBI) membranes at high temperature. The tested anode and cathode catalysts were analyzed by transmission electron microscopy (TEM) and X-ray diffraction (XRD) to determine the morphology degree of electrocatalyst agglomeration. The XRD results showed that the Pt nanoparticle (Nps) size increased from 2.2 to 6.3 and 9.4 nm before and after single cell life tests for anode and cathode, respectively. The energy dispersive X-ray established the platinum deposition on supporting material. Above results indicated that the agglomeration of Pt Nps should be responsible for the ESA loss of Pt/MWCNT electrocatalyst but not membrane degradation. The further analyses on TEM data suggested that growth of platinum particles occurred via non-spherical shape mechanism on carbon at the nanometer scale.

Keywords: Durability, High temperature PEM; Fuel cells, Pt/MWCNT, Catalyst

1 INTRODUCTION

Energy and environment are of global concern, which has stimulated worldwide research on efficient and clean energy technologies. Proton exchange membrane fuel cells (PEMFC), among other types of fuel cells, have attracted much attention since their unique and favorable advantages for applications in transportation, stationary, portable and micro-power systems. In general, the PEMFC performance is primarily limited by the slow kinetics of oxygen reduction at cathode. Since the late 1980s the PEMFC performance has been significantly increased from 0.3-0.5 mW/cm² to about 1.0-1.5 W/cm² today, whereas the noble metal loading has been decreased from 4-5 mg/cm² to 0.4-0.5 mg/cm² for cathode and <0.03 mg/cm² for anode [1].It is the carbon blacks that suffer from high corrosion, which in turn triggers the agglomeration of platinum particles resulting in reduction of the active surface area and catalytic activity. This is the major mechanism of the catalyst degradation and a key challenge to fuel cell long-term durability [1]. Improved stability of both supports and catalysts is crucial for the technology.

The newest development in the field of PEMFC aims at higher operational temperatures above 100°C, with new electrolytes e.g. acid doped polybenzimidazole (H$_3$PO$_4$/PBI) membranes, which have been the main research focus at the Energy and Materials Science Group of DTU [2]. Operating features of the HT-PEMFC include no humidification, high CO tolerance, better heat utilization and possible integration with fuel processing units, as reviewed by the group. Higher operational temperatures impose, however, more challenges for the catalysts and electrodes. First of all, the doping acid has strong adsorption on the surface of catalysts and hinders electrode reactions. As a result, more active catalysts are required to improve the performance. The high temperature obviously aggravates the carbon corrosion. So far, no catalysts with reasonable activity and sufficient long-term stability are available. The stability and durability of Pt/C electrocatalyst in mid-low temperature PEMFCs with Nafion membrane have been widely investigated [2]. However, the systematic study on the durability as well as degradation of Pt/MWCNT nanocatalyst in high temperature PEMFC with single fuel cell test has not been reported. In this study, constant life tests were conducted with a series of PBI/H$_3$PO$_4$ high temperature PEMFC single cells. Constant current performance curves were recorded and electrochemical method (CV) was applied to study the ESA loss of cathode catalysts during the life tests. The transmission electron microscopy (TEM) and X-ray diffractometer (XRD) were used to detect the increase in particle size of the cathode platinum after 3200 cycle operation. The results obtained after three time experiment replicating.

2 EXPERIMENTAL

2.1 MWCNT Pretreatment

Multiwalled carbon nanotubes (MWCNTs) with outside diameters of <8 nm, length: 0.5-2.0 um and purity of >95% were purchased from Cheap Tube Inc., in USA. As-received MWCNTs were pretreated to disentangle, purify, and decorate them. Therefore One portion of

MWCNTs was treated by HNO₃ solution; 1.0 g of MWCNTs was mixed with 100 mL of a 65% HNO₃ solution under constant stirring for 20 min and then refluxed at 140 °C for 4 h. The resulting MWCNT was rinsed thoroughly to neutral with deionized water and dried at 80 °C in a vacuum oven overnight.

2.2 Catalyst preparation

Modified precipitation method [3] was used to prepare Pt/MWCNT with high loading and good dispersion on the carbon surface. The carbon slurry was prepared by vigorous mixing of carbon nanotube with distilled water. The chloroplatinic acid solution, prepared by dissolving chloroplatinic acid ($H_2PtCl_6 \cdot 6H_2O$, Metalor® technologies (UK) Ltd) into the distilled water, was added to the carbon slurry. Then for efficient dispersion, acetic acid was added to the slurry and heated up to a desired temperature up 90°C. The pH of the slurry was adjusted to the basic using diethyl amine. An excess quantity of 40% aldehyde as reducing agent was then introduced into the slurry for *in situ* liquid-phase reduction. The Pt/MWCNT slurry was filtered, several time washed and then dried at 110°C in a vacuum oven.

2.3 Physical characterization

2.3.1 XRD Analysis

From Figure 1a it can be seen that there are obvious differences before and after loading platinum on the structures of supporting material (MWCNT). Without Pt, The structure of MWCNT is mostly amorphous carbon black, while after loading Pt; the pattern for Pt/ MWCNT exhibited four characteristic diffraction peaks, at °39.6 46.7°, 68.5° and 82.6° (2θ) degrees.

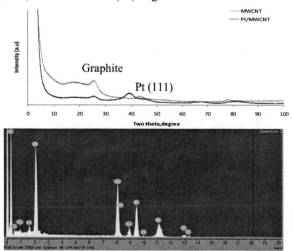

Figure 1: XRD patterns of pure MWCNT and Pt/MWCNT catalyst (a) and EDS image of Pt deposition on MWCNT used as support (b).

Crystallite size determination is performed by measuring the broadening of a particular peak in a diffraction pattern associated with a particular planar reflection from within the crystal unit cell. Using Scherrer equation, average Pt particle sizes of prepared electrocatalyst at wider Pt (111) peak is 2.2 nm. Likewise EDX analysis show that Pt successfully deposited on the carbon nanotube as support material (Figure 1b). These peaks can be attributed to the (111, crystalline facet), (200), (220) and (311) lattice planes of the Pt nanoparticles (Nps), respectively [3].

To obtain a comprehensive analysis, the Pt/ MWCNT catalyst on the electrode endured similar life test durations was characterized by XRD, and the results are shown in Figure 2. X-ray diffraction pattern of the sample was collected using a Huber G670 X-ray diffractometer (XRD) with copper rotating anode (CuK_radiation, λ = 1.54056 Å, at 40 kV/20 mA). A continuous scan rate of °5min⁻¹ from 0° to 90° of 2θ was used for all samples. After single cell series tests, the powder samples of Pt/MWCNT electrocatalyst was scraped from the surface of the anode and cathode sample by a razor blade. The characteristic diffraction peaks of the face centered cubic structure are clearly detected in all the electrode samples. The diffraction peaks at 2θ values of 39.846.3°, 67.5° and 81.4° are associated with the Pt (111), (200), (220) and (311) lattice planes, respectively (Figure 2). All the characteristic peaks of Pt in cathode and anode become sharper following the longer test time, which indicates that the Pt particle grew gradually during the lifetime test. It should be noted that the diffraction peaks of Pt in the time tested anode were broader than that tested cathodes, and its particle size had change (d= 6.3 nm) compared with before test (d= 2.2 nm), this suggests that the Pt particle growth in the cathode (d= 9.4 nm) was more serious than that in the anode [3].

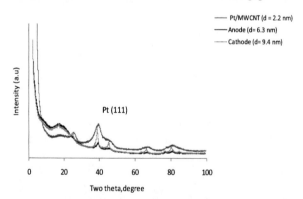

Figure 2: XRD patterns of Pt/MWCNT before and after 3200 cycles single cell test.

2.3.2 TEM analysis

Morphology and particles size distribution of Pt/MWCNT catalyst before and after single cell test was performed via High and low magnification transmission electron microscope (HMTEM and LMTEM) images were observed in a TECNAI T20 TEM equipped with a high-speed slow-scan CCD camera operated with a LaB6 filament at 200 kV. The sample deposited on holey amorphous carbon film on a Cu grid for TEM observation.

Figure 3a presents the TEM images of electrocatalyst before and after the life tests. Their particle size distributions for the electrocatalyst were obtained by manually measuring all the particles from the bright-field micrographs. Also, the diffraction rings in the selected area diffraction pattern (SAD) indicate that the Pt Nps are polycrystalline and randomly oriented, as shown in the inset of Figure 3b.

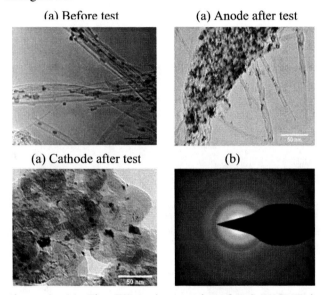

(a) Before test   (a) Anode after test

(a) Cathode after test   (b)

Figure 3: (a) The TEM photographs of Pt/MWCNT in anode and cathode before and after 3200 constant test, (b) the selected area diffraction pattern (SAD) of Pt NPs with the four rings indexed to the {111}, {200}, {220}, and {311} planes.

This result is in good agreement with crystalline fcc structure of XRD prediction.Comparing all the TEM images in figure 3, it is obvious that the size of Pt particles increases after the life tests. In anode some coalescence of particles is apparent; however, numerous small particles are still evident. Evidently, in cathode after 3200 cycles, the mass of small crystallites appear to be absent. From figure 4, it can be seen that the average mean particle size of Pt/MWCNT catalyst is 2.25 nm which is remarkably similar to those obtained in XRD results (Figure 2). At the cathode, due to the TEM images contained some agglomerated or nonspherical Pt particles histogram was not considered in the particle size distribution.

An interesting case is the study of Pt Nps growth mechanism *via* TEM analysis (Figure 3c). After normal operation through fuel cell constant test, HRTEM images revealed that Pt Nps inside the HT PEMFC have a nonspherical (rod and bent) shape typical of either longitudinal or dendrite growth (Figure 3c) [3]. Clearly, at the cathode, a marked decrease in the density of Pt Nps per cm^2 of carbon and particle growth was observed.

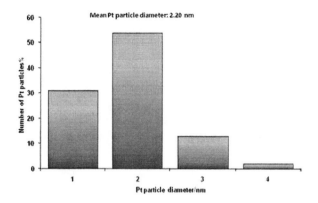

Figure 4: The histogram of Pt/MWCNT particles sizes distribution from the TEM photographs.

3   MEA INVESTIGATIONS

In fuel cell the performance loss due to either contribution of catalyst degradation or any of membrane degradation during single cell tests. Understanding features one of them would indicate performance loss origin [4].

3.1   Polarization curve

Polarization curve was obtained by a current step potentiometry with the steady state potential recorded 2 min after each current was set. All polarization curves were measured beginning with the highest currents. The fuel cell polarization at low and medium current densities was dominated by activation losses and ohmic losses. Losses were decided by the electrical resistance of the electrodes. The activation losses were decided by the cell temperature, catalyst activity, reactant concentration and the pressure. The ohmic and the resistance losses were decided by the electrical resistance of the electrodes and the resistance to the flow of ions in the electrolyte [5].

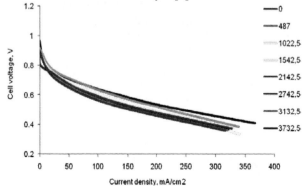

Figure 5: Polarization curves of a PBI fuel cell after different number of cycles (1cycle = 2 min) constant fuel cell test at open circuit voltage. Platinum loading of the

electrode was 0.7 mg Pt cm^{-2} and the active area of the electrode was 25 cm^2. Hydrogen and oxygen flow rates were 100 ml min^{-1} and the air flow rate was 200 ml min^{-1}.

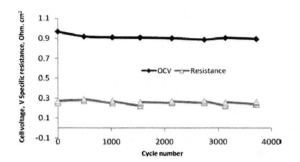

Figure 6: OCV and ohmic resistance of Pt/MWCNT catalyst after 4000 cycles.

Figure 5 shows a set of polarization curves of a Pt/MWCNT operating with hydrogen and air at 150 °C. At the anode side the hydrogen flow was set at 200 ml min^{-1}, whereas the air flow at the cathodic side was fixed at 100 ml min^{-1}. The numbers shown in the legend are accumulated number of cycles (1 cycle = 2min) at open cyclic voltage (OCV). At lower current densities, the polarization curves have the same behavior. With an increase in the test time of the Pt/MWCNT catalyst has virtually no remarkable influence on the polarization drop particularly when the current density is in a lower, though a better performance is achieved when the current density is close to the limiting value. Clearly, the performance loss occurred throughout the entire current density region, and most of the loss occurred in the first 1500 cycles. Afterwards, the performance decline rate slowed down regularly.

3.2 Open cyclic voltage (OCV), specific resistance and degradation

The impact of open circuit voltage (OCV) on the performance and degradation of PBI–H$_3$PO$_4$ membrane of PEMFC operated at 150C was investigated. The OCV showed an almost monotonous constant line (Figure 8). However, increasing OCV usually is consisting with performance loss but this outcome is not obtained in our study (Figure 5). OCV graph indicate that MEA degradation is insignificant. Likewise, since the total resistance did not change based on results from Figure 6, the proton conductive with in MEA also inconsequential, hence, OCV and resistance results show that MEA interference in catalyst performance loss insignificant and both increased catalyst activity loss due to mainly catalyst utilization (Figure 7). Finally, one may conclude that catalyst electrochemical surface area (ESA) loss (Figure 8) is likely stem from catalyst durability decrease during fuel cell constant time test.

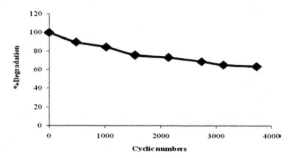

Figure 7: Performance diagram of Pt/MWCNT catalyst after 4000 cycles.

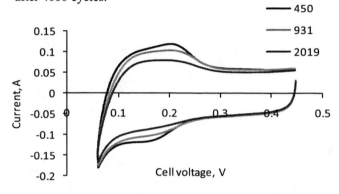

Figure 8: Cyclic voltammetry (CV) data showing catalyst electrochemical surface area (ESA) loss for the Pt/MWCNT catalyst after single cell test.

4 CONCLUSION

Durability, Preparation, characterization and of catalyst for H$_2$/O$_2$ (H$_3$PO$_4$/PBI) PEMFC –Pt/MWCNT– are presented. Our results indicate that the performance of the prepared catalyst mainly because of catalyst activity loss during single cell test but not membrane. TEM images demonstrate that the shape of Pt Nps change systematically by seeding their growth in non spherical shape (nanorod) after fuel cell test at high temperature.

REFERENCES

[1] Y. XW and Y. SY, J. Power Sources, 172, 133, 2007.
[2] Q. Li, R. He, JO. Jensen and NJ. Bjerrum, Chem. Mater. 15, 4896, 2003.
[3] D. Thompsett, In "Fuel Cell Technology," G. Hoogers (Ed.), Chap. 6, CRC Press LLC, 2003.
[4] Q. Li, JO. Jensen, RF Savinell and NJ. Bjerrum, Progress in Polymer Science, 34, 499, 2009.
[5] C. Pan, Q. Li, JO. Jensen, R. He, LN. Cleemann, MS. Nilsson, Q. Zeng and NJ. Bjerrum, J. Power Sources, 172, 278, 2007.

NANO SILVER – WHY IT IS SO HOT NOW?

K. W. Lem*[1a], J. R. Haw[1], D. S. Lee[2], C. Brumlik[3], S. Sund[4], S. Curran[5], P. Smith[6], S. Brauer[3], D. Schmidt[7], and Z. Iqbal[8]

(1) Department of Materials Chemistry & Engineering, Konkuk University, Seoul, Korea. jrhaw@ konkuk.ac.kr
(a) Corresponding Author :(O) 822-20496247;(FAX) 822-4443490; kwlem@konkuk.ac.kr; kwlem2001@yahoo.com
(2) Department of Chemical Engineering, Chonbuk National University, Jeonju, Korea. dslee@jbnu.ac.kr
(3) Nanobiz, LLC, NJ, USA. brumlik@nanobizllc.com; sambrauer@nanotechplus.net
(4) Nygard Consulting, LLC, NJ, USA. seas_nj@yahoo.com
(5) Boston Scientific, MA, USA. sean.curran2@bsci.com
(6) Knowles Electronics, IL, USA. apetesmith@aol.com
(7) Materials to Market, LLC, NJ, USA. j.david.schmidt@embarqmail.com
(8) Department of Chem. and Environmental Sci., New Jersey Institute of Technology, NJ, USA. iqbal@adm.njit.edu

ABSTRACT

Senjen and Illuminato[1] of Friends of the Earth have claimed that nanosilver (AgNP) is an extreme germ killer which presents a growing threat to public health. Volpe[2] and Height[3] of Silver Nanotechnology Working Group (SNWG) have been arguing to the USEPA[4] that AgNP poses a unique property for its size <100 nm in one dimension. They argued that the functionality of AgNP used in antimicrobial applications is not unique and not a new material. It is identical to all EPA-registered silver products for decades where any antimicrobial functionality is achieved via release of silver ions (Ag+). Nevertheless, will the "safety" of AgNP become a real "Nano-Titanic" and prevent a sustainable industrial system for this "red-hot" nanotechnology development?

Keywords: nanosilver (AgNP), life cycle, exergy, triz in dfss (dflss-g), waste minimization.

1. GENERAL APPROACH

To assess the safety of nanosilver at Konkuk University we have embarked on a Design for Lean Six Sigma based Waste Minimization research program with an international cross functional team. We have begun our study on the life cycle assessment of AgNP starting with the use of process mapping. In this paper, we will highlight our literature research findings in the life cycle assessment of AgNP. We are applying exergy analysis, TRIZ in Design for Lean Six Sigma - "Green" (DFLSS-G) to determine whether the nanosilver safety in environment is an old problem for silver, or an entirely new challenge. In any case, whether it is an old or new problem, we must deal with it seriously. As Luoma[5] stated – "Nanosilver – there is no silver lining".

2. SILVER AND NANOSILVER

Unlike most inert materials, it has been known for centuries that silver has an antiseptic effect. In ancient times, Greeks used silver vessels for drinking water storage. Koreans have also chosen silver to make their metal chopsticks. Recently, silver has found use in everyday products such as antimicrobial products, consumer products, and electronic products. The antiseptic effect gives an extra dimension in dealing with the life cycle assessment of silver.

Nanosilver can be made with different shapes such as particles, wires, and rods. Due to its enormous surface reactivity, AgNP has found utility in everyday products that require antibiotic performance, such as food contact materials, textiles and fabrics, appliances, consumer products, children's toys, infant products, 'health' supplements, cosmetics and pharmaceuticals[6].

Despite the fact that nanosilver effectively kills bacteria and thus becomes biocidal, many scientists are still not certain of its safety to humans,. Thus, safety has become a very sensitive and potentially critical issue for companies that make products containing nanosilver. A rigorous material flow analysis is needed to quantitatively assess the environmental impact of AgNP emission.

3. MATERIAL FLOW ANALYSIS

Using a material flow analysis (MFA), Johnson et al.[7] in their "anthropogenic cycling of silver in 1997" study have found that North America and Europe have the biggest share of use of silver products on a per capita basis. They found that global silver discards are approximately 57% of the silver mined and only 57% of the silver entering waste management globally is recycled. The amount of silver entering landfills globally is comparable to the amount found in tailings. Eckelman and Graedel[8] reported that more than 13 Gg of silver are emitted annually to the environment globally. The tailings and landfills make up almost three-fourths of the total emission.

4. LIFE CYCLE ASSESSMENT

The quality of our life is improved by our industrial system, but the current system is creating unintended and serious consequences for the environment at a global level. For nanotechnology, to minimize these consequences, one must be able to transform all sources of waste and toxicity into "technical" or "biological nutrients". We can then reuse them indefinitely without harm to living systems.

Figure 1 gives an overview of a silver/nanosilver product's life cycle as food and waste in industrial metabolism. The metabolization of resources should be optimized in thermodynamic terms exergy. Dewulfn and Van Langenhove [18] have previously applied exergy analysis as a quantitative tool in the thermodynamic optimization of the life cycle of plastics.

Figure 1 - A Silver/Nanosilver Product's Life Cycle as **Food** and **Waste** in Industrial Metabolism [7-16]

Adopted from Geyer (2008): Lem et al. (2009, 2006), Muller (2007); Eikelman & Graedel (2007); Johnson et al. (2006); Kobayashi(2005); Serban et al. (2004); Senge & Carstedt (2001))

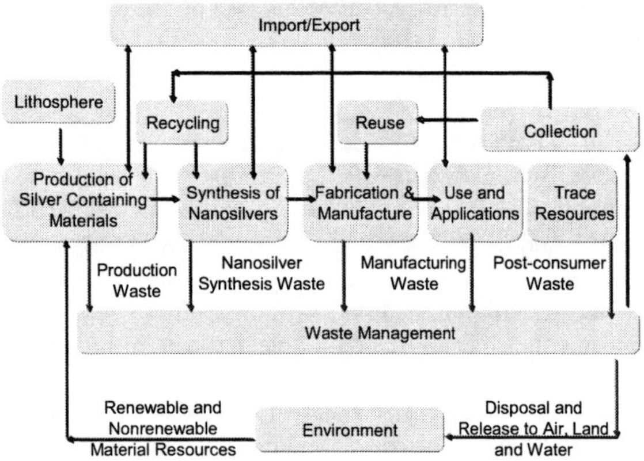

5. WASTE MINIMIZATION

Waste generation is a critical limitation for any sustainable industrial system[9]. Senge and Carstedt[16] offered a view of why industry produces waste and suggested that a synthetic process can emulate nature to reduce the waste using a cyclic industrial system. This is accomplished by focusing on three key aspects of the manufacturing process: (a) resource productivity, (b) cleaning products, and (c) re-manufacturing, recycle, and composting. One example of this approach is reycling of nylon 6 carpet[17]. This type of cyclic process has addressed and overcome the economic, technical, and logistical barriers to commercialize a closed loop recycling process and recover caprolactam from waste nylon 6 materials.

Waste generation is unavoidable so waste minimization becomes a fundamental requirement for economic feasibility. Based on the exergy ananlysis by Dewulf et al. [19], in Figure 2, Lem et al.[10] have shown that waste generation in a real process is more than just exergy loss (destroyed) in industrial metabolism.

Figure 2 - Waste Generation and Exergy Loss[18]
(Adopted from Dewulf et al, 2008)

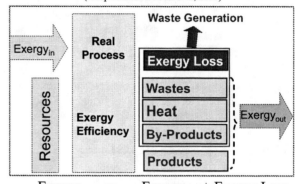

$$Exergy_{in} = Exergy_{out} + Exergy\ Loss$$

Resources = Products + [ByProducts + Heat + Wastes]+ [Exergy Loss]

6. NANOPRODUCTS

Nanoproducts can be made in two ways: top-down and bottom-up. A top-down approach is essentially tearing down of a device to gain insight into its components, materials and compositions. A bottom-up approach is the piecing together of materials to give rise to components and finally to build a device.

Figure 3 gives a roadmap of top-down and bottom-up in development of nanoproducts under a value chain of "material – properties – processing – structure – performance – applications". This roadmap helps to identify the unmet needs from materials to/from applications in nanoproduct development and manufacture.

Figure 3 – Top-Down and Bottom-Up Development [6,10]
(Lem et al., 2009; Brauer et al., 2009)

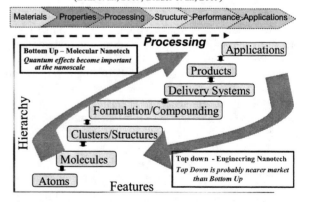

Identify the Unmet Needs from Materials to/from Applications

Commercial AgNP products are very likely produced by a bottom-up process, whereas analysis of environmental impact by AgNP products is a top-down process (see Tolaymat et al. [20]).

7. DESIGN FOR LEAN SIX SIGMA-GREEN (DFLSS-G)

We need to point out that waste minimization has been the background of the "Lean Six Sigma" methodology used to minimize wastes in manufacturing industries over the last few decades. Lean Six Sigma is a strategy that was developed to accelerate improvements in processes, products, and services, radically reduce manufacturing and/or administrative costs, and improve quality by relentlessly focusing on eliminating waste and reducing defects and variation. Design for Six Sigma (DFSS) is a tool for developing new products and processes "right from the start" for the triad in a project management - with the right quality, at the right cost, and at the right time [21-23]. Therefore, there is a need for Design for Lean Six Sigma – Green, where Green is for the environment and eco-products. Kobayashi [14] has used a product life planning methodology based on a quality function deployment (QFD) and a software tool to establish an *eco*-design concept of a product and its life cycle in multigenerational eco-products development.

8. DESIGN FOR LEAN SIX SIGMA/TRIZ

Serban et al.[15] have used a TRIZ approach to design for environment for over a product life cycle. Based on their work, we propose Design for Lean Six Sigma – Green with TRIZ to design AgNP eco-products and their life cycle as shown in Figure 4. From the mass balance of each step in the life cycle, the value generated is equal to the food resource available in each step minus the wastes at each step. Therefore, a summation of all the steps gives rise to the total value generated.

Figure 4 - DFLSS-G/TRIZ in Eco-Product Life Cycle Design

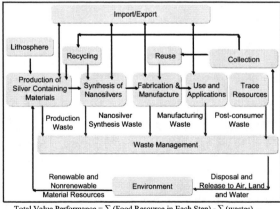

Total Value Performance = Σ (Food Resource in Each Step) - Σ (wastes) all sources

A flow chart of the procedure to be used in our study is given in Figure 5 and a TRIZ approach in Design for Lean Six Sigma – Green for AgNP products life cycle is given in Table 1. We are using the following four step iterative approaches:

Figure.5 - Flow Chart for DFLSS-G with TRIZ [24]
(Adopted from Terninko et al, 1998)

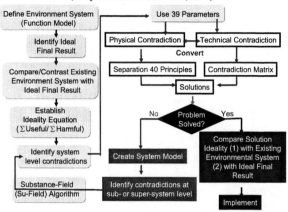

First: determine the Voice of the Environment regarding the safety of the AgNP products using two extreme sides of the debate between Friends of Earth/USEPA and Silver Nanotechnology Work Group (SNWG) to obtain a resolution regarding "Conflict". We try to answer the question - could improving one technical characteristic to solve a problem cause other technical characteristics to worsen? Once the problem is defined, we need to define the system boundaries, quantify mass flows of AgNP, and to define several emission scenarios.

Table 1 - TRIZ Approach in DFLSS-G for AgNP Products Life Cycle

DFSS Phase	TRIZ Tools	Approach	Application to AgNP Product Life Cycle
Voice of the Customer	1. Conflict Resolution, 2. Ideal Final Result, 3. Development of Measurement Systems.	Identify the Problem	Step 1: Voice of the Customer (VOE) 1. Safety of AgNP Products. 2. Define Ideality Based QFD
Concept Development	All	1. Find The Principle that Needs to be Changed 2. Then Find the Principle that is an Undesired Secondary Effect.	Step 2: Conflict resolution 1. Example - Friends of Earth/USEPA vs. Silver Nanotechnology Work Group 2. Define Functionality/ Requirements
Detailed Design	All	1. Find the Principle that Needs to be Changed, 2. Then Find the Principle that is an Undesired Secondary Effect.	3. Use of Resources 4. Search for Previously Well-Solved Problems a. Examine 39 engineering parameters/40 principles. b. IP Landscaping
Optimize	1. Conflict Resolution, 2. Trimming, 3. Subversion Analysis, 4. Problem Solving	1. Look for Analogous Solutions 2. Adapt to the Potential Solution 3. Optimize – Ideality	Step 3: Review toxicity data for environmentally relevant silver compounds. Optimize wherever possible. Review earlier search for previously well-solved problems
Validate/ Implement	1. Conflict Resolution, 2. Trimming, 3. Problem Solving	Validate potential solution	Step 4: Gap Closing - Conflict resolution/Ideality Revisit

Second: search for previously well-solved problems by looking at the 39 engineering parameters/40 principles. Antimicrobial nanoscale silver is typically embedded within substrates, mainly a polymer, where any antimicrobial functionality is achieved via release of silver ions (Ag+).

The behavior of silver in environment will be reviewed, and a mass balance model applied to calculate predicted environmental concentrations. The uncertainty of the results is assessed and predicted concentrations are compared to experimental and empirical data (an example as reported by Benn and Westerhoff [25]).

Third: compile and predict the toxicity data for environmentally relevant silver compounds for noeffect concentrations. This material flow will be optimized based on a review of our earlier search for previously well-solved problems.

Fourth: evaluate and determine the potential for risk caused by the release of silver into environment using all available experimental data and literature data.

9. FUTURE STUDIES

In our Design for Lean Six Sigma based Waste Minimization research programs, we have begun our journey to study the life cycle assessment of nanosilver starting with the use of product life cycle process mapping and Design for Lean Six Sigma with TRIZ. We are planning to have more multidisciplinary and international interactions to characterization of AgNP products and their transformations in relevant biological and environmental media, in addition to systems and life-cycle approaches to nano-safety. We endorse the nanotoxicity evaluation advocated by Yu[26] from Korea Environment and Merchandise Testing Institute (KEMTI) to have a close collaboration between toxicologists and engineers.

REFERENCES

[1] R. Senjen and I. Illuminato, "Nano & Biocidal Silver – Extreme Germ Killers Present a Growing Threat to Public Health," Friends of Earth, June 2009.

[2] J. E. Housenger, "Status of Regulating Nanoscale Particles and Prions," CPDA Mid-Year Meeting, March 11, 2009.

[3] R. Volpe, "Letter to EPA Regarding the EPA Nanosilver Scientific Advisory Panel Report," Silver Nanotechnology Working Group (SNWG), February 2010.

[4] M. J. Height, FIFRA Scientific Advisory Panel (SAP) Open Consultation Meeting, November 3 - 6, 2009.

[5] S. N. Luoma, Project on Emerging Nanotechnologies , PEN 15, September 2008.

[6] S. Brauer, K.W. Lem, and J.R. Haw, "The Markets for Soft Nanomaterials: Cosmetics and Pharmaceuticals", Presented in Nano and Green Technology Conference 2009; New York City, November 18, 2009.

[7] J. Johnson, J . Jirikowic, M. Bertram, D. Van Beers, R. B. Gordon, K. Henderson, R. J. Klee, T. Lanzono, R. Lifset, L. Oetjen, and T. E. Graedel, Environ. Sci. Technol. 39, 4655-4665, 2005.

[8] M. J. Eckelman and T.E. Graedel, Environ. Sci. Technol., 41, 6283-6289, 2007.

[9] S. Evans, M. N. Bergendahl, M. Gregory and C. Ryan, "Towards a sustainable industrial system - With recommendations for education, research, industry and policy," University of Cambridge Institute for Manufacturing, 2009.

[10] K.W. Lem, J.R. Haw, S. Sund, S. Curran, C. Brumlik, P. Smith , S. Brauer, and D. Schmidt, "Waste Minimization in Commercialization of Nanotechnology" Seminar presented Chonbuk National University, Jeonju, Jeollabuk-do, South Korea, November 26, 2009.

[11] R. Geyer, Industrial Ecology, UCSB Bren School – Session 10 – February 13, 2008

[12] N. Muller, Ph.D. Thesis, Swiss Febderal Institute of Technology, Zurich, 2007.

[13] K. W. Lem, S. A. Curran, S. Sund, and M. Gabriel, Encyclopedia of Chemical Processing (ed. S. Lee), Marcel Dekker, NY, 2006.

[14] H. Kobayashi, Research in Engineering Design, 16 (1-2) 1-16, 2005.

[15] D. Serban, E. Man, N. Ionescu and T. Roche, "A TRIZ Approach To Design For Environment," (D. Talab and T. Roche eds.), Product Engineering, 89–100, 2004.

[16] P.M. Senge and G. Carstedt, MIT Sloan Management Review, 24-38, Winter, 2001.

[17] K.W. Lem, A. Letton, T.P.J. Izod, F.S. Lupton, F. S.; W. B. Bedwell, USP 6,214,908, USP 6,414,066

[18] J. Dewulf and H. Van Langenhove, Int. J. Energy Res., 28, 969–976, 2004.

[19] J. Dewulf, H. Van Langenhove, B. Muys, S. Bruers, B. R. Bakshi , G. F. Grubb, D. M. Paulus, and E. Sciubba, Environmental Science & Technology 42 (7), 2221, 2008.

[20] T. M. Tolaymat, A.M. El Badawy, A. Genaidy, K.G. Scheckel, T.P. Luxton, M. Suidan, Science of the Total Environment, 408(5), 999-1006, 2010.

[21] S. A. Curran, M. Gabriel, and K. W. Lem, Six Sigma and Design for Six Sigma in IRI R&D Workshop, Corning, NY, March 22, 2005.

[22] S. A. Curran, K. W. Lem, S. Sund, and M. Gabriel, Encyclopedia of Chemical Processing (ed. S. Lee), Marcel Dekker, NY, 2006.

[23] S. A. Curran, Industrial Research Institute Webinar Series on Fundamentals of Quality, September 24, 2009

[24] J. Terninko, A. Zusman and B. Zlotin, "Systematic Innovation – Introduction to TRIZ (Theory of Inventive Problem Solving), CRC Press, 1998.

[25] T.M. Benn and P . Westerhoff, Environ. Sci. Technol., 42 (18), 7025–7026, 2008.

[26] I. J. Yu. "Subchronic Inhalation Toxicity Evaluation of Silver Nanoparticles," KEMTI, 5th Korea-US Nano Forum, April 17-19, 2008.

[1] [2010 NSTI – Paper# 300] Nanotech 2010, 696 San Ramon Valley Boulevard, Suite 423, Danville, CA 94526-4022, Ph: (925) 353-5004, Fax: (925) 886-8461, swenning@nsti.org.

Surface Enhanced Raman Utilizing Magnetic Core Gold Shell Nanoparticles

K.J. Carroll[*], G. P. Glaspell,[*] N. McDowall,[*] L.W. Brown III,[*] K. Zhang,[**] A.K. Pradhan,[**] J. Anderson,[*] E.E. Carpenter[*]

[*] Department of Chemistry, Virginia Commonwealth University, Richmond, VA 23284,
ecarpenter2@vcu.edu
[**] Center for Materials Research, Norfolk state University, Norfolk, VA 23504.

ABSTRACT

In this study, we investigate a novel two-step synthesis of $Fe/SiO_2/Au$ core/shell nanoparticles by aqueous reduction using sodium borohydride. The composition, morphology, and magnetic behavior were determined by X-ray diffraction, X-ray photoelectron spectroscopy, transmission electron microscopy, and room temperature vibrating sample magnetometry. We also demonstrate their application as surface-enhanced Raman substrates. We evaluate their performance as SERS active material using pyridine as a model compound. Core/shell nanoparticles with optical and magnetic functionality offer broad opportunities for biomedical and SERS applications.

Keywords: Nanoparticles, SERS, core/shell, etc

1 INTRODUCTION

Raman spectrometry has been successfully used for the identification and characterization of organic materials. However, due in part to the *low* scattering efficiency Raman spectroscopy has not been fully utilized in sensor applications. Surface enhanced Raman scattering (SERS) on the other hand has been shown to produce enhancement factors as high as 10^{14}-10^{15}, which is capable of detecting single molecules.[1] SERS is experiencing a resurgence of interest due in part to nanotechnology. Noble metal nanoparticles when excited develop a major surface plasmon along the surface of the particle resulting in enhanced Raman Scattering. The surface enhanced Raman (SER) effect is currently explained using chemical (CM) and electromagnetic (EM) mechanisms.[2] Chemical effects provide enhancement via an increase in the molecular polarizability of the adsorbate due its interaction with the metal surface. Maximum chemical enhancement is observed when the substrate forms a monolayer on the metal of interest. This phenomenon can be modulated by molecules adsorbed on the surface allowing for the development of sensors. As opposed to other sensors where specific moieties on the analyte contribute to the signal, with SERS each different type of moiety would allow for a different modulation of the SERS thus allowing for simultaneous detection of different analytes. Furthermore, SERS is capable of identifying structural features of various analytes.

Kneip and coworkers have shown that SERS spectra of various analytes were measured down to 10^{-7} M (this corresponds to a sensitivity of less than 1 pg) in colloidal gold solutions. Nie has also shown that not all particles are SER active and have reported that the aggregation of individual particles can lead to intense SER enhancement through the formation of "hot spots" (an intense plasmon resonance that exists between two particles in close proximity).[1] A "hot spot" can provide enhancement factors as high as 10^{13} to 10^{14}; sensitive enough for single molecule detection. A coupling of the particle resonances and the confined region between particles causes the observed enhancement. It is therefore possible that inducing a magnetic field may also create hotspots between adjacent nanoparticles further enhancing their sensitivity.

The last several years have seen an increasing interest in designing nanoparticles with multifunctional properties. Over the past year, nanoparticles have been designed such that they combine both optical and magnetic properties with promising applications to biomedical, engineering, and sensor design.[3-8] Specifically, Kim and coworkers for the first time claim to have designed a novel one step synthesis of $Ag@Fe_2O_3$ nanoparticles specifically for SERS-based detection of chemical species.[8] In their process, they synthesize Fe_2O_3 nanoparticles and then redisperse them for silver coating. We however report on $Fe/SiO_2/Au$ core/shell nanoparticles by aqueous reduction using sodium borohydride. The concept of utilizing a SiO_2 shell may isolate the Au from any possible interference due to the Fe core.

Raman spectroscopy is capable of providing a "fingerprint" by which molecules can be identified; the signals are often too weak to be practical for sensor applications. We propose coupling the surface enhancement observed with noble metals with the magnetic properties of an iron core. The magnetic core provides the capability to recollect distributed sensors after attachment of the target molecule, thereby concentrating the molecules of interest and increasing detection sensitivity. Since Raman spectroscopy utilizes vibrational modes that are specific to individual molecules, we believe that we should be able to identify various analytes of interest on the nanoparticle surface.

2 SYNTHESIS AND CHARACTERIZATION

2.1 Synthesis of Fe NPs

Fe NPs were synthesized by using a previously reported procedure.[9-10] Typically, 4.6 mM of iron (II) sulfate heptahydrate and 0.46 mM trisodium citrate dihydrate were added to 2L of DI-H_2O under magnetic stirring in a 4L Erlenmeyer flask at room temperature. The mixture was allowed to stir for 5 minute to ensure dissolution. Then, 1.56 g of sodium borohydride was added and allowed to stir for 10 minutes. The solution was quenched with ethanol several times and magnetically separated using a rare-earth magnet. After washing, the remaining ethanol was decanted and the particles were placed in a vacuum oven at room temperature to dry.

2.2 Synthesis of Fe@SiO_2 NPs

A total of 1.50 g of Fe NPs were added to 20 mL of DI-H_2O and 200 mL ethanol under magnetic stirring in an Erlenmeyer flask. Then 1 mL of TEOS was slowly added for 30 minutes using an addition funnel and then allowed to react for 6 hours.

2.3 Synthesis of Fe@SiO_2@Au NPs

To the Fe@SiO_2 NP solution 1 mL of 0.490 M $HAuCl_4$ was added. Then 0.204 g of sodium borohydride was added to the solution and the reaction was allowed to react for 10 minutes. The solution was quenched with ethanol several times and magnetically separated. After washing the particles were placed in a vacuum oven at room temperature to dry prior to analysis.

2.4 Heat Treatment

The dried Fe@SiO_2@Au nanoparticles were heated in an oven under N_2 atmosphere to 400 °C for 30 minutes with a 10 °C/min ramp rate.

2.5 Characterization

Microstructure analysis of the NPs was conducted using high-resolution transmission electron microscopy. HRTEM analysis was performed on the nanoparticles with a field-emission transmission electron microscope (TEM, JEOL, JEM 2010, accelerating voltage 200 KV). The nanoparticles were sonicated and vacuum dried on the copper grid before placing in the TEM chamber. The crystal structures were analyzed by a X-ray diffraction (XRD) spectrometry (Panalytical X'pert pro) equipped with a Cu Kα radiation source (λ=0.15406 nm).

For X-ray photoelectron spectroscopy (XPS) analysis, the samples were transferred to the sample chamber of a Thermo Scientific ESCALAB 250 microprobe with a focused monochromatic Al Kα radiation source (E_b= 1486.6 eV). The depth profiling experiments were conducted with a high energy Ar^+ ion beam. All XPS spectra were referred by aliphatic C 1s (284.5 eV). The magnetic properties were measured at room temperature using a Lakeshore Cryotonics Inc. Model 7300 VSM.

3 RESULTS AND DISCUSSION

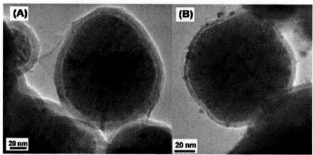

Figure 1. TEM images of (A) Fe/SiO_2 nanoparticles and (B) Fe/SiO_2/Au nanoparticles.

Figure 1 shows typical TEM images of the as-prepared 80 nm Core/shell Fe/SiO2 NPs with the iron core being 75 nm and the SiO_2 shell being 5 nm and the Fe/SiO_2/Au NPs with Au islands of 4-5 nm.

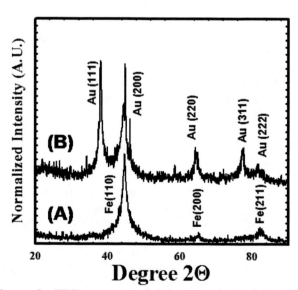

Figure 2. XRD patterns of (A) as-synthesized Fe/SiO_2 nanoparticles and (B) as-synthesized Fe/SiO_2/Au nanoparticles.

The XRD patterns of the as-prepared core/shell nanoparticles are presented in figure 2. All the diffraction peaks of Au can be indexed to face-centered cubic (fcc) crystal structure while the Fe can be indexed to body centered cubic (bcc). In the figure, the data is presented with the miller indices from the JCPDS reference powder diffraction files α-Fe (01-089-7194) and Au (01-04-0784). From the diffraction data, it can be seen that there are no Fe oxide impurities.

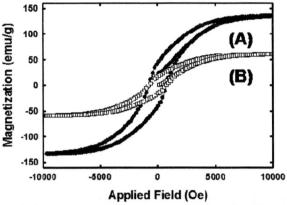

Figure 3. Room temperature Hysteresis loops for (A) as-synthesized Fe nanoparticles and (B) as-synthesized Fe/SiO$_2$/Au nanoparticles.

Room temperature hysteresis loops are shown in figure 3. The as-synthesized saturation magnetizations Fe and Fe/SiO$_2$/Au nanoparticles were found to be 148 emu/g and 58 emu/g, respectively. From the TGA data (Figure 4) it can be seen that roughly 10 percent of unreacted citrate is adsorbed onto the Fe/SiO2/Au nanoparticles and using a mass correction the corrected magnetization was calculated to be roughly 69 emu/g.

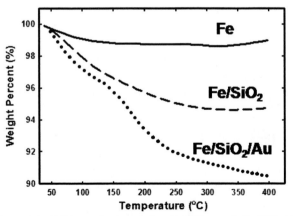

Figure 4. TGA analysis of Fe, FeSiO$_2$, and Fe/SiO$_2$/Au nanoparticles.

Using the TGA the Fe/SiO2/Au samples were heat treated to 400 °C using N$_2$ gas. As mentioned above, about 10 percent weight loss was seen when the samples were heated. It is hypothesized that this reduction is due to the loss of organics that were adsorbed onto the surface of the particles along adsorbed H$_2$O.

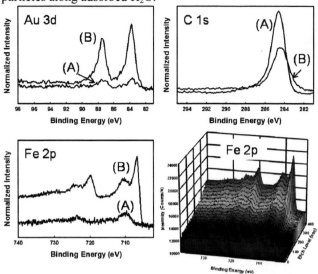

Figure 5. XPS spectra of before (A) and after (B) annealing of Fe/SiO$_2$/Au nanoparticles.

To confirm this XPS analysis was conducted on the pre- and post- annealed samples. Figure 5 depicts the results of the XPS analysis. From the C 1s spectrum a decrease in the intensity of the C-C peak is seen after the samples are annealed under N$_2$ gas. This decrease is coupled with an increase in the intensity of the Au 3d peak, which supports the hypothesis that organics were on the surface obstructing the Au surface. In addition, an increase in the Fe 2p peak is seen. A high-resolution depth profile analysis shows that as the sample is etched the intensity of the Fe 2p increases. More importantly, an increase in the metallic Fe (BE=709 eV) peak correlates with the XRD analysis.

The aggregation of nanoparticles occurs naturally in solution due in part to the reduced surface energy after aggregation, it is also possible by utilizing magnetic fields and magnetic nanoparticles. While it is difficult to collect SERS on magnetic nanoparticles consisting solely of Fe, Co or Ni, it becomes possible with core-shell nanoparticles. To test this hypothesis, we show through experimentation that by utilizing the Fe/SiO$_2$/Au core/shell nanoparticles the nanoparticles can be concentrated along magnetic field gradients which offers the distinct advantage of piloted movement of the target analyte. The observed SERS enhancement could arise from a combination of two particular scenarios (Figure 6 A). The first scenario is indicative of the analyte of interest localized by the magnetic field providing an increase in the observed Raman signal. The other scenario is related to the aggregation of gold particles. The aggregation of gold particles leads to "hot spots" showing increased SERS response.

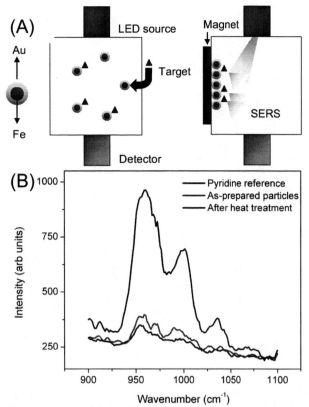

Figure 6. (A) Utilizing magnetic core shell particle to pre-concentrate the target analyte for improved sensitivity. (B) SERS of the pyridine reference and Fe/SiO$_2$/Au core/shell nanoparticles before and after heat treatment. The SERS is dominated by two intense bands at about 1,008 and 1,036 cm^{-1}, corresponding to the ring breathing and to the triangular-deformation modes, respectively.

In order to determine the affect of magnetic pre-concentration on the SERS response, we collected Raman spectra of pyridine, a compound commonly used in SERS studies, using a Raman Systems R-3000 spectrometer equipped with a 785 nm laser. The pyridine solution also contained 5 ml of ethanol which served as our internal intensity standard. In a typical experiment, magnetic SERS particles were then added to the standard pyridine solution and left to set for up to 20 minutes. In the presence of an external magnetic field all the particles were collected to one side of a cuvette and the SERS spectra was collected and is indicated in Figure 6. It is significant to note that the Raman spectra of the heat treated particles did in fact increase after the magnetic field was applied. Previously we alluded to two possible scenarios for enhancement. First, pyridine is chemisorbed onto the modified magnetic particles and after the magnetic field is applied the increased signal is proportional to the higher concentration of localized pyridine. Secondly, the increased SERS signal stems from "hot spots" generated by the aggregation of the modified magnetic particles. Since we are using a 785 nm laser which is somewhat removed from the plasmon resonance of the gold surface, typically around 520 nm, it is

reasonable to assume the first scenario likely holds true. Future experiments will incorporate the use of Nd:YAG, 532 nm, which operates closer to the plasmon resonance of gold in order to determine if "hot spots" are indeed forming.

4 CONCLUSION

In conclusion, we reported on a novel two-step synthesis of Fe/SiO$_2$/Au core/shell nanoparticles by aqueous reduction using sodium boroydride. Based on TEM, XRD, XPS, and TGA characterization, we demonstrated that the as prepared particles are not SERS active and require heat treatment to remove citrate from the surface. We have also demonstrated using Raman spectroscopy that after heat treatment the nanoparticles are SERS-active and show practical applications, including piloted movement of absorbed analytes, for SERS based detection.

5 ACKNOWLEDGEMENTS

The authors would like to thank the VCU nanomaterials core characterization facility for instrument time as well as the NSF MRI grant CHE-0820945.

6 REFERENCES

[1]   Nie, S.; Emory, S. R. *Science* **1997**, *275*, 1102.

[2]   Glaspell, G. P.; Zuo, C.; Jagodzinski, P. W. *Journal of Cluster Science* **2005**, *16*, 39.

[3]   Levin, C. S.; Hofmann, C.; Ali, T. A.; Kelly, A. T.; Morosan, E.; Nordlander, P.; Whitmire, K. H.; Halas, N. J. *ACS Nano* **2009**, *3*, 1379.

[4]   Radwan, F. R.; Carroll, K. J.; Carpenter, E. E. *J. Appl. Phys.* **2009**, *in press*.

[5]   Wang, H.; Brandl, D. W.; Le, F.; Nordlander, P.; Halas, N. J. *Nano Letters* **2006**, *6*, 827.

[6]   Noh, M. S.; Jun, B.-H.; Kim, S.; Kang, H.; Woo, M.-A.; Minai-Tehrani, A.; Kim, J.-E.; Kim, J.; Park, J.; Lim, H.-T.; Park, S.-C.; Hyeon, T.; Kim, Y.-K.; Jeong, D. H.; Lee, Y.-S.; Cho, M.-H. *Biomaterials* **2009**, *30*, 3915.

[7]   Jun, B.-H.; Noh, M. S.; Kim, J.; Kim, G.; Kang, H.; Kim, M.-S.; Seo, Y.-T.; Baek, J.; Kim, J.-H.; Park, J.; Kim, S.; Kim, Y.-K.; Hyeon, T.; Cho, M.-H.; Jeong, D. H.; Lee, Y.-S. *Small* **2009**, *9999*, NA.

[8]   Kim, K.; Jang, H. J.; Shin, K. S. *Analyst* **2009**, *134*, 308.

[9]   Carroll, K. J.; Pitts, J. A.; Zhang, K.; Pradhan, A. K.; Carpenter, E. E. *J. Appl. Phys.* **2010**, accepted.

[10]   Carroll, K. J.; Hudgins, D. M.; Brown, L. W.; Yoon, S. D.; Heiman, D.; Harris, V. G.; Carpenter, E. E. *J. Appl. Phys.* **2010**, accepted.

Highly Efficient Functionalization of Gold Nanoparticles: Strong Coupling Applications.

Diane Djoumessi Lekeufack[*], Arnaud Brioude[*], Joel Bellessa[**], Li De Zeng[**], Pierre Stadelmann[***], Anthony W. Coleman[*], Philippe Miele[*]

[*]Laboratoire des Multimatériaux et Interfaces (UMR CNRS 5615), Université Lyon1, Université de Lyon, 43 Bd du 11 Novembre 1918, Villeurbanne Cedex, France.
[**]Laboratoire de Physique de la Matière condensée et Nanostructures, CNRS UMR 5586, Université Lyon1, Université de Lyon, 43 Bd du 11 Novembre 1918, Villeurbanne Cedex, France.
[***]EPFL-SB-CIME, Bâtiment MXC 136, Station 12, CH -1015 Lausanne

ABSTRACT

Fine metal nanoparticles have unique properties different from those of the corresponding bulk material. Properties of these nanoparticles are preserved if the general tendency to aggregation can be overcome, as when a suitable stabilization can be developed. Generally, the stabilization of colloids is performed by surface modification process which brings new properties to the particles. This can be done either to enhance double layer interaction or to adsorb macromolecules that can form a physical barrier against other approaching particles.

Keywords: metal nanoparticles, functionalization, layer-by layer self-assembled, polyelectrolytes, organic dyes, ligand exchange.

1 INTRODUCTION

Nanoparticles and nanostructured metal posses a strong well-defined surface plasmon resonance (SPR) who gives them a number of unique optical properties. They have been widely used for many applications such as catalysis [1-2], medicine [3-5] electronic [6-7], and optics [8-10]. The optical properties of silver and gold nanoparticles are tunable throughout the visible and near-infrared region of the spectrum as a function of nanoparticles size, shape, aggregation state, and local environment. This can be explaining by the fact that metallic nanoparticles can absorb and scatter electromagnetic radiation with wavelength greater than the particle size. The intrinsic properties of metal nanoparticles are largely influenced by several parameters such as its size, shape, composition, refractive index of the surrounding medium, and structure. For example, the dependence between the refractive index of the surrounding medium can be used to obtain quantitative information on the environment from the optical absorption spectrum of the metals nanoparticles.

The coupling of organic molecules and metallic nanostructures brings new opportunities to develop highly efficient photonic devices that combine the best features of those two specific materials. Indeed, the interactions between the electronic transitions of organic components and the plasmon modes of noble metal nanoparticles are responsible of emerging mixed states thus creating new optical and electronic properties. As one increasingly important class of nano-structures, the immobilizations of dye molecules onto nanoparticles have interesting properties for chemical, biological and optical applications.

Manipulating the SPR is considered to be a key issue leading to the development of technologies in the areas of photonics, optoelectronics and plasmonics. Additionally, molecules adsorbed to the surface of gold and silver nanoparticles undergo enhanced surface-enhanced Raman scattering (SERS) [11-12]effects, due to the coupling of the plasmon band of the irradiated metal (i.e, the collective oscillation of the conduction band electrons upon absorption in the visible for the particular metals, due to their dielectric constant) with the molecules electronics states .

There are many examples that exploit nanoparticles interfacial properties such as ligands exchange reaction, layer by layer assembly, and DNA linked assembly. The layer by layer (LBL) technique is generally uses for surface fonctionalization. It uses sequential adsorption of oppositely charged materials from solution onto various charged supports. The process started by adsorbing a charged polymer onto an oppositely charged surface, thereby reversing the surface charge. Further layer can be added by the alternate deposition of oppositely charged polyelectrolytes. This method was introduced by Iler [13] in the mid 1960s by absorbing particles onto solid substrates. In 1990s, Decher and al [14-15] extend Iler's work to a combination of linear polycations and polyanions. The LBL method has attracted increasing attention for preparing layered thin film with tailored properties on the nanometer scale. One of the key advantages of the LBL method is that it is simple experimentally and because water is the main solvent used, the technique is considerably more environmentally friendly than techniques that use organic solvents.

Here we present the modification of gold nanoparticles by organic molecules. In the first time we describe a surface modification procedure involving ligand exchange of citrate stabilized gold nanoparticles with TDBC and layer-by layer polyelectrolytes assembly around gold nanoparticles. The surface modification of the gold nanoparticles has been proven by means of UV-Visible and Infrared spectroscopy.

2 EXPERIMENTAL SECTION

2.1 J-agregates stabilized gold nanoparticles: strong couplage.

Molecular organic compounds such as J-aggregates (JA) support excitonic states which are electrically neutral electrons/holes pairs created by the absorption of photons. JA exhibits characteristic optical absorption called J-band which is red shifted from the monomer band, ultra-short radiative lifetime and nonlinear optical susceptibility. These properties are interesting for applications in imaging materials, optoelectronic devices and dye-sensitive solar cells [16-18].

We have developed a facile route to prepare monodisperses Au@TDBC NPs allowing the direct and efficient coating of pretreated citrate-stabilized gold NPs with TDBC, without supplementary adding of salts and bases during the coating and at room temperature. Citrate-stabilized gold nanoparticles are obtained by the Fren's method [19]. The size of particles is tune to optimize the strong coupling between the electronic transitions of organic components and the plasmon modes of gold NPs.

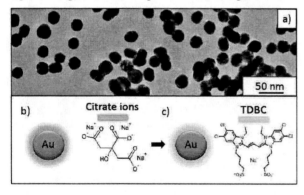

Figure 1: a) TEM images of Nps of different sizes coated with TDBC molecules, b) Citrate ions molecule, c) TDBC molecules.

The coating was realized by a ligand exchange reaction between citrate and TDBC, during the mixing of the NPs colloidal solution and the TDBC solution. Since the citrate is linked to the gold NPs surface by the oxygen atom, it can be easily replaced by the TDBC molecules which adsorption on the gold nanoparticles will be done by the nitrogen atom. Through ligand exchange reaction, the dye molecules are absorbs uniformly on the NPs. The absorbed dyes exhibit both monomer and J-aggregates absorption

bands. Strong couplage have been studied by using NPs with different size i.e. with different resonances energies (figure 2).

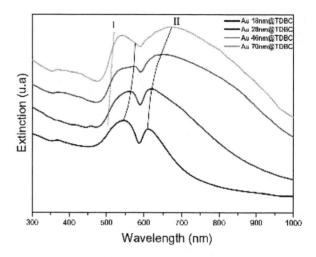

Figure 2: Extinction spectra of Au Nps with different sized coated with TDBC.

Four interesting parts can be distinguished on those spectra. The first denoted by the dotted line 1 correspond to the bare Au NPs adsorption. It is clearly shows that some particles are not coated with TDBC. This remark confirms the observations made by microscopy. The second part noted II localized at 586nm is directly related to the absorption wavelength of TDBC. Two mixed states of resonance are created.

As we can see in the extinction spectra, few particles are coated with TDBC and some are aggregated. This clearly indicates that the coating is not perfect but do not avoid the strong coupling to occur. At the contrary to previous studies, we do not manage to improve this coating by adding salts (NaCl, KCl) that is known to promote adsorption of dye molecules. The reason is that the additive salts amounts in the TDBC solution can modify the stability and the citrate ions adsorption on gold NPs.

2.2 Polyelectrolytes coated gold nanoparticles

In this work we used polyethylenimine (PEI) as cationic polyelectrolyte and poly (styrene sulfonate) sodium (PSS) as anionic polyelectrolytes. The 20nm diameter CTAB-stabilized gold nanoparticles used here are obtained prepared using the Gole et al method [20]. We have built thin films around gold nanoparticles by sequential adsorption of polyanions and polycations in aqueous medium.

Gold nanoparticles were mixed with a solution containing cationic polyelectrolytes, and a layer of polycation is adsorbed. Since the adsorption is carried out at a relatively high concentration of polyelectrolyte, a number of ionic groups remain exposed to the interface with the solution, and thus the surface charge is effectively reversed. After rinsing with deionized water, the particles were mixed with a solution containing the anionic polyelectrolyte.

The following experimental conditions were found, 10mL of the as prepared gold nanoparticles were mixed with absolute ethanol at the proportion 5:1 (nanoparticles: ethanol) and washed with water by centrifugation three times to remove the excess of CTAB. The solution is then redisperse in 5mL of Milli-Q water. For the polyelectrolytes deposition, 3mL of the colloid solution are added to 3mL of the aqueous polyelectrolyte solution of the desired concentration. The resulting solution was then left undisturbed. After 24 hours, the solution is then centrifuged three times during 20min at 13000g to eliminate the excess of polyelectrolytes. The supernatant is removed and the pellet is redispersed in milli-Q water for UV-visible and IR spectrum and TEM record.

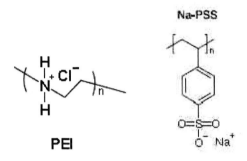

Figure 3: Polyelectrolytes structural formula.

2.3 results and discussion

For the PEI (seen figure 3) which has coexisting primary, secondary and tertiary amine groups, bonding to the gold nanoparticles occurs trough the amine functionalities. It has been reported that amine bind to gold nanoparticles through the amines functionality via weak covalent bond [21-23]. The absorption spectrum is red-shift and the broadened (figure 4), this are mainly due to the change in the refractive index (since the refractive index of PEI is 1.65, water 1.33) as a result in the absorption of PEI onto the gold nanoparticles. These suggest that CTAB are displaced from the surface of the gold nanoparticles surface upon the PEI adsorption. Despite the CTAB stabilized the gold nanoparticles, it can be adsorbed/desorbed from the gold nanoparticles surface.

For the Au@PEI@PSS, the environment of the particle changes again. Although the refractive index of the surrounding medium is decreasing (1.395 for PSS), we observe a red shift of the optical band of the Au@PEI@PSS (green and blue curves) with respect of that of Au@PEI (red curve). This red shift can be explained by increasing of the polyélectrolytes layers thickness.

Concerning the Au@PEI@PSS@PEI system, the blue shift observed in the case is attributed to the change of the refractive index of the surrounding medium.

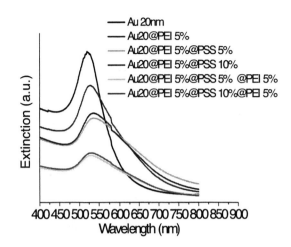

Figure 4: UV-Visible absorption spectra change with PEI and PSS functionalized gold nanoparticles.

The functionalization of gold nanoparticles by PEI (Au@PEI) was also determined by FTIR (figure 5). The main characteristic band of PEI is determined by measuring amine absorption band. Figure 5 show the FT-IR of solutions of PEI and PEI stabilized gold nanoparticles. In these spectra we can distinguished the N-H stretching band at 3300Cm^{-1}, N-H deformation band of secondary amine at 1620 Cm^{-1} and , C-N stretching band at 1050Cm^{-1}. We also have the C-H stretching band (2930 Cm^{-1} and 2860 Cm^{-1}) and C-H deformation band (1500 Cm^{-1}) of aliphatic –CH$_2$-.

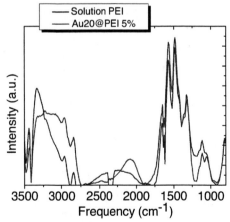

Figure 5: FTIR spectra of PEI solution and PEI functionalized gold nanoparticles.

Comparing these two spectra, we observed that in the case of Au@PEI, the absorbance peak at 3300 Cm^{-1} corresponding to the N-H stretching is broadened. This is due to interactions between the amine group and the gold nanoparticles. The vibrations of the molecules between 800 and 1700 Cm^{-1} corresponding to the C-C, CH$_2$ and the C-N bonds are unchanged, this means that these groups do not participate to the bond Au NPs-PEI.

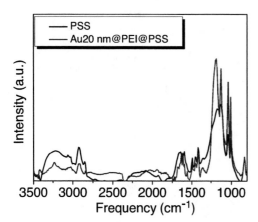

Figure 6: FTIR spectra of PSS solution and Au@PEI@PSS.

For the FT-IR characterization, the main characteristic band of PSS is determine by the absorption bands between $1000 Cm^{-1}$ and $1300 Cm^{-1}$ corresponding to the SO_3^- group. The SO_3^- group antisymmetric and symmetric vibrational adsorption peaks can be assigned to the peaks at 1195 and $1039 Cm^{-1}$ respectively. Peaks at 1128 and $1011 Cm^{-1}$ is attributed to the in-plane skeleton vibration of benzene ring and in-plane bending vibration of benzene ring. The spectral changes observe in the Au@PEI@PSS spectrum, clearly demonstrate the interaction between the PEI and the PSS.

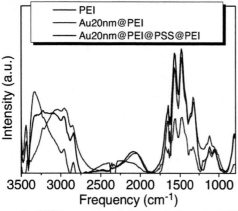

Figure 7: FTIR spectra of PEI solution, Au@PEI and Au@PEI@PSS@PEI.

The above results show that we can make a deposit layer by layer polyélectrolytes on gold nanoparticles. We also note that the absorption spectrum obtain after deposition depends on the nature of the polyélectrolytes since they don't have the same refractive index. Since the refractive index of the surrounding medium is increase. For both Au@PEI and Au@@PEI@PSS@PEI, (5% and 10%) there is a red-shift of 5nm in the resonance band. In this case, the shift observed in the optical band of gold nanoparticles is due to the increase of the average refractive index of the surrounding medium of the nanoparticles i.e. to

the coating of nanoparticles by a layer of PEI. This value corresponds to the expected value obtained by theoretical data.

These results shown us that gold nanoparticles surface can be easily modified par different molecules such as organics dyes and polyelectrolytes. Adsorption of these molecules onto gold nanoparticles surface is principally due to the presence of atoms having high affinity with gold.

REFERENCES

[1]. Narayanan, R and El-Sayed, M. A., Langmuir, 21, 2027, 2005.
[2]. Thompson, D. T., Nanotoday, 2, 40, 2007.
[3]. Salata, O.J. J., Nanobiotechnol., 2, 1, 2004,
[4]. Salem, A. K. et Saerson, P. C and Leong, K. W., Natrure Mater., 2, 668. 2003,
[5]. Pissuwan, D. J. et Velenzuela, S. M and Cortie. M. B., Trends in Biotechnol, 24,62, 2008.
[6]. Huang, Y. et Duan, X and Lieber, C. M., Science, 291, 630, 2001
[7]. Mallouk, T. E and Kovtyukhova., N. I., Chem. Eur. J., 8, 4354, 2002.
[8]. Ramakrishna, G., et al., J. Am. Chem. Soc., 130, 5032, 2008.
[9]. Polavarapu, L., et al., Appl. Phys. Lett., 92, 263110, 2008.
[10]. West, R. et Wang, Y and Goodson, T., J. Phys. Chem. B., 107, 3419. 2003.
[11]. Nie, S. et Emory, S. R., Science, 275, 1102. 1997.
[12]. Jiang, J., et al., J. Phys. Chem. B., 107, 9964. 2003.
[13]. Iler, R., J. Colloid Interface Sci., 21, 569, 1966.
[14]. Decher, G and Hong, J-D., Ber. Bunsenges. Phys. Chem., 95, 1430, 1991.
[15]. G., Decher., Science, 227, 1232, 1997.
[16]. Thomas, K. G. et Kamat, P. V., 36, 88, 2003.
[17]. Hranisavljevic, N., Dimitrijevic, N. M. et Wurtz, G. A., Wiederrecht, G. P., J. Am. Chem. Soc., 124, 4536, 2002.
[18]. Fofang, N. T., et al. 2008, Nano Letters, Vol. 8, p. 3481.
[19]. Frens, G., Nature Phys. Sci., 20, 241, 1973
[20]. Gole, A and Murphy, C. M., Chem. Mater.,16, 3633, 2004.
[21]. Leff, D.V. et Brandt, L and Heath, J. R., Langmuir, 12, 4723, 1996.
[22]. X Y Chen, J R Li and L Jiang, Nanotechnology, 11, 108 2000.
[23]. Brown, L. O and Hutchison, J. E., J. Am. Chem. Soc., 121, 882, 1999.

Electron instabilities in nanoclusters, nanostructures and inhomogeneous nanomaterials: bottom up approach

A. N. Kocharian*, G. W. Fernando**, and K. Palandage** and J. W. Davenport***

* California State University, Los Angeles, CA, USA, armen.kocharian@calstatela.edu
** Department of Physics, University of Connecticut, Storrs, CT, USA
*** Center for Functional Nanomaterials, Brookhaven National Laboratory, Upton, NY, USA

ABSTRACT

Spatial inhomogeneities in nanomaterials with strong local electronic correlations are creating a new world vision in condensed matter physics. These nanomaterials with intrinsic inhomogeneities have electronic and magnetic properties that are profoundly different from the conventional materials. Few prominent examples are the high temperature superconductors (HTSCs), colossal magnetoresistance (CMR) and multiferroic materials. The behavior of electrons in contrasting bipartite and nonbipartite topologies are quite different. The electron pairs in ensemble of clusters in a real space are complying to the Bose-Einstein and Fermi-Dirac statistics can be condensed or fully polarized in inhomogeneous media. Phase diagrams display a number of inhomogeneous, coherent and incoherent nanoscale phases seen recently in high T_c cuprates, manganites and CMR nanomaterials using scanning tunelling microscopy.

Keywords: high T_c superconductivity, incoherent pairing, spontaneous phase separation, crossover, charge and spin (pseudo)gaps, magnetism

1 INTRODUCTION

Despite tremendous experimental and theoretical efforts, there is still no microscopic theory that can yield comprehensive support for pairing correlations, phase separation and pseudogap phenomena in clusters, small nanoparticles, transition metal oxides and high-Tc superconductors (HTSCs). The concentrated materials with strong local electronic correlations, intrinsic inhomogeneities seen in scanning tunelling microscopy (STM) measurements [1]–[6] have properties that are quite different from the conventional solids. The HTSCs, multiferroics and colossal magnetoresistance (CMR) materials [7]–[9] have been under scrutiny for nearly three decades about a role of Coulomb interaction in mechanisms of electron instabilities and pairing. Strong intraatomic (local) interactions can contain the key to some of the perplexing physics observed in the these materials using STM. The test of these conjunctures proposed early on by Anderson [10] for large thermodynamic systems is currently unavailable for two and three dimensional Hubbard-like systems. Motivated by this chal-lenging problem and discovery of novel materials, we have embarked on a study of finite size two and three dimensional Hubbard clusters using the exact diagonalization technique in quantum statistical mechanics. From this perspective, exact studies of electron charge and spin instabilities and quantum critical points (QCPs) at various inter-site couplings $U > 0$ and cluster topologies in the ground state and finite temperatures can provide important answers for the understanding of charge/spin inhomogeneities and local deformations for the mechanism of pairings and magnetism in "large" concentrated systems in the absence of long range order. The exact calculations of thermodynamic properties go beyond the reach of approximate schemes. At sufficient low temperatures, the charge and spin redistribution in an ensemble of clusters can produce incipient inhomogeneities typical for nano and heterostructured materials. Pairing instabilities, phase separation and inhomogeneities in certain regions of phase space in simple Hubbard clusters have been observed in our previous studies (see Refs. listed in [11]–[15]). In our opinion, these regions of phase space have much more intriguing clues to offer, when other variables such as the temperature, chemical potential and magnetic field are included.

Magnetic inhomogeneities seen in various transition metal oxides at the nanoscale level, widely discussed in the literature, are crucial for the understanding of spin pairing instabilities, origin of ferromagnetism [16], superconductivity and ferroelectricity. Important aspects of spontaneous phase separation and electron instabilities that we found in a real space [13] strongly depend on both, the Coulomb repulsion U and cluster topology. For instance in bipartite geometries, the charge separation leads to coherent pairing also at small and moderate U values, while Nagaoka type ferromagnetic instabilities for spins occurs at large U [16]. In frustrated geometries, spontaneous transitions lead to coherent pairing and saturated ferromagnetism for all U depending on the sign of the hopping term t (energy spectrum). In what follows, we identify these phenomena as electron instabilities in a phase space defined by suitable variables. We observe such instabilities due to competition between high and low spin states. Such phenomena appear to be generic provide important clues to long standing mystery tied to spontaneous phase separation in the HTSCs and CMRs.

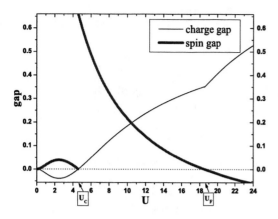

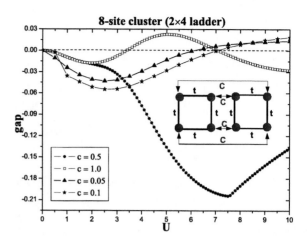

Figure 1: Charge Δ^c and spin Δ^s gaps versus U in an ensemble of squares at $\langle N \rangle \approx 3$ and $T = 0$. Phase A: Charge and spin pairing gaps of equal amplitude at $U \leq U_c$ describe bose condensation of electrons similar to BCS-like coherent pairing with a single energy gap. Phase B: Mott-Hubbard-like insulator at $U_c < U < U_F$ leads to $S = \frac{1}{2}$ spin liquid behavior. Phase C: Parallel (triplet) spin pairing ($\Delta^s < 0$) at $U > U_F$ displays $S = \frac{3}{2}$ saturated ferromagnetism.

These exact results have direct practical applications in search of novel assembled nanoclusters, inhomogeneous nanomaterials.

2 BASIC METHODOLOGY

2.1 Model

A key aspect of our study is the Hubbard model

$$H = -t \sum_{\langle ij \rangle \sigma} c_{i\sigma}^{+} c_{j\sigma} + U \sum_{i} n_{i\uparrow} n_{i\downarrow} \qquad (1)$$

with nearest neighbor hopping t and on-site Coulomb interaction $U > 0$. Focusing on this simple model gives unprecedented advantages to analyze (with great accuracy) the many body correlation effects, which are nontrivial. In our previous publications, we have outlined most of the details pertaining to the method (see Refs. [11]–[15]).

2.2 Canonical gaps

Depending on the value of the Coulomb repulsion U, Hubbard model for small clusters exhibits different pairing behavior which is evident when suitable gaps are defined and their properties examined. In a particular doping region, with one hole off half filling, when the chemical potential μ lies in an interval $[\mu_+(T), \mu_-(T)]$ with $\mu_- > \mu_+$, charge pairing is found. The boundaries of the interval are defined as $\mu_+(T) = E(N+1) - E(N)$

Figure 2: Charge gaps for the 2x4 cluster at $T = 0$ for various couplings c between the ensemble of coupled square clusters. The doping level is one electron off half-filling and the couplings t within the squares are set to 1, as indicated. There is an effective electron-electron attraction in the negative gap regions which corresponds to the coherent state with the charge and spin pairing gaps of equal amplitude, $\Delta^s = -\Delta^c$.

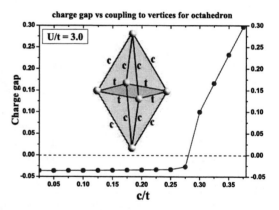

Figure 3: The zero temperature charge gap at one hole off half filling in an octahedron (at $U/t =3$) as a function of the coupling strength c/t to the vertices. The negative charge gap up to $c/t =0.28$ displays a necessary condition for charge pairing instability in deformed octahedron structures. This also underlines the role of vertex coupling and the quasi two-dimensional character of pairing which may be related to HTSC perovskites.

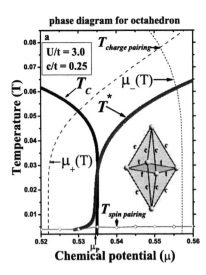

phase diagram for octahedron

$U/t = 3.0$
$c/t = 0.25$

$T_{charge\ pairing}$

T_c

$\mu_-(T)$

T^*

$\mu_+(T)$

$T_{spin\ pairing}$

μ_p

Temperature (T)

Chemical potential (μ)

Figure 4: A part of the phase diagram for the octahedron ($U/t = 3$) in Fig. 3 at $c/t = 0.25$ showing the region of instability in the chemical potential with charge and spin susceptibility peak positions (T_c and T^*) as well as spin and charge pairing temperatures. The spin rigidity of the pairs in the plane is similar in both cases although the octahedron spin susceptibilities are, of course, affected by the spins attached to the vertices.

and $\mu_-(T) = E(N) - E(N - 1)$ where $E(N)$ denotes the average canonical energy in an N electron state at temperature T, and the charge gap is defined as $\Delta^c(T) = \mu_+ - \mu_-$. The gap at $T \geq 0$, which depends on U, has to be negative for a spontaneous (charge) phase separation to occur. Analogously, we calculate a canonical spin gap as the difference in the average energies between the two cluster configurations with spin S and $S + 1$ states, $\Delta^s(T) = E(S + 1) - E(S)$, where $E(S)$ is the average canonical energy in the spin sector at fixed N.

2.3 Pairing instabilities

We define critical parameters for the level crossing degeneracies or corresponding QCPs from the condition $\Delta^{c,s}(U_c, \mu) = 0$. The sign of the gap manifests the regions for electron charge and spin instabilities, such as the electron-electron ($\Delta^c < 0$), electron-hole ($\Delta^c > 0$) pairings in the charge sector or the parallel ($\Delta^s < 0$) and opposite ($\Delta^s > 0$) spin pairings in the spin sector. The relationship between the charge gap Δ^c, and its corresponding spin counterpart Δ^s, is important in identifying the stability of phases in various cluster topologies [13]. A positive excitation gap indicates phase stability and (smooth) transition or crossover, while a negative excitation gap in the many body ground state shows an energy instability for spontaneous phase transition.

2.4 Grand canonical gaps

In the grand canonical approach, using exact analytical expressions for the grand canonical potential and partition functions, we can monitor the charge $\chi_c(\mu)$ and zero field spin $\chi_s(\mu)$ susceptibilities as a function of the chemical potential μ and temperature. The energy difference in terms of μ between the two consecutive susceptibility peaks can also serve as a natural order parameter. The distances between the peaks in double peak structures in the charge and spin density of states determine *pseudogap*. In equilibrium, the critical temperatures T_c and T^* are defined as the temperature at which the separation between the two consecutive charge or zero spin susceptibility peaks vanish. The energy (pseudo)gaps obtained from the maxima according are positive and, therefore, the transitions in the grand canonical method are always smooth.

3 DISCUSSION

3.1 Bipartite clusters

Here we carry out numerical calculations for bipartite linked 4-site clusters. Fig. 1 illustrates the charge and spin gaps at small and moderate U in ensemble of square clusters (see details Ref. [13]). The vanishing of gaps at QCPs, $U_c = 4.584$ and $U_F = 18.583$, indicates energy level crossings and electron instabilities for charge and spin, respectively. Similar QCPs and instabilities we find also in Fig. 2 for coupled 2x4 clusters, where the hopping term or coupling c between the two squares was allowed to be different from the coupling within a given square. The pairing fluctuations that are seen for the 4-site cluster exist even for these ladders at optimal doping, and most of the trends observed for the 4-site clusters remain valid here. Thermal and quantum fluctuations in the density make it energetically more favorable to form charge pairs.

3.2 Frustrated clusters

The tetrahedron cluster has a topology of square with the next nearest neighbor coupling may be regarded as a primitive unit of frustrated system [13]. Here we consider an ensemble of octahedron (or square pyramids) clusters, which are a building blocks of perovskite structure in HTSC cuprates [12]. The parental (undoped) material is an insulator. However, when excess apex oxygen is introduced, hole carriers are supplied into CuO_2 planes, and material shows superconductivity. Figure 3 shows the charge gap at fixed $U = 3$ and $\langle N \rangle \approx 4$ under the variation of the coupling term c between the plane and the apex atoms in $Bi_2Sr_2CaCu_2O_{8+\delta}$. This picture gives surprisingly plausible evidence for understanding the detrimental role of excess electron on charge pairing for possible distortions of pyramidal crystalline structure in perovskites [6]. The negative charge gap, identical to the spin gap, exists only below this QCP, $c/t = 0.28$ for level crossing degeneracy. Calculated electron distribution, as a function of c, shows that electron charge residing on the apical site does not contribute to the pairing whenever c is less than c_0. In Fig. 3, the distortion of the octahedron reproduces a charge pairing gap modulation seen in STM imaging experiments [6]. Thus we find a direct correlation between the size of the energy gap characterizing the local superconducting state and a modulation of the atomic positions seen in HTSCs [6]. These results illustrate the relevance of two dimensionality as observed in the perovskite structures, and the coupling values in this region lead to charge pairing, for one hole off half filling, as seen in the bipartite clusters; spin pairing is seen to occur at a much lower temperature as evident from the phase diagram in Fig. 4.

3.3 Nagaoka Ferromagnetism

For large U in the (repulsive) Hubbard model at one hole off half filling, there is a well-known Nagaoka theorem known for bipartite and nonbipartite topologies at $T = 0$, which claims that in higher dimensions, such systems would exhibit fully saturated ferromagnetism. Our exact results, obtained for non zero temperatures, are supporting the above statement [16], [13]. Cluster frustration appears to play a significant role. For $t = +1$, there is no spin saturation (which has also been verified at large U). For $t = -1$, there is a saturated ferromagnetic state ($S_z^{max} = 5/2$), which is insulating, at sufficiently large U (in units of $|t|$) as shown in the figure. Therefore, in these frustrated systems with spin degeneracies in the energy spectrum, (ferromagnetic) Nagaoka saturation is observed only for one type of hopping ($t = -1$).

4 KEY RESULTS

Pairing instabilities and inhomogeneities found from exact diagonalization in small bipartite and non-bipartite topologies provide novel insights into several mysterious many body problems in condensed matter physics and ultracold fermionic atoms. Rigorous Nagaoka type criteria are formulated for spontaneous phase separation, electron instabilities and magnetism driven by interaction strength, geometrical frustration, inter-site couplings (connectivity), transverse magnetic field, etc. The following is a list of properties, resulting from our exact Hubbard cluster studies:

(i) Vanishing charge and spin pairing gaps in the ground state can be directly linked to the QCPs [13].

(ii) Negative charge gap at a critical set of parameters is leading to electron charge pairing and spontaneous phase separation below T_c^P.

(iii) The pseudogap, inhomogeneities, incoherent pairing behavior below T^* but above spin pairing T_s^P similar to what is seen in the NMR and STM experiments [5].

(iv) Formation of coherent state of rigidly bound opposite spin and (doubled) paired charge below T_s^P.

(v) Phase diagrams describes pair modulation and multitude of fascinating phases, seen recently in high T_c cuprates, manganites and CMR nanomaterials using scanning tunelling microscopy [1]–[6].

(vi) The presence of a dormant magnetic state above the spin pairing temperature, observed recently [3].

(vii) Negative spin pairing gap is leading to electron spin pairing, spontaneous phase separation and Nagaoka ferromagnetism at zero and non zero temperatures [13].

Our exact calculations in finite size clusters have strong impact and immediate applications to nano systems (nanotechnology) can motivate further studies of electronic and magnetic instabilities with correlated electrons in contrasting cluster topologies. These ideas could be exploited in the nanoscience frontier by synthesizing clusters or nanomaterials with unique magnetic and electronic properties.

REFERENCES

[1] Y. Kohsaka, et al., Science, 315, 1380 2007.
[2] T. Valla, et al., Science, 314, 1914, 2006.
[3] H. E. Mohottala, et al., Nature Materials, 5, 377, 2006.
[4] M. C. Boyer, et al., Nat. Phys., 3, 802, 2007.
[5] K. K. Gomes, et al., Nature, 447, 569, 2007.
[6] J.A. Slezak, et al., Proc. Natl. Acad. Sci. USA **105** (2008) 3203.
[7] J. M. Tranquada, et al., Nature 375, 561, 1995.
[8] E. L. Nagaev, Physics - Uspechi 39, 781, 1996.
[9] E. Dagotto, Science, 309, 257, 2005.
[10] P. W. Anderson, Science, 235, 1196, 1987.
[11] A. N. Kocharian, G.W. Fernando, C. Yang, Spin and charge pairing instabilities in nanoclusters and nanomaterials pairing, in Scanning probe microscopy in nanoscience and nanotechnology, Ed. B. Bhushan, Springer, Ch 15, pp. 507-570, 2010.
[12] G. W. Fernando, A. N. Kocharian, K. Palandage, and J. W. Davenport, Phys. Rev., B80, 014525, 2009.
[13] A. N. Kocharian, G. W. Fernando, K. Palandage, and J. W. Davenport, Phys. Rev., B78, 075431, 2008.
[14] G. W. Fernando, A. N. Kocharian, K. Palandage, T. Wang, and J. W. Davenport, Phys. Rev., B75, 085109, 2007.
[15] A. N. Kocharian, G. W. Fernando, K. Palandage, and J. W. Davenport, Phys. Rev., B74, 024511, 2006.
[16] Y. Nagaoka, Phys. Rev., 147, 392, 1966.

Ultrasonically sprayed nanostructured SnO$_2$ thin films for highly sensitive hydrogen sensing

L.A. Patil

Nanomaterials Research Lab., Department of Physics, Pratap College, Amalner, 425401, India

ABSTRACT

Nanostructured SnO$_2$ thin films were prepared by ultrasonic spray pyrolysis technique. Aqueous solution (0.05M) of SnCl$_4$·5H$_2$O in double distilled water was chosen as the starting solution for the preparation of the films. The stock solution was delivered to nozzle with constant and uniform flow rate of 70 ml/h by Syringe pump SK5001. Sono-tek spray nozzle, driven by ultrasonic frequency of 120 kHz, converts the solution into fine spray. The aerosol produced by nozzle was sprayed on glass substrate heated at 150°C. The sensing performance of the films was tested for various gases such as LPG, hydrogen, ethanol, carbon dioxide and ammonia. The sensor (30 min) showed high gas response (S = 3040 at 350 °C) on exposure of 1000ppm of hydrogen and high selectivity against other gases. Its response time was short (2 s) and recovery was also fast (12 s). To understand reasons behind this uncommon gas-sensing performance of the films, their structural, microstructural, and optical properties were studied using X-ray diffraction, electron microscopy (SEM and TEM), and UV–vis spectroscopy, respectively. The results are interpreted.

Keywords: Ultrasonic spray pyrolysis technique, Nanostructured SnO$_2$ thin films, Hydrogen sensing, Gas response, Selectivity Response time, Recovery time

1. INTRODUCTION

There has been considerable interest in recent years in nanostructured gas-sensing materials in thin film form [1-13]. In comparison with conventional sintered bulk gas sensors, nanostructured thin film materials have good response, optimum operating temperature and selectivity. In addition, sensors based on thin film gas-sensing materials are essential to the development and fabrication of integrated gas sensors. For industrial applications, a low cost method is vital; hence, an ultrasonic spray pyrolysis technique is often preferred. The ultrasonic spray method(uses nebulizer) ensures generation of ultra fine droplets of narrow size distribution in the range of a few micrometers, whereas an ordinary spray method (uses spray gun) gives rise to droplets of varied sizes ranging from micrometers to a few tenths of a millimeter. Hence, the ultrasonic spray method, in present article, may be the best approach to obtain highly uniform spot-free films. Nanostructured SnO2 based hydrogen sensors prepared by ultrasonic spray pyrolysis technique are reported in present article. The most important features of the present

investigation are that the sensors show high hydrogen response, selectivity, fast response and quick recovery.

2. EXPERIMENTAL

2.1. Experimental set up to prepare nanostructured thin films

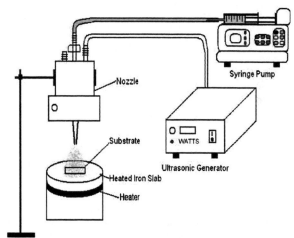

Fig. 1: Ultrasonic spray pyrolysis set up

Fig. 1 shows an ultrasonic spray pyrolysis set up to prepare nanostructured thin films. High frequency (120 kHz) from ultrasonic generator (120 kHz) is applied to piezoelectric transducer associated in nozzle which converts the precursor solution into the smoke or spray. This spray was used to prepare nanostructured thin films. The stock solution (0.05M) of SnCl$_4$·5H$_2$O in double distilled was delivered to nozzle with constant and uniform flow rate of 70 ml/h by Syringe pump SK5001. The aerosol produced by nozzle was sprayed onto the glass substrate heated at 150°C. A pyrolytic reaction takes place and nanocrystalline metal oxide was produced. The films were prepared by spraying for different spraying time intervals (5 min, 10 min, 20 min and 30 min) onto a glass substrate. Different spraying time intervals would give films of different thicknesses. As prepared thin films were fired at 500°C for 30 min. Gas-sensing properties of the films were tested.

2.2. Material characterizations
2.2.1 Structural properties: X-ray diffraction studies

Crystallite sizes of the samples with spraying time of 5min and 30 min were calculated (From fig 2) using Scherer formula which was found to be 14nm and 16.66

nm, respectively, indicating their nanocrystalline nature. XRD corresponding to 30 min film has sharper peaks as compared to 5min film. Crystallite size has a direct dependence on spraying time interval and thickness of the film.

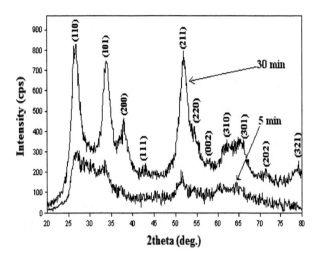

Figure 2. XRDs of SnO$_2$ thin films

2.2.2 Transmission electron microscopy (TEM).

Fig. 3 shows the TEM [CM 200 Philips (200kV HT)] of SnO$_2$ powder obtained by scratching the thin film (30 min). It is clear from TEM image that the grains are nanocrystalline in nature which are smaller than 7 nm with narrow size distribution. Surface to volume ratio of such nanoparticles is very high. Therefore, these particles are highly reactive to exposed gases. The interaction between the gases and the sensitive layer is limited to the surface itself [5]. This is one of the reasons that nanostructured films give high gas response.

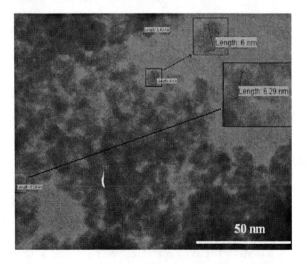

Fig. 3. TEM SnO$_2$ (30 min) thin film.

2.2.3 Absorption spectra

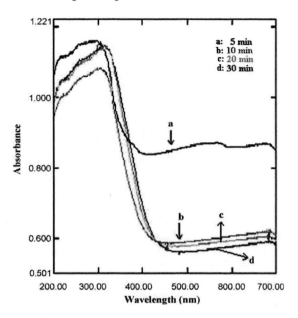

Fig.4. Absorption spectra SnO$_2$ thin films.

2.3 Gas sensing performance
2.3.1 Gas response and Selectivity

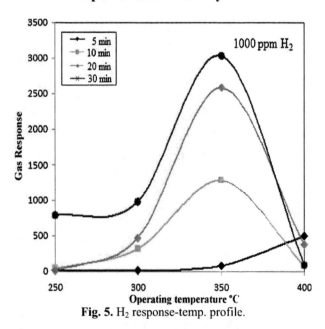

Fig. 5. H$_2$ response-temp. profile.

Hydrogen sensing performance of nanostructured SnO$_2$ thin film sensors was measured by using static gas sensing system [8]. It is clear from the Fig. 5 that the gas response increases with operating temperature, reaches to maximum (S=3040) at 350°C and falls with further increase in operating temperature.

2.3.2 Selectivity

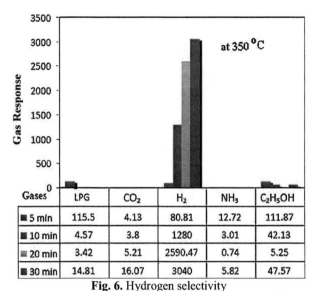

Fig. 6. Hydrogen selectivity

Fig. 6 shows selectivity of nanostructured SnO_2 thin films to H_2 against CO_2, LPG, NH_3 and ethanol gases at 350 ^{0}C.

2.3.3 Response and recovery of the sensor

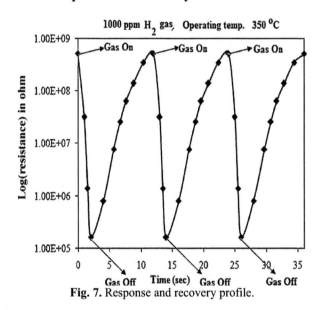

Fig. 7. Response and recovery profile.

Response and recovery of pure-SnO2 based thin film (30 min) was studied on exposure of 1000ppm hydrogen. Variation of resistance (of this film) with time was measured by using Agilent data acquisition/switch unit (Model 34970A). Typical hydrogen on-off cycle was used to monitor the flow of hydrogen on the film surface. After every new cycle, the chamber was refreshed with air. Fig. 11 represents response and recovery profile of the sensor. The sensor responded quickly (~2 s) and the recovered back fastly (~12 s). The negligible quantity of the surface reaction products and their high volatility explain the quick response to H2 and fast recovery to its initial chemical status.

2.4 References

[1] J. Hotovy, D. Bue, S. Hascik, O. Nennewitz, Characterization of NiO thin films deposited by reactive sputtering, Vacuum 50 (1998) 41–44.

[2] C.C. Chai, J. Peng, B.P. Yan, Characterization of _-Fe2O3 thin films deposited by atmospheric pressure CVD onto alumina substrates, Sens. Actuators, B 34 (1996) 412–416.

[3] M. de la, L. Olvera, R. Asomoza, SnO2 and SnO2:Pt thin films used as gas sensors, Sens. Actuators, B 45 (1997) 49–53.

[4] L.A. Patil*, M.D. Shinde, A.R. Bari, V.V. Deo, Highly sensitive ethanol sensors based on nanocrystalline SnO2 thin films, Current Applied Physics, (2010) 1-6

[5] I. Simon, N. Barson, Micromachined metal oxide gas sensors: opportunities to improve sensor performance, Sens. Actuators, B 73 (2001) 1–26.

[6] L.A. Patil *, M.D. Shinde, A.R. Bari, V.V. Deo, Novel trapping system for size wise sorting of SnO2 nanoparticles synthesized from pyrolysis of ultrasonically atomized spray for gas sensing, Sensors and Actuators B 143 (2009) 316–324

[7] D.R. Patil, L.A. Patil, Cr2O3-modified ZnO thick film resistors as LPG sensors, Talanta 77 (2009) 1409–1414

[8] L.A. Patil *, M.D. Shinde, A.R. Bari, V.V. Deo, Highly sensitive and quickly responding ultrasonically sprayed nanostructured SnO2 thin films for hydrogen gas sensing, Sens. Actuators, B 143 (2009) 270–277

[9] Y.H. Choi, S.H. Hong, H2 sensing properties in highly oriented SnO2 thin films, Sens. Actuators, B 125 (2007) 504–509.

[10] R.S. Niranjan, Y.K. Hwang, D.K. Kim, S.H. Jhung, J.S. Chang, I.S. Mulla, Nanostructured tin oxide: synthesis and gas-sensing properties, Mater. Chem. Phys. 92 (2005) 384–388.

[11] V.A. Chaudhary, I.S. Mulla, K. Vijayamohanan, Selective hydrogen sensing properties of surface functionalized tin oxide, Sens. Actuators, B 55 (1999) 154–160.

[12] K.S. Yoo, S.H. Park, J.H. Kang, Nano-grained thin film indium tin oxide gas sensor for H2 detection, Sens. Actuators, B 108 (2005) 159–164.

[13] H.R. Kim, K.I. Choi, J.H. Lee, S.A. Akbar, Highly sensitive and ultra-fast responding gas sensors using self-assembled hierarchical SnO2 sphere, Sens. Actuators, B 136 (2009) 138–143.

Synthesis of highly doped silicon and germanium nanoparticles in a low-pressure plasma-reactor for thermoelectric and solar applications

H. Grimm[*], N. Petermann[**], A. Gupta[***] and H. Wiggers[****]

[*]Institute of Combustion and Gas Dynamics, University of Duisburg-Essen
47048 Duisburg, Germany, helge.grimm@uni-due.de
[**]Institute of Combustion and Gas Dynamics, University of Duisburg-Essen
47048 Duisburg, Germany, nils.petermann@uni-due.de
[***]Institute of Combustion and Gas Dynamics, University of Duisburg-Essen
47048 Duisburg, Germany, anoop.gupta@uni-due.de
[****]Institute of Combustion and Gas Dynamics and CeNIDE, University of Duisburg-Essen
47048 Duisburg, Germany, hartmut.wiggers@uni-due.de

ABSTRACT

Microwave-plasma induced thermal decomposition of gaseous precursors like monogermane (GeH_4) and monosilane (SiH_4) is used to synthesize semiconductor nanoparticles for thermoelectric and solar applications. P- and n-doping can be achieved by mixing silane and germane with diborane (B_2H_6) and phosphine (PH_3), respectively. Adjusting the fraction of the respective precursor in the gas phase is used to control the dopant concentration. Steep temperature-time profile and kinetic control of particle formation enable for synthesizing spherical and loosely agglomerated nanoparticles with very high concentration of dopant beyond the solubility limit. The mean particle diameter can be tuned between 3 and more than 50 nm by variation of microwave-power, plasma-gas-flow, reactor pressure and precursor concentration. This enables us for producing tailor-made nanosized materials with respect to energy conversion in thermoelectrics and heterojunction solar-cells.

Keywords: silicon, doped, nanoparticles, synthesis, plasma

1 INTRODUCTION

Silicon is the most important semiconductor material and due to its very good availability and it is in the focus of a lot of solar or thermoelectric applications like heterojunction solar-cells [1] or thermoelectric generators [2,3]. There are several methods to synthesize e.g. silicon nanoparticles like solution phase synthesis [4,5], ion implantation [6] or multilayer CVD [7]. However, the gas phase synthesis has a significant scale-up possibility for industrial production. With this synthesis approach, nanoparticles with very high purity can be synthesized and it is a very cost effective method. With highly doped silicon nanoparticles one can ensure very good ohmic contacts on solar cells [8] and mixing of highly doped silicon and germanium nanoparticles is investigated to achieve high performance SiGe thermoelectric devices [9]. The structure of the nanopowder used for thermoelectric applications can be preserved during sintering to reduce thermal conductivity while the electrical conductivity of sintered samples can be adjusted with respect to dopant concentration.

2 EXPERIMENTAL

Doped as well as undoped silicon and germanium nanoparticles are synthesized in a low-pressure microwave plasma reactor. The pressure in the reactor is controlled in the range between 5 and 500 mbar (absolute) by a PID controller. Monosilane (SiH_4) or monogermane (GeH_4) is mixed with argon (Ar) and hydrogen (H_2) before being fed into the reactor. For optional n- or p-type doping, precursors like phosphine (PH_3) or diborane (B_2H_6) were added to the gas mixture, respectively. The precursor gas mixture is guided through a microwave-supported plasma ignited within a quartz gas tube that is located in the center of a microwave resonator. The microwave source is operated at a frequency of 2.45 GHz and enables a maximum microwave power of 6 kW. The composition of the precursor mixture can be adjusted in a wide range and enables for the synthesis of nanomaterials with doping concentrations ranging between 10^{16} and 10^{21} cm^{-3}. The gas temperature measured from the emission of the plasma by means of optical emission spectroscopy typically ranges between 900 and 1300 K depending on pressure, microwave power, and gas composition. The as-prepared particles are separated from the gas flow by means of a filter device and can be harvested under inert conditions to prevent them from oxidation. Figure 1 shows a brief sketch of the experimental setup used for the synthesis of the nanoparticles.

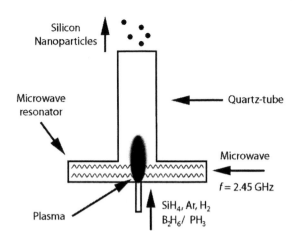

Figure 1: Schematic representation of the experimental setup for the synthesis of semiconductor nanomaterials.

3 RESULTS AND DISCUSSION

The size, shape and crystallinity of nanoparticles were analyzed by transmission electron microscopy (TEM) using a Philips CM12 at an accelerating voltage of 120 kV. The TEM specimens were prepared by ultrasonic dispersion of the nanoparticles in cyclohexane or methanol and dipping the carbon-coated Cu grids into the dispersion with subsequent drying. The mean size of the nanoparticles was investigated by means of gas-adsorption according to Brunauer, Emmett and Teller (BET) [10] using a Quantachrome Nova 2200 with N_2 as adsorption gas. The crystallinity and size of the nanoparticles were determined with a PANalytical X-ray diffractometer (X'Pert PRO) with Cu-K_α radiation and a subsequent Rietveld refinement with help of the MAUD program [11].

The nanoparticles exhibit very high crystallinity and almost every sphere consists of one single crystal (see fig. 2 and 3).

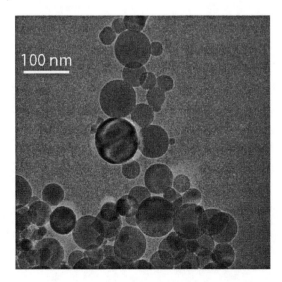

Figure 2: Single crystalline silicon nanoparticles from microwave plasma synthesis.

The results from BET analyses regarding the mean diameter of the particles comply with measurements from TEM and XRD (data not shown). With respect to synthesis conditions, parcels ranging from 3 to more than 50 nm in size are accessible. While low pressure (about 10 mbar) and low precursor concentration in the range of a few thousand ppm result in very small particles with a few nm in diameter, higher pressure as well as higher precursor concentration are necessary for the formation of bigger ones. Due to a limited residence time of the gas mixture within the hot reaction zone, the maximum particle size levels around 100 nm.

When the particles come into contact with the atmosphere the surface of the nanoparticles oxidizes. As an example, we show the HRTEM image of silicon nanoparticles as shown in figure 3. The lattice fringes clearly demonstrate that the particles are highly crystalline and that they are covered with an amorphous oxide layer. Elemental analysis of the as-prepared materials ensured that — despite the oxygen covering the particle surface — the chemical composition of the product is identical to what is expected from the gas-mixture introduced into the reactor.

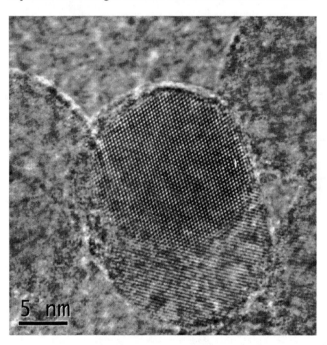

Figure 3: High resolution TEM image illustrating the crystallinity of the particles.

4 SUMMARY

We synthesized silicon and germanium nanoparticles by thermal decomposition of silane or germane in a microwave induced plasma reactor. The nanoparticles can be doped with phosphine or diborane by adding the dopant into the gas flow and the reactor enables for a quantitative decomposition of the precursor material. The nanoparticles were characterized by using BET, TEM and XRD. These particles form an oxide layer on their surface under ambient conditions. The mean diameter of the particles can be controlled by varying the microwave power, the pressure and the temperature in the reactor. The doping concentration can be precisely controlled by adjusting the gas flow of dopant inside the reactor chamber.

ACKNOWLEDGMENTS

The authors gratefully want to acknowledge the financial support by the European Union and the Ministry of Economic Affairs and Energy of the State North Rhine-Westphalia in Germany (Objective 2 Programme: European Regional Development Fund, ERDF). The authors would like to thank Z. A. Li for his support with HRTEM measurements.

REFERENCES

[1] K. v. Maydell, E. Conrad and M. Schmidt, Progress in Photovoltaics: Research and Applications 14, 289-295, 2006

[2] G. Jeffrey Snyder and Eric S. Toberer, Nature materials 7, 105-114, 2008.

[3] M. S. Dresselhaus et al., Adv. Mater. 19, 1043-1053, 2007

[4] J. Zou, R. K. Baldwin, K. A. Pettigrew, S. M. Kauzlarich, NanoLett. 4, 1181, 2004

[5] Richard A. Bley and Susan M. Kauzlarich, Journal of the American Chemical Society 118 (49), 12461-12462, 1996

[6] Y. Q. Wang, R. Smirani and G. G. Ross, Nano Letters 4 (10), 2041-2045, 2004

[7] M. Zacharias, J. Heitmann, R. Scholz, U. Kahler, M. Schmidt and J. Blasing, Appl. Phys. Lett. 80, 661, 2002

[8] A. R. Stegner, R. N. Pereira, K. Klein, R. Lechner, R. Dietmueller, M. S. Brandt, M. Stutzmann, H. Wiggers, Phys. Rev. Lett. 100, 026803, 2008

[9] G Savelli et al., J. Micromech. Microeng. 18, 104002, 2008

[10] S. Brunauer, P. Emmet, and E. Teller, J. Am. Chem. Soc. 60, 309, 1938

[11] L. Lutterotti, M. Bortolotti, G. Ischia, I. Lonardelli and H. R. Wenk, Z. Kristallogr. Suppl., 26, 125, 2007

Cleaning 10nm ceramic particles by the supersonic particle beam

Inho Kim*, Kwangseok Hwang* and Jin W. Lee*

*Department of Mechanical Engineering, POSTECH
Hyoja 31, Pohang, South Korea, jwlee@postech.ac.kr

ABSTRACT

Cryogenic aerosol beam using micron-sized aerosol particles has long been successfully used to remove contaminant particles down to 50nm, and supersonic particle beam using particles smaller than 100nm lowered the limit of cleaning down to 20nm size. In this study, the supersonic particles beam technique was improved so that the cleaning limit could be lowered to 10nm size range for ceramic contaminant particles. Two different kinds of particles were generated using Ar, CO_2 gas with or without He carrier gas. Cleaning performance for 10nm contaminant particles was sensitive to the combined condition of the beam particle size and velocity. Optimum condition for particle size and velocity were sought by varying the nozzle contour and stagnation temperature and pressure. The best removal efficiency was about 90% for 10nm Al_2O_3 particle, which it is the best performance reported to date.

Keywords: supersonic particle beam, cryogenic particle beam, nano particle cleaning, supersonic nozzle

1 INTRODUCTION

Particle contamination seriously affects the manufacturing yield of submicron scale devices. Device feature for DRAM/flash memory is expected to decrease continuously, reaching 25nm by 2015, and together with it the critical defect size to decrease to 12.5nm by 2015 [1]. Various nanotechnology based devices with feature dimensions in the nanometer size range will also be marketed in due time, which may accelerate the decrease of the required killer defect size. Since the use of drag force becomes less efficient as the contaminant size is decreased, it is generally agreed that conventional techniques should work poorly for submicron particles, and state-of-the art of various cleaning technique stays around 50-90nm.[2, 3]

One promising technique applicable in the nanometer range is the cryogenic aerosol technique, where contaminated surface is bombarded by fine particles of volatile material at high velocity. Contaminant particles adhered on a surface can be removed when the energy transferred from the bullet particles is sufficient to overcome the adhesion energy between the contaminant and substrate. Lee et al.[4] showed that Argon bullet particles could remove contaminant particles effectively, and other studies reported the applicability of Argon aerosol technique to nano contaminant cleaning. Argon,

nitrogen, carbon dioxide and water are the most common cleaning agents used, and each offers advantages and disadvantages over the others. In current technology gas is expanded through a simple nozzle like a cylindrical hole. During the cooling process part of the gas becomes liquid, and the nozzle expansion atomizes the liquid into fine droplets. Solidification follows through further expansion. Typical particle size generated in this way is a few microns and the velocity about 100m/s, which is effective for cleaning down to 50nm but cleaning efficiency drops very rapidly for smaller particles thereof.[5,6]

Recently, Yi et al.[7] showed by MD simulation that far smaller bullet particles have to be used for removing nano-sized contaminants. Removal efficiency for nano-sized contaminants depended more on the velocity of the bullet particle, and concluded that even at the same kinetic energy level a smaller particle moving at a higher velocity should give a better removal. It was also shown that when the bullet particle was too big compared to the target contaminant, by a factor of 10 or more, the fragmented atoms/molecules of the bullet particle after collision may even surround the contaminant particle, preventing it from leaving the surface. Thus the use of smaller bullets moving at high velocities is expected to have extra advantages in cleaning narrow trenches and in reducing the damage potential. Lee et al.[4, 8] succeeded in removing 20nm ceramic and metal particles using bullet particles of 100nm or smaller diameter.

This study aims to explore the possibility of using the same bullet particles smaller than 100nm for removing nano-sized contaminants down to 10nm. Argon and CO_2 particles were generated by homogeneous nucleation and growth during supersonic expansion through a Laval nozzle. Particle size and velocity were varied by varying the nozzle contour, gas composition, and stagnation pressure and temperature. Cleaning experiments were done for a flat surface contaminated with a variety of particles of size down to 10nm.

2 EXPERIMENT

The experimental system is schematically shown in Fig.1. Gas mixture is passed through a liquid-N2 dewar and cooled to a required temperature by a temperature controller. Pre-cooled gas mixture is then introduced into a stagnation chamber, and then expands through a Laval nozzle into a vacuum environment. During a supersonic expansion through the Laval nozzle, tiny condensation nuclei are formed and grow in size, where the final size can be controlled by the stagnation pressure and temperature, back pressure of the vacuum chamber, and the shape of nozzle.

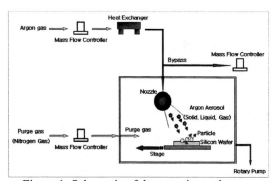

Figure 1: Schematic of the experimental setup.

In order to generate bullet particles of require size and velocity, four different gas mixtures - Ar, Ar/He, CO_2, CO_2/He - were used. Nozzle contour and nozzle dimensions were so determined that the Mach number at the nozzle exit is higher than 5.0 over the whole range of experimental conditions. Nozzles are nearly conical but rapidly expanding near the exit in order to prevent shock formation.

Silicon wafers coated with Al_2O_3 particles were exposed to the bullet particle beam for variable period of time up to 2-3minutes. Size of the contaminant particles was in the range of 5~80nm with nominal size of 10nm. Particles were dispersed in alcohol using an ultrasonic stirrer, and then dripped and spin-coated on a Si wafer. Coated wafers were dried with dry N_2 gas flowing for 1-7 days. SEM pictures of the wafer surfaces were taken before and after cleaning. Distance between the nozzle and the wafer and the angle of the particle beam were variable. Chamber pressure was also varied in the range of 10~200torr.

2.1 Cleaning with Ar and Ar/He

Cleaning experiment was first tried for Al_2O_3 particles on a flat Si surface. When pure argon was expanded at 1500torr, cleaning effect was almost zero. When the pressure was raised to 1850torr which was high enough for removing 20nm contaminants, particle removal was almost negligible (Fig.2). Aerosol particles generated at 1500torr and 1850torr are formed via homogeneous nucleation, but the size or the velocity is not good enough to remove 10nm contaminant particles. In order to vary the bullet size and velocity over a wide range via variation in stagnation pressure, mixtures of Ar and He were used, where Argon partial pressure was controlled so that no liquid formation occurs before the nozzle throat and particles are formed entirely by homogeneous nucleation and growth downstream of the throat. In order to generate bullet particles of required size and velocity, three different mixtures of 1:1, 2:8, 1:9 Ar/He were used, and stagnation pressure and temperature were varied in the range of 3000~37500 torr and 118~127K, respectively.

(a)

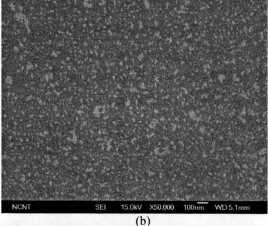

(b)

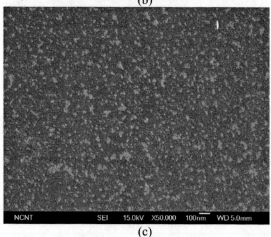

(c)

Figure 2: SEM images (a) before and after cleaning of 10nm Al_2O_3 particle using pure Argon at (b) 1500torr and (c) 1850torr, respectively.

Al_2O_3 particles were not cleaned when a 1:1 Ar/He mixture was expanded at 3000torr and 127K. (Fig. 3(a)) When 2:8 mixture gas was used at 8000torr and 118K, large Al_2O_3 particles were removed, but very small particles remained uncleaned - 10nm Al_2O_3 particles are still observed after cleaning (Fig. 3(b)). When 1:9 Ar/He

NSTI-Nanotech 2010, www.nsti.org, ISBN 978-1-4398-3401-5 Vol. 1, 2010

mixture was used at 37500torr and 123K, Al_2O_3 particles were completely removed (Fig. 3(c)). The use of Helium carrier increased particle velocity through increased sonic speed and also enhanced particle growth by removing condensation heat through collisions on growing bullets.

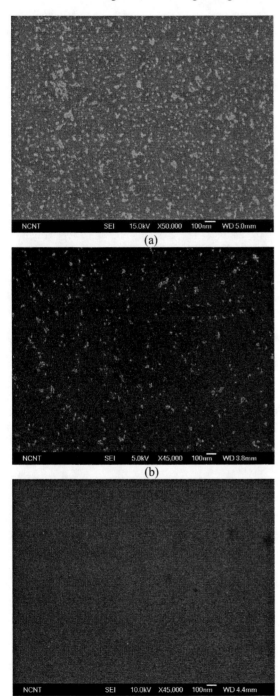

Figure 3: SEM images after cleaning of >10nm Al_2O_3 particles using (a) 1:1 Ar/He at 3000torr/127K, (b) 2:8 Ar/He at 8000torr/118K, and (c) 1:9 Ar/He at 37500torr/123K.

2.2 Cleaning with CO2 and CO2/He

In order to further increase the bullet particle velocity, CO_2 particles were generated starting from room temperature, almost three times as high as that used for Ar particle generation. To control the bullet particle size and velocity, stagnation pressure and CO_2/He mixture gas were used. When pure CO_2 gas was expanded at 10bar, cleaning effect was almost zero. When the stagnation pressure was raised to 30bar, a substantial fraction of the contaminant particles were removed (Fig. 4(a)). When the stagnation pressure was raised further to 50bar, Al_2O_3 particles were completely cleaned (Fig. 4(b)).

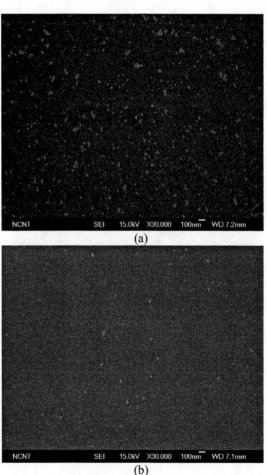

Figure 4: SEM images after cleaning of >10nm Al_2O_3 particle using pure CO_2 at (a) 30bar and (b) 50bar.

In order to further increase the velocity of the bullet particles, 1:1 CO_2/He mixture gas was used at various stagnation pressures. When this 1:1 mixture gas was used at 30bar, large particles were perfectly cleaned, but there still remained 10nm Al_2O_3 particles uncleaned (Fig. 5(a)). Al_2O_3 particles of any size were completely cleaned only when 1:1 CO_2/He mixture gas was expanded at 50bar (Fig. 5(b)). Cleaning result was a little better with CO_2/He mixture than with pure CO_2 gas.

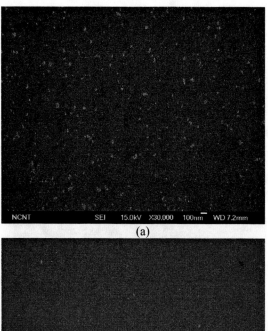

(a)

(b)

Figure 5: SEM images after cleaning of >10nm Al_2O_3 particle using 1:1 CO_2/He at (a) 30bar and (b) 50bar.

3 CONCLUSIONS

Nano-particles on a flat surface were cleaned by bombardment with supersonic particle beam. Particle beams were generated by supersonic expansion through contoured Laval nozzles, such that the final particle velocity was in the supersonic range. Particle size was varied by controlling the stagnation pressure/temperature, nozzle contour and gas composition. Cleaning was almost complete for various ceramic particles down to 10nm, which was attributed mostly to the small size and increased velocity of the bullet particles.

4 ACKNOWLEDGMENT

This work was supported by the Korea Science and Engineering Foundation (KOSEF) grants funded by the Korea government (MOST) (No.ROA-2008-000-20045-0).

REFERENCES

[1] SEMATECH, International Technology Roadmap for Semiconductors (ITRS), 2007

[2] K. Bakhtari, R.O. Guldiken, P. Makaram, A.A. Busnaina and J.G. Park, "Experimental and Numerical Investigation of Nanoparticle Removal Using Acoustic Streaming and the Effect of Time", Electrochem.Soc., 153(9),G846, 2006

[3] H.K. Lim, D.D. Jang, D.S. Kim, and J.W. Lee, "Correlation between Particle Removal and Shock-wave Dynamics in the Laser Shock Cleaning Process", Applied Phys., 97(5), 1, 2005

[4] J.W. Lee, K.H. Hwang, K.H Lee, M.Y. Yi and M.J. Lee, "Removing 20nm particles using a supersonic argon particle beam generated with a contoured Laval nozzle", Adhesion Science and Technology, 23, 769, 2009

[5] J.M. Lauerhass, J.F. Weygand, and G.P. Thomes, "Advanced cryogenic aerosol cleaning: application to damage-free cleaning of sensitive structured wafers", IEEE/SEMI Advanced Semiconductor Manufacturing Conference, 2005

[6] H.Lin, K. Chioujones, J. Lauerhaas, T. Freebern and C. Yu, "Damage-free Cryogenic Aerosol Clean Processes", IEEE Trans. Semiconductor manufacturing,20(2), 2007

[7] M.Y. Yi, D.S. Kim, J.W. Lee and J. Koplik, "Molecular dynamics (MD) simulation on the collision of a nano-sized particle onto another nano-sized particle adhered on a flat substrate, Aerosol Science and Technology, 36(12), 1427, 2005

[8] J.W. Lee, K. Kang, K. Lee, M. Yi, and M. Lee, "Removing 20nm ceramic particles using a supersonic Argon particle beam from a contoured Laval nozzle", Thin Solid Films, 517(14), 3866, 2009

Formation Mechanism of Copper-Silver Metal Nanoparticles Encapsulated in Carbon Shell

H. Y. Kang[*], H. Paul Wang[*], C. H. Huang[*], C. K. Tsai[*], C. J. G. Jou[**], U-Ser Jeng[***] and Chun-Hun Su[***]

[*]National Cheng Kung University, Department of Environmental Engineering
No. 1, University Rd., Tainan 701, Taiwan, wanghp@mail.ncku.edu.tw
[**]National Kaohsiung First University of Science & Technology, Kaohsiung, Taiwan,
george@ccms.nkfust.edu.tw
[***]National Synchrotron Radiation Research Center, HsinChu, Taiwan,
usjeng@nsrrc.org.tw

ABSTRACT

Small angle X-ray scattering (SAXS) is a very useful method in studies of electron density distributions of nanoparticles. The formation mechanism of nanosize copper-silver encapsulated in the inert carbon shell (Cu/Ag@C) has been studied by in-situ synchrotron SAXS spectroscopy. The SAXS spectra have been determined during the temperature-dependent formation of the nanosize Cu/Ag@C. A detailed structural changes corresponding to the growth of the Cu/Ag nanoparticles as the Ag^+- and Cu^{2+}- β-cyclodextrin (β–CD) complexes are carbonized at the temperature range of 363-513 K. By controlling the molar ratios of OH on β–CD to metal, the desired size (i.e., ranged from 7 to 40 nm) can be synthesized. A rapid growth of the Cu/Ag nanoparticles occurs at 393-423 K. The diameter of the complete growth of the Cu/Ag@C nanoparticles is in the range of 20-26 nm with the carbon shell thicknesses of 3-5 nm.

Keywords: core-shell nanoparticle, copper, silver, SAXS

1 INTRODUCTION

Carbon coatings are of great interests because of their high thermal and chemical stability [1]. The carbon coated Cu/Ag bimetal (Cu/Ag@C) nanoparticles can modify the functional sites in nanoreactors, enhance efficiency of a dye-sensitized solar cell [2], or improve heat dissipation of LED or CPU. In addition, Cu/Ag nanoparticles are strong antimicrobial agents for the disinfection of infectious microorganisms in wastewater [3-5]. When coated with diamond-like carbons, Cu/Ag nanoparticles is biocompatible with many materials and can resist from corrosion [6].

A very simple method for synthesis of nanosize-controllable metals (i.e., Cu or Ag) within the carbon shells has been developed by Huang et al [7]. However, the formation mechanism of the Cu/Ag bimetal nanoparticles has not yet been revealed. In the present work, in-situ small angle X-ray scattering (SAXS) spectroscopy was used to monitor the changes of nanoparticle sizes during carbonization of the Ag^+- and Cu^{2+}-β-cyclodextrin (β-CD) complexes. Molecule structural information of the Cu/Ag@C nanoparticles was also studied by X-ray absorption (extended X-ray absorption fine structure (EXAFS) and X-ray absorption near edge structure (XANES)) spectroscopy.

2 EXPERIMENTAL

The nanosize Cu/Ag (1:1, 1:2, and 2:1) nanoparticles encapsulated in the carbon shell (Cu/Ag@C) were prepared by carbonization of Ag^+- or Cu^{2+}- and β-cyclodextrin (β–CD) complexes at the temperature range of 303-573 K. Images of the Cu/Ag@C nanoparticles were determined by high resolution transmission electron microscopy (TEM) (JEOL JEM-3010). The chemical structures of the nanoparticles sampels were identified by X-ray diffraction (XRD) (Bruker, D8 Advanced) with a Cu Kα radiation scanned from 10 to 80° (2θ) at a scan rate of 3°/min. The crystalline sizes of Ag in the Cu/Ag@C nanoparticles were calculated using the Debye-Scherrer equation.

The in-situ SAXS measurements were performed on the beamline at the Taiwan National Synchrotron Radiation Research Center operated with a 1.5 GeV storage ring [8]. The beam was selectively monochromated by a double Si(111) crystal monochromater with a high energy resolution ($\Delta E/E=2\times10^{-4}$) under the energy of 10 keV.

The K-edge XANES and EXAFS spectra of copper were also collected. The beam energy was calibrated using the absorption edge of α-Cu foil at energy of 8979 eV. The X-ray absorption data were analyzed using the UWAXFS 3.0 [9] and FEFF 7.0 [10] simulated programs. The fitted pre-edged background was extrapolated throughout the data range, and subtracted and normalized to decline the effect of sample thickness. The Fourier transform was performed on k^3-weighted EXAFS oscillation in the range 3.0-11.2 Å^{-1}. Empirical fits of model compounds have an error of ± 0.01 Å in radius and ± 10% in coordination number (CN) for the first shell atoms.

3 RESULTS AND DISCUSSION

The TEM images in Fig. 1 show that the nanosize Cu/Ag (1:1) encapsulated in the carbon shell have a diameter of about 20 nm. Copper oxides are not found in the Cu/Ag@C observed by energy-dispersive X-ray (EDX) spectroscopy. Note that Ag is enriched in the outer core in the Cu/Ag/C nanoparticles. This can be attributed to the lower surface free energy of Ag than that of Cu which makes Ag more easily to enrich on the surface.

Figure 2 shows the XRD spectra of the Cu/Ag@C nanoparticles having Cu/Ag ratios of 1:1, 1:2, and 2:1 in the carbonization process at the temperature range of 363-513 K. The broadened bands at 10-40° indicate the presence of amorphous carbon. It is also clear that nanostructures of Cu and Ag are formed at 423 K. Note that no Cu/Ag alloy or oxides are observed.

The sizes of Ag nanoparticles have been calculated with the Debye-Scherrer equation as shown in Figure 3. The particle diameter of 20 nm for Cu/Ag nanoparticles (Cu/Ag =1:1) is consistent with the observations from the TEM images. A rapid growth of the Cu/Ag nanoparticles is found as the temperatures were increased from 363 to 423 K. Between 423-523 K, the growth rate of the nanoparticles is relatively reduced, and their final particle diameters are 20, 26 and 26 nm for the Cu/Ag@C having the Cu/Ag ratios of 1:1, 2:1, and 1:2, respectively.

Figure 4 shows the in-situ SAXS patterns as a function of temperature during the formation of the Cu/Ag@C nanoparticles. Similar to the observations from XRD, a rapid growth of the nanoparticles is observed at 423 K. At 513 K, the scattering pattern can be fitted with a Gaussian distribution centered at 20 nm with a standard deviation of 5.0 nm.

The XANES spectra of copper in the Cu/Ag@C nanoparticles is shown in Figure 5. The Cu/Ag@C samples contain nanosize Cu, and essentially no copper oxide is observed. Figure 6 shows the EXAFS spectra of the Cu/Ag@C fitted with >99% reliability. The Cu-Cu bond is 2.55 Å with a coordination number (CN) of 5.3 and a Debye-Waller factor of 0.0045.

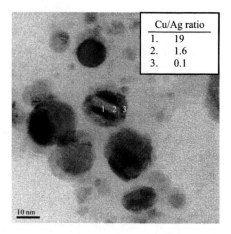

	Cu/Ag ratio
1.	19
2.	1.6
3.	0.1

Figure 1: TEM images and EDX spectrum (at points of 1, 2, and 3) of the nanosize Cu/Ag@C (Cu/Ag= 1:1).

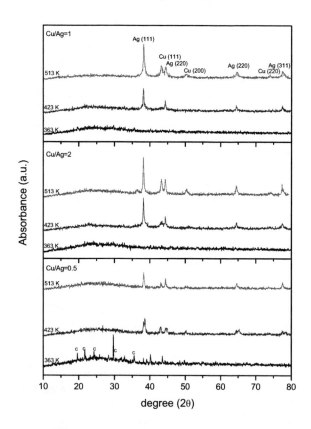

Figure 2: Temperature dependent XRD spectra of the Cu/Ag@C nanoparticles (metal ratios of Cu/Ag = 1, 2, and 0.5) at the temperature range of 363-513 K.

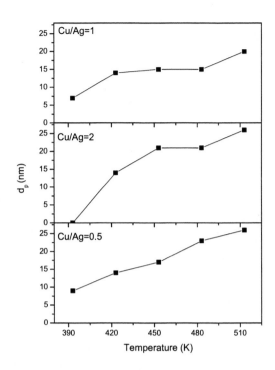

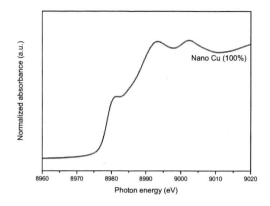

Figure 5: The least-square fitted XANES spectra of copper in the Cu/Ag@C nanoparticles.

Figure 3: Temperature dependent sizes of Ag in the Cu/Ag@C nanoparticles determined by XRD using the Scherrer's equation (metal ratios of Cu/Ag = 1, 2, and 0.5).

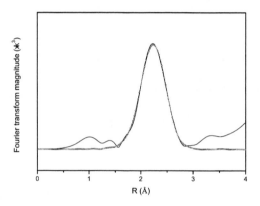

Figure 6: EXAFS spectra of copper in the Cu/Ag@C nanoparticles.

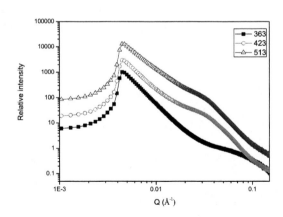

Figure 4: Temperature dependent SAXS patterns of the Cu/Ag@C nanoparticles at the temperature range of 363-513 K.

4 CONCLUSION

The temperature-dependent formation of the Cu/Ag bimetal nanoparticles encapsulated in the carbon shell has been investigated at the temperature range of 363-513 K. It is clear that nanostructures of Cu and Ag are formed at 423 K observed by XRD and in-situ SAXS. The EDX spectra indicate that neither oxygen nor metal oxide is found in the Cu/Ag@C. Note that Ag is enriched in the outer core in the Cu/Ag@C nanoparticles. The TEM images confirm that the nanoparticles have sizes of 20-26 nm with the carbon shell thicknesses of 3-5 nm which are similar to the observations from in-situ SAXS and XRD.

ACKNOWLEDGEMENT

The financial supports of the Taiwan National Science Council, Bureau of Energy, the Excellence Project of the National Cheng Kung University, and National Synchrotron Radiation Research Center are gratefully acknowledged.

REFERENCES

[1] J. Robertson, Materials Science and Engineering: R: Reports, 37, 129, 2002.

[2] F. L. Chen, I. W. Sun, H. P. Wang, C. H. Huang, J. Nanomater, 2009.

[3] Q. L. Feng, J. Wu, G. Q. Chen, F. Z. Cui, T. N. Kim, J. O. Kim, Journal of Biomedical Materials Research, 52, 662, 2000.

[4] G. Borkow, J. Gabbay, Current Medicinal Chemistry, 12, 2163, 2005.

[5] Y. S. E. Lin, R. D. Vidic, J. E. Stout, C. A. McCartney, V. L. Yu, Water Res., 32, 1997, 1998.

[6] G. Thorwarth, B. Saldamli, F. Schwarz, P. Jürgens, C. Leiggener, R. Sader, M. Haeberlen, W. Assmann, B. Stritzker, Plasma Processes and Polymers, 4, S364, 2007.

[7] C. H. Huang, H. P. Wang, J. E. Chang, E. M. Eyring, Chem. Commun., 4663, 2009.

[8] D. G. Liu, C. H. Chang, C. Y. Liu, S. H. Chang, J. M. Juang, Y. F. Song, K. L. Yu, K. F. Liao, C. S. Hwang, H. S. Fung, P. C. Tseng, C. Y. Huang, L. J. Huang, S. C. Chung, M. T. Tang, K. L. Tsang, Y. S. Huang, C. K. Kuan, Y. C. Liu, K. S. Liang, U. S. Jeng, J. Synchrot. Radiat., 16, 97, 2009.

[9] E. A. Stern, M. Newville, B. Ravel, Y. Yacoby, D. Haskel, Physica B , 208-209, 117, 1995.

[10] S. I. Zabinsky, J. J. Rehr, A. Ankudinov, R. C. Albers, M. J. Eller, Phys. Rev. B, 52, 2995, 1995.

The Use of Taguchi Method to Determine Factors affecting the size of fine attapulgite particles generated by Rapid Expansion of Supercritical Solution with and without cosolvents

Iman Asghari[*], Feridun Esmaeilzadeh[**]

[*]Chemical and Petroleum Engineering Department, Shiraz University, 7134851154, Iran , asghari.iman@gmail.com
[**]Corresponding Author: Chemical and Petroleum Engineering Department, Shiraz University, 7134851154, Iran, esmaeil@shirazu.ac.ir

Abstract

The rapid expansion of a supercritical solution (RESS) process is a promising method for the production of small, uniform and solvent-free particles of solutes. The RESS containing a non-volatile solute leads to loss of solvent power by the fast expansion of the supercritical solution through an adequate nozzle, which can cause solute precipitation. The nozzle configuration plays an important role in RESS method and has a great effect on the size and morphology of the precipitated attapulgite particles. In this work, a RESS apparatus was set up and attapulgite submicron particles were prepared successfully with the supercritical carbon dioxide. The experiments were carried out to investigate the effect of parameters such as extraction temperature (308-328K), extraction pressure (150-200 bar), spraying distance (1-5 cm) ,the nozzle configuration besides the presence of cosolvents . Taguchi method creates an orthogonal array to accommodate these requirements. The selection of a suitable orthogonal array depends on the number of control factors and their levels. In this work our experiments carried out based on L18 orthogonal array. For the Taguchi design and subsequent analysis, the software named as Qualitek-4 (version 4.82.0) was used. The RESS process could produce ultrafine spherical particles (0.09 to 2.1 μm) of attapulgite as reflected by SEM observations. However the average size of the unprocessed particles were (1-100 μm) with average of 15 μm.

1 Introduction

There are a large number of clay minerals, but those with which we are concerned in drilling fluids can be categorized into three types. The first type is needle-shaped, non-swelling clays like attapulgite or sepiolite. Attapulgus clay is usually called attapulgite which comprises 80% to 90% of the commercial product. Attapulgite has a fibrous texture. It is used in drilling mud solely for its suspending qualities. The natural fine crystal size and needle shape causes it to build a "brush heap" structure in suspension and thereby exhibit high colloidal stability even in the presence of high electrolyte concentration. Attapulgite, a naturally occurring mineral, is a crystalline hydrated magnesium alumino-silicate with a three dimensional chain structure that gives it unique colloidal and sorptive properties. Attapulgite is primary among the fuller's earth family of minerals. It was first mined in Quincy, Florida in 1893. American attapulgite is only found in southwestern Georgia and northern Florida. Attapulgite's performance is enhanced by specific thermal activation and particle size control. It is used as a viscosifier in saltwater drilling fluids where bentonite becomes ineffective [1-2]. In process

industries, for a variety of operations the particle size is a critical parameter. This is because finer and larger particles have different rates of settling in a liquid medium, and the difference is accentuated if the liquid medium is denser. Smaller particles have a higher total surface area for a given particle loading. This indicates that the strength increases with increasing surface area. . It is obvious that the submicron processed attapulgite improves the rheological properties which lead to reduce the operating cost of drilling. In other words, the processed attapulgite particles can better absorb water molecules due to increasing surface area and result in better suspension solution which leads to improve the performance of drilling mud. Drilling fluid properties can be extremely sensitive to the presence of slimes (ultrafine particles) which increase the viscosity of the slurry [3-4].The technology for producing fine particles with narrow particle size distribution (PSD) is one of the most crucial technologies in several fields such as material, pharmaceutical, food, polymer, and fine chemical industries. Conventional techniques for particle size reduction include mechanical comminuting (crushing, grinding, and milling), recrystallization of the solute particles from solution by using liquid anti-solvents, freeze-drying and spray-drying. Although the mechanical method is simple and economical, it is not suitable for thermally labile compounds or materials hard to crush. The other processes have a problem leading to the environmental pollution due to excessive use and disposal of solvent[5]. Supercritical fluid technology has been widely used for various applications such as extraction, reaction, chromatography and material processing. Carbon dioxide is the most widely used supercritical fluid, because it has a relatively low critical temperature (31.2 °C) and a moderate critical pressure (73.8 bar). In addition, it is inexpensive, leaves no toxic residue, and is not flammable. As fine particle formation processes using supercritical fluids, RESS (Rapid Expansion of Supercritical Solution), SAS (Supercritical Anti-Solvent) and PGSS (Particles from Gas-Saturated Solution) have been known[6]. The rapid expansion of a supercritical solution (RESS) process is an attractive technology for the production of small, uniform and solvent-free particles. It is to be noticed that only few examples were published on application of RESS to microspheres generation. Pre-exfoliated nanoclays were prepared using supercritical carbon dioxide as solvent by several authors C. Detrembleur et al.[7]. It should mention that there are a few papers in field of clay/polymer nanocomposites in recent years were published [8]. In this work, a RESS apparatus was set up successfully and was used to produce attapulgite fine particles which can be used as a viscosifier in drilling mud. Several authors have reviewed the applications of RESS on the preparation of fine and ultra fine particles. For instance Formation of anthracene fine particles by Nagahama et al [9]. To our knowledge, no research paper is available on micronization of attapulgite by supercritical carbon dioxide. It is obvious that the submicron processed attapulgite improves the rheological properties which lead to reduce the operating cost of drilling. The experiments were carried out to investigate the effect of extraction temperature (308–328 K) and pressure (15–20MPa), spray distance (1-5 cm), nozzle diameter (500-1200 μm) and the presence of cosolvents(Isopropyl alcohol and acetone) on the size and morphology of the precipitated attapulgite particles. Precipitated attapulgite particles were analyzed their size by scanning electron microscopy (SEM) .Before the SEM analysis, the process either original samples must be coated by a sputter-coater . Our current research is aimed towards an improved understanding of the relationship between process parameters and particle characteristics and to explore new areas of application for nanoscale particles. Taguchi method creates an orthogonal array to accommodate these requirements. The selection of a suitable orthogonal array depends on the number of control factors and their levels. Taguchi design is preferred because it reduces the number of experiments significantly. With the selection of L18 orthogonal array, using five mentioned parameters and their levels(3 levels), the number of experiments required can be drastically reduced to 18. It means that 18 experiments with different combinations of the factors should be conducted in order to study the main effects, which in the classical combination method using full factorial experimentation would require $3^5 = 243$ number of experiments to capture the influencing parameters. These control factors include Temperature, Pressure, Nozzle Diameter, Spraying Distance, and Cosolvents. All control factors have three levels. For the Taguchi design and subsequent analysis, the software named as Qualitek-4 (version 4.82.0) was used.

2 EXPERIMENTAL

2.1 Materials

The solute used during this study, attapulgite, was prepared from PDF company (Pars Drilling Fluid company,Tehran-Iran) and Carbon dioxide (99.9 %< purity) was purchased from Abughadareh Gas Chemical Company(Shiraz-Iran). The solvents, Isoporopanol (Isopropyl alcohol) and acetone was purchased from Merck.

2.2 RESS Set up

The RESS pilot plant is shown in Fig. 1,2. At first, the gaseous CO2 from a cylinder capsule was passed through a filter and then entered into a refrigerator to make liquid CO2. The liquid CO2 was then pumped by a reciprocating oil-free water-free high pressure manual pump into a vertical surge tank. The pressurized CO2 then entered into an extraction vessel (Vessel 1) which is containing with cosolvent. The attapulgite were charged with glass wool and specific packing in another extraction column (Vessel 2), and was held it in the desired conditions for two hours to ensure complete equilibrium has been obtained. The equilibrated solution was then expanded into a nozzle. The precipitated attapulgite particles were collected on the stub and analyzed by a SEM to monitor the particle size and its

morphology. A new type of nozzles was designed and fabricated to achieve ultrafine nanosize particles.

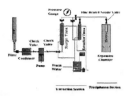

Fig.1. The schematic diagram of the experimental apparatus for the RESS process

Fig.2.The SCF apparatus was set up in Shiraz University

2.3 Particle size and morphology

The morphology and size of the precipitated attapulgite particles were examined by using scanning electron microscopy (SEM) (S360- CAMBRIDGE). In brief, prior to examine the samples by a SEM the precipitated attapulgite particles were collected on the conductive stubs which were then coated by a sputter-coater (SC-7640-Polaron) with Pd-Pt under the presence of argon (99.9% < purity) at the room temperature for a period of 100 s under an accelerating voltage of 20 KV . The mean particle size of the precipitated particles was calculated by SigmaScan Pro Image analysis.

2.4 Nozzle configuration

The structure of the new nozzle is illustrated in Fig. 3. The nozzle comprises two parts, shell part and inside part. In this design, the clearance between the inside part and the nozzle acts as a circular orifice so that supercritical fluids can spray from a narrow exit in a few microns. Also the swirled channel enhances the chance of spherical form particles formation which could be helpful to improve the particle morphology design. In the entrance of the nozzle, SC-CO$_2$ is introduced through the spiral channel, so the fluids can be swirled out of the nozzle. In this kind of nozzle instead of the usual diameter, the effective nozzle diameter has been defined.

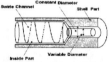

Fig.3. Schematic diagram of the used nozzle

2.5 Image analysis

As can be seen in Fig.4 the rapid expansion of supercritical solutions (RESS) was successfully used to produce attapulgite submicron particles. Attapulgite was micronized and a great size reduction of attapulgite particles in comparison with the original one was observed.

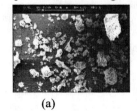
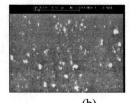

(a)   (b)

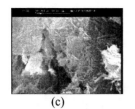
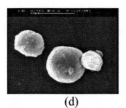

(c)   (d)

Fig.4.(a,c) Unprocessed Attapulgite (b,d)Precipitated Attapulgite by RESS process

3 Results and Discussion

In this work, the influence of RESS parameters such as extraction temperature (308-328 K), extraction pressure (150-200 bar), spraying distance (1-5cm) and effective nozzle diameter (500-1200 μm) and cosolvent (– =without cosolvent, SOL 1=Isopropanol, SOL 2=Acetone) were investigated on the mean particle size of the micronized attapulgite particles which are expressed below.   In addition, the Taguchi method was used to arrange the experimental conditions of micronization of attapulgite particles based on L-18 array as shown in Table 2.

RUN	P	T	ND	SD	COSOLVENT	Average particle size(μm)
1	150	35	500	1	–	1.00
2	150	45	900	3	SOL1	0.75
3	150	55	1200	5	SOL2	1.5
4	175	35	500	3	SOL1	0.25
5	175	45	900	5	SOL2	0.65
6	175	55	1200	1	–	1.7
7	200	35	900	1	SOL2	0.09
8	200	45	1200	3	–	0.85
9	200	55	500	5	SOL1	0.55
10	150	35	1200	5	SOL1	1.25
11	150	45	500	1	SOL2	0.3
12	150	55	900	3	–	2.1
13	175	35	900	5	–	1.45
14	175	45	1200	1	SOL1	0.17
15	175	55	500	3	SOL2	0.7
16	200	35	1200	3	SOL2	0.5
17	200	45	500	5	–	0.45
18	200	55	900	1	SOL1	0.35

Table1.L-18 Taguchi Orthogonal Array

The results of the experiments show that the presence of cosolvent is the most effective parameter in the size reduction of attapulgite particles. However the extraction pressure and extraction temperature have an important role on the size and morphology of the precipitated attapulgite particles. On the other hand, higher spraying distance and nozzle diameter cause to increase particle size distribution of the precipitated attapulgite particles. Analysis of variance (ANOVA) was used to determine the optimum conditions and most significant process parameters for the micronization of attapulgite particles. For Taguchi design the statistical software namely Qualitek-4 was applied. (fig.5.) (Dark blue=Cosolvent, Green=Pressure , Blue=Temperature , Red=Nozzle Diameter ,Violet=Spraying Distance). The results of analysis show that the optimum pressure, temperature, nozzle diameter and spraying distance and cosolvent are 200 bar, 45^{0}C , 500 μm, and 1 cm and with the presence of Isopropanol respectively.   It has to be noticed that the impression of cosolvent is very important. Anova table is drafted below (Table2).As can be observed in Fig.6 the difference between the condition of factors and their levels in table L-18 is definitely proven. because in high

temperatures the attapulgite particles are agglomerated. On the contrary by increasing the pressure the attapulgite particles become spherical (Fig.6.)

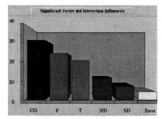

Fig.5. Significant Factor Influences for precipitated attapulgite particles

FAC	DOF	Sum of Sqrs	Var iance	F-Ratio	Pure Sum	(%)
P	2	1.0408	0.704	52.464	1.381	**24.57**
T	2	1.186	0.593	44.206	1.159	**20.63**
ND	2	0.684	0.342	25.486	0.657	**11.69**
SD	2	0.437	0.218	16.307	0.41	**7.30**
CO	2	1.81	0.905	67.439	1.783	**31.72**
Other/ Error	7	0.093	0.013			**4.06**
Total	17	5.621				**100%**

Table2. Anova Table for L-18 attapulgite

(a)L-18#12   (b) L-18#17

Fif.6.The SEM images of attapulgite (#12&#17 L-18)

3.1 Effect of Cosolvent (CO)

Attapulgite solubility is low in SC-CO2, because CO2 is a nonpolar compound. Therefore, a polar co-solvent such as Isopropylalcohol and acetone can be used together with CO2, to increase the solvating power of CO2. Increasing co-solvent concentration caused a change of morphology of the attapulgite particles. Similar result has been reported by N. Yildiz et al [10]. In other words the average particle size of attapulgite is decreased in the presence of cosolvent. However the type of solvent which is used as a cosolvent is related to the chemistry of solute. In this research because of structure of attapulgite Isopropylalcohol as a polar protic solvent is most effective. Acetone as a polar aprotic solvent is useful as well.

3.2 Effect of Extraction Pressure (P)

The extraction pressure was studied from 15 to 20MPa. The solubility of solid in supercritical fluid can be suddenly changed by slight shift of pressure, which can cause the variation of the characteristics of recrystallized particles due to a direct effect on supersaturation. As results show the micronization of attapulgite particles is too sensitive to a slight increase of pressure. The reason for this is that with increasing the pressure, the solubility of attapulgite particles increases sharply. In fact, an increase in the solute solubility results in higher supersaturations in the fluid upon expansion. According to the classical theory of nucleation, higher supersaturation causes higher nucleation rate and the particle volume is inversely proportional to the nucleation rate, our above results appear to agree with simple theoretical predictions. Similar results were reported by A.Z. Hezave and other authors [11-12].

3.3 Effect of extraction temperature (T)

The extraction temperature for attapulgite in this study ranges from 35 to 55 $^{\circ}$C. Increasing the extraction temperature leads to a decrease in the density of CO_2 and a concurrent increase in the solute's sublimation pressure. The decrease of the solvent density causes a decrease of the solvent strength. On the other hand, a concurrent increase in the solute's vapor pressure is responsible for an increase in the aspirin solubility. The net effect of these two competing factors results in an increase in the saturated attapulgite concentration in the supercritical fluid [13]. Therefore, high extraction temperature induces high attapulgite solubility (i.e., high solute concentration) at a constant extraction pressure. At lower extraction temperature, the supersaturation and nucleation rate are lower since attapulgite concentration is lower. Thus crystals initialized during the expansion might preferably grow to become larger and possibly produce one-dimensional (i.e., needle-shaped) particles along the flow direction under the shear forces. However, continuously increasing temperature (at 55^{0}C) caused to increase of the average particle size. Because increasing the temperature leads to increase the attapulgite concentration. During the expansion, high attapulgite concentration brings about the increase of the particle size, as a consequence of coagulation among particles.

3.4 Effect of nozzle diameter (ND)

It was expected that the nozzle diameter and its dimension in the RESS process affected the particle formation. In this study, the effect of the effective nozzle diameter was investigated (500-1200 μm) on the size and morphology of the precipitated attapulgite particles. Cleary, the average particles size reduces by decreasing the nozzle diameter. In other words, the particle size increases and the PSD becomes broad with increasing nozzle diameter (nozzle1=500, nozzle2=900, nozzle3=1200). Same result was reported by Wang et al[12].

3.5 Effect of spraying distance (SD)

The spraying distance can have a pronounced effect on the characterization of the attapulgite particles, since the nucleation and growth process continues in post-expansion region [14]. In this study, three spraying distances (1, 3,5cm) from the tip of the nozzle were tested. The results show only a little change in the spraying distance from 1 cm to 5cm causes to increase the average particle size. Similar results were reported by Yildiz et al[10].

4 Conclusion

The rapid expansion of supercritical solutions (RESS) was successfully used to produce attapulgite submicron particles. Attapulgite was micronized and a great size reduction of attapulgite particles in comparison with the original one was observed. The precipitated particles of attapulgite were in the range of 10 nm to 5 μm. The obtained results shows a good performance of these used nozzles in comparison with the capillary nozzle.We conclude that presence of cosolvent is an essential factor for micronization of solute in RESS process with SC-CO_2 due to the Carbon dioxide's weak solubility power . The extraction pressure and temperature also play main role in the precipitated attapulgite particle size and its morphology.

In addition, the optimum Pressure, Temperature, Nozzle Diameter, Spraying Distance and Cosolvent for micronization of attapulgite particles are 200 bar, 318 K , 500 μm , 1 cm with the presence of Isopropanol respectively.

Acknowledgment
The authors are grateful to faculty of Nanotechnology, Nanochemical engineering department of Shiraz University and also would like to thank Dr Shahabodin Ayatollahi for his collaboration to use SigmaScan Pro Image analysis software and Ms Kaveh the Operator of Scanning Electron Microscopy apparatus in shiraz university and Pars drilling fluid(PDF) company and the staff members of it: Mr Mojtaba Kalhor & Ms Niloofar Arsanjani for supporting this research.

References
[1] H.C.H.Darly George R.Gray , composition and properties of drilling and completion fluids , Gulf Professional publication ,5th eition.

[2] F. Bergaya, B.K.G. Theng and G. Lagaly Handbook of Clay Science

[3] Nanoparticles: From Theory to Application. Edited by Gu¨nter Schmid Copyright 2004 WILEY-VCH Verlag GmbH & Co. KGaA, Weinheim ISBN: 3-527-30507-6.

[4] Hsueh CH, Becher PF. Effective viscosity of suspensions of spheres. J Am Ceram Soc 2005; 88:1046–9.

[5] C.N.R. Rao, A. Govindaraj, S.R.C. Vivekchand, Inorganic nanomaterials: current status and future prospects, Annu. Rep. Prog. Chem., Sect. A Inorg. Chem. 102 (2006) 20–45.

[6] Y..Ping sun ,Supercritical fluid technology materials science and engineering synthesis , properties , and applications ,Clemson University Clemson, South Carolina.

[7] L. Urbanczyk, C. Calberg, F. Stassin, M. Alexandre, R. Jérôme, C. Jérôme, C. Detrembleur, Synthesis of PCL/clay masterbatches in supercritical carbon dioxide,49(2008) 3979-3986.

[8]S.Horsch, G.Serhatkulu, E.Gulari, R.M. Kannan , Supercritical CO_2 dispersion of nano-clays and clay/polymer nanocomposites, Journal of Polymer,47(2006) 7485-7496.

[9] K. Nagahama, G.T. Liu, Supercritical Fluid Crystallization of Solid Solution. The 4th International Symposium on Supercritical Fluids, 11–14 May, Sendai (Japan), 1997,43–46.

[10] N. Yildiz, S. Tuna, O, Doker, A, C,Alimli. Micronization of salicylic acid and taxol (paclitaxel) by rapid expansion of supercritical fluids (RESS), J. of Supercritical Fluids 41 (2007) 440–451.

[11] A.Z. Hezave, F. Esmaeilzadeh, Micronization of drug particles via RESS process, The Journal of Supercritical Fluids (2008), doi:10.1016/j.supflu.2009.09.006.

[12] J. Wang, J. Chen, Y. Yang, Micronization of titanocene dicholoride by rapid expansion of Super- critical solution and its ethylene polymerization, J. of SupercriticalFluids 33 (2005) 159-172

[13] Z. Huang, W.D. Lu, S. Kawi, Y.C. Chiew, J. Chem. Eng. Data 49 (2004) 1323.

[14] Helfgen, B. ; Hils, P. ; Holzknecht, C. ; T¨urk, M. ; Schaber, K. Simulation Of particle formation during the rapid expansion of supercritical solution, Aerosol Sci. (2001), 32, 295–319.

Manufacture of ultrafine drug particles via the Rapid Expansion of Supercritical Solution (RESS) Process using Taguchi approach

Iman Asghari[*], Feridun Esmaeilzadeh[**],

[*]Chemical and Petroleum Engineering Department, Shiraz University, 7134851154, Iran ,asghari.iman@gmail.com
[**]Corresponding Author: Chemical and Petroleum Engineering Department, Shiraz University, 7134851154, Iran, esmaeil@shirazu.ac.ir

Abstract

The poor water solubility of many drugs is a challenge in pharmaceutical research. Recently, there have been great interests in finding environmentally freindly methods producing fine particles of pharmaceutical products for applications in pharmaceutical engineering. A promising method to improve the bioavailability of pharmaceutical agents is the Rapid Expansion of Supercritical Solutions. Deferasirox (DFS), a tridentate chelator, requires two molecules for iron (III) coordination. The bioavailability (the percentage of the drug absorbed compared to its initial dosage) is limited by this insolubility. . The RESS experiments were carried out in a broad temperature range (35-55 °C) and pressures at 150–200 bar. The effects of extraction pressure, extraction temperature, nozzle diameter, and spraying distance based on Taguchi design on the shape and size of the particles formed are discussed. Our results show that extraction pressure and extraction temperature can significantly affect the morphology and size of the precipitated particles. For the Taguchi design and subsequent analysis, the software named as Qualitek-4 was used.

Key Words: Nanoparticles, Iron Chelator, Deferasirox, Rapid Expansion of Supercritical Solution, Taguchi Method

1 Introduction

In recent years, significant effort has been devoted to develop drug formulation and delivery systems for issues such as targeting and controlled release [1–3]. In fact, the solubility is a serious limitation in drug development and related requirements for bioavailability and normal absorption pattern. According to the established statistics, about a third of the drugs listed in the United States Pharmacopeia are poorly water-soluble or insoluble and more than 40% of new drug development has failed because of poor biopharmaceutical properties [4-5]. The nanosizing of drug particles has been identified as a potentially effective and broadly applicable approach.For example, smaller-diameter particles correspond to a faster dissolution rate, thus potentially higher activity and easier absorption. Other distinct advantages include tissue or cell specific targeting of drugs, longer circulating capacity in the blood, higher stability against enzymatic degradation, and the reduction of unwanted side effects [6-7]. Several conventional techniques are used for reduction of particle size such as crushing, grinding, milling, spray drying, freeze-drying and recrystallization of the solute particles from solutions using liquid antisolvents. But all these techniques have got several disadvantages. Some substances are unstable under milling conditions, and in recrystallization process the product is contaminated with

solvent. In addition, there are thermal and chemical degradation of products due to high temperatures, high-energy requirements, large amount of solvent use, solvent-disposal problems, and broad particle size distributions. Due to their several drawbacks, the use of supercritical fluids has increased rapidly over the last few years and several processes for particle formation have been studied [8].Supercritical fluid processing techniques have been applied to the particle formation in drug formulation [9-10]. The methods of fine particles formation using supercritical fluids are: rapid expansion of supercritical solutions (RESS), anti-solvent processes (gas anti-solvent (GAS), supercritical anti-solvent (SAS), aerosol solvent extraction system (ASES), solution enhanced dispersion by supercritical fluids (SEDS)) and particles from gas saturated solutions/suspensions (PGSS) [11]. The RESS process consists of extraction and precipitation unit. A substance is solubilised in a supercritical fluid (SCF) at the extraction unit, than the supercritical solution is suddenly depressurised in a nozzle causing fast nucleation and fine particle formation. Due to the rapid expansion of supercritical solution through a nozzle, the large decrease in density, and hence decreasing the SCF solvating power. The solute becomes supersaturated and then precipitaded. The driving force of the nucleation process is supersaturation. Higher supersaturation leads to increase the nucleation rate and tends to decrease the particle size. Advantages of RESS process are that nano or microparticles are produced, providing a solvent-free product and controllable particle size. The morphology and size distribution of the precipitated material is related to pre-expansion and expansion conditions, extraction parameters, spray distance and nozzle design [12]. Carbon dioxide is commonly used as a supercritical fluid because it is non-toxic, non-flammable, and cheap. It has a low critical temperature and pressure (T_c = 31.1 °C and P_c = 73.8 bar) that allow for low temperature processing. From a pharmaceutical point of view, supercritical carbon dioxide has several advantages, including being solvent-free, and being able to be used in a single-stage process and at moderate processing temperatures. Hereditary hemochromatosis (HH) is an autosomal recessive disorder characterized by progressive iron overload through increased intestinal absorption. Phlebotomy, the preferred treatment, can prevent or reverse some complications of iron overload, such as hepatic damage; however, compliance is variable and some patients are poor candidates because of underlying medical disorders and/or poor venous access. Thus, if an oral iron chelator such as deferasirox (Exjade) proves to be tolerable and effective, HH patients will have an alternative treatment option. Iron-chelation therapy is essential in iron-overloaded patients.

Iron chelation is a key feature in the management of transfusion dependent anaemias— such as b-thalassaemia major, b-thalassaemia intermedia, sickle-cell disease and myelodysplastic syndrome—to prevent end-organ damage and improve survival. Exjade is supplied as a tablet that is dispersed in water or juice. Deferasirox (Exjade, ICL670, Novartis Oncology) belongs to a novel class of tridentate iron chelators, the N-substituted bis-hydroxyphenyl triazole. Two molecules of deferasirox are needed to form a soluble complex with one Fe^{3+} ion. The active is a white to slightly yellow and not hygroscopic powder. It has a good permeability and it is practically insoluble in water and in acid medium, the solubility increasing with pH. Therefore, the particle size is likely to be important to the rate and possibly to the extent of absorption [13-15]. Poor aqueous solubility represents a major hurdle in achieving adequate oral bioavailability for a large percentage of drug compounds in drug development nowadays. Nanosizing refers to the reduction of the active pharmaceutical ingredient (API) particle size down to the sub-micron range, with the final particle size typically being 100–200 nm. The reduction of particle size leads to a significant increase in the dissolution rate of the API, which in turn can lead to substantial increases in bioavailability [16]. Several authors have reviewed the applications of RESS on the preparation of fine and ultra fine particles. Formations of anthracene fine particles by Nagahama et al [17].Krober et al in 2000 have reported an investigation of RESS for synthesis of small organic particle [18]. To our knowledge, no research paper is available on micronization of deferasirox by supercritical carbon dioxide. The experiments were carried out to investigate the effect of extraction temperature (308–328 K) and pressure (15–20MPa), spray distance (1-5 cm), nozzle diameter (500-1200 μm) on the size and morphology of the precipitated deferasirox particles.

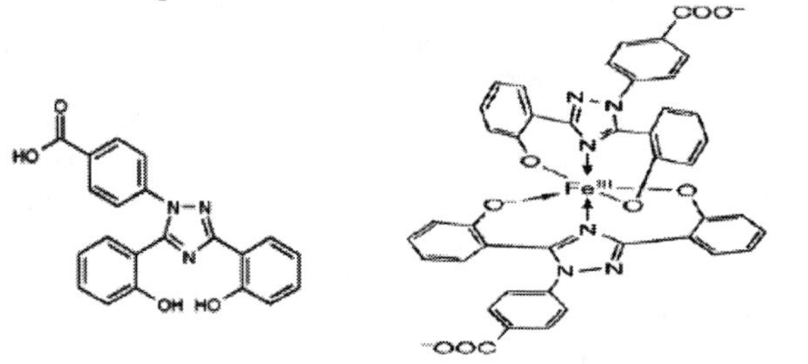

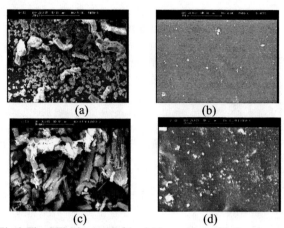

Fig.1.(a)Molecular structure of Deferasirox(b)Function of Deferasirox

2 Experimental

2.1. Materials

The solute used during this study, Deferasirox, was prepared from Pharmaceutical Arasto company and Carbon dioxide (99.9 %< purity) was purchased from Abughadareh Gas Chemical Company.

2.2. Particle characterization

Precipitated deferasirox particles were analyzed their size by scanning electron microscopy (SEM) (S360-CAMBRIDGE).Before the SEM analysis, the process either oroginal samples must be coated by a sputter-coater (SC-7640-Polaron) with Pd-Pt under the presence of argon

(99.9% < purity) at the room temperature for a period of 100 s under an accelerating voltage of 20 KV. The SEM images of original and precipitated deferasirox are given in Fig.2.

Fig.2.The SEM images of (a,c) Unprocessed Deferasirox Particles(b)Processed Deferasirox

2.3. Experimental technique

The RESS pilot plant is shown in Fig. 3. At first, the gaseous CO_2 from a cylinder capsule was passed through a filter and then entered into a refrigerator to make liquid CO_2.The liquid CO_2 was then pumped by a reciprocating high pressure pump into a surge tank The surge tank dampened the pressure fluctuations produced by operation of the pump. at the outlet of the surge tank a bourdon gauge in the range of 0-250 bar was placed. The pressurized CO_2 then entered into an extraction vessel. It should mention that the surge tank and the extraction vessel are surrounded by a regulating hot water jacket. The basket which is packed by sample and glass wool was placed into an extraction vessel. For each condition the extractor vessel was held for 2 hours to ensure equilibrium has been obtained. The equilibrated solution was then expanded by a preheated fine needle valve into a nozzle. The precipitated deferasirox particles were collected on the stub and analyzed by a SEM to monitor the particle size and its morphology. A new type of nozzle(Fig.4) were designed and fabricated to achieve ultrafine nanosize particles.

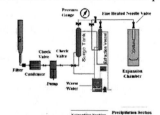

Fig3. The schematic diagram of the experimental apparatus for the RESS process

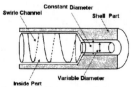

Fig.4. Schematic diagram of the used nozzle

3 Results and discussions

In this work, the influence of RESS parameters such as extraction temperature (308-328 K), extraction pressure (150-200 bar), spraying distance (1-5cm) and effective nozzle diameter (500-1200 μm) were investigated on the mean particle size of the micronized deferasirox particles which are expressed below. In addition, the Taguchi method was used to arrange the experimental conditions of micronization of deferasirox particles based on L-9 array as shown in Table 1. Anova table for deferasirox based on Qualitek-4 software is drafted below (Table2).The measurements of average particle size of deferasirox are done by SigmaScan Pro Image analysis.

Run number	P	T	SD	ND	Average Size(μm)
1	150	35	1	500	0.47
2	150	45	3	900	0.69
3	150	55	5	1200	1.64
4	175	35	3	1200	0.82
5	175	45	5	500	0.40
6	175	55	1	900	0.94
7	200	35	5	900	1.40
8	200	45	1	1200	1.00
9	200	55	3	500	1.25

Table1. Taguchi L-9 array

For Taguchi design the statistical software namely Qualitek-4 was applied. (fig5). (Dark Blue=T, Green=P, Blue=ND, Red=SD).

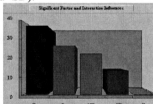

Fig.5.Influence of paparameters

Factor	DOF	Sum of Square	Variance	Percent (%)
P	2	.372	.186	26.829
T	2	.52	.26	37.513
SD	2	.182	.091	13.171
ND	2	.312	.156	22.478
ERROR	-	-	-	-
TOTAL	8	1.388		100%

Table2.Anova Table for Deferasirox

The results of analysis show that the optimum pressure, temperature, nozzle diameter and spraying distance are 175 bar, 318 K , 500 μm and 1 cm, respectively.

3.1 Effect of extraction pressure

Generally, the variation of the extraction pressure brings about a change in the concentration of Deferasirox.. The pressure studied here varies from 150 to 200 bar. An increase in extraction pressure from 150 bar to 175 is observed to induce decrease in the average particle size. Similar results were reported by A.Z. Hezave, F.

Esmaeilzadeh [19]. The reason for this is that an increase in the solute solubility results in higher supersaturations in the fluid upon expansion. According to the classical theory of nucleation, higher supersaturation causes higher nucleation rate and the particle volume is inversely proportional to the nucleation rate, our above results appear to agree with simple theoretical predictions. Similar results have also been reported for other organic solutes. But the finest average particle size and the smallest particle size distribution are observed at 175bar.In other words a further increase of extraction pressure till 200 bar yields a marked increase of the particle size, which may suggest a decoupling of the two processes of nucleation and growth. Perhaps at high deferasirox concentrations (i.e., high extraction pressure), the particle growth may be dominant or a particle may include several nuclei during the growth process. Therefore large particles seem to be readily produced and a broad particle size distribution may be obtained. Similar results have been reported for formation of aspirin by Z.Huang et al [20].As can be seen in Fig.6. , the morphology of the original particles was changed from irregular shape to spherical state at high pressure.

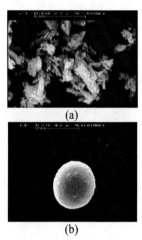

(a)

(b)

Fig.6.The SEM images of (a) original deferasirox (b) Processed deferasirox

3.2 Effect of extraction temperature

The extraction temperature for deferasirox in this study ranges from 308 to 328° K. increasing the extraction temperature leads to decrease in the density of CO2 and concurrent increase in the solute's vapour pressure. The decrease of the solvent density causes a decrease of the solvent strength. On the other hand, a concurrent increases in the solute's vapour pressure leads to an increase in the deferasirox solubility. An increase of extraction temperature from 308 to 318K leads to the increase of the supersaturation and nucleation rate as a result of increased solute concentration. The increase of nucleation rate leads to decrease in the deferasirox particle growth time, and consequently the smaller particle was obtained. However, continuously increasing temperature (at 328 K) caused to increase of the average particle size. Because increasing the temperature leads to increase the deferasirox concentration. During the expansion, high deferasirox concentration brings about the increase of the particle size, as a consequence of coagulation among particles. In the literature, similar results have also been reported in the literature [21].

3.3 Effect of nozzle diameter

It was expected that the nozzle diameter and its dimension in the RESS process highly affected the particle formation. In this study, the effect of the effective nozzle diameter was investigated (500-1200 μm). The results of the experiments show the average particles sizes can be reduced by lowering the diameter of nozzle. In other words, the particle size increases with increasing the nozzle diameter. Several studies have been reported that the change of the nozzle diameter can play an important role for processing materials [22].

3.4 Effect of spraying distance

The spraying distance can have a pronounced effect on the characterization of the particles, since the nucleation and growth process continues in post-expansion region. In this study, three spraying distances (1, 3, 5 cm) from the tip of the nozzle were tested. The results show only a little change in the spraying distance from 1 cm to 5cm causes an increase in the average particle size. Similar results were reported by Yildiz et al [23].

4 Conclusions

The rapid expansion of supercritical solutions (RESS) was successfully used to produce deferasirox submicron particles. Deferasirox was micronized and a great size reduction of deferasirox particles in comparison with the original one was observed. The obtained results show a good performance of these used nozzles in comparison with the capillary nozzle used. We conclude the extraction temperature and extraction pressure has a main role in the particle size and the morphology of deferasirox particles. However by increasing the nozzle diameter and spraying distance, the particle size increases. In addition, the optimum Pressure, temperature, nozzle diameter and spraying distance for micronization of deferasirox particles are 175 bar, 318 K , 500 μm and 1 cm, respectively.

Acknowledgment

The authors thank the Shiraz University for supporting this research. The authors would like to thank Dr.Shahab Ayatollahi and Dr.Mahmood Hoormand for their cooperation.

References

[1] R.M. Mainardes, L.P. Silva, Drug delivery systems: past, present and future, Curr. Drug Targets 5 (2004) 449.

[2] B.E. Rabinow, Nanosuspensions in drug delivery, Nat. Rev. Drug Discov. 3 (2004) 785.

[3] E.M.M. Del Valle, M.A. Galan, Supercritical fluid technique for particle engineering: drug delivery applications, Rev. Chem. Eng. 21 (2005) 33.

[4] C.A. Lipinski, Poor aqueous solubility—an industry wide problem in drug discovery, Am. Pharm. Rev. 5 (2002) 82.

[5] [10] R.H. Muller, C. Jacobs, O. Kayser, Nanosuspensions as particulate drug formulations in therapy rationale for development and what we can expect for the future, Adv. Drug Deliv. Rev. 47 (2001) 3.

[6] C. Leuner, J. Dressman, Improoving drug solubility for oral delivery using solid dispersions, Eur. J. Pharm. Biopharm. 50 (2000) 47.

[7] O. Kayser, A. Lemke, N. Hernandez-Trejo, The impact of nanobiotechnology on the development of new drug delivery systems, Curr. Pharm. Biotechnol. 6 (2005) 3.

[8] D. Kayrak, U. Akman, O¨ . Hortac¸cu, Micronization of ibuprofen by RESS, J. Supercrit. Fluids 26 (2003) 17–31.

[9] L.A. Stanton, F. Dehghani, N.R. Foster, Improving drug delivery using polymers and supercritical fluid technology, Aust. J. Chem. 55 (2002) 443.

[10] J. Fages, H. Lochard, J.J. Letourneau, M. Sauceau, E. Rodier, Particle generation for pharmaceutical applications using supercritical fluid technology, Powder Technol. 141 (2004) 219.

[11] J. Jung, M. Perrut, Particle design using supercritical fluids: literature and patent survey, J. Supercrit. Fluids 20 (2001) 179–219.

[12] P. Hirunsit, Z. Huang, T. Srinophakun, M. Charoenchaitrakool, S. Kawi, Particle formation of ibuprofen supercriticalCO2 system from rapid expansion of superciritical solutions (RESS): a mathematical model, Powder Technol. 154 (2005) 83–94.

[13]Hershko C, Konijn AM, Nick HP, Breuer W, Cabantchik ZI, Link G: ICL670A: a new synthetic oral chelator: evaluation in hypertransfused rats with selective radioiron probes of hepatocellular and reticuloendothelial iron stores and in iron-loaded rat heart cells in culture. *Blood* 2001, 97(4):1115-1122.

[14]Nick H, Acklin P, Lattmann R et al. Development of tridentates iron chelators: from desferrithiocin to ICL670. Curr Med Chem 2003; 10: 1065–1076.

[15] Piga A, Gaglioti C, Fogliacco E et al. Comparative effects of deferiprone and deferoxamine on survival and cardiac disease in patients with thalassemia major: a retrospective analysis. Haematologica 2003; 88: 489– 496.

[16] A. Noyes,W. Whitney, The rate of solution of solid substances in their own solutions, J. Am. Chem. Soc. 19 (1897) 930–934.

[17] K. Nagahama, G.T. Liu, Supercritical Fluid Crystallization of Solid Solution. The 4th International Symposium on Supercritical Fluids, 11–14 May, Sendai (Japan), 1997,43–46.

[18] H. Kro¨ ber, U. Teipel, H. Krause, The Formation of Small Organic Particles Using Supercritical Fluids. Proceedings of the 5th International Symposium on Supercritical Fluids,8–12 April, Atlanta (USA), 2000.

[19] A.Z. Hezave, F. Esmaeilzadeh, Micronization of drug particles via RESS process, The Journal of Supercritical Fluids (2008), doi:10.1016/j.supflu.2009.09.006.

[20] Z. Huang, G.B. Sun, Y.C. Chiew, S. Kawi, Formation of ultrafine aspirin particles through rapid expansion of supercritical solutions (RESS), Powder Technol. 160 (2005) 127–134.

[21] Z. Huang, W.D. Lu, S. Kawi, Y.C. Chiew, J. Chem. Eng. Data 49 (2004) 1323.

[22] J.W. Tom, P.G. Debenedetti, R.J. Jerome, J. Supercrit. Fluids 7 (1994) 9.

[23] N. Yildiz, S. Tuna, O, Doker, A, C¸Alimli. Micronization of salicylic acid and taxol (paclitaxel) by rapid expansion of supercritical fluids (RESS), J. of Supercritical Fluids 41 (2007) 440–451.

Synthesis and Charging Properties of Cobalt-Based Nanoparticles in Silicon Dioxide

Jong-Hwan Yoon

Department of Physics, College of Natural Sciences, Kangwon National University, Chuncheon, Gangwon-do 200-701, Korea, jhyoon@kangwon.ac.kr

ABSTRACT

Cobalt silicide (CoSi) nanocrystal (NC) layer distributed within narrow spatial region is synthesized by thermal annealing of a sandwich structure comprised of a thin cobalt (Co) film sandwiched between two silicon-rich oxide (SiO$_x$) layers. It is shown that the size of the CoSi NCs can be controlled by varying the Co film thickness and Si concentration, an increase in the size with increasing thickness and concentration. Capacitance-voltage (C-V) measurements on a test metal/oxide/semiconductor (MOS) structure with floating gate based on CoSi NCs of 3.8 nm in diameter and 1.4×10^{12} cm^{-2} in density are shown to have C-V characteristics suitable for nonvolatile memory applications, including a large C-V memory window of about 10 V for sweep voltages between -15 V and +8.

Keywords: cobalt silicide nanoparticles, nonvolatile memory

1 INTRODUCTION

Recently, there have been considerable research efforts devoted to realizing nonvolatile memory (NVM) devices with smaller size, faster operating speed and larger storage capacity. One approach to realize such NVM devices is to use a floating-gate transistor which consists of discrete charge traps instead of a continuous conducting layer used in many conventional devices [1]. The memory devices with floating gate based on discrete charge traps offer particular advantages over conventional floating gate structures, including the fact that charges trapped at discrete sites are more stable than in a conventional conductive floating gate for lateral leakage path in the gate oxide. This improvement in the stability of charges trapped at floating gate makes it possible that devices can be scaled to smaller dimensions by reducing the tunnel oxide layer thickness.

Various discrete-charge traps, such as semiconducting [2 - 5] and pure metal nanocrystals [6 - 9], have been studied for possible applications to NVM devices. In particular, recent works [10, 11] have demonstrated that metal nanocrystals have additional advantages compared to those of semiconducting nanocrystals, namely, an enhancement in the charge storage capacity and retention time due to a higher density of states near Fermi level and a higher work function, respectively. On the other hand, as metal silicides [12, 13] have physical properties similar to those of pure metals they have also received particular attention.

In this work, we report the direct growth of cobalt silicide (CoSi) nanocrystals, which are metallic material with a work function of about 4.7 eV [13], by thermal annealing of a simple sandwich structure consisting of an ultra thin Co layer sandwiched between two silicon-rich oxide (SiO$_x$) layers. It is observed that CoSi NCs grow into well-defined boundaries and approximately spherical shape within the SiO$_2$ matrix, increasing with Co film thickness. The MOS structures with floating gate based on CoSi NCs produced by this method are shown to have C-V characteristics suitable for nonvolatile memory applications.

2 EXPERIMENTAL DETALS

CoSi NCs were formed by thermally annealing a sandwich structure comprised of a thin Co film sandwiched between two silicon-rich oxide (SiO$_x$) layers, as depicted in Fig. 1. SiO$_x$ layers were produced by plasma-enhanced chemical vapor deposition (PECVD) at a substrate temperature of 300 °C using fixed flow rates of SiH$_4$ and N$_2$O, and ultra-thin Co layer was prepared by conventional thermal evaporation. The sandwich structures were formed by alternate deposition of SiO$_x$ and Co layers on (100) oriented p-type silicon wafers as required. Nucleation and growth of CoSi NCs were achieved by thermal annealing of the sandwich structures at elevated temperature in a quartz-tube furnace using high purity nitrogen gas (99.999 %) as an ambient.

The microstructure of CoSi NCs was investigated by transmission electron microscopy (TEM) using a JEOL JEM 2010 instrument operating at 200 kV. The chemical composition of NCs was analyzed by energy dispersive X-ray spectroscopy (EDS) using an energy dispersive spectrometer attached to the TEM instrument.

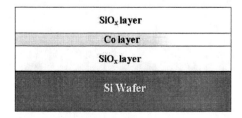

Figure 1: Schematic of layered sample structure for the formation of cobalt-based nanoparticles.

For the EDS analysis, the electron beam was focused to a spot as small as 1.5 nm in size. The MOS capacitors for memory properties were fabricated by evaporating Al through a mask with circular holes of area 0.03 mm^2 as a control electrode. Capacitance-voltage (C-V) measurements on the MOS capacitors were performed at 300 K using a Keithley 590 capacitance meter (1 MHz frequency) and a Keithley 230 programmable voltage source.

3 RESULTS AND DISCUSSION

First, the formation of CoSi nanoparticles using the layered structures shown in Fig. 1 was examined. Figure 2 shows a representative cross-sectional transmission electron microscopic (XTEM) image, which was taken for a layered structure comprised of Co film of 0.2 nm thickness sandwiched between two $SiO_{1.67}$ layers of 5 nm after annealing at 650 °C for 1 h. The XTEM image clearly shows the presence of nanoparticles with well-defined boundaries and approximately spherical shape. Especially, the image shows that the spatial distribution of the nanoparticles is much narrow, thereby demonstrating the efficacy of the technique for making nanocrystal floating gate structures for nonvolatile memory devices. In the present case, the average diameter of the nanoparticles is about 3.0 nm with standard deviation of 0.47 nm.

To identify exact phase of the Co-based nanoparticles shown in Fig. 2, energy dispersive x-ray spectroscopy (EDS) was performed on each individual nanoparticles. Figures 3(a) and (b) present representative EDS spectra obtained outside and inside nanoparticles, respectively. As seen in Figure 3, little or no Co is observed outside of nanoparticle but it is clear that the nanoparticles contain Si and Co atoms, supporting the premise that they are cobalt silicide particles. The Co-based nanoparticles shown in Fig. 2 were formed by annealing at a temperature of 650 °C where they are expected to be at a stable phase [14]. Furthermore, the high-magnification TEM image of nanoparticles shown in Fig. 5(b) reveals a regular lattice structure consistent with it being a single crystal phase. These results demonstrate the fact that the nanoparticles are cobalt silicide (CoSi) nanocrystals (NCs).

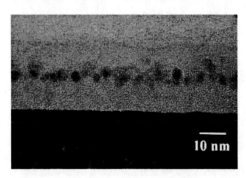

Figure 2: Cross-sectional transmission electron microscope image of the sandwich structure of $SiO_{1.67}$/Co/ $SiO_{1.67}$ after annealing for 1 h at 650 °C.

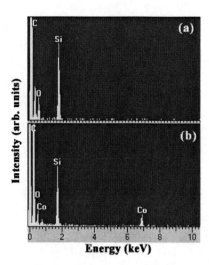

Figure 3: EDS spectra taken (a) outside and (b) inside of Co-based nanoparticle shown in Fig. 2. Note that no Co signal is observed outside of the nanocrystals.

Figure 3 shows cross-sectional transmission electron microscopic images of a layered structure with different Co thickness. The layered structure was comprised of Co films (0.2 nm, 0.5 nm, and 0.8 nm in thickness) sandwiched between two $SiO_{1.51}$ layers after annealing at 650 °C for 2 h. The image also shows the presence of NCs distributed within narrow spatial region. Furthermore, it is clear from the image that the size of the NCs increases as Co layer is thickened. The mean diameters estimated for 0.2 nm, 0.5 nm, and 0.6 nm Co thicknesses were about 2.2 nm, 2.9 nm, and 4.7 nm, respectively. In the case of 0.2 nm thickness, in particular, the diameter is smaller than that in the case shown in Fig. 2 despite the same Co thickness. This fact demonstrates that the size of CoSi NCs is also associated with concentration of Si atoms.

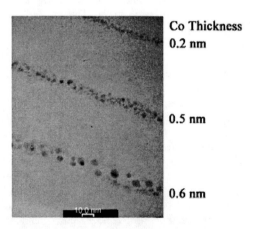

Figure 4: Cross-sectional transmission electron microscope image of $SiO_{1.51}$ layer embedded Co layers with different thickness after annealing at 650 °C for 2 h.

NSTI-Nanotech 2010, www.nsti.org, ISBN 978-1-4398-3401-5 Vol. 1, 2010

In order to explore the potential of CoSi nanocrystal floating gates for memory applications, a well-controlled metal/oxide/semiconductor (MOS) sample containing CoSi NCs between tunnel and control oxide layers was prepared for capacitance-voltage (*C-V*) measurements. The MOS structure with CoSi NCs was formed with the following layers: tunnel $SiO_2/SiO_{1.51}/Co/SiO_{1.51}/control$ SiO_2. The thickness of tunnel and control oxide layer was 5 nm and 7 nm, respectively, and the thickness of $SiO_{1.51}$ and Co layer was 3.5 nm and 0.6 nm, respectively. This structure was subsequently annealed at 650 °C for 2 h to form CoSi NCs, with the resulting microstructure shown in Figure 5, which represents the XTEM images of the sample after annealing. Figure 5(a) and (b) show the overall and high-magnification images, respectively. These images clearly show the formation of well-defined CoSi NC monolayer distributed within the narrow region, and the regular lattice structure consistent with it being a single crystal phase [Fig. 5(b)]. The average size and areal density of the CoSi NCs are about 3.8 nm and 1.4×10^{12} cm^{-2}, respectively. The size is smaller than that for 0.6 nm Co thickness shown in Fig. 4. The difference might be likely to result from a smaller concentration of Si atoms due to the relatively thinner layer of SiO_x in the case of Fig. 5. The MOS capacitors for *C-V* characteristics were fabricated by evaporating Al gate electrodes on the control oxide layer.

Figures 6(a) and (b) show the *C-V* curves measured for the MOS capacitors without and with NCs, respectively. The MOS capacitors without NCs were also fabricated with the same layer structure as the case with CoSi NCs except for the absence of Co layer. The gate voltage was swept from accumulation region (-15 V) to inversion region (8 V) before being swept back from inversion (8V) to accumulation (-15V). As seen in Fig. 6, there is clearly a distinct difference in the memory window for the two cases.

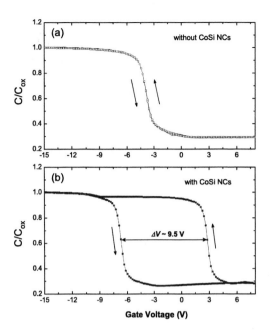

Figure 6: *C-V* hysteresis loops of MOS capacitors a) without, and b) with CoSi nanocrystals.

The MOS capacitor with CoSi NCs exhibits large memory window of approximately 10 V, while the MOS capacitor without CoSi NCs does not reveal significant memory window. This difference demonstrates that the large memory window is closely associated with the presence of the CoSi NCs. The width of *C-V* hysteresis loop, i.e., memory window, is due to the charge (Q) trapped at NCs, increasing the memory window with increasing the charge trapped at the NCs. In the case of metallic NCs, because of a high density of states near the Fermi level, several electrons per NC can be trapped [10]. We can estimate the average number, n, of electrons trapped at a CoSi NC by using the simple relation $n = (\Delta V_{FB} \cdot C_{acc})/(en_0 S)$, where ΔV_{FB} is the flat band voltage shift due to the electron charge trapped at NCs, C_{acc} is the accumulation region capacitance, n_0 is the areal density of NCs, e is the electron charge, and S is the gate electrode area. At a gate voltage of +8 V, which causes a fully electron charged state of the NC, ΔV_{FB}=4.5 V and C_{OX}= 35 pF, n_0=1.4×10^{12} cm^{-2}, and S=0.03 mm^2, so that for the sample shown in Fig. 4 the number of electrons (n) trapped in each NC is about 2.4. This value is a little bit smaller than those reported by others [6]. The difference is likely to be due to the difference in either the work function or size of NCs. As a consequence, the large memory window of the MOS capacitor with CoSi NC floating gate, as shown in Fig. 6(b), is thus attributed to a large number of electrons trapped per NC due to the metallic properties of the CoSi NCs.

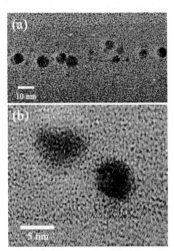

Figure 5: (a) High-resolution cross-sectional TEM images of a sample prepared using a $SiO_2/SiO_{1.51}/Co /SiO_{1.51}/SiO_2$ structure with a Co film of 0.6 nm thickness, and (b) the high-magnification TEM image of the CoSi NCs, showing regular lattice structure.

4 SUMMARY

A simple method for fabricating metallic cobalt silicide (CoSi) nanocrystals (NCs) with narrow spatial distribution has been demonstrated and shown to produce structures for nonvolatile memory applications. The method is to form CoSi NCs by thermally annealing a simple sandwich structure consisting of an ultrathin cobalt (Co) layer sandwiched between two silicon-rich oxide (SiO_x) layers, The CoSi NC size is shown to be controlled by thickness of Co layer and Si concentration in SiO_x. Capacitance-voltage (*C-V*) measurements on the test metal/oxide/semiconductor capacitor based on CoSi nanocrystal floating gate with average diameter of 3.8 nm and areal density of 1.4×10^{12} cm^{-2} in density are shown to have *C-V* characteristics suitable for future nonvolatile memory applications, including a large memory window of 10 V for sweep voltages between -15 V and +8 V.

ACKNOWLEDGEMENT

This work was supported by the Korea Research Foundation Grant funded by the Korean Government (MOEHRD, Basic Research Promotion Fund, KRF-2008-313-C00303).

REFERENCES

[1] D. Kahng and S. M. J. Sze, Bell Syst. Tech. 46, 1283, 1967.

[2] S. Tiwari, F. Rana, K. Chan, H. Hanafi, A. Hartstein, E. F. Crabbe, and K. Chan, Appl. Phys. Lett. 68, 1377, 1996.

[3] Y. Shi, K. Saito, H. Ishikuro, and T, Hiramoto, J. Appl. Phys. 84, 2358, 1998.

[4] S. H. Hong, M. C. Kim, P. S. Jeong, S. H. Choi, Y. S. Kim, and K. J. Kim, Appl. Phys. Lett. 92, 093124, 2008.

[5] T. Z. Lu, M. Alexe, R. Scholz, V. Talelaev, and M. Zacharias, Appl. Phys. Lett. 87, 202110, 2005.

[6] C. Lee, J. Meteer, V. Narayanan, and E. Kan, J. Electron Meter. 34, 1, 2005.

[7] S. K. Samanta, W. J. Yoo, and G. Samudra, Appl. Phys. Lett. 87, 113110, 2005.

[8] D. Zhao, Y. Zhu, and J. Liu, Solid State Electronics 50, 268, 2006.

[9] D. U. Lee, M. S. Lee, J. H. Kim, and E. K. Kim, Appl. Phys. Lett. 90, 093514, 2007.

[10] Z. Liu, C. Lee, V. Narayanan, C. Pei, and E. C. Kan, IEEE Trans. Electron Dev. 49, 1606, 2002.

[11] Z. Liu, C. Lee, V. Narayanan, C. Pei, and E. C. Kan, IEEE Trans. Electron Dev. 49, 1614, 2002.

[12] J. Yuan, G. Z. Pan, Y. L. Chao, and J. C. S. Woo, Mater. Res. Soc. Proc. 824, B7.11.1, 2005.

[13] E. M. Gullikson, A. P. Mills, and J. M. Phillips, Surface Science 195, L150, 1989.

[14] H. Miura, E. Ma, and C. V. Thompson, J. Appl. Phys. 70, 4287, 1991.

Microwave-assisted Superparamagnetic FePt Synthesis for MRI Application

Fan-Chun Ding[*], Ming-Jui Tsai[**], Tsiao-Yu Tsai[*], Jai-Lin Tsai[**], Jian-Ren Lai[*] and Ping-Shan Lai[*]

[*] Department of Chemistry, National Chung Hsing University
No. 250, Kuo-Kuang Road, Taichung 402, Taiwan, pslai@email.nchu.edu.tw
[**] Department of Materials Science and Engineering, National Chung Hsing University
No. 250, Kuo-Kuang Road, Taichung 402, Taiwan, tsaijl@dragon.nchu.edu.tw

ABSTRACT

Superparamagnetic iron oxide (SPIO) nanoparticles have demonstrated their practicability as MRI T_2-shortening agents for non-invasive cell labeling, drug delivery or tumor detections in clinical practice. Recently, superparamagnetic chemically disordered face-centered cubic (fcc) FePt nanoparticles has also been demonstrated as superior negative contrast agents for MRI. Superparamagnetic FePt nanoparticles, which show high saturation magnetization (M_S) compared to SPIO, are expected to be a high performance nanomagnet for magnetic medicine. However, the conventional complicated synthesis procedure and hydrophobicity has limited the potential usage in vitro, in vivo and in clinic. In this work, the simple microwave heating method was utilized for the controlled synthesis of FePt nanoparticles using Fe(acac)3 and Pt(acac)2 as the main reactants in tetra-ethylene glycol. The advantages of this process are its simplicity, the short reaction time and easy preparation. The structure and composition of the FePt nanoparticles were characterized by XRD, TEM and magnetic measurements.

By varying the Fe/Pt ratio, power and irradiation time, the microwave assisted synthesis has shown a significant advantage for the rapid production of monodisperse fcc FePt nanoparticle. The prepared FePt nanoparticles can be further capped by cysteine to improve their water solubility. Moreover, magnetic resonance relaxometry reveals that Cys-capped FePt nanoparticles have a high T2-shortening effect that can induce sufficient cell MRI contrast.

Keywords: Microwave-assisted synthesis, superparamagnetic FePt, MR imaging

1 INTRODUCTION

Superparamagnetic iron oxide (SPIO) nanoparticles have been demonstrated their practicability as MRI T_2-shortening agents for non-invasive cell labeling, drug delivery or tumor detections in clinical practice [1, 2]. However, low intracellular labeling efficiency and nanotoxicity of SPIO has limited their potential usage.[3, 4] Recently, bifunctional contrast agents for both optical and MR imaging have been developed to function as good imaging probes *in vitro* and *in vivo* [5-7] and the nanotoxicity of SPIO is suppressed by particular surface modification [8]. Thus, proper modifications of SPIO have been considered as a promising strategy to improve the aforementioned problems.

Long range ordered $L1_0$ FePt films have been studied extensively and thought to be a promising candidate for ultra-high-density magnetic recording media due to its high magneto-crystalline anisotropy (K_u). The magnetization thermal instability or superparamagnetic effect is delayed due to their very small critical grain size of around (3-5nm). However, when FePt nanoparticles are disordered with face centered cubic (fcc) structure, they exhibit superparamagnetic property under critical size. Superparamagnetic FePt nanoparticles, which show high saturation magnetization (M_S) compared to SPIO, are expected to be a high performance nanomagnet for magnetic medicine [9, 10]. In particular, the potential of FePt nanoparticles as a MRI contrast agent has been demonstrated [11]. Thus, the T2 contrast agent using superparamagnetic FePt nanoparticles seems to be a promising nanomaterial for MRI.

Microwave-assisted heating is becoming a common method for rapid synthesis. Hence, in this study we utilized the simple microwave heating method for the synthesis of superparamagnetic FePt and evaluated their potential application for MRI application.

2 EXPERIMENTAL PROCEDURE

2.1 Synthesis of FePt

The synthesis of fcc FePt nanoparticles involving simultaneous chemical reduction of $Pt(acac)_2$ and $Fe(acac)_3$ by microwave heating at high temperature in tetra-ethylene glycol (TEG, b.p. 327.3℃) solution phase was followed as described previously [12]. In brief, $Fe(acac)_3$,$Pt(acac)_2$, oleylamine and oleic acid were mixed with TEG in an inert atmosphere of argon gas. All chemicals were weighed, placed into reaction flasks. The mixture was sonicated at 60 ℃ for 1 h before transferring into the microwave apparatus. The reaction was done after heating by microwave apparatus (900 W) for 30-50 min under N_2. In the post reaction treatment, the black, solid product was separated by centrifugation, washed with ethanol and dried under vacuum at room temperature. The synthetic conditions of

FePt synthesis including molar ratio (Fe(acac)$_3$ / Pt(acac)$_2$) and heating time were shown in Table 1.

Table 1: Synthetic conditions of FePt synthesis

Molar ratio of Fe(acac)$_3$: Pt(acac)$_2$	Heating time (min)	Power (W)
2:1	30	900
	40	
	50	

2.2 Characterization of FePt

After purification, the product was characterized by X-ray powder diffraction (XRD) analysis, transmission electron microscopy (TEM), vibration sample magnetometer (VSM) and T_2 enhancing relaxivity analysis (Varian 300 MHz NMR). The iron content of the FePt was determined using an atomic absorbance spectrophotometer (Dionex ICS-90, USA)

3 RESULTS

Superparamagnetic FePt NPs were synthesized under 900w microwave heating from 30 min to 50 min in this study. Fig. 1 shows the X-ray diffraction (XRD) patterns of FePt nanocrystals prepared with different time by microwave. The XRD pattern for superparamagnetic FePt nanocrystals shows the FePt (111) and (202/200) peak (40.9°; 69.7°). These results indicate that the sample is FePt disordered fcc phase.

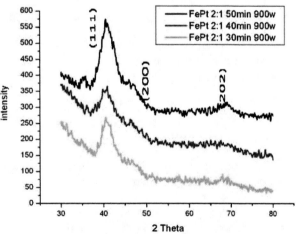

Fig. 1: X-ray diffraction patterns of FePt nanocrystals (precursor's molar ratio; Fe(acac)$_3$: Pt(acac)$_2$ is 2:1) in different heating time under 900w microwave heating.

Fig. 2 shows the VSM curves of FePt NPs. It is indicated that the prepared FePt revealed different saturation magnetization with different heating time. FePt prepared with 50 min microwave treatment at 900W exhibited the better superparamagnetic property and that

prepared with 40 min microwave treatment shows the largest saturation magnetization. There is no correlation among feed ratio, heating time and heating energy. However, comparing to the traditional synthetic procedure, the FePt synthesis using microwave-assistance decreases the reaction time in 1 h.

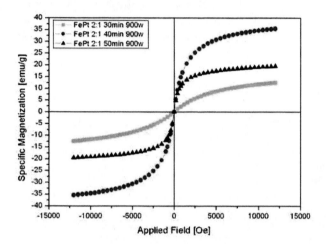

Fig. 2: VSM curves of FePt (precursor's molar ratio; Fe(acac)$_3$: Pt(acac)$_2$ is 2:1) treated with different time under 900w microwave heating.

Next, the prepared FePt nanoparticles were capped by cysteine to improve their water solubility. Fig. 3 shows The TEM image shows the morphology of FePt nanoparticles (precursor's molar ratio; Fe(acac)$_3$: Pt(acac)$_2$ is 2:1) which heating 40 minutes by 900w microwave. The average diameter is about 3 nm (average diameter : 2.99nm). The particle diameter of FePt was not increased with the molar ratio of Fe(acac)$_3$ and Pt(acac)$_2$ (2:1, 2.5:1, 3:1) (data not shown). Besides, *in vitro* results indicated the FePt nanoparticles might be taken up by cells via endocytosis. No significant cytotoxicity was observed after 500 M FePt nanoparticles incubation for 72h in HeLa cells (over 70% cell survival) (data not shown).

The T_1 and T_2 values obtained from cysteine-capped FePt dispersed in pure water was evaluated by Varian 300 MHz NMR. The inverse relaxation times of our results were almost linearly proportional to the concentration of FePt nanoparticle as previous report [11]. Consequently, the R_1 (longitudinal relaxivities) and R_2 (transverse relaxivities) values of dispersed in pure water were determined to be 5.7 and 396.1 for FePt nanoparticles. The higher the R_2/R_1 relaxivity ratio, the higher is the T_2 effect and the signal decrease on T_2-weighted images. For FePt nanoparticles, the R_2/R_1 relaxivity ratio was found to be 69.5. Resovist (Schering AG, Berlin, Germany), a carboxydextran-coated SPIO commercial MRI contrast agent for liver-specific MRI, shows a high relaxivity ratio $R_2/R_1 \sim 9.6$ ($R_1 = 19.4$ s^{-1} mM^{-1} and $R_2 = 185.8$ s^{-1} mM^{-1}). Comparing with Resovist, FePt nanoparticles exhibited a superior T_2-shortening ability.

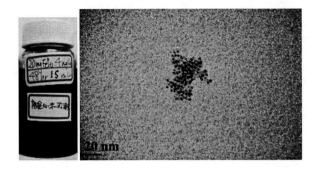

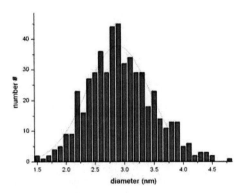

Fig. 3: Photo and TEM image of FePt-cys nanoparticles (precursor's molar ratio; Fe(acac)$_3$:Pt(acac)$_2$ is 2:1) which heating 40 min by 900w microwave.

By varying the Fe/Pt ratio, power and irradiation time, the microwave assisted synthesis has shown a significant advantage for the rapid production of monodisperse fcc FePt nanoparticle. The prepared FePt nanoparticles can be further capped by cysteine to improve their water solubility. Moreover, magnetic resonance relaxometry reveals that Cys-capped FePt nanoparticles have a high T2-shortening effect with diameter ~3 nm that can induce sufficient cell MRI contrast.

4   CONCLUSION

The advantages of this microwave heating process are theirs simplicity, the shorter reaction time and easy preparation. The longitudinal and transverse proton relaxation times obtained superparamagnetic FePt nanoparticles was measured. The R_2/R_1 relaxivity ratio of superparamagnetic FePt NPs was found to be 7.24 times larger than that of conventional SPIO-based MRI contrast agent. FePt synthesis with Fe/Pt ratio, power and microwave irradiation time has been investigated and the prepared superparamagnetic FePt can play as a T2 contrast agent for MRI application.

5   ACKNOWLEDGEMENTS

This research was supported by grants from the National Science Council of the Republic of China (NSC 98-2113-M-005-007-MY2).

REFERENCES

[1] J.V. Frangioni, New technologies for human cancer imaging. Journal of Clinical Oncology 26(24) (2008) 4012-4021.

[2] N. Nasongkla, E. Bey, J.M. Ren, H. Ai, C. Khemtong, J.S. Guthi, S.F. Chin, A.D. Sherry, D.A. Boothman, J.M. Gao, Multifunctional polymeric micelles as cancer-targeted, MRI-ultrasensitive drug delivery systems. Nano Letters 6(11) (2006) 2427-2430.

[3] N. Lewinski, V. Colvin, R. Drezek, Cytotoxicity of nanoparticles. Small 4(1) (2008) 26-49.

[4] T.R. Pisanic, J.D. Blackwell, V.I. Shubayev, R.R. Finones, S. Jin, Nanotoxicity of iron oxide nanoparticle internalization in growing neurons. Biomaterials 28(16) (2007) 2572-2581.

[5] C.F. Tu, Y.H. Yang, M.Y. Gao, Preparations of bifunctional polymeric beads simultaneously incorporated with fluorescent quantum dots and magnetic nanocrystals. Nanotechnology 19(10) (2008) 105601.

[6] H.M. Liu, S.H. Wu, C.W. Lu, M. Yao, J.K. Hsiao, Y. Hung, Y.S. Lin, C.Y. Mou, C.S. Yang, D.M. Huang, Y.C. Chen, Mesoporous silica nanoparticles improve magnetic labeling efficiency in human stem cells. Small 4(5) (2008) 619-626.

[7] W.J.M. Mulder, R. Koole, R.J. Brandwijk, G. Storm, P.T.K. Chin, G.J. Strijkers, C.D. Donega, K. Nicolay, A.W. Griffioen, Quantum dots with a paramagnetic coating as a bimodal molecular imaging probe. Nano Letters 6(1) (2006) 1-6.

[8] A.K. Gupta, M. Gupta, Cytotoxicity suppression and cellular uptake enhancement of surface modified magnetic nanoparticles. Biomaterials 26(13) (2005) 1565-1573.

[9] S.H. Sun, C.B. Murray, D. Weller, L. Folks, A. Moser, Monodisperse FePt nanoparticles and ferromagnetic FePt nanocrystal superlattices. Science 287(5460) (2000) 1989-1992.

[10] S. Maenosono, S. Saita, Theoretical assessment of FePt nanoparticles as heating elements for magnetic hyperthermia. Ieee Transactions on Magnetics 42(6) (2006) 1638-1642.

[11] S. Maenosono, T. Suzuki, S. Saita, Superparamagnetic FePt nanoparticles as excellent MRI contrast agents. Journal of Magnetism and Magnetic Materials 320(9) (2008) L79-L83.

[12] K.E. Elkins, T.S. Vedantam, J.P. Liu, H. Zeng, S.H. Sun, Y. Ding, Z.L. Wang, Ultrafine FePt nanoparticles prepared by the chemical reduction method. Nano Letters 3(12) (2003) 1647-1649.

Application of Sugimoto model on particle size prediction of colloidal TiO$_2$ Nanoparticles

H. Mehranpour[1], M. Askari[1], M. Sasani Ghamsari[2], H. Farzalibeik[1],

[1]Department of Material Science and Engineering, Sharif University of Technology, 11155-9466, Tehran, Iran.
askari@sharif.edu, h_mehranpour@alum.sharif.edu,
[2]Solid State Lasers Research Group, Laser & Optics Research School, NSTRI, 11365-8486, Tehran, Iran.
msghamsari@yahoo.com

Abstract:

In this paper, Sugimoto model has been employed to study the nucleation and the growth of TiO$_2$ nanoparticles. On the base of this model some parameters of the nucleation and the growth of nanoparticles such as supply rate of the solute (Q$_0$), molar volume of the solid (*V*m), and mean volumic growth rate of the stable nuclei during the nucleation period ($\dot{v}$) were determined. For this approach, TiO$_2$ nanoparticles were synthesized in two stages in an aqueous solution by using HNO$_3$ and TTIP and TIPO as starting materials. In this experiment, it was obviously observed that the changing of Ti^{+4} concentrations was followed with LaMer diagram. Typical obtained results were $\dot{v}$=2.93×10^2 nm^3.s^{-1} and r$_0$=8.2nm under the standard conditions at 70°C. The maximum supersaturating ratio (*S*m) was found to be 3.07. It has been concluded that the size of TiO$_2$ nanoparticles can be predict by Sugimoto model. The LaMer diagram was used as a fundamental principle of monodispersed particles formation.

Keywords: Titanium Dioxide, Sugimoto model, Nucleation, Monodispersed particles, LaMer diagram.

Introduction

Nanostructured titanium dioxide (TiO$_2$) is widely applied as a white pigment for paints or cosmetics, a support in catalysis, and a photocatalyst [1]. The performance of TiO$_2$ nanocrystals are strongly influenced by the crystallinity, morphology and the particle size. During three last decades many attempts have been done to prepare nanocrystalline TiO$_2$ particles with improved their properties [2]. It is well known that the preparation method has a critical role in performance of obtained TiO$_2$ nanomaterials [3]. Among all methods which are employed to synthesis TiO$_2$ nanocrystals, sol-gel is the best due to its controllability and low cost [4, 5]. The formation of nanostructures by sol-gel method consist a chemical process which is initiated by the reactions of molecules or ions that serve as precursors to nuclei. Nuclei undergo growth or further reactions to form nanostructures. Detailed understanding of the associated mechanism with nuclei formation and growth is important in controlling the size, shape and the physical properties of nanostructures [6, 7]. Many researches have been focused on the preparation and study on nucleation and growth of TiO$_2$ nanoparticles. For example, the synthesis of uniform anatase TiO$_2$

nanoparticles by gel–sol method and mechanisms in size control of uniform nanoparticles has investigated by Sugimoto [8,9]. In 1999, Sugimoto pointed out that the nucleation and growth of silver nanoparticles can be conformed to the LaMer theory [10]. Using Lamer mechanism, Sugimoto developed the nucleation model and verified it experimentally and described the nucleation and growth of AgCl and AgBr [10,11]. In the present work, we have tried to evaluate the nucleation and growth of TiO$_2$ nanocrystals on the base of Sugimoto approach. At first we need to present a brief description about LaMer Theory. In that theory the formation of sulfur sols has been described on the base of the decomposition of sodium thiosulfate in hydrochloric acid [10,11]. The essence of the mechanism was illustrated in Fig. 1; the time variation of nucleating species concentration (sulfur in this case) has three steps. The concentration of elemental sulfur builds up slowly until some critical concentration is reached, or more specifically a critical supersaturation level, C/C_{eq} where C_{eq} is the solubility of sulfur in the solution (stage I in Fig. 1).

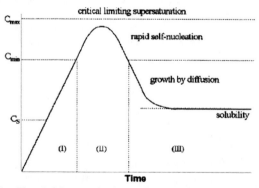

Fig.1 .The LaMer mechanism of nucleation of sulfur. The (theoretical) curve shows sulfur concentration as a function of time.

At this point, "self-nucleation" (i.e., homogeneous nucleation) occurs at a rate that has been called "effectively infinite". This step has pointed out as stage II in figure 1. That stage is the origin of the idea of "burst" nucleation.

The burst of nucleation immediately lowers the supersaturation level of monomers in the solution; as a result, nucleation essentially stops at this time. Growth then occurs by diffusion of sulfur atoms throughout the solution, stage III

NSTI-Nanotech 2010, www.nsti.org, ISBN 978-1-4398-3401-5 Vol. 1, 2010

[10,11]. The Sugimoto nucleation model can be briefly explained as follow.

When C reaches to C_{max}, the steady state of a mass balance between the supply rate of solute (R_S), consumption rates for nucleation (R_N), and growth rate of the generated nuclei (R_G) is established, and the following relationship can be given as [12,13,14]:

$$R_N + R_G - R_S = 0 \qquad (1$$

This mass balance equation may be replaced by a corresponding differential equation for the number of the generated stable nuclei, n_∞, as:

$$v_0 \frac{dn_\infty}{dt} + \dot{v} n_\infty - Q V_m = 0 \qquad (2$$

Where v_0 the minimum particle volume of the stable nuclei is, $\dot{v}$ is the mean volume growth rate of the stable nuclei, Q is the supply rate of solute in mol per unit time, and V_m is the molar volume of the solid. The stable nuclei are sufficiently large to be grown to the final product particles, and they are defined as nuclei whose each particle volume is over the maximum particle volume of the stationary nuclei. One may regard Q as being kept constant at Q_0 during the short nucleation period, and $\dot{v}$ may be regarded as a constant. In the initial time (t=0) n_∞ is equal to 0 and C reaches to C_{max}. Therefore the n_∞ can be treated as a function of time as follows [12,13,14]:

$$n_\infty = \frac{Q_0 V_m}{\dot{v}} \left[1 - \exp\left(\frac{-\dot{v} t}{v_0}\right) \right] \qquad (3$$

When $t \to \infty$ the final number of generated stable nuclei is equal to:

$$n_\infty = \frac{Q_0 V_m}{\dot{v}} \qquad (4$$

On the other hand, since particles are known to grow by diffusion control, the linear growth rate (dr/dt) in the growth stage may be given by [9,10,15]:

$$\frac{dr}{dt} = \frac{D V_m}{r} \left[C - C_\infty exp\left(\frac{2\gamma V_m}{rRT}\right) \right] \qquad (5$$

In this equation D is the diffusivity of the solute including free ions and their halide complexes, C is the molality of the solute, C_∞ is the solubility of the bulk solid in terms of total molality of solute, and γ is the specific surface energy of the solid. By considering the initial particle radius as radius of the stable nuclei during the nucleation stage, the below equation is established between r_0, D and $\dot{v}$:

$$\dot{v} = 4\pi r_0 D V_m C_\infty \left[S_m - exp\left(\frac{2\gamma V_m}{r_0 RT}\right) \right] \qquad (6$$

In this experiment the radius and the formation free energy of nucleus can be determined by Gibbs-Thomson relations [12,13,14].

In our work, the application of Sugimoto model for formation of TiO$_2$ nanoparticles was evaluated. At first, it was shown that the nucleation and growth of TiO$_2$ nanoparticles is conform with LaMer diagram and after that the parameters were obtained by Sugimoto Model for this approach we use the size of final particles from TEM photo. Comparison of our work with results of other researches shows that our results were obtained in high accuracy.

Experimental

In this investigation, the raw materials contain Titanium isopropoxide (TTIP) (98%), Triethnolamine (TIPO), Ethanol, HNO3 (98%) which have been purchased from Merck. The established controlled conditions for the preparation of TiO$_2$ nanoparticles were employed as follows. First, a stock solution of Ti^{4+} was prepared by mixing titanium isopropoxide (TIPO:Ti[OCH(CH$_3$)$_2$]$_4$) with triethanol amine (TEOA : N(CH$_2$CH$_2$OH)$_3$) at a molar ratio of TIPO: TEOA=1 : 2 under dry air to form a stable Ti^{4+} compound against the hydrolysis reaction at room temperature. The process followed by addition of doubly distilled water to make an aqueous stock solution that the concentration of Ti^{4+} is near to 5×10^{-4} mol^{-1}. Then, 10 ml of the stock solution was mixed with the same volume of doubly distilled water. The pH was controlled by addition of HClO$_4$ or NaOH solutions. The final solution (pH=9.6) was placed in a screw-capped Pyrex bottle and aged at 100°C for 36 h. Finally, the resulting highly viscous gel was mixed with 80 ml (2×10^{-3} molar) nitric acid and stirred at 25°C for 3 h to dissolve the gel and so prepared a dark solution with pH=1. The solution was set in pool water at 70°C.

Stage1. Solution was heated to achieve super saturation of Ti in solution. Each 4 min a sample has been taken from prepared solution and analyze by Atomic Adsorption.

Stage2. Using condenser to stop evaporation of solution and Each 4 min a sample has been taken again among this process TiO$_2$ particles was separated from the resulting Ti^{4+} suspension by centrifugation. This work has been repeated several times and every time the concentration of the Ti^{4+} was measured by the atomic adsorption. The products were washed with distilled water and observed using a transmission electron microscope using Germany ZEISS Em-900.

Results

3.1 *Formation of TiO$_2$ particles*

The basic alkoxide precursor used in this study, can be hydrolyzed and condensed, where an oxygen link between two Ti atoms is formed through several steps that can be written as follows:

$$\text{Ti-OR + H2O} \leftrightarrow \text{Ti-OH + ROH (1)}$$

Depends on the pH of the solution, the present of titanium (IV) compounds in aqueous solutions are TiO^{2+}, TiOH^{3+}, Ti(OH)$_2$$^{2+}$, and TiO(OH)$^+$ [15,16]. In this experiment the pH=1, and the TiO^{2+} is the dominant complex in solution. Therefore, TiO_2 can be nucleate by the below equations:

$$...Ti-OH + OR-Ti... \leftrightarrow ...Ti-O-Ti... + ROH \quad (2)$$

$$...Ti-OH + OH-Ti... \leftrightarrow ...Ti-O-Ti... + HOH \quad (3)$$

Thus, the growth of particles occurs by addition of Ti^{4+} to the existing particles. It was shown that the TiO_2 surfaces has enough number of O– or –OH sites which are available in appropriate positions to create octahedral coordination for the incoming Ti^{4+} ions. Moreover, after the attachment of Ti^{4+} ion along the planes, equal number of new O– or –OH sites are created for further attachment of Ti^{4+} ions [17].

3.2 *Nucleation of unstable particles under the standard conditions*

The time variation of Ti^{4+} supersaturation was recorded clearly and illustrated in Fig. 2. It is like the LaMer diagram. This diagram is divided in two stages under the standard conditions. In the first stage, when the liquid in the solution was evaporated, the concentration of titanium increases. It is believed that the evaporation of liquid has directly affected on changing the mass balance and so, the titanium concentration increased. The concentration of titanium increase until the C=3510ppm which can be considered as the maximum supersaturation of Ti^{+4} in solution, at this time the unstable particles are create, this particles can be solute or remain until the starting of the second stage. The radius of these particles can be found by the Gibbs-Thomson equation. In contrast, in the second stage after maximum supersaturation the concentration of titanium decrease and it was also observed that TiO_2 particles were precipitated and could separate after centrifuging the samples. Also, it can be found that the obtained supersaturation ratio (S) from the analysis of the [Ti]-curve leads to determination of the nucleation start point that is equal to 37.5 min. From the curve of the supersaturation, the maximum supersaturation ratio (S_m) can be measured as high as 3.19.

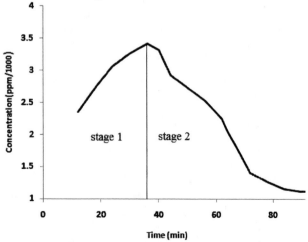

Fig2. Time evolutions of Ti^{4+} concentration in the standard system, stage1: [Ti] concentration increasing, stage 2: [Ti] concentration decreasing

The equilibrium nucleus radius r^* and ΔG^* at the maximum supersaturation (unstable particles), r_m^* and ΔG_m^*, are calculated from Gibbs-Thomson equation and are r_m=4.01nm and ΔG_m^*=1.16×10^{-16} J (=343kT), where V_m =20.9 cm^3 mol^{-1} and γ=350 mJ/m^{-2} [18].

3.3 *stable particle size and growth*

In the second stage, the concentration of Ti^{+4} decreases that is due to formation of TiO_2 particles and are precipitated during this stage. The end of the Ti^{+4} reduction, was observed after about 90 min, as shown in Fig.2. After that, the concentration of Ti^{4+} was constant at the level of C = 1180ppm. When reduction starts, at first Ti ions decrease rapidly but after a while (about 5 min) it is observed that Ti ions reduce with a less speed than before. It is believe that at this time the nucleation period of particles finished and the stable particles was created and after that this particles were grown to compose the final particles.

At this level, Gibbs-Thomason equation was used again to obtain the size of stable particles which is equal to r_0=8.2nm. TEM analysis was used to check the agreement between theory and experiment. It has been found that the synthesized TiO_2 particles in this time have a size of r =10nm which are nearly to our theoretical size. These particles are the particles which turn into final particles.

Fig. 3 shows a TEM image of TiO_2 particles prepared under the standard conditions and after 100 min. it can be found that the obtained mean particle diameter is equal to 0.200 mm and the final particle density was obtained which is n_∞ =4.8×10^{13} dm^{-3} [12,13,14].

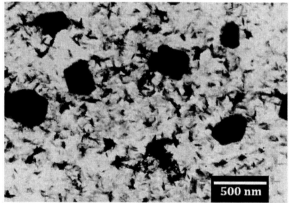

Fig3.The TEM image of uniform TiO_2 particles prepared under the standard condition.

On the other hand The supply rate of the monomer in the nucleation period, Q_0, is obtained from the product of the tangential slope of the curve of [Ti] vt t at [Ti]^{max} in Fig. 3 and hence Q_0= d[Ti]/dt=7.03×10^{-8} mol dm^{-3} s^{-1}. The amount of mean volumic growth rate of the stable nuclei obtained $\dot{v}$= $Q_0 V_m/n_\infty$=2.93×10^2 nm^{3}s^{-1}.

The growth rate of stable nuclei examine with TEM photos which was got every 10 min after the nucleation of stable particles. We expected that the size of particles will be around of 25, 35 and 45 nm in approximately 10, 15 and 20 min after supersaturation respectively. Our TEM images shows that the particle dimensions have good adaptability with our theoretical calculations. And finally, from the equation (6) diffusion coefficient of [Ti] is D = 6.18×10^{-5} cm^2s^{-1}. This D value is in a reasonable range for diffusion of ions or complexes in an aqueous solution. Moreover, this coefficient is approximately equal with the diffusion coefficient of Ti ions which and was obtained by Shingyouchi [19].

Conclusion

Tadao Sugimoto formulated a new model for studying the nucleation and growth of nano particles. The base of this model is on the nucleation and growth of particles which controlled by diffusion and the particles which their growth follows the LaMer diagram. With this model he studied the nucleation and growth of AgCl and AgBr nano particles in standard condition. At the present work an effort was made to study on nucleation and growth of TiO2 nano particles nano particles with this model the important factors obtain as follows:

S_m=3.19, r_m=4.01nm and $\Delta G_m{}^*$=1.16×$10^{-16}J$, r_0=8.2nm, Q_0=7.03×10^{-8} mol dm^{-3} s^{-1}, $\dot{v}$= 2.93×10^2 nm^3s^{-1} and D = 6.18×10^{-5} cm^2s^{-1}.

It is believe that with these parameters we can predict the size of TiO_2 nano particles in different time in growth part.

References

[1] Y. Gao, Y. Masuda, W. Seo , H. Ohta, K. Koumoto, Ceramics International, 30 , 1365–1368, 2004.

[2] R. Lee penn and Jillian F. Banfield, Geochimica et Cosmochimica Acta, Vol. 63, No. 10, 1549–1557, 1999

[3] B. Li, X. Wang, M. Yan, L. Li, Materials Chemistry and Physics, 78, 184–188, 2002.

[4] S. Mahshid, M. Askari, M. Sasani Ghamsari, J. Materials Processing Technology, 189, 296-300, 2007.

[5] S. Mahshid, M. Askari, M. Sasani Ghamsari, N. Afshar, S. Lahuti, J. Alloys and Compounds, 478, 586-589, 2009.

[6] G. Zhang, B. K. Roy, L. F. Allard, J. Chow, J. Am. Ceram. Soc., 91, 3875–3882, (2008).

[7] R. Lee Penn and Jillian F. Bbanfield, J. Colloid and Interface Science, 259, 43–52, 2003.

[8] T. Sugimoto, X. Zhou, A. Muramatsu, J. Colloid and Interface Science, 259, 53-61, 2003.

[9] T. Sugimoto, X. Zhou, A. Muramatsu, J. Colloid and Interface Science, 259, 43–52, 2003.

[10] T. Sugimoto, "Monodispersed particles", Elsevier Science B.V., 1- 368, 2001.

[11] E. E. Finney, Richard G. Finke, J. Colloid and Interface Science, 317, 351–374, 2008.

[12] T. Sugimoto, F. Shiba, T. Sekiguchi, H. Itoh, Colloids and Surfaces, A: Physicochemical and Engineering Aspects 164 , 183–203, 2000.

[13] T. Sugimoto, F. Shiba, Colloids and Surfaces A: Physico chemical and Engineering Aspects, 164, 205–215, 2000

[14] T. Sugimoto, J. Colloid and Interface Science, 309, 106–118, 2007.

[15] C. J. Brinker, G. W. Scherer," sol-gel science: the physics and chemistry of sol-gel processing" Academic Press Limited, 235-300, 1990.

[16] T. Sugimoto, X. Zhou, A. Muramatsu, J. Colloid and Interface Science,259, 339-346, 2002.

[17] I. C. Baek, M. Vithal, J. A. Chang, J. Yum, Md.K. Nazeeruddin, M. Grätzel, Y. Chung, S. Seok, , Electrochemistry Communications, 11, 909–912 , 2009.

[18] A. S. Barnard and P. Zapol, J. Phys. Chem. 108, 18435-18440, B 2004,

[19] K. Shingyouchi, A. Makishima, M. Tutumi, S. Takenouchi, J. Non-Crystalline Solids, 100, 383-387, 1988.

Preparation of nitrogen-doped mesoporous TiO$_2$ with a room-temperature ionic liquid

Tung-Li Hsiung[1], H. Paul Wang[1,2*] and Yu-Lin Wei[3]

[1]Department of Environmental Engineering, National Cheng Kung University, Tainan, Taiwan
[2]Sustainable Environment Research Center, National Cheng Kung University, Tainan, Taiwan
[3]Department of Environmental Science and Engineering, Tunghai University, Taichung, Taiwan

ABSTRACT

The nitrogen-doped mesoporous TiO$_2$ (meso N-TiO$_2$) have been prepared by the sol-gel method with room temperature ionic liquids (RTIL) (C$_4$mimBF$_4$) as the template. By ^{1}H NMR, interactions between titanium and ionic liquid are observed. The meso N-TiO$_2$ (10-50 nm) possesses the anatase phase with a pore opening of 5-8 nm observed by TEM. The band gap of the meso N-TiO$_2$ is 2.47 eV as determined by DR-UV/VIS. The meso N-TiO$_2$ has a remarkable activity in photocatalytic degradation of methylene blue under visible light (λ >400 nm). The extended X-ray absorption fine structure (EXAFS) data show that the Ti-O and Ti-Ti bond distances in the meso N-TiO$_2$ are 1.96 and 3.02 Å, respectively, suggesting that the meso N-TiO$_2$ may be distorted with insertion of nitrogen into the TiO$_2$ matrix.

Keywords: ionic liquid, visible light, photocatalysis, mesoporous TiO$_2$, EXAFS

1 INTRODUCTION

TiO$_2$ has played an important roles in many environmental applications such as photocatalytic degradation of toxic organics, reduction of NO to N$_2$, and splitting of H$_2$O [1-4]. Room temperature ionic liquids (RTILs), so called "green solvents", can be used in separation, electrochemistry, and catalysis because of its nonvolatile, nonflammable, and thermally stable properties [5-7]. Recently, it is found the RTILs can act as a template for synthesizing novel mesoporous materials by the sol-gel methods [8-10]. The high surface area mesoporous TiO$_2$ (meso TiO$_2$) possesses better photocatalytic activities for degradation of chlorinated organic compounds such as 4-chlorophenol [11,12]. However, TiO$_2$ can only absorb UV light in range of <387 nm, corresponding 5-8% of the solar spectrum. Anion (such as N, S, and P) doped TiO$_2$ has been found very active in visible light photocatalysis [13-16].

By extended X-ray absorption fine structure (EXAFS) spectrascopy, it was found that copper oxide clusters in the micropores involved in catalytic reduction of NO and oxidation of 2-chlorophenol [17,18]. These molecule-scale data turn out to be very useful in revealing nature of catalytic active species and reaction mechanisms involved. Thus, the main objective of this work was to study chemical structure of nitrogen-doped mesoporous TiO$_2$ (meso N-TiO$_2$) (synthesized using RTIL as the template) by EXAFS spectroscopy.

2 EXPERIMENTAL

The cation (1-butyl-3-methyl-imidazolium ion, [C$_4$mim]$^+$) of the RTIL ([C$_4$mim][BF$_4$]) was synthesized by refluxing of 0.6 mol of 1-chlorobutane (99.4%, TEDIA) and 0.6 mol of 1-methylimidazale (99%, Acros) at 353 K for 72 h and washed three times with 10 mL of ethyl acetate (HPLC/Spectro, Tedia). The anion solution was prepared by mixing of 0.6 mol of NaBF$_4$ (98%, Acros) and 400 mL of acetone (UST/NP grade, Pharmca) for 24 h. The nitrogen doped mesoporous TiO$_2$ (meso N-TiO$_2$) was synthesized using RTIL as the template. Generally, 15 g of NH$_4$OH (28-30%, J.T. Baker) were adding slowly into a mixture solution which contains 6 g of titanium tetrabutoxide (Ti(OBu)$_4$), 90 mL of isopropenol (90 mL) and 30 g of the RTIL with stirring for 5 h. The mixture was then heated in a teflon lined autoclave at 375 K for 24 h. The as-synthesized meso N-TiO$_2$ was filtered, washed with aceton and calcined at 773 K for 0.5 h. The nitrogen doped TiO$_2$ (N-TiO$_2$) photocatalyst was also synthesized with a similar procedure without the RTIL.

The ^{1}H NMR (nuclear magnetic resonance) chemical shifts of the RTIL were determined on a Bruker Avance 300 spectrometer with dimethylsulfoxide (DMSO-d$_6$) as the solvent and internal standard (acquisition time = 1.373 s, actual pulse repetition time = 2 s, number of scans = 32, and excitation pulse-angle = 30°). Topologies of the photocatalysts were determined by high resolution transmission electron microscopy (JEM-2100F, JEOL). Diffuse reflectance UV/VIS spectra (DR UV-VIS) of the photocatalysts were also determined on a UV-VIS spectrophotometer (HITACHI U-3010) with a scan speed of 120 nm/min (200-700 nm).

Ti K-edge (4.966 keV) XANES spectra of the photocatalysts were collected at 298 K on the Wiggler beamline (17C) at the Taiwan National Synchrotron Radiation Research Center. The electron storage ring was operated at energy of 1.5 GeV (ring current = 120-200 A). A Si(111) double-crystal monochromator was used for selection of energy with an energy resolution ($\triangle E/E$) about 1.9×10^{-4} (eV/eV). Beam energy was calibrated by the adsorption edge of a Ti foil at an energy of 4966 eV. The isolated EXAFS data were normalized to the edge jump and converted to the wavenumber scale. The structural parameters that were varied during the refinements include the bond distance and variance. Coordination numbers of Ti were systematically varied in the course of the analysis within a given fitting range. Fitting of data to model compounds was performed using FEFFIT from UWXAFS 3.0 in combination with multiple scattering code FEFF 8.0 programs [19]. The FEFFIT was used to determine the best fitting results with a minimum (<0.01 Å^{-1}) of the Debye-Waller factors. The Fourier transform was performed on k^3-weighted EXAFS oscillations in the range of 3.7-14.2 Å^{-1}.

Photocatalytic degradation of methylene blue (MB, $C_{16}H_{18}N_3S$) was determined on a home-made photoreactor with a total reflection system [20]. Typically, 0.03 g of the photocatalysts were suspended in the MB (7.8×10^{-6} moleL^{-1}) aqueous solution (50 mL) with magnetic stirring. A 20 W tungsten lamp (Newport, model 63205) with water filter (Spectra-physics, model 61945) and 400 nm cut-off window were used as the visible light source. The concentration of MB was determined by UV-VIS spectroscopy (at 665 nm).

3   RESULTS AND DISCUSSION

To form the mesopore TiO$_2$ structure, the RTIL ([C_4min$^+$][BF$_4^-$]) acts as a template. The ^{1}H NMR chemical shifts of the [C_4mim][BF$_4$] and Ti(OBu)$_4$/[C_4mim][BF$_4$] are shown in Figure 1. As the titanium oxide precursor (Ti(OBu)$_4$ is mixed with the RTIL, little perturbation is found at δ0.88, 1.24, 1.75, 3.83, and 4.14 for ^{1}H in the long chain of the [C_4mim]$^+$. The ^{1}H on the cationic five-ring (H$_A$ and H$_B$) possesses upfield shifts (H$_A$ δ9.03 → 8.97; H$_B$ δ7.72, 7.65 → 7.50, 7.47) in the mixture of Ti(OBu)$_4$ and [C_4mim][BF$_4$], suggesting chemical interactions between titanium and cationic five-ring of the ionic liquid.

The TEM images and electron diffraction patterns of N-TiO$_2$ and meso N-TiO$_2$ are shown in Figure 2. The particle sizes of the photocatalysts are in range of 10-50 nm. Note that the band gap absorptions of the N-TiO$_2$ and meso N-TiO$_2$ are 2.54 and 2.47 eV, respectively as observed by DR-UV/VIS spectroscopy. The select area electron diffraction (SAED) patterns indicate that the N-TiO$_2$ and meso N-TiO$_2$ have mainly the anatase phase. It is also clear that the meso N-TiO$_2$ has pore openings of 5-8 nm in the TEM bright field images.

EXAFS spectra of the photocatalysts were recorded and analyzed in the range of 3.7-14.2 Å^{-1}. Table 1 shows the structural parameters determined by the best fitting of the EXAFS spectra. In all EXAFS data analyzed, the Debye-Waller factors ($\triangle\sigma^2$) are less than 0.01 Å^{-1}. The 1st (Ti-O) and 2nd (Ti-Ti) bond distances of the nanosize N-TiO$_2$ and meso N-TiO$_2$ are 1.96 and 3.02-3.04 Å, respectively, which are very different from those of the bulky TiO$_2$ (1.95 and 3.07 Å). Insertion of nitrogen in the TiO$_2$ matrix as well as more surface Ti atoms may be the factors that cause the distortion of chemical structure of TiO$_2$. Note that

coordination numbers of nanosizesize particle are generally decreased [21, 22]. As expected, a less coordination numbers of the 1st and 2nd shells for the N-TiO$_2$ (5.2 and 2.4) and meso N-TiO$_2$ (2.2 and 1.4) photocatalysts are also found if compared with those of the bulky TiO$_2$ (5.5 and 3.3).

Figure 3 shows the photocatalytic degradation of MB (7.8×10^{-6} moleL^{-1}) on the N-TiO$_2$ photocatalysts under the visible light radiation. Without a photocatalyst, about 26% of MB can be degraded under radiation for 15.5 hours. On the N-TiO$_2$ photocatalyst, about 73% of MB is degraded within 14 hours. Interestingly, on the meso N-TiO$_2$ photocatalyst, in the early stage, the conversion of MB is as high as 50%. After radiation of seven hours, the meso N-TiO$_2$ has a conversion of 87%.

4   CONCLUSIONS

The meso N-TiO$_2$ visible-light photocatalyst can be synthesized with the sol-gel method using the ionic liquid (C_4mimBF$_4$) as the template. By ^{1}H NMR, interactions between Ti and the cationic five-ring of the ionic liquid are observed. The meso N-TiO$_2$ possesses an anatase phase with pore openings of 5-8 nm. The refined EXAFS data show that the Ti-O and Ti-Ti bond distances in the meso N-TiO$_2$ are 1.96 and 3.02 Å, respectively, suggesting distortion of the TiO$_2$ framework caused by facts of the increasing surface Ti atoms and insertion of nitrogen in the TiO$_2$ matrix. The meso N-TiO$_2$ has a very high photocatalytic reactivity, for instance, under the radiation of visible light for seven hrs, a conversion of 87% in degradation of MB can be obtained.

REFERENCES

[1] M.A. Fox, M.T. Dulay, Chem. Rev., 93, 341-357, 1993.
[2] H. Park, Y.C. Lee, B.G. Choi, Y.S. Choi, J.W. Yang, W.H. Hong, Small, 6, 290-295, 2010.
[3] N.G. Petrik, G.A. Kimmel, J. Phys. Chem. C, 113, 4451-4460, 2009.
[4] M.R. Hoffmann, S.T. Martin, W. Choi, D.W. Bahnemann, Chem. Rev., 95, 69-96, 1995.
[5] M.J. Earle, K.R. Seddon, Pure Appl. Chem., 72, 1391-1398, 2000.
[6] J.F. Huang, P.Y. Chen, I.W. Sun, S.P. Wang, Inogr. Chim. Acta, 320, 7-11, 2001.
[7] T. Welton, Chem. Rev., 99, 2071-2083, 1999.
[8] Y. Zhou, M. Antonietti, J. Am. Chem. Soc., 125, 14960-14961, 2003.
[9] K. Yoo, H. Choi, D.D. Dionysiou, Chem. Commun., 2000-2001, 2004.
[10] M. Estruga, C. Domingo, X. Domenech, J.A. Ayllon, J. Colloid Interf. Sci., 344, 327-333, 2010.
[11] K.S. Yoo, H. Choi, D.D. Dionysiou, Catal. Commun., 6, 259-262, 2005.
[12] K.S. Yoo, T.G. Lee, J. Kim, Micropor. Mesopor. Mat., 84, 211-217, 2005.

[13] R. Asahi, T. Morikawa, T. Ohwaki, K. Aoki, Y. Taga, Science, 293, 269-291, 2001.

[14] C. Burda, Y. Lou, X. Chen, A.C.S. Samia, J. Stout, J.L. Gole, Nano Lett., 3, 1049-1051, 2003.

[15] J.L. Gole, J.D. Stout, C. Burda, Y. Lou, X. Chen, J. Phys. Chem. B, 108, 1230-1240, 2004.

[16] V. Pore, M. Heikkilä, M. Ritala, M. Leskelä, S. Areva, J. Photoch. Photobio. A, 177, 68-75, 2006.

[17] Y.-J. Huang, H.P. Wang, J.-F. Lee, Chemosphere, 50, 1035-1041, 2003.

[18] K.-S. Lin, H.P. Wang, Langmuir, 16, 2627-2631, 2000.

[19] E.A. Stern, M. Newville, B. Ravel, T. Yacoby, D. Haskel, Physica B, 209, 117-120, 1995.

[20] S.-H. Liu, H.P. Wang, Int. J. Hydrogen Energ., 27, 859-862, 2002.

[21] H. Wilmer, M. Kurtz, K.V. Klementiev, O.P. Tkachenko, W. Grünert, O. Hinrichsen, A. Birkner, S. Rabe, K. Merz, M. Driess, C. Wöll, M. Muhler, Phys. Chem. Chem. Phys., 5, 4736-4742, 2003.

[22] V. Luca, S. Djajanti, R.F. Howe, J. Phy. Chem. B, 102, 10650-10657, 1998.

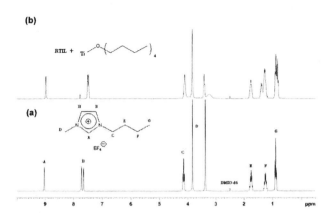

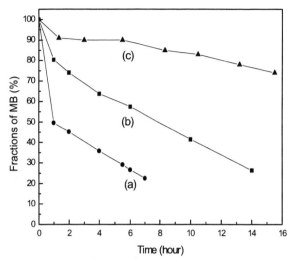

Figure 1: ^{1}H-NMR spectra of (a) [C$_{4mim}$][BF$_4$] and (b) the mixture of [C$_{4mim}$][BF$_4$] and Ti(OBu)$_4$.

Figure 3: Time dependence for photocatalytic degradation of MB on (a) meso N-TiO$_2$, (b) N-TiO$_2$ and (c) no photocatalyst.

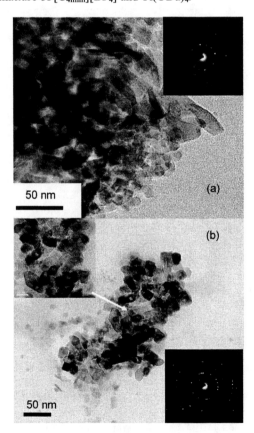

Figure 2: TEM images and electron diffraction patterns of (a) N-TiO$_2$ and (b) meso N-TiO$_2$.

	Shell	R (Å)	CN	σ^2(Å^2)
Meso N-TiO$_2$	Ti-O	1.96	2.2	0.0043
	Ti-Ti	3.02	1.4	0.0050
N-TiO$_2$	Ti-O	1.96	5.2	0.0067
	Ti-Ti	3.04	2.4	0.0039
TiO$_2$	Ti-O	1.95	5.5	0.0074
	Ti-Ti	3.07	3.3	0.0057

R: bond distance
CN: coordination number
σ^2: Debye-Waller factor

Table 1: Speciation data (by EXAFS) of meso N-TiO$_2$ and N-TiO$_2$

Preparation of nanosize copper recycled from an electroplating wastewater

C.-H. Huang[*], H. Paul Wang[**] and I.-W. Sun[***]

[*]Department of Environmental Engineering, National Cheng Kung University
Tainan, Taiwan, p5894109@mail.ncku.edu.tw
[**] Department of Environmental Engineering and Sustainable Environmental Research Center, National
Cheng Kung University, Tainan, Taiwan, wanghp@mail.ncku.edu.tw
[***] Department of Chemistry, National Cheng Kung University, Tainan, Taiwan,
iwsun@mail.ncku.edu.tw

ABSTRACT

To better utilize copper recovered from an electroplating (EP) wastewater, a simple method of preparing valuable nanoparticles has been developed in the present work. Experimentally, about 40-50% of copper in the wastewater can be chelated with β-cyclodextrin (β-CD). The β-CD-Cu^{2+} complex was dried at 373 K for 24 hours and carbonized at 573 K for at least two hours to obtain the nanosize metallic copper (Cu) encapsulated in the carbon shells (Cu@C). The images observed by transmission electron microscopy (TEM) also indicate that the size of Cu in the Cu@C is in the range of 10-60 nm. The component fitted X-ray absorption near edge structure (XANES) spectra suggest that the chelated Cu^{2+} is reduced to form Cu as the hydroxyl groups of the β-CD are oxidized to H_2O vapor during carbonization. By extended X-ray absorption fine structure (EXAFS) spectroscopy, it is found that the average coordination number (CN) of Cu in the Cu@C nanoparticles decreases along with the decrease of the size of Cu, i.e. the CN of the nanosize Cu (23 nm) is reduced to 8 comparing to 12 of the bulky Cu.

Keywords: copper, electroplating wastewater, nanoparticles, XANES, EXAFS

1 INTRODUCTION

Electroplating (EP) wastewater generally contains high concentrations of copper (typically 300-3,000 ppm) [1, 2]. To recover copper from the EP wastewater, chemical precipitation, adsorption, membrane filtration, electro-coagulation, and bio-treatment have been widely used [3-5]. However, these processes are not economically and environmentally attractive. In the separate experiments, we found that Cu^{2+} in the chemical mechanical polishing (CMP) wastewater was chelated with cyclodextrins [6]. Remarkably, Cu(II) ions chelated with β-CD molecules to which hydroxyl groups have been appended may be reduced and encapsulated in the carbon shell during carbonization of the CD-Cu^{2+} complex at the temperature of 573 K. Nanosize-controllable Cu@C particles can, therefore, be synthesized via this simple method.

The Cu@C nanoparticles have been used in enhanced efficiency of a dye-sensitized solar cell (DSSC). A small amount (0.08%) of the Cu@C nanoparticles dispersed in an ionic liquid ([bmim$^+$][PF$_6^-$]) which is used as a DSSC electrolyte can increase it electrolyte conductivity by several fold [7]. Furthermore, deposition of the nanosize copper (Cu) encapsulated carbon thin film onto the counter electrode of a DSSC can increase the efficiency by 50% if compared with the relatively expensive Pt electrode [8].

Molecule-scale data of select elements in the complex matrix can be obtained by X-ray absorption spectroscopy [9]. By X-ray absorption near edge structure (XANES) or extended X-ray absorption fine structure (EXAFS) spectroscopy, it has been found that the main As species in ground water of the blackfoot disease area are As_2O_5 and As(V)-HS [10]. The nanosize CuO in the CMP wastewater can be extracted and forms CuO clusters in the mesopores of MCM-41 as observed by XANES [11]. The CuO cluster in the micropores of ZSM-5 or ZSM-48 also played important roles in catalytic decomposition of 2-chlorophenol and reduction of NO [12, 13]. Thus, the main objective of this work was to investigate the feasibility of recovery of copper from an EP wastewater for preparing relatively high-value Cu@C core-shell nanoparticles. In addition, the Cu nanoparticles in the size range of 23-57 nm were also synthesized under the pH values of 1.0-2.3. Speciation of copper in the Cu@C formation process was also studied by XANES and EXAFS.

2 EXPERIMENTAL SECTION

The EP wastewater was sampled from an ultimate wastewater mixing tank of a copper electroplating process in Taiwan. The wastewater generally contains about 400 ppm of copper. β-CD (WAKO, 97%) was used to chelate copper ions in the EP wastewater or $CuSO_4 \cdot 5H_2O$ (RDH, 99%). At the CD/Cu molar ratio of 0.063, the CD-Cu^{2+} complex was dried at 373 K for 24 hours, washed by distilled water and filtrated with a 0.2 μm filter to remove unchelated copper, and carbonized under high purity nitrogen (99.99%) at 573 K for at least two hours to yield the nanosize Cu encapsulated in the carbon shell (Cu@C).

Concentration of copper in the EP wastewater, CD-Cu^{2+} complex, and Cu@C was analyzed by induced coupling plasma-optical emission spectroscopy (ICP-OES) (JOBIN YVON, Ultima 2000). Chemical structure of the samples was characterized by X-ray diffraction spectroscopy (XRD) (BRUKER, D8 Advanced) scanned from 10 to 80° (2θ) at a scan rate of 3°/min using CuKα radiation (40 kV, 40 mA). Images of the nanosize Cu@C were studied by transmission electron microscopy (TEM) (PHILIPS, CM-200) having an accelerating voltage of 200 kV. The EXAFS spectra of copper in the samples were collected on the Wiggler BL17C beamline at the Taiwan National Synchrotron Radiation Research Center. The EXAFS data were analyzed using the UWXAFS 3.0 and FEFF 8.0 simulation programs [14]. The component fittings of model compounds in the complexation solutions were deduced by using the least-square fittings of the XANES spectra.

3 RESULTS AND DISCUSSION

About 369 ppm of copper and a small amount of zinc (41 ppm), manganese (1.8 ppm), nickel (0.3 ppm), and lead (0.02 ppm) were found in the EP wastewater. Copper in the EP wastewater was chelated with β-CD at the β-CD/Cu ratio of 0.063. At the pH values of 1.0-2.3, 6.5-44% of copper can be recovered from the EP wastewater (see Table 1). Note that under the pH value of 1.0, a high-purity Cu@C can be obtained by carbonization of the β-CD-Cu^{2+} complex.

The component fitted XANES spectra in Fig. 1 show that the main copper compound in the EP wastewater is CuSO$_4$. After chelation of Cu^{2+} with β-CD in the EP wastewater and dried at 373 K, about 6% of β-CD-Cu^{2+} complex are formed. As the pH value of the EP wastewater is decreased to 2.0, the fraction of the β-CD-Cu^{2+} complex is increased to 34%. At the pH value of 1.0, the fraction of β-CD-Cu^{2+} complex is 50% and a small amount of Cu$_2$O (2%) is also found.

Figure 2 shows the XRD patterns of carbonized samples at the β-CD/Cu ratio of 0.063. Note that CuSO$_4$ is the main copper compound in the EP wastewater (Fig. 2(a)). After carbonization of the β-CD-Cu^{2+} complex at 573 K, metallic copper encapsulated in the carbon shells is observed in Fig. 2(b)-(d). Copper in the β-CD-Cu^{2+} complex is reduced to yield Cu as the hydroxyl groups of the β-CD are oxidized to H$_2$O vapor during carbonization. Note that the XRD peaks of Cu become broadened, suggesting the existence of nanosize Cu. The average crystalline sizes of Cu calculated with the Scherrer equation [15] using the full width at half maximum of the Cu(111) (2θ = 43.3°) peak are 23-57 nm. It is clear that smaller Cu in the Cu@C can be prepared from the CD-Cu^{2+} complex under a low pH value.

Table 1: Copper in the electroplating wastewater chelated with β-CD for synthesis of Cu@C nanoparticles.

Copper source	CD/Cu molar ratio	pH value at chelation	% chelated[a]	Purity (%) of product Cu@C[b]
EP	0	2.3	-	-
EP	0.063	2.3	6.5	19
EP	0.063	2.0	36	78
EP	0.063	1.0	44	99

[a]: determined by ICP-OES; [b]: determined by XANES.

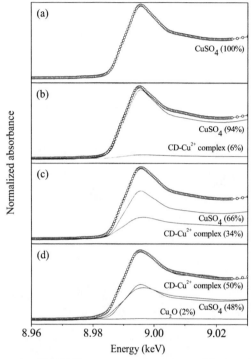

Figure 1: XANES spectra of copper (a) in the EP wastewater and chelated with β-CD at the β-CD/Cu molar ratio of 0.063 under the pH values of (b) 2.3, (c) 2.0, and (d) 1.0.

The images of the nanosize Cu encapsulated in the amorphous carbon shells can be observed by TEM shown in Fig. 3. Without the perturbation of other metal ions in the EP wastewater, in the CuSO$_4$ solution, the as-synthesized Cu@C nanoparticles have a narrow size distribution between 10-20 nm (shown in Fig. 3(a)). On the contrary, Cu recovered from the EP wastewater has a relatively wide size distribution of 10-60 nm (Fig. 3(b)). Under a low pH value of 1.0, the size distribution of Cu is in the range of 10-30 nm.

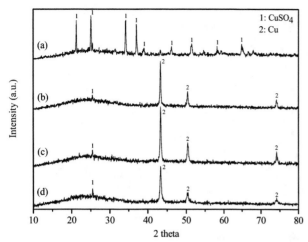

Figure 2: XRD patterns of copper (a) in the EP wastewater and the Cu@C synthesized by carbonization of the β-CD-Cu²⁺ complexes (at the β-CD/Cu molar ratio of 0.063) obtained from chelation of copper in the EP wastewater under the pH values of (b) 2.3 (c) 2.0, and (d) 1.0.

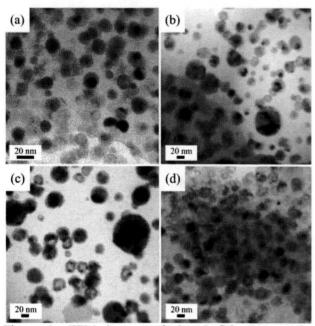

Figure 3: TEM images of the Cu@C nanoparticles synthesized from (a) CuSO₄ and EP wastewater under the pH values of (b) 2.3, (c) 2.0, and (d) 1.0 at the β-CD/Cu molar ratio of 0.063.

Generally, surface to total atoms ratios are negligible in the bulky state but significantly effective in a nanometer scale. Surface atoms have a less CN compared to the inner atoms. The average CN of a select metal may decrease along with the decrease of its crystalline sizes. By EXAFS spectroscopy (see Table 2), it is found that the average CN of Cu in the Cu@C nanoparticles (36 nm) decreases from $12 \rightarrow 9$ and further decreases to 8 for the smaller Cu@C (23 nm).

Table 2: Structure parameters of copper in Cu@C nanoparticles recovered from the electroplating wastewater.

Copper source	CD/Cu molar ratio	Cu size (nm)	pH	R (Å)	CN	σ^2 (Å²)
Cu foil	-	-	-	2.53	12.3	0.008
EP	0.063	57	2.3	2.53	11.6	0.009
EP	0.063	36	2.0	2.53	9.4	0.009
EP	0.063	23	1.0	2.53	7.7	0.008

4 CONCLUSIONS

About 40-50% of copper in the EP wastewater can be recovered by chelation with β-CD for preparation of valuable Cu@C nanoparticles. Size controllable Cu@C nanoparticles can be synthesized at the pH value range of 1.0-2.3. The average CN of Cu in the Cu@C nanoparticles decreases along with the decrease of its particle sizes. The CN of the nanosize Cu (23 nm) is reduced ($12 \rightarrow 8$) compared to its bulk state.

This work exemplifies a valuable core-shell nanoparticles can be recovered from a wastewater. In addition, a better understanding of the nanoparticles in molecular scale is also obtained.

5 ACKNOWLEDGMENT

This research was supported by the Taiwan National Science Council, Bureau of Energy and the Excellence Project of the National Cheng Kung University. We also thank Professor J.-F. Lee of the Taiwan National Synchrotron Radiation Research Center for his experimental assistances.

REFERENCES

[1] L. Monser and N. Adhoum, Sep. Purif. Technol., 26, 137, 2002.
[2] S. Venkatesan and K. Begum, Asia-Pac. J. Chem. Eng., 3, 387, 2008.
[3] T. A. Kurniawan, G. Y. S. Chan, W. H. Lo and S. Babel, Chem. Eng. J., 118, 83, 2006.
[4] I. Heidmann and W. Calmano, J. Hazard. Mater., 152, 934, 2008.
[5] Y. J. Li, X. P. Zeng, Y. F. Liu, S. S. Yan, Z. H. Hu and Y. M. Ni, Sep. Purif. Technol., 31, 91, 2005.
[6] C. H. Huang, H. P. Wang, J. E. Chang and E. M. Eyring, Chem. Comm., 31, 4663, 2009.
[7] F. L. Chen, I. W. Sun, H. P. Wang and C. H. Huang, J. Nanomater., 472950, 2009.
[8] C. H. Huang, H. P. Wang and C. Y. Liao, J. Nanosci. Nanotechnol., 10, 1, 2010.
[9] H. L. Huang, H. P. Wang, G. T. Wei, I. W. Sun, J. F. Huang and Y. W. Yang, Environ. Sci. Technol., 40, 4761, 2006.

[10] H. C. Wang, H. P. Wang, C. Y. Peng, H. L. Liu and H. L. Huang, B. Environ. Contam. Tox., 71, 798, 2003.

[11] H. L. Huang and H. P. Wang, J. Electron Spectrosc., 144, 307, 2005.

[12] K. S. Lin and H. P. Wang, J. Phys. Chem. B, 105, 4956, 2001.

[13] Y. J. Huang, H. P. Wang and J. F. Lee, Chemosphere, 50, 1035, 2003.

[14] E. A. Stern, M. Newville, B. Ravel, Y. Yacoby and D. Haskel, Physica B, 209, 117, 1995.

[15] H. Borchert, E. V. Shevehenko, A. Robert, I. Mekis, A. Kornowski, G. Grubel and H. Weller, Langmuir, 21, 1931, 2005.

Synthesis of nanocrystalline $(Zn_{1-x}Co_x)Al_2O_4$ solid solution: structural and optical properties

A. Fernández-Osorio [*], D. Hernández- Mendoza, E. Pineda-Villanueva

[*] Facultad de Estudios Superiores Cuautitlán,
Universidad Nacional Autónoma de México, CP 54740, México
ana8485@servidor.unam.mx

ABSTRACT

New inorganic pink pigments based on nanocrystalline $(Zn_{1-x}Co_x)Al_2O_4$ solid solution, were prepared using the hydrothermal method, obtaining nanoparticles below 8 nm. The spinel $ZnAl_2O_4$ was chosen as host hue to introduce Co(II) ions, because zinc aluminate is colorless and presents great thermal stability.

The most straightforward way to obtain blue colors in bulk ceramics is by means of cobalt, Co(II) ions exhibit, when tetrahedrally coordinated, highly saturated blue shades, nevertheless these nanoparticles show a brilliant pink color due to partial substitution of Al(III) ions in octahedral sites. The samples characterization was performed by X-ray diffraction and diffuse reflectance spectroscopy. The morphology of the nanoparticles was studied by high-resolution transmission electron microscopy.

The effects of Co(II) concentration and heating temperature on the spinel structure and the optical properties of the solid solution were systematically studied.

Keywords: nanoparticles, spinel pigments, solid solutions, optical properties

1 INTRODUCTION

During the last decades, much attention has been focused on the rational synthesis of nanocrystals ranging in size from 1 to 100 nm, which can exhibit very interesting size-dependent electrical, optical, magnetic and chemical properties compared with those of bulk counterparts [1].

Zinc aluminate ($ZnAl_2O_4$), naturally ocurring as the mineral ghanite, is a member of the spinel family. At present, zinc aluminate is used as a catalyst for the dehydration of satured alcohols to olefins, methanol and higher alcohol synthesis, preparation of polymethylbenzenes, synthesis of styrenes from acetophenons and isomerisation of alkenes.

Zinc aluminate have high thermal stability, high mechanical resistance, hydrophobicity, and low surface acidity. It is widely used as a high temperature material, catalysts and catalyst support, optical coating and as ceramic pigment when is doped with transitional cations [2-4]

Besides, as a wide band gap semiconductor (3.8 eV) $ZnAl_2O_4$ can be used as transparent conductor, dielectric and optical material [5].

The crystal structure of spinel ($MgAl_2O_4$) has been determined by Bragg [6] and Nishikawa [7] in 1915. It has the space group Fd3m and has a cubic structure made of eight molecular units (AB_2O_4). The A cation is divalent and the B cations trivalent, one unit cell of a compound with the spinel structure is built up from 32 oxygen ions arranged in a fcc-lattice, giving 64 tetrahedral and 32 octahedral sites. The divalent A cations occupy eight tetrahedral sites and the trivalent B cations occupy 16 octahedral sites. This distribution of the cations is designated as normal, however, this is thermodyamically not always the most stable situation, since the configurational entropy counteracts the site preference energy. Therefore, in spinels A and B cations may interchange sites via diffusion, eventually leading to the situation where all the A cations are in octahedral sites, this structure is designated as inverse.

All distribution between the two extremes are possible and the degree of inversion is given by the inversion parameter γ, which is defined as the fraction of A cations in octahedral sites.

$ZnAl_2O_4$ has the chemical formula of AB_2O_4 with the normal spinel structure [8].

The main aspect of spinels is the presence of two metallic cations, A^{2+} and B^{3+}, in tetrahedral and octahedral sites, respectively. The manner in wich such sites are occupied depend on the calcining temperature [9]

To prepare $ZnAl_2O_4$ powders, scientists have put forward various synthetic routes, generally including the traditional method of solid-solid reaction, wet chemical route such as co-precipitation, sol-gel and hydrothermal synthesis [10].

Recently, a new application of the spinels as ceramic nanopigments has been explored, owing to their high mechanical resistance, high thermal stability, low temperature sinterability and the easy incorporation of chomophore ions into the spinel lattice, allowing for

NSTI-Nanotech 2010, www.nsti.org, ISBN 978-1-4398-3401-5 Vol. 1, 2010

different types of doping, thus producing ceramic pigments with different colors [11].

The traditional and unavoidable source of blue in a ceramic pigment is the cobalt ion. All the blue ceramic pigments known currently (except vanadium-zircon) classified with the DCMA number, contain cobalt to some extent: the Co_2SiO_4 olivina (DCMA 5-08-2), the $(Co,Zn)_2SiO_4$ willmenite (DCMA 7-10-2), and the cobalt spinels $CoAl_2O_4$ (DCMA 13-26-2), Co_2SnO_4 (DCMA 13-27-2), $(Co,Zn)Al_2O_4$ (DCMA 13-28-2) and $Co(Al,Cr)_2O_4$ (DCMA 13-28-2). Co^{2+} cation is known as a chomophore which usually develops blue color, the coloring performance depends very much on the coordination of Co^{2+} ions.

The $ZnAl_2O_4$ is a normal spinel, Zn^{2+} ions occupy tetrahedral sites, while Al^{3+} cations occupy octahedral positions. When Co^{2+} is added to this structure, substituting Zn^{2+} in bulk systems usually occupy tetrahedral sites and develop blue color.

E.F. da Costa et al [12] observed that the $Zn_{1-x}Co_x Al_2O_4$ (x = 0-1) system develops violet, blue and cyan, centered in the blue colors.

Accordingly, the present work aims at prepare the nanocrystalline $Zn_{1-x}Co_x Al_2O_4$ (x = 0-1) system by the hydrothermal route, and the optical properties of the samples were analyzed.

2 EXPERIMENTAL

The $Zn_{1-x}Co_x Al_2O_4$ (x=0, 0.2, 0.4, 0.6, 0.8, 1.0) system was synthesized using the hydrothermal method, in this experiment, $AlCl_3 \cdot 6H_2O$ (99% Sigma-Aldrich), $ZnCl_2$ (98% Sigma-Aldrich) and $CoCl_2 \cdot 6H_2O$ (99% Sigma-Aldrich) were used as starting materials. Firstly, stoichiometric amounts of start materials were dissolved in deionized water, mixed well with each other, the pH was adjusted via dropping 28 wt% NH_4OH solution to 10-11, obtaining a white gel, which was washed with deionized water and then dried at room temperature for one week. Subsequently dried and grinded the gel, was calcined in air at 600°C 2h.

Six samples were prepared according to the stoichiometry:
$ZnAl_2O_4$, $Zn_{0.8}Co_{0.2} Al_2O_4$, $Zn_{0.6}Co_{0.4} Al_2O_4$, $Zn_{0.4}Co_{0.6} Al_2O_4$, $Zn_{0.2}Co_{0.8} Al_2O_4$ and $CoAl_2O_4$

Phases in the resultant powders were identified by X-ray diffraction, the patterns were obtained at room temperature with Cu K_α radiation (λ=1.5406 Å) by using a D5000 Siemens diffractometer. UV-Vis electronic absorption spectra was measured in diffuse reflectance mode in the 200–1200 nm wavelength range with an Ocean Optics HR4000 fiber optic spectrometer.

High-resolution transmission electron micrographs (HR-TEM) were obtained in a JEOL 2010 FASTEM analytical microscope operating at 200 kV.

3 RESULTS AND DISCUSSION

Fig.1 shows the XRD patterns of the compositions synthesized for the $Zn_{1-x}Co_xAl_2O_4$ solid solution calcined at 700°C/2h. At this temperature for all the values of x studied, the characteristic peaks of the spinel structure were noticed.

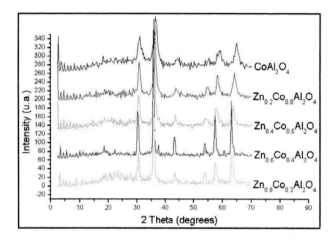

Figure 1. XRD diffraction patterns of powder samples heat treated at 700°C

The XRD patterns showed that zinc aluminate structure was formed as a single phase in all samples. In these samples, Co ions seem to be completely incorporated into the spinel lattice, in agreement with the atomic radii for divalent cations Co and Zn, both in tetrahedral coordination.

All reflections of the XRD pattern of the white powders of $ZnAl_2O_4$ (Fig.2) can be readily indexed to a spinel zinc aluminum oxide (JCPDS, 74-1136) which indicates the high purity of the nanocrystalline $ZnAl_2O_4$. This zinc aluminate has a cubic crystalline structure with a unit symmetry described by space group Fd3m and lattice parameter a = 8.083 Å.

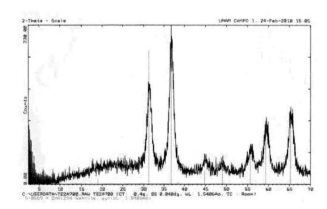

Figure 2. XRD pattern of $ZnAl_2O_4$ nanoparticles

It can be observed that the diffraction peaks are markedly broadened, which is indicative of a fine crystallite. In all samples, no peaks of impurity are observed in the XRD patterns.

In order to determine the Co_3O_4 nanoparticles average size, the peak broadening method using the classical Scherrer equation over the (3 1 1), (2 2 0), (4 2 2), (5 1 1) and (4 4 0) reflections was employed.

$ZnAl_2O_4$ nanoparticles have an average size of 6.4 ± 1.2 nm

Fig. 3 illustrates the UV-vis spectra in the range from 220-800 nm, for the $Zn_{1-x}Co_xAl_2O_4$ samples heated 700°C/2h.

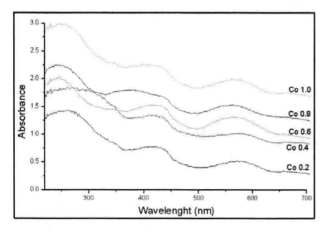

Figure 3. UV-vis absorbance spectra of $Zn_{0.8}Co_{0.2}Al_2O_4$ $Zn_{0.6}Co_{0.4}Al_2O_4$, $Zn_{0.4}Co_{0.6}Al_2O_4$, $Zn_{0.2}Co_{0.8}Al_2O_4$ and $CoAl_2O_4$

The electronic absorption spectra obtained exhibits three wide absorption bands attributed to Co^{2+} in octahedral coordination and particularly to the CF transitions:
$^4T_{1g}$ $(^4F) \rightarrow {}^4T_{2g}(^4F)$ at 560 nm, $^4T_{1g}(^4F) \rightarrow {}^4A_{2g}(^4F)$ at 420 nm and $^4T_{1g}(^4F) \rightarrow {}^4T_{1g}(^4P)$ at 250 nm.
Nano-sized $Zn_{1-x}Co_xAl_2O_4$ system is expected to have a normal spinel, with Co^{2+} situated at the tetrahedral site, however the optical spectra of the nano-sized samples are coherent with a hexa-coordinated Co^{2+} ion. It is completely different from the optical spectra of the bulk system, which exhibits the expected intense bands in the 500-600 nm range attributable to the transitions of Co^{2+} in 4-fold coordination [13].
All samples present similar spectrum, with absorption bands attributed to Co^{2+} in octahedral sites, which implies a change in the structure, inverse spinel is noticed independently on composition.

Typical HR-TEM micrograph obtained from sample $Zn_{0.8}Co_{0.2}Al_2O_4$ is shown in figure 4.

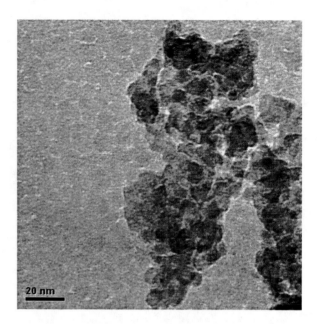

Figure 4. HRTEM micrograph of $Zn_{0.8}Co_{0.2}Al_2O_4$

In this micrograph, nanocrystallites with dimensions close to those determined by X-ray diffraction patterns can be observed.

4 CONCLUSIONS

Nanocrystalline pink powders of $Zn_{1-x}Co_xAl_2O_4$ (x= 0; 0.2; 0.4; 0.6; 0.8 and 1.0) were sinthesized by the hydrothermal method, obtaining nanoparticles with average particle size of 6.4 ± 1.2 nm. The UV-vis absorbance spectra of the samples $Zn_{1-x}Co_xAl_2O_4$ display a three wide absorption bands centered at approximately 560, 420 and 250 nm, which are attributed to Co^{2+} in octahedral coordination. It is completely different from the optical spectra of the bulk system, which exhibits the expected intense bands in the 500-600 nm range attributable to the transitions of Co^{2+} in 4-fold coordination.

REFERENCES

[1] A. Punnoose, H. Magnone, M. S. Seehra, Phys. Rev. B 64, 17420 (2001)

[2] Z. Li, S. Zhang, W. E. Lee, J. Eur.Ceram.Soc. 27, 3407, 2007

[3] G. Aguilar-Rios, M. Valenzuela, P. Salas, Appl. Catal. A 127, 65, 1995

[4] G. Lorenzi, G. Baldi, D. Bennedetto, J. Eur.Ceram.Soc. 26, 377, 2006

[5] A. R. Phani, M. Passacantando, S. Santucci, Mater. Chem. Phys. 68, 66, 2001

[6] W.H. Bragg, Phil. Mag. 30, 305, 1915

[7] S. Nishikawa, Proc. Math. Phys. Soc. Tokyo 8, 199, 1915

[8] H.St.O´Neil and W.A. Dollase, Phys. Chem. Minerals 20, 541, 1994

[9] K.E. Sickafus, J.M. Willis, J. Amer. Ceram. Soc. 82 (12) 3279, 1999

[10] X. Duan, D. Yuan, X. Wang, J. Sol-Gel Sci. Technol.35, 221, 2005

[11] M. Llusar, A. Fores, J.A. Badenes, J. Eur. Ceram. Soc. 21(8), 1121, 2001

[12] E.F. da Costa, K.C. de Souza, R. Zamian, Dyes and Pigments 81, 187, 2009

[13] P.M.T. Cavalcante, M. Dondi, G. Guarini, Dyes and Pigments 80, 226, 2009

Synthesis of Nanocrystalline VOx/MgO Aerogels and Their Application for Destructive Adsorption of CF₂Cl₂

E. V. Ilyina[*], I. V. Mishakov[*], A. A. Vedyagin[*], A. F. Bedilo[*] and K. J. Klabunde[**]

[*]Boreskov Institute of Catalysis, Novosibirsk 630090, Russia, chebka@bk.ru
[**]Department of Chemistry, Kansas State University, Manhattan, KS, USA, kenjk@k-state.edu

ABSTRACT

The developed method for synthesis of binary VOx/MgO aerogels by hydrolysis of the V and Mg alkoxides mixed together followed by gelation and supercritical drying makes it possible to obtain aerogels with exceptionally high surface areas (~ 1200 m^2/g) and uniform vanadium distribution. Their dehydration yields nanocrystalline VOx/MgO with nanoparticle size ca. 5 nm. The synthesized nanocrystalline VOx/MgO oxides were used as destructive sorbents for decomposition of halocarbons. The doping of MgO aerogels with a small amount of VOx (1-2%) results in significant acceleration of the reaction with CF₂Cl₂ causing deeper MgO transformation to MgF₂ and shortening the induction period.

Keywords: aerogels, destructive sorption, halocarbons, nanocrystalline oxides, VO$_x$/MgO

1 INTRODUCION

Aerogels are materials obtained by a sol-gel process followed by supercritical drying. Under supercritical conditions, the liquid-vapor interface that produces the collapse of the initial gel framework during conventional drying is eliminated. The obtained materials have high surface areas and pore volumes. The method is known to be versatile and allows controlling texture, composition and structural properties of the materials [1, 2].

The best known method for synthesis of nanocrystalline MgO aerogels with high surface area is the modified aerogel procedure developed by Klabunde et al. [3, 4]. Various nanocrystalline aerogel-prepared (AP) metal oxides prepared by this method have high surface areas, small crystallite sizes, unusual morphology, enhanced adsorption properties and have been extensively studied for their use as destructive adsorbents [4-6], decontamination of vegetative bacterial cells and spores [7] and in various catalytic and photocatalytic applications [8-10].

We have recently published first communications on the synthesis of the co-gelled VO$_x$/MgO aerogels and their catalytic performance in oxidative dehydrogenation of propane [11-13]. This paper is devoted to further elaboration and simplification of the synthesis of mixed V-Mg hydroxide and oxide aerogels and their application for destructive sorption of CF₂Cl₂ known to be an ozone-depleting hazard.

2 EXPERIMENTAL

The method used for synthesis of mixed V-Mg(OH)$_x$ hydroxide aerogels is generally similar to the one earlier used for preparation of nanocrystalline AP-MgO. Magnesium ribbon (Aldrich, 99.9%, 0.15 mm, 1 g), absolute methanol (30 ml), toluene (20 ml) and vanadium triisopropoxide oxide VOOCH(CH₃)₃ (Aldrich, 99%) were used as precursors. The amount of the vanadium precursor was varied depending on the desired vanadium concentration and was equal to 0.47 ml for the samples containing 10% V. All vanadium concentrations reported in this paper are normalized to V₂O₅. The concentrations were varied from 0 to 26 wt.%.

First, magnesium metal was dissolved in methanol to form magnesium methoxide solution. Then, a second solvent used as a gel stabilizer (toluene, ethanol) was added to the magnesium methoxide solution in methanol or just more methanol was added to give the same concentration of the solution. The obtained magnesium methoxide solution was slowly hydrolyzed with water (1.5 ml) to obtain a clear gel. Then, the desired amount of vanadium triisopropoxide oxide was added dropwise to the freshly prepared gel. The obtained mixed gel was placed into an autoclave (Parr Instruments) equipped with a stirrer and heated in nitrogen atmosphere to 260°C with 1.3 °C/min heating rate. The final pressure and temperature were varied to study their effect on the properties of the obtained aerogels. After reaching the desired temperature the solvents were vented and the autoclave was purged with nitrogen for 20 min. As our ultimate goal was to prepare catalysts operating at elevated temperatures, the obtained V-Mg(OH)$_x$ hydroxide aerogels were calcined up to 550°C using different procedures to convert them to the corresponding binary oxides.

The CF₂Cl₂ reaction with nanocrystalline VO$_x$/MgO aerogels was studied in a TEOM (Tapered Element Oscillating Microbalance) microanalyzer. This instrument is based on the mass measurement using a relation between the oscillation frequency of the microreactor acting as a pendulum and the mass of the studied sample. The composition of gases at the outlet of the TEOM 150 PMA microanalyzer was analyzed using QMS-200 quadrupole mass spectrometer.

High-resolution transmission electron microscopy (HRTEM) images were obtained using JEM-2010CX instrument with 100 kV acceleration voltage and 1.4 E resolution. The X-ray diffraction (XRD) patterns were registered using a X'TRA (Thermo ARL) diffractometer (Bragg-Brentano geometry, CuK_α radiation, energy dispersed detector, step scan mode). Data registration was made in the 2θ angle range from $5°$ to $80°$ with step of $0.05°$ and accumulation time of 5 s in each point. The average crystallite sizes were calculated using the Scherrer equation.

3   RESULTS AND DISCUSSION

3.1   Synthesis of V-Mg(OH)x and VOx/MgO aerogels

Sol-gel synthesis of mixed oxide aerogels based on hydrolysis of either mixed alkoxides or a homogeneous solution of two or more individual metal alkoxides followed by supercritical drying in an autoclave allows excellent control of mixing yielding homogeneous materials with high surface areas and small crystallite size [14]. However, different hydrolysis rates of alkoxides of different metals often make it difficult to synthesize homogeneous aerogels without phase segregation. The approaches suggested to overcome this problem include selection of alkoxide precursors with matching reactivity, prehydrolysis or the less reactive alkoxide, chemical modification of a more reactive one, etc [14].

In our case, vanadium triisopropoxide oxide was hydrolyzed substantially faster than magnesium methoxide. So, when we added water to the mixture of the alkoxides, we obtained a precipitate consisting mostly of amorphous hydrated vanadia. Such precipitation could be avoided if magnesium methoxide was prehydrolyzed with a small excess of water before the vanadium precursor was added. This procedure yielded clear homogeneous gels and was used for synthesis of all aerogel materials described in this paper.

Previously it was shown that the surface area of magnesium hydroxide aerogels could be stabilized by adding aromatic solvents [6]. We found toluene to be a better second solvent for stabilization of the mixed 10% V-Mg(OH)$_x$ gel than an alcohol. Typical surface area of the mixed aerogels obtained after supercritical drying was about 1150 m^2/g. 10% V-Mg(OH)$_x$ sample was even higher than that of the reference AP-Mg(OH)$_2$ prepared by a similar procedure. Even when the vanadium concentration was increased to 26%, the surface area of the hydroxide aerogel still exceeded 1000 m^2/g.

As the final goal of this study was to prepare a nanocrystalline mixed VOx/MgO oxide with high surface area, we explored the effect of dehydration conditions on the properties of obtained aerogels. Dehydration of Mg(OH)$_2$ aerogels is typically carried out by heating in vacuum to 500°C or higher temperatures [4-6]. When we applied this approach to dehydration of AP V-Mg(OH)$_x$ samples, the surface area of the synthesized mixed oxides never exceeded 300 m^2/g. For example, dehydration of 10% V-Mg(OH)$_x$ hydroxide aerogel by calcination under vacuum to 600°C led to the surface area decrease from 1100 to 220 m^2/g. Also, heating in a vacuum installation is a complex procedure, and it was desirable to develop a simpler method that could yield samples with high surface area and small particle size. Therefore, we studied several other methods for dehydration of V-Mg(OH)$_x$ hydroxide aerogels to nanocrystalline VOx/MgO mixed oxides.

The same 10% V-Mg(OH)$_x$ hydroxide aerogels with the surface area ca. 1100 m^2/g was used for determination of the optimal dehydration conditions. By far the simplest dehydration method is calcination in air in a muffle furnace. Actually, this method proved to be rather efficient. The surface area of the mixed oxide prepared by such simple calcination of our hydroxide aerogel after calcination at 550°C was equal to 325 m^2/g and was higher than the literature data on the VO$_x$/MgO materials prepared by impregnation even when AP-MgO with high surface area was used as the support [8].

To make further improvement of the dehydration procedure we studied the dehydration kinetics in a flow reactor with a spring balance using a linear heating with 1°C/min rate. There are two major weight loss regions between 100 and 200 °C and between 280 and 380°C. The first one corresponds to desorption of physically adsorbed water. The second peak corresponds to the decomposition of the mixed V-Mg(OH)$_x$ hydroxide to the corresponding mixed oxide. So, our special attention was given to this temperature range because variation of the sample heating rate in this region could have a major effect on the properties of the final mixed oxide.

The developed procedure for dehydration of V-Mg(OH)$_x$ hydroxide aerogels in a flow reactor with intense air flow and slow heating can be used for easily controlled and reproducible synthesis of nanoscale AP-VO$_x$/MgO mixed oxides with the surface area exceeding 400 m^2/g. Actually, the absolute highest surface area of 10% VO$_x$/MgO mixed oxide aerogel equal to 470 m^2/g after calcination at 550°C was obtained for the sample heated in air flow in a tubular flow reactor. This result exceeds the surface area of the sample prepared using the vacuum dehydration by more than a factor of two and appears to be the highest reported surface area for VO$_x$/MgO materials.

The XRD analysis of VO$_x$/MgO samples with different vanadium concentrations showed the presence of nanocrystalline MgO phase with the average crystallite size 5-6 nm together with some Mg(OH)$_2$. The latter can be due to partial rehydration of MgO nanoparticles exposed to the atmosphere after the heat treatment. It is remarkable that no vanadium oxide or magnesium vanadate phase was observed even in the sample with the highest vanadium concentration 26%.

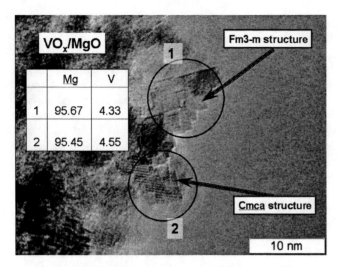

Figure 1. HRTEM image of 5% VOx-MgO aerogel (1 – Fm3m structure, 2 – Cmca structure).

Figure 1 presents high-resolution TEM image of 5% VO$_x$/MgO aerogel. The image clearly shows that the sample is nanocrystalline and consists of interconnected nanocrystals with typical dimensions below 5 nm. However, the analysis of electron diffraction images showed that the structure is changed after the vanadium introduction. The observed structure of VO$_x$/MgO nanocrystals can be attributed to two different crystalline groups denoted as 1 and 2 in Figure 1:

1 – Face-centered cubic lattice – Fm3m;
2 – Orthorhombic lattice – Cmca.

These data suggest that the growth of the vanadium concentration in VO$_x$/MgO aerogels is accompanied by gradual transformation of the face-centered cubic lattice (Fm3m) typical for MgO into the structure with orthorhombic symmetry (Cmca) typical for magnesium vanadate Mg$_3$(VO$_4$)$_2$.

The chemical composition in different regions of the sample marked in Figure 1 was studied using an EDX attachment. The data shown in Figure 1 and obtained for other aerogels clearly show that the vanadium concentrations in atomic per cents measured by EDX match the calculated bulk concentrations for all nanoparticles in all studied aerogel samples. So, the vanadium distribution in VO$_x$/MgO aerogels is homogeneous independent of the vanadium concentration in the studied concentration range. This proves that the aerogel preparation procedure yields samples where vanadium is evenly distributed in the structure of the MgO aerogel matrix.

3.2 CF$_2$Cl$_2$ destructive sorption over VO$_x$/MgO aerogels

The investigation of the weight gain kinetics using a TEOM microanalyzer with 100% contact of the sample with the halocarbon allowed us to obtain more detailed information on the reaction mechanism. The data on the reaction of AP-MgO with CF$_2$Cl$_2$ obtained in the TEOM

microanalyzer (Fig. 2) are presented for comparison. One can see that at 350°C fast MgO conversion begins in 15 minutes after the reaction start. Meanwhile, this induction period is shortened to 11 min for 1% VO$_x$/MgO.

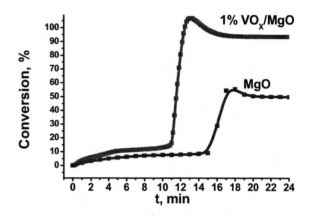

Figure 2. Kinetics of CF$_2$Cl$_2$ reaction with MgO and 1% VO$_x$/MgO aerogels at 350 °C.

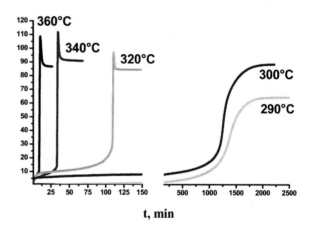

Figure 3. Kinetics of 1% VO$_x$/MgO reaction with CF$_2$Cl$_2$ at different temperatures.

Two temperature ranges can be distinguished for the kinetic curves of the CF$_2$Cl$_2$ destructive sorption on 1% VO$_x$/MgO (Fig. 3). At high temperatures (320-360° C) the reaction is characterized by a prolonged induction period when the sample weight does not change much. After the induction period ends, fast reaction accompanied by the weight takes place followed by some weight loss after reaching the maximum weight. Al lower temperatures the induction period becomes longer, whereas the reaction rate remains approximately constant. At even lower temperature it is impossible to distinguish the induction period because the sample weight grows slowly, and the weight growth kinetics has the S-shape typical for many solid-state reactions. For AP-MgO such behavior is observed at 325°C or lower temperatures. After the vanadium introduction this temperature is decreased. For instance, for 1% VO_x/MgO this change is observed at 300°C.

Figure 4 illustrates the dependence of the induction period length on the vanadium concentration in the sample. At 1% vanadium concentration the induction period is equal to 11-12 minutes at 350°C. Further increase of the vanadium concentration makes the induction period several times shorter. Most likely, such dependence results from the fact that vanadia both participates in the halocarbon decomposition and acts as a catalyst initiating the reaction with MgO matrix.

The complete mass-spectrum of the reaction products was analyzed to select the main representative peaks of the products that were continuously monitored during the reaction. These selected peaks included the peaks with m/z = 63 ($COCl_2$), 44 (CO_2), 84 (CF_2Cl_2) and 117 (CCl_4). The results obtained by the mass-spectrometric analysis show that CCl_4 and CO_2 were the main products obtained over VO_x/MgO. After the end of the induction period intense CF_2Cl_2 absorption by the sorbent takes place with the evolution of water, CO_2, CCl_4 and a small amount of phosgene. The kinetic data (Fig. 4) indicate that the CO_2 evolution takes place at the same time as the main changes of the sample weight. Meanwhile, the CCl_4 is released somewhat later, and the length of this delay grows with the increase of the vanadium concentration in the sample. The obtained data prove substantial participation of added vanadium in the halocarbon decomposition.

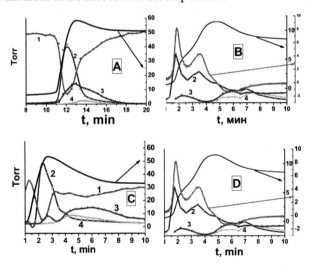

Figure 4. Kinetics of the product formation during reaction of X%VO_x/MgO aerogels with CF_2Cl_2 at 350°C. A) X =1; B) X = 8; C) X = 15; D) X = 26. Gaseous products: 1 – CF_2Cl_2, 2 – CO_2, 3 – CCl_4, 4 – $COCl_2$.

4 CONCLUSIONS

The developed aerogel method for synthesis of binary vanadium-magnesium oxides and hydroxides yields nanocrystalline materials in a wide range of vanadium concentrations. The conditions used for the aerogel hydroxide dehydration and conversion to binary oxide were optimized to obtain materials with the surface area as high as 450 m^2/g. Unlike the traditional impregnation methods, the aerogel synthesis method leads to materials with a homogeneous vanadium distribution in the MgO structure according to the EDX data.

No magnesium vanadate phase was observed in the samples with the vanadium concentration 26% or lower. Vanadium ions are introduced into the MgO nanocrystals changing their symmetry from the initial cubic (Fm3m) to orthorhombic (Cmca). Samples with introduced vanadium showed substantially higher activity in the destructive sorption of CF_2Cl_2.

Acknowledgment. This study was supported by the Russian Foundation for Basic Research (Grant 10-03-00806) and the US Civilian Research and Development Foundation (Grant RUE1-2893-NO-07).

REFERENCES

[1]. M. Schneider and A. Baiker, Catal. Rev.-Sci.Eng., 37, 515, 1995.

[2]. J. A. Schwarz, C. Contescu, and A. Contescu, Chem. Rev., 95, 477, 1995.

[3]. S. Utamapanya, K. J. Klabunde, and J. R. Schlup, Chem. Mater., , 3, 175, 1991.

[4]. K. J. Klabunde, J. Stark, O. Koper, C. Mohs, D. G. Park, S. Decker, Y. Jiang, I. Lagadic, and D. J. Zhang, J. Phys. Chem., 100, 12142, 1996.

[5]. R. Richards, W. F. Li, S. Decker, C. Davidson, O. Koper, V. Zaikovski, A. Volodin, T. Rieker, and K. J. Klabunde, J. Am. Chem. Soc., 122, 4921, 2000.

[6]. A. F. Bedilo, M. J. Sigel, O. B. Koper, M. S. Melgunov, and K. J. Klabunde, J. Mater. Chem., 12, 3599, 2002.

[7]. P. K. Stoimenov, R. L. Klinger, G. L. Marchin, and K. J. Klabunde, Langmuir, 18, 6679, 2002.

[8]. V. V. Chesnokov, A. F. Bedilo, D. S. Heroux, I. V. Mishakov, and K. J. Klabunde, J. Catal., , 218, 438, 2003.

[9]. I. V. Mishakov, A. F. Bedilo, R. M. Richards, V. V. Chesnokov, A. M. Volodin, V. I. Zaikovskii, R. A. Buyanov, and K. J. Klabunde, J. Catal., 206, 40, 2002.

[10]. J. Wang, S. Uma, and K. J. Klabunde, Appl. Catal. B -Environ., 48, 151, 2004.

[11]. I. V. Mishakov, A. A. Vedyagin, A. F. Bedilo, V. I. Zailovskii, and K. J. Klabunde, Catal. Today, , 144, 278, 2009.

[12]. I. V. Mishakov, E. V. Ilyina, A. F. Bedilo, and A. A. Vedyagin, React. Kinet. Catal. Lett., 97, 355, 2009.

[13]. E. V. Ilyina, I. V. Mishakov and A. A. Vedyagin, Inorg. Mater., 45, 1267, 2009.

[14]. J. B. Miller and E. I. Ko, Catal. Today, 35, 269, 1997.

Microwave-Assisted Synthesis of Water Soluble ZnSe@ZnS Quantum Dots

Sonia J. Bailon-Ruiz[*], Oscar Perales-Perez[*],[**] and Surinder P. Singh[**]

[*] Department of Chemistry, University of Puerto Rico, Mayaguez, sonia.bailon@uprm.edu
[**] Department of Engineering Science & Materials, University of Puerto Rico, Mayaguez,
oscarjuan.perales@upr.edu, surinder.singh@upr.edu

ABSTRACT

Water-soluble quantum dots of ZnSe@ZnS were synthesized under microwave irradiation at different temperatures in presence of thioglycolic acid (TGA). X-ray diffraction analyses suggested the development of a ZnSe-ZnS structure and an average crystallite size of 4.1 ± 0.4 nm. High-resolution transmission electron microscopy evidenced the formation of spherical nanocrystals. UV-vis analyses of nanocrystals synthesized at 120°C clearly showed the exciton peak centered at 290 nm. The corresponding photoluminescence spectrum evidenced a sharp and intense emission peak (FWHM ~ 32 nm) centered at 380 nm; the emission peak was broadened (FWHM ~ 47 nm) when the synthesis temperature was increased up to 140 °C. The strong emission in the nanocrystals synthesized at 120°C was attributed to the suitable development of the ZnS shell onto the ZnSe core.

Keywords: quantum dots, nanocrystals, semiconductors, microwave-assisted synthesis.

1 INTRODUCTION

Semiconductors nanocrystals, also known as quantum dots (QD's), are the focus of an intensive research because of the capability of tuning their optical properties by controlling crystal-size, composition or shape at the nanoscale. Compared to organic dyes, QDs are characterized by their narrow emission spectra, large quantum yield, chemical stability against photo-bleaching and surface functionality. These features make QDs dots very promising candidates for photodynamic therapy agents, biomarkers and also as components in optoelectronics devices. Regarding the use of "traditional" CdSe-based QDs for biomedical applications, their potential cytotoxicity makes necessary the search of alternative and less-toxic Cd-free nanomaterials [1-3]. Among different Zn-based QDs, highly luminescent ZnSe nanocrystals have been synthesized from organometallic diethylzinc precursor in a trioctylphospine oxide (TOPO)-hexadecylamine (HAD) mixture [4]. However, the precursor salts are extremely toxic, expensive and pyrophoric. ZnSe QDs exhibiting a hydrophobic surface have also been synthesized from precursor ZnO at temperatures as high as 350°C [5]. As an attempt to enhance the photoluminescence (PL) intensity and the quantum yield even more, the formation of QDs consisting of a ZnS-shell surrounding a ZnSe-core has been suggested [6]. In this regard, the synthesis of core-shell ZnSe@ZnS QD's in presence of thiols becomes a very promising alternative route to organometallic reactions due to: (i) high synthesis temperatures are not necessary; (ii) the precursors salts are not dangerous and easy to handle; (iii) the core-shell arrangement allows the achievement of an excellent photo-stability of the water-soluble nanocrystals; and, (iv) easy surface functionalization in aqueous media [7,8]. Microwave-assisted synthesis is an attractive method to produce the targeted core-shell arrangements because of its high heating rate and homogeneous heating capability which promotes the formation of crystalline and high-optical quality QD's [8,9]. This method is also energy efficient as compared to conventional hydrothermal approaches [9]. Under these considerations, the present study addressed the systematic evaluation of the microwave-assisted synthesis of water-soluble ZnSe@ZnS quantum dots using thioglycolic acid (TGA) as the source of sulfide ions.

2 MATERIALS AND METHODS

2.1 Materials

Zinc chloride (99.999% trace metals basis), elemental selenium powder (99.99% trace metals basis), sodium bisulfite (> 98%) and Thioglycolic acid (TGA, ≥ 98.0 %) were purchased from SIGMA-ALDRICH. All materials were used without further purification.

2.2 Synthesis of ZnSe@ZnS QDs

Selenide ions solution was prepared by reducing elemental selenium powder with sodium bisulfite. Zinc and selenium solutions were mixed in a microwave digestion vessel in presence of TGA at pH 7.0 and heated for one hour at 100°C, 120°C and 140°C. The molar ratio of Zn/TGA/Se was optimized and fixed at 1/ 4.8 / 0.04 for all tests. Once the reaction was completed, the obtained suspension was cooled down to room temperature. The formed nanocrystals were coagulated with 2-propanol, centrifuged and resuspended in deionized water.

2.3 Characterization Techniques

Samples were structurally characterized by X-ray diffraction (XRD) in a Siemens Powder Diffractometer D5000 using Cu-Kα radiation. The high-resolution TEM (HRTEM) study was performed in a FEI-TITAN unit operated at 200 KV. UV-vis absorption measurements were carried out using a Beckman DU 800 Spectrophotometer. A Shimadzu RF-5301 spectrofluorometer with spectral bandwith 5-5 and a 150W xenon lamp was used to collect the photoluminescence (PL) spectra at room temperature. Inductive Coupled Plasma Mass Spectrometer (ICP-MS) Agilent 7500ce was used to confirm the presence of Zn and Se in synthesized QD's.

3 RESULTS AND DISCUSSION

3.1 Structural and Morphological Characterization

Figure 1 shows the XRD pattern of the powder synthesized in presence of TGA at 120°C and pH 7.0. The as-synthesized sample exhibited a cubic structure of zinc blende. No diffraction peaks for isolated ZnSe (JCPDS N° 80-0021) or ZnS (JCPDS N° 77-2100) were observed, which indicates the presence of only one type of crystalline phase. Qian *et al.*, [9] suggested that the location of the XRD lines between the ones corresponding to pure ZnSe and ZnS phases should be attributed to the formation of ZnSe-ZnS core-shell like structure. Xie *et al.* mentioned that the thickness of the shell could also affect the position of the diffraction peaks [10]. In turn, the broadening of the diffraction peaks suggested the presence of very small nanocrystals; the average crystallite size, estimated according to the Debye-Scherrer's relationship, was 4.1 ± 0.4 nm. The corresponding lattice parameter "*a*" was estimated at 0.543 nm. This value is within the 0.567 nm (ZnSe) - 0.538 nm (ZnS) range.

The formation of nanosize crystals can be explained by the extremely fast nucleation rate during the formation of the ZnSe core favored by the direct reaction between available Zn and selenide species. The subsequent formation of the ZnS shell would have taken place by reaction of the excess of Zn ions with sulfide species generated by the thermal decomposition of the TGA thiol. According to Qian *et al.*, [9] the presence of atmospheric oxygen in the reaction vessel used during the microwave irradiation would have enhanced the decomposition of the HS⁻ groups in TGA to sulfide anions. Furthermore, the adsorption of non-reacted thiol species onto the QDs surface should favor the establishment of a net negative surface charge that should have prevented the nanocrystals from aggregation [7, 9].

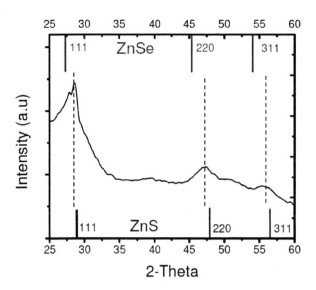

Figure 1: XRD pattern of ZnSe@ZnS powder synthesized in presence of TGA. The pH value and reaction temperature were 7.0 and 120°C, respectively. Expected diffraction lines for pure ZnSe and ZnS phases are also shown.

The HRTEM image of figure 2 corresponds to ZnSe@ZnS core-shell QDs. The image evidenced the crystallinity of the QDs even at such a small size (around 4nm). The lattice spacing for the nanocrystal in the inset was about 0.31 nm, corresponding to the (111) plane of cubic ZnSe core (JCPDS N° 80-0021).

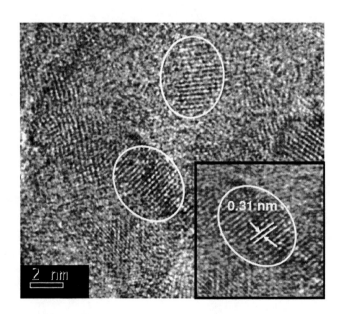

Figure 2: HRTEM image of ZnSe@ZnS quantum dots synthesized at 120°C and pH 7.0.

3.2 UV-Vis and Photoluminescence Measurements.

Figure 3 shows the UV-Vis absorption spectra of ZnSe@ZnS nanocrystals synthesized at different reaction temperatures. The promotion of crystal growth by increasing of the reaction temperature was suggested by the red-shift of the corresponding exciton peak.

The optical band gap energy (E_g) for each sample was estimated from Tauc's relationship and plotting the variation of $(\alpha h\upsilon)^2$ with $h\upsilon$. Here, $h\upsilon$ and α are the photon energy and the absorption coefficient, respectively, and are obtained from the corresponding UV-vis absorption spectra [11]. As figure 4 shows, the intercept of the linear portion of the plot with the energy axis gives the corresponding band gap value. Accordingly, the band gap for the nanocrystals synthesized at 100°C, 120°C and 140°C were estimated at 4.82 eV, 4.80 eV and 4.06 eV, respectively. These E_g values are larger than the one for bulk ZnSe (2.7 eV) and ZnS (3.67 ev) [6,9], which is indicative of the quantum confinement effect.

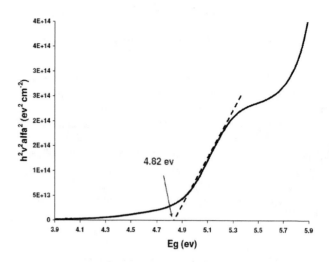

Figure 4: Determination of the band gap value for core-shell ZnSe@ZnS quantum dots synthesized at 100°C and pH 7.

As seen, an intense band edge emission with a FWHM of 32 nm became evident at 380nm in the PL spectrum of the sample synthesized at 120°C. The formation of the ZnS shell onto the ZnSe core should have improved the luminescent properties of these nanocrystals in comparison to bare ZnSe [7]. The luminescence intensity went down and the emission peak broadened when QD's were synthesized at 140°C. This drop in photoluminescence can be attributed to a much larger coverage of the ZnSe by ZnS [12]. Dabbousi *et al.* mentioned that the growth of ZnS shell onto the ZnSe core could be coherent at low coverage. However, as far as the thickness of the shell is increased, the strain due to the lattice mismatch between the core and the shell would lead to the formation of dislocations and low-angle grain boundaries, relaxing the structure and causing the ZnS growth to proceed incoherently. These defects would promote non-radiative recombinations within the ZnS shell that caused the observed decrease in the luminescence intensity [12]. On the other hand, the quenching in photoluminescence observed in QDs' synthesized at 100°C can be attributed to an incomplete or inadequate development of the ZnS layer due to the poor generation of free sulfide ions at such a low temperature. Moreover, as suggested by Deng *et al.*, the presence of structural defects in tinier crystals could also be involved with the observed drop in luminescence [7].

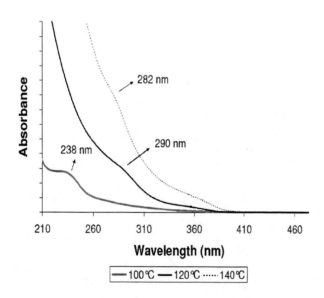

Figure 3: Absorption spectra of core-shell ZnSe@ZnS quantum dots synthesized at pH 7.0 and various reaction temperatures.

The room-temperature PL spectra of QDs synthesized at various reaction temperatures and Zn/TGA/Se molar ratio of 1/ 4.8 / 0.04, are shown in figure 5. The spectra were recorded using an excitation wavelength of 300nm.

UV-Vis and PL measurements evidenced the Stokes' shift between the absorption and emission maximum. The magnitude of the Stoke's shift for samples synthesized at 100°C, 120°C and 140°C were 150 nm, 90 nm and 140 nm, respectively. Stoke's shift can be a consequence of the increased delocalization of the electron and the consequent lowering of the corresponding excited state energy, both caused by the formation of the ZnS shell [12]. The dependence of the magnitude of the Stoke's shift with the thickness of ZnS layer has also been reported for CdSe@ZnS QDs [12].

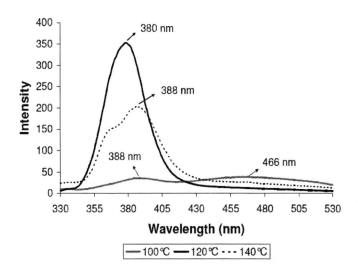

Figure 5: Photoluminescence spectra of core-shell ZnSe@ZnS quantum dots synthesized at 100°C, 120°C and 140°C. The excitation wavelength was 300 nm.

4 CONCLUSIONS

Crystalline and water-soluble quantum dots have been synthesized under microwave irradiation. X-ray diffraction and optical characterization suggested the formation of a ZnSe@ZnS core-shell structure. Luminescence of these core-shell nanocrystals was strongly dependent on the synthesis temperature; the emission intensity reached a maximum when QDs' were synthesized at 120°C. This strong emission was attributed to the adequate formation of a ZnS shell on the ZnSe core.

5 ACKNOWLEDGEMENTS

The support from Drs. E. Melendez and F. Roman, Faculty at the department of Chemistry-UPRM, is acknowledged. The authors also thank to Dr. P. M. Voyles, Faculty at the Department of Materials Science and Engineering-University of Wisconsin at Madison, and Yarilyn Cedeño, doctoral student at UPRM, for their assistance with TEM analyses. This material is based upon work supported by The National Science Foundation under Grant No. HRD 0833112 (CREST program).

REFERENCES

[1] N. Pradhan, and X. Peng, *J. Am. Chem. Soc.* 129, 3339. 2007.
[2] P. Juzenas, W. Chei, Y. Sun, M. Neto, R. Generalov, N. Generalova, and I. Christensen, *Adv. Drug Deliv. Rev.* 60, 1601-1602. 2008.
[3] W. W. Yu, E. Chang, R. Drezek, and V. L. Colvin, *Biochem. Biophys. Res. Commun.* 348, 781-782. 2006.
[4] M. A. Hines and P. Guyot-Sionnest, *J. Phys. Chem. B.* 102, 3655-3657. 1998.
[5] L. S. Li, N. Pradhan, Y. J. Wang and X. G. Peng, *Nano Lett.* 4, 2261-2264. 2005.
[6] C. Hwang and I. Cho, *Bull. Korean Chem. Soc.* 26 (11), 1776-1779. 2005.
[7] Z. Deng, F. L. Lie, S. Shen, I. Ghosh, M. Mansuripur and A. J. Muscat, *Langmuir.* 25, 434-435. 2009.
[8] S. J. Bailon-Ruiz, O. Perales-Perez, S. P. Singh and P. M. Voyles, *Mater. Res. Soc. Symp. Proc.* 1207 (N10-61), 1-2. 2010.
[9] H. Qian, X. Qiu, L. Li and J. Ren, *J. Phys. Chem.* B110, 9034-9037. 2006.
[10] R. Xie, U. Kolb, J. Li, T. Basche and A. Mews, *J. Amer. Chem. Society.* 127,7484. 2005.
[11] J. Tauc, "Amorphous and liquid semiconductors", New York , 159, 1974.
[12] B.O. Dabbousi, J. Rodriguez-Viejo, F.V. Mikulec, J.R. Heine, H. Mattoussi, R. Ober, K. F. Jensen, and M. G. Bawendi, *J. Phys. Chem.* B101, 9473. 1997.

Hydrothermal Growth of ZnO Nanostructures Using Zinc Thin Films as Seed Layer

Jung-Yen Yang[1], Hui-Wen Cheng[2], Tsung-Chieh Cheng[3] and Yiming Li[1,2,*]

[1]National Nano Device Laboratories, Hsinchu 300, Taiwan
[2]Department of Electrical Engineering and Institute of Communication Engineering,
National Chiao Tung University, Hsinchu 300, Taiwan
[3]Department of Mechanical Engineering,
National Kaohsiung University of Applied Sciences, Kaohsiung, Taiwan
[*]ymli@faculty.nctu.edu.tw

ABSTRACT

Zinc oxide nanostructure is selectively synthesized by a simple reaction of zinc acetate dehydrate. With hexamethyltetramine (HMT) under mild hydrothermal conditions, we use zinc thin films as seed layer. Scanning electron microscopy (SEM) result shows that zinc thin films were oxidized to hexagonal ZnO seeds in the absence of zinc salt and base under hydrothermal conditions. The effect of a mixed water/ethanol solvent with various ethanol contents on the morphology, nanostructure, and compositional properties of the synthesized zinc oxide is examined. SEM studies show that the ZnO pillar and wire structures could be obtained at low ethanol contents, while high ethanol contents are favorable for the formations of ZnO thin film structures through coalescence of adjacent rods. X-ray diffraction (XRD) studies show that the phenomenon of crystal tilt appeared as the ethanol content increases. Results indicate that the ethanol could acts as a shape inducing molecule. Mechanism of the growth of different ZnO nanostructures using zinc thin films as seed layer and various ethanol contents in a mixed water/ethanol solvent are also discussed.

Keywords: Hydrothermal, Zinc oxide, Seed, Mixed solvent, Nanostructures

1 INTRODUCTION

Zinc oxide is an excellent compound semiconductor for electronic and photonic applications due to its wide direct band-gap (3.37 eV). ZnO have been widely applied in high technology fields such as photodetectors [1], surface acoustic wave devices [2], photonic crystals [3], light-emitting diodes [4], photoelectrodes [5], gas sensors [6] and transparent conductive coating [7].

ZnO nanostructures can be grown via solution [8] or from gaseous route [9]. The routes mentioned above for the synthesis of ZnO nanostructures, the hydrothermal synthesis [10] is one of the most promising routes due to its low-cost and adjustable growth conditions. In recent years, Several ZnO nanostructures such as nanowires, rods, nanotubes, nanobelts, and nanoflowers have been synthesized via hydrothermal solution processes [11-17].

Seeds are needed to facilitate ZnO nanostructures growth on a substrate. Generally, ZnO seeds are prepared either by thermal decomposition of zinc salts or pre-synthesized ZnO seeds deposited on the substrate. However, uniformity of the ZnO seeds on the substrate is poor via the methods mentioned above. In the paper, ZnO nanostructures were prepared from the mixed water/ethanol solution of zinc acetate dihydrate and hexamethyltetramine (HMT) under hydrothermal conditions using pre-deposited zinc as seeds for ZnO growth. The morphology, crystal, structure, and mechanism of the ZnO nanostructures were investigated.

2 EXPERIMENTAL

Hydrothermal synthesis of ZnO nanowire arrays was carried out by suspending a Zn-deposited (100 nm) wafer upside-down in a PTFE bottle filled with 50 mL mixed water/ethanol solvent with various ethanol contents containing zinc acetate dihydrate (25 mM) and hexamethyltetramine (25 mM). The PTFE bottle was put in a pre-heated oven at 90°C for 5 h. The wafer was then removed from solution, rinsed with deionized water, and dried with an air stream.

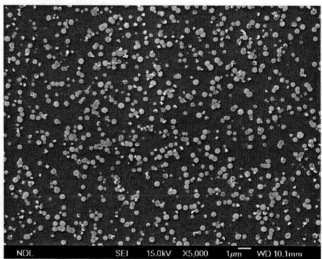

Figure 1: SEM image of hexagonal ZnO seeds grown from zinc thin films by hydrothermal route.

3 RESULTS AND DISCUSSION

To prepare the ZnO seeds for the hydrothermal growth of ZnO nanostructures, Zn-deposited wafer was immersed in water at 90°C for 1 h in the absence of precursors. Figure 1 shows the surface morphology of the treated wafer. Hexagonal particles appear on the substrate, indicating the zinc was oxidized to zinc oxide upon hydrothermal treatment. The ZnO particles were used as seeds for subsequent growth of ZnO nanostructures.

Figure 1 shows the surface morphology of ZnO nanostructures grown at various ethanol contents. ZnO nanowires were grown on the substrate, as shown in Fig. 2(a), which consist of both thick and thin nanowires. With the addition of ethanol, thin nanowires disappear and dense nanowires array is formed, as shown in Fig. 2(b). With further addition of ethanol content, no wires but hexagonal plates morphology with different thickness were found (Fig.

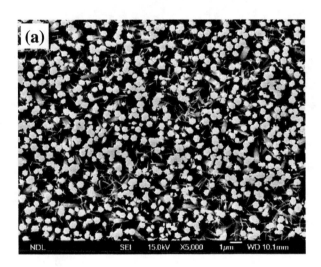

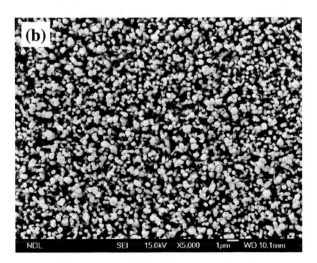

Figure 2: SEM images of ZnO thin films deposited at various water/ethanol ratios (V/V, mL): (a) 50/0, (b) 40/10, (c) 30/20, (d) 20/20, and (e) 10/40.

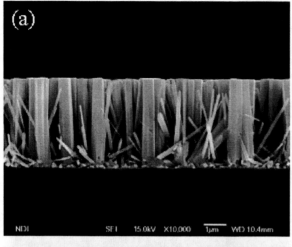

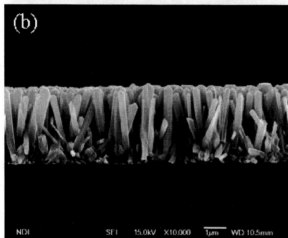

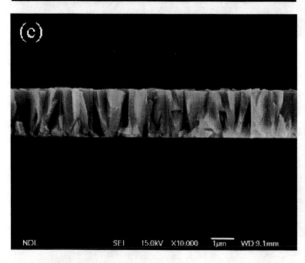

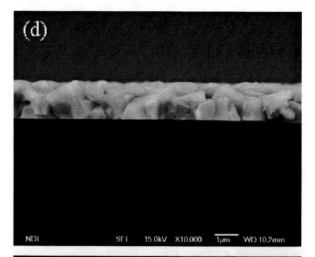

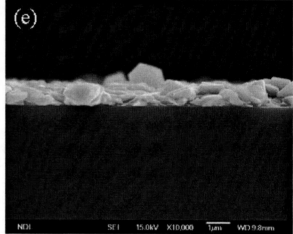

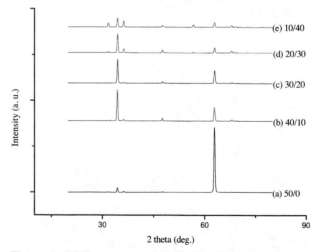

Figure 3: Cross-sectional SEM images of ZnO thin films deposited at various water/ethanol ratios (V/V, mL): (a) 50/0, (b) 40/10, (c) 30/20, (d) 20/20, and (e) 10/40.

Figure 4: XRD patterns of ZnO thin films deposited at various water/ethanol ratios (V/V, mL): (a) 50/0, (b) 40/10, (c) 30/20, (d) 20/20, and (e) 10/40.

2(c)-(e)). Figure 3 shows the cross-sectional SEM images of ZnO nanostructures grown at various ethanol contents. Both thin and thick nanowires were found, and the top of the nanowires is smooth (Fig. 3(a)). After addition of ethanol, the top of the nanowires exhibits pyramid-like morphology (Fig. 3(b)). With further addition of ethanol content, isolated nanowires coalesce to form thin films with different densities, as shown in Fig. 3(c)-(e). The growth and decline of the measured XRD peaks also reveal that the crystal tilting occurs as addition of ethanol. The XRD θ-2θ scan spectra shown in Figure 4, the strongest diffraction peak is switched from (103) to (002) with increasing ethanol content. Based on the observation of ZnO crystal growth and morphology evolution, a rational mechanism for the ZnO morphology evolution is proposed. When the Zn-coated wafer was put into the growth solution, the metal surface is oxidized to $Zn(OH)_2$, and the $Zn(OH)_2$ is subsequent dehydrated upon heating to form ZnO. The chemical reactions to form ZnO nanorods are formulated as follows:

$$Zn + 2H_2O \rightarrow Zn(OH)_2 + H_2 \qquad (1)$$
$$Zn(OH)_2 \rightarrow ZnO + H_2O \qquad (2)$$

Oxidation of zinc leads to the formation ZnO seeds in the initial stage. The ZnO seeds facilitate the formation of loose ZnO nanowires array. The thin nanowires dissolve to form tiny nuclei during crystal growth process, and the nuclei recrystallize to form thick nanowires. The dissolution-recrystallization process thus results in the formation of thick ZnO nanowires. The ethanol content in the mixed solvent also plays an important role on the morphology evolution of the ZnO nanostructures. Wang et al. indicate that the increasing ethanol contents can accelerate crystal tilting [18]. Therefore, the increasing ethanol contents lead to crystal coalescence and thicknesses reduction of the ZnO thin films.

4 CONCLUSIONS

The hydrothermal growth of ZnO nanostructures using zinc thin film as seed layer has been investigated. ZnO seeds could be in situ formed on the substrate. The effect using mixed solvent for the hydrothermal growth of ZnO nanostructures thin film was also investigated. The ZnO wire structures could be obtained at low ethanol contents, while high ethanol contents were favorable for the formations of ZnO thin film structures through coalescence of adjacent wires. The oxidation-dissolution-recrystallization mechanism has been proposed to illustrate the morphology evolution of ZnO nanostructures.

ACKNOWLEDGEMENT

This work was supported in part by Taiwan National Science Council (NSC) under Contract NSC-97-2221-E-009-154-MY2.

REFERENCES

[1] S. Liang, H. Sheng, Y. Liu, Z. Huo, Y. Lu and H. Chen, "ZnO Schottky ultraviolet photodetectors", J. Cryst. Growth, 225, 110, 2001.

[2] W. C. Shih and, M. S. Wu, "Growth of ZnO films on GaAs substrates with a SiO$_2$ buffer layer by RF planar magnetron sputtering for surface acoustic wave applications", J. Cryst. Growth, 137, 319, 1994.

[3] Y. F. Chen, D. Bagnall and T. F. Yao, "ZnO as a novel photonic material for the UV region", Mater. Sci. Eng. B 75, 190, 2000.

[4] N. Saito, H. Haneda, T. Sekiguchi, N. Ohashi, Sakaguchi, I. and K. Koumoto, "Low-Temperature Fabrication of Light-Emitting Zinc Oxide Micropatterns Using Self-Assembled Monolayers", Adv. Mater., 14, 418, 2002.

[5] J. Y. Lee, Y. S. Choi, J. H. Kim, M. O. Park and S. Im, Thin Solid Films, "Optimizing n-ZnO/p-Si heterojunctions for photodiode applications", 403, 553, 2002.

[6] A. Mitra, A. P. Chatterjee and H. S. Maiti, Mater. Lett. "ZnO thin film sensor", 35, 33, 1998.

[7] Gordon, R. G. MRS Bull., "Criteria for choosing transparent conductors", 25, 52. 2000.

[8] M. Izaki and T. Omi, "Transparent zinc oxide films prepared by electrochemical reaction", Appl. Phys. Lett., 68, 2439, 1996.

[9] S. Fay, L. Feitknecht, R. Schluchter, U. Kroll, E. Vallat-Sauvain and A. Shah, "Rough ZnO layers by LP-CVD process and their effect in improving performances of amorphous and microcrystalline silicon solar cells", Sol. Energy Mater. Sol. Cells, 90, 2960, 2006.

[10] C. S. Cundy and P. A. Cox, "The hydrothermal synthesis of zeolites: History and development from the earliest days to the present time", Chem. Rev., 103, 663, 2003.

[11] Z. R. Tian, J. A. Voigt, J. Liu, B. Mckenzie and M. J. Mcdermott, "Biommetic arrays of oriented helical ZnO nanorods and columns", J. Am. Chem. Soc., 124, 12954, 2002.

[12] Z. R. Tian, J. A. Voigt, J. Liu, B. Mckenzie, M. J. Mcdermott, M. A. Rodriguez, H. Konishi and H. F. Xu, "Complex and oriented ZnO nanostructures", Nature Mater., 2, 821, 2003.

[13] L. Vayssieres, K. Keis, S. E. Lindquist and A. Hagfeldt, "Purpose-built anisotropic metal oxide material: 3D highly oriented microrod array of ZnO", J. Phys. Chem. B, 105, 3350, 2001.

[14] L. Vayssieres, K. Keis, A. Hagfeldt and S. E. Lindquist, "Three-dimensional array of highly oriented crystalline ZnO microtubes", Chem. Mater., 13, 4395, 2001.

[15] L. Vayssieres, "Growth of arrayed nanorods and nanowires of ZnO from aqueous solutions", Adv. Mater., 15, 464, 2003.

[16] L. E. Greene, M. Law, J. Goldberger, F. Kim, J. C. Johnson, Y. F. Zhang, R. J. Saykally and P. D. Yang, "Low-Temperature Wafer-Scale Production of ZnO Nanowire Arrays", Angew. Chem., Int. Ed., 42, 3031, 2003.

[17] J. Zhang, L. D. Sun, C. S. Liao and C. H. Yan, "A simple route towards tubular ZnO", Chem. Commun., 3, 262, 2002.

[18] M. S. Wang, E. J. Kim, E. W. Shin, J. S. Chung, S. H. Hahn and C. H. Park, J. Phys. Chem. C, "Low-temperature solution growth of high-quality ZnO thin films and solvent-dependent film texture", 112, 1920, 2008.

Synergistic effect of Cobalt on Nitrogen and Carbon codoped Anatase Titania for Photodegradation under Visible Light

N.N. Binitha*,**, Z. Yaakob*, P.P Silija*, M.R. Resmi**, P.V. Suraja*

* Department of Chemical and Process Engineering, Faculty of Engineering and Built Environment, National University of Malaysia, 43600 UKM Bangi, Selangor, Malaysia, Ph: +60 (3) 8921 6400. Fax: +60 (3) 8921 6148
**Department of Chemistry, Sree Neelakanta Government Sanskrit College Pattambi, Palakkad-679306, Kerala, India Ph: +91 466-2212223. Fax: +91 466-2212223 Email:binithann@yahoo.co.in

ABSTRACT

This work reports on the synergistic effect of cobalt on Carbon and Nitrogen codoped anatase titania photocatalysts prepared via sol–gel method using urea as the nitrogen source. Visible light activity is achieved as result of anion doping and further improvement is manifested in the photocatalytic activity with the cobalt doping. The photocatalysts prepared are characterized using various analytical techniques. Diffuse reflectance spectral (DRS) analysis shows visible light absorption. Efficient dispersion of cobalt and anatase phase stability is clear from X-ray diffraction (XRD) patterns. Photodegradability under visible light is investigated using methylene blue which is a widely used dye in photodegredation studies. Titania upon C & N codoping shows good activity in very dilute solutions whereas with cobalt doping, the activity is retained even at higher dye concentrations. Reaction parameters are optimized to improve the photodegradation capacity of the prepared systems. Present systems show excellent activity even within 15 minutes and are found to be reusable as well.

Keywords: synergistic effect, cobalt doping, anion doped TiO$_2$, sol-gel method, photodegradation

1 INTRODUCTION

A dye is a colored substance that has an affinity to the substrate to which it is being applied and makes the world more beautiful. However, the release of coloured wastewaters poses a serious environmental problem and a public health concern. Colour removal, especially from textile wastewaters, has been a big challenge over the last decades, and till date there is no single and economically attractive treatment procedure available that can effectively decolourise dyes [1]. Titanium dioxide (TiO$_2$) has attracted much attention for the past few decades as the most promising photocatalyst because of its excellent capacity in the decomposition of pollutants in water and air by means of photodegredation, However, two major drawbacks in TiO$_2$ photocatalysis are the low quantum efficiency due to the high recombination rate of photoinduced electron–hole pairs (e$^-$–h$^+$) and the poor absorption ability in the visible-light region.

In order to improve the photocatalytic performance of TiO$_2$,there has been a large number of studies modifying TiO$_2$ catalyst in various ways. Some of the strategies adapted by researchers in this field include depositing noble metal on titania [2,3], doping with metal [4,5,6], doping nonmetal [7-10] and compounding TiO$_2$ with other materials [11,12]. Doping with metal and doping nonmetal were the most feasible methods for improving the photocatalytic performance of TiO$_2$.The doping of metal atoms can suppress the recombination of photo-induced electron-hole pairs so as to increase the photo quantum efficiency [9,9]. On the other hand, the nonmetal atoms can incorporate into the lattice structure of TiO$_2$, decreasing the band gap, and thus giving rise to a good response to the visible light [7-10].

Our objective was the preparation of a doped TiO$_2$ system where there is low band gap, so that visible light can be used effectively and the recombination of electron - hole pairs is low which will result in good quantum efficiency. Recently we had developed a highly active carbon and nitrogen codoped TiO$_2$. The system is found give 100% degradation of the dye pollutant within a short span of 10 minutes. But the main handicap we have faced with that catalyst was its decreased activity in solutions where the dye concentration is appreciably high. This motivated us to study the synergistic effect of transition metal on this carbon and nitrogen codoped system. The transition metal chosen here is cobalt. The effect of percentage of metal loading on the anion doped system is also investigated. The photoactivity is investigated using methylene blue (MB) degradation.

2 EXPERIMENTAL

2.1 Modified TiO$_2$ Preparation

Carbon and nitrogen codoped system is prepared following the reported procedure [13]. Cobalt loading is done on the anion doped system by using cobalt nitrate as the precursor. The detailed preparation method is as follows. Cobalt loaded Nitrogen and Carbon codoped Anatase TiO2 catalysts using Titanium (IV) isopropoxide (Sigma Aldrich) as titanium precursor, Cobalt nitrate ((Hamburg Chemical GmbH) and urea (Merck) as cobalt and nitrogen source was prepared using sol-gel method at room temperature. 18.6 ml Titanium (IV) isopropoxide and 50 mL ethyl alcohol (Merck) were mixed (solution a), to this solution b which contains 50 mL ethyl alcohol, 12.5 mL glacial acetic acid and 6.25 mL distilled water was added. The resultant transparent solution was stirred for 3h. A mixture of Urea - water - alcohol with molar ratio of urea:ethanol:water, 1:5:0.5 and Cobalt nitrate solution to get the required loading of Co was then slowly added to the above solution until the N/Ti molar content of 20 was satisfied. It was again stirred for 3h, aged for 2 days and dried in an air oven at 80˚C. Ground into fine powder and calcined at 300˚C for 5h to obtain nitrogen doped TiO_2. The systems are designated as Ti, NT,1CoNT, 2CoNT, 3CoNT and 4CoNT whereas the numbers indicates the wt% of Co loading.

2.2 Photocatalyst Characterization

XRD patterns of the samples were recorded for 2θ between 3 and 80° on a Bruker AXS D8 Advance diffractometer employing a scanning rate of 0.02°/S with Cu $K\alpha$ radiation (λ=1.5418). The FTIR spectra were recorded in NICOLET6700 FT-IRThermoscientific in the region 400–4,000 cm^{-1}. Diffuse Reflectance Ultraviolet - Visible spectroscopy (UV–Vis DRS) of powder catalyst samples was carried out at room temperature using a Varian, Cary spectrophotometer in the range of 200 to 800 nm.

2.3 Photoactivity Studies

Photodegradation of methylene blue is investigated as a model pollutant. The pollutant degrading capacity of the Cobalt loaded Nitrogen and Carbon codoped Anatase TiO2 catalysts was studied using a Rayonet type Photoreactor with visible light having 16 tubes of 8W (Associate Technica, India). In the reactor, 5 quartz tubes of 150 ml capacity, is concentrically arranged to get uniform illumination for all the systems. 50 ml of methylene blue (25mg/l) was placed in the quartz tube, containing a definite amount of the catalyst and is irradiated with visible light under continuous stirring. The methylene blue concentration was analyzed using a colorimeter (ESICO Microprocessor photo colorimeter model 1312) at a wavelength of 665 nm.

3 Results and Discussion

3.1 Characterization of the photocatalysts

XRD pattern of anion codoped sample is found to be amorphous in nature. We analyzed the cobalt incorporated samples to know about the crystallinity. All metal incorporated C and N codoped systems are found to contain anatase as the only phase. The absence of amorphous phase indicates that the metal incorporation accelerated the transition from amorphous to crystalline; of anion doped TiO_2 system. TiO_2 without any dopant also exists as anatase. The crystallite size of the systems is calculated using Scherrer equation [14]. Cobalt doping results in reduction in the crystallite size when compared to the TiO_2 system without the presence of any dopant. But increase in the percentage of metal dopant results in increase in the crystallite size, but still is less when compared to that of undoped system. Reduction in crystallite size indicates increased surface area. Efficient dispersion of Co is manifested from the absence of peaks for oxides of cobalt.

System	Crystallite size (nm)	MB Degradation (%)
Ti	7.9891	14.5454
NT	-	87.0370
1CoNT	4.2040	94.5454
2CoNT	4.6338	96.3636
3CoNT	5.8567	89.0909
4CoNT	6.5585	85.5454

Table 1: Crytstallite size and photodegradation results over different systems.

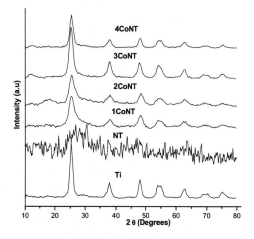

Fig. 1 XRD patterns of the prepared photocatalysts

The photocatalyst samples are found to be coloured, indicating absorption of visible light. The brownish yellow colour of NT remains unaffected after loading 1% Co. In 2CoNT the colour fades, but still remains as brownish yellow whereas 3CoNT and 4CoNT are green in colour. DRS analysis of the samples is done which shows absorption throughout the visible range. The incorporation of cobalt strengthens the visible absorption. This absorption feature suggests that these Co-doped TiO_2 can be activated by visible light. The reflectance spectra of the samples are given in figure 2. This broad-energy feature reflects the heterogeneity of Co, C and N impurities and associated charge-neutrality defects present on the samples.

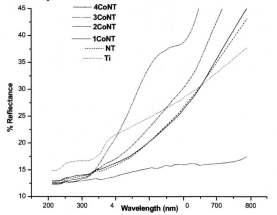

Fig. 2 UV-VIS diffuse reflectance spectroscopy of prepared photocatalysts

3.2 Photodegradation of dye pollutant

In order to assess the photocatalytic performance of Co loaded C, N codoped TiO_2 for their application in the pollutant removal processes, MB degradation is carried out. The reaction parameters are optimized to get better activity. The system selected for optimization studies is 2CoNT.

3.2.1. Effect of initial MB concentration

The effect of initial MB concentration on MB photodegradation is investigated (fig. 3). When we analyzed the activity of the catalysts using 10mg/L dye solution, all systems except that without any doping resulted in complete conversion. Thus we increased the concentration of MB to 25mg/L for further studies to get a comparison on the performance of different systems. The degradation result of 2CoNT for MB degradation over a range of concentration is studied and the result is given in Fig. 3. MB photodegradation decreases with increase in the initial concentration of MB, but still the activity is appreciable even for 100mg/L dye concentration for a short period of exposure to visible light. For NT system, the

activity drops sharply with increasing dye concentrations [14] and is found to be effective mainly in dilute solutions within this short span of exposure to visible light. For concentration variation study the other two parameters were kept constant i.e. loading of photocatalyst (1g/L) and reaction time (15 min).

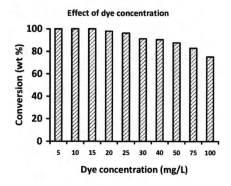

Fig. 3 Influence of dye concentration on % degradation in presence 0.1g/L catalyst for 15 min. irradiation

3.2.2 Effect of photocatalyst dose on degradation

Fig. 4 shows the effect of catalyst dose on MB degradation. It is apparent from the figure that with increase in the catalyst dose, the photodegradation increases up to dose of 0.05g. Further increase in the dose up to 0.1g does not affect MB degradation reaction. There is a marginal initial increase in the efficiency. This is because initial concentration of MB solution and light intensity is constant throughout the dose study. As the photocatalyst dose was increased from 0.01g to 0.05 g, the number of active sites increases, the extent of light absorption increases which in turn increases photoactivity. Further increase in photocatalyst dose to 0.15 g increases the opacity of solution, which decreases the penetration of light inside the solution with a slight decrease in the photodegradation of MB [15]. Thus the selected optimum catalyst dose is 1g/L dye.

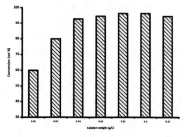

Fig.4 Effect of photocatalyst dose on the photocatalytic degradation of 25mg/l MB, for 15 min. irradiation

3.2.3 Effect of Time

The reaction is done continuously and the degraded solutions are analyzed with an interval of 5 min. It is found that the conversion reaches maximum within 15 minutes and then remains constant. Fig. 5 explains the influence of time on photodegradation of MB.

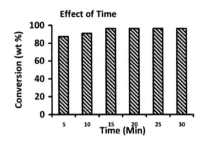

Fig. 5 Effect of reaction time on the degradation of 25mg/L MB with 0.1g/L catalyst.

3.2.4 Effect of metal loading

Effect of percentage (wt %) of metal loading on C and N codoped TiO_2 is investigated in the above selected optimum conditions. It is found that metal loading improved the activity of anion doped TiO_2. The activity increases and reaches maximum at a metal loading of 2% which then decreases. The activity of 4% metal loaded system 4CoNT is found to be even lower than that of anion doped system without any metal. The probable reason for the decreased activity with increase in metal loading may be because excess-doped metal introduce electron–hole recombination centers in the structure. Thus among the different systems 2CoNT, the 2% cobalt loaded, carbon and nitrogen codoped system is showing maximum photoactivity. The results are shown in table 1. In order to make sure about the synergistic effect of cobalt loading on the anion codoped TiO_2, degradation is done on 2% cobalt doped system without any anion doping. System preparation is done exactly as before, but eliminating the step for the addition of urea. The cobalt doped TiO_2 shows a degradation of 20.37%, which is very low when compared to the anion codoped systems.

3.2.5. Recycle ability

The reusability of TiO_2 photocatalyst was one of the key steps to make heterogeneous photocatalysis technology for practical applications. We did the reusability tests of the best system 2CoNT for 6 repeated runs and found to show appreciable activity for 3 cycles which then decreases down sharply. The results are shown in fig. 6.

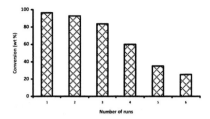

Reusability study

Fig. 5 Effect of reaction time on the degradation of 25mg/L MB with 0.1g/L catalyst.

3 CONCLUSIONS

In the present study, the successful preparation of a novel, improved photocatalyst, cobalt loaded carbon and nitrogen codoped TiO_2 is shown. Cobalt containing doped systems is found to be anatase in nature and is more photocatalytically active compared to carbon and nitrogen codoped TiO_2 without any metal doping, which is amorphous. The system, 2% cobalt loaded carbon and nitrogen codpoed TiO_2 showed excellent photoactivity for methylene blue degradation over a wide range of dye concentration within a short period of time. Reusability is another advantage of this doped system. The increased activity is assumed to be due to the synergistic effect of cobalt doping on the carbon and nitrogen codoped TiO_2. Further studies on the nature of dopants on TiO_2 and on the mechanism of photodegradation are required.

REFERENCES

[1]/ A.B. Santos, F.J. Cervantes and J.B. Lier, "Review paper on current technologies for decolourisation of textile wastewaters: perspectives for anaerobic biotechnology," Bioresour. Technol., 98, 2369–2385, 2007.

[2]/ A. Sclafani, M.N. Mozzanegay and J.M. Herrmanny, "Influence of silver deposits on the photocatalytic activity of titania," J. Catal., 168, 117–120, 1997.

[3]/ S. Sakthivel, M.V. Shankar, M. Palanichamy, B. Arabindoo, D.W. Bahnemann and V. Murugesan, "Enhancement of photocatalytic activity by metal deposition: characterisation and photonic efficiency of Pt, Au and Pd deposited on TiO_2 catalyst," Water Res., 38, 3001–3008 , 2004.

[4]/ W. Choi, A. Termin and M.R. Hoffmann, "The role of metal ion dopants in quantum-sized TiO2: correlation between photoreactivity and charge carrier recombination dynamics," J. Phys. Chem., 98, 13669–13679, 1994.

[5]/ J. Lin and J.C. Yu, "An investigation on

photocatalytic activities of mixed TiO_2-rare earth oxides for the oxidation of acetone in air," J. Photochem. Photobiol. A-chem., 116, 63–67, 1998.

[6]/ B.F. Xin, Z.Y. Ren, P. Wang, J. Liu, L.Q. Jing and H.G. Fu, "Study on the mechanisms of photoinduced carriers separation and recombination for Fe^{3+}-TiO_2 photocatalysts," Appl. Surf. Sci., 253, 4390–4395, 2007.

[7]/ R. Asahi, T. Morikawa, T. Ohwaki, K. Aoki and Y. Tage, "Visible-light photocatalysis in nitrogen-doped titanium oxides," Science, 293, 269–271, 2001.

[8]/ S.U.M. Khan, M. Al-shahry and W.B. Ingler jr., "Efficient photochemical water splitting by a chemically modified n-TiO_2," Science, 297, 2243–2245, 2002.

[9]/ T. Ohno, M. Akiyoshi, T. Umebayashi, K. Asai, T. Mitsui and M. Matsumura, "Preparation of S-doped TiO_2 photocatalysts and their photocatalytic activities under visible light," Appl. Catal. A-Gen., 265, 115–121, 2004.

[10]/ L. Lin, W. Lin, J.L. Xie, Y.X. Zhu, B.Y. Zhao and Y.C. Xie, "Photocatalytic properties of phosphor-doped titania nanoparticles," Appl. Catal. B Environ.,75, 52–58, 2007.

[11]/ H.M. zhang, X. Quan, s. Chen, and H.M. Zhao, "Fabrication and characterization of Silica/Titania nanotubes composite membrane with photocatalytic capability," Environ. Sci. Technol., 40, 6104–6109, 2006.

[12]/ W.D. Wang, C.G. Silva and J.L. Faria, "Photocatalytic degradation of chromotrope 2r using nanocrystalline TiO_2/activated-carbon composite catalysts," Appl. Catal. B-Environ., 70, 470–478, 2007.

[13]/ P.P. Silija, Z. Yaakob, N.N. Binitha, P.V. Suraja, S. Sugunanc, M.R. Resmi, "A facile sol gel method for the preparation of N-doped TiO_2 photocatalyst for pollutant degradation" ICN 2010.

Photoluminescence characteristics of GaN nanoparticles

Y. Chen, K. S. Kang*, N. Jyoti, Jaehwan Kim

Center for EAPap Actuator, Dept. of Mechanical Engineering, Inha University,

253 Yonghyun-Dong, Nam-Ku, Incheon 402-751, South Korea

ABSTRACT

GaN nanoparticles have been fabricated with sol-gel method using $Ga(NO_3)_3$ as a presusor, HNO_3 as a solvent, NH_3OH as a pH adjusting agent, and citric acid as a chelating agent. After adding citric acid the solution became clear. The solution was poured to Petri dish and dried at 400 °C for 4 h. The resulting Ga_2O_3 was gray colored powder. The gallium oxide power was put to the Al-crucible and annealed at 900 °C for 1 h. The resulting nanoparticles have been investigated using energy dispersive spectroscopy (EDS), x-ray diffeactometer, transmission electron microscopy (TEM), and fluorometer. The TEM images show that the GaN nanoparticle size was approximately 8-13 nm. X-ray diffraction patterns provide characteristic GaN diffraction peaks. There are two sharp photoluminescence (PL) peaks centered at 371 nm and 483 nm and broad PL peak between 400-600 nm.

*Author to whom correspondence should be addressed: Email: kkang@inha.ac.kr

1. Introduction

Nanoparticles, nanorods, and nanowires comprised of metlas and semiconductors exhibit size dependent electronic, magnetic, photocatalytic, and optical property. They have been considered as a next generation optoelectronic devices. Especially GaN has been extensively devoted on research to develop light emitting devices, high power electronic devices, and Schottky diodes.

The major commercial optoelectronic devices are based on the GaN thin films. Various deposition technologies including metal-organic chemical vapor deposition (MOCVD), molecular beam epitaxy, and chemical solution deposition are employed to fabricate high quality GaN thin films [1-3]. Recently, GaN nanoparticles have been fabricated with various fabrication techniques, such as MOCVD technique, synthesis GaN nanoparticles in silica matrix, and pyrolysis at high temperature [4-6]. The conversion of Ga_2O_3 or $GaO(OH)$ to GaN has been developed.

2. Experimental

$Ga(NO_3)_3$ was dissolved in nitric acid. The pH was adjusted to 8.7 with adding ammonium hydroxide. Citric acid was slowly added to the solution until the solution became clear. The solution was dried and annealed at 400 °C for 4 h. The resulting powder had gray color. The Ga_2O_3 power was annealed in tube furnace at 900 °C with ammonia gas flow. The resulting particle color became orange. The resulting nanoparticles were analyzed with energy dispersive spectroscope (EDS), x-ray diffeactometer, transmission electron microscope (TEM) and fluorometer.

3. Results and discussion

Figure 1 shows the fabrication processes of the Ga_2O_3 and GaN and pictures of sol-solution, Ga_2O_3, and GaN. Figure 2 shows the EDS of GaN nanoparticles. The three gallium peaks are clearly appeared, and strong nitrogen peak located in far left side. The cupper peaks are due to the cupper grid. This result clearly shows the gallium oxide was converted to GaN. This EDS spectrum confirms no detectable presence of oxygen in these nanoparticles.

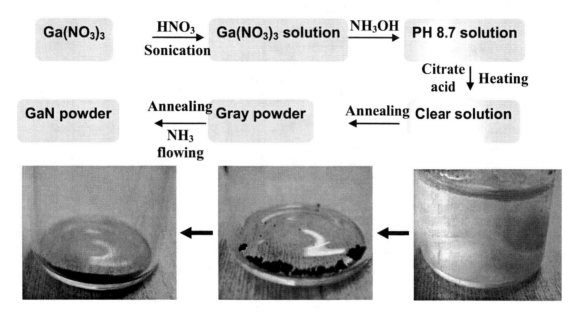

Figure 1. Schematic view of Ga_2O_3 and GaN fabrication processes and pictures of sol-solution, Ga_2O_3, and GaN.

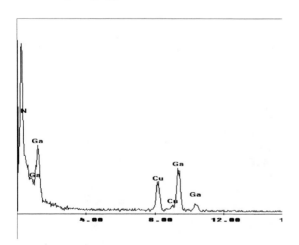

Figure 2. Energy dispersive spectrum of GaN.

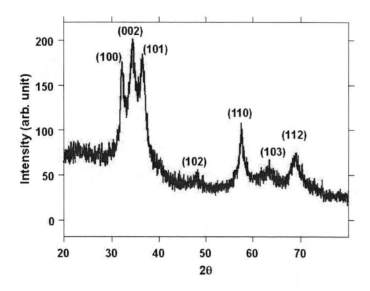

Figure 3. XRD-pattern of the GaN nanoparticles.

Figure 3 shows the XRD pattern of the GaN nanoparticles. The diffraction peaks at 2θ = 32.4, 34.6, 36.9, 48.3, 57.9, 63.6, and 69.2° indicate that the GaN has wurtzite structure. Relatively strong diffraction peaks reveal that the high crystallinity of the GaN nanoparticles.

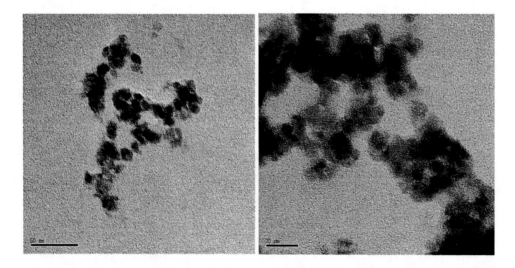

Figure 4. HRTEM images of GaN nanoparticles. The scale bar sizes are 50 nm (left) and 20 nm (right).

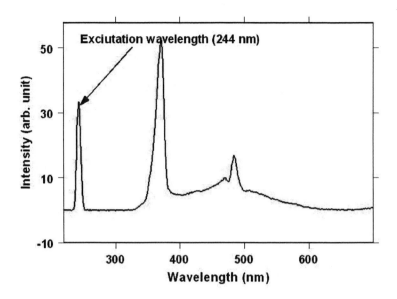

Figure 5. PL spectrum of GaN nanoparticles.

Figure 4 shows the TEM images. The measured particle sizes were approximately 8-13 nm. The GaN nanoparticles were excited with 244 nm of deep UV. Characteristic PL spectra was shown in figure 5. The major sharp luminescence peak was 371 nm. The second sharp emission peak appeared at 483 nm. Broad emission peak showed between 400 - 600 nm. This result indicates that the broad emission peak may cause the defect structure.

4. Conclusions

GaN nanoparticles were fabricated with simple sol-gel technology. The EDS and XRD results show that the Ga_2O_3 is converted to GaN. The TEM images show that the GaN nanoparticle sizes are approximately 8-13 nm. There are three characteristic luminescence peaks of the GaN nanoparticles, such as sharp tow peaks centered at 371 and 483 nm and broad peak between 400 nm – 600 nm.

Acknowledgement

This work was supported by the Korea Research Foundation (KRF) and Creative Research Initiative (EAPap Actuator) in South Korea.

References

[1] S. Y. Kim, H. W. Jang, J. L. Lee, Appl. Phys. Lett. 82 (2003) 61.

[2] D. Kisailus, J. H. Choi, F. F. Lange, J. Mater. Res. 17 (2002) 2540.

[3] S. J. Yang, T. W. Kang, T. W. Kim, K. S. Chung, J. Mater. Res 17 (2002) 1019.

[4] Y. Azuma, M. Shimada, K. Okuyama, Chem. Vap. Depos. 10 (2004) 11.

[5] Y. Yang, V. J. Leppert, S. H. Risbud, B. Twamley, P. P. Power, H. W. H. Lee, Appl. Phys. Lett. 74 (1999) 2262.

[6] J. F. Janik, R. L. Wells, Chem. Mater. 8 (1996) 2708.

Fuel Cell Catalysts Based On Core-Shell Nano-particles

Honglin Xu[*], Christopher Zalitis[*], Margarita Rivera-Hernandez[*,**], Tim Albrecht[*], Anthony Kucernak[*]

[*] Imperial College London, Department of Chemistry, Exhibition Road, London, SW7 2AZ, UK;
[**] Autonomous National University of Mexico, Physics Institute, Coyoacan, 04510, Mexico, D.F., Mexico

ABSTRACT

In this work, we used platinum black to improve the dispersion mechanisms onto highly oriented pyrolytic graphite substrates by using different solvents in order to develop a methodology to disperse core-shell Pt nanoparticles. With this technique, we found that pure isopropyl alcohol exhibited better dispersion onto graphite in comparison with a mixed solution. Although SEM and STM images revealed in general a poor dispersion, some aggregates and few individual particles scattered onto the graphite surface were observed.

Keywords: fuel cell, catalysts, Pt, nanoparticles

1 INTRODUCTION

Reducing the costs of fuel cell components is a key issue in the commercialization and broad application of fuel cell technology. Particular focus is on the catalyst: Platinum is still the most effective material for both the fuel and air electrodes in PEMFC systems [1]. However Pt is one of the most expensive metals on earth and extensive use of the technology in the future will only aggravate the situation [2]. Thus, strategies of how to use the limited Pt resources more efficiently are currently being researched extensively. One option is to increase the surface area of the catalyst, since it is the surface, rather than the bulk atoms, which is catalytically active. Nanoparticles are thus a viable alternative to bulk metals, since they exhibit a large surface-to-volume ratio. This increases the fraction of (Pt) surface atoms and ensures a higher efficiency, relative to the amount of catalyst material used. Nevertheless, a significant amount of Pt still remains in the centre of the particle and does not take part in the catalytic process. A potential alternative is to employ the so-called "core-shell" nanoparticles, where the bulk is composed of a suitable inexpensive material (e.g. Co) [3] and only the surface is covered with Pt metal. Such catalysts would require a minimal amount of Pt, while still providing sufficiently high catalytic activity. The investigation of such core-shell nanoparticles, including their physical and chemical properties is thus of key interest to the fuel cell community.

2 EXPERIMENTAL

Platinum black nanoparticles (JOHNSON MATTHEY) were ultrasonicated in either isopropyl alcohol or a 1:1 mixed solution of isopropyl alcohol (C_3H_8O, AnalaR NORMAPUR) and butyl acetate ($C_6H_{12}O_2$, SIGMA-ALDRICH). Two solutions were employed in this study. The concentration of 1mg: 1ml was employed for the pure alcohol while a 1mg: 1ml was used for the mixture. The 1 x 1 x 0.2 cm of Highly Oriented Pyrolytic Graphite (Agar Scientific) was used as a substrate material. Graphite was cleaved before each experiment prior to the deposition of a drop of solution. Samples were left to dry for 2 hours before each characterization. Scanning electron microscopy (SEM) images were obtained with LEO 420 equipment. Finally, an Agilent Scanning Tunnelling Microscope (STM) was used in the constant current mode to investigate the particle dimensions and the aggregation state of the system (Figure 1). The tunnelling conditions include: the tunnelling current 0.200nA, the bias voltage 0.800V and the scan rate 814nm/s.

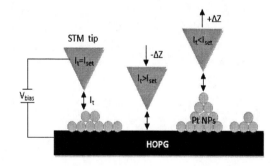

Figure 1: The constant-current mode in STM.

3 RESULTS

From Figure 2, a relatively even dispersion on the graphite surface is observed. It is possible to distinguish clusters from 100 nm to 5 microns in diameter. At higher magnification, smaller clusters can be seen although the resolution of the SEM microscope is still too low to detect individual particles. On the other hand, high resolution images, Figure 4, by STM show individual particles between the aggregates. A profile analysis of the features showed a distribution between 1.5 to 2.5 nm in height.

As the platinum concentration increased, the aggregation became more important regardless of the solvent employed. For instance, when the platinum concentration was doubled, the size of the aggregates increased up to 8 microns in diameter as seen in Figure 3. From this image, we can also notice a low surface coverage between the aggregates which complicate the imaging by STM.

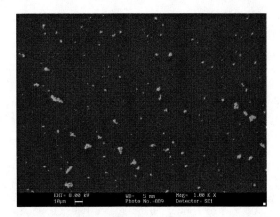

Figure 2: SEM image of Platinum black particles onto HOPG (Pt: solvent=1mg:1ml).

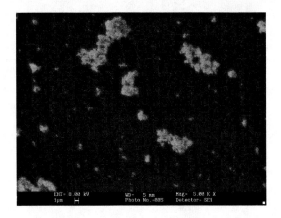

Figure 3: SEM image of Platinum black particles onto HOPG at higher concentration (Pt: solvent=2mg: 1ml).

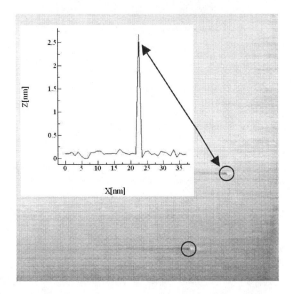

Figure 4: The platinum black particles (in red cycles) on HOPG as seen by STM (0.200nA, 0.800V, 814nm/s). The inset shows the profile of one Pt particle. Image size is 29 x 29 nm.

4 CONCLUSIONS

The results show that pure isopropyl alcohol provides better dispersion than the mixed solvent. On the other hand, lower concentrations improve the dispersion onto HOPG in comparison with higher concentrations.

Although these experiments are at an early stage, the current results demonstrate that because of the high hydrophobicity of HOPG, the method of direct dispersion onto a graphite surface is not very satisfying since platinum black particles tend to agglomerate during the solvent evaporation. Future work needs to be done in order to improve the dispersion ability of the solvent. This will reduce the mean diameter of the aggregates and hopefully increase efficiency of the catalyst.

5 FUTURE WORK

1. We will continue using the dispersion technique by using the drop technique with different solvents and mixtures. The different concentration will also be investigated.

2. We will investigate the particle deposition in situ by using the liquid cell of the STM. In this case organic and aqueous solutions will be employed.

3. The electrochemical properties of Pt black nanoparticles and Pt / Co core/shell nanoparticles will be investigated and compared. These include the in-situ electrochemical STM imaging and cyclic voltammetry.

4. Finally, fuel cell test will be performed to investigate the catalytic performance under or close to operating conditions.

REFERENCES

[1] A.J. Appleby, in: R. Dudley, W. O'Grady, S. Srinivasan (Eds.), *The Electrochemical Society Softbound Proceedings Series*, PV 79-2, Princeton, NJ, 1979.

[2] J. K. Nørskov, J. Rossmeisl, A. Logadottir, and L. Lindqvist. Origin of the Overpotential for Oxygen Reduction at a Fuel-Cell Cathode. *The Journal of Physical Chemistry B*, 108, 17886-17892, 2004.

[3] A. Sarkar, A. Vadivel Murugan, A. Manthiram. Pt-Encapsulated Pd−Co Nanoalloy Electrocatalysts for Oxygen Reduction Reaction in Fuel Cells. *Langmuir* Article ASAP, 2009, DOI: 10.1021/la902756j.

CHARACTERIZATION AND SYNTHESIS OF NANO MAGNETIC PARTICLES

Mrs.Meera Thiyagarajan Assistant Professor/Physics,
Vivekanandha Institute of Engineering and Technology ,Tiruchengode.
sn_meera@yahoo.com

ABSTRACT

Magnetic particlesfind wide applications and specifically memory device.The emerging new devices are due to large memory capacity of the nano magnetic particles.The super paramagnetism is the basic reason for the large storage of data.The synthesis of nano magnetic particles is by sol gel technique in the laboratory.The particle size distribuyion and the shape are determined by TEM ,the samples are characterized by XRD.The high resolving TEM,SEM,AFM shows the confirmation presence of nano magnetic particles.These nano magnetic particles are fit for magnetic storage of data.Our study shows the present memory devices can be replaced by these new storage devices using nano magneticmaterials.

Keywords .Data storage,super para magnetism

When the size of ferro/ferri magnetic materials is comparable or less than the single domain size, these particles essentially behave like paramagnetic materials with a notable exception of large susceptibility .The transform in the property of a materia from ferromagnetism to superparamagnetism. ■ The reason for this dissimilar behavior is enormous surface energy which originates due to the large fraction of exposed atoms at surface. ■ The surface energy is responsible for variation of property in the Nanoscale. The clear understanding of surface area to volume ratio is inevitably facilitate understanding surface energy.

■

Types of Magnetism
Dia magnetism
- Para maganetism
- Ferro magnetism
- Antiferro magnetism and
- Ferrimagnetism..

Super paramagnetism
A superparamagnet is an assembly of giant magnetic moment which are not interacting and the magnetic moment can fluctuate when the thermal energy, kBT, is larger than the anisotropy energy. superparamagnetic particles exhibit no remanence no coercivity ie., there is no

hysteresis in the magnetization

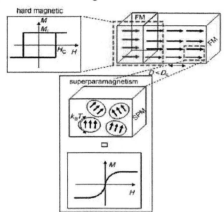

1 METHODOLOGY
FeCo−SiO2 nanocomposite films on silica glass substrates were prepared by the
sol−gel method and characterized by X-ray diffraction, transmission electron
microscopy, Rutherford backscattering spectrometry, extended X-ray absorption
fine structure spectroscopy, and magnetic susceptibility measurements. FeCo
alloy nanoparticles with average sizes around 10 nm were obtained which are

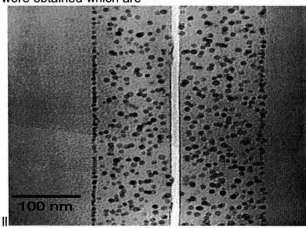

100 nm

dispersed in the silica matrix and show superparamagnetic behavior. The experimental conditions of the sol−gel preparation influence the thickness and homogeneity of the films. The magnetic properties are also affected by preparation conditions.

Magnetic anisotropy

- The term magnetic anisotropy is used to describe the dependence of the internal energy on the direction of the spontaneous magnetization, creating easy and hard directions of magnetization.

The total magnetization of a system will prefer to lie along the easy axis. Magnetocrystalline anisotropy ,Shape anisotropy,Strain anisotropy Surface anisotropy.

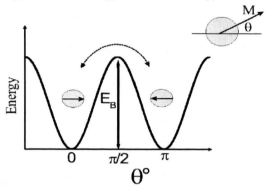

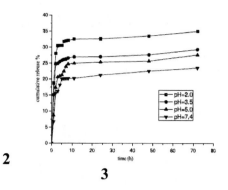

2

3

- ## Applications of Nano Magnetic Particles

Nanomagnetic materials find many applications out of which a few are listed here relevant to the current scenario.

Ultra High Density Data Storage

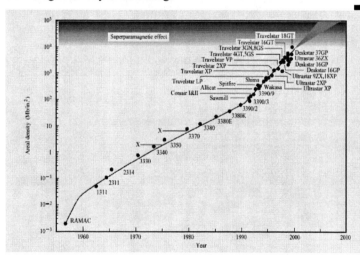

Ferro liquids

Drug delivery

Ferro Fluids

Dampers in stepper motors

Shock absorbers

Heat transfer in loud speakers

Almost every computer disk drive uses a magnetic fluid rotary seal for contaminant exclusion

DRUG DELIVERY SYSTEM.

REFERENCES

- **References**
- J. Frenkal and Dorfman, 1930, Spontaneous and Induced Magnetisation in Ferromagnetic Bodies, Nature, 3173, 126, 274-275.
- 1. a. Subankar Bedanta, 2006, PhD Thesis, Supermagnetism in magnetic nanoparticle systems, University of Duisburg Essen, Duisburg, Germany.
- Shanta R Bhattarai et al, 2008, N-hexanoyl chitosan stabilized magnetic nanoparticles: Implication for cellular labeling and magnetic resonance imaging, Journal of Nanobiotechnology, 6:1.
- 2.a. Won Seok Seo et al,2006,FeCo/graphitic-shell nanocrystals as advanced magnetic-resonance-imaging and near-infrared agents, Nature materials,5,971-976.
- Markus Zahn, 2001, Magnetic fluid and nanoparticle applications to nanotechnology, Journal of Nanoparticle Research 3:73-78.
- F H Chen et al, 2008, The grafting and release behavior of doxorubincin from Fe3O4@SiO2 core–shell structure nanoparticles via an acid cleaving amide bond: the potential for magnetic targeting drug delivery - Nanotechnology 19, 165103.
- Rudolf Hergt et al, 1998, Physical Limits of Hyperthermia Using Magnetite Fine Particles, IEEE Trans. on Magn, 34, 5, 3745-3754.
- Chul-Ho Jun et al, 2006, Demonstration of a magnetic and catalytic Co@Pt nanoparticle as a dual-function nanoplatform, Chem, Commun, 15, 1619-1621.
- A.D.McNaught and A.Wilkinson, 1997, IUPAC.Compendium of Chemical Terminology, 2nd ed.(the "Gold Book"), Blackwell Scientific Publications, Oxford.
- Guozhong Cao, 2007, Nanostructures & Nanomaterials Synthesis, Properties & Applications, Imperial College Press, London, UK.
- C.P. Bean and J.D Livingston, 1959, Superparamagnetism, J.Appl.Phys, 30, 4, 120S-129S.
- An-Hui Lu, E.L. Salabas and Ferdi Schuth, 2007, Magnetic Nanoparticles:Synthesis, Protection, Functionalization, and Application,Angew.Chem.Int.Ed,46,1222-1244.

Early Stages of Particle Formation by In-situ UV-Vis and HRS

B. Schürer,[1,2] D. Segets[1] and W. Peukert[1,2]

[1]Instiute of Particle Technology, University Erlangen-Nuremberg, Erlangen, Germany
[2]Cluster of Excellence - Engineering of Advanced Materials and Erlangen Graduate School in Advanced Optical Technologies (SAOT), University of Erlangen-Nuremberg, Erlangen, Germany

ABSTRACT

We present a new approach for the study of early states of particle formation that can be applied for precipitation processes that occur on timescales from milliseconds to several minutes. A T-shaped micro-mixer is used to precipitate nanoparticles at steady state conditions. The nucleation and growth of the nanoparticles occurs along the path through a transparent quartz capillary. Hyper Rayleigh scattering (HRS) and UV/Vis absorbance spectra are recorded simultaneously. By changing the measurement position along the capillary, different reaction stages of the nanoparticle formation process can be studied in detail. As the precipitation is performed at steady state the integration time for recording signals is not limited by the kinetics of the precipitation process. The HRS intensities and the particle size information can be used to calculate the nucleation and growth rates of the particle formation process.

Keywords: UV-Vis, Hyper Rayleigh Scattering, ZnO, nanoparticle, precipitation

1 INTRODUCTION

The early stages of nanoparticle precipitation are crucial for the morphology and size development as well as for the properties of the final nanoparticles. However, the precipitation process from cluster formation to nanoparticle growth and ripening is not well understood. So far primarily Small Angle X-Ray Scattering (SAXS) and X-ray Absorption Spectroscopy (XAS) have been used to monitor cluster formation and early stages of precipitation [1–3].

In the recent years optical Second Harmonic Generation (SHG) has proven its high potential to probe colloidal nano- and microparticles in-situ with millisecond time resolution. In literature SHG from nanoscaled structures [4] and nanoparticles is also often referred to as Hyper Rayleigh scattering which is typically used for the study of solved molecules. It has been shown that the HRS signal from nanoparticles is highly sensitive to their size, structure and shape [5]. The Second Harmonic (SH) scattered light is shifted to half of the excitation wavelength. Therefore it can easily be separated from the fundamental excitation light by means of filters and monochromators. With its high time

resolution HRS has the potential for online monitoring of dynamic processes such as growth, ripening and agglomeration of nanoparticles [5,6]. For materials with a high nonlinear response even the formation of molecular clusters can be detected [7].

Recently we have applied HRS to monitor different stages in the synthesis of ZnO nanoparticles. For the ripening process of ZnO nanocrystals the size dependence of HRS was determined quantitatively by a combination of HRS and spectrophotometry. Time-resolved measurements allowed the calculation of nucleation, growth, and ripening rates out of the nonlinear signals [6].

In this contribution we present a new approach for the study of early states of particle formation that can be applied also for precipitation processes that occur on a very short timescale and for nanoparticles that exhibit only a weak nonlinear signal.

2 EXPERIMENTAL SECTION

Ethanol was used for the preparation of all reactant solutions. Before the synthesis, a 0.1 M zinc acetate dihydrate precursor stock solution was prepared and mixed with an equimolar amount of lithium hydroxide at 293 K resulting in an instantaneous particle formation.

The absorbance spectra were measured with a fiber coupled UV-Visible spectrophotometer (Ocean Optics, USB2000+ with DH2000 as light source).

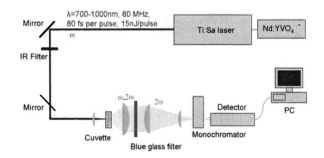

Figure 1: Transmission setup for Hyper Rayleigh scattering experiments.

For monitoring the nanoparticle formation a transmission setup as it is shown in Fig. 1 is used. For excitation of the Hyper Rayleigh scattering signal a pulsed Ti:Sa femtosecond laser (80 MHz) with a pulse length of about 80 fs and a wavelength of 800 nm is used. For the generation of the HRS signal the beam is focused into a capillary that contains the sample. With a second lens after the probe the scattered light is collected. A blue color filter that transmits only light shorter than 650 nm is used to block the transmitted and linear scattered laser light. The filtered light is then focused into the slit of a monochromator for a further separation of the HRS signal from other background radiation as two photon fluorescence. The HRS light is detected by a photomultiplier tube and recorded by a computer. The precipitation of nanoparticles is either done in the capillary of a flow cell setup as it is shown in Fig. 2.

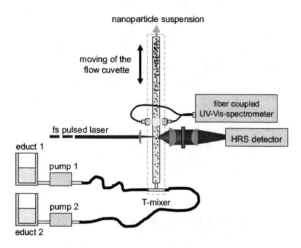

Figure 2: Schematic setup for simultaneous time resolved UV-Vis and Hyper Rayleigh scattering experiments at steady state condition.

A T-shaped micro-mixer is used accomplish a fast mixing of the reactants and to precipitate the nanoparticles at steady state conditions. After the mixing zone the cross sectional area of the capillary is increased and a laminar flow through is formed. The synthesis of nanoparticles occurs along the flow through the capillary. The precipitation setup is mounted on a moveable stage. For monitoring the growth of the nanoparticles HRS and UV/Vis absorbance spectra are recorded simultaneously. By changing the measurement position along the capillary, different reaction stages can be studied in detail.

3 RESULTS AND DISCUSSION

The precipitation of ZnO particles was carried out for different reactant concentrations and at different flow rates. HRS signals and absorbance spectra were measured simultaneously at different positions along the capillary. The measurement position was converted into an average reaction time using the flow rate and the cross section of the capillary. In Fig. 3 the smoothed HRS signal intensities

along the capillary for a flow rate of 5 ml/min and a Zn^{2+} concentration of 0.043 mol/l are shown.

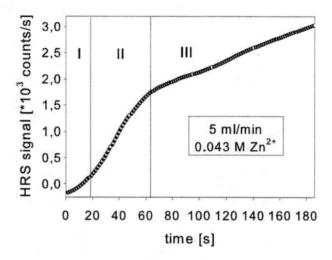

Figure 3: Smoothed HRS signal intensity for the early stages of particle formation.

Three stages can be distinguished. After the initial stage I there is a steady increase of the HRS signal intensity (stage II). In the third stage the slope of the curve changes and a linear increase of HRS signal intensity can be observed. We attribute the first stage to an initial stage of the reaction. Here a supersaturation is build up and primary nuclei are formed. After the build up of the supersaturation, a strong increase of the HRS signal is found in the second stage. The rise in the HRS signal is attributed to an increase of the number of nuclei (nucleation step) and due to the growth of the initial particles. After the formation of primary particles we assume that the number concentration does not change significantly and that the primary particles are growing. This corresponds to the third stage, which is observed after about 60 s (compare Fig. 3).

Absorbance spectra are simultaneously recorded at a distance of 2 cm after the HRS detection zone. From the absorbance spectra, ZnO particle size distributions are calculated by the algorithm that is explained in detail in reference [8]. The respective absorbance spectra are shown in Figure 4.

In the first 10 s of the precipitation, the absorption spectroscopy peak is not clearly formed and only a small and broad peak is observed. A first clear peak forms after 34 s. The absorbance peak increases with reaction time and shifts to longer wavelengths. The absolute value of the absorbance peak also increases and reaches the detection limit of the spectrophotometer after about 40 s.

NSTI-Nanotech 2010, www.nsti.org, ISBN 978-1-4398-3401-5 Vol. 1, 2010

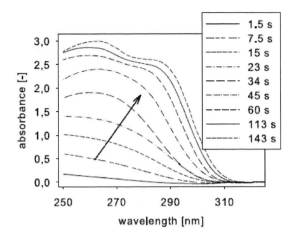

Figure 4: Absorbance spectra from ZnO nanoparticles along the axis of the glass capillary; the reactions times are calculated from the measurement position and the flow rate.

However, for the particle size correlation, the onset of the absorbance peak is most important, thus enabling the processing of the absorbance spectra even at higher levels of the peak absorbance. The calculated particle size distributions for different measurement times are shown in Fig. 5.

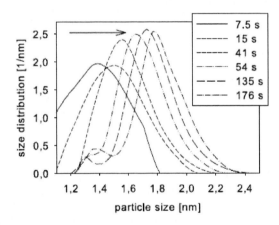

Figure 5: Volume density distribution of the ZnO nanoparticles at different reaction times calculated from the corresponding absorbance spectra.

Furthermore, the dependence of the ZnO hyperpolarizability on the particle size was determined using different measurement positions along the pipe. From the hyperpolarizabilites, the number density of the ZnO particles was calculated for different reaction times. During phase II, a strong increase in the particle number density was found which then proceeds to an almost constant particle number density in stage III. This result indicates that the ZnO precipitation process can indeed be divided into a nucleation and a particle growth stage. The particle

size distributions and the number densities are further on used to determine the nucleation and growth rates of the ZnO precipitation process.

4 CONCLUSION

We have precipitated ZnO nanoparticles using a steady state flow cell setup. With this approach the measurement time is not limited by the reaction kinetics of the precipitation process and the signal-to-noise ratio can be improved significantly. This enables the study of particles that exhibit only a weak nonlinear response. For the precipitation of ZnO particles we were able to characterize particles as small as 1.6 nm. The mean particle sizes and size distributions were calculated from the absorbance spectra. The combination of absorbance and HRS experiments can be used for the quantitative determination of nucleation and growth rates.

ACKNOWLEDGMENT

The authors gratefully acknowledge the funding of the German Research Council (DFG), which, within the framework of its `Excellence Initiative' supports the Cluster of Excellence `Engineering of Advanced Materials' (www.eam.uni-erlangen.de) and the Erlangen Graduate School in Advanced Optical Technologies (SAOT) at the University of Erlangen-Nuremberg.

REFERENCES

[1] F. Meneau, G. Sankar, N. Morgante, R. Winter, C. R. A. Catlow, G. N. Greaves, J. M. Thomas, Faraday Discuss. 122, 203, 2002.

[2] J. Rockenberger, L. Träger, A. Kornowski, T. Vossmeyer, A. Eychmüller, J. Feldhaus, H. Weller, J. Phys. Chem. B 101, 2691, 1997.

[3] P.-P.E.A. de Moor, T.P.M. Beelen, R.A. van Santen, Micropor. Mat. 9, 117, 1997.

[4] L. Martinez Tomalino, A. Voronov, A. Kohut, W. Peukert, J. Phys. Chem. B 112, 6338, 2008.

[5] K. Das, A. Uppal, P.K. Gupta, Chemical Physics Letters 426, 155, 2006.

[6] D. Segets, L. Martinez Tomalino, J. Gradl, W. Peukert, J. Phys. Chem. C 113, 11995, 2009.

[7] M.C. D'Arrigo, Cruickshank F. R., Pugh D., Sherwood J. N., Wallis J. D., Mackenzie C., Hayward D., Phys. Chem. Chem. Phys. 8, 3761, 2006.

[8] D. Segets, J. Gradl, R.K. Taylor, V. Vassilev, W. Peukert, ACS Nano 3, 1703, 2009.

Green Chemistry Used in The Synthesis and Characterization of Silver and Gold Nanoparticles

Ramón Vallejo[*], Paulina Segovia[*a, **], Nora Elizondo[*a], Odilón Vázquez[*b], Víctor Coello[**], Víctor Castaño[***]

[*a] Facultad de Ciencias Físico-Matemáticas, [*b] Instituto de Ingeniería Civil, Universidad Autónoma de Nuevo León, San Nicolás de los Garza, N. L., CP 66451, México, Tel.: (52-81) 83294030, Fax.: (52-81) 83522954, nelizond@yahoo.com

[**] CICESE, Monterrey, Km 9.5 Nva. Carretera Aeropuerto, PIIT, Apodaca, N. L., México, C.P. 66600, vcoello@cicese.mx

[***] Centro de Física Aplicada y Tecnología Avanzada, Universidad Nacional Autónoma de México, Boulevard Juriquilla 3001, Querétaro 76230, México, castano@fata.unam.mx

ABSTRACT

Gold (Au) and silver (Ag) nanoparticles have a diversity of interesting properties between which they emphasize the electrical ones, optical, catalytic and the applications in biomedicine like antibacterial and antiviral, same that depend on their morphology and size. The nanoparticles were synthesized using polyol and green methods. We made a comparison of these methods in order to investigate the influence of reaction parameters on the resulting particle size and its distribution. In the first method we use polyol process with poly (vinylpyrrolidone) (PVP) acting as a stabilizer and ethylenglycol as a reductor. The green method is an ecological synthesis technique. In this article, we made use of chemical compounds of plants in order to obtain ascorbic acid as reductor agent and saponins with surfactant properties. Here, we show that green method allows synthesize metallic nanoparticles and reduces the temperature requirement which is in contrast to the obtained with the polyol method.

Keywords: characterization, synthesis, metallic nanoparticles, green chemistry, polyol method

INTRODUCTION

Metallic nanoparticles are of great interest because of the modification of properties observed due to size effects, modifying the catalytic, electronic, and optical properties of the monometallic nanoparticles.[1-3] For this purpose, many colloidal methods of synthesis have been approached to obtain metallic nanoparticles, such as homogeneous reduction in aqueous solutions,[4] or phase transfer reactions,[5] with sodium citrate, hydrazine, $NaBH_4$, and lithium trietilborohydride ($LiBEt_3H$) as reducing agents, each of them yielding products with different physicochemical and structural characteristics.[6] Among these, the polyol method has been reported to produce small nanoparticles as the final product, easily changing composition and surface modifiers. This technique does not require an additional reducing agent since the solvent by itself reduces the metallic species. However, besides the stoichiometry and order of addition of reagents in the synthesis process, one of the most important parameters in the preparation is the temperature. Modifications in temperature influence the reaction by changing the stabilization of the nanoparticles formed and the surface modifiers, e.g., PVP, and the nucleation rate of the reduced metallic atoms.[7]

Gold (Au) and silver (Ag) nanoparticles have a diversity of interesting properties between which they emphasize the electrical ones, optical, catalytic and the applications in biomedicine like antibacterial and antiviral, same that depend on their morphology and size.

Characterization of these systems has been a difficult process where researchers have employed indirect measurements to identify the localization of the elements within the nanoparticles. A novel approach to study this kind of particles is based on the use of a high angle annular dark field (HAADF) technique, in a transmission electron microscope (TEM), which allows the observation of the elements due to atomic number, densities, or the presence of strain fields due to differences in lattice parameters. structure, the presence of surfactants or any other surface modifier besides the size of the particle and also by near-field scanning optical microscopy (NSOM) we determine the size of the particles.[8, 10]

The nanoparticles were synthesized using polyol and green methods. We made a comparison of these methods in order to investigate the influence of reaction parameters on the resulting particle size and its distribution. In the first method we use polyol process with poly (vinylpyrrolidone) (PVP) acting as a stabilizer and ethylenglycol as a reductor [11, 12]. Such procedure yield different morphologies of metal nanoparticles (including gold and silver) [13-15]. The green method is an ecological synthesis technique. There, we made use of chemical compounds of plants like rosa berberifolia and geranium manculatum in order to obtain ascorbic acid as reductor agent. Ascorbic acid ($C_6H_8O_6$) is an abundant component of plants which reaches a concentration of over 20 milimols in chloroplasts and occurs in all cell compartments including the cell wall. Additionally the acid has functions in photosynthesis as an enzyme cofactor (including synthesis of ethylene, gibberellins and anthocyanins) and in the control of cell growth [16]. We use for the synthesizes other kinds of plants also with compounds that have surfactant properties like saponins. Here, we show that green method reduces the temperature requirement which is in contrast to the obtained with the polyol method. The use of these natural components allows synthesize gold and silver nanoparticles.

EXPERIMENTAL SECTION

The polyol method was followed to obtain nanoparticles passivated with poly(vinylpyrrolidone) (PVP). Hydrogen tetrachloroaurate ($HAuCl_4$) (III) hydrate (99.99%), silver nitrate($AgNO_3$) (99.99%), and poly (N-vinyl-2-pyrrolidone) (PVP-K30, MW = 40000) were purchased from Sigma Aldrich, and 1,2-ethylenediol (99.95%) was purchased from Fischer Chemicals; all the materials were used without any further treatment. A 0.4 g sample of Poly (N-vinyl-2-pyrrolidone) (PVP) was dissolved in 50 mL of 1,2-ethylenediol (EG) under vigorous stirring, heating in reflux, until the desired temperature was reached (working temperatures ranged from 140 to 190 $^{\circ}$C in increments of 10 $^{\circ}$C). For the monometallic nanoparticles, a 0.1 mM aqueous solution of the metal precursor was added to the EG-PVP solution, with continuous agitation for 3 h in reflux.

The green method is an ecological synthesis technique. There, we made use of chemical compounds of plants like rosa berberifolia and geranium manculatum in order to obtain ascorbic acid as reductor agent from the extracts of these plants. Ascorbic acid ($C_6H_8O_6$) is an abundant component of plants, which reaches a concentration of over 20 milimols in chloroplasts and occurs in all cell compartments including the cell wall. We use for the synthesizes also cactus extracts with compounds that have surfactant properties like saponins.

The extracts were prepared as of 1 to 40 grams of the mentioned plants. Then from 10 to 50 milliliters of the extracts of these plants respectively were dissolved in water or in ethanol under vigorous stirring, heating in reflux, until the desired temperature was reached, when were drastical changes of the color of the solutions from yellow to dark brown in the case of silver nanoparticles and in the case of gold nanoparticles synthesizes the color of the solutions changed from pink to brown. For the gold and silver nanoparticles, a 0.1 mM aqueous solution of the metal precursor was added to the solutions with extracts, with continuous agitation for 30 minutes to 1 h in reflux in a working temperatures range from 60 to 100 Celsius.

The synthesis of colloidal metallic nanoparticles was carried out taking into account the optimization of the conditions of nucleation and growth. For this reason the variation of parameters like the concentration of the metallic precursors, reductor agent, amount of stabilizer, temperature and time of synthesis were realized.

For the electron microscopy analysis of the metallic nanoparticles, samples were prepared over carbon coated copper TEM grids. HAADF images were taken with a JEOL 2010F microscope in the STEM mode, with the use of a HAADF detector with collection angles from 50 mrad to 110 mrad. also by near-field scanning optical microscopy (NSOM) we determine the size of the particles. UV-vis spectra were obtained using a 10 mm path length quartz cuvette in a Cary 5000 equipment.

RESULTS AND DISCUSSION

The use of these natural components allows synthesize metallic nanoparticles. In the green method, gold and silver nanoparticles were prepared by the same reduction of $HAuCl_4$ and $AgNO_3$ respectively using extracts of plants, ascorbic acid as reducing agent obtained from geranium manculatum leaves, rosa berberifolia petals and like surfactants and simultaneous reducing agents in some cases were used cactus extracts.

The reaction time for the polyol method was around 3 hours and the nanoparticles were synthesized between 140 and 190 °C. In the green method the reaction time is reduced from 3 hours to 30 minutes until 1 hour at 60°C.

In accordance with the studies of UV visible spectroscopy, whose plasmons are in figure 1 for Au and Ag synthesized nanoparticles the results shown an absorption energy in 547 nm and 415 nm respectively.
The TEM characterization reveal the formation of nanoparticles of these metals, independent of the employed method, with a size distribution between 20 and 120 nm for gold (see figure 2) and between 10 and 27 nm for the silver. The NSOM showed that the size of gold nanoparticles syntheised was of 15 nm with very narrow distribution.

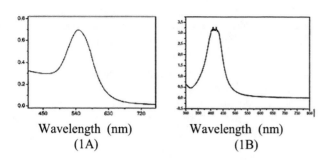

Wavelength (nm)
(1A)

Wavelength (nm)
(1B)

Figure 1. UV-Visible absorption spectrum of the Au (1A) and Ag (1B) nanoparticles synthesized by polyol and green chemistry respectively.Formatting dimensions for manuscripts.

Figure 2. TEM images of Au nanoparticles synthesized by polyol method.

CONCLUSION

In this original work, we show that green method reduces the temperature requirement, which is in contrast to the obtained with the polyol method. The use of these natural components allows synthesize metallic nanoparticles with very narrow distribution.

ACKNOWLEDGMENT

Authors would like to acknowledge to Facultad de Ciencias Físico Matemáticas de la Universidad Autónoma de Nuevo León, México.

REFERENCES

[1]. Thomas, J. M.; Raja, R.; Johnson, B. F. G.; Hermans, S.; Jones, M. D.; Khimyak, T. *Ind. Eng. Chem. Res.* **2003**, *42*, 1563.

[2]. Bronstein, L. M.; Chernyshov, D. M.; Volkov, I. O.; Ezernitskaya, M. G.; Valetsky, P. M.; Matveeva, V. G.; Sulman, E. M. *J. Catal.* **2000**, *196*, 302.

[3]. Chushak; Y. G.; Bartell, L. S. *J. Phys. Chem B* **2003**, *107*, 3747.

[4]. Liz-Marzan, L. M.; Philipse, A. P. *J. Phys. Chem.* **1995**, *99*, 15120.

[5]. Han, S. W.; Kim, Y.; Kim, K. *J. Colloid Interface Sci.* **1998**, *208*, 272.

[6]. Schmid, G. *Clusters and Colloids, From theory to Applications*; VCH Publishers: Weinheim, Germany, 1994.

[7]. Turkevich, J.; Stevenson, P.; Hillier, J. *Discuss. Faraday Soc.* **1951**, *11*, 55.

[8]. Alvarez, M. M.; Khoury, J. T.; Schaaff, G.; Shafigullin, M. N.; Vezmar, I.; Whetten, R. L. *J. Phys. Chem. B* **1997**, *101*, 3706.

[9]. Henglein, A. *J. Phys. Chem. B* **2000**, *104*, 2201.

[10]. H. K. Park, Y. T. Lim, J. K. Kim, H. G. Park and B. H. Chung, Ultramicroscopy, **2008**, *108,10*, 1115.

[11].G. Cao, Nanostructures and Nanomaterials, synthesis, properties and applications, Imperial College Press (2004).

[12]. Y. Xia, P. Yang, Y. Sun, Y. Wu, B. Mayers, B. Gates, Y. Yin, F. Kim, H.Yan, Adv. Mater., 15, 353 (2003).

[13]. C. Burda, X. Chen, R. Narayanan and M. A. El-Sayed, Chem. Rev., 2005, 105, 1025.

[14]Y. Xia and N. J. Halas, Mater. Res. Soc. Bull., 2005, 30, 338.

[15]. N. L. Rosi and C. A. Mirkin, Chem. Rev., 2005, 105, 1547.

[16]. Nicholas Smirnoff, Glen L. Wheeler , Critical Reviews in Biochemistry and Molecular Biology, 2000, Vol 35, No. 4, Pages 291-314.

Polarity sensitive colloidal gold nanoparticles

Faheem Amin[1], Dmytro Yushchenko[2], Feng Zhang[1], Zulqurnain Ali[1], Wolfgang J. Parak[1, 3]*

[1] Fachbereich Physik, Philipps Universität Marburg, Renthof 7, Marburg 35032,Germany.
[2] Max Planck Institute for biophysical Chemistry, Emeritus Group Laboratory of Cellular Dynamics, Am Faßberg 11, D-37077 Göttingen
[3] Wissenschaftliches Zentrum für Materialwissenschaften (WZMW), Philipps Universität Marburg, Marburg, Germany

* Corresponding author: wolfgang.parak@physik.uni-marburg.de

ABSTRACT

Water soluble colloidal nanoparticles possess huge surface negative charge and this provides their colloidal stability [1]. The surface charge, therefore, ultimately demonstrate the overall behavior of the nanoparticles irrespective of core material. We demonstrate fluorescence based dual emission polarity sensor that is based upon hydroxyflavone derivative [2]. The NH_2 group at position 7 facilitates the conjugation to the carboxylic acid groups on the particle surface. Two important parameters can be addressed in parallel by using the sensor i) polarity and ii) Hydration [3]. The fluorescence read-out is either single peak (hydration, due to interaction of water molecules to the ketone group) or dual peak emission (polarity of environment). The influence of external factors such as alcohols, polyamines, detergents etc. has been investigated.

Keywords Nanoparticles, Polarity, Sensor, Fluorescence

EXPERIMENTAL

Dodecanethiol stabilized Au nanoparticles were synthesized by conventional two phase method with slight modification (see figure 1) [4]. Amphiphilic polymer is constituted of backbone (Isobutylene-alt-maleic anhydride, average M_w ~6,000 g/mol, Sigma, #531278) (see figure 2a); and sidechain (dodecylamine powder, ≥98%, Fluka, # 44170) (see figure 2b) [5]. The two constituents are mixed in molar concentrations 20mmol and 15mmol respectively for backbone and sidechain (see figure 2c). The 25% free rings can be utilized later for further funtionalization. The polymer was modified with 2% FE-dye molecules (see figure 3) and then was wrapped around the particle surface to make them soluble in water [1]. The wrapping phenomenon is self triggered and is really efficient. The solvent (chloroform) was then evaporated completely and the particles were redissolved in buffer SBB 12. The anhydride rings open and we get negatively charged carboxylic groups on the surface. The highly negative charged surface ensures the colloidal stability of particles in buffer due to steric hinderance.

FIGURES AND CAPTIONS

Figure 1: Dodecanethiol stabilized Au nanoparticles.

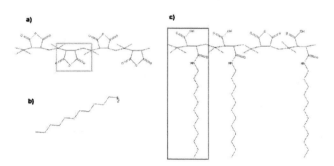

Figure 2: Amphiphilic polymer a) hydrophilic backbone consisting anhydride rings (open upon exposure to buffer) b) hydrophobic sidechains (ensure intercalation with dodecanethiol on the paticle surface) and c) polymer with 75% anhydride rings modified with side chains.

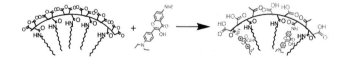

Figure 3: Amphiphilic polymer modified with 2% FE dye. The remaining rings open once dissolved in water.

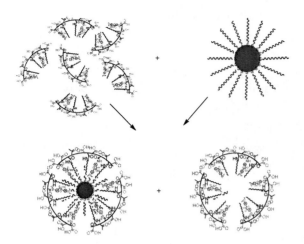

Figure 4: Polymer coating of Au nanoparticles with FE-modified amphiphilic polymer. The free polymer forms empty micelles and can be separated through either GEL electrophoresis or HPLC chromatography.

Au nanoparticles with FE dye exposed to the surrounding media via polyethylene glycol (PEG spacers) were prepared. The FE-PEG conjugates were prepared and and purified through phase separation techniques. The Boc-group on the PEG was then deprotected through trifluoroacetic acid. The conjugation of modified PEG was mediated through coupling agent (1-ethyl-3-(3-dimethylaminopropyl) carbodiimide hydrochloride, Sigma-Aldrich # E7750). The ratio of EDC was kept to 32000/NP, based upon previous findings [6] whereas for PEG the ratio was kept to 500/NP.

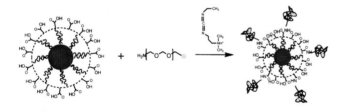

Figure 5: functionalization of plain polymer coated Au nanoparticles with FE-modified PEG spacers via EDC chemistry to keep the dye exposed to medium (Buffer).

RESULTS

Fluorescence measurements of particles with dye embedded inside polymer shell and dye located outside polymer shell has been done. It was found that the dye embedded inside polymer shell can sense the interactions on the particle surface whereas the dye located in the media is unable to sense the changes occurring on the particle surface due to influence of certain stimuli (alcohol, plyamine, detergent). It was found that the dye is sensitive to environmental changes only if it located inside polymer shell. For dye

located outside shell the readout is single emission peak due to hydration effect.

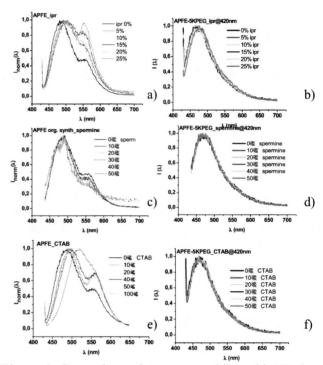

Figure 6: Comparison of Au nanoparticles with FE dye embedded in polymer shell [a), c) and e)] and FE dye exposed to buffer [b), d) and f)] under the influence of isopropanol (a, b), spermine (c, d) and CTAB (e, f).

DISCUSSION

FE incorporated Au nanoparticles can be used to sense the polarity changes in the environment. The effect of different factors has been studied successfully. The effort to locate dye outside polymer shell keeping the sensitivity of fluorophore intact failed. The dehydration effect is quite dominant on the dye molecule once in water. PEG spacers hinder the approach of small molecules to the surface of particles. This was further verified by saturating FE-incorporated particles with PEG spacers of different legths. It was found that longer the PEG spacer is better is the hindrance.

FUTURE PERSPECTIVES

FE dye with ketone group protected with a benzyl group can resolve the hydration issue. Although hydration is an important parameter in the biological membrane still it is more worthy to know the polarity of the environment. Since the trafficking of micro- as well as macro-molecules is polarity dependent. Several studies has already been devoted but no published report is available with particles.

REFERENCES

1) T. Pellegrino et al., Hydrophobic Nanocrystals Coated with an Amphiphilic Polymer Shell: A General Route to Water Soluble Nanocrystals, *Nano Lett.*, 2004, 4 (4), pp 703-707.

2) A. S. Klymchenko and A. P. Demchenko, Multiparametric probing of intermolecular interactions with fluorescent dye exhibiting excited state intramolecular proton transfer, *Phys. Chem. Chem. Phys.*, 2003, 5, 461-468.

3) G. Duportail et al., Neutral fluorescence probe with strong ratiometric response to surface charge of phospholipid membranes, *FEBS* Letters 508 (2001) 196-200.

4) M. Brust et al., Synthesis of Thiol-derivatised Gold Nanoparticles in a Two-phase Liquid-Liquid System, *J. Chem. Soc., Chem. Commun.*, 1994, pp 801.

5) Cheng-An J. Lin et al., Design of an Amphiphilic Polymer for Nanoparticle Coating and Functionalization, *Small* , 2008, 4 (3), pp 334-341.

6) F. Zhang et al., Ion and pH Sensing with Colloidal Nanoparticles: Influence of Surface Charge on Sensing and Colloidal Properties, *ChemPhysChem*, 2010, 11 (3), pp 730-735.

The Reaction of Nano-scale Iron Fluidized in Microwave Radiation Field to Treat Chlorobenzene

C-L Lee[1], Y-S Wang[2], C-R Wu[2], H-P Wang[3], C.-J.G. Jou[2*]

[1]*Department of Leisure and Recreation Management, Diwan University, Taiwan*
[2]Department of Safety, Health and Environmental Engineering,
National Kaohsiung First University of Science and Technology, Taiwan.
[3] Department of Environmental Engineering, National Cheng Kung University, Taiwan.

*Corresponding Author: Chih-Ju G. Jou, Ph.D.
E-mail address: george@ccms.nkfust.edu.tw
Tel.: 886-7-601-1000 Ext. 2316; Fax: 866-7-601-1061

Abstract- Microwave (MW) is applied to enhance chlorobenzene (CB) removal using micron iron (μ-Fe) particles, nanoscale iron particles freshly prepared in our lab (nP-Fe^0), and commercial nanoscale iron particles (nC-Fe^0) as the dielectric media. The results show that when the CB solution is irradiated with 250 W MW for 150 sec, better CB removal can be achieved than without MW irradiation. The MW radiation increases the iron oxidization rate, surface activity, and hence the CB removal rate. The MW-induced iron particles are capable of removing 13.6 times (61.2% vs. 4.5% for μ-Fe), 2.8 times (76.5% vs. 27.5% for nP-Fe^0) and 3.6 times (65.4% vs. 18.1% for nC-Fe^0) more CB, and the CB activation energy is decreased 16.8 kJ/mol (μ-Fe), 3.2 kJ/mol (nP -Fe^0) and 3.5 kJ/mol (nC-Fe^0). Using the microwave induced Nano-scale Iron Fluidized is effective in decomposing toxic organic substances as demonstrated in this laboratory study.

Keywords: Microwave-Induced; Zero valent Iron; Chlorobenzene (CB); Activation energy (Ea)

1. INTRODUCTION

Nano-scale zero-valent iron (ZVI) particles are effective in decomposing chlorine-containing organic compounds found in wastewater or groundwater [1, 2]. These organic compounds include pentachlorophenol, azo-dyestuff [3, 4], pesticides [5], polychlorinated biphenyls [6], herbicides [7] and aromatic nitric compounds [8]. Nano-particles have higher Miller Index or more stepped surface [9], and higher surface energies than micro-scale particles [10]. The factors that affect the capacity of ZVI particles to decompose organic compounds include the chemical properties, and surface activity of ZVI particles [11, 12].

Microwave (MW) is an electro-magnetic radiation with frequencies ranging from 300 MHz to 300 GHz [13]. The MW absorbed by a solution will cause the polar molecules in the solution to rotate rapidly leading to obvious thermal effect that will reduce the activation energy of the organic chemicals dissolved in the solution and weaken their chemical bonds [14]. Selecting an appropriate medium to absorb the MW energy is important to integrating the MW technology. For example, the MW radiation may be combined with granular activated carbon or ZVI for treating pentachlorophenol [15, 16] to improve the efficiency of TiO2 photocatalyst [17]. However, little information is available in literature on the microwave irradiation technique integrated with nano-scale ZVI particles to accelerate the dechlorination of chlorinated organic compounds in the liquid system.

The objectives of this study are to: (1) compare the efficiencies of different type of micron iron (μ-Fe), nanoscale iron particles freshly prepared in our lab (nP-Fe0), and commercial nanoscale iron particles (nC-Fe0) to treat aqueous chlorobenzene solution with and without MW irradiation; and (2) study the effect of particle size and surface activity on the removal of chlorobenzene dissolved in aqueous solution in the presence of MW.

2. MATERIALS AND METHODS

2.1 MATERIALS

The working solution containing 100-mg/L chlorobenzene was prepared by dissolving 905 μL of 99.9% pure chlorobenzene (GR Reagent, TEDIA, USA) in 99.9% methanol (GR Reagent, TEDIA, USA) to form a 2000-mg/L stock solution. Two hundred micro liters (μL) of the stock solution was then diluted with de-ionized water (18.2 MΩ, Millipore Co., USA) to form the 100-mg/L chlorobenzene working solution. Micro Fe was purchased from Riedel-deHaën (99.9%, < 212 μm) and Commercial Fe0 nanoparticles were obtained from Conyuan Biochemical Technology Ltd. Company (99.9%, < 60 nm).

2.2 METHODS AND ANALYSIS

A modified household microwave oven operated at 2.45 GHz with a max power of 650 W was used for generating the microwave energy. It was slightly modified by installing a programmable PID (Proportional-Integral-Derivative) controller. The MW energy was controlled by programming the PID controller for maintaining a constant pre-selected MW energy level. The sample was held in 40 mL low-energy-loss boron-silica serum bottles with Teflon coated screwed caps for carrying out the experiment.

The study was initiated by placing 40 mL of 100 mg/L CB solution in the bottle. After adding 1 g of μ-Fe, nP-Fe0 or

NSTI-Nanotech 2010, www.nsti.org, ISBN 978-1-4398-3401-5 Vol. 1, 2010

nC-Fe0, the sample was stirred in a constant-temperature shaker at 100 rpm to complete the CB decomposition study at various constant temperatures, i.e. 25, 40, 50 and 60 oC for different reaction periods, i.e. 30, 60 ... and 240 min. If the microwave energy was used, it was set at 250 W to irradiate the prepared samples according to the following operating conditions: MW irradiation time = 10 sec, MW interruption irradiation time = 120 sec, total irradiation time = 150 sec, number of cycle = 15.

Qualitative analyses was done with a HP 6890 gas chromatography (GC) coupled with an HP 5973 mass selective detector (MSD). An HP capillary column (HP-5MS, 30 m × 0.25 mm × 0.25 μm) was used for the identification of intermediates and degradation products. The carrier gas (He) flow rate was kept at 1.5 ml/min (constant flow). The oven temperature was programmed from 35 $^{\circ}$C (maintain for 1 min) to 130 $^{\circ}$C at a ramp rate of 5 $^{\circ}$C /min. IR (Raytek, Raynger 3i) was used to measure temperature variations of the media during the reaction process.

3. RESULTS AND DISCUSSION

3.1 EFFECT OF SURFACE AREA ON CB DEGRADATION

The degradation of chlorinated organic solvents by ZVI is a surface-mediated reaction; there is enough evidence to suggest that adsorption of chlorinated organic compounds at the Fe0 surface indeed occurs [18]. The rate of degradation of chlorinated solvents by Fe0 is a pseudo-first-order reaction with respect to the contaminant residual concentration [2, 12].
Fig. 1 indicates that at room temperature (25 oC), the chlorobenzene removal efficiencies after 240 min are 4.5% for μ-Fe, 27.5% for nP-Fe0 and 18.1% for nC-Fe0 particles with reaction rate constants of 2.0×10^{-4}, 1.3×10^{-3} and 9.0×10^{-4} min-1, respectively.

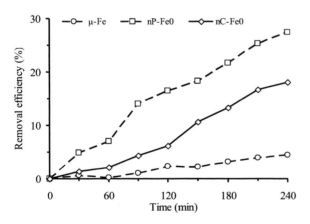

Fig. 1. CB removal efficiency for μ-Fe, nP-Fe0, and nC-Fe0 (at 25 oC, iron/liquid rate was 1 g/50 mL and CB concentration 100 mg/L).

3.2 EFFECT OF MW RADIATION ON CB DEGRADATION

When microwave irradiation is combined with some suitable microwave absorbents as the catalysts in the integrated system, the microwave can weaken the chemical bond intensities of various molecules, and reduce the activation energy of the reaction system. Commercial nano iron particles have smaller surface area to absorb more MW than the nano particle particles prepared in the laboratory. However, the former tend to agglomerate in the solution and thus absorb less MW energy to result in less efficiency of CB decomposition. In contrast, the prepared nano iron particles are used directly in wet form; they are thus evenly suspended in the CB solution. This leads to larger surface areas to absorb more MW energy than the particles that have settled to the bottom as in the case of the commercial nano-scale iron particles. Hence, the un-dried, laboratory prepared nano-scale iron particles (nP-Fe0) exhibit better CB removal efficiency than either commercial nano-scale iron particles (nC-Fe0) or micro-scale iron particles (μ-Fe0). Fig. 2 shows that applying 250 W MW for 150 sec will remove 76.5% CB for nP-Fe0, 65.4% CB for nC-Fe0, and 61.2% CB for μ-Fe0. The activated energy levels are 32.7 kJ/mol for μ-Fe, 18.6 kJ/mol for nP-Fe0, and 18.3 kJ/mol for nC-Fe0.

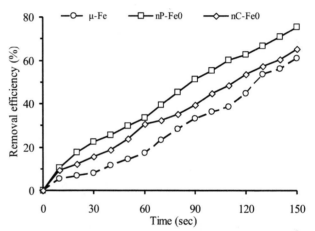

Fig. 2. CB removal efficiency for the various iron particles (irradiated with 250 W MW energy for 150 sec).

3.3 REACTION MECHANISMS

Analyses of the product analyses after CB reduction by adding 1.0 g nC-Fe0 at 25 oC for 240 min show that the major product is benzene. Analyzed using GC/MSD, the tail gas collected at the completion of the MW treatment contains benzene as the major end-product. Further FTIR analysis results show strong absorption peaks between 2320 and 2380 cm-1, and obvious weak absorption peaks in 665~670, 3598~3630, and 3703~3730 cm-1, these observations suggest that the unknown species is CO2. The chloride concentration in the original solution is 2.9 ppm; it increases to 6.3 ppm after the MW treatment.

On the other hand, the non-uniform absorption of MW energy by zero-valent iron particles may produce local hot spots with surface temperature reaching about 435 oC that is much higher than the surrounding temperature. Hence, the high-temperature oxidation can easily occur causing mineralization of benzene to form CO2. The major role of zero-valent iron is to react with CB and to generate electrons leading to the reduction and dechlorination of CB with the formation of Fe2+ or Fe+3-compounds and H2. When the MW penetrates the solution to reach the surface of zero-valent iron particles, it increases the zero-valent iron oxidation rate by creating more active sites on the surface to CB decomposition. Additionally, contacts between the high-temperature iron particle surface and H2 gas will induce iron reduction that proceeds simultaneously with CB dechlorination, and benzene mineralization as schematically shown in Fig. 3.

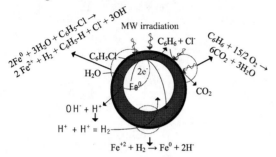

Fig. 3. Schematic diagram shown the MW irradiation reaching the surface of zero-valent iron particles to initiate reductive reactions for decomposing chlorobenzene.

4. CONCLUSION

Three types of zero-valent iron particle, µ-Fe, nP-Fe0, and nC-Fe0, are used as the microwave absorption media to enhance the removal of chlorobenzene. Laboratory data demonstrate that the MW energy enhances the CB removing 13.6 times (61.2% vs. 4.5% for µ-Fe), 2.8 times (76.5% vs. 27.5% for nP-Fe0) and 3.6 times (65.4% vs. 18.1% for nC-Fe0) more CB, and the CB activation energy decreases to 16.8 kJ/mol (µ-Fe), 3.2 kJ/mol (nP-Fe0) and 3.5 kJ/mol (nC-Fe0). The MW radiation penetrates the CB solution to reach Fe0 surface for increasing iron oxidizes, and surface activity, thus enhancing the CB removal. Additionally, the ZVI is highly oxidative, its surface activity is one of the important parameters to affect the removal of contaminant. Better CB removal is also contributed by the reduction of CB activation energy in the presence of MW irradiation.

REFERENCES

[1] C.J. Lin, S.L. Lo, Effects of iron surface pretreatment on sorption and reduction kinetics of trichloroethylene in a closed batch system, Water Res 39 (2005) 1037-1046.
[2] C.J. Clark II, P.S.C. Rao, M.D. Annable, Degradation of perchloroethylene in cosolvent solutions by zero-valent iron, J Hazard Mater 96 (2003) 65-78.
[3] J.L. Chena, S.R. Al-Abed, J.A. Ryanb, Z. Li, Effects of pH on dechlorination of trichloroethylene by zero-valent iron, J Hazard Mater 83 (2001) 243-254.
[4] A. Ghauch, C. Gallet, A. Charef, J. Rima, M. Martin-Bouyer, Reductive degradation of carbaryl in water by Zero-valent iron, Chemosphere 42 (2001) 419-424.
[5] V. Janda, P. Vasek, J. Bizova, Z. Belohlav, Kinetic models for volatile chlorinated hydrocarbons removal by zero-valent iron, Chemosphere 54 (2004) 917-925.
[6] C.B. Wang, W.X. Zhang, Synthesizing nano-scale iron particles for rapid and complete dechlorination of TCE and PCBs, Environ Sci Technol 31 (1997) 2154-2156.
[7] T. Dombek, E. Dolan, J. Schultz, D. Klarup, Rapid reductive dechlorination of atrazine by zero-valent iron under acidic conditions, Environ Pollut 111 (2001) 21-27.
[8] S.Y. Oh, P.C. Chiu, B.J. Kim, D.K. Cha, Enhanced reduction of perchlorate by elemental iron at elevated temperatures, J Hazard Mater 129 (2006) 304-307.
[9] C.J. Lin, S.L. Lo, Y.H. Liou, Degradation of aqueous carbon tetrachloride by nanoscale zerovalent copper on a cation resin, Chemosphere 59 (2005) 1299-1307.
[10] S. Choe, Y.Y. Chang, K.Y. Hwang, J. Khim, Kinetics of reductive denitrification by nano-scale zero-valent iron, Chemosphere 41 (2001) 1307-1311.
[11] S.F. Cheng, S.C. Wu, The enhancement methods for the degradation of TCE by zero-valent metals, Chemosphere 41 (2000) 1263-1270.
[12] C. Su, A.W. Puls, Kinetics of trichloroethene reduction by zero valent iron and Tin: pretreatment effect, apparent activation energy, and intermediate products, Environ Sci Technol 33 (1999) 163-168.
[13] H.S. Tai, C.J.G. Jou, Immobilization of chromium-contaminated soil by means of microwave energy, J Hazard Mater 65 (1999) 267-275.
[14] Z. Zhang, Y. Shan, J. Wang, H. Ling, S. Zang, W. Gao, Z. Zhao, H. Zhang, Investigation on the rapid degradation of congo red catalyzed by activated carbon powder under microwave irradiation, J Hazard Mater 147 (2007) 325-333.
[15] C.J. Jou, Degradation of pentachlorophenol with zero valence iron coupled with microwave energy, J Hazard Mater 152 (2008) 699-702.
[16] C.J. Jou, C.R. Wu, Granular activated carbon coupled with microwave energy for treating pentachlorophenol-containing wastewater, Environ Prog 27 (2008) 111-116.
[17] C.J.G.. Jou, C.L. Lee, C.H. Tsai, H.P. Wang, Microwave energy enhanced photocatalyses with titanium dioxide (TiO2) for degradation of trichloroethylene, Environ Eng Sci 25 (2008) 975-980.
[18] D. Schäfer, R. Köber, A. Dahmke, Competing TCE and cis-DCE degradation kinetics by zero-valent iron-experimental results and numerical simulation, J Contam Hydrol 65 (2003) 183-202.
[19] H.L. Lien, W.X. Zhang, Nano-scale Pd/Fe bimetallic particles: Catalytic effects of palladium on hydrodechlorination, Appl Catal B-Environ 77 (2007) 110-116.

Effect of C Content and Calcination Temperatures on the Characteristics of C-doped TiO$_2$ Photocatalyst

Fang-Yin Chen *, and Yao-Tung Lin **

* Dept. of Soil and Environmental Sciences, National Chung Hsing University
Taichung 402, Taiwan ROC, ivyipurple@hotmail.com
** Dept. of Soil and Environmental Sciences, National Chung Hsing University
Taichung 402, Taiwan ROC, yaotung@nchu.edu.tw

ABSTRACT

In this work, glucose was applied as C source, the ratio of glucose to TBT was 0.02 to 3.00. The range of calcination temperatures is from 400℃ to 600℃. XRD pattern shows that anatase phase is dominant. Anatase decreases with increasing in glucose/TBT ratio under the same calcination temperature. Band gap was narrowed down from 3.2 to 2.9 eV. Base on the identification of soft X-ray, results of C K edge absorption spectrum revealed that photon energy at 282.9 eV assigned to Ti-C structure. Result shows that oxygen sites were substituted by carbon atoms and C-Ti-O formed.

Keywords: C-doped TiO$_2$, C Content, Calcination Temperatures, Vis-responsible photocatalyst

1 INTRODUCTION

Photocatalytic process of TiO$_2$ is novel technology for gas sensors [1], pigment [2], solar-energy conversion [3]. Nevertheless, pure TiO2 only adsorbs UV light which is less than 5 % in the solar light spectrum because of its wide band gap of 3.2 or 3.0 eV in anatase or rutile crystalline phase, respectively. In order to utilize wider range of solar energy, studies had an attention on modifying the characteristics of TiO$_2$ by doping with nonmetal atoms such as N [4, 5], C [6-9] or transition metals such as Cr [10], and Fe [11]. However, the photoactivity of doping with transition metals is poor due to the thermal instability, or the growth of carrier trapping [12]. Therefore, doping TiO$_2$ with non-metal elements has received much attention.

Janus et al. (2006) improved the photoactivity by doping TiO$_2$ with C due to reduction of the recombination rate of photogenerated electron-hole pairs by the electron scavenger, C, doped into TiO$_2$ structure. Beside, C-doped TiO$_2$ was capable of lowering the band gap, shifting the optical response to the visible light region and decreasing the recombination rate of photogenerated electron-hole pairs [7].

In this study, we had focused on the sol gel synthesis of C-doped TiO$_2$ at various reaction temperatures. All samples were characterized using XRD, and UV-Vis. Eventually, soft X-ray spectroscopy were used to demonstrate the electronic structure of C-doped TiO$_2$.

2 EXPERIMENT

In this work, the characteristics of C-doped TiO$_2$ at various C contents and calcination temperatures were studied. The molar ratio of glucose to TBT was 0.02 to 3.0. Samples were calcined from 400 to 600^{0}C. C-doped TiO$_2$ was prepared by sol-gel method. Chemicals include 0.05 mole of titanium n-butoxide (TBT), glucose and ethanol. TBT is the catalyst precursor while glucose is the carbon source. C-doped TiO$_2$ powder was obtained after drying. The desired calcination temperatures was in the range from 400 to 600^{0}C. The molar ratio of glucose to TBT is from 0.02 to 3.0.

The crystalline structure of prepared C-doped TiO$_2$ photocatalyst was indentified using X-ray diffraction with Cu-K radiation at a scan rate of 0.05^0 s^{-1} (PANalytical X'Pert Pro MRD, USA). Soft x-ray absorption spectroscopy was used to identify the electronic and crystal structure. Ti L edge (2p), O K edge (1s), and C K edge (1s) were investigated. The experiments were performed at National Synchrotron Radiation Research Center (NSRRC) in Taiwan.

3 RESULTS

3.1 Effect of Calcination Temperature

Samples calcined at 400, 500, and 600^0 C were investigated. The effect of calcination temperature on the structure and optical properties of C-doped TiO$_2$ was listed at the Table 1. Based on the results of XRD analysis, it is seen that anatase content decreases while the calcination temperature increases from 400 to 600^{0}C. Anatase content was about 83 ~ 100 % at 400^{0}C. However, it drops to 30 ~ 73 % as the calcination temperature rises to 600^{0}C.

Particle size was investigated via XRD and DLS analysis. Based on XRD, anatase which is more active phase than rutile was the dominant component for samples from 400 to 600^0 C. Nevertheless, the phase transformation from anatase to rutile took place at 600^0 C. Based on the Scherrer equation, the results showed that particle size increases as the reaction temperature increases due to the vapor molecule accumulation and the development of rutile phase which has larger particle size than anatase. Crystal size was in the range from 9 to 38 nm while size was 175 ~

954 nm by DLS analysis. The difference was due to the aggregation of prepared particles during the process of DLS measurement. Result suggests that size grows as calcination temperature increases.

Table 1. Effect of Calcination temperatures

Sample	Calcination Temp (°C)	Phase composition A % / R %	C content (%)	Size (nm) XRD	Size (nm) DLS	Band gap (eV)
P25	-	81 / 19	-	22	-	3.20
G002T4	400	100 / 0	0.1	16	740	3.13
G002T5	500	100 / 0	0.2	20	463	2.97
G002T6	600	73 / 27	0.2	38	560	2.94
G02T4	400	100 / 0	0.2	13	669	2.95
G02T5	500	86 / 14	0.3	19	514	2.94
G02T6	600	61 / 39	0.2	34	502	2.94
G05T4	400	87 / 13	0.1	12	954	2.95
G05T5	500	81 / 19	0.2	20	318	2.92
G05T6	600	47 / 53	0.2	34	260	2.94
G10T4	400	100 / 0	0.5	10	397	2.94
G10T5	500	78 / 22	0.3	16	197	2.96
G10T6	600	47 / 53	0.2	27	175	2.94
G20T4	400	83 / 17	0.3	9	228	2.96
G20T5	500	74 / 26	0.2	17	162	2.99
G20T6	600	30 / 70	0.2	28	191	2.93
G30T4	400	84 / 16	0.3	9	221	3.03
G30T6	500	39 / 69	0.1	31	185	2.93
G30T5	600	76 / 24	0.4	19	164	2.97

3.2 Effect of Carbon Content

The effect of carbon content on the characteristics of C-doped TiO_2 was summarized in Table 2. Samples were synthesized at various amounts of glucose. Based on results of XRD analysis, two phases including anatase and rutile were dominant for C-doped TiO_2 samples. The anatase content decreases with increase in the molar ratio of glucose to TBT precursor under the same calcinaiton temperature. In the other words, the rutile phase shows gradually as the amount of glucose increases. With respect to particle size, insignificant influence was observed by the molar ratio of glucose to TBT under the same calcination temperature.

The band gap narrows from 3.20 to 2.92 eV. Band gap of samples calcinated at 600^0 C was smallest among all samples calcinated at 400 and 500^0 C. Substitutional carbon atoms introduce new states (C 2p) close to the valence band edge of TiO_2 (i.e. O 2p states). As a result the valence band edge shifts to higher energy compared with the reference TiO_2 and the band gap narrows. The energy shift of the valence band depends on the overlap of carbon states and O 2p states. A higher doping concentration of carbon results in higher energy shift due to significant overlap of carbon and oxygen states and this leads to narrower band gap in the compound [13]. Band gap of G05T5 is 2.92 eV which is the smallest.

Table 1. Effect of C content

Sample name	Calcination Temp (°C)	Phase composition A % / R %	C content (%)	Size (nm) XRD	Size (nm) DLS	Band gap (eV)
P25	-	81 / 19	-	22	-	3.20
G002T4		100 / 0	0.1	16	740	3.13
G02T4		100 / 0	0.2	13	669	2.95
G05T4	400	87 / 13	0.1	12	954	2.95
G10T4		100 / 0	0.5	10	397	2.94
G20T4		83 / 17	0.3	9	228	2.96
G30T4		84 / 16	0.3	9	221	3.03
G002T5		100 / 0	0.2	20	463	2.97
G02T5		86 / 14	0.3	19	514	2.94
G05T5	500	81 / 19	0.2	20	318	2.92
G10T5		78 / 22	0.3	16	197	2.96
G20T5		74 / 26	0.2	17	162	2.99
G30T5		76 / 24	0.4	19	164	2.97
G002T6		73 / 27	0.2	38	560	2.94
G02T6		61 / 39	0.2	34	502	2.94
G05T6	600	47 / 53	0.2	34	260	2.94
G10T6		47 / 53	0.2	27	175	2.94
G20T6		30 / 70	0.2	28	191	2.93
G30T6		39 / 69	0.1	31	185	2.93

GXTY: X means the molar ratio of glucose to TBT, Y means the calcination temperature

3.3 X-ray absorption

C K edge absorption spectra are showed in Figure 1. The C 1s spectrum revealed two components at 285.9 and 288.8 eV. The smaller component at photon energy of 285.9 eV could be attributed to C 1s electrons. Ren et al. (2007) observed only one kind of carbonate species with binding energies of 288.6 eV and revealed that carbon may substitute for some of the lattice titanium atoms and form a Ti – O – C structure [14]. The results indicated that the C 1s XPS peak (288.8 eV) could be assigned to Ti-O-C structure in carbon-doped TiO2 by substituting some of the lattice titanium atoms by carbon.

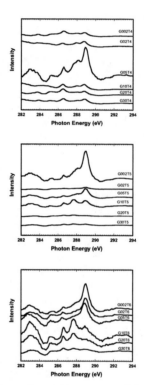

Figure 1. X-ray absorption spectra of the C K edge

4 CONCLUSION

1. The rutile content increases as the temperature rises up. The phase transformation temperature is 600^0 C.
2. Band gap narrows because the substitutional carbon atoms that introduce new states (C2p) close to the valence band edge of TiO_2 (i.e. O2p states).
3. Carbon may substitute for some of the lattice titanium atoms and form a Ti – O – C structure.

REFERENCES

1. Guidi, V., et al., "Preparation of nanosized titania thick and thin films as gas-sensors". Sensors and Actuators B: Chemical, 197-200, 1999.

2. Feldmann, C., "Preparation of Nanoscale Pigment Particles". Advanced Materials, 1301-1303, 2001.

3. Kamat, P.V. and N.M. Dimitrijevic, "Colloidal semiconductors as photocatalysts for solar energy conversion.", Solar Energy, 83-98, 1990

4. Yin, S., et al., "Low temperature synthesis of TiO2-xNy powders and films with visible light responsive photocatalytic activity.", Solid State Communications, 132-137, 2006.

5. Yin, S., et al., "Visible-light-induced photocatalytic activity of TiO2-xNy prepared by solvothermal process in urea-alcohol system." Journal of the European Ceramic Society, 2735-2742, 2006

6. Khan, S.U.M., M. Al-Shahry, and W.B. Ingler, Jr., "Efficient Photochemical Water Splitting by a Chemically Modified n-TiO2." Science, 2243-2245, 2002.

7. Janus, M., et al., "Carbon-modified TiO2 photocatalyst by ethanol carbonisation." Applied Catalysis B-Environmental, 272-276, 2006.

8. Xu, C., et al., "Photocatalytic effect of carbon-modified n-TiO2 nanoparticles under visible light illumination." Applied Catalysis B-Environmental, 312-317, 2006.

9. Xu, C., et al., "Nanotube enhanced photoresponse of carbon modified (CM)-n-TiO2 for efficient water splitting." Solar Energy Materials and Solar Cells, 938-943, 2007.

10. Yamashita, H., et al., "Characterization of metal ion-implanted titanium oxide photocatalysts operating under visible light irradiation." Journal of Synchrotron Radiation, 451-452, 1999.

11. Ranjit, K.T. and B. Viswanathan, "Synthesis, characterization and photocatalytic properties of iron-doped TiO2 catalysts." Journal of Photochemistry and Photobiology A: Chemistry, 79-84, 1997.

12. Yamashita, H., et al., "Preparation of Titanium Oxide Photocatalysts Anchored on Porous Silica Glass by a Metal Ion-Implantation Method and Their Photocatalytic Reactivities for the Degradation of 2-Propanol Diluted in Water." The Journal of Physical Chemistry B, 10707-10711, 1998

13. Wang, H. and J.P. Lewis, "Second-generation photocatalytic materials:anion-doped TiO2." JOURNAL OF PHYSICS: CONDENSED MATTER, 421-434, 2006.

14. Xiao, Q., et al., "Solar photocatalytic degradation of methylene blue in carbon-doped TiO2 nanoparticles suspension." Solar Energy, 706-713, 2008.

Formation and Stabilization of Gold Clusters and Nanoparticles in Mordenites

A. Pestryakov[1], N. Bogdanchikova[2], I. Tuzovskaya[1, 2], M. Avalos[1] and M.H. Farías[1]

[1]Tomsk Polytechnic University, Tomsk 634050, Russia, pestryakov2005@yandex.ru
[2]Centro de Nanociencias y Nanotecnología, Universidad Nacional Autónoma de México,
Apdo. Postal 14, C. P. 22800, Ensenada, México, bogdanchikova@gmail.com

ABSTRACT

Changes in electronic state of gold supported on mordenites during long-term storage in air have been studied by UV-visible spectroscopy method. Supporting gold on mordenites by ion-exchange method with further reduction with H_2 led to formation of neutral and charged metal clusters and nanoparticles. Prolonged storage of the gold catalysts in air caused intensive redox processes. Ionic states of gold were partly reduced, perhaps, under the action of light, whereas metallic particles, in contrast, were oxidized by atmospheric oxygen. After one-year storage both as-prepared and reduced samples came to more or less similar stationary state.

Cluster contribution and their stability depend strongly on acidity of zeolite (determined by SiO_2/Al_2O_3 molar ratio). Strong Brønsted acid sites favor the stabilization of gold clusters and their oxidation during prolonged storage in air because of higher redox potential of oxygen in acid media. Oxidation of gold clusters and nanoparticles can be considered as the main cause of gold catalyst deactivation during prolonged storage.

Keywords: gold, clusters, nanoparticles, oxidation, long-term storage

1 INTRODUCTION

Supported gold nanoparticles are well-known as active catalysts for some processes, for example, low-temperature CO oxidation, water gas-shift reaction, hydrocarbon oxidation, etc. [1-3]. Among all other catalysts, gold has unique properties since Au can efficiently catalyze CO oxidation and some other processes at low temperatures; and only nanosized species of gold (< 5 nm) exhibit high catalytic activity, while bigger gold particles are catalytically inert. However, along with high catalytic activity, gold nanocatalysts have a serious problem – fast deactivation during usage and storage [3-5]. The researchers suggested different explanations of this phenomena: aggregation of the metal particles, formation of carbonates or other adsorbed complexes, redox processes [2,3,6-10].

In our previous studies we suggested that gold clusters with size < 1 nm are responsible for low-temperature CO oxidation, while bigger Au nanoparticles catalyze this process at higher temperature (> 250°C) [11-13]. Formation of clusters always accompanies synthesis of the metal nanoparticles, but their study is complicated by difficulties in detection of small supported clusters with diameter < 1 nm by standard direct structural methods such as XRD and TEM [14]. According to our results [15], cluster/nanoparticle ratio as well as gold electronic state exerts a significant effect on catalytic properties of supported metals. Perhaps, changes in cluster/nanoparticle contribution and variations in redox properties of supported gold are the reason for gold catalyst deactivation during storage. UV-visible spectroscopy is a very fast and informative method for investigation of supported metal catalysts as it can detect both gold clusters and nanoparticles.

The aim of the present paper is the investigation of conditions of formation of gold clusters and nanoparticles in mordenites and their changes during prolonged storage in air. A comparative study was made varying three parameters: (1) mordenite SiO_2/Al_2O_3 molar ratios (MR), (2) reducing pretreatment, and (3) time of storage in air.

2 EXPERIMENTAL

Mordenites in protonated form with SiO_2/Al_2O_3 molar ratio 15 (M15) and 206 (M206) supplied by TOSOH corporation, Tokyo, Japan were used for the catalyst preparation. Gold containing zeolites were prepared by ion exchange procedure. The solution of $[Au(NH_3)_4](NO_3)_3$ complex for ion exchange was prepared by reaction of $HAuCl_4$ with NH_4OH and NH_4NO_3 [15]. After ion exchange the samples were washed with distilled water, dried at 60°C and heated in H_2 at 50°C. In order to reduce the initial states of gold the samples were treated with hydrogen at 100°C during 2.5 h. The samples were tested by UV-visible spectroscopy after 1 day, 1 week and 1, 6 and 12 months of storage.

UV-Visible diffuse reflectance measurements were made on a CARY 300 SCAN (VARIAN) spectrophotometer with a standard diffuse reflectance unit at room temperature. Spectra of pure supports were subtracted.

3 RESULTS AND DISCUSSION

The spectra of AuM15 sample showed three characteristic absorption bands (Fig. 1). According to some studies the absorption bands in the regions 250-260 and

320-340 nm could be attributed to neutral or charged gold clusters (charge transfer bands) [15]. The maximum wavelength at 530 nm of the spectra can be attributed to the plasmon resonance of Au nanoparticles [15-17]. Spectrum of fresh as-prepared sample exhibited very weak peak at 550 nm while signals of gold clusters (250 and 340 nm) were pronounced (Fig. 1A). According to our previous studies (XPS, FTIR, XRD, TEM data) the most part of gold in the as-prepared samples is in ionic form [15].

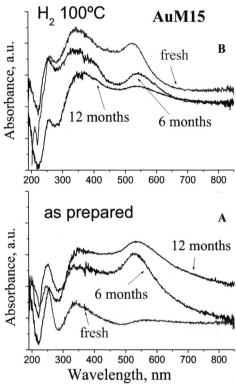

Figure 1. UV-visible spectra of AuM15 sample as-prepared and reduced with H_2 at 100^0C before and after long-term storage in air

Prolonged storage of as prepared sample in air for 6 and 12 months (Fig. 1A) led to a significant rise of intensity of the band at 550 nm belonging to gold neutral nanoparticles and a relative decrease of contribution of gold clusters. Moreover, total unstructured absorption of these samples was also increased. Obviously, the observed effects were caused by partial reduction of ionic gold. Reduction of gold ions under the action of light was noted earlier [3].

For AuM15 sample reduced at 100 °C we observed the opposite pattern. Spectrum of the fresh reduced sample contained very pronounced bands at 260, 360 and 540 nm (Fig. 1B). After long-term storage of the sample intensity of plasmon resonance bands at 540 and 250 nm as well as unstructured absorption were decreased significantly. The intensity of the signal at 350 nm remained more or less the same. The observed effects can be explained by partial oxidation of the reduced sample by oxygen during storage.

Thereby, storage of the gold catalysts in air led to intensive redox processes of gold particles. Ionic states of gold were partly reduced while metallic particles, on the contrary, were oxidized. It is interesting to note, that after 12-month storage both as-prepared and reduced samples came to more or less similar state (Fig. 1A-B).

Spectrum of the fresh as-prepared AuM206 sample demonstrated only one pronounced signal at 270 nm attributed to gold clusters (Fig. 2A). Prolonged storage in air led to appearance of intensive plasmon resonance band at 550 nm. Spectra of the sample after storage for 6 and 12 months did not differ noticeably. Only the unstructured absorption in the range of 600-900 nm slightly increased after one-year storage. Thus, as-prepared AuM206 sample reached stationary state faster as compared with AuM15.

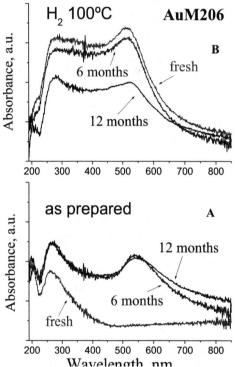

Figure 2. UV-visible spectra of AuM206 sample as-prepared and reduced with H_2 at 100^0C before and after long-term storage in air

Reduced AuM206 sample demonstrated a tendency similar to AuM15 catalyst. Spectrum of the fresh sample contained pronounced signals at 260 nm (gold clusters) and at 550 nm (Au nanoparticles) (Fig. 2B). After prolonged storage in air we observed decreasing of the intensities of both peaks due to partial oxidation of metallic particles. Final states of both as-prepared and reduced samples after one-year storage were also alike.

Thus, general trends in changes electronic states of gold species in mordenites with different acidity (MR) during long-term storage were the same. As prepared samples were partly reduced with storage, probably, under the action of light, while reduced catalysts were partly oxidized by oxygen in air. Finally, both samples came to more or less similar states corresponding to redox equilibrium.

Differences in gold states on the studied mordenites were observed mainly in the range of clusters. AuM15 sample demonstrated at least 2 types of clusters (bands at 250 and 350 nm) with different sizes and effective charge [15]. Spectrum of AuM206 exhibited only one cluster type (peak at 260 nm); and contribution of the cluster states of gold was lower as compared with AuM15.

SiO_2/Al_2O_3 molar ratio of mordenites influences mainly the concentration and strength of Brønsted acid sites on zeolite surface. Acidity of M15 is much higher as compared with M206 [18]. According to our studies, Brønsted acid sites favor formation of metal clusters in zeolites [15,18]. Analysis of UV-visible spectra of the studied samples revealed that cluster states of gold in mordenites are more resistant to redox processes as compared with metal nanoparticles. Moreover, redox potential of oxygen is higher in acid media [19]. Therefore, the surface proton acidity of zeolite has to favor oxidation of gold particles. Indeed, the rate of gold oxidation in AuM15 was much higher than in AuM206 (Figs 1B-2B).

The obtained results showed that long-term storage of gold catalysts led to drastic changes in gold state. It is interesting to investigate similar effects during short-term storage of gold samples.

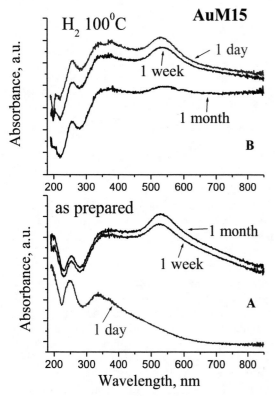

Figure 3. UV-visible spectra of AuM15 sample as-prepared and reduced with H_2 at 100^0C after short-term storage in air

Fig. 3 demonstrates the principal trend of changes in gold structural and electronic states during short-term storage (up to 1 month) to be similar to the one for long-term storage (Fig. 1). As-prepared (oxidized) sample was partially reduced while reduced catalyst was oxidized. However, after 1-month storage, in contrast with 1-year one, spectra of as-prepared and reduced samples differed noticeably. It means that longer time is necessary to bring gold species to a stationary state. 1-day storage did not change absolutely form of spectra of the fresh samples.

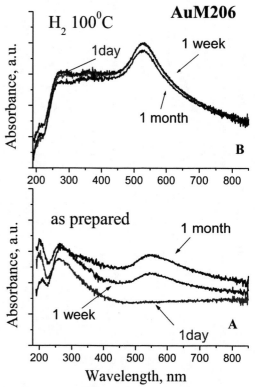

Figure 4. UV-visible spectra of AuM206 sample as-prepared and reduced with H_2 at 100^0C after short-term storage in air

For AuM206 catalyst, the character of changes during short-tem storage was also alike (Fig. 4). However, the rate of redox processes in AuM206 differed significantly from AuM15 sample. Reduction process of as-prepared AuM206 catalyst was relatively fast (Fig. 4A), while oxidation of the reduced sample for a month was not so marked (Fig. 4B). Obviously, relatively slow oxidation of AuM206 sample during short-term storage is also explained by less proton acidity of M206 zeolite as compared with M15.

Comparison with Fig. 2 shows that even 6-month storage was not enough for significant oxidation of the reduced AuM206 catalyst, while reduction of the as-prepared catalyst was relatively fast.

Thus, the obtained results showed that during prolonged storage supported gold catalysts undergo strong both reducing and oxidizing action of the medium. Initially ionic gold states can be partially reduced, probably, under the action of light, whereas neutral gold species can be oxidized by oxygen. But what is a probable cause of gold catalyst deactivation during storage? Usually pre-reduced supported gold catalysts are utilized for catalytic processes as Au^{3+} ions are inactive [11-13]. Therefore, real gold catalysts are initially reduced and during storage in air they are oxidized by oxygen. Oxidation of gold clusters and nanoparticles can be considered as the main cause of gold catalyst deactivation during prolonged storage.

Oxidation of gold clusters and nanoparticles was studied in some researches [6-10]. Authors of Ref. [10] supposed that Au particles smaller than about 30 Å are reactive to air, leading to oxidation up to 15% of Au atoms, whereas larger particles are not oxidized. These oxidized Au atoms in small particles are suggested to be active for CO oxidation. However, our studies revealed inactivity of Au^{3+} ions, whereas partly charged $Au_n^{\delta+}$ clusters were suggested to be active sites for low temperature CO oxidation [11-13].

In any case, the oxidizability of gold particles depends strongly on support type. For example, the most effective supports for gold catalysts, such as TiO_2, Fe_2O_3, CeO_2 and other metal oxides of variable valence strongly interact with the supported metal [2,3]. They favor formation of nanosized gold particles, which partially charge gold atoms bound with the support. These partly charged gold states may be active sites of the catalysts. However, in our opinion, the effective charge value is a determinant for catalytic activity of gold. Deep oxidation of gold including the catalyst prolonged storage in air leads to deactivation of active gold species.

4 CONCLUSIONS

Prolonged storage in air of as-prepared gold catalysts and reduced at 100°C leads to intensive redox processes. During storage in air ionic states of gold in as prepared samples undergo the reduction, perhaps, under the action of light, whereas metallic particles in gold catalysts reduced at 100°C, in contrast, are oxidized by atmospheric oxygen. After one-year storage both as-prepared and reduced samples came to more or less similar stationary state corresponding to redox equilibrium.

Cluster contribution and their stability depend strongly on acidity of zeolite (determined by SiO_2/Al_2O_3 molar ratio). Strong Brønsted acid sites favor the stabilization of nanosized gold species and their oxidation during prolonged storage in air because of higher redox potential of oxygen in acid media.

Oxidation of gold clusters and nanoparticles can be considered as the main cause of gold catalyst deactivation during prolonged storage.

ACKNOWLEDGEMENTS

The authors would like to express their gratitude to E. Flores, J.A. Peralta, P. Casillas, I. Gradilla, M. Sainz, C.Gonzalez, M. Vega and J. Palomares for valuable technical assistance with the experimental work. This research was supported by CONACYT grant No 79062, PAPIT-UNAM IN100908 (Mexico) and by RFBR grant 09-03-00347-a (Russia).

REFERENCES

[1] M. Haruta, N. Yamada, T. Kobayashi, S. Iijima, J. Catal., 115, 301, 1989

[2] A.S.K. Hashmi, G.J. Hutchings, Angewandte Chemie, 45, 7896, 2006.

[3] G.C. Bond, C. Louis, D.T. Thompson, Catalysis by Gold , Imperial College Press, London, 2006

[4] C.H. Kim, L.T. Thompson, J. Catal., 230, 66, 2005

[5] Y. Denkwitz, Z. Zhao, U. Hörmann, U. Kaiser, V. Plzak, R.J. Behm, J. Catal., 251, 363, 2007

[6] B.E. Salisbury, W.T. Wallace, R.L. Whetten, Chem. Phys., 262, 131, 2000

[7] B. Yoon, H. Hakkinen, U. Landman, J. Phys. Chem. A, 107, 4066, 2003

[8] D.C. Lim, I. Lopez-Salido, R. Dietsche, M. Bubek, Y.D. Kim, Surface Sci., 600, 507, 2006

[9] J. Radnik, C. Mohr, P. Claus, Phys. Chem. Chem. Phys., 5, 172, 2003.

[10] J.T. Miller, A.J. Kropf, Y. Zha, J.R. Regalbuto, L. Delannoy, C. Louis, E. Bus, J.A. van Bokhoven, J. Catal., 240, 222, 2006

[11] A. Simakov, I. Tuzovskaya, A. Pestryakov, N. Bogdanchikova, V. Gurin, M. Avalos, M.H. Farías, Appl. Catal. A, 331C, 121, 2007

[12] A.N. Pestryakov, N. Bogdanchikova, A. Simakov, I. Tuzovskaya, F. Jentoft, M. Farias, A. Díaz, Surface Sci., 601, 3792, 2007

[13] I. Tuzovskaya, A. Simakov, A. Pestryakov, N. Bogdanchikova, V. Gurin, M. Farías, H. Tiznado, M. Avalos, Catal. Commun., 8, 977, 2007

[14] N. Bogdanchikova, A. Pestryakov, M.H. Farias, J. A. Diaz, M. Avalos, J. Navarrete, Solid State Sciences, 7, 908, 2008

[15] I. Tuzovskaya, N. Bogdanchikova, A. Simakov, V. Gurin, A. Pestryakov, M. Avalos, M. H. Farías, Chem. Phys., 338, 23, 2007

[16] T. Morris, K. Kloepper, S. Wilson, and G. Szulczewski, J. Colloid and Interface Sci., 254, 49, 2002

[17] A. Tompos, J.L. Margitfalvi, E.Gy. Szabó, Z. Pászti, I. Sajó, G. Radnóczi, J. Catal. 266, 207, 2009

[18] N. Bogdanchikova, V. Petranovskii, S. Fuentes, E. Paukshtis, Y. Sugi, A. Licea-Claverie, Mater. Sci. Ing., A276, 236, 2000

[19] A.J. Bard, R. Parsons, J. Jordan, Standard Potentials in Aqueous Solutions, Marcel Dekker, New York, 1985

SAPHIR – EUROPEAN PROJECT
Safe, integrated and controlled production of high-tech multifunctional materials and their recycling

Christophe GOEPFERT, CILAS
Frédéric SCHUSTER, CEA

Industrial needs in terms of multifunctional components are increasing. Several sectors are concerned, ranging from mature high volume markets like automotive applications, high added value parts like space & aeronautic components or even emerging activities like new technologies for energy. Domains with a planetary impact like environment, new products and functions for health and safety of people are also concerned. Nanotechnologies could play a key role in promoting innovation in design and realisation of multifunctional products for the future, either by improving usual products or by creating new functions and new products. Nevertheless, this huge evolution of the industry of materials can only happen if the main technological and economic challenges are solved with reference to the societal acceptance. It concerns the mastering, over the whole life cycle of the products, of the potential risks, by an integration of the elaboration channels, while taking recycling into account.

At the present time, the manufacture of nanostructured products is not integrated so that each step entails a loss of time and money as well as potential release of nanoparticles in the atmosphere. SAPHIR aims at implementing direct production of nanoparticles through the development of a global integrated concept (from the synthesis to the final products) with a responsible approach. Indeed, the main breakthrough is to connect individual processes so that the handling of the produced powders is avoided. By means of safe recovery and conditioning systems, such as suspensions or nanostructured granulates, nanoparticles can be manipulated without risk. The transformation of these suspensions or granulates through conventional and emerging processes needs to be adapted, as far as most of these processes have only been developed for the manufacture of micro-structured components.

Description:

There is a huge number of developments and applications involving nanoparticles. The nanotech industry is moving from research to production with over 500 consumer nano-products which are already available. Nanotechnology is not tomorrow's technology anymore; nanoproducts are focused on today's market opportunities. Some previous limitations have been solved thanks to improvements in particular in the technic of dispersion of nanoparticles and the decrease of the production costs.

Nanoparticles production quantity increases (world wide market: 700-1000 billion Euros) with the consequence that more and more workers are exposed while toxicity aspects are still unknown.

NIOSH (National Institute for Occupational Safety and Health) recommends a prudent approach for manufacturing and using nanoparticles in industry. Employers should take precautions to minimize workers' exposure until more information is available.

At present no regulation exists which refers specifically to the production and the application of nanomaterials or nanoparticles. A lot more knowledge has to be generated on how nanomaterial based processes and products may interfere with human health and the environment, before any regulation in this field can be established.

In the meantime we have to apply the principle of precaution.

The development of nanotechnologies appears to be a real innovative solution in order to improve existing technologies and create new products of the future. However, at the present time, the manufacture of nanostructured products is not integrated so that each step entails a loss of time and money as well as potential release of nanoparticles in the atmosphere. SAPHIR aims at implementing direct production of nanoparticles through the development of a global integrated concept (from the synthesis to the final products) with a responsible approach as shown in fig 1.

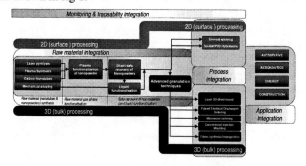

Fig 1: global integrated concept

Indeed, the main breakthrough is to connect individual processes so that the handling of the produced powders is avoided. Thanks to safe recovery and conditioning systems, suspensions or nanostructured granulates processing, nanoparticles can be manipulated without risk. The transformation of these suspensions or granulates through conventional and emerging processes needs to be adapted, as far as most of these processes have only been developed for the manufacture of micro-structured components.

The general objective of the SAPHIR project is the safe, integrated and controlled production of high-tech multifunctional nano-structured products including their recycling, and ensuring competitiveness.

It means that:

i) all along the production sequence; no nanoparticle release will be encountered. This includes synthesis, recovery (direct liquid recovery), conditioning (advanced granulation), processing and handling;

ii) ii) the whole production sequence will consist in linking in a safe way existing or emerging elementary processes (direct plasma spraying, laser 3D direct manufacturing, and powder consolidation techniques);

iii) iii) the production sequence will be controlled by innovative systems covering process efficiency, product reliability, global safety production, & traceability; recyclability issues will be addressed by proposing routes for the next future

First, the functionalization of nanoparticles will improve the properties of the final products as well as their "processability" for the next steps by a coating process (plasma treatment or fluidized bed technology are evaluated).

Then, the conditioning in suspension and the granulation will considerably reduce the risk of emission of nanoparticles in the air.

Granulation is performed by:

i) spray drying where the powder is dispersed in a liquid (water or organic solvent) and then sprayed into a chamber with hot air where the drops (granules) are shaped and dried.

ii) freeze granulation, a recently developed technology in which drops are freezed in a vessel containing liquid nitrogen or by

iii) mechanical alloying to combine alloying with CNT or nanopowders produced in a separate way.

These emerging combinations will allow the elaboration of new generation of metallic matrix composites (MMC), ductile ceramics and ceramic matrix composites (CMC), plastic matrix composites (PMC).

In order to guarantee the reliability of the chain, we have to focus on the development of on-line monitoring instrumentation (fig 2). Two parts will be developed in this field: process monitoring and traceability.

This double approach will allow to eliminate corrective actions, to do "the right things the first time" thus insuring a safe and competitive production

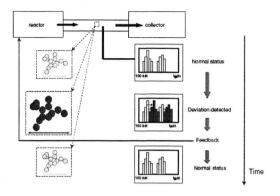

Fig 2: process monitoring

Work is done to address bulk and surface applications:

Bulk materials:
Existing processes as SPS, HIP, microwave, laser 3D processes were analyzed. SPS and HIP were successfully proven to produce nanostructured parts; the structure of these parts was studied and the properties evaluated.

Work was performed successfully towards of application of MWCNT in polymer matrices for automotive field and bipolar plates in fuel cells in the energetic field.

Main processes that are studied and developed during the project are extrusion and injection moulding; fibre synthesis/processing and direct growth of carbon nanotubes on carbon textures.

The aim of this work is to improve, in function of thermal, optical or mechanical properties, the final product.

Surface coating:
Work is performed in the field of surface engineering to improve properties for different applications.

Titanium oxide coatings were prepared by different technologies (sol-gel, thermal spray methods) using suspensions and their photocatalytic activity was studied. Hard metal coatings were prepared by thermal spraying and laser cladding, work focusses here on the in-situ formation of nanoparticles during the coating process. Performed studies contributed significantly to the understanding of the processes. As well titanium oxide coatings were prepared from powders and their photocatalytic behaviour was studied. Thermal spray coating was addressed by plasma spraying TiO_2 P25 as a suspension. Induction plasma technology was used and attention was given to the phases and to the atomisation of the suspension in the plasma torch.

SIGNIFICANT NEW ACCOMPLISHEMENTS

After the third year, the SAPHIR project now shows clearly a large number of very valuable results in 3 main domains which are:

The manufacturing of high performance raw material like ODS nanostructured powders, nanosized ZnO particles (see fig. 3) or Al/nano SiC nanostructured powders.

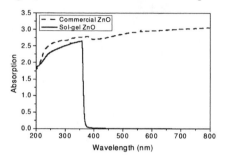

Fig 3: UV blocking capacity

The safe oriented processes which avoid any contacts with workers. Indeed, some developments have been achieved in the manufacturing process itself which is the granulation of the nanopowders (see fig. 4) in order to guarantee no release of nanoparticles at all in the air.

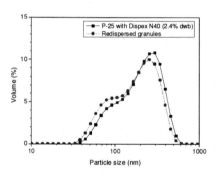

Fig 4: properties of re dispersed granules

The safe recovery system (see fig. 5) has been successfully tested in the phase vapour synthesis application and different forms of paste have been tested for the collection of particles.

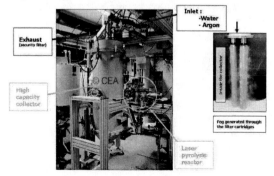

Fig 5: Recovery system

Significant progress has been also made in the field of the monitoring of processes with, in particular, windows monitoring (see fig. 6) coupling LIBS (Laser Induced Breakdown Spectroscopy) technology to the unit of nanopowder synthesis at CEA Saclay.

Fig 6: windows monitoring

Some very encouraging first results were also obtained on RFPM (Radio Frequency Plasma Monitoring) technology. Applications in surface engineering for example in the nanostructured coatings field for building applications, promising results of photocatalysis were confirmed both by the sol-gel technology and the technology of suspension plasma spraying. Bulk applications were also tested and gave good results as on bipolar plates for PEMFC using carbon nanotubes (see fig. 7).

Fig 7: CNT-containing bipolar plates

Conclusion:

The final goal is to avoid any contact between operators and material.

With regard to potential health and environmental risks dry powders of nanoparticulate materials are to be assessed as most critical, because they can easily form aerosols during production and handling processes, which might lead to human exposure or environmental contamination.

Considering these issues the solution would consist in developing fully automated tools which allow confining the production line in each of it's steps: Raw material synthesis, functionalisation, dry or wet recovery, granulation, consolidation and finally the processing in surface or bulk applications (fig 8).

Fig 8: synthesis and recovery of nanoparticles in a safe way

These elements are the foundations of the factory of the future enabling the production of new nanomaterials in conditions of maximum security.

Preparation of ZnAl$_2$O$_4$ Catalytic Support by Combustion Reaction: Influence of Type of Heating Source

E. Leal[*], K. M. S. Viana[**], R. H. G. A. Kiminami[***], D. C. Barbosa[****], S. M. P. Meneghetti[****], A. C. F. M. Costa[*], H. L. Lira[*]

[*] Department of Materials Engineering – UFCG, Avenue Aprígio Veloso - 882, Bodocongó, 58109-970, Campina Grande, PB, Brazil, elvialeal@gmail.com
[**] Faculty of Materials Engineering – UFPA, Marabá, PA, Brazil, kalinesouto@yahoo.com.br
[***] Department of Materials Engineering – UFSC, São Carlos, SP, Brazil, ruth@power.ufscar.br
[****] Department of Chemistry – UFAL, Maceió, AL, Brazil, smpm@qui.ufal.br

ABSTRACT

The aim of this work is to evaluate the esterification activity of ZnAl$_2$O$_4$ catalysts obtained by different ways of heating (plate, muffle furnace and microwave oven) during the combustion synthesis using urea as reducing agent, and nitrates of zinc and aluminum as oxidants. All samples were characterized by XRD, particle size distribution, nitrogen adsorption (BET/BJH) and SEM, and then, they were forwarded to the catalytic tests on the biodiesel esterification. XRD results showed that all catalysts presented the formation of a cubic structure of ZnAl$_2$O$_4$ spinel. The heating way had a great influence on the structural and textural features of ZnAl$_2$O$_4$, such as crystallite size and surface area, whose values were of 27, 35 and 16 nm, and 15, 9 and 57 m^2/g for the catalysts I, II and III, respectively. All catalysts presented morphology in irregular plates shape (SEM) and formation of particle agglomerates with a narrow size distribution. For all catalysts tested, very low reactions profits were observed.

Keywords: zinc aluminate, combustion, characterization, esterification.

1 INTRODUCTION

Oxide spinels comprise a very large group of structurally related compounds many of which are of considerable technological significance. Among them, zinc aluminate (ZnAl$_2$O$_4$), with typical normal spinel structure AB$_2$O$_4$, is of interest due to its combination of desirable properties such as high mechanical resistance, high thermal stability, low temperature sinterability, low surface acidity and better diffusion [1].

These powders of zinc aluminate find applications in industrial ceramics due to high performance and refinement. It also has proved interesting for catalytic applications, such as cracking processes and synthesis of alcohols and biodiesels [2, 3].

Recently, ZnAl$_2$O$_4$ has been prepared by solid state-reaction [4] or wet chemical routs such as co-precipitation [5], sol–gel [6] and hydrothermal method [1]. Among the different synthesis methods used to obtain ZnAl$_2$O$_4$ with nanosized particles, combustion reaction stands out as a promising technique, because it is a self-sustaining method that allows reaching temperatures that ensure the crystallization and phase formation in short period of time [7].

This work reports the preparation of ZnAl$_2$O$_4$ catalyst by combustion reaction using three kind of heating source (a spiral resistance, a muffle furnace and a microwave oven). The effect of heating source on the structure and morphology of the produced powders and their catalytic activity on the esterification process in biodiesel preparation were investigated.

2 EXPERIMENTAL

To obtain the ZnAl$_2$O$_4$ catalysts were used the following materials: aluminum nitrate [Al(NO$_3$)$_3$.9H$_2$O], zinc nitrate [Zn(NO$_3$)$_2$.6H$_2$O] and urea [CO(NH$_2$)$_2$], all of them with 98% of purity. The proportion of each reagent was calculated based on the theory of propellants and explosives, in order to produce the stoichiometric ratio between oxidizer/reducer equal to 1 ($\Phi e = 1$), where Φe is the elementar stoichiometric coefficient [8]. The mixture was placed in a vitreous silica crucible and subjected to different heating sources (I - a spiral resistance; II - a muffle furnace; and, III - a microwave oven) so that promote an ignition followed by combustion.

During the synthesis carried out in the spiral resistance and the muffle furnace were measured the parameters of combustion flame time and temperature, using a stopwatch (Vitese) and an infrared pyrometer (Raytek RAYR3I $\pm$ 2°C), respectively. Unable to determine the temperature and time of flame of the synthesis reactions carried out in the microwave oven, since it is formed by a Faraday cage, these were performed in a power of 100% during 10 minutes.

The samples were characterized by X-ray diffraction in a Shimadzu LAB 6000 diffractometer in a scan 2Θ range of 20-80°, using CuKα radiation (λ=1,5418Å). The values of surface area, diameter and volume pore were obtained by nitrogen adsorption, using a Quantachrome NovaWin2 system (NOVA 3200). Surface area was determined using the BET model while to determine the radius and volume pore was used the BJH theory. The average particle size was calculated from BET data using the equation (1) [9]. The agglomerate size distribution was determined using a CILAS 1064 LD particle size analyzer. The morphology of the powders also was analyzed by scanning electronic microscopy in a Philips XL30 FEG SEM microscope.

$$D_{BET} = \frac{6}{D_t . S_{BET}} \qquad (1)$$

where,

D_{BET} represents the spherical diameter (nm); D_t is the theoretical density (g/cm^3); S_{BET} corresponds to the surface area (m^2/g).

ZnAl$_2$O$_4$ catalysts were employed in esterification reactions and the percentage of biodiesel formed (FAME = fatty acid methyl ester) was determined by gas chromatography. The catalysts were tested using methanol and fatty acids derived from soybean oil, in the following reaction conditions: molar ratio of methanol/fatty acid/catalyst = 400/100/1, reaction time of 1 hour and temperature of 160 °C.

3   RESULTS

Fig. 1 shows the X-ray diffractograms of the ZnAl$_2$O$_4$ catalysts prepared by combustion reaction using the different heating sources.

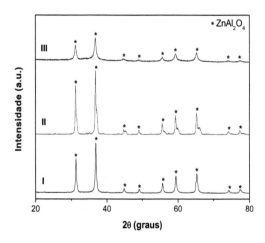

Figure 1: X-ray diffractograms of ZnAl$_2$O$_4$ catalysts. I – spiral resistance; II - muffle furnace; and, III - microwave oven.

The observed diffraction peaks in all the recorded XRD patterns correspond to those of the standard patterns of cubic ZnAl$_2$O$_4$ spinel (JCPDS, No. 05-0669) under the form of the guanine mineral. These peaks can be indexed as (220), (311), (400), (331), (422), (511), (440), (620) and (533) diffractions.

The crystallite size from XRD was calculated from X-ray line broadening of the (311) diffraction line located 2Θ = 36.87° and using the Scherrer equation [10]. The crystallite size of the catalysts I, II and III was calculated to be 27, 35 and 16 nm, respectively. The ratios of peak

intensities of all prepared powders were not the same. Comparing the XRD spectra with the flame temperatures and times measured during the synthesis performed in the spiral resistance and in the muffle furnace, whose values were of 895 and 941°C, and 27 and 21 seconds, respectively, it is observed that the higher the flame temperature reached in synthesis, more intense the peaks in the diffractograms spectra (Fig. 1). This shows that the higher the combustion temperature the greater the energy supplied to the growth of crystals.

Despite not have been performed the measurement of the flame temperature in the synthesis performed in the microwave oven, it was possible to observe through XRD spectra of the catalyst III the presence of peaks with broad bases and low intensity, that probably is a strong indicative that the combustion temperature generated in the microwave oven synthesis was lower than the temperatures generated in the synthesis performed in the spiral resistance and in the muffle furnace, which led to a crystallite size 41% and 54% nether to the crystallite sizes obtained to the catalysts I and II, respectively.

Fig. 2 shows the curves of the equivalent spherical diameter of the agglomerates as a function of the cumulative mass for the ZnAl$_2$O$_4$ catalyst.

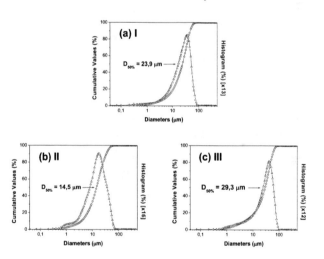

Figure 2: Granulometric distribution of the ZnAl$_2$O$_4$ catalysts. (a) I – spiral resistance; (b) II - muffle furnace; and, (c) III - microwave oven.

These curves suggest the formation of particle agglomerates with a narrow median size distribution. It was also observed that the catalysts produced in the spiral resistance (I) and microwave oven (III) showed an agglomerates distribution a little narrower and with a slight displacement of the curve to the right, indicating a cumulative mass increase in the median diameter of agglomerates ($D_{50\%}$), whose values were of 23.9 and 29.3 µm, respectively. The other hand, the median diameter of agglomerates ($D_{50\%}$) referring to the catalyst produced in the muffle furnace (II) was the lower, with a value of 14.5 µm.

This difference in the agglomerates sizes is also a result of the combustion temperature achieved during synthesis, since higher temperatures favor the formation of denser agglomerates, constituted by pre-sintered particles and without interparticle porosity, which can lead to the formation of more compacts powders. The lower temperatures favor the formation of less dense agglomerates and with interparticle porosity. Costa et al. [7] when evaluated the influence of glycine and urea on the synthesis of $ZnAl_2O_4$ catalyst by combustion reaction using a hot plate as heating source (480°C) observed a wide agglomerates size distributions, with median diameters ($D_{50\%}$) of 12.5 and 28.2 μm, respectively. Comparing our value referring to the catalyst (I) obtained in the spiral resistance, which was 23.9 μm, with the value obtained from the authors above regarding the $ZnAl_2O_4$ synthesized with urea, whose value was of 28.2 μm, it can be said that variations in synthesis conditions have great influence in the agglomerates size.

Table 1 shows the surface area, pore radius, pore volume and particle size referring to $ZnAl_2O_4$ catalysts. It can be observed that the catalysts obtained in the spiral resistance (I) and in the muffle furnace (II) presented a low surface area when compared with the catalyst obtained in the microwave oven (III), whose value was 57 m^2/g. Probably, this is explained in fact this heating source (microwave oven) have provided less energy for the particles growth, since size was 23 nm. It is observed that the larger the particle size the smaller the surface area presented. All catalysts (I, II and III) presented a mesoporous characteristic, with pore radius ranging from 17.3 to 19.2 Å. The pore volume had a small variation between the values 0.014 and 0.024 cm^3/g, with minimum belonging to the catalysts I and II, and maximum to the catalyst III.

Catalisador	S_{BET} (m^2/g)	R_P (Å)	V_P (cm^3/g)	D_{BET} (nm)
I	15	19.2	0.014	87
II	9	19.0	0.014	151
III	57	17.3	0.024	23

Table 1: Surface area (S_{BET}), pore radius (R_P), pore volume (V_P) and particle size (D_{BET}) to $ZnAl_2O_4$ catalysts.

Fig. 3 shows the nitrogen adsorption/desorption isotherms of the $ZnAl_2O_4$ catalysts. According to the IUPAC classification [12], all $ZnAl_2O_4$ catalysts showed type IV isotherm profiles, suggesting a characteristic of mesoporous material (pores with ratio ranging from 10 to 250 Å). Regarding the hysteresis, these catalysts also showed hysteresis loop of type H3 mixed with H2, indicating the presence as much of pores of slit-shaped originating in the particle aggregates, as pores open and closed with strangulations typical of an irregular morphology of bottle type [13]. Zawadzki et al. [14] synthesized $ZnAl_2O_4$ by microwave assisted solvothermal

conditions using zinc acetate, aluminium isopropoxide and 1,4-butanediol as precursors, and observed through textural analysis a N_2 adsorption/desorption isotherm similar to presented in this study, i.e. also of type IV and with a well-developed hysteresis loop of type H2, suggesting a material with basically mesoporous characteristic. Despite this similarity, the authors also observed a surface area, pore volume and pore radius of 133 m^2/g, 0.18 cm^3/g and 24 Å, respectively. I.e. values of surface area and pore volume higher than those obtained in this work.

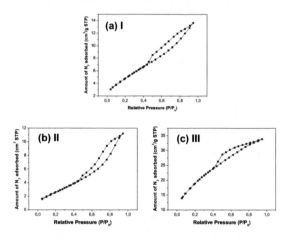

Figure 3: N_2 adsorption/desorption isotherms for the $ZnAl_2O_4$ catalysts. (a) I – spiral resistance; (b) II - muffle furnace; and, (c) III - microwave oven.

The SEM micrograph in Fig. 4 shows the morphology of the $ZnAl_2O_4$ catalysts according to heating sources used in the combustion reaction.

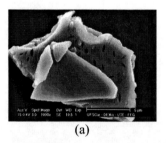

(a)

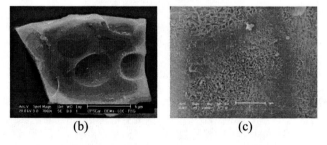

(b)   (c)

Figure 4: SEM micrographs of $ZnAl_2O_4$ catalysts. (a) I – spiral resistance; (b) II - muffle furnace; and, (c) III - microwave oven.

The micrograph referring to the catalyst obtained in spiral resistance (Fig. 4a) reveals the formation of agglomerates in the irregular plates form with rigid aspect (pre-sintered particles), sizes larger and smaller than 5 mm, and presence of large pores in slit form and uniform sizes, which probably were caused by the difficult gases release during the synthesis. The micrograph referring to the catalyst obtained in muffle furnace (Fig. 4b) shows the presence of a large agglomerate in the irregular rigid block form with approximately 10 μm. Although present a rigid aspect, this appears being composed by linked fine particles. The micrograph of the catalyst obtained in the microwave oven (Fig. 4c) shows the formation of a large agglomerate in plate form (rigid aspect) and presence of sections with fibrous aspect, providing a greater presence of pores with irregular shapes and sizes. According to Singh et al. [11], this non-uniformity in the shape could be related to the non-uniformity in the distribution of temperature and mass flow in the combustion flame.

Table 2 shows the reaction yield obtained from the esterification of fatty acids with methanol, in the presence of $ZnAl_2O_4$ catalysts. The presented results are compared to the reaction performed without catalyst, since the esterification reaction is self-catalytic. Thus, taking the results evaluated by the difference of profit between the reactions performed with and without catalyst, it was observed very low reactions profits for all catalysts. It was observed that the catalyst III despite having presented the larger surface area, this resulted in a lower reaction profit when compared to reaction without catalyst. Therefore, we assume that these low values of profit was not only a consequence of the low surface area values, but probably had interference from the nonuniformity of size and shape of particles or agglomerates, which leads to nonuniformity in size and shape of pores.

Reaction	FAME (%)
I	48
II	47
III	41
without Catalyst	45

Table 2: Profit obtained in the esterification reaction of fatty acids obtained from soybean oil referring to the different catalysts.

4 CONCLUSIONS

According to the results obtained in this work, it can conclude that changes in the $ZnAl_2O_4$ synthesis conditions, as the heating source used, led to significant changes in structural and morphological characteristics of the material. The synthesis performed in the microwave oven was most efficient to obtain powder with higher surface area and smaller crystallite size. The catalysts were not efficient in the conditions of esterification reactions tested, taking their morphological characteristics decisive interference in this behavior.

5 ACKNOWLEDGEMENTS

The authors are grateful to RENAMI / CNPq, CNPq / Multi-user case: (402561/2007-4), PROCAD / NF-CAPES, PRO-Engineering-CAPES, CNPq and CAPES for the financial support.

REFERENCES

[1] M. Zawadzki, Solid State Sciences 8, 14, 2006.
[2] T. El-Nabarawy, A.A. Attia and N. Alaya, Materials Letters 24, 319, 1995.
[3] C.T. Alves, H.M.C. Andrade, L.S. Bastos, L.S. Macêdo, S.A.B.V. Melo and E.A. Torres, In: 8th World Congress of Chemical Engineering, Montreal, 2009,
[4] S.K. Sampath and J.F. Cordaro, Journal of the American Ceramic Society 81, 649, 1998.
[5] G.F. Hetting, H. Worl, H.H. Weiter and Z. Anorg. Allg. Chem. 283, 207, 1956.
[6] M.A. Valenzuela, P. Bosh, G. Aguilar-Rios, A. Montoya and I. Schifter, Journal of Sol-Gel Science and Technology 8, 107, 1997.
[7] A.C.F.M. Costa, D.A. Vieira, R.P.T. Lula, R.H.G.A. Kiminami and L. Gama, In: 17° CBECIMat, 2339, 2006.
[8] S.R Jain, K.C. Adiga and V. Pai Verneker, Combustion and Flame, 40, 71, 1981.
[9] J.S. Reed, Principles of Ceramics Processing, 2a ed., 1994.
[10] H. Klung and L. Alexander, In X-ray diffraction procedures, Wiley, New York, EUA, p. 491, 1962.
[11] V. Singh, R.P.S. Chakradhar, J.L. Rao and D.-K. Kim, Journal of Luminescence, 128, 394, 2008.
[12] IUPAC, International Union of Pure and Applied Chemistry, Handbook, v. 2, n° 2, 1976.
[13] S.J. Gregg and K.S.W. Sing, Adsorption, Surface Area and Porosity, 2nd ed., Academic Press, London, 1982.
[14] M. Zawadzki, W. Staszak, F.E. López-Suárez, M.J. Illán-Gómez and A. Bueno-López, Applied Catalysis A: General, 371, 92, 2009.

Catalytic Degradation of Substituted Phenols in Water by Chemically Synthesized Bimetallic Nanoalloys

R. Nazir[*], M. Mazhar[**] and J. H. Zaidi[***]

[*]Department of Chemistry, Quaid-i-Azam University, Islamabad-45320, Pakistan,
Applied Chemistry Research Centre, PCSIR Laboratories Complex, Lahore, Pakistan
rabiapcsir@yahoo.com

[**]Department of Chemistry, University of Malaya, Lembah Pantai, 50603-Kualalumpur, Malaysia,
mazhar42pk@yahoo.com

[***]Department of Chemistry, Quaid-i-Azam University, Islamabad-45320, Pakistan
javid_zaidi@yahoo.com

ABSTRACT

Stable oxide free nanoalloys of Fe-Pd, Fe-Mg$_2$ and Fe-Mo, synthesized by simple chemical approach, were found to have good catalytic activity towards degradation of three substituted phenols i.e. 2-chlorophenol (2-CP), 4-chlorophenol (4-CP) and 4-nitrophenol (4-NP). The room temperature studies carried out in the absence of sunlight had shown 100% degradation of 4-CP and 4-NP to environmental friendly degradation products by all the three nanoalloys as affirmed by GC-MS, while in case of 2-CP exceptions were found for Fe-Pd and Fe-Mo. Among the three nanoalloys, Fe-Mg$_2$ is the most effective and economic alternative to the existing catalysts as it results in 100% degradation of all the substituted phenols at ambient conditions of temperature and pressure and without the use of any other external source.

Keywords: Fe-Pd, Fe-Mg$_2$, Fe-Mo, substituted phenols, degradation

1 INTRODUCTION

Water contamination from various organic pollutants like substituted phenols is the burning issue of today that is posing concern owing to its environmental toxicity and non-biodegradable nature. These persistent pollutants are causing long term health hazard [1-4] and therefore, it is necessary to eliminate them from ecosystem. Potential amount of work has been done to degrade these pollutants into less toxic byproducts and nanoparticles owing to their high reactivity are known to have better catalytic activity compared to bulk materials [5,6]. Iron being very strong reductant [7] is predominantly used for this purpose [2] but several studies in cleanup of environmental pollutants have proved that bimetallic nanoparticles are effective than zero-valent metals [5,8] due to their easy diffusibility and high activity [9]. When Pd is used in conjunction with Fe [10,11] good dechlorinating activity is observed [12] and with substantial reduction in reaction time when nano Pd/Fe is employed [7]. In addition to Fe/Pd other bimetallic systems were also tested in search of cost effectiveness but no appreciable activity was found in case of Fe-Pt, Fe-Au and Fe-Cu [13,14]. Slightly improved results were achieved by optimizing the Ru/Fe wt.% to 1.5 [14] and using Ca-Fe composite at 300°C [15]. Significant enhancement in dechlorination activity was observed by introducing Ni-Fe bimetallic system [8,16]. All these procedures end up with dechlorination of the desired reactant in case of chlorinated pollutants, while for nitro compounds such oxidation/reduction procedures are not well-known because of their non-effectiveness in properly degrading the reactant [17]. In order to enhance the degradation activity of chlorinated compounds and especially nitro compounds various other treatments were executed. Most of which require some necessary requirements to be fulfilled like presence of some specific solvent [18], additives [19] or external source like UV [20,21], ultrasound [19], day light [22], microwave [4] etc. or combination of these [1,23,24] in addition to being lengthy, tedious and costly [20,25]. Considering these aspects the present research was designed to synthesize nanoalloys that can cost effectively degrade these contaminants into environmental friendly products without employing tiresome procedures. Hence, offering an easy approach for treatment of these carcinogenic pollutants.

2 EXPERIMENTAL

2.1 Materials and Methods

All the reagents used in this study were of analytical grade and purchased from Fluka and Panreac. Phenols were supplied by Sigma-Aldrich while Anhydrous Sodium sulphate was purchased from Fluka.

2.2 Synthesis of Bimetallic Nanocatalysts

Stable bimetallic nanoalloys Fe-Pd [26], Fe-Mg$_2$ [27] and Fe-Mo were prepared as mentioned in previous studies by thermal decomposition of the homogeneous mixture of the bipyridine complexes of the component metals under inert atmosphere at 500°C. The resulting materials were characterized using EDXRF, powder XRD, AFM and TEM. EDXRF confirms the Fe:Pd and Fe:Mo metal ratio to be 1:1 except in case of Fe-Mg in which it is 1:2. The particle size as affirmed by AFM and TEM was in range of 15-30nm for Fe-Pd, 30nm for Fe-Mg$_2$ and 48-66nm for Fe-Mo. XRPD confirms the *fcc* structure of Fe-Pd while Fe-Mg$_2$ and Fe-Mo nanoalloys are amorphous in nature.

2.3 Experimental Procedure

Stock solutions (1000ppm) of the phenols were prepared using acetone as solvent. From this 20ppm standard solution of all the phenols were prepared in distilled water. Batch experiments were conducted by taking 0.1g of nanoalloy in a Schlenk tube connected with inert vacuum line and covered with a black paper to avoid sunlight interference in the reaction. To this 5.0ml of distilled water was added and stirred for 10min in order to well disperse the particles in water. After that 5.0ml of standard solution (20ppm) of the phenol was added and contents were stirred for 48hrs at room temperature and then cannula filtered. The filtrate was taken in separating funnel and 0.1ml of 3M H$_2$SO$_4$ was added. After that the contents were chloroform extracted with vigorous shaking for 5min. The chloroform extract was then treated with 1.0g anhydrous sodium sulphate, filtered and volume was leveled up to 10ml with chloroform prior to GC-MS analysis [12].

For blank experiment same procedure was adopted, except that in place of phenols distilled water was used. Standards of phenols for GC-MS analysis were made by taking 5.0ml of standard solution (20mg/l) of the phenol with 5.0ml of distilled water and 0.1ml of 3M H$_2$SO$_4$ and extracted with (5ml×2) chloroform. After treating the extract with 1.0g anhydrous sodium sulphate volume was leveled up to 10ml and this was analyzed using GC-MS.

2.4 Analysis

The catalytic activity of the as—prepared samples was checked by analyzing the chloroform extracts with help of Agilent GC-MS model 6890N. The temperature program was as follows: Initial value 60°C, ramp at 40°C/min-100°C, then 2°C/min−150°C and finally 30°C/min-250°C. Highly pure grade (99.999%) helium gas was used as carrier gas with the constant flow rate of 1.2ml/min. Temperature of inlet and detector were 220°C and 280°C, respectively. Column used was DB-5. For measurements samples were withdrawn from syringe and a volume of 1.0µL solution was injected.

3 RESULTS AND DISCUSSION

The three nanoalloys Fe-Pd, Fe-Mg$_2$ and Fe-Mo were used as catalyst for studying the degradation of three phenols (2-CP, 4-CP and 4-NP), taken as representative from the huge number of organic pollutants, in aqueous system in absence of sunlight. The results obtained for each of the sample were tabulated (Table 1) along with peak retention time (R_t) and corresponding m/z values. The results obtained have indicated that all the alloys show good catalytic activity towards *degradation* of chloro- and nitro-phenols (Figure 1).

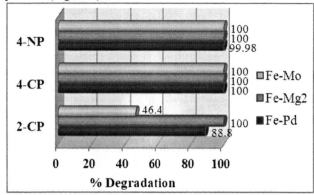

Figure 1: %age degradation of 2-CP, 4-CP and 4-NP

Based on GC-MS results and previous researches the following mechanism was proposed. In case of alloys, Fe is known to play primary role by producing hydrogen by water reduction (equation 1). Other metals (M = Pd, Mg or Mo) involved in the degradation along with Fe act as hydrogen collector [28], thus enhancing the reaction rate by providing catalytic sites to break H——H bond [9] resulting in formation of H˙ as depicted in equation 2. The hydrogen produced as a result plays an active part in degradation of CPs and NPs as presented in Scheme 1-3.

$$Fe + 2\,HOH \longrightarrow Fe^{+2} + H_2 + 2\,OH^- \qquad (1)$$

$$M + H_2 \longrightarrow H^{\cdot} + H^{\cdot} \qquad (2)$$

Schematic representation of degradation mechanism of 2-CP is presented in Scheme 1. It is obvious from the mechanism layout that 2-CP first degrades to phenol **(1)** by chloride abstraction with help of H˙ resulting in cleavage of C–Cl bond and elimination of HCl. **(1)** is degraded further following two paths.

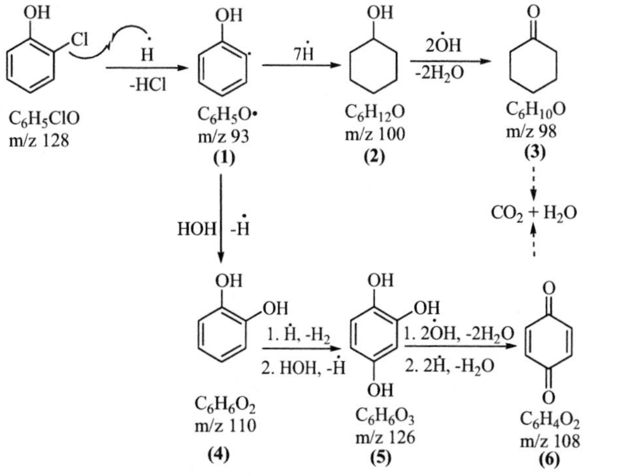

(1) Phenol (3) Cyclohexanone (5) Benzene-1,2,4-triol

(2) Cyclohexanol (4) Pyrocatechol (6) Cyclohexa-2,5-diene-1,4-dione

Scheme 1: Proposed mechanism of degradation of 2-CP

In first route reduction of (1) by H˙ results in (2), oxidation of which by ˙OH gives (3) that undergoes ring opening leading to formation of aliphatic intermediates that ultimately degrades to CO_2 and H_2O. In second path, (1) attacks water molecule and forms (4) along with H˙. This radicle attacks para position of (4) followed by attack on water molecule to give (5) which on oxidation assisted by ˙OH and reduction by H˙ results in (6) that further degrade to CO_2 and H_2O by series of ring opening intermediates.

The degradation path followed by 4-CP (Scheme 2) is not much different from the 2-CP and starts with formation of (7) that degrades to environmentally friendly products by two possible path ways. First path is identical to that of 2-CP while the second path proceeds with attack of (7) on water resulting in (10). This then oxidizes to (11) by ˙OH and reduced to (12) H˙. After that ring opening takes place and through various intermediate stages the final degradation to CO_2 and H_2O takes place.

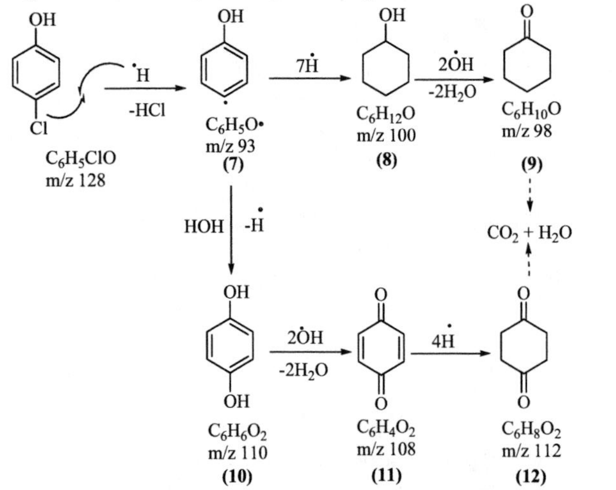

(7) Phenol   (9) Cyclohexanone   (11) Cyclohexa-2,5-diene-1,4-dione

(8) Cyclohexanol   (10) Hydroquinone   (12) Cyclohexane-1,4-dione

Scheme 2: Proposed mechanism of degradation of 4-CP

Degradation of 4-NP follows the path as given in Scheme 3. In this case OH˙ plays important role in degradation as against H˙ which is the main specie that has active role in degradation of chlorophenols'.

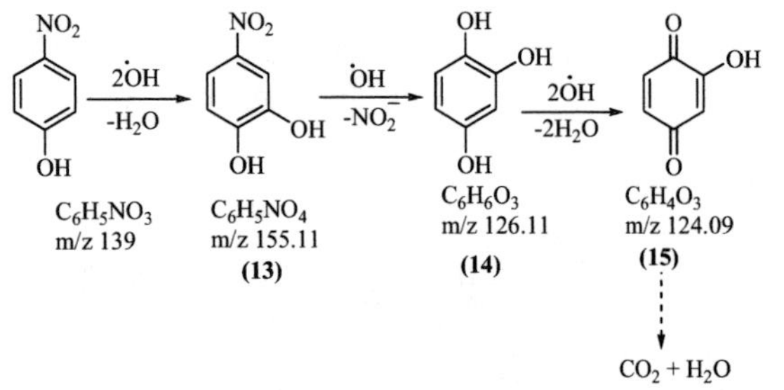

(13) 4-nitrobenzene-1,2-diol   (14) Benzene-1,2,4-triol

(15) 2-hydroxycyclohexa-2,5-diene-1,4-dione

Scheme 3: Proposed mechanism of degradation of 4-NP

The ˙OH may be generated by following route [29]

˙OH, hence produced, attacks the H on ortho position of 4-NP to produce water and 4-NP radical, that combines with another ˙OH to give (13). Further attack of ˙OH on para position of (13) results in elimination of nitro group and formation of (14) which then degrades to (15) and finally to CO_2 and H_2O through series of ring opening reactions.

Hence, the degradation studies carried out in absence of sunlight have shown that all the alloys show good catalytic activity towards *degradation* of chloro- and nitro-phenols (Fig. 1) to environmentally friendly products. Of the three nanoalloys used in this study, remarkable activity is observed in case of Fe-Mg$_2$ towards 2-CP while in case of Fe-Pd and Fe-Mo %age degradation is 88.8% and 46.4%, respectively. 100% degradation of 4-CP is observed for all the nanoalloys. Considering this, it can be safely concluded that alloys having cheap and widely abundant metal like Mg can efficiently replace alloys having expensive metals like Pd which is known for its dechlorinating activity [7,10-12]. Similarly, in the case of 2-CP, Fe-Pd shows less catalytic activity as compared to the Fe-Mg$_2$ which has 100% degradation activity towards 2-CP. The complete degradation of CPs to CO_2 and H_2O even in absence of sunlight and without utilizing expensive techniques is an important aspect of this study. This is in contrast to earlier studies that *only* resulted in dechlorination of the harmful contaminants [12,15] and in some cases even lead to the formation of more harmful products [7,30]. This high catalytic activity of Fe-Mg$_2$ alloy can be attributed to higher redox potential of Mg in comparison to other metals that are alloyed with Fe. The same trend is also shown in degradation of 4-NP that is almost 100% for all the alloys. Thus, Fe-Mg$_2$ is better *economical alternative* to Fe-Pd for the complete and safe degradation of substituted phenols because of abundant nature of its' constitutes (Fe and Mg most abundant metals of earth crust) and hence easily available and inexpensive in addition to being environmental friendly.

4   CONCLUSION

Three nanoalloys Fe-Pd, Fe-Mg$_2$, Fe-Mo prepared by simple chemical approach are found to be highly active towards degradation of carcinogenic water pollutants i.e. 2-chlorophenol, 4-chlorophenol and 4-nitrophenol into environment friendly products like water and carbon dioxide at room temperature and even in absence of sunlight or any other aid. These nanoalloys show maximum activity for 4-CP, while Mg containing alloy show good catalytic activity towards degradation of 2-CP, 4-CP and 4-NP. Fe-Mg$_2$ due to its remarkable catalytic activity towards all the substituted phenols is better economical and efficient alternative to already existing catalysts.

REFERENCES

[1] J. Bandara, J. A. Mielczarski, A. Lopez and J. Kiwi, Appl. Catal., B 34, 321, 2001.

[2] R. Cheng, J. Wang, W. Zhang, Front. Environ. Sci. Engin. China, 2, 103, 2008.

[3] A. P. Annachhatre, S. H. Gheewala, Biotechnol. Adv. 14, 35, 1996.

[4] T. Lai, J. Liu, K. Yong, Y. Shu, C. Wang, J. Hazard. Mater. 157, 496, 2008.

[5] X. Li, D. W. Elliot, W. Zhang, Crit. Rev. Solid State Mater. Sci. 31, 111, 2006.

[6] R. Cheng, J. Wang, W. Zhang, J. Hazard. Mater. 144, 334, 2007.

[7] W. Zhang, C. Wang, H. Lien, Catal. Today 40, 387, 1998.

[8] B. Schrick, J. L. Blough, A. D. Jones, T. E. Mallouk, Chem. Mater. 14, 5140, 2002.

[9] W. Liu, J. Biosci. Bioeng. 102, 1, 2006.

[10] H. Lien, W. Zhang, Appl. Catal., B 77, 110, 2007.

[11] E. V. Golubina, E. S. Lokteva, V. V. Lunin, N. S. Telegina, A. Yu. Stakheev, P. Tundo, Appl. Catal., A 302, 32, 2006.

[12] Y. Liu, F. Yang, P. L. Yue, G. Chen, Water Res. 35, 1887, 2001.

[13] F. He, D. Zhao, Appl. Catal., B 84, 533, 2008.

[14] C. J. Lin, S. L. Lo, Y. H. Liou, J. Hazard. Mater. 116, 219, 2004.

[15] X. Ma, M. Zheng, W. Liu, Y. Qian, J. Hazard. Mater., B 127, 156, 2005.

[16] Y. Tee, E. Grulke, D. Bhattacharyya, Ind. Eng. Chem. Res. 44, 7062, 2005.

[17] Y. M. Slokar, A. M. Le Marechal, Dyes Pigm. 37, 335, 1998.

[18] T. T. Bovkun, Y. Sasson, J. J. Blum, J. Mol. Catal. A: Chem. 242, 68, 2005.

[19] O. Hamdaoui, E. Naffrechoux, Ultrason. Sonochem. 16, 15, 2009.

[20] M. H. Priya, G. Madras, J. Photochem. Photobiol., A 179, 256, 2006.

[21] K. Khuanmar, W. Wirojanagud, P. Kajitvichanukul, S. Maensiri, J. Appl. Sci. 7, 1968, 2007.

[22] J. Bandara, U. Klehm, J. Kiwi, Appl. Catal., B 76, 73, 2007.

[23] J. Hofmann, U. Freier, M. Wecks, A. Demund, Top. Catal. 33, 243, 2005.

[24] N. San, A. Hatipoğlu, G. Koçtürk, Z. Çinar, J. Photochem. Photobiol., A 146, 189, 2002.

[25] X. L. Hoa, M. H. Zhou, L. C. Lei, J. Hazard. Mater. 141, 475, 2007.

[26] R. Nazir, M. Mazhar, M. J. Akhtar, M. R. Shah, N. A. Khan, M. Nadeem, M. Saddique, M. Mahmood, N. M. Butt, Nanotechnology doi. 10.1088/0957-4484/19/18/185608, 2008.

[27] R. Nazir, M. Mazhar, J. Akhtar, M. Nadeem, M. Siddique, M. R. Shah, S. K. Husanain, J. Alloys Compd. 479, 97, 2009.

[28] F. Cheng, Q. Fernando, N, Korte, Environ. Sci. Technol. 31, 1074, 1997.

[29] Q. Dai, L. Lei, X. Zhang, Sep. Purif. Technol. 61, 123, 2008.

[30] W. Ahn, S. A. Sheeley, T. Rajh, D. M. Cropek, Appl. Catal., B 74, 103, 2007.

Table 1: GC-MS results for phenols showing representative m/z values for the corresponding peak retention time, R_t

Nanoalloy	2-Chlorophenol			4-Chlorophenol			4-Nitrophenol		
	R_t	m/z	Species	R_t	m/z	Species	R_t	m/z	Species
Fe-Pd	3.375	93	**(1)**	2.911	98	**(9)**	5.273	155	**(13)**
	3.541	98	**(3)**	3.305	100	**(8)**	11.773	139	4-NP
	3.909	100	**(2)**	4.699	108	**(11)**	25.053	124	**(15)**
				10.446	110		26.550	126	**(14)**
Fe-Mg$_2$	3.357	93	**(1)**	2.727	93	**(7)**	5.268	155	**(13)**
	3.541	98	**(3)**	2.911	98	**(9)**	25.043	124	**(15)**
	3.901	100	**(2)**	3.305	100	**(8)**	26.557	126	**(14)**
				4.699	108				
				10.446	110				
Fe-Mo	3.533	98	**(3)**	2.911	98	**(9)**	5.270	155	**(13)**
	3.901	100	**(2)**	3.305	100	**(8)**	25.130	124	**(15)**
	5.980	128	2-CP	4.804	108	**(11)**	26.461	126	**(14)**
	10.954	110	**(4)**						
	26.550	126	**(5)**						

The influences of nanoscale titanium dioxide particle size and crystal structure on light absorbance

Mingqiang Zou[*], Fuqiang yang[*], Jixiang Ma[*], Yan Chen[*], Lin Cao[**] and Feng Liu[*]

[*]Chinese Academy of Inspection and Quarantine
No. A3, Gaobeidian North Road, Chaoyang District, Beijing, China, mingqiangz@sina.com
[**]University of Science and Technology Beijing, Beijing, China, lincaoustb@yahoo.com.cn

ABSTRACT

Nanoscale titanium dioxide is an important physical sun-block material and is widely used in sunscreen. In our test, the influences of the particle size and crystal structure on light absorbance were examined. The experiment shows that the absorbance (Abs) of nanoscale titanium dioxide in UV-ray district is much higher than it in visible light district under the same concentration (0.04 g/L). The Abs of nanoscale rutile keeps increasing, reaches the highest at 360 nm, and then starts to decrease as the wavelength becomes shorter. The nanoscale anatase has the similar light absorbance characteristic, with the highest Abs peak at 320 nm. In the UVA district, the rutile has stronger absorbance ability than the anatase, while in UVB and UVC, there is a similar absorbance between rutile and anatase. Compared with the bulk titanium dioxide, the nanoscale titanium dioxide is capable of absorbing the UV-ray more effectively and let the visible light pass through more easily. Since the anatase is of the stronger activity than rutile and thus probably harms the skin more easily, the nanoscale rutile should be the best selection used for sunscreen. However, when the particle size of rutile is too small, for example, smaller than 10 nm, its Abs tends to be much higher in UVB and UVC district, but lower in UVA and visible light district. Therefore the appropriate particle size and crystal structure should be considered when nanoscale titanium dioxide is applied for sun-block effect. The detailed test optimizations are being carried out in our lab.

Keywords: nanoscale titanium dioxide, particle size, crystal structure, light absorbance

1 INTRODUCTION

Nanoscale titanium dioxide has been widely used in cosmetic sunscreens for more than 25 years. In 2008, a survey of nanomaterials sold in EU cosmetic products amongst member companies of Colipa and national trade associations identified: One of the most frequently used materials is titanium dioxide. The test aims to examine the influences of the particle size and crystal structure of nanoscale titanium dioxide on the light absorbance. And then try to find the appropriate particle size and crystal structure for sunscreen application. Particle shape and element-doped is excluded in this text.

In this study we get six kinds of titanium dioxide samples. One is synthesized in the laboratory and four are bought in the market and one is abstracted from the sunscreen. The synthesis of nanoscale titanium dioxide in liquid phase reactions has been of great interest over the past decades [1-2]. Here we use hydrothermal methods to generate the 6~7nm nano-TiO$_2$ rutile samples. The four kinds of titanium dioxide are provided by Hehai Nanometer Science & Technology Co. Ltd. It is a key point to control the temperature at a low state to avoid the titanium dioxide phase changing and grain growing when abstract the titanium dioxide form the sunscreen.

2 MATERIALS AND METHODS

2.1 Material preparation

Nanoscale rutile synthesized in the laboratory: The nanoscale rutile synthesis was performed using a modified procedure as described by Cheng etc [3]. Briefly, under continuous stirring, 23 mL titanium tetrachloride slowly add into the ice water (400ml) to prepare the TiCl$_4$ suspension solution, then hydrochloric acid was used to adjust the pH <1. Next, 0.8 mol NaCl was added into the mixture and a clear solution was formed after stirring for 5~6 hours at room temperature. After heated to 90℃ and refluxed for 2h, the nanoscale TiO$_2$ particles were recovered from this solution. Finally, the pellet was dialyzed several times till pure enough and then was frozen-dry to get the sample Nano-R-1.

Four kinds of titanium dioxide bought in the market: They were as marked rutile (Bulk-R), nano-rutile (≤30nm, Nano-R-2), anatase (Bulk-A) and nano-anatase(≤30nm, Nano-A), respectively.

The sample abstracted from sunscreen: The sunscreen was added into the water and heated at 90℃ till the organic component was destroyed. Next, normal hexane was added to the mixture. The mixture was heated till it uniformity, then centrifugated, washed several times, and dried in an oven maintained at 60℃. The resulting powder 0912 was finally obtained.

2.2 Crystal structure and particle size measure

All nanoscale titanium dioxide powders were characterized by X-ray diffraction (XRD, D/max-2500/PC, Rigaku, Tokyo, Japan). The crystal structure can be given by the position of diffraction peak (27.4° of rutile type 110 crystal face; 25.3° of anatase type 101 crystal face). The background of the diffraction profile was determined after smoothing by Savitsky-Golay method [4]. Then separate the $K_{\alpha 1}$ and $K_{\alpha 2}$ doublet to get the full width at half maximum (FWHM) by Rachinger's algorithm [5]. At last the true FWHM can be determined without instrument broadening [6]. Then particle sizes were estimated according to the Debye-Scherrer formula:

$$D = \frac{K\lambda}{\beta \cos\theta} \qquad (1)$$

Where: D is the average crystallite size, K is the Scherrer Constant (here use 0.89), λ is the wavelength of X-ray (0.154nm for Cu $K_{\alpha 1}$ radiation), β is the true FWHM, θ is the diffraction angle.

2.3 Absorbance test

All the samples are tested by spectrophotometer (U3310, Hitachi, Tokyo, Japan) at the concentration of 0.04g/L except the sample which was abstracted from sunscreens (due to sample preparation limited, the sunscreen used titanium dioxide was tested at a higher concentration). The measurement parameters are: Scan Speed: 120 nm/min; Sampling Interval: 0.20 nm; Slit Width: 1 nm; Baseline Correction: User 1; Path Length: 10.0 mm.

3 RESULTS AND DISCUSSION

3.1 The crystal structure and particle size of these samples

The crystal structure and particle size of these samples are shown as follow. All the particle size is estimated by Debye-Scherrer formula.

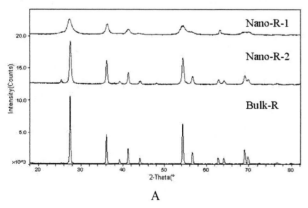

A

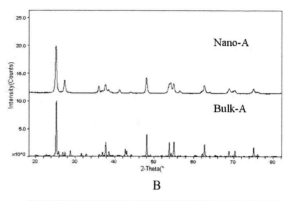

B

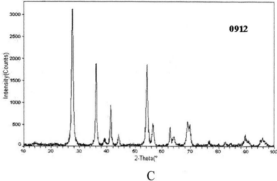

C

Figure 1: The X-ray diffraction patterns of the six titanium dioxide samples

Name	Crystal structure	Particle size
Nano-R-1	Rutile	7nm
Nano-R-2	Rutile	24nm
Bulk-R	Rutile	≥100nm
Nano-A	Anatase	22nm
Bulk-A	Anatase	≥100nm
0912	Rutile	13nm

Table 1: The crystal structure and particle size of these samples

Fig 1.A is the X-ray diffraction pattern of the three rutile samples. Nano-R-1 is prepared in lab. Nano-R-2 and Bulk-R are bought in the market. The particle size of these samples are 7nm, 24nm and bigger than 100nm separately. Fig 1.B is the X-ray diffraction pattern of two anatase samples both are bought in the market. One is 22nm and the other is bigger than 100nm. Fig 1.C is the X-ray diffraction pattern of the titanium dioxide which abstracted from the sunscreen. It is rutile and the particle size is 13nm.

3.2 The light absorbance of different crystal structure

Fig.2 is the Abs curve of Nano-R-2, Bulk-R, Nano-A and Bulk-A. The concentration is 0.04g/L. From this figure we can conclude:

1. The Abs of the two bulk titanium dioxide is very alike. They have UV-ray absorption little than visible light absorption.

2. The Abs curve of nanoscale rutile and nanoscale anatase is also alike. They both have a higher absorption in UV-ray district higher than it in visible light district. As the wavelength become shorter the Abs of nanoscale titanium dioxide quiken up increase, the Abs of rutile reach the highest at 360nm, and then become reducing when the wavelengh get shorter. The anatase has the highest Abs at 320nm, and then become shorter too. This characteristic is very alike.

3. However, the Abs curves of the two crystal structures still have some difference. First, they have different absorbance peak point, rutile is 360nm and anatase is 320nm. Second, in the UVA (320~400nm) district the nanoscle rutile has stronger absorption ability than the nanoscale anatase, while in UVB and UVC there is a similar absorbance between rutile and anatase. Considering the long-term harm to human skin of UVA and the higher photoactivity of anatase [7], the nanoscale rutile is safer and better than nanoscale anatase for sunscreen application.

higher than that of bulk. Thus we can conclude that the nanoscale material has higher transimission ability in visible light district and higher absorption ability in UV-ray district than that of bulk material. The nanoscale rutile is more effective for sunscreen application.

4. However, when the particle size of rutile is too small, as shown in figure 3, the Nano-R-1 is 7 nm, it will has much higher Abs in UVB and UVC district but lower Abs in UVA district. Compare the Nano-R-1 and Nano-R-2, we can see that the Nano-R-1 has a much higher absorption peak, but the absorbance area is much narrower than Nano-R-2. These two curves crossed at 318nm, so when the wavelength is longer than 318nm, the Abs of Nano-R-1 is lower than that of Nano-R-2. Furthermore, when the wavelength is longer than 350nm, the material can absorb litter light.

It is well know that the UVA is the main UV-ray on the surface of earth, it cause the most damage to our skin. Thus, when choose sunscreen used rutile, the particle size should not be too small. The best choice is that the material has an effective absorbance area overlay UVA (320nm~400nm) and UVB (290nm~320nm).

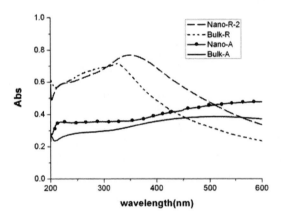

Figure 2: The Abs curve of rutile and anatase titanium dioxide samples

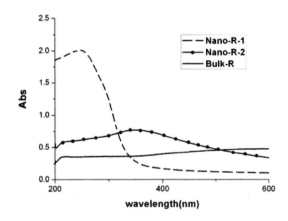

Figure 3: The Abs curve of rutile with different particle size

3.3 The light absorbance of different particle size

Fig 3 is the Abs curve of the rutile of different particle size. The concentration is 0.04g/L.

From this figure we can conclude:

1. Different particle size rutile has different light absorption ability.

2. The absorption area is blue shift as the particle size become smaller.

3. The smaller the particle is, the higher the absorbance peak is. The nanoscale rutile has a higher absorbance peak than that of bulk rutile. Compare the Abs of the Nano-R-2 and Bulk-R we can see that the tow lines are crossed at 508nm, when the wavelength longer than 508nm, the bulk has higher Abs than that of nanoscale rutile, when the wavelength shorter than 508nm, the Abs of nanoscale

3.4 The light absorbance of the sunscreen used titanium dioxide

Fig 4 is the Abs of the titanium dioxide sample which is abstracted from the sunscreen, the XRD shows it is rutile with the particle size about 13nm. The concentration of the sample is higher than 0.04g/L, while the Abs curve still can be compared with the other samples' Abs curves. From fig 4 we can see:

1. The Abs of this sample in UV-ray district is higher than the Abs in visible light district.

2. The absorbance peak of this sample is 296nm, which is in the UVB range. It has an effective absorbance when the wavelength is shorter than 360nm. Compare with Nano-R-2 we can see that if the particle size is bigger, the Abs area will be wider, but at the price of lower Abs peak. Thus, when use the bigger particle size rutile, it should be add more material in the sunscreen to get a high SPF.

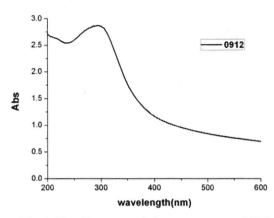

Fig 4: The Abs curve of the sunscreen used TiO$_2$

4   CONCLUSIONS

Six different kinds of titanium dioxide have been studied. The influences of nanoscale titanium dioxide particle size and crystal structure on light absorbance also been discussed. Considering the light absorbance ability and photoactivity, rutile is a better choice. Besides, the particle size is also very important for sunscreen used titanium dioxide. The nanoscale rutile has a stronger UV-ray absorption ability than bulk rutile and the smaller the particle is, the higher the absorbance peak it has. But as the particle size gets smaller the absorbance area is blue shift. Thus, if not considering the surface modification and element-doped, in order to get the most effective absorbance, the particle size should not be too big or too small. From our data, the particle size around 10~30 nm of rutile might be the best choice. While, different company may choose different particle size titanium dioxide to produce their special sunscreen.

.

REFERENCES

[1] Chemseddine A. and Mortiz, Nanostructure titania: Control over nanocrystal structure, size shap and organization, Eur. J. Inorg, 334, 1317, 1999.

[2] Trentler T. J., Denler, T. E., Bertone, J. F. etc., Synthesis of TiO2 nanocrystals by nonhydrolytic solution-based reactions, J.Am. Chem. Commun, 121, 1613, 1999.

[3] Humin Cheng, Jiming Ma, Zhenguo Zhao, and Limin Qi, Hydrothermal Preparation of Uniform Nanosize Rutile and Anatase Particles, Chem. Mater. 7, 663, 1995.

[4] Savitzky A, Golay M, Smoothing and Differentiation of Data by Simplified Least Squares Procedures, Anal Chem 36, 1627, 1964.

[5] Rachinger W. A, A Correction for the α1 α2 Doublet in the Measurement of Widths of X-ray Diffraction Lines, J. Sci. Instrum., 25, 254, 1948.

[6] Jones F.W., The Measurement of Particle Size by the X-Ray Method, Proc. Roy. Soc. London A 166, 16, 1938.

[7] Christie M. Sayes, Rajeev Wahi and Preetha A. Kurian etc, Correlating Nanoscale Titania Structure with Toxicity: A Cytotoxicity and Inflammatory Response Study with Human Dermal Fibroblasts and Human Lung Epithelial Cells, Toxicol Sci., 92, 174, 2006.

Size-Controllable Synthesis and Characterization of Wide Band Gap Semiconductor Oxide Nanoparticles

A.Sharma*, A. Sharma**, R. Sharma***

*Electrical Engineering Department, AKG College of Engineering, Uttar Pradesh Technical University,NH-24 Bypass, Ghaziabad,UP India; **Nanotechnology Lab, Electrical Engineering Department, MP A&T University, Udaipur, Raj 313001 India; ***Center of Nanobiotechnology, TCC and Florida State University, Tallahassee, FL 32304 USA

ABSTRACT

A search of new class of semiconductors with metal oxides of specific sizes motivated us to synthesize nanoparticles. The preparation of nanocomposites using dispersed nanoparticles into solid hosts is nowadays an attracting scientific art and technological interest. The properties of these hybrid nanocomposite systems are different and often improved, compared with their isolated counterparts. In this study we improved the synthesis of controlled-size semiconductor oxide nanoparticles based on the impregnation and decomposition of transition-metal 2-ethylhexanoate-based precursors in mesoporous hosts aiming to prepare MO2 (M = Ti, Ce, and Sn). The pore size of the host material was enough to control the size of the guest material synthesized within it. The linear mass gain for each cycle in the synthesis process was an advantage of this method, because it allowed the control of nanoparticle growth via a layer-by-layer assembly. We report the results of X-ray diffraction, Compton profiling, transmission electron microscopy, and Raman spectroscopy for SnO2, TiO2 and CeO2 nanocrystals. Raman spectroscopy dependence on nanocrystal sizes allowed nanocrystals to their use by this technique as prompt characterization tool to estimate nanocrystal size.

Keywords: semiconductor, oxide nanoparticles

1 INTRODUCTION

Researches on nanostructured semiconductor materials are the basis for electronics, optoelectronics, sensors, catalysis. Size-induced characteristics give novel properties to aggregate, forming microsized clusters of nanocrystals as integrated chemical system (ICS) as nanoreactors. ICS is heterogeneous and multiphase system (host and guest) designed and arranged (nanocrystals anchored in the pores) for specific functions and/or to carry out specific chemical reactions or processes-photocatalytic activity and photoluminescence intensity of TiO_2 nanoparticles to make high dispersion of Ti ions and/or the coordinative unsaturation of surface Ti ions. Wide band gap semiconductor oxides are SnO_2, TiO_2 and CeO_2 with high photocatalytic activity and high surface to volume ratio. If we consider a spherical particle, the surface to volume ratio exhibits $1/L$ dependence, where L is the nanoparticle diameter. Nanocrystal synthesis and characterization is focus in this paper.

2. NANOCRYSTAL SYNTHESIS

Metallo-organic decomposition (*MOD*) using layered solids, zeolites, carbon nanotubes. Nanocrystals anchored in porous Vycor glass (PVG, Corning code 7930) with an α-NbPO$_5$ at optimal temperature, atmosphere, time makes inorganic powders and thin films in cycles known as "one impregnation-decomposition cycle – IDC".

3. EXAMPLES OF SYNTHESIS

The integrated chemical systems (nanocrystals anchored on porous hosts) are following as illustration:

3.1. α-Nbpo$_5$ Glass-Ceramic Monolith is obtained from the glass system 6Li$_2$O–18Nb$_2$O$_5$–43CaO–33P$_2$O$_5$ has channel-like pores 130 nm shown in Figure 1a and the SEM image.

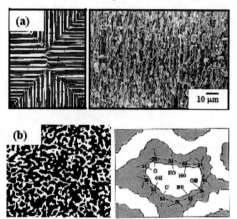

Figure 1. (a) (left) of the popsicle-like structure of porous α-NbPO$_5$ and scanning electron microscope (SEM) image (right) of the host with average pore diameter around 130 nm. (b) The porous structure of PVG (right) and the details of the chemical environmental on the pore surface (left).

3.2. Porous Vycor Glass Monolith (10x10x1) mm^3 with PVG plates in a 2 mol L^{-1} HCl-acetone become rich in silanol (Si–OH) and siloxane (Si–O–Si) groups on surface with pore diameter distribution (2 to 20 nm).

3.3. Impregnation of Single-Source Precursor into α-Nbpo$_5$ and PVG hosts is done for metallo-organic precursor titanium di-(n-propoxy)-di-(2-ethylhexanoate) [hereafter Ti(OnPr)$_2$ (hex)$_2$ or Ce(III) 2-ethylhexanoate [Ce(hex)$_3$] or Sn(II) 2-ethylhexanoate [Sn(hex)$_2$] impregnated with 50 mL, 1.0 mol L^{-1}, hexane as solvent for 24 h, at room temperature.

3.4. Thermal Decomposition at 600 °C for 8 h for Ce(hex)$_3$ at 750 °C for 8 h for Sn(hex)$_2$ and Ti(OnPr)$_2$(hex)$_2$ at ambient atmosphere using a 10 °C min^{-1} heating rate removes MO$_2$@PVG (M = Sn, Ti and Ce) and TiO$_2$@α-NbPO$_5$ plates at 200 °C (impregnation–decomposition cycle or IDC) make free MO$_2$ (M= Sn, Ti and Ce) nanocrystals.

4. EXAMPLES OF NANOCRYSTALS

4.1.Tin Dioxide is a wide band gap semiconductor with

chemical, optical and electronic properties is useful in catalysis, sensors for fuels and toxic gases, in Li-based batteries. The wide band gap (3.5 eV) makes it optical transparent in the visible region.

4.2. SnO₂ into Porous Hosts shown in Fig. 2a and 2b exhibit a specific diffraction peaks. How size is significant?

The crystallite size, L, is given by: $$L = \frac{k \cdot \lambda}{B \cdot \cos\theta} \quad (1)$$

where k is a geometry dependent constant (about 0.9 for spherical shape), λ is the X-ray wavelength, θ is the Bragg diffraction angle (in rad) and B is full-width at half maximum (FWHM) of the peak after correcting the instrumental broadening which is given by: $$B = \sqrt{B_m^2 - B_s^2} \quad (2)$$

where B_m and B_s stand for the measured FWHM of the sample and a standard. The SnO₂ crystallite sizes are listed in Table 1 showing pyrolysis of the precursor inside the monoliths makes smaller crystal sizes.

Table 1. Average crystallite sizes estimated by using Scherrer´s model for the semiconductors formed by pyrolysis of the precursors inside and outside (free) host.

Semiconductor	Average crystallite size, L (nm)		
	Free	in PVG [3]	in α-NbPO₅ [3]
SnO₂	54 [1]	6	14
CdS	24 [2]	5	_ _ _ _ _

Take home points: The linear mass gain is important. Monoliths preserve their porosity and capacity of interacting with single source precursors.

5. Titanium oxide TiO₂ has applications in photocatalysis, optics sensors, with small crystal size and high surface to volume ratio (anatase form). Temperature influences at different heat treatments can be estimated according to the equation: (3)

$$\chi = \frac{1}{1 + 0.8 \cdot \left(\frac{I_A}{I_R}\right)}$$

where χ is the TiO₂^R mass fraction in the powders, while I_A and I_B are the X-ray integrated intensities of the (101) reflection of TiO₂^A and the (110) reflection of TiO₂^R.

5.3. Tio₂@PVG is obtained with 10 IDC (upper trace) is shown in Figure 3a along with the patterns for bulk anatase TiO₂ (lower trace) with average nanocrystal diameter is 5.0 nm. The silicon threshold in the EELS spectra is 103 and 106 eV for Si and SiO₂.

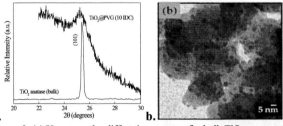

Figure 2. (a) X-ray powder diffraction patterns for bulk TiO₂ anatase and TiO₂@PVG obtained with 10 IDC. (b) TEM bright field image of TiO₂@PVG obtained with 3 IDC. The average nanocrystal size was found to be 5.0 nm.

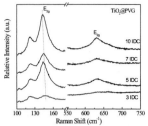

Figure 3. Raman spectra of different TiO₂ nanocrystals size dispersed into PVG obtained from different numbers of IDC.

The Raman spectra characterize by broad bands several peaks from TiO₂ nanocrystals(144 and 635 cm⁻¹, 451 and 615 cm⁻¹ of the rutile phase) based on frequency and relative intensity associated with the matrix. The lowest frequency E₂g mode experiences an upshift and becomes broader as the number of IDC increases (Figure 3) size-induced phenomena in nanostructured materials. Raman profile is based on the following equation:

$$I(\omega) \alpha \int_{BZ} \frac{|C(0,q)|^2 L^3 q}{[\omega - \omega(q)]^2 + (\gamma/2)^2} \quad (4)$$

where γ is the line width of the Raman peak and $\omega(q)$ is the phonon dispersion curve for the bulk counterpart. For spherical nanocrystals $|C(0,q)|^2 = \exp\left(-q^2 L^2 / 16\pi^2\right)$ where L is the nanocrystal average diameter. Both γ and L are fitting parameters. The phonon confinement is evident for the lowest frequency E₉ mode located at $\omega_0 = 144$ cm⁻¹ for bulk anatase-TiO₂. For simplicity, phonon dispersion can be of this mode using a linear chain model:

$$\omega(q) = \omega_0 + 20[1 - \cos(0.3768q)] \quad (5)$$

As the nanocrystal size gets smaller the phonon confinement involves large q values and the maximum of the Raman peak follows the same trend of $\omega(q)$ that is positive in the case of the E₂g mode for bulk TiO₂. Figure 4 shows Raman profile of the E₂g mode according to Eq. (4) for diameters of 15, 10 and 5 nm and considering $\Gamma_0 = 16$ cm⁻¹.

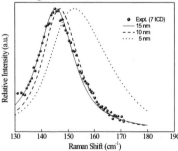

Figure 4. Size dependent Raman scattering profile of the lowest frequency E₂g mode for TiO₂. The solid points stand for the experimental data obtained for a TiO₂@PVG sample obtained with 7 IDC. The solid, dashed and dotted lines were obtained by plotting Eq. (4) for nanocrystal diameters of 15, 10 and 5 nm, respectively. The linewidth used was 16 cm⁻¹.[1].

The precursor promotes coverage of the porous surface due to its interaction with the Si–OH sites present in the PVG matrix. The interaction of metallo-organic molecule with Si–OH sites is important and Si–O–Ti bonds in decomposition during nanocrystal formation. The nanocrystal size changes with transition from anatase to rutile with influence of temperature, nucleation and growth processes at the nanocrystal surface. The anchoring process plays a fundamental role in stabilizing the nanosized-TiO$_2$ anatase phase.

5.4. Tio2@α-Nbpo5 and role of template: Peaks in XRD pattern of TiO$_2^A$ present a small 2θ variation.. Raman responses of rutile and anatase phases are distinct. TiO$_2$@α-NbPO$_5$ (A→R) transition Raman spectrum shows bands at 144 cm^{-1} [E$_g$(ν$_6$); vs], 399 cm^{-1} (B$_{lg}$; m), 519 cm^{-1} (B$_{lg}$; m) and 639 cm^{-1} [E$_g$(ν$_1$); m] while TiO$_2^R$ exhibits bands at 143 cm^{-1} (B$_{lg}$; w), 447 cm^{-1} (E$_g$; s) and 612 cm^{-1} (A$_{lg}$; s), assigned to phonon lattice modes (where: vs, very strong intensity; s, strong intensity; m, mid-intensity; and w, weak intensity) as shown in Figure 5a. Band at 144 cm^{-1} is intense and no peak of α-NbPO$_5$ host suggests the formation of TiO$_2^A$ phase at 1000 °C for 8 h byRaman band at 144 cm^{-1} [Figure 5b]. Role of the template is important in the structural phase of the TiO$_2$ nanocrystals.

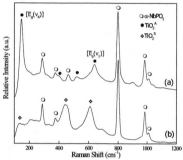

Figure 5. Raman spectra of (a) TiO$_2$@α-NbPO$_5$and (b) TiO$_2$@α-NbPO$_5$ with additional thermal treatment at 1000 °C for 8 h. [1].

For illustration, Figure 6 shows Raman spectra of TiO$_2^A$ nanocrystals. The [E$_g$(ν$_6$)] peak from TiO$_2^A$ changes from 145 cm^{-1} (free) to 148 cm^{-1} (inside the α-NbPO$_5$) and line widths modified from 19 cm^{-1} to 26 cm^{-1} suggest a smaller nanocrystal size L when the TiO$_2$ is prepared inside α-NbPO$_5$ compared with free TiO$_2$ (600 °C for 8 h), which presents L = 32 nm for TiO$_2^A$ and L = 18 nm for TiO$_2^R$. The stabilization of TiO$_2^A$ in α-NbPO$_5$ depends on interaction with the surface sites of the pores of α-NbPO$_5$. The smaller the L, the larger the fraction of surface sites that are interacting with surface sites of the pores of α-NbPO$_5$, reducing the number of nucleus sites, and causing, as a consequence, an increase in the (A→R) transition temperature. Reduction of L leads to increases the growth rate. Example: TiO$_2^A$@α-NbPO$_5$. In impregnation-decomposition cycles, a total thermal annealing of 80 h at 750 °C treatment temperature promote the kinetic coalescence of crystallites. The Raman spectrum of α-NbPO$_5$ exhibits a very strong band at 805 cm^{-1} which is assigned to niobyl groups (Nb=O) from the α-NbPO$_5$ phase; the predominant phase of this porous glass-ceramic [27]. Table 2 shows IDC presents a {I[ν(Nb=O)^{805}/I[E$_g$(ν$_6$)^{144}] = 1.0} ratio smaller than {I[ν(Nb=O)^{805}/I[E$_g$(ν$_6$)^{144}] = 1.7} to suggest larger amount of TiO$_2^A$ present in the pores. FWHM of band [E$_g$(ν$_6$)] of TiO$_2^A$ in size evaluation depends on growth of L and mass increment via impregnation-decomposition cycles. Average TiO$_2$ nanocrystal inside α-NbPO$_5$ or PVG is different due to the different pore dimensions, i.e., a larger pore diameter

allows more solution to get in, increasing the precursor amount, thus leading to larger crystal sizes.

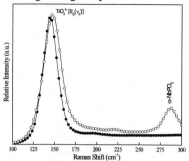

Figure 6. Raman spectra of TiO$_2^A$ obtained via decomposition of (full circle) free Ti(OnPr)$_2$(hex)$_2$ and (open circle) Ti(OnPr)$_2$(hex)$_2$@α-NbPO$_5$, in the region between 100-300 cm^{-1} for analysis of band width of the [E$_g$(ν$_6$)] phonon lattice mode of TiO$_2^A$. [1].

Table 2. {I[ν(Nb=O)^{805}/I[E$_g$(ν$_6$)^{144}]} ratio, position and FWHM of Raman bands of the [E$_g$(ν$_6$)] phonon lattice mode of the TiO$_2^A$ for: TiO$_2^A$@α-NbPO$_5$ from 10 IDC, TiO$_2^A$@α-NbPO$_5$ from 5 IDC and TiO$_2^A$@α-NbPO$_5$ from 5 IDC+5TT. Reference[1,2]

System	{I[ν(Nb=O)^{805}/I[E$_g$(ν$_6$)^{144}]} ratio	Position (cm^{-1})	FWHM (cm^{-1})
TiO$_2$@α-NbPO$_5$ from 5 IDC	1.7	147	48
TiO$_2$@α-NbPO$_5$ from 5 IDC+5TT	1.5	146	40
TiO$_2$@α-NbPO$_5$ from 10 IDC	1.0	144	26

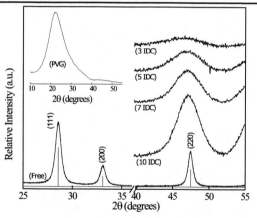

Figure 7. X-ray diffraction patterns for CeO$_2$@PVG nanocrystals obtained with 3, 5, 7 and 10 IDC. The X-ray pattern at the bottom is for free CeO$_2$ nanocrystals. The JCPDS standard for bulk CeO$_2$ is also shown as vertical lines [2]

6.1. Cerium oxide (CeO$_2$) is used as an active catalyst in vehicle emission systems for oxidation of pollutant gases, as a solid oxide fuel cell electrolyte material, as an electrode material for gas sensors for monitoring environmental pollution, as silicon-on-insulator structures , and as high T$_C$-superconductors. A route for enhancing the CeO$_2$ catalyst activity is to increase the surface to volume ratio, which can be achieved with smaller CeO$_2$ nanoparticles.

6.2. X-ray patterns of Ceo$_2$@PVG CeO$_2$ nanocrystals dispersed into PVG (obtained with 3, 5, 7, and 10 IDC) showed vertical lines, peak line widths broad suggestive of small size of the nanocrystals. The inset in Figure 7 shows

amorphous state. The shift of the diffraction peak to lower 2θ values indicates a lattice expansion for small CeO₂ particles. These line widths are much broader [≅ 3°] than those expected for bulk crystalline materials, where the diffraction peaks are as narrow as 0.15°. A typical TEM bright field image of the CeO₂@PVG nanocrystals obtained with 7 IDC is shown in Figure 8a. TEM image of the pristine PVG matrix has a different image (Figure 8b).

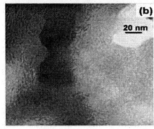

Figure 8. (a) TEM image of CeO₂@PVG nanocrystals obtained with 7 IDC and (b) of PVG. The inset to panel (a) represents particle size distribution. [2].

For illustration, TEM images show average nanocrystal size 2.3, 3.0 and 4.0 nm ± 0.9 nm for 3, 5, and 7 IDC samples similar with X-ray diffraction data. CeO₂ has fluorite structure Fm3m and shows vibrational spectra with one infrared active phonon (T₁ᵤ symmetry) and one Raman active phonon (T₂g symmetry). The T₂g Raman mode frequency downshifts and becomes broader as the number of IDC decreases, i.e., as the nanocrystal size decreases shown in Figure 9.

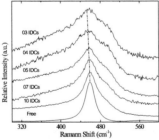

Figure 9. Raman spectra of CeO₂@PVG obtained with 3, 4, 5, 7 and 10 IDC. The Raman spectrum of free (outside the matrix) CeO₂ nanocrystals is also shown. For better comparison, the spectra are normalized to their maximum intensities. [2]

The TEM and Raman spectroscopy indicate strain in Eq. (4). Linear chain model describes the effective phonon dispersion relation and behaviour of the T₂g vibrational mode as ω(q) given by: $\omega(q) = 464 - 32\left[1 - \cos(0.5411q/2)\right]$ (6)

The phonon dispersion is responsible for involving large q wavevectors in the light scattering process. Figure 10a we plot the expected Raman profile of CeO₂ gives nanocrystal diameter as a function of the number of IDC. The nanocrystal average size versus the number of IDC presents a linear

behavior. For low diameter nanocrystals the signal-to-noise ratio is poor and the fitting of the Raman profile is difficult.

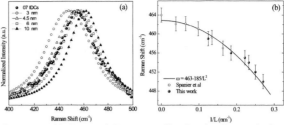

Figure 10. (a) Expected Raman profile for the T₂g mode in CeO₂ nanocrystal. (b) Raman frequency as a function of reciprocal nanocrystal size 1/L for CeO₂ nanocrystals.

The changes in lattice parameters and different strain level along the nanoparticle induce small changes in the position of the diffraction peak and broaden the Raman peak. This process leads to non-stoichiometric cerium oxide (CeO₂₋ₓ) nanocrystals. In the case of Raman scattering, the presence of oxygen vacancies along the nanocrystal also relaxes the q ≅ 0 selection rule which fundamentally has the same effect on the frequency and line width as the nanocrystal size. Raman spectroscopy gives true average diameter size of the nanoparticle distribution. Figure 10b shows the frequency plot of the main Raman peak as a function of the reciprocal nanocrystal size 1/L. A fit to the experimental data gives a quadratic dependence of frequency (in units of cm⁻¹) on 1/L, i.e.:

$$\omega(L) = 463 - 185/L^2,\qquad (7)$$

where L (in units of nm) is the average size of the nanocrystals, considering a spherical shape. CeO₂ nanocrystals can be obtained at 600 °C over 12 h final in size 10 nm. The nanocrystal growth does not occur via coalescence. PVG plays a role in controlling the coalescence process. In the first IDC the precursor promotes coverage of the porous surface, establishing some nucleation sites due to interaction with the Si–OH sites present in the matrix. The metallo-organic molecule is large enough to interact with all nearby Si–OH sites and, during the decomposition process, the dispersed nanocrystals are formed on the nucleation sites and they remain anchored to the porous matrix. The exceeding interfacial free energy is minimized by nanoparticle growth (forbidden coalescence) instead of creating new nucleation sites.

7 CONCLUSION

CeO₂, TiO₂ and SnO₂ nanocrystals have distinct size controllable properties prepared by metallo-organic decomposition, bottom-up approach in nanotechnology. Size-induced Raman spectra peaks of nanocrystals by phonon confinement model supports the average nanocrystal size measurement in TiO₂ and CeO₂ nanocrystals based on size restriction imposed by the pores and anchored through Ti–O–Si linkages. Space restriction and chemical anchoring properties prevent the coalescence process, thus allow control of the nanocrystal size via a linear mass increment.

8 REFERENCES

1. Mazali, I.O et al. Journal of Physics and Chemistry of Solids 66 (2005):37-46.
2. Mazali, I.O et al. Journal of Physics and Chemistry of Solids 68 (2007), 622-627.

Effect of grain size on Neél temperature, magnetic and electrical properties of NiFe$_2$O$_4$ nanoparticle synthesized by chemical co-precipitation technique

S. Manjura Hoque[*], M. A. Hakim[*], Indrijit Saha[**] and Per Nordblad[***]

[*]Materials Science Division, Atomic Energy Centre, Dhaka-1000, Bangladesh,
manjura_hoque@yahoo.com
[**]Department of Physics, Jahangirnagar University, Bangladesh
[***]Solid State Physics, Dept. of Engg. Physics, Uppsala University, Sweden

ABSTRACT

Nanocrystalline nickel ferrite has been prepared by chemical co-precipitation technique. X-ray diffraction patterns of as-dried and samples calcined at various temperatures have been studied. It has been observed that the sample in the as-dried condition is fully single phase and no extra peak could be observed in XRD patterns. The grain size has been obtained from Scherrer's formula and found as 7 nm in the as-dried condition. When the samples were calcined at higher temperatures subsequent grain growth has taken place. Samples calcined at 600°C led to the grain size as 16 nm. Further calcinations at 1200°C led to the grain size above 50 nm. Frequency dependence of real and imaginary part of initial permeability has been presented for the samples calcined at different temperatures. At lower firing temperatures i.e. when the grain sizes are smaller, the Neél temperature is lower. The coercivity of the samples in the as-prepared condition is almost zero, which is suggestive of superparamgnetic behavior at room temperature. Temperature and frequency dependence of resistivity and dielectric constant have been measured for the samples sintered at 1000-1350°C. Dispersion in the dielectric constants was exhibited by all the samples in the studied frequency range.

Keywords: Nickel Ferrite, superparamagnetism, coercivity, permeability, Neél temperature

1 INTRODUCTION

Ultrafine particles of nanometer dimensions located in the transition region between molecules and microscopic structures have revolutionized the recent advances in biology, physics, and chemistry. For biomedical applications the use of particles, which does not have any remanance with the withdrawal of applied field and possess superparamagnetic behavior at room temperature is being increasingly introduced [1,2]. NiFe$_2$O$_4$ and Fe$_3$O$_4$ nanoparticles is well known for their better biocompatibility and successful self-heating temperature rising characteristics. Furthermore, applications in biology and medical diagnosis and therapy require the magnetic particles to be stable in water at neutral pH and physiological salinity. The colloidal stability of this fluid will depend first, on the dimensions of the particles, which should be sufficiently small so that precipitation due to gravitation forces can be avoided. The main reason for considering NiFe$_2$O$_4$ nanoparticles as a hyperthermia agent is that these are expected to have urgently required magnetic properties for hyperthermia such as soft magnetism and small magnetic degradation at high frequency.

The exchange interaction in spinel ferrites where the antiparallel alignment of magnetic moments of A-site with B-site is mediated by oxygen ions is called superexchange interaction. The strength of the antiparallel coupling between the metal ions depends on the A-O-B bond angle with the strength being the maximum for an angle of 180°. The superexchange interaction is quite significant in ferrites due to the presence of metal ions in the A- and B-sites. Interesting magnetic properties such as magnetization, Curie temperature etc. of various ferrites have been observed to depend on the superexchange interaction strength, which is determined by the site occupancies of metal ions in the A- and B- sites. The size-dependent magnetic properties of ferrites have been investigated by many researchers for particles sizes less than 25 nm where the saturation magnetization decreases with particle size reduction due to surface spin effects. The much reduced M$_s$ in the nano-particle sample implies that outside a core of ordered moments, those on the surface layers are in a state of frozen disorder [3].

At lower firing temperatures i.e. when the grain sizes are smaller, the Neél temperature is lower. The decrease in Neél temperature with grain size can be explained on the basis of the finite size scaling theory. The finite size effect is predominant when the grain sizes are very small which results in a decrease in Neél temperature with grain size reduction. The small particles have a significant fraction of atoms on the surface and their exchange interaction should be weaker because of the lower coordination. They will have a reduced average Neél temperature compared to that of the interior atoms, which accounts for the decrease in Neél temperature at lower sintering temperature [3]. Real part of initial permeability, increases with the increase of grain growth. When the grain size is smaller the permeability is lower. The presence of small grain size interferes with wall motion, which decreases permeability and increases the stability region of real part of initial permeability. At higher frequencies, loss component of permeability are found to be lower if domain wall motion is inhibited and the magnetization is forced to change by rotation.

The purpose of the present study is to synthesize NiFe$_2$O$_4$ nanoparticle by chemical co-precipitation technique and then sintered at various temperature to avail a wide spectrum of particle size for different application starting from particle size, which will be below 10 nm and thus opens a possibility to work with hyperthermia and targeted drug delivery up to the samples sintered at high temperature for conventional ferromagnetic applications.

2 EXPERIMENTALS

We have used a standard co-precipitation technique to produce fine particles of $NiFe_2O_4$. The analytical grade of $Fe(NO_3)_3.9H_2O$, $Ni(NO_3)_2.6H_2O$ were mixed in required molar ratio and added to 6M NaOH solution with constant stirring at room temperature. The mixture was heated to 80°C with constant stirring. When the reaction was completed the precipitate was filtered and washed 10 times with distilled water. Finally, the precipitate was collected and heated at 90°C for 36 hours. X-ray diffraction patterns of as-dried powder yielded single phase ferrite. The powder was given the shape of pellets and sintered at various temperatures in the range of 200-1350°C. Curie temperature has been measured by using impedance analyzer and laboratory built furnace. Microstructure has been studied using Scanning Electron microscope. Magnetization measurements have been accomplished by VSM.

3 RESULTS AND DISCUSSION

In Fig. 1, X-ray diffraction patterns of as-dried and samples calcined at different temperatures in the range of 200-1350°C have been presented. In the as-dried condition single phase $NiFe_2O_4$ has formed with no extra peak. With the increase of sintering temperature peak width decreases which reflects the coarsening of particles. The size of the particles has been determined using Scherrer's formula from the FWHM of 311 peak and presented in Fig. 2. In the as-prepared condition, the grain size has been obtained as 7 nm. The grain size increases nonlinearly with the increase of sintering temperature. Lattice parameter presented in Fig. 3, is slightly higher than the conventional value which increases with sintering temperature. The increase in lattice parameter could be due to the deviation in cation distribution from the inverse spinel structure of $NiFe_2O_4$ [1].

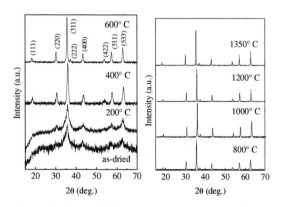

Figure 1. X-ray diffraction patterns of the as-dried and samples annealed at different temperatures.

In Fig. 4, microstructure of the as-dried and samples sintered at 1200, 1250 and 1350° C has been presented with magnification of ×15000. It can be observed from the figure that the microstructure of the sample sintered at 1200° C is heterogeneous. Grain size of different ranges exist in the microstructure. With the increase of sintering

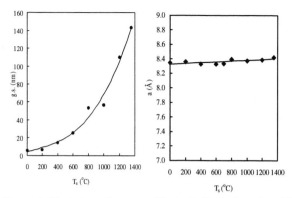

Figure 2. Variation of g.s. with sintering temperature

Figure 3. Variation of a with sintering temperature

temperature, microstructure becomes more homogeneous. It has been noticed with lower magnification (figure not shown) that there are more intergranular porosity for the sample sintered at 1200°C whereas more intrgranular porosity can be observed for the sample sintered at 1250 and 1300°C because of grain growth pores cannot migrate to the grain boundaries and therefore trapped within the grains. Intragranular porosity is more deleterious for magnetic properties since it acts as pinning centre for domain wall movement. Therefore, when there is exaggerated grain growth, there is deterioration of magnetic properties.

as-dried   1200°C

1250°C   1300°C

Figure 4. SEM micrographs of the sample in the as-dried and samples sintered (×15 k) at different temperatures.

In Fig. 5 temperature dependence of real part initial permeability has been presented in the temperature range of 600-1350°C. With the increase of sintering temperature, magnetic coupling increases. Magnetic coupling is the weakest for the sample sintered at 600°C. With the increase of sintering temperature improvement of magnetic coupling occurs. This has been reflected in the increase of Curie Temperature. Curie temperature estimated from the figures can be explained by finite scaling theory when the grain sizes are smaller. At lower firing temperatures i.e. when the grain sizes are smaller, the Neél temperature is lower.

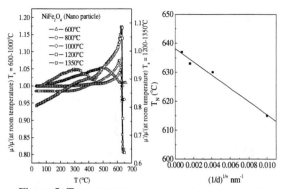

Figure 5. Temperature dependence of μ'/μ' (at room temperature)

Figure 6. Finite size scaling analysis of Neél temperature versus grain size

As shown in Fig. 6, the finite size effect is predominant when the grain sizes are very small which results in a decrease in Neél temperature with grain size reduction. The small particles have a significant fraction of atoms on the surface and their exchange interaction should be weaker because of the lower coordination. They will have a reduced average Neél temperature compared to that of the interior atoms, which accounts for the decrease in Neél temperature at lower sintering temperature.

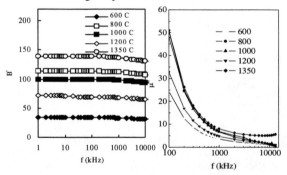

Figure 7. Frequency spectrum of real (a) and imaginary part (b) of permeability for the sample sintered at different temperature

In Fig. 7(a) and (b) frequency dependence of real and imaginary part of complex initial permeability has been presented. Real part permeability, increases with the increase of grain growth as expected i.e. when the grain size is smaller the permeability is lower. The presence of small grain size interferes with wall motion, which decreases permeability and increases the stability region of real part of initial permeability. The loss component represented by imaginary part of initial permeability decreases with frequency up to the measured frequency of this study of 13 MHz. At higher frequencies, losses are found to be lower if domain wall motion is inhibited and the magnetization is forced to change by rotation.

In Fig. 8, Field dependence of magnetization has been recorded for the samples in the as-prepared condition and sintered at 1350°C. The coercivity of the samples in the as-prepared condition is almost zero, which is suggestive of superparamgnetic behavior at room temperature [4]. Sample sintered at 1350°C shows shape of normal hysteresis loop of $NiFe_2O_4$ sample in which hysteresis loop is open. Similar effect was found by Chander et. al. [5] for

$Fe_{2.9}Zn_{0.1}O_4$. But the saturation magnetization is somewhat lower than bulk $NiFe_2O_4$ which we obtained in our previous study as 50.3 emu/g [6].

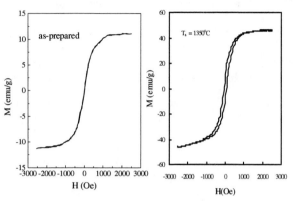

Figure 8. Field dependence of magnetization of as-prepared sample and sintered at 1350°C

The saturation magnetization in Fig 8 has been obtained as 13 emu/g for the as-prepared condition and 46 emu/g for the sample sintered at 1350°C. The much reduced M_s in the nano-particle sample implies that outside a core of ordered moments, those on the surface layer are in a state of frozen disorder. With the reduction of particle size more atoms remain on the surface, which are not exchange-coupled. As a result, the value of magnetization decreases. The effect is more prominent when the particle sizes are very small, however with the increase of particle size the saturation magnetization increases since core of the ordered moments become predominant.

In Fig. 9, temperature dependence of resistivity has been presented. Both the resistivity and the activation energy decreases with increasing sintering temperatures. When polycrystalline ferrites are considered, the bulk resistivity arises from a combination of crystallite resistivity and the resistivity of crystallite boundaries. The boundary resistivity is much greater than that of the crystallite resistivity. Thus the boundary have the greatest influence on the d.c. resistivity [7].The decrease of resistivity are also related to the decrease of porosity at higher sintering temperature since pores are non-conductive, which increases resistivity of the material [7]. The resistivity increases with the increase of porosity at lower sintering temperature because charge carrier on their way face the pores. Decrease of activation energy with the increase of sintering temperatures may be attributed to the fact that at a high sintering temperature, partial reduction of Fe^{3+} to Fe^{2+} takes place and these places act as donor centre. The conduction mechanism is due to hopping of electron of the types $Fe^{2+} \Leftrightarrow Fe^{3+}$. The value of resistivity has increased by an order of magnitude for the sample annealed at 1200 and 1350°C than the sample prepared by conventional ceramic technique. However, the value of resistivity is quite high at 1000°C due to under firing.

In Fig. 10, The real part ε' of the dielectric constant for samples sintered at higher temperature are about two orders of magnitude smaller than those of the $NiFe_2O_4$ prepared from chemicals of analytical grade [7]. The variation of dielectric constant, has been presented as a function of frequency up to f = 13 MHz. From Fig. 9 and 10, it can be seen that the higher value of dielectric constants are associated with lower resistivity. Dispersion in the dielectric constants was exhibited by all the samples in the studied frequency range. To explain dielectric dispersion in ferrites, grain and grain boundaries were assumed to be two different layers each having a different conductivity but same dielectric constant. As the frequency rises from a low value the bulk resistivity and dielectric constant fall and become asymptotic to lower values at high frequencies. This variation has the characteristic of relaxation and is attributed to the granular structure of ferrites, in which crystallites are separated by boundaries having much higher resistivities than the crystallites. Thus the structure behaves as a compound dielectric. At low frequencies the impedance of the crystallites is negligible compared to that of the boundary [7]. The dielectric constant approaches to the value, which is analogous to calculating dielectric properties from measurements on a specimen between the plates of capacitor, using a dielectric length 1 / n times the actual value. At very high frequencies the boundary capacitance becomes short circuited with the boundary resistance and the bulk dielectric properties approach those of crystallites.

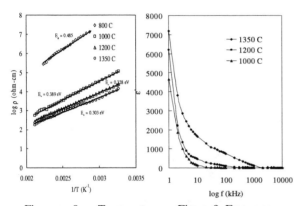

Figure 8. Temperature dependence of resistivity of the sample sintered at different temperature.

Figure 9. Frequency dependence of ε′ for the sintered samples

In Fig. 11, temperature dependence of dielectric properties have been presented for the frequencies of 10 and 100 kHz. Polycrystalline and poly-phase aggregates exhibit an interfacial or space charge polarization arising from differences between the conductivity of various phases present. This polarization resulting from heterogeneity is particularly important for compositions such as ferrites and semiconductors in which the electrical conductivity is appreciable; also for polycrystalline and polyphase materials at higher temperatures. The time constant for this interfacial polarization, and consequently the frequency at which it becomes important, is proportional to the product of resistivities of different phases. For most dielectrics this product is so large that the interfacial polarization is negligible even at low frequencies. For semiconductors and semiconducting ferrites this resistivity product is not so large, and the effect is important at room temperature. For other dielectrics this factor increase as the temperature is raised. At the same time, the conduction losses and ion jump losses also increase. These can in one sense be

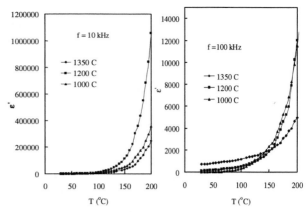

Fig. 11. Temperature dependence of real part dielectric constant for the sample sintered at different temperatures.

considered local space charge polarization resulting from mobility barriers. As the conductivity increases at higher temperatures, the apparent dielectric constant also increases, the dielectric losses increase even more rapidly, and the effectiveness of insulation decreased. To minimize this increase, compositions having the lowest possible conductivity should be used. In our results, the variation of dielectric constant with temperature for the samples sintered at different temperatures follow this general behavior.

CONCLUSIONS

Grain size of the $NiFe_2O_4$ nanoparticle prepared by chemical co-precipitation technique has been found as 6 nm in the as-dried condition with subsequent growth at higher sintering temperatures. Variation of Néel temperature with grain size can be explained by finite size scaling theory at low temperature which fits well with the experiment. Coercivity of the samples in nano level is almost zero, which is suggestive of superparamgnetic behavior at room temperature.

ACKNOWLEDGEMENT

The support provided by the International Science Programme, Uppsala University, Sweden is greatly acknowledged.

REFERENCES

1. Dong-Hyun Kim, David E. Nikles, Duane T. Johnson, Christopher S. Brazel, Journal of Magnetism and Magnetic Materials 320, 2390, 2008.
2. Seongtae Bae, Sang Won Lee, Y. Takemura, E. Yamashita, J. Kunisaki, Shayne Zurn, and Chul Sung Kim, IEEE Transactions on Magnetics, 42(10), 3566, 2006.
3. C.N. Chinnasamy, A. Narayanasamy, N. Ponpandiana, R. Justin Joseyphusa, B. Jeyadevanb, K. Tohjib, K. Chattopadhyay, Journal of Magnetism and Magnetic Materials 238, 281, 2002.
4. S. Rana, A. Gallo, R.S. Srivastava, R.D.K. Misra, Acta Biometerialia 3, 233, 2007.
5. Subhash Chander et. al. Pranama – J. Phys. 3, 617, 2003
6. S. Manjura Hoque, Md. Amanullah Choudhury, Md. Fakhrul Islam . J. Mag. Mag. Mat. 252 (2002) 292-303.
7. S. Manjura Hoque, M. Amanullah Choudhury, M. A. Hakim and M. Fakhrul Islam, Proceedings of the 2nd International Conference on Structure, Processing and Properties of Materials, p. 705-712, 2004.

Preparation and Characterization of Nanocrystalline GaP for Advanced Light Emissive Device Structures

S. Pyshkin*, J. Ballato**, G. Chumanov***, N. Tsyntsaru* and E. Rusu****

*Institute of Applied Physics, Academy of Sciences, Academy St. 5, MD2028, Kishinev, Moldova, spyshkin@yahoo.com, ashra_nt@yahoo.com
**Center for Optical Materials Science & Engineering Technologies (COMSET), Clemson University, Technology Dr. 91, Anderson, SC 29625, USA, jballat@clemson.edu
***Department of Chemistry, Clemson University, Clemson, SC 29634, gchumak@clemson.edu
**** Institute of Electronic Engineering and Industrial Technologies, Academy of Sciences, Academy St. 3/3, MD2028, Kishinev, Moldova, rusue@lises.asm.md

ABSTRACT

Presented here are recent efforts to enhance the quality of GaP nanoparticles for light emissive devices. GaP nanoparticles have been prepared by mild aqueous synthesis methods and their properties characterized by photoluminescence (PL), Raman light scattering (RLS), x-ray diffraction (XRD) and transmission electron microscopy (TEM). Due to pronounced quantum confinement effects the optical properties of GaP nanoparticles are considerably different from those of bulk crystals or particles whose dimensions exceed that of the Bohr exciton (app. 10nm). In this work, uniform GaP nanoparticles with bright luminescence at room temperature in a broad band with maximum at 3 eV have been obtained using yellow phosphorus, low temperature synthesis, and ultrasonic treatment that are found to improve the quality of the nano-suspension. The results of the nanoparticles characterization by the above mentioned methods clearly testify their high quality.

Keywords: GaP nanoparticles, mild aqueous synthesis

1 INTRODUCTION

This work reviews our recent efforts to advance the quality of GaP nanoparticles for light emissive devices based on polymer/GaP nanocomposites.

The choice of materials is based on their complementary behavior since each is a candidate for use in light emitters, waveguides, converters, accumulators and other planar, fiber or discrete micro-optic elements. While bulk and thin film GaP has been successfully commercialized for many years, its application in nanocomposites as a new optical medium has only received attention recently.

The fabrication and properties of the GaP/polymer nanocomposites for advanced light emissive device structures is discussed in another talk presented at the 2010 Nanotech Conference [1].

These papers presented to the 2010 Nanotech Conference have been fulfilled in framework of the STCU (www.stcu.int) 4610 Project "Light Emissive Device Structures" and are a part of the US/Moldova Program on advanced light-emissive sources on the base of transparent robust fluoropolymer and GaP nanoparticles [2-10].

2 EXPERIMENTAL PROCEDURE AND DISCUSSION

Nanoparticles of GaP have been prepared by mild aqueous synthesis [11-13] at different temperatures, modifications and compositions of the reacting components.

Noted here are only essential details of the aqueous syntheses as the base for comparison of nanoparticles prepared at different temperatures, modifications and compositions of the reacting components.

NaOH pellets were dissolved in distilled water. Ga_2O_3, red or yellow phosphorus powder and I_2 were mixed and added to the NaOH solution. The mixed solution was then placed into an autoclave and heated there in an oven for 8 hours at 200 or 125°C. After the completion of heating the autoclave was taken out of the oven and cooled. The obtained powder was filtered, washed with ethanol, and dried or ultrasonicated in the bath with a special solvent for separation in dimensions and preparation of a suspension for any nanocomposite. The dried powders were then characterized using standard methods of XRD, TEM, Raman scattering and photoluminescence. For comparison we used also industrial and specially grown and aged GaP single crystals [2-10].

The instruments for Raman light scattering and luminescence included spectrographs interfaced to a liquid nitrogen-cooled detector and an argon ion laser or lamp excitation sources. The spectra of Raman scattering was obtained at room temperature by excitation with 514.5 nm radiations and calibrated with the relevant etalons. Luminescence was excited by UV light of the lamps or the N_2 laser nanosecond pulses at wavelength 337 nm and measured at room temperature [9].

Figure 1 shows the TEM images of GaP nanoparticles obtained by the aqueous synthesis. One can see GaP

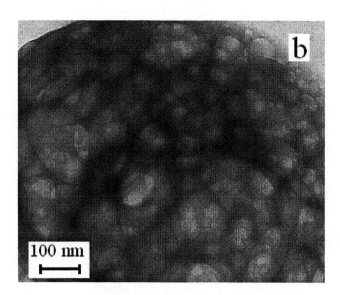

Figure 1: TEM images of GaP nanoparticles obtained by the aqueous synthesis. **a**. Thoroughly ultrasonicated and dried nanopowder. **b**. Initial clusters with the dimensions of the order of 100 nm.

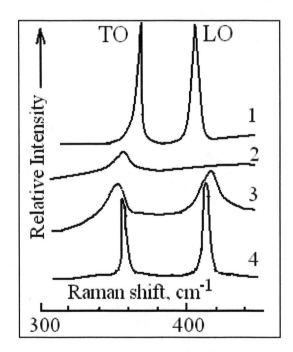

Figure 2: Raman light scattering from GaP nanoparticles of different treatment (spectra 2-4) and in comparison with perfect GaP bulk crystals (spectrum 1). 2. Not thoroughly treated powder of nanoparticles prepared using red phosphorus at 200°C. 3. Thoroughly treated GaP nanoparticles prepared using red phosphorus at 200°C. 4. Nanoparticles prepared on the base of yellow P by low temperature syntheses.

nanoparticles, having characteristic dimensions less than 10 nm. The washed, thoroughly ultrasonicated and dried nanopowder contains mainly single nanoparticles **(Fig. 1a)**,

obtained from the initial clusters with the dimensions of the order of 100 nm **(Fig. 1b)**.

Figure 2 shows spectra of Raman light scattering from GaP nanoparticles prepared on the base of yellow or red P by mild aqueous synthesis at increased or low temperatures and ultrasonically treated.

The characteristic GaP Raman lines from the doped aged GaP single crystals as well as from the nanoparticles prepared using yellow P by low temperature were narrow and intense **(Fig. 2, spectrum 1** and **4** respectively)** whereas, they were weak and broad from the any nanoparticles prepared at high temperatures **(Fig. 2, spectra 2** and **3)**. The especially weak and broad spectrum exhibits not thoroughly washed powder (please see **spectrum 2**; spectra 1-3 are taken from [13]).

In **Figure 3** one can see X-ray diffraction from GaP nanoparticles prepared at different conditions using red or yellow phosphorus **(spectra 1-3)** in comparison with the diffraction from perfect GaP single crystal **(spectrum 4)**. The nanoparticles obtained by low temperature aqueous synthesis using yellow phosphorus develop clear and narrow characteristic lines like those obtained from perfect GaP bulk single crystals taken from our unique collection of long-term (more than 40 years) ordered GaP single crystals **(Fig. 3, spectra 1** and **4)**. Contrary to that, nanoparticles prepared on the base of red phosphorus or not in the best conditions show broad and weak characteristic lines **(Fig. 3, spectra 2** and **3)**.

The luminescence was absent in newly-made industrial and our freshly prepared crystals but it was bright in the 40 years aged crystals **(Fig. 4, spectrum 1**; the features of luminescence in the perfect 40 years aged crystals please see in [3, 5, 7, 10]). Initial results on luminescent properties of GaP nanoparticles [13] confirm the preparation of 10 nm GaP nanoparticles with clear quantum confinement effects but the luminescent spectrum was not bright enough and its maximum was only slightly shifted to UV side against the 2.24 eV forbidden gap at room temperature (**Fig. 4, spectrum 2**). The nanoparticles obtained from the reaction with yellow P at low (125°C) temperature exhibit bright broad band spectra considerably shifted to UV side (**Fig. 4, spectrum 3, 4**). Note that the original powder contains only a part of GaP particles with nearly 10 nm dimension, which develop quantum confinement effect and the relevant spectrum of luminescence, so the spectrum of luminescence consists of this band with maximum at 3 eV and of the band characterizing big particles with the maximum close to the edge of the forbidden gap in GaP (**Fig. 4, spectrum 3**), but the thorough ultrasonic treatment gives an opportunity to get the pure fraction of nanoparticles with the **spectrum 4** having the maximum at 3 eV.

recombination of non-equilibrium current carriers, initially luminescence of fresh undoped crystals could be observed only at the temperatures 80K and below. Now luminescence is clearly detected in the region from 2.0 eV and until 3.0 eV at room temperature (see **Figure 4, spectrum 1**). Taking into account that the indirect forbidden gap is only 2.25 eV, it is suggested that this considerable extension of the region of luminescence to the high energy side of the spectrum as well as a pronounced increase of its brightness are connected with a very small concentration of defects, considerable improvement of crystal lattice, high transparency of perfect crystals, low probability of phonon emission at rather high temperature and participation of direct band-to-band electron transitions.

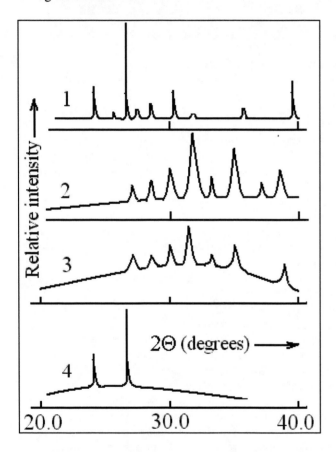

Figure 3 : X-ray diffraction from GaP nanoparticles. 1. Yellow phosphorus, best performance of low temperature syntheses, well-treated powder. 2. Yellow P, not the best performance and powder treatment. 3. Red phosphorus, the best result.

This work also continues monitoring of properties of GaP single crystals grown 40 years ago [2-10, 13]. We investigate their optical and mechanical properties in present in comparison with the data obtained in the 1960s, 1970s, 1980s and 1990s. Note that due to a significant number of defects and a highly intensive non-radiative

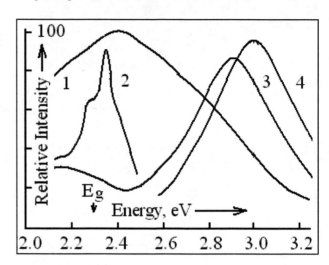

Figure 4 : Luminescence of GaP nanoparticles prepared at different conditions (spectra 2-4) and in comparison with the luminescence of perfect GaP bulk single crystals (1). Please see explanations in the text below.

Our unique collection of the long-term ordered perfect GaP single crystals gives opportunities to find deep fundamental analogies in properties of the perfect single crystals and nanoparticles [13, 14] as well as to predict and to realize in nanoparticles and perfect bulk crystals new interesting properties and applications. More detailed analyses and discussion of these results will be published.

3 CONCLUSIONS

Nanoparticles of GaP have been prepared using yellow P by mild aqueous low temperature synthesis. The spectra of PL, RLS, and XRD together with TEM images of the nanoparticles prepared under different conditions have been compared with each other as well as with those from bulk single crystals, that gives a good opportunity to find after the relevant investigation of different regimes and components for hydrothermal reactions the best performance in framework of this type of the nanoparticles synthesis. The uniform GaP nanoparticles having after severe ultrasonic treatment and a number of other operations improving the quality of the nano-suspension a bright luminescence at room temperature in a broad band with maximum at 3 eV have been used in preparation of the nanocomposites described in [1].

ACKNOWLEDGEMENTS

The authors are very grateful to the US Dept. of State, Institute of International Exchange, Washington, DC, the US Air Force Office for Scientific Research, the US Office of Naval Research Global, Civilian R&D Foundation (CRDF), Arlington, DC, Science & Technology Center in Ukraine (STCU), Clemson University, SC, University of Central Florida, FL, Istituto di elettronica dello stato solido, CNR, Rome, Italy, Universita degli studi, Cagliari, Italy, Joffe Physico-Technical Institute, St. Petersburg State Technical University, Russia and Academy of Sciences of Moldova for support and attention to our extended (1963-2009) research effort.

Our special gratitude to Prof. I. Luzinov, Dr. D. Van DerVeer, Dr. B. Zdirko and post graduate student J. Heckel from Clemson University, SC, to Dr. A. Siminel and post graduate students S. Bilevschii and A. Racu from Institute of Applied Physics, Moldova for their contribution in elaboration of nanotechnology, PL and XRD measurements.

REFERENCES

[1] S.L. Pyshkin, J. Ballato, I. Luzinov, and B. Zdyrko, "Fabrication and Characterization of the GaP/ Polymer Nanocomposites for Advanced Light Emissive Device Structures", will be published in Proceedings of the Nanotech Conference and Expo 2010, June 21-25, Anaheim, CA

[2] S. Pyshkin, J. Ballato, M. Bass, G. Chumanov and G. Turri, "Properties of the Long-term Ordered Semiconductors", The 2009 TMS Annual Meeting and Exhibition, San Francisco, Feb 15-19, Suppl. Proc., Vol. 3, pp 477-484.

[3] S. Pyshkin, J. Ballato, M. Bass and G. Turri, "Evolution of Luminescence from Doped Gallium Phosphide over 40 Years" J. Electronic Materials, Springer, Vol. 38, #5, pp 640-646 (2009).

[4] S.L. Pyshkin, J. Ballato, G. Chumanov, "Raman light scattering from long-term ordered GaP single crystals", J. Opt. A: Pure Appl. Opt. 9 (2007) 33–36, IOP Publ. House, London

[5] S.L. Pyshkin, J. Ballato, M. Bass, G. Turri, "Luminescence of Long-Term Ordered Pure and Doped Gallium Phosphide", (invited), Symposium "Recent Developments in Semiconductor, Electro Optic and Radio Frequency Materials", TMS 2007 Annual Meeting & Exhibition, Orlando, FL, Feb - March 2007; J. Electronic Materials, Springer, Vol. 37, #4, pp388-395 (2008)

[6] S. L. Pyshkin, R. Zhitaru, J. Ballato, "Modification of Crystal Lattice by Impurity Ordering in GaP", Int. Symposium on Defects, Transport and Related Phenomena, Proc. MS & T 2007 Conf., pp303-310, Sept 16 –20, 2007, Detroit, MI

[7] S.L. Pyshkin, J. Ballato, M. Bass, G. Turri, "New Phenomena in Luminescence of Gallium Phosphide", (invited), TMS 2007 Annual Meeting, Symposium: Advances in Semiconductor, Electro Optic and Radio Frequency Materials, March 9–13, New Orleans, LA, J. Electron. Mater, 37, #4, pp 388-395 (2008).

[8] Sergei L. Pyshkin, John Ballato, Michael Bass, George Chumanov and Giorgio Turri, "Time-dependent evolution of crystal lattice, defects and impurities in $CdIn_2S_4$ and GaP", The16th Int Conference on Ternary and Multinary Compounds (ICTMC16), Berlin, Sept 15-19, 2008; Phys. Status Solidi, C 6, No. 5, pp 1112–1115 (2009).

[9] S. Pyshkin, R. Zhitaru, J. Ballato, G. Chumanov, M. Bass, "Structural Characterization of Long Term Ordered Semiconductors", Proc. of the 2009 MS&T Conference, Pittsburgh, October 24-29, Int. Symposium "Fundamentals & Characterization", Session "Recent Advances in Structural Characterization of Materials", pp 698-709.

[10] Sergei Pyshkin, John Ballato, Andrea Mura, Marco Marceddu, "Luminescence of the GaP:N Long-Term Ordered Single Crystals", Suppl. Proceedings of the 2010 TMS Annual Meetings (Seattle, WA, USA, February, 2010, vol.3, pp 47-54.

[11] Shanmin Gao, Jun Lu, Nan Chen, Yan Zhao and Yi Xie, "Aqueous synthesis of III–V semiconductor GaP and InP exhibiting pronounced quantum confinement", J. Chemical Communications, pp 3064–3065 (2002)

[12] Ung Thi Dieu Thuy, Tran Thi Thuong Huyen, Nguyen Quang Liem, Peter Reiss, "Low temperature synthesis of InP nanocrystals", J. Materials Chemistry and Physics, Vol. 112 pp 1120–1123 (2008)

[13] S.L. Pyshkin , J. Ballato, G. Chumanov, J. DiMaio and A.K. Saha, "Preparation and Characterization of Nanocrystalline GaP", Symposium "Nanoelectronics and Photonics", 2006 NSTI Nanotech Conference, Boston, May 7-11, Technical Proceedings of the Conference, Vol. 3, pp 194-197

[14] S. Pyshkin and J. Ballato, "Long-term convergence of bulk- and nano-crystal properties", will be published in Proceedings of the Materials Science & Technology 2010 Conference, October 17-21, Houston, Texas.

Fabrication of Size-tunable Gold Nanoparticles Using Plasmid DNA as a Biomolecular Reactor

J. Samson[*], I. Piscopo[**], A. Yampolsky[*], P. Nahirney[***], A. Parpas[*], A. Toschi[****], and C.M. Drain[*,*****]

[*] Hunter College & Graduate Center of CUNY, New York, NY, USA, jacopo.samson@gmail.com
[**] EM consultant, EM Consulting, Huntington, NY, USA, irene.piscopo@gmail.com
[***] University of Victoria, Victoria, BC, Canada, pnahirney@gmail.com
[****] NYU Langone Medical Center, New York, NY, USA, nycalfredo@gmail.com
[*****] The Rockefeller University, New York, NY, USA, cdrain@hunter.cuny.edu

ABSTRACT

Gold nanoparticles (AuNPs) have been exploited for a wide range of potential applications, including drug delivery systems, catalysts, optical sensors and antimicrobial agents [1-5]. However, the harsh conditions employed in several synthetic approaches has forced researchers to investigate milder routes [6]. Biological macromolecules such as proteins [7], viruses [8], and plasmid DNA [9] have been shown to be successful candidates to ensure a milder pathway in the formation of AuNPs. Many of the aforementioned methodologies employing biological precursors nevertheless present other drawbacks such as lack of size tunability, broad dispersity, and poor shape control partially due to the tendency of cationic gold to disproportionate in aqueous solutions [10], as well as the difficulties in stabilizing metallic NPs. Combining plasmid DNA as a biomolecular reactor with a kinetically based approach, we have been able to stabilize and control the size of AuNPs.

Keywords: biomolecule, reactor, nanoparticle, plasmid, narrow-dispersity

1 MATERIALS AND METHODS

1.1 Plasmid DNA

pcDNA 3.1(+)/GFP (plasmid) was kindly donated by Dr. Michele Pagano, NYU Langone Medical Center. In order to determine its size as well as that of the other two plasmids utilized, a sample of each was linearized with ECORI restriction enzyme. After running a 0.8% agarose gel, the following sizes were deduced: GFP: approx. 6kbp (kilobase pairs); pMSCV: approx. 7kbp; pBABEc: approx. 6kbp.

1.2 NP Synthesis

Plasmid DNA suspensions (35 ng/uL in TE buffer, pH~8) and control samples were incubated with gold metal salts dissolved in acetone (0.5 wt. % chloro-trimethyl phosphine-gold (I), referred as gold phosphine solution) in a volume to volume ratio of 6:1, respectively in the dark at 70°C using a Lab-Line Multi block heater. In order to ensure a dark environment, a double layered aluminum foil was placed on top of the heater.

1.3 Transmission electron microscopy (TEM)

A 7 µL drop of DNA suspension was placed on a 300 mesh carbon coated copper grids, purchased from TED Pella, and allowed to dry for 5 minutes in the dark. The remaining liquid was removed using a filter paper. The control samples were prepared in the same way with the absence of DNA. The spherical shape of particles was determined by eucentric tilting (over at least an 80° range) over the particles from sample 12 h and 15 h. All data were collected at 120 kV on a Tecnai TEM at the eucentric position ensuring that all measurements and electron diffraction data were accurate for both collection and comparison. The electron diffraction patterns were collected in the microprobe or nanoprobe mode depending on the size of the area to be analyzed.

1.4 Gel Electrophoresis

0.8% agarose gels were freshly prepared by dissolving 0.4 grams of agarose, purchased from Sigma, in TRIS-acetate-EDTA buffer (1x). 2.5 µL of ethidium bromide were added when the mixture was still liquid. After the gel solidified, 10 µL of specific samples were loaded in each of the 10 wells and the gel was run at 85-90 V for 75 minutes.

1.5 UV-Vis Spectroscopy

Beckman Coulter DU800 spectrophotometer was used to collect the spectra. 50 µL of each sample was loaded into the cell (8 mm path length) before collecting the corresponding UV-visible spectrum. Backgrounds were collected before each sample determination.

2 RESULTS

The method described in section 1.2 yielded in a linear fashion the formation of approximately 6, 8, 9, 11, 18, and 21 nm-sized AuNPs after 1, 2, 4, 7, 12, and 15 hours of plasmid/metal cation incubation, respectively (Table 1, Figs. 1-3). Control experiments with TRIS-EDTA buffer (TE, pH~8) and gold phosphine solution were run in the absence of plasmid for 1, 2, 4, and 7 hours at 70°C (Figs. 1 & 2). Control experiments with nanopure water (pH~5), PBS buffer (pH~7.3), TRIS buffer (pH~6), EDTA (pH~5) and gold phosphine solution were also run in the absence of plasmid for 1 and 2 hours at 70 °C (Fig. 4).

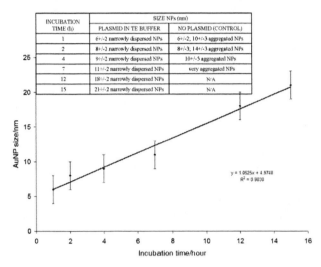

Table 1 Summary of the AuNP sizes yielded at different incubation times at 70°C in the dark in the presence and absence of plasmid DNA. The NP size using the plasmid DNA reactor is linearly correlated with time. In absence of plasmid, a correlation cannot be determined since all AuNPs are aggregated.

3 DISCUSSION

Since the one hour UV-Vis control spectra of both TE and TRIS buffers were similar to the spectrum corresponding to the 1h DNA/gold incubation, we believe that TRIS may also act as an initial seeding entity [12] (Figs. 1, 2 & 4). This is also confirmed by comparing the control experiments performed with pure water, PBS and EDTA buffers at the same conditions where absorption at ~510 nm was not observed. The 2h and the 4h UV-Vis control spectra of TE and TRIS buffers showed a progressive collapse of the peak corresponding to the gold NP absorption as well as an exponentially increasing red shift, confirming that the plasmid DNA is responsible for the size control as well as the narrow-distribution of the fabricated AuNPs (Figs. 1 & 2). Furthermore, in the absence of plasmid DNA, gold aggregated nanoparticles were observed after 7h of incubation when TE buffer control was imaged (Fig. 2). Overall, the controls run without DNA after two hours

showed a progressive collapse of the peak corresponding to the gold NP absorption. The enucleating role of the plasmid therefore was deduced from comparing the UV-Vis spectra with the analysis of gel electrophoresis (Fig. 1, 2 & 6). This comparison resulted in the conclusion that a substantial modification of the initial plasmid DNA topology in solution (Fig. 6, lanes 2 & 10) has no significant effect on the resulting intensity and produces no shift, although a certain ratio of initial plasmid condensation states is preferred to optimize the size-distribution (Fig. 6, Panel B).

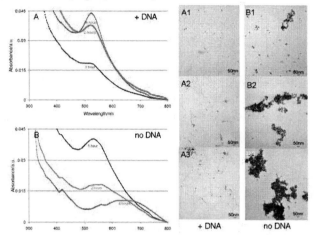

Fig. 1 A. UV-Vis spectra and TEM images of DNA-containing samples incubated with gold phosphine solution at 70°C in the dark for 1 (A1), 2 (A2) and 4 hours (A3), respectively. B. UV-Vis spectra and corresponding TEM images of TE buffer controls incubated with gold phosphine solution at 70°C in the dark for 1 (B1), 2 (B2), and 4 hours (B3), respectively.

Fabrication of narrowly dispersed particles was obtained after a progressive degradation of the DNA that becomes noticeable after ~2h of incubation time with gold phosphine (Fig. 6, lanes 2 and 5). After 7h incubation of the metal salt with the DNA, a maximized amount of narrowly dispersed particle production was observed. This trend was revealed by comparison of the absorption peaks shown in Fig. 2A and 2B as well as by TEM analysis of the AuNPs at 12 and 15 hours of incubation (Fig. 3). These degraded DNA segments maintained particle dispersion and high stability preventing re-aggregation over a period of several weeks (Fig. 7).

The size tunability of the AuNPs was later verified by TEM measurements to display correlation between UV-Vis data and corresponding incubation times (Table 1, Figs. 1 & 2). The metallic nature of the AuNPs was confirmed by superimposing the ED pattern obtained from the experimental samples with that of the gold standard (Fig. 5). The spherical shape of the AuNPs was determined by eucentric tilting. In order to ensure experimental reproducibility, the concentration of DNA used to yield the AuNPs was maintained constant. This was necessary since, upon dilution of the DNA concentration by 50 fold, a progressive broadening of the UV-Vis spectra of the NP was observed. Furthermore, the UV spectra of the 50-fold diluted DNA sample was nearly identical to the spectra of the TE buffer/Au controls, suggesting that a stoichiometric balance between metal salts and DNA substrate is necessary to ensure the NPs will be narrowly dispersed.

The above data showed that the samples containing plasmid DNA began exhibiting a trend of narrowing distribution at two hours of incubation time with respect to the corresponding control samples (Fig. 1 & 2). The overall trend of the UV spectrometry also exhibited a progressive red shift in correspondence with increasing particle size; this observation is in congruence with previous studies examining UV correlations with NP size [11]. Nucleation of the nanoparticles initiating inside the DNA minor and major grooves, which favors cation binding, may explain the size control of the AuNPs (Fig. 8). A TRIS-assisted initial conucleation is not excluded (Figs. 1B & 4) but to yield narrowly dispersed AuNPs with larger sizes, the plasmid is essential.

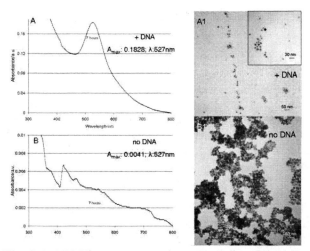

Fig. 2 A. UV-Vis spectrum and corresponding TEM image (A1) of plasmid DNA samples incubated with gold phosphine solution for 7 hours at 70°C in the dark; Inset panel: higher magnification of sample in A1. B. UV-Vis spectrum and corresponding TEM image (B1) of TE buffer control incubated with gold phosphine solution under the same conditions.

4 CONCLUSIONS

The results demonstrate an easy method to synthesize size tunable spherical AuNPs by exploiting plasmid DNA as a nanoreactor. These results are fully reproducible. Since the degraded plasmid segments act as capping agent, the stabilization of the AuNPs in solution is achieved (Fig. 7). Since linear plasmid DNA does not yield the same results, circular plasmid DNA is essential for maintaining a narrow dispersed NP outcome. Finally, the employment of mild synthetic conditions and few procedural steps makes this method environmentally friendlier than most of the current methods.

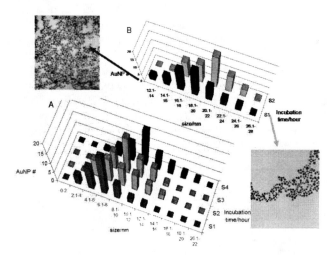

Fig. 3 A. Histogram data of particle size corresponding to plasmid samples incubated for 1 (S1, blue), 2 (S2, purple), 4 (S3, red), and 7 hours (S4, black). After 4 hours there is a progressive narrowing of the distribution and an increase in particle size that is in agreement with the Uv-Vis red shift (Figs. 1 & 2). B. Histogram data of particle size corresponding to DNA samples incubated for 12 (S1, dark blue) and 15 hours (S2, pink). TEM images, corresponding to the DNA/Au samples incubated for 12 (dark blue) and 15 hours (pink); it is possible to observe narrow distributed AuNPs confirming that the trend of nearly mono-distribution is maintained up to incubation times of 15 hours.

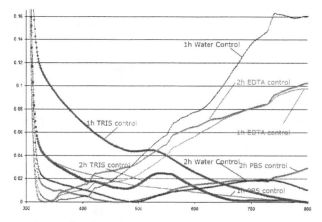

Fig. 4 Uv-Vis spectra of control experiments incubated for 1 and 2 hours at 70°C in the dark (water: green; TRIS: light blue; PBS: purple; EDTA: orange). The UV-Vis spectra of the TRIS controls (corresponding to the incubation times of 1 and 2 hours) fully resemble the spectra of the TE buffer controls corresponding to the same incubation times (Fig. 1B) as well as the 1h incubation DNA/gold (Fig. 1A).

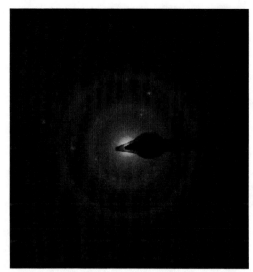

Fig. 5 ED pattern of gold standard solution (blue) superimposed on the experimental ED pattern obtained from AuNPs analysis (red)

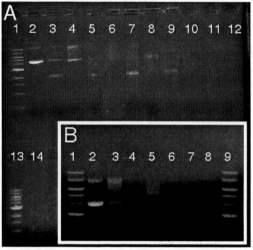

Fig. 6 0.8% agarose gel electrophoresis of the plasmid DNA incubated with gold phosphine solution at different times. The gels were run on different days (A: same day; Inset B: day after). A→ Lane 1: 1 kb ladder; Lane 2: Naked plasmid; Lane 3: Naked plasmid with rearranged initial topologies (pcR); Lane 4: plasmid after 30 min incubation time; Lane 5: plasmid after 30 min incubation time; Lane 6: plasmid after 1 hour incubation time; Lane 7: pcR after 1 hour incubation time; Lane 8: plasmid after 2 hour incubation time. Lane 9: pcR after 2 hour incubation time; Lane 10: plasmid after 4 hour incubation time; Lane 11: pcR after 4 hour incubation time; Lane 12: plasmid after 7 hour incubation time; Lane 13: 1 kb ladder; Lane 14: pcR after 7 hour incubation time. Inset B→ 1: 1 kb ladder; Lane 2: Naked plasmid; Lane 3: plasmid after 30 min incubation time; Lane 4: pcR after 30 min incubation time; Lane 5: plasmid after 1 hour incubation time; Lane 6: pcR after 1 hour incubation time. Lane 7: plasmid after 2 hour incubation time. Lane 8: pcR after 2 hour incubation time; Lane 9: 1 kb ladder.

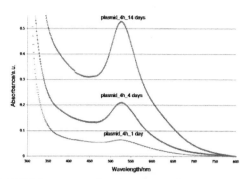

Fig. 7: UV-Vis spectra showing the peaks corresponding to the gold absorption of AuNPs. These NPs were prepared in the presence of the plasmid, incubated for 4 hours, and left undisturbed in the dark for up to two weeks indicating that the AuNPs are stable in aqueous solution. The absence of a red shift confirms that the particles in solution neither re-aggregate nor collapse through time.

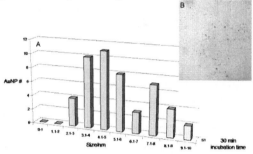

Fig. 8 A. Histogram illustrating NP size distribution after 30 min of Au salt incubation time with the plasmid. B. TEM image showing that particles are still bound to the plasmid strands. The bimodal distribution suggests that the major and minor grooves of the plasmid may act as enucleating sites and the overall function of the plasmid is that of a biomolecular reactor.

REFERENCES

[1] S. Yugang, *Science Magazine*, **298**, 2176, 2002
[2] J. Wagner, J.M. Kohler, *ACS Nano Lett.*, **5**, 685, 2005
[3] N. Jana, L. Gearheart, *Langmuir*, **17**, 6782, 2001
[4] S. Shanmugam, B. Viswathan, T.K. Varadajan, *Mat. Chem. Phys.* **95**, 51, 2006
[5] H. Hiramatsu, *Chem. Materials*, **16**, 2509, 2004
[6] J. Campelo, T. Conesa, M. Gracia, M. Juardo, R. Luque, J. Marinas, A. Romero, *Green Chem.* 2008, **10**, 853
[7] P. Ravindra, *JMSEB*, **163**, 93, 2009
[8] J. Slocik, D. Wright, *J. Mater. Chem.*, **15**, 749, 2004
[9] J. Samson, A. Varotto, P. Nahirney, A. Toschi, C. Drain, *ACS Nano*, **3**, 339, 2009
[10] J. Kimling, B. Okenve, V. Kodaidis, H. Ballot, A. Plech, *J. Phys. Chem. B*, **110**, 15700, 2006
[11] W. Haiss, D. Fernig, *Anal. Chem.*, **79**, 4215, 2007
[12] R. Guo, L. Zhang, Z. Zhu, X. Jiang, *Langmuir*, **24**, 3459, 2008

Synthesis and Optical Properties of L-cystine Capped ZnS:Co Nanoparticles

P. Kumbhakar*, R. Sarkar, and A.K. Mitra

Department of Physics, National Institute of Technology, Durgapur, 713209, India
Tel : +91-343-2546808, Fax:+91-343-2547375
*E-mail: nitdgpkumbhakar@yahoo.com & pathik.kumbhakar@phy.nitdgp.ac.in.

ABSTRACT

L-cystine capped Co^{2+} doped ZnS nanoparticles have been prepared through a soft chemical route, namely the chemical co-precipitation method. The nanostructures of the prepared samples have been analyzed using energy dispersive analysis of X-rays (EDAX), high resolution transmission electron microscope (HRTEM), Fourier transform infrared spectra (FTIR), UV-visible-NIR and photoluminescence (PL) spectrophotometers. The average sizes of the as prepared undoped and doped nanoparicles are found to be ~2.4 nm and 4.5 nm, respectively. Room-temperature PL spectrum of the undoped L-cystine capped sample exhibits emission in the blue region with a sharp spectral band peaked at ~315 nm under 250 nm UV excitation. In the L-cystine capped Co^{2+} doped ZnS samples quenching (~ 96%) of 315nm light emission takes place under same excitation wavelength.

Keywords: Cystine capped, cobalt doping, zinc sulphide, nanostructures, quenching, photoluminescence

1 INTRODUCTION

IV-VI nanocrystalline semiconductors are of great interest for fundamental studies as well as in technical applications, such as in light emitting devices, fluorescence biosensor, and solar cells [1-12]. Amongst other semiconductor materials ZnS is the most important for application as a phosphor for photoluminescence (PL), electroluminescence (EL) and cathodoluminescence (CL) devices. We have previously reported that Mn doping in ZnS leads to yellow-orange light emission under excitation by UV light [2]. Quantum confinement effect in semiconductor nanoparticles (NPs) modifies the electronic structure of the NPs when the sizes of the nanoparticles are comparable to that of Bohr excitonic radius of those materials. There are several methods for preparation ZnS NPs, such as chemical precipitation method [3-10], micro-emulsion method [11]. Recent studies suggest that the NPs may have molecular biological applications, such as in DNA detection studies [12-13]. In most of the methods stabilizers used to inhibit particle growth included reverse micelles [14], organic compounds [15] and DNA [16]. In a recent investigation we have reported a enhancement of visible light emission from uncapped Co^{2+} doped ZnS nanoparticles [17]. Various metal ions and biological

samples coming from industrial effluents are threatened to all human beings. So the determination of these toxic elements by organic fluorophore based sensor is of tremendous interest and important for analytical chemists. Photoluminescence (PL) emissions from semiconductors NPs depend on the doping concentration as well as on the capping agent, in case of capped samples. The optical properties of cystine capped nanocrystalline semiconductor have been studied extensively in recent years for their non toxic behavior and are used for the synthesis of water soluble nano fluorescent probe for chemical sensing [18-21]. Previous studies on Cystine capped ZnS shown striking differences in the pH-dependent changes in optical spectra [18]. Koneswaran *et al.* [19] reported the synthesis of the L-cystine-capped ZnS quantum dots as fluorescence sensor for the Cu^{2+} ion through both dynamic and static quenching mechanism. Borse *et al.* [20] shows the effect of iron and nickel metal ion doping on quenching of blue light emitted by mercaptoethanol capped ZnS nanoparticles. However, there are only few reports available in the literature on ZnS doped with Co^{+2} and capped with cystine. Cystine is water soluble amino acid can be used as a capping agent for ZnS NPs. The surface modification of ZnS NPs with cystine prevents the agglomeration of the NPs [21]. Hence, synthesis and characterization of the optical properties of NPs of ZnS doped with Co^{+2} capped with cystine is interesting.

Here we have adopted chemical precipitation method being simpler than all other available methods for preparation of ZnS NPs. The average sizes of the cystine capped ZnS and ZnS:Co NPs have been estimated as 2.4 nm and 4.5 nm from the HRTEM measurements. Photoluminescence (PL) spectra of the prepared ZnS NPs have also been recorded at different excitation wavelengths. However, the peak emission takes place at excitation wavelength of 250 nm and the peak PL emission takes place at 315 nm for cystine capped undoped samples, whereas for the Co^{+2} doped sample, ~96% quenching of the PL emission at 315 nm takes place under the same excitation wavelength.

2 EXPERIMENTAL DETAILS

Nanoparticles of ZnS have been prepared by chemical co-precipitation method using AR grade (Mark & SD) chemical. The synthesis is carried out at room temperature of $20^{0}C$. At first, 10ml saturated solution of both zinc nitrate and sodium sulfide in methanol are prepared. Zinc nitrate solution is vigorously stirred using a magnetic stirrer upto 1h, then calculated amount of L-cystine solution and cobalt salt are mixed with solution of Zinc nitrate. The white precipitate is separated from the reaction mixture by centrifugation for 5min at 10,000 rpm and washed several times with methanol to remove all sodium particles. The wet precipitate is then dried for characterization of optical, and nanostructural properties.

The optical transmission/absorption spectra of the sample dispersed in methanol were recorded using a UV-VIS spectrophotometer (Hitachi, U-3010). The formation of ZnS or ZnS:Co nanoparticle was confirmed using transmission electron microscope (HRTEM, JEOL JEM 2100) using carbon coated copper grid. Scanning electron microscope (SEM with EDXA, Hitachi S-3000N), has been used for compositional analysis of the prepared ZnS nanoparticles. FTIR spectra were recorded in an FTIR spectrometer (Nicolet iS 10) after preparing palette with the KBr. The photoluminescence (PL) spectrum of the ZnS

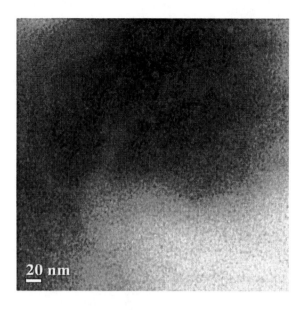

Figure 2: HRTEM micrograph of the cystine caped ZnS sample.

mean particle size of the ZnS:Co and undoped ZnS samples are of 4.5 nm and 2.4 nm respectively. The particle size distributions of undoped cystine capped sample have been shown in Figure 3.

FTIR spectra are recorded for all the samples by preparing palettes of the powder samples with the KBr in the range of 400 cm^{-1} to 4000 cm^{-1} (not shown here). Some characteristic peak appeared at 3469, 2543, 1394, 1130, 1002, and 612 cm^{-1} with some other associated peaks in all the samples and the presence of cystine was confirmed in all the samples [22]. The broad band at 3469 cm^{-1} appeared due to valence vibration of the occluded water but it also corresponds to the region where the asymmetric and symmetric stretching modes for a coordinated NH$_2$ group appear. Bands appeared around 900-1500 cm^{-1} may be due to oxygen stretching frequency (OH stretching) and binding frequency. The additional weak peaks and shoulders are appeared may be due to the nanostructural formation in the samples. The measured absorption characteristics are shown in Figure 4. The prepared nanopowders are first dispersed in methanol and then the UV-visible optical absorption characteristics of the cystine capped ZnS and ZnS: Co NPs are collected. The synthesis of ZnS NPs is clearly evident from the Figure 4.

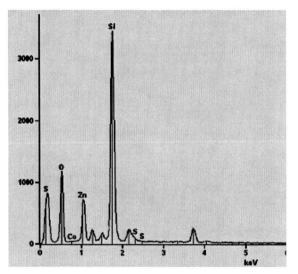

Figure 1: EDAX spectrum of ZnS:Co sample.

nanoparticles dispersed in methanol has been recorded using a spectrofluorimeter (Perkin Elmer, LS 55)

3 RESULTS AND DISCUSSIONS

Figure 1 shows the typical representative EDAX spectrum of a cystine capped ZnS:Co sample, which confirmed the compositions of the sample. To obtain the particle size and information about the nanostructure

HRTEM measurement was performed using copper grid. Figure 2 show the HRTEM micrograph of the prepared cystine capped ZnS sample for examples, the

The small excitonic absorption peaks due to the ZnS nanoparticles appeared at 280 nm, which reflects the band gap of the particle. For obtaining the absorption characteristics of all the samples, at first the transmittance

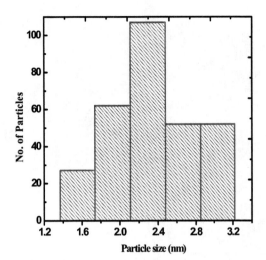

Figure 3. The particle size distribution corresponding to HRTEM image of Figure 2.

(T) at different wavelength (λ) are measured and then absorbance (α) at the corresponding wavelengths λ are calculated using the Beer-Lambert's relation;

$$\alpha = \ln(T^{-1})/d, \qquad (1)$$

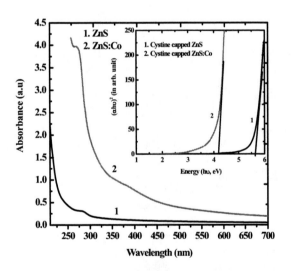

Figure 4: UV-visible absorption characteristics of the prepared samples. Inset shows the calculation of band gap of the samples.

where, d is the path length. The absorption edge for single crystal ZnS is very sharp and is determined by the nature of the electronic transition between the valence band and conduction band.

From the above Figure 4, it can be observed that the cystine capped Co^{2+} doped ZnS exhibits a shift of

absorption peak to longer wavelength and the peak becomes sharper. The relation between the incident photon energy ($h\nu$) and the absorption coefficients (α) is given by the following relation.

$$(\alpha h\nu)^{1/s} = C(h\nu - E_g), \qquad (2)$$

where, C is a constant and E_g is the bandgap of the material and the exponent s depends on the type of the transition. For direct and allowed transition $s = 1/2$, indirect transition, $s = 2$, and for direct forbidden, $s = 3/2$. For calculating the direct bandgap value $(\alpha h\nu)^2$ versus $h\nu$ is plotted and it is shown in inset of Figure 4. By extrapolating the straight portion of the graph on $h\nu$ axis at $\alpha = 0$, the bandgap value is calculated and it is 5.60 eV and 4.20 eV for ZnS and ZnS:Co samples, which is higher than that of bulk ZnS [2]. This blue shift of the bandgap takes place because of the quantum confinement effect [1]. By using the method described in [23], the calculated average size of the undoped sample is 2.5 nm, which agrees quite well with its value obtained from TEM. The sizes of ZnS:Co doped samples are slightly larger; e.g., the average size of the doped sample is 4.5 nm.

Figure 5 show the PL spectra at 250 nm excitation of cystine capped ZnS, and ZnS:Co NPs measured at room temperature. The intense PL emission from cystine capped

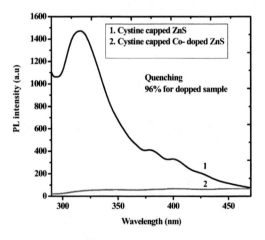

Figure 5: The PL spectra at 250 nm excitation of cystine capped ZnS and cystine capped ZnS:Co samples.

undoped ZnS nanoparticles with a peak at 315 nm along-with two other weak peaks at 380nm and 400nm are observed. The PL intensity of 315 nm peaks quenches by ~96% in the Co^{+2} doped ZnS samples. The origin of the blue-luminescence from undoped ZnS NPs has been studied by different groups [1, 2, 9, 24-30]. Becker and Bard [28] have attributed the blue emission band at 428 nm to S^{-2} vacancies. Koneswaran et al. [19] reported the synthesis of the L-cystine-capped ZnS quantum dots as fluorescence sensor for the Cu^{2+} ion through both dynamic and static quenching mechanism. Borse et al. [20] suggested that the

PL emission can be quenched by the iron and nickel doping. Wang et al. [25] explained the quenching of the PL intensity of conjugated macromolecules by bipyridinium derivatives in aqueous media due to charge dependence. Quenching of PL intensity in ZnS colloidal particles by Cd^{2+} ions was investigated by Weller et al. [26]. They have attributed this behavior to the formation of co-colloids. In the cystine capped co-doped ZnS nanoparticles, luminescent mechanism and the processes are very complex. Cobalt act as electron trapping centers which results into non radiative recombination and so the quenching of the PL intensity in the cystine capped ZnS:Co NPs have been observed in the present work.

4 CONCLUSION

The synthesis of cystine capped undoped and Co^{+2} doped ZnS nanoparticles of sizes 2.4 nm and 4.5 nm respectively, are reported by chemical precipitation method. This method being simpler and with low cost chemical components is suitable for the industrial large scale production. Photoluminescence (PL) emissions measured at room temperature from the prepared nanoparticles are reported. It is found that the cystine capped undoped sample exhibits the PL emission in the blue region. It is also observed that the PL intensity of the cystine capped ZnS:Co nanoparticles quenches by 96% than that of the cystine capped undoped ZnS nanoparticles, under the same excitation wavelength of 250 nm. In the Co^{2+} -doped ZnS nanocrystallites, luminescent mechanism and processes are very complex. However, cobalt act as electron trapping centers which results into non radiative recombination and thus the PL intensity of cystine capped ZnS:Co nanoparticles quenches. The prospects of the luminescent materials are directly decided by their fluorescence efficiency. The synthesis of L-cystine capped and Co^{2+} doped ZnS nanoparticles can easily be carried out at low cost and also can be used in analytical sensor applications.

ACKNOWLEDGEMENT

This work has been financially supported by DST Grant No. SR/FTP/PS-67/2008, Govt. of India and the Faculty Research Grant of N.I.T, Durgapur, Govt. of India. Authors are grateful to Dr. A. Patra of Chemistry Dept, NIT, Durgapur for providing the FTIR measurement facility.

REFERENCES

[1] R. N. Bhargava, D. Gallagher, Phys. Rev. Lett., 72, 416, 1994.

[2] R. Sarkar, C. S. Tiwary, P. Kumbhakar, A. K. Mitra. Physica E, 40, 3115, 2008.

[3] J. Yu, H. Lui, Y. Wang, W. Y. Jia, J. Lumin., 79 191, 1998.

[4] A. A. Khosravi, M. Angel Kundu, L. Jatwa, S. K. Deshpande, Appl. Phys. Lett., 67, 2702, 1995.

[5] P. Yang, M. K. Lu, D. Xu, D.Yuan, G. Zhou, J. Phys. Chem. Solids, 62, 1181, 2001.

[6] T. Ishihara, J. Takahasi, T. Goto, Phys. Rev. B, 42, 11099, 1990.

[7] J. Hung, Y. Yang, S. Xue, B. Yang, S. Lui, J. Shen, Appl. Phys. Lett., 70, 2335, 1997.

[8] R. N. Bhargava, D. Gallagher, T. Welker, J. Lumin., 60, 275, 1994.

[9] S. Sapara, D. D. Sharma, Phys. Rev. B, 69, 125304 2004.

[10] C. Jin, J. Yu, L. Sun, K. Dou, S. Hou, J. Zhao, Y. Chen, S. Huang, J. Lumin., 66, 315, 1996.

[11] L.M. Gan, B. Liu, C.H. Chew, S.J. Xu, S.J. Chua, G.L. Loy, G.Q. Xu, Langmuir, 13, 6427, 1997.

[12] A.P. Alivisatos, K.P. Johnsson, X. Peng, T.E. Wilson, C.P. Loweth, M.P. Bruchez, Jr., P.G. Schultz, Nature (London), 382, 609, 1996.

[13] C. A. Mirkin, R.L. Letsinger, R.C. Mucic, J.J. Storhoff, Nature, 382, 607, 1996.

[14] M. Mayer, C. Wallberg, K. Kurihara, J.H. Fendler, J. Chem. Soc., Chem. Commun., 90, 1984.

[15] C. B. Murray, D. J. Norris, M. G. Bawendi, J. Am. Chem. Soc., 115, 8706, 1993.

[16] J. L. Coffer, S. R. Bigham, R. F. Pinizzotto, H. Yang, Nanotechnology, 3, 69, 1992.

[17] R. Sarkar, C.S. Tiwary, P. Kumbhakar, A.K. Mitra, Physica B, 404, 3855, 2009

[18] W. Bae, R. Abdullah, D, Henderson, R. K. Mehra, Biochem. Biophys. Res. Commun., 237, 16, 1997.

[19] M. Koneswaran, R. Narayanaswamy, Sensor and Actuators B, 139, 104, 2009.

[20] P. H. Borse, N. Deshmukh, R.F. Shinde, S.K. Date, S.K. Kulkarni, J. Mat. Sci., 34, 6087, 1999.

[21] A.R. Kortan, R. Hull, R.L. Opila, M.G. Bawendi, M.L. Steigerwald, P.J. Carroll, L.E. Brus, J. Am. Chem. Soc., 112, 1327–1332, 1990.

[22] B. S. Rema Devi, R. Raveendran, A. V. Vaidyan, Ind. Acad. Sc., J. Phys. 68, 679, 2007.

[23] M. Chattopadhyay, P. Kumbhakar, C. S. Tiwary, R. Sarkar, A. K. Mitra and U. Chatterjee, J. Appl. Phys., 105, 024313, 2009.

[24] W. Chen, Z. Wang, Z. Lin, L. Lin, J. Appl. Phys., 82, 3111, 1997.

[25] D. Wang, J. Wang, D. Moses, G. Bazan, A. J. Heeger, Langmuir, 17, 1262, 2000.

[26] H. Weller, U. Koch, M. Gutierrez, A. Henglien, Ber, Bunsenges. Phys. Chem., 88, 649, 1984.

[27] K. Manzoor, S. R. Vadera, N. Kumar, T.R.N. Kutty, Mater. Chem. Phys., 82, 718, 2003.

[28] W.G. Becker, A.J. Bard, J. Phys. Chem., 87, 4888, 1983.

[29] N. Murase, R. Jagannathan, Y. Kanematsu, M. Watanabe, A. Kurita, H. Hirata, T. Yazawa, T. Kushida, J. Phys. Chem. B, 103, 754, 1999.

[30] S. Yanagida, M. Yoshida, T. Shiragami, C. Pac, H. Mori, H. Fujita, J. Phys. Chem., 94, 3104, 1990.

Magnetite Particle Size Dependence on the Co-precipitation Synthesis Method for Protein Separation

Raúl Terrazas Reza, Carlos A Martínez Pérez, A Martínez Martínez and Perla E García-Casillas[*]

Instituto de Ingeniería y Tecnología, Universidad Autónoma de Ciudad Juárez, Ave. Del Charro #610 norte, C.P. 32320, Cd. Juárez Chihuahua, México.
[*]pegarcia@uacj.mx or perlaelviagarcia@yahoo.com

ABSTRACT

In this work, the synthesis of magnetite nanoparticles by a three variant chemical co-precipitation methods that involves reflux and aging conditions as well as slow and rapid injection techniques was investigated. The synthesized magnetite nanoparticles showed a spherical shape with an average particle size directly influenced by the synthesis technique. Average particle size from 16 nm, 27 nm and 200 nm were obtained. All three types of magnetite nanoparticles had a superparamagnetic behavior and their saturation magnetization is influenced by the particle size. Values of 56, 67 and 78 emu/g were obtained for the 16 nm, 27 nm and 200 nm magnetite particles, respectively. The protein adsorption was improved by using a surface coating. The silica-aminosilane coating showed the best performance followed by the aminosilane coating. The silica coating showed the lowest adsorption for all three particle sizes.

Keywords: Nanoparticles, magnetite, chemical co-precipitation, protein separation, polymer coating

1 INTRODUCTION

The size of magnetic particles plays a key role for their applications and nanoparticles have shown extraordinary results compared to micro and macro scale particles [1-2]. In the nanometric scale, each particle contains a simple magnetic domain and shows a superparamagnetic behavior, which gives a rapid response to magnetization and demagnetization when an external magnetic field is applied and removed, respectively [2-4]. This kind of magnetic response is desired in applications where high precision is required to direct them easily and quickly. It offers the possibility of manipulating the magnetic nanoparticles by applying a magnetic field which makes them very attractive [5-7]. Many of these types of nanoparticles are metallic, which are not stable and are easily oxidized, therefore their applications are limited. Metallic oxides nanoparticles can overcome this limitation. Magnetite (Fe_3O_4) has been widely used in the field of magnetic materials [8-9]. In recent years, considerable effort has been focused in the design and controlled synthesis of this material with certain shape and particle size. These properties are crucial in the performance of the magnetic nanoparticles [10-12]

The synthesis of magnetic particles with the desired shape, size, and uniform size distribution, is a very important step that needs to be first achieved before any further research and use. A uniform particle size distribution is highly desired because magnetic properties show a strong dependence on the nanometric scale [10-16].

Based on the fact that different synthesis methods leads to particles with different size and morphology, this research is focus on the influence of the co-precipitation methods for the synthesis of magnetite particles and the processing parameters. Fast versus slow nucleation rates and nucleation with refluxing and aging conditions were evaluated. In order to use the magnetic nanoparticles for protein separation, they were coated with different types of materials obtaining specific superficial properties and making them suitable for attaching biological materials to their surface [17-22]. Different superficial functional groups were used: aldehyde (-CHO), hydroxyl (-OH) and amine ($-NH_2$).

2 EXPERIMENTAL PROCEDURE

The magnetite nanoparticles were coated with silica by dissolving 0.1910g of sodium metasilicate in 10 ml of water and adding 0.020 g of nanoparticles. The mixture was ultrasonically agitated during 30 min, and heated up to 80°C. The pH of the solution was lowered to 6 using HCl and a composite was obtained. The obtaining powder was centrifuged and washed several times until a pH of 7 in order to remove the residual ions; the powder was dried at 35°C during 24 h.

The coating of the aminosilane shell was obtained by mixing 100 μL of N-(2-aminoethyl)-3aminopropyltrimetoxysilane (AEAPS), 0.02 g of magnetic nanoparticles and 25 μL of deionized water in 2.5 mL of methanol. The mixture was exposed to ultrasonic agitation for 30 minutes and then1.5 mL of glycerol was added. The solution was heated up to 85°-90°C, and a high mechanical agitation was kept during 6 h. The obtained precipitate was washed with water and methanol 4 times in each case. 15 mL of glutaraldehide were added and the mixture was

NSTI-Nanotech 2010, www.nsti.org, ISBN 978-1-4398-3401-5 Vol. 1, 2010

agitated ultrasonically for 45 min. Finally, the precipitate was washed with water and dried at 20°-35°C for 24 h. A combined coating of silica and aminosilane was also applied. The nanoparticles were coated in two steps: silica coating was applied first followed by the aminosilane shell using the methodologies already described.

The protein immobilization over magnetite was evaluated using four different surface conditions: pure magnetite –no coating- (M sample), magnetite with silica coating (M-S sample), magnetite with aminosilane coating (M-A sample) and magnetite with silica and aminosilane coatings (M-S-A sample). Bovine Serum Albumin (BSA) was used as standard protein. A BSA solution with a concentration of 2 µg/µL was prepared. During each protein adsorption assay, a calibration curve was run using BSA concentrations going from 0 to 2.0 µg/µL, with increments of 0.2 µg/µL. The protein immobilization was achieved by mixing 10 µg of nanoparticles with different initial concentrations of BSA in a total volume of 300 µL, going from 0 to 1.33 µg/µL. The incubating conditions were under room temperature, with vigorous continuous agitation (Vortex®) and for a period of time of 30 min. Protein adsorption was determined using the Bradford colorimetric method at 595 nm visible light wavelength.

X-Ray Diffraction (XRD) was used to confirm the magnetite (Fe_3O_4) phase in each synthesis method. The size distribution and morphology of nanoparticles were characterized by Field Emission Scanning Electron Microscope (FE-SEM), and the elemental chemical analysis was performed by Energy Dispersive X-ray Spectroscopy (EDS). The magnetic properties were obtained using a Vibrating Sample Magnetometer (VSM). Protein adsorption on magnetite nanoparticles was studied by determining the amount adsorbed using a Visible Light Spectrometer.

3 RESULTS AND DISCUSSION

3.1 Suggested mechanism of magnetite nanoparticles formation

The following chemical reaction mechanism occurred in the three methods: trivalent iron ion hydrolyzes forming (FeOOH) (eq. 1) as pH increases; divalent iron ion forms $Fe(OH)_2$ (eq. 2) under alkaline conditions. Both chemical species reacted with each other at pH values of around 10 to 11, forming magnetite according to Cornell et al. equation (eq. 3). The overall chemical reaction is represented by equation 4:

$$(Fe(H_2O)_6)^{+3} \longrightarrow FeOOH + 3H^+ + 4H_2O \qquad (1)$$

$$Fe^{+2} + 2OH^- \longrightarrow Fe(OH)_2 \qquad (2)$$

$$2FeOOH + Fe(OH)_2 \longrightarrow Fe_3O_4 + 2H_2O \qquad (3)$$

$$2Fe^{+3} + Fe^{+2} + 8OH^{-1} \longrightarrow FeO.Fe_2O_3 + 4H_2O \qquad (4)$$

The formation of all the intermediate species were not characterized, but the use of divalent and trivalent iron precursors (Fe^{+3} and Fe^{+2}) and the pH conditions kept during reaction support the suggested mechanism based on discussed theory [1].

The chemical co-precipitation method with slow injection differs from the rapid injection method on the addition rate of ammonium hydroxide, which has a direct effect in how fast the pH change to favor the magnetite formation. In the chemical co-precipitation method with reflux and aging, the desired pH condition was obtained through the decomposition of urea at high temperature, and is represented by the following chemicals reactions [14,15].

$$(NH_2)_2CO + H_2O \longrightarrow 2NH_3 + CO_2 \qquad (5)$$

$$NH_3 + H_2O \longrightarrow NH_4^+ + OH^- \qquad (6)$$

3.2 Chemical, morphological characterization and properties.

The XRD pattern of the three synthesized nanoparticles is shown in figure 1. According to these results, it is observed that all three sample spectra are similar among them, and they are similar to the pure magnetite XRD spectrum. Pure magnetite shows six characteristics diffraction peaks, which were obtained in the three methodologies. Miller index notation is used. An inverse spinel crystal structure was obtained.

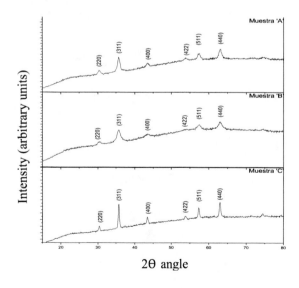

Figure 1: XRD spectra of (C) co-precipitation with reflux and aging method, (B) chemical co-precipitation with slow and (A) fast injection methods.

The obtained XRD patterns showed differences in the peak width, which is due to the different crystallite size. Wide peaks indicate the small crystals presence while narrow ones indicate larger crystals. Magnetite made with a slow and reflux method showed similarity in peaks width; however sample made with fast injection method showed smaller values. The average values of the Full Width at Half Maximum (FWHM) for the three samples is shown in table 2. It can be observed that the sample obtained by co-precipitation with reflux and aging method showed bigger crystal size than those observed in samples obtained by chemical co-precipitation with slow and fast injection methods, which showed similar values between them.

Table 2. FWHM values for the three magnetite samples.

Position 2Θ	FWHM (2Θ)		
	Co-precipitation with reflux and aging variant	Co-precipitation with slow injection variant	Co-precipitation with fast injection variant
30.362	0.6888	0.7872	0.2460
35.765	0.3936	0.3936	0.2460
43.473	0.7872	0.6888	0.1968
53.946	0.7872	0.9840	0.3936
57.512	0.6888	0.7872	0.2952
63.166	0.7200	0.8400	0.4200

The morphology of the Magnetite nanoparticles was observed by FE-SEM. All three samples exhibit spherical particle shape. The difference observed among samples was regarding the particle size. Magnetite obtained by co-precipitation with reflux and aging method showed higher particle size than magnetite obtained by chemical co-precipitation with slow and fast injection methods, the last one showed the smallest size.

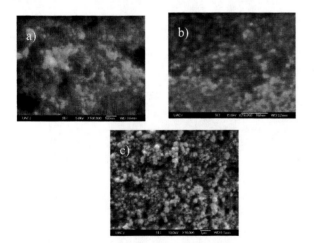

Figure 2: FESEM micrograph of magnetite obtained by a) co-precipitation with slow and b) fast injection methods, and c) co-precipitation with reflux and aging method

The figure 3 shows the average particle size and size distribution for each one of the samples. According to these results, magnetite obtained by fast injection method showed the smallest average particle size and more uniformity on particle size distribution than magnetite obtained by slow injection and with reflux and aging method, as it is established by the standard deviation values.

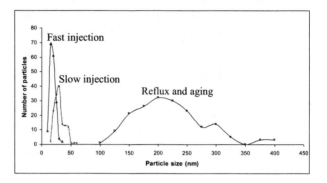

Figure 3: Particle size distribution

Magnetite obtained by reflux and aging showed the highest saturation magnetization of 78.21 emu/g and the magnetite obtained by rapid injection exhibits the smallest value (55.98 emu/g); magnetite obtained by slow injection showed a value of 64.33 emu/g. The hysteresis loop shows that all three samples present a superparamagnetic behavior, which is suitable for protein separation in biomedical applications. As result of the different method used, three different nanoparticles sizes werer obtained. Figure 4 shows the behavior of the saturation magnetization (M_s) as function of particle size and coatings. In all cases, the coating disminisht the M_s. However the aminosilane coating has a lower effect than silica.

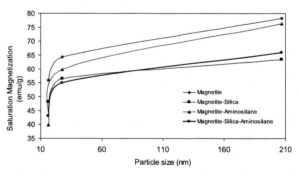

Figure 4: Saturation magnetization dependence over magnetite particle size

The coating of the particles was confirmed by FTIR analysis (figure 5). No significant differences were found between the three types of magnetite particles. The magnetite particles (M) show the characteristic band of Fe-O bond at 590 approximately but when these particles are coated with silica a new band appears at 1072 cm^{-1} due to Si-O bond (M-S) [19]; this band is present when the particles were coated with silica (M-S) and silica-aminosilane (M-S-A). In the FTIR spectra of M-A and M-S-A samples, bands at 3309 and at 1654 cm^{-1} were observed, which are attributed to the amine group (-NH$_2$) [12]. The band at 2943 cm^{-1} is due to the stretching of C-H of the methyl group (-CH$_2$, -CH$_3$). When the particles have a silica shell (Silica-Aminosilane Coating-MAS sample), a new band is shown at 802 cm^{-1} due to Si-O-Si bond [23].

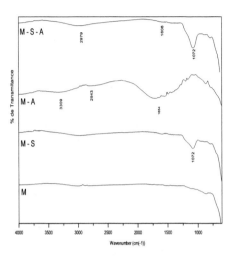

Figure 5: FTIR of magnetite (M), magnetite with silica coating (M-S), magnetite with aminosilane (M-A) and magnetite with silica and aminosilane coating (M-S-A)

The amount of Albumin Serum Bovine (BSA) protein used to immobilize the nanoparticles surface is influenced by the type of coating present, and in all cases it is superior to that observed on pure magnetite. Magnetite-aminosilane (MA) and Magnetite-silica-aminosilane (MSA), had higher capability for protein adherence than silica coating and pure magnetite (no coating). MSA showed a maximum protein retention capacity for all three magnetite particle sizes. Figure 6 shows the relationship between protein adsorption, surface coating and particle size. The best combination (regarding functionality) for the higher protein adsorption is small magnetite particles (16nm) with silica-aminosilane coating.

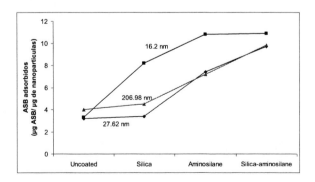

Figure 6: Protein adsorption capacity as function of particle size and coating type.

4 CONCLUSION

o The nucleation rate and growth speed during magnetite synthesis have a direct influence on the particle size. Under reflux and aging conditions, the formation of large particles (200 nm) is favored but under rapid injection conditions, small particles were obtained (16 nm). More uniformity in particle size was observed when small particles were obtained. The standard deviation values of the average particle size were 4 nm for 16 nm particle size, 8 nm for 27nm and 59 nm for the largest particles (200 nm).

o The magnetite nanoparticles coatings improved their protein adsorption capability with no detriment on their superparamagnetic behavior.

o The type of functional group present on the coating influences the protein adsorption. The silica-aminosilane coating showed the highest protein adsorption capacity in all magnetite nanoparticles studied.

o The best combination to obtain the highest protein adsorption is when an average particle size of 16nm coated with silica-aminosilane coating is used.

REFERENCES

[1]. Cornell, R.M., Schwertmann, U. 2003. The iron oxides: Structure, Properties, Reactions, Occurrences and Uses. Segunda edición. Willey-VCH GmbH & Co. Egaa.

[2]. Cao, Guozhong. 2004. Nanostructures & nanomaterials: Synthesis, Properties & Aplications. Imperial Collage Press.

[3]. Cullity, B. D., 1971. Introduction to Magnetic Materials. Addison-Wesley Publishing Company, Inc.

[4]. Jiles, 1991. Introduction to Magnetism and Magnetic Materials. Ed. Chapman & Hall. 1era edición. Londres, Inglaterra.

[5]. Rossi, L., Quach, A., Rosenzweig, Z., 2004. Glucose oxidase-magnetite nanoparticles bioconjugate for glucose sensing. Springer-Verlag. Anal Bioanal Chem 380: 606-613. DOI 10.1007/s00216-004-2770-3.

[6]. Schmid, G. 2004. Nanoparticles: From Theory to Application. Wiley-VCH Verlag GmbH / Co. KGaA.

[7]. Wang, J., Peng, Z., Huang, Y., Chen, Q. 2004. Growth of magnetite nanorods along its easy-magnetization axis of [1 1 0]. Elsevier Journal of Crystal Growth 263 616-619.

[8]. Schöth, F., Lu, A., Salabas, E.L., 2007. Magnetic Nanoparticles: Synthesis, Protection, Functionalization, and Application. Angewandte Chemie, Wiley-VCH Verlag GmbH & Co. KGaA, Winheim. DOI: 10.1002/anie.200602866

[9]. Osaka, T., Matsunaga, T., Nakanishi, T., Arakaki, A., Niwa, D., Iida, H., 2006. Synthesis of magnetic nanoparticles and their application to bioassays. Springer-Verlag. Anal Bioanal Chem 384: 593-600. DOI 10.1007/s00216-005-0255.7.

[10]. Gnanaprakash, G., Mahadevan, S., Jayakumar, T., Kalyanasundaram, P., Philip, John., RAj, Baldev., (2007). Effect of initial pH and temperature of iron salt solutions on formation of magnetite nanoparticles. Elsevier. Materials chemistry and Physics 103168-175.

[11]. Zhang, B., Xing, J., Liu, H., 2007. Preparation and application of magnetic microsphere carriers. Higher Education Press and Springer-Verlag. Front. Chem. Eng. China, 1(1): 96-101. DOI 10.1007-s11705-007-0019-3.

[12]. Günzler, H., Gremlich, H., 2002. IR Spectroscopy. Wiley-VCH Verglag GmbH. Alemania.

[13]. Kim, K. 2007. Synthesis and characterization of magnetite nanopowders. Current Applied Physics. DOI: 10.1016/j.cap.2007.04.021.

[14]. Lian, S., Wang, E., Kang, Z., Bai, Y., Gao, L., Jiang, M., Hu, C., Xu, L. 2004. Synthesis of magnetite nanorods and porous hematite nanorods. Elsevier Solid State Communications 129 485-490.

[15]. Yang, P., Yu, J., Wu, T., Liu, G., Chang, T., Lee, D., Cho, D., 2004. Urea decomposition method to synthesize hidrotalcites. Chinese Chemical Letters. Vol. 15, No. 1, pp 90-92. http://www.imm.ac.cn/journal/ccl.html

[16]. Xu, L., Hu, C., Jiang, M., Wang, E., Kang, Z., Lian, S. (2003). Convenient synthesis of single crystalline magnetic Fe_3O_4 naorods. Solid State Communications 127 (2003) 605-608.

[17]. Kobayashi, B., Saeki, S., Yoshida, M., Nagao, D., Konno, M., 2007. Synthesis of spherical submicron-sized magnetite/silica nanocomposite particles. Springer Science+Business Media, LLC. DOI 10.1007/s10971-007-1648-1.

[18]. Liu, Xianquiao., Ma, Zhiya., Xing, Jianmin., Liu, Huizhou. 2004. Preparation and characterization of amino-silane modified superparamagnetic silica nanospheres. Journal of Magnetism and magnetic Materials 270 1-6.

[19]. Xing, J., Zhang, B., Lang, Y., Liu, H. 2008. Synthesis of amino-silane modified magnetic silica adsorbents and application for adsorption of flavonoids from *Glycyrrhiza uralensis Fisch.*

[20]. Yu, C., Tam, K., Lo, C., Tsang, S., 2007. Functionalized silica coated magnetic nanoparticles with biological species for magnetic separation. IEEE Transactions on magnetics, Vol. 43, No. 6. DOI 10.1109/TMAG.2007.894203.

[21]. Yamaura, M., Camilo, R., Sampaio, L., Macedo, M., Nakamura, M., Toma, H., 2004. Preparation and characterization of (3-aminopropyl) triethoxysilane-coated magnetite nanoparticles. Journal of Magnetism and Magnetic Materials 279 (2004) 210-217.

[22]. Ma, M., Zhang, Y., Yu, W., Shen, H., Zhang, H., Gu, N., 2002. Preparation and characterization of magnetite nanoparticles coated by aminosilane. Colloids and Surfaces A: Physicochem Eng. Aspects: 212 (2003) 219-226. Elsevier Science B.V. PH: S0927-7757(02)00305-9.

[23].Y. Ivanova, Ts. Gerganova, H. M. H. V. Fernandes*, I. M. Miranda, E. Kashchieva Journal of the University of Chemical Metallurgy, Vol.41, No.3, 311-316 (2006).

Combustion synthesis and magnetic characterizations of Mn substituted Ni–Zn ferrite nanoparticles

D. R. Mane*, D. D. Birajdar*, Sagar E. Shirsath**, R. H. Kadam*

*Materials Science Research Laboratory, Shrikrishna Mahavidyalaya Gunjoti,
Osmanabad, MS, India, ram111612@yahoo.co.in
**Department of Physics, Dr. Babasaheb Ambedkar Marathwada University, Aurangabad:
431 004, MS, India, shirsathsagar@hotmail.com

ABSTRACT

The ferrite samples with a general formula $Ni_{0.6-x}Mn_xZn_{0.4}Fe_2O_4$ (where x = 0.0-0.6 in steps of x = 0.2) were synthesized by sol–gel auto combustion method using nitrates of respective metal ions. The synthesized samples were annealed at 600 0C for 4 h. The phase purity of the samples was investigated by X-ray diffraction technique. Nano size of the particle was confirmed by TEM measurement. An analysis of X-ray diffraction patterns reveals the formation of single phase cubic spinel structure. The crystal lattice constant advances gradually with increasing x from 8.391 Å to 8.451 Å. Cation distribution of constituent ions shows linear dependence of Mn substitution. Based on the cation distribution obtained from XRD data, structural parameters such as lattice parameters, ionic radii of available sites and the oxygen parameter 'u' have been calculated. An initial increase followed by a subsequent decrease of saturation magnetization with increase in x is observed.

Keywords: Ferrites, sol-gel, nano-particle

1 INTRODUCTION

Ni–Zn and Mn-Zn ferrites consists an important category of ceramic magnetic materials with a wide spectrum of technological applications, in devices that in the broadest sense can be characterized as transformers, inductors or absorbers [1, 2]. There is also interest in the magnetic properties and magneto-electronic transport behavior of Ni–Mn Ferrites. Mn-Zn ferrites are used at low frequencies (< 500 kHz) because of their low resistivity [3]. Ni–Zn ferrite has good dielectric properties, a low coercivity, high resistivity [4], and saturation magnetization is similar to that of magnetite. Ferrite particles with such properties can be realized by the sol–gel auto-combustion method; for example, the ultrafine and uniform Ni–Zn ferrite and Mn–Zn/SiO2 nanoparticles have been successfully synthesized by this method [5, 6]. It has a low production cost and can be used as a core material in transformers. The present research directed to a systematic study of structural and magnetic characterization of $Ni_{0.6-x}Mn_xZn_{0.4}Fe_2O_4$ (where x = 0.0-0.6 in steps of x = 0.2).

2 EXPERIMENTAL PROCEDURE

The powders were synthesized by sol–gel auto-combustion. Analytical grade zinc nitrate ($Zn(NO_3)_2 \cdot 6H_2O$), nickel nitrate ($Ni(NO_3)_2 \cdot 6H_2O$), manganese nitrate ($Mn(NO_3)_2 \cdot 6H_2O$) and iron nitrate ($Fe(NO_3)_3 \cdot 9H_2O$) were used as starting materials. Reaction procedure was carried out in air atmosphere without protection of inert gases. The molar ratio of metal nitrates to citric acid was taken as 1:4. The as prepared powder then annealed at 600 ^{0}C for 4 h (after confirmation by TGA). Part of the powder was X-ray examined by Phillips X-ray diffractometer (Model 3710) using Cu-K$_\alpha$ radiation (λ=1.5405Å). The scanning step was 4.26^0/min and scanning rate was 0.02^0. Magnetic measurements were performed using the commercial PARC EG&G vibrating sample magnetometer VSM 4500 with maximum applied magnetic fields up to 0.7 T.

3 RESULTS AND DISCUSSION

The X-ray diffraction pattern (Fig. 1) shows the formation of single phase cubic spinel structure without any impurity peak.

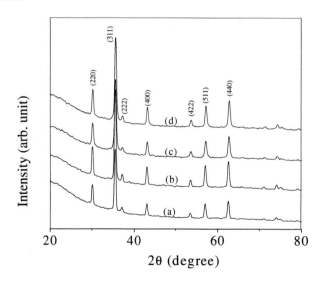

Figure 1: X-ray diffraction pattern (a) x = 0.0, (b) x = 0.2, (c) x = 0.4 and (d) x = 0.6.

The lattice constant 'a' was calculated using inter-planar spacing (d) values and Miller indices (hkl) values, with an accuracy of ± 0.002Å. The values of lattice constant 'a' are shown in Fig. 2. It is clear that lattice constant 'a' increases as the Mn substitution increases. This can be related to the fact that the Mn^{2+} ion has a radius of 0.80 Å, which is larger than the Ni^{2+} (0.69 Å). When the larger Mn ions enter the lattice, the unit cell expands while preserving the overall cubic symmetry. The X-ray density (ρ_{x-ray}) was determined using the relation $\rho_{x-ray} = 8M/Na^3$ where, M is the molecular weight of the sample, N is Avogadro's number and 'a' is lattice parameter. The ρ_{x-ray} values decreases with increase in Mn content x (Fig. 2).

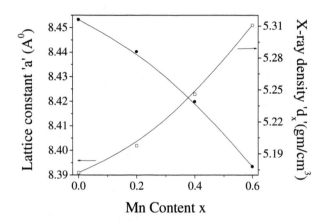

Figure 2: Variation of lattice constant and X-ray density with Mn content x.

The bulk density (ρ_{th}) measured using the formula $\rho_B = m/\pi r^2 h$, where m is mass, r is the radius and h is the height of the pellet.

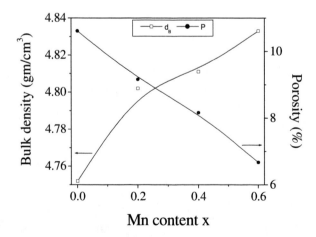

Figure 3: Variation of bulk density and porosity with Mn content x.

It is observed from Fig. 3 that the bulk density is increasing with increasing Mn substitution. The effect of the addition of Mn^{2+} on the microstructure of ferrites has been reported by Yan and Johnson [7], who found that Mn^{2+} may act as a slight accelerator of grain growth and also favor densification. Increase in bulk density may also be due to the difference in the specific gravity of MnO (5.37 g cm^{-3}), NiO (6.72 g cm^{-3}) and ZnO (5.60 g cm^{-3}) [8]. This combination of factors contributed to the reduction of the material's porosity, which consequently improved the magnetic induction of the ferrite. The percentage porosity of each sample was calculated by using the relation,

$$Porosity (P) = (\rho_{x-ray} - \rho_{th})/ \rho_{x-ray} \qquad (1)$$

Fig. 3 shows the variation of percentage porosity and particle size with composition. It is observed that porosity decreases with increasing amount of Mn. The particle size is significantly dependent on Mn substitution. The average particle size increases with increasing Mn substitution (Fig. 4).

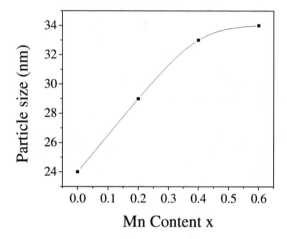

Figure 4: Variation of particle size with Mn content x.

The cation distribution in the present system was obtained from the analysis of X-ray diffraction patterns. In this method the observed intensity ratios were compared with the calculated intensity ratios. In the present study Bertaut method [9] is used to determine the cation distribution. This cation distribution is based on the premise that nearly 80% of Mn ions occupy the tetrahedral position (i.e., the A site), while the remaining occupy the octahedral position (i.e., the B site) [10]. It is known that Zn and Ni ions occupy A and B sites, respectively. Although iron and manganese ions exist at both A and B sites, they have preference for the B and the A sites, respectively [11].

Table 1 shows the results of the cation distribution analysis for $Ni_{0.6-x}Mn_xZn_{0.4}Fe_2O_4$. The most noteworthy feature of the cation distribution is that the cations exhibit nearly linear composition dependencies across the join. It can be seen that, the tetrahedral Fe^{3+} ions increase, whereas the octahedral Fe^{3+} ions decrease as the join is traversed from the manganese end number to the nickel end number.

The content of Mn^{2+} ions in the tetrahedral is quite small and remains constant. Ni^{2+} is well known to be a predominantly octahedral divalent cation, while Mn^{2+} is a predominantly tetrahedral divalent cation. Therefore, replacement of manganese by nickel will lead to the Fe^{3+} ions shifting from octahedral sites to the tetrahedral sites.

Table 1
Cation distribution of $Ni_{0.6-x}Mn_xZn_{0.4}Fe_2O_4$

| x | Cation distribution | |
	A-site	B-site
0.0	$(Zn_{0.4}Fe_{0.3})$	$[Ni_{0.6}Fe_{1.4}] O_4$
0.2	$(Mn_{0.165}Zn_{0.4}Fe_{0.435})$	$[Mn_{0.035}Ni_{0.4}Fe_{1.565}] O_4$
0.4	$(Mn_{0.301}Zn_{0.4}Fe_{0.299})$	$[Mn_{0.099}Ni_{0.2}Fe_{1.701}] O_4$
0.6	$(Mn_{0.413}Zn_{0.4}Fe_{0.187})$	$[Mn_{0.187}Fe_{1.813}] O_4$

The mean ionic radius of the tetrahedral A- 'r_A' and octahedral B-site 'r_B' can be calculated by using cation distribution and modifying the relation discussed elsewhere [12]. The variation of mean ionic radius of the A-site (r_A) and of the B-site (r_B) with Mn^{2+} is shown in Fig. 5. The r_A increases with increase in Mn^{2+} content x, this is due the increase in Mn^{2+} ions of larger ionic radii replaces Fe^{3+} ions of smaller ions at A site. The slow increase in 'r_B' is due to a combine effect of increasing fraction of Mn^{2+} and Fe^{3+} at B site which replaces Ni^{2+}.

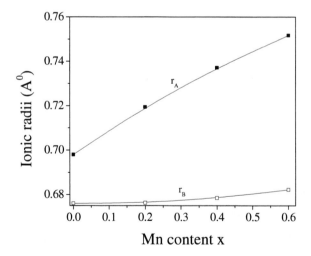

Figure 5: Variation of ionic radii of A site (r_A) and B-site (r_B) with Mn content x.

Using the values of 'a', the radius of oxygen ion $R_O = 1.32$ Å and r_A in the following expression, the oxygen positional parameter 'u' can be calculated,

$$r_A = \left(u - \frac{1}{4}\right) a\sqrt{3} - R_0 \qquad (2)$$

Fig. 7 shows decreasing value of oxygen positional parameter 'u' from 0.3875 - 0.3867 Å. In most oxidic spinels the oxygen ions are apparently larger then the

metallic ions, and in spinel like structure the oxygen positional parameter has a value in the neighborhood of 0.375 Å for which the arrangement of O^{2-} ions equals exactly a cubic closed packing but in actual spinel lattice, this ideal pattern is slightly deformed.

The theoretical lattice parameter 'a_{th}' was calculated using the values of tetrahedral and octahedral reading (r_A, r_B) and is given by the following relation,

$$a_{th} = \frac{8}{3\sqrt{3}}\left[(r_A + R_0) + \sqrt{3}(r_B + R_0)\right] \qquad (3)$$

where, r_A and r_B are radii of tetrahedral (A) site and octahedral [B] site, R_0 is radius of oxygen i.e. ($R_0 = 1.32$ Å). Theoretical lattice constant increases from 8.421 Å (x = 0.0) to 8.498 Å (x = 0.6) (Fig. 6). It is noticed that a_{th} is higher than lattice constant 'a', this deviation may be due to the formation of Fe^{2+} ions, which have an ionic radius greater than Fe^{3+}.

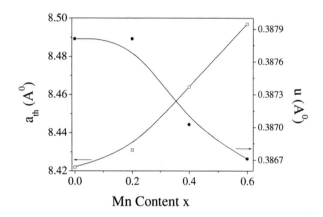

Figure 6: Variation of theoretical lattice constant 'a_{th}' and oxygen positional parameter 'u' with Mn content x.

Magnetic properties

The magnetic hysteretic loops of $Ni_{0.6-x}Mn_xZn_{0.4}Fe_2O_4$ ferrites are given in Fig. 7. Variation of saturation magnetization with composition of the series has been studied. The observed variations can be explained on the basis of cations distribution and exchange interaction between iron and between manganese ions at tetrahedral -A and octahedral -B sites. When manganese ions are introduced at the expense of nickel ions, some of the Fe ions migrate from A to the B sites in view of the site preferences for different ions. This increases the Fe ion concentration at B sites. As a result, the magnetic moment of B sublattice increases for manganese concentrations (x ≤ 0.2). However, as manganese concentration increases, the iron ions left at A site being small in number, the A–B interaction experienced by B site iron ions decreases. Also, the increased number of Fe ions at the B site increases the B–B interaction, resulting in spin canting [13]. Consequently, the magnetization of B sublattice decreases. Although the magnetic moment of Mn^{2+} is the same as that of the Fe^{3+} (5 μ_B), the exchange interaction between manganese and iron ions being small, there will be canting of spins of Fe^{3+} and Mn^{2+} ions at the A site.

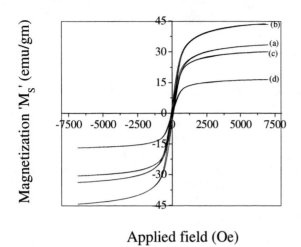

Applied field (Oe)

Figure 4: Variation of magnetization with applied field, (a) x = 0.0, (b) x = 0.2, (c) x = 0.4, (d) x = 0.6,

The increase in the manganese content in the sample therefore decreases the magnetic moment of the A sublattice. The reason for decrease in magnetization may also be the fact that low Mn concentrations reduce the number of spins occupying the A sublattices, causing the net magnetization to increase. As the Mn content increases, the exchange interactions are weakened and the B spins are no longer held rigidly parallel to the few remaining A spins. The decrease in the B-sublattice moment, interpreted as a spin departure from colinearity, causes the effect known as canting.

REFERENCES

[1]   S. Gubbala, H. Nathania, K. Koizolb and R.D.K. Misra, Physica B Condensed Mater 348, 317, 2004.

[2]   V.A.M. Brabers, in: K.H.J. Buschow (Ed.), Handbook of Magnetic Materials, North-Holland, Amsterdam, 1995, pp. 189.

[3]   S. Son, M. Taheri, E. Carpenter, V. G. Harris, M. E. McHenry, J. Appl. Phys. 91, 7589, 2002.

[6]   L. Nalbandian, A. Delimitis, V.T. Zaspalis, E.A. Deliyanni, D.N. Bakoyannakis and E.N. Peleka, 114, 465, 2008.

[4]   Parvatheeswara Rao, B.; Subba Rao, P.S.V.; Rao, K.H., IEEE Trans. Magn. 33, 4454, 1997.

[5]   A. T. Raghavender, Damir Pajic, Kreso Zadro, Tomislav Milekovic, P. Venkateshwar Rao, K. M. Jadhav and D. Ravinder, J. Magn. Magn. Mater. 316, 1, 2007.

[6]   S. F. Yan, W. Ling and E. L. Zhou, J. Cryst. Growth 273, 226, 2004.

[7]   M.F. Yan and D. Johnson Jr., J. Am. Ceram. Soc., 61, 342, 1978.

[8]   D. R. Lide, "CRC Handbook of Chemistry and Physics", 76th ed., CRC Press, London, (1995).

[9]   L. Weil, E. F. Bertaut, L. Bochirol, J. Phys. Radium, 11 (1950) 208-212.

[10]   Aria Yang, Vincent G. Harris, Scott Calvin, Xu Zuo, and Carmine Vittoria, IEEE Trans. Magn. 40, 2802, 2004.

[11]   J. Smit and H. P. J. Wijn, "Ferrites" -Philips Technical Library, 149, Eindhoven, Netherland, (1959),

[12]   Ram Kripal Sharma, Varkey Sebastian, N. Lakshmi, K. Venugopalan, V. Raghavendra Reddy, Ajay Gupta, Phys. Rev. B, 75, 144419, 2007.

[13]   S. V. Kakatkar, S. S. Kakatkar, R. S. Patil, A. M. Sankpal, S. S. Suryawanshi, D. N. Bhosale and S. R. Sawant, Phys. Stat. Sol. (b). 198, 853, 1996.

Luminescent mechanism of bulk and nano ZnS:Mn^{2+} phosphors

M.M. Duvenhage, O.M. Ntwaeaborwa, H.C. Swart and M.J.H. Hoffman

Department of Physics, University of the Free State,
P.O. Box 339, Bloemfontein, ZA9300, South Africa, hoffmamj@ufs.ac.za

ABSTRACT

ZnS nanophosphors, with or without different concentrations of Mn^{2+} activator ions, were synthesized by using the chemical precipitation method. Micron sized ZnS:Mn^{2+} powder was obtained from Phosphor Technology in the United Kingdom. Structure, morphology and particle sizes of the samples were determined by using XRD, TEM and SEM. The average particle size was ~3 nm in diameter. The band gap of ZnS nanophosphors, estimated from the UV-Vis data, was 4.1 eV. Luminescent properties of the samples were determined by photoluminescence (PL) measurements. Two emission peaks at 450 nm and 590 nm were observed from the micron sized sample. The 590 nm peak was red-shifted by 10 nm in the case of ZnS:Mn^{2+} nanophosphors used in this study. In this work a luminescent mechanism for both bulk (micron sized) and nano sized ZnS:Mn^{2+} is proposed, and the red-shifting is explained using the configuration coordinate model for Mn^{2+}.

Keywords: ZnS:Mn^{2+}, nanophosphors, luminescence, luminescent mechanism

1 INTRODUCTION

Although ZnS:Mn^{2+} is a well known phosphor used in display panels, solar cells and also in biological imaging together with CdS [1],[2], the excitation and emission mechanism of the nano sized ZnS is still not well understood. Most of the reported mechanisms attribute the 450 nm peak to emission from shallow traps in the band gap [3],[4] and the 590 – 600 nm peak to the $^4T_1$ - $^6A_1$ transition of Mn^{2+} [3],[4],[5]. None of these mechanisms differentiate between emission processes of bulk and nano scale ZnS or explain the red–shifting of the 590 nm peak. Some reports explain the mechanism in terms of the configurational coordinate model for a Mn^{2+} impurity in ZnS [6],[7],[8], without explaining the red shift in the peak position. In this work a luminescent mechanism for both bulk and nano ZnS:Mn^{2+} is proposed. This mechanism accounts for a blue shift in the excitation and absorption spectra of nano ZnS:Mn^{2+}, a PL peak that were observed at 450 nm , a peak at 590 nm for the bulk sample and one at 600. This red shift in the peak can be explained by the configuration coordinate model for Mn^{2+}. Understanding the mechanism could lead to development of high quality nanoscale ZnS:Mn^{2+} phosphor for a variety of lighting applications.

2 EXPERIMENTAL SETUP

2.1 Synthesis

ZnS:Mn^{2+} sols were prepared by using the chemical precipitation method described by Lu et al. [9]. The synthesis was carried out using Zn(CH$_3$COO)$_2$.2H$_2$O, Mn(CH$_3$COO)$_2$ and Na$_2$S as starting materials. Zn(CH$_3$COO)$_2$.2H$_2$O and Mn(CH$_3$COO)$_2$ were dissolved in ethanol with stirring at 150°C. A 1:1 ratio of ethanol to deionized water solution of Na$_2$S was added to the Zn^{2+} and Mn^{2+} solution drop by drop with vigorous stirring at 75°C. The resulting white precipitate was centrifuged and washed using a mixture of toluene and ethanol in a 2:1 volume ratio. The precipitate was dried at 120°C for two hours.

2.2 Characterization

The structure and morphology of the samples were determined by XRD and TEM and SEM using a Siemens Diffractometer D5000 equipped with a Cu Kα source and JEM 2100F TEM and JEM 7500F SEM. The bandgap of ZnS was determined from the UV-Vis data collected by a Shimadzu UV-1700 PharmaSpec UV-Vis spectrophotometer. The PL data was collected using a 15 W Xenon flash lamp and 325 nm HeCd (26mW) laser as excitation sources.

3 RESULTS AND DISCUSSION

The XRD pattern of ZnS:Mn^{2+} nanoparticles is shown in Figure 1. The three diffraction peaks indexed as (111), (220) and (311) (JCPDS 5-566) match the lattice planes of the zincblende (sphalerite) ZnS crystal structure. Manganese impurities did not contribute to any additional diffraction peaks or shifts in the peak position and it indicates that the Mn^{2+} ions were well dispersed in the ZnS matrix [10] and their concentration was relatively low. The particle sizes were calculated using Scherrer's equation [11] by making use of the (111) diffraction peak. The average particle size was determined as 3 ± 1 nm in diameter. The TEM image in Figure 2 shows a cluster of agglomerated nanoparticles as indicated by the circle. The SEM image of the sample showed a similar cluster of agglomerated nanoparticles. The average particle size estimated from the TEM image was ~2-4 nm in diameter, which is consistent with the particle size calculated from the XRD data. The band gap of synthesized nano ZnS was determined from the UV absorption data and found to be 4.1 eV. This band gap

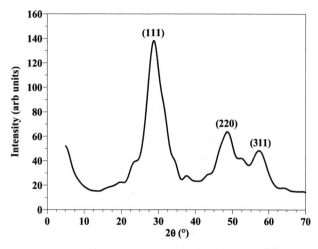

Figure 1: XRD pattern of synthesized ZnS:Mn^{2+} 5 mol%.

Figure 2: A TEM image of ZnS:Mn^{2+} with average particle size of 2 – 4 nm in diameter.

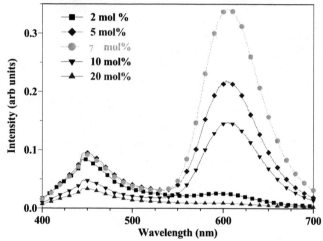

Figure 3: PL spectra for synthesized ZnS:Mn^{2+} with different doping concentrations of Mn^{2+}.

is blue shifted from the band gap of bulk ZnS (3.6 eV) [17]. The band gap of ZnS nanoparticles has thus enlarged probably due to quantum confinement effects. Figure 3 shows the PL excitation spectra for synthesized ZnS:Mn^{2+} with different doping concentrations of Mn^{2+}. Two peaks are observed at 450 and 600 nm. The 450 nm peak corresponds to excitonic emission of ZnS and the 600 nm peak corresponds to the characteristic $^4T_1 \rightarrow {}^6A_1$ transition of Mn^{2+} ions [12]. There was an intensity increase with an increase in Mn^{2+} from 2 – 5 mol% and a decrease in intensity for the 10 and 20 mol% Mn^{2+} doping. This may be due to concentration quenching effects [13]. Figure 4 shows possible electron transitions that could lead to the orange emission from the Mn^{2+} ions. These transitions were explained in terms of Tanabe-Sugano diagrams for the d^5 level and the Ligand field theory by different researchers [14]-[18]. It is known that Mn^{2+} has a d^5 electron structure with a tetrahedral symmetry. When doped into ZnS, it occupies the sites of Zn^{2+}. The ground state of Mn^{2+} is $^6A_1(^6S)$ and the first excited state is ^4G. Under tetrahedral symmetry, the excited stated can be split into $^4T_1$, $^4T_2$, $^4A_1$ and ^4E multiplets whose energies were reported by McClure [17]. The d electron states of the Mn^{2+} ion acts as efficient luminescent centers while interacting strongly with the host crystal's s-p electronic states at which the external electronic states are normally directed [18]. Figure 4 describes (a) the excitation mechanism of bulk and nano ZnS (b) the emission mechanism of ZnS and ZnS:Mn^{2+}(c) the emission mechanism of ZnS:Mn^{2+} using the configurational coordinate model.

Figure 4 (a): The energy dispersion for a bulk semi-conductor and a nanoparticle is described in ref [3]. In the case of a nanoparticle the band gap (E_g) is blue shifted. The band gap for bulk ZnS is 3.6 eV [19] and from UV-Vis measurements the band gap for ZnS nanoparticles was determined as 4.1 eV. The insert (i) in Figure 4 shows the excitation spectra of bulk and nanoparticles of ZnS. This corresponds to excitation over the band gap. A hole is created in the valence band by the incident photon and the electron is excited over the band gap to the conduction

band. In the case of bulk ZnS the energy of this excitation is 3.6 eV and for ZnS nanoparticles it is 4.1 eV, that is the excitation energy of the ZnS nanoparticles is blue shifted.

Figure 4(b): The excited electron immediately transfers to the shallow traps (dashed line in Figure 4). This shallow trap-state is delocalized over the entire nanocrystal, because of the quite small effective mass of the electron. Because the hole has a much higher effective mass it will initially remain in the valence band and will be trapped on a longer time scale [20]. The emission spectra of ZnS:Mn^{2+} (Figure 3) show that the interaction between ZnS and Mn^{2+} resulted in two emissions peaks at 450 (minor) and 600 nm (major). The minor peak at 450 nm (blue emission) can be attributed to the hole trapping and recombination with electrons by defect states (zinc or sulphur vacancies) in ZnS. The orange emission at 600 nm for nanoparticles can be attributed to $^4T_1 \rightarrow {}^6A_1$ transitions of Mn^{2+} ions. For the process denoted by A, two possible routes for this emission are possible. For A1, an excited electron in the conduction band can relax to the vacancy level and then to the $^4T_1$ level of Mn^{2+} non-radiatively, followed by radiative transition to the ground state ($^6A_1$) and recombination with a hole trapped in the ground state. The radiative transition is accompanied by emission of orange photons at 600 nm. Since the orange emission at 600 nm is more intense than the blue (violet) emission at 450 nm, this suggests that non-radiative relaxation to the $^4T_1$ level of Mn^{2+} was faster than hole capture and recombination with electrons by defects states of ZnS. For A2, an excited electron in the conduction band can first relaxes non-radiatively to the $^4T_1$ level of Mn^{2+} and finally to the $^6A_1$ level emitting orange photons in the process.

Figure 4(c): The emission peak of nano-sized ZnS:Mn^{2+} was observed to be red shifted by 10 nm from that of the bulk spectrum. So far, there have been many reports about the origin of this red shift in the emission. Cruz et al. [21] reported that the red shift might be caused by a large density of surface states in the nanoparticles, or by strong electron-phonon coupling in the nanoparticles. They also report that it might be possible that the size-dependent crystal field effect is responsible for this red shift. Li et al. [22] reported that the red shift may come from the quantum confinement effect in nanoparticles which leads to a change of the crystal field surrounding the Mn^{2+} ions. Shionoya and Yen [23] reported that when a metal ion occupies a certain position in a crystal, the crystal field strength that affects the ion increases as the space containing the ion becomes smaller. Therefore for increases in the field, the transition energy between the $^4T_1$ and $^6A_1$ levels of Mn^{2+} is predicted to decrease (shift to longer wavelengths). From these reports it can be seen that a change in the crystal field will most likely cause a red shift in the emission of nanoparticles. The quantum confinement effect (that arises from the very small particle sizes) will change the crystal field. From XRD data it can also be seen that the bulk sample is much more crystalline than the nano samples. It can therefore be concluded that a change in the crystal field would cause a change in the parabola offset in the configurational coordinate diagram. The offset is given by $\Delta Q = Q_0 - Q'_0$ where Q_0 is the equilibrium distance of the excited state, Q'_0 is the equilibrium distance of the ground state and ΔQ is the parabola offset. The configurational coordinate diagram also describes the Stokes'shift (difference in absorption and emission energy) [23]. When there is a displacement between the potential wells of the ground and excited states, there will be a difference in the Stokes' shift. The parabola offset will therefore also increase. In figure 4(c) the configurational coordinate model for Mn^{2+} [8],[24] is used. The same process as in figure 4(a) occurs, but the difference between the emission of the bulk and nanoparticles is shown. The insert (ii) in Figure 4 shows the emission spectra for bulk and nanoparticles of ZnS:Mn^{2+}. For bulk particles there is only a slight offset in the parabola of the excited state. An electron relaxes non-radiatively from the conduction band to the $^4T_1$ state. This electron then relaxes non-radiatively to the bottom of the $^4T_1$ state. The excess energy will be released in the form of heat to the lattice. The electron will then recombine with a trapped hole in the ground state ($^6A_1$) and emission of orange photons at 590 nm takes place. In

the case of nanoparticles the parabola offset is greater than that of bulk particles. The same process of electron relaxation and recombination will occur, but in this case emission at 600 nm takes place.

4. Conclusion

ZnS:Mn^{2+} nanoparticles were synthesized using the chemical precipitation method. The structure and particle morphology were analyzed by using XRD, TEM and SEM. The average particles size was found to be 3 $\pm$1 nm. The band gap of ZnS nanoparticles was determined as 4.1eV and it was blue shifted from that of bulk ZnS (3.6eV). For the Mn^{2+} doped samples two PL peaks were observed at 450 nm and at 600 nm. The 450 nm peak corresponds to excitonic emission of ZnS and the 600 nm peak corresponds to the characteristic $^4T_1 \rightarrow {}^6A_1$ transition of Mn^{2+} ions. These transitions were explained in terms of Tanabe-Sugano diagrams for the d^5 level and the Ligand field theory. The blue shift in the excitation spectra of nano sized ZnS can be explained according to the energy dispersion diagram for bulk and nano semiconductors, while the red shift in the emission spectra can be explained according to the configurational coordinate model of Mn^{2+}.

Acknowledgements: The Physics Department at the Nelson Mandela Metropolitan University for the use of their PL system, department of Geography at the University of the Free State for the help with the XRD measurements, the South African National Research Foundation (NRF) and the University of the Free State for financial assistance.

REFERENCES

[1] B.S. Rema Devi, R. Raveendran and A.V. Vaidyan, Pramana – J. Phys., **68** (4) (2007) 679.

[2] S. Santra, H. Yang, J.T. Stanley, P.H. Holloway, B.M. Moudgil, G. Walter and R.A. Mericle, Chem. Commun.(2005) 3144.

[3] B. Bhattacharjee, D. Ganguli, K. Iakoubovskii, A. Stesmans and S. Chaudhuri, Bull. Mater. Sci., **25** (3) (2002) 179.

[4] R. Sarkar, C.S. Tiwary, P. Kumbhakar, S. Basu and A.K. Mitra, Physica E, **40** (2008) 3199.

[5] M. Tanaka, J. Qi and Y. Masumoto, J. Lumin., **87 – 89** (2000) 474.

[6] F.G. Anderson, W.M. Dennis and G.F. Imbusch, J. Lumin., **90** (2000) 28.

[7] H. Hanzawa, M. Kobayashi, O. Matsuda, K. Murase and W. Giriat, Phys. Stat. Sol., **175** (a), (1999) 721.

[8] T.A. Prokof'ev, B.A. Polezhaev and A.V. Kovalenko, J. Appl. Spec., **72** (6) (2005) 869.

[9] S.W. Lu, B.I.Lee, Z.L. Wang, W. Tong, B.K. Wagner, W. Park and C.J. Summers, J. Lumin., **92** (2001) 74.

[10] W.Q. Peng, S.C. Qu, G.W. Cong, Z.G. Wang, J. Crystal Growth, **279** (2005) 456

[11] B.D. Cullity, Elements of X-Ray Diffraction, Addison-Wesley, Reading, MA, 1978

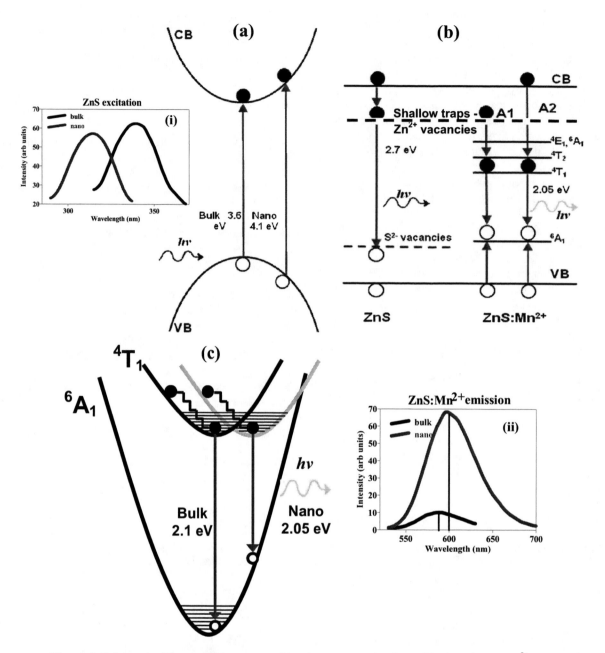

Figure 4: Schematic diagram of the proposed luminescent mechanism of ZnS and ZnS:Mn^{2+}.

[12] R.N. Bhargava, D. Gallagher, X. Hong and N. Nurmikko, Phys. Rev. Lett., **72** (1994) 416

[13] K.H. Yoon, J.K. Ahn and J.Y. Cho, J. Mater. Sci., **36** (2001) 1376.

[14] M. Tanaka, J. Lumin., **100** (2002) 166.

[15] M. Jain, J.L. Roberts, in: M. Jain (Ed.), Diluted Magnetic Semiconductors, World Scientific, Singapore, 1991, 31.

[16] Y. Tanabe and S. Sugano, J. Phys. Soc. Japan, **9** (1954) 753.

[17] D.S. McClure, J. Chem. Phys., **39** (1963) 2850.

[18] H. Yang, J. Zhao, L. Song, L. Shen, Z. Wang, L. Wang and D. Zhang, Mater. Lett., **57** (2003) 2290.

[19] C. Kittel, Introduction to Solid State Physics, 6th Ed., John Wiley, New York, (1986) p185.

[20] S.C. Ghosh, C. Thanachayanont and J. Dutta, ECTI-CON, Pattaya, Thailand, (2004) 146.

[21] A.B. Cruz, Q. Shen and T. Toyoda, Thin Solid Films, **499** (2006) 106.

[22] G.H. Li, F.H. Su, B.S. Ma, K. Ding, S.J. Xu and W. Chen, Phys. Stat. Sol., **241** (2004) 3252.

[23] S. Shionoya and W.M. Yen, Phosphor Handbook, CRC Press LLC, Boca Raton, (1999) p 36, 169-170.

[24] H. Hanzawa, M. Kobayashi, O. Matsuda, K. Murase and W. Giriat, Phys. Stat. Sol., (a) **175** (1999) 721.

From Micro to Nano Magnetic Spheres: Size-Controllable Synthesis, Multilayer Coatings, and Biomedical Applications

Ken Cham-Fai Leung[*], Shouhu Xuan[*] and Yixiang J. Wang[**]

[*]Center of Novel Functional Molecules, Department of Chemistry, The Chinese University of Hong Kong, Shatin, NT, Hong Kong SAR, cfleung@cuhk.edu.hk (KCFL)
[**]Department of Diagnostic Radiology and Organ Imaging, Prince of Wales Hospital, The Chinese University of Hong Kong, Shatin, NT, Hong Kong SAR

ABSTRACT

Size-controllable monodispersed ferrite microspheres to nanoparticles have been synthesized with a range of particle sizes from 10 to 1500 nm. The particle's grain size can also be tuned from 5.9 to 21.5 nm. Some of these spheres have been coated efficiently with silica, poly(aniline), dendrimer, etc., to yield well-defined core/shell-like structures. These magnetic microspheres have been functionalized successfully with organic moieties such as crown ethers, acting as a tool for the magnetic separation of discrete functional gold nanoparticles. Moreover, ultrasmall magnetic nanoparticles have been employed as both *in vitro* and *in vivo* magnetic resonance imaging contrast agents.

Keywords: contrast agent, magnetic resonance imaging, magnetite, microsphere, nanoparticle

1 INTRODUCTION

Magnetic material has become a particularly important research area and is attracting a great deal of attentions because of their potential technological applications, such as in ferrofluids, colored pigments, high-density magnetic recording media, chemical sensors, electrophotographic developers, biological assays and magnetic resonance imaging [1-9]. High-quality nanocrystals with well-defined shapes and controllable sizes play an important role in nanotechnology because the chemical and physical properties of nanomaterials are strongly related to their mean size and size distribution. To date, therefore, a range of techniques have been developed for fabrication of magnetic nanostructures with sophisticated nanostructures, among which spherical nanoparticles are of particular interest owning to their shape and size-dependent properties [10-18].

Spinel ferrites (MFe_2O_4; M=Fe, Mn, Zn, Co, and Ni) exhibit interesting magnetic, magnetoresistive, and magneto-optical properties that are useful for a broad range of applications [19-22]. In principle, the magnetic properties of the ferrite can be systematically varied by changing the identity of the divalent M^{2+} cation or by partial substitution while maintaining the basic crystal structure. Besides the composition, the size and crystallinity of the crystals also have great influence on the properties of the ferrite nanomaterials. For this reason, various approaches, including co-precipitation, hydrothermal, sol-gel, reversed and normal micelles, sonochemical reactions, ball milling, etc., are being adopted to produce spinel ferrites to realize the desired properties [23-26]. In order to attain reliable and high performance for biomedical applications, ferrite particles should possess narrow size distribution, large surface area, high magnetic saturation, good dispersion in liquid media, well-defined shape and surface texture [27-29]. Therefore, the synthesis of monodispersed ferrite nanospheres with tunable sizes and magnetic properties has been intensively pursued for their technological and fundamental scientific importance.

2 RESULTS AND DISCUSSION

Up to now, the general method that was employed for the preparation of monodispersed ferrite nanospheres, was usually based on a wet chemistry-based synthetic route such as co-precipitation [30]. Another effective method for the size and shape-controllable synthesis of the ferrite nanoparticles is *via* a high-temperature thermal decomposition route, by using organometallic complex, such as metal acetylacetonates, carbonyls, etc., as the precursors along with some aliphatic acids, amines and diols [31-36]. In this work, a modified solvothermal method [37] was developed to construct a series of superparamagnetic ferrite microspheres/nanoparticles which comprised of tunable particle sizes, grain sizes, and magnetic properties. The as-prepared ferrite particles can be coated with different functional groups or layer of materials such as silica, poly(aniline), dendrimers, etc. The superparamagnetic properties equipped with functional coatings render the particles ideal candidates for practical applications such as magnetic separation, magnetic resonance imaging and drug delivery.

2.1 Synthesis

When $FeCl_3 \cdot xH_2O$ and $MCl_2 \cdot yH_2O$ (M = Mn, Zn, Co, or Ni) were used as the metal source, the grain size and particle size of the secondary structures can be tuned rationally by using different amounts of ethylene glycol

(EG), diethylene glycol (DEG), sodium acrylate (Na acrylate), and sodium acetate (NaOAc) (Figure 1) [38,39].

By way of examples, monodispersed ferrite microspheres with different grain sizes that have an average particle size about 280 nm, can be synthesized by controlling the ratios of Na acrylate/NaOAc from 1.5/0 to 1.0/0.5, 0.5/1.0, 0.3/1.2, and 0.1/1.4 in EG at a temperature of 200 °C for 10 h. These processes afford the ferrite microspheres which possess a grain size from 5.9 to 6.9, 8.3, 10.1, and 13.5 nm, respectively [38]. When Na acrylate was introduced to the $FeCl_3$/EG solution, a significant amount of $Fe(OOCCH=CH_2)_3$ complex was formed. Then, in the presence of EG, the complex was transformed into Fe_3O_4 nucleates under hydrothermal treatment [37]. Meanwhile, the freshly formed nanocrystals are unstable because they contain high surface energy, and thus they have a great tendency to aggregate rapidly. Because of the strong coordinations between iron(III) ions and carboxylate, the as-formed Fe_3O_4 nanograins possessed a coating of polyacrylate. By increasing the amount of Na acrylate, more polyacrylate will bind to the primary Fe_3O_4 nanocrystals' surface. This results in a decrease of the size of Fe_3O_4 nanograins. Consequently, the main driving force for the aggregation of nanograins is generally attributed to the reduced high surface energy through both the attachment among the primary nanoparticles and the polyacrylate. Additionally, the grain size tunability may also be the result of slight differences in alkalinity caused by varying the amounts of NaOAc. The hydrolysis rate of $FeCl_3$ could be accelerated by increasing alkalinity so as to promote the formation of larger Fe_3O_4 nanocrystals.

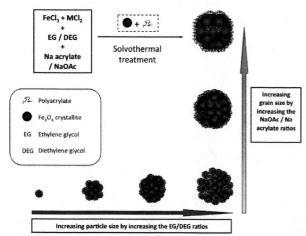

Figure 1: Schematic representation of the formation of the monodispersed ferrite particles with tunable grain size and particle size. The particle sizes are also tunable when Na acrylate is replaced by PVP.

On the other hand, by varying the EG/DEG ratios in the system instead of Na acrylate/NaOAc ratios, the particle sizes can be tuned. EG plays both roles as a reducing agent and a solvent during the formation of the particles. Since the reaction temperature in the sealed container is 200 °C,

which is slightly higher than the boiling point of EG (196 °C), therefore, EG may exist as a superfluid. At this temperature, Fe_3O_4 nanograins and polyacrylate assemble to form large secondary structural microspheres. When DEG (b.p. = 240 °C) was introduced together with EG, the size of the particles can be reduced. The diameter of the as-prepared Fe_3O_4 particles are ranging from 6 to 60, 120, and 170 nm when the EG/DEG ratio changes from 0/20 to 5/15, 10/10, and 20/0, respectively. These particles are superparamagnetic which were characterized by vibrating sample magnetometer (VSM).

From the above reaction conditions, if Na acrylate was replaced by poly(vinylpyrrolidone) (PVP), the sizes of ferrite nanoparticles/microspheres can be tuned with size from 20 to 300 nm, by simply varying the EG/DEG ratios. For example, the EG/DEG ratios of 1/19, 5/15, 10/10, and 20/0 lead to the formation of Fe_3O_4 nanoparticles/microspheres with average sizes of 20, 90, 165, and 300 nm, respectively, at a temperature of 200 °C for 10 h [39]. Additionally, if the reaction suspension was heated under a continuous flow of N_2 with mechanical stirring, the as-prepared particles are superparamagnetic.

2.2 Coating

Core/shell-structured composite materials which combine with the advantageous properties of both materials of the core and the shell, have attracted increasing interest due to their unique physicochemical properties. Various types of bifuntional or trifunctional core/shell materials, such as electronic/optical, electronic/catalytic, magnetic/electronic, etc., could be technically obtained. In our work, ferrite particles were coated with functional shells to give magnetic core/shell composite materials.

Among these shell materials, poly(aniline) (PANI) has received more attention for its unique electrical properties. In our group, Fe_3O_4@PANI microspheres with blackberry-like morphology have been successfully obtained by using an *in situ* surface polymerization method in the presence of aniline and PVP [40]. Naked Fe_3O_4 particles (350 nm) were coated with a thin silica shell with amine functional group, prior to PANI coating. The thickness of the PANI shells (30-60 nm) depends on the polymerization time, the initial concentration of aniline, and the reaction temperature. Furthermore, a more uniform PANI coating (25 nm) apart from the rough blackberry-like morphology has been obtained for a carboxylic acid-modified Fe_3O_4 microspheres (160 nm) [41]. The uniform Fe_3O_4@PANI microspheres are then used as templates for supporting gold nanoparticles (4 nm). For the synthesis of the Fe_3O_4@PANI@Au composites, nitrogen atoms on the Fe_3O_4@PANI microspheres' periphery were protonated such that the citrate-coated, negatively charged Au seeds could be immobilized by electrostatic interactions [41].

Silica has been frequently used to coat ferrite particles, unambiguously for biological applications [30]. This is because silica has good biocompatibility and hydrophilicity, and can prevent aggregation of particles in liquids and improve their physicochemical stability. Silica can be effectively coated on hydroxyl functionalized ferrite particles by a sol-gel method using silica precursors such as tetraethyl orthosilicate (TEOS) and aminopropyl triethoxysilane (APTES).

Despite inorganic coating shells, organic shells are practical to coat ferrite particles to provide specific functions and properties. In our work, poly(arylether) dendrons have been successfully attached onto the surface of Fe_3O_4 microspheres [42]. The dendron attachment/detachment processes on ferrite particles are controlled by water concentration via dynamic covalent chemistry, wherein the hydrophilicity and hydrophobicity of the ferrite particles' surface can be reversibly tuned [42]. Organic, supramolecular molecules such as crown ethers were attached on $Fe_3O_4@SiO_2$ microspheres for magnetic separation purposes [43]. Crown ethers such as dibenzo[24]crown-8 that are attached on $Fe_3O_4@SiO_2$ microspheres, can be recognized by dibenzyl ammonium derivatives upon pH changes [44]. Discrete functional gold nanoparticles that are linked with dibenzyl ammonium groups, can be magnetically separated and purified by crown ether-coated ferrite microspheres. The pH-switchable property of the recognition unit allows the detachment of the final product from the ferrite microspheres [43].

2.3 Properties and Applications

Magnetic separation by using functional ferrite nanoparticles/microspheres is attractive. In our work, the $Fe_3O_4@PANI@Au$ composite materials are effective co-catalysis for reduction, that can be magnetically recyclable [41]. Furthermore, one of the important applications of ferrite is the use as contrast agents for magnetic resonance imaging (MRI). As MRI contrast agents, superparamagnetic ferrite should be employed wherein superparamagnetic materials behave as magnets only if external magnetic field is applied.

$Fe_3O_4@SiO_2$ nanoparticles (8 nm) appeared to be the preferred material for magnetic labeling of stem cells without the addition of any transfecting agent [30]. This is because (1) they provide a strong change in signal per unit of metal, in particular on $T2^*$-weighted images; (2) they are composed of biodegradable iron, which is biocompatible and can thus be reused/recycled by cells by using normal biochemical pathways for iron metabolism; (3) they can be easily detected by light and electron microscopy. After biodegradation, iron oxide nanoparticles are turned into soluble iron ions. Our amine-functionalized $Fe_3O_4@SiO_2$ nanoparticles (8 nm) have been employed as contrast agents for *in vitro* and *in vivo* MRI. For *in vivo* MRI, nanoparticle-labeled stem cells were implanted to rabbit's brain (Figure 2) and spinae muscle, demonstrating a long-lasting, durable MRI labeling efficacy after 8-12 weeks [30].

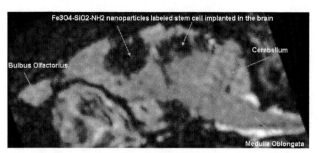

Figure 2: *In vivo* MRI. Amine-functionalized $Fe_3O_4@SiO_2$ nanoparticle-labeled stem cells were implanted into rabbit's brain.

The $T2$ relaxation rates of some of our as-synthesized superparamagnetic ferrite particles are summarized in Table 1. From the results, the ferrite microspheres/nanoparticles with a larger grain size yields a higher $T2$ relaxation rate. The $r2$ relaxivities of commercially available Fe_3O_4-based MRI contrast agents range from 33.4 for VSOP-C184 and 38 for Supravist to 120 for Feridex and 189 for Resovist [45]. It is well-known that the effects of crystallization nature of the ferrite are related to the $T2$ relaxation rate [46]. Also, $r2$ relaxivity of ferrites is closely related to their particle size, and that larger particles tend to have higher $r2$ relaxivity. The grain sizes of some of our ferrite microspheres/nanoparticles are close to that of VSOP-C184 and Supravist. From this aspect, because of the difference in crystallization nature, our ferrite microspheres/nanoparticles demonstrate stronger relaxivities than the VSOP-C184 and Supravist.

Particles	Particle size (nm)	Grain size (nm)	$T2$ relaxation rate ($s^{-1}mmol^{-1}$)
Fe_3O_4	280	5.9	65.3
Fe_3O_4	280	10.1	80.4
Fe_3O_4	280	13.5	84.6
Fe_3O_4	90	10-20	93.7
$ZnFe_2O_4$	90	7-10	30.4
$Fe_3O_4@SiO_2$	10	3-4	18.9
$Fe_3O_4@SiO_2$	8	3-4	43.5

Table 1: $T2$ relaxation rates of some of our as-synthesized superparamagnetic ferrite particles.

3 CONCLUSIONS

Superparamagnetic ferrite nanoparticles/microspheres have been successfully prepared with tunable particle sizes and grain sizes. Coatings of these particles with selected inorganic and organic shells are feasible. It has been demonstrated that these particles are ideal contrast materials for *in vitro* and *in vivo* MRI.

REFERENCES

[1] K. Raj and R. Moskowitz, J. Magn. Magn. Mater., 85, 233, 1990.

[2] A. G. Hu, G. T. Yee and W. B. Lin, J. Am. Chem. Soc., 127, 12486, 2005.

[3] T. Sen, A. Sebastianelli and I. J. Bruce, J. Am. Chem. Soc., 128, 7130, 2006.

[4] B. R. Martin, D. J. Dermody, B. D. Reiss, M. M. Fang, L. A. Lyon, M. J. Natan and T. E. Mallouk, Adv. Mater., 11, 1021, 1999.

[5] L. Huo, W. Li, L. Lu, H. Cui, S. Xi, J. Wang, B. Zhao, Y. Shen and Z. Lu, Chem. Mater., 12, 790, 2000.

[6] P. E. Matijevic, L. L. Hench and D. B. Ulrich, In Colloid Science of Composites System Science of Ceramic Chemical Processing Ed. Wiley: New York, 463, 1986.

[7] A. H. Lu, E. L. Salabas and F. Schuth, Angew. Chem., Int. Ed., 46, 1222, 2007.

[8] U. Y. Jeong, X. W. Teng, Y. Wang, H. Yang and Y. N. Xia, Adv. Mater., 19, 33, 2007.

[9] S. Laurent, D. Forge, M. Port, A. Roch, C. Robic, L. V. Elst and R. N. Muller, Chem. Rev., 108, 2064, 2008.

[10] T. He, D. R. Chen and X. L. Jiao, Chem. Mater., 16, 737, 2004.

[11] X. L. Hu, J. M. Gong, L. Z. Zhang and J. C. Yu, Adv. Mater., 20, 4845, 2008.

[12] Y. G. Sun and Y. N. Xia, Science, 298, 2176, 2002.

[13] M. Yin, C. K. Wu, Y. B. Lou, C. Burda, J. T. Koberstein, Y. M. Zhu and S. O'Brien, J. Am. Chem. Soc., 127, 9506, 2005.

[14] X. G. Peng, J. Wickham and A. P. Alivisatos, J. Am. Chem. Soc., 120, 5343, 1998.

[15] Z. Y. Tang, N. A. Kotov and M. Giersig, Science, 297, 237, 2002.

[16] F. Bai, D. S. Wang, Z. Y. Huo, W. Chen, L. P. Liu, X. Liang, C. Chen, X. Wang, Q. Peng and Y. D. Li, Angew. Chem., Int. Ed., 46, 6650, 2008.

[17] J. Q. Zhuang, H. M. Wu, Y. G. Yang and Y. C. Cao, J. Am. Chem. Soc., 129, 14166, 2007.

[18] J. Q. Zhuang, H. M. Wu, Y. G. Yang and Y. C. Cao, Angew. Chem., Int. Ed., 47, 2208, 2008.

[19] M. Sugimoto, J. Am. Ceram. Soc., 82, 269, 1999.

[20] J. Park, J. Joo, S. Kwon, Y. Jang and T. Hyeon, Angew. Chem., Int. Ed., 46, 4630, 2007.

[21] X. Wang, J. Zhuang, Q. Peng and Y. D. Li, Nature, 437, 121, 2005.

[22] N. Bao, L. Shen, Y. Wang, P. Padhan and A. Gupta, J. Am. Chem. Soc., 129, 12374, 2007.

[23] Z. J. Zhang, Z. L. Wang, B. C. Chakoumakos and J. S. Yin, J. Am. Chem. Soc., 120, 1800, 1998.

[24] Y. Lee, J. Lee, C. J. Bae, J. G. Park, H. J. Noh, J. H. Park and T. Hyeon, Adv. Funct. Mater., 15, 503, 2005.

[25] C. Liu, B. S. Zou, A. J. Rondinone and Z. J. Zhang, J. Phys. Chem. B, 6, 1141, 2000.

[26] K. E. Mooney, J. A. Nelson and M. J. Wagner, Chem. Mater., 16, 3155, 2004.

[27] S. Mitragotri and J. Lahann, Nat. Mater., 8, 15, 2009.

[28] J.-H. Park, G. von Maltzahn, L. Zhang, A. M. Derfus, D. Simberg, T. J. Harris, E. Ruoslahti, S. N. Bhatia and M. J. Sailor, Small, 5, 694, 2009.

[29] A. Shavel and L. M. Liz-Marzán, Phys. Chem. Chem. Phys., 11, 3762, 2009.

[30] H.-H. Wang, Y.-X. J. Wang, K. C.-F. Leung, D. W. T. Au, S. Xuan, C.-P. Chak, S. K. M. Lee, H. Sheng, G. Zhang, L. Qin, J. F. Griffith and A. T. Ahuja, Chem. Eur. J., 15, 12417, 2009.

[31] S. Sun and H. Zeng, J. Am. Chem. Soc., 124, 8024, 2002.

[32] S. Sun, H. Zeng, D. B. Robinson, S. Raoux, P. M. Rice, S. X. Wang and G. Li, J. Am. Chem. Soc., 126, 273, 2004.

[33] H. Zeng, P. M. Rice, S. X. Wang and S. Sun, J. Am. Chem. Soc., 126, 11458, 2004.

[34] Q. Song and Z. J. Zhang, J. Am. Chem. Soc., 126, 6164, 2004.

[35] E. Tirosh, G. Shemer and G. Markovich, Chem. Mater., 18, 465, 2006.

[36] N. Bao, L. Shen, Y. Wang, J. Ma, D. Mazumdar and A. Gupta, J. Am. Chem. Soc., 131, 12900, 2009.

[37] H. Deng, X. Li, Q. Peng, X. Wang, J. Chen and Y. Li, Angew. Chem., Int. Ed., 44, 2782, 2005.

[38] S. Xuan, Y.-X. J. Wang, J. C. Yu and K. C.-F. Leung, Chem. Mater., 21, 5079, 2009.

[39] S. Xuan, F. Wang, Y.-X. J. Wang, J. C. Yu and K. C.-F. Leung, J. Mater. Chem., in revision.

[40] S. Xuan, Y.-X. J. Wang, K. C.-F. Leung and K. Shu, J. Phys. Chem. C, 112, 18804, 2008.

[41] S. Xuan, Y.-X. J. Wang, J. C. Yu and K. C.-F. Leung, Langmuir, 25, 11835, 2009.

[42] K. C.-F. Leung, S. Xuan and C.-M. Lo, ACS Appl. Mater. Interfaces, 1, 2005, 2009.

[43] C.-P. Chak, S. Xuan, P. M. Mendes, J. C. Yu, C. H. K. Cheng and K. C.-F. Leung, ACS Nano, 3, 2129, 2009.

[44] K. C.-F. Leung, C.-P. Chak, C.-M. Lo, W.-Y. Wong, S. Xuan and C. H. K. Cheng, Chem. Asian J., 4, 364, 2009.

[45] C. Corot, P. Robert, J.-M. Idee and M. Port, Adv. Drug Delivery Rev., 58, 1471, 2006.

[46] B. A. Larsen, M. A. Haag, N. J. Serkova, K. R. Shroyer, C. R. Stoldt, Nanotechnology, 19, 265202, 2008.

Transport of large Particles and Macromolecules in Flow Through Inorganic Membrane

A. O. Imdakm and T. Matsuura

Department of Chemical Engineering, University of Ottawa, 161 Louis Pasteur St., Ottawa, Ont., Canada KIN 6N

Abstract

The influence of transport of large particles and macromolecules on membrane separation efficiency and permeation rate limitation has been observed in many separation processes of microscopic scale. In this paper a Monte Carlo simulation model is developed to study this phenomenon. The membrane pore space is described by a three-dimensional network model of inter-connected cylinders pores. The effective radii of these pores are distributed according to experimentally reported pore size distribution. The paths of the injected particles throughout the pore space are determined by transition probability function proportional to the pore flux. Pores are considered completely blocked when particles of effective radii comparable with the pore size pass through them. Thus, the membrane capability of transport can ultimately vanish and, therefore, this process is a percolation process and it is characterized by a percolation threshold below which the membrane losses its macroscopic connectivity. When the model applied to this problem, the predictions are in excellent agreement with the available experimental data.

Keywords: Ultrafiltration; Network model; Suspensions; Macromolecules; Permeability Reduction; Size Exclusion; SE.

1 Introduction

In recent years, there is a wide spread of commercial development in separation processes utilizing membranes in large area of industrial interest. In many of such transport operations in chemical and biochemical industries involve separation of large particles and macromolecules in restricted environment. For example, reverse osmosis (RO), micro-filtration (MF), ultra-filtration (UF), and deep-bed filtration (DBF). The efficiency of such process, which is the point of interest of this study, depends crucially on the availability of an open membrane pore space through which mass transport takes place. Thus, if such transporting paths (pore spaces) are reduced during the separation process, the permeation rate, which is considered as one of the measures of membrane performance, is reduced. As a result, the whole separation operation is also reduced to a certain degree depending on the degree of pore space reduction.

In many separation processes as in MF/UF, the most important issue is the transport of large particles and macromolecules whose effective radii are comparable to the pore size of the membrane. These transporting particles and macromolecules can be the component of the feed fluid or sometimes they can also arise due to the release of material deposited on the membrane pore walls or due to chemical reaction between the flowing fluid and membrane pore wall, which may take place because of the change in transporting fluid ionic environments. When emulsions droplets are present in flow of stable emulsions such as those in oily wastewater, paints, starch, and skim milk, significant number of membrane pores can also be blocked by emulsions droplets, the sizes of which are comparable to membrane pore sizes. This type of membrane pore plugging by solid particles or emulsions droplets of effective sizes comparable to the size of the membrane pores is known as pore plugging by **size-exclusion (SE)**. This mechanism of membrane pore space reduction is found to be the most dominant mechanism of flux rate reduction in many cases [1,3,4,5]. The capture of particles or emulsions droplets by pore wall because of electrical interaction between them is another mechanism of pore space reduction, but pore plugging by size exclusion mechanism still plays the most fundamental role in pore space reduction. Interesting reader may referred to the excellent reviews of Belfort et al. [2] and M. Sahimi et al. [5] for the theoretical models describing the transport processes in flow through porous media, their validity and limitation. These models in general ignore the detail descriptions of the membrane pore space (morphology), and the phyisco-chemical interaction between the transporting particles and membrane pore wall.

In this study an attempt is made to predict the performance of ultrafiltration membranes with the assumption that the pore space reduction is only due to size exclusion, SE, mechanism. SE is believed to be one of the primary reasons for the membrane pore space reduction and a thorough study based on SE will give some insight into the fouling phenomena by membrane pore space reduction. The calculation carried out in this study is also based on the parameters describing the pore size distribution obtained by Wang et al. [6]. Agreement of calculation and experimental data is further examined.

2 Model Formulations

2.1 Membrane pore space

The membrane pore space is described by simple-cubic (**L x L x L, L=12**) network model in which the bonds of the network represent the pore-throat. Each pore is assumed to be a cylindrical tube of constant length l and an effective diameter D_{pore}. The initial effective diameter of these pores are distributed according to a lognormal probability density function with geometric mean pore size μ_{pore} and standard deviation σ_{pore} as reported by Wang et al. [6]. The geometrical mean pore size and standard deviation are listed in **Table-1.** The sizes of the sites, which represent the pore bodies, are

$$\frac{df(D_{pore})}{dD_{pore}} =$$

$$\frac{1}{D_{pore}(ln\sigma_{pore})\sqrt{2\pi}} \exp\left[\frac{(lnD_{pore}-ln\mu_{pore})^2}{2(ln\sigma_{pore})^2}\right] \quad (1)$$

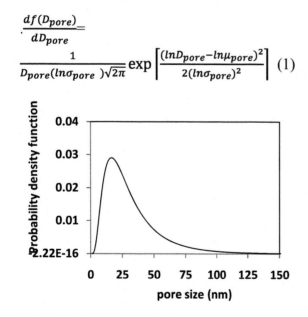

Figure 1: Probability density function for membranes 0.5 A..

assumed to be large enough in order to be able to contain the transporting particles. The pore length

was estimated based on the mean pore length determined microscopically, Wang et al. [6]. Creping flow within each pore is assumed, as such, the volumetric flow rate q_i of pore i is defined as $q_i = g_i\Delta p_i$ where Δp_i is pressure drop across the pore and g_i is the hydraulic conductance of the pore, which is proportional to R_{pore}^4. The boundary conditions are constant pressure drop as given experimentally, Wang et al. [6], imposed in the z-direction, and periodic boundary condition in the X and Y direction.

Therefore, steady state pressure distribution in the entire network can be computed. From the pressure distribution, one can calculate network permeability, average flow velocity, flow rate in each pore, the time needed for the suspended particles to pass through the pore, and cross-flux (permeate) using experimentally reported surface porosity shown in Table-1.

Thickness (μm) "Calcinations effect"		μ_{pore} (nm)	σ_{pore}	Surface Porosity (%)
Before	After			
97	85	25.7	1 .95	3.15

Table 1: Properties of Sepiolite Membrane (0.5 A) as given by Wang et al. [6].

2.2 Particles sizes

The macromolecules solutes, which are injected into the network, are polyethylene glycol PEG, (A), and polyethylene oxide PEO, (B), (C), (D), and (E) . They are assumed to be spheres with radii estimated from the **Einstein-Stock** equation depending on their molecular weight, Mw, (see table-2).

Molecular weight	Macromolecules radius r_px10^9 (m)
35,000 (A)	5.6825
100,000 (B)	8.9888
200,000 ©	13.5022
300,000 (D)	17.1305
400,000 (E)	20.282

Table 2: Radius of injected molecules, r_p. based on Einstein-Stocks equation.

Thus, for PEG

$$r_p = 16.73x10^{-12}M_w^{0..557}, \quad (2)$$

and for PEO

$$r_p = 10.44x10^{12}M_w^{0.587} \qquad (3)$$

2.3 Particles Movement

Suppose $V_{pore}\tau$ of fluid comes into the membrane within time interval Δt, where V_{pore} is the total volume of the network bonds at initial condition (i.e., at t=0), and τ is the number of pore volumes being injected into the network. The time interval Δt can be given as $V_{pore}\tau/Q$, where Q is the volumetric flow rate of the fluid coming into the network, and Δt can be defined as the time needed to inject such amount of constant volume of fluid ($V_{pore}\tau$). Q decreases with time because of pore plugging by SE that increases with time t. Therefore, Δt will increase with an increase in time until the system reaches steady state condition (i.e., no more pore plugging). The number of particles, M, being injected into the network within the time interval Δt is assumed to be constant (constant solute concentration) as reported experimentally (Wang et al. [6]), and it can be calculated given solute concentration (ppm), $V_{pore}\tau$, r_p, and particles density, ρ_p. Once the particle is injected at the network entrance (Z=o), it moves in the Z-direction to the next node, this because of the constant pressure drop imposed in the Z-direction, or it may cause pore plugging by SE (if $r_p \geq R_{pore}$) (rejection).When the particle arrived at anode, it moves to the one of the unplugged pores attached to this node with probability proportional to the fraction of flow rate departing from the node through this pore. Once the pore is chosen, the effective radius, r_p, of the particle is compared with the radius of the selected pore, R_{pore}. If $r_p \geq R_{pore}$, the pore entrance considered totally blocked by the solid particle (setting its radius to zero), this is equivalent to removing that bond from the network. If $r_p < R_{pore}$, the particle is allowed to pass through the pore to the next node. It is clear that this process is a percolation process. Once a set of plugged pores are identified, a new flow filed is computed, taking into consideration the presence of all the solid particles and their effect on the flow field.

3 Results and Discussion

In this study, an extensive Monte Carlo simulation of this process described above is carried out; it is obvious that the sharp permeability reduction observed experimentally (Wang et al. [6]) is due to the significant pore plugging mainly by SE.

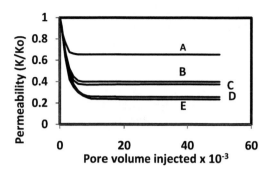

Figure 2: Normalized permeability vs. pore volume injected.

This is may due the fact that the injected particles are of size r_p that is comparable to the mean pore size, μ_{pore}, and as such, a large fraction of pores are plugged by SE and in particular at network entrance. Figure 2 shows clearly the permeability reduction observed in the early stage of the ultra- filtration. It decreases quickly due to SE in the first stage of filtration and reaches steady state after some time.

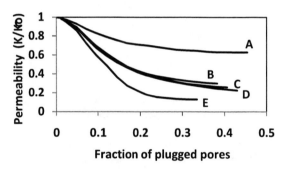

Figure 3: Normalized permeability of Figure 2 vs. fraction of plugged pores.

The results are carried out for operating pressure of 55 psig. The permeability changes with the fraction of pores plugged is given in Figures 3. This figure show that permeability decrease quickly until about 20% of the pores are blocked particularly in close vicinity to the network entrance and then level off slowly due to small fraction of pore plugging deep down into the network. Figure 4 show the flux rate of pure water and the flux decline due to the injection of solutes of different molecular weights and under different operating pressures. Experimental results are also shown on the same figure. The agreement between the experimental and calculated data is excellent in a pressure range 35-115 psig.

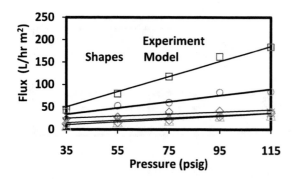

Figure 4: Comparison of predicted flux rates vs. pressure with the experimental data. □ pure water, ○ A, ◇ B&C, x D, and △ E , Wang et al. [6].

The number of particles, M, injected into the pore network during the time interval Δt, not all of them will pass through the network. Some of them get rejected at the network entrance, other will cause pore plugging within the network (and may cause also particles trapping within the plugged pores, and the rest will eventually pass through the network. As the system approach steady state, the particles which may cause pore plugging decrease. When the steady reached, all the particles injected will take relatively a long path and time period far larger than the given time interval Δt before they can pass through the network. If the number of particles passing through the network during the time interval Δt becomes M_{out}, thus after steady state is reached, $1 - M_{out}/M$ is the separation of particles by the network, The separation plotted versus the molecular weight of the macromolecules solutes in Figure 5 does not show any applied pressure dependence. The results confirmed clearly by reported experimental data.

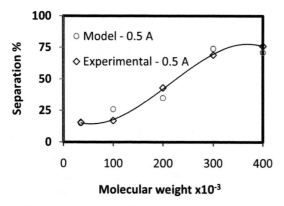

Figure 5: Comparison of steady state separation with experimental data , Wang et al. [6].

4 Conclusions

From the results presented above, we can draw the following conclusion:

- The Monte Carlo simulation model based the transport through three-dimensional network model of interconnected pores can describe the experimental ultra-filtration data both qualitatively and quantitatively *without resorting to any adjustable parameters.*
- According to the network model prediction, the membrane pore is blocked mainly by size exclusion, SE, mechanism, which results in the *sever reduction* of the permeability (flux rate) during the first a few seconds.

5 References

[1] Baghdikian S, Y,, Sharma M. M., ana Handy L. L., "Flow of clay suspensions through porous media." SPE 16257, Denver, CO., (1987).

[2] Belfort, G., R. Davis. And A. L. Zydeny, "The behavior of suspensions and macromolecular solutions in crossflow microfiltration," J. Membr. Sci., 96 1(1994).

[3] Imdakm, A. O., and M. Sahimi, "Transport of large particles in flow through porous media." Phys. Rev., A 36, 3099 (1987).

[4] Imdakm, A. O., and M. SAhimi, "computer simulation of particle transport processes in flow through porous media," Chem. Eng. Sci. 46, 1977 (1991).

[5] Sahimi M., and A. O. Imdakm, "The effect of morphological disorder on hydrodynamic dispersion in flow through porous media," J. Phys., A21, 3833 (1988).

[6] Wang Q. K., T. Matsuura, C. y. Feng, M. R. Weir, C. Detellier, E. Rutadinka, and R. L. Van Mao, "The sepiolite membrane for ultrafilteration,"J. Membr. Sci. 184, 153 (2001).

A facile low-temperature route for preparing monodisperse Mn_3O_4 nanopolyhedrons from amorphous MnO_2 nanoparticles

Shuping Zhang*, Wei Liu**, Jie Ma*, and Yan Zhao***

* College of Science, University of Shanghai for Science &Technology, majie@usst.edu.cn
**College of Environment & Construction, University of Shanghai for Science &Technology
***School of Medical Instrument and Food Engineering, University of Shanghai for Science &Technology

ABSTRACT

A simple hydrothermal route has been utilized for preparing monodisperse tetragonal Mn_3O_4 nanocrystal from new-made amorphous MnO_2 nanomaterials. The MnO_2 nanomaterials with its mother solution are directly used as precursor without any pre-separation or pre-disposal process. The as-obtained samples are characterized by XRD, FTIR, SEM, TEM, SAED, and VSM. The two stages reaction procedures are clarified with the hydrothermal treatment time increasing. First stage is the formation of MnOOH nanorods and Mn_3O_4 nanoparticles mixture. Another is the transformation from MnOOH nanorods to monodisperse Mn_3O_4 polyhedron nanoparticles. The VSM results indicate the excellent superparamagnetic character of as-obtianed Mn_3O_4 polyhedron nanoparticles at room-temperature.

Keywords: Nanomaterial; Mn_3O_4; Hydrothermal; MnO_2; superparamagnetic.

1. INTRODUCTION

Manganese oxides and oxyhydroxides have been paid more attentions as important multifunctional and extensive utilized-prospects materials. Manganese oxides nano-materials show excellent supercapacitive characteristics with high values of specific capacitance, catalysis, magnetic properties[1-3]. Moreover, manganese oxides are in abundance and low cost as well as environmentally benign nature[4]. Mn_3O_4 shows an effective catalysis for selective reduction of nitrobenzene and the decomposition of NO_x [5]. Mn_3O_4 is also used to produce manganese zinc ferriteas soft magnetic materials and lithium manganese oxides as rechargeable electrode materials[6]. MnOOH has been used for adsorption of chemical and radioactive wastes and for oxidation of non-biodegradable

For preparation of manganese oxides, various method have been developed and exploited, such as hydrothermal [8], sol-gel[9], electro-deposition[10], solvothermal[11], etc. As we known, though $KMnO_4$ as a solvothermal precursor has been widely adopted for synthesizing Mn_3O_4 nanomaterials, most reports have just focused on obtained a single targeted product and neglected reaction intermediate processes[11]. Consequently, the ambiguous reaction mechanism is necessary to be elucidated

For clarifying the reaction procedures in water-ethanol solvothermal system, we divide the whole reaction processes into room-temperature mixture reaction stage and solvothermal treatment stage, rather than directly treatment of the mixture of $KMnO_4$ and water-ethanol in hydrothermal system. By this alternative route, not only reaction intermediate products including amorphous MnO_2 nanoparticles, and MnOOH nanorods are acquired, but also it is disclosed MnO_2 could be directly utilized as precursor to prepare lower valance manganese oxides or oxyhydroxides nanomaterials under appropriate reducer assistance. The samples obtained at different stages were characterized by XRD, FTIR, SEM, TEM and SAED. Furthermore, the room-temperature magnetic properties of typical samples were measured by VSM。

2. EXPERIMENTAL

MnO_2 amorphous nanomaterial was synthesized according to the reference[12]. For preparing Mn_3O_4 nanomaterial, the solution was not re-newly made by fresh Mn_3O_4 amorphous nanomaterial, but directly used the turbid mixture contained amorphous MnO_2 in hydrothermal approach. Take a typical example, 20 ml above-mentioned mixture was transferred into Teflon-lined stainless-steel 25 ml capacity autoclave. The autoclave was sealed and maintained constantly at 160 °C for 24 h in a digital-controlled constant temperature oven. Then autoclave was cooled to room-temperature naturally. The yellow-brownish precipitate was centrifuged and washed extensively with de-ionized water and ethanol in turn many times. Finally, the obtained products were dried under vacuum at 60 °C for storage and further characterization.

The product were characterized by X-ray powder diffraction (XRD, Bruker D8, Germany) equipped with graphite monochromatized Cu Ka radiation (λ = 1.54056 Å), employing a scanning rate of 0.02° s^{-1}, in the 2θ ranges from 10° to 70°. Infrared spectra (4000–400 cm^{-1}) were recorded by a Nicolet 5DX FTIR spectrometer with 1 cm^{-1} resolution. The micro-morphologies and nanostructure of products were inspected by scanning electron microscopy (SEM, Philip XL30, Holland), transmission electron microscopy (TEM, Hitachi, H800EM, Japan) with selected area electron diffraction (SAED). The room-temperature magnetic properties of typical samples are characterized by vibrating sample magnetometer (VSM, Lake Shore 735VSM, USA) in the range of -5000 -5000 Oe.

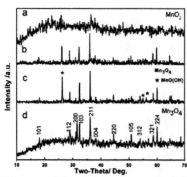

Fig.1 The XRD patterns of a) at room-temperature; b) obtained via hydrothermal at 160 °C for 12 h; c) via hydrothermal at 160 °C for 12 h without CTAB; d) via hydrothermal at 160 °C for 24 h.

3. RESULTS AND DISCUSSIONS

The XRD patterns of the samples obtained by different routes are investigated and laid out in Figure 1. Figure 1a shows the amorphous material sample obtained via solutions route at room-temperature. The material is confirmed as MnO_2 by element analysis and FTIR spectra shown in figure 2a, which is according with the literature[12]. The pattern shown in figure 1b exposes the sample obtained via hydrothermal at 160 °C for 12 h is composed of monoclinic phase γ-MnOOH (JCPDS 41-1379) and tetragonal phase Mn_3O_4 (JCPDS 24-0734 hausmannite-mine type). The diffraction peaks of γ-MnOOH are signed by "★" in the pattern. When hydrothermal process was lasted at 160 °C for 24h, the pure Mn_3O_4 nanomaterial was demonstrated by sample's XRD pattern shown the figure 1d. The cell parameters of the product are corrected as a = 5.749(2) Å and c = 9.455(1) Å. Above results suggest that the reaction from MnO_2 to Mn_3O_4 nanomaterial can be accomplished via hydrothermal at 160 °C for 24 h. Meanwhile, The series of XRD patterns also explored that this transformational processes from MnO_2 to Mn_3O_4 nanomaterial can be divide two stages including the transfer from MnO_2 to γ-MnOOH intermediate, and then to Mn_3O_4 nanomaterial. In other words, the Mn^{4+} is firstly reduced to Mn^{3+}, and then part of Mn^{3+} is further reduced to Mn^{2+} in the procedures. For investigating the function of additive CTAB, the hydrothermal process was also performed at similar conditions except without CTAB at 160 °C for 12 h. The XRD pattern of the sample indicate the product is made up of γ-MnOOH and Mn_3O_4 nanomaterials (**figure 1c**), which is also approved by the FTIR spectrum shown in figure 2c.

FTIR spectra were inspected and listed in figure 2. When the sample was obtained at room-temperature, the spectrum was resembled with that of MnO_2 shown in figure 2a. The spectrum displayed a strong absorbance band at 523 cm-1 due to Mn-O vibrations of MnO_2. The bands at 2931 and 2860 cm[-1] indicated some CTAB absorbed by the amorphous MnO_2 nanomaterial. The peaks at 1080 to 1160 cm[−1] were attributed to the OH bending modes. Another peak at 2089 cm[-1] considered as a combination band of the

OH stretching mode at 2680 cm[-1] and the excited lattice mode at 592 cm[-1] was also observed. The belonged to γ-MnOOH. The peaks below 627 cm[−1] could be attributed to the Mn–O vibrations. The results indicated that the γ-MnOOH was existed in the sample, which was consistent with some studies of Kohler et al [13] and Sharma et al [14]. Meanwhile, the characteristic FTIR absorption peaks of the Mn_3O_4 were appeared in the spectrum with the bands at 634, 531 and 416 cm[-1]. The FTIR spectra of the as-obtained sample by hydrothermal treatment 24 h with CTAB was presented in figure 2d. Three main absorption peaks were inspected in the range of 400-650 cm[-1], the band 634 cm[-1] was characteristic of Mn-O stretching modes in tetrahedral sites, whereas vibration frequency at 531 cm[-1] was attributed to the distortion vibration of Mn-O in an octahedral environment. The third vibration band at 416 cm[-1] corresponded to the vibration of manganese species (Mn^{3+}) in an octahedral site, and 1632 and 1402 cm[-1] bands were normally attributed to O-H bending vibrations combined with Mn atoms [15].

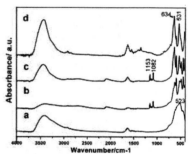

Fig. 2 The FTIR spectra of a) at room-temperature; b) via hydrothermal at 160 °C for 12h; c) via hydrothermal at 160 °C for 12h without CTAB; d) via hydrothermal at 160 °C for 24 h.

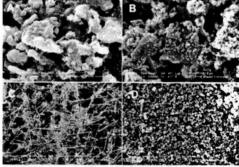

Fig. 3 The SEM images of a. a) via hydrothermal for 12 without CTAB; b) at room-temperature; c) via hydrothermal 12h; d) via hydrothermal 24 h.

Figure 3 showed SEM micrographs of the products obtained via different routes. Figure 3A exhibited the sample was consisted of seriously agglomerating nanoparticles and nanorods, which was synthesized via hydrothermal treatment without CTAB at 160 °C for 12 h. The obtained MnO_2 sample revealed loose or agglomerated nano-multiceps with size around 30-50 nm (figure 3B). When the mixture was treated via hydrothermal with CTAB at 160 °C for 12 h, the micrographs in figure 3C exposed the sample was made up of γ-MnOOH nanorods and Mn₃O₄ nanoparticles, which was proved by SAED patterns shown in figure 4. Compared figure 3c to 3a, with some CTAB assistance, more regular product with lower agglomeration was synthesized. It should be contributed to the interaction inhibited the crystal aggregation between CTAB and crystal surface. With the hydrothermal treatment enhanced to 24 h, the sample was composed of uniform mono-disperse regular polyhedral nanoparticles with the size ca.50-60 nm shown in figure 3D.

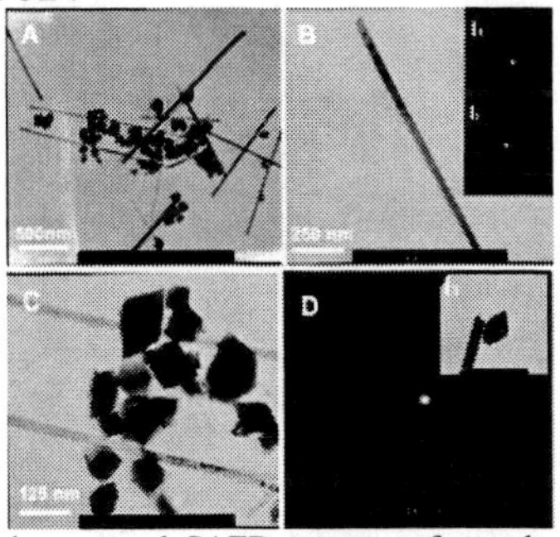

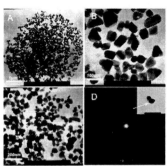

Fig. 4 The TEM images and SAED patterns of samples obtained via hydrothermal for 12h. A) Low magnification image; B) the single nanorod with its SAED patterns signed as I_1 and I_2; C) High magnification image; D the SAED pattern of the nanoparticle with its corresponding TEM image inserted

To further investigate crystal status of samples obtained hydrothermal treatment, the TEM and SAED were applied to inspect samples and typical micrographs and ED pattern were shown in figure 4 and figure 5, respectively. Figure 4 A and 4C showed two different magnification TEM images of the samples obtained at 12h, the sample was constitutes of nanorods with diameter ca.40 nm and polyhedron nanoparticles with the size about 60-80nm accordingly with figure 3C. The ED patterns and corresponding TEM image were exhibit in the figure 4 B and 4D, respectively. Two different crystal facet ED pattern of nanorods were shown in figure 4b as inserts I_1 and I_2, which is accordance with monoclinic phase γ-MnOOH calculated by crystal facet interspacing and correspond the diffraction direction of ED are as [1-2 1] and[011], respectively. The obvious contrast diffracted SAED pattern of representative nanoparticle indicated nanoparticles was belonged to the different structure with the nanorods, and its corresponding TEM image was inserted into figure 4D. By calculated of patterns, the sample was the tetragonal phase Mn₃O₄ structure. The diffraction direction of electron beam is along [001]. Above SAED patterns and TEM images further demonstrated that the sample was composed of different shape γ-MnOOH and Mn₃O₄.

The TEM image shown in figure 5 further indicated uniform polyhedron structure nanoparticles with the narrow size distribution between 50- 60nm. The typical SAED pattern was shown in figure 5D and its responding representation particle was inserted in the picture. The electron beam was ascertained as [010] direction and the d values of various crystal facets accord with them of tetragonal phase Mn₃O₄.

Fig. 5 The TEM images and SAED patterns of samples obtained via hydrothermal for 24h. A) ,B) and C) are different magnification image of the sample; D) the SAED pattern of the nanoparticle with its corresponding TEM image inserted in the picture.

Based on above inspected results, the pure Mn₃O₄ nanoparticles growth procedures can be demonstrated by the scheme 1 and described as follow: 1) The amorphous loose MnO₂ nano-multiceps formation. 2) During the hydrothermal treatment earlier stage, the mixture of MnO₂ is gradually turned into γ-MnOOH nanorods and Mn₃O₄ nanoparticles. 3) In the hydrothermal treatment later stage, the γ-MnOOH is gradually reduced to Mn₃O₄. The firstly stage is easily accomplished at room-temperature, for KMnO₄ shows strong oxidizability and oxidizes ethanol into aldehyde with the MnO₂ formation. For the CTAB as a surfactant is readily absorbed on surface of MnO₂ crystal nucleus, the crystal MnO₂ grow very slowly and loose structure is finally formed at low-temperature. The intention of keeping the reaction at room-temperature for 24h is that no residues KMnO₄ are insured. In the second stages, the formation of γ-MnOOH and Mn₃O₄ nanometerial should be attributed to the infirm oxidability and reducibility of MnO₂ and ethanol. Under hydrothermal treatment conditions at 160 °C for 12 h, A part of amorphous MnO₂ particles are further reduced to γ-MnOOH by ethanol with the crystal shape transference from nano-multiceps to regular nanorods. Meanwhile, other parts of MnO₂ particles are directly reduced to Mn₃O₄ nanoparticles with the crystal phase and appear transformation. In the finally stage, with the treatment time changed from 12 h to 24h, the γ-MnOOH is further reduced and changed to Mn₃O₄ in water-ethanol system. Meanwhile, nanorods completely disappear and form the single component nanoparticles. The formation of Mn₃O₄ nanoparticles in the hydrothermal system are reasonable illuminated by a dissolution–recrystallization[16] and cleavage-decomposition mechanism[17]. Compared to the sample via directly used KMnO₄ and water-ethanol (1:1) as hydrothermal precursors at 160 °C for 12 without CTAB, the above indirectly reaction process is more facilely obtained lower aggregative nanomaterial.

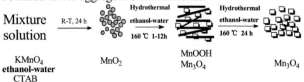

Scheme 1 The reaction processes for forming of Mn₃O₄ nanoparticles via hydrothermal treatment.

For further distinguishing as-obtained MnO_2, intermediate product, and pure Mn_3O_4 nanomaterials, the room-temperature magnetic properties of them are characterized by VSM and obtained hysteresis loops are shown in figure 6. The M–H curves show that the loops of MnO_2 and mixture intermediate are linear with the field with coercivity about 17.107 and 20.059 Oe, respectively. Contrarily, the loop of the Mn_3O_4 nanomaterial reveals an obvious small bend magnetic field in the range of -1500 to 1500 Oe with coercivity about 9.6798 Oe. Compared to the MnO_2 precursor, the coercivity of pure Mn_3O_4 nanoparticles is declined to half, however, the coercivity of the mixture intermediate is slightly increased. This result should be attributed to the transformation of crystal structure, valence of Mn and shape of samples. Because the no hysteresis shown in Mn_3O_4 sample M–H behavior is like in the case of paramagnetism and the magnetization never gets saturated even at very high applied field, this state of the ferromagnetism is called 'superparamagnetism[18], indicating the superparamagnetic character of the Mn_3O_4 samples. When ferromagnetic systems loose their multidomain character, they are then termed as single domain with their magnetic spins aligned along one of the easy directions of magnetization. Compared with the report of Ozkaya et al [19], the bend existed in the loop of Mn_3O_4 nanomaterials should be contributed to different size and shape come from crystallization condition of the product.

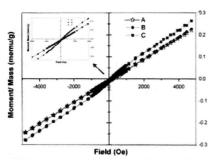

Fig. 6 The room-temperature magnetic properties of samples. A) the mixture of MnOOH nanorods and Mn_3O_4 nanoparticles; B) amorphous MnO_2 nanoparticles; C) Mn_3O_4 nanoparticles. Inset: the magnified pattern of the low field region.

4. CONCLUSIONS

In this work, Mn_3O_4 nanoparticles have been successfully obtained via a facile hydrothermal treatment, in which new-made amorphous MnO_2 nanoparticles with its mother-solution are directly used as precursor. This work firstly shows that MnO_2 can be transferred into monodisperse Mn_3O_4 polyhedron nanoparticles with the size around 60-80 nm via a suitable hydrothermal process, and reveals MnOOH nanorods is important intermediate. During the formation of Mn_3O_4 nanoparticles, the crystal phase growth and shape change are reasonable illuminated by a dissolution–recrystallization mechanism. The as-

prepared Mn_3O_4 nanoparticles show super- paramagnetic character due to the small size effect.

ACKNOWLEDGEMENTS
The authors are thankful to the National '11th-five' Science & Technology Plan Key Project Foundation (No.2007BA62B04), the Shanghai International Cooperation Foundation (No. 073458014) and the Shanghai excellent young Foundation for financial support.

REFERENCES
[1] Zhang H, Cao G, Wang Z, Yang Y, Shi Z, Gu Z. Nano Letters 8, 2664, 2008
[2] Subramanian V, Zhu H, Vajtai R, Ajayan PM, Wei B. J. Phys. Chem. B 109, 20207, 2005
[3] Cui X, Liu G, Lin Y. Nanomedic Nanotech Bio Medic 1, 130, 2005
[4] Bakenov Z, Wakihara M, Taniguchi I.. J Solid State Electrochem 12, 57, 2008
[5] Han YF, Chen F, Zhong Z, Ramesh K, Chen L, Jian D, Ling WW. Chem Eng J 134, 276, 2007
[6] Pan'kov VV. Ceram. Int. 14, 87, 1988
[7] Rabiei S, Miser DE, Lipscomb J A, Saoud K, Gedevanishvili S, Rasouli F. J Mater Sci 40: 4995, 2005
[8] Shen X F, Ding Y S, Hanson J C, Aindow M, Suib S L. J Am Chem Soc. 128, 4570, 2006
[9] Ching S, Petrovay DJ, Jorgensen ML, Suib SL. (Inorg Chem 36, 883, 1997
[10] Nakayama M, Tagashira H. Langmuir 22, 3864, 2006
[11] Zhang W, Yang Z, Liu Y, Tang S, Han X, Chen M. J. Crystal Growth 263, 394, 2004
[12] Subramanian V, Zhu H, Wei B. Chem Phys Lett 453, 242, 2008
[13] Kohler T, Armbrustor T, Libowitzky E. J. Solid State Chem. 133, 486, 1997
[14] Sharma P K, Whittingham MS. Mater. Lett. 48: 319, 2001
[15]Wang XL, Yuan AB, Wang YQ. J Power Sources 2, 1007. 2007
[16] Ma J, Wu Q, Ding Y. J Nanopart Res 10: 775, 2008
[17] Chen Y, Zhang Y, Yao QZ, Zhou G T, Fu S, Fan H. J Solid State Chem 180, 1218, 2007
[18] Liu C, Rondinone A J, Zhang ZJ. Pure Appl Chem 72, 37, 2000
[19] Ozkaya T, Baykal A, Kavas H, Köseoğlu Y, Toprak MS. Physica B 403: 3760, 2008

Preparation of Silver Nanoparticles via Ionomer Clusters

P. Atorngitjawat[*] and K. Pipatpanyanugoon[**]

[*]Faculty of Science, Burapha University, Chonburi, Thailand, pornpena@buu.ac.th
[**]Faculty of Engineering, Burapha University, Chonburi, Thailand, kasemp@buu.ac.th

ABSTRACT

This work focuses on preparing silver nanoparticles by using a novel technique, which is simple, fast and economical. Silver nanoparticles were produced by using a sulfonic acid polystyrene ionomer as a template, and these were then characterized by UV-visible spectrophotometer, wide angle X-ray diffraction, and transmission electron microscopy. Two different levels of acid groups of ionomers were used, 30 and 50 mole %. The results show that silver nanoparticles prepared from an ionomer containing 30 mole % of acid groups provided a smaller particle diameter, about 14.8 nm. Furthermore, using partially neutralized ionomers as a template, an even smaller diameter of silver nanoparticles was achieved. Particles 10.1 nm and 14.3 nm were observed from the partially neutralized ionomers containing 30 and 50 mole% sulfonated groups respectively.

Keywords: silver nanoparticles, ionomer clusters, ionomer templates, sulfonated ionomers, nanoparticle preparation

1 INTRODUCTION

Nanotechnology has played a key role in both academia and industry for the past two decades. This is due to the unique properties and potential applications for nano-scale materials, especially metallic nanoparticles. Metallic nanoparticles can be used for catalytic, optoelectronic and biomedical applications. Particularly, silver nanoparticles have exhibited antimicrobial properties and also been used as biosensing materials for surface enhanced Raman spectroscopy and metal-enhanced fluorescence [1-4]. A variety of procedures have attempted to synthesize nanoparticles. Nanoparticles of silver have been prepared via a microemulsion of water in supercritical CO_2 [5-6] and via a rapid expansion of a supercritical solution into a liquid solvent [7-9]. Oleic acid has been used as a stabilizer for silver nanoparticle synthesis under various pH conditions via the solvent exchange method, in which silver nitrate was reduced and transferred into organic solvent by reorientation of oleic acid [10].

The ion solvent exchange method provides a simple and reproducible technique to prepare silver nanoparticles. For example, the formation of silver nanoparticles embedded in Nafion® membranes [4,11], natural rubbers [12], polysaccharides [13], liquid crystalline polymers [14] and dendrimers [15] are based on a standardized ion exchange process. By utilizing ionomers that contain a nanoscale morphology of ionic clusters [16], which are typically viewed as roughly spherical aggregates about 3-15 nm in size [17], this allows the formation of metallic nanoparticles using the cavities of ionic clusters. Therefore, in this work, silver nanoparticles were prepared by reducing silver nitrate with sodium borohydride within the clusters of the ionomers. UV-visible spectroscopy (UV-vis), wide angle X-ray diffraction (XRD), and transmission electron microscopy (TEM) were employed to characterize the formation and the morphology of silver nanoparticles.

2 EXPERIMENT

2.1 Materials

Silver nitrate ($AgNO_3$), sodium borohydride ($NaBH_4$), sulfuric acid, sodium hydroxide (NaOH) and methanol were purchased from Qrec and used without further purification. General purpose polystyrene (PS) with number average molecular weight of about 130,000 was obtained from Siam Polystyrene Co., Ltd. Propionic anhydride was obtained from Fluka. Sulfonic acid polystyrene (SPS) was synthesized from propionyl sulfate according to a procedure reported in a previous article [18]. The degree of sulfonation of SPS was determined by titration with a standard 0.05 M NaOH solution in methanol. Sulfonation levels of the SPS were 30 and 50 mol%, which will be referred to as SPS30 and SPS50, respectively. The partially neutralized SPS was obtained via an ion exchange process with NaOH [19]. The partially neutralized versions of SPS30 and SPS50 will be referred to as SPS30Na and SPS50Na, respectively.

2.2 Preparation of Silver Nanoparticles

In a typical ion exchange/silver reduction process, 1 g of SPS was dissolved in 20 mL of methanol before adding 5 mL of 1 M $AgNO_3$ solution with continue stirring and heated to 60 °C for 1 hr. Then 20 mL of 1 M of $NaBH_4$ solution was added into the sample. Silver particles were precipitated and rinsed with water and methanol several times. The sample was dried in the oven at 90 °C for 48 hrs. The silver particles prepared via ionomers of SPS30, SPS50, SPS30Na and SPS50Na are denoted as AgH30, AgH50, AgNa30 and AgNa50, respectively.

2.3 Characterization

UV-vis absorption spectra of the silver colloid solution were acquired using a UV-4100 spectrometer. Silver powders were measured using a Bruker D8 X-ray diffractometer with a Cu Kα radiation monochromatic filter in the 2θ range of 30-80°. The morphology of synthesized silver nanoparticles was observed on a Philips Tecnai 20 transmission electron microscopy. The samples were prepared by dropping the silver colloid on carbon-coated Cu grids and allowing them to dry in air at 60 °C for 24 hrs. They were observed at an accelerating voltage of 100 kV and magnification of 620,000x.

3 RESULTS AND DISCUSSION

UV- vis. The absorption spectra of AgH30, AgH50, AgNa30 and AgNa50 were similar. Provided in Figure 1 is an absorption spectra of AgNa30, which serves as a typical example. The absorption band λ_{max}, which ranged from 400 to 480 nm, corresponds to the surface plasmon resonance of the nanoparticles, and indicate that the silver nanoparticles are well dispersed in a colloidal suspension. Some degree of unsymmetrical spectra were observed, which may be due to a slight distribution in the size of the silver particles.

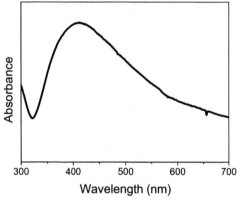

Figure 1: UV-vis absorption spectra of AgNa30.

XRD. The formation and size of solid silver nanoparticles were obtained from the X-ray diffraction analysis. As provided in Figure 2, the XRD pattern of the four prominent peaks at 2θ values of about 38, 44, 64 and 78° matches well with that of the bulk FCC silver in the JCPDS database and in the literature [4, 9]. These peaks can be attributed to the (111), (200), (220) and (311) planes. The broadness of the peaks is due to the particles being nanoscale. Peak broadening can be used to estimate the average silver particle size, based on the Debye-Scherrer equation [20]:

$$D = \frac{K\lambda}{\beta \cos \theta} \qquad (1)$$

where D is an average particle diameter, K is a constant related to the crystal shape (equal to 1 here), β is the corrected band broadening parameter, λ is wavelength of the x-ray source and θ is the diffraction angle.

According to Debye-Scherrer equation, the average diameters for silver nanoparticles prepared via AgH30, AgH50, AgNa30 and AgNa50 are 14.8, 17.7, 10.1 and 14.3 nm, respectively.

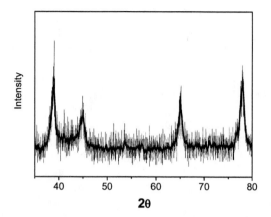

Figure 2: Wide angle X-ray diffraction pattern for silver nanoparticles of AgNa30.

TEM. To obtain further information on the particle size of the nanoparticles, TEM was employed. Figure 3 shows the TEM micrograph of silver nanoparticles of AgNa30. A spherical shape and excellent distribution in the colloid can be observed. The size of silver nanoparticles obtained from TEM image agrees well with the result from XRD.

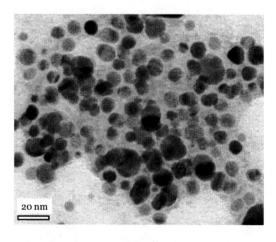

Figure 3: TEM image of silver nanoparticles, prepared from AgNa30.

4 SUMMARY

Silver nanoparticles were prepared by the ion solvent exchange method via the clusters of the ionomers acting as a media. The average diameter for silver nanoparticles prepared from AgH30, AgH50, AgNa30 and AgNa50 is 14.8, 17.7, 10.1 and 14.3 nm, respectively. At the same degrees of sulfonation, the size of silver nanoparticles prepared from partial neutralized ionomers is smaller than that of without neutralization.

ACKNOWLEDGMENT

The authors thank the Thailand Research Fund (MRG5180223) and Faculty of Engineering, Burapha University for financial support.

REFERENCES

[1]   J-M. Zen, G.C. Chen, R.F. Fan, and A.J. Bard, Chem. Phys. Lett.23, 169, 1990.

[2]   Y. Wang, A. Suna, J. McHugh, E.F. Hilinshi, P.A. Lucas, R.D. Johnson, J. Chem. Phys. 92, 6927, 1990.

[3]   A. Albu-yaron, and L. Arcan, Thin Solid Films. 185, 181, 1990.

[4]   Y.P. Sun, P. Atorngitjawat, Y. Lin, *et al.* Journal of membrane science.245, 211, 2004.

[5]   M. Ji, X.Y. Chen, C.M. Wai, and J.L. Fulton, J. Am. Chem. Soc. 121, 2631, 1999.

[6]   H. Ohde, J.M. Rodriguez, X.R. Ye, C.M. Wai, Chem. Commun. 23, 2353, 2000.

[7]   R.C. Petersen, D.W. Matson, and R.D. Smith, J. Am. Chem. Soc. 108, 2100, 1986.

[8]   C.A. Eckert, B.L. Knutson, and P.G. Debenedetti, Nature. 383, 313, 1996.

[9]   Y.P. Sun, P. Atorngitjawat, and M.J. Meziani, Langmuir. 17, 5707, 2001.

[10]   D. Seo, W. Yoon, S. Park, and J. Kim, Colloids and Surfaces A. 313-314, 158, 2008.

[11]   A. Sachdeva, S. Sodaye, A.K. Pandey, and A. Goswami, Anal. Chem. 78, 7169, 2006.

[12]   N.H.H. Abu Bakar, J. Ismail, M. Abu Bakar, Materials Chemistry and Physics. 104, 276, 2007.

[13]   N.E. Kotelnikova, T. Paakkari, R. Serimaa, G. Wegener, V.P. Kotelnikov, V.N. Demidov, A.V. Schukarev, E. Windeisen, and H. Knozinger, Macromol. Symp. 114, 165, 1997.

[14]   E.B. Barmatov, D.A. Pebalk, and M.V. Barmatova, Langmuir. 20, 10868, 2004.

[15]   R.W.J. Scott, O.M. Wilson, and R.M. Crooks, J. Phys. Chem. B. 109, 692, 2005.

[16]   S. Schlick, "Ionomers: Characterization, theory, and Applications," CRC Press, 1996.

[17]   A. Eisenberg, B. Hird, and R.B. Moore, Macromolecules. 23, 4098, 1990.

[18]   P. Atorngitjawat, R.J. Klein, and J. Runt, Macromolecules. 39, 1815, 2006.

[19]   P. Atorngitjawat, and J. Runt, Macromolecules. 40, 991, 2007.

[20]   H.P. Klug, L.E. Alexander, "X-ray Diffraction Procedures", John Wiley and Son, 1959.

Preparation and Application of Copper Oxide Nanoparticles

M. H. Chang and Clifford Y. Tai[*]

[*]Department of Chemical Engineering, National Taiwan University
Taipei, Taiwan, cytai@ntu.edu.tw

ABSTRACT

Copper oxide nanoparticles were synthesized in this study using a spinning disk reactor (SDR). The precursor of copper oxide was first synthesized by a liquid-liquid reaction of $CuSO_4$ and Na_2CO_3 solutions in a continuous mode using SDR. Then the precursor was calcined up to 500°C to obtain copper oxide nanoparticles. The effects of operating variables, including rotation speed, reactant concentration, and liquid flow rate, on the size of copper oxide nanoparticles were investigated. Smaller copper oxide particles were obtained under lower reactant concentration and higher disk rotation speed. The size of obtained primary particles of copper oxide was between 20-30 nm observed under a transmission electron microscope as compared to the number mean size of 40-50 nm determined by a particle size analyzer. Then a CuO-water nanofluid was prepared and the effective thermal conductivity of the CuO-water nanofluid increased with an increase in CuO content up to 0.40 vol%.

Keywords: spinning disk reactor, nanofluid, copper oxide, nanoparticle

1 INTRODUCTION

Nanofluid is a system that is composed of nanoparticles dispersed in a heat transfer fluid in order to enhance the thermal conductivity, specific heat capacity, and electrical conductivity of the fluid. One of the methods reported in the literature, for synthesizing nanofluid was called VEROS (vacuum evaporation on running oil substrate) [1]. Magnetic materials such as iron, cobalt, and nickel were straightly evaporized into silicon oil to obtain nanofluid, in which the size of nanoparticles was about 2.5 nm. A similar method using evaporation/condensation technique, called SANSS (submerged arc nanoparticle synthesis system), was reported by Jwo et al. [2] to prepare CuO-water nanofluid using a high temperature arc between 6000-12000 °C, and the thermal conductivity of the nanofluid can be improved by 9.6 % as the CuO content of 0.4 vol%. Lee et al. [3] has prepared Al_2O_3 and CuO nanoparticles using the condensation method and then dispersed them in a mixing chamber to obtain nanofluid. The thermal conductivities of CuO-water nanofluid were were improved by 12 % as the CuO content of 3.4 vol%. Although the thermal conductivity of the fluid can be increased by these methods, the production rate need to be improved and aggregation of

particles need to be solved. Furthermore, these processes were time and energy consuming, and the manufacturing processes were difficult to scale-up.

One possible method for mass production of nanoparticles is the HiGee technique, which has been developing for more than ten years. Two types of equipments, i.e., rotating packed-bed reactor (RPBR) and spinning disk reactor (SDR), have been applied in this regard [4-6]. For using the SDR in crystallization, reactant solutions were spread onto a spinning disk to achieve a uniform and high supersaturation through micromixing, thus small and uniform particles were obtained via precipitation. In our laboratory, uniform nanoparticles of different chemicals, such as barium carbonate [7], magnesium hydroxide [8], and silver nanoparticles [9,10], has been successfully synthesized using this technique, and the production rate was greatly improved, i.e., 33 kg/day for silver nanoparticles.

In this research, copper oxide particles were synthesized by using an SDR, with the expectation of a product of the uniform and nano-scale particles, and the effects of operating variables were also investigated. The produced copper oxide particles were further used to prepare nanofluid, and the thermal conductivity was measured using the transition hot-wire method.

2 EXPERIMENTAL

2.1 Preparation of the Precursor and CuO Nanoparticles

The experimental setup of continuous operating is shown in Fig. 1. It consists of a liquid feeding system (1~8), a spinning disk reactor (9 and 10), a slurry outlet (11) and a collection tank (12). The liquid feeding system contains two storage tanks (1 and 2), from which liquid reactants are pumped into the reactor chamber through flowmeters (5 and 6). Two liquid distributors (7 and 8) perpendicular to the spinning disk (9) were designed to distribute the reactants uniformly on the disk. The main part of the spinning disk reactor is a stainless-steel disk (9), which is 20 cm in diameter and driven by a variable-speed motor (M). The spinning disk is enclosed in a cylindrical acrylic-chamber (10). A funnel-like outlet (11) is located at the bottom of the chamber for guiding the slurry into the collection vessel (12).

At the beginning of an experiment, one-liter $CuSO_4 \cdot 5H_2O$ and Na_2CO_3 solutions with concentrations ranging from 0.01 M to 0.40 M were charged separately into tanks 1 and 2. The two reactant solutions were pumped onto the center of the spinning disk at a rotation speed N through the two distributors at a controlled flow rate (L_1, L_2). Then the liquid velocity was accelerated due to centrifugal force, causing it to spread over the disk surface and forming a thin film where the precursor of copper oxide precipitated. The slurry left the disk, hit the inner wall of the chamber, flowed down along the wall past the funnel-like outlet, and was finally collected in the vessel. The product slurry was centrifuged at 11,000 rpm for 10 minutes, washed with deionized water and acetone, and finally dried in vacuum for one day at room temperature. Then the produced precursor powder was subjected to a calcination process up to 500°C to produce CuO powder.

2.2 Characterization of the Precursor and CuO nanoparticles

Powder samples were then analyzed with an X-ray diffractometer (XRD, Philips, X'PERT) to determine their crystal structures, and their morphologies were observed with a transmission electron microscope (JEOL, JEM1010). To determine the particle size distribution, CuO powder was dispersed in water agitated with a sonicator (Misonix, XL2020), using sodium hexametaphosphate (Kokusan Chemical Works, Ltd.) as dispersant, and was analyzed with a dynamic light scattering analyzer (Malvern, nano ZS). The functional groups of the precursor of copper oxide were also determined by using a Fourier transform infrared spectroscope (FT-IR, BIO-RAD FTS3000). The weight change of the precursor during the calcination process was analyzed by a thermalgravimetric analyzer (TGA, PerkinElmer Pyris 1 TGA) under nitrogen environment.

2.3 Preparation of CuO Nanofluid and The Measurement of Thermal Conductivity

To prepare CuO-water nanofluid, NaHMP was used as the dispersant with the concentration ranging from 0.2 to 2 g/100mL. The produced CuO powder of desired amount was agitated for 10 min in a jacketed vessel containing NaHMP solution kept at room temperature using sonicator with a power intensity of 165 W. Effective thermal conductivity of nanofluid was measured by a Thermal Properties Analyzer (Decagon, KD2), which consists of a handheld readout and a 6 cm stainless needle sensor that can be inserted into the fluid. The measurement theory was based on transient hot-wire method [3].

3 RESULTS AND DISCUSSION

For synthesizing copper oxide, the effects of operating variables were investigated, including reactant concentration, reactant flow rate, and rotation speed of SDR. Characteristics of the precursor and copper oxide were also

determined. The produced copper oxide nanoparticles were further used for preparing nanofluid, and the thermal conductivity was measured and compared with that calculated by theoretical model.

3.1 Effects of Reactant Concentration on the CuO particle size

In order to investigate the effects of the reactant concentration on the CuO particle size, other operating variables were fixed: the liquid flow rates of $CuSO_4 \cdot 5H_2O$ and Na_2CO_3 solution set at 0.2 L/min, the rotation speed of the spinning disk at 4000rpm, and the [$CuSO_4 \cdot 5H_2O$]/[Na_2CO_3] ratio at 1/1 according to the stoichiometric ratio of the reaction. The results were shown in Table 1. When the concentration of copper sulfate, [$CuSO_4 \cdot 5H_2O$], was smaller than 0.10 M, the change in particle size was less significant. Then the mean size of particles increased dramatically with increasing reactant concentrations. As the concentration of copper sulfate increased from 0.10 M to 0.40 M, the volume mean size of CuO increased nearly threefold, from 63.3 to 167.2 nm, and the number mean size was almost doubled, from 48.3 nm to 93.0 nm. The reason was conjectured as follows: under all the controlled reactant concentrations, the nucleation took place were at an extremely fast rate, and resulted in uniform, tiny primary particles. However, under higher concentrations, the tiny nuclei collided with each other more easily to yield more agglomerates.

3.2 Effects of Reactant Flow Rate on the CuO Particle size

In this experiment, the concentration of $CuSO_4 \cdot 5H_2O$ and Na_2CO_3 solution fixed at 0.1 M, and the rotation speed at 4000 rpm, the liquid flow rates of both reactant solutions were changed simultaneously from 0.2 L/min to 5.0 L/min to see how they affect the CuO particle size. The results indicated that when the liquid flow rate was increased from 0.2 L/min to 3.0 L/min, both the volume and number mean size of copper oxide changed very little, with a variation within 4 nm from the average, which was 61.3 nm and 47.0 nm respectively. For a further increase in the flow rate to 5.0 L/min, the volume mean size of CuO particles jumped from 59.2 to 74.6 nm and the number mean size increased from 44.5 to 51.5 nm. It was understood that an increase in liquid flow rate resulted in a thicker liquid film on the spinning disk, and thus, a worse mixing between the two reactant solutions. The poor mixing caused an incomplete reaction on the disk surface. As a result, the residual reactants further reacted in the collection vessel, where the crystals grew to form larger particles with a wider distribution.

3.3 Effects of Rotation Speed on the CuO Particle Size

As to the effect of rotation speed of the disk was concerned, the mean sizes of CuO particles obtained under different rotation speeds of the disk were measured, and the flow rates of both reactant solutions were kept at 1.5 L/min. When the rotation speed increased from 1000 rpm to 4000 rpm, the volume and number mean size of copper oxide remained almost constant, about 60 and 46 nm, respectively. As the disk rotation speed was below 1000 rpm, i.e., at 500 rpm, the volume mean size of CuO increased to 79.3 nm. Lower rotation speed of the disk caused poorer agitation and less disturbances within the liquid film, and thus, the lower mixing efficiency. As a result, the particle size distribution of CuO product was wider, just like the case of high flow rate of reactants.

3.4 Analysis of the Precursor and CuO particles

To determine the chemistry of calcination process leading the decomposition of precursors to form copper oxide, the precursors were subject to FT-IR and TGA analyses. The FT-IR pattern of the produced precursor of CuO was shown in Fig. 2. Both OH and CO_3 bondings were observed, predicting the precursor to be $Cu_2(OH)_2CO_3$ compound. As to the TGA analysis, the precursor sample was heated to 600 °C with a heating rate of 10 °C/min, and the final weight percentage remained almost constant at temperatures above 500 °C was 74.5 %, which was closed to the theoretical molecular weight ratio, $2CuO/Cu_2(OH)_2CO_3$, of 72 % predicted by the following equation (1). This small difference could be attributed to some impurity which was included in the precursor.

$$Cu_2(OH)_2CO_{3(s)} \xrightarrow{\Delta} 2CuO_{(s)} + CO_2 \uparrow + H_2O \qquad (1)$$

Figure 3(a) shows the results of XRD analysis of the CuO product, which indicates a tenorite structure. The TEM micrograph shown in Fig. 3(b) revealed that the morphology of CuO was spherical, 20-30 nm in diameter, with some aggregates of 50-100 nm. The mean particle size measured by a dynamic light scattering analyzer was 40-50 nm and 60-70 nm for number and volume, respectively. The particle size distribution is shown in Fig. 4, and the percentage of CuO particle smaller than 100 nm is greater than 90 vol%.

3.5 Thermal Conductivity of CuO Nanofluid

The effective thermal conductivities under CuO content ranging from 0.01 to 0.40 vol% were measured and compared with that by theoretical calculation using Maxwell model [11]. The thermal conductivity of the nanofluid was increased as the CuO content increased, and the best result of this research was more than 10% improvement as CuO content was 0.40 vol%. In addition, the thermal conductivity of all the samples of nanofluid were higher than that of calculated, which was 1.01 as CuO content of 0.40 vol%. It was probably due to the simplicity of the model which neglected the effects of motion, interaction, and size of particles.

4 CONCLUSION

Copper oxide nanoparticles were successfully synthesized using the spinning disk reactor. The effects of operating variables on particle size were investigated. Smaller copper oxide particles were obtained under lower reactant concentration and higher disk rotation speed. Spherical copper oxide nanoparticles with size 20-30 nm were observed under an electron microscope, and their volume mean size was about 60-70nm measured by a dynamic light scattering analyzer. For the prepared CuO nanofluid, the thermal conductivities of all the samples were higher than that calculated from theoretical model, showing that CuO nanoparticles has great potential for the heat transfer application.

5 TABLES AND ILLUSTRATIONS

[CuSO$_4$ · 5H$_2$O] (M)	volume mean size (nm)	number mean size (nm)
0.01	56.7	44.2
0.05	60.3	45.7
0.10	63.3	48.3
0.20	94.3	63.7
0.30	122.3	75.3
0.40	167.2	93.2

Table 1: Effect of reactant concentration on CuO particle size. Other operating conditions: $L_1=L_2=0.2$ L/min, N= 4000rpm, [CuSO$_4$ · 5H$_2$O]/ [Na$_2$CO$_3$]= 1

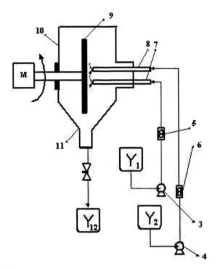

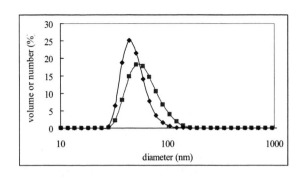

Figure 4 CuO particle size distribution on volume basis (■) and number basis (◆).

1,2: storage tank   3,4: pump   5,6: flowmeter
7,8: liquid distributor   9: spinning disk   10: chamber
11: outlet   12: collection tank   M: motor

Figure 1 Experimental Setup.

ACKNOWLEDGMENTS

The authors wish to thank the National Science Council of Taiwan for financial support of this work.

REFERENCES

[1] H. Akoh, Y. Tsukasaki, S. Yatsuya, and A. Tasaki, J. Cryst. Growth 45, 495, 1978.

[2] C. S. Jwo, T. P. Teng, and J. Chang, J. Alloy. Compd. 434-435, 569, 2007.

[3] S. Lee, S. U. S. Choi, S. Li, and J. A. Eastman, Trans. Am. Soc. Mech. Eng. 121, 280, 1999.

[4] C. Ramshaw and R. H. Mallinson, US Patent 4383255, 1981.

[5] J. F. Chen, Y. H. Wang, F. Guo, X. M. Wang, and C. Zheng, Ind. Eng. Chem. Res. 39, 948, 2000.

[6] L. M. Cafiero, G. Baffi, A. Chianese, and R. J. J. Jachuck, Ind. Eng. Chem. Res. 41, 5240, 2002.

[7] C. Y. Tai, C. T. Tai, and H. S. Liu, Chem. Eng. Sci. 61, 7479, 2006.

[8] C. Y. Tai, C. T. Tai, M. H. Chang, and H. S. Liu, Ind. Eng. Chem. Res. 46, 5536, 2007.

[9] C. Y. Tai, Y. H. Wang, and H. S. Liu, AIChE J. 54, 445, 2008.

[10] C. Y. Tai, Y. H. Wang, C. T. Tai, and H. S. Liu, Ind. Eng. Chem. Res. 48, 10104, 2009.

[11] J. C. Maxwell, "A Treatise on Electricity and Magnetism," Vol. 1, Clarendon Press: Oxford, 1881.

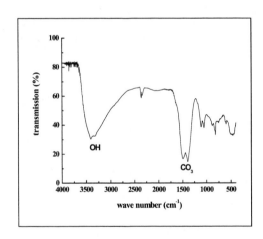

Figure 2 FT-IR pattern of the precursor of copper oxide.

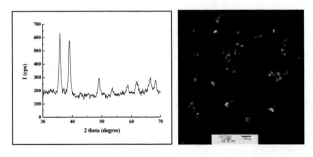

Figure 3 (a)XRD pattern; (b) TEM micrograph of produced CuO nanoparticles.

Nanopowders of TiO$_2$ obtained by combustion reaction: effect of fuels

E. P. Almeida[*], J. P. Coutinho*, N. L. Freitas*, R. H. G. A. Kiminami**, H. L. Lira* and
A. C. F. M. Costa*

[*]Universidade Federal de Campina Grande, Doutorado em Engenharia de Processos, Av. Aprígio Veloso,
882-Bodocongó, 58429-900,Campina Grande - PB, Brasil
, sther_almeida@yahoo.com.br
[**]Universidade Federal de São Carlos, Departamento de Engenharia de materiais, São Carlos – SP, Brasil,
ruth@power.ufscar.br

ABSTRACT

The aim of this work is to evaluate the effect of the different fuels in the combustion reaction synthesis. The structural and morphological characteristics of the TiO$_2$ powders were investigated. The powders were prepared according to the propellants theory by using a vitreous silica container and titanium isopropoxide as precursor. The fuels oxalic dihydrazide, aniline, monohydrated citric acid and carbohydrazide were used in a stoichiometric ratio as reducing agent. The powders were characterized by X-ray diffraction, nitrogen adsorption by BET, particle size distribution and scanning electron microscopy. The results of the X-ray diffraction showed that all fuels studied produced rutile as majority phase with presence of anatase as second phase. Carbohydrazide give the large amount of rutile phase, great peaks of X-ray diffraction, indicating the formation of powders with large crystallite size when comparison with the others fuels.

Keywords: TiO$_2$, fuels, combustion reaction, nanopowders.

1 INTRODUÇÃO

Titanium oxide is a polymorphic material, which can be found in three different crystalline phases at atmospheric pressure: rutile, anatase and brookite. The rutile phase is formed at high temperature (>1000°C), and the anatase phase is formed at low temperatures (close to 450°C) [1]. In general the brookite phase is unstable and has low interest. From the three titania crystalline phases, anatase is the most studied due to the possibility to be used in applications such as semiconductors and photocatalyst [2]. Anatase can be used to discolor residual water or to oxidize organic components [3,4]. Rutile phase can be used to applications such as pigment in paint, environmental purification, hydrogen gas generation due to their photoactivity, in the electronic industry as capacitors and electrical circuits and temperature condenser [5,6].

The use of chemical methods to prepare nanoparticles, with desirable physical, chemical and structural properties has been of great importance, due to the molecular stability and good chemical homogeneity that can be reached with materials. The combustion reaction process can offer advantages such as is quickly and simple, do not need calcinations step and has low energy consumption during the synthesis. Also, the combustion reaction method synthesizes products of high purity, nanometric particles with good chemical homogeneity [7,8].

The study of different fuels in the combustion reaction to prepare some ceramic systems has been published in several works [9]. To obtain specifically TiO$_2$ it can be cited the use of citric acid, glycine, oxalic dihydrazide and urea [10,11] and others. All these fuels contain nitrogen, but with different reducer capacity, organic chain size, valence and, so, gas quantity generated by them, which define the temperature and time of the combustion flame reached during the synthesis. High temperatures favored the crystallization and sintering of the powder and as higher the amount of gas produced higher is the energy dissipation, that is, smaller will be the amount of energy available to sintering and crystallization, which generate low combustion temperatures and in some case favored the formation of secondary phase, once, part of the heat that must go to the phase formation was lost to the surrounding area. In this way, the aim of this work is to evaluate the effect of fuels (Eac, Ean, Eox and Ecb) in the combustion temperature and in the morphology and structure of the TiO$_2$ powder prepared by combustion reaction.

2 MATERIALS AND METHODS

To prepare TiO$_2$ by combustion reaction it was used aniline (C$_6$H$_7$N) – Ean, monohydrated citric acid (C$_6$H$_8$O$_7$.H$_2$O) – Eac, carbohydrazide (CO(N$_2$H$_3$) – Ecb, and oxalic dihydrazide (CH$_6$N$_4$O) - Eox as fuels and reducer agents, and titanium isopropoxide [Ti[OCH(CH$_3$)$_2$]$_4$ as source of cations and oxidizer agent. All reagents present 98% of purity. The mixture was added to a silica vitreous crucible and submitted to a direct heating by a spiral electrical resistance (temperature close to 600°C). Following, it was added the titanium isopropoxide to the fuel, with releasing a great amount of volatile gases (product of reagents decomposition). In the moment of gas release decrease, the mixture was transferred to a pre-heated oven at 700°C (Model: EDG3P – 3000) until ignition and subsequently combustion. The measurement of maximum temperature of the combustion flame was done by a infrared pyrometer (Raytec Model-Raynger 3i) and the time of combustion flame by a digital chronometer (Model: Technos). After powder preparation with each fuel it was de-agglomerate in a mortar and passed in a sieve 325 mesh (45μm) to subsequent characterization.

The X-ray diffraction data were obtained from a diffractometer Shimadzu model 6000, with Kα Cu radiation. The phase identification was done from the X-ray diffraction data and using a computational software (Pmgr) from Shimadzu and JCPDF data. The crystallinity was estimated with software and using the Lorentz coefficient of correction. The surface area and the adsorption/desorption isotherms were obtained from gas adsorption and using an ASAP 2000, from Micromeritics. The particle size was calculated by Braunauer, Emmet and Teller theory (BET). The morphology of the TiO_2 powder was observed by scanning electron microscopy (SEM) model XL30 FEG, from Philips.

3 RESULTS

Fig. 1 show the intensity and the aspect of the flame reached during the combustion reaction to the TiO_2 preparation and using Ean, Eac and Eox as fuels. The color of the combustion flame is a strong indicative to evaluate the heating produced by combustion during the synthesis and can be high or low.

| a –Ean | b – Eac | c – Ecb | d – Eox |

Fig. 1: Intensity and aspect of the flame burning.

From Fig. 1, it was observed that each fuel produce a characteristics combustion flame, in relation to the color, intensity or brightness color of the flame. So, it can be observed that aniline fuel (Fig. 1a) produced an intense flame, with brilliant aspect in the interior of the crucible and in the top extremity of the flame, also present dark color in the center of the flame. Eac (Fig. 1b) also produce an intense flame, but apparently less strong than Ean and with light aspect. Ecb (Fig. 1c) produced a spread and tiny flame, without brightness and clear aspect only in the extremity. The wideness of the flame give to a loss of heat to the environment and this can affect the combustion temperature reached during the synthesis. Eox (Fig. 1d) produced a clear flame with bright color close in the top extremity of the crucible, but with low intensity and dark aspect in the center and in the interior of the crucible.

Table 1 presents maximum temperature and time of the flame, also the combustion flame color observed during the combustion reaction. It can be observed that Ean fuel produce the biggest combustion temperature and an intense yellow color in the top and bottom of the flame (light grey

in the Fig. 1a), and red color close to the center of the flame (dark grey in Fig. 1a), however the time of combustion in this case was smaller when compared with others fuels. The Eox and Eac present intense yellow flame with long time of flame when compared with others fuels. The Ecb show a spread red flame and small time of combustion when compared with Eac and Eox. By analogy between the flame color, their aspect and period of time, with the measured temperature it was observed a good correlation, once the greatest temperature in the combustion was reached with Ean and the smallest temperature was reached with Ecb with a difference of 201°C.

Table 01 – Temperature and time of combustion to the fuels use in the TiO_2 synthesis.

Fuels	Temperature (°C)	Time (seg)	Color
Ean	900	50	Intense yellow
Ecb	699	65	Red
Eac	798	172	Yellow
Eox	852	195	Intense yellow

Fig. 2 - presents the X-ray patterns of the TiO_2 powders prepared by combustion reaction with the following fuels: Ean, Eac, Ecb and Eox.

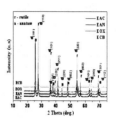

Fig. 2 - X-ray powder diffraction of TiO_2.

It can be observed that for all fuels, except for aniline, TiO_2 powder with rutile crystalline structure (JCPDF 65-0190) was obtained as major phase and traces of anatase as secondary phase (JCPDF 21-1272). Ean produces TiO_2 with anatase as major phase and rutile as secondary phase. This results is due to the Ean has bigger organic chain, greater valence and strong lateral groups or firmly bond when compared with other fuels.

Table 2 presents the results of crystallinity and crystallite size calculated from the X-ray diffraction data to TiO_2 powders prepared by combustion reaction and using Ean, Eac, Ecb and Eox.

Table 2 – Crystallinity and crystallite size of TiO2 powders prepared by combustion reaction with Ean, Eac, Ecb and Eox.

Fuels	Cristalinity (%)	Crystallite Size - (nm)
Ean	94	43
Ecb	93	26
Eac	81	46
Eox	82	52

From Table 2 it was observed that the fuels produces powders with high crystallinity (above 80%) and crystallite size below 100m, and characterize as nanometric powder. Ecb fuel produces powder with smallest crystallite size (26nm) when compared with other fuels. This is a consequence of small temperature reached by combustion flame during the synthesis. It was observed that the Ean give the formation of a powder with crystallite size slightly inferior to the Eac and Eox. This is due to the superior flame time produced during the synthesis with Eac and Eox, which was close to 180 s and close to 30% superior to the flame time reached with Ean.

Oliveira et al (2007) when prepared TiO_2 doped with Ni by combustion reaction and using urea as fuel, to be used as yellow pigment, show the presence of rutile phase and crystallite size between 26-38nm [12]. The researchers related only the presence of rutile as major phase and anatase as secondary phase and crystallite size of 65 and 44nm to glycine and urea fuels. By comparison these results with that one obtained in this study, it can be observed that only the Ecb produced crystallite size similar to the related by Almeida et al(2007) [13] and the others fuels produced two times crystallite size. This shows clearly, the influence of the physical-chemical character of the fuel in the structural characteristics of the TiO_2 powder prepared by combustion reaction.

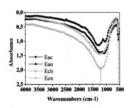

Fig. 3 – Infrared spectroscopy of the TiO_2 powder.

Fig. 3 presents results of the infrared spectroscopy of the TiO_2 powder prepared by combustion reaction and with Ean, Eac, Ecb and Eox as fuels. It can be observed similar behavior for the powders produced with different fuels, that is, for all conditions the desirable phase was the same (TiO_2). For all powders prepared it was observed a strong band in the range of 1259.95 cm^{-1} characteristics of the O-Ti-O axial vibration. It was not possible to observe bands bellow 500cm^{-1}, due to limit of detection of the equipment, but it can be observed a tendency of absorption band formation close to 551cm^{-1} that correspond to the O-Ti-O vibration. This absorption band between 500 and 800cm^{-1} related with O-Ti-O also was reported by Phonthammachai et al. [14] when studied the structural and rheological aspects of TiO_2 prepared by sol-gel method.

Table 3 present the surface area, particle size and powder diameter of TiO_2 prepared by combustion reaction using Ean, Eac, Ecb and Eox as fuels. It can be observed that for all fuels do not occurred significant changing in the

values of surface area. It was observed a small reduction of 16% when compared the powder prepared with Ecb with Eac.

Table 3 – Surface area, particle size and pore diameter of the TiO_2.

Fuels	Surface área BET (m^2/g) *	Particle size BET (nm) *	Pore diameter Dp (nm)	D_{BET}/ D_{DRX}
Eac	3.89	362	8.0	14
Ean	4.38	322	10.8	8
Ecb	4.63	304	6.8	7
Eox	4.12	342	11.8	7

* calculated from surface area (BET)
Theoretical density = 4.26 g/cm^3 [15]

This is probable due to a small flame temperature and small combustion time (699°C and 65s, respectively) reached during the synthesis with Ecb in relation to the Eac (798°C and 172s, respectively). From the ratio between crystallite size calculated by DRX and particle size calculated from the values of surface area (D_{BET}/D_{DRX}) it was observed that the powder prepared with Eac more agglomerated in relation to the others fuels, and this increasing of agglomerate state (particle with pre-sintering start) give the condition to a reduction in the surface area.

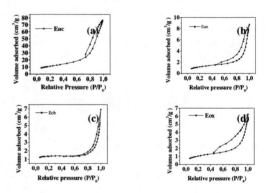

Fig. 4 – Nitrogen adsorption/desorption isothermal toTiO_2.

The results from textural characterization of TiO_2 powders prepared by combustion reaction using Ean, Eac, Ecb and Eox as fuels are represented by the N_2 adsorption/desorption isothermals in the Fig. 4. For all powders the isothermal profiles were type IV and V with a loop of hysteresis type H3, according to the IUPAC classification [16]. These isothermal profiles characterize the powders as mesopores materials (pore dimensions between 2-50 nm) and the hysteresis type H3 indicate that the pores are in shape of cone or parallel plates. Despite the powders presented similar isothermal profiles, the powder prepared with Eac presented a greatest widening curve due to the greater average diameter. Valente et al [17] when study the method to modify TiO_2 by sol-gel process with the addition of cerium oxide also showed that TiO_2 presented isothermal profile type II but hysteresis loop type H3, similar to the reported in this study.

Fig. 5 show the morphology of the powders prepared by combustion reaction using Ean, Eac, Ecb and Eox as fuels from scanning electron microscopy.

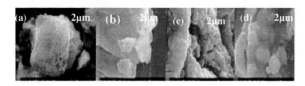

Fig. 5 – Images from SEM of the TiO₂ powders.

From the images it was observed that the TiO₂ powders prepared with all fuels presented the formation of agglomerates with pre-sintering particles, except for the powder prepared with carbohydrazide that presented a morphology of soft agglomerates (formed by interparticles weak forces), non-uniform and with irregular shape (Fig. 5a). Apparently these agglomerates are soft and present aspects more fragile when compared with the morphology presented by the powders prepared with Eox, Eac and Ean (Fig. 5b, 5c and 5d, respectively) that showed agglomerates with rigid aspect, that is, apparently dense.

4 CONCLUSIONS

The flame time and temperature of combustion reached with different fuels contributed to significant differences in the structural and morphological characteristics of the powders. All studied fuels used in the combustion reaction prepared TiO₂ powder with presence of rutile phase and anatase. However, Ean contributed to the formation of anatase as major phase and rutile as secondary phase. Eac produced powder with smallest crystallite size, but with greatest agglomerate size and smallest surface area. All fuels presented similar ratio between surface area and particle size. The pores presented values between 6.8 and 11.8 nm. All fuels produced powders formed by polycrystalline and agglomerates particles.

ACKNOWLEDGEMENTS

The authors would like to thank the Brazilian institutions CAPES and RENAMI-CNPq for their financial support of this research.

REFERENCES

[1] L. Castañeda, J. C. Alons, A. Ortiz, E. Andrade, J. M. Saniger, Rañuelos. Spray pyrolysis deposition and characterization of titanium oxide thin films. Mat. Chem. and Phys. 77, p.938–944, 2003.

[2] G. Bertoni, E. Beyers, J. Verbeeck, M. Mertens, P. Cool, E. F. Vansant, G.V. Tendeloo, Quantification of crystalline and amorphous content in porous TiO₂ samples from electron energy loss spectroscopy. Ultramicroscopy, 106, p.630–635, 2006.

[3] M. Sökmen, A. Özkan, Decolourising textile wastewater with modified titania: the effects of inorganic anions on the photocatalysis J. Photochemistry and Photobiology A: Chemistry., 147, p. 77-81, 2002.

[4] J. Chen, M. Liu, L. Zhang,.L. Jin, Application of nano TiO₂ towards polluted water treatment combined with electro-photochemical method. W. Res. 37, p. 3815-3820, 2003.

[5] M. A. Fox, M. T. Dulay, Heterogeneous photocatalysis. Chem. Rev. 93, p. 341- 357, 1993.

[6] M. R. Hoffman, S. T. Martin, W. Choi, D.W. ahnemann, Environmental applications of semiconductor photocatalysis, Chem. Rev. 95,69–96, 1995.

[7] D. A. Fumo, Cimentos em betões refratários LCC e ULCC: Síntese mecanismos de hidratação e interação com os microenchedores. Aveiro, Portugal, p.157. Tese (Doutorado em Ciências e Engenharia Cerâmica), Universidade de Aveiro, 1997.

[8] S. S. Manoharan, V. Prasad, S. V. Subramanyam, K. C. Patil, Combustion synthesis and properties of fine particle La2CuO4 and La1,8Sr0,2CuO4. Physica C, v. 190, p. 225-228, 1992.

[9] N. L. Freitas, E. P. Almeida, J. P. Coutinho, H. L. Lira, A. C. F. M. Costa, Síntese de TiO₂ por Reação de combustão usando dihidrazida oxálica como combustível. XVII COBEQ, Recife-PE, 2008.

[10] E. P. Almeida, C. H. Silva, J. P. Coutinho, N. L. Freitas, R. H. G. A. Kiminami, H. L. Lira, A. C. F. M. Costa, Synthesis of TiO₂ by combustion reaction: evaluation of different fuels. PTECH 2007.

[11] M. H. Priya, G. Madras, Photocatalytic degradation of nitrobenzenes with combustion synthesized nano-TiO2. J. Photochemistry and Photobiology A: Chem. 178, 1-7, 2006

[12] J. B. L. Oliveira, A. C. F. M. Costa, T. Weger, L. Gama, Yellow Pigment prepared by reaction combustion. on Powder technology, PTECH 2007 .

[13] E. P. Almeida, C. H. Silva, J. P Coutinho, N. L. Freitas, R. H. G. A. Kiminami, H. L. Lira, A. C. F. M. COSTA, Synthesis of TiO2 by combustion reaction: evaluation of different fuels. PTECH 2007.

[14] N. Phonthammachai, T. Chairassameewong, E. Gulari, A.M. Jamieson, S. Wongkasemjit, Structural and rheological aspect of mesoporous nanocrystalline TiO2 synthesized via sol–gel process. Microporous and Mesoporous Mater., p. 66, 261–271,2003.

[15] J. S. Reed, Principles of Ceramics Processing. 2. ed. USA: Jonh Wiley e Sons, Inc. 1995.

[16] S. J. Greeg, K. S. W. Sing, "Adsorption, Surface and Porosity", 2ª Edition, Academic Press, London, 1982.

[17] J. P. S. Valente, A. B. Araújo, D. F. Bozano, P. M. Padilha, A. O. Florentino, Síntese e caracterização textural do catalisador CeO₂/TiO₂ obtido via sol-gel: fotocatálise do composto modelo hidrogenoftalato de potássio. Revista Eclética Química, 004, 7-13, 2005.

Preparation and Characterization of BiFeO₃ Nanosystems as Visible Light Driven Photo-degradation of Organic Compounds

Chang Hengky[1,2], Steve Dunn[3], Valdew Singh[1], Leonard Loh[1]

[1] *Nanotechnology and Materials Science, Biomedical Engineering Group*
School of Engineering (Manufacturing), Nanyang Polytechnic, Singapore, 569830, Singapore
Hengky_Chang@nyp.gov.sg

[2] *Nanotechnology Centre, Department of Materials, School of Applied Sciences*
Cranfield University, Bedfordshire, MK43 0AL, United Kingdom

[3] *Materials Department, School and Engineering and Materials, Queen Mary, University of London,*
E1 4NS, UK

ABSTRACT

BiFeO3 (BFO) nanopowders were synthesized at relatively low temperatures via a self combustion method with the aid of citric acid as the fuel. The effect of different fuel concentration, calcinations temperature and duration is performed. The resultant BFO nanopowders are characterized in terms of their structural, morphological and optical properties. The synthesized BFO nanopowders are of very high purity without any secondary phases and exhibited a good absorption in the visible-light regime, which resulted in the efficient photocatalytic activity for decomposition of organic compounds.

1 INTRODUCTION

The use of renewable or low energy sources for the degradation of organic pollutants has generated broad interest in both scientific research and potential applications in recent years. Research on semiconductor photocatalysis came sharply into focus after the discovery that TiO_2 photochemical electrodes could split water using ultraviolet light [1]. Following that, the photocatalytic oxidation of organic contaminants using TiO_2-based semiconductors as a photocatalyst has been extensively investigated due to their excellent photochemical stability, high-efficiency, low cost, and non-toxicity. However, the photo-efficiency of TiO_2 is severely limited by its large band gap (3.2 eV), and the associated restriction of the absorption in the visible-light region. Only 4% of terrestrial radiation is suitable for the photoexcitation of TiO_2 [2]. This renders the overall process impractical.

Perovskite – type BiFeO3 (BFO) materials have attracted much interest due to their multiferroic properties at room temperature [3,4]. In particular, BFO thin films have been investigated intensively in their various applications, including capacitors [5], nonvolatile memory [6], and magnetoelectric devices [7]. In addition to these electronic applications, the photocatalytic properties of BFO powders under visible-light illumination have also been reported recently [8]. Although the photocatalytic activity of BFO powders under visible light is ascribed mainly to a small band-gap energy, attempts to elucidate other factors, such as

their weak ferromagnetic properties, are still being made [9]. Despite the fact that some work in this area has been undertaken, various research avenues such as synthesis, characterization, and the effects of magnetic fields remain to be explored.

BFO powders have been prepared using a variety of synthetic methods, such as sol–gel techniques [10], combustion [11], and hydrothermal processes [12]. From the perspective of synthesis, it is important to obtain pure materials using temperatures as low as possible. This point may be especially important for BFO materials because some secondary phases such as $Bi_2Fe_4O_9$, Bi_{12} $(Bi_{0.5}Fe_{0.5})$ $O_{19.5}$, and $Bi_{25}Fe_1O_{39}$ are usually formed during the synthesis [13,14]. In this study, pure BFO nanopowders were synthesized at very low temperatures via a self combustion method and follow by calcinations with the use of citric acid as fuel, and the absorptions property of the resultant BFO nanopowder were evaluated.

2 EXPERIMENTAL

2.1 Chemicals and materials

A commercially available reagents grade Bismuth Nitrate Pentahydrate ($Bi(NO_3)_3 \cdot 5H_2O$) and Iron Nitrate Nonahydrate ($Fe(NO_3)_3 \cdot 9H_2O$) (purity > 99%, International Laboratory) are used as raw materials. Nitric acid (HNO_3) with 70% concentration and De-Ionised (DI) water with resistivity of 18.3 MΩ were used to dissolve the salts raw materials above. All the chemicals and materials mentioned were used as purchased without further purification.

2.2 BFO nanopowders synthesis technique

Bismuth Nitrate Pentahydrate ($Bi(NO_3)_3 \cdot 5H_2O$) and Iron Nitrate Nonahydrate ($Fe(NO_3)_3 \cdot 9H_2O$) were used in stoichiometric proportions of 1:1 molar ratio to create 0.2 M solution. A solution containing of 0.2 mol of nitric acid (HNO_3) (25ml of 70% HNO_3) is mixed with 25 ml of DI water. Bismuth Nitrate Pentahydrate ($Bi(NO_3)_3 \cdot 5H_2O$) salts were dissolved in the nitric acid solution mentioned earlier and Iron Nitrate Nonahydrate ($Fe(NO_3)_3 \cdot 9H_2O$) were just dissolved in the DI water only. Once dissolved, both Bismuth and Iron ionic solutions were mixed together and citric acid with molar ration of Citric/Metal of 1:1 was added into the Bi-Fe ionic solution and stirred until complete dissolution. The Bi-Fe with citric acid solution were then heated up to 250°C until the solution dried and follow by auto ignition of the combustion and formed dark brownish flakes and left heated for another ½ hour. The dark brownish flakes were calcined in the furnace at varying temperatures of 450°C, 550°C, 650°C for 6 hours and varying soaking timing at 450°C of 6, 9, 12 and 24 hours and to evaluate the phase evolution of the BFO powders.

The calcined BFO powders were in the form of small flakes aggregates, and the calcined powder were then subjected to high energy planetary ball milling process at 400 RPM rotational speed for ½ hour to breakdown the agglomerated BFO nanopowders to obtained fine powders.

2.3 BFO nanopowders characterization

The synthesized BFO nanopowders size and morphology was characterized using FESEM (JEOL-JSM7500F). XRD patterns were obtained using PANAlytical X'Pert Pro MPD advanced powder X-ray diffractometer (using Cu Kα = 1.54056 A° radiation) with scanning range of 2 theta from 20° to 70°. The absorption optical property of the resultant BFO nanopowders was characterized using Perkin Elmer model Lambda 950 UV-Vis-NIR spectrophotometer with integrating sphere. BFO nanopowders were spread thinly on the surface of quartz glass with the help of ethanol to prepare samples for absorption testing.

3 RESULTS AND DISCUSSION

The synthesized BFO nanopowders X-Ray Difraction patterns at varying parameters are shown below:

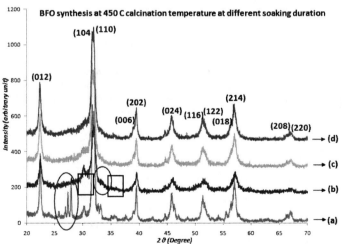

Fig. 1. XRD pattern of synthesized BFO nanopowders calcined at 450°C for (a) 6 hours, (b) 9 hours, (c) 12 hours and (d) 24 hours.

Figure 1(a) presents the XRD pattern of the BFO nanopowders calcined at 450°C for 6 hours duration, and it was revealed that under this condition, there are Bi_2O_3 and Fe_2O_3 phases exist as shown by the circled and square box area respectively in figure 1(a). As the soaking duration increased to 9 hours as shown by figure 1(b), the Bi_2O_3 phase started to disappear and some Fe_2O_3 phases still existed. At figure 1(c) and 1(d) where soaking time were 12 hours and 24 hours at respectively, it is revealed that the BFO nanopowders obtained are highly crystallized and of high purity and exhibit a single-phase perovskite structure according to JCPDS card no. 71-2494. Non-perovskite phases such as Bi2Fe4O9 and Bi2O3/Fe2O3 are not detected in XRD spectra. The obvious peak-splitting shows that the nanoparticles are rhombohedral, consistent with the structure of BFO ceramics,[8 from DOI: 10.1002/adma.200602377].

Another experiment carried out for BFO nanopowder synthesis was calcination at different temperature to see the effect of varying calcinations temperature on the phases of the BFO nanopowders obtained. From figure 2 where BFO brownish flakes after self combustion process, they were subjected to different calcinations temperature of 450°C, 550°C and 650°C for 6 hours duration, it is revealed that the BFO phases were gradually vanished as the calcinations temperature increases from 450°C to 650°C.

Also, it is observed that as the calcination temperature increased, the resultant BFO materials is go towards more complete oxide systems as the peaks of Bi_2O_3 and Fe_2O_3 became more obvious as shown by the square box area in the figure 2. 6 hours of calcination soaking time at different calcinations temperature was enough to see the evolution of resultant BFO nanopowders phase instead of longer calcinations time such as 24 hours which is the optimum synthesis parameter.

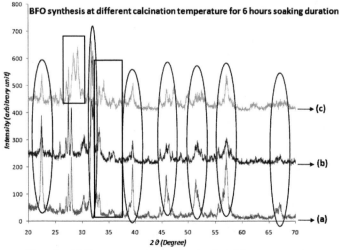

Fig. 2. XRD pattern of synthesized BFO nanopowders calcined for 6 hours at (a) 450°C, (b) 550°C and (C) 650°C.

The synthesized BFO nanopowders microstructure of optimum synthesized conditions (calcinations at 450°C for 24 hours) was characterized by FESEM and shown as follows:

(a)

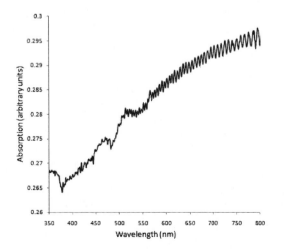

(b)

Fig. 3. FESEM micrograph of synthesized BFO nanopowder at (a) 10,000X magnifications, and (b) 25,000X magnifications

It was observed that the resultant BFO nanopowders are in spherical shape and ultra fine size with observed estimated size in the range of around 100 nm. Some nanowires morphology are observed as well, but it is of minority content as indicated by a circled area in figure 3(b). The BFO nanopowders are in agglomerated form, and further efforts need to break down the agglomerations such as using sonication to obtained better dispersed nanoparticles.

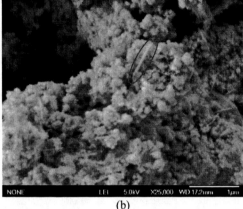

Fig. 4. UV-Vis absorption spectroscopy of BFO nanopowders

As a feasible potential application of BFO nanopowders in photocatalytic degradation activity of organic compounds, the absorption characteristics of the BFO nanopowders were investigated using UV–Vis spectroscopy. As can be seen in Fig. 4, the absorption spectrum shows absorption peaks at 480 nm and 530 nm wavelength, suggesting that BFO nanopowders can absorb remarkable amounts of visible light.

In figure 5 shows the optical band gap of the BFO nanopowders calculated using Tauc's relationship formula of $\alpha h\nu = A(h\nu - E_g)^n$ and n is $\frac{1}{2}$ since BFO is a direct bandgap semiconductor. To obtain optical band gap (E_g) value, a curve of $(\alpha h\nu)^2$ are plotted against $(h\nu)$ and extrapolation of the linear region of the graph intersecting x-axis at 0 value gives the value of the optical band gap, E_g. The calculated value from the absorption spectrum is 1.9 eV, which is smaller band gap value with that of another report [8] and smaller than the BFO thin film of 2.5 eV as reported at other report [15] . The smaller band gap of BFO nanoparticles indicates a possibility of utilizing more visible light for photocatalysis.

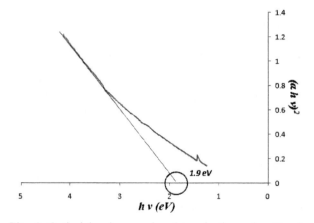

Fig. 5. Optical band gap value determination using Tauc's relationship formula of the synthesized BFO nanopowders

4 CONCLUSION

The Bismuth Ferric Oxide perovskite structure nanopowders is successfully synthesized via self combustion method and its process parameters have been optimized to obtain high purity, high crystalline BFO nanopowders. From the micrograph study, the synthesized BFO nanopowders were confirmed obtained in the nanoscale size.

The resultant nanopowders optical property shown obvious absorption peaks at around 480 nm and 530 nm wavelength, hence the synthesized BFO nanopowders could be potentially used as photocatalytic agent at visible light range. The estimated optical band gap calculated using Tauc's relationship is 1.9 eV, confirming the potential applications of the synthesized BFO nanopowders as photocatalyst under visible light region for degradation of organic compounds pollutants as smaller band gap indicates a possibility of utilizing more visible light for photocatalysis. Future works following this work would be photocatalytic activity evaluation of the synthesized BFO powder on Rhodamine

Blue dyes which is commonly used dyes in the textile industry.

5 ACKNOWLEDGEMENT

The author would like to take this opportunity to express his sincere gratitude to Nanyang Polytechnic Nanomaterials Laboratory, Biomedical Engineering Hub and Dr. Steve Dunn for his invaluable guidance and support in the capacity as the author supervisor, as well as fellow post-graduate students and lab officers at Nanotechnology Centre, Department of Materials, School of Applied Sciences, Cranfield University for the assistance given during this course of work.

6 REFERENCES

[1] A. Fujishima and K. Honda, Nature (London) 37, 238 (1972)

[2] Linsebigler, A.L., Lu, G & Yates, J.T. Jr Photocatalysis on TiO₂ surfaces: principles, mechanisms, and selected results. Chem. Rev. 95, 735-758 (1995)

[3] J. R. Teague, R. Gerson, and W. J. James, ''Dielectric Hysteresis in Single Crystal BiFeO3,'' Solid State Commun., 8 [13] 1073–4 (1970).

[4] I. Sosnowskat, T. Peterlin-Neumaier, and E. Steichele, ''Spiral Magnetic Ordering in Bismuth Ferrite,'' J. Phys. C, 15 [23] 4835–46 (1982).

[5] K. Y. Yun, M. Noda, and M. Okuyama, ''Prominent Ferroelectricity of BiFeO3 Thin Films Prepared by Pulsed-Laser Deposition,'' Appl. Phys. Lett., 83[19] 3981–3 (2003).

[6] N. Hur, S. Park, P. A. Sharma, J. S. Ahn, S. Guha, and S.-W. Cheong, ''Electric Polarization Reversal and Memory in a Multiferroic Material Induced by Magnetic Fields,'' Nature, 429, 392–5 (2004).

[7] V. E. Wood and A. E. Austin, Magnetoelectric Interaction Phenomena in Crystals. Gordon and Breach Science Publisher, New York, 1975.

[8] F. Gao, X. Chen, K. Yin, S. Dong, Z. Ren, F. Yuan, T. Yu, Z. Zou, and J.-M.Liu, ''Visible-Light Photocatalytic Properties of Weak Magnetic BiFeO3 Nanoparticles,'' Adv. Mater., 19 [19] 2889–92 (2007).

[9] F. Gao, Y. Yuan, K. F. Wang, X. Y. Chen, F. Chen, and J.-M. Liu, ''Preparation and Photoabsorption Characterization of BiFeO3 Nanowires,'' Appl.Phys. Lett., 89, 102506–8 (2006).

[10] S. Ghosh, S. Dasgupta, A. Sen, and H. S. Maiti, ''Low-Temperature Synthesis of Nanosized Bismuth Ferrite by Soft Chemical Route,'' J. Am. Ceram. Soc., 88 [5] 1349–52 (2005).

[11] J.-B. Li, G. H. Rao, J. K. Liang, Y. H. Liu, J. Luo, and J. R. Chen, ''Magnetic Properties of Bi(Fe1-xCrx)O3Synthesized by a Combustion Method,'' Appl. Phys. Lett., 90, 162513–6 (2007).

[12] J. Luo and P. A. Maggard, ''Hydrothermal Synthesis and Photocatalytic Activities of SrTiO3-Coated Fe₂O₃ and BiFeO₃,'' Adv. Mater., 18 [4] 514–7 (2006).

[13] M. W. Lufaso, T. A. Vanderah, I. M. Pazosa, I. Levina, R. S. Roth, J. C.Nino, V. Provenzano, and P. K. Schenck, ''Phase Formation, Crystal Chemistry, and Properties in the System Bi2O3–Fe2O3–Nb2O5,'' J. Solid State Chem., 179[12]3900–10 (2006).

[14] M. I. Morozov, N. A. Lomanova, and V. V. Gusarov, ''Specific Features of BiFeO3 Formation in a Mixture of Bismuth(III) and Iron(III) Oxides,'' Russ. J.Gen. Chem., 73 [11] 1772–6 (2003).

[15] K. Takahashi, N. Kida, M. Tonouchi, *Phys. Rev. Lett.* 2006, *96*, 117 402.

Magnetite Nanoparticles as Moderators of Extreme Conditions During *Bradyrhizobium japanicum* Growth

M. R. GHALAMBORAN,[†*] J. J. RAMSDEN[†]

[†] Centre for Microsystems and Nanotechnology, Cranfield University, Bedfordshire, MK43 0AL, UK.

[*] Corresponding author:
E-mail: m.r.ghalamboran.s06@cranfield.ac.uk; Tel: +44(0)1234750111; Fax: +44(0)1234751346

ABSTRACT

The aim of this study was to investigate whether magnetite nanoparticles and pH simultaneously affect *Bradyrhizobium* growth rate (BGR) in liquid media, with a view to diminishing sensitivity to high pH and, hence, enhancing survival in extreme conditions. Treatments were different concentrations of nanoparticles and different pH. Statistical design was a complete randomized block (CRB) in a factorial 3×6 experimental arrangement with four replications, the first factor being the different concentrations of nanoparticles and the second factor the different pH values. The most important variables were bacterial growth rate (BGR), growth rate constant (GRC), mean generation time (MGT), number of generations (NG) of bacteria before the death phase, oxidation reaction potential (ORP), and water activity (a_w). Nanoparticles in alkaline and acidic pH could increase GRC and NG, and decrease MGT. The nanoparticles could buffer available oxygen, so that the ORP and the a_w were optimized with respect to survival in extreme conditions. The combined effects of pH and nanoparticles could enhance NG, GRC and decrease MGT, beyond a certain value increased concentrations of nanoparticles in the media had the opposite effects on the physiological growth indices.

Keywords: magnetite nanoparticles, *Bradyrhizobium japanicum*, viability, extreme conditions.

1 INTRODUCTION

Development of the cultivation of soybean in semiarid and arid lands depends on preserving the ability of *Bradyrhizobium japanicum* growth under environmental stress; e.g., alkaline pH is an extreme inhibitor for *B. japanicum*. Since secretions of *Bradyrhizobium* are alkaline it needs to grow at neutral pH [1, 2, 9]. In addition the rhizobium–legume symbiosis can lead to the release of reactive oxygen species (ROS: O_2^-, H_2O_2, and $HO^.$). These ROS may play a positive role in deterring bacterial invasion [8] and helping to limit infection in plant root cells by either crosslinking cell wall glycoproteins or degrading the cell wall, but in their absence in the lost host ROS will damage the bacterial cell wall [8]. Recently, iron nanoparticles are being investigated as a new generation of environment-improving technologies [14, 21]. Previous studies indicated that magnetite nanoparticles can sequester oxidizing materials secreted by the bacteria during growth, hence enhancing growth, and furthermore can acting as oxygen scavengers and as buffers against environmental stress [6, 7]. As *Bradyrhizobium japanicum* is an aerobic bacterium, therefore its survival depends on oxidative phosphorylation [4], consequently variation of ORP in the medium affects bacterial growth [15], in other words enzymes like nitrogenase can be inactivated when sensitive groups like sulfhydryls are oxidized [16], and oxygen derivatives have degraded bacterial cell walls in the absence of host cells [8]. Moreover, the products of oxygen reduction are extremely toxic for the survival of *B. japanicum*. Thus the ORP in a culture medium is a useful index for bacterial growth rate. Previous studies have revealed a positive relationship between pH and ORP [15, 8]. Most chemical and biological reactions are actually redox ones reactions in which electron transfers; that is, occur, they can be characterized by pH and ORP measurements. However, magnetite nanoparticles will accept electrons and they thereby prevent the products of oxygen reduction from increasing and thus encourage microbial growth [19]. The question we seek to answer is whether magnetite nanoparticles, which might have been added to media for detoxification or to soil for remediation proposes [9], alter the growth of the symbiotic *B. japanicum* and the ability of crops like soybean to fix nitrogen. We have examined this by adding nanoparticles to pure *B. japanicum* cultures.

2 MATERIALS AND METHODS

2.1. Bacterial strains and culture media

The Histic strain of *Bradyrhizobium japanicum* was obtained from the Soil and Water Institute, Tehran. The bacterial cultures were prepared using the procedures described by Ghalamboran et al. (2009). Briefly, 50 ml of YMB [18] in a 250 ml flask were autoclaved (121 °C, 15–20 min, 103 kPa), and after cooling one loop full of the Histic strain was introduced to the flask, which was incubated on a rotary shaker at 150 rpm at 28–30ºC for 7–8 days. At the early stationary growth phase, 1 ml of broth culture (YMB) was transferred to new 250 ml flasks of YMB (50 ml) broth with varying concentrations of added magnetite nanoparticles (0, 40, 80 µg ml^{-1}) and pH (4, 5, 6, 7, 8, 9). The pH was adjusted by adding 1 M NaOH or HCl after the nanoparticles, prior to autoclaving. The inoculated flasks were held on a rotary shaker at 150 rpm at 28–30 °C for 7 days, then every day 1 ml from each flask was removed separately and eight times sequentially tenfold diluted in test tubes. To count the bacteria, twice 100 µl from each of these eight tubes was removed and spread on two 9 cm Petri dishes containing YMA solid medium [18]. The Petri dishes were immediately transferred to an incubator and kept at 28–30 ºC for 6–8 days. Counting the number of colony-forming units N_f on the surface of the Petri dishes was started after 5–6 days.

2.2. Physiological indices of bacterial growth

The bacterial growth indices (GRC, NG, and MGT) were calculated during the logarithmic phase [6].

2.3. Preparation of magnetite nanoparticles

The magnetite nanoparticles were prepared by co-precipitating ferric and ferrous salts in sodium hydroxide solution according to the procedure described in the previous study [6].

2.4. Water activity (a_w)

Bacterial growth and division depend on the water activity [16]. The amount of water available to the bacteria can be reduced by interaction with solute molecules (the solvation effect) or by adsorption to the surfaces of chaotropic solids [19]. The water activity is the ratio of the vapour pressure p of water in a material (liquid medium) to the vapour pressure p_0 of pure water at the same temperature [11], i.e.:

$$a_w = \rho / \rho_0 \qquad (1)$$

The a_w was measured by placing 5 ml of the liquid medium in a small 10 ml chamber of a water activity meter. The average of three measurements was recorded. This variable was measured at the beginning of the experiment (before inoculation) and 7 days after inoculation.

2.5. Oxidation reaction potential (ORP)

The ORP was determined by detecting the concentration ratio of a selected ion in the reduced form to the same ion in the oxidized form. An ORP electrode measured the redox potential according to the Nernst half-cell potential equation [4]. The ORP was measured at the beginning of the experiment (before inoculation) and 7 days after inoculation.

3 RESULTS AND STATISTICAL ANALYSIS

3.1. Population of viable bacterial cells

Fig. 1 shows the growth of the bacteria in the presence of different concentration of nanoparticles.

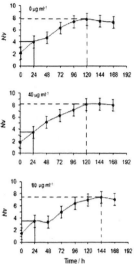

Figure 1. Variation of log number of viable *B. japanicum* cells per ml (N_v) in the presence of different concentrations of magnetite nanoparticles at standard pH (6.8).

Fig. 2 shows the growth of bacteria at different pH. Note the need of the population to periodically recover (no growth take place) at pH extremes.

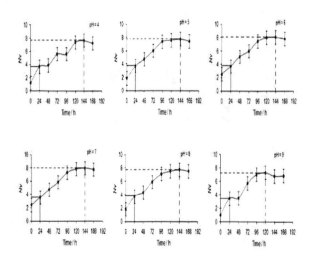

Figure 2. Variation of log number of viable *B. japanicum* cells per ml (N_v) at different pH in the absence of nanoparticles.

According to the analysis of variance (Table 1) the population of the viable cells was affected by both nanoparticles and pH levels.

Table 1. Analysis of variance of the physiological growth indices of *B. japanicum* Histic in the presence of nanoparticles.

Source of Variation	df[†]	N_v	MGT / h	GRC / h^{-1}	NG
Replication (R)	3	85	0.16	1.73	0.33
Nanoparticles (N)	2	11[b]	7.16[b]	0.004[b]	69.1[b]
pH	5	3.1[b]	2.1 b	0.0006[b]	11.43[b]
N × pH	10	0.3[n.s.]	1.75[b]	0.0005[b]	9.92[b]
R × N × pH	51	0.2	0 .254	3.145	0.60

[†] df is the number of degrees freedom; [b] F test indicates significance at $P < 0.01$, and [n.s.] indicates no significance.

3.2.   Mean generation time (MGT)

The magnetite nanoparticles significantly decrease the MGT compared with the control (without nanoparticles) treatment at $P < 0.01$, also the MGT was affected by pH levels at $P < 0.01$, (Table 1). The MGT increased more in the acid region than in the alkaline.

3.3.   Growth rate constant (GRC)

The GRC was affected by nanoparticles at $P < 0.01$ (Table 1), and pH at $P < 0.01$ (Table 1). Minimum of the GRC occurred at pH=4 and at alkaline pH ≥ 8.

3.4.   Number of generations (NG)

The nanoparticles could significantly increase NG at $P < 0.01$ (Table 1). Also pH significantly changed NG at $P < 0.01$.

3.5.   Comparison of nanoparticles

According to Table 2, all physiological growth indices were significant at $P < 0.01$, and just the NG between C_1 (0 µg ml^{-1}) and C_3 (80 µg ml^{-1}) was significant at $P < 0.05$.

Table 2. Comparison of the effects of the different concentrations of magnetic nanoparticles based on Duncan's test.

Treatments	MGT / hr	GRC / hr^{-1}	NG
C_1-C_2	1.1[b]	0.02[b]	2.8[b]
C_1-C_3	0.6[b]	0.2[b]	0.94[a]
C_2-C_3	1.1[b]	0.02[b]	3.8[b]

[a], [b] F test-indicated significance at $P < 0.05$ and $P < 0.01$, respectively.

3.6. Water activity (a_w)

The analysis of variance (Table 3) indicated that a_w was affected separately by nanoparticles and pH, while their combination had no significant effect. Using the magnetite

nanoparticles increases a_w. Moreover the a_w increased after 7 days, and its rate at alkaline pH was greater than at the control (standard pH). Maximum a_w occurred at pH= 8 and 9 after 7 days.

Table 3. Analysis of variance of the water activity (a_w) in the batch culture.

Sources of Variation	df[†]	Zero time	After 7 days
Replication (R)	3	1.2	4.2
Nanoparticles (N)	2	1.5[b]	1.14[a]
pH	5	2.5[b]	1.03[a]
N × pH	10	2.8[n.s.]	3.71[n.s.]
R × N × pH	51	2.9	2.34

[a], [b] F test indicated significance at $P < 0.05$ and $P < 0.01$, respectively; [n.s.] indicated non-significant difference.

3.7. Oxidation reaction potential (ORP)

The ORP was affected by nanopartcles, pH, and their combination (Table 4).

Table 4. Analysis of variance of the oxidation reaction potential (ORP)

Sources of Variation	df[†]	Zero time	After 7 days
Replication (R)	3	122	405
Nanoparticles (N)	2	1717.04[b]	234557.1[b]
pH	5	111587.5[b]	130344.4[a]
N × pH	10	437.4[b]	336[a]
R × N × pH	51	93.61	123.5

[a], [b] F test-indicated significances at $P < 0.05$ and $P < 0.01$, respectively.

4 DISCUSSION

According to the above results, ORP increased during *Bradyrhizobium japanicum* Histic growth without nanoparticles, possibly due to releasing ROS [8], while in the presence of the magnetite nanoparticles the secretions of *Bradyrhizobium* were presumably sequestered by the nanoparticles, hence the nanoparticles decreased the ORP.

Previous studies indicated that *B. japanicum* has an antioxidant defence system (AD), which involves enzymatic activities (catalases, superoxide dismutases, peroxidases) and together ROS and AD contribute to the redox balance, the modulation of which is probably crucial for physiological regulation [15]. Clearly, variation of pH could affect the enzymatic activities, and then the redox potential of culture medium will be changed [6, 7, 14, 15]. Extreme conditions like acid or alkaline pH are an obstacle to redox balance stability. Besides, survival and increasing bacterial growth depend on antioxidant defence activity in extreme conditions [15]. Thus any factor that can increase AD activity, result in improved survival and physiological indices. In this study treatments with nanoparticles significantly increased viability of *B. japanicum* cells. Based on the results, one of the roles of the nanoparticles is inferred to be a pH buffering effect, and another is decreasing ORP.

Furthermore, not only were the magnetite nanoparticles mild reducing reagents at slightly acid pH but also at extremes pH(4, 8 and 9), the presence of nanoparticles significantly regulated water activity. These nanoparticles acting as catalysts could convert superoxides to peroxides and oxygen molecules and produce water molecules ($O_2^- + O_2^- + 2H^+ \rightarrow H_2O_2 + O_2$ and $H_2O_2 + 2H^+ \rightarrow 2H_2O$); likewise they are likely to be able to absorb water molecules temporarily [3, 15, 16, 17], and then they prevent from decreasing available oxygen during bacterial culture.

Although some constituents in the YMB like K_2HPO_4 are used to buffer variation of pH during growth, our results show that it could not act effectively in extreme conditions, while the nanoparticles were potent in alkaline pH and they could apparently moderate extreme conditions for bacterial growth. In fact nanoparticles have the capacity of preserving oxygen molecules by neutralization of the ROS [3]. The present results are consistent with the application of nanoparticles for free radical scavenging [3, 15]. Preserving optimum values of parameterslike the a_w and the ORP at extreme pH by nanoparticles could enhanced N_f, NG, GRC and decreased the MGT. On the other hand the application of a high concentration of magnetite naoparticles has inversed effects on the optimum physiological indices of the bacterial growth, because the high density of nanoparticles probably increases osmotic pressure in the liquid media, and thus water activity declined and decreased the survival of the bacterial cells.

5 CONCLUSION

Feeble *B. japanicum* growth under extreme condition can be revived by adding magnetite nanoparticles. The optimum concentration was 40 µg ml^{-1} of nanoparticles. Furthermore, useful enhancement of growth even at standard pH (6.8) could be obtained by adding nanoparticles. Evidence from water activity and ORP suggested that sequestration of secreted ROS is the most significant effect of the nanoparticles. Therefore, cultivating legume plants like soybean under extreme conditions depends on enhancing the natural defence systems for rhizobium growth, supplementing the natural defence systems (like producing catalysts or polysaccharides in the bacterial cell wall for reducing the ROS). Nevertheless, results from farms are not satisfactory. Nowadays under conditions of intensive agriculture implying the use of a high level of fertilizer in order to maximize the yield of crops is a serious problem leading to extreme pH in the fields. Moreover our results indicate that nanoparticles may play an important role in the regulation of bacterial growth in such extreme conditions, consequently allowing legume plants to be developed in infertile soils and poor lands.

REFERENCES

[1] C. Appunu, Dhar, B. Symbiotic effectiveness of acid-tolerant B*radyrhizobium* strains with soybean in low pH soil. African J. Biotechnol. (2006), 5, 842-845.

[2] A.B. Arun, Sridhar, K.R. Growth tolerance of rhizobia isolated from sand dune legumes of the southwest coast of India. Engng. Life Sci. (2005), 5, 134-138.

[3] C. Buzea, Pacheco, I. I., Robbie, K. Nanomaterials and nanoparticles: Sources and toxicity. Biointerphases. (2007), 2, MR17-MR71.

[4] C.-N. Chang, Bangcheng, H., Chao, A. Applying the nernst equation to simulate redox potential variations for biological nitrification and denitrification processes. Environ. Sci. Technol. (2004), 38, 1807-1812.

[5] J. M. Frustaci, Sangwan, I., O'Brian, M. R. Arerobic growth and respiration of a δ-aminolevulinic acid synthase (hem*A*) mutant of Bradyrhizobium iaponicum. J. Bacteriol. (1991), 173, 1145-1150.

[6] M. R. Ghalamboran, Ramsden, J. J. Magnetite nanoparticles enhance growing rate of *Bradyrhizobium japanicum*. Proc. Nanotechnol. Conf., May 3-7, (2009); Houston, Texas.

[7] M. R. Ghalamboran, Ramsden, J. J., Ansari F. Growth rate enhancement of Bradyrhizobium japanicum due to magnetite nanoparticles. J. Bionanosci. (2009); 3, 1-6.

[8] K.E. Gibson, Kobayashi, H., and Walker, C. G. Molecular determinants of a symbiotic chronic infection. Rev. Genet. (2008), 42, 413-441.

[9] P.H. Graham, Viteri, S.E., Mackie, F., Vargas, A.T., Placios, A. Variation in acid soil tolerance among strains of *Rhizobium* and *Bradyrhizobium*. Field Crops Res. (1982), 5, 121-128.

[10] P.H. Graham, Draeger K., Ferrey M.L, Hammer B.E., Martinez R.E. Naarons S.R., Quinto C. Acid pH tolerance in strains of *Rhizobium* and *Bradyrhizobium*, and initial studies on the basis for acid tolerance of *Rhizobium* tropici. J. Microbiol. (1994), 40, 198.

[11] W. D. Grant. Life at low water activity. Phil. Trans. R. Soc. Lond. B. Biol. Sci. (2004), 359, 1249–1267.

[12] C. Mace, Desrocher, S. Gheorghiu, F. Kane, A. Pupeza, M. Cernik, M. Kvapil, P. Venkatakrishnan, R. Zhang, W. X. Nanotechnology and groundwater remediation: A step forward in technology understanding. Remediation J. (2006), 16, 23-33.

[13] M. K. Meghvansi, Prasad, K., and Mahna, S.K. Identification of pH tolerant *Bradyrhizobium japanicum* strain and their symbiotic effectiveness in soybean in low nutrient soil. African J. Biotechnol. (2005), 4 , 663-666.

[14] H. Park, Park, Y. M., Oh, S. K., You, K. M., and Lee, S. H. Enhanced reduction of nitrate by supported nanoscale zero-valent iron prepared in ethanol-water solution. Environ. Technol. (2009), 30, 261-267.

[15] N. Pauly, Pucciariello, C., Mandon, K., Innocenti, G., Jamet, A., Baudouin, E., Hérouart, D., Frendo, P., and Puppo, A. Reactive oxygen and nitrogen species and glutathione: key players in the legume-rhizobium symbiosis. J. Exp. Botany. (2006), 57, 1769-1776.

[16] L.M. Prescott, Harley, J.P., and Klein, D.A. Microbiology. New York: McGraw-Hill (2002), pp.113-125.

[17] D. Schubert, Dargusch, R., Raitano, J., Chan, S. W. Cerium and yttrium oxide nanoparticles are neuroprotective. Biochemi. And Biophysi. Res. Comm. (2006), 342, 86-91.

[18] P. Somasegaran, Hoben, H.J., Handbook for Rhizobia. New York: Springer-Verlag (1994).

[19] E. Tombàcz, Hajdú, A., Illés. Water in contact with magnetite nanoparticles, as seen from experiments and computer simulations Langmuir. Langmuir. (2009), 25, 13007-13014.

[20] P. M. Wiggins. Enzyme reactions and two-state water. J. Biol. Phys. Chem. (2002), 2, 25-37.

[21] W. X. Zhang. Nanoscale iron particles for environmental remediation. J. Nanoparticles Res. (2003), 5, 323-332.

NanoZeolites – small particles, small pores, big impact!

Wayne Daniell, Jürgen Sauer, Andreas Kohl

NanoScape AG
Am Klopferspitz 19, Planegg-Martinsried, D-82152, Germany
E-mail: daniell@nanoscape.de; http://www.nanoscape.de

ABSTRACT

Zeolites are well known porous materials established in applications which employ their large surface area, controllable porosity and tuneable surface chemistry. Through a reduction in their particle size by an order of magnitude, significant increases in product performance have been achieved and, more notably, new applications have been accessed which until now were unsuitable for classical zeolite materials.

Keywords: zeolites, porous, encapsulation, triggered release

1 INTRODUCTION

Zeolites were first discovered at the end of the 18th century and came to prominence in the 1960's when synthetically produced zeolites were employed in large industrial applications as molecular sieves for filtration and purification purposes, and as catalysts in the cracking of crude oil. However, recent advances in hydrothermal synthesis techniques [1] and procedures to modify the surfaces of the crystalline particles [2] have led to the development of a new form of these materials – the NanoZeolites, with particle size in the lower sub-micron range (e.g. NanoZeolite-FAU, see Fig.1).

Figure 1: TEM micrograph of crystalline NanoZeolite Faujasite (mean particle size ca. 50nm)

These porous, nanocrystalline materials have breathed new life into an established material class and possess significantly improved properties (such as higher specific surface area), leading to performance enhancement in many known applications.

Furthermore, the availability of nanoscale zeolites has opened up new applications, which were until now unserviceable with microcrystalline particles.

2 NANOZEOLITES

Advancements in synthesis and purification techniques have taken the mean particle size of zeolites down from the few microns range to around 100 nm (an order of magnitude). Furthermore, narrower particle size distributions can be achieved in comparison with classical microcrystalline zeolites which are currently commercially available (see Figure 2). This not only means a more homogeneous material and an increase in the surface area (per gram of particles), but also increases the kinetics of adsorption/desorption processes.

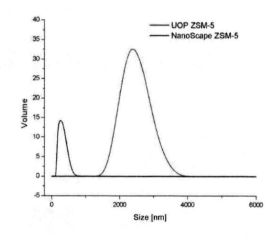

Figure 2: Comparison of particle size distribution of NanoZeolite MFI with a commercial microcrystalline MFI zeolite

This property can be put to good use in the form of thin film adsorber coatings in industrial dehumidification and/or energy recovery systems [3]. The use of nanoscale zeolites means: a reduction in weight (less adsorber material required to coat surfaces effectively); smaller cross-sections (due to thinner films) and thereby lower air stream resistance; enhanced adsorption/desorption kinetics; and a higher overall efficiency of the complete system (through savings in energy). The use of an LTA zeolite structure (with pore diameter 3-5 Å) furthermore imposes a high selectivity towards moisture adsorption, and reduces risk of bacterial contamination. Operational lifetime of the systems is also lengthened due to the better adhesion of the nanoparticles onto the substrate (typically aluminium foil or stainless steel).

3. OPEN-PORE APPLICATIONS

NanoZeolites lend themselves to applications in which the porous material is applied either as a thin-film coating or as an additive dispersed within a matrix. The so-called "open pore" applications include using the zeolite as a filter, adsorber or catalyst. A combination of the pore size and pore-surface chemistry can be employed to effectively and selectively filter out, adsorb, desorb or catalytically convert gas or liquid phase molecules. When coated onto the surfaces of various substrates (aluminium, steel, ceramic, carbon fibres), NanoZeolite thin films have significantly contributed to improving the performance of air purification filters and exhaust-gas catalysts. On fibre mats, these nanoparticle coatings have increased the active surface area without causing a significant loss in porosity (which would reduce airflow rate). For water purification applications NanoZeolites have proven suitable for both polymer and ceramic membranes. In the former case, the inorganic material is used as an additive to an organic polymer matrix; whereas for ceramic membranes a coating is employed, which modifies the effective pore size of the ceramic material.

4. ENCAPSULATION

NanoZeolites can also be employed as hosts for functional molecules such as dyes or catalysts (see Fig. 3). Through a suitable encapsulation process (e.g. impregnation, grafting etc.), the guest molecules are brought into the pore structure of the NanoZeolite and, depending on the size of molecule (in relation to pore dimensions) and the pore wall – guest molecule interactions, reside either in the pore channels or the central cavities of the unit cell. These encapsulated guest molecules exhibit more stability than when solvated and an increased resistance to chemical, mechanical and thermal influences. Furthermore, through encapsulation the application range of the functional molecule can be extended, e.g. organic dyes dispersed in aqueous systems; photochromic dyes in polymers with high flexural modulus. Through this technique the use of (and reliance upon) organic solvents can be reduced, and the lifetime of certain molecules significantly prolonged.

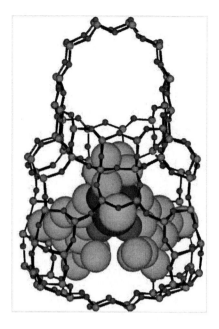

Figure 3: Functional molecule encapsulated within the pore structure of a NanoZeolite host

For certain applications (e.g. catalysts, sensors) it is not the encapsulation, protection or transport of the molecule that is of importance, but its release into a surrounding medium. A slow release can be achieved through careful selection of parameters such as: the strength of host-guest interaction (physical/chemical); and the steric confinements imposed upon the guest through encapsulation. For larger, bulkier molecules a nanoscale mesoporous host is often more appropriate, with pore diameters reaching up to 100Å (see Fig. 4).

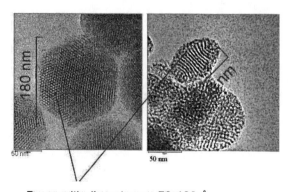

Pores with diameter ca. 50-100 Å
available for loading with functional molecules

Figure 4: TEM micrographs of nanometre sized mesoporous materials

A controlled or triggered release can be achieved through use of any external stimulus ("trigger"). This can be temperature, pressure or a burst of radiation (e.g. UV light).

Thermal triggers typically focus on the increased vibrations which occur within the zeolite framework when heat is applied. The oscillations of the cage structure can cause a pore mouth to expand to a size large enough to allow the guest molecule to exit the pores. A change in pressure can also force a molecule out through a pore opening, though more commonly the effect is that of varying the equilibrium adsorption state of weakly bound molecules. Attachment of guest molecules to the pore channel walls through special linkers, or the fitting of "flaps" over the pore mouth openings, allows the use of UV light as a trigger. There is scope here not only to control the speed of release but also the wavelength range over which the trigger activates. This leads to a highly sensitive and extremely controllable release mechanism.

REFERENCES

[1] T. Bein and V. P. Valtchev, "Synthesis and stabilisation of nanoscale zeolite particles" EP 1334 068 B1

[2] J. Sauer and A. Kohl, "Coated molecular sieve" WO 2008/000457 A2

[3] J. Sauer, T. Westerdorf, and H. Klingenburg "Humidity and/or heat-exchange device for example plate heat-exchanger, sorption rotor, adsorption dehumidifying rotor or the similar" WO 2006/079448 A1

Adhesion of Water on Flat Polymer Surfaces and Superhydrophobic Surfaces

K. Y. Law,[*] H. Zhao and B. Samuel

Xerox Research Center Webster, Xerox Innovation Group
Webster, NY, USA, klaw@xeroxlabs.com

ABSTRACT

The interactions between water and surfaces of varying hydrophobicity have been studied using a high-sensitivity microelectromechanical balance system (tensiometer). The surfaces studied include solution coated polymer films, commercial plastic sheets, self-assembled polymer layers on Si-wafer, fluorinated textured surfaces on Si-wafer, CVD polymerized films, and rose petals. Static contact angle data indicate that these surfaces varying from hydrophilic to superhydrophobic, from smooth and flat to rough and textured. While the wetting force of water on these surfaces was found to correlate well to the advancing water contact angle, the same could not be said about the pull off force, which is a measure of adhesion between the water droplet and the surface of interest. Rather the pull off force was shown to correlate well to the receding contact angle. Evidence is provided that the adhesion between water and the rough surface can be as strong as the flat polymer surface when the water droplet is in the Wenzel state on the rough surface. Finally, adhesion of water was found to play a role to the quality of the detachment of the water droplet from a surface. A minute amount of water residue could be found on certain surfaces after pull-off. The correlation between the quality of detachment and the receding contact angle is discussed.

Keywords: wetting, adhesion, contact angle, sliding angle, superhydrophobicity, polymer surfaces, rose petals

1 INTRODUCTION

The interplay between the adhesion of imaging materials (ink and toner) on various print surfaces versus the cohesive forces within the patches or droplets of these imaging materials play a crucial role in print hardware design and print process development [1,2]. If the imaging material adheres strongly on the print surface, particularly when the work-of-adhesion is stronger than the cohesive force, patches or droplets of image materials may break and a small amount of imaging materials may leave behind the print surface after contact. This will result in image quality defect, an undesirable effect in printing [3]. In this work, we attempted to measure the attractive and adhesive interactions of organic liquids, such as hexadecane and molten solid ink, on a microelectromechanical balance system (tensiometer) and then correlate them with surface property data. Our initial scoping experiments were not successful due to difficulties in handling low surface tension liquids. For the sake of experimental procedure development and to find out if there are correlations between attractive/adhesive forces and the surface properties, we use water in our model investigation. The surface tension of water is at 72.5 mN/m and is three times higher than those of hexadecane and solid ink. Here we report the static and dynamic (advancing, receding and sliding) contact angle measurements of water on 20 different surfaces. The correlation between the force data from the tensiometer and the surface property data are presented and discussed.

2 EXPERIMENTAL

The surfaces studied in this work include solution coated polymer films (Table 1, **1 - 3**, **6** and **13**), commercial plastic sheets (**4**, **5**, **14**), self-assembled polymer layers on Si-wafer (**10**, **11**), CVD polymerized films (**7 – 9**, **12**) [4], fluorinated textured surfaces on Si-wafer (Table 2, **15 – 18**), [5] and rose petals (**19**, **20**). Contact angle and sliding angle measurements were performed on a goniometer model OCA20 from Dataphysics using DI water as the test liquid. The size of the drop was controlled to be ~ 5 μL. For dynamic measurements, advancing angle was measured by adding a small volume of water to the surface; and the receding angle was measured by slowly removing water from the drop (0.15 μL/s). Sliding angle measurement was done by tilting the base unit at a rate of 1°/sec with a ~ 10 μL droplet using titling base unit TBU90E. The sliding angle is defined as the angle where the test liquid droplet starts to slide (or move).

The interaction of water with different surfaces was studied on a high-sensitivity microelectromechanical balance system or a tensiometer (Dataphysics DCAT 21, Germany). A hydrophilic platinum/iridium ring suspended with a 5 mg deionized water droplet was connected to the cantilever, and the force of this balance system was initialized to zero initially. The substrate was then placed on the sample table which was raised up at a speed of 0.03 mm/s toward the water droplet and the surface was considered "touched" when the force change was larger than 0.3 mg (or 2.9 μN). At the point of contact, the snap-in force was recorded. The substrate surface continued to move up for another 0.1 mm. After that, the surface and the water droplet retracted or separated at a speed of 0.01 mm/s. At the point of separation, the pull-off force was recorded. The entire procedure was performed in lab ambient. The relation between the interactive forces (both attraction and separation) and the distance (or time) were recorded automatically and analyzed by the software.

3 RESULTS AND DISCUSSION

3.1 Contact Angle Measurement of Water on Different Surfaces

Flat Surfaces. The surfaces of 14 different polymer surfaces obtained from different sources were studied by both static and dynamic contact angle measurement techniques. Table 1 summarizes the surfaces/materials studied and their static contact angle, sliding angle, advancing contact angle and receding contact angle. The results show that these surfaces vary from hydrophilic for **1** to highly hydrophobic for **14**. It is also important to note that, some surfaces are very smooth (on Si wafer) and others may have a different degree of roughness due to their coating or fabrication process. Thus surfaces **1 – 14** represent a snap shot of surfaces of all traits.

Table 1. Water contact angle data on flat polymer surfaces.

	Surfaces	θ [a]	θ_A [b]	θ_R [c]	α [d]
1	Polyurethane (PU)	70.5°	85°	48.9°	51°
2	PU – 2% Silclean	98.2°	104.3°	76.3°	31°
3	PU – 8% Silclean	104.3°	105.9°	88.1°	23°
4	Polyimide	80.1°	82.5°	56.1°	26.4°
5	Plexiglass	86.5°	93.9°	77.3°	29.1°
6	Polycarbonate	92.4°	98.2°	68.1°	59.2°
7	i-CVD Silicone	87.9°	91.2°	62.2°	90° [e]
8	i-CVD Fluorosilicone	115.9°	118°	90.3°	18.2°
9	i-CVD PTFE	127.7°	134.9°	73.6°	90° [e]
10	SAM OTS	109°	117.4°	94.6°	13°
11	SAM FOTS	107.3°	116°	95°	13.6°
12	Perfluoroacrylate polymer	113°	113.1°	61°	90° [e]
13	Hydrophobic sol gel coating	112.2°	111.6°	92.4°	5.6°
14	PTFE	117.7°	126.6°	91.9°	64.3°

(a) static contact angle; (b) advancing contact angle, (c) receding contact angle, (d) sliding angle; estimated errors are all < 3°, and (e) upper limit of the tilting unit, water drop does not slide even at 90°.

Rough/Textured Surfaces. In an attempt to generalize the conclusion in this work, we extend our study to superhydrophobic textured surfaces and two natural rough surfaces in the front and back side of the rose petals. The superhydrophobic textured surfaces was made using photolithography, followed by surface modification with a fluorosilane and the preparative procedure has been outlined elsewhere [5]. The rose petals were home grown flower picked from Law's garden in Penfield, New York. The contact angle measurement data for these surfaces are summarized in Table 2. The results indicate that water droplets are in the Cassie Baxter state on surfaces **15 – 19**,

whereas the water drop is in the Wenzel state on surface **20**, the back side of the rose petal [6].

Table 2. Water contact angle data on rough surfaces.

No	Si textured surface (pillar diameter/spacing)	θ [a]	θ_A [b]	θ_R [c]	α [d]
15	3 µm/4.5 µm	149°	160°	130.8°	20°
16	3 µm/6 µm	156.2°	161.3°	142.6°	10.1°
17	3 µm/9 µm	154.1°	159.9°	148.9°	5.7°
18	3 µm/12 µm	156.2°	160.8°	151.8°	3.4°
19	Rose petal - front	144.7°	150.7°	131.6°	6.1°
20	Rose petal - back	132.4°	136.6°	85.7°	90° [e]

a - e, same as Table 1.

3.2 Adhesion Measurements of Water on Different Surfaces

The interactive forces between water and various surfaces were studied on a tensiometer. Figures 1a-1d depict a series of photographs showing the inside view of the chamber of the tensiometer.

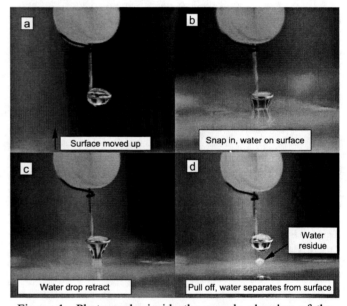

Figure 1 Photographs inside the sample chamber of the microelectromechanical balance system: (a) measuring surface moves upward towards the water droplet, (b) surface makes contact to the water droplet, (c) surface starts to retract separating from the water droplet, and (d) surface separates from the water droplet leaving a small drop of water residue on the surface.

The figure also stepwise illustrates how the experiment was performed. Firstly, 5 mg of deionized water was placed on the platinum/iridium ring, which was connected to the cantilever. The force of this balance system was initialized to zero and the surface of the substrate was moved upwards at a speed of 0.03 mm/s as shown in Figure 1a. As the surface and the water droplet approached and "touched", the attractive, wetting force was detected and measured as the snap-in force (Figure 1b). After the snap-in, the surface continued to move ~ 0.1 mm closer to the ring probe. At this point, the surface was retracted (at a speed of 0.01 mm/s) from the water droplet and the maximum force was recorded (Figure 1c). Depending on the nature of the surface, a very small drop of water residue might leave behind after the water droplet is pulled off from the surface (Figure 1d). At the point of separation, the pull-off force was recorded. The relation between the interactive forces and the distance (or time) was recorded automatically by the tensiometer. A typical output for the entire experiment is given in Figure 2.

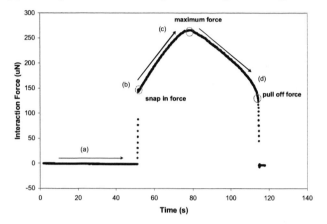

Fig. 2 A typical experiment output from the tensiometer

Among the data points in Figure 2, the snap-in force and pull-off force are most relevant to this work. The former measures the wetting force when the water droplet first interacts with the surface. The pull-off force is a little bit more complicated. The force recorded at pull-off will depend on how or when the water droplet is separated from the surface. When the separation is clean where no water residue is observed after the separation, the pull-off force is the adhesion force between the water drop and the surface at vertical separation. On the other hand, when water residue is observed on the surface after pull-off, it implies that the adhesion between the surface and the water droplet is stronger than the cohesive force within the water droplet. What's recorded then is the force when the water droplet breaks. The magnitude of this force is related to the work-of-adhesion between water and the surface, cohesive force of the water droplet (surface tension) and the contact area when the water droplet breaks. Detailed analysis of this force is beyond the scope of this investigation. In this work, our main concern is the snap-in force and the pull-off force for surfaces 1 – 20.

3.3 Correlation of Water Adhesion to Contact Angle Data

Figure 3 depict the plot of the snap-in force as a function of the advancing contact angle and a fairly good correlation is obtained. The snap-in, wetting force is shown to decrease

as the advancing contact angle increases and becomes negligible when the surface is superhydrophobic. The result makes sense because the motion of "snapping in" is very similar to the advancing of the contact line during the advancing contact measurement.

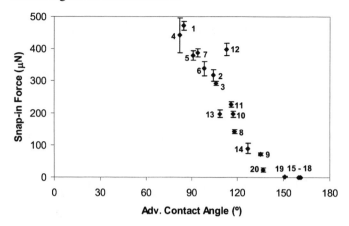

Fig. 3 Plot of snap-in force measured on tensiometer versus advancing contact angle.

Numerous attempts were made, but fail to correlate the pull-off force with contact angle parameters that are known to relate to adhesion, such as sliding angle, contact angle hysteresis and $(\cos\theta_R - \cos\theta_A)$. On the other hand, a good correlation is obtained when we plot the pull-off force versus the receding contact angle (Figure 4). The pull-off, adhesion force between water and the surface is shown to decrease monotonously as the receding contract angle increases.

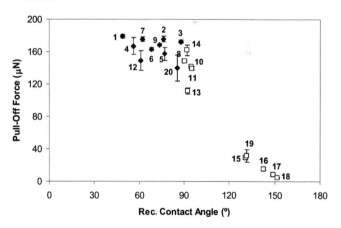

Fig. 4 Plot of the pull-off force for all surfaces measured on the tensiometer versus receding contact angle.

Another silent feature of the data in Figure 4 lies in the labeling. As noted earlier, water drop may break off differently during the pull-off: □ denotes to a clean break where no water residue was observed and ■ indicates the presence of water residue after pull-off. A glance of the plot in Figure 4 reveals a remarkable trend. For surfaces with receding angle > ~ 90°, the water-surface separation is clean, whereas residual water drops are observed for surfaces with receding angle $< \sim 90°$. While this may be intuitively expected because surfaces with low receding contact angle should have high adhesion. What's amazing is the clean transition from bad separation to good separation.

4 CONCLUDING REMARKS

This work shows with great generality that the wetting force correlates well to the advancing contact angle and the adhesion force correlates to the receding contact angle, the lower the receding angle the stronger the adhesion force towards the liquid. In case of water, this work also demonstrates that when the receding angle is $< 90°$, the adhesion force becomes stronger than the cohesion of the water droplet. This results in breakage of the water droplet during pull-off. As noted in the introductory section, water was used as a model liquid to study the tool to investigate the intricate relationship between surface property and adhesion; our next attention will be focused on applying the tool and the understanding to study low tension liquids.

5 ACKNOWLEDGEMENTS

The authors thank Dr. Varun Sambhy for providing us with the OTS surface and the polyurethane films.

REFERENCES

[1] For an introduction to electrophotographic printing, please see, L. B. Schein, Electrophotography and Development Physics, Springer-Verlag, New York, 1987.
[2] For examples of fusing apparatus and process, please see R. Moser, US Patents, 5,601,926, 1997 and 4,197, 445, 1980; T. Mitsuya, D. Hara, R. Yabuki and H. Ueki, Proceedings of IS&T's NIP20, 215, 2004; and S. Hasebe, Proceedings of IS&T's NIP24, 519, 2008.
[3] M. Esterman, S. Dargan, J. S. Arney and B. K. Thorn, Proceedings of IS&T's NIP24, 526, 2008.
[4] W. E. Tenhaeff and K. K. Gleason, Adv. Funct. Mater., 18, 979, 2008; T. P. Martin, K. K. S. Lau, K. Chan, Y. Mao, M. Gupta, W. S. O'Shaughnessy and K. K. Gleason, Surface & Coating Tech., 201, 9400, 2007.
[5] K. Y. Law, H. Zhao and V. Sambhy, NSTI Nanotech Conference & Expo 2009, Technical Proceedings, 3, 177.
[6] L. Feng, Y. Zhang, J. Xi, Y. Zhu, N. Wang, F. Xia and L. Jiang, Langmuir, 24, 4114, 2008.

Biofouling Control Coatings Utilizing VOC-free Biodegradable Chitothane and Nanotubes Containing Non-toxic Natural Product Repellents

Ronald R. Price* and Jonathan Matias**

* Glen Muir Technologies Inc, Stevensville, MD, glenmuirtek@gmail.com
** Poseidon Ocean Sciences, New York, NY, matias@poseidon.com

ABSTRACT

Control of biological fouling is a complex problem if non-toxic means are employed, as bioflims consist of a broad range of species and thus require a range of control strategies. Currently the standard for control of biofouling in the marine and aquaculture industries consists of the use of copper based biocides. A novel new approach combines the idea of ablative coatings in conjunction with encapsulated synthetic analogs of natural fouling-repellent compounds. We report the application of a polysaccharide-based coating that is crosslinked with a water soluble diisocyanate which will ablate on fouling and two novel compounds found to control a wide range of marine and fresh water fouling organisms.

Keywords: biofouling, chitothane, nanotubes, non-toxic

1 Introduction

A range of industries from commercial shipping to power generation and aquaculture facillities depend on an efficient and cost effective means to control biofouling. Uncontrolled fouling on marine surfaces increases fuel use which results in higher direct operating costs in addition to producing increased emissions from marine fuels. In power generation facillities, fouling requires additional maintenance which increases the costs of power generation that results in higher rates for electricity. In aquaculture, the failure to control fouling results in the rapid deterioration of netting, holding tanks and water circulation equipment and also increases disease and parasitism of the species being cultured.

Current approaches consist of the use of polymer based coatings that contain copper as an oxide or more commonly as a copper thiocyanate. Some coatings utilize non-metallic biocides such as Sea Nine 211 or other herbicide compounds that are highly toxic to marine species. Due to increased regulation of metallic and toxic biocides it has become necessary to look for low-toxicity or non-toxic means for the control of biofouling. To optimize biofouling control using non- or low-toxicity materials, it is necessary to craft a total system that is not only biomimitic but is capable of being utilized in a commercial setting at a reasonable cost [1].

Chitosan is the second most abundant natrural polymer in nature. Many marine species such as shrimp and crabs utilize protective shells made of chitin, thus, as a marine-based polymer, many species of fouling organisims produce enzymes that break down chitin into sugars that can then be directly utilized as food.

Treatment of the naturally occuring chitin with hydrocloric acid and sodium hydroxide will modify it to form chitosan which is soluble in a 5% acetic acid solution. By crosslinking the chitosan with modified diisocyanates it is possible to form chitothane coatings. These coatings exhibit a range of degradation rates depending on the degree of crosslinking and thus serve as a replacement for the self-polishing coatings that were developed for tributyltin and copper containing paints.

Designed to ablate only when subjected to chitinase enzymes produced by marine fouling organisms, these coatings exhibit their best performance when challenged by fouling species that produce natural chitin degradating enzymes. Thus the greatest self-polishing rate occurs in an environment where fouling pressure is high. The coating does not readily degrade when fouling pressure is low which can extend the lifetime of the coating under some conditions.

Many currently available commercial biocides and natural repellent compounds unfortunately degrade the performance of the base polymer systems in paints when included at the high rates necessary to achieve a long working life. This poses a problem for the formulator who must utilize large amounts of biocide to obtain a long-lasting, effective coating. To solve this problem both microcapsule-based formulations and nanostructured biocide delivery systems have been developed; however, the utilization of large microcapsules results in increased surface roughness and thus increases drag. On the other hand, nanostructured tubules based on halloysite clay form coatings that are easily sprayable yet result in minimal surface roughness. In addition, as 1:1 aluminosilicate clays, they pose no environmental risk to the aquatic environment or to workers who apply or remove such coatings.

To control fouling it is necessary to find compounds that can disrupt the ability of bacteria and other marine species to colonize a surface. By disrupting their means of inter-species communication and their ability to sense the surface they wish to colonize, the fouling cycle may be disrupted, thereby resulting in a fouling free surface. Two compounds capable of acting in such a manner have been developed:

MR-08 is effective against a range of fouling species, and analogs of 2-FNPK[2] are also capable of acting not only as a repellent for marine species but also as an anti-bacterial control against many common bacteria including e-coli.

By combining an ablate-on-fouling coating with a delivery system that acts as a depot to protect the coating from degradation of physical properties such as adheasive strength and scratch resistance, it becomes possible to load the coating with active agent to high levels while still maintaining the physical properties of the coating. This is due to the ability of the halloysite nanotubes to form a composite with the chitothane polymer, thereby increasing abrasion resistance. In addition to providing protection for the delivery system, the chitosan protects the active agent(s) from early degradation by the chemistry of the coating itself or the external environment, thus extending the performance of the active agents. The utilization of natural products that are non-toxic also reduces the build up of resistance in target species which is often induced by toxic agents and thus should result in a longer useful lifetime for the coating.

2 Materials and Methods

Chitosan (Sigma-Aldrich); the polyisocyanate was synthesized in our lab; MR-08 (Poseidon Ocean Sciences); and 2-FNPK (Naval Research Laboratory) were utilized in the production of the coating.

Chitosan was solubilized in a 5% acetic acid solution in deionized water. Chitosan was added at a rate of 3%-5% by weight and was subjected to a high shear mixing environment for 30 min. after which it was filtered through a 300 thread count pillow case to remove any remaining shell or other contaminates from the chitosan solution.

The blocked polyisocyanate is produced by heating deionized water to 78°C and saturating that solution with sodium metabisulfite then cooling that solution to 22°C and adding dropwise tolune diisocyanate at a 1:2 ratio to the sulfite solution. Once reacted (stirred for 1 hr.), acetone is added sufficient to precipitate the resulting blocked isocyanate from the water solution which is then filtered through No. 4 Wattman filter paper utilizing a buchner funnel. Once recovered it is again rinsed with fresh acetone and allowed to air dry.

The active agents (MR-08 and 2-FNPK) are incorporated into halloysite nanoclay that has been surface treated to facilitate the inclusion of the active agent(s). The treated clay is allowed to adsorb the chemical and then subjected to a 30mm/Hg vacuum which entrapps the active agent. Following the entrapment the clay and active agent are then added to the chitosan solution at a rate between 30% and 50% by weight utilizing a high sheer electric laboratory mixer. Following this step the blocked isocyanate is added at a rate of between 1% and 5% by weight of the chitosan in solution. The surfaces to be protected are then coated and allowed to air dry. As an alternative heat may be applied to increase the reaction rate.

3 Results

Test panels were exposed either in the marine environment or in a laboratory fish tank that contained a wide variety of fouling species derived from the local marine environment. Following exposure for a period of months the panels were withdrawn for visiual testing of the extent of fouling coverage.

After several months of exposure the panels were removed from the water and examined for fouling. MR-08 was tested in the Indian Ocean at Tuticorin, India [Fig. 2] and 2-FNPK was tested in our laboratory fouling tanks [Fig 3].

Neither compound was found to control marine algae or grasses but both offer excellent performance against both marine bacterial fouling and larger fouling species such as barnacles and briazoans. The addition of 1% by weight zinc oxide to the 2s-FNPK coating eliminated algal fouling [Fig 3].

References

[1] Novel Biodegradable Biofoulijng Control Coatings and Method of Formation, Ronald, R.Price, Eric R. Welsh, Linda Passaro, US Patent Application Publication US 2009/0162312.

[2] Process for the Preparation of 21-Furyl-N-Pentylketone and Longer Chain Analogs., Linda Passaro US Patent Application Publication uas 2006/0094888.

Figures

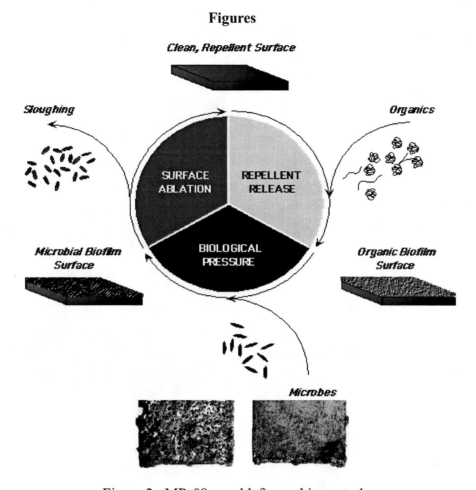

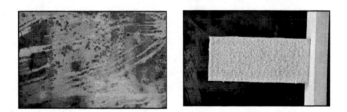

Figure 2: MR-08 panel left panel is control

Figure 3: Analog of 2-FNPK (4 months) Right Panel MR-08 (5 months exposure).

Nanoscale Silver Coatings on Polymer Fibers

M. Amberg[*], E. Körner[*], D. Hegemann[*] and M. Heuberger[*]

[*]Empa, Swiss Laboratories for Materials Testing and Research, *Advanced* **Fibers**
CH-9014 St. Gallen, Switzerland, manfred.heuberger@empa.ch

ABSTRACT

The low pressure plasma coating technology offers new possibilities when applied to fibers or textile fabrics. Two important strengths of plasma coatings are i) typically less than 200nm coating thickness, which does not affect the fiber mechanical properties, and, ii) the availability of a wide spectrum of different coating materials and morphologies.

This paper presents two different examples of plasma coatings, both are silver containing; we report some of their properties and sketch the respective application windows. The first coating is a homogeneous plasma silver metallization while the second example is a highly cross-linked plasma polymer containing variable amounts of silver nano particles.

Keywords: fibers, plasma, metallization, polymerization, nano-composite

1 PLASMA COATING OF FIBERS AND FABRICS

Nanometer thick metal coatings are widely used on different polymer substrates like foils; e.g. for anti-statics, optical effects, diffusion barriers and more [1]. The physical and chemical properties of such plain coatings are well understood.

We investigate the characteristics and properties of nanometer metal films and metal nano particle containing coatings on fibers and textiles. Fibers present a special geometrical challenge to the coating process, particularly if the process is a directed one [2]. For the same reason it is also difficult to reach shadowed areas of fiber material in a woven fabric.

A convenient way to produce homogeneous nanometer coatings on fibers and textiles is using a low-pressure plasma deposition [3]. In this type of plasma one ionizes a gas of selected chemical composition in an alternating electric field (i.e. RF or microwave). A reduced pressure in the process reactor is used to insure a useful diffusion length of the chemical radicals and control the extension of the plasma zone. The substrate temperature is moderately raised to 50-60°C, which represents a mild condition; thus not deteriorating the fiber mechanical properties.

Plasma coating reactors come in various shapes and geometries to accommodate different types of substrates. The here used reactors are of two types i) batch processing fabric stripes, or, ii) endless fiber or yarn coater, which delivers the substrate fiber from air to air via an orifice and a differential pumping configuration. Fig1 below shows the example of the cylindrical plasma zone inside an endless fiber coater.

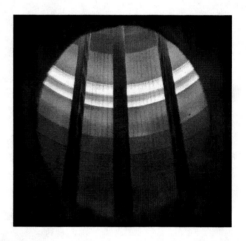

Fig1: Snapshot of synthetic fibers passing through a low-pressure plasma zone. A cylindrical target allows a homogeneous coating of the fibers.

The plasma deposition rates are limited to a typical range of 1-100nm per minute. Therefore, plasma coatings are well suited to produce thin coatings in a range 10-200nm. In order to increase the dwell time of the fiber inside the plasma zone, one can use configurations with multiple passes. The example shown in Fig1 provides a maximum of 64 passes.

It is important to note that the nanometer thickness of such coatings ensures that the "textile character" of a fiber or fabric remains intact. This is an important asset for the subsequent textile processing (e.g. weaving, stitching...) and also for the haptics of the final textile.

We present two different types of plasma processes to produce silver containing coatings on fibers. These are i) plasma sputter process to produce homogeneous metal films and ii) a combination of plasma sputter process with a plasma polymerization process to produce a cross linked polymer layer containing silver nano-particles (i.e. Ag nano-composite).

2 HOMOGENEOUS SILVER COATINGS

At a plasma base pressure of nominally 1Pa we introduced argon, which is ionized by irradiation of RF. The silver material is present inside the plasma reactor in form of a so-called silver target – a several mm thick plate of silver with magnets mounted in such a way that the ions are accelerated in circular trajectories towards the target surface (i.e. magnetron), where they sputter silver atoms and clusters into the adjacent space.

When a fiber is passed through the plasma zone a condensation of the metal clusters on the substrate surface takes place. There is a finite surface mobility of silver atoms on the fiber surface. A continuous rotation of the fiber along its axis is however needed to achieve a homogeneous metal coating all around, as the one shown in Fig2 below.

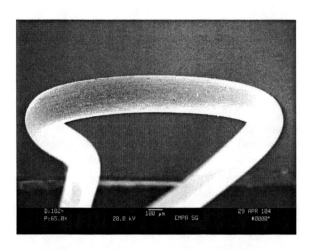

Fig2: SEM image of a polyester monofilament (PES 150f1) coated with a highly flexible, homogeneous silver coating of 250nm thickness.

It is important to note that the smoothness of such thin metal coating is directly dependent on the quality of surface cleaning prior to the silver deposition. An argon/oxygen plasma zone at the entrance of the plasma reactor is used to remove residual surface contaminations. A prominent source of fiber surface contamination stems from processing oils (~0.1 wt%) of so-called "avivage" or "appreture". These substances often contain a number of components that are used for fiber lubrication and anti-statics during and after the melt-spinning process [2].

The same plasma sputtering process can also be applied to a multifilament fiber or yarn. Again, a homogeneous coating can be achieved via spreading and long-axis rotation of individual filaments in the plasma zone. Fig3 below shows an example of a polyester multifilament fiber processed in a silver plasma sputtering process. The silver quantity (per length) is now distributed over a larger surface area, thus reducing the nominal film thickness by a factor of roughly 5x over the monofilament equivalent.

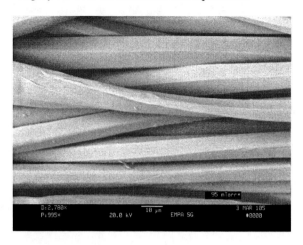

Fig3: SEM image of a polyester multifilament (PES 150f48) coated with a highly flexible, homogeneous silver coating of nominally 50nm thickness.

The electrical conductivity of such silver-coated fibers was investigated as a function of the silver quantity (per length) [4]. It was found that plasma metalized fibers can reliably cover a wide range of different conductivities and possible applications starting from EM-shielding or power supply ($Z < 1\Omega/cm$) to signal conduction ($Z < 100\ \Omega / cm$), and anti-static applications ($Z < 10^8\ \Omega / cm$). A comparision with the electrical properties of bulk silver is illustrative. Fig4 depicts how the specific resistance of silver coated fibers varies with the deposited amount of silver per unit length. For convenience, results are related to the theoretical limit (dotted line) of a silver wire with the same Ag quantity per length.

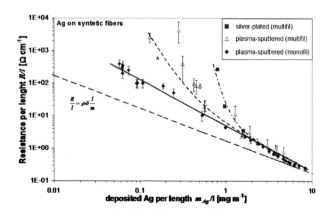

Fig4: Resistance of silver coated fibers versus the amount of silver deposited. The dashed line corresponds to the theoretical minimum, which would be reached for a film of ideal morphology (i.e. no defects).

All silver coated fibers have a resistance exceeding the theoretical minimum, which is indicative of small defects in the film. We note that at higher silver thickness these defects play a vanishingly smaller role, as expected. Furthermore, the monofilament outperforms the multifilament at lower silver quantities – again this is because defects play a more important role at the 5x smaller film thickness on the multifilament. For comparison, we have also plotted the resistivity of conventional electrochemically plated silver fibers that exhibit a much rougher silver film due to the different deposition technique [5]. At lower silver quantities, the plated silver fibers have therefore a comparably higher specific resistivity.

3 NANO-COMPOSITE SILVER/POLYMER COATINGS

A second type of silver containing plasma coating can be realized by the introduction of carbon containing monomer gases (e.g. ethylene) into the plasma chamber during the argon plasma sputtering process [6,7]. The pressure inside the reactor is thus raised to 10Pa. In order to accelerate argon ions towards the target under these conditions, an asymmetric electrode setup can be used as shown in Fig5 [8]. The different sizes of the electrodes effectively produces a dynamically induced DC bias voltage between the electrodes that is used to accelerate the ions towards the metal target. Simultaneously, the monomers fed into the reactor will be ionized and the produced hence chemical radicals in the plasma zone will deposit onto any substrate to form a highly cross-linked polymer coating. The Ag clusters that are simultaneously condensing on the substrate form silver nano particles according to the growth mode that is dictated by the local conditions (i.e. interfacial energies and surface diffusion kinetics).

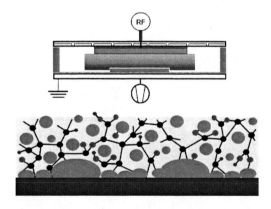

Fig5: An asymmetric plasma reactor setup can be used to simultaneously sputter silver metal and deposit a highly cross-linked polymer matrix. The growth mechanism of the silver nano-particles is governed by interface energetics and surface diffusion.

The number and size distribution of the thus produced nano-particles can be controlled by plasma input power and gas composition. It is to note that the silver contained in these coatings can be made in a highly oxidized form, exhibiting a very large specific surface area. These combined properties are responsible for a bust-like release of silver I ions from such coatings when exposed to an aqueous environment. As shown in Fig6 below, a fast initial silver ion release is a unique characteristic for this form of nano-composite.

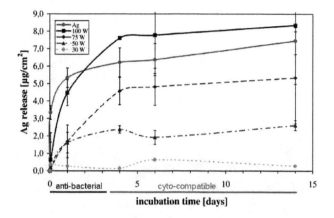

Fig6: The silver ion release kinetics of Ag nano-composite coatings depends on the size distribution of the silver nano-particles, their oxidation state as well as the water and proton conductivity of the polymer host matrix.

Furthermore, we have carried out bacterial colonialization tests with such surfaces that revealed a strong antibacterial effect for the first 24 hours. Once the silver containing supernatant was exchanged against fresh solution, it was found that these surfaces provide the grounds for cells to proliferate. The cell compatibility of these nano-composites was further enhanced by introducing nitrogen containing monomers into the plasma reactor, which led to the formation of amines and amides in the cross-linked polymer substrate. Such nitrogen rich coatings were found to promote cell growth comparable to a typical cell growth substrate [9].

4 DICUSSION AND CONCLUSIONS

We have presented two examples of silver containing nano-coatings on fibers and fabrics. These coatings are produced using the low-pressure plasma technology that allows control over chemistry and morphology of the coatings in ways that are hardly possible with other methods.

The example of homogeneous silver coatings has demonstrated that it is possible to coat literally any fiber material with a flexible, well-adherent metal film. We have undertaken a number of chemical and mechanical tests with the coated fibers and fabrics to assure a high electro-mechanical quality. In addition to the electrical characterization mentioned in this paper we have also performed extensive tests for wash fastness and sweat fastness of these homogeneous silver coatings – in view of possible application of such fibers in wearable electronics. Future developments on the basis of the here presented fiber will include smart additional layers to add better chemical inertness or functional electronic coatings (e.g. insulation, organic semiconductors, sensors, and also multiple electrodes on a single fiber).

The silver nano-composite layers are now further explored as tunable antibacterial coatings on (textile) implants. Particularly the fact that we are able to install an application window with an initial anti-microbial burst, followed by a surface exhibiting cyto-compatible conditions is an interesting way to approach the common problem of post-implantation infection, which causes a significant amount of cost and patient discomfort.

REFERENCES

[1] J. Reece Roth, Industrial Plasma Engineering, Vol. **2**, IOP, Bristol, 2001, p451.

[2] M. Amberg et al, J. Adhesion Sci Technol, **24** 123-134, (2010).

[3] Plasma technologies for textiles, ed. R. Shishoo, Woodhead, Cambridge, (2007).

[4] M. Amberg et al., Plasma Process. Polym, **5**, 874-880, (2008).

[5] D. Hegemann et al., Mater. Technol., **24**, 41-45 (2009).

[6] H. Biederman et al., Pure Appl. Chem., **60**, 607-618 (1988).

[7] D. Hegemann et al., Progr. Organic Coat., **58**, 237-240 (2007).

[8] E. Körner et al., Plasma Process. Polym., **6**, 119-125 (2009).

[9] S. Lischer et al, Biomaterials, **submitted**, (2010).

Tribological Studies of Conventional Microcrystalline and Engineered Near-Nanocrystalline WC-17Co HVOF Coatings

G.C. Saha*, A. Mateen* and T.I. Khan*

*Department of Mechanical & Manufacturing Engineering, University of Calgary, 2500 University Drive N.W., Calgary, AB, T2N 1N4, Canada, gsaha@ucalgary.ca

ABSTRACT

Tungsten carbide based cermet composite coatings possess the unique surface protection properties enabling their increasing use in the oil sands industry to combat severe sliding and high stress abrasion wear and other types of slurry abrasion and erosion. Thermal spraying such as the use of high velocity oxy-fuel (HVOF) technique has increased the wear life of coatings based on WC-Co system when compared with high velocity air plasma (HVAP) technique. The ability to deposit HVOF coatings with a nanocrystalline structure provides even higher wear resistance and fracture toughness over those achieved using conventional microcrystalline coatings of the same composition. In this research, conventional microcrystalline and engineered near-nanocrystalline WC grains were used to obtain two powder feedstocks of WC-17Co composition. The chemical analysis of the powders by XRD and EDS showed that the novel near-nanocrystalline powder had an engineered particle formation with 'duplex' cobalt layer surrounding WC core, comparing a typical spray-dried microcrystalline WC-17Co particle. Characterization of the coatings by XRD and SEM suggested that in case of near-nanocrystalline WC-17Co coating an extensive decarburization of WC during the spraying of the powder may had been prevented due to the use of the duplex Co coated WC-17Co powder, a novelty in this research. Tribological properties of the coatings were tested by atomic force microscopy, Vickers microhardness, fracture toughness, and two-body pin-on-plate (ASTM G133-05) abrasive wear tests. The near-nanocrystalline WC-17Co coating had shown significantly better performance than the conventional microcrystalline WC-17Co coating.

Keywords: Cermets; Near-nanocrystalline WC-17Co powder; HVOF thermal spraying; Two-body abrasive wear; Fracture toughness

1 INTRODUCTION

Tungsten carbide cobalt coating has been extensively used for wear resistant applications. Tungsten carbide having a Vickers hardness of about 2242 adds to the hardness of the composite while cobalt acts as a softer binder and hence adds to the toughness of the composite. High velocity oxy-fuel (HVOF) process has advantages of low dwell time, lower process temperature and high impingent velocity and, therefore, results in a higher adhesion and higher phase retention during coating [1]. Even then there are chances of decarburization of WC and formation of lower carbides such as W_2C, $(W, Co)_6C$ etc. W_2C has a higher brittleness and therefore degrades the quality of the coating. To minimize this degradation of WC, a 'duplex' Co coated cermet powder was used in this study. Nanocrystalline materials have the advantage of increased hardness and toughness simultaneously which otherwise are exclusive properties [2]. Recent work by Khan *et al.* [3] showed that near-nanocrystalline WC-Co coatings are more resistant to abrasive wear than microcrystalline coatings and that HVOF sprayed micro- and near-nanocrystalline coatings have superior abrasive wear resistance compared to heat treated and untreated steels [4]. Other researchers have found that during HVOF spraying the nanocrystalline WC decarburizes to a higher extent than the microcrystalline WC [5]. Correspondingly, a higher wear rate was reported during sliding in nanocrystalline coatings than in microcrystalline coatings. On the other hand, Stewart *et al.* and others reported that the abrasive wear resistance of nanocrystalline WC-Co coatings was 1.4 to 3.1 times better than that of the coatings obtained by spraying microcrystalline powders [6,7].

In this study, a novel duplex Co coated near-nanocrystalline WC-17Co powder was used for HVOF coating and the hardness, fracture toughness and two-body abrasive wear tests were conducted and compared with those of the conventional microcrystalline WC-17Co cermet powder sprayed coating. The extent of WC decarburization was also investigated and correlated with the wear behavior of the coatings.

2 EXPERIMENTAL PROCEDURE

Two different cermet powders were used as feedstock in the investigation: engineered near-nanocrystalline WC-17Co (provided by MesoCoat Inc., Euclid, OH) and microcrystalline WC-17Co. These were near-nanocrystalline WC-17Co (provided by Sulzer Metco, Westbury, NY). The chemical analysis of the coating powders was carried out using a Noran 8-channel energy dispersive spectrometer (Table 1). Spray deposition of the micro- and near-nanocrystalline powders was carried out at Hyperion Technologies Inc. in Calgary, Canada, under equivalent conditions using a Sulzer Metco (Westbury, NY) Diamond Jet Hybrid DJ2700 HVOF torch. Methane was used as a fuel gas utilizing a DJ2700 aircap. The deposition parameters are shown in Table 2. The ratio between methane-to-oxygen in the fuel mixture was maintained at 0.68 to create an optimum crystalline phase structure within the final coatings. A traverse speed of the torch across the substrate of approximately 0.2 m/s was used and approximately 5 µm thickness/pass was deposited. Coatings of 210 µm thick were produced for microscopic examination and coatings of 450 µm thick were deposited for two-body abrasive wear test analysis. The coating thickness measurement was performed according to ASTM E376-06 using a thickness gauge 3000FX model.

Table 1: Coating powder characteristics

Characteristics	Coating powder	
	Micro WC-17Co	Near-nano WC-17Co
Element, wt.%	Balance W, 17-Co, 5-C, 1-Others	Balance W, 18.5-Co, 5.6-C, 1.11-O
Manufacturing route	Agglomerated /sintered	Clad/conversion
Shape	Mostly spherical	Mostly spherical

Table 2: Deposition parameters used in the HVOF spraying

Deposition parameter	Parametric value
Shroud gas (air), slpm*	1742
Oxygen, slpm	1346
Methane, slpm	918
Carrier gas (nitrogen), slpm	60
Powder feed rate, g/min	38
Spray angle, degrees	90

*Standard litre per minute

A hardened AISI 1118 steel 'wear plate' with a microstructure and properties of that of a water-quenched and tempered C-Mn steel was received from an oilfield in Fort McMurray in Alberta, Canada and used to produce samples of cross-section 9 x 7 x 5 mm. The steel had a chemical composition in wt.% of 0.18-C and 1.4-Mn.

The wear tests of the coated and uncoated samples were carried out under 10, 20, 40 and 60 N loads. The wear rates were studied, as per ASTM G133-05 standard, for varying sliding distance. The microhardness values of the coatings and uncoated steel were determined using a microhardness tester (Micromet II, Buehler) using 300 g load. Atomic force microscopy (AFM, Park system XE100) and scanning electron microscopy (JEOL,JSM-8200) were carried out for microstructural characterization. XRD analysis using Rigaku Multiflex (Rigaku, Japan, model Rigaku multiflex ZD3609N), using a CuKα radiation source with a step size of 0.02°, a step time of 5 s and a 2θ scan window from 10° to 90° was used. XRD study was conducted in order to determine the phase changes that took place during the process of coatings. The fracture toughness for micro- and near-nanocrystalline coatings were calculated using Rockwell hardness tester equipped with a diamond cone and an applied load of 150 Kg.

3 RESULTS AND DISCUSSION

The SEM micrographs in Figures. 1(a) and (b) show differences between the dispersed WC grain sizes in the microcrystalline and near-nanocrystalline WC-17Co as-received powders, respectively. The near-nanocrystalline powder used in this study had a mean WC grain size of 427 nm with a standard deviation of 120 nm. The conventional microcrystalline powder particle size was in the range 15-35 µm with a WC grain size of 3-5 µm.

(a)

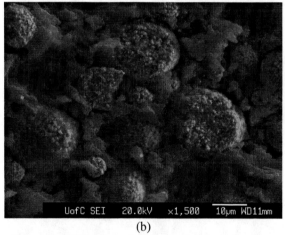
(b)

Figure 1: SEM micrographs showing cross-sections through: (a) microcrystalline and (b) near-nanocrystalline particles.

XRD analysis taken from the sprayed coatings showed a significant difference in phase composition compared to the powders before HVOF spraying, see Figure 2. The presence of W_2C and W phases is clearly seen in both types of coatings, whilst peaks of metallic Co is present in near-nanocrystalline coating. The proportion of WC transformed to W_2C and W phases was higher for the microcrystalline coating compared with the near-nanocrystalline coating. Furthermore, peaks of W_3C and mixed carbides such as η-Co_3W_3C were also observed for the microcrystalline coating.

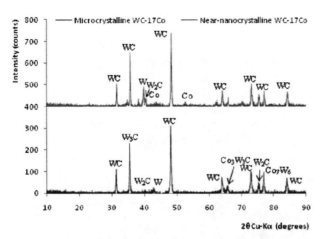

Figure 2: XRD analysis of the sprayed powders.

The microstructures of the coatings were examined using the SEM, and the micrographs of the microcrystalline and near-nanocrystalline as-sprayed coatings are shown in Figures 3(a) and (b), respectively. The near-nanocrystalline coating appeared to be denser and free of micro-voids or porosity compared to the microcrystalline coating.

(a)

(b)

Figure 3: SEM micrographs showing the sprayed: (a) microcrystalline and (b) near-nanocrystalline coatings.

The AFM surface profile measurements taken from the sprayed coatings showed that the near-nanocrystalline WC-17Co coating resulted in a lower surface roughness value compared to the microcrystalline coating (see Table 3). The porosity measurements for the coatings are given in Table 3. It is seen that the near-nanocrystalline coating resulted in a lower porosity than that of the conventional coating.

Table 3: A comparison of characteristics for the microcrystalline and near-nanocrystalline coatings

Characteristic	Coating	
	Micro WC-17Co	Near-nano WC-17Co
Surface roughness, Ra, μm	3.98 ± 0.46	4.17 ± 0.75
Porosity, %	> 1	2.29 ± 0.16
Vickers microhardness, VHN	1440	1237
Fracture toughness, MPa.m$^{1/2}$	25.12 ± 0.8	19.04 ± 1.1

The microhardness and fracture toughness values for the coatings are shown in Table 3. The change in microhardness as a function of depth through the steel substrate and the coated surface is shown in Fig. 4. An almost uniform hardness value of 348 VHN was obtained for the hardened steel. The hardness values were significantly higher for coatings giving a value of 1237 VHN for the microcrystalline coating and a value of 1440 VHN for the near-nanocrystalline coating. Both coatings were sprayed same conditions and a hardness increase of more than 16% for the near-nanocrystalline over the microcrystalline coatings was achieved.

The Anstis *et al* [8] formula was developed by researchers to determine the indentation fracture toughness of coatings. The measurement of the cracks and substituting values into the formula gives a quantitative value for fracture toughness:

$$K_{IC} = 0.016[E/H]^{1/2}Pc^{-3/2} \qquad (1)$$

Where K_c is the fracture toughness [MPa*m$^{1/2}$], E is the elastic modulus [GPa], H is the hardness [GPa], P is the indentation load [N], c is the crack length measured from the indentation center [μm]. In Table 3, the fracture toughness calculated for the microcrystalline WC-17Co coating based on crack model (Equation 1) gave a value of 19.04 MPa*m$^{1/2}$ and was in consistent with earlier published data [9]. More importantly, the duplex near-nanocrystalline WC-17Co coating gave a toughness value of 25.12 MPa*m$^{1/2}$. This represented a 31% increase in fracture toughness for the near-nanocrystalline coatings compared to the microcrystalline coating.

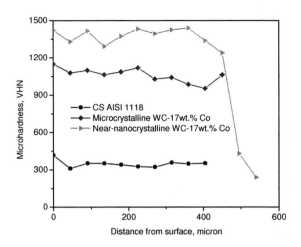

Figure 5: Microhardness depth profiles of hardened steel and micro- and near-nanocrystalline coating surfaces.

The sliding wear rate of the hardened steel and coated surfaces was compared using a range of applied loads using a sliding speed of 1.68 m/min, as shown in Fig. 6. It is evident from the wear test graphs that the best wear resistance was recorded for the near-nanocrystalline coating followed by the microcrystalline coating. The worst wear resistance was shown by the uncoated CS AISI 1118 steel. Comparing the steady state wear rates for the coated and uncoated materials, a value of 5.40 mm³/m was obtained for the hardened steel, and a lower value of 3.34 mm³/m was recorded for the microcrystalline WC-17wt.%Co coating. The wear rate of the near-nanocrystalline coating was extremely low with a value of 0.32 mm³/m, which was approximately 10 times more wear resistant than the conventional microcrystalline coating.

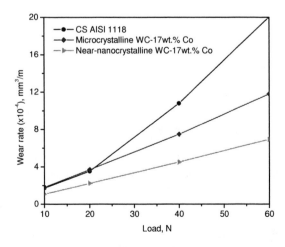

Figure 6: Wear rate as a function of load for hardened steel and microcrystalline and near-nanocrystalline coatings.

4   CONCLUSIONS

Wear is not a material property. It is the response of a material towards a particular system [10]. Therefore, one material may have different response to different environments. Wear mechanism and mode also depend upon the structure of the system where it takes place. In this research, two-body abrasive wear behavior of microcrystalline and engineered near-nanocrystalline WC-17Co coatings was studied. It was observed that the wear rate changed with changing the sliding distance and load. The abrasive wear mechanism was studied and the wear rate was correlated with properties such as coating microhardness, WC grain structure, distribution of Co content in the coating composition, fracture toughness, surface roughness as well as coating phase formation/retention during the HVOF thermal spraying. Sliding velocity, temperature, wear medium and wear environment were kept constant during the tests.

The wear rate changed with different sliding distances. It was recorded by optical microscopy that the grain structure of the coatings changed during sliding, thereby creating an atmosphere in which strain hardening was taking place. Significant strain hardening was also observed when microhardness tests were conducted near the grooves in the near-nanocrystalline coating. SEM images of the grooves revealed that the wear surface was fractured in the case of microcrystalline coating, while the near-nanocrystalline coating surface was plastically deformed.

ACKNOWLEDGEMENTS
The authors would like to thank the Natural Sciences and Engineering Research Council of Canada (NSERC) for providing financial support to carry out this research.

REFERENCES

[1] B. Wielage, A. Wank, H. Pokhmurska, T. Grund, C. Rupprecht, G. Reisel and E. Friesen, Surf. Coat. Technol. 201, 2032-2037, 2006.
[2] R. Valiev, Nature 419, 887-889, 2002.
[3] T.I. Khan, G.C. Saha and L.B. Glenesk, Surf. Engg. 2009, In Press.
[4] G.C. Saha, T.I. Khan and L.B. Glenesk, J. Nanos. Nanotechnol. 9, 4316-4323, 2009.
[5] P.H. Shipway, D.G. McCartney and T. Sudaprasert, Wear 259, 820-827, 2005.
[6] D.A. Stewart, P.H. Shipway and D.G. McCartney, Wear 225-229, 789-798, 1999.
[7] J.M. Guilemany, S. Dosta and J.R. Miguel, Surf. Coat. Technol. 201, 1180-1190, 2006.
[8] G.R. Anstis, P. Chantikul, B.R. Lawn and D.B. Marshall, J. American Ceramic Society 64(9), 533-538, 1981.
[9] D. Han and J.J. Mecholsky Jr., J. Mater. Sci. 25, 4949-4956, 1990.
[10] K. Kato, J. of engineering tribology 216, 349-355, 2002.

A generalized approach for the surface engineering of nanoparticles in suspension for highly efficient hybrid capture of bio-molecules

T. Sen[*&**] and I. J. Bruce[**]

[*] Current address: Centre for Materials Science, School of Forensic and Investigative Sciences, University of Central Lancashire, Preston, United Kingdom, tsen@uclan.ac.uk
[**] Department of Biosciences, University of Kent, Canterbury, United Kingdom, i.j.bruce@kent.ac.uk

ABSTRACT

Surface engineering of nanoparticles with hydrophilic surfaces has been carried out in suspension by a strict control of the surface ad-layer of water surrounding the nanoparticles using a generalized strategy called tri-phasic reverse emulsion (TPRE). This approach produces an optimal density of surface amine groups with monolayer patterns on the nanoparticles surfaces. Surface functionalized nanoparticles when chemically conjugated to oligonucleotide ($5'NH_2$-dC_6-dT_{25}), the oligonucleotide conjugated bio-nanoparticles exhibited exceptional hybridization efficiency to fluorescence labeled complementary oligonucleotide (dA_{25}). This strategy has overcome the problem associated with the hydrolysis and condensation of aminosilane molecules during the surface functionalization of nanoparticles in suspension and can have improved efficiency in applications ranging from medicine to pharmaceutics to materials science.

Keywords: Surface functionalization, Aminosilane, nanoparticles, DNA hybridization, Tri-phasic reverse emulsion, magnetic bio-separation

1 INTRODUCTION

Engineering atomic and molecular nanostructures on flat surfaces is extensively reported [1,2] whereas that in the case of nanoparticles in suspension is not fully understood as the process can be difficult to control due to the aggregation of nanoparticles in suspension, solvation of nanoparticles and an uncontrolled reaction of reactant molecules in the suspension. Surface functionalization of flat surfaces by aminosilane is a controlled process due to the elimination of water from the surface by a simple drying step [3-6] whereas in suspension phase they are difficult to control due to a series of consecutive reactions of aminosilane molecules in water to form polymeric species (see reaction scheme 1). Under heterogeneous conditions e.g. in the presence of a solid support, the reactive monomers could react with native surface –OH groups to form a monolayer of –NH$_2$ functionalities or they can undergo self polymerization before they condense onto the surface. The oligomers or polymers can further react on the surfaces to form –NH$_2$ functionalized multilayer. The

rate constants (k_1, k_2, k_3, k_4, k_4', k_5) of these various reactions control the functionalization of solid surfaces. In order to produce a uniform monolayer of –NH$_2$ functionalized surfaces, ideally k_1 should be equal to k_4 and other rate constants (k_2, k_3, k_4' and k_5) should be zero.

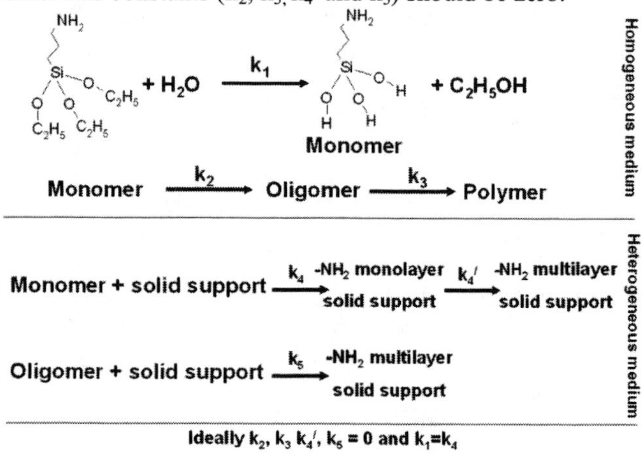

Scheme 1 Hydrolysis and condensation reaction of aminosilane molecules in water under homogeneous and heterogeneous (in the presence of solid surfaces) condition

Moreover, the flat surfaces have only two dimensions so that the silane molecules only interact through the z-axis onto the surfaces. Bein and co-workers [4] reported that 3-aminopropyl triethoxy silane (APTS) molecules formed a monolayer (surface amine density of 5.3×10^{14} silanes/cm^2; thickness of 7Å) on Quartz surfaces under gas phase reaction. Atomic layer deposition (ALD) technique under gas phase is also reported [7, 8] to be a controlled process where the water molecules were excluded from the reaction. Gas phase deposition by ALD technique or surface engineering of flat surfaces by controlled water condition is well studied; however, surface engineering of hydrophilic nanoparticles in suspension using aminosilanes is not fully studied.

Surface functionalization using aminosilane is reported to be an essential step for the chemical conjugation of bio-molecules. Applications in nano and nanobiotechnology [9, 10] rely on surface functionalization step using aminosilane. However, no one has ever reported the mechanism of formation of aminosilane layers on the nanoparticles surfaces and its importance in improving the efficiency in such applications. Sen *et al* have reported [11] the difficulty for the functionalization of nanoparticles in suspension using aminosilane. Herein, we report that tri-phasic reverse emulsion (TPRE) approach can be used as a generalized strategy for surface engineering of hydrophilic nanoparticles is suspension for various applications such as highly efficient hybrid capture of bio-molecules.

In the actual process, amino molecules were added to a tri-phasic (nanoparticles-surface water-organic solvent) reverse emulsion of hydrated magnetic nanoparticles in an organic solvent (toluene), in the presence of a common biocompatible non-ionic surfactant, (Triton X100). The bulk water was magnetically separated from a core-shell silica magnetite suspension and the hydrated nanoparticles were dispersed in toluene in the presence of Triton X100. As aminosilane is soluble but do not hydrolyse or self condense in toluene, hence, it will remain unreacted in the continuous toluene phase. Aminosilane can only hydrolyse and subsequently condense onto the surfaces of the nanoparticles where there is water present in the system. In the present system this water is present only as adsorbed water on the surfaces of the nanoparticles. This permits the aminosilane to react in a controlled fashion and form an ordered uniform layer of aminosilicate (Scheme 2). In this approach therefore the process that degrade the quality of the nanoparticles and the functionalisation i.e. aggregation of nanoparticles and self condensation of APTS monomers to oligo/polymers are eliminated.

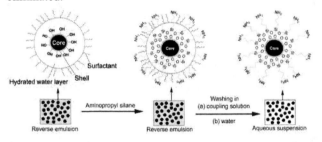

Scheme 2 Surface engineering of core-shell nanoparticles in suspension using TPRE approach

2 EXPERIMENTAL

2.1 TPRE approach of surface engineering of nanoparticles

150mg of either core-shell silica-magnetite nanoparticles (I) and diamagnetic silica spheres (II) were collected either magnetic separation or centrifugation. 30ml of toluene and 5gm of triton X100 were added and the mixture shaken to form a tri-phasic reverse emulsion. APTS was added to the emulsion to a final concentration of 2% (w/v) and allowed to react in a 100mL glass reactor fitted with condenser at 50°C in an oil bath for 5 hrs with stirring. The amount of surface water was controlled by washing stepwise with water miscible solvent (dry tetrahydrofuran). The suspension was washed with coupling solution (0.8% v/v glacial acetic acid in dry methanol) three times and

stored at RT in the same solution. Surface amine densities were determined by colorimetric assay using 4-nitrobenzaldehyde [12]. The total amine densities were determined by combustion (CHN) analysis.

2.2 Covalent coupling of single stranded oligonucleotides (5'-NH$_2$ dC$_6$ dT$_{25}$) to nanoparticles

(1×)SSC and (13×)SSC buffers were prepared by diluting a stock solution of (20×)SSC buffer (175.3g NaCl, 88.2g sodium citrate, 1L H$_2$O, pH 7.4) with distilled, deionized water, adjusted to pH 7.4 and autoclaved before use. Glutaraldehyde solutions were prepared immediately before use. 2mg of aminosilanized nanoparticles were washed (×3) with 1ml of coupling buffer (1×SSC buffer, pH 7.3) for 2 minutes at 18°C. After removal of the supernatant, 0.5ml of a 5% v/v glutaraldehyde solution in coupling buffer were added and the suspension incubated for 3 hours with end-over-end rotation at 18°C. The material was subsequently washed (×3) with 1ml coupling buffer to remove excess glutaraldehyde. 1ml of a 3.3μM solution of 5'-amine modified oligo-dT$_{25}$ were added and the mixture left incubating overnight whilst shaking. The oligo-modified nanoparticles were then washed once with coupling buffer and placed in 0.8ml of NaBH$_3$CN solution (0.03% w/v in coupling buffer) for 30 minutes at 18°C. After this the material was washed (×3) with 0.8ml of coupling buffer and finally resuspended in 200 μl of the same.

2.3 DNA hybrid capture experiments (model assay)

1mg of oligo-dT$_{25}$ modified nanoparticles was washed twice with 0.5ml of water and resuspended in and heated to 80°C for 4 minutes. 1ml of a 1.5μM solution of 5'-fluorescein modified oligo-dA$_{25}$ in (13×)SSC /0.05%BSA was added to the particles and the suspension incubated with gentle shaking for 30 minutes at 18°C. The supernatant was removed and kept for analysis. After washing (×3) with 1ml of 13×SSC, 200μl of water was added to the particles and the suspension heated to 85°C for 4 minutes to disassociate the annealed/hybridized oligo-dA$_{25}$ sequences. The supernatant was removed and analyzed by fluorescence spectrophotometer (excitation 460nm and emission 515nm).

3 RESULTS AND DISCUSSION

Transmission electron microscopy (TEM) showed both shell-core silica-magnetite (I) and model silica (II) materials were spherical in morphology (figure 1a). The diameter of shell-core silica-magnetite was measured to be around 40nm whereas the diameter of model silica nanoparticles was measured to be around 400nm. The silica nanoparticles (II) were used as a diamagnetic analog of superparamagnetic core shell silica-magnetite.

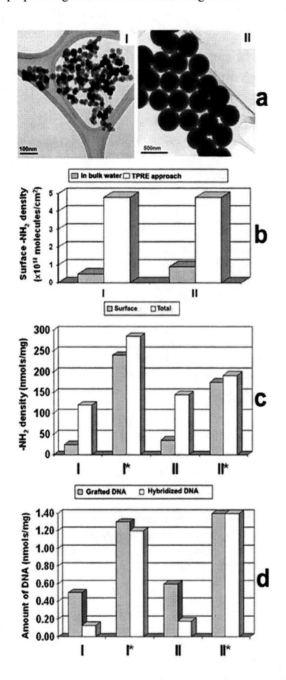

Figure 1 (a) TEM images of core-shell silica-magnetite (I) and silica spheres (II). (b) Surface amine density of amine functionalized nanoparticles of shell-core silica-magnetite (I) and spherical silica (II) measured by colorimetric assay. (c) Surface to total amine density of nanoparticles (I, II), * indicate TPRE approach. (d) Amount of oligonucleotide (5'NH$_2$-dC$_6$-dT$_{25}$) grafted on nanoparticles (I, II) and the hybridization efficiency of fluorescent labeled complementary oligonucleotide (dA$_{25}$).

The Brunauer, Emmett, Teller (BET) surface area of materials I and II were measured to be 30 and 22m^2/g. Both shell-core silica-magnetite (I) and model spherical silica (II) behaved similarly in Salmon sperm DNA binding and elution experiment (data not shown) indicating that both surfaces were identical in nature hence pure silica nanoparticles could be used as model nanoparticles.

Functionalized nanoparticles (I and II) by TPRE approach exhibited a high surface amine density (4.8×10^{14}molecules/cm^2) compared to nanoparticles functionalized in water (Fig. 1b). A high value (>80%) of surface to total amine density was observed in materials functionalized by TPRE approach compared to functionalization in water (20%). The high value of surface to total amine density by TPRE approach is a direct proof of a controlled surface engineering and the surface amine density values (4.8 ×10^{14}/cm^2) were close to the surface amine density of a monolayer (5.3×10^{14}/cm^2) on flat surfaces [4].

The orientation of the surface amine groups (shown in scheme 2) was tested by applying them in biology (Figure 2c and 2d) by attaching oligonucleotide (5'NH$_2$-dC$_6$-dT$_{25}$) and their efficiency in capturing fluorescent labeled complementary target oligonucleotide (dA$_{25}$) by hybridization mechanism. Surface engineered nanoparticles prepared by TPRE approach exhibited a high value (85%) of 5'NH$_2$-dC$_6$-dT$_{25}$ attachment efficiency with respect to the initial concentration of 5'NH$_2$-dC$_6$-dT$_{25}$ during the reaction whereas materials functionalized by bulk water phase exhibited only 43% attachment of 5'NH$_2$-dC$_6$-dT$_{25}$. The hybrid capture efficiency of oligonucleotide (5'NH$_2$-dC$_6$-dT$_{25}$) attached bio-nanoparticles in a solution of complementary fluorescence labeled oligonucleotide (dA$_{25}$) was observed to be very different in TPRE approach compared to nanoparticles functionalized in water. TPRE approach provided up to 100% capture of oligonucleotide (dA$_{25}$) by hybridization mechanism compared to 30% capture in water phase. This high performance of hybridization efficiency is due to the proper orientation of grafted oligonucleotide (5'NH$_2$-dC$_6$-dT$_{25}$) to the surface of the nanoparticles so that it is available to capture the complementary oligonucleotide (dA$_{25}$). These results suggest that the proper orientation of surface –NH$_2$ groups is very important and is achieved by TPRE approach. Maxwell *et al* [13] reported that the orientation of grafted bio-molecules in gold nanoparticles surfaces is related to the performance and detection of target bio-molecules for the applications in biosensors.

^{29}Si CPMAS solid state NMR spectra (Figure 2) of surface engineered diamagnetic silica core materials and polymerized silica obtained from APTS indicate the presence of various silicon environments due to two distinct chemical shift regions: -90 to -110ppm {Q Si sites{Si*(OSi)$_n$(OH)$_{4-n}$, n can have values from 1 to 3}and -50 to -70ppm {T Si sites {H$_2$NCH$_2$CH$_2$CH$_2$Si*(OSi)$_n$(OH)$_{3-n}$ with n can have values from 1 to 3}. Core silica sphere was observed to have three different silicon environments

(Figure 2; types 1, 2, 3) only in Q Si-sites and self polymerized APTS have two different silicon environments (Figure 2; types 4, 5) only in T Si-sites. The surface engineered silica nanoparticles exhibited both Q and T Si sites. TPRE approach provided only one type of T site (type 4) whereas functionalization in bulk water provided two types of T Si-sites (types 4 and 5). The absence of type 5 Si-site on surfaces support the model structure (shown in scheme 2) proposed in TPRE approach. The absence of other types of bonding such as amino end of APTS to surface silica (N-Si) as previously reported by ALD technique [7,8]. Vanblaaderen *et al* [14] reported the one step synthesis of amino-functionalized monodispersed colloidal organo-silica spheres using APTS but the surface amine groups were observed to be various types {$H_2NCH_2CH_2CH_2Si^*(OSi)_n(OH)_{3-n}$ with n values from 1 to 3} similar to uncontrolled functionalization in bulk water phase. It is also observed that the relative amount of type 3 to type 1 Si-species decreased after surface functionalization. The change is very prominent in the case of surface engineered nanoparticles prepared by TPRE approach and this is due to the simultaneous conversion of surface Si-species from type 3 to 2 to 1.

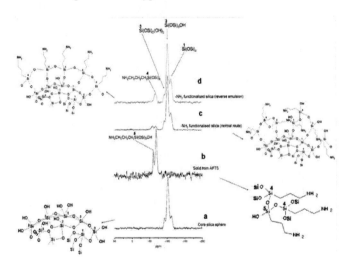

Figure 2 ^{29}Si CPMASNMR spectra of (a) Diamagnetic core silica sphere; (b) Polymerized APTS; (c) Amine functionalized silica sphere engineered in bulk water; (d) Amine functionalized silica sphere engineered by TPRE approach.

The organization of amino silane molecules around the nanoparticles could also have an effect on the resultant particle sizes. The surface engineered magnetic nanoparticles by TPRE approach exhibited no change in particle sizes (see Figure 3) whereas average particle sizes shifted from 50nm to 145nm when surface engineering of nanoparticles was carried out in water. In the case of TPRE approach, the density (4.8×10^{14} molecules/cm^2) and an ordered orientation of surface functional groups around the nanoparticles supports the formation of silane monolayer (7Å i.e. 0.7nm thickness) identical to the flat surfaces

measured by ellipsometry and therefore there is no change in particle size. Surface functionalization of nanoparticles in water exhibited an increase (ca. 100nm) in size and this may due to the presence of polymerized layer of silane or an interparticle connectivity through polymerized amino silane molecules as presented in scheme 2.

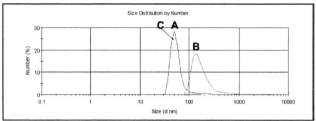

Figure 3 Particle size distribution of magnetic nanoparticles (I) before (A) and after surface engineering in water (B) and by TPRE approach (C)

REFERENCES

[1] Y. Yin and A. P. Alivisatos, Nature, 437, 664, 2005.

[2] J. V. Barth, G. Costantini and K. Kern, Nature, 437, 671, 2005.

[3] S. R. Wasserman, Y. T. Tao and G. M. Whitesides, Langmuir, 5, 1074, 1989.

[4] D. G. Kurth and T. Bein, Angew. Chem., Int. Ed. Engl., 31, 336, 1992.

[5] S. J. Oh, S. J. Cho, C. K. Kim and J. W. Park, Langmuir, 2002, 18, 1764.

[6] T. Nakagawa, T. Tanaka, D. Niwa, T. Osaka, H. Takeyama and T. Matsunaga, J. Biotechnol., 2005, 116, 105.

[7] S. Ek, E. I. Iiskola and L. Niinisto, J. Phys. Chem. B, 108, 9650. 2004.

[8] S. Ek, E. I. Iiskola, L. Niinisto, J. Vaittinen, T. T. Pakkanen and A. Root, Chem. Commun., 2032, 2003.

[9] L. Stelter, J. G. Pinkernelle, R. Michel, R. Schwartlander, N. Raschzok, M. H. Morgul, M. Koch, T. Denecke, J. Ruf, H. Baumler, A. Jordan, B. Hamm, I. M. Sauer, U. Teichgraber, Mol. Imag. Biol. 12, 25, 2010.

[10] T. Tanaka, R. Sakai, R. Kobayashi, K. Hatakeyama, T. Matsunaga, Langmuir, 25, 2956, 2009.

[11] T. Sen and I. J. Bruce, Chemical Communications DOI: 10.1039/b817354K. 2009.

[12] J. H. Moon, J. H. Kim, K. Kim, T. H. Kang, B. Kim, C. H. Kim, J. H. Hahn, J. W. Park, J. W. Langmuir, 13, 4305, 1997.

[13] D. J. Maxwell, J. R. Taylor and S. M. Nie, J. Am. Chem. Soc., 124, 9606, 2002.

[14] A. Vanblaaderen and A. Vrij, J. Colloid Interface Sci., 156, 1, 1993.

Fabrication of superhydrophobic nanostructured films by Physical Vapour Deposition

M. Yoldi, J.A. García, R.J. Rodríguez, A. Martínez, R. Bueno, M. Rico, J. Osés

Centre of Advanced Surface Engineering, R&D Department, AIN
E-31191 Cordovilla, SPAIN, myoldi@ain.es

ABSTRACT

This work reports on recent advances in the deposition of non-wettable coatings with high water contact angles (WCAs). We propose a simple and easily-controlled method for fabricating different types of nanoestructured films of Al_2O_3 with hydrophobic properties by Physical Vapour Deposition (PVD).

The films have been deposited over three types of hydrophilic substrates with very different values of roughness and free surface energy. The results of this study showed that the hydrophobicity of the Al_2O_3 coatings do not depend on the starting conditions of the substrate.

GD-OES and FESEM showed a 50-250 nm surface flake structure of Al_2O_3. This surface flake structure, which is responsible of the hydrophobic effects, is optimized in thicker films.

These Al_2O_3 hydrophobic films presented similar values of contact angle in water to the classic polymeric films based on methyl and fluorocarbons groups.

Keywords: Superhydrophobic surface, nanostructured films, aluminum oxide, Physical Vapour Deposition, wettability

1 INTRODUCTION

The wettability of a solid surface is a very important property that depends not only on the surface energy but also on the geometry structures on the surface. When the contact angle (CA) is smaller than 90°, the surface is hydrophilic; when it is higher than 90°, the surface is hydrophobic. Superhydrophilic surfaces (CA smaller than 5°) and superhydrophobic surfaces (CA higher than 150°) present very interesting properties not only in fundamental research but also in many practical applications, such as surfaces for satellite dishes, solar energy panels, photovoltaics, exterior architectural glass and green houses or heat transfer surfaces in air conditioning equipment.

Due to the small contact area between the surface and water, chemical reactions or bond formation with water are limited on a hydrophobic surface. Because of that, non-wettable surface prevent the adherence of frost or snow, the deposition of dust, oxidation processes or electrical conduction.

Many surfaces in nature exhibit superhydrophobic properties, such as the winds of butterflies [1] or the leaves of some plants [2]. Moreover, the best-known example of a non-wettable surface is the lotus plant [4]. Electron microscopy of the surface of lotus leaves shows a micrometer structure covered with nanocrystallites. Several studies conclude that this combination of micrometer-scale and nanometer-scale roughness is responsible for the hydrophobic properties [5]. However, other natural examples do not exhibit two-length scales and many studies calling into a question this requirement [6, 7].

To obtain superhydrophobic surfaces, surface energy and surface roughness are two dominant factors. When surface energy is lowered, by adding methyl or fluorocarbons groups, the hydrophobicity property is enhanced. On the other side, the ability to control the morphology of a surface on micron and nanometer scale is the key to achieve non-wetting effect [8, 9, 10].

Most of the superhydrophobic surfaces that are described in the literature are based on activated polymeric compounds [11]. Although good results have been obtained with this type of coatings, the surface energy of these coatings increases with time and the hydrophobic properties of the material fade. The main objective of this study is to prepare inorganic films with low free surface energy, low hysteresis and low tilt angle, which present highly and permanent hydrophobic properties. As the tribological properties play an important role in many applications requiring water-repellent properties, another important objective is to analyze the adhesion and the friction properties of these coatings.

2 EXPERIMENTAL

Three different types of ceramic substrates have been used: S06, S12 and S13. Before coating, the roughness of the substrates was determined by optical interferometry, using a WYKO RST 500 interferometric profilometer. Five measures were developed along each sample to determine the averaged roughness, R_a.

The hydrophilic character of these substrates was quantified by determining the contact angle of the surfaces in water, using as EASY DROOP equipment [13]. Ten measures were developed along each sample, five on the left side and five on the right one, to determine the contact angle.

Before coating, the substrates were cleaned with absolute ethanol.

The polymeric fluorine-based coatings were developed by a plasma polymerization (PP) process, using an

EUROPLASMA equipment. Two different procedures were applied, which are denoted as PP1 and PP2.

The Al_2O_3 coatings were prepared by Physical Vapour Deposition (PVD) in a commercial METAPLAS MZR 323 PVD chamber equipped with 3 magnetron sputtering sources. The characteristic parameters of the PVD process (dose, bias, temperature, pressure, flow of O_2) have been studied and optimized to obtain films of Al_2O_3 with the desire combination of micro- and nanostructure. The optimization of these parameters of control allows obtaining different effects on the final film properties, from hydrophobic to superhydrophobic. Once these parameters of control were optimized, two different processes were studied depending on the time of deposition, in order to study the influence of the thickness in the micro- and nanostructure of the final surface. They are denoted PVD1 and PVD2.

Table 1 summarizes the nomenclature used to denote the samples.

Sample	Substrate	Treatment	Procedure	Nomenclature
1	S06	PP	1	S06PP1A
2	S06	PP	1	S06PP1B
3	S06	PP	2	S06PP2A
4	S06	PP	2	S06PP2B
5	S12	PP	1	S12PP1A
6	PVD	PP	1	S12PP1B
7	S12	PP	2	S12PP2A
8	S12	PP	2	S12PP2B
9	S13	PP	1	S13PP1A
10	S13	PP	1	S13PP1B
11	S13	PP	2	S13PP2A
12	S13	PP	2	S13PP2B
13	S06	PVD	1	S06PVD1A
14	S06	PVD	1	S06PVD1B
15	S06	PVD	2	S06PVD2A
16	S06	PVD	2	S06PVD2B
17	S06	PVD	2	S12PVD1A
18	S12	PVD	1	S12PVD1B
19	S12	PVD	1	S12PVD2A
20	S12	PVD	2	S12PVD2B
21	S12	PVD	2	S12PVD1A
22	S13	PVD	1	S13PVD1B
23	S13	PVD	1	S13PVD2A
24	S13	PVD	2	S13PVD2B

Table 1 Nomenclature of the samples

The composition of the films was studied by GD-OES, with a Jobin-Yvon HD1000 Glow Discharge analyzer [14]. Changes in the chemical composition profile of the coatings have been also analyzed. The morphology of these new coatings has been studied by FE-SEM, using a Hitachi S-4800 FE-SEM microscope.

The wettability of the surfaces was studied by determining the angle contact in water with an EASY DROP equipment.

The mechanical and tribological characterization of the coatings includes the determination of the film adhesion (scratch test), thickness, and wear and friction coefficient.

3 RESULTS AND DISCUSSION

Figure 1 shows the roughness and the contact angles in water of the three types of substrates that have been used in this study.

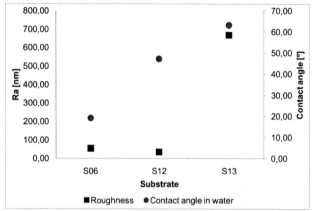

Figure 1: Roughness and contact angles in water of the substrates

This plot shows that surfaces S06 and S12 are smooth, with R_a of 53.61±8.75 nm and 34.75±2.80 nm, respectively. The surface roughness of S13 is one order of magnitude larger, with a R_a of 667.52±192.02 nm. This plot also reveals that S06 is the most hydrophilic substrate, with a contact angle in water of 19.11±4.26°. S12 shows 47.20±1.73° and S13, 63.17±3.27°.

The visual inspection of the samples showed that the PVD coatings, as well as the polymeric films, are transparent.

The scratch tests revealed a good adherence between the PVD coatings and the three types of substrates. No significant differences were observed between them.

GD-OES analysis confirmed that the chemical composition of the PVD coatings is Al_2O_3. The changes in the chemical composition profile allow determining the thickness of the coatings. Al_2O_3 films of 50 nm were observed for the shorter PVD deposition (PVD1) and 250 nm films for the longer process (PVD2)

The investigation by FE-SEM microscopy of cross-sections of the samples revealed the thickness of the coatings. These values were in good agreement with the ones obtained by GD-OES.

Figure 2 shows the averaged contact angles of water of the coatings. It shows that the hydrophobic character of the final surface do not depend on the properties of the substrate, because similar values are obtained for S06, S12 and S13. However, it reveals that PVD2 process is more efficient than PVD1, because thicker Al_2O_3 coatings are more hydrophobic that the thinner ones. FESEM micrographs showed that the surface flake structure, which is responsible for the hydrophobic properties, is optimized in thicker samples. Al_2O_3 coatings prepared by PVD2 process showed similar values of wettability to the polymeric films of reference.

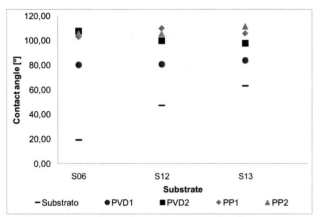

Figure 2: Averaged contact angles in water

FE-SEM micrographs showed a surface flake structure of the PVD coatings. Figure 3 shows the micrograph of a sample prepared by the longer PVD procedure. The thickness of this sample is 300 nm. As it can be observed in figure 3, this flake structure combines a micro and nano scale. According to the bibliography, this micro/nano structure in the surface is supposed to be responsible of the hydrophobic properties [12].

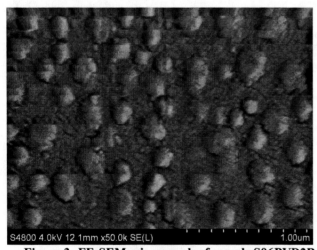

Figure 3: FE-SEM micrograph of sample S06PVD2B

However, this flake structure is not well-defined in all the samples. Figure 4 shows the micrograph of a sample prepared by the shorter PVD procedure. The thickness of the sample is 50 nm.

Comparing figures 3 and 4, we can confirm that the surface flake structure is more defined in thicker samples than in the thinner ones. These results are in good accordance with the contact angles in water that are summarized in figure 2, and explained why thicker Al_2O_3

coatings are more hydrophobic that the thinner ones: as the flake structure of the Al_2O_3 is more defined in thicker coatings, they are more hydrophobic. Compare figure 3 and 4.

Figure 3: FE-SEM micrograph of sample S06PVD2B

Figure 2 summarized the averaged values of wettability of the samples. However, the samples prepared by PVD process presented an important variation in the values of contact angles in water, up to 140º (see figure 5). The cleaning procedure, the position of the samples in the PVD chamber, the uniformity of the plasma and the rate of O_2 flow during the process should be analyzed in following studies, in other to optimize the wettability of the Al_2O_3 coatings.

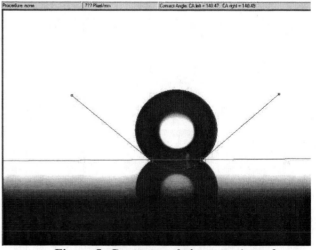

Figure 5: Contact angle in water (sample S13PVD2B)

4 CONCLUSIONS

Hydrophobic Al_2O_3 coatings with contact angles up to 140 ° were deposited by PVD over three different types of ceramic substrates.

The contact angles of these Al_2O_3 coatings do not depend either on the roughness neither on the surface energy of the substrates.

These coatings present a surface flake structure which combines a micro and a nanoscale. The quality of this structure determines the hydrophobicity of the coatings, and it is optimized in thicker films.

The results of this study might open a new pathway to the fabrication of superhydrophobic transparent conducting surfaces that can be used in several fields such as engineering, photovoltaic, decorative, among others.

ACKNOWLEDGEMENTS

The authors thank Peter Martens, from EUROPLASMA, for the development of the polymeric coatings by plasma polymerization technique.

REFERENCES

[1] T. Wagner, C. Neinhuis and W. Barthlott, "Wettability and contaminability of insect wings as a function of their surface sculptures", Acta Zool. 77, 213-25, 1996.

[2] C.Neinhuis and W. Barthlott, "Characterization and distribution of water-repellent, self-cleaning plant surfaces", Ann. Bot. 79, 667-77, 1997.

[3] R.M. Hill, "Superspreading", Curr. Opin. Colloid Interface Sci. 3 , 247-54, 1998.

[4] W. Barthlott and C. Neinhuis, "Purity of the sacred lotus, or scape from contamination in biological surfaces", Planta 202, 1-8, 1997.

[5] A. Nakajima, K. Hashimoto and T. Watanabe, "Recent studies on superhydrophobic films", Monatsh Chem 131, 31-41, 2001.

[6] E. Martines, K. Seunarine. H Morgan, N. Gadegaard, C.D.W. Wilkinson and M.O. Riehle, "Superhydrophobicity and superhydrophobicity of regular nanopatterns", Nano Lett. 5, 2097-2103, 2005.

[7] T. Onda, S. Shibuichi, N. Satoh and K. Tsujii "Super-water-repelent fractal surfaces", Langmuir 12, 2125-2127, 1996.

[9] T.L.Sun, L. Feng, X.F. Gao and L. Jiang, ""Bioinspired surfaces with special wettability", Acc. Chem. Res. 38, 644-52

[10] D. Quere, "Non-sticking drops", Rep. Prog. Phys. 68 (2005) 2495-2532, 2005.

[11] M. Ma and R.M.Hill, "superhydrophobic surfaces", Curr. Opin. Colloid Interface Sci. 11, 193-202, 2006.

[12] X. Zhang, M. Honkanen, M. Järn, J. Peltonen, V. Pore, E. Levänen and T. Mäntylä, "Thermal stability of the structural features in the superhydrophobic bohemite films on austenitic stainless steels", App. Surface Sci. 254, 5129-5133, 2008.

[13] SESSILE DROOP, Static Contac Angle. www.kruss.de

[14] GD-OES Horiba Jobi-Yvon HD 1000 http://www.horiba.com

Development of Durable Nanostructured Superhydrophobic Self-Cleaning Surfaces on Glass Substrates

Y. D. Jiang[*], M. Kitada[**], M. White[*], T. Fitz[***], and A. T. Hunt[*]

[*]nGimat Co., Atlanta, GA, USA, [*]yjiang@nGimat.com, mwhite@nGimat.com, ahunt@nGimat.com
[**]Honda R&D Co., Ltd., 4630 Shimotakanezawa Haga-machi Tochigi Fref., 321-3393 Japan, masayoshi_kitada@n.t.rd.honda.co.jp
[***]Honda R&D Americas, Inc., Raymond, OH, USA, tfitz@oh.hra.com

ABSTRACT

High quality SiO_2 based transparent superhydrophobic coatings with double-roughness microstructure and high durability were deposited onto glass substrates by the Combustion Chemical Vapor Deposition (CCVD) technique and the combination of an innovative device and the CCVD. A contact angle of higher than $165°$, a rolling angle of $<5°$, a haze of $<0.5\%$, and an increased transmittance by 2% higher than bare glass have been achieved by the CCVD technique. By the combination of the innovative device and the CCVD, a contact angle of $137°$ has been achieved after being wiped by a standard automobile wiper for 60,000 times with coating's morphology keeping the same.

Keywords: nanostructure, superhydrophobic, durable, coating, CCVD

1 INTRODUCTION

Studies of superhydrophobic self-cleaning surfaces have been attracting increasing interest in recent years for both fundamental research and practical applications. The applications of superhydrophobic self-cleaning surfaces include architectural glass, automotive glass, solar panels, shower doors, and nanochips, etc. [1-3]. Wide usage of self-cleaning surfaces will result in huge energy savings by removing the need for washing, scrubbing, and chemical polishing of windows, ceramics, and other surfaces. In addition to a reduction in cleaning requirements, these superhydrophobic surfaces have additional benefits, such as improved safety when driving in severe rain and snow, and improved efficiency in solar cells.

The development of superhydrophobic self-cleaning surfaces was first inspired by the observation of natural cleanness of lotus leaves [4] and other plant leaves [5]. The typical superhydrophobic self-cleaning effect in nature is found from lotus leaves. Scanning electron microcopy (SEM) studies on lotus leaves [5,6] revealed that a microscopically rough surface consisting of an array of randomly distributed micropapillae with diameters ranging from 5 to 10 μm. These micropapillae are covered with waxy hierarchical structures in the form of branch-like nanostructures with an average diameter of about 125 nm. Motivated by the lotus leaf, many techniques have been being developed to create nanostructures mimicking the lotus effect from many organic and inorganic materials [7,8]. However, existing hydrophobic coatings either have low transmittance not suitable for windows and solar panels or not as hydrophobic as lotus leaves or low durability.

In this work, an open atmosphere CCVD technique and the combination of an innovative device and CCVD were employed to deposit SiO_2 based superhydrophobic coatings onto glass substrates. The coatings' morphological, hydrophobic, and other physical properties are presented.

2 EXPERIMENTAL PRECEDURES

2.1 Superhydrophobic coatings by the CCVD Process

In the CCVD process, shown in Figure 1 [9], precursors are dissolved in a solvent, acting also as the combustible fuel. This solution is atomized to form submicron droplets by the proprietary Nanomiser™ device. The resulting vaporous fog is then convected by an oxygen containing stream to a flame. The heat from the flame provides the energy required to flash vaporize the ultrafine droplets and for the precursors to react and vapor deposit on the substrates. The CCVD technique uses a wide range of inexpensive, soluble precursors that do not need to have a high vapor pressure. The key advantages of the CCVD technique include:

- Open-atmosphere processing
- High quality at low cost
- Wide choice of substrates and
- Continuous production capability

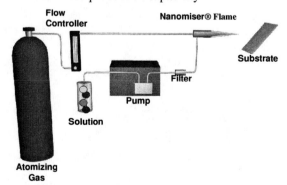

Figure 1. Schematic representation of the CCVD system

In this work, the coatings were fabricated by either the CCVD technique or a combination of an innovative device and CCVD. Prior to depositions, the glass substrates were ultrasonically cleaned in organic solvents such as isopropanol, rinsed in deionized water, and blown dry using nitrogen. The substrate was then mounted on a metal chunk or the top of a back heater. Key process parameters include deposition temperature, glass surface temperature, and coating composition.

2.2 Analytical techniques

Equilibrium, receding, and advancing CAs were measured by a CA measuring system (G10, Kruss USA). Equilibrium CAs were measured using deionized water droplets of approximately 1 – 2 mm in diameter. If not indicated specifically, all the CAs in the following sessions are equilibrium water CAs. Three data points were tested on all samples.

The coating's morphology was observed by SEM (Hitachi s-800 and s-4800). Transmittance and reflectance in the visible range were measured by a spectrometer (PERKIN-ELMER Lambda 900 UV/VIS/NIR spectrometer). Surface roughness (as root mean square, RMS) was evaluated by an optical profilometer (Burleigh Instruments, Inc.). Haze in the visible range was characterized by a haze meter (BYK Gardner Haze Meter). Coating's durability was tested by measuring the CA before and after wiping the coating by a standard automobile wipe for certain times.

3 RESULTS AND DISCUSSION

Superhydrophobic coatings have been grown by many techniques such as CVD [7] and sol-gel [8]. These techniques can require costly starting materials, and/or are time consuming and have low throughput. The open atmosphere CCVD and its combination with an innovative device offer an attractive alternative to grow durable transparent superhydrophobic coatings on glass and plastic substrates with good yield and high throughput potential. SiO$_2$ is chosen as primary coating material because of its low cost, ease to make, and refractive index match between the coating and the substrate, which reduces reflectivity of the coated specimens.

To achieve low haze and high transmittance, the feature size of the coating must be much smaller than the visible wavelength to reduce large light scattering. Figure 2 shows the SEM image of a typical SiO$_2$ coating on glass substrate by the CCVD technique. The coating has a rough surface and double surface roughness, in which coarse features are composed of nanostructures of 30 to 200 nm. The double roughness morphology is similar to the topology of lotus leaves but smaller. The nanometer sized hierarchical structure is essential to simultaneously achieve low haze, increased transparency, and superhydrophobicity.

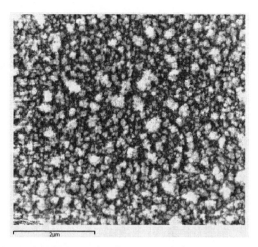

Figure 2. SEM image of a typical CCVD SiO$_2$ based superhydrophobic coating on glass substrate

A CA of over 165° was achieved on as-deposited samples with a rolling angle of less than 5°. The haze of the samples is less than 0.5%. For reference, the bare glass substrate treated with the same fluorinate silane has a CA of approximately 110° with a haze of about 0.2%. The transmittance and reflectance of bare glass are in the range of 91.8 to 92.6% and 7.4 to 8.5%, respectively while those of the CCVD SiO$_2$ coated samples are in the range of 93.9 to 94.5% and 5.6 to 6.2, respectively. The SiO$_2$ coated glass is hyper-transparent with an increased transmission of about 2% higher and a reduced reflectance of about 2% lower than bare glass substrate, suggesting the CCVD SiO$_2$ coatings reduce reflection in the visible range and are of anti-reflection, which will benefit many applications, especially solar cells, lighting and imaging.

In addition to CA, durability is another important factor for many practical applications of self-cleaning surfaces. Durability tests were conducted by moving the samples across a defined distance on a polishing cloth surface. Force was determined by the weight of the samples themselves. CA was measured after each two passes across the abrasion surface. In total, twenty passes were completed for each sample. In a previous report [10], coating's durability was improved considerably by a post-deposition treatment process. In this report, coatings' composition was studied to improve their durability. Figure 3 shows the contact angle of samples with different compositions as a function of abrasion pass. For the sample with composition A, its initial contact angle was 170°. It decreased rapidly in the first 10 passes of abrasion. After 20 passes the contact angle decreased to 150°. It was noticed that after the abrasion test, the rolling angle increased significantly. For the sample with composition B, its initial contact angle was 170°. After 20 passes of abrasion it decreased to 160°, which is even higher than the sample with a post-deposition treatment [10]. Therefore, in addition to processing conditions and post-deposition treatment, coating's composition plays a critical role in its hydrophobility and durability. Figure 4 shows the SEM image of a superhydrophobic coating after the abrasion test. It can be seen that the coating's surface became much smoother and lost its double roughness, which leads to decreased contact angle and increased rolling angle. These results suggest that initial superhydrophobic surfaces are not strong enough to keep their superhydrophobicity to meet a number of practical applications.

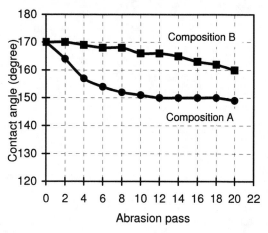

Figure 3. Contact angle of SiO_2 based coatings as a function of abrasion pass with different compositions

Figure 4. SEM image of a superhydrophobic coating after 20 passes of abrasion test

To further improve durability for practical applications, *n*Gimat and its commercialization partner Honda R&D designed and manufactured an innovative device (WRIN). This WRIN device and related process embed ultrafine particles into the substrate surface and forms robust structures. Figure 5 shows plan view and cross sectional SEM images of a hydrophobic coating by the WRIN device. It can be seen that larger dual-sized semi-spherical and irregular features with a size from 0.2 to 1.5 μm were embedded into the glass surface. Smaller features with size 200 nm or less were distributed between the larger features. The cross sectional SEM image (Figure 5 (b)) clearly shows an embedded particle which deformed the glass surface. A defused interface layer was formed between the embedded

particle and the glass substrate. Energy dispersive X-ray (EDX) spectra were collected from point 1 and point 2 in Figure 5 (b). The EDX spectra clearly show a distinct intermediate layer between the embedded particle and the glass substrate, which provide strong bonding and therefore, improved abrasion resistance.

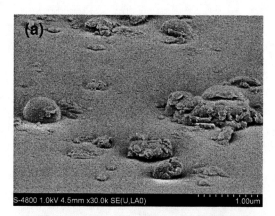

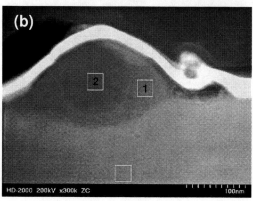

Figure 5. (a) Plan view (75° tilt) and (b) cross sectional SEM images of a sample deposited by the WRIN device. EDX spectra collected from (b) point 1 and (b) point 2 showed a change in composition.

To further increase the distribution density of the deposited pillars and hence, coating's hydrophobicity, after the larger pillars were deposited by the WRIN device, smaller features with a size of 50 nm or less were deposited by the CCVD technique. These smaller features cover both the larger pillars and the valleys between the larger pillars to form double-roughness structure. The samples deposited by the combination of the WRIN device and the CCVD technique were subjected to abrasion tests by a standard automobile wiper. After being wiped for 60,000 times and re-treated, the contact angle of the sample decreased from 151° to 137°. The morphology of the sample after abrasion tests is shown in Figure 6. The smaller features by the CCVD process on the larger pillars by the WRIN device were wiped off. The larger pillars and the smaller features in the valleys between the larger pillars are kept almost the same before abrasion tests, showing the high robustness of the larger pillars and their protection to the smaller features with less robustness. The overall durability of the double-sized super-hydrophobic structure has been improved significantly. More optimization and tests will be conducted to optimize both the durability and hydrophobicity that will ultimately meeting the practical requirements.

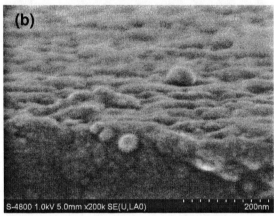

Figure 6. SEM images of a sample deposited by the combination of the WRIN device and CCVD technique, (a) no tilt and (b) tilt 75°

4 SUMMARY

As a summary, transparent superhydrophobic surfaces with high performance have been successfully prepared on glass substrates by the CCVD technique and the combination of the WRIN device and the CCVD technique. A contact angle of higher than 165°, a rolling angle of <5°, a haze of <0.5%, and an increased transmittance by 2% higher than bare glass have been achieved by the CCVD technique. By the combination of the innovative device and the CCVD, a contact angle of 137° has been achieved after abrasion tests by a standard automobile wiper for 60,000 times with coating's morphology keeping the same.

This work was supported by the Deportment of Energy (DOE) through Grant No. DE-FG02-04ER84007 and Honda R&D Co., Ltd.

REFERENCES

[1] A. Nakajima, K. Hashimoto, and T. Watanabe, Chem. Monthly 132, 31, 2001.

[2] N. A. Patankar, Langmuir 20, 8209, 2004.

[3] D. Quere, A. Lafuma, and J. Bico, Nanotechnology 14, 1109, 2003.

[4] W. Barthlott and C. Neinhuis, Planta 202, 1 (1997)

[5] L. Feng, S. H. Li, Y. s. Li, H. J. Li, L. J. Zhang, J. Zhai, Y. L. Song, B. Q. Liu, L. Jiang, and D. B. Zhu, Adv. Mater. 14, 1857, 2002.

[6] B. He, J. Lee, and N. A. Patankar, Colloids and Surfaces A: Physicochem. Eng. Aspects 248, 101, 2004.

[7] Y. Y. Wu, H. Sugimura, Y. Inoue, and O. Takai, Chem. Vap. Deposition 8, 47, 2002.

[8] H. M. Shang, Y. Wang, S. J. Limmer, T. P. Chou, K. Takahashi, and G. Z. Cao, Thin Solid Film 472, 37, 2005.

[9] A. T. Hunt, W. B. Carter, and J. K. Cochran, Jr., App. Phys. Lett. 63, 266, 1993.

[10] Y. D. Jiang, Y. Smalley, H. Harris, and A. T. Hunt, Nanotech 2008, Boston, MA, USA, June 1-5, 2008

Superhydrophobic Coating Using Metallic Nanorods For Aerospace Applications

Ganesh K Kannarpady[a*], Yoann Lefrileux[b], Thomas Laurent[b], Justin Woo[a], Khedir R Khedir[a], Hidetaka Ishihara[a], Steve Trigwell[c], Charles Ryerson[d], Alexandru S Biris[a**]

[a] Nanotechnology Center, University of Arkansas at Little Rock
2801 South University Avenue, Little Rock, AR, 72204
[b] Ecole d'Ingenieurs du CESI-EIA,
86 route de Breuty, Domaine universitaire, 16400 La Couronne, France
[c] Applied Science and Technology, ASRC Aerospace, ASRC-24
Kennedy Space Center, Orlando, FL, 32899
[d] Terrestrial and Cryospheric Sciences Branch Cold Regions
Research & Engineering Laboratory Engineer Research and Development
Center U.S. Army Corps of Engineers, Hanover, NH 03755-1290

ABSTRACT

Ice formation on the body of airplanes poses several problems like increasing drag and decreasing lift. It can also cause engine stoppage due to its accumulation on carburetor/fuel nozzles and the engine's air source. A superhydrophobic coating that has a high ice adhesion redactor factor can be a solution to avoid ice formation. In this paper, we are presenting a modified technique to generate Al nanorods followed by Teflon coating to obtain the contact angle of 155° with water droplet. A combination of anodization and glancing angle sputtering deposition (GLAD) technique was used to fabricate nanorods of Al. To our knowledge this is the first time GLAD sputtering technique was coupled with aluminum lattice templates obtained by the anodization for the growth of nanorods. The nominal size of the nanorods was 30-80 nm in diameter. A conformal coating of teflon was applied using RF sputtering. The water droplets bounced off from the surface indicating possible applications in anti-icing coating for aerospace.

Keywords: Anti-icing coating, superhydrophobic surface, Al nanorods, glancing angle deposition technique (GLAD), aluminum lattice membrane.

1. INTRODUCTION

Mimicking naturally occurring superhydrophobic behavior in some of the plant leaves such as in lotus, on a metallic surface has been a great challenge for researchers. If a metallic surface can be successfully tailored to exhibit superhydrophobicity, that can find applications in various fields such as aerospace, power transmission lines, wind turbines, etc. There are several reports of metallic surface, with a surface modification, showing fairly high contact angle with a water droplet [1-3]. However, it is a great challenge to develop a super hydrophobic coating for aerospace applications as the coating must be strong enough to sustain rugged conditions like abrasion from impacting water droplets in rain in flight. As well, such coatings should be suitable to operate effectively with other chemicals common in the aerospace environment.

Glancing angle deposition technique has been effectively used to generate nanorods of metals with various size and shape [4-8]. Although, GLAD can be used with various thin film deposition techniques such as e-beam evaporation, thermal evaporation, the use of GLAD with sputtering deposition technique gives a new dimension to the fabrication of the nanorods. The size, shape, spacing and distribution of nanorods can be tailored by controlling a partial pressure of Ar during sputtering deposition that are not possible with the other methods of thin film deposition techniques. Instead of GLAD sputtering technique, some researchers used normal sputtering to generate nanorods using aluminum lattice templates [9]. The anodization of aluminum surface along with subsequent chemical etching leaves aluminum lattice templates that act as nucleation center for the growth of the nanorods.

In this paper we are presenting a combination of two techniques, aluminum lattice template and GLAD sputtering, to generate a nano surface roughness in aluminum. An ultrathin coating of Teflon on aluminum nanorod surface was used to reduce the surface energy. While aluminum lattice template had been extensively used for the generation of metallic nanorods, to our knowledge it is the first time the GLAD technique was coupled with aluminum lattice templates to generate metallic nanorods. The variation of contact angle of water droplet with aluminum surface at different stages of surface modification has been discussed.

2. EXPERIMENTAL PROCEDURE

2.1 Fabrication of aluminum lattice template

Aluminum sheet (99.99% pure) of 1 mm thick, obtained from Alfa Aeser, was annealed at 500°C for 4 hours. The annealed sheet of aluminum was then cleaned and electro-polished to get a smooth shining surface. Anodization was carried out in 0.3M/L oxalic acid solution at a constant cell potential of 40V for 45 min. at 278 K. The anodized aluminum surface was subjected to chemical etching for 20 min. in a mixture of phosphoric acid and chromic acid. The resulted aluminum lattice template surface is shown in Fig. 1.

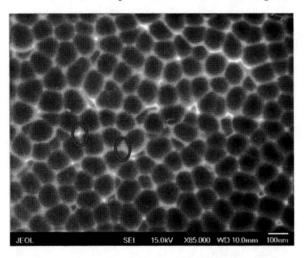

Figure 1: SEM image of the aluminum lattice template.

2.2 Generation of aluminum nanorods

Aluminum nanorods were grown on aluminum lattice templates using a DC magnetron sputtering GLAD technique as shown in Fig. 2. As seen in the schematic diagram, the substrate normal was making 85° angle with the target normal. The chamber was pumped down to a base pressure of 5×10^{-7} torr. The plasma was generated using Ar gas at 2.5 mTorr pressure. The sample-substrate distance was 6". The nanorods were deposited at power density of 3.5 W/cm^2. The substrate was rotated at 10 rpm using a DC stepper motor. To reduce the surface energy of the nanorods, a thin coating of Teflon was deposited using a RF magnetron sputtering source. The substrate-target distance for this deposition was 2" and the deposition was carried out in a normal deposition mode with power density of 6W/cm^2. The thickness of Teflon coating was less than 10 nm.

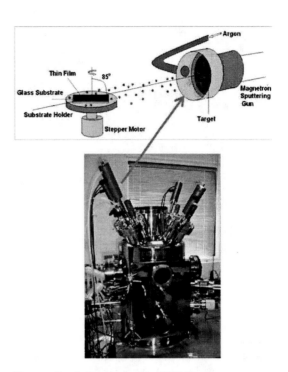

Figure 2: In-house built RF/DC magnetron sputtering deposition system along with a schematic of glancing angle deposition technique.

2.3 Morphology and contact angle measurement

Surface morphology of aluminum lattice template and aluminum nanorods were studied using scanning electron microscope (JEOL700). The static contact angle of water with Teflon coated nanorod aluminum surface was measured using sessile drop method (VCA Optima).

3. RESULTS AND DISCUSSION

The aluminum nanorods grown on aluminum lattice templates using GLAD sputtering technique are shown in Fig. 3. As shown in the figure, the aluminum nanorods of the size in the range, 30-80 nm are distributed uniformly over the substrate. It is interesting to note that the size and shape of the nanorods show a large deviation from the size and shape of aluminum lattice templates. The aluminum lattice templates show mostly hexagonal structures as seen in Fig. 1, in agreement with the reports on aluminum lattice templates [10]. The aluminum nanorods grown on aluminum lattice templates exhibit mostly near-circular structures. Moreover, each aluminum nanorod looks like a bunch of agglomerated smaller nanorods as evident from the multiple facets on each nanorod.

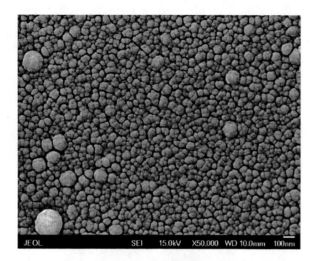

Figure 3: SEM image of Aluminum nanorods generated on the nano-templates using glancing angle sputtering technique.

The growth of nanorods on aluminum lattice templates using normal sputtering deposition technique has been discussed by many researchers [9]. In such a case, the shape and size of the nanorods closely follow the shape and size of aluminum lattice templates. The template pits generally serve as nucleation centers for the nanorod growth. The separation of nanorods in such a scenario is mostly governed by the width of the template wall. The width of the template wall depends on the cell potential and temperature during anodization process [11]. However, the growth of nanorods on aluminum lattice templates using GLAD has not been discussed so far in the literature. Form the preliminary results that we have presented here, the growth of the nanorods are mostly governed by the height of the template wall. During GLAD, the substrate normal was tilted at an angle of 85° with the target normal as shown in Fig. 2. As the growth of nanorods on a normal surface during GLAD mainly governed by the shadowing effect of the material that is deposited in the beginning [12], in the present scenario hypothetically the template walls provide nucleation centers for the growth rather than template pits. In other words, the template walls shadow the template pits. In addition, the junction of different hexagonal structures (marked by red circles in Fig. 2) has larger area than the rest of the template walls. Hence, it can be assumed that these junctions are preferred for the growth of nanorods over the rest of the template walls. These preferred centers coupled with the substrate rotation during the deposition, facilitate a near-circular structure of the aluminum nanorods.

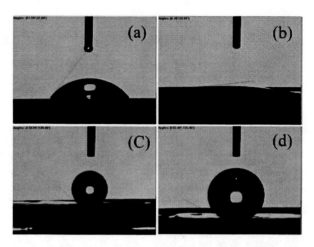

Figure 4: The contact angle of water droplet with, (a) as-received aluminum surface, (b) aluminum lattice templates, (c) aluminum nanorods, coated with Teflon, grown on as-received aluminum surface, (d) aluminum nanorods, coated with Teflon, grown on aluminum lattice templates.

Figure 5: Shape of water droplets on modified aluminum surface (the surface corresponding to Fig. 4(d)). The central circular groove is the modified surface.

Figs. 4(a)-(d) show the variation of contact angle of water droplet on aluminum surface at different stages of surface modification. The as-received aluminum surface shows contact angle of ~51° with water droplet. This is in agreement with the value reported in the literature [13]. The aluminum lattice templates, obtained by anodization and subsequent etching, show about 6-10° contact angle with water droplet. The formation of aluminum lattice templates increased the surface roughness and hence the roughness factor r (the ratio of actual over apparent surface area). This in turn reduced the contact angle according to the Wenzel's law of wettability [14]. In Fig. 4(C), the shape and contact angle of water droplet on aluminum nanorods, coated with the thin layer of Teflon, on as-received aluminum surface. The observed contact angle of 139° is in agreement with our previously reported work [15]. Finally, Fig. 4(d) shows a near spherical shape of water droplet on the aluminum nanorods, coated with a thin layer of Teflon, grown on aluminum lattice templates. The measured contact angle was ~155°, the highest among all the other surfaces discussed above.

Fig. 5 shows lotus like behavior of aluminum nanorod surface (corresponding to the surface that is discussed in Fig. 4(d) above). These water droplets are larger than that of the water droplets shown in Figs. 4(a)-(d). It was observed that a small tilt to the substrate, bounced off all the water droplets shown in Fig. 5 confirming the superhydrophobic behavior of the sample. From the Cassie theory [16] of superhydrophobicity, it is understood that a superhydrophobic surface needs to have a low surface energy rough surface and trapped air in the gap of rough structures. While the generation of the nanorods increased surface roughness, the coating of thin layer of Teflon on the nanorod surface reduced the surface energy. The reduction in the surface energy ensured the repulsion of the water drops and the air trapped in the gap of the nanorods ensured floating of water droplets and thereby preventing the water from sliding inside the gap. The generation of nanorods on aluminum lattice membrane created a two-stage roughness. The first stage roughness was provided by the pits of templates and the second stage roughness was provided by the gap between aluminum nanorods. This may be the reason for a significant improvement in the observed contact angle of 155° (Fig. 4(d)) with water droplet as opposed to 139° that was observed on Teflon coated aluminum nanorods on the as-received aluminum surface (Fig. 4(c)).

4. CONCLUSIONS

The observation of superhydrophobic behavior on nanorod aluminum surface grown on aluminum template is highly promising for anti-icing applications in various fields such as aerospace. The combination of anodization and GLAD technique for the growth of nanorods has shown a potential of generating a two-stage roughness that is required for obtaining very high contact angle with water. With further investigation, it is possible to modify aluminum like aerospace compatible lightweight materials to exhibit a very high ice adhesion reduction factor.

ACKNOWLEDGEMENTS

Financial support from U.S. Army (ERDC Cooperative Agreement Number: W912HZ-09-02-0008) and Arkansas Science & Technology Authority (Grant # 08-CAT-03) is greatly acknowledged.

REFERENCES

[1]. DK Sarkar, M Farzaneh, RW Paynter, Mater Lett., 62, 1226-1229, 2008.

[2]. Z. Youfa, Y. Xinquan, Z. Quanhui, C. Feng, L. Kangning, Appl. Surf.Sci., 256(6), 1883-1887, 2010.

[3]. X. Wu, G. Shi, J. Phys.Chem. B, 110(23), 11247-52, 2006.

[4]. W. K. Choi, L. Li, H.G. Chew, F. Zheng, Nanotechnology, 18(38), 385302-1-6, 2007.

[5]. H. Alouach, G. J. Mankey, J. Mat. Res., 19(12), 3620-3625, 2004

[6]. R. Krishnan, T. Parker, S. Lee, T.M. Lu, Nanotechnology, 20(46), 465609, 2009.

[7]. P. I. Wang, T. C. Parker, T. Karabacak, G. C. Wang, T. M. Lu, Nanotechnology, 20(8), 2009

[8]. T. Karabacak, P. I. Wang, G. C. Wang, T. M. Lu, Thin Solid Films, 493(1-2), 293-296, 2005.

[9]. P. Lee, O. Lee, S. Hwang, S. Jung, S. E. Jee, K. H. Lee, Chem.Mater., 17, 6181-6185, 2005.

[10] H. Zheng, J. Zhong, W. Wang, Y. Zheng, C. Ma, Thin Solid Films, 516, 4983-4987, 2008.

[11] U. Sahaym, M. G. Norton, J.Mater.Sci., 43, 5395-5429, 2008.

[12] Y. P. Zhao, D. X. Ye, G. C. Wang, T.M. Lu, Proceedings of SPIE Nanotubes and Nanowires, 5219, 59-73, 2003.

[13]. A. Gajewski, Int. J. Heat and Mass Transfer, 51 (19-20), 4628-4636, 2008.

[14]. R.N. Wenzel, Ind.Eng.Chem. 28, 988-994, 1936

[15] G. K. Kannarpady, R. Sharma, B. Liu, S. Trigwell, C. Ryerson, A.S. Biris, Appl. Surf. Sci., 256(6), 1679-1682.

[16] A.B.D. Cassie, S. Baxter, Trans. Farady Soc., 40, 546-551, 1944.

*gkkannarpady@ualr.edu; Tel: 501-569-8067; Fax: 501-683-7601
**asbiris@ualr.edu; Tel: 501-683-7456; Fax: 501-683-7601

Durable Super-Hydrophobic Nano-Composite Films

Jeff Chinn[*], Fred Helmrich[*], Rolf Guenther[*], Mark Wiltse[*], Kendall Hurst[**], and Robert W. Ashurst[**]

[*]Integrated Surface Technologies, Inc.
1455 Adams Dr., Ste 1125, Menlo Park, CA 94025, info@insurftech.com

[**]Dept. of Chemical Engineering, Auburn University, Auburn, AL 36849

ABSTRACT

A durable super-hydrophobic nano-particle based coating based on an ALD/CVD deposition process of alumina has been characterized. This new conformal coating is ideal for protecting printed circuit boards from water damage. Products coated with this composite film exhibit excellent resilience when exposed to aqueous solutions. The mechanical durability of this film was improved by using silane surface reactions which improved inter-particle and surface adhesions. Under water erosion tests, an improvement of >400x over non-reinforced films has been obtained. The shear failure of the film is tuned to allow electrical connections to be made through the coating. This allows many existing products, such as printed circuit boards, to be protected with a non-wetting surface coating without product redesign.

Keywords: super-hydrophobic, super-oleophobic, nano-composite, conformal coatings, water-safe electronics.

1 INTRODUCTION

For decades naturally occurring non-wetting and super-hydrophobic surfaces such as lotus leaves have captivated the imagination of engineers in their efforts to overcome failures and problems arising from water damage in unwanted places. With the recent development of a durable super-hydrophobic coating, protection methods against water damage are shifting from traditional physical barriers by encapsulation, or from costly packaging techniques with rubberized seals around ingress openings. Currently, printed circuit boards are covered with a protective film or "conformal coating" consisting of an organic material such as silicone, epoxy or urethane. The added thickness of these conformal coatings becomes problematic and must be accounted for in product packaging, because of the compact and sleek design requirements of popular portable consumer products. In addition, the added film causes contact resistance and inter-connectivity problems between electronic assemblies. Thus, the critical connectors are not coated and are often the failure points from shorts and electrical leakages caused by accidental exposures to liquids.

The mechanical structures required to create the properties of a super-hydrophobic coating are on the scale of nano-meters (10^{-9} meters) and thus require careful engineering [1,2]. As a result of the required ultra-fine structure, these nano-structured protective films are often fragile and have limited durability. In the past, the various techniques used to create the nano-structures depended on the targeted material and their application over large areas and complex 3D shapes tended to limit their applicability, which also was costly.

In this paper, a non-wetting, super-hydrophobic coating which can be economically applied to large surfaces is briefly described. The mechanical durability developed to provide a commercially acceptable coating for protecting electronics is shown in Fig. 1. The ability of these types of durable coatings is not limited to PCB's and could be applied to other applications which will be discussed in the future.

Fig 1: LEFT: A cell phone PCB immersed in water. The entire boards gets wet. RIGHT: Also immersed in water, a cell phone PCB that has been coated with a super-hydrophobic film. This coating, normally invisble, acts as a virtual force field that pushes water away from the surface, yielding a silvery appearance under water that protects the board from water damage.

2 EXPERIMENTAL AND APPARATUS

The coating technology developed is called "Vapor Particle Deposition" (VPD) for creating nano-particle composite films. The VPD method uses a sub-atmospheric gas-phase flow-through reactor which is suitable for large batch processing. The nano-composite structure is created using a hybrid atomic layer deposition (ALD) / chemical vapor deposition (CVD) process in which super-saturated vapor conditions are created [3]. During the process (patent pending), a metal organic precursor is oxidized in such a way that the proper film roughness and film coverage over the substrate are achieved. Subsequently, these nano-particles are encapsulated into a glass-like matrix to improve the film's durability. The final step includes a surface modification treatment to create a low surface energy state.

Nano-scale roughness of alumina were formed by the following oxidation reactions:

$$2Al(CH_3)_3 + 3H_2O \longrightarrow Al_2O_3 + 6CH_4 \quad or\ by$$

$$2AlOH^* + 2Al(CH_3)_3 \longrightarrow 2[Al\text{-}O\text{-}Al(CH_3)_2]^* + 2CH_4$$

$$2[Al\text{-}O\text{-}Al(CH_3)_2]^* + 3H_2O \longrightarrow Al_2O_3 + 2AlOH^* + 4CH_4$$

where * are surface bound species.

Alkyl bridge silanes are used to create silsesquioxane structures to strengthen the composite film and improve the mechanical durability of the film [4,5]. These were formed by the oxidation of silanes as described by the reactions of:

$$SiCl_4 + 2H_2O \longrightarrow SiO_2^* + 4HCl$$

$$C_2H_4Cl_6Si_2 + 6H_2O \longrightarrow SiO_x\text{-}(CH_2)_2\text{-}SiO_x^* + 6HCl + 3H_2O$$

$$C_6H_{12}Cl_6Si_2 + 6H_2O \longrightarrow SiO_x\text{-}(CH_2)_2\text{-}SiO_x^* + 6HCl + 3H_2O$$

The critical process parameters of the VPD process include the chemical doses and timing between the metal organic and oxidation precursors in the ALD/CVD reaction, which affect the surface diffusion and the chemical reaction. If the chemical injections are spatially partitioned to prevent gas phase interactions, an ALD type surface reaction occurs with insufficient roughness and structure formation. Conversely, a simultaneous injection of precursors results in a gas-phase CVD reaction in which "nano-dust" is formed with very poor surface coverage. The key process challenge was to allow sufficient and controlled intermixing so that the super-saturated condition was precisely tailored to create nano-particles uniformly through-out the reactor volume, leading to sufficient surface coverage of particles over the parts to be coated.

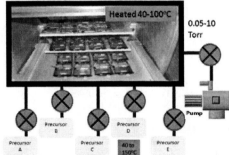

Fig 2: Schematic of the VPD coating chamber. Precursor vapors are introduced sequentially or concurrently to create composite layered films. A multi-shelf chamber configuration is available for large batch processing.

The coating equipment used was the RPX-540 manufactured by Integrated Surface Technologies (IST) (www.insurftech.com). The RPX-540 can deliver up to 5 different chemical precursors to create a custom nano-composite film. The entire coating system is precisely temperature controlled to ensure vapor delivery to the process chamber without condensation. Precursor

chemistries can be heated from 40-150°C and the process chamber to 100°C. The process timing control is based on National Instruments' LABVIEW®. The process chamber can be configured to accommodate a variety of media, including printed circuit boards, small objects like hearing aids, or glass and plastic substrates. Batches of materials up to 2000-in² of flat board material can be processed in a single batch in about 30 minutes. The super-hydrophobic coating described for printed circuit board water protection is currently sold as "Repellix™". Low surface energy monolayers can be also processed, typically used as MEMS anti-stiction or imprint release layers or for a variety of adhesion promoters or bio-compatible anti-fouling coatings.

3 RESULTS AND DISCUSSIONS

Super-hydrophobic composite films were prepared under a variety of process conditions with varying roughness, thickness, and linking (glue) schemes. Since these films consist of "glued reinforced" particles or roughness, it was observed that film roughness alone (rms) was inadequate for predicting the performance of these films. An improved classification method which includes aerial surface density (or coverage) was determined a key part of film quality. Coverage between 25-60% was required. Future publications on this metrology characterization are forthcoming.

3.1 Film Characterization

Films were routinely characterized utilizing AFM, SEM, and Contact Goniometry. Additional metrology included TEM, FTIR, XPS, mechanical scanning wear by Hysitron, water erosion, optical transmission and TGA-MS. IST also perfomed biocompatibility tests based on ISO-10993 FDA guidelines and the material was deemed to be biocapatible. Film roughness ranged from 30-700nm rms (Fig. 3). The as-deposited composite is normally super-hydrophilic (water contact angle (CA) <10°). After surface modification with a perfluoronated silane, the films exhibited super-hydrophobic (CA: >160°) and super-oleophobic (olive oil >160°) properties (Fig. 4). No change on contact angle was observed even under immersion conditions for >9 months. The surface energy of perfluoronated silicon was ~20 dyne-cm and the rough composite film was <5 dyne-cm. Typical film thicknesses ranged from 50-200nm.

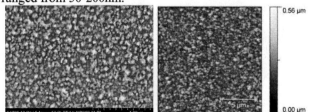

Fig 3: SEM and AFM of the super-hydrophobic nano-composite film. The rms roughness was ~164nm, average particle size ~30-35nm with an aerial surface coverage of ~42%.

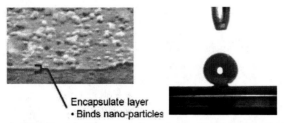

Encapsulate layer
• Binds nano-particles

Fig 4: LEFT: An electron micrograph showing the glueing encapsulant for improved interparticle and particle-to-surface adhesion. RIGHT: After surface modification, the water contact angle is greater than 160°.

3.2 Electrical Performance

This unique material is a non-wetting film that can be deposited over large areas or over entire assemblies, even over sensitive electrical connection points, creating a boundary layer of air. This feature provides protection so that manufacturers do not have to re-engineer the packaging enclosures of their products to accommodate for any thickness and stiffness. It is essential that the film permits electrical contact when mating with electrical connectors. To test contact resistance, a classic 4-point probe technique was employed. Molex brand flex ribbon connectors with 10 pins and 0.5mm pitch were coated with a super-hydrophobic film. The connectors were unmated during coating, and mated once for initial testing. A challenge current of 1mA was used and the voltage across each contact point was measured. The measured values differed only slightly from the nominal (manufacturer supplied value).

Various surface mount 40 pin socket connectors shown in Fig. 5 were also tested before and after the non-wetting Repellix coating was deposited. The chart in Fig. 6 shows the typical contact resistance of uncoated and coated connectors. No statistical significance was observed.

Fig 5: Typical multi-pin surface mounted connectors that are used in many consumer electronic products.

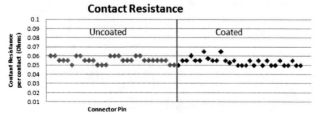

Fig 6: A representative chart of measured contact resistance of an array of connectors compare before and after coating results.

The durability of the nano-composite film is tuned by the inclusion of various levels of silane oxides which serve as inter-particle and surface adhesion "glues". This accounts for the cohesion of the film and the durability to accommodate light handling during the assembly process. During the connector insertion, the composite film is mechanically abraded, allowing physical contact to be made (Fig. 7). While a very small area of the sliding connector surface is exposed, surface tensions of aqueous liquids will not wet the contacting surfaces. If the glue oxide is made too thick, the mechanical force of the connector could be insufficient to reliably penetrate the composite film. In Fig. 8, the thick composite film shows areas in which good and poor contacting surfaces are achieved.

Fig 7: Optical micrograhs of the super-hydrophobic film which has been scraped away by the contacting barbs of the connector. The surface tension keeps the connecting pads from being wetted by liquids.

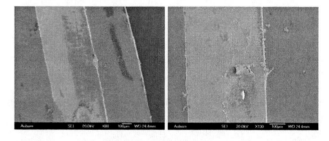

Fig 8: Durability of the connecting surface is controlled by the glueing agent process and the thickness of the film. Left: the non-wetting nano-composite film which is displaced by the mating connector revealing the base metal for good electrical connection. Right: A more durable coating which is not removed by the mating connector, leading to high resistivity and therefore electrical connection problems.

3.3 Durability Characterization

The mechanical durability of the nano-composites was tested using a Hysitron TI-950 Triboindenter. Wear patterns are created by raster scanning the nano-composite film with a given force predefined by the user. The scan can consist of a single pass or multiple passes over the same area within one test. By applying a known force and selecting the number of passes over which this force is applied, the amount of material that is removed during a scan can be measured post-testing, using the probe in an AFM contact mode as shown in Fig. 9.

Another technique used was a water droplet erosion test in which a stream of water droplets was used to erode a fixed area. The impact and shear force on the film can be calculated from the incident angle, impact height, rate, drop volume and time [6]. Shown in Fig. 10, an improvement of >400x can be obtained with increased inter-particle silanation.

The nano-composite films were also subjected to environmental stress testing and neutral pH salt fog. The particle film itself was resilient to these stresses and its super-hydrophobicity remained unchanged.

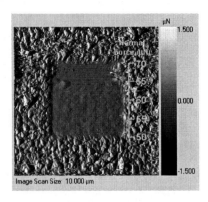

Fig 9: A 10x10μm topographical SPM image showing the 5x5μm worn area and surrounding surface. In this case, the film failed at a normal force of roughly 62μN.

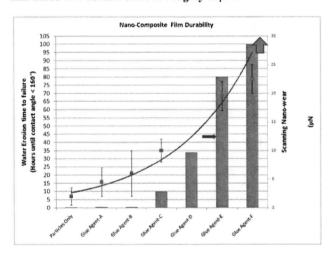

Fig 10: Chart showing the film failure time from water erosion tests overlaid with the failure loads from the TriboIndenter ScanningWear Tests. A film with a Scanning Wear failure of 25μN failed at 100 hours under water erosion.

3.4 Water Resistance Performance

The water resistance performance was tested by exposing a number of products to a variety of harsh liquid exposures. In Fig. 11, comparisons of several cell phones were tested in a rain-simulation spray chamber. Additional tests were performed by immersing them into Gatorade. The Gatorade electrolyte solution consisting of potassium phosphate / citric acid instantaneously shorted any uncoated products, while Repellix coated devices were resistant up to 30 minutes.

Depending on the Repellix film thickness, the particular device packaging and ingress points, coated PCB's can still be subjected to electro-chemical induced corrosion around high voltage field points such as power-supply lines. However, overall, the coated devices significantly out performed the uncoated devices.

Phone Vendor	Brand A	Brand B	Brand C
Style	Candy Stick	Slider	Flip
Control	1 minute	2-3 minutes	1 minutes
Repellix	20 minutes	20-30 minutes	2 minutes
Sample Size	20	5	1
Phones Passed	19	5	1

Fig 11: Table shows performance of three (3) brands of cell phones which were exposed in a "rain chamber". In all observed cases, the Repellix coated electronics outperformed all uncoated products.

4 CONCLUSIONS

An improved durable particle based nano-composite film in which bridge oxides are used to improve the mechanical durability has been characterized. This water resistant conformal coating is ideally suited for electronic applications to protect printed circuit boards used in cell phones and other portable devices from water damage. Further improvements are expected and the application to other devices and surfaces is expected.

5 ACKNOWLEDGEMENTS

Support for this work was provided in part by the National Science Foundation under grant 0911783.

REFERENCES

[1] L. Gao, A. Fadeev, T. McCarthy, "Superhydrophobicity and Contact-Line Issues", MRS Bulletin, Vol. 33, Aug. 2008, p747

[2] Tuteja el al, "Design of Superoleophobic Surfaces", Science, Vol. 318, 7Dec2007, p 1618

[3] M. Swihart, "Vapor-phase Synthesis of Nanoparticles", Current Opinion in Colloid and Interface Science 8 (2003), 127-133.

[4] Mabry, J. M.; Vij, A.; Viers, B. D.; Grabow, W. W.; Marchant, D.; Ruth, P. N.; Vij, I. "Hydrophobic Silsesquioxane Nanoparticles and Nanocomposite Surfaces.", ACS Symposium Series, The Science and Technology of Silicones and Silicone-Modified Materials, Clarson, S. J.; Fitzgerald, J. J.; Owen, M. J.; Van Dyke, M. E. (Eds.), 2006,

[5] Yüce, M. Y.; Demirel, A. L.; Menzel, F. "Tuning the Surface Hydrophobicity of Polymer/Nanoparticle Composite Films in the Wenzel Regime by Composition", Langmuir, 2005, 21, 5073.

[6] Water Drop Height: 0.20 meters, Volume 0.046cc, Velocity: 1.979 m/sec, Rate: 2.3 Drops/sec, Incident Angle: 45°, (Estimated Impact Force: 901μN)

Tunable Wetting of Titanium and Gold Based Wicking Materials -For Uses under High Accelerations

C. Ding[*], P. Bozorgi[**], C.D. Meinhart[**], and N.C. MacDonald[**]

[*]California Nanosystems Institute, UCSB
Room 2229, ESB, Santa Barbara, CA 93106, USA, ding@engr.ucsb.edu
[**]Mechanical Engineering Dept., University of California, Santa Barbara, CA, USA

ABSTRACT

Heat pipes adopt wicking materials to achieve some unique functions that thermal siphons could not achieve [1]. Based on titanium micromachining, gold plating and surface modification techniques, we fabricated pillar arrays, and modify the dimension or parameters to investigate their tunable wicking behaviors. Our study shows that water completely wets the micro/nano-scale titania structures fabricated with micromachining and surface modification techniques, and the spreading trends of water were found to be diffusive. Wicking material under different acceleration conditions were also investigated and we found the bitextured titania structures (BTS) undergoes twice larger inertia force than the monostructures with same dimensions.

Keywords: Bitextured nanostructure titania, wetting, acceleration, cooling, titanium.

1 INTRODUCTION

Wicking materials are widely used in heat pipes which have long been utilized for cooling microprocessor chips [2, 3], nuclear power systems, space based radar systems, etc.. Heat pipes have been of particular useful in cooling highly power intensive chips because of their high thermal conductivity and the capacity to carry high amounts of heat. The working fluid evaporates after absorbing heat from the hot end and flows towards the cold end, condensing thereby. A very fine wick structure on the internal wall of the heat pipe brings the condensate back from the cold end towards the hot end by capillary actions.

A flat heat pipe has wicking materials on the interior walls of its chamber to automatically pump the cooling fluids from the heat sink side to the evaporator side when transporting heat. The flat geometry allows micro/nano-scale machining techniques to be applied for forming unique wicking materials such as nanowires, carbon nanotubes, biporous copper, micromachined structures, etc. The wicking material is usually designed to provide high capillary force on the spreading liquid. The capillary force can be derived using virtual energy method when a working fluid contacts with a solid surface. Although capillary can be well defined, the friction force between fluid and solid surfaces is relatively difficult to measure and analytically to define for complicated structures. For some application

such as the wicking material in spacecraft based cooling systems may undergo a certain range of accelerations which may induce another term of force due to inertia. Wetting behaviors of wicks can be dramatically changed because of inertia forces depending how the device is aligned according to acceleration direction.

1.1 Wetting and Roughness

When a droplet of fluid such as water contacts with a smooth and flat surface (Fig. 1A), hydrophobicity or hydrophilicity is governed by the relative reaction of the forces at the triple line arising from the three interfacial energies (or call surface tensions), γ_{SV}, γ_{SL}, and γ, taking place at the solid-vapor, solid-liquid, and liquid-vapor interfaces, respectively. Projecting these three surface tensions acting on a specific contact line gives the Young's equation:

$$\cos\theta_e = \frac{\gamma_{SV} - \gamma_{SL}}{\gamma} \tag{1}$$

where θ_e is the equilibrium contact angle.

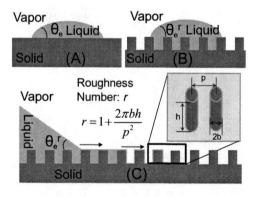

Figure 1: Flow of the impregnating film at the front boundary. Figure A shows the Young's contact angle; Figure B and C show the static and dynamic cases of the wetting on solid surfaces.

In reality it usually differs from this ideal situation in which we have smooth and clean surface. Solid surfaces are usually rough or carry undulations, pores, surface defects. Using virtual energy method together with Young's equation we can derived to get the Wenzel's relation [4]:

$$\cos \theta_e^{\ r} = r \cos \theta_e \qquad (2)$$

where $\theta_e^{\ r}$ is the apparent contact angle. The above equation comes with a limit of $-1 \le r \cos \theta_e \le 1$. The Wenzel's relation tells us that the surface roughness can enhance both the nonwetting (hydrophobic) and wetting (hydrophilic) ability of liquid on solid surfaces. When the Young's contact angle on flat/smooth surface is less than $90°$, roughness will reduce the apparent contact angle leading to superhydrophilic/superwetting case [5-7].

1.2 Statics and dynamics

In the previous subsection we show that the roughness modifies the value of apparent contact angle, and superwetting can be possibly achieved by tuning the roughness. A simple idea is to use micromachined pillars to obtain large surface roughness and optimize the wetting by tuning the pillar dimensions. We denote such a surface by its pillar roughness r which is defined in Fig. 1. Provided that the case in Fig. 1, the pillars guide the liquid within the pillar arrays and form in a manner similar to wicking but more accurately hemiwicking which is intermediate between spreading and imbibition [8]. Assuming that the top surface areas can also get wet during the progression of the liquid film the surface energy change is then only due to the suppression of the liquid-vapor interfaces. Using virtual energy method again, the capillary force is given by [9, 10]:

$$F_{cp} = \gamma(r-1) = \frac{2\pi\gamma bh}{p^2} \qquad (3)$$

The capillary force is then constant and depends only on the surface roughness. This force is generally resisted by viscous force due to the flow of liquid. When the wicking material works under accelerations, then the inertial force will get involved when the liquid wets or de-wets. The net driving force for wetting of fluids on wicks will be a sum of the following three forces:

$$F_{net} = F_{cp} - F_{friction} \pm F_{inertia} \qquad (4)$$

where the inertial force may help increase of decrease the net force depending on how the wetting direction is aligned with the acceleration direction. The resistance force to the flow generally comes from two domains: 1) velocity gradients over a distance h due to the friction between the bottom surface and the moving fluid; 2) the perturbation to the flow by the pillars arrays. These two resistant force shall be both present during the flow, but one may dominate the other for different designs (for example, the pillar height difference) of the micro patterns. No accurate general rule is available at this point to predict the flow behavior because of the wide variety of surface properties. However, balancing the driving force with the viscous force, the dynamic behavior of the flow on rough solids was found to obey the famous Washburn relation [11] which stipulates the travel path of the flow increases as the square root of time: $x = (Dt)^{1/2}$ where D is the dynamic coefficient of the flow. A high dynamical coefficient is preferred for the wicking material (pillar arrays) used in heat pipes for ensuring acceptable operations. It'll be interesting to find out the dynamic coefficient experimentally, and provide insightful information for optimizing it by tuning the design parameters.

2 FBARICATION

2.1 Dry Etching and Surface Oxidization

Based on titanium micromachining and surface modification techniques, we fabricated BTS with two main designs: $5\mu m/5\mu m$ (Dia./Gap) and $100\mu m/50\mu m$ (Dia./Gap) with an etch depth of $50\mu m$. The fabrication process for making BTS (Fig. 2 and Fig. 3C) begins with the SiO$_2$-masked deep etching of a 300 μm thick Ti substrate using Inductively Coupled Plasma (ICP) etch in an Ar/Cl$_2$ ambience. The SiO$_2$ layer is approximately 3 μm thick. After deep etching is complete, the sample is oxidized in a 30% solution of hydrogen peroxide at ~83^{0}C for 15-20minutes to obtain the BTS and the bottom floor surface. The surfaces show nano-scale walls (30~50nm thick) and pores (150~200nm in diameter). The nanostructures are self assembled on high-aspect-ratio Ti pillar arrays when oxidized in H$_2$O$_2$ solution.

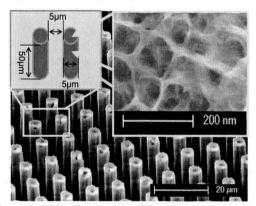

Figure 2: SEM photograph of the BTS with $5\mu m/5\mu m$ (Dia./Gap). The height of the pillars is ~$50\mu m$.

2.2 Gold Plating

The sample (Fig. 3C) is further processed for gold plating. After forming BTS on the Ti pillars, a seed layer (20nm Ti/150nm Au) was deposited on all surfaces then it was dipped into a non-cyanide based solution for Au plating at 55^{0}C. A current 3mA/cm^2 was applied during plating. The sample was taken out for SEM study after plating for 7600 seconds (Fig. 3D). The minimum gap was reduced from 50μm to 5μm after 7600 seconds plating.

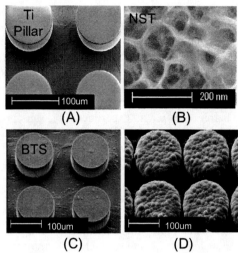

(A) (B)

(C) (D)

Figure 3: SEM photographs of Ti pillars (A), nanostructure titania-NST (B), bitextured titania structure-BTS (C), and BTS with plated gold (D).

3 EXPERIMENTS AND RESULTS

3.1 Wetting Experiments and Analysis

DI water was used as a working fluid for its high merit number [12] for heat pipe systems. The experiment was done by bringing the wicks into contact with a drop of water (~40μL), and the flow is recorded with a camera (Nikon D90). The position of the front liquid interface was extracted by decomposing the video into 24 frames/second. The drop completely wetted the nano/micro structured surface at the end, and the contact angle goes to zero. A curve of the form $x^2 = Dt$ is well fitted to the collected distance-vs-time data shown in Fig. 4. The experimental data generate the dynamic coefficients, D, for different wetting experiments. By differentiating the fitted curve we get the wetting speed:

$$\frac{dx}{dt} = \frac{D}{2}\frac{1}{x} \qquad (5)$$

The velocity plot of each experiment is shown in Fig. 5.

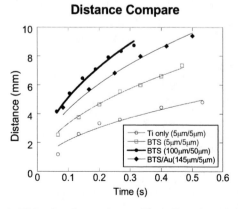

Distance Compare

Figure 4: This plot shows the capillary flow front interface path as a function of time.

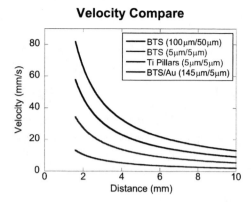

Velocity Compare

Figure 5: Velocity plot of equation (5) for four wicks.

As shown in Fig. 5, the NST enhanced the wetting due to the increased surface roughness for the same design parameter (5μm Dia./5μm gap). However a different design parameter (100μm Dia./50μm gap) gives over ~160% improvement. Further tuning of the design parameter by gold plating reduced the velocity by a small amount.

3.2 Acceleration Tests and Discussions

The wetting performances of the wicking material under acceleration conditions are tested with Ideal Aerosmith acceleration table (Model: 1291BR). A schematic of the testing setup with this testing system is shown in Fig. 6. The sample is fixed at one end of the extension arm to be 0.25m away from the axis of the rotation. A drop of 30μL of DI water is used to completely wet the wicks from the distal end to the near-to-rotation-center end of the wicking substrate. A transparent cover (3.5cm in diameter, 1cm in height) was fixed on top of the sample to reduce evaporation when the device is under acceleration and jogging. The angular acceleration rate is set to be 300 °/s², and this rate is used in each test for comparison. We chose the sample as shown in Fig. 3A and 3C for experiments for their relatively lower capillary pressure due to their larger dimension both in diameter and gap.

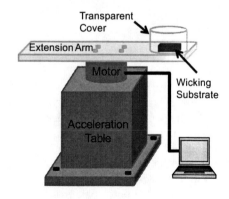

Figure 6: Schematic of the acceleration testing systems.

The goal of the experiments is to see if they can keep the wick wet under 10g jogging acceleration at radial direction when wetting from the distal end of the extension arm to the center. The sample was accelerated at an increasing and decreasing angular acceleration rate both at $300°/s^2$ to get a specific jogging acceleration in radial direction, and then kept at certain jogging speed for 10 seconds.

The wicking substrate with pure Ti pillars was found to be able to undergo a radial acceleration of ~6.3g where g stands for gravity acceleration, above this rate dry area was found as shown in Fig. 7A. The sample as shown in Fig. 3C with BTS was found to be able to undergo up to a radial acceleration of 12.13g. Part of the water volume is found gone during spin due to evaporation or flying so that the total volume of water is reduced. A well-sealed device would help avoid reduction of water volume.

From the analysis of in section 1.2 we know that when a drop of water is put contact with BTS wicks the spreading volume of water undergoes both capillary force and viscous force. When the device is under zero acceleration the capillary force helps increase the wetting speed while the viscous force functions as to decrease the speed. Experimental wetting results in Fig. 5 show that a design in favor of increasing speed is possible by forming BTS or changing the dimensions of the micromachined wicking structures. When the wicking substrate is under acceleration as in Fig. 6 both the capillary force and viscous force help to be in favor of fighting with de-wetting flows. With this specific situation shown in Fig. 6 BTS not only helps increase capillary force but may also helps increase viscous force for avoiding de-wetting.

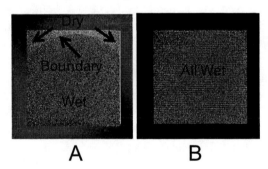

Figure 7: Acceleration (10g) test result comparison between Ti mono-structure (A) and BTS (B). Figure A shows the dry-out around the two top corners. Figure B shows that all areas stay wet due to BTS.

4 CONCLUSIONS

Complete wetting of DI water on micromachined titanium structures is achieved, and the spreading trends of water were found to be diffusive. Wetting on titanium based microstructures was improved by oxidizing the titanium pillars and forming bitextured titania structures. The nanoscale titania structure greatly enhances the wetting dynamics by ~160%. A dimension change from 5μm/5μm (Dia./gap) to 100μm/100μm provides much higher wetting

speed. This indicates that the wetting velocity of wicking material can be tuned dramatically by adopting nanostructures and dimension changes. Two wicking material were investigated under different acceleration conditions for studying of de-wetting behaviors, and we found all the bitextured titania structures (BTS) can keep wet under much larger radial acceleration (12.13g) than the monostructures (6.3g) with same dimensions.

This work is supported by the TGP program under the Microsystems Technology Office at the Defense Advanced Research Projects Agency. A portion of this work was done in the UCSB nanofabrication facility, part of the NSF funded NNIN network. The authors would like to thank Timothy J. Reed and Chris Burgner for their help on testing the wicks under accelerations.

REFERENCES

[1] P. D. Dunn and D. A. Reay, "Heat pipe," Physics in Technology, vol. 4, pp. 187-201, 1973.

[2] U. Vadakkan, G. M. Chrysler, and S. Sane, "Silicon/water vapor chamber as heat spreaders for microelectronic packages," IEEE, 2005, pp. 182-186.

[3] H. Xie, H. Xie, A. Ali, and R. Bhatia, "The use of heat pipes in personal computers," ITHERM, 1998

[4] R. N. Wenzel, "Resistance of Solid Surfaces to Wetting by Water," Industrial & Engineering Chemistry, vol. 28, pp. 988-994, 1936.

[5] X. Feng and L. Jiang, "Design and creation of superwetting/antiwetting surfaces," Adv. Mater, vol. 18, pp. 3063-78, 2006.

[6] C. Extrand, S. Moon, P. Hall, and D. Schmidt, "Superwetting of Structured Surfaces," Langmuir, vol. 23, pp. 8882-8890, 2007.

[7] J. Yuan, X. Liu, O. Akbulut, J. Hu, S. Suib, J. Kong, and F. Stellacci, "Superwetting nanowire membranes for selective absorption," Nature Nanotechnology, vol. 3, p. 332, 2008.

[8] D. Quéré,, "Wetting and Roughness," Annual Review of Materials Research, vol. 38, pp. 71-99, 2008.

[9] C. Ishino, M. Reyssat, E. Reyssat, K. Okumura, and D. Quere, "Wicking within forests of micropillars," EPL, vol. 79, p. 56005, 2007.

[10] C. Ding, P. Bozorgi, N. Srivastava, M. Sigurdson, C. D. Meinhart, and N. C. MacDonald, "Super Wetting of Micro&Nano Structured Titania Surfaces," in Transducers, Denver, CO, 2009.

[11] E. W. Washburn, "The Dynamics of Capillary Flow," Physical Review, vol. 17, p. 273, 1921.

[12] W. G. Anderson, J. H. Rosenfeld, D. Angirasa, and Y. Me, "Evaluation of Heat Pipe Working Fluids in the Temperature Range 450 to 700 K," STAIF-04

CONTACT
*C. Ding, Public, tel: +1-805-893-5341;
changsong_ding@yahoo.com

Self Assembly of Colloidal Gold Nanoparticles on Diblock Copolymer Thin Film Templates

S.M. Adams and R. Ragan*

University of California, Irvine, CA, USA, rragan@uci.edu

ABSTRACT

Self-assembly of metallic nanoparticles were investigated for the development of field-enhanced chemical and biological detection devices with the capacity to achieve single-molecule level detection resulting from surface enhanced Raman scattering, associated with closely spaced noble metal nanostructures. Using chemical self-assembly, we attached monodisperse, colloidal gold nanoparticles on self-organized polymer templates, patterning arrays of nanoparticle clusters with sub-10 nanometer interparticle spacing in order to engineer enhanced optical fields. Poly(methyl methacrylate) domains in phase-separated polystyrene-b-poly(methyl methacrylate) diblock copolymer thin films were chemically modified with surface amination for varied arrangements. Chemically synthesized sub-20 nm diameter gold nanoparticles were attached to the amine-functionalized surfaces using EDC linking chemistry with thioctic acid ligand-bound to the nanoparticle surface.

Keywords: self assembly, nanoparticle, sensor, thin film

1 INTRODUCTION

Metallic nanoparticles exhibit a feature in which a surface plasmon resonance with wavelength approaching that of incident light is sustained through oscillation of the metal's conduction electrons. With closely spaced metal nanoparticles, the surface plasmon resonance is observed as a dipole due to the collective electronegative nature with respect to the positive ionic background [1]. These nanoparticle plasmonic resonance effects are desirable to enhance detection capabilities of molecular biosensors. The development of metal nanoparticle arrays with low interparticle spacing for use as field enhanced detection of chemical and biological components with high sensitivity has been a subject of great interest in the past several decades [2,3]. Comparison of state of the art biosensor technology design indicates significant benefit with nanoparticle array sensor technology. Current methods for optical sensor detection generally rely on fluorescent

tagging, whereas the application of metal nanoparticle arrays have displayed a strong optical signal from plasmon scattering effects the equivalent of 500,000 fluorescein molecules, permitting detection of biomolecules up to near single-molecule detection [4-6].

Self-assembly techniques have been increasingly considered for their potential application in the development of nanoscale surface patterning as an alternative to conventional lithography techniques, enabling ordered patterning at scales previously unachievable with increased yield and cost-effectiveness using simple thermodynamic and chemical synthesis techniques [7,8]. In this study, metallic nanoparticles were investigated for the development of field-enhanced chemical and biological detection devices with the capacity to achieve single-molecule level detection resulting from surface enhanced Raman scattering (SERS) associated with closely spaced noble metal nanostructures [3]. Localized surface plasmon resonance (LSPR) sensors likewise benefit from ordered metal nanoparticles on surfaces, providing increased shift in minimum of reflectivity with biological binding event (figure 1). Furthermore, strong scattering from the interacting surface plasmons of metal nanoparticles in the patterned array is also applicable to enhancement of photovoltaic technology, in which the dipole interactions increase the optical path of incident light in the absorber layers of photovoltaic solar cells [9].

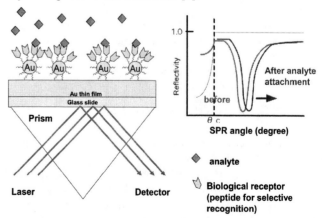

Figure 1: Refractive index of surface changes when analyte molecule binds to biological receptor on nanoparticle surface yielding a shift in SPR minimum reflectance angle.

Use of chemical self-assembly techniques for developing nanoparticle arrays also provide significant benefits. As enhanced optical [10] and catalytic [11] properties have been observed with sub-10 nm inter-particle spacing, adaptation of bottom-up chemical assembly techniques for template design would likewise be optimal for development for its capacity for such low inter- particle dimensions. Nanoparticles prepared from colloidal solution can be produced with diverse size [12], shape [13], and interfacial features [14]. By using surface chemistry techniques to instigate nanoparticle attachment from colloidal solution, metal nanoparticles with desired physical characteristics can be implemented in template design. Using chemical self-assembly, the surface template can likewise be varied by simple changes in the chemical design to optimize for particle spacing and alignment. Moreover, the method of processing relies on a simple batch of chemical reactions, which would be both reproducible when optimized as well as provide a relatively low cost fabrication technique in comparison to traditional lithographic methods. These design characteristics provide generous motivation for development of self-assembled nanoparticle templates.

2 EXPERIMENTAL MODEL

With chemical self-assembly, we have attached monodisperse, colloidal gold nanoparticles on self-organized polymer templates, patterning arrays of nanoparticle clusters with sub-10 nanometer interparticle spacing in order to engineer enhanced optical fields. Poly (methyl methacrylate) (PMMA) domains in phase-separated polystyrene-b-poly(methyl methacrylate) (PS-*b*-PMMA) diblock copolymer thin films were chemically modified with surface amination for both lamellar and hexagonal arrangements. Chemically synthesized sub-20 nm diameter gold nanoparticles were attached to the amine-functionalized surfaces using EDC linking chemistry with thioctic acid ligand-bound to the nanoparticle surface (figure 2) [7].

2.1 Surface Functionalization

Sodium citrate stabilized Au nanoparticle colloids with particle size of 20 and 10 nm in diameter were prepared from reduction of $HAuCl_4$. The surface of the Au nanoparticles was chemically functionalized with carboxylic acid surface functional groups by ligand attachment of DL-6,8-thioctic acid (Sigma- Aldrich) as depicted in figure 2a [7]. This method of functionalizing the

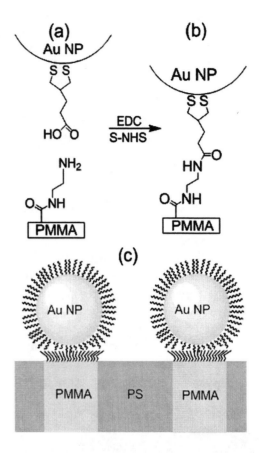

Figure 2: Schematic representation of assembly method: (a) Nanoparticles functionalized with thioctic acid and PMMA with primary amine; (b) Crosslinking chemistry of nanoparticles to amine-treated surfaces; and (c) Au nanoparticles selectively assembled on PMMA regions of the PS-b-PMMA template [7].

nanoparticle from a colloidal solution permits the attachment of particles of various sizes and shapes, as long as carboxyl ligand can be attached.

The surface of the PMMA regions of the PS-*b*-PMMA thin film diblock copolymer, prepared by spin coating on a Si substrate followed by thermal annealing, were functionalized with primary amine functional groups by reacting with dilute ethylenediamine [7]. Controlled surface patterning of the aminated regions is accomplished by thin film self-assembly relative to copolymer fractional molecular weight and degree of polymerization [15].

2.2 Nanoparticle Self-Assembly

The carboxyl-functionalized Au nanoparticles were selectively attached to the amine-functionalized PMMA regions of the diblock copolymer (figure 2c) by the use of a chemical crosslinker, designed to join carboxylic acids with primary amines. In this case, 1- ethyl-3-[3-dimethylaminopropyl] carbodiimide hydrochloride (EDC) linking chemistry with N-hydroxy sulfosuccinimide (S-NHS) [16] is used to form a covalent bond is formed between the metallic nanoparticles and patterned copolymer surface (figure 2b), yielding a self-assembled patterned array of nanoparticles on the polymer thin film [7].

3 RESULTS AND DISCUSSION

Atomic force and scanning electron microscopy was used to demonstrate that Au nanoparticles were preferentially immobilized on poly(methyl methacrylate) (PMMA) domains of the copolymer templates. AFM topography images are shown of templates exposed to TA-Au colloids with (figure 3a) and without (figure 3b) EDC and S-NHS presence to promote chemical crosslinking. Gold nanoparticle distribution on the sample with EDC/S-NHS treatment exhibited high levels of selectivity for attachment on regions where the PMMA has been selectively etched [7].

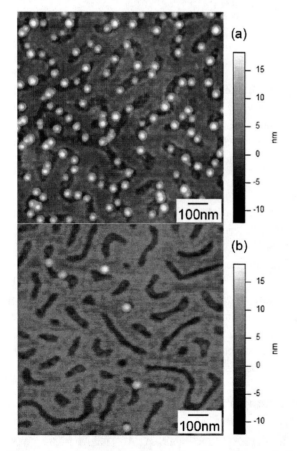

Figure 3. Amine-treated polymer surface with attached nanoparticles with (a) and without (b) EDC crosslinker treatment [7].

By controlling the size and spacing of the reactive polymer domains and the relative nanoparticle size, nanoparticle arrays are fabricated with controlled clustering arrangements (figure 4). The spacing between nanoparticle clusters is directly affected by the interdomain spacing (λ) of the copolymer matrix. While the number of particles in the nanoparticle cluster is a function of the difference between nanoparticle size (d) and PMMA domain size (D). Interparticle distances within nanoparticle clusters arranged on the patterned surface are controlled by the relative ionic strength of the nanoparticle colloid used for attachment. This clustering technique is especially useful for arranging nanoparticles in dimer formations, which provide significant increase in detection limits for Raman enhanced molecules in SERS systems.

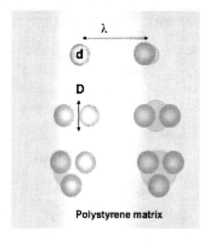

Figure 4. Schematic of metal nanoparticle cluster array assembly for nanoparticles of diameter d, on polymer domain size D, and interdomain spacing λ.

Analysis of the preliminary SERS enhancement measurement on the gold nanoparticle array showed an enhancement factor of the order of 10^7 for the detection of molecular pyridine (figure 5). As there would be 10^4 more pyradine molecules per cubic cm of solution than square cm of a closely packed monolayer of pyradine on the surface, this indicates that far fewer molecules need be present for detection. This is beneficial for developing sensitive signal enhancement and supports further efforts to analyze and optimize this self-assembly-based sensor design.

with primary amines. In this case, 1- ethyl-3-[3-dimethylaminopropyl] carbodiimide hydrochloride (EDC) linking chemistry with N-hydroxy sulfosuccinimide (S-NHS) [16] is used to form a covalent bond is formed between the metallic nanoparticles and patterned copolymer surface (figure 2b), yielding a self-assembled patterned array of nanoparticles on the polymer thin film [7].

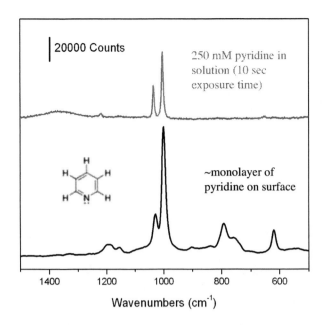

20000 Counts

250 mM pyridine in solution (10 sec exposure time)

~monolayer of pyridine on surface

Wavenumbers (cm⁻¹)

Figure 5. Preliminary SERS measurement of nanoparticle attached diblock copolymer surface.

4 CONCLUSIONS

By altering the surface chemistry of colloidal nanoparticles with carboxylic acid functional groups and a nanoscale patterned chemical templates with primary amine functional groups, we have selectively attached the nanoparticles as an array structure using crosslinking chemistry techniques. Through incorporating chemical assembly techniques with fluidic colloids, inexpensive production of arrays of gold nanostructures suitable for biomolecular sensors with exceedingly low detection limits and enhanced photovoltaics are shown to be achievable.

REFERENCES

[1]   T. Jensen, L. Kelly, A. Lazarides, and G. C. Schatz, J. Clust. Sci., 10, 295-317, 1999.

[2]   S. Nie and S. R. Emory, Science, 275, 1102-1106, 1997.

[3]   K. Kneipp, Y. Wang, H. Kneipp, L. T. Perelman, I. Itzkan, R. R. Dasari, and M. S. Feld, Phys. Rev. Lett., 78, 1667-1670, 1997.

[4]   M. Bally, M. Halter, J. Vörös, and H. M. Grandin, Surf. Interface Anal., 38, 1442-1458, 2006.

[5]   P. Bao, A. G. Frutos, C. Greef, J. Lahiri, U. Muller, T. C. Peterson, L. Warden, and X. Xie, Anal. Chem., 74, 1792-1797, 2002.

[6]   G. Raschke, S. Kowarik, T. Franzl, C. Sonnichsen, T. A. Klar, J. Feldmann, A. Nichtl, and K. Kurzinger, Nano Lett., 3, 935-938, 2003.

[7]   J. H. Choi, S. M. Adams, and R. Ragan, Nanotechnology, 20, 065301-065306, 2009.

[8]   W. A. Lopes and H. M. Jaeger, Nature, 414, 735-738, 2001.

[9]   K. Nakayama, K. Tanabe, and H. A. Atwater, Appl. Phys. Lett., 93, 121904, 2008.

[10]   P. J. Schuck, D. P. Fromm, A. Sundaramurthy, G. S. Kino, and W. E. Moerner, Phys. Rev. Lett., 94, 017402, 2005.

[11]   M. Valden, X. Lai, and D. W. Goodman, Science, 281, 1647-1650, 1998.

[12]   T. S. Ahmadi, Z. L. Wang, T. C. Green, A. Henglein, and M. A. Elsayed, Science, 272, 1924-1926, 1996.

[13]   S. E. Skrabalak, L. Au, X. Li, and Y. Xia, Nat. Protoc., 2, 2182-2190, 2007.

[14]   H. Chen, X. Kou, Z. Yang, W. Ni, and J. Wang, Langmuir, 24, 5233-5237, 2008.

[15]   U. Y. Jeong, H. C. Kim, R. L. Rodriguez, I. Y. Tsai, C. M. Stafford, J. K. Kim, C. J. Hawker, and T. P. Russell, Adv. Mater., 14, 274-276, 2002.

[16]   Z. Grabarek and J. Gergely, Anal. Biochem., 185, 131-135, 1990.

Non-Silicone Based Nano Thickness Coating to Improve Protein Compatibility

Rosa Yeh, Michael T.K. Ling

Baxter Healthcare Corporation,
25212 W. Illinois Route 120, RLT-14
Round Lake, IL 60073, USA
rosa_yeh@baxter.com (847) 270-3422
michael_ling@baxter.com (847)270-4418

ABSTRACT

Medical devices have relied to a large extent on silicone oil surface treatment. A drawback of silicone oil is that it may induce aggregation of proteins in vivo or in vitro. The conformational change of proteins may lead the immune system to consider the aggregated protein as a foreign object, thus stimulating an immune response (Jones, Kaufmann, & Middaugh, 2005). Because of this concern, market demand for silicone-free medical devices has increased. The purpose of this study is to demonstrate technical concepts of silicone-free surface coating that could be used to design a protein compatible and lubricious surface for drug delivery devices.

A proprietary acrylate derivative coating material was developed and its effectiveness as a silicone oil coating material replacement for a syringe system was demonstrated. The materials, which can be synthesized by plasma polymerization for nano thickness coating, have excellent protein drug compatibility. Medical devices coated with the newly developed materials dramatically reduce the loss of expensive protein drugs because of its low protein absorption. In addition, the coating materials show better protein compatibility than silicone oil. The coating minimizes the risk of immunogenicity related to protein-silicone oil interactions.

This acrylate derivative coating is stable and provides good lubricity and water vapor barrier.

Keywords: nanocoating, silicone replacement, protein drug compatible coating, medical devices, syringe

1 INTRODUCTION

Silicone oil may induce aggregation of proteins. The conformational change of proteins may lead the immune system to consider the aggregated protein as a foreign object, thus stimulating an immune response (Jones, Kaufmann, & Middaugh, 2005). Because of this concern, market demand for silicone-free medical devices has increased. The purpose of this study is to demonstrate technical concepts of silicone-free surface coating that could be used to design a protein compatible surface.

2 METHOD

Using plasma polymerization technology (Figure 1), a few proprietary monomers containing acrylate derivatives were polymerized and covalently bonded onto a plastic substrate to form a nano-scale thin film.

The coating was applied to a Cyclic Olefin Copolymer (COC) (Figure 2) syringe material. After coating, the coated surfaces were characterized by contact angle measurement, protein adsorption, coating stability, syringe and piston sliding force, coated surface morphology, water vapor transmission rate (WVTR), and protein stability tests.

Radioactive iodine (I^{125}) labeled human serum albumin (HSA) and erythropoietins (EPO) were used for the protein adsorption study. 1.2 ml of I^{125} labeled protein solution was injected into a 3ml syringe (Figure 3). After incubation and rinsed, the amount of protein bound to the container surface was read by a gamma counter.

3 RESULTS

An acrylate derivative coating was successfully applied to the interior surface of a Cyclic Olefin Copolymer (COC) syringe. Water contact angle was measured on the coated and non-coated COC surfaces. The result showed that surface properties of the substrates changed dramatically from hydrophobic to hydrophilic after coating as shown in Figure 4. HSA and EPO protein adsorption were reduced by 50 times and 20 times respectively, when compared to a neat or silicone oil coated surface. The stability of the coating was evaluated by subjecting the test article to 20-40 kGy of gamma irradiation, followed by 3 weeks of aging at 60°C. Two types of COC syringes were evaluated in this study. There is no indication of increase in HSA adsorption after 3 week of aging at 60°C as shown in Figure 5.

The main concern of high piston sliding force in the absence of silicone oil coating was addressed via testing (Figure 6). It was demonstrated that the sliding force of a piston on a silicone oil coated syringe versus an acrylate derivative coated syringe barrel were comparable.

Another concern was a possible increase in water loss through the coating interface between the piston and barrel. A water loss study for a 3 ml ISO standard COC syringe suggested that the water vapor barrier for the coated syringe is better than or equivalent to the silicone oil coated syringe, Table 1.

4   CONCLUSIONS

A proprietary acrylate derivative coating material was developed and its effectiveness as a silicone oil coating material replacement for a syringe system was demonstrated. The materials, which can be synthesized by plasma polymerization, have excellent protein drug compatibility. Medical devices coated with the newly developed materials dramatically reduce the loss of expensive protein drugs because of its low protein absorption. In addition, the coating materials show better protein compatibility than the non-coated materials and silicone coated material. The coating may minimize the potential risk of immunogenicity related to protein-silicone oil interactions.

This proprietary acrylate derivative coating is stable and provides good lubricity and water vapor barrier.

5   FIGURES AND TABLE

Figure 3: Protein adsorption study.

Non-coated surface   Coated surface

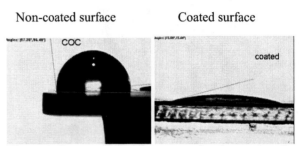

Figure 4: Contact angle of non-coated and coated surface.

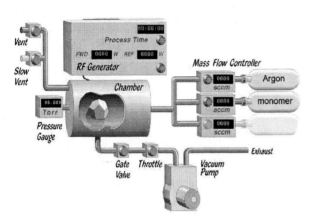

Figure 1: Plasma polymerization system.

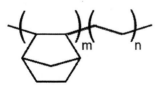

Figure 2: Chemical structure of COC.

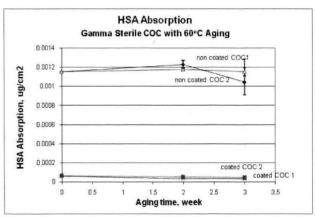

Figure 5: Polyolefin syringe HSA adsorption as a function of aging time.

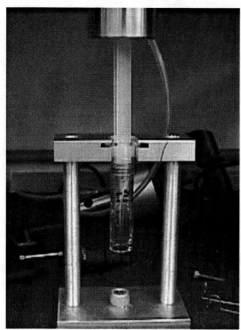

Figure 6: Syringe sliding force test.

Syringe	WVTR, g/day
Acrylic derivative coated syringe	5.68E-05
Silicone oil coated - syringe	8.64E-05

Table 1: Water vapor transmission rate (WVTR) for the acrylic derivative coated versus silicone oil coated syringe.

REFERENCE

[1] Jones, L. S., Kaufmann, A., & Middaugh, C. R. (2005). Silicone oil induced aggregation of proteins. *Journal of Pharmaceutical Sciences, 94 (4)*, 918-927.

Transparent Conducting Films of Antimony-Doped Tin Oxide with Uniform Mesostructure Assembled from Preformed Nanocrystals

Vesna Müller,[*] Jiri Rathousky,[**] Matthias Rasp,[*] Benedikt Schütz,[*] Markus Niederberger,[***] and Dina Fattakhova-Rohlfing[*]

[*]Department of Chemistry and Biochemistry, University of Munich (LMU), Butenandtstrasse 5-13, 81377 Munich, Germany, Dina.Fattakhova@cup.uni-muenchen.de
[**]J. Heyrovsky Institute of Physical Chemistry, Academy of Sciences of the Czech Republic, Dolejskova 3, 182 23 Prague 8, Czech Republic, jiri.rathousky@jh-inst.cas.cz
[***]Department of Materials, ETH Zürich, Wolfgang-Pauli-Strasse 10, 8093 Zürich, Switzerland, markus.niederberger@mat.ethz.ch

ABSTRACT

Transparent conducting films of antimony-doped tin oxide (ATO) with a uniform 3D-mesostructure are prepared by self-assembly of crystalline ATO nanoparticles with various antimony content directed by commercially available Pluronic copolymers. The mesostructure of the films calcined at 300 – 500 °C exhibits periodicity of 14 nm, surface area of 200 - 300 m²/cm³ and a well-developed accessible porosity of 45 - 55 %. The high crystallinity of the nanoparticles serving as building blocks enables to obtain the fully crystalline inorganic frameworks with sufficient electric conductivity already at temperatures as low as 300 °C. Ferrocene molecules covalently immobilized in the conducting mesoporous matrix show significantly enhanced electrochemical response proportional to the electrode surface area. The high electric conductivity and the uniform accessible mesoporosity combined with a simple and generally applicable preparation procedure make the developed ATO films attractive as the nanostructured transparent electrodes for various optoelectronic applications.

Keywords: Transparent conducting materials, antimony-doped tin oxide, nanoparticles, mesoporous films

1 INTRODUCTION

The recent rapid progress in nanoscience and nanotechnology has lead to the development of completely new nanoscale materials. Novel morphologies in already known materials have substantially extended their application potential. Transparent conducting oxides are known as thin dense films, being indispensible for touch panel displays, OLEDs or non-silicon solar cells. However, other types of optoelectronic devices, especially those involving immobilized species or functional layers, require conducting networks rather than flat electrode interfaces. Such 3D-electrode architectures with very high interface area are suitable as conducting hosts accommodating functional guest species, which allows a substantially more efficient electron transfer and consequently increased efficiency of the devices.

Some years ago we reported the preparation of 3D-networks of indium tin oxide (ITO) and demonstrated their potential as efficient electrode systems [1, 2]. Of great interest is the extension of such architectures to other 3D-organized transparent conductors replacing the rare and expensive indium and providing a different surface chemistry or energy level properties. One widely used system is extrinsically doped tin oxide, such as fluorine- or antimony-doped tin oxide (ATO). Very recently, two communications reporting mesoporous ATO electrodes were published [3, 4] and their potential for, e.g., fabrication of efficient electrochemiluminescence biosensors was demonstrated [3]. However, the manufacturing of 3D-transparent electrodes still remains a non-trivial task, hindering the full exploitation of their application potential. All the reported methods are practically solely based on self-assembly of sol-gel derived metal oxide precursors assisted by amphiphilic structure-directed agents (the so called Evaporation-Induced Self-Assembly process). As the formed inorganic frameworks are amorphous, they have to be treated at higher temperatures to induce the crystallization indispensible for electron conductivity. Such phase transformation without the collapse of mesostructure was achieved for ATO films only using special amphiphilic copolymers of the KLE type [4] with outstanding templating properties and the increased thermal robustness, which however are not commercially available. Easily available and rather cheap copolymers of the Pluronic family, successfully used for preparation of mesoporous crystalline metal oxides, have failed due to the different sol-gel chemistry of tin. The formation of a crystalline mesoporous structure of tin oxide requires a special post-synthetic treatment at carefully controlled humidity and temperature for several days, which is rather time-consuming.

Consequently, the development of a simple and generally applicable procedure for the preparation of transparent conducting 3D-frameworks using easily available components is highly desirable. One passable solution might be the replacement of amorphous metal oxide precursor by crystalline building blocks, which could form the mesoporous crystalline network at much milder conditions. The pre-formed nanocrystals could also bring the additional advantage of a fine adjustment of the doping level and thus the electric conductivity, which is often better controlled in the particle synthesis. Nevertheless, the use of nanocrystalline building blocks for the preparation of mesoporous materials has been reported only for a few oxides, probably due to the fact that it is rather difficult to prepare stable and agglomeration-free nanoparticle dispersions.

Recently we have reported the preparation of crystalline monodisperse ATO nanoparticles 3-4 nm in size by a non-aqueous approach, which utilizes cheap and easily available metal precursors, being completed after heating the precursor mixture for several hours at 100-150°C in a closed glass vessel, without the need of an autoclave [5]. The obtained nanocrystals exhibit a high electric conductivity and are dispersible at high concentration in both water and a number of organic solvents. Consequently, these nanocrystals are suitable building blocks for assembly into a mesostructure. In this communication we present a procedure enabling the assembly of such crystals to mesoporous conducting films, this process being directed by commercially available Pluronic copolymers. The parameters influencing the self-assembly properties have been analyzed in detail, as well as the texture properties, crystallinity and conductivity of the obtained films.

2 RESULTS AND DISCUSSION

In spite of the apparent simplicity of the approach based on the directed assembly of nanoparticles to mesostructures (Scheme 1), their use as the building blocks is far from being straightforward because of a number of strictly required particle properties. In addition to a suitable morphology and a good dispersibility of the particles themselves, their dispersions should be compatible with the added structure-directing agent. Moreover, a suitable type of interaction should exist between particles and polymer molecules acting as a driving force for their self-assembly to the desired supramolecular structures. We have found that these interactions depend to a large extend on the bulk as well as surface chemical composition of the particles and on the solvent. Even if the ATO particles are well dispersible in polar solvents such as water or ethanol, the addition of Pluronic copolymers causes the precipitation of the particles. Tetrahydrofurane was found to be so far the only solvent, in which ATO nanoparticles form stable colloidal dispersions in the presence of Pluronic copolymers. The antimony percentage in the ATO nanoparticles is another important parameter governing the dispersibility and the self-assembly of particles. To form stable colloidal dispersions in the presence of the Pluronic

copolymer, the percentage of antimony should exceed 6 mol % (in the investigated range from 0 to 40 mol %).

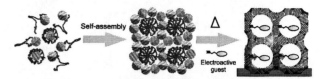

Scheme 1. The formation of mesoporous crystalline electrode layers by the self-assembly of crystalline nanoparticles directed by amphiphilic molecules. The arrows (right) indicate the direct electron flow to redox species immobilized in the 3D-conducting framework.

The surface composition of the ATO nanoparticles was found to be of primary importance for their self-assembling properties and the thermal stability of the formed mesostructures. When the as-synthesized particles were used for the film preparation, only a thermally unstable lamellar structure was formed. According to the thermogravimetry data, such particles still contain ca. 50 % of adsorbed benzyl alcohol and the benzyl alcoholate moieties, whose removal requires washing in organic solvents. Finally, as shown below, only washed ATO nanoparticles with Sb content exceeding 6 % could be assembled in the presence of the Pluronic copolymer to thermally stable periodic mesostructured films. The observed influence of the composition of the particles and their surface chemistry on their self-assembly behavior suggests an important role of the coordination of the Pluronic block-copolymer molecules, presumably of their hydrophilic polyethyleneoxide blocks, on the particle surface. Although we can only speculate about the structure of the formed coordination complexes due to the lack of the currently available experimental data, it will be surely influenced by the presence of the surface-adsorbed organic moieties as well as the surface charge and composition of the particles. As was demonstrated by the zeta-potential measurements, the point of zero charge (pzc) of the ATO particles changes linearly with the Sb content. This altering surface charge and the replacement of the surface Sn atoms by Sb ones, as demonstrated by the Raman measurements [5], could be one of the reasons of the remarkably different self-assembly behavior of Sb-doped particles compared to the undoped ones.

In the final step of the mesoporous ATO films assembly, the Pluronic-containing colloidal dispersion in THF of washed ATO nanocrystals with doping levels from 6 to 30 mol % (assigned further as ATO_6 etc) was coated on various substrates and calcined in air at temperatures varying from 300 °C (Pluronic combustion temperature) to 500 °C in order to remove the template and to sinter the nanoparticles. The obtained data show that 300 °C is enough to remove all the organics from the material. Scanning electron microscopy (SEM) images of the films (Fig. 1b) prove the very good uniformity of the films and their crack-free character. Transmission electron microscopy (TEM) images (Fig. 1a) show that the nanoparticles with different doping levels form similar films with highly porous uniform 3D-architectures. The wormhole mesostructure contains ca. 100 nm large domains with some indication of pore ordering. The Fourier-transform spectra of the films calcined at 300 – 400 °C show two rings corresponding to the mesostructure periodicity of 14±1 and 8±0.5 nm, respectively (Fig. 1a, inset). The electron microscopy data are in a good agreement with small-angle XRD patterns, showing two reflections due to the pore organization with a similar mesostructure periodicity (Fig. 2a). The first reflection corresponding to a mesostructure d-spacing of ca. 14 nm slightly shifts to the higher angles due to a shrinkage of mesostructure in a direction perpendicular to the substrate. In contrast to sol-gel derived films, the mesostructure shrinkage of the films assembled from nanoparticles is only marginal, indicating high temperature stability of the mesoporous framework. The second ring/reflection, corresponding to higher order of periodicity, becomes weaker with the increase in the calcination temperature due to the increasing distortion in the mesostructure ordering.

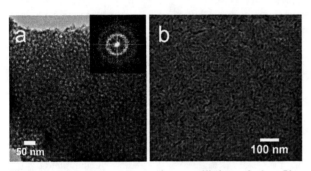

Figure 1. Mesostructure and crystallinity of the films assembled from ATO_20 nanoparticles and calcined at 400 °C.: (a) TEM image, inset: the Fourier transform spectra; (b) SEM image (top view).

High-resolution TEM images of the mesoporous ATO films show the complete crystallinity of the pore walls, being composed of similarly sized nanocrystals. The d-spacing of the crystalline fringes and WAXS patterns of the film (Fig. 2b) correspond solely to the cassiterite SnO_2 phase without any traces of antimony oxides, indicating that no phase separation occurred during the temperature treatment. The crystal size determined from the line broadening of the WAXS patterns (Fig. 2b) slightly grows from 3.0 nm (size of the particles used for the film assembly [5]) to 4.3 nm after calcination at 500 °C. The sintering of nanoparticles within the pore walls leads to a slight mesostructure shrinkage and thus some deterioration of the mesostructure periodicity. The typical film thickness was 210 ±10 nm.

The texture properties and the accessibility of the internal surface of the ATO thin films were studied by krypton adsorption at ca 77 K. The Kr adsorption isotherms

on the films calcined at 300 to 500 °C (Fig. 2c) show that all the materials exhibit well-developed mesoporosity without any pore-blocking. While the variation in the calcination temperature does not essentially change the character of the porosity, there are some quantitative changes in the texture characteristics. The specific surface area related to the total film volume slightly decreases with increasing calcination temperature, but remains remarkably high achieving 300, 270 and 200 m^2/cm^3 for calcination temperatures of 300, 400 and 500 °C, respectively. The porosity equals 45, 30 and 55 % for the calcination temperatures as above. The pore width slightly increases with increasing calcination temperature from ca 6 nm (for 300 and 400 °C) to ca 8 nm (for 500 °C).

The Kr adsorption results in combination with the electron microscopy and X-ray diffraction data indicate a high thermal stability of the mesostructure assembled from the pre-formed nanocrystals.

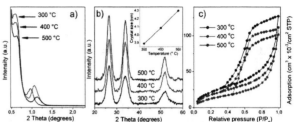

Figure 2. Characterization of the mesoporosity and crystallinity of the films assembled from ATO_20 nanoparticles and treated at different temperatures: (a) small angle X-ray diffraction, (b) wide angle X-ray diffraction and (c) Kr adsorption isotherms. The inset in the Fig. 2b shows dependence on the calcination temperature of the crystal size estimated from the peak broadening.

The variation of the electric conductivity of the mesoporous ATO films with the Sb doping level in the precursor nanoparticles and the calcination temperature was determined. The film conductivity is the highest when the film is assembled from particles containing 8-10 % of Sb (Fig. 3a). These results are in good agreement with the conductivity of the nanoparticles themselves, which achieves a maximum at 4-10 % of Sb [5]. Conductivity of the films reasonably increases with increasing calcination temperature, which is due to the particle sintering, a slight increase in the crystal size and the related increase in the electron mobility. The highest conductivity of 0.8 S cm^{-1} was obtained for the film assembled from the ATO_8 nanoparticles and calcined at 500 °C. This is practically the same conductivity which was obtained for the parent ATO_8 nanoparticles in pressed pellets treated at the same temperature (1 S cm^{-1}) [5], being quite high for a film with ca. 50 % of porosity.

To test of the applicability of mesoporous ATO films as electrode layers, they were deposited on flat conducting ITO substrates in order to minimize ohmic losses. Ferrocene moieties were covalently anchored via peptide bonds to their surface as standard redox probes. Fig. 3b shows an example of electrode performance of the film assembled from the ATO_10 nanoparticles calcined at different temperatures.

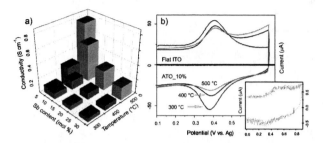

Figure 3. (a) Variation of the electric conductivity of the mesoporous films assembled from ATO nanocrystals with Sb content in the precursor nanoparticles and the calcination temperature of the prepared films. (b) The cyclic voltammograms of the ferrocenecarboxylic acid covalently anchored on the mesoporous ATO_10 layers calcined at 300 °C, 400 °C and 500 °C and on flat ITO electrode (black and inset). Scan rate is 50 mV s^{-1}.

The voltammograms of all samples exhibit Gaussian-shaped peaks of reversible ferrocene oxidation and reduction, characteristic for modified electrodes with immobilized redox species. Both double layer charging and Faradaic currents of the ferrocene oxidation and reduction on the mesoporous electrodes drastically increase compared to those on the flat ITO electrode modified in the same way. The surface coverage of the electrochemically accessible ferrocene molecules obtained by the integration of peak-area corresponds to $1.9 \cdot 10^{-11}$ mol cm^{-2} for the flat ITO electrode and $9.4 \cdot 10^{-10}$ mol cm^{-2} (50-times), $6.5 \cdot 10^{-10}$ mol cm^{-2} (35-times) and $5.5 \cdot 10^{-10}$ mol cm^{-2} (29-times) for the mesoporous ATO electrodes calcined at 300 °C, 400 °C and 500 °C, respectively. The numbers in brackets indicate corresponding increase in surface coverage compared to the flat ITO electrode. This increase in the Faradaic charge is in excellent agreement with the specific surface areas of the films obtained by Kr adsorption measurements. Consequently, the mesoporous ATO films calcined at lower temperatures are suitable for the application as transparent electrode layers with the very high surface area, enabling direct electrochemical addressing of each redox molecule immobilized on its surface. The surface area of the mesoporous ATO layers seems to play a more important role in enhancing the charge collection efficiency compared to their specific conductivity. This can be due to the fact the total serial resistance of the "sandwich" electrodes composed of porous conducting layer on a dense conducting substrate is sufficiently low to provide an efficient electron transport. The transparency of the films assembled from the particles with different Sb content is above 93 % of transmission, which makes them promising candidates for electrode layers for the optoelectronic applications.

3 CONCLUSIONS

Transparent conducting layers of antimony-doped tin oxide with a uniform mesoporosity and crystalline framework can be easily manufactured by a directed assembly of pre-formed ATO nanocrystals using commercially available and cheap polymers of the Pluronic family [6]. The crystallinity of the nanoparticles serving as building blocks enables to obtain the fully crystalline inorganic frameworks with sufficient electric conductivity already at temperatures as low as 300 °C without any need for the elaborate post-synthetic treatment. Another attractive feature of this approach is the high thermal stability of the obtained mesostructure exhibiting practically no shrinkage due to the temperature treatment, which is a typical disadvantage of amorphous sol-gel derived precursors. The use of preformed nanocrystals, whose preparation can be controlled with a high precision, ensures a high reproducibility. Owing to the open and accessible character of porosity, the high surface area and the uniform pore size, the obtained mesoporous frameworks are promising host materials for the accommodation of functional redox moieties. The high potential of the obtained mesoporous layers as nanostructured transparent electrodes with the high surface area was demonstrated on an example of ferrocene molecules which were covalently immobilized in the conducting matrix, showing the significantly enhanced electrochemical response proportional to the electrode surface area.

ACKNOWLEDGEMENT

The authors are grateful to the German Research Foundation (DFG, grant No. FA 839/1-1) and the Grant Agency of the Czech Republic (grant No. 104/08/0435-1) for the financial support.

REFERENCES

[1] D. Fattakhova Rohlfing, T. Brezesinski, J. Rathouský, A. Feldhoff, T. Oekermann, M. Wark, B. Smarsly, Adv. Mater. 18, 2980, 2006.

[2]. D. Fattakhova-Rohlfing, T., B. Smarsly, J. Rathouský, Superlattices and Microstructures 44, 686, 2008.

[3] K. Hou, D. Puzzo, M. G. Helander, S. S. Lo, L. D. Bonifacio, W. Wang, Z.-H. Lu, G. D. Scholes, G. A. Ozin, Adv. Mater. 2009, 21, 2492.

[4] Y. Wang, T. Brezesinski, M. Antonietti, B. Smarsly, ACS Nano 2009, 3, 1373.

[5] V. Müller, M. Rasp, G. Štefanić, S. Ba, S. Günther, J. Rathousky, M. Niederberger, D. Fattakhova-Rohlfing, Chem. Mater. 21, 5229, 2009.

[6] V. Müller, M. Rasp, J. Rathousky, M. Niederberger, D. Fattakhova-Rohlfing:, Small, 2010, 6, 633.

Correlation Between the Oxygen Content and the Structure of AlCrSiO$_x$N$_{(1-x)}$ Thin Films Deposited by Pulsed DC Magnetron Sputtering

H. Najafi[*], A. Karimi[*] and M. Morstein[**]

[*] Institute of Condensed Matter Physics (ICMP), Ecole Polytechnique Fédérale de Lausanne (EPFL),
CH-1015, Lausanne, Switzerland, hossein.najafi@epfl.ch, ayat.karimi@epfl.ch
[**] Platit AG, Advanced Coating Systems, CH-2545 Selzach, Switzerland, m.morstein@platit.com

ABSTRACT

Al$_{52}$Cr$_{40}$Si$_8$(N$_{1-x}$O$_x$) thin films were prepared using pulsed DC magnetron sputtering. Two series of films were deposited at 400ºC and 650ºC by changing the O/(O+N) in the films from 0% (pure nitrides) to 100% (pure oxides). The obtained films were characterized by various techniques including XRD, XPS, SEM, TEM and nanoindentation. XRD results revealed that in the nitride region, films are crystallized in the cubic (B1) structure, which survives more at 650ºC rather than 400ºC against the oxygen incorporation. In the transition region between nitrides and oxides, the structure collapses to completely amorphous network and hardness was decreased to 12 GPa. In the oxide region, the formation of nanocrystallites of α-(Al,Cr)$_2$O$_3$ was observed in particular by the appearance of (113) and (116) reflections in XRD and TEM. The films hardness in this region was increased again to 30-35 GPa.

Keywords: oxynitride, XPS, nanoindentation, amorphous network, TEM

1 INTRODUCTION

Thin transition metal oxynitride films have led to another class of thin films, which appear to be a simple combination of oxides and nitrides but present new properties and thus allow for outstanding applications [1-3]. Recent years have seen a growing interest in multi element oxynitride films due to their ability to display a variety of properties that cannot be obtained by single-phase films [4]. There are several reasons for choosing multi-element oxynitride films, some of them are diffusion barrier resulting in improved oxidation resistance and chemical stability, thermal barrier, dielectric properties, solid solution hardening and grain size hardening. In this contribution, our aim is to investigate the structural properties of ternary oxynitride thin films of Al-Cr-Si-O-N, covering from pure nitrides to pure oxides. Al-Cr-O-N hard thin films are attracting much attention as the replacement to Ti-Al-O-N films because of the higher solubility of AlN into CrN (i.e. maximum 77%) in the cubic structure (B1) in comparison with the solubility of AlN into TiN (i.e. maximum 65%) system [5]. In our previous research work [6], we studied Al-Cr-Si-O-N system in the hexagonal

crystal structure (B4), whereas currently we are studying this system in the cubic structure. For this purpose, an Al$_{52}$Cr$_{40}$Si$_8$ target was used to deposit films varying from pure nitrides to pure oxides by changing the O/(O+N) ratio in the films. Then the change in microstructure as well as mechanical properties with the oxygen content are discussed.

2 EXPERIMENTAL

The oxynitride thin films were deposited using pulsed DC (f = 100 kHz, t = 2 μs) magnetron sputtering of an Al$_{52}$Cr$_{40}$Si$_8$ target. The [100]-oriented silicon wafer was employed as substrate and a mixed gas of Ar, N$_2$ and O$_2$ was applied for sputtering. In order to track the properties of the obtained films from oxygen incorporation, two series of films were deposited at 400ºC and 650ºC by changing the O$_2$ / (O$_2$+N$_2$) ratio in reactive gas from 0% to 100%. The flow rates (sccm) of Ar, N$_2$ and O$_2$ gases were controlled via mass flow controller, and the relative partial pressure of reactive gases over the total gas flow was maintained at 30%. Other deposition parameters consisting base pressure in the chamber, working pressure, DC input power and substrate bias were fixed at 3×10^{-3} Pa, 4.5×10^{-1} Pa, 150 W and -50 V, respectively. The films thickness was kept almost the same around 2 μm. The crystallographic structure was investigated by X-ray diffraction using monochoromatised CuKα radiation in grazing incidence (4º) geometry. The chemical composition of the achieved films was evaluated via surface X-ray photoelectron spectroscopy (XPS). The XPS data were collected by Axis Ultra (Kratos analytical, Manchester, UK) under ultra-high vacuum condition (<10^{-3} Pa), using a monochromatic Al K_a X-ray source (1486.6 eV). Gold (Au 4$f_{7/2}$) and copper (Cu 2$p_{3/2}$) lines at 84.0 and 932.6eV, respectively, were used for calibration, and the adventitious carbon 1s peak at 285eV as an internal standard to compensate for any charging effects. Both high resolution scanning electron microscopy (SEM-XLF30-SFEG) operated at 5 kV and and transmission electron microscopy (TEM), using Philips CM-20 equipment working at 200 kV, were employed in order to track the microstructures of the selected films. To prepare a suitable sample for TEM, films were thinned by mechanical polishing on diamond pads and then subjected to ion-milling for final thinning up to transparency to the electron beam. Hardness of samples was determined by the nanoindentor XP with a Berkovitch-type pyramidal diamond tip, indenting the films under continuous stiffness measurement method to a maximum depth of 1000 nm, but the values corresponding to the penetration depth of 100 - 200 nm were assigned as film hardness.

3 RESULTS AND DISCUSSION

3.1 Chemical Composition

The atomic composition of the Al-Cr-Si-O-N films was determined from the XPS measurements. Fig.1a illustrates the chemical composition of films, which were deposited at 650°C versus oxygen fraction ($O_2/(O_2+N_2)$) in the reactive gas. It is noteworthy that there are three different film structures with respect to the oxygen ratio in the reactive gas: nitride films with presence of oxygen in the nitride lattice (cation/anion = 1), films in the transition mode from nitride to oxide structure with no particular lattice, and oxide films with the presence of nitrogen in the oxide lattice (cation/anion = 0.66). The variations of Al/Cr ratio with different oxygen fraction were also plotted in Fig. 1a. It is noticeable that Al/Cr ratios are constant around 1 at lower oxygen fractions (nitride mode), which is lower than the composition of target (i.e. Al/Cr = 1.3). As the oxygen content of the films is raised, the Al/Cr ratio increases from 1 to 1.8 (oxide mode). This could be related to target poisoning that chromium oxide is formed more than aluminum oxide and sputtering of chromium oxide is more difficult than aluminum oxide.

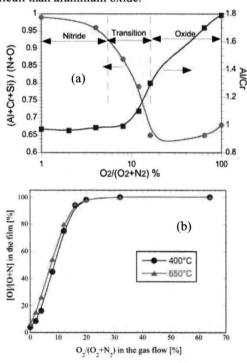

Fig. 1. Variations of cation/anion and Al/Cr ratios of the films (a) and [O]/[O+N] ratio in the films (b) as a function of oxygen fraction ($O_2 / (O_2+N_2)$) in the gas flow.

It is noticeable from Fig. 1b that the incorporation of oxygen into the films increases much faster than the fraction of oxygen in the reactive gas, so that the oxides can be formed from the oxygen fraction of about $O_2 / (O_2+N_2) = 20\%$.

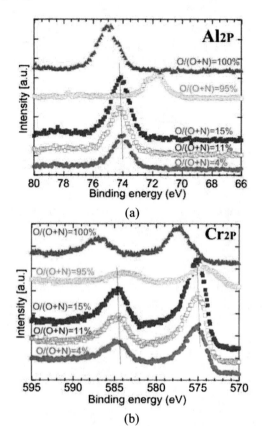

Fig. 2. Al_{2p} (a) and Cr_{2p} (b) XPS spectra from nitrides to oxides.

Fig. 2a-b shows the chemical environment of the Al and Cr species based on XPS chemical bonding analysis of the Al_{2p} and Cr_{2p} peaks. The trends of the peak shifts and peak intensities revealed that oxides increase in the film on the expense of nitrides as oxygen was increased. The position and the width of the peaks implied that it contained contributions from both nitride and oxide. Al_{2p} and Cr_{2p} had a peaks at 74 eV and 575 eV respectively that broadened and shifted to higher energies for increasing oxygen content except for one which corresponds to the transition region that causes the peak to shift to lower energies. This could confirm that chemical environment in the transition region is totally different in comparison to others. In fact at the transition region, quantitative XPS data indicates that stoichiometry of film is much like oxides. As N concentration increases in the expense of oxygen, N may start substituting for O in this bonding configuration. As the valence of N (-3) is substantially more negative than the oxygen valence of -2, this dictates a formal Al valence shifts to a value greater than +3 in the ionic limit of bonding, then the binding energy of electron decreases in XPS measurement [7].

3.2 Phase Content and Crystalline Structure

Fig. 3a-b exhibits the grazing incidence XRD spectra of two different series of films deposited at 650°C and 400°C. The results reveal that in both temperatures, nitride films of (Al,Cr)N crystallize in the rock salt cubic type of structure (B1) with (111) and (200) peaks being the highest. One can note that the film structure considerably depends on the oxygen fraction. With the increasing of oxygen fraction in the films, the preferred orientation is changed from (111) to (200) as confirmed also by Bragg-Brentano (θ-2θ) which are not shown here. Furthermore, it is noteworthy that cubic structure survives more at 650°C than 400°C against the oxygen incorporation in the films. This can be described by higher atomic diffusion and crystallinity at high temperatures. In the transition region, films are completely amorphous, as also confirmed by TEM. The oxide films deposited at 650°C exhibit diffused peaks at 43.5° and 57.5° corresponding to (113) and (116) of α-(Al,Cr)₂O₃, respectively (pdf n° 43-1484).

Films deposited at 400°C in oxide region display the existence of $(Al,Cr)_2O_3$ in spite of low substrate temperature. These findings are consistent with those reported in [8] confirming the role of Cr in low-temperature growth of Al_2O_3. It is also in accordance with the results of mechanical properties presented in the next sections (Sec. 3.4)

3.3 Microscope Study

The morphology of the films deposited at 650°C with different composition has been observed along the plan view SEM and reported in Fig. 4a-c. In the nitride region a granular surface of small crystallites with a grain size of approximately 20 nm was observed. It is noteworthy that at higher oxygen concentrations, for all films initially, the structure collapses into a completely amorphous network followed by the formation of nanocrystallites. This is related to the fact that the mobility of oxygen in the growth surface of amorphous film is not sufficient to form oxide lattice.

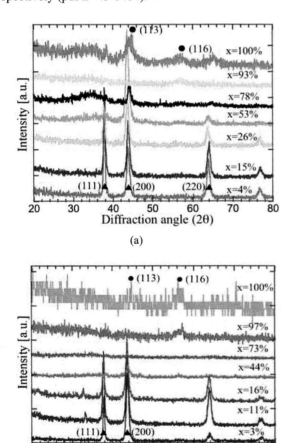

(a)

(b)

Fig. 3. Grazing incidence XRD patterns of oxynitride films for different concentrations of oxygen x=O/(O+N), deposited at a) 650°C and b) 400°C. The symbols ▲ and ● are c-(Al,Cr)N and (Al,Cr)₂O₃ respectively.

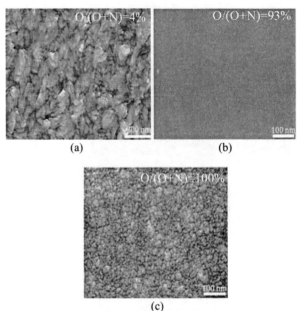

(a) (b)

(c)

Fig. 4: Plane view SEM micrographs showing microstructure of films with different ratio of O/(O+N) in the film:a) O/(O+N) = 4% b) O/(O+N) = 93% c) O/(O+N) =100%

The TEM cross-sectional image and electron diffraction pattern of the films are shown in Fig. 5. The films containing less oxygen (O/(O+N) = 4%) had a nano-columnar structure (Fig. 5a) and diffraction pattern could be indexed using the cubic (Al,Cr)N rings. According to the Fig. 5b, in the transition region, film was found to be completely amorphous since the diffraction pattern recorded on the film is made of diffuse rings. As depicted in Fig. 5c, films with oxide structure become well-oriented nano-crystalline films and the diffraction pattern could be indexed to (113) and (116) which is in good agreement with the XRD results.

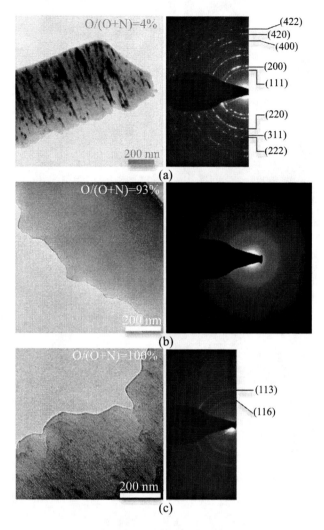

(a)

(b)

(c)

Fig. 5. Cross-sectional TEM micrographs showing the microstructure and diffraction patterns of the films with different ratio of O/(O+N) in the film: a) O/(O+N)= 4%,b) O/(O+N) = 93% c) O/(O+N) = 100% ,

3.4 Mechanical Properties

The variations of nano-hardness of the films, as a function of oxygen fraction in the reactive gas ($O_2/(O_2+N_2)$) are presented in Fig. 6. The existence of three different modes of films, which correspond to nitride region, transition region and oxide region are obvious as also observed in the chemical analysis and XRD results. For all specimens, as the oxygen fraction increases from 0% to 8%, both hardness and modulus of films alter slightly, this can be dependent to the formation of metal vacancies together with change of chemical bonding in the nitride lattice. Significant decline of hardness from 36 GPa to 16GPa was observed upon increasing the oxygen fraction in the flow gas from 8% to 12%. It is notable that the mechanical properties remain low until the oxygen fraction of 20%. Low mechanical properties of the films in transition mode can be explained by the fact that those are completely amorphous and do not exhibit the peaks of nitride or oxide

phases. By increasing the fraction of oxygen in the reactive gas from 24% to 100% (oxide region), hardness again takes higher values (33-35 GPa), corresponding to the formation of oxides. In the oxide region, films deposited at 400°C exhibit elevated mechanical properties in spite of low temperature for oxide formation. This can be explained by the role of Cr for low-temperature growth of $(Al,Cr)_2O_3$ [8].

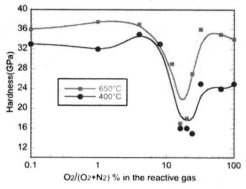

Fig. 6. Hardness of the films as a function of oxygen fraction in the reactive gas for two series of films deposited at 400°C and 650°C.

4 CONCLUSION

Both structural and mechanical properties of oxynitride thin films of Al-Cr-Si-O-N were probed versus oxygen concentration. The nitride lattice survives to incorporating of 53% oxygen, and the oxide lattice starts to crystallize when the oxygen content passes beyond 93%. In between, amorphous oxynitride films are formed.

ACKNOWLDGEMENTS
The authors are thankful to the Swiss Commission for Technology and Innovation (CTI) for financial support. They would like also to thank CIME-EPFL for providing electron microscopy facilities.

REFERENCES

[1] J. Ye, S. Ulrich, C. Ziebert, M. Stüber, Thin Solid Films, 517, 1151, 2008.

[2] C. C. Lung, C. L. Shu, W. T. Ko, Solid-State Electronics, 50, 103, 2006.

[3] W. Liu, X. P. Su, S. Zhang, H. B. Wang, J. H. Liu, L. Q. Yan, Vacuum, 82, 1280, 2008.

[4] A. Karimi, M. Morstein, T. Cselle, Surf. Coat. Technol, In press, 2010.

[5] Y. Makino, ISIJ Int. 38, 925, 1998.

[6] H. Najafi, A. Karimi, M. Morstein, Proceeding of 37th international conference on metallurgical coatings & thin films, USA, 2010.

[7] J. R. Smith, P. C. J. Graat, D. A. Bonnell, R. H. French, Mat. Res. Soc. Symp. Proc., 2001.

[8] J. M. Andersson, E. Wallin, U. Helmersson, U. Kreissig, E. P Munger, Thin Solid Films, 513, 57, 2006.

DSSC Photoanodes Fabricated with Mesoporous TiO₂ Hollow Spheres

Chang-Yu Liao[*], H. Paul Wang[*,**], and Hong-Ping Lin[***]

[*]Department of Environmental Engineering, National Cheng Kung University,
Tainan 70101, Taiwan, wanghp@mail.ncku.edu.tw
[**]Sustainable Environment Research Center, National Cheng Kung University,
Tainan 70101, Taiwan
[***]Department of Chemistry, National Cheng Kung University, Tainan 70101, Taiwan

ABSTRACT

To enhance solar energy efficiency of a dye-sensitized solar cell (DSSC), high crystallinity mesoporous TiO₂ hollow spheres (MHS-TiO₂) have been coated on the DSSC photoanode. The MHS-TiO₂ was prepared by calcination of the Ti(OBu)₄ coated mesoporous carbon hollow spheres at temperature of 873 K. Observations by XRD, FE-SEM, TEM, and HR-TEM suggest that the MHS-TiO₂ is composed of mainly fine anatase nano structures. The specific absorption edge and band gap of the MHS-TiO₂ determined by diffuse reflectance ultraviolet-visible spectroscopy are 200-400 nm and 3.44 eV, respectively. By X-ray absorption near edge structure (XANES) and extended X-ray absorption fine structure (EXAFS), it is found that the MHS-TiO₂ possesses Ti-O and Ti-(O)-Ti bond distances of 1.98 and 3.07 Å with coordination numbers of 2.9 and 0.7, respectively. The current-voltage (I-V) curves of the DSSC utilizing the MHS-TiO₂ photoanodes were determined with an AM 1.5G solar simulator. Notably, the photovoltaic performances of the DSSC are 0.714 V in the open-circuit voltage, 2.52 mA·cm⁻² in the short-circuit current density, and 0.666 in the fill factor, and a better conversion efficiency (1.2%) is obtained.

Keywords: DSSC, TiO₂, mesoporous TiO₂, EXAFS

1 INTRODUCTION

Mesoporous silica materials, coated with carbon jackets and modified by metal oxide skin layers, are widely utilized for super capacitors in electron storage [1] and photocatalysts in environmental pollution control [2]. Very recently, the use of titanium containing meso- and microporous molecular sieves and zeolites and the Ti/MCM-41 in photoanodes of a dye-sensitized solar cell (DSSC) for photovoltaic demonstration has been documented by Atienzar and coworkers [3, 4].

The DSSC photoanodes play the major key functions on loading of sufficient sunlight-absorbing dyes and effective transportation of photoexcited electrons generated on the surface of metal oxide thin films [5]. Nanostructured anatase TiO₂ possesses high surface area and abundant mesopores while immobilized on the transparent conducting oxide (TCO) glass in the form of a thin film, is

desired for photoanode materials in DSSCs [6]. However, little effort is dedicated to use the mesoporous TiO₂ hollow spheres (MHS-TiO₂) in DSSC photoanodes.

Speciation data such as bond distance, coordination number (CN), and chemical identity of select elements involved in the disorder matrix can be determined by extended X-ray absorption fine structure (EXAFS) spectroscopy. With the benefits of EXAFS, migration routes and extraction pathways of copper and zinc species can be elucidated during electrokinetic treatments of copper-rich sludges and recovery of nanosize zinc from phosphor wastes with an ionic liquid [7, 8]. The decrease of the CN of Cu-Cu with decreasing sizes of Cu in the Cu@C deposited on the counter electrode of a DSSC, of which having 80% enhancement in the conversion efficiency by use of smaller Cu@C particles (e.g., 20→7 nm) has also been found [9].

The informative X-ray absorption near edge structure (XANES) data can also indentify the oxidation states, coordination geometry, and local bonding environment of hard X-ray absorbing atoms existed in complicated solid matrixes. Using XANES, Hsiung and coworkers have reported that the A₂ active species were the mainly surface photoactive sites in nanosize TiO₂ during photocatalytic degradation of methylene blue and 2-chlorophenol [10, 11]. By XANES, it was found that dispersion of Cu@C in the DSSC electrolyte may improve its performances [12, 13]. Thus, the main objective of the present work was to study chemistry structure of the MHS-TiO₂ employed in photoanodes of a DSSC by EXAFS and XANES. Solar conversion efficiency of DSSCs having the MHS-TiO₂ photoanode was also measured on a sunlight simulator.

2 EXPERIMENTAL

Detailed synthesis procedures of the MHS-TiO₂ has been reported [14]. Briefly, 0.4 g of titanium(IV) n-butoxide (Ti(OBu)₄) (Acrös) were added into an acidic ethanolic solution containing 20 g of ethanol (95%) and 1 g of concentrated nitric acid to yield a titanium-to-carbon ratio (Ti/C) of about 0.02 mol·mol⁻¹. The acidic ethanolic titanium precursor was vigorously mixed with 0.4 g of mesoporous carbon hollow sphere templates for 1 d at 298 K. The titanium xerogel encapsulated mesoporous carbon hollow sphere was obtained by drying the mixture at 333 K for solvent removal. The MHS-TiO$_2$ was thus obtained by calcining the titanium xerogel encapsulated mesoporous carbon hollow sphere at 873 K for 8 h for the removal of carbon templates.

The crystalline structure of the MHS-TiO$_2$ was characterized on an X-ray diffraction (XRD) spectrometer (Rigaku Ultima$^+$) scanned from 10 to 80° (2θ) at a scanning rate of 3°·minute^{-1}. The surface topologic images of the MHS-TiO$_2$ were analyzed by field-emission scanning electron (FE-SEM) microscopy (Philips, XL-40FEG) under an accelerating voltage of 10 kV. The sectional morphologic images of the MHS-TiO$_2$ were determined by high-resolution transmission electron microscopy (HR-TEM) (FEI E.O, Tecnai F20 G2). The nano-beam diffraction (NBD) patterns of the MHS-TiO$_2$ were also measured. The diffuse reflectance ultraviolet-visible (DR UV-vis) spectrum of the MHS-TiO$_2$ was measured on a spectrophotometer (Varian, Cary 100 Conc) at 200-800 nm.

The EXAFS and XANES spectra of the MHS-TiO$_2$ were collected on the Wiggler beamline (17C) at the Taiwan National Synchrotron Radiation Research Center. The electron storage ring was operated at an energy of 1.5 GeV. A Si(111) double-crystal monochromator was used for energy selection at an energy resolution (ΔE/E) of 1.9×10^{-4} (eV/eV). The beam energy was calibrated against the absorption edge of a titanium foil at 4966 eV. The EXAFS spectrum of the MHS-TiO$_2$ was refined and analyzed using the UWXAFS 3.0 and FEFF 8.0 simulation programs [15, 16]. The Fourier transform coupled with a Bessel function was performed to convert the k^3-weighted EXAFS oscillation from k-space (3.6-12.3 Å^{-1}) to R-space. The many-body factor (S$_0^2$) was fixed at 0.9 to reduce the fit variables during the simulation.

The TiO$_2$ slurry was prepared by mixing MHS-TiO$_2$ (3 g), Triton X-100 (Sigma-Aldrich) (0.1 mL), acetylacetone (Fluka) (0.2 mL), polyethylene glycol (FW = 20000, Fluka) (1.5 g) in water (10 mL). The MHS-TiO$_2$ photoanode was made by coating the viscous TiO$_2$ slurry onto a fluorine-doped tin oxide (FTO) glass (8 Ω/square, Solaronix), having an area of 0.25 cm^2 (0.5 cm×0.5 cm) which was, dried and calcined at 723 K for 30 min. The as-prepared photoanode was immersed in the N3 (Solaronix) ethanolic solution (3×10^{-4} M) for 24 h. Three drops of H$_2$PtCl$_6$ (Alfa Aesar) ethanolic solution (5×10^{-3} M) were spread onto a FTO glass which was calcined at 658 K for 10 min to yield the counter electrode. The iodide/triiodide (I$^-$/I$_3^-$) electrolyte was composed of 1,2-dimethyl-3-propylimidazolium iodide (DMPII, Solaronix) (0.6 M), lithium iodide (LiI, Aldrich) (0.1 M), iodide (I$_2$, Riedel-de Haën) (0.05 M), and 4-tert-butylpyridine (TBP, Aldrich) (0.5 M) in acetonitrile (Mallinckrodt). The photoanode and counter electrode were separated by a 60 μm-thick thermal foil (SO-SX1170, Solaronix). The I$^-$/I$_3^-$ electrolyte was injected into the cell by means of the capillary force. An epoxy gel was then used to seal the cell. The assembled DSSC was tested on a 300 W solar simulator (91160A, Oriel) combined with an AM 1.5G filter (59044, Oriel). The power density of

illumination of the class A simulator was calibrated and fixed to 100 mW·cm^{-2} with a reference solar cell and meter (91150, Oriel). The current-voltage (I-V) data of the DSSC was recorded and analyzed with the I-V test station program (Oriel).

3 RESULTS AND DISCUSSION

The XRD pattern of the MHS-TiO$_2$ is shown in Fig. 1. The crystal polymorph of the MHS-TiO$_2$ corresponds very well with the anatase TiO$_2$ (PDF#21-1272). The crystalline size of the MHS-TiO$_2$ calculated with Scherrer equation using the FWHM (full-width at half-maximum) of the (101) peak at 25.3° is about 8 nm. Peaks broadening arisen at 36-39 and 53-56° (2θ) are also evident for the existence of the TiO$_2$ crystalline.

The FE-SEM image of the MHS-TiO$_2$ is shown in Fig. 2. It is clear that the MHS-TiO$_2$ having a hollow structure is constructed with TiO$_2$ aggregates (30-50 nm) (see Fig. 2). The HR-TEM image of the MHS-TiO$_2$ is shown in Fig. 3. It seems that the sizes of the MHS-TiO$_2$ having the wall thickness of 20-60 nm are 400-600 nm in diameter. The NBD pattern of the MHS-TiO$_2$ shows distinct diffraction peaks at (101), (004), (200), (105), and (211), which can be assigned to anatase phase. By HR-TEM, it is also worth to note that the MHS-TiO$_2$ are composed of TiO$_2$ nanocrystallines sized in the range of 5-10 nm (see Fig. 3).

The DR UV-vis spectrum of the MHS-TiO$_2$ is shown in Fig. 4. The absorption wavelength of the MHS-TiO$_2$ is extended from 400 nm to about 500 nm. The (F(R)*hv)^2 as a function of the band gap denoting the Tauc plot is also shown in the inseted DR UV-vis spectrum. The F(R) and hv represent the Kebulk-Munk function and photon energy, respectively obtained from the dividing irradiative wavelength by a constant of 1240. The absorption band gap of the MHS-TiO$_2$ is determined by the fitting of the linear part of the Tauc plot to a straight line and extrapolated to the x-axis, herein denotes the band gap. The direct band gap of 3.44 eV for the MHS-TiO$_2$ is consistent with that of an anatase reported in the literature [17].

The XANES and EXAFS spectra of the MHS-TiO$_2$ are shown in Fig. 5. The main features of four-fold tetrahedral (TiO$_4$), five-fold square pyramid ((Ti=O)O$_4$), and six-fold octahedral (TiO$_6$) titanium oxide structures arisen from 1s-3d dipole forbidden electron transitions in the pre-edge XANES can be attributed to A$_1$ (4969.5 eV), A$_2$ (4970.5 eV), and A$_3$ (4971.5 eV) (see Fig. 5(a)) [18]. Wu and co-workers have reported that the intense features at 4970-4975 eV yielded from 3d-4p dipole forbidden electron transitions can also be assigned to A$_3$ and B features [19]. In addition, two distinct features at 4980-5010 eV of the post-edge XANES are generally referred to C and D traits which are induced by the 3s-np dipole allowed electron transitions [20]. By Gaussian-Lorentzian curve fitting, the conjugated A$_1$-A$_3$ signals can be deconvoluted and semi-quantified. It is found that the fractions of A$_1$-A$_3$ in the MHS-TiO$_2$ are 22-55% (see Fig. 5(b)). Note that the A$_2$

species have a high activity in photocatalytic degradation of toxic pollutants [10, 11].

To better understand the molecule-scale information of the MHS-TiO$_2$, its EXAFS spectrum was also collected and refined. Fig. 5(b) shows the EXAFS oscillation in terms of FT magnitude and $\chi(k)*k^3$ in R and k domain, respectively. The experimental spectrum which was theoretically simulated by FEFF 8.0 program and extracted speciation data are summarized in Table 1. The MHS-TiO$_2$ possesses Ti-O and Ti-(O)-Ti average bond distances of 1.98 and 3.07 Å with CNs of 2.9 and 0.7 in the first and second shells, respectively. Notably, the reduced CNs in the shells of Ti-O ($6 \rightarrow 2.9$) and Ti-(O)-Ti in the MHS-TiO$_2$ compared to the bulk anatase with little perturbation in the average bond distances are also found. Similar observation of the nanosize effect on the CNs of copper and its alloys encapsulated in the carbon shells has also been reported [9].

To comprehend the electron transportation characteristic of the MHS-TiO$_2$, photovoltaic performances of the DSSC having the MHS-TiO$_2$ photoanodes were studied on an AM 1.5G solar simulator. Figure 6 shows the current-voltage curves of the DSSC with MHS-TiO$_2$ photoanode under light-illuminated and dark conditions. The photovoltaic data of the DSSC utilizing the MHS-TiO$_2$ photoanode in the open-circuit voltage, short-circuit current density, and fill factor are 0.714 V, 2.52 mA·cm^{-2}, and 0.666 consecutively, which yielding a conversion efficiency of 1.2% illuminated by an AM 1.5G solar power (100 mW·cm^{-2}).

4   CONCLUSIONS

The DSSC possessing MHS-TiO$_2$ photoanodes have been fabricated and examined. The MHS-TiO$_2$ contains mainly anatase phase with the crystalline sizes of 5-10 nm, which has an UV-vis absorption threshold at about 500 nm and has a band gap of 3.44 eV. The FE-SEM and HR-TEM observations also indicate that the MHS-TiO$_2$ has the wall thicknesses of 20-60 nm and sphere diameter of 400-600 nm. By EXAFS, molecule-scale information of the MHS-TiO$_2$ can be revealed, which indicates the Ti-O and Ti-(O)-Ti average bond distances of 1.98 and 3.07 Å with CNs of 2.9 and 0.7 in the first and second shells, respectively. The XANES spectroscopic data also indicate that fraction of the photoactive A$_2$ sites is 23% in the MHS-TiO$_2$. The I-V curve of the DSSC having the MHS-TiO$_2$ photoanodes demonstrates 0.714 V, 2.52 mA·cm^{-2}, and 0.666 in open-circuit voltage, short-circuit current density, and fill factor consecutively, apparently yields a conversion efficiency of 1.2%. This work exemplifies the application of mesoporous TiO$_2$ hollow spheres in DSSC photoanodes for enhanced electricity generation from the solar irradiation.

5   ACKNOWLEDGEMENTS

The financial supports of the Taiwan National Science Council, Bureau of Energy, the Excellence Project of the National Cheng Kung University and National Synchrotron Radiation Research Center are gratefully acknowledged. We also thank the experimental assistances on the measurements of EXAFS and XANES supported by Dr. J.-F. Lee of the NSRRC.

REFERENCES

[1]   C.W. Huang and H. Teng, J. Electrochem. Soc. 155, A739, 2008.

[2]   Z.F. Bian, J. Ren, J. Zhu, S.H. Wang, Y.F. Lu and H.X. Li, Appl. Catal. B-Environ. 89, 577, 2009.

[3]   P. Atienzar, S. Valencia, A. Corma and H. Garcia, ChemPhysChem 8, 1115, 2007.

[4]   P. Atienzar, M. Navarro, A. Corma and H. Garcia, ChemPhysChem 10, 252, 2009.

[5]   B. O'Regan and M. Grätzel, Nature 353, 737, 1991.

[6]   C.J. Barbé, F. Arendse, P. Comte, M. Jirousek, F. Lenzmann, V. Shklover and M. Grätzel, J. Am. Ceram. Soc. 80, 3157, 1997.

[7]   S.H. Liu, H.P. Wang, C.H. Huang and T.L. Hsiung, J. Synchrot. Radiat. 17, 202, 2010.

[8]   H.L. Huang, H.P. Wang, E.M. Eyring and J.E. Chang, Environ. Chem. 6, 268, 2009.

[9]   C.H. Huang, H.P. Wang, J.E. Chang and E.M. Eyring, Chem. Commun. 31, 4663, 2009.

[10] T.L. Hsiung, H.P. Wang and H.C. Wang, Radiat. Phys. Chem. 75, 2042, 2006.

[11] T.L. Hsiung, H.P. Wang and H.P. Lin, J. Phys. Chem. Solids 69, 383, 2008.

[12] C.Y. Liao, H.P. Wang, F.L. Chen, C.H. Huang and Y. Fukushima, J. Nanometer. 698501, 2009.

[13] F.L. Chen, I.W. Sun, H.P. Wang and C.H. Huang, J. Nanometer. 472950, 2009.

[14] C.Y. Chang-Chien, C.H. Hsu, T.Y. Lee, C.W. Liu, S.H. Wu, H.P. Lin, C.Y. Tang and C.Y. Lin, Eur. J. Inorg. Chem. 24, 3798, 2007.

[15] E.A. Stern, M. Newville, B. Ravel, Y. Yacoby, and D. Haskel, Physica B 209, 117, 1995.

[16] A.L. Ankudinov, B. Ravel, J.J. Rehr and S.D. Conradson, Phys. Rev. B 58, 7565, 1998.

[17] H. Tang, F. Levy and H. Berger, Phys. Rev. B 52, 7771, 1995.

[18] F. Farges, G.E. Brown and J.J. Rehr, Phys. Rev. B 56, 1809, 1997.

[19] Z.Y. Wu, G. Ouvrard, P. Gressier and C.R. Natoli, Phys. Rev. B 55, 10382, 1997.

[20] V. Luca, S. Djajanti and R.F. Howe, J. Phys. Chem. B 102, 10650, 1998.

[21] P.T. Hsiao and H.S. Teng, J. Am. Ceram. Soc. 92, 888, 2009.

	Shells	CN[a]	R (Å)[b]	σ^2 (Å²)[c]
MHS-TiO$_2$	Ti-O	2.9	1.98	0.008
	Ti-(O)-Ti	0.7	3.07	0.001
Anatase TiO$_2$ standar[d]	Ti-O	6.00	1.95	0.006
	Ti-(O)-Ti	4.00	3.03	0.005

[a]: coordination number; [b]: bond distance; [c]: Debye-Waller factor; [d]: speciation data are adapted from the ref. [21].

Table 1: Speciation data of the MHS-TiO$_2$ studied by extended X-ray absorption fine structural spectroscopy .

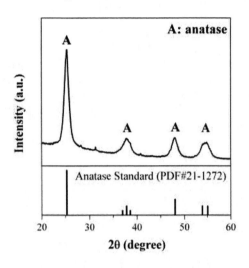

Figure 1: X-ray diffraction pattern of the MHS-TiO$_2$.

Figure 2: Field-emission scanning electron microscopic image of the MHS-TiO$_2$.

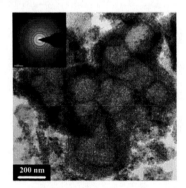

Figure 3: High resolution transmission electron microscopic image of the MSH-TiO$_2$.

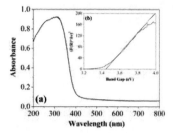

Figure 4: Diffuse reflectance ultraviolet-visible spectrum of the MHS-TiO$_2$. Its Kubelka-Munk spectrum is also inserted in the upper corner.

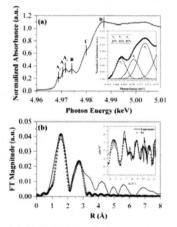

Figure 5: The (a) XANES (and the component fitting) and (b) EXAFS (and the k^3-weighted χ(k) curve) spectra of the MHS-TiO$_2$.

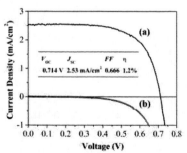

Figure 6: Current-voltage (I-V) curves of the DSSC photoanode containing the MHS-TiO$_2$ under the (a) light illumination and (b) dark conditions.

Electronic properties control of single-walled carbon nanotube using spontaneous redox process of MnO_4^- and carbon nanotube

Dong-Wook Shin[1], Jong Hak Lee[1], and Ji-Beom Yoo[1,*]

[1]Department of Sungkyunkwan Advanced Instituted of Nanotechnology (SAINT),
Sungkyunkwan University, 300, Chunchun-dong, Jangan-gu, Suwon, 440-746, Korea
[*]Corresponding author: jbyoo@skku.edu

ABSTRACT

Single-walled carbon nanotube (SWCNT) thin film is three dimensional interconnected networks of quantum wires and a large number of factors can affect their optical and electrical properties. In order to obtain the low sheet resistance of SWCNT thin film that has the high transmittance at 550 nm, electronic properties of SWCNTs film are modified through chemical treatment and enhanced carrier injection. We investigated the effectiveness of electronic properties with single-walled carbon nanotube through spontaneous redox-reaction between the metal oxide ion such as MnO_4^- and carbon nanotube. The sheet resistance of SWCNTs film decorated with MnO_2 has lower than that of as-prepared SWCNTs film due to the reduction of MnO_2 particle on the surface of SWCNT. The effect of doping on electronic structure of SWCNTs was investigated by four point probe (FPP), Raman spectroscopy, and UV vis-NIR absorption spectrometer.

Keywords: single-walled carbon nanotube, MnO_2, transparent conducting film, spontaneous redox-reaction, doping

1 INTRODUCTION

Transparent conducting films (TCFs) fabricated by single-walled carbon nanotube networks with low sheet resistance and high transmittance has been exploited for the wide range of applications, including next generation displays, stretchable and flexible electronic devices, optical devices, sensors [1] and solar cells [2]. However, further improvement in the sheet resistance of SWCNT films is needed to make them an attractive alternative to indium tin oxide (ITO). For improvement of the sheet resistance with transparent conducting films containing SWCNTs resulting from the percolation effect [3] and the heterogeneous contact resistance of the random networks, removal of solubilizer for the dispersion of SWCNTs such as surfactant and organic solvent, chemical doping and enhanced carrier injection have been needed to decrease the contact resistance between the heterogeneous junctions and enhance the percolation effect.

Strategies for doping SWCNT films with n-type and p-type to decrease the contact resistance between the heterogeneous junctions of SWCNT networks were to use dopants such as alkali metals, alkaline-earth metals, Brønsted acid, and spontaneous redox-reaction between the metal ion (Au and Pt ion) and carbon nanotube. Chemical doping of SWCNT films with several alkali metals, such as Li, K and Cs, and alkaline-earth metals, such as Sr, has resulted in n-type doping behavior [4-8]. In contrast, the decrease of the sheet resistance in SWCNTs films due to the spontaneous reduction of Au and Pt ions on the sidewalls of SWCNTs has been used for p-type doping [9-11]. Similarly, the influence of a Brønsted acid, e.g. H_2SO_4, HNO_3, and HCl, in SWCNTs has been used as the dopant for p-type doping [12-15].

In particular, attaching metal nanoparticles to nanotube sidewalls is of interest for obtaining nanotube/nanoparticle hybrid materials with useful properties, for forming metal nanowires on nanotube templates, and for enhancing the conductivity of SWCNTs film. Spontaneous redox-reaction between of metal ions, such as Au and Pt ions, and carbon nanotube can be created Au and Pt nanoparticles on the surface of the SWCNTs and ejected the electrons in SWCNTs as result of the p-type doping. Similarly, decoration of the metal oxide on the surface of SWCNTs due to spontaneous redox-reaction was suggested because the sheet resistance of TCFs containing SWCNTs decreased. Here, we investigated the effectiveness of electronic properties with single-walled carbon nanotube through spontaneous redox-reaction between the metal oxide ion such as MnO_4^- and carbon nanotube. The change of the sheet resistance in TCF containing SWCNTs, bleaching the transition peaks related to van Hove singularities, and the suppression in Raman spectra were checked through the control of the reaction time in the aqueous $KMnO_4$ solution.

2 EXPERIMENT DETAILS

2.1 Dispersion of SWCNTs

The SWCNTs with a mean diameter of 1.2–1.6 nm and a purity of 90 vol%, which were synthesized by an arc discharge process, were purchased from Hanwha Nanotech. In order to obtain a well-dispersed SWCNTs solution, 10 mg of an SWCNT powder was dispersed in a 1 wt% aqueous solution of sodium dodecylbenzene sulfonate (SDBS) surfactant (10 g of SDBS in 990 ml of deionized water). The mixture was sonicated in a horn-type sonicator (VC 505, Sonics & Materials Inc.), and all sonication processes were performed in ambient conditions for 10 min. And then, the SWCNTs solution was centrifuged in 12000 *g for 1 h, and 80% of the upper solution was used to this experiments.

2.2 Fabrication of TCF containing SWCNTs

The SWCNT film used in this experiment was fabricated on various substrates using a filtration-wet transfer (FWT) method [15-17]. In order to transfer the SWCNT film from the filter to the polyethylene terephthalate (PET) substrate, a conformal contact was first made between the filter and PET. After clipping them together, the filter and PET were placed in deionized water. Initially, lateral wetting occurred at the interface of the filter due to the water flow through the channels in the filter. After complete wetting, the filter was separated from the set and a SWCNT film was formed on the PET substrate. The entire process took only about 80 s. In case of substrates like Si wafer and glass, SWCNTs film was transferred due to the change of the surface energy with Si wafer and glass through the treatment of octadecyl trichlorosilane (OTS) solution for 5 min.

2.3 KMnO₄ solution treatment

SWCNTs film fabricated on various substrates was treated in the aqueous KMnO$_4$ solution (1 mM) for different treatment time, and all treatment processes were performed in ambient conditions. After the immersion in the aqueous KMnO$_4$ solution for 1 h to 18 h, MnO$_2$-doped SWCNTs was rinsed by deionized water for 1 min.

2.4 Analysis of MnO₂-doped SWCNTs

UV–vis–NIR absorption spectroscopy (Shimadzu UV-3600) was used to characterize the change in intensity in the van Hove singularity-related transition peaks, E_{11}^{S}, E_{22}^{S} and E_{11}^{M} through the doping of SWCNT films. Raman spectroscopy (Kaiser Optical System Model RXN 1) was also used to characterize the optical properties of the doped or de-doped SWCNTs. Several excitation energies of 1.96 eV (632.8 nm, helium–neon gas laser) were used. The SWCNT film used for the Raman measurement was fabricated on SiO₂ using a filtration-wet-transfer (FWT) method and the data was collected on numerous spots (five spots) on the sample. All Raman spectra were normalized to the intensity of the G-band to allow a visual comparison between the samples. The change in sheet resistance of the as-prepared SWCNT film and MnO$_2$-modified SWCNT film were measured by a four-point probe resist meter (AIT CMT-SR1000N).

3 RESULTS AND DISCUSSION

In order to identify the effect of the doping in the reduction of MnO$_2$ for the decrease in sheet resistance of the SWCNT film immersed in the aqueous KMnO$_4$ solution,

this study examined the change in sheet resistance of the MnO$_2$-doped SWCNT film as the treatment of SWCNTs film from 1 h to 18 h. Figure 1 shows a sheet resistance of the as-prepared SWCNT film and MnO$_2$-doped SWCNTs film as a function of treatment of MnO$_2$. The sheet resistance of the SWCNT film decreased about 45 % after immersing the as-prepared SWCNTs film in the aqueous KMnO$_4$ solution for 1 h. The ratio of the change of sheet

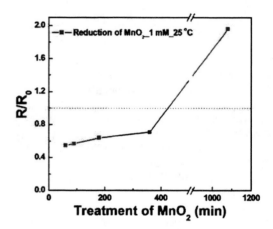

Figure 1. Treatment time of MnO$_2$ versus sheet resistance for transparent conducting film generated from as-prepared SWCNTs film and MnO$_2$-doped SWCNTs film as the increase of the treatment time

resistance with the MnO2-doped SWCNT film decreased with increasing treatment time from 1 h to 18 h, and for 18 h, the sheet resistance of MnO$_2$-doped SWCNTs film increased by a factor of 2. This indicates that the increase of the contact resistance between heterogeneous junctions due to a number of the creation of the MnO$_2$ particle on the surface of SWCNT was resulted from the increase of the sheet resistance with MnO$_2$-doped SWCNTs film. Therefore, we can control the electronic properties of SWCNTs film through the control of the treatment time in the KMnO$_4$ solution.

Figure 2 shows the UV–vis–NIR absorption spectra of the as-prepared SWCNT film and doped SWCNT film after immersion in the aqueous KMnO$_4$ solution for 1 h to 18 h. The van Hove singularities-related transition peaks correspond to the intrinsic excitonic transitions of semiconducting (E_{11}^{S}, E_{22}^{S} and E_{33}^{S}) and metallic (E_{11}^{M} and E_{22}^{M}) SWCNTs from the valence band to the conduction band can be observed in these absorption spectra. These peaks were gradually bleached after immersion in the aqueous KMnO$_4$ solution for 1h to 18 h. Kim et al examined SWCNTs doped with various concentrations of AuCl₃ and reported that the van Hove singularity transitions disappeared gradually in the UV–vis–NIR absorption spectra with increasing p-doping concentration and the work function of the SWCNTs increased [9]. In this absorption spectra, the

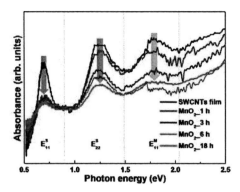

Figure 2. UV–vis–NIR optical absorption spectra with increasing treatment time of MnO_2. The arrows indicate the suppression of intensity in transition peaks related to van Hove singularity through p-doped SWCNTs. E_{ii}^{S} and E_{ii}^{M} indicate the van Hove singularity transitions between the i th energy levels of the semiconducting and metallic SWCNT.

decrease of intensity with transition peaks related to the van Hove singularities, such as the E_{11}^{S}, E_{22}^{S} and E_{11}^{M} bands (red arrow and blue arrow), in the spectra of MnO2-doped SWCNT films as the increase of the treatment time suggest that the Fermi level was shifted further down below the first van Hove singularity in the valence band of the semiconductor and metallic nanotube. After the immersion in the aqueous $KMnO_4$ solution for 1 h, the transition peak of E_{11}^{S} was almost bleached. In case of the transition peak of E_{22}^{S} and E_{11}^{M}, intensity with transition peak was gradually decreased as the increase of the treatment time.

The strong and non-monotonic dependence of the SWCNT Raman spectra on the laser excitation energy E_{laser} demonstrated Raman scattering associated with a resonance

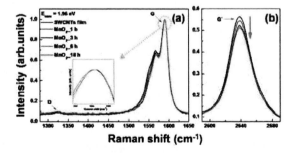

Figure 3. Raman spectra of as-prepared SWCNTs film and MnO$_2$-doped SWCNT films at an excitation energy of 1.96 eV. Inset of (a) shows the unchanged peak position in the G band. All Raman spectra were normalized by the G-band intensity. The arrow shows the decrease of G`-band intensity after immersion in the aqueous $KMnO_4$ solution.

process for the excitation laser energy E_{laser} with the optical transition energy E_{ii} between the van Hove singularities in the valence band and the conduction band [18]. The gradual decrease of intensity in G`-band (figure 3(b)) as the increase of the treatment time was observed due to the transfer of the electrons from the SWCNTs to metal oxide ions. This phenomenon was related to the p-type doping. Inset of figure 3(a) showed the peak position of the G band. Although we can suggest another evidence of p-type doping through the blueshift in G-band, we did not observe the blueshift in the G-band. The non-covalent interaction between the SWCNTs and dopants was attributed to the negligible change in the intensity of the D band.

4 CONCLUSIONS

Transparent conducting films containing SWCNTs were fabricated on various substrates using the filtration wet transfer method. Decoration of the metal oxide on the surface of SWCNTs due to spontaneous redox-reaction was suggested because the sheet resistance of TCFs containing SWCNTs decreased. The sheet resistance of the SWCNT film decreased about 45 % after immersing the as-prepared SWCNTs film in the aqueous $KMnO_4$ solution for 1 h. MnO$_2$-doped SWCNTs film chemically doped with an electron acceptor showed the suppression of the van Hove singularity in the UV–vis–NIR absorption spectra and the gradual decrease of intensity in G`-band. That is, after treatment of MnO$_2$, the Fermi level was shifted further down below the first van Hove singularity with the valence band of the semiconductor and metallic nanotube in the original position on the electronic structure with the as-prepared SWCNTs. In case of 18 h, however, the sheet resistance of MnO$_2$-doped SWCNTs film increased by a factor of 2. This indicates that the increase of the contact resistance between heterogeneous junctions due to a number of the creation of the MnO$_2$ particle on the surface of SWCNT was resulted from the increase of the sheet resistance with MnO$_2$-doped SWCNTs film.

REFERENCES

[1] Cattanach K, Kulkarni R D, Kozlov M and Manohar S K, Nanotechnology, 17, 4123, 2006
[2] ContrerasM A, Barnes T, Lagemaat J, Rumbles G, Coutts T J, Weeks C, Glatkowski P, Levitsky I, Peltola J and Britz D A, J. Phys. Chem. C, 111, 14045, 2007
[3] Hu L, Hecht D S and Gruner G, Nano Lett., 4, 2513, 2004
[4] Lee R S, Kim H J, Fischer J E, Thess A and Smalley R E, Nature, 388, 255, 1997
[5] Grigorian L, Sumanasekera G U, Loper A L, Fang S, Allen J L and Eklund P C, Phys. Rev. B 58, R4195, 1998
[6] Kong J, Zhou C, Yenilmez E and Dai H, Appl. Phys.

Lett., 77, 3977, 2000

[7] Chen B, Li B and Chen L, Appl. Phys. Lett., 93, 043104, 2008

[8] Kim B H, Park T H, Baek T H, Lee D S, Park S J, Kim J S and Park Y W, J. Appl. Phys., 103, 096103, 2008

[9] Kim K K et al, J. Am. Chem. Soc., 130, 12757, 2008

[10] Kong B-S, Jung D-H, Oh S-K, Han C-S and Jung H-T, J. Phys. Chem. C, 111, 8377, 2007

[11] Choi H C, Shim M, Bangsaruntip S, and Dai H, J. Am. Chem. Soc., 124, 9058, 2002

[12] Sankapal B R, Setyowati K and Chen J, Appl. Phys. Lett., 91, 173103, 2007

[13] Nascimento G M, Hou T, Kim Y A, Muramatsu H, Hayashi T, Endo M, Akuzawa N and DresselhausM S, Nano Lett., 8, 4168, 2008

[14] Graupner R, Abraham J, Vencelova A, Seyller T, Hennrich F, Kappes M M, Hirsch A and Ley L, Phys. Chem. Chem. Phys., 5, 5472, 2003

[15] Shin D-W, Lee J H, Kim Y-H, Yu S M, Park S-Y, and Yoo J-B, Nanotechnology, 20, 475703, 2009

[16] Shin J-H, Shin D W, Patole S P, Lee J H, Park S M and Yoo J B, J. Phys. D: Appl. Phys., 42, 045305, 2009

[17] Lee J H, Shin D W, Makotchenko V G, Nazarov A S, Fedorov V E, Kim Y H, Choi J-Y, Kim J M and Yoo J B, Adv. Mater. 21, 4383, 2009

[18] Rao A M, Richter E, Bandow S, Chase B and Eklund P C, Science, 275, 187, 1997

NSTI-Nanotech 2010, www.nsti.org, ISBN 978-1-4398-3401-5 Vol. 1, 2010

Dry native protein assays on substrates by non-contact Laser-Induced-Forward Transfer LIFT process

S. Genov[1], I. Thurow[1], D. Riester[2], T. Hirth[1,3], K. Borchers[3], A. Weber[1,3], G. E. M. Tovar[1,3]

[1] University of Stuttgart, Institute for Interfacial Engineering, Nobelstrasse 12, 70569 Stuttgart, Germany
[2] Fraunhofer-Institute for Laser Technology, Steinbachstrasse 15, 52074 Aachen, Germany
[3] Fraunhofer Institute for Interfacial Engineering and Biotechnology, Nobelstrasse 12, 70569 Stuttgart, Germany, guenter.tovar@igb.fraunhofer.de

ABSTRACT

Laser-Induced-Forward Transfer (LIFT) is a non-contact method for transferring bioactive substances such as proteins in a microstructured way from a transfer layer on a LIFT target to more or less any surface of a 2D or 3D solid work piece. The quality of the biological transfer layer on the target, i.e. its homogeneity and the bioactivity of the ingredients to be transferred, is crucial for the efficiency of the process and quality of the transferred pattern. Here, aqueous trehalose solutions containing green fluorescent protein (GFP) or the ECM protein laminin type 1 were used in a spin-coating/drying process to render dry thin native protein trehalose films on titanium-coated COP foils. The homogeneity of the layer thicknesses was characterized by spectroscopic ellipsometry and the homogeneity of the GFP distribution within the layer was monitored by fluorescence scan analysis. After application of the LIFT process, the transferred proteins were tested in their bioactivity by fluorescence scan analyses and cell assays.

Keywords: native proteins, trehalose, LIFT, homogenous thin layers, micro array, fibroblast adhesion

1 INTRODUCTION

Micro-structured protein films on solid surfaces with retained protein activity are necessary for a variety of bio-technological and biomedical applications.[1] There are multiple possibilities for generating protein films or patterned protein layers: (electro-)spraying, dip-coating, spin-coating, micro-spotting, micro-contact printing and ink-jet printing. A promising non-contact process to generate micro-scale native protein patterns on 3D and 2D substrates is based on Laser-Induced Forward Transfer (LIFT).[2] This technique uses a laser pulse for jetting a substance from a planar so-called target onto a substrate (Figure 1). The target is coated with a light-absorbing layer and the substances to be transferred. The absence of a dead volume makes this method of printing very efficient and saves valuable resources. LIFT overcome the risk of clogging of nozzles or pins and has recently been adapted for transfer processes of a variety of biomolecules and cells.[3-10]

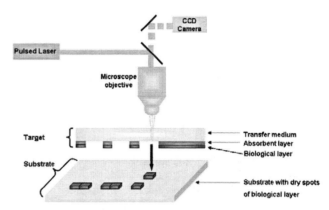

Figure 1: Drawing of the LIFT process in principle according to Barron et al.[11]

Targets comprise a solid transfer medium (glass or polyurethane) which is coated with a light absorbent layer (titanium or silver) and a biological layer. A laser pulse via the transfer medium of the target induces vaporization of the absorbent layer and the local vapor jet propels a small portion of the biological layer from the target onto the substrate. To date, little attention was paid on optimizing the target coating with regard to the homogeneity of the protein distribution and layer-thickness, alternative transfer media or the suitability for storage of "ready-to-use" targets.

In this study we chose titanium-coated cyclo-olefin polymer (COP) foil as transfer substrate due to its adaquate optical and chemical properties and its possible importance for future scale-up and automation. For the biological layer we prepared protein containing films by spin-coating. film thickness was examined by means of ellipsometric spectroscopy. In order to observe the protein distribution within the biological layer the self-fluorescent green fluorescent protein (GFP) was applied. The quality of the micro-structured patterns transferred by the LIFT process was evaluated by protein activity tests and cell-assays.

2 MATERIALS AND METHODS

2.1 Titanium coated COP foil

For sample preparation cyclo-olefin polymer foils (COP foils, Zeonor®, Zeon Europe GmbH) were coated with a titanium mirror with a thickness of 100 nm. COP foils were sputtered in a vacuum chamber at 6 µbar with titanium in presence of argon gas (physical vapour deposition).

2.2 Spin-coating

Titanium coated COP foils were incubated in 2% alkaline detergent solution (Hellmanex®, Hellma) for 45 min at 40 °C. Subsequently, the prepared foils were cut in circular pieces with a diameter of 16 mm. The protein-trehalose solution was degassed in a vacuum chamber (Heraeus vacutherm®), for 10 min to a pressure of 100 mbar. To achieve protein-trehalose films a spin-coating device was applied (P6700, Specialty Coating Systems Inc. Indianapolis, USA). 25 µl were deposited on the COP foils and spin-coated for 60 sec at given speeds at 23 °C.

2.3 Evaluation of layer thickness and protein distribution

For layer thickness evaluations 600 mM aequous trehalose solution containing 50 µg mL^{-1} GFP (rTuboGFP, 1 mg mL^{-1}, evrogen, Russia) were spin-coated at 1000 rpm, 3000 rpm and 5000 rpm. Layer thickness was determined by spectroscopic ellipsometry [12, 13] using a SENTECH SE 800 ellipsometer and the analysis software Spectra Ray2 V.5.1, SENTECH Instruments. The determined refractive indices of the organic layers varied from 1.52 to 1.57. The resulting layer thickness was determined to be 171 nm ± 9.9 nm at 5000 rpm, 283 nm ± 11.0 nm at 3000 rpm and 532 nm ± 9.5 nm at 1000 .

2.4 LIFT process

To transfer proteins by LIFT a pulsed laser device (Fraunhofer Institute for Laser Technique ILT, Aachen) was applied. The inserted Q3 laser (CryLas GmbH, Berlin) was used at λ = 355 nm, pulse width >1 ns, frequency 15 kHz and laser powers 5 mW and 10 mW. The target-substrate distance was adjusted to 70 µm.

2.5 Adhesion of primary fibroblast cells

LIFT generated spots of ECM protein laminin type I (Sigma-Aldrich Chemie Gmbh, Munich Germany) were evaluated by exploiting the adhesion behavior of fibroblast cells. Laminin type 1 was transferred in an arrayed format onto nitrocellulose coated glass substrates. The microarrays were blocked by spray-coating with Stable Guard (SurModics, USA, via Diarect AG, Freiburg, Germany). Fibroblasts were isolated from human foreskin. And pre-cultivated in T-25 culture flasks (Greiner BioOne, Frickenhausen, Germany) in DMEMr (DMEM + 10 % FCS + 1 % Gentamicin (Invitrogen, Karlsruhe, Germany). Incubation of fibroblasts with laminin microarrays was done in DMEM (without FCS) for 2h. In order to avoid unspecific

adhesion of the cells to the nitrocellulose, the substrates were shaken every 10 min.

3 RESULTS AND DISCUSSION

3.1 Layer thickness, spin-coating reproducibility and protein distribution

Spectroscopic ellipsometry was applied to evaluate the film thickness homogeneity of spin-coated trehalose-protein films on titanium coated COP foil.

Figure 2: Bar chart of mean layer thicknesses resulting from spin-coating processes at 5000 rpm, 3000 rpm, 1000 rpm. The layers consist of GFP (50 µg mL-1) in dried trehalose (600 mM). Three sample replicates for each speed. shows the coating thicknesses of GFP-trehalose films on COP substrates each with an area of 15 mm x 15 mm. Samples were coated at three speeds 5000 rpm, 3000 rpm and 1000 rpm. The layer thicknesses were measured at nine measuring points per sample. At a speed of 3000 rpm the mean resulting layer thickness was 283 nm which is approximately half of the film thickness of 532 nm resulting at 1000 rpm. At 5000 rpm the layer thickness was 171 nm and therefore about half the thickness of 3000 rpm films. Error bars show the double standard deviation and represent the homogeneity of the layer thickness of each sample. Standard deviations of the film thickness were < 7.1 % of the mean at 5000 rpm < 4.0 % at 3000 rpm and < 2.3 % at 1000 rpm. The reproducibility of the spin-coating technique was evaluated by comparing three independently prepared samples for each speed (samples 1 -3). The overall standard deviations of three samples (27 measuring points) were 5.8 %, 3.9 %, 1.8 % of the mean for 5000 rpm, 3000 rpm and 1000 rpm, respectively. The bulging border areas of the sample pieces where observed to become wider with decreasing speed and were not integrated into analysis.

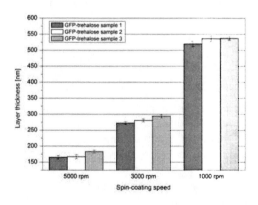

Figure 2: Bar chart of mean layer thicknesses resulting from spin-coating processes at 5000 rpm, 3000 rpm, 1000 rpm. The layers consist of GFP (50 µg mL-1) in dried trehalose (600 mM). Three sample replicates for each speed.

The results demonstrate that spin-coating created protein trehalose films of homogenous layer thickness. The layer thickness was controlled by spin-coating speed. In order to achieve high protein load per area but minimal width of bulging borders we chose 3000 rpm speed for layer preparation, as far as not stated otherwise.

3.2 Homogeneity of protein distribution

It has been reported that immobilization of proteins and disaccharides in amorphous (i.e. glassy) state protected the proteins from chemical and conformational degradation [14]. During the spin-coating process the solvent (water) evaporates quickly from deposited protein trehalose solutions resulting in amorphous dry coatings [15]. We aimed to investigate whether homogenous protein distribution is achieved within such amorphous trehalose layers. Therefore, self-luminescent protein GFP was incorporated into trehalose layers at 3000 rpm. Protein distribution in dried trehalose films was investigated by means of fluorescence detection: The coated area was divided into rows A-E and columns 1-10. Figure 3 shows the mean fluorescence intensities and standard deviations for each row for two independently generated samples in a bar chart. The overall standard deviation was 17.4 % (sample 1) and 18.5 % (sample 2).

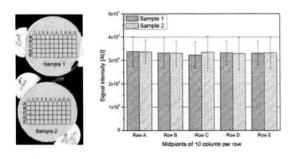

Figure 3: Left: Fluorescence scan image of a d = 16 mm target piece. Right: Midpoint bar chart: Signal intensity of self-luminescent native protein GFP (50 µg mL^{-1}) embedded in dried trehalose (600 mM) film.

3.3 Analyses after LIFT process

LIFT targets were prepared as described using solutions of GFP (1 mg mL^{-1}) and trehalose (600 mM). in order to print protein microarrays via the LIFT process. The biological layers were generated at 1000 rpm spin-coating speed resulting in layers of approximately 550 nm (see section 3.1). GFP was used as indicator protein to visualize native proteins within the transferred spots and to optimize the spot morphology by adaption of the laser parameters. The LIFT process was enforced with two different laser power settings: P = 5 mW and 10 mW. Protein spots were generated applying either three laser shots per deposited spot or one laser shot per deposited spot.

Figure 4 (1a-2b) shows details from the fluorescence scan from 5 x 5 arrays generated by various combinations of the mentioned parameters. As only native GFP molecules emit fluorescence the measured signals proofed that GFP was transferred and deposited in native conformation.

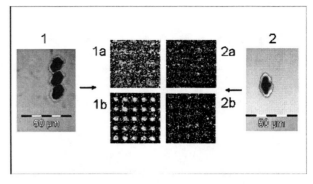

Figure 4: Microscope images of the target and fluorescence scan images of an acceptor glass substrate): Ablation marks from three laser shots at 5 mW (1) and one laser shot at 10 mW (2) and GFP deposited at 10 mW, accumulation of three shots per spot (1a), 5 mW, accumulation of three shots per spot (1b), 10 mW, one shot per spot (2a), 5 mW, one shot per spot (2b).

The presented experiment further revealed that three shots at 5 mW (1b) produced well defined, distinct spots in an arrayed format, while three shots per spot at 10 mW resulted in laminar protein distribution (1a). Printing with one laser shot per spot failed to deliver considerable spots. Atomic force microscopy (AFM) was applied to evaluate the ablation of the biological layer from the COP target during the LIFT process. The AFM image seen in Figure 5 shows well-defined cylindric ablation mark with 40 µm diameter and 600 nm depth caused by the laser induced vapor jet. This results indicates that the biological layer (layer thickness approximately 550 nm, see above) was totally removed from the target.

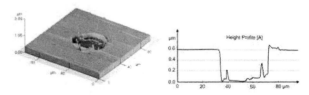

Figure 5: Left: AFM image showing ablation of the protein trehalose layer on COP foil target after one laser shot with 6 mW laser power during the LIFT process. Right: Height profile showing ablation of the protein trehalose layer on target after the LIFT process.

Electron spectroscopy for chemical analysis (ESCA) detected titanium and titanium oxide at the bottom of the laser induced cavity (data not shown), affirming the above stated interpretation

3.4 Cell assay

LIFT targets were prepared as described using solutions of laminin type 1 (50 µg mL^{-1}) and trehalose (600 mM) in order to print protein microarrays for cell adhesion assays. The distance between the deposited protein spots was adjusted to 400 µm.

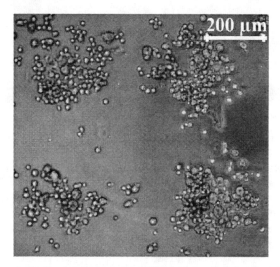

Figure 6: Optical micrograph of fibroblast cells adsorbed to LIFT-transferred laminin type 1 spots on a nitrocellulose coated glass substrate.

The optical micrograph in Figure 6 shows fibroblast cells adhering preferable to four distinct spots on the nitrocellulose covered target. This behaviour is attributed to the micropatterned laminin successfully transferred by the LIFT process.

5. CONCLUSIONS

This study demonstrates that spin-coating is a well suited method for preparation of dry thin films of trehalose and proteins with uniform protein distribution on circular targets. The resulting film thickness was controlled via the spin-coating speed and proofed to be homogeneous all over the coated area (standard deviation per sample < 7.1%). Spin-coated dry thin native protein trehalose films on titanium-coated cyclo-olefin polymer (COP) foil were applied as targets for laser-based protein-transfer by Laser Induced Forward Transfer (LIFT). The titanium coated COP foil proofed to be well suited as transfer medium for the LIFT process. The laser power and the number of pulses were adapted such that arrays of circular spots with diameters of approximately 100 µm were generated on the LIFT substrate. Green fluorescent protein GFP was detected in its native self-luminescent state after being transferred by LIFT and LIFT-generated laminin microarrays mediated microstructured fibroblast adhesion.

Thus, we conclude that dry trehalose protein films are a versatile tool for storing proteins in their native state for subsequent processing, e.g. by means of LIFT. The use of COP foil as transfer medium for the LIFT process enables adaption of the process towards up-scaling and further automation due to the materials mechanical flexibility and processibility.

4 ACKNOWLEDGEMENTS

The authors thank Joachim Mayer, Fraunhofer IGB for AFM images and the Federal German Ministry BMWI (grant ID 16IN0491), the Fraunhofer Gesellschaft and the state Baden-Württemberg for financial support.

REFERENCES

[1] C. D. Heyes, G. U. Nienhaus, Zeitschrift fuer Physikalische Chemie, 221, 75, 2007.

[2] P. K. Wu, B. R. Ringeisen, J. Callahan, M. Brooks, D. M. Bubb, H. D. Wu, A. Pique´, B. Spargo, R. A. McGill, D. B. Chrisey, Thin Solid Films, 398/399, 607, 2001.

[3] J. M. Fernandez-Pradas, M. Colina, P. Serra, J. Dominguez, J. L. Morenza, Thin Solid Films, 453/454, 27, 2004.

[4] B. Hopp, T. Smausz, Z. Antal, N. Kresz, Z. Bor, D. Chrisey, Journal of Applied Physics, 96, 3478, 2004.

[5] B. Hopp, T. Smausz, N. Kresz, N. Barna, Z. Bor, L. Kolozsvari, D. B. Chrisey, A. Szabo, A. Nogradi, Tissue Engineering, 11, 1817, 2005.

[6] G. Kecskemeti, N. Kresz, T. Smausz, B. Hopp, A. Nogradi, Applied Surface Science, 247, 83, 2005.

[7] I. Zergioti, A. Karaiskou, D. Papazoglou, C. Fotakis, D. Kapsetaki, D. Kafetzopulous, Applied Physics Letters, 86, 16390, 2005.

[8] M. Duocastella, M. Colina, J. M. Fernandez-Pradas, P. Serra, J. L. Morenza, Applied Surface Science, 253, 7855, 2007.

[9] C. B. Arnold, P. Serra, A. Piqué, Materials Research Society BULLETIN, 32, 23, 2007.

[10] V. Dinca, M. Fasari, D. Kafetzopulous, A. Popescu, M. Dinescu, C. Fotakis, Thin Solid Films, 516, 6504, 2008.

[11] J. A. Barron, P. Wu, H. D. Ladouceur, B. R. Ringeisen, Biomedical Microdevices, 2, 139, 2004.

[12] F. Höök, J. Vörösb, M. Rodahla, R. Kurratb, P. Bönid, J. J. Ramsdene, M. Textorb, N. D. Spencerb, P. Tengvallc, J. Golda, B. Kasemo, Colloids and Surfaces B: Biointerfaces, 24, 155, 2000.

[13] R. Kurrat, B. Wälivaara, A. Marti, M. Textor, P. Tengvall, J. J. RAMSDEN, N. D. Spencer, Colloids and Surfaces B: Biointerfaces, 11, 187, 1998.

[14] F. Franks, R. H. M. Hatley, S. F. Mathias, BioPharm International, 4, 38, 1991.

[15] J. H. Crowe, J. F. Carpenter, L. M. Crowe, Annual Reviews Physics & Chemistry, 60, 73, 1998.

Novel Metal Oxide Deposition Method to Fabricate Nanostructures

Mansouri, Robab*,**; Sandooghchian, Mehri**; Mansouri, Mohammad Reza**

* Department of Physic, University of Payame Noor, kooye Golestan, Shiraz, Iran,
robabmansouri@yahoo.com

** Kerman Semiconductor Company, Flat no.7, building no25, Sarve Sharghi BLVD., Sa'adat Abad,
Tehran-Iran, info@kermansem.com

Abstract

ZnO is well promised for applying in semiconductor nanodevices. In recent years, wide types of various nanostructures were interested and investigated. In this work, we constructed new designed CVD set up (Chemical Vapor Deposition) and then studied about ZnO nanostructures grown on a SnO_2 deposited on a Si wafer, which was deposited by a new method of deposition, to find more sensitivity and selectivity than other nanostructures for applying in gas sensors. SnO_2 is one of the nanomaterials, which is well promised to apply in gas sensors and ZnO too. Likely by applying both of them on a Si wafer, to achieve more sensitivity and selectivity.

Key Words: metal oxide, deposition, ZnO naopillar

1. Introduction

Semiconducting metal oxide have been extensively studied due to their simplicity, sensitivity and applicability in many field. Zinc oxide (ZnO), a n-type metal oxide semiconductor sensing material with a wide band gap (Eg = 3.37 eV at 300 K) and the exciton binding energy of 60 meV, much larger than those of other semiconductors, has been under extensive research owing to its high chemical stability, low cost, and good flexibility in fabrication. Most of the studies up to now are focused on sintered particle or thin-film based devices. Several groups have recently proposed that the nanowires, nanotetrahedral or nanoribbons of semiconducting oxides are very promising sensors, diodes and transistors and some of their results have shown that the devices based on one-dimensional nanostructures have great potential in overcoming the fundamental limitations.

To enhance these properties of nanostructures, we determined to prepare some depositions, such as stannous oxide, zirconium oxide and cobaltous oxide and then growth various nanostructures on them. Here we fabricated only nanostructures of zinc oxide on SnO_2 deposited layer.

Deposition of metals and metal oxides render new properties to nanostructures and were interested by scientists. For example, by proper deposition, new properties are rendered to a nanostructure.

Depositions have been executed by sputtering method so far. It has own advantages and disadvantages on quality of layer. It needs a strong vacuum and it is extremely extortionate specially for wide areas, for example, by this method it is impossible and very expensive to deposit 1 m^2.

By this project, we innovated new approach of deposition of metal oxide layer and executed with stannous oxide on type n of Si wafer to fabricate ZnO nanostructures first, and developed later by other metals. By this method, we could fabricate aligned nanowire of ZnO. Early aligned ZnO nanowire was fabricated on gold-deposited Si wafer and it was expensive.

This new method gives some advantages like Easy process, cheap process, low waste, and fabricating some new nanostructures, for example aligned nanowire of ZnO by deposition of gold has been fabricated in pervious. It gives us an opportunity to have various depositions with cheap and easy process.

2. Experiment

At the out set we prepared Stannous oxide layer on type n of 100 Si wafer by the following explained method. Then we innovated and constructed a new designed CVD set up and after the very zinc oxide nanostructures were fabricated on the very deposited Si wafer by new constructed CVD set up. At terminally stage the sample was investigated by SEM (scanning electron microscope), EDX (electron diffraction X-ray) and chemical analysis. The each single of steps are explained individually by detail.

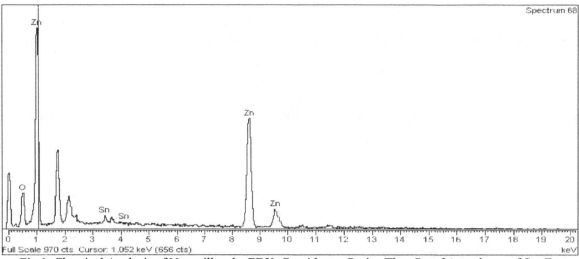

Fig 1: Chemical Analysis of Nanopillars by EDX. Consider on Peaks, They Proof Attendance of Sn, Zn and O On Si Wafer.

2.1 Deposition detail

The deposition is executed by chemical approach. Early, $SnCl_2$ is dissolved into sufficient acetone to make suitable concentration into100 ml volumetric flask. Then Si wafer 100 type n is put on the hotplate and heated up to 300 °C. Later the solution is spread on the wafer by proper pipette and dried by heat. This action is repeated for several times. The very wafer is placed into furnace which its temperature is carried out to 900°C for 2 hours. It causes chlorine is removed and oxygen constitute instead of it. The EDX chemical analysis shows that the layer contains Sn and Cl was completely removed and deposition is complete. The detail of process will be explained following.

3.1. Fabricating of nanostructures

Fabricating of nanostructures was followed by developing and constructing novel Chemical Vapor Deposition setting. The new designed CVD (Chemical Vapor Deposition) set up is somehow differing from the ones constructed before. This setting additional to fabricating some nanostructures can produce powder form of nanostructures in mass form about 0.5-1 gram. Setting would have been capability to produce 1-2 kilograms per run by scaling it up. Additional it has other advantages, which is brought following:

- Well-ordered nanostructures.
- Low energy waste.
- Low cost fabricating of nanostructures procedure.
- Saving time.
- Fabricating of some novel nanostructures.
- High yield

The setup, which was modified several times during this research project, was turned out to be unique enough to be registered as a patent.

This set up includes four main parts, which are:

- A furnace, which its unique design allow us omit thermal gradient. As a result, that is thermally isotropic.

- A Quartz cell, which is lied into the chamber of the furnace on two thin walled alumina tubes in a way that the cell is almost suspended in the

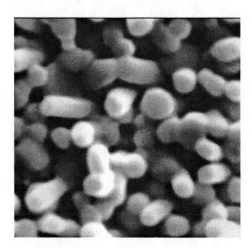

Fig 2: The SEM Image of Prepared Sample Which shows its Top View. Consider on perpendicularity of nanopillars of ZnO, which grows on SnO_2 deposited layer on Si wafer. Also this image shows unclearly hexagonal cross section of grown nanostructures.

chamber. Its unique design renders us to have more and well-ordered nanostructures.

- A narrower Quartz tube, acting as fabrication chamber, into which are placed an n-type (100) Si wafer and mixture of powders. This tube is accommodated into the long quartz chamber.
- Gas supplier includes inlet and outlet.

The quartz cell is put into the furnace which is heated before up to about 900°C. After putting cell into the furnace, the heat is carried up to 950°C and temperature is fixed for 30 minutes. After reaching to the predetermined temperature the gas mixture is introduced. The gas flow includes proper mixture of nitrogen about 1000 ml/min and air about 200 ml/min, is introduced into cell and fabricating of nanostructures is launched to grow at predetermined temperature. This situation is fixed for 30 min. At the end of 30 minutes, the air supplying is cut off and the quartz cell is withdrawn out of furnace and wafers are removed to investigate.

3. Results and Conclusion

By considering on EDX (electron diffraction X ray) results, fig 1, shows that the sample contain Zn, O and Sn, and this is not far from of our expectation. According to our executed test existence of Zn,O and Sn is reasonable. The deposition had been executed by $SnCl_2$ solution and existence of Sn is reasonable but absence of Cl is fuzzy. It is assumed that the $SnCl_2$ solution is evaporated on Si wafer and ignition causes to regenerate $SnCl_2$ molecules to SnO_2 molecules and this is proven by EDX image (fig. 1).

$$SnCl_2 + O_2 \longrightarrow SnO_2 + Cl_2$$

Additional chemical analysis for definition of nanostructures was executed. Zinc oxide becomes yellow, when it is ignited. The very sample is ignited and the nanostructures which were deposited on SnO_2 layer and the ignited sample become yellow and confirms the EDX result and existence of zinc oxide.

By considering on taken SEM images (figs. 2, 3, 4 & 5), it is found that nanostructures of zinc oxide are all aligned pudgy hexagonal which are grown on deposited SnO_2 layer. Fig 2 shows conspicuously that the nanostructures are completely aligned however, aligned ZnO nanostructures has been grown by sputtering deposition of gold so far. It is very expensive method and is work only for small areas. Figs 3 & 4 show the launching point of growing of nanostructures. As it is seen they come straight up. Also fig 4 shows this result. The dimension of nanpillars as shown in fig 4, is about 200 nm. Fig 5 reveals that the very nanostructures are hexagonal.

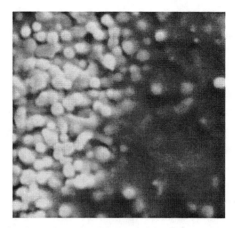

Fig 3: The SEM Image of Prepared Sample, Which Shows Its Top View. Consider on Launching of Growing of Nanopillars. That is arising point.

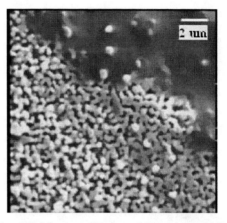

Fig 4: The SEM Image of Prepared Sample, Which Shows Its Top View. Consider on Launching of Growing of Nanopillars. The diameters of nanopillars are about 200 nm.

Fig 5: The SEM Image of Prepared Sample from its Top View. Consider on Hexagonal Cross sections of Nanopillars.

As well as known, SnO$_2$ has orthorhombic crystal structure and during regeneration of deposited layer of SnCl$_2$ solution Sn molecules is laid on Si atoms in flat style by its hexagonal bases. It is supposed that very layer which bounded with Si atoms forms a pattern to growth. The hexagonal of SnO$_2$ layer act as a pattern to align growing and it is proven by SEM images in figs 2and 5.

It seems that the SnO$_2$ layer remain on Si wafer surface during the growth because EDX analysis shows Sn peak. Addition, after removing the nanostructures from wafer by ultrasonic the SnO$_2$ metallic surface of layer is appeared.

However, this new method arise some advantages, which are brought following:

- Easy and cheap process

- Low waste

- Fabricating some new nanostructures, for example aligned nanowire of ZnO by deposition of gold has been fabricated in pervious.

4. Acknowledgment

These schievment would not have been possible without sustained efforts of Narges Arampasand, Morteza Bayati, Ali Sanee and R. Rahimi.

5. References

[1]. Ki-Won Kim, Pyeong-Seok Cho, Sun-Jung Kim, Jong-Heun Lee, Chong-Yun Kang, Jin-Sang Kim, Seok-Jin Yoon "The selective detection of C$_2$H$_5$OH using SnO$_2$–ZnO thin film gas sensors prepared by combinatorial solution deposition" Sensors and Actuators B **123** (2007) 318–324.

[2]. Ji. Haeng Ya. Gyoong, Man. Choa; "Selective CO gas Delection of CuO- and ZnO doped Sno$_2$ gas sensor"Sensors and Actuators B **75** (2001) 56-61.

Effect of nano-sized silica particle for high insulation resistance on the surface of non-oriented electrical steel

J.W. Kim*, M.S. Kwon and C.H. Han

* Electrical Steel Research Group, Technical Research Laboratories, POSCO
1, Goedong-dong, Nam-gu, pohang, Kyungbuk 790-300, Korea, Pohang P.O.Box 36

ABSTRACT

The effect of a nano-sized silica particle on the surface insulation resistivity of non-oriented electrical steel sheet was investigated. Several kinds of nano-sized colloidal silica were mixed to the chromium-free coating solution, which is consisted of mixture of metal phosphate and polyester resin.

To explain the mechanism of insulation current in the coating layer, the series micro-dielectric effect is introduced. In well-dispersed nano-sized particle in the coating layer, the successive micro-dielectric results in the electrical insulation current. As a result of successive micro-dielectric, the value of insulation current was increased. Here Franklin insulation tester was employed to determine the properties of surface resistivity associated with the conduction of an electrical current.

As an experimental result, it was found that the size and solid content of a nano-sized silica particle were strongly influenced on the properties of insulation current. It is increased by decreasing the particle size of silica and increasing solid content of silica. It is, however, only increased to a certain extent of solid content and then it begins to decrease because the coalescence and cohesion phenomena are broken out. Such as a high percent of silica content, therefore, micro-dielectric effect should be decreased with increasing the total surface area of silica due to clustered silica acts as one large silica particle.

Keywords: Chromium-free, Nano-sized colloidal silica, Surface insulation current, Series micro-dielectric effect

1 INTRODUCTION

Electrical steels are used in the magnetic cores of such machines as motors, generators and transformers. Intensive effects have been directed at developing improved steels with lower iron loss for use in energy efficient machines. The eddy current loss component of the core can be reduced by use of insulating coatings to ensure that the eddy current is restricted to individual laminations. The selection of the most suitable coating solution for a specific application has to be made with reference to the insulation current, punchability, corrosion resistance properties, weldability, heat resistance, stacking factor, chemical resistance, burn-out characteristics, resistance to compression, coating thickness, surface roughness and scratch resistance properties.

A coating solution which forms an insulation coating of a non-oriented electrical steel sheet generally is classified into three types. There are inorganic coating solution, organic coating solution and organic-inorganic hybrid coating solution.

The inorganic coating solution (So called C-4 type) is composed mainly of inorganic materials such as phosphate which enables the formation of a coating film with high heat resistance, weldability, and layering properties. However, due to the high hardness of the insulation coating film, a mold of punching machine is more quickly damaged than when using a coating film containing an organic material.

Consequently, the inorganic coating solution is not desirable from the aspect of punching processability. The organic coating solution (So called C-3 type) is consisted mostly of an organic material and is very superior in terms of punchability and adhesion properties. Even when the thickness of the film is increased, adhesion is excellent, and thus the organic coating solution is chiefly used for large iron cores requiring interlayer insulating properties. However, the weldability of the organic coating film is not good because resin-decomposing gas is generated upon welding. Moreover, it is not suitable for burn-out process (So call SRA [Stress Relief Annealing] process) because of a weak heat resistance.

The interest of improving heat resistance and insulation properties have been required the development of an organic-inorganic hybrid coating system (So called C-5, C-6 type) using both an organic and an inorganic material. A phosphate (H_3PO_4) or chromate (CrO_3) is mainly used for the inorganic materials. In the case of a coating film using such hybrid composition, the heat resistance and insulation properties are strongly influenced by the inorganic material. At the same time, the lubrication effect is increased by the properties of organic material.

The method of forming an insulation coating film by using an organic-inorganic coating is well-known. In the case where chromate is used, a hydrogen bond (O-H) is formed between chromate and an iron oxide (FeO) layer of a non-oriented electrical steel sheet, thus attaining excellent coating properties, including adhesion and punchability, and furthermore, good coating properties may be obtained even after SRA process. However, due to the composition of such a conventional coating solution essentially contained chromium trioxide, unfavorable effects on the human health and environmental problems are caused. Attributable to the above problems, the use of heavy metals, including hexavalent chromium(Cr nj), is limited under the current strict control of environmental restrictions such as RoHS (restriction of the use of hazardous substances) and REACH (Registration, Evaluation, Authorization and Restriction of Chemicals) among EU countries. For reason of the regulation of using hexavalent chromium, many researchers have forced on the development of chromium-free coating solutions for electrical steel in these days [1-3]. In order to improve on the low corrosion resistance and adhesion properties resulting from the absence of the chromate, a metal-phosphate (ex: aluminum phosphate (Al $(H_2PO_4)_3$), calcium phosphate (Ca $(H_2PO_4)_2$), zinc phosphate (Zn $(H_2PO_4)_2$) and so on) are introduced.

In several decades, colloidal silica particles have been used in a wide range of applications including composite materials [4]. Many research groups have reported that nano-sized colloidal silica has unique properties: improvement of the mechanical properties, progression of the chemical and thermal resistance and enhance the gas barrier, even more, increase of the insulation current. In all cases, the benefits of smaller sized particles can only be realized if the particles are well dispersed, either in a solvent or in other solid matrices [5-6].

Accordingly, in the current work, the various types of well dispersed nano-sized colloidal silica are applied to improve the insulation properties of non-oriented electrical steel. The Franklin tester (ASTM A-717) was used to evaluate the insulation current of coated samples. The mechanism of insulation current on the surface of the electrical steel was investigated. The effect of nano-sized colloidal silica, especially for the particle size and solid content of silica, on the insulation properties is elucidated in detail.

2 EXPERIMENTAL PROCEDURE

A 50A1300 JIS grade steel sheet produced by POSCO was used for the experiment samples. The grooved bar coater (Coater # 4 [wire diameter: 0.1mm] made by R.D.S Webster, N.Y) was utilized to coat the sample having 50mm in width and 150mm in length. Coating solutions were prepared with the following compositions: 25~40wt. % of metal phosphate (Mixture of aluminum phosphate, $Al(H_2PO_4)_3$ and zinc phosphate, $Zn(H_2PO_4)_2$), 25~40wt. % of emulsion-type polyester and 10~50wt. % of 4 different size of colloidal silica. Here, industrial grade-colloidal silica is purchased by Nissan-Chemical Industries, LTD and each component is well dispersed by turbine-type stirrer at least 60 min. The size range of colloidal silica is between 7nm and 88 nm, and the range of solid content of silica in the coating solution is up to 15.5 wt. %.

The as-prepared samples are coated by various kinds of coating solution. After that, the coated samples are baked at 650f in the radiant tube furnace for 20 seconds. Here, coating thickness is measured by the SEM (Scanning electron microscope) & FIB (Focused Ion Beam) techniques. The Franklin insulation tester was employed in order to determine the insulation current associated with the

conduction of an electrical current through the surface of the samples. Here, the value of insulation current (mA) can be converted to a surface insulation resistivity per one side of sheet (Ь Є /sheet) as described equation 1 (ASTM Standards: A 717/A 717M).

$$\Omega cm^2 / sheet = 6.45[(\frac{1000}{mA}) - 1] \qquad (1)$$

To explain the behavior and morphology of nano-sized silica in coating layer, EDS (Energy dispersive spectroscopy), EPMA (Electron Probe Micro Analyzer) and GDS (Glow Discharge Spectroscopy) technique are utilized.

3 RESULT AND DISCUSSION

Electrical steels in the magnetic cores of such electrical applications are usually coated with an insulation coating in order to improve the performance of the electrical steel in terms of reduced iron loss (Including eddy current loss), punchability and welding properties and corrosion resistance. In order to minimize iron loss in proportion to the square of the thickness of the electrical steel, an insulation coating is formed on both sides of metal surfaces. As following the equation 2, usually the eddy current loss, Pec can be expressed as the following form. Here Bo, t, f, and ρ are maximum permeability, steel thickness, frequency and density, respectively.

$$P_{ec} = \frac{10^{-9} \pi^2 B_o^2}{6} \cdot \frac{t^2 \cdot f^2}{\rho} \qquad (2)$$

An equation 3 shows the relationship between the insulation current, r and the increasing rate of iron loss by eddy current loss, C (%). This equation means that the higher insulation current, the smaller increasing rate of iron loss. Increasing insulation current leads to increasing the efficiency of electrical machines. Here, K, f, ρ, Bo, a, t, and Wo are stacking factor, frequency, density, permeability, width of electrical steel, thickness and iron loss at standard f and Bo, respectively.

$$C (\%) = \frac{1.65 \cdot f \cdot B_o^2 \cdot a^2 \cdot k \cdot t}{\rho \cdot w_o \cdot r} \times 10^{-3} \qquad (3)$$

A number of papers already explained the effect of nano-sized particle on the insulation properties [4-6]. These papers are described that insulation property is strongly affected by a barrier effects of particle in the coating layer. Here, the barrier effect is increased by increasing the amount of nano-sized particle. In this study, to confirm the behavior of nano-sized silica on the surface and in the layer of coating, various size of colloidal silica is employed.

First of all, a various analysis techniques are utilized to explain the behavior of silica particle on the surface of coating. Here the coating solution is composed of 40 wt. %

of metal phosphate, 40 wt. % of polyester resin and 20 wt. % of colloidal silica. Solid content and particle size of the colloidal silica are 6.2 wt. % and 45 nm, respectively.

Figure 1 shows the SEM image (Included surface roughness) (a), EDS analysis pick (b) and EPMA mapping images (c) of the coating surface, respectively. From the SEM image & surface roughness, EDS analysis pick and EPMA mapping images, we found that coated sample has macroscopically very smooth surface and surface roughness is approximately 0.2~0.3 ʊ . Furthermore, it can be clearly confirmed by the presence of the silica particle (SiO_2) which is well-dispersed on the surface of coating.

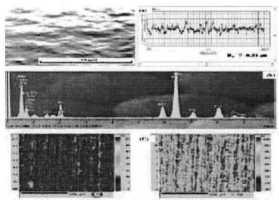

Figure 1 : SEM image & surface roughness (a), EDS spectrum (b) and EPMA analysis (c) of coating surface.

Figure 2 shows the cross section of coating layer fabricated by FIB & SEM techniques (a) and GDS analysis graphic (b). As a result of SEM image and GDS analysis graphic, the structure of the coating layer, is formed compactly without cracking, and the silica particle is well-distributed in the layer of coating. Here, the coating layer be composed of metal-phosphate, polyester resin and silica particle (Solid content: 6.2 wt. %; particle size: 45 nm).

As already noted above, in order to enhance the barrier effect, the colloidal silica is added into the coating solution. The main factor of insulation properties is the size and amount of nano-sized particle. However, no one has considered the relationship between the barrier effect and electric intensification (Especially current). That is why, in this study, a series micro-dielectric effect is introduced to explain the insulation mechanism of the particle of coating surface.

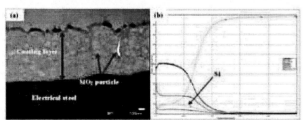

Figure 2 : Cross section of coating layer (a) and GDS analysis graphic (b)

Figure 3 illustrates the schematic sketch of series micro-dielectric effect, which explains the electric field intensifying due to the conductive nano-sized particles in coating layer. Under the DC voltage, separations of positive and negative charges within the coating layer occur respectively at the top and bottom of the surface. A micro-dielectric is caused if the nano-sized particles are well-distribution in the coating layer. As it were, local electric intensification is generated between the particle-to-polymer layer (or metal phosphate), particle-to-particle and particle-to-(+/-) electrode. This phenomenon takes place consecutively in throughout coating layer. It means that the conductive nano-sized particle acts as the barrier of current. Thereafter, Successive micro-dielectric in the coating layer results in the electrical insulation current. As a result of successive micro-dielectric, the electrical insulation may be increased within well-dispersed particle in the coating layer.

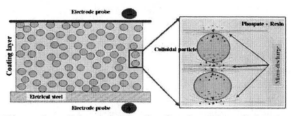

Figure 3 : Schematic sketch of series micro-dielectric effect model

Main parameters having an effect on the electrical properties in field of electrical steel sheet are as following: 1. Coating thickness, 2. Volume insulation resistivity (Ω-cm) of materials, 3. Structure of coating layer. Especially it can be strongly influenced by the volume insulation resistivity of materials. Normally the insulation current is increased with increasing a volume insulation resistivity of materials. Therefore, in this study, the size and amount of silica particle as a factor of the coating component are only considered.

First of all, the effect of silica size on the insulation properties is investigated. Figure 4 shows the insulation current at 4 different sizes of silica which is chosen to 7, 15, 45 and 88 nm. All other parameters kept a constant value: coating thickness of 0.7 ʊ and solid content of silica particle of 6.2 wt. %. As shown in the Figure 4, under basic coating component consisting 50 wt. % of metal phosphate and 50 wt. % of polyester resin, the value of insulation current indicated in Franklin tester is about 100 mA in average of 5 sheets. It means that 900 mA of current are charged through the surface of coating between the contact probes of Franklin tester and electrical steel sheet. However, if the colloidal silica is well-dispersed in coating layer, the value of insulation current is increased with decreasing the size of silica particle. At the 7 nm of silica size, the value of insulation current is approximately 500 mA. It means that here 500 mA of current is restricted by a layer of coating entrapped the silica particle. It can be explained as series micro-dielectric effect. As explained before, the micro-dielectric is generated between the particle-to-polymer layer, particle-to-particle and particle-to-electrode in coating layer. As increasing total surface area of silica particle, the effect of micro-dielectric should be increased.

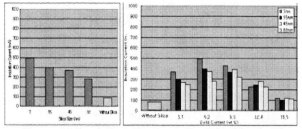

Figure 4 : Insulation current vs. Particle size of silica at the same coating thickness / Figure 5 : Insulation current vs. solid content of silica at the same coating thickness.

Figure 5 shows the relationship between the value insulation current and the solid content of silica at particle size of 7 nm to 88 nm and coating thickness of 0.7 ὐ . As mentioned above, the series micro-dielectric effect increases with increasing solid content of silica, because total surface area of silica increases rapidly. The value of insulation current is, however, only increased up to 6.2 ~ 9.3 wt% of silica particle and then the value insulation current begin to decrease. The reason for the mentioned behavior should be related with dispersion of silica particle in the coating layer. Figure 6 show the cross section of coating layer by SEM photos at high amount of silica (Solid content of SiO_2: 12.4 wt. %; particle size: 88 nm). It can be visually found that the nano-sized silica particle has a tendency to coalesce together (So called, clustered silica). Under the presences of clustered silica, it should be decreased total surface area of silica particle. The clustered silica, which acts as one large silica particle, might be increased with increased the solid content of silica. As explained in Figure 4, the effect of micro-dielectric is decreased with decreasing the total surface area of SiO2. Also the micro-crack or interfacial micro-pores incorporated among particles should affect a micro-dielectric effect. Unfortunately, it is impossible to investigate.

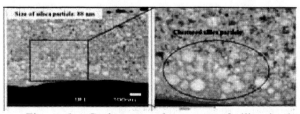

Figure 6 : Coalescence phenomena of silica in the coating layer.

4   CONCLUSION

In this study, the effect of a nano-sized silica particle on the insulation properties of non-oriented electrical steel was

investigated. Several kinds of nano-sized colloidal silica was added into coating solution which is consisted of metal phosphate (Mixture of Al $(H_2PO_4)_3$ and Zn $(H_2PO_4)_2$) and polyester resin. With a various analysis techniques, it can be confirmed that nano-sized silica particles are well-dispersed in the layer and surface of coating.

The series micro-dielectric effect was introduced to explain the mechanism of insulation current. As a result of successive micro-dielectric, electrical insulation current was increased in well-dispersed nano-sized silica particles within the coating layer.

As an experimental result, the value of insulation current was increased by decreasing the particle size of silica. Moreover, it was increased by increasing solid content of silica. It was, however, only increased to a certain extent of solid content and then it began to decrease by reason of the clustered silica generated by coalescence and cohesion phenomena among particles.

REFERENCES

[1] David Snell, Alan Coombs, "Novel coating technology for non-oriented electrical steels," Journal of magnetism and magnetic materials, 215-216 (133-135), 2000.

[2] M. Lindenmo, A. Coombs and D. Snell, "Advantages, properties and types of coating on non-oriented electrical steels," J. magnetism and magnetic materials, 215-216(79-82), 2000.

[3] D.C.L. Vasconcelos, R.L. Or´efice, W.L. Vasconcelos, "Processing, adhesion and electrical properties of silicon steel having non-oriented grains coated with silica and alumina sol–gel," Materials Science and Engineering A 447, 77–82, 2007.

[4] R.J. Hunter, Introduction to Modern Colloidal Science, Oxford Univ. Press, New York, 1993.

[5] Xiaolin Chen, Yonghong Cheng, Bo Yue, Hengkun Xie, "Study of epoxy/mica insulation deterioration in generator stator using ultra-wide band partial discharge testing technique," Polymer Testing 25, 724–730, 2006.

[6] W. S. Zhao, Q. G. Meng and Z. L. Wang, "The application of research on powder mixed EDM in rough machining," J. Mater. Process Technol, Vol. 129, 30-33, 2002.

[7] Lucian Dascalescu, Adrian Samuila and Robert Tobazéon, "Dielectric behaviour of particle-contaminated air-gaps in the presence of corona," J. Electrostatics, Vol. 36, 253-275, 1996.

[8] Tatsushi Matsuyama and Hideo Yamamoto, "Charge relaxation process dominates contact charging of a particle in atmospheric conditions," J. Phys. D: Appl. Phys. 28, 2418-2423, 1997.

----
[1] Nanotech 2010, 696 San Ramon Valley Boulevard, Suite 423, Danville, CA 94526-4022, Ph: (925) 353-5004, Fax: (925) 886-8461, swenning@nsti.org.

Rhodamine B degradation efficiency of differently annealed titanium dioxide nanotubes

M.V. Diamanti, F. Stercal, MP. Pedeferri

CMIC Dept, Politecnico di Milano, Via Mancinelli 7, 20131 Milan, Italy

Tel: +39 02 23993144, Fax: +39 02 23993180, E-mail address: mariavittoria.diamanti@polimi.it

ABSTRACT

Titanium dioxide is employed as photocatalytic substrate in heterogeneous catalysis, specially in the development of Advanced Oxidation Processes (AOP) which exploit the synergetic use of UV irradiation and oxidizing compounds to increase the degradation efficiency of hazardous chemical compounds [1]. Its main uses range from the mineralization of organic substances in air and in wastewaters, for environmental clean-up and deodorizing, to the removal of metal ions from wastewaters, from photovoltaic cells to the creation of antifouling, antibacterial, self-cleaning and antifogging surfaces [2]. When titanium undergoes anodic oxidation in fluoride-containing electrolytes, nanotubular amorphous oxides are obtained [3]: these oxides can acquire photoactivity through annealing treatments, which modify the amorphous structure of the oxide by inducing the formation of anatase crystals [4] and may lead to oxide doping depending on the annealing atmosphere. This paper presents an investigation over the photoefficiency of nanotubular TiO_2 films in the degradation of organic pollutants: rhodamine B was chosen as model reactant [5]. The effects of thermal treatments in several atmospheres on the oxides photoactivity are investigated.

First, specimens were characterized by means of Scanning electron microscopy (to investigate the attained morphology: fig. 1), X-ray diffraction (to achieve information on the oxide crystal structure) and Glow-discharge optical emission spectroscopy (to analyze the oxide elemental composition). The degradation of the organic compound was proved to follow a pseudo-first order kinetics, and a 90% mineralization of the rhodamine B solution was achieved after 23 hour irradiation of a 6 cm^2 nominal area specimen with UV-Vis light. While at a first glance no marked influence of nanotubes length was detected on the photocatalytic efficiency of the TiO_2 layer, an interesting influence was noticed when the adsorption extent of rhodamine B was taken into account: in fact, the active sites saturation was reached at growing concentrations with increasing nanotubes length, which was proved by a sudden change in the degradation kinetics with initial organic compound concentration. Finally, the effect of the annealing atmosphere was investigated by using air, pure nitrogen, and mixtures of: 20% oxygen 80% argon, 20% argon 80% nitrogen, and 50% oxygen 50% nitrogen; the presence of nitrogen was intended to dope the oxide and therefore to shift its activity towards visible light [6]. Only negligible effects were found on the UV-Vis photoactivity of the specimens (fig. 2); conversely, in all cases a detrimental effect was noticed on the photocatalytic activity of the specimens irradiated with pure visible light, specially when annealing was performed in atmosphere totally devoid of oxygen. A beneficial effect was exerted by the presence of nitrogen on the oxide activity under visible light radiation.

Submission track: Advanced materials - Nanostructured coatings, surfaces and films

REFERENCES

[1] S. Chiron, A. Fernandez-Alba, A. Rodriguez, E. Garcia-Calvo, Water Res. 34 (2000), 366-377.
[2] O. Carp, C.L. Huisman, A. Reller, Prog. Solid State Chem. 32 (2004) 33-177.
[3] J.M. Macak, K. Sirotna, P. Schmuki, Electrochimica Acta 50 (2005), 3679-3684.
[4] MP. Pedeferri, C. E. Bottani, D. Cattaneo, A. Li Bassi, B. Del Curto, M.F. Brunella, Proceedings of NSTI Nanotech 2005, Anaheim 8-12 May 2005, Vol.2, pp.332-334.
[5] X. Zhao, Y. Zhu, Environ. Sci. Technol. 40 (2006), 3367-3372.
[6] R. Asahi, T .Morikawa, T. Ohwaki, K. Aoki, Y. Taga, Science 293 (2001), 269-271.

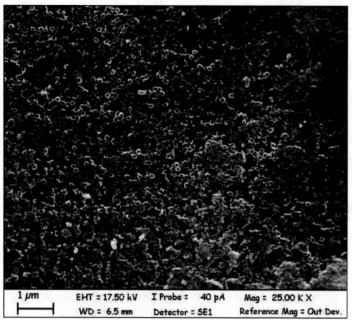

Figure 1 – SEM investigation of the surface morphology of titanium anodized in the chosen fluoride-based solutions: formation of amorphous TiO$_2$ nanotubes.

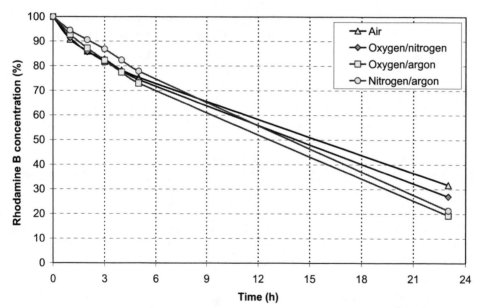

Figure 2 – Photocatalytic degradation in time of rhodamine B in presence of differently annealed substrates irradiated with UV-Vis light source. Resulting degradation extents and kinetics are comparable for all annealing treatments.

Control of Solid Surface Properties by Electron Beam Processing

I.V.Vasiljeva, S.V.Mjakin

Engineering Technology Center RADIANT, Ltd.
10 Kurchatova str. 194223 St-Petersburg, Russia, radiant@skylink.spb.ru

ABSTRACT

Electron beam modification of various solids (oxides, sulfides, ceramics, polymers) is shown to provide a control over their surface properties from the content of certain atoms and functional groups to overall hydrophilicity and reactivity depending on the process parameters. Particularly, the content of hydroxyl groups on the surface can be adjusted by the appropriate absorbed dose selection promoting either hemisorption of physically adsorbed water molecules or dehydroxylation and dehydration. The developed approach is successfully applied to enhance the exploration performances of such materials as fused silica useful for chromatographic capillaries, constructive ceramics, adsorbents, electroluminescent phosphors, polymeric substrates and coatings due to the improved compatibility between the irradiated components of these compounds.

Keywords: electron beam, absorbed dose, surface functional groups, modification, compatibilization

1 Introduction

Electron beam processing is a promising approach to the control over the surface properties of various solids. The main advantages of electron beam technologies over the conventional chemical methods for surface pretreatment include the direct impact onto chemical bonds in the surface layer, possibility for the activation and control of adsorption-desorption processes, precise adjustment of the process parameters and absence of any chemical or radioactive contamination.

The effect of electron beam treatment upon various materials (oxides, sulfides, ceramics, polymers) is studied to clarify the general features of chemical reactions involving surface functional groups in correlation with the process parameters. The irradiation was carried out using a resonance transforming electron accelerator RTE-1V produced by the Efremov Institute of Electrophysical Apparatus (St-Petersburg, Russia) with the beam energy 500-900 keV and absorbed dose 50-500 kGy. The changes in the exposed surface properties were characterized by the adsorption of acid-base indicators with different pK_a values (distribution of Lewis and Brensted sites according to their acidity strength), FTIR spectroscopy, contact wetting angle and surface tension measurements for boundaries with various liquids.

2 Results and discussion

2.1. Modification of oxides

Generally, electron beam processing of various oxides in air provides the control over the content of various hydroxyl groups and specific chemical bonds on their surface. For initially non- or weakly hydroxylated oxides (e.g. aerosils) irradiation at relatively small doses (50-100 kGy) leads to the dissociation of physically adsorbed water molecules into $H^{\bullet}$ and $OH^{\bullet}$ radicals interacting with the element-oxygen bridging groups to form hydroxyls [1]:

At higher absorbed doses (150-200 kGy) the surface of such materials undergoes dehydroxylation due to the disruption of relatively weak M-O-H bonds and overall dehydration.

In contrast, for some oxides featuring with initially hydroxylated surface (e.g. FeO, CuO) the lowest doses (about 50 kGy) promote dehydroxylation due to the disruption of weak O-H bonds in Brensted acidic groups yielding oxygen bridges (Fig 1b). The further dose growth provides hydroxylation and dehydration processes according to the mechanism described above.

Oxides with higher cation charges such as Fe_2O_3 and Al_2O_3 also undergo dehydroxylation at low doses (~50 kGy) but acidic OH-groups are recovered at 200-300 kGy due to the presence of water molecules strongly bonded with Lewis acid sites and high metal-oxygen bond strengths requiring significant energies for disruption (Fig. 1c).

Upon the optimization of electron beam processing parameters the considered effects were successfully used for the following applications:

- modification of fillers for cements to impart them with enhanced strength performances and frost resistance due to the activation of solidification processes [2];

- pretreatment of oxide additives for ceramic materials providing up to 50% increase in their bent and compression strength due to the formation of additional hydroxyls at the surface promoting chemical binding of the particles through oxygen bridges upon annealing [1].

a

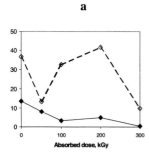

b

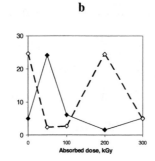

c

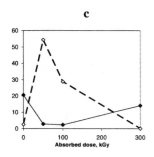

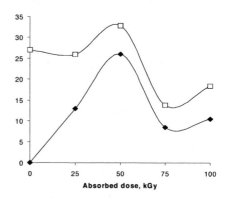

Figure 1. Variations in the content (μmol/g) of Lewis basic (——, bridging O atoms or ions, pK_a -0.3) and Brensted acidic (-----, hydroxyls with pK_a 2.5) sites on the surface of Aerosil-175 (a), CuO (b), Fe_2O_3 (c)

Figure 2. Effect of absorbed dose at electron beam (700 keV) pretreatment of fused silica glass on the content of acidic Brensted centers (hydroxyls) with pK_a 2.5 (, μmol/cm^2) on the surface of fused silica glass samples after electron beam processing and efficiency of methacryloxypropyl-trimethoxysilane grafting characterized by the optical density ($D \cdot 10^3$) of the specific absorption band 2954.7 cm^{-1} (♦)

2.2. Functionalization of fused silica

Electron beam pretreatment of fused silica glass used for the production of chromatographic capillaries allowed a controllable functionalization of the glass surface. Particularly, the content of Brensted acidic centers (silanol groups) on the irradiated fused silica surface is found to follow an "oscillatory" trend as a function of the absorbed dose below 100 kGy at electron beam processing due to the alternating reactions of hydroxylation (probably as a result of Si-O-Si bond disruption and interaction with radiolyzed physically adsorbed water) and thermal dehydration/dehydroxylation at radiation heating. The highest content of acidic hydroxyls with pK_a 2.5 (measured using the adsorption of the acid-base indicator o-nitroaniline with the corresponding pK_a value) achieved upon electron beam processing with the energy 700 keV and absorbed dose 50 kGy provides the best conditions for the subsequent immobilization of methacryloxypropyl-trimethoxysilane (MOPTMS) layer as confirmed by the highest intensity of the absorption band at 2954.7 cm^{-1} intrinsic to stretching vibrations of CH-groups according to FTIR spectroscopy data (Fig. 2).

This approach allows imparting the initial fused silica glass surface with a stable, uniform and reproducible chemical functionality. Electron beam initiated grafting of MOPTMS provides the enhancement of the processes for the production of fused silica glass capillaries for electrochromatography and electrophoresis at the stage of an intermediate bifunctional layer formation required for the subsequent deposition of specific polymer coatings [3].

2.3. Enhancement of adsorbents

The controllable formation of specific functional groups on the surface upon electron beam processing is useful for the modification of their selective adsorption properties towards various compounds.

Electron beam treatment of natural sand taken from St-Petersburg suburban area Lahta (Russia) imparted it with adsorption activity towards Mn^{2+} ions growing with the increase of absorbed dose up to the saturation at 200-500 kGy (Fig. 3) due to the formation of Brensted acidic and basic hydroxyls as well as Lewis basic sites.

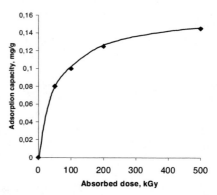

Figure 3. Adsorption capacity of sand towards Mn cations as function of the absorbed dose at electron beam pretreatment with the energy 700 keV

NSTI-Nanotech 2010, www.nsti.org, ISBN 978-1-4398-3401-5 Vol. 1, 2010

In another experiment commercial charcoal Dratnasirs 50-325 (produced from coconuts) with the average particle size 20 µm, specific surface 1150 m²/g was irradiated at 700 keV and tested on the adsorption of Cu^{2+} ions as a model heavy metal pollutant in water. A certain increase of adsorption capacity was observed at a relatively low absorbed dose 50 kGy followed by a drastic decrease at higher doses (Fig. 4). The adsorption capacity was found to correlate with the content of Lewis basic and Brensted acidic centers capable of binding Cu^{2+} ions according to the donor-acceptor and cation exchange mechanisms correspondingly.

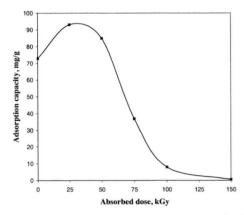

Figure 4. Adsorption capacity of charcoal as function of the absorbed dose at electron beam pretreatment

2.4. Enhancement of electroluminescent phosphors

Electron-beam processing of ZnS:Cu based electroluminescent phosphors (containing the optimal content of Cu doping additive about 0.3% wt.) provided a significant increase of their brightness, further improved upon an additional pretreatment of the initial charge mixture for their synthesis at a higher absorbed dose (Table 1). The observed effect is determined by the following factors:

- improvement of the doping additive diffusion into the material (particularly in the case of the initial charge mixture irradiation);
- decomposition of ZnS:Cu solid solutions;
- generation of specific defects promoting electroluminescent electron transitions;
- elimination of defects negatively affecting the electroluminescence. Particularly, adverse "electron traps" formed by sulfur vacancies and Lewis acidic (Zn, Cu ions) centers are shielded and "repaired" by electron beam processing at optimal parameters resulting in the disruption of Zn-S and Cu-S bonds and yielding of -SH and -OH groups according to the mechanism described above.

In addition, electron beam irradiation provides a certain storability enhancement for such phosphors and a possibility for the control over their emission spectra color due to the variation of the processing parameters.

Table 1. Relative changes in electroluminescence brightness of ZnS-Cu based phosphors prepared using 900 keV electron beam processing at different stages of the synthesis

Processing conditions	Electro-luminescence brightness, %
Reference (no electron beam processing)	100
Phosphor synthesized from non-irradiated charge mixture and irradiated by electron beam at optimal absorbed dose 50 kGy	145
Phosphor synthesized from a charge mixture preliminarily irradiated at absorbed dose 500 kGy	160
Phosphor synthesized from a charge mixture preliminarily irradiated at absorbed dose 500 kGy and irradiated at 50 kGy after the synthesis	190

2.5. Modification of polymers

In our studies electron beam processing was successfully applied for the modification of the following polymeric materials:

- *Poly(ethylene terephthalate) (PET)* useful in numerous applications from supports for specific coatings and layers in various technical devices to bottles for beverages.

Electron beam processing of commercial PET films (45 and 100 µm thickness produced by AO "SVEMA" enterprise (Shostka, Ukraine) and Du Pont, USA) was performed with the energy 500 keV and absorbed dose in the range 25-500 kGy. The changes in PET properties were characterized by absorption of acid-base indicators, UV-vis and FTIR spectroscopy and contact angle measurements at boundaries with different liquids (water, isooctane, toluene).

Generally the analysis of surface sites distribution indicated that electron beam treatment of PET leads to surface reactions according to the mechanisms similar to those described above for oxides. The initial PET surface is predominantly occupied with Lewis basic centers with pK_a –0.3 (Fig. 5) formed by oxygen atoms in C=O containing ester groups. Processing at relatively low absorbed doses (up to 100 – 150 kGy) results in their transformation into hydroxyls (pK_a 2.5) [4].

The increase of absorbed dose to 200 – 300 kGy resulted in the opposite trend towards the surface dehydroxylation due to the removal of adsorbed water upon radiation heating.

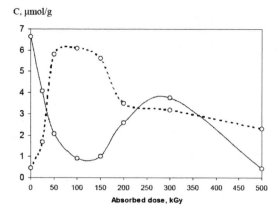

Figure 5. Content of adsorption centers with pK_a –0.3 (—) and 2.5 (---) as function of absorbed dose.

Mutual transformations were also observed for the centers with pKa 5.0 and 0.8 (Fig. 6) probably due to the oxidation of methylene and methine groups with ozone resulting from the interactions of accelerated electrons with the air oxygen:

$$CH_n + O_3 \rightarrow CH_{n-1}-OH + O_2 \; (n = 1–3)$$

or such reactions as

$$H_2O \rightarrow H^{\bullet} + OH^{\bullet}$$

and

$$\underset{\|}{\overset{O}{R-C-O-R'}} + H^{\bullet} + OH^{\bullet} \rightarrow \underset{\|}{\overset{O}{R-C-OH}} + R'-OH$$

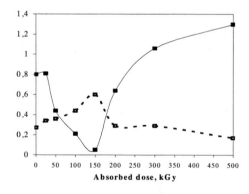

Figure 6. Content of adsorption centers with pK_a 5.0 (—) and 0.8 (---) as function of absorbed dose.

These data were confirmed by FTIR spectroscopy indicating the transformation of methylene and methine groups into hydroxyls and carboxyls at 50-150 kGy and contact wetting angle measurements showing the polymer surface hydrophilization under these conditions followed by a certain hydrophobization at higher doses. The considered approach afforded a successful grafting of N-vinyl pyrrolidone on PET film surface [1] unattainable without electron beam pretreatment.

- Cyan ester of poly(vinyl alcohol) (CEPA)

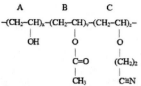

used as a binder with a high dielectric constant in composites for various electronic devices such as electroluminescent displays. Electron beam irradiation at optimal processing parameters provided a significant increase of the dielectric constant for this polymer (Fig. 7) due to a partial oxidation of the polymer yielding highly polar C=O groups [5].

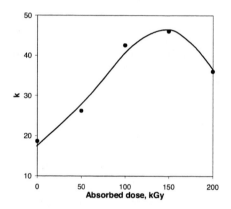

Figure 7. Dielectric constants of CEPA as function of the absorbed dose at electron beam processing with the energy 700 keV.

REFERENCES

[1] Ed. Sergey V. Mjakin, Maxim M. Sychov, Inna V. Vasiljeva, "Electron beam modification of solids: mechanisms, common features and promising applications", Nova Science Publishers, Inc., 2009.

[2] L.B. Svatovskaya, A.M. Sychova, V.M. Sychov, V.A. Chernakov, A.V. Khitrov, T.S. Titova, M.M. Sychov, I.V. Vasylieva, S.V. Mjakin, Innovations and Developments in Concrete Materials and Construction: Proc. of the International conf., Scotland, Dundee, 2002, P.61.

[3] A.Ju.Shmykov, S.V.Mjakin, I.V.Vasiljeva, V.N.Filippov, M.E.Vylegzhanina, T.E.Sukhanova, V.E.Kurochkin, Applied Surface Science, 255, 639, 2009.

[4] I.V.Vasiljeva, S.V.Mjakin, A.V.Makarov, A.N.Krasovsky, A.V.Varlamov, Applied Surface Science, 252, 8768, 2006.

[5] M.M. Sychov, V.G. Korsakov, A. L. Zagranichek, S.V. Mjakin, I.V. Vasiljeva, A.G. Rodionov, L.L. Ejenkova. Proc. of Eurodisplay 2007 Int. Conf., Moscow, 2007, P.372.

Fabrication and assembly of nanowrinkle patterns on metallic surface: a nonlithography method for nano fabrication

Z. Zhang, C. Guo and Q. Liu*

National Center for Nanoscience and Technology, China
Beiyitiao No.11, Zhongguancun, Beijing 100190, China,
zhangzw@nanoctr.cn, guocf@nanoctr.cn, *liuq@nanoctr.cn

ABSTRACT

Motivated by wrinkles, which are ubiquitous in nature, artificial patterns with micro periods were also fabricated in stiff/soft bilayer. Since its great importance in both fundamental research and practical use, ordered periodic nanostructures fabricated by wrinkling has fascinated researchers. In this paper, the wrinkle patterns with periods from micro- to nano-scale (sub 500nm) were fabricated by choosing the appropriate metal film and optimizing the film thicknesses. The influence of metal film morphology on wrinkle pattern was studied, which is seldom mentioned in existing works. Furthermore, the wrinkle patterns were assembled by the prefabricated structures produced via nanoimprint, deposition with masks and laser scanning. Our research results show that the wrinkling of polymer/metal bilayer is a potential nano fabrication method combining conventional bottom-up with top-down nanofabricating technique.

Keywords: wrinkle, assembly, nano fabrication, nonlithography

1 INTRODUCTION

With the development of modern technology, the fabrication of periodic microstructures has been attracting persistent attention [1]. Although kinds of techniques such as contact printing techniques[2, 3], electron beam lithography (EBL)[4, 5], focused ion beam (FIB) processing[6] have been extensively pursued to fabricate nanoscale features, novel techniques is still developed to satisfy increasing demands of low cost, large area nanofabrication. Motivated by wrinkle, the ubiquitous phenomenon in nature, spontaneous formation of wrinkle patterns has been utilized to fabricate periodic microstructure. Various applications of periodic structures via wrinkling, such as flexible electronics [7, 8], tunable diffraction grating [9], biocompatible cell and particle alignment [10], advanced metrology methods [11-13], have been developed.

The wrinkle pattern has a well defined wavelength, which is approximately proportional to the thickness of buckled film. Thus a periodical structure with wavelength on micro/nano scale can be fabricated by wrinkling of thin films. Compare to the existing micro/nano fabrication method, the fabrication method via wrinkling has the follow

advantages. First, the wrinkle can be simply generated by thermal stress, stretching and swelling. Second, the wrinkle wavelength can be controlled by the film thickness and materials properties. Last, the wrinkle can be realized in a large area easily. However, to our best knowledge, the nano fabrication via wrinkling has not been realized yet, because there are problems needed to be solved.

The larger wrinkle wavelength is the first problem for nanofabrication. The previous works have significantly reduced the wavelength from macroscopic scale to micro scale [14-18]. Recently, the wrinkle wavelength has been reduced to submicron scale (ca. 500 nm) on metallic surface via thermal wrinkling [19, 20]. Unfortunately, the smaller wrinkle pattern on metallic surface has not been reported yet, which hinders the application of wrinkle from micron scale to nano scale.

Another barrier of nanofabrication via wrinkling is the randomness of wrinkle pattern. The isotropic stress induced by annealing, swelling etc. usually produce a chaos and complicated patterns, which is hard to utilize and most treated as a nuisance. The ordered wrinkle patterns are desired for device use. The effective and low cost assembly method is still desired urgently.

Not confined to a technique issue, nano fabrication via wrinkling makes the conventional theory of wrinkle phenomena face a great challenge. The small wavelength demand an ultra thin metal film, which shows a discontinuous morphology composed of grains. The influence of metal film morphology on wrinkle pattern is seldom mentioned.

As a consequence, the wrinkling method is a promising nanofabrication, but there are still problems need to be resolved. In this paper, we have fabricated the wrinkle patterns with periods from micro- to nano-scale (sub 500nm) by material selection and film thickness optimization. Several assembly methods have been proposed to obtain ordered wrinkle patterns. A brief theory analysis has been given based on conventional wrinkle models.

2 EXPERIMENTAL DETAILS

2.1 Sample preparation and wrinkle process

The wrinkle experiments were preformed on the polymer/metal bilayer system. The polystyrene (PS) was chose to be the polymer layer due to its mechanic property and inexpensiveness. A toluene solution of the PS (Mw=100,000) was spin-coated onto a cleaned silicone wafer, and then was annealed at 60 °C for 12 h to remove the residual solvent and to relieve the stress induced by spin-coating. The PS films were controlled to various thicknesses by adjusting the solution concentration or/and the spin-coating speed and determined by profilegraph (Dektak 150, Veeco). Various kinds of metal were deposited on PS film by RF magnetron sputtering (ULVAC, ACS-4000-C4). The bilayer system was heated to a temperature which is higher than the glass transition temperature of PS (100 ℃) in a vacuum oven. The thermal stress during the heating process generated the wrinkle patterns on the sample surfaces.

2.2 Characterization of wrinkle patterns

The surface morphology of wrinkled sample was examined by optical microscopy (OM, OLYMPUS BX51), field emission scanning electron microscopy scanning electron microscopy (SEM, HITACHI S4800), and atomic force microscopy (AFM). The AFM images were all acquired in the tapping mode with a VEECO Dimensions 3100 instrument and a Nanoscope Ⅳa controller. All measurements by AFM were performed at ambient conditions in an acoustic damping box. The software packages 5r30 and 6r13 were used for data analysis.

3 RESULTS AND DISCUSSIONS

3.1 The wavelength reduced by optimizing material parameters

A simplified prediction of wrinkle wavelength has been given as $\lambda \sim t_m(E_m/E_p)^{1/3}$. Most of achievements in wrinkle wavelengths are based on thinning the films. And the submicron metallic wrinkle pattern has been successfully fabricated based on PS/Au bilayer by thermal wrinkling However the wrinkle wavelength is about 500 nm[19] ,while the thickness of Au film has reached to 5 nm, which is nearly the lower limit of continuous film, and the PS film is about 120 nm. The thinner PS film will cause the sharply increase of the critical stress to wrinkle. Thus reducing the thickness for smaller wrinkle pattern is effective but hard to be improved. As another influence factor, the lower modulus of metal film can also lead to smaller wavelength in a lower level than film thickness. In Fig. 1, the wrinkle patterns of some soft metal supported on PS film were fabricated by thermal wrinkling, which verified that thermal wrinkling is a versatile microstructure fabrication on various metals.

Among kinds of soft metal, we choose Sn as the metal film due to its low modulus (41GPa for bulk Sn) and high quality of film. Fig 2 shows the wrinkle patterns of Sn with different thickness on 56 nm PS film. The thickness of Sn film is controlled by the deposited time. It is obvious the wrinkle have an increasing wavelength as the increase of

deposited time. It should be noted that the Sn film not only affects the wrinkle wavelength, but also the morphology of the wrinkle. For an excessively thick Sn film, such as deposited for 300, 400, 500 s, the wrinkle pattern have an island-like structure with low amplitudes, because the thick film has a high stiffness. However, for an excessively thin Sn film, the wrinkle gets hazy and looses its periodicity, because the poor continuity of .the film. The Sn film deposited for 100 s, has a best labyrinth pattern and smallest wavelength. The thickness is determined to be about 5 nm by AFM.

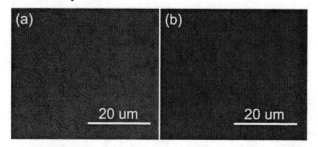

Figure 1: OM images of wrinkles of 20 nm Bi (a) and Sn (b) on 120 nm PS film. The wrinkles were fabricated by annealing the samples at 120 °C for 2 h.

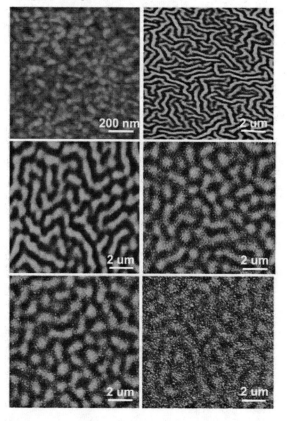

Figure 2: Wrinkles of different Sn film on 56 nm PS. The Sn films are deposited for 60(a), 100(b), 200(c), 300(d), 400(e) and 500(f) s. The wrinkles were fabricated by annealing the samples at 120 °C for 2 h. The color scale is 50 μm.

Besides the metal film, the polymer film is also seriously affects the wrinkle wavelength, especially when the polymer film is comparable to the metal film. Fig 3 shows the wrinkle patterns of 5 nm Sn film supported on the PS films with different thicknesses. A downtrend of wrinkle wavelength is obvious as decreasing the thickness of PS film. Thus wrinkle wavelength has been reduced to below 200 nm, which is much lower than existing reported metallic wrinkle wavelength, by using the appropriate metal and optimizing the thicknesses of both metal and polymer films.

3.2 The assembly method of wrinkle patterns

Reducing the wavelength is the first step of nanofabrication by wrinkling. The controllability of wrinkle pattern is important and necessary. Many methods have been proposed to assemble the wrinkle [17, 18].Here, we present some novel methods to obtain ordered wrinkle pattern. In Fig. 4, some ordered wrinkle patterns have been fabricated via different methods. All these methods are prefabricating patterned structures before annealing. The wrinkle pattern will have some favored direction related to the prefabricated structures. Fig 4(a) (b) show ordered wrinkle patterns assembled via thermal nano-imprint. A template is needed to imprint structures on PS films. The wrinkles are assembled to be perpendicular to the edges of prefabricated structures, because the stress relaxation perpendicular to the edges. The wrinkle pattern in Fig 4(c) is obtained via selective area film deposition. After a PS/Sn bilayer is produced, the left part is covered by a mask. A continuative Sn film deposition will cause the different thickness of covered and uncovered parts. The edge of thicker part will produce a pinning effect, which cause the increase of perpendicular stress in the adjacent area. An ordered wrinkle pattern paralleled to the edges appears. Another wrinkle assembly via laser scanning is shown in Fig 4(d). The ablation of Sn film via laser scanning produces break of films. The break will relieve the perpendicular stress, thus the wrinkles are perpendicular to the scanned line.

All these methods have been proved to be effective assembly method, although there are many defects. The defects come from the sensitiveness of stress distribution, the defects of prefabricated structures and the geometrical limit. Development of novel, simple assembly method is still needed

3.3 The influence of metal film morphology on wrinkle pattern

The conventional wrinkling theory is based on a model, in which the metal film is treated as a flat, continuous layer. When the metal film is limited to a very thin thickness due to the demands of small wavelength, the detailed morphology of film is a network of grains. The influence of film morphology on wrinkle pattern can not be studied from

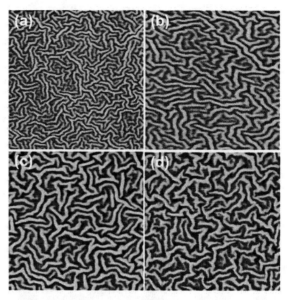

Figure 3: AFM images (5 μm×5 μm) of wrinkles on 5 nm Sn film supported on different PS films. The thicknesses of PS films are 23 nm (a), 56 nm (b), 71 nm (c), 165 nm (d). The color scale is 50 μm.

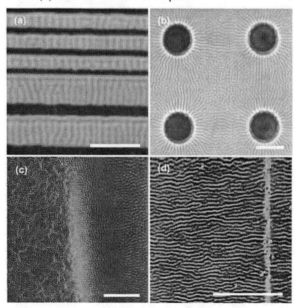

Figure 4: OM images (a, b) and AFM images (c, d) of wrinkle patterns assembled by various method. The scale bar is 5 μm.

the conventional model. Here we have studied the wrinkle on the bilayers with same thickness but different morphology.

The 20 nm Sn film with different morphology was fabricated by once 400 s (Fig. 5(b)), twice 200 s (Fig. 5(d)), and four 100 s (Fig. 5(f)) deposition respectively [21]. The SEM images show that the four 100 s deposited film has the finest and homogenous grains while the 400 s deposited film has the largest and inhomogeneous grains. The wrinkle patterns of these 20 nm Sn films supported on 310 nm PS are shown in Figs. 5(a), 5(c) and 5(e), respectively. All these wrinkles have no significant change in wavelength because the same thickness and materials of films, while the amplitude is different. The film with finest and homogenous grains is most likely to a continuous layer. Thus the corresponding wrinkle is clearest and has largest amplitude. However, the multi-times deposited films have both larger grains and fine grains, which damaged the uniformity. The existing of giant grains will bury the wrinkle structure, thus make the wrinkle hazy and reduce the amplitudes. In order to obtain a clear, well developed wrinkle pattern, the thin film with fine and homogenous morphology is preferred.

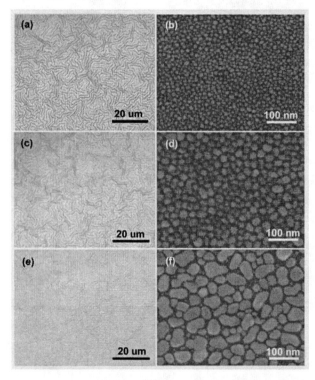

Fig. 5: The influence of film morphology on wrinkle patterns: (a), (c), (e) wrinkle patterns of 20 nm Sn film on 310 nm PS film after heating at 120 °C for 4 h. (b), (d), (f), the film morphology of Sn film in (a), (c),(e) respectively.

4 SUMMARY

In summary, we have fabricated the wrinkle patterns with periods from micro to nano scale (sub 500nm) by choosing the appropriate metal film and optimizing the film thicknesses. Various methods of wrinkle assembly have been presented and show its effect in controlling the wrinkle direction. The controllability of wrinkles gives possibility of nanofabrication via wrinkling, although many efforts are still needed. Furthermore the influence of film morphology on wrinkle pattern has been studied, which indicated a homogenous thin film with fine grains is preferred for high quality wrinkle pattern.

The authors gratefully acknowledge the financial support from the National Natural Science Foundation of China (NSFC) (10974037), the National Basic Research Program of China (NBRPC) (2010CB934102, 2006CB932603).

REFERENCES

[1] J. Genzer and J. Groenewold, Soft Matter, 2, 310, 2006.
[2] J. L. Wilbur, A. Kumar, H. A. Biebuyck, E. Kim and G. M. Whitesides, Nanotechnology, 7, 452, 1996
[3] H. W. Li, D.-J. Kang, M. G. Blamire and W. T. S. Huck, Nano Lett., 2, 347, 2002
[4] Y. F. Chen, K. W. Peng and Z. Cui, Microelectron. Eng., 73/74, 662, 2004
[5] W. Chen and H. Ahmed, Appl. Phys. Lett., 62, 1499,1993
[6] W. H. Li, D. H. Kang, M. G. Blamire and W. T. S. Huck, Nanotechnology, 14, 220, 2003
[7] M. Watanabe, H. Shirai and T. Hirai, J. Appl. Phys., 92, 4631, 2002.
[8] S. P. Lacour, S. Wagner, Z. Huang and Z. Suo, Appl. Phys. Lett., 82, 2404, 2003.
[9] C. Harrison, C. M. Stafford, W. H. Zhang and A. Karim, Appl. Phys. Lett., 85, 4016, 2004
[10] K. Efimenko, M. Rackaitis, E. Manias, A. Vaziri, L. Mahadevan and J. Genzer, Nat. Mater., 4, 293 2005.
[11] C. M. Stafford, C. Harrison, K. L. Beers, A. Karim, E. J. Amis, M. R. Vanlandingham, H. C. Kim, W. Volksen, R. D. Miller and E. E. Simonyi, Nat. Mater., 3, 545, 2004.
[12] J. Y. Chung, T. Q. Chastek, M. J. Fasolka, H. W. Ro, and C. M. Stafford, ACS Nano, 3, 844, 2009.
[13] E. P. Chan, K. A. Page, S. H. Im, D. L. Patton, R. Huang and C. M. Stafford, Soft Matter, 5, 4638,2009.
[14] M.-W. Moon, S. H. Lee, J.-Y. Sun, K. H. Oh, A. Vaziri, and J. W. Hutchinson, Proc. Natl. Acad. Sci. U.S.A., 104, 1130, 2007.
[15] A. Chiche, C. M. Stafford and J. T. Cabral, Soft Matter 4, 2360 2008.
[16] B. Kolaric, H. Vandeparre, S. Desprez, R. A. L. Vallée and P. Damman, Appl. Phys. Lett., 96, 043119, 2010.
[17] N. Bowden, S. Brittain, A. G. Evans, J. W. Hutchinson and G. M. Whitesides, Nature, 393, 1466, 1998.
[18] P. J. Woo, K. Y. Suh, S. Y. Park and H. H. Lee, Adv. Mater., 14, 1383, 2002.
[19] Z. Zhang, C. Guo, S. Cao, L. Bai, Y. Xie and Q. Liu, Jpn. J. Appl. Phys., 48, 090208, 2009.
[20] C.-C. Fu, A. Grimes, M. Long, C. G. L. Ferri, B. D. Rich, S. Ghosh, S. Ghosh, L. P. Lee, A. Gopinathan, and M. Khine, Adv. Mater., 21, 1, 2009.
[21] C. F. Guo, J. Zhang, J. Miao, Y. Fan and Q.ian Liu, Optics Express, 18, 2621, 2010

Simple Fabrication of Hierarchical Structures on a Polymer Surface

Bahador Farshchian, Jeong Tae Ok, Steven Maxwell Hurst, and Sunggook Park*

*Department of Mechanical Engineering and Center for Bio-Modular Multiscale Systems
Louisiana State University, Baton Rouge, LA70803
Tel: 225-578-0279, Fax: 225-578-5924, email: sunggook@me.lsu.edu

ABSTRACT

The ability to fabricate hierarchical structures has drawn significant interest as a means of achieving superhydrophobic surfaces. Water repellency has potential industrial importance, such as self-cleaning, anti-oxidation and minimization of the resistance against liquid flow in microfluidic systems. In this paper we introduce a simple and fast technique to fabricate hierarchical structures on polymer surface in order to achieve a superhydrophobic surface. In the first step microstructures were fabricated on surface of poly(methyl methacrylate) using a sand paper. In order to make nanostructures on top of microstructures O_2 reactive ion etching was employed. In the last step the surface was coated with fluorinated silane molecules to lower its surface energy. Static and dynamic contact angle measurement showed water contact angles were higher than 150° and sliding angles were 2° on the fabricated surface.

Keywords: hierarchical structures, superhydrophobic surface, self cleaning, Cassie's state, Wenzel's state.

1 INTRODUCTION

The ability to fabricate hierarchical structures has drawn significant interest as a means of achieving superhydrophobic surfaces. It is well proven that in addition to surface chemistry topography of a surface has a significant influence on the wettability. Some of the plant leaves have shown superhydrophobic properties in a way that water droplet can easily roll off and carry away the dust and debris on their surface, a mechanism which is called as self-cleaning [1]. This property has been demonstrated for lotus, rice, taro, india canna leaves and is known as lotus effect [2]. It was first believed that this behavior is because of microscale textures and chemistry of the surfaces of those leaves but further investigation of those leaves under scanning electron microscope (SEM) revealed that there are branch-like nanostructures on top of microstructures of those surfaces. For example, the surface of a lotus leaf is composed of fine, branched nanostructures (120 nm) on top of micropapilla (5-9 μm) [3]. In fact, the combination of hierarchical structures on the surface of those leaves and their chemistry is the cause of their water repellency. Water repellency has potential industrial importance, such as self cleaning, anti-oxidation, minimization of the resistance

against liquid flow in microfluidic systems. Ou et al. [4, 5] fabricated a microchannel which had a superhydrophoic bottom and systematically investigated the effect of topography on drag reduction and velocity profiles using pressure drop measurements and microparticle image velocimetry. They reported a pressure drop reduction of over 40% and an apparent slip length exceeding 20 μm.

The two major models to predict water contact angles on rough surfaces are the Wenzel and Cassie models. In the Wenzel's theory a droplet deposited on a rough surface fills completely the grooves of the rough surface. The Wenzel's contact angle can be calculated using equation 1 [6].

$$cos\ \theta^* = r\ cos\ \theta \qquad (1)$$

where θ^* and θ are contact angles on rough and flat surfaces, respectively, and r is surface roughness. The key parameter controlling Wenzel's contact angel is the roughness defined as the ratio of true surface area to the projected area. Therefore r is always larger than 1. Based on this equation, if the contact angle on flat surface is less than 90°, the contact angel on a rough surface will be smaller than that for the corresponding flat surface while for a true contact angle higher than 90° an increase in surface roughness increases the contact angle.

In the Cassie's state air pockets are assumed to be trapped under the surface of liquid droplet. As a result the liquid doesn't wet completely the surface area under the liquid droplet and a composite surface consisting of air and substrate forms under the droplet. If we assume $\emptyset_s$ as the surface fraction of solid under droplet then contact angle can be calculated using equation 2 [6].

$$cos\ \theta^* = -1 + \emptyset_s\ (cos\ \theta + 1) \qquad (2)$$

The Cassie's contact angle monotonously increases as $\emptyset_s$ decreases, implying that $\emptyset_s$ should be as small as possible. But it should be noted that reducing $\emptyset_s$ decreases roughness as well causing to reach the critical roughness below which Wenzel's state is energetically more preferable.

There are two requirements for a surface to be superhydrophobic. The surface should have a very high contact angle, larger than 150°, and the sliding angle should be less than 10° [7]. The tilting angle at which the droplet rolls off the substrate is called as the sliding angle. In case of superhydrophobic surfaces the droplet can easily roll off by tilting the surface for a small amount. It is well known that both Cassie and Wenzel's states can make high contact angle droplets but it is only Cassie's state which can cause a low sliding angle. In Wenzel's state droplets wet the surface completely and therefore it cannot easily move on the surface. In Cassie's state the droplet is in touch with solid in less area, thus reducing the dragging force (friction) applied on the droplet. As a result, the droplet can easily move on the surface at a low sliding angle. Therefore, it can be concluded that only Cassie state can fulfill the two requirements for a superhydrophobic surface.

Various micro- and nanofabrication techniques have been employed to artificially produce superhydrophobic surfaces. Sun et al. [8] has reported creating of a superhydrophobic surface using a natural template of lotus leaf and replicating it by casting. Jeong et al. [9] have investigated the wettability of nanoengineered dual-roughness surfaces fabricated by UV-assisted capillary force lithography. Lee et al. [10] have fabricated multiscale structures on high density polyethylene (HDPE) by heat and pressure-driven imprinting methods using patterned anodized alumina as replication template.

In this paper we introduce a simple and fast technique to fabricate hierarchical structures on polymers surface in order to achieve a superhydrophobic surface. In the this technique we make a superhydrophobic surface on poly(methyl methacrylate) (PMMA) surface by scratching it by sand paper followed by reactive ion etching (RIE) and silane coating processes. Scanning electron microscope (SEM) images of surfaces fabricated using abovementioned technique proved formation of hierarchical structures on PMMA. Static and dynamic contact angle of water droplet on fabricated surfaces were measured. We also observed bouncing of water droplet on the fabricated surface.

2 EXPERIMENTAL PROCEDURES

2.1 Fabrication of superhydrophobic surface on PMMA.

Figure 1 shows schematics of the process steps to achieve a hierarchical structure with superhydrophobic property in polymer substrates. PMMA sheet with thickness of 2.9 mm were purchased from United States plastic corp. and cut in 10×30 mm^2 pieces. For comparison, four pieces of PMMA sheets which underwent different treatments were prepared. The first piece (S0) is considered as a reference sample and is used to measure contact angle on plane PMMA. A commercial sand paper with 240 grit number was used to sand surfaces of second and third PMMA pieces (S1 and S2). Then oxygen plasma etching process was done to treat surface of third and fourth sample (S2 and S3). The RIE power and oxygen pressure were set at 150 W and 150 mTorr respectively and the surface of samples S3 and S4 were treated for 20 minutes. Finally samples S1, S2 and S3 were treated for 20 minutes with a fluorinated silane molecule (heptadecafluoro-1, 1, 2, 2-

tetrahydrodecyl) trichlorosilane ($C_{10}H_4Cl_3F_{17}Si$) in the vapor phase using a home built vacuum chamber in order to reduce surface energy of the samples. In table 1, the type of surface treatment on each sample is summarized.

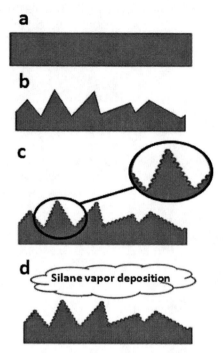

Figure 1- Schematic illustration of the process used to fabricate hierarchical structure on surface of PMMA to achieve a superhydrophobic surface. (a) Plane PMMA. (b) PMMA sanded by sand paper. (c) oxygen plasma etching of PMMA. (d) vapor deposition of silane molecules on surface.

Table 1- Type of surface treatment on different PMMA samples.

Sample	Treatment
S0	No treatment (plane PMMA)
S1	Sanded + silane coated
S2	Sanded + oxygen plasma + silane coated
S3	Oxygen plasma + silane coated

2.2 Contact angle measurements.

A contact angle analyzer (FTA 125, First Ten Ångstrom, Inc. Portsmouth, Virginia) was used to measure static contact angles of water droplets on the samples prepared in this study. The samples were placed on a stage and adjusted to an appropriate level. A droplet is dispensed on the sample by pushing a button of the pipette. The image is captured immediately after the droplet is placed on the sample for accurate measurement. The software automatically analyzes the drop shape and computes the contact angle. For each sample contact angle measurements were done 3 times with water droplets of a target volume of 5 µL. The averaged values of contact angles were used for each data point. Sliding angle measurements were done for samples which had a static contact angle higher than 150°. In order to measure sliding angle water droplet with volume of 5 μL were placed on sample with zero degree of tilting and then sample was tilted manually. The angle at which droplet slides on surface was measured using a goniometer. 3 measurements were done to define sliding angle of each sample.

2.3 Recording of bouncing droplet.

To record bouncing of water droplets on the sample S2 water droplets with volume of 8 – 10 μL were released from height of 3.4 mm to 8.4 mm respect to the sample surface. The high speed camera (Kodak Ektapro 1000HRC), which has a maximum frame rate of 1000 fr/s and minimum exposure time of 50 μs, was used to capture images during droplet impact on the surface.

3 RESULTS AND DISSCUSION

In figure 2 water droplets on the surfaces of plane PMMA (S0), PMMA sanded with sand paper and coated with silane (S1), PMMA sanded with sanded paper, etched using oxygen plasma and coated with silane (S2) and PMMA etched using oxygen plasma and coated with silane (S3) are shown.

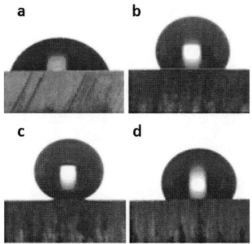

Figure 2- water droplet on surface of (a) S0, (b) S1, (c) S2 and (d) S3 samples.

Static contact angles of water droplet on samples S0, S1, S2 and S3 were measured as 70.5±1.6°, 119.1±1.9°, 165.3±3.2° and 115.7±1.0° respectively. As can be seen the contact angle for plane PMMA without any silane coating (S0) is the lowest among all samples while for the sample (S2) which has a hierarchical structure and treated with silane molecule it is the highest. Scanning electron microscope image of sample S2 is shown in figure 3. In fact sanding PMMA produces mostly microstructures on the sample surface while the subsequent O_2 plasma etching process bring about the formation of nanostructures on top

of sanded microstructures. Despite the presence of hierarchical micro/nanostructures, the surface (S2) does not produce superhydrophobicity without combining a fluorinated silane coating. Coating the sample with fluorinated silane molecules further reduces their surface energy. A combination of formation hierarchical structures on surface of PMMA and silnae coating on sample surface will cause water droplets to follow the Cassie's model where it just stays on top of protrusions on sample surface. Having silane molecules on surface of sample will cause the water droplet not to fill cavities of samples and be repelled by silane molecules. In this state water is in touch with the solid substrate in much lower area compared to the case it is placed on plane PMMA. Our sliding angle measurements showed water slides on surface at 2° which is due to low friction of water droplet and the substrate. The contact angle of water droplet on sample S1 and S3 which have only microstructures and nanostructures and are silane coated are lower than sample S2. The results corroborate the fact that having hierarchical structures in combination of appropriate surface chemistry is an important means in order to minimize surface fraction of solid substrate, increase roughness and increase contact angle.

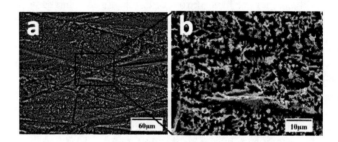

Figure 3- SEM images of sample S2 which is sanded by sanded by sand paper, etched by oxygen plasma and silane coated.

One efficient way to show water-repellency on superhydrophobic surface is the droplet impact experiment. The impact of microliter water droplets in the range of injection droplet volume of 8 – 10 μL on the superhydrophobic surface (sample S2) was investigated for different initial impact speeds (v_0: 12.4 – 33.6 cm/s), which were varied by changing the dropping needle height from 3.4 mm to 8.4 mm. The Weber number, a dimensionless number indicative of the ratio between the kinetic energy of the droplet and surface energy, were in the range of 0.28 – 2.04 depending on the needle height. Figure 4 shows image sequences for one of the examples of rebounding behavior of water droplet, which was released at 8.4 mm needle height. For all the Weber numbers, water droplets successively bounce off the surface. The restitution coefficient of droplet, e, gradually decreases from 0.73 to 0.28 as every rebound occurs (or the number of rebound increases), where e was obtained by dividing initial impact speed into take-off speed. The bouncing continues until all the energy in the droplet was dissipated by internal flow of the bouncing droplet and surface friction [11]. The impact experiments on superhydrophobic PMMA sufaces (sample S2) indicate that the surface exerts an excellent anti-sticking character with water droplets.

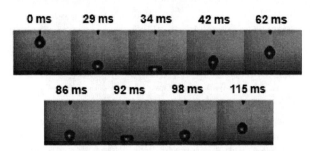

Figure 4- image sequences for rebounding water droplet (droplet volume V0: 9.3 µL) on superhydrophobic PMMA surface (sample S2, needle height: 8.4 mm, initial impact speed v0: 33.6 cm/s and We: 2.04).

4 CONCLUSIONS

In this paper we have introduced a cheap, fast and easy technique to produce hierarchical structures in order to achieve superhydrophic surface on PMMA. In this technique hierarchical structures are produced on surface of PMMA by sanding PMMA sheet with sand paper and etching it by oxygen plasma. After fabrication of hierarchical structures on polymers they are coated with a fluorinated silane. This simple technique was able to successfully produce a superhydrophobe surface on PMMA. Having a hierarchical structure on PMMA will decrease the surface fraction of solid in contact with water droplets (Cassie's model). Presence of silane molecules on surface also repels water preventing it from wetting inside the grooves causing the static contact angle to increase more than 150° and water to slide on surface at sliding angle 2°.

ACKNOWLEDGMENTS
This research was supported by National Science Foundation CAREER Award (CMMI-0643455). We also thank Profs. Dimitris E. Nikitopoulos and Michael C. Murphy for fruitful discussion.

REFERENCES

[1] Z. Yang, C. Chiu, J. Yang and J. Yeh, "Investigation and Application of an Ultrahydrophobic Hybrid-Structured Surface with Anti-sticking Character", Journal of Micromechanics and Microengineering, 19, 2009.

[2] Z. Guo and W. Liu, "Biomimic from Superhydrophobic Plant Leaves in Nature: Binary Structure and Unitary Structure", Plant Science ,172, 1103-1112, 2007.

[3] X. F and L. Jiang, "Design and Creation of Superwetting/Antiwetting Surfaces", Advanced Materials, 18, 3036-3078, 2006.

[4] J. Ou, B. Perot, and J. P. Rothstein, "Laminar Drag Reduction in Microchannels Using Ultrahydrophobic Surfaces", Physics of fluids, 16, 4635-4643, 2004.

[5] J. Ou and J. P. Rothstein, "Direct Velocity Measurements of the Flow Past Drag-reducing Ultrahydrophobic Surfaces", Physics of fluids, 17,103606, 2005.

[6] M. Callies and D. Quere, "On Water Repellency", Soft Materials, 1, 55-61, 2005.

[7] B. Bhushan, Y. Jung, and K. Koch, "Micro-, Nano- and Hierarchical Structures for Superhydrophobicity, Self-cleaning and Low Adhesion", Philosophical Transactions of the Royal Society, 367, 1631-1672, 2009.

[8] M. Sun, C. Luo, L. Xu, H. Ji, Q. Ouyang, D. Yu, and Y. Chen, "Artificial Lotus Leaf by Nanocasting", Langmuir, 21, pp. 8978-8981, 2005.

[9] H. E. Jeong , M. K. Kwak , C. I. Park, K. Y. Suh, "Wettability of Nanoengineered Dual-roughness Surfaces Fabricated by UV-assisted Capillary Force Lithography", Journal of Colloid and Interface Science, 339, 202–207, 2009.

[10] Y. Lee, S. H. Park, K. B. Kim, and J. K. Lee, "Fabrication of Hierarchical Structures on a Polymer Surface to Mimic Natural Superhydrophobic Surfaces", Advanced Materials, , 19, 2330–2335, 2007.

[11] Y. Kwon, N. Patankar, J. Choi and J. Lee, "Design of Surface Hierarchy for Extreme Hydrophobicity", Langmuir Article, 25, 6129-6136, 2008.

The Nanophysics of TiO₂/Au Model Catalyst as a Key to Understanding the High Efficiency of Real Au/TiO₂ Catalyst and Technological Consequences

I.V. Tvauri[*], D.F. Remar[*], A.M. Turiev[*], N.I. Tsidaeva[*], K. Fukutani[**], T.T. Magkoev[*]

[*]Nanophysics Research Centre, University of North Ossetia, Vatutina 46,
Vladikavkaz 362025, Russian Federation, ateia@mail.ru
[**]Institute of Industrial Science, The University of Tokyo, 4-6-1, Komaba,
Meguro-ku, Tokyo 153-8505, Japan

ABSTRACT

Carbon monoxide oxidation on the surface of in-situ prepared direct and inverse model Au/TiO₂ catalyst have been studied in ultra-high vacuum condition by a set of surface science analytical techniques. It is found that a metal/oxide perimeter interface is an active area for CO to CO₂ transformation. Measurements with isotopically labeled $^{18}O_2$ reveal that carbon dioxide is formed over Au/titania via the participation of oxygen specie of titanium oxide. Efficiency of the process is higher for the system consisting of reduced titania (TiO$_x$, x<2) compared to TiO₂. Results offer a number of possibilities to improve the performance of real Au/TiO₂ catalyst.

Keywords: surface nanostructure, thin films, molecular reaction on surface, gold, titanium oxide.

1 INTRODUCTION

About a decade ago it was found that catalytic activity for a range of oxidation reactions, (most notably carbon monoxide oxidation) of a mixture of Au ultra-small particles with titanium oxide powder considerably exceeds that exhibited by Au or titania on their own [1]. The explanation of high catalytic activity of titania/Au is still controversial. On one hand it is considered that the catalytic efficiency is due to the unique electronic and/or structural properties of oxide supported nanometer-scale Au particles which differ from those of the bulk Au, on the other – to the effect of the Au/titania perimeter interface [2]. To overcome this controversy it would be informative to study the catalytic behavior of the system winch emphasizes one of the two above-mentioned parameters, either an Au ultra-small cluster or titania/Au perimeter interface effect. The real catalyst is hardly suitable for thus purpose, however the corresponding model system might be a good candidate. In relation to this the aim of the present study is to investigate the CO+O₂ interaction on the surface of the system obtained by the controlled growth of titanium oxide sub monolayer to multilayer thin film islands on the surface of atomically clean Au (111). The advantage of this system is that carbon monoxide does not adsorb molecularly on titania and Au (111) [3] under ultra high vacuum (UHV) at a substrate temperature higher than 200 K, thus allowing to attribute any possible CO oxidation effect to the titania/Au interface. This system can be viewed as an inverse model catalyst due to complementary nature to real Au/TiO₂ catalyst [4].

To further elucidate the effect of the titania as well as the metal oxide perimeter interface in catalytic CO oxidation additional set of measurements on CO/Au/TiO$_x$ (x<2) systems consisting of titania films oxidized by isotopically labeled oxygen $^{18}O_2$ have been carried out. Mass-spectroscopic analysis of the formed carbon dioxide specie unequivocally reveal essential role of oxygen of titania in catalytic CO oxidation. Efficiency of the process is notably higher for reduced titanium oxide film (TiO$_x$, x<2) compared to TiO₂ [5].

2 EXPERIMENTAL

Since gaseous molecules are very sensitive to the influence of electron or ion impact, on which surface probe techniques of electron and ion spectroscopy are based, investigation of CO+O₂ interaction have been carried out by reflection adsorption infrared spectroscopy (RAIRS) – an unrestrictive high-sensitive technique which is well suited for the study of the systems under consideration, i.e., dipole gas molecules on the metal substrate and thin film. For this purpose conventional Fourier-transform spectrometer adapted to measure normal molecular vibrations at grazing incidence and detection beam angles was used. Additional techniques used mainly for characterization of the preparation of titania films were Auger electron spectroscopy (AES) using double-pass cylindrical mirror analyzer with coaxial electron gun, X-ray photoelectron spectroscopy (XPS, Mg K$_\alpha$ anode) and low energy ion scattering (LEIS, He$^+$ ions) with the aid of a hemispherical analyzer, low energy electron diffraction (LEED) with the rear-view four-grid optics and work function measurements by Anderson method. Mass-spectroscopic analysis of carbon dioxide specie were carried out by thermal desorption spectroscopy (TDS) with the aid of quadrupole mass-spectrometer. For LEIS the chamber was equipped with a differentially pumped ion gun using He$^+$ ions at a primary energy of 1 keV. All measurements have been carried out in ultra high vacuum chamber with the base pressure normally not exceeding 2×10^{-10} Torr. The Au (111) surface was obtained by the well-known procedure of growing thick film (50 ML) of Au on Mo (110) at a substrate held at 800 K [4]. Gold of high purity (99.99%) was evaporated from a conical W/Re filament source. It was verified that the film surface prepared in this way is identical to that of bulk Au (111). Titanium oxide was grown by thermal evaporation of metallic titanium (99.996% purity) in oxygen ambient at a substrate held at 700 K. The evaporation rate of Ti was controlled by the work function change of Mo (110) crystal as well as by Mo MNV (188 eV) Auger signal attenuation [5, 6]. One monolayer equivalent (MLE) of Ti is defined as 2.34×10^{15} cm^{-2}. Adsorption of carbon monoxide and oxygen were carried out by backfilling the UHV chamber through the leak valve with the respective spectroscopically pure gas at a partial pressure not exceeding 10^{-6} Torr. The value of exposure of 1 L was defined as 10^{-6} Torr s. In more detail experimental conditions are described elsewhere [5, 7, 8].

3 RESULTS AND DISCUSSION

Initially the Au film of about 50 ML thick was grown by thermal evaporation of metallic Au on the surface of Mo (110) bulk crystal. No impurities as well as Mo (110) substrate signals were detected by AES and XPS after the formation of the Au film. LEED exhibited hexagonal spots characteristic for (111) surface structure. Titanium oxide layers were grown by reactive evaporation of Ti in an oxygen atmosphere at a partial pressure of 10^{-6} Torr at a substrate held at elevated temperature of 700 K. According to LEIS data titanium oxide does not fully cover the Au surface but likely grows via three-dimensional islands beginning from the submonolayer coverage. An observed LEED pattern is and evidence for the formation of TiO$_2$ (100) structure with some local reconstruction of the surface. Formation of oxide with TiO$_2$ stoichiometry is corroborated with the XPS data. As has been shown previously [6] the Ti 2p photoelectron line is very sensitive to the degree of Ti oxidation. The corresponding spectra are shown in Fig. 1. Curves 1 and 2 correspond to the oxide films obtained by the above mentioned procedure with the difference that the latter one was not exposed to oxygen after stopping the Ti atom flux. The curve 1 indicating a "deeper" oxidation of Ti is close to the spectrum of the bulk TiO$_2$ crystal [6], whereas the curve 2 is an evidence of sub-stoichiometric oxide film.

The RAIRS measurements of carbon monoxide adsorption separately on pure Au (111) and TiO$_2$ films held at 200 K have shown that CO does not easily adsorb on these surfaces. The spectrum of the system obtained after exposing the titanium oxide film (30 MLE) corresponding to curve 1 of Fig. 1 to about 100 L of carbon monoxide consists of only one low intensity absorption line at a wavenumber of 2180 cm^{-1} (see Fig. 2, inlay). No lines have been detected upon exposing the Au (111) to CO up to exposure value of 400 L (wavenumber range probed: 1800 – 2200 cm^{-1}). In addition, no C 1s lines have been observed by XPS. However quite noticeable IR absorption line at

2114 cm^{-1} has been detected when CO was adsorbed on the surface consisting of titania clusters on Au (111). The corresponding spectrum for titania 5 MLE at CO exposure of 100 L is shown in Fig. 2 (curve 1). Exposure of about 100 L corresponds to CO saturation. Up to this exposure the IR intensity gradually increases with the total blue shift of intramolecular vibration frequency of 8 cm^{-1}. The latter can be attributed to increasing dipole-dipole interaction with CO coverage increase. Above 100 L exposure the CO IR intensity and line position do not noticeably change. Therefore one can consider that the interface area created by the contact of Au (111) and titanium oxide becomes an active place for CO adsorption while titania and Au (111) separately are poor substrates for CO bonding at this temperature. Such a dramatic change of the adsorption capability is an evidence of considerable change of the electronic and/or structural properties of both titania and Au when they are in contact. According to literature data on

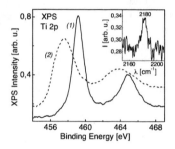

Figure 1: Ti 2p photoelectron spectra of thick (30 MLE) titanium oxide film on Au (111) annealed (1) and unannealed (2) in an oxygen ambient after stopping the Ti evaporation flux. Inlay: Infrared absorption line (intensity I versus wave-number λ) observed after exposing the titanium oxide film, corresponding to curve (1), to 100 L of carbon monoxide at a substrate held at 200 K.

titanium dioxide growth on different metal substrates the stoichiometry as well as the electronic structure of Ti and O ions of the oxide layer which is in direct contact with the metal is different from that of TiO$_2$. Moreover the electronic state of the topmost layer of the metallic substrate also changes. The Au atoms at the oxide/Au perimeter interface acquire some anionic character. Thus adsorption of the titanium oxide on Au (111) results in a formation of a ternary Ti-O-Au system at the interface with a chemisorptive properties which are notably different from those of TiO$_2$ and Au separately [8].

Upon subsequent inlet of the molecular oxygen into the UHV chamber the intensity of the CO absorption line gradually decreases with a slight blue shift (total detected shift is about 10 cm^{-1}) (see Fig. 2, curves 1 and 7). Possible explanations of such IR intensity depletion are 1) oxygen adsorption stimulated CO desorption; 2) formation of CO$_2$ and its subsequent desorption; 3) change of the CO molecular axis tilt in the way that intramolecular vibrations are no longer accessible by RAIRS in the configuration used. However, the latter one might not be the case since XPS measurements before and after oxygen inlet revealed that in the latter case the C 1s intensity is by the order of magnitude lower than in the former case. (It should be noted that the C 1s line intensity decrease is not accompanied by the change of its binding energy which indicates that CO does not dissociate on the surface). This means that the IR line intensity depletion is due to CO surface concentration decrease, rather than to molecular reorientation on the surface. The reason for CO concentration decay is CO_2 formation and its subsequent desorption.

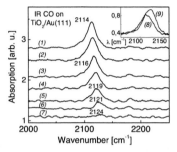

Figure 2: Change of the infrared absorption spectra during exposure of the CO/titania/Au (111) system to oxygen at a substrate held at 200 K. Carbon monoxide initial exposure and oxygen partial pressure are 100 L and 10^{-6} Torr, respectively. Oxygen exposure time (min): 1-0; 2-5; 3-10; 4-15; 5-20; 6-25; 7-30. Inlay: Comparison of IR absorption bands of CO on bare (8) and oxygen predosed 5 MLE titania/Au (111) held at 200 K

In the previous studies it was shown that CO, as well as NO, coadsorbed with atomic oxygen on a number of substrates exhibit an increase of the vibrational frequency caused by molecular-oxygen interaction. An observed blue shift of CO vibrational mode in this case (see Fig. 2, inlay) can be viewed as an evidence that oxygen is accommodated by the surface prior to interaction with preadsorbed CO. Accommodation by the surface results in the formation of some anionic species at the oxide/Au interface which cause an observed increase of the CO intramolecular vibrational frequency and which are active for CO oxidation. In addition, coexistence of a negatively charged species on the surface in the vicinity of a polar molecules like CO and NO results in an effective increase of their dynamical dipole moment [9]. The latter one leads to the enhanced infrared light absorption which results in the increase of IR line intensity. In relation to this, the observed increase of the CO IR intensity for the reversed order of adsorption (see Fig. 2, curves 8 and 9) is a further evidence of appearance of anionic species upon oxygen adsorption on titania/Au. Thus oxygen and carbon monoxide adsorption on titania/Au (111) results in the change of the electronic state of the interface in addition to the change of the electronic state of both titania and Au when they form the interface. Such a complex transformation of the electronic state of the system resulting from titania/Au (111) formation and subsequent

adsorption of oxygen and carbon monoxide leads to an adsorption and catalytic behavior of the interface which are dramatically different from those exhibited by titania and Au on their own. Also taking into account a rather high CO IR intensity decay rate (see Fig. 2) one can conclude that the titania/Au perimeter interface is a place for efficient carbon monoxide and oxygen chemical interaction.

Above consideration relates to saturation coverage of CO which can be achieved at an exposure of 100 L. Similar CO IR intensity degradation was also observed for lower CO initial coverage. Moreover, it was found that the rate of the CO IR line intensity decay is higher at lower initial CO coverage. In order to find the corresponding dependence the four types of CO/oxide/Au systems differing only by the initial CO exposure have been investigated measuring the CO IR intensity versus oxygen exposure time. In this case the scan area was restricted to 2070 – 2150 cm^{-1} with the increase of the number of scans to increase signal to noise ratio. The corresponding plots for titania 5 MLE and different CO initial exposures are shown in Fig. 3 (curves 1 to 4). It is seen that the lower the CO initial coverage the higher is its decay rate upon oxygen exposure. Assuming that this decay is due to CO_2 formation and its subsequent desorption the initial slopes of the corresponding curves (slope angles (α) of the initial part of the curves (see Fig. 3) are proportional to the $CO+O_2$ interaction efficiency (σ), σ = tg (α). This value plotted versus the CO initial exposure is shown in Fig. 3 (inlay). It is seen that at lower concentration the CO molecules more efficiently interact with oxygen: The reaction rate varies by a factor of two or even higher depending on the CO initial coverage. Possible explanation of this might be the fact that the reactant species may need a free space on the surface (specifically at the oxide/Au perimeter interface) for the reaction to occur.

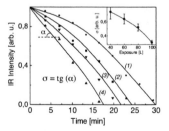

Figure 3: Dependence of the CO IR intensity line on the oxygen exposure time for different initial CO exposures of titania/Au (111) system (L): 1-100; 2-80; 3-60; 4-40. Inlay: Dependence of the value of σ characterizing the CO surface concentration decay rate upon the initial CO exposure

It may turn out that this space is necessary because the reaction between species proceeds via the mobile precursor. At higher CO coverage when the concentration of free sites is lower, the mobility of the species is suppressed due to site blocking effect so that the reaction rate decreases, and vice versa. As some CO adsorption sites become unoccupied due to CO_2 formation and desorption, the reaction rate increases as follows from the corresponding curves shown in Fig. 3.

Results with titania formed by oxidation in an isotopically labeled oxygen ($^{18}O_2$) ambient corroborate above conclusion of the essential role of the titania surface in CO oxidation. Reduced (TiO_x, $x<2$) and totally oxidized TiO_2 films were prepared by the technique mentioned above. The formed Au/TiO_2 and Au/TiO_x systems after cooling down to 80K were exposed to carbon monoxide at an exposure of 100 L. Afterwards molecular oxygen $^{16}O_2$ was supplied into the UHV chamber to a partial pressure of 10^{-6} Torr. The thermal desorption spectra of particles recorded by the quadrupole mass analyzer at a temperature sweep rate of 2 K/s, tuned to the mass of 46 m/z ($C^{16}O^{18}O$), are shown in Fig. 4. Appearance of this desorption peak is a clear indication of direct participation of oxygen of titania in catalytic oxidation of CO over Au/TiO_2 and Au/TiO_x. Notably higher intensity of the desorption peak in the latter case is an indication of higher efficiency of reduced titania film in CO oxidation, despite lower oxygen concentration compared to fully oxidized titanium oxide.

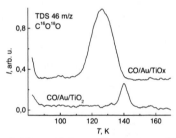

Figure 4: Thermal desorption spectra of isotopically labeled carbon dioxide specie ($C^{16}O^{18}O$) formed over Au/TiO_2 and Au/TiO_x ($x<2$).

In this sense one should attribute the main effect of CO oxidation not to the absolute concentration of available oxygen atoms but mainly to their specific electronic state in the oxide as well as at the metal/oxide perimeter interface. Repeated cycles of the above adsorption-desorption process of oxygen and carbon mono- and dioxide does not change the Ti 2p photoelectron line both for the reduced and totally oxidized titania (see Fig. 1), whereas one should expect further reduction because of the oxygen loss due to $C^{18}O^{16}O$ formation. Such stability of the stoichiometry of the titania during catalytic CO oxidation can be explained assuming oxygen exchange between the gas phase and titania, maintaining oxygen concentration of the oxide at an initial level. Higher efficiency of the oxidation process for the reduced titania film indicates that higher delocalization and lower bonding energy of oxygen of oxide is a key factor governing the catalytic effect.

4   CONCLUSIONS

Reflection absorption infrared spectroscopy in combination with other surface sensitive techniques of electron and ion spectroscopy (XPS, AES, LEED, LEIS, TDS) was used to study catalytic CO oxidation effect on the model direct and inverse Au/titania supported catalysts in ultra-high vacuum condition. It was found that in both cases the metal/oxide perimeter interface is a key factor for adsorption and CO oxidation. Efficiency of the effect is higher for lower CO concentration on the surface, pointing at a crucial role of the free sites for catalytic CO oxidation. For technological perspective the results obtained offer the following steps to gain the performance of the Au/TiO_2 catalysts: 1) to produce smaller gold particles to increase the Au/TiO_2 perimeter interface length; 2) to supply the reaction $CO+O_2$ mixture to the catalyst at lower pressure to enhance the oxidation rate.

ACKNOWLEDGMENTS

The work was partly supported by Russian Fund for Basic Research (grant # 10-02-00558-a) and Japanese Society for the Promotion of Science (grant # 09-02-92109-YaF_a) and by Russian Agency of Education under contract # 2.1.1/3938

REFERENCES

[1] Haruta M., "Size- and support-dependency in the catalysis of gold",Catal. Today, 36, 153–164, 1997.

[2] R. Meyer, C. Lemire, Sh.K. Shaikhutdinov, H.-J. Freund, "Surface Chemistry of Gold Catalysis", Gold Bulletin, 37, 72-124, 2004.

[3] Outka, R. J. Madix, "The Oxidation of Carbon Monoxide on the Au(110) Surface", Surf. Sci., 179, 351, 1987.

[4] F.P. Leisenberger, S. Surnev, G. Koller, M.G. Ramsey, F.P. Netzer, "Investigations of CO adsorption on the V-oxide-Pd (111) model „inverse catalyst"", Surf. Sci., 444, 211, 2000.

[5] T.T. Magkoev, G.G. Vladimirov, D. Remar, A.M.C. Moutinho, "Comparative study of metal adsorption on the metal and the oxide surfaces" Solid. State Commun, 122, 341-346, 2002.

[6] Q. Guo, W.S. Oh, D.W. Goodman, "Titanium Oxide Films Grown on Mo (110)", Surf. Sci., 437, 49-60, 1999.

[7] T.T. Magkoev, D. Rosenthal, S.L.M. Schroeder, K. Christmann, "Effect of the titanium oxide stoichiometry on the efficiency of CO oxidation on the Au/TiO_x system surface", Techn. Phys. Lett., 26, 894-896, 2000.

[8] T.T. Magkoev, K. Christmann, A.M.C. Moutinho, Y. Murata, "Alumina vapour condensation on Mo(110) surface and adsorption of copper and gold atoms on the formed oxide layer", Surf. Sci., 515, 538-552, 2002.

[9] L. Zhang, R. Persaud, T.E. Madey, "Ultrathin Metal Films on a Metal Oxide Surface: Growth of Au on TiO_2", Phys. Rev. B., 56, 10549, 1997.

Production Technology for Nanofiber Filtration Media

M. Maly and S. Petrik

Elmarco s.r.o.
V Horkach 76/18, CZ-46007 Liberec, Czech Republic, stanislav.petrik@elmarco.com

ABSTRACT

The theoretical background and technical capabilities of the free liquid surface (nozzle-less) electrospinnig process is described. The process is the basis of both laboratory and industrial production machines known as Nanospider™ and developed by Elmarco s.r.o. Technical capabilities of the machines (productivity, nanofiber layer metrics, and quality) are described.

Comparison with competing/complementary technologies is given, e.g. nozzle electrospinning, nano-meltblown, and islets-in-the sea. Application fields for nanofiber materials produced by various methods are discussed. Consistency of the technology performance and production capabilities are demonstrated using an example of polyamide nanofiber air filter media.

Keywords: nanofibers, electrospinning, production, technology, filtration

1 INTRODUCTION

Electrospinning methods for creating nanofibers from polymer solutions have been known for decades [1, 2]. The nozzle-less (free liquid surface) technology opened new economically viable possibilities to produce nanofiber layers in a mass industrial scale, and was developed in the past decade [3]. Hundreds of laboratories are currently active in the research of electrospinning process, nanofiber materials, and their applications. Nanofiber nonwoven-structured layers are ideal for creating novel composite materials by combining them with usual nonwovens. The most developed application of this kind of materials is air filtration [4]. Liquid filters and separators are being developed intensively with very encouraging results. Also well known are several bio-medical applications utilizing nanofiber materials, often from biocompatible/degradable polymers like PLA, gelatine, collagen, chitosan. These developing applications include wound care, skin-, vessel-, bone- scaffolds, drug delivery systems and many others. [3, 5]. Inorganic/ceramic nanofibers attract growing interest as materials for energy generation and storage (solar and fuel cells, batteries), and catalytic materials [6-10].

To fully explore the extraordinary number of application opportunities of nanofibers, the availability of reliable industrial-level production technology is essential. This paper intends to demonstrate that the technology has matured to this stage.

2 THEORETICAL BACKGROUND

The electrospinning process is an interesting and well-characterized physical phenomenon and has been an attractive subject for theoretical investigations of several groups [9, 11-17, 1, 2]. Most work concentrates on the essentials of the process – the nanofiber formation from a liquid polymer jet in a (longitudinal) electric field. It has been theoretically described and experimentally proven that the dominant mechanism is whipping elongation occurring due to bending instability [13, 16, 17]. Secondary splitting of the liquid polymer streams can occur also [1], but the final thinning process is elongation.

In Figure 1, the schematic of bending mechanism derived from physical model (a) is compared with a stroboscopic snapshot (b) [18].

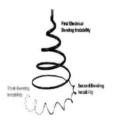

Figure 1. The path of an electrospinning jet (a – schematic, b – stroboscopic photograph).
(Courtesy of Darrell Reneker, University of Akron)

A comprehensive analysis (electrohydrodynamic model) of the fiber formation mechanisms published by Hohman et al. [16, 17] describes the regions of individual kinds of instability observed during the process. It has predicted and experimentally proven that there is a domain of the process variables where bending instability dominates.

The efforts to scale up the electrospinning technology to an industrial production level used to be based on multiplication of the jets using multi-nozzle constructions [1]. However, the number of jets needed to reach economically acceptable productivity is very high, typically thousands. This brings into play many challenging tasks, generally related to reliability, quality consistency, and machine maintenance (especially cleaning). The nozzle-less electrospinning solves most of these problems due to its mechanical simplicity, however, the process itself is more complex because of its spontaneous multi-jet nature. The Lukas' et al. [19] study focused on the process of multi-jet generation from a free liquid surface in an electric field. They showed that the process can be analyzed using Euler's equations for liquid surface waves

$$\nabla\left(\rho\frac{\partial\Phi}{\partial t}+p\right)=0$$

(1)

where Φ is the scalar velocity potential, p is the hydrostatic pressure, and ρ is the liquid density. They derived the dispersion law for the waves in the form

$$\omega^2=(\rho g+\gamma k^2-\varepsilon E_0^2 k)\frac{k}{\rho}$$

(2)

where E_0 is electric field strength, γ – surface tension.

When a critical electric field intensity is reached (E_c), ω^2 is turned to be negative, ω is then a purely imaginary value, and hence, the amplitude of the liquid surface wave

$$\xi=Ae^{qt}\exp(ikx)$$

(3)

exponentially grows, which leads to an instability. Critical field strength can then be expressed

$$E_c=\sqrt[4]{4\gamma\rho g/\varepsilon^2}$$

(4)

From this equation, they derived the expression for the critical spatial period („wavelength") – the average distance between individual jets emerging from the liquid surface (Figure 2).

$$\lambda_c=2\pi/k_c=2\pi a$$

(5)

and

$$\lambda=12\pi\gamma/[2\varepsilon E_0^2+\sqrt{(2\varepsilon E_0^2)^2-12\gamma\rho g}]$$

(6)

a is the capillary length

$$a=\sqrt{\gamma/\rho g}$$

(7)

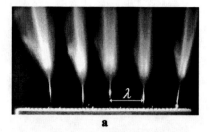
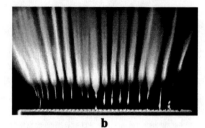

Figure 2. Free liquid surface electrospinning of Polyvinyalcohol at 32 kV (a) and 43 kV (b). *(Courtesy of David Lukas, Technical University of Liberec)*

3 TECHNICAL REALIZATION AND DISCUSSION

The simplest realization of the nozzle-less electrospinning head is in Figure 3. A rotating drum is dipped into a bath of liquid polymer. The thin layer of polymer is carried on the drum surface and exposed to a high voltage electric field. If the voltage exceeds the critical value (4), a number of electrospinning jets are generated. The jets are distributed over the electrode surface with periodicity given by equation (6). This is one of the main advantages of nozzle-less electrospinning: the number and location of the jets is set up naturally in their optimal positions. In the case of multi-needle spinning heads, the jet distribution is made artificially. The mismatch between "natural" jet distribution and the real mechanical structure leads to instabilities in the process, and to the production of nanofiber layers which are not homogenous.

Several types of rotating electrodes for free liquid surface electrospinning for industrial machines have been developed (Figure 3b). However, the drum type is still one of the most productive.

Figure 3. Free liquid surface electrospinning from a rotating electrode (a), and various types of spinning electrodes (b).

Research and development centers are very active in their efforts to further improve productivity of the manufacturing process. Novel methods for the production of sub-micron fibers are being developed. The most advanced methods alternative to electrospinning are "Fine Hole" meltblown and "Islets-in-the-sea. The individual methods can be considered to be complementary rather than competing. This is especially true with respect to the fiber diameter distribution and fiber layer uniformity. Individual methods will likely find different areas of application. More productive Nano-meltblown and Islets-in-the-sea technologies compromise fiber diameter and homogeneity and will likely be used in production cost sensitive applications like hygiene nonwovens, while high quality electrospinning technologies will be used in products where their high added value and need for low amounts of the material can be easily implemented (air and liquid filtration, biomedicine).

The nozzle-less principle using rotating electrodes has been developed into a commercially available industrial scale. A photograph of a modular Nanospider [TM] machine is in Figure 4.

Figure 4. Nozzle-less production electrospinning line (Nanospider[TM]).

In addition to productivity (or throughput) of the production line, individual industrial applications require certain production consistency. We will illustrate the nozzle-less electrospinning technology performance with the example of air filtration media composed of a regular cellulose substrate and a thin nanofiber layer made from Polyamide 6. The product can be characterized by a number of parameters, like fiber diameter distribution (mean value and its standard deviation), basis weight of the nanofiber layer, etc. For the particular application, functional product parameters are more important. Typical values are the initial gravimetric filtration efficiency (IGE), differential filtration efficiency, and pressure drop, measured according to the norms widely accepted within industry.

The correlation between nanofiber diameter and basis weight of the nanofiber layer with differential filtration efficiency is illustrated in Figure 5. To obtain various basis weights, substrate speed was varied from 0.2 m/min to 4 m/min for each series of samples. Polymer solution parameters (concentration, etc.) together with electric field intensity determine the range of nanofiber diameter. Nanofiber diameter distribution has been measured using a scanning electron microscope (SEM). Basis weight values were obtained either by using an analytical scale Mettler (higher values), or by extrapolation from its known dependence on substrate velocity (lower ones).

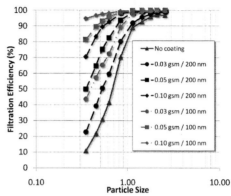

Figure 5. Filtration efficiency of nanofiber media samples

Pressure drop and initial gravimetric filtration efficiency have been chosen as representative product parameters. They were measured using NaCl aerosol at the following settings: air flow speed: 5 m/min, sample area 100 cm^2, flow rate 50 l/min.

In Figure 6, results of long-term stability and reproducibility of the IGE are presented. It can be seen that the individual runs differ within the standard deviation of the process, and the mean value of the filtration efficiency does not exhibit any significant shift after 16 hours of machine run. Similar consistency exhibits the value of the basis weight of the nanofiber layer.

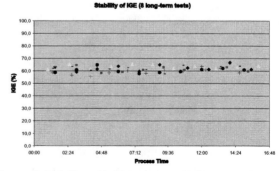

Figure 6. Stability of initial gravimetric filtration efficiency of the media produced at the industrial nozzle-less electrospinning equipment.

Production capacity of the industrial electrospinning line for Polyamide 6 is illustrated in Figure 7.

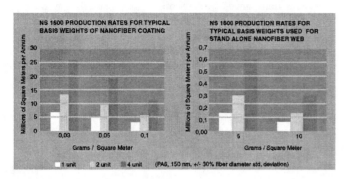

Figure 7. Production capacity of the nozzle-less electrospinning line with Polyamide 6.

4 CONCLUSION

High-quality low-cost production of nanofiber layers is essential to support the enormous amount of research results being obtained at many universities and research centers. The described nozzle-less electrospinning technology has matured to a level where large scale production use is common, and can be modified for practically all known polymers soluble in organic solvents and water, as well as for polymer melts. This opens commercial opportunities for hundreds of ideas developed in the academic sphere.

REFERENCES

1. V. Kirichenko,Y. Filatov, A. Budyka, *Electrospinning of Micro- and Nanofibers: Fundamentals in Separation and Filtration Processes,* Begell House, USA (2007)
2. S. Ramakrishna, K. Fujihara, W. Teo, T. Lim, and Z. Ma, *An Introduction to Electrospinning and Nanofibers,* World Scientific, Singapore (2005)
3. Jirsak, O., Sanetrnik, F., Lukas, D., Kotek, V., Martinova, L., Chaloupek, J., "A method of nanofibers production from a polymer solution using electrostatic spinning and a device for carrying out the metod", The Patent Cooperation Treaty WO 2005/024101, (2005)
4. T. Jaroszczyk, S. Petrik, K. Donahue, *"The Role of Nanofiber Filter Media in Motor Vehicle Air Filtration"*, 4th Biennial Conference on Emissions Solutions in Transportation, AFS, Ann Arbor, MI, Oct 5-8 (2009)
5. Proceedings of „Nanofibers for the 3rd Millenium – Nano for Life" Conference, Prague, March 11-12 (2009)
6. L. Kavan, M. Grätzel, *Electrochem. Solid-State Lett.,* **5**, A39, (2002)
7. J. Duchoslav, L. Rubacek, *Electrospun TiO2 Fibers as a Material for Dye Sensitizied Solar Cells,* In: Proc. of NSTI Nanotech Conference, Boston, MA, June 1-5 (2008)
8. L. Rubacek, J. Duchoslav, *Electrospun Nanofiber Layers for Applications in Electrochemical Devices,* In: Proc. of NSTI Nanotech Conference, Boston, MA, June 1-5 (2008)
9. M. Bognitzki, W. Czado, T. Frese, A. Schaper, M. Hellwig, M. Steinhard, A. Greiner and J. H. Wendorff, *Adv. Mater.* **13**, 70 (2001)
10. H. Guan, C. Shao, S. Wen, B. Chen, J. Gong, X. Yang, *Inorg. Chem. Commun.* **6**, 1302 (2003)
11. G. F. Taylor and M. D. Van Dyke, *Proc. R. Soc. London, Ser. A* **313**, 453 (1969)
12. J. Doshi and D. H. Reneker, "Electrospinning process and applications of electrospun fibers," *J. Electrost.* **35**, 151 (1995)
13. C.J. Thompson, G.G. Chase, A.L. Yarin and D.H. Reneker, „Effects of parameters on nanofiber diameter determined from electrospinning model", *Polymer,* **48**, 6913 (2007).
14. Y. M. Shin, M. M. Hohman, M. P. Brenner and G. C. Rutledge Electrospinning: A whipping fluid jet generates submicron polymer fibers, *Appl. Phys. Lett.* **78**, 1149 (2001)
15. J. H. Yu, S. V. Fridrikh and G. C. Rutledge, The role of elasticity in the formation of electrospun fibers, *Polymer* **47**, 4789 (2006)
16. M. M. Hohman, M.Shin, G. C. Rutledge and M. P. Brenner, Electrospinning and electrically forced jets. I. Stability theory, *Phys. Fluids* **13**, 2201 (2001)
17. M. M. Hohman, M.Shin, G. C. Rutledge and M. P. Brenner, Electrospinning and electrically forced jets. II. Applications, *Phys. Fluids* **13**, 2221(2001)
18. D. Reneker, Personal communication (2009)
19. D. Lukas, A. Sarkar and P. Pokorny, *J. Appl. Phys.* **103**, 084309 (2008)

Atmospheric Plasma Surface Modification of Polytetrafluoroethylene Films

B.E. Stein, V. Rodriguez-Santiago, A.A. Bujanda, K.E. Strawhecker, D.D. Pappas

U.S. Army Research Laboratory, Aberdeen Proving Ground, MD 21005, ben.stein@arl.army.mil

ABSTRACT

This work details the use of atmospheric-pressure dielectric barrier discharge plasmas to enhance the adhesion of thin metal coatings to surfaces of polytetrafluoroethylene (PTFE) films. Prolonged exposure to the energetic species in a He-O$_2$ plasma discharge leads to a decrease in water contact angle, indicating either a chemical restructuring of the PTFE surface, and/or changes in the morphology of the surface. Our initial results show a maximum decrease in water contact angle (WCA) of approximately 25° after only 10 seconds of He-O$_2$ plasma exposure. Scanning electron microscopy imaging performed on plasma-treated surfaces reveal changes in morphology, indicating physical etching and roughness changes caused by the plasma. Preliminary scratch test results indicate improved adhesion of metal films to the surface of the polymer.

Keywords: atmospheric plasma, polytetrafluoroethylene, metal deposition

1 INTRODUCTION

Polytetrafluoroethylene (PTFE) is a widely studied and used material due to its inert nature and chemical stability. However, these same attributes also render PTFE surfaces difficult to modify and functionalize. Developing a treatment method that would allow for the surface functionalization of PTFE with the same ease and variety as other industrial polymers like polyethylene would be a useful and valuable technology for a number of industries. Primary interests would focus on augmenting and enhancing PTFE's desired features like bacterial resistance, biocompatibility, high-temperature stability, and hydrophilicity [1, 2].

Numerous methods exist for polymer surface modification including wet-chemical and mechanical treatments, flame exposure, and exposure to ions, plasmas, and UV radiation, among others [3, 4]. Specifically, atmospheric-pressure (AP) plasma treatments have proven as successful as the above processes in modifying the surface chemistry of a variety of polymers [1, 2], but also possess the added advantage of a quick and scalable process without the need for vacuum systems. Previous studies have focused on the effects of atmospheric plasmas [1,5,6], atmospheric pressure plasma assisted liquid deposition [7], and the results of sputter-coating on as-received PTFE substrates [8]. In this study we demonstrate the effects of plasma-induced surface modification of PTFE on the adhesive behavior of a commercial epoxy and a metallic thin film.

2 EXPERIMENTAL

2.1 Materials

Polytetrafluoroethylene sheets (Goodfellow, 0.1 mm thick) were cleaned before use by soaking them in ethanol for 15 minutes, then allowing them to air dry.

The polyurethane adhesive used for the peel tests was Dev-Con 2 (Devthane), which was kept at low temperature in a standard freezer for 5 minutes immediately before mixing to slow the polymerization process and allow for an extended working time.

2.2 Experimental Procedures

Plasma surface modification was performed using an atmospheric pressure dielectric barrier discharge system, with the bottom electrode in a roller configuration (Fig.1). The system (Sigma Technologies, model APC2000) is equipped with two electrodes with a total working area of approximately 0.065 m^2. For plasma treatment, the carrier and reactive gases were injected into the electrodes at atmospheric pressure and allowed to diffuse, forming a filamentary glow discharge. The system includes two aluminum slotted electrodes, which have a working distance of 1.2 mm from the rotating alumina ground electrode.

Helium was used to initiate and generate the plasma at atmospheric pressure before the reactive gas was introduced to the system, and for this study the He flow rate was set at 12,000 sccm. Oxygen was used as a reactive gas at flow rates from 50 to 1000 sccm. The operating frequency of the applied voltage was 90 kHz with power densities varying between 6.2 kW/m^2 and 12.4 kW/m^2. Plasma exposure times of the polymer surfaces ranged between 2 seconds and 2 minutes.

For the metallization process, a gold-palladium alloy was used and sputtered on the polymer surface using a Hummer XP Sputter Coater. Samples were placed inside the sputtering chamber and evacuated to ~30mTorr. The sputter chamber was then backfilled with Ar to a pressure of 55mTorr (7.33 Pa). A DC voltage of 1.3 kV was then supplied to the electrode, yielding sputtering plasma.

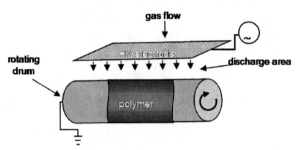

Figure 1. Atmospheric plasma system setup.

2.3 Analytical Techniques

Changes in sample hydrophilicity were assessed by observing water contact angles, using the sessile-drop method, with standard water contact angle (WCA) procedures (ASTM D5946-04). Contact angles were calculated using the Young-Laplace fitting technique coupled with the auto-fitting software CAM 2008 (KSV Instruments).

Near-surface compositional depth profiling was performed using a Kratos Axis Ultra 165 X-ray photoelectron spectroscopy (XPS) system equipped with a hemispherical analyzer. A 100 W monochromatic Al Kα (1486.7 eV) beam irradiated a 1 mm x 0.5 mm sampling area with a take-off angle of 90°. The pressure in the XPS chamber was held between 10^{-9} and 10^{-10} Torr. Elemental high resolution scans for C1s, O1s, and F1s were taken at a pass energy of 20 eV.

Changes in surface morphology were observed using scanning electron microscopy (SEM). A field-emission electron microscope (FEI NOVA NanoSEM 600) was operated in a low-vacuum detection mode, with chamber pressures held between 50-90 Pa. The low-vacuum mode of operation allows for beam currents up to 15KeV without damaging the sample at working distances between 8-15 mm.

Peel testing was carried out according to ASTM D-1876, using the Dev-con 2 polyurethane adhesive. Samples were cut into 0.0254m by 0.2286m strips, with 0.0762m tabs, and were peeled at a rate of 0.254m/min, using an MTS Synergie-1000lbs (500N load cell).

Scratch testing was performed on the metalized films to determine their adhesive strength using a nanoindenter (Nanoindenter XP, MTS Systems Corp), with a Berkovich tip.

3 RESULTS
3.1 Surface Chemistry and Morphology

Water contact angle measurements (Figs. 2a,b) showed a maximum drop in contact angle of approximately 25°, with the greatest effect occurring after a few seconds of exposure, then leveling out after that, indicating that atmospheric plasma treatments are effective even at short exposure times (Figure 3). A likely explanation for the leveling of the WCA is that under the plasma conditions used, no further sites are activated thus a saturation point is reached within a few seconds of exposure.

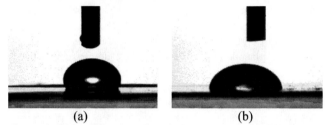

(a)   (b)

Figure 2: WCA micrographs of (a) untreated and (b) plasma treated PTFE.

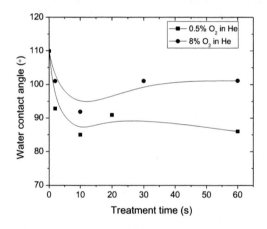

Figure 3: Effects of plasma exposure on WCA.

XPS analysis of plasma treated surfaces revealed a minor increase of about 2% in the atomic concentration of surface oxygen (O 1s) as shown in Figure 4, however, the major peaks were assigned to C 1s and F 1s (not shown). This result is not surprising given the high strength of the C-F bond due to its partial ionic character. Nonetheless, the increase in surface oxygen groups is still detectable, and indicates that the plasma treatment is indeed capable of breaking C-F bonds on the PTFE surface at atmospheric pressure. Further plasma parameter optimization will be carried out to achieve higher oxygen levels.

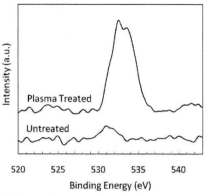

Figure 4: XPS results showing the O 1s peak of untreated and plasma treated PTFE.

Scanning electron microscopy did not reveal any significant changes in surface morphology (Figures 5a, b), indicating that plasma effects on the surface, such as wettability, are chemical rather than morphological. This is supported by the XPS results.

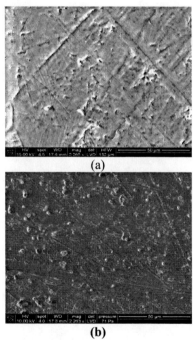

Figure 5: SEM images of (a) untreated and (b) plasma treated PTFE films.

3.2 Adhesion Strength

Adhesion tests were carried out as a means to verify the effectiveness of the plasma modification via peel tests with a polyurethane adhesive. Figure 6 shows the peel strength of untreated and plasma treated films. The results reveal two orders of magnitude increase in peel strength compared to control samples. This supports the argument that the plasma is responsible for the mild oxidation of the PTFE surface by enhancing the bonding between the surface and the adhesive.

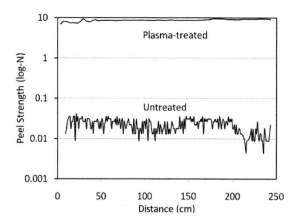

Figure 6: Increase in adhesive strength of PTFE sheets using a polyurethane adhesive.

Based on these results, the increased adhesion between the polymer surface and a metallic film was investigated by scratch tests. Au-Pd alloy films (50 nm) were sputtered on the surface of untreated and plasma treated PTFE samples. An applied load ramp from 0.05 mN to a maximum of 5 mN was used to scratch the surface of the metallized polymer surface. Scratch profiles and post-profiles (not shown) indicated plastic and elastic components attributed

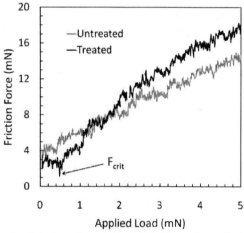

Figure 7. Friction force as a function of applied load during scratch tests for untreated and plasma treated samples, and the corresponding critical force (F_{crit}).

to the mechanical properties of the bulk polymer. In order to obtain a value for a critical load corresponding to the failure of the metal/polymer interface, the friction force as a function of applied load during the scratch was inspected as shown in Figure 7. We defined the critical load here as the applied load at which the slope of Figure 7 changes. For the plasma treated sample, F_{crit} was 0.55 mN whereas for the untreated sample a change in slope was undetected so F_{crit} was assumed to be negligible. These preliminary results suggest that the force required to penetrate the metallic film is less for the untreated sample, arguably due to a weaker adhesion to the polymer surface. Nonetheless, more testing is required to fully characterize and quantify the adhesive behavior and strength.

4 CONCLUSIONS

In this study we used an atmospheric plasma treatment to modify the surface of PTFE. Chemical analysis showed minor chemical modification by the addition of oxygen chemical groups. However, peel testing showed a two orders-of-magnitude increase in the adhesion strength of PTFE to a polyurethane adhesive. Scratch testing was used to investigate the adhesion of a thin Au/Pd film to the plasma-modified surface. Preliminary results indicate that the plasma treatment does improve the metal adhesion to the surface. Future studies will include further exploration of various plasma parameters including applied power, gas composition, and exposure time, as well as aging studies to determine the stability of the newly modified surfaces over time. Once an optimized plasma profile is developed, masks will be employed to selectively modify the surface and create a variety of patterns.

5 ACKNOWLEDGEMENTS

This research was supported in part by an appointment to the Postgraduate Research Participation Program at the U.S. Army Research Laboratory administered by the Oak Ridge Institute for Science and Education through an interagency agreement between the U.S. Department of Energy and USARL.

REFERENCES

1. Sarra-Bournet, C, S Turgeon, D Mantovani, and G Laroche. "A study of atmospheric pressure plasma discharges for surface functionalization of PTFE used in biomedical applications." *Journal of Physics D: Applied Physics* 39 (2006): 3461-3469.

2. Desmet, Tim, Rino Morent, Nathalie De Geyter, Christophe Leys, Etienne Schacht, and Peter Dubruel. "Nonthermal Plasma Technology as a Versatile Strategy for Polymeric Biomaterials Surface Modification: A Review." *BioMacromolecules* 10, no. 9 (2009): 2351-2379.

3. Wertheimer, M R, A C Fozza, and A Hollander. "Industrial processing of polymers by low-pressure plasmas: the role of VUV radiation." *Nuclear Instruments and Methods in Physics Research B* 151 (1999): 65-75.

4. Truica-Marasescu, F, S Guimond, and M R Wertheimer. "VUV-induced nitriding of polymer surfaces: Comparison with plasma treatments in nitrogen." *Nuclear Instruments and Methods in Physics Research B* 208 (2003): 294-299.

5. Fang, Zhi, Lili Hao, Hao Yang, Xiangqian Xie, Yuchang Qiu, and Kuffel Edmund. "PTFE Surface modification by filamentary and homogeneous dielectric barrier discharges in air." *Applied Surface Science* 255 (2009): 7279-7285.

6. Pappas, Daphne D., Andres A. Bujanda, Joshua A. Orlicki, and Robert E. Jensen. "Chemical and morphological modification of polymers under a helium–oxygen dielectric barrier discharge." *Surface & Coatings Technology*, no. 203 (2008): 830-834.

7. Zettsu, Nobuyuki, Hiroto Itoh, and Kazuya Yamamura. "Surface functionalization of PTFE sheet through atmospheric plasma liquid deposition approach." Surface and Coatings Technology 202 (2008): 5284-5288.

8. Slepicka, P, Z Kolska, J Nahlik, V Hnatowicz, and V Svorcik. "Properties of Au nanolayers on PET and PTFE." *Surface and Interface Analysis* 41, no. 9 (August 2009): 741-745.

Bio-Molecules Immobilization on Nanowire Electronic Devices

*W.C. Maki, N.N. Mishra, M. Fellegy, B. Filanoski, E. Cameron and *G.K. Maki

Center for Advanced Microelectronics and Biomolecular Research,
University of Idaho, Post Falls, ID USA, nmishra@cambr.uidaho.edu
*Integrated Molecular Sensors, Sylvania OH, USA,
wusimaki@intergratedmolecularsensors.com

ABSTRACT

Nano-scale Field Effect Transistor based devices have shown great promise in bio-molecular detection. However, immobilization of bio-molecules on nano-scaled devices is a challenge. We investigated methods for immobilizing oligonucleotide and peptide nucleic acid on nanowire devices. Oligonucleotide and peptide nucleic acid were covalently linked on the nanowire surface which is covered by native silicon dioxide. Standard silanization chemistry combined with a bi-functional linker approach was used in the immobilization process. Using this approach, negatively charged target nucleic acids were closely attached to the nanowire. Streptavidin-Alkaline phosphatease conjugate was used as indicator to confirm immobilized oligonucleotide on the surface. Biotin labeled oligonucleotides with complimentary sequences was used to report the presence of PNA on the surface.

Keywords: nanowire transistor, surface cleaning, oligonucleotide, peptide nucleic acid, immobilization.

1 INTRODUCTION

Nanowire field effect transistors (nano-FET) have been developed and investigated in bio-detection for the last decade [1, 2, 3, 4 and 5]. It has shown promises for improving biosensor characteristics such as sensitivity, simplicity and economic viability. Compared with traditional bio-detection techniques such as optical spectroscopy, mass spectroscopy or immune precipitation, advantages of electronic nano-FET bio-detection are: direct detection without labeling, ultra-sensitivity without amplification and simple device implementation without the requirement of expensive equipment.

It has been reported that nano-electronic bio-detection achieved much higher detection sensitivity,

at least on the order of 10^3, fM verses pM and nM [6,7,8] Our previous work demonstrated electronic detection of methylated DNA, bacterial 16S rRNA and toxins on nano-FET biosensors [9, 10, 11 and 12]. However, there are challenges in the development of nano-FET biosensors. One of these challenges is immobilization of bio-molecules on nano-scaled sensing surface and the distance between captured bio-molecules and nanowire, which is critical in electronic detection. Since the detection mechanism is based on molecular charge on the sensing surface, which affects electronic properties of semiconductor nanowire and generates detectable signal. These issues must be addressed in the development and application of nano-FET biosensors. Investigations and studies of silicon nanowire surface modification have been reported and showed improvement [13, 14, and 15].

Herein, we report the investigation of immobilization oligonucleotide and peptide nucleic acid (PNA) on silicon chips and nanowire device surface. Oligonucleotide and PNA were covalently linked on the nanowire surface which is covered by 1-2 nm native silicon dioxide. To reduce the distance between oligonucleotide and nanowire, a low concentration acetic acid solution was used to wash the silanized surface. Chemiluminescence detection method was used to confirm immobilized oligonucleotide and PNA.

2 METHODS

2.1 Cleaning and Activating Silicon Chips

Slicon chips with a 25nm thin layer of silicon dioxide were used to investigate cleaning and immobilization methods. A set of six nanowire devices was used to demonstrate device surface modification. Two methods were investigated. In method-1, silicon chips were rinsed with ethanol and sonicated in 2% Hellmanex solution for 5 minutes.

After rinsing with ddH₂O chips were sonicated in ddH$_2$O for another 3 minutes. In method-2, silicon chips were cleaned using Plasma. In both methods, further cleaning was performed in a acid solution of HCl:Meth (1:1) for 30 minutes, rinsed with ddH$_2$O, and dried in Argon gas.

To activate the silicon dioxide surface, chips were immersed in 5% APTES ethanol solution and silanization at 50°C for 3 hrs, or applied 300 ul APTES on a set of nano-device surfaces in an Argon gas chamber for 1 hr. After rinsing with 1 mM or 1% AcOH and then water, chips or nanowire devices were dried in argon gas and store in desiccators.

2.2 Immobilization of Oligonucleotide and Peptide Nucleic Acid

Oligonucleotide and PNA were immobilized on an activated surface through a bi-functional linker uccinimidyl-4-(N-maleimidomethyl)cyclohexane-1-carboxylate (sulfo-SMCC). This oligonucleotide contains a thio-group at 5'end for covalently linking to the surface, and biotin at 3'end for chemiluminescence detection. To immobilize PNA, a cystein residue was designed at the N-terminal of PNA, which provides thio-group for covalent linkage.

4 mg/ml sulfo-SMCC was prepared in 100mM NaHCO$_2$ buffer, pH 9.0. The chips were soaked in above solution for 1 hr at room temperature (RT), then, washed with ddH$_2$O and dried in Argon gas. In the modification of nanowire devices, 100ul of sulfo-SMCC solution was applied onto a set of nano-device surface, incubated in a sealed moisture chamber for one hr.

A free SH group is required in both oligonucleotide and PNA for immobilization. A reduction reaction was carried out in the presence of 50mM DTT in 50ul 1uM oligonucleotide for 15 minutes in RT. After reduction, the sample was purified through a Sephadex G-25 spin column to remove DTT molecules. A drop of purified sample was then applied onto a sulfo-SMCC modified surface, and incubated in a sealed moisture chamber overnight. Unbound molecules were washed away with TE buffer and ddH$_2$O. After drying in Argon gas, these modified chips can be store in dry at 4°C for several months.

Immobilization of oligonucleotide and PNA on a nanowire device was performed using the same method described above.

2.3 Characterization of Immobilized Bio-molecules

The chemiluminescence detection method was used for characterization of immobilized oligonucleotide and PNA on the chip surface. To prevent non-specific binding, chips and nanowire devices were blocked using a blocking buffer containing 1mM PEG and 3% BAS for 3 hrs. Strepavidin and Alkaline phosphatease conjugate (SA-AP) was diluted 1:1,000 in a TBS buffer. Chips and devices were incubated in SA-AP solution for 30 minutes at RT. After washing with a TBS buffer (3 x 3minutes), chemiluminescence detection was performed in the presence of substrate AutoGlow 450. Images were taken using a CCD camera.

To determine the presence of PNA on the chip surface, biotin labeled oligonucleotides with complimentary sequences were used as the reporter. Hybridization was performed at 33°C for 3 hrs. Unbound oligonucleotides were washed away. To investigate the specificity of hybridization, three PNAs and three oligonucleotides were used in the tests. Chemiluminescence signal was detected and recorded using CCD camera.

3 RESULTS

3.1 Surface cleaning and activation

To investigate the modification methods, Oligonucleotide and PNA immobilization were first studied on silicon chips with a thin layer of silicon dioxide. Two methods were used and compared for surface cleaning of silicon chips and nanowire devices. In method-1, acid and sonication treatment were used to clean the surface. In method-2, acid treatment combined with plasma cleaning was used. Both methods worked well for surface cleaning of chips. In the device surface cleaning, however, nanowires were damaged in the sonication process using method-1, particularly in the nanowire contact with the gold pads (data not shown). Therefore, plasma cleaning method was used.

The thickness of the salinization layer on the surface is a critical issue in the bio-molecular immobilization on nano-FET biosensors, because the distance between nanowire and molecular charge directly affects detection sensitivity. It is important that molecular charge should be close to sensing surface within 10nm. Therefore, reducing the thickness of the salinization layer on the sensing surface must be addressed. It was reported that using 1mM AcOH to rinse APTES activated surface reduced the thickness of salinization layer to less than 4nm [15]. The length of a 20 mer oligonucleotide is about 6nm. We applied an additional AcOH washing steps after sialinization reaction which brought the total distance between nanowire and oligonucleotide to 10nm. We found that results were reproducible.

3.2 Bio-molecular immobilization

Immobilization of oligonucleotide on silanized silicon chips was tested using a modified Oligo-dT$_{20}$. This oligonucleotide contains a thio-group at 5'end for immobilization, and a biotin modified residue at 3'end for chemiluminescence detection. The sequence of the oligonucleotide is 5'-SH-ttttttttttttttttttttt- biotin-3'.

Chemiluminescence detection of the presence of biotin labeled oligo-dT$_{20}$ on chip surface is shown in Figure 1.

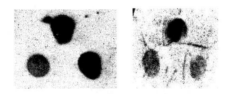

Figure 1 Images of chemiluminescence detection of biotin labeled oligo-dT$_{20}$ on silicon chips.

Alkaline phosphatease activity was detected in the areas of immobilized oligo-dT$_{20}$ on chips. Light signal generated by enzymatic reaction was captured by CCD camera with exposrue time of one minute. Longer exposure time increased the signal. However, significant diffusion was observed in the images.

3.3 Immobilization of multiplex PNA

For the investigation of immobilizing multiplex PNA, three PNA molecules and three biotin labeled oligonucleotides with complimentary sequences were used in the study. The following is the sequences of PNAs and oligonucleotides.

PNA-1   Ac-Cys--TTA-TCT-TCC-TCT-CONH2
PNA-2   Ac-Cys-TTC-ACT-TGT-TAT-CONH2
PNA-3   Ac-Cys-CAC-CAC-TAT-TAT-CONH2

Oligo-1 5'-biotinAAA AGA GGA AGA TAA -3'
Oligo-2 5'- biotinAAA ATA ACA AGT GAA –3'
Oligo-3 5'- biotinAAA ATA ATA GTG GTG -3'

Three PNAs were immobilized on each of four chips. Hybridization was performed as: a) buffer only (top left); b) buffer contains one oligonucleotide (top right); c) buffer contains two oligonucleotides (bottom left); and d) buffer contains three oligonucleotides. Results showed that the presence of all three PNAs on chips, and demonstrated specific hybridization and detection.

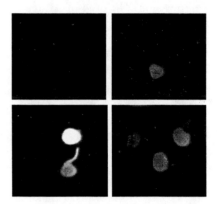

Figure 2 Chemiluminescence detection of PNAs on silicon chips

3.4 Nanowire surface modification

To investigate immobilization of PNA on nanowire devices, a set of six devices (Figure 3 A) was used in the study. The "gap" area of each device is the location of nanowire PNA immobilization was confirmed through hybridization of biotin labeled oligonucleotide which was detected using CCD imaging (Figure 3 B).

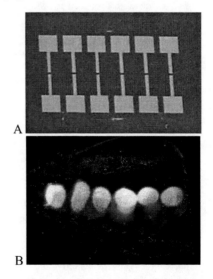

Figure 3 Chemiluminescence detection of PNA   immobilized on nano-devices

Alkaline phosphatease activity was detected on six devices as shown in Figure 3B. The image clearly showed six dots. However, the dots are larger than the actual sensing area. This is because of the exposure time for imaging was set for 1 minute which resulted in light diffusion. Short exposure times, 15, 30 and 45 seconds, were tested. No clear signal was detected due to the time required for enzymatic reaction. The peak of AP reaction is 12 - 15 minutes after applying the substrate.

4 CONCLUSION

Immobilization of oligonucleotide and PNA was investigated in the presented work. Additional AcOH washing step reduces the thickness of salinization layer and brought bio-molecules close to the sensing surface. Chemiluminescence characteriazation of immobilized molecules has shown the presence of oligonucleotide and PNA on silicon chips and nanowire devices. Multiple PNA immobilization and specific hybridization were also demonstrated.

5 ACKNOWLEDGEMENTS

This work was supported in part by NIH contract HHSH261200800064C and USDA grant CSREES 3447917058

REFERENCES

1. Stern, E. et al. Label-free immunodetection with CMOS-compatible semiconducting nanowires *Nature* **455**, 519-522 (2007).
2. Anderson, E. P. et al. A system for multiplexed direct electrical detection of DNA synthesis *Sensors and Actuators B: Chemical* **SNB-10185** (2007).
3. Patolsky, F., Zheng, G. and Lieber, C. M. Fabrication of silicon nanowire devices for ultra-sensitive, label free, real time detection of biological and chemical species *Nature Protocol* **1**, 1711-1724 (2006).
4. Hahm, J. and Lieber, C. M. Direct ultrasensitive electrical detection of DNA and DNA sequence variations using nanowire nanosensors. *Nano Letter* **4**, 51-54 (2004).
5. Uslu, F. Label free fully electronic nucleic acid detection system based on a field-effect transistor device *Biosensors and Bioelectronics* **19**, 1723-1731 (2004).
6. Wang, Z. and Jin G. Silicon surface modification with a mixed silanes layer to immobilize proteins for biosensor with imaging ellipsometry *Colloids and Surfaces B: Biointerfaces* **34**, 173-177 (2004).
7. Barbaro, M. et al. Fully electronic DNA hybridization detection by a standard CMOS biochip *Sensors and Actuators B: Chemical* **118**, 41-46 (2006).
8. Kerman, K. et al. Peptide nucleic acid modified carbon nanotube field effect transistor for ultra-sensitive real-time detection of DNA hybridization *NanoBiotechnology* **1**, 65-70 (2005).
9. N.N Mishra, P. Winterrowd, R. Nelson, S. Rastogi, W.C. Maki and G.K.Maki, " Fabrication and Characterization of Nano-Field Effect Transistor for Bio-safety" (Nano Science and Technology Institute, Nanotech 2007, May 20-24, Santa Clara, CA; ISBN 1420061836 **2**, 465-468 (2007).
10. N. N. Mishra, W. C. Maki, R. Nelson, P. Winterrowd, S. K. Rastogi, E. Cameron, B. Filanoski and G.K. Maki, "Polysilicon Nano-wires Based Nano-device on Silicon Chip: Fabrication and its Application", Nano Science and Technology Institute , NSTI-Nanotech 2008, **2**, 607-610 (2008)
11. Mishra, N., Maki, W., Nelson,R., Winterrowd, P., Rastogi, S., et al. "Ultra-Sensitive Detection of Bacterial Toxin with Silicon Nanowire Transistor" *Lab-on Chip* **8**, 868-871 (2008)
12. B., Rastogi, S. and Maki, G. "Nanowire-transistor based ultra-sensitive DNA methylation detection", *Biosensor & Bioelectronics*, **23**, 780-787 (2008)
13. Man-Fai Ng· Liping Zhou, Shuo-Wang Yang,Li Yun Sim, Vincent B. C. Tan, and Ping Wu Theoretical investigation of silicon nanowires: Methodology, geometry, surface modification, and electrical conductivity using a multiscale approach. *Physical. Review.* 76, 155435 (2007)
14. Diao, J. Ren, D. Engstrom, J. R. and Lee, K. A. Surface modification strategy on silicon nitride for developing biosensors *Analytical Biochemistry* **343**, 322-328 (2005).
15. Y. Han, D. Mayer, A. Offenhäusser and S. Ingebrandt. Surface activation of thin silicon oxides by wet cleaning and silanization *Thin Solid Films,* 510: 175-180 (2006).

Address for Correspondence: Dr. Nirankar N Mishra, CAMBR, University of Idaho, 721 Lochsa St, Post Falls, ID 83854. Tel: 208-262-2047; Fax: 208 262 2001
Email: nmishra@cambr.uidaho.edu

Dr. Wusi Maki,
*Integrated Molecular Sensors, Sylvania OH, USA, wusimaki@intergratedmolecularsensors.com

Chemical Structure of ZnO Nanorod Array on a DSSC Photoanode

Yu-Hao Chang[*], Chang-Yu Liao[*], H. Paul Wang[*,**], Yu-Lin Wei[***], Tzung-Fang Guo[****],
and Yi-Fang Chiang[*****]

[*]Department of Environmental Engineering, National Cheng Kung University,
Tainan 70101, Taiwan, wanghp@mail.ncku.edu.tw
[**]Sustainable Environment Research Center, National Cheng Kung University,
Tainan 70101, Taiwan
[***]Department of Environmental Science and Engineering, Tunghai University,
Taichung 40704, Taiwan
[****]Institute of Electro-Optical Science and Engineering, Advanced Optoelectronic Technology Center,
National Cheng Kung University, Tainan 70101, Taiwan
[*****]Institute of Electro-Optical Science and Engineering, National Cheng Kung University,
Tainan 70101, Taiwan

ABSTRACT

The main objective of the present work was to study speciation of ZnO nanorods during the photoexcited electron transfer between photosensitive dyes and the ZnO nanorod arrays on the photoanode of a dye-sensitized solar cell (DSSC) by X-ray absorption (near edge structure (XANES) and extended X-ray absorption fine structure (EXAFS)) spectroscopy. The least-square component fitted XANES spectra show that the main zinc species in the ZnO-nanorod photoanode are nanosize ZnO (88%) and $Zn(OH)_2$ (12%). An enhanced adsorption of the dyes N3 ($C_{26}H_{20}O_{10}N_6S_2Ru$) and mercurochrome ($C_{20}H_8Br_2HgNa_2O_6$) on the ZnO nanorods is found, of which 6-10% of the nanosize ZnO involve in the chemical interaction with the dyes. Their EXAFS spectra also indicate that the bond distance and coordination number of Zn-O in the ZnO nanorods is 1.95 Å (which is less than that in the ZnO seedlayer by 0.01 Å) and in the range of 4.6-4.8, respectively. The efficiencies of the DSSCs containing the N3 and mercurochrome dispersed ZnO nanorods on the photoanodes are 0.33-0.92%. The relatively less J_{sc} (photocurrent density) of the DSSC with the N3-dispersed ZnO nanorod may be due to the formation of a N3-Zn^{2+} complex. A rapid recombination of the photoexcited electrons with I_3^- in the electrolyte of the DSSC containing the N3 dispersed ZnO nanorods is also found, which causes a reduction of its FF (fill factor) value.

Keywords: DSSC, ZnO, XANES, EXAFS

1 INTRODUCTION

Dye-sensitized solar cells (DSSCs) have been attracted increasing attention in recent decades, due to their low fabrication cost, relatively simple process, and high efficiency [1, 2]. The typical DSSC is consisted of a dye adsorbed on a nanostructure films made up of metal oxide nanoparticles, an electrolyte as a charge transportation layer containing iodide/tri-iodide redox couples, and a platinum coated transparent conducting oxide glass as a counterelectrode [3-5]. Dye sensitizers adsorbed on the surfaces of photoactive metal oxide nanoparticles are used for absorbing incident sunlight. During illumination of solar irradiation, the photoexcited electrons are migrated from the dye sensitizers to the conduction band of photoactive metal oxide and transported to the transparent conducting oxide (TCO) glass. The dye sensitizers are regenerated with electrons from the platinum counterelectrode through the redox couple in the electrolyte [6, 7]. Tens of thousands of the grain boundary in a traditional DSSC may restrain electron transportation from dye sensitizer to a TCO glass. Therefore, promoting the electron transfer efficiency of photoactive metal oxide nanoparticle photoanode is an essential issue [8-10].

Zinc oxide possessing an effective band gap (3.37 eV), high electron mobility and good optical characteristic has attracted extensive attention in UV light emitters, field emissions, gas/chemical sensors, and field effect transistors [11]. Very recently, it has been found that an improvement of the electron transportation in the photoanode of a DSSC can be achieved by using single crystal and vertical ZnO nanowires or rods on a TCO glass [10, 12-14].

Coordination number (CN), bond distance, and oxidation state of select element in the complex matrix can be determined by X-ray absorption near-edge structure (XANES) and extended X-ray absorption fine structure (EXAFS) spectroscopy. By EXAFS, it was observed that copper oxide clusters played an important role in catalytic oxidation of chlorophenols (in supercritical water) and decomposition of NO [16, 17]. The photoactive sites in the nanosize TiO_2 during photocatalytic degradation of $CHCl_3$ and methylene blue have also been formed by in situ XANES [18, 19].

Speciation of zinc in the ZnO nanorod photoanode adsorbed with dye sensitizers is still lacking in the literature.

The molecule-scale understanding of photoactive species can facilitate development of new DSSC technologies. Hence, the main objective of this work was to study chemical structure of zinc in the ZnO nanorod photoanode adsorbed with dye sensitizers such as mercurochrome and N3.

2 EXPERIMENTAL

The ZnO sol-gel precursor was prepared by dissolving zinc acetate dihydrate ($Zn(CH_3COO_2)\cdot 2H_2O$) (OSAKA) in a mixture of containing 2-aminoethanol ($NH_2CH_2CH_2OH$) (WAKO) (0.75 M) and 2-methoxyethanol ($CH_3OCH_2CH_2OH$) (Fluka) at a 1:1 molar ratio of zinc acetate dihydrate to 2-aminoethanol. The solution was mixed with stirring at 333 K for 30 min to yield a clear solution which was dropped onto an ITO substrate for spin coating on a spin coater at 3000 rpm for 30 sec. The thin film was dried at 353 K and annealed at 623 K for one hour.

Zinc nitrate hexahydrate ($Zn(NO_3)_2\cdot 6H_2O$) (KANTO) (0.05 M) and hexamethylenetetramine (($CH_2)_6N_4$) (0.05 M) were used to prepare the ZnO nanorod array thin film which was self-assembled at 363 K. The as synthesized ZnO nanorod photoanode was washed with deioned water and dried at 353 K for 24 hour.

Chemical structure of the ZnO nanorod photoanode was determined by X-ray diffraction spectroscopy (D8 advance, Bruker) with a Cu Kα (1.542 Å) radiation. The microstructure of the ZnO nanorod on the photoanode was studied by scanning electron microscopy (SEM) (PHILIPS, S3000N). After dehydration at 373 K, the photoanode was immersed in 0.04 mM of N3 dye ($C_{26}H_{20}O_{10}N_6S_2Ru$) (Solaronix) or mercurochrome ($C_{20}H_8Br_2HgNa_2O_6$) (Fluka) in C_2H_5OH (99.5%) at 298 K for 24 hours in the dark condition.

XANES and EXAFS spectra of zinc (9659 eV) in the ZnO NRA photoanode were measured on the Wiggler beamline (17C) at the Taiwan National Synchrotron Radiation Research Center. The electron storage ring was operated at the energy of 1.5 GeV. A Si(111) double-crystal monochromator was used for selection of energy with an energy resolution ($\Delta E/E$). Data were collected in fluorescence mode with a Lytle detector. The photon energy was calibrated by characteristic pre-edge peaks in the absorption of metallic Zn powder (9659 eV). The absorption spectra were collected using ion chambers which were filled with helium gas.

The Pt counter electrode was prepared by spin-coating of a H_2PtCl_6 solution (a drop) on the ITO and annealed at 673 K for 15 minutes. The electrolyte of the DSSC was prepared by dissolving 0.01 M of iodine (I_2) (Riedel-de Haën), 0.5 M of lithium iodide (LiI) (ALDRICH), 0.06 M of 1,2-dimethyl-3-propylimidazolium iodide ($C_8H_{15}N_2I$) (Solaronix) and 0.5 M of 4-*tert*-butylpyridine ($C_9H_{13}N$) (ALDRICH) in acetonitrile (CH_3CN) (Mallinckrodt).

In the photocurrent measurements, the ZnO nanorod photoanode and Pt counterelectrode were separated by a 60

µm thick Surlyn (Solaronix) and heated at 373 K for 10 minutes. After infuse electrolyte into the spacer, the channels between two electrodes were sealed by epoxy resin.

The dye-sensitized ZnO nanorod photoanode was illuminated through a conducting glass by with an AM 1.5 simulated light irradiation (300W Xenon lamp and AM 1.5 filter). The current-voltage characteristics of the DSSCs were investigated on a solar simulator (Newport, 91160A).

3 RESULTS AND DISCUSSION

Figure 1 shows XRD patterns of the ZnO seedlayer and nanorod on the ITO substrate. The characteristic diffraction peaks at (100), (002) and (004) that are attributed to wurtzite (JCPDS card no. 36-1451) are observed. A sharp diffraction peak (002) at 2θ = 34.8° suggests that ZnO nanorod on the photoanode has a high crystallinity. Remarkably, the enhanced diffraction peak (002) is much more intensive than the (100) and (004) peaks, indicating that the ZnO nanorods are oriented perpendicular to the substrate surface perfectly and grown along the c-asix, which is similar to observation of Cheng and Samulski [20]. The (002) diffraction peak for the ZnO seedlayer may be attributed to the well establish ZnO nanorod in the crystal growth process.

The SEM image of the ZnO nanorod on the photoanode is shown in Figure 2. The vertically well-aligned arrays of ZnO nanorod with a high density are uniformly formed on the substrate. The length, diameter and aspect ratio of nanorods are about 900, 50 nm and 18, respectively.

XANES can provide information including oxidation state of an excited atom, coordination geometry and bonding of local environment in the complex matrix. XANES spectra of model compounds such as nanosize ZnO, $Zn(OH)_2$, and metallic Zn have also been determined. The XANES spectra of zinc were expressed mathematically in a LC XANES fit vectors, using the absorption data within the energy range of 9640-9700 eV. The least-square fitted XANES spectra of zinc in the ZnO nanorod are shown in Figures 3. The main zinc species in the ZnO seedlayer is nanosize ZnO (100%). Zinc species in the ZnO nanorod are mainly nanosize ZnO (88%) and $Zn(OH)_2$ (12%). Adsorption of N3 ($C_{26}H_{20}O_{10}N_6S_2Ru$) and mercurochrome ($C_{20}H_8Br_2HgNa_2O_6$) dyes on the ZnO nanorod photoanode causes decreasing of nanosize ZnO fractions to 83 and 80%, respectively.

The EXAFS spectra of zinc were recorded and analyzed in the k rang of 3.5-12.5 Å^{-1}. The Debye-Waller factors ($\Delta\sigma^2$) are less than 0.01 Å^{-2} in all refined EXAFS data (see Table 1). In the ZnO seedlayer and nanorod, the bond distances of Zn-O are 1.96 and 1.95 Å, respectively. The coordination numbers (CNs) of the ZnO nanorod photoanode adsorbed with N3 and mercurochrome are 4.6 and 4.8, respectively.

The photocurrent density-voltage characteristics of the dye-sensitized solar cell fabricated with the ZnO nanorod photoanode adsorbed with N3 and mercurochrome dye under an AM 1.5 simulated light irradiation are shown in Figure 4. The solar cell efficiency can be expressed by $\eta = (J_{sc} \times V_{oc} \times FF/P_{in})$, where, J_{sc} is the short-circuit current density, V_{oc} is the open-circuit voltage, FF is the fill factor, and P_{in} is the incident light power. The parameters of the ZnO nanorod photoanode in the DSSC are summarized in Table 2. The V_{oc} value of the ZnO nanorod/N3 on the photoanode of the DSSC is slightly greater than that of the mercurochrome dye. The J_{sc} of the DSSC with the ZnO nanorod adsorbed with N3 is much less than that with mercurochrome, which may be caused by the formation of a N3-Zn^{2+} complex [21]. The less FF value in the ZnO nanorod/N3 on the photoanode of the DSSC suggests that the photoexcited electrons may be more rapidly to be recombined by I_3^- in the electrolyte if compare to the DSSC using the mercurochrome dye. It seems that the mercurochrome has a better performance in the ZnO nanorod on the photoanode of the DSSC.

4 CONCLUSIONS

The main zinc species in the ZnO nanorod photoanode are nanosize ZnO (88%) and $Zn(OH)_2$ (12%). An enhanced adsorption of the dyes N3 ($C_{26}H_{20}O_{10}N_6S_2Ru$) and mercurochrome ($C_{20}H_8Br_2HgNa_2O_6$) on the ZnO nanorods is found, of which 6-10% of the nanosize ZnO involve in the chemical interaction with the dyes. The coordination numbers (CNs) of the ZnO nanorod photoanode adsorbed with N3 is less than that with mercurochrome. Because of the formation of a N3-Zn^{2+} complex, the J_{sc} of the DSSC with the ZnO nanorod adsorbed with N3 is much less than that with mercurochrome. A rapid recombination of the photoexcited electrons with I_3^- in the electrolyte of the DSSC containing the N3 dispersed ZnO nanorods is also found, which causes a reduction of its FF (fill factor) value.

5 ACKNOWLEDGEMENTS

The financial supports of the Taiwan National Science Council, Bureau of Energy, the Excellence Project of the National Cheng Kung University and National Synchrotron Radiation Research Center are gratefully acknowledged.

REFERENCES

[1] A. Hagfeldt and M. Grätzel, Chem. Rev., 95, 49, 1995.

[2] M. Grätzel, J. Photochem. Photobiol. C-Photochem. Rev., 4, 145, 2003.

[3] M. Grätzel, Prog. Photovoltaics, 8, 171, 2000.

[4] M. Grätzel, J. Photochem. Photobiol. A-Chem., 164, 3, 2004.

[5] M. Grätzel, Chem. Lett., 34, 8, 2005.

[6] A. Hagfeldt and M. Grätzel, Accounts Chem. Res., 33, 269, 2000.

[7] M. Grätzel, Inorg. Chem., 44, 6841, 2005.

[8] S.Y. Huang, G. Schlichthorl, A.J. Nozik, M. Grätzel and A.J. Frank, J. Phys. Chem. B, 101, 2576, 1997.

[9] A. Kay and M. Grätzel, Chem. Mat., 14, 2930, 2002.

[10] J.B. Baxter and E.S. Aydil, Appl. Phys. Lett., 86, 2005.

[11] X.D. Wang, C.J. Summers and Z.L. Wang, Nano Lett., 4, 423, 2004.

[12] M. Law, L.E. Greene, J.C. Johnson, R. Saykally and P.D. Yang, Nat. Mater., 4, 455, 2005.

[13] J.B. Baxter, A.M. Walker, K. van Ommering and E.S. Aydil, Nanotechnology, 17, S304, 2006.

[14] Y.H. Jiang, M. Wu, X.J. Wu, Y.M. Sun and H.B. Yin, Mater. Lett., 63, 275, 2009.

[15] I. Gonzalez-Valls and M. Lira-Cantu, Eng. Environ. Sci., 2, 19, 2009.

[16] Y.J. Huang, H.P. Wang and J.F. Lee, Chemosphere, 50, 1035, 2003.

[17] Y.J. Huang, H.P. Wang and J.F. Lee, Appl. Catal. B-Environ., 40, 111, 2003.

[18] T.L. Hsiung, H.P. Wang, Y.M. Lu and M.C. Hsiao, Radiat. Phys. Chem., 75, 2054, 2006.

[19] T.L. Hsiung, H.P. Wang and H.P. Lin HP, J. Phys. Chem. Solids, 69, 383, 2008.

[20] B. Cheng and E.T. Samulski, Chem. Commun., 986, 2004.

[21] K. Keis, J. Lindgren, S.E. Lindquist and A. Hagfeldt, Langmuir, 16, 4688, 2002.

ZnO	Shell	Bond Distance (Å)	Coordination Number	σ^2 (Å^2)
Seedlayer		1.96±0.0001	5.20	0.006
ZnO nanorod	Zn-O	1.95±0.0001	5.20	0.006
ZnO NRA /mercurochrome		1.95±0.0001	4.75	0.005
ZnO NRA/N3		1.95±0.0001	4.55	0.005

σ^2: Debye-Waller.

Table 1: Structural parameters of the ZnO nanorod photoanode absorbed with mercurochrome and N3.

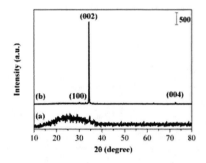

Figure 1: XRD patterns of the ZnO (a) seedlayer (magnified by 10 times) and (b) nanorod on the ITO substrate.

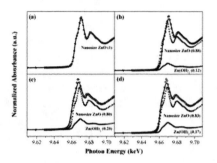

Figure 3: SEM images of the ZnO nanorod photoanode.

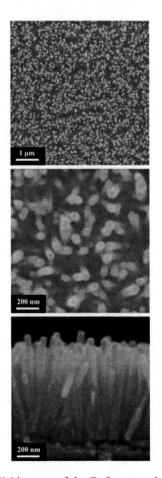

Figure 2: SEM images of the ZnO nanorod photoanode.

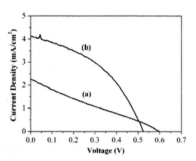

Figure 4: Current-voltage curves of the DSSCs containing the ZnO NRA photoanode absorbed with (a) N3 and (b) mercurochrome.

Photoanode of DSSCs	V_{oc} (V)	J_{sc} (mA/cm^2)	FF	η (%)
ZnO nanorod / Mercurochrome	0.52	1.03	0.42	0.92
ZnO nanorod / N3	0.60	0.57	0.24	0.33
TiO$_2$ / Mercurochrome	0.49	2.84	0.45	0.64
TiO$_2$ /N3	0.68	11.8	0.40	3.22

V_{oc}: open-circuit voltage; J_{sc}: short-circuit current density; FF: fill factor; η: efficiency.

Table 2: Structural parameters of the ZnO nanorod photoanode absorbed with mercurochrome and N3.

An experimental and theoretical study of the inhibition mechanism of organic substances in concrete

M.V. Diamanti[1], M. Ormellese[1], E.A. Pérez-Rosales[2], MP. Pedeferri[1], G. Raffaini[1], F. Ganazzoli [1]

[1] Dept. Chemistry, Materials and Chemical Engineering – Politecnico di Milano
Via Mancinelli, 7 - 20131 Milano – Italy
[2] Corrosion consultant

ABSTRACT

Corrosion inhibitors are used to prevent chloride-induced corrosion in reinforced concrete structures. Several commercial inhibitors are available, but their efficiency and their inhibitive mechanism are not well understood. In this paper the inhibiting behaviour of five organic substances in delaying chloride-induced corrosion was evaluated in alkaline solution using EIS and potentiodynamic tests. The studied substances were sodium tartrate, sodium benzoate, sodium glutamate, DMEA and TETA. Theoretical calculations based on molecular mechanics and molecular dynamics were used to establish which functional groups bind to the passive film, the strength of their binding, the functional groups contributing to their filming, and their evolution with time.

Keywords: inhibitor, concrete, chloride, mechanism, efficiency.

1 INTRODUCTION

Steel reinforcements in concrete are normally passivated due to concrete's high alkalinity. However, this passive film can be destroyed mainly by two specific conditions: concrete carbonation and chlorides presence at the metal surface. The former causes a decrease in concrete pH, changing the steel condition from passive to active, whereas the latter provokes local breakdown of the passive film, originating pits on the metal surface. In most cases, the use of high quality concrete and adequate cover are enough to prevent corrosion [1]. Corrosion inhibitors are a very attractive prevention method, due to their low cost and easy handling. They can be added to the concrete mix as additives, or applied to the concrete surface and migrate towards the reinforcement. So far, most researches have focused on determining the efficiency of some organic substances as corrosion inhibitors, either in solution or in concrete; however, little has been done to determine the interaction mechanism between these substances and the passive film [2-7]. In this research, five organic substances were studied by theoretical and experimental methodologies in order to establish how their interaction with the passive film delayed corrosion initiation.

2 EXPERIMENTAL SET-UP

Tests and simulations were performed on five organic substances: sodium tartrate, sodium benzoate, sodium glutamate, dimethylethanolamine (DMEA) and triethylenetetramine (TETA). High purity commercial products, except for TETA, with a 70% purity, were used. Computer simulations using Molecular Mechanics (MM) and Molecular Dynamics (MD) techniques were used to establish which functional groups bind to the passive film, the strength of their binding, the functional groups contributing to their filming, and their evolution with time. Electrochemical test results from electrochemical impedance spectroscopy (EIS) and potentiodynamic tests (obtained from previous research) were used to confirm the proposed mechanism.

2.1 MM and MD simulation protocol

The interaction between the inhibitors and the passive film was studied using a simulation protocol [8]. Calculations were performed with Materials Studio Modeling v.3.2.0.0 software, using the condensed-phase optimized molecular potentials for atomistic simulation studies (COMPASS) force field. A force field simply gives the potential energy of the system as a function of the coordinates of all constituent atoms. In practice, it is expressed through appropriate bonded and non-bonded energy contributions. The solvent was modeled as a dielectric medium with a distance-dependent dielectric constant. The monolayers were composed of 16, 25 and 36 molecules distributed on the surface. They were disposed as close as possible to one another to check whether the intermolecular interactions affect the adsorption energy. Only the systems containing a monolayer of 36 molecules were then subjected to MD simulations. Energy minimizations with respect to all of the variables (the atomic coordinates) were done with the conjugate gradient algorithm. MD simulations were carried out at a constant temperature (300 K), controlled by the Berendsen thermostat. The integration of the dynamic equations was performed with the Verlet algorithm using a time-step of 1 fs (10^{-15} s), and the instantaneous coordinates (or frames) were periodically saved for further analysis or geometry optimization.

2.2 Electrochemical tests

Three electrochemical tests were carried out: potentiodynamic tests (PT) to determine pitting potential at increasing chloride content and increasing inhibitor dosage; free corrosion tests (FC tests) to establish time-to-pitting initiation at increasing chloride content and increasing inhibitor dosage; electrochemical impedance spectroscopy (EIS) tests to determine adsorption parameters. Specimens were machined from commercial reinforcing carbon steel: exposed surface area was 18 cm^2. Tests were carried out in simulated concrete pore solution (SPS), consisting of saturated Ca(OH)$_2$ at pH 13.0 by addition of Na(OH) 0.06 M, at room temperature (25ºC). Potentiodynamic curves were recorded with an Autolab Potentiostat (PGSTAT 30). FC tests were performed at increasing inhibitor dosage: from 10^{-4} M up to 0.1 M. Specimens were passivated for 168 hours in SPS. Free corrosion potential was monitored daily from the immersion of specimens. When passivation condition were reached, 0.1 M NaCl was added to the solution. Potential measurements continued till initiation of corrosion. Potentiodynamic Tests (PT) were performed in a 0.1 M NaCl solution containing 0.1 M of inhibitor, starting from -1.2 V vs SCE, increasing the potential with a scan rate of 1 V/h. Pitting potential was evaluated as the potential corresponding to a sudden increase of anodic current density. EIS spectra were recorded after 48 hours of immersion of specimens in SPS, at progressive inhibitor concentrations from 10^{-4} M up to 0.1 M, every 48 hours. EIS spectra were obtained at corrosion potential (Ecorr), using a signal amplitude of ±10 mV in a frequency range 100 kHz to 5 mHz. Experiment and data collection were carried out under computer control using FRA v4.8 software. Data were analyzed with ZSimpWin v3.10.

3 RESULTS

3.1 MD & MM Results

Table 1 reports the energy calculated for the adsorption of a single inhibitor molecule on the substrate at its most stable conformation after MM and MD calculations. The adsorption energy E_{ads} was calculated as following:

$$E_{ads} = E_{tot} - E_s - E_{mol} \qquad (1)$$

where E_{tot} is the total energy of the system, E_s accounts for the energy of the γ–FeOOH slab and E_{mol} refers to the energy of the isolated inhibitor. The E_{mol} values reported in Table 1 gives the energy of each molecule due to stretching, bending, deformations with respect to the ideal values, and to the non-bonded interactions within the molecule. In the present context the values of E_{mol} must simply be taken as reference values for the isolated molecule to be used in equation (1) for calculating the adsorption energy of the inhibitor molecules subject to the surface interaction

through the total energy E_{tot} of the composite system. In fact, E_{tot} is calculated for the slab with the adsorbed inhibitor molecules through the same terms as before in their optimized, i.e., energy-minimized, geometry. Moreover, following standard procedures for MM and MD simulations, the whole slab was kept fixed at a given position in the numerical minimization of the potential energy, so that its contribution to the total energy of the system, E_s, is equal to zero, even though all of its surface atoms (slightly more than 11500) correctly interact with the adsorbate molecules and their pairwise interactions with all inhibitors atoms are taken into account. The optimized values of E_{tot} and the corresponding adsorption energies calculated with equation (1) are reported in Table 1. By comparing the resulting adsorption energies, it is possible to establish an adsorption ranking for the studied substances, where glutamate shows the strongest adsorption and DMEA the weakest one.

Table 1 – Energy for a single molecule of the inhibitors

	E_{tot} (kJ/mol)	E_{mol} (kJ/mol)	E_{ads} (kJ/mol)
Tartrate	−114	−16.39	−97
Benzoate	−111	−44.04	−67
Glutamate	−134	−34.18	−100
TETA	−323	−229.35	−94
DMEA	−122	−68.08	−54

In order to ascertain how the inhibitor molecules interact among themselves and with the substrate, monolayers containing 16, 25 and 36 molecules of inhibitor were regularly placed on the surface as close as possible to one another. Their energy was then minimized considering only isolated islands of molecules without periodic boundary conditions. Placing the molecules in a regular way allowed determining the presence of intermolecular interactions, and to check whether their overall effect is repulsive or attractive. To establish the differences in the average adsorption energy per molecule between the monolayer and the single molecule, the total energy of the system was divided by the number of molecules forming the monolayer. This value was introduced in equation (1) to obtain the adsorption energy per molecule, which is shown in Table 2. MD calculations were also performed in order to see the time evolution of the monolayer. Figure 1 shows an optimized snapshot of typical instantaneous conformations obtained. Detailed results are reported in [9].

Table 2 – Energy for a monolayer of 36 inhibitor molecules

	E_{tot} (kJ/mol)	E_{ads} (kJ/mol)
Tartrate	−111	−94
Benzoate	−117	−73
Glutamate	−133	−98
TETA	−304	−75
DMEA	−125	−57

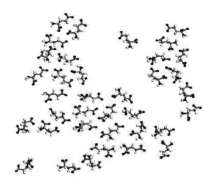

Figure 1 – Optimized snapshot of the monomolecular layer conformation for TETA

3.2 Electrochemical test results

Pitting potential around -0.25 V SCE was measured in the SPS without inhibitors. Tartrate and benzoate showed a significant inhibitive effect, since pitting potential increased up to +0.4 V SCE. TETA and DMEA increased pitting potential to +0.2 V SCE. Glutamate performances resulted to be worse than in the other cases. These results were confirmed by free corrosion tests through the measurement of time-to-pitting initiation, as reported in Figure 2: tartrate and benzoate showed best inhibition effect, while TETA and DMEA showed the worst inhibition, even at the highest inhibitor concentration. For carboxylate compounds time-to-corrosion increased with concentration.

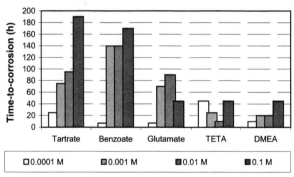

Figure 2 – Time-to-corrosion in SPS with NaCl 0.1 M

EIS spectra were used to derive adsorption parameters. Impedance response of the case study was investigated through the following equivalent electric: $R_s(Q_1R_1)(Q_2R_2)$, where R_s is solution resistance, Q_1 and R_1 represent processes occurring at high frequencies, thus corresponding to double layer capacitance and charge transfer resistance, respectively, Q_2 and R_2 are capacitance and resistance at low frequency range, describing redox processes occurring at the passive layer. The electric model adopted implies that R_1 and Q_1 describe the evolution of the interaction between substrate and inhibitor. A constant phase elements replacing

the pure capacitor was considered, in order to allow a non-ideal response of the system. The capacitance is therefore expressed in $[S \cdot s^n/cm^2]$.

Figure 3 reports the evolution of R_1 and Q_1 at increasing inhibitor concentration. As expected, the charge transfer resistance, R_1, increases as inhibitor concentration increases (i.e., with decreasing corrosion rate). Double layer capacitance, Q_1, decreases as inhibitor concentration increases, indicating the formation of a homogeneous film ($n = 0.97$: the double layer is considered as pure capacitor).

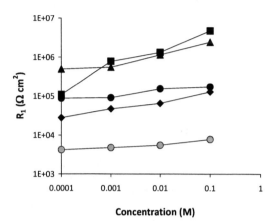

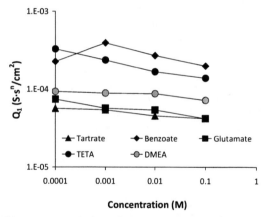

Figure 3 – Evolution of charge transfer resistance and double layer capacitance (n = 0.97)

4 DISCUSSION

Comparing the values in Tables 1 and 2, it is possible to notice that the adsorption energy tends to become slightly less negative, indicating the presence of repulsive forces between the molecules. The related change of position of the molecules (see detail in [9]) present on the substrate is due to the intermolecular interactions, either attractive or repulsive, in the presence of the attractive interactions with the surface. In fact, the presence of delocalized charges on the functional groups of different molecules may result in a repulsive force, which makes them drift apart. This is more evident in the case of the carboxylate groups present in tartrate. Conversely, the presence of groups such as hydroxyls which form hydrogen bonds with either the carboxylate or the amino groups, may produce attractive interactions among the molecules and allow them to rearrange in different ways. This is more evident in the case of TETA, but it is also clear in tartrate. These findings apply to the monolayers of glutamate and DMEA, whereas the only inhibitor where no clear trend can be discerned is benzoate. Upon comparison of Tables 1 and 2, most inhibitors present more negative adsorption energies for the isolated molecules than for the monolayer. An increase in the number of molecules forming the monolayer could imply an increase in the adsorption strength: in fact, intermolecular attractive interactions, such as the formation of hydrogen bonds, may stabilize the film, with an energy drop. On the other hand, the outcome was the opposite one: more molecules produce a slightly less stable film due to enhanced repulsive interactions. However, it should be noticed that these results may be affected by the limitations inherent to these calculations, such as the use of a dielectric medium instead of the explicit solvent and the lack of periodic boundary conditions.

The MD runs carried out starting from the initial optimized monolayers with an implicit solvent suggest a different behavior among the different inhibitors. An even surface coverage is observed with tartrate and benzoate with only some random local clustering of molecules. A significant clustering of the glutamate anions can clearly be noticed due to the intermolecular dipolar interactions and hydrogen bonds: it shows a strong surface adsorption, but unlike tartrate and benzoate it tends to form isolated clusters, leaving many exposed surface areas. DMEA shows an even surface coverage, but a weaker adsorption, whereas TETA shows strong hints of a significant molecular clustering leaving exposed surface area.

The results of molecular simulations contributed to confirm the electrochemical test results. Glutamate and tartrate showed the highest charge transfer resistance (R_1), i.e. the lowest corrosion rate. In the absence of inhibitor values close to 10^3 Ωcm^2 are reported as charge transfer resistance for carbon steel in alkaline environment [1]. Glutamate and tartrate showed the lowest double layer capacitance (Q_1), which means the highest homogeneity of the adsorbed organic layer (Figure 3). As reported in [1], capacitance values lower than 10^{-3} S/cm^2 are typical of homogeneous adsorbed monolayers.

A possible mechanism of the inhibition effect produced by the organic substances tested as inhibitors can be based on the following feature: an inhibitor molecule must include a free electron pair to enhance adsorption on passive film. From this viewpoint, aminic and carboxylic groups show this characteristic as potential inhibitive substances. On the other hand, once a molecule is adsorbed on the substrate, it must cover as much as possible the whole exposed surface. As a final requirement, such adsorbed molecules have to compete with the polar molecules, or charged ions, present in solution, such as water and chloride ions, since both can induce the detachment of the inhibitors. Therefore, an ideal inhibitor should show two basic requirements: 1) it must strongly adsorb to the surface; 2) it must cover the whole surface.

Electrochemical tests, namely FC tests and PT, have the important feature to give the final behavior as a result of the combination of the two above-discussed requirements. Tartrate and benzoate showed the highest pitting potential. The inhibition effect of these two organic compounds was also confirmed by the increase of time-to-corrosion: with the higher concentration of tartrate and benzoate corrosion occurred more that 160 h after chloride addition. This electrochemical results confirmed what obtained from MM and MD simulations.

5 CONCLUSION

The following can be concluded:
–/ a favorable interaction energy with the surface of the adsorbed molecules, with repulsive intermolecular interactions, mainly among the anions, was confirmed. The presence of repulsive lateral interactions between the inhibitors' molecules was confirmed. Different cluster-type patterns were obtained for inhibitor molecules
–/ glutamate showed the strongest adsorption energy and an enhanced tendency to clustering
–/ tartrate and benzoate showed less negative adsorption free energy and uniform distributed pattern, leading to more efficient inhibiting effect
–/ amines showed an intermediate behavior, most likely because of a balance between adsorption and clustering with respect to other substances

REFERENCES

[1] L. Bertolini, B. Elsener, P. Pedeferri, R. Polder, Corrosion of steel in concrete: prevention, diagnosis, repair, Wiley, Weinheim (2004)
[2] B. Elsener, EFC 35, Institute of Materials, London, 2001.
[3] C. Monticelli, A. Frignani and G. Trabanelli, Cem. Conc. Res., 30, p. 635, 2003.
[4] L. Dhoubi, E. Triki and A. Raharinaivo, Cem. Concr. Comp., 24, p. 35, 2002.
[5] J.M. Gaidis, Cem. Concr. Comp., 26, p. 181, 2004.
[6] H.E. Jamil, M.F. Montemor, R. Boulif, A. Shriri, M.G.S. Ferreira, Electr. Acta, 48, p. 3509, 2005.
[7] M. Ormellese, L. Lazzari, S. Goidanich, G. Fumagalli, A. Brenna, Corr. Sci, 51, 12, p. 2959, 2009.
[8] F. Ganazzoli, G. Raffaini, Phys. Chem., Chem. Phys., 7, p. 3651, 2005.
[9] M. Ormellese, E.A. Pérez, G. Raffaini, F. Ganazzoli, L. Lazzari, *Inhibition mechanism in concrete by organic substances: an experimental and theoretical study*, Proc. Int. Conf. Corrosion/09, Paper N. 09221, NACE Int., Houston, TX, 2009

Thin-Film Deposition on Nanoparticles in Low-Pressure Plasma

A. Shahravan*, and T. Matsoukas**

Department of Chemical engineering, The Pennsylvania State University
Room150, University Park, Pa 16802
*ashahravan@engr.psu.edu, **matsoukas@engr.psu.edu

ABSTRACT

We have synthesize hollow nanoparticles by a two-step process: a particle that acts as a template is coated with amorphous hydrogenated carbon in a low pressure plasma. The core particle is then dissolved to produce a hollow nanoshell. Transmission electron microscopy (TEM) images depict the thickness and the uniformity of the adherent polymer coatings. In order to determine the permeability of the shell, coated KCl particles are dissolved in water and the conductivity of the solution is measured as a function of time. These nanoshells exhibit a delayed release response which can be controlled via the deposition time of the film.

Keywords: Plasma polymerization, Hollow nanoparticles, Controlled release

1 INTRODUCTION

Because of their physical and chemical properties, nanoparticles are used in a variety of applications that range from biomedical [1],[2],[3], to optical [4],[5], to electronic fields [5]. By depositing thin films onto the particle surface it is possible to enhance the optical, mechanical, thermal, and electrical properties that are unattainable in the particles alone. Silica nanoparticles are one of the highly attractive nanostructures for a wide variety of nanotechnology applications such as catalysis, adsorption, separation, and sensing [7]. In addition to the coating process that improves the surface properties of nanoparticles, they can be used to improve the performances of organic polymers [8],[9]. By mixing the inorganic particles specially silica with polymers organic/inorganic composites form which has the rigidity and thermal stability of inorganic materials and the flexibility and ductility of organic polymers. Various synthesis routes can prepare the so-called nanocomposite structures as Hajji et al. reported [10]. Sol-gel, blending, and in situ polymerization are the general processing techniques to produce polymer/silica nanocomposites [9]. In addition to these methods, plasma deposition technique is another approach for a wide range of materials in order to modify their surfaces and also to fabricate nanocomposites.

Plasma enhanced chemical vapor deposition, PECVD, is a technique that modifies nanoparticles by coating them with a solid produced by the plasma decomposition of an organic (usually hydrocarbon) molecule. Such solids are often referred to as plasma polymers. This process offers a number of advantages that are not possible with colloidal processing: (1) the precursor used in plasma may not contain the type of functional groups normally associated with conventional polymerization; (2) Very uniform thin films from few nm to more than 1 μm can formed and the thickness can be easily controlled via the deposition time; (3) Because of the cross-linked structure, the polymer film is mechanically and chemically stable [11],[12]. By taking advantage of these properties we tailored plasma polymer coating process to produce hollow nanoparticles.

Hollow particles are of great interest for the encapsulation of materials such as drugs, dyes and cosmetics in order to release them controllably over the time [13],[14]. These are commonly produced by a two-step process that starts with the coating the surfaces of templates with a desirable thickness, followed by a chemical dissolution or calcination of the template. This synthesis process has been employed for variety of materials including ceramics, polymers, and metals [15]. In some cases, core removal through the polymer shell is so slow because of the low diffusion in the small pores of the shell, <10 nm. In this case, mesoporous polymer capsules are used to accelerated the core removal process [16],[17]. Here we report the preparation of plasma polymer hollow particles after a successful selective core dissolution process which proves the permeability of the shell.

In our previous work we developed a plasma based process for depositing plasma polymers of various thicknesses onto micro- and nanoparticles [18]. Here we show that these films can be used to produce hollow particles (nanoshells) via chemical etching of the template particle. In this paper we describe the fabrication of the hollow particles and present experimental results that reveal their permeability characteristics.

2 EXPERIMENTAL PROCEDURE

Deposition process in order to coat nanoparticles takes place in two different plasma reactors one with two par-

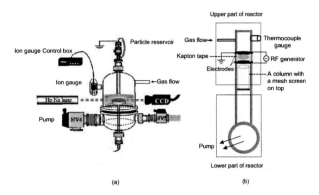

Figure 1: Schematic of the reaction setup: (a) a classic semi-batch reactor, (b) a novel tubular reactor.

allel electrodes, and one with two external ring electrodes. A schematic of the reaction setup is shown in figure 1. The major part of the setup is a pyrex reactor that allows a full visual access to the coating process. The total volume of the chamber is 1.5 liter while the plasma glow occupies approximately 0.5 liter in volume between the two electrodes. Adjustable stainless steel electrodes are spaced apart by a distance of 0.5 cm to 5 cm. The lower electrode, 4 cm in diameter, is powered and the upper electrode, 8 cm in diameter, is grounded. Both electrodes are capacitively coupled by a radio frequency (RF) power supplier at 13.56 MHz and operates in the pressure range 100 mtorr to 1000 mtorr. Two pumps, a roughing pump and a diffusion pump, are connected to the gas outlet for evacuating gases from the chamber. During operation, the precursor vapor and Ar gas enter the chamber via the gas inlet pipe. The particles are introduced into the chamber as a dry powder after pressing through a set of mesh screens. These particles acquire a charge as they fall through the plasma and become trapped in the sheath region above the bottom electrode where they remain over extended periods of time. A beam-expanded 14 mW HeNe laser illuminates the trapped particles and a charge coupled device (CCD) camera is used to capture the process. When the plasma is turned off, particles are collected from the surface of the lower electrode.

Since particles are suspended, this reactor leads to the formation of coatings with a high degree of radial uniformity. However, the amount of particles that can be processed is very small and limited by the volume of the sheath space. The second reactor was designed to circumvent this difficulty. The reactor operates with the same RF power, 10-150 W at 13.56 MHz, which is applied to the external electrodes, and is attached to the same vacuum system which we used for the parallel-plate reactor. Particles are placed on the wall of the reactor close to the charged electrode by spreading them on the surface of the Kapton tape. Double sided Kapton tape that are suitable for powder coating are used

to carry particles on their surfaces. A set of 100-400 mesh screens are used to press and separate particles on the HN tape. This Kapton tape has silicone adhesive protection and is ideal for powder coating since it does not leave any residue. Through this process we can now treat 0.1-1 g of particles per run. After the deposition process, which typically lasts 1 to 30 min, the particles are removed from the reactor and placed in 5% aqueous HF overnight to dissolve the silica and produce hollow particles. Many other materials can be used as a sacrificial template. In addition to silica we have used KCl, which offers the advantage that its dissolution kinetics can be studied easily with conductivity measurements.

Monodisperse spherical silica particles of 200 nm and 1μm diameter are obtained from Geltech Inc. By using Stober method we also synthesize silica particles as small as 20 nm diameter in our laboratory. In order to eliminate the moisture of particles, they should be alcohol washed and oven dried prior to inserting them into the reactors . KCl particles are prepared in our lab by using a constant output atomizer (model 3075). Isopropyl slcohol (IPA) is purchased from Sigma-Aldrich and is used as precursor since it produces smooth and well-defined films. The discharge is normally run for 10-40 minutes. When particles are collected either from the electrode or the surface of the Kapton tape, they are placed on TEM support films. The process conditions for this study are 6 sccm of Ar and 0.5 sccm of isopropanol leading to the total pressure of 200-500 mtorr. TEM analysis of the nanoparticles was performed on a Philips EM420T operating at 120 kV. This instrument is equipped with a CCD camera and an X-ray Energy-Dispersive Spectroscopy (EDS).

3 RESULT AND DISCUSSION

TEM images of uncoated silica and coated silica particles in plasma are given in figure 2. The size distribution of the bare silica particles are 200±12 nm. Different film thicknesses can easily be achieved by controlling the plasma processing duration since the average deposition rate is 2.5±0.5 nm/min. Figure 2 shows silica particles 200 nm in diameter coated with the amorphous hydrogenated carbon (a:H-C) films for 10 minutes.

TEM images of IPA plasma polymer hollow particles are shown in figure 3. The spherical silica particle diameter is 200±12 nm. The thickness of the film is 50±10 nm and was obtained afted 20 min of deposition. Figure 3a shows a TEM image of the particles after treatment in HF. These images demonstrate that the silica core has completely dissolved while the shells have remain intact. Apparently, the films are chemically inert to the corrosive power of hydrofluoric acid. Moreover, it is apparent that the films are permeable to small molecules since HF, H_2O, and dissolved silica (aggregates of silicic acid) can all pass through the shell. To obtain quanti-

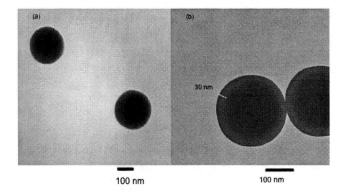

Figure 2: TEM images of: (a) untreated silica particles with diameter of 200 nm, (b) treated silica particles with a plasma coating of 30 nm

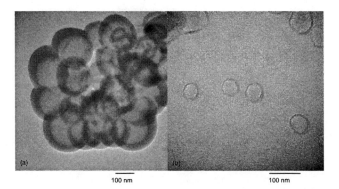

Figure 3: TEM images of: hollow particles after etching (a) silica (b) KCl core

tative information about the permeability of the shells, we switched from silica to potassium chloride. Spherical KCl particles were produced by spray-drying an aqueous solution of the salt. The resulting particles are spherical but not as monodisperse as silica, with sizes in the range 20–100 nm. The fabrication of the hollow particles was similar to that of silica except that the dissolution of the core was accomplished in water. Figure 3b shows the hollow nanoshells after the dissolution of the KCL core.

To measure the concentration of dissolved KCl in water, a conductivity meter (Thermo Orion model 105) is used. The in-vitro release of KCl from the core is illustrated in figure 4. The release profile may be divided into three stages. The first stage is marked by a very rapid increase of conductivity. Within 10 minutes of stirring the coated KCl particles in solution, the conductivity rises to about 72% of its final value. We attribute this increase to uncoated KCl particles and not uniformly coated KCl that are present in the sample. In contrast, for the control the concentration of the uncoated KCl reaches the maximum within the first 82 seconds. The next stage is characterized by a slow, continuous release that lasts over a period of several days. Within the 13 days we observe the slow increase of KCl

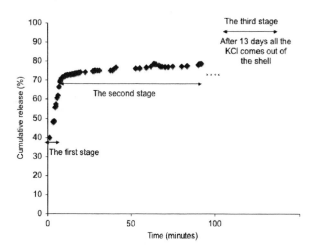

Figure 4: Release profile of KCl from hollow particles

release from 71.7% to 99% of the total amount of the salt. Beyond this point, no further increase in conductivity is observed. To confirm this result, TEM image was taken after 14 days and no KCl was observed in the hollow particles.

By comparing the uncoated KCl dissolution release in water with the coated KCl, the polymer shell around the KCl particles delayed the salt dissolution in water. This study is comparable with Shao et al. result[19]. They studied the cefradine release profile from porous hollow silica nanoparticles and showed a delayed dissolution effect. Cefradine dissolves in water from the shell within 12 hours while it takes several days for us to dissolve all of the coated KCl. But the pattern of the etching is almost the same. First they observed a fast release, then a slower stage and finally reached a constant concentration which is almost the same as our result.

4 CONCLUSION

In summary, we developed a method to synthesize hollow nanoparticles by a template-dissolution process. First, we deposit a plasma polymer by the plasma enhanced chemical vapor deposition technique. Then the template is dissolved to produce a hollow nanoshell. The hollow particle shell has a uniform thickness as it is shown in TEM images and is intact after mixing with HF acid. One of the interesting behavior of hollow particles is their permeability which was studied in this work by measuring the profile release of KCl from the hollow shell. The concentration of the coated KCl is detected by a conductivity meter and is measured over a long period of time. The hollow particles exhibit a delayed release response which can be used in encapsulation applications.

4.1 ACKNOWLEDGMENT

This work was supported by grant CBET-0651283 from the U.S. National Science Foundation, the Pennsylvania State University Materials Research Institute Nanofabrication Lab, and the National Science Foundation Cooperative Agreement No. 0335765, National Nanotechnology Infrastructure Network, with Cornell University.

REFERENCES

[1] Laura Ann Bauer, Nira S. Birenbaum and Gerald J. Meyer, J. Mater. Chem. 14, 517-526, 2004

[2] Yi Cui, Qingqiao Wei, Hongkun Park, Charles M. Lieber, Science 293, 1289-1292, 2001

[3] Paul Alivisato, Nature Biotechnology 22, 47-52, 2004

[4] P. Yang, H. Yan, S. Mao, R. Russo, J. Johnson, R. Saykally, N. Morris, J. Pham, R. He, H.-J. Choi, Advanced Functional Materials 12, 323-331, 2002

[5] Wensheng Shi, Yufeng Zheng, Hongying Peng, Ning Wang, Chun Sing Lee, Shuit-Tong Lee, Journal of the American Ceramic Society 83, 3228-3230, 2000

[6] Yi Cui, Xiangfeng Duan, Jiangtao Hu, and Charles M. Lieber, J. Phys. Chem. B 104, 5213-5216, 2000

[7] Igor I. Slowing, Brian G. Trewyn, Supratim Giri, and Victor S.-Y. Lin, Adv. Funct. Mater. 17, 1225-1236, 2007

[8] Masahiro Fujiwara, Katsunori Kojima,Yuko Tanaka, and Ryoki Nomura, J. Mater. Chem. 14 , 1195-1202, 2004

[9] Hua Zou, Shishan Wu, and Jian Shen, Chemical Reviews 108, 3893-3957, 2008

[10] Hajji P,David, L. Gerard, J. F. Pascault, J. P. Vigier, G. J. Polym. Sci., Part B: Polym. Phys. 37, 3172-3187, 1999

[11] Keijiro Nakanishi, Hitoshi Muguruma, and Isao Karube, Anal.Chem. 68, 1695-1700, 1996

[12] Donglu Shi and Peng He, Rev.Adv.Mater.Sci. 7, 97-107, 2004

[13] Albert H. L. Chow, Henry H. Y. Tong, Pratibhash Chattopadhyay, and Boris Y. Shekunov, Pharmaceutical Research 24, 411-437, 2007

[14] Kumaresh S. Soppimath, Tejraj M. Aminabhavi, Anandrao R. Kulkarni, Walter E. Rudzinski, Journal of Controlled Release 70, 1-20, 2001

[15] Sang Hyuk Im, Unyong Jeong, and Younan Xia, Nature Materials 4, 671-675, 2005

[16] Suk Bon Yoon, Kwonnam Sohn, Jeong Yeon Kim, Chae-Ho Shin, Jong-Sung Yu, and Taeghwan Hyeon, Adv. Mater. 14, 19-21, 2002

[17] Minsuk Kim, Suk Bon Yoon, Kwonnam Sohn, Jeong Yeon Kim, Chae-Ho Shin, Taeghwan Hyeon, Jong-Sung Yu, Microporous and Mesoporous Materials 63, 1-9, 2003

[18] Jin Cao and Themis Matsoukas, J. Appl. Phys. 92, 2916-2922, 2002

[19] Jian-Feng Chen, Hao-Min Ding, Jie-Xin Wang, Lei Shao, Biomaterials 25, 723727, 2004

Surface Modification and Performance Analysis of Jute Based Nanophased Green Composite

M. K. Hossain[*], M. W. Dewan[**], M. Hosur[***] and S. Jeelani[****]

Assistant Professor, hossainm@tuskegee.edu
[**]Graduate Student, mdewan3020@student.tuskegee.edu
[***]Associate Professor, hosur@tuskegee.edu
[****]Professor of ME, Director of T-CAM & Ph. D Program in Materials Science and Engineering, and
Vice President for Research and Sponsored Programs at Tuskegee University, jeelanis@tuskegee.edu
Center for Advanced Materials (T-CAM), Tuskegee University
101 Chappie James Center, Tuskegee, AL 36088, USA

ABSTRACT

Jute-based green composites can be used in consumer goods, low-cost housing, and interior of cars, civil structures, and biomedical applications due to their ease of availability, echo-friendliness, low cost, and good specific properties that are comparable to synthetic fibers. Biodegradable and green nanophased jute composites are manufactured using chemically treated jute fibers and biodegradable polymer (Biopol) for this study. The surface modification of jute fibers was accomplished by performing subsequent chemical treatments such as detergent washing, dewaxing, alkali, and acetic acid treatment. The morphology of the modified surface was examined using the scanning electron microscopy (SEM), and Fourier transform infrared spectroscopy (FTIR). Thermal performance of the treated fibers was studied using the state-of-the-art thermogravimetric analysis (TGA). It was found that the finally treated jute fiber contained only 4% moisture and the higher amount of cellulose, which is a promising result for proper fiber-surface wetting and bonding with the matrix. Biodegradable jute Composites were fabricated using traditional routes such as compression molding technique. The mechanical properties of this green composite were studied by three point bending test. The result showed that surface modification improved the thermal and mechanical properties of the jute based biocomposites. Addition of nanoclay in this biocomposite is on progress and the resultant data would be presented in the conference.

Keywords: Biodegradable composite, Biopol, Jute fiber, Surface modification, FTIR.

1. INTRODUCTION

Synthetic fibers such as carbon, glass, aramid fibers etc. are generally used to make composite materials due to their better mechanical and thermal properties than natural fibers. However, they are expensive and non-degradable. On the other hand, natural fibers like banana, cotton, sisal, coir, jutes are inexpensive, biodegradable or recyclable at reasonable cost, easily available in nature, and have good mechanical, thermal, and electrical properties. Moreover, they are renewable [1]. Jute is one kind of natural bast fiber which is found from jute plant. Among the all natural fibers, jute fibers are easily available in fabric and fiber forms with good mechanical and thermal properties [2]. Jute fibers are echo-friendly, low cost, and low-density fibers with high specific properties. Therefore, jute-based composite materials can be used in consumer goods, low-cost housing, and interior of cars, civil structures, and biomedical applications. The main elements of jute fibers are- cellulose, hemicelluloses, lignin, and pectin. Cellulose is the main element of jute fiber, which is resistant to alkali but hydrolyzed by acid. Hemicellulose works as supporting matrix agents of cellulose. Hemicellulose is hydrophilic, soluble in alkali and easily hydrolyzed in acids. Lignins are amorphous and hydrophobic in nature. It contains aromatic and aliphatic constituents. Pectins are like waxes. It gives plant flexibility [3].

Despite the advantages of cellulosic fiber, its main disadvantage is the incompatibility of phase during the mixing of the hydrophilic cellulosic fiber with the hydrophobic polymer matrix. The prime objective of the surface treatment is to increase the bonding strength between hydrophilic fiber and hydrophobic polymeric matrix [4, 5]. There are many attempts underway to control the interfacial adhesion between fiber and matrix. These methods include physical or chemical modification of fibers or the use of coupling agents. Chemical modification of the fibers such as dewaxing, treatment with alkali, acetylation, and grafting alters the surface properties that results better wetting of the fibers with the matrix [6]. Co-polyester of 3-hydroxybutyrate (HB) and 3-hydroxyvalerate (HV) is commercially known as biopol was used as a matrix system in the jute based biocomposites. Among all the biodegradable polymeric materials biopol is considered as the true bio-polymers, because it is synthesized by bacteria. Poly(3-hydroxybutyrate) is brittle while poly(3-hydroxybutyrate-co-3-hydroxyvalerate) is ductile in nature [7]. Nanoclay improves barrier, flammability resistance, thermal and mechanical properties for polymer composites at a very low filler loading. The nanoclay particle increases the modulus and decreases strain to failure of nanocomposite [8]. Dispersion of nanoclay into the composites is very important for nanophased composites.

The objective of this study is to modify the jute fabrics by chemical treatment for the better interfacial adhesion with matrix and produce biodegradable nanocomposites. The surface modification was evaluated by the morphological, mechanical, and thermal characterization of jute fibers and its composites.

2. EXPERIEMNTAL

2.1 Materials Selection

The bacterial copolyester, poly(3-hydroxybutyrate-co-3-hydroxyvalerate) or biopol granule (5 mm), with the average 3-hydroxyvalerate content of 12 mol%, supplied by the Goodfellow Company (UK) was used without additional purification. Natural jute fabric obtained from the onlinefabricstore.net (USA) was used after chemically surface modification. Alcojet detergent, ethanol (50% solution), NaOH (50% solution), and acetic acid (99%) received from the Sigma–Aldrich were used without further purification.

2.2 Surface Modification

Jute fibers are hydrophilic in nature. On the other hand, all polymers (resins) are hydrophobic in nature. Hydrophilic fiber and hydrophobic polymer result poor interaction bonding at fiber/matrix interface. Surface modifications of the fibers are necessary to improve bonding and adhesion affinity to polymer matrices and dimensional stability. Surface treatment not only decreases moisture absorption, but also increases wettability of the fibers in the matrix polymer and the interfacial bond strength.

- Detergent washed: Detergent wash is performed to remove dirt which comes with jute fabrics. Jute fabrics keep into 5% detergent solutions at 30 °C for 1 hour and it is then washed with water and dried at room temperature.
- Dewaxing: Dewaxing is performed to remove pectin and waxes come with jute fabrics during the processing. Detergent washed jute fabrics keep into 5% ethanol solution at 30 °C for 2 hours followed by washing with distilled water and drying at room temperature for 24 hours.
- Alkali treatment: It is the most important steps for surface modification. It increases the wettability of the fibers and improves adhesion affinity to

polymer matrices. Dewaxed jute fabrics keep into 5% NaOH solution for 2 hours at 30 °C.
- Alkali treated jute fabrics are washed with distilled water-acetic acid (2%) solution followed by washing with distilled water and vacuum drying.

The morphology of the modified surface was examined using SEM and the performance of the treated fibers was studied by TGA and FTIR.

2.2 Composite Fabrication

Jute biopol composites were fabricated using untreated and chemically treated jute fabrics and bipol using a hot press unit. The solution of biopol granules were prepared dissolving it into the cholorofrom in a ratio of 1:10. The solution was then dried and a thin firm was prepared using the hot press machine at 180 °C and 2 ton pressure. The thickness of the film was maintained 0.5 mm. The biopol film and jute fabric were stacked like a sandwich and jute/biopol composites were manufactured using the compression molding technique. In step 1- the composite panel was heated upto 166 °C at 1 ton pressure and held for 5 minutes. In step 2- the temperature was maintained 140 °C at a pressure of 2 ton and held for 10 minutes. After that samples were kept on the heated plate for cooling at room temperature.

2.3 Test Procedure

2.3.1 Spectral and Thermal Analysis

The FTIR spectra of parent and surface-treated jute fabrics were recorded with a Nicolet 10 DX FTIR spectrophotometer. Thermo gravimetric analysis (TGA) was carried out under nitrogen gas atmosphere on a TA Instruments (Q 500), Inc. apparatus. TGA measurements were carried out to obtain information on the thermal stability of the jute fabrics, biopol, and jute/biopol composite. The samples were cut into small pieces to maintain the sample weight within a 10–15 mg range. Then the samples were kept in a platinum sample pan, weighed and heated to 450 °C from room temperature (30 °C) at a heating rate of 10 °C/min. The real time characteristic curves were generated by Universal Analysis 2000-TA Instruments, Inc. data acquisition system.

2.3.2 Morphological Characterization

Morphology of untreated and treated jute samples were examined under a Field Emission Scanning Electron Microscope (FE-SEM Hitachi S-900) JEOL JSM 5800). An accelerating voltage was applied to accomplish desired magnification.

2.3.3 Differential Scanning Calorimetry (DSC)

DSC scans were performed using TA Instruments Q 1000 with sealed pans in the temperature range 30 °C to 200 °C at 5 °C/min heating rate under nitrogen flow. The melting and crystallization temperatures (T_m and T_c, respectively) of the biopol were taken at the corresponding peak maximum.

2.3.4 Flexure Test

Flexural tests under three-point bend configuration were performed on the Zwick Roell testing unit according to ASTM D790-02 to determine the ultimate strength, strain and young modulus of the jute/biopol composites. The machines were run under displacement control mode at a crosshead speed of 2.0 mm/min and tests were performed at room temperature.

3. RESULTS AND DISCUSSIONS

3.1 FTIR Spectroscopy

Characterizations of modified fabrics constituents are an important part of our investigation. Figure 1 illustrates the FTIR spectra of different modified jute fabrics. A broad absorption band in the region 3400–3300 cm^{-1} characteristic of hydrogen bonded O-H stretching vibration, is common to all of the spectra. The O-H stretching vibration decreases due to surface modification. Near to wave number 2950 cm^{-1} shows the C-H vibration. It is also decreases due to the surface modification. Hemicelluloses are the main concern of the surface modification that appears near to 1100 cm^{-1} wave number. The amount of hemicelluloses decreases due to alkali treatment.

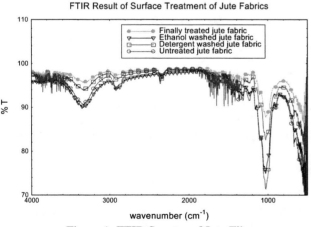

Figure 1: FTIR Spectra of Jute Fibers

3.2 Thermal Stability

From the TGA curves, it is found that the moisture content of the fabrics decreases due to the surface modification of fabrics. It is also observed that the decomposition temperature slightly increases with the surface treatment. Decomposition temperatures of the finally treated jute fiber and the neat biopol are about 310 and 270 °C, respectively (Figure 2). It is noticed that the decomposition temperature of jute/biopol composite is higher than the neat biopol but lower than finally treated jute fiber, as shown in Figure 3. From the TGA curve it is also evident that the amount of residue increases due to the surface modification. It indicates the higher percentage of cellulose content compared to the untreated fiber.

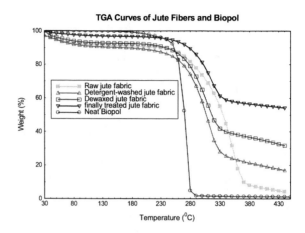

Figure 2: TGA curves of Jute and Biopol

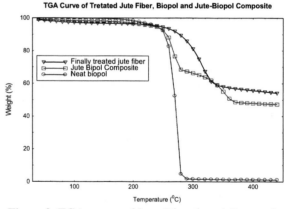

Figure 3: TGA curves of Jute, Biopol, and Composite

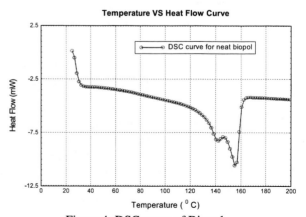

Figure 4: DSC curve of Biopol

3.3 Thermal Properties of Biopol

Thermal properties including melting and crystallization temperature of biopol were measured using differential scanning calorimetry (DSC) test. Melting temperature is important to process the thermoplastic polymer. Form the DSC result it is found that the melting temperature of neat biopol is 158 °C.

3.4 Flexural Properties

Figure 5 shows the flexural result of untreated and treated jute/biopol composites. Surface modified jute fabrics result better adhesion with the matrix. As a result, better flexural properties of the surface modified jute/biopol composite are found compared to the untreated jute/biopol composite. The results are shown in the table 1.

Jute/ Biopol Composite	Max Force (N)	Strain at F_{max} (%)	Max Stress (MPa)	Modulus (GPa)
Treated	60.03	6.13	30.30	1.33
Untreated	56.06	12.61	24.06	1.25

Table 1: Flexural test results

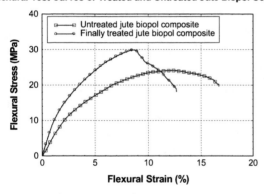

Flexural Test Curves of Treated and Untreated Jute Biopol Composite

Figure 5: Flexural stress-strain curves

3.5 Morphological Characterization

The surface morphology of the jute fabrics was investigated using SEM micrograph (Figure 6). From the micrograph it is noticed that the untreated fiber surface is smooth. Untreated jute fibers contain cellulose, hemicelluloses, pectins, and lignins. Detergent and ethanol washed jute fabrics contain a lot of granules which are waxes and other impurities on the fiber surfaces. Those impurities are removed due to the subsequent detergent and ethanol treatment. Finally treated jute fibers do not have the white granules. It indicates that the hemicellulose is removed due to the alkali treatment and results higher percentage of cellulose. The finally treated surface also rougher compared to untreated jute fibers. The rough

surface attributes to better interaction with matrix due to larger surface area.

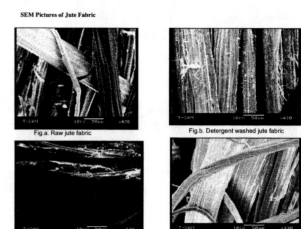

SEM Pictures of Jute Fabric

Fig.a. Raw jute fabric | Fig.b. Detergent washed jute fabric
Fig.c. Ethanol washed jute fabric | Fig.d. Finally treated jute fabric

Figure 6: SEM micrographs of Jute Fibers

4. SUMMARY AND CONCLUSIONS

In this paper, it is demonstrated that four subsequent surface modification treatments improved cellulose content which is promising for composite fabrication. FTIR results showed lower amount of hemicelluloses due to surface modifications. TGA results showed that decomposition temperature increases due to the surface treatment. SEM results also showed the improvement of surface due to the chemical modification. From flexural results it was concluded that the treated jute fiber/biopol composite results better mechanical properties compared to the untreated jute fiber/biopol composite.

REFERENCES

[1] T. Doan, S. Gao and E. Mader, Compos. Sci. Technol., 66, 952-963, 2006.
[2] M. Khan, J. Ganster and H. Fink, Compos. Part A, 40, 846-851, 2009.
[3] M. John and S. Thomas, 71, 343-364, 2008.
[4] C.K. Hong, I. Hwang, N. Kim, D.H. Park, B.S. Hwang and C. Nah, J. Ind. Eng. Chem., 14, 71-76, 2008.
[5] F. Corrales, F. Vilaseca, M. Llop, J. Girones, J.A., Mendez and P. Mutje, J. Hazard. Mater., 144, 730-735, 2007.
[6] S. Wong, R. Shanks and A. Hodzic, Compos. Sci.Technol., 64, 1321-1330, 2004.
[7] D. Snow, M. Major and L. Green, Microelectronic Engr. 30, 969, 1996.
[8] G. Huang and A. Netravali, Compos. Sci. Technol., 67, 2005-2014, 2007.

Chemical and Phase Transformations of Nanocrystalline Oxides in a Carbon Nanoreactor

A. M. Volodin[*], V. I. Zaikovskii[*], A. F. Bedilo[*] and K. J. Klabunde[**]

[*]Boreskov Institute of Catalysis, Novosibirsk 630090, Russia, volodin@catalysis.ru
[**]Department of Chemistry, Kansas State University, Manhattan, KS, USA, kenjk@k-state.edu

ABSTRACT

The application of nanocrystalline oxides as precursors for synthesis of new nanomaterials with the preservation of the dimensions of the initial nanoparticles is of considerable interest. A promising approach to the overcoming the problem of the sintering and growth of the nanoparticles during the solid-state chemical is the deposition of a coating on the surface of the precursor nanoparticles that would prevent their sintering. Current study was devoted to the investigation of the solid-state phase and chemical transformations of nanocrystalline TiO2, MgO an Al_2O_3 coated with carbon. Such coating appeared to be penetrable by the molecules present in the gas phase. The anatase phase in all synthesized C@TiO_2 nanocomposites could be completely converted to $TiOF_2$ in the reaction with halocarbons. Overall, it is possible to deposit an intelligent carbon coating on the surface of the oxide nanoparticles that would prevent their sintering and make it possible to synthesize new nanomaterials.

Keywords: carbon nanoreactor, destructive sorption, halocarbons, nanocrystalline oxides

1 INTRODUCION

Nanocrystals of common metal oxides such as MgO, CaO, ZnO, TiO2, Al2O3, and Fe2O3, have been shown to be highly efficient and active adsorbents for many toxic chemicals including air pollutants, chemical warfare agents, and acid gases [1-8]. In most cases, destructive adsorption takes place on the surface of the nanocrystals, so that the adsorbate is chemically dismantled and thereby made nontoxic. In particular, aerogel-prepared (AP) nanocrystalline MgO has been shown to have small average particle size (~ 4nm), high surface area (> 500 m^2/g) and high reactivity [1, 5]. At elevated temperatures, chlorocarbon reactions can be driven to stoichiometric proportions, especially if small amounts of transition metal catalysts are added.

The use of these nanocrystalline metal oxides is limited under conditions where liquid water or water vapor is present due to their tendency to adsorb water, and thereby be partially deactivated toward adsorption of the target pollutants. Although the target pollutants are usually subjected to conversion in the presence of water as well, relatively large amounts of water can mitigate against the adsorption of the target adsorbate decreasing the efficiency of the destructive adsorbent. We have shown that coating destructive sorbent with an intelligent carbon layer significantly improves the stability of such nanocrystalline materials in the presence of water and the possible time of their storage in air in comparison with not coated nanoscale oxides without a considerable loss of activity [9]. For practical application of nanoscale oxides as destructive sorbents, it is extremely important that the carbon coating makes such materials hydrophobic.

In the current publication we shall discuss the specific features of the reactions of nanoscale MgO and TiO_2 with chlorofluorocarbons and the effect of carbon on the performance of the destructive sorbents.

2 EXPERIMENTAL

AP-MgO with the surface area of 385 m^2/g was prepared by a sol-gel technique involving high-temperature supercritical drying described in detail earlier [4, 9]. The final preparation step was overnight evacuation at 500°C.

A flow reactor equipped with a McBain spring balance and an on-line gas chromatograph was used in the flow experiments. This made it possible to monitor both the composition of the products and the sample weight during the reaction. The sensitivity in the determination of the sample weight was 10^{-4} g. CF_2Cl_2 and CFCl_3 of "pure" grade and Ar of "ultra pure" grade were used in the experiments. The halocarbon was passed through the reactor with a volume rate of 3 L/h. MgO sample weight was 0.05-0.1 g. The samples were activated in argon flow at 500°C for 30 min.

The same flow reactor equipped with a McBain spring balance was used for "in situ" carbon formation. AP-MgO was placed in a quartz basket and placed on a spring. This allowed for measuring of the sample weight change and accurate determination of carbon percentage in the sample. Then, the sample was heated under a flow of argon, at 500 °C for 1 hour. After that butadiene used as a carbon source was added to the flow. The flow rates were: butadiene - 7.5 L/h; argon - 75 L/h. Initial carbon formation occurred at a rate of approximately 2 wt. % per hour. After the desired amount of carbon was obtained, the butadiene flow was stopped and the product was cooled to room temperature under an argon flow.

Two TiO_2 samples were used in the experiments: Degussa P-25 and Nanoactive TiO_2 (Nanoscale Materials). Before the experiments, the samples were calcined in an argon flow at 500°C for 1 h. Then, the temperature was decreased to the desired reaction temperature, and the halocarbon flow was turned on. The standard reaction time was 30 min. Then, the sample was cooled in the argon flow to room temperature. The products were analyzed using a Shimadzu GCMS-QP5000 gas chromatograph mass spectrometer equipped with a 30 m Restec capillary column with XTI-5 phase. XRD studies were performed using a Brucker D8 instrument with CuK_α monochromatic irradiation.

3 RESULTS AND DISCUSSION

3.1 Effect of carbon coating on MgO reaction with CF_2Cl_2

CF_2Cl_2 is a colorless and odorless gas, which is chemically very inert and has a high thermal stability. It appeared to react with nanoscale MgO at temperatures as low as 325°C to give a significant gain of the sample weight as evidenced by kinetic curves presented in Figure 1. The presence of an obvious induction period, after which an intense gain of the sample weight starts, is the most distinctive and interesting feature of the kinetic curves.

The induction period phenomenon is reproduced with 100% probability, and its duration is reproduced with ca. 5-10% accuracy when all conditions are kept the same. The existence of an induction period on the kinetic curve must be connected with gradual accumulation of certain active sites or defects on the MgO surface, which eventually lead to complete transformation of the nanoparticles. As soon as enough active sites are accumulated, intense bulk substitution of oxygen by fluorine starts. Note that the weight gain stops at values close to 40% indicating the end of the reaction. The maximum theoretical weight gain corresponding to complete transformation of the oxide to fluoride is 55%. This indicates that about 80% transformation is actually achieved in our experiments.

The main products resulting from the AP-MgO reaction with CF_2Cl_2 are MgF_2, CO_2 and CCl_4 [10] formed according to overall reaction (1). The former is identified as the only solid phase according to the XRD and HRTEM data. CCl_4 appears to result from CF_2Cl_2 exchange reaction with partially chlorinated MgO/MgF_2 surface.

$$2\,MgO + 2\,CF_2Cl_2 \rightarrow 2\,MgF_2 + CO_2 + CCl_4 \qquad (1)$$

We have earlier shown that carbon-coated AP-MgO has improved stability in the presence of water and the possible storage time in air in comparison with not coated nanoscale MgO [9]. Our EPR investigation of the active basic sites present on the surface of AP-MgO partially covered with carbon showed that samples containing 10% carbon or less retain significant amounts of open active sites that are available for destructive adsorption reactions [11]. So, it was interesting to study how the introduction of such

carbon coating affects the AP-MgO reactivity with respect to destructive adsorption of halocarbons.

Unexpectedly we observed that the carbon coating improves the reactivity of AP-MgO in destructive adsorption of CF_2Cl_2. Its reaction with AP-MgO is characterized by a prolonged induction period that is nearly 1 h long at 400°C (Fig. 1). The addition of only 1 wt.% carbon significantly shortens the induction period, while induction period over 10% C/AP-MgO was 3 times shorter than over AP-MgO. Similar results were obtained in the reaction with another halocarbon $CFCl_3$.

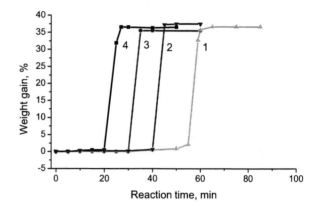

Figure 1. Kinetic weight gain curves of AP-MgO (1), 1% C/AP-MgO (2), 3% C/AP-MgO (3) and 10% C/AP-MgO (4) treated with CF_2Cl_2 at 400°C.

3.2 Phase selectivity in solid-state transformation of nano-TiO_2 to $TiOF_2$

The phase composition of the oxide nanoparticles appeared to be another parameter having a significant effect on such reactions. Relatively easily the phase composition and size of the nanoparticles can be regulated for titanium dioxide. We have studied reaction of nanocrystalline TiO_2 samples with CHF_2Cl resulting in the formation of a new chemical compound titanium oxyfluoride:

$$TiO_2 + CHF_2Cl \rightarrow TiOF_2 + CO + HCl \qquad (2)$$

The reaction was studied in the flow of the halocarbon at different temperatures. Its kinetics appeared to be significantly dependent on the size of TiO_2 nanoparticles and their crystalline phase. Figure 2 presents the XRD spectra characterizing the solid products of the reaction of Degussa P-25 TiO_2 with CHF_2Cl at different temperatures. The initial TiO_2 sample (spectrum 1) consists of a mixture of anatase (~ 75%) and rutile (~25%) with the particle size about 25 nm. The reaction with CHF_2Cl leads to the formation of a new solid product identified by XRD as $TiOF_2$ and decrease in the intensity of anatase lines (spectra 2 and 3). Note that even after complete transformation of anatase the lines of rutile remain unchanged (spectrum 3).

No reaction of rutile with CHF_2Cl was observed even when the reaction temperature was increased to 400°C.

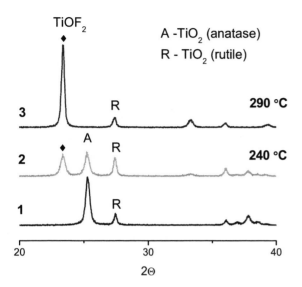

Figure 2. XRD spectra of Degussa P-25 TiO_2 before reaction (1) and after reaction with $CHFCl_2$ at 240°C (2) and 290°C (3).

The observed phase selectivity phenomenon vividly shows that the crystalline phase of titania nanoparticles has a major effect on their reactivity. This reaction can be used for preparation of pure nanoscale rutile phase for use as a photocatalyst.

Using Nanoactive TiO2 prepared by sol-gel technology it is possible to prepare sample with different particle size and phase composition by simply varying the heat treatment temperature. For example, after calcination at 500°C for 1 h the average size of nanoparticles according to the XRD data is about 15 nm, and anatase is the only identified crystalline phase. Its reaction with CHF2Cl at temperature as low as 240°C results in almost complete TiO2 transformation to TiOF2. This temperature is 50°C lower than the temperature required for complete conversion of 25 nm anatase nanoparticles. Thus, similar to reactions of halocarbons with MgO, a decrease of the TiO2 particle size allows for a significant decrease of the temperature necessary for complete conversion of solid phase nanoparticles.

According to the XRD data, the particle size of reaction product TiOF2 is similar to that of the nanoparticles of the initial TiO2. The surface area of the material also does not change much. These results prove that such reactions can be used for synthesis of new halogenated nanocrystalline materials using nanocrystalline oxide as precursors.

3.3 Effect of carbon on the solid-state transformations of TiO_2 and Al_2O_3

Current study was devoted to the investigation of the solid-state phase and chemical transformations of nanocrystalline TiO_2 and Al_2O_3 with carbon coatings obtained by CVD method and by decomposition of polyvinyl alcohol. In all cases the carbon coating acting as a carbon nanoreactor prevented the sintering of the nanoparticles and helped to stabilize their size. The dimensions of the anatase nanoparticles in such $C@TiO_2$ nanocomposites could be stabilized at 10-12 nm. Almost no rutile was observed by XRD even after calcination at 750°C (Fig. 3, curve 3). This is due to the thermodynamic stability of the anatase phase at small particle size. Meanwhile, pure TiO_2 without the carbon coating was almost completely transformed to rutile under the same conditions (Fig. 3, curve 2)..

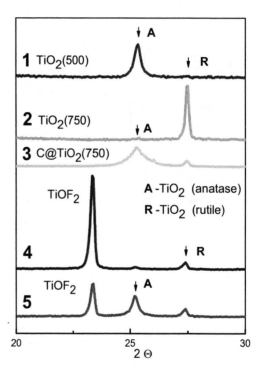

Figure 3. XRD patterns registered for nanocrystalline TiO_2 (1,2) and $C@TiO_2$ (3-5) after heating at 500° (1) and 750°C (2,3), after reaction of $C@TiO_2$ with CF_2Cl_2 at 330°C (4) and following hydrolysis at 310°C (5).

Meanwhile, such coating appeared to be penetrable by the molecules present in the gas phase. The anatase phase in all synthesized $C@TiO_2$ nanocomposites could be completely converted to $TiOF_2$ in the reaction with CHF_2Cl and CF_2Cl_2 halocarbons (Fig. 3, curve 4). The analysis of the reaction product by HRTEM coupled with EDX detected $TiOF_2$ nanoparticles with typical dimensions 10-15 nm inside the carbon coating. Their size was close to that of the initial anatase nanoparticles. Intriguingly, this reaction was phase-selective. Only anatase nanoparticles were converted, whereas rutile, if present in the precursor, was totally inactive. The synthesized $TiOF_2$ nanoparticles could be converted to an interesting nanostructured anatase by hydrolysis in humid air (Fig. 4).

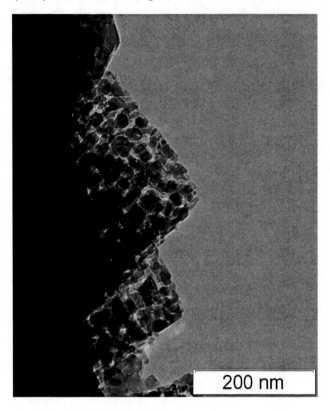

Figure 4. HRTEM image of nanostructured anatase prepared by $TiOF_2$ hydrolysis in humid air.

Similar results were also obtained for the phase transformations ($\gamma \rightarrow \theta \rightarrow \alpha$) in C@$\gamma$-$Al_2O_3$ nanocomposites. The nanoparticles inside the carbon shell are resistant to sintering until very high temperatures (1000°C for Al_2O_3). It was shown that such method can be used to synthesize a carbon film on the surface of Al_2O_3 nanoparticles without substantial changes of the initial oxide morphology. The phase transformations of such C@γ-Al_2O_3 core-shell samples from gamma to delta and, then, alfa alumina phase were studied by laser-induced luminescence, XRD, TEM and adsorption methods up to 1150°C. It was shown that the carbon coating prevents sintering of the Al_2O_3 nanoparticles and as a result prevents the corundum phase from being formed at such temperatures even in low concentrations.

4 CONCLUSIONS

We have demonstrated that nanocrystalline oxides can be used as precursors for synthesis of novel halogenated nanocrystalline materials. Of special interest are composite materials consisting of carbon-coated nanocrystalline oxides. Such permeable carbon coating not only provides high stability of the destructive sorbents under atmospheric conditions, but in some cases shown in this publication considerably increases their reactivity.

The carbon coating deposited on the oxide nanoparticles prevents their sintering and makes it possible to synthesize new nanomaterials with nanoparticle dimensions close to those of the initial oxide precursors. The preservation of the small size of the oxide nanoparticles inside such carbon nanoreactor is a key factor preventing the phase transformation in the studied C@TiO_2 and C@γ-Al_2O_3 nanocomposites.

Acknowledgment. This study was supported by the Russian Foundation for Basic Research (Grant 10-03-00806) and the US Civilian Research and Development Foundation (Grant RUE1-2893-NO-07).

REFERENCES

[1] A. Khaleel, P. N. Kapoor and K. J. Klabunde,. Nanostruct. Mater., 11, 459, 1999.

[2] E. M. Lucas, and K. J. Klabunde, Nanostruct. Mater., 12, 179, 1999.

[3] O. B. Koper, I. Lagadic, A. Volodin and K. J. Klabunde, Chem. Mater., 9, 2468, 1997.

[4] K. J. Klabunde, J. Stark, O. Koper, C. Mohs, D. G. Park, S. Decker, Y. Jiang, I. Lagadic and D. J. Zhang, J. Phys. Chem., 100, 12142, 1996.

[5] Y. X. Li and K. J. Klabunde, Langmuir, 7, 1388, 1991.

[6] G. W. Wagner, P. W. Bartram, O. Koper and K. J. Klabunde, J. Phys. Chem. B, 103, 3225, 1999.

[7] G. W. Wagner, O. B. Koper, E. Lucas, S. Decker and K. J. Klabunde, J. Phys. Chem. B, 104, 5118, 2000.

[8] S. Decker and K. J. Klabunde, J. Am. Chem. Soc., 118, 12465, 1996.

[9] A. F. Bedilo, M. J. Sigel, O. B. Koper, M. S. Melgunov, and K. J. Klabunde, J. Mater. Chem., 12, 3599, 2002.

[10] I. V. Mishakov, V. I. Zaikovskii, D. S. Heroux, A. F. Bedilo, V. V. Chesnokov, A. M. Volodin, I. N. Martyanov, S. V. Filimonova, V. N. Parmon, and K. J. Klabunde, J. Phys. Chem. B, 109, 6982, 2005.

[11] D. S. Heroux, A. M. Volodin, V. I. Zaikovski, V. V. Chesnokov, A. F. Bedilo, and K. J. Klabunde, J. Phys. Chem. B, 108, 3140, 2004.

Integration of SWCNT into Olefin polymer composites

J. W. Guan[*], B. Simard[*], Y. Wan[**], Y. Ma[**] and F. Ko[**]

[*]Steacie Institute for Molecular Sciences, National Research Council of Canada
100 Sussex Drive, Ottawa, ON, Canada, K1A 0R6, jingwen.guan@nrc-cnrc.gc.ca
[**]Advanced Fibrous Materials Laboratory, AMPEL, University of British Columbia
2355 East Mall, Vancouver BC, Canada V6T 1Z4

ABSTRACT

Due to the lack of interfacial compatibility, unmodified single walled carbon nanotubes (SWCNT) cannot be composited with polyolefin matrices. Here we show how SWCNT can be modified chemically to overcome this difficulty. The approach relies on forming a compatible polymer layer on the surface of SWCNT through *in-situ* polymerization. Here we demonstrate proof-of-principle with UHMWPE and show that the composite can be gel spun into fiber. The uniformity, interfacial interaction, preliminary mechanical tests are also reported.

Keywords: single-walled carbon nanotubes, integration, ultrahigh molecular weight polyethylene, composite, fibres

1 INTRODUCTION

Ultra high molecular weight polyethylene (UHMWPE) possesses many excellent material characteristics, including high impact resistance, high wear resistance, good sunlight and humidity resistance, good chemical resistance and biocompatibility. Therefore, UHMWPE has been and continues to be widely used as one of the major ballistic protection materials in the industry. However, to reach the required protection level the amount of UHMWPE needed can be quite significant, making the protection gear heavy and bulky which results in loss of mobility and comfort.

One of the major drawback of UHMWPE is its processability. In the past this has been circumvented by combining it with low molecular weight polyethylene (PE), polypropylene (PP), polyacrylate (PA), polycarbonate (PC), polyaniline (PANI) and other materials [1-6], but, this yielded no improvement in the anti-ballistic properties of the final composites over the neat UHMWPE.

Owing to their superlative mechanical properties, large surface areas and high aspect ratios, single-walled carbon nanotubes (SWCNT) are widely considered to be the most promising additives for the development of high performance multifunctional composite materials [7-8]. However, pristine SWCNT are relatively chemically inert and owing to strong van der walls forces they tend to agglomerate in bundles; the higher the purity and quality, the more difficult the chemistry and processing are. These difficulties are exacerbated when the matrix is chemically very stable such as poly-olefin thermoplastics. In general, to enhance the properties of a given matrix, the SWCNT must be ex-foliated ideally at the single tube level and must exhibit excellent interfacial compatibility with the matrix [9-10]. This can only be realized through surface chemical modification on the side wall of SWCNT.

Many literature reports have focused on physical blending of multi-walled CNT (MWCNT) with polyolefin matrices such as Poly-Ethylene (PE) and Poly-Propylene (PP). In all of those cases, the dispersion is poor and the CNT-polymer interfacial interaction is weak, resulting in little or no improvement, or in many cases worsening in properties [11-14]. Few studies have been published in which chemical functionalization with alkyl functional groups have been used to improve dispersion but again the SWCNT-polymer interaction remained weak [15]. There have been few reports using *in situ* polymerization on the side wall of CNT to ensure efficient interfacial interaction but as exercised the process was random and lack of control [16-18]. Therefore, to take full advantage of the properties of SWCNT, interfacial chemistry is essential and must be tailored to create an effective covalent connection between the side wall of SWCNT and the polymer matrix to optimize load transfer effectively. Here, we report on a novel approach to resolve the issues of dispersion and interfacial bonding between SWCNT and poly-olefin matrices through covalent anchoring and *in situ* polymerization propagation to produce a SWCNT-polymer coated compatibilizer. This compatibilizer can play roles of either reinforcement filler or interface eliminator between SWCNT and polyolefin matrices. This approach is not only limited to polyolefin but has been extended to other matrices such as epoxy resin and epoxy resin system, epoxy vinyl ester resin and other thermoplastic such as polycarbonate, the results of which are beyond the scope of this paper and will be published elsewhere.

2 EXPERIMENTATION

The SWCNT used in this work were produced using our reported procedures [19] with double laser system and biochar as feedstock. Subsequently, the pristine SWCNT were purified with our in-house developed WCPP [20-21] method that gives high purity and high quality materials without damaging or introducing functional groups on the tubes as shown in the SEM of Fig. 1a and the TEM of Fig. 1b.

Since this paper is focusing mainly on the integration of modified SWCNT into polyolefin, specifically in UHMWPE, the detailed chemistry of the surface modifications will be the subject of a separate paper. Briefly, WCPP purified SWCNT were modified chemically by forming C-C covalent bonds on the surface of the SWCNT. The addends were used to initiate *in-situ* polymerization with the reagents that carry the desired functional groups. The result is a core-shell like structure in which the core is a single or small bundle of SWCNT and the shell is a polymer layer of adjustable thickness. Fig. 1c and 1d show typical SEM and TEM images, respectively.

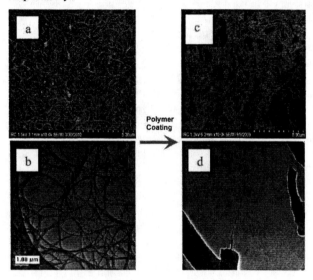

Figure 1. SEM (a), TEM (b) images of WCPP purified SWCNT; SEM (c) and TEM (d) of surfaced modified with polymer coating.

UHMWPE (MW: 3-6X10^6) was purchased from Sigma-Aldrich. A typical small size composite preparation for gel-spinning fiber is: 100 mg UHMWPE and 25 mg of 2, 6-di-tert-butyl-p-cresol (DBPC) were added to 4 g of decalin and the mixture was heated to 140 °C until a transparent solution was obtained. Ten (10) mg of surface modified SWCNT (Fig. 1c) were dispersed in tetrahydrofuran (THF) and then added to 0.9 g of decalin. Next, the THF was evaporated so that SWCNT/decalin suspension was obtained. The suspension was heated to 140 °C and mixed with the UHMWPE solution. The mixture was quickly stirred for a while and transferred to a syringe that was preheated and kept at 135-140 °C with a heating mantel.

Figure 2. Bulk quantity of SWCNT-UHMWPE composite (a) and (b); gel-spinning and fibre-drawing (c) and SWCNT integrated fibers and neat UHMWPE fibers (d).

A scale-up composite preparation as shown in Fig. 2a and 2b is as follows: 8.06 g of the greenish dry powder of Fig. 1c was dispersed in THF (*ca.* 3 L) through tip and bath-sonication to form a well dispersed suspension. The suspension was concentrated to a final volume of about 1 L by evaporating the THF through heating under stirring. The concentrated suspension was then mixed with 1.5 L of xylene (Cas: 1330-20-7, Anachemia) that contains 192 g of UHMWPE powder in an Erlenmeyer flask and the mixture was mechanically stirred (Fig. 2b). The mixture was slowly heated up under vigourous mechanical stirring and nitrogen atmosphere. The rest of the THF was slowly and continuously evaporated. When the THF is completely evaporated, the temperature of the mixture started to increase. At about 130 °C that is near the boiling point of xylene, the morphology and color of the mixture changed to become black uniformly. Within about 3 to 5 minutes, the temperature had risen to about 140 °C and the mixture became very viscous, forming one big block. At this point mechanical stirring became impractical. The heating source was quickly removed and the mixture was let to cool down to room temperature under nitrogen flow. Free xylene was decanted and the solid was filtrated with wet-strength filter paper and washed with methanol. The filtrand was dried in air and then dried at 85 °C in an oven overnight. The final loading of modified SWCNT of Fig.1c in PE composite is 4 wt%, which corresponds to a neat SWCNT loading of about 0.5 wt%.

3 RESULTS

The SWCNT-UHMWPE composite shown in Fig. 2a was used to spun fibers by the gel-spinning process. In this case, the composite was gelated in mineral oil and introduced in the syringe which was maintained at 135-140 °C. A syringe pump was used to extrude the fiber into a water bath. The gel fiber was then drawn in hot air (about 100 °C). The draw ratio is about 3 (Fig. 2c and 2d). Just like the bulk material, the spun fibres are grey or greenish in color. Compared with the neat UHMWPE fibres, the composited fibres have many more voids. As shown in Fig. 3, the core-shell structure of the compatibilizer is still clearly visible within these voids and in peripheries. The non-uniform distribution of the compatibilizer within the matrix is obviously undesirable and is the result of incomplete processing, something that can be resolved in future experiments. The incomplete processing has given overall weaker structures of the fibres. In particular, as shown in Fig 4, the obtained SWCNT-UHMWPE fibers have an average strength of 203.7 MPa and a strain of 425.8 %, while neat PE fibers made have 252.3 MPa and 762.7 % respectively. However, the reinforcement of SWCNT was embodied by the increase of modulus from about 2.1 GPa for the PE to an average value of 4.2 GPa for the composited fiber.

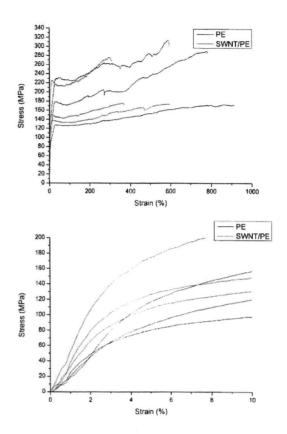

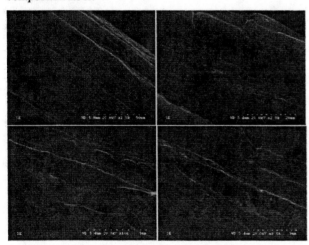

Figure 3. Close look at the morphology on SWCNT-UHMWPE fiber by SEM

As mentioned before, we are hopeful that significant improvement will be obtained upon process improvement, including higher draw ratio (in the example used here, the draw ratio is only 3). To this regard, as shown in Fig. 5, the void structure shown in Fig. 3 can be readily eliminated by simple improvement in the blending process.

Figure. 4. Mechanical property test of SWCNT-UHMWPE fibres.

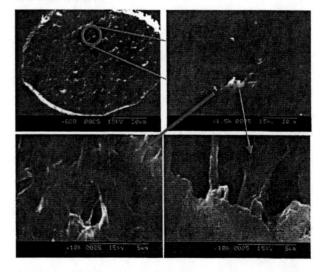

Figure 5. Views at cross section and air void of SWCN-PE fibre

4 CONCLUSIONS

We have shown that SWCNT can be chemically modified to form a core-shell like structure in which the core is the SWCNT and shell a polymer coating of various thicknesses that allows compatibilization with various matrices. In particular, we have illustrated the concept with the challenging UHMWPE matrix. Further, we showed that the composited SWCNT-UHMWPE allows for the fabrication of fibers through gel-spinning. The early experiments are very promising and we are hopeful that highly performing ballistic fiber can be developed using this approach

REFERENCES

[1] Y. Z. bin, L. Ma, R. Adachi, H. Kurosu and M. Matsuo, Polymer 42, 8125, 2001.

[2] G.R. Valenciano, A. E. Job, L. H. C. Mattoso, Polymer 41, 4757, 2000.

[3] S. B. Xie, S. M. Zhang, F. S. Wang, H. J. Liu, M. S. Yang, Polym. Eng. Sci. 45, 1247, 2007.

[4] M. Okamoto, A. Kojima, T. Kotaka, Polymer 39, 2149, 1998.

[5] K. Sakurai, A. Nakajo, T. Takahashi, S. Takahashi, T. Kawazura, T. Mizoguchi, Polymer 37, 3953, 1996.

[6] A. j. Busby, J. X. Zhang, C. J. Roberts, E. Lester, S. M. Howdle, Adv. Mater. 17, 364, 2005.

[7] S. Iijima, Nature 354, 56, 1991.

[8] Z. Spitalsky, D. Tasis, K. Papagelis, C. Galiotis, Prog. Polym. Sci. 35, 357, 2010.

[9] H. Alter, J. Appl. Polym. Sci. 9, 1525, 1965.

[10] V. Buryachenko, A. Roy, K. Lafdi, K. Anderson, S. Chellapilla, Compos. Sci. Technol. 65, 2435, 2005.

[11] J. Yeh, S. Lin, K.chen, K. Huang, J. Appl. Polym. Sci. 110, 2538, 2008.

[12] H. E. Miltner, N. Grossiord, K. Lu, J. Loos, C. E. Koning, B. V. Mele, Macromolecules 41, 5753, 2008

[13] M. Moniruzzaman, F. Du, N. Romero, K. I. Winey, Polymer 47, 293, 2006.

[14] J. Zhu, J. Kim, H. Peng, J. L. Margrave, V. N. Khabashesku, E. V. Barrera, Nano Lett. 3, 1107, 2003.

[15] A. A. Koval'chuk, v. G. Shevchenko, A. N. Shchegolikhin, P. M. Nedorezova, A. N. Klyamkina, A. M. Aladyshev, Macromolecues 41, 7536, 2008.

[16] D. McIntosh, V. N. Khabashesku, E. V. Barrera, J. Phys. Chem. C, 111, 1592, 2007.

[17] B. Yang, K. P. Pramoda, G. Xu, S. H. Goh, Adv. Func. Mater. 17, 2062, 2007.

[18] S. Park, S. W. Yoon, H. Choi, J. S. Lee, W. K. Cho, J. Kim, H. Park, W. S. Yun, C. H. Choi, Y. Do, I. S. Choi, Chem. Mater. 20, 4588, 2008.

[19] C. T. Kingston, Z. J. Jakubek, S. Dénomée, B. Simard, Carbon 42, 1657, 2004.

[20] M. B. Jakubinek, M. B. Johnson, M. A. White, J. W. Guan, B. Simard, J. Nanosci. Nanotechn. In press.

[21] E. Najafi, J. Wang, A. P. Hitchcock, J. W. Guan, S. Denommee, S. Benoit, J. Am. Chem. Soc. In press.

Role of Nanoparticles in Life Cycle of ZnO/Polyurethane Nanocomposites

Xiaohong Gu, Guodong Chen, Minhua Zhao, Stephanie. S. Watson, Paul E. Stutzman,
Tinh Nguyen, Joannie W. Chin and Jonathan W. Martin

Materials and Construction Research Division
National Institute of Standards and Technology, Gaithersburg, MD 20899, USA
xiaohong.gu@nist.gov

ABSTRACT

This study investigated the role of ZnO nanoparticles in the photodegradation of a waterborne polyurethane (PU) nanocomposite during exposure to ultraviolet (UV) radiation. The effects of loading and size of ZnO nanoparticles on the photodegradation of ZnO/PU films were evaluated. Chemical and surface morphological changes of exposed specimens were examined by Fourier transform infrared spectroscopy (FTIR) and atomic force microscopy (AFM), respectively. The results clearly showed that ZnO nanoparticles can act as a photo-catalyst or a photo-stabilizer, depending on the UV exposure conditions. Both size and loading of the ZnO nanoparticles had a strong effect on photodegradation of the nanocomposites. The polymer in the vicinity of the nanoparticles preferentially degrades when the nanoparticles act as a photo-catalyst, and is shielded when they behave as a photo-stabilizer.

Keywords: nanoparticles, nanocomposites, UV exposure, life cycle, AFM

1 INTRODUCTION

Recent developments in nanoparticle technology have initiated the use of inorganic UV absorbers such as zinc oxide (ZnO) for UV protection of polymers [1]. This approach is based on the unique UV absorbing ability of these metal oxides. Compared to the traditional organic UV absorbers, the inorganic UV absorbers offer transparency in the visible range and are non-migratory in the polymer matrix. They also offer enhancements in electrical, mechanical, and antibacterial properties for the polymers [1]. However, due to the photoreactivity of some metal oxide nanoparticles [2], their contributions to UV protection of the polymeric matrix can be complicated.

In this paper, we have investigated the role of ZnO nanoparticles in the life-cycle of a waterborne PU nanocomposite exposed to UV radiation. Nano-filled polyurethane thin films were prepared with ZnO nanoparticles having various sizes and loadings. The UV exposure of the nanocomposites was conducted on both NIST SPHERE (**S**imulated **P**hotodegradation via **H**igh **E**nergy **R**adiant **E**xposure) and an UV/Ozone (UVO) chamber. Chemical changes of the nanocomposite films

were examined by FTIR. Changes in film morphology were characterized by AFM. The photoreactivity of ZnO nanoparticles was measured by electron paramagnetic resonance *(EPR)* spectroscopy. The results clearly showed that the ZnO nanoparticles played a critical role in the life cycle of the ZnO/PU nanocomposites. They acted either as a photo-catalyst or a photo-stabilizer, depending on the UV exposure conditions. The degradation behavior of the polymers in the vicinity of the nanoparticles exposed to different conditions was discussed.

2 EXPERIEMNTALS*

2.1 Materials and Specimen Preparation

A commercial waterborne dispersion of polyurethane (PU, Bayer Material Science) and a series of waterborne dispersions of ZnO nanoparticles (BYK Additives & Instruments) were selected for preparing ZnO/PU nanocomposites. The PU dispersion was a one-component, anionic dispersion of an aliphatic polyester urethane resin in water/n-methyl-2-pyrrolidone solvent. The nominal average diameters of the ZnO nanoparticles were 20 nm, 40 nm, and 60 nm, and their respective specific surface areas were 54 m^2/g, 33 m^2/g and 18 m^2/g (as provided by the manufacturer).

ZnO/PU nanocomposite films (hereafter referred to as ZnO/PU films) were prepared by mixing the PU dispersion with different loadings and/or different sizes of ZnO nanoparticles using a mechanical stirrer (Dispermat, VMA) at 315 rad/s (3000 rpm) for 20 minutes. Three ZnO nanoparticle loadings, 1 %, 2 %, and 5 % (based on mass of the solid PU), were used for 20 nm ZnO nanoparticles. Similarly, three sizes of nanopartilces, 20 nm, 40 nm and 60 nm were used at 5 % loading. After degassing for 1 h in vacuum, the mixture was then applied to the substrates. Thin films having a thickness of approximately 5 μm were prepared by spin coating onto calcium fluoride (CaF$_2$) substrates. Thick films having a nominal thickness of approximately 100 μm were prepared by drawdown technique on a glass substrate that was pretreated with a release agent. All films were dried overnight under ambient conditions, followed by an oven post-curing at 150 °C for 10 min. Free-standing ZnO/PU films were obtained by removing the thick films from glass substrates. In addition,

PU films without ZnO nanoparticles were also prepared for comparison.

2.2 High UV Radiant Laboratory Exposure

UV exposure of ZnO/PU films was mainly conducted on the NIST SPHERE, a 2 m diameter integrating sphere-based weathering chamber. This weathering device utilizes a mercury arc lamp system that produces a collimated and highly uniform UV flux of approximately 480 W/m² in the spectral range between 295 nm and 400 nm. ZnO/PU films coated on CaF₂ were placed in two 17-window sample holders, and exposed to UV radiation at 45 °C, 0 % RH and 45 °C, 75 % RH. To study UV exposure conditions on the photo-catalytic effect of nanoparticles, ZnO/PU films were also exposed in an UV/Ozone (UVO) cleaner (Jetlight Company, Inc.). The main wavelengths of the light emission of this device are 254 nm and 185 nm.

2.3 Measurements

2.3.1 Fourier Transform Infrared Spectroscopy (FTIR)

Chemical degradation of the ZnO/PU films coated on CaF₂ was measured by transmission FTIR spectroscopy using a PIKE auto-sampling accessory (PIKE Technologies). This automated sampling device allows efficient and precise recording of transmission FTIR spectra. The auto-sampler accessory was placed in an FTIR spectrometer compartment (Thermo-Nicolet Nexus 670xl) equipped with a liquid nitrogen-cooled mercury cadmium telluride (MCT) detector. Spectra were recorded at a resolution of 4 cm⁻¹ and were averaged over 128 scans. The peak height was used to represent IR intensity, which was expressed in absorbance units.

2.3.2 Atomic Force Microscopy (AFM)

A Dimension 3100 AFM (Veeco Metrology) was used to image the morphology and the microstructure of the ZnO/PU films before and after UV exposure. The AFM was operated in tapping mode using commercial silicon probes (Olympus AC-160TS) with a resonance frequency of approximately 300 kHz.

2.3.3 Dynamic Mechanical Thermal Analysis (DMTA)

The glass transition temperatures (T_g) of the ZnO/PU films were measured on an RSA III (TA Instruments) dynamic mechanical thermal analyzer (DMTA). Measurements were performed from - 100 °C to 150 °C at a temperature ramp of 3 °C/min, with a frequency of 1.0 Hz and a strain of 0.5 %. T_g was determined from the maximum of the tan δ peak, and was the average of three measurements.

2.3.4 Electron Paramagnetic Resonance (EPR) Spectroscopy

EPR analysis was conducted to quantify the photoreactivity of ZnO nanoparticles using radical spin traps 3-aminoproxyl (AP) (Aldrich) at 500 μmol/L in deionized water and stock preparations (0.2 g/L) of metal oxide suspensions in deionized water. For each spin trap experiment, a new mixture of spin trap stock solution and ZnO suspension was prepared to monitor the photo-initiated radicals under UV irradiation at ambient temperature. Aliquots (50 μL) of the spin trap/ZnO mixture were placed into EPR capillary tubes in the EPR cavity and irradiated in situ with a Xe arc lamp at 500 W. All EPR spectra were recorded at ambient conditions with a Bruker Elexsys E500 EPR spectrometer. Control experiments were carried out to ensure that observed EPR signals did not arise from photolysis or oxidation products of the spin traps themselves as well as to monitor the EPR signal stability in the dark and under illumination.

3. RESULTS AND DISCUSSION

3.1 Effect of Nanoparticles on Chemical and Thermo-mechanical Properties of ZnO/PU Films before UV Exposure

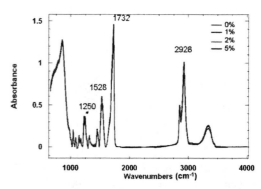

Figure 1: FTIR transmission spectra of pure PU and ZnO/PU films containing 1 %, 2 %, and 5 % ZnO loadings by mass. The specimens are spin-coated films on CaF₂ substrates.

Figure 1 displays the FTIR transmission spectra of unexposed ZnO/PU films having ZnO loadings of 0 %, 1 %, 2 %, and 5 % by mass. The ZnO/PU films having different ZnO loadings show similar absorption bands at 2928 cm⁻¹, 1732 cm⁻¹, 1528 cm⁻¹, and 1250 cm⁻¹, which are attributed to the CH₂ anti-symmetric stretching, C=O stretching in urethane and ester, N-H bending plus C-N stretching, and C-O-C stretching in ester, respectively [3, 4]. The variation in the peak intensity is due to thickness differences. The characteristic absorption of ZnO around 500 cm⁻¹ was not observed in Figure 2 due to the cut-off of the CaF₂ substrate below 750 cm⁻¹.

The effect of loadings and sizes of ZnO manoparticles on Tg of the ZnO/PU films are displayed in Table 1. A progressive increase in T_g with an increase in particle loading or a decrease in particle size was observed. Given that all systems had similar degrees of dispersion, both increase in loading or decrease in size of the nanoparticles would lead to an increase in the interfacial volume fraction for nanoparticles in the polymer matrix. The increase in T_g of the nanocomposite films suggests a reduction in the mobility of the PU polymer chains [5]. Because T_g can be used to assess the interfacial interactions between nanoparticles and polymers [6], the dependence of T_g on the loading and the size of the nanoparticles indicates that the interactions between ZnO particles and polymers matrix increased with an increase in the interfacial volume fractions in the nanocomposites.

ZnO Loading	Tg (°C)	ZnO Size	Tg (°C)
0%	93.5	None	93.5
1%	114.3	20 nm	127.4
2%	122.6	40 nm	119.8
5%	127.4	60 nm	110.8

Table 1 Effects of loadings and sizes of ZnO nanoparticles on T_g of the ZnO/PU nanocomposite films. For the system with different loadings, the size of ZnO nanoparticle is 20 nm; while for the system with different sizes, the loading of the nanoparticles is 5 %. The T_g value averaged from three specimens. The uncertainties for the data points are less than 2 %.

3.2 Role of Nanoparticles in Long-term Performance of ZnO/ PU Nanocomposites during UV Exposure

Figures 2 and 3 illustrate the effects of loading and size of ZnO nanoparticles on chemical degradation of PU films exposed on SPHERE at 45 °C, 0 % RH. The chemical degradation of the PU is typically represented by the intensity decreases of the bands at 2928 cm^{-1}, 1732 cm^{-1}, 1528 cm^{-1}, and 1041 cm^{-1}. Little change was observed for pure PU films. However, substantial changes were shown for the ZnO/PU films as a function of exposure time. The magnitude of these changes increased with an increase in loading and a decrease in size of the nanoparticles. Note that the specimens exposed to the same temperature/RH condition without UV light showed little degradation. These results indicated that the ZnO nanoparticles used in this study had a photo-catalytic effect on the degradation of PU during SPHERE exposure. This photo-catalytic effect is proposed to be an interface-related phenomenon which increases in magnitude with an increase in the nanoparticle interfacial areas. This hypothesis is further confirmed by EPR results (Figure 4), which shows the photoreactivity of

ZnO nanoparticles increased with decreasing nanoparticle size.

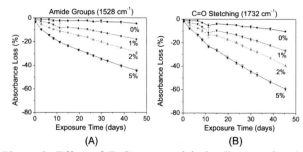

Figure 2: Effect of ZnO nanoparticle loading on chemical degradation of 20 nm ZnO/PU films during exposure on SPHERE at 45 °C, 75 % RH. (A) and (B) represent the absorbance loss of amide II band at 1528 cm^{-1} and C=O stretching band of ester groups at 1732 cm^{-1}, respectively. Each data point in this figure is the average of four specimens, and error bars represent one standard deviation.

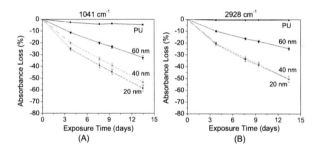

Figure 3: Effect of the nanoparticle size on chemical degradation of 5 % ZnO/PU films exposed on SPHERE at 45 °C, 75 % RH. Three sizes of ZnO particles, 20 nm, 40 nm and 60 nm, were used. (A) and (B) represent absorbance loss of C-O-C asymmetric stretching at 1041 cm^{-1} and CH$_2$ asymmetric stretching at 2928 cm^{-1}, respectively. Each data point was the average of four specimens, and error bars represent one standard deviation.

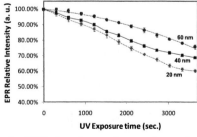

Figure 4: EPR absorbance decay plots using the EPR spin trap method for ZnO nanoparticle dispersions with different particle sizes. Error bars represent one standard deviation.

However, when UVO chamber was used as a UV exposure device, we found that the degradation rate of pure PU was the highest among the films. As shown in Figure 5, when the particle size was fixed at 20 nm, films with 5 %

loading still degraded at a faster rate than the 1 % filled system, but slower than the pure PU films. Further, all three nanoparticle sizes (20 nm, 40 nm and 60 nm) decelerated the degradation of PU films exposed to the UVO condition; that is, they acted as a UV stabilizer. Such a different phenomenon may be attributed to a different degradation mechanism for the specimens exposed to the different UV devices. The short wavelength (i.e., 254 nm and 185 nm) and the consequent atomic oxygen and ozone have caused a different degradation mechanism for specimens exposed to the UVO chamber than those on the SPHERE. In this environment, the role of ZnO nanoparticles has changed: from a photo-catalyst to a photo-stabilizer.

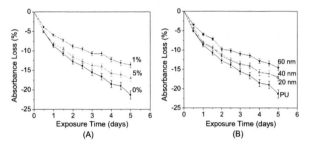

Figure 5: Effect of nanoparticle loading (A) and size (B) on absorbance loss of amide II band at 1528 cm^{-1} as a function of exposure time for ZnO/PU films exposed to UVO cleaner at ambient condition. For the system with different loadings, the size of ZnO nanoparticle is 20 nm; while for the system with different sizes, the loading of the nanoparticles is 5 %. Each data point was the average of four specimens, and error bars represent one standard deviation.

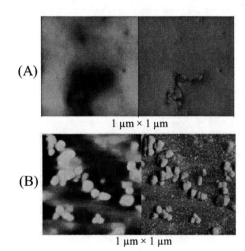

(A)

1 μm × 1 μm

(B)

1 μm × 1 μm

Figure 6: AFM height (left) and phase (right) images for ZnO/PU films exposed to (A) SPHERE at 45 °C, 0 % RH for two months, and (B) UVO chamber in ambient condition for 120 h. Lateral dimensions are 1 μm × 1 μm.

The different roles played by the ZnO nanoparticles in the chemical degradation for ZnO/PU films also resulted in

distinctive morphological changes when specimens were exposed to different UV conditions. As shown in Figure 6, for SPHERE exposure, the polymers in the regions surrounding the nanoparticles degraded faster than the bulk polymer matrix, leading to pit formation on the surface of the exposed film (Figure 6A). On the other hand, for UVO chamber exposure, the polymer regions next to the particles changed more slowly than the matrix, forming elevated domains on the film surface. It appeared that the degradation of polymer in the vicinity of the particles was preferentially accelerated when nanoparticles behave as a photo-catalyst, or was protected when nanoparticles act as a photo-stabilizer. Such differences in the local interfacial regions ultimately will result in the overall performance of nanocomposites during exposure to UV irradiation. The results of this study suggested that the nanoparticles play a critical role in the long-term performance and life cycle of nanocomposites.

4 CONCLUSIONS

The role of nanoparticles in a waterborne ZnO/PU system exposed to UV irradiation has been investigated. The effects of loading and size of ZnO nanoparticles on photodegradation of ZnO/PU films were evaluated. The results have shown that ZnO nanoparticles can act as a photo-stabilizer or a photo-catalyst in nanocomposites, depending on the exposure conditions. This study demonstrated that the role of nanoparticles in the life cycle of nanocomposites is critical but also complicated, suggesting that it is necessary to develop new methodologies to accurately predict the service life of these advanced polymer composites.

* Instruments and materials are identified in this paper to describe the experiments. In no case does such identification imply recommendation or endorsement by the National Institute of Standards and Technology (NIST).

REFERENCES

1. C. Hegedus, F. Pepe, D. Lindenmuth, and D Burgard, JCT Coatings Tech, April, 42, 2008.
2. Y.Q. Li, S.Y. Fu, Y.W. Mai, Polymer, 47, 2127, 2006.
3. T. Nguyen, et al. "Service Life Prediction: Challenge the Status Quo," Federation of Societies for Coatings Technology, Blue Bell, PA, 13, 2005.
4. J. Lemaire and N. Siampiringue, "Service Life Prediction of Organic Coatings: A Systems Approach," ACS Symposium Series 772, American Chemical Society, Oxford Press, NY, 198, 1999.
5. D.Y. Perera, Progress in Organic Coatings, 50, 247, 2004.
6. L.S. Schadler, et al. MRS Bulletin, 32, 335, 2007.

Surface tailoring of carbon fabric reinforced PES composites with PTFE filler in micro and nano- sizes

Mohit Sharma and Jayashree Bijwe[i]

Industrial Tribology Machine Dynamics and Maintenance Engineering Centre (ITMMEC)
Indian Institute of Technology, Delhi, Hauz Khas, New Delhi 110016, India

ABSTRACT

Effect of surface tailoring on mechanical and tribological properties of plasma treated carbon fabric (CF), used as reinforcement, to develop composites is reported in this paper. CF being inert does not have good adhesion with matrix. Cold remote nitrogen oxygen plasma (CRNOP) was used for surface treatment to enhance their adhesion with the matrix, polyethersulphone (PES) of higher molecular weight. Composites with twill weave carbon fabric (CF) ($\approx$67-70 wt %) were developed using impregnation technique followed by compression molding. Further, the top two layers of composites with best properties were tailored with 2wt % PTFE fillers in micron and nano-scale range. Tribo performance of all composites was evaluated on UMT tribometers by sliding of composite pin against silicon carbide (SiC) paper; improved wear resistance (W_R) and reduced Coefficient of friction (μ). Worn surfaces of the pin and disc were studied with SEM to understand wear mechanisms.

Keywords: PES-CF composites, solid lubricants, plasma treatment, wear.

1. INTRODUCTION

Carbon fabric reinforced polymers (CFRPs) are being increasingly used as advanced composites for structural materials applications, especially in aircraft industry because of very high specific strength and modulus apart from their higher thermal conductivity, corrosion resistance etc [1]. In spite of excellent properties of CF such as very high strength, modulus, thermal conductivity, lubricity etc. it suffers with serious drawback. These are inert towards matrix and show little adhesion. Various surface treatments are reported to be successful to enhance adhesion either by mechanical interlocking or grafting various functional groups which interacts with matrix and improve adhesion. Plasma treatment is a well known technique to improve adhesion by incorporating functional groups on CF [2]. It is less destructive method and allows a greater control over the number of unwanted reaction pathways. It has been reported that such treatments proved beneficial for improvement in mechanical properties of some polymers and composites and also for tribological properties in adhesive and fretting wear mode [3, 4]. However, for abrasive wear studies such efforts are not reported.

Moreover, it is well known fact that addition of solid lubricants such as PTFE in polymers and composites proves beneficial in improving friction and wear properties. However, inclusion of such inert and low surface energy filler leads to deterioration in mechanical properties significantly.

Hence it is wiser to tailor the surfaces with solid lubricants rather than bulk. Not much is reported in this regard in the literature [5]. Hence in this work, efforts were also made to tailor the surface of composite with solid lubricant PTFE in two sizes viz. micro and nano. Nano particles were tried since these are reported to be more beneficial as compared to micro-size with equal loading [6, 7].

In current work, the influence of cold remote nitrogen oxygen plasma treatment on fabric and subsequently on its PES composites on physical, mechanical and abrasive wear properties is investigated. Further efforts of surface tailoring composite by using PTFE fillers in micron and nano-scale range and their subsequent influence on abrasive wear performance are also reported.

2. FABRICATION OF COMPOSITES

PES polymer GAFONE B3500 was used as matrix material. (properties as shown in Table 1).

Melt Flow Index (MFI) (g/10min) ASTM D 1238	Weight-average molecular (M_w)*	Polydispersiy (M_w/M_n)
8.5 ± 0.2	39,825	2.64

*Supplier's data

Table 1. Melt flow rate (MFI) and molecular wt of PES.

Carbon fabric of twill weave supplied by Fiber Glast USA was used as reinforcement for development of composites by impregnation technique. Cold remote nitrogen oxygen plasma (CRNOP) was employed for surface treatment of chemically inert CF to enhance its reactivity towards matrix materials [8].

Dichloromethane (CH_2Cl_2) was used as a solvent to prepare the solution of PES (20 wt %). Twenty fabric plies were then immersed for 12 hrs in a sealed steel container filled with viscous solution of PES in CH_2Cl_2. The prepregs were taken out carefully to avoid the disturbance in weave and dried in oven for 1 hour at 100^0C in a stretched condition. Twenty such prepregs were then stacked in the mould carefully to avoid misalignment. Compression molding was done at pressure of 7.3 MPa, temperature of 380-390⁰C. The composites were then cooled in compressed condition and then cut with the help of diamond cutter as per ASTM standards for different mechanical and tribological characterizations.

The developed composites were designated as follows : C_U, C_T (subscripts U and T denotes untreated and plasma treated fabrics, respectively), P_N denotes virgin PES polymer and S_{tef} denotes the composite tailored with PTFE fillers.

2.1 Surface Tailoring

Surface tailoring of one composite (C_T, composite with best performance) was done separately by developing top two layers of composites with 2 wt % PTFE fillers (12μm and 30 nm). Prior to impregnation the fillers were well dispersed in solvent using high frequency probe sonication for 20 minutes.

3. CHARACTERIZATION OF COMPOSITES

The physical and mechanical characterizations of all composites as per ASTM standards are shown in Table 2.

Properties	P_N	C_U	C_T
Fiber weight (%) (ASTM D2584)	-	67.5	71.5
Void fraction (vol.) (ASTM D2734)	-	0.47	0.44
Density (g/cm³) (ASTM D792)	1.37	1.52	1.58
Tensile strength (MPa) (ASTM D638)	83	744	747
Tensile modulus (GPa) (ASTM D638)	3.22	65	74
Strain at break (%) (ASTM D638)	>60	1.1	1.06
Toughness (MPa) (ASTM D638)	3.2	4.1	4.4
Flexural strength (MPa) (ASTM D790)	112	692	736
Flexural modulus (GPa) (ASTM D790)	2.8	54	63
ILSS (MPa) (ASTM D2344)	-	42	46

Table 2. Physical and mechanical properties of CF-PES composites reinforced with virgin and CRNOP treated CF.

3.1 Abrasive wear studies

Abrasive wear studies were carried on tribometer (UMT-3MT), CETR, USA. The tests were carried out in Pin-on-Disc configuration (Fig. 1) on S33HE high torque rotational motion drive loading with 1-100N dimensional force sensor

with the resolution of 5mN in a single pass rotational motion.

For bedding purpose, square sized composite pin (10 mm × 10 mm × 3.5 mm) was abraded against silicon carbide (SiC) abrasive paper of 800 grade for uniform contact. The bedded sample after cleaning ultra centrifugally followed by drying was fitted in a holder and was then abraded against SiC abrasive paper of 120 grade (grit size ~118 μm) fixed on the rotary drive with the speed of 8 rpm. The total sliding distance of 1.15 m was achieved by abrading the pin in 5 rotations.   Experiments were conducted at different loads (10, 20, 30 and 40 N). After the experiment, pin was again cleaned, dried and weighed with an accuracy of 0.0001g. The experiment was repeated for two times and the average value of weight loss (difference in weight) was used for specific wear rate calculations.

The average coefficient of friction (μ) was recorded with respect to time with UMT test viewer software and average value of μ from the data on five tracks recorded for each experiment was considered. The specific wear rate K_0 was calculated from the following equation;

$$K_0 = \Delta m / \rho \, L \, d \; (m^3/Nm)\ldots\ldots\ldots (1)$$

Where Δm is the weight loss in kg, ρ the density in kg/m³, L the load in N and d the distance abraded in m. The fabric was always parallel to the abrading plane and warp fibers were parallel to the sliding direction.

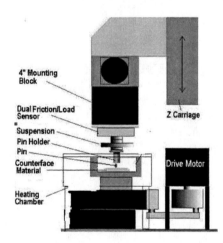

Figure 1. Schematic of Pin-on-Disc Configuration

4. RESULTS AND DISCUSSION

Mechanical properties of pristine polymer and composites are shown in Table 2. High mol wt polymer showed slightly superior properties. Composites based on treated fabric showed better properties than those with virgin/untreated CF. confirming enhancement in fiber-matrix adhesion due

NSTI-Nanotech 2010, www.nsti.org, ISBN 978-1-4398-3401-5 Vol. 1, 2010

to plasma treatment. Most of the properties such as ILSS, flexural strength and modulus, and tensile modulus showed improvement due to enhanced fiber matrix bonding. Specific wear rates (K_0) and Coefficient of friction (μ) as a function of load for selected composites are plotted in Fig. 2 (a, b). Following were salient observations.

- Both wear rate and μ decreased with load for all selected materials.
- CF reinforcement led to significant decrease in friction and wear, both.
- Treated fabric composites led to better performance.
- Surface tailoring by PTFE proved to improve both friction and wear performance in abrasive wear mode, though not significantly.
- There was marginal difference in the performance of micro and nano-sized PTFE when placed on the surface, nano being slightly more effective. Improper dispersion of nano-filler could be one of the reasons for this behavior.

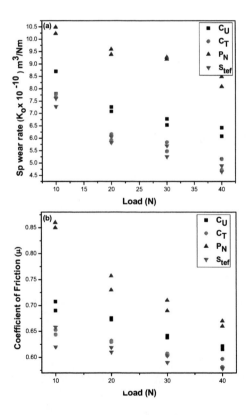

Figure 2. Specific wear rate (K_0) and coefficient of friction (μ) as a function of load for pristine polymer, P_N; composites C_U and C_T and surface tailored composite $S_{tef.}$

It was observed that μ decreased gradually with load from 10 to 40N for all the composites (from 0.9 to 0.5), which is a common trend for polymer composites [9]. At 10N load, the highest μ was recorded for virgin PES (0.84) followed by C_U (0.69), C_T (0.63) composites and surface tailored composite (0.6).

CF reinforcement (untreated and plasma treated) significantly decreased the specific wear rates. At 40N load, the K_0 reduced almost two times due to treated CF reinforcement. When the performance of treated and untreated fabric composites was compared, enhancement was around 15 % in wear resistance (W_R) due to treatment.

The trends confirmed that the plasma treatment has beneficially improved the W_R of composite and are in accordance to the mechanical strength properties.

K_0 decreased with increase in load in general and was in accordance to the Lhymn's equation. The increase in W_R due to reinforcement could be due to lubricating effect of CF [10].

4.1 SEM Studies

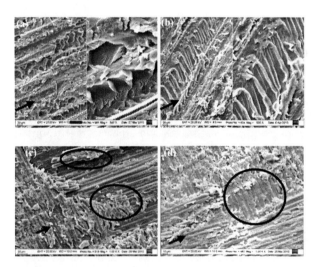

Figure 3. SEM micrographs (X1000) of abraded surfaces of composites worn under loads (a) P_N at 10N, (b) P_N at 40N (c) C_U at 10 (d) for C_T at 10N.

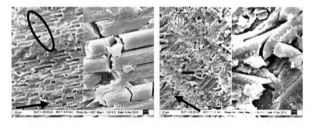

Figure 4. SEM micrographs (X1000) of abraded surfaces of composites worn under 40N load showing various failure mechanisms; (a) Micro cutting and multiple fractures to the fibers lying in perpendicular direction of abrasion; (b) fiber fracture/slicing-lengthwise, micro cracking and debonding.

SEM micrographs of the groves of virgin polymer P_N under extreme loads (10 and 40 N) are shown in Fig. 3 (a and b).

Groves of fatigue striations were observed. Such feature are reported for thermoset polymers during abrasion [11]. For amorphous polymer such as PES, failure is more by fatigue cracks (parallel multiple cracks in the direction perpendicular to stress). The wide furrows in the direction of sliding due to micro-ploughing are also visible. The extent of micro-ploughing by SiC particles and plastic deformation which was the main cause for wear was load dependent. As seen in Fig 3(b), the grooves are deeper and wider. Effect of treatment on fiber matrix adhesion can be seen in Fig. 3(c) and 3(d). Fig. 3(c) shows the less fiber matrix adhesion and most of the fibers are almost without any appreciable amount of matrix. On the contrary, treated fibers show good adhesion. This enhanced adhesion was responsible for less damage to the fibers and hence less wear. Fig. 4(a, b) shows various failure mechanics of fibers in worn composites.

5. CONCLUSIONS

Based on the studies on composites developed with carbon fabric treated with cold remote nitrogen plasma, following conclusions were drawn.

- CF reinforcement showed significant enhancement in all mechanical properties except elongation to break.
- Plasma treatment of CF proved beneficial for improvement in fiber-matrix adhesion leading to more enhancement in mechanical properties.
- CF reinforcement proved beneficial for reducing friction and wear of PES significantly. Treatment of fabric proved more beneficial in this aspect also.
- Surface tailoring of composite is a new technique and offers best combination of mechanical and tribological properties apart from saving the cost of solid lubricant which would have been otherwise required in the bulk. However, in case of abrasive wear, though some improvement was observed due to micro and nano-sized PTFE particles on surface, it was not so significant. Perhaps in adhesive wear mode, the effect would be manifold.

ACKNOWLEDGMENTS

Authors are grateful to the Council of Scientific and Industrial Research (CSIR), New Delhi, India for funding the work reported in this paper. Authors are also grateful to Prof. Brigitte Mutel from University of Lille, France for offering facility of plasma treatment to fabric and Mr. Sanjay Charati & Mr. Atul Raja from Solvay Advanced Polymers for providing PES material for this research.

REFERENCES

1. K. Friedrich, (ed.) "Advances in Composite Tribology", Composite Materials Series 8, Elsevier Amsterdam, 1993.

2. R Li., L. Yea and W. Mai, "Application of plasma technologies in fiber-reinforced polymer composites: A review of recent developments" Composites Part A. 28, 73-86, 1997.

3. F. Sua, Z. Zhang, K. Wang, W. Jiang and W. Liu, "Tribological and mechanical properties of the composites made of Carbon Fabrics modified with various methods" Composites Part A. 36, 1601–1607, 2005.

4. M. Sharma, S. Tiwari and J. Bijwe, "Optimization of materials parameters for development of Polyetherimide composites" Materials Sci.& Engg. B, Article in press, Available online November 2009.

5. J. Bijwe, W. Hufenbach, K. Kunze and A. Langkamp, " Polymeric Composites and Bearings with Engineered Tribo-surfaces" In: K. Friedrich and A. K. Schlarb (ed.) "Tribology of Polymeric Nano-composites" Tribology and Interface Engineering Series, Elsevier B. V., 55, 483-500, 2008.

6. K. Friedrich, S. Fakirov and and Z. Zhang Editor, "Tribological Characteristics of Micro and Nanoparticle Filled Polymer Composites" Polymer Composites, Springer US, 2005.

7. J. Bijwe , J. J. Rajesh , A. Jeyakumar and A. K. Ghosh, "A study of the role of solid lubricant and fibrous reinforcement in modifying the wear performance of polyethersulphone" J. of Synthetic Lubn, 17(2), 99, 2006.

8. B. Mutel, M. Bigan and H. Vezin, "Remote nitrogen plasma treatment of a polyethylene powder; Optimization of the process by composite experimental designs" Applied Surf. Sci., 239, 25–35, 2004.

9. J. Bijwe, R. Rattan, M. Fahim, "Abrasive wear performance of carbon fabric reinforced Polyetherimide composites: Influence of content and orientation of fabric" Tribo Intl, 40, 844–854, 2007.

10. C. Lhymn, K. E. Templemeyer and P. K. Davis, "The abrasive wear of short fiber composites" Composites. 16(2), 127-136, 1985.

11. B. K.Satapathy, "Performance Evaluation of Non-asbestos Fibre Reinforced Organic Friction Materials" Ph.D thesis, IIT Delhi, 2003.

[i] Corresponding author; e-mail: jbijwe@gmail.com, jbijwe@hotmail.com; Telefax: 91-11-26591280

CHARACTERIZING PARTICLE EMISSIONS FROM BURNING POLYMER NANOCOMPOSITES[&]

Marc R. Nyden, Richard H. Harris, Yeon Seok Kim, Rick D. Davis,
Nathan D. Marsh and Mauro Zammarano

Building and Fire Research Laboratory, National Institute of Standards and Technology, 100 Bureau
Drive, Mail Stop 8652, Gaithersburg, MD 20899, marc.nyden@nist.gov

ABSTRACT

The use of nanoscale materials as performance additives in polymers may pose significant health and environmental risks. Although it is unlikely that nanoadditives encapsulated in polymers will be released during normal use, there is the potential for nanoparticle aerosolization when these materials are exposed to heat and fire (either accidentally or during incineration). Furthermore, the nanoparticle morphologies and chemical compositions generated during the combustion of nanocomposite materials may be vastly different than the structures of the pristine nanoadditives due to oxidation and/or interactions with other decomposition products. The potential health and environmental impact of these morphological transformations are unknown.

In an effort to gain a better understanding of the potential hazards associated with the commercialization of polymer nanocomposites, we have undertaken an investigation into the nature of the particles released when these materials are burned. The first question we hope to answer is whether significant amounts of nanoscale additives are released (in addition to soot, which is a ubiquitous byproduct of gas phase combustion) when polymer nanocomposites are burned.

Keywords: carbon nanofibers, fire, polyurethane foam, nanocomposites.

1   INTRODUCTION

Engineered nanoparticles, such as carbon nanotubes (CNTs) and nanofibers (CNFs) are increasingly being used as fire retardants and performance additives in polymeric materials. However, because of their small size and ability to interact with biological molecules, these nanoadditives may pose significant health and environmental risks if they are released into the environment [1,2]. Although it is unlikely that encapsulated nanostructures will be released when the materials containing them are used in protected, indoor environments, this risk becomes more significant when these materials are exposed to fire (either unintended or during incineration) or other forms of intense energy (e.g., sunlight) when they are disposed of in landfills. The nanoparticulate morphologies generated in this way may be vastly different than those adopted by the pristine

nanostructures due to agglomeration and interactions with other decomposition products and may pose significant new health risks [3].

In an effort to gain a better understanding of the potential hazards associated with the commercialization of polymer nanocomposites, we have undertaken an investigation into the nature of the particles released when these materials are burned. Some of the questions we hope to answer are: 1) Do nanocomposite materials release significant amounts of nanoparticles (in addition to soot, which is a ubiquitous byproduct of gas phase combustion) when they are burned? 2) If so, under what circumstances are these nanoparticles emitted and 3) Are the size distributions, morphologies, and chemistries of the released nanoparticles different from what they were in the nanocomposite?

2   MATERIALS AND METHODS

2.1   Polyurethane foam (PUF) with and without (3.9 %) oxidized CNFs was prepared as described in reference 4. Specimens measuring approximately 76 mm x 76 mm x 10 mm thick were burned in an NBS smoke density chamber in the horizontal orientation under an incident heat flux of 50 kW/m2; in accordance with the procedure specified by NFPA 270 (or equivalently, ISO 5659-2) [5]. Particulate emissions were collected by re-circulating at 1 L/min the smoke generated from the burning PUFs through an assembly containing a 2.0 μm teflon membrane filter. Samples of the emitted particulates were suspended in de-ionized water (by sonicating for 30 minutes in a Branson 2210 sonicator*) and deposited on glass slides, which were examined by optical microscopy. Samples of the non-volatilized particulates (char) were similarly suspended in de-ionized water and subjected to microscopic study.

2.2   Measurements were made to determine whether CNFs were released when the char left after burning the CNF-containing PUF was mechanically disturbed. Samples of CNF chars were placed in test tubes and agitated using a mechanical vibrator. The atmosphere above the char was sampled continuously with a TSI Model 3007 condensation particle counter.

2.3   The mass of CNFs released from the char was also quantified. Air was circulated through the test tube during mechanical agitation of the char and bubbled into a trap

---

containing deionized water. The contents from the trap were transferred to separate containers at 30 minute intervals over a total of 3 hours. Each of these (10) time resolved suspensions was treated by adding deionized water and sodium dodecyl sulfate (SDS) to obtain a volume of 100 ml containing 0.23 % SDS by mass. These suspensions were then sonicated using a Sonics VCX 130 ultrasonic processor with a 13 mm probe for 1 hour at 50 W to exfoliate and suspend the aerosolized CNFs. Absorbance spectra of these suspensions were measured in the UV-VIS region extending from 185 nm to 1800 nm (over a 1 cm path length) using a Perkin Elmer Lambda 950 spectrometer. The concentration of CNFs in the time resolved suspensions were obtained by inversion of a linear fit (Beer's law) generated before hand from absorbance measurements made on a series of CNF suspensions of known concentration (figure 1).

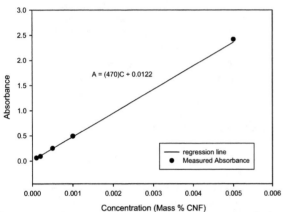

$$A = (470)C + 0.0122$$

Figure 1. Plot of the measured absorbance (at 267 nm) versus concentration for a series of CNF suspensions of known concentrations.

3 RESULTS AND DISCUSSION

In the absence of added CNFs, PUF produces little if any char when it is burned. However, when the CNFs are well dispersed in PUF they form an entangled network that prevents the collapse of the foam during burning, thereby reducing dripping and flammability. This network of CNFs is left as char at the conclusion of the burn [4]. Figure 2a is an optical micrograph obtained by imaging a suspension of a representative sample of these chars. The CNFs are clearly visible appearing more elongated than the spherical structures, which are typical of soot formed during gas phase combustion. These elongated structures are not apparent in the optical micrographs obtained from the smoke generated during the burning of CNF-containing (figure 2b) and CNF-free (figure 2c) foams. In fact, the micrographs of the volatilized component (smoke) from both foams look similar in that they are dominated by spherical structures which we attribute to soot. These observations suggest that the CNFs, which are present in the both the unburnt foam and nonvolatilized char, are destroyed in the flames (presumably, soot is destroyed as

well, but it is also generated in the gas phase combustion process). The fact that the CNFs decompose at temperatures attained in the flames was verified by heating slides prepared with the char of the CNF-containing foam. The elongated structures disappeared after heating about a minute at a temperature of about 650°C (in air), which is far below the temperatures attained in the flames above the sample, which are probably in excess of 1200 ^{O}C. The next step will be to investigate the effects of changing the combustion conditions from well-ventilated to under-ventilated, which should reduce the destructive efficiency of the flames thereby increasing the likelihood that CNFs will be released into the environment.

Figure 2. Optical micrographs of the nonvolatized char from the CNF-containing foam (a) and smoke from the CNF-containing (b) and CNF-free PUFs (c). Note the presence of the fibers in micrograph a and the absence of these structures in b and c.

The possibility that CNFs might be released when chars left behind after burning CNF-containing PUF are mechanically disturbed was also investigated. Samples of these CNF chars were placed in enclosed plastic test tubes and agitated using a mechanical vibrator. UV-VIS analysis of the suspensions obtained from the trap indicated that the rate of aersolization of the CNFs was approximately 1.4 mg/hour. The particle count measurements made during agitation of a representative char sample is compared to the background obtained in the absence of char in figure 3. The maximum peak count from agitation of the char (~10,000 counts/cm3) is an order of magnitude larger than the background signal (due to particles already present in the air), suggesting that significant amounts of sub-micron particles are released in this way. Since these chars are comprised almost exclusively of CNFs, we assume that they are responsible for the observed particle counts. However, we have not yet determined the state of aggregation of the emitted CNFs in time to include results in this report. We expect to complete our investigation of the size distributions and morphologies of the emitted CNFs in time for presentation at the meeting in June. 2010.

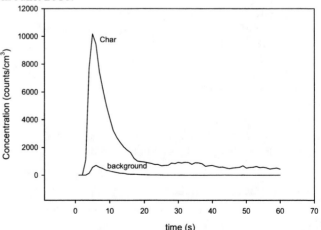

Figure 3. Comparison of particle counts measurements made in a plastic tube after disturbing the residual char to the background signal obtained in the absence of char.

4 CONCLUSIONS

Microscopic analyses of the particulate emissions produced in burning CNF containing PUF under well-ventilated conditions indicates that the CNFs were effectively destroyed in the flames. However, high levels of nanoparticles were detected when the residual char was disturbed by air currents or mechanical forces. Thus, it appears that the major hazard for CNF exposure during well-ventilated burning arises from agitation of the residual char, rather than from the fire smoke.

REFERENCES

1. Nanoscience and Nanotechnologies: Opportunies and Uncertainties; The Royal Society & The Royal Academy of Engineering; July, 2004; http://www.nanotec.org.uk/report/Nano report 2004 fin.pdf.

2. The National Nanotechnology Initiative Strategy for Nanotechnology-Related Environmental, Health, and Safety Research, Subcommittee on Nanoscale Science, Engineering, and Technology, Committee on Technology, National Science and Technology Council; Feb. 2008; http://www.nano.gov/NNI_EHS_Research_Strategy .pdf.

3. Günter Oberdörster, G., Maynard, A., Donaldson, K., Castranova, V., Fitzpatrick, J. Ausman, K., Carter, J., Karn, B., Kreyling, W., Lai, D., Olin, S., Monteiro-Riviere, N., Warheit, D., Yang, H.; Part. Fibre Toxicol. 2:8; doi:10.1186/1743-8977-2-8 [Online 6 October 2005].

4. Zammarano, M., Krämer, R.H., Harris, R., Ohlemiller, T.J., Shields, J.R., Rahatekar, S.S., Lacerda, S., Gilman, J.W.; Polym. Adv. Technol. 19; 588-595; 2008.

5. NFPA 270: Standard Test Method for Measurement of Smoke Obscuration Using a Conical Radiant Source in a Single Closed Chamber

New method for the characterization of abrasion-induced nanoparticle release into air from nanomaterials

L.Golanski, A.Guiot, D.Braganza and F.Tardif

CEA-Liten, 17 rue des Martyrs, F38054 Grenoble Cedex09, France

ABSTRACT

A new method for the characterization of abrasion induced nanoparticle release into air from nanomaterials using a normalized Taber equipment has been developed. The influence on nanoparticle releases of the abrasion process conditions and the tool type were investigated. Concentrations of nanoparticles released in the air higher than that of micrometric particles were measured by an Electrical Low Pressure Impactor (ELPI): up to 10000 particles/cm^3. The released nanoparticles were then collected on Transmission Electron Microscopy (TEM) grids and filters for further analysis in order to determine if nanoparticles are leached by abrasion in free or agglomerated form. At this stage, Cu nanoparticles were observed nearly detached from a Polymethyl Methacrylate (PMMA) nanomaterial using an SiC grid paper P1200 (ISO classification). This confirms the interest to develop such a method. No standard method for the characterization of nanoparticle release is currently available.

Keywords: Taber method, abrasion, nanoparticles, CNT, release

1. INTRODUCTION

Nanotechnologies are emerging around the world. Nanoparticles are finding new industrial applications every day. Today we are witnessing the advent of a new step of the industrial history of nanoparticles as nanoparticles developed in laboratories are about to be manufactured in mass production. Thus, economists are talking about the emergence of a new industry for the 21st century which could rank alongside automobile and microelectronics industries in terms of turnover as soon as 2015. Nevertheless, this new industry can develop dynamically only if the safety issues are solved and this for the whole life cycle of the nano products: from fabrication to the end of life through usage. Due to the variety of nanoparticles, toxicology studies will take time to define dangerousness of every kind of nanoparticle. Today in the absence of reliable data on the toxicity of nanoparticles, the only way to ensure that nanomaterials are not dangerous is to verify that they do not release their nanocharges. No standard method for characterization of nanoparticle release is currently available.

A test bench capable of measuring the release of nanoparticles from nanomaterials using a linear normalized Taber equipment has been achieved in the frame of Nanosafe 2 project [1], [2]. The purpose of this bench is to quantify whether nanomaterials after abrasion leach particles in free or agglomerated form. Our first results showed that only a few released nanoparticles were detected from nanoproducts by abrasion [2]. Another study proposes a similar method using a circular Taber equipment instead of a linear Taber [3]. This work shows that the total number of submicrometric particles or nanoparticles generated by abrasion was extremely low [3].

In this paper, efficiency of standard and non-standard abrasion tools to aerosolize nanoparticles from polymer matrix was quantified. The most efficient tool to aerosolize the nanodust by abrasion was searched. The release of nanoparticles from PMMA containing 10% wt Cu nanoparticles and from polycarbonate containing 3% wt carbone nanotubes (CNT) were quantified and analyzed.

2. EXPERIMENTAL METHOD

All manipulations are performed in a sealed glove box equipped with a HEPA filter in order to limit the initial presence of particles in the atmosphere. The background noise in the sealed glove box is quantified less than 10 part./cm^3. These special conditions are obtained by creating a vacuum at the top of the box, that will suck the air rich in particles at a rate of 150 l / min and a clean air is obtained using an absolute filter placed in the bottom of the glove box. Nanomaterials are mechanically solicited by abrasion using a standardized linear Taber equipment. The tool is rubbed on the sample in order to generate a dynamic friction.

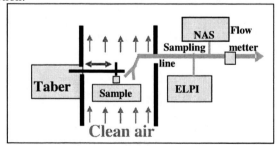

Fig. 1 Schematic of the measurement tool.

The friction is linear and cyclical on the surface of the sample. The cycle speed, number of cycles are parameters that can be modified for the abrasion process. The normal force applied can be modified as well adding mass on the sample.

The characterization of the particles released by the abrasion process requires the detection of the particle concentration and the number size distribution. One instrument was used for measuring the number size distribution: an ELPI. The ELPI allows the determination of particle number distribution in the range of 7 nm to 10 μm. This device is capable of achieving real-time measurements and displays, for each size class, changes in concentration over time. It also allows depositing released particles on filters after putting them on impactors. These filters (hydrophilic polycarbonate membrane) are then analyzed by scanning electron microscopy (SEM) to observe the morphology and size of the collected particles. Thus, it is possible to measure the size distribution and to perform physical-chemical characterization of the collected particles.

An electrostatic nanoaerosol sampler (NAS-Nanometer Aerosol Sampler,model 3089, TSI Incorporated, USA) was used in parallel with the ELPI. The released particles were collected on TEM grids in order to determine if nanoparticles are leached by abrasion in free or agglomerated form and for an elemental identification of the nanoparticles of interest by EDX analysis (Energy dispersive X-ray spectroscopy).

3. RESULTS

3.1 Influence of abrasion conditions on nanoparticle release

First, the influence of the abrasion process conditions on nanoparticle release were investigated. All the measurements shown here were performed on polycarbonate sample containing 3% wt CNT.

The abrasion was performed using a steel brush and the speed of the tool generating friction on the sample was increased. As shown on the Fig. 2 the abrasion effect is enhanced when the abrasion speed increases.

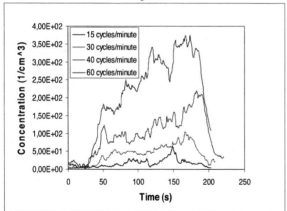

Fig. 2: Concentration of particles at different abrasion speed measured with an ELPI.

The concentration of released particles reaches a value of 370 part./cm^3 for a speed of 60 cycles per minute by comparison with 50 part./cm^3 particles for a speed of 15 cycles per minute. Further, all experiments were carried out with the maximum abrasion speed, 60 cycles per minute.

In a second step the abrasion was performed using a steel brush and the normal force applied on the sample was increased. As shown on the Fig.3b the abrasion effect is enhanced when increasing normal force without changing the size distribution of released particles (Fig.3a, 3b). Up to 1800 part/cm^3 nanoparticles were measured using an ELPI, the maximum number of nanoparticles were measured at stage 2 (Fig 3b).

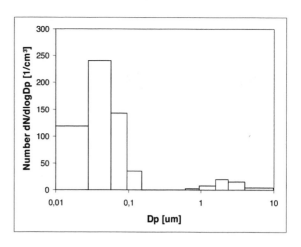

Fig.3a Effect of applied force (m = 1200g) on particle distribution.

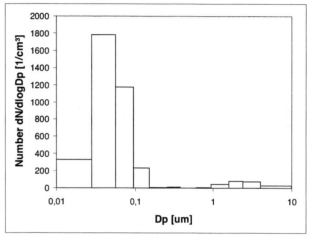

Fig.3b Effect of applied force (m = 2100 g) on particle distribution.

3.2 Quantification of nanoparticle release using different tools

An efficient tool to aerosolize the nanodust by abrasion was searched. Standard and non-standard tools were quantified. PMMA containing up to 10 wt % nanoparticles was investigated. Polycarbonate samples containing 3% wt CNT was investigated as well. Normalized standard tools efficiency- abrasive ribbons of SiC -were tested (ISO 7784-1). Non standard steel brushes efficiency has been evaluated as well.

3.2.1 Abrasion using a non-standard tool brush

First abrasion test was performed using a steel brush tool. Polycarbonate containg 3% wt of CNT was investigated. Higher concentrations of nanoparticles than micrometric particles were released in the air (Fig. 4).

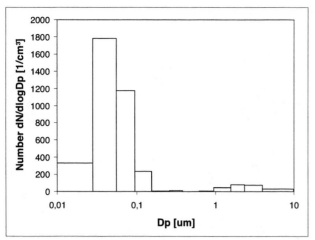

Fig.4 Size distribution from polycarbonate polymer containing 3% wt CNT (Carbone Nanotubes) abraded using a steel brush. A normal force was applied (m = 2100g).

Significant release of nanoparticles was detected: up to 1800 particles/cm³ were measured at stage 2 of the ELPI.

3.2.2 Abrasion using a standard tool

Polymethyl Methacrylate PMMA polymer containing 10 % wt of nanoparticles with a mean size around 40 nm was investigated. The influence of the type of abrasive paper (which can also simulate a sanding) on the release of nano particles in the air was quantified. Silicon carbide-type abrasive paper P120 and P1200 with variable grit size (from 125μm up to 15.3μm) were used. The influence of the abrasive grit size on the number size distribution was estimated.

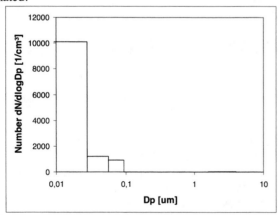

Fig.5a Size distribution from PMMA polymer containing wt 10% Cu nanoparticles abraded using a P1200 SiC abrasive paper. No normal force was applied.

Using a P1200 paper, up to 10000 particles/cm³ were measured at stage 1 of the ELPI.
Using a P120 paper with abrasive grains larger than in the case of P1200, smaller concentrations of nanoscale objects are detected while larger concentrations of sub micron objects are issued.

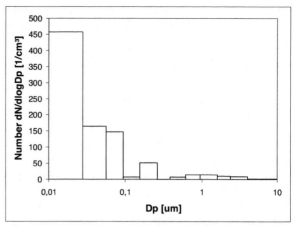

Fig.5b From PMMA polymer containing 10% wt Cu nanoparticles abraded using a P120 SiC abrasive paper. No normal force was applied.

20 time less nanoparticles can be observed on channel 1 of the ELPI in this case by comparison with results obtained using P1200 abrasive paper .
The type of abrasion tool has a high impact on the quantity of nanoparticles releases in the air from the polymer.

3.3 Analysis of released particles from samples

In order to identify which kind of particles are released, analysis on TEM grids are performed. The dust was deposited on a TEM grid using the NAS and on filters on different stages of the ELPI (as described in the section 2).

In a first step the presence of free or agglomerate nanoparticles or CNT and polymer embedded with nanoparticles were analyzed by TEM and SEM. In a second step the samples were examined with EDX (energy dispersive X-ray spectroscopy) for an elemental identification and the presence of Cu nanoparticles was confirmed.

3.3.1 Polycarbonate containing 3% wt CNT

Nanoparticles emitted from polycarbonate samples containing wt 3% CNT by abrasion was investigated. These particles were collected on the filter in stage 2 of the ELPI and analyzed by SEM. The particle number distribution performed on this sample is showed on fig. 4.

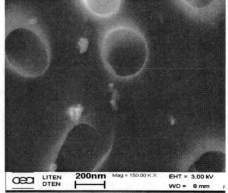

Fig.6 SEM observation of detected particles on filter out on the ELPI channel abraded using a steel brush.

At this stage particles with aerodynamic diameter between 28 and 56 nm are deposited (Fig 3b).

The presence of nanometric polymer particles on the filter can be observed. No free NTC was observed. These observations were also confirmed by TEM.

3.3.2 PMMA containing 10% wt Cu

Nanoparticles emitted by abrasion from polycarbonate samples containing 10% wt Cu was investigated further by TEM (Fig.7 and 8). No visible free Cu nanoparticles were observed when the surface was abraded using a brush. Micrometric and submicronic polymer particles were visible on TEM grids. The TEM image proves that Cu nanoparticles are still embedded in the polymer (Fig 7).

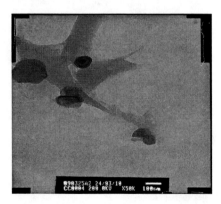

Fig.7 Observation of abraded particles on TEM grids.

When the surface was abraded using a P1200 SiC paper, nanoparticles nearly detached from PMMA polymer were observed (Fig.8).

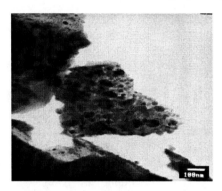

Fig.8 Observation of abraded particles on TEM grids.

In a second step Cu nanoparticles were identified by EDX (Fig.9). The size of nanoparticles from TEM image corresponds to the size expected and given by the manufacturer.

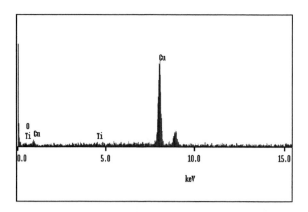

Fig.9 EDX graph of abraded polymer and nanoparticles identified on Fig.8.

4. CONCLUSION

The abrasion process itself was optimised to give the highest release and then the highest aerosolization of the nanodust.

- Increase of nanodust was obtained by increasing the abrasion speed and the normal force on the sample.

- The sandpaper P1200 friction on PMMA containing 10% wt Cu promotes the formation of nanoscale dust in higher quantity than P120.

- Abrasion performed on polycarbonate containing CNT and PMMA containing Cu using stainless brush produced nano-sized dust but no isolated CNTs or Cu nanoparticles.

- In the case where PMMA was abraded using an SiC grid paper P1200, free Cu nanoparticles were nearly detached.

At this stage standard SiC standard paper seems to be efficient to detach Cu nanoparticles from PMMA polymer. This confirms the interest to develop such a method.

[1] A. Guiot, L. Golanski and F. Tardif, "Measurement of nanoparticle removal by abrasion", Proceedings of the Nanosafe 2008 Conference, Grenoble, France, www.nanosafe2008.org, 2008.
[2] F.Tardif, A. Guiot, L.Golanski,"Measurement of nanofiller removal by abrasion", Proceedings of the Nanotech 2009 Conference, Houston, USA.
[3] M. Vorbau, L. Hillemann, M. Stintz, "Method for the characterisation of the abrasion induced nanoparticle release into air from surface coatings", J. Aerosol Science, Vol 40, Issue 3, P 209-217, 2009.

Acknowledgments

The authors would like to thank C.Cayron for the TEM and SEM measurements and to P.Tiquet for the fabrication of some polymers.

Direct Evidence of Nanoparticle Release from Epoxy Nanocomposites Exposed to UV Radiation

T. Nguyen, B. Pellegrin, C. Bernard, X. Gu, J. M. Gorham, P. Stutzman, A. Shapiro, E. Byrd, and J. Chin

National Institute of Standards and Technology, Gaithersburg, MD 20899

ABSTRACT

This study assessed the fate of SiO_2 nanoparticles (nanoSiO$_2$) in epoxy/nanoSiO$_2$ composites exposed to ultraviolet (UV) radiation. The matrix was a stoichiometric mixture of a diglycidyl ether of bisphenol A epoxy and an aliphatic tri-amine. Films of unfilled and 5 mass % SiO_2-filled amine-cured epoxy were exposed to 75 % relative humidity (RH), 50 °C, and (295-400) nm UV radiation. Photodegradation, mass loss, surface morphology of the exposed composites, and composition of the released particles were characterized. Amine-cured epoxy/nanoSiO$_2$ composites underwent rapid photodegradation, resulting in substantial mass loss, accumulation of SiO_2 nanoparticles on the composite surface, and release of SiO_2 nanoparticles. The results of this study will provide useful information to assess the potential risk of nanoSiO$_2$ in epoxy nanocomposites during outdoor use.

Keywords: epoxy, life cycle, nanocomposites, nanoparticles, release, UV.

1 INTRODUCTION

Polymer nanocomposites are being used or potentially will be used in large volumes in a wide variety of applications (1,2). Whatever the application, both the long-term performance of the nanocomposite itself and the fate of the nanoparticles in the polymer matrix play a key role in the acceptance and commercialization of these advanced products. This is because the matrix in a polymer nanocomposite undergoes degradation during service and post-service, potentially releasing nanoparticles into the environment via the effects of mechanical vibration/abrasion, rain, condensed water, and wind. Since nanoparticles have shown potential risks to human health and environment (3,4), their release during the life cycle of polymer nanocomposites could present a roadblock to innovation and commercialization of polymer nanocomposites. However, little data is available on the environmentally-induced degradation of polymer nanocomposites, the state of the embedded nanoparticles during exposure, or how they may be released during service. The lack of this type of information hinders our ability to understand the release mechanisms, predict the long-term release behavior, and develop strategies to mitigate this potentially serious problem. This study investigated the fate of SiO_2 nanoparticles embedded in an epoxy matrix exposed to ultraviolet (UV) radiation. Results showed that, when exposed to UV radiation having wavelengths similar to those of the sunlight, the epoxy matrix underwent photodegradation and subsequent mass loss, accumulation of a large amount of SiO_2 nanoparticles on the composite surface, and release of SiO_2 nanoparticles into the environment.

2. EXPERIMENTAL PROCEDURES

2.1 Materials and Nanocomposite Preparation

The matrix was a model amine-cured epoxy commonly used for protective coatings and fiber-reinforced polymer composites. It was a stoichiometric mixture of a diglycidyl ether of bisphenol A epoxy having an equivalent mass of 189 (grams of resin containing one gram equivalent of epoxide) and a tri-polyetheramine curing agent. After curing, this amine-cured epoxy forms a crosslinked network structure. The SiO_2 nanoparticles (nanoSiO$_2$) had an average diameter of 7 nm and were surface-treated with a silane material, whose composition and coverage on the nanoSiO$_2$ surface were not known. Reagent grade toluene was used for composite processing.

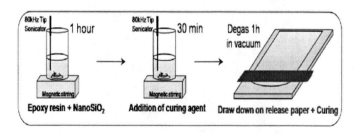

Figure 1. Process used to prepare the amine-cured epoxy/nanoSiO$_2$ composites.

Free-standing epoxy films containing 5 % mass fraction of SiO_2 nanoparticles having a dry thickness between 125 μm and 175 μm were prepared according to the steps shown in Figure 1. SiO_2 nanoparticles were first dispersed in a large amount of toluene using a tip sonicator. After adding epoxy resin, the nanoSiO$_2$ suspension was sonicated under constant stirring for an additional 1 h. Curing agent was added to the suspension, and the mixture was sonicated for 30 minutes. NanoSiO$_2$/amine/epoxy mixture was degassed for 1 h in vacuum at room temperature, followed by drawing down on a polyethylene terephthalate sheet. Films were cured at ambient conditions

(24 °C and 45 % relative humidity) for three days, followed by post-curing for 4 h at 110 °C in an air circulating oven. Unfilled epoxy films were also prepared for comparison.

2.2 UV Exposure Conditions

The UV radiation source was a 2 m integrating sphere-based environmental chamber, referred to as SPHERE (Simulated Photodegradation via High Energy Radiant Exposure) (5). This SPHERE UV chamber utilizes a mercury arc lamp system that produces a collimated and highly uniform UV flux of approximately 480 W/m^2 in the 295 nm to 400 nm range. This chamber can also precisely control the relative humidity (RH) and temperature. Specimens having 25 mm x 25 mm were mounted on a 17-window exposure cell, which was exposed in the UV chamber at 50 °C and 75 % RH. For nanoparticle release, a special sample holder (Figure 2) was employed. This holder collects particles expelled from the sample during UV exposure. It consists of a sample chamber, tubes to supply desired RH and temperature, and a container to collect released particles. A cover containing a quartz window that allows UV to irradiate the sample is used to seal the holder. In this study, a poly(tetrafluoroethylene) film was placed on the collector surface. All samples were mounted normal to the horizontal direction.

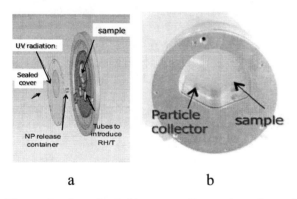

a   b

Figure 2. Sample holder to collect released particles: a) a schematic, b) a photograph with an exposed sample.

2.3. Characterization of UV-exposed Composite and Released Particles

Mass loss, surface morphology, and chemical degradation of UV-exposed composites were characterized before and after exposure. Mass loss was measured with an analytical balance, surface morphology via field emission scanning electron microscopy (FE-SEM), and chemical changes by FTIR in attenuated total reflection (ATR) mode (FTIR-ATR) and X-ray photoelectron spectroscopy (XPS). FTIR spectra were recorded at a resolution of 4 cm^{-1} using a spectrometer equipped with a liquid nitrogen-cooled mercury cadmium telluride (MCT) detector. A ZnSe prism was used for the ATR measurement. XPS analysis was

carried out using a Mg Kα X-ray source (1253.6 eV) at a 45° angle between the sample surface normal and the lens/hemispherical analyzer. Spectra were acquired at a pass energy 44.75 eV and a step size of .125 eV/step. All XP spectra were fit using 100 % Gaussian peaks, a Shirley baseline, and adjusted with the appropriate sensitivity factors. The released particles were characterized by FE-SEM and Energy Dispersive X-ray Spectroscopy (EDS).

3. RESULTS

3.1 Mass Loss

Figure 3 displays the mass loss of unfilled and 5 % nanoSiO$_2$-filled epoxy samples as a function of exposure time in the UV/50 °C/75 % RH environment. Except for a small mass increase at early exposure times, which was likely due to the moisture uptake, the mass loss in both materials was nearly linear with exposure time. The rate of mass loss of the epoxy/nanoSiO$_2$ composite was slightly greater than that of the unfilled material. After 58 d exposure, the maximum mass losses of unfilled and nanocomposite films were 1.1 % ± 0.05 % and 2.1 % ± 0.28%, respectively.

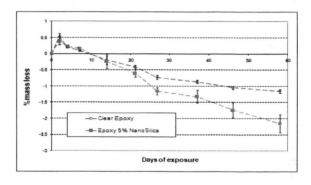

Figure 3. Mass loss vs. time for unfilled and 5 % nanoSiO$_2$-filled epoxy films exposed to UV/50 °C/75 % RH.

3.2 Surface Morphological Changes

Figure 4 shows FE-SEM images of epoxy/5 % nanoSiO$_2$ composite surface for different exposure times.

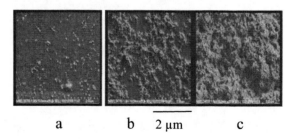

a   b   2 µm   c

Figure 4. FE-SEM images of epoxy/5 % nanoSiO$_2$ composite exposed to UV radiation for various times: a) before exposure; b) 7 d; and c) 43 d.

Before exposure, the surface contained a small amount of nanoparticles. The density of nanoparticles on the composite surface increased with exposure time. After 43 d exposure, almost the entire surface appeared to be covered with nanoparticles (Figure 4c).

3.3 Surface Composition Changes

Chemical changes in the matrix surface following exposure to UV radiation can be readily observed in the FTIR-ATR difference spectra (spectra of exposed specimen minus spectrum of unexposed specimen) displayed in Figure 5a. Various epoxy absorbance bands, e.g., 1508 cm^{-1} (benzene ring vibration), decreased in intensity, and new bands in the 1650 cm^{-1}-1750 cm^{-1} region(C=O) appeared. These changes were due to photo-oxidation by UV radiation, leading to chain scission in the epoxy and generation of various volatile products (6,7). UV radiation has been identified as the main weathering factor causing severe degradation in this particular epoxy, with temperature and RH playing minor roles (8).

Figure 5b shows FTIR-ATR intensity changes in the 1508 cm^{-1} and 1714 cm^{-1} bands, attributed to chain scission and oxidation, respectively, as a function of exposure time. Note that the intensity changes have been normalized to both the initial absorbance and that of the least-changed band (1360 cm^{-1}, due to CH$_3$) to minimize the effects of

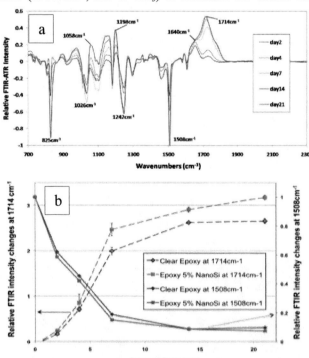

Figure 5. a) Difference FTIR-ATR spectra for different exposures times, and b) chain scission and oxidation of epoxy/nanoSiO$_2$ composite exposed to UV radiation. Each data point in Figure 5b was the average of four specimens, and the error bar represents one standard deviation.

thickness differences between samples and contact variations by the ATR probe on the sample. Both unfilled and nanoSiO$_2$-filled epoxy films underwent rapid photodegradation during UV exposure, and the presence of the SiO$_2$ nanoparticles appeared to have only a small effect on the oxidation rate of this epoxy. It should be noted that the depth of analysis in the epoxy polymer or the FTIR-ATR technique in the 800 cm^{-1}-3000 cm^{-1} range and using a ZnSe prism is between 0.5 μm and 2.5 μm from the surface. Therefore, the chemical changes observed originate from the polymer layer at or near the composite surface. Figure 5a also shows that the intensity of the 1058 cm^{-1} band, assigned to Si-O-Si bond, also increased with exposure, suggesting that the concentration of SiO$_2$ nanoparticles near the surface increased with UV exposure.

The degradation of the epoxy polymer and an increase of the SiO$_2$ material near the composite surface was consistent with XPS data from the Si (2p) (101-102.5 eV) and C (1s) (~284.5 eV) regions whose percent surface concentrations are reported in Figure 6. The surface concentration of carbon decreased from 82.2 % ± 1.7 % to 43.8 % ± 0.7 %, while that of silicon increased from 0.5 % ± 0.1 % to 10.1 % ± 0.3 % after 62 d exposure. The O (1s) region (one major component of the composite) also increased in surface concentration with exposure (not shown), probably from both the polymer oxidation and the exposed SiO$_2$. The substantial increase of SiO$_2$ material on the composite surface following UV irradiation has been confirmed by inductively-coupled plasma–optical emission spectroscopy (ICP-AES) analysis (9).

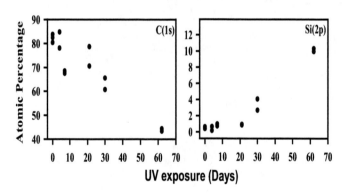

Figure 6. XPS based C and Si concentrations on the composite surface as a function of time exposed to UV radiation. Except for the 21 d where only one specimen was used, data at other exposure times consisted of two or more specimens.

Microscopic and spectroscopic data shown in Figures 4,5 and 6 strongly indicated that the increase of nanoSiO$_2$ concentration at the composite surface with UV exposure was a result of the epoxy matrix degradation. However, these results do not answer the question: were SiO$_2$ nanoparticles released from the nanocomposite during exposure to UV radiation? Analysis of particles collected at the bottom of the sample holder (i.e., Figure 2b) address this question. Figure 7 shows SEM images and EDS spectra of the particle collector surface before and after the sample was exposed to UV radiation. Before the exposure, the collector surface showed no evidence of particles (Figure 7a). After 43 d exposure, many particles were observed on the collector surface (Figure 7b), and numerous spherical nanoparticles can be seen at higher magnification (Figure 7c).

EDS spectrum obtained from the collector surface before the sample was exposed to UV radiation showed only F and C (Figure 7d), as expected for a poly(tetrafluoroethylene) film. However, EDS analysis of the collector surface after the sample was irradiated by UV for 43 days revealed the presence of Si (Figure 7e). The concentration of O also increased after the UV exposure. The increase in O concentration supports the suggestion that the spherical nanoparticles on the collector surface observed in Figure 7c were likely due to SiO_2. The results of Figure 7 provide direct evidence that SiO_2 nanoparticles or their aggregates were released from the epoxy nanocomposites to the surroundings during exposure to UV radiation. However, these preliminary results do not reveal whether these SiO_2 nanoparticles are pristine or covered with polymer molecules. Work is in progress to address this question.

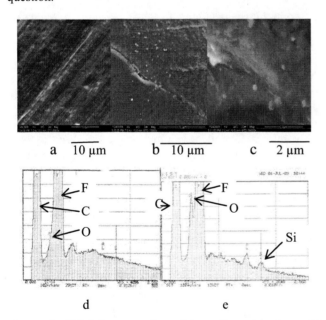

Figure 7. SEM images of particle collector surface: (a) before, and (b) after the sample was exposed to UV radiation for 43 d, (c) higher magnification of b, showing numerous spherical nanoparticles; EDS spectrum of the collector surface: before (d) and after the sample was exposed for 43 d (e), showing the presence of Si element.

Based on microscopic and spectroscopic evidence, the release of SiO_2 nanoparticles from the epoxy nanocomposites during UV irradiation condition used in this study probably followed the following sequence: Epoxy polymer on the surface was first removed through the photodegradation process, resulting in accumulation of a large concentration of SiO_2 nanoparticles on the composite surface. At a critical thickness/concentration, particles containing SiO_2 fell off the vertical surface likely by gravitation force. For nanocomposites exposed outdoors, environmental elements such as rain, condensed water, wind, mechanical vibration/abrasion, and stresses resulting from dimensional changes likely affect the release rate of nanoparticles.

CONCLUSIONS

The fate of SiO_2 nanoparticles in epoxy nanocomposites during exposure to UV radiation has been investigated. Analyses of chemical composition and morphology of the UV-exposed composite surfaces and released particles showed that the epoxy matrix underwent rapid photodegradation, resulting in accumulation of a large concentration of SiO_2 nanoparticles on the composite surface, some of which were then released into the surroundings. The results of this study will provide useful information to assess the potential risk of nanoSiO$_2$ in epoxy nanocomposites during outdoor exposure.

REFERENCES

1. F. Hussain, M. Hojjati, M. Okamoto, R. E. Gorga, J. Composite Materials 40, 1511, (2006).
2. P. M. Ajayan and J. M. Tour, Nature 447, 1066 (2007).
3. A. Poland, R. Duffin, I. Kinloch, A. Maynard, W. A. H. Wallace, A. Seaton, V. Stone, S. Brown, W. MacNee, and K. Donaldson, Nature Nanotechnology 3, 423 (2008).
4. Helland, P. Wick, A. Koehler, K. Schmid, and C. Som, Environmental Health Perspectives 115, 1125 (2007).
5. J. Chin, E. Byrd, N. Embree, J. Garver, B. Dickens, T. Fin, and J. W. Martin, Review Scientific Instruments 75, 4951 (2004).
6. J.F. Rabek, "Polymer Photodegradation: Mechanism and Experimental Methods", Chapman &Hall, NY., 1995, pp 185-216.
7. Bellinger, V. and Verdu, Oxidative Skeleton Breaking in Epoxy-Amine Networks, J. Appl. Polym. Sci. 30, 363 (1985).
8. X. Gu, et al., in "Service Life Prediction of Polymeric Materials, Global Perspectives", Eds. J. W. Martin, R. A. Ryntz, J. Chin, R. A. Dickie, Springer, NY., 2009, p. 1.
9. S. Rabb, L. Yu, C. Bernard, and T. Nguyen, in this proceedings.

An Analytical Method to Quantify Silica Nanoparticles Accumulated on the Surface of Polymer Nanocomposites Exposed to UV Radiation

Savelas A. Rabb[1], Lee L. Yu[1], Coralie Bernard[2] and Tinh Nguyen[2]

[1]Analytical Chemistry Division, Chemical Science and Technology Laboratory,
savelas.rabb@nist.gov, lee.yu@nist.gov
[2]Materials and Construction Research Division, Building and Fire Research Laboratory,
coralie.bernard@nist.gov, tinh.nguyen@nist.gov
Gaithersburg, MD, USA

ABSTRACT

Three films containing different mass fractions (0 %, 5 % and 10 %) of silica (SiO_2) nanoparticles have been subjected to controlled degradation conditions at specified humidity, temperature, and ultraviolet (UV) radiation doses over a period of 59 days. The mass fraction of extracted nanosilica is measured by inductively coupled plasma–optical emission spectroscopy (ICP-OES). Results show the presence of nanosilica on the films containing 5 % and 10 % nanosilica, regardless of the exposure conditions. However, after UV exposure for 59 days, the mass fraction of nanosilica in the extracted solutions increase by factors of 2.3 and 1.7 for the 5 % and 10 % nanosilica films, respectively. Measurement variability is lower for the 10 % nanosilica films, likely due to the larger mass of nanosilica that is released. Results of this study will assist in establishing a reliable and sensitive method for measuring the release rate of silica nanoparticles from UV-irradiated polymer nanocomposites.

Keywords: nanocomposites, release rate, ICP-OES, silica nanoparticles

1 INTRODUCTION

Nanocomposites have generated widespread interest in areas such as construction, transportation, optics and electronics [1,2]. As the polymer degrades or weakens, the nanomaterials (e.g., nanoparticles, nanotubes) used to fill the polymer matrix will be potentially released to the environment. The release of these nanomaterials is of concern as the impact on the ecosystem is not fully known [3-5]. However, current research is lacking in the accurate assessment of the release of nanomaterials upon degradation of the nanocomposites. The National Institute of Standards and Technology has developed a technique using inductively coupled plasma-optical emission spectroscopy (ICP-OES) to aid in assessing the release of SiO_2 nanoparticles from epoxy nanocomposites exposed to the weathering environments via the determination of SiO_2 nanoparticle surface accumulation. Films containing SiO_2 mass fractions of 0 %, 5 % and 10 % were investigated.

Silica nanoparticles on the surface of the degraded films were chemically extracted by using a solution containing 5 % hydrofluoric acid (HF). Silicon in the extracts was determined with ICP-OES. Results will lead to an accurate methodology of determining release rates of SiO_2 nanoparticles from epoxy nanocomposites exposed to UV radiation.

2 EXPERIMENTAL

2.1 Preparation of Nanocomposites and Procedure for UV Exposure

Epoxy films containing 0 %, 5 % and 10 % mass fraction of SiO_2 nanoparticles having 7 nm diameter were prepared according to the procedure described in Nguyen et al. [6]. Subsequently, the films were placed in the Simulated Photodegradation *via* High Energy Radiant Exposure (SPHERE) UV chamber, a 2 m integrating sphere-based environmental system. The SPHERE utilizes a mercury arc lamp system that produces a collimated and highly uniform UV flux of approximately 480 W/m^2 in the 295 nm to 400 nm range. This chamber can also precisely control the relative humidity (RH) and temperature. Specimens having 25 mm x 25 mm dimensions were mounted on a 17-window exposure cell, which was exposed in the SPHERE UV chamber at 50 °C and 75 % RH. Specimens were removed at a specified time interval for characterization.

2.2 Characterization of Nanocomposites for the Surface Accumulation of Silica Nanoparticles Using ICP-OES

The film specimens were removed from the SPHERE chamber, and the irradiated sections (8 to 9 for each SiO_2 mass fraction) were cut from each film and placed in petri dishes, separated according to the silica mass fraction. Silica nanoparticles on the surface of these sections were extracted using 5 % hydrofluoric (HF) acid for 5 min in a volume of 10 mL. After the removal of the exposed films, the extraction solutions were diluted to 24 mL. These solutions were further diluted by a factor of 50 and analyzed. The solutions contained 0.2 % NaOH to neutralize the remaining HF and mitigate the Si background.

The method of standard additions was used to compensate for matrix effects in the Si determinations of the extracted solutions of the SiO$_2$ nanoparticles from the polymer nanocomposites. Each extracted solution was split into two solutions and one solution was spiked with Si. The spike stock solutions ranged from 0.38 µg/g Si to 2.7 µg/g Si. A 0.5 g spike was taken from the Si spike stock solutions and added to a 5 g sample solution. The Si spike stock solutions were prepared from the SRM 3150 Silicon Standard Solution. Sn was used as an internal standard at 0.5 µg/g.

A PerkinElmer Optima 5300 DV ICP-OES instrument (Shelton, CT) was used for the analyses. (Identification of commercial products in this paper was done in order to specify the experimental procedure. In no case does this imply endorsement or recommendation by the National Institute of Standards and Technology.) The Si mass fractions in the samples were measured according to the parameters in Table 1. Each sample measurement was comprised of five replicates, and each solution was repeatedly measured two different times for the analysis.

Power (kW)	1.5
Plasma gas (L min^{-1})	15
Auxiliary gas (L min^{-1})	0.5
Nebulizer gas (L min^{-1})	0.6
Nebulizer	MiraMist
Spray chamber	cyclone
Viewing	axial
Sample uptake (mL min^{-1})	1.0
Analyte wavelength (nm)	Si I 251.611
Reference wavelength (nm)	Sn II 189.927
On-chip integration time (s)	0.256
Total read time (s)	8.192

Table 1: Operating conditions for ICP-OES.

All uncertainties shown for the data consist of expanded uncertainties expressed at the 95 % level of confidence and are calculated according to the principles of the *ISO Guide to the Expression of Uncertainty in Measurement* (GUM) [7]. Expanded uncertainties were determined for ICP-OES silicon measurements by using the following equations:

$$u_c = \sqrt{u_1^2 + u_2^2 + u_3^2 + \dots}$$ (1)

$$U = ku_c$$ (2)

$$\% \ U = \frac{U}{x} * 100$$ (3)

where u_i represents the individual component of uncertainty, u_c is the combined uncertainty, k is the expansion factor based on the Student's t for the chosen level of confidence, U is the expanded uncertainty, and x is the observed measurement of Si. Propagated components of uncertainty include observed measurement repeatability, observed variability in the determination of ICP-OES sensitivity, and uncertainties in the known values for the calibration standards.

3 RESULTS AND DISCUSSION

3.1 Extractions of Silica Nanoparticles Using Hydrofluoric Acid

To determine the accumulation of SiO$_2$ nanoparticles on the polymer nanocomposite surface during UV exposure, a chemical extraction method was developed to selectively attack SiO$_2$ nanoparticles on the surface of the film without extracting nanoparticles from the interior. Variable concentrations of HF (1 % to 50 %) were investigated to optimize this dissolution process using silica nanoparticles (sample sizes < 10 mg), not embedded in the polymer. To generate the different concentrations of HF, H$_2$O was added first to the nanoparticles and then concentrated HF was added to give a total of 10 mL. It was observed that concentrations of HF from 10 % to 50 % dissolved the SiO$_2$ nanoparticles in 1 min or less. The dissolutions at these concentrations are likely too rapid and could potentially attack the SiO$_2$ nanoparticles embedded in the polymer film. At 1 % HF and 5 % HF, the nanoparticles dissolved in ~15 min and ~2 min, respectively. Of the two concentrations, 5 % HF was chosen as an extraction solution for the nanoparticles as the timing of dissolution was not excessively long. Extractions for the films using 5 % HF showed successful results. Differences were observed in the Si I 251 nm intensities for the films embedded with different mass fractions of nanosilica (Figure 1), before and after exposure. Successive extractions observed either decreases to signal levels of the unexposed samples or to signal levels that were consistent with extraction from the interior of the nanocomposite (Figure 2).

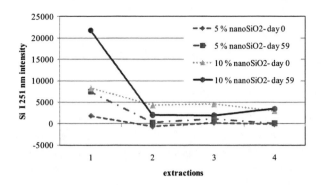

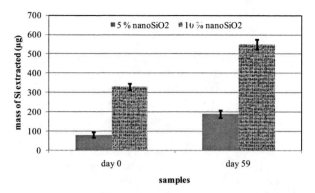

Figure 1: Observed Si I 251 nm intensity for multiple extractions of SiO$_2$ nanoparticles from the surface of the polymer nanocomposites.

Figure 2: Average mass of Si (μg) extracted from the surface of the polymer nanocomposites. 5 % nanoSiO$_2$ film and 10 % nanoSiO$_2$ film represented by the solid and patterned fill, respectively.

3.2 Determination of Surface Accumulation of Silica Nanoparticles from Polymer Nanocomposites Using ICP-OES

Current results show the presence of SiO$_2$ nanoparticles on the surface of multiple film specimens containing 5 % and 10 % SiO$_2$ nanoparticles, regardless of the UV exposure (Figure 2). This is in agreement with data from the preliminary extractions shown in Figure 1 and those obtained using atomic force microscopy. Exposure to UV radiation for 59 days increases the observed Si mass fraction in the extracted solutions by factors of 2.3 and 1.7 for the 5 % and 10 % nanosilica films, respectively. The higher masses of Si were extracted from the unexposed and exposed 10 % nanosilica films at 330 μg and 550 μg, respectively. Ratios of the Si mass extracted from the 10 % nanosilica and 5 % nanosilica films (Figure 3) show that the rate of nanosilica surface accumulation for the 5 % nanosilica films could be increasing at a faster rate over the 59 day exposure period. However, net accumulations for the 10 % nanosilica films are a factor of 2 greater than the 5 % nanosilica films (Figure 4) after the exposure period. Multiple sections of the film were analyzed to determine repeatability of nanosilica accumulation for each film. The repeatability ranges from 4.6 % to 16 % for all films. Variability of observed Si masses is lower for the 10 % nanosilica films, likely due to the larger mass of nanosilica that is released.

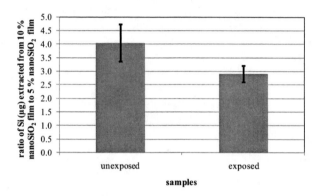

Figure 3: Ratio of Si (μg) extracted from the surface of the 10 % nanoSiO$_2$ film to the 5 % nanoSiO$_2$ film.

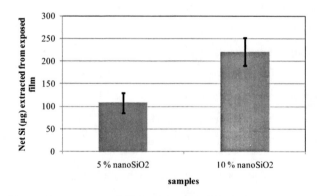

Figure 4: Net mass of Si (μg) extracted from the surface of the exposed polymer nanocomposites.

4 CONCLUSIONS

A technique using ICP-OES has been successfully developed to determine the surface accumulation of SiO$_2$ nanoparticles on polymer nanocomposites after being exposed to weathering conditions such as UV radiation. The SiO$_2$ nanoparticles are chemically extracted from the films using 5 % HF for a period of 5 min, diluted and analyzed. Results demonstrate that more nanosilica is released from the 10 % nanosilica films, regardless of UV exposure. Also, with UV exposure, the surface accumulation is greater in both nanosilica films. Measurement variability is $\leq$ 16 % with measurements from the exposed 10 % nanosilica films having the lowest variability at 4.6 %. These ICP-OES results will aid in the assessment of the release rate of SiO_2 nanoparticles during life times of polymer nanocomposites.

REFERENCES

1. F. Hussain, M. Hojjati, M. Okamoto, R. E. Gorga, J. Composite Materials 40, 1511, (2006).
2. P. M. Ajayan and J. M. Tour, Nature 447, 1066, (2007).
3. A. Barnard, Nanoscale 1, 89, (2009).
4. H. Hildebrand, D. Kuhnel, A. Potthoff, K. Mackenzie, A. Springer and K. Schirmer, Environ. Pollut. 158, 65, (2010).
5. V. K. Sharma, J. Environ. Sci. Health, Part A 44, 1485, (2009).
6. T. Nguyen, B. Pellegrin, C. Bernard, X. Gu, J. Gorham, P. Stutzman, A. Shapiro, E. Byrd and J. Chin, in this proceedings.
7. JCGM 100:2008; *Guide to the Expression of Uncertainty in Measurement*; (ISO GUM 1995 with Minor Corrections), Joint Committee for Guides in Metrology: BIPM, Sèvres Cedex, France (2008); available at http://www.bipm.org/utils/common/documents/jcgm/JCGM_100_2008_E.pdf; see also Taylor, B.N.; Kuyatt, C.E.; *Guidelines for Evaluating and Expressing the Uncertainty of NIST Measurement Results*; NIST Technical Note 1297; U.S. Government Printing Office: Washington, DC (1994); available at http://physics.nist.gov/Pubs/.

Developing a Strategy for the Spatial Localisation and Autonomous Release of Silver Nanoparticles within Smart Implants

Laura A. A. Newton[*], Duncan G. Sharp[**], Ray Leslie[*] and James Davis[***]

[*] School of Science and Technology, Nottingham Trent University,
Nottingham, NG11 8NS, UK. laura.newton@ntu.ac.uk
[**] Clinical Biochemistry, Leeds Metropolitan University,
Leeds, LS1 3HE, UK
[***] School of Engineering, NIBEC, University of Ulster, Jordanstown, Newtownabbey,
Co.Antrim, Northern Ireland, BT37 0QB

ABSTRACT

The prevalence of antibiotic resistance has increased during the last few years and is viewed as a growing problem which has fuelled copious amounts of research on the development of new antibacterial agents. Silver is well recognized as possessing antibacterial activity and so by harnessing these properties and incorporating silver nanoparticles (AgNPs) within microdevices possessing microfluidic channels has much promise. Progress in developing these types of systems has been limited due to the technological difficulties involved in controlling the inclusion of these nanoparticles within the devices. This work provides an insight into a novel electrochemical interaction that can enable not only the localisation of silver within such structures but also enables the smart release of AgNPs.

Keywords: Silver nanoparticles, Homocysteine, Glutathione, Antibacterial Agents

INTRODUCTION

Access related infections (ARIs) are among the most prevalent causes of nosocomial infections and are complicated by the fact that bacteria associated with biofilm formation on the catheter surfaces play a key role in the morbidity and pathogenesis of these infections with hospital admissions for vascular ARIs having doubled in the last decade [1-4]. The bacteria contained within the biofilm layer benefit from having an increased resistance to antimicrobial agents and, as such, are generally more difficult to treat and to eradicate [5]. While systemic antibiotics are largely effective at eliminating circulating bacteria, their ability to sterilize lines in which a biofilm has taken hold is less efficient and life threatening complications may arise. Most of these are treatable albeit at a price in terms of morbidity for the patient and financial costs incurred by the need for intravenous antibiotics and an increased length of hospital stay [6-7]. This is compounded by the alarming long term issues associated with the repeated use of antibiotic regimes in such contexts as they have been attributed to the selection of multi-resistant organisms, such as methicillin-resistant *Staphylococcus aureus* (MRSA) and vancomycin-resistant enterococci [8-9].

There has been much research into surface modifications incorporating antibacterial components that attempt to minimise the propensity for biofilm formation. In spite of this, no currently available indwelling line is entirely free from the complication of infection. The incorporation of silver particles as an antibacterial coating is however now commonplace whereby its ionized form has been shown to bind to tissue proteins and bacterial RNA/DNA resulting in structural changes that can inhibit cell replication and can result in cell distortion and death [10-12]. Silver has long been viewed as an ideal weapon to be used in the fight against pathogenic bacteria such that the impregnation of silver nanoparticles (AgNPs) within medical devices and wound dressings to prevent bacterial growth is routine[13-17]. Conventional approaches to the incorporation of silver are well suited for macroscale application but their deployment within the emerging microscale devices creates an issue. There is also some concern over there mid to long term effectiveness as the coatings, irrespective of composition, can be rendered inactive by the exopolysaccharides that constitute the biofilm. Numerous investigations have shown silver impregnated tubes to be ineffective at preventing the development of biofilms with the formation of the latter leading to capsule formation that could serve to promote re-infection [4]. It could be proffered that their main value lies more in slowing the development of the biofilm rather than in its prevention [3-5].

METHODOLOGY

Harnessing the antibacterial properties of AgNPs within microdevices possessing microfluidic channels that could be prone to bacterial colonisation creates a technological dilemma in terms of not only controlling their incorporation within the device but in producing the nano particle in a form that is readily accessible to the bacteria and thus maintain the antimicrobial action. AgNPs can be synthesized using a variety of different techniques and at present, they are mainly produced by bulk processing methods, through the chemical reduction of a silver salt (typically silver nitrate), which allows them to be subsequently implanted within macro structures [18-19]. Electrochemical methods for preparing AgNPs and nano structured surfaces have also been widely reported [18,20] and while the present investigation seeks to adapt the basis of the former, the remit was focused more on a novel approach that could enable not only the localisation of silver within such structures but also the controlled formation of AgNPs which can be either localised at the device interface or ejected into the solution channel.

The basis of the approach lies in the functionalisation of an electrode surface with silver through conventional electrodeposition methods [18,20]. The surface is then oxidised to strip the silver but, importantly, this is done in the presence of a thiol ligand. The sulfur functionality binds to the released silver ion resulting in the formation of a silver thiolate complex (Ag-S-R). This is insoluble and is effectively deposited at the electrode surface. Electrochemical reduction of the silver-thiolate bond would, under normal instances, be expected to lead to the release of the thiol back into solution. Such transformations have long been studied – particularly with thiolate monolayers on gold [21-22]. In such cases, the gold from the gold thiolate is simply re-incorporated back within the underlying gold layer [22]. In this case the normal expectation, and our original aim, would be for the silver to be re-deposited onto polycrystalline platinum – ultimately leading to a nanostructured deposit on the surface. It was found however that, depending on the thiol used, a wholly different behaviour could be obtained and rather than re-integration of the silver onto the platinum surface, electro-ejection of AgNPs can occur.

A variety of silver thiolate systems have been investigated in an effort to elucidate the underlying dynamics of the interfacial processes involved which enables the controlled ejection of AgNPs. These were facilitated through a combination of electroanalytical investigations, electrochemical quartz crystal microbalance (EQCM) studies, atomic force microscopy (AFM), electron microscopy (SEM-EDX) and x-ray photoelectron spectroscopy (XPS).

RESULTS

The preliminary investigations were peformed using an EQCM. The silver was electrodeposited onto a platninum crystal used as the working electrode (-0.4V). It could then be stripped off from the surface through the application of an oxidising potential (+0.5V) as indicated in Figure 1. The decrease in the frequency indicating the accumulation of silver at the surface and the return to the baseline the subsequent removal of the silver ion into the bulk of the solution.

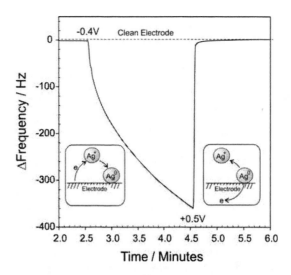

Figure 1. EQCM data highlighting the deposition and stripping of silver

The process was repeated but with a view to assessing the impact of a thiol additive (homocysteine, 10mM). Silver was electrodeposited (-0.4V) onto the crystal as before from a simple solution of silver nitrate. The silver plated crystal was then removed, rinsed and placed in a buffered (pH 7) solution of the thiol and the electrode held at 0 V to equilibrate. The imposition of the oxidising potential (+0.5V) resulted in a further decrease in the crystal frequency (Figure 2) and is in marked contrast to the behaviour observed in Figure 1.

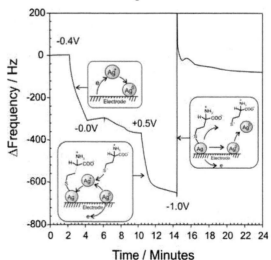

Figure 2. EQCM data highlighting the deposition and stripping of silver in the presence of homocysteine

This is attributed to the stripping of Ag^0 to Ag^+, the subsequent formation of the silver-thiolate complex (Ag-SR) and its deposition at the electrode as a consequence of the poor solubility of the resulting product. It was anticipated that the imposition of a reducing potential (-1.0V) would therefore result in the reduction of the Ag-

SR bond resulting in the release of the thiol but the retention of the silver at the electrode. Similar processes have been observed on gold substrates and the initial behaviour followed this prediction with a sharp increase in crystal frequency. However, rather than simply releasing the thiol, the frequency returned to almost that observed with a clean platinised crystal indicating that the entire silver layer had been lost.

Visual examination of both the crystal and the interfacial solution layer revealed the appearance of colloidal silver aggregates. This was corroborated by conducting elemental analysis of the both the bulk solution and the solution adjacent to the electrode interface. This is clearly contrary to what would normally be expected. The release of silver into the solution is usually mandated through an oxidising potential (c.f. Figure 1) whereas the reducing potential should have resulted in the electrodeposition of the silver consistent with the behaviour shown in Figure 1 and when repeating the above experiment in absence of the thiol. This therefore suggests that the properties of the amino acid thiol must be instrumental in removing the silver.

The experiment was repeated with 3-mercaptopropane and the results are shown in Figure 3. The EQCM profile is initially similar to that obtained with the homocysteine but there is a marked difference when the electroreduction of the Ag-SR complex is attempted. As expected, there is an increase in the frequency as the thiol is released back into solution and it then returns to the value observed before the oxidative step was applied. In contrast to the profile observed with homocysteine, it effectively terminates at this point indicating that the silver is retained at the electrode - in keeping with what would normally be expected.

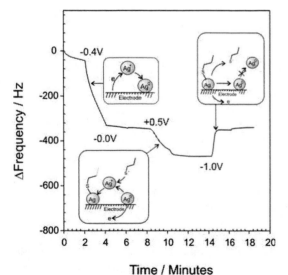

Figure 3. EQCM data highlighting the deposition and stripping of silver in the presence of 3-mercaptopropane.

The experiments were repeated with 3-mercaptopropanol and the tripeptide, glutathione. The former was found to possess an EQCM profile almost identical to that of the propane analogue while glutathione mirrored that of the homocysteine – indicating that both of the previous responses were accurate but providing two distinct pathways which could be considered to be representative of the nature of the thiol used. The main difference between the two groups is the presence of the acid-base functionality. Both homocysteine and glutathione possess the latter.

DISCUSSION

An initial assumption was that the complexing properties of the amino acid group must be responsible for the removal of the silver. However, this is flawed in two respects. The electro-oxidation of the silver in the presence of the amino acid thiol leads to an insoluble product. The greater hydrophilicity inherent to glutathione was also found to be ineffectual in mobilising the product into solution. In addition, the reducing potential (-1V) reduces the silver (I) to metallic silver thus making it unavailable for complex formation. It could be rationalised that the oxidation process leading to the Ag-SR formation leads to the precipitation of a loose aggregation of Ag-SR particles. The subsequent reduction of the Ag-SR species at the immediate surface leads to the release of the thiol which then induces the release of the overlying layer. In this respect the interfacial Ag-SR serves as electro-reducible anchor. Yet, the mercaptopropane and mercaptopropanol results clearly contradict this hypothesis as the entire silver is effectively recaptured and only the thiol is released.

The presence of the acid-base groups appears pivotal to the release mechanism but rather than through complexation, the mode of action could be through their ability to serve as conduits for the transfer of electro-reducible protons to the electrode. The underlying platinum on the crystal is an excellent catalytic substrate for the electroreduction of protons and given the application of the reducing potential (-1V) it could well be that the adherence of the metallic silver (or aggregate layer) is weakened as a consequence of the formation of an adsorbed hydrogen adlayer on the platinum.

CONCLUSIONS

The results highlight the tunable nature of the electrodeposited Ag-SR interaction. While it is clear that it is possible to control the physical ejection of the Ag-SR layer through the application of an appropriate potential, the nature of the ligand composition is shown to be crucial. The electrochemical approach provides an option for controlling not only the spatial location of the Ag-SR particles/aggregations but also enables control over their release.

REFERENCES

[1] K.A. Barrooclough, C.M. Hawley, G. Playford and D.W. Johnson, Pevention of access-related infection indialysis. Expert Review of Antiinfective Therapy, 7, 1185-1200, 2009.

[2] E.L. Lawrence and I.G. Turner, Material for urinary catheters: a review of their history and development in the UK. Medical Engineering and Physics, 27, 443-453, 2005.

[3] F. Suska, S. Svensson, A. Johansson, L. Emanuelsson, H. Karlholm, M. Ohrlander and P. Thomsen, In vivo evaluation of noble metal coatings. Journal of Biomedical Materials Research Part B: Applied Biomaterials

[4] E. Hurrel, E. Kucerova, J. Callbilla-Barron and S.J. Forsythe, Biofilm formation on enteral feeding tubes by cronobacter sakazakii, salmonella serovars and other enterobacteriaceae. International Journal of Food Microbiology, 136, 227-231, 2009.

[5] M.E. Falagas, A.M. Kapaskelis, A.D. Kouranas, O.K. Kakisi, Z. Athanassa and D.E. Karageorgopouls, Outcome of Antimicrobial therapy in documented biofilm associated infections: A review of the available clinical evidence. Drugs, 69, 1351-1361, 2009.

[6] C. Stephens, Microbiology- Breaking Down Biofilms. Current Biology, 12, R132-R134, 2002.

[7] C.A. Fux, J.W. Costerton, R.S. Stewart and P. Stroodley, Survival strategies of infectious biofilms. Trends in Microbiology, 13, 34-40, 2005.

[8] K. Hiramatsu, L. Cui, M. Kuroda and T. Ito, The emergence and evolution of methicillin-resistance staphylococcus aureus. Trends in Microbiology, 9, 486-493, 2001.

[9] G.A. Noskin, Vancomycin-resistant enterococci: Clinical, microbiologic and epidemiologic features. Journal of Laboratory and Clinical Medicine, 130, 14-20, 1997.

[10] M. Rai, A. Yadar and A. Gadi, Silver nanoparticles as a new generation of antimicrobials. Biotechnology Advances, 27, 76-83, 2009.

[11] V.K. Sharma, R.A. Yngard and Y. Lin, Silver nanoparticles: Green synthesis and their antimicrobial activities. Advances in Colloid and Interface Science, 145, 83-96, 2009.

[12] J.S. Kim, E. Kuk, K.N. Yu, J-H. Kim, S.J. Park, H.J. Lee, S.H. Kim, Y.K. Park, Y.H. Park, C-Y. Hwang, Y-K. Kim, Y-S. Lee, D.H. Jeong and M-H. Cho, Antimicrobial effects of silver nanoparticles. Nanomedicine: Nanotechnology, Biology and Medicine, 3, 95-101, 2007.

[13] J.M. Schierholz, L.J. Lucas, A. Rump and G. Pulverer, Efficacy of silver coated medical devices. Journal of Hospital Infection, 40, 257-262, 1998.

[14] D.R. Monterio, L.F. Gorup, A.S. Takamiug, A.C. Ruvollofilho, E. Rodrigues de Camargo and D.B. Barbosa, The growing importance of materials the prevent microbial adhesion: antimicrobial effects of medical devices containing silver. International Journal of Antimicrobial Agents, 34, 103-110, 2009.

[15] M.E. Rupp, T. Ftzgerald, N. Marion, V. Helget, S. Puumala, J.R. Anderson and P.D. Fegy, Effect of silver-coated urinary catheters: efficacy, cost effectiveness and antimicrobial resistance. American Journal of Infection Control, 32, 445-450, 2004.

[16] S. Gaisford, A.E. Beezer, A.H. Bishop, M. Walker and D. Parsons, An in vitro method for the quantitative determination of the antimicrobial efficacy of silver-containing wound dressings. International Journal of Pharmaceutics, 366, 111-116, 2009.

[17] M.A.A. O'Neill, G.J. Vine, A.E. Beezer, A.H. Bishop, J. Hadgraft, C. Labetoulle and M. Walker, Antimicrobial properties of silver containing wound dressings: a microcalorimetric study. International Journal of Pharmaceutics, 263, 61-68, 2003.

[18] W. Zhang, X. Qiao and J. Chen, Synthesis of silver nanoparticles – effects of concerned parameters in water/oil microemulsion. Materials Science and Engineering B, 142, 1-15, 2007.

[19] J.H. Lim and J.S. Lee, A statistical design and analysis illustrating the interactions between key experimental factors for the synthesis of Ag Nanoparticles. Colloids and Surfaces A: Physiochemical and Engineering Aspects, 322, 155-163, 2008.

[20] M. Muzur, Electrochemically prepared silver nanoflakes and nanowires. Electrochemistry Communications, 6, 400-403, 2004.

[21] M.B. Buchmann, T. Fyes and T. Sutherland, Electrochemical release from gold-thiolate electrodes for controlled release of ion channels into bilayer membranes. Bioorganic and Medicinal Chemistry, 12, 1315-1324, 2004.

[22] D.F. Yang and M. Morin, Chronoamperometric study of the reduction of chemisorbed thiols on Au(III). Journal of Electroanalytical Chemistry, 429, 1-5, 1997.

Nanoparticle release from nano-containing product, introduction to an energetic approach

L. Gheerardyn[*],[**], O. Le Bihan[*], J.-M. Malhaire[***], O. Guillon[***], C. Gibaut[***], E. Dore[**]
and M. Morgeneyer[**]

[*]Institut National de l'Environnement industriel et des RISques (INERIS), Parc Technologique Alata
BP 2, Verneuil-en-Halatte, France, ludovic.gheerardyn@ineris.fr
[**]Université de Technologie de Compiègne (UTC), Compiègne, France
[***]Renault, Guyancourt, France

ABSTRACT

The use of nanoparticles in manufactured products becomes more and more frequent and could lead to new potential risks for the workers, the consumers and the environment. This study focused on the R&D process of a nano-structured part; the stress operated during this process was reproduced into a nanosecurised facility, in order to characterize possible release of nanoparticles from the product. It appears that the release of nanoparticle is proportional to the energy applied on the surface. The particle release have been studied for three energy levels. The maximum particle concentration was around 10 p/cc for scratching operation, 380 p/cc while sawing and above 100.000 p/cc while sanding. Stress tests operated on a non nanostructured part induce also a release of nanoparticles.

Keywords: energy, abrasion, coating, nanoparticle, release

1 INTRODUCTION

Nowadays, with nanotechnology, consumer products become self-cleaning, antibacterial, UV-resistant... The effects of nanoparticles are not well known and represent a possible risk for people and the environment. The mechanical behaviour of nano-containing products has to be studied during their whole life-cycle. For this, INERIS develops new methods and tools allowing the characterization of the nanoparticles released into the air from the product.

A first study demonstrates the release of nanoparticules from both nanocomposite and standard material while drilling and sanding [1]. Abrasive wear is a very common wear mechanism. Taber-based tests, reproducing daily-life stress, have also been applied on polymer coatings containing nanoclay [2] and nanometric zinc oxide [3]. The concentration level of released nanoparticle from these products is very low and the particles remain embedded in the polymer matrix. Effects of sanding on nano-enabled paints have also been studied [4].

In this context, car manufacturer RENAULT is investigating the mechanical properties of nanomaterials, in the frame of the FOREMOST European Project. The company asked INERIS to determine whether there are nanoparticle releases from a nanostructured part during the R&D process. This metallic part, coated with inorganic fullerene, was designed for lowering the friction coefficient to reduce vehicle fuel consumption. The aim of this study is to reproduce representative operations of the process that are: unpacking, handling, roughness measurement and fracturing for microscopy observation. Accidental scenarios such as surface rubbing or the falling of the part have been considered.

2 TOOLS & METHODS

The stress tests have been operated sequentially, in order to evaluate a potential impact of each operation, in term of particle concentration. After a stress, the stabilization of the level of particle concentration was necessary to separate a possible input for the next test.

A first attempt was done on a nanostructured sample part to optimize and validate the whole experimental protocol. The main testing was led on three nanostructured sample parts, which are supposed to be the same, to evaluate the reliability of the results. Two standard parts, which are not coated with the nanostructured layer, have also been tested for particle release comparison.

2.1 Instrumentation

The choice of the devices allows informing about total concentration in number, on a range of time as short as possible to detect brief emissions, and about size distribution from 10 nm up to 20 μm.

Instruments are SMPS with long DMA (TSI 3080 coupled to 3031) and an optical counter (Grimm 1.108) for size distribution. The use in parallel of two CPC (TSI 3785 and Grimm 5.400) gives more reliability to the results when measuring low-level concentrations.

Device	Range	Sampling time
TSI 3080+3031	[10 nm ; 500 nm]	235 s
TSI 3785	[4 nm ; 1 μm]	1 s
Grimm 5.400	[4 nm ; 1 μm]	6 s
Grimm 1.108	[0.3 μm ; 20 μm]	6 s

Table 1 : Range and sampling time of the devices

2.2 Nano-safety

Throughout testing, the following security principles have been followed: the stress was applied in a nanosecurised facility which consists in a double-containment of the controlled atmosphere, by putting a glove-box within a venting hood. When they are outside of the nanosecurised facility, sample parts were systematically contained in two plastic bags, in order to avoid nanoparticle release into the ambient air.

2.3 Experiment

At the end of each experiment, the glove-box and every tool within were cleaned with a dust repellent spray. The controlled atmosphere was recycled, to reach a background noise level lower than 1 p/cc.

The tools used during the experiment were: a hand saw, a vise and a needle. Concerning the inlet flow rate, its value was equal to the devices flow consumption, which was around 4.5 lpm. A sampling point was set above the nanostructured surface to exhaust the maximum of particles released into the air. The description of each test is given in the following table.

#	Description	Aim
1	Extraction from the safety bag	Is there any nanoparticle release when handling the part?
2	Plastic bag agitation	
3	Part extraction from plastic bag	
4	Part moves in 3 dimensions + shock test	
5	Surface rubbing with glove	Accidental scenario #1
6	Falling from a height of 20 cm	Accidental scenario #2
7	Surface scratching (3 directions)	Is there any nanoparticle release during roughness measurement?
8	Surface sawing	Is there any nanoparticle release during sample preparation for microscopy?
9	Surface sanding	Extreme stress

Table 2 : Description and aim of the stress tests.

3 RESULTS

3.1 Nanostructured part

Stress tests 1 to 7 generate emission peaks that are around 10 p/cc. While sawing, emission peaks in a range from 50 to 1,000 p/cc are measured (see Fig. 1).

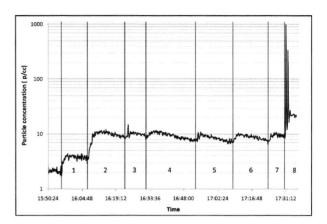

Figure 1 : Particle concentration through time while testing a single nanostructured part (except sanding).

3.2 Comparison with standard part

The same tests as those described in table 2 were applied on a standard part which does not contain nanoparticles. The particle concentration during tests 1 to 6 is in a range from 4 to 12 p/cc. In comparison with a nanostructured part (see Tab. 3), no specific emission –that is the difference between the emission peak and the background noise– was detected during the scratch test. Also, no scratch was observed on the standard part while the nanostructured layer is damaged and thus involves a particle emission. The nanostructured layer is less wear-resistant than the non nanostructured surface. Sanding results are not representative while testing the standard part because the surface was too hard for the abrasive wheel.

#	Test	Nanostructured part	Standard part
1	Extraction bag #1	+	+
2	Bag agitation	+	+
3	Part extraction	+	+
4	Handling + shock	+	+
5	Surface rubbing	+	+
6	Part falling	+	+
7	Surface scratching	+	< 1
8	Surface sawing	++	++
9	Surface sanding	++++	✕

Table 3 : Comparison of particle emissions between nanostructured and standard parts.

3.3 Surface sanding

A Dremel 400 DIGITAL has been used to force particle release by sanding the surface, even if this stress does not appear during the R&D process. Velocity of rotation was 19,000 rpm, using an aluminum oxide abrasive wheel. The machine engine is located outside of the glove-box, to avoid nanoparticle release from the brushes.

The surface sanding generates higher particle emissions than that observed during the previous stress tests. The particle concentration is above 100,000 p/cc and depends on the worn surface area.

Size distribution is unimodal with a mode included between 30 and 40 nm. Also, 95% of the measured particles have a diameter range which is 10 nm to 100 nm (see Fig. 2).

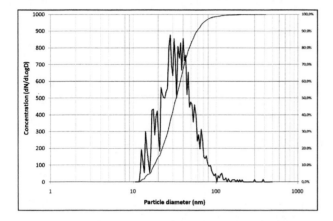

Figure 2 : Particle size distribution while sanding.

4 DISCUSSION

When handling the part, either nanostructured or not, the devices measured particle emissions in a range from 4 nm to 1 μm. In the both cases, the particle concentration level was very low. The situation is different while sawing or sanding the surface. The emission induced by the sawing test remains short-time emissions while the sanding test generates a high concentration. A comparison is made between three stress tests: scratching, sawing and sanding (Fig. 3). The maximum particle concentration is the mean value of the highest emission peak measured with CPC while testing the three nanostructured parts.

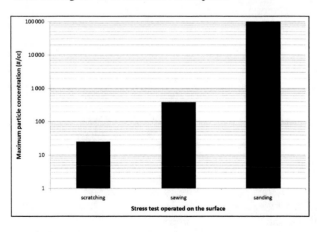

Figure 3 : Maximum particle concentration for three stress test.

The maximum particle concentration is clearly due to the brutality of the stress. Scratching, sawing and sanding represent low, medium and high energy mechanical processes respectively.

Figure 3 suggests that there is a relationship between the maximum particle concentration and the energy applied on the surface. Because of this, an estimation of the energy loss by friction, to determine an order of magnitude of these processes, is made (see table 4) with the following equations [5]:

- For scratching and sanding :

$$E_d = \mu \times F_N \times d \qquad (1)$$

- For sanding :

$$E_d = \mu \times F_N \times r \times \omega \times t \qquad (2)$$

With E_d the energy loss (J), F_N the normal load (N), d the crossed distance (m), μ the friction coefficient, r the radius of the abrasive wheel (m), ω the velocity of rotation (rad.s^{-1}) ant t the stress duration (s).

Process	Energy loss (J)
Scratching	0.01
Sawing	0.4
Sanding	2.0

Table 4 : Estimation of the energy loss.

Figure 4 represents the relationships that can be established between the maximum particle concentration and the energy loss.

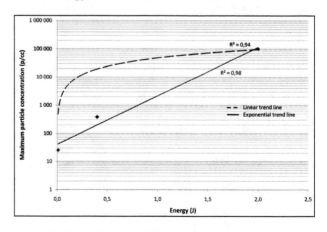

Figure 4 : Relationship between maximum particle emission and energy loss.

As a first approach, two trend lines can be defined : a linear and an exponential functions, both of them with a correlation coefficient R^2 higher than 0,90. It clearly appears that a strong correlation exists between the particle release and the energy applied on the coating.

More data are required, in order to determine which trend line is the more suitable to describe the relationship between the maximum particle concentration and the energy loss.

5 CONCLUSION

The objective of this study was to evaluate a possible nanoparticle release into the air from a metallic part coated with inorganic fullerene, during the R&D process in Renault laboratories.

The stress operated on the nanostructured parts during this process have been reproduced successfully into a controlled atmosphere. The impact of each action on the surface has been identified. Even in the case of low-level emissions, the devices made possible to detect some particle release.

When comparing with a standard part, emissions due to the nanostructured layer were identified. The nano-coating starts to be damaged while scratching whereas the standard part remains intact, due to the differences of hardness between the nanostructured coating and the standard surface. During sawing process, nanoparticle emissions duration stills short, independently of the maximum particle concentration. The surface sanding generates an aerosol which size mode is centered around 30-40 nm with a distribution width from 10 to 100 nm. However, for a normal use, the particle release is under 10 particles per stress, whereas the total particle concentration into the ambient air is above 10,000 p/cc.

A relationship between the maximum particle concentration and the energy loss during the stress test has been approximated, meaning that threshold values can be defined to forecast the mechanical behavior of the layer during specific uses. More stress tests, with different levels of energy, are required to refine the relationship between the energy loss and the maximum particle concentration.

ACKNOWLEDGEMENTS

The authors wish to thank Program 189 for supporting the PhD work of Ludovic Gheerardyn and the FOREMOST European project for the financial support of this study.

REFERENCES

[1] O. Le Bihan and K. Schierholz, "Assessment of nanoparticles emission from manufactured products : feasibility study", Nanosafe 2008.

[2] A. Guiot, L. Golanski and F. Tardif, "Measurement of nanoparticle removal by abrasion", Journal of Physics: Conference series, 170, 2009.

[3] M. Vorbau, L. Hilleman and M. Stintz, "Method for the characterization of the abrasion induced nanoparticle release into air from surface", Journal of Aerosol Science, 40, 209-217, 2009.

[4] I. K. Koponen, K. A. Jensen and T. Schneider, "Sanding dust from nanoparticle-containing paints", Journal of Physics : Conference Series, 151, 2009.

[5] P. Lallemand, "Mécanique (Berkeley : cours de physique, volume 1)", Armand Colin, 144-145, 1972.

Poly(vinyl chloride)/TiO₂ nanocomposite resin via the synthesis of TiO₂ nanoparticles in vinyl chloride followed by *in situ* suspension polymerization

H. Yoo[*] and S.-Y. Kwak[**]

Department of Materials Science and Engineering, Seoul National University,
599 Gwanak-ro, Gwanak-gu, Seoul 151-744, Republic of Korea
[*]friends1@snu.ac.kr [**]sykwak@snu.ac.kr

ABSTRACT

There are various methods established and implemented to prepare polymer/TiO₂ nanocomposites, but the choice of preparation method is limited in the case of most frequently used commodity polymers because of their enormous production and low price. *In situ* suspension polymerization is rarely reported, because suspension is a severe condition to maintain the dispersion of nanoparticles in monomer phase. However, not only the incorporation of nanoparticle but also the preservation of morphology and processing condition can be attained if *in situ* suspension polymerization is achieved in the polymer resins mainly produced by suspension polymerization such as PVC. In this study, we prepared PVC/TiO₂ nanocomposite resin via the synthesis of highly dispersible TiO₂ nanoparticles in the vinyl chloride, followed by *in situ* suspension polymerization of vinyl chloride which contained the nanoparticles. Solvothermal method is applied to the nanoparticle preparation, and the prepared TiO₂ nanoparticles have controlled sizes, shapes, and crystalline phases with the types and concentrations of precursor and oxygen donor. The film samples from prepared PVC/TiO₂ nanocomposites resins are highly transparent due to the minimized visible light scattering which is an evidence of high dispersity of TiO₂ nanoparticles.

Keywords: nanocomposite, titanium dioxide, poly(vinyl chloride), *in situ* suspension polymerization

1 INTRODUCTION

Polymer/inorganic nanocomposites have been extensively studied in the last decade. They combine the advantages of the polymer and the inorganic material, and usually contain special properties of nanofillers leading to materials with improved properties. Among the various inorganic materials, titanium dioxide (TiO₂) can be widely applied to the polymeric materials for the addition of the eco-friendly characteristics such as adsorption and catalytic degradation of various contaminations, inhibition of pathogenic microorganisms, protection of ultravisible radiation, etc. Various methods, including melt extrusion of unmodified or organically modified TiO₂, dispersion of organically modified TiO₂ in organic solvent followed by dissolution of the matrix polymer and solvent casting, or *in situ* polymerization, are established and implemented to prepare polymer/TiO₂ nanocomposites. However, the choice of nanocomposite preparation method is limited in the case of most frequently used commodity polymers such as poly(vinyl chloride) (PVC), because of their enormous production and low price. Moreover, there are considerable difficulties that TiO₂ nanoparticles are easily conglomerated and not simply broken apart by conventional methods. It is obvious that the most important factors which affect the properties of nanocomposites are the dispersion and the adhesion at the polymer and nanofillers [1]. The agglomerates of TiO₂ nanoparticles act as defects which inevitably result in the degradation of the mechanical properties in polymer/inorganic nanocomposites. The decrement of interface area between the polymer matrix and TiO₂ due to the agglomeration, moreover, leads to inferior catalytic activity of TiO₂. Therefore, it is necessary to find the preparation route which satisfies high performance-cost balance in the area of polymer/TiO₂ nanocomposites.

The majority of vinyl polymers are produced by heterogeneous polymerization, e.g. emulsion polymerization, dispersion polymerization, and suspension polymerization. *In situ* emulsion polymerization is the most important method for polymer encapsulation of inorganic particles by far [2]. On the other hand, *in situ* suspension polymerization is rarely reported, because the suspension obtained by shearing a system containing monomer, water, and suspending agent is a severe condition to maintain the dispersion of nanoparticles in monomer phase. However, not only the incorporation of nanoparticle but also the preservation of morphology and processing condition can be attained if *in situ* suspension polymerization is achieved in the polymer resins mainly produced by suspension polymerization such as PVC. In this study, we prepared PVC/TiO₂ nanocomposite resin via the synthesis of highly dispersible TiO₂ nanoparticles in the vinyl chloride (VC), the monomer of PVC, and *in situ* suspension polymerization of VC which contained the nanoparticles. Solvothermal method is applied to the nanoparticle preparation [3], and the prepared TiO₂ nanoparticles have controlled sizes, shapes, and crystalline phases with the types and concentrations of precursor and oxygen donor. The sizes and shapes of the nanoparticles are determined by high-resolution transmission electron microscopy (HR-TEM), and the size distributions of the nanoparticles are determined by dynamic light scattering (DLS) method. The prepared TiO₂ nanoparticles are finally incorporated in the

NSTI-Nanotech 2010, www.nsti.org, ISBN 978-1-4398-3401-5 Vol. 1, 2010

polymer matrix through *in situ* suspension polymerization. The film samples from prepared PVC/TiO$_2$ nanocomposites resins are highly transparent due to the minimized visible light scattering which is an evidence of high dispersity of TiO$_2$ nanoparticles. The predominant dispersions of the nanoparticles are confirmed by transmission electron microscopic and UV-visible transmittance observations of prepared PVC/TiO$_2$ nanocomposites.

2 EXPERIMENTAL

2.1 Materials

Titanium tetrabutoxide (Ti(OBu)$_4$, 98%) and titanium tetraisopropoxide (Ti(iPr)$_4$, 98%) used in the preparation of TiO$_2$ were obtained from Aldrich. Lauric acid, stearic acid, oleic acid, and linoleic acid used for the surface capping of TiO$_2$ were also from Aldrich. Ethanol, cyclohexane, and tetrahydrofuran were of the highest purity available and purchased from Dae Jung Chem. All materials were used without further purification.

Reagents used for suspension polymerization were obtained from the following sources: vinyl chloride (VC) from Korea Gas; octyl peroxyneodecanoate (BND) and dioctyl peroxydicarbonate (OPP) from Chemex Co. Ltd.; poly(vinyl alcohol) (trade name K-420^{TM}) from Kuraray Co. Ltd.; octadecyl dibutyl-4-hydroxyphenylpropionate (IR), dilauryl thiodipropionate (DL), and aluminum sulfate (AS) from Hanwha Chem. Corp. All of these chemicals were used as received.

2.2 Synthesis of TiO$_2$ nanoparticles

In the typical synthesis procedure of TiO$_2$ nanoparticles, Ti(OBu)$_4$ (2 mL), lauric acid (0.05 mol), and liquefied VC (50 g) were mixed at room temperature by mechanical stirring. After continued stirring at room temperature for 10 min, the solution was kept at 80~100 °C for 6~12 h.

Before characterization of TiO$_2$ nanoparticles, VC was slowly vented out and the residual VC was evaporated at room temperature. Organic-capped TiO$_2$ nanoparticles were easily redispersed in solvents such as cyclohexane or tetrahydrofuran. An as-obtained sol of TiO$_2$ nanoparticle materials could be stable for a few weeks. By adding an excess of ethanol, TiO$_2$ nanoparticles were precipitated at room temperature. The resulting precipitates were isolated by centrifugation and washed with ethanol to remove surfactant residuals.

2.3 Preparation of poly(vinyl chloride)/TiO$_2$ nanocomposite resin

In a high-pressure PVC polymerization reactor equipped with a temperature controller, a high-speed mechanical agitator, and a balance-controlled vacuum VC feeder, PVC/TiO$_2$ nanocomposite resin was prepared by *in-situ* suspension polymerization of VC which

contained the TiO$_2$ nanoparticles. A schematic illustration of the reaction procedure is shown in Figure 1. The experimental procedure was as follows. First, the reactor was charged with water and additives. The additives used in the suspension polymerization were K-420^{TM} (as a suspending agent), BND and OPP (as initiators), IR and DL (as antioxidants), and AS (as a scale inhibitor); these additives were combined in the reactor. After TiO$_2$ nanoparticle dispersion in vinyl monomer is added to the dispersion, polymerization was carried out at a temperature of 57.5 ± 0.5 °C and a pressure of approximately 9.0 ± 0.5 bar while agitating the suspension at a speed of *ca.* 600 to 700 rpm until there was a pressure drop of approximately 0.5 bar from the maximum pressure attained. At the end of the reaction, the remaining gas was vented out. The crude product was then washed several times with ethanol and water, filtered, and dried overnight under vacuum.

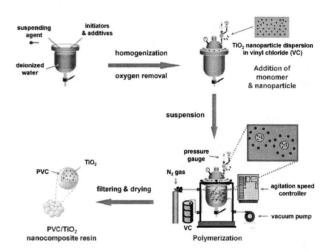

Figure 1: Schematic illustration of the synthetic pathway for the preparation of PVC/TiO$_2$ nanocomposite resin.

2.4 Characterization Techniques

Transmission electron microscopy (TEM). TEM images of TiO$_2$ nanoparticles were obtained using a JEOL JEM-3010 microscope operating at 300 kV. The samples were prepared by dropping dilute solutions of TiO$_2$ nanoparticles in tetrahydrofuran onto 300-mesh carbon-coated copper grids and immediately evaporating the solvent. TEM images of PVC/TiO$_2$ nanocomposite were obtained using a Philips CM20 microscope operating at 200 kV. Ultra-thin cross-sections of the specimens were prepared using a Leica Ultracut UCT ultramicrotome at room temperature.

Powder X-ray Diffraction (XRD). XRD patterns of TiO$_2$ nanocrystal powders were collected with a MAC Science Co. M18XHF-SRA diffractometer using Cu K_α radiation (λ=1.54 Å).

Dynamic light scattering (DLS). The dispersity of TiO$_2$ nanoparticles in hydrophobic liquid medium was estimated by DLS using an DLS-7000 spectrophotometer coupled with a GC-1000 autocorrelator (Otsuka Electronics Co., Ltd., Osaka, Japan) by utilizing an Ar laser (λ=488 nm) at a scattering angle of 90°.

Scanning electron microscopy (SEM). The morphology of PVC/TiO$_2$ nanocomposite was determined with a JEOL JSM-6700F FE-SEM operating at 5 kV. The FE-SEM samples were coated with platinum for 5 min using a JEOL JFC-1100 ion sputter coater.

UV-visible transmittance. Optical properties of PVC/TiO$_2$ nanocomposite were observed with a PerkinElmer Lambda 25 UV-visible spectrometer.

3 RESULTS & DISCUSSION

The prepared TiO$_2$ nanoparticles capped with various organic ligands have controlled sizes, shapes, and crystalline phases with the types and concentrations of precursor, oxygen donor, and organic ligand. Figure 2 are the HR-TEM images of TiO$_2$ samples prepared with titanium tetrabuthoxide as a precursor, lauric acid as an oxygen donor. In the case of L1 and L2 (Figure 2(a) and 2(b)), TiO$_2$ nanospheres were prepared. As the increment of precursor, nanorods appeared (Figure 2(c) and 2(d)). TiO$_2$ nanospheres were almost disappeared and nanorods became major products in L4. Relatively broad diffraction peaks in the X-ray diffraction patterns due to anatase structures are present for all these nanoparticles. Figure 3 is the size distribution graph of a TiO$_2$ nanosphere sample. The TiO$_2$ nanoparticles with very small size and narrow size distribution are highly dispersible in not only vinyl monomers but also various solvents such as cyclohexane or tetrahydrofuran. The dispersibility of TiO$_2$ nanoparticles on the hydrophobic medium can be evaluated by analyzing the hydrodynamic diameter of dispersed nanoparticles through DLS measurement.

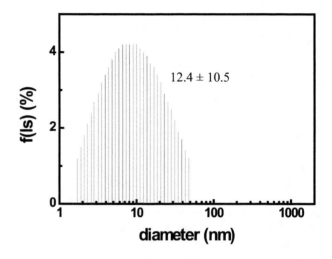

Figure 3: Size distribution graph of a prepared TiO$_2$ nanosphere sample (L1, dispersed in cyclohexane).

Sample code	L1	L2	L3	L4
Average Diameter (nm)	12.4 ± 10.5	7.2 ± 8.4	20.8 ± 54.6	40.8 ± 112

Table 1: Average diameters of the synthesized TiO$_2$ nanoparticles measured by DLS

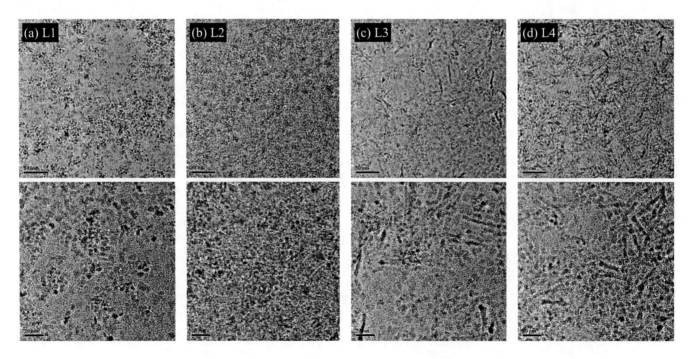

Figure 2: HR-TEM images of the prepared TiO$_2$ nanoparticle samples.

NSTI-Nanotech 2010, www.nsti.org, ISBN 978-1-4398-3401-5 Vol. 1, 2010

The preparation of PVC/TiO₂ nanocomposite resin via *in situ* suspension polymerization of VC requires no additional process step to disperse TiO₂ nanoparticles in the monomer phase, because the synthesis of TiO₂ nanoparticles and the capping of hydrophobic ligand onto the nanoparticles can be performed in liquefied VC by one step process. VC contained the nanoparticles was readily suspended in water and VC droplets containing the TiO₂ nanoparticles were subsequently polymerized to form PVC/TiO₂ nanocomposite resin. Figure 4 shows FE-SEM images of PVC/TiO₂ nanocomposite resin at various magnifications. The diameter of the grains was in the range 100–160 μm, and was thus similar to the grain size range of commercial-grade suspension PVC. The dispersion of TiO₂ nanoparticles in PVC matrix was examined by TEM, as shown in Figure 5. The TiO₂ nanoparticles were well dispersed without significant agglomeration. The degree of dispersion of nanoparticles in the polymer matrix was found to be significantly enhanced by *in situ* suspension polymerization.

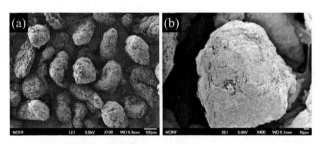

Figure 4: FE-SEM images of PVC/TiO₂ nanocomposite resin with magnifications of (a) × 100 and (b) × 400

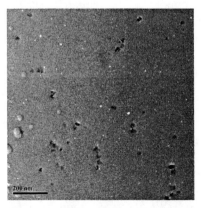

Figure 5: TEM image of PVC/TiO₂ nanocomposites resin.

The film samples from prepared PVC/TiO₂ nanocomposites resins are as transparent as pure PVC (Figure 6). PVC/TiO₂ nanocomposite are transmitting visible light (400 nm < λ < 700 nm) due to the minimized visible light scattering which is an evidence of high dispersity of TiO₂ nanoparticle. Moreover, PVC/TiO₂ nanocomposite is absorbing much of UV light which shows the potential of UV protection material (Figure 7).

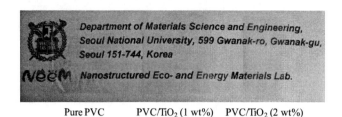

Department of Materials Science and Engineering, Seoul National University, 599 Gwanak-ro, Gwanak-gu, Seoul 151-744, Korea

Nanostructured Eco- and Energy Materials Lab.

Pure PVC   PVC/TiO₂ (1 wt%)   PVC/TiO₂ (2 wt%)

Figure 6: The film samples from pure PVC and prepared PVC/TiO₂ nanocomposites resins.

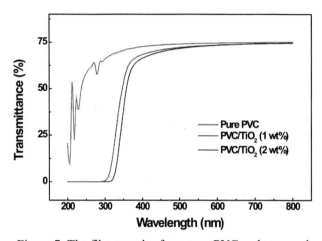

Figure 7: The film samples from pure PVC and prepared PVC/TiO₂ nanocomposites resins.

4 CONCLUSION

In this study, PVC/TiO₂ nanocomposite resin was successfully prepared via the synthesis of highly dispersible TiO₂ nanoparticles in the vinyl chloride (VC) and *in situ* suspension polymerization of VC which contained the nanoparticles. TiO₂ nanoparticles with hydrophobic surface could be synthesized in vinyl chloride directly and the sizes and shapes of TiO₂ nanoparticles were controlled in the variant of precursor concentration. The TiO₂ nanoparticles were well dispersed in PVC matrix without significant agglomeration. The films from PVC/TiO₂ nanocomposites are as transparent as pure PVC, because of the minimized visible light scattering. More investigations on their physical properties will be presented and discussed at the meeting.

REFERENCES

[1] H. Zou, S. Wu, J. Shen, *Chem. Rev.*, 108, 3893, 2008.
[2] E. Bourgeat-Lami, P. Espiard, A. Guyot, *polymer*, 36, 4385, 1995.
[3] X. L. Li, Q. Peng, J. X. Yi, X. Wang, Y. D. Li, *Chem.-Eur. J.*, 12, 2383, 2006.

Functionalization of Carbon Nanotubes
for Advanced CNT/LCP Nanocomposites

Nanda Gopal Sahoo[*], Henry Kuo Feng Cheng[*,**], Lin Li[*], Siew Hwa Chan[*], Jianhong Zhao[**]

[*]School of Mechanical and Aerospace Engineering, Nanyang Technological University, 50 Nanyang Avenue, Singapore 639798, mlli@ntu.edu.sg

[**]Singapore Institute of Manufacturing Technology, 71 Nanyang Drive, Singapore 638075

Introduction

Carbon nanotubes (CNTs) have been highly attractive materials due to their unique structure and extraordinary mechanical, electrical, and thermal properties. CNTs are ideal reinforcing fillers for fabrication of high strength and lightweight polymer composites [1]. Two key factors for development of high performance CNT/polymer composites are (i) homogeneous dispersion of CNTs in a selected polymer matrix and (ii) strong interfacial adhesion between CNTs and the polymer matrix. However, since commercial raw CNTs have a great tendency to form bundles due to strong van der Waals attraction between individual tubes, it is extremely difficult to disperse CNTs into a polymer matrix. Functionalization of CNTs is an effective way to prevent carbon nanotubes from aggregating, which can help CNTs to achieve uniform dispersion and stabilization within a polymer matrix. In addition, for an ideal functionalization of CNTs, it is not only to prevent carbon nanotubes from aggregating for homogeneous dispersion of CNTs, but more importantly functional groups attached to CNTs can act as a compatibilizer (i) to enhance interfacial adhesion between CNTs and a given polymer and (ii) to stabilize the dispersion of CNTs in the polymer matrix. As a result, overall properties such as mechanical, thermal and electrical properties of CNT/polymer composites will be enhanced significantly.

In the present study, we have used a liquid crystalline polymer (LCP) and multiwalled carbon nanotubes (MWCNTs) to develop high performance polymer composites using a melt mixing method. Recently we have selectively introduced two types of chemical moieties (i.e. carboxylic, -COOH, and hydroxyl benzoic acid groups, HBA) on the sidewalls of MWCNTs and mixed them with LCP in order to improve their dispersion in LCP and interaction with LCP [1]. At the same time, a strong interfacial adhesion between the functionalized CNTs and LCP is achieved through formation of hydrogen bonds and between them. The effects of CNT functionalization and loading on the rheological, mechanical, dynamic mechanical and thermal properties of LCP have been investigated in detail. SEM, FTIR and Raman spectroscopy were used to examine the interaction between the functionalized MWCNTs and the LCP. The strong interaction between the functionalized MWCNTs and LCP has greatly contributed to the improvement in the dispersion of MWCNTs in the polymer matrix as well as the interfacial adhesion.

Experimental

The liquid crystalline polymer (LCP) used in this study was a copolymer of 4-hydroxybenzoic acid (HBA) and ethylene terephthalate (ET), with a proportion of 80 % HBA to 20 % ET, which was manufactured by Unitika Ltd. (Japan) with a trade name of Rodrun LC 5000. The multiwalled carbon nanotubes (MWCNTs) were purchased from Iljin Nano Tech, Korea. The diameter and length of MWCNTs were about 10–20 nm and 20 mm, respectively. The synthesis of MWCNT-COC_6H_4COOH (MWCNT-HBA) was followed according to our previous report [1].

The LCP and MWCNTs were melt-mixed using a Haake microcompounder at 300 ^{o}C and a screw speed of 50 rpm. The MWCNT/LCP nanocomposite samples were injection-molded using a Haake micro-injection machine. The content of MWCNT-HBA in the LCP composites was varied to be 0.5, 1, 2, and 4 wt %. For comparison, we have also prepared the LCP composites containing 1 wt % of raw MWCNTs and 1 wt % of MWCNT-COOH, respectively.

NSTI-Nanotech 2010, www.nsti.org, ISBN 978-1-4398-3401-5 Vol. 1, 2010

Results and Discussion

The comparison between the FTIR spectra of LCP and MWCNT-HBA/LCP composites have shown that the -C=O groups of LCP chains might form the hydrogen bonds with -COOH groups of the MWCNT-HBA. The possible mechanism for the improved interfacial interaction between MWCNTs and LCP is proposed in Figure 1.

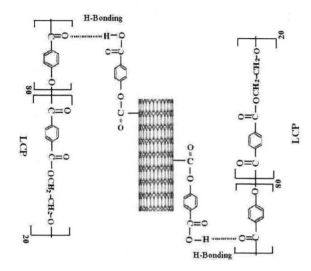

Figure 1. The functionalized MWCNT with hydroxyl benzoic acid groups (HBA) and the possible formation of hydrogen bonding with LCP chains.

At the same time, the dispersion of the MWCNT-HBA became much better than that of raw MWCNTs. This result has also been supported by the observed good dispersion of the MWCNTs in the LCP matrix using SEM. Especially, the longer chemical moieties (like HBA) containing benzene rings attached on the surface of the MWCNTs may provide stronger interactions with the LCP chains via not only H-bonding but also π-π interaction. As an important result, the stronger interaction between the HBA-functionalized MWCNTs and the LCP matrix greatly enhanced the dispersion as well as the interfacial interaction as compared to –COOH functionalized MWCNTs.

The rheological properties of MWCNT/LCP were also significantly affected by the stronger CNT-polymer interaction in the composites than that in a raw MWCNTs/LCP composite at the same CNT loading. The functional groups attached on MWCNTs contributed to the increased CNT-LCP interaction, thus leading to the increase in melt viscosity and storage modulus.

The overall mechanical performance of MWCNT/LCP composites could be improved greatly by a proper functionalization of MWCNTs. For example, the addition of 4 wt % HBA-MWCNTs to LCP could result in the considerable improvements in the tensile strength and modulus of LCP by 66 % and 90 %, respectively. The incorporation of a very small quantity of MWCNT-HBA improved the elongation at break for the composite. However, when the MWCNT-HBA loading was higher, the elongation at break was reduced. In conclusion, the presence of the HBA groups on the MWCNTs surface is likely to strengthen the interfacial interaction between the LCP matrix and the nanotubes in the composites.

References:
Advanced Functional Materials, <u>19</u>, 3962-3971(2009).

Atomistic-Based Continuum Modeling of the Interfacial Properties of Structural Adhesive Bonds

*J.M. Wernik and **S.A. Meguid

*University of Toronto, Toronto, ON, Canada, jake.wernik@utoronto.ca
**University of Toronto, Toronto, ON, Canada, meguid@mie.utoronto.ca

ABSTRACT

This work is concerned with the multiscale modeling of nano-reinforced structural adhesive bonds (SABs) that are typically used in the transport industry, with the aerospace sector being the prime target. Specifically, it is our objective to develop the appropriate formulations to allow the atomistic-based continuum modelling of SABs and the development of a representative volume element that accounts for the nonlinear behaviour of its constituents. This model is used to evaluate the load transfer and strengthening mechanism(s) and interfacial properties of the nanofiller and surrounding adhesive. The work is further extended to establish experimentally, the relative merits of the candidate nanofillers upon the mechanical properties of SABs. We experimentally explore the benefits of selected adhesives, interface properties of SABs with a view to identify candidate systems for industrial applications. The experimental results show that the addition of 2.5 wt% of carbon nanotubes (CNTs) into aerospace structural adhesive can lead to as much as a 40% increase in the tensile strength and 30% increase in the interlaminar shear strength.

1 INTRODUCTION

Most multiscale modeling techniques adopt either the discrete model or atomistic-based continuum models. In the discrete model, it is common to employ tight binding (TB) for electronic band structure, molecular dynamics (MD) for atomistics and finite element FE (or meshfree) methods for continuum. Unfortunately, most FE/MD multiscale modeling techniques suffer from one or more of the following difficulties: (i) the necessity of meshing traditional FE regions down to the atomic scale, thus leading to physical inconsistency and numerical difficulties, (ii) contamination of the solution due to wave reflection resulting from the improper description of the transition zone, and (iii) the FE/MD combination of energy mismatch and mesh incompatibility leads to erroneous non-physical effects in the transition region. A summary of the different discrete modeling techniques can be viewed in the recent review by Wernik and Meguid [1].

The other multiscale modeling approach is the atomistic-based continuum technique. It has the unique advantage of describing atomic positions, their interactions and their governing interatomic potentials in a continuum framework.

The interatomic potentials (deformation measures) introduced in the model capture the underlying atomistic structure of the different phases considered. Thus, the influence of the nanophase is taken into account via appropriate atomistic constitutive formulations. Consequently, these measures are fundamentally different from those in the classical continuum theory. The strength of atomistic-based continuum techniques lies in their ability to avoid the large number of degrees of freedom encountered in the discrete modeling techniques, whilst at the same time allow for the description of the nonlinear constitutive behaviour of the constituents. A schematic of this approach as it relates to the modeling of CNT structures is depicted in Fig. 1.

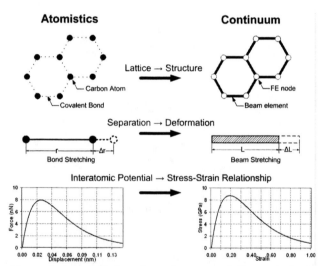

Figure 1: Atomistic-based continuum modeling technique as it relates to CNT structures.

In this paper we describe a novel approach to the development of a nonlinear atomistic-based representative volume element for the study of carbon nanotube reinforced composites.

2 ATOMISTIC-BASED CONTINUUM MODEL

A three-dimensional nonlinear representative volume element (RVE) was developed to study the nano-reinforced polymer system. The RVE consists of the carbon nanotube (CNT), the surrounding polymer matrix, and the CNT-polymer interface, as depicted in Fig. 2. Due to the nanoscale involved in simulating carbon nanotube structures, an atomistic description was incorporated on two levels. First, the carbon-carbon (C-C) covalent bonds in the CNT structure were described using the Modified Morse potential. Secondly, the atomic van der Waals interactions between the atoms in the CNT and the atoms in the polymer matrix were described using the Lennard-Jones potential. This description implies the assumption of a non-bonded interfacial region.

Representative Volume Element

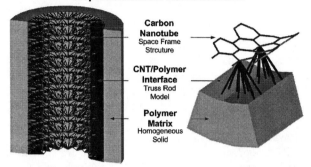

Figure 2: The components used in the development of the representative volume element.

The CNT was modeled as a space-frame structure as depicted in Fig. 3. In the space-frame model, each beam element corresponds to an individual chemical bond in the CNT. As in traditional FE models, nodes were used to connect the beam elements to form the CNT structure. In this case the nodes represent the carbon atoms and their positions are defined by the same atomic coordinates. In this way, the space-frame structure allows the mechanical behavior of the CNT to be accurately modeled in terms of the displacements of the atoms. Figure 3 also illustrates a portion of the hexagonal lattice to demonstrate the connectivity of the beam elements in the continuum representation of the CNT. The Modified Morse interatomic potential was used to derive the material models for the continuum representations of the atomic interactions in the CNT. This potential is well suited for characterizing the behavior of C-C covalent bonding.

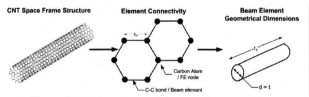

Figure 3: Carbon nanotube space frame structure.

To model the surrounding polymer matrix, a two component-epoxy adhesive (SikaDur330) was used. The stiffness of the epoxy polymer was assumed to be 4.6 GPa and Poisson's ratio was taken to be 0.3. This epoxy adhesive is commonly used in the fabrication of glass reinforced polymer composites (GFRP). The polymer was modeled as a homogeneous solid. The volume of the polymer was varied in order to investigate the effect of the CNT's volume fraction on the effective mechanical properties of the nano-reinforced polymer system.

In this study, we investigate a non-bonded configuration which implies that only van der Waals interactions are considered. In order to simulate the van der Waals interactions, we have adopted the use of a truss rod model whereby each interaction was represented by one truss rod. Each rod extends out from a carbon atom in the CNT structure to an atom in the polymer matrix. The Lennard-Jones interatomic potential was used to describe the behavior of the van der Waals interactions.

3 RESULTS AND DISCUSSION

As an intermediary step in the study of the effective properties of nano-reinforced composites, the nonlinear response of the individual carbon nanotube was investigated. Two carbon nanotube arrangements were studied: the (16,0) zigzag nanotube and the (9,9) armchair nanotube, both having diameters of approximately 1.2 nm and lengths of 10 nm. Two loading conditions were investigated to obtain both tensile and shear properties.

The results show that the zigzag nanotube can withstand a strain of 17%, while the armchair nanotube can withstand a strain of up to 22%. The respective tensile modulus for the armchair and zigzag nanotubes are 944.8 GPa and 920.2 GPa. The overall trend of the stress-strain curves agree well with others in the literature [2,3], and also show that the armchair configuration exhibits higher strength and stiffness when compared to the zigzag configuration. The shear modulus for zigzag and armchair nanotubes was 322.6 GPa and 333.8 GPa, respectively. The present study has predicted shear strengths of 68 GPa and 96 GPA for armchair and zigzag nanotubes, respectively, with corresponding failure strains of approximately 30% and 37%. These values do agree with several quoted values in literature [4,5].

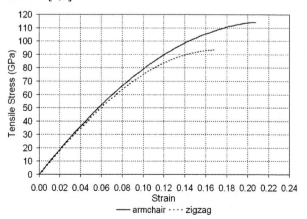

Figure 4: Tensile stress-strain curves for zigzag and armchair nanotubes

The RVE was loaded under tension and torsion to provide the respective properties. For both loading conditions, the volume of the surrounding polymer was varied to investigate the effect of CNT volume fraction on the mechanical properties. For all cases, the CNT was treated as a continuous fiber in the representative volume element. The RVE stress-strain curves confirm that the armchair nanotube acts as a better reinforcing agent when compared to the zigzag nanotube. For CNT volume fractions ranging from 1% to 10%, we see a 3- to 23-fold increase in the Young's Modulus over that of the pure polymer. This demonstrates the reinforcement capabilities of the carbon nanotubes. However, both zigzag and armchair RVEs show a clear deviation from the traditional continuum rule of mixtures. This proves that this continuum law cannot be used to predict the effective properties of nano-reinforced polymers. In the case of torsional loading we again see the same desirable increase in the shear modulus for volume fractions ranging from 1% to 10%. At 1% we see a 16% decrease in the shear modulus while at 10% there is up to an approximate 8-fold increase.

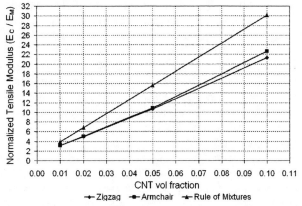

Figure 5: Normalized composite tensile modulus for different CNT volume fractions

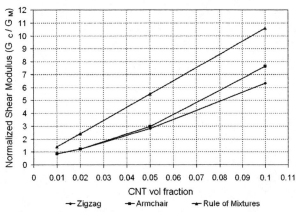

Figure 6: Normalized composite shear modulus for different CNT volume fractions

The measured values of mechanical properties of CNT-based composites reported by various researchers are lower than those predicted by theoretical models. The published experimental results [6-8] have all shown lower improvements in composite stiffness from the addition of carbon nanotubes. This discrepancy can be attributed to the assumption of a defect-free system in the present analysis as well as the use of perfectly aligned CNT fibers. In comparison, the results of the experimental work would have been hindered by the random orientation of CNTs in the polymer matrix, the inability to ensure defect-free nanotubes, as well as the possible agglomeration of CNTs. In addition, the use of multi-walled carbon nanotubes (MWCNTs) have been shown to perform less efficiently as reinforcing agents when compared with single-walled CNTs.

REFERENCES

[1] J.M. Wernik and S.A. Meguid, "Coupling Atomistics and Continuum in Solids: Status, Prospects, and Challenges," International Journal of Mechanics and Materials in Design 5, 79-110, 2009.

[2] T. Belytschko, S.P. Xiao, G.C. Schatz, R.S. Ruoff, "Atomistic Simulations of Nanotube Fracture," Physical Review B 65, 1-8,2002.

[3] J.R. Xiao, J. Staniszewski, J.W. Gillespie Jr., "Fracture and Progressive Failure of Defective Graphene Sheets and Carbon Nanotubes," Composite structures 88, 602-609, 2009.

[4] G.I. Giannopoulos, P.A. Kakavas, N.K. Anifantis, "Evaluation of the Effective Mechanical Properties of Single Walled Carbon Nanotubes Using a Spring Based Finite Element Approach," Computational Materials Science 41, 561-569, 2008.

[5] S. Gupta, K. Dharamvir, V.K. Jindal, "Elastic Moduli of Single-Walled Carbon Nanotubes and their Ropes," Physical Review B 72, 165428-1-16, 2005.

[6] D. Qian, E.C. Dickey, R. Andrews, T. Rantell, "Load Transfer and Deformation Mechanisms in Carbon Nanotube-Polystyrene Composites," Applied Physics Letters 76, 2868-2870, 2000.

[7] L.S. Schadler, S.C. Giannaris, P.M. Ajayan, "Load Transfer in Carbon Nanotube Epoxy Composites," Applied Physics Letters 73, 3842-3844, 1998.

[8] M. Yeh, T. Hsieh, N. Tai, "Fabrication and Mechanical Properties of Multi-walled Carbon Nanotubes/Epoxy Nanocomposites," Materials Science and Engineering A 483-484, 289-292, 2008.

Alignment of Si$_3$N$_4$ nanorods in Polypropylene single fibers

Vijay K. Rangari*, M.Yousuf and Shaik Jeelani.

Materials Science and Engineering, Tuskegee University Tuskegee, AL 3608

* Email: rangariv@tuskegee.edu, Fax: 334 727 2286

ABSTRACT

In the present investigation, we have aligned the Si$_3$N$_4$ nanorods in the polypropylene (PP) fibers through extrusion process to fabricate PNCs. The alignment of Si$_3$N$_4$ in polypropylene/Si$_3$N$_4$ nanocomposite filaments has been carried out in a systematic manner to produce uniform monofilaments. The nanocomposite filaments were first uniformly stretched and stabilized using a two-set godet machine. The as extruded fibers were tested for their thermal and mechanical properties. The results of nanocomposite polymer filaments were compared with neat polymer filaments. These results clearly show that the nanocomposite polymer filaments are much (307 %) higher in tensile strength and modulus (>1000%) as compared to the neat polymer filaments

Keywords: polypropylene, silicon nitride, nanocomposite fibers, mechanical properties

1 INTRODUCTION

Polymer nanocomposites (PNCs) have garnered great academic and industrial interest since their inception [1-6]. PNCs have been shown to enhance physical, thermomechanical, and processing characteristics compared to their neat polymer counterparts. These nanocomposites can be tailored according to the applications and requirements by selection of nanoparticles morphology and compatible polymer matrix. Nanostructured silicon nitride particles exhibit a high-potential for the reinforcement of polymers [4,7-10]. These covalently bonded material in its larger form exhibits high strength and high toughness. It has been used in many industrial applications, such as engine components, bearings and cutting tools [11,12]. Polypropylene fibers have been widely used in apparel, upholstery, floor coverings, hygiene medical, geotextiles, car industry, automotive textiles, various home textiles, wall-coverings [13].

A significant progress has been made in dispersion of acicular or spherical nanoparticles in polymer matrix by surface modification and other techniques. While alignment of acicular or rod shape particles in polymeric fiber and manufacturing advanced macroscopic structures remain a challenge. Most commonly use techniques for alignment are wet spinning [14], magnetic alignment [15] electrospinning [16] and melt processing [17,18]. To align

the nanoparticles while extruding, many aspects should be considered. Such as the starting materials, chemical compositions, mixing techniques, type of extruder, material loading process, extrusion temperature, material residence time within the extruder, the die and its orifice shape and size, extrusion rate, extrusion direction, surrounding air temperature and its speed, fibers cooling type and process, filament draw ratio and winding speed. Proper combination of such factors leads to the production of fine fibers.

In the recent years, much attention has been paid to the synthesis of different morphologies of compounds of silicon nitride nanoparticles. The spherical and rod shaped morphologies are studied extensively among them. Our present interest deals with the study of rod-silicon nitride nanoparticles along with PP. Both neat and nanoparticles infused PP single fibers were produced using a single screw extrusion technique. Morphological and mechanical properties are presented in this paper.

2 EXPERIMENTAL

Materials

The Si$_3$N$_4$ nanorods (light gray color, Si$_3$N$_4$ alpha, 99%, Size 100X800 nm) and spherical shaped (white color, Si$_3$N$_4$ amorphous, 98.5+%, Size 15-30 nm) particles were obtained from *Nanostructured & amorphous Materials, Inc.* Polypropylene (Microthene FP-800-00 microfine powder ~20 μm, melt flow rate 35g/10min and density ~ 0.909 g/cc) powder was purchased from Equistart Chemicals, LP, USA.

Rod shaped Si$_3$N$_4$ particles and PP powder were carefully measured in the ratio of 1:99 by weight (1wt%) and dry mixed using a mechanical blender at a speed of 2300 rpm with a circulating cold base container (5°C). The mixing process was paused for 6 minutes after 3 minutes of continuous mixing and then resumed for further mixing. This was done to avoid temperature rise, which may result in softening the PP. This technique is repeated 20 to 30 times until uniform mixing was observed. The effective time for which the mixture was blended was between 75 to 90 minutes. At this time, the mixture took a light gray appearance and it seemed that the rod Si$_3$N$_4$ particles were thoroughly mixed with the PP.

The mixture of PP and rod-Si$_3$N$_4$ nanoparticles was then further air dried in a dryer for 12 h and extruded using a *Wayne Yellow Label Table Top single screw extruder*.

3 RESULTS AND ISCUSSION

Tensile tests were carried out to study the effect of alignment of rod-Si$_3$N$_4$ particles in PP polymer on tensile strength and modulus. The single fiber tests were carried out according to the ASTM standard D 3379-75 [31] using Zwick/Roell tensile testing equipment with a 20 N load cell. Test specimens were cut each of fibers were aligned to the two ceramic grips, with a distance of separation of 102mm (gauge length). Tensile test results are shown in Figure 1. The ultimate tensile strength of 1wt% rod-Si$_3$N$_4$/ and 2wt% rod--Si$_3$N$_4$ in PP is 249.48 MPa, and 307.21 MPa respectively. Where as the neat PP is 63.58MPa. These results show that the ultimate tensile strength is increased by 383% as compared to that of neat PP.

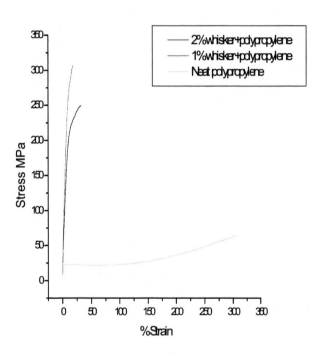

Figure 1. Tensile response of 1wt% and 2wt% of rod- Si$_3$N$_4$ Polypropylene and neat polypropylene filaments

These significant increments in tensile properties are assigned to the alignment and exfoliation of rod-Si$_3$N$_4$ particles in PP polymer. The tensile modulus was also calculated from the stress vs. strain curve of neat PP, 1wt%, and 2wt% of rod- Si$_3$N$_4$ are 0.49GPa, 4.03GPa and 5.03GPa respectively. These significant improvements of tensile

modulus were attributed to the infusion of high strength Si$_3$N4 nanoparticles and their alignment in extrusion direction of PP polymer fibers. This strength/modulus of nanocomposite fiber (307.21MPa/ 5.03.4GPa) is more than that of the neat PP (63.58MPa/0.49GPa) which indicates the exceptional load bearing capability of Si$_3$N$_4$ and their potential applications in structural composite materials. When rod-Si$_3$N$_4$ nanoparticles are increase in strength and modulus of the PP at the cost of a loss in ultimate PP polymer strain.

Morphological Studies

FESEM (*JEOL FE-JSM7000F*) studies have been carried out to study the alignment of rod-Si$_3$N$_4$ in PP fibers. To corroborate the alignment of the Si$_3$N$_4$ rods in PP matrix the polymer was etched and coated with gold palladium before the FE-SEM analysis. Etching of the specimens was carried out in a HUMMER 6.2 plasma etching system with aluminum target. The etching process was done in the argon atmosphere for a span of 1 hour. The vacuum was maintained between 60to 80 millitorr and the current flow was maintained between 10 to 15 Amps. The samples for FE-SEM experiments were prepared by etching the as extruded rod-Si$_3$N$_4$/PP fibers in argon gas for 30 minutes. Due to the etching most of the polymer that covered the matrix layer facing the target was etched out and the Si$_3$N$_4$ rods were clearly seen in the FE-SEM. The Figure 2 (a)

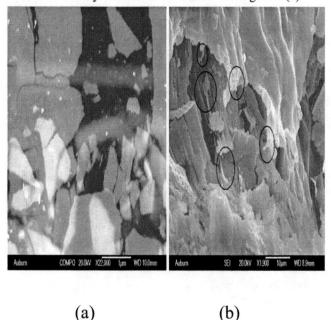

(a)   (b)

Figure 2. SEM micrographs of (a) aligned silicon nitride nanoparticles in the polypropylene matrix and b) particles pullout from the polymer matrix

very clearly shows that the rod shaped Si_3N_4 particles are well aligned in the fiber extrusion direction. These Si_3N_4 particles are surfaced only after deep etching of the surface polymer. Such alignment of the rod shaped particles proves the ability of the extrusion method to align the particles in the direction of extrusion. To find out the alignment of rod-Si_3N_4 particles deep in the fiber a higher magnification FE-SEM was carried out. Figures 2(b) micrograph shows the rod-Si_3N_4 particles are pulled out after tensile testing. These rod-Si_3N_4 particles are aligned in the fiber in different layers one below the other layers. The micrograph clearly shows that the rod shaped particles are aligned in the various levels of the fiber thickness. This phenomenon plays the leading role in enhancing the mechanical properties of composite fibers.

4 CONCLUSIONS

We have successfully aligned the silicon nitride nanorods in polypropylene single filaments through melt extrusion process. Significant enhancement in tensile properties has been observed for 2wt% rod shaped Si3N4 particles. The tensile results clearly show that the nanocomposite polymer filaments are much (307 %) higher in tensile strength and modulus (>1000%) as compared to the neat polymer filaments. The FE-SEM study proves that the rod shaped particles can be successfully aligned in the extrusion direction. These results are consistence with our previous studies on rod- Si_3N_4 nanoparticles infused in Nylon-6 polymers and CNTs infused Nylon6-polymer filaments.

5 ACKNOWLEDGEMENTS

The authors would like thank National Science Foundation for their financial support through NSF-PREM, and NSF-RISE grants.

REFERENCES

[1] Vijaya K Rangari, Mohammed Yousuf, Shaik Jeelani, Merlyn X Pulikkathara and Valery N Khabashesku. Nanotechnology 19 (2008) 245703-245712.

[2] Vijaya K Rangari, Tarig A Hassan, Quentin Mayo and Shaik Jeelani, Composite Science and Engineering, 69, (2009), 2293-2300

[3] Vijaya K. Rangari, Ghouse M. Mohammad, Shaik Jeelani, Angel Hundley, Komal Vig, Shree Ram Singh and Shreekumar Pillai, Nanotechnology, 21 (2010) 095102

[4] Vijaya K. Rangari, Mohammad Y. Shaik, Hassan Mahfuz, Shaik Jeelani, Materials Science and Engineering A 500 (2009) 92–97

[5] Hassan Mahfuz, Ashfaq Adnan, Vijaya k. Rangari and Shaik Jeelani. International Journal of Nanoscience 4, (2005) 55–72.

[6] Richard A. Vaia and John F. Maguire, Chem. Mater, 19, (2007), 2736 -2751.

[7] Zeynep Ergungor, Miko Cakmak, Celal Batur, Macromol. Symp. 185, (2002), 259–276 .

[8] Weifu Dong, Yiqun Liu, Xiaohong Zhang, Jianming Gao, Fan Huang, Zhihai Song, BangHui Tan, and Jinliang Qiao, Macromolecules,38 (2005), 4551-4553.

[9] Wei De Zhang, Lu Shen, In Yee Phang, and Tianxi Liu Macromolecules, 37, (2004)256-259.

[10] Liangwei Qu, L. Monica Veca, Yi Lin, Alex Kitaygorodskiy, Bailin Chen, Alecia M. McCall, John W. Connell, and Ya-Ping Sun, Macromolecules 38,(2005), 10328-10331.

[11] Frank L. Riley, J. Am. Ceram. Soc., 83 (2000), 245–65

[12] P.S. Songa, S. Hwangb, B.C. Sheub Cement and Concrete Research 35, (2005) 1546– 1550.

[13] Sheng Z, Richard H.A, Richard H and Baljinder K. K, Polymer Degradation and Stability, 91, (2006), 7l9-725.

[14] Vigolo, B. ; Penicaud, A.; Coulon, C. ; Sauder, C. ; Pailler, R. ; Journet, P. B.; Poulin, P. Science, 2000, 290,1331.

[15] Correa-Duarte, M.A.; Grzelczak, M.; Salugueirino-Maceira, V.; Giersig, M.; Liz- Marzán, L.M.; Farle, M.; Sierazdki, K.; Diaz, R. J. Phys. Chem. B. 2005, 109, 19060.

[16] Ko, F.; Gogotsi, Y.; Ali, A.; Naguib, N.; Ye, H.; Yang, G.; Li, C.; Willis, P. AdV. Mater 2003,15, 1161.

[17] Rangari, V.K.; Mohammad Y. S.; Mahfuz, H.; Jeelani, S. Materials Science and Engineering A 2009, 500, 92.

[18] Mahfuz, H.; Ashfaq, A.; Mohammad, M.H.; Rangari, V.K.; Jeelani, S.; Steven, D.J.; Applied Physics Letters 2006, 88, 83119

Trifunctional Epoxy – SWNTs Composites. A New Non – Covalent SWNTs Integration Approach

J. M. González-Domínguez[*], A. Ansón-Casaos[*], A. M. Díez-Pascual[**], B. Ashrafi[***], M. Naffakh[**], A. Johnston[***], M. A. Gómez[**] and M. T. Martínez[*]

[*] Instituto de Carboquímica, CSIC, c/Miguel Luesma Castan 4, 50018 Zaragoza, Spain.
jmgonzalez@icb.csic.es ; mtmartinez@icb.csic.es
[**] Departamento de Física e Ingeniería de Polímeros, Instituto de Ciencia y Tecnología de Polímeros, CSIC, c/Juan de la Cierva 3, 28006 Madrid, Spain.
[***] Institute for Aerospace Research, NRC, 1200 Montreal Road, Ottawa, Canada.

ABSTRACT

A novel non-covalent approach to integrate Single Walled Carbon nanotubes (SWNTs) into an epoxy system was addressed. It was based on the SWNT wrapping by a PEO-based Block copolymer (Pluronic F68) which provided highly debundled SWNTs. This procedure allowed performing the integration process without participation of organic solvents. The resulting nanocomposites were characterized by means of electrical and mechanical properties, and compared with different control samples. The reinforcement of the epoxy matrix with Pluronic-wrapped SWNTs provided higher electrical conductivity values and much lower percolation threshold than samples where SWNTs have been integrated without Pluronic. Storage moduli were decreased upon Pluronic incorporation in the control samples, but the presence of SWNTs compensated this plasticizing effect, maintaining the matrix stiffness. Tensile testing of the 0.5 wt% Pluronic-wrapped SWNTs sample revealed a toughness improvement of 276% with respect to the neat matrix.

Keywords: single-walled carbon nanotubes, block copolymer, wrapping, trifunctional epoxy, nanocomposites.

1 INTRODUCTION

Due to their unique physical properties, Carbon Nanotubes (CNTs) have great potential as property enhancers of polymers. However, properly dispersing CNTs within polymers, specifically epoxy matrices, is difficult. Surfactant dispersion of CNTs has been shown to be an effective non-covalent method to improve integration [1], using organic solvents. Block copolymers (BCs) have been treated as ordinary dispersants of CNTs in organic solvents and subsequently incorporated into epoxy resins in a few cases [2-4]. But it is well known that the use of solvents is harmful for the epoxy properties [5]. The solventfree non-covalent CNT-BC wrapping approach has not been used in epoxy resins reinforcement so far. Moreover, BCs exhibit a toughening effect by self-assembly nanostructure formation in epoxy resins [6] and they can also be employed to obtain CNT powders with a high content on disentangled wrapped tubes [7,8]. Our aim is to take advantage of these features to manufacture high performance nano-filled thermoset materials. A novel non-covalent solventless integration technique, in which a block copolymer is used to wrap SWNTs, has been developed combining the SWNT debundling effect with the toughness enhancement of a BC. Characterization of nanocomposites physical properties has been also addressed.

2 EXPERIMENTAL SECTION

2.1 Materials

SWNTs were produced by the arc discharge technique at the Instituto de Carboquímica - CSIC (Zaragoza, Spain), using Ni/Y as catalysts in 2/0.5 atomic ratio. In order to increase purity and reduce metal content, these SWNTs were subjected to a refluxing HNO_3 treatment (1.5M, 2h) followed by centrifugation (10000 rpm, 4h). Acid-treated SWNTs were finally filtered through a polycarbonate membrane (Millipore, 3µm pore size) and thoroughly rinsed with deionised water. These acid-treated SWNTs will be called "SWNTs" hereafter.

The epoxy matrix consists of an aerospace-grade trifunctional epoxy monomer: triglycidyl-p-aminophenol (TGAP), and 4,4'-diaminodiphenyl sulfone was used as the curing agent. Both reagents were kindly provided by Hunstman Advanced Materials (Spain) and used as received.

2.2 Wrapping SWNTs in amphiphilic BC

The wrapping process was carried out by using an amphiphilic BC. A judicious selection of the respective blocks was made attending to their relative miscibility between the epoxy matrix and the SWNTs. Polyethylenoxide (PEO) was selected as the "epoxyphilic" block while Polypropilenoxide (PPO) was chosen as the block with more affinity to SWNTs. One BC which satisfies this composition, was the commercially available Pluronic F68 (Sigma-Aldrich, PEO-PPO-PEO, $M_w = 8400$

g/mol) used as purchased. SWNTs were dispersed in aqueous Pluronic F68 solution following an experimental procedure reported elsewhere [8], making use of tip sonication, centrifugation and filtering. The resulting wrapped SWNTs exhibited a high degree of debundling and contained 30 wt% of BC. The successful integration of Pluronic-wrapped SWNTs into the TGAP+DDS system was studied in terms of curing kinetics revealing an improvement in species mobility [8].

2.3 Nanocomposites preparation

The wrapping effect of Pluronic F68 over SWNTs greatly increased the miscibility of the filler with the epoxy medium, which allowed carrying out the integration process with no participation of organic solvents. The typical experimental protocol to manufacture TGAP-DDS-SWNT blends, which consisted of hot magnetic stirring (at 60°C) and mild tip sonication, can be found in reference [8]. Different filler ratios were incorporated into the epoxy system: 0.1, 0.25, 0.5, 1 and 2 wt% of bare SWNTs or Pluronic-wrapped SWNTs. Different control samples were also manufactured by adding to the TGAP-DDS epoxy system the same amounts of BC present in the different wrapped SWNTs ratios. Epoxy-CNT blends were cast on a homemade steel dish mould sealed by teflon plates. A curing process was carried out in a Perkin Elmer hydraulic press coupled to a Greaseby Specac controlled heater, under 3 Tm of pressure during 30 min. Then, sample specimens were transferred to a Carbolite LTH4/30 oven and postcured at 200°C for 4h.

2.4 Characterization techniques

Scanning electron microscopy (SEM) experiments were performed in a scanning electron microscope (Hitachi S3400N), working in the secondary electrons mode at a high voltage 15 kV and a distance of 5 mm. Cured samples were fractured and the edge was sputtered with a 10nm gold layer prior to their observation.

Dynamical Mechanical Analyses (DMA) were carried out using a Mettler DMA 861 dynamic mechanical analyzer, in the tensile mode at frequencies of 0.1, 1 and 10 Hz. The specimen dimension was ~19.5 × 5 × 2 mm³. The measurements were performed from -100 °C to the temperature at which the specimens degraded and broke, at a heating rate of 2 °C/min.

Tensile testing was made on an MTS 858 Table Top Servohydraulic test frame equipped with hydraulic grips. Samples were cured as a dog-bone coupon following the ASTM D638 standard [9]. Each dog-bone coupon was placed in the grips and tested in displacement control at a loading rate of 0.050 inches/min (1.27 mm/min) to failure. Up to 9 coupons were measured for each sample.

Electrical conductivity was registered with a Keithley 4200-SCS source measurement unit, working at 20V. The specimens were obtained from a dish mould with 8.8 x 8.8

x 2.9 mm³ holes. Nanocomposite samples were set in a sandwich-like arrangement using two copper sheets (0.2 mm thick). Measurements were carried out in a two point probe configuration, placing each probe in different 8.8 x 8.8 mm² square surfaces of the test sample.

3 RESULTS AND DISCUSSION

3.1 SEM observation

SEM images (Figure 1) clearly show how the integration of Pluronic-wrapped SWNTs leads to a much more homogeneous distribution of the filler. SWNTs bundles become thinner than unwrapped SWNTs, which tend to be present in the matrix as agglomerates of entangled bundles.

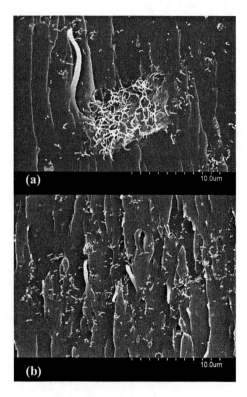

Figure 1: Typical SEM images of the nanocomposite sample containing: (a) 0.5 wt% SWNTs, and (b) 0.5 wt% Pluronic-wrapped SWNTs

3.2 Electrical conductivity

Table 1 shows electrical conductivity values for the different studied samples. The TGAP-DDS epoxy matrix exhibited a conductivity value of $9.7 \cdot 10^{-14}$ S/cm. Conductivity values for the nanocomposite samples showed higher improvements when using Pluronic-wrapped SWNTs as reinforcer. Samples without Pluronic have a less pronounced raise of conductivity values with increasing content of SWNTs, settling at 1 wt% (~10^{-7} S/cm). The maximum conductivity values are achieved with Pluronic-wrapped SWNTs at lower filler content than SWNTs.

In all cases the BC-wrapping approach provides better overall conductivity values and a much lower percolation threshold. This threshold is reached below 0.1 wt% SWNTs in the case of Pluronic-wrapped SWNTs, since the effective SWNT content is lower considering the presence of BC (30 wt%). The conductivity raise of the sample containing 0.1 wt% wrapped SWNTs was about 4 orders of magnitude with respect to the neat matrix.

Filler content (wt%)	σ (S/cm)	
	SWNTs	**Wrapped SWNTs**
0.1	$6.70 \cdot 10^{-13}$	$1.38 \cdot 10^{-09}$
0.25	$5.73 \cdot 10^{-13}$	$6.85 \cdot 10^{-09}$
0.5	$1.04 \cdot 10^{-08}$	$2.91 \cdot 10^{-08}$
1	$1.78 \cdot 10^{-07}$	$2.62 \cdot 10^{-06}$
2	$1.11 \cdot 10^{-07}$	$1.60 \cdot 10^{-06}$

Table 1: Electrical conductivity values for the different epoxy nanocomposite samples. The filler content of SWNTs stands for the actual load. In wrapped SWNTs the filler content stands for 70 wt% SWNTs and 30 wt% BC.

3.3 Dynamical Mechanical Analysis

Figure 1 depicts the storage moduli data drawn from the DMA experiments in all samples. Three series of samples are represented: nanocomposite samples containing variable amounts of bare SWNTs, nanocomposites with variable amounts of Pluronic-wrapped SWNTs and blank samples made by incorporating the same amounts of Pluronic present in each of the wrapped SWNTs samples. Nanocomposite samples based on bare SWNTs showed a progressive and moderate modulus increase with increasing filler content. This series reaches a maximum modulus value at 1 wt% bare SWNTs (~4 GPa) which represents around a 30% improvement with respect to the neat TGAP+DDS matrix. On the other hand, the presence of Pluronic F68 within the epoxy matrix visibly decreased its storage modulus in ~ 30% due to a strong plasticizing effect

caused by the BC. This effect has been previously observed by other authors working with other epoxy systems filled with PEO-based BCs [10]. This effect seems to be independent of the BC amount in the studied range. In Pluronic-wrapped SWNT samples, the storage moduli noticeably increase with respect to their respective blank samples. SWNTs compensate the effect of Pluronic BC, reaching the matrix stiffness (similar modulus value than TGAP+DDS), again with no clear dependence of the filler concentration. The DMA data are in good agreement with atomic force microscopy stiffness characterization of these nanocomposites (unpublished data).

3.4 Tensile tests

Tensile tests were carried out on the sample containing 0.5 wt% Pluronic-wrapped SWNTs (called Epoxy + wrapped SWNTs). Control samples consisted of several cases: the neat TGAP+DDS matrix (called Epoxy), the epoxy containing the same amount of BC present in the nanocomposite (called Epoxy + BC), and finally the epoxy nanocomposite containing 0.5 wt% SWNTs (called Epoxy + SWNTs). Stress (σ) – strain (ε) curves derived from the tensile analyses are represented in Figure 2.

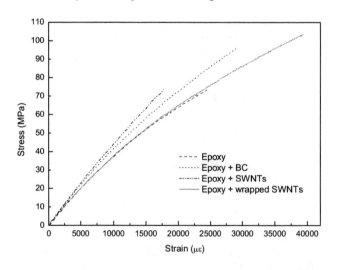

Figure 3: Representative σ – ε curves for the epoxy + 0.5 wt% Pluronic-wrapped SWNTs and its control samples

From the above curves, different mechanical parameters can be obtained. The Young's modulus (E) stands for the slope of the initial linear part of the curve (up to 5000με), while toughness (G) is estimated by calculating the area under the curve. E and G data are collected in Table 2. It is clear how the incorporation of bare SWNTs slightly increases the TGAP+DDS Young's modulus and toughness by 4.7% and 35.3% respectively. The presence of the BC in the same amount as the Epoxy + wrapped SWNTs sample (0.15 wt% BC) has a more pronounced effect over toughness, causing an increase of 70.6% with respect to the neat matrix. This is consistent with the fact that BCs are

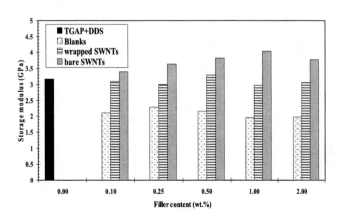

Figure 2: Storage moduli data drawn from DMA

well known to toughen epoxy matrices at a nanoscale level by formation of self-assembled nanostructures [6]. In our case, Pluronic F68 itself is causing a noticeable toughness improvement despite the low amounts in which it is present. Young's modulus for Epoxy + BC sample, however, remains unchanged with respect to the neat matrix. Finally, for the Epoxy + wrapped SWNTs sample there is huge increase of toughness with respect to the neat matrix (276%), which evidences a synergic toughening effect between bare SWNTs and Pluronic F68. While bare SWNTs or BC are not separately causing much toughening, the combination of both (in the form of Pluronic-wrapped SWNTs) provides extremely high toughening effect. The elastic properties seem to be slightly worsened in this last case, since Young's modulus decreases a 7% with respect to the neat matrix.

Sample	Tensile Parameters	
	E (GPa)	G (MPa)
Epoxy	4.3 ± 0.4	0.34 ± 0.30
Epoxy + BC	4.3 ± 0.3	0.58 ± 0.52
Epoxy + bare SWNTs	4.5 ± 0.3	0.46 ± 0.15
Epoxy + wrapped SWNTs	4.0 ± 0.2	1.28 ± 0.74

Table 2: Tensile parameters extracted from $\sigma - \varepsilon$ curves for 0.5 wt% nanocomposite sample.

3.5 Thermo-oxidative stability

Thermogravimetric analysis (TGA) was applied in both inert and air atmospheres respectively to deepen into the degradation profile and its changes with the addition of SWNTs and/or Pluronic at a fixed loading. From the TGA profile (data not shown), isoconversional kinetics calculations [11] were applied and variations of the activation energy (AE) along the degradation process were obtained. Characterization data indicated that the thermo-oxidative profile of the nanocomposites degradation is composed by a very complex thermal stage (with, at least, two steps) followed by a neat single oxidative step (when registered in air) partially overlapped. The presence of SWNTs modifies the AE profile for both the degradation and oxidative steps, suggesting a mechanism alteration. The addition of Pluronic wrapped SWNTs increases the AE at lower conversions in each stage, whereas the control experiment with Pluronic evidenced an improvement in the stability as indicated by an AE increasing in the oxidative stage. It has also been drawn from the TGA data a significant improvement in the flame retardant ability of the composites when Pluronic-wrapped SWNTs are incorporated.

4 CONCLUSIONS

Successful preparation and characterization of novel nanocomposite materials based on a trifunctional epoxy resin and SWNTs was addressed through an integration

protocol which avoids the use of organic solvents. The approach consisted of the SWNTs wrapping by an amphiphilic BC (Pluronic F68). Samples containing BC-wrapped SWNTs possessed higher overall conductivity than samples without Pluronic, with a lower effective amount of SWNTs. Mechanical properties revealed that storage moduli remain unmodified for Pluronic-wrapped SWNT samples while toughness increased a 276% for the 0.5 wt% loading. This enhancement is a consequence of a synergic effect between bare SWNTs and the BC, since both components do not separately present such improvements. These features may be ascribed to the highly homogeneous distribution of SWNTs inside the epoxy matrix when incorporated as Pluronic-wrapped, far better than bare SWNTs (as determined by SEM). Storage and Young's moduli for samples without Pluronic are higher than the neat matrix. For a 0.5 wt% SWNTs epoxy nanocomposite, Storage and Young's moduli are increased by 30% and 35% respectively. The used approach allows enhancing the storage and Young's moduli, or toughness selectively depending on the incorporation of SWNTs with or without Pluronic F68. A thermal conductivity improvement is also expected. These measurements are currently under progress.

REFERENCES

[1] D. Tasis, N. Tagmatarchis, A. Bianco, M. Prato, Chem. Rev. 106, 1105, 2006.
[2] Q. Li, M. Zaiser, V. Koutsos, Phys. Stat. Sol. A, 201, R89, 2004.
[3] J. Cho, I. M. Daniel, Scripta Mater. 58, 533, 2008.
[4] J. Cho, I. M. Daniel, D. A. Dikin, Composites: Part A 39, 1844, 2008.
[5] K. T. Lau, M. Lu, C. Lam, H. Cheung, F. L. Sheng, H. L. Li, Compos. Sci. Technol. 65, 719, 2004.
[6] L. Ruiz-Pérez, G. J. Royston, P. A. Fairclough, A. J. Ryan, Polymer 49, 4475, 2008.
[7] R. Schvartzman-Cohen, E. Nativ-Roth, Y. Levi-Kalisman, R. Yeruzalmi-Rozen, Langmuir 20, 6085, 2004.
[8] J. M. González-Domínguez, P. Castell, A. Ansón, W. K. Maser, A. M. Benito, M. T. Martínez, J. Nanosci. Nanotechnol. 9, 6104-6112, 2009.
[9] ASTM D638. Standard Test Method for Tensile Properties of Plastics. ASTM International, West Conshohocken, PA, USA, 2008.
[10] J. M. Dean, R. B. Grubbs, W. Saad, R. F. Cook, F. S. Bates, J. Polym. Sci. B: Polym. Phys. 41, 2444, 2003.
[11] S. Vyazovkin, J. Comput. Chem. 18, 393, 1997.

Synthesis of polynobornene on catalyst functionalized-MWCNTs with a "grafting from" approach: The influence of the catalyst-MWCNT spacing

P. Longo*,***, C. Costabile*,***, F. Grisi*,***, N. Latorraca*, P. Ciambelli**,***, M. Sarno**,***, C. Leone**, D. Sannino**,***

* Department of Chemistry, University of Salerno, via Ponte Don Melillo, 84084 Fisciano (SA), Italy
** Department of Chemical and Food Engineering, University of Salerno, Via Ponte don Melillo, 84084 Fisciano (SA), Italy
*** NANO_MATES, Research Centre for NANOMAterials and nanoTEchnology at University of Salerno, University of Salerno, 84084 Fisciano (SA), Italy

ABSTRACT

The synthesis of polynorbornene by ring opening metathesis polymerizations (ROMP) in the presence of 2nd generation Grubbs catalyst-functionalized nanotube initiators, is reported. A full characterization of Grubbs catalyst-functionalized nanotubes was performed by FTIR (Fourier Transform Infrared Spectroscopy) and TG-DTG-MS (Thermogravimetrical Analysis coupled with Mass Spectrometer), in order to quantify the amount of catalyst grafted to the nanotube surface.

Keywords: ROMP, 2nd generation Grubbs catalyst (G2), carbon nanotubes, thermogravimetrical analysis, activity

1 INTRODUCTION

The synthesis of polymeric composite materials, containing multi-walled carbon nanotubes (MWNTs) [1], able to combine the mechanical strength and electronic conductivity of carbon nanotubes with the solubility and processability of polymers, has emerged as one of the most important and challenging topic in the framework of the nanostructured materials [2].

The main approaches to obtain homogeneously dispersed filler in the nanocomposites aim for a covalent bound between the polymer and the carbon nanotubes. In particular, as for the "grafting from" approach, catalyst functionalized nanotubes behave as polymerization initiators and polymers can be grown directly from the nanotube surface. This method prevents a full control of the polymerization reaction, but allows high amount of grafted polymer on the nanotube surface [3].

Herein, the synthesis of polynorbornene by ring opening metathesis polymerizations (ROMP) [4] in the presence of 2nd generation Grubbs catalyst-functionalized nanotube initiators is reported. Grubbs catalyst-functionalized nanotubes have been characterized by FTIR (Fourier Transform Infrared Spectroscopy) and TG-DTG-MS (Thermogravimetrical Analysis coupled with Mass Spectrometer), in order to quantify the amount of catalyst grafted to the nanotube surface. Four initiators have been

prepared, differing for the length of the arm connecting the catalyst to the MWNTs.

2 EXPERIMENTAL

2.1 Materials and methods

Materials

Carbon nanotubes, have been prepared by catalytic ethylene chemical vapor deposition (CVD) [4-5] on Co/Fe modified Al_2O_3 powder, and –CCOH functionalized as reported in [5].

The –COOH-MWCNTs were successively functionalized by following the procedure reported by Adronov et al. [3], obtaining initiators **1-4**.

The synthesis of polynorbornene by ring opening metathesis polymerizations (ROMP) in the presence of 2nd generation Grubbs catalyst-functionalized nanotube, initiators **1-4** (Figure 1) has been performed. All initiators have a Grubbs catalyst linked to the nanotube through "chemical arms" differing for the alcoholic group connecting the catalyst to the –COOH functionality of the MWCNTs.

Methods

The Grubbs catalyst-functionalized nanotubes characterisation was performed by FTIR (Fourier Transform Infrared Spectroscopy) and TG-DTG-MS (Thermogravimetrical Analysis coupled with Mass Spectrometer) in order to quantify the amount of catalyst grafted to the nanotube surface. The activity of the **1-4** initiators were estimated.

Infrared spectroscopy has been used to characterize the functionalized MWNTs. The infrared spectra were obtained at room temperature by using a Nexus Thermo Nicolet FTIR (32 scans collected). Powdered materials (100 mg, 0.5 wt% of CNT in KBr) were pelleted (pressure 2 . 104 kg cm22) in self-supporting disks of 0.1–0.2 mm.

Simultaneous TG-DTG-MS analysis, on G2 and initiators **1-4,** was performed at 10K/mn heating rate under flowing air, with a Thermogravimetric Analyser (SDTQ

600 TA Instruments), online connected to a quadrupole mass detector (Quadstar 422, Pfeiffer Vacuum).

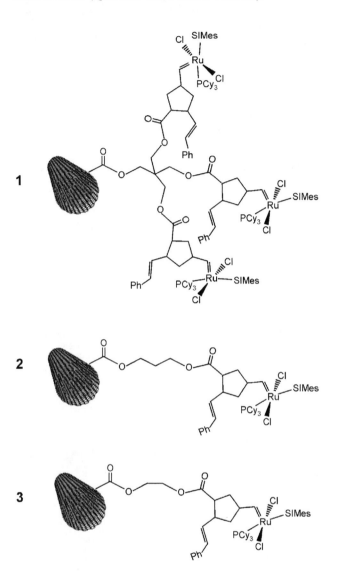

1

2

3

4

Figure 1 Scheme of the 4 initiators (**1-4**) prepared

2.2   Results and discussion

FTIR spectra of initiators **1-4** are reported in Figure 2.

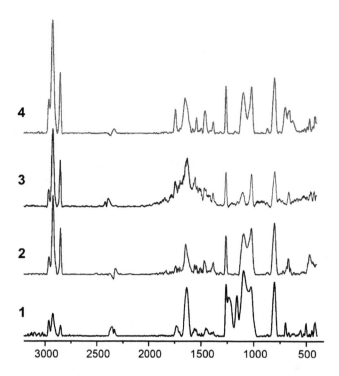

Figure 2: FTIR spectra of initiators **1-4**.

Despite the crowd of signals of FTIR spectra of all initiators, it is possible to identify some of the most characteristic signals. The spectra (Figure 2) present the C-H bond stretching at around 2960 cm^{-1}, due to methyl groups on MWCNTs and/or to –CH$_2$- groups in cyclopentane. The signals at 2125 and 2850 cm^{-1} can be attributed to the asymmetric and symmetric stretching of -CH$_2$- in cyclohexyl groups of the ruthenium coordinated PC$_{y3}$ ligands. Signals at around 1740 cm^{-1} are associated to the C=O stretching of the ester groups. CH$_2$ scissoring absorptions are observed at 1482 cm^{-1}, likely related to the aromatic ring, and at 1463 cm^{-1}, due to cyclohexyl groups. In the range 1270-900 cm^{-1} CH$_2$ rocking and wagging vibrations, mainly attributed to cyclic alkane absorption, are present for all the samples. Finally, the strong peaks at 1635 and 802, and that at 697 cm^{-1}, likely due to the abundance of aromatic groups, could be attributed, respectively, to C=C stretching of phenyl groups and to the C-H and C-C out of plane bending of the aromatic ring of the NHC ligand.

In figure 3 the TG profile of G2 was shown. We can distinguish three distinct steps of weight loss in the ranges 144-220°C, 220-375°C, 375-495°C. Moreover, figure 1c shows the ion currents of the main mass fragments (in the range 30-110 m/z). Taking into account the molecular formula of G2, the 6 most intense mass fragments (m/z = 39, 51, 65, 77, 78, 105) due to the trimethylphenyl groups release, those (m/z = 39, 41, 51, 55, 67, 77, 78, 79, 80, 81) attributed to the decomposition of tricyclohexylphosphine, and the mass fragments (m/z = 39, 51, 61, 62, 63, 65, 89, 90, 91) coming from benzylidene can be recognized.

Together with the main fragments of HCl (35 and 36 m/z) the most intense fragment (m/z = 30) from the decomposition of imidazolidinylidene, and that from CO_2 (m/z = 44) are observed.

By matching the profiles of weight losses and mass fragments of G2 we can associate the decomposition of trimethylphenyls and benzylidene to the first weight loss, and tricyclohexylphosphine to the second loss, in the same temperature range in which the mass fragments of HCl are observed. The decomposition of imidazolidinylidene fraction is associated with the presence of CO_2 mass fragments detected during the third weight loss.

the mass detector in the same temperature range. In particular, the profile of the fragment m/z = 44, formed from the oxidation of the two composite fractions, G2 and CNT, is reported in Figure 4c. We can conclude that the nG2 weight loss at 495°C is due to G2 and the chains that bind to the nanotube G2, and that it constitutes the 28.0 wt% of nG2. From molecular weight calculations we can quantify the contribution of G2 to this first weight loss and conclude that the 23.6 wt.% of nG2 is due to G2. Reminding that in this temperature range G2 loses the 79.0 % of its weight, the total amount of G2 in nG2 is equal to 29.6 wt%. Finally, the Ru amount in G2 has been calculated according to stoichiometry (Table 1).

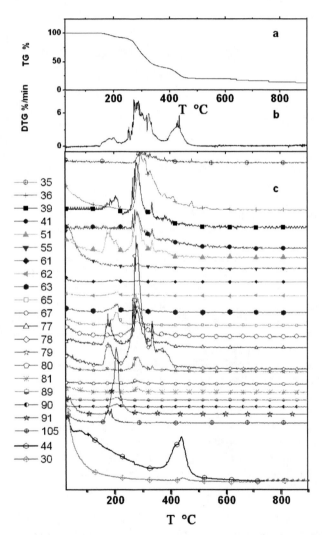

Figure 3: TG-DTG of G2 (a, b); main mass fragments m/z of G2 (c)

In Figure 4 the air flow TG profiles of 2nd generation Grubbs catalyst, run **1** and COOH-functionalized MWNTs (NT) are reported for comparison. We can see that at 495°C the oxidation of 79.6 wt.% of G2 is completed. In the TG profile of nG2 (see, in particular, the DTG profile) we can distinguish the oxidation of G2 and that of functionalized CNTs, as confirmed by the ion currents fragments given by

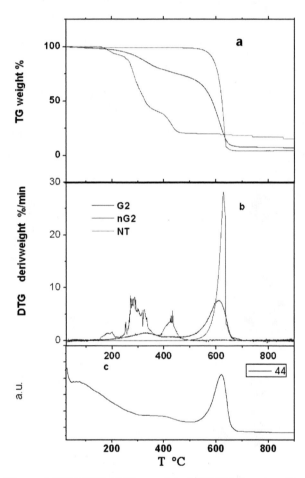

Figure 4: TG-DTG of G2, run **1** and COOH-functionalized MWNT

In Table 1 were collected also the estimation of the amount of catalyst grafted to MWCNTs by TG-DTG –MS analysis for the other three initiators **run 2-4.**

All the initiators **1-4** were able to promote the ROMP of 2-norbornene, results are reported in Table 1, as well. The activities of **1-4** depend on the length of the arm connecting the catalyst to the MWCNTs. In fact, most active initiators showed to be **1** and **2**, whereas a drop of the activity is observed for catalyst **4**. These results can be rationalized by

NSTI-Nanotech 2010, www.nsti.org, ISBN 978-1-4398-3401-5 Vol. 1, 2010

considering the lower accessibility of the monomer to the catalyst grafted to the nanotube through a shorter arm.

Table 1. Estimation of the amount of Ru grafted to MWCNT by TG-DTG-MS and catalyst activity

Samples	$mg_{catalyst}/mg_{sample}$	mol_{Ru}/mg_{sample}	Activity*
1	0.296 ± 0.025	$3.48 * 10^{-7}$	$0.94*10^2$
2	0.309 ± 0.025	$3.64* 10^{-7}$	$0.68*10^2$
3	0.247 ± 0.025	$2.91* 10^{-7}$	$0.26*10^2$
4	0.223 ± 0.025	$2.62* 10^{-7}$	$0.65*10^1$

*Activity: $kg_{polymer}/[monomer][catalyst]h$.

3   CONCLUSION

Four 2nd generation Grubbs catalyst-functionalized nanotubes initiators have been prepared, differing for the length of the arm connecting the catalyst to the MWNTs.

The amount of catalyst grafted to the nanotube surface has been quantified by TGA on-line connected with a quadrupole-mass spectrometer. All the initiators have been able to promote the ROMP of 2-norbornene. We have found that the catalytic activity depends of the length of the arm connecting the catalyst to the MWCNTs. In fact, the most active initiators are those with longer arm, while a drop of the activity is observed for the catalyst obtained connecting G2 directly to the –COOH functionalization. These results can be rationalized by considering the lower accessibility of the monomer to the catalyst grafted to the nanotube through a shorter arm.

REFERENCES

[1] S.Iijima, Nature (London) 1991, 354, 56-58; Dai, H. Surf. Sci., 500, 218, 2002.

[2] J. N Coleman. U.Khan, Y. K. Gun'ko, Adv. Mater. 18, 689, 2006; M. Moniruzzaman,; K.Winey I., Macromolecules 39, 5194, 2006.

[3] A. Adronov, Y.Liu Macromolecules, 37, 4755, 2004.

[4] Bielawski, C.W.; Grubbs, R.H. Prog. Polym. Sci., 32, 1, 2007.

[5] P. Ciambelli, D. Sannino, M. Sarno, C. Leone, U. Lafont Diam Relat Mater 16, 1144, 2007

[6] P. Longo, C. Costabile, F. Grisi, G. Siniscalchi, P. Ciambelli, M. Sarno, C. Leone, D. Sannino Nanotech 2009 - 2009 NSTI Nanotechnology Conference and Expo, 3-7 May 2009, Houston TX ISBN: 978-1-4398-1783-4. Vol. 2 pp 531-533

Quantum Dot – Fibre Composites

Andreas Zeller[*] and James H. Johnston[*]

[*]School of Chemical and Physical Sciences, Victoria University of Wellington,
PO Box 600, Wellington 6140, New Zealand, email: a.j.zeller@gmx.de, jim.johnston@vuw.ac.nz

ABSTRACT

The nanotechnology research presented here is concerned with the development and characterisation of selected quantum dot materials which can be used to produce new hybrid materials comprising such quantum dots and wool and paper fibres that exhibit tuneable optical fluorescence and electronic properties, for potential use in textiles and packaging and labelling papers. ZnS, ZnO and Cu_2O have been selected as they can be readily synthesised and fluoresce in the visible and UV regions. Wet chemical methods have been used to synthesise these quantum dots. The quantum dots have been attached to or incorporated into wool and paper fibre substrates either directly or by using linkers. The chemical bonding, optical and electronic properties of the new quantum dot – fibre composites have been characterised. The applications are being progressed.

Keywords: quantum dots, ZnS, ZnO, Cu_2O, wool, paper, photoluminescence.

1 INTRODUCTION

Quantum dots are nano-scale inorganic semiconductor materials typically 1-10 nm in size, which exhibit interesting electroluminescent, photoluminescent, electronic, magnetic and catalytic properties. Their unique properties have attracted widespread interest and have resulted in applications ranging from optoelectronic devices such as light–emitting diodes, laser diodes and solar cells, to stable inorganic fluorophores [1], fluorescent labelling in biotechnology (e.g. biochips) and medical imaging (optical detection of certain cancerous tumours) [2]. Quantum dots offer advantages in comparison to organic fluorophores as they can be less toxic, relatively resistant to photobleaching and possess a longer lifetime/stability.

The electroluminescent and photoluminescent properties are related to the very small dimension of the nanostructure of the semiconductor crystal. An inorganic semiconductor is characterised by a band gap (E_g) between its valence and its conduction band. By decreasing the size of the semiconductor crystal, the conduction and valence band become a continuum of individual orbital energy levels. The smaller the crystal, the more discrete are the energy levels. Hence the mobility of the electrons is limited within discrete (quantized) energy states of the nanocrystals. This is known as the "quantum confinement effect". Due to the quantum confinement effect, quantum dots exhibit optical

fluorescence when excited thermally, electrically or by UV light. For example, when irradiated with UV light, an electron gains sufficient energy to cross the band gap from the valence band to the conduction band and leave a hole in the valence band. After a series of heat-emitting steps, the excited electron falls back from the lower edge of the conduction band (LUMO) to the hole in the top of the valence band (HOMO), by emitting light. The emitted light is the excess energy of the radiative recombination of the electron-hole pair or exciton, which corresponds to the energy value of the band gap ($E_g = E_{LUMO} - E_{HOMO}$). Hence the wavelength of the emitted light is directly related the energy of the band gap. Changes in the band gap energy have significant effects on the optical and electrical properties of the quantum dots. A decrease in the size of the quantum dot results in an increase in the band gap energy. From this, it follows that quantum dots of the same semiconductor material but with different sizes emit light in different colours. However, light emission and absorption is not only dependent on the physical dimension of the nanostructure, other factors also determine the optical properties. For example, the presence of impurity ions (dopants) in the quantum dot crystal structure can also affect the radiative recombination of the electron-hole pair. Hence an electron or hole is trapped in the energy level within the band gap introduced by a dopant. A controlled addition of small quantities of dopants into the crystal lattice can therefore be used to change the fluorescent wavelength to within the visible range.

We have applied both methods, varying the size and incorporation of metal ion impurities, for selected quantum dot materials in order to obtain tuneable fluorescent colours in the visible range and have developed new hybrid nano-materials comprising quantum dots and paper or wool fibres. Potential applications for these include security packaging and labels for the paper industry and new photoluminescent wool fibres for product authentication in the textiles and fashion industry [3-5]. The quantum dots have been attached to wool and paper fibre substrates either directly or by using linkers which provide chemical bonding between the quantum dots and the respective fibre substrates.

2 QUANTUM DOT SYNTHESIS

By choosing solution and sol-gel chemistry methods, the structure, size and size distribution of the quantum dots can be controlled by changing the reaction concentrations, temperature, time, pH and the stabilizer agent. Stabilizing agents play an important role in the wet chemistry method. Very small particles exhibit a strong affinity to undergo agglomeration and oxidation processes due to the attractive van der Waals forces. This can be inhibited by shielding the semiconductor surface with an ionic capping agent (electrostatic stabilization) or by tailoring the particle surface with a polymeric dispersant (steric stabilization). The stability of the particle dispersion is not only dependent on stabilizer agents, but also on the temperature and solvent used in the wet chemistry method. If the interaction between the surfactant and the solvent is favourable, the nano-crystals remain isolated and do not agglomerate.

2.1 Chemical reduction method

Copper (I) oxide (Cu_2O) quantum dots with interesting optical properties have been synthesised by a chemical reduction method in the presence of an ionic and polymeric stabilizer in aqueous solution.

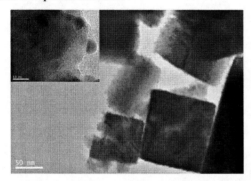

Figure 1: TEM image of regular shaped nanocubes consisting of small Cu_2O nanocrystals.

Initially, 5 nm spherical Cu_2O nanocrystals are formed by reducing aqueous copper nitrate with sodium borohydride ($NaBH_4$) at 100°C in the presence of sodium citrate ($Na_3C_6H_5O_7$) as a capping agent and poly(methacrylic acid sodium salt) (PMAA-Na) as a polymeric dispersant. These small spherical nanocrystals successively transform over time to larger cube-shaped Cu_2O nanocrystals (Figure 1).

Sodium borohydride reduces the Cu^{2+} to Cu^+ ions which then combine with oxygen due to their high surface energy to form the small 2-5 nm Cu_2O nanocrystals. In the second step, these small Cu_2O nanocrystals aggregate to form larger cubic Cu_2O particles which range in size from about 25-250 nm depending upon the synthesis conditions (Figure 1). By changing the process parameters, it has been established that the particle size distribution is dependent principally on the stirring time and sodium borohydride concentration. These Cu_2O nanocrystals exhibit strong yellow-green colour in visible light. When excited with UV light, fluorescent peaks occur at 380, 422, 486 and 530 nm,

with the latter two being responsible for the yellow-green colour. Cu_2O nanocrystals prepared in the cubic shaped form are resistant to oxidation in air and in solution. They also exhibit an interesting dual colour effect. In suspension or coated on the inner side of glass tubes, the crystals show a red colour in transmitted light and a yellow green colour in reflected light (Figure 2). This is similar to the well known Lycurgus Cup which has nanogold and nanosilver particles in the glass matrix.

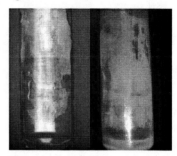

Figure 2: Colour effect of Cu_2O nanocrystals in transmitted light (left) and reflected light (right).

2.2 Sol-gel method

The sol-gel method is often used for the synthesis of metal oxide materials. It make use of the aqueous chemistry of metal salts (typically metal alkoxides and metal chlorides), which is dominated by the complexation with water molecules and their subsequent deprotonation (hydrolysis). ZnO nanocrystals have been prepared by a sol-gel method using a reducing agent, $NaBH_4$, in the presence of polyvinylpyrrolidine (PVP) with a zinc chloride solution, at a reaction temperature of 70°C. The addition of sodium borohydride (alkali) to the acidic zinc salt solution results in a rapid reaction whereby hydroxyl groups coordinate to the zinc ions forming reactive hydrolysed monomers or oligomers. These reactive intermediates are more likely to be involved in condensation reaction due to the higher nucleophilic properties of the OH^- ligand. The polymeric dispersant PVP stabilises the zinc ions in solution and controls the nucleation growth via the condensation reaction by the formation of a protective layer around the ZnO nanocrystals.

The resulting ZnO nanocrystals or quantum dots typically show a yellow-brown fluorescence under UV light with a maximum emission wave length of 550nm (Figure 3).

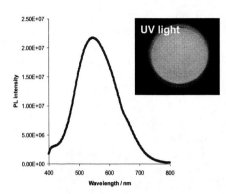

Figure 3: Optical emission spectrum of ellipsoidal shaped ZnO nano-crystals in aqueous solution.

A TEM image shows individual ZnO nanocrystals of about 5 to 10nm in size, loosely adhering to each other to form nanoclusters. (Figure 4).

Figure 4: TEM image of ZnO nanocluster (left) and SEM image of the final polycrystalline product (right). Note the different magnifications and marker bar scales.

The presence of hydroxyl ligands (OH⁻) on the surface of the nanocrystals may have a destabilising effect on the particles due to the higher ionic strength of the medium. ZnO nanocrystals undergo further particle growth via coalescence and aggregation to form approximately ellipsoid shaped crystals of about 25 nm to 1µm in size, as shown by SEM (Figure 4).

2.3 Wet chemical precipitation method

The principle of the wet chemical method is the precipitation of a solid phase from solution during a chemical reaction by controlling the reaction equilibrium between the solid phase and solvated metal ions. This method is readily used in the synthesis of II-VI quantum dots. Zinc sulfide quantum dots have been formed by the slow addition of a Na_2S solution into a $ZnCl_2$ solution in the presence of poly(methyl methacrylate) (PMMA) as a polymeric stabilizer at room temperature. This method also allows the introduction of small quantities of impurities or dopants such as Mn^{2+} and Cu^{2+} ions into ZnS lattice during the precipitation process. This method of adding controlled impurities into the semiconductor material is known as "doping". As mentioned above, the nature and level of the

dopant ion(s) can be used to control the fluorescent wavelength(s). ZnS quantum dots synthesised by this method show a strong blue fluorescence under UV light. However, ZnS quantum dots doped with Mn^{2+} exhibit a yellow-orange fluorescence and Cu^{2+} doped ZnS quantum dots show green-grey fluorescence (Figure 5).

Figure 5: ZnS quantum dots in the undoped form (left), doped with Mn^{2+} ions (centre) and Cu^{2+} ions (right)

3 QUANTUM DOT – FIBRE COMPOSITES

The combination of these quantum dots with paper and wool fibres has enabled the development of new hybrid nanomaterials with photoluminescent functionality.

3.1 Quantum dot – paper fibre composites

Paper fibres consist mainly of cellulose which makes them an excellent choice as a substrate for the formation of a quantum dot – hybrid material. This is due to the presence and the ability of the surface hydroxyl groups on the cellulose chains to bind to various materials through hydrogen bonding. ZnS quantum dots doped with manganese ions and copper ions have already been successfully attached to *Kraft* paper fibres by members of our research group [3]. The printing of doped ZnS quantum dots directly onto fibre surfaces using modified ink-jet technology has also been successfully demonstrated [3].

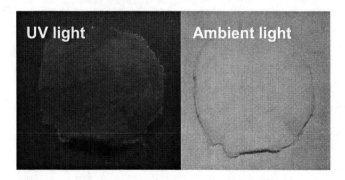

Figure 6: A ZnO paper fibre composite showing a red-orange fluorescence tinge under UV light (left), and the original off white colour under ambient light (right).

ZnO quantum dots produced by the sol-gel method described in section 2.2, can be attached to *Kraft* paper fibres. The new paper composite material shows a red-orange fluorescence tinge under UV light, whereas under ambient light it is an off white colour (Figure 6).

3.2 Quantum dot – wool fibre composites

The main component of wool is keratin, a structural protein, which is made up of long chains of various amino acids. The wool fibre surface is coated with an external fatty acid monolayer, which hinders the adsorption of quantum dots. Alcoholic alkali treatment has been shown to remove the lipid fatty acid mono layer and expose the underlying amino acids. This enables the quantum dot or linker molecule to bind to the nitrogen, oxygen and sulfur atoms on the surface proteins of the wool fibres.

The attachment of the quantum dots onto the wool fibre surface can take place by the use of linkers such as mercaptosuccinic acid. This can be realised by the formation of disulfide, peptide and hydrogen bonds between the quantum dot, linker and functional groups of the wool fibre substrates.

Mn^{2+} doped and Cu^{2+} doped ZnS quantum dots have been attached to wool fibres using mercaptosuccinic acid. Under UV light, the Mn^{2+} doped ZnS wool composite fluoresces with a red-orange colour and the Cu^{2+} doped ZnS wool composite fluoresces with a green colour (Figure 7). The photoluminescent emission spectrum for the Mn^{2+} doped ZnS wool composite shows the typical fluorescent peaks for Mn^{2+} doped quantum dots [3] (Figure 8). The peak at 460 nm is due to S^{2-} vacancies and the peak at 590-600 nm is due to the $^{4}T_{1}$-$^{6}A_{1}$ transition. The incorporation of the quantum dots onto the wool fibres essentially does not change the wavelengths of these photoluminescent emissions.

Figure 7: Photoluminescent Mn^{2+} doped ZnS quantum dot wool fibre composite (left) and Cu^{2+} doped ZnS quantum dot wool fibre composite (right), viewed under UV light.

It has been observed that the intensity of the photoluminescence emission decreases with time due to quenching effects. Wool fibre itself contains fluorescent amino acids, notably phenylalanine, tyrosine and tryptophan. The excitation energy of the quantum dots is transferred to the fluorescent amino acids via resonance energy transfer, thereby decreasing the photoluminescent intensity with time.

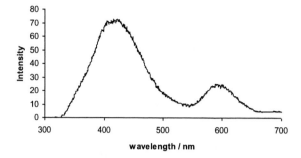

Figure 8: Photoluminescent spectrum of Mn^{2+} doped ZnS quantum dots attached to wool fibres using mercaptosuccinic acid as the linker.

4 CONCLUSION

Photoluminescent quantum dots of Cu_2O, ZnO, and Mn^{2+} and Cu^{2+} doped ZnS materials have been successfully synthesised by wet chemical methods and characterised. These have been incorporated into paper and wool fibres to produce new photoluminescent composite materials for possible security labelling and tracer applications.

REFERENCES

[1] V.M. Ustinov, A.E. Zhukov, A.Yu. Egorov and N.A. Maleev. Quantum Dot Lasers. *Oxford University Press, Oxford* (2003).
[2] M. Bruchez, M. Moronne, P. Gin, S. Weiss and A.P. Alivisatos. Semiconductor Nanocrystals as Fluorescent Biological Labels. *Science*, 281, 2013-2015 (1998)..
[3] Aaron C. Small and James H. Johnston: Novel Hybrid Materials of Cellulose and Doped ZnS Nanoparticles. Current Applied Physics 8, (2008), 512-515.
[4] Michael J. Richardson and James H. Johnston: Sorption and Binding of Nanocrystalline Gold by Merino Wool Fibres – An XPS Study. J. Colloid Interface Science. Vol 310, Iss 2, 425-430 (2007).
[5] F.M. Kelly, J.H. Johnston and M.J. Richardson: Silver Nanoparticles as Colourants and Anti-Microbial Agents For Woollen Textiles. Proceedings of the AUTEX2007: From Emerging Innovations to Global Business, Tampere Finland, (2007).

ACKNOWLEDGEMENT

The research has been carried out with the financial support of a New Zealand FRST NERF Quantum Dot – Fibre Composites Grant, VICX0702 to Professor J. H. Johnston. This is gratefully acknowledged.

Nanogold and Nanosilver Hybrid Plastics

M. Pobedinsky[*] and James H. Johnston[*]

[*] School of Chemical and Physical Sciences, Victoria University of Wellington,
PO Box 600, Wellington 6140, New Zealand. email; maria.pobedinsky@vuw.ac.nz

ABSTRACT

This paper presents an innovative development of proprietary new hybrid plastic materials functionalised by nanogold and nanosilver entities which exhibit optical and anti-microbial effects. The syntheses of these new nano-functionalised plastics have been successfully performed utilising the chemical affinity of gold and silver for nitrogen in order to bind nanogold and nanosilver entities to the nitrogen groups in polyurethane and other N-containing polymers such as nylon. The electron microscopy has shown that nanogold and nanosilver particles have been formed on the surface and within the bulk of the plastic substrates.

Keywords: gold, silver, nanoparticles, anti-microbial, polymers, polyurethane, nylon.

1 INTRODUCTION

Major developments in the properties of thermoplastic polyurethanes (TPUs) were carried out in the 1950s in the research laboratories of BF Goodrich by Schollenberger et al. [1]. Today TPUs are one of the most versatile plastic materials in the world because of their unique properties such as the high elasticity of rubber combined with toughness and durability of metal, flexibility and resistance to many environmental factors. TPU elastomers can be also moulded into any shape which makes the processing easier. TPU is utilized in many applications such as in footwear production, industrial machinery, coatings and paints, production of elastic fibres, water heaters, insulation for buildings, refrigeration, medical devices etc. [2]

The incorporation of metallic nanoparticles into polymers arouses a great deal of interest amongst researchers because of their unusual physical and chemical properties. This paper presents an innovative proprietary development of new plastic materials functionalised by nanogold and nanosilver entities which exhibit surface plasmon resonance optical effects and anti-microbial properties for applications such as air and water filters, membranes, protective textiles, wound dressings and coatings on medical implants [3], which open up new business opportunities. Nanogold and nanosilver entities have increased chemical activity resulting from their large surface to volume ratios and crystallographic surface structures compared with their bulk forms. They exhibit interesting optical properties due to surface plasmon resonance effects and anti-microbial properties, particularly

silver, due to their strong binding to the electron-donating groups in the bacterial cells [4]. There is an increasing demand for new hybrid plastics with modified surface, chemical and anti-microbial properties. Nanogold plastics offer low temperature catalyst applications. The work presented here utilises and builds on the proprietary knowhow of Johnston et al. who have developed new chemistry technology to bind nanogold and nanosilver to natural and synthetic fibres and substrates, and generate new product suites. [5]. The chemical affinity of gold and silver for nitrogen has been used to bind nanogold and nanosilver entities to the nitrogen groups in polyurethane to produce new functionalised plastics that can be moulded and shaped by conventional processes. In a comparable manner, nanogold and nanosilver have been chemically bound to other N-containing polymers such as nylon 66.

2 MATERIALS AND METHODS

2.1 Materials and reagents

All the chemicals were supplied by Sigma Aldrich. Polyurethane beads and nylon 66 beads were provided by Centre for Advanced Composite Materials and the Plastics Centre of Excellence at the University of Auckland, New Zealand.

2.2 Analysis Methods

The extent of uptake or absorption of gold and silver by the plastic substrates was determined by Atomic Absorption using a GBC 9600 Atomic Absorbance Spectrometer. The Transmission Electron Microscopy (TEM) images of the nanocrystals were acquired on JEOL 2011 TEM operating at 200 kV. For TEM analyses, the TPU samples were dissolved in DMF and a drop of the resulting solution was placed onto carbon-coated copper grids, air dried and further carbon coated. The extent of dispersion and elemental analysis of the nanoparticles into the polymeric matrix were investigated by means of a JEOL 6500 F field-emission scanning electron microscope (SEM) and energy dispersive analysis (EDS) operating in a low-vacuum mode at 15 kV and a working distance of 9 mm. The UV–Vis spectra were recorded using a Varian Cary 100 Scanning spectrometer over wavelengths of 200-900 nm.

The antimicrobial activity of nanogold and nanosilver functionalised plastics were tested against ATCC 29213 *Staphylococcus aureus*. Samples of the functionalised beads were placed on a Mueller Hinton agar plate and incubated at 35°C for 18hrs in aerob incubator.

2.3 Synthesis of Nanogold and nanosilver plastics

The uptake of dissolved gold and silver by TPU and nylon was studied by soaking samples of the plastics in different concentrations of gold or silver solutions for different times and temperatures, and analyzing the resulting solution for residual dissolved gold or silver respectively. This showed that at low concentrations all the gold and much of the silver were absorbed respectively.

The nanosilver-TPU hybrid material was then synthesized by the reduction of the absorbed Ag^+ on and within the TPU under controlled conditions. Here the TPU acted as the substrate and simultaneously as the reducing agent due to its amine functionality [5]. A similar procedure was used for synthesis of gold nanoparticles by reduction of an Au^{3+} ($AuCl_4^-$) containing solution in the presence of TPU. By varying the metal ion concentration, temperature and time, the nanoparticle size can be controlled and therefore the color can be tuned. The formation of nanogold on and within the plastics is evident by the appearance of a pink-purple colour resulting from the surface plasmon resonance scattering of light by gold nanoparticles about 20 nm in size. The shade and intensity of the colour increase with increasing nanogold content (Figure 1). For nanosilver, the surface plasmon resonance colour is yellow which similarly increases in shade and depth with increasing nanosilver content (Figure 2). An example of polyurethane containing paint shows that with increasing reaction time the colour develops from light pink to purple (Figure 3).

Figure 1: Polyurethane with increasing nanogold content.

Figure 2: Polyurethane with increasing nanosilver content.

Figure 3: Polyurethane paint with increasing reaction time.

3 RESULTS AND DISCUSSION

The morphologies of the nano-sized silver and gold particles observed by TEM are shown in Fig 4. Higher magnification reveals the characteristic spherical and polyhedral structure of silver nanoparticles in TPU. The diameter of the particles was measured at 6–20 nm (Fig. 4 a). A TEM image of typical gold nanoparticles in TPU is shown in Fig. 4 b. These are about 2-50 nm in size. The SEM and EDS analysis have confirmed the formation of nanogold and nanosilver on the polymer surface and within the bulk of these new hybrid TPU and nylon 66 plastic materials.

The UV-Vis spectrum of the nanosilver TPU sample shows a peak at 420 nm, which clearly indicates the presence of silver nanoparticles (Fig. 5a) [6]. For TPU samples containing gold nanoparticles the UV-Vis spectrum shows a broader peak with the maximum absorption at about 590 nm (Fig. 5b) consistent with the surface plasmon resonance band for gold.

The nanogold and nanosilver hybrid plastics also exhibit effective anti-microbial activity against Gram-positive bacteria *Staphylococcus aureus* (ATCC 29213) as shown by the extensive zone of inhibition (clear area) around a nanosilver TPU bead (Figure 6). Nanogold is similarly effective.

A comparable suite of nanosilver nylon and nanogold nylon materials have been prepared which exhibit the same characteristics and properties as the nanosilver and nanogold TPU hybrid plastic materials.

Examples of the nanosilver and nanogold TPU beads which have been moulded into conventional "dog bone" test strips are shown in Fig. 7. These are very uniform in colour and show that the nanogold and nanosilver entities are distributed evenly through the moulded plastic, confirming that these entities do not affect the thermoplastic forming properties of the polymer substrates.

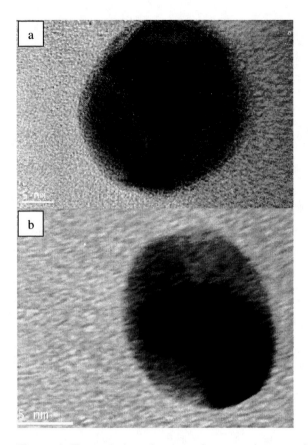

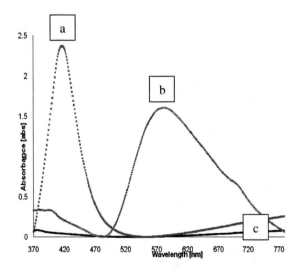

Figure 4: Transmission electron microscope images of a silver nanoparticle (a) and gold nanoparticle (b). Marker bar = 5 nm.

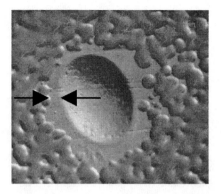

Figure 6: Zone of inhibition for *Staphylococcus Aureus* around a nanosilver TPU bead as shown by the arrows.

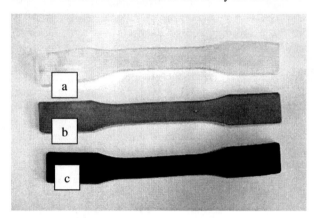

Figure 7: Molded polyurethane "dog bones"; untreated TPU (a); TPU with nanosilver (b); and nanogold (c)

4 CONCLUSIONS

New TPU and nylon hybrid plastic materials functionalised by nanogold and nanosilver entities have been successfully prepared and characterised. These can be moulded into different shapes. by conventional thermoplastic processes. They exhibit different colours due to the surface plasmon resonance effects of the nanogold and nanosilver and have highly effective anti-microbial properties, particularly the nanosilver plastics. As such these nanosilver plastics have considerable potential in commercial applications.

5 ACKNOWLEGMENT

We gratefully acknowledge support from Centre for Advanced Composite Materials and the Plastics Centre of Excellence of University of Auckland, New Zealand This research is part of the "Hybrid Plastics" FRST RFI UOAX0812 programme which provided the funding for the work.

Figure 5: UV-Visible spectra of treated TPU with silver solution (a); treated TPU with gold solution (b); and untreated TPU (c).

REFERENCES

[1] C. Hepburn, *Polyurethane elastomers (2nd ed.),* *Elsevier Science Publishers, London, 1992,*

[2] M. Szycher, Szycher's *Handbook of Polyurethane,* *CRC Press LLC, Florida, 1999.*

[3] H. J. Jeon, J. S. Kim, T. G. Kim, J. H. Kim, W. Yu, J. H. Youk, *Applied Surface Science 245, 5886-5890, 2008.*

[4] Y. Zhang, H. Peng, W. Huang, Y. Zhou, D. Yan, Journal of Colloid and Interface Science 325, 371-376, 2008.

[5] James Johnston, Kerstin Burridge, Fern Kelly and Aaron Small. *NZ and PCT Patent Specification,* *2009.*

[6] J. Prashant, T. Pradeep, *Biotechnology and bioengineering. 90 (1), 59-63, 2005.*

Wide Area Deposition of Transparent Flexible Nanocomposites

T. Druffel

Optical Dynamics Nanotechnology, 1950 Production Ct., Louisville, KY 40299, USA
tdruffel@opticaldynamics.com

ABSTRACT

Inorganic-organic nanocomposites offer a multitude of options to engineer abrasion resistance, improve UV weathering, enhance thermal conductivity, and control refractive index. Enhancements of this kind are typically only achieved if the nanoparticles are randomly dispersed as discrete entities within the polymer matrix. This constraint requires tight control of nanoparticle synthesis and chemical modification of the nanoparticle surface for appropriate polymerization within the matrix. Metal oxide nanocomposite coatings are distinguished by their high visible transparence and preservation of polymer flexibility. In this paper we discuss the deposition of nanocomposite coatings using techniques that are suitable for large and small areas. We also explore specific material properties and the behavior of these coatings under large strains.

Keywords: Nanocomposite, Thin Film, Flexible, Deposition

1 INTRODUCTION

Nanocomposites with various refractive indices and mechanical properties are produced by using different monodisperse inorganic nanoparticles in an optimized polymer matrix. A self-assembled polymer nanocomposite coating composed of metal oxide nanoparticles and a UV-cured acrylate has been developed for flexible applications requiring high visible transparence. This commercial coating includes the hardware, process and chemistry to provide a broadband anti-reflective filter spanning the visible wavelengths for independent opticians and to date there are nearly 100,000 people wearing these nanocomposite coatings in the United States and Worldwide [1]. The strain domains of these films are very similar to those of the polymer substrate and as such do not exhibit the crazing or cracking often associated with films applied through vapor deposition [2]. These coatings perform well on flexible substrates as shown in Figure 1, and phenomenal flexibility has also been demonstrated on 38-layer films on a urethane substrate strained in excess of 25% without film failure. The refractive index of the coatings can be manipulated by changing the packing density of the nanoparticles, and tight control of the particle diameters results in highly transparent films.

A recent review of thin films in solar energy discussed several applications of transparent conductors [3] that could be fulfilled with nanocomposites. Photovoltaic devices

Figure 1. Multilayer film on flexible substrate

utilizing nanoparticles have been demonstrated, and devices using metal and metal oxide nanoparticles in which electromagnetic radiation is transmitted, reflected, concentrated or absorbed can replace more expensive physical vapor deposited (PVD) materials. To move the nanocomposite technology towards sustainable and renewable requires that the films be produced across wide areas. Roll-to-roll techniques are aptly suited for deposition of nanocomposite thin films over wide area substrates. These flexible systems and deposition techniques open applications of sustainable and renewable solar applications on three dimensional surfaces.

We will discuss the chemistry, process and equipment needed to move the evaporative spin coating of nanocomposites to cover wide areas. The deposition represents a significant reduction in costs and offers a versatile process to cover a number of film depositions. The process is capable of depositing highly transparent films of distinct refractive index as shown in figure 2 which is a feasibility of the concept by depositing a nanocomposite with approximately a 45 percent nanoparticles by spray coating. The thickness of the film was measured to be several microns thick and the refractive index of the coating was determined to be approximately 1.75. The coating has been scraped from each end of the slide, and it is evident from the reflectance of the light that the center section has a higher refractive index. This demonstration shows that transparent nanocomposites can be deposited utilizing printing technique without agglomeration of the particles even at high packing densities.

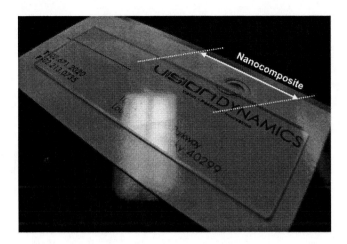

Figure 2: Demonstration of a nanocomposite coating deposited using a micro fluid dispensing valve.

2 BACKGROUND

The primary optical property of a material is its refractive index. This simple ratio is used to explain the propagation of electromagnetic waves through materials and at interfaces. The design and production of optical filters is mainly accomplished through the combination of thin films with unique refractive indices. These are typically dielectric and metal films that are almost exclusively deposited using vacuum deposition. Although these materials have been studied extensively, they have significant disadvantages related to the processing steps and limited mechanical flexibility. The nanocomposites composed of inorganic nanoparticles embedded in an organic polymer matrix directly address these issues. The challenges for nanocomposites are to alter the optical properties of a material without affecting the visible transparence of the final article. The need and methods to build nanocomposites of randomly dispersed nanoparticles for high visible transparence is discussed. The need for discrete dispersions of nanoparticles in the polymer matrix will be discussed along with methods to maintain the separation between the nanoparticles.

2.1 Nanocomposites

Nanocomposites composed of inorganic nanoparticles in a polymer matrix offer a unique platform to engineers, in such that the optical and mechanical properties of the composites can be changed. Nanocomposites with various refractive indices and mechanical properties are produced using different monodisperse inorganic nanoparticles in an optimized polymer matrix [4]. The strain domains of these films are very similar to those of the polymer substrate and as such do not exhibit the crazing or cracking often associated with films applied through vapor deposition. There is no need to use high heats to establish the coating, thus processing on plastics is straightforward. Inorganic-

organic nanocomposites can be assembled from three distinct components: crystalline nanoparticles, polymers and organometallics as shown in the Venn diagram (see Figure 3). The organometallics are the precursor chemistries for the sol-gel process and can be used solely to produce thin film filters. Starting with crystalline nanoparticles to produce an inorganic-organic nanocomposite can be done by simply incorporating a colloidal dispersion of nanoparticles into a monomer and curing. The limitation to this method is that the nanoparticles do not actively participate in the curing, and so a homogeneous material is not formed. Yu et. al. produced thin films on the order of several microns using a colloidal silica and acrylic monomer cured in the presence of heat [5]. Others have developed nanoparticle based layers into an antireflective coating by spin coating and layer by layer deposition [6,7].

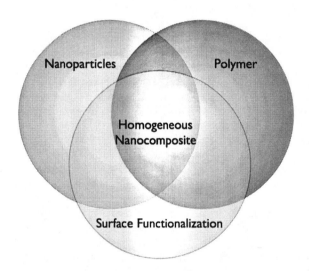

Figure 3: Venn Diagram of an inorganic-organic nanocomposite.

Steric stabilization of the nanoparticles through surfactants or polymer surfaces reduces the surface energy of the nanoparticles and can result in inferior coatings. By functionalizing the crystalline nanoparticles with a polymerizable group, chemical bonds between the nanoparticles and the polymer matrix are formed. This creates a route to produce a homogeneous dispersion of nanoparticles in a polymer matrix which is the key to producing nanocomposites that have high visible transparence while maintaining the elastic properties of the polymer.

2.2 Engineered Properties

The optical and mechanical properties of a nanocomposite are engineered by varying the volume concentration of nanoparticles in the matrix. Using spin coating techniques these properties can be engineered from the base polymer up to 65 volume percent, nearly the theoretical close packing of spheres. The spin coating method is a well understood deposition technique that produces reproducible uniform films that are spread across a substrate with considerable shear forces. When the optical diameter of the nanoparticles becomes to large, optical scattering results as the light waves are reflected from the boundary of the inorganic and organic phases.

The refractive index of a nanocomposite that is a homogeneous mixture of an inorganic metal oxide nanoparticle and an organic polymer can be modeled as a mixture of the two. The refractive index of a material is related to the square root of the permittivity and the permittivity of a composite material is equivalent to the average of the components. Thus the refractive index of a composite material is related as:

$$n^2 = \sum_i n_i^2 v_i \qquad (1)$$

Where n_i and v_i are the refractive index and volume fractions of the components [8]. Although there are numerous relations modeling the mixture of two discrete materials the following is sufficient for most modeling purposes. So by knowing the refractive index of the two materials modeling software can be used to determine the refractive index of a thin film nanocomposite.

The inorganic nanoparticles have a non-uniform refractive index across the visible spectrum, whereas the organic polymer is near constant across the spectrum. Nanocomposite films approximately 500 nm thick with refractive indices between 1.5 and 1.75 measured at 480 nm were made using ZnO nanoparticles a 5 to 50% loading of nanoparticles and a UV curable monomer as shown in figure 4. The films used ZnO dispersed in TMPTA at varying volume loadings and spun coated onto a quartz substrate using the Optical Dynamics spin coater. The source of the ZnO is a nanoparticle dispersion in methyl ethyl ketone (MEK) by Umicore (Zano MEK 067) which is reported to have 30 nm ZnO nanoparticles dispersed at 45 weight percent. The dispersion uses a surfactant to maintain the nanoparticle separation. The refractive index was determined by measuring the reflectance spectrum using a contact spectrophotometer F20 by Filmetrics and measuring the thickness and roughness using a contact profilometer XP-1 by Ambios corporation. This data was then used to determine the refractive index using a Cauchy model. It should be noted that the refractive index of the fully loaded ZnO film has been determined to be approximately 1.75, which is slightly lower than the expected 1.82. This is most likely due to the added surfactant reducing the effective refractive index of the ZnO nanoparticles. A similar study of TiO2 nanocomposite films was undertaken and resulted in a maximum refractive index of 1.88 at a volume packing of 65 volume percent. These films were used to produce thin film reflective filters with up to 38 layers and strained up to 25 percent [9].

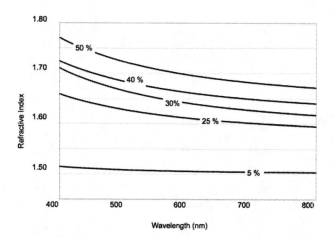

Figure 4: Refractive index dispersion of nanocomposites featuring varying volume fractions of ZnO dispersed in a UV cured polyurethane acrylate.

Another use of nanocomposite films would be as absorptive filters that maintain a high visible transparence. Examples of these would be eye protection from harmful ultra violet and infra red wave-lengths. Metal oxides also have high absorption in the UV region, this coupled with the low absorption in the visible region makes these ideal for optical applications requiring UV blocking. It is well known that titanium dioxide and ZnO have very high absorption in the ultra- violet region. The optical and UV response of 500 nm thick ZnO films of varying volume ratios of ZnO nanoparticles was measured using a UV-Vis spectrophotometer 8453 by Hewlett Packard. As anticipated the ZnO dispersions had a solid edge at about 380 nm which is in line with what is expected of the UV absorbance of ZnO. The UV absorbance ($\lambda = 340$ nm) of the ZnO was linear with the volume fraction of ZnO in the films.

A similar study of the impact of volumetric loading of nanoparticles on the modulus of a thin film resulted in a maximum modulus of near 60 percent loading [10]. In this study silica nanoparticles were dispersed at volumes ranging from 30 to 75 percent by volume and the modulus was measured using nanoindentation.

3   DEPOSITION

All of the films discussed above were spin coated onto 80 mm substrates using a piece of equipment that was specifically designed to deposit thin film nanocomposites. The coater moves up to four separate 80 mm diameter substrates through three process steps: cleaning, coating, and curing. All of the coating parameters (spin speed, air flow, temperature, and sweep) are computer controlled, with the solvent being removed from the coating chamber using a fan. After the coating is applied, the films are cured using a pulsed xenon strobe lamp. All of these parameters must be considered when designing large area deposition.

Spin coating is not an ideal candidate for large area substrates, and thus the engineer is faced with depositing high volume density films without a simply applied force to overcome the thermodynamic surface forces of the nanoparticles. The stabilization techniques used to keep the nanoparticles dispersed in a solvent may not translate into a discrete dispersion in the nanocomposite. Ideally, the functionalization would reduce the surface energy of the nanoparticles to be equivalent to the monomers creating a bulk nanocomposite monomer.

The deposition of thin films using dip coating would be an ideal application of these nanocomposites for large area substrates. A simple set-up was built to pull a glass slide out of a nanoparticle dispersion at speeds between 1 and 25 mm/s. The nanocomposite dispersion was a Cerium dioxide, which is available as a colloidal suspension from Sigma-Aldrich (Product No 289744) and a Trimethylolpropane Triacrylate. The ceria dispersion was functionalized such that acrylate groups surrounded the nanoparticles. The total volume of the nanoparticles was 40 percent. The thickness of the coating was determined to be 270 nm, with a refractive index of 1.8 measured at 480 nm. The composition was then diluted to produce a film that would be on the order of a quarter wavelength (approximately 70 nm) shown in figure 5b. This was then alternately dipped between a high (CeO2) and low (SiO2) nanocomposite to build a nine layer reflective coating (figure 5a).

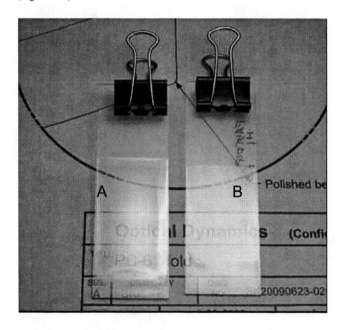

Figure 5: Dipped coated thin films of CeO2 (right b) and CeO2 and SiO2 (left a).

Roll-to-roll methods such as Gravure, knife blade and spray to deposit these coatings on flexible substrates has also been employed. These methods as well as the dip coating method mentioned earlier do not use high shear forces to aid in the homogenous dispersion of the thin films, and so the surface functionalization is important. A demonstration of a roll-to-roll mirror is shown in figure 6.

The color variations across the mirror are due to PET substrate not being flat.

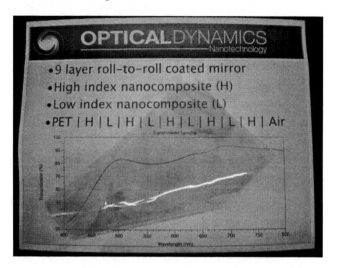

Figure 6: Nanocomposite thin film miror deposited onto a flexible PET substrate

4   CONCLUSION

The results show that metal oxide nanoparticles can be used to engineer important material properties of thin films. These films can be deposited onto very flexible substrates using methods that are very amenable to in-line wide area deposition. Furthermore, the deposition techniques are carried out at very near atmospheric conditions.

REFERENCES

[1] Druffel, T., et al., **2006**, United States Patent Application No. 20060065989

[2] Druffel, T., Geng, K., Grulke, E., **2006** *Nanotechnology*, 17(14): p. 3584.

[3] Granqvist, C.G., **2007**, *Solar Energy Materials and Solar Cells*, 91: p. 1529

[4] Krogman, K., Druffel, T., Sunkara, M., **2005**, *Nanotechnology*, 16, S338

[5] Yu, Y.-Y., Chen, W.-C., **2003**, *Materials Chemistry and Physics*, 82, p. 388

[6] Gemici, Z., Shimmer, H., Cohen, R. E., Rubner, M. F., **2008**, *Langmuir*, 24 (5), p. 2168

[7] Lee, D., Rubner, M. F., Cohen, R. E., **2006**, *Nano Letters*, 6 (10) p. 2305

[8] Yamasaki, T., Tsetse, T., 1998, Applied Physics Letters, 72, p. 1957

[9] Druffel, T., Lattis, M, Spencer, M., Buazza, O., **2010**, *Nanotechnology*, 21 in press

[10] Druffel, T., Mandzy, M., Sunkara, M., Grulke, E., 2008, Small, 4 (4) p. 459

Fabrication and Characterization of the GaP/Polymer Nanocomposites for Advanced Light Emissive Device Structures

S.L. Pyshkin[*], J. Ballato[**], I. Luzinov[***], and B. Zdyrko[***]

[*] Adjunct Professor of Clemson University, SC, permanent address: Institute of Applied Physics, Academy of Sciences, Academy St. 5, MD2028, Kishinev, Moldova, spyshkin@yahoo.com
[**] Center for Optical Materials Science & Engineering Technologies (COMSET), Clemson University, Technology Dr. 91, Anderson, SC 29625, USA, jballat@clemson.edu
[***] School of Materials Science and Engineering and Center for Optical Materials Science and Engineering Technologies, 161 Sirrine Hall, Clemson University, Clemson, SC 29634, USA, LUZINOV@exchange.clemson.edu, BOGDANZ@exchange.clemson.edu

ABSTRACT

Nanoparticles of GaP have been prepared using yellow P and a mild aqueous low temperature synthesis followed by ultrasonication. The resulting nanocomposites yielded a bright luminescence at room temperature in a broad band with the maximum ranging between 2.5-3.2 eV and showing pronounced quantum confinement effects.

Using an improved technology for the preparation of GaP nanoparticles and methods for fabricating GaP/polymer nanocomposites we expect to create a framework for novel light emissive device structures.

Keywords: Light emissive GaP/polymer nanocomposites

1 INTRODUCTION

This work continues our efforts to advance GaP/ polymer nanocomposites for light emissive devices based on GaP nanoparticles. The components of these nanocomposites demonstrate complementary behavior and each is a candidate for use in light emitters, waveguides, converters, accumulators and other planar, fiber or discrete micro-optic elements. While bulk and thin GaP films have been successfully commercialized for many years, its application in device nanocomposite structures for accumulation, conversion and transport of light energy has only received attention recently.

Nanoparticles of GaP have been prepared on the basis of yellow P using mild aqueous low temperature synthesis [1]. The spectra of photoluminescence (PL) and Raman light scattering (RLS), X-ray diffraction (XRD) and electron microscopy (TEM) of the nanoparticles prepared under different conditions have been compared with each other as well as with those from bulk single crystals. After the relevant investigation of different regimes and components for hydrothermal reactions this type of the synthesis has been chosen as an optimal one. In preparation of the nanocomposite we used the fractions of uniform GaP nanoparticles having after a thorough ultrasonic treatment and a number of other operations improving the quality of the nano-suspension a bright luminescence at room temperature in a broad band with the maximum, dependently on the concrete synthesis conditions, between 2.4 – 3.2 eV, while the value of the forbidden gap in GaP at room temperature is only 2.24 eV. Note that according to our investigations of these conditions and data on nanoparticles characterization, only a combination of low temperature synthesis, using yellow P and thorough ultrasound treatment of the reaction products leads to the maximum broad band and UV shift of luminescence.

This paper and a companion paper [1] discuss light emissive sources on the base of transparent robust fluoropolymer and GaP nanoparticles.

2 EXPERIMENTAL PROCEDURE AND DISCUSSION

The details of preparation by mild aqueous synthesis [1, 3] suitable for light emissive nanocomposites GaP nanoparticles as well as the influence of different temperatures, modifications and compositions of the reacting components on quality of the particles can be found in [1].

The dried powders were characterized using standard methods of XRD, TEM, Raman scattering and photoluminescence. For comparison we used also industrial and specially grown and aged GaP single crystals [1-11]. Polyglycidyl methacrylate (PGMA), polyglycidyl methacrylate-co-polyoligoethyleneglycol methacrylate (PGMA-co-POEGMA) and biphenyl vinyl ether (BPVE) polymers were used for preparation on GaP nanocomposites. Thickness of the polymer composite film was within 250-300 nm defined from AFM scratch experiment. The next procedures have been used in fabrication of the nanocomposites:

1) GaP powder was ultrasonicated in methylethylketone (MEK) using Branson 5210 ultrasonic bath. Then, PGMA was added to the MEK solution. GaP to polymer ratio was less than 1:10. Films were deposited via dip-coating;

2) GaP powder was dispersed in water-ethanol mixture (1:1 volume ratio) and ultrasonicated using Branson 5210 bath for 120 min. Then, polyglycidyl methacrylate-co-polyoligoethyleneglycol methacrylate (PGMA-co-

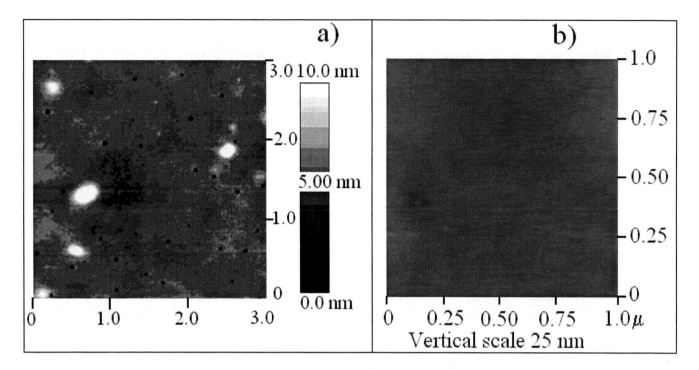

Figure 1 : AFM topography images of the GaP/PGMA (Fig. 1a) and GaP/ PGMA-co-POEGMA (Fig. 1b) nanocomposites.

POEGMA) was added in form of water-ethanol mixture (1:1 volume ratio) solution. GaP to polymer ratio was less than 1:3. Films were deposited on quartz slides via dip-coating;

3) GaP powder was dispersed in the biphenyl vinyl ether/dichloromethane (BPVE/DCM) solution, then the solution was stirred and filtered from the excess of the powder. A few mL drops of the settled solution were casted onto silicon wafer.

Surface morphology of the films was studied with AFM. AFM was performed on a Dimension 3100 (Digital Instruments, Inc.) microscope. We used the tapping mode to study the surface morphology of the composite films in ambient air.

Luminescence of the nanocomposites was excited by the N_2 laser nanosecond pulses at wavelength 337 nm and measured at room temperature.

Figure 1 shows the AFM topography images of the GaP/PGMA (Fig. 1a) and GaP/ PGMA-co-POEGMA (Fig. 1b) nanocomposites. Note that thoroughly washed, ultrasonicated and dried nanopowder obtained by mild low temperature aqueous synthesis from yellow P may be used for fabrication of quite good blue light emissive nanocomposites (**Fig. 1a** and **Fig. 2**, **spectrum 2**), but the best quality can be obtained only in fabrication of the nanocomposite from similarly prepared nanoparticles, but stored as a suspension in a suitable liquid (**Fig. 1b** and **Fig. 2, spectrum 1**).

Figure 2 shows spectra of luminescence from GaP/ PGMA-co-POEGMA nanocomposites. GaP nanoparticles have been prepared using yellow P by mild aqueous synthesis at decreased temperature, stored as the suspension

in a liquid (**spectrum 1**) or the dry powder (**spectrum 2**).

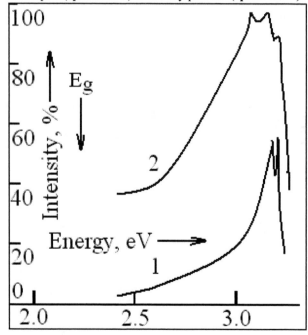

Figure 2 : Spectra of luminescence from GaP/ PGMA-co-POEGMA nanocomposites. GaP nanoparticles have been prepared using yellow P by mild aqueous synthesis at decreased or temperature and stored as the suspension in a liquid (**spectrum 1**) or the dry powder (**spectrum 2**).

The maximum UV shift of the luminescence, app. 1 eV has the nanocomposite prepared on the base of the suspension (**spectrum 1**), so, one can assume that the mostly uniform and small high quality nanoparticles with pronounced quantum confinement effect are stored in the suspension.

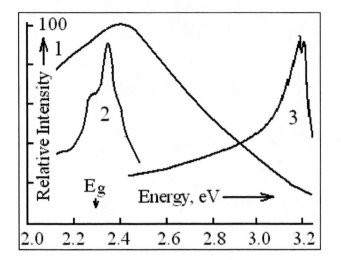

Figure 3 : Luminescence spectra from perfect, long-term ordered GaP single crystals (1) in comparison with the spectra of GaP nanoparticles and nanocomposites, prepared in 2005 – 2006 (2) and now (3).

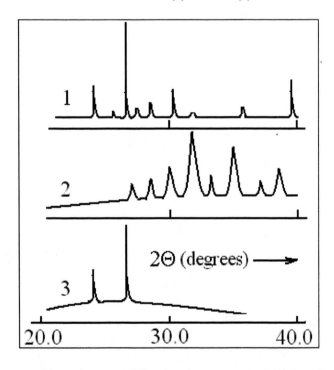

Figure 4: X-ray diffraction from GaP nanoparticles of the best to the moment performance (**spectrum 1**) in comparison with the diffraction from not thorough prepared GaP nanoparticles and perfect GaP bulk single crystal (**spectra 2 and 3 relatively**).

Figure 3 presents the luminescence spectra of our perfect, long-term (up to 50 years) ordered GaP single crystals in comparison with the spectra of GaP

nanoparticles and nanocomposites, prepared in 2005 – 2006 and now. As we note earlier [3, 6, 8, 11], the luminescence was absent at room temperature in newly-made industrial and in our freshly prepared crystals, but it was bright in the same our 40 years aged crystals and shifted to UV up to 2.4 eV at the 2.24 eV forbidden gap for GaP bulk single crystals (**Fig. 3, spectrum 1**). Our first attempts to prepare GaP nanoparticles, dated by the years 2005-2006 [3], gave their room temperature luminescence with maximum shifted only to 2.4 eV (**Fig. 3, spectrum 2**) that in comparison with the new maximum at 3.2 eV (**Fig. 3, spectrum 2**) confirms serious achievements in technology of GaP nanoparticles and GaP based nanocomposites.

Considerable changes in quality of GaP nanoparticles due to improvement of their preparation are shown in **Figure 4**, where one can see X-ray diffraction from GaP nanoparticles of the best to the moment performance (**spectrum 1**) in comparison with the diffraction from not thorough prepared GaP nanoparticles and perfect GaP bulk single crystal (**Figure 4, spectrum 2 and 3 respectively**). One can see very narrow and intense XRD lines from GaP nanoparticles and bulk single crystals of high quality contrary to broad diffraction lines from not thoroughly prepared GaP nanoparticles (**spectra 1, 3** contrary to **spectrum 2**).

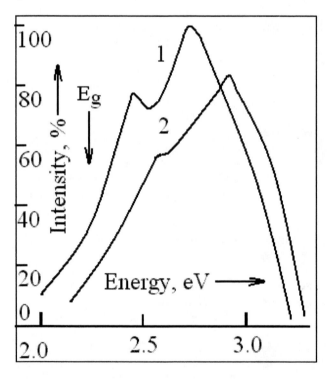

Figure 5 : Luminescence spectra of 2 GaP/BPVE nanocomposites produced on the base of 2 parties of GaP nanoparticles prepared using different conditions.

Recently we started investigation of the GaP/BPVE nanocomposite as a prospective light emissive device structure. **Figure 5** shows the luminescence spectra of 2 nanocomposites produced on the base of 2 parties of GaP nanoparticles obtained under different conditions (temperature, modification of P, etc.). The details and results of this work will be published in the nearest future.

In this paper we note that in the GaP/BPVE nanocomposite the position of the luminescent maximum can be changed between 2.5 – 3.2 eV and the brightness is 20-30 more than in the PGMA and PGMA-co-POEGMA matrixes.

3   CONCLUSIONS

Using an improved technology for preparation of GaP nanoparticles and methods of fabrication of the GaP/ polymer nanocomposites we can change within the broad limits the main parameters of luminescence.

ACKNOWLEDGEMENTS

The authors are very grateful to the US Dept. of State, Institute of International Exchange, Washington, DC, the US Air Force Office for Scientific Research, Science & Technology Center in Ukraine (STCU), Clemson University, SC, Istituto di elettronica dello stato solido, CNR, Rome, Italy, Universita degli studi, Cagliari, Italy, Joffe Physico-Technical Institute, St. Petersburg State Technical University, Russia and Academy of Sciences of Moldova for support and attention to our extended (1963-2009) research efforts.

Our special gratitude to Dr. E. Rusu and Dr. N. Tsyntsaru from Institute of Electronic Engineering and Industrial Technologies, Dr. A. Siminel, post graduate students S. Bilevschii and A. Racu from Institute of Applied Physics, Moldova, Dr. D. Van DerVeer, Clemson University, for their contribution in elaboration of GaP nanotechnology, characterization of nanoparticles and nanocomposites.

REFERENCES

[1] S.L. Pyshkin, J. Ballato, G. Chumanov, N. Tsyntsaru and E. Rusu, "Preparation and Characterization of Nanocrystalline GaP for Advanced Light Emissive Device Structures", will be published in Proceedings of the Nanotech Conference and Expo 2010, June 21-25, Anaheim, CA

[2] S. Pyshkin, J. Ballato, M. Bass, G. Chumanov and G. Turri, "Properties of the Long-term Ordered Semiconductors", The 2009 TMS Annual Meeting and Exhibition, San Francisco, Feb 15-19, Suppl. Proc., Vol. 3, pp 477-484.

[3] S.L. Pyshkin , J. Ballato, G. Chumanov, J. DiMaio and A.K. Saha, "Preparation and Characterization of Nanocrystalline GaP", Symposium "Nanoelectronics and Photonics", 2006 NSTI Nanotech Conference,

Boston, May 7-11, Technical Proceedings of the Conference, Vol. 3, pp 194-197

[4] S.L. Pyshkin, J. Ballato, G. Chumanov, "Raman light scattering from long-term ordered GaP single crystals", J. Opt. A: Pure Appl. Opt. **9** (2007) 33–36, IOP Publ. House, London

[5] S.L. Pyshkin, J. Ballato, M. Bass, G. Turri, "Luminescence of Long-Term Ordered Pure and Doped Gallium Phosphide", (invited), Symposium "Recent Developments in Semiconductor, Electro Optic and Radio Frequency Materials", TMS 2007 Annual Meeting & Exhibition, Orlando, FL, Feb - March 2007; J. Electronic Materials, Springer, Vol. 37, #4, pp 388-395 (2008)

[6] S. L. Pyshkin, R. Zhitaru, J. Ballato, "Modification of Crystal Lattice by Impurity Ordering in GaP", Int. Symposium on Defects, Transport and Related Phenomena, Proc. MS & T 2007 Conf., pp 303-310, Sept 16 –20, 2007, Detroit, MI

[7] S.L. Pyshkin, J. Ballato, M. Bass, G. Turri, (invited), "Evolution of Luminescence from Doped Gallium Phosphide over 40 Years", TMS 2008 Annual Meeting, Symposium: Advances in Semiconductor, Electro Optic and Radio Frequency Materials, March 9–13, New Orleans, LA, J. Electron. Mater, Vol. 38, #5, pp 640-646 (2009).

[8] Sergei L. Pyshkin, John Ballato, Michael Bass, George Chumanov and Giorgio Turri, "Time-dependent evolution of crystal lattice, defects and impurities in $CdIn_2S_4$ and GaP",The16th Int Conference on Ternary and Multinary Compounds (ICTMC16), Berlin, Sept 15-19, 2008; Phys. Status Solidi, C 6, No. 5, pp 1112–1115 (2009).

[9] S. Pyshkin, R. Zhitaru, J. Ballato, G. Chumanov, M. Bass, "Structural Characterization of Long Term Ordered Semiconductors", Proc. of the 2009 MS&T Conference, Pittsburgh, October 24-29, Int. Symposium "Fundamentals & Characterization", Session "Recent Advances in Structural Characterization of Materials", pp 698-709.

[10] Sergei Pyshkin, John Ballato, Andrea Mura, Marco Marceddu, "Luminescence of the GaP:N Long-Term Ordered Single Crystals", Suppl. Proceedings of the 2010 TMS Annual Meetings (Seattle, WA, USA, February, 2010, vol.3, pp 47-54.

[11] "Advanced Light Emissive Device Structures", STCU (www.stcu.int) Project 4610, 2009-2012.

Carbon Nanocomposite Electrode
for Electrostatic Precipitators

K. Alam, J. Morosko

Russ College of Engineering and Technology, Ohio University, Athens, OH 45701

e-mail: alam@ohio.edu Phone: (740) 593-1558   Fax: (740) 593-0476

D. Burton

Applied Sciences, Inc, Cedarville, OH

e-mail: dburton@apsci.com, Phone: (937) 766-2020

ABSTRACT

The electrostatic precipitator[1] (ESP) is a key pollution control component in air pollution control devices for small oil and coal-fired industrial boilers. ESPs can operate with an efficiency of 98 to 99% for the removal of mercury and fly ash from the flue gas stream. Within the electrostatic precipitator, fly ash particles are charged electrically due to a corona current from discharge electrodes (Figures 1 and 2) as the flue gas passes through the precipitator, allowing the charged particles to be collected on oppositely charged plates typically made of metal.

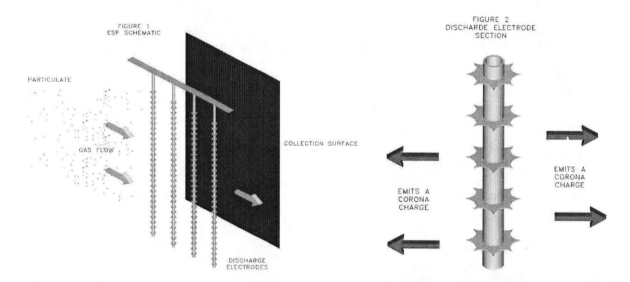

Figure 1. Conceptual view of ESP showing the discharge electrode in the gas flow

Figure 2. Detailed schematic of discharge electrode and corona field

Recent advances in the ESP technology include the development of the wet ESP[2], in which water flow on the collection electrode improves the performance of the ESP. Wet ESPs use a metal or fabric for the collection surface. They typically have higher capture of sub-micron particulates due to the better adhesion of the particles to the liquid flowing on the collection surfaces, and virtually no re-entrainment. Wet ESPs are very effective when fine particles need to be removed from the exhaust air stream, typically with aerodynamic diameters less than 2.5 µm. Wet ESPs that use metal discharge electrodes must use alloys such as stainless steels or different grades of hastelloy to protect against the highly corrosive environment. Therefore, these electrodes are very expensive. Polymer nanocomposite electrodes are a less expensive alternative for wet ESPs, particularly because of their corrosion resistance. The lower

NSTI-Nanotech 2010, www.nsti.org, ISBN 978-1-4398-3401-5 Vol. 1, 2010

weight of the polymer nanocomposite electrode also provides an advantage because the supporting structure can be designed for lower loads.

This paper discusses the development of a novel carbon composite electrode for ESPs by using an electrically conductive polymer matrix composite with carbon fiber. Graphitic carbon microfibers that have been used in composites for aero-space and sports equipment can be sufficiently conductive for the discharge electrode application. To make a conductive polymer, carbon nanofibers can also be used as an additive in small volume fractions. When the volume fraction of the nanofiber is small, the nanofiber may not provide much enhancement of the mechanical strength; but the electrically conductivity can be improved significantly.

One option for making conductive electrodes is to incorporate continuous carbon microfiber in a polymer matrix using a pultrusion process. Further improvement of electrical conductivity can be achieved by adding carbon nanofibers to the polymer resin so that there is greater continuity of electrical contact between the microfibers and nanofibers within the polymer matrix. Good results are obtained if the nanofibers are well dispersed and aligned. Analysis and simulation of the nanofiber incorporation (both dispersion and alignment) into a polymer has been carried out at Ohio University[3].

However, producing a composite with nanofibers requires careful processing steps because the stiff nanofibers must be dispersed without significant breakage. If the fibers are broken due to the mixing action, the reduction in fiber length can significantly reduce the conductivity because the shorter fibers may not provide a continuous path for the flow of current. This may degrade the electrical conductivity and make it difficult for the composite product to match the performance requirements of the metal discharge electrodes.

For the nanocomposite electrode to be a potential replacement for the traditional metal electrode, a typical test is to verify that it will produce comparable or superior voltage-current (V-I) characteristic as the current state-of-the-art ESP components. In this study, nanocomposites were produced in the form of pultruded tapes containing carbon microfibers and carbon nanofibers in a polypropylene matrix. The nanofibers (Pyrograf III, PR-19, produced by Applied Sciences, Inc., Cedarville, OH) were embedded into the tape by a thermoplastic pultrusion process with tows of carbon microfiber (Besfight G30-700). The tapes were typically 10 mm wide and 1 mm thick.

The nanocomposite tapes were then supported on a polymer tube (Figure 3) to form the electrode. The metal electrode shown in the middle was fabricated on the basis of guidelines provided by a designer and manufacturer of electrodes[4]. All the electrodes were then tested in three different small and large scale ESP chambers (Figures 4 to 7) to determine the corona discharge current. Figure 4 shows the metal electrode in a 6 inch high test chamber. Figure 5 shows a similar metal electrode in a 3 feet high test chamber. And Figures 6 and 7 show a large scale test chamber that can test electrodes that are 12 feet long. In Figure 7, the composite electrode is shown in the test configuration within the large scale test chamber.

Figure 3. Electrode configurations of nanocomposite tapes wrapped on a polymer tube. The steel electrode in the center is designed on the basis of a commercial discharge electrode.

Figure 4. ESP test chamber 1

Figure 5. ESP test chamber 2

Figure 6. ESP test chamber 3

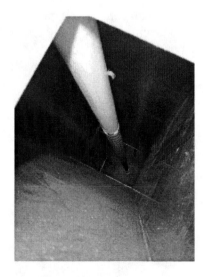

Figure 7. ESP test chamber 3 (inside view)

NSTI-Nanotech 2010, www.nsti.org, ISBN 978-1-4398-3401-5 Vol. 1, 2010

The composite tape electrodes and the commercial type metal electrodes were tested in the ESP chambers to evaluate and compare the voltage-current (V-I) behavior. A more efficient electrode will generate higher current at a specific voltage. Comparison of the V-I performance (Figure 8) of the electrodes reveals that the highest performing electrode contained carbon fiber and carbon nanofiber. The best performing composite electrode was the tape produced with carbon fiber and polypropylene that was preloaded with 20 weight percent carbon nanofiber. This demonstrates that polymer composite electrodes with carbon fiber have the potential of replacing metal electrodes in ESPs.

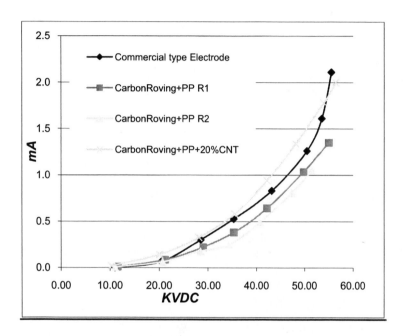

Figure 8. Comparison of the corona discharge current produced by 4 different electrodes: one commercial metal electrode, two composite electrodes (R1 & R2) with carbon microfiber, and one nanocomposite electrode.

Reference
[1] Dayley, M. and Holbert, K. Electrical Precipitators for Power Plants. Retrieved December 14, 2006, from http://www.eas.asu.edu/~holbert/wise/electrostaticprecip.html
[2] Bayless, D., Alam, M.K., Radcliff, J., and Caine, J., "Membrane-Based Wet Electrostatic Precipitation," *Fuel Processing Technology*, (85) 6-7, pp. 781-798, 2004.
[3] Chatterjee, A., Alam, M.K., and P. Klein, "Electrically Conductive Carbon Nanofiber Composites with High Density Polyethylene and Glass Fibers," Materials and Manufacturing Processes, Vol. 22, pp 62-65, 2007.
[4] Caine, J., Southern Environmental Inc. (Pensacola, FL), personal communication, 2008.

Nanopalladium – Wool: A Novel Fibre for Multifunctional Textiles

C. Fonseca-Paris* and J. H. Johnston**.

School of Chemical and Physical Sciences and the MacDiarmid Institute for Advanced Materials and Nanotechnology, Victoria University of Wellington, PO Box 600, Wellington, New Zealand, *carla.fonseca-paris@vuw.ac.nz, ** jim.johnston@vuw.ac.nz

ABSTRACT

New multifunctional textile products based on composite materials of natural wool fibres and palladium nanoparticles have been developed thereby combining the inherent properties of their precursor components in a synergistic manner through the use of chemistry and nanotechnology.

Palladium nanoparticles were successfully immobilized into wool fibres imparting their unique optical, electronic, catalytic and anti-microbial properties and their intrinsic value as a precious metal to produce an innovative new multifunctional composite textile fibre. By varying the reaction conditions employed, the size of the palladium nanoparticles and the color of the resultant nanopalladium-wool fibres have been specifically tuned, providing a range of colored textiles varying from orange, to brown and green. The size of the palladium nanoparticles varied in size from 2 to 30 nm as confirmed by TEM images. Anti-microbial testing of nanopalladium–wool fibres carried out using *Staphylococcus aureus* showed significant inhibition of bacteria growth around and underneath the nanopalladium-fibres. Preliminary catalysis tests based on hydrogenation of cyclohexene showed promising catalytic activity of the nanopalladium composites at room temperature.

Keywords: palladium, nanoparticle, wool, multifunctional textile, antimicrobial, catalyst.

1 INTRODUCTION

The development of new functional textile products based on composite materials of natural fibres and nanomaterials have become of increasing interest within both academic and industrial sectors. The inherent properties of their precursor components can be combined in a synergistic manner to create new innovative multifunctional textiles.

New Zealand is the world's largest producer and exporter of crossbred wool, and is the second-largest in the export of all wool [1]. Wool is a first class textile fibre possessing flexibility and elasticity, and acting as an insulator against hot and cold temperatures. It imparts softness to fabrics made from it and these are also naturally flame retardant and can absorb and release moisture. Keratin, the main constituent of wool fibres, is a complex protein comprising amino acids containing amine, sulphur, carboxylate and hydroxy functional groups, all of having the potential to bind to nanoparticles.

Nanoparticles can be immobilized on wool imparting their functionality such as optical, electronic, catalytic and antimicrobial properties to the textiles. In particular, nanoparticles of precious metals such as gold, silver and palladium not only combine their unique physical and chemical properties, but also represent high value, quality and wealth that can be transferred to the final textile or apparel product. When small metal particles are irradiated by light, the oscillating electric field causes the conduction band electrons to oscillate coherently resulting in a variety of brilliant hues, this phenomenon is known as Localized Surface Plasmon Resonance (LSPR) [2]. The color observed is dependent mainly on the size and shape of the nanoparticle [3]. Thus by controlling the formation of the metal nanoparticles the color may be specifically tuned. Professor James Johnston and his research group at Victoria University of Wellington [4-5] have captured the unique opportunity to use metal nanoparticles as novel colorants on wool fibres through their development and use of proprietary chemistry and nanotechnology. Their methodology is a clean process that provides a substantial improvement to the current environmentally problematic dyeing processes. Nanopalladium-wool fibres therefore provide the opportunity to develop new multi-functional textiles ideally suited for application in high end fabrics and fashions, and also for applications requiring anti-microbial and also catalytic properties. This will create new business opportunities and add significant value to New Zealand wool products and exports.

2 MATERIALS AND METHODS

2.1 Materials and reagents

All chemicals were purchased from Sigma Aldrich. Crossbred (coarse fibre) and merino (fine fibre) wools were provided by AgResearch Limited, New Zealand.

2.2 Nanopalladium-wool composites synthesis

The palladium nanoparticles were prepared by chemical reduction of a palladium chloride $PdCl_2$ solution in the presence of the wool wherein the wool acts as the reductant [4]. Merino and crossbred wool were used for all the experiments. The nanoparticles simultaneously bind to the wool fibres. By varying the reaction conditions (pH, temperature, time and $PdCl_2$ concentration) the size and shape of the nanoparticles and their distribution on and within the wool fibres can be controlled, therefore the color can be specifically tuned.

2.3 Palladium uptake

The uptake of the palladium by the wool was monitored by measuring the amount of residual palladium in solution by flame atomic absorption spectroscopy on a GBC 906AA spectrometer. Samples of the residual solution where taken at different time intervals and measured using an air-acetylene flame.

2.4 Palladium nanoparticles analysis

The palladium nanoparticles were examined by a JEOL 6500F field-emission gun scanning electron microscope (SEM) operating at 15 KV and a JEOL 2011 transmission electron microscope (TEM) operating at 200 KV. The palladium presence was confirmed by energy dispersive spectroscopy (EDS) for both techniques. All the samples were coated with carbon prior to analysis. In addition the TEM samples were embedded in resin before cut into thin films and placed on a copper grid.

2.5 In vitro antimicrobial testing

Anti-microbial testing of nanopalladium–wool fibres was carried out by means of two methods with two different bacteria.

Samples of palladium-wool composites were placed on agar plates inoculated with *Staphylococcus aureus* (ATCC 29213). The plates where incubated at 35 °C overnight in an O_2 incubator. The inhibition of the growth of the bacteria around the nanopalladium-wool fibres was clearly evident.

Samples of palladium-wool composites were transferred to Eppendorf tubes containing a culture of *Escherichia coli* (W3110) in PBS. The tubes were incubated in the dark for 7 days. A diluted aliquot from each tube was after plated onto an LB agar plate. The Plates were incubated at 37 °C overnight, and colonies were counted. Total CFU (viable bacterial numbers) were calculated.

3 RESULTS

Nanoparticles of palladium were synthesized and immobilized on and within the wool fibres under controlled conditions to effect the reduction of Pd^{2+} to Pd^0. By varying the reaction conditions employed (temperature, time, concentration and pH) the size of the palladium nanoparticles and the color of the resultant metal nanopalladium-wool fibres were controlled providing a range of colored textiles for merino and crossbred wool ranging from light brown to bright oranges and dark green (Fig 1).

Figure 1: Resulting colors of the nanopalladium-wool fibres for merino (top) and crossbred wool (bottom) for different reaction conditions.

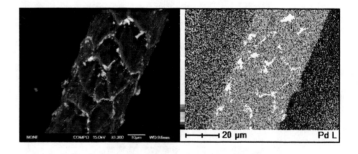

Figure 2: Backscatter SEM image of nanopalladium-crossbred wool composite (composites prepared at 100 °C, pH=2.0 with a metal load of 14.04 mg Pd/g wool) and EDS elemental map showing the location of palladium on the surface of the fibre.

The uptake studies for Pd^{2+} by the wool fibres at 50 °C shows that merino wool absorbs up to 90 % of a palladium solution of 260 ppm after soaking the fibres for 24 h. Crossbred wool fibres absorb up to 60 % of the palladium. The difference presented by the two types is wool is due to the different diameter of the fibres resulting in different specific surface areas.

SEM imaging shows the distribution of the nanoparticles in the wool fibre. The presence of the Pd nanoparticles was confirmed by elemental EDS analysis (Fig 2). The nanoparticles are located mainly at the cuticle edges of the wool fibres.

781

TEM images show nanoparticles of palladium from 2 to 30 nm in size on and within the fibre depending on the experimental conditions employed for the reduction of Pd^{+2} to Pd^0 and the type of wool employed. The nanoparticles have a polydisperse particle size distribution with larger nanoparticles located on the surface of the fibre and smaller particles within the fibre. Figure 3 shows nanoparticles of palladium within the fibres of a crossbreed wool composite ranging in size from 2 to 7 nm. The resulting color of the fibre is due to the different localized surface plasmon resonance bands of the reflected light generated by the different size of the nanoparticles on the fibre [2].

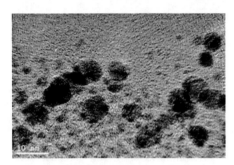

Figure 3: TEM image of Nanopalladium particles on and within the fibres of crossbred wool ranging in size from 2 to 7 nm (14.04 mg Pd/g wool composites prepared at 100 °C and pH=2.0 with a reduction time of 2.0 h). Scale bar 10 nm.

Anti-microbial testing of nanopalladium–wool fibres carried out using *S. aureus* (ATCC 29213) showed that the nanopalladium-wool was able to inhibit the growth of the bacteria around and underneath the fibres as shown on Figure 4.

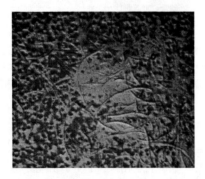

Figure 4: Antimicrobial activity of nanopalladium-wool shown by the inhibition of bacteria growth around and underneath the fibres (composites prepared at 100°C and pH=2.0 with a reduction time of 2.0 h).

The effect of the concentration of nanopalladium composites in the viability of *E. coli* cells was also measured. The bacteria cells were contacted with palladium composites prepared under the same experimental conditions (50 °C, pH=2.0) varying only the metal load. All the nanocomposites tested showed bactericidal effect. The cell survival decreases with the increase in the palladium concentration of the composite (Fig. 5). All the results are expressed as a percentage of cell survival relative to control wool (merino and crossbred). Higher concentration of palladium is lethal to the bacteria with a total CFU of cero.

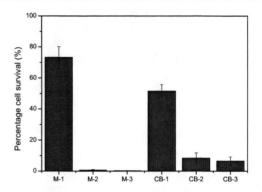

Figure 5: Percentage of *E. coli* cell survival for merino and crossbred wool at different metal loads (M-1: 0.59 mg Pd/g merino wool, M-2: Merino, 5.85 mg Pd/g merino wool, M-3: Merino, 11.70 mg Pd/g merino wool, CB-1: 0.35 mg Pd/g crossbred wool, CB-2: 3.51 mg Pd/g crossbred wool, CB-3: 7.02 mg Pd/g crossbred wool).

The nanoparticles have increased reactivity due to the large surface to volume ratio. The small size of the nanoparticles of palladium in the composites makes them ideal for catalysis applications. Palladium is widely known as a catalyst for hydrogenation reactions. These composite fibres were tested on their catalytic properties showing promising initial results. Catalysis tests based on hydrogenation of cyclohexene were carried out at low temperature. This work is currently being progressed and the results will be published at a later date.

CONCLUSIONS

New hybrid materials of wool fibre and palladium nanoparticles have been successfully produced. By varying the reaction conditions the color of the resultant metal nanopalladium-wool fibres were tuned obtaining colors ranging from orange to brown to green.

SEM, TEM and EDS analysis shows that the nanoparticles range in size from 2 to 30 nm wide, with the larger nanoparticles located mainly at the cuticle edges of the fibres. The nanoparticles are also formed inside the fibres where the particle size is smaller as shown by electron microscopy measurements.

The nanopalladium-wool fibres tested positive against *S. aureus* and *E. coli* microbes. The bactericidal effect is dependant on the concentration of the palladium contained in the fibres. The bacteria are also inhibited from growing around the vicinity of the fibres.

Nanopalladium-wool fibres therefore provide the opportunity to develop new multi-functional textiles for application in high end fabrics and fashions in the case of merino wool, whereby the high value and wealth traditionally associated with precious metals and the quality of the merino wool are combined, or carpet and interior furnishings where the anti-microbial properties are desired to prevent the proliferation and growth of microbes. This will add significant value to New Zealand wool products and exports creating new business opportunities for the wool industry.

REFERENCES

[1] The Ministry of Agriculture and Forestry.http://www.maf.govt.nz.

[2] Mie, Greifswald Ber Physk Ges, 5, 492-500, 1908.

[3] J. J. Mock, M. Barbic, D. R. Smith, D. A. Schultz, and S. Schultz, Journal of Chemical Physics V. 116, N. 15, 2002.

[4] James H. Johnston, Fern M. Kelly, Kerstin A. Burridge, and Aaron C. Small, NZ Patent Specification, 2009.

[5] M. J. Richardson and James H. Johnston, Journal of Colloid and Interface Science, 310, 2, 425-430, 2007.

ACKNOWLEDGEMENTS

This work was supported by the MacDiarmid Institute for Advanced Materials and Nanotechnology, New Zealand.

We also gratefully acknowledge the support from AgResearch Limited, New Zealand and Dr. David Ackerley from the School of Biological Science, Victoria University of Wellington, New Zealand.

Novel Nanocomposites: Hierarchically ordered porous silica for the immobilization of enzyme as bio-catalyst

T. Sen[*]

[*]Centre for Materials Science, School of Forensic and Investigative Sciences, University of Central Lancashire, Preston, United Kingdom, tsen@uclan.ac.uk

ABSTRACT

Hierarchically ordered porous silica with porosity in three different lengths (microporosity: pore diameter < 2nm; mesoporosity: 2nm < pore diameter < 50nm; macroporosity: pore diameter > 50nm) has been synthesized using polystyrene latex as colloidal template to generate macroporosity, pluoronic F127 ($EO_{107}PO_{70}EO_{107}$) as a non-ionic surfactant to generate meso and microporosity in an acidic pH in the presence of tetramethyl orthosilicate as silica source. The calcined material has been functionalized using aminopropyl triethyoxy silane (APTS) in toluene at 50^0C in the presence of surface adsorbed water followed by glutaraldehyde treatment. Lipase has been immobilized by chemical conjugation on the surface of functionalized materials in PBS buffer (pH, 7.4). Hydrolysis of 4-nitrophenyl palmitate has been studied using Lipase immobilized hierarchically ordered porous silica and the results have been compared with lipase immobilized non functionalized materials. The rate of hydrolysis was observed to be increased exponentially with time for both materials, however, functionalized materials exhibited a two fold increase in hydrolysis efficiency without loss of catalytic efficiency after 2nd catalytic cycle.

Keywords: Enzyme catalysis, lipase, hierarchically ordered porous silica, surface functionalization, hydrolysis of ester

1   INTRODUCTION

Complex nanocomposites with porous structures are widespread in nature where hierarchy of pore structure has evolved to provide either strength or a pathway for fluid transport. For instance the structure of calcium phosphate bone is highly porous such that the material is as light as possible but with optimal load bearing properties associated with the distribution of pores and the curvature of the wall structure. The silica exo-skeleton of a diatom is designed both for protection of the single-celled algae and for transport of fluid to the organism within. Nature also gives clues to the optimal structures for transport of high volumes of gases to active sites in the hierarchical pore structure of the lungs. To date, there is a vast literature on synthesizing materials with monodisperse pore sizes at different length scales, e.g. microporous (pore diameter < 2nm), mesoporous (2nm < pore diameter < 50nm) and macroporous (pore diameter > 50nm). However, there is far less on trying to combine these different length scales into one composite material. Sen *et al* reported [1, 2] such hierarchically ordered porous nanocomposites of silica structures using polystyrene latex spheres of defined sizes as a template for macropores, pluoronic tri-block copolymer as a surfactant for meso and microporosity in the presence of a co-solvent, however, the applications of such novel nanocomposites have never been reported. Recently, hierarchically ordered porous materials have been used for the applications such as scaffolds [3] and rechargeable battery [4], however, such nanocomposites have not been exploited as catalytic supports.

Lipase is a well known enzyme used extensively for the bio-catalysis. However, using lipase in solution phase bio-catalysis is not economical as the separation of lipase from the reaction mixtures is not an easy process. One of the important reactions reported using lipase is the hydrolysis of vegetable oil for the production of bio-diesel [5, 6]. Dyal *et al* reported [7] the immobilization of *candida rugosa* on magnetic nanoparticles (maghemite) for the hydrolysis of ester in heterogeneous medium.

Herein, we report the immobilization of lipase (*candida rugosa*) on novel hierarchically ordered porous silica nanocomposites by chemical conjugation with functionalized surface and their efficiency in the hydrolysis of *p*-nitro phenyl palmitate to p-nitro phenol and palmitic acid as heterogeneous bio-catalysts.

2   EXPERIMENTAL

Hierarchically ordered porous silica materials were synthesized following the previously reported protocol [1, 2]. Non cross linked polystyrene latex spheres were synthesized following the reported protocol using an emulsion polymerization technique [2]. The polystyrene latex spheres were packed into a monolithic structure by centrifugation at 6000 rpm for 1 hr and dried at 60^0C overnight. Polystyrene latex monolith was filled with a gel containing silica (tetramethyl orthosilicate) in an acidic medium (aqueous hydrochloric acid) in the presence of tri-block copolymer F127 and n-butanol. The silica gel filled polystyrene monolith composite was dried at 60^0C overnight. Templates were removed from the composites by washing with toluene followed by calcinations at 500^0C for 8hr in the presence of air.

Lipase (*candida rugosa*) immobilisation was carried out by chemical conjugation and the method has been presented schematically (see reaction scheme 1).

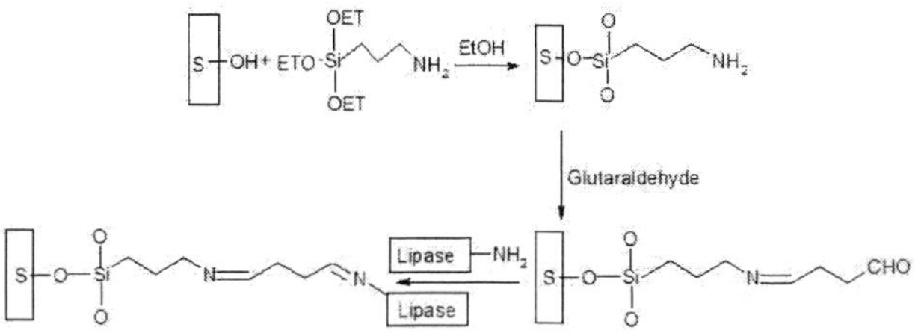

Scheme 1 Reaction scheme for the chemical conjugation of lipase on hierarchically ordered porous silica support (S).

150mg of calcined materials were thoroughly mixed with 10µg deionised water and the wetted materials were functionalised using 5% (w/v) 3-aminopropyl triethoxy silane (APTS) in 10mL toluene at 50^0C for 24 hrs in a glass reactor. Surface-adsorbed water was used for the hydrolysis of APTS on the surface of the materials [8]. The surface amine density of functionalised nanocomposite was determined by established colorimetric assay using 4-nitro benzaldehyde [9]. The amine functionalized silica nanocomposites were washed with 1 mL (3 times) of coupling buffer (1× SSC, pH 7.3) for 2 min at RT. 1×SSC buffer was prepared by diluting a stock solution of 20×SSC buffer (175.3 g of NaCl, 88.2 g of sodium citrate, and 1 L of H_2O, pH 7.4) with distilled, deionized water, adjusted to pH 7.4. After removal of the supernatant, 1 mL of a 5% (w/v) glutaraldehyde solution in coupling buffer was added and the reaction mixture was incubated for 3 h with end-over-end rotation at RT. The material was subsequently washed with 1 mL (3 times) of coupling buffer to remove excess glutaraldehyde. The materials were then washed with 1mL (3 times) of PBS buffer (pH 7.2) and stored in PBS buffer before immobilisation of Lipase.

50mg of glutaraldehyde modified silica nanocomposites were treated with 1mL of Lipase solution (2mg/mL) in PBS buffer for 15 hrs with end-over-end rotation at RT. The concentration of Lipase solutions before and after the reaction with silica nanocomposites were measured by Bradford assay [10] using UV visible spectroscopy (absorption at λ_{595nm}). The amount of lipase in the solution was calculated using a calibration curve constructed from a range of standard solutions of lipase prepared separately in PBS buffer reacting with 1.2mL of Bradford reagent and measuring λ_{595nm}. A similar reaction was carried out on calcined silica nanocomposites (non-functionalized) for the immobilisation of lipase.

Hydrolysis of ester (4-nitro phenyl palmitate) was carried out using 15mg support reacting with 1mL of ester solution (3.74 µmol ml^{-1}) prepared in a 1:1 mixture of isopropanol and reagent A (0.0667g Gum Arabic + 12mL of 250mM Tris-HCl buffer, pH 7.8 + 48mL of deionized

water + 0.267g of sodium deoxycholate) at 20^0C for 4 hrs in 1.5mL Eppendorf tube by end-over-end rotation. The products were collected in different time intervals and measured the concentration of 4-nitrophenol. Scheme 2 presents the hydrolysis of 4-nitro phenyl palmitate to palmitic acid and 4-nitro phenol.

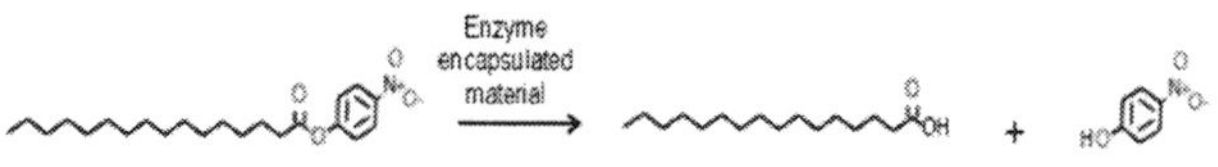

Scheme 2 Reaction scheme for the hydrolysis of 4-nitro phenyl palmitate

The hydrolysis reaction was monitored by measuring the concentration of 4-nitro phenol in the reaction solution. The concentrations of 4-nitrophenol in the reaction mixture at different time interval were determined by using a calibration curve constructed from a range of standard solutions of 4-nitrophenol prepared separately in a solution containing 1:1 mixture of isopropanol and reagent A by measuring the absorbance at λ_{410nm}.

Nanocomposites were washed in 1mL (3 times) buffer solution containing 1:1 mixture of isopropanol and reagent A. The washed materials were further used for the hydrolysis of 4-nitrophenyl palmitate under identical condition as above for testing the catalytic efficiency of lipase immobilized nanocomposites in the 2nd catalytic cycle.

3 RESULTS AND DISCUSSION

Scanning electron micrographs of the hierarchically ordered porous nanocomposites before and after lipase immobilization have been presented in figure 1. Materials exhibited uniform macroporous structure of pore sizes of around 330nm with interconnecting windows of around 100nm. The void structure and interconnecting windows remains free after lipase immobilization indicating that the pore structure remains unchanged which is essential for the transport of reactant molecules into the nanocomposites. The presence of meso and microporosity has been confirmed through low angle powder X-ray diffraction and nitrogen adsorption measurement (data not shown). BET surface area of the calcined materials was measured to be around 210m^2/g. The formation mechanism of such porous silica composites containing macro, meso and microporosity with interconnecting windows of hierarchical ordering has been reported earlier [1, 2].

Surface amine density of the APTS functionalized nanocomposites was measured to be 54nmol/mg by colorimetric assay.

Figure 2A and B presents the calibration curves of lipase in PBS buffer using Bradford reagent (Fig.2A) and 4-nitrophenol in a 1:1 mixture of isopropanol and reagent A (Fig.2B). Calibration curve 2A exhibited a straight line with a positive intercept (regression value of 0.9653) and calibration curve B exhibited straight line passing through

Figure 1 Scanning electron micrographs of hierarchically ordered porous silica. A and B represent calcined nanocomposites before surface functionalization. C and D represent the surface functionalized lipase immobilized nanocomposites

the origin with no intercept (regression value of 0.9853) indicating that the calibration curves are reliable for measuring the concentration of unknown solutions within the concentration ranges presented in figure 2A and 2B.

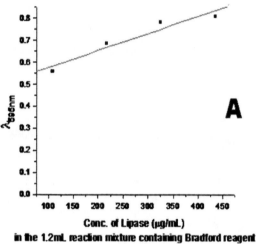

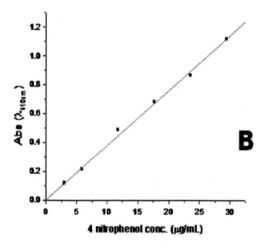

Figure 2 Calibration curves of Lipase in PBS buffer (A) and 4-nitrophenol in 1:1 mixture of isopropanol to reagent A (B)

The lipase concentration on non-functionalized and functionalized nanocomposite has been calculated using calibration curve 2A and the values were 32.3 µg/mg and 42.6 µg/mg respectively. Immobilization of lipase on functionalized nanocomposites exhibited a change in color from white to pink whereas no color-change observed in non-functionalized nanocomposites.

The kinetics of hydrolysis of 4-nitrophenyl palmitate using lipase immobilized nanocomposites (functionalized and non-functionalized) has been presented in figure 3. The rate of hydrolysis was observed to be increased exponentially with time for lipase immobilized non-functionalized and functionalized nanocomposites.

However, the catalytic efficiency of lipase immobilized functionalized nanocomposites was observed to be two fold higher in value (826 μmol/g of lipase) compared to lipase immobilized non-functionalized nanocomposite (437 μmol/g of lipase). The catalytic efficiency was unaffected on 2nd cycle on lipase immobilized functionalized nanocomposites whereas a six fold reduction was observed in non-functionalized nanocomposites.

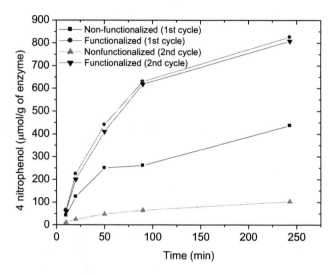

Figure 3 Kinetics of hydrolysis of 4-nitro phenyl palmitate to palmitic acid and 4-nitrophenol

The reason for the loss of catalytic efficiency of lipase immobilized non-functionalized nanocomposites could be due to the leaching of lipase from the nanocomposites into the reaction mixture in the 1st catalytic cycle. Leaching of lipase onto the non-functionalized nanocomposites perhaps due to the loosely bonded lipase to the surface by physical adsorption whereas lipase could be chemically bonded to the functionalized surface through chemical conjugation as presented in scheme 1.

In conclusion, hierarchically ordered porous silica nanocomposites have been used for the immobilization of lipase (*candida rugosa*). Lipase immobilization by chemical conjugation through an intermediate surface functionalization step is important for the fabrication of efficient bio-catalysts. Chemically conjugated novel bio-catalysts reported in this paper could be utilized in other important bio-transformation processes in pharmaceutical industry as a greener route.

REFERENCES

[1] T. Sen, J. L. Casci, G. J. T. Tiddy and M. W. Anderson, "One-Pot Synthesis of Hierarchically Ordered Porous Silica Materials on Three Orders of Length Scale" Angew. Chem. Int. Edit. 42, 4649, 2003.

[2] T. Sen, J. L. Casci, G. J. T. Tiddy and M. W. Anderson, "Synthesis and Characterization of Novel Hierarchically Ordered Porous Silica Materials" Chem. Mater. 16, 2044, 2004.

[3] C.F. Xue, J. X. Wang, T. Bu, D. Y. Zhao "Hierarchically Porous Silica with Ordered Mesostructure from Confinement Self-Assembly in Skeleton Scaffolds." Chem. Mater. 22, 494, 2010.

[4] Y. S. Hu, P. Adelhelm, B. M. Smarsly, S.Hore, M. Antonietti, J. Maier, "Synthesis of hierarchically porous carbon monoliths with highly ordered microstructure and their application in rechargeable lithium batteries with high-rate capability." Adv. Func. Mater. 17, 1873, 2007.

[5] F. Yagiz, D. Kazan, N. Akin, "Biodiesel production from waste oils by using lipase immobilized on hydrotalcite and zeolites" Chem. Eng. J., 134, 262, 2007.

[6] L. Huang, Z. Cheng., "Immobilization of lipase on chemically modified bimodal ceramic foams for olive oil hydrolysis" *Chem. Eng. J., 144*, 103, 2008.

[7] A. Dyal, K. Loos, M. Noto, S. W. Chang, C.Spagnoli, K. V. P. M. Shafi, A.Ulman, M.Cowman, R.A. Gross, "Activity of Candida rugosa lipase immobilized on gamma-Fe2O3 magnetic nanoparticles" J. *Am. Chem. Soc., 125*, 1684, 2003.

[8] T. Sen and I. J. Bruce "Surface activation of nanoparticulates in suspension by a novel strategy: tri-phasic reverse emulsion for highly efficient DNA hybridization" Chemical Communications DOI: 10.1039/b817354K. 2009.

[9] J. H. Moon, J. H. Kim, K. Kim, T. H. Kang, B. Kim, C. H. Kim, J. H. Hahn, J. W. Park, J. W. "Absolute surface density of the amine group of the aminosilylated thin layers: Ultraviolet-visible spectroscopy, second harmonic generation, and synchrotron-radiation photoelectron spectroscopy study" Langmuir, 13, 4305, 1997.

[10] M. Bradford, "A Rapid and Sensitive Method for the Quantitation of Microgram Quantities of Protein Utilizing the Principle of Protein-Dye Binding" Anal. Biochem. 72, 248, 1976.

The physical properties of zein nanoclay hybrid resin as a base for zein nanoclay nanocomposite films

J. Luecha[*], N. Sozer[**] and J.L. Kokini[***]

[*]University of Illinois, Urbana-Champaign, IL, USA, jluecha2@illinois.edu
[**] University of Illinois, Urbana-Champaign, IL, USA, nsozer@illinois.edu
[***] University of Illinois, Urbana-Champaign, IL, USA, kokini@illinois.edu

ABSTRACT

From our previous study, zein nanoclay nanocomposite films were successfully prepared by blown extrusion method. This method involved two main steps; i) formation of zein resin and ii) formation of zein film by blowing. The structure analysis of the nanocomposite films prepared by this method suggested partially exfoliated nanocomposite structure. However, the tensile and barrier properties of zein nanoclay hybrid films were not found to improve at the percentages used. The amount of nanoclay included in the zein resin had a percentage limit (< 5 wt %) during blown extrusion. The insufficient extensibility of the zein nanoclay resin caused rupture of the film during blowing after the critical level. Therefore, it is important to understand the interaction between zein and nanoclay in the resin form in order to design the proper preparation method. The objectives of this study were to understand the interaction of zein-nanoclay mixture in the resin form by characterizing the physical properties including extensional rheology of the hybrid resin using dough inflation system for the first time. It was found that the nanoclay was delaminated and dispersed within the zein matrix as shown in the TEM image. The extrusion played important role in mixing the nanoclay in zein resin but at certain amount of nanoclay content. The critical nanoclay content which improve the extensional properties of the zein hybrid resin without sacrificing the processability was below 5 wt% which was inconsistent with previous work. The measurement of extensional rheological properties of zein resin using dough inflation still requires testing parameters adjustment. These findings can be useful for adjusting the blown extrusion process to produce improved mechanical and film barrier properties. The incorporation of nanoclay within the biopolymer matrix provided a novel pathway to engineer mechanically durable and biodegradable packaging films.

Keywords: zein, nanoclay, nanocomposites, dough inflation system, extensional rheology

1 INTRODUCTION

The waste of petroleum-based packaging materials has become the worldwide major environmental concern. Therefore, the biodegradable materials derived from renewable resources have become under spot light as a potential approach to reduce the packaging material waste problems. Zein, a protein found in corn, is one of the potential bio-based materials which have been studied extensively. Similar to other biopolymers, corn zein films have poorer mechanical and barrier properties in comparison with other petroleum-based films [1]. To improve zein film properties, the polymer nanoclay nanocomposite technique was introduced to zein film preparation. The polymer nanoclay nanocomposite technique involves the intercalation/exfoliation of several layers of ~ 1 nm thick clay platelet in the polymer matrix which then improve the mechanical properties and barrier properties of the polymers by means of the high aspect ratio of dispersed nanoclay platelets in polymer matrix [2]. In previous study [3], zein nanoclay nanocomposite films were formed by extrusion blowing technique which comprises of the zein resin formation and zein bubble formation steps. It was shown that the blown extrusion could prepare zein nanoclay nanocomposite films with the partially exfoliated nanocomposite structure, however, the zein nanoclay hybrid resin failed to form bubble at the 5 wt% of nanoclay. The failure to form zein bubble after a certain nanoclay loading indicated the effect of nanoclay on the zein resin rheological properties which finally would have impact on the quality of the zein-nanoclay nanocomposite film. The role of nanoclay on rheological properties of polymer melt systems has been studied extensively in several polymers as reviewed by Shinha Ray [4]. It was believed that the rheological properties of nanocomposites strongly depend on the interaction of polymer and nanoclay based on the interfacial interactions at the nanoscale. For instance, in the uniaxial small elongational flow study, the Polylactide (PLA) nanocomposites exhibited great value of viscosity and tendency to have strong strain-induced hardening as a result of the nanoclay alignment 90 degree towards the extension direction. However, this was not hold true for every polymer [4]. Therefore, it is essential to investigate the role of nanoclay on rheological properties of zein resin in order to gain fundamental understanding of the processability of zein nanoclay nanocomposite films.

NSTI-Nanotech 2010, www.nsti.org, ISBN 978-1-4398-3401-5 Vol. 1, 2010

The dough inflation system has been useful for monitoring large biaxial extension of bread dough resembling the deformation in the real breadmaking conditions [5]. This new dough rheology testing apparatus was developed by Bogdan Dobraszczyk and Clive Roberts at the University of Reading in conjunction with a texture analyzer company, Stable Micro System [6]. The extensional flow measurement is always difficult to evaluate as a result of the high deformation level, nonuniformity within the structure and difficulties in converting the strain to the machine displacement. Therefore, the calculation of changes in sample dimension is essential [5]. The stability of the bubble studied by this system was directly related to the elongational strain hardening properties of the bread dough [6]. Strain hardening is the phenomena in which the materials can maintain their stability before failure during deformation [5]. The strain hardening is noticed as an increase in slope of the plot of stress (σ) versus Hencky strain (ε) where $\varepsilon = \ln (L/L_0)$; L= initial sample length and L_0= instantaneous sample length. For the materials in which obey a power law $\sigma = K\varepsilon^n$, the strain hardening index (n) can be obtained. It has been proved that the strain hardening obtained from dough inflation system correlated with the bubble failure strain [6]. The downsides of this system are that the large amount of sample is required; the measurement is destructive; and the well-defined and accepted method of measurement has not been established [5]. It has been known that the large extensional viscosity of the food materials plays an important role in several food processing events including extrusion, sheeting, molding, fiber formation, adhesion and bubble expansion [5]. Therefore, it is challenging to evaluate the biaxial extensional rheological behavior of the zein nanoclay hybrid resin in the similar manner of the film formation process. In this pioneer study, the observation on the extensional rheological properties of zein resin as affected by nanoclay loadings and extrusion were investigated by using dough inflation system.

2 MATERIALS AND METHODS

Zein powder (90% crude protein dry weight basis) was purchased from Sigma Aldrich (Milwaukee, WI, USA). The Nanomer ® I.34 TCN which was surface-modified nanoclay containing 25-30 wt. % methyl dihydroxy ethyl hydrogenated tallow ammonium and oleic acid were supplied from Sigma Aldrich (Milwaukee, WI, USA). Ethanol (95%) was purchased from Decon Laboratories Inc. (King of Prussia, PA).

2.1 Zein nanoclay hybrid resin preparation

The zein nanocomposite resins were prepared by first, dissolving 100 g of zein, 30 g of oleic acid and certain amount of nanoclay (0, 3, 4, 5, and 9 % w/w of zein.) in 650 ml of ethanol solution. The mixture was stirred for 10 minutes after they reached 60-65°C. Then, the solution was poured into an ice water bath for precipitation. The precipitate zein was collected and then formed into dough-like resin. The resin was hand kneaded until it was not sticky and labeled as fresh resin. Another set of resins prepared from the same manner was then passed four times through single screw extruder (Model EPL-V501, C.W. Brabender, Hackensack, NJ) and labeled as extruded resin.

2.2 Characterization

Transmission electron microscope (TEM) of the hybrid resin was carried out by Phillip CM12 TEM (Mahwah, NJ) at 120 KV.

The extensional rheological properties of zein nanoclay hybrid dough were obtained by the method described by Dobraszczyk [6]. The zein resin sheet preparation was done by rolling the zein resin to a required thickness and cutting into a certain round shape by cookie dough cutter provided by the dough inflation system. The sheet of zein resin was inflated at room temperature on the dough inflation system connected to a texture analyzer (TA.HD. Plus, Stable Micro Systems Ltd., Godalming, UK). The sheet of zein nanoclay hybrid resin was inflated by volume displacement of air. The pressure during inflation was monitored by pressure transducer and the volume of the bubble was calculated from the piston displacement. The experiment set up is shown in Figure 1. The dough inflation system can carry out several parameters including bi-axial extensibility (the length of the curve up to the point of rupture), L (mm); tenacity (the maximum pressure required during inflation of the bubble), P (mm); ; deformation energy (the energy necessary to inflate the ; strain hardening index, n; and stress (KPa) at bubble failure. The typical result of an inflating dough bubble is shown in Figure 2. The strain hardening of bread dough can be observed in stress-Hencky strain curve as its slope increases with increasing extension which indicates that the material is stiffened as the extension goes on. The more curvature of the curve indicates the greater the strain hardening of the materials [5]. Three samples of each treatment were measured.

Figure 1: Extensibility testing of zein nanoclay resin using dough inflation system.

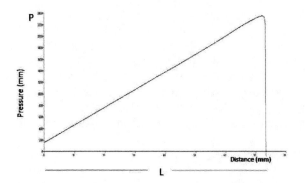

Figure 2: Typical dough inflation result of inflating dough bubble.

3 RESULTS AND DISCUSSION

The structure characterization of zein nanoclay hybrid resin was obtained from TEM as shown in Figure 3. It is clearly seen that at 9 % of nanoclay, the nanoclay platelets exfoliated in fresh zein resin. This has confirmed that the well-dispersion of nanoclay can occur at the resin formation step. We believe that the abrupt conformational change of zein protein during precipitation delaminate the stacking structure of nanoclay. A similar approach has been observed in polymer nanoclay nanocomposites prepared by in situ polymerization.

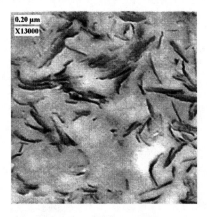

Figure 3: TEM images of zein nanoclay hybrid resin at nanoclay content of 9 wt. %.

Figure 4 and 5 show the tenacity and biaxial extensibility parameters which are obtained directly from the measurement. It is obvious that extrusion dramatically influences the biaxial extensional behavior of zein resins. The extrusion process creates entanglement of zein molecules which facilitates zein network formation that corresponds to the high elongational behavior [7]. It is seen that nanoclay improved the tenacity of the extruded zeinanoclay hybrid resins. Samples with 3 and 4% nanoclay loadings exhibited relatively higher tenacity values among other samples.This impact was not noticeable

for fresh samples (Figure 4). At high percentage of nanoclay (9 %), the hybrid resins exhibited poorer tenacity. The similar trend was also found in the bi-axial extensibility values (Figure 5). The bi-axial extensibility improved by extrusion. The inclusion of nanoclay content increased the bi-axial extensibility of both fresh and extruded zein nanocomposite resins up to 4 % then the values declined by further increase in clay content. At nanoclay content greater than 5 %, zein nanoclay hybrid resin obviously lost its extensibility which was in agreement with Luecha et al. [3] where they found that the bubble could not be formed above 5% nanoclay content [3].

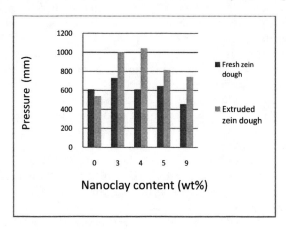

Figure 4: The tenacity of fresh and extruded zein dough at various nanoclay contents.

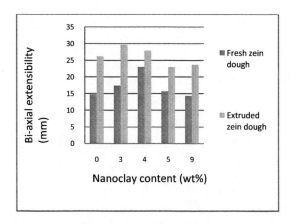

Figure 5: The biaxial extensibility of fresh and extruded zein dough at various nanoclay contents.

Figure 6 shows the stress at bubble failure. The extruded samples were capable of holding higher stress before rupture occurred. Moreover, extrusion definitely assisted the mixing of nanoclay in zein matrix and formed into strong resin as the stress at bubble failure increased extensively at the addition of nanoclay of 3 and 4 % then leveled off in the samples containing 5 and 9 %.

The strain hardening index of zein nanoclay hybrid resins are shown in Figure 7. Surprisingly, the nanoclay content and extrusion did not have distinct impact on the strain hardening index as was first expected.

The zein nanoclay hybrid resins exhibit great variation in rheological properties among samples which caused the testing parameters adjustment more difficult. The inconsistencies were highest in the fresh resin with high amount of water. Some of the fresh samples were too sticky that could not be lifted from the base of the retainer even though the oil was applied to prevent stickiness. In contrary, some of the extruded resins were very elastic that the bubbles did not fall even though the rupture had occurred. This might influence the strain hardening index calculations.

The extensional rheological properties of thermoplasticized zein as a material for extrusion blown films have been studied by Oliviero et al. [8], by using rheometer equipped with elongational tool. The mixture of zein and plasticizer were mixed at high temperature to form thermoplasticized zein material which was then blown into bubble by extrusion blowing technique. At the small deformation levels, they found that the strain hardening parameter which was obtained from the slope of the transient elongational viscosity versus time curve at the region of rapid deformation could be an indicator for blown film process. The higher the strain hardening of the thermoplasticized zein the better the blown film formation characteristics [8].

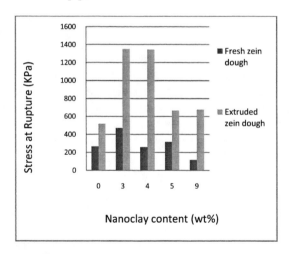

Figure 6: The stress at bubble failure of fresh and extruded zein dough at various nanoclay contents.

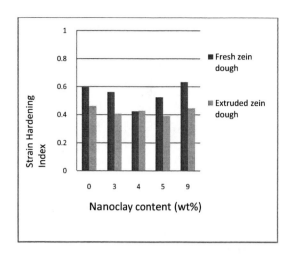

Figure 7: The strain hardening index of fresh and extruded zein dough at various nanoclay contents.

4 CONCLUSIONS

The dough inflation system was used for characterizing zein nanoclay hybrid resin for the first time. The extensional rheology of the zein resin was greatly influenced by extrusion. At nanoclay contents around 3 to 4% were the best loading range that could improve the zein resin properties as well as facilitate the blown film processing.

REFERENCES

[1] H. -M. Lai and G.W. Padua, Cereal Chem. 74: 771-775, 1997.
[2] M. Alexandre and P. Dubois. Material Science and Engr.28: 1-63, 2000.
[3] J. Luecha, N. Sozer, and J.L. Kokini, Journal of Material Science [accepted] 2010.
[4] S. Shinha Ray, Journal of Industrial Engineering Chemistry., 12, 811-842, 2006.
[5] B. Dobraszczyk, J. Smewing, M. Albertini, G. Maesmans, and J. D. Schofield. Cereal Chemistry, 80, 218-224, 2003.
[6] B. Dobraszczyk, Cereal Foods World, 42, 516-519, 1997
[7] M. Lau, "Effect of mechanical energy and moisture content on the rheological properties of zein-oleic acid dough during kneading, mixing and extrusion". Thesis (M.S.) Rutgers University, 2007.
[8] M. Oliviero, E. D. Maio, and S. Iannace. Journal of Applied Polymer Sci. 115, 277-287, 2010.

Nanogold and Nanosilver Wool: New Products for High Value Fashion Apparel and Functional Textiles

*James H. Johnston, Kerstin A. Burridge, Fern M. Kelly and Aaron C. Small

School of Chemical and Physical Sciences, Victoria University of Wellington,
PO Box 600, Wellington 6140, New Zealand
* Contact person. Email jim.johnston@vuw.ac.nz

ABSTRACT

We have developed a unique and exciting new proprietary nanotechnology and product suite of high value textiles involving the use of nanogold and nanosilver as colourants and functional components in wool. This innovatively links the high value and prestige of gold and silver with high value textiles and fashion apparel through nanoscience and technology. The products contain only pure wool and pure gold or silver respectively and are environmentally desirable. Gold and silver in bulk form exhibit their characteristic metallic colour. However, when their particle size is reduced to nano dimensions they exhibit different colours due surface plasmon resonance effects resulting from the collective oscillation of conduction band electrons. The colours depend on the size and shape of the nano entities and do not fade like traditional organic dyes. The nanogold and nanosilver entities are formed by the controlled reduction of Au^{3+} to Au^0 and of Ag^+ to Ag^0 respectively, on the surface and within the wool fibres. These entities are strongly chemically bound and do not leach out or rub off. By controlling the formation and particle size of the nanogold and nanosilver in the wool fibres, an attractive range of colours in shades of pink, purple, blue-grey, grey and yellow for nanogold wool and yellow, peach, purple and green for nanosilver wool are possible. The technology and product suite have been protected by New Zealand and PCT patent applications, and are being progressed to commercialization. Investment partners are being sought.

Keywords: gold, silver, nanoparticles, colourants, wool, textile, apparel, fashion.

1 INTRODUCTION

Nobel metals such as gold and silver have long been known to form stable colloids of nanosize particles. Such nanoparticles exhibit different colours due to surface plasmon resonance effects which result from the interaction of incoming electromagnetic radiation in the visible region with the collective plasmon oscillations at the metal surface [1]. The colour is dependent on the particle size and shape and the dielectric constant of the surrounding medium. Colloids of nanogold particles with spherical particles of about 10-20 nm in size are red. As the particle size

increases, the colour changes through darker shades of red to purple, then to blue-grey for particles up to about 80-100 nm. These colours relate to a shift and broadening of the transverse plasmon resonance absorption band at about 520 nm. Gold nanorods are typically green in colour as they show this transverse band as well as a longitudinal plasmon resonance band at about 720 nm [2, 3]. These colours are stable in sunlight and UV light and do not change providing there is no change in particle size, through the growth or reduction of discrete nanoparticles or agglomeration.

Colloids of nanosilver similarly exhibit a variety of colours. Spherical nanoparticles of silver of about 40 nm in size are yellow. Increasing the particle size progressively shifts the colour from yellow to red. By increasing the size of silver triangular prisms from about 60 nm to 100 nm to about 120 nm, the colour changes from purple to blue to green.

Interestingly, colloidal gold nanoparticles have been used to colour glass dating back to about the 17th century. Also, the Lycurgus Cup dating back to around 400AD used gold and silver nanoparticles as colourants. The science was not understood until 1869 when Michael Faraday recognized and explained the role of gold as a colourant in general terms. In 1908 Mie [4] provided a theoretical explanation by solving Maxwells equations for the absorption and scattering of electromagnetic radiation by very small metallic particles – nanoparticles.

Wool fibres have been used since historical times in fabrics and textiles for garments, furnishings and floor coverings. The wool fibres are typically about 12-40 microns in diameter and several tens of centimeters in length. They are woven into yarn that is used to make a very wide range of fabrics and textiles. The yarn or the fabric and textiles can be dyed using conventional technology to produce many different colours. *Merino* wool with its smaller fibre diameter of about 10-25 microns is highly valued for fine fabrics and fashion apparel, and commands a price premium. *Strong* wool with a fibre diameter of about 25-40 microns is used in textiles for furnishing and carpet, where higher durability is required.

Wool belongs to the group of fibrous proteins consisting of α-keratins and comprises long polyhedral inner cortical cells surrounded by flattened external cuticle cells and partially by an external fatty acid monolayer, mainly 18-methyleicosanic acid. The helix arrangement of the proteins gives wool its flexibility, elasticity and crimp properties. The fatty acid layer provides some surface hydrophobicity which can influence further processability such as dyeing, enhancing shrink resistance and anti-static properties.

Wool garments are traditionally coloured by natural dyes such as *madder (red), indigo (blue)* and synthetic dyes typically anthraquinone or azo-based compounds. Unfortunately, these dyes, particularly the natural ones are not colourfast. Hence colour changes resulting from exposure to sunlight and UV light, repeated washing and surface abrasion from wearing are issues that adversely affect the overall quality of the fabrics and textiles. Colourfastness is important in high quality fabrics and textiles for the high fashion market. This is becoming even more so due to the strong environmental awareness and move towards natural and sustainable dye formulations.

New Zealand is the world's second largest producer of *merino* wool which is renown for its fine fibre and hence its use in very high quality knitwear and fashion apparel. Similarly, New Zealand is a large producer of *strong* wool which is used in the manufacture of high quality wool carpets and hard wearing textiles.

We have innovatively captured and developed a unique and exciting new proprietary nanotechnology and product suite of high value knitwear apparel and textiles involving the use of nanogold and nanosilver as colourants and functional components in wool and other substrates [5, 6]. This links the high value and prestige of gold and silver with high quality New Zealand *merino* and *strong* wool for high value fashion apparel and textiles through nanoscience and technology.

The products contain only pure wool and pure gold or silver, making them environmentally attractive compared with current environmentally problematic dyes. Nanosilver wool and nanogold wool exhibit effective anti-microbial properties even at very low levels due to the natural anti-microbial properties of silver and gold. A key competitive advantage is that these nano entities are chemically bound to the fibres and do not wash or rub off, obviating any consumer concerns.

The technology and product suite have been protected by New Zealand and International patents (pending) [5]. *This exciting new technology and product suite is a world leading innovation and a world first. It offers attractive new investment and business opportunities for the high value fashion apparel, carpets and functional textiles markets where high quality, uniqueness, functionality and environmental attractiveness, rather than price, are the key drivers and product hallmarks.*

2   NANOGOLD - WOOL

Nanogold entities have been attached to *merino* and *strong* wool fibres using our proprietary methodology [5]. In this, we control the reduction of Au^{3+} in the $AuCl_4^-$ ion to Au^0 to produce nanogold entities of the required particle size and shape to yield the desired colour and chemically bind them to the keratin in the wool fibres, to provide stable colourfast nanogold-wool products in an attractive range of colours. An electronmicroscope image (Fig. 1) shows the nanogold particles (white dots) binding primarily to the cuticle edges of the wool fibre with a lesser distribution over the cuticle surface. The very strong extinction coefficient of nanogold due to surface plasmon resonance effects, means that very little nanogold is required to provide a strong colour. This is shown by the very small amount of nanogold on the surface of the fibre (Fig. 1).

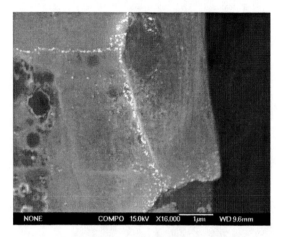

Figure 1: Backscattered electronmicroscope image of a *merino* wool fibre with nano gold on the surface and mainly at the cuticle edges.

The technology was initially developed on the laboratory scale using small quantities of *merino* and *strong* wool fibres in top or sliver form. Samples of spun yarn and woven fabric were also used but difficulties were encountered here due to the processing lubricants that were used during yarn and fabric production absorbing into the wool and interfering with the process chemistry. Further work is being carried out here. A selection *merino* wool fibres coloured with nanogold are shown in Fig. 2

The technology has been successfully scaled up to colour larger quantities of wool, preferably in the sliver or top form which have then been spun into yarn and fashioned into textiles, scarves and carpet (Figs 3, 4). No difficulties were encountered in the spinning or fabrication processes. Interestingly however, the presence of nanogold reduced the static electricity build-up which often arises during spinning in conventionally dyed wools. From this it appears that the electrical conductivity of the wool is enhanced by the nanogold which desirably dissipates the build up of static electricity.

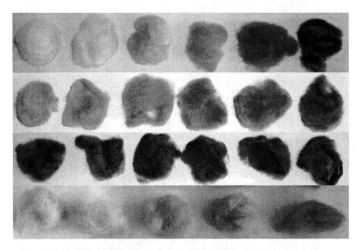

Figure 2: A selection of nanogold coloured *merino* wools

Figure 3: Purple nanogold-*strong* wool in coloured top form and spun into yarn.

X-ray photoelectron spectroscopy (XPS) measurements show that the nanogold is chemically bound to the sulfur in the cystine amino acids of the keratin protein. This is shown by the appearance of peaks in the XPS spectrum (Fig. 5) of the Au 4f electrons of gold in the nanogold-wool which can be assigned to Au-S bonds. Peaks from Au^0 and Au^{3+} are also present The sulfur 2p XPS spectrum of the nanogold wool shows a complementary reduction in the –S-S- peak compared with the –S-H peak due to the –S-S- bonds being broken and Au-S bonds being formed (Fig. 5). In the merino wool itself, the sulfur 2p peaks for the –S-S- of cystine and the –S-H of cysteine are similar in size [6].

Figure 4: Mauve nanogold-*merino* wool as a ball of yarn and woven into a scarf.

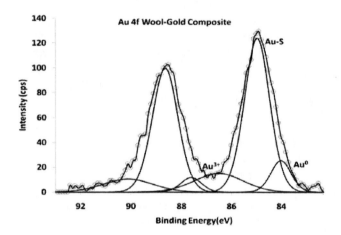

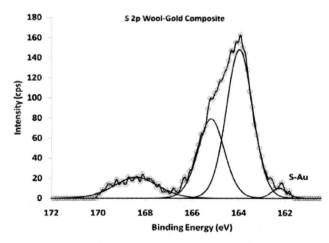

Figure 5: Top – Au 4f XPS spectrum of nanogold-wool; Au-S peaks at 84.2, 87.8 eV; Au^0 at 82.8, 86.4 eV; Au^{3+} at 85.8, 89.4 eV. Bottom – S 2p XPS spectrum of nanogold-wool; S-S at 165.2 eV, S-H at 163.8 eV, S-Au at 162.2 eV, SO$_3^-$ at 168.6 eV.

3 NANOSILVER - WOOL

Nanosilver entities have been similarly attached to *merino* and *strong* wool fibres using our proprietary methodology [5]. For this, we control the reduction of Ag^+ to Ag^0 to produce nanosilver entities of the required particle size and shape to produce the desired colour and again chemically bind them to the keratin in the wool fibres to provide nanosilver-wool products in an attractive range of colours (Fig. 6).

Figure 6: A selection of nanosilver coloured *merino* wools

These wool fibres have also been spun into yarn. As an example of an apparel garment, the nanosilver-*merino* wool has been woven into a scarf and the nanosilver-*strong* wool has been made into carpet (Fig. 7).

Figure 7: Nanosilver-*merino* wool woven into a scarf and nanosilver-*strong* wool made into carpet.

The anti-microbial activity of silver has been known for a long time. It has been shown to be effective against *Escherichia coli, Staphylococcus Aureus*, and *HIV-I*. [7] Due to their large relative surface area and crystallographic surface structure, silver nanoparticles exhibit increased chemical and microbial activity [7]. Anti-microbial testing of nanosilver-*merino* wool was carried out against *Staphylococcus aureus* (ATCC 25923) with an incubation time of 24 hrs. The anti-microbial activity was compared to that for untreated *merino* wool and silver wire. Fibres coloured with nanosilver significantly inhibited the growth of *Staphylococcus aureus* microbes, as shown by the zone of inhibition (clear area) in Fig. 8. For individual nanosilver-*merino* wool fibres, the zone of inhibition was up to 100 times the fibre diameter, whereas for silver wire it

was only 3 times the diameter. Untreated *merino* wool showed no anti-microbial activity. Nanosilver wool therefore exhibits excellent anti-microbial activity which is considerably enhanced over other forms of silver.

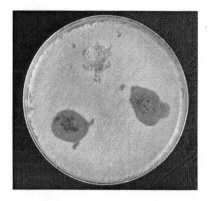

Figure 8: The anti-microbial activity of nanosilver-*merino* wool against Staphylococcus aureus (ATCC 25923) shown by the clear area around clumps of nanosilver-merino wool (bottom) and loose fibres (top).

4 CONCLUSION

High quality *merino* and *strong* wool have been successfully coloured with nanogold and nanosilver to provide a new suite of novel wool fibres for high value fabrics and textiles. A range of attractive colours have been produced and the wool made into examples of high fashion apparel and carpet. *The technology and products are being commercialised. Investment partners are being sought*

REFERENCES

[1] T.J. Norman Jr., C.D. Grant, D. Magana and J.Z. Zhang, *J. Phys. Chem. B* **106** 7005-7012 (2002).
[2] Marie-Christine Daniel and Didier Astruc, *Chem. Rev.* **104** 293-346 (2004).
[3] Marek Grzelczak, Jorge Pérez-Juste, Paul Mulvaney and Luis M. Liz-Marzán, *Chem. Soc. Rev.* 2 **37** 1783-1791 (2008).
[4] G. Mie, Ann. Physik, 25, 377, 1908.
[5] James H. Johnston, Kerstin A. Burridge, Fern M. Kelly, and Aaron C Small, NZ and PCT Patent (Pending), 2009.
[6] James H. Johnston, Kerstin A. Burridge and Fern M. Kelly, *Curr. Appl. Phys. accepted* (2009).
[7] J.M.T. Hamilton-Miller, S. Shah, *Int. J. Antimicrobial Agents* **7(2)** 97-99 (1996)

ACKNOWLEDGEMENTS

We gratefully acknowledge support from Wool Partners International Ltd and the World Gold Council (London).

Deposition of Layered N,C-(TiO$_2$/ITO) Electrodes for Solar Hydrogen Production via Photoelectrochemical Degradation of Aqueous Pollutant

K.R. Wu[*], C.H. Hung[**], C.W. Yeh[***], C.C. Wang[*], T.P. Cho[****]

[*] National Kaohsiung Marine University, Kaohsiung, Taiwan, krwu@mail.nkmu.edu.tw
[**] National Kaohsiung First University of Science and Technology, jeremyh@ccms.nkfust.edu.tw
[***] Kao Yuan University, alice@cc.kyu.edu.tw
[****] Metal Industries Research and Development Centre, tpcho@mail.mirdc.org.tw

ABSTRACT

To enhance the visible-light photoelectrochemical (PEC) response, the heterostructured TiO$_2$/ITO electrodes were codoped with nitrogen (N) and carbon (C) ions grown on ITO glass substrate as a layered N,C-(TiO$_2$/ITO)ITO (Ti/TO) electrode. Under AM 1.0 simulated solar light irradiation and 1.5 V applied bias, the layered Ti/TO electrode yields simultaneously a hydrogen production rate of 12.0 μmol/cm^2 h and pollutant degradation rate of 0.0126 1/cm^2 h, whereas the N,C-ITO electrode shows little activity of pollutant oxidation. One can concludes that the overlaid N,C-TiO$_2$ layer enhances not only the photocurrent response of the layered Ti/TO electrode at entire applied potentials, but also the flat band potential; a shift of about 0.1 V toward the cathode, which is desperately beneficial in PEC process. The PEC activities are attributable in part to the columnar film surface of the layered Ti/TO electrodes.

Keywords: solar hydrogen, TiO$_2$, ITO, PEC degradation

1 INTRODUCTION

The photoelectrochemical (PEC) splitting of water using solar energy has attracted substantial attention as a means of producing hydrogen as a clean and renewable resource [1]. Researchers have sought suitable photocatalysts for splitting water since a pioneering work of Honda and Fujishima by employing TiO$_2$ semiconductor as photoanode in 1972 [2]. As an anatase TiO$_2$ photoanode is excited by incident light with wavelengths shorter than 387 nm, electrons and holes can be generated. In short, the photogenerated holes oxidize and decompose water, even some organic or inorganic substances in aqueous solution, at the photoanode while the electrons can interact with hydrogen ions into hydrogen at the counter platinum (Pt) electrode. It has been shown that hydrogen production can be enhanced by irradiating Pt/TiO$_2$ suspensions with simultaneous degradation of azo-dyes [3]. Moreover, photoelectrolytic cleavage of biomass wastes, such as urine, ethanol and glycerol, in water is much more efficient for electricity generation [4] and hydrogen production [5].

For efficient PEC splitting water, the n-type photoanode is required to have narrow band gap around 2.0 eV, suitable negative flat band potential, good stability and high quantum efficiency [1]. Among various semiconductor photocatalysts, TiO$_2$ is one of the most popular catalysts because it is environmentally friendly and chemically stable in electrolyte solution with high quantum efficiency. However, the use of TiO$_2$ is limited by its wide band gap (~3.2 eV). Alternatively, semiconductors with smaller band gaps, such as CdS (2.4 eV), Fe$_2$O$_3$ (2.3 eV) and Cu$_2$O (2.2 eV), commonly suffer from photocorrosion in electrolyte solution and fast recombination of the photogenerated carriers [1]. It is well-known that metal or semiconductor with a high work function coupled with other semiconductor can significantly enhance the oxidation ability of photogenerated holes in the semiconductor, due to efficient carriers separation. A heterostructured film, such as TiO$_2$/ITO, TiO$_2$/WO$_3$, CdS/TiO$_2$ or TiO$_2$/SnO$_2$, have been proposed for providing a potential driving force for photogenerated charge carriers separation [6-9]. Dai *et al.* have demonstrated that the superior photocatalytic (PC) reactions of the TiO$_2$/ITO film are mostly associated with the photogenerated holes, because ITO has a higher work function (~4.7 eV) than TiO$_2$ and a Schottky barrier can form at the TiO$_2$/ITO interface, where the ITO thin film accepts electrons [7]. One should note that in heterostructured film not only the proximity of their conduction band to conduction band [10], but also the proper redox positions of heterostructure plays significant roles on charge carriers transfer and thus photocatalysis.

Recently, both N-doped In$_2$O$_3$ and ITO (N-ITO) and C-doped In$_2$O$_3$ electrodes were demonstrated to be promising photocatalysts with favorable PEC properties, especially under visible light (λ>378 nm) [11-13]. However, both ITO and N-ITO films exhibit no PC activity in degradation of methylene blue solution [12], though some PC degradation of azo dyes has been reported [14]. In our previous work, we found that an applied bias could serve as a highly efficient way to suppress hole-electron recombination of the N-doped ITO electrode, where the generated photocurrent density sharply increased with its applied bias potential. This has been ascribed mostly to better electrical conductivity and proper positions of the flat band potentials [11,15] as a result of suppression of InN and SnO$_2$ phases in the domain-structured N-ITO electrode which is prepared at a low N-doping content [13].

Titania codoping with anionic species, such as C and N, appears to have very promising visible-light PC properties on pollutant oxidation [16,17]. Yin *et al.* have shown that substitution of the O sites by N and C is responsible for visible-light PC activities [17]. To enhance the visible-light PEC response, the heterostructured TiO$_2$/ITO electrode was codoped with nitrogen and carbon ions grown on ITO glass substrate, as a layered N,C-(TiO$_2$/ITO)ITO film electrode (Ti/TO), for better durability and charged carriers transfer and thus PEC capability. We aimed to investigate the Ti/TO electrodes for solar hydrogen production with simultaneous degradation of aqueous pollutant.

Dimethyl sulfoxide (DMSO) is one of the most common phenol-free organic solvents applied in semiconductor manufacturing industries due to its superior solvent property and water miscibility. However, DMSO is unable removed effectively by most typical biological wastewater treatment units [18]. Thus, some advanced oxidation processes, e.g. O$_3$/UV, or UV/TiO$_2$-based photocatalysts are being developed for the particular purposed of reducing the DMSO concentration in the wastewater [19,20]. In this study, DMSO was chosen as the target pollutant for PEC degradation with simultaneous hydrogen production.

2 EXPERIMENTAL PROCEDURE

A closed-field unbalanced magnetron sputtering system (MIRDC, Taiwan) was used to prepare the samples. Layered TiO$_2$/ITO films were codoped with N and C ions on as-received ITO (17 Ω/sq. and 100±10 nm thick) glass substrates. In addition to the heterostructured Ti/TO film, N,C codoped TiO$_2$ (N,C-TiO$_2$, 2.2 µm) and ITO (N,C-ITO, 1.8 µm) samples were prepared under the same sputtering conditions as each counterparts of the Ti/TO were, denoted as N,C-TiO$_2$ and N,C-ITO, respectively. An as-received ITO electrode (e-ITO) with a thickness of 300 nm was used for comparison. Details were described elsewhere [12,21].

The crystal structures of the samples were analyzed using a high resolution X-ray diffractometer (XRD, Rigaku ATX-E) and Micro-PL/Raman spectroscope (Jobin-Yvon T64000). The surface topography of samples was analyzed using an atomic force microscope (AFM, SPI 3800N, Seiko. The microstructure and thickness of the films were investigated by a scanning electron microscope (SEM, JEOL JSM-6700F).

PEC oxidation tests were performed using a standard PEC three-cell system that included a sample anode, a saturated calomel electrode (SCE) as a reference electrode and a Pt wire counter electrode in DMSO or 4.0 N Na$_2$CO$_3$ solution (pH~11). The initial concentration of DMSO was 0.03 mole/L. The DMSO residual concentration was monitored by a GC (HP4890) with FID as its detector. The analytic conditions of the GC/FID for DMSO quantification were: VOCOL separation column (60-m long×0.53 mm, 5 µm, Supelco Co.), temperature ramping from 100 °C up to 220 °C, and detector temperature was set at 250 °C. The system was controlled by a potentiostat (CHI 610C). The samples were illuminated by artificial sunlight (Newport 96000 150 W solar simulator) with a light intensity of AM

1.0 (75 mW/cm^2). An applied bias of 0.5 V and 1.5 V vs. SCE was used in all PEC and electrolysis tests, since a total of applied bias of higher than 2.0 V has been reported to be supplied to drive electrolysis in a practical PEC system [22].

3 RESULTS AND DISCUSSION

3.1 Microstructural and Morphological Properties

Fig. 1 presents the XRD patterns of all three samples along with e-ITO electrode, which shows distinct diffraction patterns of polycrystalline cubic bixbyite In$_2$O$_3$ phase. All ITO preferred planes revealed in N,C-TiO$_2$ sample are about the same characteristics in the e-ITO electrode. However, N,C-ITO and Ti/TO samples exhibit a significantly different pattern against the e-ITO, exhibiting relatively large intensity ratios of (400)/(222) and (440)/(222) planes, as listed in Table 1, which are associated with the increase in resistivity of the host ITO [12]. The increase in resistance of conducting ITO substrate has been correlated to the decrease in fill factor of dye sensitized TiO$_2$ solar cells [1]. The intensity ratio that is mostly ascribed to the N dopant increases with its concentration in the ITO lattice [12]. This implies that the intensity ratio, namely resistivity, can be suppressed by doping fewer amount of N with C ions into the host ITO. In addition, a phase separation of small SnO$_2$ particles in the host crystalline ITO lattice may occur, since tin is liable to be segregated close to the surface of the lattice or along its grain boundaries under elevated deposition or post treatment conditions [23]. This can cause the increase in resistivity of the N,C-ITO films. The segregation of SnO$_2$ particles can be further examined by Raman scattering spectra in the following section.

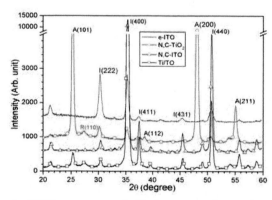

Fig. 1: XRD patterns of all three samples and e-ITO electrode; A: anatase, I: In$_2$O$_3$

Raman spectra are known to be very sensitive to local crystallinity and microstructures near the film surface. The Raman spectra shown in Fig. 2 reveal typical patterns of the anatase TiO$_2$ phase for both N,C-TiO$_2$ and Ti/TO samples. The most intense Raman peaks of the two samples are in the range of 144±0.5 cm^{-1}, which implies well-crystallized nature of the anatase TiO$_2$ phase [24]. A broad peak around 235 cm^{-1} of SnO$_2$ phase is observed in N,C-ITO sample. The same broad peak with very weak intensity is also observed in Ti/TO sample, as shown in the inset of Fig. 2. This confirms the finding of the XRD patterns discussed in Fig. 1.

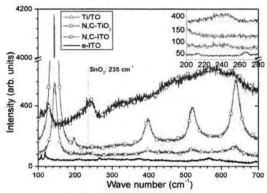

Fig. 2: Raman spectra of all three samples and e-ITO electrode

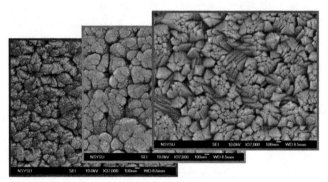

Fig. 3: Plain view of SEM images: (a) N,C-ITO, (b) Ti/TO and (c)N,C-TiO$_2$ samples

Fig. 3 displays plain SEM images of N,C-ITO, Ti/TO and N,C-TiO$_2$ samples. The N,C-ITO film reveals various shapes and sizes of prismatic grains (50-200 nm) with few small sub-grains oriented in the same direction on the surface of the film, as shown in Fig. 3(a). This grain-subgrain structure is commonly found in typical DC-sputtered ITO films [23] and on N-doped ITO films prepared at an optimal N content [12]. This implies that the N,C-ITO film still preserve a relative low resistance. As presented in Fig. 3(b), the columnar film morphology observed on Ti/TO sample is more distinct in comparison with that on N,C-TiO$_2$, as shown in Fig. 3(b) and 3(c), respectively. The AFM measurements indicate that the N,C-ITO has a RMS roughness and specific surface area ratio of 16.3 nm and 1.21, where N,C-TiO$_2$ has the corresponding values of 18.5 nm and 1.25, respectively. The Ti/TO film which consisted of a N,C-ITO film overlaid by a DC-sputtered TiO$_2$ film with a thickness of 300 nm has the values of 33.6 nm and 1.35, respectively, as shown in Table 1. One can expects that a significant increase in

surface morphology rends more reactive surface area of the sample.

Table 1: General structural properties of samples

sample	In$_2$O$_3$ intensity ratio		RMS (nm)
	(400)/(222)	(440)/(222)	
N,C-ITO	12.15	4.51	16.3
N,C-TiO$_2$	0.97	0.77	18.5
Ti/TO	38.53	5.93	33.6
e-ITO	0.99	0.65	<2.0

3.2   Photoelectrochemical Properties

Fig. 4 shows a set of current-voltage (IV) characteristics recorded on three samples and e-ITO electrode in dark and under illumination of simulated solar light. Under illumination, the photocurrent increases sharply as the applied potential reaches 0.5 V and 0.3 V vs. SCE on N,C-ITO and Ti/TO electrodes, respectively. There is no saturation of photocurrent observed in both two samples, which indicates efficient charge separation under illumination, while the saturation of photocurrent can be seen on N,C-TiO$_2$ sample. Under illumination, however, the e-ITO electrode exhibits no gain on photocurrent with a breakdown point at about 0.9 V vs. SCE. This indicates that the N and C dopants can enhance the PEC properties of pristine ITO film. That is a shift of the flat band potential from -0.70 to -0.82 V vs. SCE is achieved on N,C-ITO sample, which was estimated from the I-V measurements in Fig. 4 [25]. The relatively high photoactive response of the Ti/TO electrode is attributable to, at least in part, the synergetic effect of N,C-codoping on band gap narrowing and photosensitizing [16,21]. Moreover, the overlaid N,C-TiO$_2$ layer (300 nm thick) enhances not only the photocurrent response of the layered electrode at the entire applied potentials, but also the flat band potential from -0.82 to -0.91 V vs. SCE; a shift of negative flat band potential is desired for facilitating the PEC process [1,25].

Under irradiation of AM 1.0 simulated solar light and 1.5 V applied bias, the layered Ti/TO film electrode has the highest photocurrent density of 0.42 mA/cm^2 and hydrogen yield rate of 28.8 µmol/cm^2 h in 4.0 N Na$_2$CO$_3$ solution (pH~11). Comparatively, the N,C-ITO electrode has a photocurrent density of 0.35 mA/cm^2 and a hydrogen yield rate of 24.6 µmol/cm^2 h, respectively. However, the N,C-TiO$_2$ electrode exhibits a photocurrent density of 0.09 mA/cm^2 and a relatively high hydrogen yield rate of 25.0 µmol/cm^2 h, respectively.

The PEC and PC activities of the layered Ti/TO film were further evaluated by studying the degradation of DMSO aqueous solution with simultaneous hydrogen production under illumination of simulated solar light. Hydrogen evolution by electrolysis was undetectable at an applied bias of even higher than 2.0 V vs. SCE, which is in good agreement with the report by Mishra et al. [22]. The PEC activities of Ti/TO electrode at 1.5 V show a significant increase in the hydrogen evolution with simultaneous DMSO degradation activity, as summarized in Table 2. The hydrogen yield rate obtained at 1.5 V was 12.0 $\mu mol/cm^2$; about 40 times higher than that at 0.5 V (0.3 $\mu mol/cm^2$ h) vs. SCE, which is correlated well with the sharp increase of photocurrent density of 0.61 mA/cm^2 at 1.5 V vs. SCE (figure not shown). The degradation rate is found to fit approximately a pseudo-first-order kinetic model as a function of applied bias; the degrading rate constant of DMSO was equal to 0.0108 $1/cm^2$ h and 0.0126 $1/cm^2$ h at an applied bias of 0.5 V and 1.5 V, respectively. In other words, only a 17% increase in degradation rate was gained as the applied bias was raised from 0.5 V to 1.5 V, but a 50% increase in degradation rate was obtained from the PC (0.072 $1/cm^2$ h) to PEC reaction at 0.5 V.

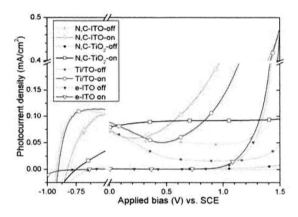

Fig. 4: Photocurrent response of three samples and e-ITO electrode

Table 2: The degradation rate and hydrogen yield rates obtained from the PEC reaction in DMSO solution

sample	degradation rate, $1/cm^2$ h		hydrogen yield rate, $\mu mol/cm^2$ h	
	0.5 V	1.5 V	0.5 V	1.5 V
Ti/TO	0.0108	0.0126	0.3	12.0

4 CONCLUSION

In summary, the layered Ti/TO electrode that combined a versatile N,C-TiO$_2$ top layer and novel bias-driven N,C-ITO intercalated layer is expectedly to perform the synergetic PEC capability for simultaneous DMSO degradation and hydrogen production under solar light illumination. In addition to a distinct columnar morphological feature, the overlaid N,C-TiO$_2$ layer enhances both the photocurrent response and the negative flat band potential of the layered Ti/TO electrode as well.

Acknowledgements

The authors would like to thank the National Science Council of Taiwan, ROC, for financial support this research under Contract No. NSC 98-2221-E-022 -004 -MY2.

REFERENCES

[1] J. Nowotny, T. Bak, M.K. Nowotny and L.R. Sheppard, Int. J. Hydrog. Energy 32, 2609, 2007.

[2] A. Fujishima and K. Honda K. Nature 238, 37, 1972.

[3] A. Patsoura, D.I. Kondarides and X.E. Verykios, Appl Catal B 64, 171, 2006.

[4] D.I. Kondarides, V.M. Daskalaki, A. Patsoura and X.E. Verykios, Catal. Lett. 122, 26, 2008.

[5] M. Antoniadou, P. Bouras, N. Strataki and P. Lianos, Int. J. Hydrog. Energy 33, 5045, 2008.

[6] H. Irie, H. Mori, K and. Hashimoto, Vacuum 74, 625, 2004.

[7] W. Dai, X. Wang, P. Liu, Y. Xu, G. Li and X. Fu, J. Phys. Chem. B 110, 13470, 2006.

[8] H. Chen, S. Chen, X. Quan, H. Yu, H. Zhao and Y. Zhang, J. Phys. Chem. C 112, 9285, 2008.

[9] J. Bai, J. Li, Y. Liu, B. Zhou and W. Cai, Appl. Catal. B 95, 408, 2010.

[10] A.J. Nozik and R. Memming, J. Phys. Chem. 100, 13061, 1996.

[11] K.R. Reyes-Gil, EA. Reyes-Garcıa and D. Raftery, J. Phys. Chem. C 111, 14579, 2007.

[12] K.R. Wu, C.W. Yeh, C.H. Hung, L.H. Cheng and C.Y. Chung, Thin Solid Films 518, 1581, 2009.

[13] Y. Sun, C.J. Murphy, K.R. Reyes-Gil, EA. Reyes-Garcia, J.P. Lilly and D. Raftery, Int. J. Hydrog. Energy 33, 5967, 2008.

[14] M.H. Habibi and N. Talebian, Dyes Pigment 73, 186, 2007.

[15] Y. Bessekhouad, D. Robert and J.-V. Weber, Catal. Today 101, 315, 2005.

[16] D. Noguchi, Y. Kawamata, and T. Nagatomo, J. Electrochem. Soc. 152, D124, 2005.

[17] S. Yin, M. Komatsu, Q. Zhang, F. Saito and T. Sato, J. Mater. Sci. 42, 2399, 2007.

[18] S. Matsui, Y. Okawa and R. Ota, Water Sci. Technol. 20, 201, 1998.

[19] J.J. Wu, M. Muruganandham and S.H. Chen, J. Hazard. Mater. 149, 218, 2007.

[20] A.M. Cojocariu, P.H. Mutin, E. Dumitriu, A. Vioux, F. Fajula and V. Hulea, Chemosphere 77, 1065, 2009.

[21] K.R. Wu and C.H. Hung, Appl. Surf. Sci. 256, 1595, 2009.

[22] P.R. Mishra, P.K. Shukla, A.K. Singh and O.N. Srivastava, Int. J. Hydrog. Energy 28, 1089, 2003.

[23] M. Kamei, Y. Shigesato and S. Takaki, Thin Solid Films 259, 38, 1995.

[24] V. Swamy, A. Kuznetsov, L.S. Dubrovinsky, R.A. Caruso, D.G. Shchukin and B.C. Muddle, Phys. Rev. 71, 184302, 2005.

[25] M. Radecka, M. Rekas, A. Trenczek-Zajac and K. Zakrzewska, J. Power Sources 181, 46, 2008.

Bimodal powder mixtures for
the powder injection molding of SiC and AlN

V. Onbattuvelli*, G. Purdy*, G. Kim*, S. Laddha** and S.V. Atre*
*School of MIME, Oregon State University, Corvallis, OR 97331, USA
**Pacific Northwest National Laboratory, Richland, WA 99354, USA

ABSTRACT

In this paper, we studied the effect of nanoparticles addition on the green density and sintering of injection molded SiC and AlN. Using a bimodal powder mixture increased the solids loading for SiC and AlN systems, respectively. The nanoparticles addition also improved the sinterability of SiC and AlN by reducing the liquid phase formation to 1800 and 1500 °C, respectively. Around 99.3% density is obtained for bimodal μ-n AlN system when pressureless sintered at 1650 °C.

Keywords: SiC, AlN, bimodal, PIM, sintering

1 INTRODUCTION

Silicon carbide (SiC) and aluminum nitride (AlN) exhibit thermal and mechanical properties (**Table 1**) of relevance to applications in electronics, aerospace, defense and automotive industries [1-4]. A successful translation of these properties into final applications depends on the fabrication of these ceramics into fully dense microstructures.

Table 1: Properties of sintered SiC and AlN ceramics

Property	SiC	AlN
Density (g/cc)	3.2	3.26
Thermal Conductivity (W/m.K)	120	170
CTE (10^{-6}/°C)	4	4.5
Vicker's Hardness (GPa)	22	10.4
Fracture Toughness (MPa.m$^{0.5}$)	4.5	2.6
Flexural Strength (MPa)	450	320
Volume Resistivity (Ω.cm)	$10^{\wedge 4}$	$>10^{\wedge 14}$

High sintered densities are required for attaining preferred values of thermal and mechanical properties. For example, She and Ueno studied the effect of varying the amount of Al_2O_3-Y_2O_3 as the sintering aids over the % porosity [5]. Similarly, Sciti and Bellosi compared the combinations of Al_2O_3-Y_2O_3 and La_2O_3-Y_2O_3, as the sintering aids for the SiC densification [6]. Sintering temperature and hold time were the commonly studied process parameters in the past. For example, Junior and Shanafield [4] and She and Ueno [5] reported the increase in the sintered density with the increase in the sintering temperature. On the other hand, Rodriguez et al [7] and Izhevskyi et al [8] noticed an increase in the sintered density with higher hold time at the sintering temperature.

In this paper, we tried to improve the green density of the feedstock by blending two different particle sizes [9, 10]. Such a bimodal mixture was selected in such a way that the smaller particles fit into the interstitial spaces between the larger particles without forcing them apart. In the present work, bimodal μ-n SiC and AlN were mixed in different ratios along with their sintering aids. The effect of varying the nanoparticle content over the solids loading was studied and composition providing the maximum solids loading was thus chosen. The formularions were then scaled up and injection molded to provide the "green" samples. Powder injection molding (PIM) was chosen as the net-shaping technique so as to fabricate complex part geometries at a higher production rate. Following the molding cylce, the polymer was removed by solvent and thermal debinding. The debound "brown" samples were then sintered and measured for densities and weight loss.

2 EXPERIMENTAL SECTION

2.1. Materials

Commercially available α-SiC, AlN, Y_2O_3 were used as the starting materials as received. Multi-component binder system comprising of paraffin wax, polypropylene, polyethylene and stearic acid was used in the current study. The binder composition was chosen in such a way to facilitate a multi-step (solvent, thermal) debinding.

2.2. Instrumentation

Torque rheometry (Intelli-Torque Plasticorder - Brabender) was performed to determine the composition (critical solids loading) of SiC and AlN feedstocks. The formulations determined were scale-up with a co-rotating twin-screw extruder (Entek Extruders, OR). Thermogravimetric analysis (TGA) (TA-Q 500) and differential scanning calorimetry (DSC) (TA-Q 1000) were done to shortlist the molding temperatures. Arburg 221M injection molding machine was used to shape the "green" SiC and AlN samples. All sintering experiments were conducted in the CPF-tube furnace systems (Thermal Technology LLC) with no external pressure. The sintered densities were measured via Archimedes apparatus. The micrographs of the samples were taken with the Qunta^{TM} –FEG (FEI) dual beam electron microscope.

3 RESULTS AND DISCUSSION

3.1 Feedstock Formulation and Scale-Up

Torque rheometry was performed to determine the powder fractions required to achieve a homogenous bimodal powder-polymer mixtures. The binder components were melted first at 160 °C and bimodal powders premixed with the sintering aids were then slowly added.

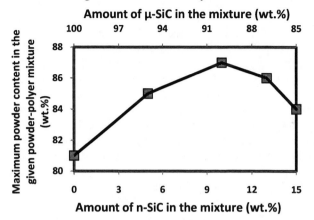

Figure 1a: Effect of n-SiC addition to μ-SiC on the maximum powder fraction in the powder-polymer mixture

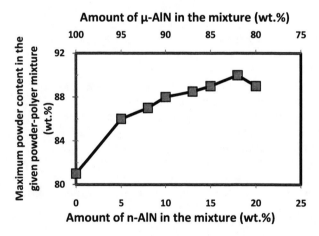

Figure 1b: Effect of n-AlN addition to μ-AlN on the maximum powder fraction in the powder-polymer mixture

There is an upper limit to powder content that is possible in a powder-polymer mixture for any particle size distribution. The particle size distribution was varied by adding nanoparticles to the μ-sized powders. The fraction of of n-particles in the bimodal μ-n powder mixture was plotted against the maximum powder content in the resulting powder-polymer mixtures as shown in the **Figures 1a** and **1b**. In both the SiC and AlN systems, a maximum limit to increasing the packing fraction of the particles in the powder-polymer mixture by nanoparticle additions was observed. Thus, a maximum powder content of 87 wt. % (65.5 vol.%) in the powder-polymer mixture was achieved with a bimodal mixture containing 90 wt.% μ- SiC and 10 wt.% n-SiC (**Figure 1a**). Similarly, a maximum powder

content of 90 wt. % (71 vol.%) in the powder-polymer mixture was achieved for the bimodal AlN mixture containing 82 wt.% μ- AlN and 18 wt.% n-AlN (**Figure 1b**). A comparative experiment involving a monomodal 100 wt.% μ-SiC and μ-AlN, resulted in a maximum powder content of only 81 wt. % (53 vol.% - SiC and 52 vol.% - AlN) in the powder-polymer mixture. In addition to the increment in the solids loading, an increase in the torque (directly proportional to the mixture viscosity) is noticed with the nanoparticle addition.

With their tendency to agglomerate, nanoparticles exhibit poor packing density [10, 11]. Hence, the current finding of increase in the powder content (solids loading) with the nanoparticle addition proves to be novel and could result in a possible change in the sintering behavior of SiC and AlN. From the critical solids loading obtained, an optimal solids loading of 59 vol.% (82 wt.%) and 64 vol.% (5 wt.%) were selected for scale up of bimodal μ-n SiC and AlN feedstocks, respectively.

3.2 Feedstock Characterization

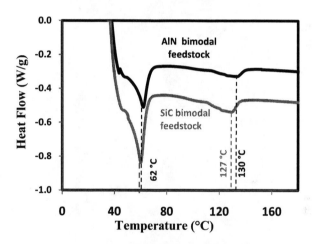

Figure 2: DSC of extruded SiC and AlN feedstocks showing two melting peaks at 62 and 130 °C

The thermal properties of the extruded feedstocks provide basic guidelines for the subsequent PIM steps. **Figure 6** shows the DSC of the feedstocks where two endothermic peaks at 62+/- 2 °C and 130+/-3 °C are observed during heating. These peaks indicate the melting of PW and SA at 62 °C and backbone polymers at 130+/-3 °C [12]. It can thus be inferred that the melt temperatures for injection molding should be higher than 135 °C.

From the TGA plot (**Figure 3**), the AlN bimodal feedstock composition can be finalized as 84.7+/- 0.2 wt.% powder and 15.7 +/- 0.2 wt.% binder. Similarly, it is 81.7+/- 0.3 wt.% powder and 18.7+/- 0.3 wt.% binder for the SiC bimodal feedstock. The two stage degradation reveals the degradation of the filler phase from 175-400°C and backbone polymers from 450-525°C. These pyrolysis data can be used for establishing the upper limit for the injection molding temperatures and thermal debinding profiles [12].

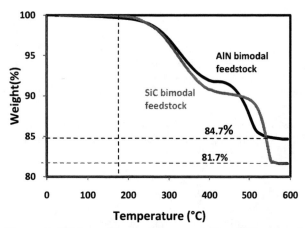

Figure 3: TGA confirming the composition of the bimodal SiC and AlN feedstocks along with their thermal profiles.

3.3 Injection Molding and Debinding

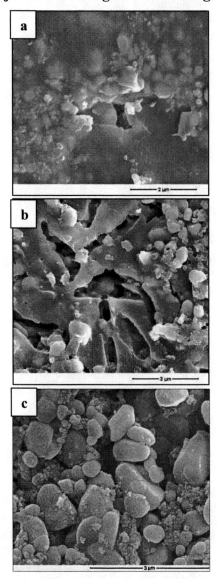

Figure 4: Microstructure of μ-n AlN sample after, (a) injection molding, (b) solvent and (c) thermal debinding

From the TGA and DSC measurements, injection molding temperature window is found to be between 135 to 170 °C. To harness the pseudoplastic behavior of the binder mixture, molten feedstock pellets were injected at 165 °C to produce the "green" samples. The binder's role of providing fluidity and shape to the green parts is complete after injection molding and is thus subjected to removal. Initially, the green samples were immersed in heptane to remove soluble binder components. A large amount of open pores, after solvent debinding, allows the degraded products to diffuse to the surface easily. Consequently, the thermal removal of insoluble binder components will be finished in a much shorter period without endangering the integrity of the resulting "brown" sample [13]. Micrographs in the **figures 4 a-c** represent the solvent and thermal debinding of injection molded AlN samples. Absence of prior studies and importance of the successful debinding necessitates a future work on studying the effect of material (particle size, solids loading and part geometry) and process parameters (solvent type, immersion time, temperature, heating rate and hold time) on the debinding kinetics.

3.4. Sintering

Figure 5 plots the effect of sintering temperature over the % relative densities of bimodal AlN and SiC samples. Around 99.3% relative density was achieved for bimodal AlN at 1650 °C. This is around 100 °C less than what is reported prior for pressureless sintering of AlN [4, 14]. The drastic decrease in the sintering temperature suggests the reaction between Y_2O_3 and Al_2O_3 from the n-AlN to form the yttrium aluminum garnet (YAG) liquid phase. The above argument is once again confirmed with the SEM micrographs revealing liquid phase formation at 1500 °C (**Figure 6a**) and densification at 1650 °C (**Figure 6b**). Prior reports by Molisani et al [14] and Watari et al [15] involving μ-Y_2O_3 and μ-AlN mentioned that the least temperature require for eutectic is 1690 °C. Thus, the reduced sintering temperature in our work can be inferred as the outcome of employing a bimodal powder mixture along with a nanosized sintering aid.

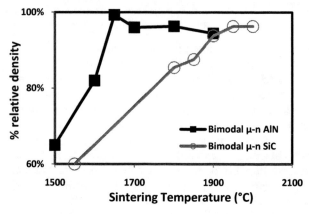

Figure 5: Effect of sintering temperature on the densification of bimodal μ-N SiC and AlN samples

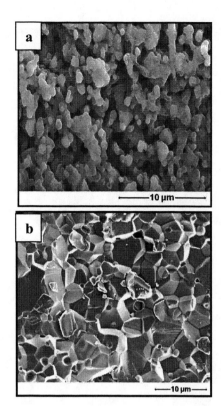

Figure 6: SEM micrographs of bimodal μ-n AlN revealing liquid phase formation at 1500 °C (a) and 99.3% relative density at 1650 °C (b).

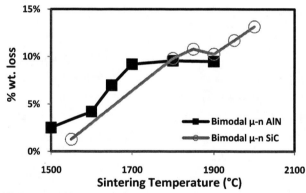

Figure 7: Effect of sintering temperature over the % wt. loss experienced during the densification of bimodal SiC and AlN samples

The bimodal SiC on contrary exhibited a steady increase in the relative density with increase in sintering temperature with the maximum being 97% at 1950 °C. Similar results in SiC were noticed in the past by She and Ueno [5] and Herrmann et al [18]. Achieving around 85% densification at 1800 °C however confirms the formation of liquid phase below 1800 °C. Employing higher hold time at this temperature might further help in understanding the sintering behavior. The inability to reach the maximum densification in SiC and decrease in the AlN density with temperature might be correlated to the % weight loss experienced in both the systems during sintering. **Figure 7** plots the effect of sintering temperature over the % wt. loss

experienced by these bimodal systems. She and Ueno [5], and Molisani et al [16] have reported similar weight loss behavior exhibited by SiC and AlN during sintering. The immediate future work in these systems will involve studying the effect of sintered density and microstructure over the mechanical and thermal properties.

4 CONCLUSIONS

Higher solids loading of 65 and 71 vol.% were obtained for SiC and AlN feedstocks by employing bimodal powder mixtures. With the addition of AlN and SiC nanoparticles, the onset of sintering was brought down to 1500 and 1800 °C, respectively. In addition to the above results, pressureless densification at relatively low temperatures (1650 °C for AlN) suggests the use of bimodal μ-n powder mixtures as a potential approach to process a broad range of difficult-to-sinter ceramic and metal powders.

REFERENCES

1. M. Flinders, D. Ray, A. Anderson and R.A. Cutler, *Journal of American Ceramic Society*, 88, 2217, 2005.
2. K. Pelissier, T. Chartier and J.M. Laurent, *Ceramics International*, 24, 371, 1998.
3. Okada, *Journal of the European Ceramic Society*, 28, 1097, 2008.
4. A.F. Junior and D. Shanafield, *Journal of Materials Science: Materials in Electronics*, 16, 139, 2005.
5. J. H. She and K.Ueno, *Materials Research Bulletin*, 34, 1629, 1999.
6. D. Sciti and A. Bellosi, *Journal of Materials Science*, 35, 3849, 2000.
7. M.C. Rodriguez, A. Munoz and A.D. Rodriguez, *Journal of European Ceramic Society*, 26, 2397, 2006.
8. V.A. Izhevskyi, A.H.A. Bressiani and J.C. Bressiani, *Journal of American Ceramic Society*, 88, 1115, 2005.
9. J.Y. Qiu, Y. Hotta and K. Watari, *Journal of American Ceramic Society*, 89, 377, 2006.
10. P. Suri, S. Atre, R. German and J. D'Souza, *Materials Science and Engineering A*, 356, 337, 2003.
11. X.L. Li, H.A. Ma, Y.J. Zheng, Y. Liu, G.H. Zuo, W.Q. Liu, J.G. Li and X. Jia, *Journal of Alloys and Compounds*, 463, 412, 2008.
12. G. Aggarwal, S.J. Park and I. Smid, *International Journal of Refractory Metals and Hard Materials*, 24, 253, 2006.
13. M.A. Omar, R. Ibrahim, M.I. Sidik, M. Mustapha and M. Mohamad, *Journal of Materials Processing Technology*, 140, 397, 2003.
14. L. Molisani, H.N. Yoshimura and H. Goldenstein, *Journal of Material Science: Material Electronics*, 20, 1, 2009.
15. K. Watari, H.J. Hwang, M. Toriyama and S. Kanzaki, Journal *of Materials Research*, 14, 1409, 1999.
16. M. Herrmann, R. Neher, K. Brandt and S. Hoehn, *Journal of European Ceramic Society*, 30, 1495, 2010.

Preparation, properties, and thermal control applications of silica aerogel infiltrated with solid-liquid phase change materials

Zhou Xiangfa[a,b*], Xiao Hanning[a*], Feng Jian[b],Zhang Changrui[b],Jiang Yonggang[b]

a Department of Materials Science and Engineering, Hunan University, Changsha 410082, China
b State Key Lab of Advanced Ceramic Fibers & Composites, College of Aerospace &Materials Engineering , National Univ.of Defense Technology, Changsha 410073, China

ABSTRACT

In the present work, silica aerogel saturated with erythritol as phase change materials (PCMs) was prepared by melt infiltration. In the novel composite, erythritol with high latent heat of fusion was used as PCM for thermal control, whereas nanoporous silica aerogel was prepared as the phase change matrix to provide structural strength and prevent leakage of melted erythritol. Nitrogen gas adsorption curves and SEM analysis indicate that the pore structure of silica aerogel was porous and connected with each other. FTIR analysis showed that the composite formation of silica aerogel and erythritol were physical, whereas DSC analysis showed that the melting point and heat storage capacity of the composite were 123.8° C and 289.92 kJ/kg, respectively. The thermal protection properties of phase change composites were designed under laboratory conditions using a thermal measurement setup of a simulated thermal environment of an aircraft. Compared with the paraffin-silica aerogel composite, the erythritol-silica aerogel composite could rapidly control the rising temperature by absorbing heat under high thermal environments.

Keywords: Aerospace materials; Electron microscopy; Sol-gel process; Thermal properties; Phase change materials

Corresponding Author's : Tel: +86-731-8822269;
E-mail address:hnxiao@hnu.cn; flucky-zhou@tom.vip.com .

1. INTRODUCTION

The control of temperature and/or heat transfer is of crucial importance in numerous technical processes and natural occurrences. Protecting a surface subjected to high heat flux could be achieved by using a thermal barrier to keep its temperature under emergency operating conditions. Latent heat absorption phenomenon associated with melting suitable PCMs could be effectively used to delay or modify temperature increase in surfaces subjected to high fluxes. A number of inorganic and organic PCMs and their mixtures have been investigated as latent heat storage materials. Sari et al. [1,2] investigated some fatty acids and their eutectic mixtures used for latent heat storage, while Alkan et al. [3] investigated a novel material for thermal storage-solid/liquid transitions in ethylene glycol distearate. Thermophysical properties of three kinds of paraffin were determined by DSC and their heat transfer processes were investigated [4]. Among the investigated PCMs, paraffin has been widely used for latent heat storage applications due to its proper thermal characteristics such as little or no super cooling, low vapor pressure, chemical stability, and self-nucleating behavior [5, 6]. However, paraffin has now been proven to suffer from some technical problems, as well as in its use and application, which include relatively low latent heats of fusion, inability to control shape and form of melted paraffin, and unevenness in heating and cooling. Hayes [7] investigated composite fabric endothermic materials (CFEMs) in devices suitable for aerospace and military use, such as heat sinks. CFEMs preferably comprise PCMs that induce capillary action and chemical adhesion to the fibers of the matrix. As a result, a greatly increased surface area for heat transfer is obtained, thus providing controlled melting and thermal dissipation of the fusion cooling agent.

However, they cannot be effectively used for applications that require cooling at very high temperatures or at prolonged periods, such as in aircraft electronics and spacecraft devices. Thus, appropriate selection of PCMs is essential for the successful design of thermal protection composites. Firstly, higher latent heat storage capacity determines the stability of the thermal performance of phase change composites requiring higher latent heat of fused PCMs. Secondly, the leakage behavior of phase change composites under service temperature—another important property—determines durability and potential use in different applications. Kaizawa et al. [8] showed that erythritol with a large latent heat of 344 kJ/kg at melting point of 117° C exhibited excellent chemical stability under repeated melting and solidifying cycles. On the other hand, the problem on leakage of melted organic PCMs could be improved by embedding a porous matrix like porous ceramics. Wirtz et al. [9, 10] demonstrated that graphite foam can effectively immobilize liquid paraffin. Harel et al. [11] investigated pore structure and connectivity of silica aerogel with high porosity and large surface area, and subsequently obtained many applications especially in aerospace materials.

The objective of the present work is to integrate the aforementioned advantages. The high heat storage capacity of the erythritol-silica composites, which was determined by melt infiltration suitable for heat-absorbing applications and pore structural properties of the silica matrix, were analyzed. Properties of the novel erythritol-silica composite were measured by means of DSC, SEM, and FTIR. The thermal protection properties of the erythritol-silica composites were measured under laboratory conditions using the simulated thermal environment of an aircraft.

2. EXPERIMENTAL PROCEDURES
2.1. Materials

Erythritol ($C_4H_{10}O_4$) with melting point of 119–123° C was obtained from Green Brand Biotechnology Co. (Shangdong, China).

2.2. Preparation of silica aerogel and silica-erythritol composite

Silica alcogels were prepared by a two-step sol-gel process using tetraethoxysilane (TEOS) as the precursor and a co-precursor diluted in methanol (MeOH) solvent. Initially, hydrolysis and condensation were carried out under acidic (oxalic acid) and basic (ammonium hydroxide (NH_4OH)) catalysis conditions. Then, the alcogels were dried under supercritical conditions in an autoclave above the critical temperature and pressure of methanol. The silica aerogel was then obtained.

Finally, to prepare the phase change composite with high heat storage capacity, silica aerogel was heated with molten erythritol by melt infiltration.

2.3. Pore characterization of silica aerogel

Pore-size distributions were measured using a nitrogen gas adsorption Brunauer-Emmett-Teller (BET) surface area analyzer (ASAP 2000). BET analysis based on the amount of N_2 gas adsorbed at various partial pressures (five points at $0.05 < P/P_0 < 0.3$; nitrogen molecular cross-sectional area=$0.162nm^2$) was used to determine surface area. A single condensation point (P/P_0=0.99) was used to determine pore size and pore volume. Pore size distributions were calculated from the desorption isotherms.

2.4. Determination of properties of the erythritol-silica composite

Microphotographs were taken from the fracture surface of the composite by SEM (JSM-6360LV). The chemical properties of the erythritol, silica, and the composite were determined by FTIR. The melting point and heat storage capacity of the erythritol and the erythritol-silica composite were determined by DSC (NETZSCH STA449C) calibrated according to indium standards in the 30–200° C range. The scanning rate is 5° C/min.

2.5. Thermal protection measurement set-up

Paraffin-silica composites were fabricated [12] and introduced for thermal protection investigation rather than the erythritol-silica. Both paraffin-silica and erythritol-silica composites (dimensions: 100×100×2mm) were introduced as solid matrices for the phase change composites. The experimental data on the two samples is illustrated in Table 1. The samples were introduced into the experimental setup (Fig. 5), and a steel plate with thickness of 2 mm was glued to the left side of the silica aerogel composite using a silicon layer tightened by screws. The composites were heated from the left side using a heater installed on the furnace to provide the required heating source. The temperature of heater surface (hot face) was detected by the displaying instrument of the furnace. Another steel plate with the same thickness was used to press PCM-silica samples from the right side. J-type thermocouple installed on the right steel plate was used to detect the temperature of the samples (cold face). Temperature varied with time was then recorded. While conducting the experiments, the top and bottom walls of the container were insulated with adiabatic materials.

3. RESULTS AND DISCUSSION
3.1. Structural properties

Figure 1 shows the microstructure of the silica and the composite. The microstructure of the silica aerogel is porous and connected to each other. Meanwhile, the microstructure of the composite shows that the erythritol was dispersed uniformly into the porous network of the silica aerogel.

Figure 2 shows the nitrogen adsorption/desorption isotherms and pore size distributions of the silica aerogel. The isotherms of the sample belong to type IV and are characteristic of porous material with capillary condensation [13]. For the samples, the capillary condensation step is at a higher relative pressure (0.85 P/P_0), which suggest bigger pore diameters. Pore size distribution shows two kinds of pores centered at approximately 53.09 nm. The silica aerogel has a high BET surface area of 552.75 m^2/g and a pore volume of 4.390cm^3/g.

3.2. Chemical properties

The FTIR spectra of the erythritol-silica composite are shown in Fig. 3. For the erythritol, the FTIR spectra displayed a broad O-H stretching band at wavenumber 3438 cm^{-1} and peaks at wavenumbers 1412 cm^{-1}, 1046cm^{-1}, and 612 cm^{-1}, which are typical of carbohydrates. Peaks at 2944 cm^{-1} and 1641 cm^{-1} were caused by the stretching vibration of the functional group of C-H. For the silica aerogel, the FTIR spectra showed an intensive peak at 1077cm^{-1} indicating the bending vibration of Si-O; meanwhile, the peaks at 878 cm^{-1} were specifically caused by the same Si-O bending vibrations. The peak at 3449cm^{-1} represents the stretching vibration of the functional group of Si-OH. For the composite, no significant new peaks were observed. Findings from the FTIR spectra illustrate that the composite is merely a physical combination of silica aerogel and erythritol.

3.3. Thermal properties

Figure 4, which presents the thermal characteristics of the erythritol and the composite, shows that the latent heat of the erythritol is 338.46 kJ/kg (Tm melting point=121.4° C). This proves that erythritol has a large latent heat of fusion; the quite large enthalpy is caused by its linear straight chain containing the hydroxyl group. The DSC curve of the composite indicates that a phase change at 123.8° C has a large heat storage capacity of 289.92 kJ/kg. The thermal characteristics of the composite are very close with the erythritol.

3.4. Thermal protection measurement

The thermal protection characteristics of the two-phase change composites were studied through the experiment illustrated in Fig. 6. Figure 6(a) represents the temperature of cold face subjected to different temperatures of the hot face; the figure shows that the temperature of cold face of the erythritol-silica is lower than the paraffin-silica composite when subjected to the same hot face temperature. Figure 6(b) represents the temperature of cold face varying with time when the temperature of the hot face was set to 600° C; the figure shows that temperature changes are not distinct with time around the phase change temperature of the composites. The erythritol-silica increased the holding time or duration of temperature compared with the

paraffin-silica composite. For the erythritol-silica composite, minimal temperature changes were observed from 800–1600 s, which indicate that the time-temperature curve slope hardly increased at temperatures around the heat absorption capacity of the composite. Thus, the erythritol-silica composite obtained better heat absorption properties.

4. CONCLUSIONS

(1) The high heat storage capacity phase change of the nanocomposite was successfully prepared by melt infiltration of the erythritol into the porous silica.

(2) There was no reaction between the erythritol and silica aerogel based on structure analysis via SEM and chemical analysis via FTIR.

(3) The heat storage capacity of the composite at the melting point, 123.8° C is 289.92 kJ/kg.

(4) The erythritol-silica composite exhibited fairly efficient thermal regulation compared with the paraffin-silica composite.

References

[1] A.Sari,K. Kaygusuz .Renew Energy; 28:939-948, 2003.

[2] A.Sari..Energy conver and manag; 47:1207-1221,2006

[3] C.Alkan,K.Kaya ,A.Sari. Mater lett; 62:1122-1125, 2008.

[4] M.Akgun, O.Aydin, K.Kaygusuz. Appl Therm Engi; 28:405-413, 2008.

[5] X.Liu,H.Liu. Energy Conver Manag; 47:2515-22, 2006.

[6] F.Frusteri, V.Leonardi, et al. Appl Therm Eng; 25:1623-1633, 2005.

[7] C.Hayes.U.S. Pat No. 4446916. 1984, 05,08.

[8]A.Kaizawa,N.Maruoka,etal.Heat mass transf; 44:763-769,2008.

[9]R.Wirtz, P.Shuo, A.Fuchs. In:The 6th ASME-JSME thermal engineering joint conference, March 2003.

[10]R.Wirtz , A.Fuchs , et al. AIAA Paper 2003-05-13.

[11] E.Harel , J.Granwehr , A.Juliette . Nat mater, 5: 4, 2006.

[12] X.F Zhou, H.N Xiao, et al.Compo Sci Tech，69：1246-1249, 2009.

[13]K.Sing .Colloids and Surfaces A, 187-188: 3-9, 2001

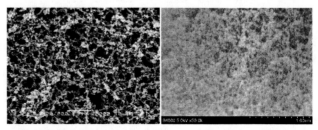

Fig.1 SEM of the microstructures of the silica aerogel and the erythritol-silica composite

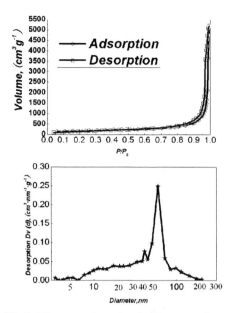

Fig.2. Nitrogen adsorption isotherms and pore size
distribution of the silica aerogel

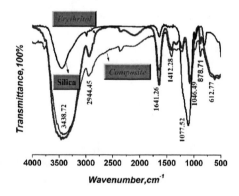

Fig. 3.FTIR spectrum of the erythritol and the silica aerogel
and the erythritol-silica composite

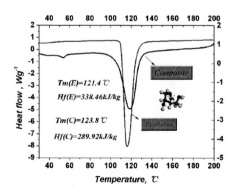

$Tm(E)=121.4\ ℃$

$Hf(E)=338.46kJ/kg$

$Tm(C)=123.8\ ℃$

$Hf(C)=289.92kJ/kg$

Fig.4. DSC Curves of the erythritol and the erythritol-silica
composite

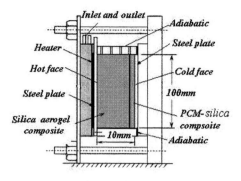

Fig.5. Thermal protection setup for the PCM-silica
composite

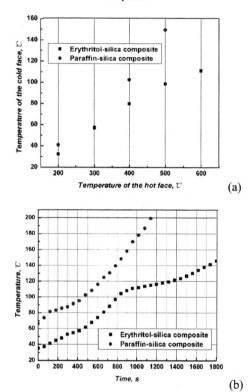

Fig.6. (a) Comparison of cold face and hot face temperature
of erythritol-silica and paraffin-silica composites

(b) Cold face temperature varied with time as hot face
temperature is 600℃

Table 1

Thermal physical parameter of the erythritol-silica and
paraffin-silica composites

Composites(dimension: 100×100×2mm)	Erythritol-silica	Paraffin-silica
Phase change temperature, ℃	123.8	56.3
Heat storage capacity, kJ(kg)$^{-1}$	289.9	165.1
Conductivity,Wm^{-1}K^{-1}	0.31	0.38
Density, kg·m^{-3}	1327.6	944.1

Multifunctional Nanocomposite Systems

J. Dinglasan, A. Das, N. Loukine, D. Anderson[*]

[*]Vive Nano, Inc.
700 Bay St., Toronto, ON., M5G 1Z6, Canada, danderson@vivenano.com

ABSTRACT

A method to produce multifunctional nanocomposite materials is presented. Polymer stabilized nanoparticles (e.g. hydroxides, oxides, metals, salts, etc.) are incorporated into a bulk inorganic matrix (zinc oxide/hydroxide, aluminum oxide/hydroxide, ferric oxide, silica, ceria, etc) through the formation of an inorganic phase around the polymer stabilized nanoparticles. This results in the precipitation of a porous nanocomposite that possesses multiple functionalities that come from the different combinations of components of the composite. Nanocomposites that can be used as catalysts, sorbents for heavy metals and arsenic, ion exchangers, and pigments have been made, and data illustrating these properties will be shown

.

Keywords: nanoparticles, nanocomposite, sorbents, ion exchange, oxidation catalyst

1 SYNTHESIS AND CHARACTERIZATION OF MULTIFUNCTIONAL NANOCOMPOSITES

The nanocomposite system is composed of two distinct components: polymer coated nanoparticles and an inorganic phase. The synthesis of the polymer coated nanoparticles is illustrated in Figure 1.

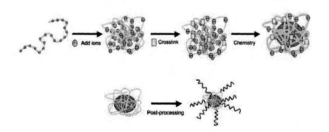

Figure 1. Schematic of Vive Nano's nanoparticle synthesis

Water dispersible nanoparticles are produced using counterion-induced collapsed polyelectrolyte chains. Polyelectrolytes are polymers having monomer units with ionizable groups such as amines, carboxylic acids, or sulfonic acids. Their conformation in aqueous solution is governed by electrostatic interactions between the charged moieties in the chain. Under conditions of high ionization and low ionic strength, the individual polyelectrolyte chains assume a highly extended, swollen coil conformation arising from repulsive electrostatic interactions along the chain. The addition of ionic species results in the screening of these repulsive interactions by the counterions, causing the highly extended polyelectrolyte coils to collapse into globules [1]. These globules contain the counterions used to induce collapse, and can range in size from ~ 1-50 nm. Crosslinking the collapsed polymer chain results in a stabilized globular structure thereby forming a template in which to synthesize nanoparticles. Nanoparticles are formed from the trapped counterions in the globules. For example, if we use Ag^+ ions to collapse the polyelectrolyte, the trapped Ag^+ ions can be reduced to form Ag^0 nanoparticles. This method is very versatile and can be used to synthesize different noble metal, metal oxide, semiconductor and alloy nanoparticles encapsulated inside a charged polymer template.

These polymer encapsulated inorganic nanoparticles can then be embedded in a porous matrix support comprising of various inorganic oxides, effectively creating a supported multi functional nanocomposite with a very high active surface area. The formation of nanocomposite can simplified into three steps and is summarized in Figure 2 below.

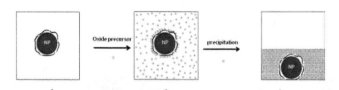

Figure 2. Steps outlining the formation of the multifunctional nanocomposite.

The polymer coated nanoparticles are first dispersed in water (1). An inorganic oxide precursor (e.g. salts) is then added to the solution of nanoparticles. These oxide precursors coordinate with the functional groups of the polymer through either ionic, donor-acceptor or hydrogen bonds (2). Finally, the insoluble oxide is formed by changing the pH of the solution to form an insoluble solid inorganic hydroxide/oxide phase wherein the nanoparticles are entrapped as part of a composite product (3).

Electron microscopy characterization done on polymer encapsulated Fe_2O_3 nanoparticles embedded in a Fe_2O_3 matrix (Fe_2O_3/PAA on Fe_2O_3) have shown that these systems have porosity of ~ 10 nm and that the nanoparticles retain their structural integrity even after nanocomposite formation. Dye adsorption tests on the nanocomposite system have also shown that the functional groups of the encapsulating polymer (i.e. from the nanoparticle) still remain accessible. A high resolution electron microscopy image (FE-SESM) of the Fe_2O_3-based nanocomposite system is shown in Figure 3.

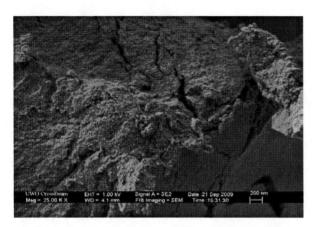

Figure 3. FE-SEM image of Fe_2O_3 nanoparticles embedded in a Fe_2O_3 matrix.

The process can provide unparalleled control over nanocomposite properties by simple and straightforward changes to the combination of nanocomposite elements and synthesis conditions. The great advantage of the proposed nanocomposite system is the mulifunctionality of the material, which is derived from:

1. The inorganic nanoparticle inside the polymer template – different types of nanoparticle can easily be synthesized;

2. The charged polymer encapsulating the inorganic nanomaterials;

3. The inorganic oxide support material that serves as a structural support as well as a means to prevent sintering of the nanoparticles.

2 MULTIFUNCTIONAL NANOCOMPOSITE PERFORMANCE

Nanocomposites were tested for their ability to remove both organic and inorganic contaminants from water solution. These results are shown in figures 4 and 5. Figure 4 shows the performance of a nanocomposite composed of Fe_2O_3/PAA nanoparticles on an Fe_2O_3 support towards the removal of metal cations (Na^+, K^+, Ca^{2+}, Mg^{2+}, Ni^{2+}, Zn^{2+}, Pb^{2+}, Al^{3+} and Sb^{3+}) from solution. The data demonstrates that under certain solution conditions these nanocomposites show some selectivity towards divalent metal cations (like Ni^{2+} and Zn^{2+}) even in the presence of other cations (like K^+ and Na^+). Figure 5 shows the removal of up to 200 mg methylene blue dye using 1 g of the same Fe_2O_3 nanoparticle-based nanocomposite [2].

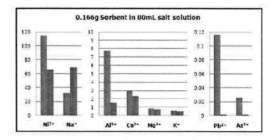

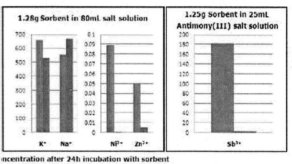

Figure 4. Removal of heavy metals, arsenic and antimony ions from aqueous solutions with Fe_2O_3/PAA on Fe_2O_3. (units in ppm)

Figure 5. Removal of methylene blue from an aqueous solution with Fe_2O_3/PAA on Fe_2O_3.

CO oxidation performance of a nanocomposite comprising (0.5%)Pt, (0.5%)Pd nanoparticles on a Ceria support is shown in Figure 6. This nanocomposite was subjected to a temperature programmed reaction protocol to compare its performance in CO oxidation against the standard (1%) Pt on Alumina support [3]. The data shows this nanocomposite exhibits improved performance over the existing standard by having lower light off temperatures for both CO and propylene oxidation.

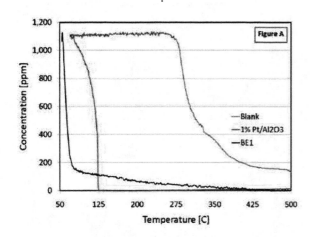

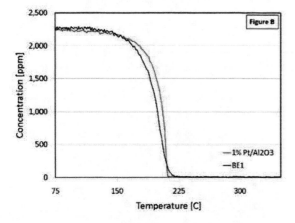

Figure 5. Catalytic converter applications of multifunctional nanocomposites. (A) CO oxidation using (0.5%)Pt (0.5%)Pt on CeO$_2$ (labeled as BE1 in the figure). (B) Propylene oxidation using the BE1.

3 REFERENCES

[1] A. V. Dobrynin and M. Rubinstein. *Prog. Polym. Sci.* 30, 1049, 2005.

[2] L. Xiong, Y. Yang, J. Mai, W. Sun, c. Zhang, D. Wei, Q. Chen and J. Ni. *Chem. Eng. J.* 313, 156, 2010.

[3] O. Shakir, A. Yezerets, N. W. Currier and W.S. Epling. *Appl. Catal. A.* 301, 365, 2009.

In-situ Generation of Paramagnetism during Polymerization

K. Mallick[*], M. Coyanis[*], M. Witcomb[**], R. Erasmus[***]
and A. Strydom[****]

[*]Advanced Materials Division, Mintek, Private Bag X3015, Randburg 2125, South Africa,
kaushikm@mintek.co.za; mabelc@mintek.co.za
[**]Microscopy and Microanalysis Unit, University of the Witwatersrand, Private Bag 3,
South Africa. michael.witcomb@wits.ac.za
[***]School of Physics, University of the Witwatersrand, Private Bag 3, WITS 2050,
South Africa. rudolph.erasmus@wits.ac.za
[****]Physics Department, University of Johannesburg, PO Box 524, Auckland Park 2006,
South Africa, amstrydom@uj.ac.za

ABSTRACT

Polyaniline and its derivatives have gained significant interest because of their unique electronic property, simple synthesis process and their environmental stability. We report on the cerium (IV) ammonium nitrate mediated synthesis of poly (amino-acetanilide), PAA, using an interfacial polymerization technique in which PAA serves as a guest of the cerium (III) ion, a paramagnetic species produced during the synthesis condition. Cerium (III) ionic species bonded with the chain nitrogen of the PAA and the supramolecular system show the paramagnetic behavior throughout the experimental temperature range 400-1.9 K.

Keywords: cerium ammonium nitrate, composite, paramagnetism, SEM, optical characterization

1 INTRODUCTION

Among the known conducting polymers, polyaniline is unique due to its doping adjustable electrical conductivity and metal-like transport property at both room and low temperatures [1]. The electrical conductivity of the polyaniline can be varied over the full range from insulator to metal by doping. Through doping, the chemical potential (Fermi level) can be moved into the region of energy of the high density of electronic states either by a redox reaction or by an acid-base reaction. Doped polyaniline and its derivatives are good conductors due to the fact that doping introduces charge carriers into the electronic structure and the attraction of an electron in one repeat unit to the nuclei in the neighboring unit leads to carrier delocalization along the polymer chain and to charge carrier mobility, which is extended into three dimensions through inter-chain electron transfer [2]. Paramagnetic behaviour in highly protonic acid doped polyaniline has also been reported at low temperatures [3]. EPR study of the camphor-sulphonic acid doped conducting polyaniline showed temperature independent Pauli susceptibility within the temperature range 300 to 50K. The Curie contribution to the electronic paramagnetic susceptibility of the heavily doped polyaniline arises from a disordered metallic state close to the metal-insulator transition. This has been observed only below 50K [3]. Paramagnetism in polyaniline has been reported by introducing paramagnetic metal nanoparticles into the polyaniline matrix [4, 5].

In the present communication we report on an *in situ* synthesis technique for the preparation of a paramagnetic PAA-cerium (III) supramolecular composite material by applying an '*in situ* polymerization and composite formation' (IPCF) technique [6] using cerium (IV) ammonium nitrate (CAN) as an oxidizing agent for polymerizing *para*-amino-acetanilide. CAN is most extensively used in synthetic organic chemistry as an oxidant (reduction potential value of +1.61V vs. NHE) [7]. During the polymerization process each step is associated with a release of electron and that electron reduces the Ce^{+4} ion to form a Ce^{+3} ion. The Ce^{+3} ion binds with the chain nitrogen of the polyaniline which causes the emergence of paramagnetism in polyaniline.

2 EXPERIMENTAL

2.1 Materials

Cerium ammonium nitrate and *para*-amino acetanilide were purchased from Sigma-Aldrich and BDH respectively. Ultra-pure water (specific resistivity >17MΩcm) was used to prepare the solution of cerium ammonium nitrate (10^{-2} mol dm^{-3}). Toluene was purchased from Merck.

2.2 Characterization Techniques

Scanning electron microscopy (SEM) studies were undertaken in a FEI FEG Nova 600 Nanolab at 5 kV. For UV-vis spectra analysis, a small portion of the solid sample was dissolved in methanol and scanned within the range 300-800 nm using a Varian, CARY, 1E, digital spectrophotometer. Raman spectra were acquired using the green (514.5nm) line of an argon ion laser as the excitation source. Light dispersion was undertaken via the single spectrograph stage of a Jobin-Yvon T64000 Raman spectrometer. Power at the sample was kept very low (0.73mW), while the laser beam diameter at the sample was ~1 μm. X-ray photoelectron spectra (XPS) were collected in a UHV chamber attached to a Physical Electronics 560 ESCA/SAM instrument. Magnetic measurements were performed under controlled temperatures using a Magnetic Properties Measurement System (MPMS) (Quantum Design, USA).

2.3 Synthesis of Ce^{+3} Ion Doped PAA

In a typical experiment 0.104g of *para*-amino acetanilide was placed in a conical flask. 10 mL of toluene was slowly added to it under a mild stirring condition. Cerium ammonium nitrate, CAN, (10 ml) having a concentration of 10^{-2} mol dm^{-3} was added drop wise to the conical flask. A dark brown colour developed at the bottom of the conical flask during the addition of the CAN solution to the toluene suspended amino acetanilide. After all the CAN was added, the dark brown precipitation was kept under static conditions for another 10 min. The whole process was carried out at room temperature (~25 °C). Subsequently, the colloidal precipitation was taken from the bottom of the flask and pipetted onto lacey, carbon-coated, copper TEM grids for SEM analysis. The rest of the solution was filtered and washed with distilled water several times. A small portion of the solid mass was used for UV-vis and Raman analysis. For the magnetic property study, the solid mass was dried and measured under controlled temperatures using a MPMS system.

3. RESULTS AND DISCUSSION

Figure 1 shows typical SEM images of the poly (amino-acetanilide). In the Raman spectrum the bands within the range of 1100 and 1700 cm^{-1} are sensitive to the PAA oxidation state. The spectrum reveals C–C deformation bands of the benzenoid ring at 1605 cm^{-1}, which is the characteristic for semiquinone rings [8, 9]. Further the 1504 cm^{-1} band corresponds to the N–H bending deformation band of Ce^{3+} functionalized amine. Two bands at 1325 and 1370 cm^{-1} correspond to C–N^{+} stretching modes created as a result of the incorporation of Ce^{3+} ions with the chain nitrogen. A strong band at 1467 cm^{-1} corresponds to the C=N stretching mode of the quinoid units. The broad band at 1230 cm^{-1} corresponds to the C–N stretching mode of single bonds, the benzenoid units. The position of the benzene C–H bending deformation band at 1171 cm^{-1} is characteristic of the reduced and semiquinone structures. The Raman spectrum indicates the presence of both benzenoid and quinoid structures present in the polymer chain. For the acetamide functional group (CH$_3$-CO-NH–) the symmetric CH$_3$ stretching mode appeared at 2925 cm^{-1} and for CH$_3$-CO– functional group the asymmetric CH$_3$

stretching mode appeared at 3040 cm^{-1} (inset Figure 2) [10].

The electronic absorption spectra of the polyaniline base, the polyaniline salt and their respective derivatives have been well documented in the literature [11, 12]. The PANI salt shows three absorption peaks at 310–360, 400–440, and above 700 nm. The absorption peak at 310–360 nm is due to the π-π^* transition of the benzenoid rings. The peak at 400–440 nm is due to the polaron–bipolaron transition, whereas the broad absorption band appearing above 700 nm is due to the benzenoid-to-quinoid excitonic transition. In this study, the UV–vis spectrum of PAA (Figure 3), which was synthesized with toluene as a solvent and using CAN as an oxidant, showed two absorption bands with peaks positions at 310 and 450 mn corresponding to the aforementioned transitions. In this case, we did not find any prominent absorption band for the excitonic transition, which probably overlapped with the strong absorption due to the polaron–bipolaron transition.

Figure 4 shows the temperature dependence of reverse magnetic susceptibility, $(M/B)^{-1}$, of Ce-PAA measured in a small field of 300 Oe. The temperature evolution is that of a paramagnetic material. There are no co-operative ordering effects down to the lowest measured temperature (2 K). At intermediate temperatures, the $(M/B)^{-1}$ values proceed linear in temperature. In an effort to characterize the magnetic species giving rise to this behaviour, we provide a qualitative measure of the Curie-Weiss law, in terms of the permeability of free space μ_0, Avogadro's number N_A, the Bohr magneton μ_B, and Boltzmann's constant k_B.

$$\chi(T) = \frac{C}{T - \theta_p}, \quad C = \frac{\mu_0 N_A \mu_B^2}{3 k_B} \mu_e^2 \qquad (1)$$

The free parameters are the Weiss temperature θ_p and μ_e, the effective moment of the magnetic species giving rise to paramagnetic behaviour. Here, this role would be assumed by cerium, which could in general take on either the magnetic Ce^{3+} ionic form, or the non-magnetic Ce^{4+} state. For purposes of illustration, since the precise molar mass of cerium is not known, the susceptibility data at intermediate temperatures are fitted well according to the Curie-Weiss law with a θ_p = -120 K, and μ_e= 2.27. Although a Weiss temperature that is negatively signed points towards an antiferromagnetic kind of interaction, caution has to be exercised in the present situation in view of the lack of an accurate cerium ion concentration level. The obtained effective moment value, on the other hand, is close to but somewhat less than that of the full J=5/2 magnetic multiplet value (J the total angular momentum) which amounts to 2.54 for the free-ion situation of Ce^{3+}. We furthermore present the magnetization at the base temperature of 2 K (inset, Fig. 4) as function of applied field. A weak curvature is observed, which arises from coupling of the applied field with the magnetic species in the sample and hence a small decrease in magnetic susceptibility at elevated fields. The magnetization is clearly in evidence of a substance containing a magnetic species, i.e. Ce^{3+}, in the paramagnetic state in this sample.

4. CONCLUSIONS

Results demonstrate that two diamagnetic reactant species form a single supramolecular paramagnetic product due to the *in situ* formation of the paramagnetic Ce^{3+} species at the reaction condition. We believe that this material holds promise for magnetic applications by tuning magnetism (nature and strength of exchange interaction between the magnetic species) through suitable concentration levels of the ionic dopant. The advantages of CAN are low toxicity, ease of handling, experimental simplicity and solubility in a number of solvents. During the polymerization process each step is associated with a release of electron and that electron reduces the Ce^{+4} ion to form a Ce^{+3} ion. The Ce^{+3} ion binds with the chain nitrogen of the polyaniline which causes the emergence of paramagnetism in polyaniline. The resultant paramagnetic organic macromolecule complex compound could be useful in the application of flexible magnetic materials and we believe this work will show the way towards further research for synthesizing organic magnetic materials.

REFERENCES

[1] K. Lee, S. Cho, S. H. Park, A. J. Heeger, C. W. Lee, S. H. Lee, Nature 441, 65 (2006).

[2] A. J. Heeger, J. Phys. Chem. B, 105, 8475 (2001).

[3] N. S. Sariciftci, A. J. Heeger, Y. Cao, Phys. Rev. B 49, 5988 (1994).

[4] T. Viswanathan, B. Berry, US Patent, US 6,764,617 B1 (2004).

[5] D. Cottevieille, US Patent, US 6,303,671 B1 (2001).

[6] K. Mallick, M. Witcomb, R. Erasmus, A. Strydom, J. Appl. Phys. 106, 074303 (2009).

[7] V. Nair, A. Deepthi, Chem. Rev. 107, 1862 (2007).

[8] M. Łapkowski, K. Berrada, S. Quillard, G. Louarn, S. Lefrant, A. Proń, Macromolecules, 28, 1233 (1995).

[9] G. Louarn, M. Łapkowski, S. Quillard, A. Proń, J. P. Buisson, S. Lefrant, J. Phys. Chem. 100, 6998 (1996).

[10] G. Socrates, Infrared and Raman Characteristic Group Frequencies: Tables and Charts, John Wiley and Sons Ltd, 2001.

[11] S. K. Pillalamarr, F. D. Blum, A. T. Tokuhiro, M. F. Bertino, Chem. Mater. 17, 5941 (2005).

[12] K. Mallick, M. Witcomb, R. Erasmus, A. Strydom, J. Appl. Poly. Sci. 116, 1587 (2010).

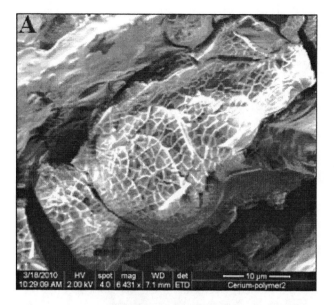

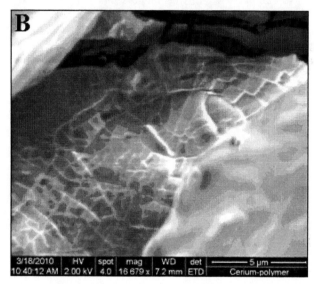

Figure 1 : SEM images of the cerium (III) ion doped poly-(amino-acetanilide).

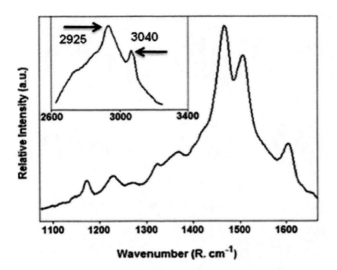

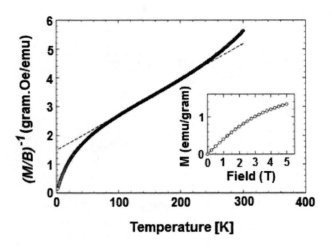

Figure 2 : Raman spectrum of the cerium doped poly-(amino-acetanilide) within the range from 1100 to 1700 cm^{-1}. The spectrum indicates the presence of both the benzenoid and the quinoid structures in the poly-(amino-acetanilide). The inset spectrum indicates the presence of acetanilide group in the polymer.

Figure 4: Temperature dependence of reverse magnetic susceptibility, $(M/B)^{-1}$, of Ce (III)-poly (amino-acetanilide), measured at 300 Oe. The inset figure is the magnetization at 2 K as function of applied field. The curvature is due to the coupling of the applied field with the magnetic species in the sample.

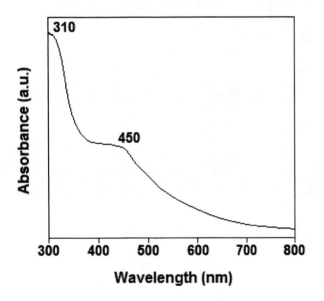

Figure 3 : UV-vis spectrum of the Ce^{3+} doped poly (amino-acetanilide).

NSTI-Nanotech 2010, www.nsti.org, ISBN 978-1-4398-3401-5 Vol. 1, 2010

Graphene and polymer composite using easy soluble expanded graphite

Jong Hak Lee[1], Xianhui Meng[2], Dong-Wook Shin[1], and Ji-Beom Yoo[1,2*]

[1] Department of Sungkyunkwan Advanced Instituted of Nanotechnology (SAINT),
Sungkyunkwan University, 300, Chunchun-dong, Jangan-gu, Suwon, 440-746, Korea
[2] Advanced Materials Science and Engineering, Sungkyunkwan University, 300 Chunchun-dong, Jangan-gu, Suwon 440746, Republic of Korea

ABSTRACT

Easy soluble expanded graphite (ESEG) was prepared from a fluorinated graphite intercalation compound (FGIC) - $C_2F \cdot nClF_3$ containing inorganic volatile intercalating agent ClF_3. The interlayer distance of ESEG was approximately 17 % higher than that of ordinary graphite. This severe expanded state, together with high specific surface area of the ESEG can give us essential material properties for dispersing easily the ESEG in the solution. ESEG suspension prepared in several organic solvents and water with common surfactants using sonication process was subjected to electron microscopy techniques. This one step exfoliation process of ESEG can allow the low cost mass production of graphene because of the very simple and short process time. In addition, well-dispersed graphene in water and organic solutions have potential use in high-performance, scalable graphene-based applications. We demonstrated that ESEG could be considered as the preferable reinforcing filler in polymer matrices.

Keywords: easily soluble expanded graphite, dispersion, mass production, graphene-polymer composite

1 INTRODUCTION

Recently, graphene have attracted enormous scientific attention on account of its extraordinary electronic and mechanical properties resulting from one-atom-thick layers and hexagonally arrayed sp2-hybridized carbon atom structure[1,2]. There are several examples of graphene layers being used in devices and composites[3,4]. However, graphene suffers from several problems, such as the difficulty in depositing a uniform film onto a substrate. Therefore, a great deal of effort has been made to form a uniform graphene film. Thus far, graphene films have been prepared by a variety of techniques, including "mechanical exfoliation"[5,6], "graphene in solution"[7,8] and "epitaxial growth"[9,10] methods. "Mechanical exfoliation" produces the highest quality graphene, which is suitable for fundamental studies. "Epitaxial growth" provides the shortest path to graphene-based electronic circuits. On the other hand, "graphene in solution" can offer lower cost and higher throughput for the production of graphene based nanocomposites as well as relatively larger sized films than other methods[11]. However, the method is relatively complicated and involves a personal-dependant procedures due to the use of several chemicals involving oxidation, funtionalization and reduction processes. Researchers most commonly use Hummer's method[12] to produce well dispersed graphene oxide (GO) solutions in account of its hydrophilic property. However, the as-prepared graphene oxide is an electrical insulator. Therefore, a chemical reduction to recover its conductivity is necessary. Nevertheless, the reduced GO still exhibits lower conductivity than pristine graphene[11,13]. Some researchers have introduced different re-intercalation and dispersion methods to circumvent the oxidation of graphene [14,15]. Although Xiaolin[14] and others[16] suggested methods for high yield and quality graphene suspensions, their method is quite complicated with a lengthy process time because the process to form a homogeneous suspension without the oxidation of graphene consists of 6~8 steps. Notwithstanding these efforts, the production of stable suspensions of graphene in water or organic solvents is an important goal for the fabrication of graphene based devices. In this study, the expanded graphite was synthesized using a fluorinated graphite intercalation compound in only one step exfoliation comprising a single intercalation and thermal expansion process within 5~6 hours. The expansion state of the easy soluble expanded graphite can allow dispersion in organic solvents or even water by ultrasonication with normal surfactants used for carbon nanotubes (CNT), such as sodium dodecylbenzene sulfonate (SDBS). After the dispersion state examination, we demonstrated that ESEG could be considered as the preferable reinforcing filler in polymer matrices.

2 EXPERIMENT

2.1 Synthesis of fluorinated graphite intercalation compound (FGIC) - $C_2F \cdot nClF_3$

A Teflon reactor was filled with 30 g of liquid ClF_3 and cooled with liquid nitrogen. Five grams of pure natural graphite (ash content < 0.05 mass %, particle size = 200–300 microns, Zaval'evsk coal field, Ukraine) was then added. The reactor was sealed hermetically, and the temperature was increased slowly to 22°C and kept at that temperature for 5 hours. The excess ClF_3 was removed in a nickel vessel cooled with liquid nitrogen until a constant mass was measured. The intercalation product (approximately 11 g) had an approximate composition of $C_2F \cdot 0.13ClF_3$. Elemental analysis data (mass %): C 44.22; F 44.79; Cl 12.49.[17]

2.2 Expansion of FGIC

The intercalation compound, $C_2F \cdot 0.13ClF_3$ (~50 mg), was decomposed thermally in a quartz reactor (V~500 ml) heated to 600-700°C. After 4 minutes, the compound was decomposed using a thermal shock process and the ESEG filled the larger volume of the reactor.

2.3 Synthesis of graphene-polymer composite

These ESEG were dispersed in 50mL freshly distilled N,Ndimethylacetamide (DMAc). The suspension was ultrasonicated for 1 hour and 2 g of diamine(ODA) 4,4-diaminodiphenylether was added, then magnetically stirred for 1 h. The suspension was transferred into a three-neck round-bottom flask equipped with a mechanical stirrer and filled with nitrogen gas. After stirring, 2.94 g of the dianhydride (s-BPDA) was added. The reaction was kept for 3 h and formed an graphene–poly(amic acid) solution.

A series of graphene–poly(amic acid) solutions with different solid contents was obtained using the same process. In order to obtain graphene–polyimide composite films, the latter were imidized by heating as follows. The graphene–polyamic acid solutions prepared were cast onto glass plates and dried in a dry air-flowing chamber. Subsequently, the dried films were cured at 110, 170, 210, and 250°C for 1 h each in an air-circulating oven to obtain solvent free films. Then, the composite films were imidized at 350°C for 1 h under vacuum and MWNT–polyimide composite films were formed.

3 RESULTS AND DISCUSSION

Before characterizing the mechanical properties of the graphene-polyimide composite, the solubility of ESEG in DMAc was examined. The photograph in Fig. 1 (a) shows that the graphene suspension in DMAc was quite stable. A few drops from this solution were collected onto the cropper–grids for HRTEM analysis. The central part of the graphene layers was mostly uniform, homogeneous and featureless. However, the folding edges provided a clear indication of the number of graphene sheets (Fig. 1 (b) and (c)). The incident electron beam was locally parallel to the folding edge. Hence, a fold exhibits one dark fringe for each graphene layer. A large number of folding edges were examined and a cluster of 3–6 graphene sheets with a thickness 1~2 nm were observed. The spacing between neighboring fringes was measured to be 3.6~3.9 Å, which is considerably larger than the planes of bulk graphitic layers (3.35 Å).[17] In a previous paper, it was demonstrated that the yield of a few layers of graphene in DMAc was approximately 93% and the yields of graphene were approximately 23%.[18]

Figure 1: photograph of ESEG suspension in DMAc (a), TEM images of dispersed graphene sheets (b), (c) and photograph of ESEG (0.1 wt%) based polyimide paste (d).

Due to the good stability of graphene-DMAc suspension, graphene-polyimide paste also had the good dispersion state. After film formation, the tensile properties of polyimide and graphene–polyimide nanocomposites were examined. Table 1 is showing the mechanical properties of neat polyimide and graphene-polyimide composites. The enhanced mechanical properties of epoxy nano-composites at low graphene content is reported.[19] At low nano-filler content graphene platelets perform significantly better than carbon nanotubes in terms of enhancing a variety of mechanical properties including tensile strength, Young's modulus, fracture toughness, fracture energy, and resistance to fatigue crack growth.[19] In our study, graphene-polyimide composite had the similar mechanical enhancement at low graphene concentration. As increase the graphene content to 0.05 wt%, the tensile strength was increase from 100kgf/cm² to 111.9kgf/cm². However, as graphene content increased further, at the graphene content about 0.1 wt%, tensile strength was decreased to the 89.8kgf/cm². There are several possible reasons for this. At the low graphene content, enhanced specific area of graphene and improved mechanical adhesion of the grpahene and matrix interface can make the graphene-polyimide composite has the good enhanced mechanical properties. As increase the reinforcing agent, however, graphene disturb the polymerization process.

Graphene (wt%)	Room temperature tensile strength (kgf/cm²)	Elongation at break (%)
0	100	20
0.05	111.9	16.03
0.1	89.8	15.7

Table 1: neat polymer and graphene-polyimide composite mechanical characterization.

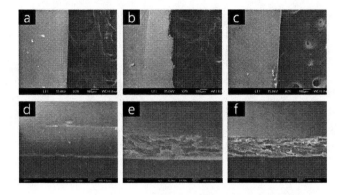

Figure 2: SEM images of the polymer and graphene composite films after fracture examination. neat polyimide (a) and (d), 0.05wt% (b) and (e), 0.1wt% (c) and (f). upper images are plane view and bottom parts are cross-sectional view.

Figure 2 is showing the SEM images of the neat polyimide and graphene-polyimide composite films after fracture experiments. The image of fracture plane could be a vital clue. As increase the graphene content, the fracture plane became more rough (Fig. 2 (b) and (e)). Until the adequate graphene content, graphene can act as a reinforcing agent. However, after adequate graphene content, graphene disturbs the polymerization of polyimide. As a result, the elongation property became poor.

4 CONCLUSION

Easy soluble expanded graphite (ESEG) was prepared from a fluorinated graphite intercalation compound (FGIC) - C2F·nClF3 containing inorganic volatile intercalating agent ClF3 was examined as the reinforcing agent. Well-dispersed graphene in DMAc solutions has the good dispersability. We demonstrated that ESEG could be considered as the preferable reinforcing filler in polymer matrices. As increase the graphene content to 0.5wt%, the tensile strength was increased to 12% higher than that of neat polyimide film.

REFERENCES

[1] Geim, A. K. & Novoselov, K. S. "The rise of graphene" Nature Mater. 6, 183–191 (2007).
[2] Novoselov, K. S. et al. "Two-dimensional gas of massless Dirac fermions in graphene." Nature 438, 197–200 (2005).
[3] Blake, P. et al. "Graphene-based liquid crystal device." Nano Lett. 8, 1704–1708 (2008).
[4] Dikin, D. A. et al. "Preparation and characterization of graphene oxide paper." Nature 448, 457–460 (2007).
[5] Novoselov, K. S. et al. "Electric field effect in atomically thin carbon films." Science 306, 666–669 (2004).
[6] Novoselov, K. S. et al. "Two-dimensional atomic crystals." Proc. Natl Acad. Sci. USA 102, 10451–10453 (2005).
[7] Li, D., Muller, M. B., Gilje, S., Kaner, R. B. & Wallace, G. G. "Processable aqueous dispersions of graphene nanosheets." Nature Nanotech. 3, 101–105 (2008).
[8] Stankovich, S. et al. "Stable aqueous dispersions of graphitic nanoplatelets via the reduction of exfoliated graphite oxide in the presence of poly(sodium 4-styrenesulfonate)." J. Mater. Chem. 16, 155–158 (2006).
[9] Berger, C. et al. "Electronic confinement and coherence in patterned epitaxial graphene." Science 312, 1191–1196 (2006).
[10] Berger, C. et al. "Ultrathin epitaxial graphite: 2D electron gas properties and a route toward graphenebased nanoelectronics." J. Phys. Chem. B 108, 19912–19916 (2004).
[11] Goki eda, et al. "Large-area ultrathin films of reduced graphene oxide as a transparent and flexible electronic material." Nature Nanotech. 3, 270-274 (2008)
[12] Hummers, W. S. & Offeman, R. E. "Preparation of graphite oxide." J. Am. Chem. Soc. 80, 1339 (1958).
[13] Stankovich, S. et al. "Synthesis of graphene-based nanosheets via chemical reduction of exfoliated graphite oxide." Carbon 45, 1558–1565 (2007).
[14] Yenny Hernandez, et al. "High-yield production of graphene by liquid-phase exfoliation of graphite." Nature Nanotech. 3, 563-568 (2008)
[15] Xiaolin li, et al. "Highly conducting graphene sheets and Langmuir Blodgett films." Nature Nanotech. 3, 538-542 (2008)
[16] Mohammad Choucair, et al. "Gram-scale production of graphene based on solvothermal synthesis and sonication." Nature Nanotech. 4, 30-33 (2009)
[17] Lee JH, Shin DW, Makotchenko VG, Nazarov AS, Fedorov VE, Kim YH, et al. "One step exfoliation synthesis of easily soluble graphite and transparent conducting graphene sheet" Adv. Mater. 21, 4384-4387 (2009)
[18] Lee JH, Shin DW, Makotchenko VG, Nazarov AS, Fedorov VE, Yoo JB. "Superior dispersion property of easily soluble graphite." Small, 1, 58-62 (2010)
[19] Mohammad A. Rafiee, Javad Rafiee, Zhou Wang, Huaihe Song, Zhong-Zhen Yu, and Nikhil Koratkar. "Enhanced mechanical properties of nano composites at low graphene content" ACS Nano, 3, 3884-3890 (2009)

Thermomechanical Properties of Epoxy Resin/MWCNTs Nanocomposites

H. Zois*, A. Kanapitsas* and C.G. Delides**

*Technological Educational Institute (T.E.I.) of Lamia, 3rd km Old National Road Lamia-Athens, 35 100 Lamia, Greece, kanapitsas@teilam.gr
**Technological Educational Institute (T.E.I.) of West Macedonia, Laboratories of Physics and Materials Technology, 50100 Kila, Kozani, Greece, kdelidis@yahoo.com

ABSTRACT

The thermomechanical properties and thermal stability of a nanocomposite system, which consist of epoxy resin (ER) as matrix and carbon nanotubes (MWCNTs) as filler, were investigated by Thermogravimetric/Differential Thermal Analysis (TGA/DTA) and Dynamic Mechanical Analysis (DMA). The dependence of the thermal properties (i.e. weight loss rate, decomposition temperature, residual mass) of the nanocomposites is associated with the filler content. The addition of carbon nanotubes improves generally, the thermal behavior of the epoxy matrix. The results were discussed in terms of the epoxy resin-MWCNTs interactions. The interactions between the polymer matrix and the filler nanoparticles seem to play an important role, especially for the higher MWCNTs concentration,

Keywords: nanocomposites, carbon nanotubes, mechanical properties, thermal stability

1 INTRODUCTION

Polymer nanocomposites show different properties than the bulk polymer matrix, due to the small size of the filler and the corresponding increase of their surface area [1, 2]. It is well known that the composite properties can change dramatically with the dispersion state, geometric shape, surface properties, particle size, and particle size distribution. Because of the recent commercial availability of nanoparticles, there is increasing interest in polymer nanocomposites. These composites have been shown to undergo substantial improvements in mechanical properties such as the strength, modulus and dimensional stability, permeability to gases, water and hydrocarbons, thermal stability, flame retardancy, chemical resistance and electrical, dielectric, magnetic and optical properties [3-8].

Epoxy resin (ER) is one of the most commonly used matrices for the preparation of polymer composites, because of its easy processability and excellent mechanical properties. Carbon nanotubes (MWCNTs) have received much attention since their discovery by Ijiima in 1992 [9]. Extensive research and development efforts have been devoted to the use of MWCNTs as nanofillers to produce a variety of high-performance polymer nanocomposites for specific technological applications. Among the advantages of MWCNTs is their high aspect ratio (as high as 1000), which can induce better adhesion with the polymeric matrix. This is an important factor for effective enhancement of the nanocomposites properties. The modification of filler particles with specific chemical groups (such as amine groups) improves the compatibility with the epoxy resin and, thus, their dispersion within the polymer matrix.

The main aim of the present work is to investigate the thermomechanical properties and thermal stability of nanocomposite materials based on epoxy resin matrix filled with amine-modified multi-walled carbon nanotubes. We also focus on the influence of filler content on the thermal and mechanical behavior of the systems studied.

2 EXPERIMENTAL

2.1 Sample Preparation

The pre-polymer D.E.R.332 used in this work is diglycidyl ether of bisphenol A (DGEBA) supplied by Fluka SA. The hardener used was triethylenetetramine (TETA) supplied by Sigma Aldrich. The amine-modified multi-walled carbon nanotubes (average diameter 9.5 nm and average length < 1 μm) was supplied by Nanocyl SA. All the components of the system are commercial products and were used without any purification.

The DGEBA/TETA/MWCNTs nanocomposites were prepared by the dispersion of the determined amount of carbon in a glass vessel. Prior to that procedure, the pre-polymer was heated at 40 °C in order to decrease its viscosity. The stoichiometric amount of TETA (14 phr) was added to the DGEBA matrix, and then the mixture was mechanically stirred for 1 h at 2000 rpm and degassed under vacuum for 15 min. Finally, the mixture was sonicated for 30 min in order to break up the MWCNTs agglomerates [10] and degassed again. The homogeneous liquid was poured in rectangular-shaped Teflon molds and the samples were cured at 60 °C for 20 h and at 150 °C for 2 h. Several specimens were prepared, with the filler content systematically varied between 0 % (pure resin) and 1 % w/w MWCNTs.

2.2 Experimental Techniques

The characterization of the ER/MWCNTs nanocomposites includes two different experimental techniques: Thermogravimetric/Differential Thermal Analysis (TGA/DTA) and Dynamic Mechanical Analysis (DMA).

A Polymer Laboratories dynamic mechanical thermal analyzer MK III operating at a frequency of 10 Hz, a strain of 4× and a scanning rate of 2°C/min was used. Measurements were performed from room temperature up to 200°C and the resultant changes in the mechanical parameters (E', E'' and $\tan\delta$) were plotted.

Thermogravimetric and Differential Thermal Analysis measurements were performed on a simultaneous thermal analyzer (TGA/DTA) STA 503 device (BAEHR ThermoAnalyse GmbH, Germany). TGA tests were conducted in temperature range from 20 to 1400 °C under dry nitrogen (N_2) atmosphere at a heating rate of 20 °C/min and the weight loss was monitored.

3 RESULTS AND DISCUSSION

The thermomechanical properties of the ER/MWCNTs nanocomposites are changed, due to the MWCNTs' contribution to crosslinking procedure and the existence of an interfacial layer between the nanoparticles and the polymer matrix.

The effect of MWCNTs on the thermal stability of the nanocomposites was studied using thermogravimetric analysis (TGA). Relative TGA and derivative (DTGA) curves for neat epoxy and the prepared ER/MWCNTs nanocomposites are presented in Fig. 1a and 1b, respectively. Similar curves have also been obtained for epoxy resin nanocomposites filled with carbon black (ER/CB) [11]. The results reveal that the samples show good thermal stability for temperatures up to around 340 °C with a maximum decomposition temperature higher than 375 °C. A large weight loss occurs between 350-550 °C. The decomposition (thermal degradation) temperature corresponds to the temperature of the peak of the derivative mass loss (DTGA) curve, i.e. the temperature of the maximum weight loss rate. The thermal degradation temperatures of the pure epoxy and the ER/MWCNTs nanocomposites are similar, within experimental errors (± 1.5 °C). The higher the filler content is, the lower the weight loss rate is (dm/dt peak magnitude decreases).

The temperature corresponding to 5 % initial mass loss, $T_{5\%}$, (onset of the TGA curve) indicates the thermal stability of the samples. The variation of $T_{5\%}$ with MWCNTs content are shown in Fig. 2. The data reveals that the pure epoxy matrix shows, generally, higher weight loss onset (i.e. the degradation starts at higher temperatures), compared to the ER/MWCNTs nano-composites. The $T_{5\%}$ temperature is slightly decreases by the addition of MWCNTs.

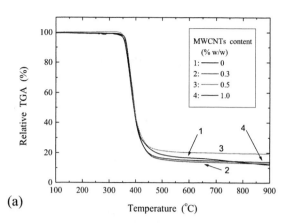

(a)

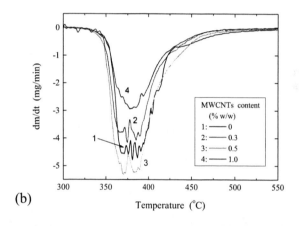

(b)

Figure 1: Comparative relative TGA (a) and DTGA (b) curves for ER/MWCNTs nanocomposites with various filler contents shown on the plots.

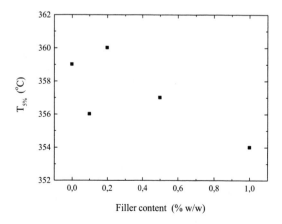

Figure 2: The temperature corresponding to 5 % initial mass loss as a function of the filler content for ER/MWCNTs nanocomposites.

Fig. 3 shows the variation of residual mass (char content) of the nanocomposites at 1000 °C as a function of the filler composition. The char content are indicative of the degradation level and the flammability resistance of the samples. The char formation mechanism is, generally, enhanced when adding filler particles, however the values of the residual mass for the ER/MWCNTs nanocomposites decrease with increasing filler content.

The reduction observed for the samples with higher MWCNTs content is probably attributed to the fact that the degradation process of the nanocomposites is hampered by the increased filler concentration, because the interactions between the ER matrix and the MWCNTs filler particles are stronger. This result is in agreement with DMA measurements and glass transition temperature studies [10]. It should also be mentioned that MWCNTs were found to have better dispersion within the polymer matrix, compared to ER/CB nanocomposites [11]. A complete comparison of the thermomechanical properties and thermal stability of ER/MWCNTs and ER/CB nanocomposites can be found in [11].

In Fig. 4, characteristic plots of DMTA measurements for ER/MWCNTs nanocomposites are presented. It was found that the storage modulus, E′, decreases significantly in the glassy state. On the contrary, it is slightly affected by the filler content in the rubbery state.

The glass transition temperature, T_g, is determined by the peak of the tanδ curves. It varies in a strange way. The glass transition temperature as a function of the filler content is presented in Fig. 5. For the samples with filler content up to 0,05 % w/w, the glass transition temperature decreases, compared to the pure polymer matrix. For the samples with higher MWCNTs content, T_g is higher compared to the pure epoxy matrix and varies slightly with increasing filler content.

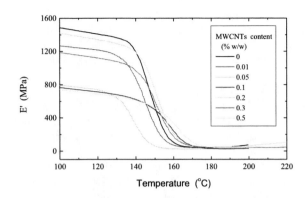

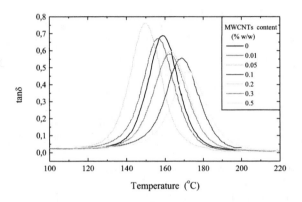

Figure 4: DMA spectra: E′ (top) and tanδ (bottom) as a function of temperature at constant frequency of 10 Hz for ER/MWCNTs nanocomposites and various filler contents shown on the plots.

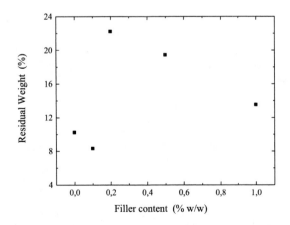

Figure 3: The residual weight at temperature 1000 °C as a function of the filler content for ER/MWCNTs nanocomposites.

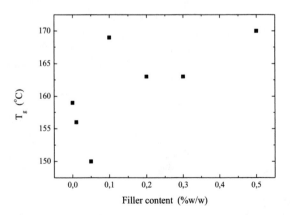

Figure 5: The glass transition temperature as a function of the filler content for ER/MWCNTs nanocomposites.

NSTI-Nanotech 2010, www.nsti.org, ISBN 978-1-4398-3401-5 Vol. 1, 2010

Filler Content (% w/w)	0	0.1	0.2	0.5	1.0
Maximum weight loss rate (mg/min)	4.8	4.6	5.0	5.2	3.0
Decomposition temperature (^{o}C)	380	379	382	378	379
$T_{5\%}$ (^{o}C)	359	356	360	357	354
Residual mass (%)	10.2	8.3	22.2	19.4	13.5
T_g (^{o}C)	159	169	163	170	-

Table 1: The characteristic values of TGA parameters for ER/CNTs nanocomposites

It is also clear that the minimum value of tanδ is observed for the sample with higher glass transition temperature. These results are, probably, attributed to the formation of MWCNTs aggregates within the epoxy matrix. More results on the effect of the nanofillers on the glass transition and the relaxation behavior of the polymer matrix will be given in a future work.

In the Table 1 above the values of the characteristic parameters of TGA measurements for ER/MWCNTs nanocomposites are summarized. It is also included the glass transition temperatures obtained from DMA measurements.

4 CONCLUSIONS

In this work, the thermal stability and dynamic mechanical behaviour of epoxy resin/multi-walled carbon nanotubes (ER/MWCNTs) composites was systematically investigated. The dependence of thermal properties on the filler content was also studied. It was found that the addition of carbon nanotubes enhances the thermal properties of the epoxy matrix. The degradation starts at higher temperatures and the char formation mechanism is enhanced when adding filler particles. For higher MWCNTs content, the interactions between the polymer matrix and the filler particles play an important role and the degradation process becomes slower. From DMA measurements it was found that the addition of MWCNTs increases the glass transition temperature of the pure polymer matrix and decreases the storage modulus in the glassy state. The interactions between the filler particles and the epoxy matrix seem to play an important role on the thermal properties and molecular dynamics mechanisms.

REFERENCES

[1] Y. Sun, Z. Zhang, K.-S. Moon and C.P. Wong, J. Polym. Sci. Part B: Polym. Phys. 42, 3849, 2004.
[2] J. Brown, I. Rhoney and R.A. Pethrick, Polym. Int. 53, 2130, 2004.
[3] R. Krishnamoorti and E.P. Giannelis, Macromolecules 30, 4097, 1997.
[4] R. Pelster and U. Simon, Colloid and Polymer Science 277, 2, 1999.
[5] E. Reynaud, T. Jouen, C. Gauthier, G. Vigier and J. Varlet, Polymer 42, 8759, 2001.
[6] A. Kanapitsas, P. Pissis and R. Kotsilkova, J. Non-Cryst. Solids 305 204, 2002.
[7] G.C. Psarras, K.G. Gatos, P.K. Karahaliou, S.N. Georga, C.A. Krontiras and J. Karger-Kocsis, eXPRESS Polym. Lett. 1, 837, 2007.
[8] Y.X. Zhou, P.X. Wu, Z.-Y. Cheng, J. Ingram and S. Jeelani, eXPRESS Polym. Lett. 2, 26, 2008.
[9] S. Ijiima, Nature 354, 56, 1991.
[10] Th. Kosmidou, A. Vatalis, C. Delides, E. Logakis, P. Pissis and G. Papanicolaou, eXPRESS Polym. Lett. 2, 364, 2008.
[11] A. Kanapitsas, H. Zois and C.G. Delides, Proc. 9th Med. Conf. Medicta 2009, Marseille (France), 2009.

Ag-TiO$_2$ coated Polyurethane Composite Nanofiber Web for Application in the Protection against Chemical and Biological Warfare Agent

Su-Yeol Ryu*, Woo Hyuk Choi**, Chi Hyung Ahn***, and Seung-Yeop Kwak*****

Department of Materials Science and Engineering, Seoul National University,
599 Gwanak-ro, Gwanak-gu, Seoul 151-744, Korea
*suyol@hanmail.net, **illogick@hanmail.net, ***piscesoul@snu.ac.kr, ****sykwak@snu.ac.kr

ABSTRACT

The objective in this work is to develop protective clothing that is effective against both chemical warfare (CW) and biological warfare (BW) agents. The protective clothing material was made of composite nanofiber web that was composed of Ag-loaded TiO$_2$ (Ag-TiO$_2$) coated onto polyurethane (PU) nanofiber by a synchronous process consisted of electrospinning and electrospray. TiO$_2$ is one of the most popular photocatalysts. Its main drawback is excitation only by UV. To extend the absorption range of TiO$_2$ into the visible range, Ag was deposited onto TiO$_2$ surface. Ag is nontoxic metal with excellent antibacterial activity. PU nanofiber web can provide deposition sites for introducing Ag-TiO$_2$. The morphologies of Ag-TiO$_2$ and Ag-TiO$_2$ coated PU composite nanofiber web were examined. Presence of Ag was confirmed by various X-ray analyses. A broad absorption covering the visible range is confirmed by UV-vis. Protective ability of the composite nanofiber web is evaluated using CW agent simulants.

Keywords: Ag-loaded TiO$_2$, electrospinning, nanofiber, chemical warfare agent, biological warfare agent

1 INTRODUCTION

Threats of chemical warfare (CW) and biological warfare (BW) agents, which have been used in military conflicts, are serious international problems. It is positively necessary to protect humankind against CW/BW agents. A number of studies have presented about development of new materials for detoxification of CW/BW agent.

TiO$_2$ is the most efficient photocatalyst because of its photoactivity, electronic property, low cost, non toxicity, and chemical stability; it have been used for decomposition of CW agents [1]. Its main drawback is poor photoactivity to visible light. To extend the photoactivity of TiO$_2$ to the visible range, there are various modification techniques to enhance photoactivity to visible light. One method of enhancing visible photoactivity is loading of silver (Ag) nanoparticles onto TiO$_2$ surface [2]. The loaded Ag on TiO$_2$ surface extends the light absorption into the visible range and enhances the surface electron excitation by plasmon resonances. Ag is a powerful natural antibiotic being with excellent antibacterial activity [3]; it can be used for

decontamination of BW agents. Moreover, nanofiber web can be acted as physical shield from CW/BW agents. High surface area and micro/nano pores of nanofiber web can provide deposition sites for introducing Ag-TiO$_2$ and good vapor release property.

In this work, we developed the chemical and biological protective clothing material that might be effective against both CW and BW agents. This protective clothing material, which was consisted of Ag-loaded TiO$_2$ (Ag-TiO$_2$) coated on the nanofiber, was developed by using of the synchronous electrospinning/electrospray process. Besides, the photocatalytic activities of the protective clothing materials under visible/UV light were examined with nerve agent simulant. This work will have significance in development of protective materials against CW/BW agents.

2 EXPERIMENTAL

2.1 Materials

Titanium dioxide (TiO$_2$, Degussa, P25), silver nitrate (AgNO$_3$, ≥99.0%, Sigma-Aldrich, ACS Reagent), sodium borohydride (NaBH$_4$, ≥99.0%, Sigma-Aldrich), ammonium hydroxide solution (NH$_4$OH$_{(aq)}$, 28-30% in H$_2$O, ≥99.99%, Sigma-Aldrich, ACS Reagent), polyurethane (PU, MW 30,000 g/mol, supplied by HYOSUNG R&DB Labs (Korea)). *N,N*-dimethylformamide (DMF, anhydrous, 99.8%, Sigma-Aldrich), tetrahydrofuran (THF, anhydrous, ≥99.0%, Sigma-Aldrich), methanol (99.8%, Sigma-Aldrich), dimethyl methylphosphonate (DMMP, 97%, Sigma-Aldrich), benzyl triethylammonium chloride (BTMAC, >99.0%, Tokyo Chemical Industry Co.) were used as received without further purification.

2.2 Preparation of Ag-TiO$_2$

In a typical procedure, 0.20 g (2.5 mmol) commercial TiO$_2$ nanoparticles (P25, Degussa, containing crystalline phases of anatase and rutile) were placed into 100 ml of deionized water, and dispersed by using an ultrasonication for 30 minutes. After ultrasonic treatment, 0.0425 g (0.25 mmol) of AgNO$_3$ was added in TiO$_2$ suspension. This solution, which was adjusted to various pH conditions, was stirred for 30 minutes. The pH condition of the solution was adjusted by adding a little NH$_4$OH aqueous solution.

Then, 0.0095 g (0.25 mmol) of NaBH$_4$ was added for rapid reduction. There is a changing color from white to brownish. The reduction yields brownish solution. This solution was vigorously stirred for 30 minutes. Brownish solution was centrifuged to wash and separate the Ag-TiO$_2$ nanoparticles from the solution. This produced was repeated for several times. Finally, the Ag-TiO$_2$ nanoparticles were dried at 298 K for 24 hours.

2.3 Characterization of Ag-TiO$_2$

The morphology of Ag-TiO$_2$ and presence of Ag nanoparticles on Ag-TiO$_2$ was detected using FE-SEM/EDS (SUPRA 55VP, Carl Zeiss) and HR-TEM (JEM-3010, JEOL). Presence of Ag nanoparticles was additionally confirmed using WXRD (M18XHF, MAC Science Co.) with Cu K$_\alpha$ irradiation at λ = 1.54Å. Further structural information for Ag-TiO$_2$ is obtained using a XPS (AXIS-HIS, KRATOS) fitted with Ma/Al K$_\alpha$ X-ray source and a sweep mode analyzer electronics. The binding energy scale was calibrated at the main C 1s peak (284.6 eV). The light absorbance of Ag-TiO$_2$ was observed by UV-vis (Lambda25, Perkin Elmer).

2.4 Preparation and Characterization of Ag-TiO$_2$ coated Polyurethane Composite Nanofiber Web

To make electrospun PU nanofibers, Polyurethane (PU) was dissolved in DMF/THF (3:1 v/v). Small amount of BTMAC (0.5 wt%), a kind of organic salt, was added into PU solution to increase the conductivity of PU solution. It is important to obtain optimum electrospinning conditions for preparation of uniform nanofibers without defects. So, various conditions were studied as follows: concentrations of polymer solution were 20-30 wt%, applied voltages were 14-20 kV, tip-to-collector distances were 9-15 cm, and flow rate of syringe pump were 0.5-1 ml/h.

Ag-TiO$_2$ coated PU composite nanofiber web was prepared by an electrospinning/electrospray synchronous process. Ag-TiO$_2$ was dispersed in methanol (1.0 wt%) by using an ultrasonication for 1 hour. Ag-TiO$_2$ dispersion was electrically sprayed onto PU nanofiber surface by synchronous electrospray during PU electrospinning at the optimum condition. Electrospray conditions were as follows: applied voltage was 20 kV, tip-to-collector distance was 5 cm, and the suspension feed rate was 5 ml/h.

Figure 1 shows a schematic diagram of the synchronous electrospinning/electrospray process. It consists of three major components: an electrospinning tip, an electrospray tip, and a collector drum.

The morphology of the composite nanofiber web was examined by FE-SEM (JSM-6700F, JEOL). Ag-TiO$_2$ content of the composite nanofiber web was confirmed by TGA (Q500, TA Instruments) at heating rate of 10 °C in N$_2$ atmosphere. To evaluate decomposition ability of the composite nanofiber web against CW agents, the catalytic

test was carried under UV light or visible light irradiation. DMMP was used as nerve agent simulant. 100 mg of the composite nanofiber web was immersed in 100 mL of DMMP aqueous solution (20 mM) contained in quartz bottle. The sample was magnetically stirred in the dark for 1 hour for stabilization. During UV irradiation, the DMMP solution was taken with fixed intervals and checked the UV-visible adsorption. The decomposition of DMMP was analyzed by UV-vis (Lambda25, Perkin Elmer) in the range 190-400 nm.

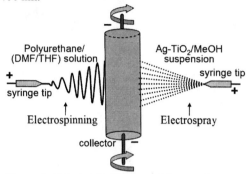

Figure 1: Schematic diagram of the synchronous electrospinning/electrospray process.

3 RESULT AND DISCUSSTION

3.1 Confirmation of Ag Loading on Ag-TiO$_2$

The morphology of Ag-TiO$_2$ samples was examined by FE-SEM and HR-TEM. Figure 2 shows the HR-TEM images. White Arrow shows the Ag nanoparticles on the surface of Ag-TiO$_2$. Size distribution analysis for Ag nanoparticles was done with 50 particles. The Ag nanoparticles on the Ag-TiO$_2$ pH 4, Ag-TiO$_2$ pH 6, and Ag-TiO$_2$ pH 8 has a size ranging from 5 to 15 nm.

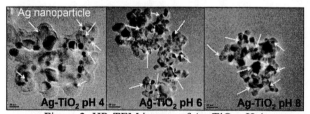

Figure 2: HR-TEM images of Ag-TiO$_2$ pH 4, Ag-TiO$_2$ pH 6, and Ag-TiO$_2$ pH 8. The arrows in figures show the Ag nanoparticles on the surface of TiO$_2$.

Figure 3 is the FE-SEM image of Ag-TiO$_2$ pH 4 nanoparticle; the size is unchanged as compared with P25. But Ag-TiO$_2$ nanoparticle shows a rough surface. Roughness of surface results from the existence of Ag nanoparticles. Presence of Ag nanoparticles is detected by EDS analysis (inset in Figure 3). The content of Ag nanoparticles on the Ag-TiO$_2$ surface is *approx.* 11.5 wt%. The Presence of Ag was additionally confirmed using various X-ray analyses.

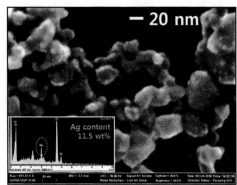

Figure 3: FE-SEM/EDS images of Ag-TiO₂ pH 4.

To confirm the presence of Ag on the Ag-TiO₂ surface, XRD and XPS analyses were performed. The XRD patterns are shown in Figure 4. All of Ag-TiO₂ samples show anatase and rutile phases like P25. Also, diffraction peaks for Ag(0) species (38.1, 44.2, 64.4, and 77.4°) are observed in all of Ag-TiO₂ samples [4]. Diffraction peak for Ag(I) species (34.2°) are not observed [2]. It might be due to the low amount of Ag(I) and the amorphous state of Ag [3].

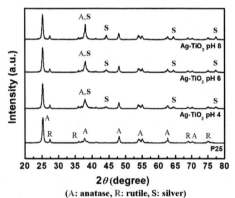

(A: anatase, R: rutile, S: silver)
Figure 4: XRD patterns of P25, Ag-TiO₂ pH 4, Ag-TiO₂ pH 6, and Ag-TiO₂ pH 8.

The XPS spectra provide the valence state of Ag on all of Ag-TiO₂ samples, as shown in Figure 5. Two individual peaks found at ~372.5 (for Ag $3d_{3/2}$) and ~366.2 eV (for Ag $3d_{5/2}$) confirm the presence of Ag in the all of Ag-TiO₂; these results agree with EDS data. The peaks of Ag $3d_{3/2}$/ $3d_{5/2}$ for Ag-TiO₂ pH 6 and pH 8 samples are slightly shifted to high binding energy. During reduction of Ag nanoparticles, the pH condition would affect the valence states of Ag on the Ag-TiO₂.

The light absorption spectra of P25 nanoparticle and Ag-TiO₂ pH 4, 6, and 8 are demonstrated in Figure 6. A band at about 450 nm corresponding to the signal of signal of Ag nanoparticles is present in all of Ag-TiO₂ samples. Compared to P25, an absorption covering the wavelength above 400 nm in the spectra of Ag-TiO₂ samples would be attributed to the enhanced photoactivities under visible light [7]. So, all of Ag-TiO₂ samples could be used in photocatalysis without UV.

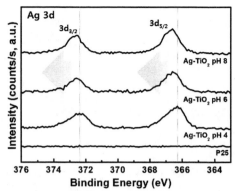

Figure 5: Ag 3d XPS spectra of P25, Ag-TiO₂ pH 4, Ag-TiO₂ pH 6, and Ag-TiO₂ pH 8.

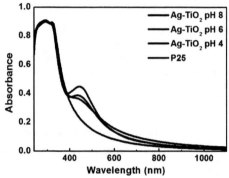

Figure 6: Light absorption of P25, Ag-TiO₂ pH 4, Ag-TiO₂ pH 6, and Ag-TiO₂ pH 8 as measured by UV-vis spectroscopy

3.2 Polyurethane Electrospinning Conditions

Before a preparation of Ag-TiO₂ coated PU composite nanofiber web, it was very important to find optimum electrospinning conditions. Figure 7 shows the FE-SEM images of electrospun PU nanofibers under the various conditions.

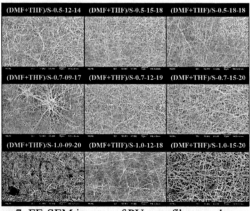

Figure 7: FE-SEM images of PU nanofibers under various electrospinning conditions. (solvent: DMF/THF (3:1 v/v), organic salt: BTMAC, flow rate: 0.5-1.0 ml/h, tip-to-collector distance: 9-15 cm, voltage: 14-20 kV)

Electrospinning stability could be improved through addition of BTMAC as an organic salt. The optimum electrospinning condition for PU nanofiber is determined to be the PU solution concentration of 25 wt% in DMF/THF (3:1 v/v), addition of BTMAC of 0.5 wt%, flow rate of 0.5-0.7 ml/h, tip-to-collector distance of 12-15 cm, and voltage of 18-20 kV. The Electrospun PU nanofibers under the optimum condition have an average diameter 180 nm.

In this work, it was considered that Ag-TiO$_2$ pH 4 nanoparticles had better photoactivities than others. So, the dispersed Ag-TiO$_2$ in methanol was electrically sprayed onto PU nanofiber surface by synchronous electrospray during PU electrospinning at the optimum condition.

Figure 8 shows the morphology of Ag-TiO$_2$ coated PU composite nanofiber web. Electrosprayed Ag-TiO$_2$ nanoparticles are coated on the surface of PU nanofiber.

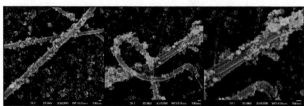

Figure 8: FE-SEM images of
Ag-TiO$_2$ coated PU composite nanofiber web.

Figure 9 shows TGA analysis for Ag-TiO$_2$ coated PU composite nanofiber web. Content of Ag-TiO$_2$ in the composite nanofiber web is *approx.* 17.2 wt% by TGA. Compared to PU nanofiber web, the composite nanofiber web exhibits no dramatic decrease of thermal stability.

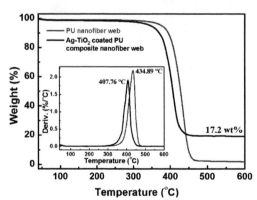

Figure 9: TGA data for Ag-TiO$_2$ coated PU
composite nanofiber web.

Photocatalytic decomposition of DMMP over PU nanofiber web and Ag-TiO$_2$ coated PU composite nanofiber web was evaluated by measuring the amount of remained DMMP in solutions under visible light or UV light irradiation. Compared to PU nanofiber web, the composite nanofiber web exhibits a faster decomposition of DMMP as shown in Figure 10. During 60 min, decomposition capacities of the composite nanofiber web are 2.63 mmol/g

under visible light irradiation and 6.74 mmol/g under UV light irradiation.

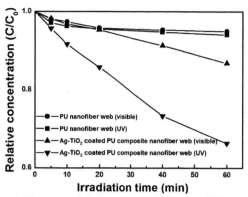

Figure 9: Decomposition study of DMMP on PU nanofiber web and Ag-TiO$_2$ coated PU composite nanofiber web, monitored by UV-vis.

4 CONCLUSION

Under various pH conditions, Ag-TiO$_2$ nanoparticles were successfully prepared by chemical reduction method. Existence of Ag on the surface of Ag-TiO$_2$ was investigated by various analyses. The valence state of Ag on the Ag-TiO$_2$ would be affected the pH condition of synthesis process. The visible light photoactivity of Ag-TiO$_2$ samples was confirmed from light absorption behavior. The Ag-TiO$_2$ coated PU composite nanofiber web was prepared by synchronous electrospray during electrospinning at the optimum condition. CW protective ability of the composite nanofiber web is evaluated by decomposition experiment for DMMP (nerve agent simulant) under visible/UV light irradiation. Decomposition capacities of the composite nanofiber web are 2.63 mmol/g (visible light irradiation) and 6.74 mmol/g (UV light irradiation).

REFERENCES

[1] Moss, J. A.; Szczepankiewicz, S. H.; Park, E.; Hoffmann, M. R. J. Phys. Chem. B 109, 19779, 2005.

[2] Zhang, H.; Wang, G.; Chen, C.; Lv, X. J.; Jinghong, U. H. Chem. Mater. 20, 6543, 2008.

[3] Hu, C.; Lan, Y.; Qu, J; Xu, X.; Wang, A. J. Phys. Chem. B 110, 4066, 2006.

[4] Zhang, F. X.; Guan, N. J.; Ji, Y. Z.; Zhang, X.; Chen, J. X.; Zeng, H. S. Langmuir 19, 8230, 2003.

[5] Arabatzis, I. M.; Stergiopoulos, T.; Bernard, M. C.; Labou, D.; Neophytides, S. G.; Falaras, P. Appl. Catal. B 42, 187, 2003.

[6] Guin, D.; Manorama, S. V.; Latha, J. N. L.; Singh, S. J. Phys. Chem. C 111, 13393, 2007.

[7] Ni~no-Martinez, N.; Martinez-Cata~non, G. A.; Aragon-Pi~na, A.; Martinez-Gutierrez, F.; Martinez-Mendoza, J. R.; Ruiz, F. Nanotechnology 19, 065711, 2008.

Development of Melt Blown Electrospinning Apparatus of Isotactic Polypropylene

Y. Lee[*], K. Watanabe[*], T. Nakamura[*], B. S. Kim[*], H. Y. Kim[**] and I. S. Kim[*]

[*]Department of Functional Machinery and Mechanics, Faculty of Textile Science and Technology, Shinshu University, Ueda, Nagano 386-8567, Japan, kim@shinshu-u.ac.jp
[**]Department of Textile Engineering, Chonbuk National University, Jeonju 561-756, Korea, khy@chonbuk.ac.kr

ABSTRACT

Melt blown electrospinning is a method of combining melt blown with melt electrospinning. In this study, we developed the multi nozzle melt blown electrospinning machine. The significant point of this method is to use air blowing force as well as electrostatic force at the same time. Also, this machine has many processing parameters, such as applied voltage, spinning distance, temperature at polymer reservoir, temperature at nozzle, air flow rate at nozzle, nozzle diameter, nozzle length, flow rate of melt polymer, molecular weight of polymer, and addictives, etc. We have successfully prepared nonwoven isotactic polypropylene fibers by controlling these parameters.

Keywords: melt blown electrospinning, polypropylene, fibers, air flow, nozzle

1 INTRODUCTION

In recent years, electrospinning is becoming of great interest because not only it can produce polymer fibers with diameters in the range of nano- to a few micrometers using polymer solutions or melts, but it also has the advantages of being simple, conventional methods with which a wide range of porous structures can be produced, and inexpensive as compared with conventional methods[1-5]. In many studies, this method has attracted considerable attention since setup is inexpensive and simple. In our previous study, we have firstly prepared a syndiotactic polypropylene (sPP) fibrous membrane from the solution via solution electrospinning at slightly elevated temperature [6]. SEM images and the frequency distribution of sPP fiber diameters of electrospun sPP fibers are shown in Figure 1. However, organic solvents used for electrospinning has serious issues which are mostly related to solvent recycling and environmental safety. Moreover, the mixed solvent system used for electrospinning of sPP is difficult for mass production. Among many studies, to figure out these problems, the use of molten polymers to produce electrospun fibers via melt-electrospinning becomes a subject of great interest. In spite of the potential benefits of melt-electrospinning, little progress has been made in the past twenty years. Fully understanding of the melt-electrospinning process, and its potential to replace solution electrospinning, has not yet been realized [8]. Meanwhile melt spinning and melt blowing are the most commonly used processes for producing the nonwoven fibers. Melt blown process is known because it has higher productivity than melt spinning process. In general, melt blown commercial products are composed of fibers with average diameters exceeding 1-2 μm [9].

In this study, we developed multi nozzle melt blown electrospinning machine. The significant point in this machine is able to use coincidentally two kinds of force, for instance, air blowing force and electrostatic force. We therefore strongly anticipate that air blowing force increases production rate and electrostatic force produces thin diameter of fibers, respectively. Here, we attempt to study the effects of air-flow rate, additive, and nozzle system on fiber diameter of the iPP nanofibers prepared by recently developed multi nozzle melt blown electrospinning apparatus.

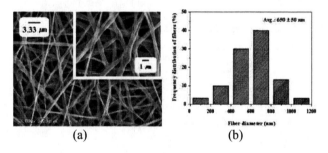

(a)   (b)

Figure 1: (a) SEM image and (b) frequency distribution of electrospun sPP fibrous membrane from solutions at a slight elevated temperature (35 °C)

2 EXPERIMENTAL

2.1 Materials

The iPP (isotactic polypropylene, 1200 melt flow rate (MFR) at 230°C) was kindly provided from Kuraray CO., LTD. (Japan). Irgatec CR 76 was obtained from Ciba Japan K.K. (Japan). LiCl was purchased from Wako (Japan). Two samples were used in this study. Sample A is prepared by mixing iPP and Irgatec CR 76 (200g/6g). Sample B is prepared by mixing iPP, Irgatec CR 76, and LiCl (200g/6g/2g).

2.2 Melt Blown Electrospinning

Conditions	Levels
Air flow rate at nozzle (*l*/min)	30, 60 , 110, 130
Applied voltage (kV)	15, 28, 33
Tip to collector distance (cm)	3.0, 8.5
Temperature at polymer reservoir (°C)	260
Air temperature at nozzle (°C)	450

Table 1: Melt blown electrospinning conditions of iPP electrospun fibers.

This apparatus has many parameters: applied voltage (kV), spinning distance (cm), temperature at polymer reservoir (°C), air temperature at nozzle (°C), air flow rate at nozzle (*l*/min), nozzle diameter (gauge), nozzle length (mm), flow rate of melt polymer (m*l*/min). In addition, polymer characteristics such as molecule weight, melt index, and melt point notably can affect the spinnability and the resulting fiber diameter. In this work, we investigated only two important parameters. Firstly, we checked the effect of air flow rate at nozzle on fiber diameter. Two different conditions of air flow rate (at 60 *l*/min and 130 *l*/min) were tested, respectively, whereas other parameters were fixed. A high-voltage power supply was used to generate a potential difference of 28 kV between the needle (with a 19 gauge, 15 mm) and a grounded metallic rotating drum placed at the distance of 8.5 cm from the tip of the needle. Polymer was melt at 260°C, and the temperature at nozzle was controlled to be 350°C by blowing of hot air. Secondly, we investigated the effect of LiCl on fiber diameter. Polypropylene has a poor electrical conductivity. Generally, this property may not be appropriate for electrospinning. LiCl can help the enhancement of electrical conductivity of PP polymer and thereby improvement of electrospinning conditions [10].

2.3 Characterization

The morphology of the melt blown electrospun fibers was examined with scanning electron microscopy (SEM, VE-8800, Keyence Co., Japan) and also using a Digital microscope (VHX-900,Keyence.co.Japan)

To compare the crystal structure of melt blown electrospun fibers verses melt spun fibers were determined using a X-ray Diffraction Spectroscopy. (XRD, Dmax2500, Rigaku Ltd., Japan) operating at 50kV and 200mA using Cu- Kα radiation. (wave length 1.5405nm) The diffracting intensities were recorded every 1 ° from 2θ scans in range 5 ° ~ 30 °.

3 RESULTS AND DISCUSSION

3.1 Effect of Air Flow Rate At Nozzle on Fiber Diameter

Melt blown electrospun iPP nonwoven fibers shown in Figure 2. As shown in this figure, we could successfully produce the melt blown electrospun iPP fiber mats (Figure 2), showing a good spinnability. We also investigated the effect of air flow rate at nozzle on fiber diameter. Figure 2 shows SEM images of melt blown electrospun fibers prepared at different air flow rates (Figure 2b: 60 *l*/min and 2c: 130 *l*/min, respectively). Table 2 summarizes averaged fiber diameter of each sample at different air flow rate, respectively. From SEM analysis, we found that the higher air flow rate at nozzle is, the thinner fiber diameter is formed. The diameter of the obtained fibers was found to be about 31.3 μm. However, we couldn't observe narrower ultrafine fiber on the nano-scale.

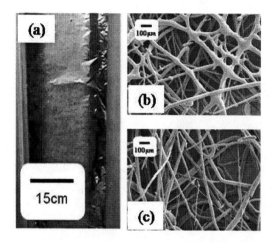

Figure 2: Melt blown electrospun iPP nonwoven fibers. (a) is Digital photograph, (b) and (c) are SEM images. Air flow rate at nozzle: (b) 60 *l*/min, (c) 130 *l*/min.

Sample	(b)	(c)
Average fiber diameter	31.3 ± 12	25.0 ± 10
Air flow rate at nozzle (*l*/min)	60	130
Applied voltage (kV)	28	28
Tip-to-collector distance (cm)	8.5	8.5

Table 2: The average diameters of melt blown iPP electrospun fibers produced from different air flow rate at nozzle.

3.2 Effect of LiCl on Fiber Diameter

As expected, the addition of small amounts of salt (about 1 wt.% in solution) was found to increase the fiber diameter and thereby decrease the jet length. [10] Consequently, to produce nanoscaled fibers, we added an additive, such as LiCl, which will increase the conductivity of molten iPP and results in an improved spinnability and thinner fiber diameter. The digital microscope images of melt blown electrospun iPP nonwoven fiber mats of sample B (with 1% LiCl) is shown in Figure 3 (left). Figure 3 (right) shows SEM image of melt blown electrospun iPP nanofiber with the smallest diameter of about 260 nm.

Figure 3: SEM images of melt blown electrospun iPP nonwoven fibers incorporating 1% LiCl.

3.3 XRD

The WAXD profiles of the melt blown electrospun iPP nonwoven fibers are shown in Figure 4. Sample A is melt electrospun iPP nonwoven fibers, sample B is melt blown iPP nonwoven fibers. It clearly shows the (110), (040), (130), and (131) crystal reflections at $2\theta=14.0°$, 17.0 °,18.5° and 22°, indicative of form I with an antichiral helical conformation [7,11]. Specifically, as seen in Figure 4, the melt blown electrospun iPP nanofibers (Figure 4A) exhibited sharper peak than that of melt blown fibers (Figure 4B), suggesting the enhanced crystalline structures.

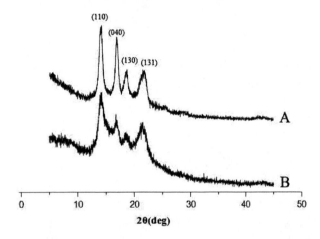

Figure 4: WAXD profiles of (A) melt electrospun iPP nonwoven fibers (B) Melt blown iPP nonwoven fibers.

3.4 Effects of Nozzle System on Fiber Diameter

Based on our previous results, melt blown electrospun iPP nonwoven fibers with the diameter of ca. 25 nm was successfully prepared by controlling the air flow rate and an addition of LiCl. But, we have faced on the difficulty in controlling the temperature of nozzle system, which will therefore give rise to poor spinnability and broader fiber diameter. Thus, we further improved the nozzle system to control the fine temperature during melt blown

electrospinning. Firstly, we changed the thickness of the walls of molten polymer tank in nozzle system for improving the thermal transfer. Secondly, we attached the heater on nozzle system for a stable heating environment. As a result, as seen in Figure 5, after changing the nozzle system, the temperature could be well-controlled, but the spinnability was not stable at higher air flow rate of 130 ℓ/min.

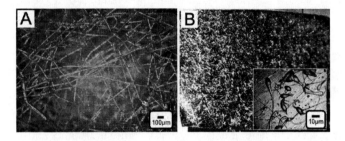

Figure 5: Digital microscope images of (A) previous nozzle system and (B) new nozzle system (air flow rate: 130ℓ/min, TCD: 8.5cm, applied voltage: 28kV, air temperature at nozzle: 230 °C).

Thirdly, the number of tips in nozzle system was also changed from 6 metal tips to 3 metal tips. In principle, when the metal tips get closer, the electric filed between the metal tips will be interrupted each other during electrical charging. As a result, such interaction will interrupt a full whipping and extension of polymer chains. The results are shown in Figure 6, and indicate that the more stable spinnability was observed than that of the previous nozzle system with 6 metal tips. In addition, Figure 7 shows SEM images of the corresponding melt blown electrospun iPP nanofibers.

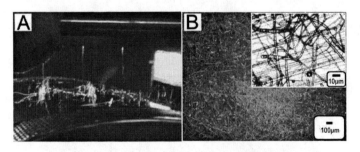

Figure 6: (A) Digital camera image of changed nozzle system and (B) digital microscope image of iPP melt blown electrospinning at the changed nozzle system. (air flow rate : 30ℓ/min, TCD : 3.0cm, applied voltage: 15kV, Air temperature at nozzle: 230 °C)

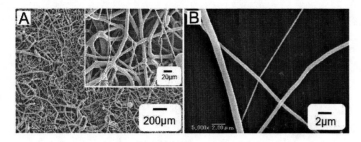

Figure 7: SEM images of iPP melt blown electrospinning at the changed nozzle system (air flow rate : 30ℓ/min, TCD : 3.0cm, applied voltage: 15kV, Air temperature at nozzle: 230 °C)

4 CONCLUSIONS

We developed the multi nozzle melt blown electrospinning machine. The crucial point of this method is to use both air blowing force and electrostatic force at the same time. Also iPP nonwoven fiber mat was successfully prepared by Melt blown electrospinning. We found that the higher air flow rate at nozzle is, the thinner fiber diameter is produced. Moreover, by adding LiCl into polypropylene, we could obtain the nanofiber with the smallest diameter (~ 260 nm) on melt blown electrospinning.

5 ACKNOWLWDGEMENTS

This work was supported by project for "Innovation Creative Center for Advanced Interdisciplinary Research Area" in Special Coordination Funds for Promoting Science and Technology from the Ministry of Education, Culture, Sports, Science and Technology of Japan. The authors acknowledge the support of Shinshu University Global COE Program "International Center of Excellence on Fiber Engineering".

REFERENCES

[1] S.A. Therona, E. Zussmana E, A.L. Yarin Polymer 45, 2017, 2004.

[2] S.H. Tan, R. Inai, M. Kotaki, S. Ramakrishna, S. Polymer 46, 6128, 2005.

[3] K. H. Lee, O.Ohsawa, S. Lee, J. C. Park, K. W. Kim , H.Y. Kim, Y. Watanabe, I. S. Kim, Sen'i Gakkaishi 64, 306, 2008.

[4] J. Doshi, D. H. J. Reneker, Electrostat 35, 151, 1995.

[5] K. Wei, T. Ohta, B.S. Kim, K.H. Lee, M.S. Khil, H. Y. Kim, I.S. Kim, Polym. Adv. Technol. in press, doi:10.1002/pat.1490.

[6] J. H. Park, B. S Kim, Y.C. Yoo, M.S. Khil, H.Y. Kim, J. Appl. Polym. Sci. 107, 2211, 2007.

[7] K. H. Lee, O. Ohsawa, K. Watanabe, I. S. Kim, S. R. Givens, B. Chase, J. F. Labolt, Macromolecules 42, 5215, 2009.

[8] Jason Lyons, Christopher Li, Frank Ko, Polymer 45, 7597, 2004.

[9] C. J. Ellison, A. Phatak, D.W. Giles, C. W. Macosko, F. S. Bates, Polymer 48, 3306, 2007.

[10] X. H. Qin, Y. Q. Wan, J. H. He, J. Zhang, J. Y. Yu, S. Y. Wang, Polymer 45, 6409, 2004.

[11] Giovanna Machado a,b, Eder J. Kinast c, Jackson D. Scholten b, Arthur Thompson a,Tiago De Vargas d, Sérgio R. Teixeira c, Dimitrios Samios, European Polymer Journal 45, 700, 2009.

Magnetic Dye-Adsorbent Catalyst: A "Core-Shell" Nanocomposite

S. Shukla[†,*], M.R. Varma[†], S. Suresh[§], and K.G.K. Warrier[†]

[†]Ceramic Technology Department, Materials and Minerals Division (MMD), National Institute for Interdisciplinary Science and Technology (NIIST), Council of Scientific and Industrial Research (CSIR), Thiruvananthapuram – 695019, Kerala, India, *satyajit_shukla@niist.res.in
[§]Department of Physics, Indian Institute of Technology, Mumbai 400076, India
*Present Address: Indo-US Science and Technology Forum (IUSSTF) Research Fellow, Ceramics Section, Energy Systems Division, Argonne National Laboratory, Argonne, IL 60439, sshukla@anl.gov

ABSTRACT

Hydrogen titanate nanotubes have been processed via hydrothermal using the commercially available nanocrystalline anatase-titania powder as a precursor. The samples have been characterized using TEM, XRD, and BET to determine the product morphology, structure, and specific surface-area, and have been utilized for a typical industrial application involving the removal of an organic-dye from an aqueous solution via surface-adsorption mechanism under the dark-condition. The dye-adsorption kinetics of hydrogen titanate nanotubes has been compared with that of the as-received anatase-titania nanoparticles. The nanotube morphology has been observed to play a significant role in enhancing the dye-adsorption kinetics under the dark-condition. In order to ease the separation of nanotubes after the dye-adsorption process, a magnetic dye-adsorbent catalyst has been developed. It consists of a nanocomposite particle having a "core-shell" structure, with a magnetic ceramic particle as a core and dye-adsorbent hydrogen titanate nanotubes as a shell. Due to the presence of nanotubes on the surface, the magnetic dye-adsorbent catalyst possesses an enhanced specific surface-area relative to that of the conventional magnetic photocatalyst. As a result, the former has been successfully utilized to remove an organic-dye from an aqueous solution via surface-adsorption mechanism, under the dark condition (energy-independent process), while simultaneously possessing the ferromagnetic property required for its separation, from the treated solution, using an external magnetic field. It is demonstrated that, under the given test-conditions, the magnetic dye-adsorbent catalyst removes almost ~100% of an organic-dye from an aqueous solution in dark in just 30 min, where as the conventional magnetic photocatalyst removes only ~40-55% of an organic-dye in 3 h via surface-adsorption mechanism. A surface-cleaning treatment has been developed for reusing the magnetic dye-adsorbent catalyst by removing the previously adsorbed organic-dye from its surface. The dye-adsorption and surface-cleaning treatment have been demonstrated by monitoring the variation in the original color of the catalyst-powder during the test.

Keywords: dye-adsorbent, magnetic photocatalyst, nanotubes, titania, hydrogen titanate

1 INTRODUCTION

Nanocrystalline anatase-titania (TiO_2) is a well known wide bind-gap (~3.2 eV) semiconductor, which has been extensively used for photocatalysis application, which involves the generation of electron/hole pairs within the particle via radiation-exposure (ultraviolet, visible, or solar), subsequently generating hydroxyl ($-OH^-$) radicals on the surface, which attack and degrade the dye molecules in an aqueous environment. Various parameters such as the nanocrystallite size, specific surface-area, nature and relative amount of phases, dopants, surface-catalyst, and surface-purity are known to affect the photocatalytic activity of nanocrystalline anatase-TiO_2. Although, TiO_2-based photocatalysts are very effective for the dye-removal application, they suffer from few major drawbacks. First, the photocatalysis is highly energy-dependent process, which makes it relatively expensive. Second, for the utilization of the visible-light, band-gap tuning is required via dopants, which also increases the cost of the photocatalysis process. Third, the separation of TiO_2-based photocatalysts after the dye decomposition, via conventional approaches such as coagulation and sedimentation, is tedious and costlier. In order to overcome these drawbacks associated with the TiO_2-based photocatalysts, we process here the nanotubes of hydrogen titanate ($H_2Ti_3O_7$) via hydrothermal and utilize them for the removal of an organic-dye via surface-adsorption mechanism under the dark-condition. Since this is an energy-independent process, the nanotubes of $H_2Ti_3O_7$ offer significant cost-saving over the anatase-TiO_2 nanoparticles. However, being nano-magnetic, the separation of $H_2Ti_3O_7$ nanotubes with the surface-adsorbed dye, from an aqueous solution, is also a major problem. To overcome this issue, we further process a "core-shell" magnetic dye-adsorbent catalyst, which involves the core of a magnetic ceramic particle and the shell of $H_2Ti_3O_7$ nanotubes. Such nanocomposite particle possesses both the magnetic and dye-adsorption properties, which overcome almost all limitations associated with the photocatalysts based on the anatase-TiO_2 nanoparticles and $H_2Ti_3O_7$ nanotubes.

2 EXPERIMENTAL

2.1 Chemicals

Sodium hydroxide (NaOH, Assay 97 %) and methylene blue (MB, >96 %) were purchased from S.D. Fine Chemicals Limited, India; 1 M hydrochloric acid (HCl, 35 wt.%) from Ranbaxy Fine Chemicals Limited, India; ammonium hydroxide (NH$_4$OH, 25 wt.%) from Merk Limited, India; and anatase-TiO$_2$ from Central Drug House (CDH) Laboratory (P) Limited, India. All chemicals and powders were used as-received without any further purification.

2.2 Processing H$_2$Ti$_3$O$_7$ Nanotubes via Hydrothermal

The hydrothermal process has been already described in detail elsewhere [1]. 3 g of as-received anatase-TiO$_2$ (CDH) was suspended in a highly alkaline aqueous solution, containing 10 M NaOH filled up to 84 vol.% of Teflon-beaker placed in a stainless-steel (SS 316) vessel of 200 ml capacity. The process was carried out with continuous stirring in an autoclave (Amar Equipment Pvt. Ltd., Mumbai, India) at 120 °C for 30 h under an autogenous pressure. The autoclave was allowed to cool naturally to room temperature and the product was separated by decanting the top solution. The product was then washed using 100 ml of 1 M HCl solution for 1 h followed by washing using 100 ml of distilled water for 1 h (termed here as the "first washing-cycle"). The product was separated from the solution using a centrifuge (R23, Remi Instruments India Ltd.), and subjected to the "second washing-cycle". In this cycle, the product was washed once using 100 ml of 1 M HCl and then multiple times (#8-9) using 100 ml of distilled water (1 h each) or till the pH of the filtrate became almost constant. The product was separated from the solution using a centrifuge and dried in an oven at 80 °C overnight.

2.3 Processing Magnetic Dye-Adsorbent Catalyst

The processing of magnetic dye-adsorbent catalyst has been described in detail elsewhere [2]. In short, a mixed cobalt ferrite (CoFe$_2$O$_4$)-hematite (Fe$_2$O$_3$) magnetic ceramic powder was first processed via polymerized complex technique. This was followed by the deposition of an insulating layer of silica (SiO$_2$) via Stober process. The SiO$_2$-deposited magnetic ceramic particles were then coated with the nanocrystalline anatase-TiO$_2$ via sol-gel to obtain the "conventional magnetic photocatalyst". The latter was subjected to a hydrothermal process, as described in the section-2.2, to obtain the "core-shell" nanocomposite particles, with the core of a magnetic ceramic particle and the shell of H$_2$Ti$_3$O$_7$ nanotubes.

2.4 Characterization

The morphologies of different samples at the nanoscale were examined using the transmission electron microscope (TEM) and high-resolution TEM (Tecnai G2, FEI, Netherlands) operated at 300 kV. The selected-area electron diffraction (SAED) patterns were obtained to confirm the crystallinity and the structure of different samples. The crystalline phases present in the samples were also determined via X-ray diffraction (XRD, PW1710, Phillips, Netherlands). The broad-scan analysis was typically conducted within the 2-θ range of 10-80° using the Cu $K\alpha$ (λ_{Cu}=1.542 Å) X-radiation. The Brunauer, Emmett, and Teller (BET) specific surface-area (Micrometrics Gemini 2375 Surface Area Analyzer, U.S.A.) was measured via nitrogen adsorption, using the multi-point method, after degassing the powders at 200 °C for 2 h. The magnetic properties of the magnetic dye-adsorbent catalyst were measured using a vibrating sample magnetometer attached to a Physical Property Measurement System.

2.5 Dye-Adsorption Measurements

The dye-adsorption experiments in the dark were conducted using the MB as a model catalytic dye-agent. A 75 ml aqueous suspension was prepared by dissolving 7.5 μmol•L^{-1} of MB dye and then dispersing specific amount of the catalyst (0.4 and 1.0 g•L^{-1} for the pure H$_2$Ti$_3$O$_7$ nanotubes and the magnetic dye-adsorbent catalyst). The suspension was stirred in the dark and 3 ml sample suspension was separated after each 30 min time interval for total 180 min stirring time. The catalyst powder was separated using a centrifuge and the solution was used to obtain the absorption spectra using the ultraviolet (UV)-visible absorption spectrometer (UV-2401 PC, Shimadzu, Japan). The normalized concentration of surface-adsorbed MB dye was calculated using the equation of the form,

$$MB_{Adsorbed}(\%) = \left(\frac{C_0 - C_t}{C_0}\right) \times 100 \qquad (1)$$

which is equivalent of the form,

$$MB_{Adsorbed}(\%) = \left(\frac{A_0 - A_t}{A_0}\right) \times 100 \qquad (2)$$

where, C_0 and C_t correspond to the MB dye concentration at the start and after string time 't' with the corresponding absorbance of A_0 and A_t. The catalyst powders were then subjected to a typical surface-cleaning treatment in a bath of specific composition [2] either in a dark or under continuous radiation-exposure (UV, solar, or visible) with continuous stirring for 8 h. The surface-cleaned samples were utilized for the next cycle of dye-adsorption under similar test-conditions.

3 RESULTS AND DISCUSSION

The TEM image and XRD pattern as obtained for the hydrothermal product is presented in Fig. 1(a) and 1(b). The corresponding SAED pattern is shown as an inset in Fig. 1(a). It is clearly seen that, the hydrothermal product has a nanotube morphology. The SAED and XRD patterns suggest that, the nanotubes are made up of $H_2Ti_3O_7$. When the as-received anatase-TiO_2 particles are subjected to the hydrothermal treatment (autoclave) under highly alkaline condition, the exfoliation of single-layer nanosheets of $Na_2Ti_3O_7$ results from the bulk anatase-TiO_2 structure, which continuously undergo the dissolution and crystallization processes. This is supported by the fact that, $Na_2Ti_3O_7$ easily adopts a layered-structure whereas anatase-TiO_2 does not. Hence, the formation of $Na_2Ti_3O_7$ is vital for the subsequent exfoliation of single-layer nanosheets. Due to their higher specific surface-area to volume ratio and the presence of dangling bonds along the two long-edges, the nanosheets of $Na_2Ti_3O_7$ have a strong drive to rollup. However, this rollup tendency is opposed by the repulsive force produced by the charge on the nanosheets created by the presence of Na^+-ions. These Na^+-ions are replaced via ion-exchange process with H^+-ions during the washing steps, which results in the formation of pure-$H_2Ti_3O_7$; thus, reducing the repulsive force for the rollup. Hence, at the completion of an ion-exchange process, the nanosheets of pure-$H_2Ti_3O_7$ immediately rollup to form the nanotubes, Fig. 1(a). The $H_2Ti_3O_7$ nanotubes exhibit very high specific surface-area as high as 180 $m^2 \cdot s^{-1}$, which is almost 18 times larger than that of the as-received anatase-TiO_2 (CDH). As a result, the nanotubes provide large number potential sites for the removal of organic-dye from an aqueous solution via surface-adsorption mechanism under the dark-condition. The variation in the amount of surface-adsorbed MB as a function stirring time in the dark, for $H_2Ti_3O_7$ nanotubes, is presented in Fig. 2(a) and compared with that of the as-received anatase-TiO_2 nanoparticles. It is clearly seen that, the nanoparticles of as-received anatase-TiO_2 exhibit only ~50-55 % of MB adsorption under the dark condition. On the other hand, the $H_2Ti_3O_7$ nanotubes show ~100% adsorption after 30 min of stirring time. The variation in the powder-color as a function of different test-conditions is shown in Fig. 2(b). After the surface-cleaning treatment, the original high dye-adsorption capacity is regained [2].

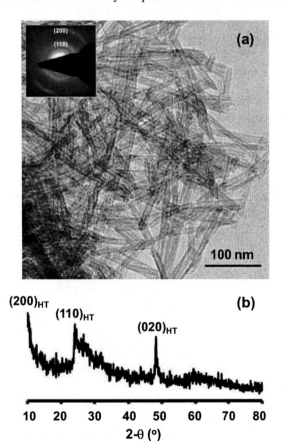

Figure 1: Typical TEM image (a) and XRD pattern (b) of hydrothermally processed $H_2Ti_3O_7$ nanotubes.

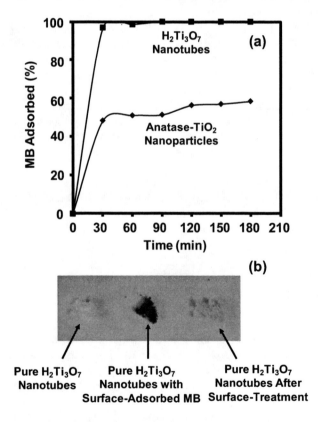

Figure 2: (a) Typical variation in the normalized concentration of surface-adsorbed MB (%) as a function of stirring time in dark. (b) Variation in the color of $H_2Ti_3O_7$ nanotubes powder as a function of different test-conditions.

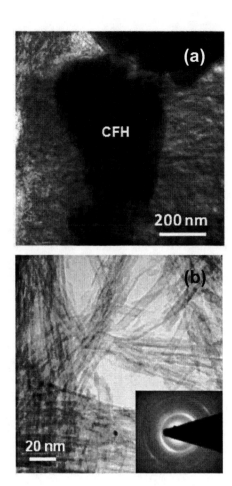

Figure 3: Typical TEM images of the magnetic dye-adsorbent catalyst at lower (a) and higher (b) magnifications.

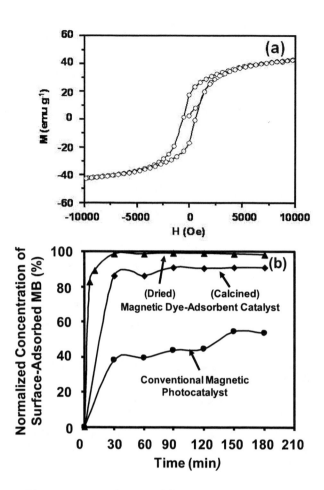

Figure 4: Magnetic (a) and dye-removal (in dark) (b) characteristics of the magnetic dye-adsorbent catalyst.

In order to enhance the separation efficiency, a magnetic ceramic particle has been incorporated within $H_2Ti_3O_7$ nanotubes. The TEM image of a typical "core-shell" nanocomposite particle with the core of a mixed cobalt ferrite and hematite (CFH) magnetic particle and the shell of $H_2Ti_3O_7$ nanotubes is presented in Fig. 3(a). The central dark-region in Fig. 3(a) represents the core magnetic ceramic particle and the region of grey-contrast surrounding it consists of $H_2Ti_3O_7$ nanotubes, which are revealed in a higher magnification image presented in Fig. 3(b), where the corresponding SAED pattern is shown as an inset. Such "core-shell" nanocomposite structure is termed here as a "magnetic dye-adsorbent catalyst" since it possesses both the magnetic and dye-adsorption properties. The presence of a hysteresis loop, Fig. 4(a), suggests the ferromagnetic nature, which makes its separation possible using an external magnetic field. In Fig. 4(b), the magnetic dye-adsorbent catalyst is observed to remove ~100% MB dye, whereas the conventional magnetic photocatalyst removes only ~40-55% of the dye. Since MB dye is cationic, very high dye-adsorption capacity can be retained for the initial MB dye concentration as high as 60 $\mu mol \cdot L^{-1}$ by adjusting the solution-pH in the basic range (~8-10) [1].

4 CONCLUSION

Hydrothermally processed magnetic dye-adsorbent catalyst consisting a nanocomposite particle, having a "core-shell" structure with a magnetic ceramic particle as a core and $H_2Ti_3O_7$ nanotubes as shell, is much superior to the nanocrystalline anatase-TiO_2, pure-$H_2Ti_3O_7$ nanotubes, and conventional magnetic photocatalyst for the removal of an organic-dye from an aqueous solution under dark-condition.

ACKNOWLEDGEMENT

Authors thank CSIR, India and IUSSTF* for the financial support (Project # NWP0010 and P81113).

REFERENCES

[1] N. Harsha, K.R. Ranya, K.B. Babitha, S. Shukla, S. Biju, M.L.P. Reddy, K.G.K. Warrier, J. Nanosci. Nanotech. (Proofs Corrected)

[2] S. Shukla, M.R. Varma, K.G.K. Warrier, M.T. Lajina, N. Harsha, C.P. Reshmi, Indian Patent Application Number 0067DEL2010 (PCT to be filed)

Nanohybrid silica/polymer subcritical aerogels

A. Fidalgo, L.M Ilharco, J.P.S Farinha and J.M.G. Martinho

*CQFM - Centro de Química-Física Molecular and IN - Institute of Nanoscience and Nanotechnology, Instituto Superior Técnico, Universidade Técnica de Lisboa, Av. Rovisco Pais 1, 1049-001 Lisboa, Portugal, Alexandra.fidalgo@ist.utl.pt

ABSTRACT

Novel nanohybrid silica/polymer aerogels were produced by the soft sol-gel process, under subcritical conditions. The inclusion of functionalized core-shell polymer nanoparticles (PNP) at the first stages of the sol-gel process builds up a structure stiff enough to withstand capillary pressure, yielding machineable monoliths with good mechanical properties. The influence of the silica precursor and of the size and content of the PNPs was analyzed.

Keywords: aerogels, hybrid nanocomposites, insulators

1 INTRODUCTION

The most common insulators (mineral wools and expanded polymers) are low-priced, but do not meet the demanding energetic and environmental requirements of modern industry. However, aerogels are conquering the market, so far only for high technology applications [1]. They are non-flammable, extremely light (densities in the range 3-500 $kg \cdot m^{-3}$), and have excellent thermal and acoustic properties (thermal conductivity in the range 0.01-0.02 $W \cdot m^{-1} \cdot K^{-1}$ and acoustic impedance 103-106 $kg \cdot m^{-2} \cdot s^{-1}$) [2]. The drawbacks of these wonder materials are their high production costs, brittleness and instability to atmospheric moisture. The present work explores the synthesis of a nanohybrid silica/polymer aerogel with improved mechanical properties, obtained at ambient pressure.

Hybrid inorganic-organic materials are the result of an intimate mixture of components, where interactions may range from van der Waals or hydrogen bonds (class *I*) to strong covalent or ionic-covalent bonds (class *II*) [3]. The sol-gel process is the best method to synthesize these materials, since the mild conditions involved are ideal to preserve the organic components [4].

Previous work proved the possibility of obtaining pure silica aerogels by subcritical drying of aged alcogels [5]. The inclusion of core-shell polymer nanoparticles (PNP) at the first stages of the sol-gel process allowed improving their mechanical properties [6-8].

In the present work, different silica precursors and PNP sizes and contents were rehearsed in view of attaining even superior properties.

2 EXPERIMENTAL

The hybrid alcogels were prepared by a two-step acid/basic catalyzed co-hydrolysis/co-condensation of a silica precursor and cross-linked core-shell particles. These consist of a trimethoxysilyl-functionalized poly(butyl methacrylate) shell and a poly(butyl methacrylate-co-butyl acrylate) core (TMS-PNP). Aqueous sodium silicate was used in alternative to tetraethoxysilane as silica precursor (Figure 1).

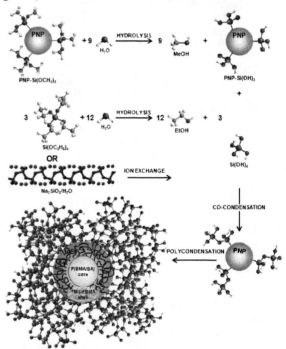

Figure 1: Reaction scheme for the synthesis of class *II* SiO$_2$/TMS-PNP hybrid materials.

The synthesis of the core-shell cross-linked colloidal polymer particles followed a two-stage emulsion polymerization technique. In the first stage, cross-linked seed particles were produced by batch emulsion co-polymerization of n-butylmethacrylate (BMA) and n-butylacrylate (BA), with ethylene glycol dimethacrylate (EGDMA) as cross-linker. The mixture was emulsified with sodium dodecyl sulfate (SDS), and initiated with potassium persulfate (KPS). It was then purged with nitrogen gas for 2 h, and heated to 353 K for an additional 2 h period. In the second stage, the seeds were used to prepare the PNPs, with

NSTI-Nanotech 2010, www.nsti.org, ISBN 978-1-4398-3401-5 Vol. 1, 2010

size and composition controlled by semi-continuous emulsion polymerization, with monomer and initiator fed under starving conditions: two mixtures (one organic and one aqueous) were independently added to the seed dispersion (purged under nitrogen) over 8 h, at constant rate, under stirring. The organic phase contained BMA, 3-trimethoxysilyl-propylmethacrylate (TMS-PMA) and EGDMA. The aqueous phase contained SDS and KPS. The reaction mixture was stirred at 353 K for an additional 2 h period before cooling to room temperature. This procedure yielded a 30 wt% solids stable dispersion of cross-linked core-shell particles with a trimethoxysilyl modified poly(butylmetacrylate) shell and a poly(butylmetacrylate-co-butylacrylate) core (TMS-PNP).

For the synthesis of the TEOS based hybrid alcogels, a mixture of TEOS, TMS-PNP and water in 2-propanol (2-PrOH) was acidified with hydrochloric acid (HCl), sealed and stirred for 1 h at 333 K, to promote the co-hydrolysis. This sol was then neutralized with ammonia (NH_4OH) and left to gel. An appropriate ageing followed at 333 K: 24 h in the mother liquor and another 24 h in a mixture of water, TEOS and 2-PrOH in the initial proportions. The aged gels were washed with 2-PrOH, and subcritically dried, at 333 K, in a *quasi*-saturated solvent atmosphere, until negligible weight loss.

For the aqueous sodium silicate based gels, the silica precursor (Na_2O:3.3 SiO_2:25.9 H_2O; d=1.37 g.cm^{-3}) was diluted to a weight percent of SiO_2 in solution of 3-10%, and ionic exchanged with Amberlite®, with different Amberlite /Na^+ mass ratios and contact periods, in order to obtain silicic acid with different pH values (1-3). The hybrid hydrogels were then obtained by catalytic condensation of silicic acid with ammonia, as used for the TEOS based alcogels. The same ageing, washing and drying procedures were used to prepare the sodium silicate based hybrid aerogels.

Hybrid silica/polymer aerogels with different TMS-PNP contents (weight percent, roughly estimated assuming full conversion of TEOS into SiO_2) and average diameters (50, 100 and 150 nm) were prepared. For comparison, pure silica aerogels were synthesized following the same procedures.

The chemical and physical properties of the dry gels have been analyzed by volume shrinkage upon drying, envelope density determinations, scanning electron microscopy (SEM), nitrogen sorption isotherms, and diffuse reflectance infrared (DRIFT) spectroscopy. The mechanical properties were measured by unidirectional compression tests, and the thermal conductivities were determined using a modified Lee's disk method.

3 RESULTS AND DISCUSSION

The inorganic aerogel is translucent, whereas the hybrid ones are opaque, even for very low polymer contents (as low as 0.5 wt%). Clearly, the presence of TMS-PNP

modifies the aerogel's pore network, since the light dispersion becomes so altered.

In Figure 2 the SEM microphotographs of fracture surfaces of the inorganic and a 3 wt% TMS-PNP (100 nm diameter) hybrid silica/polymer aerogels are compared. All the samples show a bicontinuous distribution of interconnected particles and macropores, irrespective of the silica precursor and PNP content or average size, but it is clear that the inclusion of TMS-PNP (in contents as low as 0.5 wt%) is sufficient to impart a drastic change to the aerogel's microstructure: it becomes more granular, with a more uniform distribution of smaller macropores.

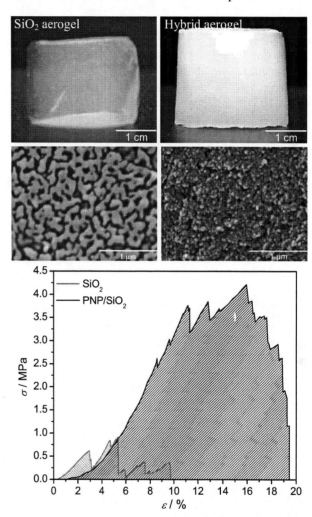

Figure 2: Photographs, corresponding cross-sectional SEM microphotographs and stress-strain curves obtained from unidirectional compression tests of a pure silica sample and a hybrid aerogel with 3 wt% PNP (100 nm).

The changes induced in the mechanical behavior of the silica aerogels by incorporation of TMS-PNP were evaluated by unidirectional compression tests. The stress-strain (σ-ε) curves in Figure 2 refer to the pure silica and the 3 wt% TMS-PNP 100 nm hybrid aerogels. The brittle nature of the silica aerogel remains upon the introduction of

TMS-PNP, as seen from the serrations along the compression curve, resulting from successive fracture of the cells' walls [9]. However, even low polymer contents promote a clear improvement in the mechanical properties of the material: the Young's modulus increases, the maximum compression strength and the corresponding strain become 3 to 5 times higher. Moreover, the improvement in the mechanical behavior is noticeable, since the hybrid aerogel is able to absorb a much higher energy (roughly the area under the σ-ε curve) up to the maximum compression strength.

The effect of the polymer content on the pore morphology of the aerogels was characterized by analysis of the N_2 adsorption-desorption isotherms, shown in Figure 3: they remain type *IV* over the entire range of polymer content, with the hysteresis loops usually associated with capillary condensation in mesopores [10]. The hysteresis is clearly type H1 for all the samples, characteristic of a uniform distribution of cylindrical mesopores. However, for the inorganic aerogel the adsorption branch is not so steep, resulting in a broader hysteresis, which points to a less defined mesopore shape.

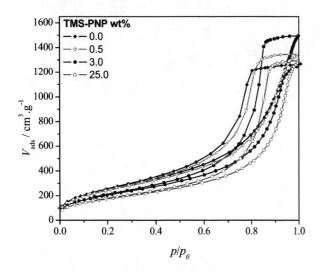

Figure 3: N_2 Adsorption-desorption isotherms (at 77 K) of the inorganic and selected hybrid aerogels with different 100 nm TMS-PNP contents.

The plateau observed at high relative pressure shows the existence of two distinct evolution trends: for low polymer contents (up to ~3 wt%), the total pore volume of the hybrid samples increases above that of the corresponding inorganic aerogel; higher polymer contents result in a decrease of the total pore volume, eventually reaching values below that of the inorganic sample.

Concurrently, for TMS-PNP contents between 0 and 3 wt%, the envelope density, ρ_e, of the aerogels decreases (from 420 to 357 kg·m^{-3}) with the increase of polymer content. This decrease reflects an increase in the percent porosity, P_r, from 80 to 83% (calculated as $100 \times (1-\rho_e/\rho_s)$, taking the skeletal density, ρ_s, as 2100 kg m^{-3} [11]).

Simultaneously, there is an increase of the average mesopore diameter (from 8.4 to 11.5 nm), estimated by application of the Barrett-Joyner-Halienda (BJH) method to the N_2 desorption isotherm, assuming a cylindrical pore shape [10]. On the other hand, the specific surface area (inferred from the N_2 sorption isotherms by a BET analysis of the amount of gas adsorbed at relative pressures between 0.05 and 0.3, using N_2 cross sectional area of 16.2 Å^2 [10]) decreases with increasing TMS-PNP content up to 3 wt%, from 936 to 766 m^2.g^{-1}.

The effect of the TMS-PNP average diameter on the pore morphology of the hybrid aerogels is shown in Figure 4, where the BJH pore size distribution is included in the inset.

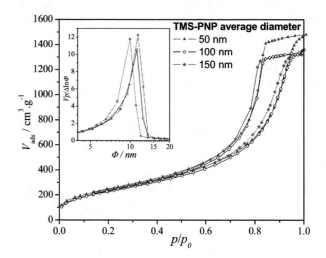

Figure 4: N_2 sorption isotherms of hybrid aerogels containing 5 wt% TMS-PNP with different sizes.

The total pore volume is similar for the 50 and 100 nm TMS-PNP hybrids, but increases when the polymer nanoparticles size is 150 nm. This increase, which corresponds to a decrease in ρ_e to 200 kg·m^{-3} (corresponding to a porosity of 89%) appears to be associated with an increase in the specific surface area and average mesopore diameter, as indicated in Table 1.

TMS-PNP size	S_{BET} (m^2/g)	V_p (cm^3/g) (p/p$_0$=0.99)	Φ_{BJH} (nm)
50nm	873±3	2.041514	8.15
100 nm	850±2	2.064906	8.68
150 nm	907±3	2.262190	9.32

Table 1: Surface area (S_{BET}), total pore volume (V_p) and average mesopore diameter (Φ_{BJH}) for SiO$_2$/5 wt% TMS-PNP hybrid aerogels with different PNP average sizes.

From these results, it is noticeable that all the synthesized aerogels possess a hierarchical bimodal pore structure, containing mesopores within the individual particles, and macropores in the interstitial voids between them. The presence of TMS-PNP, irrespectively of its content and size, seems to promote homogeneity in the aerogel's pore network, both in the macro and mesopore ranges. A slight decrease in density (increase in porosity) may be achieved by synthesizing the aerogels with a small amount of TMS-PNP (up to ~3 wt%) with an average diameter of 150 nm. This is not due to an increase in macroporosity (which seems to decrease) but rather to the increase in the average dimensions of the mesopores, which acquire a more cylindrical shape. The same trends are observed for aerogels prepared from aqueous sodium silicate solution, with envelope densities ranging between 480 (for the inorganic aerogel) and 280 kg·m^{-3} (for the hybrid with 3 wt% of the 100 nm TNS-PNP).

The hybrid materials were further characterized, at the molecular structure level, by diffuse reflectance infrared (DRIFT) spectroscopy. The spectra of the inorganic aerogel and of the hybrid containing 3 wt% TMS-PNP are compared in Figure 5.

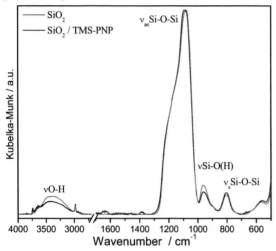

Figure 5: DRIFT spectra, normalized to the most intense band (at ~1100 cm^{-1}) of pure silica (red), and hybrid aerogel containing 3 wt% (blue) of TMS-PNP.

Both spectra present the typical features of a sol-gel silica network. The broad band centered at ~3300 cm^{-1} is assigned to the stretching mode of O-H groups involved in different hydrogen bonding. They are correlated with silanol (SiOH) groups, since there is no molecular water in these samples (the H-O-H deformation mode, expected at ~1650 cm^{-1} is absent). Another silanol band, assigned to the Si-O(H) stretching mode, appears at 950 cm^{-1}. The feature at ~800 cm^{-1} is assigned to the symmetric Si-O-Si stretching mode. The corresponding antisymmetric mode is the strongest band in the spectrum, with maximum absorption at ~1100 cm^{-1}. The similarity of this band shape in the two spectra indicates that the silica structure is not significantly altered by the presence of the polymer. The absence of the adsorbed water band and the low relative intensities of the silanol related bands are good indications that all the aerogels are relatively hydrophobic.

4 CONCLUSIONS

Nanohybrid silica/TMS-PNP aerogels were successfully produced under subcritical conditions. Aqueous sodium silicate proved to be an excellent silica precursor with cost benefits, and the size of the polymer nanoparticles was a key parameter of the aerogel's mechanical resistance. These hybrid aerogels take advantage of the synergy between organic and inorganic properties to improve the mechanical behavior of pure inorganic aerogels, still preserving their unique properties. The method developed is a green chemistry process that avoids the expensive steps of the current technologies.

REFERENCES
[1] J. E. Fesmire, Cryogenics 46, 111, 2006.
[2] L.W. Hrubesh, T.M. Tillotson, J.F. Poco *in* "Better Ceramics Through Chemistry", (Eds: C.J. Brinker D.E. Clark, D.R. Ulrich, B.J. Zelinski), Elsevier, New York 1988.
[3] C. Sanchez, B. Julián, P. Belleville, M. Popall, J. Mater. Chem. 15, 3559, 2005.
[4] N. Leventis, A. Palczer, L. McCorkle, G. Zhang, C. Sotiriou-Leventis J. Sol-Gel Sci. Tecnol. 35, 99, 2005.
[5] A. Fidalgo, M.E. Rosa, L.M. Ilharco, Chem. Mater. 15, 2186, 2003.
[6] A. Fidalgo, J.P.S Farinha, J.M.G. Martinho, M.E. Rosa, L.M. Ilharco, Chem. Mater. 19, 2603, 2007.
[7] L.M. Ilharco, A. Fidalgo, J.P.S. Farinha, J.M.G. Martinho, M.E. Rosa, J. Mater. Chem. 17, 2195, 2007.
[8] J.M.G. Martinho, L.M. Ilharco, J.P.S. Farinha, A.M. Fidalgo, P.O. Martinho, WO 2006/107226 A1; US2008/0188575 A1.
[9] L.J. Gibson, M.F. Ashby, "Cellular Solids: Structure and Properties" (2nd Ed.), Cambridge University Press, Cambridge 1997.
[10] S.J. Gregg, K.S.W. Sing, "Adsorption, Surface Area and Porosity" (2nd Ed.), Academic Press, London 1982.
[11] T. Woignier, J. Phalippou, J. Non-Cryst. Solids 93, 17, 1987.

The magnetic properties optimization in diluted magnetic oxide semiconductor nanoparticles obtained by a modified co-precipitation approach in intrazeolitic networks

R. BOSÎNCEANU*, F. IACOMI

*Department of Physics,"Al. I. Cuza" University, Bd. Carol I
Iasi, Romania*

Abstract

Ferromagnetism in diluted magnetic semiconductors (DMS) has been a subject of great scientific and technological interest for the past few years due to its implications for spintronics, advanced magneto-optics and sensors applications. Our research involves a modified co-precipitation approach that allows controlled growth of Fe3O4/γ-Fe2O3 nanocrystals within the superparamagnetic and single domains limits by creating the precipitating conditions with the help of hydrolase reaction of the urea. XRD confirms the formation of the ferrite and the partial loss of host crystallinity observed in X-ray diffraction patterns thereby suggests that the process of cluster formation involves aggregation within the large cages of the zeolite and local destruction of the pores network. Room-temperature M-H measurements verifies the influence of synthesis conditions and crystal size on the magnetic properties of ferrite nanocrystals. Distinct absorption maxima in diffuse reflectance UV−visible spectra and sharp exciton peaks in low-temperature excitation spectra verified the presence of quantum-confined Fe3O4/γ-Fe2O3.

Keywords: magnetic properties optimization, diluted magnetic oxide semiconductor, Superparamagnetism.

1. Introduction

Magnetic properties of nanoparticles are subject to intense research activity driven by a fundamental interest in the novel physical properties of the nanoscale system and also potential industrial application of nanostructured materials. This letter reports on the magnetic properties of well-dispersed diluted magnetic oxide quantum systems within an aluminous-silicious matrix. The use of the A zeolite as a self-assembeled nanotemplate enables us to synthesize composite materials containing magnetic oxide nanoparticles at different temperature for optimization of magnetic properties. Comparable inorganic methods for the synthesis of nanoscale mixed-metal oxides require heating at high temperatures in order to produce the desired oxide composition and microstructure [1] and [2]. The development of such mixed-metal oxide and zeolite-based nanocomposites is targeting the functionalization into device technologies for sensors and spintronics applications.

The materials challenge is to magnetize functional non-magnetic materials by synthesis dilute concentrations of magnetic species, thereby harnessing the benefits of a magnetic response without significantly affecting the desirable physical properties of the non-magnetic host

3. Experimental Part

3.1. Starting materials

All the chemicals were of reagent grade and used without further purification. The following materials were used for the loading $Fe_3O_4/\gamma - Fe_2O_3$ thereby : A zeolite with pore size 0.4 nm and 2.5 nm obtained in Chemical Science and Technology Laboratory of Faculty of Chemistry from the "Al.I.Cuza" University of Iasi, ferric chloride hexahydrate ($FeCl_3 * 6H_2O$, >99%), ferrous chloride tetrahydrate ($FeCl_2 * 4H_2O$, >99%), hydrochloric acid (HCl, $1N$) were obtained from KEBO, distilled water and urea (46.66%).

3.2. The construction of the ion exchange isotherms for Fe^{+3}/Na-A with a cation exchange capacity of 5.5 meq/g. The procedure carried out was described by Barros et al. (1995) [3]. Fig.1 presents the ion-exchange isotherms experimentally obtained.

One observes from Fig.1 is that the isothermal behaviour is very similar with the type "d" [4] for that from 298K and the type "a" [4] for the rest.

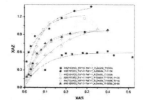
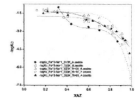

Fig1. Schematic representation of ion-exchange isotherms for A zeolite to (T=298K, 313K, 333K) and in treated A zeolite with urea (R=6, 10, 60).

Fig.2. Kielland plots for A zeolite to (T=313K, 333K) and treated A zeolite with urea (R=6, 10, 60) for Fe+3-Na+1.

However, it is strongly hydrated at 298K, steric problems inhibit in some extent the retention of ferric ions. Possibly, at this temperature, Fe^{+3} cations are only located in the large cages due to the low X_{AZ} values obtained for them. It is also observed at 313K and 333K some overexchanged values ($X_{AZ} > 1$), which is probably due to some multilayer adsorption. A densely occupied monolayer will act in some degree as an extension of the zeolite, and will be able to attract further cations from the solution phase. The adsorption phenomenon in zeolite ion-exchange isotherms was already seen (Barros et al., 2003) [5] indicating that X_AZ values greater than 1 are possible to occur and the hypothesis of experimental errors should be completed discarded.

The functions XAZ=f(XAS) are fited by the Table Curve program.

Construction of Kielland Plots.
The Kielland quotients were obtained according to the equation: log Kc'= Co + 2C1XAZ', where Co=logK_B^A, 2C1 is Kielland coefficient and Kc' is the corrected selectivity coefficient. XBZ was considered to be XBZ = 1- XAZ. ZB is the charge of the ferric ion, +3, and, supposing that only sodium ions are exchanged, ZA was considered to be +1. The experimental points of the isotherms were fitted to equations and the functions XAZ=f(XAS) calculated with the help of the software package Table Curve and activity coefficients with PHRQPITZ program. In this way, the curve points were normalized and they were used to calculate Kc. Afterwards, the Kielland plots were constructed for isotherms.

With the ion exchange data it was possible to calculate the selectivity coefficients and the respective Kielland plot of each isotherm. The mathematical algorithm performed to obtain the Kielland plot improves the visualization of the straight lines. It indicates that the exchange sites have the same energy. Nevertheless, the mineral analysed display several exchange sites of different energies. As the Kielland plots are linear, we believe that only one kind of site acts in the exchange process with Fe^{+3}. The straight lines present positive values of C1, indicating that "W" is exothermic and that the Fe^{+3} ions tend to cluster. The degree of exchange increases as the temperature increases, because it apparently inhibits this tendency to cluster. It is observed that the higher temperature, the more regular the distribution of the Fe^{+3} ions through the sites. Thus, the zeolite is able to shelter a greater quantity of these ions, elevating the inflexion point.

All Kielland plots show non-linear shape for the Fe^{+3} cations investigated. This feature was already expected since NaA framework has sites with distinct energies (Barros, 2003) [5].

3.3. The determination of the optimum urea/ferric ratio.

The change of the pH of each solution with R=06, 10 and 60 with reaction time are shown in fig.3.

The pH rate rise rapidly with R and the final pH of the solution became high. The rate of the pH rise slowed down at pH 4-5 in the solution with R= 6 and 10. These phenomena were not observed in the solution with R=60.

The products formation to distinct pH values from the solutions with R=6 and 60 is shown in fig.4 (a) and (b).

From the pH measurements results of products, β-FeO(OH) was produced at pH 1-2 and magnetite at pH 4-

6. α-FeO(OH) was recognized from the solutions with R=6.

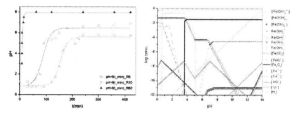

Fig.3. The relation between change of pH and the reaction time for R=urea/ Fe^{+3} mole ratio.

Fig. 5. Thermodynamic calculations of the concentrations of Fe^{+2} and Fe^{+3} species formed at different solution pHs [1].

Magnetite and α-FeO(OH) were produced at the same time, after that, α-FeO(OH) gradually decreased and was finally transformed into magnetite. α-FeO(OH) was not observed from the solution with R=60. Magnetite was precipitated rapidly since the rate of the rise of pH was quick with an increase in R.

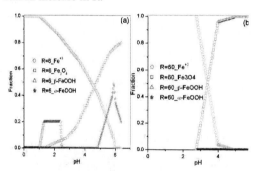

Fig.4. The formation of Fe^{+3}, Fe$_3$O$_4$, β-FeOOH and α-FeOOH to distinct pH values from solutions with (a) R=6 and (b) R= 60.

When R value was small, R=6, the rate of the rise of pH was slow and the hydrolysis of urea was a rate-determining step in the formation of magnetite, and the pH of the solution was kept at a constant value of pH about 4.6 during precipitation of magnetite.

When R value was large, the rate of rise of pH was rapid and the pH value quickly reached 7-8, being optimum ratio for the proposal goal.

3.4. Chemical conditions:
chemical composition of the zeolitic reaction media, pH, the redox potential, the concentration of the different salts used, the boundary conditions. In this way it is possible to establish, on a rational basis, the chemical conditions as well as the boundary conditions in which magnetite could be formed reproducibly as a pure phase [1].

3.5. The modified co-precipitation of $Fe_3O_4/\gamma - Fe_2O_3$ into zeolites

The chemical activation process of zeolites realized on chemical and thermal way, thereby: a) one gram of 200 mesh Na-A dried probe is calcinate at a 550-600 ℃ temperature in air or nitrogen current for the removal of the network-forming organic molecules. The activated zeolite probes were washed with distilled water and dried at 105 °C temperature in reaction column, fig.6.

When the ion-exchange isotherm indicates the preset concentration of ferric ions it is stoped the urea-ferric chloride mixture injection and concomitant begin the automated ferrous chloride injection at 98 °C thereby it is realized the ferrous ions absorption and according fig 3. the urea hydrolysis rising to the optimal pH value.

The co-precipitation direction it is realized according diagrams from fig.4 (a) and (b). For the urea excess hydrolysis and the anions removal it is needed the treatment of 30 minutes with steams at 120 °C temperature.

The modified zeolite is dried 20 minutes with air at 105 °C, extracted from column with air jet and transported into vessel for maturing through cyclone and subsequent 20 minutes calcination at 900 °C.

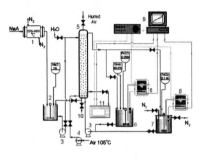

Fig.6. Schematic representation of the batch reactor for the modification of zeolites with magnetic oxides: 1- furnace at 550-600°C,, 2-chemical treatment vessel NaCl 1N, 3- electric water pump, 4- electric air blower, 5-reaction column, 6- mixing vessel of solution with R=(urea/Fe^{+3})=6,10,20,60, 7- mixing vessel for FeCl$_2$ 0.1M, 8- masterflex 7521 pump with pH-meter, 9-black star interface, 10-Sewage,11-Microwave Oven.

3.6. Characterization

The crystalline structure of the parent zeolites and modified zeolites were studied by X-ray diffraction analyses. Powder XRD patterns were recorded in the large angle (2θ=30-80°) region with DRON 2 diffractometer using a CuK$_{\alpha 1}$ radiation (U: 25KV, I: 20mA, U$_d$: 40V) with wavelength of 1,5405 Å and a step size of 0.01° in laboratory of structural analysis of faculty of physics by univ. "Al.I.Cuza" Iaşi.

DRUV-VIS spectra of the samples were recorded (200-700nm) on an SPEKOL ® 1300 UV VIS spectral photometer.

Magnetic measurements are carried out by vibrating magnetometer at the room temperature (293K) were recorded in Center of Applied Research in Physics and Advanced Technologies (CARPATH), Romania.

EPR measurements were performed at room temperature with a RADIOPAN spectrometer in X band.

4. Results and discussions

X-ray powder diffraction patterns for *I_ γ-FeO(OH)/A* and *II_ γ-FeO(OH)/γ-Fe$_2$O$_3$/A* samples showed

that the zeolite crystallinity was maintained during synthesis, Fig.7.(a) and (b).

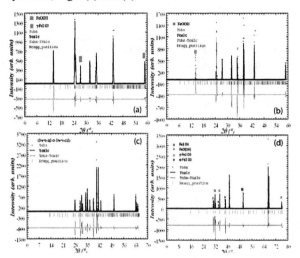

Fig.7. The XRD Spectra of
(a)- γ-FeO(OH)/γ-Fe$_2$O$_3$/A 0.9 mEgFe^{+3}/1gA
(b)- γ-FeO(OH)/A 0.15 mEgFe^{+3}/1gA
(c)-(Fe$_2$O)$^{+4}$/A 1.5 mEgFe^{+3}/1gA
(d)-Fe$_3$O$_4$/γ-FeO(OH)/γ-Fe$_2$O$_3$/α-Fe$_2$O$_3$/A 2.2 mEgFe^{+3}/1gA

In the case of *III_(Fe$_2$O)$^{+4}$*/A sample, *Fig7.(c)*, X-ray absorption spectroscopic measurements could prove the existence of small, slightly disordered iron (III) oxide nanoparticles and *IV_Fe$_3$O$_4$/γ-FeO(OH)/γ-Fe$_2$O$_3$/α-Fe$_2$O$_3$/A* sample, *Fig.7.(d)*, XRD confirms the formation of the ferrite and the partial loss of host crystallinity observed in X-ray diffraction patterns thereby suggests that the process of cluster formation involves aggregation within the large cages of the zeolite A and local destruction of the pores network.

The optical absorption spectra give the quantization effect of band gap. The DRUV-VIS , fig.8, of loaded samples show excitonic peak at ~250 nm, whereas the bulk FeO(OH), Fe2O3 and calcined FeO(OH), show strong maxima at 535, 560, and 515 nm, respectively [6, 7]. However, it is interesting to note that FeA composite also absorbs nearly in the same region (~250 nm) as that of the loaded samples [8, 9]. Thus, the occurrence of these exciton peaks in this range can be interpreted for both framework trivalent iron and iron oxide nanoparticles (non-framework) inside the mesopores. Hence, must be taken in deducing the information regarding the nature of the species on the basis of DRUV-VIS spectra alone. In present DRUV-VIS study, we have not observed the formation of iron oxides particles on the outer surface of A zeolite, unless iron salt concentration (0.01 M) is increased, which gives the absorption in the range of 350-550 nm taking into account, the observation of Abe et al. [8, 9].

We report the first observations of ferromagnetism above room temperature for dilute (<4 at.%) Fe-embedded A zeolite, Fig.9.(d). The Fe is found to carry an average magnetic moment of 0.16 μ$_B$ per ion. Our ab initio calculations find a valance state of Fe^{2+}and that the magnetic moments are ordered ferromagnetically, consistent with the experimental findings. We have obtained room-temperature ferromagnetic ordering in the powder form of the nanocomposite material. The unique

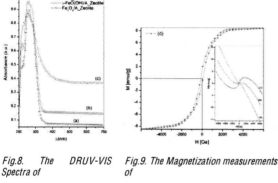

Fig.8. The DRUV-VIS Spectra of

(a)-$\gamma - FeO(OH)/\gamma - Fe_2O_3/A$,
(b)-γ-$FeO(OH)/A$,
(c)-$(Fe_2O)^{+4}/A$ Zeolite.

Fig.9. The Magnetization measurements of

(a)-γ-$FeO(OH)/\gamma$-Fe_2O_3/A _0.9 mEgFe^{+3}/1gA
(b)- γ-$FeO(OH)/A$ _0.15 mEgFe^{+3}/1gA
(c)- $(Fe_2O)^{+4}/A$ 1.5 mEgFe^{+3}/1gA
(d)- Fe_3O_4/γ-$FeO(OH)/$ γ-$Fe_2O_3/$ α-Fe_2O_3/A_2.2 mEgFe^{+3}/1gA

feature of our sample preparation was the low-temperature processing. When standard high-temperature ($T > 700$ °C) methods were used, samples were found to exhibit clustering and were not ferromagnetic at room temperature, *Fig.9.(a), (b) and (c)*. This capability to fabricate ferromagnetic Fe-embedded A zeolite semiconductors promises new spintronic devices as well as magneto-optic components.

The electrical conductivity of zeolites in their cation forms and with semiconductor clusters was studied as a function of temperature Fig.10.

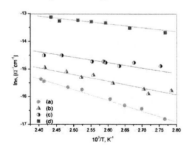

Fig.10. The electrical comductivity as a function of temperature and the iron charge

The formation of semiconductor clusters into the zeolite channels determines the increase in electrical conductivity and the sharp decrease in the activation energy of electrical conductivity, Table 1.

Table 1. The values of activation energy of electrical conductivity.

Sample	*(a)*	*(b)*	(c)	(d)
E_a, eV	3.02	1.86	1.51	1.05

The differences in ΔE_a values, calculated from the dependence $\sigma = f(10^3/T)$, can be related to the substitutional or interstitial Na impurities. The changes in electrical properties are due to transformations in the density of electronic levels as a function of the size, known as quantum size effects.

5. Conclusions

In the present study, the superparamagnetic iron oxide nanocomposite with particle size in the order of 200-400 mesh was targeted to satisfy its use in the spintronics domain. In most reports, however, the experimental conditions have been reached through a "trial and error" approach, with little consideration for theoretical basis of the solution chemistry of the reaction system. A rational approach has been adopted to synthesize the magnetic phases of $Fe_3O_4/\gamma - Fe_2O_3$ in the zeolitic cavities and channels based on the quantitative analysis of different reaction equilibriums. The results of the thermodynamic modeling [1], together with the available kinetic information of relevant reactions, have been used for the selection of the optimum experimental conditions.

Where traditional electronics are based on control of charge carriers (n- or p-type), practical magnetic semiconductors would also allow control of quantum spin state (up or down). This would theoretically provide near-total spin polarization (as opposed to iron and other metals, which provide only ~50% polarization), which is an important property for spintronics applications, eg. spin transistors.

References

[1] R. Bosîceanu, N. Sulițanu, JOAM, vol.10, (2008), p 3482-5486.

[2] WEIDENTHALER Claudia; ZIBROWIUS Bodo; SCHIMANKE Julia; YACHUN MAO; MIENERT Bernd; BILL Eckhard; SCHMIDT Wolfgang, Microporous and mesoporous materials ISSN 1387-1811, vol.84,n1-3,pp.302-317, 2005.

[3] Barros, M.A.S.D.; Machado, N.R.C.F. and Sousa Aguiar, E. F., Construção de Isoterma de Troca Iônica em Zeólitas Naturais, In. Encontro Regional de Catálise do Estado de São Paulo, 3, 1995. Campinas. Anais Campinas, 47-50 (1995).

[4] Barrer, R.M. and Klinowski, J., Theory of Isomorphous Replacement in Aluminosilicates, Phil. Trans. R. Soc. Lond., 285, 637-80 (1977).

[5] Barros et al.. 2003. Boletim Tecnico Cientifico do CEPENE 11 (1): 63-90.

[6] S.K. Badamali, P. Selvam, Stud. Surf. Sci. Catal. 113 (1998) 749.

[7] S.K. Badamali, A. Sakthivel, P. Selvam, Catal. Lett. 65 (200C 153.

[8] T. Abe, Y. Tachibana, T. Uematsu, M. Iwamoto, J. Chem. Soc., Chem. Commun. (1995) 1617.

[9] M. Iwamoto, T. Abe, Y. Tachibana, J. Moi. Catal. A 155 (2000) 143.

* Corresponding author: r1bosinceanu@yahoo.com

Evaluations of PP-g-GMA and PP-g-HEMA as a Compatibilizer for Polypropylene/Clay Nanocomposites

Byungwook Nam[1], Dong Hak Kim[2] and Younggon Son[3]

[1] Department of Applied Chemical Engineering, Korea University of Technology and Education, Cheonan, Chungnam 330708, S. Korea
[2] Department of Chemical Engineering, Soonchunhyang University, Asan, Chungnam 336745, S. Korea
[3] Advanced Materials Science and Engineering, College of Engineering, Kongju National University, Cheonan, Chungnam 331717, S. Korea

ABSTRACT

In this study, we investigated the performances of a hydroxyl ethyl methacrylate grafted PP (PP-g-HEMA) and a glycidyl methacrylate grafted PP (PP-g-GMA) as a compatibilizers in PP/clay nanocomposites. The compatibilizers were prepared by melt grafting with a radical initiator. Since the PP-g-MA is successfully and widely used in the PP/clay nanocomposites, we also studied three PP-g-MAs containing different amount of MA and having different molecular weights for a comparison. PP/clay nanocomposites compatibilized by the PP-g-HEMA and the PP-g-GMA show similar level of the clay interlayer distances with those of the PP-g-MAs. We also investigated the effect of molecular weights of the compatibilizers. In general, the compatibilizer of lower molecular weight was observed to exhibit lower performance as a compatibilizer. It is observed that increase of polar group content in the modified PP (PP-g-HEMA, PP-g-GMA and PP-g-MA) always accompanies the molecular weight reduction which deteriorates the mechanical properties. Thus, we prepared the PP-g-HEMA and PP-g-GMA by incorporation of a styrene comonomer. The compatibilizers (PP-g-HEMA-*co*-styrene and PP-g-GMA-*co*-styrene) thus obtained show good performance as a compatibilizer in the PP/clay nanocomposites. We observed that the PP/clay composites containing the PP-g-HEMA-*co*-styrene and the PP-g-GMA-*co*-styrene have very well balanced mechanical properties.

Keywords: PP/clay nanocomposites, PP-g-HEMA-*co*-styrene, PP-g-GMA-*co*-styrene

1 INTRODUCTION

In this study, we investigated a hydroxy ethyl methacrylate grafted PP (PP-g-HEMA) and a glycidyl methacrylate grafted PP (PP-g-GMA) as a compatibilizer in the PP/clay nanocomposites. Additionally, the effects of molecular weights of the PP-g-HEMA and the PP-g-GMA were also investigated. As far as we know, PP-g-HEMA has never been tried as a compatibilizer in the PP/clay nanocomposite. Though a few studies have reported the performance of PP-g-GMA [1], there has been no study on the effect of the molecular weight of the PP-g-GMA. It has been reported that the incorporation of styrene as a comonomer improves the grafting yield of GMA and MA onto PP [2-4]. Thus, we fabricated PP-g-HEMA and PP-g-GMA by incorporation of styrene comonomer and evaluated their efficiencies as a compatibilizer. Since PP-g-MA has been successfully and widely used as a compatibilizer for the PP/clay nanocomposites, it is used as a reference in this study.

2 EXPERIMENTAL

Materials
Two different PPs were used in this study. Commercial homo PP, used for fabrication of PP-g-HEMA and PP-g-GMA, was from Samsung-Total Chemicals. Metallocene catalyzed poly(ethylene-*co*-octene) was used as an elastomer in PP/clay nanocomposites to increase the impact strength. Organoclay (trade name: Cloisite 20A) based on dimethyl, dehydrogenated tallowquaternary ammonium was purchased from Southern Clay Products, Inc. Hydroxy ethyl methacrylate (HEMA), a glycidyl methacrylate (GMA), styrene and 1,3-bis-(t-butyl peroxy-isopropyl) benzene were from the Aldrich Chemicals.

Preparations of the compatibilizers
Preparations of the compatibilizers and the nanocomposites were carried out in a twin screw extruder. The powdery PP was tumble-mixed with a polar monomer (HEMA or GMA), 1,3-bis-(t-butylperoxy-isopropyl) benzene as an initiator and antioxidant (Iganox 1010) in a sealed plastic bag. The mixture was immediately put into a hopper of the twin screw extruder for compounding. The content of the polar monomer varied from 1.2 to 2.0 phr. The ratio of polar monomer/initiator was set to 10. The compatibilizers incorporated by the styrene comonomer were also prepared. In this paper, a code for the compatibilizer will be used as follows:
The code PP-g-HEMA2.0St2.0 represents a compatibilizer prepared from the mixture of PP of 100 g, HEMA of 2.0 g, styrene of 2.0 g and an initiator 0.2 g. In

the same rule, PP-g-GMA1.2 represents the compatibilizer prepared from the mixture of PP of 100 g, GMA of 1.2 g and an initiator of 0.12 g.

Preparations of the nanocomposites

Since the nanocomposites prepared by masterbatch method provide better micro scale dispersion [5-9], all composites were prepared by a two-step masterbatch method. The first step was blending the compatibilizers and clay at a ratio of 7/3 wt./wt. (compatibilizer/Cloisite20A), creating a masterbatch. In the second dilution step, each of the masterbatch samples was dry mixed with predetermined amounts of PP and elastomer and subsequently metered to the screws. The amounts of the masterbatch, PP, and elastomer were 50, 35, and 15 (by weight), respectively. Therefore, the percentages of clay, compatibilizer, PP, and elastomer in the nanocomposites were 15, 35, 35, and 15, respectively.

Characterizations

Wide angle X-ray data were collected on a Rigaku D/MAX-IIIC X-ray diffractometer (Cu Kα radiation, wavelength = 1.5418) with accelerating voltage of 40 kV. Diffraction spectra were obtained over a 2θ range of 1.28 – 10°.

Tensile properties and Flexural properties were tested using a universal mechanical testing machine (Model Hounsfield H25KS). Notched Izod impact tests were carried out at a room temperature. Melt flow index (MFI) was measured at 230°C and 2.16 Kg load (ASTM D1238).

3   RESULT & DISCUSSION

It is known that achieving the homogeneous dispersion of the silicate layers (well exfoliated or at least intercalated structure) within the polymer matrix is a key factor to achieve good physical properties of the polymer/clay nanocomposites. Thus, we investigated the level of the polymer intercalation by the XRD. Fig. 1 shows the XRD patterns for the PP/clay nanocomposites. The vertical line in Fig. 1 represents the position of the d001 peak of the organo clay (Cloisite 20A). Based on the XRD data, the d001 spacings (clay interlayer distance) are calculated and the values are shown in each curve. As can be seen, all composites except the PP/PP-g-MA2.6/C20A show the intercalated structures. The PP/PP-g-MA2.6/C20A does not show the clay characteristic peak implying the exfoliated structure. The d001 peaks of the other nanocomposites are observed to shift to lower angles compared to the peak of the organo clay, showing the clay interlayer distances are enlarged from 2.4 nm to around 2.8 nm.

Since the performance of PP-g-MA as a compatibilizer is affected by the level of the MA contents and the molecular weight, we investigated three PP-g-MAs containing different MA contents and molecular weights. It is known that PP-g-MA having higher MA content provides

better dispersion of the silicate layers. Thus, it is understandable that PP/PP-g-MA2.6/C20A, which has the highest MA content among the PP-g-MAs investigated, shows the well exfoliated structure. On the other hand, the other composites compatibilized by PP-g-MAs (PP-g-MA0.5 and PP-g-MA0.8), PP-g-GMA1.2 and PP-g-HEMA1.2 provide the intercalated structures. As Fig. 1 shows, those four compatibilizers significantly increase the clay interlayer distance in the composites. The PP-g-GMA1.2 and PP-g-HEMA1.2 show similar levels of the interlayer distances to those of PP-g-MA0.5 and PP-g-MA0.8.

Selective physical properties of the PP/clay nanocomposites are shown in Table 1. When looking at the properties of the nanocomposites compatibilized by the PP-g-MAs, it is seen that the stiffness (tensile strength and flexural modulus) tends to increase with the MA content. This is because the dispersion of clay interlayers increases as the MA content increases. Due to economic reasons, the grafting of MA onto PP is generally carried out in an extruder with aid of radical initiators. To increase the MA content, the amounts of MA and the initiator should generally be increased. The radical initiator enhances not only the radical formation but also the chain scission of the polymer. Therefore, the molecular weight of PP-g-MA produced by the melt grafting method decreases with the MA content. Additionally, the maleated polymer becomes more brittle as the MA content increases. Though PP/PP-g-MA2.6/C20A exhibits the exfoliated nanostructure due to the highest polarity, the miscibility between PP and PP-g-MA2.6 is expected to be low because the high level of the polar MA group prevents a good mixing of PP-g-MA2.6 with the non polar PP, which may deteriorate the impact property of the nanocomposite [10] Though the stiffness of PP/PP-g-MA2.6/C20A is highest among composites investigated, its toughness (elongation at break and Izod impact strength) is poor because of the two reasons mentioned above. It is noteworthy that MFIs of PP/PP-g-MA2.6/C20A are much lower than those of PP/PP-g-HEMA1.2/C20A and PP/PP-g-GMA1.2/C20A, implying that the melt processibilities of PP/PP-g-MA2.6/C20A are poor compared with those of PP/PP-g-HEMA1.2/C20A and PP/PP-g-GMA1.2/C20A. This is a somewhat unexpected result because the MFI of PP-g-MA2.6 is higher than those of PP-g-HEMA1.2 and PP-g-GMA1.2. It is well known that rheological properties of the nanocomposites are very sensitive to the dispersion of nano-fillers. It has been frequently reported that the complex viscosity of the nanocomposites increases (especially at low frequency), the Newtonian plateau at low frequency weakens, and the yielding phenomenon becomes more pronounced as the level of clay dispersion increases. Therefore, MFI (sensitive property to low shear rate) of the nanocomposites compatibilized by PP-g-MA of the highest MA content is lowest as a consequence of the well-dispersed nanostructure.

In order to be used as a commercial plastic, property balance is important. Thus, PP-g-MA2.6 (with higher MA content) is regarded to be an inappropriate compatibilizer for PP/clay nanocomposite. It should be noted that there is a proper level of MA content in PP-g-MA to have the nanocomposite of well-balanced properties. In this study, the proper level of MA content is around 0.8 wt. % since the property balance of PP/PP-g-MA0.8/C20A is best among three nanocomposites compatibilized by the PP-g-MA. Many studies reported similar levels of MA content to this study [11]. Thus, we will compare the performance of PP-g-HEMA and PP-g-GMA with PP-g-MA0.8 afterward.

The physical properties of PP/PP-g-HEMA1.2/C20A are well balanced, as can be seen in Table 1. PP/PP-g-HEMA1.2/C20A is superior to PP/PP-g-MA0.8/C20A in most properties; especially, MFI. PP/PP-g-GMA1.2/C20A exhibits a similar level of physical properties with those of PP/PP-g-MA0.8/C20A. Thus, it can be concluded that the PP-g-HEMA and PP-g-GMA provide good performances as a compatibilizer in PP/clay nanocomposites.

The effects of the polar group content and molecular weights of PP-g-HEMA and PP-g-GMA were also investigated. Fig. 2 is the XRD patterns for the PP/clay nanocomposites compatibilized by the PP-g-HEMA and PP-g-GMA having various molecular weights and polar group contents. The physical properties of the corresponding nanocomposites are shown in Table 2. It is seen that the interlayer distance of PP/PP-g-HEMA/C20A and PP/PP-g-GMA/C20A is affected by the polar group content. PP/PP-g-HEMA1.5/C20A and PP/PP-g-GMA1.5/C20A exhibit slightly larger interlayer distance than those of PP/PP-g-HEMA1.2/C20A and PP/PP-g-GMA1.2/C20A. This is due to the compatibilizing effect of more polar PP derivative. However, PP/PP-g-HEMA2.0/C20A and PP/PP-g-GMA2.0/C20A show the smallest interlayer distance in spite of the highest polar group content. This is an inconsistent result with that of PP-g-MA.

Intercalation ability of a compatibilizer onto clay interlayers is affected by the interaction with clay and the viscosity of the compatibilizer. Higher polarity tends to increase the intercalation ability due to the better interaction with the organosilicates. Higher molecular weight tends to increase the intercalation ability due to the higher viscosity and stress during melt mixing since the higher stress can easily break the clay tactoids. As mentioned earlier, the molecular weight of the polar monomer grafted PP produced by the melt grafting method decreases with the amount of the grafted polar group. Therefore, two opposite effects exit in PP-g-HEMA and PP-g-GMA containing higher amounts of HEMA and GMA. In these polymers, the effect of molecular weight (thus viscosity) is inferred to be more dominant than the polarity.

The stiffness of PP/PP-g-HEMA1.5/C20A and PP/PP-g-GMA1.5/C20A is slightly higher than that of the PP/PP-g-HEMA1.2/C20A and the PP/PP-g-GMA1.2/C20A due to the better dispersion of the clay interlayers by the higher polarity. However, their toughness is decreased with the amount of the polar group. The processibility is also decreased with the amount of the polar group due to the same reason mentioned earlier. The physical properties of PP/PP-g-HEMA2.0/C20A and PP/PP-g-GMA2.0/C20A are observed to be the lowest. Again, it should be noted that the highest amount of a polar group does not guaranty the highest performance of the nanocomposite.

We observed that increasing the amount of the polar group in the modified PP derivative always accompanies the drastic decrease in molecular weight. It is reported that the incorporation of styrene as a comonomer improves the grafting yield of MA and GMA onto PP [2-4]. It is also expected that the incorporation of styrene comonomer prevents the chain scission in a certain degree, and consequently the polar group grafted PP having the high molecular weight and high grafting yield can be obtained. Therefore, we prepared the PP-g-HEMA and PP-g-MA by the incorporation of the styrene.

In this study, we evaluated the performance of PP-g-HEMA and PP-g-GMA as a compatibilizer in the PP/clay nanocomposites. PP-g-HEMA is first tried as a compatibilizer in the PP/clay nanocomposites. We observed that PP-g-HEMA and PP-g-GMA have similar levels of the compatibilizing efficiency with PP-g-MA. PP-g-HEMA gives slightly better efficiency than PP-g-GMA. In most previous studies on PP/clay nanocomposites, an improvement of the clay dispersion and increased stiffness of the nanocomposite by using the polar monomer grafted PPs have been highlighted and focused on. However, losing the toughness is always accompanied when the polar monomer grafted PPs are employed. We solved this problem by an incorporation of the styrene comonomer in the melt grafting step.

4 CONCLUSION

We evaluated the performance of PP-g-HEMA and PP-g-GMA as a compatibilizer in PP/clay nanocomposites. It was observed that PP-g-HEMA and PP-g-GMA show similar levels of compatibilizing effect with the PP-g-MA. It was also observed that an increase of the polar group content leads to a decrease of the molecular weight in the polar group grafted PP, which deteriorates the physical properties of the PP/clay nanocomposites. It was found out that the preparations of PP-g-HEMA and PP-g-GMA by incorporation of styrene monomer solve this problem. PP-g-HEMA-co-styrene and PP-g-GMA-co-styrene thus obtained show very good performances as a compatibilizer in the PP/clay nanocomposites.

REFERENCES

[1] Lofez-Quintanilla ML, Sanchez-Valdes S, Ramos De Valle LF, Medellin-Rodriguez FJ, (2006) *J Appl Polym Sci* 100:4748

[2] Cartier H, Hu GH,(1998) *J Polym Sci A* 36:1053

[3] Cartier H, Hu GH, (1998) *J Polym Sci A* 36:2763

[4] Sun Y, Hu GH, Lambla M, (1995) *J Appl Polym Sci* 57:1043

[5] Treece MA, Zhang W, Moffitt RD, Oberhauser JP, (2007) *Polym Eng Sci* 47:898

[6] Betega De Paiva L, Morales AR, Guimaraes TR, (2007) *Mat Sci Eng A* 447:261

[7] Lertwimonum W, Vergnes B, (2006) *Poly Eng Sci* 46:314

[8] Modesti M, Lorenzetti A, Bon D, Besco S, (2005) *Polymer* 46:10237

[9] Modesti M, Lorenzetti A, Bon D, Besco S, (2006) *Polymer Degradation and Stability* 91:672

[10] Kurokawa Y, Yasuda YH, Kashiwagi M, Oya A, (1997) *J Mat Sci Lett* 16:1670

[11] Dubnikova IL, Berezina SM, Korlev YM, Kim GM, Lomakin SM, (2007) *J Appl Polym Sci* 105:385

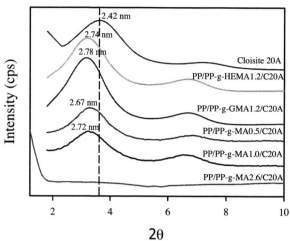

Figure 1: X-ray diffraction patterns of the PP/clay nanocomposites. Vertical line represents the position of the d_{001} peak of the organo clay (Cloisite 20A). Numbers shown in each peak represent the clay interlayer distances of the composites.

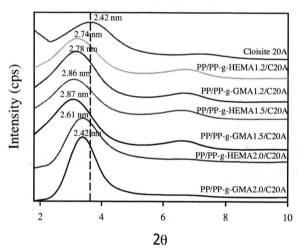

Figure 2: X-ray diffraction patterns of the PP/clay nanocomposites compatibilized by PP-g-HEMA and PP-GMA with different polar group contents. Vertical line represents the position of the d_{001} peak of the organo clay (Cloisite 20A). Numbers shown in each peak represent the clay interlayer distances of the composites.

Table 1: Physical properties of PP/clay nanocomposites.

Samples properties	PP/PP-g-HEMA1.2/C20A	PP/PP-g-GMA1.2/C20A	PP/PP-g-MA2.6/C20A	PP/PP-g-MA0.5/C20A	PP/PP-g-MA0.8/C20A
MFI (g/10min) of the nanocomposite	8.5	4.9	0.7	2.6	4.5
MFI (g/10min) of the compatibilizer	15	18	170	50	60
Tensile Strength (Kgf/cm²)	248	257	290	260	230
Elongation at break (%)	380	110	10	40	210
Flexural modulus (Kgf/cm²)	15200	16600	25300	17500	14500
Izod impact Strength (Kgf cm/cm)	21.3	20.7	3.5	5.9	20.5

Table 2: Physical properties of PP/PP-g-HEMA/clay and PP/PP-g-GMA/clay nanocomposites.

Samples Properties*	PP/PP-g-HEMA1.2/C20A	PP/PP-g-GMA1.2/C20A	PP/PP-g-HEMA1.5/C20A	PP/PP-g-GMA1.5/C20A	PP/PP-g-HEMA2.0/C20A	PP/PP-g-GMA2.0/C20A
MFI of the nanocomposite	8.5	4.9	5.1	2.6	5.7	5.2
MFI of the compatibilizer	15	18	49	42	82	85
Tensile Strength	248	257	250	263	216	232
Elongation at break	380	110	184	101	97	115
Flexural modulus	15200	16600	18700	19400	15600	16000
Izod impact Strength	21.3	20.7	8.2	11.2	6.4	9.1

*The units are same as Table 1.

Synthesis and Application of Nano-Size Polymer Colloids

C. J. Tucker,* T. H. Kalantar, D. M. Meunier, J. K. Harris

The Dow Chemical Company, Midland, Michigan, USA, *CJTucker@Dow.com

ABSTRACT

Over the last 50+ years The Dow Chemical Company has sponsored numerous research programs involving the preparation and use of nano-size colloidal systems. The nano-colloids used range from soft (liquid microemulsions) to solid (polymeric) materials and from organic to inorganic. Application areas include catalysis, polymer latices, delivery systems, consumer and personal care products, reaction media and nanoporous media. In this paper a fundamental study and two related application areas will be reviewed. The first topic covered is fundamentals of microemulsion formulation, followed by their application in the preparation of nano-sized polymer colloids. Also reviewed will be the use of amphipathic block copolymers to form highly stable unilamellar and multilamellar vesicles and their use as delivery systems for sensitive functional ingredients.

Keywords: microemulsion, nanoparticle poragen, vesicle, diblock surfactant, encapsulation

1 INTRODUCTION

The work described in this paper has it's origins in efforts at The Dow Chemical Company to develop a fundamental understanding of the formulation and application of surfactants and microemulsions in a wide variety of applications ranging from cleaning products to reaction media [1]. As part of this work, formulation strategies were developed to form microemulsions which were stable over broad temperature ranges and oil/water volume ratios. Efficient surfactants and cosurfactants were also identified and synthesized. Using this fundamental understanding, microemulsion approaches were utilized for the synthesis of nano-sized polymer colloids as small as 5 nm. These colloids were subsequently dispersed in a polymer films for the preparation of nanoporous films for use in microelectronic devices. Studies designed to identify more efficient surfactants led to the development of ethylene oxide/butylene oxide block copolymers. In addition to their utility for the preparation of microemulsions, we have found that these block copolymers form highly stable vesicles. These vesicles have shown promise in the stabilization and delivery of actives in a number of applications.

2 MICROEMULSION FORMULATION

Microemulsions have been used in a broad range of applications ranging from drug delivery to enhanced oil recovery. Work has focused on the development of stable microemulsion formulations requiring the minimum amount of surfactant while maintaining phase stability under use conditions. In order to study the efficiency of surfactants for the preparation of microemulsions we characterized the efficiency of surfactant systems in bicontinuous (Type III) microemulsions for equal volumes of water and oil. This characterization can typically be done through the generation of "fish" diagrams such as that shown in Figure 1 [2, 3]. The surfactant efficiency is defined by the minimum surfactant concentration at which a single phase microemulsion is obtained.

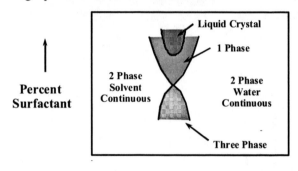

Figure 1: "Fish" diagram used for determining surfactant efficiency.

Fish diagrams can also be used to study the effect of oil structure, oil/water ratio, electrolyte content, temperature and other factors on phase behavior. An example of one such diagram is shown in Figure 2. In this study it was shown that it is possible to produce microemulsions using as little as 0.5wt% surfactant at equal volumes of oil and water.

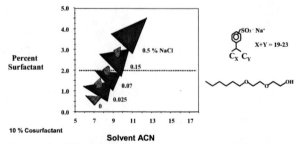

Figure 2 : "Fish" diagram (top portion) used for model system prepared using 10% diethylene glycol hexyl ether. Numbers in chart represent percent NaCl present

Using the "fish" diagram approach, numerous surfactants have been characterized for their efficiency in forming microemulsions. In addition to commercially available surfactants experimental surfactants such as ethylene oxide/butylene oxide (EO/BO) block copolymers have also been characterized. Figure 3 shows surfactant efficiency for a series of EO/BO copolymers with equivalent EO content in a water-hydrocarbon system. In this series, surfactant efficiency increases with molecular weight up to a total polymer molecular weight of ~4500. Presumably this increase in efficiency is due to an increasing tendency of the polymers of higher polymer MW's to reside at the oil-water rather than the air-water interface.

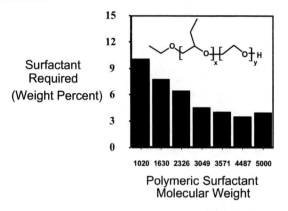

Figure 3: Efficiency vs. MW (EO/BO Polymeric Surfactants).

This fundamental understanding of surfactants and microemulsion phase behavior was used to develop polymerization conditions for the preparation of polymer nanoparticles such as those described in the Section 3. In addition the understanding of the relationship between surfactant structure and interfacial activity has led to the development of surfactants for other colloidal systems such as the EO/BO block copolymer vesicles described in Section 4.

3 POLYMER NANOPARTICLE TEMPLATES FOR POROUS FILMS

A continuing need in the microelectronics industry is, as interconnect linewidths decrease, for better interlayer insulating materials with ever lower dielectric constant κ to enable increased device densities and faster processing. SiLK[TM], a polyarylene interlayer dielectric developed by Dow, has dielectric constant κ=2.65; efforts to develop lower κ polyarylene materials have focused on decreasing κ by introducing porosity. This section will overview a key approach taken in these laboratories to develop the desired porosity [4, 5, 6].

Requirements for the porosity include: uniformly small pore size and pore distribution, closed porosity, and the porous material must exhibit good fracture toughness and a high pore yield.

A variety of attempts were made to introduce porosity by including a polymeric poragen with the matrix monomer, B-staging, spin coating, and heating to cure and devolatilize the poragen. Linear and star-shaped polymers with ethynyl functionality, as well as densely side-chain functionallized polyacrylates were employed to this end [4,7].

A key issue with these poragens is that pore generation is highly process sensitive; the result is that pore size, shape and density are highly dependent on the cure temperature profile. In retrospect, this is not surprising as pore formation depends on self-assembly of the poragen in a polymer matrix in which it may not have been fully soluble.

It was hypothesized that an approach employing a pre-organized spherical poragen, in which one poragen would template one pore, would allow good control of pore shape, size, and size distribution, with low process sensitivity.

A microemulsion polymerization approach [8, 9, 10] was employed to prepare the desired poragens [11, 12, 13]. Key poragen requirements include ionic and elemental purity (C, H, O only), a low volume swell factor (VSF) to avoid particle-particle overlap, ability to yield spherical pores, robustness to thermal processing, with low residue yield and volume (~weight) average diameter (D_v) controllable from 5-20 nm. These requirements pointed to the use of styrene-based poragens made with non-ionic components only.

The microemulsion polymerization method also offered a great deal of synthetic flexibility in terms of allowing flexibility of monomer content, surface functionality, etc. This would allow tuning of properties such as the thermal devolatilization profile, and minimization of devolatilization residue.

A typical polymerization [11] was conducted in semi-continuous mode as follows: a styrene-divinylbenzene plus a functional monomer mix was continuously fed into an aqueous solution of a nonionic surfactant at 30 °C. The redox initiator (hydrogen peroxide/ascorbic acid) was co fed. The product was precipitated with an equal volume of methyl ethyl ketone followed by centrifugation. The resulting solid was resuspended and washed repeatedly. The total residual metal content was 1.3 ppm and the residue on thermal treatment at 500 °C under N_2 was 0.37 wt%.

The nanoparticles were characterized by size exclusion chromatography (SEC) using viscometric detection [11]. The retention volume at each time increment was related to a Log(IV*M) using the universal calibration [14]; M could then be calculated at each retention volume from the measured intrinsic viscosity (IV). The VSF was also be determined from these data. From the polymer density ρ, weight average molecular weight M, and Avogadro's number L, D_v was calculated as follows:

$$D_v = 2*\left(\frac{0.75M}{\pi L \rho}\right)^{1/3}$$

The Mark-Houwink plot had slope $\alpha \sim 0$ across the molecular weight distribution, showing that the suspended polymer nanoparticles are spherical (Figure 4) [15].

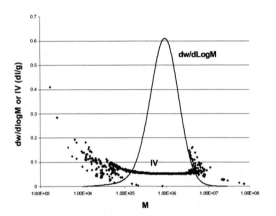

Figure 4. Molecular weight distribution and Mark-Houwink plot for a 15.4 nm diameter polymer nanoparticle.

By this method, the particles have a "collapsed" (unswollen) D_v=15.4 nm. Dynamic light scattering in THF shows a mean particle diameter, after correcting for solvent swelling (VSF=2.10) and the M_z/M_w ratio, of 17.5 nm. SEC using MALLS detection yielded a collapsed particle diameter of 16.6 nm. The particle sizes determined by these different methods are in very good agreement. Results using these methods typically also correlate well with hydrodynamic chromatography and TEM results.

The isolated dry particles form a dusty solid that may be swollen and dispersed in appropriate solvents. The proton NMR spectrum shows signals typical for a crosslinked styrenic polymer; a resonance typical of poly(ethylene oxide) is present also. This resonance did not disappear on repeated re-precipitation, suggesting that a small amount of the nonionic surfactant grafts to the particle surface during nanoparticle formation.

Typically, the poragens are B-staged with arylene monomers in solution to yield a prepolymer containing the poragen [11, 13]. The resulting solution is diluted and spin coated on a silicon wafer, then cured to 430 °C resulting in a nanoporous film (Figure 5) with approximately spherical pores 8~10 nm diameter, in high yield.

Porous films made using 12 and 7 nm poragens exhibited successively smaller pore diameters of 8-10 and 5-6 nm respectively on film formation, as shown by SAXS. Pore morphology was relatively insensitive to the cure profile.

Spherical polymer particles as small as 4.9 nm (M_w=28,500) have been made by this method. The synthetic process has been successfully scaled from a lab scale of a few grams to multiple kilograms in a pilot plant.

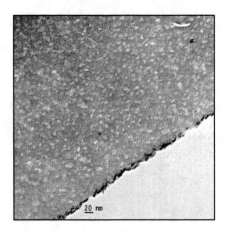

Figure 5. Porous polyarylene film prepared using a 12 nm poragen. κ = 2.1-2.2.

4 EO/BO BLOCK COPOLYMER VESICLES

Surfactant materials can self-assemble into a wide variety of meso-scale structures depending on surfactant composition, extent of hydration and aqueous environment [16]. One such structure is a vesicle; a hollow, spherical nano-particle composed of a bilayer surfactant wall and an aqueous core. These structures are naturally produced from phospholipids and are the basic template of biological cells. However, formation of vesicles from phospholipids can be a laborious and complicated process [17]. Work was initiated in the 1970's to explore whether or not these structures could be useful for encapsulation of actives, but they proved to be too fragile for most commercial applications [18].

Figure 6. Cryo-SEM micrograph of $BO_{12}EO_{12}$ MLV in 1.0 wt % aqueous solution taken 10 weeks after formation. The particle is ~7 μm in diameter.

Surfactants prepared from ethylene oxide/ butylene oxide diblock copolymers have been shown to spontaneously produce vesicles on hydration [19]. Proper selection of block lengths ensures vesicle formation, while the hydration regime can produce vesicles that range from

~80 nm to 1 mm in diameter from the same copolymer composition [20]. Vesicles may be multi-lamellar (MLV) or uni-lamellar (ULV) structures depending on synthesis conditions. Like phospholipid-based structures, post-structure formation processing can be used to convert MLV's into ULV's.

Research on the potential of these materials for actives encapsulation has shown that both hydrophilic and hydrophobic materials can be sequestered [21]. These structures are particularly robust, easy to load and easy to process, thus offering significant potential for use in many encapsulation applications. Work done by Jordan *et al.* show that the materials can encapsulate, protect and deliver vitamin A and other selected organic acids [22].

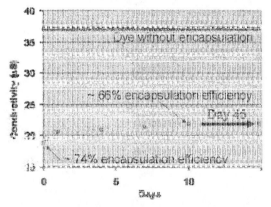

Figure 7: Release profile of an ionic dye in $BO_{12}EO_{12}$ MLV in 1.0 wt % aqueous solution as measured by conductivity.

5 CONCLUSIONS

Fundamental work done over a number of years in the areas of microemulsion phase behavior and surfactant structure activity relationships has been applied to a number of technology areas of interest to The Dow Chemical Company. Nanoparticle poragens made by microemulsion polymerization have found application in low κ dielectric film formation, and vesicles composed of novel block copolymer surfactants can be formed spontaneously, and have utility in personal care applications.

REFERENCES

[1] J. Klier, C. Tucker, T. Kalantar, D. Green, *Advanced Materials* **2000**, *12*, 1751.

[2] M. Bourrel, R. Schechter, *Microemulsions and Related Systems*. Marcel Dekker, New York, 1988.

[3] M. Kahlweit, et al. *J. Colloid Interface Science* **1987**, *118*, 436.

[4] Silvis, H. C.; Hahnfeld, J. L.; Niu, Q. J.; Radler, M. J.; Godschalx, J. P.; Stokich, T. M.; Lyons, J. W.; Kalantar, T. H.; Hefner, R. E., Jr.; Yontz, D. J.; Landes, B. G.; Ouellette, K. B.; Kern, B.; Marshall,

J. G.; Bouck, K. J.; Syverud, K.; Leff, M.; Bruza, K. J. *PMSE Preprints* **2002**, *87* 427.

[5] Niu, Q. J.; Strittmatter, R. J.; Hahnfeld, J. L.; Silvis, H. C.; Landes, B. G.; Waeterloos, J.; Meyers, G. F.; Kalantar, T. H. *PMSE Preprints* **2004**, *90* 101.

[6] Martin, S.J.; Godschalx, J.P; Mills, M.E.; Shaffer, E.O.; Townsend, P.H. *Advanced Materials* **2000**, *12*, 1769.

[7] Niu, Q. J.; Martin, S. J.; Godschalx, J. P.; Townsend, P. H. Molecular bottle-brushes as templating agents for nanoporous SiLK* dielectric films. Abstracts of Papers, 224th ACS National Meeting, Boston, MA, United States, August 18-22, 2002, PMSE-118.

[8] Xu, X. J.; Siow, K. S.; Wong, M. K.; Gan, L. M. *Langmuir* **2001**, *17*, 4519.

[9] Larpent, C.; Tadros, T. F. *Colloid and Polymer Science* **1991**, *269*, 1171.

[10] Larpent, C.; Bernard, E.; Richard, J.; Vaslin, S. *Macromolecules* **1997**, *30*, 354.

[11] Kalantar, T. H.; Niu, Q. J.; Tucker, C. J.; Domke, C. H. US 2004054111 A1 20040318.

[12] Mubarekyan, E.; Kalantar, T. H.; Silvis, H. C.; Niu, Q. J. US 2004253442 A1 20041216.

[13] So, Y.-H.; Niu, Q. J.; Townsend, P. H.; Martin, S. J.; Kalantar, T. H.; Godschalx, J. P.; Bruza, K. J.; Bouck, K. J. WO 2003070813 A1 20030828.

[14] See for example: http://www.viscotek.com/the-gpc-visc.aspx

[15] W. Funke, W.; Walther, K. *Polymer Journal* **1985**, *17*, 179.

[16] Israelachvili, J. N. *Intermolecular and Surface Forces*, 2nd Ed., Academic Press, San Diego CA, 1992, 345-372.

[17] Fendler, J. H. *Membrane Mimic Chemistry: Characterization and Application of Micelles, Microemulsions, Bilayers, vesicles, Host-Guest Systems and Polyions*; John Wiley and Sons, NY, NY, 1982, 25-26.

[18] Bangham, A.D.; Standish, M. M.; Watkins, J. C. *Journal of Molecular Biology*, **1965**, 13, 238.

[19] Harris, J. K.; Bruening M.; Rose, G. D. *Langmuir*, **2002**, *18*, 5337.

[20] Harris, J. K. Doctoral Thesis, Michigan State University, 2003.

[21] Harris, J. K.; Zhang, X; Jordan, S. L.; Rose, G. D.; Garcia, J. M. WO 2008024703 A1 20080228.

[22] Zhang, X.; Jordan, S. L.; Harris, J. K.; Rose, G. D. Encapsulation technology for skin care applications using multi-lamellar vesicles of ethylene oxide/butylene oxide diblock copolymers. Abstracts of Papers, 234th ACS National Meeting, Boston, MA, United States, August 19-23, 2007.

ZnO:Polydiacetylene Films as Chromic Sensors

J.L. Zunino III*, A. Patlolla**, D.P. Schmidt*, Z. Iqbal**

*U.S. Army Armaments Research Development and Engineering Center's (ARDEC)
Materials, Manufacturing & Prototype Technology Division
Picatinny Arsenal, New Jersey, USA
james.zunino@us.army.mil; daniel.p.schmidt@us.army.mil

**New Jersey Institute of Technology
Department of Chemistry and Environmental Science, Newark, NJ 07102
anithapatlolla@gmail.com; iqbal@adm.njit.edu

ABSTRACT

Films of the conjugated polydiacetylenes (PDAs) can function as chromatic temperature, chemical and stress sensors, and can be prepared by spin-coating from monomer solutions followed by polymerization using UV- or γ- irradiation. PDAs thus have the potential for advanced applications as sensors

Nanocomposite ZnO:PDA and ZnO/ZrO$_2$:PDA thin films have been prepared and their structural properties have been characterized by a variety of techniques including Fourier transform infrared spectroscopy, Raman spectroscopy, differential thermal calorimetry (DSC) and extended x-ray absorption fine structure (EXAFS) measurements. We have conducted testing and calibration of the thermal and stress sensing properties of the composite films using optical density measurements. It was also observed that the fluorescent red phase transforms to a non-flourescent blue phase in the presence of trace amounts of chemical vapors. Results on the sensing of trace amounts of vapors of tri-nitrotoluene (TNT) and dinitrotoluene (DNT) by optical densitometry and fluorescence, and efforts to enhance the chemical sensitivity of TNT detection by addition of single wall carbon nanotubes to the polydiacetylene (PDA), will also be discussed.

Keywords: polydiacetylenes, sensors, chromic, Army, thermal

1 INTRODCUTION

Materials that change color in response to external stimuli are of great interest to the U.S. Army. Chromic materials refer to those materials which radiate, lose color, or change properties induced by external stimuli. Different stimuli result in different responses in the material being affected. "Chromic" is a suffix that means color, so chromic materials are named based on the stimuli energy affecting them, for example:

photochromic - light

thermochromic – heat
piezorochromic – pressure
solvatechromic – liquid
electrochromic – electricity/voltage conflicts

Polydiacetylenes (PDAs) are a series or conjugated polymers which can undergo thermochromic transitions when exposed to stimuli. This chromic change is caused by the shortening of the conjugation length of the polymeric backbone [1]. Furthermore, the thermochromic transition temperature is a function of the side groups on the polymeric backbone structures. By changing the side groups, repeatable response to set stimuli are possible. This allows these materials to function as sensors. There are several applications of these PDAs as chromic sensors for Army applications.

The U.S. Army ARDEC is interested in chromic materials for sensing applications. As part of the Active Coatings Technologies Program, a research team was established to investigate chromic materials for active sensor systems and develop active/smart paints and coatings.

Films of these conjugated polydiacetylenes (PDAs) can function as chromatic temperature, chemical, and stress sensors, and are of particular interest to the Army.

2 TECHNICAL DISCUSSION

2.1 Materials Development

The conjugated PDA films are prepared by spin-coating monomer solutions followed by polymerization using UV- or γ- irradiation. The polymer backbone of PDAs is comprised of alternating double bonded (ene)-triply bonded (yne) groups which are responsible for intriguing temperature, stress and chemically-induced chromatic (blue, non-flourescent to red, fluorescent) transitions. Figure 1 illustrates basics of thermochromism in PDAs.

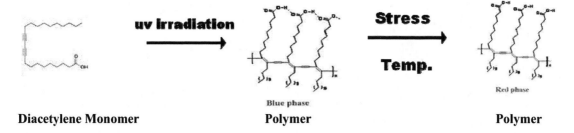

| Diacetylene Monomer | Polymer | Polymer |

Figure 1. Illustration of monomer to polymer transition and Blue to Red Phase transition of PDAs.

Research on the properties of PDAs in the past has primarily focused on their large non-linear optical response originating from their long conjugated backbones [2]. More recently interest has shifted to the chromatic transitions exhibited by the PDAs [3]. The PDA monomers polymerize rapidly under external stimuli, such as uv-irradiation, and some of the polymers undergo chromatic phase transitions due to changes in the backbone structure and conformation [4]. The monomers are normally colorless and become increasingly blue with polymerization. Color in PDAs is a result of π to π^* electronic transitions associated with the C≡C-C≡C diacetylene backbone. Irreversible blue to red chromatic transitions occur in many PDAs under external stimuli. This also means that these PDAs will undergo phase changes in two stable states, the low temperature blue state and the high temperature red state.

Although these thermochromic polydiacetylene possess the desired properties of interest to the U.S. Army, in order for them to be acceptable solutions for military applications, the overall cost must be acceptable versus the added utility they will provide. If the PDAs and resulting films, coatings, paints, etc. are too expensive, difficult to apply, or causes other additional cost/time burdens, the Army will not implement them. Therefore, cost effective monomer and polymers to develop these PDAs were analyzed, tested, and selected.

The Army research team has found that ZnO uniquely forms a weak complex with acidic diacetylenes containing carboxylic groups: 10, 12-pentacosadiynoic acid (PCDA), 10, 12-docosadiynedioic acid (DCDA), and 10, 12-tricosadiynoic acid (TCDA), resulting in reversal of the chromatic blue to red transition. They have also observed that this reversibility is slowed down by mixing the PDA with an alloy of ZnO with ZrO_2, thus allowing the mixed oxide-PDA films to function as elapsed time-temperature indicators.

Three commercially available thermochromic PDA monomers were used as the active component in paint formulations developed for the Army. The PDA monomers selected have the following chemical structures:

- $CH_3(CH_2)_{11}$-C≡C-C≡C-$(CH_2)_8$-COOH
  (10,12 pentacosadiynoic acid-PCDA)

- $CH_3(CH_2)_9$-C≡C-C≡C-$(CH_2)_8$-COOH
  (10,12 tricosadiynoic acid-TCDA)

- HOOC-$(CH_2)_8$-C≡C-C≡C-$(CH_2)_8$-COOH
  (10,12 docosadiynedioic acid -Bis-1)

PCDA and TCDA were found to show a blue to red transition with brighter contrast in the polymer phase compared with Bis-1. PCDA and TCDA were therefore incorporated in most of the formulations used. Bis-1 was added to either PCDA or TCDA in the higher temperature formulations to fine-tune the transition temperature.

The team discovered that nanosized Zinc oxide (ZnO) or a heat treated oxide alloy of ZnO and zirconium oxide (ZrO_2) could be mixed into this formulation to control the reversibility of the chromatic transition.

2.2 Understanding effects of ZnO on Polydiacetylenes

Nanocomposite ZnO:PDA and ZnO/ZrO_2:PDA thin films were prepared and their structural properties were characterized by a variety of techniques including FTIR, Raman spectroscopy, DSC and EXAFS measurements. FTIR was used to understand the role that pure ZnO has in affecting PDA thermochromisms. Figure 2 illustrates the effect.

Preliminary EXAFS results on poly-PDCA metal oxide nanocomposites show changes in the mean square disorder in Zn-Zn distances in the ZnO-poly-PCDA complex relative to pure ZnO (Figure 3). This is consistent with the bonding interactions between ZnO and poly-PCDA and supports the results indicated by the FTIR spectra.

Raman spectroscopy is another method used to characterize and understand the effects of ZnO, ZrO_2, ZnO/ZrO_2 added to the PDAs being used. Figure 4 shows Raman Spectra of both the pure and mixed oxides of interest. A peak at 425 cm^{-1} of the ZnO in mixed ZnO/ZrO_2 shows a loss of intensity caused by the alloying. PDAs head group interactions with mixed oxides are different than those of pure ZnO, helping to understand the effects the oxides have and how to tailor the response of the materials to different stimuli.

Figure 5 and Figure 6 show the C≡C effect on the PDA backbone induced by addition of ZnO acting with the –COOH of the head groups (see at ~2260 cm^{-1} in Figure 6). Pure ZrO_2 does not affect the irreversibility of the chromatic transition, but thermally alloying ZrO_2 with ZnO was found to substantially slow down the rate of conversion of the red to the blue phase. Initial x-ray studies indicated no change in crystal structure of ZnO on thermal treatment with ZrO$_2$.

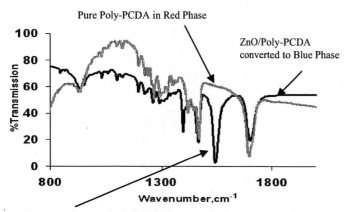

Pure Poly-PCDA in Red Phase

ZnO/Poly-PCDA converted to Blue Phase

Infrared line @ 1540 cm^{-1} in presence on ZnO is assigned to backbone C=C mode activated by ZnO interaction with head groups

Figure 2. FTIR showing role of ZnO in PDA chromism.

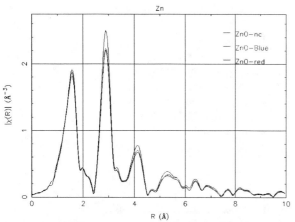

Figure 3. EXAFS spectra results of ZnO showing plots of ZnO nanocrystal (nc), Red Phase & Blue Phase.

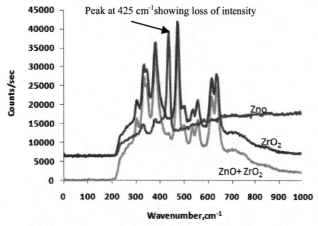

Peak at 425 cm^{-1}showing loss of intensity

Figure 4. Raman Spectra of pure and mixed oxides.

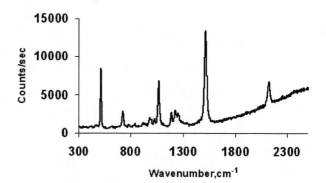

Figure 5. Pure Poly-PCDA in Red Phase (Raman 785 nm excitation).

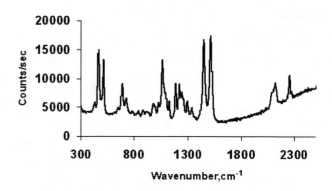

Figure 6. ZnO-poly PCDA in Red Phase converted to Blue Phase (Raman 785 nm excitation).

Both reversible and irreversible systems are possible. ZnO interactions can also take place with the head groups to repair hydrogen bonds and thus re-from the Blue Phase. Furthermore, reversing of the Red Phase to Blue Phase can be slowed down by using a mixed oxide of zinc and zirconium, ZnO/ZrO$_2$, in compositions near Zr$_{1-x}$Zn$_x$O$_y$ where x=0.4. This interaction is believed to lead to the reversibility of the red phase and sizeable increase of the chromatic transition temperature in the poly-PCDA-ZnO nanocomposites. By understanding the effects of ZnO and ZrO$_2$ and their interactions with the head groups of PDAs, the research team is able to control the reversibility and irreversibility of the chromic materials being developed.

2.3 Chromaticity of ZnO- and ZrO$_2$ Containing Polydiacetylene Nanocomposites for Sensing

The effect of ZnO and ZrO$_2$ on PDA and the resulting measurable chromaticity of nanocomposite ZnO:PDA has led to the development ZnO:PDA films as chromatic sensors. By adjusting the formulations these materials can be modified to respond to thermal stimuli at a desired "trigger" temperature. The ability allows for thermal indicating paints, coatings, and films to serve as sensors for thermal exposure monitoring, as well as other applications.

While the "trigger" temperature will provide an easy "go/no go" indication of an item's exposure to thermal conditions, the "soak" time or duration of exposure is important as well. As mentioned above, the addition of nanosized ZnO in pure form and alloyed with ZrO_2 can cause the red phase to revert to the blue phase at variable rates. Therefore paint formulations with added ZnO and mixtures of ZnO with ZrO_2 can be used to monitor the cumulative time of exposure in certain temperature ranges by measuring the chromaticity of the material. An optical densitometer can be used to accurately measure the resulting chromaticity. Chromaticity is a quantitative measure of the vividness or dullness of a color.

This provides both accurate visual cumulative time information as well as quantitative cumulative time information allowing ZnO:PDAs to function as thermal indicating paints, coatings, films for thermochromic sensing applications.

Cumulative time of exposure in multiple temperature ranges can be sensed and optically detected by mixing small amounts of the time-temperature sensitive diacetylene monomers, and ZnO:PDAs into the coating matrices. This allows for the thermal indicating paints, coatings, sensors, etc. to "remember" prior day exposures and monitor thermal cumulative thermal dwell or soak times.

Thermal indicating paints are being investigated by the Army for munitions monitoring, bake-off testing, and other applications where soldiers' safety can be increased by providing early warning indicators of potential thermal exposures and off-gassing conditions.

It was also observed that fluorescent Red Phase transforms to a non-flourescent Blue Phase in the presence of trace amounts of certain chemical vapors. There have been successful results on the sensing of these trace amounts of vapors, including tri-nitrotoluene (TNT) and dinitrotoluene (DNT), by optical densitometry and fluorescence.

TNT, DNT, and related stimulants (toluene) expresses the ability to open up the hydrogen bonds of the Blue Phase head groups and causing a predictable and controlled Red Phase transition. An effort to enhance the chemical sensitivity of TNT detection by infiltrating blue PCDA into high surface area microporous zeolite and addition of single wall carbon nanotubes is underway (Figure 7).

This may allow stand-off detection for soldiers, first responders, etc. of IEDs and other potential dangers.

5. CONCLUSIONS

Polydiacetylene polymers that undergo chromic transitions when exposed to stimuli have great potential for cost effective sensing applications. Since the chromic transition is a function of the modification of head groups on the polymeric backbone structures, by changing these groups, repeatable response to set stimuli are possible. Understanding the effects of ZnO and ZrO_2 interactions with the head groups allows control of reversibility, irreversibility, and "triggered" response to desired stimuli.

This allows for the development of cost effective active or smart coatings, paints, films, etc. that can be utilized as sensors, indicators and warnings. The research being performed will directly and indirectly support the warfighter.

6. ACKNOWLEDGEMENTS

Support was provided by the U.S. Army Armaments Research Development & Engineering Command (ARDEC), Program Executive Office Ammunition (PEO Ammo), Logistics Research & Engineering Directorate (LRED), Active Coatings Technology Program, Tactical Ammunition Accountability, and the New Jersey Institute of Technology (NJIT).

REFERENCES

[1] A. Patlolla, Q. Wang, A. Frenkel, J. Zunino III, D. Skelton, Z. Iqbal, "Studies of the blue to red phase transition in polydiacetylene nanocomposites and blends," Materials Research Society Proceedings, Vol. 1190, (2009).

[2] D. B. Wolfet, M.S. Paley, D.O. Frazier, Proc. SPIE-Int. Soc. Opt. Eng. 150, 3123 (1997).

[3] R. W. Carpick, D.Y. Sasaki, M.S.Marcus, M.A. Eriksson and A.R. Burns, J.Physics: Condensed Matter 16, R679 (2004).

[4] J. Zunino III and Z. Iqbal, "Thermal indicating paints for ammunition health monitoring," SPIE, Proceedings SPIE 7646, (2010).

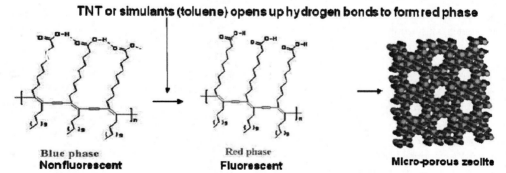

Figure 7. Opening of Blue Phase head groups by TNT or toluene will cause the Red Phase transition in zeolite at low concentrations.

Directed assembly of magnetic nanoparticles in polymers: the formation of anisotropic plastic films containing aligned nanowires.

Despina Fragouli[*], Raffaella Buonsanti[**,***], Giovanni Bertoni[****], Claudio Sangregorio[*****], Francesca Villafiorita[**,***], Carola La Tegola[**], Bruno Torre[****], Roberto Cingolani[*,****], Athanassia Athanassiou[*,**]

[*] Center for Bio-Molecular Nanotechnologies (CBN) of IIT@NNL-UniLe-Lecce, via Barsanti c/o STAMMS, Arnesano 73010 (LE), Italy, despina.fragouli@iit.it, athanassia.athanassiou@iit.it, roberto.cingolani@iit.it

[**] National Nanotechnology Laboratory (NNL) CNR-Istituto di Nanoscienze, via per Arnesano km 5, 73100 Lecce, Italy

[***] Scuola Superiore ISUFI, Università del Salento, Distretto Tecnologico ISUFI, via per Arnesano km 5, 73100 Lecce, Italy, raffaella.buonsanti@unisalento.it

[****] Fondazione Istituto Italiano di Tecnologia (IIT), via Morego 30, I-16163 Genova, Italy, giovanni.bertoni@iit.it, bruno.torre@iit.it

[*****] INSTM Research Unit-Dipartimento di Chimica, Università di Firenze, via della Lastruccia 3, 50019 Sesto F.no, Firenze, Italy, claudio.sangregorio@unifi.it

ABSTRACT

We present a facile technique for the preparation of plastic films containing aligned magnetic nanowires with anisotropic behavior. Solutions of magnetic nanoparticles mixed with polymers are drop casted on a substrate under the application of a weak magnetic field, and upon solvent evaporation the nanoparticles are assembled following the field's direction forming thus aligned nanowires into the polymer. The application of the magnetic field in different time intervals reveals the kinetics of the nanowires formation in the film, while microscopy techniques reveal their morphology and their positioning in the film. The resulting nanocomposites exhibit magnetic and mechanical anisotropy. The efficiency and universality of the method is proved by the assembly of the magnetic nanoparticles with diverse polymers, heterostructures and organic molecules resulting in composite materials for a wide range of applications. Finally, the method is successfully applied during the photopolymerization process, allowing the formation of patterns with aligned nanowires.

Keywords:, field directed assembly, magnetic nanoparticles, nanowires, nanocomposites, anisotropic behavior

1 INTRODUCTION

Nanoparticles (NPs) are intensively investigated for their unique properties, and as building blocks for new mesoscopic forms of matter. Assembling NPs in 1D arrays forming nanowires (NWs) is particularly challenging because highly anisotropic objects must be formed starting from essentially isotropic NPs. Magnetic NPs have a large domain of applications, for many of which high density and spatially oriented arrays of polymer embedded magnetic NWs has a critical role, due to their unique properties arising from their anisotropic morphology.

The assembly of magnetic NPs under controllable external magnetic fields (MFs) [1,2,3] during evaporation of a nanocrystal solution is an attractive option for the fabrication of 1D magnetic NWs supported on a film, due to its simplicity, effectiveness, and speed. The nanocomposite films of NWs attract particular attention because of their synergistic and hybrid properties derived from individual components [4]. Specifically, the resulting films demonstrate unique properties, since the physical properties of the polymers are combined with the highly anisotropic properties of the 1D NWs guiding to novel materials with directionally enhanced properties [5,6].The application of MFs for the alignment of fillers in polymers was mainly used in the carbon nanotubes composites, by applying high MFs (>10T) [6]. For the NWs formation in polymers, due to the magnetic orientation of particles, are used mostly micrometer sized particles [7,8], while the use of NPs resulted to formation of oriented supra-aggregates [5]. This work demonstrates that using polymers and MFs, 1D NWs with structural evolution from dots to wires are developed in a controlled manner directly in a polymer matrix. Besides, the nanocomposite films of the NWs may be formed directly upon photopolymerization resulting in patterned structures, while they may be of diverse polymer/NPs combinations. Depending on their individual properties, they exhibit anisotropic behavior of various physical properties, or response to external stimuli, opening thus new perspectives for applications.

2 EXPERIMENTAL METHODS

2.1 Samples Preparation:

Iron oxide colloidal nanocrystal spheres of 10 nm diameter were synthesized by modifying a wet-chemical synthetic approach [9]. Fe(CO)$_5$ was used as the precursor, while oleic acid, oleylamine and hexadecane- 1,2-diol were used as both reactants and capping molecules. Chloroform solutions containing 1%-5%wt of iron oxide and various polymers such as: acrylate copolymer poly(ethylmethacrylate-co-methylacrylate (PEMMA), polystyrene (PS), regioregular poly(3-hexylthiophene) (P3HT), poly (9-vinylcarbazole) (PVK), etc. were drop casted on glass substrates. The system was either subjected or not to a homogeneous MF (~160mT), produced by two permanent magnets, applied parallel to the substrate during the deposition and evaporation process. For the kinetics study the films were exposed to the MF for different time intervals. After the removal of the MF all films were left to dry overnight. The magnetic study has been performed on magnetic films in which NWs are present in the whole volume. For the formation of responsive films 8% wt photochromic molecules, 2H-1-benzopyran-2,2'-(2H)-indole, were dissolved in the chloroform solutions containing PEMMA and 1%wt of NPs. For the photopolymerization process one drop of chloroform solution containing 10%wt of NPs, 0.5%wt of photoinitiator (PI) (Irgacure 1700, Ciba) and 89.5%wt of methylmethacrylate was deposited on glass substrate under magnetic field. After 15min a small area of the drop (0.3 cm^2) was irradiated for 90min at 355nm with fluence F=10.5mJ/cm^2 (Nd:YAG laser, repetition rate=10Hz, Quanta-Ray GCR-190, Spectra Physics). After the photopolymerization, each sample was washed with methanol and then dried in ambient dark condition.

2.2 Samples Characterization:

The films were studied at low magnification with an optical microscope (Olympus BX41). Higher magnifications were obtained with a scanning electron microscope, SEM, (JEOL JSM-6490LA) detecting the backscattered electrons (BSE), and a 100kV transmission electron microscope, TEM, (JEOL JEM-1011) in bright field mode (BF), imaging thin film sections with thickness about 500 nm -120 nm, which were cut with a Leica EM UC6 Ultramicrotome. Freestanding films were characterized by holographic microscopy (Lyncèe Tec DHM 1000) in transmission mode. Sample topography and magnetic data are acquired by means of MFP-3D atomic force microscope, AFM, (Asylum Research, Santa Barbara CA) using magnetic force microcopy technique, MFM, in order to resolve magnetic structure with nanometer resolution. Magnetic forces are detected using non-contact cantilevers covered with a thin magnetic film on the tip side, and signal is detected in a modified dynamic-AFM mode.[10] The magnetic properties were studied with a Quantum Design Ltd. SQUID magnetometer. The temperature dependence of zero-field-cooled (ZFC)-field cooled (FC) magnetizations were collected in a static field of 10mT after cooling the samples down to 2 K in a zero magnetic field (ZFC) or in the same probe field of 10mT (FC). The angular dependence of the magnetization was measured at 300 K by rotating the

sample by 20° steps, while keeping the 5 mT applied field in the plane of the film. All the measurements were performed applying the field in the plane of the slab, first parallel and then perpendicular to the direction of the aligned NWs in the polymer matrix. All data were corrected for the diamagnetic contributions of the polymer and of the substrate which were separately measured and which were found negligible.

The mechanical properties, and the viscoelastic behavior of the films were studied with the dynamical mechanical analysis, DMA, (Q800 TA Instruments). The stress/strain evaluation was obtained by ramping a controlled force (stress) of 1N/min at constant temperature of 25°C through breaking.

The irradiation of the films containing photochromic molecules was conducted with a Nd:YAG laser (LDF-80-P, Alphalas) of energy 20μW, repetition rate=10Hz, and irradiation time 260sec at 355nm, and 40μW for 33min at 532nm.

3 RESULTS AND DISCUSSION

3.1 Formation Details:

The starting material is a polymer/NPs solution casted on a substrate, which upon application of a weak MF during solvent evaporation results to nanocomposite films, incorporating aligned NW arrays. In particular the magnetic NPs of the casted solution, follow the direction of the magnetic lines of the field during the evaporation of the solvent, resulting to their assembly and to the subsequent formation of the magnetic NWs in the polymer (figure 1a,c). On the contrary, casted solutions without the MF application result in a nanocomposite film containing amorphous aggregates (figure 1b).

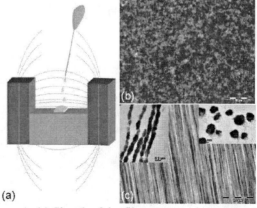

Figure 1: (a) Sketch of the films preparation upon MF. Microscope images of a 1%wt iron oxide/PEMMA film formed (b) without MF; and (c) upon application of a MF; insets: TEM images of film slices cut parallel (left) and perpendicular (right) to the NWs direction.

As shown at the TEM images of figure 1, the aligned structures are composed by the assembled NPs. The combined topographic and magnetic study with the MFM, showed that the NPs maintain their magnetic properties proving thus the formation of magnetic NWs (figure 2).

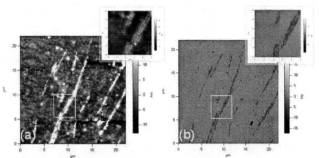

Figure 2: (a) Topography; and (b) magnetic image of a film slice, with thickness 200nm. The insets correspond to the area indicated with the yellow rectangular.[10]

The control of the NWs dimensions (max length 15μm, diameter ca. 80nm) and of their localization across the polymer is achieved by varying the NPs concentration, and the duration of the applied MF. Concerning the former, NPs concentration lower than 0.7%wt results in short elongated structures randomly aligned, whereas concentrations above 5%wt permit the formation of aligned NWs(figure 3).

Figure 3: (a) Microscope and TEM image of a 0.5%wt NPs/PEMMA film. (b) Image of a 5%wt NPs/PEMMA film

Concerning the latter, microscopy studies show that the NWs start forming on the liquid/air interface while the increase of MF application time gives longer NWs deeper in the volume of the films (figure 4), phenomenon attributed to the viscosity gradient change during evaporation [9]. (figure 4).

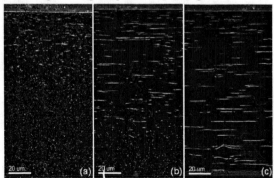

Figure 4: SEM images of cross sections of 1%wt iron oxide/PEMMA films formed after the MF application of 1min (a); 5min (b); and 15min (c) during solvent evaporation.

3.2 Anisotropic Properties:

A magnetic study of the films demonstrates that the NWs embedded in the polymer maintain the room temperature superparamagnetic behavior of the pristine NPs, since it is observed thermal

reversibility above ca. 140 K in the magnetization recorded as a function of temperature after ZFC and FC procedures. Moreover, the magnetizations of the temperature dependence of the structured films after application of a static MF parallel and perpendicular to the NW alignment direction shows a strong directional dependence with the temperature, with the magnetization higher in the parallel orientation. The anisotropy is better evidenced by measuring the angular dependence of the magnetic moment, which shows an increase of ca. 90% upon rotating the film from the perpendicular to the parallel orientation (figure 5). [9]

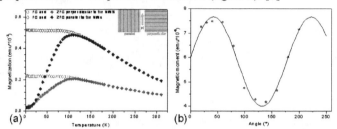

Figure 5: a) ZFC and FC magnetization curves measured with the direction of the external MF parallel and perpendicular to the aligned NWs as indicated in the inset. b) Angular dependence of the magnetic moment measured at 5 mT and 300 K. The film used is a 1%wt iron oxide/PEMMA.

Additionally the films also exhibit mechanical anisotropy. In particular, by studying the stress/strain behavior of films, applying a stretching force parallel and perpendicular to the NWs, it is shown that the films have different behavior. In detail, when the stress force is applied parallel to the aligned NWs, the strain of the film is much lower than in the case that the stress is perpendicular to the NWs alignment. These results demonstrate that the films in the direction of the alignment of the NWs are much more resistant.(figure 6)

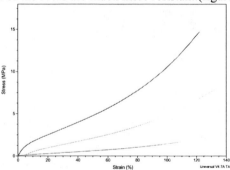

Figure 6: Stress strain curves of 1%wt iron oxide/PEMMA films without NWs (grey line), with NWs aligned parallel to the applied force (blue line) and with NWs aligned perpendicular to the force (red line).

3.3 Further Applications:

The method followed for the NWs formation in the polymer matrix may be applied to various NP/polymer combinations. NW plastic films are obtained by mixing iron oxide NPs with diverse polymers e.g. acrylates, conductive polymers etc., while the use of bigger iron oxide NPs (diameter 18nm), or colloidal heterostructures such as iron oxide/titanium dioxide, or iron oxide/gold spheres, form NWs in the polymer matrices (figure 7).

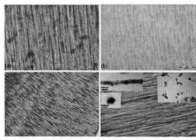

Figure 7: Microscope images of (a) 1%wt iron oxide/PS; (b) 1%wt iron oxide/P3HT; (c) 1% wt iron oxide/PVK; (d) 1%wt iron oxide-titanium dioxide/PEMMA.

Moreover, the mixing of the abovementioned solutions with other responsive molecules guide to the formation of responsive anisotropic magnetic films. Specifically, polymer/iron oxide solution was mixed with the photochromic molecule spiropyran (SP) and dried under MF. The SP is a molecule which upon UV irradiation is proved to photoisomerize to merocyanine (MC). MC reverts back to the SP form photochemically, using visible light irradiation [11], and this transformation between the isomers alters reversibly the macroscopic volume of their host polymer matrices [11]. The nanocomposite film was irradiated with UV and green light, and a holographic microscopy study was conducted after each irradiation. Figure 8a demonstrates 3D phase image of the film, where it is shown that the NWs are responsible for the phase modification, with apparent aligned stripes parallel to each other in the focused area. These stripes contain all the aligned NWs at the focal plane of examination. The line profile analysis of the whole phase image (figure 8b) before and after UV-vis irradiation, and the calculation of the mean width of the observed stripes reveals that they follow a reversible behaviour (figure 8c) attributed to the reversible behaviour of the photochromic molecules.

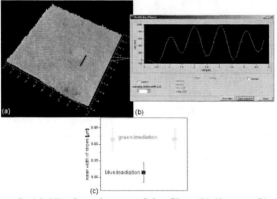

Figure 8: (a) 3D phase image of the film; (b) line profile of a selected area of (a); (c) the mean width of the stripes upon UV and green irradiation.

Finally, we achieved to form patterned nanocomposite films containing aligned NWs upon photopolymerization with pulsed UV irradiation. Figure 8a shows the microscope image of a monomer/PI/NPs film before the photopolymerization process. After irradiation of a small area, is formed a polymer film with embedded aligned NWs. The rest of the film contains amorphous aggregates after washing. (figure 8b)

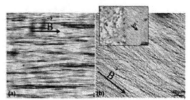

Figure 8: Microscope images of (a) the monomer/NPs film; (b) the photopolymerized film. Inset: an edge of the film showing the photopolymerized area (left) and the non-polymerized area (right).

4 CONCLUSIONS

We present a simple method for the formation and alignment of magnetic NWs in polymer matrices. The resulting plastic films exhibit magnetic and mechanical anisotropy and may be of various polymer/NPs combinations. The NWs can be formed and positioned in the films during the drying process of the NPs/polymer solutions under MF, or during photopolymerization process. They are either free-standing or may have the desired shape, making easier their incorporation to devices. In addition, the analytical results presented may be used as model systems for the reproduction of similar nanocomposite systems, or for the theoretical exploration of the forces and interactions taking place in the formation procedure. Finally, the possibility of having NWs of different lengths, aligned in various layers, can imply different physical behavior in the same film and opens the way for the application of this advanced multilayer structure in various devices.

REFERENCES

[1] Y. Sahoo, M. Cheon, S. Wang, H. Luo, E. P. Furlani and P. N. Prasad, J. Phys. Chem. B 108, 3380, 2004.
[2] Y. Lalatonne, L. Motte, V. Russier, A. T. Ngo, P. Bonville and M. P. Pileni, J. Phys. Chem. B 108, 1848, 2004.
[3] J. I. Park, Y. W. Jun, J. S. Choi and J.Cheon, Chem. Commun. 5001, 2007.
[4] G. Schmidt and M. M. Malwitz Curr. Opin. Colloid Interface Sci. 8, 103, 2003.
[5] J. Jestin, F. Cousin, I. Dubois, C. Ménager, R. Schweins, J. Oberdisse and F. Boué, Adv. Mater. 20, 2533, 2008.
[6] T. Kimura, H. Ago, M. Tobita, S. Ohshima, M. Kyotani and M. Yumura Adv. Mater. 14, 1380, 2002.
[7] J. E. Martin, E. Venturini, J. Odinek and R. A. Anderson, Phys. Rev. E 61(3), 2818, 2000.
[8] F. Fahrni, M. W. J. Prins, L. J. van Ijzendoorn, J. Magn. Magn. Mater. 321, 1843, 2009.
[9] D. Fragouli, R. Buonsanti, G. Bertoni, C. Sangregorio, C. Innocenti, A. Falqui, D. Gatteschi, P. D. Cozzoli, A. Athanassiou and R. Cingolani ACS NANO, DOI: 10.1021/nn901597a, 2010.
[10] D. Fragouli, B. Torre, G. Bertoni, R. Buonsanti, R. Cingolani and A. Athanassiou, MRT, DOI: 10.1002/jemt.20845, 2010.
[11] A. Athanassiou, M. Kalyva, K. Lakiotaki, S. Georgiou, C. Fotakis, Adv. Mater. 17, 988, 2005.

Soft plasma polymer coatings based on atomic polymerization

V. Cech, R. Trivedi, S. Lichovnikova, and L. Hoferek

Institute of Materials Chemistry, Brno University of Technology
Purkynova 118, CZ-61200 Brno, Czech Republic, cech@fch.vutbr.cz

ABSTRACT

Atomic polymerization (plasma polymerization) may be used to form soft polymer coatings of controlled physicochemical properties. Such coatings are utilized to construct functionally gradient and multilayered nanostructures of continuously or quasi-continuously varying physicochemical properties. Tailored nanostructures have high application potential for optical devices, photonic crystals, dielectric coatings, chemical sensors, separation membranes, biocompatible coatings, and polymer composites or nanocomposites of controlled interphase – composites without interfaces.

Keywords: thin films, soft coatings, plasma polymerization, nanocomposites

1 ATOMIC POLYMERIZATION

Plasma polymerization is a thin film-forming process based on atomic polymerization [1]. Monomer molecules are activated and fragmented during plasma process producing free radicals, electrons, and ions, and the highly reactive radicals recombine at substrate surface forming the thin film. The low-temperature plasma is a gentle but powerful tool for coating of substrates retaining their bulk properties. The substrate (glass, metal, ceramics, polymer, or composite) can be of planar, fibrous, or particulate form. The fragmentation of monomer molecule is controlled by the energy used for plasma process and the molecule can be fragmented in smaller species or even single atoms enabling the material construction at nanoscale synthesis [2]. The film thickness of deposited films can be from nanometers to microns.

2 SOFT PLASMA POLYMER COATINGS

Plasma-polymerized organosilicones constitute a class of materials with a rich and varied scientific background. This class of materials possesses a special characteristic, which distinguishes it from other plasma polymers – the ability to vary and control the degree of its organic/inorganic character (i.e., the carbon content) and the polymer cross-linking by the appropriate choice of fabrication variables [3,4]. This allows one to control many physicochemical properties over wide ranges resulting in an extraordinary potential for useful applications, which are only now beginning to be tapped. The organosilicon plasma polymers are widely recognized for their potential in optical, mechanical, and electronic applications. Mostly hard coatings are developed as protective layers, but we aimed at soft coatings using pulsed plasma. A reduction of plasma energy (power), but operated in several orders of magnitude, enabled us to control chemical composition and structure of soft plasma polymer coatings resulting in a wide range of mechanical, optical, and surface properties.

2.1 Chemical Properties

The elemental composition (Fig. 1) of soft plasma polymer coatings of hexamethyldisiloxane (pp-HMDSO), vinyltriethoxysilane (pp-VTES), and pure tetravinylsilane or TVS in a mixture with oxygen gas (pp-TVS/O_2) can be varied by deposition condition (RF power: 0.1 – 10 W, flow rate: 0.1 to 10 sccm, process pressure was 1 – 10 Pa).

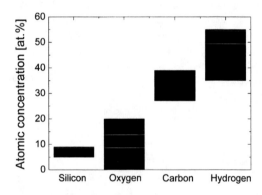

Figure 1: Elemental composition of soft plasma polymer coatings can be varied by deposition conditions.

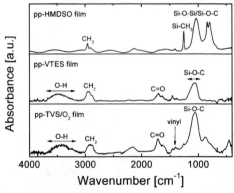

Figure 2: Infrared spectra characterize chemical structure of deposited films.

Chemical structure of coatings can be controlled by deposition conditions as well (Fig. 2). The functional group (e.g., ethoxy, vinyl) in monomer molecule can be preserved or eliminated and the new chemical groups ((e.g., hydroxyl, carbonyl) can be formed using suitable deposition conditions.

2.2 Mechanical Properties

The plasma polymer coating is deposited in a form of hydrogenated amorphous carbon-silicon (a-SiC:H) or carbon-silicon oxide (a-SiOC:H) alloy. Mechanical and optical properties of coatings are influenced by the bond energy and the level of polymer cross-linking. The Young's modulus of a-SiC:H alloy increases due to higher cross-linking at enhanced power (Fig. 3a). However, an incorporation of oxygen atoms into plasma polymer network reduces polymer cross-linking of a-SiOC:H alloy for RF power > 1 W. Similar behavior can be found for hardness (Fig. 3b).

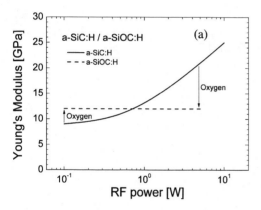

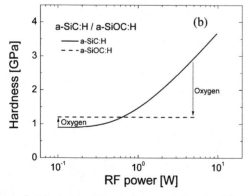

Figure 3: Mechanical properties controlled by RF power and amount of oxygen incorporated into plasma polymer network, (a) Young's modulus, (b) hardness.

2.3 Optical Properties

Dispersion curves for the refractive index and extinction coefficient are controlled by RF power and amount of oxygen in plasma polymer network. The refractive index increases with enhanced power due to increasing optical density of polymer (Fig. 4a) but decreases due to stronger Si-O-C bonding species that are formed instead of weaker Si-C bonding groups (Fig. 4b).

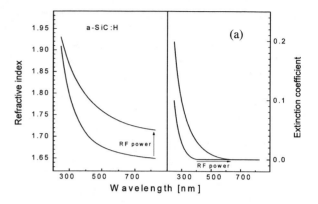

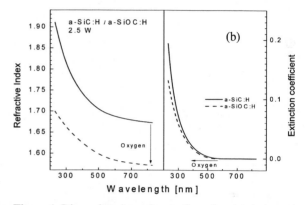

Figure 4: Dispersion dependences for refractive index and extinction coefficient (a) controlled by RF power and corresponding to a-SiC:H alloy, (b) controlled by amount of oxygen in polymer network of a-SiOC:H alloy.

2.4 Surface Properties

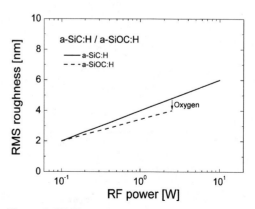

Figure 5: RMS roughness of polymer coatings.

Surface morphology of soft plasma polymer coatings is influenced by RF power, amount of oxygen, and film thickness (Fig. 5).

Wettability of polymer coatings can be influenced by RF power and functional groups at coating surface. The total surface free energy (Fig. 6a) consists of dispersion and polar components. The dispersion component increases due to decreasing concentration of vinyl groups (pp-TVS) in polymer with enhanced power (Fig. 6b). The polar component increases due to increasing concentration of polar groups (OH, C=O) (Fig. 6c).

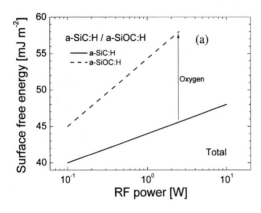

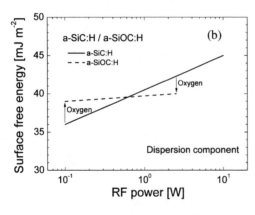

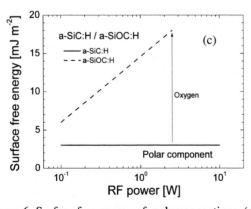

Figure 6: Surface free energy of polymer coatings, (a) total, (b) dispersion, and (c) polar component.

3 APPLICATION

Next progress was aimed at construction of functionally nanostructured films in a form of multilayers, with sharp of diffusive interfaces, and gradient coatings of physicochemical properties continuously varying from material A to material B. The deposited multilayered coatings were subjected to ellipsometric measurements that enabled us to distinguish individual layers in the layered structure and determine the layer thickness and its optical constants. Such tailored nanostructures may be used for optical filters, photonic crystals, and polymer composites or nanocomposites with controlled interphase.

3.1 Optical Devices

Soft plasma polymer coatings can be used to construct nanostructures from two polymers A and B using a layer-by-layer rotating system (Fig. 7a), or the coatings can be stack up varying properties of the individual layer stepwise (Fig. 7b), or the nanostructure of continuously varying properties can be constructed (linear gradient in Fig. 7c) to be tailored for specific application. Mechanical properties (Young's modulus, hardness) can be varied across the nanostructure in a similar way.

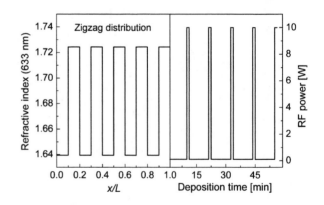

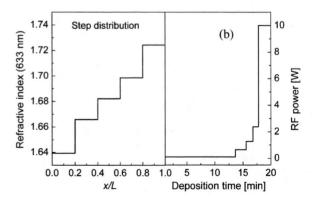

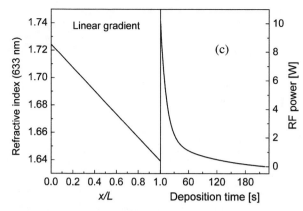

Figure 7: Refractive index across polymer coating and corresponding time-dependent RF power used to deposit the nanostructure, (a) zigzag distribution, (b) step distribution, and (c) linear gradient.

3.2 Polymer Composites

Soft plasma polymer coatings can be used to improve performance of polymer composites and enable to produce a new type of composites without interfaces. Soft plasma polymers were applied for nanocoating of glass fibers (GF) used as reinforcements for polyester composites and the functional nanocoating (pp-TVS/O2) overcame the industrial sizing (Fig.8).

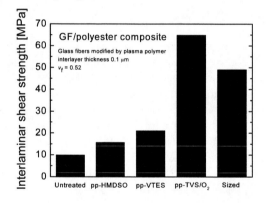

Figure 8: Interlaminar shear strength of GF/polyester composite reinforced by untreated, plasma coated, and industrially sized glass fibers.

3.3 Hybrid Coatings

Hybrid coatings may be deposited at higher powers to form soft grains in stiffer matrix (Fig. 9). The softer grains improve toughness of coatings and decrease the internal stress.

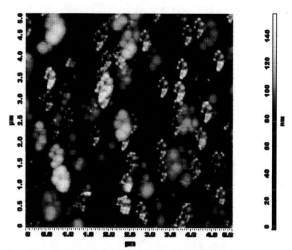

Figure 9: Hybrid coating including soft grains.

Acknowledgements: This work was supported in part by the Czech Ministry of Education, grant no. ME09061, grant no. MSM0021630501, and the Academy of Sciences of the Czech Republic, grant no. KAN101120701.

REFERENCES

[1] H. Yasuda, J. Polym. Sci., Macromol. Rev. 16 (1981) 199.

[2] V. Cech, Plasma polymer films: from nanoscale synthesis to macroscale functionality, in "Handbook of Nanostructured Thin Films and Coatings" (Ed. S. Zhang) Vol. 1, pp. 481-527, CRC Press, New York 2010.

[3] A. M. Wrobel and M. R. Wertheimer, Plasma-polymerized organosilicones and organometallics, in "Plasma deposition, treatment, and etching of polymers" (Ed. R. d'Agostino) pp.163-268, Academic Press, New York 1990.

[4] Y. Segui, Plasma Deposition from Organosilicon Monomers. in Proc. NATO ASI Plasma Processing of Polymers (Eds. R. d'Agostino, P. Favia, F. Fracassi) pp.305-319, Kluwer Academic Publ., Acquafredda di Maratea 1997.

Structural Control of Porous Polymeric Materials through Magnetic Gelation

M. Furlan, M. Morbidelli, M. Lattuada*

ETH Zürich, Institute for Chemical and Bioengineering
HCI F135, Zurich 8093, Switzerland, marco.lattuada@chem.ethz.ch

ABSTRACT

In this work, we introduce a novel procedure to create porous polymeric materials using a process that we named magnetic gelation. The method starts with the preparation of magnetic polymer nanoparticles, composed of magnetite nanocrystals dispersed into a polymer matrix via miniemulsion polymerization. Due to the superparamagnetic behavior of the nanocrystals embedded in the polymer matrix, the nanoparticles prepared in this manner develop strong and reversible dipolar interactions only in the presence of an external magnetic field. Both the size of the nanoparticles and the amount of magnetite nanocrystals encapsulated inside them has been varied, so as to tune the strength of the magnetic interactions. Dispersions containing a few volume percentages of these magnetic nanoparticles, which are stabilized by means of electrostatic interactions, have been partially destabilized through the addition of controlled amounts of electrolytes and their self-assembly and gelation behavior in the presence and in the absence of an external magnetic field have been investigated. In the absence of magnetic fields, the particles self-assemble into random fractal clusters, which eventually percolate to form a colloidal gel. Instead, when an external magnetic field is applied, the particles align themselves in columnar structures in the direction of the field. By tuning the strength of magnetic interactions, different extent of anisotropy in the final material are obtained. The materials obtained through this magnetic gelation process have been hardened though partial fusion of the nanoparticles and characterized by means of electron microscopy.

Keywords: Magnetic Nanoparticles, Self-Assembly, Porous Materials, Gelation, Anisotropic Materials, Structure Control.

1 Introduction

Self-assembly is one the most used bottom-up approached for the preparation of materials with advanced functionalities [1]–[3]. Many efforts have been dedicated in the last ten years to design nanoparticles that can spontaneously arrange into desired structures. Spherical nanoparticles are probably the most commonly investigated type of building block, because an enormous variety of materials can be easily prepared in nanoparticulate form, with good control over size and surface functionality [2], [4]–[6]. Among the different types of particles, magnetic nanoparticles are especially interesting since they display anisotropic interactions and superparamagnetism [4], [7]. Thanks to the strong dipolar interactions that they can develop, magnetic nanoparticles have the tendency to align into strong-like structures in the presence of na applied magnetic field [8].

In this work, we introduce a novel approach, from now on referred to as magnetic gelation, which allows us to produce polymer porous monoliths with controlled anisotropic structures. Our strategy consists in extending the conventional colloidal gelation process, which generates a highly porous network of particles, to the case of magnetic nanoparticles assembled in the presence of a magnetic field. First of all, magnetic composite nanoparticles have been prepared by encapsulating magnetite nanocrystals through miniemulsion polymerization inside a polymer matrix. In this way, a suspension of electrostatically stabilized magnetic polymer colloids is produced, which are then swollen with additional monomers and initiator. Subsequently, monoliths were generated by destabilization of the magnetic nanoparticles suspension through an ionic strength increase in the presence of a magnetic field. The combination of isotropic attractive Van der Waals interactions, repulsive electrostatic interactions and magnetic dipolar interactions lead to the formation of a percolated network with controlled anisotropic structure. Mechanical stability was brought to these gels by a post-polymerization step, forming covalent bonds between the nanoparticles and leading to a rigid monolithic structure, according to a procedure developed by Butté et al. [9]. The effect of the strength of dipolar interaction on the anisotropy of the material has been investigated. Scanning electron microscopy have been used to assess the degree of anisotropy achieved.

2 Experimental Procedures

The synthesis of Fe_3O_4 follows the coprecipitation method developed by Massart [10], with the only difference that ricinoleic acid was used as ligand to stabilize the nanocrystals. The synthesis of monolaurylmaleate surfactant follows the method developed by Kozuka et al. [11].

100nm magnetic polymer colloids with 50 w% of magnetite were prepared by dissolving 0.5 g sodium monolaurylmaleate surfactant in 48 ml of water and by mixing 7.98 g styrene, 2.9 g diethyl ether 0.42 g divinylbenzene, 0.25 g hexadecane and 8.4 g magnetite. The two solutions were then mixed using a magnetic stirrer and ultrasonicated in an ice bath to obtain a miniemulsion. Then, 0.12 g KPS in 2 ml water were added before conducting the polymerization at 70°C for four hours. After that, the reaction mixture was cooled down to room temperature and filtered through a filter paper. The final dry fraction of the latex was 19.02%.

The preparation of 150nm magnetic polymer colloids with 20 w% of magnetite followed a similar recipe, but using 36 ml of water and 21.2 g styrene, 2.35 g divinylbenzene, 0.49 g hexadecane and 6 g magnetite. Directly after sonification, the emulsion was diluted so that the organic content decreased from 40% to 20% by adding a proper amount of a solution containing 0.023 g sodium monolaurylmaleate surfactant in 60 ml of water.

The gelation process follows the method developed by Gauckler et al. [12]. A stock swelling solution was prepared by mixing 9 g styrene, 1 g divinylbenzene and 0.1 g AIBN. The swelling of the latex was carried out as follows: an amount of the above solution equal to the 20% or 40% in weight of the dry fraction of the latex was added dropwise and mixed for 4 hours. Afterwards, 1 ml of the swollen latex was put inside the magnetic field and it was mixed with 0.5 ml urea solution(4 M) and 0.5 ml urease solution (960 units/ml), while the gel samples for magnetic torque measurements were prepared by mixing 0.5 ml swollen latex, 0.25 ml urea solution (4 M) and 0.25 ml urease solution (960 units/ml). The final weight percentage of the gel were 11.7% for the gel obtained from Latex 1, 9.5% for the gel obtained from Latex 2 and 12% for the gel obtained from latex 3. After the gelation the temperature was risen to 70°C and the postpolymerization was carried out for 24 hours in the presence of the magnetic field by means of a heating jacket.

3 Results and Discussion

The aim of this work is to develop a new method for the production of porous polymeric materials with anisotropic structures. In order to achieve this goal, and take advantage of the existing technology in the preparation of magnetic nanoparticles, we have chosen to exploit the natural tendency of colloidal particles to form percolating networks, called gels, when their colloidal stability is compromised in the presence of strong attractive interactions and their particle volume fraction is in the range of 1-30% volume percentage [13]. Among the different types of mechanisms that can be used to

Figure 1: (a) SEM picture of a gel from latex 1 obtained in the absence of a magnetic field, showing no particles alignment.

induce the formation of colloidal gels, we have used the screening of electrostatic interactions on charged-stabilized colloidal nanoparticles, achieved by means of an ionic strength increase. By tuning the amount of charge stabilizing the nanoparticles, and the amount of electrolytes added, the kinetics of the gelation process can been regulated. The traditional colloidal gelation that has been utilized in this work has however been modified in two manners. First of all, we have used superparamagnetic nanoparticles, which can respond to the application of an external magnetic field by developing strong dipolar interactions. The second modification of conventional gelation process is the use of the reactive gelation process, developed by Butté et al. [9]. Since colloidal gels are usually very soft materials, the polymer latex nanoparticles are swollen with additional monomer and oil-soluble initiator prior to gelation. Once the gel is formed, the additional monomers is converted to polymer in order to link the nanoparticles with strong covalent bonds.

In order to avoid irreproducibility in the time required to reach the gel point and to improve the control of the gelation process a destabilization method developed by Gauckler's group [12] was used. This method uses the enzyme-catalyzed decomposition of urea, yielding one Ca^{2+} ion and two NH_4^+ ions per molecule of urea, i.e. producing one double charged ion and two single charged ions from one neutral molecule, leading to an increase of the ionic strength and therefore to the electric double layer compression.

For the experiments carried out in this work two latexes were prepared that differ in magnetite content and and particle size. All latexes were prepared via miniemulsion polymerization as explained in the Experimental section. The particle size and magnetite content (i.e. weight fraction of magnetite inside a particle) of the two latexes are different. The first latex (latex 1) has an average diameter of 100 nm with a magnetite content of 50%. This latex has been produced with the aid of a volatile solvent. The solvent facilitates the dispersion of the high amounts of magnetite in the monomer phase and evaporates during polymerization. The second latex (latex 2) has an average diameter of 150 nm with a magnetite content of 20%, and was produced by a first creating a monomer-magnetite miniemulsion in water with a disperse phase weight fraction of 40%, followed by a dilution as described in the experimental section.

(a) low magnification

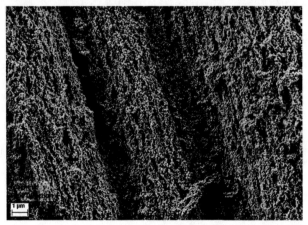

(b) high magnification

Figure 2: (a) and (b): SEM pictures of gel 2 at low and high magnification, respectively. Particles alignment is clearly visible in the SEM pictures. The direction of the applied magnetic field is indicated by the arrow.

Both latexes 1 and 2 were gelled in the presence of a magnetic field with a magnetic flux density of 1 T, yielding gels 1 and 2, respectively. For comparison purposes, Figure 1 shows the SEM picture of a typical gel obtained using latex 1 in the absence of a magnetic field. The structure of such a gel shows the expected disordered porous network of particles, without any anisotropy or specific particles alignment.

On the other hand, Figures 2(a) and 2(b) show two

(a) low magnification

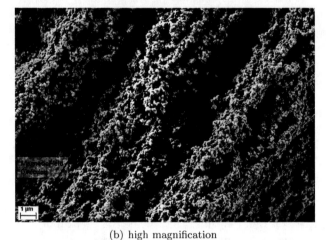

(b) high magnification

Figure 3: (a) and (b): SEM pictures of gel 3 at low and high magnification, respectively. Particles alignment is clearly visible in the SEM pictures. The direction of the applied magnetic field is indicated by the arrow.

SEM pictures of gel 1, obtained from nanoparticles containing a high amount of magnetite (50w%). The SEM figures clearly reveal the presence of particles aligned into parallel fiber-like structures in the direction of the applied magnetic field. The strings are anyway not made of perfectly aligned particles, because of both the scrambling effect caused by diffusion and because particles are all of exactly the same size with the same amount of magnetite. The anisotropy of this material is anyway especially apparent in the low magnification picture, where the broken portion of the material shows an aligned side surface and a random front surface. The front surface was perpendicular to the magnetic field during gelation and has a random structure similar to that of the gel obtained in the absence of a magnetic field (see Figure 1).

Figures 3(a) and 3(b) show two SEM pictures of gel 2. Once again, clear anisotropic structures of particles aligned in strings are observed. 3(a) obtained at low magnification clearly indicates how the structure of the magnetic gels resembles qualitatively the structure of wood fibers. The structure of gel 2 can be qualitatively be compared to that of gel 1, shown in Figure 2. Gel 1 nanoparticles contain 50% magnetite and have a diameter of about 100 nm while gel 2 nanoparticles contain 20% magnetite with a diameter of 150 nm, giving comparable amounts of magnetite per particle. Both gels present qualitatively similar anisotropic structures with aligned strings in the direction of the applied magnetic field, despite the different characteristics of the latexes they are made of.

4 Conclusions

In this work we have introduced a novel and reliable method, which we named magnetic gelation, to produce porous polymeric monoliths with controllable anisotropic structure. This method is based on a bottom up approach composed of four steps. The first steps is the production of magnetite nanocrystals stabilized with ricinoleic acid via salts coprecipitation in alkali solution. These nanocrystals are then dispersed, in a second step, in styrene so that after miniemulsion polymerization a stable magnetic latex is obtained. The suspension of magnetic nanoparticles is then swollen with additional monomer and initiator. The third step is the most important of the process, because is the step were the monolith is obtained and its structure, in particular its anisotropy, is controlled. The latex is gelled in a magnetic field by increasing the electrolyte concentration via an enzyme catalyzed reaction. When the gelation process is carried out in the presence of a magnetic field, the alignment of the magnetic nanoparticles in the direction of the applied magnetic field can be controlled by tuning the intensity of the field. In this manner, the structure of the final porous monolith can be tuned by from fully isotropic and random, when diffusion becomes the dominating mechanism, to strongly anisotropic in one direction, with particles mostly aligned in bundles of strings. In addition to the intensity of the applied magnetic field, the anisotropy of the structure can also be controlled by varying the size of the particles and the amount of magnetite dispersed into them. All these factors affect the strength of the dipole-dipole interactions experienced by the nanoparticles. In the final fourth step, after gelation occurred, the structure is glued together by converting the remaining monomer and thus chemically crosslinking the bonds among the particles. The structure of the monoliths has been qualitatively characterized by SEM analysis. It has been shown that the produced monoliths have anisotropic structure in the direction of the applied magnetic field, whereas in the perpendicular direction no alignment is observed and a random structure similar to that of a a gel prepared outside a magnetic field is obtained.

5 Acknowledgments

The authors acknowledge the support by the Electron Microscopy of ETH Zurich (EMEZ) and Frank Krumeich for the SEM pictures. The financial support from the Swiss National Science Foundation (SNF), with grant number 200021-116687, is gratefully acknowledged.

REFERENCES

[1] Hornyak, G. L., Tibbals, H. F., Dutta, J., and Moore, J. J., "Introduction to Nanoscience & Thechnology", CRC Press, 2009.

[2] Katz, E. and Willner, I., Angewandte Chemie-International Edition, 43, 6042-6108, 2004.

[3] Nie, Z. H. and Petukhova, A. and Kumacheva, E., Nature Nanotechnology, 5, 15-25, 2010.

[4] Gupta, A. K. and Gupta, M., Biomaterials, 26, 3995-4021, 2005.

[5] Antonietti, M. and Landfester, K., Progress in Polymer Science, 27, 689-757, 2002.

[6] Landfester, K., Angewandte Chemie-International Edition, 48, 4488-4507, 2009.

[7] Lu, A. H. and Salabas, E. L. and Schuth, F., Angewandte Chemie-International Edition, 46, 1222-1244, 2007.

[8] Jeong, U. and Teng, X. W. and Wang, Y. and Yang, H. and Xia, Y. N., Advanced Materials, 19, 33-60, 2007.

[9] Marti, N. and Quattrini, F. and Butte, A. and Morbidelli, M., Macromolecular Materials and Engineering, 290, 221-229, 2005.

[10] Massart, R., *IEEE* Transactions on Magnetics, 17, 1247-1248, 1981.

[11] Kozuka K. et al., US Patent No. 3980622: Polymerizable emulsifying agent and application thereof, 1976.

[12] Gauckler, L.J. and Graule, Th. and Baader, F., Materials Chemistry and Physics, 61, 78-102, 1999.

[13] Poon, W. C. K. and Haw, M. D., Advances in Colloid and Interface Science, 73, 71-126, 1997.

Polymer Nanocomposites: An Advanced Material for Functional Fibers

M. Joshi

Department of Textile Technology
Indian Institute of Technology, IIT Delhi 110016, India
Email: mangala@textile.iitd.ernet.in
Tel. No. 91 11 26596623, Fax No. 91 11 26581103

ABSTRACT

Polymer nanocomposites have recently gained a great deal of attention because of the much superior properties in terms of   increased strength and modulus, improved heat resistance, decreased gas permeability and flame retardance at very low loadings of < 5-wt % of nanofillers, as compared to conventional composites. Nanocomposite fibres that contain nanoscale embedded rigid particles as reinforcements show improved high temperature mechanical property, thermal stability, useful optical, electrical, barrier or other functionality such as improved dyeability, flame retardance, antimicrobial property etc. This paper discusses the recent developments in this area and also presents some highlights of the research and development activity undertaken at Textile Department, IIT Delhi, in this area.

Key words: polymer nanocomposites, functional fibers, nanoclay, POSS

1 INTRODUCTION

Textile Industry is likely to be hugely impacted by Nanotechnology, as research involving nanotechnology based finishes and coatings to improve performances or to create new functionality on textile materials has gained a big impetus / momentum across the globe. The current R&D trends show a clear shift to nanomaterials as a new tool to improve properties and gain multifunctionalities for textile materials. Organized nanostructures as exhibited by either fibers, nanocoatings, nanofinishing, nanofibers and nanocomposites seem to have the potential to revolutionize the textile industry with new functionality such as self cleaning surfaces, conducting textiles, antimicrobial properties, controlled hydrophilicity or hydrophobicity, protection against fire, UV radiation, etc. All this can be achieved without affecting the bulk properties of fibers and fabrics. The market for textiles based on nanotechnology is expected to reach US$115 billion by 2012 [1].

Polymer nanocomposites are the advanced new class of materials with an ultrafine dispersion of nanofillers or nanoparticles in a polymeric matrix, where at least one dimension of nanofillers is smaller than about 10nm. The volume and influence of the interfacial interactions increases exponentially with decreasing filler /reinforcement size and thus forms an additional separate phase known as interphase, which is distinct from the dispersed and continuous phases and hence influences the composite properties to a much greater extent even at low nanofiller loading (< 5%). Therefore, their properties are much superior to conventional composites. The interest in polymer nanocomposites further arises from the fact that, they are light weight as compared to conventional composites because of the low filler loadings, are usually transparent as scattering is minimized because of the nanoscale dimension involved and are still processable in many different ways including production of fibers with nanoscale fillers embedded in the polymer matrix. With these improved set of properties, they show promising applications in developing advanced textile materials such as Nanocomposite fibers, nanofibers and other nanomaterial incorporated coated textiles for applications in medical, defense, aerospace and other technical textile applications such as filtration, protective clothing besides a range of smart and intelligent textiles.

Polymers nanocomposites thus offer tremendous potential when produced in fiber form and offer properties that leapfrog those of currently known commodity synthetic fibres. Nanocomposite fibres that contain nanoscale embedded rigid particles as reinforcements show improved high temperature mechanical property, thermal stability, useful optical, electrical, barrier or other functionality such as improved dyeability, flame retardance, antimicrobial property, etc. These novel biphasic nanocomposites fibres in which dispersed phase is of nanoscale dimension, will make a major impact in tire reinforcement, electro-optical devices and other applications such as medical textiles, protective clothing etc. The work on spinning of nanocomposites started about seven years ago and several research groups across the world are exploring the synthesis, fiber processing, structure- property characterization and correlation and molecular modeling of these unique new composites fibers. Polymeric nanocomposite fibres have been mostly spun through three basic methods of fiber spinning - Melt spinning, Solution spinning and Electrospinning. Although, most of the research reports on polymeric nanocomposites is where it has been studied in form of films or moulded specimens and very few reports on their spinning into fiber form [1, 2-4].

2 RESEARCH AT IIT DELHI

2.1 Nanocomposite Fibers

At Textile Department at IIT Delhi, India we have investigated nanocomposite fibers based on all the three major types of nanofillers viz layered silicate nanoclays (MMT), carbon nanotubes (CNT), nanofibers/ nanographite as well as hybrid nanostructured materials such as POSS.

2.1.1 Polypropylene/NanoclayNanocom-posite Fibers

Compatibilized polypropylene/nanoclay composite filaments were produced by melt intercalation route using twin screw compounder coupled to a fiber take up device and drawing machine and characterized to study the effect of the compatibilizer and the role of nanoclay in improving the properties. The compatibilizer used was maleic anhydride grafted Polypropylene (PP-g-MA).

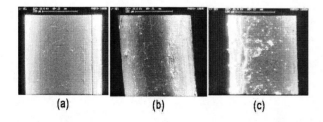

(a)   (b)   (c)

Figure1: Scanning Electron Micrographs of polypropylene/nanoclay composite filaments (a) Neat PP (b) 0.5% clay loading (c) 1 % clay loading (surface view)

Clay loadings of up to 1 wt % with up to 3 wt % of the compatibilizer were studied. The dyeability properties of these filaments showed that nanocomposite filaments took up disperse dyes unlike the neat PP filaments which have to be dope dyed. There was a significant improvement in

tensile, thermal, dynamic mechanical and creep resistance properties of PP/nanoclay composite filaments over neat PP filaments at very low loadings of the organomodified clay [5, 6].

2.1.2 HDPE/ POSS Nanocomposite Fibers

Another development was making high performance fibers based on polymeric nanocomposites based on a novel class of hybrid nanostructured filler, Polyhedral Oligomeric Silsesquioxane (POSS). The system chosen for the study is the simple 'octamethyl POSS', a molecular silica as the nanofiller and HDPE as the polymeric matrix. At comparatively very low loadings (0.25-0.5 wt %), molecularly dispersed POSS actually gives a lubricating effect and facilitates the drawing of filaments, which results in higher tensile strength and modulus. With increase in POSS concentration beyond 1 wt %, POSS existing as nanocrystals/aggregates starts hindering the orientation of HDPE chains leading to a gradual fall in tensile strength and modulus. Incorporation of POSS also modifies the thermal degradation behaviour of HDPE and broadens the temperature range of thermal degradation. The HDPE-POSS nanocomposite filaments also exhibit better UV resistance than neat HDPE filaments, which may be attributed to the scattering/reflective action of POSS [7, 8].

2.1.3 Polyurethane/Nanoclay Nano-composite Fibers

An attempt to explore the feasibility of producing filaments from polyurethane (PU) /clay nanocomposites and compare their structure and properties vis-a-vis neat PU filaments has been carried out as a part of doctoral thesis by our research group. This work reports the production of filaments from neat polyurethane and polyurethane/clay nanocomposite by dry-jet-wet spinning; a technique being used for the first time for this system [9, 10]. An organomodified nanoclay was used as a filler and thermoplastic polyurethane as the matrix. Dispersed nanoclay in PU matrix has induced both external morphological changes as well as internal micromorphology. Nanoclay dispersion reduces the stretchability by enhancing the void content of the nanocomposite filaments. Modulus and tenacity are enhanced significantly in the presence of nanoclay at low concentrations; nevertheless elongation and elastic recovery are marginally affected. Thermal studies suggest that a significant improvement in thermal stability of PU/clay nanocomposites filaments is due to hybridization with inorganic nanoclay. High thermal shrinkage at low clay concentration indicated high orientation due to good dispersion and exfoliation of clay in filaments. Boiling water shrinkage and water swelling also indicated high orientation and reduced swelling due to incorporation of clay. Fire retardant properties studied by cone calorimetry shows excellent fire retardant properties at low clay content (0.25 wt %). Dyeability properties of nanocomposite fibres also get significantly enhanced in the presence of nanoclay. Weatherability resistance of PU/clay filaments are significant only at higher clay concentration (> 1 wt %).

2.2 Future Trends

Nanotechnology has thus emerged as the 'key' technology, which has revitalized the material science and has the potential for development and evolution of a new range of improved materials including polymers and textiles. However there are many challenges in the development of these products, which need to be intensively researched so that the wide range of applications envisaged can become a commercial reality. An excellent dispersion and stabilization of the nanoparticles in the polymer matrix is crucial to achieving the desired nano effects. The tendency to agglomerate due to extremely high surface area is the major problem facing the effective incorporation of nanoadditives in coatings/finishing as well as in nanocomposite preparation. Surface engineering of nanoparticles and combining them with functional surface-active polymers can bring the nanoparticles onto fibers/textiles without loosing their superb, nanoscopic properties.

REFERENCES

[1] M. Joshi, The Impact of Nanotechnology on Polyesters and Polyamides' Chapter in the book titled "Advances in Polyesters and Polyamides",Woodhead Publishing Co. Ltd. Cambridge, UK, 2008.

[2] L. Razafimahefa, S. Chlebicki, I. Vroman and E. Devaux, Dyes and Pigments, 66, 55, 2005.

[3] E. Giza, K. Hiroshi, T. Kikutani and N. O. Kui, J. Macromolecular Sci. Physics, B 39, 545, 2000.

[4] G. H. Guan, C. C. Li and D. Zhang, J. Appl. Polym. Sci. 95, 1443, 2005.

[5] M. Joshi, M. Shaw and B. S. Butola, Fibres and Polym. (Korea) 5, 1, 2004.

[6] M. Joshi and V. Viswanathan, J. Appl. Polym. Sci.102, 2164, 2006.

[7] M. Joshi, B. S. Butola, G. Simon and N. Kukalevab, Macromolecules, 39, 1839, 2006.

[8] M. Joshi and B. S. Butola, J. Appl. Polym. Sci. 105, 978, 2007.

[9] G. V. R. Reddy, Studies on Polyurethane/clay nanocomposite fibers, PhD Thesis, IIT Delhi, 2008, Supervisors: M. Joshi and B. L. Deopura.

[10] G. V. R. Reddy, B. L. Deopura and M. Joshi, J. Appl. Polym. Sci. 116, 843, 2010.

Polymer Nanoparticles as Formulation Agents

H. Pham, F.Li, J.B. Goh, J. Dinglasan and D. Anderson[*]

[*]Vive Nano, Inc.
700 Bay St., Suite 1100, Toronto, ON. M5G1Z6 Canada, danderson@vivenano.com

ABSTRACT

An approach to produce polymer nanoparticles with tailored characteristics to accommodate different nanoformulation applications is presented. Polymer nanoparticles are made from the collapse of polyelectrolytes in the presence of counterions and/or different solvents. The size, hydrophobic/hydrophilic character, as well as surface interaction characteristics of collapsed polyelectrolyte nanoparticles can be controlled by varying the degree of crosslinks and the degree of chemical modification of the parent polyelectrolyte chain. Depending on the characteristics of the polymer nanoparticle (i.e. size, hydrophobic/hydrophilic character, degree of crosslinking, etc.) different molecules like dyes, pigments, agricultural and pharmacological active ingredients, and aromatics can be loaded into the polymer nanoparticles. Polymer nanoparticles made with this approach are typically smaller in size (10-50 nm) than those achievable using other methods. Data illustrating the different characteristics of the polymer nanoparticles and the interactions of the polymer nanoparticles with different active ingredients will be presented.

Keywords: polymer nanoparticles, nano formulations, active ingredients, hydrophobic/hydrophilic character

1 SYNTHESIS AND CHARACTERIZATION OF POLYMERIC NANOPARTICLES

Water dispersible nanoparticles are produced using counterion-induced collapsed polyelectrolyte chains. Polyelectrolytes are polymers having monomer units with ionizable groups such as amines, carboxylic acids, or sulfonic acids. Their conformation in aqueous solution is governed by electrostatic interactions between the charged moieties in the chain. Under conditions of high ionization and low ionic strength, the individual polyelectrolyte chains assume a highly extended, swollen coil conformation arising from repulsive electrostatic interactions along the chain. The addition of ionic species results in the screening of these repulsive interactions by the counterions, causing the highly extended polyelectrolyte coils to collapse into globules. These globules contain the counterions used to induce collapse, and can range in size from ~ 1-50 nm [1]. Crosslinking the collapsed polymer chain results in a stabilized globular structure even after removal of the ions used to collapse the polyelectrolyte. This process is summarized in figure 1.

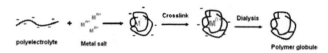

Figure 1. -ion induced collapse of polyelectrolytes to form polymer nanoparticles/globules.

Transmission Electron Microscopy (TEM) and Atomic force microscopy (AFM) was used to characterize the structure and size of these polymer nanoparticles. Figure 2 shows the TEM images of polymer nanoparticles that have been dialyzed to remove most of the collapsing ions. Figure 3A shows an AFM image of $Al(OH)_3$ collapsed polyacrylic nanoparticles and figure 3B shows the same nanoparticles that were dialyzed against acid to remove the Al^{3+} ions.

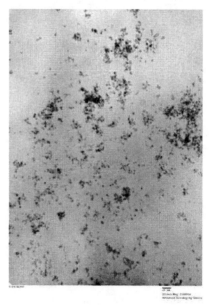

Figure 2. TEM image of polyacrylic acid nanoparticles (note: contrast comes from residual metal ions (Fe^{3+}) not entirely removed by dialysis)

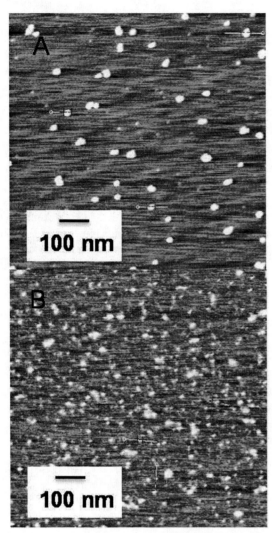

2 RELATIVE POLARITY OF DIFFERENT POLYMER NANOPARTICLES

The relative hydrophobicity of the polymer nanoparticle microenvironment can be characterized using an environment-sensitive fluorescent probe such as pyrene. The intensity ratio of the first and third vibronic bands (I1/I3) in the emission spectra of the pyrene monomer is very sensitive to the monomer's microenvironment, and can be used as a metric to gauge the hydrophobic nature of different polymer nanoparticles produced using the methods described here. As shown in figure 4, we measured the hydrophobic character of the nanoparticles as a function of the solution pH and the polymer used to make the polymer nanoparticles. At pH 3-6, a polymer nanoparticle microenvironment similar to o-dichlorobenzene [2] was achieved using polymer nanoparticles from poly(methacrylic acid) (PMAA) or poly(methacrylic acid-co- ethyl acrylate) (P(MAA-co-EA)), while a less hydrophobic microenvironment similar to dioxane was achieved from Zn^{2+}- collapsed polyacrylic acid nanoparticles. A microenvironment similar to glycerol was achieved by using Na^+- collapsed polyacrylic nanoparticles. Similarly, in the pH range 6-10 different microenvironments were achieved using different polymers make the nanoparticles. A microenvironment similar to methylene chloride was achieved from PMMA or P(MAA-co-EA) nanoparticles while a less hydrophobic microenvironment similar to glycerol was achieved from Na^+ collapsed polyacrylic acid nanoparticles.

Figure 3. A) Al^{3+} collapsed polyacrylic acid nanoparticles. (B) Dialyzed nanoparticles to remove Al^{3+}.

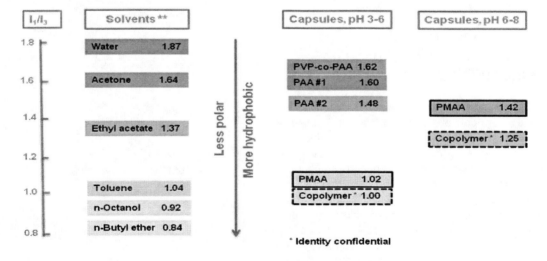

** D.C. Dong, M. A. Winnik* *Photochemistry and Photobiology* **1982**, 35, 17-21.

Figure 4. . Summary of relative hydrophobicty of the polymeric nanoparticles using pyrene as a microenvironment sensitive probe.

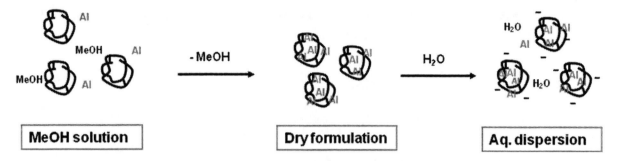

Figure 5. . Proposed mechanism for loading hydrophobic active ingredients into polymer nanoparticles.

AIs	Solubility in water (mg/mL)	AI wt% loaded *	Polymer capsules	Improved solubility In water (mg/mL) *
Imazethapyr	1.40	50	PMAA	6.7
Trifluralin	2.2×10^{-4}	33	PMAA	1.0
		33	PAA/PVP	1.0
		50	Copolymer	6.5

* **Data shown as tested – does not represent upper limit.**

Table 1.

3 LOADING OF ACTIVE INGREDIENTS INTO POLYMER NANOPARTICLES

To illustrate the utility of these polymeric nanoparticles in formulations, we have loaded several types of active ingredients into these nanoparticles. A simplified version of the loading protocol is illustrated in figure 5. Polymeric nanoparticles are dispersed in a solvent that is able to disperse both the hydrophobic active ingredient of interest and the polymeric nanoparticles. As the solvent is removed, the active ingredients associate with the polymeric nanoparticles until the point where a dry formulation is formed. This dry formulation is then redispersed in water to form an aqueous dispersion of the polymer nanoparticle/active ingredient formulation. Once in aqueous solution, the hydrophobic active ingredient remains associated with the polymer nanoparticles.

Using this method, the solubility of hydrophobic active ingredients in water can be greatly improved without the use of any additional surfactants and solvents. We have demonstrated this for an active ingredient that has parts per trillion (ppt) solubility in water by improving its solubility in water by 6 orders of magnitude (data not shown due to confidentiality restrictions). Table 1 shows two of the different active ingredients we have investigated.

REFERENCES

[1] A. V. Dobrynin and M. Rubinstein. *Prog. Polym. Sci.* 30, 1049, 2005.
[2] D. C. Dong and M. A. Winnik. *Photochem. Photobiol.* 17, 35, 1982.

Preparation of nickel doped multi-functional micro-patternable polydimethylsiloxane nanocomposite polymer with characterization of its magnetic, electrical and mechanical properties.

A. Khosla [*], J. L. Korčok [**], M. Haiducu [***], B.L. Gray [*], D. B. Leznoff [**], M. Parameswaran [***]

[*] Micro instrumentation Lab, School of Engineering Science, Simon Fraser University, 8888 University Drive, Burnaby, British Columbia, Canada V5A 1S6.

[**] Department of Chemistry, Simon Fraser University, 8888 University Drive, Burnaby, British Columbia, Canada V5A 1S6.

[***] IMMR, School of Engineering Science, Simon Fraser University, 8888 University Drive, Burnaby, British Columbia, Canada V5A 1S6.

ABSTRACT

We present the preparation and characterization of a novel polydimethylsiloxane (PDMS) multifunctional nanocomposite with electric and magnetic properties. The composite is fabricated by ultrasonic agitation of nickel nanoparticles in the PDMS matrix. The prepared nanocomposite is micromolded using conventional soft lithography techniques down to a feature size of 20µm. Microstructures including coils and cantilevers are molded against a poly(methyl methacrylate) mold fabricated via Deep UV LIGA. Nickel polydimethylsiloxane nanocomposites containing up to 45.5 % Nickel nanoparticles by weight in a polydimethylsiloxane matrix demonstrate the largest saturation magnetization of 21.0 emu/g as measured by SQUID magnetometry. It is also observed that the Young Modulus is a linear function of filler loading up to 50 wt% and indicates a material with superior mechanical properties as compared to undoped polydimethylsiloxane. The insulator–conductor transition for the nickel-polydimethylsiloxane composite occurs at 40 wt.%.

Keywords: Polydimethylsiloxane, Nanocomposite, Young's Modulus, MEMS, Deep UV LIGA, conductive polymer, magnetic polymer.

1 INRODUCTION

Micro total analysis systems (µTAS), or lab-on-a-chip systems, are microsystems that may contain multiple components for sample preparation, chemical reaction, fluid control, analyte separation and detection, and data acquisition, to accomplish complex tasks such as the detection of disease markers or environmental toxins in small (e.g., nanoliter) samples of fluid [1]. While many materials have been employed to fabricate microfluidic devices for µTAS, polydimethylsiloxane (PDMS), a silicone based elastomer, has been widely used because of its biocompatability, low toxicity, high oxidative and thermal stability, optical transparent, low permeability to water, low electrical conductivity, and ease of micropatterning [2]. In order to realize handheld PDMS based µTAS, all of the required components needed by the microsystem must be miniaturized and interconnected into a complete functional system. However, fluid control and data acquisition components may require electrical functionality in addition to fluidic functionality, which may be problematic in PDMS-based microsystems, as adhesion of metals to PDMS is poor and often leads to development of microcracks and electrical failure [3]. Hence, there is a need to develop PDMS-based active materials of similar flexibility to the undoped and insulating PDMS, that can also be easily micromicromolded using similar soft lithography techniques, to provide robust system electrical routing. In addition, for lab on a chip systems, the ability to manipulate fluid on chip (fluid control) via on-chip microvalves and pumps is usually required [4], with magnetically actuated valves and pumps a good choice due to the high energy density of magnetic actuation schemes [5]. While PDMS is inherently electrically insulating and non magnetic, these properties can be modified by the introduction of conducting and/or magnetic nanoparticles in the polymer matrix. Previously we reported fabrication of electrically conducting PDMS by incorporating multiwalled carbon nanotubes in PDMS matrix [6]. We now report on a different multifunctional material that posses both electronic and magnetic properties that we characterize, as well as improved mechanical properties, for utilization as electronic routing and magnetic actuation for PDMS-based micro total analysis systems.

2 PREPARATION OF MATERIALS AND MICROFABRICATION

2.1 Materials

PDMS (Sylgard 184 Elastomer Kit) was purchased from Dow Corning. Nickel nanoparticles with purity of 99.7+% and a diameter of 30-50 nm were obtained from Nanostructure & Amorphous Materials, Incorporated USA. Poly (methyl methacrylate) (PMMA) which is a deep UV photo-patternable polymer was bought from Plaskolite Incorporated USA. All the materials were used as purchased.

2.2 Electric and magnetic PDMS preparation

PDMS polymer consists of a base elastomer and curing agent [7] that was manually stirred in 10:1 ratio by weight for 2 minutes. Nickel nanoparticles were ultrasonically dispersed in the prepared PDMS polymer matrix in varying weight percentage by using a VC 750 (Sonic Inc.) programmable ultrasonic processor. The ultra sonic processor was operated in pulse mode (10 seconds on and 15 seconds off cycle) for 2 minutes, which provided mixing by repeatedly allowing the sample to settle back under the probe after each burst. The prepared nanocomposite was placed into a vacuum chamber to remove air bubbles for 30 min prior to micromolding.

2.3 Poly(methyl methacrylate) (PMMA) mould microfabrication

PMMA is an excellent choice for a micromolding master due to its high mechanical strength and good dimensional stability. However, the use of PMMA as a positive tone resist or a mold usually involves a highly energetic and very expensive radiation source. One classical example is the use of synchrotron light (x-ray) for LIGA [8-11]. LIGA is a German acronym, standing for **Li**thographie, **G**alvanoformung, **A**bformung which translates to lithography, electroplating, and molding, for which the high cost x-ray source is required for the lithographic aspects of the process. However, it has been demonstrated that PMMA patterning with a latent image can be accomplished using low-pressure mercury (germicidal) lamps [12-14], thus reducing drastically the cost associated with PMMA lithography. We utilize this low-cost, deep-UV lithography process for the fabrication of our molds.

Figure 1 shows the basic fabrication sequence. The chosen structural material for the micromold was OPTIX®, a commercial acrylic purchased from Plaskolite. The irradiation source was a Stratalinker 2400UV crosslinker equipped with five low-pressure mercury lamps able to provide a nominal power of $4mW/cm^2$ and a deep-UV spectrum with a maximum peak at 253.7 nm. In photolithography, collimated light is required for high aspect ratio patterning. However, the Stratalinker provides uncollimated illumination. To improve the resulting aspect ratio a plastic grate made of 12.5 by 12.5 mm squares, each with a height of 13 mm, was set at a distance of 2 cm above the substrates during exposure. This way, only the light that hits the bottom plane of the grate at angles less than 46° with respect to the perpendicular on the surface are allowed through. Further, to even out the irradiation so that no square patterns are formed as a result of positioning the grate above the substrates, the latter were set on top of a rotational stage driven by a dc motor, which was fed by a 9 Volt battery. The rotational frequency of the stage was approximately 0.125 Hz.

Prior to the following succession of steps, the acrylic was cut into 3 x 3 inch squares, cleaned with water and methanol, and then dried with a stream of nitrogen gas. A hard metal mask of 100nm of gold was sputtered on the PMMA substrate using a Corona Vacuum System sputterer. After metal deposition, Shipley 1813 photoresist was spun on at 4000rpm for 30 seconds. Because the PMMA may warp if baked above 100°C, the photoresist baking temperature was kept below 75°C. The sample was exposed through a contact mask using an i-line source and developed in MF-319. The hard mask was etched using TFA gold etchant from Transene Company, Inc. Following the gold etch, the photoresist was removed using a 60 second blanket exposure followed by another development in MF-319, thus avoiding damage to the OPTIX® by strong organic solvents such as acetone or photoresist stripper typically used to remove photoresist. At this point, the samples were exposed to deep UV using the hard mask for patterning, using a dose of 6650 Joules.

The samples were immersed in a developer bath consisting of 7:3 isopropyl alcohol (IPA) to de-ionized water (DI water) at 28°C for 1 hour. The temperature was maintained constant throughout the development and the bath was given a slight manual agitation. The development of the samples was quenched in an ultrasonic IPA bath at room temperature (~18°C) for 10 seconds, followed by an IPA rinse for another 10 seconds. This quenching was necessary to prevent the re-deposition of the partially dissolved PMMA. After removing the gold mask using TFA gold etchant , rinsing with DI water, and blowing the samples dry with nitrogen gas, the depth of the etched channels were measured using an Alpha-step 500 profilometer and found to be 100μm deep and a width of 20μm.

2.4 Micromolding

Micromolding was accomplished using standard soft lithography methods as is typically utilized for undoped

PDMS [15]. The Ni-PDMS nanocomposites with different doping levels were prepared as defined in section 2.2 and poured on the PMMA molds prepared as discussed in section 2.3. The excess PDMS on the PMMA mould was scraped off using a glass slide and then de-gassed to remove any air bubbles for 30 minutes. The samples were baked on a hot plate at a temperature of 75 °C for 1 h followed by cooling to room temperature. The structures were then peeled off from the mold and were ready for testing. Figure 2 shows examples of resulting Ni-PDMS microfabricated structures

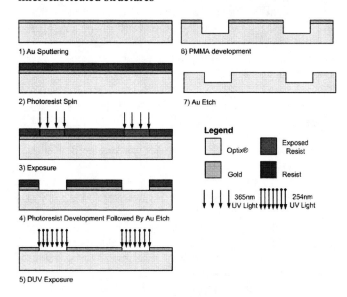

Figure 1: Deep-UV exposure of PMMA: fabrication process steps

3 EXPERIMENTAL RESULTS AND DISCUSSION

3.1 Magnetic characterization

Magnetic properties of Ni-PDMS nanocomposites were analyzed by SQUID magnetometry and were compared to the pure nickel nanoparticles. Magnetometry measurements were carried out on a Quantum Design MPMS-XL-7S SQUID magnetometer equipped with an Evercool liquid helium dewar. Ni-PDMS samples, as well as the pure (30-50 nm) Ni nanoparticles, were packed in polycarbonate capsules and mounted in low-background diamagnetic straws. For each sample, magnetization (M) vs. field (H) hysteresis measurements were obtained at 300 K in fields between +2 T and -2 T, taking data every 25-200 G at low fields (< 0.2 T) and 1000-5000 G at higher fields. The pure Ni nanoparticles were found to have a saturation magnetization, M_{sat}, of 46.0 emu/g. The coercivity, H_c, is 105 G, indicating that the material is a soft magnet. The shapes of the hysteresis loops, fields at which saturation occurs, and coercive fields are nearly identical for all Ni-

PDMS composites, verifying that the incorporation of the

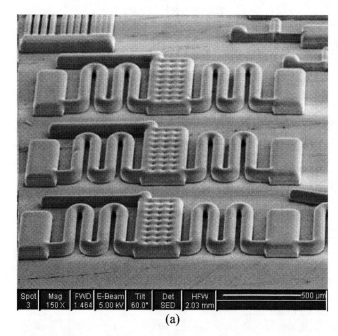

(a)

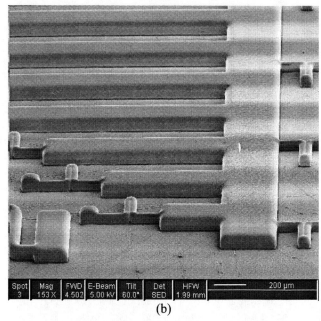

(b)

Figure 2: Scanning Electron Microscopy (SEM) images (a) Bridges (b)Micromolded cantilevers

nanoparticles into PDMS films did not change their fundamental magnetic properties. As expected, M_{sat} values per gram of nanocomposite correlate linearly with the weight percent of Ni in PDMS (Figure 3), with the 45.5 % sample having the largest saturation magnetization, at 21.0 emu/g as compared to the pure Ni Nanoparticles which had a saturation magnetization of 45.7 emu/g (shown in Figure 3 as the 100 wt% point).

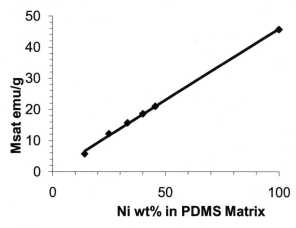

Figure 3: Saturation magnetization (M_{sat}) per gram of composite as a function of mass percent of Ni in the Ni-PDMS composites.

3.2 Mechanical characterization

Young's Modulus measurements were done using an Instron +3340 Series Single Column System. A total of ten test specimens were fabricated with varying weight percentage of nickel nanoparticles in PDMS (eg. 0, 14.06, 25, 30 ...55), with each sample having a gage length of 75 mm, a width of 25 mm, and a thickness of 1mm. Tests were carried out employing the ASTM (E8-04) Standard Test Method. After the sample was loaded into the testing apparatus and the strain gage attached to the specimen, all variables were cleared and the speed of the grips was set to 15 mm/minute. Each specimen was pulled until failure while data was collected by a computer acquisition system. The measured Young's Modulus of pure PDMS was found to be 0.59 MPa, which lies well within the range of earlier reported values of 0.36-0.87 MPa [16]. Figure 4 shows results of Young's Modulus for all fabricated nanocomposites. TheYoung's Modulus increases with increasing Wt% of nickel nanoparticles in the PDMS matrix. Thus, doping the PDMS with nickel nanoparticles results in materials with higher Young's Modulus. However, after 60 Wt% the nanocomposite could not be easily micromolded because of increased viscosity at high fill levels. It was observed that the Young Modulus's is an approximately linear function of filler loading up to 50 wt%, and and can be particularly enhanced. The curve levels off slightly above 50 wt%, although still increases. The results obtained are in accordance with the experimental values previously reported by other researchers on nanocomposite polymers doped with 200nm silver [17].

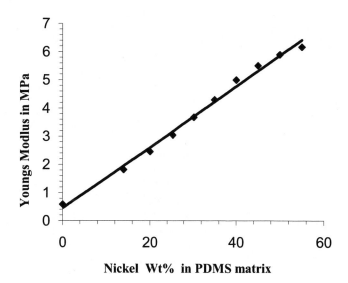

Figure 4: Young Modulus's versus nickel weight percentage (Wt%) in PDMS matrix

3.3. Electrical characterization

The direct current (DC) electrical properties of 10 mm x 5mm x 1mm films of the Ni-PDMS nanocomposite were measure using an HP3748A digital multimeter, which was set to operate in a four probe mode to eliminate the contact resistance. Figure 4 shows the resulting resistivity of nickel nanoparticle doped PDMS with varying weight percentage. It can be clearly seen that the resistivity decreases with increasing nickel nanoparticle weight percentage. It is observed that up to 40 weight percent the resistivity decreases slowly. However, after this point, which is the percolation threshold, the resistivity decreases rapidly and can be explained in terms of percolation theory [18]. At

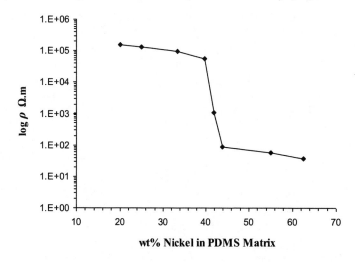

Figure 5: Resisitivity versus nickel wt % in PDMS matrix

lower concentrations the nickel nanoparticle percolation paths are not set up and as the concentration is increased the percolation paths are set up by the conducting nickel nanoparticles. At this point the nickel nanoparticles control the conductivity of the nanocomposite. The percolation threshold was observed to be equal to 41 wt% for the Ni-PDMS nanocomposite.

4   CONCLUSION

The electrical, magnetic and mechanical properties of micropatternable nickel nanoparticle (30-50nm) doped polydimethylsiloxane have been characterized as a function of weight percentage of nickel nanoparticles in the PDMS matrix. It is observed that a percolation threshold occurs at approximately 40 weight percentage and the highest saturation magnetization occurs at 45.2 weight percentage. It was also observed that beyond 60 weight percentage, the nanocomposite becomes difficult to micropattern. The fabricated nanocomposite has superior maechanical properties, with a higher Young's Modulus than undoped PDMS; the relationship between Young's Modulus and nanoparticle doping was found to be approximately linear. We have been successfully able to micromold the prepared naocomposite down to features sizes of 50µm (limited by the photomask resolution limit), using a fabrication process similar to unfilled PDMS [19]. The multifunction material can be not only used to solve the signal routing problem in µTAS but also shows promise as a magnetically actuated material for fluid control components such as microvalves and pumps. Further work will involve incorporationg this novel material in developing microheaters, microvalves and micropumps for all polymer in lab on a chip devices.

ACKNOWLEDGEMENTS

The authors would like to acknowledge NSREC for Discovery Grant funding; CMC Microsystems for providing us with tools for mask design; and CFI and BCKDF for providing funds for the polymer microstructuring facility. SERC (DBL, JLK) and the SQUID magnetometer facility was also funded by CFI and BCKDF. We would also like to thank Bill Woods for help with microfabrication.

REFERENCES

[1] George M. Whitesides, NATURE, Vol 442, 27 July 2006.

[2] Younan Xia and George M. Whitesides, Annu Rev Mater Sci 28 (1998), pp. 153–184

[3] A.J Gallant.; D.A Zeze; D Wood,.; M.C Petty, D. Dai; J.M Chamberlain, Infrared and Millimeter Waves, 2007 and the 2007 15th International Conference on Terahertz Electronics. IRMMW-THz., vol., no., pp.982-983, 2-9 Sept. 2007

[4] Petra S Dittrich, Kaoru Tachikawa, Andreas Manz, Analytical Chemistry, Vol. 78, No. 12. (1 June 2006), pp. 3887-3908.

[5] Nicole Pamme, Critical Review: Lab Chip, 2006, 6, 24 – 38

[6] A. Khosla and B. L. Gray, Materials Letters Volume 63, Issues 13-14, 31 May 2009, Pages 1203-1206.

[7] http://www.dowcorning.com/applications/search/products/Details.aspx?prod=01064291&type=PROD&bhcp=1.

[8] Becker, E.W., et al., Microelectronic Engineering, 1986. 4(1): p. 35-36.

[9] Ehrfeld, W. and D. Münchmeyer, Nuclear Instruments and Methods in Physics Research A, 1991. 303(3): p. 523-531.

[10] Ehrfeld, W. and H. Lehr, Radiation Physics and Chemistry, 1995. 45(3): p. 349-365.

[11] Achenbach, S., Klymyshyn D, Haluzan D, Mappes T, Wells G, Mohr J., Microsystem Technologies, 2006. 13(3-4): p. 343-347.

[12] M Haiducu, M Rahbar, I G Foulds, RW Johnstone, D Sameotoand M Parameswaran Journal of Micromechanics and Microengineering, 2008. 18(11): p. 115029-115036.

[13] Johnstone, R., I.G. Foulds, and M. Parameswaran,Journal of Vacuum Science and Technology B, 2008. 26(2): p. 682-685.

[14] Johnstone, R.W., I.G. Foulds, and M. Parameswaran. Proceedings of the Canadian Conference on Electrical and Computer Engineering (2007). CCECE 2007. Vancouver, BC, Canada: IEEE

[15] S. Jaffer and B.L. Gray, J Micromechanics Microengineering 18 (2008), p. 035043.

[16] http://www.mit.edu/~6.777/matprops/pdms.htm

[17] P McCluskey,.; Nagvanshi, M.; Verneker, V.R.P.; Kondracki, P.; Finello, D., Proceedings of 3rd International Conference on Adhesive Joining and Coating Technology in Electronics Manufacturing, 1998, p 282-6.

[18] B.J. Last and D.J. Thouless, Phys Rev Lett 27 (1971), pp. 1719–1721.

[19] S.M. Westwood, S. Jaffer and B.L. Gray, J Micromechanics Microengineering 18 (2008), p. 064014 (9 pp.)

Environmentally Friendly Synthesis of Polypyrrole within Polymeric Nanotemplates-mechanism of polypyrrole synthesis

X.Li and C.Maladier-Jugroot[*]

[*]Department of Chemistry & Chemical Engineering,
Royal Military college of Canada, Kingston, ON, Canada,
summer.li@rmc.ca, cecile.maladier-jugroot@rmc.ca

ABSTRACT

This paper describes an innovative way to synthesize polypyrrole using self-assembly nano-templates poly (isobutylene-alt-maleic anhydride) (IMA) or poly (styrene-alt-maleic anhydride) (SMA). The reaction occurs in neutral pH water solution, requires no further electrochemical induction or chemical oxidizers as opposed to currently used methods. To understand the reaction mechanism, the reaction path was investigated theoretically by *ab initio* Density Functional Theory (DFT) simulations. It was found that water has a crucial role in the activation and elongation steps of the polymerization. This study is expected to have an important impact on a novel environmental friendly synthesis of conducting polymers.

Keywords: polypyrrole, amphiphilic alternating copolymer, confinement effect, computational chemistry

1 INTRODUCTION

In recent years, polypyrrole has been intensively studied due to its conducting property and its applications in organic light-emitting diodes (LED) [1], display [2], to charge dissipating films [3] and volatile organic light sensors [4][5]. To the best of our knowledge, up to now all the syntheses reported in literature involve electrochemical oxidation or chemical oxidation [6-8]. Lately, our group discovered an innovative way to synthesize polypyrrole using self-assembly nano- templates poly (isobutylene-alt-maleic anhydride) (IMA) [9] or poly (styrene-alt-maleic anhydride) (SMA) [10]. The templates are amphiphilic copolymers with hydrophobic core and hydrophilic outer layer in water solution. When immersed in the template system, pyrrole molecules polymerize in the hydrophobic cavity, with no external inductions. The reaction only takes place in a confined environment where the cavity is 2nm. This phenomenon, the confinement effect, has been widely reported in biological systems. Recent works [11,12] show that confined environment results in different dynamic and thermodynamic properties of fluids inside of the small cavities and therefore one could apply it to induce synthesis that would not happen in bulk environment. With that in mind, it is therefore important to investigate the reaction mechanism to provide insight information on synthesis within nanotubes. The aim of this paper is to use theoretical calculations to follow the reaction pathway. We performed a thorough *ab initio* quantum chemical study over the reaction potential energy surface corresponding to different structures. The detailed reaction pathway is described including the transition states (TS), intermediate products and the energies will be calculated accordingly.

2 METHODS

All results have been found using the Gaussian 03 software package. All ground states and transition state calculations are done at the DFT level using the hybrid functional B3LYP and basis set 6-31+G. The imaginary frequencies of each TS were animated to ensure that the transition state corresponded to the correct reaction path.

2.1 Activation of Pyrrole

To determine the role of water on the activation step, two simulations were performed. The first simulation involved the increase of the N-H bond of pyrrole to obtain the activated form with protonations of either on alpha or on beta carbon. The second simulation was performed in the presence of water close to the N-H group of pyrrole. Water was aligned linearly with pyrrole through N-H. A relaxed scan was performed increasing N-H bond distance at 0.05A increment 30 steps. The scan was done and compared at three levels: B3LYP/6-31+G (d), B3LYP/6-31+G and B3LYP/6-31G. Transition state was calculated through QST3 method at B3LYP/6-31+G level. Products and reactants were optimized at B3LYP/6-31+G.

2.2 Dimerization

Five conformations were proposed for dimerization with different binding arrangements (table 1). The most stable structure was chosen for further calculations. The binding of two activated monomers was performed in relaxed scan where C_α-C_α distance was decreased at 0.1A increments 30 steps at B3LYP/6-31+G level. Further geometry rearrangement and proton shuffling were completed through 2D scans at HF/3-31G. All transition sates were searched through QST3 method at B3LYP/6-31+G level. The stable intermediates were all optimized at the level of B3LYP/6-31+G.

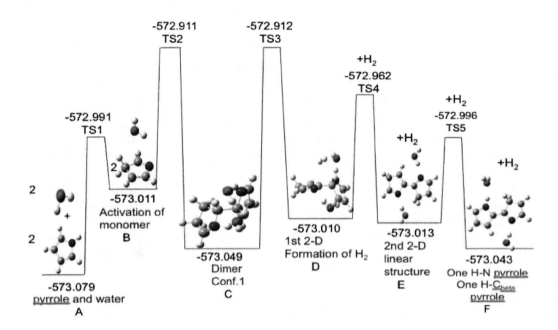

Figure 1: reaction pathway. All intermediates are marked from A-F. The transition states are labeled from TS1-TS5. The energies are in Hartree.

2.3 Rearrangement

The final structural rearrangement of the pyrrole dimer was performed so that one pyrrole is reprotonated on its nitrogen while the other pyrrole stays activated at beta-carbon. The ground state and transition state calculations were done at B3LYP/6-31+G level.

3. RESULTS

The complete reaction pathway is mapped out through computation simulations (Figure 1). The reaction is mediated by water and the final product is linear and conjugated.

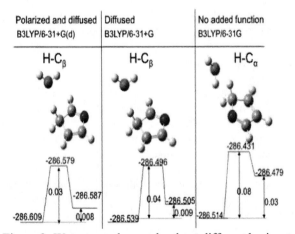

Figure 2: Water-pyrrole scan by three different basis sets

3.1 Activation of Pyrrole

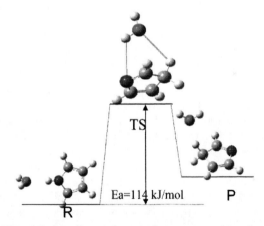

Figure 3: reaction pathway of water-pyrrole activation. In this step, water facilitates proton transfer from nitrogen to beta carbon. B3LYP/6-31+G basis set was used here.

When water is present during the reaction, pyrrole can act both as a proton donor (the proton attached to the nitrogen) to water and as a proton acceptor (to the pi-electron cloud). A study on water-pyrrole complex indicated that the proton donor form is energetically favored by 6.72 kcal/mol [13]. Therefore in the initial step water was linearly linked to N-H of pyrrole.

A relax scan was performed where N-H bond increases at 0.05A increment. Three different levels of basis sets were used to calculate for comparison: B3LYP/6-31+G (d), B3LYP/6-31+G and B3LYP/6-31G. Both B3LYP/6-31+G (d) and B3LYP/6-31+G basis sets result in protonation at beta-carbon, where the lower basis set B3LYP/6-31G protonates the alpha-carbon instead which gave energetically less favored reaction path (Figure 2). The structures from B3LYP/6-31+G (d) and B3LYP/6-31+G are similar and are close in energy. For simplicity only B3LYP/6-31+G was chosen as the level of theory in this study.

To compare two conformations, H-C_β and H-C_α, both transition states were searched by QTS3 method [14]. The result shows that the protonated beta conformation is kinetically (100kJ/mol lower) and thermodynamically (54.6kJ/mole lower) favored. Furthermore, a control group was done with no water molecule presented. It shows that only protonated-alpha conformation was obtained. Water was the only agent present in reaction system, so that it is reasonable to believe water directs proton transferring during the reaction. We can then calculate the stabilization energy by water using equation (1)

$$\Delta E(\text{stabilization})=E(\text{water-pyrrole})-E(\text{water})-E(\text{pyrrole})(1)$$

The stabilization energy -40 kJ/mol shows that the presence of water molecules not only facilitates proton transfer, but also was needed to obtain an activated pyrrole molecule in a beta-carbon conformation.

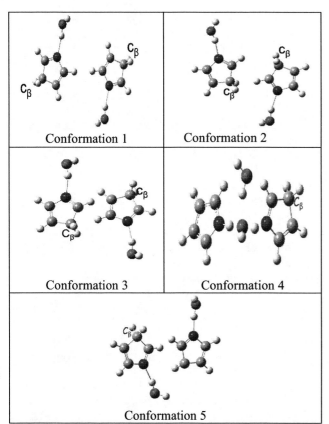

Table 1: Five conformations proposed for dimerization. 1 gave the lowest activation energy and the most stable product.

Previous studies [15, 16] have indicated that pyrrole could reversibly convert to H-C_β and H-C_α through tautomerization in the absence of water with the alpha pyrrole (H-C_α) tautomer being more stable kinetically and thermodynamically. In contrast, our work shows the H-C_β conformation is to be more favorable (Figure 2).

3.2 Dimerization of Pyrrole

The dimerization of activated pyrrole monomers was studied subsequently and the reaction is shown in figure 1, step B-C. Five possible conformations were proposed based on possible binding arrangements (Table 1). In all structures, C_α's are the polymerization sites, given that the reaction occurs between C_α-C_α. Conformations 1, 2 and 3 include two H-C_β pyrrole monomers with different geometry arrangements. Specifically, conformation 1 and 2 align two monomers show an inversion point: 1 has both H-C_β facing outward and 2 positions both H-C_β inward. Conformation 3 lines two pyrrole monomers in the same direction so that one H-C_β faces outward and another one faces inward. Conformations 4 and 5 also included two pyrroles, however one is activated and the other one is not.

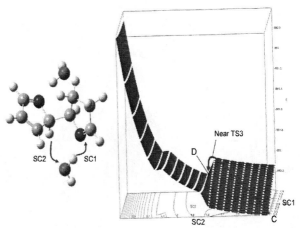

Figure 4: first 2-D scan. The two dimensions of scan are shown in red arrows on the left. The reaction pathway is marked on the right in red arrow. The reaction goes from structure C to D, with the pre-transition state of TS3.

All five structures underwent relaxed scans where C_α-C_α distance was decreased at 0.1A increments 30 steps at B3LYP/6-31+G level. The most stable product, conformation 1, with the lowest activation energy was selected for further characterization. Following the C_α-C_α bond formation of the dipyrrole, a second proton transfer occurs from H-C_β to adjacent pyrrole resulting in two H-C_β's on one pyrrole, with the second pyrrole remaining unprotonated (Figure 1, step B-C). This calculated dimer C is structurally different from stand-alone polypyrrole where its nitrogen atoms are unprotonated and the C_α centers are protonated. With that in mind, further proton transfer was studied.

Using dimmer C as a starting structure, a two-dimensional relax scan was performed. The N-H distance was decreased along the first dimension (0.2 A increments, 20 scans) and C-H distance was increased along the second as shown in figure 4 (0.2A, 20 scans). Due to the complexity of the calculation, the 2-D scan was performed at HF/3-31G level. The potential energy surface obtained during the scan is illustrated in figure 4. The formation of H_2 gas was indicated and the structure is further stabilized in energy. The lowest energy path from reactant C to product D was chosen as the reaction pathway (Figure 4). The saddle point was searched by QST3 method at B3LYP/6-31+G level. The second 2-D scan was done in the same fashion with omission of the H_2 molecule, but on the C-H and N-H bonds on the opposite side of dimer (figure 1, step D-E). During the second 2-D scan, proton shuffled again resulted a completely linear structure where both pyrroles are protonated at beta-carbon. The reaction pathway was also selected based on the lowest energy sequence, and the saddle point was chosen as the near transition state. Which was then optimized using QST3 at the B3LYP/6-31+G level (Figure 5).

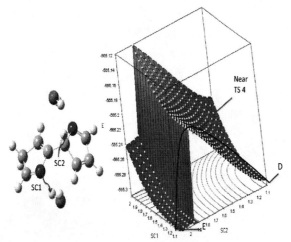

Figure 5: second 2-D scan. The two dimensions of scan are shown in red arrows on the left. The reaction pathways marked on the right in black arrow from D-E.

3.3   Rearrangement of Pyrrole Dimers

Comparison between structure E and the polypyrrole indicated that there is further proton shuffling since both nitrogen atoms remained unprotonated in structure E (Figure 1). The final product was achieved by protonating one nitrogen and keeping the other pyrrole activated at the beta carbon (Figure 1. Step E-F). The final structure is further stabilized in energy and the dimerization is complete. As a result, polymerization will proceed in the fashion that the incoming monomers will add to the activated beta-carbon pyrrole at C_α centers and the proton will transfer back to the nitrogen from C_β after dimerization.

CONCLUSION

This paper described a detailed study of the polymerization of pyrrole in aqueous environment. Understanding the details of the novel synthesis in great details would provide us profound insight to control the structure ar the nanoscale. It is our interests, from both theoretical and practical point of view, to understand the reaction pathway, intermediate reactivity and calculate the energies accordingly. The theoretical characterization of the synthesis and the number steps involved showed the essential role of water in this confined polymer template for the activation as well as polymerization of pyrrole. The synthesis is template-directed and the soft templates used here provide a confined space. Future work will study the confinement effect with larger cavity spacing to help determine the role of space restriction in directing the synthesis. This study is expected to have an important impact on a novel environmentally friendly synthesis of nanostructures. It will contribute to knowledge in utilizing nanotemplates to synthesize novel materials.

REFERENCES

[1] Yu G., Heeger, A. *J., Synth. Met.* **1997**, 85, 1183.
[2] Roussel F., Chan-Yu-King R., Buisine J.M., *Eur. Phys. J. E*, **2003,** 11, 293-300
[3] Lerch K., Jonas F., Linke M., J. *Chim. Phys. Phys.-Chim.Biol.* **1998**, 95, 1506
[4] MaQuade D, Pullen A., Swager T., *Chem. Rev.,* **2000,** 100, 2537-2574.
[5] Sotzing G. A., Briglin S., Grubbs R., Lewis N., *Anal.Chem.* **2000**, 72, 3181.
[6] S. Sadki. et. al., *Chem. Soc. Rev.*, **2000** 29, 283-293
[7] S.J. Hawkins. et. al., *J. Mater. Chem.*, **2000**, 10, 2057-2062.
[8] M. Zhou., et. al., *J. Phys. Chem. B* **1999**, 103, 8443-8450
[9] A.S.W. Chan, M.N. Groves, C. Malardier-Jugroot, Molecular Simulation, accepted for publication, **2009**
[10] C. Malardier-Jugroot, et. al., *Langmuir*, **2005**, 21(22), 10179.
[11] V. Kocherbitov., *J.Phys.Chem.* C **2008,** 112, 16893-16897.
[12] Y.Lipatov. et al., *Polym. Bull.* **2008**, 739-747
[13] M.J. Tubergen, Anne M. Andrews *J.phys.chem.* **1993**,   97, 7451-7457.
[14] C. Peng, P. Y. Ayala, H. B. Schlegel and M. J. Frisch, *J. Comp. Chem.*, **1996**, 17, 49
[15] S.M. Bachrach. *J.Org.Chem.* **1993**, 58, 5414-5421
[16] M. Martoprawiro, et al *J. Phys. Chem.* A **1999**, 103,3923-3934

Directing Cell Migration by Electrospun Fibers

Y. Liu[*], A. Franco[*], L. Huang[**], R. Clark[***], and M. Rafailovich[*]

[*]Department of Materials Science and Engineering, Stony Brook University, Stony Brook, NY, USA, yinliu@ic.sunysb.edu
[**]Department of Condensed Matter Physics and Materials Science, Brookhaven National Laboratory, Upton, NY, USA, lhuang@bnl.gov
[***]Department of Biomedical Engineering, Stony Brook University, Stony Brook, NY, USA, rclark@notes.cc.sunysb.edu

ABSTRACT

In this study, aligned poly (methyl methacrylate) (PMMA) fibers designed as guidance structure were produced by electrospinning and tested in en masse cell migration assays. We have shown that migration of fibroblasts on the planar surface results in a radial outward trajectory, and a spatially dependent velocity distribution decreases exponentially in time towards the single cell value. If the cells are plated on the surface of aligned electrospun fibers above 1 micro in diameter, they become polarized along the fiber. The velocity of the cells on the fibrous scaffold is lower than that on the planar surface. Contrary to expectations, the rate at which cells migrate within the construct is mostly dependent on the orientation of the fibers and nearly independent on the size of the holes. These results indicate that electrospun fibers can be used to regulate cell migration and therefore provide controlled cell delivery systems.

Keywords: electrospinning, polymer fibers, cell migration, wound healing

1 INTRODUCTION

Cell migration into the wound is an essential part of all wound healing in mammals [1, 2]. Even though cell migration mechanisms and regulation of cell migration have been studied extensively on numerous substrates, the surfaces were mostly planar [3]. The research on fibers may be more relevant to understand the process involved in the wound healing process, since in vivo cells migrate on the extracellular matrix (ECM), which is fibrillar [4].

In this study, we chose to study human dermal fibroblasts migration on flat and fibrillar poly (methyl methacrylate) (PMMA) substrates. Here we show that the cell migration rate can dependent on the microstructure of the morphology of the substrate, such that on the flat surfaces the speed decreases with increasing incubation time, whereas on the aligned fibrous scaffold, cells can migrate for very long time with constant velocity. These results are critical important to fulfill our long-term goal, which is to develop an artificial implant as a conduit for cell migration and skin regeneration after skin injury.

2 MATERIALS AND METHODS

2.1 Fabrication and Characterization of PMMA Aligned Scaffold

Solutions of PMMA (M_w=120 kDa, M_w/M_n=3, Sigma-Aldrich Inc., St. Louis, MO) were made in toluene (Fisher Scientific, Pittsburgh, PA) at a concentration of 30 mg/ml and spun cast onto the clean glass coverslips using a photoresist spinner with a frequency of 25,000 PRM. The film thickness was determined by ellipsometry to be approximately 100 nm. The polymer-coated coverslip were then annealed at 120°C in a vacuum of 10^{-7} Torr for 12 h to remove the residual solvent. Similar substrates were also prepared as targets for the electrospun fibers. This served as a two folds purpose: (a) the small amount of residual solvent plasticized the film and allowed the fibers to adhere; (b) placing the fibers onto a chemical identical substrate allowed for a direct comparison of the cell structure on flat and fibrillar surfaces.

The procedure to fabricate PMMA fibers is shown in figure 1. In the electrospinning process, PMMA solutions were loaded in a 5 ml glass syringe (Popper and Sons Inc., New Hyde Park, NY) attached to a syringe pump (KDS200, KD Scientific Inc., New Hope, PA) which provided a steady solution flow rate of 20 μl/min. A high-voltage power supply (Gamma High Voltage Research, Ormond Beach, FL) was employed to apply a potential of 15 kV between a 25 gauge blunt end needle (I.D. 0.26 mm, Popper and Sons Inc.) and a metal target which was horizontally placed 10 cm away from the tip of the needle a counter electrode. Highly aligned PMMA fibrous scaffolds were obtained by collecting the fibers with rotating drum at a speed of 6750 r/min. In further to fabricate the cross-aligned PMMA fibrous scaffold, the sample was removed and rotated to 90° and the procedure was repeated for the next layer. The substrates used for measuring the two dimensional (2D) migration consisted of a single layer of parallel fibers. The substrate used for measuring three dimensional (3D) migration consisted of two layers of fibers, oriented at 90° to each other and

The fiber diameter distribution of PMMA scaffolds was calculated by analyzing scanning electron microscope (SEM) images with Image Tool. For cell seeding, fiber-carrying coverslips were sterilized with ultraviolet (UV) light for 20 min. Serum free Dulbecco's Modified Eagle medium (DMEM) containing 30 μg/ml intact human plasma fibronectin (Fn) (Calbiochem, San Diego, CA) was added into each sample and incubated at 37°C for 2 h.

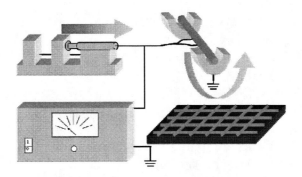

Figure 1: Schematic illustration of the procedure of fabricating PMMA cross-aligned fibrous scaffold. PMMA solution was loaded in the syringe and pushed out by syringe pump. When a high voltage (5-8kV) is applied between the needle and the collector, the fibers will be deposited on the rotating drum.

2.2 Cell Culture and Membrane Staining

Primary human dermal fibroblasts were obtained from Clonetics (San Diego, CA) and used between passages 11 and 12. The cells were cultured in DMEM supplemented with 10% fetal bovine serum (HyClone, Logan, UT) and an antibiotic mix of penicillin strepmycin, and L-Glutamine (full-DMEM), in a 37°C, 5% CO_2, 95% humidity incubator (Napco Scientific Company, Tualatin, OR).

The standard phase contrast microscopy limited the visualization of the cells on the fibers. To visualize the cells under fluorescent microscopy in real time, cell membrane were stained with DiD (Invitrogen, Carlsbad, CA). The DiD lipid-labeling procedure was obtained by a modification of the manufacture's instruction. Briefly, cells were washed well with phosphate buffer saline (PBS) and re-suspended at density of 1×10^6 cells/ml in serum free DMEM containing DiD at a concentration of 3.5 μg/ml, incubated for 30 min at 37°C. The labeled suspension tubes were centrifuged and the supernatant was removed.

2.3 Cellular F-actin Cytoskeleton and Nucleus Organization

To observe cells under fluorescent microscope, cells were fixed, permeabilized, followed by staining with Alexa Fluor 488 Phalloidin (Invitrogen, Carlsbad, CA) and propidium iodide (PI, Sigma Chemical Co., St. Louis, MO) for actin cytoskeleton and nucleus, respectively. The morphology of the cells was visualized with Nikon Diaphot-TMD inverted microscope or Leica TCS SP2 laser scanning confocal microscope (LSCM, Leica microsystem Inc., Bannock burn, IL).

2.4 Assessment of Cell Migration Velocity on Different PMMA Substrates

The agarose drop assay is a method that is frequently used to analyze cell mobility [5]. This method is performed by packing cells at high density in a small drop of agarose placed on the substrate to be studied. Migratory properties of cells are evaluated by measuring the ability of the cells to migrate out of the agarose drop. Fibroblasts to be tested were stained with DiD as described above, and then re-suspended in a volume of 0.2% (w/v) agarose solution to obtain the final cell density of 1.5×10^7 cells/ml. Droplets of 1.25 μl the cells suspension were delivered with a sterile micropipette into the wells with different substrates. The wells were then placed at 4°C for 15 min to allow the agarose to solidify. After cooling, the agarose droplets were covered with full-DMEM gently.

After incubation for a predetermined time, time-lapse images of the cells were recorded every 15 min for up to 60 min with a MetaMorph-operated CoolSNAP™HQ camera attached to a Nikon Diaphot-TMD inverted microscope fitted with a 37°C incubator stage and a 10× objective lens. For each sample, the migration speed was determined from the time-lapse images using MetaMorph, which tracked the distance covered by the center of a nucleus very 15 min over a period of 1 h. The same measurement was done after 6, 16 and 24 h of cell culture. Each measurement represents the average and standard deviation for 20 cells, with 3 duplicates and the entire curve was repeated at least 4 times.

3 RESULTS AND DISCUSSION

3.1 Cell Migrating Track on PMMA Spun Cast Film and Aligned Fibrous Scaffold

Previously, we showed that the appearance of the cells on an electrospun fiber mat, whose diameter was smaller than one micron, was similar to that of the cells cultured on a flat film. In addition to fiber diameter, orientation is another crucial parameter which can determine cell migration. To detect the effect of fiber orientation, the agarose droplet was placed on a PMMA thin film and scaffold composed of aligned fibers with diameter of 8 μm. From figure 2, we can clearly see that cells appeared to migrate radially on PMMA thin film. However, cells migrated along the direction of the electrospun fibers.

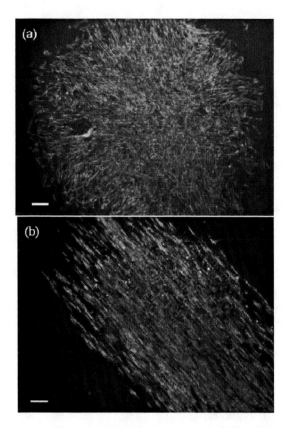

Figure 2: Fluorescence microscope overlap images of the live cells, stained with DiD (red), onto the image of the cells incubated for 24 h, fixed and stained for F-actin with Alexa Fluor (green), where the edge of the droplet is clearly marked: the agarose droplet deposited on Fn-coated PMMA (a) film and (b) aligned fiber. Scale bar=0.5 mm.

3.2 Cell Migration Velocity on PMMA Spun Cast Film and Aligned Fibrous Scaffold

Average speed of en masse cell migration was measured using time-lapse imaging and MetaMorph software. In order to measure cell migration velocity reproducibly, we chose to focus on cells at the leading edge of the corona, since it was easier to determine their position. The average cell migration speed, of the fibroblasts at the leading edge of the corona formed on different substrates, is plotted as a function of incubation time in figure 3. From the figure, we can see that on the flat film surface (figure 3, black line), the migration velocity decreased exponentially with incubation time. On the aligned fibrous scaffold, cell migration velocity remained constant in time and of the same magnitude as that of the single cell, or the asymptotic value reached by the cells on the Fn-coated flat surface.

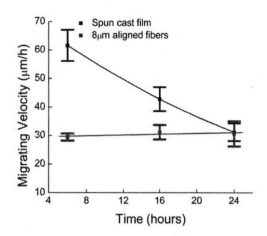

Figure 3: Cell migration velocity as a function of incubation time. Each measurement represents the average and standard deviation for 20 cells, with 3 duplicates and the entire curve was repeated at least 4 times.

3.2 Cell Migrating on PMMA Cross-aligned Fibrous Scaffold

A true 3D scaffold consists of alternating layers of fibers. The cell migration results on cross-aligned scaffold are shown in figure 4. From the figure, we could see that most of the cells were oriented in the cross hatched pattern of the matrix.

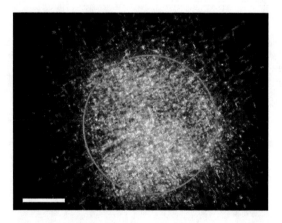

Figure 4: Fluorescent microscope image of live cells, stained with DiD, migrating from an agarose droplet onto Fn-adsorbed PMMA cross-aligned fibrous scaffold after 24 hours incubation. The ring surrounds the periphery of the original droplets. Scale bar=100 μm.

In figure 5 we show a SEM image of the cells that were fixed and dried on the fibers. From the figure we can see that even though most of the cytoplasm of the cells was centered on one fiber, a large number of filopodia extended out to the neighboring fiber. In this manner, the cells could sense the cell density in their environment.

We could also see that when the cells migrated on the 3D cross-aligned fibers, cells will not "flow" and migrate through the spaces between the fibers. Rather, cells will adhere to the fibers and either remain on a straight course or follow the fiber on the layer above or below. In figure 5 we also show two cells on a cross hatched fiber mat. The arrows point to one cell, which is oriented entirely on one fiber and migrating in a straight course. The second set of arrows point to another cell, whose cytoplasm is bent at 90° as the cell began to migrate to a lower layer and follows the fiber oriented at 90°. Since the two cells are in close proximity, they are trying to move away from each other, which become possible only through the 90° rotation of one of the cells.

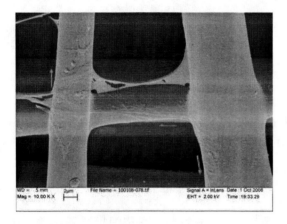

Figure 5: SEM image of cells migrating on the cross-aligned PMMA scaffold. Note that the cells extended between adjacent fibers.

Figure 6: Cartoon of cells plated on a scaffold composted of two layers of aligned fibers oriented at 90° to each other. The cells migrated by following the contours of the fibers. At a junction, the cells can either continue straight on one fiber or follow the fiber oriented at 90° into the next layer.

4 CONCLUSION

The migration of cells on 3D electrospun scaffolds is summarized in figure 6. Cells prefer fibers to the flat surface of identical chemical composition [6]. As a result, when an agarose droplet containing dermal fibroblast cells

are plated on the surface of the aligned electrospun fibers fiber mat, where the diameter of the fibers is larger than 1 micrometer, the cells migrate along the fibers. In contrast to cells migrating from a droplet placed on a flat film, the cells migrate with constant velocity, similar to that observed for single cell migration. On a flat surface, the cell migration is radial, as cells try to reduce the cell density. The velocity is initially fast and eventually reduces to that of the single cell velocity as the cell density decreases. On fibers, the cell density is constant, as is the cell migration velocity. When placed on a three dimensional mat, such as the one shown in figure 6, the cells will migrate from one level to another following the contour of the fibers. Hence the migration is unaffected the by size of the "holes" formed by the fibers.

5 ACKNOWLEDGMENTS

This work was supported by the National Science Foundation Materials Research Science (NSF-MRSEC) program.

REFERENCES

[1] D. A. Lauffenburger, A. F. Horwitz, "Cell migration: a physically integrated molecular process," Cell, 84, 359-369, 1996.
[2] T. J. Mitchison, L. P. Cramer, "Actin-based motility and cell locomotion, Cell, 371-379, 1996.
[3] A. Curtis, C. Wilkinson, "Topographical control of cells," Biomaterials, 18, 1573-1583, 1997.
[4] P. Friedl, E. B. Brocker, "The biology of cell locomotion within three-dimensional extracellular matrix," Cell Mol. Life Sci., 57, 41-64, 2000.
[5] J. Varani, W. Orr, P. A. Ward, "Comparison of migration patterns of normal and malignant cells in 2 assay systems," Am. J. Pathol., 90, 159-171, 1978.
[6] Y. Liu, A. Franco, L. Huang, D. Gerdappe, R. Clark, M. Rafailovich, "Control of cell migration in two and three dimensions using substrate morphology," Exp. Cell Res. 315, 2544-2557, 2009.

Design and multiobjective optimization of a novel reactive extrusion process for the production of nanostructured PA12/ PDMS blends

F. Pla, R. Rached, S. Hoppe, C. Fonteix

Laboratoire Réactions et Génie des Procédés (LRGP)-CNRS-Nancy Université
1 rue Grandville - BP 20451 - 54001 NANCY Cedex (France)

ABSTRACT

This work deals with the manufacture, by reactive extrusion, of Polyamide 12 (PA12)/Polydimethylsiloxane (PDMS) nanostructured blends, with the purpose of improving the mechanical properties of PA12. The originality of the process is based on the simultaneous anionic synthesis of both PA12 and a compatibilizer (PA12-PDMS-PA12) in the presence of the dispersed phase (PDMS).The process was studied owing to an adapted experimental strategy which allowed elaborating empirical models able to predict the effect of the main operating variables on the corresponding blends properties. These models were then used in a multiobjective optimization procedure which allowed, determining the best operating conditions for the production of blends for which all specific properties were maximized except the diameter of the dispersed phase which was minimized.
This work led to a significant improvement of the mechanical properties of PA12, particularly its impact energy at low temperature.

Keywords: nanostructured blends, PA12, PDMS, reactive extrusion, multiobjective optimization

1 INTRODUCTION

The concept of physically combining two or more polymers via blending to enhance mechanical properties of materials, has received a lot of attention [1], [2].

Traditionally, the elaboration of these blends is carried out by mixing, at the melt state, the polymers in presence of a compatibilizer [3]. Usually, under these conditions, this leads to a micrometric scale dispersion of the minor phase and it is known that at this scale of dispersion the mechanical properties of the final materials are not optimal, whatever the method of compatibilization. By contrast, this work sets out a new process of elaboration of polymers blends, through a typical example: the manufacture, by reactive extrusion, of PA12/PDMS nanostructured blends. In this process, the continuous phase *(PA12)* is synthesized in the extruder in the presence of: (i) the dispersed phase *(PDMS)*, (ii) a new and particularly efficient catalytic system *(aluminium hydride* as catalyst *and MDI caprolactam* as activator and (iii) a macro-activator *(functionalized PDMS: α,ω–dicarbamoyloxy-caprolactam-PDMS)* well suited to develop *in situ* a tri-block (PA12-PDMS-PA12) copolymer acting as compatibilizer. The originality of the process is based on the <u>simultaneous synthesis</u> of both PA12 and the compatibilizer by activated anionic polymerization of lauryl lactam (LL). The final goal is to obtain blends in which the elastomer particles *(PDMS)* are dispersed and stabilized at a nano-scale and with mechanical properties better than those of PA12.

This research comprises (i) the synthesis, the kinetic study and the modeling of the activated anionic polymerization of LL (AAPL). This was preceded by the synthesis of both the macro-activator and a catalytic system giving rise to complete conversion of LL in a short and controlled reaction time, (ii) the complete design, modeling and optimization of the reactive extrusion process.

2 STUDY AND MODELING OF THE ANIONIC POLYMERIZATION

Prior to the study of this polymerization, the catalytic system and the macro-activator were synthesized. The efficiency of the catalytic system was tested according to methods described in previous papers [4], [5], which clearly showed that the total conversion of LL can be achieved, at 220°C, in a reaction time comparable to classical residence times in the extruder (3-4 minutes). This result was confirmed owing to experiments carried out in the extruder and followed by the construction of a kinetic model for the syntheses of both the matrix (PA12) and the interfacial agent (PA12-PDMS-PA12) [5].

2.1 Elementary chemical reactions

The chemistry of AAPL is complex in nature. However, the process is of high interest, particularly, owing to (i) its fast reaction rate and (ii) the interesting mechanical properties of the resulting polymer. AAPL involves several reversible and irreversible reactions in which the active species are consumed and regenerated. The general reaction scheme responsible for the polymer chains formation has long been established. It comprises:
- an initiation step (reaction between the activator and the catalyst),
- a propagation step consisting of a proton transfer followed

by a nucleophilic addition,
- side reactions, mainly Claisen's reactions characterized by the formation of enols and condensation reactions.

2.2 Main assumptions

The establishment of the model requires the use of several assumptions to enhance the speed of convergence. These assumptions can be summarized as following:
- initiation is very fast and is considered as irreversible,
- the reaction giving rise to enols will not be taken into account because it is analytically similar to the proton transfer reaction. Therefore, its kinetic parameter will be included into that of the proton transfer,
- proton transfer, propagation and Claisen's reactions rate are supposed to be independent of the chain-length,
- the reactor is isothermal and ideally mixed.

2.3 Mathematical model

Thanks to the above mechanisms and assumptions, the model is based on material balances expressed for each species (activator, catalyst, monomer, active species and macromolecules). It consists of a system of differential algebraic equations involving 18 parameters which were determined through the minimization of the maximum likelihood criterion, J, with the experimental data:

$$J = \sum_{data} (X^{experimental} - X^{simulated})^2 \qquad (1)$$

where X is LL conversion. This minimization was obtained using an evolutionary algorithm previously developed in our laboratory [6], [7]. These 18 parameters were identified thanks to batch polymerizations carried out at 180°C and 220°C with the same initial concentrations of monomer: $[M]_0 = 4.55$ mol.l^{-1}, catalyst: $[C]_0 = 3.10^{-2}$ mol.l^{-1} and activator: $[A]_0 = 3.10^{-2}$ mol.l^{-1}). Monomer conversion, X, was determined for reaction times varying between zero and around 15 minutes, through a Soxhlet extraction of the residual monomer during 24 hours, followed by weighting the resulting polymer sample previously dried under vacuum at 80°C during 24 hours.

2.4 Validation of the model

Figure 1 shows the time evolution of X for experiments carried out (i) at 180 and 220°C with the same recipe ([M]$_0$=4.55mol.l^{-1} and [C]$_0$=[A]$_0$=4.55.10^{-2}mol.l^{-1}), (ii) at 200°C for two different activator and catalyst concentrations.

The results clearly show the good agreement between simulated and experimental data. Moreover, as expected, X is higher when the temperatures and the catalytic system concentrations are increased.

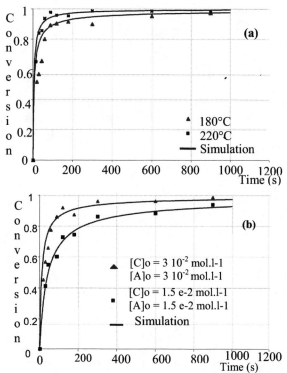

Figure 1: Effect of temperature (a) and catalytic system concentration (b) on lauryl lactame conversion

3 BLENDS ELABORATION

3.1 Preliminary experiments

Blends were synthesized in a conical intermeshing counter-rotating twin screw extruder. Melting of LL, PDMS, functionalized PDMS and activator was carried out in an electrically heated flask mechanically stirred under a nitrogen atmosphere. This mixture and the catalyst were pumped separately by controlling the mass flow into the extruder hopper. Connections between the flask and the extruder were accomplished by a heated metal tube. The resulting materials were then pelletized, dried at 80°C and injection-molded in the form of test bars for tensile and strength measurements. Preliminary experiments allowed to analyze the effects of the macro-activator concentration, [MA], the [PDMS]/[MA] ratio and PDMS number average molar mass, Mn, on blends morphology and mechanical properties. The results clearly showed that the average diameter of the dispersed phase decreased while the corresponding impact energy increased significantly when [MA] and Mn were increased.

3.2 Process modeling

An experimental design was then carried out in order to elaborate a model able to predict the effect of the main process operating variables [*Screw rotation speed (N)*,

PDMS feed rate (Q), concentration (F) and number average molar mass (Mn)], on the corresponding blends properties [LL conversion (X), average diameter of the dispersed phase (D_n), Young modulus (E), melting temperature of the matrix (T_m) and Charpy impact energy (E_{imp})].

In order to obtain a model with correct specification and well determined parameters, a series of 30 runs were carried out. Given the fact that the expected responses do not vary in a linear manner with the selected variables, a D-optimal plan, where the response could be modeled in a quadratic manner, was selected to enable the quantification of the prediction of the responses.

An optimal number of 20 experiments were obtained using an evolutionary algorithm [6], [7].

The rotatability method was then used to complete the experimental design in order to increase the degree of freedom and to choose 6 more experiments to validate the model.

4 repetition experiments were also performed in order to validate the statistical significance of the chosen experimental plan.

The proposed mathematical equations were complete second degree polynomials using the normalized values of N, F, Q, and Mn, respectively.

The different steps used for the calculation of the model coefficients (a_i), together with their confidence intervals are summarized here only for Young's modulus. The final polynomial equations of the other properties were obtained using the same method.

By applying a multiple regression analysis on 20 experimental data, the coefficients of equation (1) were estimated.

The quality of the estimations needs to be checked out. The confidence interval for each coefficient with a risk of 5% was then estimated. If the confidence interval of a coefficient is large and include zero, the parameter must be discarded (its effect being negligible). So, a total of 7 coefficients were deleted which led to a model containing 8 coefficients.

The final structure of the model was then defined by:

$$E = 520 - 34x_2 - 189x_3 - 113x_4 + 53x_1x_2 - 60x_2x_4 + 128x_3^2 + 132x_4^2 \quad (1)$$

It was then necessary to check out the absence of correlation between the coefficients of the model.

The degrees of correlation between two coefficients were evaluated using the variance-covariance matrix. In the present case no correlation between the different coefficients was found.

The validity of the polynomial equation was finally verified by use of Fischer-Snedecor's test based on the ratio of the variances applied to identification, validation and repetition experiments.

Figure 2 shows, as examples, the validation of the model equations obtained for D_n and E_{impact} for which a fairly good agreement was observed between simulated and experimental data.

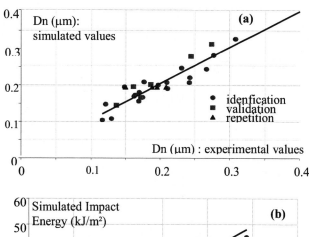

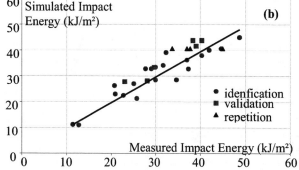

Figure 2: Validation of the model: simulation versus experiments (a): Average particles diameters; (b): Impact energy

3.3 Multiobjective optimization

The model thus obtained was then used in a multiobjective optimization procedure which allowed, owing to the use of the concept of domination, obtaining a set of non-dominated solutions called Pareto's domain.

A specific decision engineering tool was then set up to classify all these solutions taking into account the preferences defined by a decision maker in order to determine the best operating conditions for the production of nanostructured blends characterized by a high impact energy at low temperature.

The concept of domination used to select a solution, is defined in such manner that a solution (compromise or vector) z_1 dominates another solution z_2 if it is better or equal for all criteria and strictly better for at least one criterion [8]. The input variables constitute Pareto's domain while the output criteria constitute Pareto's front. In the present study, this domain was defined by an infinite number of solutions obtained using an evolutionary algorithm [7].

Among the different methods developed to classify the non-dominated solutions, we used the Net Flow Method [9] based on a decision-maker's choice defined for each property by 4 parameters: weight and preference, indifference and veto thresholds. Table 1 gives the values of the 4 parameters used in this investigation. The selection of a proper level of each threshold is not a trivial task. It is based on a good knowledge of the process and on the range spanned by each criterion of Pareto's front.

Decision-maker's choice	T_m	D_n	X	E_{impact}	E
Weight	0.10	0.30	0.15	0.30	0.15
Indifference threshold	0.5	0.005	0.005	2	5
Preference threshold	1	0.01	0.01	5	20
Veto threshold	2.6	0.03	0.03	10	40

Table 1. Decision-maker's choice for each criterion

The final objective of this work was to maximize T_m, X, E_{impact} and E, and to minimize Dn. According to the decision maker's choice, an example of visualization of the results is given through an input-input graph (Figure 3a): F, versus N and an output-output graph (Figure 3b): $E_{impact)}$, versus Dn. This example clearly shows that N covers all the range of its use, while (F) is limited in a restricted range. On each figure are represented 5000 non-dominated solutions constituting Pareto's domain -figure 3(a)- and Pareto's front -figure (3b)-, respectively. The 20% best and the preferred solutions are also indicated in grey and by a red triangle respectively.

These results show that, for the proposed decision-maker's choice, the recommended operating conditions are: N=75.6 rpm, F=3.43%, Q=20.1 g/min and Mn=62,700 g/mol. The corresponding values of the blends properties are: X=1, D_n=128 nm, T_m=167.6°C, E=485 mPa and E_{impact}=43 kJ/m². This proves that the chosen decision-maker's preferences allowed well to minimise Dn and to maximize E_{impact}. This was a deliberate choice to obtain a nanostructured blend with high impact energy.

To validate these results, an experimental run was carried out using the following operating conditions: N=80 rpm, F=3.30 g/min., Q=20 g/min. and Mn=62700 g/mol. The corresponding blend properties were: X=0.95, D_n=115 nm, T_m=167.4°C, E=490 mPa and E_{impact}=42 kJ/m². They correspond to an actual improvement of pure PA12 properties which are: E_{impact}=7 kJ/m² (at 23°C) and 6 kJ/m² (at -30°C), T_m=178°C and E=1100 mPa.

4 CONCLUSION

The objective of this work was to determine the optimal operating conditions of a reactive extrusion process for the production of nanostructured PA12/PDMS blends. In this process, the matrix (PA12) and the compatibilizer (PA12-PDMS-PA12) were simultaneously synthesized in the presence of the dispersed phase (PDMS). The optimization is a multiobjective procedure using: (i) Pareto's concept, to determine a large number of non-dominated solutions, (ii) ranking of these solutions according to decision-makers'

preferences. This method allowed selecting the optimal operating conditions giving rise to the best properties of the corresponding blends and confirmed the importance of their nanostructured morphology.

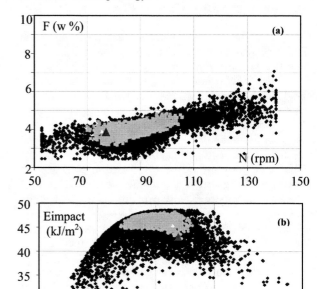

Figure 3: Ranked set of non dominated solutions:
(a) F versus N ; (b) E_{impact} versus Dn

REFERENCES

[1] R.P. Kambour, *NATO* ASI Series, E : Applied Sci., 331-47, 1985

[2] M. Hara, J.A. Sauer, *J. Macromol. Science, Reviews in Macromol. Chem. and Phys.*, C38 (2), 327-362, 1998

[3] K. Dedeker, G. Groeninck, J. Appl. Polym. Sci., 73 (6), 889-898, 1999

[4] R. Rached, S. Hoppe, A. Jonquieres, P. Lochon, F. Pla J.Appl. Polym Sci., 102, 3, 2818-2831, (2006)

[5] R. Rached, S. Hoppe, C. Fonteix, C. Schrauwen, F. Pla. Chem. Eng. Sci. 60, 2715-2727, (2005)

[6] F. Bicking, C. Fonteix, J.P. Corriou, I. Marc, *Recherche opérationnelle /Operations res.*, 28, 23-36, 1994

[7] C. Fonteix, F. Bicking, I.Marc, Intern. J.Systems Sci., 26, 1919-1933, 1995

[8] R. Viennet, C. Fonteix, I. Marc, In: *Lecture Notes in Computer Science, "Artificial Evolution"*, Springer Verlag Ed., 120-127, (1996).

[9] L.N. Kiss, K. Zaras, C. Fonteix, R. Dominique, ASOR Bulletin, 21 (2), 2-8, (2002)

EFFECT OF SIZE ON PROPERTIES OF NANO-STRUCTURED POLYMERS –TRANSITION FROM MACROSCALING TO NANOSCALING

K. W. Lem*[1a], J. R. Haw[1], D. S. Lee[2], C. Brumlik[3], S. Sund[4], S. Curran[5], P. Smith[6], S. Brauer[3], D. Schmidt[7], and Z. Iqbal[8]

(1) Department of Materials Chemistry & Engineering, Konkuk University, Seoul, Korea. jrhaw@ konkuk.ac.kr
(a) Corresponding author :(O) 822-20496247 ;(FAX) 822-4443490; kwlem@konkuk.ac.kr; kwlem2001@yahoo.com
(2) Department of Chemical Engineering, Chonbuk National University, Jeonju, Korea. dslee@jbnu.ac.kr
(3) Nanobiz, LLC, NJ, USA. brumlik@nanobizllc.com; sambrauer@nanotechplus.net
(4) Nygard Consulting, LLC, NJ, USA. seas_nj@yahoo.com
(5) Boston Scientific, MA, USA. sean.curran2@bsci.com
(6) Knowles Electronics, IL, USA. apetesmith@aol.com
(7) Materials to Market, LLC, NJ, USA. j.david.schmidt@embarqmail.com
(8) Department of Chem. and Environmental Sci., New Jersey Institute of Technology, NJ, USA. iqbal@adm.njit.edu

ABSTRACT

The growth in investment on the effect of size in nanoscience and nanotechnology has been astounding. Of the $12.4B spent in 2006 worldwide Nanotechnology research funding [1], at least 50 % was spent on the effect of size for the development of nanomaterials and devices. The focus of nanoscience is to understand the influence of size on the material properties, while nanotechnology focuses on using size effects to create structures, devices and systems with novel properties and functions.

Based on literature data, we have demonstrated that a modified Hall-Petch relation is useful to relate the observed "effective" nanosize to physical properties. Several viscoelastic models have been developed to explain the role surface molecular dynamics plays in thermal mechanical behavior. The polymers examined including polystyrene, polymethyl methacrylate, and Metafuse[TM].

Keywords: nanoscaling, tunable properties, hall-petch relation, transitions, interfacial dynamic

1. INTRODUCTION

Materials such as metals, ceramics, and polymers exhibit characteristic bulk properties, as well as characteristic response to sizes on the nanoscale. Metals are strong and ductile, but optically opaque and reflective. They exhibit high thermal and electrical conductivity. Ceramics are compounds of metallic & non-metallic elements such as oxides, carbides, nitrides, sulfides. They are brittle, glassy, elastic, and non-conducting. Polymers are of long chain organic molecules with many repeated units. They are soft, ductile, low strength, low density; and can be optically translucent or transparent. They are typically thermal and electrical insulators.

Fundamental structure changes with size have led to the size dependent material properties which are substantially different from their counterparts in bulk[2]. The extent of delocalization of valence electrons can vary with the size of the system, and quantum effects become relevant for sizes less than 10 nm. Material properties become tunable by size; notably, coordination number imperfection, surface relaxation behavior and surface, nanosolidification in physical properties, superplasticity in mechanical properties, melting and thermal diffusivity in thermal properties, acoustic phonon hardening and optical phonon softening behavior, quantum confinement effects in optical properties, work function and dielectric suppression in electrical properties, and magnetic modulation in magnetic properties. For example, the bandgap of semiconductors such as ZnO, CdS, and Si, changes with size, and. magnetic materials such as Fe, Co, Ni, Fe3O4, etc., exhibit size dependent magnetic memory properties.

2. EFFECT OF SIZE ON PROPERTIES

The Hall-Petch relation in Eq. 1 where b = - 0.5 has been used to describe the effect of grain size on the yield or flow stress for metal systems empirically[3]

$$\sigma_y = \sigma_o + K_y * d^b \qquad (1)$$

Where σ_y is the yield stress for a polycrystal, σ_o is the yield stress for a single crystal or a polycrystal with an infinitely large grain size, d is the average grain size, and K_y is a Hall-Petch material specific constant. Grain boundaries in metal systems play a critical role obstructing the movement of dislocations, thereby influencing the yield or flow stress. When the grain size is very large or extremely small, Hall-Petch relation at b= - 0.5 fails to describe the effect of grain size on yield or flow stress, and toughness for metal systems. Consequently, a new relationship must be developed to predict grain size effects at the nanoscale.

3. SIZE EFFECT ON NANOSCALING

Nanoscale, material properties vary as a function of the size compared with the same materials on a macroscale. Sun in 2007[2] has given an extensive review on evaluation of the bond-order-legnth-strength (BOLS) mechanism on the effect of size on the usual behavior of material properties. An inverse Hall–Petch relation was reported by Nieh and Wadsworth[4] for grain sizes below a certain threshold, d_c as seen in Figure 1. Conrad and Jung[5] have identified the following three discernible regions in relationship of the effect of grain size from mm to nm on the flow stress and plastic deformation kinetics of Au particles at 300 °K:

1. Region I (d $>\sim 0.5$ μm), Grain size hardening in accord with the Hall–Petch Relation;
2. Region II (d $\approx$ 10–500 nm) Grain size hardening in accord with the Hall–Petch Relation;
3. Regime III (d $<\sim 10$ nm). Grain size softening with the flow stress proportional to b = 1 in Relation (Inverse Hall-Petch Relation).

They reported that the general behavior of Au nanoparticles is similar to that reported previously for Cu and Ag nanoparticles. Sülleiová et al.[6] have confirmed findings of these three regions in their work on nanocrystalline copper. They reported that the Hall-Petch equation is applicable to the range from mono-crystal to the grain size of 1 μm, where the exponent, b, has values from -0.5 to 1. However, at grain size ranging from 10 to 100 nm b is less than -0.5. Finally for the grain sizes < 10 nm, b can change its value. At this level of grain size, the material has extremely high volume fraction of grain boundaries, exceeding 50 %[6].

Figure 1 - Macroscopic to Nanoscopic Transition
Adopted from Nieh and Wadsworth[4]; Sülleiová et al.[6]

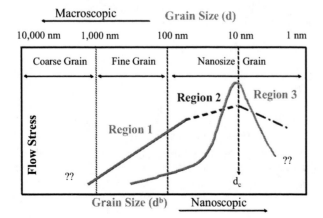

TUNABLE PROPERTIES BY SIZE

In most of nanotechnology studies, the size range of greatest interest is typically from 100nm down to approximately 0.2 nm, because in this size range properties of the materials become tunable. Three primary reasons for this tunable behavior are believed to be:

(1) Increases in relative surface area vs. bulk volume,
(2) Increases in % of atoms/molecules on the surface,
(3) The dominance of quantum effects.

The increased surface area per unit mass and % of atoms on the surface result in a corresponding increase in chemical reactivity. Thus, some nanomaterials become useful as catalysts to improve the efficiency of fuel cells and batteries, or in "soft" nanotechnology for cosmetics and pharmaceuticals[7].

The effect of size on other properties such as surface tension or 'stickiness' and thermal transition temperatures are also important, because these characteristics directly affect physical and chemical properties.

4. EQUATIONS FOR NANOSCALING

Based on the data reported in the literature to date, an equation similar to Hall and Petch relation (Eq. 1) has been reported to predict size dependent nanomaterial properties as shown in Eq. 2.

$$A = A_o + K * d^b \qquad (2)$$

Where A is the size-dependent material properties, Ao and K are material specific constants independent of size. They may be affected by temperature, time, pressure, and other service parameters. The exponent, b, is a constant for a given size range.

We must point out at this juncture that equations obtained empirically may oversimplify underlying scientific principles. Such equations should be used with caution. However this empirical approach may find its usefulness when the data is difficult to obtain. It may provide at least an order of magnitude estimate to predict trends of tunable material properties.

Some examples of such predictions are given in the literature. Nanda[8] highlighted that Pawlow in 1909 had predicted a melting point depression of nano metal particles. Melting point is inversely proportional to the particle size (d) using a thermodynamic model. Similar results were reported by Kang et al.[9] in their phonon spectrum study. In a simulation based on a statistical melting criterion, they found that a lowering in nano metal particles' melting point is inversely proportional to the particle size (d). Kelsall et al.[10] reported that the linear strain of a nanomaterial is inversely proportional to particle size (d) as the nearest-neighbor distance decreases with decreasing size. Surface cohesion energy can be actually more stable than the volume cohesion energy in nanoparticles due to the fact that cohesion energy decreases as diameter of nanoparticles decrease. Yaghmaee and Shokri[11] have proposed Eq. (3) illustrating that the cohesive energy of nanoparticle is inverse proportional to the nanoparticle' diameter.

$$E_p = E_b \left(1 - \frac{1}{fv} \frac{d_a}{d_p} \right) \tag{3}$$

Where E_p is the cohesive energy for particle, and E_b is the cohesive energy in bulk, f_v is the package factor, d_a is the diameter of the atom in neutral state, and d_p is the diameter of the particle.

5. POLYMER HYBRIDS - METAFUSE ™

DuPont's MetaFuse™ product line is based on nanocrystalline metal/polymer hybrid materials that have lightweight engineering plastics components in complex shapes with the stiffness of soft metals such as magnesium or aluminum. The technology is based on deposition of nanocrystalline metal (such as nickel with average grain size of 20 nm) on the surface of molded engineering plastics. According to Ray[12], at this range of the grain size, the nanocrystalline metal particles increase the mechanical properties including yield strength, hardness, and wear resistance because it follows a Hall-Petch relation (see Eq. (1)). Interestingly, Sun [2] reported that the onset of modulus increase in nickel nanoparticles is ~ 2 – 3 nm, at which the inverse Hall-Petch relation may prevail.

Polymers are viscoelastic which means they exhibit both viscous and elastic responses when undergoing deformation. Viscoelastic behaviors are dynamic in nature, characterized by their time dependence, loading history dependence, and lack of a one-to-one correspondence between stress and strain. All polymers exhibit a glass transition temperature (Tg) and this can be considered to be one of the most important polymeric material properties. Below the glass transition temperature, the materials act as hard, rigid glasses. Above Tg, they are soft and flexible. In general, any factor that affects segmental mobility will affect Tg. This includes factors that influence the nature of the moving segment, chain stiffness or steric hindrance, and those that influence the free volume available for segmental motion. All of these factors come into play for nanoscaled polymer structures.

6. NANOSCALING IN POLYMERS

In the last two decades, literature studies have reported that the dynamic properties of polymer materials at a nanosize thickness (typically <100nm thick) are found to be very different to those found in the bulk materials. Fu et al.[13] reported recently that the Young's modulus of poly(vinyl alcohol) (PVA) nanofibers increased dramatically when their diameter became very small. The onset of the change is at approximately 140 nm. They report that the Young's modulus increases from 35 GPa at 160 nm diameter to about 530 GPa at 20 nm diameter. In their recent dielectric dynamic study of isotactic PMMA, polyvinylacetate, poly(2-vinylpyridine), and hyperbranched aromatic polyesters, Serghei et al.[14] observed that branched polymers (hyperbranched aromatic polyesters) appear to experience a more pronounced change in measured mean relaxation time for the dynamic glass transition compared with linear polymers (isotactic PMMA). In addition, the onset of the change for the branched polymers was observed to take place at much larger film thicknesses than that of the linear polymers. They reported onset temperatures of isotactic PMMA at 6 nm, polyvinylacetate at 16 nm, poly(2-vinylpyridine) at 36 nm, and hyperbranched aromatic polyesters at 200 nm film thickness.

The root cause of these anomalies is believed to be interfacial properties. This hypothesis has been substantiated by both experiments and computer simulation. Most researchers have found that these anomalies become profound with decreasing film thickness. Thus, the dependence of dynamic properties of nanothin polymer films qualitatively follows the relation described by Eq. (2). Yet quantitative understanding of the role of the interface still remains an elusive and controversial topic[15].

7. NANOTHIN POLYMER FILMS - T_G

Kim et al.[16] reported that the Tg increases with decreasing film thickness for the nanothin films of polymers that have attractive interaction such as PMMA and P2VP. The Tg of nanothin polystyrene (PS) films was reported to be lower than the Tg of bulk films when the thickness of the film was reduced. At approximately 10 nm film thickness, the Tg of PS films can be decreased by as much as 25 °K compared to the Tg in the bulk [17]. Singh et al. [18] have observed that the decrease in Tg depends on both film thickness and polystyrene molecular weight.

Table 1 gives a summary of the equations reported in the literature for Tg of as a function of film thickness in form of Eq. (2).

Table 1 – PS Films Thickness on Tg in Eq. (2)

A_o	K	b	Source
$A_o = T_{g, bulk}$	$K = - (T_{g, bulk}) \times (a)^{1.8}$, where a = 32 angstroms	b = -1.80	Forrest and Dalnoki-Veress [17]
$A_o = T_{g, bulk}$ = (366 – 113000/Mw)	$K = - (T_{g, bulk}) \times (g)$, where $g = (1/14.4) \times (Rg)^{1.78}$ and $Rg = \{s^2\}^{1/2}$ = RMS radius of gyration	b = -1.78	Singh et al. [18]

Table 2 gives a summary of regression results based on the data and equation (by keeping b = -1.78) as reported by Singh et al.[18]. As seen on Table 2, The value Ao is very close to the bulk Tg of polystyrene (~92°C or 365°K), whereas the material constant, K, is related to the radius of gyration of polystyrene. K shows a molecular weight and molecular weight distribution dependence as expected.

Table 2: Regression Results for Tg as a Function of PS Films Thickness in Eq. (2)

Polymer	A_o	K	b	Source
Polystyrene	$A_o = T_{g, bulk}$ = (366 – 113000/Mw)	K = - ($T_{g, bulk}$) x (g), where g = (1/14.4) x (Rg)$^{1.78}$ and Rg = $\{s^2\}^{1/2}$ = RMS radius of gyration	b = -1.78	Singh et al. [18]
Mw/Mn = 2.00; Mw = 239,700	A=367.5	K = -3203.5	b = -1.78	R^2 = 0.98
Mw/Mn = 1.05; Mw = 212,400	A=365.7	K = -2029.1	b = -1.78	R^2 = 0.98
Mw = 157,100	A=365.0	K = -5087.7	b = -1.78	R^2 = 0.89
Mw = 212,400	A=365.9	K = -1970.0	b = -1.78	R^2 = 0.95
Mw = 560,900	A=365.9	K = -3677.2	b = -1.78	R^2 = 0.97

8. FUTURE STUDIES

In this paper, we showed how a modified Hall-Petch relation was found to be useful to predict the effect of size on the change of nanomaterial properties. It provides a "quick and dirty" means to qualitatively evaluate nanoscaling trends and forecast material properties. Nonetheless, we need to be cautious because these types of relations are empirical and often oversimplify the underlying science. Ideally, size effect on dynamic interface properties and the role molecular dynamics at the surface interface should be quantified from the scientific first principles. As the size of matter is reduced to <100 nm, and material properties become tunable, specific quantum effects need to be identified that play a critical role in significantly changing a material's optical, magnetic, electrical, thermal and mechanical properties.

The studies from Serghei et al.[14], Qi[15] and Sun[2] should be continued with a focus to determine the role of molecular dynamic at the surfaces. The techniques developed by Han[19] (especially the plots of log (storage modulus) vs. log (loss modulus)), and Nagai et al. [20] (nanorheological measurement by AFM) should be employed to study the effect of size on dynamic flow behavior and molecular mobility. This will provide an understanding of the role of molecular characteristics and chemical nature in transitions from the bulk to the interface. The focus on the molecular characteristics include molecular weight, molecular weight distribution, degree of long chain branching, influence of ionic, hydrogen and other secondary bonding. Chemical nature is emphasized on flexible homopolymers, miscible and immiscible blends, composites, block and graft copolymers, liquid crystalline polymers, laminates, nanocellular plastics, and polymer

hybrids. Our current research focus is on pursuing this elucidation of the relationships of polymer properties to chemistry, molecular weight, branching at the nanoscale.

REFERENCES

[1] V. Mamikunian, "Investor Enthusiasm for Nanotech Opportunities in Electronics,"Lux Research Inc. 3-15, 2007
[2] C. Q. Sun, Progress in Solid State Chemistry, 35, 1-159, 2007.
[3] W.D. Callister, Jr. and D. G. Rethwisch, "Materials Science and Engineering – An Introduction," 8th Edition, Wiley & Sons, 2010.
[4] T.G. Nieh and J. Wadsworth, *Scripta Metall. Mater.* **25**, 955, 1991.
[5] H. Conrad, K. Jung, Materials Science and Engineering A 406, 78–85, 2005
[6] K. Sülleiová, M.Besterci, and T.Kvackaj, METAL 2009, 5, 19–21, 2009
[7] S. Brauer, K.W. Lem, and J.R. Haw, Presented in Nano and Green Technology Conference 2009, New York City, November 17-19, 2009.
[8] K.K. Nanda, PRAMANA - Journal of Physics, 72 (4), 617-628, 2009.
[9] K. Kang, S. J Qin, C. Wang, Physica E41, 817–821, 2009.
[10] R.W. Kelsall, I.W. Hamley and M Geoghegan; "Nanoscale Science and Technology," Wiley & Sons, 2005.
[11] M. S. Yaghmaee and B. Shokri, Smart Mater. Struct. 16, 349–354, 2007.
[12] M. R. Day, "Nanometal Polymer Hybrid," Advanced Materials & Processes, 25-27, April 2008.
[13] Q. Fu, Y. Jin, X. F. Song, J. Y. Gao, X. B. Han, X.G. Jiang, Q. Zhao, and D.P. Yu, Nanotech. 21, 095703, 2010.
[14] A. Serghei, M. Tress, and F. Kremer, J. Chem. Physics 131, 154904, 2009.
[15] D. P. Qi, "On near-free-surface dynamics of thin polymer films," Ph.D. Dissertation, University of Waterloo, Waterloo, Ontario, Canada, 2009.
[16] C. J. Kim, A. Facchetti, and T. J. Marks, JACS, 131, 9122-9132, 2009.
[17] J. A. Forrest and K. Dalnoki-Veress, Advances in Colloid and Interface Science, 94, 167-196, 2001.
[18] L. Singh, P. J. Ludovice, C.L. Henderson, Thin Solid Films 449, 231–241, 2004.
[19] C. D. Han, "Rheology and Processing of Polymeric Materials (Vol. 1 & 2)", Oxford University Press, 2007.
[20] S. Nagai, S. Fujinami, K. Nakajima, and T. Nishi, Composite Interfaces , 16, 13–25, 2009.

[1] [2010 NSTI – Paper# 301] Nanotech 2010, 696 San Ramon Valley Boulevard, Suite 423, Danville, CA 94526-4022, Ph: (925) 353-5004, Fax: (925) 886-8461, swenning@nsti.org.

Preparation and characterization of Sulfonated High Impact Polystyrene (SHIPS) ion exchange nanofiber by electrospinning

Taek Sung Hwang, Eun Jung Choi, Noh Seok Kwak

Department of Applied Chemistry and Biological Engineering, College of Engineering, Chungnam National University, 79 Daehangno, Yuseong-gu, Daejeon 305-764, South Korea, tshwang@cnu.ac.kr

ABSTRACT

High impact polystyrene (HIPS) nanofiber was prepared by electrospinning overcoming brittleness of polystyrene (PS). HIPS nanofiber can be crosslinked after electrospinning with divinylbenzene as a crosslinking agent and heat. After thermal crosslinking, HIPS nanofiber was sulfonated by sulfuric acid. FT-IR, X-ray photoelectron spectroscopy (XPS), water uptake, ion exchange capacity (IEC), morphology and contact angle were used to characterize sulfonated HIPS nanofiber. According to XPS analysis, SO_3H group was increased with sulfonation time and water uptake was increased with sulfonation time, due to the hydrophilic group (SO_3H) introduced in the HIPS nanofiber.

Keywords: High impact Polystyrene (HIPS), nanofiber, electrospinning, Sulfonation, ion exchange capacity (IEC)

1 INTRODUCTION

The study of nanofibers has been a subject of intensive research because of their unique properties and widespread applications in many areas. Among the various methods used to prepare nanofibers, electrospinning has been very attractive in the past decade because it is a simple and versatile approach for producing a matrix of fibers with diameters ranging from a few nanometers to micrometers. To date, over 100 different polymers have been electrospun into nanofibers. Because of the very high aspect ratio, specific surface area, and porosity, electrospun fibers are used for various applications, including tissue engineering, wound dressings, drug delivery, and vascular grafts [1].

More recently fibers have been used for ion exchange and have the advantages of simplification of the preparation, fabrication, contact efficiency and physical requirements of strength and dimensional stability [2].

In this study, the electrospinning of solutions of HIPS dissolved in DCE and DMF was completed and nanometer fibers of HIPS were prepared by electrospinning. We obtained high capacity HIPS cation exchange nanofiber by sulfonation and evaluated the performance of HIPS cation exchange nanofiber. The chemical properties such as water uptake, ion exchange capacity (IEC), morphology and contact angle were evaluated FT-IR spectral analysis, XPS apparatus, via titration, scanning electron microscopy (SEM) observation and contact angle meter.

2 EXPERIMENTAL

2.1 Materials

The High Impact Polystyrene (HIPS) used in the study were obtained from Kumho petrochemical Co. Ltd. The 1,2-Dichloroethane (DCE, 99%) and N,N-Dimethylformamide (DMF, 99%) used as a solvent was obtained from Samchun Co. Ltd. The thermal initiator, benzoyl peroxide (BPO, 75%) from Lancaster Co. Ltd., was purified by recrystallization from water. As a crosslinking agent, divinylbenzene was high-purity grades purchased from Aldrich Co. Ltd. Sulfuric acid were from Duksan Co. Ltd.

2.2 Preparation of Sulfonated High Impact Polystyrene (SHIPS) Nanofiber

HIPS nanofiber was prepared by electrospinning of its solutions in 1,2-Dichloroethane (DCE) and N,N-Dimethylformamide (DMF). The process parameters studied in this work included the concentration of the polymer solution, the solvent (DCE:DMF) mixing ratios, the applied voltage, the flow rate. Figure 1 shows the schematic diagram of electrospining apparatus. Uniform HIPS nanofiber were prepared by this electrospinning with the following optimal process parameters: the concentration of HIPS in DCE and DMF was 23 wt%, the solvent (DCE:DMF) ratios was 30:70, the applied voltage was 15 kV, the flow rate was 1.5 ml/hr. The HIPS nanofiber was crosslinked by divinylbenzene at 90 °C and was treated by immersion in 95 % sulfuric acid (Aldrich) for 40-120 min at 60 °C. Finally the sulfonated HIPS nanofiber was treated to stepped dilution rinse (70%, 50%, 30% sulfuric acid) and followed by rinsing with deionized water [3]. Table 1 and Figure 2 show the reaction conditions and crosslinking and sulfonation mechanism used to prepare the HIPS cation exchange nanofiber.

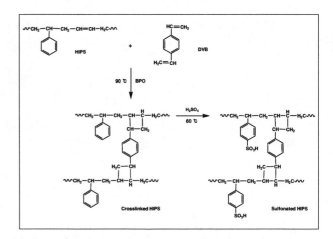

Figure 2: Mechanism of crosslinking and sulfonation on HIPS ion exchange nanofiber.

2.3 Characterization

The chemical structures of HIPS cation exchange nanofiber was characterized by Fourier transform infrared spectrometry (FT-IR).

X-Ray photoelectron spectroscopy (XPS) with Al Ka (1486.6 eV) X-rays evaluated the amount and state of nitrogen and sulfur ions doped into HIPS cation exchange nanofiber.

The swelling behavior of the sulfonated HIPS caion exchange nanofiber was measured by water uptake (WU). The water uptake was calculated as the ratio of the weight of the absorbed water to the dry membrane weight using the following equation.

$$WU(\%) = \frac{W_{wet} - W_{dry}}{W_{dry}} \times 100 \qquad (1)$$

Here, W_{wet} and W_{dry} are the wet and dry weights of the samples, respectively.

The ion exchange capacity (IEC) of the sulfonated HIPS nanofber were determined by Fisher's titration method. The experimental ion-exchange capacity (IEC) values were calculated from the following equation.

$$IEC(meq/g \cdot dry) = \frac{(V_{NaOH} \times N_{NaOH}) - (V_{HCl} \times N_{HCl})}{\text{Weight of sample}} \qquad (2)$$

V_{NaOH} is the volume of the NaOH standard solution, V_{HCl} is the volume of consumed HCl standard solution, and

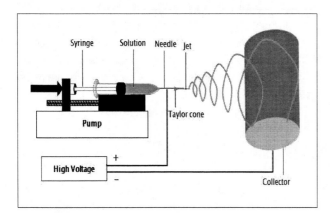

Figure 1: Schematic diagram of electrospining apparatus.

Solvent Mixing ratio (DCE:DMF)	Condition of electropinning				Crosslinking agent (wt%)	Sulfonation time (min)
	Conc. (%)	Volt (kV)	Flow rate (ml/hr)	TCD (cm)		
30:70	23	15	1.5	15	5 7.5 10	40 80 120 160 200

Table 1: Preparation conditions of SHIPS ion exchange nanofiber.

N_{NaOH} and N_{HCl} are the concentrations of the NaOH and HCl solutions, respectively.

The surface morphology of nanofiber before and after sulfonation reactions was observed by scanning electron microscopy (SEM).

The contact angle measurements were performed at room temperature ($25\,^{\circ}C$) using a commercial contact angle meter. All measurements were performed with deionized water.

3 RESULTS AND DISCUSSION

To confirm the change in the chemical structure of the HIPS ion exchange nanofiber, FT-IR spectra of the HIPS nanofiber before and after the sulfonation reaction were analyzed. Figure 3 shows the FT-IR spectra of HIPS and SHIPS, which were prepared with varying sulfonation times. In the sulfonated HIPS nanofiber, the absorption bands assigned to sulfonic acid groups were observed at 1250~1150 and 1060~1030 cm^{-1}. The strong band in the frequency range of 1250~1150 cm^{-1} can be ascribed to the stretching vibration of S=O, and the absorption band around 1060~1030 cm^{-1} is assigned to the symmetric (O=S=O) stretching band.

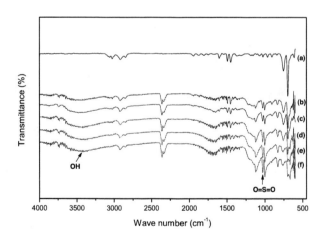

Figure 3: FT-IR spectra of HIPS nanofiber and sulfonated HIPS ion exchange nanofiber on 7.5% DVB content at sulfonation time (min): (a) 0, (b) 40, (c) 80, (d) 120, (e) 160 and (f) 200.

The composition of the surface layers of sulfonated HIPS cation exchange nanofiber was investigated by XPS. Figure 4 shows XPS wide scan spectra of sulfonated HIPS cation

exchange nanofiber having various sulfonation time. All nanofiber show three primary characteristic peaks corresponding to O1s, C1s, and S2p at binding energies of 532.5 eV, 284.9 eV, and 169.7 eV respectively.

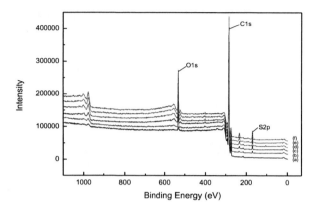

Figure 4: XPS spectra of HIPS nanofiber and sulfonated HIPS ion exchange nanofiber on 7.5% DVB content at sulfonation time (min): (a) 0, (b) 40, (c) 80, (d) 120, (e) 160 and (f) 200.

The water uptake is the most important property for ion exchange nanofiber in terms of their application. As shown in Figure 5, the water uptake decreased with increasing DVB contents. The maximum water uptake was 75.6 %, which was found for nanofiber prepared with 5 wt% of DVB and 200 min of sulfonation time. From this result, it can be concluded that DVB contents as crosslinking agent was an important factor for the water uptake.

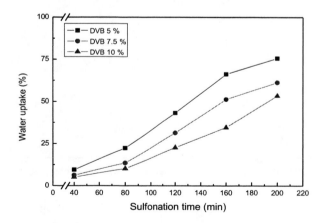

Figure 5: Effect of DVB contents on the water uptake.

The ion-exchange capacity (IEC) of the sulfonated HIPS nanofiber prepared with different sulfonation time and

DVB content as crosslinking agent were investigated by titration methods. Figure 6 shows the ion exchange capacity for HIPS nanofiber with different sulfonation reaction time and DVB content. The highest IEC value occurred for the sulfonated HIPS nanofiber prepared with 7.5 wt% of DVB and 200 min of sulfonation reaction time, having 2.53 meq/g.

The relationship between the hydrophilicity of nanofiber and the sulfonation reaction time is shown in Figure 8. It can be found that the contact angle decreases, implying the hydrophilicity of the nanofiber increases with the increase of sulfonation time.

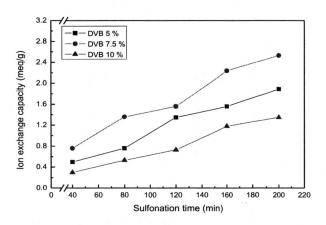

Figure 6: Effect of DVB contents on the ion exchange capacity.

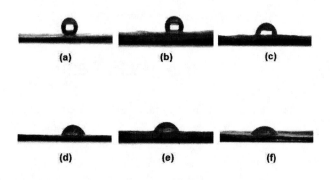

Figure 8: Contact angle of sulfonated HIPS ion exchange nanofiber (original magnification = 1000x) at sulfonation time (min): (a) 0, (b) 40, (c) 80, (d) 120, (e) 160 and (f) 200.

The morphology of the nanofiber influences the sulfonation reaction. The SEM images of HIPS and SHIPS nanofiber prepared with various sulfonation reaction times are shown in Figure 7. As shown in Figure 7, the SHIPS nanofiber surface is rougher than the HIPS nanofiber due to the increased sulfonation time.

REFERENCES

[1] Y. Lin, Y. Yao, X.Yang, N. Wei, X. Li, P. Gong, R. Li, D. Wu, "Preparation of Poly(ether sulfone) Nanofibers by Gas-Jet/Electrospinning," Journal of Applied Polymer Science, 107, 909, 2008.

[2] L. Dominguez, K.R. Benak, "Design of High Efficiency Polymeric Cation Exchange Fibers," Polymers for Advanced Technologies, 12, 197, 2001.

[3] H. An, C. Shin, G.G. Chase, "Ion exchanger using electrospun polystyrene nanofiber," Jouranl of Membrane Science, 283, 84, 2006.

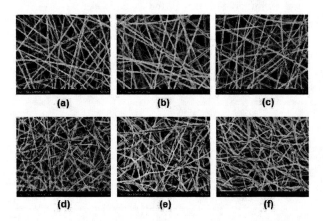

Figure 7: SEM images of sulfonated HIPS ion exchange nanofiber (original magnification = 1000x) at sulfonation time (min): (a) 0, (b) 40, (c) 80, (d) 120, (e) 160 and (f) 200.

Mechanical and Moisture Resistance Performance of Silver Nanoparticle Reinforced Fish Skin Gelatin Films

H. Harding[*], A. Ludwick[*], T. Samuel[**], S. Young[*], and H. Aglan[*]

[*]Department of Mechanical Engineering, Tuskegee University
218 Foster Hall, Tuskegee University, Tuskegee, AL 36088, aglanh@tuskegee.edu
[**]Department of Pathobiology, Tuskegee University, Tuskegee, AL, USA

ABSTRACT

Films from the biodegradable polymer gelatin have been studied for biomedical applications not only because of the film forming properties of gelatin, but also because of the availability and economical advantages of gelatin. In this work, nanostructured silver reinforced biocompatible films for controlled drug release were developed. Cross-linked and uncross-linked films from cold water fish skin gelatin were prepared. The cross-linking agent was 1,4-butanediol diglycidyl ether (1% (v/v)). Silver nanoparticles at loadings of 0.3% (w/v) were included in both types of films. A plasticizer was used for all neat and silver-loaded films. Cross-linking of the gelatin enhanced the mechanical properties and moisture absorbing properties of the films even though the uncross-linked film is completely soluble in water while the cross-linked is not. However, the strain to failure of the cross-linked films decreased in comparison to the uncross-linked films. Studies of biomedical applications are underway.

Keywords: silver nanoparticles, fish skin gelatin, tearing energy, moisture resistance

1 INTRODUCTION

Gelatin is a heterogeneous mixture of proteins extracted from mainly skin, tendons, ligaments, and bones. The structure of the gelatin contains repeating sequences of glycine-X-Y where X and Y are most frequently proline and hydroxyproline [1]. These sequences are the building blocks for the triple helical structures of the gelatin. Cross-linking of gelatin films is done in order to strengthen the mechanical properties and decrease the solubility of the films for use in biomedical applications. In a recent study, the solubility of fish-skin gelatin films was decreased by increasing the amount of the cross-linking agent 1-ethyl-3-(3-dimethylaminopropyl) carbodiimide [2].

In this work, gelatin extracted from cold water fish skin was examined. The objective was to formulate and manufacture fish skin gelatin films containing silver nanoparticles. The microstructure and processing conditions of these films were formulated with silver nanoparticles at a concentration of 0.3% (w/v). The tearing energy and moisture uptake properties of these gelatin films with silver nanoparticles were characterized.

2 MATERIALS AND EXPERIMENTAL

Gelatin (CAS # 9000-70-8) from cod, pollock, and haddock (cold water fish skin), with an average molecular weight of 60 kDa, was purchased from Sigma. This gelatin is a heterogeneous mixture of proteins extracted by a proprietary process, developed by the supplier. It remains unchanged for many years in sealed containers at room temperature, gelling below $35\,^{\circ}$ C. The isoelectric point of this gelatin is approximately 6. The plasticizer D-sorbitol was purchased from Fisher Scientific (BP439) (CAS# 50-70-4). The gelatin was cross-linked using 1,4-butanediol diglycidyl ether (CAS# 110-63-4) in a standard buffer solution of pH 10, both purchased from Sigma Aldrich.

Silver nano powder (less than or equal to 100 nm in diameter) was purchased from Sigma-Aldrich Inc. (Product #576832). These nanoparticles have an organic coating for dispersion in polar solvents and a high surface area ($5.0\,m^2/g$).

2.1 Processing of Neat and Nanoreinforced Films (Neat and 0.3%)

Gelatin powder was dissolved in distilled water while being mixed on a magnetic stirring plate (7.5 grams per 60 ml of water; 12.5%). The ambient temperature was about $20\,^{\circ}$ C $\pm\,5\,^{\circ}$ C and it took approximately 15 minutes for the gelatin to dissolve. Sorbitol was added into the gelatin solution (0.3 g sorbitol per 1 g of gelatin) in order to make the films soft [3]. For the nanoreinforced films, the silver nanoparticles were mixed into the 12.5% gelatin solution and the system was mixed overnight. The concentration of the silver nanoparticles was 0.3% (w/v). About 30 mL were poured into Teflon molds lined with Bytac ® contact paper and dried at $37\,^{\circ}$ C for 7 h in a convection oven.

After the films cooled to ambient temperature, a thin razor blade was used to carefully lift the outer edges of the film from the mold. All of the films were stored at ambient temperature in plastic sheets with 6 mm holes for ventilation, for at least a week before testing.

2.2 Processing of Cross-linked Neat and Nanoreinforced Films (crNeat and cr0.3%)

A method for cross-linking gelatin films was modified from a recent study involving gelatin-chitosan scaffolds which were cross-linked with 1,4-butanediol diglycidyl ether [4]. First, a 12.5% (w/v) solution of gelatin was fabricated by dissolving gelatin in a standard buffer solution of pH 10. The solution was mixed for about 45 minutes; then 2.625 g of sorbitol was added into the solution and mixed for 10 minutes at room temperature. For the nanoreinforced films, silver nanoparticles were added at 0.3% (w/v) to the solution. The solution was electronically mixed at 20°C $\pm$ 5°C overnight. The cross-linking agent, 1,4-butanediol diglycidyl ether, was added to form a 1% (v/v) solution and it was electronically mixed at 50°C for 4 hr with an air-cooled condenser. Finally, the solution was poured into Teflon molds and it was dried at 37°C for 7 h in a convection oven.

In this work, the selection of a pH of 10 for the cross-linking reaction was based on convenience. No clear colorless commercial pH 9 buffer was available. Hence a commercial colorless pH 10 buffer was used. Recently, 1,4-butanediol diglycidyl ether was used to cross-link a gelatin, chitosan, and hydroxypropyl hybrid at a pH of 9 for a biomedical application [4]. Cross-linking occurred when the epoxide groups of the agent, 1,4-butanediol diglycidyl ether reacted with the amine groups of gelatin [5]. The extent of cross-linking is unknown since not all of the epoxy groups were cross-linked. However, a study of cross-linking a gelatin film with 1,4-butanediol diglycidyl ether showed at least 50% of the film was cross-linked at a pH of 10 [6]. The result of the current work at pH 10 can be assumed to be similar.

2.3 Mechanical Testing of Gelatin Films

The tensile tests were performed on an electromechanical material testing system with a 25 N load cell. The uncross-linked samples were cut into rectangles with 6.5 mm width and a gauge length of 25.4 mm. The thickness varied from 0.3 - 0.6 mm. The samples were tested at a speed of 5.0 mm/min. The crNeat and cr0.3% films were tested with a 44.5 N load cell. The samples were cut into dog bones with a 4 mm width and a gauge length of 37 mm. At least 5 samples of all films were tested. The tearing energy tests were conducted on notched rectangular samples with 12.7 mm width and a gauge length of 25.4 mm.

2.4 Instrumentation

Fracture surfaces of notched samples from tearing energy experiments were examined with a Hitachi S-2150 scanning electron microscope (SEM). Thermal scans of the films were done on a TA Instruments DSC Q1000.

3 RESULTS AND DISCUSSION

3.1 Mechanical Properties of Neat and Cross-linked Neat Gelatin Films

A comparison of the Neat and crNeat films is shown in Figure 1. The crNeat films had higher ultimate strengths than the Neat films but lower strength to failure [7-8]. The average ultimate strength was increased by 126% and the average strain to failure was considerably reduced by 96% after cross-linking. Strain-hardening is observed in the stress-strain curve of the crNeat film but not in the uncross-linked film. The crNeat film stress increased linearly to about 6 MPa, then decreased to below 5 MPa, and finally rapidly increased until failure near 131% strain. The Neat film stress linearly increased to about 2 MPa and then gradually increased until failure at 300% strain. Cross-linking has caused strain hardening to occur in the crNeat stress-strain behavior. The crNeat films are brittle compared to the Neat film as seen by the lower strain to failure.

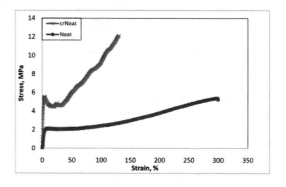

Figure 1: Stress vs. strain behavior of the neat and cross-linked neat films

3.2 Mechanical Properties of Silver-filled Gelatin Films

The stress-strain behavior of 0.3% and cr0.3% films is shown in Figure 2. There was no significant difference between the strain to failure nor the ultimate strength of the 0.3% and cr0.3% films. Strain-hardening was observed in the stress-strain curve of the cr0.3% film. The stress of the cr0.3% film increased linearly to about 3.5 MPa, then decreased to below 3 MPa, and finally rapidly increased until failure near 140% strain. Cross-linking of gelatin films has been observed to increase the tensile strength or have no effect [9]. Interestingly, the strain to failure has been determined to either increase, decrease, or remain constant, after cross-linking [9].

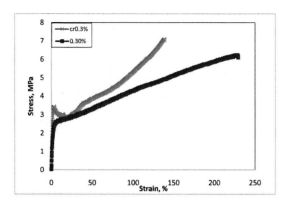

Figure 2: Stress vs. strain behavior of the silver-filled gelatin films

	Cut Depth (mm)	Tearing Energy (KJ/m2)	Tear Strength (KN/m)
0.3%	4.5	31.3	2.0
	4	33.8	2.0
	4	34.4	2.6
cr0.3%	5	19.4	1.1
	4.9	18.9	1.1
	5	20.9	1.4

Table 1: Tearing Energy Data for Gelatin Films.

3.3 Tearing Energy of Silver-Filled Gelatin Films

Elastomeric materials have a characteristic property of resistance to tear propagation, known as the tearing energy. A higher tearing energy means the material has a stronger resistance to tearing and is a tougher material. It is important in the mechanical design of materials to calculate the tearing energy to determine if the material has an acceptable tearing energy for a given application. Equations for the tearing energy have been developed and used for many years. In a recent study [10], the following tearing energy equation was used:

$$T = 2KaW_0 \qquad (1)$$

where a was the initial cut length, W_0 the strain energy, and $K = \pi/\sqrt{\lambda}$, λ was the extension ratio. The strain energy was calculated by computing the area under the load-displacement curve and dividing it by the volume of the sample. Another factor of performance of polymers is the tear strength [10]. It is the maximum force required to tear a sample with elastic properties. The equation is:

$$T_s = F/d \qquad (2)$$

where F is peak load and d is the sample thickness.

The load vs. displacement curves of three typical notched 0.3% samples films are plotted in Figure 3. The load rapidly increased to about 7 N and then more gradually reached an average maximum load of 9.6 ± 0.6 N before the tear propagated to complete fracture. The areas under the curves were approximated by Microcal Origin software since the relationship was non-linear. The cut depth, tearing energy, and tear strength values were calculated for each of the films as seen in Table 1.

In the cross-linked samples, the load vs. displacement curves, plotted in Figure 4, increased more gradually and the average maximum load reached was lower at 7.1 ± 0.8 N.

It should be noted that the average tearing energy and tear strength of the uncross-linked films was higher than the cross-linked by 40% and 47%, respectively. This was attributed to the higher resistance to tear propagation observed in the uncross-linked films, thus requiring more tearing energy and tear strength (Table 1). No tearing energy values were found for comparison in the literature. Only the initial tensile strengths to propagate a tear in notched bovine gelatin films has been reported in the literature [11].

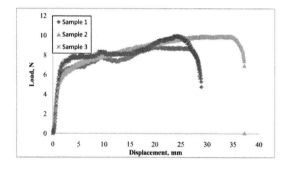

Figure 3: Load - Displacement behavior of notched uncross-linked silver-filled gelatin films.

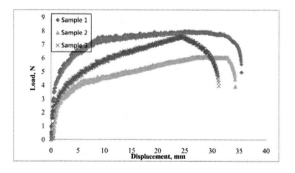

Figure 4: Load - Displacement behavior of notched cross-linked silver-filled gelatin films.

3.4 Moisture Uptake of Gelatin Films

A comparison of the moisture uptake of the cross-linked and the uncross-linked films is shown in Figure 5.

Initially, the moisture uptake properties were not affected by the cross-linking; however, after about 150 hrs, the weight of the cross-linked films was significantly greater than the uncross-linked films. At this point, the cross-linked films absorbed moisture more rapidly than the uncross-linked films. This observation could be related to the differential scanning calorimetry (DSC) patterns noted for the cross-linked and uncross-linked samples. Broader endotherms interpreted as due to water absorption were observed in the DSC scans of the cross-linked films compared to the uncross-linked films.

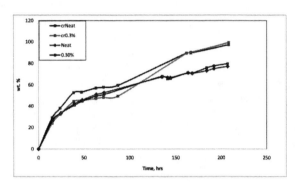

Figure 5: Moisture uptake of cross-linked silver-filled gelatin films compared to uncross-linked films with time.

4 CONCLUSIONS

After cross-linking, the mechanical properties of the uncross-linked films were increased by 126%; the strain to failure was decreased. Strain-hardening was observed in the stress-strain behavior of the cross-linked neat films. The addition of silver nanoparticles slightly increased the ultimate strength of the uncross-linked system and decreased the ultimate strength of the cross-linked system.

The tearing energy was found to be higher for the uncross-linked silver-filled films than for the cross-linked silver-filled films by about 40%. This data was correlated to the SEM observations made from the fracture surface morphologies. It was determined that the mechanism of resistance to tear propagation was tearing ridges. The uncross-linked silver-filled films had coarse tearing ridges and more pull-out material than the cross-linked silver-filled films.

It was found that the moisture uptake properties of the cross-linked films were greater than the uncross-linked after about 150 hrs. At about 208 hrs the percent moisture uptake of the cross-linked silver-filled films increased by about 25% compared to the uncross-linked.

ACKNOWLEDGEMENTS

This work was supported by Alabama A&M Research Institute GRSP contract # AAMURI – 08-C-0016 and NSF- CREST HRD-0317741.

REFERENCES

1. Choi, Y., Song, K., Park, M., Nam, Y., *Study on gelatin-containing artificial skin: Preparation and characteristics of novel gelatin-alginate sponge.* Biomaterials, 1999. **20**: p. 409-417.

2. Piotrowska, B., Sztuka, K., Kolodziejska, I., Dobrosielska, E., *Influence of transglutaminase or 1-ethyl -3-(3-dimethylaminopropyl) carbodiimide (EDC) on the properties of fish-skin gelatin films.* Food Hydrocolloids, 2008. **22**: p. 1362-1371.

3. Carvalho, R.A., Sobral, P.J.A., Thmazine, M., Habitante, A.M.Q.B., Gimenez, B., Gomez-Guillen, M.C., Montero, P., *Development of edible films based on differently processed Atlantic halibut (Hippoglossus) skin gelatin.* Food Hydrocolloids, 2008: p. 1117-1123.

4. Wang, S., Liu, W., Han, B., Yang, L., *Study on a hydroxypropyl chitosan-gelatin based scaffold for corneal stroma tissue engineering.* Applied Surface Science, 2009. **255**: p. 8701-8705.

5. Zeeman, R., van Wachem, P., van Luyn, M., Hendriks, M., Cahalan, P., Feijen, J., *Crosslinking and modification of dermal sheep collagen using 1,4-butanediol diglycidyl ether.* Journal of Biomedical Materials Research 1999. **46**: p. 424-433.

6. Zeeman, R., Dijkstra, P., van Wachem, P., van Luyn, M., Hendriks, M., Cahalan, P., Feijen, J., *The kinetics of 1,4-butanediol diglycidyl ether crosslinking of dermal sheep collagen.* Journal of Biomedical Materials Research Part A, 1999. **51**(4): p. 541-548.

7. Bigi, A., Panzavolta, S., Rubini, K., *Relationship between triple-helix content and mechanical properties of gelatin films.* Biomaterials, 2004. **21**: p. 5675-5680.

8. Boanini, E., Rubini, K., Panzavolta S., Bigi, A., *Chemico-physical characterization of gelatin films modified with oxidized alginate.* Acta Biomaterialia, 2009. **6**(2): p. 383-388.

9. Chiou, B., Avena-Bustillos R., Bechtel, P., Jafri, J., Narayan, R., Imam, S., Glenn, G., Orts, W., *Cold water fish gelatin films: Effects of cross-linking on thermal, mechanical, barrier, and biodegradation properties.* European Polymer Journal, 2008. **44**(11): p. 3748-3753.

10. Aglan, H., Calhoun, M., Allie, L., *Effect of UV and Hygrothermal Aging on the Mechanical Performance of Polyurethane Elastomers.* Journal of Applied Polymer Science, 2008. **108**: p. 558-564.

11. Rivero, S., Garcia, M.A., Pinotti, A., *Correlations between structural, barrier, thermal and mechanical properties of plasticized gelatin films.* Innovative Food Science and Emerging Technologies, 2009. **11**(2): p. 369-375.

Experiments to Investigate Electromagnetic Shielding Performance of Polyaniline-Coated Thin Films

B.R. Kim[*] and H.K. Lee[**]

[*]Department of Civil and Environmental Engineering, KAIST
371-1 Guseong-dong, Yuseong-gu, Daejeon 305-701, South Korea, bong-ida@kaist.ac.kr
[**]Department of Civil and Environmental Engineering, KAIST
371-1 Guseong-dong, Yuseong-gu, Daejeon 305-701, South Korea, leeh@kaist.ac.kr

ABSTRACT

The electromagnetic wave shielding/absorbing is of increasing important to prevent the performance degradation of electrical system/equipment and health threats from harmful electromagnetic wave [1-4]. Therefore, studies of electromagnetic wave shielding/absorbing materials are required to improve the electromagnetic environment. Many studies on electromagnetic wave shielding/absorbing materials have been performed [5-15]. Recently, the investigation of the electromagnetic wave transmission, reflection, and absorption characteristics of polyaniline-coated thin films was conducted by Kim et al. [1]. In this paper, the investigation on electromagnetic shielding/absorbing characteristics of polyaniline-coated thin films conducted by Kim et al. [1] is introduced and reviewed.

Keywords: polyaniline, thin Films, electromagnetic wave, shielding effectiveness, experimental method

1 INTRODUCTION

The development of digital communication technologies in modern industry has led to the use of high-performance electrical apparatus and the use of them has been causing artificial electromagnetic environmental pollution called electromagnetic radiation [1, 16]. In these electromagnetic environment, electromagnetic wave shielding/ absorbing is of increasing important to prevent the performance degradation of electrical system/equipment and health threats from harmful electromagnetic wave [1-4]. Therefore, studies of electromagnetic wave shielding/absorbing materials are required to improve the electromagnetic environment.

Many studies on electromagnetic wave shielding/ absorbing materials have been performed [5-15]. As representative shielding materials, the metallic materials such as steel, copper, and aluminum have been widely used due to their high conductivity and dielectric constant [1, 3, 17-20]. However, these metallic materials have disadvantages due to limited mechanical flexibility, heavy weight, corrosion, and poor processibility [1, 3, 19-21]. In this reason, the researches on novel materials have been carried out as an alternative to these metallic materials [1,

22-24]. In particular, the investigation of the electromagnetic wave transmission, reflection, and absorption characteristics of polyaniline-coated thin films was recently conducted by Kim et al. [1]. In Kim et al. [1], the optical transmittance, sheet resistance, and electromagnetic interference shielding efficiency of the polyaniline-coated thin films were systematically investigated. In this paper, the investigation on electromagnetic shielding/absorbing characteristics of polyaniline-coated thin films conducted by Kim et al. [1] is introduced and reviewed.

2 EXPERIMENTAL PROGRAM

Electromagnetic interference shielding efficiency (SE) is defined as a measure of the reduction or attenuation in the electromagnetic field strength at a point in space caused by the insertion of a shield between the source and that point and can be expressed as summation of the initial reflection loss (SE_R), absorption loss (SE_A), and internal reflection loss (SE_B) [1, 16]:

$$SE = SE_R + SE_A + SE_B. \qquad (1)$$

In Kim et al. [1], the electromagnetic interference SE was measured in a frequency range of 30 MHz~1.5 GHz according to ASTM D 4935-99 [25] which is the suitable method for planar material. The test setup and specimen set (reference and load specimens) are illustrated in Figure 1 and 2.

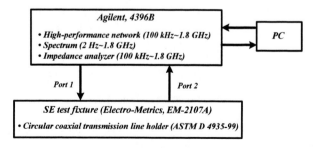

Figure 1: Test setup for electromagnetic interference SE (cf. Kim et al. [1])

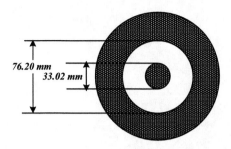

Reference specimen

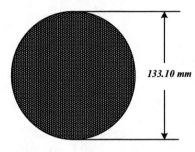

Load specimen

Figure 2: Detailed dimension of test specimen set (reference and load specimens) in accordance with ASTM D 4935-99 [25] (cf. Kim et al. [1])

According to ASTM D 4935-99 [25], the electromagnetic interference SE was calculated as

$$\text{SE(dB)} = 10\log\frac{P_1}{P_2} = 20\log\frac{V_1}{V_2} \qquad (2)$$

where, P_1 (or V_1) and P_2 (or V_2) are the received powers (or respective voltage levels) of the load and reference specimens, respectively [1, 25]. Details of test method for electromagnetic interference shielding efficiency can be found in Kim et al. [1]

The polyaniline-coated thin films for test specimens were manufactured based on the newly proposed synthetic method [26] in Elpani, Co. The thicknesses of polyaniline coated in thin films were 229, 366, 640, and 823 nm, respectively [1]. Details of synthesis of polyaniline and manufacturing of polyaniline-coated thin films can be found in Lee et al. [26], Lee et al. [27], and Kim et al. [1].

3 RESULT SUMMARY

The investigation of electromagnetic interference shielding performance of polyaniline-coated thin films was carried out by Kim et al. [1] and was reviewed in this paper. According to Kim et al. [1]'s experimental investigation, the overall electromagnetic interference SE values were obtained as 14.28%, 30.68%, 46.50%, and 52.67% with respect to the thicknesses 229, 366, 640, and 823 nm, respectively. It is noted that the research of Kim et al. [1] considered the optical transmittance as well as electromagnetic interference SE.

Kim et al. [1] also investigated the reflection loss and absorption loss of polyaniline-coated thin films. The result of Kim et al. [1] was clear that the reflection and absorption loss increased as the thickness of polyaniline coated in thin films increased. It was emphasized in Kim et al. [1] that the electromagnetic wave absorption characteristic of the polyaniline-coated thin films the unique one that distinguishes from the typical metallic materials (cf. Kim et al. [1] and Avloni et al. [13]). Although the metallic materials are superior electromagnetic interference shielding materials with very high electromagnetic interference SE values, the primary electromagnetic interference shielding mechanism of them is surface reflection of the electromagnetic wave due to their high conductivity [1]. The scattered reflection of electromagnetic wave may also cause electromagnetic environmental pollution together with previously existing electromagnetic wave.

As well as electromagnetic interference SE, the optical transmittance and sheet resistance of polyaniline-coated thin filim were investigated in Kim et al [1]. In Kim et al. [1], the applicability of an SE formula [28-30] was also checked by comparing the experimentally measured SE values with calculated SE values. Kim et al. [1]'s results showed the potential use of polyaniline-coated thin films as an electromagnetic interference shielding/absorbing materials that satisfies both optical transmission and electromagnetic interference SE requirements for transparent bodies such as windows or doorways of buildings. Details of experimental results of polyaniline-coated thin films can be found in Kim et al. [1].

4 FORTHCOMING RESEARCH

The authors investigated the electromagnetic interference shielding performance of polyaniline-coated thin films in the previous research [1]. In a forthcoming research, for serviceability of polyaniline-coated films, extensive studies through far-field test and near-field test will be carried out. Also, the authors will conduct the study on polyaniline-coated film systems having the more superior electromagnetic shielding/ absorbing and optical transparent characteristics.

ACKNOWLEDGEMENT

This work was supported by the IT R&D program of MKE/KEIT [2008-F-044-02, Development of new IT convergence technology for smart building to improve the environment of electromagnetic waves, sound and building] and the MOCT R&D program [C008R3020001-08R040200230, Development of a new interior decoration system for absorbing harmful electromagnetic waves based on carbon nanotubes (CNTs)]. The first author would like to thank to researchers in Elpani Co. Ltd.

REFERENCES

[1] Kim, B.R., Lee, H.K., Kim, Eunmi, and Lee, Suck-Hyun, "Intrinsically electromagnetic radiation shielding/absorbing characteristics of polyaniline-coated transparent thin film," *Synthetic Metal*, submitted for publication, 2010.

[2] Satheesh Kumar, K.K., Geetha, S., and Trivedi, D.C., "Freestanding conducting polyaniline film for the control of electromagnetic radiations," *Current Applied Physics*, 5, 603-608, 2005.

[3] Niu, Y., "Electromagnetic interference shielding with polyaniline nanofibers composite coatings," *Polymer Engineering and Science*, 48, 355-359, 2008.

[4] Kim, B.R., Nam, I.W., Lee, H.K., Choi, S.M., and Sim, J.B., "Electromagnetic wave shielding characteristics of carbon nanotube (CNT) coated film," *The 22th KKCNN Symposium on Civil Engineering*, Chiangmai, Thailand, 229-232, Oct. 31-Nov. 2, 2009.

[5] Chiou, J.M., Zheng, Q., and Chung, D.D.L., "Electromagnetic interference shielding by carbon fiber reinforced cement," *Composites*, 20, 379-381, 1996.

[6] Sau, K.P., Chaki, T.K., Chakraborty, A., and Khastgir, D., "Electromagnetic interference shielding by carbon black and carbon fibre filled rubber composites," *Plastics Rubber & Composites Processing & Applications*, 26, 291-297, 1997.

[7] Nagasawa, C., Kumagai, Y., Urabe, K., and Shinagawa, S., "Electromagnetic shielding particleboard with nickel-plated wood particles," *Journal of Porous Materials*, 6, 247-254, 1999.

[8] Chung, D.D.L., "Electromagnetic interference shielding effectiveness of carbon materials," *Carbon*, 39, 279-285, 2001.

[9] Cao, J. and Chung, D.D.L., "Colloidal graphite as an admixture in cement and as a coating on cement for electromagnetic interference shielding," *Cement and Concrete Research*, 33, 1737-1740, 2003.

[10] Lee, S.H., Shim, J.W., Park, D.C., and Jung, M.Y., "Properties and shielding efficiency of electromagnetic wave absorbing inorganic paint using carbon," *Journal of the Architectural Institute of Korea*, 19, 69-76, 2003. (in Korean)

[11] Fan, Z., Luo, G., Zhang, Z., Zhou, L., and Wei, F., "Electromagnetic and microwave absorbing properties of multi-walled carbon nanotubes/polymer composites," *Materials Science & Engineering B. Solid-state materials for advanced technology*, 132, 85-89, 2006.

[12] Soto-Oviedo, M.A., Araújo, O.A., Faez, R., Rezende, M.C., and De Paoli, M.A., "Antistatic coating and electromagnetic shielding properties of a hybrid material based on polyaniline/organoclay nanocomposite and EPDM rubber," *Synthetic Metals*, 156, 1249-1255, 2006.

[13] Avloni, J., Ouyang, M., Florio, L., Henn, A.R., and Sparavigna, A., "Shielding effectiveness evaluation of metalized and polypyrrole-coated fabrics," *Journal of Thermoplastic Composite Materials*, 20, 241-254, 2007.

[14] Jou, W.S., Cheng, H.Z., and Hsu, C.F., "The electromagnetic shielding effectiveness of carbon nanotubes polymer composites," *Journal of Alloys and Compounds*, 434-435, 641-645, 2007.

[15] Kwon, S.H. and Lee, H.K., "A computational approach to investigate electromagnetic shielding effectiveness of steel fiber reinforced mortar," *CMC: Computers, Materials, & Continua*, 12, 197-222, 2009.

[16] Hemming, L.H., *Architectural Electromagnetic Shielding Handbook: A Design and Specification*, The Institute of Electrical and Electronics Engineers (IEEE), Inc., New York, 1992.

[17] Lee, C.Y., Song, H.G., Jang, K.S., Oh, E.J., Epstein, A.J., and Joo, J., "Electromagnetic interference shielding efficiency of polyaniline mixtures and multilayer films," *Synthetic Metal*, 102, 1346-1349, 1999.

[18] Luo, X.C.H. and Chung, D.D.L., "Electromagnetic interference shielding using continuous carbon-fiber carbon-matrix and polymer-matrix composites," *Composites Part B: Engineering*, 30, 227-231, 1999.

[19] Yuping, D., Shunhua, L., and Hongtao, G., "Investigation of electrical conductivity and electromagnetic shielding effectiveness of polyaniline composites," *Science and Technology of Advanced Materials*, 6, 513-518, 2005.

[20] Vulpe, S., Nastase, F., Nastase, C., Stamatin, I., "PAN-PAni nanocomposites obtained in thermocentrifugal fields," *Thin Solid Films*, 495, 113-117, 2006.

[21] Yuexian, S., Hongli, W., and Yuansuo, Z., "Electric and electromagnetic shielding properties of highly conducting polyaniline films," *2002 3rd International Symposium on Electromagnetic Compatibility*, 582-585, 2002.

[22] Taka, T., "EMI shielding measurements on poly(3-octyl thiophene) blends," *Synthetic Metals*, 41, 1177-1170, 1991.

[23] Trivedi, D.C. and Dhawan, S.K., "Shielding of electromagnetic interference using polyaniline," *Synthetic Metals*, 59, 267-272, 1993.

[24] Pomposo, J.A., Rodríguez, J., and Grande, H., "Polypyrrole-based conducting hot melt adhesives for EMI shielding applications," *Synthetic Metals*, 104, 107-111, 1999.

[25] ASTM D 4935-99, *Standard test Method for Measuring the Electromagnetic Shielding Effectiveness of Planer Materials*, 1999.

[26] Lee, S.H., Lee, D.H., Lee, K., and Lee, C.W., "High-performance polyaniline prepared via polymerization in a self-stabilized dispersion," *Advanced Functional Materials*, 15, 1495-1500, 2005.

[27] Lee, K., Cho, S., Park, S.H., Heeger, A.J., Lee, C.W., and Lee, S.H., "Metallic transport in polyaniline," *Nature*, 441, 65-68, 2006.

[28] Shaklette, L.W. and Colaneri, N.F., "EMI shielding measurements of conducting polymer blends," *Instrumentation and Measurement Technology Conference, IMTC-91, Conference Record, 8th IEEE*, 72-78, 1991.

[29] Colaneri, N.F. and Shaklette, L.W., "EMI shielding measurements of conductive polymer blends," *IEEE Transactions on Instrumentation and Measurement*, 41, 291-297, 1992.

[30] Mäkelä, T., Pienimaa, S., Taka, T., Jussila, S., and Isotalo, H., "Thin polyaniline films in EMI shielding," *Synthetic Metals*, 85, 1335-1336.

Fabrication of poly(alizarin red)-carbon nanofibers composite film and its application in developing nitric oxide microsensor

Dongyun Zheng[*,**], Chengguo Hu[*], Shengshui Hu[*,**]

[*]College of Chemistry, Wuhan University, Wuhan 430072, PR China, sshu@whu.edu.cn
[**]State Key Laboratory of Transducer Technology, Chinese Academy of Sciences, Beijing 100084, PR China, sshu@whu.edu.cn

ABSTRACT

In this work, a simple method for the stable dispersion of carbon nanofibers (CNFs) in water by alizarin red and controllable surface addition onto carbon fiber microelectrodes (CFE) via electropolymerization is reported. We have characterized these poly(alizarin red S)-carbon nanofiber (PARS-CNF) composite films, with techniques including scanning electron microscopy (SEM), infrared spectroscopy (IR) and voltammetry. These investigations showed that the films have loose three-dimensional structure with plenty of dispersed CNFs. This PARS-CNF composite-modified CFE (PARS-CNF /CFE) exhibited good electrocatalytic ability to the electrochemical oxidation of nitric oxide (NO). We also successfully applied the PARS-CNF/CFE system to the determination of NO released from rat liver cells. The results were satisfied, which indicated that this electrode may have wide practical applications in the determination of NO.

Keywords: alizarin red, carbon nanofibers, electropolymerization, carbon fiber electrode, nitric oxide, sensor

1 INTRODUCTION

Carbon nanofibers (CNFs) are characterized as having high degree of orientation on their graphitic basal planes parallel to the fiber axis and possessing excellent mechanical strength and electrical conductivity [1]. They are also recognized as one of the promising materials based on its nanostructure and particular properties, being expected in various applications including catalysts or catalyst supports, selective adsorption agents, energy storage devices such as lithium ion second battery or electric double layer capacitor and chemical sensing on molecular scale. Moreover, compared with carbon nanotubes (CNTs), CNFs have larger surface-active groups-to-volume ratio, better mechanical stability, and easier mass production. Especially, CNFs process more edge sites on the out wall than CNTs. Based on these advantages; the acid treated CNFs as a biocompatible electron conductor have been widely used in electroanalysis. CNFs can be modified onto the electrode surface by

methods like casting, deposition, chemical vapor deposition and adsorption. These methods, however, generally lack film controllability and are not suitable for electrodes with small dimensions or special shapes.

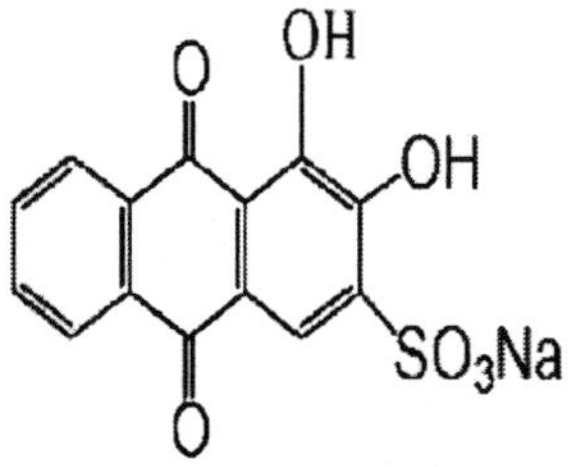

Fig. 1 Molecular structure of alizarin red.

As Wu reported [2], due to their special molecular structure (Fig. 1), AR molecules can interact with MWNTs through noncovalent interactions (such as van der Waals interactions, and π–π stacking) to dissolve the MWNTs into aqueous solution. In this work, we improved the solubility of CNFs in water with AR through the non-covalent interactions such as π–π stacking and strong adsorption between AR and CNFs. Via successive cyclic sweeping CFEs in CNF-AR solution, a composite film of PAR-CNF was electropolymerized onto the electrode surface. The thin CNF-AR film was characterized by various methods. This method for depositing CNFs onto a surface is not cumbersome and the thickness of the resulting PAR-CNF film is conveniently controlled by the cycle number of the cyclic voltammetric sweeps. This composite film exhibited good catalytic ability to the electrochemical oxidation of nitric oxide (NO). The present work not only provided a more sensitive electrochemical method for the determination of NO, but also foresaw the promising applications of in situ electropolymerization method in developing high-performance electrochemical microsensors.

2 EXPERIMENTAL

2.1.Chemicals and Apparatus

CNFs were provided by World Precision Instruments Inc. (WPI Inc.) and made hydrophilic by sonicating for 4 h in the mixture of H_2SO_4 (98%) and HNO_3 (36–38%) with the volume ratio of 3:1 followed by thorough rinsing with water. The treated CNFs were dried in vacuum oven and then kept in desiccator for future uses. AR was purchased from Shanghai Reagent Company. CNF suspensions were prepared by sonicating CNFs in 5 mg/mL alizarin red aqueous solution for 10 h to form a final concentration of 0.5 mg/mL. Other reagents were of analytical grade and were used as received. Aqueous solutions were prepared with doubly distilled water.

NO-saturated solution was prepared as described previously [3], using a value of 1.8 mmol·L^{-1} (mM) for its concentration at saturation. Shortly, NO gas was generated by slowly dropping a 6 M H_2SO_4 solution into a rockered flask containing saturated $NaNO_2$ solution, NO being formed in this disproportional reaction. The gas generated was forced to bubble in 30% NaOH solution twice and in water once in order to trap any NO_2 formed as a result of oxidation of NO from traces of oxygen. Before the addition of H_2SO_4, all apparatus were degassed meticulously with nitrogen for 30 min to exclude O_2, as NO is rapidly destroyed by O_2. NO-saturated solution was obtained by bubbling NO pure gas through 6 mL deoxygenated distilled water for 30 min and kept under NO atmosphere until use. NO standard solutions were prepared by making serial dilutions of the saturated NO solution as previously reported [4]. NO solutions were made fresh and kept in a glass flask with a rubber plug, stored in a light-free place.

All electrochemical measurements were performed on a CHI 660A electrochemical analyzer (Shanghai Chenhua Co., China) in a conventional three-electrode arrangement, equipped with a platinum wire counter electrode, a saturated calomel electrode (SCE) reference electrode and a PAR-CNF/CFE working electrode. Scanning electron microscopy (SEM) was performed on a scanning electron microscope (Quanta 200, Fei, Holland). An inverted microscope (XDP-1) was used to control the length of the CFE.

2.2 Preparation of Modified Electrodes

CFE was prepared as follows: carbon fibers (7.8 μm in diameter, Kureha Chemical Industry, Tokyo, Japan) were firstly cleaned by sonicating in acetone, alcohol and doubly distilled water, each for 2 min, and allowed to dry in air. The cleaned carbon fibers were connected to copper wires for conducting by silver glue and sealed in capillary by fusing the tips on an alcohol burner. The length of carbon fibers exposed outside the pipettes was controlled to be about 0.5 cm. Following this, the protruded carbon fibers were carefully cut to be 800–1000 μm in length under the microscope for sensing. Finally, the tail of capillary was also sealed with AB glues leaving a part of the copper wire extended outside. The obtained CFE was compared by sweeping from -0.2 to 0.8 V in 1.0 M KCl containing 5.0 mM $K_3[Fe(CN)_6]$ to make sure that all the employed CFE possessed the same effective surface area. Prior to surface modification, CFE was pretreated by sweeping from -0.8 to

1.0 V in 0.1 M H_2SO_4 for 20 cycles by cyclic voltammetry.

The fabrication of PAR-CNF/CFE was carried out by potentially sweeping CFE in the range of 0 V~2.2 V at a scan rate of 50 mV/s for 25 cycles in aqueous solutions containing 5 mg/mL alizarin red and 0.5 mg/mL CNFs without any supporting electrolyte.

PAR-modified CFEs (PAR/CFEs) were prepared by scanning a CFE in aqueous vanillin solution (5 mg/mL) with cyclic voltammetry in the range of 0 V-2.2 V at a scan rate of 50 mV/s for 25 cycles.

Prior to use, all of the above modified electrodes were first washed with water and then swept in the potential range of 0 V to 1.2 V in 0.1 M phosphate buffer solution (PBS, pH 7.4) at a scan rate of 50 mV/s until stable cyclic voltammograms were obtained.

3   RESULTS AND DISCUSSION

3.1. In Situ Electropolymerization of CNFs and AR on CFEs

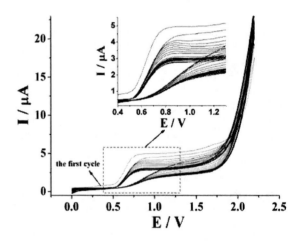

Fig. 2. The in situ electropolymerization process of CNFs and alizarin red on the surface of CFE. Scan rate: 50 mV/s.

The cyclic voltammograms for the surface modification of CFEs by PAR-CNF composite films via electropolymerization are shown in Fig. 2. The oxidation peak current of the AR monomer at about 0.8 V decreases with an increasing number of electropolymerization cycles (inset of Fig. 1), demonstrating the successful deposition of poly (alizarin red) films on the CFE.

AR is a binary weak acid and its acid ionization constants are $pK_1 = 5.5$, $pK_2 = 11.0$, respectively. They exist in different forms at different conditions (Fig. 3A). As the condition is neutral in this work, we deduce the possible electropolymerization mechanism of AR onto the surface of CFE is as Fig. 3B:

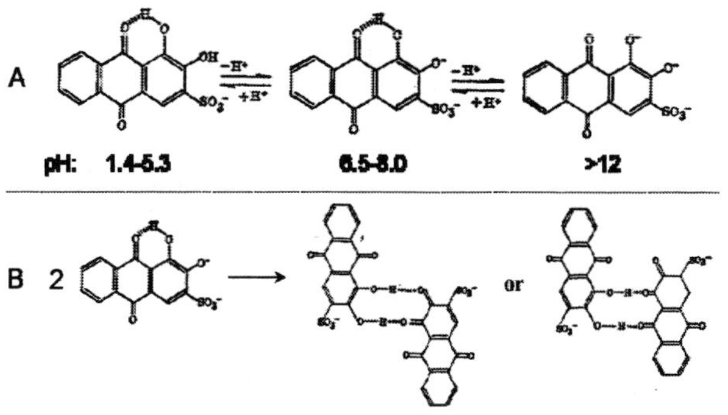

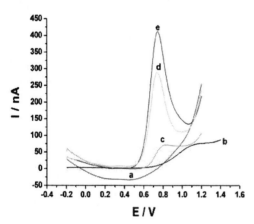

Fig. 3. A: Different forms of AR at different conditions; B: The possible electropolymerization mechanism of AR onto the surface of CFE.

3.2. Electrochemical response of nitric oxide at PAR-CNF/CFE

Fig. 4. Square wave voltammagramms of bare CFE (curve b), PAR/CFE (c) and PAR-CNF/CFE (a, d and e) in the absence of NO (a), and in the presence of 18 μM NO (b, c and d) and 36 μM NO (e) in 0.1 M phosphate buffer solution (pH 7.4).

Fig. 4 shows the square wave voltammogramms of different electrodes in the absence and presence of NO. When the concentration of NO was 18 μM, only a small and wide oxidation peak at 1.1 V with a current of 17.3 nA is observed (curve b) at bare CFE. The oxidation current is enlarged to 37.1 nA at PAR/CFE (curve c), along with the apparent negative shift of the oxidation potential to 0.8 V. However, a sharp oxidation peak at 0.74 V with a current of 273.6 nA is observed at PAR-CNF/CFE (curve d) and there is no electrochemical signal in the absence of NO (curve a). Moreover, the peak current increased to 353.4 nA when the NO concentration was added to 36 μM. These results suggest that the peak corresponds to the electrochemical oxidation of NO and the PAR-CNF film can catalyze the electrochemical oxidation of NO on CFE. This is probably

attributed to the synergetic enhancement effect of PAR and CNFs for the electrochemical oxidation of NO. That is to say, the formation of a conductive porous structure of CNFs on CFE and the surface coating of CNFs by PAR produce a three-dimensional reaction network with plenty of electro-active centers, which possesses apparent advantages for enhancing the accumulation and improving the electrochemical response of NO in comparison with the compact PAR film on CFE.

3.3. Amperometric detection of nitric oxide at PAR-CNF/CFEs

Electrochemical response of NO microsensor based PAR-CNFs composite film was examined. The results of amperometric detection for the consecutive addition of different concentrations of NO show that the response of the microsensor is rapid and has an excellent reproducibility. The current exhibits a linear relationship with the concentration of NO in the range of 36 nM ~ 78 μM and the detection limit is estimated to be 5 nM (S/N=3).

3.4. Interference study

Interferents	Concentration (mol/L)	Signal change (%)
glucose	9×10^{-4}	-1.89
caffein	9×10^{-4}	-3.84
DL-valine	9×10^{-4}	-4.98
L-glutamic acid	9×10^{-4}	+2.76
L-aspartic acid	9×10^{-4}	-2.06
Glycine	9×10^{-4}	-1.03
cholesterol	9×10^{-4}	-1.73
L-serine	9×10^{-4}	-0.82
barbitone	9×10^{-4}	+0.32
L-arginine	4.5×10^{-4}	+2.08
L-tyrosine	4.5×10^{-4}	+4.64
sucrose	4.5×10^{-4}	+0.41
L-cystine	4.5×10^{-4}	+1.62
epinephrine	9×10^{-5}	+6.49
UA	9×10^{-5}	+5.79
AA	9×10^{-5}	+8.77
DA	9×10^{-5}	+15.21
NO_2^-	9×10^{-5}	+20.56

Table 1 Influence of some materials on the peak current of 9×10^{-6} M NO.

Table 1 shows the experimental results of the interference study. The addition of 9.0×10^{-4} M glucose, caffein, DL-valine, L-glutamic acid, L-aspartic acid, Glycine, cholesterol, L-serine and barbitone, 4.5×10^{-4} M L-arginine, L-tyrosine, sucrose and L-cystine did not interfere with the detection of 9.0×10^{-6} M NO (signal change <5%), which shows that PAR-CNF/CFE has a selective reactivity to NO, and that this selectivity is not easily perturbed. Hydrogen bonding between NO and the oxygen-containing groups on PAR might be responsible for the sensitive and selective response of the PAR-CNF composite film to NO. However, 9×10^{-5} M epinephrine, uric acid (UA), ascorbic acid (AA), dopamine (DA) and NO_2^- had some interferences on the determination of 9×10^{-6} M NO, which may be because they can be oxidized easily at this operation potential.

3.5. Measurement of NO released from rat liver cells

In biological systems, NO is biosynthesized from L-Arg via a specific biochemical pathway that includes nitric oxide synthase (NOS). Amperometric measurements at the operational potential of 0.85 V were carried out in a stirred deaerated 0.1 mol/L PBS. When the stimulator, 50 µL 0.3 M L-Arg was injected into 5 mL of 0.1 M PBS (pH 7.4) with rat liver cells and L-NNA (an inhibitor), an immediate current increase is observed, belonging to the oxidation of high concentration of L-Arg. However, when 50 µL 0.3 M L-Arg was added to the PBS containing only rat liver cells, an additional delayed current response appears at about 80 s after the appearance of L-Arg signal, which is in good agreement with a previous work [5] and attributed to NO released from the liver cells by the stimulation of L-Arg. This result foresees the promising application of PAR-CNF/CFE in detecting NO in real samples.

ACKNOWLEDGEMENTS

This research is supported by the National Nature Science Foundation of China (Nos. 30770549, 20805035 and 90817103).

REFERENCES

[1] A. Oberlin, M. Endo, T. Koyama, "Filamentous growth of carbon through benzene decomposition", J. Cryst. Growth. 32, 335-349, 1976.

[2] K.B. Wu, S.S. Hu, "Deposition of a thin film of carbon nanotubes onto a glassy carbon electrode by electropolymerization", Carbon 42, 3237-3242, 2004.

[3] M.N. Friedemann, S.W. Robinson, G.A. Gerhardt, "o-Phenylenediamine-Modified Carbon Fiber Electrodes for the Detection of Nitric Oxide", Anal. Chem. 68, 2621-2628, 1996.

[4] S. Trevin, F. Bedioui, J. Devynck, "New electropolymerized nickel porphyrin films: Application to the detection of nitric oxide in aqueous solution", J. Electroanal. Chem. 408, 261-265, 1996.

[5] Y.F. Peng, C.H. Hu, D.Y. Zheng, S.S. Hu, "A sensitive nitric oxide microsensor based on PBPB composite film modified carbon fiber microelectrode", Sens. Actuators, B, Chem. 133, 571-576, 2008.

Characterization of Carbon Nanotube-reinforced Polyethylene Nanocomposite Produced by Cryogenic Ball-milling Process

G. Terife and K. A. Narh

Department of Mechanical & Industrial Engineering
New Jersey Institute of Technology, Newark, NJ 07102-1982

ABSTRACT

The feasibility of using the cryogenic ball-milling process as an environmentally friendly method to produce polymer/CNT nanocomposites was investigated. Linear Low density Polyethylene was used as the matrix material, and 1wt % of Multi-walled carbon nanotubes was used as reinforcement; the influence of the milling time and balls size was evaluated. The morphology of the nanocomposite, and the degree of dispersion of the MWCNTs were studied using SEM, visual inspection and light transmission microscopy; ropes as well as aggregates of MWCNTs were observed, and wetting of the nanotubes by the matrix was also evidenced. An increase of up to 28% in the elastic modulus (determined by tensile testing) with respect to the matrix, was obtained. DSC analysis showed evidence of increase in the degree of crystallization, a result of the nucleating capability of the CNTs in the matrix.

Keywords: Carbon nanotubes, nanocomposites, cryogenic ball milling, mechanical properties.

INTRODUCTION

Any polymer composite where a fiber is used as reinforcement should fulfill some basic requirements so that a significant enhancement in the properties of the matrix can be achieved. First, the aspect ratio (the ratio of the length to the diameter of the fiber) has to be large, second the fiber has to form an intimate contact with the matrix, so when stress is applied, it can be efficiently transferred from the matrix to the reinforcement, third the strength of the fiber has to be much greater than the strength of the matrix, and finally the fiber has to be well dispersed and distributed throughout the matrix. [1]. Carbon nanotubes fulfill most of the reinforcement criteria described above. The combination of high aspect ratio, small size, very low density, and more importantly, excellent physical properties, such as extremely high mechanical strength and stiffness, high electrical and thermal conductivity, make carbon nanotubes (CNTs) perfect candidates as ideal reinforcing fillers in high strength, lightweight polymer nanocomposites with high performance and multi-functions.

Besides the filler properties and geometry (i.e. aspect ratio), other factors to be considered when analyzing the enhancement of the mechanical properties in nanocomposites are: degree of dispersion, distribution and alignment, as well as polymer-CNT interactions and interface characteristics.

CNTs tend to form bundles or ropes, which then form highly entangled and stable aggregates. In order to produce a good polymer/CNT nanocomposite, one in which a significant enhancement of the properties of the matrix is achieved, it is necessary to break the CNT aggregates at least into isolated bundles or ropes; however, the ideal nanocomposite will have individual CNTs uniformly distributed throughout the matrix. A good dispersion and distribution will result in a more efficient stress transfer and in a more uniform stress distribution, avoiding stress concentration.

The objective of the present work was to evaluate the efficiency of the cryogenic mixing process to produce Polyethylene/MWCNTs nanocomposites with enhanced mechanical properties. The mechanical properties of the nanocomposites and unfilled matrix were determined by tensile testing, and the change in elastic modulus was compared with matrix was has been reported in the literature by other authors.

It is worth pointing out at this stage, that the cryogenic device used in work could only operate at cryogenic temperatures. Hence, it was not the goal of the authors to compare the results with measurements made at room temperatures using the same device; it was simply one of the several devices the authors have been using in their efforts to obtain the best procedure for efficient mixing of nanoparticles with polymeric matrices. Another device that is currently being used by the authors for declustering of carbon nanotubes is the magnetic assisted impact coating device, by Aveka Inc.; the study is being conducted at room temperature, and the results will be published at a later stage.

EXPERIMENTAL

Materials

The nanocomposites were prepared using as the matrix, Linear Low Density Polyethylene (LLP8555.25) supplied by Exxon Mobil. The LLDPE was used in powder form with the average particle diameter of 378 μm (determined in dry state using laser diffraction particle size analyzer, Beckman Coulter LS230). MWCNTs, purchased from Cheap Tubes Inc., used as reinforcement, were said to be of 95 wt% purity with outside diameter in the range 20-30 nm and length 10-30 m. The CNTs were incorporated into the matrix as received; no surface treatment or further purification was performed.

Sample Preparation

The mixing of the LLDPE powder with 1wt% of MWCNTs was done in batches of 1.5 g, at liquid nitrogen temperatures, using a cryogenic milling device (6850 Freezer/Mill from SPEX CertiPrep Group). The mixing action of the apparatus results only from the movement of metallic balls inside four vials, due to the magnetic field generated by the coil surrounding the vials.

The influence of the mixing time and ball size on the mechanical properties of the nanocomposites was studied. The two ball sizes used were 1/2" and 3/16" in diameter; in order to keep the mass ratio of material to balls constant at 3/50 throughout the study, 3 large balls or 58 small balls were used. The total mixing times were 12, 18, 24 and 30 minutes. The following parameters were kept constant: frequency at 10 Hz, 10 minutes of pre-cooling, and 2 minutes of cooling or rest between every 3 minutes of continuous milling.

Nanocomposites Characterization

The morphology of the recovered, cryogenically mixed PE/MWCNT powder, was evaluated using Scanning Electron Microscopy (SEM) (*LEO 1530vp*). The homogeneity of the nanocomposites was investigated by light transmission microscopy of thin films; the films were prepared by melting the nanocomposites between glass slides using a hot plate, set at 200°C. The optical micrographs were acquired with a Nikon microscope (model Eclipse E200) using an objective lens with 4x magnification.

In order to evaluate the mechanical properties of the nanocomposites, the recovered material was compression molded at 150°C and 2000psi (using a *CAVEN* press). Rectangular test specimens of 40x10mm and 0.5mm thick were cut from sheets of 40x50mm and 0.5mm thick.

The tensile tests were performed using an Instron universal testing machine (Instron 5567); with a 500N load cell, crosshead speed of 30mm/min, and initial distance between grips of 20mm. Five specimens of each sample were tested. It is important to note that even though the dimensions of test specimens did not fulfill any particular standard (ASTM or ISO), all the calculations were done using ASTM D638 standard as a reference.

The actual content of MWCNTs in the nanocomposites was determined through Thermogravimetric analysis (TGA) using a TA Instrument (model Q50); about 7mg of each sample was heated at 10°C/min from room temperature to 550°C. The specimens were held in Aluminum pans, and the system was purged with nitrogen.

RESULTS AND DISCUSSION

Morphology of the nanocomposites

SEM inspection of the mixed powder recovered from the cryogenic ball miller showed that the CNT aggregates were broken into smaller aggregates, and ropes. Figure 1 shows, at two levels of magnification, a characteristic SEM image of cryogenically mixed powder using large balls for 12min. Even though small aggregates of nanotubes can be observed at both magnification levels (Figure 1), at the higher magnification (Figure 1b), ropes of MWCNTs can be seen attached to a LLDPE particle. Since the nanotube ropes are fused into the PE particle and the CNTs were coated by the polymer (as demonstrated in previous work by the authors [1]), it can be concluded that through cryogenic ball-milling it is possible to achieve good wetting of the filler.

Through optical microscopy (Figure 2), small CNT aggregates were observed, but their sizes decreased with increasing mixing time. For a longer mixing time better dispersion and distribution was obtained (see Figure 2, for example).

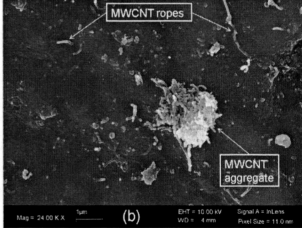

Figure 1: SEM images of the PE/MWCNT cryogenically mixed powder using 1/2" for 12min. (b) is a magnification of the boxed region in (a).

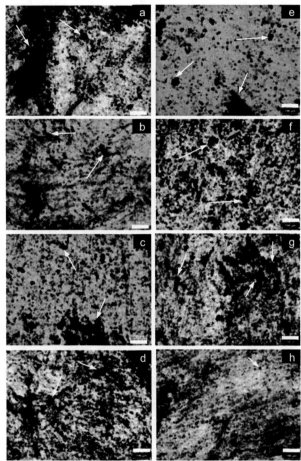

Figure 2: Optical micrographs of the nanocomposites cryogenically mixed using large balls for: (a) 12min, (b) 18min, (c) 24 min, (d) 30 min, and small balls for: (e) 12min, (f) 18min, (g) 24 min, and (h) 30 min. The scale bar in all the images is 100μm, the arrows indicate aggregates.

Mechanical properties of the nanocomposites

In order to indentify the influence that the mixing parameters have on the mechanical properties of the nanocomposites, normalized value of the tensile properties were determined according to the following expression:

$$Y_N = \frac{Y}{1 - \blacksquare_T} \qquad (1)$$

Where Y_N is the normalized tensile property, Y is the nominal value (measured values of E: Elastic modulus, or σ_y: yield strength) and φ_{CNT} is the volume fraction of MWCNTs calculated from the mass fraction determined through TGA (densities of PE and the CNT used were 0.936 g/cm³ and 2.1 g/cm³, respectively; values reported by suppliers).

Figures 3 and 4 show the normalized values of the yield strength (σ_{yN}), and elastic modulus (E_N) as a function of the mixing time and ball size.

Even though the SEM images show good wetting of the nanotubes by the LLDPE, the presence of agglomerated

CNTs limits the stress that can be transferred from the matrix to the filler; therefore the yield strength of the nanocomposites was not significantly improved by incorporating 1wt% of MWCNTs (see Figure 3). On the other hand, the yield strain of the nanocomposites tends to be lower than the yield strain of the unfilled LLDPE (results not shown due to space constraints), because CNTs are more rigid than the polymeric matrix, they restrain the deformation of the matrix [2].

Figure 3 shows that the elastic modulus of the nanocomposites is increased between 8 and 28% with respect to the matrix. The enhancement in Young's modulus results not only from the mechanical reinforcement that CNTs impart, but also from a higher degree of crystallinity, which was demonstrated through DSC analysis in previous work [1]. However, as it can be seen in Figure 3, there is no clear correlation between the Young's modulus and the mixing time; which can be attributed to the presence of CNT aggregates of different sizes and shapes (Figure 2), causing the aspect ratio of the filler to be less than expected. This will affect not only the mechanical properties of the filler but also how it interacts with the matrix [3].

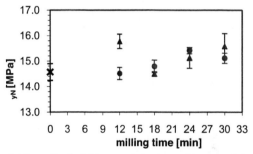

Figure 3: Normalized yield strength (σ_{yN}) as a function of the mixing time for: × unfilled LLDPE a LLDPE/1wt%MWCNTs cryogenically mixed using ▲ large balls, and ● small balls.

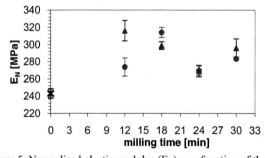

Figure 5: Normalized elastic modulus (E_N) as a function of the mixing time for: × unfilled LLDPE, and LLDPE/1wt%MWCNTs cryogenically mixed using ▲ large balls, and ● small balls.

The change in elastic modulus of the cryogenically mixed nanocomposites was compared to experimental values reported in the literature by other authors. Figure 5 shows a summary of the change in elastic modulus as a function of the content of nanotubes for different Polyethylene grades reinforced with MWCNTs. From the information presented in

Figure 5, it can be seen that even though CNT aggregates are still present in the cryogenic mixed nanocomposites, the elastic modulus enhancement obtained is, in many cases, superior to what has been reported by other authors. The sample preparation methods used by these authors are summarized below.

Gorrasi et al. [4] produced LLDPE/MWCNTs nanocomposites, by centrifugal ball milling in solid state at room temperature. The MWCNTs used had diameters of 10-15nm and length up to 10μm. They observed a 20% increase in elastic modulus by incorporating 1 and 2 wt% of MWCNTs.

Xiao et al. [5] studied the mechanical properties of LDPE reinforced with MWCNTs with diameters ranging between 10 and 20nm, and lengths between 1and 5μm. The nanocomposites were prepared through mechanical mixing at 140°C. The results of their tensile tests showed that the elastic modulus increases with the nanotube content.

Kanagaraj et al. [6] studied the mechanical properties of injection molded tensile specimens of HDPE/MWCNT nanocomposites. The nanotubes were functionalized with chemical groups such as carboxyl, carbonyl, and hydroxyl, through acid treatment. The mixing was done in water; pellets of HDPE were added to an aqueous suspension of the MWCNTs, which was heated and magnetically stirred to produce coated polymer pellets. The results of tensile tests show that the elastic modulus increases linearly with increasing MWCNTs contents.

Wang et al. [7] prepared nanocomposites of UHMWPE and functionalized MWCNTs (with diameter of 20-40nm and length of 0.5-50μm) through solution mixing. The mechanical properties were determined in tension using gel spun fibers. They reported an increase in elastic modulus of between 5 and 14%, with respect to the unfilled matrix, depending on the content of MWCNTs.

Ruan et al. [8] reported 38% increase in modulus with respect to the matrix of hot-drawn films of nanocomposites, consisting of UHMWPE reinforced with 1 wt% of MWCNTs The specimens were prepared through solution mixing.

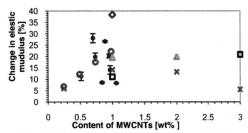

Figure 4: Change in Elastic modulus as a function of CNT content. ● LLDPE-MWCNTs cryogenically mixed, △ LLDPE-MWCNTs energy ball milling [4], □ LDPE-MWCNTs mechanical mixing [5], ◆ HDPE-MWCNTs (functionalized) [6], ✖ UHMWPE-MWCNTs (functionalized), ■ UHMWPE-MWCNTs hot-drawn films [8]

By comparing the change in elastic modulus with respect to the matrix, for different PE/MWCNTs nanocomposites, it is possible to conclude that (1) for similar content of CNTs, fuctionalization of the nanotubes does not significantly improve the reinforcement capability of the nanotubes, (2) as the nanotube content is increased the elastic modulus tends to be higher, and (3) by using the cryogenic ball milling device to mix the nanocomposites it is possible to achieve greater enhancement in elastic modulus than other reported methods, even at a lower level of nanotube concentration

CONCLUSIONS

The cryogenic ball-milling process allows production of PE/CNT nanocomposites with enhanced elastic modulus. When compared to other techniques it was evident that the increase in modulus that can be achieved through cryogenic mixing is superior to what has been reported in the literature.

Acknowledgments

This work was funded by the National Science Foundation (Award: CMMI-0802947). The authors acknowledge the collaboration of Dr. Edward Dreizin and Dr. George Collins.

REFERENCES

1. Terife, Graciela and Narh, Kwabena A., "Creating Polymer-Carbon nanotubes nanocomposites by cryomilling," *SPE ANTEC Conference Proceedings*, Chicago, p. 349-353, (2009).
2. Callister, William D., (2007), Materials Science and Engineering, New York, John Wiley & Sons, Inc, 832.
3. Moniruzzaman, Mohammad and Winey, Karen I., (2006), Polymer Nanocomposites Containing Carbon Nanotubes, *Macromolecules*, 39(16), 5194-5205.
4. Gorrasi, Giuliana, Sarno, Maria, Bartolomeo, Antonio Di, Sannino, Diana, Ciambelli, Paolo and Vittoria, Vittoria, (2007), Incorporation of carbon nanotubes into polyethylene by high energy ball milling: Morphology and physical properties, *Journal of Polymer Science Part B: Polymer Physics*, 45(5), 597-606.
5. Xiao, K. Q., Zhang, L. C. and Zarudi, I., (2007), Mechanical and rheological properties of carbon nanotube-reinforced polyethylene composites, *Composites Science and Technology*, 67(2), 177-182.
6. Kanagaraj, S., Varanda, Fátima R., Zhil'tsova, Tatiana V., Oliveira, Mónica S. A. and Simões, José A. O., (2007), Mechanical properties of high density polyethylene/carbon nanotube composites, *Composites Science and Technology*, 67(15-16), 3071-3077.
7. Wang, Yanping, Cheng, Ruiling, Liang, Linli and Wang, Yimin, (2005), Study on the preparation and characterization of ultra-high molecular weight polyethylene-carbon nanotubes composite fiber, *Composites Science and Technology*, 65(5), 793-797.
8. Ruan, S. L., Gao, P., Yang, X. G. and Yu, T. X., (2003), Toughening high performance ultrahigh molecular weight polyethylene using multiwalled carbon nanotubes, *Polymer*, 44(19), 5643-5654.

Electronic Properties and Transitions States involved in the Pyrrole Oligomerization

A. Cuán*, C.M. Cortés-Romero**, M.T. Ramírez-Silva***, M.A. Romero-Romo*, M.E. Palomar-Pardavé*

* Materials Department, Metropolitan Autonomous University- Azcapotzalco, Mexico City 02200, MEXICO, acuan@correo.azc.uamm.mx, mmrr@correo.azc.uamm.mx, mepp@correo.azc.uamm.mx
** IATC, 76802 San Juan del Río, Querétaro. México, carlos.cortes@cpimex.com
*** Chemistry Department, Metropolitan Autonomous University-Iztapalapa, Mexico City 09340, MEXICO, mtrs@xanum.uam.mx

ABSTRACT

The electrochemical pyrrole polymerization mechanism at initial stages is theoretically studied, through sequential monomer units coupling to obtain 2 to 9 oligomer formation. The transitions states, involved in the initial stages, give the sequential growing of oligomers and the electronic properties dependence. The relevance of these oligomers lies on their solubility in water as well as on the influence in uniformity of the final polymer film. The dependence of the growing chain and electronic properties has been studied by the Polarizable Continuum Method (PCM) and then water solvent effect is taking into account.

Keywords: pyrrole, transitions states, oligomerization, solvent effect, DFT study.

1 INTRODUCTION

The technological impact of polymer materials with conducting properties has been the aim of many academic and industrial studies to gain a better understanding of the phenomena involved in the synthesis and characterization of such materials. This is the case of polypyrrole, which is a polymer widely employed in industry [1], not only due to its inherent properties that make it amply usable in batteries, heavy duty capacitors, conductive textiles and fabrics, mechanical actuators, electrochemical sensors [2,3], but also due to its oxidative capability and water solubility. Furthermore, polypyrrole is commercially available and industrially produced [1, 4, 5]. The literature that has reported reaction mechanisms consider that such steps as electron transfer, proton transfer and direct pyrrole radicals formation may be involved during polymerization depending on the synthesis method used [4]. In the case of pyrrole electrochemical polymerization, it is assumed that the latter mechanism is the most likely reaction pathway. Additionally, the use of this method has shown that some operating variables like the nature of the electrolyte [6], the sort of solvent [7], the temperature [8] and pH [9] influence the process by promoting or controlling the interaction among the reaction intermediates.

The reaction mechanism involved during electrochemical polymerization appears to entail some difficulties [10-13]. The main one is related to the determination of the different reaction stages and polymerization's rate controlling step [12,13]. Apart from the poor solubility of certain pyrrole oligomers, there is the non-crystalline nature of the product that makes structure characterization and analysis of physical properties a difficult issue./ In this regard, quantum-chemical calculations provide additional information to elucidate the structure and stability of the species involved in the reaction mechanisms

2 METHODOLOGY

Calculations have been performed with *Gaussian03* [14] using the hybrid B3LYP functional [15] using the hybrid B3LYP (for singlet state) and UB3LYP (for triplet and doublet states) functional, with 6-311+(d, p) basis set in vacuo and with the Polarizable Continuum Method (PCM) [16] accounting for the solvent (water) effect.Geometry optimization calculations have been carried out for all the complexes involved (reactants, intermediates and products) and the frequencies were computed by using analytical second derivatives to check that the stationary points of the involved systems were corresponding to a minimum in the Potential Energy Surface (PES)./ For transitions state (TS) an initial guess was given and then the TS was relaxed in the direction of the reaction coordinate to verify that it connected the reactants and the product, by using IRC formalism implemented in Gaussian program. Frequency analyses were carried out to make sure first order saddle point for TS.

3 DISCUSSION AND RESULTS

To study the reaction mechanism of pyrrole dimmer it has made a scan of the dihedral angle N-C-C-N (ϕ) as it is shows in the Figure 1. The dimer formed displays different conformations around the Potential Energy Surface (PES) when torsion of the dihedral angle, -N-C-C-N- (ϕ) is carried out, vide Figure 1. The profile of the PES is rather similar to early reported by Yurtsever et at. [17]; where different levels of theory were compared. The relative dihedral angle rotation freedom may produce two more stable conformations which correspond to the nitrogens lying in a semi-trans (namely, trans-) and semi-cis conformation (namely, cis-), the dihedral angle φ is about 28.5 and 44.9 degrees, respectively; and they are denoted as the trans-Prod1, and cis-Prod2 in Figure 1c. This result agrees with reported in [17]./ The notation cis- or trans- in this work is not a properly cis- trans- isomerization but we use this notation for the sake of simplicity and clarity.

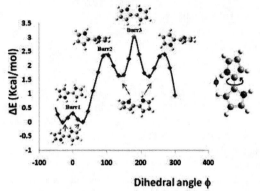

Figure 1. Graphical representation of the PES for the dihedral angle torsion, -N-C-C-N- (φ) showed on the right side of the picture. This graph displays the energy associated to the different model structural conformations. Level of theory B3LYP/6311+(d,p).

From the PES, vide Figure 1; the maximums in the profile correspond to perpendicular and planar conformations contrary to semi-empirical and smaller bases set than 4-31 are used [17], in which cases those conformations resulted as minima in the PES [11,17, 18], therefore a caution has to be taken when those levels of calculations are used. Then for the 6311+(d,p) where taken only the minimal conformations. The analysis of the differences among the structural changes along the PES, in Figure 1, shows that trans- and cis- conformations are minimums as the frequency calculation showed only positive values and also they are situated at minimum positions within the PES, being trans- more stable than cis- phase by 2.33 Kcal mol^{-1}. According to the picture, reaching the cis- conformation from trans- conformation requires a large energy supply of 2.37 Kcal mol^{-1} (Barr2), Barrier 1 (Barr1) indicates the amount of energy to the rocking mode between trans—trans passing by the planar conformation and is about 0.11 Kcal mol^{-1}; barrier 3 (Barr3) is about 1 Kcal mol^{-1} and indicates the amount of energy to the rocking mode but now between cis—cis conformation passing by the planar conformation. Based on the computed results, the energy barrier for converting the cis- into trans- structural conformation is not so high, and then both phases can be exit when the polymer is growing, although to reach cis- to trans- required less energy than the vice-versa.

It has been reported two mainly mechanisms to dimmer formation, where when the reaction mechanism in which the pyrrole radical cation, formed by imposing a potential to the electrode surface, reacts directly with a neutral molecule [19], vide Figure 2, or the chemical process goes through the reaction of two radical cations (Py$^{•+}$) [20], vide

Figure 3, in which case, both yielding the same products, namely, cis-prod and trans-prod. The latter system has two associated stages a single state in which the electrons could be coupled, and triplet state in which the electrons remain uncoupled, for triplet states UB3LYP/631+(d,p) level of theory was used. Now, for the first mechanism, the charge and mass balance of the overall process is similar to the one previously discussed, but the chemical process differs by the sequential steps and the intermediate species involved, see Figure 2 and 3. Since in this mechanism there are only one electron and one positive charge, then a doublet state is associated to the intermediate formed/ and for this study also UB3LYP/631+(d,p) level of theory was used.

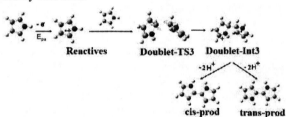

Figure 2. Reaction mechanism for one radical cation with a neutral molecule.

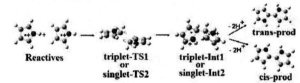

Figure 3. Reaction mechanism of two radical cations.

Then in the Table 1 shows the energy values obtained for both mechanism and also it is taking into account the solvent water effect. So, the non-interacting system for each mechanism followed was considered as a reference.

Now it has been discussed the case of the radical cation-radical cation interaction, [(Py$^{•+}$) + (Py$^{•+}$)]. Table 1 shows the ΔE for the different intermediate systems involved and their degree of stabilization, *in vacuo*. The ΔE for the singlet and triplet states were calculated from the difference in energy between products and reagents, i. e., [(•+Py—Py•+)] − [(Py•+) + (Py•+)], vide Table 1, in which for the singlet state yielded 31.9 kcal mol^{-1} for cis-int1 intermediate and 32.4 kcal mol^{-1} for trans-int1 intermediate over the reactants energy. Of course, to reach these intermediates a Transition State (TS) are involved for each one. The structure obtained for the TS is shown in the Figure 4(a), for the •+Py—Py•+ approaching. The calculated activation energy for Singlet-TS2 corresponds to 71.57 Kcal mol^{-1}, for trans- and 56.88 Kcal mol^{-1} for cis- conformations. Analysing its corresponding Singlet-TS2 geometry, the arrows in the Figure 3 indicate the displacements associated to the imaginary frequency. The C$_1$-C$_1'$ distance in both involved transitions states, one to trans- one to cis- conformations is 2.20 Å, same as that reported by Lacroix et al. [21] and their activation energy value by [21] is about 74.5 Kcal mol^{-1} at AM1//AM1 and 60.5 Kcal mol^{-1} at B3LYP//AM1 levels of theory in gas-phase; but it is not specified for which specific state.

For the triplet state only one intermediate Cis-int1 was stabilized, Trans-int1 was not stabilized, a repulsion effect for trans- isomer was observed. Then, the energy obtained for the Triplet-TS1 in cis- conformation is 120.71 kcal mol^{-1} and the energy stabilization to reach Triplet-Int1 (cis-conformation) is about 26.71 kcal mol^{-1}, with activation energy of 94.0 kcal mol^{-1}. All the differences energy values for the associated intermediates are higher than for the non-interacting system, which means that high activation energy is required, thus, this interaction might not be favored at vacuo conditions. That is not the case for the pyrrole radical cation interacting with a neutral molecule as can be seen in the Table 1, where only one associated transition state was obtained and it corresponds to cis-conformation. The Doublet-TS3 looks like the Cis conformation obtained the two radicals cations interactions. Nevertheless this doublet-TS3 yielding the cis- trans- conformation intermediates and now a stabilization of 12.0 and 20.0 kcal mol^{-1} with respect to the non-interacting system is obtained for cis and trans, respectively.

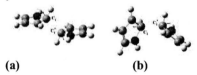

(a)   (b)

Figure 4. Transitions states geometries, for the pyrrole-pyrrole interaction. (a) to trans conformation and (b) to cis conformation. The pictures displays the Schematical displacements associated to the imaginary frequencies of the corresponding TS. The C_1-C_1' distance in the TS is 2.20 Å. Level of theory B3LYP/6311+(d,p).

When water in taking into account as a solvent, by PCM, it is showed meaningful changes, as can be compared in the Table 1. The activations barriers diminish enormously does not matter which mechanism is about, but also, the profile in inverted and now the radical cation interactions results stabilized more than radical cation-neutral molecule by about 30 kcal mol^{-1}. It has to be noted that both cis-, trans- conformations are stabilized in the same proportion which means lower selectivity by water presence, as they have also similar energy activation (Singlet-TS2) also. The triplet state again is not stabilized. For the radical cation interacting with a neutral molecule seems more selective to trans- conformation, compare Doublet-TS3 with Doublet-Int3 in the last column of the Table. This obtained result gives an insight of the mechanism profile and some preferences that could be governed in a reaction oligomerization, which certainly gives the final properties of the product.

Systems	In vacuo		With solvent effect (water)	
Non-interacting	$(Py^{\bullet+}) + (Py^{\bullet+})$	$(Py^{\bullet+}) + Py(N)$	$(Py^{\bullet+}) + (Py^{\bullet+})$	$(Py^{\bullet+}) + Py(N)$
Reference system	0.00	0.00	0.00	0.00
Interacting	$(^{\bullet+}Py{-}Py^{\bullet+})$	$(Py^{\bullet+}){-}Py(N)$	$(^{\bullet+}Py{-}Py^{\bullet+})$	$(Py^{\bullet+}){-}Py(N)$
Triplet-**TS1**	120.71 (Cis) ----- (Trans)	----	59.17 (Cis) ------- (Trans)	----
Triplet-**Int1**	94.0 (Cis) ---- (Trans)	----	26.0 (Cis) ---- (Trans)	----
Singlet-**TS2**	56.88 (Cis) 71.57 (Trans)	----	9.79 (Cis) 11.03 (Trans)	----
Singlet-**Int2**	31.9 (Cis) 32.4 (Trans)	----	-35.7 (Cis) -37.4 (Trans)	----
Doublet-**TS3**	----	14.18 (Cis) ----- (Trans)	----	2.67 (Cis) 25.8 (Trans)
Doublet-**Int3**		-12.0 (Cis) -20.0 (trans)		1.0 (Cis) -6.0 (trans)

Table 1. Energy difference values, in kcal mol^{-1}, for intermediate systems and transitions states (TS) involved in the reaction mechanism for the two processes, in vacuo and with solvent effect. Non-interacting system is taking as a reference state. Calculated values are at B3LYP/6311+G(d,p) level of theory. Vide Figures 2 and 3 for identified the involved systems.

Chain lenght	Eg, eV		
	In vacuo	Solvent effect	Experimental (23)
1	5.55	5.95	5.96
2	4.82	4.79	4.37
3	4.17	4.20	3.61
4	3.87	3.91	3.48
5	3.63	3.67	2.82
6	4.07	4.10	2.77
7	3.43	3.46	2.45
8	3.37	3.40	2.28
9	3.34	3.37	2.10

Table 2. $|E_{HOMO}-E_{LUMO}|$ (E_g) energy difference for the different oligomers, calculated in vacuo and with water as solvent effect. Comparison of the PCM calculated energy gap (Eg) for Py and the sequential formed oligomers until 9 units at in vacuo and with water (solvent) effect. Applied level of theory: B3LYP/6-311+(d,p). Eg= $|E_{HOMO}-E_{LUMO}|$, Chain lenght=number of coupled Py units (oligomers)

When the oligomer increases, some of the electronic properties changes as the $|E_{HOMO}-E_{LUMO}|$ (E_g) energy difference. In the Table 2, the growing of Py-oligomer is presented and depicted schematically in Figure 3. The comparison involves experimental literature reported values. It can be observed that for the initial monomer and until the 4 units Py-oligomer, the trend in theoretical calculated E_g for both at in vacuo and with water conditions decreases, in line with the experimental energy values, vide Table 2. However, from the 5 units Py-oligomer on there is a more pronounced difference between these E_g's being more notorious for the 6 units Py oligomer where the theoretical energy suddenly increases to start lowering again for the further Py-oligomer. Within these calculations, the properties of water such as dipolar momentum and conductivity were accounted for, but no strong contribution to the calculated E_g is observed as compared to vacuo conditions, vide Figure 5.

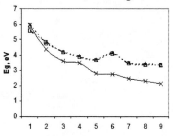

Figure 5. Calculated energy Eg, at in vacuo conditions (□) and in the presence of water (Δ) and that experimental literature reported (X) (17), as a function of the chain length of the Py-oligomer. Level of theory: B3LYP/6-311+(d,p). $Eg = |E_{HOMO} - E_{LUMO}|$.

4 CONCLUSION

The mechanism involving one neutral and one radical cation interaction yields better stabilization in vacuo than when the water is presented, contrary to the two radicals cations interaction, where water presence stabilized the transitions states and intermediates. The obtained results showed that when water is taking into account, by PCM, the reaction mechanism stabilized the transitions states involved along the reaction and gives low selectivity to the products, as cis- trans- conformation have similar stabilization.

The electronic properties of the pyrrole molecule and its consecutive oligomers until a 9 units Py-oligomer were calculated finding proper fit to those experimental reported values. The stabilization of E_g for all the molecules decreases as the oligomer chain length increases which is ascribed to the distribution through the HOMO-LUMO of the stepwise products. No apparent energy contribution from water, used as solvent, on the formed stepwise Py-oligomer is obtained. This indicates that water does not affect the final oligomer properties, but it contributes to Py activation energies by enhancing the polarization of the reactant molecule as compared to vacuo conditions in line with experimental reported results.

ACKNOLEDGEMENTS

AC thanks the financial support from SEP-CONACYT through the project 80361 and MEPP gratefully acknowledge financial support from CONACYT through the projects 49775-Y (24658).

REFERENCES

[1] B. Scrosati, Applications of Electroactive Polymers, Chapman & Hall, London, 1993.

[2] G.A. Álvarez-Romero, A. Morales-Pérez, A. Rojas-Hernández, M. Palomar- Pardavé, M. T. Ramírez-Silva, Electroanalysis, 16, 1236, 2004.

[3] G.A. Álvarez-Romero, M.E. Palomar-Pardavé, M.T. Ramírez-Silva, Anal. Bioanal. Chem., 387, 1533, 2007.

[4] D.L. Wise, G.E. Winek, D.J. Trantolo, T.M. Cooper, D.J. Gresser, in Electrical and Optical Polymer System, Marcel Dekker, Inc., New York, Vol. 17, 1998.

[5] J. Rodríguez, H.J. Grande, T.F. Otero, in Handbook of Organic Conductive Molecules and Polymers, ed. H. S. Nalwa, John Wiley & Sons, New York, 415, 1997.

[6] G.A. Álvarez-Romero, E. Garfias-García, M.T. Ramírez-Silva, C. Galán-Vidal, M. Romero-Romo, M. Palomar-Pardavé, App. Surf. Sci., 252, 5783, 2006.

[7] K. Imanishi, M. Satoh, Y. Yasuda. R. Tsushima, S. Aoki, J. Electroanal. Chem. 242, 203, 1988.

[8] M. Satoh, K. Kaneto, K. Yoshino, Synth. Met. 14, 289, 1986.

[9] M. Satoh, K. Imanishi, K. Yoshino, J. Electroanal. Chem., 317, 139, 1991.

[10] C.M. Jenden, R.G. Davidson and T.G. Turner, Polymer, 34, 1649, 1993.

[11] E. Chaînet, M. Billon, Synth. Met., 99, 21, 1999.

[12] J.A. Cobos-Murcia, L. Galicia, A. Rojas-Hernández, M.T. Ramírez-Silva, R. Álvarez-Bustamante, M. Romero-Romo, G. Rosquete-Pina, M. Palomar-Pardavé, Polymer, 46, 9053, 2005.

[13] E. Garfias-García, M. Romero-Romo, M.T. Ramírez-Silva, J. Morales, M. Palomar-Pardavé, J. Electroanal. Chem.,613, 67, 2008.

[14] M.J. Frisch et al, *Gaussian 03, Version* B.3; *Gaussian, Inc.,* Pittsburgh, PA, 2003.

[15] C. Lee, W. Yang, R.G. Parr, Phys. Rev. B, 37 (1988) 785.

[16] S. Miertus, E. Scrocco, J. Tomasi, J. Chem. Phys. 55 (1981) 117.

[17] M. Yurtsever, E. Yurserver, Synthetic Metals 98, 221-227, 1999.

[18] M. Takakubo, J. Electroanal. Chem., 258, 303, 1989.

[19] S. Asavapiriyanont, G.K. Chandler, G.A. Gunawardena, D. Pletcher, J. Electroanal. Chem. 177, 229, 1984.

[20] B.L. Funt, A. F. Diaz, Organic Electrochemistry: an introduction and a guide, Marcel Dekker, New York, (1991) 1337.

[21] J. C. Lacroix, F. Maurel, and P.C. Lacaze. J. Am. Chem. Soc., 123, 1989-1996, 2001.

Amphiphilic Polymeric Core-Shell Particles:

Novel Synthetic Strategy and Potential Applications of the Particles

Pei Li*

*Department of Applied Biology and Chemical Technology
The Hong Kong Polytechnic University, Hung Hom, Kowloon,
Hong Kong, P. R. China, bcpeili@polyu.edu.hk

ABSTRACT

We have developed a novel and commericially viable route to a wide range of highly uniform, amphiphilic core-shell particles in nano- to micro-scaled sizes. Novel feature of this synthetic approach is that it combines graft copolymerization, *in situ* self-assembly of the resulting amphiphilic graft copolymers and emulsion polymerization in a one-step synthesis. This versatile methodology allows us to design and tailor-made particles for specific applications through selection of appropriate amino-containing water-soluble polymers and hydrophobic monomers.

Keywords: nanoparticles, core–shell, amphiphilic, well-define, uniform

INTRODUCTION

Core–shell polymeric particles that consist of two or more very different chemical components and have particle diameters in nano- to micro-size ranges often exhibit improved physical and chemical properties over their single-component counterparts in applications as diverse as biomedicine and surface coatings. In particular, amphiphilic particles that consist of well-defined hydrophobic cores and hydrophilic shells have attracted much attention because of their potential applications in diagnostics, bio-separation, drug delivery, gene therapy, enzyme immobilization, coatings, and catalysis [1, 2]. They are also of interest from a fundamental point of view in colloid and interface science [3-5].

Various synthetic approaches have been reported to synthesize amphiphilic core–shell particles based on polymerization and assembly methods. They include: (1) Graft copolymerization of a hydrophilic monomer onto a reactive seeded particle via either conventional or living radical polymerization [6, 7]. (2) Copolymerization of a reactive macro-monomer with a hydrophobic monomer [8].

(3) Emulsion polymerization in the presence of block copolymer [9] or comb-like copolymer [10] containing controlled free radicals moieties. (4) Ab initio emulsion polymerization by self-assembly using controlled radical polymerization [11]. (5) Self-assembly of amphiphilic block copolymers, followed by crosslinking the core or shell via covalent or ionic bonding [12]. (6) Stepwise deposition of polyelectrolytes onto charged particle surface [13].

METHODOLOGY

We have developed another route to synthesize well-defined, amphiphilic core–shell particles and hydrophilic microgels [14-17], based on the aqueous-phase redox reaction between alkyl hydroperoxide and amine functional groups of a water-soluble polymer. Initiation comes about when the amine-containing water-soluble polymers interact with a small amount of catalyst in water at between 70–80 °C to generate free radicals on the polymer backbone. The macroradicals subsequently initiate the graft polymerization of the hydrophobic monomer. The amphiphilic macro-radicals generated can *in situ* self-assemble to form polymeric micelle-like microdomains, which become loci for the subsequent polymerization of the monomer: a type of emulsion polymerization. Well-defined amphiphilic core–shell particles with diameters between 50 and 300 nm, thus can be produced in the absence of surfactant.

FEATURES

Novel feature of this synthetic approach is that it combines graft copolymerization, *in situ* self-assembly of the resulting amphiphilic graft copolymers and emulsion polymerization in a one-step synthesis. This versatile methodology allows us to design and tailor-made particles for specific applications through selection of appropriate amino-containing water-soluble polymers and hydrophobic monomers. Special features of this synthetic route and products include:

- The particles are easy to synthesize in high solids content (up to 30%) without using surfactant.
- The particles have well-defined core-shell nanostructure ranging from nano-to micro-scale with narrow particle size distribution.
- The process uses aqueous-based Chemistry, which is environmentally benign.
- The core property of the particle can be varied (e.g. hard, soft, temperature-sensitive and hollow).
- The shell component can use a wide range of amine containing water-soluble polymers such as synthetic and biopolymers
- Surface functionalities and properties can be easily altered.

This process offers a commercially viable route to a wide variety of novel core-shell particles with different sizes, compositions, structures and functions (examples are shown in Figure 1). Applications of these unique particles in gene and drug deliveries, enzyme immobilization, bioseparation, water treatment, antibacterial and functional coatings have been demonstrated. This presentation will show the synthetic methodology that allows for tuning of the internal core and external shell compositions and properties for selected applications.

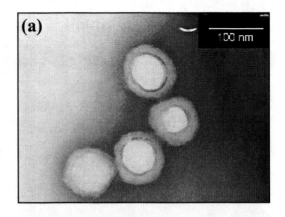

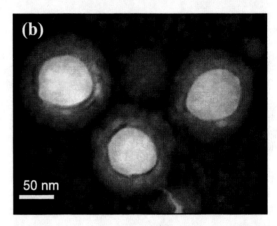

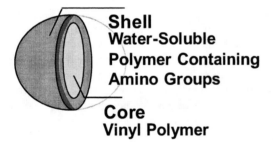

Shell
Water-Soluble
Polymer Containing
Amino Groups

Core
Vinyl Polymer

Hydrophobic core + hydrophilic shell
(Amphiphilic core-shell particle)

Hydrophilic core and shell
(Core-shell microgel or smart microgel)

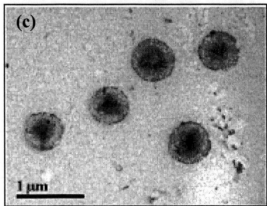

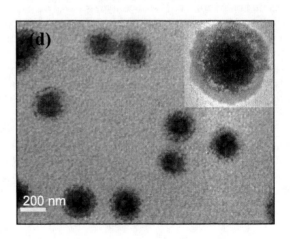

Figure 1. TEM images of particles (a) poly(methyl methacrylate) cores with polyethyeneimine shells; (b) polystyrene cores and polyethyeneimine shells; (c) poly(*n*-butyl acrylate) cores and chitosan shells: (d) poly(*N*-isopropylacrylamide) cores with chitosan shells.

POTENTIAL APPLICATIONS

These core–shell particles have been applied in various fields. For example, particles with PMMA cores (~ 90 nm in diameter) and hairy poly(ethylenimine) shells have been successfully used as carriers in gene and drug deliveries [18-20]. Particles with biopolymer shells such as chitosan and soft poly(n-butyl acrylate) cores have been applied onto cotton fabrics as a biocompatible antibacterial finishing [21, 22]. Particles with PMMA cores coated with either polyethylenimine or chitosan shells (~150 to 500 nm in diameter) have been applied as effective nanosorbents to remove heavy metals and organic contaminants in wastewater treatment [23]. The core-shell particles can be used for multiple times via a process of regeneration. For application in enzyme immobilization, novel one-step immobilization of celluase onto a non-porous nanoparticle has been developed [24]. The core-shell nanoenzyme particles show many advantages over the conventional immobilization methods. Furthermore, modification of the core-shell particles to hollow particles allows us to create novel materials such as nanotubes, nanofibers and rods in nano- and micro-size ranges [25, 26]. Magnetic-responsive, biocompatible core-shell particles with magnetic nanoparticles encapsulated inside the core have also been prepared [27] for usage in bio-separation, targeted drug and gene delivery and immobilization of enzyme.

ACKNOWLEDGEMENT

I would like to thank my talent students and researchers for their contributions to this work. I also gratefully acknowledge the Hong Kong Polytechnic University, the Research Grant Council and Innovation and Technology Fund of the Hong Kong SAR for their financial support on this research.

REFERENCES

[1] H. Kawaguchi, Prog. Polym. Sci. 25, 1171-1210, 2000.

[2] C. Pichot, Current Opinion in Colloid & Interf. Sci. 9, 213-221, 2004.

[3] F. Caruso, Adv. Mater. 13, 11-22, 2001.

[4] E. M. Coen, R. A. Lyons, R. G. Gilbert, Macromolecules 29, 5128-5135, 1996.

[5] L. Vorwerg, R. G. Gilbert, Macromolecules 33, 6693-6703, 2000.

[6] S. Tsuji, H. Kawaguchi, Langmuir 20, 2449-2455, 2004.

[7] F. D'Agosto, M. Charreyre, C. Pichot, R. Gilbert, G., J. Polym. Sci. A Polym. Chem. 41, 1188-1195, 2003.

[8] T. Basinska, S. Slomkowski, S. Kazmierski, A. Dworak, M. Chehimi, M. , J. Polym. Sci. A Polym. Chem. 42, 615-623, 2004.

[9] J. Nicolas, A. V. Ruzette, C. Farcet, P. Gérard, S. Magnet, B. Charleux, Polymer 48, 7029-7040, 2007.

[10] M. Save, M. Manguian, C. Chassenieux, B. Charleux, Macromolecules 38, 280-289, 2004.

[11] C. J. Ferguson, R. J. Hughes, D. Nguyen, B. T. T. Pham, R. G. Gilbert, A. K. Serelis, C. H. Such, B. S. Hawkett, Macromolecules 38, 2191-2204, 2005.

[12] Q. Zhang, E. E. Remsen, K. L. Wooley, J. Am. Chem. Soc. 122, 3642-3651, 2000.

[13] A. J. Khopade, F. Caruso, Langmuir 19, 6219-6225, 2003.

[14] P. Li, J. Zhu, P. Sunintaboon, F. W. Harris, Langmuir 18, 8641-8646, 2002.

[15] J. Zhu, P. Li, J. Polym. Sci. A Polym. Chem. 41, 3346-3353, 2003.

[16] M. F. Leung, J. Zhu, F. W. Harris, P. Li, Macromol. Rapid Commun. 25, 1819-1823, 2004.

[17] W. Li, P. Li, Macromol. Rapid Commun. 28, 2267-2271, 2007.

[18] M. Feng, P. Li, J. Biomed. Mater. Res. 80, 184, 2007.

[19] J. Zhu, A. Tang, L. P. Law, M. Feng, K. M. Ho, D. K. L. Lee, F. W. Harris, P. Li, Bioconjugate Chem 16, 139-146, 2005.

[20] M. Feng, K. L. D. Lee, P. Li, Int. J. Pharm. 311, 209, 2006.

[21] W. Ye, M. F. Leung, J. Xin, T. L. Kwong, D. K. L. Lee, P. Li, Polymer 46, 10538-10543, 2005.

[22] W. Ye, J. Xin, P. Li, K. L. D. Lee, T. L. Kwong, J Appl Polym Chem 102, 1787, 2006.

[23] P. Li Wastewater treatment process using core-shell particles 7323110, 2008.

[24] K. M. Ho, X. Mao, L. Gu, P. Li, Langmuir 24, 11036-11042, 2008.

[25] P. Sunintaboon, K. M. Ho, P. Li, S. Z. D. Cheng, F. W. Harris, J. Am. Chem. Soc. 128, 2168-2169, 2006.

[26] C. H. Lee, K. M. Ho, F. W. Harris, S. Z. D. Cheng, P. Li, Soft Matter 5, 4914-4921, 2009.

[27] K. M. Ho, P. Li, Langmuir 24, 1801-1807, 2008.

Green Chemistry Derived Nanocomposite of Silver-modified Titania for Disinfectant

Jingbo Liu[1], Iliana Medina-Ramirez[2], Zhiping Luo[3], Sajid Bashir[1] and Ray Mernaugh[4]

1: Texas A&M University-Kingsville, Kingsville, Texas 78363, kfjll00@tamuk.edu; br9@tamuk.edu;
2: Universidad Autonoma de Aguascalientes, Aguascalientes, Ags., 20100, Mexico,
iemedina@correo.uaa.mx;
3: Texas A&M University, College Station, Texas 77843, luo@mic.tamu.edu;
4: Biochemistry, Vanderbilt University, Nashville, TN 37232, r.mernaugh@vanderbilt.edu.

ABSTRACT

A significant accomplishment of this study centers on developing nanocomposite materials via green synthesis techniques for inactivation of prokaryotes like *Escherichia coli (E coli)*. The motivations of this study are to: (1) implement fundamental research in nanochemistry for biological application; (2) understand the mechanism(s) of inactivation of bacteria; and (3) to integrate advanced instrumentation to aid understanding of antibacterial effects of novel nanoparticles. The originality of this study focuses on use of bottom-up synthesis to produce Ag-modified TiO_2 nanocomposite particles with decreased band gap energy in the visible light range. In addition, several characterization techniques were used to study the nanostructure of these nanocomposites.

Keywords: Nanocomposite of Ag-TiO_2, Colloidal Synthesis, Characterization, Antibacterial Performance

1 INTRODUCTION

Titania (TiO_2) is one common photocatalyst, which has been extensively studied for applications in the field of catalysis and disinfection [1, 2]. While the importance silver (Ag) as a bactericide has been well-documented [3,4], the mode of action remains to be explored. Other approaches include the use of Ag modified TiO_2 nanoparticles (Ag-TiO_2 NPs) as photocatalyst [5], however, its use as a disinfectant request ultraviolet light. This research suggests that Ag-TiO_2 NPs can be prepared in a cost-efficient and environment-friendly manner [6]. Modification of TiO_2 using Ag resulted in reduced band gap energy as measured by ultra-violet visible light (UV-VIS) spectroscopy. The decreased band gap will allow us to eliminate bacteria using visible light instead of UV radiation. Significance of this study was to develop environmental-friendly synthesis of nanomaterials, and characterization, including use of several state-of-the-art instrumentation to analyze nanostructure of nanocomposite for in-depth study of inactivation of bacteria.

2 EXPERIMENTAL

All chemicals unless otherwise specified were obtained from Sigma-Aldrich (St Louis, MO), and solvents from VWR International (West Chester, PA). The reagents were reagent grade and were used without further purification. Doubly-distilled and 0.2 micron filtered (Milli-Q) water was used in the dissolution of our compounds. A number of formulations were attempted to optimize the final composition of Ag-TiO_2. Various reducing agents such as ascorbic acid, sodium citrate, sodium borohydride and dimethylamine borane (DMAB) were used for the complete chemical reduction of silver ion. In addition, the nanostructure of the Ag-TiO_2 was systematically studied using several analytical techniques, including transmission electron microscopy equipped with X-ray energy dispersive spectroscopy, UV-VIS spectroscopy, X-ray powder diffraction, and zeta-potential to determine the particle size and its distribution, crystalline phase structure and surface stability. Furthermore, antibacterial effect of Ag-TiO_2 colloid was conducted under visible light.

2.1 Materials

Starting material of titanium (IV) butoxide ($Ti(O^nBu)_4$) was added into acetic acid (CH_3COOH) aqueous solution with continuous agitation. The acetic acid was used to control the hydrolysis rate of TiO_2 in our experiments. Arabic gum (AG, a dispersing agent) of 3.00 mass % solutions was prepared and stirred for min at 60 °C. Into this AG solution, the second starting material $AgNO_3$ (molar ratio of Ag: Ti was controlled at 1:20), which was dissolved in 30.00 ml of Milli-Q water, was added. The Ag^+ cation was then reduced using the above reducing agents. The concentration of Ag was kept at 1.06×10^{-3} mol/L.

2.2 Characterization

A Tecnai F20-G^2 transmission electron microscopy (TEM) (FEI Company, Hillsboro, Oregon) was used to characterize the nanoparticles and its size distribution. X-ray energy dispersive spectroscopy (EDS) technique was also used to obtain elemental composition Ag-TiO_2 nanocomposite. An atomic-resolved lattice fringe of these nanocomposite materials was characterized by high resolution TEM (HRTEM). The samples were dispersed in absolute ethanol and deposited onto the carbon-coated copper (Cu) grid. An X-ray powder diffraction (XRD) was employed to determine the crystalline phase and average crystallite size of the nanocomposite. The XRD patterns were collected using a Rigaku multiflex diffractometer (Rigaku Americas Incorporation, Houston, TX), equipped with a Cu target at 40 kV and 44 mA. The scanning range was varied between 30 and 80 degrees at a scanning rate of 2°/min. The phase structure was identified using the Rigaku Jade 7.0 software package. A ZetaPALS (Brookhaven Instruments Corporation, New York, USA) was utilized to measure the particle size distribution and surface energy to evaluate the stability of Ag-TiO$_2$ nanocomposite. The ZetaPALs utilizes dynamic light scattering to measure the electrophoretic mobility of negatively-charged colloid.

2.3 Inactivation of Bacteria

Pure cultures of *Escherichia coli (E. coli*, TG$_1$) were maintained as frozen glycerol stocks in the Department of Biochemistry at Vanderbilt University. *E. coli* cultures were grown on Bacto-Tryptone (2×YT) agar plates or in 20 ml of sterile 2×YT or Luria-Bertani (LB) broth. The cells were grown overnight, aerobically on a rotatory shaker at 37 °C. Cells were harvested from overnight culture by centrifugation for 10 min at 4 °C and 6000 rpm. The cells were re-suspended and diluted to a final concentration of 1×10^7 cfu/ml in a 0.05M KH$_2$PO$_4$ and 0.05M K$_2$HPO$_4$ (pH 7.0) buffer solution.

3   RESULTS AND DISCUSSION

Synthesis of Ag-TiO$_2$ nanocomposite was developed in this study to prevent the aggregation of the particles during the synthesis process. The primary objectives of this work were to characterize the samples derived from colloidal chemistry and to attempt to prevent particles from agglomeration with a surfactant capable to improve their dispersion in the final product.   This process has the advantage to improve the wetting and dispersion of the surface of the Ag-TiO$_2$ composite. A continuous external TiO$_2$ thin layer was then deposited onto the Ag core successively. The present work has successfully overcome the negative attributes of the existing products and procedure of Ag-TiO$_2$ fabrication (shown in *Figure 1*).

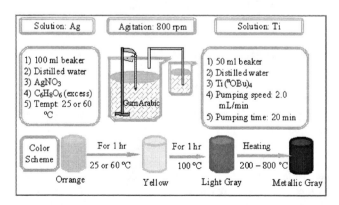

Figure 1: Fabrication of the Ag-TiO$_2$ via Bottom-up Green Chemical Approach.

3.2 Nanostructural Characterization

The X-ray powder patterns for Ag-TiO$_2$ nanocomposite are depicted in Figure 1. Results clearly indicate the presence of silver (Ag) in the form of highly crystalline face-centered cubic (*fcc*) structure (PDF 01-089-3722 with lattice constants of 4.0855 Å and 90 °). The diffraction from TiO$_2$ crystals indicates an anatase structure was obtained (PDF 03-065-5714 with lattice constants of 3.7855/9.514 Å and 90 °). ***Figure 2*** display four specimens with identical spectra, which indicates that the reproducibility was excellent. The insert was the Ag XRD pattern, which did not show in the spectra of Ag-TiO$_2$ composite due to its low concentration.

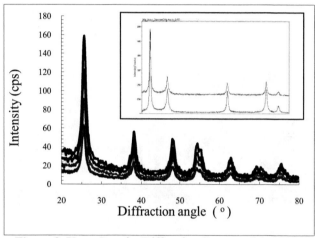

Figure 2: X-ray Powder Diffraction Spectra of Ag-TiO$_2$.

Results of TEM morphological and nanostructural studies are summarized in ***Figure 3***. The results indicate that mono-dispersive and highly crystalline Ag-TiO$_2$ nanocomposites were obtained. The composite's appearance is near-spherical with the average size ranged from 20 to 50 nm in diameter. Particle agglomeration occurred and was presumably due to the high surface tension of the ultrafine nanocomposite particles. The fine particle size will result in a large surface area that will, in turn, enhance the nanocomposite catalytic activity. The Ag and TiO$_2$ distribution indicates that Ag was used as a seed to allow for TiO$_2$ deposition. TiO$_2$ formed continuous and successive membranes around the Ag core, which was consequently subjected to decreased band gap energy.

X-ray energy dispersive spectroscopic (EDS) analysis indicate that nanocomposite consisted mainly of Ag, Ti, and O (***Figures 4***). The spectra (***left insert***) of L$_\alpha$ for the Ag metal at 2.984 eV and L$_n$ at 2.806 eV is well-indexed with the standard metallic Ag. Similarly, the spectra (***right insert***) Ti occurred at 4.510 eV and resulted from a K$_\alpha$ electron configuration and another one occurred at 0.452 eV and resulted from an L$_\alpha$ configuration. An insignificant peak at 4.930 eV caused by K$_\beta$ was also detected.

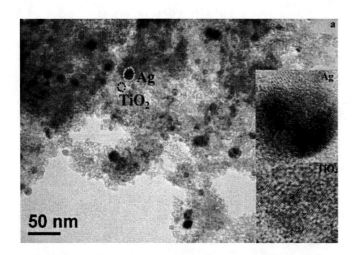

Figure 3: TEM image of Ag-TiO₂ nanocomposite (top insert: Ag fringe; bottom insert: TiO₂ fringe).

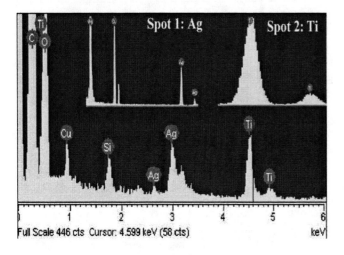

Figure 4: EDX spectra of Ag-TiO₂ nanocomposite (left insert: Ag spectrum; right insert: TiO₂ spectrum).

In order to evaluate the stability and the electrokinetic behavior of the Ag-TiO₂ colloid, the zeta-potential (known as effective surface potential, (denoted as ζ) was measured by imposing an electrical field (2 V/cm) across the colloid particles. The measured ζ allowed an estimation of the degree of aggregation. It was found that zetapotential essentially ranged from (minus) -20 to -85 mV according to the various fabrication parameters of the TSNC. With increasing pH, the absolute ζ values decreased and were essentially constant at basic solution. The large zetapotential of negative charges minimized particle aggregation due to electrostatic repulsion. In the finer particle region (10 - 20 nm), the zetapotential was from (minus) -55 to -45 mV. The less negative zetapotential ranging from -25 to -15 mV can be attributed to the crystal growth to form larger particle with size of 50 to 70 nm.

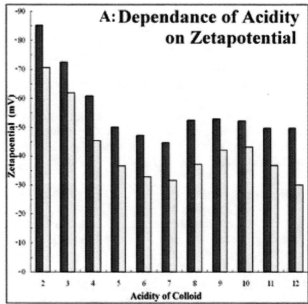

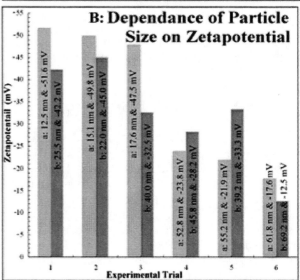

Figure 5: The electrokinetics analysis of Ag-TiO₂ nanocomposite to evaluate the surface energy (**A**); the particle size distribution (**B**) of Ag-TiO₂ nanocomposite corresponding to the zeta potential.

3.3 Inactivation of Bacteria

The nanocomposite optical absorption properties were consistent with noble metal nanoparticles. The GS derived Ag-TiO₂ were successful in the removal of biological impurities from drinking and underground water supplies. *E. coli* were inactivated using Ag-TiO₂ under visible light at ambient temperature and pressure. The results of the study indicated that nanocomposites could be specifically designed to prevent growth of bacteria in water (***Figure 6 A to C***). Significance of work is development of Ag-TiO₂ for inactivation of bacteria.

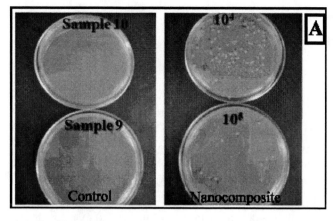

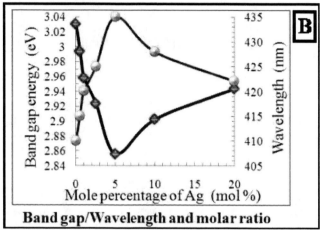

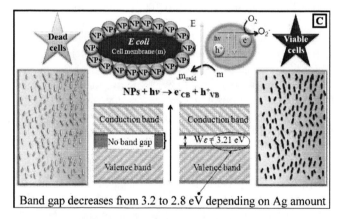

Figure 6: The inactivation of *E coli* using Ag-TiO₂ nanocomposite (*A*); UV-VIS spectroscopic study of the band gap decrease of TiO₂ modified by Ag (*B*); and hypothesis of inactivation mechanism of *E coli* (*C*).

4   CONCLUSION

The colloidal suspensions of the nanocomposites are composed of metal and ceramic (Ag-TiO₂). They were found to be extremely stable over a prolonged time period, and were identified using zetapotential data at various pH with negative surface charges. Morphologically, nanocomposites were found to be composed of near-spherical particles that were highly crystalline. The nanocomposites were mono-dispersed with particles varying in size from 15.8 to 22.5 nm, depending upon nanocomposite solution pH. Nanocomposite elemental composition studies indicated that the molar ratio of Ag and Ti was approximately 1 to 20. The binding energies and energy differences of Ag and Ti were well indexed with their associated standard spectra. Nanocomposite optical absorption properties were consistent with noble metal nanoparticles. The 'green chemistry' derived Ag-TiO₂ composites are applicable for the removal of biological impurities from drinking and underground water supplies. *E. coli*, a gram-negative model microorganism was effectively inactivated using the nanocomposite under visible light at ambient temperature and pressure. The results of the study indicated that nanocomposite could be specifically designed to prevent growth of bacteria in water.

ACKNOWLEDGEMENT

The authors are thankful to the Academia Mexicana de Ciencias (AMC), Fundación México Estados Unidos para la Ciencia (FUMEC), Texas A&M University-Kingsville (TAMUK), the College of Arts and Sciences Research and Development Fund, (160310-00014) and Robert A. Welch Departmental grant, (AC0006, at TAMUK) for the financial support. Technical support and facility access were provided by the South Texas Environmental Institute and also by the Department of Chemistry, the Microscope and Imaging Center and the Center of Materials Characterization Facility at TAMU College Station. In addition, technical support from biochemistry at Vanderbilt University is also duly acknowledged to allow us in conducting the advanced instrumentation analysis. The US NSF MRI program was further acknowledged to allow the use of Ultima III Rigaku X-ray powder diffraction (TAMU-Corpus Christi).

REFERENCES

[1] R. Kelsall, I. W. Hamley, M. Geoghegan, *Nanoscale Science and Technology*, John Wiley & Sons, Inc. New Jersey, USA, Ch 1, 2005.

[2] Y. Yao, Y. Ohko, Y. Sekiguchi, A. Fujishima, Y. Kubota, *Journal of Biomedical Materials Research Part B: Applied Biomaterials*, 85B, 453, 2008.

[3] A. Borras, Á. Barranco, J. P. Espinós, J. Cotrino, J. P. Holgado, A. R. González-Elipe, *Plasma Processes and Polymers*, 4, 515, 2007.

[4] F. Zhang, R. Jin, J. Chen, C. Shao, W. Gao, L. Li, N. Guan, Journal of Catalysis, 232, 424, 2005.

[5] A. Kubacka, M. Ferrer, A. Martínez-Arias, M. Fernández-García, *Applied Catalysis B: Environmental*, 84, 87, 2008.

[6] F. Sayılkan, M. Asiltürk, N. Kiraz, E. Burunkaya, E. Arpaç, H. Sayılkan, *Journal of Hazardous Materials*, 162, 1309, 2009.

Magneto-liposomes: Stability of magnetic nanoparticles in suspension for drug delivery

T. Sen[*], S. J. Sheppard[*&***], T. Mercer[**] and A. Elhissi[***]

[*]Centre for Materials Science, School of Forensic and Investigative Sciences, University of Central Lancashire, Preston, United Kingdom, tsen@uclan.ac.uk
[**]School of Computing Engineering and Physical Sciences, tmercer1@uclan.ac.uk
[***]School of Pharmacy, aelhissi@uclan.ac.uk

ABSTRACT

Superparamagnetic iron oxide nanoparticles (SPIONs) of diameter 30-50 nm have been synthesized and stabilized in suspension by one-pot fabrication of magneto-liposomes of diameter around 80nm. Magneto-liposomes have been prepared by mixing with optimum ratio of prefabricated SPIONs and multi lamellar vesicles (MLVs) under strong ultrasonic vibration. Magneto-liposome nanocomposites are stable in aqueous suspension at physiological pH (~7) for days. Surfaces of magneto-liposomes have been characterized by measuring zeta potential at physiological pH and Salmon sperm DNA binding/elution experiments. Stable magneto-liposome exhibited nearly neutral surfaces whereas bared SPIONs exhibited positive and MLVs exhibited negative surface charge at physiological pH (~7). SPIONs have been uniformly coated with liposomes evidenced from high binding and elution of Salmon sperm DNA unlike to bared SPIONs. Pro-drug (Mitomycin-C) entrapment into stable magneto-liposomes at physiological pH (~7) is under study and the results will be presented in the conference (NSTI 2010).

Keywords: SPIONs, liposome, magneto-liposomes, surface charge, drug loading

1 INTRODUCTION

Superparamagnetic iron oxide nanoparticles (SPIONs) have become increasingly important materials for the quick, easy, sensitive and reliable separation of specific bio-molecules and for magnetic hyperthermia agents in medical diagnostics and therapeutics [1-4]. The surface properties and the interactions between nanoparticles in suspension are not well understood. One of the major problems of working on such nanoparticles is the aggregation. Surface modification of such nanoparticles in suspension raised an important question "Does the self-assembled coating of magnetic nanoparticles cover individual particles or agglomerates?"[5]. Recently Sen *et al* [6] have reported the stability of magnetic nanoparticles using commercial dispersing agents, however, they lacked bio-compatibility for *in vivo* purposes such as drug delivery, magnetic hyperthermia and magnetic contrasting.

Liposomes (vesicles formed by amphilphilic phospholipid molecules in water) are well known for drug delivery for long times [7-9] and several review papers have been published [10-12]. Drugs can be entrapped either in the inner aqueous phase or in the lipid bilayers, depending on their hydrobicity/hydrophilicity ratio. Similarly SPIONs are also well known in the field of drug delivery [13]. The advantage of using SPION as a drug carrier is that they can be transported through the vascular system and can be concentrated at a particular point of the body with the aid of magnetic field [14]. Nanoparticles (diameter < 100nm) based drug carrier are promising candidate for drug delivery as they can diffuse through cell membrane. Size, morphology and surface charge are three important parameters of drug loaded nanoparticles and their behaviour in the blood stream when injected intravenously. Gupta *et al* [15] have reported that diameter of nanoparticles ranging from 10 to 100 nm are most effective for drug delivery because they can evade reticuloendothelial system (RES). Recently, Gabizon *et al* [16] have reported that incorporating polyethylene glycol (PEG) to liposome bilayer results in inhibition of liposome uptake by the reticulo-endothelial system and significant prolongation of liposome residence time in the blood stream. Fabrication of bio-compatible SPIONs of diameter 10-100 nm with specific surface charge and hydrophilicity is a challenge for drug delivery.

Sasaki *et al* [17] reported the entrapment of pro-drug Mitomycin-C in liposome and Tokunaga *et al* [18] have reported the release in to the blood stream through intravenous injection. Mitomycin-C is a potent antibiotic-antineoplastic drug for ocular surgery, however, suffers due to ocular toxicity. Chetoni *et al* [19] have reported that liposomal preparation containing Mitomycin-C was capable of reducing the corneal healing rate and drug toxicity of a corneal lesion in a rabbit model. Recently, Zalipsky *et al* [20] have reported that Mitomycin-C loaded STEALTH liposome (SL) has enhanced antitumor activity compared to pure Mitomycin-C or pure Doxorubicin or Doxorubicin loaded SL.

Liposome coated magnetic nanoparticles (magneto-liposomes) are new class of nanocomposites for drug delivery [21-24]. Herein, we report the one-pot fabrication of stable magneto-liposomes and their surface characteristics at physiological pH (~7) along with the preliminary data on entrapment of Mitomycin-C *in vitro*.

2 EXPERIMENTAL

Bared SPIONs were synthesized following Massart method [25] by co-precipitation of an aqueous solution of ferrous and ferric chloride in the presence of ammonium hydroxide.

Multilamellar liposomes (MLVs) solution was prepared by mixing 270mg of phospholipid (SPC) in 1ml chloroform in a 500mL round bottom flask. The flask containing the phospholipid solution was attached to a rotator evaporator placed on 35^0C water bath. Upon evaporation, a thin film of lipid was observed to be formed in the round bottom flask. The flask containing lipid film was flashed with nitrogen gas in order to remove chloroform residue. The film was dissolved in 27mL of deionized water and stirred manually for 10 min and annealed for 2 hr at room temperature before use for the fabrication of magneto-liposomes.

Magneto-liposomes suspensions were prepared by mixing SPIONs of various amount (0.25mg/mL to 15mg/mL) in 7mL MLVs and placed under strong ultrasonic vibration (titanium horn) for 2, 4, 6 and 8 min.

Bared SPIONs were characterized by X-ray diffraction, magnetometry, scanning and transmission electron microscopy. All materials (bared SPIONs and magneto-liposomes) were characterized by laser particle size analyzer (Nanosizer) and salmon sperm binding and elution experiment [26].

3 RESULTS AND DISCUSSION

The synthesized iron oxide materials exhibited X-ray diffraction pattern of multiple peaks within the 2θ range 30 to 75^0 (data not shown). Indexing peaks corresponding to the fingerprint peaks (220, 311, 400, 440) of pure magnetite (Fe_3O_4). Figure 1 presents the magnetic data of the materials at room temperature. Almost closed loop with negligible coercivity was observed indicating the superparamagnetic nature of the iron oxide materials (SPIONs).

Figure 2 presents the particle size distribution of the bared iron oxide nanoparticles (SPIONs), multilamellar vesicles (MLVs), small unilamellar vesicles (SLVs) and magneto-liposomes (liposome coated SPIONs). Bared SPIONs in suspension exhibited a mean diameter of 60nm whereas MLVs exhibited a bimodal size distribution of sizes around 300nm and 1000nm. SLVs exhibited the smallest size (30nm); whereas SPION coated liposomes (magntoliposomes) was observed to be around 80nm in diameter. The mean size distribution of magneto-liposomes positioned in between bared SPIONs and MLVs with unimodal distribution indicating that a thin layer (~10nm) of phospholipids was formed around SPIONs as a shell.

Electron micrographs (Figure 3) exhibited that bared superparamagnetic iron oxide materials (SPIONs) were 30-50nm in diameter. The size of magneto-liposomes can be characterized by Cryo-TEM, however, currently we don't have the facility to obtain such information.

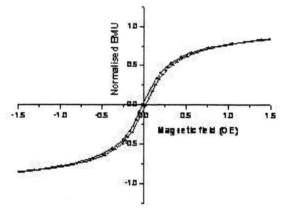

Figure 1 Magnetic susceptibility data of iron oxide materials

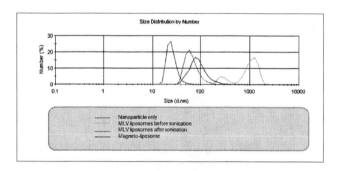

Figure 2 Particle size distributions of liposome, SPIONs and magneto-liposomes

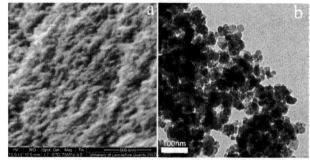

Figure 3 Scanning electron micrograph (a) and transmission electron micrograph (b) of bared SPIONs.

Based on the size of the MLVs, SUVs, SPIONs and magneto-liposomes, formation mechanism of vesicles, SUVs and magneto-liposomes are presented graphically in figure 4. The thickness of phospholipids bilayers can be considered as 8 nm based on the head to tail distance (~4nm) of a typical phospholipid molecule (see figure 4a). The diameter of MLV varies from few hundred nm to few microns (see figure 2) but upon ultrasonication, MLVs can produce LUV or SUV of smaller in diameter (see figure 4b). Based on the hydrodynamic diameter of SPIONs (~60nm), the minimum size of magneto-liposomes can be around 76nm with SPIONs core and single bilayer shell. The size of our magneto-liposomes (~80nm) supports the formation of single bilayer phospholipid coated SPIONs (see figure 4c).

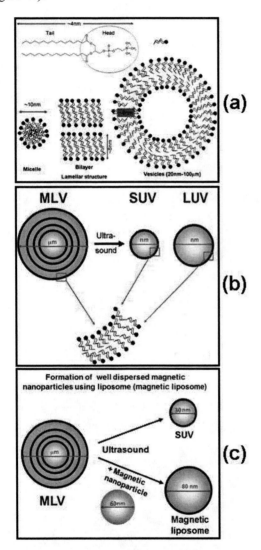

Figure 4 Graphical representation of the formation of liposomes using phospholipids (a), small (SUV) and large (LUV) unilamellar vesicles from multilamellar vesicles (b) and magneto-liposomes (c)

The suspension stability of SPIONs and magneto-liposomes prepared using a mixture of SPIONs (0.5mg/mL to 15mg/mL) in prefabricated MLV solution has been presented in figure 5. Bared SPIONs are unstable in suspension with (Figs.5B and 5C) or without ultrasonication (Fig. 5A). Magneto-liposomes with increasing concentration of SPIONs in suspension increased the suspension stability (Fig.5G). Further increase of SPIONs concentration (> 10 mg/mL) in MLV solution

produced an unstable magneto-liposomes suspension (Fig.5H). Ultrasonication time had an effect on the stability of magneto-liposomes (data not shown) and the optimum time was observed to be 8 min for the formation of stable magneto-liposomess.

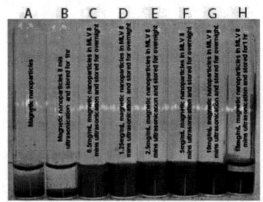

Figure 5 Pictorial presentation of bared SPIONs (A), bared SPIONs after ultrasonication batch SS068(B,C), magneto-liposomes of various batches SS067 (C), SS065 (D), SS063 (E), SS069 (F), SS070(G), SS073(H).

The surface characteristics of bared SPIONs and magneto-liposomes have been studied by Salmon sperm DNA binding/elution experiment and zeta potential measurements (see figure 6). The binding of DNA at physiological pH (~7) under high salt concentration (NaCl) was observed to be quite high in all materials (Fig. 6a), however, elution of DNA from the surface was observed to be different (see Figs. 6a and 6b). The elution of DNA passes through a maxima with increasing concentration of SPIONs during the fabrication of magneto-liposomes. The best elution results were observed with 5 and 10mg/mL of magnetite in magneto-liposomes suspensions. This is a direct indication of a different surface of magneto-liposomes compared with bared SPIONs. Bared SPIONs were reported [26] to be good in binding DNA and poor in elution from the surface.

Zeta potential measurement (Fig. 6c) indicates that SPIONs were positively charged at physiological pH (~7) whereas MLV or SUVs were negatively charged. Magneto-liposomes containing a range of concentration of SPIONs exhibited a systematic change in their zeta potential values. The zeta potential data indicates that increasing the concentration of SPIONs in magneto-liposomes suspension, the surface charge cross over took place from negative to positive values.

In conclusion, a series of stable magneto-liposomes were synthesized in suspension at physiological pH (~7) by simple one-pot method under strong ultrasonication. The size of the magneto-liposomes was measured to be around 80 nm and they were observed to be stable in suspension for days. The drug entrapment into bared SPIONs and magneto-liposomes at physiological pH (~7) is under study. The initial results indicate that entrapment of Mitomycin-C is found to be related to the stability of the magneto-liposomes in suspension. Detailed data will be presented in NSTI 2010 conference followed by a full manuscript publication in a relevant journal.

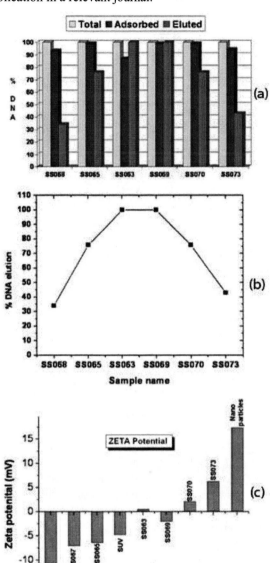

Figure 6 DNA Salmon sperm binding and elution results (a and b) and zeta potential (c) of bared SPIONs, MLVs, SUVs and magneto-liposomes.

REFERENCES

[1] Q. A. Pankhurst, J. Connolly, S. K. Jones and J. Dobson, J. Phys. D: Appl. Phys., 36, R167, 2003.

[2] R. Langer and D. A. Tirrell, Nature, 428, 487, 2004.

[3] J. M. Perej, Nat. Nanotechnol., 2, 525, 2007.

[4] T. Sen, A. Sebastianelli and I. J. Bruce, J. Am. Chem. Soc., 128, 7130, 2006

[5] T. Prozorov, R. Porozorov and A. Gadanken Adv. Mat., 10, 1529, 1998.

[6] T. Sen, S. Magdassi, G. Nizri, I. J. Bruce, Micro. Nano. Lett., 1, 37, 2006.

[7] M. Mezei, V. Gulasekharam, Life Sciences, 26, 2473, 1980.

[8] R. L. Juliano, Trends Pharmacol. Sci., 2, 39, 1981.

[9] G. Poste, C. Cucana, A, Raz, P. Bugelski, R. Kirsh, I. J. Fidler, Cancer Res., 42, 1412, 1982.

[10] T. M. Allen, Drugs, 54, 8, 1997.

[11] M. L. Matteucci, D. E. Thrall, Vet. Radiol., Ultrasoun. 41, 100, 2000.

[12] W. T. Al-Jamal and K. Kostarelos, Nanomedicine, 2, 85, 2007.

[13] S. Goodwin, C. Peterson, C, Hoh, C.J. Bittner, J. Mag. Mag. Mater., 194, 132, 1999.

[14] P. Tatraj, M. D. P. Morales, S. V. Verdaguer, T. G. Carreno, C. J. Serna, J. Phys. D. Appl. Phys. 36, R182, 2003.

[15] A. K. Gupta, M. Gupta, Biomaterials, 26, 3995, 2005.

[16] A. A. Gabizon, H. Shmeeda, S. Zalipsky, J. Liposome. Res., 16, 175, 2006.

[17] H. Sasaki, Y. Takakura, M. Hashida, T, Kimura, H. Sezaki, J. Pharmacobio-Dynamics, 7, 120, 1984.

[18] Y. Tokunaga, T. Iwasa, Y. Tokunaga, T. Iwasa, J. Fujisaki, S. Sawai, A. Kagayama, Chem. Pharm. Bull., 36, 3557, 1988.

[19] P. Chetoni, S. Burgalassi, D. Monti, M. Najarro, E. Boldrini, J. Drug Deliv. Sci. Technol., 17, 43, 2007.

[20] S. M. Zalipsky, S., M. Saad, J. Drug Target., 15, 518, 2007.

[21] De Cuyper, M., S. J. H. Soenen, K. Coenegrachts, L. Ter Beek, Anal. Biochem., 376, 266-273, 2007.

[22] K. J. Mart, K. P. Liem, S. J. Webb, " Pharm. Res., 26, 1701, 2009.

[23] S. J. H. Soenen, J. Cocquyt, L. Defour, P. Saveyn,P. Van Der Meeren, M. De Cuyper, Mater. Manuf. Process., 23, 611, 2008.

[24] S. J. H. Soenen, E. Illyes, D. Vercauteren, K. Braeckmans, Z. Majer, S. C. De Smedt, M. De Cuyper, Biomaterials, 30, 6803, 2009.

[25] R. Massart, E. Dubois, V. Cabuil, and E. Hasmonay, J. Magn. Mater.,149, 1, 1995.

[26] I. J. Bruce, J. Taylor, M. Todd, M. J. Davies, E. Borioni, C. Sangregorio and T. Sen, J. Magn. Magn. Mater., 284, 145, 2004.

Nanoemulsion Formation and Characterization of Neutraceutical Component with the Sonication and Self-Assembly Methods

A. J. Choi, C. G. Park, S. H. Han, and C. B. Park

Department of Herbal Crop Research, Rural Development Administration, Chungbuk, Korea,
aejini77@korea.kr

ABSTRACT

In this research, we investigated the optimum condition for the preparation of O/W nanoemulsion containing surfactants, neutraceutical ingredients extracted from *Angelica gigas* Nakai and oil phase in ternary phase diagrams of systems, and characterized the stability and formation of nanoemulsions. O/W nanoemulsions of neuraceutical ingredients could be prepared by the ultra-sonication process and by self-assembly method with a particle size of 20-100nm and having a good stability during storage. The self-assembly method have an advantage with using low energy, but have a defect of using with many amounts of synthetic surfactants. Therefore the ultra-sonication method may be more powerful tool to prepare the nanoemulsion than the self-assembly method, it might be due to the formation capacity of nanoemulsion phase.

Keywords: nanoemulsion, *Angelica gigas* Nakai, neutraceutical ingredients self-assembly, ultra-sonication

INTRODUCTION

Nanotechnology focuses on the characterization, fabrication, and manipulation of biological and nonbiological structures smaller than 100nm. Structures on this scale have been shown to have unique and novel functional properties[1]. In recent years, much attention has focused on lipid-based formulations to improved the oral bioavailability of lipophilic functional components. In fact, the most popular approach is the incorporation of the active lipophilic component into inert lipid vehicles such as oil, surfactant dispersions, microemulsions, nanoemulsions, and liposomes[2-4]. One of the promising nanotechnologies is nanoemulsions, which have also potential advantages over macroemulsions offering sustained controlled release, improved bioavailability and high stability for nutraceutical components[5]. *Angelica gigas* Nakai (also known as Cham-Danggui in Korea) is a perennial plant belonging to the Umbelliferae family, and the root has been traditionally used in Korean folk medicine containing neutraceutical components such as decursin and decursinol angelate[6].

MATERIALS AND METHODS

1. Materials

Polyoxyethylene sorbitan monooleate (Tween 80) was supplied from IlshinWells (Seoul, Korea, commercially available as Almax-9080). Medium chain triglyceride(MCT) supplied from Wellga (Yangsan, Korea).

2. Preparation of *Angelica gigas* Nakai Extracts

Neutraceutical ingredients such as decursin and decursinol angelate was extracted from *Angelica gigas* Nakai, medicinal herb in Korea. The neutraceutical ingredients of *Angelica gigas* Nakai was extracted with 100% MeOH (5% w/v) for 2hr at room temperature. The extracts was concentrated with evaporation at 50℃.

3. Preparation of nanoemulsion

The composition of three-component nanoemulsion system included *Angelica gigas* Nakai Extracts(AE)+MCT, Tween 80 and water. An oil-in-water emulsion was prepared by self-assembly or ultra-sonication method[7].

4. Particle size measurements

The mean droplet size and size distribution were determined by laser light scattering (Nanotrac ^{TM}250, Microtrac Inc., PA, U.S.A) at 25°C.

RESULTS AND DISCUSSIONS

1. Ternary phase diagram

The phase diagrams indicating the behavior of the systems composed of AE+MCT, Tween 80 and water with self-assembly and ultra-sonication method, respectively, and area of nanoemulsion existence are shown Fig. 1 and 2. Area enclosed within the solid line represents the extent of nanoemulsion formation, which was a white region and below 100nm. In place of a gray region, was formed with microemulsion, above 100nm and an extensive region, 2-phase regions were formed at high surfactant (degree of 60-

80 wt %), low AE concentrations, together with very large, cloudy and separation.

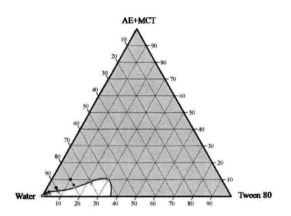

Fig. 1. Pseudo-ternary phase diagram of nanoemulsions formed by self-assembly method on the system AE+MCT/Tween 80/Water
(AE; *Angelica gigas* Nakai Extracts, White region; Nanoemulsion <100nm)

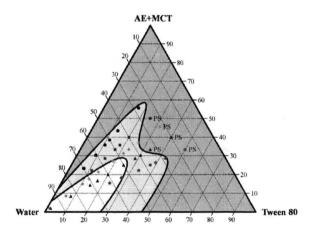

Fig. 2. Pseudo-ternary phase diagram of nanoemulsions formed by ultra-sonication method on the system AE+MCT/Tween 80/Water
(AE; *Angelica gigas* Nakai Extracts, PS; phase separation, White region; Nanoemulsion <100nm, Gray region; Microemulsion >100nm)

2. Stability of nanoemulsions during storage

The stability of nanoemulsion measured by particle size over 14 days period of storage at 25 °C. The nanoemulsions prepared in this study maintained stable form having particle size, below 100nm. However, microemulsions were not stable during storage, and had phase separation and flocculation.

REFERENCES

1. J. Weiss, P. Takhistov, & J. Mcclements (2006) Functional materials in food nanotechnology. *Journal of Food Science*, 71(9), 107-116.
2. S. A. Charman, M. C. Charman, T. D. Rogge, F. J. Wilson, C. W. Dutko, & C. W. Pouton (1992) Self-emulsifying drug delivery systems formulation and biopharmaceutic evaluation of an investigational lipophilic compound. *Pharmaceutical Research*, 9(1), 87-93.
3. B. K. Kang, J. S. Lee, S. K. Chon, S. Y. Jeong, S. H. Yuk, G. Khang, H. B. Lee, & S. H. Cho (2004) Development of self-microemulsion formulation using delivery systems (SMEDDS) for oral bioavailability enhancement of simvastatin in beagle dogs. *International Journal of Pharmaceutics*, 274, 65-73.
4. K. Bouchemal, S. Briancon, E. Perrier, & H. Fessi, (2004) Nano-emulsion formulation using spontaneous emulsification: solvent, oil and surfactant optimization. *International Journal of Pharmaceutics*, 280, 241-151.
5. J. Flangan, & H. Singh (2006) Microemulsions: A potential system for bioactives in food. *Critical Reviews in Food Science & Nutrition*, 46, 221.
6. S. D. Sarker, & L. Nahar (2004) Natural medicine: the genus Angelica. *Current Medicinal Chemistry* ,11, 1479-1500.
7. C. T. Kim, Y. J. Cho, C. J. Kim, I. H. Kim, Y. H. Kim, & A. J. Choi (2007) Research report of Korea Food Research Institute, No. E071000-07085. Development of process technology for functional food nanocomposite. In: Report of the Development of Food Nanotechnology, December 31, Kyeonggi, Korea, 123-266.

Rural Development Administration, Department of Herbal Crop Research, Bisanri 80, Eumseong, Chungbuk, 369-873, Republic of Korea, Ph: +82-43-871-5576, Fax: +82-43-871-5569, aejini77@korea.kr.

Control of the morphology of nanoparticles resulting from dynamic optimization of a fed-batch emulsion copolymerization process

B. Benyahia, M.A. Latifi, C. Fonteix, and F. Pla

Laboratoire Réactions et Génie des Procédés (LRGP) - CNRS-Nancy Université
1 rue Grandville - BP 20451 - 54001 NANCY Cedex (France)

ABSTRACT

This paper deals with the design and control of the morphology of core-shell nanoparticles elaborated by fed-batch emulsion copolymerization of styrene and butyl-acrylate in the presence of a chain transfer agent (n-dodecyl mercaptan). A mathematical model was elaborated and validated. It consists of a system of differential algebraic equations deduced from population balance and involving 49 unknown kinetic and thermodynamic parameters, many of them being impossible to be accurately estimated, due to the lack of experimental data. A method based on the sensitivity analysis allowed us to determine a subset of the 21 most influential parameters. The 28 non estimable parameters were taken from the literature. The model was then used to optimize the best profile of the pre-emulsion feed rate to control (i) the composition and average molar masses of the copolymer, (ii) the instantaneous glass transition temperature, corresponding to a core-shell morphology adapted to special end-use properties.

Keywords: emulsion copolymerization, modeling, dynamic optimization, core-shell nanoparticles, morphology control

1 INTRODUCTION

Emulsion polymerization is an important industrial process used to produce a large variety of polymers for multiple uses (e.g. paints, adhesives, coatings, varnishes...). Moreover, it has significant advantages over bulk and solution polymerization processes such as heat removal capacity and viscosity control. These advantages result mostly from the multiphase and compartmentalized nature of the emulsion polymerization which allows the production of polymers of high molecular weights with high polymerization rates, delivering a high versatility to product qualities. However, the complexity of emulsion polymerization systems, arising from factors such as their multiphase nature, nonlinear behavior and sensitivity to disturbances, induces more intense difficulties on modeling and makes the development of optimization procedures of emulsion polymerization reactions a very challenging task. Moreover, the production of polymers with specified end-use properties is one of the key issues in polymer industry. The desired end-use properties are usually carried out by using optimization approaches where many conflicting objective functions are frequently involved. This is known as multiobjective optimization problems increasingly encountered in chemical processes [1-4]. The optimal solutions are therefore not unique but constitute sets of non dominated compromises (Pareto's front) which show trade-offs among the whole objectives. A decision making approach is then used to rank Pareto's solutions in order to select the best compromise to be implemented. This communication deals with modeling and dynamic multiobjective optimization of batch and fed-batch emulsion copolymerization of styrene and butyl-acrylate in the presence of a chain transfer agent (CTA): n-dodecyl mercaptan. The objective is to optimize the operating variables in order to produce core-shell particles with a specific glass transition temperature profile and high conversion rate.

2 PROCESS MODEL

There are many research contributions on modeling emulsion polymerization processes, starting with the conventional Harkins' model which identifies three stages: nucleation, particles growth and the end of polymerization. The models available in the literature have different degrees of complexity depending upon their scope and application. The most representative have been reviewed [5], [6].

2.1 Main assumptions

The establishment of a model requires generally the use of several assumptions to enhance the speed of convergence. In this work, some of these assumptions are made without providing justification, as they are readily accepted and validated in the classical literature. Others which must be given with the necessary explanations are summarized as follows:

- Due to the high surfactant concentration used in this work, only micellar nucleation is considered,
- All reactions in the aqueous phase are neglected except initiation and inhibition,
- The chain transfer agent is subject to diffusional limitations mainly in the droplet-aqueous phase interface,
- The growing particles and the monomer droplets are considered to be monodisperse,
- The reactor is perfectly mixed and isothermal.

2.2 Kinetic Scheme

According to these assumptions, the model is based on the following elementary chemical reactions:
- *in the aqueous phase*: initiation, inhibition, nucleation and radical absorption
- *in the organic phase (particles)*: propagation, terminations by combination and disproportionation, inhibition, transfer to monomers, transfer to chain transfer agent and radical desorption.

2.3 Mathematical model

Using this scheme, the development of the kinetic model comprises the writing of reactions rates, mass balance of the various species (initiator, monomers, solvent, CTA, macroradicals and macromolecules), balance of the moments of order 0, 1 and 2 of the degree of polymerization distribution (DPD) of both macroradicals and macromolecules and influence of temperature on kinetic constants. The population balance is based on the assumption that the fraction of particles containing j free radicals follows Poisson's law. The model takes also into account the main phenomena involved in the process (radicals desorption, gel and glass effects...). It consists of a system of differential algebraic equations involving 49 parameters to be estimated.

2.4 Parametric identification

A first step, prior to the parameters identification is to evaluate the estimability of these parameters and to determine the subset of potentially estimable. Due to the model structure and possible lack of measurements, the estimation of some parameters appeared to be impossible regardless the amount of available data. The main limitations to the parameters estimability are their weak effect on the measured outputs and the correlation between their effects. Moreover, this estimation can lead to significant degradation in the predictive capability of the model. The development of an effective solution to the parameters selection requires establishing a methodology based on the magnitude of the individual effect of each parameter on the measured outputs [7]. This approach has been applied to the 49 parameters of the model leading to a subset of 21 parameters. The aim of the model was to correctly predict simultaneously the global conversion (X_{ove}), the fraction of residual styrene (Fr_2), the number- and weight-average molecular weights ($\overline{M_n}, \overline{M_w}$) and the average particles diameters (d_p). The model parameters were determined through the minimization of the maximum likelihood criterion, J, with the experimental data.

$$J = \sum_{k=1}^{5} N_k \cdot \ln\left(\sum_{l=1}^{N_k} \left(x_k(t_{kl}) - \hat{x}_k(t_{kl}, \theta) \right)^2 \right) \quad (1)$$

where N_k is the number of measurements of the variables x_k, t_{kl} is the lth time of measurement of the variable x_k and $\hat{x}_k$ is the value of x_k predicted by the model using the values θ of the unknown parameters. In this relation, the five variables x_k were: X_{ove}, $\overline{M_n}, \overline{M_w}$, d_p and Fr_2.

2.5 Associated results

The measured data were obtained from several batch runs carried out in a 1-liter jacketed reactor, using 1g of initiator, 60 g of styrene, 60g of butyl-acrylate and various CTA concentrations and temperatures [8],[9]. Global conversion, residual monomers, Mn and Mw and Tg were determined by gravimetry, GC using a Delsi Nermag DN 200 chromatograph, SEC using a Waters Millipore equipment, DSC using a Pyris 1 Perkin Elmer apparatus, respectively. Figure 1a shows the time evolution of X_{ove}, for experiments carried out at 60 and 70 °C, each for two different CTA concentrations. As expected, when the temperature is increased the conversion rate is higher. On the other hand, the effect of CTA on the global conversion is quite clear in spite of the weak differences between experimental and simulated values due to the CTA concentrations used in this work. Figure 1b presents the

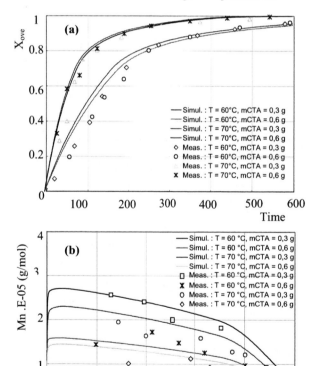

Figure 1: Effect of temperature and CTA concentration on: (a): Overall conversion; (b): Number average molecular weight

evolution of Mn versus X_{ove} for the same runs. As expected, Mn decreases when CTA concentration increases and decreases when the temperature is increased. The same observations were obtained for Mw. On the other hand, smaller particles were produced when the temperature increased. Nevertheless, CTA has a weak effect on the particles average diameters. Moreover, due to the difference between the reactivity ratios of each monomer, styrene is consumed faster than butyl-acrylate and the copolymer composition drifts till the total consumption of styrene. Globally the results show an acceptable agreement between simulated and experimental data.

2.6 Model validation

The model was then validated on new runs realized in batch mode. As shown in the two examples given in figure 2, good agreement was again observed between simulated and experimental data.

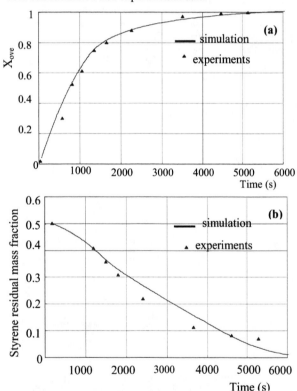

Figure 2: Validation of the model in batch mode:
(a): Overall conversion, (b): Styrene residual mass fraction

The model was also validated with runs carried out in fed-batch mode (figure 3).

3   MULTIOBJECTIVE OPTIMIZATION

The final objective of this work was to produce core-shell particles with specific end-use properties depending on the glass transition temperature profile.

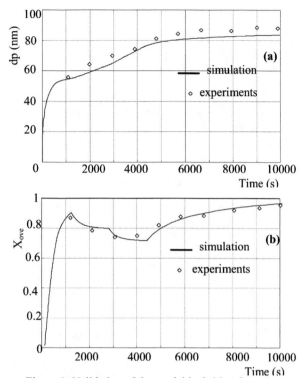

Figure 3: Validation of the model in fed-batch mode:
(a)   Average particles diameters,(b) Overall conversion

Considering that the two monomers used have different reactivity ratios and that the corresponding polymers have different Tg (-54°C for PBu and 100 °C for PS), the key feature of the optimization problem is to determine the optimal feed rate profiles which control the polymerization reactions in order to produce particles with a designed morphology and glass transition temperature.

Two objective functions, f_1 and f_2, given in equations (2), have been selected for the optimization. The first one aims at minimizing the error between the glass transition temperature and the desired profile, while the second aims at minimizing the final conversion.

$$Min f = [f_1, f_2]$$

$$f_1 = \frac{1}{t_{fc} - t_0} \int_0^{t_{fc}} |T_g - T_{g1}| dt + \frac{1}{t_{fs} - t_{fc}} \int_{t_{fc}}^{t_{fs}} |T_g - T_{g2}| dt$$

$$f_2 = -X(t_f)$$

$$st. \quad \dot{x} = f(x(t), u(t), p, t) ; \quad x(t_0) = x_0$$

$$\frac{1}{t_{fc} - t_0} \int_0^{t_{fc}} (0.9 - X(t))^2 dt \le \varepsilon^2$$

$$u_{inf} \le u(t) \le u_{sup}$$

(2)

where T_g is the glass transition temperature, T_{g1} the desired glass transition temperature for the core, T_{g2} the desired glass transition temperature for the shell, t_{fc} and t_{fs} the times necessary to obtain the corresponding core and shell respectively, $X(t_f)$ is the conversion at the end of the process and $\boldsymbol{u}$ the control vector (feeds and time periods).

The feed profiles and the time periods which maximize the conversion at the end of the copolymerization and minimize the difference between the measured and a designed profile of glass transition temperature, were determined by means of a multi-objective optimization approach based on Pareto's approach. The set of non-dominated solutions (Pareto's front), obtained by the use of an evolutionary algorithm [10], is given in figure 4.

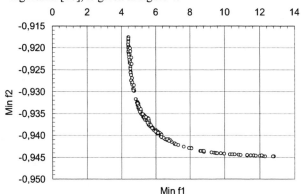

Figure 4: Set of non-dominated solutions (Pareto's front)

Multiattribute utility theory (MAUT) was then used as a decision making tool to rank Pareto's solutions. The resulting best solution implemented within the real system is given in figure 5.

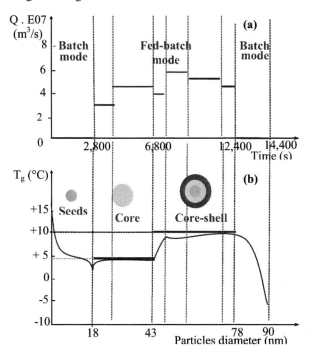

Figure 5: Feed rate (a) and corresponding instantaneous glass transition temperature profiles (b)

4   CONCLUSION

In this work, a dynamic model has been developed and validated for the batch and fed-batch emulsion copolymerization of styrene and butyl-acrylate in the presence of a chain transfer agent (n-dodecyl mercaptan). After its validation, this model has then been used in a multiobjective optimization problem designed to determine the optimal feed profiles necessary to produce, with a high conversion, core-shell latex particles with specific glass transition temperature. A decision support approach was used to determine this optimal solution.

REFERENCES

[1]   C. Fonteix, S. Massebeuf, F. Pla, L. Nandor Kiss, Eur. J. Oper. Res., 153, 350-359, 2004

[2]   S. Garg and S. K. Gupta, Macromol. The. Simul. 8, 46-53, 1999

[3]   K. Mitra., S. Majundar, S. Raha, Ind. Eng. Chem., 43, 6055-6063, 2004

[4]   D. Sakar, S. Rohani, A. Jutan, AIChE J., 53, 5, 1164-1174, 2007

[5]   J. Gao, and A. Penlidis, Prog. Polym. Sci. 27, 403-535, 2002

[6]   S. C. Thickett and R. G. Gilbert, Polymer 48, 6965-6991, 2007

[7]   B. Benyahia, M. A. Latifi, C. Fonteix, F. Pla, S. Nacef Chemical Engineering Science, 65, (2), 850-869, 2010

[8]   B. Benyahia, PhD thesis, National Polytechnic Institute of Lorraine, Nancy-University, 2009

[9]   B. Benyahia, M. A. Latifi, C. Fonteix, F. Pla, S. Nacef, CHISA 2008, 18th International Congress of Chemical and Process Engineering, Prague, Czech Republic, 24-28 August, 2008

[10] S. Massebeuf, C. Fonteix, S. Hoppe, F. Pla, J. Appl. Polym. Sci. 87,14, 2383-2396, 2003

NSTI-Nanotech 2010, www.nsti.org, ISBN 978-1-4398-3401-5 Vol. 1, 2010

Index of Authors

Index of Keywords

NSTI-Nanotech 2010, www.nsti.org, ISBN 978-1-4398-3401-5 Vol. 1, 2010